Upgrade with BASE

British Aerospace (Systems and Equipment) Ltd. – BASE – has a wide variety of products and technologies which are available for both new–build and upgrade applications in military and civil aircraft programmes:

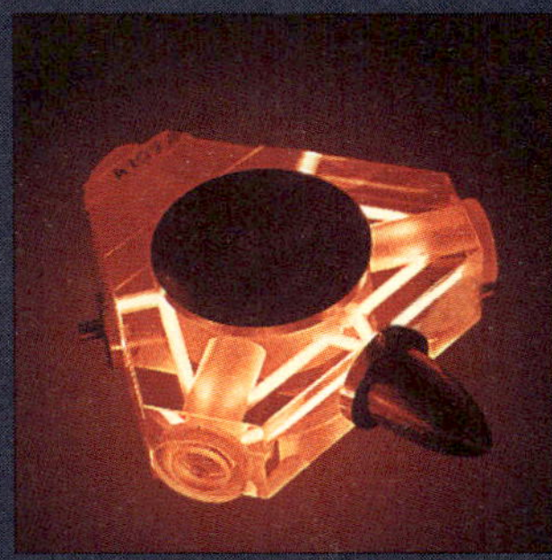

LINS – Laser Inertial Navigation Unit

LINS 300 offers high reliability, superior performance and low cost of ownership for precise navigation requirements. LINS 764G meets the latest embedded GPS inertial navigation system specification (EGI), and is a candidate for a wide range of UK and European upgrade opportunities. (See editorial entry).

TERPROM – Digital Terrain System

Gives highly accurate, drift-free terrain referenced navigation, predictive ground and obstacle proximity warning, stealthy terrain-following, enhanced target acquisition and covert weapon aiming, all with a significant reduction in crew workload and no reliance on external navigation aids. (See editorial entry).

Aircraft Radomes

The BASE Reinforced and Microwave Plastics facility is an industry leader in the design, development, testing and qualification of radomes for aircraft and missiles, for new-build and upgrade applications. (See editorial entry).

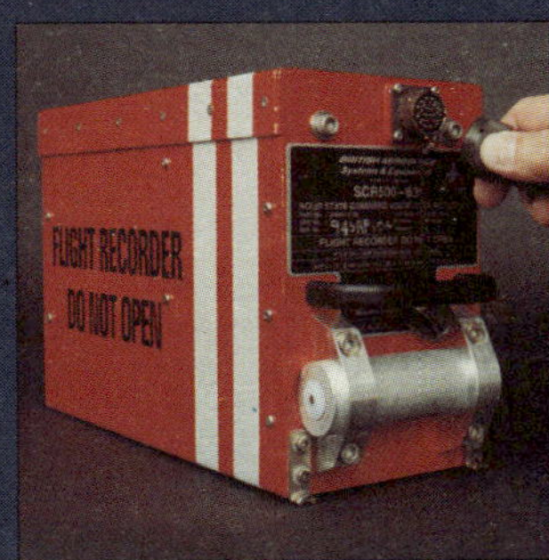

SCR500 – Flight Recorders

The latest digital recording and solid-state memory techniques make the SCR500 the lightest and most versatile range of recorders for commercial aircraft and helicopter applications:

- *Solid-state reliability, 3 year warranty, no routine maintenance.*
- *Complies with EUROCAE ED 55, ED 56A and TSO - C123 and TSO - C124 draft. ARINC 757/557 interchangeability.*
- *PC-based downloader/ground support equipment.*
- *Configuration options include CVR, FDR and combined CVR/FDR.*
- *Software to level C for future HUMS compatibility.*
- *Worldwide British Aerospace support and dealer network.*

VITS 1000 – Video Image Tracking System

The first in a new family of high performance electro-optical tracking systems developed by BASE for the civil and military, retro-fit and new build, helicopter and fixed wing aircraft markets:

- *Precision tracks video images of targets even in low contrast, cluttered scenes.*
- *Capable of integration with new or existing systems through standard interfaces.*
- *Processes TV or IR image data.*
- *Capable of tracking high-dynamic targets.*

–Your Upgrade Partner

British Aerospace
(Systems and Equipment) Ltd.
Clittaford Road, Southway, Plymouth,
Devon, England PL6 6DE
Telephone: 44 1752 695695
Telex: 45708
Fax: 44 1752 695500

See Editorial Entry on page 237-254

Jane's

AIRCRAFT UPGRADES

Formerly Jane's Civil and Military Aircraft Upgrades

Edited by Simon Michell

Third Edition
1995-96

JAWA VOLUME II

ISBN 0 7106 1281 8
"Jane's" is a registered trade mark

In the USA and its dependencies
Jane's Information Group Inc, 1340 Braddock Place, Suite 300, Alexandria, VA 22314-1651, USA

I(T)P An International Thomson Publishing Company

British Library Cataloguing-in-Publication Data.
A catalogue record for this book is available from the British Library.

Printed and bound in Great Britain by Biddles Limited, Guildford and King's Lynn

Contents

ADMINISTRATION

Publishing Director: Robert Hutchinson

Managing Editor: Ruth Jowett

Yearbook Editorial Production Manager: Ruth Simmance

Production Editor: Jane Stimson

EDITORIAL OFFICES

Jane's Information Group Limited, Sentinel House,
163 Brighton Road, Coulsdon, Surrey CR5 2NH, UK

Tel: 0181 700 3700 International +44 181 700 3700
Telex: 916907 Janes G
Fax: 0181 763 2582 International +44 181 763 2582
e-mail:yearbook@janes.co.uk

SALES OFFICES

Send enquiries to International Sales Manager:
Fabiana Angelini (UK/MoD, Europe)
David Eaton-Jones (Middle East, Asia Pacific, Africa, Austria, Germany, Greece, Switzerland, Eastern Europe)
Jane's Information Group Limited, UK address as above

Tel Enquiries: 0181 700 3700 International: +44 181 700 3700
Fax Enquiries: 0181 763 1006 International: +44 181 763 1006
Fax Orders: 0181 763 0276 International: +44 181 763 0276

Send USA enquiries to:
Joe McHale, Senior Vice-President Product Sales,
Jane's Information Group Inc, 1340 Braddock Place, Suite 300,
Alexandria, Virginia 22314-1651

Tel: +1 703 683 3700
Telex: 6819193
Fax: +1 703 836 0029

ADVERTISEMENT SALES OFFICES

Advertisement Sales Manager: Richard West

Australia: Brendan Gullifer, Havre & Gullifer (Pty) Ltd, Level 50, 101 Collins Street, Melbourne 3000

Tel: +61 3 9650 1100
Fax: +61 3 9650 6611

Benelux: Richard West, Jane's Information Group (see UK/Rest of World)

Brazil: L Bilyk, Brazmedia International S/C Ltda, Alameda Gabriel Monterio da Silva, 366 CEP, 01442 São Paulo

Tel: +55 11 853 4133
Telex: 32836 BMED BR
Fax: +55 11 852 6485

France: Patrice Février, Jane's Information Group – France, BP 418, 35 avenue MacMahon, F-75824 Paris Cedex

Tel: +33 1 45 72 33 11
Fax: +33 1 45 72 17 95

Germany and Austria: Janet Scott, Jane's Information Group (see UK/Rest of World)

Hong Kong: Jeremy Miller, Major Media Ltd, Room 1402, 14F Capitol Centre, 5-19 Jardine's Bazaar, Causeway Bay

Tel: +852 890 3110
Fax: +852 576 3397

Israel: Oreet Ben-Yaacov, Oreet International Media, 15 Kinneret Street, IL-51201 Bene-Berak

Tel: +972 3 570 6527
Fax: +972 3 570 6526

Italy and Switzerland: Ediconsult Internazionale Srl, Piazza Fontane Marose 3, I-16123 Genoa, Italy

Tel: +39 10 583684
Telex: 281197 EDINT I
Fax: +39 10 566578

Japan: Intermart/EAC Inc, 1-7 Akasaka 9-chome, Minato-Ku, Tokyo 107

Tel: +81 3 5474 7835
Fax: +81 3 5474 7837

Korea, South: Young Seoh Chinn, JES Media International, 6th Floor Donghye Building, 47-16 Myungil-Dong, Kangdong-Gu, Seoul 134-070

Tel: +82 2 481 3411
Fax: +82 2 481 3414

Scandinavia: Gillian Thompson, The Falsten Partnership, 11 Chardmore Road, Stamford Hill, London N16 6JA, UK

Tel: +44 181 806 2301
Fax: + 44 181 806 8137

Singapore, Indonesia, Malaysia, Philippines, Taiwan and Thailand: Hoo Siew Sai, Major Media (Singapore) Pte Ltd, 6th Floor, 52 Chin Swee Road, Singapore 0316

Tel: +65 738 0122
Telex: RS 43370 AMPLS
Fax: +65 738 2108

South Africa: Janet Scott, Jane's Information Group (see UK/Rest of World)

Spain: Jesus Moran Iglesias, Varex SA, Modesto Lafuente 4, E-28010 Madrid

Tel: +34 1 448 7622
Fax: +34 1 446 0198

UK/Rest of World: Janet Scott, Jane's Information Group, Sentinel House, 163 Brighton Road, Coulsdon, Surrey CR5 2NH

Tel: 0181 700 3700 International +44 181 700 3700
Telex: 916907 Janes G
Fax: 0181 763 0643 International +44 181 763 0643

USA and Canada: Kimberley S Hanson, Advertising Sales and Marketing Director, Jane's Information Group Inc, 1340 Braddock Place, Suite 300, Alexandria, Virginia 22314-1651

Tel: +1 703 683 3700
Telex: 6819193
Fax: +1 703 836 0029

USA West Coast: Anne Marie St John-Brooks, Regional Advertising Sales Manager, Jane's Information Group Inc, 1523 Rollins Road, Burlingame, California 94010

Tel: +1 415 259 9982
Fax: +1 415 259 9751

Administration: UK: Tara Betts
USA and Canada: Maureen Nute

How To Use This Book

In order to help our readers we have added pages at the beginning of this edition, the first being this page on *'How to use this book'*. It gives an overview of what is contained in each section and an explanation of the *'Entry Tags'* we have added.

There is another new page, containing *New Entries*, which will enable the reader to find out quickly what is new in this edition.

These new pages and the Foreword and Contents will be printed on coloured paper from this year to make it easier for readers to find the page they need.

Jane's Aircraft Upgrades (JAU) contains full *Jane's All The World's Aircraft (JAWA)* type descriptions of commercial and military aircraft, both fixed- and rotary-wing, that are no longer in production or are coming to the end of their production cycles but are still in operational service worldwide. This contrasts with *Jane's All The World's Aircraft* which covers all aircraft either in full production or in their design or development stages.

This book is organised in alphabetical country order, with each country's section containing information on airframe manufacturers as well as companies that offer retrofit packages to upgrade, convert, modify, refurbish and repaint these aircraft. These entries are also in alphabetical order.

Each manufacturer's entry lists aircraft built by that company or by a predecessor of the company. Each airframe entry gives full details of the development history, listing the various versions still in service. This is followed by a full description of the aircraft's design, structure, flying controls, equipment avionics, power plant and, where applicable, armament. The fullest data possible on the aircraft's dimensions, both internal and external, weights and loadings and performance are also listed. An UPGRADES heading is located directly after the VERSIONS heading. This contains all significant upgrade activity and refers the reader to the relevant section in the book. For example, on reading the MiG-21 entry, you will notice the upgrade section lists companies that have carried out work, are developing upgrade packages or have prepared bids for upgrade work on the aircraft as well as countries that have either put out requests for proposals (RFPs) or have a requirement for upgrade work. The reader is referred either to the VERSIONS heading of that aircraft entry, a separate description which follows within the same manufacturers entry (in this case Mikoyan) or an entry which appears in another section of the book. The exact location of every entry can be gained by referring to the index. Information is also given on where each aircraft is in service. Military aircraft have an OPERATORS heading which lists each air arm still operating the aircraft along with the number in service. At the front of the book are two fleet lists each consisting of four columns. The first column of the first list contains the aircraft type, the second column contains the power plant which has been installed on that aircraft, the third column contains the number of that aircraft in the airline/fleet and the fourth column contains the name of the airline/fleet which operates that aircraft. The second fleet list contains the same information with the power plant appearing in the first column.

To help users of this title evaluate the published data, Jane's Information Group has divided entries into three categories:-

● ***VERIFIED*** The editor has made a detailed examination of the entry's content and checked its relevancy and accuracy for publication in the new edition to the best of his ability.

● ***UPDATED*** During the verification process, significant changes to content have been made to reflect the latest position known to Jane's at the time of publication.

● ***NEW ENTRY*** Information on new equipment and/or systems appearing for the first time in the title.

We have also added dates to the pictures included in this edition. All pictures will be dated with the first year of publication. This year's pictures will be dated ***1995*** and next year's ***1996*** and so on.

Alphabetical list of advertisers

ELBIT'S AIRCRAFT UPGRADES TURN OUTDATED PLATFORMS INTO FIRST-CLASS FIGHTERS

With across-the-board expertise in aircraft retrofits and advanced avionics, Elbit is in a unique position to provide cost-effective solutions to assure that current platforms match future requirements. Elbit upgrades "rejuvenate" aircraft through structural improvements, advanced cockpits, and next-generation capabilities assured by the following high performance systems:

■ Weapon delivery and navigation ■ Mission computers ■ Sight helmet ■ Day and night vision ■ AACMI ■ Multifunction displays (color and monochromatic) ■ Stores management ■ Radar ■ EW ■ Training systems.

Elbit's expertise is based on 25 years of experience in 50 upgrade programs all over the globe for A-4, F-4, F-5 A/B, F-5 E/F, F-15, F-16, L-39 and Mig-21 aircraft.

Team with Elbit. Together we can reach new heights.

Elbit A STRONG SENSE OF DIRECTION

Elbit Ltd. P.O.Box 539, Haifa 31053, Israel. Tel: (972) 4-315315. Fax: (972)4-550002/551623.

Northrop Grumman F-5E Tiger IV demonstrator aircraft cockpit suite *(Northrop Grumman)*

Jane's

AIRCRAFT UPGRADES

1995-96

Jane's Information Group Limited, Sentinel House, 163 Brighton Road, Coulsdon, Surrey CR5 2NH, UK
Jane's Information Group Inc, 1340 Braddock Place, Suite 300, Alexandria, VA 22314-1651, USA

YOUR
F-5 PLUS
Now A Fighter Fit For Front-Line Missions
When IAI's LAHAV takes charge of your F-5 modernization program, you commit your fleet to the most comprehensive upgrading program world wide - already proven in active service. Unmatched experience and know-how are applied to deliver a configuration that responds to your strategy: an air superiority fighter, multi-role combat aircraft, or lead-in trainer.
Our modular integrated package includes:
Advanced avionic system
Improved cockpit design
State-of-the-art radar
Electronic warfare suite
Improved flexibility in carrying capability
Your customized F-5 Plus will take off with state-of-the-art capabilities. Contact LAHAV today.
IAI LAHAV DIVISION
Military Aircraft Group
ISRAEL AIRCRAFT INDUSTRIES LTD
The Power to Empower Your Fleet
Israel Ben Gurion Int'l Airport 70100. Tel: 972-3-935-3163/5.
Fax: 972-3-935-3687. U.S.A. Tel: 1-703-875-3777. Fax: 1-703-875-3740.
Europe Tel: 33-1-46.40.47.47. Fax: 33-1-47.45.73.22.

YOUR
MIG-21
2000
Now A Modern Fighter Ranking With The Best
When IAI's LAHAV takes charge of your MIG-21 modernization program, you commit your fleet to the world leader in fighter upgrading. Unmatched experience and know-how are applied to deliver a combat aircraft ready to challenge any operational mission.
Our package includes:
Advanced avionic systems
State-of-the-art navigation systems with embedded GPS
Accurate homing and positioning system
One-piece windshield for enhanced pilot visibility
Carrying capability for Western and Eastern armaments
Your customized F-5 Plus will take off with state-of-the-art capabilities. Contact LAHAV today.
IAI LAHAV DIVISION
Military Aircraft Group
ISRAEL AIRCRAFT INDUSTRIES LTD
The Power to Empower Your Fleet
Israel Ben Gurion Int'l Airport 70100. Tel: 972-3-935-3163/5.
Fax: 972-3-935-3687. U.S.A. Tel: 1-703-875-3777. Fax: 1-703-875-3740.
Europe Tel: 33-1-46.40.47.47. Fax: 33-1-47.45.73.22.

Foreword

Introduction

In early 1995, it was decided that the name of the third edition of this reference work should be changed from *Jane's Civil and Military Aircraft Upgrades* to *Jane's Aircraft Upgrades*. At the same time the book became part of a three-volume family of Jane's products and was subtitled, *Jane's All The World's Aircraft Volume II*. Volume I is the original *Jane's All The World's Aircraft*. Volume III is a new title, *Jane's World Air Forces*. The strategy behind this grouping is to create a family of books, which will become the most accurate and complete source of aircraft information available.

This third edition contains more than 30 new full aircraft descriptions including the Bell 47 helicopter, Ilyushin Il-86, Jetstream 31 and Zlin 50 as well over 20 new companies which specialise in aircraft upgrading including Elbit of Israel, Hunting Aviation of South Africa, Racal Avionics of the UK and Greenwich Air Services of the USA. In all there are 134 new entries in this third edition. I have also included 300 new photographs and line drawings. Those who own a copy of the second edition of this book, will notice that the Upgrades heading in each aircraft description has been relocated so that it follows the Versions heading, and that a new Operators heading has been created to give details of where each military aircraft is in service.

Military Drawdown and Fiscal Squeeze

Since the publication of the second edition of this volume there has been no change in the prevailing geopolitical retrenchment of the two Superpowers. The USA continues its policy of drawdown, closing bases all over the world and Russia has all but left central Europe and the Baltics. With the exception of France and the Tiger economies of South-east Asia practically every country has reduced its military spending drastically.

Despite the fact that economic statisticians continue to point out that the Western economies have experienced prolonged growth with exceedingly low inflation, there remains a profound sense of economic gloom. Germany has begun to shift investment to overseas production facilities with negative results to its unemployment figures. France continues to struggle with its franc fort policy which has also led to increased unemployment. The UK is still gripped by domestic market stagnation and Japan faces further crises in its banking infrastructure. However, it appears that these problems have reached their nadirs and that economic indicators will start on an upward trend, especially if the USA's economy continues to grow at current rates. This will hopefully lead to a significant upturn in the aerospace market spend. Passenger travel has already started to increase and new orders are slowly gathering speed. However, the combination of the military and commercial downturns has had a dramatic affect on the aerospace industry as decreased markets have led to drastic production overcapacity.

1995, like 1994, has seen continued merger activity on a grand scale. The merger of Lockheed with Martin Marietta has created a huge conglomerate with a staff of 170,000 personnel and annual sales of $23 billion. Raytheon Aircraft Company, with production of the Hawker aircraft now installed at its Wichita site combined with its total integration of Beech Aircraft, is well on its way to becoming a very significant player. It remains to be seen what Raytheon Aircraft will do next.

Europe continues to rationalise its aircraft production. The UK companies Avro and Jetstream signed a joint marketing agreement in June with the Franco-Italian company ATR. Although this is by no means a merger it does represent a significant but not sufficient reaction to the reality of Europe's regional aircraft overcapacity. A continued consequence of the combination of military and commercial cutbacks in the early 1990s, is the drop in new aircraft orders and therefore the need to keep existing airframes airworthy, and in line with new safety and environmental regulations as well as new technology. Retrofit activity therefore continues to grow in both military and commercial aircraft fleets.

Northrop Grumman F-5E Tiger IV demonstrator aircraft *(Northrop Grumman)*

MILITARY AIRCRAFT UPGRADE ACTIVITY 1995

Fixed-Wing Fighters

A-4 Skyhawk

The A-4M Skyhawk aircraft, supplied to the Argentine Air Force by the US Navy, has finally been kick-started, following the takeover of the management of Argentina's Fabrica Militares Aviation (FMA) by Lockheed Martin. The newly privatised and renamed company, Lockheed Aircraft Argentina, will commence operations with two major modernisation programmes. One worth $200 million for an unspecified Argentine Air Force aircraft and the other worth $280 million for the A-4M upgrade. Lockheed Martin will carry out the first 18 upgrades at its facilities in Ontario, California and Lockheed Aircraft Argentina will carry out the work on the second batch. The upgrade itself consists of installing a variant of the Westinghouse APG-66 radar (designated ARG-1) as well as AlliedSignal LED displays.

F-4 Phantom II

Although the F-4, like the A-4, is a relatively old aircraft, it is still in service with 10 air forces worldwide and subject to various upgrade programmes. Mitsubishi had completed over 80 of the planned 100 F-4EJ to F-4EJKai upgrades for the Japanese Air Force by 1995. This programme, like the Argentine A-4M, is based on a version of the Westinghouse APG-66 in this case the J/APG-66J. Further features include a Litton J/ASN-4 inertial navigation system (INS), head-up display (HUD), and J/APR-4Kai radar warning receiver (RWR). Daimler-Benz Aerospace (DASA) is scheduled to complete its F-4F ICE by the end of 1996. This programme consists of, amongst other things, replacing the existing Westinghouse AN/APG-120 radar with an all-digital multimode Hughes AN/APG-65, built under licence by DASA. Greece, South Korea and Turkey operate fleets of F-4 Phantom II aircraft and all three of these countries have upgrade requirements. Greece has released an interim request for proposal (RFP) for a structural and avionics upgrade. Known bidders for this are DASA, Israeli Aircraft Industries (IAI) and Rockwell. Turkey has selected IAI for its F-4 upgrade but no deal has yet been signed as it seems that negotiations concerning workshare and technology transfer agreements have stalled. IAI is also bidding for the South Korean requirement to upgrade its fleet of F-4D/E and RF-4C aircraft.

F-5 Freedom Fighter and Tiger II

The F-5A/B Freedom Fighter, F-5E/F Tiger II and RF-5A/E represent one of the largest upgrade markets past, present and future. The world's most advanced companies have developed, manufactured and implemented structural and avionics packages on what is one of the most ubiquitous fighter aircraft in service. In March 1995 Northrop, the design authority, was selected by the USAF's San Antonio Air Logistics Center (SA-ALC) to offer its structural update programme (SUP) worldwide. Northrop also celebrated the first flight on 20 April 1995 of its F-5E Tiger IV advanced avionics demonstrator. IAI, not surprisingly, has proven expertise in this field and is actively seeking a contract to upgrade the South Korean F-5 aircraft. Bristol Aerospace of Canada was contracted in June by the Canadian Government to market all 72 of Canada's CF-5s including the aircraft that the company had already upgraded. Bristol Aerospace is currently manufacturing new wings for the Norwegian Air Force's F-5A/B Freedom Fighters. These aircraft have already undergone the Tiger-PAWS programme for avionics and weapon systems improvements, developed by Sierra Technologies of New York, and completed on the Norwegian aircraft in June 1994.

MiG-21

The MiG-21, following the F-5, represents the next big upgrade scramble. There are over 3,000 MiG-21 aircraft in service around the world and many of these, especially the MiG-21bis versions are prime targets for retrofit work. IAI flew its MiG-21-2000 demonstrator on 24 May 1995 (see photograph in addenda). Mikoyan/Sokol exhibited their MiG-21-93 at Paris and Moscow. It is still not known, however, whether the Russian aircraft is a flying demonstrator or a static display. Aerea of Italy has collaborated with Mikoyan in a programme to develop a modular weapons storage upgrade for both NATO and Russian weapons. Actual upgrade contract awards include the Indian Government contract with Mikoyan. This programme covers 100 aircraft with an option for 70 more. The new equipment is to include the Kopyo multifunctional coherent pulse Doppler radar with slot antenna, head-up display (HUD), multifunctional display (MFD), hands-on throttle and stick (HOTAS), helmet-mounted target designator, inertial reference system (IRS), navigation computer, IFF system and radar warning receiver (RWR). Significantly, the system design allows the use of western equipment. Romania has awarded a contract to Elbit in collaboration with IAI to upgrade 100 of its MiG-21s. IAI is also contracted to upgrade Cambodia's fleet of 19 MiG-21 aircraft. Other countries with current MiG-21 upgrade requirements include the Czech Republic and Hungary. The Hungarian Air Force is already equipping its MiG-21 aircraft with western IFF and radio sets.

Mirage F1

The Spanish Air Force has a requirement to upgrade its fleet of Mirage F1-CE/EE/C aircraft to extend their operational lives until 2010. No prime contractor has yet been selected although CASA of Spain looks to be a very likely candidate for the contract. The core of the upgrade will be a radar altimeter, GPS, onboard computer, stores management system, colour VHF camera, and an ECM suite. The aircraft are also to be fitted with the Barax detector/jammer, HAVE-QUICK II communication units and IFF/SIF Mode 4 Crypto. South Africa, which operates a fleet of 25 Mirage F1s has become very active in designing systems for this aircraft. Grinaker Avitronics has developed an EW upgrade package which provides radar warning using RWS-130 radar warning receiver, a chaff dispensing system and an active ECM system using an SPJ-200 CW jammer. Aerosud of South Africa has conducted over 50 hours of flight tests with a re-engined Mirage F1 as part of a development programme which is being undertaken in collaboration with Rusjet of Russia and is aimed at the Mirage F1, MiG-21 and Mirage III. This engine is also undergoing flight trials in an Atlas Cheetah. Another engine collaboration which includes a South African company, is the programme to develop an upgrade for the Atar 8/9K50 turbojet which is in use in the Mirage F1 as well as the Atlas Cheetah, Mirage 50 and Super Etendard. This team consists of Snecma of France, ITP of Italy and Simera of South Africa. Main elements of this programme are to develop a modification to the core in order to decrease acceleration time by one third. This will improve maintainability by increasing the time between overhaul from 900 hours to 1,200 hours.

Transport, Tanker and Patrol Aircraft

C-130

Versions of the C-130 Hercules aircraft are in military service with 57 countries worldwide and many of these aircraft are the early A, B, E and H models which are prime targets for relifing and avionic systems upgrades.

Sabena Technics of Belgium successfully flew its C-130H avionics upgrade programme (AUP) Hercules on 22 June 1995 (see photograph in addenda). The contract is for 12 aircraft and covers the design, development and installation of Honeywell Defense Avionics Systems Division (DASD) integrated vehicle mission management system (IVMMS) consisting of a new autopilot, with air data sensor unit (ADSU); flight management system (FMS), with cockpit display; electronic flight instrument system (EFIS); new weather radar; altitude heading reference system (AHRS); inertial navigation system (INS); and GPS. Snow Aviation of Ohio is currently installing a two-man cockpit upgrade on a C-130A. This is a four-part retrofit package consisting of an electrical system, communication, navigation and identification systems, engine indication and crew alerting system as well as flat glass displays and HOTAS installation. Snow Aviation is hoping to market this upgrade for installation on other large aircraft including the Boeing 727-200. The Spanish Government gave the go-ahead in July 1995 to extend the lives of its fleet of seven C-130H and five KC-130H aircraft, and in June 1995 Westinghouse announced that it had been awarded a contract to upgrade the Portuguese Air Force fleet of C-130H with its AN/APN-241 navigation and weather radar. Marshall Aerospace has recently undertaken an extensive study for the UK MoD, to determine what refurbishment would be necessary to keep its fleet of C-130K tanker aircraft operational with acceptable availability rates.

DHC-5

The Egpytian Air Force received the fifth DHC-5 Buffalo aircraft, completely refurbished by Greenwich Air Services, in July 1995. This work was financed under the US Foreign Military Sales programme and included stripping the aircraft down, overhauling all the components, installing new avionics and refurbishing the interior. Greenwich Air Services teamed with LAB, of France, and used LAB's facilities in Dinard to fulfil the contract.

HS 748

Despite the Andovers being withdrawn from the RAF in the early part of the year, the HS 748 is still in service with 12 air forces worldwide. The Royal Australian Air Force has initiated the Project AIR 5232 programme to update the systems on six of the eight RAAF navigation training aircraft. Improved systems are to include: ground-mapping radar (similar to the one in the RAF Dominie T. Mk 1), off-the-shelf navigation computer and simulated radar warning receiver (RWR). The programme also includes installation of GPS as part of the RAAF Project 5195 to install GPS in all RAAF, RAN and Army tactical aircraft. The GPS system will allow fusion of radar, GPS and inertial navigation system data from the existing Litton LTN-92 inertial navigation system (INS). The first aircraft is planned for redelivery in 1997.

VC-10

FR Aviation and its sister company Flight Refuelling are still on target to complete the conversion of 13 Royal Air Force VC-10 C. Mk 1(K) tanker/passenger conversions during 1996. Both these companies are subsidiaries of the Cobham Group which acquired the US company, Sargent Fletcher, in September 1994. The programme covers installation of two-point air-to-air refuelling equipment giving the aircraft dual passenger and in-flight refuelling capabilities. The contract also covers upgrading the communications and navigation equipment.

Still in the field of in-flight refuelling, the Turkish Government has finally awarded Boeing a contract to modify seven KC-135A to KC-135R configuration. There has been a great deal of opposition to this $315 million contract in Turkey but it is unlikely that a better deal could have been negotiated. In 1995 the official Islamic Republic News Agency of Iran reported that it had developed an air-to-air refuelling system for its fleet of MiG-29 aircraft. Although no details have been given, it is believed to have been designed for its fleet of Boeing 707 aircraft which were purchased during the Shah's reign.

In August of this year, the first of two Dutch KDC-10 tanker conversions started its flight tests and is likely to be delivered during 1995. This programme has suffered cost overruns and delays as the original in-service dates were April and June. KLM Engineering which is acting as prime contractor blames the delays on problems in installing the fuel control systems as well as difficulties with the remote aerial refuelling operator (RARO) developed by McDonnell Douglas.

MILITARY HELICOPTER UPGRADES

Attack, patrol and utilty helicopters

The British Army took redelivery of its AH-1 Gazelle helicopters, retrofitted with GEC-Marconi Defence Systems Rewarder radar warning receiver (RWR) in June, and Army Lynx helicopters fitted with the same system are also scheduled to return to service this year. Racal Avionics is nearing completion of the Puma navigation upgrade contract awarded to it in February 1993. This contract consists of the installation of the Supertans system and a new GPS receiver to offer P(Y) code rather than the C/A (coarse acquisition) code together with TACAN and VOR/ILS/MKR. Elements of the instrumentation are being upgraded to enhance instrument flight rules (IFR) and cleared flight operations along airways by installing dual EHSIs that use LCD technology. The system includes a GPS sensor as well as omnidirectional radio. The display is night vision goggle (NVG) compatible. Racal Avionics was also awarded a very significant contract in July 1995 to equip all UK Royal Navy front-line helicopters with a navigation/mission computer known as the control/display and navigation unit (CDNU), which hosts a P(Y) code GPS module. A total of 150 aircraft are covered by this contract including Sea King Mks 4/5/6 and Lynx HMA 8 aircraft. The Lynx HMA 8 CDNU is to be interfaced with a central tactical system (CTS). The Lynx HMA 8 upgrade itself is due to be completed by the year 2000. This programme covers 49 aircraft and is being carried out by Westland Helicopters and the Royal Navy at Fleetlands.

Westlands is also currently upgrading the Belgian fleet of five Sea King Mk 48, the Royal Norwegian Sea King Mk 43B and the Royal Australian Navy fleet of Sea King helicopters. In early 1995, the UK MoD issued an invitation to tender for the missions systems update (MSU), for the Royal Navy's fleet of 10 airborne early warning (AEW) to bring them up to AEW7 standard. These aircraft are due to re-enter service in the year 2000. Pakistan is currently upgrading six of its Sea King helicopters with GEC Avionics AQS 928G/206 dipping sonar system. Mann Aviation of the UK is manufacturing kits which will be installed by the Pakistani Navy at PN Base Mehran, Karachi. Pakistan hopes to have completed this programme by December 1996.

The Italian Army is in the process of upgrading its fleet of 60 A 129 Mangusta helicopters with the Marconi RALM 01 laser warning system, Lockheed Sanders AN/ALQ-144(V)E infra-red countermeasures set and NVG-compatible cockpit displays. In a separate and complementary contract, 15 Mangustas are to be fitted with the Martin Marietta/OTOBREDA M197 three-barrel 20 mm Gatling gun as well as provision for air-to-air Stingers.

The British Army order for 67 WAH-64D Apache helicopters, which itself is based on the US AH-64D Longbow Apache upgrade, has had a knock-on affect for the proposed AH-1(4B)W integrated weapons system (IWS) upgrade which was shelved after the UK MoD did not select the AH-1 Cobra Venom.

Apart from the AH-64D Longdow Apache upgrade, 227 of which are due to enter service in 1997, the USA is currently upgrading 380 OH-58A/C to Kiowa Warrior configuration for the US Army as well as continuing work on the AH-1W MLU for the US Marine Corps. This programme includes

McDonnell Douglas Helicopter Company AH-64D Longbow Apache upgrade launching a Hellfire missile *(McDonnell Douglas)*

First Lockheed L-1011-200 TriStar undergoing commercial freighter conversion at Marshall Aerospace facilities in Cambridge, UK *(Marshall Aerospace)*

installation of night targeting systems (NTS), INS and GPS, and four-blade rotor systems.

COMMERCIAL AIRCRAFT UPGRADE ACTIVITY

Re-engining

Upgrade activity in the field of aircraft re-engining, continues with companies such as Snow Aviation of Columbus, Ohio developing a programme to re-engine commercial L-100 Hercules with four Allison AE2100 turboprop engines. As well as the proposal to replace the three Pratt & Whitney JT8D turbofan engines of the Boeing 727-200 with two IAE V2500 turbofans. The Alenia subsidiary, Dee Howard completed its contract with United Parcel Service (UPS) to re-engine 44 Boeing 727-100 aircraft with Rolls Royce Tay 651 turbofans in August 1995. Professional Aviation of South Africa and Basler Turbo Conversions of the USA continue to re-engine the Douglas DC-3 with Pratt & Whitney Canada PT6A turboprop engines. Basler re-engines with PT6A-67 engines and Professional Aviation uses the PT6A-65.

Noise reduction

Commercial aircraft noise reduction programmes are offered by Nordam with its Boeing 737-200 high gross weight hush kit and its newly developed Boeing 737-200 low gross weight hush kit, which received its Federal Aviation Authority (FAA) Supplemental Type Certificated (STC) in November 1994. The Quiet Nacelle Corporation (QNC) plans to exploit the work it has performed in developing its hush kit for the US Air Force Boeing 707 'Open Skies' aircraft by offering the hush kit for commercial operation. QNC also expects to gain certification for its BAC One-Eleven hush kit in 1996. The company also commenced flight tests for DC-8-50 hush kit in mid-1995. Avacelle is still testing its Boeing 707 noise reduction nacelle and expects to commence production this year. A new company, Noises Off, is developing a noise reduction system which will cancel random jet engine noise at its source, by generating an identical new noise that is 180° out of phase with the original noise. This system is being designed to be applicable to all types of jet engined aircraft. A cabin noise reduction system has been developed by Ultra Electronics of the UK for retrofitting on Saab 340 aircraft. The package consists of installing loudspeakers inside the cabin trim. Once installed, the system can reduce cabin noise by up to 10 dB.

Cargo conversions

Hamilton Aviation has added the sliding door installation to its cargo conversion capabilities. The company has already delivered converted Boeing 727-200 and McDonnell Douglas DC-9-30 cargo conversions with a standard cargo door, 9 *g* bulkhead, freight handling system and class E compartment. The aircraft are also subject to heavy C level maintenance checks as well as the installation of GPS. The company announced its multiple STC for the sliding door cargo conversion in 1994. This will enable it to be fitted to both narrow and wide-body commercial jet aircraft. The first installation will be on a Boeing 727-200 followed by a McDonnell Douglas DC-10-30 aircraft. A major feature of the sliding door is that it can be opened for loading in all weather conditions. Marshall Aerospace of Cambridge, UK, has delivered its first commercial Lockheed L-1011 TriStar freighter conversion to American International Airways which has ordered a total of eight aircraft. The programme entails removal of passenger interior, strengthening the floor, installation of cargo door and fixed crash bulkhead. Daimler-Benz Aerospace Airbus have signed an agreement with Simera of South Africa for Simera to install freighter conversions on Airbus A300 and A310 aircraft. The work will be carried out at Jan Smuts airport using a complete kit and tooling supplied by Daimler-Benz Aerospace Airbus. The main features of the Simera programme are installation of a main deck cargo door, 9 *g* safety barrier net, fire detection system, smoke curtain, window blanking as well as reinforcing the floor structure.

CONCLUSION

MARKET TRENDS AND MARKET OPPORTUNITIES

The global upgrade market for military aircraft has entered a new phase. There is now a very mature market for systems enhancement, up-arming, life extension and role conversion of western-built combat aircraft. This market will continue to grow for the reasons mentioned in the introduction. Fighter aircraft, such as the F-4 Phantom II, F-5 Freedom Fighter/Tiger II, F-16 Fighting Falcon, Sepecat Jaguar, Mirage III/5/50 and F1 and Tornado will be operating for many years to come. Air forces will continue to install these tried and tested aircraft with the latest technology. There are already plans to equip hundreds of fighter aircraft built in the 1970s with the very latest avionics suites. There is now a proliferation of companies such as Rockwell in the USA, Sabca in Belgium, Smiths Industries in the UK, Sextant Avionique in France, DASA in Germany, IAI and Elbit in Israel as well as companies in Asia such as Singapore Aerospace, which have the capabilities to re-equip these ageing western aircraft with complete modular avionics suites including head-up displays, multifunction displays, radar, mission computers, weapons control computers, communications, VOR/ILS, TACAN, GPS, ECM, IFF and electro-optic equipment including FLIR and laser designators.

The next phase in the combat aircraft upgrade market has just begun. This is the development of the same modular suites for installation in the aircraft manufactured by the former Soviet Union. Upgrade packages exist in embryonic form as explained earlier and contracts have already been signed with established air forces. However, an area of particular interest, is those countries which have recently gained sovereign status as a consequence of the break-up of the Soviet Union. There are now a core of countries in central Europe, the Baltics and South-east Asia which, for a combination of reasons including lack of funds and realignment of economic/strategic dependence, will form air arms by mixing second-hand aircraft with new-build. It is very likely that these air forces will resemble the nascent Croatian Republic Air Force which has managed to build up a force of former Soviet equipment such as MiG-21s and Mi-8/17 helicopters, despite the arms embargo. It is very likely that these new air forces will re-equip their aircraft with a mixture of western and Russian systems.

The commercial aviation market is still grappling with the next generation landing system. The decision to replace the existing instrument landing system (ILS) with a microwave landing system (MLS) or a developed global positioning system (GPS) will be the single most significant event in the commercial passenger aircraft retrofit market. The installation of traffic collision avoidance systems (TCAS) also represents a growth area, as does the development and installation of digital telephone systems on board passenger aircraft.

Acknowledgements

The compilation of this third edition of *Jane's Aircraft Upgrades* is the result of a large group of people who have advised, supported and encouraged me throughout the continued development of this title. I would especially like to thank, Mike Gething, editor of *Jane's Defence Systems Modernisation* who has liaised closely with me and has generously shared his knowledge and expertise. Paul Beaver has, once again supplied me with invaluable advice and information. Paul Portnoi and Danny Pratt of *Jane's World Airlines* supplied me with the fleet lists which appear in the prelim pages of the book. The efforts of Peter Cooper, Peter March, Peter Steinemann and Paul Tompkins who have supplied many photographs are also greatly appreciated.

The efforts of all these people however, would have come to nothing without the dedication and professionalism of the in-house production team headed by Ruth Simmance who is responsible for the production cycle, ably assisted by Sulann Staniford, Jane Stimson, Marc Firmager and Rebecca Irons who organised the pictures, read the proofs and kept everything under control. Sarah Erskine, Kathryn Jones, Carol Offer and Christine Varndell were responsible for typesetting the book and Dave Ward's team of Jack Brenchley and Judy Kent were responsible for the page layout. Chrissie Richards liaised with the in-house production team and has been a great help in putting the book together. Finally thanks must go to Robert Hutchinson, the publisher, for his continued support and to Ruth Jowett, the Managing Editor of the Aviation/Space Group.

MO
707

New Entries in this Edition

Country	Manufacturer	Entry
Belgium	Sabca	Sabca (Northrop) F-5E/F Upgrade Programme
Belgium	Sabena	Sabena Technics (Lockheed) C-130 Life Extension Programme
Belgium	Sabena	Sabena Technics (Lockheed) C-130H
Belgium	Sabena	Sabena Technics (Mcdonnell Douglas) DC-10-30 Cargo Conversion
Belgium	Sabena	Sabena Technics (Swearingen/Fairchild) Merlin IIIA Cabin Refurbishment
Belgium	Sabena	Sabena Technics SA
Brazil	Neiva	Neiva N621 Universal
Canada	Canadair	Canadair Challenger
Canada	Conair	Conair (Cessna) Caravan C208B
Canada	De Havilland	DHC-8 Dash 8M
Chile	Enaer	Enaer T-35 Pillán (Devil)
Czech Republic	Zlin	Moravan Inc
Czech Republic	Zlin	Zlin 42 M
Czech Republic	Zlin	Zlin 50
Finland	Valmet	Valmet Aviation Industries Inc
Finland	Valmet	Valmet L-70 Miltrainer
France	Aerospatiale	Aerospatiale (Fouga) CM 170 Magister and Super Magister
France	Aerospatiale	Aerospatiale (Fouga) CM 175 Zephyr
France	Aerospatiale	Aerospatiale N 262 and Frégate
France	Dassault	Dassault Mirage 2000 Upgrades
France	Mudry	Avions Mudry et Cie
France	Mudry	Mudry CAP 21
France	Sagem	Elkan Upgrade Programme
Greece	Hai	Hellenic Aerospace Industry Ltd
Indonesia	IPTN	IPTN/Boeing (Boeing) B737-200 Surveiller Upgrade
International	EHI	EH Industries EH 101 Upgrade
International	EHI	EH Industries Limited
International	Panavia	Tornado Production
International	Transall	Transall C-160 Gabriel
International	Transall	Transall C-160H Astarte
Israel	Elbit	(Mcdonnell Douglas) F-4 Phantom 2000 Upgrade
Israel	Elbit	Elbit (Aero Vodochody) L-39 Upgrade
Israel	Elbit	Elbit (Mikoyan) MiG-21 Upgrade
Israel	Elbit	Elbit (Northrop) F-5 Upgrade
Israel	Elbit	Elbit Ltd
Israel	IAI	IAI (Mikoyan) Cambodian Air Force MiG-21BIS Upgrade
Israel	IAI	Lahav Division
Italy	Aerea	Aerea SpA
Italy	IAI	Aerea (Mikoyan) MiG-21 Universal Stores Carriage Systems Upgrade
Japan	Fuji	Fuji T-5
Japan	Kawasaki	Kawasaki C-1
Netherlands	Fokker	FAS (Boeing) 737 Modification Capabilities
Netherlands	Fokker	FAS (Westland) Lynx Avionics Upgrade Evaluation
New Zealand	RNZAF	RNZAF (Lockheed) P-3K Orion Project Rigel Upgrade
Russia	Ilyushin	Ilyushin Il-86 Command Post
Russia	Ilyushin	Ilyushin Il-86
Russia	MiG	Mikoyan MiG-29 Export Upgrade Options
Russia	Sokol	Sokol Government-Owned Aircraft Plant
Russia	Sokol	Sokol/Mikoyan (Mikoyan) MiG-21-93 Upgrade
Russia	Sukhoi	Sukhoi Su-22M-5 Upgrade
Russia	Tupolev	Tupolev Tu-160SC
South Africa	Aerosud	Aerosud/SAAF Generic Avionics System
South Africa	Hunting	Hunting Aviation (Pty) Limited
Sweden	SAAB-Scania	SAAB 340B AEW & C
Sweden	SAAB-Scania	SAAB AJS 37 Viggen (Thunderbolt)
Sweden	SAAB-Scania	SAAB SK60 (SAAB 105) Re-Engine Programme
Switzerland	Pilatus	Pilatus PC-6 Turbo-Porter (Agricultural Versions)
UK	Base	Base/Honeywell (British Aerospace) F/A 2 Sea Harrier Navigation Upgrade
UK	Fairey Hydraulics Ltd	Fairey (Lockheed) L-1011 Tristar Cargo Door Actuator
UK	FLS Aerospace	FLS (Bae/Hawker) 125 Dominie Overhaul and Conversion
UK	FLS Aerospace	FLS (Boeing 747) Pylon Retrofit
UK	FLS Aerospace	FLS Exterior Recoating
UK	FLS Aerospace	FLS Interior Refurbishment
UK	FRA	FR Aviation (BAE/BAC) VC10 C. MK 1(K) Tanker Conversion
UK	FRA	FR Aviation (BAE/English Electric) Canberra Life-Extension Programme
UK	FRA	FR Aviation (Dassault) Falcon 20 Stores Carriage Upgrade
UK	FRA	FR Aviation (North American) F-100D Super Sabre Life-extension
UK	Jetstream Aircraft Ltd	Jetstream 31
UK	Jetstream Aircraft Ltd	Jetstream Aircraft Ltd (Wholly owned subsidiary of British Aerospace plc)
UK	Mann	Mann (Westland) Sea King Upgrade
UK	Mann	Mann Aviation Flight Systems Ltd
UK	Marshall Aerospace	Marshall (Lockheed) C-130K Refurbishment Study
UK	Pilatus Britten-Norman	Pilatus Britten-Norman BN-2B (AL.1) Upgrade
UK	Racal	Racal (BAe/Hawker Dominie T. Mk 2 Navigation Upgrade
UK	Racal	Racal Avionics (Aerospatiale) SA 330 Puma Navigation Update (PNU)
UK	Racal	Racal Avionics Ltd (Part of the Racal Electronics Group)
UK	Shorts	Shorts S312 Tucano
UK	Smiths Industries Aerospace	Smiths Industries (Lockheed) C-130 Navigation Upgrade
UK	Ultra	Ultra (SAAB) 340 Active Noise Control Retrofit System
USA	Bell	Bell 205
USA	Bell	Bell 47G
USA	Boeing	Boeing (Federal Express) 727RE 'Quiet 727'
USA	Boeing	Boeing 707 Military Conversions
USA	Boeing	Boeing 747-400 Performance Improvement Package (PIP)
USA	Cessna	Cessna 305A Bird Dog
USA	Chrysler	CTAS (Boeing) E-6B Command Post Modification Programme
USA	E-Systems	E-Systems Flight Calibration Conversion
USA	E-Systems	E-Systems Greenville Division
USA	Hamilton Aviation	Hamilton Aviation Commercial Jet Cargo Door Conversion
USA	Hamilton Aviation	Hamilton Aviation Sliding Door Installation
USA	Isolair	Isolair (Mil) Terminator II
USA	Lockheed Martin	RNZAF/Lockheed (Lockheed) P-3K Orion Kestrel Relife Programme
USA	Mcdonnell Douglas	Mcdonnell Douglas Apache Upgrades
USA	Nordam	Nordam (Boeing) 737-200 Low Gross Weight Hush Kit
USA	Northrop Grumman	Northrop F-5 Structural Update Programme (SUP)
USA	Northrop Grumman	Northrop Grumman F-5E Tiger IV Advanced Avionics Demonstrator
USA	Rockwell	Rockwell (North American) T-2 Buckeye
USA	Rockwell	Rockwell/Boeing B-1B Lancer Conventional Weapons Upgrade
USA	Snow	Snow Aviation (Boeing) 727-200 Re-Engine Programme
USA	Snow	Snow Aviation Modern Technology Cockpit MTC
USA	Soloy	Soloy Dual Pac (Cessna) 208B Grand Carvan
USA	Soloy	Soloy (Enaer) T-35DT Turbo Pillán

Glossary of aerospace terms in this book

AAM Air-to-air missile.
AATH Automatic approach to hover.
AC Alternating current.
ACARS Automatic communications and recording system.
ACCP Automatic configuration control processor.
ACE Actuator control electronics.
ADI Attitude/director indicator.
ACLS (1) Automatic carrier landing system; (2) Air cushion landing system.
ACMI Air combat manoeuvring instrumentation.
ACN Aircraft classification number (ICAO system for aircraft pavements).
ADAC Avion de décollage et atterrissage court (STOL).
ADAV Avion de décollage et atterrissage vertical (VTOL).
ADC (1) US Air Force Aerospace Defense Command (no longer active); (2) air data computer.
ADF Medium frequency automatic direction-finding (equipment).
ADG Accessory-drive generator.
ADI Attitude/director indicator.
aeroplane (N America, airplane) Heavier-than-air aircraft with propulsion and a wing that does not rotate in order to generate lift.
AEW Airborne early warning.
AFB Air Force Base (USA).
AFCS Automatic flight control system.
AFRP Aramid fibre-reinforced plastics.
afterburning Temporarily augmenting the thrust of a turbofan or turbojet by burning additional fuel in the jetpipe.
AGM Air-to-ground missile.
AGREE Advisory Group on Reliability in Electronic Equipment.
Ah Ampère-hours.
AHRS Attitude/heading reference system.
AIDS Airborne integrated data system.
aircraft All man-made vehicles for off-surface navigation within the atmosphere, including helicopters and balloons.
AIS Advanced instrumentation subsystem.
airstair Retractable stairway built into aircraft.
ALCM Air-launched cruise missile.
AM Amplitude modulation.
AMAD Airframe-mounted accessory drive.
anhedral Downward slope of wing seen from front, in direction from root to tip.
ANVIS Aviator's night vision system.
AOA Angle of attack (see 'attack' below).
AP Ammonium perchlorate.
APFD Autopilot flight director.
approach noise Measured 1 n mile from downwind end of runway with aircraft passing overhead at 113 m (370 ft).
APS Aircraft prepared for service; a fully equipped and crewed aircraft without usable fuel and payload.
APU Auxiliary power unit (part of aircraft).
ARINC Aeronautical Radio Inc, US company whose electronic box sizes (racking sizes) are the international standard.
ARV Air recreational vehicle or air reconnaissance vehicle, according to context.
ASE (1) Automatic stabilisation equipment; (2) Aircraft survivability equipment.
ASI Airspeed indicator.
ASIR Airspeed indicator reading.
ASM Air-to-surface missile.
aspect ratio Measure of wing (or other aerofoil) slenderness seen in plan view, usually defined as the square of the span divided by gross area.
ASPJ Advanced self-protection jammer.
AST Air Staff Target (UK).
ASV (1) Air-to-surface vessel; (2) Anti-surface vessel.
ASW Anti-submarine warfare.
ATA Air Transport Association of America.
ATC Air traffic control.
ATDS Airborne tactical data system.
ATHS Airborne target handover (US, handoff) system.
ATR Airline Transport Radio ARINC 404 black box racking standards.
attack, angle of (alpha) Angle at which airstream meets aerofoil (angle between mean chord and free-stream direction). Not to be confused with angle of incidence (which see).
augmented Boosted by afterburning.
autogyro Rotary-wing aircraft propelled by a propeller (or other thrusting device) and lifted by a freely running autorotating rotor.
AUW All-up weight (term meaning total weight of aircraft under defined conditions, or at a specific time during flight). Not to be confused with MTOGW (which see).
avionics Aviation electronics, such as communications radio, radars, navigation systems and computers.
AVLF Airborne very low frequency.
AWACS Airborne warning and control system (aircraft).

bar Non-SI unit of pressure adopted by this yearbook pending wider acceptance of Pa. 1 bar 10^5 Pa. ISA pressure at S/L is 1,013.2 mb or just over 1 bar. ICAO has standardised hectopascal for atmospheric pressure, in which ISA S/L pressure is 101.32 hPa.
bare weight Undefined term meaning unequipped empty weight.
basic operating weight MTOGW minus payload (thus, including crew, fuel and oil, bar stocks, cutlery etc).
BCAR British Civil Airworthiness Requirements (see JAR).
bearingless rotor Rotor in which flapping, lead/lag and pitch change movements are provided by the flexibility of the structural material and not by bearings. No rotor is rigid.
Beta mode Propeller or rotor operating regime in which pilot has direct control of pitch.
birotative Having two components rotating in opposite directions.
BITE Built-in test equipment.
bladder tank Fuel (or other fluid) tank of flexible material.
BLC Boundary-layer control.
bleed air Hot high-pressure air extracted from gas-turbine engine compressor or combustor and taken through valves and pipes to perform useful work such as pressurisation, driving machinery or anti-icing by heating surfaces.
blisk Blade plus disc (of turbine engine) fabricated in one piece.
blown flap Flap across which bleed air is discharged at high (often supersonic) speed to prevent flow breakaway.
BOW Basic operating weight (which see).
BPR Bypass ratio.
BRW Brake release weight, maximum permitted weight at start of T-O run.
BTU Non-SI unit of energy (British Thermal Unit) 0.9478 J.
bulk cargo All cargo not packed in containers or on pallets.
bus Busbar, main terminal in electrical system to which battery or generator power is supplied.
BVR Beyond visual range.
bypass ratio Airflow through fan duct (not passing through core) divided by air flow through core.

C^3 Command, control and communications.
C^3CM Command, control, communications and countermeasures.
CAA Civil Aviation Authority (UK).
cabin altitude Height above S/L at which ambient pressure is same as inside cabin.
CAD/CAM Computer-assisted design/computer-assisted manufacture.
CAM Cockpit-angle measure (crew field of view).
canards Foreplanes, fixed or controllable aerodynamic surfaces ahead of CG.
CAN 5 Committee on Aircraft Noise (ICAO) rules for new designs of aircraft.
CAR Civil Airworthiness Regulations.
carbonfibre Fine filament of carbon/graphite used as strength element in composites.
CAS (1) Calibrated airspeed, ASI calibrated to allow for air compressibility according to ISA S/L; (2) close air support; (3) Chief of the Air Staff (also several other aerospace meanings).
CBR California bearing ratio, measure of ability of airfield surface (paved or not) to support aircraft.
CBU Cluster bomb unit.
CCV Control configured vehicle.
CEAM Centre d'Expériences Aériennes Militaires.
CEAT Centre d'Essais Aéronautiques de Toulouse.
CEP Circular error probability (50/50 chance of hit being inside or outside) in bombing, missile attack or gunnery.
CEV Centre d'Essais en Vol.
CFRP Carbonfibre-reinforced plastics.
CG Centre of gravity.
chaff Thin slivers of radar-reflective material cut to length appropriate to wavelengths of hostile radars and scattered in clouds to protect friendly aircraft.
chord Distance from leading-edge to trailing-edge measured parallel to longitudinal axis.
CKD Component knocked down, for assembly elsewhere.
clean (1) In flight configuration with landing gear, flaps, slats etc retracted; (2) Without any optional external stores.
c/n Construction (or constructor's) number.
COINS Computer operated instrument system.
combi Civil aircraft carrying both freight and passengers on main deck.
comint communications intelligence.
composite material Made of two constituents, such as filaments or short whiskers plus adhesive forming binding matrix.
CONUS Continental USA (ie, excluding Hawaii, etc).
convertible Transport aircraft able to be equipped to carry passengers or cargo.
core Gas generator portion of turbofan comprising compressor(s), combustion chamber and turbine(s).
CPU Central processing unit.
C/R Counter-rotating (propellers).
CRT Cathode-ray tube.
CSAS Command and stability augmentation system (part of AFCS).
CSD Constant-speed drive (output shaft speed held steady, no matter how input may vary).
CSRL Common strategic rotary launcher (for air-launched missiles of various types).
CTD Colour tactical display.

DADC Digital air data computer.
DADS Digital air data system.
daN Decanewtons (Newtons force × 10); thus, torque measured in daN-metres.
DARPA Defense Advanced Research Projects Agency.
databus Electronic highway for passing digital data between aircraft sensors and system processors, usually MIL-STD-1553B or ARINC 419 (one-way) and 619 (two-way) systems.
dB Decibel.
DC Direct current.
DECU Digital engine control unit.
derated Engine restricted to power less than potential maximum (usually such engine is flat rated, which see).
design weight Different authorities have different definitions; weight chosen as typical of mission but usually much less than MTOGW.
DF Direction-finder, or direction-finding.
DGAC Direction Générale à l'Aviation Civile.
dibber bomb Designed to cause maximum damage to concrete runways.
dihedral Upward slope of wing seen from front, in direction from root to tip.
DINS Digital inertial navigation system.
disposable load Sum of masses that can be loaded or unloaded, including payload, crew, usable fuel etc; MTOGW minus OWE.
DLC Direct lift control.
DME UHF distance-measuring equipment; gives slant distance to a beacon; DME element of Tacan.
dog-tooth A step in the leading-edge of a plane resulting from an increase in chord (see also saw-tooth).
Doppler Short for Doppler radar — radar using fact that received frequency is a function of relative velocity between transmitter or reflecting surface and receiver; used for measuring speed over ground or for detecting aircraft or moving vehicles against static ground or sea.
double-slotted flap One having an auxiliary aerofoil ahead of main surface to increase maximum lift.
dP Maximum design differential pressure between pressurised cabin and ambient (outside atmosphere).
DRA Defence Research Agency, Farnborough and Bedford.
drone Pilotless aircraft, usually winged, following preset programme of manoeuvres.
DS Directionally solidified.

EAA Experimental Aircraft Association (divided into local branches called Chapters).
EAS Equivalent airspeed, RAS minus correction for compressibility.
EB welding Electron beam wealding.
ECCM Electronic counter-countermeasures.
ECM Electronic countermeasures.
EFIS Electronic flight instrument(ation) system, in which large multifunction CRT displays replace traditional instruments.
EGT Exhaust gas temperature.
ehp Equivalent horsepower, measure of propulsive power of turboprop made up of shp plus addition due to residual thrust from jet.
EICAS Engine indication (and) crew alerting system.
EIS Entry into service.
ekW Equivalent kilowatts, SI measure of propulsive power of turboprop (see ehp).
elevon Wing trailing-edge control surface combining functions of aileron and elevator.
ELF Extreme low frequency.
elint electronics intelligence.
ELT Emergency locator transmitter, to help rescuers home on to a disabled or crashed aircraft.
EMP Electromagnetic pulse of nuclear or electronic origin.
EO Electro-optical.
EPA Environmental Protection Agency.

EPDS Electrical power distribution system.
EPNdB Effective perceived noise decibel, SI unit of EPNL.
EPNL Effective perceived noise level, measure of noise effect on humans which takes account of sound intensity, frequency, character and duration, and response of human ear.
EPU Emergency power unit (part of aircraft, not used for propulsion).
ESA European Space Agency.
ESM (1) Electronic surveillance (or support) measures; (2) Electronic signal monitoring.
ETOPS Extended-range twin (engine) operations, routing not more than a given flight time (120 or 180 minutes) from a usable alternative airfield.
EVA Extra-vehicular activity, ie outside spacecraft.
EW Electronic warfare.
EWSM Early warning support measures.

FAA Federal Aviation Administration.
factored Multiplied by an agreed number to take account of extreme adverse conditions, errors, design deficiencies or other inaccuracies.
FADEC Full authority digital engine (or electronic) control.
FADS Flexible air data system.
FAI Fédération Aéronautique Internationale.
fail-operational System which continues to function after any single fault has occurred.
fail-safe Structure or system which survives failure (in case of system, may no longer function normally).
FAR Federal Aviation Regulations.
FAR Pt 23 Defines the airworthiness of private and air taxi aeroplanes of 5,670 kg (12,500 lb) MTOGW and below.
FAR Pt 25 Defines the airworthiness of public transport aeroplanes exceeding 5,670 kg (12,500 lb) MTOGW.
FBL Fly-by-light (which see).
FBW Fly-by-wire (which see).
FDS Flight director system.
feathering Setting propeller or similar blades at pitch aligned with slipstream to give resultant torque (not tending to turn shaft) and thus minimum drag.
FEL Fibre elastomeric rotor head.
fence A chordwise projection on the surface of a wing, used to modify the distribution of pressure.
fenestron Helicopter tail rotor with many slender blades rotating in short duct.
ferry range Extreme safe range with zero payload.
FFAR Folding-fin (or free-flight) aircraft rocket.
FFVV Fédération Française de Vol à Voile (French gliding authority).
field length Measure of distance needed to land and/or takeoff; many different measures for particular purposes, each precisely defined.
flaperon Wing trailing-edge surface combining functions of flap and aileron.
flat-four Engine having four horizontally opposed cylinders; thus, flat-twin, flat-six etc.
flat rated Propulsion engine capable of giving full thrustor power for take-off to high airfield height and/or high ambient temperature (thus, probably derated at S/L).
FLIR Forward-looking infra-red.
FLOT Forward line of own troops.
fly-by-light Flight control system in which signals pass between computers and actuators along fibre optic leads.
fly-by-wire Flight control system with electrical signalling (ie, without mechanical interconnection between cockpit flying controls and control surfaces).
FM Frequency modulation.
FMCS Flight management computer system.
FMS (1) Foreign military sales (US DoD); (2) Flight management system.
FOL Forward operating location.
footprint (1) A precisely delineated boundary on the surface, inside which the perceived noise of an aircraft exceeds a specified level during take-off and/or landing; (2) Dispersion of weapon or submunition impact points.
Fowler flap Moves initially aft to increase wing area and then also deflects down to increase drag.
FQIS Fuel quantity indicating system.
free turbine Turbine mechanically independent of engine upstream, other than being connected by rotating bearings and the gas stream, and thus able to run at its own speed.
frequency agile (frequency hopping) Making a transmission harder to detect by switching automatically to a succession of frequencies.
FSCP Flap and slat control processor.
FSD Full-scale development.
FSW Forward-swept wing.
FY Fiscal year; 1 October to 30 September in USA government affairs (1 October 1992 starts FY 1993).

g Acceleration due to mean Earth gravity, ie of a body in free-fall; or acceleration due to rapid change of direction of flight path.
gallons Non-SI measure; 1 Imp gallon (UK) 4.546 litres, 1 US gallon 3.785 litres.
GCI Ground-controlled interception.
GfK Glassfibre-reinforced plastics (German).
GFRP Glassfibre-reinforced plastics.
glide ratio Of a sailplane, distance travelled along track divided by height lost in still air.
glideslope Element giving descent path guidance in ILS.
glove (1) Fixed portion of wing inboard of variable sweep wing; (2) additional aerofoil profile added around normal wing for test purposes.
GPS Global Positioning System, US military/civil satellite-based precision navaid.
GPU Ground power unit (not part of aircraft).
GPWS Ground-proximity warning system.
green aircraft Aircraft flyable but unpainted, unfurnished and basically equipped.
gross wing area See wing area.
GS Glideslope, of ILS.
GSE Ground support equipment (such as special test gear, steps and servicing platforms).
GTS Gas-turbine starter (ie starter is miniature gas turbine).
gunship Helicopter designed for battlefield attack, normally with slim body carrying pilot and weapon operator only.

h Hour(s).
hardened Protected as far as possible against nuclear explosion.
hardpoint Reinforced part of aircraft to which external load can be attached, eg weapon or tank pylon.
HASC US House Armed Services Committee.
HBPR High-bypass ratio (engine).
hectopascal (hPa) Unit of pressure, Pa × 100.
HDD Head-down display.
helicopter Rotary-wing aircraft both lifted and propelled by one or more power-driven rotors turning about substantially vertical axes.
HF High frequency.
HMD Helmet-mounted display; hence HMS sight.
Hocac Hands on cyclic and collective.
homebuilt Experimental category aircraft built/assembled from plans or kits.
hot and high Adverse combination of airfield height and high ambient temperature, which lengthens required TOD.
Hotas Hands on throttle and stick.
hovering ceiling Ceiling of helicopter (corresponding to air density at which maximum rate of climb is zero), either IGE or OGE.
HP High pressure (HPC, compressor; HPT, turbine).
hp Horsepower.
HSDB High-speed databus.
HSI Horizontal situation indicator.
HUD Head-up display (bright numbers and symbols projected on pilot's aiming sight glass and focused on infinity so that pilot can simultaneously read display and look ahead).
HUDWAC Head-up display and weapon aiming computer.
HUNTER Head-up navigation and targeting equipment for retrofit.
HVAR High-velocity aircraft rocket.
Hz Hertz, cycles per second.

IAMS Integrated armament management system.
IAS Indicated airspeed, ASIR corrected for instrument error.
IATA International Air Transport Association.
ICAO International Civil Aviation Organisation.
ICDU Intelligent control and display unit.
ICNIA Integrated communications, navigation and identification avionics.
IDG Integrated-drive generator.

IFF Identification friend or foe.
IFR Instrument flight rules (ie, not VFR).
IGE In ground effect: helicopter performance with theoretical flat horizontal surface just below it (eg mountain).
ILS Instrument landing system.
IMC Instrument meteorological conditions. Meteorological conditions too poor for pilot to fly without reference to blind-flying instruments.
IMK Increased manoeuvrability kit.
Imperial gallon 1.20095 US gallons; 4.546 litres.
IMS Integrated multiplex system.
INAS Integrated nav/attack system.
incidence Strictly, the angle at which the wing is set in relation to the fore/aft axis. Wrongly used to mean angle of attack (which see).
inertial navigation Measuring all accelerations imparted to a vehicle and, by integrating these with respect to time, calculating speed at every instant (in all three planes) and by integrating a second time calculating total change of position in relation to starting point.
INEWS Integrated electronic warfare system.
INS Inertial navigation system.
integral construction Machined from solid instead of assembled from separate parts.
integral tank Fuel (or other liquid) tank formed by sealing part of structure.
intercom Wired telephone system for communication within aircraft.
INTPS Integrated navigation and tactical plotting system.
inverter Electric or electronic device for inverting (reversing polarity of) alternate waves in AC power to produce DC.
IOC Initial operational capability.
IP (1) Intermediate pressure; (2) initial point in attack manoeuvre.
IPC, IPT Intermediate-pressure compressor, turbine.
IR Infra-red.
IRAN Inspect and repair as necessary.
IRLS Infra-red linescan (builds TV-type picture showing cool regions as dark and hot regions as light).
IRS Inertial reference system.
IRST Infra-red search and track.
ISA International Standard Atmosphere.
ISIS (1 Boeing Vertol) Integral spar inspection system; (2 Ferranti) integrated strike and interception sight.
ITE Involute throat and exit (rocket nozzle).
IVSI Instantaneous VSI.

J Joules, SI unit of energy.
JAR Joint Airworthiness Requirements, agreed by all major EC countries (JAR.25 equivalent to FAR.25).
JASDF Japan Air Self-Defence Force.
JATO Jet-assisted take-off (actually means rocket-assisted).
JCAB Japan Civil Airworthiness Board.
JDA Japan Defence Agency.
JGSDF Japan Ground Self-Defence Force.
JMSDF Japan Maritime Self-Defence Force.
joined wing Tandem wing layout in which forward and aft wings are swept so that the outer sections meet.
JPATS Joint Primary Aircraft Training System.
J-STARS US Air Force/Navy Joint Surveillance Target Attack Radar System in Boeing E-8A.
JTIDS Joint Tactical Information Distribution System.

kbit One thousand bits of memory.
Kevlar Aramid fibre used as basis of high-strength composite material.
km/h Kilometres per hour.
kN Kilonewtons (the Newton is the SI unit of force; 1 lbf 4.448 N).
knot 1 n mile per hour.
Krueger flap Hinges down and then forward from below the leading-edge.
kVA Kilovolt-amperes.
kW Kilowatt, SI measure of all forms of power (not just electrical).

LABS Low-altitude bombing system designed to throw the bomb up and forward (toss bombing).
LANTIRN Low-altitude navigation and targeting infra-red, night.
LARC Low-altitude ride control.
LBA Luftfahrtbundesamt (German civil aviation authority).
lbf Pounds of thrust.
LCD Liquid crystal display, used for showing instrument information.
LCN Load classification number, measure of 'flotation' of aircraft landing gear linking aircraft weight, weight distribution, tyre numbers, pressures and disposition.
LED Light-emitting diode.
Lidar Light detection and ranging (laser counterpart of radar).
LITVC Liquid-injection thrust vector control.
LLTV Low-light TV (thus, LLLTV, low-light level).
load factor (1) percentage of maximum payload; (2) design factor for airframe.
LOC Localiser (which see).
localiser Element giving steering guidance in ILS.
loiter Flight for maximum endurance, such as supersonic fighter on patrol.
longerons Principal fore-and-aft structural members (eg in fuselage).
Loran Long-range navigation; family of hyperbolic navaids based on ground radio emissions, now mainly Loran C.
LOROP Long-range oblique photography.
low observables Materials and structures designed to reduce aircraft signatures of all kinds.
lox Liquid oxygen.
LP Low pressure (LPC, compressor; LPT, turbine).
LRMTS Laser ranger and marked-target seeker.
LRU Line-replaceable unit.

m Metre(s), SI unit of length.
MAC (1) US Air Force Military Airlift Command; (2) mean aerodynamic chord.
Mach number The ratio of the speed of a body to the speed of sound (1,116 ft; 340 m/s in air at 15°C) under the same ambient conditions.
MAD Magnetic anomaly detector.
Madar Maintenance analysis, detection and recording.
Madge Microwave aircraft digital guidance equipment.
marker, marker beacon Ground beacon giving position guidance in ILS.
mb Millibars, bar × 10^{-3}.
MBR Marker beacon receiver.
MEPU Monofuel emergency power unit.
METO Maximum except take-off.
MF Medium frequency.
MFCD Multifunction colour display.
MFD Multifunction (electronic) display.
mg Milligrammes, grammes × 10^{-3}.
MLS Microwave landing system.
MLW Maximum landing weight.
mm Millimetres, metres × 10^{-3}.
MMH Monomethyl hydrazine.
MMO Maximum operating Mach number.
MMS Mast-mounted sight.
MNPS Minimum navigation performance specification.
Mo Maximum permitted operating Mach number.
monocoque Structure with strength in outer shell, devoid of internal bracing (semi-monocoque, with some internal supporting structure).
MoU Memorandum of understanding.
MPA Man-powered aircraft.
MPCD Multipurpose colour display.
MPD Multipurpose display.
mph Miles per hour.
MRW Maximum ramp weight.
MSIP US armed forces multistaged improvement programme.
MTBF Mean time between failures.
MTBR Mean time between removals.
MTI Moving-target indication (radar).
MTOGW Maximum take-off gross weight (MRW minus taxi/run-up fuel).
MTTR Mean time to repair.
MZFW Maximum zero-fuel weight.

N Newton, SI unit of force (4.448 lb f).
NACA US National Advisory Committee for Aeronautics (now NASA).
Nadge NATO air defence ground environment.
NAS US Naval Air Station.
NASA National Aeronautics and Space Administration.
NASC US Naval Air Systems Command (also several other aerospace meanings).
NATC US Naval Air Training Command or Test Center (also several other aerospace meanings).
NBAA US National Business Aircraft Association.
NDB Non-directional beacon.
NDT Non-destructive testing.
NGV Nozzle guide vane.
NH_4ClO_4 Ammonium perchlorate.
n mile nautical mile, 1.8532 km, 1.15152 miles.
NOAA US National Oceanic and Atmospheric Administration.
NOE Nap-of-the-Earth (low-flying in military aircraft, using natural cover of hills, trees etc).
NOS Night observation surveillance.
NO_x Generalised term for oxides of nitrogen.
Ns Newton-second (1 N thrust applied for 1 second).
NVG Night vision goggles.

OAT Outside air temperature.
OBOGS Onboard oxygen generating system.
OCU Operational Conversion Unit.
OEI One engine inoperative.
offset Workshare granted to a customer nation to offset the cost of an imported system.
OGE Out of ground effect; helicopter hovering, far above nearest surface.
Omega Long-range hyperbolic radio navaid.
OMI Omni-bearing magnetic indicator.
omni Generalised word meaning equal in all directions (as in omni-range, omni-flash beacon).
on condition maintenance According to condition rather than at fixed intervals.
Opeval Operational evaluation.
Optronics Combination of optics and electronics in viewing and sighting systems.
OSTIV Organisation Scientifique et Technique Internationale du Vol à voile (international gliding authority).
OTH Over-the-horizon.
OTPI On-top position indicator (indicates overhead of submarine in ASW).
OWE Operating weight empty. MTOGW minus payload, usable fuel and oil and other consumables.

PA system Public or passenger address.
pallet (1) for freight, rigid platform for handling by forklift or conveyor; (2) for missile, mounting and electronics box outside aircraft.
payload Disposable load generating revenue (passengers, cargo, mail and other paid items), in military aircraft loosely used to mean total load carried of weapons, cargo or other mission equipment.
PCU Pilot's control unit.
PD radar Pulse Doppler.
PDCS Performance data computer system.
penaids Penetration aids, such as jammers, chaff or decoys to help aircraft fly safely through hostile airspace.
PFA Popular Flying Association (UK).
PFCS Primary flight computer system.
phased-array Radar in which the beam is scanned electronically in one or both axes without moving the antenna.
PHI Position and heading (or homing) indicator.
plane A lifting surface (eg wing, tailplane).
plug door Door larger than its frame in pressurised fuselage, either opening inwards or arranged to retract parts prior to opening outwards.
plume The region of hot air and gas emitted by a helicopter jetpipe.
pneumatic de-icing Covered with flexible surfaces alternately pumped up and deflated to throw off ice.
port Left side, looking forward.
power-by-wire Using electric power alone (not electro-hydraulic) to drive control surfaces and perform other mechanical tasks.
power loading Aircraft weight (usually MTOGW) divided by total propulsive power or thrust at T-O.
prepreg Glassfibre cloth or rovings pre-impregnated with resin to simplify layup.
pressure fuelling Fuelling via a leakproof connection through which fuel passes at high rate under pressure.
pressure ratio In gas-turbine engine, compressor delivery pressure divided by ambient pressure (in supersonic aircraft, divided by ram pressure downstream of inlet).
primary flight controls Those used to control trajectory of aircraft (thus, not trimmers, tabs, flaps, slats, airbrakes or lift dumpers etc).
primary flight display Single screen bearing all data for aircraft flight path control.
propfan A family of new technology propellers characterised by multiple scimitar-shaped blades with thin sharp-edged profile. Single and contra-rotating examples promise to extend propeller efficiency up to about Mach 0.8. See also UDF.
pulse Doppler Radar sending out pulses and measuring frequency-shift to detect returns only from moving target(s) in background clutter.
pylon Structure linking aircraft to external load (engine nacelle, drop tank, bomb etc). Also used in conventional sense in pylon racing.

radius In terms of performance, the distance an aircraft can fly from base and return without intermediate landing.
RAE See DRA.
RAI Registro Aeronautico Italiano.
RAM Radar absorbent material.
ram pressure Increased pressure in forward-facing aircraft inlet, generated by converting (relative) kinetic energy to pressure.
ramp weight Maximum weight at start of flight (MTOGW plus taxi/run-up fuel).
range Too many definitions to list, but essentially the distance an aircraft can fly (or is permitted to fly) with specified load and usually whilst making allowance for specified additional manoeuvres (diversions, standoff, go-around etc).
RANSAC Range surveillance aircraft.
RAS Rectified airspeed, IAS corrected for position error.
raster Generation of large-area display, eg TV screen, by close-spaced horizontal lines scanned either alternately or in sequence.
RAT Ram air turbine.
redundant Provided with spare capacity or data channels and thus made to survive failures.
refanned Gas-turbine engine fitted with new fan of higher BPR.

RFP Request(s) for proposals.
rigid rotor (see bearingless rotor).
RLD Rijksluchtvaartdienst. Netherlands civil aviation department.
RMI Radio magnetic indicator; combines compass and navaid bearings.
R/Nav Calculates position, distance and time from groups of airways beacons.
RON Research octane number.
rotor-kite Rotary-wing aircraft with no internal power, lifted by a freely running autorotating rotor and towed by an external vehicle.
roving Multiple strands of fibre, as in a rope (but usually not twisted).
RPV Remotely piloted vehicle (pilot in other aircraft or on ground).
RSA Réseau du Sport de l'Air.
ruddervators Flying control surfaces, usually a V tail, that control both yaw and pitch attitude.
RVR Runway visual range.

s Second(s).
SAC US Air Force Strategic Air Command.
safe-life A term denoting that a component has proved by testing that it can be expected to continue to function safely for a precisely defined period before replacement.
salmon (French saumon) Streamlined fairings, usually at wingtip of sailplane, serving same function as end plate and acting also as tip-skid.
SAR (1) Search and rescue; (2) synthetic aperture radar.
SAS Stability augmentation system.
SATS (1) Small airfield for tactical support; (2) Small Arms Target System.
saw-tooth Same as dog-tooth.
SCAS Stability and control augmentation system.
SCNS Self-contained navigation system.
second-source Production of identical item by second factory or company.
semi-active Homing on to radiation reflected from target illuminated by radar or laser energy beamed from elsewhere.
SENSO ASW sensor operators.
service ceiling Usually height equivalent to air density at which maximum attainable rate of climb is 100 ft/min.
servo A device which acts as a relay, usually augmenting the pilot's efforts to move a control surface or the like.
SFAR Special Federal Aviation Regulation(s).
sfc Specific fuel consumption.
SGAC Secrétariat Général a l'Aviation Civile (now DGAC).
shaft Connection between gas-turbine and compressor or other driven unit. Two-shaft engine has second shaft, rotating at different speed, surrounding the first (thus, HP surrounds inner LP or fanshaft).
shipment One item or consignment delivered (by any means of transport) to customer.
Shoran Short-range navigation (radio).
shp Shaft horsepower, measure of power transmitted via rotating shaft.
sideline noise EPNdB measure of aircraft landing and taking off, at point 0.25 nm (2- or 3-engined) or 0.35 nm (4-engined) from runway centreline.
sidestick Control column in the form of a short handgrip beside the pilot.
SIF Selective identification facility.
sigint Signals intelligence.
signature Characteristic 'fingerprint' of all electromagnetic radiation (radar, IR etc).
single-shaft Gas turbine in which all compressors and turbines are on common shaft rotating together.
S/L Sea level.
SLAR Side-looking airborne radar.
snap-down Air-to-air interception of low-flying aircraft by AAM fired from fighter at a higher altitude.
soft target Not armoured or hardened.
specific fuel consumption Rate at which fuel is consumed divided by power or thrust developed, and thus a measure of engine efficiency. For jet engines (airbreathing, ie not rockets) unit is mg/Ns, milligrams per Newton-second; for shaft engines unit is μ/J, micrograms (millionths of a gram) per Joule (SI unit of work or energy).
specific impulse Measure of rocket engine efficiency; thrust divided by rate of fuel/oxidant consumption per second, the units for mass and force being the same so that the answer is expressed in seconds.
SPILS Stall protection and incidence-limiting system.
spool One complete axial compressor rotor; thus a two-shaft engine may have a fan plus an LP spool.
SSB Single-sideband (radio).
SSR Secondary surveillance radar.
SST Supersonic transport.
st Static thrust.
stabiliser Fin (thus, horizontal stabiliser tailplane).
stall strips Sharp-edged strips on wing leading-edge to induce stall at that point.
stalling speed Airspeed at which aircraft stalls at 1*g*, ie wing lift suddenly collapses.
standard day ISA temperature and pressure.
starboard Right side, looking forward.
static inverter Solid-state inverter of alternating wave-form (ie, not rotary machine) to produce DC from AC.
stick-pusher Stall-protection device that forces pilot's control column forward as stalling angle of attack is neared.
stick-shaker Stall-warning device that noisily shakes pilot's control column as stalling angle of attack is neared.
STOL Short take-off and landing. (Several definitions, stipulating allowable horizontal distance to clear screen height of 35 or 50 ft or various SI measures.)
store Object carried as part of payload on external attachment (eg bomb, drop tank).
strobe light High-intensity flashing beacon.
substrate The underlying layer on which something (such as a solar cell or integrated circuit) is built up.
supercritical wing Wing of relatively deep, flat-topped profile generating lift right across upper surface instead of concentrated close behind leading-edge.
sweepback Backwards inclination of wing or other aerofoil, seen from above, measured relative to fuselage or other reference axis, usually measured at quarter-chord (25%) or at leading-edge.
synchronous satellite Geostationary.

t Tonne, 1 Megagram, 1,000 kg.
tabbed flap Fitted with narrow-chord tab along entire trailing-edge which deflects to greater angle than main surface.
tabs Small auxiliary surfaces hinged to trailing-edge of control surfaces for purposes of trimming, reducing hinge moment (force needed to operate main surface) or in other way assisting pilot.
TAC US Air Force Tactical Air Command.
Tacan Tactical air navigation UHF military navaid giving bearing and distance to ground beacons; distance element (see DME) can be paired with civil VOR.
TACCO Tactical commander, ASW aircraft.
taileron Left and right tailplanes used as primary control surfaces in both pitch and roll.
tailplane Main horizontal tail surface, originally fixed and carrying hinged elevator(s) but today often a single 'slab' serving as control surface.
TANS Tactical air navigation system; Decca Navigator or Doppler-based computer, control and display unit.
TAS True airspeed, EAS corrected for density (often very large factor) appropriate to aircraft height.
TBO Time between overhauls.
t/c ratio Ratio of the thickness (aerodynamic depth) of awing or other surface to its chord, both measured at the same place parallel to the fore-and-aft axis.
TET Turbine entry temperature (of the gas); also turbine inlet temperature (TIT), inter-turbine temperature (ITT) and turbine gas temperature (TGT).
TFR Terrain-following radar (for low-level attack).
thickness Depth of wing or other aerofoil; maximum perpendicular distance between upper and lower surfaces.
tilt-rotor Aircraft with fixed wing and rotors that tilt up for hovering and forward for fast flight.
T-O Take-off.
TOD Take-off distance.
TOGW Take-off gross weight (not necessarily MTOGW).
T-O noise EPNdB measure of aircraft taking off, at point directly under flight path 3.5 nm from brakes-release (regardless of elevation).
ton Imperial (long) ton 1.016 t (Mg) or 2,240 lb, US (short) ton 0.9072 t or 2,000 lb.
track Distance between centres of contact areas of main landing wheels measured left/right across aircraft (with bogies, distance between centres of contact areas of each bogie).
transceiver Radio transmitter/receiver.
transponder Radio transmitter triggered automatically by a particular received signal as in civil secondary surveillance radar (SSR).
TRU Transformer rectifier unit.
TSFC Thrust specific fuel consumption of jet engine (turbojet, turbofan, ducted propfan or ramjet).
TSO Technical Standard Order (FAA).
turbofan Gas-turbine jet engine generating most thrust by a large-diameter cowled fan, with small part added by jet from core.
turbojet Simplest form of gas turbine comprising compressor, combustion chamber, turbine and propulsive nozzle.
turboprop Gas turbine in which as much energy as possible is taken from gas jet and used to drive reduction gearbox and propeller.
turboshaft Gas turbine in which as much energy as possible is taken from gas jet and used to drive high-speed shaft (which in turn drives external load such as helicopter transmission).
TVC Thrust vector control (rocket).
TWT Travelling-wave tube.
tyre sizes In simplest form, first figure is rim diameter (in or mm) and second is rim width (in or mm). In more correct three-unit form, first figure is outside diameter, second is maximum width and third is wheel diameter.

UAV Unmanned air vehicle.
UBE, Ubee Ultra bypass engine, alternative terminology (Boeing) for UDF.
UDF Unducted fan, one form of advanced propulsion system in which gas-turbine blading directly drives large fan (propfan) blades mounted around the outside of the engine pod. (GE registered abbreviation.)
UHF Ultra-high frequency.
unfactored Performance level expected of average pilot, in average aircraft, without additional safety factors.
upper surface blowing Turbofan exhaust vented over upper surface of wing to increase lift.
usable fuel Total mass of fuel consumable in flight, usually 95 to 98 per cent of system capacity.
useful load Usable fuel plus payload.
US gallon 0.83267 Imperial gallon; 3.785 litres.

variable geometry Capable of grossly changing shape in flight, especially by varying sweep of wings.
VCID Voice command interactive device.
V_D Maximum permitted diving speed.
VDU Video (or visual) display unit.
vernier Small thruster, usually a rocket, for final precise adjustment of a vehicle's trajectory and velocity.
VFR Visual flight rules.
VHF Very high frequency.
VLF Very low frequency (area-coverage navaid).
V_{MO} Maximum permitted operating flight speed (IAS, EAS or CAS must be specified).
VMS Vehicle management system.
V_{NE} Never-exceed speed (aerodynamic or structural limit).
VOR VHF omnidirectional range (VHF radio beacons providing bearing to or from beacon).
vortex generators Small blades attached to wing and tail surfaces to energise local airflow and improve control.
vortillon Short-chord fence ahead of and below leading-edge.
VSI Vertical speed (climb/descent) indicator.
VTOL Vertical take-off and landing.

washout Inbuilt twist of wing or rotor blade reducing angle of incidence towards the tip.
WDNS Weapon delivery and navigation system.
wheelbase Minimum distance from nosewheel or tailwheel (centre of contact area) to line joining mainwheels (centres of contact areas).
wing area Total projected area of clean wing (no flaps, slats etc) including all control surfaces and area of fuselage bounded by leading- and trailing-edges projected to centreline (inapplicable to slender-delta aircraft with extremely large leading-edge sweep angle). Described as gross wing area; net area excludes projected areas of fuselage, nacelles, etc.
wing loading Aircraft weight (usually MTOGW) divided by wing area.
winglet Small auxiliary aerofoil, usually sharply upturned and often sweptback, at tip of wing.
WIU Weapons interface unit.

zero-fuel weight MTOGW minus usable fuel and other consumables, in most aircraft imposing severest stress on wing and defining limit on payload.
zero/zero seat Ejection seat designed for use even at zero speed on ground.
ZFW Zero-fuel weight.
μg Microgrammes, grammes × 10^{-6}.

JANE'S WORLD AIRLINES FLEET LISTS

The following fleet lists of world airlines were extracted from the *Jane's World Airlines* database which is compiled by Paul Portnoi. The list includes information verified up to June 1995. Where there is a zero or no number in the third column, the aircraft are either on order/option, or the exact number in fleet is not known.

AIRCRAFT

The following list contains aircraft which are currently produced as well as aircraft which are out of production

Boeing 737-200 passenger aircraft

AIRFRAME	POWER PLANT	NO. IN FLEET	FLEET NAME
A300-600	GE CF6-80C2A3	11	LUFTHANSA GERMAN AIRLINES - DEUTSCHE LUFTHANSA AG
A300-600	P&W JT9D-7R4H1	11	SAUDIA - SAUDI ARABIAN AIRLINES
A300-600F	GE CF6-80C24F	6	FEDERAL EXPRESS CORPORATION
A300-600R	GE CF6-50C2A5	2	AIR AFRIQUE
A300-600R	GE CF6-50C2A5	1	AIR FINLANDIA
A300-600R	GE CF6-80C2A5	7	CHINA EASTERN AIRLINES
A300-600R	GE CF6-80C2A5	4	CHINA NORTHWEST AIRLINES
A300-600R	GE CF6-80C2A5	6	EMIRATES
A300-600R	GE CF6-80C2A5	0	IRAN AIR THE AIRLINE OF THE ISLAMIC REPUBLIC OF IRAN
A300-600R	GE CF6-80C2A5	5	KUWAIT AIRWAYS CORPORATION
A300-600R	GE CF6-80C2A5	3	OLYMPIC AIRWAYS S.A.
A300-600R	P&W PW4158	2	AIR LIBERTE
A300-600R	P&W PW4158	6	CHINA AIRLINES
A300-600R	P&W PW4158	3	CHINA NORTHERN AIRLINES
A300-600R	P&W PW4158	9	EGYPTAIR
A300-600R	P&W PW4158	10	GARUDA INDONESIA PT
A300-601	GE CF6-80C2A1	6	THAI AIRWAYS INTERNATIONAL
A300-605R	GE CF6-80C2A5	35	AMERICAN AIRLINES
A300-605R	GE CF6-80C2A5	4	MONARCH AIRLINES LTD
A300-605R	GE CF6-80C2A5	2	THAI AIRWAYS INTERNATIONAL
A300-622	P&W PW4158	5	KOREAN AIR
A300-622R	P&W PW4158	16	KOREAN AIR
A300-622R	P&W PW4158	8	THAI AIRWAYS INTERNATIONAL
A300B2-1	GE CF6-50C2R	13	AIR INTER
A300B2-101	GE CF6-50C	4	AIR FRANCE
A300B2-101	GE CF6-50C	8	INDIAN AIRLINES LTD
A300B2-200	GE CF6-50C2	2	ALITALIA
A300B2-203	GE CF6-50C2	5	IRAN AIR THE AIRLINE OF THE ISLAMIC REPUBLIC OF IRAN
A300B2-203	GE CF6-50C2	3	VASP - VIACAO AEREA SAO PAULO S.A.
A300B2-K3C	GE CF6-50C2R	1	AIR INTER
A300B2-K3C	GE CF6-50C2R	9	JAPAN AIR SYSTEM CO LTD (NIHON AIR SYSTEM)
A300B2-K3C	GE CF6-50C2R	4	SOUTH AFRICAN AIRWAYS (SAA)
A300B4	GE CF6-50	2	AIR OPS OF EUROPE
A300B4-100	GE CF6-50C	1	TRANSLIFT AIRWAYS
A300B4-100	GE CF6-50C2	4	ALITALIA
A300B4-100	P&W JT9D-59A	8	IBERIA - LINEAS AEREAS DE ESPANA, S.A.
A300B4-100	P&W JT9D-59A	3	PREMAIR A/S
A300B4-100	P&W JT9D-59A	2	ZAS AIRLINES OF EGYPT
A300B4-103	GE CF6-50C	6	CONTINENTAL AIRLINES
A300B4-103	GE CF6-50C2	8	KOREAN AIR
A300B4-103	GE CF6-50C2	6	OLYMPIC AIRWAYS S.A.
A300B4-103	GE CF6-50C2	2	PHILIPPINE AIR LINES, INC - PAL
A300B4-103	GE CF6-50C2	7	THAI AIRWAYS INTERNATIONAL
A300B4-200	GE CF6-50C2	8	ALITALIA
A300B4-200	GE CF6-50C2	1	EUROPEAN AIRLIFT
A300B4-200	GE CF6-50C2	1	MALAYSIA AIRLINES
A300B4-200	GE CF6-50C2	9	PAKISTAN INTERNATIONAL AIRLINES (PIA)
A300B4-200	P&W JT9D-59A	6	CHINA AIRLINES
A300B4-203	GE CF6-50C2	3	AIR AFRIQUE
A300B4-203	GE CF6-50C2	2	AIR CHARTER
A300B4-203	GE CF6-50C2	11	AIR FRANCE
A300B4-203	GE CF6-50C2	3	AIR INDIA
A300B4-203	GE CF6-50C2	5	AIR JAMAICA
A300B4-203	GE CF6-50C2	1	AIR SINAI
A300B4-203	GE CF6-50C2	15	CONTINENTAL AIRLINES
A300B4-203	GE CF6-50C2	5	EGYPTAIR
A300B4-203	GE CF6-50C2	2	FINNAIR OY
A300B4-203	GE CF6-50C2	1	GHANA AIRWAYS
A300B4-203	GE CF6-50C2	2	INDIAN AIRLINES LTD
A300B4-203	GE CF6-50C2	10	PHILIPPINE AIR LINES, INC - PAL
A300B4-203	GE CF6-50C2	4	QANTAS AIRWAYS LTD
A300B4-203	GE CF6-50C2	3	SEMPATI AIR TRANSPORT

AIRFRAME	POWER PLANT	NO. IN FLEET	FLEET NAME
A300B4-203	GE CF6-50C2	3	SOUTH AFRICAN AIRWAYS (SAA)
A300B4-203	GE CF6-50C2	2	SULTAN AIR
A300B4-203	GE CF6-50C2	5	THAI AIRWAYS INTERNATIONAL
A300B4-203	GE CF6-50C2	1	TUNIS AIR - SOCIETE TUNISIENNE D EL'AIR
A300B4-203	GE CF6-50C2	2	VIASA, VENEZOLANA INTERNACIONAL DE AVIACION SA
A300B4-220	P&W JT9D-59A	9	GARUDA INDONESIA PT
A300B4-2C	GE CF6-50C2R	4	AIR INTER
A300B4-2C	GE CF6-50C2R	8	JAPAN AIR SYSTEM CO LTD (NIHON AIR SYSTEM)
A300B4-622R	P&W PW4158	12	JAPAN AIR SYSTEM CO LTD (NIHON AIR SYSTEM)
A300B4K-100	GE CF6-50C2	1	EUROPEAN AIRLINES SA
A300B4K-103	GE CF6-50C2	1	FAUCETT PERU
A300B4K-200	GE CF6-50C2	2	EUROPEAN AIRLINES SA
A300C4-203	GE CF6-50C2	1	SOUTH AFRICAN AIRWAYS (SAA)
A300F4-203	GE CF6-50C2	2	KOREAN AIR
A310	P&W PW4152	2	AIR CLUB INTERNATIONAL
A310-200	GE CF6-80A3	7	AIR FRANCE
A310-200	GE CF6-80A3	4	FEDERAL EXPRESS CORPORATION
A310-200	GE CF6-80A3	10	KLM ROYAL DUTCH AIRLINES
A310-200	GE CF6-80A3	13	LUFTHANSA GERMAN AIRLINES - DEUTSCHE LUFTHANSA AG
A310-200	GE CF6-80A3	2	MIDDLE EAST AIRLINES - AIRLIBAN
A310-200	GE CF6-80A3	7	TURKISH AIRLINES, INC (TURK HAVA YOLLARI A.O)
A310-200	GE CF6-80C2A2	4	HAPAG-LLOYD FLUGGESELLSCHAFT.MBH
A310-200	GE CF6-80C2A2	2	THAI AIRWAYS INTERNATIONAL
A310-200	P&W JT9D-7R4	2	YEMEN AIRLINES
A310-200	P&W JT9D-7R4E1	7	DELTA AIR LINES, INC
A310-200	P&W JT9D-7R4E1	2	SABENA BELGIAN WORLD AIRLINES
A310-200	P&W JT9D-7R4E1	2	SILKAIR (SINGAPORE) PRIVATE LIMITED
A310-200	P&W JT9D-7R4E1	4	SINGAPORE AIRLINES LIMITED
A310-203	GE CF6-80A3	2	AIR ALGERIE
A310-203	GE CF6-80A3	3	CYPRUS AIRWAYS LTD
A310-203C	GE CF6-80A3	1	MARTINAIR HOLLAND NV
A310-204	GE CF6-80C2A2	1	CYPRUS AIRWAYS LTD
A310-222	P&W JT8D-7R4E1	3	CHINA NORTHWEST AIRLINES
A310-222	P&W JT9D-7R4E1	4	NIGERIA AIRWAYS
A310-222	P&W JT9D-7R4E1	2	QATAR AIRWAYS
A310-222	P&W JT9D-7R4H	5	KUWAIT AIRWAYS CORPORATION
A310-300	NYK	0	BIMAN BANGLADESH AIRLINES
A310-300	GE CF6-50C2A2	1	LAP - LINEAS AEREAS PARAGUAYAS
A310-300	GE CF6-80C2A2	5	AEROFLOT RUSSIAN INTERNATIONAL AIRLINES
A310-300	GE CF6-80C2A2	4	AIR AFRIQUE
A310-300	GE CF6-80C2A2	4	AIR FRANCE
A310-300	GE CF6-80C2A2	8	AIR INDIA
A310-300	GE CF6-80C2A2	5	CHINA EASTERN AIRLINES
A310-300	GE CF6-80C2A2	2	CSA - CESKOSLOVENSKE AEROLINIE
A310-300	GE CF6-80C2A2	10	EMIRATES
A310-300	GE CF6-80C2A2	3	HAPAG-LLOYD FLUGGESELLSCHAFT.MBH
A310-300	GE CF6-80C2A2	3	KENYA AIRWAYS LTD
A310-300	GE CF6-80C2A2	12	LUFTHANSA GERMAN AIRLINES - DEUTSCHE LUFTHANSA AG
A310-300	GE CF6-80C2A2	1	MIDDLE EAST AIRLINES - AIRLIBAN
A310-300	GE CF6-80C2A2	5	ROYAL JORDANIAN
A310-300	GE CF6-80C2A2	1	SUDAN AIRWAYS
A310-300	GE CF6-80C2A2	5	TAP AIR PORTUGAL
A310-300	GE CF6-80C2A2	7	TURKISH AIRLINES, INC (TURK HAVA YOLLARI A.O)
A310-300	GE CF6-80C2A2	1	YEMEN AIRLINES
A310-300	GE CF6-80C2A8	3	KUWAIT AIRWAYS CORPORATION
A310-300	GE CF6-80C2A8	2	SAETA
A310-300	P&W JT9D-7R4	0	YEMEN AIRLINES
A310-300	P&W JT9D-7R4E1	1	SABENA BELGIAN WORLD AIRLINES
A310-300	P&W PW4152	3	AEROCANCUN
A310-300	P&W PW4152	3	AEROLINEAS ARGENTINAS
A310-300	P&W PW4152	2	AIR NIUGINI
A310-300	P&W PW4152	4	AUSTRIAN AIRLINES, OSTERREICHISCHE LUFTVERKEHRS AG
A310-300	P&W PW4152	13	DELTA AIR LINES, INC
A310-300	P&W PW4152	0	ECUATORIANA
A310-300	P&W PW4152	0	IRAQI AIRWAYS
A310-300	P&W PW4152	1	JES AIR
A310-300	P&W PW4152	1	OASIS INTERNATIONAL AIRLINES
A310-300	P&W PW4152	17	SINGAPORE AIRLINES LIMITED
A310-300	P&W PW4152	1	VIETMAN AIRLINES
A310-300	P&W PW4156	4	BALAIRCTA
A310-300	P&W PW4156	2	TAROM
A310-300	P&W PW4158	1	AIR LIBERTE
A310-304	GE CF6-80C2A2	1	LLOYD AEREO BOLIVIANO SAM
A310-304	GE CF6-80C2A2	1	SOMALI AIRLINES
A310-307	GE CF6-80C2A8	6	PAKISTAN INTERNATIONAL AIRLINES (PIA)
A310-322	P&W JT9D-7R4E1	5	SWISSAIR (SOCIETE ANONYME SUISSE POUR LA NAVIGATION AERIENNE)
A319	CFMI CFM56-5A	0	LUFTHANSA GERMAN AIRLINES - DEUTSCHE LUFTHANSA AG
A319	CFMI CFM56-5B2	0	SWISSAIR (SOCIETE ANONYME SUISSE POUR LA NAVIGATION AERIENNE)
A319	CFMI CFM56-5B6	0	AIR CANADA
A319	CFMI CFM56-5B6	0	AIR INTER
A320	IAE V2500-A5	0	AERO LLOYD
A320-100	CFMI CFM56-5A1	6	AIR FRANCE
A320-100	CFMI CFM56-5A1	6	AIR INTER
A320-100	CFMI CFM56-5A1	5	BRITISH AIRWAYS
A320-100	CFMI CFM56-5A1	7	VIETMAN AIRLINES
A320-200	CFMI CFM56-5A1	34	AIR CANADA
A320-200	CFMI CFM56-5A1	19	AIR FRANCE
A320-200	CFMI CFM56-5A1	28	AIR INTER
A320-200	CFMI CFM56-5A1	2	AIR MALTA CO. LTD.
A320-200	CFMI CFM56-5A1	1	AIR NIPPON
A320-200	CFMI CFM56-5A1	18	ALL NIPPON AIRWAYS
A320-200	CFMI CFM56-5A1	14	ANSETT AUSTRALIA
A320-200	CFMI CFM56-5A1	5	BRITISH AIRWAYS
A320-200	CFMI CFM56-5A1	3	CALEDONIAN AIRWAYS
A320-200	CFMI CFM56-5A1	4	CANADA 3000 AIRLINES LIMITED
A320-200	CFMI CFM56-5A1	12	CANADIAN - CANADIEN
A320-200	CFMI CFM56-5A1	0	FIRST EUROPEAN AIRWAYS
A320-200	CFMI CFM56-5A1	14	GULF AIR
A320-200	CFMI CFM56-5A1	22	IBERIA - LINEAS AEREAS DE ESPANA, S.A.

AIRFRAME	POWER PLANT	NO. IN FLEET	FLEET NAME
A320-200	CFMI CFM56-5A1	3	KUWAIT AIRWAYS CORPORATION
A320-200	CFMI CFM56-5A1	33	LUFTHANSA GERMAN AIRLINES - DEUTSCHE LUFTHANSA AG
A320-200	CFMI CFM56-5A1	0	MIDWAY AIRLINES
A320-200	CFMI CFM56-5A1	50	NORTHWEST AIRLINES
A320-200	CFMI CFM56-5A1	4	ONUR AIR
A320-200	CFMI CFM56-5A1	2	ROYAL JORDANIAN
A320-200	CFMI CFM56-5A1	6	TAP AIR PORTUGAL
A320-200	CFMI CFM56-5A1	7	TUNIS AIR - SOCIETE TUNISIENNE D EL'AIR
A320-200	CFMI CFM56-5A1	4	VIETMAN AIRLINES
A320-200	CFMI CFM56-5A3	4	EXCALIBUR AIRWAYS LTD
A320-200	CFMI CFM56-5A3	6	LINEAS AEREAS COSTARRICENSES, S.A. - LACSA
A320-200	CFMI CFM56-5A3	7	MONARCH AIRLINES LTD
A320-200	CFMI CFM56-5B4	0	AUSTRIAN AIRLINES, OSTERREICHISCHE LUFTVERKEHRS AG
A320-200	CFMI CFM56-5B4	1	SWISSAIR (SOCIETE ANONYME SUISSE POUR LA NAVIGATION AERIENNE)
A320-200	IAE V2500-A1	3	ADRIA AIRWAYS
A320-200	IAE V2500-A1	4	AIR 2000 LTD
A320-200	IAE V2500-A1	2	AIR FOYLE LIMITED
A320-200	IAE V2500-A1	2	AIR LANKA
A320-200	IAE V2500-A1	3	AIRTOURS INTERNATIONAL AIRWAYS LTD
A320-200	IAE V2500-A1	16	AMERICA WEST AIRLINES, INC
A320-200	IAE V2500-A1	4	BALKAN
A320-200	IAE V2500-A1	2	CHINA AIRLINES
A320-200	IAE V2500-A1	8	CYPRUS AIRWAYS LTD
A320-200	IAE V2500-A1	7	DRAGONAIR (HONG KONG DRAGON AIRLINES LIMITED)
A320-200	IAE V2500-A1	7	EGYPTAIR
A320-200	IAE V2500-A1	3	EUROCYPRIA AIRLINES
A320-200	IAE V2500-A1	30	INDIAN AIRLINES LTD
A320-200	IAE V2500-A1	14	MEXICANA
A320-200	IAE V2500-A1	6	PREMAIR A/S
A320-200	IAE V2500-A1	5	REGION AIR
A320-200	IAE V2500-A1	7	SOUTH AFRICAN AIRWAYS (SAA)
A320-200	IAE V2500-A1	3	TRANSASIA AIRWAYS
A320-200	IAE V2500-A1	6	TRANSLIFT AIRWAYS
A320-200	IAE V2500-A5	2	AMERICA WEST AIRLINES, INC
A320-200	IAE V2527-A5	27	UNITED AIRLINES, INC
A321	CFMI CFM56-5B	0	AIR FRANCE
A321	IAE V2500-A5	0	AERO LLOYD
A321	IAE V2530-A5	0	TRANSASIA AIRWAYS
A321-100	(TBD)	0	ALL NIPPON AIRWAYS
A321-100	CFMI CFM56-5B	2	AIR INTER
A321-100	CFMI CFM56-5B	5	ALITALIA
A321-100	CFMI CFM56-5B	0	AUSTRIAN AIRLINES, OSTERREICHISCHE LUFTVERKEHRS AG
A321-100	CFMI CFM56-5B	0	KUWAIT AIRWAYS CORPORATION
A321-100	CFMI CFM56-5B	2	SWISSAIR (SOCIETE ANONYME SUISSE POUR LA NAVIGATION AERIENNE)
A321-100	IAE V2530-A5	10	LUFTHANSA GERMAN AIRLINES - DEUTSCHE LUFTHANSA AG
A330-300	GE CF6-80E1	3	AER LINGUS
A330-300	GE CF6-80E1	5	AIR INTER
A330-300	P&W PW4164	1	LTU - LUFTTRANSPORT UNTERNEHMEN GMBH
A330-300	P&W PW4164	0	NORTHWEST AIRLINES
A330-300	P&W PW4168	5	MALAYSIA AIRLINES
A330-300	RR Trent 768	1	DRAGONAIR (HONG KONG DRAGON AIRLINES LIMITED)
A330-300	RR Trent 768	0	GARUDA INDONESIA PT
A330-300	RR Trent 768	0	TRANS WORLD AIRLINES, INC
A330-300	RR Trent 772	3	CATHAY PACIFIC AIRWAYS LTD
A330-300	TBD	0	KOREAN AIR
A330-320	P&W PW4164	1	THAI AIRWAYS INTERNATIONAL
A340-200	CFMI CFM56-5C2	3	AIR FRANCE
A340-200	CFMI CFM56-5C2	2	AUSTRIAN AIRLINES, OSTERREICHISCHE LUFTVERKEHRS AG
A340-200	CFMI CFM56-5C2	4	CATHAY PACIFIC AIRWAYS LTD
A340-200	CFMI CFM56-5C2	0	IBERIA - LINEAS AEREAS DE ESPANA, S.A.
A340-200	CFMI CFM56-5C2	7	LUFTHANSA GERMAN AIRLINES - DEUTSCHE LUFTHANSA AG
A340-200	CFMI CFM56-5C2	0	PHILIPPINE AIR LINES, INC - PAL
A340-200	CFMI CFM56-5C2	4	SABENA BELGIAN WORLD AIRLINES
A340-300	CFMI CFM56-5C2	5	AIR FRANCE
A340-300	CFMI CFM56-5C2	0	AIR LANKA
A340-300	CFMI CFM56-5C2	0	CHINA EASTERN AIRLINES
A340-300	CFMI CFM56-5C2	0	CHINA SOUTHERN AIRLINES
A340-300	CFMI CFM56-5C2	4	GULF AIR
A340-300	CFMI CFM56-5C2	8	LUFTHANSA GERMAN AIRLINES - DEUTSCHE LUFTHANSA AG
A340-300	CFMI CFM56-5C2	3	TURKISH AIRLINES, INC (TURK HAVA YOLLARI A.O)
A340-300	CFMI CFM56-5C3	2	AIR MAURITIUS
A340-300	CFMI CFM56-5C3	4	TAP AIR PORTUGAL
A340-300	CFMI CFM56-5C3	4	VIRGIN ATLANTIC AIRWAYS
A340-300	CFMI CFM56-5C4	0	AIR CANADA
A340-300	CFMI CFM56-5C4	0	CATHAY PACIFIC AIRWAYS LTD
A340-300	CFMI CFM56-5C4	1	KUWAIT AIRWAYS CORPORATION
A340-300	CFMI CFM56-5C4	0	SINGAPORE AIRLINES LIMITED
A340-300	TBD	0	ALL NIPPON AIRWAYS
AIAe RJ100 Avroliner	Lyc LF507-1F	0	CROSSAIR
AIAe RJ100 Avroliner	Lyc LF507-1F	8	SAM - SOCIEDAD AERONAUTICA DE MEDELLIN
AIAe RJ100 Avroliner	Lyc LF507-1F	10	TURKISH AIRLINES, INC (TURK HAVA YOLLARI A.O)
AIAe RJ70 Avroliner	Lyc LF507-1F	4	AIR MALTA CO. LTD.
AIAe RJ70 Avroliner	Lyc LF507-1F	6	BUSINESS EXPRESS (THE DELTA CONNECTION)
AIAe RJ85 Avroliner	Lyc LF507-1F	4	CROSSAIR
AIAe RJ85 Avroliner	Lyc LF507-1F	7	LUFTHANSA CITYLINE
AIAe RJ85 Avroliner	Lyc LF507-1F	0	MALMO AVIATION
AIAe RJ85 Avroliner	Lyc LF507-1F	1	PELITA AIR SERVICE
AS (Nord) 262	TU Bastan VIC1	1	EQUATORIAL - INTERNATIONAL AIRLINES OF SAO TOME E PRINCIPE
AS (Sud) Caravelle 10B1R	P&W JT8D-7B	3	ISTANBUL AIRLINES - ISTAIR
AS (Sud) Caravelle 10B3	P&W JT8D-7	2	SYRIAN ARAB AIRLINES
AS (Sud) Caravelle 11R	P&W JT8D-7B	1	AEROSUCRE COLOMBIA, S.A.
AS 330J Puma	TU Turmo IVC	2	MYANMA AIRWAYS (BURMA AIRWAYS CORPORATION)
AS 332L Super Puma	TU Makila 1A	3	MAERSK AIR
AS 350B Ecureuil	TU Arriel 1B	4	ERA AVIATION, INC
AS 350B2 Ecureuil	TU Arriel 1D1	1	GROENLANDSFLY A/S
AS 365N2 Dauphin 2	TU Arriel 1C2	2	MAERSK AIR
AS SA316B Alouette III	TU Artouste III	1	SCHREINER AIRWAYS, BV
AS SA316B Alouette III	TU Artouste IIIB1	1	SCHREINER AIRWAYS, BV
AS SA365C2 Dauphin 2	TU Arriel 1A2	1	SCHREINER AIRWAYS, BV

AIRFRAME	POWER PLANT	NO. IN FLEET	FLEET NAME
AS SA365N Dauphin 2	TU Arriel 1C	4	SCHREINER AIRWAYS, BV
AS SE313B Lama	TU Artouste IIIB	2	AIR AFFAIRES GABON
AS/BAe Concorde 101	RR Olympus 593-610	7	AIR FRANCE
AS/BAe Concorde 101	RR Olympus 593-610	7	BRITISH AIRWAYS
ATR 42	P&WC PW120	3	AIR CALEDONIE
ATR 42	P&WC PW120	2	AIR MAURITIUS
ATR 42	P&WC PW120	47	AMR EAGLE
ATR 42	P&WC PW120	7	CIMBER AIR DENMARK A/S
ATR 42	P&WC PW120	6	CITYFLYER EXPRESS LTD
ATR 42	P&WC PW120	42	CONTINENTAL EXPRESS
ATR 42	P&WC PW120	17	EUROWINGS
ATR 42	P&WC PW120	1	HOLMSTROM AIR
ATR 42	P&WC PW120	8	INTER-CANADIAN
ATR 42	P&WC PW120	1	REGIONAL AIRLINES
ATR 42	P&WC PW120	2	ROYAL AIR MAROC
ATR 42	P&WC PW120	2	SOCIETE NOUVELLE AIR GUADELOUPE
ATR 42	P&WC PW120	2	SOCIETE NOUVELLE AIR MARTINIQUE
ATR 42	P&WC PW120	8	TRANS STATES AIRLINES
ATR 42	P&WC PW121	10	BAR HARBOR AIRLINES (CONTINENTAL/DELTA EXPRESS)
ATR 42	P&WC PW121	37	BRITT AIRWAYS, INC (CONTINENTAL EXPRESS)
ATR 42	P&WC PW121	2	ETHIOPIAN AIRLINES CORPORATION
ATR 42	P&WC PW121	4	OLYMPIC AVIATION, S.A.
ATR 42	P&WC PW121	12	ROCKY MOUNTAIN AIRWAYS (CONTINENTAL EXPRESS)
ATR 42	P&WC PW121	2	THAI AIRWAYS INTERNATIONAL
ATR 42-300	P&WC PW120	1	AEROMAR
ATR 42-300	P&WC PW120	1	AIR INTER GABON, SA.
ATR 42-300	P&WC PW120	5	AIR LITTORAL
ATR 42-300	P&WC PW120	3	AIR TAHITI
ATR 42-300	P&WC PW120	9	AVIANOVA
ATR 42-300	P&WC PW120	14	BRIT'AIR
ATR 42-300	P&WC PW120	15	CANADIAN REGIONAL
ATR 42-300	P&WC PW120	2	CROATIA AIRLINES
ATR 42-300	P&WC PW120	2	TACV - TRANSPORTES AEREOS DE CABO VERDE
ATR 42-300	P&WC PW120	8	TAT EUROPEAN AIRLINES
ATR 42-300	P&WC PW120	13	TRANS WORLD EXPRESS, INC
ATR 42-300	P&WC PW120	1	TRANS-JAMAICAN AIRLINES LTD
ATR 42-300	P&WC PW120	5	TRANSASIA AIRWAYS
ATR 42-300	P&WC PW120	1	TUNINTER
ATR 42-320	P&WC PW120	4	ACES
ATR 42-320	P&WC PW120	1	AIR SAINT-PIERRE
ATR 42-320	P&WC PW120	2	COMAIR (COMMERCIAL AIRWAYS (PTY) LTD)
ATR 42-320	P&WC PW121	6	AEROMAR
ATR 42-320	P&WC PW121	2	AIR BOTSWANA
ATR 42-320	P&WC PW121	1	AIR MALAWI
ATR 42-320	P&WC PW121	2	AIRLINK AIRLINE (PTY) LTD
ATR 42-320	P&WC PW121	1	IRAN ASSEMAN AIRLINES
ATR 42-320A	P&WC PW120	0	CSA - CESKOSLOVENSKE AEROLINIE
ATR 42-500	P&WC PW127E	0	AIR TAHITI
ATR 42-500	TBD	0	AEROMAR
ATR 72	P&WC PW124		AIR GABON
ATR 72	P&WC PW124	2	AIR TAHITI
ATR 72	P&WC PW124	2	BANGKOK AIRWAYS
ATR 72	P&WC PW124	6	BINTER
ATR 72	P&WC PW124	3	JAT - JUGOSLOVENSKI AEROTRANSPORT
ATR 72	P&WC PW124	6	KARAIR OY
ATR 72	P&WC PW124	4	LOT - POLSKIE LINIE LOTNICZE
ATR 72	P&WC PW124	6	TAT EUROPEAN AIRLINES
ATR 72	P&WC PW124	2	THAI AIRWAYS INTERNATIONAL
ATR 72	P&WC PW124	3	TRANS STATES AIRLINES
ATR 72	P&WC PW124	2	TUNINTER
ATR 72	P&WC PW124B	2	CITYFLYER EXPRESS LTD
ATR 72	P&WC PW124B	1	LAO AVIATION
ATR 72	P&WC PW124B	6	OLYMPIC AVIATION, S.A.
ATR 72	P&WC PW125	2	AIR CAPE
ATR 72	P&WC PW125	4	CSA - CESKOSLOVENSKE AEROLINIE
ATR 72-101	P&WC PW125	2	BRIT'AIR
ATR 72-200	P&WC PW124	33	AMR EAGLE
ATR 72-200	P&WC PW124	6	EUROWINGS
ATR 72-200	P&WC PW124	11	TRANSASIA AIRWAYS
ATR 72-201	P&WC PW123	1	AIR LITTORAL
ATR 72-202	P&WC PW124B	1	KAMPUCHEA AIRLINES
ATR 72-202	P&WC PW124B	3	VIETNAM AIRLINES
ATR 72-210	P&WC PW124	12	ATLANTIC SOUTHEAST AIRLINES, INC (THE DELTA CONNECTION)
ATR 72-210	P&WC PW124	4	IRAN ASSEMAN AIRLINES
ATR 72-210	P&WC PW127	3	AVIANOVA
ATR 72-210	P&WC PW127	2	CONTINENTAL EXPRESS
ATR 72-210	P&WC PW127	4	EUROWINGS
ATR 72-300	P&WC PW124	4	CORSE MEDITERRANEE
Agusta A109	Allison 250-C20R1	2	NATIONAL OVERSEAS AIRLINE
Agusta A109 II	Allison 250-C20	1	OLYMPIC AVIATION, S.A.
Agusta-Bell 204B	Lyc T53-11A	1	AIR FAST INDONESIA
Antonov 12	ZMKB AI-20K	54	AEROFLOT RUSSIAN INTERNATIONAL AIRLINES
Antonov 12	ZMKB AI-20K	1	TAAG, LINHAS AEREAS DE ANGOLA (ANGOLA AIRLINES)
Antonov 12	ZMKB AI-20M	2	AIR CHINA INTERNATIONAL
Antonov 12	ZMKB AI-20M	5	SAMARA AIRLINES
Antonov 12	ZMKB AI-20M	1	VOLGA DNEPR CARGO AIRLINE
Antonov 12 (F)	ZMKB AI-20	5	IRAQI AIRWAYS
Antonov 12 (F)	ZMKB AI-20K	3	BALKAN
Antonov 12 (F)	ZMKB AI-20M	2	CAAC
Antonov 12 (F)	ZMKB AI-20M	2	CHINA AIR CARGO
Antonov 124-100 Ruslan	ZMKB D-18T	3	AIR FOYLE LIMITED
Antonov 124-100 Ruslan	ZMKB D-18T	6	HEAVYLIFT CARGO AIRLINES
Antonov 124-100 Ruslan	ZMKB D-18T	6	VOLGA DNEPR CARGO AIRLINE
Antonov 2	Aviadvigatel ASh-621R	1	VIETNAM AIRLINES
Antonov 24	ZMKB AI-24	8	AIR KORYO
Antonov 24	ZMKB AI-24	31	CAAC
Antonov 24	ZMKB AI-24	2	CHINA NORTHWEST AIRLINES
Antonov 24	ZMKB AI-24	5	CHINA SOUTHERN AIRLINES
Antonov 24	ZMKB AI-24	1	IRAQI AIRWAYS

AIRFRAME	POWER PLANT	NO. IN FLEET	FLEET NAME
Antonov 24	ZMKB AI-24	6	KEMEROVSKOE AVIAPREDPRIYATIYE
Antonov 24	ZMKB AI-24		NOVOKUZNETSKOE AVIAPREDPRIYATIYE
Antonov 24	ZMKB AI-24		OMSKOYE AVIAPREDPRIYATIYE
Antonov 24	ZMKB AI-24	6	SIBERIA AIRLINE
Antonov 24	ZMKB AI-24	1	SYRIAN ARAB AIRLINES
Antonov 24	ZMKB AI-24	6	TOMSKOYE AVIAPREDPRIYATIYE
Antonov 24	ZMKB AI-24		TURKMENISTAN AIRLINE
Antonov 24	ZMKB AI-24	7	ULAN-UDESKOYE AVIAPREDPRIYATIYE
Antonov 24	ZMKB AI-24	5	VIETNAM AIRLINES
Antonov 24B	ZMKB AI-24	4	MONGOLIAN AIRLINES - MIAT
Antonov 24B	ZMKB AI-24 II	8	BELAVIA
Antonov 24B	ZMKB AI-24VT	1	AIR GUINEE
Antonov 24RV	ZMKB AI-24	3	AEROFLOT RUSSIAN INTERNATIONAL AIRLINES
Antonov 24RV	ZMKB AI-24	1	AIR MALI SA
Antonov 24RV	ZMKB AI-24	2	ARIANA AFGHAN AIRLINES
Antonov 24RV	ZMKB AI-24	7	CUBANA
Antonov 24RV	ZMKB AI-24	3	LAO AVIATION
Antonov 24RV	ZMKB AI-24	2	LITHUANIAN AIRLINES
Antonov 24RV	ZMKB AI-24	8	MONGOLIAN AIRLINES - MIAT
Antonov 24RV	ZMKB AI-24 II	14	TAROM
Antonov 24RV	ZMKB AI-24VT	1	AIR GUINEE
Antonov 24RV	ZMKB AI-24VT	10	BELAVIA
Antonov 24RV	ZMKB AI-24VT	3	KAMPUCHEA AIRLINES
Antonov 24T	ZMKB AI-24	2	IRAQI AIRWAYS
Antonov 24TV	ZMKB AI-24	1	IRAQI AIRWAYS
Antonov 24V	ZMKB AI-24	12	AEROFLOT RUSSIAN INTERNATIONAL AIRLINES
Antonov 24V	ZMKB AI-24	5	CUBANA
Antonov 24V	ZMKB AI-24	2	IRAQI AIRWAYS
Antonov 24V	ZMKB AI-24	2	LITHUANIAN AIRLINES
Antonov 24V	ZMKB AI-24	7	VIETNAM AIRLINES
Antonov 24V-2	ZMKB AI-24	14	BALKAN
Antonov 26	ZMKB AI-24	3	AEROCARIBBEAN
Antonov 26	ZMKB AI-24	1	AEROFLOT RUSSIAN INTERNATIONAL AIRLINES
Antonov 26	ZMKB AI-24	2	ARIANA AFGHAN AIRLINES
Antonov 26	ZMKB AI-24	6	KEMEROVSKOE AVIAPREDPRIYATIYE
Antonov 26	ZMKB AI-24	3	LITHUANIAN AIRLINES
Antonov 26	ZMKB AI-24	3	MONGOLIAN AIRLINES - MIAT
Antonov 26	ZMKB AI-24	5	NOVOKUZNETSKOE AVIAPREDPRIYATIYE
Antonov 26	ZMKB AI-24		OMSKOYE AVIAPREDPRIYATIYE
Antonov 26	ZMKB AI-24	5	SIBERIA AIRLINE
Antonov 26	ZMKB AI-24	4	SYRIAN ARAB AIRLINES
Antonov 26	ZMKB AI-24	5	TOMSKOYE AVIAPREDPRIYATIYE
Antonov 26	ZMKB AI-24		TURKMENISTAN AIRLINE
Antonov 26	ZMKB AI-24VT	1	KAMPUCHEA AIRLINES
Antonov 26 (F)	ZMKB AI-24	24	CAAC
Antonov 26 (F)	ZMKB AI-24	24	CAAC
Antonov 26 (F)	ZMKB AI-24VT	4	BELAVIA
Antonov 26B (F)	ZMKB A-24VT	4	BELAVIA
Antonov 28	PZL Rz TVD-10S	1	AEROFLOT RUSSIAN INTERNATIONAL AIRLINES
Antonov 30	ZMKB AI-24	4	AEROFLOT RUSSIAN INTERNATIONAL AIRLINES
Antonov 30	ZMKB AI-24	1	MONGOLIAN AIRLINES - MIAT
Antonov 32	ZMKB AI-20M	1	VOLGA DNEPR CARGO AIRLINE
Antonov 32	ZMKB AI-24	2	AERONICA - AEROLINEAS NICARAGUENSES, SA
Antonov 32	ZMKB AI-24	1	SIBERIA AIRLINE
B707-300	P&W JT3D-3B	4	IRAN AIR THE AIRLINE OF THE ISLAMIC REPUBLIC OF IRAN
B707-300	P&W JT3D-3B	1	LLOYD AEREO BOLIVIANO SAM
B707-300	P&W JT3D-7	4	CHINA SOUTHWEST AIRLINES
B707-300	P&W JT3D-7	1	EUROPEAN AIRLIFT
B707-300	P&W JT3D-7 (Q)	2	AIR CHINA INTERNATIONAL
B707-300B	P&W JT3D-3B	1	AFRICAN AIRLINES INTERNATIONAL LTD
B707-300B	P&W JT3D-3B	1	ECUATORIANA
B707-300B	P&W JT3D-3B	1	SOMALI AIRLINES
B707-300B	P&W JT3D-3B	1	SUDAN AIRWAYS
B707-300B	P&W JT3D-3B (Q)	3	AEROVIAS NACIONALES DE COLOMBIA S.A. AVIANCA
B707-300B	P&W JT3D-3B (Q)	2	ECUATORIANA
B707-300B	P&W JT3D-7	2	AIR ZIMBABWE
B707-300B	P&W JT3D-7 (Q)	1	AFRICAN AIRLINES INTERNATIONAL LTD
B707-300B (F)	P&W JT3D-3B	1	ETHIOPIAN AIRLINES CORPORATION
B707-300BC	P&W JT3D-3B	1	NATIONAL OVERSEAS AIRLINE
B707-300C	P&W JT3D-3B	1	APISA - AEROTRANSPORTES PERUANOS INTERNCIONALES SA
B707-300C	P&W JT3D-3B	3	CARIBBEAN INTERNATIONAL AIRWAYS
B707-300C	P&W JT3D-3B	1	DOMINICANA DE AVIACION
B707-300C	P&W JT3D-3B	1	IRAQI AIRWAYS
B707-300C	P&W JT3D-3B	3	LIBYAN ARAB AIRLINES
B707-300C	P&W JT3D-3B	2	NATIONAL OVERSEAS AIRLINE
B707-300C	P&W JT3D-3B	1	NIGERIA AIRWAYS
B707-300C	P&W JT3D-3B	1	OKADA AIR LTD
B707-300C	P&W JT3D-3B	2	PAKISTAN INTERNATIONAL AIRLINES (PIA)
B707-300C	P&W JT3D-3B	1	PAN AVIATION
B707-300C	P&W JT3D-3B	2	ROYAL AIR MAROC
B707-300C	P&W JT3D-3B	3	ROYAL JORDANIAN
B707-300C	P&W JT3D-3B	1	SAETA
B707-300C	P&W JT3D-3B	3	SEAGREEN AIR TRANSPORT
B707-300C	P&W JT3D-3B	2	SUDAN AIRWAYS
B707-300C	P&W JT3D-3B	1	TAAG, LINHAS AEREAS DE ANGOLA (ANGOLA AIRLINES)
B707-300C	P&W JT3D-3B	3	TAROM
B707-300C	P&W JT3D-3B	4	TRANSBRASIL S.A. LINHAS AEREAS
B707-300C	P&W JT3D-3B	1	VIETNAM AIRLINES
B707-300C	P&W JT3D-3B	2	YEMEN AIRLINES
B707-300C	P&W JT3D-3B (Q)	2	AERONAVES DEL PERU, S.A.
B707-300C	P&W JT3D-3B (Q)	1	AIR RWANDA
B707-300C	P&W JT3D-3B (Q)	3	BUFFALO AIRWAYS
B707-300C	P&W JT3D-3B (Q)	1	CHALLENGE AIR CARGO
B707-300C	P&W JT3D-3B (Q)	4	DAS AIR CARGO (Dairo Air Services)
B707-300C	P&W JT3D-3B (Q)	2	FAST AIR CARRIER
B707-300C	P&W JT3D-3B (Q)	1	HEAVYLIFT CARGO AIRLINES
B707-300C	P&W JT3D-3B (Q)	1	IRAQI AIRWAYS
B707-300C	P&W JT3D-3B (Q)	2	KUWAIT AIRWAYS CORPORATION
B707-300C	P&W JT3D-3B (Q)	1	LADECO SA/LINEA AEREA DE COBRE/LADECO AIRLINES

AIRFRAME	POWER PLANT	NO. IN FLEET	FLEET NAME
B707-300C	P&W JT3D-3B (Q)	8	MIDDLE EAST AIRLINES - AIRLIBAN
B707-300C	P&W JT3D-3B (Q)	1	PAN AVIATION
B707-300C	P&W JT3D-3B (Q)	1	PRIMERAS LINEAS URUGUAYAS DE NAVEGACION AEREA - PLUNA
B707-300C	P&W JT3D-3B (Q)	2	SAUDIA - SAUDI ARABIAN AIRLINES
B707-300C	P&W JT3D-3B (Q)	2	TAAG, LINHAS AEREAS DE ANGOLA (ANGOLA AIRLINES)
B707-300C	P&W JT3D-3B (Q)	5	TAMPA SA
B707-300C	P&W JT3D-7	2	SUDAN AIRWAYS
B707-300C	P&W JT3D-7 (Q)	1	SCIBE AIRLIFT
B707-300C (F)	P&W JT3D-3B (Q)	8	BURLINGTON AIR EXPRESS
B707-300C (F)	P&W JT3D-3B (Q)	1	ECUATORIANA
B707-300C (F)	P&W JT3D-3B (Q)	4	FLORIDA WEST AIRLINES INC
B707-300C (F)	P&W JT3D-3B (Q)	1	LAN-CHILE
B707-300C (F)	P&W JT3D-3B (Q)	1	SEAGREEN AIR TRANSPORT
B707-300C (F)	P&W JT3D-3B (Q)	6	TMA - TRANS MEDITERRANEAN AIRWAYS
B707-300C (F)	P&W JT3D-7	1	EGYPTAIR
B707-300F	P&W JT4A-12	1	ZAIREAN AIRLINES
B707-400	RR Conway 508	1	ZAIREAN AIRLINES
B720B	P&W JT3D-1	2	MIDDLE EAST AIRLINES - AIRLIBAN
B720B	P&W JT3D-3B	1	PAN AVIATION
B727-100	P&W JT8D-7	5	AEROVIAS NACIONALES DE COLOMBIA S.A. AVIANCA
B727-100	P&W JT8D-7	2	BALTIC INTERNATIONAL AIRLINES
B727-100	P&W JT8D-7	5	EXPRESS ONE
B727-100	P&W JT8D-7	1	SAETA
B727-100	P&W JT8D-7	2	SCIBE AIRLIFT
B727-100	P&W JT8D-7	1	TAME - TRANSPORTES AEREOS MILITARES ECUATORIANOS, CA
B727-100	P&W JT8D-7	8	TRANS WORLD AIRLINES, INC
B727-100	P&W JT8D-7A	1	AEROVIAS NACIONALES DE COLOMBIA S.A. AVIANCA
B727-100	P&W JT8D-7A	1	EVERGREEN INTERNATIONAL AIRLINES, INC
B727-100	P&W JT8D-7A	1	LINEAS AEREAS COSTARRICENSES, S.A. - LACSA
B727-100	P&W JT8D-7A	1	SAHSA - SERVICIO AEREO DE HONDURAS, SA
B727-100	P&W JT8D-7A	1	SAM - SOCIEDAD AERONAUTICA DE MEDELLIN
B727-100	P&W JT8D-7A	1	ZAIREAN AIRLINES
B727-100	P&W JT8D-7B	4	ACES
B727-100	P&W JT8D-7B	2	AEROPERU
B727-100	P&W JT8D-7B	5	AEROVIAS NACIONALES DE COLOMBIA S.A. AVIANCA
B727-100	P&W JT8D-7B	4	AMERICANA DE AVIACION
B727-100	P&W JT8D-7B	3	AVENSA
B727-100	P&W JT8D-7B	8	EXPRESS ONE
B727-100	P&W JT8D-7B	1	FAUCETT PERU
B727-100	P&W JT8D-7B	2	IRAN AIR THE AIRLINE OF THE ISLAMIC REPUBLIC OF IRAN
B727-100	P&W JT8D-7B	1	KABO AIR
B727-100	P&W JT8D-9	3	CONTINENTAL AIRLINES
B727-100	P&W JT8D-9A	1	DOMINICANA DE AVIACION
B727-100	P&W JT8D-9A	1	KABO AIR
B727-100	P&W JT8D-9A	2	LLOYD AEREO BOLIVIANO SAM
B727-100	P&W JT8D-9A	2	ROYAL NEPAL AIRLINES CORPORATION
B727-100	P&W JT8D-9A	1	SAETA
B727-100	P&W JT8D-9A	1	SAN
B727-100	P&W JT8D-9A	2	TAME - TRANSPORTES AEREOS MILITARES ECUATORIANOS, CA
B727-100 (F)	P&W JT8D-7	4	DHL AIRWAYS
B727-100 (F)	P&W JT8D-7	6	EXPRESS ONE
B727-100 (F)	P&W JT8D-7B	4	AMERIJET INTERNATIONAL
B727-100 (F)	P&W JT8D-7B	3	KALITTA AMERICAN INTERNATIONAL AIRWAYS
B727-100C	P&W JT8D-7	1	ARIANA AFGHAN AIRLINES
B727-100C	P&W JT8D-7	1	EVERGREEN INTERNATIONAL AIRLINES, INC
B727-100C	P&W JT8D-7	1	EXPRESS ONE
B727-100C	P&W JT8D-7	2	REEVE ALEUTIAN AIRWAYS, INC
B727-100C	P&W JT8D-7A	1	AEROPERU
B727-100C	P&W JT8D-7A	1	EVERGREEN INTERNATIONAL AIRLINES, INC
B727-100C	P&W JT8D-7B	2	AMERIJET INTERNATIONAL
B727-100C	P&W JT8D-7B	2	BRADLEY AIR SERVICES LIMITED (FIRST AIR)
B727-100C	P&W JT8D-7B	1	BRADLEY AIR SERVICES LIMITED (FIRST AIR)
B727-100C	P&W JT8D-7B	1	BURLINGTON AIR EXPRESS
B727-100C	P&W JT8D-7B	1	BURLINGTON AIR EXPRESS
B727-100C	P&W JT8D-7B	1	DOMINICANA DE AVIACION
B727-100C	P&W JT8D-7B	19	EMERY WORLDWIDE
B727-100C	P&W JT8D-7B	1	EVERGREEN INTERNATIONAL AIRLINES, INC
B727-100C	P&W JT8D-7B	10	KELOWNA FLIGHTCRAFT AIR CHARTER LTD
B727-100C	P&W JT8D-7B	6	TAESA
B727-100C	P&W JT8D-7B	1	VARIG - VIACAO AEREA RIO-GRANDENSE S.A.
B727-100C	P&W JT8D-9	1	ARIANA AFGHAN AIRLINES
B727-100C	P&W JT8D-9	3	CONTINENTAL MICRONESIA, INC
B727-100C	P&W JT8D-9A	1	LLOYD AEREO BOLIVIANO SAM
B727-100C	P&W JT8D-9A	4	VARIG - VIACAO AEREA RIO-GRANDENSE S.A.
B727-100C (F)	P&W JT8D-7	3	DHL AIRWAYS
B727-100C (F)	P&W JT8D-7	1	EVERGREEN INTERNATIONAL AIRLINES, INC
B727-100C (F)	P&W JT8D-7A	1	FEDERAL EXPRESS CORPORATION
B727-100C (F)	P&W JT8D-7B	3	DHL AIRWAYS
B727-100C (F)	P&W JT8D-7B	2	EVERGREEN INTERNATIONAL AIRLINES, INC
B727-100C (F)	P&W JT8D-7B	17	UNITED PARCEL SERVICE COMPANY UPS
B727-100F	P&W JT8D-7	6	EUROPEAN AIR TRANSPORT N.V./S.A.
B727-100F	P&W JT8D-7B	1	ARROW AIR INC
B727-100F	P&W JT8D-7B	6	EMERY WORLDWIDE
B727-100F	P&W JT8D-7B	67	FEDERAL EXPRESS CORPORATION
B727-100F	P&W JT8D-7B	3	ITAPEMIRIN TRANSPORTES AEREOS SA
B727-100F	P&W JT8D-9A	1	ITAPEMIRIN TRANSPORTES AEREOS SA
B727-100QF	RR Tay 651-54	34	UNITED PARCEL SERVICE COMPANY UPS
B727-100QF	RR Tay 654-54	3	STAR AIR A/S
B727-200	P&W JT8D-15	3	AVIOGENEX
B727-200	P&W JT8D-15	1	FAUCETT PERU
B727-200	P&W JT8D-15	6	KIWI INTERNATIONAL AIR LINES, INC
B727-200	P&W JT8D-15	5	MIAMI AIR INTERNATIONAL INC
B727-200	P&W JT8D-15	1	POLYNESIAN AIRLINES
B727-200	P&W JT8D-15	4	TUR EUROPEAN AIRWAYS
B727-200	P&W JT8D-15A	2	ARROW AIR INC
B727-200	P&W JT8D-17	2	ACES
B727-200	P&W JT8D-17	9	AMERICAN TRANS AIR, INC
B727-200	P&W JT8D-17	1	SERVIVENSA
B727-200	P&W JT8D-17	3	SYRIAN ARAB AIRLINES

AIRFRAME	POWER PLANT	NO. IN FLEET	FLEET NAME
B727-200	P&W JT8D-17A	10	AMERICAN TRANS AIR, INC
B727-200	P&W JT8D-17A	1	SAM - SOCIEDAD AERONAUTICA DE MEDELLIN
B727-200	P&W JT8D-1A	2	CARNIVAL AIR LINES INC
B727-200	P&W JT8D-7	3	ARIANA AFGHAN AIRLINES
B727-200	P&W JT8D-7	6	DHL AIRWAYS
B727-200	P&W JT8D-7	2	KEY AIRLINES INC
B727-200	P&W JT8D-7B	2	AIR CHARTER
B727-200	P&W JT8D-7B	7	FLORIDA WEST AIRLINES INC
B727-200	P&W JT8D-7B	2	ISTANBUL AIRLINES - ISTAIR
B727-200	P&W JT8D-7B	2	KABO AIR
B727-200	P&W JT8D-7B	4	KIWI INTERNATIONAL AIRLINES INC
B727-200	P&W JT8D-7B	15	NORTHWEST AIRLINES
B727-200	P&W JT8D-7B	1	ROYAL AIR MAROC
B727-200	P&W JT8D-7B	12	SHUTTLE INC (dba USAIR SHUTTLE)
B727-200	P&W JT8D-7R	2	SAM - SOCIEDAD AERONAUTICA DE MEDELLIN
B727-200	P&W JT8D-9	2	AIR ALGERIE
B727-200	P&W JT8D-9	1	LIBYAN ARAB AIRLINES
B727-200	P&W JT8D-9A	1	AVENSA
B727-200	P&W JT8D-9A	2	CARNIVAL AIR LINES INC
B727-200	P&W JT8D-9A	1	KIWI INTERNATIONAL AIR LINES, INC
B727-200	P&W JT8D-9A	1	MONGOLIAN AIRLINES - MIAT
B727-200	P&W JT8D-9A	6	OLYMPIC AIRWAYS S.A.
B727-200 Advanced	P&W JT8D-15	1	AEROVIAS NACIONALES DE COLOMBIA S.A. AVIANCA
B727-200 Advanced	P&W JT8D-15	9	AIR ALGERIE
B727-200 Advanced	P&W JT8D-15	4	AIR CHARTER
B727-200 Advanced	P&W JT8D-15	7	AIR FRANCE
B727-200 Advanced	P&W JT8D-15	5	AIR JAMAICA
B727-200 Advanced	P&W JT8D-15	22	AMERICAN AIRLINES
B727-200 Advanced	P&W JT8D-15	5	ANSETT AUSTRALIA
B727-200 Advanced	P&W JT8D-15	1	BRADLEY AIR SERVICES LIMITED (FIRST AIR)
B727-200 Advanced	P&W JT8D-15	25	CONTINENTAL AIRLINES
B727-200 Advanced	P&W JT8D-15	7	CONTINENTAL MICRONESIA, INC
B727-200 Advanced	P&W JT8D-15	2	CYPRUS TURKISH AIRLINES / KIBRIS TURK HAVA YOLLARI
B727-200 Advanced	P&W JT8D-15	133	DELTA AIR LINES, INC
B727-200 Advanced	P&W JT8D-15	5	IRAN AIR THE AIRLINE OF THE ISLAMIC REPUBLIC OF IRAN
B727-200 Advanced	P&W JT8D-15	0	IRAN ASSEMAN AIRLINES
B727-200 Advanced	P&W JT8D-15	5	ISTANBUL AIRLINES - ISTAIR
B727-200 Advanced	P&W JT8D-15	8	LIBYAN ARAB AIRLINES
B727-200 Advanced	P&W JT8D-15	25	NORTHWEST AIRLINES
B727-200 Advanced	P&W JT8D-15	3	OLYMPIC AIRWAYS S.A.
B727-200 Advanced	P&W JT8D-15	6	ROYAL AIR MAROC
B727-200 Advanced	P&W JT8D-15	2	SUN COUNTRY AIRLINES
B727-200 Advanced	P&W JT8D-15	1	TAN - TRANSPORTES AEREOS NACIONALES, SA
B727-200 Advanced	P&W JT8D-15	4	TURKISH AIRLINES, INC (TURK HAVA YOLLARI A.O)
B727-200 Advanced	P&W JT8D-15	75	UNITED AIRLINES, INC
B727-200 Advanced	P&W JT8D-15/15A	5	USAIR, INC
B727-200 Advanced	P&W JT8D-15A	0	KIWI INTERNATIONAL AIRLINES INC
B727-200 Advanced	P&W JT8D-17	1	AEROLINEAS ARGENTINAS
B727-200 Advanced	P&W JT8D-17	7	AEROVIAS NACIONALES DE COLOMBIA S.A. AVIANCA
B727-200 Advanced	P&W JT8D-17	4	AVENSA
B727-200 Advanced	P&W JT8D-17	2	EMIRATES
B727-200 Advanced	P&W JT8D-17	6	IRAQI AIRWAYS
B727-200 Advanced	P&W JT8D-17	15	NORTHWEST AIRLINES
B727-200 Advanced	P&W JT8D-17	3	SUN COUNTRY AIRLINES
B727-200 Advanced	P&W JT8D-17	1	SUN COUNTRY AIRLINES
B727-200 Advanced	P&W JT8D-17	1	TAME - TRANSPORTES AEREOS MILITARES ECUATORIANOS, CA
B727-200 Advanced	P&W JT8D-17	13	VASP - VIACAO AEREA SAO PAULO S.A.
B727-200 Advanced	P&W JT8D-17	5	YEMEN AIRLINES
B727-200 Advanced	P&W JT8D-17R	2	AEROPERU
B727-200 Advanced	P&W JT8D-17R	3	LLOYD AEREO BOLIVIANO SAM
B727-200 Advanced	P&W JT8D-17R	26	MEXICANA
B727-200 Advanced	P&W JT8D-17R	2	SAFAIR (PTY) LTD
B727-200 Advanced	P&W JT8D-17R	3	SYRIAN ARAB AIRLINES
B727-200 Advanced	P&W JT8D-9	14	DELTA AIR LINES, INC
B727-200 Advanced	P&W JT8D-9	21	IBERIA - LINEAS AEREAS DE ESPANA, S.A.
B727-200 Advanced	P&W JT8D-9	5	VIASA, VENEZOLANA INTERNACIONAL DE AVIACION SA
B727-200 Advanced	P&W JT8D-9/9A	86	AMERICAN AIRLINES
B727-200 Advanced	P&W JT8D-9A	2	AVENSA
B727-200 Advanced	P&W JT8D-9A	53	CONTINENTAL AIRLINES
B727-200 Advanced	P&W JT8D-9A	1	DOMINICANA DE AVIACION
B727-200 Advanced	P&W JT8D-9A	8	JAT - JUGOSLOVENSKI AEROTRANSPORT
B727-200 Advanced	P&W JT8D-9A	3	KOREAN AIR
B727-200 Advanced	P&W JT8D-9A	8	TUNIS AIR - SOCIETE TUNISIENNE D EL'AIR
B727-200 Advanced (F)	P&W JT8D-15	6	UNITED PARCEL SERVICE COMPANY UPS
B727-200 Advanced (F)	P&W JT8D-17	2	UNITED PARCEL SERVICE COMPANY UPS
B727-200 Advances	P&W JT8D-15	4	OKADA AIR LTD
B727-200/200 Advanced	P&W JT8D-9A	41	TRANS WORLD AIRLINES, INC
B727-200A	P&W JT8D-17	2	ROYAL JORDANIAN
B727-200A	P&W JT8D-7B	1	TAESA
B727-200C	P&W JT8D-9A	1	EXPRESS ONE
B727-200F	P&W JT8D-15	3	TURKISH AIRLINES, INC (TURK HAVA YOLLARI A.O)
B727-200F	P&W JT8D-15A	3	AMERIJET INTERNATIONAL
B727-200F	P&W JT8D-15A	1	ITAPEMIRIN TRANSPORTES AEREOS SA
B727-200F	P&W JT8D-17R	1	ITAPEMIRIN TRANSPORTES AEREOS SA
B727-200F	P&W JT8D-7	6	EUROPEAN AIR TRANSPORT N.V./S.A.
B727-200F	P&W JT8D-7B	1	BRADLEY AIR SERVICES LIMITED (FIRST AIR)
B727-200F	P&W JT8D-7B	3	EMERY WORLDWIDE
B727-200F	P&W JT8D-7B (Q)	2	TNT INTERNATIONAL AVIATION SERVICES
B727-200F	P&W JT8D-9A	1	AMERIJET INTERNATIONAL
B727-200F Advanced	P&W JT8D-15	1	ANSETT AUSTRALIA
B727-200F Advanced	P&W JT8D-15	56	FEDERAL EXPRESS CORPORATION
B727-200F Advanced	P&W JT8D-15A	2	ARROW AIR INC
B727-200F Advanced	P&W JT8D-15A	6	FEDERAL EXPRESS CORPORATION
B727-200F Advanced	P&W JT8D-17	5	FEDERAL EXPRESS CORPORATION
B727-200F Advanced	P&W JT8D-17A	5	FEDERAL EXPRESS CORPORATION
B727-200F Advanced	P&W JT8D-17R	7	FEDERAL EXPRESS CORPORATION
B727-200F Advanced (RE)	P&W JT8D-217C/-17A	11	FEDERAL EXPRESS CORPORATION
B727-200RE	P&W JT8D-217C/-17	2	SUN COUNTRY AIRLINES
B737-100	P&W JT8D-7A	13	CONTINENTAL AIRLINES

AIRFRAME	POWER PLANT	NO. IN FLEET	FLEET NAME
B737-100	P&W JT8D-7A	2	FAR EASTERN AIR TRANSPORT
B737-100	P&W JT8D-9A	1	AMERICA WEST AIRLINES, INC
B737-100	P&W JT8D-9A	1	FAUCETT PERU
B737-100	P&W JT8D-9A	1	SIERRA PACIFIC AIRLINES, INC
B737-200	P&W JT8D	2	LAP - LINEAS AEREAS PARAGUAYAS
B737-200	P&W JT8D-15	2	AIR CHARTER
B737-200	P&W JT8D-15	19	AIR FRANCE
B737-200	P&W JT8D-15	1	AVIATECA
B737-200	P&W JT8D-15	4	AVIOGENEX
B737-200	P&W JT8D-15	37	BRITISH AIRWAYS
B737-200	P&W JT8D-15	4	GB AIRWAYS
B737-200	P&W JT8D-15	3	IRAN AIR THE AIRLINE OF THE ISLAMIC REPUBLIC OF IRAN
B737-200	P&W JT8D-15	1	LITHUANIAN AIRLINES
B737-200	P&W JT8D-15	11	OLYMPIC AIRWAYS S.A.
B737-200	P&W JT8D-15	1	SAHARA INDIA AIRLINES
B737-200	P&W JT8D-15	20	SAUDIA - SAUDI ARABIAN AIRLINES
B737-200	P&W JT8D-15	3	THAI AIRWAYS INTERNATIONAL
B737-200	P&W JT8D-15	3	TRANSAVIA AIRLINES CV
B737-200	P&W JT8D-17	1	AIR GABON
B737-200	P&W JT8D-17	1	FAUCETT PERU
B737-200	P&W JT8D-17	4	MALEV - HUNGARIAN AIRLINES
B737-200	P&W JT8D-17	1	SIERRA PACIFIC AIRLINES, INC
B737-200	P&W JT8D-17A	1	ETHIOPIAN AIRLINES CORPORATION
B737-200	P&W JT8D-7	1	TACA INTERNATIONAL AIRLINES SA
B737-200	P&W JT8D-7	1	TRANSAVIA AIRLINES CV
B737-200	P&W JT8D-7	6	VASP - VIACAO AEREA SAO PAULO S.A.
B737-200	P&W JT8D-7A	1	AIR FAST INDONESIA
B737-200	P&W JT8D-7A	1	AIRFAST INDONESIA
B737-200	P&W JT8D-7A	1	FAR EASTERN AIR TRANSPORT
B737-200	P&W JT8D-7B	1	SAHSA - SERVICIO AEREO DE HONDURAS, SA
B737-200	P&W JT8D-7B	45	UNITED AIRLINES, INC
B737-200	P&W JT8D-9	1	AIR FAST INDONESIA
B737-200	P&W JT8D-9	1	AIR MADAGASCAR
B737-200	P&W JT8D-9	5	CARIBBEAN INTERNATIONAL AIRWAYS
B737-200	P&W JT8D-9	1	LAM - LINHAS AEREAS DE MOCAMBIQUE
B737-200	P&W JT8D-9	1	TAN - TRANSPORTES AEREOS NACIONALES, SA
B737-200	P&W JT8D-9A	4	AEROLINEAS ARGENTINAS
B737-200	P&W JT8D-9A	1	ARKIA ISRAELI AIRLINES
B737-200	P&W JT8D-9A	2	CANAIR CARGO LTD
B737-200	P&W JT8D-9A	5	CARNIVAL AIR LINES INC
B737-200	P&W JT8D-9A	3	COMAIR (COMMERCIAL AIRWAYS (PTY) LTD)
B737-200	P&W JT8D-9A	24	CONTINENTAL AIRLINES
B737-200	P&W JT8D-9A	5	EURALAIR INTERNATIONAL
B737-200	P&W JT8D-9A	3	FAR EASTERN AIR TRANSPORT
B737-200	P&W JT8D-9A	2	FAUCETT PERU
B737-200	P&W JT8D-9A	4	INDIAN AIRLINES LTD
B737-200	P&W JT8D-9A	2	JETALL
B737-200	P&W JT8D-9A	1	LADECO SA/LINEA AEREA DE COBRE/LADECO AIRLINES
B737-200	P&W JT8D-9A	4	NWT AIR (NORTHWEST TERRITORIAL AIRWAYS LTD)
B737-200	P&W JT8D-9A	3	SEMPATI AIR TRANSPORT
B737-200	P&W JT8D-9A	1	SIERRA PACIFIC AIRLINES, INC
B737-200	P&W JT8D-9A	3	SULTAN AIR
B737-200	P&W JT8D-9A	1	TACA INTERNATIONAL AIRLINES SA
B737-200	P&W JT8D-9A	1	TRANSMILE AIR SERVICES
B737-200	P&W JT8D-9A	1	TURKS & CAICOS AIRWAYS LIMITED
B737-200 (F)	P&W JT8D-15	2	SOUTH AFRICAN AIRWAYS (SAA)
B737-200 Advanced	P&W JT8D-15	10	AIR ALGERIE
B737-200 Advanced	P&W JT8D-15	1	AIR MADAGASCAR
B737-200 Advanced	P&W JT8D-15	1	AIR NAMIBIA
B737-200 Advanced	P&W JT8D-15	13	AIR NEW ZEALAND
B737-200 Advanced	P&W JT8D-15	2	AIR ZAIRE
B737-200 Advanced	P&W JT8D-15	3	AIR ZIMBABWE
B737-200 Advanced	P&W JT8D-15	1	ALOHA AIRLINES
B737-200 Advanced	P&W JT8D-15	15	AMERICA WEST AIRLINES, INC
B737-200 Advanced	P&W JT8D-15	1	CAMEROON AIRLINES
B737-200 Advanced	P&W JT8D-15	1	CANADIAN - CANADIEN
B737-200 Advanced	P&W JT8D-15	8	COMPANIA PANAMENA DE AVIACION, SA (COPA)
B737-200 Advanced	P&W JT8D-15	5	CROATIA AIRLINES
B737-200 Advanced	P&W JT8D-15	10	DELTA AIR LINES, INC
B737-200 Advanced	P&W JT8D-15	3	EAST WEST AIRLINES
B737-200 Advanced	P&W JT8D-15	4	LADECO SA/LINEA AEREA DE COBRE/LADECO AIRLINES
B737-200 Advanced	P&W JT8D-15	1	LAN-CHILE
B737-200 Advanced	P&W JT8D-15	35	LUFTHANSA GERMAN AIRLINES - DEUTSCHE LUFTHANSA AG
B737-200 Advanced	P&W JT8D-15	8	NIGERIA AIRWAYS
B737-200 Advanced	P&W JT8D-15	4	ROYAL AIR MAROC
B737-200 Advanced	P&W JT8D-15	6	RYANAIR LTD
B737-200 Advanced	P&W JT8D-15	8	SABENA BELGIAN WORLD AIRLINES
B737-200 Advanced	P&W JT8D-15	1	SAHSA - SERVICIO AEREO DE HONDURAS, SA
B737-200 Advanced	P&W JT8D-15	3	SOBELAIR - SOCIETE BELGE DE TRANSPORTES AERIENS
B737-200 Advanced	P&W JT8D-15	11	SOUTH AFRICAN AIRWAYS (SAA)
B737-200 Advanced	P&W JT8D-15	2	TRANSAERO AIRLINES
B737-200 Advanced	P&W JT8D-15	2	TRANSAVIA AIRLINES CV
B737-200 Advanced	P&W JT8D-15/15A	47	USAIR, INC
B737-200 Advanced	P&W JT8D-15A	46	DELTA AIR LINES, INC
B737-200 Advanced	P&W JT8D-15A	2	EAST WEST AIRLINES
B737-200 Advanced	P&W JT8D-15A	3	TRANSAERO AIRLINES
B737-200 Advanced	P&W JT8D-17	3	AIR ALGERIE
B737-200 Advanced	P&W JT8D-17	9	AIR NIPPON
B737-200 Advanced	P&W JT8D-17	2	AIR SINAI
B737-200 Advanced	P&W JT8D-17	9	ALL NIPPON AIRWAYS
B737-200 Advanced	P&W JT8D-17	3	ARKIA ISRAELI AIRLINES
B737-200 Advanced	P&W JT8D-17	14	CANADIAN - CANADIEN
B737-200 Advanced	P&W JT8D-17	2	CAYMAN AIRWAYS LTD
B737-200 Advanced	P&W JT8D-17	7	CHINA SOUTHERN AIRLINES
B737-200 Advanced	P&W JT8D-17	2	DELTA AIR LINES, INC
B737-200 Advanced	P&W JT8D-17	3	EGYPTAIR
B737-200 Advanced	P&W JT8D-17	2	EL AL ISRAEL AIRLINES, LTD
B737-200 Advanced	P&W JT8D-17	14	INDIAN AIRLINES LTD
B737-200 Advanced	P&W JT8D-17	9	JAPAN TRANSOCEAN AIR

AIRFRAME	POWER PLANT	NO. IN FLEET	FLEET NAME
B737-200 Advanced	P&W JT8D-17	2	KENYA AIRWAYS LTD
B737-200 Advanced	P&W JT8D-17	1	LAN-CHILE
B737-200 Advanced	P&W JT8D-17	3	LINEAS AEREAS COSTARRICENSES, S.A. - LACSA
B737-200 Advanced	P&W JT8D-17	2	TAAG, LINHAS AEREAS DE ANGOLA (ANGOLA AIRLINES)
B737-200 Advanced	P&W JT8D-17	2	TAAG, LINHAS AEREAS DE ANGOLA (ANGOLA AIRLINES)
B737-200 Advanced	P&W JT8D-17	1	TACA INTERNATIONAL AIRLINES SA
B737-200 Advanced	P&W JT8D-17	8	TAP AIR PORTUGAL
B737-200 Advanced	P&W JT8D-17	1	TUNIS AIR - SOCIETE TUNISIENNE D EL'AIR
B737-200 Advanced	P&W JT8D-17	17	UNITED AIRLINES, INC
B737-200 Advanced	P&W JT8D-17	17	VARIG - VIACAO AEREA RIO-GRANDENSE S.A.
B737-200 Advanced	P&W JT8D-17A	4	AIR CHINA INTERNATIONAL
B737-200 Advanced	P&W JT8D-17A	2	LAN-CHILE
B737-200 Advanced	P&W JT8D-17A	1	UGANDA AIRLINES
B737-200 Advanced	P&W JT8D-17A	3	XIAMEN AIRLINES
B737-200 Advanced	P&W JT8D-9A	3	AER LINGUS
B737-200 Advanced	P&W JT8D-9A	6	AEROLINEAS ARGENTINAS
B737-200 Advanced	P&W JT8D-9A	1	AIR NIPPON
B737-200 Advanced	P&W JT8D-9A	8	ALOHA AIRLINES
B737-200 Advanced	P&W JT8D-9A	6	AMERICA WEST AIRLINES, INC
B737-200 Advanced	P&W JT8D-9A	19	CANADIAN - CANADIEN
B737-200 Advanced	P&W JT8D-9A	3	CHINA AIRLINES
B737-200 Advanced	P&W JT8D-9A	1	CONTINENTAL AIRLINES
B737-200 Advanced	P&W JT8D-9A	2	EAST WEST AIRLINES
B737-200 Advanced	P&W JT8D-9A	2	FAR EASTERN AIR TRANSPORT
B737-200 Advanced	P&W JT8D-9A	3	LADECO SA/LINEA AEREA DE COBRE/LADECO AIRLINES
B737-200 Advanced	P&W JT8D-9A	3	PRIMERAS LINEAS URUGUAYAS DE NAVEGACION AEREA - PLUNA
B737-200 Advanced	P&W JT8D-9A	3	SEMPATI AIR TRANSPORT
B737-200 Advanced	P&W JT8D-9A	50	SOUTHWEST AIRLINES COMPANY
B737-200 Advanced	P&W JT8D-9A	2	TUNIS AIR - SOCIETE TUNISIENNE D EL'AIR
B737-200 Advanced	P&W JT8D-9A	7	UNITED AIRLINES, INC
B737-200 Advanced	P&W JT8D-9A	17	USAIR, INC
B737-200 Advanced (ER)	P&W JT8D-17	7	CANADIAN - CANADIEN
B737-200 Heavy Advanced	P&W JT8D-17	2	BRAATHENS S.A.F.E. A/S
B737-200A	P&W JT8-15	3	TAESA
B737-200A	P&W JT8D-17	1	RYANAIR LTD
B737-200C	P&W JT8D-15	1	ALOHA AIRLINES
B737-200C	P&W JT8D-15A	4	TAT EUROPEAN AIRLINES
B737-200C	P&W JT8D-7B	1	TNT INTERNATIONAL AVIATION SERVICES
B737-200C	P&W JT8D-9A	2	AEROLINEAS ARGENTINAS
B737-200C	P&W JT8D-9A	2	CANADIAN - CANADIEN
B737-200C	P&W JT8D-9A	2	L'AEROPOSTALE
B737-200C	P&W JT8D-9A	1	TRANSMILE AIR SERVICES
B737-200C Advanced	P&W JT8D-15	1	AIR ALGERIE
B737-200C Advanced	P&W JT8D-15	2	CAMEROON AIRLINES
B737-200C Advanced	P&W JT8D-15	2	IRAQI AIRWAYS
B737-200C Advanced	P&W JT8D-15	1	KAMPUCHEA AIRLINES
B737-200C Advanced	P&W JT8D-15	1	LINA CONGO
B737-200C Advanced	P&W JT8D-15	2	ROYAL AIR MAROC
B737-200C Advanced	P&W JT8D-15	4	SABENA BELGIAN WORLD AIRLINES
B737-200C Advanced	P&W JT8D-15	1	YEMEN AIRLINES
B737-200C Advanced	P&W JT8D-17	1	AIR GUINEE
B737-200C Advanced	P&W JT8D-17	2	AIR TANZANIA
B737-200C Advanced	P&W JT8D-17	8	ALASKA AIRLINES, INC
B737-200C Advanced	P&W JT8D-17	1	ALOHA AIRLINES
B737-200C Advanced	P&W JT8D-17	2	CANADIAN - CANADIEN
B737-200C Advanced	P&W JT8D-17	5	MARKAIR, INC
B737-200C Advanced	P&W JT8D-17	1	TAAG, LINHAS AEREAS DE ANGOLA (ANGOLA AIRLINES)
B737-200C Advanced	P&W JT8D-17	1	TACA INTERNATIONAL AIRLINES SA
B737-200C Advanced	P&W JT8D-17	1	TUNIS AIR - SOCIETE TUNISIENNE D EL'AIR
B737-200C Advanced	P&W JT8D-17A	2	YEMEN AIRLINES
B737-200C Advanced	P&W JT8D-7	2	SUDAN AIRWAYS
B737-200C Advanced	P&W JT8D-9	1	AIR ALGERIE
B737-200C Advanced	P&W JT8D-9	1	LAM - LINHAS AEREAS DE MOCAMBIQUE
B737-200C Advanced	P&W JT8D-9A	4	CANADIAN - CANADIEN
B737-200C Advanced (F)	P&W JT8D-17	2	VASP - VIACAO AEREA SAO PAULO S.A.
B737-200F	P&W JT8D-15	2	GERMAN CARGO SERVICES
B737-200QC	P&W JT8D-9A	1	COMPANIA PANAMENA DE AVIACION, SA (COPA)
B737-200QC	P&W JT8D-9A	0	SAF SERVICE AERIEN FRANCAIS, SA
B737-200QC	P&W JT8D-9A	2	TRANSMILE AIR SERVICES
B737-300	CFMI CFM56-3	0	NATIONAL OVERSEAS AIRLINE
B737-300	CFMI CFM56-3B1	14	AIR CHINA INTERNATIONAL
B737-300	CFMI CFM56-3B1	6	AIR FRANCE
B737-300	CFMI CFM56-3B1	1	AIR PACIFIC LIMITED
B737-300	CFMI CFM56-3B1	2	ALOHA AIRLINES
B737-300	CFMI CFM56-3B1	20	AMERICA WEST AIRLINES, INC
B737-300	CFMI CFM56-3B1	21	ANSETT AUSTRALIA
B737-300	CFMI CFM56-3B1	4	BRITISH MIDLAND AIRWAYS LTD
B737-300	CFMI CFM56-3B1	0	CAAC
B737-300	CFMI CFM56-3B1	20	CHINA SOUTHERN AIRLINES
B737-300	CFMI CFM56-3B1	13	CHINA SOUTHWEST AIRLINES
B737-300	CFMI CFM56-3B1	73	CONTINENTAL AIRLINES
B737-300	CFMI CFM56-3B1	18	DELTA AIR LINES, INC
B737-300	CFMI CFM56-3B1	8	GARUDA INDONESIA PT
B737-300	CFMI CFM56-3B1	9	JAT - JUGOSLOVENSKI AEROTRANSPORT
B737-300	CFMI CFM56-3B1	4	JET AIRWAYS
B737-300	CFMI CFM56-3B1	15	KLM ROYAL DUTCH AIRLINES
B737-300	CFMI CFM56-3B1	2	LAUDA AIR LUFTFAHRT AKTIENGESELLSCHAFT
B737-300	CFMI CFM56-3B1	31	LUFTHANSA GERMAN AIRLINES - DEUTSCHE LUFTHANSA AG
B737-300	CFMI CFM56-3B1	4	MARKAIR, INC
B737-300	CFMI CFM56-3B1	4	MONARCH AIRLINES LTD
B737-300	CFMI CFM56-3B1	6	PAKISTAN INTERNATIONAL AIRLINES (PIA)
B737-300	CFMI CFM56-3B1	12	PHILIPPINE AIR LINES, INC - PAL
B737-300	CFMI CFM56-3B1	133	SOUTHWEST AIRLINES COMPANY
B737-300	CFMI CFM56-3B1	11	TAESA
B737-300	CFMI CFM56-3B1	1	TAGB - TRANSPORTES AEREOS DA GUINE BISSAU
B737-300	CFMI CFM56-3B1	8	TRANSBRASIL S.A. LINHAS AEREAS
B737-300	CFMI CFM56-3B1	2	VASP - VIACAO AEREA SAO PAULO S.A.
B737-300	CFMI CFM56-3B1/3B2	10	AIR EUROPA (AIR ESPANA, SA)
B737-300	CFMI CFM56-3B1/3B2/3C1	101	UNITED AIRLINES, INC

AIRFRAME	POWER PLANT	NO. IN FLEET	FLEET NAME
B737-300	CFMI CFM56-3B2	1	AIR CALEDONIE INTERNATIONAL
B737-300	CFMI CFM56-3B2	3	AIR CHINA INTERNATIONAL
B737-300	CFMI CFM56-3B2	1	AIR FOYLE LIMITED
B737-300	CFMI CFM56-3B2	2	AIR HOLLAND CHARTER BV
B737-300	CFMI CFM56-3B2	1	AIR MADAGASCAR
B737-300	CFMI CFM56-3B2	15	AMERICA WEST AIRLINES, INC
B737-300	CFMI CFM56-3B2	3	AVIATECA
B737-300	CFMI CFM56-3B2	1	BRITISH MIDLAND AIRWAYS LTD
B737-300	CFMI CFM56-3B2	4	CHINA SOUTHWEST AIRLINES
B737-300	CFMI CFM56-3B2	4	CONDOR FLUGDIENST GMBH
B737-300	CFMI CFM56-3B2	7	DEUTSCHE BA LUFTFAHRTGESELLSCHAFT MBH
B737-300	CFMI CFM56-3B2	3	EUROBELGIAN AIRLINES
B737-300	CFMI CFM56-3B2	1	EXCALIBUR AIRWAYS LTD
B737-300	CFMI CFM56-3B2	10	GERMANIA FLUGGESELLSCHAFT KOLN
B737-300	CFMI CFM56-3B2	1	LADECO SA/LINEA AEREA DE COBRE/LADECO AIRLINES
B737-300	CFMI CFM56-3B2	13	MAERSK AIR
B737-300	CFMI CFM56-3B2	6	SABENA BELGIAN WORLD AIRLINES
B737-300	CFMI CFM56-3B2	4	SILKAIR (SINGAPORE) PRIVATE LIMITED
B737-300	CFMI CFM56-3B2	1	SOBELAIR - SOCIETE BELGE DE TRANSPORTES AERIENS
B737-300	CFMI CFM56-3B2	3	SULTAN AIR
B737-300	CFMI CFM56-3B2	1	SUN EXPRESS
B737-300	CFMI CFM56-3B2	4	TACA INTERNATIONAL AIRLINES SA
B737-300	CFMI CFM56-3B2	8	TAP AIR PORTUGAL
B737-300	CFMI CFM56-3B2	5	TEA BASEL
B737-300	CFMI CFM56-3B2	2	TEA FRANCE
B737-300	CFMI CFM56-3B2	8	TRANSAVIA AIRLINES CV
B737-300	CFMI CFM56-3B2	3	TRANSBRASIL S.A. LINHAS AEREAS
B737-300	CFMI CFM56-3B2	2	TRANSMED AIRLINES
B737-300	CFMI CFM56-3B2	3	TURKMENISTAN AIRLINE
B737-300	CFMI CFM56-3B2	58	USAIR, INC
B737-300	CFMI CFM56-3B2	25	VARIG - VIACAO AEREA RIO-GRANDENSE S.A.
B737-300	CFMI CFM56-3B2	1	VIETNAM AIRLINES
B737-300	CFMI CFM56-3B2	9	VIVA AIR
B737-300	CFMI CFM56-3B2/3C1	16	QANTAS AIRWAYS LTD
B737-300	CFMI CFM56-3C	4	CHINA HAINAN AIRLINES
B737-300	CFMI CFM56-3C1	1	AIR MALAWI
B737-300	CFMI CFM56-3C1	5	AIR MALTA CO. LTD.
B737-300	CFMI CFM56-3C1	0	AIR NEW ZEALAND
B737-300	CFMI CFM56-3C1	5	EUROBELGIAN AIRLINES
B737-300	CFMI CFM56-3C1	2	LAM - LINHAS AEREAS DE MOCAMBIQUE
B737-300	CFMI CFM56-3C1	4	MALEV - HUNGARIAN AIRLINES
B737-300	CFMI CFM56-3C1	2	OMAN AIR
B737-300	CFMI CFM56-3C1	1	SAETA
B737-300	CFMI CFM56-3C1	2	SUN EXPRESS
B737-300	CFMI CFM56-3C1	5	TAROM
B737-300	CFMI CFM56-3C1	10	YUNNAN AIRLINES
B737-300/400/500	CFMI CFM56-3C	0	UNITED AIRLINES, INC
B737-300F	CFMI CFM56-3B2	1	SINGAPORE AIRLINES LIMITED
B737-300F	CFMI CFM56-3C1	2	MALAYSIA AIRLINES
B737-300LR	CFMI CFM56-3B2	33	USAIR, INC
B737-300QC	CFMI CFM56-3B1	2	LUFTHANSA GERMAN AIRLINES - DEUTSCHE LUFTHANSA AG
B737-300QC	CFMI CFM56-3B2	1	POLYNESIAN AIRLINES
B737-300QC	CFMI CFM56-3C1	3	FALCON CARGO AB
B737-400	CFMI CFM56	0	CHINA HAINAN AIRLINES
B737-400	CFMI CFM56-3	0	EAST WEST AIRLINES
B737-400	CFMI CFM56-3	0	MALEV - HUNGARIAN AIRLINES
B737-400	CFMI CFM56-3B1	2	JET AIRWAYS
B737-400	CFMI CFM56-3B1	0	SAHSA - SERVICIO AEREO DE HONDURAS, SA
B737-400	CFMI CFM56-3B1	2	TAN - TRANSPORTES AEREOS NACIONALES, SA
B737-400	CFMI CFM56-3B2	6	AER LINGUS
B737-400	CFMI CFM56-3B2	2	ALOHA AIRLINES
B737-400	CFMI CFM56-3B2	3	CARNIVAL AIR LINES INC
B737-400	CFMI CFM56-3B2	1	JAPAN AIRLINES COMPANY LTD - JAL
B737-400	CFMI CFM56-3B2	12	KLM ROYAL DUTCH AIRLINES
B737-400	CFMI CFM56-3B2	54	USAIR, INC
B737-400	CFMI CFM56-3C1	2	AIR BELGIUM
B737-400	CFMI CFM56-3C1	7	AIR BERLIN
B737-400	CFMI CFM56-3C1	1	AIR CHARTER
B737-400	CFMI CFM56-3C1	2	AIR NAURU
B737-400	CFMI CFM56-3C1	7	AIR UK (LEISURE) LIMITED
B737-400	CFMI CFM56-3C1	1	AIR VANUATU
B737-400	CFMI CFM56-3C1	24	ALASKA AIRLINES, INC
B737-400	CFMI CFM56-3C1	16	ASIANA AIRLINES
B737-400	CFMI CFM56-3C1	7	BRAATHENS S.A.F.E. A/S
B737-400	CFMI CFM56-3C1	36	BRITISH AIRWAYS
B737-400	CFMI CFM56-3C1	6	BRITISH MIDLAND AIRWAYS LTD
B737-400	CFMI CFM56-3C1	2	CORSAIR
B737-400	CFMI CFM56-3C1	3	EUROBELGIAN AIRLINES
B737-400	CFMI CFM56-3C1	8	FUTURA INTERNATIONAL AIRWAYS
B737-400	CFMI CFM56-3C1	7	GARUDA INDONESIA PT
B737-400	CFMI CFM56-3C1	2	GB AIRWAYS
B737-400	CFMI CFM56-3C1	9	HAPAG-LLOYD FLUGGESELLSCHAFT.MBH
B737-400	CFMI CFM56-3C1	4	ICELANDAIR
B737-400	CFMI CFM56-3C1	1	ISTANBUL AIRLINES - ISTAIR
B737-400	CFMI CFM56-3C1	2	JAPAN TRANSOCEAN AIR
B737-400	CFMI CFM56-3C1	2	LAUDA AIR LUFTFAHRT AKTIENGESELLSCHAFT
B737-400	CFMI CFM56-3C1	4	LOT - POLSKIE LINIE LOTNICZE
B737-400	CFMI CFM56-3C1	7	LUFTHANSA GERMAN AIRLINES - DEUTSCHE LUFTHANSA AG
B737-400	CFMI CFM56-3C1	2	LUXAIR S.A.
B737-400	CFMI CFM56-3C1	39	MALAYSIA AIRLINES
B737-400	CFMI CFM56-3C1	6	MARKAIR, INC
B737-400	CFMI CFM56-3C1	7	OLYMPIC AIRWAYS S.A.
B737-400	CFMI CFM56-3C1	18	QANTAS AIRWAYS LTD
B737-400	CFMI CFM56-3C1	6	ROYAL AIR MAROC
B737-400	CFMI CFM56-3C1	3	SABENA BELGIAN WORLD AIRLINES
B737-400	CFMI CFM56-3C1	2	SAHARA INDIA AIRLINES
B737-400	CFMI CFM56-3C1	3	SOBELAIR - SOCIETE BELGE DE TRANSPORTES AERIENS
B737-400	CFMI CFM56-3C1	1	SOLOMON AIRLINES
B737-400	CFMI CFM56-3C1	7	THAI AIRWAYS INTERNATIONAL

AIRFRAME	POWER PLANT	NO. IN FLEET	FLEET NAME
B737-400	CFMI CFM56-3C1	4	TRANSBRASIL S.A. LINHAS AEREAS
B737-400	CFMI CFM56-3C1	2	TRANSMED AIRLINES
B737-400	CFMI CFM56-3C1	27	TURKISH AIRLINES, INC (TURK HAVA YOLLARI A.O)
B737-400	CFMI CFM56-3C1	2	UKRAINIAN AIRLINES
B737-400	CFMI CFMI56-3C1	1	ROYAL TONGAN AIRLINES
B737-500	CFMI CFM56-3B1	10	AER LINGUS
B737-500	CFMI CFM56-3B1	1	AIR PACIFIC LIMITED
B737-500	CFMI CFM56-3B1	12	BRITISH MIDLAND AIRWAYS LTD
B737-500	CFMI CFM56-3B1	14	CHINA SOUTHERN AIRLINES
B737-500	CFMI CFM56-3B1	5	CSA - CESKOSLOVENSKE AEROLINIE
B737-500	CFMI CFM56-3B1	1	ESTONIAN AIR
B737-500	CFMI CFM56-3B1	3	EURALAIR INTERNATIONAL
B737-500	CFMI CFM56-3B1	4	MAERSK AIR
B737-500	CFMI CFM56-3B1	6	SABENA BELGIAN WORLD AIRLINES
B737-500	CFMI CFM56-3B1	12	SAS - SCANDINAVIAN AIRLINES SYSTEM
B737-500	CFMI CFM56-3B1	25	SOUTHWEST AIRLINES COMPANY
B737-500	CFMI CFM56-3B1	5	TAESA
B737-500	CFMI CFM56-3B1	2	TUNIS AIR - SOCIETE TUNISIENNE D EL'AIR
B737-500	CFMI CFM56-3B1/3C1	32	LUFTHANSA GERMAN AIRLINES - DEUTSCHE LUFTHANSA AG
B737-500	CFMI CFM56-3B1/3C1	57	UNITED AIRLINES, INC
B737-500	CFMI CFM56-3C1	1	AIR AUSTRAL
B737-500	CFMI CFM56-3C1	16	AIR FRANCE
B737-500	CFMI CFM56-3C1	0	AIR NIPPON
B737-500	CFMI CFM56-3C1	3	ASIANA AIRLINES
B737-500	CFMI CFM56-3C1	3	BALKAN
B737-500	CFMI CFM56-3C1	20	BRAATHENS S.A.F.E. A/S
B737-500	CFMI CFM56-3C1	5	EGYPTAIR
B737-500	CFMI CFM56-3C1	5	HAPAG-LLOYD FLUGGESELLSCHAFT.MBH
B737-500	CFMI CFM56-3C1	6	LOT - POLSKIE LINIE LOTNICZE
B737-500	CFMI CFM56-3C1	2	LUXAIR S.A.
B737-500	CFMI CFM56-3C1	9	MALAYSIA AIRLINES
B737-500	CFMI CFM56-3C1	5	ROYAL AIR MAROC
B737-500	CFMI CFM56-3C1	2	TURKISH AIRLINES, INC (TURK HAVA YOLLARI A.O)
B737-500	CFMI CFM56-3C1	3	XIAMEN AIRLINES
B737-600	CFMI CFM56-7	0	SAS - SCANDINAVIAN AIRLINES SYSTEM
B737-700	CFMI CFM56-7	0	GERMANIA FLUGGESELLSCHAFT KOLN
B737-700	CFMI CFM56-7	0	MAERSK AIR
B737-700	CFMI CFM56-7	0	SOUTHWEST AIRLINES COMPANY
B737-800	CFMI CFM56-7	0	HAPAG-LLOYD FLUG GMBH
B747	NYK	0	QATAR AIRWAYS
B747	P&W JT9D-7	2	ACS OF CANADA - AIR CHARTER SYSTEMS
B747	P&W JT9D-7J	1	AIR CLUB INTERNATIONAL
B747-100	P&W JT9D-3A	3	UNITED AIRLINES, INC
B747-100	P&W JT9D-7	3	AIR CANADA
B747-100	P&W JT9D-7	13	AIR FRANCE
B747-100	P&W JT9D-7	11	TOWER AIR, INC
B747-100	P&W JT9D-7A	3	AER LINGUS
B747-100	P&W JT9D-7A	15	BRITISH AIRWAYS
B747-100	P&W JT9D-7A	2	CORSAIR
B747-100	P&W JT9D-7A	3	EVERGREEN INTERNATIONAL AIRLINES, INC
B747-100	P&W JT9D-7A	1	JAPAN ASIA AIRWAYS (NIHON ASIA KOKU)
B747-100	P&W JT9D-7A	2	KALITTA AMERICAN INTERNATIONAL AIRWAYS
B747-100	P&W JT9D-7A	2	NORTHWEST AIRLINES
B747-100	P&W JT9D-7A	1	OKADA AIR LTD
B747-100	P&W JT9D-7A	8	TRANS WORLD AIRLINES, INC
B747-100	P&W JT9D-7A	15	UNITED AIRLINES, INC
B747-100	P&W JT9D-7AH	1	VIRGIN ATLANTIC AIRWAYS
B747-100	P&W JT9D-7F	1	IRAN AIR THE AIRLINE OF THE ISLAMIC REPUBLIC OF IRAN
B747-100	RR RB211-524C2-19	8	SAUDIA - SAUDI ARABIAN AIRLINES
B747-100 (F)	P&W JT9D-7F	2	EVERGREEN INTERNATIONAL AIRLINES, INC
B747-100 (F) (SCD)	P&W JT9D-7A	15	UNITED PARCEL SERVICE COMPANY UPS
B747-100 (LR)	P&W JT9D-7A	2	JAPAN AIRLINES COMPANY LTD - JAL
B747-100 (SCD)	P&W JT9D-7J	1	EL AL ISRAEL AIRLINES, LTD
B747-100B (SR)	P&W JT9D-7A	3	JAPAN AIRLINES COMPANY LTD - JAL
B747-100B (SR/EUD)	P&W JT9D-7A	2	JAPAN AIRLINES COMPANY LTD - JAL
B747-100F (SCD	P&W JT9D-7A	1	ARKIA ISRAELI AIRLINES
B747-100F (SCD)	P&W JT9D-7A	2	KALITTA AMERICAN INTERNATIONAL AIRWAYS
B747-100F (SCD)	P&W JT9D-7A	2	SOUTHERN AIR TRANSPORT, INC
B747-100SF	P&W JTD9-7A	1	AHK AIR HONG KONG
B747-100SR (F)	P&W JT8D-7A	1	EVERGREEN INTERNATIONAL AIRLINES, INC
B747-100SR (F)	P&W JT8D-7A	1	EVERGREEN INTERNATIONAL AIRLINES, INC
B747-200	GE CF6-50E	2	AIR FRANCE
B747-200	GE CF6-50E2	6	THAI AIRWAYS INTERNATIONAL
B747-200	P&W JT9D-7A	3	TRANS WORLD AIRLINES, INC
B747-200	P&W JT9D-7F	2	IRAN AIR THE AIRLINE OF THE ISLAMIC REPUBLIC OF IRAN
B747-200	P&W JT9D-7J	4	EL AL ISRAEL AIRLINES, LTD
B747-200	P&W JT9D-7J	1	SAUDIA - SAUDI ARABIAN AIRLINES
B747-200	P&W JT9D-7Q	2	CORSAIR
B747-200	P&W JT9D-7Q	6	GARUDA INDONESIA PT
B747-200	P&W JT9D-7Q	4	SINGAPORE AIRLINES LIMITED
B747-200	RR RB211-524C2	5	CATHAY PACIFIC AIRWAYS LTD
B747-200	RR RB211-524D4	5	AIR NEW ZEALAND
B747-200	RR RB211-524D4	1	AIR PACIFIC LIMITED
B747-200	RR RB211-524D4	2	CATHAY PACIFIC AIRWAYS LTD
B747-200	RR RB211-524D4U	13	BRITISH AIRWAYS
B747-200 (SCD)	GE CF6-50E2	1	AIR GABON
B747-200 (SCD)	GE CF6-50E2	1	SABENA BELGIAN WORLD AIRLINES
B747-200 (SCD)	P&W JT9D-7Q	1	AEROVIAS NACIONALES DE COLOMBIA S.A. AVIANCA
B747-200B	GE CF6-50E2	7	ALITALIA
B747-200B	GE CF6-50E2	6	ALL NIPPON AIRWAYS
B747-200B	GE CF6-50E2	4	LUFTHANSA GERMAN AIRLINES - DEUTSCHE LUFTHANSA AG
B747-200B	GE CF6-50E2	6	PHILIPPINE AIR LINES, INC - PAL
B747-200B	P&W JT9D-7A	2	CONTINENTAL AIRLINES
B747-200B	P&W JT9D-7A	1	IBERIA - LINEAS AEREAS DE ESPANA, S.A.
B747-200B	P&W JT9D-7A	1	JAPAN AIR CHARTER
B747-200B	P&W JT9D-7A	1	JAPAN AIRLINES COMPANY LTD - JAL
B747-200B	P&W JT9D-7A	6	PAKISTAN INTERNATIONAL AIRLINES (PIA)
B747-200B	P&W JT9D-7A/7J	5	VIRGIN ATLANTIC AIRWAYS
B747-200B	P&W JT9D-7AW	1	CHINA AIRLINES

AIRFRAME	POWER PLANT	NO. IN FLEET	FLEET NAME
B747-200B	P&W JT9D-7AW	1	JAPAN ASIA AIRWAYS (NIHON ASIA KOKU)
B747-200B	P&W JT9D-7F	3	CONTINENTAL AIRLINES
B747-200B	P&W JT9D-7F	5	NORTHWEST AIRLINES
B747-200B	P&W JT9D-7FW	3	MIDDLE EAST AIRLINES - AIRLIBAN
B747-200B	P&W JT9D-7J	4	AIR INDIA
B747-200B	P&W JT9D-7J	1	OLYMPIC AIRWAYS S.A.
B747-200B	P&W JT9D-7J	7	UNITED AIRLINES, INC
B747-200B	P&W JT9D-7Q	6	AEROLINEAS ARGENTINAS
B747-200B	P&W JT9D-7Q	5	AIR INDIA
B747-200B	P&W JT9D-7Q	2	CHINA AIRLINES
B747-200B	P&W JT9D-7Q	2	IBERIA - LINEAS AEREAS DE ESPANA, S.A.
B747-200B	P&W JT9D-7Q	8	JAPAN AIRLINES COMPANY LTD - JAL
B747-200B	P&W JT9D-7Q	1	JAPAN ASIA AIRWAYS (NIHON ASIA KOKU)
B747-200B	P&W JT9D-7Q	12	NORTHWEST AIRLINES
B747-200B	P&W JT9D-7Q	3	OLYMPIC AIRWAYS S.A.
B747-200B	P&W JT9D-7Q	3	PHILIPPINE AIR LINES, INC - PAL
B747-200B	P&W JT9D-7Q3	1	IBERIA - LINEAS AEREAS DE ESPANA, S.A.
B747-200B	P&W JT9D-7R4G2	3	JAPAN AIRLINES COMPANY LTD - JAL
B747-200B	P&W JT9D-7R4G2	3	KOREAN AIR
B747-200B	P&W JT9D-7R4G2	3	NORTHWEST AIRLINES
B747-200B	P&W JT9D-7R4G2	5	SOUTH AFRICAN AIRWAYS (SAA)
B747-200B	P&W JT9D-7R4G2	2	UNITED AIRLINES, INC
B747-200B	RR RB211-524D4	2	MALAYSIA AIRLINES
B747-200B	RR RB211-524D4	3	QANTAS AIRWAYS LTD
B747-200B (F)	P&W JT9D-7AW	1	JAPAN AIRLINES COMPANY LTD - JAL
B747-200B (LR)	P&W JT9D-7AW	10	JAPAN AIRLINES COMPANY LTD - JAL
B747-200B (SCD)	GE CF6-50E2	10	AIR FRANCE
B747-200B (SCD)	GE CF6-50E2	5	ALITALIA
B747-200B (SCD)	GE CF6-50E2	10	LUFTHANSA GERMAN AIRLINES - DEUTSCHE LUFTHANSA AG
B747-200B (SCD)	GE CF6-50E2	2	PAKISTAN INTERNATIONAL AIRLINES (PIA)
B747-200B (SCD)	GE CF6-50E2	3	VARIG - VIACAO AEREA RIO-GRANDENSE S.A.
B747-200B (SCD)	P&W JT9D-70A	1	AIR MADAGASCAR
B747-200B (SCD)	P&W JT9D-7F	1	ROYAL AIR MAROC
B747-200B (SCD)	P&W JT9D-7J	4	KUWAIT AIRWAYS CORPORATION
B747-200B (SCD)	P&W JT9D-7OA	1	PHILIPPINE AIR LINES, INC - PAL
B747-200B (SCD)	P&W JT9D-7Q	1	CAMEROON AIRLINES
B747-200B (SCD)	P&W JT9D-7Q	1	PHILIPPINE AIR LINES, INC - PAL
B747-200B (SCD)	P&W JT9D-7Q	1	SOUTH AFRICAN AIRWAYS (SAA)
B747-200B (SCD)	P&W JT9D-7Q3	3	IBERIA - LINEAS AEREAS DE ESPANA, S.A.
B747-200B (SCD)	P&W JT9D-7R4G2	3	AIR CHINA INTERNATIONAL
B747-200B (SCD)	RR RB211-524D4	2	QANTAS AIRWAYS LTD
B747-200B (SCD)	RR RB211-524DX	3	BRITISH AIRWAYS
B747-200C	P&W JT9D-7F	1	EVERGREEN INTERNATIONAL AIRLINES, INC
B747-200C	P&W JT9D-7J	1	CAL - CARGO AIR LINES LTD
B747-200C	P&W JT9D-7J	3	EL AL ISRAEL AIRLINES, LTD
B747-200C	P&W JT9D-7Q	3	AIR CANADA
B747-200C (F)	P&W JT9D-7A	1	KOREAN AIR
B747-200C (F)	P&W JT9D-7F	1	EVERGREEN INTERNATIONAL AIRLINES, INC
B747-200C (SCD)	GE CF6-50E2	2	MARTINAIR HOLLAND NV
B747-200C (SCD)	P&W JT9D-7FW	3	IRAQI AIRWAYS
B747-200CRAF	P&W JT9D-7J	1	EVERGREEN INTERNATIONAL AIRLINES, INC
B747-200F	GE CF6-50E2	1	MARTINAIR HOLLAND NV
B747-200F	P&W JT9D-7	1	EL AL ISRAEL AIRLINES, LTD
B747-200F	P&W JT9D-7A	5	FEDERAL EXPRESS CORPORATION
B747-200F	P&W JT9D-7F	1	IRAN AIR THE AIRLINE OF THE ISLAMIC REPUBLIC OF IRAN
B747-200F	P&W JT9D-7Q	1	AHK AIR HONG KONG
B747-200F	P&W JT9D-7Q	2	EL AL ISRAEL AIRLINES, LTD
B747-200F	P&W JT9D-7Q	1	FEDERAL EXPRESS CORPORATION
B747-200F	P&W JT9D-7Q	1	JAPAN UNIVERSAL SYSTEMS TRANSPORT - JUST
B747-200F	P&W JT9D-7Q	1	KOREAN AIR
B747-200F	P&W JT9D-7R4G2	1	AIR CHINA INTERNATIONAL
B747-200F	RR RB211-524D4	4	CATHAY PACIFIC AIRWAYS LTD
B747-200F	RR RB211-524D4	2	SAUDIA - SAUDI ARABIAN AIRLINES
B747-200F (SCD)	GE CF6-50E2	10	AIR FRANCE
B747-200F (SCD)	GE CF6-50E2	2	ALITALIA
B747-200F (SCD)	GE CF6-50E2	1	ARKIA ISRAELI AIRLINES
B747-200F (SCD)	GE CF6-50E2	4	CARGOLUX
B747-200F (SCD)	GE CF6-50E2	3	GERMAN CARGO SERVICES
B747-200F (SCD)	GE CF6-50E2	5	LUFTHANSA GERMAN AIRLINES - DEUTSCHE LUFTHANSA AG
B747-200F (SCD)	GE CF6-50E2	7	NIPPON CARGO AIRLINES CO. LTD
B747-200F (SCD)	P&W JT9D-7A	6	CHINA AIRLINES
B747-200F (SCD)	P&W JT9D-7A	2	KOREAN AIR
B747-200F (SCD)	P&W JT9D-7AW	1	JAPAN AIRLINES COMPANY LTD - JAL
B747-200F (SCD)	P&W JT9D-7F	5	NORTHWEST AIRLINES
B747-200F (SCD)	P&W JT9D-7Q	6	JAPAN AIRLINES COMPANY LTD - JAL
B747-200F (SCD)	P&W JT9D-7Q	4	KOREAN AIR
B747-200F (SCD)	P&W JT9D-7Q	1	NORTHWEST AIRLINES
B747-200F (SCD)	P&W JT9D-7R4G2	2	JAPAN AIRLINES COMPANY LTD - JAL
B747-200F (SCD)	P&W JT9D-7R4G2	2	NORTHWEST AIRLINES
B747-200SF	P&W JT9D-7J	1	EVERGREEN INTERNATIONAL AIRLINES, INC
B747-300	GE CF6-50C2B1	2	THAI AIRWAYS INTERNATIONAL
B747-300	GE CF6-50E2	3	KLM ROYAL DUTCH AIRLINES
B747-300	GE CF6-80C2B1	3	VARIG - VIACAO AEREA RIO-GRANDENSE S.A.
B747-300	P&W JT9D-7R4G2	1	JAPAN ASIA AIRWAYS (NIHON ASIA KOKU)
B747-300	P&W JT9D-7R4G2	2	KOREAN AIR
B747-300	P&W JT9D-7R4G2	2	SOUTH AFRICAN AIRWAYS (SAA)
B747-300	P&W JT9D-7R4G2	2	SWISSAIR (SOCIETE ANONYME SUISSE POUR LA NAVIGATION AERIENNE)
B747-300	RR RB211-524C2	6	CATHAY PACIFIC AIRWAYS LTD
B747-300	RR RB211-524D4	6	QANTAS AIRWAYS LTD
B747-300	RR RB211-524D4-19	11	SAUDIA - SAUDI ARABIAN AIRLINES
B747-300 (LR)	P&W JT9D-7R4G2	9	JAPAN AIRLINES COMPANY LTD - JAL
B747-300 (SCD)	GE CF6-50E2	2	AIR FRANCE
B747-300 (SCD)	GE CF6-50E2	2	AIR FRANCE
B747-300 (SCD)	GE CF6-50E2	2	VARIG - VIACAO AEREA RIO-GRANDENSE S.A.
B747-300 (SCD)	P&W JT9D-7R4G2	2	EGYPTAIR
B747-300 (SCD)	P&W JT9D-7R4G2	1	MALAYSIA AIRLINES
B747-300 (SCD)	P&W JT9D-7R4G2	3	SWISSAIR (SOCIETE ANONYME SUISSE POUR LA NAVIGATION AERIENNE)
B747-300 (SR)	P&W JT9D-7R4G2	4	JAPAN AIRLINES COMPANY LTD - JAL
B747-300C	GE CF6-50E2	2	SABENA BELGIAN WORLD AIRLINES

AIRFRAME	POWER PLANT	NO. IN FLEET	FLEET NAME
B747-300C	GE CF6-80C2B1	2	AIR INDIA
B747-300M	P&W JT9D-7R4G2	2	ANSETT AUSTRALIA
B747-300M	P&W JT9D-7R4G2	1	KOREAN AIR
B747-300M	P&W JT9D-7R4G2	9	SINGAPORE AIRLINES LIMITED
B747-300M (SCD)	GE CF6-50E2	10	KLM ROYAL DUTCH AIRLINES
B747-300M (SCD)	P&W JT9D-7R4G2	3	SINGAPORE AIRLINES LIMITED
B747-400	GE CF6-80C2	16	ALL NIPPON AIRWAYS
B747-400	GE CF6-80C2	2	GARUDA INDONESIA PT
B747-400	GE CF6-80C2	5	KLM ROYAL DUTCH AIRLINES
B747-400	GE CF6-80C2	3	PHILIPPINE AIR LINES, INC - PAL
B747-400	GE CF6-80C2	2	VIRGIN ATLANTIC AIRWAYS
B747-400	GE CF6-80C2B1F	7	AIR FRANCE
B747-400	GE CF6-80C2B1F	5	ASIANA AIRLINES
B747-400	GE CF6-80C2B1F	4	CANADIAN - CANADIEN
B747-400	GE CF6-80C2B1F	10	EVA AIRWAYS
B747-400	GE CF6-80C2B1F	0	JAPAN AIR SYSTEM CO LTD (NIHON AIR SYSTEM)
B747-400	GE CF6-80C2B1F	24	JAPAN AIRLINES COMPANY LTD - JAL
B747-400	GE CF6-80C2B1F	10	LUFTHANSA GERMAN AIRLINES - DEUTSCHE LUFTHANSA AG
B747-400	GE CF6-80C2B1F	2	MALAYSIA AIRLINES
B747-400	GE CF6-80C2B1F	1	ROYAL AIR MAROC
B747-400	GE CF6-80C2B1F	7	THAI AIRWAYS INTERNATIONAL
B747-400	P&W PW4056	3	AIR CANADA
B747-400	P&W PW4056	5	AIR CHINA INTERNATIONAL
B747-400	P&W PW4056	4	AIR INDIA
B747-400	P&W PW4056	5	CHINA AIRLINES
B747-400	P&W PW4056	2	EL AL ISRAEL AIRLINES, LTD
B747-400	P&W PW4056	17	KOREAN AIR
B747-400	P&W PW4056	8	MALAYSIA AIRLINES
B747-400	P&W PW4056	10	NORTHWEST AIRLINES
B747-400	P&W PW4056	28	SINGAPORE AIRLINES LIMITED
B747-400	P&W PW4056	24	UNITED AIRLINES, INC
B747-400	RR RB211-524G	5	AIR NEW ZEALAND
B747-400	RR RB211-524G	28	BRITISH AIRWAYS
B747-400	RR RB211-524G	18	QANTAS AIRWAYS LTD
B747-400	RR RB211-524G	4	SOUTH AFRICAN AIRWAYS (SAA)
B747-400	RR RB211-524G2	1	CATHAY PACIFIC AIRWAYS LTD
B747-400	RR RB211-524H2	14	CATHAY PACIFIC AIRWAYS LTD
B747-400 (SCD)	GE CF6-80C2B1F	7	AIR FRANCE
B747-400 (SCD)	GE CF6-80C2B1F	7	LUFTHANSA GERMAN AIRLINES - DEUTSCHE LUFTHANSA AG
B747-400 (SCD)	GE CF6-80C2B1F	2	MALAYSIA AIRLINES
B747-400 (SCD)	P&W PW4056	3	AIR CHINA INTERNATIONAL
B747-400C	GE CF6-80C2B1F	1	AIR FRANCE
B747-400C (SCD)	GE CF6-80C2B1F	3	ASIANA AIRLINES
B747-400C (SCD)	P&W PW4056	1	KOREAN AIR
B747-400D	GE CF6-80C2B1F	8	JAPAN AIRLINES COMPANY LTD - JAL
B747-400F	GE CF6-80C2B1F	2	CARGOLUX
B747-400F	GE CF6-80C2B1F	0	KLM ROYAL DUTCH AIRLINES
B747-400F	P&W JT9D-7R4G2	3	SINGAPORE AIRLINES LIMITED
B747-400F	RR RB211-524H2	1	CATHAY PACIFIC AIRWAYS LTD
B747-400F (SCD)	GE CF6-80C2B1F	1	ASIANA AIRLINES
B747-400M	GE CF6-80C2B1F	1	KUWAIT AIRWAYS CORPORATION
B747-400M (SCD)	GE CF6-80C2B1F	11	KLM ROYAL DUTCH AIRLINES
B747SP	P&W JT9D-7A	4	CHINA AIRLINES
B747SP	P&W JT9D-7A	2	KOREAN AIR
B747SP	P&W JT9D-7A	7	UNITED AIRLINES, INC
B747SP	P&W JT9D-7F	4	IRAN AIR THE AIRLINE OF THE ISLAMIC REPUBLIC OF IRAN
B747SP	P&W JT9D-7FW	1	AIR MAURITIUS
B747SP	P&W JT9D-7FW	1	AIR NAMIBIA
B747SP	P&W JT9D-7FW	1	CORSAIR
B747SP	P&W JT9D-7FW	5	SOUTH AFRICAN AIRWAYS (SAA)
B747SP	P&W JT9D-7FW	2	SYRIAN ARAB AIRLINES
B747SP	P&W JT9D-7J	4	AIR CHINA INTERNATIONAL
B747SP	RR RB211-524C2-19	3	SAUDIA - SAUDI ARABIAN AIRLINES
B747SP	RR RB211-524D4	2	QANTAS AIRWAYS LTD
B747SR	GE CF6-45A2	14	ALL NIPPON AIRWAYS
B757-200	P&W PW2037	4	AMERICAN TRANS AIR, INC
B757-200	P&W PW2037	84	DELTA AIR LINES, INC
B757-200	P&W PW2037	2	FAR EASTERN AIR TRANSPORT
B757-200	P&W PW2037	33	NORTHWEST AIRLINES
B757-200	P&W PW2037	2	ROYAL AIR MAROC
B757-200	P&W PW2037	6	SHANGHAI AIRLINES
B757-200	P&W PW2037/2039	88	UNITED AIRLINES, INC
B757-200	P&W PW2040	6	AEROMEXICO
B757-200	P&W PW2040	3	AMERICAN TRANS AIR, INC
B757-200	P&W PW2040	16	CONDOR FLUGDIENST GMBH
B757-200	P&W PW2040	4	ETHIOPIAN AIRLINES CORPORATION
B757-200	RR RB211-535C	2	AVENSA
B757-200	RR RB211-535C	34	BRITISH AIRWAYS
B757-200	RR RB211-535C	3	LTE INTERNATIONAL AIRWAYS SA
B757-200	RR RB211-535E4	2	AEROVIAS NACIONALES DE COLOMBIA S.A. AVIANCA
B757-200	RR RB211-535E4	15	AIR 2000 LTD
B757-200	RR RB211-535E4	1	AIR BELGIUM
B757-200	RR RB211-535E4	3	AIR EUROPA (AIR ESPANA, SA)
B757-200	RR RB211-535E4	2	AIR HOLLAND CHARTER BV
B757-200	RR RB211-535E4	4	AIR TRANSAT
B757-200	RR RB211-535E4	6	AIRTOURS INTERNATIONAL AIRWAYS LTD
B757-200	RR RB211-535E4	14	AMERICA WEST AIRLINES, INC
B757-200	RR RB211-535E4	1	AMERICAN TRANS AIR, INC
B757-200	RR RB211-535E4	5	ARKIA ISRAELI AIRLINES
B757-200	RR RB211-535E4	1	BRITANNIA AIRWAYS LTD
B757-200	RR RB211-535E4	10	BRITISH AIRWAYS
B757-200	RR RB211-535E4	0	CAAC
B757-200	RR RB211-535E4	19	CHINA SOUTHERN AIRLINES
B757-200	RR RB211-535E4	7	CHINA SOUTHWEST AIRLINES
B757-200	RR RB211-535E4	1	CONDOR FLUGDIENST GMBH
B757-200	RR RB211-535E4	5	CONTINENTAL AIRLINES
B757-200	RR RB211-535E4	7	EL AL ISRAEL AIRLINES, LTD
B757-200	RR RB211-535E4	8	IBERIA - LINEAS AEREAS DE ESPANA, S.A.
B757-200	RR RB211-535E4	12	LTU - LUFTTRANSPORT UNTERNEHMEN GMBH

AIRFRAME	POWER PLANT	NO. IN FLEET	FLEET NAME
B757-200	RR RB211-535E4	9	LTU SUD INTERNATIONAL AIRWAYS
B757-200	RR RB211-535E4	8	MONARCH AIRLINES LTD
B757-200	RR RB211-535E4	1	MYANMA AIRWAYS (BURMA AIRWAYS CORPORATION)
B757-200	RR RB211-535E4	1	NORTH AMERICAN AIRLINES
B757-200	RR RB211-535E4	2	ROYAL NEPAL AIRLINES CORPORATION
B757-200	RR RB211-535E4	5	TAESA
B757-200	RR RB211-535E4	3	TRANSAERO AIRLINES
B757-200	RR RB211-535E4	1	TURKMENISTAN AIRLINE
B757-200	RR RB211-535E4	34	USAIR, INC
B757-200	RR RB211-535E4	3	VENUS AIRLINES SA
B757-200	RR RB211-535E4	3	XIAMEN AIRLINES
B757-200	RR RB211-535E4B	77	AMERICAN AIRLINES
B757-200	RR RB211-535E5	3	TRANSAVIA AIRLINES CV
B757-200ER	P&W PW2040	2	LADECO SA/LINEA AEREA DE COBRE/LADECO AIRLINES
B757-200ER	RR RB211-535E4	1	AIR SEYCHELLES
B757-200ER	RR RB211-535E4	19	BRITANNIA AIRWAYS LTD
B757-200ER	RR RB211-535E4	3	CALEDONIAN AIRWAYS
B757-200ER	RR RB211-535E4	8	CANADA 3000 AIRLINES LIMITED
B757-200ER	RR RB211-535E4	1	GUYANA AIRWAYS CORP
B757-200ER	RR RB211-535E4	3	ICELANDAIR
B757-200ER	RR RB211-535E4	3	ROYAL BRUNEI AIRLINES
B757-200F	P&W PW2040	1	ETHIOPIAN AIRLINES CORPORATION
B757-200F	RR RB211-535E4	1	GULF AIR
B757-200PF	P&W PW2040	45	UNITED PARCEL SERVICE COMPANY UPS
B757-200PF	RR RB211-535E4	3	CHALLENGE AIR CARGO
B757-200PF	RR RB211-535E4	10	UNITED PARCEL SERVICE COMPANY UPS
B767-200	GE CF6-80A	8	AMERICAN AIRLINES
B767-200	GE CF6-80A	6	ANSETT AUSTRALIA
B767-200	GE CF6-80A	2	BRITANNIA AIRWAYS LTD
B767-200	GE CF6-80A	13	DELTA AIR LINES, INC
B767-200	GE CF6-80A	1	TACA INTERNATIONAL AIRLINES SA
B767-200	GE CF6-80A	3	TRANSBRASIL S.A. LINHAS AEREAS
B767-200	GE CF6-80A2	2	DELTA AIR LINES, INC
B767-200	GE CF6-80C2	25	ALL NIPPON AIRWAYS
B767-200	GE CF6-80C2B2	4	EVA AIRWAYS
B767-200	P&W JT9D-7R4D	12	AIR CANADA
B767-200	P&W JT9D-7R4D	2	JAPAN AIRLINES COMPANY LTD - JAL
B767-200	P&W JT9D-7R4D	11	UNITED AIRLINES, INC
B767-200	P&W JT9D-7R4G2	2	EL AL ISRAEL AIRLINES, LTD
B767-200ER	GE CF6-80A	1	VIETNAM AIRLINES
B767-200ER	GE CF6-80A2	6	AIR NEW ZEALAND
B767-200ER	GE CF6-80A2	22	AMERICAN AIRLINES
B767-200ER	GE CF6-80A2	5	BRITANNIA AIRWAYS LTD
B767-200ER	GE CF6-80A2	2	LAN-CHILE
B767-200ER	GE CF6-80A2	1	TRANSBRASIL SA - LINHAS AEREAS
B767-200ER	GE CF6-80C2	2	MALEV - HUNGARIAN AIRLINES
B767-200ER	GE CF6-80C2B2	3	BRITISH AIRWAYS
B767-200ER	GE CF6-80C2B2	12	USAIR, INC
B767-200ER	GE CF6-80C2B2	6	VARIG - VIACAO AEREA RIO-GRANDENSE S.A.
B767-200ER	GE CF6-80C2B4	2	AIR MAURITIUS
B767-200ER	GE CF6-80C2B4	1	AIR SEYCHELLES
B767-200ER	GE CF6-80C2B4	2	BRITANNIA AIRWAYS LTD
B767-200ER	GE CF6-80C2B4	2	LOT - POLSKIE LINIE LOTNICZE
B767-200ER	P&W JT9D-7R4D	2	EL AL ISRAEL AIRLINES, LTD
B767-200ER	P&W JT9D-7R4D	10	TRANS WORLD AIRLINES, INC
B767-200ER	P&W JT9D-7R4D	1	TRANSBRASIL S.A. LINHAS AEREAS
B767-200ER	P&W JT9D-7R4D	8	UNITED AIRLINES, INC
B767-200ER	P&W JT9D-7R4D1	8	AIR CANADA
B767-200ER	P&W JT9D-7R4E	3	EGYPTAIR
B767-200ER	P&W JT9D-7R4E	3	ETHIOPIAN AIRLINES CORPORATION
B767-200ER	P&W JT9D-7R4E	7	QANTAS AIRWAYS LTD
B767-200ER	P&W JT9D-7R4E4	4	AIR CHINA INTERNATIONAL
B767-200ER	P&W PW4052	2	AIR CHINA INTERNATIONAL
B767-200ER	P&W PW4056	2	AEROVIAS NACIONALES DE COLOMBIA S.A. AVIANCA
B767-200ER	P&W PW4056	2	AIR ZIMBABWE
B767-200ER	P&W PW4056	2	LAM - LINHAS AEREAS DE MOCAMBIQUE
B767-200ER	P&W PW4056	0	SAS - SCANDINAVIAN AIRLINES SYSTEM
B767-200ER	P&W PW4056	1	SOUTH AFRICAN AIRWAYS (SAA)
B767-200ER	P&W PW4056	2	TRANSBRASIL SA - LINHAS AEREAS
B767-200ER	P&W PW4060	1	AEROMEXICO
B767-200ER	P&W PW4060	2	BALKAN
B767-300	NYK	0	UKRAINIAN AIRLINES
B767-300	GE CF6-80	0	AEROFLOT RUSSIAN INTERNATIONAL AIRLINES
B767-300	GE CF6-80A2	25	DELTA AIR LINES, INC
B767-300	GE CF6-80C2B2	31	ALL NIPPON AIRWAYS
B767-300	GE CF6-80C2B2F	3	AIR ALGERIE
B767-300	GE CF6-80C2BF	4	ASIANA AIRLINES
B767-300	GE CF6-80C2E2	1	WORLD AIR NETWORK
B767-300	P&W JT9D-7R4D	16	JAPAN AIRLINES COMPANY LTD - JAL
B767-300	P&W PW4000	0	SHANGHAI AIRLINES
B767-300	P&W PW4056	3	AIR CHINA INTERNATIONAL
B767-300	P&W PW4060	4	TRANS WORLD AIRLINES, INC
B767-300	RR RB211-524H	0	YUNNAN AIRLINES
B767-300ER	GE CF6-80C2	5	AIR NEW ZEALAND
B767-300ER	GE CF6-80C2	2	ALITALIA
B767-300ER	GE CF6-80C2	0	CONTINENTAL AIRLINES
B767-300ER	GE CF6-80C2	0	KLM ROYAL DUTCH AIRLINES
B767-300ER	GE CF6-80C2	2	LEISURE INTERNATIONAL AIRWAYS LTD
B767-300ER	GE CF6-80C2	2	LOT - POLSKIE LINIE LOTNICZE
B767-300ER	GE CF6-80C2	4	VARIG - VIACAO AEREA RIO-GRANDENSE S.A.
B767-300ER	GE CF6-80C2B4	1	AIR PACIFIC LIMITED
B767-300ER	GE CF6-80C2B4	0	BRITANNIA AIRWAYS LTD
B767-300ER	GE CF6-80C2B4	18	GULF AIR
B767-300ER	GE CF6-80C2B6	2	AIR EUROPE SPA
B767-300ER	GE CF6-80C2B6	6	ALL NIPPON AIRWAYS
B767-300ER	GE CF6-80C2B6	35	AMERICAN AIRLINES
B767-300ER	GE CF6-80C2B6	1	AUSTRALIAN ASIAN AIRLINES
B767-300ER	GE CF6-80C2B6	11	CANADIAN - CANADIEN
B767-300ER	GE CF6-80C2B6	6	CHINA SOUTHERN AIRLINES

AIRFRAME	POWER PLANT	NO. IN FLEET	FLEET NAME
B767-300ER	GE CF6-80C2B6	15	QANTAS AIRWAYS LTD
B767-300ER	GE CF6-80C2B6F	3	ASIANA AIRLINES
B767-300ER	GE CF6-80C2B6F	5	EVA AIRWAYS
B767-300ER	GE CF6-80C2B6F	1	SOBELAIR - SOCIETE BELGE DE TRANSPORTES AERIENS
B767-300ER	GE CF6-80C2B7F	2	AIRTOURS INTERNATIONAL AIRWAYS LTD
B767-300ER	P&W JT9D-7R4G2	2	EGYPTAIR
B767-300ER	P&W PW4050	12	SAS - SCANDINAVIAN AIRLINES SYSTEM
B767-300ER	P&W PW4056	1	LAUDA AIR LUFTFAHRT AKTIENGESELLSCHAFT
B767-300ER	P&W PW4056	6	ROYAL BRUNEI AIRLINES
B767-300ER	P&W PW4056	1	TACA INTERNATIONAL AIRLINES SA
B767-300ER	P&W PW4060	2	AER LINGUS
B767-300ER	P&W PW4060	2	AEROMEXICO
B767-300ER	P&W PW4060	2	AIR CANADA
B767-300ER	P&W PW4060	1	AIR EUROPE SPA
B767-300ER	P&W PW4060	3	AIR FRANCE
B767-300ER	P&W PW4060	2	ASIANA AIRLINES
B767-300ER	P&W PW4060	6	CONDOR FLUGDIENST GMBH
B767-300ER	P&W PW4060	14	DELTA AIR LINES, INC
B767-300ER	P&W PW4060	2	LAN-CHILE
B767-300ER	P&W PW4060	3	LAUDA AIR LUFTFAHRT AKTIENGESELLSCHAFT
B767-300ER	P&W PW4060	4	LTU - LUFTTRANSPORT UNTERNEHMEN GMBH
B767-300ER	P&W PW4060	4	LTU SUD INTERNATIONAL AIRWAYS
B767-300ER	P&W PW4060	5	MARTINAIR HOLLAND NV
B767-300ER	P&W PW4060	2	SPANAIR
B767-300ER	P&W PW4060	1	TAESA
B767-300ER	P&W PW4060	1	TRANSBRASIL S.A. LINHAS AEREAS
B767-300ER	P&W PW4060	23	UNITED AIRLINES, INC
B767-300ER	PW PW4060	1	REGION AIR
B767-300ER	RR RB211-524H	23	BRITISH AIRWAYS
B767-300ER (F)	GE CF6-80C2	1	UNITED PARCEL SERVICE COMPANY UPS
B767-300F	GE CF6-80C2B7F	0	ASIANA AIRLINES
B777	TBD	0	GARUDA INDONESIA PT
B777	TBD	0	VIRGIN ATLANTIC AIRWAYS
B777-200	GE GE90-B4	0	CONTINENTAL AIRLINES
B777-200	P&W PW4084	0	ALL NIPPON AIRWAYS
B777-200	P&W PW4084	0	JAPAN AIR SYSTEM CO LTD (NIHON AIR SYSTEM)
B777-200	P&W PW4084	0	JAPAN AIRLINES COMPANY LTD - JAL
B777-200	RR Trent 875	0	THAI AIRWAYS INTERNATIONAL
B777-200/200ER	GE GE90-B4	0	BRITISH AIRWAYS
B777-200/200ER	RR Trent 871	0	CATHAY PACIFIC AIRWAYS LTD
B777-200/200ER	RR Trent 871	0	EMIRATES
B777-200A	P&W PW4073	1	UNITED AIRLINES, INC
B777-200ER	GE GE90-B3	0	CHINA SOUTHERN AIRLINES
B777-200ER	GE GE90-B3	0	LAUDA AIR LUFTFAHRT AKTIENGESELLSCHAFT
B777-200ER	GE GE90-B4	0	EURALAIR INTERNATIONAL
B777-200ER	RR Trent 877	0	TRANSBRASIL S.A. LINHAS AEREAS
B777-200ER	RR Trent 884	0	EMIRATES
B777-200ER	TBD	0	KOREAN AIR
BAe (AW) 650-222 Argosy	RR Dart 532-2	2	SAFE AIR LTD
BAe (BAC) One-Eleven	RR Spey 511-14	3	DELTA AIR TRANSPORT NV
BAe (BAC) One-Eleven 200	RR Spey 506-14	9	KABO AIR
BAe (BAC) One-Eleven 200	RR Spey 506-14	2	OKADA AIR LTD
BAe (BAC) One-Eleven 200	RR Spey 511-14	2	LADECO SA/LINEA AEREA DE COBRE/LADECO AIRLINES
BAe (BAC) One-Eleven 300	RR Spey 511-14	2	LADECO SA/LINEA AEREA DE COBRE/LADECO AIRLINES
BAe (BAC) One-Eleven 300	RR Spey 511-14	4	OKADA AIR LTD
BAe (BAC) One-Eleven 400	RR Spey 511-14	2	GAS AIR
BAe (BAC) One-Eleven 400	RR Spey 511-14	4	KABO AIR
BAe (BAC) One-Eleven 400	RR Spey 511-14	11	OKADA AIR LTD
BAe (BAC) One-Eleven 400	RR Spey 511-14	1	PELITA AIR SERVICE
BAe (BAC) One-Eleven 400	RR Spey 512-14	4	MAERSK AIR LIMITED
BAe (BAC) One-Eleven 400	RR Spey 512-14DW	1	GAS AIR
BAe (BAC) One-Eleven 475F	RR Spey 511-14	1	TAROM
BAe (BAC) One-Eleven 500	RR Spey 511-14DW	7	TAROM
BAe (BAC) One-Eleven 500	RR Spey 512-14DW	8	AUSTRAL
BAe (BAC) One-Eleven 500	RR Spey 512-14DW	5	BRITISH WORLD AIRLINES
BAe (BAC) One-Eleven 500	RR Spey 512-14DW	3	CYPRUS AIRWAYS LTD
BAe (BAC) One-Eleven 500	RR Spey 512-14DW	2	MAERSK AIR LIMITED
BAe (BAC) One-Eleven 500	RR Spey 512-14DW	4	RYANAIR LTD
BAe (DH) 114 Riley Heron	Lyc IO-540-K1C5	2	AIR FIJI
BAe (DH) 114 Riley Heron	Lyc IO-540-K1C5	1	SUNFLOWER AIRLINES LTD
BAe (HS) 121 Trident 2E	RR Spey 511-5W	21	CAAC
BAe (HS) 121 Trident 2E	RR Spey 512-5W	4	CAAC
BAe (HS) 125-600B	RR Viper 601-22A	1	PELITA AIR SERVICE
BAe (HS) 125-700B	Ga TFE731-3-1H	1	KUWAIT AIRWAYS CORPORATION
BAe (HS) 125-800	Ga TFE731-3R-1H	1	MAERSK AIR
BAe (HS) 748	RR Dart 536-2	1	ROYAL TONGAN AIRLINES
BAe (HS) 748 LFD Srs 2A	RR Dart 534-2	3	BRADLEY AIR SERVICES LIMITED (FIRST AIR)
BAe (HS) 748 Srs 1	RR Dart 534-2	3	EMERALD AIRWAYS LTD
BAe (HS) 748 Srs 1A	RR Dart 533-2	2	AIR BVI LTD
BAe (HS) 748 Srs 2A	RR Dart 514-2	7	EMERALD AIRWAYS LTD
BAe (HS) 748 Srs 2A	RR Dart 534-2	1	AIR AUSTRAL
BAe (HS) 748 Srs 2A	RR Dart 534-2	1	AIR BVI LTD
BAe (HS) 748 Srs 2A	RR Dart 534-2	3	AIR FAST INDONESIA
BAe (HS) 748 Srs 2A	RR Dart 534-2	4	AIR INUIT (1985) LTD
BAe (HS) 748 Srs 2A	RR Dart 534-2	2	AIR SAINT-PIERRE
BAe (HS) 748 Srs 2A	RR Dart 534-2	9	BOURAQ INDONESIA AIRLINES PT
BAe (HS) 748 Srs 2A	RR Dart 534-2	5	BRADLEY AIR SERVICES LIMITED (FIRST AIR)
BAe (HS) 748 Srs 2A	RR Dart 534-2	4	CALM AIR (CANADIAN PARTNER)
BAe (HS) 748 Srs 2A	RR Dart 534-2	7	MOUNT COOK AIRLINE
BAe (HS) 748 Srs 2A	RR Dart 534-2	1	SATENA
BAe (HS) 748 Srs 2A	RR Dart 534-2	2	TACV - TRANSPORTES AEREOS DE CABO VERDE
BAe (HS) 748 Srs 2A	RR Dart 534-2	1	TAME - TRANSPORTES AEREOS MILITARES ECUATORIANOS, CA
BAe (HS) 748 Srs 2A	RR Dart 543-2	2	AIR SENEGAL, SOCIETE NATIONALE DE TRANSPORT AERIEN
BAe (HS) 748 Srs 2A	RR Dart 543-2	2	ROYAL NEPAL AIRLINES CORPORATION
BAe (HS) 748 Srs 2A (SCD)	RR Dart 534-2	1	TAME - TRANSPORTES AEREOS MILITARES ECUATORIANOS, CA
BAe (HS) 748 Srs 2B	RR Dart 535-2	1	BRADLEY AIR SERVICES LIMITED (FIRST AIR)
BAe (HS) 748 Srs 2B	RR Dart 535-2	1	SUN AIR (BOP AIR PTY LTD dba)
BAe (HS) 748 Srs 2B	RR Dart 536-2	2	AIR MADAGASCAR
BAe (HS) 748 Srs 2B	RR Dart 536-2	1	AIR MARSHALL ISLANDS - AMI

AIRFRAME	POWER PLANT	NO. IN FLEET	FLEET NAME
BAe (HS) 748 Srs 2B	RR Dart 536-2	6	BOURAQ INDONESIA AIRLINES PT
BAe (HS) 748 Srs 2B	RR Dart 536-2	1	SATENA
BAe (HS) 748 Super 2B	RR Dart 536-2	1	CAMEROON AIRLINES
BAe (HS) 748 Super 2B	RR Dart 536-2	4	LIAT (1974) LTD.
BAe (HS) 748 Super 2B	RR Dart 536-2	2	MAKUNG AIRLINES, CO LTD
BAe (HS) 748-200 Srs 2	RR Dart 534-2	4	AIR CREEBEC INC
BAe (HS) 748-2A Cargo	RR Dart 534-2	1	AIR MANITOBA
BAe (HS) 748-2A Combi	RR Dart 534-2	4	AIR MANITOBA
BAe (HS) 748-333 Srs 2A	RR Dart 535-2	1	TAGB - TRANSPORTES AEREOS DA GUINE BISSAU
BAe (Vickers) Frt master	RR Dart 510	8	BRITISH WORLD AIRLINES
BAe (Vickers) Merchantman	RR Tyne 506-10	2	HUNTING CARGO AIRLINES
BAe (Vickers) Viscount	RR Dart 506	1	ZAIREAN AIRLINES
BAe (Vickers) Viscount	RR Dart 525	1	MANDALA AIRLINES PT
BAe (Vickers) Viscount	RR Dart 525F	4	BOURAQ INDONESIA AIRLINES PT
BAe (Vickers) Viscount	RR Dart 525F	1	INTERCONTINENTAL DE AVIACION
BAe 146-100	Lyc ALF502R-5	1	AIR BOTSWANA
BAe 146-100	Lyc ALF502R-5	4	AIR CHINA INTERNATIONAL
BAe 146-100	Lyc ALF502R-5	1	AIR MALDIVES
BAe 146-100	Lyc ALF502R-5	1	AIR UK LTD
BAe 146-100	Lyc ALF502R-5	3	CHINA EASTERN AIRLINES
BAe 146-100	Lyc ALF502R-5	3	CHINA NORTHWEST AIRLINES
BAe 146-100	Lyc ALF502R-5	2	DRUK AIR CORP
BAe 146-100	Lyc ALF502R-5	1	JERSEY EUROPEAN
BAe 146-100	Lyc ALF502R-5	2	NATIONAL AIR CHARTER P.T.
BAe 146-100	Lyc ALF502R-5	1	QANTAS AIRWAYS LTD
BAe 146-100A	Lyc ALF502R-5	2	AIR WISCONSIN (UNITED EXPRESS)
BAe 146-200	Lyc ALF502R-5	3	AIR ATLANTIC LTD (CANADIAN PARTNER)
BAe 146-200	Lyc ALF502R-5	1	AIR UK LTD
BAe 146-200	Lyc ALF502R-5	1	AIR ZIMBABWE
BAe 146-200	Lyc ALF502R-5	1	AVIACION DE CHIAPAS (AVIACSA)
BAe 146-200	Lyc ALF502R-5	7	DELTA AIR TRANSPORT NV
BAe 146-200	Lyc ALF502R-5	2	EUROWINGS
BAe 146-200	Lyc ALF502R-5	2	HAMBURG AIRLINES LUFTFAHRTGESELLSCHAFT.MBH
BAe 146-200	Lyc ALF502R-5	4	JERSEY EUROPEAN
BAe 146-200	Lyc ALF502R-5	2	LAN-CHILE
BAe 146-200	Lyc ALF502R-5	5	MALMO AVIATION
BAe 146-200	Lyc ALF502R-5	2	MANX AIRLINES LTD
BAe 146-200	Lyc ALF502R-5	4	MERIDIANA SPA
BAe 146-200	Lyc ALF502R-5	1	PELITA AIR SERVICE
BAe 146-200	Lyc ALF502R-5	3	QANTAS AIRWAYS LTD
BAe 146-200A	Lyc ALF502R-5	5	AIR NOVA
BAe 146-200A	Lyc ALF502R-5	5	AIR WISCONSIN (UNITED EXPRESS)
BAe 146-200A	Lyc ALF502R-5	5	AIRBC LTD
BAe 146-200A	Lyc ALF502R-5	6	ANSETT AUSTRALIA
BAe 146-200A	Lyc ALF502R-5	1	ATLANTIC AIRWAYS (FAROE ISLANDS)
BAe 146-200A	Lyc ALF502R-5	5	BUSINESS EXPRESS (THE DELTA CONNECTION)
BAe 146-200A	Lyc ALF502R-5	17	USAIR, INC
BAe 146-200A	Lyc ALF502R-5	3	WESTAIR COMMUTER AIRLINES (MESA AIRLINES)
BAe 146-200QC	Lyc ALF502R-5	2	AIR JET
BAe 146-200QC	Lyc ALF502R-5	1	ANSETT NEW ZEALAND
BAe 146-200QT	Lyc ALF502R-5	3	AIR FOYLE LIMITED
BAe 146-200QT	Lyc ALF502R-5	2	ANSETT AUSTRALIA
BAe 146-200QT	Lyc ALF502R-5	2	EUROWINGS
BAe 146-200QT	Lyc ALF502R-5	1	MALMO AVIATION
BAe 146-200QT	Lyc ALF502R-5	3	PAN AIR LINEAS AEREAS SA
BAe 146-200QT	Lyc ALF502R-5	12	TNT INTERNATIONAL AVIATION SERVICES
BAe 146-300	Lyc ALF502R-5	3	AER LINGUS
BAe 146-300	Lyc ALF502R-5	0	AEROTAXI SUD
BAe 146-300	Lyc ALF502R-5	10	AIR UK LTD
BAe 146-300	Lyc ALF502R-5	6	ANSETT AUSTRALIA
BAe 146-300	Lyc ALF502R-5	8	ANSETT NEW ZEALAND
BAe 146-300	Lyc ALF502R-5	3	BRITISH WORLD AIRLINES
BAe 146-300	Lyc ALF502R-5	8	CHINA NORTHWEST AIRLINES
BAe 146-300	Lyc ALF502R-5	1	CROSSAIR
BAe 146-300	Lyc ALF502R-5	2	JERSEY EUROPEAN
BAe 146-300	Lyc ALF502R-5	5	MAKUNG AIRLINES, CO LTD
BAe 146-300	Lyc ALF502R-5	2	QANTAS AIRWAYS LTD
BAe 146-300	Lyc ALF502R-5	2	TAS AIRWAYS SPA
BAe 146-300	Lyc ALF502R-5	5	THAI AIRWAYS INTERNATIONAL
BAe 146-300A	Lyc ALF502R-5	5	AIR WISCONSIN (UNITED EXPRESS)
BAe 146-300QT	Lyc ALF502R-5	4	AIR FOYLE LIMITED
BAe 146-300QT	Lyc ALF502R-5	2	MALMO AVIATION
BAe 146-300QT	Lyc ALF502R-5	1	PAN AIR LINEAS AEREAS SA
BAe 146-300QT	Lyc ALF502R-5	9	TNT INTERNATIONAL AVIATION SERVICES
BAe ATP	P&WC PW125	5	MERPATI NUSANTARA AIRLINES PT
BAe ATP	P&WC PW126	2	BIMAN BANGLADESH AIRLINES
BAe ATP	P&WC PW126	14	BRITISH AIRWAYS
BAe ATP	P&WC PW126	3	BRITISH MIDLAND AIRWAYS LTD
BAe ATP	P&WC PW126	13	MANX AIRLINES LTD
BAe ATP	P&WC PW126	3	SATA AIR ACORES - SERVICO ACOREANO DE TRANSPORTES AEREOS, E.P.
BAe Jetstream 31	Ga TPE331-10UF-511H	1	MAERSK AIR LTD
BAe Jetstream 31	Ga TPE331-10UF-513H	1	AIR FOYLE LIMITED
BAe Jetstream 31	Ga TPE331-10UF-513H	1	REGIONAIR
BAe Jetstream 31	Ga TPE331-10UG-513H	5	AIRBC LTD
BAe Jetstream 31	Ga TPE331-10UG-513H	3	EASTERN AUSTRALIA AIRLINES
BAe Jetstream 31	Ga TPE331-10UG-513H	25	EXPRESS AIRLINES I, INC (NORTHWEST AIRLINK)
BAe Jetstream 31	Ga TPE331-10UG-513H	17	JETSTREAM INTERNATIONAL AIRLINES
BAe Jetstream 31	Ga TPE331-10UG-513H	21	MESA AIRLINES INC
BAe Jetstream 31	Ga TPE331-10UR-513H	10	SUN-AIR OF SCANDINAVIA A/S
BAe Jetstream 31	Ga TPE331-14GR/HR	34	WESTAIR COMMUTER AIRLINES (MESA AIRLINES)
BAe Jetstream 31	Ga TPE331-5-252D	12	CCAIR, INC (USAIR EXPRESS)
BAe Jetstream 41	Ga TPE331-14GR/HR	0	AMR EAGLE
BAe Jetstream 41	Ga TPE331-14GR/HR	1	BRITISH MIDLAND AIRWAYS LTD
BAe Jetstream 41	Ga TPE331-14GR/HR	10	MANX AIRLINES LTD
BAe Jetstream 41	Ga TPE331-14GR/HR	2	SUN-AIR OF SCANDINAVIA A/S
BAe Jetstream 41	Ga TPE331-14GR/HR	0	TRANS STATES AIRLINES
BAe Jetstream Super 31	Ga TPE331-10UR-513H	50	AMR EAGLE
BAe Jetstream Super 31	Ga TPE331-12UAR-701H	1	EASTERN AUSTRALIA AIRLINES
BAe Jetstream Super 31	Ga TPE331-12UAR-703H	9	REGIONAL AIRLINES

AIRFRAME	POWER PLANT	NO. IN FLEET	FLEET NAME
BAe Jetstream Super 31	Ga TPE331-12UR	22	TRANS STATES AIRLINES
BE 1300C Airliner	P&WC PT6A-41	1	ALPHA AIR
BE 18	P&W R-985	4	BERING AIR, INC
BE 1900 Airliner	P&WC PT6A-65B	13	AIR MIDWEST, INC (USAIR EXPRESS)
BE 1900 Airliner	P&WC PT6A-65B	16	BAR HARBOR AIRLINES (CONTINENTAL/DELTA EXPRESS)
BE 1900 Airliner	P&WC PT6A-65B	20	MESA AIRLINES INC
BE 1900C Airliner	P&WC PT6A-65B	3	AIR NAMIBIA
BE 1900C Airliner	P&WC PT6A-65B	1	ALPHA AIR
BE 1900C Airliner	P&WC PT6A-65B	21	BUSINESS EXPRESS (THE DELTA CONNECTION)
BE 1900C Airliner	P&WC PT6A-65B	9	CONQUEST AIRLINES
BE 1900C Airliner	P&WC PT6A-65B	5	CONTINENTAL EXPRESS
BE 1900C Airliner	P&WC PT6A-65B	1	INTEROT AIR SERVICE
BE 1900C Airliner	P&WC PT6A-65B	7	LIBERTY EXPRESS (USAIR EXPRESS)
BE 1900C Airliner	P&WC PT6A-65B	13	PENNSYLVANIA COMMUTER AIRLINES, INC (USAIR EXPRESS)
BE 1900C-1 Airliner	P&WC PT6A-65B	1	ALPHA AIR
BE 1900C-1 Airliner	P&WC PT6A-65B	2	BUSINESS EXPRESS (THE DELTA CONNECTION)
BE 1900C-1 Airliner	P&WC PT6A-65B	18	COMMUTAIR (USAIR EXPRESS)
BE 1900C-1 Airliner	P&WC PT6A-65B	10	CONTINENTAL EXPRESS
BE 1900C-1 Airliner	P&WC PT6A-65B	14	ROCKY MOUNTAIN AIRWAYS (CONTINENTAL EXPRESS)
BE 1900C-1 Airliner	P&WC PT6A-65B	4	STATESWEST AIRLINES, INC (USAIR EXPRESS)
BE 1900C-1 Airliner	P&WC PT6A-65B	1	TURKS & CAICOS AIRWAYS LIMITED
BE 1900D Airliner	P&WC PT6A-65D	6	LIBERTY EXPRESS (USAIR EXPRESS)
BE 1900D Airliner	P&WC PT6A-67D	8	COMMUTAIR (USAIR EXPRESS)
BE 1900D Airliner	P&WC PT6A-67D	91	MESA AIRLINES INC
BE 1900D Airliner	P&WC PT6A-67D	13	MIDWEST EXPRESS
BE 200 Airliner	P&WC PT6A-41	3	AIR SCHEFFERVILLE, INC
BE 65 Excalibur Queenaire	Lyc IO-720-A1B	1	SUNFLOWER AIRLINES LTD
BE 99 Airliner	P&WC PT6A-20	15	BAR HARBOR AIRLINES (CONTINENTAL/DELTA EXPRESS)
BE 99 Airliner	P&WC PT6A-27	1	AIR SCHEFFERVILLE, INC
BE A36 Bonanza	Cont IO-550-B	9	LUFTHANSA GERMAN AIRLINES - DEUTSCHE LUFTHANSA AG
BE A36 Bonanza	Cont IO-550-B	6	SAUDIA - SAUDI ARABIAN AIRLINES
BE Baron 58	Cont IO-520-C	3	AIR AFFAIRES GABON
BE Baron 58	Cont IO-520-C	1	AIR PROVENCE
BE Baron 58	Cont IO-520-C	1	AIRKENYA AVIATION LIMITED
BE Baron 58	Cont IO-540-C	19	LUFTHANSA GERMAN AIRLINES - DEUTSCHE LUFTHANSA AG
BE Baron 95-B55	Cont IO-470-L	2	ROYAL AIR MAROC
BE Baron 95-C55	Cont IO-520-C	1	AIR FIJI
BE Baron 95-E55	Cont IO-520-C	2	AIRKENYA AVIATION LIMITED
BE C99 Airliner	P&WC PT6A-36	14	BAR HARBOR AIRLINES (CONTINENTAL/DELTA EXPRESS)
BE C99 Airliner	P&WC PT6A-36	8	BRITT AIRWAYS, INC (CONTINENTAL EXPRESS)
BE Duke A60	Lyc TIO-541-E1B4	1	KELOWNA FLIGHTCRAFT AIR CHARTER LTD
BE F33A Bonanza	Cont IO-520-BA	59	LUFTHANSA GERMAN AIRLINES - DEUTSCHE LUFTHANSA AG
BE King Air 200	P&WC PT6A-41	3	TAT EUROPEAN AIRLINES
BE King Air 90	P&WC PT6A-28	1	AVENSA
BE King Air 90	P&WC PT6A-6	2	AIR PROVENCE
BE King Air A100	P&WC PT6A-28	2	SAUDIA - SAUDI ARABIAN AIRLINES
BE King Air A90 / B90	P&WC PT6A-20	1/1	AIR PROVENCE
BE King Air B100	GA TPE331-6-252B	1	AIRLINK AIRLINE (PTY) LTD
BE King Air B200	P&WC PT6A-41	4	PROVINCIAL AIRWAYS LTD
BE King Air E90	P&WC PT6A-28	1	AIR AFFAIRES GABON
BE King Air F90	P&WC PT6A-28	1	AIR AFFAIRES GABON
BE Queen Air 65-B80	Lyc IGSO-540-A1D	1	AIR FAST INDONESIA
BE Super King Air 200	P&WC PT6A-41	1	AIR CREEBEC INC
BE Super King Air 200	P&WC PT6A-41	1	AIR FOYLE LIMITED
BE Super King Air 200	P&WC PT6A-41	2	AIR IVOIRE
BE Super King Air 200	P&WC PT6A-41	2	AIR PROVENCE
BE Super King Air 200	P&WC PT6A-41	2	CALM AIR (CANADIAN PARTNER)
BE Super King Air 200	P&WC PT6A-41	1	INTEROT AIR SERVICE
BE Super King Air 200	P&WC PT6A-41	2	RATIOFLUG GMBH
BE Super King Air 200	P&WC PT6A-41	2	ROYAL AIR MAROC
BE Super King Air 200	P&WC PT6A-42	2	JAPAN AIR SYSTEM CO LTD (NIHON AIR SYSTEM)
BE Super King Air 200C	P&WC PT6A-41	1	AIR INUIT (1985) LTD
BE Super King Air 200C	P&WC PT6A-41	1	ERA AVIATION, INC
BE Super King Air 200C	P&WC PT6A-41	2	LAM - LINHAS AEREAS DE MOCAMBIQUE
BE Super King Air 200C	P&WC PT6A-42	1	AIR AFFAIRES GABON
BE Super King Air B200	P&WC PT6A-42	1	INTEROT AIR SERVICE
BE T34 Mentor	P&WC PT6A-25	2	LUFTHANSA GERMAN AIRLINES - DEUTSCHE LUFTHANSA AG
BE Twin Bonanza	Lyc GO-480-F4A6	1	ABOITIZ AIR TRANSPORT
BN-2 Islander	Lyc O-540-E4C5	1	TRANS-JAMAICAN AIRLINES LTD
BN-2A Islander	Lyc IO-540-E4C5	2	SERVICIOS AEREOS ASTRO SA
BN-2A Islander	Lyc IO-540-E4C5	3	TRANS ISLAND AIRWAYS LIMITED
BN-2A Islander	Lyc IO-540-K1B5	2	BALI AIR
BN-2A Islander	Lyc O-540-E4B5	1	ARKIA ISRAELI AIRLINES
BN-2A Islander	Lyc O-540-E4B5	3	MOUNT COOK AIRLINE
BN-2A Islander	Lyc O-540-E4C5	2	AIR ATLANTIQUE
BN-2A Islander	Lyc O-540-E4C5	3	AIR BVI LTD
BN-2A Islander	Lyc O-540-E4C5	2	AIR INTER GABON, SA.
BN-2A Islander	Lyc O-540-E4C5	1	AIR SEYCHELLES
BN-2A Islander	Lyc O-540-E4C5	3	FORMOSA AIRLINES
BN-2A Islander	Lyc O-540-E4C5	1	LIAT (1974) LTD.
BN-2A Islander	Lyc O-540-E4C5	2	SOLOMON AIRLINES
BN-2A Islander	Lyc O-540-E4C5	5	SUNFLOWER AIRLINES LTD
BN-2A Islander	Lyc O-540-E4C5	2	TURKS & CAICOS AIRWAYS LIMITED
BN-2A Trislander	Lyc IO-540-E4C5	2	TAIWAN AIRLINES
BN-2A Trislander	Lyc O-540-E4C5	2	AERO COZUMEL, SA
BN-2A Trislander	Lyc O-540-E4C5	7	AIR SAINT THOMAS
BN-2A Trislander	Lyc O-540-E4C5	2	AIR TUNGARU CORP
BN-2A Trislander	Lyc O-540-E4C5	9	AURIGNY AIR SERVICES LTD
BN-2A Trislander	Lyc O-540-E4C5	3	BALI AIR
BN-2A Trislander	Lyc O-540-E4C5	2	TRANS-JAMAICAN AIRLINES LTD
BN-2A-26 Islander	Lyc IO-540-E4C5	3	TAIWAN AIRLINES
BN-2A-26 Islander	Lyc O-540-E4C5	1	LOGANAIR LTD
BN-2A-26 Islander	Lyc O-540-E4C5	5	MAYA AIRWAYS LTD
BN-2A-3 Islander	Lyc IO-540-K1B5	2	IRAN ASSEMAN AIRLINES
BN-2A-8 Islander	Lyc O-540-E4C5	4	AIR MOOREA
BN-2A-8 Islander	Lyc O-540-E4C5	1	POLYNESIAN AIRLINES
BN-2B-20 Islander	Lyc IO-540-K1B5	5	CHINA SOUTHERN AIRLINES
BN-2B-20 Islander	Lyc IO-540-K1P5	1	AIR FIJI
BN-2B-26 Islander	Lyc IO-540-E4C5	1	TAIWAN AIRLINES

AIRFRAME	POWER PLANT	NO. IN FLEET	FLEET NAME
BN-2B-26 Islander	Lyc O-540-E4C5	4	LOGANAIR LTD
BN-2T Islander	Allison 250-B17C	1	AIR MAURITANIE
Bell 204B Iroquois	Lyc T53-11A	3	AIR FAST INDONESIA
Bell 206B JetRanger	Allison 250-C20	2	AIR MAURITIUS
Bell 206B JetRanger	Allison 250-C20	1	DHL AIRWAYS
Bell 206B JetRanger	Allison 250-C20B	1	NATIONAL OVERSEAS AIRLINE
Bell 206B JetRanger II	Allison 250-C20	3	AIR FAST INDONESIA
Bell 206B JetRanger II	Allison 250-C20	2	ERA AVIATION, INC
Bell 206B JetRanger II	Allison 250-C20	1	GROENLANDSFLY A/S
Bell 206B JetRanger II	Allison 250-C20	1	TRANSAFRIK
Bell 206B JetRanger III	Allison 250-C20	6	ERA AVIATION, INC
Bell 206B JetRanger III	Allison 250-C20	3	GROENLANDSFLY A/S
Bell 206L LongRanger	Allison 250-C20	5	ERA AVIATION, INC
Bell 206L-1 LongRanger II	Allison 250-C28B	1	DHL AIRWAYS
Bell 212	P&WC PT6T-3 TwinPac	1	AIR FAST INDONESIA
Bell 212	P&WC PT6T-3B TwinPac	14	ERA AVIATION, INC
Bell 212	P&WC PT6T-3B TwinPac	6	GROENLANDSFLY A/S
Bell 222	Allison LTS101-650C1	1	PAN AVIATION
Bell 412	P&WC PT6T-3 TwinPac	2	AIR FAST INDONESIA
Bell 412	P&WC PT6T-3B TwinPac	16	ERA AVIATION, INC
C 172 Skyhawk	Cont IO-520-D	2	PROVINCIAL AIRWAYS LTD
C 180	Cont IO-520-D	1	AIR SCHEFFERVILLE, INC
C 185	Cont IO-520-D	1	HOLMSTROM AIR
C 185A Skywagon	Cont IO-520-D	2	MOUNT COOK AIRLINE
C 185F Skywagon II	Cont IO-520-D	6	MOUNT COOK AIRLINE
C 207 Bonair conversion	Cont IO-550	1	MOUNT COOK AIRLINE
C 207 Stationair 8 II	Cont IO-520-F	4	BERING AIR, INC
C 208 Caravan I	P&WC PT6A-114	3	SATENA
C 208 Caravan I	P&WC PT6A-114A	35	EMPIRE AIRLINES
C 208A Caravan I	P&WC PT6A-114	10	FEDERAL EXPRESS CORPORATION
C 208B Caravan I	P&WC PT6A-114A	3	SCENIC AIRLINES INC
C 208B Caravan I	P&WC PT6A-114A	2	TRANSMILE AIR SERVICES
C 208B Caravan I Super	P&WC PT6A-114	155	FEDERAL EXPRESS CORPORATION
C 208B Caravan I Super	P&WC PT6A-114A	1	AIR MANITOBA
C 208B/3 Caravan I Super	P&WC PT6A-114	51	FEDERAL EXPRESS CORPORATION
C 210N Turbo Centurion II	Cont TSIO-520-R	1	SIERRA PACIFIC AIRLINES, INC
C 310L	Cont IO-470-V	1	AIR ATLANTIQUE
C 310Q II	Cont IO-470-VO	1	AIR TCHAD, SOCIETE DE TRANSPORTES AERIENS
C 310R II	Cont IO-520-M	2	AIR ATLANTIQUE
C 310R II	Cont IO-520-M	2	HAZELTON AIRLINES
C 310R2	Cont IO-520-MB	1	CROATIA AIRLINES
C 337 Super Skymaster	Cont IO-360-C	2	NATIONAL OVERSEAS AIRLINE
C 337B Super Skymaster	Cont IO-360-C/D	1	ARKIA ISRAELI AIRLINES
C 340	Cont TSIO-520-N	1	KELOWNA FLIGHTCRAFT AIR CHARTER LTD
C 401	Cont TSIO-520-E	1	AMERIJET INTERNATIONAL
C 402 II	Cont TSIO-520-VB	4	ALPHA AIR
C 402A	Cont TSIO-520-E	1	NATIONAL AIR CHARTER P.T.
C 402B	Cont TSIO-520-E	1	AIR ATLANTIQUE
C 402B	Cont TSIO-520-E	2	KELOWNA FLIGHTCRAFT AIR CHARTER LTD
C 402B	Cont TSIO-520-E	2	THERON AIRWAYS
C 402B	Cont TSIO-520-VB	1	MUK AIR
C 402B II	Cont TSIO-520-E	1	FLANDERS AIRLINES NV
C 402C	Cont TSIO-520-E	1	AIR ATLANTIQUE
C 402C Utililiner II	Cont TSIO-520-VB	1	AIR GUYANE
C 402C Utililiner II	Cont TSIO-520-VB	3	BAHAMASAIR
C 402C Utililiner II	Cont TSIO-520-VB	1	BIG SKY AIRLINES
C 402C Utililiner II	Cont TSIO-520-VB	3	CROATIA AIRLINES
C 402C Utililiner II	Cont TSIO-520-VB	6	LAM - LINHAS AEREAS DE MOCAMBIQUE
C 402C Utililiner II	Cont TSIO-520-VB	1	SUN-AIR OF SCANDINAVIA A/S
C 404 Titan	Cont GTSIO-520-M	3	AIR ATLANTIQUE
C 404 Titan	Cont GTSIO-520-M	1	THERON AIRWAYS
C 404 Titan II	Cont GTSIO-520-M	2	AIR INTER GABON, SA.
C 406	Cont TSIO-520	1	AIR ATLANTIQUE
C 421B Golden Eagle	Cont GTSIO-520-H	1	NATIONAL OVERSEAS AIRLINE
C 421C Golden Eagle	Cont GTSIO-520-L	1	NATIONAL OVERSEAS AIRLINE
C 500 Citation	P&WC JT15D-1A	1	SUN-AIR OF SCANDINAVIA A/S
C 500 Citation I	P&WC JT15D-1	1	AIR ATLANTIQUE
C 500 Citation I	P&WC JT15D-1A	2	EURALAIR INTERNATIONAL
C 500 Citation IA	P&WC JT15D-1A	1	SUN-AIR OF SCANDINAVIA A/S
C 501 Citation I/SP	P&WC JT15D-1A	1	AMERIJET INTERNATIONAL
C 501 Citation I/SP	P&WC JT15D-1A	1	SUN-AIR OF SCANDINAVIA A/S
C 525 CitationJet	RR/Williams Int FJ44	1	EURALAIR INTERNATIONAL
C 550 Citation II	P&WC JT15D-4	1	OMAN AIR
C 550 Citation II	P&WC JT15D-4	2	SAUDIA - SAUDI ARABIAN AIRLINES
C 550 Citation II	P&WC JT15D-4	1	SUN AIR (BOP AIR PTY LTD dba)
C 560 Citation V	P&WC JT15D-5A	2	EURALAIR INTERNATIONAL
C Citation I/SP	P&WC JT15D-1A	1	CIMBER AIR DENMARK A/S
C F337E Super Skymaster	Cont IO-360-C	1	ARKIA ISRAELI AIRLINES
C F406 Caravan II	P&WC PT6A-112	1	TRANSMILE AIR SERVICES
C T210L Tbo Centurion II	Cont TSIO-520-H	1	PAN AVIATION
C T210M Tbo Centurion II	Cont TSIO-520-R	1	THERON AIRWAYS (PTY) LTD
C T337GP Skymaster	Cont TSIO-360-C	1	ARKIA ISRAELI AIRLINES
C T337HP Skymaster II	Cont TSIO-360-C	1	ARKIA ISRAELI AIRLINES
C U206G Stationair 6 II	Cont IO-520-F	2	TRANS-JAMAICAN AIRLINES LTD
C550 CiTation II	P&WC JT15D-4	1	LAUDA AIR LUFTFAHRT AKTIENGESELLSCHAFT
CASA 212-200 Aviocar	Ga TPE331-10-501C	1	AIR FAST INDONESIA
CASA 212-200 Aviocar	Ga TPE331-10-501C	4	LAM - LINHAS AEREAS DE MOCAMBIQUE
CASA 212-200 Aviocar	Ga TPE331-10R-511C	1	AIR TUNGARU CORP
CASA 212-200 Aviocar	Ga TPE331-10R-511C	5	SATENA
CASA 212-300 Aviocar	Ga TPE331-10R-511C	2	SATENA
CASA/IPTN CN235-10	GE CT7-7A	4	BINTER
CASA/IPTN CN235-200	GE CT7-9C	5	BINTER
CASA/IPTN CN235-QC	GE CT7-9	1	BINTER
CL 600 Challenger	Lyc ALF502L-2C	0	RATIOFLUG GMBH
CL 600 Regional Jet	GE CF34-3A	9	AIR CANADA
CL 600 Regional Jet	GE CF34-3A	4	AIR LITTORAL
CL 600 Regional Jet	GE CF34-3A	1	BRIT'AIR
CL 600 Regional Jet	GE CF34-3A	17	COMAIR
CL 600 Regional Jet	GE CF34-3A	15	LUFTHANSA CITYLINE

AIRFRAME	POWER PLANT	NO. IN FLEET	FLEET NAME
CL 600 Regional Jet	GE CF34-3A	0	YORK AIRCRAFT LEASING INC.
CL 600 Regional Jet 100ER	GE CF34-3A	8	SKYWEST AIRLINES INC (DELTA CONNECTION)
CL 600 Regional Jet 100LR	GE CF34-3A	7	LAUDA AIR LUFTFAHRT AKTIENGESELLSCHAFT
CL 601-3A Challenger	GE CF34-3A	1	SCIBE AIRLIFT
CL 601-3A ER Challenger	GE CF34-3A	1	THAI AIRWAYS INTERNATIONAL
CV 240	P&W R-2800	1	AEROCHAGO AIRLINES, S.A.
CV 240	P&W R-2800	3	KITTY HAWK GROUP, INC.
CV 340	P&W R-2800	1	AEROCARIBE
CV 440 Metropolitan	P&W R-2800	1	AEROCARIBE
CV 440 Metropolitan	P&W R-2800	5	KITTY HAWK GROUP, INC.
CV 580	Allison 501-D13	3	AIR CAPE
CV 580	Allison 501-D13	4	AIR INUIT (1985) LTD
CV 580	Allison 501-D13	2	AIR NIAGARA EXPRESS INC
CV 580	Allison 501-D13	4	AIR RESORTS AIRLINES
CV 580	Allison 501-D13	1	AVENSA
CV 580	Allison 501-D13	10	CANAIR CARGO LTD
CV 580	Allison 501-D13	4	ERA AVIATION, INC
CV 580	Allison 501-D13	16	KELOWNA FLIGHTCRAFT AIR CHARTER LTD
CV 580	Allison 501-D13	2	REGIONAL AIR (PTY) LTD
CV 580	Allison 501-D13	7	SIERRA PACIFIC AIRLINES, INC
CV 580	Allison 501-D13	2	TURKS & CAICOS AIRWAYS LIMITED
CV 580 (F)	Allison 501-D13	11	EUROPEAN AIR TRANSPORT N.V./S.A.
CV 5800	Allison 501-D22G	2	KELOWNA FLIGHTCRAFT AIR CHARTER LTD
CV 580F (SCD)	Allison 501-D13	4	JETALL
CV 600	RR Dart 542-6	2	SERVICIOS AEREOS ASTRO SA
CV 600F	RR Dart 542-4	8	KITTY HAWK GROUP, INC.
CV 640 (F)	RR Dart 542-4	11	ZANTOP INTERNATIONAL AIRLINES, INC
Convair 440	P&W R-2800	2	FOUR STAR AIR CARGO
DA Falcon 10	Ga TFE731-2-1C	1	EURALAIR INTERNATIONAL
DA Falcon 20C	GE CF700-2C	1	LIBYAN ARAB AIRLINES
DA Falcon 20E	GE CF700-2D2	3	IRAN ASSEMAN AIRLINES
DA Falcon 20F	GE CF700-2D2	1	IRAN ASSEMAN AIRLINES
DA Falcon 20F	GE CF700-2D2	2	IRAQI AIRWAYS
DA Falcon 20F	GE CF700-2D2	2	SYRIAN ARAB AIRLINES
DA Falcon 20F-5	Ga TFE731-5AR-2C	1	EURALAIR INTERNATIONAL
DA Falcon 50	Ga TFE731-3-1C	3	IRAQI AIRWAYS
DA Falcon 50	Ga TFE731-3-1C	1	LIBYAN ARAB AIRLINES
DA Falcon 900	Ga TFE731-5AR-1C2	2	SAUDIA - SAUDI ARABIAN AIRLINES
DC-10	GE CF6-50C2B	0	ACS OF CANADA - AIR CHARTER SYSTEMS
DC-3	P&W R-1830	1	AIR FAST INDONESIA
DC-3	P&W R-1830	2	CARIBBEAN INTERNATIONAL AIRWAYS
DC-3	P&W R-1830	1	COMAIR (COMMERCIAL AIRWAYS (PTY) LTD)
DC-3	P&W R-1830	1	KELOWNA FLIGHTCRAFT AIR CHARTER LTD
DC-3	P&W R-1830	1	REGIONAL AIR (PTY) LTD
DC-3	P&W R-1830	4	SERVIVENSA
DC-3C	P&W R-1830	3	AEROCARIBBEAN
DC-3C	P&W R-1830	10	AIR ATLANTIQUE
DC-3C	P&W R-1830	4	AIR NORTH
DC-3C	P&W R-1830	2	AIRKENYA AVIATION LIMITED
DC-3C	P&W R-1830	5	FOUR STAR AIR CARGO
DC-3C	P&W R-1830	3	SAHSA - SERVICIO AEREO DE HONDURAS, SA
DC-4	P&W R-2000	1	AIR NORTH
DC-6A	P&W R-2800	2	AIR ATLANTIQUE
DC-6A	P&W R-2800	11	NORTHERN AIR CARGO INC
DC-6A	P&W R-2800	3	TRANS-AIR-LINK CORPORATION
DC-6A (F)	P&W R-2800	1	DOMINICANA DE AVIACION
DC-6B	P&W R-2800	1	NORTHERN AIR CARGO INC
DC-6BF	P&W R-2800	1	AESA AIRLINES
DC-6BF Swingtail	P&W R-2800	2	NORTHERN AIR CARGO INC
DC-7CF	CW R-3350	1	TRANS-AIR-LINK CORPORATION
DC-9	P&W JT8D	1	SATENA
DC-9	P&W JT8D-7	2	FLORIDA WEST AIRLINES INC
DC-9	P&W JT8D-7B	3	KITTY HAWK GROUP, INC.
DC10-10	GE CF6-6D	6	CONTINENTAL AIRLINES
DC10-10	GE CF6-6D	6	CONTINENTAL MICRONESIA, INC
DC10-10	GE CF6-6D	31	UNITED AIRLINES, INC
DC10-10	GE CF6-6D1A	4	PREMAIR A/S
DC10-10	GE CF6-6D1A	2	SUN COUNTRY AIRLINES
DC10-10	GE CF6-6K	46	AMERICAN AIRLINES
DC10-10	GE CF6-6K	6	HAWAIIAN AIRLINES
DC10-10ER	GE CF6-6K2	3	AMERICAN AIRLINES
DC10-10F	GE CF6-6D1A	2	FEDERAL EXPRESS CORPORATION
DC10-10F (CF)	GE CF6-6D	11	FEDERAL EXPRESS CORPORATION
DC10-15	GE CF6-50C2F	2	AEROPERU
DC10-15	GE CF6-50C2F	6	MEXICANA
DC10-30	GE CF6-50C	3	AIR AFRIQUE
DC10-30	GE CF6-50C	1	AIR ZAIRE
DC10-30	GE CF6-50C	4	AOM FRENCH AIRLINES
DC10-30	GE CF6-50C	7	BRITISH AIRWAYS
DC10-30	GE CF6-50C	1	ECUATORIANA
DC10-30	GE CF6-50C	6	GARUDA INDONESIA PT
DC10-30	GE CF6-50C	8	IBERIA - LINEAS AEREAS DE ESPANA, S.A.
DC10-30	GE CF6-50C	3	KOREAN AIR
DC10-30	GE CF6-50C	2	NIGERIA AIRWAYS
DC10-30	GE CF6-50C	6	NORTHWEST AIRLINES
DC10-30	GE CF6-50C	1	PRIMERAS LINEAS URUGUAYAS DE NAVEGACION AEREA - PLUNA
DC10-30	GE CF6-50C	5	VIASA, VENEZOLANA INTERNACIONAL DE AVIACION SA
DC10-30	GE CF6-50C1	3	JAT - JUGOSLOVENSKI AEROTRANSPORT
DC10-30	GE CF6-50C2	9	AMERICAN AIRLINES
DC10-30	GE CF6-50C2	5	AOM FRENCH AIRLINES
DC10-30	GE CF6-50C2	6	BIMAN BANGLADESH AIRLINES
DC10-30	GE CF6-50C2	2	CALEDONIAN AIRWAYS
DC10-30	GE CF6-50C2	3	CANADIAN - CANADIEN
DC10-30	GE CF6-50C2	3	CONDOR FLUGDIENST GMBH
DC10-30	GE CF6-50C2	13	CONTINENTAL AIRLINES
DC10-30	GE CF6-50C2	3	FINNAIR OY
DC10-30	GE CF6-50C2	1	GHANA AIRWAYS CORPORATION
DC10-30	GE CF6-50C2	2	JAPAN AIR SYSTEM CO LTD (NIHON AIR SYSTEM)
DC10-30	GE CF6-50C2	11	LUFTHANSA GERMAN AIRLINES - DEUTSCHE LUFTHANSA AG

AIRFRAME	POWER PLANT	NO. IN FLEET	FLEET NAME
DC10-30	GE CF6-50C2	4	MALAYSIA AIRLINES
DC10-30	GE CF6-50C2	1	PHILIPPINE AIR LINES, INC - PAL
DC10-30	GE CF6-50C2	2	SABENA BELGIAN WORLD AIRLINES
DC10-30	GE CF6-50C2	8	UNITED AIRLINES, INC
DC10-30	GE CF6-50C2	8	VARIG - VIACAO AEREA RIO-GRANDENSE S.A.
DC10-30	GE CF6-50C2	4	WORLD AIRWAYS INC
DC10-30	GE CF6-50C2B	2	AOM FRENCH AIRLINES
DC10-30	GE CF6-50C2B	2	CUBANA
DC10-30	GE CF6-50C2B	1	NORTHWEST AIRLINES
DC10-30	GE CF6-50C2R	1	AFRICAN SAFARI AIRWAYS LTD
DC10-30	GE CF6-50C2R	5	AIR FRANCE
DC10-30	GE CF6-50C2R	2	EUROPEAN AIRLIFT
DC10-30ER	GE CF6-502CB	3	THAI AIRWAYS INTERNATIONAL
DC10-30ER	GE CF6-50C2	5	CANADIAN - CANADIEN
DC10-30ER	GE CF6-50C2B	1	FINNAIR OY
DC10-30ER	GE CF6-50C2B	1	NORTHWEST AIRLINES
DC10-30F	GE CF6-50C2	1	AEROFLOT RUSSIAN INTERNATIONAL AIRLINES
DC10-30F (AF)/(CF)	GE CF6-50C2	22	FEDERAL EXPRESS CORPORATION
DC10-30F (CF)	GE CF6-50C2	2	VARIG - VIACAO AEREA RIO-GRANDENSE S.A.
DC10-40	P&W JT9D-20	21	NORTHWEST AIRLINES
DC10-40	P&W JT9D-59A	1	JAPAN AIR CHARTER
DC10-40	P&W JT9D-59A	15	JAPAN AIRLINES COMPANY LTD - JAL
DC10-40	P&W JT9D-59A	4	JAPAN ASIA AIRWAYS (NIHON ASIA KOKU)
DC8-33F	P&W JT4A-9	1	APISA - AEROTRANSPORTES PERUANOS INTERNCIONALES SA
DC8-33F	P&W JT4A-9	2	TRANSAFRIK
DC8-50	P&W JT3D-3B (Q)	2	CARGO D'OR
DC8-51F	P&W JT3D-3B (Q)	2	KALITTA AMERICAN INTERNATIONAL AIRWAYS
DC8-52	P&W JT3D-3B (Q)	1	KALITTA AMERICAN INTERNATIONAL AIRWAYS
DC8-53F	P&W JT3D-3B	1	ARCA COLOMBIA
DC8-54	P&W JT3D-3B (Q)	1	BUFFALO AIRWAYS
DC8-54	P&W JT3D-3B (Q)	2	KALITTA AMERICAN INTERNATIONAL AIRWAYS
DC8-54	P&W JT3D-3B (Q)	2	ZANTOP INTERNATIONAL AIRLINES, INC
DC8-54F	P&W JT3D-3B (Q)	1	AERONAVES DEL PERU, S.A.
DC8-54F	P&W JT3D-3B (Q)	3	AFRICAN INTERNATIONAL AIRWAYS
DC8-54F	P&W JT3D-3B (Q)	1	AIR ZAIRE
DC8-54F	P&W JT3D-3B (Q)	1	EMERY WORLDWIDE
DC8-55	P&W JT3D-3B	2	AFFRETAIR (PRIVATE) LTD
DC8-55	P&W JT3D-3B	1	KABO AIR
DC8-55	P&W JT3D-3B (Q)	1	FLASH AIRLINES
DC8-55	P&W JT3D-3B (Q)	6	KALITTA AMERICAN INTERNATIONAL AIRWAYS
DC8-55	P&W JT3D-3B (Q)	1	LINEAS AEREAS COSTARRICENSES, S.A. - LACSA
DC8-55F	P&W JT3D-3B	1	ARCA COLOMBIA
DC8-55F	P&W JT3D-3B (Q)	1	ACS OF CANADA - AIR CHARTER SYSTEMS
DC8-61	P&W JT3D-3B (Q)	1	AEROPERU
DC8-61	P&W JT3D-3B (Q)	10	AIRBORNE EXPRESS (ABX AIR, INC dba)
DC8-61	P&W JT3D-3B (Q)	3	BUFFALO AIRWAYS
DC8-61F	P&W JT3D-3B (Q)	1	AERONAVES DEL PERU, S.A.
DC8-61F	P&W JT3D-3B (Q)	2	AIR TRANSPORT INTERNATIONAL
DC8-61F	P&W JT3D-3B (Q)	5	KALITTA AMERICAN INTERNATIONAL AIRWAYS
DC8-62	P&W JT3D-3B (Q)	2	AEROPERU
DC8-62	P&W JT3D-3B (Q)	6	AIRBORNE EXPRESS (ABX AIR, INC dba)
DC8-62	P&W JT3D-3B (Q)	1	AOM FRENCH AIRLINES
DC8-62	P&W JT3D-3B (Q)	2	BUFFALO AIRWAYS
DC8-62	P&W JT3D-3B (Q)	3	RICH INTERNATIONAL AIRWAYS INC
DC8-62	P&W JT3D-7	1	AEROPERU
DC8-62CF	P&W JT3D-3B (Q)	2	INTERNATIONAL CARGO EXPRESS (ICX dba)
DC8-62CF	P&W JT3D-7 (Q)	1	AIR MARSHALL ISLANDS - AMI
DC8-62F	P&W JT3D-3B (Q)	2	AIR TRANSPORT INTERNATIONAL
DC8-62F	P&W JT3D-3B (Q)	1	ARROW AIR INC
DC8-62F	P&W JT3D-3B (Q)	7	EMERY WORLDWIDE
DC8-62F	P&W JT3D-3B (Q)	1	EVERGREEN INTERNATIONAL AIRLINES, INC
DC8-62F	P&W JT3D-7 (Q)	9	ARROW AIR INC
DC8-62F	P&W JT3D-7 (Q)	2	CARGOSUR - COMPANIA DE EXPLOITATION DE AVIONES CARGUEROS SA
DC8-63	P&W JT3D-3B (Q)	1	BUFFALO AIRWAYS
DC8-63	P&W JT3D-3B (Q)	1	RICH INTERNATIONAL AIRWAYS INC
DC8-63	P&W JT3D-7	1	SURINAM AIRWAYS
DC8-63	P&W JT3D-7 (Q)	2	AER TURAS TEORANTA
DC8-63	P&W JT3D-7 (Q)	1	AIR STARLINE AG
DC8-63	P&W JT3D-7 (Q)	13	AIRBORNE EXPRESS (ABX AIR, INC dba)
DC8-63AF (F)	P&W JT3D-7 (Q)	1	AIR MARSHALL ISLANDS - AMI
DC8-63F	P&W JT3D-7	4	BURLINGTON AIR EXPRESS
DC8-63F	P&W JT3D-7 (Q)	2	AIR TRANSPORT INTERNATIONAL
DC8-63F	P&W JT3D-7 (Q)	4	ARROW AIR INC
DC8-63F	P&W JT3D-7 (Q)	12	EMERY WORLDWIDE
DC8-63F	P&W JT3D-7 (Q)	2	KALITTA AMERICAN INTERNATIONAL AIRWAYS
DC8-63F	P&W JT3D-7 (Q)	1	SAUDIA - SAUDI ARABIAN AIRLINES
DC8-71F	CFMI CFM56-2	24	UNITED PARCEL SERVICE COMPANY UPS
DC8-71F	CFMI CFM56-2B1	3	BURLINGTON AIR EXPRESS
DC8-71F	CFMI CFM56-2C	1	CARGOSUR - COMPANIA DE EXPLOITATION DE AVIONES CARGUEROS SA
DC8-71F	CFMI CFM56-2C	9	EMERY WORLDWIDE
DC8-71F	CFMI CFM56-2C1	9	AIR TRANSPORT INTERNATIONAL
DC8-71F	CFMI CFM56-2C1	3	FLAGSHIP EXPRESS
DC8-71F	CFMI CFM56-2C1	2	LAN-CHILE
DC8-71F	CFMI CFM56-2C1	3	SOUTHERN AIR TRANSPORT, INC
DC8-72	CFMI CFM56-2C5	1	SAUDIA - SAUDI ARABIAN AIRLINES
DC8-73	CFMI CFM56-2C5	0	DHL AIRWAYS
DC8-73F	CFMI CFM56-2	1	AIR INDIA
DC8-73F	CFMI CFM56-2	2	EVERGREEN INTERNATIONAL AIRLINES, INC
DC8-73F	CFMI CFM56-2	28	UNITED PARCEL SERVICE COMPANY UPS
DC8-73F	CFMI CFM56-2C	8	EMERY WORLDWIDE
DC8-73F	CFMI CFM56-2C1	2	SOUTHERN AIR TRANSPORT, INC
DC8-73F	CFMI CFM56-2C5	5	GERMAN CARGO SERVICES
DC9-10	P&W JT8D-7	7	TRANS WORLD AIRLINES, INC
DC9-14	P&W JT8D-7A	1	BRITISH MIDLAND AIRWAYS LTD
DC9-14	P&W JT8D-7A	1	NORTHWEST AIRLINES
DC9-14	P&W JT8D-7B	6	MIDWEST EXPRESS
DC9-14	P&W JT8D-7B	16	NORTHWEST AIRLINES
DC9-15	P&W JT8D-7A	1	BRITISH MIDLAND AIRWAYS LTD
DC9-15	P&W JT8D-7A/7B	13	INTERCONTINENTAL DE AVIACION

AIRFRAME	POWER PLANT	NO. IN FLEET	FLEET NAME
DC9-15	P&W JT8D-7B	11	AEROCALIFORNIA
DC9-15	P&W JT8D-7B	2	AIRBORNE EXPRESS (ABX AIR, INC dba)
DC9-15	P&W JT8D-7B	1	EVERGREEN INTERNATIONAL AIRLINES, INC
DC9-15	P&W JT8D-7B	3	GREAT AMERICAN AIRWAYS
DC9-15	P&W JT8D-7B	2	MIDWEST EXPRESS
DC9-15	P&W JT8D-7B	4	NORTHWEST AIRLINES
DC9-15F	P&W JT8D-7A	1	EVERGREEN INTERNATIONAL AIRLINES, INC
DC9-15F	P&W JT8D-7A	1	KALITTA AMERICAN INTERNATIONAL AIRWAYS
DC9-15F	P&W JT8D-7B	5	EMERY WORLDWIDE
DC9-15F	P&W JT8D-7B	1	KALITTA AMERICAN INTERNATIONAL AIRWAYS
DC9-21	P&W JT8D-11	2	SAS - SCANDINAVIAN AIRLINES SYSTEM
DC9-30	P&W JT8D-17	2	BINTER
DC9-30	P&W JT8D-7A	35	AIR CANADA
DC9-30	P&W JT8D-9A	2	ADRIA AIRWAYS
DC9-30	P&W JT8D-9A	34	CONTINENTAL AIRLINES
DC9-31	P&W JT8D-15	7	NORTHWEST AIRLINES
DC9-31	P&W JT8D-17	3	AEROMEXICO
DC9-31	P&W JT8D-17	3	AEROPOSTAL
DC9-31	P&W JT8D-7	1	NORTHWEST AIRLINES
DC9-31	P&W JT8D-7B	18	AIRBORNE EXPRESS (ABX AIR, INC dba)
DC9-31	P&W JT8D-7B	1	AVENSA
DC9-31	P&W JT8D-7B	23	NORTHWEST AIRLINES
DC9-31	P&W JT8D-7B/9A	68	USAIR, INC
DC9-31	P&W JT8D-9	26	NORTHWEST AIRLINES
DC9-31	P&W JT8D-9	18	TRANS WORLD AIRLINES, INC
DC9-31	P&W JT8D-9A	1	AIR ARUBA
DC9-32	P&W JT8D-11	1	BRITISH MIDLAND AIRWAYS LTD
DC9-32	P&W JT8D-15	10	NORTHWEST AIRLINES
DC9-32	P&W JT8D-15	5	SUN AIR (BOP AIR PTY LTD dba)
DC9-32	P&W JT8D-17	15	AEROMEXICO
DC9-32	P&W JT8D-17	2	SERVIVENSA
DC9-32	P&W JT8D-7	4	IBERIA - LINEAS AEREAS DE ESPANA, S.A.
DC9-32	P&W JT8D-7B	7	AEROCALIFORNIA
DC9-32	P&W JT8D-7B	16	AIRBORNE EXPRESS (ABX AIR, INC dba)
DC9-32	P&W JT8D-7B	4	MIDWEST EXPRESS
DC9-32	P&W JT8D-7B	1	SERVIVENSA
DC9-32	P&W JT8D-9	3	BRITISH MIDLAND AIRWAYS LTD
DC9-32	P&W JT8D-9	6	MERPATI NUSANTARA AIRLINES PT
DC9-32	P&W JT8D-9/-15	14	TRANS WORLD AIRLINES, INC
DC9-32	P&W JT8D-9A	34	ALITALIA
DC9-32	P&W JT8D-9A	14	AVIACO
DC9-32	P&W JT8D-9A	5	MIDWEST EXPRESS
DC9-32	P&W JT8D-9A	10	NORTHWEST AIRLINES
DC9-32F	P&W JT8D-7	2	EVERGREEN INTERNATIONAL AIRLINES, INC
DC9-32F	P&W JT8D-7B	3	AIRBORNE EXPRESS (ABX AIR, INC dba)
DC9-33	P&W JT8D-9	1	TRANS WORLD AIRLINES, INC
DC9-33F	P&W JT8D-9	6	AIRBORNE EXPRESS (ABX AIR, INC dba)
DC9-33F	P&W JT8D-9A	1	EVERGREEN INTERNATIONAL AIRLINES, INC
DC9-33RC(F)	P&W JT8D-9A	3	EVERGREEN INTERNATIONAL AIRLINES, INC
DC9-34	P&W JT8D-15	3	TRANS WORLD AIRLINES, INC
DC9-34	P&W JT8D-17	3	AVIACO
DC9-34	P&W JT8D-17	3	IBERIA - LINEAS AEREAS DE ESPANA, S.A.
DC9-34CF	P&W JT8D-17	1	AEROPOSTAL
DC9-34F	P&W JT8D-17	2	AVIACO
DC9-41	P&W JT8D-11	10	AIRBORNE EXPRESS (ABX AIR, INC dba)
DC9-41	P&W JT8D-11	12	NORTHWEST AIRLINES
DC9-41	P&W JT8D-11	25	SAS - SCANDINAVIAN AIRLINES SYSTEM
DC9-41	P&W JT8D-15	6	AIRBORNE EXPRESS (ABX AIR, INC dba)
DC9-41	P&W JT8D-15	10	JAPAN AIR SYSTEM CO LTD (NIHON AIR SYSTEM)
DC9-41	P&W JT8D-15	3	TRANS WORLD AIRLINES, INC
DC9-41	P&W JT8D-17	5	FINNAIR OY
DC9-50	P&W JT8D-17	1	GHANA AIRWAYS CORPORATION
DC9-51	P&W JT8D-17	6	AEROPOSTAL
DC9-51	P&W JT8D-17	4	AVENSA
DC9-51	P&W JT8D-17	2	EUROFLY SPA
DC9-51	P&W JT8D-17	12	FINNAIR OY
DC9-51	P&W JT8D-17	13	HAWAIIAN AIRLINES
DC9-51	P&W JT8D-17	6	MERIDIANA SPA
DC9-51	P&W JT8D-17	34	NORTHWEST AIRLINES
DC9-51	P&W JT8D-17	12	TRANS WORLD AIRLINES, INC
DC9-51	P&W JT8D-17A	3	AEROPOSTAL
DHC-2 Beaver	P&W R-985	2	BRADLEY AIR SERVICES LIMITED (FIRST AIR)
DHC-3 Otter	P&W R-1340	2	AIR SCHEFFERVILLE, INC
DHC-5A Buffalo	GE CT64-820-4	1	ETHIOPIAN AIRLINES CORPORATION
DHC-6 Twin Otter	PW&C PT6A-20	1	TAGB - TRANSPORTES AEREOS DA GUINE BISSAU
DHC-6 Twin Otter 100	P&WC PT6A-20	1	CALM AIR (CANADIAN PARTNER)
DHC-6 Twin Otter 103	P&WC PW6A-20	2	ALOHA ISLAND AIR
DHC-6 Twin Otter 200	P&WC PT6A-20	4	AIRBC LTD
DHC-6 Twin Otter 200	P&WC PT6A-20	1	AIRKENYA AVIATION LIMITED
DHC-6 Twin Otter 200	P&WC PT6A-20	1	CALM AIR (CANADIAN PARTNER)
DHC-6 Twin Otter 200	P&WC PT6A-20	5	ERA AVIATION, INC
DHC-6 Twin Otter 200	P&WC PT6A-27	1	AIR INUIT (1985) LTD
DHC-6 Twin Otter 200	P&WC PT6A-27	1	SEAGREEN AIR TRANSPORT
DHC-6 Twin Otter 210	P&WC PT6A-27	2	SUNFLOWER AIRLINES LTD
DHC-6 Twin Otter 300	P&WC PT6A-27	10	ACES
DHC-6 Twin Otter 300	P&WC PT6A-27	2	AIR BURUNDI
DHC-6 Twin Otter 300	P&WC PT6A-27	1	AIR CALEDONIE INTERNATIONAL
DHC-6 Twin Otter 300	P&WC PT6A-27	1	AIR FIJI
DHC-6 Twin Otter 300	P&WC PT6A-27	1	AIR GUYANE
DHC-6 Twin Otter 300	P&WC PT6A-27	2	AIR INTER GABON, SA.
DHC-6 Twin Otter 300	P&WC PT6A-27	4	AIR INUIT (1985) LTD
DHC-6 Twin Otter 300	P&WC PT6A-27	4	AIR MADAGASCAR
DHC-6 Twin Otter 300	P&WC PT6A-27	1	AIR MOOREA
DHC-6 Twin Otter 300	P&WC PT6A-27	2	AIR NIPPON
DHC-6 Twin Otter 300	P&WC PT6A-27	1	AIR RWANDA
DHC-6 Twin Otter 300	P&WC PT6A-27	1	AIR SENEGAL, SOCIETE NATIONALE DE TRANSPORT AERIEN
DHC-6 Twin Otter 300	P&WC PT6A-27	4	AIR SEYCHELLES
DHC-6 Twin Otter 300	P&WC PT6A-27	1	AIR TAHITI
DHC-6 Twin Otter 300	P&WC PT6A-27	1	AIR TCHAD, SOCIETE DE TRANSPORTES AERIENS

AIRFRAME	POWER PLANT	NO. IN FLEET	FLEET NAME
DHC-6 Twin Otter 300	P&WC PT6A-27	6	AIRKENYA AVIATION LIMITED
DHC-6 Twin Otter 300	P&WC PT6A-27	8	ALOHA ISLAND AIR
DHC-6 Twin Otter 300	P&WC PT6A-27	1	ARKIA ISRAELI AIRLINES
DHC-6 Twin Otter 300	P&WC PT6A-27	5	BRADLEY AIR SERVICES LIMITED (FIRST AIR)
DHC-6 Twin Otter 300	P&WC PT6A-27	3	ERA AVIATION, INC
DHC-6 Twin Otter 300	P&WC PT6A-27	5	ETHIOPIAN AIRLINES CORPORATION
DHC-6 Twin Otter 300	P&WC PT6A-27	1	FARNER AIR TRANSPORT AG
DHC-6 Twin Otter 300	P&WC PT6A-27	2	GROENLANDSFLY A/S
DHC-6 Twin Otter 300	P&WC PT6A-27	1	GUYANA AIRWAYS CORP
DHC-6 Twin Otter 300	P&WC PT6A-27	4	JAPAN TRANSOCEAN AIR
DHC-6 Twin Otter 300	P&WC PT6A-27	2	LESOTHO AIRWAYS CORPORATION
DHC-6 Twin Otter 300	P&WC PT6A-27	6	LIAT (1974) LTD.
DHC-6 Twin Otter 300	P&WC PT6A-27	2	LINA CONGO
DHC-6 Twin Otter 300	P&WC PT6A-27	6	MALAYSIA AIRLINES
DHC-6 Twin Otter 300	P&WC PT6A-27	1	MAYA AIRWAYS LTD
DHC-6 Twin Otter 300	P&WC PT6A-27	11	MERPATI NUSANTARA AIRLINES PT
DHC-6 Twin Otter 300	P&WC PT6A-27	1	MOUNT COOK AIRLINE
DHC-6 Twin Otter 300	P&WC PT6A-27	2	NORLANDAIR (ICELAND)
DHC-6 Twin Otter 300	P&WC PT6A-27	6	NORONTAIR
DHC-6 Twin Otter 300	P&WC PT6A-27	1	OMAN AIR
DHC-6 Twin Otter 300	P&WC PT6A-27	2	PAKISTAN INTERNATIONAL AIRLINES (PIA)
DHC-6 Twin Otter 300	P&WC PT6A-27	1	POLYNESIAN AIRLINES
DHC-6 Twin Otter 300	P&WC PT6A-27	2	ROSS AVIATION
DHC-6 Twin Otter 300	P&WC PT6A-27	9	ROYAL NEPAL AIRLINES CORPORATION
DHC-6 Twin Otter 300	P&WC PT6A-27	2	ROYAL TONGAN AIRLINES
DHC-6 Twin Otter 300	P&WC PT6A-27	1	SAUDIA - SAUDI ARABIAN AIRLINES
DHC-6 Twin Otter 300	P&WC PT6A-27	24	SCENIC AIRLINES INC
DHC-6 Twin Otter 300	P&WC PT6A-27	2	SOCIETE NOUVELLE AIR GUADELOUPE
DHC-6 Twin Otter 300	P&WC PT6A-27	1	SOLOMON AIRLINES
DHC-6 Twin Otter 300	P&WC PT6A-27	2	SURINAM AIRWAYS
DHC-6 Twin Otter 300	P&WC PT6A-27	2	TACV - TRANSPORTES AEREOS DE CABO VERDE
DHC-6 Twin Otter 300	P&WC PT6A-27	2	TURKS & CAICOS AIRWAYS LIMITED
DHC-6 Twin Otter 300	P&WC PT6A-27	11	WIDEROE'S FLYVESELSKAP A/S
DHC-6 Twin Otter 300	P&WC PT6A-27	3	WINDWARD ISLANDS AIRWAYS INTERNATIONAL NV
DHC-6 Twin Otter 300	P&WC PT6A-27	2	XINJIANG AIRLINES
DHC-6 Twin Otter 310	P&WC PT6A-27	1	MOUNT COOK AIRLINE
DHC-7 Dash 7-100	P&WC PT6A-50	2	ADRIA AIRWAYS
DHC-7 Dash 7-100	P&WC PT6A-50	1	AIR GUINEE
DHC-7 Dash 7-100	P&WC PT6A-50	3	AIR NIUGINI
DHC-7 Dash 7-100	P&WC PT6A-50	10	ARKIA ISRAELI AIRLINES
DHC-7 Dash 7-100	P&WC PT6A-50	1	BRADLEY AIR SERVICES LIMITED (FIRST AIR)
DHC-7 Dash 7-100	P&WC PT6A-50	5	BRITISH AIRWAYS
DHC-7 Dash 7-100	P&WC PT6A-50	2	BRITISH MIDLAND AIRWAYS LTD
DHC-7 Dash 7-100	P&WC PT6A-50	5	BRITT AIRWAYS, INC (CONTINENTAL EXPRESS)
DHC-7 Dash 7-100	P&WC PT6A-50	5	CONTINENTAL EXPRESS
DHC-7 Dash 7-100	P&WC PT6A-50	3	GROENLANDSFLY A/S
DHC-7 Dash 7-100	P&WC PT6A-50	4	PARADISE ISLAND AIRLINES
DHC-7 Dash 7-100	P&WC PT6A-50	5	PELITA AIR SERVICE
DHC-7 Dash 7-100	P&WC PT6A-50	5	PIEDMONT AIRLINES INC (USAIR EXPRESS)
DHC-7 Dash 7-100	P&WC PT6A-50	5	ROCKY MOUNTAIN AIRWAYS (CONTINENTAL EXPRESS)
DHC-7 Dash 7-100	P&WC PT6A-50	1	ROSS AVIATION
DHC-7 Dash 7-100	P&WC PT6A-50	2	TYROLEAN AIRWAYS
DHC-7 Dash 7-100	P&WC PT6A-50	8	WIDEROE'S FLYVESELSKAP A/S
DHC-7 Dash 7-100	P&WC PT6A-50	4	YEMEN AIRLINES
DHC-8 Dash 8	P&WC PW120	1	AIR INUIT (1985) LTD
DHC-8 Dash 8-100	P&WC PW120	11	AIR ATLANTIC LTD (CANADIAN PARTNER)
DHC-8 Dash 8-100	P&WC PW120	1	AIR CREEBEC INC
DHC-8 Dash 8-100	P&WC PW120	14	AIR NOVA
DHC-8 Dash 8-100	P&WC PW120	16	AIR ONTARIO, INC
DHC-8 Dash 8-100	P&WC PW120	12	AIRBC LTD
DHC-8 Dash 8-100	P&WC PW120	2	AIRES (AEROVIAS DE INTERGRACION REGIONAL SA)
DHC-8 Dash 8-100	P&WC PW120	3	ANSETT NEW ZEALAND
DHC-8 Dash 8-100	P&WC PW120	2	BRITISH AIRWAYS
DHC-8 Dash 8-100	P&WC PW120	2	CANADIAN REGIONAL
DHC-8 Dash 8-100	P&WC PW120	4	CCAIR, INC (USAIR EXPRESS)
DHC-8 Dash 8-100	P&WC PW120	5	EASTERN AUSTRALIA AIRLINES
DHC-8 Dash 8-100	P&WC PW120	4	GREAT CHINA AIRLINES
DHC-8 Dash 8-100	P&WC PW120	3	HAMBURG AIRLINES LUFTFAHRTGESELLSCHAFT.MBH
DHC-8 Dash 8-100	P&WC PW120	23	HORIZON AIR
DHC-8 Dash 8-100	P&WC PW120	0	INTEROT AIR SERVICE
DHC-8 Dash 8-100	P&WC PW120	8	LIAT (1974) LTD.
DHC-8 Dash 8-100	P&WC PW120	25	MESABA AIRLINES (NORTHWEST AIRLINK)
DHC-8 Dash 8-100	P&WC PW120	16	PENNSYLVANIA COMMUTER AIRLINES, INC (USAIR EXPRESS)
DHC-8 Dash 8-100	P&WC PW120	43	PIEDMONT AIRLINES INC (USAIR EXPRESS)
DHC-8 Dash 8-100	P&WC PW120	3	SUNSTATE AIRLINES
DHC-8 Dash 8-100	P&WC PW120	2	TYROLEAN AIRWAYS
DHC-8 Dash 8-100	P&WC PW121	0	STATESWEST AIRLINES, INC (USAIR EXPRESS)
DHC-8 Dash 8-100	P&WC PW121	6	TYROLEAN AIRWAYS
DHC-8 Dash 8-100	P&WC PW123	2	ERA AVIATION, INC
DHC-8 Dash 8-102	P&WC PW120	4	NORONTAIR
DHC-8 Dash 8-103B	P&WC PW121	4	WIDEROE'S FLYVESELSKAP A/S
DHC-8 Dash 8-200	P&WC PW120A	0	MESA AIRLINES INC
DHC-8 Dash 8-300	P&WC PT6A-27	2	AIRES (AEROVIAS DE INTERGRACION REGIONAL SA)
DHC-8 Dash 8-300	P&WC PW123	6	AIR ONTARIO, INC
DHC-8 Dash 8-300	P&WC PW123	6	AIRBC LTD
DHC-8 Dash 8-300	P&WC PW123	2	ALM ANTILLEAN AIRLINES
DHC-8 Dash 8-300	P&WC PW123	7	BAHAMASAIR
DHC-8 Dash 8-300	P&WC PW123	3	BANGKOK AIRWAYS
DHC-8 Dash 8-300	P&WC PW123	2	BRITISH AIRWAYS
DHC-8 Dash 8-300	P&WC PW123	14	CANADIAN REGIONAL
DHC-8 Dash 8-300	P&WC PW123	5	CONTACTAIR
DHC-8 Dash 8-300	P&WC PW123	6	GREAT CHINA AIRLINES
DHC-8 Dash 8-300	P&WC PW123	2	HAMBURG AIRLINES LUFTFAHRTGESELLSCHAFT.MBH
DHC-8 Dash 8-300	P&WC PW123	0	MESA AIRLINES INC
DHC-8 Dash 8-300	P&WC PW123	5	SCHREINER AIRWAYS, BV
DHC-8 Dash 8-300	P&WC PW123	2	TABA - TRANSPORTES AEREOS DA BACIA AMAZONICA SA
DHC-8 Dash 8-300	P&WC PW123B	1	AIR CREEBEC INC
DHC-8 Dash 8-300	P&WC PW123E	9	S A EXPRESS AIRLINES (PTY) LTD
DHC-8 Dash 8-300A	P&WC PW123	3	TYROLEAN AIRWAYS

AIRFRAME	POWER PLANT	NO. IN FLEET	FLEET NAME
DHC-8 Dashe 8-300	P&WC PW123	1	REGION AIR
Dornier 228	Ga TPE331-5-252D	1	AIR MALAWI
Dornier 228-100	Ga TPE331-5-252D	1	HOLMSTROM AIR
Dornier 228-100	Ga TPE331-5A-252D	1	AIRLINK AIRLINE (PTY) LTD
Dornier 228-200	Ga TPE331-5-252D	3	JAPAN AIR COMMUTER CO LTD
Dornier 228-200	Ga TPE331-5-252D	7	OLYMPIC AVIATION, S.A.
Dornier 228-200	Ga TPE331-5-252D	1	SUCKLING AIRWAYS
Dornier 228-200	Ga TPE331-5A-252D	1	AIR TAHITI
Dornier 228-201	Ga TPE331-5-252D	1	AIR HUDIK AB
Dornier 228-201	Ga TPE331-5-252D	2	AIR MARSHALL ISLANDS - AMI
Dornier 228-201	Ga TPE331-5-252D	2	DRUK AIR CORP
Dornier 228-201	Ga TPE331-5-252D	2	HOLMSTROM AIR
Dornier 228-201	Ga TPE331-5-252D	2	JAGSON AIRLINES
Dornier 228-201	Ga TPE331-5-252D	4	VAYUDOOT
Dornier 228-201K	Ga TPE331-5-252D	1	REGIONAIR
Dornier 228-202	Ga TPE331-5-252D	3	FORMOSA AIRLINES
Dornier 228-202	Ga TPE331-5-252D	13	PRECISION AIRLINES
Dornier 228-202	Ga TPE331-5-252D	1	RATIOFLUG GMBH
Dornier 228-202	Ga TPE331-5-252D	1	SATA AIR ACORES - SERVICO ACOREANO DE TRANSPORTES AEREOS, E.P.
Dornier 228-202	Ga TPE331-5-252D	2	SOMALI AIRLINES
Dornier 228-202K	Ga TPE331-5-252D	1	FLEXAIR, BV
Dornier 228-202K	Ga TPE331-5-252D	3	PELANGI AIR, SDN BHD
Dornier 228-202K	Ga TPE331-5-252D	1	SILKAIR (SINGAPORE) PRIVATE LIMITED
Dornier 228-202K	Ga TPE331-5-252D	3	SOCIETE NOUVELLE AIR GUADELOUPE
Dornier 228-202K	Ga TPE331-5-252D	4	SUCKLING AIRWAYS
Dornier 228-212	Ga TPE331-5-252D	2	AIR CALEDONIE
Dornier 228-212	Ga TPE331-5-252D		TAHITI CONQUEST AIRLINES - TCA
Dornier 228-212	Ga TPE331-5A-252D	1	AIR MOOREA
Dornier 228-212	Ga TPE331-5A-252D	4	FORMOSA AIRLINES
Dornier 228-212	Ga TPE331-5A-252D	1	PELANGI AIR, SDN BHD
Dornier 27Q-5	Lyc GO-480-B1A6	1	TAGB - TRANSPORTES AEREOS DA GUINE BISSAU
Dornier 28D Skyservant	Lyc IGSO-540-A1E	1	TAGB - TRANSPORTES AEREOS DA GUINE BISSAU
Dornier 328	P&WC PW119B	0	AIR CALEDONIE
Dornier 328	P&WC PW119B	10	HORIZON AIR
Dornier 328	P&WC PW119B	20	JETSTREAM INTERNATIONAL AIRLINES
Dornier 328	P&WC PW119B	0	TAHITI CONQUEST AIRLINES - TCA
Dornier 328-110	P&WC PW119B	1	LONE STAR AIRLINES
EMB 110 Bandeirante	P&WC PT6A-27	1	NORDESTE
EMB 110 Bandeirante	P&WC PT6A-27	14	TABA - TRANSPORTES AEREOS DA BACIA AMAZONICA SA
EMB 110 Bandeirante	P&WC PT6A-34	2	STELLAIR T.A.
EMB 110C Bandeirante	P&WC PT6A-27	3	BRASIL CENTRAL
EMB 110C Bandeirante	P&WC PT6A-27	3	NORDESTE
EMB 110P Bandeirante	P&WC PT6A-27	4	BRASIL CENTRAL
EMB 110P Bandeirante	P&WC PT6A-27	2	NORDESTE
EMB 110P Bandeirante	P&WC PT6A-27	5	RIO-SUL SERVICOS AEROS REGIONAIS SA
EMB 110P Bandeirante	P&WC PT6A-27	4	TAM - TRANSPORTES AEREOS REGIONAIS SA
EMB 110P1 Bandeirante	P&WC PT6A-34	2	AIR AFFAIRES GABON
EMB 110P1 Bandeirante	P&WC PT6A-34	8	AIRES (AEROVIAS DE INTERGRACION REGIONAL SA)
EMB 110P1 Bandeirante	P&WC PT6A-34	2	AIRLA
EMB 110P1 Bandeirante	P&WC PT6A-34	11	ATLANTIC SOUTHEAST AIRLINES, INC (THE DELTA CONNECTION)
EMB 110P1 Bandeirante	P&WC PT6A-34	3	BRASIL CENTRAL
EMB 110P1 Bandeirante	P&WC PT6A-34	7	EAGLE AIRWAYS LTD
EMB 110P1 Bandeirante	P&WC PT6A-34	2	FLANDERS AIRLINES NV
EMB 110P1 Bandeirante	P&WC PT6A-34	5	NORDESTE
EMB 110P1 Bandeirante	P&WC PT6A-34	1	REGIONAIR
EMB 110P1 Bandeirante	P&WC PT6A-34	1	SUN-AIR OF SCANDINAVIA A/S
EMB 110P1 Bandeirante	P&WC PT6A-34	2	SUNSTATE AIRLINES
EMB 110P2 Bandeirante	P&WC PT6A-27	1	NORDESTE
EMB 110P2 Bandeirante	P&WC PT6A-34	1	AIR BURKINA
EMB 110P2 Bandeirante	P&WC PT6A-34	1	FLANDERS AIRLINES NV
EMB 110P2 Bandeirante	P&WC PT6A-34	4	MUK AIR
EMB 110P2 Bandeirante	P&WC PT6A-34	2	SUNAIR (BOP AIR PTY LTD dba)
EMB 110P6 Bandeirante	P&WC PT6A-34	1	AIR VANUATU
EMB 120 Brasilia	P&WC PW115	3	LUXAIR S.A.
EMB 120 Brasilia	P&WC PW118	14	BRITT AIRWAYS, INC (CONTINENTAL EXPRESS)
EMB 120 Brasilia	P&WC PW118	2	TAT EUROPEAN AIRLINES
EMB 120 Brasilia	P&WC PW118	3	WIDEROE NORSKAIR A/S
EMB 120ER Brasilia	P&WC PW118A	29	SKYWEST AIRLINES INC (DELTA CONNECTION)
EMB 120RT Brasilia	P&WC PW118	1	AIR ARUBA
EMB 120RT Brasilia	P&WC PW118	9	AIR LITTORAL
EMB 120RT Brasilia	P&WC PW118	60	ATLANTIC SOUTHEAST AIRLINES, INC (THE DELTA CONNECTION)
EMB 120RT Brasilia	P&WC PW118	20	BRITT AIRWAYS, INC (CONTINENTAL EXPRESS)
EMB 120RT Brasilia	P&WC PW118	40	COMAIR
EMB 120RT Brasilia	P&WC PW118	38	CONTINENTAL EXPRESS
EMB 120RT Brasilia	P&WC PW118	9	DELTA AIR TRANSPORT NV
EMB 120RT Brasilia	P&WC PW118	41	MESA AIRLINES INC
EMB 120RT Brasilia	P&WC PW118	2	NORDESTE
EMB 120RT Brasilia	P&WC PW118	5	RIO-SUL SERVICOS AEROS REGIONAIS SA
EMB 120RT Brasilia	P&WC PW118	15	WESTAIR COMMUTER AIRLINES (MESA AIRLINES)
EMB 120RT Brasilia	P&WC PW118A	2	SUNAIR (BOP AIR PTY LTD dba)
EMB 145	Allison AE 3007A	0	REGIONAL AIRLINES
EMB 145	Allison AE 3007A	0	TRANSBRASIL S.A. LINHAS AEREAS
Eurocopter 332L S.Puma	TU Makila 1A	2	ERA AVIATION, INC
Eurocopter 332L S.Puma	TU Makila 1A1	8	ERA AVIATION, INC
Eurocopter 350B-1	TU Arriel 1B	5	ERA AVIATION, INC
Eurocopter 350B-2	TU Arriel 1D1	8	ERA AVIATION, INC
Eurocopter AS355F2	Allison 250-C20F	2	OLYMPIC AVIATION, S.A.
Eurocopter BO 105 C	Allison 250-C20	5	PELITA AIR SERVICE
Eurocopter BO 105 CBS	Allison 250-C20	31	ERA AVIATION, INC
Fairchild Ind. F-27	RR Dart 514-7	1	AERO COZUMEL, SA
Fairchild Ind. F-27	RR Dart 514-7	2	BRITT AIRWAYS, INC (CONTINENTAL EXPRESS)
Fairchild Ind. F-27	RR Dart 532-7	2	STELLAIR T.A.
Fairchild Ind. F-27A	RR Dart 528-7	1	AERO COZUMEL, SA
Fairchild Ind. F-27F	RR Dart 529-7E	2	AEROCARIBE
Fairchild Ind. F-27F	RR Dart 529-7E	2	EMPIRE AIRLINES
Fairchild Ind. F-27J	RR Dart 532-7	1	AIRES (AEROVIAS DE INTERGRACION REGIONAL SA)
Fairchild Ind. F-27J	RR Dart 532-7	4	CATA
Fairchild Ind. FH-227B	RR Dart 532-7	6	TABA - TRANSPORTES AEREOS DA BACIA AMAZONICA SA
Fairchild Ind. FH-227B	RR Dart 532-7L	2	AEROCARIBE

AIRFRAME	POWER PLANT	NO. IN FLEET	FLEET NAME
Fairchild Ind. FH-227B	RR Dart 536-7P	2	IRAN ASSEMAN AIRLINES
Fairchild Ind. FH-227D	RR Dart 532-7L	2	AEROCARIBE
Fairchild Metro 23	Ga TPE331-12U-701G	2	AIRLA
Fairchild Metro 23	Ga TPE331-12U-701G	2	CHINA HAINAN AIRLINES
Fairchild Metro 23	Ga TPE331-12U-701G	1	HAZELTON AIRLINES
Fairchild Metro 23	Ga TPE331-12U-701G	7	KENDELL AIRLINES (AUSTRALIA) PTY LIMITED
Fairchild Metro 23	Ga TPE331-12U-701G	4	LONE STAR AIRLINES
Fokker 100	RR Tay 620-15	1	AIR GABON
Fokker 100	RR Tay 620-15	2	AIR IVOIRE
Fokker 100	RR Tay 620-15	9	AIR UK LTD
Fokker 100	RR Tay 620-15	75	AMERICAN AIRLINES
Fokker 100	RR Tay 620-15	4	AVIACION DE CHIAPAS (AVIACSA)
Fokker 100	RR Tay 620-15	2	CORSE MEDITERRANEE
Fokker 100	RR Tay 620-15	6	KLM ROYAL DUTCH AIRLINES
Fokker 100	RR Tay 620-15	10	MEXICANA
Fokker 100	RR Tay 620-15	10	SWISSAIR (SOCIETE ANONYME SUISSE POUR LA NAVIGATION AERIENNE)
Fokker 100	RR Tay 620-15	14	TAT EUROPEAN AIRLINES
Fokker 100	RR Tay 650-15	5	AIR INTER
Fokker 100	RR Tay 650-15	4	AIR LITTORAL
Fokker 100	RR Tay 650-15	4	BRITISH MIDLAND AIRWAYS LTD
Fokker 100	RR Tay 650-15	10	CHINA EASTERN AIRLINES
Fokker 100	RR Tay 650-15	5	DEUTSCHE BA LUFTFAHRTGESELLSCHAFT MBH
Fokker 100	RR Tay 650-15	6	IRAN AIR THE AIRLINE OF THE ISLAMIC REPUBLIC OF IRAN
Fokker 100	RR Tay 650-15	12	KOREAN AIR
Fokker 100	RR Tay 650-15	3	MERPATI NUSANTARA AIRLINES PT
Fokker 100	RR Tay 650-15	3	MIDWAY AIRLINES
Fokker 100	RR Tay 650-15	1	PELITA AIR SERVICE
Fokker 100	RR Tay 650-15	6	PORTUGALIA - COMPANIA PORTUGESA DE TRANSPORTES AEREOS SA
Fokker 100	RR Tay 650-15	1	ROYAL SWAZI NATIONAL AIRWAYS CORPORATION
Fokker 100	RR Tay 650-15	4	SEMPATI AIR TRANSPORT
Fokker 100	RR Tay 650-15	2	TABA - TRANSPORTES AEREOS DA BACIA AMAZONICA SA
Fokker 100	RR Tay 650-15	2	TAIWAN AIRLINES
Fokker 100	RR Tay 650-15	17	TAM - TRANSPORTES AEREOS REGIONAIS SA
Fokker 100	RR Tay 650-15	7	TRANSWEDE
Fokker 100	RR Tay 650-15	40	USAIR, INC
Fokker 27-200	RR Dart 532-7R	1	MESABA AIRLINES (NORTHWEST AIRLINK)
Fokker 50	P&WC PW124	5	BUSY BEE OF NORWAY A/S
Fokker 50	P&WC PW124	4	LUXAIR S.A.
Fokker 50	P&WC PW125	6	AER LINGUS
Fokker 50	P&WC PW125	2	AIR ZIMBABWE
Fokker 50	P&WC PW125	11	ANSETT AUSTRALIA
Fokker 50	P&WC PW125	6	AUSTRIAN AIR SERVICES
Fokker 50	P&WC PW125	1	CROSSAIR
Fokker 50	P&WC PW125	4	ICELANDAIR
Fokker 50	P&WC PW125	3	KENYA AIRWAYS LTD
Fokker 50	P&WC PW125	10	KLM CITYHOPPER
Fokker 50	P&WC PW125	25	LUFTHANSA CITYLINE
Fokker 50	P&WC PW125	9	MAERSK AIR
Fokker 50	P&WC PW125	11	MALAYSIA AIRLINES
Fokker 50	P&WC PW125	2	NORDESTE
Fokker 50	P&WC PW125	2	PELANGI AIR, SDN BHD
Fokker 50	P&WC PW125	10	PHILIPPINE AIR LINES, INC - PAL
Fokker 50	P&WC PW125	7	RIO-SUL SERVICOS AEROS REGIONAIS SA
Fokker 50	P&WC PW125	22	SAS - SCANDINAVIAN AIRLINES SYSTEM
Fokker 50	P&WC PW125	2	SUDAN AIRWAYS
Fokker 50	P&WC PW125B	9	AIR UK LTD
Fokker 50	P&WC PW125B	2	FORMOSA AIRLINES
Fokker 50	P&WC PW125B	3	NORWEGIAN AIR SHUTTLE A/S
Fokker 50	P&WC PW125B	2	ROYAL BRUNEI AIRLINES
Fokker 50	P&WC PW127B	6	AEROVIAS NACIONALES DE COLOMBIA S.A. AVIANCA
Fokker 70	RR Tay 620	2	AIR LITTORAL
Fokker 70	RR Tay 620	0	ALITALIA
Fokker 70	RR Tay 620	0	AUSTRIAN AIRLINES, OSTERREICHISCHE LUFTVERKEHRS AG
Fokker 70	RR Tay 620	2	BRITISH MIDLAND AIRWAYS LTD
Fokker 70	RR Tay 620	0	MALEV - HUNGARIAN AIRLINES
Fokker 70	RR Tay 620	0	MERPATI NUSANTARA AIRLINES PT
Fokker 70	RR Tay 620	2	MESA AIRLINES INC
Fokker 70	RR Tay 620	0	PELITA AIR SERVICE
Fokker 70	RR Tay 620	2	SEMPATI AIR TRANSPORT
Fokker 70	RR Tay 620	3	TYROLEAN AIRWAYS
Fokker 70	RR Tay 620-15	1	SILKAIR (SINGAPORE) PRIVATE LIMITED
Fokker F27-100	RR Dart 514-7	1	AIR UK LTD
Fokker F27-100	RR Dart 514-7	6	BUSY BEE OF NORWAY A/S
Fokker F27-100	RR Dart 514-7	1	FLANDERS AIRLINES NV
Fokker F27-100	RR Dart 514-7	2	INDIAN AIRLINES LTD
Fokker F27-100	RR Dart 514-7	4	VAYUDOOT
Fokker F27-100	RR Dart 514-7	3	WDL AVIATION (KOLN)
Fokker F27-100	RR Dart 532-7	1	JERSEY EUROPEAN
Fokker F27-100 (SCD)	RR Dart 514-7	2	FLANDERS AIRLINES NV
Fokker F27-100 (SCD)	RR Dart 514-7	1	FLANDERS AIRLINES NV
Fokker F27-200	RR Dart 528-7E	1	AIR COMORES
Fokker F27-200	RR Dart 528-7E	10	AIR UK LTD
Fokker F27-200	RR Dart 528-7E	1	AIRKENYA AVIATION LIMITED
Fokker F27-200	RR Dart 528-7E	1	EXPRESSO AEREO
Fokker F27-200	RR Dart 532-7	1	BRASIL CENTRAL
Fokker F27-200	RR Dart 532-7	1	REGIONAIR
Fokker F27-200	RR Dart 532-7	2	SEMPATI AIR TRANSPORT
Fokker F27-200	RR Dart 532-7E	13	PAKISTAN INTERNATIONAL AIRLINES (PIA)
Fokker F27-200	RR Dart 532-7R	4	BUSY BEE OF NORWAY A/S
Fokker F27-200	RR Dart 532-7R	4	COMAIR (COMMERCIAL AIRWAYS (PTY) LTD)
Fokker F27-200	RR Dart 536-7P	1	LLOYD AEREO BOLIVIANO SAM
Fokker F27-400	RR Dart 532-7	1	INDIAN AIRLINES LTD
Fokker F27-400	RR Dart 532-7	2	SCHREINER AIRWAYS, BV
Fokker F27-400	RR Dart 532-7	1	SEMPATI AIR TRANSPORT
Fokker F27-400	RR Dart 532-7R	1	FARNER AIR TRANSPORT AG
Fokker F27-400	RR Dart 532-7R	1	FARNER AIR TRANSPORT HUNGARY
Fokker F27-400	RR Dart 536-7	1	LIBYAN ARAB AIRLINES
Fokker F27-400M	RR Dart 536-7R	7	AIR ALGERIE
Fokker F27-400M (F)	RR Dart 532-7R	2	TAAG, LINHAS AEREAS DE ANGOLA (ANGOLA AIRLINES)

AIRFRAME	POWER PLANT	NO. IN FLEET	FLEET NAME
Fokker F27-500	RR Dart 532-7	15	AIR FRANCE
Fokker F27-500	RR Dart 532-7	2	AIR SINAI
Fokker F27-500	RR Dart 532-7	5	AIR UK LTD
Fokker F27-500	RR Dart 532-7	1	CHANNEL EXPRESS (AIR SERVICES) LTD
Fokker F27-500	RR Dart 532-7	2	EGYPTAIR
Fokker F27-500	RR Dart 532-7	10	EMPIRE AIRLINES
Fokker F27-500	RR Dart 532-7	1	EUROPEAN AIRLIFT
Fokker F27-500	RR Dart 532-7	2	EXPRESSO AEREO
Fokker F27-500	RR Dart 532-7	24	FEDERAL EXPRESS CORPORATION
Fokker F27-500	RR Dart 532-7	9	JERSEY EUROPEAN
Fokker F27-500	RR Dart 532-7	8	MERPATI NUSANTARA AIRLINES PT
Fokker F27-500	RR Dart 532-7	2	RIO-SUL SERVICOS AEROS REGIONAIS SA
Fokker F27-500	RR Dart 532-7	15	SAF SERVICE AERIEN FRANCAIS, SA
Fokker F27-500	RR Dart 532-7	1	STAR AIR A/S
Fokker F27-500	RR Dart 532-7	1	SUDAN AIRWAYS
Fokker F27-500	RR Dart 532-7	7	TAM - TRANSPORTES AEREOS REGIONAIS SA
Fokker F27-500	RR Dart 532-7R	3	EAST WEST AIRLINES
Fokker F27-500	RR Dart 532-7R	2	FARNER AIR TRANSPORT AG
Fokker F27-500	RR Dart 532-7R	1	TAAG, LINHAS AEREAS DE ANGOLA (ANGOLA AIRLINES)
Fokker F27-500	RR Dart 535-7R	3	PENNSYLVANIA COMMUTER AIRLINES, INC (USAIR EXPRESS)
Fokker F27-500	RR Dart 536-7	6	MERPATI NUSANTARA AIRLINES PT
Fokker F27-500	RR Dart 536-7	4	OMAN AIR
Fokker F27-500	RR Dart 552-7R	2	RATIOFLUG GMBH
Fokker F27-600	RR Dart 528-7E	1	AIR UK LTD
Fokker F27-600	RR Dart 532-7	2	AIR JET
Fokker F27-600	RR Dart 532-7	1	AIR TCHAD, SOCIETE DE TRANSPORTES AERIENS
Fokker F27-600	RR Dart 532-7	2	BRASIL CENTRAL
Fokker F27-600	RR Dart 532-7	8	CUBANA
Fokker F27-600	RR Dart 532-7	1	FARNER AIR TRANSPORT AG
Fokker F27-600	RR Dart 532-7	8	FEDERAL EXPRESS CORPORATION
Fokker F27-600	RR Dart 532-7	1	FLANDERS AIRLINES NV
Fokker F27-600	RR Dart 532-7	1	LINA CONGO
Fokker F27-600	RR Dart 532-7	5	MYANMA AIRWAYS (BURMA AIRWAYS CORPORATION)
Fokker F27-600	RR Dart 532-7	3	RATIOFLUG GMBH
Fokker F27-600	RR Dart 532-7	1	SEMPATI AIR TRANSPORT
Fokker F27-600	RR Dart 532-7	2	TAAG, LINHAS AEREAS DE ANGOLA (ANGOLA AIRLINES)
Fokker F27-600	RR Dart 532-7P	1	LLOYD AEREO BOLIVIANO SAM
Fokker F27-600	RR Dart 532-7R	1	STAR AIR A/S
Fokker F27-600	RR Dart 536-3B	1	SOMALI AIRLINES
Fokker F27-600	RR Dart 536-7	12	LIBYAN ARAB AIRLINES
Fokker F27-600	RR Dart 536-7B	1	AIR IVOIRE
Fokker F27-600	RR Dart 536-7R	1	IRAN ASSEMAN AIRLINES
Fokker F27-600	RR Dart 536-7R	1	LESOTHO AIRWAYS CORPORATION
Fokker F27-600	RR Dart 536-7R	5	WDL AVIATION (KOLN)
Fokker F27-600CF	RR Dart 536-7R	2	AIR TANZANIA
Fokker F28-1000	RR Spey 555-15	3	AEROLINEAS ARGENTINAS
Fokker F28-1000	RR Spey 555-15	2	AEROPERU
Fokker F28-1000	RR Spey 555-15	1	AIR GABON
Fokker F28-1000	RR Spey 555-15	2	AIR LITTORAL
Fokker F28-1000	RR Spey 555-15	7	AIR NIUGINI
Fokker F28-1000	RR Spey 555-15	4	ANSETT AUSTRALIA
Fokker F28-1000	RR Spey 555-15	11	CANADIAN REGIONAL
Fokker F28-1000	RR Spey 555-15	12	HORIZON AIR
Fokker F28-1000	RR Spey 555-15	1	IRAN ASSEMAN AIRLINES
Fokker F28-1000	RR Spey 555-15	1	LINA CONGO
Fokker F28-1000	RR Spey 555-15	1	MYANMA AIRWAYS (BURMA AIRWAYS CORPORATION)
Fokker F28-1000	RR Spey 555-15	2	PELITA AIR SERVICE
Fokker F28-1000	RR Spey 555-15	10	TAT EUROPEAN AIRLINES
Fokker F28-1000	RR Spey 555-15H	4	DELTA AIR TRANSPORT NV
Fokker F28-1000	RR Spey 555-15N	3	SAS - SCANDINAVIAN AIRLINES SYSTEM
Fokker F28-1000C	RR Spey 555-15	1	AIR AUSTRAL
Fokker F28-1000C	RR Spey 555-15	1	AIR IVOIRE
Fokker F28-2000	RR Spey 555-15	2	AIR GABON
Fokker F28-2000	RR Spey 555-15	4	TAT EUROPEAN AIRLINES
Fokker F28-2000	RR Spey 555-15N	1	GHANA AIRWAYS CORPORATION
Fokker F28-3000	RR Spey 555-15	2	ANSETT AUSTRALIA
Fokker F28-3000	RR Spey 555-15H	1	DELTA AIR TRANSPORT NV
Fokker F28-3000	RR Spey 555-15H	1	ROYAL SWAZI NATIONAL AIRWAYS CORPORATION
Fokker F28-3000C	RR Spey 555-15H	1	SATENA
Fokker F28-4000	RR Spey 555-15	2	AIR NIUGINI
Fokker F28-4000	RR Spey 555-15H	2	AIR IVOIRE
Fokker F28-4000	RR Spey 555-15H	2	BIMAN BANGLADESH AIRLINES
Fokker F28-4000	RR Spey 555-15H	4	IRAN ASSEMAN AIRLINES
Fokker F28-4000	RR Spey 555-15H	26	MERPATI NUSANTARA AIRLINES PT
Fokker F28-4000	RR Spey 555-15H	5	TAT EUROPEAN AIRLINES
Fokker F28-4000	RR Spey 555-15H/-15P	2	MYANMA AIRWAYS (BURMA AIRWAYS CORPORATION)
Fokker F28-4000	RR Spey 555-15P	1	AIR BURKINA
Fokker F28-4000	RR Spey 555-15P	1	AIR MAURITANIE
Fokker F28-4000	RR Spey 555-15P	7	ANSETT AUSTRALIA
Fokker F28-4000	RR Spey 555-15P	1	COMAIR (COMMERCIAL AIRWAYS (PTY) LTD)
Fokker F28-4000	RR Spey 555-15P	3	DELTA AIR TRANSPORT NV
Fokker F28-4000	RR Spey 555-15P	1	GHANA AIRWAYS CORPORATION
Fokker F28-4000	RR Spey 555-15P	4	KLM CITYHOPPER
Fokker F28-4000	RR Spey 555-15P	3	LIBYAN ARAB AIRLINES
Fokker F28-4000	RR Spey 555-15P	5	PELITA AIR SERVICE
Fokker F28-4000	RR Spey 555-15P	16	SAS - SCANDINAVIAN AIRLINES SYSTEM
Fokker F28-4000	RR Spey 555-15P	1	TAME - TRANSPORTES AEREOS MILITARES ECUATORIANOS, CA
Fokker F28-4000	RR Spey 555-15P	19	USAIR, INC
GA 7 Cougar	Lyc O-320-D1D	1	ALPHA AIR
GAF Nomad N22B	Allison 250-B17B	1	MOUNT COOK AIRLINE
GAF Nomad N24A	Allison 250-B17B	1	MOUNT COOK AIRLINE
Grumman G-21 Goose	P&W R-985	1	PAN AVIATION
Gulfstream I	RR Dart 529-8D	1	KELOWNA FLIGHTCRAFT AIR CHARTER LTD
Gulfstream I	RR Dart 529-8X	5	AIR PROVENCE
Gulfstream I	RR Dart 529-8X	7	TAS AIRWAYS SPA
Gulfstream II	RR Spey 511-8	1	LIBYAN ARAB AIRLINES
Gulfstream II	RR Spey 511-8	1	PELITA AIR SERVICE
Gulfstream II	RR Spey 511-8	4	SAUDIA - SAUDI ARABIAN AIRLINES
Gulfstream III	RR Spey 511-8	1	PELITA AIR SERVICE

AIRFRAME	POWER PLANT	NO. IN FLEET	FLEET NAME
Gulfstream III	RR Spey 511-8	3	SAUDIA - SAUDI ARABIAN AIRLINES
Gulfstream III	RR Spey 511-8	1	SUNAIR (BOP AIR PTY LTD dba)
Gulfstream IV	RR Tay 611-8	6	SAUDIA - SAUDI ARABIAN AIRLINES
HAL/BAe (HS) 748 Srs 2	RR Dart 533-2	5	VAYUDOOT
HAL/BAe (HS) 748 Srs 2A	RR Dart 533-2	5	VAYUDOOT
HAL/Dornier 228-201	Ga TPE331-5-252D	5	VAYUDOOT
Handley Page Herald	RR Dart 532-9	9	CHANNEL EXPRESS (AIR SERVICES) LTD
Handley Page Herald 401	RR Dart 532-9	1	AEROSUCRE COLOMBIA, S.A.
Harbin Yunshuji Y-12	P&WC PT6A-20	1	AIR FIJI
Harbin Yunshuji Y-12	P&WC PT6A-27	3	LAO AVIATION
Harbin Yunshuji Y-12 II	P&WC PT6A-27	1	GUIZHOU AIRLINES
Harbin Yunshuji Y-12 II	P&WC PT6A-27	5	MONGOLIAN AIRLINES - MIAT
Harbin Yunshuji Y-12II	P&WC PT6A-27	4	CHINA SOUTHWEST AIRLINES
Hiller UH-12E	VO-540	3	FORMOSA AIRLINES
Hughes 369D	Allison 250-C20B	1	NATIONAL OVERSEAS AIRLINE
IAI Arava 102	P&WC PT6A-34	1	CATA
IAI Westwind 1124	GA TFE731-3-1G	1	ARKIA ISRAELI AIRLINES
IAI Westwind 1124	GA TFE731-3-1G	1	KELOWNA FLIGHTCRAFT AIR CHARTER LTD
IPTN N-250-100	Allison AE 2100	0	BOURAQ INDONESIA AIRLINES PT
IPTN N-250-100	Allison AE 2100	0	MERPATI NUSANTARA AIRLINES PT
IPTN N-250-100	Allison AE 2100	0	SEMPATI AIR TRANSPORT
IPTN/CASA 212-100 Aviocar	Ga TPE331-5-251C	4	PELITA AIR SERVICE
IPTN/CASA 212-200 Aviocar	Ga TPE331-10-511C	11	MERPATI NUSANTARA AIRLINES PT
IPTN/CASA 212-200 Aviocar	Ga TPE331-10-511C	8	PELITA AIR SERVICE
IPTN/CASA CN235-10	GE CT7-7A	14	MERPATI NUSANTARA AIRLINES PT
IPTN/CASA CN235-100	GE CT7-9C	0	CONQUEST AIRLINES
IPTN/CASA CN235-200	GE CT7-9C	0	MERPATI NUSANTARA AIRLINES PT
IPTN/Eurocopter NBO-105CB	Allison 250-C20	28	PELITA AIR SERVICE
IPTN/NAS SA330G Puma	TU Turmo IVA/IVC	4	PELITA AIR SERVICE
IPTN/NAS SA330J Puma	TU Turmo IVC	11	PELITA AIR SERVICE
IPTN/NAS SA332C S.Puma	TU Makila 1A	2	PELITA AIR SERVICE
Ilyushin 14	SH ASh-82T	1	AEROCARIBBEAN
Ilyushin 14	SH ASh-82T	5	AIR KORYO
Ilyushin 14	SH ASh-82T	18	CAAC
Ilyushin 18	ZMKB AI-20	2	AIR KORYO
Ilyushin 18	ZMKB AI-20M	9	CAAC
Ilyushin 18D	ZMKB AI-20	5	AIR CARGO
Ilyushin 18D	ZMKB AI-20	2	AIR KORYO
Ilyushin 18D	ZMKB AI-20	2	BALKAN
Ilyushin 18D	ZMKB AI-20M	4	AEROCARIBBEAN
Ilyushin 18D	ZMKB AI-20M	55	AEROFLOT RUSSIAN INTERNATIONAL AIRLINES
Ilyushin 18D	ZMKB AI-20M	2	TAROM
Ilyushin 18D	ZMKB AI-20M	3	VIETNAM AIRLINES
Ilyushin 18E	ZMKB AI-20	2	BALKAN
Ilyushin 18V	ZMKB AI-20	2	BALKAN
Ilyushin 18V	ZMKB AI-20K(M)	2	TAROM
Ilyushin 18V (F)	ZMKB AI-20K	1	AEROCARIBBEAN
Ilyushin 62	KKBM NK-8-4	61	AEROFLOT RUSSIAN INTERNATIONAL AIRLINES
Ilyushin 62	KKBM NK-8-4	3	AEROFLOT RUSSIAN INTERNATIONAL AIRLINES
Ilyushin 62	KKBM NK-8-4	2	CSA - CESKOSLOVENSKE AEROLINIE
Ilyushin 62	KKBM NK-8-4	1	TAROM
Ilyushin 62M	Aviadvigatel D-30	89	AEROFLOT RUSSIAN INTERNATIONAL AIRLINES
Ilyushin 62M	Aviadvigatel D-30	27	AEROFLOT RUSSIAN INTERNATIONAL AIRLINES
Ilyushin 62M	Aviadvigatel D-30	1	AIR INDIA
Ilyushin 62M	Aviadvigatel D-30	6	AIR KORYO
Ilyushin 62M	Aviadvigatel D-30	10	CUBANA
Ilyushin 62M	Aviadvigatel D-30	2	TAAG, LINHAS AEREAS DE ANGOLA (ANGOLA AIRLINES)
Ilyushin 62M	Aviadvigatel D-30KU	2	TAROM
Ilyushin 62M / MK	Aviadvigatel D-30	4	AEROFLOT RUSSIAN INTERNATIONAL AIRLINES
Ilyushin 62MK	Aviadvigatel D-30	6	AEROFLOT RUSSIAN INTERNATIONAL AIRLINES
Ilyushin 76	Aviadvigatel D-30	19	AEROFLOT RUSSIAN INTERNATIONAL AIRLINES
Ilyushin 76	Aviadvigatel D-30		BARNAULSKOYE AVIAPREDPRIYATIYE
Ilyushin 76M	Aviadvigatel D-30	31	AEROFLOT RUSSIAN INTERNATIONAL AIRLINES
Ilyushin 76M	Aviadvigatel D-30	4	SYRIAN ARAB AIRLINES
Ilyushin 76M (F)	Aviadvigatel D-30	12	IRAQI AIRWAYS
Ilyushin 76MD	Aviadvigatel D-30	7	AEROFLOT RUSSIAN INTERNATIONAL AIRLINES
Ilyushin 76MD (F)	Aviadvigatel D-30	20	IRAQI AIRWAYS
Ilyushin 76T	Aviadvigatel D-30	31	AEROFLOT RUSSIAN INTERNATIONAL AIRLINES
Ilyushin 76TD	Aviadvigatel D-30	16	AEROFLOT RUSSIAN INTERNATIONAL AIRLINES
Ilyushin 76TD	Aviadvigatel D-30	2	CAIRO CHARTER AND CARGO
Ilyushin 76TD	Aviadvigatel D-30	1	HEAVYLIFT CARGO AIRLINES
Ilyushin 76TD	Aviadvigatel D-30	4	VOLGA DNEPR CARGO AIRLINE
Ilyushin 76TD	Aviadvigatel D-30KP	4	DACONO AIR
Ilyushin 76TD	Aviadvigatel D-30KP	3	SAMARA AIRLINES
Ilyushin 86	KKBM NK-86	64	AEROFLOT RUSSIAN INTERNATIONAL AIRLINES
Ilyushin 86	KKBM NK-86	19	AEROFLOT RUSSIAN INTERNATIONAL AIRLINES
Ilyushin 86	KKBM NK-86	2	CHINA NORTHERN AIRLINES
Ilyushin 86	KKBM NK-86	6	SIBERIA AIRLINE
Ilyushin 86	KKBM NK-86	1	TRANSAERO AIRLINES
Ilyushin 86	KKBM NK-86	22	VNUKOVO AIRLINES
Ilyushin 96-300	Aviadvigatel PS-90A	4	AEROFLOT RUSSIAN INTERNATIONAL AIRLINES
Ilyushin 96M	P&W PW2337	0	AEROFLOT RUSSIAN INTERNATIONAL AIRLINES
L1011 TriStar 1	RR RB211-22B	8	AIR OPS OF EUROPE
L1011 TriStar 1	RR RB211-22B	4	ALL NIPPON AIRWAYS
L1011 TriStar 1	RR RB211-22B	14	AMERICAN TRANS AIR, INC
L1011 TriStar 1	RR RB211-22B	1	CALEDONIAN AIRWAYS
L1011 TriStar 1	RR RB211-22B	12	CATHAY PACIFIC AIRWAYS LTD
L1011 TriStar 1	RR RB211-22B	32	DELTA AIR LINES, INC
L1011 TriStar 1	RR RB211-22B	2	DRAGONAIR (HONG KONG DRAGON AIRLINES LIMITED)
L1011 TriStar 1	RR RB211-22B	1	FAUCETT PERU
L1011 TriStar 1	RR RB211-22B	11	RICH INTERNATIONAL AIRWAYS INC
L1011 TriStar 1	RR RB211-22B	10	TRANS WORLD AIRLINES, INC
L1011 TriStar 1 (F)	RR RB211-22B	1	TRADEWINDS INTERNATIONAL AIRLINES (WRANGLER AVIATION dba)
L1011 TriStar 1/100	RR RB211-22B	5	BRITISH AIRWAYS
L1011 TriStar 1/150	RR RB211-22B	3	AIR CANADA
L1011 TriStar 100	RR RB211-22B	5	AIR TRANSAT
L1011 TriStar 100	RR RB211-22B	4	CALEDONIAN AIRWAYS
L1011 TriStar 100	RR RB211-22B	2	CATHAY PACIFIC AIRWAYS LTD
L1011 TriStar 100	RR RB211-22B	4	LTU - LUFTTRANSPORT UNTERNEHMEN GMBH

AIRFRAME	POWER PLANT	NO. IN FLEET	FLEET NAME
L1011 TriStar 100	RR RB211-22B	6	TRANS WORLD AIRLINES, INC
L1011 TriStar 100	RR RB211-22B-02	1	AIR LANKA
L1011 TriStar 200	RR RB211-524B	1	DELTA AIR LINES, INC
L1011 TriStar 200	RR RB211-524B-02	2	AIR LANKA
L1011 TriStar 200	RR RB211-524B4	1	LTU - LUFTTRANSPORT UNTERNEHMEN GMBH
L1011 TriStar 250	RR RB211-524B	6	DELTA AIR LINES, INC
L1011 TriStar 300	RR RB211-524B-02	17	SAUDIA - SAUDI ARABIAN AIRLINES
L1011 TriStar 50	RR RB211-22B-02	1	AIR LANKA
L1011 TriStar 500	RR RB211-524B	17	DELTA AIR LINES, INC
L1011 TriStar 500	RR RB211-524B4	2	AIR LANKA
L1011 TriStar 500	RR RB211-524B4	2	LTU - LUFTTRANSPORT UNTERNEHMEN GMBH
L1011 TriStar 500	RR RB211-524B4	5	ROYAL JORDANIAN
L1011 TriStar 500	RR RB211-524B4	2	SAUDIA - SAUDI ARABIAN AIRLINES
L1011 TriStar 500	RR RB211-524B4	5	TRINIDAD & TOBAGO (BWIA INTERNATIONAL) AIRWAYS CORPORATION
L1011 TriStar 500	RR RB211-524B4-02	1	TAAG, LINHAS AEREAS DE ANGOLA (ANGOLA AIRLINES)
L1011 TriStar 500	RR RB211-524B4-02	7	TAP AIR PORTUGAL
L1011 Tristar (F)	RR RB211-524B-02	0	KALITTA AMERICAN INTERNATIONAL AIRWAYS
L1049 S Constellation	WR R-3350	2	AEROCHAGO AIRLINES, S.A.
L1329 JetStar 6	P&W JT12A-8	2	PAN AVIATION
L1329 JetStar 731	Ga TFE731-3	1	PAN AVIATION
L1329 JetStar 8	P&W JT12A-8	1	SEAGREEN AIR TRANSPORT
L1329 JetStar II	Ga TFE731-3	6	IRAQI AIRWAYS
L182 Hercules	Allison T56A-11	4	ABOITIZ AIR TRANSPORT
L188 Electra	Allison 501-D13	6	FRED OLSEN AIR TRANSPORT LTD
L188 Electra	Allison 501-D13	7	HUNTING CARGO AIRLINES
L188 Electra	Allison 501-D13	3	REEVE ALEUTIAN AIRWAYS, INC
L188 Electra	Allison 501-D13	2	TAN - TRANSPORTES AEREOS NACIONALES, SA
L188 Electra (F)	Allison 501-D13	17	ZANTOP INTERNATIONAL AIRLINES, INC
L188C Electra	Allison 501-D13	3	AIR ATLANTIQUE
L188C Electra	Allison 501-D13	5	CHANNEL EXPRESS (AIR SERVICES) LTD
L188C Electra	Allison 501-D13	3	MANDALA AIRLINES PT
L188C Electra	Allison 501-D13	1	TRANSAFRIK
L382C Hercules	Allison T56A-15	2	YEMEN AIRLINES
L382E/F Hercules	Allison 501-D22A	1	SOUTHERN AIR TRANSPORT, INC
L382G Hercules	Allison 501-D22A	2	AIR ALGERIE
L382G Hercules	Allison 501-D22A	2	AIR CHINA INTERNATIONAL
L382G Hercules	Allison 501-D22A	1	AIR GABON
L382G Hercules	Allison 501-D22A	2	CHINA AIR CARGO
L382G Hercules	Allison 501-D22A	2	ETHIOPIAN AIRLINES CORPORATION
L382G Hercules	Allison 501-D22A	1	HEAVYLIFT CARGO AIRLINES
L382G Hercules	Allison 501-D22A	1	ICAS INTER CIEL SERVICE
L382G Hercules	Allison 501-D22A	1	NWT AIR (NORTHWEST TERRITORIAL AIRWAYS LTD)
L382G Hercules	Allison 501-D22A	3	PELITA AIR SERVICE
L382G Hercules	Allison 501-D22A	9	SAFAIR (PTY) LTD
L382G Hercules	Allison 501-D22A	17	SOUTHERN AIR TRANSPORT, INC
L382G Hercules	Allison 501-D22A	2	TAAG, LINHAS AEREAS DE ANGOLA (ANGOLA AIRLINES)
L382G Hercules	Allison 501-D22A	6	TRANSAFRIK
Learjet 24D / 25C	GE CJ610-6	1/1	AIR PROVENCE
Learjet 25	GE CJ610-8A	1	THERON AIRWAYS (PTY) LTD
Learjet 25D	GE CJ610-8A	1	AERONAVES DEL PERU, S.A.
Learjet 25D	GE CJ610-8A	2	FLAGSHIP EXPRESS
Learjet 31A	Ga TFE731-2-3B	4	SINGAPORE AIRLINES LIMITED
Learjet 35	Ga TPE731-2-2B	1	RATIOFLUG GMBH
Learjet 35A	Ga TFE731-2-2B	1	ERA AVIATION, INC
Learjet 35A	Ga TFE731-2-2B	2	TAESA
Learjet 35A	Ga TPE731-2-2B	1	AIR AFFAIRES GABON
Learjet 45	Ga TFE731-20	0	SINGAPORE AIRLINES LIMITED
Learjet 55	Ga TPE731-3A-2B	2	RATIOFLUG GMBH
Learjet 55	Ga TPE731-3AR-2B	1	CHINA NORTHERN AIRLINES
Learjet 60	P&WC PW305A	1	LAUDA AIR LUFTFAHRT AKTIENGESELLSCHAFT
Let 410A	P&WC PT6A-27	1	ECUATO GUINEANA DE AVIACION
Let 410UVP	Mot M-601D	2	FARNER AIR TRANSPORT HUNGARY
Luftschiff WDL 1B Airship	Cont IO-360-D	2	WDL AVIATION (KOLN)
MD-11	GE CF6-80C2	5	CHINA EASTERN AIRLINES
MD-11	GE CF6-80C2	6	GARUDA INDONESIA PT
MD-11	GE CF6-80C2	1	GHANA AIRWAYS CORPORATION
MD-11	GE CF6-80C2D1BF	4	FINNAIR OY
MD-11	GE CF6-80C2D1F	17	AMERICAN AIRLINES
MD-11	GE CF6-80C2D1F	6	EVA AIRWAYS
MD-11	GE CF6-80C2D1F	7	KLM ROYAL DUTCH AIRLINES
MD-11	GE CF6-80C2D1F	4	THAI AIRWAYS INTERNATIONAL
MD-11	GE CF6-80C2D1F	6	VARIG - VIACAO AEREA RIO-GRANDENSE S.A.
MD-11	GE CF6-80C2D1F	4	VASP - VIACAO AEREA SAO PAULO S.A.
MD-11	P&W PW4360	11	DELTA AIR LINES, INC
MD-11	P&W PW4360	5	KOREAN AIR
MD-11	P&W PW4460	4	CHINA AIRLINES
MD-11	P&W PW4460	5	JAPAN AIRLINES COMPANY LTD - JAL
MD-11	P&W PW4460	0	JAT - JUGOSLOVENSKI AEROTRANSPORT
MD-11	P&W PW4460	4	LTU - LUFTTRANSPORT UNTERNEHMEN GMBH
MD-11	P&W PW4460	13	SWISSAIR (SOCIETE ANONYME SUISSE POUR LA NAVIGATION AERIENNE)
MD-11	P&W PW4460	4	WORLD AIRWAYS INC
MD-11C	GE CF6-80C2D1F	6	ALITALIA
MD-11CF	P&W PW4462	4	MARTINAIR HOLLAND NV
MD-11F	GE CF6-80C2D1F	1	CHINA EASTERN AIRLINES
MD-11F	GE CF6-80C2D1F	13	FEDERAL EXPRESS CORPORATION
MD-11F	P&W PW4462	0	MARTINAIR HOLLAND NV
MD-80	P&W JT8D-217	1	ADRIA AIRWAYS
MD-81	P&W JT8D-209	2	AUSTRAL
MD-81	P&W JT8D-209	7	AUSTRIAN AIRLINES, OSTERREICHISCHE LUFTVERKEHRS AG
MD-81	P&W JT8D-209	8	JAPAN AIR SYSTEM CO LTD (NIHON AIR SYSTEM)
MD-81	P&W JT8D-209	17	SWISSAIR (SOCIETE ANONYME SUISSE POUR LA NAVIGATION AERIENNE)
MD-81	P&W JT8D-217	5	CONTINENTAL AIRLINES
MD-81	P&W JT8D-217	12	JAPAN AIR SYSTEM CO LTD (NIHON AIR SYSTEM)
MD-81	P&W JT8D-217	31	SAS - SCANDINAVIAN AIRLINES SYSTEM
MD-81	P&W JT8D-217	19	USAIR, INC
MD-82	P&W JT8D-217	10	AEROMEXICO
MD-82	P&W JT8D-217	4	ALASKA AIRLINES, INC
MD-82	P&W JT8D-217	4	ALASKA AIRLINES, INC
MD-82	P&W JT8D-217	2	ALM ANTILLEAN AIRLINES

AIRFRAME	POWER PLANT	NO. IN FLEET	FLEET NAME
MD-82	P&W JT8D-217	1	ALM ANTILLEAN AIRLINES
MD-82	P&W JT8D-217	234	AMERICAN AIRLINES
MD-82	P&W JT8D-217	6	AUSTRIAN AIRLINES, OSTERREICHISCHE LUFTVERKEHRS AG
MD-82	P&W JT8D-217	2	BALAIRCTA
MD-82	P&W JT8D-217	2	CHINA EASTERN AIRLINES
MD-82	P&W JT8D-217	18	CONTINENTAL AIRLINES
MD-82	P&W JT8D-217	8	MERIDIANA SPA
MD-82	P&W JT8D-217	8	NORTHWEST AIRLINES
MD-82	P&W JT8D-217	1	OASIS INTERNATIONAL AIRLINES
MD-82	P&W JT8D-217	29	TRANS WORLD AIRLINES, INC
MD-82	P&W JT8D-217	12	USAIR, INC
MD-82	P&W JT8D-217	2	VENUS AIRLINES SA
MD-82	P&W JT8D-217A	39	ALITALIA
MD-82	P&W JT8D-217A	36	CONTINENTAL AIRLINES
MD-82	P&W JT8D-217C	36	ALITALIA
MD-82	P&W JT8D-217C	5	FAR EASTERN AIR TRANSPORT
MD-82	P&W JT8D-217C	9	KOREAN AIR
MD-82	P&W JT8D-217C	2	SPANAIR
MD-82	P&W JT8D-219	15	AERO LLOYD
MD-82	P&W JT8D-219	2	ALASKA AIRLINES, INC
MD-82	P&W JT8D-219	4	CONTINENTAL AIRLINES
MD-82	P&W JT8D-219	9	FINNAIR OY
MD-82	P&W JT8D-219	4	RENO AIR
MD-82	P&W JT8D-219	15	SAS - SCANDINAVIAN AIRLINES SYSTEM
MD-82 / MD/SAIC MD-82	P&W JT8D-217	11	CHINA EASTERN AIRLINES
MD-82 / MD/SAIC MD-82	P&W JT8D-217	24	CHINA NORTHERN AIRLINES
MD-83	P&W JT8D-217	1	AEROLINEAS ARGENTINAS
MD-83	P&W JT8D-219	3	AERO LLOYD
MD-83	P&W JT8D-219	4	AEROCANCUN
MD-83	P&W JT8D-219	3	AEROMEXICO
MD-83	P&W JT8D-219	3	AEROPOSTAL
MD-83	P&W JT8D-219	10	AEROVIAS NACIONALES DE COLOMBIA S.A. AVIANCA
MD-83	P&W JT8D-219	1	AIR ARUBA
MD-83	P&W JT8D-219	6	AIR LIBERTE
MD-83	P&W JT8D-219	7	AIRTOURS INTERNATIONAL AIRWAYS LTD
MD-83	P&W JT8D-219	28	ALASKA AIRLINES, INC
MD-83	P&W JT8D-219	26	AMERICAN AIRLINES
MD-83	P&W JT8D-219	7	AOM FRENCH AIRLINES
MD-83	P&W JT8D-219	2	AUSTRIAN AIRLINES, OSTERREICHISCHE LUFTVERKEHRS AG
MD-83	P&W JT8D-219	2	BALAIRCTA
MD-83	P&W JT8D-219	5	CENTENNIAL AIRLINES
MD-83	P&W JT8D-219	2	CONTINENTAL AIRLINES
MD-83	P&W JT8D-219	1	EUROFLY SPA
MD-83	P&W JT8D-219	1	FAR EASTERN AIR TRANSPORT
MD-83	P&W JT8D-219	5	FINNAIR OY
MD-83	P&W JT8D-219	1	NORTH AMERICAN AIRLINES
MD-83	P&W JT8D-219	5	OASIS INTERNATIONAL AIRLINES
MD-83	P&W JT8D-219	13	RENO AIR
MD-83	P&W JT8D-219	12	SPANAIR
MD-83	P&W JT8D-219	9	TRANSWEDE
MD-83	P&W JT8D-219	9	TRINIDAD & TOBAGO (BWIA INTERNATIONAL) AIRWAYS CORPORATION
MD-83	P&W JT8D-219	2	TUR EUROPEAN AIRWAYS
MD-83	P&W JT8D-219	1	VENUS AIRLINES SA
MD-83	P&W JT8D-219	2	ZAS AIRLINES OF EGYPT
MD-83 / SAIC/MD-83	P&W JT8D-217	16	TRANS WORLD AIRLINES, INC
MD-87	P&W JT8D-217	4	BALAIRCTA
MD-87	P&W JT8D-217	8	JAPAN AIR SYSTEM CO LTD (NIHON AIR SYSTEM)
MD-87	P&W JT8D-217	16	SAS - SCANDINAVIAN AIRLINES SYSTEM
MD-87	P&W JT8D-217	2	TRANSWEDE
MD-87	P&W JT8D-217C	24	IBERIA - LINEAS AEREAS DE ESPANA, S.A.
MD-87	P&W JT8D-219	2	AERO LLOYD
MD-87	P&W JT8D-219	2	AEROMEXICO
MD-87	P&W JT8D-219	3	FINNAIR OY
MD-87	P&W JT8D-219	1	ZAS AIRLINES OF EGYPT
MD-87	P&W JT8D-219B	2	GREAT AMERICAN AIRWAYS
MD-87ER	P&W JT8D-219	2	AUSTRIAN AIRLINES, OSTERREICHISCHE LUFTVERKEHRS AG
MD-87SR	P&W JT8D-217	3	AUSTRIAN AIRLINES, OSTERREICHISCHE LUFTVERKEHRS AG
MD-88	P&W JT8D-217	13	AVIACO
MD-88	P&W JT8D-219	4	AEROLINEAS ARGENTINAS
MD-88	P&W JT8D-219	10	AEROMEXICO
MD-88	P&W JT8D-219	2	AIR ARUBA
MD-88	P&W JT8D-219	117	DELTA AIR LINES, INC
MD-88	P&W JT8D-219	2	MIDWEST EXPRESS
MD90-30	IAE V2500-D5	0	GREAT CHINA AIRLINES
MD90-30	IAE V2525-D5	3	DELTA AIR LINES, INC
MD90-30	IAE V2525-D5	0	JAPAN AIR SYSTEM CO LTD (NIHON AIR SYSTEM)
MD90-30	IAE V2525-D5	0	SAS - SCANDINAVIAN AIRLINES SYSTEM
Mil Mi-17	Klimov TV3-117MT-3	3	KAMPUCHEA AIRLINES
Mil Mi-8	Klimov TV2-117A		BARNAULSKOYE AVIAPREDPRIYATIYE
Mil Mi-8	Klimov TV2-117A	2	KAMPUCHEA AIRLINES
Mil Mi-8	Klimov TV2-117A	3	MONGOLIAN AIRLINES - MIAT
Mil Mi-8	Klimov TV2-117A		NOVOKUZNETSKOE AVIAPREDPRIYATIYE
Mil Mi-8	Klimov TV2-117A		TURKMENISTAN AIRLINE
Mil Mi-8	Klimov TV2-117A		ULAN-UDESKOYE AVIAPREDPRIYATIYE
Mil Mi-8T	Klimov TV2-117A	1	KAMPUCHEA AIRLINES
Mitsubishi MU-2F	Ga TPE331-10-501A	1	SUN-AIR OF SCANDINAVIA A/S
NAMC YS-11-100	RR Dart 543-10J/K	5	AIR NIPPON
NAMC YS-11-200	RR Dart 542-10J/K	5	JAPAN TRANSOCEAN AIR
NAMC YS-11-200	RR Dart 543-10J/K	14	AIR NIPPON
NAMC YS-11A Combi	RR Dart 542-10K	2	REEVE ALEUTIAN AIRWAYS, INC
NAMC YS-11A-200	RR Dart 542-10K	11	AIRBORNE EXPRESS (ABX AIR, INC dba)
NAMC YS-11A-500	RR Dart 542-10J/K	6	JAPAN AIR COMMUTER CO LTD
NAMC YS-11A-500	RR Dart 542-10J/K	15	JAPAN AIR SYSTEM CO LTD (NIHON AIR SYSTEM)
NAMC YS-11A-500	RR Dart 543-10J/K	7	AIR NIPPON
NAMC YS-11A-500	RR Dart 543-10J/K	7	ALL NIPPON AIRWAYS
PA-23-250 Aztec B	Lyc IO-540-A1D5	2	PROVINCIAL AIRWAYS LTD
PA-23-250 Aztec C	Lyc IO-540-C4B5	1	AIR SAINT-PIERRE
PA-23-250 Aztec D	Lyc IO-540-C4B5	1	AIR FAST INDONESIA
PA-23-250 Aztec D	Lyc IO-540-C4B5	1	NORLANDAIR (ICELAND)

AIRFRAME	POWER PLANT	NO. IN FLEET	FLEET NAME
PA-23-250 Aztec D	Lyc IO-540-C4B5	1	SOLOMON AIRLINES
PA-23-250 Aztec D	Lyc IO-540-C4B5	3	TRANS ISLAND AIRWAYS LIMITED
PA-23-250 Aztec E	Lyc IO-540-C4B5	1	AIR RWANDA
PA-23-250 Aztec F	Lyc IO-540-C4B5	2	TURKS & CAICOS AIRWAYS LIMITED
PA-28-140 Cherokee	Lyc IO-360	10	PROVINCIAL AIRWAYS LTD
PA-28-151/161 Cherokee	Lyc IO-360	2	PROVINCIAL AIRWAYS LTD
PA-28-180	Lyc IO-360	1	ZAIREAN AIRLINES
PA-28-181 Archer II	Lyc O-360-A4M	8	SAUDIA - SAUDI ARABIAN AIRLINES
PA-31 Cheyenne IIIA	P&WC PT6A-6	4	CAAC
PA-31-310 Navajo	Lyc TIO-540-A1A	1	AIR NORTH
PA-31-310 Navajo	Lyc TIO-540-A1A	7	PROVINCIAL AIRWAYS LTD
PA-31-310 Navajo	Lyc TSIO-540-A2C	1	AIRKENYA AVIATION LIMITED
PA-31-310 Navajo B	Lyc TIO-540-A2C	1	GAS AIR
PA-31-310 Navajo C	Lyc TIO-540-A1A	1	HOLMSTROM AIR
PA-31-325 Navajo C/R	Lyc TIO-540-F2BD	1	GAS AIR
PA-31-350 Chieftain	Lyc TIO-540-J2BD	1	AIR SAINT-PIERRE
PA-31-350 Chieftain	Lyc TIO-540-J2BD	3	AIRLINK AIRLINE (PTY) LTD
PA-31-350 Chieftain	Lyc TIO-540-J2BD	4	ARKIA ISRAELI AIRLINES
PA-31-350 Chieftain	Lyc TIO-540-J2BD	4	BERING AIR, INC
PA-31-350 Chieftain	Lyc TIO-540-J2BD	1	CALM AIR (CANADIAN PARTNER)
PA-31-350 Chieftain	Lyc TIO-540-J2BD	7	HAZELTON AIRLINES
PA-31-350 Chieftain	Lyc TIO-540-J2BD	3	IRAN ASSEMAN AIRLINES
PA-31-350 Chieftain	Lyc TIO-540-J2BD	2	MOUNT COOK AIRLINE
PA-31-350 Chieftain	Lyc TIO-540-J2BD	2	NORLANDAIR (ICELAND)
PA-31-350 Chieftain	Lyc TIO-540-J2BD	3	PROVINCIAL AIRWAYS LTD
PA-31-350 Chieftain	Lyc TIO-540-J2BD	1	SUNAIR (BOP AIR PTY LTD dba)
PA-31-350 Navajo	LY TIO-540-J2BD	6	SCENIC AIRLINES INC
PA-31-350 T1020	Lyc TIO-540-J2B	3	BERING AIR, INC
PA-31-350 T1040	P&WC PT6A-27	1	AIRKENYA AVIATION LIMITED
PA-31T Cheyenne II	P&WC PT6A-28	1	AIR MAURITANIE
PA-34-200 Seneca	Lyc IO-360-C1E6	1	PROVINCIAL AIRWAYS LTD
PA-34-200T Seneca II	Cont TSIO-360-E	4	THERON AIRWAYS
PA-34-300 Seneca	Lyc IO-360-C1E6	1	FLANDERS AIRLINES NV
PA-42-720 Cheyenne IIIA	P&WC PT6A-6	7	LUFTHANSA GERMAN AIRLINES - DEUTSCHE LUFTHANSA AG
PZL Mielec (Antonov) 2	Aviadvigatel ASh-621R		BARNAULSKOYE AVIAPREDPRIYATIYE
PZL Mielec (Antonov) 2	Aviadvigatel ASh-621R	2	ESTONIAN AIR
PZL Mielec (Antonov) 2	Aviadvigatel ASh-621R		KEMEROVSKOE AVIAPREDPRIYATIYE
PZL Mielec (Antonov) 2	Aviadvigatel ASh-621R		KYRGYZSTAN AIRLINE
PZL Mielec (Antonov) 2	Aviadvigatel ASh-621R	47	MONGOLIAN AIRLINES - MIAT
PZL Mielec (Antonov) 2	Aviadvigatel ASh-621R		NOVOKUZNETSKOE AVIAPREDPRIYATIYE
PZL Mielec (Antonov) 2	Aviadvigatel ASh-621R		OMSKOYE AVIAPREDPRIYATIYE
PZL Mielec (Antonov) 2	Aviadvigatel ASh-621R		TOMSKOYE AVIAPREDPRIYATIYE
PZL Mielec (Antonov) 2	Aviadvigatel ASh-621R		ULAN-UDESKOYE AVIAPREDPRIYATIYE
Partenavia AP68TP Victor	Lyc IO-360-A1B6	1	AIR CAPE
Partenavia P68C	Lyc IO-360-A1B6	4	LAM - LINHAS AEREAS DE MOCAMBIQUE
Partenavia PN68 Observer	Lyc IO-360-A1B6	1	SUN-AIR OF SCANDINAVIA A/S
Piaggio P166-DL2	Lyc IGSO-540-A1H	4	IRAQI AIRWAYS
Pilatus PC-6 Turbo Porter	P&WC PT6A-27	2	FARNER AIR TRANSPORT AG
Pilatus PC-6 Turbo Porter	P&WC PT6A-27	3	MOUNT COOK AIRLINE
Pilatus PC-6 Turbo Porter	P&WC PT6A-27	3	SATENA
Pilatus PC-6B Tbo. Porter	P&WC PT6A-20	1	ROYAL NEPAL AIRLINES CORPORATION
ROMBAC One-Eleven 561RC	RR Spey 511-14DW	6	TAROM
ROMBAC One-Eleven 561RC	RR Spey 512-14DW	4	RYANAIR LTD
Rockwell Grand Commander	Lyc IGSO-540-B1A	1	ARKIA ISRAELI AIRLINES
Rockwell Shrike Commander	Lyc IO-540-E1B5	11	IRAN ASSEMAN AIRLINES
Rockwell Turbo Cmdr 690B	Ga TPE331-5-251K	2	CATA
Romaero One-Eleven 2500	RR Tay 650-15	0	KIWI INTERNATIONAL AIRLINES INC
SA 227AC Metro III	Ga TPE331-11U-611G	2	EMPIRE AIRLINES
SA226 Metro II	Ga TPE331-10UA-511G	3	PROVINCIAL AIRWAYS LTD
SA226 Metro II	Ga TPE331-3UW-303G	3	EUROPEAN AIR TRANSPORT N.V./S.A.
SA226 Metro III	Ga TPE331-11U-611G	4	EAGLE AIRWAYS LTD
SA226AC Metro III	Ga TPE331-11U-611G	9	COMAIR
SA226AT Merlin IV	Ga TPE331-3UW-303G	1	PROVINCIAL AIRWAYS LTD
SA226AT Merlin IV	Ga TPE331-3UW-304G	2	FLANDERS AIRLINES NV
SA226AT Merlin IVA	Ga TPE331-10UA-511G	2	JETALL
SA226TC Metro II	Ga TPE331-10UA-511G	4	BIG SKY AIRLINES
SA226TC Metro II	Ga TPE331-10UA-511G	4	JETALL
SA226TC Metro II	Ga TPE331-3UW-303G	2	AIRLINK AIRLINE (PTY) LTD
SA226TC Metro II	Ga TPE331-3UW-303G	1	JETALL
SA226TC Metro II	Ga TPE331-3UW-304G	2	KENDELL AIRLINES (AUSTRALIA) PTY LIMITED
SA227AC Metro III	Ga TPE331-11U-601G	1	NORLANDAIR (ICELAND)
SA227AC Metro III	Ga TPE331-11U-611G	1	AIR ATLANTIQUE
SA227AC Metro III	Ga TPE331-11U-611G	1	AIRLA
SA227AC Metro III	Ga TPE331-11U-611G	10	CHAUTAUQUA AIRLINES
SA227AC Metro III	Ga TPE331-11U-611G	22	HORIZON AIR
SA227AC Metro III	Ga TPE331-11U-611G	6	LONE STAR AIRLINES
SA227AC Metro III	Ga TPE331-11U-611G	26	MESABA AIRLINES (NORTHWEST AIRLINK)
SA227AC Metro III	Ga TPE331-11U-612G	2	TRANS STATES AIRLINES
SA227AC Metroliner	Ga TPE331-11U-611G	25	SKYWEST AIRLINES INC (DELTA CONNECTION)
SA227AT Expediter	Ga TPE331-11U-601G	11	UNITED PARCEL SERVICE COMPANY UPS
SA227AT Expediter I	Ga TPE331-11U-611G	10	DHL AIRWAYS
SA227AT Merlin IV	Ga TPE331-3UW-303G	1	PROVINCIAL AIRWAYS LTD
SA227AT Merlin IVC	Ga TPE331-11U-611G	1	JETALL
SA227AT Metro III	Ga TPE331-11U-611G	3	JETALL
SAC Yunshuji Y-8	WJ 6	2	CHINA EASTERN AIRLINES
Saab 2000	Allison AE 2100	1	AIR MARSHALL ISLANDS - AMI
Saab 2000	Allison AE 2100	0	AMR EAGLE
Saab 2000	Allison AE 2100	0	BUSINESS EXPRESS (THE DELTA CONNECTION)
Saab 2000	Allison AE 2100	0	CALM AIR (CANADIAN PARTNER)
Saab 2000	Allison AE 2100	5	CROSSAIR
Saab 2000	Allison AE 2100	3	DEUTSCHE BA LUFTFAHRTGESELLSCHAFT MBH
Saab 2000	Allison AE 2100	0	KENDELL AIRLINES (AUSTRALIA) PTY LIMITED
Saab 2000	Allison AE 2100	1	REGIONAL AIRLINES
Saab 340A	GE CT7-5A2	16	AMR EAGLE
Saab 340A	GE CT7-5A2	6	BAR HARBOR AIRLINES (CONTINENTAL/DELTA EXPRESS)
Saab 340A	GE CT7-5A2	6	BRIT'AIR
Saab 340A	GE CT7-5A2	17	BUSINESS EXPRESS (THE DELTA CONNECTION)
Saab 340A	GE CT7-5A2	5	CHAUTAUQUA AIRLINES
Saab 340A	GE CT7-5A2	19	COMAIR

AIRFRAME	POWER PLANT	NO. IN FLEET	FLEET NAME
Saab 340A	GE CT7-5A2	10	CROSSAIR
Saab 340A	GE CT7-5A2	5	DEUTSCHE BA LUFTFAHRTGESELLSCHAFT MBH
Saab 340A	GE CT7-5A2	29	EXPRESS AIRLINES I, INC (NORTHWEST AIRLINK)
Saab 340A	GE CT7-5A2	5	FINNAVIATION OY
Saab 340A	GE CT7-5A2	3	FORMOSA AIRLINES
Saab 340A	GE CT7-5A2	9	KENDELL AIRLINES (AUSTRALIA) PTY LIMITED
Saab 340A	GE CT7-5A2	2	LAPA - LINEAS AEREAS PRIVADAS ARGENTINIAS, SA
Saab 340A	GE CT7-5A2	3	SKYWAYS, AB
Saab 340B	GE CT7-5A2	4	HAZELTON AIRLINES
Saab 340B	GE CT7-9B	4	AER LINGUS
Saab 340B	GE CT7-9B	100	AMR EAGLE
Saab 340B	GE CT7-9B	22	BUSINESS EXPRESS (THE DELTA CONNECTION)
Saab 340B	GE CT7-9B	4	CHAUTAUQUA AIRLINES
Saab 340B	GE CT7-9B	4	CHINA SOUTHERN AIRLINES
Saab 340B	GE CT7-9B	0	CHINA SOUTHWEST AIRLINES
Saab 340B	GE CT7-9B	15	CROSSAIR
Saab 340B	GE CT7-9B	6	EXPRESS AIRLINES I, INC (NORTHWEST AIRLINK)
Saab 340B	GE CT7-9B	1	FINNAVIATION OY
Saab 340B	GE CT7-9B	6	FORMOSA AIRLINES
Saab 340B	GE CT7-9B	2	JAPAN AIR COMMUTER CO LTD
Saab 340B	GE CT7-9B	1	JAPAN AIR SYSTEM CO LTD (NIHON AIR SYSTEM)
Saab 340B	GE CT7-9B	11	KLM CITYHOPPER
Saab 340B	GE CT7-9B	7	REGIONAL AIRLINES
Saab 340B	GE CT7-9B	4	SKYWAYS, AB
Saab 340Bplus	GE CT7-5A2	0	AMR EAGLE
Saab 340Bplus	GE CT7-5A2	1	CALM AIR (CANADIAN PARTNER)
Saab 340Bplus	GE CT7-5A2	1	HAZELTON AIRLINES
Saab 340Bplus	GE CT7-5A2	0	KENDELL AIRLINES (AUSTRALIA) PTY LIMITED
Saab 340Bplus	GE CT7-5A2	1	REGIONAL AIRLINES
See notes	NYK		NATIONAL AIR CHARTERS
See notes	NYK		SUN D'OR INTERNATIONAL AIRLINES
Shijiazhuang Yunshuji Y-5	HS 5	12	CHINA NORTHWEST AIRLINES
Shorts 330-200	P&WC PT6A-45B	1	CANAIR CARGO LTD
Shorts 330-200	P&WC PT6A-45B	4	OLYMPIC AVIATION, S.A.
Shorts 330-200	P&WC PT6A-45R	4	BANGKOK AIRWAYS
Shorts 330-200	P&WC PT6A-45R	1	MUK AIR
Shorts 330-200	P&WC PT6A-45R	1	OLYMPIC AVIATION, S.A.
Shorts 330-200	P&WC PT6A-45R	1	SKYWAYS, AB
Shorts 330-200	P&WC PT6A-45R	2	SUNSTATE AIRLINES
Shorts 360	P&WC PT6A-6	2	AIR UK LTD
Shorts 360	P&WC PT6A-6	1	AURIGNY AIR SERVICES LTD
Shorts 360	P&WC PT6A-6	8	BUSINESS EXPRESS (THE DELTA CONNECTION)
Shorts 360	P&WC PT6A-6	2	CHINA EASTERN AIRLINES
Shorts 360	P&WC PT6A-6	3	CHINA SOUTHERN AIRLINES
Shorts 360	P&WC PT6A-6	2	CITYFLYER EXPRESS LTD
Shorts 360	P&WC PT6A-6	10	PENNSYLVANIA COMMUTER AIRLINES, INC (USAIR EXPRESS)
Shorts 360	P&WC PT6A-6	3	SUNSTATE AIRLINES
Shorts 360	P&WC PT6A-65R	1	BRITISH MIDLAND AIRWAYS LTD
Shorts 360	P&WC PT6A-65R	3	CANADIAN REGIONAL
Shorts 360	P&WC PT6A-65R	2	MUK AIR
Shorts 360 Advanced	P&WC PT6A-6	1	AIR HUDIK AB
Shorts 360 Advanced	P&WC PT6A-6	4	JERSEY EUROPEAN
Shorts 360 Advanced	P&WC PT6A-65AR	2	BANGKOK AIRWAYS
Shorts 360-100	P&WC PT6A-6	3	LOGANAIR LTD
Shorts 360-100 Advanced	P&WC PT6A-6	5	LOGANAIR LTD
Shorts 360-200	P&WC PT6A-6	25	AMR EAGLE
Shorts 360-300	P&WC PT6A-6	4	AMR EAGLE
Shorts 360-300	P&WC PT6A-6	9	CCAIR, INC (USAIR EXPRESS)
Shorts 360-300	P&WC PT6A-6		CHINA ASIA AIRLINES
Shorts 360-300	P&WC PT6A-6	2	HAZELTON AIRLINES
Shorts 360-300	P&WC PT6A-6	2	PHILIPPINE AIR LINES, INC - PAL
Shorts 360-300	P&WC PT6A-67R	2	LIBERTY EXPRESS (USAIR EXPRESS)
Shorts Belfast	RR Tyne 515-101W	5	HEAVYLIFT CARGO AIRLINES
Sikorsky S-58ET	P&WC PT6T-3 TwinPac	2	AIR FAST INDONESIA
Sikorsky S-58ET	P&WC PT6T-6 TwinPac	2	AIR FAST INDONESIA
Sikorsky S-61N	GE CT58-140-1	4	GROENLANDSFLY A/S
Sikorsky S-76A II	Allison 250-C30S20	4	PELITA AIR SERVICE
Transall C-160NG	RR Tyne 522	3	PELITA AIR SERVICE
Transall C-160P	RR Tyne 522	3	PELITA AIR SERVICE
Transall C-160P	RR Tyne 522	2	SAF SERVICE AERIEN FRANCAIS, SA
Tupolev 134	Aviadvigatel D-30	36	AEROFLOT RUSSIAN INTERNATIONAL AIRLINES
Tupolev 134	Aviadvigatel D-30		KYRGYZSTAN AIRLINE
Tupolev 134-3	Aviadvigatel D-30	8	AEROFLOT RUSSIAN INTERNATIONAL AIRLINES
Tupolev 134A	Aviadvigatel D-30	69	AEROFLOT RUSSIAN INTERNATIONAL AIRLINES
Tupolev 134A	Aviadvigatel D-30	2	BALTIC INTERNATIONAL AIRLINES
Tupolev 134A	Aviadvigatel D-30	19	BELAVIA
Tupolev 134A	Aviadvigatel D-30	5	CSA - CESKOSLOVENSKE AEROLINIE
Tupolev 134A	Aviadvigatel D-30	6	ESTONIAN AIR
Tupolev 134A	Aviadvigatel D-30	8	LITHUANIAN AIRLINES
Tupolev 134A	Aviadvigatel D-30	7	SAMARA AIRLINES
Tupolev 134A	Aviadvigatel D-30	5	VIETNAM AIRLINES
Tupolev 134A-3	Aviadvigatel D-30	5	BALKAN
Tupolev 134A-3	Aviadvigatel D-30	2	KAMPUCHEA AIRLINES
Tupolev 134A-3	Aviadvigatel D-30	1	LITHUANIAN AIRLINES
Tupolev 134A-3	Aviadvigatel D-30-III	6	MALEV - HUNGARIAN AIRLINES
Tupolev 134B-3	Aviadvigatel D-30	2	AIR KORYO
Tupolev 134B-3	Aviadvigatel D-30	4	SYRIAN ARAB AIRLINES
Tupolev 134B-3	Aviadvigatel D-30	4	VIETNAM AIRLINES
Tupolev 154	KKBM NK-8-2	44	AEROFLOT RUSSIAN INTERNATIONAL AIRLINES
Tupolev 154	KKBM NK-8-2	2	AERONICA - AEROLINEAS NICARAGUENSES, SA
Tupolev 154	KKBM NK-8-2		BARNAULSKOYE AVIAPREDPRIYATIYE
Tupolev 154	KKBM NK-8-2	3	KEMEROVSKOE AVIAPREDPRIYATIYE
Tupolev 154	KKBM NK-8-2		KYRGYZSTAN AIRLINE
Tupolev 154	KKBM NK-8-2	5	NOVOKUZNETSKOE AVIAPREDPRIYATIYE
Tupolev 154	KKBM NK-8-2		OMSKOYE AVIAPREDPRIYATIYE
Tupolev 154	KKBM NK-8-2	12	SIBERIA AIRLINE
Tupolev 154	KKBM NK-8-2	3	TOMSKOYE AVIAPREDPRIYATIYE
Tupolev 154	KKBM NK-8-2		TURKMENISTAN AIRLINE
Tupolev 154	KKBM NK-8-2	1	ULAN-UDESKOYE AVIAPREDPRIYATIYE

AIRFRAME	POWER PLANT	NO. IN FLEET	FLEET NAME
Tupolev 154	KKBM NK-8-2	33	VNUKOVO AIRLINES
Tupolev 154A	KKBM NK-8-2	45	AEROFLOT RUSSIAN INTERNATIONAL AIRLINES
Tupolev 154B	KKBM NK-8-2	101	AEROFLOT RUSSIAN INTERNATIONAL AIRLINES
Tupolev 154B	KKBM NK-8-2	4	AEROFLOT RUSSIAN INTERNATIONAL AIRLINES
Tupolev 154B	KKBM NK-8-2	3	AIR KORYO
Tupolev 154B	KKBM NK-8-2	17	BELAVIA
Tupolev 154B	KKBM NK-8-2U	1	TAROM
Tupolev 154B-1	KKBM NK-8-2	70	AEROFLOT RUSSIAN INTERNATIONAL AIRLINES
Tupolev 154B-1	KKBM NK-8-2	5	TAROM
Tupolev 154B-2	KKBM NK-8-2	296	AEROFLOT RUSSIAN INTERNATIONAL AIRLINES
Tupolev 154B-2	KKBM NK-8-2	1	AIR KORYO
Tupolev 154B-2	KKBM NK-8-2	17	BALKAN
Tupolev 154B-2	KKBM NK-8-2	10	MALEV - HUNGARIAN AIRLINES
Tupolev 154B-2	KKBM NK-8-2	2	TAROM
Tupolev 154B-2	KKBM NK-8-2U	7	SAMARA AIRLINES
Tupolev 154C	KKBM NK-8-2	5	AEROFLOT RUSSIAN INTERNATIONAL AIRLINES
Tupolev 154C	KKBM NK-8-2	3	AEROFLOT RUSSIAN INTERNATIONAL AIRLINES
Tupolev 154C	KKBM NK-8-2	1	VNUKOVO AIRLINES
Tupolev 154M	Aviadvigatel D-30	30	AEROFLOT RUSSIAN INTERNATIONAL AIRLINES
Tupolev 154M	Aviadvigatel D-30	30	AEROFLOT RUSSIAN INTERNATIONAL AIRLINES
Tupolev 154M	Aviadvigatel D-30	2	ARIANA AFGHAN AIRLINES
Tupolev 154M	Aviadvigatel D-30	5	BALKAN
Tupolev 154M	Aviadvigatel D-30	2	CAIRO CHARTER AND CARGO
Tupolev 154M	Aviadvigatel D-30	10	CHINA NORTHWEST AIRLINES
Tupolev 154M	Aviadvigatel D-30	5	CHINA SOUTHWEST AIRLINES
Tupolev 154M	Aviadvigatel D-30	4	CSA - CESKOSLOVENSKE AEROLINIE
Tupolev 154M	Aviadvigatel D-30	4	CUBANA
Tupolev 154M	Aviadvigatel D-30	6	IRAN AIR THE AIRLINE OF THE ISLAMIC REPUBLIC OF IRAN
Tupolev 154M	Aviadvigatel D-30	1	LAO AVIATION
Tupolev 154M	Aviadvigatel D-30	1	MONGOLIAN AIRLINES - MIAT
Tupolev 154M	Aviadvigatel D-30	2	NOVOKUZNETSKOE AVIAPREDPRIYATIYE
Tupolev 154M	Aviadvigatel D-30	7	SIBERIA AIRLINE
Tupolev 154M	Aviadvigatel D-30	5	SICHUAN AIRLINES
Tupolev 154M	Aviadvigatel D-30	3	SYRIAN ARAB AIRLINES
Tupolev 154M	Aviadvigatel D-30	9	UNION AIRLINES
Tupolev 154M	Aviadvigatel D-30	5	XINJIANG AIRLINES
Tupolev 154M	Aviadvigatel D-30KU-154	3	BELAVIA
Tupolev 154M	Aviadvigatel D-30KU-154	3	SAMARA AIRLINES
Tupolev 204	Aviadvigatel PS-90-A	3	VNUKOVO AIRLINES
Xian Yunshuji Y-7	WJ 5A-1	3	AIR CHINA INTERNATIONAL
Xian Yunshuji Y-7	WJ 5A-1	9	CHINA EASTERN AIRLINES
Xian Yunshuji Y-7	WJ 5A-1	5	CHINA NORTHERN AIRLINES
Xian Yunshuji Y-7	WJ 5A-1	5	CHINA SOUTHERN AIRLINES
Xian Yunshuji Y-7	WJ 5A-1	1	LAO AVIATION
Xian Yunshuji Y-7-100	WJ 5A-1	3	AIR CHINA INTERNATIONAL
Xian Yunshuji Y-7-100	WJ 5A-1	3	CHINA AIR CARGO
Xian Yunshuji Y-7-100	WJ 5A-1	6	CHINA NORTHERN AIRLINES
Xian Yunshuji Y-7-100	WJ 5A-1	7	CHINA NORTHWEST AIRLINES
Xian Yunshuji Y-7-100	WJ 5A-1	3	CHINA SOUTHWEST AIRLINES
Xian Yunshuji Y-7-100	WJ 5A-1	2	GUIZHOU AIRLINES
Xian Yunshuji Y-7-100	WJ 5A-1	5	SICHUAN AIRLINES
Xian Yunshuji Y-7-100	WJ 5A-1	4	UNION AIRLINES
Xian Yunshuji Y-7-100	WJ 5A-1	2	WUHAN AIR LINES
Xian Yunshuji Y-7-100	WJ 5A-1	2	ZHONGYUAN AIRLINES
Yakovlev 40	ZMKB AI-25	25	AEROFLOT RUSSIAN INTERNATIONAL AIRLINES
Yakovlev 40	ZMKB AI-25	2	ARIANA AFGHAN AIRLINES
Yakovlev 40	ZMKB AI-25		BARNAULSKOYE AVIAPREDPRIYATIYE
Yakovlev 40	ZMKB AI-25	8	BELAVIA
Yakovlev 40	ZMKB AI-25	8	CUBANA
Yakovlev 40	ZMKB AI-25	4	ESTONIAN AIR
Yakovlev 40	ZMKB AI-25		KYRGYZSTAN AIRLINE
Yakovlev 40	ZMKB AI-25	1	NATIONAL OVERSEAS AIRLINE
Yakovlev 40	ZMKB AI-25	3	SAMARA AIRLINES
Yakovlev 40	ZMKB AI-25	6	SYRIAN ARAB AIRLINES
Yakovlev 40	ZMKB AI-25	5	VIETNAM AIRLINES
Yakovlev 40	ZMKB AI-25	1	VOLGA DNEPR CARGO AIRLINE
Yakovlev 42	ZMKB D-36	4	AEROFLOT RUSSIAN INTERNATIONAL AIRLINES
Yakovlev 42	ZMKB D-36	12	LITHUANIAN AIRLINES
Yakovlev 42D	ZMKB PS-90AT	3	DACONO AIR

Notes

NYK = Not yet confirmed

TBD = To be determined

POWER PLANT

This list contains aircraft which are currently produced as well as aircraft which are out of production

General Electric CT7 Turboprop engine

POWER PLANT	AIRFRAME	NO. IN FLEET	FLEET NAME
Allison 250-B17B	GAF Nomad N22B	1	MOUNT COOK AIRLINE
Allison 250-B17B	GAF Nomad N24A	1	MOUNT COOK AIRLINE
Allison 250-B17C	BN-2T Islander	1	AIR MAURITANIE
Allison 250-C20	Agusta A109 II	1	OLYMPIC AVIATION, S.A.
Allison 250-C20	Bell 206B JetRanger	2	AIR MAURITIUS
Allison 250-C20	Bell 206B JetRanger	1	DHL AIRWAYS
Allison 250-C20	Bell 206B JetRanger II	3	AIR FAST INDONESIA
Allison 250-C20	Bell 206B JetRanger II	2	ERA AVIATION, INC
Allison 250-C20	Bell 206B JetRanger II	1	GROENLANDSFLY A/S
Allison 250-C20	Bell 206B JetRanger II	1	TRANSAFRIK
Allison 250-C20	Bell 206B JetRanger III	6	ERA AVIATION, INC
Allison 250-C20	Bell 206B JetRanger III	3	GROENLANDSFLY A/S
Allison 250-C20	Bell 206L LongRanger	5	ERA AVIATION, INC
Allison 250-C20	Eurocopter BO 105 C	5	PELITA AIR SERVICE
Allison 250-C20	Eurocopter BO 105 CBS	31	ERA AVIATION, INC
Allison 250-C20	IPTN/Eurocopter NBO-105CB	28	PELITA AIR SERVICE
Allison 250-C20B	Bell 206B JetRanger	1	NATIONAL OVERSEAS AIRLINE
Allison 250-C20B	Hughes 369D	1	NATIONAL OVERSEAS AIRLINE
Allison 250-C20F	Eurocopter AS355F2	2	OLYMPIC AVIATION, S.A.
Allison 250-C20R1	Agusta A109	2	NATIONAL OVERSEAS AIRLINE
Allison 250-C28B	Bell 206L-1 LongRanger II	1	DHL AIRWAYS
Allison 250-C30S20	Sikorsky S-76A II	4	PELITA AIR SERVICE
Allison 501-D13	CV 580	3	AIR CAPE
Allison 501-D13	CV 580	4	AIR INUIT (1985) LTD
Allison 501-D13	CV 580	2	AIR NIAGARA EXPRESS INC
Allison 501-D13	CV 580	4	AIR RESORTS AIRLINES
Allison 501-D13	CV 580	1	AVENSA
Allison 501-D13	CV 580	10	CANAIR CARGO LTD
Allison 501-D13	CV 580	4	ERA AVIATION, INC
Allison 501-D13	CV 580	16	KELOWNA FLIGHTCRAFT AIR CHARTER LTD
Allison 501-D13	CV 580	2	REGIONAL AIR (PTY) LTD
Allison 501-D13	CV 580	7	SIERRA PACIFIC AIRLINES, INC
Allison 501-D13	CV 580	2	TURKS & CAICOS AIRWAYS LIMITED
Allison 501-D13	CV 580 (F)	11	EUROPEAN AIR TRANSPORT N.V./S.A.
Allison 501-D13	CV 580F (SCD)	4	JETALL
Allison 501-D13	L188 Electra	6	FRED OLSEN AIR TRANSPORT LTD
Allison 501-D13	L188 Electra	7	HUNTING CARGO AIRLINES
Allison 501-D13	L188 Electra	3	REEVE ALEUTIAN AIRWAYS, INC
Allison 501-D13	L188 Electra	2	TAN - TRANSPORTES AEREOS NACIONALES, SA
Allison 501-D13	L188 Electra (F)	17	ZANTOP INTERNATIONAL AIRLINES, INC
Allison 501-D13	L188C Electra	3	AIR ATLANTIQUE
Allison 501-D13	L188C Electra	5	CHANNEL EXPRESS (AIR SERVICES) LTD
Allison 501-D13	L188C Electra	3	MANDALA AIRLINES PT
Allison 501-D13	L188C Electra	1	TRANSAFRIK
Allison 501-D22A	L382E/F Hercules	1	SOUTHERN AIR TRANSPORT, INC
Allison 501-D22A	L382G Hercules	2	AIR ALGERIE
Allison 501-D22A	L382G Hercules	2	AIR CHINA INTERNATIONAL
Allison 501-D22A	L382G Hercules	1	AIR GABON
Allison 501-D22A	L382G Hercules	2	CHINA AIR CARGO
Allison 501-D22A	L382G Hercules	2	ETHIOPIAN AIRLINES CORPORATION
Allison 501-D22A	L382G Hercules	1	HEAVYLIFT CARGO AIRLINES
Allison 501-D22A	L382G Hercules	1	ICAS INTER CIEL SERVICE
Allison 501-D22A	L382G Hercules	1	NWT AIR (NORTHWEST TERRITORIAL AIRWAYS LTD)
Allison 501-D22A	L382G Hercules	3	PELITA AIR SERVICE
Allison 501-D22A	L382G Hercules	9	SAFAIR (PTY) LTD
Allison 501-D22A	L382G Hercules	17	SOUTHERN AIR TRANSPORT, INC
Allison 501-D22A	L382G Hercules	2	TAAG, LINHAS AEREAS DE ANGOLA (ANGOLA AIRLINES)
Allison 501-D22A	L382G Hercules	6	TRANSAFRIK
Allison 501-D22G	CV 5800	2	KELOWNA FLIGHTCRAFT AIR CHARTER LTD
Allison AE 2100	IPTN N-250-100	0	BOURAQ INDONESIA AIRLINES PT
Allison AE 2100	IPTN N-250-100	0	MERPATI NUSANTARA AIRLINES PT
Allison AE 2100	IPTN N-250-100	0	SEMPATI AIR TRANSPORT
Allison AE 2100	Saab 2000	1	AIR MARSHALL ISLANDS - AMI
Allison AE 2100	Saab 2000	0	AMR EAGLE
Allison AE 2100	Saab 2000	0	BUSINESS EXPRESS (THE DELTA CONNECTION)

POWER PLANT	AIRFRAME	NO. IN FLEET	FLEET NAME
Allison AE 2100	Saab 2000	0	CALM AIR (CANADIAN PARTNER)
Allison AE 2100	Saab 2000	5	CROSSAIR
Allison AE 2100	Saab 2000	3	DEUTSCHE BA LUFTFAHRTGESELLSCHAFT MBH
Allison AE 2100	Saab 2000	0	KENDELL AIRLINES (AUSTRALIA) PTY LIMITED
Allison AE 2100	Saab 2000	1	REGIONAL AIRLINES
Allison AE 3007A	EMB 145	0	REGIONAL AIRLINES
Allison AE 3007A	EMB 145	0	TRANSBRASIL S.A. LINHAS AEREAS
Allison LTS101-650C1	Bell 222	1	PAN AVIATION
Allison T56A-11	L182 Hercules	4	ABOITIZ AIR TRANSPORT
Allison T56A-15	L382C Hercules	2	YEMEN AIRLINES
Aviadvigatel ASh-621R	Antonov 2	1	VIETNAM AIRLINES
Aviadvigatel ASh-621R	PZL Mielec (Antonov) 2		BARNAULSKOYE AVIAPREDPRIYATIYE
Aviadvigatel ASh-621R	PZL Mielec (Antonov) 2	2	ESTONIAN AIR
Aviadvigatel ASh-621R	PZL Mielec (Antonov) 2		KEMEROVSKOE AVIAPREDPRIYATIYE
Aviadvigatel ASh-621R	PZL Mielec (Antonov) 2		KYRGYZSTAN AIRLINE
Aviadvigatel ASh-621R	PZL Mielec (Antonov) 2	47	MONGOLIAN AIRLINES - MIAT
Aviadvigatel ASh-621R	PZL Mielec (Antonov) 2		NOVOKUZNETSKOE AVIAPREDPRIYATIYE
Aviadvigatel ASh-621R	PZL Mielec (Antonov) 2		OMSKOYE AVIAPREDPRIYATIYE
Aviadvigatel ASh-621R	PZL Mielec (Antonov) 2		TOMSKOYE AVIAPREDPRIYATIYE
Aviadvigatel ASh-621R	PZL Mielec (Antonov) 2		ULAN-UDESKOYE AVIAPREDPRIYATIYE
Aviadvigatel D-30	Ilyushin 62M	89	AEROFLOT RUSSIAN INTERNATIONAL AIRLINES
Aviadvigatel D-30	Ilyushin 62M	27	AEROFLOT RUSSIAN INTERNATIONAL AIRLINES
Aviadvigatel D-30	Ilyushin 62M	1	AIR INDIA
Aviadvigatel D-30	Ilyushin 62M	6	AIR KORYO
Aviadvigatel D-30	Ilyushin 62M	10	CUBANA
Aviadvigatel D-30	Ilyushin 62M	2	TAAG, LINHAS AEREAS DE ANGOLA (ANGOLA AIRLINES)
Aviadvigatel D-30	Ilyushin 62M / MK	4	AEROFLOT RUSSIAN INTERNATIONAL AIRLINES
Aviadvigatel D-30	Ilyushin 62MK	6	AEROFLOT RUSSIAN INTERNATIONAL AIRLINES
Aviadvigatel D-30	Ilyushin 76	19	AEROFLOT RUSSIAN INTERNATIONAL AIRLINES
Aviadvigatel D-30	Ilyushin 76		BARNAULSKOYE AVIAPREDPRIYATIYE
Aviadvigatel D-30	Ilyushin 76M	31	AEROFLOT RUSSIAN INTERNATIONAL AIRLINES
Aviadvigatel D-30	Ilyushin 76M	4	SYRIAN ARAB AIRLINES
Aviadvigatel D-30	Ilyushin 76M (F)	12	IRAQI AIRWAYS
Aviadvigatel D-30	Ilyushin 76MD	7	AEROFLOT RUSSIAN INTERNATIONAL AIRLINES
Aviadvigatel D-30	Ilyushin 76MD (F)	20	IRAQI AIRWAYS
Aviadvigatel D-30	Ilyushin 76T	31	AEROFLOT RUSSIAN INTERNATIONAL AIRLINES
Aviadvigatel D-30	Ilyushin 76TD	16	AEROFLOT RUSSIAN INTERNATIONAL AIRLINES
Aviadvigatel D-30	Ilyushin 76TD	2	CAIRO CHARTER AND CARGO
Aviadvigatel D-30	Ilyushin 76TD	1	HEAVYLIFT CARGO AIRLINES
Aviadvigatel D-30	Ilyushin 76TD	4	VOLGA DNEPR CARGO AIRLINE
Aviadvigatel D-30	Tupolev 134	36	AEROFLOT RUSSIAN INTERNATIONAL AIRLINES
Aviadvigatel D-30	Tupolev 134		KYRGYZSTAN AIRLINE
Aviadvigatel D-30	Tupolev 134-3	8	AEROFLOT RUSSIAN INTERNATIONAL AIRLINES
Aviadvigatel D-30	Tupolev 134A	69	AEROFLOT RUSSIAN INTERNATIONAL AIRLINES
Aviadvigatel D-30	Tupolev 134A	2	BALTIC INTERNATIONAL AIRLINES
Aviadvigatel D-30	Tupolev 134A	19	BELAVIA
Aviadvigatel D-30	Tupolev 134A	5	CSA - CESKOSLOVENSKE AEROLINIE
Aviadvigatel D-30	Tupolev 134A	6	ESTONIAN AIR
Aviadvigatel D-30	Tupolev 134A	8	LITHUANIAN AIRLINES
Aviadvigatel D-30	Tupolev 134A	7	SAMARA AIRLINES
Aviadvigatel D-30	Tupolev 134A	5	VIETNAM AIRLINES
Aviadvigatel D-30	Tupolev 134A-3	5	BALKAN
Aviadvigatel D-30	Tupolev 134A-3	2	KAMPUCHEA AIRLINES
Aviadvigatel D-30	Tupolev 134A-3	1	LITHUANIAN AIRLINES
Aviadvigatel D-30	Tupolev 134B-3	2	AIR KORYO
Aviadvigatel D-30	Tupolev 134B-3	4	SYRIAN ARAB AIRLINES
Aviadvigatel D-30	Tupolev 134B-3	4	VIETNAM AIRLINES
Aviadvigatel D-30	Tupolev 154M	30	AEROFLOT RUSSIAN INTERNATIONAL AIRLINES
Aviadvigatel D-30	Tupolev 154M	30	AEROFLOT RUSSIAN INTERNATIONAL AIRLINES
Aviadvigatel D-30	Tupolev 154M	2	ARIANA AFGHAN AIRLINES
Aviadvigatel D-30	Tupolev 154M	5	BALKAN
Aviadvigatel D-30	Tupolev 154M	2	CAIRO CHARTER AND CARGO
Aviadvigatel D-30	Tupolev 154M	10	CHINA NORTHWEST AIRLINES
Aviadvigatel D-30	Tupolev 154M	5	CHINA SOUTHWEST AIRLINES
Aviadvigatel D-30	Tupolev 154M	4	CSA - CESKOSLOVENSKE AEROLINIE
Aviadvigatel D-30	Tupolev 154M	4	CUBANA
Aviadvigatel D-30	Tupolev 154M	6	IRAN AIR THE AIRLINE OF THE ISLAMIC REPUBLIC OF IRAN
Aviadvigatel D-30	Tupolev 154M	1	LAO AVIATION
Aviadvigatel D-30	Tupolev 154M	1	MONGOLIAN AIRLINES - MIAT
Aviadvigatel D-30	Tupolev 154M	2	NOVOKUZNETSKOE AVIAPREDPRIYATIYE
Aviadvigatel D-30	Tupolev 154M	7	SIBERIA AIRLINE
Aviadvigatel D-30	Tupolev 154M	5	SICHUAN AIRLINES
Aviadvigatel D-30	Tupolev 154M	3	SYRIAN ARAB AIRLINES
Aviadvigatel D-30	Tupolev 154M	9	UNION AIRLINES
Aviadvigatel D-30	Tupolev 154M	5	XINJIANG AIRLINES
Aviadvigatel D-30-III	Tupolev 134A-3	6	MALEV - HUNGARIAN AIRLINES
Aviadvigatel D-30KP	Ilyushin 76TD	4	DACONO AIR
Aviadvigatel D-30KP	Ilyushin 76TD	3	SAMARA AIRLINES
Aviadvigatel D-30KU	Ilyushin 62M	2	TAROM
Aviadvigatel D-30KU-154	Tupolev 154M	3	BELAVIA
Aviadvigatel D-30KU-154	Tupolev 154M	3	SAMARA AIRLINES
Aviadvigatel PS-90-A	Tupolev 204	3	VNUKOVO AIRLINES
Aviadvigatel PS-90A	Ilyushin 96-300	4	AEROFLOT RUSSIAN INTERNATIONAL AIRLINES
CFMI CFM56	B737-400	0	CHINA HAINAN AIRLINES
CFMI CFM56-2	DC8-71F	24	UNITED PARCEL SERVICE COMPANY UPS
CFMI CFM56-2	DC8-73F	1	AIR INDIA
CFMI CFM56-2	DC8-73F	2	EVERGREEN INTERNATIONAL AIRLINES, INC
CFMI CFM56-2	DC8-73F	28	UNITED PARCEL SERVICE COMPANY UPS
CFMI CFM56-2B1	DC8-71F	3	BURLINGTON AIR EXPRESS
CFMI CFM56-2C	DC8-71F	1	CARGOSUR - COMPANIA DE EXPLOITATION DE AVIONES CARGUEROS SA
CFMI CFM56-2C	DC8-71F	9	EMERY WORLDWIDE
CFMI CFM56-2C	DC8-73F	8	EMERY WORLDWIDE
CFMI CFM56-2C1	DC8-71F	9	AIR TRANSPORT INTERNATIONAL
CFMI CFM56-2C1	DC8-71F	3	FLAGSHIP EXPRESS
CFMI CFM56-2C1	DC8-71F	2	LAN-CHILE
CFMI CFM56-2C1	DC8-71F	3	SOUTHERN AIR TRANSPORT, INC
CFMI CFM56-2C1	DC8-73F	2	SOUTHERN AIR TRANSPORT, INC
CFMI CFM56-2C5	DC8-72	1	SAUDIA - SAUDI ARABIAN AIRLINES
CFMI CFM56-2C5	DC8-73	0	DHL AIRWAYS

POWER PLANT	AIRFRAME	NO. IN FLEET	FLEET NAME
CFMI CFM56-2C5	DC8-73F	5	GERMAN CARGO SERVICES
CFMI CFM56-3	B737-300	0	NATIONAL OVERSEAS AIRLINE
CFMI CFM56-3	B737-400	0	EAST WEST AIRLINES
CFMI CFM56-3	B737-400	0	MALEV - HUNGARIAN AIRLINES
CFMI CFM56-3B1	B737-300	14	AIR CHINA INTERNATIONAL
CFMI CFM56-3B1	B737-300	6	AIR FRANCE
CFMI CFM56-3B1	B737-300	1	AIR PACIFIC LIMITED
CFMI CFM56-3B1	B737-300	2	ALOHA AIRLINES
CFMI CFM56-3B1	B737-300	20	AMERICA WEST AIRLINES, INC
CFMI CFM56-3B1	B737-300	21	ANSETT AUSTRALIA
CFMI CFM56-3B1	B737-300	4	BRITISH MIDLAND AIRWAYS LTD
CFMI CFM56-3B1	B737-300	0	CAAC
CFMI CFM56-3B1	B737-300	20	CHINA SOUTHERN AIRLINES
CFMI CFM56-3B1	B737-300	13	CHINA SOUTHWEST AIRLINES
CFMI CFM56-3B1	B737-300	73	CONTINENTAL AIRLINES
CFMI CFM56-3B1	B737-300	18	DELTA AIR LINES, INC
CFMI CFM56-3B1	B737-300	8	GARUDA INDONESIA PT
CFMI CFM56-3B1	B737-300	9	JAT - JUGOSLOVENSKI AEROTRANSPORT
CFMI CFM56-3B1	B737-300	4	JET AIRWAYS
CFMI CFM56-3B1	B737-300	15	KLM ROYAL DUTCH AIRLINES
CFMI CFM56-3B1	B737-300	2	LAUDA AIR LUFTFAHRT AKTIENGESELLSCHAFT
CFMI CFM56-3B1	B737-300	31	LUFTHANSA GERMAN AIRLINES - DEUTSCHE LUFTHANSA AG
CFMI CFM56-3B1	B737-300	4	MARKAIR, INC
CFMI CFM56-3B1	B737-300	4	MONARCH AIRLINES LTD
CFMI CFM56-3B1	B737-300	6	PAKISTAN INTERNATIONAL AIRLINES (PIA)
CFMI CFM56-3B1	B737-300	12	PHILIPPINE AIR LINES, INC - PAL
CFMI CFM56-3B1	B737-300	133	SOUTHWEST AIRLINES COMPANY
CFMI CFM56-3B1	B737-300	11	TAESA
CFMI CFM56-3B1	B737-300	1	TAGB - TRANSPORTES AEREOS DA GUINE BISSAU
CFMI CFM56-3B1	B737-300	8	TRANSBRASIL S.A. LINHAS AEREAS
CFMI CFM56-3B1	B737-300	2	VASP - VIACAO AEREA SAO PAULO S.A.
CFMI CFM56-3B1	B737-300QC	2	LUFTHANSA GERMAN AIRLINES - DEUTSCHE LUFTHANSA AG
CFMI CFM56-3B1	B737-400	2	JET AIRWAYS
CFMI CFM56-3B1	B737-400	0	SAHSA - SERVICIO AEREO DE HONDURAS, SA
CFMI CFM56-3B1	B737-400	2	TAN - TRANSPORTES AEREOS NACIONALES, SA
CFMI CFM56-3B1	B737-500	10	AER LINGUS
CFMI CFM56-3B1	B737-500	1	AIR PACIFIC LIMITED
CFMI CFM56-3B1	B737-500	12	BRITISH MIDLAND AIRWAYS LTD
CFMI CFM56-3B1	B737-500	14	CHINA SOUTHERN AIRLINES
CFMI CFM56-3B1	B737-500	5	CSA - CESKOSLOVENSKE AEROLINIE
CFMI CFM56-3B1	B737-500	1	ESTONIAN AIR
CFMI CFM56-3B1	B737-500	3	EURALAIR INTERNATIONAL
CFMI CFM56-3B1	B737-500	4	MAERSK AIR
CFMI CFM56-3B1	B737-500	6	SABENA BELGIAN WORLD AIRLINES
CFMI CFM56-3B1	B737-500	12	SAS - SCANDINAVIAN AIRLINES SYSTEM
CFMI CFM56-3B1	B737-500	25	SOUTHWEST AIRLINES COMPANY
CFMI CFM56-3B1	B737-500	5	TAESA
CFMI CFM56-3B1	B737-500	2	TUNIS AIR - SOCIETE TUNISIENNE D EL'AIR
CFMI CFM56-3B1/3B2	B737-300	10	AIR EUROPA (AIR ESPANA, SA)
CFMI CFM56-3B1/3B2/3C1	B737-300	101	UNITED AIRLINES, INC
CFMI CFM56-3B1/3C1	B737-500	32	LUFTHANSA GERMAN AIRLINES - DEUTSCHE LUFTHANSA AG
CFMI CFM56-3B1/3C1	B737-500	57	UNITED AIRLINES, INC
CFMI CFM56-3B2	B737-300	1	AIR CALEDONIE INTERNATIONAL
CFMI CFM56-3B2	B737-300	3	AIR CHINA INTERNATIONAL
CFMI CFM56-3B2	B737-300	1	AIR FOYLE LIMITED
CFMI CFM56-3B2	B737-300	2	AIR HOLLAND CHARTER BV
CFMI CFM56-3B2	B737-300	1	AIR MADAGASCAR
CFMI CFM56-3B2	B737-300	15	AMERICA WEST AIRLINES, INC
CFMI CFM56-3B2	B737-300	3	AVIATECA
CFMI CFM56-3B2	B737-300	1	BRITISH MIDLAND AIRWAYS LTD
CFMI CFM56-3B2	B737-300	4	CHINA SOUTHWEST AIRLINES
CFMI CFM56-3B2	B737-300	4	CONDOR FLUGDIENST GMBH
CFMI CFM56-3B2	B737-300	7	DEUTSCHE BA LUFTFAHRTGESELLSCHAFT MBH
CFMI CFM56-3B2	B737-300	3	EUROBELGIAN AIRLINES
CFMI CFM56-3B2	B737-300	1	EXCALIBUR AIRWAYS LTD
CFMI CFM56-3B2	B737-300	10	GERMANIA FLUGGESELLSCHAFT KOLN
CFMI CFM56-3B2	B737-300	1	LADECO SA/LINEA AEREA DE COBRE/LADECO AIRLINES
CFMI CFM56-3B2	B737-300	13	MAERSK AIR
CFMI CFM56-3B2	B737-300	6	SABENA BELGIAN WORLD AIRLINES
CFMI CFM56-3B2	B737-300	4	SILKAIR (SINGAPORE) PRIVATE LIMITED
CFMI CFM56-3B2	B737-300	1	SOBELAIR - SOCIETE BELGE DE TRANSPORTES AERIENS
CFMI CFM56-3B2	B737-300	3	SULTAN AIR
CFMI CFM56-3B2	B737-300	1	SUN EXPRESS
CFMI CFM56-3B2	B737-300	4	TACA INTERNATIONAL AIRLINES SA
CFMI CFM56-3B2	B737-300	8	TAP AIR PORTUGAL
CFMI CFM56-3B2	B737-300	5	TEA BASEL
CFMI CFM56-3B2	B737-300	2	TEA FRANCE
CFMI CFM56-3B2	B737-300	8	TRANSAVIA AIRLINES CV
CFMI CFM56-3B2	B737-300	3	TRANSBRASIL S.A. LINHAS AEREAS
CFMI CFM56-3B2	B737-300	2	TRANSMED AIRLINES
CFMI CFM56-3B2	B737-300	3	TURKMENISTAN AIRLINE
CFMI CFM56-3B2	B737-300	58	USAIR, INC
CFMI CFM56-3B2	B737-300	25	VARIG - VIACAO AEREA RIO-GRANDENSE S.A.
CFMI CFM56-3B2	B737-300	1	VIETNAM AIRLINES
CFMI CFM56-3B2	B737-300	9	VIVA AIR
CFMI CFM56-3B2	B737-300F	1	SINGAPORE AIRLINES LIMITED
CFMI CFM56-3B2	B737-300LR	33	USAIR, INC
CFMI CFM56-3B2	B737-300QC	1	POLYNESIAN AIRLINES
CFMI CFM56-3B2	B737-400	6	AER LINGUS
CFMI CFM56-3B2	B737-400	2	ALOHA AIRLINES
CFMI CFM56-3B2	B737-400	3	CARNIVAL AIR LINES INC
CFMI CFM56-3B2	B737-400	1	JAPAN AIRLINES COMPANY LTD - JAL
CFMI CFM56-3B2	B737-400	12	KLM ROYAL DUTCH AIRLINES
CFMI CFM56-3B2	B737-400	54	USAIR, INC
CFMI CFM56-3B2/3C1	B737-300	16	QANTAS AIRWAYS LTD
CFMI CFM56-3C	B737-300	4	CHINA HAINAN AIRLINES
CFMI CFM56-3C	B737-300/400/500	0	UNITED AIRLINES, INC
CFMI CFM56-3C1	B737-300	1	AIR MALAWI
CFMI CFM56-3C1	B737-300	5	AIR MALTA CO. LTD.

POWER PLANT	AIRFRAME	NO. IN FLEET	FLEET NAME
CFMI CFM56-3C1	B737-300	0	AIR NEW ZEALAND
CFMI CFM56-3C1	B737-300	5	EUROBELGIAN AIRLINES
CFMI CFM56-3C1	B737-300	2	LAM - LINHAS AEREAS DE MOCAMBIQUE
CFMI CFM56-3C1	B737-300	4	MALEV - HUNGARIAN AIRLINES
CFMI CFM56-3C1	B737-300	2	OMAN AIR
CFMI CFM56-3C1	B737-300	1	SAETA
CFMI CFM56-3C1	B737-300	2	SUN EXPRESS
CFMI CFM56-3C1	B737-300	5	TAROM
CFMI CFM56-3C1	B737-300	10	YUNNAN AIRLINES
CFMI CFM56-3C1	B737-300F	2	MALAYSIA AIRLINES
CFMI CFM56-3C1	B737-300QC	3	FALCON CARGO AB
CFMI CFM56-3C1	B737-400	2	AIR BELGIUM
CFMI CFM56-3C1	B737-400	7	AIR BERLIN
CFMI CFM56-3C1	B737-400	1	AIR CHARTER
CFMI CFM56-3C1	B737-400	2	AIR NAURU
CFMI CFM56-3C1	B737-400	7	AIR UK (LEISURE) LIMITED
CFMI CFM56-3C1	B737-400	1	AIR VANUATU
CFMI CFM56-3C1	B737-400	24	ALASKA AIRLINES, INC
CFMI CFM56-3C1	B737-400	16	ASIANA AIRLINES
CFMI CFM56-3C1	B737-400	7	BRAATHENS S.A.F.E. A/S
CFMI CFM56-3C1	B737-400	36	BRITISH AIRWAYS
CFMI CFM56-3C1	B737-400	6	BRITISH MIDLAND AIRWAYS LTD
CFMI CFM56-3C1	B737-400	2	CORSAIR
CFMI CFM56-3C1	B737-400	3	EUROBELGIAN AIRLINES
CFMI CFM56-3C1	B737-400	8	FUTURA INTERNATIONAL AIRWAYS
CFMI CFM56-3C1	B737-400	7	GARUDA INDONESIA PT
CFMI CFM56-3C1	B737-400	2	GB AIRWAYS
CFMI CFM56-3C1	B737-400	9	HAPAG-LLOYD FLUGGESELLSCHAFT.MBH
CFMI CFM56-3C1	B737-400	4	ICELANDAIR
CFMI CFM56-3C1	B737-400	1	ISTANBUL AIRLINES - ISTAIR
CFMI CFM56-3C1	B737-400	2	JAPAN TRANSOCEAN AIR
CFMI CFM56-3C1	B737-400	2	LAUDA AIR LUFTFAHRT AKTIENGESELLSCHAFT
CFMI CFM56-3C1	B737-400	4	LOT - POLSKIE LINIE LOTNICZE
CFMI CFM56-3C1	B737-400	7	LUFTHANSA GERMAN AIRLINES - DEUTSCHE LUFTHANSA AG
CFMI CFM56-3C1	B737-400	2	LUXAIR S.A.
CFMI CFM56-3C1	B737-400	39	MALAYSIA AIRLINES
CFMI CFM56-3C1	B737-400	6	MARKAIR, INC
CFMI CFM56-3C1	B737-400	7	OLYMPIC AIRWAYS S.A.
CFMI CFM56-3C1	B737-400	18	QANTAS AIRWAYS LTD
CFMI CFM56-3C1	B737-400	6	ROYAL AIR MAROC
CFMI CFM56-3C1	B737-400	3	SABENA BELGIAN WORLD AIRLINES
CFMI CFM56-3C1	B737-400	2	SAHARA INDIA AIRLINES
CFMI CFM56-3C1	B737-400	3	SOBELAIR - SOCIETE BELGE DE TRANSPORTES AERIENS
CFMI CFM56-3C1	B737-400	1	SOLOMON AIRLINES
CFMI CFM56-3C1	B737-400	7	THAI AIRWAYS INTERNATIONAL
CFMI CFM56-3C1	B737-400	4	TRANSBRASIL S.A. LINHAS AEREAS
CFMI CFM56-3C1	B737-400	2	TRANSMED AIRLINES
CFMI CFM56-3C1	B737-400	27	TURKISH AIRLINES, INC (TURK HAVA YOLLARI A.O)
CFMI CFM56-3C1	B737-400	2	UKRAINIAN AIRLINES
CFMI CFM56-3C1	B737-500	1	AIR AUSTRAL
CFMI CFM56-3C1	B737-500	16	AIR FRANCE
CFMI CFM56-3C1	B737-500	0	AIR NIPPON
CFMI CFM56-3C1	B737-500	3	ASIANA AIRLINES
CFMI CFM56-3C1	B737-500	3	BALKAN
CFMI CFM56-3C1	B737-500	20	BRAATHENS S.A.F.E. A/S
CFMI CFM56-3C1	B737-500	5	EGYPTAIR
CFMI CFM56-3C1	B737-500	5	HAPAG-LLOYD FLUGGESELLSCHAFT.MBH
CFMI CFM56-3C1	B737-500	6	LOT - POLSKIE LINIE LOTNICZE
CFMI CFM56-3C1	B737-500	2	LUXAIR S.A.
CFMI CFM56-3C1	B737-500	9	MALAYSIA AIRLINES
CFMI CFM56-3C1	B737-500	5	ROYAL AIR MAROC
CFMI CFM56-3C1	B737-500	2	TURKISH AIRLINES, INC (TURK HAVA YOLLARI A.O)
CFMI CFM56-3C1	B737-500	3	XIAMEN AIRLINES
CFMI CFM56-5A	A319	0	LUFTHANSA GERMAN AIRLINES - DEUTSCHE LUFTHANSA AG
CFMI CFM56-5A1	A320-100	6	AIR FRANCE
CFMI CFM56-5A1	A320-100	6	AIR INTER
CFMI CFM56-5A1	A320-100	5	BRITISH AIRWAYS
CFMI CFM56-5A1	A320-100	7	VIETMAN AIRLINES
CFMI CFM56-5A1	A320-200	34	AIR CANADA
CFMI CFM56-5A1	A320-200	19	AIR FRANCE
CFMI CFM56-5A1	A320-200	28	AIR INTER
CFMI CFM56-5A1	A320-200	2	AIR MALTA CO. LTD.
CFMI CFM56-5A1	A320-200	1	AIR NIPPON
CFMI CFM56-5A1	A320-200	18	ALL NIPPON AIRWAYS
CFMI CFM56-5A1	A320-200	14	ANSETT AUSTRALIA
CFMI CFM56-5A1	A320-200	5	BRITISH AIRWAYS
CFMI CFM56-5A1	A320-200	3	CALEDONIAN AIRWAYS
CFMI CFM56-5A1	A320-200	4	CANADA 3000 AIRLINES LIMITED
CFMI CFM56-5A1	A320-200	12	CANADIAN - CANADIEN
CFMI CFM56-5A1	A320-200	0	FIRST EUROPEAN AIRWAYS
CFMI CFM56-5A1	A320-200	14	GULF AIR
CFMI CFM56-5A1	A320-200	22	IBERIA - LINEAS AEREAS DE ESPANA, S.A.
CFMI CFM56-5A1	A320-200	3	KUWAIT AIRWAYS CORPORATION
CFMI CFM56-5A1	A320-200	33	LUFTHANSA GERMAN AIRLINES - DEUTSCHE LUFTHANSA AG
CFMI CFM56-5A1	A320-200	0	MIDWAY AIRLINES
CFMI CFM56-5A1	A320-200	50	NORTHWEST AIRLINES
CFMI CFM56-5A1	A320-200	4	ONUR AIR
CFMI CFM56-5A1	A320-200	2	ROYAL JORDANIAN
CFMI CFM56-5A1	A320-200	6	TAP AIR PORTUGAL
CFMI CFM56-5A1	A320-200	7	TUNIS AIR - SOCIETE TUNISIENNE D EL'AIR
CFMI CFM56-5A1	A320-200	4	VIETMAN AIRLINES
CFMI CFM56-5A3	A320-200	4	EXCALIBUR AIRWAYS LTD
CFMI CFM56-5A3	A320-200	6	LINEAS AEREAS COSTARRICENSES, S.A. - LACSA
CFMI CFM56-5A3	A320-200	7	MONARCH AIRLINES LTD
CFMI CFM56-5B	A321	0	AIR FRANCE
CFMI CFM56-5B	A321-100	2	AIR INTER
CFMI CFM56-5B	A321-100	5	ALITALIA
CFMI CFM56-5B	A321-100	0	AUSTRIAN AIRLINES, OSTERREICHISCHE LUFTVERKEHRS AG
CFMI CFM56-5B	A321-100	0	KUWAIT AIRWAYS CORPORATION

POWER PLANT	AIRFRAME	NO. IN FLEET	FLEET NAME
CFMI CFM56-5B	A321-100	2	SWISSAIR (SOCIETE ANONYME SUISSE POUR LA NAVIGATION AERIENNE)
CFMI CFM56-5B2	A319	0	SWISSAIR (SOCIETE ANONYME SUISSE POUR LA NAVIGATION AERIENNE)
CFMI CFM56-5B4	A320-200	0	AUSTRIAN AIRLINES, OSTERREICHISCHE LUFTVERKEHRS AG
CFMI CFM56-5B4	A320-200	1	SWISSAIR (SOCIETE ANONYME SUISSE POUR LA NAVIGATION AERIENNE)
CFMI CFM56-5B6	A319	0	AIR CANADA
CFMI CFM56-5B6	A319	0	AIR INTER
CFMI CFM56-5C2	A340-200	3	AIR FRANCE
CFMI CFM56-5C2	A340-200	2	AUSTRIAN AIRLINES, OSTERREICHISCHE LUFTVERKEHRS AG
CFMI CFM56-5C2	A340-200	4	CATHAY PACIFIC AIRWAYS LTD
CFMI CFM56-5C2	A340-200	0	IBERIA - LINEAS AEREAS DE ESPANA, S.A.
CFMI CFM56-5C2	A340-200	7	LUFTHANSA GERMAN AIRLINES - DEUTSCHE LUFTHANSA AG
CFMI CFM56-5C2	A340-200	0	PHILIPPINE AIR LINES, INC - PAL
CFMI CFM56-5C2	A340-200	4	SABENA BELGIAN WORLD AIRLINES
CFMI CFM56-5C2	A340-300	5	AIR FRANCE
CFMI CFM56-5C2	A340-300	0	AIR LANKA
CFMI CFM56-5C2	A340-300	0	CHINA EASTERN AIRLINES
CFMI CFM56-5C2	A340-300	0	CHINA SOUTHERN AIRLINES
CFMI CFM56-5C2	A340-300	4	GULF AIR
CFMI CFM56-5C2	A340-300	8	LUFTHANSA GERMAN AIRLINES - DEUTSCHE LUFTHANSA AG
CFMI CFM56-5C2	A340-300	3	TURKISH AIRLINES, INC (TURK HAVA YOLLARI A.O)
CFMI CFM56-5C3	A340-300	2	AIR MAURITIUS
CFMI CFM56-5C3	A340-300	4	TAP AIR PORTUGAL
CFMI CFM56-5C3	A340-300	4	VIRGIN ATLANTIC AIRWAYS
CFMI CFM56-5C4	A340-300	0	AIR CANADA
CFMI CFM56-5C4	A340-300	0	CATHAY PACIFIC AIRWAYS LTD
CFMI CFM56-5C4	A340-300	1	KUWAIT AIRWAYS CORPORATION
CFMI CFM56-5C4	A340-300	0	SINGAPORE AIRLINES LIMITED
CFMI CFM56-7	B737-600	0	SAS - SCANDINAVIAN AIRLINES SYSTEM
CFMI CFM56-7	B737-700	0	GERMANIA FLUGGESELLSCHAFT KOLN
CFMI CFM56-7	B737-700	0	MAERSK AIR
CFMI CFM56-7	B737-700	0	SOUTHWEST AIRLINES COMPANY
CFMI CFM56-7	B737-800	0	HAPAG-LLOYD FLUG GMBH
CFMI CFMI56-3C1	B737-400	1	ROYAL TONGAN AIRLINES
CW R-3350	DC-7CF	1	TRANS-AIR-LINK CORPORATION
Cont GTSIO-520-H	C 421B Golden Eagle	1	NATIONAL OVERSEAS AIRLINE
Cont GTSIO-520-L	C 421C Golden Eagle	1	NATIONAL OVERSEAS AIRLINE
Cont GTSIO-520-M	C 404 Titan	3	AIR ATLANTIQUE
Cont GTSIO-520-M	C 404 Titan	1	THERON AIRWAYS
Cont GTSIO-520-M	C 404 Titan II	2	AIR INTER GABON, SA.
Cont IO-360-C	C 337 Super Skymaster	2	NATIONAL OVERSEAS AIRLINE
Cont IO-360-C	C F337E Super Skymaster	1	ARKIA ISRAELI AIRLINES
Cont IO-360-C/D	C 337B Super Skymaster	1	ARKIA ISRAELI AIRLINES
Cont IO-360-D	Luftschiff WDL 1B Airship	2	WDL AVIATION (KOLN)
Cont IO-470-L	BE Baron 95-B55	2	ROYAL AIR MAROC
Cont IO-470-V	C 310L	1	AIR ATLANTIQUE
Cont IO-470-VO	C 310Q II	1	AIR TCHAD, SOCIETE DE TRANSPORTES AERIENS
Cont IO-520-BA	BE F33A Bonanza	59	LUFTHANSA GERMAN AIRLINES - DEUTSCHE LUFTHANSA AG
Cont IO-520-C	BE Baron 58	3	AIR AFFAIRES GABON
Cont IO-520-C	BE Baron 58	1	AIR PROVENCE
Cont IO-520-C	BE Baron 58	1	AIRKENYA AVIATION LIMITED
Cont IO-520-C	BE Baron 95-C55	1	AIR FIJI
Cont IO-520-C	BE Baron 95-E55	2	AIRKENYA AVIATION LIMITED
Cont IO-520-D	C 172 Skyhawk	2	PROVINCIAL AIRWAYS LTD
Cont IO-520-D	C 180	1	AIR SCHEFFERVILLE, INC
Cont IO-520-D	C 185	1	HOLMSTROM AIR
Cont IO-520-D	C 185A Skywagon	2	MOUNT COOK AIRLINE
Cont IO-520-D	C 185F Skywagon II	6	MOUNT COOK AIRLINE
Cont IO-520-F	C 207 Stationair 8 II	4	BERING AIR, INC
Cont IO-520-F	C U206G Stationair 6 II	2	TRANS-JAMAICAN AIRLINES LTD
Cont IO-520-M	C 310R II	2	AIR ATLANTIQUE
Cont IO-520-M	C 310R II	2	HAZELTON AIRLINES
Cont IO-520-MB	C 310R2	1	CROATIA AIRLINES
Cont IO-540-C	BE Baron 58	19	LUFTHANSA GERMAN AIRLINES - DEUTSCHE LUFTHANSA AG
Cont IO-550	C 207 Bonair conversion	1	MOUNT COOK AIRLINE
Cont IO-550-B	BE A36 Bonanza	9	LUFTHANSA GERMAN AIRLINES - DEUTSCHE LUFTHANSA AG
Cont IO-550-B	BE A36 Bonanza	6	SAUDIA - SAUDI ARABIAN AIRLINES
Cont TSIO-360-C	C T337GP Skymaster	1	ARKIA ISRAELI AIRLINES
Cont TSIO-360-C	C T337HP Skymaster II	1	ARKIA ISRAELI AIRLINES
Cont TSIO-360-E	PA-34-200T Seneca II	4	THERON AIRWAYS
Cont TSIO-520	C 406	1	AIR ATLANTIQUE
Cont TSIO-520-E	C 401	1	AMERIJET INTERNATIONAL
Cont TSIO-520-E	C 402A	1	NATIONAL AIR CHARTER P.T.
Cont TSIO-520-E	C 402B	1	AIR ATLANTIQUE
Cont TSIO-520-E	C 402B	2	KELOWNA FLIGHTCRAFT AIR CHARTER LTD
Cont TSIO-520-E	C 402B	2	THERON AIRWAYS
Cont TSIO-520-E	C 402B II	1	FLANDERS AIRLINES NV
Cont TSIO-520-E	C 402C	1	AIR ATLANTIQUE
Cont TSIO-520-H	C T210L Tbo Centurion II	1	PAN AVIATION
Cont TSIO-520-N	C 340	1	KELOWNA FLIGHTCRAFT AIR CHARTER LTD
Cont TSIO-520-R	C 210N Turbo Centurion II	1	SIERRA PACIFIC AIRLINES, INC
Cont TSIO-520-R	C T210M Tbo Centurion II	1	THERON AIRWAYS (PTY) LTD
Cont TSIO-520-VB	C 402 II	4	ALPHA AIR
Cont TSIO-520-VB	C 402B	1	MUK AIR
Cont TSIO-520-VB	C 402C Utililiner II	1	AIR GUYANE
Cont TSIO-520-VB	C 402C Utililiner II	3	BAHAMASAIR
Cont TSIO-520-VB	C 402C Utililiner II	1	BIG SKY AIRLINES
Cont TSIO-520-VB	C 402C Utililiner II	3	CROATIA AIRLINES
Cont TSIO-520-VB	C 402C Utililiner II	6	LAM - LINHAS AEREAS DE MOCAMBIQUE
Cont TSIO-520-VB	C 402C Utililiner II	1	SUN-AIR OF SCANDINAVIA A/S
GA TFE731-3-1G	IAI Westwind 1124	1	ARKIA ISRAELI AIRLINES
GA TFE731-3-1G	IAI Westwind 1124	1	KELOWNA FLIGHTCRAFT AIR CHARTER LTD
GA TPE331-6-252B	BE King Air B100	1	AIRLINK AIRLINE (PTY) LTD
GE CF34-3A	CL 600 Regional Jet	9	AIR CANADA
GE CF34-3A	CL 600 Regional Jet	4	AIR LITTORAL
GE CF34-3A	CL 600 Regional Jet	1	BRIT'AIR
GE CF34-3A	CL 600 Regional Jet	17	COMAIR
GE CF34-3A	CL 600 Regional Jet	15	LUFTHANSA CITYLINE
GE CF34-3A	CL 600 Regional Jet	0	YORK AIRCRAFT LEASING INC.
GE CF34-3A	CL 600 Regional Jet 100ER	8	SKYWEST AIRLINES INC (DELTA CONNECTION)

POWER PLANT	AIRFRAME	NO. IN FLEET	FLEET NAME
GE CF34-3A	CL 600 Regional Jet 100LR	7	LAUDA AIR LUFTFAHRT AKTIENGESELLSCHAFT
GE CF34-3A	CL 601-3A Challenger	1	SCIBE AIRLIFT
GE CF34-3A	CL 601-3A ER Challenger	1	THAI AIRWAYS INTERNATIONAL
GE CF6-45A2	B747SR	14	ALL NIPPON AIRWAYS
GE CF6-50	A300B4	2	AIR OPS OF EUROPE
GE CF6-502CB	DC10-30ER	3	THAI AIRWAYS INTERNATIONAL
GE CF6-50C	A300B2-101	4	AIR FRANCE
GE CF6-50C	A300B2-101	8	INDIAN AIRLINES LTD
GE CF6-50C	A300B4-100	1	TRANSLIFT AIRWAYS
GE CF6-50C	A300B4-103	6	CONTINENTAL AIRLINES
GE CF6-50C	DC10-30	3	AIR AFRIQUE
GE CF6-50C	DC10-30	1	AIR ZAIRE
GE CF6-50C	DC10-30	4	AOM FRENCH AIRLINES
GE CF6-50C	DC10-30	7	BRITISH AIRWAYS
GE CF6-50C	DC10-30	1	ECUATORIANA
GE CF6-50C	DC10-30	6	GARUDA INDONESIA PT
GE CF6-50C	DC10-30	8	IBERIA - LINEAS AEREAS DE ESPANA, S.A.
GE CF6-50C	DC10-30	3	KOREAN AIR
GE CF6-50C	DC10-30	2	NIGERIA AIRWAYS
GE CF6-50C	DC10-30	6	NORTHWEST AIRLINES
GE CF6-50C	DC10-30	1	PRIMERAS LINEAS URUGUAYAS DE NAVEGACION AEREA - PLUNA
GE CF6-50C	DC10-30	5	VIASA, VENEZOLANA INTERNACIONAL DE AVIACION SA
GE CF6-50C1	DC10-30	3	JAT - JUGOSLOVENSKI AEROTRANSPORT
GE CF6-50C2	A300B2-200	2	ALITALIA
GE CF6-50C2	A300B2-203	5	IRAN AIR THE AIRLINE OF THE ISLAMIC REPUBLIC OF IRAN
GE CF6-50C2	A300B2-203	3	VASP - VIACAO AEREA SAO PAULO S.A.
GE CF6-50C2	A300B4-100	4	ALITALIA
GE CF6-50C2	A300B4-103	8	KOREAN AIR
GE CF6-50C2	A300B4-103	6	OLYMPIC AIRWAYS S.A.
GE CF6-50C2	A300B4-103	2	PHILIPPINE AIR LINES, INC - PAL
GE CF6-50C2	A300B4-103	7	THAI AIRWAYS INTERNATIONAL
GE CF6-50C2	A300B4-200	8	ALITALIA
GE CF6-50C2	A300B4-200	1	EUROPEAN AIRLIFT
GE CF6-50C2	A300B4-200	1	MALAYSIA AIRLINES
GE CF6-50C2	A300B4-200	9	PAKISTAN INTERNATIONAL AIRLINES (PIA)
GE CF6-50C2	A300B4-203	3	AIR AFRIQUE
GE CF6-50C2	A300B4-203	2	AIR CHARTER
GE CF6-50C2	A300B4-203	11	AIR FRANCE
GE CF6-50C2	A300B4-203	3	AIR INDIA
GE CF6-50C2	A300B4-203	5	AIR JAMAICA
GE CF6-50C2	A300B4-203	1	AIR SINAI
GE CF6-50C2	A300B4-203	15	CONTINENTAL AIRLINES
GE CF6-50C2	A300B4-203	5	EGYPTAIR
GE CF6-50C2	A300B4-203	2	FINNAIR OY
GE CF6-50C2	A300B4-203	1	GHANA AIRWAYS
GE CF6-50C2	A300B4-203	2	INDIAN AIRLINES LTD
GE CF6-50C2	A300B4-203	10	PHILIPPINE AIR LINES, INC - PAL
GE CF6-50C2	A300B4-203	4	QANTAS AIRWAYS LTD
GE CF6-50C2	A300B4-203	3	SEMPATI AIR TRANSPORT
GE CF6-50C2	A300B4-203	3	SOUTH AFRICAN AIRWAYS (SAA)
GE CF6-50C2	A300B4-203	2	SULTAN AIR
GE CF6-50C2	A300B4-203	5	THAI AIRWAYS INTERNATIONAL
GE CF6-50C2	A300B4-203	1	TUNIS AIR - SOCIETE TUNISIENNE D EL'AIR
GE CF6-50C2	A300B4-203	2	VIASA, VENEZOLANA INTERNACIONAL DE AVIACION SA
GE CF6-50C2	A300B4K-100	1	EUROPEAN AIRLINES SA
GE CF6-50C2	A300B4K-103	1	FAUCETT PERU
GE CF6-50C2	A300B4K-200	2	EUROPEAN AIRLINES SA
GE CF6-50C2	A300C4-203	1	SOUTH AFRICAN AIRWAYS (SAA)
GE CF6-50C2	A300F4-203	2	KOREAN AIR
GE CF6-50C2	DC10-30	9	AMERICAN AIRLINES
GE CF6-50C2	DC10-30	5	AOM FRENCH AIRLINES
GE CF6-50C2	DC10-30	6	BIMAN BANGLADESH AIRLINES
GE CF6-50C2	DC10-30	2	CALEDONIAN AIRWAYS
GE CF6-50C2	DC10-30	3	CANADIAN - CANADIEN
GE CF6-50C2	DC10-30	3	CONDOR FLUGDIENST GMBH
GE CF6-50C2	DC10-30	13	CONTINENTAL AIRLINES
GE CF6-50C2	DC10-30	3	FINNAIR OY
GE CF6-50C2	DC10-30	1	GHANA AIRWAYS CORPORATION
GE CF6-50C2	DC10-30	2	JAPAN AIR SYSTEM CO LTD (NIHON AIR SYSTEM)
GE CF6-50C2	DC10-30	11	LUFTHANSA GERMAN AIRLINES - DEUTSCHE LUFTHANSA AG
GE CF6-50C2	DC10-30	4	MALAYSIA AIRLINES
GE CF6-50C2	DC10-30	1	PHILIPPINE AIR LINES, INC - PAL
GE CF6-50C2	DC10-30	2	SABENA BELGIAN WORLD AIRLINES
GE CF6-50C2	DC10-30	8	UNITED AIRLINES, INC
GE CF6-50C2	DC10-30	8	VARIG - VIACAO AEREA RIO-GRANDENSE S.A.
GE CF6-50C2	DC10-30	4	WORLD AIRWAYS INC
GE CF6-50C2	DC10-30ER	5	CANADIAN - CANADIEN
GE CF6-50C2	DC10-30F	1	AEROFLOT RUSSIAN INTERNATIONAL AIRLINES
GE CF6-50C2	DC10-30F (AF)/(CF)	22	FEDERAL EXPRESS CORPORATION
GE CF6-50C2	DC10-30F (CF)	2	VARIG - VIACAO AEREA RIO-GRANDENSE S.A.
GE CF6-50C2A2	A310-300	1	LAP - LINEAS AEREAS PARAGUAYAS
GE CF6-50C2A5	A300-600R	2	AIR AFRIQUE
GE CF6-50C2A5	A300-600R	1	AIR FINLANDIA
GE CF6-50C2B	DC-10	0	ACS OF CANADA - AIR CHARTER SYSTEMS
GE CF6-50C2B	DC10-30	2	AOM FRENCH AIRLINES
GE CF6-50C2B	DC10-30	2	CUBANA
GE CF6-50C2B	DC10-30	1	NORTHWEST AIRLINES
GE CF6-50C2B	DC10-30ER	1	FINNAIR OY
GE CF6-50C2B	DC10-30ER	1	NORTHWEST AIRLINES
GE CF6-50C2B1	B747-300	2	THAI AIRWAYS INTERNATIONAL
GE CF6-50C2F	DC10-15	2	AEROPERU
GE CF6-50C2F	DC10-15	6	MEXICANA
GE CF6-50C2R	A300B2-1	13	AIR INTER
GE CF6-50C2R	A300B2-K3C	1	AIR INTER
GE CF6-50C2R	A300B2-K3C	9	JAPAN AIR SYSTEM CO LTD (NIHON AIR SYSTEM)
GE CF6-50C2R	A300B2-K3C	4	SOUTH AFRICAN AIRWAYS (SAA)
GE CF6-50C2R	A300B4-2C	4	AIR INTER
GE CF6-50C2R	A300B4-2C	8	JAPAN AIR SYSTEM CO LTD (NIHON AIR SYSTEM)
GE CF6-50C2R	DC10-30	1	AFRICAN SAFARI AIRWAYS LTD

POWER PLANT	AIRFRAME	NO. IN FLEET	FLEET NAME
GE CF6-50C2R	DC10-30	5	AIR FRANCE
GE CF6-50C2R	DC10-30	2	EUROPEAN AIRLIFT
GE CF6-50E	B747-200	2	AIR FRANCE
GE CF6-50E2	B747-200	6	THAI AIRWAYS INTERNATIONAL
GE CF6-50E2	B747-200 (SCD)	1	AIR GABON
GE CF6-50E2	B747-200 (SCD)	1	SABENA BELGIAN WORLD AIRLINES
GE CF6-50E2	B747-200B	7	ALITALIA
GE CF6-50E2	B747-200B	6	ALL NIPPON AIRWAYS
GE CF6-50E2	B747-200B	4	LUFTHANSA GERMAN AIRLINES - DEUTSCHE LUFTHANSA AG
GE CF6-50E2	B747-200B	6	PHILIPPINE AIR LINES, INC - PAL
GE CF6-50E2	B747-200B (SCD)	10	AIR FRANCE
GE CF6-50E2	B747-200B (SCD)	5	ALITALIA
GE CF6-50E2	B747-200B (SCD)	10	LUFTHANSA GERMAN AIRLINES - DEUTSCHE LUFTHANSA AG
GE CF6-50E2	B747-200B (SCD)	2	PAKISTAN INTERNATIONAL AIRLINES (PIA)
GE CF6-50E2	B747-200B (SCD)	3	VARIG - VIACAO AEREA RIO-GRANDENSE S.A.
GE CF6-50E2	B747-200C (SCD)	2	MARTINAIR HOLLAND NV
GE CF6-50E2	B747-200F	1	MARTINAIR HOLLAND NV
GE CF6-50E2	B747-200F (SCD)	10	AIR FRANCE
GE CF6-50E2	B747-200F (SCD)	2	ALITALIA
GE CF6-50E2	B747-200F (SCD)	1	ARKIA ISRAELI AIRLINES
GE CF6-50E2	B747-200F (SCD)	4	CARGOLUX
GE CF6-50E2	B747-200F (SCD)	3	GERMAN CARGO SERVICES
GE CF6-50E2	B747-200F (SCD)	5	LUFTHANSA GERMAN AIRLINES - DEUTSCHE LUFTHANSA AG
GE CF6-50E2	B747-200F (SCD)	7	NIPPON CARGO AIRLINES CO. LTD
GE CF6-50E2	B747-300	3	KLM ROYAL DUTCH AIRLINES
GE CF6-50E2	B747-300 (SCD)	2	AIR FRANCE
GE CF6-50E2	B747-300 (SCD)	2	AIR FRANCE
GE CF6-50E2	B747-300 (SCD)	2	VARIG - VIACAO AEREA RIO-GRANDENSE S.A.
GE CF6-50E2	B747-300C	2	SABENA BELGIAN WORLD AIRLINES
GE CF6-50E2	B747-300M (SCD)	10	KLM ROYAL DUTCH AIRLINES
GE CF6-6D	DC10-10	6	CONTINENTAL AIRLINES
GE CF6-6D	DC10-10	6	CONTINENTAL MICRONESIA, INC
GE CF6-6D	DC10-10	31	UNITED AIRLINES, INC
GE CF6-6D	DC10-10F (CF)	11	FEDERAL EXPRESS CORPORATION
GE CF6-6D1A	DC10-10	4	PREMAIR A/S
GE CF6-6D1A	DC10-10	2	SUN COUNTRY AIRLINES
GE CF6-6D1A	DC10-10F	2	FEDERAL EXPRESS CORPORATION
GE CF6-6K	DC10-10	46	AMERICAN AIRLINES
GE CF6-6K	DC10-10	6	HAWAIIAN AIRLINES
GE CF6-6K2	DC10-10ER	3	AMERICAN AIRLINES
GE CF6-80	B767-300	0	AEROFLOT RUSSIAN INTERNATIONAL AIRLINES
GE CF6-80A	B767-200	8	AMERICAN AIRLINES
GE CF6-80A	B767-200	6	ANSETT AUSTRALIA
GE CF6-80A	B767-200	2	BRITANNIA AIRWAYS LTD
GE CF6-80A	B767-200	13	DELTA AIR LINES, INC
GE CF6-80A	B767-200	1	TACA INTERNATIONAL AIRLINES SA
GE CF6-80A	B767-200	3	TRANSBRASIL S.A. LINHAS AEREAS
GE CF6-80A	B767-200ER	1	VIETNAM AIRLINES
GE CF6-80A2	B767-200	2	DELTA AIR LINES, INC
GE CF6-80A2	B767-200ER	6	AIR NEW ZEALAND
GE CF6-80A2	B767-200ER	22	AMERICAN AIRLINES
GE CF6-80A2	B767-200ER	5	BRITANNIA AIRWAYS LTD
GE CF6-80A2	B767-200ER	2	LAN-CHILE
GE CF6-80A2	B767-200ER	1	TRANSBRASIL SA - LINHAS AEREAS
GE CF6-80A2	B767-300	25	DELTA AIR LINES, INC
GE CF6-80A3	A310-200	7	AIR FRANCE
GE CF6-80A3	A310-200	4	FEDERAL EXPRESS CORPORATION
GE CF6-80A3	A310-200	10	KLM ROYAL DUTCH AIRLINES
GE CF6-80A3	A310-200	13	LUFTHANSA GERMAN AIRLINES - DEUTSCHE LUFTHANSA AG
GE CF6-80A3	A310-200	2	MIDDLE EAST AIRLINES - AIRLIBAN
GE CF6-80A3	A310-200	7	TURKISH AIRLINES, INC (TURK HAVA YOLLARI A.O)
GE CF6-80A3	A310-203	2	AIR ALGERIE
GE CF6-80A3	A310-203	3	CYPRUS AIRWAYS LTD
GE CF6-80A3	A310-203C	1	MARTINAIR HOLLAND NV
GE CF6-80C2	B747-400	16	ALL NIPPON AIRWAYS
GE CF6-80C2	B747-400	2	GARUDA INDONESIA PT
GE CF6-80C2	B747-400	5	KLM ROYAL DUTCH AIRLINES
GE CF6-80C2	B747-400	3	PHILIPPINE AIR LINES, INC - PAL
GE CF6-80C2	B747-400	2	VIRGIN ATLANTIC AIRWAYS
GE CF6-80C2	B767-200	25	ALL NIPPON AIRWAYS
GE CF6-80C2	B767-200ER	2	MALEV - HUNGARIAN AIRLINES
GE CF6-80C2	B767-300ER	5	AIR NEW ZEALAND
GE CF6-80C2	B767-300ER	2	ALITALIA
GE CF6-80C2	B767-300ER	0	CONTINENTAL AIRLINES
GE CF6-80C2	B767-300ER	0	KLM ROYAL DUTCH AIRLINES
GE CF6-80C2	B767-300ER	2	LEISURE INTERNATIONAL AIRWAYS LTD
GE CF6-80C2	B767-300ER	2	LOT - POLSKIE LINIE LOTNICZE
GE CF6-80C2	B767-300ER	4	VARIG - VIACAO AEREA RIO-GRANDENSE S.A.
GE CF6-80C2	B767-300ER (F)	1	UNITED PARCEL SERVICE COMPANY UPS
GE CF6-80C2	MD-11	5	CHINA EASTERN AIRLINES
GE CF6-80C2	MD-11	6	GARUDA INDONESIA PT
GE CF6-80C2	MD-11	1	GHANA AIRWAYS CORPORATION
GE CF6-80C24F	A300-600F	6	FEDERAL EXPRESS CORPORATION
GE CF6-80C2A1	A300-601	6	THAI AIRWAYS INTERNATIONAL
GE CF6-80C2A2	A310-200	4	HAPAG-LLOYD FLUGGESELLSCHAFT.MBH
GE CF6-80C2A2	A310-200	2	THAI AIRWAYS INTERNATIONAL
GE CF6-80C2A2	A310-204	1	CYPRUS AIRWAYS LTD
GE CF6-80C2A2	A310-300	5	AEROFLOT RUSSIAN INTERNATIONAL AIRLINES
GE CF6-80C2A2	A310-300	4	AIR AFRIQUE
GE CF6-80C2A2	A310-300	4	AIR FRANCE
GE CF6-80C2A2	A310-300	8	AIR INDIA
GE CF6-80C2A2	A310-300	5	CHINA EASTERN AIRLINES
GE CF6-80C2A2	A310-300	2	CSA - CESKOSLOVENSKE AEROLINIE
GE CF6-80C2A2	A310-300	10	EMIRATES
GE CF6-80C2A2	A310-300	3	HAPAG-LLOYD FLUGGESELLSCHAFT.MBH
GE CF6-80C2A2	A310-300	3	KENYA AIRWAYS LTD
GE CF6-80C2A2	A310-300	12	LUFTHANSA GERMAN AIRLINES - DEUTSCHE LUFTHANSA AG
GE CF6-80C2A2	A310-300	1	MIDDLE EAST AIRLINES - AIRLIBAN
GE CF6-80C2A2	A310-300	5	ROYAL JORDANIAN

POWER PLANT	AIRFRAME	NO. IN FLEET	FLEET NAME
GE CF6-80C2A2	A310-300	1	SUDAN AIRWAYS
GE CF6-80C2A2	A310-300	5	TAP AIR PORTUGAL
GE CF6-80C2A2	A310-300	7	TURKISH AIRLINES, INC (TURK HAVA YOLLARI A.O)
GE CF6-80C2A2	A310-300	1	YEMEN AIRLINES
GE CF6-80C2A2	A310-304	1	LLOYD AEREO BOLIVIANO SAM
GE CF6-80C2A2	A310-304	1	SOMALI AIRLINES
GE CF6-80C2A3	A300-600	11	LUFTHANSA GERMAN AIRLINES - DEUTSCHE LUFTHANSA AG
GE CF6-80C2A5	A300-600R	7	CHINA EASTERN AIRLINES
GE CF6-80C2A5	A300-600R	4	CHINA NORTHWEST AIRLINES
GE CF6-80C2A5	A300-600R	6	EMIRATES
GE CF6-80C2A5	A300-600R	0	IRAN AIR THE AIRLINE OF THE ISLAMIC REPUBLIC OF IRAN
GE CF6-80C2A5	A300-600R	5	KUWAIT AIRWAYS CORPORATION
GE CF6-80C2A5	A300-600R	3	OLYMPIC AIRWAYS S.A.
GE CF6-80C2A5	A300-605R	35	AMERICAN AIRLINES
GE CF6-80C2A5	A300-605R	4	MONARCH AIRLINES LTD
GE CF6-80C2A5	A300-605R	2	THAI AIRWAYS INTERNATIONAL
GE CF6-80C2A8	A310-300	3	KUWAIT AIRWAYS CORPORATION
GE CF6-80C2A8	A310-300	2	SAETA
GE CF6-80C2A8	A310-307	6	PAKISTAN INTERNATIONAL AIRLINES (PIA)
GE CF6-80C2B1	B747-300	3	VARIG - VIACAO AEREA RIO-GRANDENSE S.A.
GE CF6-80C2B1	B747-300C	2	AIR INDIA
GE CF6-80C2B1F	B747-400	7	AIR FRANCE
GE CF6-80C2B1F	B747-400	5	ASIANA AIRLINES
GE CF6-80C2B1F	B747-400	4	CANADIAN - CANADIEN
GE CF6-80C2B1F	B747-400	10	EVA AIRWAYS
GE CF6-80C2B1F	B747-400	0	JAPAN AIR SYSTEM CO LTD (NIHON AIR SYSTEM)
GE CF6-80C2B1F	B747-400	24	JAPAN AIRLINES COMPANY LTD - JAL
GE CF6-80C2B1F	B747-400	10	LUFTHANSA GERMAN AIRLINES - DEUTSCHE LUFTHANSA AG
GE CF6-80C2B1F	B747-400	2	MALAYSIA AIRLINES
GE CF6-80C2B1F	B747-400	1	ROYAL AIR MAROC
GE CF6-80C2B1F	B747-400	7	THAI AIRWAYS INTERNATIONAL
GE CF6-80C2B1F	B747-400 (SCD)	7	AIR FRANCE
GE CF6-80C2B1F	B747-400 (SCD)	7	LUFTHANSA GERMAN AIRLINES - DEUTSCHE LUFTHANSA AG
GE CF6-80C2B1F	B747-400 (SCD)	2	MALAYSIA AIRLINES
GE CF6-80C2B1F	B747-400C	1	AIR FRANCE
GE CF6-80C2B1F	B747-400C (SCD)	3	ASIANA AIRLINES
GE CF6-80C2B1F	B747-400D	8	JAPAN AIRLINES COMPANY LTD - JAL
GE CF6-80C2B1F	B747-400F	2	CARGOLUX
GE CF6-80C2B1F	B747-400F	0	KLM ROYAL DUTCH AIRLINES
GE CF6-80C2B1F	B747-400F (SCD)	1	ASIANA AIRLINES
GE CF6-80C2B1F	B747-400M	1	KUWAIT AIRWAYS CORPORATION
GE CF6-80C2B1F	B747-400M (SCD)	11	KLM ROYAL DUTCH AIRLINES
GE CF6-80C2B2	B767-200	4	EVA AIRWAYS
GE CF6-80C2B2	B767-200ER	3	BRITISH AIRWAYS
GE CF6-80C2B2	B767-200ER	12	USAIR, INC
GE CF6-80C2B2	B767-200ER	6	VARIG - VIACAO AEREA RIO-GRANDENSE S.A.
GE CF6-80C2B2	B767-300	31	ALL NIPPON AIRWAYS
GE CF6-80C2B2F	B767-300	3	AIR ALGERIE
GE CF6-80C2B4	B767-200ER	2	AIR MAURITIUS
GE CF6-80C2B4	B767-200ER	1	AIR SEYCHELLES
GE CF6-80C2B4	B767-200ER	2	BRITANNIA AIRWAYS LTD
GE CF6-80C2B4	B767-200ER	2	LOT - POLSKIE LINIE LOTNICZE
GE CF6-80C2B4	B767-300ER	1	AIR PACIFIC LIMITED
GE CF6-80C2B4	B767-300ER	0	BRITANNIA AIRWAYS LTD
GE CF6-80C2B4	B767-300ER	18	GULF AIR
GE CF6-80C2B6	B767-300ER	2	AIR EUROPE SPA
GE CF6-80C2B6	B767-300ER	6	ALL NIPPON AIRWAYS
GE CF6-80C2B6	B767-300ER	35	AMERICAN AIRLINES
GE CF6-80C2B6	B767-300ER	1	AUSTRALIAN ASIAN AIRLINES
GE CF6-80C2B6	B767-300ER	11	CANADIAN - CANADIEN
GE CF6-80C2B6	B767-300ER	6	CHINA SOUTHERN AIRLINES
GE CF6-80C2B6	B767-300ER	15	QANTAS AIRWAYS LTD
GE CF6-80C2B6F	B767-300ER	3	ASIANA AIRLINES
GE CF6-80C2B6F	B767-300ER	5	EVA AIRWAYS
GE CF6-80C2B6F	B767-300ER	1	SOBELAIR - SOCIETE BELGE DE TRANSPORTES AERIENS
GE CF6-80C2B7F	B767-300ER	2	AIRTOURS INTERNATIONAL AIRWAYS LTD
GE CF6-80C2B7F	B767-300F	0	ASIANA AIRLINES
GE CF6-80C2BF	B767-300	4	ASIANA AIRLINES
GE CF6-80C2D1BF	MD-11	4	FINNAIR OY
GE CF6-80C2D1F	MD-11	17	AMERICAN AIRLINES
GE CF6-80C2D1F	MD-11	6	EVA AIRWAYS
GE CF6-80C2D1F	MD-11	7	KLM ROYAL DUTCH AIRLINES
GE CF6-80C2D1F	MD-11	4	THAI AIRWAYS INTERNATIONAL
GE CF6-80C2D1F	MD-11	6	VARIG - VIACAO AEREA RIO-GRANDENSE S.A.
GE CF6-80C2D1F	MD-11	4	VASP - VIACAO AEREA SAO PAULO S.A.
GE CF6-80C2D1F	MD-11C	6	ALITALIA
GE CF6-80C2D1F	MD-11F	1	CHINA EASTERN AIRLINES
GE CF6-80C2D1F	MD-11F	13	FEDERAL EXPRESS CORPORATION
GE CF6-80C2E2	B767-300	1	WORLD AIR NETWORK
GE CF6-80E1	A330-300	3	AER LINGUS
GE CF6-80E1	A330-300	5	AIR INTER
GE CF700-2C	DA Falcon 20C	1	LIBYAN ARAB AIRLINES
GE CF700-2D2	DA Falcon 20E	3	IRAN ASSEMAN AIRLINES
GE CF700-2D2	DA Falcon 20F	1	IRAN ASSEMAN AIRLINES
GE CF700-2D2	DA Falcon 20F	2	IRAQI AIRWAYS
GE CF700-2D2	DA Falcon 20F	2	SYRIAN ARAB AIRLINES
GE CJ610-6	Learjet 24D / 25C	1/1	AIR PROVENCE
GE CJ610-8A	Learjet 25	1	THERON AIRWAYS (PTY) LTD
GE CJ610-8A	Learjet 25D	1	AERONAVES DEL PERU, S.A.
GE CJ610-8A	Learjet 25D	2	FLAGSHIP EXPRESS
GE CT58-140-1	Sikorsky S-61N	4	GROENLANDSFLY A/S
GE CT64-820-4	DHC-5A Buffalo	1	ETHIOPIAN AIRLINES CORPORATION
GE CT7-5A2	Saab 340A	16	AMR EAGLE
GE CT7-5A2	Saab 340A	6	BAR HARBOR AIRLINES (CONTINENTAL/DELTA EXPRESS)
GE CT7-5A2	Saab 340A	6	BRIT'AIR
GE CT7-5A2	Saab 340A	17	BUSINESS EXPRESS (THE DELTA CONNECTION)
GE CT7-5A2	Saab 340A	5	CHAUTAUQUA AIRLINES
GE CT7-5A2	Saab 340A	19	COMAIR
GE CT7-5A2	Saab 340A	10	CROSSAIR

POWER PLANT	AIRFRAME	NO. IN FLEET	FLEET NAME
GE CT7-5A2	Saab 340A	5	DEUTSCHE BA LUFTFAHRTGESELLSCHAFT MBH
GE CT7-5A2	Saab 340A	29	EXPRESS AIRLINES I, INC (NORTHWEST AIRLINK)
GE CT7-5A2	Saab 340A	5	FINNAVIATION OY
GE CT7-5A2	Saab 340A	3	FORMOSA AIRLINES
GE CT7-5A2	Saab 340A	9	KENDELL AIRLINES (AUSTRALIA) PTY LIMITED
GE CT7-5A2	Saab 340A	2	LAPA - LINEAS AEREAS PRIVADAS ARGENTINIAS, SA
GE CT7-5A2	Saab 340A	3	SKYWAYS, AB
GE CT7-5A2	Saab 340B	4	HAZELTON AIRLINES
GE CT7-5A2	Saab 340Bplus	0	AMR EAGLE
GE CT7-5A2	Saab 340Bplus	1	CALM AIR (CANADIAN PARTNER)
GE CT7-5A2	Saab 340Bplus	1	HAZELTON AIRLINES
GE CT7-5A2	Saab 340Bplus	0	KENDELL AIRLINES (AUSTRALIA) PTY LIMITED
GE CT7-5A2	Saab 340Bplus	1	REGIONAL AIRLINES
GE CT7-7A	CASA/IPTN CN235-10	4	BINTER
GE CT7-7A	IPTN/CASA CN235-10	14	MERPATI NUSANTARA AIRLINES PT
GE CT7-9	CASA/IPTN CN235-QC	1	BINTER
GE CT7-9B	Saab 340B	4	AER LINGUS
GE CT7-9B	Saab 340B	100	AMR EAGLE
GE CT7-9B	Saab 340B	22	BUSINESS EXPRESS (THE DELTA CONNECTION)
GE CT7-9B	Saab 340B	4	CHAUTAUQUA AIRLINES
GE CT7-9B	Saab 340B	4	CHINA SOUTHERN AIRLINES
GE CT7-9B	Saab 340B	0	CHINA SOUTHWEST AIRLINES
GE CT7-9B	Saab 340B	15	CROSSAIR
GE CT7-9B	Saab 340B	6	EXPRESS AIRLINES I, INC (NORTHWEST AIRLINK)
GE CT7-9B	Saab 340B	1	FINNAVIATION OY
GE CT7-9B	Saab 340B	6	FORMOSA AIRLINES
GE CT7-9B	Saab 340B	2	JAPAN AIR COMMUTER CO LTD
GE CT7-9B	Saab 340B	1	JAPAN AIR SYSTEM CO LTD (NIHON AIR SYSTEM)
GE CT7-9B	Saab 340B	11	KLM CITYHOPPER
GE CT7-9B	Saab 340B	7	REGIONAL AIRLINES
GE CT7-9B	Saab 340B	4	SKYWAYS, AB
GE CT7-9C	CASA/IPTN CN235-200	5	BINTER
GE CT7-9C	IPTN/CASA CN235-100	0	CONQUEST AIRLINES
GE CT7-9C	IPTN/CASA CN235-200	0	MERPATI NUSANTARA AIRLINES PT
GE GE90-B3	B777-200ER	0	CHINA SOUTHERN AIRLINES
GE GE90-B3	B777-200ER	0	LAUDA AIR LUFTFAHRT AKTIENGESELLSCHAFT
GE GE90-B4	B777-200	0	CONTINENTAL AIRLINES
GE GE90-B4	B777-200/200ER	0	BRITISH AIRWAYS
GE GE90-B4	B777-200ER	0	EURALAIR INTERNATIONAL
Ga TFE731-2-1C	DA Falcon 10	1	EURALAIR INTERNATIONAL
Ga TFE731-2-2B	Learjet 35A	1	ERA AVIATION, INC
Ga TFE731-2-2B	Learjet 35A	2	TAESA
Ga TFE731-2-3B	Learjet 31A	4	SINGAPORE AIRLINES LIMITED
Ga TFE731-20	Learjet 45	0	SINGAPORE AIRLINES LIMITED
Ga TFE731-3	L1329 JetStar 731	1	PAN AVIATION
Ga TFE731-3	L1329 JetStar II	6	IRAQI AIRWAYS
Ga TFE731-3-1C	DA Falcon 50	3	IRAQI AIRWAYS
Ga TFE731-3-1C	DA Falcon 50	1	LIBYAN ARAB AIRLINES
Ga TFE731-3-1H	BAe (HS) 125-700B	1	KUWAIT AIRWAYS CORPORATION
Ga TFE731-3R-1H	BAe (HS) 125-800	1	MAERSK AIR
Ga TFE731-5AR-1C2	DA Falcon 900	2	SAUDIA - SAUDI ARABIAN AIRLINES
Ga TFE731-5AR-2C	DA Falcon 20F-5	1	EURALAIR INTERNATIONAL
Ga TPE331-10-501A	Mitsubishi MU-2F	1	SUN-AIR OF SCANDINAVIA A/S
Ga TPE331-10-501C	CASA 212-200 Aviocar	1	AIR FAST INDONESIA
Ga TPE331-10-501C	CASA 212-200 Aviocar	4	LAM - LINHAS AEREAS DE MOCAMBIQUE
Ga TPE331-10-511C	IPTN/CASA 212-200 Aviocar	11	MERPATI NUSANTARA AIRLINES PT
Ga TPE331-10-511C	IPTN/CASA 212-200 Aviocar	8	PELITA AIR SERVICE
Ga TPE331-10R-511C	CASA 212-200 Aviocar	1	AIR TUNGARU CORP
Ga TPE331-10R-511C	CASA 212-200 Aviocar	5	SATENA
Ga TPE331-10R-511C	CASA 212-300 Aviocar	2	SATENA
Ga TPE331-10UA-511G	SA226 Metro II	3	PROVINCIAL AIRWAYS LTD
Ga TPE331-10UA-511G	SA226AT Merlin IVA	2	JETALL
Ga TPE331-10UA-511G	SA226TC Metro II	4	BIG SKY AIRLINES
Ga TPE331-10UA-511G	SA226TC Metro II	4	JETALL
Ga TPE331-10UF-511H	BAe Jetstream 31	1	MAERSK AIR LTD
Ga TPE331-10UF-513H	BAe Jetstream 31	1	AIR FOYLE LIMITED
Ga TPE331-10UF-513H	BAe Jetstream 31	1	REGIONAIR
Ga TPE331-10UG-513H	BAe Jetstream 31	5	AIRBC LTD
Ga TPE331-10UG-513H	BAe Jetstream 31	3	EASTERN AUSTRALIA AIRLINES
Ga TPE331-10UG-513H	BAe Jetstream 31	25	EXPRESS AIRLINES I, INC (NORTHWEST AIRLINK)
Ga TPE331-10UG-513H	BAe Jetstream 31	17	JETSTREAM INTERNATIONAL AIRLINES
Ga TPE331-10UG-513H	BAe Jetstream 31	21	MESA AIRLINES INC
Ga TPE331-10UR-513H	BAe Jetstream 31	10	SUN-AIR OF SCANDINAVIA A/S
Ga TPE331-10UR-513H	BAe Jetstream Super 31	50	AMR EAGLE
Ga TPE331-11U-601G	SA227AC Metro III	1	NORLANDAIR (ICELAND)
Ga TPE331-11U-601G	SA227AT Expediter	11	UNITED PARCEL SERVICE COMPANY UPS
Ga TPE331-11U-611G	SA 227AC Metro III	2	EMPIRE AIRLINES
Ga TPE331-11U-611G	SA226 Metro III	4	EAGLE AIRWAYS LTD
Ga TPE331-11U-611G	SA226AC Metro III	9	COMAIR
Ga TPE331-11U-611G	SA227AC Metro III	1	AIR ATLANTIQUE
Ga TPE331-11U-611G	SA227AC Metro III	1	AIRLA
Ga TPE331-11U-611G	SA227AC Metro III	10	CHAUTAUQUA AIRLINES
Ga TPE331-11U-611G	SA227AC Metro III	22	HORIZON AIR
Ga TPE331-11U-611G	SA227AC Metro III	6	LONE STAR AIRLINES
Ga TPE331-11U-611G	SA227AC Metro III	26	MESABA AIRLINES (NORTHWEST AIRLINK)
Ga TPE331-11U-611G	SA227AC Metroliner	25	SKYWEST AIRLINES INC (DELTA CONNECTION)
Ga TPE331-11U-611G	SA227AT Expediter I	10	DHL AIRWAYS
Ga TPE331-11U-611G	SA227AT Merlin IVC	1	JETALL
Ga TPE331-11U-611G	SA227AT Metro III	3	JETALL
Ga TPE331-11U-612G	SA227AC Metro III	2	TRANS STATES AIRLINES
Ga TPE331-12U-701G	Fairchild Metro 23	2	AIRLA
Ga TPE331-12U-701G	Fairchild Metro 23	2	CHINA HAINAN AIRLINES
Ga TPE331-12U-701G	Fairchild Metro 23	1	HAZELTON AIRLINES
Ga TPE331-12U-701G	Fairchild Metro 23	7	KENDELL AIRLINES (AUSTRALIA) PTY LIMITED
Ga TPE331-12U-701G	Fairchild Metro 23	4	LONE STAR AIRLINES
Ga TPE331-12UAR-701H	BAe Jetstream Super 31	1	EASTERN AUSTRALIA AIRLINES
Ga TPE331-12UAR-703H	BAe Jetstream Super 31	9	REGIONAL AIRLINES
Ga TPE331-12UR	BAe Jetstream Super 31	22	TRANS STATES AIRLINES
Ga TPE331-14GR/HR	BAe Jetstream 31	34	WESTAIR COMMUTER AIRLINES (MESA AIRLINES)

POWER PLANT	AIRFRAME	NO. IN FLEET	FLEET NAME
Ga TPE331-14GR/HR	BAe Jetstream 41	0	AMR EAGLE
Ga TPE331-14GR/HR	BAe Jetstream 41	1	BRITISH MIDLAND AIRWAYS LTD
Ga TPE331-14GR/HR	BAe Jetstream 41	10	MANX AIRLINES LTD
Ga TPE331-14GR/HR	BAe Jetstream 41	2	SUN-AIR OF SCANDINAVIA A/S
Ga TPE331-14GR/HR	BAe Jetstream 41	0	TRANS STATES AIRLINES
Ga TPE331-3UW-303G	SA226 Metro II	3	EUROPEAN AIR TRANSPORT N.V./S.A.
Ga TPE331-3UW-303G	SA226AT Merlin IV	1	PROVINCIAL AIRWAYS LTD
Ga TPE331-3UW-303G	SA226TC Metro II	2	AIRLINK AIRLINE (PTY) LTD
Ga TPE331-3UW-303G	SA226TC Metro II	1	JETALL
Ga TPE331-3UW-303G	SA227AT Merlin IV	1	PROVINCIAL AIRWAYS LTD
Ga TPE331-3UW-304G	SA226AT Merlin IV	2	FLANDERS AIRLINES NV
Ga TPE331-3UW-304G	SA226TC Metro II	2	KENDELL AIRLINES (AUSTRALIA) PTY LIMITED
Ga TPE331-5-251C	IPTN/CASA 212-100 Aviocar	4	PELITA AIR SERVICE
Ga TPE331-5-251K	Rockwell Turbo Cmdr 690B	2	CATA
Ga TPE331-5-252D	BAe Jetstream 31	12	CCAIR, INC (USAIR EXPRESS)
Ga TPE331-5-252D	Dornier 228	1	AIR MALAWI
Ga TPE331-5-252D	Dornier 228-100	1	HOLMSTROM AIR
Ga TPE331-5-252D	Dornier 228-200	3	JAPAN AIR COMMUTER CO LTD
Ga TPE331-5-252D	Dornier 228-200	7	OLYMPIC AVIATION, S.A.
Ga TPE331-5-252D	Dornier 228-200	1	SUCKLING AIRWAYS
Ga TPE331-5-252D	Dornier 228-201	1	AIR HUDIK AB
Ga TPE331-5-252D	Dornier 228-201	2	AIR MARSHALL ISLANDS - AMI
Ga TPE331-5-252D	Dornier 228-201	2	DRUK AIR CORP
Ga TPE331-5-252D	Dornier 228-201	2	HOLMSTROM AIR
Ga TPE331-5-252D	Dornier 228-201	2	JAGSON AIRLINES
Ga TPE331-5-252D	Dornier 228-201	4	VAYUDOOT
Ga TPE331-5-252D	Dornier 228-201K	1	REGIONAIR
Ga TPE331-5-252D	Dornier 228-202	3	FORMOSA AIRLINES
Ga TPE331-5-252D	Dornier 228-202	13	PRECISION AIRLINES
Ga TPE331-5-252D	Dornier 228-202	1	RATIOFLUG GMBH
Ga TPE331-5-252D	Dornier 228-202	1	SATA AIR ACORES - SERVICO ACOREANO DE TRANSPORTES AEREOS, E.P.
Ga TPE331-5-252D	Dornier 228-202	2	SOMALI AIRLINES
Ga TPE331-5-252D	Dornier 228-202K	1	FLEXAIR, BV
Ga TPE331-5-252D	Dornier 228-202K	3	PELANGI AIR, SDN BHD
Ga TPE331-5-252D	Dornier 228-202K	1	SILKAIR (SINGAPORE) PRIVATE LIMITED
Ga TPE331-5-252D	Dornier 228-202K	3	SOCIETE NOUVELLE AIR GUADELOUPE
Ga TPE331-5-252D	Dornier 228-202K	4	SUCKLING AIRWAYS
Ga TPE331-5-252D	Dornier 228-212	2	AIR CALEDONIE
Ga TPE331-5-252D	Dornier 228-212		TAHITI CONQUEST AIRLINES - TCA
Ga TPE331-5-252D	HAL/Dornier 228-201	5	VAYUDOOT
Ga TPE331-5A-252D	Dornier 228-100	1	AIRLINK AIRLINE (PTY) LTD
Ga TPE331-5A-252D	Dornier 228-200	1	AIR TAHITI
Ga TPE331-5A-252D	Dornier 228-212	1	AIR MOOREA
Ga TPE331-5A-252D	Dornier 228-212	4	FORMOSA AIRLINES
Ga TPE331-5A-252D	Dornier 228-212	1	PELANGI AIR, SDN BHD
Ga TPE731-2-2B	Learjet 35	1	RATIOFLUG GMBH
Ga TPE731-2-2B	Learjet 35A	1	AIR AFFAIRES GABON
Ga TPE731-3A-2B	Learjet 55	2	RATIOFLUG GMBH
Ga TPE731-3AR-2B	Learjet 55	1	CHINA NORTHERN AIRLINES
HS 5	Shijiazhuang Yunshuji Y-5	12	CHINA NORTHWEST AIRLINES
IAE V2500-A1	A320-200	3	ADRIA AIRWAYS
IAE V2500-A1	A320-200	4	AIR 2000 LTD
IAE V2500-A1	A320-200	2	AIR FOYLE LIMITED
IAE V2500-A1	A320-200	2	AIR LANKA
IAE V2500-A1	A320-200	3	AIRTOURS INTERNATIONAL AIRWAYS LTD
IAE V2500-A1	A320-200	16	AMERICA WEST AIRLINES, INC
IAE V2500-A1	A320-200	4	BALKAN
IAE V2500-A1	A320-200	2	CHINA AIRLINES
IAE V2500-A1	A320-200	8	CYPRUS AIRWAYS LTD
IAE V2500-A1	A320-200	7	DRAGONAIR (HONG KONG DRAGON AIRLINES LIMITED)
IAE V2500-A1	A320-200	7	EGYPTAIR
IAE V2500-A1	A320-200	3	EUROCYPRIA AIRLINES
IAE V2500-A1	A320-200	30	INDIAN AIRLINES LTD
IAE V2500-A1	A320-200	14	MEXICANA
IAE V2500-A1	A320-200	6	PREMAIR A/S
IAE V2500-A1	A320-200	5	REGION AIR
IAE V2500-A1	A320-200	7	SOUTH AFRICAN AIRWAYS (SAA)
IAE V2500-A1	A320-200	3	TRANSASIA AIRWAYS
IAE V2500-A1	A320-200	6	TRANSLIFT AIRWAYS
IAE V2500-A5	A320	0	AERO LLOYD
IAE V2500-A5	A320-200	2	AMERICA WEST AIRLINES, INC
IAE V2500-A5	A321	0	AERO LLOYD
IAE V2500-D5	MD90-30	0	GREAT CHINA AIRLINES
IAE V2525-D5	MD90-30	3	DELTA AIR LINES, INC
IAE V2525-D5	MD90-30	0	JAPAN AIR SYSTEM CO LTD (NIHON AIR SYSTEM)
IAE V2525-D5	MD90-30	0	SAS - SCANDINAVIAN AIRLINES SYSTEM
IAE V2527-A5	A320-200	27	UNITED AIRLINES, INC
IAE V2530-A5	A321	0	TRANSASIA AIRWAYS
IAE V2530-A5	A321-100	10	LUFTHANSA GERMAN AIRLINES - DEUTSCHE LUFTHANSA AG
KKBM NK-8-2	Tupolev 154	44	AEROFLOT RUSSIAN INTERNATIONAL AIRLINES
KKBM NK-8-2	Tupolev 154	2	AERONICA - AEROLINEAS NICARAGUENSES, SA
KKBM NK-8-2	Tupolev 154		BARNAULSKOYE AVIAPREDPRIYATIYE
KKBM NK-8-2	Tupolev 154	3	KEMEROVSKOE AVIAPREDPRIYATIYE
KKBM NK-8-2	Tupolev 154		KYRGYZSTAN AIRLINE
KKBM NK-8-2	Tupolev 154	5	NOVOKUZNETSKOE AVIAPREDPRIYATIYE
KKBM NK-8-2	Tupolev 154		OMSKOYE AVIAPREDPRIYATIYE
KKBM NK-8-2	Tupolev 154	12	SIBERIA AIRLINE
KKBM NK-8-2	Tupolev 154	3	TOMSKOYE AVIAPREDPRIYATIYE
KKBM NK-8-2	Tupolev 154		TURKMENISTAN AIRLINE
KKBM NK-8-2	Tupolev 154	1	ULAN-UDESKOYE AVIAPREDPRIYATIYE
KKBM NK-8-2	Tupolev 154	33	VNUKOVO AIRLINES
KKBM NK-8-2	Tupolev 154A	45	AEROFLOT RUSSIAN INTERNATIONAL AIRLINES
KKBM NK-8-2	Tupolev 154B	101	AEROFLOT RUSSIAN INTERNATIONAL AIRLINES
KKBM NK-8-2	Tupolev 154B	4	AEROFLOT RUSSIAN INTERNATIONAL AIRLINES
KKBM NK-8-2	Tupolev 154B	3	AIR KORYO
KKBM NK-8-2	Tupolev 154B	17	BELAVIA
KKBM NK-8-2	Tupolev 154B-1	70	AEROFLOT RUSSIAN INTERNATIONAL AIRLINES
KKBM NK-8-2	Tupolev 154B-1	5	TAROM
KKBM NK-8-2	Tupolev 154B-2	296	AEROFLOT RUSSIAN INTERNATIONAL AIRLINES

POWER PLANT	AIRFRAME	NO. IN FLEET	FLEET NAME
KKBM NK-8-2	Tupolev 154B-2	1	AIR KORYO
KKBM NK-8-2	Tupolev 154B-2	17	BALKAN
KKBM NK-8-2	Tupolev 154B-2	10	MALEV - HUNGARIAN AIRLINES
KKBM NK-8-2	Tupolev 154B-2	2	TAROM
KKBM NK-8-2	Tupolev 154C	5	AEROFLOT RUSSIAN INTERNATIONAL AIRLINES
KKBM NK-8-2	Tupolev 154C	3	AEROFLOT RUSSIAN INTERNATIONAL AIRLINES
KKBM NK-8-2	Tupolev 154C	1	VNUKOVO AIRLINES
KKBM NK-8-2U	Tupolev 154B	1	TAROM
KKBM NK-8-2U	Tupolev 154B-2	7	SAMARA AIRLINES
KKBM NK-8-4	Ilyushin 62	61	AEROFLOT RUSSIAN INTERNATIONAL AIRLINES
KKBM NK-8-4	Ilyushin 62	3	AEROFLOT RUSSIAN INTERNATIONAL AIRLINES
KKBM NK-8-4	Ilyushin 62	2	CSA - CESKOSLOVENSKE AEROLINIE
KKBM NK-8-4	Ilyushin 62	1	TAROM
KKBM NK-86	Ilyushin 86	64	AEROFLOT RUSSIAN INTERNATIONAL AIRLINES
KKBM NK-86	Ilyushin 86	19	AEROFLOT RUSSIAN INTERNATIONAL AIRLINES
KKBM NK-86	Ilyushin 86	2	CHINA NORTHERN AIRLINES
KKBM NK-86	Ilyushin 86	6	SIBERIA AIRLINE
KKBM NK-86	Ilyushin 86	1	TRANSAERO AIRLINES
KKBM NK-86	Ilyushin 86	22	VNUKOVO AIRLINES
Klimov TV2-117A	Mil Mi-8		BARNAULSKOYE AVIAPREDPRIYATIYE
Klimov TV2-117A	Mil Mi-8	2	KAMPUCHEA AIRLINES
Klimov TV2-117A	Mil Mi-8	3	MONGOLIAN AIRLINES - MIAT
Klimov TV2-117A	Mil Mi-8		NOVOKUZNETSKOE AVIAPREDPRIYATIYE
Klimov TV2-117A	Mil Mi-8		TURKMENISTAN AIRLINE
Klimov TV2-117A	Mil Mi-8		ULAN-UDESKOYE AVIAPREDPRIYATIYE
Klimov TV2-117A	Mil Mi-8T	1	KAMPUCHEA AIRLINES
Klimov TV3-117MT-3	Mil Mi-17	3	KAMPUCHEA AIRLINES
LY TIO-540-J2BD	PA-31-350 Navajo	6	SCENIC AIRLINES INC
Lyc ALF502L-2C	CL 600 Challenger	0	RATIOFLUG GMBH
Lyc ALF502R-5	BAe 146-100	1	AIR BOTSWANA
Lyc ALF502R-5	BAe 146-100	4	AIR CHINA INTERNATIONAL
Lyc ALF502R-5	BAe 146-100	1	AIR MALDIVES
Lyc ALF502R-5	BAe 146-100	1	AIR UK LTD
Lyc ALF502R-5	BAe 146-100	3	CHINA EASTERN AIRLINES
Lyc ALF502R-5	BAe 146-100	3	CHINA NORTHWEST AIRLINES
Lyc ALF502R-5	BAe 146-100	2	DRUK AIR CORP
Lyc ALF502R-5	BAe 146-100	1	JERSEY EUROPEAN
Lyc ALF502R-5	BAe 146-100	2	NATIONAL AIR CHARTER P.T.
Lyc ALF502R-5	BAe 146-100	1	QANTAS AIRWAYS LTD
Lyc ALF502R-5	BAe 146-100A	2	AIR WISCONSIN (UNITED EXPRESS)
Lyc ALF502R-5	BAe 146-200	3	AIR ATLANTIC LTD (CANADIAN PARTNER)
Lyc ALF502R-5	BAe 146-200	1	AIR UK LTD
Lyc ALF502R-5	BAe 146-200	1	AIR ZIMBABWE
Lyc ALF502R-5	BAe 146-200	1	AVIACION DE CHIAPAS (AVIACSA)
Lyc ALF502R-5	BAe 146-200	7	DELTA AIR TRANSPORT NV
Lyc ALF502R-5	BAe 146-200	2	EUROWINGS
Lyc ALF502R-5	BAe 146-200	2	HAMBURG AIRLINES LUFTFAHRTGESELLSCHAFT.MBH
Lyc ALF502R-5	BAe 146-200	4	JERSEY EUROPEAN
Lyc ALF502R-5	BAe 146-200	2	LAN-CHILE
Lyc ALF502R-5	BAe 146-200	5	MALMO AVIATION
Lyc ALF502R-5	BAe 146-200	2	MANX AIRLINES LTD
Lyc ALF502R-5	BAe 146-200	4	MERIDIANA SPA
Lyc ALF502R-5	BAe 146-200	1	PELITA AIR SERVICE
Lyc ALF502R-5	BAe 146-200	3	QANTAS AIRWAYS LTD
Lyc ALF502R-5	BAe 146-200A	5	AIR NOVA
Lyc ALF502R-5	BAe 146-200A	5	AIR WISCONSIN (UNITED EXPRESS)
Lyc ALF502R-5	BAe 146-200A	5	AIRBC LTD
Lyc ALF502R-5	BAe 146-200A	6	ANSETT AUSTRALIA
Lyc ALF502R-5	BAe 146-200A	1	ATLANTIC AIRWAYS (FAROE ISLANDS)
Lyc ALF502R-5	BAe 146-200A	5	BUSINESS EXPRESS (THE DELTA CONNECTION)
Lyc ALF502R-5	BAe 146-200A	17	USAIR, INC
Lyc ALF502R-5	BAe 146-200A	3	WESTAIR COMMUTER AIRLINES (MESA AIRLINES)
Lyc ALF502R-5	BAe 146-200QC	2	AIR JET
Lyc ALF502R-5	BAe 146-200QC	1	ANSETT NEW ZEALAND
Lyc ALF502R-5	BAe 146-200QT	3	AIR FOYLE LIMITED
Lyc ALF502R-5	BAe 146-200QT	2	ANSETT AUSTRALIA
Lyc ALF502R-5	BAe 146-200QT	2	EUROWINGS
Lyc ALF502R-5	BAe 146-200QT	1	MALMO AVIATION
Lyc ALF502R-5	BAe 146-200QT	3	PAN AIR LINEAS AEREAS SA
Lyc ALF502R-5	BAe 146-200QT	12	TNT INTERNATIONAL AVIATION SERVICES
Lyc ALF502R-5	BAe 146-300	3	AER LINGUS
Lyc ALF502R-5	BAe 146-300	0	AEROTAXI SUD
Lyc ALF502R-5	BAe 146-300	10	AIR UK LTD
Lyc ALF502R-5	BAe 146-300	6	ANSETT AUSTRALIA
Lyc ALF502R-5	BAe 146-300	8	ANSETT NEW ZEALAND
Lyc ALF502R-5	BAe 146-300	3	BRITISH WORLD AIRLINES
Lyc ALF502R-5	BAe 146-300	8	CHINA NORTHWEST AIRLINES
Lyc ALF502R-5	BAe 146-300	1	CROSSAIR
Lyc ALF502R-5	BAe 146-300	2	JERSEY EUROPEAN
Lyc ALF502R-5	BAe 146-300	5	MAKUNG AIRLINES, CO LTD
Lyc ALF502R-5	BAe 146-300	2	QANTAS AIRWAYS LTD
Lyc ALF502R-5	BAe 146-300	2	TAS AIRWAYS SPA
Lyc ALF502R-5	BAe 146-300	5	THAI AIRWAYS INTERNATIONAL
Lyc ALF502R-5	BAe 146-300A	5	AIR WISCONSIN (UNITED EXPRESS)
Lyc ALF502R-5	BAe 146-300QT	4	AIR FOYLE LIMITED
Lyc ALF502R-5	BAe 146-300QT	2	MALMO AVIATION
Lyc ALF502R-5	BAe 146-300QT	1	PAN AIR LINEAS AEREAS SA
Lyc ALF502R-5	BAe 146-300QT	9	TNT INTERNATIONAL AVIATION SERVICES
Lyc GO-480-B1A6	Dornier 27Q-5	1	TAGB - TRANSPORTES AEREOS DA GUINE BISSAU
Lyc GO-480-F4A6	BE Twin Bonanza	1	ABOITIZ AIR TRANSPORT
Lyc IGSO-540-A1D	BE Queen Air 65-B80	1	AIR FAST INDONESIA
Lyc IGSO-540-A1E	Dornier 28D Skyservant	1	TAGB - TRANSPORTES AEREOS DA GUINE BISSAU
Lyc IGSO-540-A1H	Piaggio P166-DL2	4	IRAQI AIRWAYS
Lyc IGSO-540-B1A	Rockwell Grand Commander	1	ARKIA ISRAELI AIRLINES
Lyc IO-360	PA-28-140 Cherokee	10	PROVINCIAL AIRWAYS LTD
Lyc IO-360	PA-28-151/161 Cherokee	2	PROVINCIAL AIRWAYS LTD
Lyc IO-360	PA-28-180	1	ZAIREAN AIRLINES
Lyc IO-360-A1B6	Partenavia AP68TP Victor	1	AIR CAPE
Lyc IO-360-A1B6	Partenavia P68C	4	LAM - LINHAS AEREAS DE MOCAMBIQUE

POWER PLANT	AIRFRAME	NO. IN FLEET	FLEET NAME
Lyc IO-360-A1B6	Partenavia PN68 Observer	1	SUN-AIR OF SCANDINAVIA A/S
Lyc IO-360-C1E6	PA-34-200 Seneca	1	PROVINCIAL AIRWAYS LTD
Lyc IO-360-C1E6	PA-34-300 Seneca	1	FLANDERS AIRLINES NV
Lyc IO-540-A1D5	PA-23-250 Aztec B	2	PROVINCIAL AIRWAYS LTD
Lyc IO-540-C4B5	PA-23-250 Aztec C	1	AIR SAINT-PIERRE
Lyc IO-540-C4B5	PA-23-250 Aztec D	1	AIR FAST INDONESIA
Lyc IO-540-C4B5	PA-23-250 Aztec D	1	NORLANDAIR (ICELAND)
Lyc IO-540-C4B5	PA-23-250 Aztec D	1	SOLOMON AIRLINES
Lyc IO-540-C4B5	PA-23-250 Aztec D	3	TRANS ISLAND AIRWAYS LIMITED
Lyc IO-540-C4B5	PA-23-250 Aztec E	1	AIR RWANDA
Lyc IO-540-C4B5	PA-23-250 Aztec F	2	TURKS & CAICOS AIRWAYS LIMITED
Lyc IO-540-E1B5	Rockwell Shrike Commander	11	IRAN ASSEMAN AIRLINES
Lyc IO-540-E4C5	BN-2A Islander	2	SERVICIOS AEREOS ASTRO SA
Lyc IO-540-E4C5	BN-2A Islander	3	TRANS ISLAND AIRWAYS LIMITED
Lyc IO-540-E4C5	BN-2A Trislander	2	TAIWAN AIRLINES
Lyc IO-540-E4C5	BN-2A-26 Islander	3	TAIWAN AIRLINES
Lyc IO-540-E4C5	BN-2B-26 Islander	1	TAIWAN AIRLINES
Lyc IO-540-K1B5	BN-2A Islander	2	BALI AIR
Lyc IO-540-K1B5	BN-2A-3 Islander	2	IRAN ASSEMAN AIRLINES
Lyc IO-540-K1B5	BN-2B-20 Islander	5	CHINA SOUTHERN AIRLINES
Lyc IO-540-K1C5	BAe (DH) 114 Riley Heron	2	AIR FIJI
Lyc IO-540-K1C5	BAe (DH) 114 Riley Heron	1	SUNFLOWER AIRLINES LTD
Lyc IO-540-K1P5	BN-2B-20 Islander	1	AIR FIJI
Lyc IO-720-A1B	BE 65 Excalibur Queenaire	1	SUNFLOWER AIRLINES LTD
Lyc LF507-1F	AIAe RJ100 Avroliner	0	CROSSAIR
Lyc LF507-1F	AIAe RJ100 Avroliner	8	SAM - SOCIEDAD AERONAUTICA DE MEDELLIN
Lyc LF507-1F	AIAe RJ100 Avroliner	10	TURKISH AIRLINES, INC (TURK HAVA YOLLARI A.O)
Lyc LF507-1F	AIAe RJ70 Avroliner	4	AIR MALTA CO. LTD.
Lyc LF507-1F	AIAe RJ70 Avroliner	6	BUSINESS EXPRESS (THE DELTA CONNECTION)
Lyc LF507-1F	AIAe RJ85 Avroliner	4	CROSSAIR
Lyc LF507-1F	AIAe RJ85 Avroliner	7	LUFTHANSA CITYLINE
Lyc LF507-1F	AIAe RJ85 Avroliner	0	MALMO AVIATION
Lyc LF507-1F	AIAe RJ85 Avroliner	1	PELITA AIR SERVICE
Lyc O-320-D1D	GA 7 Cougar	1	ALPHA AIR
Lyc O-360-A4M	PA-28-181 Archer II	8	SAUDIA - SAUDI ARABIAN AIRLINES
Lyc O-540-E4B5	BN-2A Islander	1	ARKIA ISRAELI AIRLINES
Lyc O-540-E4B5	BN-2A Islander	3	MOUNT COOK AIRLINE
Lyc O-540-E4C5	BN-2 Islander	1	TRANS-JAMAICAN AIRLINES LTD
Lyc O-540-E4C5	BN-2A Islander	2	AIR ATLANTIQUE
Lyc O-540-E4C5	BN-2A Islander	3	AIR BVI LTD
Lyc O-540-E4C5	BN-2A Islander	2	AIR INTER GABON, SA.
Lyc O-540-E4C5	BN-2A Islander	1	AIR SEYCHELLES
Lyc O-540-E4C5	BN-2A Islander	3	FORMOSA AIRLINES
Lyc O-540-E4C5	BN-2A Islander	1	LIAT (1974) LTD.
Lyc O-540-E4C5	BN-2A Islander	2	SOLOMON AIRLINES
Lyc O-540-E4C5	BN-2A Islander	5	SUNFLOWER AIRLINES LTD
Lyc O-540-E4C5	BN-2A Islander	2	TURKS & CAICOS AIRWAYS LIMITED
Lyc O-540-E4C5	BN-2A Trislander	2	AERO COZUMEL, SA
Lyc O-540-E4C5	BN-2A Trislander	7	AIR SAINT THOMAS
Lyc O-540-E4C5	BN-2A Trislander	2	AIR TUNGARU CORP
Lyc O-540-E4C5	BN-2A Trislander	9	AURIGNY AIR SERVICES LTD
Lyc O-540-E4C5	BN-2A Trislander	3	BALI AIR
Lyc O-540-E4C5	BN-2A Trislander	2	TRANS-JAMAICAN AIRLINES LTD
Lyc O-540-E4C5	BN-2A-26 Islander	1	LOGANAIR LTD
Lyc O-540-E4C5	BN-2A-26 Islander	5	MAYA AIRWAYS LTD
Lyc O-540-E4C5	BN-2A-8 Islander	4	AIR MOOREA
Lyc O-540-E4C5	BN-2A-8 Islander	1	POLYNESIAN AIRLINES
Lyc O-540-E4C5	BN-2B-26 Islander	4	LOGANAIR LTD
Lyc T53-11A	Agusta-Bell 204B	1	AIR FAST INDONESIA
Lyc T53-11A	Bell 204B Iroquois	3	AIR FAST INDONESIA
Lyc TIO-540-A1A	PA-31-310 Navajo	1	AIR NORTH
Lyc TIO-540-A1A	PA-31-310 Navajo	7	PROVINCIAL AIRWAYS LTD
Lyc TIO-540-A1A	PA-31-310 Navajo C	1	HOLMSTROM AIR
Lyc TIO-540-A2C	PA-31-310 Navajo B	1	GAS AIR
Lyc TIO-540-F2BD	PA-31-325 Navajo C/R	1	GAS AIR
Lyc TIO-540-J2B	PA-31-350 T1020	3	BERING AIR, INC
Lyc TIO-540-J2BD	PA-31-350 Chieftain	1	AIR SAINT-PIERRE
Lyc TIO-540-J2BD	PA-31-350 Chieftain	3	AIRLINK AIRLINE (PTY) LTD
Lyc TIO-540-J2BD	PA-31-350 Chieftain	4	ARKIA ISRAELI AIRLINES
Lyc TIO-540-J2BD	PA-31-350 Chieftain	4	BERING AIR, INC
Lyc TIO-540-J2BD	PA-31-350 Chieftain	1	CALM AIR (CANADIAN PARTNER)
Lyc TIO-540-J2BD	PA-31-350 Chieftain	7	HAZELTON AIRLINES
Lyc TIO-540-J2BD	PA-31-350 Chieftain	3	IRAN ASSEMAN AIRLINES
Lyc TIO-540-J2BD	PA-31-350 Chieftain	2	MOUNT COOK AIRLINE
Lyc TIO-540-J2BD	PA-31-350 Chieftain	2	NORLANDAIR (ICELAND)
Lyc TIO-540-J2BD	PA-31-350 Chieftain	3	PROVINCIAL AIRWAYS LTD
Lyc TIO-540-J2BD	PA-31-350 Chieftain	1	SUNAIR (BOP AIR PTY LTD dba)
Lyc TIO-541-E1B4	BE Duke A60	1	KELOWNA FLIGHTCRAFT AIR CHARTER LTD
Lyc TSIO-540-A2C	PA-31-310 Navajo	1	AIRKENYA AVIATION LIMITED
Mot M-601D	Let 410UVP	2	FARNER AIR TRANSPORT HUNGARY
P&W JT12A-8	L1329 JetStar 6	2	PAN AVIATION
P&W JT12A-8	L1329 JetStar 8	1	SEAGREEN AIR TRANSPORT
P&W JT3D-1	B720B	2	MIDDLE EAST AIRLINES - AIRLIBAN
P&W JT3D-3B	B707-300	4	IRAN AIR THE AIRLINE OF THE ISLAMIC REPUBLIC OF IRAN
P&W JT3D-3B	B707-300	1	LLOYD AEREO BOLIVIANO SAM
P&W JT3D-3B	B707-300B	1	AFRICAN AIRLINES INTERNATIONAL LTD
P&W JT3D-3B	B707-300B	1	ECUATORIANA
P&W JT3D-3B	B707-300B	1	SOMALI AIRLINES
P&W JT3D-3B	B707-300B	1	SUDAN AIRWAYS
P&W JT3D-3B	B707-300B (F)	1	ETHIOPIAN AIRLINES CORPORATION
P&W JT3D-3B	B707-300BC	1	NATIONAL OVERSEAS AIRLINE
P&W JT3D-3B	B707-300C	1	APISA - AEROTRANSPORTES PERUANOS INTERNCIONALES SA
P&W JT3D-3B	B707-300C	3	CARIBBEAN INTERNATIONAL AIRWAYS
P&W JT3D-3B	B707-300C	1	DOMINICANA DE AVIACION
P&W JT3D-3B	B707-300C	1	IRAQI AIRWAYS
P&W JT3D-3B	B707-300C	3	LIBYAN ARAB AIRLINES
P&W JT3D-3B	B707-300C	2	NATIONAL OVERSEAS AIRLINE
P&W JT3D-3B	B707-300C	1	NIGERIA AIRWAYS
P&W JT3D-3B	B707-300C	1	OKADA AIR LTD

POWER PLANT	AIRFRAME	NO. IN FLEET	FLEET NAME
P&W JT3D-3B	B707-300C	2	PAKISTAN INTERNATIONAL AIRLINES (PIA)
P&W JT3D-3B	B707-300C	1	PAN AVIATION
P&W JT3D-3B	B707-300C	2	ROYAL AIR MAROC
P&W JT3D-3B	B707-300C	3	ROYAL JORDANIAN
P&W JT3D-3B	B707-300C	1	SAETA
P&W JT3D-3B	B707-300C	3	SEAGREEN AIR TRANSPORT
P&W JT3D-3B	B707-300C	2	SUDAN AIRWAYS
P&W JT3D-3B	B707-300C	1	TAAG, LINHAS AEREAS DE ANGOLA (ANGOLA AIRLINES)
P&W JT3D-3B	B707-300C	3	TAROM
P&W JT3D-3B	B707-300C	4	TRANSBRASIL S.A. LINHAS AEREAS
P&W JT3D-3B	B707-300C	1	VIETNAM AIRLINES
P&W JT3D-3B	B707-300C	2	YEMEN AIRLINES
P&W JT3D-3B	B720B	1	PAN AVIATION
P&W JT3D-3B	DC8-53F	1	ARCA COLOMBIA
P&W JT3D-3B	DC8-55	2	AFFRETAIR (PRIVATE) LTD
P&W JT3D-3B	DC8-55	1	KABO AIR
P&W JT3D-3B	DC8-55F	1	ARCA COLOMBIA
P&W JT3D-3B (Q)	B707-300B	3	AEROVIAS NACIONALES DE COLOMBIA S.A. AVIANCA
P&W JT3D-3B (Q)	B707-300B	2	ECUATORIANA
P&W JT3D-3B (Q)	B707-300C	2	AERONAVES DEL PERU, S.A.
P&W JT3D-3B (Q)	B707-300C	1	AIR RWANDA
P&W JT3D-3B (Q)	B707-300C	3	BUFFALO AIRWAYS
P&W JT3D-3B (Q)	B707-300C	1	CHALLENGE AIR CARGO
P&W JT3D-3B (Q)	B707-300C	4	DAS AIR CARGO (Dairo Air Services)
P&W JT3D-3B (Q)	B707-300C	2	FAST AIR CARRIER
P&W JT3D-3B (Q)	B707-300C	1	HEAVYLIFT CARGO AIRLINES
P&W JT3D-3B (Q)	B707-300C	1	IRAQI AIRWAYS
P&W JT3D-3B (Q)	B707-300C	2	KUWAIT AIRWAYS CORPORATION
P&W JT3D-3B (Q)	B707-300C	1	LADECO SA/LINEA AEREA DE COBRE/LADECO AIRLINES
P&W JT3D-3B (Q)	B707-300C	8	MIDDLE EAST AIRLINES - AIRLIBAN
P&W JT3D-3B (Q)	B707-300C	1	PAN AVIATION
P&W JT3D-3B (Q)	B707-300C	1	PRIMERAS LINEAS URUGUAYAS DE NAVEGACION AEREA - PLUNA
P&W JT3D-3B (Q)	B707-300C	2	SAUDIA - SAUDI ARABIAN AIRLINES
P&W JT3D-3B (Q)	B707-300C	2	TAAG, LINHAS AEREAS DE ANGOLA (ANGOLA AIRLINES)
P&W JT3D-3B (Q)	B707-300C	5	TAMPA SA
P&W JT3D-3B (Q)	B707-300C (F)	8	BURLINGTON AIR EXPRESS
P&W JT3D-3B (Q)	B707-300C (F)	1	ECUATORIANA
P&W JT3D-3B (Q)	B707-300C (F)	4	FLORIDA WEST AIRLINES INC
P&W JT3D-3B (Q)	B707-300C (F)	1	LAN-CHILE
P&W JT3D-3B (Q)	B707-300C (F)	1	SEAGREEN AIR TRANSPORT
P&W JT3D-3B (Q)	B707-300C (F)	6	TMA - TRANS MEDITERRANEAN AIRWAYS
P&W JT3D-3B (Q)	DC8-50	2	CARGO D'OR
P&W JT3D-3B (Q)	DC8-51F	2	KALITTA AMERICAN INTERNATIONAL AIRWAYS
P&W JT3D-3B (Q)	DC8-52	1	KALITTA AMERICAN INTERNATIONAL AIRWAYS
P&W JT3D-3B (Q)	DC8-54	1	BUFFALO AIRWAYS
P&W JT3D-3B (Q)	DC8-54	2	KALITTA AMERICAN INTERNATIONAL AIRWAYS
P&W JT3D-3B (Q)	DC8-54	2	ZANTOP INTERNATIONAL AIRLINES, INC
P&W JT3D-3B (Q)	DC8-54F	1	AERONAVES DEL PERU, S.A.
P&W JT3D-3B (Q)	DC8-54F	3	AFRICAN INTERNATIONAL AIRWAYS
P&W JT3D-3B (Q)	DC8-54F	1	AIR ZAIRE
P&W JT3D-3B (Q)	DC8-54F	1	EMERY WORLDWIDE
P&W JT3D-3B (Q)	DC8-55	1	FLASH AIRLINES
P&W JT3D-3B (Q)	DC8-55	6	KALITTA AMERICAN INTERNATIONAL AIRWAYS
P&W JT3D-3B (Q)	DC8-55	1	LINEAS AEREAS COSTARRICENSES, S.A. - LACSA
P&W JT3D-3B (Q)	DC8-55F	1	ACS OF CANADA - AIR CHARTER SYSTEMS
P&W JT3D-3B (Q)	DC8-61	1	AEROPERU
P&W JT3D-3B (Q)	DC8-61	10	AIRBORNE EXPRESS (ABX AIR, INC dba)
P&W JT3D-3B (Q)	DC8-61	3	BUFFALO AIRWAYS
P&W JT3D-3B (Q)	DC8-61F	1	AERONAVES DEL PERU, S.A.
P&W JT3D-3B (Q)	DC8-61F	2	AIR TRANSPORT INTERNATIONAL
P&W JT3D-3B (Q)	DC8-61F	5	KALITTA AMERICAN INTERNATIONAL AIRWAYS
P&W JT3D-3B (Q)	DC8-62	2	AEROPERU
P&W JT3D-3B (Q)	DC8-62	6	AIRBORNE EXPRESS (ABX AIR, INC dba)
P&W JT3D-3B (Q)	DC8-62	1	AOM FRENCH AIRLINES
P&W JT3D-3B (Q)	DC8-62	2	BUFFALO AIRWAYS
P&W JT3D-3B (Q)	DC8-62	3	RICH INTERNATIONAL AIRWAYS INC
P&W JT3D-3B (Q)	DC8-62CF	2	INTERNATIONAL CARGO EXPRESS (ICX dba)
P&W JT3D-3B (Q)	DC8-62F	2	AIR TRANSPORT INTERNATIONAL
P&W JT3D-3B (Q)	DC8-62F	1	ARROW AIR INC
P&W JT3D-3B (Q)	DC8-62F	7	EMERY WORLDWIDE
P&W JT3D-3B (Q)	DC8-62F	1	EVERGREEN INTERNATIONAL AIRLINES, INC
P&W JT3D-3B (Q)	DC8-63	1	BUFFALO AIRWAYS
P&W JT3D-3B (Q)	DC8-63	1	RICH INTERNATIONAL AIRWAYS INC
P&W JT3D-7	B707-300	4	CHINA SOUTHWEST AIRLINES
P&W JT3D-7	B707-300	1	EUROPEAN AIRLIFT
P&W JT3D-7	B707-300B	2	AIR ZIMBABWE
P&W JT3D-7	B707-300C	2	SUDAN AIRWAYS
P&W JT3D-7	B707-300C (F)	1	EGYPTAIR
P&W JT3D-7	DC8-62	1	AEROPERU
P&W JT3D-7	DC8-63	1	SURINAM AIRWAYS
P&W JT3D-7	DC8-63F	4	BURLINGTON AIR EXPRESS
P&W JT3D-7 (Q)	B707-300	2	AIR CHINA INTERNATIONAL
P&W JT3D-7 (Q)	B707-300B	1	AFRICAN AIRLINES INTERNATIONAL LTD
P&W JT3D-7 (Q)	B707-300C	1	SCIBE AIRLIFT
P&W JT3D-7 (Q)	DC8-62CF	1	AIR MARSHALL ISLANDS - AMI
P&W JT3D-7 (Q)	DC8-62F	9	ARROW AIR INC
P&W JT3D-7 (Q)	DC8-62F	2	CARGOSUR - COMPANIA DE EXPLOITATION DE AVIONES CARGUEROS SA
P&W JT3D-7 (Q)	DC8-63	2	AER TURAS TEORANTA
P&W JT3D-7 (Q)	DC8-63	1	AIR STARLINE AG
P&W JT3D-7 (Q)	DC8-63	13	AIRBORNE EXPRESS (ABX AIR, INC dba)
P&W JT3D-7 (Q)	DC8-63AF (F)	1	AIR MARSHALL ISLANDS - AMI
P&W JT3D-7 (Q)	DC8-63F	2	AIR TRANSPORT INTERNATIONAL
P&W JT3D-7 (Q)	DC8-63F	4	ARROW AIR INC
P&W JT3D-7 (Q)	DC8-63F	12	EMERY WORLDWIDE
P&W JT3D-7 (Q)	DC8-63F	2	KALITTA AMERICAN INTERNATIONAL AIRWAYS
P&W JT3D-7 (Q)	DC8-63F	1	SAUDIA - SAUDI ARABIAN AIRLINES
P&W JT4A-12	B707-300F	1	ZAIREAN AIRLINES
P&W JT4A-9	DC8-33F	1	APISA - AEROTRANSPORTES PERUANOS INTERNCIONALES SA
P&W JT4A-9	DC8-33F	2	TRANSAFRIK

POWER PLANT	AIRFRAME	NO. IN FLEET	FLEET NAME
P&W JT8-15	B737-200A	3	TAESA
P&W JT8D	B737-200	2	LAP - LINEAS AEREAS PARAGUAYAS
P&W JT8D	DC-9	1	SATENA
P&W JT8D-11	DC9-21	2	SAS - SCANDINAVIAN AIRLINES SYSTEM
P&W JT8D-11	DC9-32	1	BRITISH MIDLAND AIRWAYS LTD
P&W JT8D-11	DC9-41	10	AIRBORNE EXPRESS (ABX AIR, INC dba)
P&W JT8D-11	DC9-41	12	NORTHWEST AIRLINES
P&W JT8D-11	DC9-41	25	SAS - SCANDINAVIAN AIRLINES SYSTEM
P&W JT8D-15	B727-200	3	AVIOGENEX
P&W JT8D-15	B727-200	1	FAUCETT PERU
P&W JT8D-15	B727-200	6	KIWI INTERNATIONAL AIR LINES, INC
P&W JT8D-15	B727-200	5	MIAMI AIR INTERNATIONAL INC
P&W JT8D-15	B727-200	1	POLYNESIAN AIRLINES
P&W JT8D-15	B727-200	4	TUR EUROPEAN AIRWAYS
P&W JT8D-15	B727-200 Advanced	1	AEROVIAS NACIONALES DE COLOMBIA S.A. AVIANCA
P&W JT8D-15	B727-200 Advanced	9	AIR ALGERIE
P&W JT8D-15	B727-200 Advanced	4	AIR CHARTER
P&W JT8D-15	B727-200 Advanced	7	AIR FRANCE
P&W JT8D-15	B727-200 Advanced	5	AIR JAMAICA
P&W JT8D-15	B727-200 Advanced	22	AMERICAN AIRLINES
P&W JT8D-15	B727-200 Advanced	5	ANSETT AUSTRALIA
P&W JT8D-15	B727-200 Advanced	1	BRADLEY AIR SERVICES LIMITED (FIRST AIR)
P&W JT8D-15	B727-200 Advanced	25	CONTINENTAL AIRLINES
P&W JT8D-15	B727-200 Advanced	7	CONTINENTAL MICRONESIA, INC
P&W JT8D-15	B727-200 Advanced	2	CYPRUS TURKISH AIRLINES / KIBRIS TURK HAVA YOLLARI
P&W JT8D-15	B727-200 Advanced	133	DELTA AIR LINES, INC
P&W JT8D-15	B727-200 Advanced	5	IRAN AIR THE AIRLINE OF THE ISLAMIC REPUBLIC OF IRAN
P&W JT8D-15	B727-200 Advanced	0	IRAN ASSEMAN AIRLINES
P&W JT8D-15	B727-200 Advanced	5	ISTANBUL AIRLINES - ISTAIR
P&W JT8D-15	B727-200 Advanced	8	LIBYAN ARAB AIRLINES
P&W JT8D-15	B727-200 Advanced	25	NORTHWEST AIRLINES
P&W JT8D-15	B727-200 Advanced	3	OLYMPIC AIRWAYS S.A.
P&W JT8D-15	B727-200 Advanced	6	ROYAL AIR MAROC
P&W JT8D-15	B727-200 Advanced	2	SUN COUNTRY AIRLINES
P&W JT8D-15	B727-200 Advanced	1	TAN - TRANSPORTES AEREOS NACIONALES, SA
P&W JT8D-15	B727-200 Advanced	4	TURKISH AIRLINES, INC (TURK HAVA YOLLARI A.O)
P&W JT8D-15	B727-200 Advanced	75	UNITED AIRLINES, INC
P&W JT8D-15	B727-200 Advanced (F)	6	UNITED PARCEL SERVICE COMPANY UPS
P&W JT8D-15	B727-200 Advances	4	OKADA AIR LTD
P&W JT8D-15	B727-200F	3	TURKISH AIRLINES, INC (TURK HAVA YOLLARI A.O)
P&W JT8D-15	B727-200F Advanced	1	ANSETT AUSTRALIA
P&W JT8D-15	B727-200F Advanced	56	FEDERAL EXPRESS CORPORATION
P&W JT8D-15	B737-200	2	AIR CHARTER
P&W JT8D-15	B737-200	19	AIR FRANCE
P&W JT8D-15	B737-200	1	AVIATECA
P&W JT8D-15	B737-200	4	AVIOGENEX
P&W JT8D-15	B737-200	37	BRITISH AIRWAYS
P&W JT8D-15	B737-200	4	GB AIRWAYS
P&W JT8D-15	B737-200	3	IRAN AIR THE AIRLINE OF THE ISLAMIC REPUBLIC OF IRAN
P&W JT8D-15	B737-200	1	LITHUANIAN AIRLINES
P&W JT8D-15	B737-200	11	OLYMPIC AIRWAYS S.A.
P&W JT8D-15	B737-200	1	SAHARA INDIA AIRLINES
P&W JT8D-15	B737-200	20	SAUDIA - SAUDI ARABIAN AIRLINES
P&W JT8D-15	B737-200	3	THAI AIRWAYS INTERNATIONAL
P&W JT8D-15	B737-200	3	TRANSAVIA AIRLINES CV
P&W JT8D-15	B737-200 (F)	2	SOUTH AFRICAN AIRWAYS (SAA)
P&W JT8D-15	B737-200 Advanced	10	AIR ALGERIE
P&W JT8D-15	B737-200 Advanced	1	AIR MADAGASCAR
P&W JT8D-15	B737-200 Advanced	1	AIR NAMIBIA
P&W JT8D-15	B737-200 Advanced	13	AIR NEW ZEALAND
P&W JT8D-15	B737-200 Advanced	2	AIR ZAIRE
P&W JT8D-15	B737-200 Advanced	3	AIR ZIMBABWE
P&W JT8D-15	B737-200 Advanced	1	ALOHA AIRLINES
P&W JT8D-15	B737-200 Advanced	15	AMERICA WEST AIRLINES, INC
P&W JT8D-15	B737-200 Advanced	1	CAMEROON AIRLINES
P&W JT8D-15	B737-200 Advanced	1	CANADIAN - CANADIEN
P&W JT8D-15	B737-200 Advanced	8	COMPANIA PANAMENA DE AVIACION, SA (COPA)
P&W JT8D-15	B737-200 Advanced	5	CROATIA AIRLINES
P&W JT8D-15	B737-200 Advanced	10	DELTA AIR LINES, INC
P&W JT8D-15	B737-200 Advanced	3	EAST WEST AIRLINES
P&W JT8D-15	B737-200 Advanced	4	LADECO SA/LINEA AEREA DE COBRE/LADECO AIRLINES
P&W JT8D-15	B737-200 Advanced	1	LAN-CHILE
P&W JT8D-15	B737-200 Advanced	35	LUFTHANSA GERMAN AIRLINES - DEUTSCHE LUFTHANSA AG
P&W JT8D-15	B737-200 Advanced	8	NIGERIA AIRWAYS
P&W JT8D-15	B737-200 Advanced	4	ROYAL AIR MAROC
P&W JT8D-15	B737-200 Advanced	6	RYANAIR LTD
P&W JT8D-15	B737-200 Advanced	8	SABENA BELGIAN WORLD AIRLINES
P&W JT8D-15	B737-200 Advanced	1	SAHSA - SERVICIO AEREO DE HONDURAS, SA
P&W JT8D-15	B737-200 Advanced	3	SOBELAIR - SOCIETE BELGE DE TRANSPORTES AERIENS
P&W JT8D-15	B737-200 Advanced	11	SOUTH AFRICAN AIRWAYS (SAA)
P&W JT8D-15	B737-200 Advanced	2	TRANSAERO AIRLINES
P&W JT8D-15	B737-200 Advanced	2	TRANSAVIA AIRLINES CV
P&W JT8D-15	B737-200C	1	ALOHA AIRLINES
P&W JT8D-15	B737-200C Advanced	1	AIR ALGERIE
P&W JT8D-15	B737-200C Advanced	2	CAMEROON AIRLINES
P&W JT8D-15	B737-200C Advanced	2	IRAQI AIRWAYS
P&W JT8D-15	B737-200C Advanced	1	KAMPUCHEA AIRLINES
P&W JT8D-15	B737-200C Advanced	1	LINA CONGO
P&W JT8D-15	B737-200C Advanced	2	ROYAL AIR MAROC
P&W JT8D-15	B737-200C Advanced	4	SABENA BELGIAN WORLD AIRLINES
P&W JT8D-15	B737-200C Advanced	1	YEMEN AIRLINES
P&W JT8D-15	B737-200F	2	GERMAN CARGO SERVICES
P&W JT8D-15	DC9-31	7	NORTHWEST AIRLINES
P&W JT8D-15	DC9-32	10	NORTHWEST AIRLINES
P&W JT8D-15	DC9-32	5	SUN AIR (BOP AIR PTY LTD dba)
P&W JT8D-15	DC9-34	3	TRANS WORLD AIRLINES, INC
P&W JT8D-15	DC9-41	6	AIRBORNE EXPRESS (ABX AIR, INC dba)
P&W JT8D-15	DC9-41	10	JAPAN AIR SYSTEM CO LTD (NIHON AIR SYSTEM)
P&W JT8D-15	DC9-41	3	TRANS WORLD AIRLINES, INC

POWER PLANT	AIRFRAME	NO. IN FLEET	FLEET NAME
P&W JT8D-15/15A	B727-200 Advanced	5	USAIR, INC
P&W JT8D-15/15A	B737-200 Advanced	47	USAIR, INC
P&W JT8D-15A	B727-200	2	ARROW AIR INC
P&W JT8D-15A	B727-200 Advanced	0	KIWI INTERNATIONAL AIRLINES INC
P&W JT8D-15A	B727-200F	3	AMERIJET INTERNATIONAL
P&W JT8D-15A	B727-200F	1	ITAPEMIRIN TRANSPORTES AEREOS SA
P&W JT8D-15A	B727-200F Advanced	2	ARROW AIR INC
P&W JT8D-15A	B727-200F Advanced	6	FEDERAL EXPRESS CORPORATION
P&W JT8D-15A	B737-200 Advanced	46	DELTA AIR LINES, INC
P&W JT8D-15A	B737-200 Advanced	2	EAST WEST AIRLINES
P&W JT8D-15A	B737-200 Advanced	3	TRANSAERO AIRLINES
P&W JT8D-15A	B737-200C	4	TAT EUROPEAN AIRLINES
P&W JT8D-17	B727-200	2	ACES
P&W JT8D-17	B727-200	9	AMERICAN TRANS AIR, INC
P&W JT8D-17	B727-200	1	SERVIVENSA
P&W JT8D-17	B727-200	3	SYRIAN ARAB AIRLINES
P&W JT8D-17	B727-200 Advanced	1	AEROLINEAS ARGENTINAS
P&W JT8D-17	B727-200 Advanced	7	AEROVIAS NACIONALES DE COLOMBIA S.A. AVIANCA
P&W JT8D-17	B727-200 Advanced	4	AVENSA
P&W JT8D-17	B727-200 Advanced	2	EMIRATES
P&W JT8D-17	B727-200 Advanced	6	IRAQI AIRWAYS
P&W JT8D-17	B727-200 Advanced	15	NORTHWEST AIRLINES
P&W JT8D-17	B727-200 Advanced	3	SUN COUNTRY AIRLINES
P&W JT8D-17	B727-200 Advanced	1	SUN COUNTRY AIRLINES
P&W JT8D-17	B727-200 Advanced	1	TAME - TRANSPORTES AEREOS MILITARES ECUATORIANOS, CA
P&W JT8D-17	B727-200 Advanced	13	VASP - VIACAO AEREA SAO PAULO S.A.
P&W JT8D-17	B727-200 Advanced	5	YEMEN AIRLINES
P&W JT8D-17	B727-200 Advanced (F)	2	UNITED PARCEL SERVICE COMPANY UPS
P&W JT8D-17	B727-200A	2	ROYAL JORDANIAN
P&W JT8D-17	B727-200F Advanced	5	FEDERAL EXPRESS CORPORATION
P&W JT8D-17	B737-200	1	AIR GABON
P&W JT8D-17	B737-200	1	FAUCETT PERU
P&W JT8D-17	B737-200	4	MALEV - HUNGARIAN AIRLINES
P&W JT8D-17	B737-200	1	SIERRA PACIFIC AIRLINES, INC
P&W JT8D-17	B737-200 Advanced	3	AIR ALGERIE
P&W JT8D-17	B737-200 Advanced	9	AIR NIPPON
P&W JT8D-17	B737-200 Advanced	2	AIR SINAI
P&W JT8D-17	B737-200 Advanced	9	ALL NIPPON AIRWAYS
P&W JT8D-17	B737-200 Advanced	3	ARKIA ISRAELI AIRLINES
P&W JT8D-17	B737-200 Advanced	14	CANADIAN - CANADIEN
P&W JT8D-17	B737-200 Advanced	2	CAYMAN AIRWAYS LTD
P&W JT8D-17	B737-200 Advanced	7	CHINA SOUTHERN AIRLINES
P&W JT8D-17	B737-200 Advanced	2	DELTA AIR LINES, INC
P&W JT8D-17	B737-200 Advanced	3	EGYPTAIR
P&W JT8D-17	B737-200 Advanced	2	EL AL ISRAEL AIRLINES, LTD
P&W JT8D-17	B737-200 Advanced	14	INDIAN AIRLINES LTD
P&W JT8D-17	B737-200 Advanced	9	JAPAN TRANSOCEAN AIR
P&W JT8D-17	B737-200 Advanced	2	KENYA AIRWAYS LTD
P&W JT8D-17	B737-200 Advanced	1	LAN-CHILE
P&W JT8D-17	B737-200 Advanced	3	LINEAS AEREAS COSTARRICENSES, S.A. - LACSA
P&W JT8D-17	B737-200 Advanced	2	TAAG, LINHAS AEREAS DE ANGOLA (ANGOLA AIRLINES)
P&W JT8D-17	B737-200 Advanced	2	TAAG, LINHAS AEREAS DE ANGOLA (ANGOLA AIRLINES)
P&W JT8D-17	B737-200 Advanced	1	TACA INTERNATIONAL AIRLINES SA
P&W JT8D-17	B737-200 Advanced	8	TAP AIR PORTUGAL
P&W JT8D-17	B737-200 Advanced	1	TUNIS AIR - SOCIETE TUNISIENNE D EL'AIR
P&W JT8D-17	B737-200 Advanced	17	UNITED AIRLINES, INC
P&W JT8D-17	B737-200 Advanced	17	VARIG - VIACAO AEREA RIO-GRANDENSE S.A.
P&W JT8D-17	B737-200 Advanced (ER)	7	CANADIAN - CANADIEN
P&W JT8D-17	B737-200 Heavy Advanced	2	BRAATHENS S.A.F.E. A/S
P&W JT8D-17	B737-200A	1	RYANAIR LTD
P&W JT8D-17	B737-200C Advanced	1	AIR GUINEE
P&W JT8D-17	B737-200C Advanced	2	AIR TANZANIA
P&W JT8D-17	B737-200C Advanced	8	ALASKA AIRLINES, INC
P&W JT8D-17	B737-200C Advanced	1	ALOHA AIRLINES
P&W JT8D-17	B737-200C Advanced	2	CANADIAN - CANADIEN
P&W JT8D-17	B737-200C Advanced	5	MARKAIR, INC
P&W JT8D-17	B737-200C Advanced	1	TAAG, LINHAS AEREAS DE ANGOLA (ANGOLA AIRLINES)
P&W JT8D-17	B737-200C Advanced	1	TACA INTERNATIONAL AIRLINES SA
P&W JT8D-17	B737-200C Advanced	1	TUNIS AIR - SOCIETE TUNISIENNE D EL'AIR
P&W JT8D-17	B737-200C Advanced (F)	2	VASP - VIACAO AEREA SAO PAULO S.A.
P&W JT8D-17	DC9-30	2	BINTER
P&W JT8D-17	DC9-31	3	AEROMEXICO
P&W JT8D-17	DC9-31	3	AEROPOSTAL
P&W JT8D-17	DC9-32	15	AEROMEXICO
P&W JT8D-17	DC9-32	2	SERVIVENSA
P&W JT8D-17	DC9-34	3	AVIACO
P&W JT8D-17	DC9-34	3	IBERIA - LINEAS AEREAS DE ESPANA, S.A.
P&W JT8D-17	DC9-34CF	1	AEROPOSTAL
P&W JT8D-17	DC9-34F	2	AVIACO
P&W JT8D-17	DC9-41	5	FINNAIR OY
P&W JT8D-17	DC9-50	1	GHANA AIRWAYS CORPORATION
P&W JT8D-17	DC9-51	6	AEROPOSTAL
P&W JT8D-17	DC9-51	4	AVENSA
P&W JT8D-17	DC9-51	2	EUROFLY SPA
P&W JT8D-17	DC9-51	12	FINNAIR OY
P&W JT8D-17	DC9-51	13	HAWAIIAN AIRLINES
P&W JT8D-17	DC9-51	6	MERIDIANA SPA
P&W JT8D-17	DC9-51	34	NORTHWEST AIRLINES
P&W JT8D-17	DC9-51	12	TRANS WORLD AIRLINES, INC
P&W JT8D-17A	B727-200	10	AMERICAN TRANS AIR, INC
P&W JT8D-17A	B727-200	1	SAM - SOCIEDAD AERONAUTICA DE MEDELLIN
P&W JT8D-17A	B727-200F Advanced	5	FEDERAL EXPRESS CORPORATION
P&W JT8D-17A	B737-200	1	ETHIOPIAN AIRLINES CORPORATION
P&W JT8D-17A	B737-200 Advanced	4	AIR CHINA INTERNATIONAL
P&W JT8D-17A	B737-200 Advanced	2	LAN-CHILE
P&W JT8D-17A	B737-200 Advanced	1	UGANDA AIRLINES
P&W JT8D-17A	B737-200 Advanced	3	XIAMEN AIRLINES
P&W JT8D-17A	B737-200C Advanced	2	YEMEN AIRLINES
P&W JT8D-17A	DC9-51	3	AEROPOSTAL

POWER PLANT	AIRFRAME	NO. IN FLEET	FLEET NAME
P&W JT8D-17R	B727-200 Advanced	2	AEROPERU
P&W JT8D-17R	B727-200 Advanced	3	LLOYD AEREO BOLIVIANO SAM
P&W JT8D-17R	B727-200 Advanced	26	MEXICANA
P&W JT8D-17R	B727-200 Advanced	2	SAFAIR (PTY) LTD
P&W JT8D-17R	B727-200 Advanced	3	SYRIAN ARAB AIRLINES
P&W JT8D-17R	B727-200F	1	ITAPEMIRIN TRANSPORTES AEREOS SA
P&W JT8D-17R	B727-200F Advanced	7	FEDERAL EXPRESS CORPORATION
P&W JT8D-1A	B727-200	2	CARNIVAL AIR LINES INC
P&W JT8D-209	MD-81	2	AUSTRAL
P&W JT8D-209	MD-81	7	AUSTRIAN AIRLINES, OSTERREICHISCHE LUFTVERKEHRS AG
P&W JT8D-209	MD-81	8	JAPAN AIR SYSTEM CO LTD (NIHON AIR SYSTEM)
P&W JT8D-209	MD-81	17	SWISSAIR (SOCIETE ANONYME SUISSE POUR LA NAVIGATION AERIENNE)
P&W JT8D-217	MD-80	1	ADRIA AIRWAYS
P&W JT8D-217	MD-81	5	CONTINENTAL AIRLINES
P&W JT8D-217	MD-81	12	JAPAN AIR SYSTEM CO LTD (NIHON AIR SYSTEM)
P&W JT8D-217	MD-81	31	SAS - SCANDINAVIAN AIRLINES SYSTEM
P&W JT8D-217	MD-81	19	USAIR, INC
P&W JT8D-217	MD-82	10	AEROMEXICO
P&W JT8D-217	MD-82	4	ALASKA AIRLINES, INC
P&W JT8D-217	MD-82	4	ALASKA AIRLINES, INC
P&W JT8D-217	MD-82	2	ALM ANTILLEAN AIRLINES
P&W JT8D-217	MD-82	1	ALM ANTILLEAN AIRLINES
P&W JT8D-217	MD-82	234	AMERICAN AIRLINES
P&W JT8D-217	MD-82	6	AUSTRIAN AIRLINES, OSTERREICHISCHE LUFTVERKEHRS AG
P&W JT8D-217	MD-82	2	BALAIRCTA
P&W JT8D-217	MD-82	2	CHINA EASTERN AIRLINES
P&W JT8D-217	MD-82	18	CONTINENTAL AIRLINES
P&W JT8D-217	MD-82	8	MERIDIANA SPA
P&W JT8D-217	MD-82	8	NORTHWEST AIRLINES
P&W JT8D-217	MD-82	1	OASIS INTERNATIONAL AIRLINES
P&W JT8D-217	MD-82	29	TRANS WORLD AIRLINES, INC
P&W JT8D-217	MD-82	12	USAIR, INC
P&W JT8D-217	MD-82	2	VENUS AIRLINES SA
P&W JT8D-217	MD-82 / MD/SAIC MD-82	11	CHINA EASTERN AIRLINES
P&W JT8D-217	MD-82 / MD/SAIC MD-82	24	CHINA NORTHERN AIRLINES
P&W JT8D-217	MD-83	1	AEROLINEAS ARGENTINAS
P&W JT8D-217	MD-83 / SAIC/MD-83	16	TRANS WORLD AIRLINES, INC
P&W JT8D-217	MD-87	4	BALAIRCTA
P&W JT8D-217	MD-87	8	JAPAN AIR SYSTEM CO LTD (NIHON AIR SYSTEM)
P&W JT8D-217	MD-87	16	SAS - SCANDINAVIAN AIRLINES SYSTEM
P&W JT8D-217	MD-87	2	TRANSWEDE
P&W JT8D-217	MD-87SR	3	AUSTRIAN AIRLINES, OSTERREICHISCHE LUFTVERKEHRS AG
P&W JT8D-217	MD-88	13	AVIACO
P&W JT8D-217A	MD-82	39	ALITALIA
P&W JT8D-217A	MD-82	36	CONTINENTAL AIRLINES
P&W JT8D-217C	MD-82	36	ALITALIA
P&W JT8D-217C	MD-82	5	FAR EASTERN AIR TRANSPORT
P&W JT8D-217C	MD-82	9	KOREAN AIR
P&W JT8D-217C	MD-82	2	SPANAIR
P&W JT8D-217C	MD-87	24	IBERIA - LINEAS AEREAS DE ESPANA, S.A.
P&W JT8D-217C/-17	B727-200RE	2	SUN COUNTRY AIRLINES
P&W JT8D-217C/-17A	B727-200F Advanced (RE)	11	FEDERAL EXPRESS CORPORATION
P&W JT8D-219	MD-82	15	AERO LLOYD
P&W JT8D-219	MD-82	2	ALASKA AIRLINES, INC
P&W JT8D-219	MD-82	4	CONTINENTAL AIRLINES
P&W JT8D-219	MD-82	9	FINNAIR OY
P&W JT8D-219	MD-82	4	RENO AIR
P&W JT8D-219	MD-82	15	SAS - SCANDINAVIAN AIRLINES SYSTEM
P&W JT8D-219	MD-83	3	AERO LLOYD
P&W JT8D-219	MD-83	4	AEROCANCUN
P&W JT8D-219	MD-83	3	AEROMEXICO
P&W JT8D-219	MD-83	3	AEROPOSTAL
P&W JT8D-219	MD-83	10	AEROVIAS NACIONALES DE COLOMBIA S.A. AVIANCA
P&W JT8D-219	MD-83	1	AIR ARUBA
P&W JT8D-219	MD-83	6	AIR LIBERTE
P&W JT8D-219	MD-83	7	AIRTOURS INTERNATIONAL AIRWAYS LTD
P&W JT8D-219	MD-83	28	ALASKA AIRLINES, INC
P&W JT8D-219	MD-83	26	AMERICAN AIRLINES
P&W JT8D-219	MD-83	7	AOM FRENCH AIRLINES
P&W JT8D-219	MD-83	2	AUSTRIAN AIRLINES, OSTERREICHISCHE LUFTVERKEHRS AG
P&W JT8D-219	MD-83	2	BALAIRCTA
P&W JT8D-219	MD-83	5	CENTENNIAL AIRLINES
P&W JT8D-219	MD-83	2	CONTINENTAL AIRLINES
P&W JT8D-219	MD-83	1	EUROFLY SPA
P&W JT8D-219	MD-83	1	FAR EASTERN AIR TRANSPORT
P&W JT8D-219	MD-83	5	FINNAIR OY
P&W JT8D-219	MD-83	1	NORTH AMERICAN AIRLINES
P&W JT8D-219	MD-83	5	OASIS INTERNATIONAL AIRLINES
P&W JT8D-219	MD-83	13	RENO AIR
P&W JT8D-219	MD-83	12	SPANAIR
P&W JT8D-219	MD-83	9	TRANSWEDE
P&W JT8D-219	MD-83	9	TRINIDAD & TOBAGO (BWIA INTERNATIONAL) AIRWAYS CORPORATION
P&W JT8D-219	MD-83	2	TUR EUROPEAN AIRWAYS
P&W JT8D-219	MD-83	1	VENUS AIRLINES SA
P&W JT8D-219	MD-83	2	ZAS AIRLINES OF EGYPT
P&W JT8D-219	MD-87	2	AERO LLOYD
P&W JT8D-219	MD-87	2	AEROMEXICO
P&W JT8D-219	MD-87	3	FINNAIR OY
P&W JT8D-219	MD-87	1	ZAS AIRLINES OF EGYPT
P&W JT8D-219	MD-87ER	2	AUSTRIAN AIRLINES, OSTERREICHISCHE LUFTVERKEHRS AG
P&W JT8D-219	MD-88	4	AEROLINEAS ARGENTINAS
P&W JT8D-219	MD-88	10	AEROMEXICO
P&W JT8D-219	MD-88	2	AIR ARUBA
P&W JT8D-219	MD-88	117	DELTA AIR LINES, INC
P&W JT8D-219	MD-88	2	MIDWEST EXPRESS
P&W JT8D-219B	MD-87	2	GREAT AMERICAN AIRWAYS
P&W JT8D-7	AS (Sud) Caravelle 10B3	2	SYRIAN ARAB AIRLINES
P&W JT8D-7	B727-100	5	AEROVIAS NACIONALES DE COLOMBIA S.A. AVIANCA
P&W JT8D-7	B727-100	2	BALTIC INTERNATIONAL AIRLINES

POWER PLANT	AIRFRAME	NO. IN FLEET	FLEET NAME
P&W JT8D-7	B727-100	5	EXPRESS ONE
P&W JT8D-7	B727-100	1	SAETA
P&W JT8D-7	B727-100	2	SCIBE AIRLIFT
P&W JT8D-7	B727-100	1	TAME - TRANSPORTES AEREOS MILITARES ECUATORIANOS, CA
P&W JT8D-7	B727-100	8	TRANS WORLD AIRLINES, INC
P&W JT8D-7	B727-100 (F)	4	DHL AIRWAYS
P&W JT8D-7	B727-100 (F)	6	EXPRESS ONE
P&W JT8D-7	B727-100C	1	ARIANA AFGHAN AIRLINES
P&W JT8D-7	B727-100C	1	EVERGREEN INTERNATIONAL AIRLINES, INC
P&W JT8D-7	B727-100C	1	EXPRESS ONE
P&W JT8D-7	B727-100C	2	REEVE ALEUTIAN AIRWAYS, INC
P&W JT8D-7	B727-100C (F)	3	DHL AIRWAYS
P&W JT8D-7	B727-100C (F)	1	EVERGREEN INTERNATIONAL AIRLINES, INC
P&W JT8D-7	B727-100F	6	EUROPEAN AIR TRANSPORT N.V./S.A.
P&W JT8D-7	B727-200	3	ARIANA AFGHAN AIRLINES
P&W JT8D-7	B727-200	6	DHL AIRWAYS
P&W JT8D-7	B727-200	2	KEY AIRLINES INC
P&W JT8D-7	B727-200F	6	EUROPEAN AIR TRANSPORT N.V./S.A.
P&W JT8D-7	B737-200	1	TACA INTERNATIONAL AIRLINES SA
P&W JT8D-7	B737-200	1	TRANSAVIA AIRLINES CV
P&W JT8D-7	B737-200	6	VASP - VIACAO AEREA SAO PAULO S.A.
P&W JT8D-7	B737-200C Advanced	2	SUDAN AIRWAYS
P&W JT8D-7	DC-9	2	FLORIDA WEST AIRLINES INC
P&W JT8D-7	DC9-10	7	TRANS WORLD AIRLINES, INC
P&W JT8D-7	DC9-31	1	NORTHWEST AIRLINES
P&W JT8D-7	DC9-32	4	IBERIA - LINEAS AEREAS DE ESPANA, S.A.
P&W JT8D-7	DC9-32F	2	EVERGREEN INTERNATIONAL AIRLINES, INC
P&W JT8D-7A	B727-100	1	AEROVIAS NACIONALES DE COLOMBIA S.A. AVIANCA
P&W JT8D-7A	B727-100	1	EVERGREEN INTERNATIONAL AIRLINES, INC
P&W JT8D-7A	B727-100	1	LINEAS AEREAS COSTARRICENSES, S.A. - LACSA
P&W JT8D-7A	B727-100	1	SAHSA - SERVICIO AEREO DE HONDURAS, SA
P&W JT8D-7A	B727-100	1	SAM - SOCIEDAD AERONAUTICA DE MEDELLIN
P&W JT8D-7A	B727-100	1	ZAIREAN AIRLINES
P&W JT8D-7A	B727-100C	1	AEROPERU
P&W JT8D-7A	B727-100C	1	EVERGREEN INTERNATIONAL AIRLINES, INC
P&W JT8D-7A	B727-100C (F)	1	FEDERAL EXPRESS CORPORATION
P&W JT8D-7A	B737-100	13	CONTINENTAL AIRLINES
P&W JT8D-7A	B737-100	2	FAR EASTERN AIR TRANSPORT
P&W JT8D-7A	B737-200	1	AIR FAST INDONESIA
P&W JT8D-7A	B737-200	1	AIRFAST INDONESIA
P&W JT8D-7A	B737-200	1	FAR EASTERN AIR TRANSPORT
P&W JT8D-7A	B747-100SR (F)	1	EVERGREEN INTERNATIONAL AIRLINES, INC
P&W JT8D-7A	B747-100SR (F)	1	EVERGREEN INTERNATIONAL AIRLINES, INC
P&W JT8D-7A	DC9-14	1	BRITISH MIDLAND AIRWAYS LTD
P&W JT8D-7A	DC9-14	1	NORTHWEST AIRLINES
P&W JT8D-7A	DC9-15	1	BRITISH MIDLAND AIRWAYS LTD
P&W JT8D-7A	DC9-15F	1	EVERGREEN INTERNATIONAL AIRLINES, INC
P&W JT8D-7A	DC9-15F	1	KALITTA AMERICAN INTERNATIONAL AIRWAYS
P&W JT8D-7A	DC9-30	35	AIR CANADA
P&W JT8D-7A/7B	DC9-15	13	INTERCONTINENTAL DE AVIACION
P&W JT8D-7B	AS (Sud) Caravelle 10B1R	3	ISTANBUL AIRLINES - ISTAIR
P&W JT8D-7B	AS (Sud) Caravelle 11R	1	AEROSUCRE COLOMBIA, S.A.
P&W JT8D-7B	B727-100	4	ACES
P&W JT8D-7B	B727-100	2	AEROPERU
P&W JT8D-7B	B727-100	5	AEROVIAS NACIONALES DE COLOMBIA S.A. AVIANCA
P&W JT8D-7B	B727-100	4	AMERICANA DE AVIACION
P&W JT8D-7B	B727-100	3	AVENSA
P&W JT8D-7B	B727-100	8	EXPRESS ONE
P&W JT8D-7B	B727-100	1	FAUCETT PERU
P&W JT8D-7B	B727-100	2	IRAN AIR THE AIRLINE OF THE ISLAMIC REPUBLIC OF IRAN
P&W JT8D-7B	B727-100	1	KABO AIR
P&W JT8D-7B	B727-100 (F)	4	AMERIJET INTERNATIONAL
P&W JT8D-7B	B727-100 (F)	3	KALITTA AMERICAN INTERNATIONAL AIRWAYS
P&W JT8D-7B	B727-100C	2	AMERIJET INTERNATIONAL
P&W JT8D-7B	B727-100C	2	BRADLEY AIR SERVICES LIMITED (FIRST AIR)
P&W JT8D-7B	B727-100C	1	BRADLEY AIR SERVICES LIMITED (FIRST AIR)
P&W JT8D-7B	B727-100C	1	BURLINGTON AIR EXPRESS
P&W JT8D-7B	B727-100C	1	BURLINGTON AIR EXPRESS
P&W JT8D-7B	B727-100C	1	DOMINICANA DE AVIACION
P&W JT8D-7B	B727-100C	19	EMERY WORLDWIDE
P&W JT8D-7B	B727-100C	1	EVERGREEN INTERNATIONAL AIRLINES, INC
P&W JT8D-7B	B727-100C	10	KELOWNA FLIGHTCRAFT AIR CHARTER LTD
P&W JT8D-7B	B727-100C	6	TAESA
P&W JT8D-7B	B727-100C	1	VARIG - VIACAO AEREA RIO-GRANDENSE S.A.
P&W JT8D-7B	B727-100C (F)	3	DHL AIRWAYS
P&W JT8D-7B	B727-100C (F)	2	EVERGREEN INTERNATIONAL AIRLINES, INC
P&W JT8D-7B	B727-100C (F)	17	UNITED PARCEL SERVICE COMPANY UPS
P&W JT8D-7B	B727-100F	1	ARROW AIR INC
P&W JT8D-7B	B727-100F	6	EMERY WORLDWIDE
P&W JT8D-7B	B727-100F	67	FEDERAL EXPRESS CORPORATION
P&W JT8D-7B	B727-100F	3	ITAPEMIRIN TRANSPORTES AEREOS SA
P&W JT8D-7B	B727-200	2	AIR CHARTER
P&W JT8D-7B	B727-200	7	FLORIDA WEST AIRLINES INC
P&W JT8D-7B	B727-200	2	ISTANBUL AIRLINES - ISTAIR
P&W JT8D-7B	B727-200	2	KABO AIR
P&W JT8D-7B	B727-200	4	KIWI INTERNATIONAL AIRLINES INC
P&W JT8D-7B	B727-200	15	NORTHWEST AIRLINES
P&W JT8D-7B	B727-200	1	ROYAL AIR MAROC
P&W JT8D-7B	B727-200	12	SHUTTLE INC (dba USAIR SHUTTLE)
P&W JT8D-7B	B727-200A	1	TAESA
P&W JT8D-7B	B727-200F	1	BRADLEY AIR SERVICES LIMITED (FIRST AIR)
P&W JT8D-7B	B727-200F	3	EMERY WORLDWIDE
P&W JT8D-7B	B737-200	1	SAHSA - SERVICIO AEREO DE HONDURAS, SA
P&W JT8D-7B	B737-200	45	UNITED AIRLINES, INC
P&W JT8D-7B	B737-200C	1	TNT INTERNATIONAL AVIATION SERVICES
P&W JT8D-7B	DC-9	3	KITTY HAWK GROUP, INC.
P&W JT8D-7B	DC9-14	6	MIDWEST EXPRESS
P&W JT8D-7B	DC9-14	16	NORTHWEST AIRLINES
P&W JT8D-7B	DC9-15	11	AEROCALIFORNIA

POWER PLANT	AIRFRAME	NO. IN FLEET	FLEET NAME
P&W JT8D-7B	DC9-15	2	AIRBORNE EXPRESS (ABX AIR, INC dba)
P&W JT8D-7B	DC9-15	1	EVERGREEN INTERNATIONAL AIRLINES, INC
P&W JT8D-7B	DC9-15	3	GREAT AMERICAN AIRWAYS
P&W JT8D-7B	DC9-15	2	MIDWEST EXPRESS
P&W JT8D-7B	DC9-15	4	NORTHWEST AIRLINES
P&W JT8D-7B	DC9-15F	5	EMERY WORLDWIDE
P&W JT8D-7B	DC9-15F	1	KALITTA AMERICAN INTERNATIONAL AIRWAYS
P&W JT8D-7B	DC9-31	18	AIRBORNE EXPRESS (ABX AIR, INC dba)
P&W JT8D-7B	DC9-31	1	AVENSA
P&W JT8D-7B	DC9-31	23	NORTHWEST AIRLINES
P&W JT8D-7B	DC9-32	7	AEROCALIFORNIA
P&W JT8D-7B	DC9-32	16	AIRBORNE EXPRESS (ABX AIR, INC dba)
P&W JT8D-7B	DC9-32	4	MIDWEST EXPRESS
P&W JT8D-7B	DC9-32	1	SERVIVENSA
P&W JT8D-7B	DC9-32F	3	AIRBORNE EXPRESS (ABX AIR, INC dba)
P&W JT8D-7B (Q)	B727-200F	2	TNT INTERNATIONAL AVIATION SERVICES
P&W JT8D-7B/9A	DC9-31	68	USAIR, INC
P&W JT8D-7R	B727-200	2	SAM - SOCIEDAD AERONAUTICA DE MEDELLIN
P&W JT8D-7R4E1	A310-222	3	CHINA NORTHWEST AIRLINES
P&W JT8D-9	B727-100	3	CONTINENTAL AIRLINES
P&W JT8D-9	B727-100C	1	ARIANA AFGHAN AIRLINES
P&W JT8D-9	B727-100C	3	CONTINENTAL MICRONESIA, INC
P&W JT8D-9	B727-200	2	AIR ALGERIE
P&W JT8D-9	B727-200	1	LIBYAN ARAB AIRLINES
P&W JT8D-9	B727-200 Advanced	14	DELTA AIR LINES, INC
P&W JT8D-9	B727-200 Advanced	21	IBERIA - LINEAS AEREAS DE ESPANA, S.A.
P&W JT8D-9	B727-200 Advanced	5	VIASA, VENEZOLANA INTERNACIONAL DE AVIACION SA
P&W JT8D-9	B737-200	1	AIR FAST INDONESIA
P&W JT8D-9	B737-200	1	AIR MADAGASCAR
P&W JT8D-9	B737-200	5	CARIBBEAN INTERNATIONAL AIRWAYS
P&W JT8D-9	B737-200	1	LAM - LINHAS AEREAS DE MOCAMBIQUE
P&W JT8D-9	B737-200	1	TAN - TRANSPORTES AEREOS NACIONALES, SA
P&W JT8D-9	B737-200C Advanced	1	AIR ALGERIE
P&W JT8D-9	B737-200C Advanced	1	LAM - LINHAS AEREAS DE MOCAMBIQUE
P&W JT8D-9	DC9-31	26	NORTHWEST AIRLINES
P&W JT8D-9	DC9-31	18	TRANS WORLD AIRLINES, INC
P&W JT8D-9	DC9-32	3	BRITISH MIDLAND AIRWAYS LTD
P&W JT8D-9	DC9-32	6	MERPATI NUSANTARA AIRLINES PT
P&W JT8D-9	DC9-33	1	TRANS WORLD AIRLINES, INC
P&W JT8D-9	DC9-33F	6	AIRBORNE EXPRESS (ABX AIR, INC dba)
P&W JT8D-9/-15	DC9-32	14	TRANS WORLD AIRLINES, INC
P&W JT8D-9/9A	B727-200 Advanced	86	AMERICAN AIRLINES
P&W JT8D-9A	B727-100	1	DOMINICANA DE AVIACION
P&W JT8D-9A	B727-100	1	KABO AIR
P&W JT8D-9A	B727-100	2	LLOYD AEREO BOLIVIANO SAM
P&W JT8D-9A	B727-100	2	ROYAL NEPAL AIRLINES CORPORATION
P&W JT8D-9A	B727-100	1	SAETA
P&W JT8D-9A	B727-100	1	SAN
P&W JT8D-9A	B727-100	2	TAME - TRANSPORTES AEREOS MILITARES ECUATORIANOS, CA
P&W JT8D-9A	B727-100C	1	LLOYD AEREO BOLIVIANO SAM
P&W JT8D-9A	B727-100C	4	VARIG - VIACAO AEREA RIO-GRANDENSE S.A.
P&W JT8D-9A	B727-100F	1	ITAPEMIRIN TRANSPORTES AEREOS SA
P&W JT8D-9A	B727-200	1	AVENSA
P&W JT8D-9A	B727-200	2	CARNIVAL AIR LINES INC
P&W JT8D-9A	B727-200	1	KIWI INTERNATIONAL AIR LINES, INC
P&W JT8D-9A	B727-200	1	MONGOLIAN AIRLINES - MIAT
P&W JT8D-9A	B727-200	6	OLYMPIC AIRWAYS S.A.
P&W JT8D-9A	B727-200 Advanced	2	AVENSA
P&W JT8D-9A	B727-200 Advanced	53	CONTINENTAL AIRLINES
P&W JT8D-9A	B727-200 Advanced	1	DOMINICANA DE AVIACION
P&W JT8D-9A	B727-200 Advanced	8	JAT - JUGOSLOVENSKI AEROTRANSPORT
P&W JT8D-9A	B727-200 Advanced	3	KOREAN AIR
P&W JT8D-9A	B727-200 Advanced	8	TUNIS AIR - SOCIETE TUNISIENNE D EL'AIR
P&W JT8D-9A	B727-200/200 Advanced	41	TRANS WORLD AIRLINES, INC
P&W JT8D-9A	B727-200C	1	EXPRESS ONE
P&W JT8D-9A	B727-200F	1	AMERIJET INTERNATIONAL
P&W JT8D-9A	B737-100	1	AMERICA WEST AIRLINES, INC
P&W JT8D-9A	B737-100	1	FAUCETT PERU
P&W JT8D-9A	B737-100	1	SIERRA PACIFIC AIRLINES, INC
P&W JT8D-9A	B737-200	4	AEROLINEAS ARGENTINAS
P&W JT8D-9A	B737-200	1	ARKIA ISRAELI AIRLINES
P&W JT8D-9A	B737-200	2	CANAIR CARGO LTD
P&W JT8D-9A	B737-200	5	CARNIVAL AIR LINES INC
P&W JT8D-9A	B737-200	3	COMAIR (COMMERCIAL AIRWAYS (PTY) LTD)
P&W JT8D-9A	B737-200	24	CONTINENTAL AIRLINES
P&W JT8D-9A	B737-200	5	EURALAIR INTERNATIONAL
P&W JT8D-9A	B737-200	3	FAR EASTERN AIR TRANSPORT
P&W JT8D-9A	B737-200	2	FAUCETT PERU
P&W JT8D-9A	B737-200	4	INDIAN AIRLINES LTD
P&W JT8D-9A	B737-200	2	JETALL
P&W JT8D-9A	B737-200	1	LADECO SA/LINEA AEREA DE COBRE/LADECO AIRLINES
P&W JT8D-9A	B737-200	4	NWT AIR (NORTHWEST TERRITORIAL AIRWAYS LTD)
P&W JT8D-9A	B737-200	3	SEMPATI AIR TRANSPORT
P&W JT8D-9A	B737-200	1	SIERRA PACIFIC AIRLINES, INC
P&W JT8D-9A	B737-200	3	SULTAN AIR
P&W JT8D-9A	B737-200	1	TACA INTERNATIONAL AIRLINES SA
P&W JT8D-9A	B737-200	1	TRANSMILE AIR SERVICES
P&W JT8D-9A	B737-200	1	TURKS & CAICOS AIRWAYS LIMITED
P&W JT8D-9A	B737-200 Advanced	3	AER LINGUS
P&W JT8D-9A	B737-200 Advanced	6	AEROLINEAS ARGENTINAS
P&W JT8D-9A	B737-200 Advanced	1	AIR NIPPON
P&W JT8D-9A	B737-200 Advanced	8	ALOHA AIRLINES
P&W JT8D-9A	B737-200 Advanced	6	AMERICA WEST AIRLINES, INC
P&W JT8D-9A	B737-200 Advanced	19	CANADIAN - CANADIEN
P&W JT8D-9A	B737-200 Advanced	3	CHINA AIRLINES
P&W JT8D-9A	B737-200 Advanced	1	CONTINENTAL AIRLINES
P&W JT8D-9A	B737-200 Advanced	2	EAST WEST AIRLINES
P&W JT8D-9A	B737-200 Advanced	2	FAR EASTERN AIR TRANSPORT
P&W JT8D-9A	B737-200 Advanced	3	LADECO SA/LINEA AEREA DE COBRE/LADECO AIRLINES

POWER PLANT	AIRFRAME	NO. IN FLEET	FLEET NAME
P&W JT8D-9A	B737-200 Advanced	3	PRIMERAS LINEAS URUGUAYAS DE NAVEGACION AEREA - PLUNA
P&W JT8D-9A	B737-200 Advanced	3	SEMPATI AIR TRANSPORT
P&W JT8D-9A	B737-200 Advanced	50	SOUTHWEST AIRLINES COMPANY
P&W JT8D-9A	B737-200 Advanced	2	TUNIS AIR - SOCIETE TUNISIENNE D EL'AIR
P&W JT8D-9A	B737-200 Advanced	7	UNITED AIRLINES, INC
P&W JT8D-9A	B737-200 Advanced	17	USAIR, INC
P&W JT8D-9A	B737-200C	2	AEROLINEAS ARGENTINAS
P&W JT8D-9A	B737-200C	2	CANADIAN - CANADIEN
P&W JT8D-9A	B737-200C	2	L'AEROPOSTALE
P&W JT8D-9A	B737-200C	1	TRANSMILE AIR SERVICES
P&W JT8D-9A	B737-200C Advanced	4	CANADIAN - CANADIEN
P&W JT8D-9A	B737-200QC	1	COMPANIA PANAMENA DE AVIACION, SA (COPA)
P&W JT8D-9A	B737-200QC	0	SAF SERVICE AERIEN FRANCAIS, SA
P&W JT8D-9A	B737-200QC	2	TRANSMILE AIR SERVICES
P&W JT8D-9A	DC9-30	2	ADRIA AIRWAYS
P&W JT8D-9A	DC9-30	34	CONTINENTAL AIRLINES
P&W JT8D-9A	DC9-31	1	AIR ARUBA
P&W JT8D-9A	DC9-32	34	ALITALIA
P&W JT8D-9A	DC9-32	14	AVIACO
P&W JT8D-9A	DC9-32	5	MIDWEST EXPRESS
P&W JT8D-9A	DC9-32	10	NORTHWEST AIRLINES
P&W JT8D-9A	DC9-33F	1	EVERGREEN INTERNATIONAL AIRLINES, INC
P&W JT8D-9A	DC9-33RC(F)	3	EVERGREEN INTERNATIONAL AIRLINES, INC
P&W JT9D-20	DC10-40	21	NORTHWEST AIRLINES
P&W JT9D-3A	B747-100	3	UNITED AIRLINES, INC
P&W JT9D-59A	A300B4-100	8	IBERIA - LINEAS AEREAS DE ESPANA, S.A.
P&W JT9D-59A	A300B4-100	3	PREMAIR A/S
P&W JT9D-59A	A300B4-100	2	ZAS AIRLINES OF EGYPT
P&W JT9D-59A	A300B4-200	6	CHINA AIRLINES
P&W JT9D-59A	A300B4-220	9	GARUDA INDONESIA PT
P&W JT9D-59A	DC10-40	1	JAPAN AIR CHARTER
P&W JT9D-59A	DC10-40	15	JAPAN AIRLINES COMPANY LTD - JAL
P&W JT9D-59A	DC10-40	4	JAPAN ASIA AIRWAYS (NIHON ASIA KOKU)
P&W JT9D-7	B747	2	ACS OF CANADA - AIR CHARTER SYSTEMS
P&W JT9D-7	B747-100	3	AIR CANADA
P&W JT9D-7	B747-100	13	AIR FRANCE
P&W JT9D-7	B747-100	11	TOWER AIR, INC
P&W JT9D-7	B747-200F	1	EL AL ISRAEL AIRLINES, LTD
P&W JT9D-70A	B747-200B (SCD)	1	AIR MADAGASCAR
P&W JT9D-7A	B747-100	3	AER LINGUS
P&W JT9D-7A	B747-100	15	BRITISH AIRWAYS
P&W JT9D-7A	B747-100	2	CORSAIR
P&W JT9D-7A	B747-100	3	EVERGREEN INTERNATIONAL AIRLINES, INC
P&W JT9D-7A	B747-100	1	JAPAN ASIA AIRWAYS (NIHON ASIA KOKU)
P&W JT9D-7A	B747-100	2	KALITTA AMERICAN INTERNATIONAL AIRWAYS
P&W JT9D-7A	B747-100	2	NORTHWEST AIRLINES
P&W JT9D-7A	B747-100	1	OKADA AIR LTD
P&W JT9D-7A	B747-100	8	TRANS WORLD AIRLINES, INC
P&W JT9D-7A	B747-100	15	UNITED AIRLINES, INC
P&W JT9D-7A	B747-100 (F) (SCD)	15	UNITED PARCEL SERVICE COMPANY UPS
P&W JT9D-7A	B747-100 (LR)	2	JAPAN AIRLINES COMPANY LTD - JAL
P&W JT9D-7A	B747-100B (SR)	3	JAPAN AIRLINES COMPANY LTD - JAL
P&W JT9D-7A	B747-100B (SR/EUD)	2	JAPAN AIRLINES COMPANY LTD - JAL
P&W JT9D-7A	B747-100F (SCD	1	ARKIA ISRAELI AIRLINES
P&W JT9D-7A	B747-100F (SCD)	2	KALITTA AMERICAN INTERNATIONAL AIRWAYS
P&W JT9D-7A	B747-100F (SCD)	2	SOUTHERN AIR TRANSPORT, INC
P&W JT9D-7A	B747-200	3	TRANS WORLD AIRLINES, INC
P&W JT9D-7A	B747-200B	2	CONTINENTAL AIRLINES
P&W JT9D-7A	B747-200B	1	IBERIA - LINEAS AEREAS DE ESPANA, S.A.
P&W JT9D-7A	B747-200B	1	JAPAN AIR CHARTER
P&W JT9D-7A	B747-200B	1	JAPAN AIRLINES COMPANY LTD - JAL
P&W JT9D-7A	B747-200B	6	PAKISTAN INTERNATIONAL AIRLINES (PIA)
P&W JT9D-7A	B747-200C (F)	1	KOREAN AIR
P&W JT9D-7A	B747-200F	5	FEDERAL EXPRESS CORPORATION
P&W JT9D-7A	B747-200F (SCD)	6	CHINA AIRLINES
P&W JT9D-7A	B747-200F (SCD)	2	KOREAN AIR
P&W JT9D-7A	B747SP	4	CHINA AIRLINES
P&W JT9D-7A	B747SP	2	KOREAN AIR
P&W JT9D-7A	B747SP	7	UNITED AIRLINES, INC
P&W JT9D-7A/7J	B747-200B	5	VIRGIN ATLANTIC AIRWAYS
P&W JT9D-7AH	B747-100	1	VIRGIN ATLANTIC AIRWAYS
P&W JT9D-7AW	B747-200B	1	CHINA AIRLINES
P&W JT9D-7AW	B747-200B	1	JAPAN ASIA AIRWAYS (NIHON ASIA KOKU)
P&W JT9D-7AW	B747-200B (F)	1	JAPAN AIRLINES COMPANY LTD - JAL
P&W JT9D-7AW	B747-200B (LR)	10	JAPAN AIRLINES COMPANY LTD - JAL
P&W JT9D-7AW	B747-200F (SCD)	1	JAPAN AIRLINES COMPANY LTD - JAL
P&W JT9D-7F	B747-100	1	IRAN AIR THE AIRLINE OF THE ISLAMIC REPUBLIC OF IRAN
P&W JT9D-7F	B747-100 (F)	2	EVERGREEN INTERNATIONAL AIRLINES, INC
P&W JT9D-7F	B747-200	2	IRAN AIR THE AIRLINE OF THE ISLAMIC REPUBLIC OF IRAN
P&W JT9D-7F	B747-200B	3	CONTINENTAL AIRLINES
P&W JT9D-7F	B747-200B	5	NORTHWEST AIRLINES
P&W JT9D-7F	B747-200B (SCD)	1	ROYAL AIR MAROC
P&W JT9D-7F	B747-200C	1	EVERGREEN INTERNATIONAL AIRLINES, INC
P&W JT9D-7F	B747-200C (F)	1	EVERGREEN INTERNATIONAL AIRLINES, INC
P&W JT9D-7F	B747-200F	1	IRAN AIR THE AIRLINE OF THE ISLAMIC REPUBLIC OF IRAN
P&W JT9D-7F	B747-200F (SCD)	5	NORTHWEST AIRLINES
P&W JT9D-7F	B747SP	4	IRAN AIR THE AIRLINE OF THE ISLAMIC REPUBLIC OF IRAN
P&W JT9D-7FW	B747-200B	3	MIDDLE EAST AIRLINES - AIRLIBAN
P&W JT9D-7FW	B747-200C (SCD)	3	IRAQI AIRWAYS
P&W JT9D-7FW	B747SP	1	AIR MAURITIUS
P&W JT9D-7FW	B747SP	1	AIR NAMIBIA
P&W JT9D-7FW	B747SP	1	CORSAIR
P&W JT9D-7FW	B747SP	5	SOUTH AFRICAN AIRWAYS (SAA)
P&W JT9D-7FW	B747SP	2	SYRIAN ARAB AIRLINES
P&W JT9D-7J	B747	1	AIR CLUB INTERNATIONAL
P&W JT9D-7J	B747-100 (SCD)	1	EL AL ISRAEL AIRLINES, LTD
P&W JT9D-7J	B747-200	4	EL AL ISRAEL AIRLINES, LTD
P&W JT9D-7J	B747-200	1	SAUDIA - SAUDI ARABIAN AIRLINES
P&W JT9D-7J	B747-200B	4	AIR INDIA

POWER PLANT	AIRFRAME	NO. IN FLEET	FLEET NAME
P&W JT9D-7J	B747-200B	1	OLYMPIC AIRWAYS S.A.
P&W JT9D-7J	B747-200B	7	UNITED AIRLINES, INC
P&W JT9D-7J	B747-200B (SCD)	4	KUWAIT AIRWAYS CORPORATION
P&W JT9D-7J	B747-200C	1	CAL - CARGO AIR LINES LTD
P&W JT9D-7J	B747-200C	3	EL AL ISRAEL AIRLINES, LTD
P&W JT9D-7J	B747-200CRAF	1	EVERGREEN INTERNATIONAL AIRLINES, INC
P&W JT9D-7J	B747-200SF	1	EVERGREEN INTERNATIONAL AIRLINES, INC
P&W JT9D-7J	B747SP	4	AIR CHINA INTERNATIONAL
P&W JT9D-7OA	B747-200B (SCD)	1	PHILIPPINE AIR LINES, INC - PAL
P&W JT9D-7Q	B747-200	2	CORSAIR
P&W JT9D-7Q	B747-200	6	GARUDA INDONESIA PT
P&W JT9D-7Q	B747-200	4	SINGAPORE AIRLINES LIMITED
P&W JT9D-7Q	B747-200 (SCD)	1	AEROVIAS NACIONALES DE COLOMBIA S.A. AVIANCA
P&W JT9D-7Q	B747-200B	6	AEROLINEAS ARGENTINAS
P&W JT9D-7Q	B747-200B	5	AIR INDIA
P&W JT9D-7Q	B747-200B	2	CHINA AIRLINES
P&W JT9D-7Q	B747-200B	2	IBERIA - LINEAS AEREAS DE ESPANA, S.A.
P&W JT9D-7Q	B747-200B	8	JAPAN AIRLINES COMPANY LTD - JAL
P&W JT9D-7Q	B747-200B	1	JAPAN ASIA AIRWAYS (NIHON ASIA KOKU)
P&W JT9D-7Q	B747-200B	12	NORTHWEST AIRLINES
P&W JT9D-7Q	B747-200B	3	OLYMPIC AIRWAYS S.A.
P&W JT9D-7Q	B747-200B	3	PHILIPPINE AIR LINES, INC - PAL
P&W JT9D-7Q	B747-200B (SCD)	1	CAMEROON AIRLINES
P&W JT9D-7Q	B747-200B (SCD)	1	PHILIPPINE AIR LINES, INC - PAL
P&W JT9D-7Q	B747-200B (SCD)	1	SOUTH AFRICAN AIRWAYS (SAA)
P&W JT9D-7Q	B747-200C	3	AIR CANADA
P&W JT9D-7Q	B747-200F	1	AHK AIR HONG KONG
P&W JT9D-7Q	B747-200F	2	EL AL ISRAEL AIRLINES, LTD
P&W JT9D-7Q	B747-200F	1	FEDERAL EXPRESS CORPORATION
P&W JT9D-7Q	B747-200F	1	JAPAN UNIVERSAL SYSTEMS TRANSPORT - JUST
P&W JT9D-7Q	B747-200F	1	KOREAN AIR
P&W JT9D-7Q	B747-200F (SCD)	6	JAPAN AIRLINES COMPANY LTD - JAL
P&W JT9D-7Q	B747-200F (SCD)	4	KOREAN AIR
P&W JT9D-7Q	B747-200F (SCD)	1	NORTHWEST AIRLINES
P&W JT9D-7Q3	B747-200B	1	IBERIA - LINEAS AEREAS DE ESPANA, S.A.
P&W JT9D-7Q3	B747-200B (SCD)	3	IBERIA - LINEAS AEREAS DE ESPANA, S.A.
P&W JT9D-7R4	A310-200	2	YEMEN AIRLINES
P&W JT9D-7R4	A310-300	0	YEMEN AIRLINES
P&W JT9D-7R4D	B767-200	12	AIR CANADA
P&W JT9D-7R4D	B767-200	2	JAPAN AIRLINES COMPANY LTD - JAL
P&W JT9D-7R4D	B767-200	11	UNITED AIRLINES, INC
P&W JT9D-7R4D	B767-200ER	2	EL AL ISRAEL AIRLINES, LTD
P&W JT9D-7R4D	B767-200ER	10	TRANS WORLD AIRLINES, INC
P&W JT9D-7R4D	B767-200ER	1	TRANSBRASIL S.A. LINHAS AEREAS
P&W JT9D-7R4D	B767-200ER	8	UNITED AIRLINES, INC
P&W JT9D-7R4D	B767-300	16	JAPAN AIRLINES COMPANY LTD - JAL
P&W JT9D-7R4D1	B767-200ER	8	AIR CANADA
P&W JT9D-7R4E	B767-200ER	3	EGYPTAIR
P&W JT9D-7R4E	B767-200ER	3	ETHIOPIAN AIRLINES CORPORATION
P&W JT9D-7R4E	B767-200ER	7	QANTAS AIRWAYS LTD
P&W JT9D-7R4E1	A310-200	7	DELTA AIR LINES, INC
P&W JT9D-7R4E1	A310-200	2	SABENA BELGIAN WORLD AIRLINES
P&W JT9D-7R4E1	A310-200	2	SILKAIR (SINGAPORE) PRIVATE LIMITED
P&W JT9D-7R4E1	A310-200	4	SINGAPORE AIRLINES LIMITED
P&W JT9D-7R4E1	A310-222	4	NIGERIA AIRWAYS
P&W JT9D-7R4E1	A310-222	2	QATAR AIRWAYS
P&W JT9D-7R4E1	A310-300	1	SABENA BELGIAN WORLD AIRLINES
P&W JT9D-7R4E1	A310-322	5	SWISSAIR (SOCIETE ANONYME SUISSE POUR LA NAVIGATION AERIENNE)
P&W JT9D-7R4E4	B767-200ER	4	AIR CHINA INTERNATIONAL
P&W JT9D-7R4G2	B747-200B	3	JAPAN AIRLINES COMPANY LTD - JAL
P&W JT9D-7R4G2	B747-200B	3	KOREAN AIR
P&W JT9D-7R4G2	B747-200B	3	NORTHWEST AIRLINES
P&W JT9D-7R4G2	B747-200B	5	SOUTH AFRICAN AIRWAYS (SAA)
P&W JT9D-7R4G2	B747-200B	2	UNITED AIRLINES, INC
P&W JT9D-7R4G2	B747-200B (SCD)	3	AIR CHINA INTERNATIONAL
P&W JT9D-7R4G2	B747-200F	1	AIR CHINA INTERNATIONAL
P&W JT9D-7R4G2	B747-200F (SCD)	2	JAPAN AIRLINES COMPANY LTD - JAL
P&W JT9D-7R4G2	B747-200F (SCD)	2	NORTHWEST AIRLINES
P&W JT9D-7R4G2	B747-300	1	JAPAN ASIA AIRWAYS (NIHON ASIA KOKU)
P&W JT9D-7R4G2	B747-300	2	KOREAN AIR
P&W JT9D-7R4G2	B747-300	2	SOUTH AFRICAN AIRWAYS (SAA)
P&W JT9D-7R4G2	B747-300	2	SWISSAIR (SOCIETE ANONYME SUISSE POUR LA NAVIGATION AERIENNE)
P&W JT9D-7R4G2	B747-300 (LR)	9	JAPAN AIRLINES COMPANY LTD - JAL
P&W JT9D-7R4G2	B747-300 (SCD)	2	EGYPTAIR
P&W JT9D-7R4G2	B747-300 (SCD)	1	MALAYSIA AIRLINES
P&W JT9D-7R4G2	B747-300 (SCD)	3	SWISSAIR (SOCIETE ANONYME SUISSE POUR LA NAVIGATION AERIENNE)
P&W JT9D-7R4G2	B747-300 (SR)	4	JAPAN AIRLINES COMPANY LTD - JAL
P&W JT9D-7R4G2	B747-300M	2	ANSETT AUSTRALIA
P&W JT9D-7R4G2	B747-300M	1	KOREAN AIR
P&W JT9D-7R4G2	B747-300M	9	SINGAPORE AIRLINES LIMITED
P&W JT9D-7R4G2	B747-300M (SCD)	3	SINGAPORE AIRLINES LIMITED
P&W JT9D-7R4G2	B747-400F	3	SINGAPORE AIRLINES LIMITED
P&W JT9D-7R4G2	B767-200	2	EL AL ISRAEL AIRLINES, LTD
P&W JT9D-7R4G2	B767-300ER	2	EGYPTAIR
P&W JT9D-7R4H	A310-222	5	KUWAIT AIRWAYS CORPORATION
P&W JT9D-7R4H1	A300-600	11	SAUDIA - SAUDI ARABIAN AIRLINES
P&W JTD9-7A	B747-100SF	1	AHK AIR HONG KONG
P&W PW2037	B757-200	4	AMERICAN TRANS AIR, INC
P&W PW2037	B757-200	84	DELTA AIR LINES, INC
P&W PW2037	B757-200	2	FAR EASTERN AIR TRANSPORT
P&W PW2037	B757-200	33	NORTHWEST AIRLINES
P&W PW2037	B757-200	2	ROYAL AIR MAROC
P&W PW2037	B757-200	6	SHANGHAI AIRLINES
P&W PW2037/2039	B757-200	88	UNITED AIRLINES, INC
P&W PW2040	B757-200	6	AEROMEXICO
P&W PW2040	B757-200	3	AMERICAN TRANS AIR, INC
P&W PW2040	B757-200	16	CONDOR FLUGDIENST GMBH
P&W PW2040	B757-200	4	ETHIOPIAN AIRLINES CORPORATION
P&W PW2040	B757-200ER	2	LADECO SA/LINEA AEREA DE COBRE/LADECO AIRLINES

POWER PLANT	AIRFRAME	NO. IN FLEET	FLEET NAME
P&W PW2040	B757-200F	1	ETHIOPIAN AIRLINES CORPORATION
P&W PW2040	B757-200PF	45	UNITED PARCEL SERVICE COMPANY UPS
P&W PW2337	Ilyushin 96M	0	AEROFLOT RUSSIAN INTERNATIONAL AIRLINES
P&W PW4000	B767-300	0	SHANGHAI AIRLINES
P&W PW4050	B767-300ER	12	SAS - SCANDINAVIAN AIRLINES SYSTEM
P&W PW4052	B767-200ER	2	AIR CHINA INTERNATIONAL
P&W PW4056	B747-400	3	AIR CANADA
P&W PW4056	B747-400	5	AIR CHINA INTERNATIONAL
P&W PW4056	B747-400	4	AIR INDIA
P&W PW4056	B747-400	5	CHINA AIRLINES
P&W PW4056	B747-400	2	EL AL ISRAEL AIRLINES, LTD
P&W PW4056	B747-400	17	KOREAN AIR
P&W PW4056	B747-400	8	MALAYSIA AIRLINES
P&W PW4056	B747-400	10	NORTHWEST AIRLINES
P&W PW4056	B747-400	28	SINGAPORE AIRLINES LIMITED
P&W PW4056	B747-400	24	UNITED AIRLINES, INC
P&W PW4056	B747-400 (SCD)	3	AIR CHINA INTERNATIONAL
P&W PW4056	B747-400C (SCD)	1	KOREAN AIR
P&W PW4056	B767-200ER	2	AEROVIAS NACIONALES DE COLOMBIA S.A. AVIANCA
P&W PW4056	B767-200ER	2	AIR ZIMBABWE
P&W PW4056	B767-200ER	2	LAM - LINHAS AEREAS DE MOCAMBIQUE
P&W PW4056	B767-200ER	0	SAS - SCANDINAVIAN AIRLINES SYSTEM
P&W PW4056	B767-200ER	1	SOUTH AFRICAN AIRWAYS (SAA)
P&W PW4056	B767-200ER	2	TRANSBRASIL SA - LINHAS AEREAS
P&W PW4056	B767-300	3	AIR CHINA INTERNATIONAL
P&W PW4056	B767-300ER	1	LAUDA AIR LUFTFAHRT AKTIENGESELLSCHAFT
P&W PW4056	B767-300ER	6	ROYAL BRUNEI AIRLINES
P&W PW4056	B767-300ER	1	TACA INTERNATIONAL AIRLINES SA
P&W PW4060	B767-200ER	1	AEROMEXICO
P&W PW4060	B767-200ER	2	BALKAN
P&W PW4060	B767-300	4	TRANS WORLD AIRLINES, INC
P&W PW4060	B767-300ER	2	AER LINGUS
P&W PW4060	B767-300ER	2	AEROMEXICO
P&W PW4060	B767-300ER	2	AIR CANADA
P&W PW4060	B767-300ER	1	AIR EUROPE SPA
P&W PW4060	B767-300ER	3	AIR FRANCE
P&W PW4060	B767-300ER	2	ASIANA AIRLINES
P&W PW4060	B767-300ER	6	CONDOR FLUGDIENST GMBH
P&W PW4060	B767-300ER	14	DELTA AIR LINES, INC
P&W PW4060	B767-300ER	2	LAN-CHILE
P&W PW4060	B767-300ER	3	LAUDA AIR LUFTFAHRT AKTIENGESELLSCHAFT
P&W PW4060	B767-300ER	4	LTU - LUFTTRANSPORT UNTERNEHMEN GMBH
P&W PW4060	B767-300ER	4	LTU SUD INTERNATIONAL AIRWAYS
P&W PW4060	B767-300ER	5	MARTINAIR HOLLAND NV
P&W PW4060	B767-300ER	2	SPANAIR
P&W PW4060	B767-300ER	1	TAESA
P&W PW4060	B767-300ER	1	TRANSBRASIL S.A. LINHAS AEREAS
P&W PW4060	B767-300ER	23	UNITED AIRLINES, INC
P&W PW4073	B777-200A	1	UNITED AIRLINES, INC
P&W PW4084	B777-200	0	ALL NIPPON AIRWAYS
P&W PW4084	B777-200	0	JAPAN AIR SYSTEM CO LTD (NIHON AIR SYSTEM)
P&W PW4084	B777-200	0	JAPAN AIRLINES COMPANY LTD - JAL
P&W PW4152	A310	2	AIR CLUB INTERNATIONAL
P&W PW4152	A310-300	3	AEROCANCUN
P&W PW4152	A310-300	3	AEROLINEAS ARGENTINAS
P&W PW4152	A310-300	2	AIR NIUGINI
P&W PW4152	A310-300	4	AUSTRIAN AIRLINES, OSTERREICHISCHE LUFTVERKEHRS AG
P&W PW4152	A310-300	13	DELTA AIR LINES, INC
P&W PW4152	A310-300	0	ECUATORIANA
P&W PW4152	A310-300	0	IRAQI AIRWAYS
P&W PW4152	A310-300	1	JES AIR
P&W PW4152	A310-300	1	OASIS INTERNATIONAL AIRLINES
P&W PW4152	A310-300	17	SINGAPORE AIRLINES LIMITED
P&W PW4152	A310-300	1	VIETMAN AIRLINES
P&W PW4156	A310-300	4	BALAIRCTA
P&W PW4156	A310-300	2	TAROM
P&W PW4158	A300-600R	2	AIR LIBERTE
P&W PW4158	A300-600R	6	CHINA AIRLINES
P&W PW4158	A300-600R	3	CHINA NORTHERN AIRLINES
P&W PW4158	A300-600R	9	EGYPTAIR
P&W PW4158	A300-600R	10	GARUDA INDONESIA PT
P&W PW4158	A300-622	5	KOREAN AIR
P&W PW4158	A300-622R	16	KOREAN AIR
P&W PW4158	A300-622R	8	THAI AIRWAYS INTERNATIONAL
P&W PW4158	A300B4-622R	12	JAPAN AIR SYSTEM CO LTD (NIHON AIR SYSTEM)
P&W PW4158	A310-300	1	AIR LIBERTE
P&W PW4164	A330-300	1	LTU - LUFTTRANSPORT UNTERNEHMEN GMBH
P&W PW4164	A330-300	0	NORTHWEST AIRLINES
P&W PW4164	A330-320	1	THAI AIRWAYS INTERNATIONAL
P&W PW4168	A330-300	5	MALAYSIA AIRLINES
P&W PW4360	MD-11	11	DELTA AIR LINES, INC
P&W PW4360	MD-11	5	KOREAN AIR
P&W PW4460	MD-11	4	CHINA AIRLINES
P&W PW4460	MD-11	5	JAPAN AIRLINES COMPANY LTD - JAL
P&W PW4460	MD-11	0	JAT - JUGOSLOVENSKI AEROTRANSPORT
P&W PW4460	MD-11	4	LTU - LUFTTRANSPORT UNTERNEHMEN GMBH
P&W PW4460	MD-11	13	SWISSAIR (SOCIETE ANONYME SUISSE POUR LA NAVIGATION AERIENNE)
P&W PW4460	MD-11	4	WORLD AIRWAYS INC
P&W PW4462	MD-11CF	4	MARTINAIR HOLLAND NV
P&W PW4462	MD-11F	0	MARTINAIR HOLLAND NV
P&W R-1340	DHC-3 Otter	2	AIR SCHEFFERVILLE, INC
P&W R-1830	DC-3	1	AIR FAST INDONESIA
P&W R-1830	DC-3	2	CARIBBEAN INTERNATIONAL AIRWAYS
P&W R-1830	DC-3	1	COMAIR (COMMERCIAL AIRWAYS (PTY) LTD)
P&W R-1830	DC-3	1	KELOWNA FLIGHTCRAFT AIR CHARTER LTD
P&W R-1830	DC-3	1	REGIONAL AIR (PTY) LTD
P&W R-1830	DC-3	4	SERVIVENSA
P&W R-1830	DC-3C	3	AEROCARIBBEAN
P&W R-1830	DC-3C	10	AIR ATLANTIQUE

POWER PLANT	AIRFRAME	NO. IN FLEET	FLEET NAME
P&W R-1830	DC-3C	4	AIR NORTH
P&W R-1830	DC-3C	2	AIRKENYA AVIATION LIMITED
P&W R-1830	DC-3C	5	FOUR STAR AIR CARGO
P&W R-1830	DC-3C	3	SAHSA - SERVICIO AEREO DE HONDURAS, SA
P&W R-2000	DC-4	1	AIR NORTH
P&W R-2800	CV 240	1	AEROCHAGO AIRLINES, S.A.
P&W R-2800	CV 240	3	KITTY HAWK GROUP, INC.
P&W R-2800	CV 340	1	AEROCARIBE
P&W R-2800	CV 440 Metropolitan	1	AEROCARIBE
P&W R-2800	CV 440 Metropolitan	5	KITTY HAWK GROUP, INC.
P&W R-2800	Convair 440	2	FOUR STAR AIR CARGO
P&W R-2800	DC-6A	2	AIR ATLANTIQUE
P&W R-2800	DC-6A	11	NORTHERN AIR CARGO INC
P&W R-2800	DC-6A	3	TRANS-AIR-LINK CORPORATION
P&W R-2800	DC-6A (F)	1	DOMINICANA DE AVIACION
P&W R-2800	DC-6B	1	NORTHERN AIR CARGO INC
P&W R-2800	DC-6BF	1	AESA AIRLINES
P&W R-2800	DC-6BF Swingtail	2	NORTHERN AIR CARGO INC
P&W R-985	BE 18	4	BERING AIR, INC
P&W R-985	DHC-2 Beaver	2	BRADLEY AIR SERVICES LIMITED (FIRST AIR)
P&W R-985	Grumman G-21 Goose	1	PAN AVIATION
P&WC JT15D-1	C 500 Citation I	1	AIR ATLANTIQUE
P&WC JT15D-1A	C 500 Citation	1	SUN-AIR OF SCANDINAVIA A/S
P&WC JT15D-1A	C 500 Citation I	2	EURALAIR INTERNATIONAL
P&WC JT15D-1A	C 500 Citation IA	1	SUN-AIR OF SCANDINAVIA A/S
P&WC JT15D-1A	C 501 Citation I/SP	1	AMERIJET INTERNATIONAL
P&WC JT15D-1A	C 501 Citation I/SP	1	SUN-AIR OF SCANDINAVIA A/S
P&WC JT15D-1A	C Citation I/SP	1	CIMBER AIR DENMARK A/S
P&WC JT15D-4	C 550 Citation II	1	OMAN AIR
P&WC JT15D-4	C 550 Citation II	2	SAUDIA - SAUDI ARABIAN AIRLINES
P&WC JT15D-4	C 550 Citation II	1	SUN AIR (BOP AIR PTY LTD dba)
P&WC JT15D-4	C550 CiTation II	1	LAUDA AIR LUFTFAHRT AKTIENGESELLSCHAFT
P&WC JT15D-5A	C 560 Citation V	2	EURALAIR INTERNATIONAL
P&WC PT6A-112	C F406 Caravan II	1	TRANSMILE AIR SERVICES
P&WC PT6A-114	C 208 Caravan I	3	SATENA
P&WC PT6A-114	C 208A Caravan I	10	FEDERAL EXPRESS CORPORATION
P&WC PT6A-114	C 208B Caravan I Super	155	FEDERAL EXPRESS CORPORATION
P&WC PT6A-114	C 208B/3 Caravan I Super	51	FEDERAL EXPRESS CORPORATION
P&WC PT6A-114A	C 208 Caravan I	35	EMPIRE AIRLINES
P&WC PT6A-114A	C 208B Caravan I	3	SCENIC AIRLINES INC
P&WC PT6A-114A	C 208B Caravan I	2	TRANSMILE AIR SERVICES
P&WC PT6A-114A	C 208B Caravan I Super	1	AIR MANITOBA
P&WC PT6A-20	BE 99 Airliner	15	BAR HARBOR AIRLINES (CONTINENTAL/DELTA EXPRESS)
P&WC PT6A-20	BE King Air A90 / B90	1/1	AIR PROVENCE
P&WC PT6A-20	DHC-6 Twin Otter 100	1	CALM AIR (CANADIAN PARTNER)
P&WC PT6A-20	DHC-6 Twin Otter 200	4	AIRBC LTD
P&WC PT6A-20	DHC-6 Twin Otter 200	1	AIRKENYA AVIATION LIMITED
P&WC PT6A-20	DHC-6 Twin Otter 200	1	CALM AIR (CANADIAN PARTNER)
P&WC PT6A-20	DHC-6 Twin Otter 200	5	ERA AVIATION, INC
P&WC PT6A-20	Harbin Yunshuji Y-12	1	AIR FIJI
P&WC PT6A-20	Pilatus PC-6B Tbo. Porter	1	ROYAL NEPAL AIRLINES CORPORATION
P&WC PT6A-25	BE T34 Mentor	2	LUFTHANSA GERMAN AIRLINES - DEUTSCHE LUFTHANSA AG
P&WC PT6A-27	BE 99 Airliner	1	AIR SCHEFFERVILLE, INC
P&WC PT6A-27	DHC-6 Twin Otter 200	1	AIR INUIT (1985) LTD
P&WC PT6A-27	DHC-6 Twin Otter 200	1	SEAGREEN AIR TRANSPORT
P&WC PT6A-27	DHC-6 Twin Otter 210	2	SUNFLOWER AIRLINES LTD
P&WC PT6A-27	DHC-6 Twin Otter 300	10	ACES
P&WC PT6A-27	DHC-6 Twin Otter 300	2	AIR BURUNDI
P&WC PT6A-27	DHC-6 Twin Otter 300	1	AIR CALEDONIE INTERNATIONAL
P&WC PT6A-27	DHC-6 Twin Otter 300	1	AIR FIJI
P&WC PT6A-27	DHC-6 Twin Otter 300	1	AIR GUYANE
P&WC PT6A-27	DHC-6 Twin Otter 300	2	AIR INTER GABON, SA.
P&WC PT6A-27	DHC-6 Twin Otter 300	4	AIR INUIT (1985) LTD
P&WC PT6A-27	DHC-6 Twin Otter 300	4	AIR MADAGASCAR
P&WC PT6A-27	DHC-6 Twin Otter 300	1	AIR MOOREA
P&WC PT6A-27	DHC-6 Twin Otter 300	2	AIR NIPPON
P&WC PT6A-27	DHC-6 Twin Otter 300	1	AIR RWANDA
P&WC PT6A-27	DHC-6 Twin Otter 300	1	AIR SENEGAL, SOCIETE NATIONALE DE TRANSPORT AERIEN
P&WC PT6A-27	DHC-6 Twin Otter 300	4	AIR SEYCHELLES
P&WC PT6A-27	DHC-6 Twin Otter 300	1	AIR TAHITI
P&WC PT6A-27	DHC-6 Twin Otter 300	1	AIR TCHAD, SOCIETE DE TRANSPORTES AERIENS
P&WC PT6A-27	DHC-6 Twin Otter 300	6	AIRKENYA AVIATION LIMITED
P&WC PT6A-27	DHC-6 Twin Otter 300	8	ALOHA ISLAND AIR
P&WC PT6A-27	DHC-6 Twin Otter 300	1	ARKIA ISRAELI AIRLINES
P&WC PT6A-27	DHC-6 Twin Otter 300	5	BRADLEY AIR SERVICES LIMITED (FIRST AIR)
P&WC PT6A-27	DHC-6 Twin Otter 300	3	ERA AVIATION, INC
P&WC PT6A-27	DHC-6 Twin Otter 300	5	ETHIOPIAN AIRLINES CORPORATION
P&WC PT6A-27	DHC-6 Twin Otter 300	1	FARNER AIR TRANSPORT AG
P&WC PT6A-27	DHC-6 Twin Otter 300	2	GROENLANDSFLY A/S
P&WC PT6A-27	DHC-6 Twin Otter 300	1	GUYANA AIRWAYS CORP
P&WC PT6A-27	DHC-6 Twin Otter 300	4	JAPAN TRANSOCEAN AIR
P&WC PT6A-27	DHC-6 Twin Otter 300	2	LESOTHO AIRWAYS CORPORATION
P&WC PT6A-27	DHC-6 Twin Otter 300	6	LIAT (1974) LTD.
P&WC PT6A-27	DHC-6 Twin Otter 300	2	LINA CONGO
P&WC PT6A-27	DHC-6 Twin Otter 300	6	MALAYSIA AIRLINES
P&WC PT6A-27	DHC-6 Twin Otter 300	1	MAYA AIRWAYS LTD
P&WC PT6A-27	DHC-6 Twin Otter 300	11	MERPATI NUSANTARA AIRLINES PT
P&WC PT6A-27	DHC-6 Twin Otter 300	1	MOUNT COOK AIRLINE
P&WC PT6A-27	DHC-6 Twin Otter 300	2	NORLANDAIR (ICELAND)
P&WC PT6A-27	DHC-6 Twin Otter 300	6	NORONTAIR
P&WC PT6A-27	DHC-6 Twin Otter 300	1	OMAN AIR
P&WC PT6A-27	DHC-6 Twin Otter 300	2	PAKISTAN INTERNATIONAL AIRLINES (PIA)
P&WC PT6A-27	DHC-6 Twin Otter 300	1	POLYNESIAN AIRLINES
P&WC PT6A-27	DHC-6 Twin Otter 300	2	ROSS AVIATION
P&WC PT6A-27	DHC-6 Twin Otter 300	9	ROYAL NEPAL AIRLINES CORPORATION
P&WC PT6A-27	DHC-6 Twin Otter 300	2	ROYAL TONGAN AIRLINES
P&WC PT6A-27	DHC-6 Twin Otter 300	1	SAUDIA - SAUDI ARABIAN AIRLINES
P&WC PT6A-27	DHC-6 Twin Otter 300	24	SCENIC AIRLINES INC
P&WC PT6A-27	DHC-6 Twin Otter 300	2	SOCIETE NOUVELLE AIR GUADELOUPE

POWER PLANT	AIRFRAME	NO. IN FLEET	FLEET NAME
P&WC PT6A-27	DHC-6 Twin Otter 300	1	SOLOMON AIRLINES
P&WC PT6A-27	DHC-6 Twin Otter 300	2	SURINAM AIRWAYS
P&WC PT6A-27	DHC-6 Twin Otter 300	2	TACV - TRANSPORTES AEREOS DE CABO VERDE
P&WC PT6A-27	DHC-6 Twin Otter 300	2	TURKS & CAICOS AIRWAYS LIMITED
P&WC PT6A-27	DHC-6 Twin Otter 300	11	WIDEROE'S FLYVESELSKAP A/S
P&WC PT6A-27	DHC-6 Twin Otter 300	3	WINDWARD ISLANDS AIRWAYS INTERNATIONAL NV
P&WC PT6A-27	DHC-6 Twin Otter 300	2	XINJIANG AIRLINES
P&WC PT6A-27	DHC-6 Twin Otter 310	1	MOUNT COOK AIRLINE
P&WC PT6A-27	DHC-8 Dash 8-300	2	AIRES (AEROVIAS DE INTERGRACION REGIONAL SA)
P&WC PT6A-27	EMB 110 Bandeirante	1	NORDESTE
P&WC PT6A-27	EMB 110 Bandeirante	14	TABA - TRANSPORTES AEREOS DA BACIA AMAZONICA SA
P&WC PT6A-27	EMB 110C Bandeirante	3	BRASIL CENTRAL
P&WC PT6A-27	EMB 110C Bandeirante	3	NORDESTE
P&WC PT6A-27	EMB 110P Bandeirante	4	BRASIL CENTRAL
P&WC PT6A-27	EMB 110P Bandeirante	2	NORDESTE
P&WC PT6A-27	EMB 110P Bandeirante	5	RIO-SUL SERVICOS AEROS REGIONAIS SA
P&WC PT6A-27	EMB 110P Bandeirante	4	TAM - TRANSPORTES AEREOS REGIONAIS SA
P&WC PT6A-27	EMB 110P2 Bandeirante	1	NORDESTE
P&WC PT6A-27	Harbin Yunshuji Y-12	3	LAO AVIATION
P&WC PT6A-27	Harbin Yunshuji Y-12 II	1	GUIZHOU AIRLINES
P&WC PT6A-27	Harbin Yunshuji Y-12 II	5	MONGOLIAN AIRLINES - MIAT
P&WC PT6A-27	Harbin Yunshuji Y-12II	4	CHINA SOUTHWEST AIRLINES
P&WC PT6A-27	Let 410A	1	ECUATO GUINEANA DE AVIACION
P&WC PT6A-27	PA-31-350 T1040	1	AIRKENYA AVIATION LIMITED
P&WC PT6A-27	Pilatus PC-6 Turbo Porter	2	FARNER AIR TRANSPORT AG
P&WC PT6A-27	Pilatus PC-6 Turbo Porter	3	MOUNT COOK AIRLINE
P&WC PT6A-27	Pilatus PC-6 Turbo Porter	3	SATENA
P&WC PT6A-28	BE King Air 90	1	AVENSA
P&WC PT6A-28	BE King Air A100	2	SAUDIA - SAUDI ARABIAN AIRLINES
P&WC PT6A-28	BE King Air E90	1	AIR AFFAIRES GABON
P&WC PT6A-28	BE King Air F90	1	AIR AFFAIRES GABON
P&WC PT6A-28	PA-31T Cheyenne II	1	AIR MAURITANIE
P&WC PT6A-34	EMB 110 Bandeirante	2	STELLAIR T.A.
P&WC PT6A-34	EMB 110P1 Bandeirante	2	AIR AFFAIRES GABON
P&WC PT6A-34	EMB 110P1 Bandeirante	8	AIRES (AEROVIAS DE INTERGRACION REGIONAL SA)
P&WC PT6A-34	EMB 110P1 Bandeirante	2	AIRLA
P&WC PT6A-34	EMB 110P1 Bandeirante	11	ATLANTIC SOUTHEAST AIRLINES, INC (THE DELTA CONNECTION)
P&WC PT6A-34	EMB 110P1 Bandeirante	3	BRASIL CENTRAL
P&WC PT6A-34	EMB 110P1 Bandeirante	7	EAGLE AIRWAYS LTD
P&WC PT6A-34	EMB 110P1 Bandeirante	2	FLANDERS AIRLINES NV
P&WC PT6A-34	EMB 110P1 Bandeirante	5	NORDESTE
P&WC PT6A-34	EMB 110P1 Bandeirante	1	REGIONAIR
P&WC PT6A-34	EMB 110P1 Bandeirante	1	SUN-AIR OF SCANDINAVIA A/S
P&WC PT6A-34	EMB 110P1 Bandeirante	2	SUNSTATE AIRLINES
P&WC PT6A-34	EMB 110P2 Bandeirante	1	AIR BURKINA
P&WC PT6A-34	EMB 110P2 Bandeirante	1	FLANDERS AIRLINES NV
P&WC PT6A-34	EMB 110P2 Bandeirante	4	MUK AIR
P&WC PT6A-34	EMB 110P2 Bandeirante	2	SUNAIR (BOP AIR PTY LTD dba)
P&WC PT6A-34	EMB 110P6 Bandeirante	1	AIR VANUATU
P&WC PT6A-34	IAI Arava 102	1	CATA
P&WC PT6A-36	BE C99 Airliner	14	BAR HARBOR AIRLINES (CONTINENTAL/DELTA EXPRESS)
P&WC PT6A-36	BE C99 Airliner	8	BRITT AIRWAYS, INC (CONTINENTAL EXPRESS)
P&WC PT6A-41	BE 1300C Airliner	1	ALPHA AIR
P&WC PT6A-41	BE 200 Airliner	3	AIR SCHEFFERVILLE, INC
P&WC PT6A-41	BE King Air 200	3	TAT EUROPEAN AIRLINES
P&WC PT6A-41	BE King Air B200	4	PROVINCIAL AIRWAYS LTD
P&WC PT6A-41	BE Super King Air 200	1	AIR CREEBEC INC
P&WC PT6A-41	BE Super King Air 200	1	AIR FOYLE LIMITED
P&WC PT6A-41	BE Super King Air 200	2	AIR IVOIRE
P&WC PT6A-41	BE Super King Air 200	2	AIR PROVENCE
P&WC PT6A-41	BE Super King Air 200	2	CALM AIR (CANADIAN PARTNER)
P&WC PT6A-41	BE Super King Air 200	1	INTEROT AIR SERVICE
P&WC PT6A-41	BE Super King Air 200	2	RATIOFLUG GMBH
P&WC PT6A-41	BE Super King Air 200	2	ROYAL AIR MAROC
P&WC PT6A-41	BE Super King Air 200C	1	AIR INUIT (1985) LTD
P&WC PT6A-41	BE Super King Air 200C	1	ERA AVIATION, INC
P&WC PT6A-41	BE Super King Air 200C	2	LAM - LINHAS AEREAS DE MOCAMBIQUE
P&WC PT6A-42	BE Super King Air 200	2	JAPAN AIR SYSTEM CO LTD (NIHON AIR SYSTEM)
P&WC PT6A-42	BE Super King Air 200C	1	AIR AFFAIRES GABON
P&WC PT6A-42	BE Super King Air B200	1	INTEROT AIR SERVICE
P&WC PT6A-45B	Shorts 330-200	1	CANAIR CARGO LTD
P&WC PT6A-45B	Shorts 330-200	4	OLYMPIC AVIATION, S.A.
P&WC PT6A-45R	Shorts 330-200	4	BANGKOK AIRWAYS
P&WC PT6A-45R	Shorts 330-200	1	MUK AIR
P&WC PT6A-45R	Shorts 330-200	1	OLYMPIC AVIATION, S.A.
P&WC PT6A-45R	Shorts 330-200	1	SKYWAYS, AB
P&WC PT6A-45R	Shorts 330-200	2	SUNSTATE AIRLINES
P&WC PT6A-50	DHC-7 Dash 7-100	2	ADRIA AIRWAYS
P&WC PT6A-50	DHC-7 Dash 7-100	1	AIR GUINEE
P&WC PT6A-50	DHC-7 Dash 7-100	3	AIR NIUGINI
P&WC PT6A-50	DHC-7 Dash 7-100	10	ARKIA ISRAELI AIRLINES
P&WC PT6A-50	DHC-7 Dash 7-100	1	BRADLEY AIR SERVICES LIMITED (FIRST AIR)
P&WC PT6A-50	DHC-7 Dash 7-100	5	BRITISH AIRWAYS
P&WC PT6A-50	DHC-7 Dash 7-100	2	BRITISH MIDLAND AIRWAYS LTD
P&WC PT6A-50	DHC-7 Dash 7-100	5	BRITT AIRWAYS, INC (CONTINENTAL EXPRESS)
P&WC PT6A-50	DHC-7 Dash 7-100	5	CONTINENTAL EXPRESS
P&WC PT6A-50	DHC-7 Dash 7-100	3	GROENLANDSFLY A/S
P&WC PT6A-50	DHC-7 Dash 7-100	4	PARADISE ISLAND AIRLINES
P&WC PT6A-50	DHC-7 Dash 7-100	5	PELITA AIR SERVICE
P&WC PT6A-50	DHC-7 Dash 7-100	5	PIEDMONT AIRLINES INC (USAIR EXPRESS)
P&WC PT6A-50	DHC-7 Dash 7-100	5	ROCKY MOUNTAIN AIRWAYS (CONTINENTAL EXPRESS)
P&WC PT6A-50	DHC-7 Dash 7-100	1	ROSS AVIATION
P&WC PT6A-50	DHC-7 Dash 7-100	2	TYROLEAN AIRWAYS
P&WC PT6A-50	DHC-7 Dash 7-100	8	WIDEROE'S FLYVESELSKAP A/S
P&WC PT6A-50	DHC-7 Dash 7-100	4	YEMEN AIRLINES
P&WC PT6A-6	BE King Air 90	2	AIR PROVENCE
P&WC PT6A-6	PA-31 Cheyenne IIIA	4	CAAC
P&WC PT6A-6	PA-42-720 Cheyenne IIIA	7	LUFTHANSA GERMAN AIRLINES - DEUTSCHE LUFTHANSA AG
P&WC PT6A-6	Shorts 360	2	AIR UK LTD

POWER PLANT	AIRFRAME	NO. IN FLEET	FLEET NAME
P&WC PT6A-6	Shorts 360	1	AURIGNY AIR SERVICES LTD
P&WC PT6A-6	Shorts 360	8	BUSINESS EXPRESS (THE DELTA CONNECTION)
P&WC PT6A-6	Shorts 360	2	CHINA EASTERN AIRLINES
P&WC PT6A-6	Shorts 360	3	CHINA SOUTHERN AIRLINES
P&WC PT6A-6	Shorts 360	2	CITYFLYER EXPRESS LTD
P&WC PT6A-6	Shorts 360	10	PENNSYLVANIA COMMUTER AIRLINES, INC (USAIR EXPRESS)
P&WC PT6A-6	Shorts 360	3	SUNSTATE AIRLINES
P&WC PT6A-6	Shorts 360 Advanced	1	AIR HUDIK AB
P&WC PT6A-6	Shorts 360 Advanced	4	JERSEY EUROPEAN
P&WC PT6A-6	Shorts 360-100	3	LOGANAIR LTD
P&WC PT6A-6	Shorts 360-100 Advanced	5	LOGANAIR LTD
P&WC PT6A-6	Shorts 360-200	25	AMR EAGLE
P&WC PT6A-6	Shorts 360-300	4	AMR EAGLE
P&WC PT6A-6	Shorts 360-300	9	CCAIR, INC (USAIR EXPRESS)
P&WC PT6A-6	Shorts 360-300		CHINA ASIA AIRLINES
P&WC PT6A-6	Shorts 360-300	2	HAZELTON AIRLINES
P&WC PT6A-6	Shorts 360-300	2	PHILIPPINE AIR LINES, INC - PAL
P&WC PT6A-65AR	Shorts 360 Advanced	2	BANGKOK AIRWAYS
P&WC PT6A-65B	BE 1900 Airliner	13	AIR MIDWEST, INC (USAIR EXPRESS)
P&WC PT6A-65B	BE 1900 Airliner	16	BAR HARBOR AIRLINES (CONTINENTAL/DELTA EXPRESS)
P&WC PT6A-65B	BE 1900 Airliner	20	MESA AIRLINES INC
P&WC PT6A-65B	BE 1900C Airliner	3	AIR NAMIBIA
P&WC PT6A-65B	BE 1900C Airliner	1	ALPHA AIR
P&WC PT6A-65B	BE 1900C Airliner	21	BUSINESS EXPRESS (THE DELTA CONNECTION)
P&WC PT6A-65B	BE 1900C Airliner	9	CONQUEST AIRLINES
P&WC PT6A-65B	BE 1900C Airliner	5	CONTINENTAL EXPRESS
P&WC PT6A-65B	BE 1900C Airliner	1	INTEROT AIR SERVICE
P&WC PT6A-65B	BE 1900C Airliner	7	LIBERTY EXPRESS (USAIR EXPRESS)
P&WC PT6A-65B	BE 1900C Airliner	13	PENNSYLVANIA COMMUTER AIRLINES, INC (USAIR EXPRESS)
P&WC PT6A-65B	BE 1900C-1 Airliner	1	ALPHA AIR
P&WC PT6A-65B	BE 1900C-1 Airliner	2	BUSINESS EXPRESS (THE DELTA CONNECTION)
P&WC PT6A-65B	BE 1900C-1 Airliner	18	COMMUTAIR (USAIR EXPRESS)
P&WC PT6A-65B	BE 1900C-1 Airliner	10	CONTINENTAL EXPRESS
P&WC PT6A-65B	BE 1900C-1 Airliner	14	ROCKY MOUNTAIN AIRWAYS (CONTINENTAL EXPRESS)
P&WC PT6A-65B	BE 1900C-1 Airliner	4	STATESWEST AIRLINES, INC (USAIR EXPRESS)
P&WC PT6A-65B	BE 1900C-1 Airliner	1	TURKS & CAICOS AIRWAYS LIMITED
P&WC PT6A-65D	BE 1900D Airliner	6	LIBERTY EXPRESS (USAIR EXPRESS)
P&WC PT6A-65R	Shorts 360	1	BRITISH MIDLAND AIRWAYS LTD
P&WC PT6A-65R	Shorts 360	3	CANADIAN REGIONAL
P&WC PT6A-65R	Shorts 360	2	MUK AIR
P&WC PT6A-67D	BE 1900D Airliner	8	COMMUTAIR (USAIR EXPRESS)
P&WC PT6A-67D	BE 1900D Airliner	91	MESA AIRLINES INC
P&WC PT6A-67D	BE 1900D Airliner	13	MIDWEST EXPRESS
P&WC PT6A-67R	Shorts 360-300	2	LIBERTY EXPRESS (USAIR EXPRESS)
P&WC PT6T-3 TwinPac	Bell 212	1	AIR FAST INDONESIA
P&WC PT6T-3 TwinPac	Bell 412	2	AIR FAST INDONESIA
P&WC PT6T-3 TwinPac	Sikorsky S-58ET	2	AIR FAST INDONESIA
P&WC PT6T-3B TwinPac	Bell 212	14	ERA AVIATION, INC
P&WC PT6T-3B TwinPac	Bell 212	6	GROENLANDSFLY A/S
P&WC PT6T-3B TwinPac	Bell 412	16	ERA AVIATION, INC
P&WC PT6T-6 TwinPac	Sikorsky S-58ET	2	AIR FAST INDONESIA
P&WC PW115	EMB 120 Brasilia	3	LUXAIR S.A.
P&WC PW118	EMB 120 Brasilia	14	BRITT AIRWAYS, INC (CONTINENTAL EXPRESS)
P&WC PW118	EMB 120 Brasilia	2	TAT EUROPEAN AIRLINES
P&WC PW118	EMB 120 Brasilia	3	WIDEROE NORSKAIR A/S
P&WC PW118	EMB 120RT Brasilia	1	AIR ARUBA
P&WC PW118	EMB 120RT Brasilia	9	AIR LITTORAL
P&WC PW118	EMB 120RT Brasilia	60	ATLANTIC SOUTHEAST AIRLINES, INC (THE DELTA CONNECTION)
P&WC PW118	EMB 120RT Brasilia	20	BRITT AIRWAYS, INC (CONTINENTAL EXPRESS)
P&WC PW118	EMB 120RT Brasilia	40	COMAIR
P&WC PW118	EMB 120RT Brasilia	38	CONTINENTAL EXPRESS
P&WC PW118	EMB 120RT Brasilia	9	DELTA AIR TRANSPORT NV
P&WC PW118	EMB 120RT Brasilia	41	MESA AIRLINES INC
P&WC PW118	EMB 120RT Brasilia	2	NORDESTE
P&WC PW118	EMB 120RT Brasilia	5	RIO-SUL SERVICOS AEROS REGIONAIS SA
P&WC PW118	EMB 120RT Brasilia	15	WESTAIR COMMUTER AIRLINES (MESA AIRLINES)
P&WC PW118A	EMB 120ER Brasilia	29	SKYWEST AIRLINES INC (DELTA CONNECTION)
P&WC PW118A	EMB 120RT Brasilia	2	SUNAIR (BOP AIR PTY LTD dba)
P&WC PW119B	Dornier 328	0	AIR CALEDONIE
P&WC PW119B	Dornier 328	10	HORIZON AIR
P&WC PW119B	Dornier 328	20	JETSTREAM INTERNATIONAL AIRLINES
P&WC PW119B	Dornier 328	0	TAHITI CONQUEST AIRLINES - TCA
P&WC PW119B	Dornier 328-110	1	LONE STAR AIRLINES
P&WC PW120	ATR 42	3	AIR CALEDONIE
P&WC PW120	ATR 42	2	AIR MAURITIUS
P&WC PW120	ATR 42	47	AMR EAGLE
P&WC PW120	ATR 42	7	CIMBER AIR DENMARK A/S
P&WC PW120	ATR 42	6	CITYFLYER EXPRESS LTD
P&WC PW120	ATR 42	42	CONTINENTAL EXPRESS
P&WC PW120	ATR 42	17	EUROWINGS
P&WC PW120	ATR 42	1	HOLMSTROM AIR
P&WC PW120	ATR 42	8	INTER-CANADIAN
P&WC PW120	ATR 42	1	REGIONAL AIRLINES
P&WC PW120	ATR 42	2	ROYAL AIR MAROC
P&WC PW120	ATR 42	2	SOCIETE NOUVELLE AIR GUADELOUPE
P&WC PW120	ATR 42	2	SOCIETE NOUVELLE AIR MARTINIQUE
P&WC PW120	ATR 42	8	TRANS STATES AIRLINES
P&WC PW120	ATR 42-300	1	AEROMAR
P&WC PW120	ATR 42-300	1	AIR INTER GABON, SA.
P&WC PW120	ATR 42-300	5	AIR LITTORAL
P&WC PW120	ATR 42-300	3	AIR TAHITI
P&WC PW120	ATR 42-300	9	AVIANOVA
P&WC PW120	ATR 42-300	14	BRIT'AIR
P&WC PW120	ATR 42-300	15	CANADIAN REGIONAL
P&WC PW120	ATR 42-300	2	CROATIA AIRLINES
P&WC PW120	ATR 42-300	2	TACV - TRANSPORTES AEREOS DE CABO VERDE
P&WC PW120	ATR 42-300	8	TAT EUROPEAN AIRLINES
P&WC PW120	ATR 42-300	13	TRANS WORLD EXPRESS, INC
P&WC PW120	ATR 42-300	1	TRANS-JAMAICAN AIRLINES LTD

POWER PLANT	AIRFRAME	NO. IN FLEET	FLEET NAME
P&WC PW120	ATR 42-300	5	TRANSASIA AIRWAYS
P&WC PW120	ATR 42-300	1	TUNINTER
P&WC PW120	ATR 42-320	4	ACES
P&WC PW120	ATR 42-320	1	AIR SAINT-PIERRE
P&WC PW120	ATR 42-320	2	COMAIR (COMMERCIAL AIRWAYS (PTY) LTD)
P&WC PW120	ATR 42-320A	0	CSA - CESKOSLOVENSKE AEROLINIE
P&WC PW120	DHC-8 Dash 8	1	AIR INUIT (1985) LTD
P&WC PW120	DHC-8 Dash 8-100	11	AIR ATLANTIC LTD (CANADIAN PARTNER)
P&WC PW120	DHC-8 Dash 8-100	1	AIR CREEBEC INC
P&WC PW120	DHC-8 Dash 8-100	14	AIR NOVA
P&WC PW120	DHC-8 Dash 8-100	16	AIR ONTARIO, INC
P&WC PW120	DHC-8 Dash 8-100	12	AIRBC LTD
P&WC PW120	DHC-8 Dash 8-100	2	AIRES (AEROVIAS DE INTERGRACION REGIONAL SA)
P&WC PW120	DHC-8 Dash 8-100	3	ANSETT NEW ZEALAND
P&WC PW120	DHC-8 Dash 8-100	2	BRITISH AIRWAYS
P&WC PW120	DHC-8 Dash 8-100	2	CANADIAN REGIONAL
P&WC PW120	DHC-8 Dash 8-100	4	CCAIR, INC (USAIR EXPRESS)
P&WC PW120	DHC-8 Dash 8-100	5	EASTERN AUSTRALIA AIRLINES
P&WC PW120	DHC-8 Dash 8-100	4	GREAT CHINA AIRLINES
P&WC PW120	DHC-8 Dash 8-100	3	HAMBURG AIRLINES LUFTFAHRTGESELLSCHAFT.MBH
P&WC PW120	DHC-8 Dash 8-100	23	HORIZON AIR
P&WC PW120	DHC-8 Dash 8-100	0	INTEROT AIR SERVICE
P&WC PW120	DHC-8 Dash 8-100	8	LIAT (1974) LTD.
P&WC PW120	DHC-8 Dash 8-100	25	MESABA AIRLINES (NORTHWEST AIRLINK)
P&WC PW120	DHC-8 Dash 8-100	16	PENNSYLVANIA COMMUTER AIRLINES, INC (USAIR EXPRESS)
P&WC PW120	DHC-8 Dash 8-100	43	PIEDMONT AIRLINES INC (USAIR EXPRESS)
P&WC PW120	DHC-8 Dash 8-100	3	SUNSTATE AIRLINES
P&WC PW120	DHC-8 Dash 8-100	2	TYROLEAN AIRWAYS
P&WC PW120	DHC-8 Dash 8-102	4	NORONTAIR
P&WC PW120A	DHC-8 Dash 8-200	0	MESA AIRLINES INC
P&WC PW121	ATR 42	10	BAR HARBOR AIRLINES (CONTINENTAL/DELTA EXPRESS)
P&WC PW121	ATR 42	37	BRITT AIRWAYS, INC (CONTINENTAL EXPRESS)
P&WC PW121	ATR 42	2	ETHIOPIAN AIRLINES CORPORATION
P&WC PW121	ATR 42	4	OLYMPIC AVIATION, S.A.
P&WC PW121	ATR 42	12	ROCKY MOUNTAIN AIRWAYS (CONTINENTAL EXPRESS)
P&WC PW121	ATR 42	2	THAI AIRWAYS INTERNATIONAL
P&WC PW121	ATR 42-320	6	AEROMAR
P&WC PW121	ATR 42-320	2	AIR BOTSWANA
P&WC PW121	ATR 42-320	1	AIR MALAWI
P&WC PW121	ATR 42-320	2	AIRLINK AIRLINE (PTY) LTD
P&WC PW121	ATR 42-320	1	IRAN ASSEMAN AIRLINES
P&WC PW121	DHC-8 Dash 8-100	0	STATESWEST AIRLINES, INC (USAIR EXPRESS)
P&WC PW121	DHC-8 Dash 8-100	6	TYROLEAN AIRWAYS
P&WC PW121	DHC-8 Dash 8-103B	4	WIDEROE'S FLYVESELSKAP A/S
P&WC PW123	ATR 72-201	1	AIR LITTORAL
P&WC PW123	DHC-8 Dash 8-100	2	ERA AVIATION, INC
P&WC PW123	DHC-8 Dash 8-300	6	AIR ONTARIO, INC
P&WC PW123	DHC-8 Dash 8-300	6	AIRBC LTD
P&WC PW123	DHC-8 Dash 8-300	2	ALM ANTILLEAN AIRLINES
P&WC PW123	DHC-8 Dash 8-300	7	BAHAMASAIR
P&WC PW123	DHC-8 Dash 8-300	3	BANGKOK AIRWAYS
P&WC PW123	DHC-8 Dash 8-300	2	BRITISH AIRWAYS
P&WC PW123	DHC-8 Dash 8-300	14	CANADIAN REGIONAL
P&WC PW123	DHC-8 Dash 8-300	5	CONTACTAIR
P&WC PW123	DHC-8 Dash 8-300	6	GREAT CHINA AIRLINES
P&WC PW123	DHC-8 Dash 8-300	2	HAMBURG AIRLINES LUFTFAHRTGESELLSCHAFT.MBH
P&WC PW123	DHC-8 Dash 8-300	0	MESA AIRLINES INC
P&WC PW123	DHC-8 Dash 8-300	5	SCHREINER AIRWAYS, BV
P&WC PW123	DHC-8 Dash 8-300	2	TABA - TRANSPORTES AEREOS DA BACIA AMAZONICA SA
P&WC PW123	DHC-8 Dash 8-300A	3	TYROLEAN AIRWAYS
P&WC PW123	DHC-8 Dashe 8-300	1	REGION AIR
P&WC PW123B	DHC-8 Dash 8-300	1	AIR CREEBEC INC
P&WC PW123E	DHC-8 Dash 8-300	9	S A EXPRESS AIRLINES (PTY) LTD
P&WC PW124	ATR 72		AIR GABON
P&WC PW124	ATR 72	2	AIR TAHITI
P&WC PW124	ATR 72	2	BANGKOK AIRWAYS
P&WC PW124	ATR 72	6	BINTER
P&WC PW124	ATR 72	3	JAT - JUGOSLOVENSKI AEROTRANSPORT
P&WC PW124	ATR 72	6	KARAIR OY
P&WC PW124	ATR 72	4	LOT - POLSKIE LINIE LOTNICZE
P&WC PW124	ATR 72	6	TAT EUROPEAN AIRLINES
P&WC PW124	ATR 72	2	THAI AIRWAYS INTERNATIONAL
P&WC PW124	ATR 72	3	TRANS STATES AIRLINES
P&WC PW124	ATR 72	2	TUNINTER
P&WC PW124	ATR 72-200	33	AMR EAGLE
P&WC PW124	ATR 72-200	6	EUROWINGS
P&WC PW124	ATR 72-200	11	TRANSASIA AIRWAYS
P&WC PW124	ATR 72-210	12	ATLANTIC SOUTHEAST AIRLINES, INC (THE DELTA CONNECTION)
P&WC PW124	ATR 72-210	4	IRAN ASSEMAN AIRLINES
P&WC PW124	ATR 72-300	4	CORSE MEDITERRANEE
P&WC PW124	Fokker 50	5	BUSY BEE OF NORWAY A/S
P&WC PW124	Fokker 50	4	LUXAIR S.A.
P&WC PW124B	ATR 72	2	CITYFLYER EXPRESS LTD
P&WC PW124B	ATR 72	1	LAO AVIATION
P&WC PW124B	ATR 72	6	OLYMPIC AVIATION, S.A.
P&WC PW124B	ATR 72-202	1	KAMPUCHEA AIRLINES
P&WC PW124B	ATR 72-202	3	VIETNAM AIRLINES
P&WC PW125	ATR 72	2	AIR CAPE
P&WC PW125	ATR 72	4	CSA - CESKOSLOVENSKE AEROLINIE
P&WC PW125	ATR 72-101	2	BRIT'AIR
P&WC PW125	BAe ATP	5	MERPATI NUSANTARA AIRLINES PT
P&WC PW125	Fokker 50	6	AER LINGUS
P&WC PW125	Fokker 50	2	AIR ZIMBABWE
P&WC PW125	Fokker 50	11	ANSETT AUSTRALIA
P&WC PW125	Fokker 50	6	AUSTRIAN AIR SERVICES
P&WC PW125	Fokker 50	1	CROSSAIR
P&WC PW125	Fokker 50	4	ICELANDAIR
P&WC PW125	Fokker 50	3	KENYA AIRWAYS LTD
P&WC PW125	Fokker 50	10	KLM CITYHOPPER

POWER PLANT	AIRFRAME	NO. IN FLEET	FLEET NAME
P&WC PW125	Fokker 50	25	LUFTHANSA CITYLINE
P&WC PW125	Fokker 50	9	MAERSK AIR
P&WC PW125	Fokker 50	11	MALAYSIA AIRLINES
P&WC PW125	Fokker 50	2	NORDESTE
P&WC PW125	Fokker 50	2	PELANGI AIR, SDN BHD
P&WC PW125	Fokker 50	10	PHILIPPINE AIR LINES, INC - PAL
P&WC PW125	Fokker 50	7	RIO-SUL SERVICOS AEROS REGIONAIS SA
P&WC PW125	Fokker 50	22	SAS - SCANDINAVIAN AIRLINES SYSTEM
P&WC PW125	Fokker 50	2	SUDAN AIRWAYS
P&WC PW125B	Fokker 50	9	AIR UK LTD
P&WC PW125B	Fokker 50	2	FORMOSA AIRLINES
P&WC PW125B	Fokker 50	3	NORWEGIAN AIR SHUTTLE A/S
P&WC PW125B	Fokker 50	2	ROYAL BRUNEI AIRLINES
P&WC PW126	BAe ATP	2	BIMAN BANGLADESH AIRLINES
P&WC PW126	BAe ATP	14	BRITISH AIRWAYS
P&WC PW126	BAe ATP	3	BRITISH MIDLAND AIRWAYS LTD
P&WC PW126	BAe ATP	13	MANX AIRLINES LTD
P&WC PW126	BAe ATP	3	SATA AIR ACORES - SERVICO ACOREANO DE TRANSPORTES AEREOS, E.P.
P&WC PW127	ATR 72-210	3	AVIANOVA
P&WC PW127	ATR 72-210	2	CONTINENTAL EXPRESS
P&WC PW127	ATR 72-210	4	EUROWINGS
P&WC PW127B	Fokker 50	6	AEROVIAS NACIONALES DE COLOMBIA S.A. AVIANCA
P&WC PW127E	ATR 42-500	0	AIR TAHITI
P&WC PW305A	Learjet 60	1	LAUDA AIR LUFTFAHRT AKTIENGESELLSCHAFT
P&WC PW6A-20	DHC-6 Twin Otter 103	2	ALOHA ISLAND AIR
PW PW4060	B767-300ER	1	REGION AIR
PW&C PT6A-20	DHC-6 Twin Otter	1	TAGB - TRANSPORTES AEREOS DA GUINE BISSAU
PZL Rz TVD-10S	Antonov 28	1	AEROFLOT RUSSIAN INTERNATIONAL AIRLINES
RR Conway 508	B707-400	1	ZAIREAN AIRLINES
RR Dart 506	BAe (Vickers) Viscount	1	ZAIREAN AIRLINES
RR Dart 510	BAe (Vickers) Frt master	8	BRITISH WORLD AIRLINES
RR Dart 514-2	BAe (HS) 748 Srs 2A	7	EMERALD AIRWAYS LTD
RR Dart 514-7	Fairchild Ind. F-27	1	AERO COZUMEL, SA
RR Dart 514-7	Fairchild Ind. F-27	2	BRITT AIRWAYS, INC (CONTINENTAL EXPRESS)
RR Dart 514-7	Fokker F27-100	1	AIR UK LTD
RR Dart 514-7	Fokker F27-100	6	BUSY BEE OF NORWAY A/S
RR Dart 514-7	Fokker F27-100	1	FLANDERS AIRLINES NV
RR Dart 514-7	Fokker F27-100	2	INDIAN AIRLINES LTD
RR Dart 514-7	Fokker F27-100	4	VAYUDOOT
RR Dart 514-7	Fokker F27-100	3	WDL AVIATION (KOLN)
RR Dart 514-7	Fokker F27-100 (SCD)	2	FLANDERS AIRLINES NV
RR Dart 514-7	Fokker F27-100 (SCD)	1	FLANDERS AIRLINES NV
RR Dart 525	BAe (Vickers) Viscount	1	MANDALA AIRLINES PT
RR Dart 525F	BAe (Vickers) Viscount	4	BOURAQ INDONESIA AIRLINES PT
RR Dart 525F	BAe (Vickers) Viscount	1	INTERCONTINENTAL DE AVIACION
RR Dart 528-7	Fairchild Ind. F-27A	1	AERO COZUMEL, SA
RR Dart 528-7E	Fokker F27-200	1	AIR COMORES
RR Dart 528-7E	Fokker F27-200	10	AIR UK LTD
RR Dart 528-7E	Fokker F27-200	1	AIRKENYA AVIATION LIMITED
RR Dart 528-7E	Fokker F27-200	1	EXPRESSO AEREO
RR Dart 528-7E	Fokker F27-600	1	AIR UK LTD
RR Dart 529-7E	Fairchild Ind. F-27F	2	AEROCARIBE
RR Dart 529-7E	Fairchild Ind. F-27F	2	EMPIRE AIRLINES
RR Dart 529-8D	Gulfstream I	1	KELOWNA FLIGHTCRAFT AIR CHARTER LTD
RR Dart 529-8X	Gulfstream I	5	AIR PROVENCE
RR Dart 529-8X	Gulfstream I	7	TAS AIRWAYS SPA
RR Dart 532-2	BAe (AW) 650-222 Argosy	2	SAFE AIR LTD
RR Dart 532-7	Fairchild Ind. F-27	2	STELLAIR T.A.
RR Dart 532-7	Fairchild Ind. F-27J	1	AIRES (AEROVIAS DE INTERGRACION REGIONAL SA)
RR Dart 532-7	Fairchild Ind. F-27J	4	CATA
RR Dart 532-7	Fairchild Ind. FH-227B	6	TABA - TRANSPORTES AEREOS DA BACIA AMAZONICA SA
RR Dart 532-7	Fokker F27-100	1	JERSEY EUROPEAN
RR Dart 532-7	Fokker F27-200	1	BRASIL CENTRAL
RR Dart 532-7	Fokker F27-200	1	REGIONAIR
RR Dart 532-7	Fokker F27-200	2	SEMPATI AIR TRANSPORT
RR Dart 532-7	Fokker F27-400	1	INDIAN AIRLINES LTD
RR Dart 532-7	Fokker F27-400	2	SCHREINER AIRWAYS, BV
RR Dart 532-7	Fokker F27-400	1	SEMPATI AIR TRANSPORT
RR Dart 532-7	Fokker F27-500	15	AIR FRANCE
RR Dart 532-7	Fokker F27-500	2	AIR SINAI
RR Dart 532-7	Fokker F27-500	5	AIR UK LTD
RR Dart 532-7	Fokker F27-500	1	CHANNEL EXPRESS (AIR SERVICES) LTD
RR Dart 532-7	Fokker F27-500	2	EGYPTAIR
RR Dart 532-7	Fokker F27-500	10	EMPIRE AIRLINES
RR Dart 532-7	Fokker F27-500	1	EUROPEAN AIRLIFT
RR Dart 532-7	Fokker F27-500	2	EXPRESSO AEREO
RR Dart 532-7	Fokker F27-500	24	FEDERAL EXPRESS CORPORATION
RR Dart 532-7	Fokker F27-500	9	JERSEY EUROPEAN
RR Dart 532-7	Fokker F27-500	8	MERPATI NUSANTARA AIRLINES PT
RR Dart 532-7	Fokker F27-500	2	RIO-SUL SERVICOS AEROS REGIONAIS SA
RR Dart 532-7	Fokker F27-500	15	SAF SERVICE AERIEN FRANCAIS, SA
RR Dart 532-7	Fokker F27-500	1	STAR AIR A/S
RR Dart 532-7	Fokker F27-500	1	SUDAN AIRWAYS
RR Dart 532-7	Fokker F27-500	7	TAM - TRANSPORTES AEREOS REGIONAIS SA
RR Dart 532-7	Fokker F27-600	2	AIR JET
RR Dart 532-7	Fokker F27-600	1	AIR TCHAD, SOCIETE DE TRANSPORTES AERIENS
RR Dart 532-7	Fokker F27-600	2	BRASIL CENTRAL
RR Dart 532-7	Fokker F27-600	8	CUBANA
RR Dart 532-7	Fokker F27-600	1	FARNER AIR TRANSPORT AG
RR Dart 532-7	Fokker F27-600	8	FEDERAL EXPRESS CORPORATION
RR Dart 532-7	Fokker F27-600	1	FLANDERS AIRLINES NV
RR Dart 532-7	Fokker F27-600	1	LINA CONGO
RR Dart 532-7	Fokker F27-600	5	MYANMA AIRWAYS (BURMA AIRWAYS CORPORATION)
RR Dart 532-7	Fokker F27-600	3	RATIOFLUG GMBH
RR Dart 532-7	Fokker F27-600	1	SEMPATI AIR TRANSPORT
RR Dart 532-7	Fokker F27-600	2	TAAG, LINHAS AEREAS DE ANGOLA (ANGOLA AIRLINES)
RR Dart 532-7E	Fokker F27-200	13	PAKISTAN INTERNATIONAL AIRLINES (PIA)
RR Dart 532-7L	Fairchild Ind. FH-227B	2	AEROCARIBE
RR Dart 532-7L	Fairchild Ind. FH-227D	2	AEROCARIBE

POWER PLANT	AIRFRAME	NO. IN FLEET	FLEET NAME
RR Dart 532-7P	Fokker F27-600	1	LLOYD AEREO BOLIVIANO SAM
RR Dart 532-7R	Fokker 27-200	1	MESABA AIRLINES (NORTHWEST AIRLINK)
RR Dart 532-7R	Fokker F27-200	4	BUSY BEE OF NORWAY A/S
RR Dart 532-7R	Fokker F27-200	4	COMAIR (COMMERCIAL AIRWAYS (PTY) LTD)
RR Dart 532-7R	Fokker F27-400	1	FARNER AIR TRANSPORT AG
RR Dart 532-7R	Fokker F27-400	1	FARNER AIR TRANSPORT HUNGARY
RR Dart 532-7R	Fokker F27-400M (F)	2	TAAG, LINHAS AEREAS DE ANGOLA (ANGOLA AIRLINES)
RR Dart 532-7R	Fokker F27-500	3	EAST WEST AIRLINES
RR Dart 532-7R	Fokker F27-500	2	FARNER AIR TRANSPORT AG
RR Dart 532-7R	Fokker F27-500	1	TAAG, LINHAS AEREAS DE ANGOLA (ANGOLA AIRLINES)
RR Dart 532-7R	Fokker F27-600	1	STAR AIR A/S
RR Dart 532-9	Handley Page Herald	9	CHANNEL EXPRESS (AIR SERVICES) LTD
RR Dart 532-9	Handley Page Herald 401	1	AEROSUCRE COLOMBIA, S.A.
RR Dart 533-2	BAe (HS) 748 Srs 1A	2	AIR BVI LTD
RR Dart 533-2	HAL/BAe (HS) 748 Srs 2	5	VAYUDOOT
RR Dart 533-2	HAL/BAe (HS) 748 Srs 2A	5	VAYUDOOT
RR Dart 534-2	BAe (HS) 748 LFD Srs 2A	3	BRADLEY AIR SERVICES LIMITED (FIRST AIR)
RR Dart 534-2	BAe (HS) 748 Srs 1	3	EMERALD AIRWAYS LTD
RR Dart 534-2	BAe (HS) 748 Srs 2A	1	AIR AUSTRAL
RR Dart 534-2	BAe (HS) 748 Srs 2A	1	AIR BVI LTD
RR Dart 534-2	BAe (HS) 748 Srs 2A	3	AIR FAST INDONESIA
RR Dart 534-2	BAe (HS) 748 Srs 2A	4	AIR INUIT (1985) LTD
RR Dart 534-2	BAe (HS) 748 Srs 2A	2	AIR SAINT-PIERRE
RR Dart 534-2	BAe (HS) 748 Srs 2A	9	BOURAQ INDONESIA AIRLINES PT
RR Dart 534-2	BAe (HS) 748 Srs 2A	5	BRADLEY AIR SERVICES LIMITED (FIRST AIR)
RR Dart 534-2	BAe (HS) 748 Srs 2A	4	CALM AIR (CANADIAN PARTNER)
RR Dart 534-2	BAe (HS) 748 Srs 2A	7	MOUNT COOK AIRLINE
RR Dart 534-2	BAe (HS) 748 Srs 2A	1	SATENA
RR Dart 534-2	BAe (HS) 748 Srs 2A	2	TACV - TRANSPORTES AEREOS DE CABO VERDE
RR Dart 534-2	BAe (HS) 748 Srs 2A	1	TAME - TRANSPORTES AEREOS MILITARES ECUATORIANOS, CA
RR Dart 534-2	BAe (HS) 748 Srs 2A (SCD)	1	TAME - TRANSPORTES AEREOS MILITARES ECUATORIANOS, CA
RR Dart 534-2	BAe (HS) 748-200 Srs 2	4	AIR CREEBEC INC
RR Dart 534-2	BAe (HS) 748-2A Cargo	1	AIR MANITOBA
RR Dart 534-2	BAe (HS) 748-2A Combi	4	AIR MANITOBA
RR Dart 535-2	BAe (HS) 748 Srs 2B	1	BRADLEY AIR SERVICES LIMITED (FIRST AIR)
RR Dart 535-2	BAe (HS) 748 Srs 2B	1	SUN AIR (BOP AIR PTY LTD dba)
RR Dart 535-2	BAe (HS) 748-333 Srs 2A	1	TAGB - TRANSPORTES AEREOS DA GUINE BISSAU
RR Dart 535-7R	Fokker F27-500	3	PENNSYLVANIA COMMUTER AIRLINES, INC (USAIR EXPRESS)
RR Dart 536-2	BAe (HS) 748	1	ROYAL TONGAN AIRLINES
RR Dart 536-2	BAe (HS) 748 Srs 2B	2	AIR MADAGASCAR
RR Dart 536-2	BAe (HS) 748 Srs 2B	1	AIR MARSHALL ISLANDS - AMI
RR Dart 536-2	BAe (HS) 748 Srs 2B	6	BOURAQ INDONESIA AIRLINES PT
RR Dart 536-2	BAe (HS) 748 Srs 2B	1	SATENA
RR Dart 536-2	BAe (HS) 748 Super 2B	1	CAMEROON AIRLINES
RR Dart 536-2	BAe (HS) 748 Super 2B	4	LIAT (1974) LTD.
RR Dart 536-2	BAe (HS) 748 Super 2B	2	MAKUNG AIRLINES, CO LTD
RR Dart 536-3B	Fokker F27-600	1	SOMALI AIRLINES
RR Dart 536-7	Fokker F27-400	1	LIBYAN ARAB AIRLINES
RR Dart 536-7	Fokker F27-500	6	MERPATI NUSANTARA AIRLINES PT
RR Dart 536-7	Fokker F27-500	4	OMAN AIR
RR Dart 536-7	Fokker F27-600	12	LIBYAN ARAB AIRLINES
RR Dart 536-7B	Fokker F27-600	1	AIR IVOIRE
RR Dart 536-7P	Fairchild Ind. FH-227B	2	IRAN ASSEMAN AIRLINES
RR Dart 536-7P	Fokker F27-200	1	LLOYD AEREO BOLIVIANO SAM
RR Dart 536-7R	Fokker F27-400M	7	AIR ALGERIE
RR Dart 536-7R	Fokker F27-600	1	IRAN ASSEMAN AIRLINES
RR Dart 536-7R	Fokker F27-600	1	LESOTHO AIRWAYS CORPORATION
RR Dart 536-7R	Fokker F27-600	5	WDL AVIATION (KOLN)
RR Dart 536-7R	Fokker F27-600CF	2	AIR TANZANIA
RR Dart 542-10J/K	NAMC YS-11-200	5	JAPAN TRANSOCEAN AIR
RR Dart 542-10J/K	NAMC YS-11A-500	6	JAPAN AIR COMMUTER CO LTD
RR Dart 542-10J/K	NAMC YS-11A-500	15	JAPAN AIR SYSTEM CO LTD (NIHON AIR SYSTEM)
RR Dart 542-10K	NAMC YS-11A Combi	2	REEVE ALEUTIAN AIRWAYS, INC
RR Dart 542-10K	NAMC YS-11A-200	11	AIRBORNE EXPRESS (ABX AIR, INC dba)
RR Dart 542-4	CV 600F	8	KITTY HAWK GROUP, INC.
RR Dart 542-4	CV 640 (F)	11	ZANTOP INTERNATIONAL AIRLINES, INC
RR Dart 542-6	CV 600	2	SERVICIOS AEREOS ASTRO SA
RR Dart 543-10J/K	NAMC YS-11-100	5	AIR NIPPON
RR Dart 543-10J/K	NAMC YS-11-200	14	AIR NIPPON
RR Dart 543-10J/K	NAMC YS-11A-500	7	AIR NIPPON
RR Dart 543-10J/K	NAMC YS-11A-500	7	ALL NIPPON AIRWAYS
RR Dart 543-2	BAe (HS) 748 Srs 2A	2	AIR SENEGAL, SOCIETE NATIONALE DE TRANSPORT AERIEN
RR Dart 543-2	BAe (HS) 748 Srs 2A	2	ROYAL NEPAL AIRLINES CORPORATION
RR Dart 552-7R	Fokker F27-500	2	RATIOFLUG GMBH
RR Olympus 593-610	AS/BAe Concorde 101	7	AIR FRANCE
RR Olympus 593-610	AS/BAe Concorde 101	7	BRITISH AIRWAYS
RR RB211-22B	L1011 TriStar 1	8	AIR OPS OF EUROPE
RR RB211-22B	L1011 TriStar 1	4	ALL NIPPON AIRWAYS
RR RB211-22B	L1011 TriStar 1	14	AMERICAN TRANS AIR, INC
RR RB211-22B	L1011 TriStar 1	1	CALEDONIAN AIRWAYS
RR RB211-22B	L1011 TriStar 1	12	CATHAY PACIFIC AIRWAYS LTD
RR RB211-22B	L1011 TriStar 1	32	DELTA AIR LINES, INC
RR RB211-22B	L1011 TriStar 1	2	DRAGONAIR (HONG KONG DRAGON AIRLINES LIMITED)
RR RB211-22B	L1011 TriStar 1	1	FAUCETT PERU
RR RB211-22B	L1011 TriStar 1	11	RICH INTERNATIONAL AIRWAYS INC
RR RB211-22B	L1011 TriStar 1	10	TRANS WORLD AIRLINES, INC
RR RB211-22B	L1011 TriStar 1 (F)	1	TRADEWINDS INTERNATIONAL AIRLINES (WRANGLER AVIATION dba)
RR RB211-22B	L1011 TriStar 1/100	5	BRITISH AIRWAYS
RR RB211-22B	L1011 TriStar 1/150	3	AIR CANADA
RR RB211-22B	L1011 TriStar 100	5	AIR TRANSAT
RR RB211-22B	L1011 TriStar 100	4	CALEDONIAN AIRWAYS
RR RB211-22B	L1011 TriStar 100	2	CATHAY PACIFIC AIRWAYS LTD
RR RB211-22B	L1011 TriStar 100	4	LTU - LUFTTRANSPORT UNTERNEHMEN GMBH
RR RB211-22B	L1011 TriStar 100	6	TRANS WORLD AIRLINES, INC
RR RB211-22B-02	L1011 TriStar 100	1	AIR LANKA
RR RB211-22B-02	L1011 TriStar 50	1	AIR LANKA
RR RB211-524B	L1011 TriStar 200	1	DELTA AIR LINES, INC
RR RB211-524B	L1011 TriStar 250	6	DELTA AIR LINES, INC
RR RB211-524B	L1011 TriStar 500	17	DELTA AIR LINES, INC

POWER PLANT	AIRFRAME	NO. IN FLEET	FLEET NAME
RR RB211-524B-02	L1011 TriStar 200	2	AIR LANKA
RR RB211-524B-02	L1011 TriStar 300	17	SAUDIA - SAUDI ARABIAN AIRLINES
RR RB211-524B-02	L1011 Tristar (F)	0	KALITTA AMERICAN INTERNATIONAL AIRWAYS
RR RB211-524B4	L1011 TriStar 200	1	LTU - LUFTTRANSPORT UNTERNEHMEN GMBH
RR RB211-524B4	L1011 TriStar 500	2	AIR LANKA
RR RB211-524B4	L1011 TriStar 500	2	LTU - LUFTTRANSPORT UNTERNEHMEN GMBH
RR RB211-524B4	L1011 TriStar 500	5	ROYAL JORDANIAN
RR RB211-524B4	L1011 TriStar 500	2	SAUDIA - SAUDI ARABIAN AIRLINES
RR RB211-524B4	L1011 TriStar 500	5	TRINIDAD & TOBAGO (BWIA INTERNATIONAL) AIRWAYS CORPORATION
RR RB211-524B4-02	L1011 TriStar 500	1	TAAG, LINHAS AEREAS DE ANGOLA (ANGOLA AIRLINES)
RR RB211-524B4-02	L1011 TriStar 500	7	TAP AIR PORTUGAL
RR RB211-524C2	B747-200	5	CATHAY PACIFIC AIRWAYS LTD
RR RB211-524C2	B747-300	6	CATHAY PACIFIC AIRWAYS LTD
RR RB211-524C2-19	B747-100	8	SAUDIA - SAUDI ARABIAN AIRLINES
RR RB211-524C2-19	B747SP	3	SAUDIA - SAUDI ARABIAN AIRLINES
RR RB211-524D4	B747-200	5	AIR NEW ZEALAND
RR RB211-524D4	B747-200	1	AIR PACIFIC LIMITED
RR RB211-524D4	B747-200	2	CATHAY PACIFIC AIRWAYS LTD
RR RB211-524D4	B747-200B	2	MALAYSIA AIRLINES
RR RB211-524D4	B747-200B	3	QANTAS AIRWAYS LTD
RR RB211-524D4	B747-200B (SCD)	2	QANTAS AIRWAYS LTD
RR RB211-524D4	B747-200F	4	CATHAY PACIFIC AIRWAYS LTD
RR RB211-524D4	B747-200F	2	SAUDIA - SAUDI ARABIAN AIRLINES
RR RB211-524D4	B747-300	6	QANTAS AIRWAYS LTD
RR RB211-524D4	B747SP	2	QANTAS AIRWAYS LTD
RR RB211-524D4-19	B747-300	11	SAUDIA - SAUDI ARABIAN AIRLINES
RR RB211-524D4U	B747-200	13	BRITISH AIRWAYS
RR RB211-524DX	B747-200B (SCD)	3	BRITISH AIRWAYS
RR RB211-524G	B747-400	5	AIR NEW ZEALAND
RR RB211-524G	B747-400	28	BRITISH AIRWAYS
RR RB211-524G	B747-400	18	QANTAS AIRWAYS LTD
RR RB211-524G	B747-400	4	SOUTH AFRICAN AIRWAYS (SAA)
RR RB211-524G2	B747-400	1	CATHAY PACIFIC AIRWAYS LTD
RR RB211-524H	B767-300	0	YUNNAN AIRLINES
RR RB211-524H	B767-300ER	23	BRITISH AIRWAYS
RR RB211-524H2	B747-400	14	CATHAY PACIFIC AIRWAYS LTD
RR RB211-524H2	B747-400F	1	CATHAY PACIFIC AIRWAYS LTD
RR RB211-535C	B757-200	2	AVENSA
RR RB211-535C	B757-200	34	BRITISH AIRWAYS
RR RB211-535C	B757-200	3	LTE INTERNATIONAL AIRWAYS SA
RR RB211-535E4	B757-200	2	AEROVIAS NACIONALES DE COLOMBIA S.A. AVIANCA
RR RB211-535E4	B757-200	15	AIR 2000 LTD
RR RB211-535E4	B757-200	1	AIR BELGIUM
RR RB211-535E4	B757-200	3	AIR EUROPA (AIR ESPANA, SA)
RR RB211-535E4	B757-200	2	AIR HOLLAND CHARTER BV
RR RB211-535E4	B757-200	4	AIR TRANSAT
RR RB211-535E4	B757-200	6	AIRTOURS INTERNATIONAL AIRWAYS LTD
RR RB211-535E4	B757-200	14	AMERICA WEST AIRLINES, INC
RR RB211-535E4	B757-200	1	AMERICAN TRANS AIR, INC
RR RB211-535E4	B757-200	5	ARKIA ISRAELI AIRLINES
RR RB211-535E4	B757-200	1	BRITANNIA AIRWAYS LTD
RR RB211-535E4	B757-200	10	BRITISH AIRWAYS
RR RB211-535E4	B757-200	0	CAAC
RR RB211-535E4	B757-200	19	CHINA SOUTHERN AIRLINES
RR RB211-535E4	B757-200	7	CHINA SOUTHWEST AIRLINES
RR RB211-535E4	B757-200	1	CONDOR FLUGDIENST GMBH
RR RB211-535E4	B757-200	5	CONTINENTAL AIRLINES
RR RB211-535E4	B757-200	7	EL AL ISRAEL AIRLINES, LTD
RR RB211-535E4	B757-200	8	IBERIA - LINEAS AEREAS DE ESPANA, S.A.
RR RB211-535E4	B757-200	12	LTU - LUFTTRANSPORT UNTERNEHMEN GMBH
RR RB211-535E4	B757-200	9	LTU SUD INTERNATIONAL AIRWAYS
RR RB211-535E4	B757-200	8	MONARCH AIRLINES LTD
RR RB211-535E4	B757-200	1	MYANMA AIRWAYS (BURMA AIRWAYS CORPORATION)
RR RB211-535E4	B757-200	1	NORTH AMERICAN AIRLINES
RR RB211-535E4	B757-200	2	ROYAL NEPAL AIRLINES CORPORATION
RR RB211-535E4	B757-200	5	TAESA
RR RB211-535E4	B757-200	3	TRANSAERO AIRLINES
RR RB211-535E4	B757-200	1	TURKMENISTAN AIRLINE
RR RB211-535E4	B757-200	34	USAIR, INC
RR RB211-535E4	B757-200	3	VENUS AIRLINES SA
RR RB211-535E4	B757-200	3	XIAMEN AIRLINES
RR RB211-535E4	B757-200ER	1	AIR SEYCHELLES
RR RB211-535E4	B757-200ER	19	BRITANNIA AIRWAYS LTD
RR RB211-535E4	B757-200ER	3	CALEDONIAN AIRWAYS
RR RB211-535E4	B757-200ER	8	CANADA 3000 AIRLINES LIMITED
RR RB211-535E4	B757-200ER	1	GUYANA AIRWAYS CORP
RR RB211-535E4	B757-200ER	3	ICELANDAIR
RR RB211-535E4	B757-200ER	3	ROYAL BRUNEI AIRLINES
RR RB211-535E4	B757-200F	1	GULF AIR
RR RB211-535E4	B757-200PF	3	CHALLENGE AIR CARGO
RR RB211-535E4	B757-200PF	10	UNITED PARCEL SERVICE COMPANY UPS
RR RB211-535E4B	B757-200	77	AMERICAN AIRLINES
RR RB211-535E5	B757-200	3	TRANSAVIA AIRLINES CV
RR Spey 506-14	BAe (BAC) One-Eleven 200	9	KABO AIR
RR Spey 506-14	BAe (BAC) One-Eleven 200	2	OKADA AIR LTD
RR Spey 511-14	BAe (BAC) One-Eleven	3	DELTA AIR TRANSPORT NV
RR Spey 511-14	BAe (BAC) One-Eleven 200	2	LADECO SA/LINEA AEREA DE COBRE/LADECO AIRLINES
RR Spey 511-14	BAe (BAC) One-Eleven 300	2	LADECO SA/LINEA AEREA DE COBRE/LADECO AIRLINES
RR Spey 511-14	BAe (BAC) One-Eleven 300	4	OKADA AIR LTD
RR Spey 511-14	BAe (BAC) One-Eleven 400	2	GAS AIR
RR Spey 511-14	BAe (BAC) One-Eleven 400	4	KABO AIR
RR Spey 511-14	BAe (BAC) One-Eleven 400	11	OKADA AIR LTD
RR Spey 511-14	BAe (BAC) One-Eleven 400	1	PELITA AIR SERVICE
RR Spey 511-14	BAe (BAC) One-Eleven 475F	1	TAROM
RR Spey 511-14DW	BAe (BAC) One-Eleven 500	7	TAROM
RR Spey 511-14DW	ROMBAC One-Eleven 561RC	6	TAROM
RR Spey 511-5W	BAe (HS) 121 Trident 2E	21	CAAC
RR Spey 511-8	Gulfstream II	1	LIBYAN ARAB AIRLINES
RR Spey 511-8	Gulfstream II	1	PELITA AIR SERVICE

POWER PLANT	AIRFRAME	NO. IN FLEET	FLEET NAME
RR Spey 511-8	Gulfstream II	4	SAUDIA - SAUDI ARABIAN AIRLINES
RR Spey 511-8	Gulfstream III	1	PELITA AIR SERVICE
RR Spey 511-8	Gulfstream III	3	SAUDIA - SAUDI ARABIAN AIRLINES
RR Spey 511-8	Gulfstream III	1	SUNAIR (BOP AIR PTY LTD dba)
RR Spey 512-14	BAe (BAC) One-Eleven 400	4	MAERSK AIR LIMITED
RR Spey 512-14DW	BAe (BAC) One-Eleven 400	1	GAS AIR
RR Spey 512-14DW	BAe (BAC) One-Eleven 500	8	AUSTRAL
RR Spey 512-14DW	BAe (BAC) One-Eleven 500	5	BRITISH WORLD AIRLINES
RR Spey 512-14DW	BAe (BAC) One-Eleven 500	3	CYPRUS AIRWAYS LTD
RR Spey 512-14DW	BAe (BAC) One-Eleven 500	2	MAERSK AIR LIMITED
RR Spey 512-14DW	BAe (BAC) One-Eleven 500	4	RYANAIR LTD
RR Spey 512-14DW	ROMBAC One-Eleven 561RC	4	RYANAIR LTD
RR Spey 512-5W	BAe (HS) 121 Trident 2E	4	CAAC
RR Spey 555-15	Fokker F28-1000	3	AEROLINEAS ARGENTINAS
RR Spey 555-15	Fokker F28-1000	2	AEROPERU
RR Spey 555-15	Fokker F28-1000	1	AIR GABON
RR Spey 555-15	Fokker F28-1000	2	AIR LITTORAL
RR Spey 555-15	Fokker F28-1000	7	AIR NIUGINI
RR Spey 555-15	Fokker F28-1000	4	ANSETT AUSTRALIA
RR Spey 555-15	Fokker F28-1000	11	CANADIAN REGIONAL
RR Spey 555-15	Fokker F28-1000	12	HORIZON AIR
RR Spey 555-15	Fokker F28-1000	1	IRAN ASSEMAN AIRLINES
RR Spey 555-15	Fokker F28-1000	1	LINA CONGO
RR Spey 555-15	Fokker F28-1000	1	MYANMA AIRWAYS (BURMA AIRWAYS CORPORATION)
RR Spey 555-15	Fokker F28-1000	2	PELITA AIR SERVICE
RR Spey 555-15	Fokker F28-1000	10	TAT EUROPEAN AIRLINES
RR Spey 555-15	Fokker F28-1000C	1	AIR AUSTRAL
RR Spey 555-15	Fokker F28-1000C	1	AIR IVOIRE
RR Spey 555-15	Fokker F28-2000	2	AIR GABON
RR Spey 555-15	Fokker F28-2000	4	TAT EUROPEAN AIRLINES
RR Spey 555-15	Fokker F28-3000	2	ANSETT AUSTRALIA
RR Spey 555-15	Fokker F28-4000	2	AIR NIUGINI
RR Spey 555-15H	Fokker F28-1000	4	DELTA AIR TRANSPORT NV
RR Spey 555-15H	Fokker F28-3000	1	DELTA AIR TRANSPORT NV
RR Spey 555-15H	Fokker F28-3000	1	ROYAL SWAZI NATIONAL AIRWAYS CORPORATION
RR Spey 555-15H	Fokker F28-3000C	1	SATENA
RR Spey 555-15H	Fokker F28-4000	2	AIR IVOIRE
RR Spey 555-15H	Fokker F28-4000	2	BIMAN BANGLADESH AIRLINES
RR Spey 555-15H	Fokker F28-4000	4	IRAN ASSEMAN AIRLINES
RR Spey 555-15H	Fokker F28-4000	26	MERPATI NUSANTARA AIRLINES PT
RR Spey 555-15H	Fokker F28-4000	5	TAT EUROPEAN AIRLINES
RR Spey 555-15H/-15P	Fokker F28-4000	2	MYANMA AIRWAYS (BURMA AIRWAYS CORPORATION)
RR Spey 555-15N	Fokker F28-1000	3	SAS - SCANDINAVIAN AIRLINES SYSTEM
RR Spey 555-15N	Fokker F28-2000	1	GHANA AIRWAYS CORPORATION
RR Spey 555-15P	Fokker F28-4000	1	AIR BURKINA
RR Spey 555-15P	Fokker F28-4000	1	AIR MAURITANIE
RR Spey 555-15P	Fokker F28-4000	7	ANSETT AUSTRALIA
RR Spey 555-15P	Fokker F28-4000	1	COMAIR (COMMERCIAL AIRWAYS (PTY) LTD)
RR Spey 555-15P	Fokker F28-4000	3	DELTA AIR TRANSPORT NV
RR Spey 555-15P	Fokker F28-4000	1	GHANA AIRWAYS CORPORATION
RR Spey 555-15P	Fokker F28-4000	4	KLM CITYHOPPER
RR Spey 555-15P	Fokker F28-4000	3	LIBYAN ARAB AIRLINES
RR Spey 555-15P	Fokker F28-4000	5	PELITA AIR SERVICE
RR Spey 555-15P	Fokker F28-4000	16	SAS - SCANDINAVIAN AIRLINES SYSTEM
RR Spey 555-15P	Fokker F28-4000	1	TAME - TRANSPORTES AEREOS MILITARES ECUATORIANOS, CA
RR Spey 555-15P	Fokker F28-4000	19	USAIR, INC
RR Tay 611-8	Gulfstream IV	6	SAUDIA - SAUDI ARABIAN AIRLINES
RR Tay 620	Fokker 70	2	AIR LITTORAL
RR Tay 620	Fokker 70	0	ALITALIA
RR Tay 620	Fokker 70	0	AUSTRIAN AIRLINES, OSTERREICHISCHE LUFTVERKEHRS AG
RR Tay 620	Fokker 70	2	BRITISH MIDLAND AIRWAYS LTD
RR Tay 620	Fokker 70	0	MALEV - HUNGARIAN AIRLINES
RR Tay 620	Fokker 70	0	MERPATI NUSANTARA AIRLINES PT
RR Tay 620	Fokker 70	2	MESA AIRLINES INC
RR Tay 620	Fokker 70	0	PELITA AIR SERVICE
RR Tay 620	Fokker 70	2	SEMPATI AIR TRANSPORT
RR Tay 620	Fokker 70	3	TYROLEAN AIRWAYS
RR Tay 620-15	Fokker 100	1	AIR GABON
RR Tay 620-15	Fokker 100	2	AIR IVOIRE
RR Tay 620-15	Fokker 100	9	AIR UK LTD
RR Tay 620-15	Fokker 100	75	AMERICAN AIRLINES
RR Tay 620-15	Fokker 100	4	AVIACION DE CHIAPAS (AVIACSA)
RR Tay 620-15	Fokker 100	2	CORSE MEDITERRANEE
RR Tay 620-15	Fokker 100	6	KLM ROYAL DUTCH AIRLINES
RR Tay 620-15	Fokker 100	10	MEXICANA
RR Tay 620-15	Fokker 100	10	SWISSAIR (SOCIETE ANONYME SUISSE POUR LA NAVIGATION AERIENNE)
RR Tay 620-15	Fokker 100	14	TAT EUROPEAN AIRLINES
RR Tay 620-15	Fokker 70	1	SILKAIR (SINGAPORE) PRIVATE LIMITED
RR Tay 650-15	Fokker 100	5	AIR INTER
RR Tay 650-15	Fokker 100	4	AIR LITTORAL
RR Tay 650-15	Fokker 100	4	BRITISH MIDLAND AIRWAYS LTD
RR Tay 650-15	Fokker 100	10	CHINA EASTERN AIRLINES
RR Tay 650-15	Fokker 100	5	DEUTSCHE BA LUFTFAHRTGESELLSCHAFT MBH
RR Tay 650-15	Fokker 100	6	IRAN AIR THE AIRLINE OF THE ISLAMIC REPUBLIC OF IRAN
RR Tay 650-15	Fokker 100	12	KOREAN AIR
RR Tay 650-15	Fokker 100	3	MERPATI NUSANTARA AIRLINES PT
RR Tay 650-15	Fokker 100	3	MIDWAY AIRLINES
RR Tay 650-15	Fokker 100	1	PELITA AIR SERVICE
RR Tay 650-15	Fokker 100	6	PORTUGALIA - COMPANIA PORTUGESA DE TRANSPORTES AEREOS SA
RR Tay 650-15	Fokker 100	1	ROYAL SWAZI NATIONAL AIRWAYS CORPORATION
RR Tay 650-15	Fokker 100	4	SEMPATI AIR TRANSPORT
RR Tay 650-15	Fokker 100	2	TABA - TRANSPORTES AEREOS DA BACIA AMAZONICA SA
RR Tay 650-15	Fokker 100	2	TAIWAN AIRLINES
RR Tay 650-15	Fokker 100	17	TAM - TRANSPORTES AEREOS REGIONAIS SA
RR Tay 650-15	Fokker 100	7	TRANSWEDE
RR Tay 650-15	Fokker 100	40	USAIR, INC
RR Tay 650-15	Romaero One-Eleven 2500	0	KIWI INTERNATIONAL AIRLINES INC
RR Tay 651-54	B727-100QF	34	UNITED PARCEL SERVICE COMPANY UPS
RR Tay 654-54	B727-100QF	3	STAR AIR A/S

POWER PLANT	AIRFRAME	NO. IN FLEET	FLEET NAME
RR Trent 768	A330-300	1	DRAGONAIR (HONG KONG DRAGON AIRLINES LIMITED)
RR Trent 768	A330-300	0	GARUDA INDONESIA PT
RR Trent 768	A330-300	0	TRANS WORLD AIRLINES, INC
RR Trent 772	A330-300	3	CATHAY PACIFIC AIRWAYS LTD
RR Trent 871	B777-200/200ER	0	CATHAY PACIFIC AIRWAYS LTD
RR Trent 871	B777-200/200ER	0	EMIRATES
RR Trent 875	B777-200	0	THAI AIRWAYS INTERNATIONAL
RR Trent 877	B777-200ER	0	TRANSBRASIL S.A. LINHAS AEREAS
RR Trent 884	B777-200ER	0	EMIRATES
RR Tyne 506-10	BAe (Vickers) Merchantman	2	HUNTING CARGO AIRLINES
RR Tyne 515-101W	Shorts Belfast	5	HEAVYLIFT CARGO AIRLINES
RR Tyne 522	Transall C-160NG	3	PELITA AIR SERVICE
RR Tyne 522	Transall C-160P	3	PELITA AIR SERVICE
RR Tyne 522	Transall C-160P	2	SAF SERVICE AERIEN FRANCAIS, SA
RR Viper 601-22A	BAe (HS) 125-600B	1	PELITA AIR SERVICE
RR/Williams Int FJ44	C 525 CitationJet	1	EURALAIR INTERNATIONAL
SH ASh-82T	Ilyushin 14	1	AEROCARIBBEAN
SH ASh-82T	Ilyushin 14	5	AIR KORYO
SH ASh-82T	Ilyushin 14	18	CAAC
TBD	A330-300	0	KOREAN AIR
TBD	A340-300	0	ALL NIPPON AIRWAYS
TBD	ATR 42-500	0	AEROMAR
TBD	B777	0	GARUDA INDONESIA PT
TBD	B777	0	VIRGIN ATLANTIC AIRWAYS
TBD	B777-200ER	0	KOREAN AIR
TU Arriel 1A2	AS SA365C2 Dauphin 2	1	SCHREINER AIRWAYS, BV
TU Arriel 1B	AS 350B Ecureuil	4	ERA AVIATION, INC
TU Arriel 1B	Eurocopter 350B-1	5	ERA AVIATION, INC
TU Arriel 1C	AS SA365N Dauphin 2	4	SCHREINER AIRWAYS, BV
TU Arriel 1C2	AS 365N2 Dauphin 2	2	MAERSK AIR
TU Arriel 1D1	AS 350B2 Ecureuil	1	GROENLANDSFLY A/S
TU Arriel 1D1	Eurocopter 350B-2	8	ERA AVIATION, INC
TU Artouste III	AS SA316B Alouette III	1	SCHREINER AIRWAYS, BV
TU Artouste IIIB	AS SE313B Lama	2	AIR AFFAIRES GABON
TU Artouste IIIB1	AS SA316B Alouette III	1	SCHREINER AIRWAYS, BV
TU Bastan VIC1	AS (Nord) 262	1	EQUATORIAL - INTERNATIONAL AIRLINES OF SAO TOME E PRINCIPE
TU Makila 1A	AS 332L Super Puma	3	MAERSK AIR
TU Makila 1A	Eurocopter 332L S.Puma	2	ERA AVIATION, INC
TU Makila 1A	IPTN/NAS SA332C S.Puma	2	PELITA AIR SERVICE
TU Makila 1A1	Eurocopter 332L S.Puma	8	ERA AVIATION, INC
TU Turmo IVA/IVC	IPTN/NAS SA330G Puma	4	PELITA AIR SERVICE
TU Turmo IVC	AS 330J Puma	2	MYANMA AIRWAYS (BURMA AIRWAYS CORPORATION)
TU Turmo IVC	IPTN/NAS SA330J Puma	11	PELITA AIR SERVICE
VO-540	Hiller UH-12E	3	FORMOSA AIRLINES
WJ 5A-1	Xian Yunshuji Y-7	3	AIR CHINA INTERNATIONAL
WJ 5A-1	Xian Yunshuji Y-7	9	CHINA EASTERN AIRLINES
WJ 5A-1	Xian Yunshuji Y-7	5	CHINA NORTHERN AIRLINES
WJ 5A-1	Xian Yunshuji Y-7	5	CHINA SOUTHERN AIRLINES
WJ 5A-1	Xian Yunshuji Y-7	1	LAO AVIATION
WJ 5A-1	Xian Yunshuji Y-7-100	3	AIR CHINA INTERNATIONAL
WJ 5A-1	Xian Yunshuji Y-7-100	3	CHINA AIR CARGO
WJ 5A-1	Xian Yunshuji Y-7-100	6	CHINA NORTHERN AIRLINES
WJ 5A-1	Xian Yunshuji Y-7-100	7	CHINA NORTHWEST AIRLINES
WJ 5A-1	Xian Yunshuji Y-7-100	3	CHINA SOUTHWEST AIRLINES
WJ 5A-1	Xian Yunshuji Y-7-100	2	GUIZHOU AIRLINES
WJ 5A-1	Xian Yunshuji Y-7-100	5	SICHUAN AIRLINES
WJ 5A-1	Xian Yunshuji Y-7-100	4	UNION AIRLINES
WJ 5A-1	Xian Yunshuji Y-7-100	2	WUHAN AIR LINES
WJ 5A-1	Xian Yunshuji Y-7-100	2	ZHONGYUAN AIRLINES
WJ 6	SAC Yunshuji Y-8	2	CHINA EASTERN AIRLINES
WR R-3350	L1049 S Constellation	2	AEROCHAGO AIRLINES, S.A.
ZMKB A-24VT	Antonov 26B (F)	4	BELAVIA
ZMKB AI-20	Antonov 12 (F)	5	IRAQI AIRWAYS
ZMKB AI-20	Ilyushin 18	2	AIR KORYO
ZMKB AI-20	Ilyushin 18D	5	AIR CARGO
ZMKB AI-20	Ilyushin 18D	2	AIR KORYO
ZMKB AI-20	Ilyushin 18D	2	BALKAN
ZMKB AI-20	Ilyushin 18E	2	BALKAN
ZMKB AI-20	Ilyushin 18V	2	BALKAN
ZMKB AI-20K	Antonov 12	54	AEROFLOT RUSSIAN INTERNATIONAL AIRLINES
ZMKB AI-20K	Antonov 12	1	TAAG, LINHAS AEREAS DE ANGOLA (ANGOLA AIRLINES)
ZMKB AI-20K	Antonov 12 (F)	3	BALKAN
ZMKB AI-20K	Ilyushin 18V (F)	1	AEROCARIBBEAN
ZMKB AI-20K(M)	Ilyushin 18V	2	TAROM
ZMKB AI-20M	Antonov 12	2	AIR CHINA INTERNATIONAL
ZMKB AI-20M	Antonov 12	5	SAMARA AIRLINES
ZMKB AI-20M	Antonov 12	1	VOLGA DNEPR CARGO AIRLINE
ZMKB AI-20M	Antonov 12 (F)	2	CAAC
ZMKB AI-20M	Antonov 12 (F)	2	CHINA AIR CARGO
ZMKB AI-20M	Antonov 32	1	VOLGA DNEPR CARGO AIRLINE
ZMKB AI-20M	Ilyushin 18	9	CAAC
ZMKB AI-20M	Ilyushin 18D	4	AEROCARIBBEAN
ZMKB AI-20M	Ilyushin 18D	55	AEROFLOT RUSSIAN INTERNATIONAL AIRLINES
ZMKB AI-20M	Ilyushin 18D	2	TAROM
ZMKB AI-20M	Ilyushin 18D	3	VIETNAM AIRLINES
ZMKB AI-24	Antonov 24	8	AIR KORYO
ZMKB AI-24	Antonov 24	31	CAAC
ZMKB AI-24	Antonov 24	2	CHINA NORTHWEST AIRLINES
ZMKB AI-24	Antonov 24	5	CHINA SOUTHERN AIRLINES
ZMKB AI-24	Antonov 24	1	IRAQI AIRWAYS
ZMKB AI-24	Antonov 24	6	KEMEROVSKOE AVIAPREDPRIYATIYE
ZMKB AI-24	Antonov 24		NOVOKUZNETSKOE AVIAPREDPRIYATIYE
ZMKB AI-24	Antonov 24		OMSKOYE AVIAPREDPRIYATIYE
ZMKB AI-24	Antonov 24	6	SIBERIA AIRLINE
ZMKB AI-24	Antonov 24	1	SYRIAN ARAB AIRLINES
ZMKB AI-24	Antonov 24	6	TOMSKOYE AVIAPREDPRIYATIYE
ZMKB AI-24	Antonov 24		TURKMENISTAN AIRLINE
ZMKB AI-24	Antonov 24	7	ULAN-UDESKOYE AVIAPREDPRIYATIYE
ZMKB AI-24	Antonov 24	5	VIETNAM AIRLINES

POWER PLANT	AIRFRAME	NO. IN FLEET	FLEET NAME
ZMKB AI-24	Antonov 24B	4	MONGOLIAN AIRLINES - MIAT
ZMKB AI-24	Antonov 24RV	3	AEROFLOT RUSSIAN INTERNATIONAL AIRLINES
ZMKB AI-24	Antonov 24RV	1	AIR MALI SA
ZMKB AI-24	Antonov 24RV	2	ARIANA AFGHAN AIRLINES
ZMKB AI-24	Antonov 24RV	7	CUBANA
ZMKB AI-24	Antonov 24RV	3	LAO AVIATION
ZMKB AI-24	Antonov 24RV	2	LITHUANIAN AIRLINES
ZMKB AI-24	Antonov 24RV	8	MONGOLIAN AIRLINES - MIAT
ZMKB AI-24	Antonov 24T	2	IRAQI AIRWAYS
ZMKB AI-24	Antonov 24TV	1	IRAQI AIRWAYS
ZMKB AI-24	Antonov 24V	12	AEROFLOT RUSSIAN INTERNATIONAL AIRLINES
ZMKB AI-24	Antonov 24V	5	CUBANA
ZMKB AI-24	Antonov 24V	2	IRAQI AIRWAYS
ZMKB AI-24	Antonov 24V	2	LITHUANIAN AIRLINES
ZMKB AI-24	Antonov 24V	7	VIETNAM AIRLINES
ZMKB AI-24	Antonov 24V-2	14	BALKAN
ZMKB AI-24	Antonov 26	3	AEROCARIBBEAN
ZMKB AI-24	Antonov 26	1	AEROFLOT RUSSIAN INTERNATIONAL AIRLINES
ZMKB AI-24	Antonov 26	2	ARIANA AFGHAN AIRLINES
ZMKB AI-24	Antonov 26	6	KEMEROVSKOE AVIAPREDPRIYATIYE
ZMKB AI-24	Antonov 26	3	LITHUANIAN AIRLINES
ZMKB AI-24	Antonov 26	3	MONGOLIAN AIRLINES - MIAT
ZMKB AI-24	Antonov 26	5	NOVOKUZNETSKOE AVIAPREDPRIYATIYE
ZMKB AI-24	Antonov 26		OMSKOYE AVIAPREDPRIYATIYE
ZMKB AI-24	Antonov 26	5	SIBERIA AIRLINE
ZMKB AI-24	Antonov 26	4	SYRIAN ARAB AIRLINES
ZMKB AI-24	Antonov 26	5	TOMSKOYE AVIAPREDPRIYATIYE
ZMKB AI-24	Antonov 26		TURKMENISTAN AIRLINE
ZMKB AI-24	Antonov 26 (F)	24	CAAC
ZMKB AI-24	Antonov 26 (F)	24	CAAC
ZMKB AI-24	Antonov 30	4	AEROFLOT RUSSIAN INTERNATIONAL AIRLINES
ZMKB AI-24	Antonov 30	1	MONGOLIAN AIRLINES - MIAT
ZMKB AI-24	Antonov 32	2	AERONICA - AEROLINEAS NICARAGUENSES, SA
ZMKB AI-24	Antonov 32	1	SIBERIA AIRLINE
ZMKB AI-24 II	Antonov 24B	8	BELAVIA
ZMKB AI-24 II	Antonov 24RV	14	TAROM
ZMKB AI-24VT	Antonov 24B	1	AIR GUINEE
ZMKB AI-24VT	Antonov 24RV	1	AIR GUINEE
ZMKB AI-24VT	Antonov 24RV	10	BELAVIA
ZMKB AI-24VT	Antonov 24RV	3	KAMPUCHEA AIRLINES
ZMKB AI-24VT	Antonov 26	1	KAMPUCHEA AIRLINES
ZMKB AI-24VT	Antonov 26 (F)	4	BELAVIA
ZMKB AI-25	Yakovlev 40	25	AEROFLOT RUSSIAN INTERNATIONAL AIRLINES
ZMKB AI-25	Yakovlev 40	2	ARIANA AFGHAN AIRLINES
ZMKB AI-25	Yakovlev 40		BARNAULSKOYE AVIAPREDPRIYATIYE
ZMKB AI-25	Yakovlev 40	8	BELAVIA
ZMKB AI-25	Yakovlev 40	8	CUBANA
ZMKB AI-25	Yakovlev 40	4	ESTONIAN AIR
ZMKB AI-25	Yakovlev 40		KYRGYZSTAN AIRLINE
ZMKB AI-25	Yakovlev 40	1	NATIONAL OVERSEAS AIRLINE
ZMKB AI-25	Yakovlev 40	3	SAMARA AIRLINES
ZMKB AI-25	Yakovlev 40	6	SYRIAN ARAB AIRLINES
ZMKB AI-25	Yakovlev 40	5	VIETNAM AIRLINES
ZMKB AI-25	Yakovlev 40	1	VOLGA DNEPR CARGO AIRLINE
ZMKB D-18T	Antonov 124-100 Ruslan	3	AIR FOYLE LIMITED
ZMKB D-18T	Antonov 124-100 Ruslan	6	HEAVYLIFT CARGO AIRLINES
ZMKB D-18T	Antonov 124-100 Ruslan	6	VOLGA DNEPR CARGO AIRLINE
ZMKB D-36	Yakovlev 42	4	AEROFLOT RUSSIAN INTERNATIONAL AIRLINES
ZMKB D-36	Yakovlev 42	12	LITHUANIAN AIRLINES
ZMKB PS-90AT	Yakovlev 42D	3	DACONO AIR

Notes
NYK = Not yet confirmed
TBD = To be determined

ARGENTINA

LAA

LOCKHEED AIRCRAFT ARGENTINA

Avenida Fuerza Aérea Argentina Km 5½, 5103 Córdoba
Telephone: 54 (51) 690594
Fax: 54 (51) 690698
Telex: 51965 AMCOR AR

BOARD OF MANAGEMENT:
Eng Hilario Francisco Luciano (Commercial) (President)
Eng Osvaldo Ricardo Giraudo (Production) (Vice-President)
Eng Alberto Osvaldo Buthet (Technical)
Jorge Alberto Carrizo (Finance)
Dr Juan Luis Smekens (Industrial Relations)

Original FMA (Military Aircraft Factory) came into operation 10 October 1927 as central organisation for aeronautical research and production; underwent several name changes before reverting 1968 to original title as component of Area de Material Córdoba (AMC) of Argentine Air Force. Principal activities are aircraft design, manufacture, maintenance and repair, major current product being IA 63 Pampa jet trainer; laboratories, factories and other aeronautical division buildings occupy total covered area of 253,000 m² (2,723,275 sq ft); AMC had workforce of about 2,100 in early 1995, of whom about 1,300 engaged in design and manufacturing; Córdoba facility also accommodates Centro de Ensayos en Vuelo (Flight Test Centre), a separate division also controlled by Argentine Air Force, at which all aircraft produced in Argentina undergo certification testing.

Argentine government has passed legislation to privatise most national defence companies, including FMA; this required Argentine Air Force to convert FMA into a joint stock company from April 1992, Air Force buying 30 per cent of the shares to establish itself as FMA's holding and management authority; government would sell remaining 70 per cent to private investors at stock exchange valuation. In pursuance of this, as per Argentine Ministry of Defence (MoD) resolution, AMC control of FMA handed over to Planning Secretary of MoD on 20 December 1993. Argentine MoD and Lockheed Aircraft Argentina SA (LAASA) signed concession agreement on 15 December 1994, allocating management of FMA to Lockheed from 1 March 1995. This was connected to a $200 million deal to refurbish and upgrade 36 A-4 Skyhawk aircraft and a $200 million contract from the MoD to maintain Argentine Air Force aircraft for the next five years. Lockheed took over the management of FMA on 1 July 1995. The name was subsequently changed to Lockheed Aircraft Argentina (LAA).

UPDATED

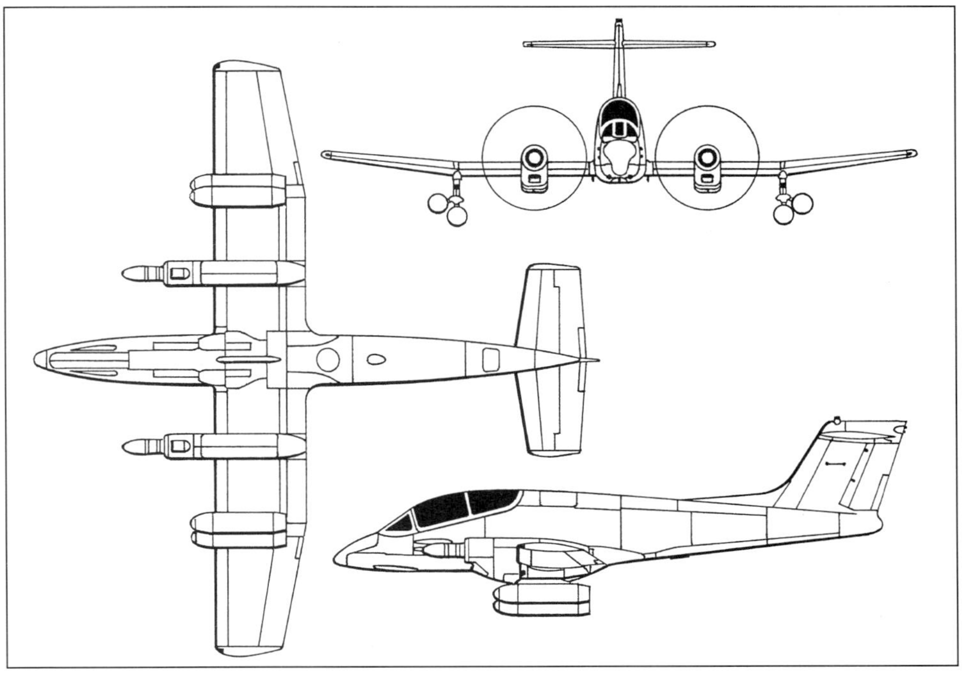

LAA (FMA) IA 58A Pucará twin-turboprop close support, reconnaissance and counter-insurgency aircraft *(Jane's/Mike Keep)*

1994

LAA (FMA) IA 58A Pucará aircraft of the Argentine Air Force

1995

FMA IA 58 PUCARÁ

TYPE: Twin-turboprop close support, reconnaissance and counter-insurgency aircraft.

PROGRAMME: Four versions of this twin-turboprop light attack aircraft were built as follows:

VERSIONS: **IA 58A:** Initial (two-seat) production version. Total of 60 ordered originally for the Fuerza Aérea Argentina (FAA), which later ordered 48 more, partly to replace about 24 aircraft lost in 1982 during fighting in the South Atlantic. Deliveries began in the Spring of 1976, and the last example was completed in 1986. Some early production aircraft converted to single-seat configuration, with extra fuselage fuel tank in place of rear seat. Six IA 58As were delivered to the Fuerza Aérea Uragaya, and a further 40 were made available for export in 1986. Four sold to Sri Lankan Air Force 1993.

IA 58B: Prototype only.

IA 58C: Improved single-seat version. Programme suspended.

IA 66: Prototype only.

OPERATORS: The IA 58 Pucará is in service with the following air forces: Argentina (75), Columbia (3), Sri Lanka (4) and Uruguay (6).

DESIGN FEATURES: Cantilever low-wing monoplane. Wing section NACA 64_2A215 at root, NACA 64_1 at tip. Dihedral 7° on outer panels. Incidence 2°. No sweepback. Fixed incidence tailplane and elevators mounted near top of fin. Curved dorsal fin.

FLYING CONTROLS: All-dural electrically controlled hydraulically actuated trailing-edge slotted flaps, inboard and outboard of each engine nacelle, modified Frise ailerons of duralumin, with magnesium alloy trailing-edges actuated by push/pull rods. No slats. Balance tab in starboard aileron, electrically operated trim tab in port aileron. Rudder and elevators actuated by push/pull rods, and each fitted with electrically operated inset trim tab.

STRUCTURE: The wing is a conventional two-spar semi-monocoque fail-safe structure of duralumin, with 075 ST upper and 024 ST lower skins. The fuselage is a conventional semi-monocoque fail-safe structure of duralumin frames and stringers, built in forward, central and rear main sections. Upper part of nosecone opens upward for access to avionics and equipment. The tail unit is a cantilever semi-monocoque structure of duralumin; two-spar rudder and elevators have magnesium alloy trailing-edges.

LANDING GEAR: Hydraulically retractable tricycle type, with emergency mechanical back-up, all units retract forward, steerable nose unit (33° left and right) into fuselage, main units into engine nacelles. Kronprinz Ring-Feder shock-absorber in each unit. Single wheel on nose unit, twin wheels on main units, all with Type III tubeless tyres size 7.5-10. Tyre pressure 3.10 bars (45 lb/sq in) on all units. Hydraulic disc brakes on mainwheels only. Parking and emergency brake. No anti-skid units. Landing gear suitable for grass strip operation. Provision for 80 m (262 ft) take-off run using three JATO bottles attached to underfuselage pylon.

POWER PLANT: Two 729 kW (978 shp) Turbomeca Astazou XVIG turboprops, each driving a Ratier-Forest 23LF-379 three-blade variable-pitch fully feathering metal propeller with spinner. Water injection system, flow rate 2.5 litres (0.66 US gallons; 0.55 Imp gallons)/min for two AMC (Area de Material Córdoba) fuselage tanks combined capacity 772 litres (204 US gallons; 170 Imp gallons) and one AMC self-sealing tank in each wing combined capacity 508 litres (134 US gallons; 111 Imp gallons). Overall usable internal capacity 1,280 litres (338 US gallons; 281 Imp gallons). Gravity refuelling point for all tanks on top of fuselage aft of cockpit. Fuel system includes two accumulator tanks, permitting up to 30 seconds of inverted flight. A long-range auxiliary tank, usable capacity 318 or 1,100 litres (84 or 290 US gallons; 70 or 242 Imp gallons) can be attached to the fuselage centreline pylon, and a 318 litre (84 US gallon; 70 Imp gallon) auxiliary on each underwing pylon. Possible external fuel loads are therefore 318, 636, 954, 1,100 or 1,736 litres (84, 168, 252, 290 or 458 US gallons; 70, 140, 210, 242, or 382 Imp gallons); maximum internal and external usable fuel capacity is 3,016 litres (796 US gallons; 663 Imp gallons). Oil capacity 11 litres (2.9 US gallons; 2.4 Imp gallons).

ACCOMMODATION: Pilot and co-pilot in tandem on Martin-Baker AP06A zero/zero ejection seats beneath single AMC moulded Plexiglas canopy which is hinged at rear and opens upward. Rear (co-pilot) seat elevated 25 cm (10 in) above front seat. Rearview mirror for each crew member. Teleflex tinted and bulletproof windscreen with wiper. Armour plating in cockpit floor, resistant to 7.62 mm ground fire from 150 m (500 ft). Dual controls and blind-flying instrumentation standard. Cockpits heated and ventilated by mixture of engine bleed and external air.

SYSTEMS: Air conditioning, de-icing and anti-*g* systems optional. Hydraulic system, with air/oil reservoir and two engine-driven pumps each delivering 4 litres (1 US gallon; 0.9 Imp gallon)/min at 175 bars (2,538 lb/sq in), actuates landing gear, flaps, nosewheel steering and mainwheel brakes. Independent pneumatic (compressed air) system on each engine to supply water injection, fuel system, inverted flight accumulators, auxiliary fuel tank transfer, and (port engine only) canopy sealing system. Electrical system includes two 28.5 V 5 kW Turbomeca engine-driven DC starter/generators; two 250 VA Flite-Tronics static inverters (main and standby), fed from DC emergency busbar, for 115/26 V single-phase AC power at 400 Hz; and a 24 V 36 Ah SAFT Voltabloc 4006A Ni/Cd battery. No APU. Main oxygen system uses 5 litre (1.3 US gallon; 1.1 Imp gallon) Bendix liquid oxygen bottle and

LAA (FMA) IA 58A Pucará (ex-Argentine Air Force) twin-turboprop aircraft *(Peter J. Cooper)*

1994

LAA (FMA) IA 58A Pucará twin-turboprop close-support aircraft

1995

lox converter; separate gaseous oxygen supply for emergency use; l'Hotellier fire detection and extinguishing system, with Graviner extinguisher bottle.

AVIONICS AND SYSTEMS: Standard avionics include Delta VOR/ILS, Smith magnetic compass, Sperry gyro compass and dual artificial horizons, Bendix DFA-73A-1, ADF receiver, Bendix RTA-42A VHF com transceiver, SunAir RE-800 HF com transceiver, Bendix RNA-34 VOR/LOC/glide slope receiver, SunAir ACU-810 HF coupler/amplifier, Delta audio amplifier and intercom. Optional avionics include ECM, weather radar, IFF, inertial navigation system, Machmeter, and VHF-FM tactical communications system. Standard equipment includes dual Pioneer airspeed and vertical speed indicators, dual Kollsman altimeters, dual Air Precision turn and bank indicators, dual Bendix accelerometers, dual attitude indicators (plus standby), dual bearing/distance/heading indicators, flap position indicator, dual landing gear position indicators, Air Precision chronometer, dual Jaeger engine rpm and AMC torque indicators, dual Brion Leroux propeller pitch indicators, dual Jaeger fuel and oil temperature indicators, dual Faure Herman fuel flow and AMC fuel quantity indicators, heated pitot intake, three pitot static ports, GE landing light in leading-edge of each underwing pylon, AMC taxying light on nosewheel strut, fin-tip anti-collision light, wingtip navigation lights, instrument panel lights and warning lights.

ARMAMENT: Two 20 mm Hispano DCA-804 cannon, each with 270 rds in underside of forward fuselage; and four 7.62 mm FN Browning M2-30 machine guns, each with 900 rds, in sides of fuselage abreast of cockpit. Alkan 115E ejector pylon on centreline beneath fuselage, capacity 1,000 kg (2,205 lb); Alkan 105E pylon capacity 500 kg (1,102 lb), beneath each wing outboard of engine nacelle. Maximum external stores load with full internal fuel is 1,500 kg (3,307 lb), including gun and rocket pods, bombs, cluster bombs, incendiaries, mines, torpedoes, air-to-surface missiles, camera pod(s) or auxiliary fuel tank(s). Maximum external weapons load when carrying drop tanks on the fuselage or wing stations is 1,000 kg (2,205 lb). Typical loads can include 12 125 kg bombs; seven launchers each with 19 2.75 in rockets; a 12.7, 20 or 30 mm gun pod and two 318 litre drop tanks; six 125 kg bombs and 16 5 in rockets; six launchers each with 74 mm cartridges, plus onboard ECM; twelve 250 lb napalm bombs; three 500 kg delayed action bombs. Alternative weapon stores include two twin 7.62 mm machine gun pods three launchers each containing 19 2.75 in rockets. SFOM 83A3 reflector sight permits weapon release at any desired firing angle; optional Bendix AWE-1 programmer allows release in step or ripple modes of single weapons, pairs or salvos.

DIMENSIONS, EXTERNAL:
Wing span 14.50 m (47 ft 6⅞ in)
Wing chord:
centre-section (constant) 2.24 m (7 ft 4¼ in)
at tip 1.60 m (5 ft 3 in)
Wing aspect ratio 6.9
Length overall 14.253 m (46 ft 9⅛ in)
Length of fuselage 13.657 m (44 ft 10⅛ in)
Fuselage: Max width 1.32 m (4 ft 4 in)
Max depth 1.95 m (6 ft 4¾ in)
Height overall 5.362 m (17 ft 1⅛ in)
Tailplane span 4.70 m (15 ft 5 in)
Wheel track (c/l of shock-absorbers) 4.40 m (13 ft 9¼ in)
Wheelbase 3.885 m (12 ft 9 in)
Propeller diameter 2.59 m (8 ft 6 in)

DIMENSIONS, EXTERNAL:
Cockpit: Length 2.85 m (9 ft 4¼ in)
Max width 0.81 m (2 ft 8 in)
Max height 1.25 m (4 ft 1¼ in)
Floor area 2.90 m^2 (31.2 sq ft)
Volume 2.74 m^3 (96.8 cu ft)

AREAS:
Wings gross 30.30 m^2 (326.1 sq ft)
Ailerons (total) 2.00 m^2 (21.53 sq ft)
Trailing-edge flaps (total) 3.54 m^2 (38.10 sq ft)
Fin, excl dorsal fin 38.85 m^2 (41.76 sq ft)
Rudder, incl tab 1.15 m^2 (12.38 sq ft)
Tailplane 4.60 m^2 (49.51 sq ft)
Elevators (total incl tabs) 2.612 m^2 (28.11 sq ft)

WEIGHTS AND LOADINGS:
Weight empty, equipped 4,020 kg (8,862 lb)
Max fuel load: internal 1,000 kg (2,205 lb)
external 1,359 kg (2,997 lb)
Max external stores load 1,500 kg (3,307 lb)
Max T-O weight 6,800 kg (14,991 lb)
Max zero-fuel weight 4,546 kg (10,222 lb)
Max landing weight 5,600 kg (12,345 lb)
Max wing loading 224.4 kg/m^2 (45.97 lb/sq ft)
Max power loading 4.666 kg/kW (7.66 lb/shp)

PERFORMANCE (at AUW of 5,500 kg: 12,125 lb except where indicated):
Max critical Mach number at max T-O weight 0.77
Never-exceed speed at max T-O weight Mach 0.63 (405 kts; 750 km/h; 466 mph)
Max cruising speed at 6,000 m (19,680 ft) 259 kts (480 km/h; 298 mph)
Econ cruising speed 232 kts (430 km/h; 267 mph)
Max speed for landing gear extension (all weights) 150 kts (278 km/h; 172 mph)
Stalling speed, flaps and landing gear down, AUW of 4,790 kg (10,560 lb) 78 kts (143 km/h; 89 mph)
Max rate of climb at S/L 1,080 m (3,543 ft)/min
Service ceiling 10,000 m (32,800 ft)
Min ground turning radius 6.50 m (21 ft 4 in)
T-O run 300 m (985 ft)
T-O to 15 m (50 ft) 705 m (2,313 ft)
Landing from 15 m (50 ft), landing weight of 5,100 kg (11,243 lb) 603 m (1,978 ft)
Landing run, landing weight as above 200 m (656 ft)
Attack radius at T-O weight of 6,500 kg (14,330 lb), 10% reserves of initial fuel:
with 1,500 kg (3,307 lb) of external weapons:
lo-lo-lo 121 n miles (225 km; 140 miles)
lo-lo-hi 175 n miles (325 km; 202 miles)
hi-lo-hi 189 n miles (350 km; 217 miles)
with 1,200 kg (2,645 lb) of external weapons:
lo-lo-lo 216 n miles (400 km; 248 miles)
lo-lo-hi 310 n miles (575 km; 357 miles)
hi-lo-hi 350 n miles (650 km; 404 miles)
with 800 kg (1,764 lb) of ordnance and 450 l (119 US gallons; 99 Imp gallons of external fuel):
lo-lo-lo 310 n miles (575 km; 357 miles)
lo-lo-hi 445 n miles (825 km; 512 miles)
hi-lo-hi 526 n miles (975 km; 606 miles)
Ferry range at 5,485 m (18,000 ft) with max internal and external fuel 2,002 n miles (3,710 km; 2,305 miles)
g limits +6/–3

VERIFIED

AUSTRALIA

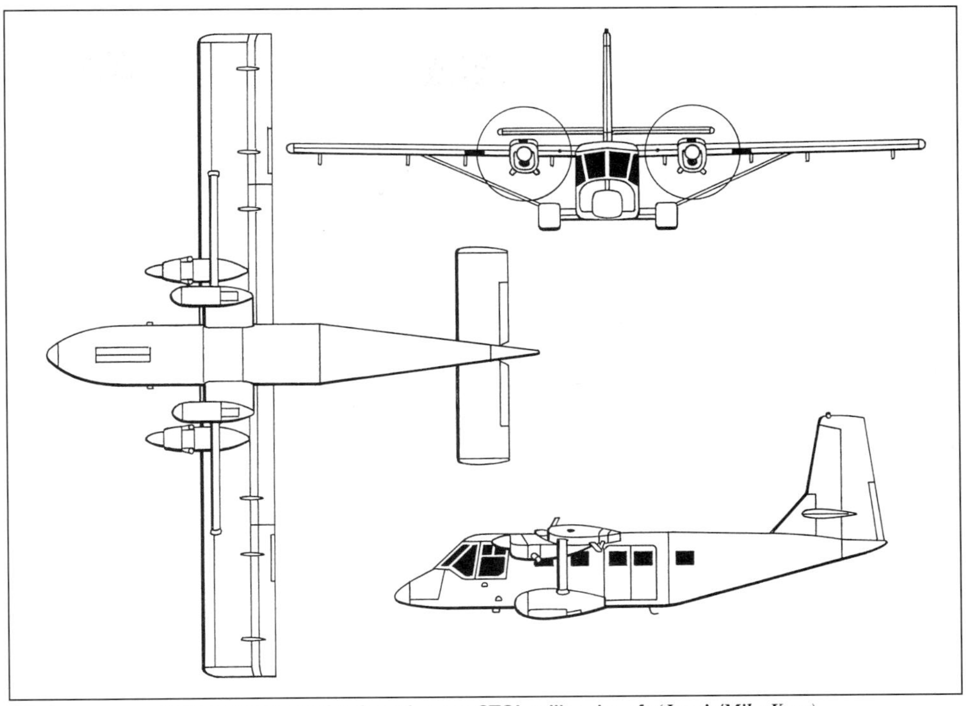

ASTA (GAF) Nomad twin-turboprop STOL utility aircraft *(Jane's/Mike Keep)* *1994*

ASTA (GAF) Nomad of the Royal Australian Army Air Corps *1994*

ASTA

AEROSPACE TECHNOLOGIES OF AUSTRALIA LTD

CORPORATE OFFICE: Private Box 226, Port Melbourne, Victoria 3207
Telephone: 61 (3) 647 3111
Fax: 61 (3) 646 4381 or 2253
Telex: AA 34851
CHAIRMAN: Leigh Masel
MANAGING DIRECTOR AND ACTING CEO: Richard Santo
MEDIA COMMUNICATIONS: Andrea Edwards
BUSINESS UNITS:

ASTA Airport, Private Bag 9, Avalon Airport, Lara, Victoria 3212
Telephone: 61 (52) 279 555
Fax: 61 (52) 823 335
AIRPORT DEVELOPMENT MANAGER: Dennis Chant

ASTA Defence, Private Bag 4, Avalon Airport, Lara, Victoria 3212
Telephone: 61 (52) 279 444
Fax: 61 (52) 823 345
GENERAL MANAGER: Colin Giles

ASTA Engineering, PO Box 376, Port Melbourne, Victoria 3207
Telephone: 61 (3) 647 3111
Fax: 61 (3) 645 2582
GENERAL MANAGER: Michael Barker

ASTA Components, Private Bag 4, Port Melbourne, Victoria 3207
Telephone: 61 (3) 647 3111
Fax: 61 (3) 645 3424
GENERAL MANAGER: Geoff Steele

ASTA Aircraft Services Pty Ltd (ASTAAS), Private Bag 2, Avalon Airport, Lara, Victoria 3212
Telephone: 61 (52) 279 111
Fax: 61 (52) 823 892
MANAGING DIRECTOR: David Main

Pacific Aerospace Corporation Ltd (PAC): See New Zealand section

Established 1986 as private enterprise organisation to succeed former Government Aircraft Factories (GAF). Group specialises in design, development, manufacture, assembly, maintenance and modification of aircraft, target drones and guided weapons; component subassembly, final assembly, modification, repair and test flying of jet and other aircraft undertaken at Avalon airfield.

ASTA Airport created October 1990 following July 1990 purchase of Avalon airfield; ASTA Defence created 1 July 1992 by combining former Military Aircraft Services, Systems and General Aviation business units. ASTA group has 75.1 per cent stake in Pacific Aerospace Corporation (see New Zealand section); HAECO of Hong Kong has 9.9 per cent interest in ASTAAS.

Recent and current aircraft activity includes component manufacture, forward fuselage fitting out, final assembly and flight test of RAAF AF/ATF-18A Hornets; 40 per cent of Australian production work on RAAF Pilatus PC-9/A programme (completed October 1991); and assembly of eight of 16 Seahawks for Royal Australian Navy. Subcontract work includes components for Boeing 747 (Krueger flaps), 757 (inspar wing ribs and rudders) and 777 (rudders); Airbus A320 (flap shrouds); Sikorsky Black Hawk/Seahawk (rotor blade components); McDonnell Douglas MD-80 series; and McDonnell Douglas MD-11 (CFM56 nacelle components). ASTA also a risk-sharing contractor to Aerospatiale to design/develop/manufacture carbonfibre main and central landing gear doors and floor support panels for Airbus A330/A340.

UPDATED

ASTA (GAF) NOMAD

TYPE: Twin-turboprop STOL utility aircraft.

PROGRAMME: The N2 Nomad prototype flew for the first time on 23 July 1971; Australian type certificates for the N22 and N24 initial production versions, were issued in May 1975 and October 1977 respectively.

The N22B and N24A were type certificated in August 1975 and May 1978 respectively, and in December 1978 these versions were awarded US FAR Pt 135 Appendix A certification (Transport category). Type certificates have also been issued in various European, Asian, South American and Pacific countries.

Nomad production was suspended in late 1984 after completion of 170 aircraft; 148 of these had been sold by early 1985.

VERSIONS: **N22B:** Short fuselage civil version, for up to 13 passengers and/or cargo. Total of 33 delivered by January 1985, including Commuter/Cargo, Medicmaster and Surveymaster versions, and two Floatmasters (see following paragraph).

Floatmaster: Version of N22B with twin Wipline 9812 floats or 9500 amphibious gear.

Missionmaster: Short fuselage military version, in service for forward area support and surveillance, and as a light personnel and equipment transport. Delivered to, or ordered by, Australian Army (23), Papua New Guinea Defence Force (4), Philippine Air Force (15+), and Royal Thai Air Force (22).

N24A: Civil version with 1.14 m (3 ft 9 in) plug inserted in cabin. Accommodation for up to 17 passengers; increased forward baggage capacity. Total of 40 delivered, including Commuterliner, Medicmaster and Cargomaster versions. Five in service with Royal Thai Navy N24A (SLAR).

Searchmaster B: Basic coastal patrol version. Bendix RDR 1400 search radar in nose radome. Delivered to, or ordered by, Indonesian Navy (12), Marshall Islands (2), Northern Territory Air Work (3) and Queensland Air (1).

Searchmaster L: More sophisticated coastal patrol version, operated by Indonesian Navy (6), Northern Territory Air Work (2), Executive Air (3) and Reprographics (1). Litton APS-504(V) 2 search radar in under-nose 'guppy' radome. Litton APS-504(V) 5 USCS (7).

OPERATORS: Versions of the ASTA (GAF) Nomad are in service with the air arms of the following countries: Australia (N22/24 23), Indonesia (Searchmaster B/L 18), Papua New Guinea (N22 4), Philippines (Mission Master 16) and Thailand (N24A 5).

DESIGN FEATURES: Braced high-wing monoplane. Basic NACA 23018 wing section, modified to incorporate increased nose radius and camber. Dihedral 1° from roots. Incidence 2°. No sweepback. Small stub wings at cabin floor level support the main landing gear fairings from which a single strut on each side braces the main wing. The tail unit is a cantilever all-metal structure. Ventral fin on floatmaster.

FLYING CONTROLS: Full-span double-slotted trailing-edge flaps. All-metal ailerons, which droop with the flaps and transfer their motion progressively to slot-lip ailerons as the flaps extend, resulting in full-span flap. Controls actuated manually by cables and pushrods. One-piece all-moving tailplane with inset trim and anti-balance tab. Tailplane and rudder actuated manually by cables. Trim tab in rudder.

STRUCTURE: The wing is a two-spar fail-safe torsion-box structure of riveted light alloy. The fuselage is a conventional semi-monocoque riveted light alloy structure of stringers and frames.

LANDING GEAR: Retractable tricycle type with electrical retraction by means of a single actuator in the fuselage. Main wheels retract forward into streamline fairings at outer end of stub-wings. Dual hydraulically operated single-disc brakes on main units. No anti-skid units. Wipaire Wipline 9812 twin-float gear on Floatmaster.

POWER PLANT: Two 313 kW (420 shp) Allison 250-B17C turboprop engines, each driving a Hartzell three-blade constant-speed fully feathering reversible-pitch metal propeller with Beta control. Standard fuel capacity 1,018 litres (269 US gallons; 224 Imp gallons) plus 20 litres (5.3 US gallons; 4.4 Imp gallons) unusable in flexible bag tanks. Provision for internal auxiliary tanks for ferry purposes. An additional fuel capacity of 335 litres (88.5 US gallons; 73.7 Imp gallons) is provided by two operational integral tanks, one in each wingtip. Gravity refuelling via overwing point above each pair of tanks. Oil capacity 8.5 litres (2.3 US gallons; 1.9 Imp gallons) per engine.

ACCOMMODATION (N22B): Flight deck accommodates a crew

ASTA (GAF) Nomad Searchmaster 'B' of the Indonesian Naval Air Arm

1994

of two on side by side seats, but certification covers single pilot operation in countries where this applies. Access to flight deck by forward-opening door on each side. Main cabin has individual seats for up to 12 passengers, at 74 cm (29 in) pitch, with continuous seat tracks and readily removable seats which allow rapid rearrangement of the cabin to suit alternative loads. Access to main cabin via double doors on port side, with single emergency exit on starboard side. Baggage compartments in nose (with door on each side - left-hand door only on later units) and optionally in rear of fuselage (with internal and external access). Whole interior, including flight deck, is heated and ventilated.

ACCOMMODATION (N24A): Flight deck accommodation and access as for N22B. Lengthened main cabin, with similar internal provision to N22B for up to 16 passengers, and access via double port-side doors as in N22B. Enlarged nose baggage compartment. Rear baggage compartment of same capacity as N22B. Ventilation and heating system with individual adjustable outlets.

ARMAMENT AND OPERATIONAL EQUIPMENT (Missionmaster/Searchmaster): Provision for four underwing hardpoints capable of accepting up to 227 kg (500 lb) loads, including gun and rocket pods. The nose bay can accommodate surveillance and night vision aid equipment. Removable seat armour and self-sealing fuel tanks can be fitted.

ASTA (GAF) Searchmaster L maritime patrol aircraft

1995

DIMENSIONS, EXTERNAL:
- Wing span: 16.51 m (54 ft 2 in)
- Length overall: N22B: 12.57 m (41 ft 3 in)
  - N24A: 14.35 m (47 ft 1 in)
- Height overall: 5.54 m (18 ft 2 in)
- Wheel track: 2.90 m (9 ft 6 in)
- Wheelbase: N22B: 3.73 m (12 ft 3 in)
  - N24A: 4.45 m (14 ft 7 in)

AREA:
- Wings, gross: 30.10 m² (324.0 sq ft)

WEIGHTS AND LOADINGS (landplane versions):
- Manufacturer's basic weight empty:
  - N22B: 2,092 kg (4,613 lb)
  - N24A: 2,377 kg (5,241 lb)
- Typical weight empty, equipped:
  - Cargomaster: 2,269 kg (5,002 lb)
  - Missionmaster: 2,290 kg (5,050 lb)
  - Medicmaster: N22B: 2,304 kg (5,080 lb)
    - N24A: 2,449 kg (5,399 lb)
  - Surveymaster: 2,786 kg (6,143 lb)
  - Commuterliner: N24A: 2,407 kg (5,306 lb)
  - Searchmaster: B: 2,413 kg (5,321 lb)
    - L: 2,675 kg (5,897 lb)
- Max fuel load (usable), N22B and N24A:
  - standard: 803 kg (1,770 lb)
  - extended range: 1,066 kg (2,350 lb)
- Max T-O weight:
  - N22B (all versions): 3,855 kg (8,500 lb)
  - Searchmaster L: 4,127 kg (9,100 lb)
  - N24A: 4,264 kg (9,400 lb)
- Max zero-fuel weight: N22B: 3,742 kg (8,250 lb)
  - N24A: 4,150 kg (9,150 lb)
- Max landing weight: N22B: 3,855 kg (8,500 lb)
  - N24A: 4,173 kg (9,200 lb)
- Max floor loadings:
  - main cabin: N22B: 732 kg/m² (150 lb/sq ft)
    - N24A: 488 kg/m² (100 lb/sq ft)
  - rear main cabin: N24A: 342 kg/m² (70 lb/sq ft)
  - rear fuselage: N22B, N24A: 244 kg/m² (50 lb/sq ft)
- Max wing loading: N22B: 127.9 kg/m² (26.2 lb/sq ft)
  - N24A: 141.6 kg/m² (20.0 lb/sq ft)
- Max power loading:
  - N22B: 6.16 kg/kW (10.12 lb/shp)
  - N24A: 6.81 kg/kW (11.19 lb/shp)

PERFORMANCE (landplane versions, at max T-O weight, ISA at S/L, except where indicated):
- Normal cruising speed:
  - N22B, N24A: 168 kts (311 km/h; 193 mph)
- Stalling speed, power off, flaps up, at AUW of 3,402 kg (7,500 lb):
  - N22B, N24A: 65 kts (121 km/h; 75 mph)
- Stalling speed, power off, flaps down, at AUW of 3,402 kg (7,500 lb):
  - N22B, N24A: 47 kts (88 km/h; 55 mph)
- Max rate of climb at S/L, both engines, T-O rating for 5 min: N22B: 445 m (1,460 ft)/min
  - N24A: 390 m (1,280 ft)/min
  - N22B (ISA + 20°C): 396 m (1,300 ft)/min
  - N24A (ISA + 20°C): 325 m (1,066 ft)/min
- Rate of climb at S/L, one engine out, max continous rating:
  - N22B: 73 m (240 ft)/min
  - N24A: 67 m (220 ft)/min
  - N22B (ISA + 20°C): 52 m (170 ft)/min
  - N24A (ISA + 20°C): 49 m (160 ft)/min
- Service ceiling, both engines, climbing at 30.5 m (100 ft)/min, max cruise rating:
  - N22B at 3,630 kg (8,000 lb) AUW: 6,400 m (21,000 ft)
  - N24A at 4,082 kg (9,000 lb) AUW: 6,100 m (20,000 ft)
- Min ground turning radius:
  - N22B, N24A: 11.66 m (38 ft 3 in)
- Runway LCN at max T-O weight: N22B: 2.3
  - N24A: 2.5
- T-O run: N22B (FAR 23): 223 m (730 ft)
  - N24A (FAR 23): 296 m (970 ft)
  - N22B (STOL): 183 m (600 ft)
  - N22B (FAR 23), ISA + 20°C: 296 m (970 ft)
  - N24A (FAR 23), ISA + 20°C: 366 m (1,200 ft)
  - N22B (STOL), ISA + 20°C: 213 m (700 ft)
- T-O to 15 m (50 ft):
  - N22B (FAR 23): 360 m (1,180 ft)
  - N24A (FAR 23): 521 m (1,710 ft)
  - N22B (STOL): 320 m (1,050 ft)
  - N22B (FAR 23), ISA + 20°C: 463 m (1,520 ft)
  - N24A (FAR 23), ISA + 20°C: 610 m (2,000 ft)
- Landing from 15 m (50 ft), AUW of 3,630 kg (8,000 lb):
  - N22B (FAR 23): 408 m (1,340 ft)
  - N22B (STOL): 216 m (710 ft)
  - N22B (FAR 23), ISA + 20°C: 353 m (1,160 ft)
- Landing from 15 m (50 ft), AUW of 4,082 kg (9,000 lb):
  - N24A (FAR 23): 408 m (1,340 ft)
  - N24A (FAR 23), ISA + 20°C: 439 m (1,440 ft)
- Landing run; AUW of 3,630 kg (8,000 lb):
  - N22B (FAR 23): 212 m (695 ft)
  - N22B (STOL): 76 m (250 ft)
  - N22B (FAR 23), USA + 20°C: 204 m (670 ft)
- Landing run, AUW of 4,082 kg (9,000 lb):
  - N24A (FAR 23): 238 m (780 ft)
  - N24A (FAR 23), ISA + 20°C: 256 m (840 ft)
- Max range at 90% power, standard fuel, reserves for 45 min hold: N22B, N24A at S/L
  - 580 n miles (1,074 km; 668 miles)
  - N22B, N24A at 3,050 m (10,000 ft): 730 n miles (1,352 km; 840 miles)

UPDATED

BELGIUM

FN HERSTAL

FNNH SA

Voie de Liége 33, B-4040 Herstal
Telephone: 32 (41) 408111
Fax: 32 (41) 408679
MANAGING DIRECTOR: Daniel Huet
DIRECTOR, SALES: Philippe Glaessens
DIRECTOR, TECHNICAL: Lucien Beciani
MANAGER, PUBLICITY: Chantal Regibeau

FN Herstal originates from 1889 when the Fabrique Nationale d'Armes de Guerre was founded to manufacture rifles for the Belgian Army. The company is a subsidiary of the French Giat Industries group. FN Herstal is currently active in the development and manufacture of airborne weapon systems, weapon support systems as well as infantry weapons and ammunition.

VERIFIED

FN Herstal modular weapons kit consisting of twin FN 7.62 mm twin Mag machine gun pod and FN 2.75 MRL rocket launcher *(FNNH)*

1994

FN HERSTAL MODULAR WEAPONS RETROFIT KIT

TYPE: Airborne weapons upgrade kit (rotary- and fixed-wing).

PROGRAMME: FN Herstal has installed 300 fixed-wing and 250 rotary-wing aircraft with modular weapons kits worldwide.

DESIGN FEATURES: The modular kit consists of a combination of different elements including: an FN .50 calibre heavy machine gun pod (HMP), a combined .50 calibre (HMP) and a four-tube rocket launcher (HMP-RL), an FN range of 7-, 12- and 19-tube rocket launchers (LAU-7H, LAU-12H, LAU-19H, LAU-7A, LAU-19A), an FN 7.62 mm twin MAG machine gun pod (TMP-5). All FN rocket launchers are compatible with FFAR or W-R motors and can be installed on both fixed- and rotary-wing aircraft. Tube calibre is 70 mm (2.75 in). The system is installed in a symmetric configuration to counter any twist effect experienced during weapons firing. Recoil force on any combination of weapons is constantly less than the system weight. FN Herstal has also developed an external mounting system (EMA) in collaboration with Bell Helicopter to enhance even distribution of weapons stores weight and firing load reaction. All systems are installed with constant bore sighting and gravity emergency release mechanism.

UPDATED

FN Herstal HMP fitted on Embraer EMB-312 aircraft *(FNNH)*

1995

FN HERSTAL CREW-SERVED WEAPONS SYSTEM RETROFIT

TYPE: Pintle-mounted 7.62 or .50 calibre machine gun retrofit (rotary-wing).

DESIGN FEATURES: Option 1. Installation of FN .50 calibre M3M machine gun on floor pintle or FN .50 calibre M3M machine gun on external pintle, with elastic cradle for recoil attenuation, perfect balance and ease of firing. Gun grips are also fixed on support cradle.

Option 2. Installation of FN 7.62 mm calibre MAG M58 machine gun on choice of three pintle mountings: floor, external and window.

VERIFIED

SABCA

SOCIETE ANONYME BELGE DE CONSTRUCTION AERONAUTIQUES

Chaussée de Haecht 1470, B-1130 Brussels
Telephone: 32 (2) 729 5511
Fax: 32 (2) 216 1570
Telex: 21 237 SABCA B
CHAIRMAN: J. Detemmerman
DEPUTY DIRECTOR/GENERAL MANAGER: Remo Pellichero
DIRECTOR, MARKETING AND SALES: M. Humblet
OTHER WORKS: Aéroport de Gosselies-Charleroi, B-6041 Gosselies
Telephone: 32 (71) 254211
Fax: 32 (71) 344214
Telex: 51 251 SABGO B

Founded in 1920, Sabca is the major aerospace company in Belgium. Since the Second World War, it has participated in various European aircraft programmes. In Brussels, Sabca is manufacturing structures and parts for the Lockheed F-16, Alpha Jet, Mirage III/5/F.1 and Atlantique 2, Northrop F-5, Airbus A330/340, Fokker 60 and 100, Eurocopter AS 330 Puma and Ariane launchers. Servo controls are produced for the F-16, Ariane launchers and the Hermes space shuttle.

In Charleroi, Sabca, assembled and tested F-16s for Belgium and Denmark. It also modified Belgian Air Force and USAF F-16s and assembled Agusta A 109A helicopters for Belgium including installation and integration of HeliTOW 2 weapon systems and various avionic systems. Sabca's Electronic Division produces airborne electronic components and ground equipment as well as maintaining Doppler equipment.

For many years Sabca has been responsible for the maintenance and overhaul of Belgian and other armed forces military aircraft, their electronic components and accessories, as well as commercial aircraft and helicopters. It is currently integrating ECM devices on Belgian aircraft.

Sabca is a member of various European consortia. Dassault Aviation has a 53 per cent shareholding in the company and Fokker has a shareholding of over 42 per cent. The company's works occupy a total area of approximately 82,000 m^2 (882,640 sq ft) and employs an average of 1,500 personnel.

Dassault Belgique was acquired by Sabca in late 1990. Construction of a new plant for the Sabca Limburg NV subsidiary began in 1990. This new factory has been operational since mid-1991 and produces components in high-tech composite materials for aircraft and aerospace applications.

UPDATED

SABCA (DASSAULT) MIRAGE 5 MIRSIP/ ELKAN UPGRADE PROGRAMME

TYPE: Combat aircraft upgrade.

PROGRAMME: Sabca is prime contractor for the Belgian Air Force's Mirage System Improvement Programme (MIRSIP). In addition to life-extending modifications and maintenance.

The original MIRSIP plan called for upgrading 15 single-seat Mirage 5BAs of No. 8 Squadron and five two-seat Mirage 5BD operational trainers, to be redelivered by the end of 1994. The aircraft were originally to be operated by the Belgian Air force but restructuring has now made the aircraft surplus to requirements. The first upgraded Mirage flew in December 1992. In July 1994 the Chilean government agreed to purchase the upgraded aircraft as well as an additional five aircraft which will not be upgraded. Deliveries to the Chilean Air Force are to take place between January and October 1995.

DESIGN FEATURES: The aircraft are to be fitted with canard control surfaces and pressure refuelling (both Dassault systems), and a new weapons delivery, navigation and reconnaissance system (Sagem Uliss 92 INS, Sagem Tercor terrain contour matching system with 3D digital map database, Thomson-CSF laser rangefinder, Vinten 3150/2768 colour video recording system and GEC Ferranti

Sabca (Dassault) Mirage 5BA upgraded aircraft renamed Elkan (Guardian) by Chilean Air Force *(Sabca)*

1995

HUD) operating via a MIL-STD-1533B databus. (See also Dassault and Sagem entries in French section.) Before delivery, additional complementary modifications will be implemented to the radio navigation system to adapt it to existing ground stations peculiar to Chile.

UPDATED

Sabca (Dassault) Mirage 5BD upgraded aircraft now serving with the Chilean Air Force *(Sabca)*

1995

SABCA (WESTLAND) SEA KING UPGRADE

TYPE: Combat helicopter upgrade.

PROGRAMME: Sabca is to install modification kits, produced by Westland Helicopters, into the Belgian Air Force's fleet of five Sea King Mk 48 helicopters. Programme to begin in 1995.

DESIGN FEATURES: The kits will include a Bendix RD1500B search radar, FLIR 2000FD forward-looking infra-red and a Racal RNS252 navigation system integrated with the Canadian Marconi CMA 3012 global positioning system.

VERIFIED

SABCA (NORTHROP) F-5E/F UPGRADE PROGRAMME

TYPE: Fixed-wing combat aircraft upgrade.

PROGRAMME: Sabca was selected by the Indonesian government in early 1995 to act as systems integrator for Indonesian Air Force F-5E/F Freedom Fighter upgrade programme. Sabca to carry out initial engineering and integration work on prototype F-5E single-seat aircraft and F-5F two-seat operational trainer, including flight testing in Belgium. Remaining aircraft to be upgraded by Indonesian Air Force in Indonesia from kits supplied by Sabca. The programme which is set to last three years includes a four-year spares support programme and structural repair work.

DESIGN FEATURES: Package includes the following preselected equipment: GEC-Marconi Avionics (GMAv) HUD/WAC and Litton LN-93 INS, GEC-Marconi Sky Guardian RWR, MIL-STD-1553 databus, new air data computer, stores management system, mission camera and HOTAS controls. Plans to include new pulse Doppler radar were dropped due to limited funding. However the existing ESCO (formerly Emerson) APG-59(V)3 is currently being upgraded to (V)5 standard.

NEW ENTRY

SABENA

SABENA TECHNICS SA

Brussels National Airport, B-1930 Zaventem
Telephone: 32 (2) 723 4958
Fax: 32 (2) 723 5997/5995
Telex: 63697 SABCOM B
CHAIRMAN: P. Godfroid
EXECUTIVE VICE-PRESIDENT, TECHNICAL: J. Delance
GENERAL MANAGER, SALES AND CONTRACTS: J. M. Rivier
PUBLIC RELATIONS: Miss Manuela Hollanders

Sabena Technics has covered facilities of over 143,000 m^2 (1,538,680 sq ft) and a workforce of over 2,500 personnel. Responsibilities include passenger aircraft heavy maintenance, engines, APU and components overhaul, engineering and engine test nacelle manufacture. The company is also responsible for passenger cargo conversions as well as military life extension programmes for the Belgian Air Force C-130 aircraft. The company has carried out work on the following types of aircraft: B707, B727, B737, B747, DC-10 and A310. In May 1995 Swissair announced that it had acquired a 49.5 per cent stake in Sabena Technics.

NEW ENTRY

Sabena Technics (Lockheed) C-130 Hercules of the Belgian Air Force undergoing the outer wing replacement programme *(Sabena)*

1995

SABENA TECHNICS (LOCKHEED) C-130 LIFE EXTENSION PROGRAMME

TYPE: Military transport life extension.

PROGRAMME: In order to increase the lives of Belgian Air Force C-130 Hercules by 15 years, Sabena Technics was contracted to re-wing the aircraft. The programme was initiated in 1990 and completed May 1994. Number of man-hours required per aircraft was 2,000.

CUSTOMERS: Belgian Air Force.

DESIGN FEATURES: Programme consisted of replacing the outer wings of the 12 Belgian Air Force C-130H Hercules transport aircraft. The wings were manufactured by Lockheed (Georgia USA).

NEW ENTRY

SABENA TECHNICS (LOCKHEED) C-130H AVIONICS UPGRADE PROGRAMME (AUP)

TYPE: Military transport aircraft avionics upgrade.

PROGRAMME: Contract awarded late 1992 to Sabena as prime contractor. Honeywell Defense Avionics Systems Division (DASD) selected as main subcontractor to install its Integrated Vehicle Mission Management System (IVMMS). Fabrisys subcontracted for rewiring work. Prototype ground testing commenced December 1994. Prototype test flying to commence first quarter 1995.

CUSTOMERS: Programme covers 12 Belgian Air Force C-130H Hercules transport aircraft.

COSTS: Total programme budget is approximately $38 million. Half of this sum covers the development work and prototyping.

DESIGN FEATURES: System incorporates flight management calculations and control including: computer air release point for parachuting of men and equipment/provisions (CARP), search and rescue (SAR), take-off and landing data, navigation and communication radio management, Jeppesen database, performance computations, flight planning. System also provides data transfer of: data prepared on a mission planning ground station (MPGS), Jeppesen database updates, test data.

AVIONICS: IVMMS consists of installation of: new autopilot with air data sensor unit (ADSU), flight management system (FMS) with cockpit display unit (CDU), electronic flight instrument system (EFIS), new weather radar, attitude heading reference system (AHRS), inertial navigation system (INS) and global positioning system (GPS).

NEW ENTRY

SABENA TECHNICS (SWEARINGEN/ FAIRCHILD) MERLIN IIIA CABIN REFURBISHMENT

TYPE: Military VIP/calibration aircraft interior upgrade.

PROGRAMME: Five aircraft of the Belgian Air Force were refurbished. The programme was initiated in January 1994 and completed in December 1994.

DESIGN FEATURES: Programme consisted of the installation of seat tracks over the entire cabin length, new floor panels, emergency lighting system, new insulation blankets behind sidewall linings, new window and sidewall linings, new carpets throughout aircraft and new toilet system and seat. Programme also included refurbishing forward and aft partitions and the relocation of reading lights and fresh air and oxygen outlets.

NEW ENTRY

SABENA TECHNICS (MCDONNELL DOUGLAS) DC-10-30 CARGO CONVERSION

TYPE: Passenger to cargo conversion.

PROGRAMME: Three Federal Express aircraft converted from mixed cabin to full cargo configuration. The programme was initiated in November 1991 and completed in November 1992 and required three months per aircraft.

DESIGN FEATURES: Programme included: conversion from 4 zone passenger type to 3 zone freighter type, removal of cabin air recirculation system, zone temperature control modification, removal of PA and entertainment system and the removal of all bins, galleys, toilets and coatrooms. Conversion also included installation of sidewall and ceiling gilliners, smoke barrier and 9*g* barrier net, courier seat, anti-skid coating on cargo floors, complete cargo kits as well as passenger door deactivation and window plug installation.

NEW ENTRY

Sabena Technics (Lockheed) C-130 Hercules of the Belgian Air Force with new avionics suite *(Sabena)*
1995

Sabena Technics (Swearingen/Fairchild) refurbished Merlin IIIA of the Belgian Air Force *(Sabena)*
1995

Sabena Technics (McDonnell Douglas) DC-10-30 undergoing cargo conversion *(Sabena)*
1995

BOSNIA-HERZEGOVINA

SOKO

VAZDUHOPLOVNA INDUSTRIJA SOKO DD

Radõc bb, 88000 Mostar
Telephone: 38 (88) 53749, 21692
Fax: 38 (88) 423049, 423205
Telex: 46322

Soko (Falcon) Aircraft Industry founded 14 October 1950, initially as subcontractor to existing aircraft industry, including Ikarus, to which Soko was informal successor; produced parts for indigenous S-49C, Aero 2, 212 and 213, Soviet-designed Yak-2 and Yak-9, and rebuilt Republic F-84G Thunderjets; prime contractor on Type 552 and G-2A Galeb trainers, plus Jastreb attack aircraft; manufactured Kraguj counter-insurgency type; Galeb/Jastreb built since 1963 and exported to Zambia and Libya; Galeb 3 developed late 1969, but not built. Soko privatised March 1991, 3,000 of its 3,500 workers having held 20 per cent of $110 million capital.

Claimed independence of Bosnia-Herzegovina from former Yugoslavia recognised by European Community 7 April 1992. Successor state of Yugoslavia transferred many Soko activities to Utva (which see for G-2 Galeb, G-4 Super Galeb and J-1 Jastreb); Mostar factory ceased production 1 May 1992; some partly complete G-4 Super Galebs abandoned on production line; infrastructure damaged by civil war; possession of Mostar area in dispute.

International collaboration included subcontract work on de Havilland Dash 8 (emergency exits); Embraer EMB-120; Airbus A300/310/330/340 (rudder shell), A320 (cargo compartment) and A330/340 (fuselage panels) — in co-operation with Aerospatiale. SOCEA-SOGERMA, Dornier/DASA and Korean Air; ATR 42 (frames); McDonnell Douglas MD-80 and MD-11 in co-operation with Alenia and IAI; Eurocopter Super Puma (sponsons and rotor blades); Dassault Aviation; and Tupolev Tu-204 (structural elements and assemblies) in association with UAPK. On 29 July 1990, Soko delivered to Aviaexport of Moscow a half-scale Ilyushin Il-114, produced in collaboration with Utva, including electrical, pressure and hydraulic systems, for wind tunnel tests. Efforts continued until 1992 to join international collaborative ventures including design and manufacture of structural assemblies and components in metals and composites. Mostar plant included 140,000 m² (1,507,000 sq ft) of covered floor area.

Other activities included fabrication of wings and forward fuselage, and assembly of G-4 Super Galeb; licence manufacture of Eurocopter Gazelle, including some subassemblies for French assembly line; final assembly of Orao attack aircraft (see Soko/Avioane in International section) and wing production for both assembly lines; manufacture of front fuselage and wings of Utva Lasta; and production of aircraft ground and test equipment, rocket pods and bomb racks.

VERIFIED

BRAZIL

EMBRAER

EMPRESA BRASILEIRA DE AERONÁUTICA SA

Av Brig Faria Lima 2170, Caixa Postal 343, 1227901 São José dos Campos, SP
Telephone: 55 (123) 25 1711
Fax: 55 (123) 21 8466
Telex: (391) 1233589 EBAE BR
CHAIRMAN: José Elisland Baya de Barros
CHIEF EXECUTIVE OFFICER: Ozires Silva
COMMERCIAL DIRECTOR: Paulo Rubens Lancia Cury
TECHNICAL DIRECTOR: Horácio Aragonés Forjaz
PRESS OFFICER: Mario B. de M. Vinagre
US SUBSIDIARY:
Embraer Aircraft Corporation, 276 Southwest 34th Street, PO Box 21623, Fort Lauderdale, Florida 33315
Telephone: 1 (305) 524 5755 and 5744
Telex: (230) 522318 EMBRAER FORT LAUDERDALE
PARIS OFFICE:
Embraer Aviation International, BP 74, Aéroport du Bourget, Zone d'Aviation d'Affaires, F-93350 Le Bourget, France
Telephone: 33 48 35 94 20
Telex: 213498F EBAE PAR

Embraer, which was founded on 19 August 1969 and commenced operations on 2 January 1970, celebrated its silver anniversary in 1994. The Brazilian government owns 63.81 per cent of the voting shares, 36.19 per cent of the subscribed capital being held by private shareholders. Embraer has a factory area of 275,198.55 m² (2,962,210 sq ft), and had built a total of 4,614 aircraft by 1 January 1994.

Since August 1974, Embraer has had an agreement with Piper Aircraft Corporation to manufacture various Piper types under licence. A 1987 agreement with McDonnell Douglas provides for the supply of 200 sets of wing flaps, in composite material, for the MD-11 airliner, with a further 100 sets on option. Deliveries began on 29 October 1988 and are continuing.

Current programmes include the EMB-120 Brasilia in its various versions (EMB-120 Cargo, EMB-120 Combi, EMB-120QC (Quick-Change), all of them based on the current production model - the EMB-120ER Advanced Brasilia); the EMB-312F Tucano, for France; the EMB-312H Super Tucano for the Brazilian Air Force and to compete, through association with Northrop, in the JPATS (joint primary aircraft training system) competition that will select the trainer that will equip both the US Navy and Air Force from the mid-1990s; the AMX attack jet for Brazil and Italy, and the EMB-145 jet transport, which is due to fly early this year (1995) and has letters of intent from 14 customers in eight countries covering 136 aircraft. Also current are subcontract programmes with McDonnell Douglas (MD-11 composite outboard flaps) and Boeing (machined parts for the 747, 767 and 777). The Embraer privatisation was completed on 7 December 1994 with the sale of 55.4 per cent of the voting stock. A consortium led by the bank Bozano Simonsen acquired 45.44 per cent of the stock. The remaining 9.96 per cent of the stock was acquired by private investors and two investment banks. The new owners are obliged to inject $36 million in fresh money to re-capitalise the company.

UPDATED

Embraer EMB-110 Bandeirante C-95 *(Jane's/Mike Keep)*

1994

EMBRAER EMB-110 BANDEIRANTE (PIONEER)

Brazilian Air Force designations: C-95, EC-95, R-95 and SC-95

TYPE: Twin-turboprop general purpose transport.

PROGRAMME: Bandeirantes have been sold to more than 80 operators in 36 countries worldwide, and by 1 September 1990 the 490 aircraft then delivered had logged more than 5.2 million flying hours. Production ended with the 500th aircraft, a C-95C for the Brazilian Air Force, which was delivered during 1990. With its delivery, that service had received 156 Bandeirantes (60 C-95, 20 C-95A and 28 C-95B transports; two EC-95Bs for navaid calibration; 1 XC-95B for aircraft rain research; five search and rescue SC-95Bs; six aerial photogrammetry R-95s; 12 dihedral-tailed, EFIS equipped C-95Cs and 22 EMB-111s).

VERSIONS: **EMB-110P1:** Stretched, passenger/cargo convertible version of the EMB-110P.

EMB-110P1A: Updated version of P1 with the following improvements: 10° tailplane dihedral, to reduce vibration and noise mass balance relocated to centreline of elevator; mass balance on elevator tab; duplicated control rods to elevator; passenger seats attached to floor only; improved acoustic internal lining; improved emergency exit seals; improved main cabin door seal; new cabin carpet with foam insulation; front intake for venting system. First two P1As delivered in December 1983 to Provincetown-Boston Airlines, USA. Replaced P1 as standard version from c/n 439 onwards. Two delivered to Colombian Air Force in December 1993.

EMB-110P1K: Military utility, cargo and paradropping version. No tailplane dihedral. Brazilian Air Force designation SC-95A.

EMB-110P1K SAR: Search and rescue version of P1K, equipped for inland and overwater search, paradropping and aeromedical evacuation. Maximum T-O weight 6,000 kg (13,230 lb). Accommodation for up to six stretchers plus seats for observers and space for inflatable dinghies and other rescue equipment. Two bubble windows in each side of cabin. Independent oxygen system for medevac missions. Brazilian Air Force designation SC-95A/B.

EMB-110P2: Third-level commuter transport version, carrying up to 21 passengers. First flown on 3 May 1977. Detailed description applies mainly to this version.

EMB-110P2: Replaced former P2 as third-level commuter transport version, carrying up to 21 passengers. Incorporates same changes as PIA.

EMB-110P1A/41 and EMB-110P2A/41: Replaced P1/41 and P2/41 from 1983, certificated under SFAR Pt 41 for a maximum T-O weight of 5,900 kg (13,010 lb). Power plant and dimensions unchanged.

EMB-111: Maritime version of the 110.

OPERATORS: Versions of the EMB-110 are in service with the air arms of the following countries: Brazil (120+), Colombia (2) and Peru (3).

DESIGN FEATURES: Cantilever low-wing monoplane. Wing section NACA 23016 (modified) at root, NACA 23012 (modified) at tip, sweepback 0° 19' 48'' at quarter chord. Dihedral 7° at 28 per cent chord. Incidence 3°. Two upward hinged doors, one on each side of nose, provide access to avionics. Ventral fin. Wing and tail de-icing system optional.

FLYING CONTROLS: Statically balanced Frise ailerons and double-slotted flaps. Trim tab in port aileron. Trim tabs in rudder and port elevator.

STRUCTURE: The wing is an all-metal two-spar structure, of 2024-T3 and -T4 aluminium alloy, with detachable glassfibre wingtips. Glassfibre wing/fuselage fairing. All-metal ailerons and flaps. The fuselage is an all-metal semi-monocoque structure of 2024-T3 aluminium alloy. The tail unit is a cantilever all-metal structure, with sweptback vertical surfaces. Glassfibre dorsal fin.

LANDING GEAR: Hydraulically retractable tricycle type, with single wheel and ERAM oleo-pneumatic (nitrogen) shock-absorber in each unit. Mainwheel tyre size 670 × 270-12 (10 ply rating), pressure 5.86-6.20 bars (85-90 lb/sq in). Steerable forward-retracting nosewheel unit has tyre size 6.50-8, pressure 4.27-4.96 bars (62-68 lb/sq in).

POWER PLANT: Two 559 kW (750 shp) Pratt & Whitney Canada PT6A-34 turboprop engines, each driving a Hartzell HC-B3TN-3C/T10178H-8R constant-speed three-blade metal propeller with autofeathering and full reverse-pitch capability. Four integral fuel tanks in wings,

Embraer EMB-110P1 Bandeirante twin-turboprop corporate jet aircraft

1995

with total capacity of 1,720 litres (453 US gallons; 378 Imp gallons). Oil capacity 8.7 litres (2.28 US gallons; 1.9 Imp gallons). Gravity refuelling point on top of each wing. Optional de-icing system for engine air inlets and propellers.

ACCOMMODATION: Pilot and co-pilot side by side on flight deck. Seats for up to 21 passengers in main cabin of P2, at 74 cm (29 in) pitch. P1 has quick-change cabin seating up to 18 persons. Crew/passenger door at front and passenger/baggage door at rear, both on port side; emergency exit over wing on each side, and opposite crew/passenger door on starboard side. Crew/passenger door can also be used as an emergency exit. Cabin floor stressed for uniformly distributed loads of up to 488 kg/m² (1,000 lb/sq in). Baggage compartment at rear of cabin, with total capacity of 2.0 m³ (70.6 cu ft). Flush type toilet in compartment at rear of cabin. Toilet/lavatory standard. Windscreen de-icing optional.

SYSTEMS: Air cycle air conditioning system with cooling capacity of 25,000 BTU/h and engine bleed heating. Hydraulic system, pressure 207 bars (3,000 lb/sq in), for landing gear actuation, dual independent braking systems, nosewheel steering and parking brake. Electrical system utilises two starter/generators, giving 200 A continuously or 24 V 34 Ah Ni/Cd battery with two 250 VA static inverters to supply 115/26 V 400 Hz AC power. External power receptacle on port side of fuselage. Oxygen system for crew and passengers (standard in P2, optional in P1), using oxygen cylinder in rear of fuselage with capacity of 3.3 m³ (115 cu ft) at 128 bars (1,850 lb/sq in) pressure.

AVIONICS AND EQUIPMENT: Collins Pro Line, Collins Microline and King Silver Crown II avionics packages. Pro Line package includes two VHF-20 com transceivers: VIR-30A VHF nav VOR/ILS marker beacon receiver, second VIR-30A; PN-101 pictorial navigation system with 331A-3G pilot's HSI: 331-3G co-pilots VOR/ILS indicator; ADF receiver; 332C-10 RMI with NAV 1/ADF 1 on single needle and NAV2/ADF 2 on double needle; two 387C-4 audiocontrol panels with two 356F-3 speaker amplifiers and eight speakers (flight deck and cabin); PA system; pilot's interphone and radio master switch. Options to Pro Line package include second ADF-60; second 332C-10 RMI; DME with 339F-12A indicator and NAV 1/HOLD/NAV 2 switching; TDR-90 transponder; second PN-90; second PN-101 with 331 A-3G co-pilot's HSI (instead of co-pilot's AIM-200-DC-FM directional gyro and 331H-3G VOR/ILS indicator); and ALT-50 radar altimeter with 339H-4 indicator. Microline package includes two VHF-251 com transceivers; VIR-351 VHF nav receiver with Dorne & Margolin DMH 21-1 nav adaptor plus Collins GLS-350 glide slope receiver; second VIR-351 with GLS-350 glide slope receiver only; ADF-650A ADF, including RCR-650A receiver; Aeronetics 7137 RMI with Aeronetics 7100 pilot's dual RMI converter; MKR-350 pilot's marker beacon receiver; PN-101 pilot's pictorial navigation system with 331A3G HSI; IND-351A co-pilot's VOR/ILS indicator (coupled to NAV 2); two 387C-4 audio control panels with two 356F-3 speaker amplifiers and eight speakers (flight deck and cabin); PA system; pilot's interphone; and radio master switch. Options to Microline package include ANS-351 R/Nav (only if aircraft is equipped also with DME-451): second ADF-650A; one or two DME-451, each with TCR-451 tranceiver and IND-451 indicator; one or two TDR-950 transponders; second Aeronetics 7137 RMI for co-pilots with one or two ADF adaptors as appropriate; second PN-101 with 331A-3G co-pilot's HSI (instead of IND-351A and AIM-200-DC-FM directional gyro); second MKR-350; King KI 207 pilot's VOR/ILS indicator repeater, coupled to NAV 2; second KI 207 for co-pilot, coupled to NAV 1; ALT-50A radio altimeter with 339H-4 indicator; AP-106 autopilot with pilot's FD-112V flight director (instead of 331A-A-3G HSI and RAI-303 artificial horizon on pilot's side; manual electric trim for pitch control, with command for both pilots (Collins 334D-67 trim servo, as used in autopilot); and NAV 1/NAV 2 transfer switch. Silver Crown II package includes two KY 196 VHF com transceivers; two KN 53 VHF nav receivers; KR 87 ADF receiver; KN 582 RMI with NAV 1/ADF 1 on single needle and NAV 2/ADF 2 on double needle; KNR 633 RMI converter; KR 21 pilot's marker beacon receiver; KCS 55A compass system with KI 525A pilot's HSI; KN 72 converter; KI 204 co-pilot's VOR/ILS indicator, coupled to NAV 2; two Collins 387C-4 audio control panels with two 356F-3 speaker amplifiers and eight speakers (flight deck and cabin); PA system; pilot's interphone; and radio master switch. Options to Silver Crown II package include one or two KNS 81 R/Nav (only if aircraft is also equipped with KN 63 DME systems, each with KDI 572 indicator; one or two KT 76A transponders; second KNI 582 system; second KCS 55A with KI 525A co-pilot's HSI (instead of directional gyro and KNI 204); KN 72 navconverter (required when ordering second KCS 55A without second KNS 81); KI 204 pilot's VOR/ILS indicator, coupled to NAV 2; second KI 204 for co-pilot, coupled to NAV 1; second KR 21 for co-pilot; KRA 405 radio altimeter, with KNI 415 indicator, and two KY 196E wide bandwidth VHF com transceivers (instead of both KY 196).

Embraer EMB-111 Bandeirante of the Brazilian Air Force. This version fitted with search, rescue and weather avoidance radar

1995

DIMENSIONS, EXTERNAL:

Wing span	15.33 m (50 ft 3½ in)
Wing chord: at root	2.33 m (7 ft 7¾ in)
at tip	1.37 m (4 ft 6 in)
Wing aspect ratio	8.10
Length overall	15.10 m (49 ft 6½ in)
Length of fuselage	14.59 m (47 ft 10½ in)
Height overall	4.92 m (16 ft 1¾ in)
Fuselage: Max width	1.72 m (5 ft 7¾ in)
Tailplane span	7.54 m (24 ft 9 in)
Wheel track	4.94 m (16 ft 2½ in)
Wheelbase	5.10 m (16 ft 8¾ in)
Propeller diameter	2.36 m (7 ft 9 in)
Distance between propeller centres	4.80 m (15 ft 9 in)
Propeller ground clearance	0.276 m (10¾ in)
Passenger door (rear, port): Height	1.35 m (4 ft 5¼ in)
Width	0.85 m (2 ft 9½ in)
Crew/passenger door (fwd, port):	
Height	1.42 m (4 ft 8 in)
Width	0.63 m (2 ft 1 in)
Passenger and crew emergency exits (three, each):	
Height	0.80 m (2 ft 7½ in)
Width	0.63 m (2 ft 1 in)

DIMENSIONS, INTERNAL:

Cabin: Max length	9.53 m (31 ft 3¼ in)
Width	1.60 m (5 ft 3 in)
Height	1.60 m (5 ft 3 in)
Floor area	12.00 m² (129.2 sq ft)
Volume	20.40 m³ (720.4 cu ft)

AREAS:

Wings, gross	29.10 m² (313.23 sq ft)
Ailerons (total)	2.16 m² (23.25 sq ft)
Flaps (total)	4.90 m² (52.74 sq ft)
Fin, excl dorsal fin	3.81 m² (41.01 sq ft)
Dorsal fin	0.82 m² (8.83 sq ft)
Ventral fin	0.80 m² (8.61 sq ft)
Rudder, incl tab	1.69 m² (18.19 sq ft)
Tailplane	5.51 m² (59.31 sq ft)
Elevators, incl tabs	4.31 m² (46.39 sq ft)

WEIGHTS AND LOADINGS (A: P2, B: P141 and P2/42, C: P1 in passenger configuration):

Weight empty, equipped: A	3,516 kg (7,751 lb)
B, commercial	3,590 kg (7,915 lb)
B, cargo	3,393 kg (7,480 lb)
Max payload: A	1,681 kg (3,706 lb)
B, commercial	1,561 kg (3,443 lb)
B, cargo	1,712 kg (3,774 lb)
C	1,633 kg (3,600 lb)

Max T-O weight: A 5,670 kg (12,500 lb)
B 5,900 kg (13,010 lb)
Max ramp weight: B 5,930 kg (13,073 lb)
Max landing weight: A 5,670 kg (12,500 lb)
B 5,700 kg (12,566 lb)
Max zero-fuel weight: A, B 5,540 kg (12,015 lb)
Max wing loading: A 195.52 kg/m² (40.04 lb/sq ft)
B 202.61 kg/m² (41.50 lb/sq ft)
Max power loading: A 5.07 kg/kW (8.33 lb/shp)
B 5.27 kg/kW (8.67 lb/shp)

PERFORMANCE (at max T-O weight, ISA, except where indicated. A: P2, B: P1/41 and P2/42):
Max level speed at 2,440 m (8,000 ft):
A 248 kts (460 km/h; 286 mph)
Max cruising speed at 2,440 m (8,000 ft):
A 223 kts (413 km/h; 257 mph)
B 222 kts (411 km/h; 256 mph)
Econ cruising speed at 3,050 m (10,000 ft):
A 181 kts (335 km/h; 208 mph)
B 184 kts (341 km/h; 212 mph)
Stalling speed at max landing weight:
A 69 kts (128 km/h; 80 mph) CAS
Max rate of climb at S/L: A 545 m (1,788 ft)/min
B 500 m (1,640 ft)/min
Rate of climb at S/L, one engine out:
A 131 m (430 ft)/min
B 113 m (370 ft)/min
Time to 3,050 m (10,000 ft): A 6 min
Time to 4,575 m (15,000 ft): A 10 min
Service ceiling: A 6,860 m (22,500 ft)
B 6,550 m (21,500 ft)
Service ceiling, one engine out: A 3,385 m (11,100 ft)
B 3,050 m (10,000 ft)
T-O distance:
A, FAR 23.135A 675 m (2,215 ft)
B, FAR 23.135/SER 41 A 807 m (2,650 ft)
Landing distance (non-factored) at max landing weight:
A 850 m (2,790 ft)
B 868 m (2,850 ft)
Range (long-range cruising speed, 45 min reserves):
A 1,080 n miles (2,001 km; 1,244 miles)
B 1,060 n miles (1,964 km; 1,220 miles)

VERIFIED

Embraer EMB-111P Bandeirante of the Gabonese Air Force

1994

EMBRAER EMB-121A XINGU I

TYPE: Twin-turboprop general purpose transport and advanced training aircraft.

PROGRAMME: The prototype Xingu (PP-ZXI) flew for the first time on 10 October 1976. A second airframe was built for static testing, and a third for fatigue testing. First production Xingu (PP-ZCT) was flown 20 May 1977, and Brazilian CTA certification was awarded in May 1979. British CAA certification was granted in July of the same year. Xingu production totalled 100 by 1983. Six VU-9s were delivered to the Grupo de Transporto Especial (Special Transport Group) of the Brazilian Air Force for service with the 6th Air Transport Squadron at Brasilia. Other customers included the CSE in the UK, Sabena, Aerovias Atlantico (Columbia), the government of the Rivers State (Nigeria), and Brazilian civil operators. The French Air Force and Navy ordered 41 Xingus (25 and 16 respectively), for aircrew training and liaison duties, all of which have been delivered. A developed version with PT6A-135 engines was also produced as the **EMB-121A1 Xingu II** and existing Xingu Is can be retrofitted to this standard.

UPGRADES: **Embraer.** EMB-121A1 Xingu II. See following entry.

OPERATORS: Versions of the EMB-121A Xingu I are in service with the air arms of Brazil (6).

The following description applies to the Xingu I:

DESIGN FEATURES: Cantilever low-wing monoplane. Utilises basically same wing as EMB-110P2 (which see), but with reduced span and modified tips. Sweepback 0° at 28 per cent chord. Ventral fin. Pneumatic de-icing boots on ailerons and trailing-edges. Pneumatic de-icing boots on fin and tailplane leading-edges.

FLYING CONTROLS: Slotted ailerons and double-slotted trailing-edges. Fixed incidence tailplane. Trim and balance tabs in rudder and each elevator.

STRUCTURE: The fuselage is an all-metal semi-monocoque safe-life structure of 2024-T3/T351/T3511 aluminium alloys, with circular cross-section except for nose and tailcones; made up of stretched and bent C frames, extruded stringers, bulkheads and stressed skin panels. Rear bulkhead and all-skin panels chemically milled. Wing/fuselage attachment strengthened by a machine rib on each side of carry-through spars. Flight deck floor is clad 2024-T3 sheet; cabin floor panels are sandwich plates with a balsa core. The tail unit is a Cantilever metal T tail (2024-T351 aluminium alloy) with glassfibre fin leading-edge and dorsal fin.

LANDING GEAR: ERAM hydraulically retractable tricycle type with oleo-pneumatic shock-absorber in each unit. Single ERAM forward-retracting mainwheels with Goodyear do Brasil 26.3 × 8.43 × 12 in tyres. Steerable, forward-retracting twin Oldi nosewheels with Goodyear do Brasil 15.75 × 4.45 × 8 in tyres. Tyre pressure 5.52-5.86 bars (80-85 lb/sq in) on all units. ERAM single-disc hydraulic brakes.

POWER PLANT: Two 507 kW (600 shp) Pratt & Whitney Aircraft of Canada PT6A-28 turboprop engines each driving a Hartzell HC-B3TN-3D/T10178HB-8R three-blade constant-speed metal propeller with autofeathering and full reverse-pitch capability. Four integral fuel tanks in wings with total capacity of 1,666 litres (439.55 US gallons; 366 Imp gallons). Gravity refuelling point on top of each wing. Oil capacity 8.3 litres (2.16 US gallons; 1.8 Imp gallons).

ACCOMMODATION: Normal flight crew of two, but certificated for single-pilot operation. Cabin seats up to nine passengers. Downward hinged door on port side, aft of wing, with built-in airstairs. Emergency exit over wing on starboard side. Baggage compartments in nose (unpressurised, with external access via door on each side) and at rear of cabin (pressurised, with internal access). Toilet/lavatory and galley standard. Entire accommodation pressurised and air conditioned.

SYSTEMS: Hamilton Standard air cycle air conditioning system, maximum capacity 20,000 BTU/h for cooling and 40,000 BTU/h for heating. Pressurisation system (maximum differential 0.41 bars; 6 lb/sq in) maintains S/L cabin environment up to 4,270 m (14,000 ft) and 2,440 m (8,000 ft) environment up to 8,335 m (27,350 ft). Hydraulic system, pressure 207 bars (3,000 lb/sq in), for landing gear extension/retraction, nosewheel steering and brake actuation. Emergency hand pump for back-up gear extension. Primary electrical system is 28 V DC, supplied by two 300 A starter/generators; 40 Ah Ni/Cd battery for assisted starting and emergency power. Main and standby 600 VA static inverters provide 115/26 V AC power at 400 Hz. Electric anti-icing of windscreen, engine air intakes, pitot static ports and propellers; pneumatic boot de-icing of wing and tail leading-edges. High-pressure 127 bars (1,850 lb/sq in) oxygen system for crew and passengers.

AVIONICS: Standard avionics include dual RCA AVC-110A VHF com transceivers, dual RCA AVN-220A VOR/ILS/ marker beacon receivers, dual Collins ADF-60A ADF, one VOR/ILS OBS indicator (co-pilot), dual RCA AVA-310 audio panels, one Sperry C-14 gyromagnetic compass with dual Sperry RD-44 course indicator (pilot) (SFIM CG130 compass in aircraft for French Air Force and Navy), one AIM-200 DC FM direction gyro (co-pilot) one Sperry GH-14-330 gyro horizon (pilot), one AIM-500 DC FM gyro horizon (co-pilot), one Collins TDR-90 transponder with Smiths 01-200-105 encoding altimeter, one Collins DME-40 DME, two telex TEL-66C microphones, two pairs of telex A1210 earphones, dual Flite-Tronics PC-15 BC(D) static converters and one SunAir ASB-100A HF transceiver. Optical avionics are available in three standard packages, with dual Collins DF-206 ADF, SunAir ASB-100A HF/AM/SSB, dual RCA AVA-310 and Sperry SPZ-200 autopilot common to all three. In addition, package No. 1 offers dual RCA AVC-110A VHF com, one RCA, AVN-220A VHF nav and one Sperry C-14 gyromagnetic compass; package No 2 offers dual RCA AVC-110A VHF com, dual RCA AVN-220A VHF nav, dual Sperry C-14 gyromagnetic compasses and Garret Rescu/88L ELT; package No. 3 offers dual Collins VHF 20A com, dual Collins VIR-30A VHF nav, dual Sperry C-14 gyromagnetic compasses, Garret Rescu/88L ELT, Collins ALT-50 radio altimeter, and dual Sperry STARSIVC flight directors for pilot and co-pilot. Bendix RDR-1200 or RCA Primus 40 weather radar, cabin music and PA system are optional extras on all three packages; Collins TDR-90 transponder on packages 2 and 3; Colllins DME-40 on package 3; and pilot's Sperry STARS IVB or IVC on package 2.

EQUIPMENT: Standard equipment comprises maximum, permissible airspeed indicator, eight day clock, chronometer, cabin rate of climb indicator, cabin altitude and differential pressure indicator, annunciator panel, heated stall warning system, dual heated pitot tubes and heated static ports, external power sockets, wing ice light, dual landing lights, dual taxi lights, dual anti-collision strobe lights, dual map lights, cabin dome lights, instrument lighting system, low profile glareshield, and hand type cabin fire extinguishers. Optional equipment includes three-light strobe system, fire extinguishing system, de-icing and anti-icing system, toilet, and a range of galley equipment.

DIMENSIONS, EXTERNAL:
Wing span 14.05 m (46 ft 1¼ in)
Wing chord: at fuselage c/l 2.47 m (8 ft 1¼ in)
at root 2.33 m (7 ft 7¾ in)
at structural tip 1.49 m (4 ft 10¾ in)
Wing aspect ratio 7.18
Length overall 12.25 m (40 ft 2¼ in)
Length of fuselage 11.16 m (36 ft 7½ in)
Fuselage: Max width 1.86 m (6 ft 1¼ in)
Height overall 4.84 m (15 ft 10½ in)
Tailplane span 5.58 m (18 ft 3¾ in)
Wheel track 5.24 m (17 ft 2¼ in)
Wheelbase 2.90 m (9 ft 6¼ in)
Propeller diameter 2.36 m (7 ft 9 in)
Propeller ground clearance 0.29 m (11 ½ in)
Distance between propeller centres 5.10 m (16 ft 8¾ in)
Passenger door (rear port): Height 1.32 m (4 ft 4 in)

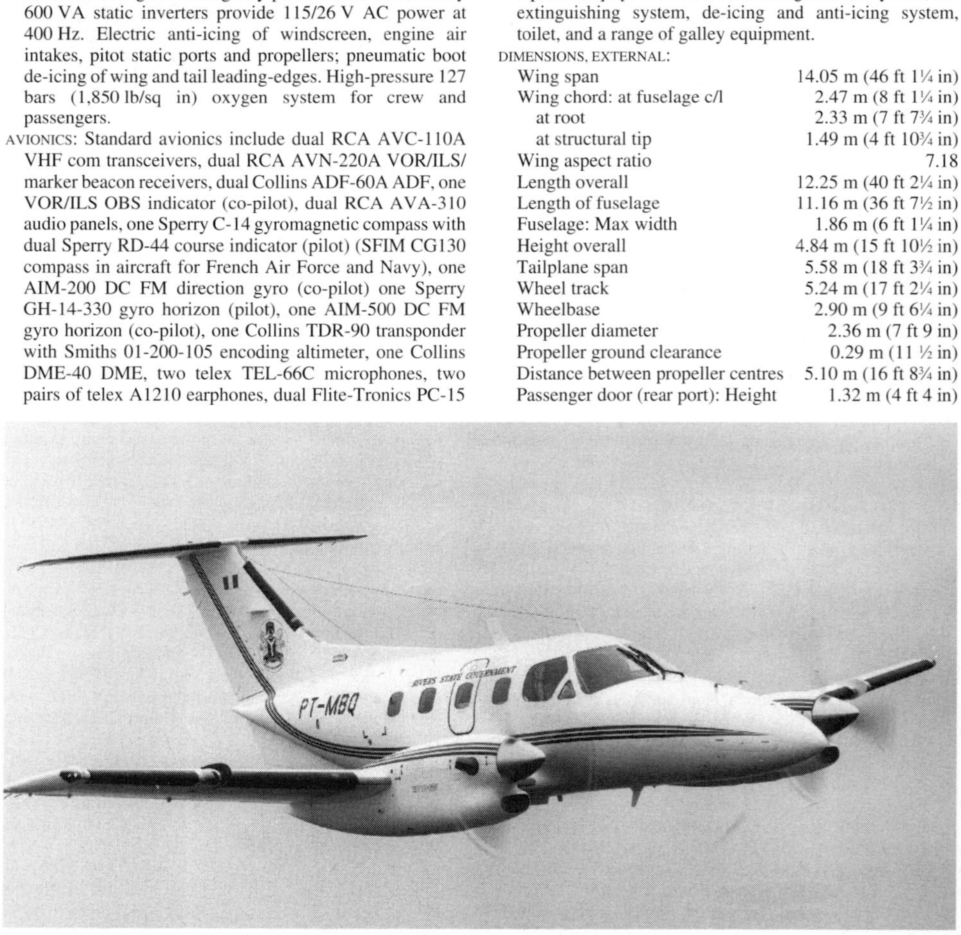

Embraer EMB-121A Xingu I corporate propjet

1994

Width 0.63 m (2 ft 0¾ in)
Height to sill 1.245 m (4 ft 1 in)
Emergency exit (overwing, stbd):
Height 0.96 m (3 ft 1¾ in)
Width 0.51 m (1 ft 8 in)

DIMENSIONS, INTERNAL:
Cabin, incl flight deck: Max length 5.18 m (17 ft 0 in)
Max width 1.74 m (5 ft 8½ in)
Max height 1.52 m (4 ft 11¾ in)
Passenger cabin, incl rear baggage compartment:
Length 3.57 m (11 ft 8½ in)
Floor area 4.24 m² (45.64 sq ft)
Volume 6.9 m³ (243.7 cu ft)
Baggage compartment volume:
Nose 0.30 m³ (10.6 cu ft)
Rear 0.71 m³ (25.1 cu ft)

AREAS:
Wing, gross 27.50 m² (296.0 sq ft)
Ailerons (total) 1.84 m² (19.81 sq ft)
Trailing-edge flaps (total) 4.90 m² (52.74 sq ft)
Vertical tail surfaces (total, excl dorsal fin) 4.00 m² (43.06 sq ft)
Rudder, incl tab 1.30 m² (13.99 sq ft)
Dorsal fin 0.54 m² (5.81 sq ft)
Ventral fin 0.94 m² (10.12 sq ft)
Horizontal tail surfaces (total) 5.84 m² (62.86 sq ft)
Elevators, incl tabs 2.17 m² (23.36 sq ft)

WEIGHTS AND LOADINGS:
Weight empty, equipped 3,620 kg (7,984 lb)
Max payload (one pilot) 860 kg (1,896 lb)
Max T-O weight 5,670 kg (12,500 lb)
Max ramp weight 5,700 kg (12,565 lb)
Max zero-fuel weight 4,660 kg (10,273 lb)
Max landing weight 5,340 kg (11,772 lb)
Max cabin floor loading 488 kg/m² (100 lb/sq ft)
Max wing loading 206.2 kg/m² (42.2 lb/sq ft)
Max power loading 5.59 kg/kW (9.19 lb/shp)

PERFORMANCE (at max T-O weight, ISA, except where indicated):
Max cruising speed 3,350 m (11,000 ft) 243 kts (450 km/h; 280 mph)
Econ cruising speed at 6,100 m (20,000 ft) 197 kts (365 km/h; 227 mph)
Stalling speed at max T-O weight, flaps up 96 kts (178 km/h; 111 mph)
Stalling speed at max landing weight, flaps down 72 kts (134 km/h; 83 mph)
Max rate of climb at S/L 426 m (1,400 ft)/min
Service ceiling 7,925 m (26,000 ft)
Service ceiling, one engine out 3,960 m (13,000 ft)
T-O to 15 m (50 ft) 865 m (2,840 ft)
Landing from 15 m (50 ft) at max landing weight 850 m (2,790 ft)
Range with 780 kg (1,720 lb) payload at 6,100 m (20,000 ft), 45 min reserves 1,225 n miles (2,270 km; 1,410 miles)
Range with max fuel and 610 kg (1,345 lb) payload at 6,100 m (20,000 ft), ISA 45 min reserves 1,270 n miles (2,352 km; 1,461 miles)

VERIFIED

EMBRAER EMB-121A1 XINGU II

TYPE: Twin-turboprop general purpose transport and advanced trainer aircraft.

PROGRAMME: A first example of this version of the Xingu (PP-ZCT, the re-engined version of the Xingu I) flew for the first time on 4 September 1981. It retains the same airframe as the Xingu I, but has more powerful Pratt & Whitney Canada PT6A-135 turboprop engines, four-blade propellers, increased fuel capacity and a small strake added on each side of the tailcone.

An improved standard interior for new production Xingu IIs (available as a retrofit on earlier Xingus) was introduced in 1984. This features an improved air conditioning system, new noise-suppressing materials, redesigned leather seats, new carpeting and toilet installations, foldout tables on each side of the cabin which can be joined in the centre to make a larger surface, and Collins Pro Line radio communications equipment. TAWA ordered 16 of these aircraft. Details are the same as the Xingu I except for the following.

OPERATORS: Versions of the EMB-121 Xingu II are in service with the French Air Force (25) and French Navy (17).

AVIONICS: Standard avionics include dual Collins VHF-20 com transceivers, dual Collins VIR-30 nav receivers, one Collins ADF-60A, one Collins TDR-90 transponder, Collins CLT-21/31/61/91 control heads, one Collins DME-40, one Bendix RDR-1150 HP weather radar, two Sperry C-14 gyro compasses, two Sperry VG-14 vertical gyros, two Collins RMI-36s, dual Baker audio/interphone and PA system, J. E. T. standby attitude indicator, dual Sperry AD-550B ADIs, dual Sperry RD-550B HSIs, and a Dorne & Margolin DMELT-8, one emergency locator transmitter. Optional avionics include second ADF, transponder and DME; Bendix checklist and nav interface unit; one or two Collins ANS-31C R/nav; Sperry SPZ-4000 AFCS; Sperry digital air data computer; Global GNS-500 VLF/Omega; one or two Collins ALT-50 radio altimeters; and a tape recorder/FM receiver.

WEIGHTS AND LOADINGS: As detailed for Xingu I except:
Weight empty, equipped 3,500 kg (7,716 lb)
Max payload 900 kg (1,984 lb)
Max fuel load 1,308 kg (2,884 lb)
Max power loading 5.07 kg/kW (8.33 lb/shp)

PERFORMANCE (at max T-O weight, ISA):
Max operating speed 252 kts (466 km/h; 290 mph) CAS
Max cruising speed at 3,050 m (10,000 ft) 251 kts (465 km/h; 289 mph)
Econ cruising speed at 3,050 m (10,000 ft) 205 kts (380 km/h; 236 mph)
Stalling speed, power off:
flaps up 96 kts (178 km/h; 111 mph) CAS
flaps down 76 kts (141 km/h; 88 mph) CAS
Max rate of climb at S/L 548 m (1,800 ft)/min
Rate of climb, one engine out 103 m (340 ft)/min
Service ceiling 8,535 m (28,000 ft)
Service ceiling, one engine out 3,290 m (10,800 ft)
T-O run 580 m (1,903 ft)
T-O to 15 m (50 ft) 760 m (2,495 ft)
Landing from 15 m (50 ft) 890 m (2,920 ft)
Landing run 560 m (1,835 ft)
Min ground turning radius 10.73 m (35 ft 2½ in)
Runway LCN 10
Range at 6,100 m (20,000 ft), 45 min reserves:
with max payload 880 n miles (1,630 km; 1,012 miles)
with max fuel 1,230 n miles (2,278 km; 1,415 miles)

VERIFIED

Embraer EMB-121A1 Xingu II of the French Navy ***(Peter J. Cooper)***

1994

EMB-121A1 Xingu II of the French Air Force (foreground) and EMB-121A1 Xingu II of the French Navy (background) ***(Embraer)***

1995

NEIVA

INDÚSTRIA AERONÁUTICA NEIVA S/A
(Subsidiary of Embraer)
Caixa Postal 10, 18608-900 Botucatu, SP
Telephone: 55 (149) 21-2122
Fax: 55 (149) 22-1285
Telex: 142 736 SOAN BR
PRESIDENT: Paulo Cury
ENGINEERING MANAGER: Luiz Carlos Benetti

UPDATED

NE-821 Carajá a Schafer Comanchero 500B built under licence by Neiva

1993

NEIVA NE-821 CARAJÁ

TYPE: High-powered version of Schafer turboprop conversion of Piper Navajo Chieftain.

PROGRAMME: Completed in 1992. Responsibility for the EMB-820C was transferred to Neiva in mid-1983, and in 1984 four of the last five Embraer assembled examples were converted to Comanchero 500B eight-seat executive configuration, in which form they are designated NE-821 Carajá. The first Carajá made its initial flight on 9 March 1984, and deliveries began in November of that year. Brazilian certification covers the use of either PT6A-27 or PT6A-34 engines, flat rated in both cases at 410 kW (550 shp) and driving Hartzell three-blade constant-speed propellers with reverse pitch and automatic synchronisation. The most recent examples have been delivered with the -34 power plant. Usable fuel capacity is 1.314 litres (347 US gallons; 289 Imp gallons).

CUSTOMERS: Total of 39 delivered 1 January 1991.

WEIGHTS:

Weight empty, equipped	2,300 kg (5,070 lb)
Max payload	784 kg (1,728 lb)
Max T-O weight	3,629 kg (8,000 lb)
Max ramp weight	3,651 kg (8,050 lb)
Max landing weight	3,447 kg (7,600 lb)
Max zero-fuel weight	3,084 kg (6,800 lb)

PERFORMANCE:

Max cruising speed at 3,050 m (10,000 ft), max T-O weight, ISA	232 kts (430 km/h; 267 mph)
Normal cruising speed	221 kts (410 km/h; 255 mph)
Econ cruising speed	174 kts (323 km/h; 200 mph)
Stalling speed	80 kts (148 km/h; 92 mph)
Max rate of climb at S/L	740 m (2,425 ft)/min
Rate of climb at S/L, one engine out	170 m (560 ft)/min
Max operating altitude	7,315 m (24,000 ft)
Service ceiling, one engine out	5,335 m (17,500 ft)
T-O to 15 m (50 ft)	540 m (1,772 ft)
Landing from 15 m (50 ft)	563 m (1,847 ft)
Max range, 45 min reserves	945 n miles (1,750 km; 1,087 miles)

VERIFIED

Neiva N621 Universal two/three-seat basic trainer

1995

NEIVA N621 UNIVERSAL

Brazilian Air Force designation: T-25 and T-25A

TYPE: Two/three-seat basic trainer.

PROGRAMME: The Universal was designed by Eng Joseph Kovacs to meet a Brazilian Air Force requirement. Initial design started in January 1963, and the prototype (PP-ZTW) flew for the first time on 29 April 1966.

The first production T-25 was flown on 7 April 1971, and 132 were delivered to the Brazilian Air Force between Autumn 1971 and early 1975. Production took place at Botucatu and São José dos Campos, with wingtips, tailplanes and elevators manufactured under subcontract by Motortec in Rio de Janeiro. The Brazilian Air Force ordered eight more in 1978, and an additional order for 20 was being negotiated in early 1979. Ten T-25s were supplied to the Chilean Army; these were transferred subsequently to the Chilean Air Force.

OPERATORS: The T-25 is in service with the armed forces of the following countries: Brazil (120), Paraguay (5+).

DESIGN FEATURES: Cantilever low-wing monoplane. Wing section NACA 63_2A315 at root, NACA $63_1$212 at tip. Dihedral 6°. Incidence 2°. The tail unit is a cantilever all-metal structure.

FLYING CONTROLS: All-metal dynamically balanced slotted ailerons. All-metal split flaps. Electrically actuated tab in port elevator.

STRUCTURE: The wing is a single-spar structure of riveted aluminium alloy. The centre-fuselage is a welded steel tube with skin panels. Semi-monocoque tailcone of riveted aluminium alloy.

LANDING GEAR: Retractable tricycle type. Hydraulic retraction, main units inward, nosewheel rearward. ERAM oleo shock-absorbers. Mainwheels fitted with Goodyear tyres size 6.50-8 and Goodyear or OLDI disc brakes. Nosewheel steerable and fitted with Goodyear tyre size 6.00-6. Tyre pressure 2.28 bars (33 lb/sq in) on main units, 1.79 bars (26 lb/sq in) on nose unit.

POWER PLANT: One 224 kW (300 hp) Lycoming IO-540-K1D5 flat-six engine, driving a Hartzell HC-C2YK-4/C8475-A2 non-feathering two-blade constant-speed metal propeller. Six aluminium fuel tanks in wings, total capacity 332 litres (87.6 US gallons; 73 Imp gallons). Refuelling points above wings. Oil capacity 11.5 litres (3 US gallons; 2.5 Imp gallons).

ACCOMMODATION: Two seats side by side, with full dual controls, and optional third seat at rear. Large rearward-sliding transparent canopy. Baggage compartment aft of rear seat.

SYSTEMS: Electrically actuated hydraulic system, pressure 103 bars (1,500 lb/sq in), for flaps and landing gear. Manual emergency pump. 28 V electrical system.

AVIONICS AND EQUIPMENT: 140-channel Brazilian-made VHF radio, ADF and VOR/LOC. Complete IFR instrumentation.

ARMAMENT: Two underwing hardpoints for 7.62 mm machine gun pods.

DIMENSIONS, EXTERNAL:

Wing span	11.00 m (36 ft 1 in)
Wing chord: at root	2.00 m (6 ft 6½ in)
at tip	1.08 m (3 ft 6½ in)

Neiva N621 Universal trainer aircraft, in service with the air forces of Brazil and Paraguay

1995

Wing aspect ratio	7.1
Length overall	8.60 m (28 ft 2½ in)
Height overall	3.00 m (9 ft 9¾ in)
Tailplane span	3.95 m (12 ft 11½ in)
Wheel track	2.65 m (8 ft 8¼ in)
Wheelbase	2.33 m (7 ft 7¾ in)
Propeller diameter	2.13 m (7 ft 0 in)
Propeller ground clearance	0.37 m (1 ft 2½ in)
DIMENSIONS, INTERNAL:	
Cabin: Length	2.20 m (7 ft 2½ in)
Max width	1.25 m (4 ft 1 in)
Max height	1.25 m (4 ft 1 in)
Floor area	3.0 m² (32 sq ft)
Volume	4.00 m³ (141 cu ft)
Baggage compartment volume	0.35 m³ (12.5 cu ft)
AREAS:	
Wings, gross	17.20 m² (185.14 sq ft)
Ailerons (total)	1.47 m² (15.82 sq ft)
Trailing-edge flaps (total)	1.34 m² (14.42 sq ft)
Fin	0.82 m² (8.83 sq ft)
Rudder	0.90 m² (9.69 sq ft)
Tailplane	1.72 m² (18.51 sq ft)
Elevators, incl tab	1.35 m² (14.53 sq ft)
WEIGHTS AND LOADINGS (A: Aerobatic, U: Utility):	
Weight empty, equipped: A, U	1,150 kg (2,535 lb)
Max T-O weight: A	1,500 kg (3,306 lb)
U	1,700 kg (3,747 lb)
Max wing loading: A	88.2 kg/m² (18.1 lb/sq ft)
U	100.0 kg/m² (20.5 lb/sq ft)
Max power loading: A	6.70 kg/kW (11.02 lb/hp)
U	7.59 kg/kW (12.49 lb/hp)
PERFORMANCE (at max T-O weight. A: Aerobatic, U: Utility):	
Never-exceed speed:	
A, U	269 kts (500 km/h; 310 mph)
Max level speed at S/L:	
A	162 kts (300 km/h; 186 mph)
U	160 kts (296 km/h; 184 mph)
Max cruising speed (75% power) at S/L:	
A	153 kts (285 km/h; 177 mph)
U	151 kts (280 km/h; 174 mph)
Stalling speed:	
flaps up: A	63.5 kts (117 km/h; 73 mph)
U	66 kts (122 km/h; 76 mph)
flaps down: A	56.5 kts (104 km/h; 65 mph)
U	59.5 kts (110 km/h; 68.5 mph)
Max rate of climb at S/L: A	400 m (1,312 ft)/min
U	320 m (1,050 ft)/min
Service ceiling: A	6,100 m (20,000 ft)
U	5,000 m (16,400 ft)

NEW ENTRY

CANADA

AIRTECH

AIRTECH CANADA

Peterborough Municipal Airport, PO Box 415, Peterborough, Ontario K9J 6Z3
Telephone: 1 (705) 743 9483
Fax: 1 (705) 749 0841
Telex: 06-962912

PRESIDENT: John O'Dwyer
GENERAL MANAGER: Bernard LaFrance
CHIEF ENGINEER: James C. Mewett
PRESS RELATIONS: Alison M. Mewett

Airtech specialises in retrofitting versions of the de Havilland Canada Otter and Beaver with more powerful Polish built engines, offering increased climb rates and considerably greater fuel economy, at lower power settings, than the engines which they replace. It has also designed, manufactured, tested and installed modifications (in particular, auxiliary fuel tanks and medevac equipment) for various types of aircraft including the Cessna 401, 414 and 421, Piper PA-31 and PA-42, Mitsubishi MU-2 and Fairchild Metro IIB.

There has been no recent news of the DC-3/2000 programme.

VERIFIED

AIRTECH CANADA DHC-2/PZL-3S BEAVER

TYPE: STOL utility aircraft re-engine programme.

PROGRAMME: Airtech Canada introduced a conversion of the DHC-2 Beaver, with the PZL-3S engine, at the request of operators who wanted an increase in power to provide improved performance and safer operation from short airstrips. Four such conversions had been completed by early 1989.

POWER PLANT: One 447 kW (6,000 hp) PZL-3s seven-cylinder air-cooled radial engine, driving a PZL US132000A four-blade constant-speed propeller. Fuel capacity 359 litres (96 US gallons; 79 Imp gallons), or 523 litres (138 US gallons; 115 Imp gallons) with auxiliary in each wingtip.

WEIGHTS AND LOADINGS:	
Weight empty	1,419 kg (3,129 lb)
Useful load with full fuel	636 kg (1,402 lb)
Max T-O weight	2,313 kg (5,100 lb)
PERFORMANCE:	
Normal operating speed and max cruising speed	126 kts (233 km/h; 145 mph) IAS
Max rate of climb at S/L	488 m (1,600 ft)/min
T-O to 15 m (50 ft)	305 m (1,000 ft)

VERIFIED

AIRTECH CANADA DHC-3/1000 OTTER

TYPE: STOL utility aircraft re-engine programme.

PROGRAMME: Airtech Canada refitted eight de Havilland Canada DHC-3 Otters with Polish PZL-3S radial engines.

Following the first flight of a prototype on 25 August 1983, the Otter conversion was then offered with a 746 kW (1,000 hp) Polish ASz-62IR engine instead of the 447 kW (600 hp) PZL-3S. 10th conversion completed in May 1993, installed by third party. Kit available.

VERIFIED

AIRTECH CANADA MEDEVAC INTERIOR UPGRADES

TYPE: Fixed- and rotary-wing aircraft medical evacuation conversions.

PROGRAMME: Airtech Canada is responsible for the conversion of aircraft interiors for use by emergency medical services. Installations for the following aircraft are available: C401A, C411, C414, C421, C500, King Air 10, MU-2, SA26AT S76, PA-31, PA-31-350, PA-42.

DESIGN FEATURES: The installations are DOT (DAR) approved and include berths, dual oxygen and vacuum, dual incubator power outlets and patient restraints. Airtech is participating in upgrading the Ontario Ministry of Health's dedicated helicopter fleet.

VERIFIED

Airtech Canada DHC-3/1000 Otter with PZL-3S radial engine

1994

Airtech Canada medevac installation with dual oxygen and vacuum and incubator power outlets

1994

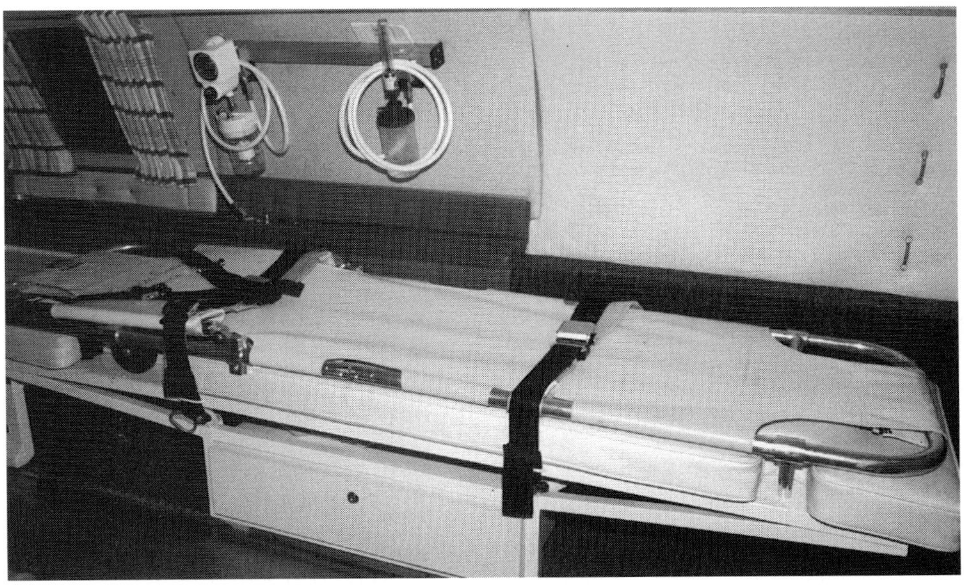

Airtech Canada medevac installation with stretcher

1994

BELL

BELL HELICOPTER TEXTRON
(a Division of Textron Canada Ltd)

12800 rue de l'Avenir, St Janvier, Mirabel, Quebec J7J 1R4
Telephone: 1 (514) 437 3400
Fax: 1 (514) 437 6010

PRESIDENT: Mr Dell Young

Memorandum of Understanding to start helicopter industry in Canada signed 7 October 1983. 34,560 m² (372,000 sq ft) factory opened late 1985 and employs 1,300 people. US civil production of 206B JetRanger and 206L LongRanger transferred to Canada by early 1987; then Model 212 in August 1988 and Model 412 in February 1989. The 230 was introduced and certificated in Canada in 1992. Over 900 (all models) delivered by early 1994 about half of each helicopter made in Canada (dynamic systems supplied by Bell Fort Worth). Production rate from 15 to 25 per month.

UPDATED

Bell Model 206L-3 LongRanger III general purpose light helicopter

1994

Bell Model 206L-3 LongRanger III with Allison 250-C30P turboshaft engines

1994

BELL MODEL 206L-3 LONGRANGER III

TYPE: Stretched turbine-powered general purpose light helicopter.

PROGRAMME: Announced 25 September 1973; production in Canada began January 1987.

CUSTOMERS: More than 1,100 LongRangers produced by January 1990; Canadian deliveries totalled 120 by January 1990.

DESIGN FEATURES: Cabin length increased to make room for club seating and extra window; Bell Noda-Matic transmission to reduce vibration; improvements introduced on LongRanger II include new freewheel unit, modified shafting and increased-thrust tail rotor; main rotor rpm 394; rotor brake optional.

FLYING CONTROLS: Hydraulic fully powered cyclic and collective controls and foot powered tail rotor control; tailplane with highly cambered inverted aerofoil section and stall strip produces appropriate nose-up and nose-down attitude during climb and descent; optional autostabiliser, autopilot and IFR systems. Model 206L-3 has endplate fins on tailplane; single-pilot IFR with Collins AP-107H autopilot; optional SFENA autopilot with stabilisation and holds for heading, height and approach.

STRUCTURE: Conventional light alloy structure with two floor beams and bonded honeycomb sandwich floor; transmission mounted on two beams and deck joined to floor by three fuselage frames; main rotor blades have extruded aluminium D-section leading-edge with honeycomb core behind covered by bonded skin; tail rotor blades have bonded skin without honeycomb core.

POWER PLANT: One 485 kW (650 shp) Allison 250-C30P turbobshaft (maximum continuous rating 415 kW; 557 shp). Transmission rated at 324 kW (435 shp) for take-off, with a continuous rating of 276 kW (370 shp); 340 kW (456 shp) transmission optional. Rupture resistant fuel system, capacity 416 litres (110 US gallons; 91 Imp gallons).

ACCOMMODATION: Redesigned rear cabin, more spacious than JetRanger with 5 cm (2 in) more headroom. With a crew of two, standard cabin layout accommodates five passengers in two canted rearward-facing seats and three forward-facing seats. An optional executive cabin layout has four individual passenger seats. Port forward passenger seat has folding back to allow loading of a 2.44 × 0.91 × 0.30 m (8 × 3 × 1 ft) container, making possible carriage of such items as survey equipment, skis and other long components. Double doors on port side of cabin provide opening 1.52 m (5 ft 0 in) wide, for straight-in loading of stretcher patients or utility cargo; in ambulance or rescue role two stretcher patients and two ambulatory patients/attendants may be carried.

SYSTEMS: Hydraulic system; electrical power from starter/generator and 17 Ah battery.

AVIONICS: Standard Collins MicroLine suite includes dual nav/com, ADF, DME, transponder and marker beacon receiver. R/Nav, radio altimeter and encoding altimeter optional.

EQUIPMENT: Optional kits include emergency floating gear, 907 kg (2,000 lb) cargo hook, and engine bleed air ECS.

DIMENSIONS, EXTERNAL:	
Main rotor diameter	11.28 m (37 ft 0 in)
Tail rotor diameter	1.65 m (5 ft 5 in)
Length: overall rotors turning	13.02 m (42 ft 8½ in)
fuselage, incl tailskid	10.44 m (34 ft 3 in)
Height: over tailfin	2.90 m (9 ft 6¼ in)
to top of rotor head	3.14 m (10 ft 3¾ in)
Fuselage: Max width	1.32 m (4 ft 4 in)
Stabiliser span	1.98 m (6 ft 6 in)
Width over skids	2.34 m (7 ft 8¼ in)
DIMENSIONS, INTERNAL:	
Cabin volume	2.35 m³ (83 cu ft)
AREAS:	
Main rotor disc	99.89 m² (1,075.2 sq ft)
Tail rotor disc	2.14 m² (23.04 sq ft)
WEIGHTS AND LOADINGS:	
Weight empty, standard: to 1991	998 kg (2,200 lb)
from 1992	1,024 kg (2,258 lb)
Max external load	907 kg (2,000 lb)
Max T-O weight: normal	1,882 kg (4,150 lb)
external load	1,927 kg (4,250 lb)
Max disc loading: normal	18.85 kg/m² (3.86 lb/sq ft)
external load	19.29 kg/m² (3.95 lb/sq ft)
Max power loading: transmission for T-O normal	5.81 kg/kW (9.54 lb/shp)
transmission for T-O, external load	5.95 kg/kW (9.77 lb/shp)
PERFORMANCE (at max normal T-O weight ISA):	
Never-exceed speed (VNE):	
at S/L	130 kts (241 km/h; 150 mph)
at 1,525 m (5,000 ft)	133 kts (246 km/h; 153 mph)
Max cruising speed at 1,525 m (5,000 ft)	110 kts (203 km/h; 126 mph)
Max rate of climb at S/L	408 m (1,340 ft)/min
Service ceiling at max cruise power	6,100 m (20,000 ft)
Hovering ceiling: IGE	5,030 m (16,500 ft)
OGE	1,645 m (5,400 ft)
Range, no reserves:	
at S/L	320 n miles (592 km; 368 miles)
at 1,525 m (5,000 ft)	360 n miles (666 km; 414 miles)

VERIFIED

BELL MODEL 412HP

TYPE: Four-blade, twin-engined utility helicopter.

PROGRAMME: Model 412 announced 8 September 1978; FAR Pt 29 VRF approval received 9 January 1981, IFR 13 February 1981; production (SP version) transferred to Canada February 1989; first delivery (civil) 18 January 1981.

VERSIONS: **Model 412SP:** Special Performance version with increased maximum T-O weight, new seating options and 55 per cent greater standard fuel capacity. Superseded by 412HP early 1991.

Military 412: Announced by Bell June 1986; fitted with Lucas Aerospace chin turret and Honeywell Head Tracker helmet sight similar to that in AH-1S; turret carries 875 rounds, weighs 188 kg (414 lb) and can be removed in under 30 minutes; firing arcs ±110° in azimuth, +15° and −45° in elevation; other armament includes twin dual FN Herstal 7.62 mm gun pods, single FN Herstal 0.50 in pod, pods of seven or 19 2.75 in rockets, M240E1 pintle-mounted door guns, FN Herstal four-round 70 mm rocket launcher and a 0.50 in gun or two Giat M621 20 mm cannon pods.

Model 412EP: Current version. See *Jane's All the World's Aircraft 1995-96.*

Model 412HP: Improved transmission giving better OGE hover; FAR Pt 29 certification 5 February 1991, first delivery (c/n 36020) later that month. *Now standard current model, to which detailed description applies.*

Agusta Griffon: Agusta developed military version (see Italian section); capable of medevac, armed support, transport, SAR and patrol.

IPTN: Indonesia's IPTN (which see) has licence to produce up to 100 Model 412SPs with progressive increase in manufacture.

OPERATORS: Versions of the 412 are in service with the armed forces of the following countries: Bahrain (3), Botswana (5), Canada (100), Colombia (1), Guatemala (6), Guyana (1), Honduras (9), Indonesia (28+), Italy (16), Lesotho (3), Netherlands (3), Norway (18), Peru (2), Slovenia (3), South Korea (3), Sri Lanka (4), Thailand (1), Uganda (2), Venezuela (2) and Zimbabwe (2).

DESIGN FEATURES: Four-blade main rotor, blades retained within central metal star fitting by single elastomeric bearings. Blades can be folded; rotor brake standard; two-blade tail rotor; main rotor rpm 314.

FLYING CONTROLS: Fully powered hydraulic controls; gyroscopic stabiliser bar above main rotor; automatic variable incidence tailplane; IFR versions have large fin above cabin to improve roll-yaw responses during manual instrument flying.

STRUCTURE: Main rotor blade spar unidirectional glassfibre with 45° wound torque casting of glassfibre cloth. Nomex rear section core with trailing-edge of unidirectional glassfibre; leading-edge protected by titanium abrasion strip and replaceable stainless steel cap at tip; lightning protection mesh embedded; provision for electric de-icing heater elements; main rotor hub of steel and light alloy; all-metal tail rotor.

LANDING GEAR: Optional high skid, emergency pop-out float or non-retractable tricycle gear.

POWER PLANT: Pratt & Whitney Canada PT6T-3BE Turbo Twin-Pac, rated at 1,342 kW (1,800 shp) for 5 minutes for T-O and 1,193 kW (1,600 shp) maximum continuous. In event of engine failure, remaining engine can deliver up to 764 kW (1,025 shp) for 2½ minutes, or 723 kW (970 shp) for 30 minutes. Transmission rating 1,022 kW (1,370 shp) for T-O, 828 kW (1,110 shp) maximum continuous. Optional 30 kW (40 shp) for accessory drives from main gearbox. Seven interconnected rupture resistant fuel cells, with automatic shut-off valves (breakaway fittings), have a combined capacity of 1,249 litres (330 US gallons; 275 Imp gallons). Two 76 or 310 litre (20 or 82 US gallon; 16.7 or 68.3 Imp gallon) auxiliary fuel tanks, in any combination, can increase maximum total capacity to 1,870 litres (494 US gallons; 411 Imp gallons). Single-point refuelling on starboard side of cabin.

EQUIPMENT: Optional equipment includes a cargo sling and rescue hoist.

DIMENSIONS, EXTERNAL:	
Main rotor diameter	14.02 m (46 ft 0 in)
Tail rotor diameter	2.59 m (8 ft 6 in)

Main rotor blade chord: at root 0.40 m (1 ft 4 in)
at tip 0.22 m (8½ in)
Tail rotor blade chord 0.29 m (11½ in)
Length: overall, rotors turning 17.07 m (56 ft 0 in)
fuselage, excl rotors 12.92 m (42 ft 4¾ in)
Height: to top of rotor head 3.29 m (10 ft 9½ in)
overall, tail rotor turning 4.32 m (14 ft 2¼ in)
Stabiliser span 2.86 m (9 ft 4½ in)
Width over skids 2.59 m (8 ft 6 in)
Door sizes as Model 212

AREAS:
Main rotor disc 154.40 m² (1,661.9 sq ft)
Tail rotor disc 5.27 m² (56.75 sq ft)

WEIGHTS AND LOADINGS:
Weight empty, equipped: VFR 2,950 kg (6,505 lb)
IFR 3,002 kg (6,620 lb)
Max external hook load 2,041 kg (4,500 lb)
Max T-O and landing weight 5,397 kg (11,900 lb)
Max disc loading 34.96 kg/m² (7.16 lb/sq ft)
Max power loading 4.02 kg/kW (6.61 lb/shp)

PERFORMANCE (at max T-O weight, ISA):
Max cruising speed:
at S/L 122 kts (226 km/h; 140 mph)
at 1,525 m (5,000 ft) 124 kts (230 km/h; 143 mph)
Service ceiling, one engine out, 30 min power rating 2,070 m (6,800 ft)
Hovering ceiling: IGE 3,110 m (10,200 ft)
OGE 1,585 m (5,200 ft)
Range at 1,525 m (5,000 ft), long-range cruising speed, standard fuel, no reserves 402 n miles (745 km; 463 miles)

UPDATED

Bell Model 412SP of the Royal Norwegian Air Force

1994

Bell Model 412SP twin-engined utility helicopter of the Royal Norwegian Air Force

1995

BRISTOL AEROSPACE

BRISTOL AEROSPACE LTD

PO Box 874, 660 Berry Street, Winnipeg, Manitoba R3C 2S4
Telephone: 1 (204) 775 8331
Fax: 1 (204) 885 3195
Telex: 0757774
PRESIDENT: K. F. Burrows

VERIFIED

NORTHROP F-5 UPGRADE PROGRAMME

Canadian Forces designation: CF-116

TYPE: Trainer aircraft upgrade. Lead-in trainer for CF-188 Hornets.

PROGRAMME: Two 'prototypes' ordered late 1988; CF fleet fitment ordered November 1990; prototype first flown 14 June 1991; life extension changes completed on 23 CF-5s by January 1992; first service re-entry Spring 1992 with No. 419 Squadron at CFB Cold Lake, Alberta; completion due end 1995. In early 1994 the Canadian government reduced the total number of upgrades from 46 to 36.

CUSTOMERS: Canadian Forces (13 single-seat and 23 two-seat); eight of the two-seat aircraft have also received structured upgrades; also contracted by CASA in 1991 to undertake structural improvements on 23 Spanish Air Force F-5Bs. Contract completed. In 1991, work began on structural life extension elements for the Norwegian Air Force including the manufacture of new dorsal longeron

Bristol Aerospace CF-5A avionics upgrade

1994

New Bristol Aerospace CF-5 instrumentation including GEC-Marconi Avionics HUD

1994

Northrop F-5A of the Norwegian Air Force being delivered to Bristol Aerospace facilities ***(Bristol Aerospace)***

1995

kits. The company is also manufacturing new wings for the Royal Norwegian Air Force. Company has also carried out F-5 upgrades for Thailand, USAF, USN and Venezuela.

COSTS: Canadian fleet fitment contract worth C$69.73 million (1990) including installation, a new database and a training simulator.

DESIGN FEATURES: Life extension modifications include repair, overhaul and re-skinning of wings and fin; replacement of dorsal longeron, rear fuselage formers and landing gear, complete rewiring and repairing.

AVIONICS: New suite includes MIL-STD-1553 databus interfaced GEC-Marconi Avionics HUD and mission computer/display processor. Litton INS, Magnavox AN/ARC-164 VHF com, J. E. T. standby attitude indicator, Conrac angle of attack sensor, Honeywell radar altimeter and Ferranti video camera.

UPDATED

CAE

CAE AVIATION LTD

Formerly Northwest Industries Limited Inc
(a Division of CAE Inc)

PO Box 9864, Edmonton International Airport, Edmonton, Alberta T5J 2T2
Telephone: 1 (403) 890 6300
Fax: 1 (403) 890 6550
PRESIDENT: L. H. Prokop
VICE-PRESIDENT, BUSINESS DEVELOPMENT: B. C. McKenzie
VICE-PRESIDENT, AIRCRAFT PROGRAMS: T. Nord

CAE is one of Canada's largest aircraft maintenance, repair, overhaul and modification centres for military and commercial aircraft, including the Lockheed C-130 Hercules Dassault Falcon, Lockheed T-33 and Canadair CL-41 (CT-114). In addition to its major in-plant aircraft programmes, mobile repair parties are stationed at CAE Aviation in support of CF-5 and CF-18 aircraft of the Canadian Forces. The manufacturing shops produce structural, mechanical and electronic components for its aircraft overhaul and modification programmes and, under subcontract, for North America's principal aerospace manufacturers.

Depot level inspection and repair, as well as sampling inspections, will continue to be carried out on CFT-33A Silver Star jet trainers.

UPDATED

CAE AVIATION LTD (CANADAIR) CL-41 TUTOR UPGRADE

TYPE: Two-seat jet counter-insurgency and armament training aircraft upgrade.

PROGRAMME: Programme to rewire the CL-41 (CT-114) Tutor jet trainers was started in 1994. This activity, together with a depot level inspection and repair, avionics update, windscreen replacement and operation load monitoring programmes to be combined into the complete fleet starting in early 1994.

UPDATED

Canadair CL-41 Tutor aircraft (CT-114) undergoing avionics update, rewiring and repair ***(CAE Aviation)***

1995

CAE AVIATION LTD (LOCKHEED) C-130 HERCULES UPGRADE

TYPE: Tactical transport and multimission aircraft upgrade.

PROGRAMME: In 1987 CAE Aviation Ltd completed a major structural upgrade of the CF fleet of 22 Lockheed C-130E Hercules. In addition two new C-130H-84 and two used C-130H-73 aircraft, purchased by the CF, underwent extensive avionics upgrading, modification, repair and repaint to achieve commonality with the remainder of the fleet. A progressive structural inspection (PSI) programme for CFC-130s, begun in Autumn 1987, will result in the complete fleet being recycled through CAE every three years. CAE Aviation has been contracted by the Canadian Department of National Defence for the Avionics update of the Canadian Air Force's Fleet of C-130 Hercules aircraft. The programme is for the design, development and installation of a modernised suite of avionics. The contract is to be extended to cover all 30 aircraft in the fleet.

UPDATED

CAE Aviation Ltd C-130 Outer Wing Improvement Programme showing installation of modified outer wing
1994

CANADAIR

CANADAIR, BOMBARDIER INC

400 Côte-Vertu, Dorval, Québec H4S 1Y9
POSTAL ADDRESS: PO Box 6087, Station Centreville, Montreal, Quebec H3C 3G9
Telephone: 1 (514) 855 300
Fax: 1 (514) 855 7903
PRESIDENT: Robert E. Brown
PRESIDENT, BOMBARDIER BUSINESS AIRCRAFT: Bryant T. Moss
PRESIDENT, CANADAIR AMPHIBIOUS AIRCRAFT DIVISION: James C. Cherry
PRESIDENT, BOMBARDIER REGIONAL AIRCRAFT DIVISION: Pierre Lortie
EXECUTIVE VICE-PRESIDENT: Roland Gagnon
PRESIDENT, CANADAIR: David Huddleston
VICE-PRESIDENTS/GENERAL MANAGERS:
William R. Dawes (Surveillance Systems Division)
Keith Garner (Canadair Challenger Inc)
PUBLIC RELATIONS: Catherine Chase

Canadair has manufactured civil and military aircraft since 1944. It has also been employed in the research, design, development and production of missile components unmanned surveillance systems and a variety of non-aerospace products. Headquarters for Canadair, consolidating all Canadair aerospace activities, and for the Bombardier Aerospace Group - North America, is a new nine storey office tower alongside Dorval Airport. It currently has a manufacturing plant in the St Laurent complex at Cartierville Airport, and another has been expanded at Montreal (Dorval) International Airport, Mirabel, houses the CF-18 Hornet programme and other military aircraft services.

Canadair became a wholly owned Bombardier subsidiary 23 December 1986. Bombardier Aerospace Group - North America encompasses three leading aircraft manufacturers - Canadair and de Havilland in Canada and Learjet in the USA. These operate in close association with the Bombardier Shorts Group in Northern Ireland. Four Canadair business divisions — the Business Aircraft Division; Amphibious Aircraft Division; Manufacturing Division; Defence Systems Division and along with a new marketing organisation, Bombardier Regional Aircraft Division — are responsible for marketing and supporting Canadair products.

The Challenger entered production during 1978, and programme go-ahead for the Regional Jet was given in March 1989. Manufacture of the fifth series of CL-215 tanker/utility amphibious aircraft (c/n 1081-1125) ended in 1989; certification of the CL-215T turboprop took place March 1991. The CL-415 turboprop programme was launched in October 1991. Major subcontracts concern nose barrel assemblies for the McDonnell Douglas F/A-18C Hornet; six major fuselage components for 600 Airbus A330/A340 aircraft (for Aerospatiale) inboard wing leading-edge assemblies for these aircraft (for BAe); and rear fuselage sections for the Boeing 767. Engineering support for the CF's CF-18 Hornet fleet is provided under a renewable contract (begun in 1986) by a team comprising Canadair, CAE Electronics (Montreal) and CAE Aviation (Edmonton), with Canadair as prime contractor. Production of aircraft spares and the modification, repair and overhaul of aircraft are also included in the current work programme.

UPDATED

CANADAIR CL-41

RCAF designation: CT-114 Tutor

TYPE: Two-seat jet counter-insurgency and armament training aircraft.

PROGRAMME: Two prototypes of the Canadair designed CL-41 two-seat side by side jet basic trainer were built as a private venture, each powered by a 2,400 lb st Pratt & Whitney JT12A-5 turbojet. The first of these flew for the first time on 13 January 1960.

VERSIONS: **CL-41A**: Basic training version, ordered into production for the RCAF in September, 1961. The total order of 190 aircraft, was each powered by a General Electric J85/CJ610-1B turbojet, and the first of these was delivered to the RCAF on 29 October 1963. Production was at the rate of one aircraft every five days and more than 110 were delivered by mid-June 1965. Equipment includes dual blind-flying panels, an automatic emergency escape system for the crew and provision for light armament and external stores. Production ended in 1966.

CL-41G: Countery-insurgency and armament training aircraft adapted from CL-41A. First flown in June 1964, the prototype had logged more than 200 hours of demonstration and test by Spring 1965. Production configuration has uprated General Electric J85-J4 engine, six stores suspension points capable of carrying up to 3,500 lb (1,590 kg) of gun pods, bombs, rockets, Sidewinder air-to-air missiles, auxiliary fuel tanks and soft-field landing gear. Available to customer's specification with variations in avionics, protective armour plate and so on.

UPGRADES: **CAE Aviation Ltd**: Life extension programme. See separate entry.

OPERATORS: Versions of the CL-41 are in service with the armed forces of the following countries: Canada (144), Malaysia (in storage).

The following details refer to the CL-41G:

DESIGN FEATURES: Cantilever low-wing monoplane. Wing section NACA 63A014 at root, NACA 63A212 at tip. Aspect ratio 6.0. Chord at root 2.46 m (8 ft 1 in), at tip 1.23 m (4 ft 0½ in). Dihedral 2°. Incidence 1° at root –½° at tip. Taxi light in nosecone. Cantilever tailplane mounted on tip of fin.

STRUCTURE: Each wing has an aluminium alloy box primary structure, with detachable tip, formed by single main spar and stressed skin, supplemented by two auxiliary spars providing attachments for leading- and trailing-edge assemblies. Chemical milling and honeycomb sandwich utilised. The fuselage is an aluminium alloy semi-monocoque structure with four extruded main longerons and two primary longitudinal beams. The tailplane is a cantilever aluminium alloy structure with tailplane mounted at tip of fin. Metal honeycomb used in elevator.

FLYING CONTROLS: Internally sealed light alloy ailerons with honeycomb trailing-edges. Spring tab and electrically actuated trim tab in port aileron; spring tab and geared tab in starboard aileron. Hydraulically actuated single-slotted flaps of light alloy and honeycomb sandwich, with Jarry Hydraulics actuators. Door-type airbrake on each side of rear fuselage. Trim tab in port elevator, geared tab in starboard elevator, ground adjustable tab in rudder.

LANDING GEAR: Hydraulically retractable tricycle type. Main units retract inward into wings, nose unit forward. Dowty oleo-pneumatic main units with Type III 6.50 × 8 tyres, pressure 3.72 kg/cm² (53 lb/sq in). Jarry Hydraulics steerable nose oleo with Type III tyre size 5.00 × 5.

POWER PLANT: One 1,340 kg (2,950 lb st) General Electric J85-J4 turbojet engine. Internal fuel in five interconnected flexible cells in fuselage, with total capacity of 1,170 litres (309 US gallons; 258 Imp gallons). Single refuelling point aft of cockpit on starboard side of fuselage. Provision for up to six external fuel tanks, each of 181 litres (48 US gallons; 40 Imp gallons).

ACCOMMODATION: Crew of two side by side, under electrically operated rearward-opening and jettisonable canopy. Integrated escape system with Weber ejection seats operable down to ground level. Personal baggage stowage in air intake fairings.

SYSTEMS: Cabin cooling turbine and heat exchanger operated by engine bleed air. Hydraulic system pressure 105 kg/cm² (1,500 lb/sq in) for landing gear and doors, nosewheel steering, flaps and airbrakes. Pressure regulated pneumatic supply from engine bleed air for canopy seal, hydraulic reservoir pressurisation and anti-*g* suits. Electrical system

Canadair CL-41 Tutor of the Canadian Air Force *(Peter March)*
1994

includes 300 A generator, two 22 Ah Ni/Cd batteries, two 750 kVA inverters and 115/26 V transformer. No de-icing system on wings.

ARMAMENT: Up to 1,590 kg (3,500 lb) of gun pods, bombs, rockets and air-to-air missiles, as specified by customer.

ELECTRONICS AND EQUIPMENT: Blind-flying instrumentation standard. Electronics provides for selection from Collins 618 VHF, Collins DF-203 radio compass, AN/ARC-552 UHF, AN/ARC-504 emergency UHF, AN/ARN-501 TACAN, AN/APX-46 IFF and AN/AIC-502 intercom. Also permits fitment of PHI-5C-1 position and homing indicator.

DIMENSIONS, EXTERNAL:
Wing span 11.13 m (36 ft 5 9/10 in)
Length overall 9.75 m (32 ft 0 in)
Height overall 2.84 m (9 ft 3¾ in)
Tailplane span 4.16 m (13 ft 7 in)
Wheel track 4.02 m (13 ft 2¼ in)
Wheelbase 3.38 m (11 ft 1 in)

AREAS:
Wings, gross 20.44 m^2 (220 sq ft)
Ailerons (total) 1.72 m^2 (18.5 sq ft)
Flaps (total) 3.04 m^2 (32.6 sq ft)
Fin 1.07 m^2 (11.47 sq ft)
Rudder 0.56 m^2 (6.0 sq ft)
Tailplane 2.92 m^2 (31.4 sq ft)
Elevators 0.92 m^2 (9.9 sq ft)

WEIGHTS AND LOADINGS:
Weight empty: CL-41G 2,400 kg (5,296 lb)
Max T-O weight: CL-41A 3,355 kg (7,397 lb)
CL-41G 5,000 kg (11,000 lb)
Max landing weight: CL-41A 3,175 kg (7,000 lb)
CL-41G 4,040 kg (8,900 lb)

PERFORMANCE (CL-41G, without external stores):
Max level speed at 8,700 m (28,500 ft) with 50% fuel 407 kts (755 km/h; 470 mph)
Max permissible diving speed 478 kts (885 km/h; 550 mph)
Stalling speed, with 50% fuel 71 kts (133 km/h; 83 mph)
Service ceiling 12,800 m (42,200 ft)
T-O to 15 m (50 ft) 532 m (1,750 ft)
Landing from 15 m (50 ft) 710 m (2,330 ft)

VERIFIED

CANADAIR CL-215

TYPE: Twin-engined multipurpose amphibian.

PROGRAMME: The Canadair CL-215 is intended primarily for firefighting but is adaptable to a wide variety of other duties. Operating from small airstrips, lakes, ocean, bays and so on, it made its first flight on 23 October 1967 and its first water take-off on 2 May 1968. DoT certification in the Utility and Restricted categories was obtained on 7 March 1969, followed by FAA certification in the Restricted category on 15 May of the same year.

A total of 125 was built and deliveries have been made to the governments of France (15); Greece (16); Italy (5); Spain (30); Thailand (2); Venezuela (2); Yugoslavia (5); and eight Canadian provinces (Alberta 4, Manitoba 5, Newfoundland 4, Northwest Territories 2, Ontario 9, Quebec 19, Saskatchewan 4 and Yukon 2). All aircraft are capable of a variety of roles including firefighting. Spain equipped eight for SAR and coastal patrol, and Thailand (2). The Venezuelan pair were configured as passenger transport aircraft.

Production of the first 125 aircraft was completed in batches of 30, 20, 15, 15 and 45. The last C-215 (for Greece) was completed and delivered 3 May 1990.

The firefighting installation consists of two internal tanks, two retractable probes and two drop doors; plus the associated operating systems. It attacks fires with either water or chemical retardants ground-loaded at airports or with fresh or salt water scooped from a suitable body of water as the aircraft skims across the surface. Water can be mixed with small amounts of concentrate to produce effective extinguishing foam.

The aircraft carries a maximum water or retardant load of 5,346 litres (1,412 US gallons; 1,176 Imp gallons). The tanks can be ground filled in 2 minutes, or scoop filled in 10 seconds while the aircraft planes at 70 knots (130 km/h; 81 mph). Pick-up distance in still air, from 15 m (50 ft) above the surface on approach to 15 m (50 ft) above the surfacing during climb-out, is 1,200 m (3,935 ft).

Single CL-215s have frequently made over 100 drops totalling more than 534,600 litres (141,230 US gallons; 117,600 Imp gallons) in one day. Full loads have been scooped up from the Mediterranean in wave heights of up to 2 m (6 ft). In 1983 a Yugoslav CL-215 made 225 drops totalling 1,202,950 litres (317,760 US gallons; 264,590 Imp gallons) on fires in one day.

A lightweight integrated liquid spray system has also been certificated, and four have been purchased for the Yugoslav CL-215s. The system which does not interfere with the primary role of firefighting, is available for retrofit. Uses include the application of oil dispersants and pesticides. Tests conducted at Canadair have shown that the CL-215 can be used to extinguish oil fires by airdropping a suitable foaming agent. Another type of foam agent, mixed onboard the aircraft after scooping, has proved particularly effective against forest fires. All CL-215s in Canada are now fitted with foam injection systems.

UPGRADES: **Canadair:** CL-215T. See separate entry.

OPERATORS: Versions of the CL-215 are in service with the armed forces of the following countries: Greece (14) and Spain (21).

DESIGN FEATURES: Cantilever high-wing monoplane. No dihedral. All-metal one-piece fail-safe structure, with front and rear spars at 16 and 49 per cent chord. Provision for de-icing on tail unit leading-edges.

FLYING CONTROLS: Hydraulically operated all-metal single-slotted flaps, supported by four external hinges on interspar ribs on each wing. Trim tab and geared tab in port aileron, rudder/aileron interconnect tab in starboard aileron. Elevators and rudder fitted with dynamic balance, trim tab (port elevator only), spring tabs and geared tabs.

STRUCTURE: Wing spars of conventional construction, with extruded caps and webs stiffened by vertical members. Aluminium alloy skin with riveted span-wise extruded stringers, is supported at 762 mm (30 in) pitch by interspar ribs. Leading-edge consists of aluminium alloy skin attached to pressed nose-ribs and spanwise stringers. Detachable glassfibre wingtips. The fuselage is an all-metal single-step flying boat hull of conventional fail-safe construction. The tail unit is a cantilever all-metal fail-safe structure with horizontal surfaces mounted midway up fin. Structure of aluminium alloy sheet, honeycomb panels, extrusions and fittings.

LANDING GEAR: Hydraulically retractable tricycle type. Fully castoring, self-centring twin-wheel nose unit retracts rearward into hull and is fully enclosed by doors. Fifth series aircraft are fitted with nosewheel steering, and a retrofit kit is available for earlier aircraft. Main gear support structures retract into wells in sides of hull. A plate mounted on each main gear assembly encloses bottom of wheel well. Mainwheel tyre pressure of 5.31 bars (77 lb/sq in); nosewheel tyre pressure 6.55 bars (95 lb/sq in). Hydraulic disc brakes. Non-retractable stabilising floats are each carried on a pylon cantilevered from wing box structure, with break-away provision.

Canadair CL-215, water bomber in service with France's Sécurité Civile *1994*

POWER PLANT: Two 1,566 kW (2,100 hp) Pratt & Whitney R-2800-CA3 18-cylinder radial engines, each driving a Hamilton Standard constant-speed fully feathering three-blade propeller, with 43E60 hub and Type 6903A blades. Two fuel tanks, each of eight flexible cells, in wing spar box with total usable capacity of 5,910 litres (1,561 US gallons; 1,300 Imp gallons). Gravity refuelling through two points above each tank. Oil in two tanks, with total capacity of 272.75 litres (72 US gallons; 60 Imp gallons), aft of engine firewalls.

ACCOMMODATION (water bomber version): Crew of two side by side on flight deck. Dual controls standard. Two 2,673 litre (706 US gallon; 588 Imp gallon) water tanks in main fuselage compartment, with retractable pickup probe in each side of hull bottom. Water drop doors in each side of hull bottom. Flush doors on port side of fuselage forward and aft of wings. Emergency exit on starboard side aft of wing trailing-edge. Emergency hatch above starboard cockpit. Mooring hatch in upper surface of nose. Side-facing canvas folding seats for eight people are located in the forward cabin area.

ACCOMMODATION (other roles): When configured for patrol and search and rescue missions, aircraft has additional observers. Navigator's station, immediately behind flight deck includes search radar display. Observers' stations in rear fuselage have sliding seats which can be positioned alongside blister windows. Toilet in rear of cabin; galley installed. Additional seats and/or stretchers available. In passenger transport configuration, up to 26 forward-facing seats can be fitted in a fully furnished interior with toilet and galley. Cargo tiedown fittings for loads up to 2,268 kg (5,000 lb). Provision for extra cabin windows, to a maximum of 14.

SYSTEMS: Hydraulic system, pressure 207 bars (3,000 lb/sq in), utilises two engine-driven pumps maximum flow rate 45.5 litres (12 US gallons; 10 Imp gallons)/min to actuate nosewheel steering, landing gear, flaps, water drop doors, pick-up probes and wheel brakes. Unpressurised air/oil, reservoir. Electrically driven third pump provides hydraulic power for emergency actuation of landing gear and brakes and closure of water doors. Electrical system includes two 400 VA 115 V 400 Hz static inverters (800 VA in SAR version), two 28 V 200 A DC engine-driven generators, one 36 Ah lead-acid battery, and one air-cooled petrol engine-driven 28 V 200 A generator GPU.

AVIONICS AND EQUIPMENT: Standard installation includes dual VHF transceivers single VHF/FM com, dual VOR/ILS receivers, dual ADF, two marker beacon receivers, ATC transponder and ELT. Optional avionics include HF, DME and radio altimeter. A search radar is optional on the SAR version.

DIMENSIONS, EXTERNAL:
Wing span 28.60 m (93 ft 10 in)
Wing chord (constant) 3.54 m (11 ft 7½ in)
Wing aspect ratio 8.2
Length overall 19.82 m (65 ft 0½ in)
Beam (max) 2.59 m (8 ft 6 in)
Length/beam ratio 7.5
Height overall: on land 8.98 m (29 ft 5½ in)
on water 6.88 m (22 ft 7 in)
Draught: wheels up 1.12 m (3 ft 8 in)
wheels down 2.03 m (6 ft 8 in)
Tailplane span 10.97 m (36 ft 0 in)
Wheel track 5.28 m (17 ft 4 in)
Wheelbase 7.23 m (23 ft 9 in)
Propeller diameter 4.34 m (14 ft 3 in)
Forward door: Height* 1.37 m (4 ft 6 in)
Width 1.03 m (3 ft 4 in)
Height to sill 1.68 m (5 ft 6 in)
Rear door: Height 1.12 m (3 ft 8 in)
Width 1.03 m (3 ft 4 in)
Height to sill 1.83 m (6 ft 0 in)
Water drop door: Length 1.60 m (5 ft 3 in)
Width 0.81 m (2 ft 8 in)
Emergency exit: Height 0.91 m (3 ft 0 in)
Width 0.51 m (1 ft 8 in)
** incl 25 cm (10 in) removable sill*

DIMENSIONS, INTERNAL:
Cabin, excl flight deck: Length 9.38 m (30 ft 9½ in)
Max width 2.39 m (7 ft 10 in)
Max height 1.90 m (6 ft 3 in)
Floor area 19.69 m^2 (212 sq ft)
Volume 35.59 m^3 (1,257 cu ft)

AREAS:
Wings, gross 100.33 m^2 (1,080 sq ft)
Ailerons, (total) 8.05 m^2 (86.6 sq ft)
Flaps (total) 22.39 m^2 (241.0 sq ft)
Fin 17.2 m^2 (185.5 sq ft)
Rudder, incl tabs 6.02 m^2 (64.75 sq ft)
Tailplane 28.4 m^2 (306.0 sq ft)
Elevators (total incl tabs) 7.888 m^2 (84.8 sq ft)

WEIGHTS AND LOADINGS:
Manufacturer's weight empty 12,337 kg (27,200 lb)
Typical operating weight, empty 12,855 kg (28,340 lb)
Max fuel weight 4,245 kg (9,360 lb)
Max payload: Water bomber 5,443 kg (12,000 lb)
Utility version 3,747 kg (8,260 lb)
Max T-O weight (land) 19,731 kg (43,500 lb)

Canadair CL-215 in service with the government of Quebec

1994

Max T-O weight (water)	17,100 kg (37,700 lb)
Max zero-fuel weight	18,597 kg (41,000 lb)
Max landing weight on land or water	16,780 kg (37,000 lb)
Max cabin floor loading	732 kg/m² (150 lb/sq in)
Max wing loading	196.66 kg/m² (40.3 lb/sq in)
Max power loading	6.23 kg/kW (10.36 lb/hp)

PERFORMANCE:

Cruising speed (max recommended power) at AUW of 18,595 kg (41,000 lb) at 3,050 m (10,000 ft) 157 kts (291 km/h; 181 mph)
Stalling speed 15° flap, AUW of 19,731 kg (43,500 lb) 79 kts (145 km/h; 90 mph)
Stalling speed 25° flap, AUW of 16,780 kg (37,000 lb): power off 66 kts (123 km/h; 76 mph)
Max rate of climb at S/L at AUW of 19,731 kg (43,500 lb): at max continuous power 305 m (1,000 ft)/min
Rate of climb at S/L, one engine out, at AUW of 17,100 kg (37,700 lb) at T-O power 30.5 m (100 ft)/min
T-O to 15 m (50 ft):
from land at AUW of 19,731 kg (43,500 lb) 811 m (2,660 ft)
from water at AUW of 17,100 kg (37,700 lb) 800 m (2,625 ft)
Landing from 15 m (50 ft) at AUW of 16,780 kg (37,000 lb): on land 768 m (2,520 ft)
on water 835 m (2,740 ft)
Range with 1,587 kg (3,500 lb) payload:
at max cruise speed 925 n miles (1,714 km; 1,065 miles)
at long-range cruise power 1,130 n miles (2,094 km; 1,301 miles)

UPDATED

Canadair CL-215 of the Royal Thai Navy *(Canadair)*

1995

Canadair CL-215T amphibian retrofitted CL-215 with Pratt & Whitney Canada PW123AF turboprops *(Canadair)*

1995

CANADAIR CL-215T

TYPE: Twin-turboprop multipurpose amphibian aircraft and upgrade (CL-215).

PROGRAMME: In January 1987 Bombardier Inc announced its intention to develop this turboprop version of the CL-215.

VERSIONS: **CL-215T:** Turboprop version of the CL-215 with a number of improvements. These include modern turboprop engines, an upgraded and air conditioned cockpit, pressure refuelling, powered flight controls, wingtip endplates and auxiliary fins. C/N 1056 and subs are the best candidates for upgrade to CL-215T configuration.

Two CL-215s were modified as CL-215T prototypes and underwent an extensive flight test programme. The first of these (C-FASE) made its initial flight 8 June 1989, followed on 20 September 1989 by the second (C-FAWQ), which was fitted with powered ailerons. Restricted Category approval (for firefighting) was obtained 31 March 1991 followed on 23 December 1991 by Utility Category. Retrofit kits for existing CL-215s became available in 1991. FAA certification in the Restricted category was awarded in March 1993. The government of Quebec was the launch customer for the retrofit programme, having supplied the prototype airframes. There were 15 conversion kits, including powered ailerons, ordered by the Spanish government on 3 August 1989 for aircraft operated by No. 43 Squadron of the Spanish Air Force. To date, 13 retrofitted aircraft have been delivered to Spain along with two complete kits, which are being installed during 1995.

DESIGN FEATURES: Cantilever high-wing monoplane. No dihedral. Metal endplates, which improve lateral stability and permit use of full engine ratings.

FLYING CONTROLS: Hydraulically operated all-metal single-slotted flaps, supported by four external hinges on interspar ribs on each wing. Powered ailerons, with Jacottet actuators. Trim tab and geared tab in port aileron, rudder/aileron interconnect tab in starboard aileron. Hydraulically powered elevators and rudder fitted with dynamic balance, trim tab (port elevator only), spring tabs and geared tabs, for manual reversion.

STRUCTURE: The wing is an all-metal one-piece fail-safe structure, with front and rear spars at 16 and 49 per cent chord. Spars of conventional construction, with extruded caps and webs stiffened by vertical members. Aluminium alloy skin, with riveted spanwise extruded stringers, is supported at 762 mm (30 in) pitch by interspar ribs. Leading-edge consists of aluminium alloy skin attached to pressed nose-ribs and spanwise stringers. All-metal single-step flying boat hull of conventional fail-safe construction. The tail unit is a cantilever all-metal fail-safe structure with horizontal surfaces mounted midway up fin. Structure of aluminium alloy sheet, honeycomb panels, extrusions and fittings. 'Arrowhead' auxiliary fin on each half of tailplane.

LANDING GEAR: Hydraulically retractable tricycle type. Self-centring twin-wheel nose unit retracts rearward into hull and is fully enclosed by doors. Nosewheel steering standard on fifth series CL-215s and CL-215Ts. Main gear support structures retract into wells in sides of hull. A plate mounted on each main gear assembly encloses bottom of wheel well. Mainwheel tyre pressure 5.31 bars (77lb/sq in); nosewheel tyre pressure 6.55 bars (95 lb/sq in). Hydraulic disc brakes. Non-retractable stabilising floats are each carried on a pylon cantilevered from wing box structure, with breakaway provision.

POWER PLANT: Two 1,775 kW (2,380 shp) Pratt & Whitney

Canadair CL-215T turboprop amphibian dropping water *(Canadair)*

1995

Canada PW123AF turboprops, on damage-tolerant mounts, each driving a Hamilton Standard 14SF-17 four-blade constant-speed fully feathering reversible-pitch propeller. Two fuel tanks, each of eight identical flexible cells, in wing spar box with total usable capacity of 5,796 litres (1,531 US gallons; 1,275 Imp gallons). Single-point pressure refuelling (rear fuselage, starboard side), plus gravity point above each tank. Pneumatic/electric intake de-icing system.

ACCOMMODATION: Normal crew of two side by side on flight deck, with dual controls. Additional stations in maritime patrol/SAR versions for flight engineer, navigator and two observers. For water bomber cabin installation, see Equipment paragraph. With water tanks removed, transport configurations can include layout for 30 passengers plus toilet, galley and baggage area, with seat pitch of 79 cm (31 in). Combi layout offers cargo at front, full firefighting capability, plus 11 seats at rear. Other quick-change interiors available for medevac (12 stretchers and two medical attendants), utility/paratroop (up to 14 fold-up troop-type canvas seats in cabin, in two inward facing rows), all-cargo or other special missions according to customer's requirements. Flush doors to main cabin on port side of fuselage forward and aft of wings. Emergency exit on starboard side aft of wing trailing-edge. Crew emergency hatch in flight deck roof on starboard side. Mooring hatch in upper surface of nose. Provision for additional cabin windows.

SYSTEMS: Casey vapour-cycle air conditioning system and Janitrol heater. Hydraulic system, pressure 207 bars (3,000 lb/sq in) utilises two engine-driven pumps (maximum flow rate 45.5 litres, 12 US gallons; 10 Imp gallons) to actuate nosewheel steering, landing gear, flaps, water drop doors, pick-up probes, flight controls and wheel brakes. Pressurised air/oil reservoir. Electrically driven third pump provides hydraulic power for emergency actuation of landing gear and brakes and closure of water doors. Electrical system includes two 800 VA 115 V 400 Hz static inverters, two 28 V 400 A DC engine-driven starter/generators and two 40 Ah Ni/Cd batteries. Ice protection system optional.

AVIONICS: Standard installation includes dual VHF transceivers, single VHF/FM com. Dual VOR/ILS receivers, dual ADF, two marker beacon receivers, ATC transponder and ELT. Optional avionics include autopilot, VLF/Omega nav system, search radar and colour weathe radar.

EQUIPMENT (Waterbomber): Two integral water tanks in main fuselage compartment near CG (combined capacity 5,346 litres; 1,410 US gallons; 1,176 Imp gallons), plus seven inward-facing seats. Tanks filled by two hydraulically actuated scoops aft of hull step, fillable also on ground by hose adaptor on each side of fuselage. Two water doors capable of independent opening in hull bottom. Onboard foam concentrate reservoirs and mixing system. Optional spray kit can be coupled with firefighting tanks for large scale spraying of oil dispersants and insecticides. In a typical firefighting mission with a fire 85 n miles (157 km; 98 miles) from the CL-215T's base, and a water source 10 n miles (18.5 km; 11.5 miles) from the fire, and 45 minutes fuel reserves, aircraft could make 26 water scoop and drop circuits before having to return to base to refuel. Water tanks can be scoop-filled completely (on smooth water, ISA, conditions) in an on-water distance of only 564 m (1,850 ft); partial water loads can be scooped on smaller water bodies. Minimum safe water depth for scooping is only 1.40 m (4 ft 7 in).

EQUIPMENT (other versions): Stretcher kits, passenger or troop seats, cargo tiedowns, searchlight and other equipment according to customer requirements. Provisions for two underwing pylon attachment points for stores.

DIMENSIONS, EXTERNAL:	
Wing span	28.60 m (93 ft 10 in)
Wing chord (constant)	3.54 m (11 ft 7½ in)
Wing aspect ratio	8.16
Length overall	19.82 m (65 ft 0½ in)
Beam (max)	2.59 m (8 ft 6 in)
Length/beam ratio	7.5
Height overall: on land	8.98 m (29 ft 5½ in)
on water	6.88 m (22 ft 7 in)
Draught: wheels up	1.12 m (3 ft 8 in)
wheels down	2.03 m (6 ft 8 in)
Tailplane span	10.97 m (36 ft 0 in)
Wheel track	5.28 m (17 ft 4 in)
Wheelbase	7.23 m (23 ft 9 in)
Propeller diameter	3.97 m (13 ft 0¼ in)
Propeller/fuselage clearance	0.59 m (1 ft 11¼ in)
Propeller/water clearance	1.30 m (4 ft 3¼ in)
Propeller/ground clearance	2.77 m (9 ft 1 in)
Forward door: Height*	1.37 m (4 ft 6 in)
Width	1.03 m (3 ft 4 in)
Height to sill	1.68 m (5 ft 6 in)
Rear door: Height	1.12 m (3 ft 8 in)
Width	1.03 m (3 ft 4 in)
Height to sill	1.83 m (6 ft 0 in)
Water drop door: Length	1.60 m (5 ft 3 in)
Width	0.81 m (2 ft 8 in)
Emergency exit: Height	0.91 m (3 ft 0 in)
Width	0.51 m (1 ft 8 in)

* *incl 25 cm (10 in) removable sill*

DIMENSIONS, INTERNAL:	
Cabin, excl flight deck: Length	9.38 m (30 ft 9½ in)
Max width	2.39 m (7 ft 10 in)
Max height	1.90 m (6 ft 3 in)
Floor area	19.69 m² (212 sq ft)
Volume	35.59 m³ (1,257 cu ft)

AREAS:	
Wings, gross	100.33 m² (1,080.0 sq ft)
Ailerons (total)	8.05 m² (86.6 sq ft)
Flaps (total)	22.39 m² (241.0 sq ft)
Fin	17.2 m² (185.5 sq ft)
Rudder, incl tabs	6.02 m² (64.75 sq ft)
Tailplane	28.4 m² (306.0 sq ft)
Elevators (total, incl tabs)	7.88 m² (84.8 sq ft)

Canadair CL-215T turboprop amphibian operated by Quebec's Fonds du service aérien gouvernement *(Canadair)*

1994

WEIGHTS AND LOADINGS (A: water bomber, B: utility, land or water based):

Typical operating weight empty:
A 12,265 kg (27,040 lb)
B 12,000 kg (26,460 lb)
Max internal fuel weight: A, B 4,649 kg (10,250 lb)
Max payload: A (disposable) 5,443 kg (12,000 lb)
B 4,432 kg (9,770 lb)
Max ramp weight: A (land) 19,800 kg (43,650 lb)
A (water), B 17,236 kg (38,000 lb)
Max T-O weight:
A, production (land) 19,890 kg (43,850 lb)
A, (water), B (land and water) 17,100 kg (37,700 lb)
Max touchdown weight for water scooping:
A 16,420 kg (36,200 lb)
Max flying weight after water scooping (A and B, land and water) 20,525 kg (45,250 lb)
Max landing weight 16,783 kg (37,000 lb)
Max zero-fuel weight: A 18,257 kg (40,250 lb)
B 16,511 kg (36,400 lb)
Max wing loading: A (land) 196.7 kg/m² (40.30 lb ft²)
A (water), B (land and water) 170 kg/m² (34.91 lb/sq ft)
Max power loading: A (land) 5.56 kg/kW (9.14 lb/shp)
A (water), B (land and water) 4.82 kg/kW (7.92 lb/shp)

PERFORMANCE (estimated at weights shown):

Max cruising speed at 3,050 m (10,000 ft), AUW of 14,741 kg (32,500 lb) 203 kts (376 km/h; 234 mph)
Long-range cruising speed at 3,050 m (10,000 ft), AUW of 14,741 kg (32,500 lb) 145 kts (269 km/h; 167 mph)
Patrol speed at S/L, AUW of 15,876 kg (35,000 lb) 130 kts (241 km/h; 150 mph)
Stalling speed:
15° flat, AUW of 20,865 kg (46,000 lb) 80 kts (148 km/h; 92 mph) CAS
25° flap, AUW of 16,783 kg (37,000 lb) 65 kts (121 km/h; 75 mph) CAS
Max rate of climb at S/L, AUW of 20,865 kg (46,000 lb) 419 m (1,375 ft)/min
T-O distance at S/L, ISA:
land, AUW of 19,731 kg (43,500 lb) 823 m (2,700 ft)
water, AUW of 17,168 kg (37,850 lb) 814 m (2,670 ft)
Landing distance at S/L, ISA:
land, AUW of 16,783 kg (37,000 lb) 671 m (2,200 ft)
water, AUW of 16,783 kg (37,000 lb) 664 m (2,180 ft)
Scooping distance at S/L, ISA (incl safe clearance heights) 1,189 m (3,900 ft)
Ferry range with 500 kg (1,100 lb) payload 1,300 n miles (2,410 km; 1,500 miles)
Design *g* limits (15° flap) −1.0/+3.25

UPDATED

CANADAIR CHALLENGER

Canadian Forces designations: CC-144 and CE-144A

TYPE: Twin-turbofan business, cargo and regional transport.

PROGRAMME: First flight of first of three prototypes (C-GCGR-X) 8 November 1978; first flight production Challenger 600 with Lycoming ALF 502L-2 turbofans 21 September 1979; first customer delivery 30 December 1980; prototype Challenger 601 with GE CF34s 10 April 1982; first 601-1A delivered 6 May 1983; first 601-3A 6 May 1987 and first 601-3A/ER 19 May 1989; 601-S launched June 1989 but not committed to production.

VERSIONS: **Challenger 600:** Total 83 built after certification in 1980 (76 since retrofitted with winglets); 16 delivered to Canadian Department of National Defence as CC-144 (7) and CE-144A (6), and CX-144 (3) for coastal patrol. Production completed.

Challenger 601-1A: First production version (66 built) to have CF34 engines; first flight 17 September 1982. Production completed.

Challenger 601-3A: Standard version with 'glass cockpit' and upgraded CF34s; first flight 28 September 1986; Canadian and US certification 21 and 30 April 1987; also certificated for Cat. II and in 22 other countries; improvements include CF34-3A engines flat rated to 21°C, and fully integrated digital flight guidance and flight management systems.

Challenger 601-3R: Extended-range version. Available as retrofit. For full details see *Jane's All the World's Aircraft 1995-96.*

UPGRADES: **Bombardier - Canadair Defense Systems Division:** Currently upgrading three Challenger 600 aircraft for Electronic Support and Training Missions. These three aircraft will supplement the three other aircraft which are in use with the Canadian Air Force. First three aircraft were modified by Lockheed and Canadair.

Canadair: 76 Challenger 600 have been retrofitted with winglets, after the introduction of a modification which was certificated in 1984.

Lockheed/Canadair: CE-144. Three Challenger 600 aircraft extensively reconfigured for Electronic Support and Training Missions.

OPERATORS: Versions of the Challenger in service with the armed forces of Canada (16), China (2) and Germany (6).

CUSTOMERS: Total 84 Challenger 600 and 66 601-1A delivered; total 121 Challenger 601-3A delivered between 6 May 1987 and 5 January 1993; total 288 Challengers of all versions delivered by early 1994.

DESIGN FEATURES: Advanced wing section; quarter-chord sweep 25°; thickness/chord ratio 14 per cent at root, 12 per cent at leading-edge sweep break and 10 per cent at tip; dihedral 2° 33′; incidence at root 3° 30′; fuselage circular cross-section, pressurised.

FLYING CONTROLS: Fully powered hydraulic controls; electrically actuated variable incidence tailplane; two-segment spoilers (outboard airbrake panels, inboard lift dumpers); two-segment double-slotted flaps.

STRUCTURE: Two-spar wing torsion box; chemically milled fuselage skin panels with riveted frames and stringers form damage-tolerant structure; multispar fin and tailplane.

LANDING GEAR: Hydraulically retractable tricycle type, with twin wheels and Dowty oleo-pneumatic shock-absorber on each unit. Mainwheels retract inward into wing centre-section, nose unit forward. Nose unit steerable and self-centring. Mainwheels have Goodyear 25 × 6.75 tyres, pressure 14.20 bars (206 lb/sq in); nosewheels have B. F. Goodrich 18 × 4.4 tyres, pressure 10.41 bars (151 lb/sq in). ABS (Aircraft Braking Systems) hydraulically operated multiple-disc carbon brakes with fully modulated anti-skid system. Minimum ground turning radius 12.19 m (40 ft 0 in).

POWER PLANT: Two 33.36 kN (7,500 lb st) Textron Lycoming ALF 502L-2 or L-3 turbofans in Challenger 600; General Electric CF34-1A in Challenger 601 or CF34-3A in 601-3A, each rated at 40.66 kN (9,140 lb st) with automatic power reserve, or 38.48 kN (8,650 lb st) without APR. One engine is pylon mounted on each side of rear fuselage, fitted with cascade type fan-air thrust reversers. Integral fuel tank in centre-section (capacity 2,839 litres; 750 US gallons; 624 Imp gallons), one in each wing (each 2,725 litres; 720 US gallons; 600 Imp gallons) and auxiliary tank beneath cabin floor; total capacity 9,278 litres (2,451 US gallons; 2,041 Imp gallons). Optional tank in tailcone, capacity 696.5 litres (184 US gallons; 153 Imp gallons). Pressure and gravity fuelling and defuelling. Oil capacity 13.6 litres (3.6 US gallons; 3 Imp gallons).

ACCOMMODATION: Pilot and co-pilot side by side on flight deck with dual controls. Blind-flying instrumentation standard. Interiors are installed to customer's specifications. A maximum of 19 passenger seats is approved. Typical installations include toilet, buffet, bar and wardrobe. Medevac version can carry up to seven stretcher patients, an infant incubator, a full complement of medical staff and comprehensive intensive care equipment. The baggage compartment, with its own loading door, is accessible in flight. Downward opening door on port side, forward of wing, from c/n 24 onwards. (Upward opening on earlier aircraft.) Door is power assisted on 601-3A. Overwing emergency exit on starboard side. Entire accommodation heated, ventilated and air conditioned.

SYSTEMS: Sundstrand pressurisation and AiResearch air conditioning systems, maximum pressure differential 0.64 bars (9.3 lb/sq in) in Challenger 600, 0.62 bars (9.0 lb/sq in) in Challenger 601 and 601-3A. Three independent hydraulic systems, each of 207 bars (3,000 lb/sq in). No. 1 system powers flight controls (via servo-actuators positioned by cables and pushrods); No. 2 system for flight controls and brakes; No. 3 system for flight controls, landing gear extension/retraction, brakes and nosewheel steering. Nos. 1 and 2 systems each powered by an engine-driven pump, supplemented by an AC electric pump; No. 3 system by two AC pumps. Two 30 kVA engine-driven generators supply primary 115/200 V three-phase AC electric power at 400 Hz. Three transformer-rectifiers to convert AC power to 28 V DC; one 43 Ah Ni/Cd battery. Alternative primary power provided by

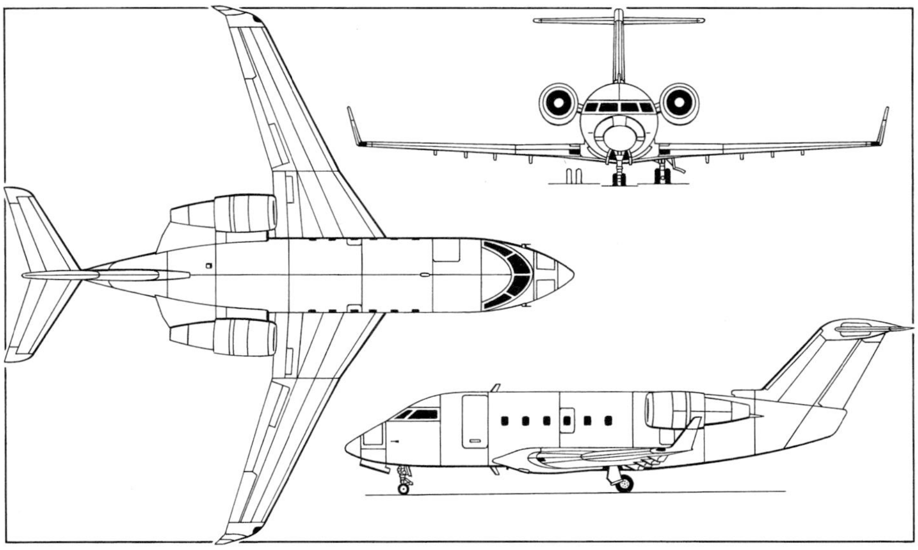

Canadair Challenger 601-3A (two General Electric CF34-3A turbofans) ***(Jane's/Mike Keep)*** *1995*

Canadair Challenger 601-3A twin-turbofan business, cargo and regional aircraft ***(Canadair)*** *1995*

Canadair Challenger 600 (CE-144) in service with the Canadian Defence Forces Air Command ***(Canadair)***

1995

APU and an air-driven generator, the latter being deployed automatically if the engine-driven generators and APU are inoperative. Stall warning system, with stick shakers and stick pusher. Garrett GTCP-100E gas turbine APU (to be certificated for in-flight operation) for engine start, ground air conditioning and other services. Electric anti-icing of windscreen, flight deck side windows and pitot heads; Sundstrand bleed air anti-icing of wings, tailplane, engine intake cowls and guide vanes. Gaseous oxygen system, pressure 124 bars (1,800 lb/sq in). Continuous-element fire detectors in each engine nacelle and APU; two-shot extinguishing system for engines, single-shot system for APU.

AVIONICS (except 601-3A): Standard avionics include dual Collins VHF-20A com, dual Collins VIR-30A VOR/ILS/marker beacon receiver, dual Honeywell SPZ 600 flight directors, SPZ 600 autopilot with dual servos, dual TDR-90 transponders, dual DME (with HSI display), one ALT-55B radio altimeter, dual C-14 compasses, standby compass, one ADF-60, dual intercom system, comparator warning system RCA Primus 400 weather radar, HF and other antennae. Provision, at customer's option, for HF com, third VHF com, second ADF, VLF nav, INS, GPWS, ELT, flight data recorder and cockpit voice recorder.

AVIONICS (601-3A): Honeywell digital avionics include SPZ-8000 five-tube electronic flight instrument system (EFIS); single multifunction display (MFD); dual Honeywell laser inertial reference systems (LIRS); dual flight management systems; digital automatic flight control system, with dual channel fail-operational autopilot and flight director; Mach trim and auto trim; dual digital air data system; Honeywell four-colour digital weather radar; radio altimeter; Collins Pro Line II nav/coms, including dual VHF com; dual VHF nav; dual DME; dual ATC transponders; dual ADF; dual HF com; cockpit voice recorder; standby instruments (artificial horizon, airspeed indicator and altimeter). Systems certificated for Cat. II operations. Space provisions for flight data recorder, ELT, VLF/Omega, GPWS, and full provisions for third LIRS.

EQUIPMENT (Medevac version): Includes a cardiopulmonary resuscitation unit; a physio control lifepack comprising a heart defibrilator, ECG and cardioscope; an ophthalmoscope; respirators and resuscitators; an infant monitor; X-ray viewer; cardiostimulator; foetal heart monitor; and an anti-shock suit.

DIMENSIONS, EXTERNAL:
Wing span: 600 18.85 m (61 ft 10 in)
601, over winglets 19.61 m (64 ft 4 in)
Wing chord: at fuselage c/l 4.89 m (16 ft 0½ in)
at tip 1.27 m (4 ft 1.9 in)
Wing aspect ratio: 600 7.4
601 8.5
Length overall 20.85 m (68 ft 5 in)
Fuselage: Max diameter 2.69 m (8 ft 10 in)
Height overall 6.30 m (20 ft 8 in)
Tailplane span 6.20 m (20 ft 4 in)
Wheel track (c/l of shock struts) 3.18 m (10 ft 5 in)
Wheelbase 7.99 m (26 ft 2½ in)
Passenger door (port, fwd): Height 1.78 m (5 ft 10 in)
Width 0.91 m (3 ft 0 in)
Height to sill 1.61 m (5 ft 3½ in)
Baggage door (port, rear): Height 0.84 m (2 ft 9 in)
Width 0.71 m (2 ft 4 in)
Height to sill 1.61 m (5 ft 3½ in)
Overwing emergency exit (stbd):
Height 0.91 m (3 ft 0 in)
Width 0.51 m (1 ft 8 in)

DIMENSIONS, INTERNAL:
Cabin: Length, incl galley, toilet and baggage area, excl flight deck 8.61 m (28 ft 3 in)
Max width 2.49 m (8 ft 2 in)
Width at floor level 2.18 m (7 ft 2 in)
Max height 1.85 m (6 ft 1 in)
Floor area 18.77 m² (202 sq ft)
Volume 32.6 m³ (1,150 cu ft)

AREAS:
Wings, gross (excl winglets on 601/601-3A) 48.31 m² (520.0 sq ft)
Ailerons (total) 1.39 m² (15.0 sq ft)
Trailing-edge flaps (total) 7.80 m² (84.0 sq ft)
Fin 9.18 m² (98.8 sq ft)
Rudder 2.03 m² (21.9 sq ft)
Tailplane 6.45 m² (69.4 sq ft)
Elevators (total) 2.15 m² (23.1 sq ft)

WEIGHTS AND LOADINGS:
Manufacturer's weight empty:
600 8,464 kg (18,660 lb)
601 9,049 kg (19,950 lb)
601-3A 9,292 kg (20,485 lb)
Typical operating weight empty:
600 10,562 kg (23,285 lb)
601 11,151 kg (24,585 lb)
601-3A 11,197 kg (24,685 lb)
Max fuel: 600 6,754 kg (14,890 lb)
601, 601-3A 7,559 kg (16,665 lb)
601, 601-3A with extended-range option 8,119 kg (17,900 lb)
Max payload: 600 2,365 kg (5,215 lb)
601 2,229 kg (4,915 lb)
601-3A 2,184 kg (4,815 lb)
Payload with max fuel: 600 1,395 kg (3,075 lb)
601 907 kg (2,000 lb)
601-3A 862 kg (1,900 lb)
Max T-O weight: 600 18,642 kg (41,100 lb)*
601, 601-3A 19,550 kg (43,100 lb)
601-3A with extended-range option 20,230 kg (44,600 lb)
Max ramp weight: 600 18,710 kg (41,250 lb)
601, 601-3A 19,618 kg (43,250 lb)
601-3A with extended-range option 20,300 kg (44,750 lb)
Max landing weight:
600, 601, 601-3A 16,329 kg (36,000 lb)
Max zero-fuel weight: 600 12,927 kg (28,500 lb)
601 14,061 kg (31,000 lb)
601-3A 13,381 kg (29,500 lb)**
Max wing loading: 600 385.9 kg/m² (79.04 lb/sq ft)
601, 601-3A 404.7 kg/m² (82.88 lb/sq ft)
Max power loading: 600 279.41 kg/kN (2.74 lb/lb st)
601 254.03 kg/kN (2.49 lb/lb st)
601-3A 240.81 kg/kN (2.36 lb/lb st)

* *18,710 kg (41,250 lb) with optional winglets*
** *14,061 kg (31,000 lb) available optionally*

PERFORMANCE (at max T-O weight except where indicated):
Max cruising speed:
600, 601, 601-3A 459 kts (851 km/h; 529 mph)
Normal cruising speed:
600, 601, 601-3A 442 kts (819 km/h; 509 mph)
Long-range cruising speed:
600 401 kts (743 km/h; 462 mph)
601, 601-3A 424 kts (786 km/h; 488 mph)
Time to initial cruise altitude: 600 25 min
601 21 min
601-3A 20 min
Max operating altitude (all) 12,500 m (41,000 ft)
Service ceiling, one engine out (all) 7,315 m (24,000 ft)
Balanced T-O field length (ISA at S/L):
600 1,737 m (5,700 ft)
601, 601-3A 1,645 m (5,400 ft)
Landing distance at S/L at max landing weight:
600, 601-3A 1,006 m (3,300 ft)
601 1,082 m (3,550 ft)
Min ground turning radius (all) 20.27 m (66 ft 6 in)
Range with max fuel and five passengers, NBAA IFR reserves (200 n mile; 370 km; 230 mile alternate) at long-range cruising speed:
600 (basic) 2,800 n miles (5,186 km; 3,222 miles)
600 (with fuselage tank) 3,123 n miles (5,784 km; 3,594 miles)
601 3,440 n miles (6,371 km; 3,959 miles)
601-3A 3,430 n miles (6,356 km; 3,950 miles)
Design *g* limit: 600 +2.7
601, 601-3A +2.6

OPERATIONAL NOISE LEVELS (FAR Pt 36):
T-O: 600 84.7 EPNdB
601 79.9 EPNdB
601-3A 79.4 EPNdB
Sideline: 600 89.5 EPNdB
601 84.8 EPNdB
601-3A 85.9 EPNdB
Approach: 600 91.6 EPNdB
601, 601-3A 89.4 EPNdB

NEW ENTRY

CANADAIR REGIONAL JET FLIGHT MANAGEMENT SYSTEM (FMS) UPGRADE

TYPE: Twin-turbofan regional transport aircraft FMS upgrade.

PROGRAMME: In February 1994, Regional Jet announced that it was introducing the Collins FMS 4050 Flight Management System on its new-build Regional Jets. This system is also available as a retrofit. For full details of Canadair Regional Jet see *Jane's All the World's Aircraft 1995-96.*

DESIGN FEATURES: The Collins FMS 4050 will enhance the aircraft's navigational capabilities by allowing the operator to preset company routes.

UPDATED

CONAIR

CONAIR AVIATION LTD

PO Box 220, Abbotsford, British Columbia V2S 4N9
Telephone: 1 (604) 855 1171
Fax: 1 (604) 855 1017
PRESIDENT: K. B. Marsden
VICE-PRESIDENT: J. K. Dunkel
MARKETING MANAGER: Robert M. Stitt
MANAGER, CORPORATE COMMUNICATIONS: Lorna Thomassen

Operates largest private fleet of fire control aircraft in world with 90 aircraft (52 fixed-wing and 38 helicopters). Conair also operates oil spill control, insect control and forest fertilisation as well as fisheries surveillance. Company also designs and manufactures associated systems, including fire control belly tanks for various helicopters and Douglas DC-6B; Grumman and Canadian built S-2 Trackers (see Firecat and Turbo Firecat), spray systems for Lockheed C-130, Douglas DC-6B, Fokker F27 and Alenia G222. C-130 conversion includes 7,560 litre (1,997 US gallon; 1,663 Imp gallon) modular spray system; six delivered by 1991, to 356th Tactical Airlift Squadron of USAF.

UPDATED

Conair (Fokker) F27 Firefighter dropping fire retardant

1994

CONAIR (FOKKER) F27 FIREFIGHTER

TYPE: Twin-turboprop firefighter conversion.

PROGRAMME: Conair has modified three Fokker F27 Mk 600 commuter transports for firefighting roles, for which it received Canadian DoT type approval on 5 June 1986. It was the world's first turboprop conversion dedicated to forest fire suppression and resource protection. There are two F27 Firefighters currently owned and operated by the French Sécurité Civile.

DESIGN FEATURES: The modification includes installing a Conair delivery system which can carry 6,364 litres (1,681 US gallons; 1,400 Imp gallons) of long-term retardant. A 455 litre (120 US gallon; 100 Imp gallon) foam injection system is also available. The converted aircraft are readily adaptable to other functions such as transporting cargo and fire crews, infra-red fire detection and mapping, aerial survey aerial spraying and pararescue operations. Unnecessary internal items such as cabin insulation, bulkheads, pressurisation equipment and galleys are deleted. Modern avionics are installed, and the eight-compartment retardant delivery system 15 fitted ventrally as an integral part of the fuselage. The tank is blended to the fuselage with Kevlar fairings, and can be loaded at a rate of 1,514 litres (400 US gallons; 333 Imp gallons)/min. Door sequencing is computer controlled, and the entire vent system is integral with the modified fuselage floor so that the aircraft's cargo-carrying capabilites are retained. The aircraft is crewed by two pilots, and seating for 19 support crew members is retained, together with the large forward (port) freight door.

WEIGHTS AND LOADINGS:

Operating weight empty	10,646 kg (23,471 lb)
Max payload	6,731 kg (14,840 lb)
Max fuel	4,152 kg (9,153 lb)
Max T-O weight	20,411 kg (45,000 lb)
Max landing weight	18,143 kg (40,000 lb)

PERFORMANCE (at max T-O weight except where indicated):

Never-exceed speed (V_{NE})	259 kts (480 km/h; 298 mph)
Max cruising speed	230 kts (426 km/h; 265 mph)
Normal drop speed	125 kts (232 km/h; 144 mph)
Min control speed	80 kts (149 km/h; 92 mph)
Stalling speed, flaps down, power off	77 kts (143 km/h; 89 mph)
Max rate of climb at S/L	366 m (1,200 ft)/min
Rate of climb, at S/L, one engine out, AUW of 14,060 kg (31,000 lb)	177 m (580 ft)/min
Service ceiling	7,620 m (25,000 ft)
Service ceiling, one engine out, AUW of 14,060 kg (31,000 lb)	5,640 m (18,500 ft)
T-O to 10.7 m (35 ft)	1,600 m (5,250 ft)
Landing from 15 m (50 ft) at max landing weight	987 m (3,240 ft)
Min field length	1,525 m (5,000 ft)
Max endurance	3 h 24 min

VERIFIED

CONAIR HELITANKERS

TYPE: Rotary-wing helitanker conversions.

PROGRAMME: Conair has developed a growing number of helicopter-mounted fire control systems known as helitankers. The semi-monocoque belly mounted tanks feature individually operated, full length rigid doors which may be opened in various combinations over a wide range of airspeeds to permit variable retardant line lengths and drop concentrations. A self-loading hover-fill system allows the tank to be filled while the helicopter hovers above a remote water source, and an offload feature allows the water payload to be pumped to a portable ground reservoir for the use of ground-based firefighters. A foam injection system permits the fire suppressing qualities of a water payload to be greatly enhanced. A reversible pump allows single-point foam loading injection into the tank and single-point offloading.

CUSTOMERS: Helitanker system sales to date have included 31 Bell 205/212s, 16 Aerospatiale AS 350B$_1$ Ecureuils, 10 SA 315B Lamas and two AS 330/332 Pumas.

DESIGN FEATURES: System capacities are 1,360 litres (359 US

Conair (Fokker) F27 Firefighter. The World's first turboprop conversion dedicated to forest fire suppression

1994

Conair (Bell) 212 Helitanker self-loading its fire control tank ***(Conair)***

1995

gallons; 299 Imp gallons) for the Bell 205/212; 900 litres (238 US gallons; 198 Imp gallons) for the Lama; 800 litres (211 US gallons; 176 Imp gallons) for Ecureuil; and 2,355 litres (622 US gallons; 517.5 Imp gallons) for the Puma. The Puma system features an 800 litre (211 US gallon; 176 Imp gallon) two-door belly tank, and a 1,296 litre (342 US gallon; 285 Imp gallon) fuselage main tank with two internal doors for reloading the external tank via a 261 litre (69 US gallon; 57 Imp gallon) chute. Foam tank capacity is 173 litres (46 US gallons; 38 Imp gallons). The Bell 205 and 212 helitankers are available with a rappelling system to deliver firefighters to remote fire sites.

UPDATED

CONAIR FIRECAT

TYPE: Fixed-wing aircraft fire control conversion.

PROGRAMME: In production.

CUSTOMERS: Total deliveries 32 by mid-1994.

DESIGN FEATURES: Based on Grumman S-2A (S2F-1) or de Havilland Canada CS2F-1/2 Trackers. Firefighting conversion includes a four-compartment retardant tank allowing choice of drop patterns, cabin floor raised 20.3 cm (8 in) to accommodate tank, reworked/updated flight instrument panels, fitting larger landing gear wheels with low-pressure tyres, removing 1,361 kg (3,000 lb) of military equipment, inspecting and repairing wing spars and complete rewiring. Options include hydraulic or pneumatic discharge system with microcomputer control of drop sequence and 173 litre (45.6 US gallon; 38 Imp gallon) foam injection system.

FLYING CONTROLS: As Tracker.

STRUCTURE: As Tracker.

POWER PLANT: Two 1,110 kW (1,475 hp) Wright 982C9HE2 (R-1820-82) Cyclone nine-cylinder air-cooled radial engines, each driving a Hamilton Standard three-blade propeller. Total internal fuel capacity; 1,968 litres (520 US gallons; 433 Imp gallons).

ACCOMMODATION: Minimum crew of one pilot.

WEIGHTS AND LOADINGS:

Operating weight empty	6,895 kg (15,200 lb)
Max payload	3,300 kg (7,275 lb)
Max fuel	1,418 kg (3,126 lb)
Max T-O weight	11,793 kg (26,000 lb)
Max landing weight	11,113 kg (24,500 lb)

PERFORMANCE (at max T-O weight):

Never-exceed speed (VNE)	240 kts (444 km/h; 276 mph)
Max level speed at 1,220 m (4,000 ft)	220 kts (408 km/h; 253 mph)
Normal cruising speed	180 kts (333 km/h; 207 mph)
Normal drop speed	120 kts (222 km/h; 138 mph)
Stalling speed, flaps down, power off	82 kts (152 km/h; 95 mph)
Max rate of climb at S/L	366 m (1,200 ft)/min
Rate of climb at S/L, OEI	107 m (350 ft)/min
Service ceiling	6,100 m (20,000 ft)
T-O to 15 m (50 ft)	915 m (3,000 ft)
Landing from 15 m (50 ft)	762 m (2,500 ft)
Min field length	915 m (3,000 ft)
Endurance with max payload	4 h 30 min

UPDATED

Conair 2,355 litre Helitanker system on an Aerospatiale AS 330 Puma

1994

CONAIR TURBO FIRECAT

TYPE: Fixed-wing aircraft fire control and turboprop conversion.

PROGRAMME: First flight 7 August 1988; Canadian certification 22 December 1989.

CUSTOMERS: First delivery to French Sécurité Civile August 1988; total of 10 for delivery by mid-1995. Conair built its own demonstrator in 1992.

DESIGN FEATURES: Turboprops to extend life of aircraft and its fire control role; other advantages include improved performance, higher speed for greater productivity, better manoeuvring in mountainous terrain, turbine reliability, reduced operating and maintenance costs and fuel availability.

FLYING CONTROLS: As Tracker with enhanced longitudinal control feedback.

STRUCTURE As Tracker.

POWER PLANT: Two 1,062 kW (1,424 shp) Pratt & Whitney Canada PT6A-67AF turboprops, each driving a Hartzell HC-B5MA-3BXI/M11296SX five-blade propeller. Total fuel capacity of 2,936 litres (775.6 US gallons; 645.8 Imp gallons), consisting of 1,972 litres (521 US gallons; 433.8 Imp gallons) internally and 964 litres (254.6 US gallons; 212 Imp gallons) in two underwing pylon tanks. Single-point refuelling station in starboard engine nacelle.

ACCOMMODATION: Minimum crew of one pilot.

AVIONICS: Include angle of attack and stall warning system.

EQUIPMENT: Four-compartment retardant tank, maximum capacity 3,455 litres (913 US gallons; 760 Imp gallons); maximum discharge rate 3,955 litres (1,045 US gallons; 870 Imp gallons)/s.

DIMENSIONS, INTERNAL:

Retardant tank: Length 3.25 m (10 ft 8 in)
Width 1.27 m (4 ft 2 in)
Depth 0.86 m (2 ft 10 in)

WEIGHTS AND LOADINGS: As for Firecat except:

Operating weight empty 6,884 kg (15,177 lb)
Max fuel 2,339 kg (5,158 lb)
Max T-O weight 12,473 kg (27,500 lb)

PERFORMANCE (at max T-O weight except where indicated):

As for Firecat except:
Normal cruising speed 220 kts (408 km/h; 253 mph)
Rate of climb at S/L, OEI:
at max T-O weight 61 m (200 ft)/min

UPDATED

CONAIR (CESSNA) CARAVAN C208B SURVEILLANCE AIRCRAFT

TYPE: Fixed-wing aircraft surveillance conversion.

PROGRAMME: Conair has integrated a surveillance package into a standard Cessna Caravan (C208B). The first aircraft was delivered in June 1994; a second aircraft in May 1995. Integration of similar systems for other aircraft types are being pursued.

DESIGN FEATURES: The features of the modification include replacement of the existing cargo door with an automated roll-up door, a controllable deployment platform for day and night imaging systems, a fuselage-mounted universal camera system protected with an electrically driven sliding door and a package of communications and remote tracking systems. A console in the cabin contains all system controls as well as a video monitor and two VCRs, permitting an operator to accomplish all surveillance tasks from one position. The floor-mounted universal camera system is positioned adjacent to the operator, allowing adjustments to the camera and still frame photographs to be taken with ease. Conair designed, performed and certified the complete conversion to FAR 23 standards. Other modifications incorporated at the same time to enhance the aircraft's effectivity included a single-point refuelling system, navigation system upgrades, replacement of a cabin window with one that opens, additional storage provisions and an upgraded communication system for communication between the cabin, cockpit and ground resources.

ACCOMMODATION: Two crew, one equipment operator, one to three observers.

WEIGHTS AND LOADINGS:

Standard empty weight (modified passenger version) ≈2,727 kg (6,000 lb)

PERFORMANCE: As for basic aircraft. New limitations are:

Max speed: during deployment of platform 105 kts (194 km/h; 122 mph
with platform deployed 125 kts (232 km/h; 145 mph)
with aerial camera door open or in transit 125 kts (232 km/h; 145 mph)
with cabin window open 125 kts (232 km/h; 145 mph)

NEW ENTRY

Conair Turbo Firecat fire control aircraft conversion of a Grumman S-2 Tracker *(Conair)*

1995

Conair Turbo Firecat dropping fire retardant

1994

Conair (Cessna) Caravan C208B surveillance aircraft

1995

DE HAVILLAND

DE HAVILLAND INC

Garrett Boulevard, Downsview, Ontario M3K 1Y5
Telephone: 1 (416) 633 7310
Fax: 1 (416) 375 4546
Telex: 0622128
PRESIDENT: Kenneth G. Laver
VICE-PRESIDENT, OPERATIONS: Serge Perron
VICE-PRESIDENT, ENGINEERING: C. Gerard
MANAGER, PUBLIC RELATIONS: Colin S. Fisher

Established 1928 as The de Havilland Aircraft of Canada Ltd, subsidiary of The de Havilland Aircraft Company Ltd, both subsequently absorbed by Hawker Siddeley Group. Ownership transferred to Canadian government 26 June 1974. Purchased by Boeing Company 31 January 1986 and made a division of Boeing of Canada Ltd. Sale to Bombardier Inc (51 per cent) and Province of Ontario (49 per cent), finalised 9 March 1992. Supported by help from Ontario and federal governments, all government support conditionally repayable.

VERIFIED

de Havilland DHC-1 Chipmunks of the Canadian air force. The DHC-1 has since been withdrawn from the Canadian Air Force but remains in service with the Sri Lankan and Thai air forces

1994

DHC-1 CHIPMUNK

TYPE: Two-seat primary trainer.

PROGRAMME: Designed by de Havilland Canada, the Chipmunk was also built by de Havilland in Great Britain. First flown 22 May 1946.

OPERATORS: The Chipmunk is still in service with the Royal Air Force (60+), the Royal Army Air Corps (20+) and the air forces of Sri Lanka (5) and Thailand (8+).

DESIGN FEATURES: Cantilever low-wing monoplane. High-lift wing section based on NACA 2415 and USA 35B with increased camber towards tips. Aspect ratio 6.82. Taper ratio 2.1:1. Incidence 2°. Flap area 2.03 m (22 sq ft). Total aileron area 1.3 m (14 sq ft). Gross wing area 16.01 m (172.5 sq ft).

FLYING CONTROLS: Slotted ailerons with trim-tab adjustable on ground. Slotted flaps inboard of ailerons. Rudder tab adjustable on ground, elevator tabs controllable from cockpits.

STRUCTURE: All-metal single-spar wing structure with stressed-skin. D-section leading-edge and fabric covering aft of spar. Auxiliary spar carries flaps and ailerons. Metal framed fabric covered ailerons with metal trim tab. All-metal flaps inboard of ailerons. The fuselage is an all-metal semi-monocoque structure. The tail unit is a cantilever monoplane type. Aerodynamically balanced rudder and elevators with fabric covering and metal trim tab in each.

LANDING GEAR: Fixed tailwheel type. Rubber-in-compression shock-absorber struts. Tailwheel on levered-suspension unit with air/oil springing. Track 2.66 m (8 ft 9 in).

POWER PLANT: One 145 hp D. H. Gipsy Major four-cylinder in-line inverted air-cooled engine. Two-blade fixed-pitch wood or Fairey-Reed metal (optional) airscrew. Fuel capacity 114 litres (30 US gallons; 25 Imp gallons) in two wing tanks. Oil capacity 11.4 litres (3.3 US gallons; 2.75 Imp gallons).

ACCOMMODATION: Tandem cockpit with dual controls under sliding one-piece 'bubble' canopy.

DIMENSIONS, EXTERNAL:

Wing span	10.46 m (34 ft 4 in)
Length	7.75 m (25 ft 5 in)
Height	2.134 m (7 ft 0 in)

WEIGHTS AND LOADINGS:

Weight empty	526 kg (1,158 lb)
Total disposable load	350 kg (772 lb)
Weight loaded	875 kg (1,930 lb)
Wing loading	54.7 kg/m^2 (11.2 lb/sq ft)
Power loading	6.15 kg/hp (13.8 lb/hp)

PERFORMANCE (fixed-pitch airscrew):

Max speed at S/L	120 kts (223 km/h; 139 mph)
Cruising speed	107 kts (200 km/h; 124 mph)
Max diving speed	172 kts (320 km/h; 200 mph)
Rate of climb at S/L	275 m (900 ft)/min
Service ceiling	5,242 m (17,200 ft)
Absolute ceiling	5,913 m (19,400 ft)
Stalling speed (flaps down)	43 kts (80 km/h; 50 mph)
Stalling speed (flaps up)	47 kts (88 km/h; 55 mph)
T-O run (zero wind)	137 m (450 ft)
Endurance	2.3 h

PERFORMANCE (Fairey-Reed airscrew):

Max speed at S/L	121 kts (225 km/h; 140 mph)
Speed at 1,525 m (5,000 ft)	116 kts (216 km/h; 134 mph)
Cruising speed	107 kts (200 km/h; 124 mph)
Rate of climb at S/L	275 m (900 ft)/min
Service ceiling	5,243 m (17,200 ft)
Absolute ceiling	5,913 m (19,400 ft)
T-O distance to 15 m (50 ft)	265 m (870 ft)
Landing distance from 15 m (50 ft)	283 m (930 ft)
Still air range	421 n miles (780 km; 485 miles)

VERIFIED

DHC-2 BEAVER

US Navy designation: U-6

TYPE: Piston engine or turboprop STOL utility transport aircraft.

PROGRAMME: The prototype DHC-2 Beaver flew for the first time in August 1947, and type certification was received from the Canadian Department of Transport on 12 March 1948. The Beaver has been in service with the USAF and RAF but now is mainly operated by commercial operators.

VERSIONS: **DHC-2 Mk I:** Initial version of the Beaver.

DHC-2 Mk I Beaver Amphibian: Amphibious version of the Beaver.

DHC-2 Turbo-Beaver Mk III: Operators of the DHC-2 Beaver utility aircraft can convert them to Mk III Turbo-Beaver by means of a retrofit kit. DHC produced 60 Turbo-Beavers during the mid-1960s as the last variant of the DHC-2 production run, and believes that many operators of the R-985 piston-engined version may wish to upgrade them to Mk III standard. The Canadian DoT has drawn attention to the danger of cylinder head assembly failures in older radial engines if cylinder bores have been chromium plated to prolong life, and the R-985 in the original DHC-2 comes into this category.

The retrofit kit is based on the use of a PT6A-20 turboprop engine, which can be purchased as part of the kit or supplied by other sources and can operate on aviation kerosene instead of avgas. It also includes a Hartzell reversible-pitch propeller, a 76 cm (30 in) fuselage plug permitting the installation of two additional seats (making eight in addition to the pilot), increased fuel capacity, and strengthening of the wings, bracing struts and tail unit. Engine TBO can be raised from 1,000 to 1,400 hours for the R-985 to 3,500 hours for the PT6A-20 with a possible maximum in certain circumstances of 5,000 hours.

UPGRADES: **Airtech:** DHC-2/PZL-3S Beaver. See separate entry.

de Havilland: DHC-2 Turbo-Beaver Mk III. See Versions.

Wipaire: Super Beaver. See separate entry.

OPERATORS: Versions of the DHC-2 Beaver are in service with the armed forces of the following countries: Colombia (9), Republic of Korea (5), Turkey, Uganda (1), US Navy (3).

DESIGN FEATURES: Braced high-wing monoplane with single streamline-section bracing strut each side. Wing section NACA 64A series type CL, 0.4. Aspect ratio 9.2. Constant-chord of 1.59 m (5 ft 2½ in). Dihedral 2°. Incidence 0°. Sweepback at quarter-chord 0°.

FLYING CONTROLS: NACA slotted ailerons and hydraulically operated NACA slotted flaps. Trim tabs on ailerons adjustable on ground.

STRUCTURE: The wing is an aluminium alloy two-cell box structure with built-up front and rear spars and lateral stringers. NACA slotted ailerons, of aluminium alloy construction, droop 15° with NACA slotted flaps also of aluminium alloy. The front fuselage is constructed of welded steel tubes covered with dural skin. Centre portion of aluminium alloy channel members and stressed skin. Rear portion is a conventional aluminium alloy stressed-skin semi-monocoque structure.

TAIL UNIT: Cantilever aluminium alloy stressed skin structure with single-piece fixed-incidence tailplane. Controllable elevator and rudder trim tabs.

LANDING GEAR: Non-retractable tailwheel tyre. Rubber-in-compression shock-absorption on main units. DH shock-absorber strut on steerable tailwheel. Main units have Goodyear 95-2902 wheels and 8.50 × 10 6 ply tyres and tubes, pressure 1.75 kg/cm^2 (25 lb/sq in). Tail unit has Goodyear 512129M wheel, Dunlop 5.50 × 4TC tyre and Goodyear 5.00 × 4 tube pressure 2.46 kg/cm^2 (35 lb/sq in). Goodyear disc brakes. Alternatively, can be fitted with Edo 58-4580 floats, Bristol 348-4580 amphibious floats, fixed skis of DH design or combination wheel-ski gear of DH design. With wheel/ski gear, the changeover from wheels to skis or vice versa is accomplished normally from the cockpit, but an electric pump unit is available as optional equipment.

POWER PLANT: One 450 hp Pratt & Whitney R-958 Wasp Junior* nine-cylinder radial air-cooled engine, driving a two-blade Hamilton Standard 2D30-237 controllable-pitch propeller, diameter 2.59 m (8 ft 6 in). Fuel in the three metal or rubber cell tanks under cabin floor. Front and centre tanks each have a capacity of 132 litres (34.8 US gallons; 29 Imp gallons), rear tank capacity is 95 litres (25.2 US gallons; 21 Imp gallons). Total standard capacity is 359 litres (94.8 US gallons; 79 Imp gallons). Provision for two 82 litre (21.6 US gallon; 18 Imp gallon) tanks in wingtips to raise total capacity to 523 litres (138.1 US gallons; 115 Imp gallons). Refuelling point for standard tanks on port side of fuselage. Oil capacity 20.5 litres (5.4 US gallons; 4.5 Imp gallons).

* *Standard engine is Wasp Junior SB-3. Alternative models are the military R-958-AN-2, -4, -6B, -8, -10, -12 12B or 14BM1. Engines in the USAF U-6A (now withdrawn) were R-958-AN1, -3, 14B, -39 and 39A.*

British built de Havilland DHC-1 Chipmunk in service with the RAF *(Peter March)*

1994

de Havilland DHC-2 Beaver with floats

1994

ACCOMMODATION: Pilot's compartment with pilot on port side and removable seat on starboard side. Dual rubber and Y-type control column with throw-over wheel. Door with automobile-type sliding window on each side. Cabin seats up to seven passengers (six when both seats occupied in pilot's compartment). Cabin heating. Floor stressed for freight carrying. Lightweight collapsible bush seats are interchangeable with cargo attachments. Two cabin doors, one on each side, are wide enough to roll a 45 gallon petrol drum into cabin on its side. Hatches in rear wall of cabin enable long pieces of freight, such as 10 ft drilling rods, to be loaded and stowed. Baggage space at back of cabin, with separate locker aft for emergency rations and so on.

SYSTEMS: Manual low-pressure hydraulic system for brakes and flaps. Engine-driven 24 V DC 50 A or 100 A generator. Outside receptacle for ground APU.

ELECTRONICS AND EQUIPMENT: Radio to operator's requirements. Blind-flying instrumentation is not standard. Provision for navigation lights, instrument lighting, anchor riding light and cabin lights, agricultural spraying or dusting equipment and so on.

DIMENSIONS, EXTERNAL:

Wing span	14.64 m (48 ft 0 in)
Length overall: Landplane	9.24 m (30 ft 4 in)
Seaplane	9.98 m (32 ft 9 in)
Height over tail: Landplane	2.75 m (9 ft 0 in)
Seaplane	3.18 m (10 ft 5 in)
Tailplane span	4.83 m (15 ft 10 in)
Wheel track	3.10 m (10 ft 2 in)
Wheelbase	6.94 m (22 ft 9 in)
Float track (c/l of floats)	2.92 m (9 ft 6¾ in)
Cockpit door (each side): Height	1.02 m (3 ft 4 in)
Width	0.43 m (1 ft 5 in)
Height to sill	1.42 m (4 ft 8 in)
Cabin door (each side): Height	1.02 m (3 ft 4 in)
Width	0.99 m (3 ft 3 in)
Height to sill	1.24 m (4 ft 1 in)

DIMENSIONS, INTERNAL:

Cabin, incl cockpit: Length	2.74 m (9 ft 0 in)
Max width	1.22 m (4 ft 0 in)
Max height	1.30 m (4 ft 3 in)
Floor area	2.93 m² (31.5 sq ft)
Volume	3.40 m³ (120 cu ft)
Baggage hold (in rear fuselage)	0.40 m³ (14 cu ft)

AREAS:

Wings, gross	23.2 m² (250 sq ft)
Ailerons (total)	2.20 m² (24.6 sq ft)
Trailing-edge flaps (total)	1.86 m² (20.0 sq ft)
Fin	1.49 m² (16.0 sq ft)
Rudder, incl tab	0.87 m² (9.4 sq ft)
Tailplane	2.36 m² (25.4 sq ft)
Elevators, incl tab	2.14 m² (23.0 sq ft)

WEIGHTS AND LOADINGS:

Basic operating weight: Landplane	1,361 kg (3,000 lb)
Seaplane	1,506 kg (3,316 lb)
Max T-O weight: Landplane	2,313 kg (5,100 lb)
Seaplane	2,309 kg (5,090 lb)
Max landing weight: Landplane	2,313 kg (5,100 lb)
Max wing loading: Landplane	99.6 kg/m² (20.4 lb/sq ft)
Max power loading: Landplane	5.1 kg/hp (11.3 lb/hp)

PERFORMANCE (landplane at max T-O weight):

Max level speed at S/L	121 kts (225 km/h; 140 mph)
Max permissible diving speed	156 kts (290 km/h; 180 mph)
Max cruising speed at S/L	117 kts (217 km/h; 135 mph)
Econ cruising speed at S/L	108 kts (201 km/h; 125 mph)
Stalling speed	52 kts (97 km/h; 60 mph)
Rate of climb at S/L	311 m (1,020 ft)/min
Service ceiling	5,490 m (18,000 ft)
T-O run	170 m (560 ft)
T-O run to 15 m (50 ft)	310 m (1,015 ft)
Landing run from 15 m (50 ft)	305 m (1,000 ft)
Landing run	152 m (500 ft)
Range with max fuel, 45 min reserve	676 n miles (1,252 km; 778 miles)
Range with max payload, 45 min reserve	419 n miles (777 km; 483 miles)

UPDATED

DHC-3 OTTER

US Navy designation: NU-1B

TYPE: Single-engined STOL utility transport.

PROGRAMME: The Otter which flew for the first time on 12 December 1951 received its Canadian certification of airworthiness as both a landplane and seaplane in November 1952, and thereupon became the first single-engined aircraft to qualify for approval in accordance with ICAO Cat. D airworthiness requirements.

The following description applies to the standard DHC-3 Otter:

DESIGN FEATURES: Braced high-wing monoplane, with single streamline-section bracing strut each side. Wing section NACA 64A series Type CL, 0.4. Aspect ratio 8.97. Constant chord of 1.98 m (6 ft 6 in). Dihedral 2°. Incidence 2° 30′. Sweepback at quarter-chord 0°.

FLYING CONTROLS: NACA slotted ailerons and slotted flaps. Trim tabs on ailerons adjustable on ground. Controllable trim tabs in rudder and elevator.

STRUCTURE: The wing is an aluminium alloy two-cell box structure with built-up front and rear spars and lateral stringers. Ailerons of aluminium alloy construction droop 26° with flaps, also of aluminium alloy. The fuselage is a conventional all-metal semi-monocoque stressed-skin structure. The tail unit is a cantilever all-metal structure. Adjustable-incidence tailplane mounted halfway up fin, which is integral with fuselage.

LANDING GEAR: Non-retractable tailwheel type. Main units have rubber-in compression shock-absorption. Goodyear 530884M mainwheels, tyres and tubes, size 11.00 × 12, pressure 1.97 kg/cm² (28 lb/sq in). Goodyear 511500M-1 tailwheel, tyre and tube, size 6.00 × 6, pressure, 2.53 kg/cm² (36 lb/sq in). Goodyear disc brakes. Alternatively, can be fitted with Edo 55-7170 floats, Bristol 324 amphibious floats, fixed skis of DH design, or combination wheel-skis of DH design. In the wheel/ski version, the change from wheels to skis or vice versa is accomplished normally from the cockpit by a manually operated hydraulic pump, but an electric pump unit is available as optional equipment.

POWER PLANT: One 600 hp Pratt & Whitney R-1340-S1H1-G or S3H1 nine-cylinder radial air-cooled engine, driving a three-blade Hamilton Standard 3D40 Hydromatic propeller, diameter 3.30 m (10 ft 10 in). Fuel in three rubber cell tanks under cabin floor. Front tank has capacity of 233 litres (61 US gallons; 51 Imp gallons), centre tank 389 litres (102 US gallons; 85 Imp gallons) and rear tank 192 litres (50 US gallons; 42 Imp gallons). Total fuel capacity 814 litres (213 US gallons; 178 Imp gallons). Three refuelling points on port side of fuselage. Oil capacity 41 litres (10.8 US gallons; 9 Imp gallons).

ACCOMMODATION: Pilot (port) and co-pilot or passenger side by side on flight deck. Dual rudder pedals and W-type control column with throw-over wheel. Door on each side of flight deck and one in bulkhead to cabin. Cabin normally accommodates nine passengers on folding seats, with provision for a 10th seat as an optional extra. All passenger seats quickly removable. Alternatively, six stretchers and four passengers, may be carried. There is a door on each side of cabin and the floor is reinforced for cargo-carrying. Cargo drop hatch, camera hole or paratroop exit on floor.

SYSTEMS: Manual low-pressure hydraulic system for brakes

de Havilland DHC-3 Otter with floats

1994

and flaps. Engine-driven 24 V DC 50 A or 100 A generator. Outside receptacle for ground APU.

ELECTRONICS AND EQUIPMENT: Navigation lights, controllable-intensity instrument lights, a 250 W sealed landing light in the port wing leading-edge and cabin lights are provided. Radio and navigation equipment as specified by the operator.

DIMENSIONS, EXTERNAL:

Wing span	17.69 m (58 ft 0 in)
Length overall	12.80 m (41 ft 10 in)
Height over tail: Landplane	3.83 m (12 ft 7 in)
Seaplane	4.57 m (15 ft 0 in)
Tailplane span	6.46 m (21 ft 2 in)
Wheel track	3.42 m (11 ft 2 in)
Wheelbase	8.49 m (27 ft 10 in)
Float base (c/l of floats)	3.20 m (10 ft 6 in)
Cockpit door (each side): Height	1.12 m (3 ft 8 in)
Width	0.66 m (2 ft 2 in)
Height to sill	1.85 m (6 ft 1 in)
Cabin door (port): Height	1.14 m (3 ft 9 in)
Width	1.18 m (3 ft 10½ in)
Height to sill	1.24 m (4 ft 1 in)
Cabin door (starboard): Height	1.14 m (3 ft 9 in)
Width	0.76 m (2 ft 6 in)
Height to sill	1.24 m (4 ft 1 in)

DIMENSIONS, INTERNAL:

Cabin, excl flight deck: Length	5.00 m (16 ft 5 in)
Max width	1.58 m (5 ft 2 in)
Max height	1.50 m (4 ft 11 in)
Floor area	7.34 m² (79 sq ft)

AREAS:

Wings, gross	34.84 m² (375 sq ft)
Ailerons (total)	2.44 m² (26.3 sq ft)
Trailing-edge flaps (total)	9.10 m² (98.0 sq ft)
Fin	3.08 m² (33.2 sq ft)
Rudder, incl tab	2.51 m² (27.0 sq ft)
Tailplane	3.62 m² (39.0 sq ft)
Elevators, incl tab	4.27 m² (46.0 sq ft)

WEIGHTS AND LOADINGS:

Basic operating weight: Landplane	2,010 kg (4,431 lb)
Seaplane	2,219 kg (4,892 lb)
Skiplane	2,110 kg (4,652 lb)
Wheel-ski version	2,147 kg (4,734 lb)
Max T-O weight: Landplane	3,629 kg (8,000 lb)
Seaplane	3,614 kg (7,967 lb)
Max landing weight: Landplane	3,629 kg (8,000 lb)
Max wing loading: Landplane	104 kg/m² (21.3 lb/sq ft)
Max power loading: Landplane	6.03 kg/hp (13.3 lb/hp)

PERFORMANCE:

Max speed at S/L:	
Landplane	133 kts (246 km/h; 153 mph)
Max speed at 1,525 m (5,000 ft):	
Landplane	138 kts (257 km/h; 160 mph)
Seaplane	133 kts (246 km/h; 153 mph)
Skiplane	137 kts (254 km/h; 158 mph)
Max permissible speed in dive	167 kts (309 km/h; 192 mph)
Max cruising speed at S/L:	
Landplane	114 kts (212 km/h; 132 mph)
Econ cruising speed at S/L:	
Landplane	105 kts (195 km/h; 121 mph)
True cruising speed (400 hp) at 1,525 m (5,000 ft):	
Landplane	120 kts (222 km/h; 138 mph)
Seaplane	111 kts (207 km/h; 129 mph)
Skiplane	115 kts (214 km/h; 133 mph)
Stalling speed: Landplane	50 kts (93 km/h; 58 mph)
Max rate of climb at S/L:	
Landplane	259 m (850 ft)/min
Rate of climb at S/L (500 hp):	
Seaplane	198 m (650 ft)/min
Skiplane	210 m (690 ft)/min
Service ceiling (S1H1-G engine):	
Landplane	5,730 m (18,800 ft)
Seaplane	5,455 m (17,900 ft)
Skiplane	5,640 m (18,500 ft)
Service ceiling (S3H1-G engine):	
Landplane	5,300 m (17,400 ft)
Seaplane	5,000 m (16,400 ft)
Skiplane	5,210 m (17,100 ft)
T-O run: Landplane	192 m (630 ft)
T-O to 15 m (50 ft): Landplane	352 m (1,155 ft)
Seaplane	586 m (1,925 ft)
Landing from 15 m (50 ft): Landplane	268 m (880 ft)
Seaplane	366 m (1,200 ft)
Landing run: Landplane	134 m (440 ft)
Max range:	
Landplane	820 n miles (1,520 km; 945 miles)
Seaplane	742 n miles (1,375 km; 855 miles)
Range with 953 kg (2,100 lb) payload:	
Landplane	761 n miles (1,410 km; 875 miles)
Max endurance: Landplane	8.6 h
Seaplane	7.9 h

NOTE: Range and endurance include allowances for 10 minute warm up, take-off, climb to 1,525 m (5,000 ft) and fuel reserves for 45 minutes at cruise power.

VERIFIED

DHC-4 CARIBOU

TYPE: Twin-engined all-weather STOL utility transport.

PROGRAMME: The Caribou was developed with the co-operation of the Canadian Department of Defence Production and an order for one prototype was placed by the Royal Canadian Air Force. Construction began in 1957 and the prototype flew for the first time on 30 July 1958.

The original DHC-4 Caribou obtained US Type Approval on 23 December 1960, at a gross weight of 11,793 kg (26,000 lb). The DHC-4A was approved on 11 July 1961, at a maximum gross weight of 12,928 kg (28,500 lb).

Five Caribou were delivered to the US Army for evaluation in 1959, under the designation YAC-1. As a result the Caribou was adopted by the US Army, which subsequently ordered 159 aircraft under the designation CV-2 (originally AC-1). The 134 aircraft still in service on 1 January 1967 were transferred to the USAF under an inter-service agreement.

VERSIONS: **CV-2A.** Equivalent to the DHC-4.

C-7A. (formerly CV-2B): Equivalent to the DHC-4A, with higher AUW. Delivery began in 1963. Change of designation followed transfer from US Army to USAF.

One of these Caribou was converted into a flying command post by Collins Radio Company, for operation with the US 1st (Air) Cavalry Division in Vietnam. Nine separate communications positions along one side of the cabin provided commanders with ready contact with ground unit as well as long-range communications.

DHC-4 Caribou: Standard utility version.

UPGRADES: **Newcal:** DHC-4T. See separate entry.

OPERATORS: Versions of the DHC-4 Caribou are in service with the armed forces of the following countries: Australia (14), Cameroon (1), Liberia (2), Malaysia (13), Thailand (2).

DESIGN FEATURES: Cantilever high-wing monoplane. Wing section NACA 64-3A417.5 throughout one-piece centre-section, varying to NACA 63-2A615 near tips of outer panels. Dihedral on outer panels 5°. Incidence 3° inboard, 0° outboard. Sweepback at quarter-chord 0°. Variable incidence tailplane.

FLYING CONTROLS: Full-span double-slotted flaps, outer trailing portions operated independently as ailerons. Trim tabs on ailerons. Spring tabs on rudder and elevators.

STRUCTURE: The wing is an all-metal two-spar fail-safe structure. The fuselage is an all-metal fail-safe semi-monocoque structure. Rear portion is upswept aft of wings, with upward and inward hinging door forming underside of rear fuselage. The tail unit is a cantilever all-metal structure. Glassfibre fairings at top and bottom of rudder.

SYSTEMS: Wing has Goodrich flush-mounted inflatable de-icing boots in four sections. Tail unit has Goodrich inflatable de-icer boots.

LANDING GEAR: Retractable tricycle-type. Hydraulic retraction, main units forward, nose unit rearward. Main gear, produced by Jarry Hydraulics, shortens as it retracts. Dual wheels on all units. Hydraulically steerable nosewheel. Goodyear wheels and tyres, size 11.00 × 12 on main units 7.50 × 10 on nose unit. Tyre pressures (nominal) 2.81 kg/m² (40 lb/sq in). Goodyear four-cylinder single-disc brakes.

POWER PLANT: Two 1,450 hp Pratt & Whitney R-2000-7M2 14-cylinder two-row radial air-cooled engines, driving Hamilton Standard Type 43D50-7107A three-blade fully feathering Hydromatic propellers. Fuel in two wing tanks, each of 10 cells, with total capacity of 3,137 litres (828.7 US gallons; 690 Imp gallons). Refuelling point in upper surface of each wing. Oil capacity 227 litres (60 US gallons; 50 Imp gallons).

ACCOMMODATION: Flight compartment seats two side by side. Civil version accommodates 30 passengers, who enter by two doors, at rear of cabin on each side (an airstair type can be provided) or via the large rear-loading ramp, formed by electrically lowering the sloping under-surface of the rear fuselage. The military version carries 32 troops on inward-facing folding seats or 26 fully equipped paratroops, or, in an ambulance role, up to 22 litter patients, four sitting casualties and four attendants. Typical freight loads are 3 tons of cargo or two fully loaded jeeps. The floor is stressed to support distributed loads of 975 kg/m² (200 lb/sq ft) and has tie-down fittings on a 0.50 m (20 in) grid over the entire area. Tie-down rings can be fitted at 36 points on the side walls.

SYSTEMS: Flaps, brakes, landing gear retraction and nose-wheel steering actuated by 210 kg/cm² (3,000 lb/sq in) hydraulic system. No pneumatic system. Two engine-driven generators, 24 V 300 A DC/115 V 400 Hz AC.

ELECTRONICS AND EQUIPMENT: To customer's requirements. Blind-flying instrumentation standard.

de Havilland DHC-4 Caribou

DIMENSIONS, EXTERNAL:

Wing span	29.15 m (95 ft 7½ in)
Wing chord at root	3.60 m (11 ft 10 in)
Wing chord at tip	1.72 m (5 ft 7¾ in)
Wing apsect ratio	10
Length overall	22.13 m (72 ft 7 in)
Height over tail	9.70 m (31 ft 9 in)
Tailplane span	11.00 m (36 ft 0 in)
Wheel track	7.05 m (23 ft 1½ in)
Wheelbase	7.82 m (25 ft 8 in)
Passenger door (each side): Height	1.40 m (4 ft 7 in)
Width	0.76 m (2 ft 6 in)
Main cargo door (rear fuselage):	
Height	1.90 m (6 ft 3 in)
Width	1.86 m (6 ft 1½ in)
Height to sill	1.16 m (3 ft 9 ½ in)

DIMENSIONS, INTERNAL:

Cabin, excl flight deck: Length	8.76 m (28 ft 9 in)
Max width	2.21 m (7 ft 3 in)
Max height	1.90 m (6 ft 3 in)
Floor area	16.35 m² (176 sq ft)
Volume	32.57 m³ (1,150 cu ft)

AREAS:

Wings, gross	84.72 m² (912 sq ft)
Ailerons (total)	8.45 m² (91 sq ft)
Trailing-edge flaps (total)	26.47 m² (285 sq ft)
Fin	11.80 m² (127 sq ft)
Rudder, incl tab	7.80 m² (84 sq ft)
Tailplane	13.37 m² (144 sq ft)
Elevators, incl tab	7.99 m² (86 sq ft)

WEIGHTS AND LOADINGS:

Basic operating weight (incl two crew)	8,283 kg (18,260 lb)
Max payload	3,965 kg (8,740 lb)
Normal max T-O weight	12,928 kg (28,500 lb)
Max permissible weight for ferry missions	14,197 kg (31,300 lb)
Max zero-fuel weight	12,250 kg (27,000 lb)
Max landing weight	12,928 kg (28,500 lb)
Normal max wing loading	152.3 kg/m² (31.2 lb/sq ft)
Normal max power loading	4.45 kg/hp (9.83 lb/hp)

PERFORMANCE (at normal max T-O weight):

Max level speed at 1,980 m (6,500 ft)	188 kts (347 km/h; 216 mph)
Max diving speed	208 kts (386 km/h; 240 mph)
Max and econ cruising speed at 2,285 m (7,500 ft)	158 kts (293 km/h; 182 mph)
Stalling speed	59 kts (109 km/h; 68 mph)
Rate of climb at S/L	413 m (1,355 ft)/min
Service ceiling	7,560 m (24,800 ft)
Service ceiling, one engine out	2,680 m (8,800 ft)
T-O run	221 m (725 ft)
T-O to 15 m (50 ft)	361 m (1,185 ft)
Landing from 15 m (50 ft)	376 m (1,235 ft)
Landing run	204 m (670 ft)
Range with max fuel	1,135 n miles (2,103 km; 1,307 miles)
Range with max payload	210 n miles (390 km; 242 miles)

NOTES: Ranges are for long-range cruising speed at 2,285 m (7,500 ft), with allowances for warm-up, taxi, take-off, climb, descent, landing and 45 minutes reserve.

VERIFIED

DHC-5 BUFFALO

US Navy designation: UC-8A

TYPE: Twin-turboprop STOL utility transport.

PROGRAMME: First flown on 9 April 1964 and was developed from the piston-engined DHC-4 Caribou. Development costs were shared equally by the US Army, the Canadian government and de Havilland Canada.

A total of 126 Buffalos of all versions was built before production ended in December 1986. Final production version was the DHC-5D, of which deliveries began in 1976. Customers included the Abu Dhabi Defence Force, Brazilian Air Force, Cameroon Air Force, Canadian Armed Forces, Chilean Air Force, Ecuadorean Air Force, Egyptian Air Force (with LAPES), Ethiopian Air Force, Kenyan Air Force, Mauritanian Air Force, Mexican Air Force and Navy, Omani Police Wing, Peruvian Air Force, Sudan Air Force, Tanzanian Air Force, Togolese Air Force, US Army, Zaïre Air Force and Zambian Air Force.

UPGRADES: **Greenwich Aviation:** See addenda.

OPERATORS: Versions of the DHC-5 Buffalo in service with the armed forces of the following countries: Brazil (20), Cameroon (4), Canada (14), Ecuador (2), Egypt (9), Kenya (6), Mauritania (2), Mexico (1), Peru (13), Sudan (3), Tanzania (5), Togo (2), UAE (5), United States (1), Zaire (3), and Zambia (5).

The following description applies to the military DHC-5D:

DESIGN FEATURES: Cantilever high-wing monoplane. Wing section NACA 64-3A417.5 (mod) at root, NACA 63-2 (mod) at tip. Dihedral 0° inboard of nacelles, 5° outboard. Incidence 2° 30′. Sweepback at quarter-chord 1° 40′. Fixed incidence T tailplane.

FLYING CONTROLS: Full-span double-slotted aluminium alloy flaps, outboard sections functioning as rear ailerons. Aluminium alloy slot-lip spoilers, forward of inboard flaps, are actuated by Jarry Hydraulics unit. Spoilers coupled to manually operated ailerons for lateral control, uncoupled for symmetrical ground operation. Electrically actuated trim tab in starboard aileron. Geared tab in each aileron. Rudder/aileron interconnect tab in port aileron.

STRUCTURE: Conventional fail-safe two-spar box structure of high strength aluminium alloys. The fuselage is a fail-safe non-pressurised structure of high strength aluminium alloy. Longitudinal keel members support cargo floor. The tail unit is a cantilever structure of high strength aluminium alloy, with fixed incidence T tailplane.

SYSTEMS: Outer wing leading-edges fitted with electrically controlled flush pneumatic rubber de-icing boots.

LANDING GEAR: Retractable tricycle type, with twin wheels on each unit. Hydraulic actuation, nose unit retracting aft, main units forward. Jarry hydraulics oleo-pneumatic shock-absorbers. Goodrich mainwheels and tyres size 37 × 15-12, pressure 3.10 bars (45 lb/sq in) for STOL assault role, 4.14 bars (60 lb/sq in) as STOL transport. Goodrich nosewheels and tyres, size 8.9 × 12.5, presssure 3.17 bars (46 lb/sq in). Goodrich multidisc anti-skid brakes.

POWER PLANT: Two General Electric CT64-820-4 turboprop engines, each flat rated at 2,336 kW (3,133 shp) and driving a Hamilton Standard 63E60-25 three-blade constant-speed reversible fully feathering metal propeller with Beta control. Propellers have integral hydraulic systems and electric de-icing of spinners, cuffs and blades. Fuel in one integral tank in each inner wing, combined capacity 4,841 litres (1,279 US gallons; 1,065 Imp gallons), and 10 interconnected rubber bag tanks in each outer wing, capacity 3,137 litres (829 US gallons; 690 Imp gallons). Overall fuel capacity 7,978 litres (2,108 US gallons; 1,755 Imp gallons). Refuelling points above wings and in starboard side of fuselage for pressure refuelling. Total oil capacity 45.5 litres (12 US gallons; 10 Imp gallons).

ACCOMMODATION: Crew of three comprising pilot, co-pilot and crew chief. Main cabin can accommodate roll-up troop seats or folding forward-facing seats for 41 combat equipped troops, 35 paratroops or 24 stretchers and six seats. Provision for toilet in forward part of cabin. Door on each side at rear of cabin. Loading height with rear-loading door up and ramp down 2.90 m (9 ft 6 in). Tie-down points in 508 mm (20 in) grid, with additional tie-downs at sides of cabin.

SYSTEMS: Garrett bleed air cabin heating and cooling system. Two independent hydraulic systems, each of 207 bars (3,000 lb/sq in), actuate landing gear, flaps spoilers, rudders, brakes, nosewheel steering, winch and APU starting, 3,45 bar (50 lb/sq in) pneumatic system for engine starting, de-icing and environmental control. Two Lucas Aerospace engine-driven variable frequency three-phase 20 kVA AC generators with 28 V DC and 400 Hz conversion subsystems. Solar T-62T-40-5 gas-turbine APU in port engine nacelle provides electric (10 kVA generator), hydraulic and pneumatic power for environmental control, hydraulic operation of cargo winch, electrical systems and other utility functions.

de Havilland DHC-5 Buffalo

1994

AVIONICS AND EQUIPMENT: Full IFR instrumentation and weather radar standard, as are 34 troop seats, cargo buffer rail, cargo winch, roller conveyors, parachute anchor cables and retrieval system, pendulum release system. Brooks and Perkins palletised loading system with integral side rail restraint, tiedown straps and chains and crew oxygen system. Optional items include seven forward-facing troop seats, airframe de-icing, toilet, cabin oxygen system, self-sealing fuel cells, and a low altitude parachute extraction system (LAPES). With LAPES, aircraft is capable of delivering up to 2,268 kg (5,000 lb) with pinpoint accuracy, even where no airstrip exists. In a typical LAPES mission, aircraft descends to about 60 m (200 ft) where a 4.6 m (15 ft) drogue parachute is deployed. This slows down aircraft, and pilot immediately reduces height to 1.8 m (6 ft) over drop zone. A mechanical release is then activated, enabling drogue to pull out main parachute which in turn extracts cargo, carried on a sled which skids quickly to a stop on reaching ground. Software for LAPES is manufactured by Irvin Industries Canada Ltd, sled and release mechanism by Metric Systems of Fort Walton Beach, Florida.

DIMENSIONS, EXTERNAL:

Wing span	29.26 m (96 ft 0 in)
Wing chord: at root	3.59 m (11 ft 9¼ in)
at tip	1.19 m (3 ft 11 in)
Wing aspect ratio	9.75
Length overall	24.08 m (79 ft 0 in)
Height overall	8.73 m (28 ft 8 in)
Tailplane span	9.75 m (32 ft 0 in)
Wheel track	9.29 m (30 ft 6 in)
Wheelbase	8.48 m (27 ft 10 in)
Propeller diameter	4.42 m (14 ft 6 in)
Propeller/fuselage clearance	0.97 m (3 ft 2½ in)
Propeller ground clearance	0.97 m (3 ft 2½ in)
Cabin doors (each side): Height	1.68 m (5 ft 6 in)
Width	0.84 m (2 ft 9 in)
Height to sill	1.17 m (3 ft 10 in)
Emergency exits (each side, below wing leading-edge):	
Height	1.02 m (3 ft 4 in)
Width	0.66 m (2 ft 2 in)
Height to sill	approx 1.52 m (5 ft 0 in)
Rear cargo loading door and ramp:	
Height	6.33 m (20 ft 9 in)
Width	2.34 m (7 ft 8 in)
Height to ramp hinge	1.17 m (3 ft 10 in)

DIMENSIONS, INTERNAL:

Cabin, excl flight deck:	
Length, cargo floor	9.58 m (31 ft 5 in)
Width at floor	2.36 m (7 ft 9 in)
Max width	2.67 m (8 ft 9 in)
Max height (aft wings)	2.08 m (6 ft 10 in)
Height forward of rear spar	1.98 m (6 ft 6 in)
Floor area	22.48 m² (242 sq ft)
Volume (rectangular)	44.74 m³ (1,580 cu ft)

AREAS:

Wings, gross	87.8 m² (945 sq ft)
Ailerons (total)	3.62 m² (39 sq ft)
Trailing-edge flaps (total, incl ailerons)	26.01 m² (280 sq ft)
Spoilers (total)	2.34 m² (25.2 sq ft)
Fin	8.55 m² (92 sq ft)
Rudder	5.57 m² (60 sq ft)
Tailplane	14.07 m² (151.5 sq ft)
Elevator, incl tabs	7.57 m² (81.5 sq ft)

WEIGHTS AND LOADINGS (A: STOL assault mission from unprepared airfield; B: STOL transport mission, firm smooth airfield surface):

Operational weight empty incl crew and 680 kg (1,500 lb) allowance for options and avionics:	
A, B	11,412 kg (25,160 lb)
Max payload: A	5,370 kg (11,840 lb)
B	8,164 kg (18,000 lb)
Max normal fuel: A, B	6,212 kg (13,696 lb)
Max unit load for airdrop: A, B	2,721 kg (6,000 lb)
Manoeuvring limit load factor: A	3.0
B	2.5
Max T-O weight: A	18,597 kg (41,000 lb)
B	22,316 kg (49,200 lb)
Landing weight: A	17,735 kg (39,100 lb)
B	21,273 kg (46,900 lb)
Max zero-fuel weight: A	16,782 kg (37,000 lb)
B	19,731 kg (43,500 lb)
Max uniform cabin floor loading:	
A, B	976 kg/m² (200 lb/sq ft)
Max wing loading: A	211.8 kg/m² (43.4 lb/sq ft)
B	254.4 kg/m² (52.1 lb/sq ft)
Max power loading: A	3.98 kg/kW (6.54 lb/shp)
B	4.78 kg/kW (7.85 lb/shp)

PERFORMANCE (at max T-O weight except where indicated. A: STOL assault mission from unprepared airfield, B: STOL transport mission from firm smooth airfield surface):

Max cruising speed at 3,050 m (10,000 ft):	
A**	252 kts (467 km/h; 290 mph)
B*	227 kts (420 km/h; 261 mph)
Stalling speed, 40° flap:	
A at 17,690 kg (39,000 lb) AUW	67 kts (124 km/h; 77 mph)

B at 21,273 kg (46,900 lb) AUW
73 kts (135 km/h; 84 mph)
Max rate of climb at S/L, normal rated power:
A 710 m (2,330 ft)/min
B 555 m (1,820 ft/min)
Rate of climb at S/L, one engine out:
A, max power 205 m (675 ft)/min
B, military power 116 m (380 ft)/min
† Service ceiling, normal rated power:
A 9,450 m (31,000 ft)
B 8,380 m (27,000 ft)
Service ceiling, one engine out:
A, military power 5,575 m (18,300 ft)
B*, max power 3,810 m (12,500 ft)
STOL T-O run: A** 289 m (950 ft)
B 701 m (2,300 ft)
STOL T-O to 15 m (50 ft), mid-CG:
A** 381 m (1,250 ft)
B 838 m (2,750 ft)
STOL landing from 15 m (50 ft):
A** 346 m (1,135 ft)
B 613 m (2,010 ft)
STOL landing run: A** 168 m (550 ft)
B 259 m (850 ft)
Range at 3,050 m (10,000 ft):
A, max payload 225 n miles (416 km; 259 miles)
B, max payload 600 n miles (1,112 km; 691 miles)
A, B, zero payload
1,770 n miles (3,280 km; 2,038 miles)
Max range at 7,620 m (25,000 ft):
A, B with ferry tanks
3,300 n miles (6,115 km; 3,800 miles)

† *Recommended max operating altitude of 7,620 m (25,000 ft)*
* *at 21,200 kg (46,737 lb) AUW*
** *with 5,443 kg (12,000 lb) payload*

VERIFIED

de Havilland DHC-6 Twin Otter twin-turboprop STOL aircraft

1994

DHC-6 TWIN OTTER

Canadian designation: CC-138
US Air Force designation: UV-18B

TYPE: Twin-turboprop STOL transport.

PROGRAMME: Design of DHC-6 Twin Otter transport aircraft started in January 1964, and construction started in November of that year. The first of these (CF-DHC-X), powered by two 432 kW (579 bhp) PT6A-6 engines, flew in May 1965.

The fourth and subsequent aircraft of the inital series 100 version were fitted with PT6-A engines and the first delivery of a production aircraft, to Ontario Department of Lands and Forests, was made in July 1966, shortly after the Twin Otter received FAA Type Approval. All series were certificated to FAR 23 Pt 135.

In all, Twin Otters are in military service with 12 nations, de Havilland Canada developed for this market the **Series 300** specialised military version (see separate entry).

The Twin Otter is used as a photo survey aircraft in China, Sudan and Switzerland. China and Kenya also each have one Twin Otter modified for geophysical survey work. A firefighting Twin Otter carrying up to 10 smoke jumpers was purchased in late 1983 by the Forest Service of the US Department of Agriculture. In 1984 the Norwegian company, Widerøe's Flyveselkap A/S modified a Twin Otter for the dispersal of oil spills. The aircraft was fitted with two aluminium retardant tanks, each with a capacity of 1,500 litres (409.5 US gallons; 341 Imp gallons) and a 5.50 m (18 ft) long sprayboom mounted transversely under the fuselage aft of the wings. Instead of using electrical or wind-driven pumps, the retardant was expelled by pressurising the tanks with engine bleed air.

Four standard versions of the Twin Otter have been built: Series 100 (115 built); Series 200 (115 built) and Series 300 (which see). In total 844 Twin Otters were manufactured for customers in over 80 countries.

Series 300 deliveries began in Spring 1969 with the 231st Twin Otter off the line. Available with a short nose, as a floatplane. Ten of the aircraft supplied to Peru were fitted with floats, for operation by Group Aé No. 42 of the Peruvian Air Force based at Iquitos.

OPERATORS: Versions of the DHC-6 Twin Otter series 100 and 200 are in service with the armed forces of the following countries: Argentina (7), Chile (5), France, Haiti (1), Norway (3) and Paraguay (1).

DESIGN FEATURES: Strut braced high-wing monoplane. Wing section NACA 6A series mean line; NACA 0016 (mod) thickness distribution. Dihedral 3°. No sweepback. Fixed incidence tailplane.

FLYING CONTROLS: All-metal double-slotted full-span trailing-edge flaps. All-metal ailerons which also droop for use as flaps. Electrically actuated tab in port aileron; geared trim tabs in port and starboard ailerons. Manually operated trim tabs in rudder and elevators. A geared tab is fitted to the rudder to lighten control forces, and a tab fitted to the starboard elevator is linked to the flaps to control longitudinal trim during flap retraction and extension.

STRUCTURE: All-metal safe-life structure, each wing being attached to the fuselage by two bolts at the front and rear spar fitting and braced by a single streamline section strut on each side. Light alloy riveted construction is used throughout except for the upper skin panels which have spanwise corrugated stiffeners bonded to them. The fuselage is a conventional semi-monocoque safe-life structure, built in three sections. Primary structure of frames, stringers and skin of aluminium alloy. Windscreen and cabin windows of acrylic plastics. Cabin floor is of low density aluminium faced sandwich construction and is designed to accommodate distributed loads of up to 976 kg/m (200 lb/sq in). The tail unit is a cantilever all-metal structure of high strength aluminium alloys. Fin and fixed incidence tailplane are bolted to rear fuselage.

LANDING GEAR: Non-retractable tricycle type, with single wheel on each unit. Fully steerable nosewheel. Urethane compression-block shock-absorption on main units. Oleo-pneumatic nosewheel shock-absorber. Goodyear mainwheel tyre size 11.00-12 pressure 2.62 bars (38 lb/sq in).

Goodyear nosewheel tyre size 8.9-12.50, pressure 2.28 bars (33 lb/sq in). Goodrich independent, hydraulically operated disc brakes on mainwheels. Alternatively, high-flotation wheels and tyres, for operation in soft field conditions, also available at option size 15-0-12.0 for nosewheel and mainwheels. Provision for alternative wheel/ski landing gear. Twin-float gear for short-nose Srs 300, with added wing fences and small auxiliary fins.

POWER PLANT: Two 486/462 kW (6.52 ehp); (620 shp) Pratt & Whitney Canada PT6A-27 turboprop engines, each driving a Hartzell HC-B3TN-3DY three-blade reversible-pitch fully feathering metal propeller with Beta control (zero-pitch propeller on floatplane). Two underfloor fuel tanks (eight cells), total capacity of 1,446 litres (381.9 US gallons; 318 Imp gallons). Refuelling point for each tank on port side of fuselage. Oil capacity 9.1 litres (2.4 US gallons; 2 Imp gallons) per engine. Optional electric de-icing system for propellers and air intakes.

ACCOMMODATION: Side by side for one or two pilots on flight deck, access to which is by a forward-opening car type door on each side or via the passenger cabin. Dual controls standard. Windscreen demisting and defrosting standard. Cabin divided by bulkhead in top main passenger for freight compartment and baggage compartment. Seats for up to 20 passengers in main cabin. Standard interior is 20-seat commuter layout with Douglas track, carpets, double windows, individual air vents and reading lights, and airstair door. Optional layouts include 18- or 19-seat commuter versions, and 13/20-passenger utility version with foldaway seats and double cargo doors with ladder. Access to cabin by door on each side of rear fuselage; airstair door on port side. Optional double door for cargo on port side instead of airstair door. Compartments in nose and aft of main cabin, each with upward-hinged door on port side, for 136 kg (300 lb) and 227 kg (500 lb) of baggage respectively; rear baggage hold accessible from cabin in emergency. Emergency exits near front of cabin on each side. Heating of flight deck and passenger cabin by engine bleed air; ventilation via a ram air intake on the port side of the fuselage nose. Oxygen system for crew and passengers optional. Executive, survey or ambulance interiors

de Havilland DHC-6 Twin Otter with floats

1994

de Havilland DHC-6 Twin Otter of the British Antartic Survey

1994

optional. Tie-down cargo rings standard for the freighter role.

SYSTEMS: Hydraulic system, pressure 103.5 bars (1,500 lb/sq in), for flaps, brakes nosewheel steering and (where fitted) ski retraction mechanism. A hand pump in the crew compartment provides emergency pressure for standby or ground operation if the electric pump is inoperative. Accumulators smooth the system pressure pulses and provide pressure for parking and emergency braking. Optional low-pressure pneumatic system (1.24 bars; 18 lb/sq in) for operation of autopilot or wing and tail de-icing boots, if fitted. Primary electrical system is 28 V DC with one 24 A starter/generator on each engine. One 40 Ah 20-cell Ni/Cd battery (optional a 36 Ah lead-acid battery) for emergency power and engine starting. Separate 3.6 Ah battery supplies independent power for engine starting relays and ignition. 250 VA main and standby static inverters provide 400 Hz AC power for instruments and avionics. External DC receptacle aft of port side cabin door permits operation of complete system on the ground. Optional boot de-icing equipment on wing. Optional pneumatic-boot de-icing of tailplane leading-edge.

AVIONICS AND EQUIPMENT: Blind-flying instrumentation standard. Navigation and communications equipment, including weather radar, according to requirements.

DIMENSIONS, EXTERNAL:

Wing span	19.81 m (65 ft 0 in)
Wing chord (constant)	1.98 m (6 ft 6 in)
Wing aspect ratio	10
Length overall: Landplane	15.77 m (51 ft 9 in)
Seaplane	15.09 m (49 ft 6 in)
Height overall: Landplane	5.94 m (19 ft 6 in)
Seaplane from waterline	6.04 m (19 ft 10 in)
Tailplane span	6.30 m (20 ft 8 in)
Wheel track: Landplane	3.71 m (12 ft 2 in)
Wheelbase: Landplane	4.53 m (14 ft 10½ in)
Length of floats: Seaplane	9.65 m (31 ft 8 in)
Width over floats: Seaplane	5.18 m (17 ft 0 in)
Seaplane track (c/l of floats)	4.06 m (13 ft 4 in)
Propeller diameter	2.59 m (8 ft 6 in)
Passenger door (port): Height	1.27 m (4 ft 2 in)
Width	0.76 m (2 ft 6 in)
Height to sill	1.32 m (4 ft 4 in)
Passenger door (stbd): Height	1.15 m (3 ft 9½ in)
Width	0.77 m (2 ft 6¼ in)
Height to sill	1.32 m (4 ft 4 in)
Baggage compartment door (nose):	
Mean height	0.69 m (2 ft 3¼ in)
Width	0.76 m (2 ft 6 in)
Height to sill	1.32 m (4 ft 4 in)
Baggage compartment door (port, rear):	
Max height	0.97 m (3 ft 2 in)
Width	0.65 m (2 ft 1 ½ in)
Height to sill	1.32 m (4 ft 4 in)

DIMENSIONS, INTERNAL:

Cabin, excl flight deck, galley and baggage compartment:	
Length	5.64 m (18 ft 6 in)
Max width	1.61 m (5 ft 3¼ in)
Max height	1.50 m (4 ft 11 in)
Floor area	7.45 m² (80.2 sq ft)
Volume	10.87 m³ (384 cu ft)
Baggage compartment (nose):	
Volume	1.08 m³ (38 cu ft)
Baggage compartment (rear):	
Length	1.88 m (6 ft 2 in)
Volume	2.49 m³ (88 cu ft)

AREAS:

Wings, gross	39.02 m² (420 sq ft)
Ailerons (total)	3.08 m² (33.2 sq ft)
Trailing-edge flaps (total)	10.42 m² (112.2 sq ft)
Fin	4.46 m² (48.0 sq ft)
Rudder, incl tabs	3.16 m² (34.0 sq ft)
Tailplane	9.92 m² (100 sq ft)
Elevator, incl tabs	3.25 m² (35.0 sq ft)

WEIGHTS AND LOADINGS:

Typical operating weight (20-seat commuter, incl two crew and 50 kg (130 lb) of avionics)	3,363 kg (7,415 lb)
Max payload for 100 n miles (185 km; 115 miles)	1,941 kg (4,280 lb)
Max T-O weight	5,670 kg (12,500 lb)
Max landing weight:	
Wheels and skis	5,579 kg (12,300 lb)
Floats	5,690 kg (12,544 lb)

PERFORMANCE (at max T-O weight, ISA):

Max cruising speed at 3,050 m (10,000 ft)	182 kts (338 km/h; 210 mph)
Stalling speed: flaps up	74 kts (138 km/h; 86 mph)
flaps down	58 kts (108 km/h 67 mph)
Max rate of climb S/L	488 m (1,600 ft)/min
Rate of climb at S/L, one engine out	104 m (340 ft)/min
Service ceiling	8,140 m (26,700 ft)
Service ceiling, one engine out	3,530 m (11,600 ft)
T-O run: STOL	213 m (700 ft)
CAR Pt 3	262 m (860 ft)
T-O to 15 m (50 ft): STOL	366 m (1,200 ft)
CAR Pt 3	457 m (1,500 ft)
Landing from 15 m (50 ft): STOL	320 m (1,050 ft)
CAR Pt 3	591 m (1,940 ft)
Landing run: STOL	157 m (515 ft)
CAR Pt 3	290 m (950 ft)
Range at long-range cruising speed with 1,134 kg (2,500 lb) payload	700 n miles (1,297 km; 806 miles)
Range at long-range cruising speed with 862 kg (1,900 lb) payload with tanks	920 n miles (1,704 km; 1,509 miles)

VERIFIED

DHC-6-300M and MILITARY TWIN OTTER

TYPE: Twin-turboprop STOL transport, counter-insurgency and reconnaissance aircraft.

PROGRAMME: Announced on 16 July 1982, the specialised military Twin Otter was produced in three basic configurations.

VERSIONS: **Military transport (DHC-6-300M):** Basic military version with standard interior for 15 troops; convertible to 20-passenger/cargo, paratroop or ambulance layout.

Counter-insurgency (DHC-6-3M): Based on the military transport version; provision for cabin-mounted machine gun, armour protection and external ordnance on four underwing hardpoints.

Maritime reconnaissance (DHC-6-300MR): Developed from military transport, from which it is distinguishable externally by chin-mounted radome, and a searchlight pod on the starboard outer wing pylon. Prototype/demonstrator, (C-GFJQ-X), first flown in 1982, is in this configuration.

OPERATORS: Versions of the DHC-6 300 Twin Otter are in service with the armed forces of the following countries: Argentina (1), Benin (1), Canada (7), Chile (11), Ecuador (5), Ethiopia (2), France, Peru (18), Senegambia (1), Sudan (1), Uganda (1), United States (8).

DESIGN FEATURES: Power plant, systems, dimensions and areas as described for civil Twin-Otter except for provision of in-flight operable folding door, aluminium alloy wheels with heavy-duty tyres, wing structural hardpoints (two per wing), wingtip tanks and auxiliary finlets on tailplane. Self-sealing fuel cells optional.

ACCOMMODATION: Side by side seats for one or two pilots on flight deck. Dual controls, and windscreen de-icing/defrosting standard. Cabin divided by bulkhead into main passenger or freight compartment and baggage compartment. Cabin of military transports is equipped to accommodate 15 troops on lightweight, side-facing, folding tubular seats. Utility seating arrangement allows full or partial freight carrying capacity without having to remove seats. Provisions for 20 forward-facing utility seats in lieu of troop seats. In-flight operable folding door and ladder for passenger exit and entry. Complete provisions for the following optional equipment: six standard aluminium poles or 'cot type' *(folding leg)* military litters; one utility jump seat in starboard rear cabin area; floor-mounted parachute static line anchor cable. Cabin of counter-insurgency version has provision for 20 folding seats in the forward port side corner, machine gun in cabin and flight crew armour protection. Cabin of maritime version offers provisions for 20 forward-facing seats (or 15 side-facing troop seats), two observers' seats in forward section, six litters, four observers' bubble windows, camera installation, one console and operator seat, rear baggage compartment toilet, or two optional cabin fuel tanks.

AVIONICS: Standard avionics in all versions include crew interphone system, cabin PA system, dual VHF com, single HF, single VHF-FM, autopilot flight director, VLF, single VOR 1 with glide slope, single VOR 2 with marker beacon receiver, dual ADF, single DME, single transponder, radio altimeter, compass and dual RMIs.

ARMAMENT AND OPERATIONAL EQUIPMENT (Military transport and counter-insurgency versions): Optional items for military transport include paratroop jump controls and provisions for installing most Wild or Zeiss aerial survey cameras in rear cabin area, with a floor opening for a navigation drift sight. Standard mission equipment on counter-insurgency version includes four underwing hardpoints, each capable of taking an Alkan 663 NATO standard carrier of 250 kg (551 lb) capacity. Each carrier can be used for carriage of gun and/or rocket pods; co-pilot operated controls are also provided. Options for counter-insurgency version include NATO standard pylons, CRV7 rocket pods, optical projection gun/rocket control panel, 0.50 in side firing Browning machine gun, (pintle-mounted), two 7.62 mm pods and a 0.5 in machine gun pod.

ARMAMENT AND OPERATIONAL EQUIPMENT: (Maritime reconnaissance version): Chin-mounted radome housing a Litton AN/APS-504(V)2 360° search radar. Four underwing hardpoints, as on other military versions. One of these is for Spectrolab HIPAS II 50 million candlepower searchlight, with co-pilot operated controls; remaining three can be used for gun and/or rocket pods, as listed for counter-insurgency version. Tactical operator's station and console containing radar display and controls; VLF navigation controls; and a hand-held 70 mm high resolution camera

de Havilland DHC-6-300M. In service with the Alaskan National Guard

1994

with a data annotation component fed with navigation subsystem outputs. Optional items include crew rest stations, flare chutes, IR linescan, VLF/Omega.

WEIGHTS AND LOADINGS (A: military transport, B: counter-insurgency version, C: maritime version, D: maritime version with two auxiliary fuel tanks in cabin):

Operational weight empty: A 3,430 kg (7,561 lb)
B 4,143 kg (9,133 lb)
C 4,306 kg (9,494 lb)
D 4,440 kg (9,788 lb)
Max payload:
A (100 n miles + IFR reserves) 2,540 kg (5,600 lb)
B 1,436 kg (3,167 lb)
C, D 1,351 kg (2,979 lb)
Max fuel load: A (standard) 1,171 kg (2,583 lb)
A (long-range), C 1,447 kg (3,190 lb)
B (self-sealing tanks) 1,068 kg (2,354 lb)
D 2,581 kg (5,690 lb)
Max T-O weight: A, C, D 6,350 kg (14,000 lb)
B 5,670 kg (12,500 lb)
Max landing weight: A, C, D 6,123 kg (13,500 lb)
B 5,579 kg (12,300 lb)
Max wing loading: A, C, D 161.1 kg/m² (33.0 sq ft)
B 145.5 kg/m² (29.8 sq ft)
Max power loading: A, C, D 6.8 kg/kW (11.29 lb/shp)
B 6.13 kg/kW (10.08 lb/shp)

PERFORMANCE (at max T-O weight, ISA, except where indicated):

Max level speed at S/L:
A, B 170 kts (315 km/h; 196 mph)
Max cruising speed at 3,050 m (10,000 ft):
A, at 95% of MTOGW 177 kts (328 km/h; 204 mph)
B 182 kts (337 km/h; 209 mph)
C, D, at 95% of MTOGW
168 kts (311 km/h; 193 mph)
Econ cruising speed at 3,050 m (10,000 ft):
A, at 95% of MTOGW 150 kts (278 km/h; 173 mph)
B 148 kts (274 km/h; 170 mph)
C, D, at 95% of MTOGW
148 kts (274 km/h; 170 mph)
Stalling speed at max landing weight, flaps down, engines idling: all versions 58 kts (108 km/h; 67 mph) CAS
Max rate of climb at S/L: A 414 m (1,360 ft)/min
B 487 m (1,600 ft)/min
C, D 369 m (1,210 ft)/min
Rate of climb at S/L, one engine out:
A 67 m (220 ft)/min
B 104 m (340 ft)/min
C, D 43 m (140 ft)/min
Service ceiling: A 7,620 m (25,000 ft)
B 8,105 m (26,600 ft)
C, D 7,070 m (23,200 ft)
Service ceiling, one engine out: A 2,895 m (9,500 ft)
B 3,535 m (11,600 ft)
C, D 2,315 m (7,600 ft)
T-O run (conventional): A, C, D 358 m (1,176 ft)
T-O run STOL: A 219 m (720 ft)
B 213 m (700 ft)
T-O to 15 m (50 ft) (conventional):
A, C, D 597 m (1,960 ft)
T-O to 15 m (50 ft) (STOL): A, B 366 m (1,200 ft)
Landing from 15 m (50 ft) (conventional):
A, C, D 628 m (2,060 ft)
Landing from 15 m (50 ft) (STOL):
A, B 320 m (1,050 ft)
Landing run (conventional): A 376 m (1,233 ft)
C, D 345 m (1,133 ft)
Landing run (STOL):A 192 m (630 ft)
B 157 m (515 ft)
Range with max payload:
A 120 n miles (222 km; 138 miles)
B, with allowance for 7 min and VFR reserves
430 n miles (796 km; 495 miles)
D, at long-range cruising speed at 3,050 m (10,000 ft), allowance for 7 min and VFR reserves:
190 n miles (352 km; 219 miles)
Range with max fuel, allowance for 7 min and VFR reserves: A 850 n miles (1,575 km; 978 miles)
B 720 n miles (1,334 km; 829 miles)
D, at long-range cruising speed at 3,050 m (10,000 ft)
1,460 n miles (2,705 km; 1,681 miles)

VERIFIED

DHC-7 DASH 7

TYPE: Four-engined short/medium-range quiet STOL transport.

PROGRAMME: Dash 7 quiet STOL airliner project was begun by de Havilland in Canada late 1972, following a worldwide market survey of short-haul transport requirements.

Two preproduction aircraft were built, the first of these (C-GNBX-X) flying on 27 March 1975 and the second (C-GNCA-X) on 26 June 1975. A third airframe was built for structural testing and a fourth for fatigue testing. The first production Dash 7 (C-GOIW, c/n 3), flew on 30 March 1977.

Certification by Canadian Department of Transport to FAR 25 was received on 2 May 1977; STOL performance is approved under conventional FAR 25 and FAR 121 regulations. In addition, certification has been given for 7° 30′ glide slope and 10.7 m (35 ft) landing reference height adopted by the FAA for STOL aircraft. The first Dash 7 to enter service was c/n 4, with Rocky Mountain Airways (USA), on 3 February 1978. Deliveries of the Dash 7 by October 1986 totalled 105 to 34 customers in 22 countries.

VERSIONS: **Series 100:** Passenger aircraft.

Series 101: All-cargo aircraft.

Series 150: Version with higher weight and greater fuel capacity.

Series 151: Version with higher weight and greater fuel capacity.

Dash 7 IR: Ice reconnaissance version (which see)

OPERATORS: Versions of the DHC-7 DASH 7 are in service with the armed forces of the following countries: Venezuela (1), United States (3).

The following description applies to the Series 100 and standard Series except where indicated:

DESIGN FEATURES: Cantilever high-wing monoplane, with 4° 30′ dihedral from centre section. Wing section NACA 63A418 (modified) at root, NACA 63A415 (modified) at tip. Sweepback 3° 12′ at quarter-chord. Incidence 3° at root. Fixed incidence tailplane.

FLYING CONTROLS: Double-slotted flaps, extending over approximately 80 per cent of trailing-edge, are actuated mechanically for take-off, by irreversible screwjacks, and hydraulically for landing. Two inboard ground spoilers/lift dumpers and two outboard air spoilers in each upper surface, forward of flaps, also actuated hydraulically. Outboard sections can be operated symmetrically, or differentially in combination with the cable operated ailerons. Servo-tab in each aileron. Tailplane has one-piece operated horn-balance elevator with spring tabs. Fore and trailing serially hinged rudders, actuated hydraulically.

STRUCTURE: Conventional all-metal two-spar bonded skin/stringer structure. The fuselage is an all-metal stressed skin pressurised structure of bonded skin/stringer construction. Basically circular, with flattened profile under floor level. The tail unit is a cantilever all-metal T tail, with large dorsal fin.

SYSTEMS: Pneumatic boot de-icing of leading-edges outboard of the inner nacelles. Pneumatic boot de-icing of tailplane leading-edges and elevator horns.

LANDING GEAR: Menasco retractable tricycle type, with twin wheels on all units. Oleo-pneumatic shock-absorbers. Hydraulic retraction, main units forward into inboard engine nacelles, steerable nose unit rearward into fuselage. Goodrich mainwheel and tyres size 30 × 9.00-15, pressure 7.38 bars (107 lb/sq in); nosewheel tyres size 6.50-10, pressure 5.31 bars (77 lb/sq in). Larger, low pressure tyres optional, with pressures of 4.83 bars (70 lb/sq in) on main units, 4.76 bars (69 lb/sq in) on nose unit. Goodrich anti-skid hydraulic braking system for all units. Small retractable tailskid under rear fuselage.

POWER PLANT: Four Pratt & Whitney Aircraft Canada PT6A-50 turboprop engines, each flat rated 835 kW (1,220 shp) and driving a Hamilton Standard 24PF-305 constant-speed fully feathering reversible-pitch four-blade propeller, with Beta control, of slow-turning type (1,320 rpm) to reduce noise level. Propeller blades are of GRP, with forged aluminium spars and foam cores. Fuel in two integral tanks in each wing, total capacity 5,602 litres (1,478 US gallons; 1,232 Imp gallons). Single pressure refuelling/defuelling point on underside of rear fuselage, aft of pressure dome. Pneumatic de-icing of engine air intakes; electric de-icing for propellers. Oil capacity 23 litres (6 US gallons; 5 Imp gallons).

ACCOMMODATION: Flight crew of two, plus one or two cabin attendants. Dual controls standard. Seats for 50 passengers at 81 cm (32 in) pitch, in pairs on each side of centre aisle, with generous provision for underseat carry-on baggage. Outward opening airstair door at rear on port side. Emergency exits on each side at front of cabin and on starboard side at rear. Baggage compartment in rear fuselage (capacity 998 kg; 2,200 lb), with external access on starboard side and internal access from cabin. Galley coat rack and toilet at rear of cabin. Optional arrangements include movable bulkhead for mixed freight/passenger loads with large forward freight door on port side. Up to five standard pallets can be accommodated in an all-cargo role. Quick-change cargo handling system available optionally. Entire accommodation pressurised and air conditioned.

SYSTEMS: Cabin pressure differential 0.294 bars (4.26 lb/sq in). Two air cycle systems, driven by engine bleed air, for cabin air conditioning. Two independent hydraulic systems, each of 207 bars (3,000 lb/sq in). No. 1 system actuates flaps, rudder, wing spoilers and mainwheel brakes. No. 2 system actuates landing gear, nosewheel and back-up mainwheel brakes, parking brakes, nosewheel steering, rudder and outboard wing spoilers. Primary power provided by four Phoenix 28 V 250 A 7.5 kW starter/generators. Power of 115/200 V three-phase AC at 400 Hz from four 10 kVA Lucas brushless generators for propeller and windsceen de-icing and standby fuel pumps. Lucas static inverters supply constant frequency 400 Hz loads, including engine instrumentation and navigation systems. Ni/Cd batteries for engine starting. APU for cabin air conditioning and electrics, and engine starting available optionally.

AVIONICS AND EQUIPMENT: Standard avionics include crew interphone system; cabin PA system; flight data recorder;

de Havilland DHC-7 Dash 7 four-engine short/medium-range quiet STOL aircraft

1994

flight compartment voice recorder; emergency locator transmitter; two independent VHF communications systems; two independent VHF (VOR/ILS) radio navigation systems; one LF (ADF) radio navigation system; one transponder; one DME; one RCA Primus 40 weather radar; one marker beacon receiver; Sperry SPZ-700 autopilot/flight director system, incorporating Z-500 flight computer and ADC-200 central air data computer; Sperry STARS ADI and HSI; Sperry AA-215 radio altimeter and two Sperry C-14 slaved gyro compasses and VG-14 vertical gyros. Provision for variety of optional avionics to customer's requirements. Standard options include Collins 618M-3 com transceiver, Collins 51RV-4D nav receiver, Collins 51Z-4 glide slope/market beacon receiver, and Collins 621A-6 transponder.

DIMENSIONS, EXTERNAL:
Wing span 28.35 m (93 ft 0 in)
Wing chord: at root 3.81 m (12 ft 6 in)
at tip 1.68 m (5 ft 6 in)
means aerodynamic 2.99 m (9 ft 9¾ in)
Wing aspect ratio 10
Length overall 24.58 m (80 ft 7¾ in)
* Height overall 7.98 m (26 ft 2 in)
Tailplane span 9.45 m (31 ft 0 in)
Fuselage: Max diameter 2.79 m (9 ft 2 in)
Wheel track 7.16 m (23 ft 6 in)
Wheelbase 8.38 m (27 ft 6 in)
Propeller diameter 3.43 m (11 ft 3 in)
Propeller ground clearance (inboard engines) 1.60 m (5 ft 3 in)
Min propeller/fuselage 0.75 m (2 ft 5½ in)
Passenger door (rear, port): Height 1.75 m (5 ft 9 in)
Width 0.76 m (2 ft 6 in)
* Height to sill 1.09 m (3 ft 7 in)
Emergency exit doors (fwd, each):
Height 0.91 m (3 ft 0 in)
Width 0.51 m (1 ft 8 in)
* Height to sill 1.55 m (5 ft 1 in)
Emergency exit door (rear, stbd):
Height 1.35 m (4 ft 5 in)
Width 0.61 m (2 ft 0 in)
* Height to sill 1.09 m (3 ft 7 in)
Baggage hold door (rear, stbd):
Height 1.02 m (3 ft 4 in)
Width 0.84 m (2 ft 9 in)
* Height to sill 1.47 m (4 ft 10 in)
Cargo door (fwd, port, optionally):
Height 1.78 m (5 ft 10 in)
Width 2.31 m (7 ft 7 in)
* Height to still approx 1.22 m (4 ft 0 in)
* *will vary with configuration and loading configurations*

DIMENSIONS, INTERNAL:
Cabin excl flight deck: Length 12.04 m (39 ft 6 in)
Max width 2.59 m (8 ft 6 in)
Floor width 2.13 m (7 ft 0 in)
Max height 1.94 m (6 ft 4½ in)
Height under wing 1.85 m (6 ft 1 in)
Volume 54.1 m^3 (1,910 cu ft)
Baggage compartment (rear fuselage):
Max length 2.30 m (7 ft 6½ in)
Volume 6.8 m^3 (240 cu ft)

AREAS:
Wings, gross 79.90 m^2 (860.0 sq ft)
Ailerons (total) 2.26 m^2 (23.22 sq ft)
Trailing-edge (total) 27.33 m^2 (294.20 sq ft)
Spoilers (total) 3.63 m^2 (39.04 sq ft)
Vertical tail surfaces (total, excl dorsal fin) 15.79 m^2 (170.0 sq ft)
Horizontal tail surfaces (total) 20.16 m (217.0 sq ft)

WEIGHTS AND LOADINGS:
Basic weight empty (standard 50-passenger layout) 12,247 kg (27,000 lb)
Operating weight empty 12,560 kg (27,690 lb)
Max payload (50 passengers or cargo) 5,130 kg (11,310 lb)
Max usable fuel (standard tanks) 4,502 kg (9,925 lb)
Max fuel for extended range 6,717 kg (14,810 lb)
Max T-O weight 19,958 kg (44,000 lb)
Max zero-fuel weight 17,690 kg (39,000 lb)
Max landing weight 19,050 kg (42,000 lb)
Max cabin floor loading 366.3 kg/m^2 (975 lb/sq ft)
Max wing loading 249.8 km/m^2 (51.17 lb/sq ft)
Max power loading 5.98 kg/kW (9.82 lb/shp)

PERFORMANCE (at max T-O weight, FAR Pt 25, at S/L, ISA except where indicated):
Max cruising speed at 2,440 m (8,000 ft) at AUW of 18,597 kg (41,000 lb) 231 kts (428 km/h; 266 mph)
Max cruising speed at 4,575 m (15,000 ft) at AUW of 18,597 kg (41,000 lb) 227 kts (420 km/h; 261 mph)
En route rate of climb, flaps and landing gear up:
4 engines, max climb power 372 m (1,220 ft)/min
3 engines, max continuous power 220 m (720 ft)/min
Service ceiling at AUW of 18,597 kg (41,000 lb):
4 engines, max climb power 6,400 m (21,000 ft)
3 engines, max continuous power 3,855 m (12,650 ft)
FAR Pt 25 T-O field length, 25° flap, AUW of 18,597 kg (41,000 lb) 689 m (2,260 ft)
T-O field length at 3,050 m (10,000 ft), 15° flap 1,829 m (6,000 ft)
FAR Pt 25 STOL landing field length at max landing weight 45° flap 594 m (1,950 ft)
Landing field length at 3,050 m (10,000 ft) at 18,915 kg (41,700 lb) landing weight, 45° flap 823 m (2,700 ft)
Min ground turning radius 8.84 m (29 ft 0 in)
Runway LCN with 32 × 11.50-15 low-pressure tyres, rigid, 30 in relative stiffness 16.2
Range at 4,575 m (15,000 ft) with 50 passengers and baggage at long-range cruising speed, IFR reserves 690 n miles (1,279 km; 795 miles)
Max range at 4,575 m (15,000 ft) with standard fuel and 2,948 kg (6,500 lb) payload, long-range cruising speed 1,170 n miles (2,168 km; 1,347 miles)

OPERATIONAL NOISE LEVELS (FAR Pt 36 at S/L, ISA + 10°C, confirmed):
T-O 80.5 EPNdB
Approach on 3° glide slope 91.4 EPNdB
Sideline 82.8 EPNdB

VERIFIED

de Havilland DHC-7 Dash 7 IR ice reconnaissance aircraft

1994

DHC-7 DASH 7IR

TYPE: Four-engined short/medium-range quiet STOL ice reconnaissance aircraft.

PROGRAMME: Somewhat later than originally planned the Dash 7 IR (for ice reconnaissance) entered service with the Canadian Department of Environment in Spring 1986. This one-off aircraft registered C-GCFR, is a specially equipped non-standard example of the Dash 7 Series 150, intended for use in surveying sea ice and icebergs in the shipping and oil drilling regions of the Labrador coast and the Gulf of St Lawrence, where it supplements two Lockheed Electras already used for this purpose by the DoE's Atmospheric Environment Service.

DESIGN FEATURES: Non-standard features of the Dash 7 IR include a special dorsal observation cabin just aft of the flight deck, and a Canadian Astronautics Ltd SLAR 100 side-looking radar mounted in a fairing on the port side of the fuselage, to locate ice in shipping lanes and drilling areas. Other mission equipment includes a laser profilometer to measure ice formation contours, photographic mapping equipment, and a datalink between the aircraft and ships and drilling rigs in the patrol area.

VERIFIED

DHC-8 DASH 8M

Canadian Forces designations: CC-142 and CT-142
US Air Force designation: E-9A

TYPE: Multimission medium utility aircraft.

VERSIONS: **CC-142 and CT-142:** Passenger/cargo transport and navigation trainer versions of Dash 8M-100 respectively; both have long-range fuel tanks, low-pressure tyres, high-strength floors and mission avionics; Canadian Dept of National Defence operates two CC-142s and four CT-142s, latter with mapping radar in extended nose. For full description of de Havilland Dash 8 see *Jane's All the World's Aircraft 1995-96*.

E-9A: US Air Force Dash 8-100 as missile range control aircraft; relays telemetry, voice and drone and fighter control data and simultaneously observes range with radar; avionics include large electronically steered phased-array radar in fuselage side and AN/APS-128D surveillance radar in ventral dome.

OPERATORS: The DHC-8 Dash 8M is in service with the armed forces of the following countries: Canada (8) and United States (2).

NEW ENTRY

de Havilland Dash 8 E-9A in service with the USAF *(de Havilland)*

1995

IMP

IMP AEROSPACE LTD

Suite 400, 2651 Dutch Village Road, Halifax, Nova Scotia B3L 4T1
Telephone: 1 (902) 873 2250
Fax: 1 (902) 873 2003
Telex: 019 22504
MARKETING MANAGER: M. J. Garvey

Created 1970 from former Fairey Canada company when latter acquired by IMP Group. Major programmes concern weapon system engineering, outfitting, upgrading and maintenance support of maritime aircraft used by Canadian Forces.

VERIFIED

IMP (GRUMMAN) S-2E TRACKER UPGRADE

TYPE: Twin-turboprop anti-submarine aircraft.

PROGRAMME: Order to re-engine and upgrade 12 S-2E Tracker ASW aircraft for Brazilian Air Force (FAB designation **P-16E)** placed 1989. Conversion includes 1,230 kW (1,650 shp) P&WC PT6A-67CF turboprops, Hartzell five-blade propellers and modification of fuel, hydraulic, pneumatic and electrical systems. First flight of 'prototype' was 14 June 1990; redelivered to Brazil 17 December 1990 for trials on aircraft carrier *Minas Gerais*. Having received the first of the 12 upgraded S-2Es, the Brazilian government cancelled the contract for the remaining 11 aircraft. The first and only aircraft converted by IMP performed well in trials. IMP retain S-2 Tracker upgrade capabilites.

VERIFIED

Brazilian Air Force P-16E Tracker converted by IMP *(IMP Aerospace)*
1994

IMP (LOCKHEED) CP-140A ARCTURUS

TYPE: Arctic and maritime surveillance aircraft variant of Lockheed P-3 Orion (see US section).

PROGRAMME: Outfitting and completion of the three aircraft for non-ASW surface surveillance and search and rescue duties with Canadian Forces. Completed and delivered 1992-93. Details of avionics in Lockheed P-3 entry.

CUSTOMERS: Canadian Forces (CFB Greenwood, Nova Scotia): three aircraft.

COSTS: C$254 million (1992).

DESIGN FEATURES: Last three P-3C airframes, from Lockheed California production line; bought 'green' and outfitted by IMP with cockpit avionics. Texas Instruments AN/APS-134 Plus surveillance radar and interior equiment and furnishing.

EQUIPMENT: SAR equipment includes liferafts and survival kit, airdroppable (SKAD).

VERIFIED

IMP (Lockheed) CP-140A Arcturus arctic and maritime surveillance aircraft *(IMP Aerospace)*
1995

IMP (SIKORSKY) SEA KING CONVERSION (1)

TYPE: Anti-submarine helicopter conversion (Sea King).

PROGRAMME: Six Canadian Forces CH-124A Sea Kings converted for surface surveillance over Gulf from HMCS *Athabaskan* and *Protecteur,* one converted by IMP and remainder at Canadian Forces Base Shearwater, Nova Scotia with IMP kits. Programme commenced August 1990.

AVIONICS: Tracor SLIPAR (short light pulse alerting receiver) laser warning system, E-Systems AN/APR-39 radar warning receiver, Loral AN/AAR-47 passive missile approach warning system. Sanders AN/ALQ-144 infra-red jammer, Tracor AN/ALE-37 chaff dispenser and M-130 flare dispenser.

ARMAMENT: One C-9 machine gun mounted in cabin doorway.

UPDATED

IMP (SIKORSKY) SEA KING CONVERSION (2)

TYPE: Anti-submarine helicopter conversion.

PROGRAMME: Mission system avionics modification of six Canadian Forces CH-124A Sea Kings to CH-124B, for compatibility with frigates and destroyers equipped with Canadian towed-array sonar system (CANTASS), under programme known as HELTAS (helicopter towed-array support); redelivered between August 1992 and May 1993. Seventh Sea King equipped with MAD and Tac/Nav of CH-124B plus a 99-channel sonobuoy receiver, HAPS processor, and three competing types of acoustic recorder. Conversion completed November 1993.

DESIGN FEATURES: Intended to evaluate equipment of, and provide training platforms for, Canadian versions of EH 101. Canadian government has since cancelled the EH 101 order.

ACCOMMODATION: Crew of four (two pilots and two navigators, latter acting as TACCO and SENSO).

AVIONICS: Existing dipping sonar, analog Tac/Nav, sonobuoy receiver and relay systems deleted: AN/ASQ-504 internal MAD retained. New equipment includes AN/ASN-123 Tac/Nav. AN/UYS-503 sonobuoy processor and 31-channel AN/ARR-75 sonobuoy receiver, plus 20 sonobuoys.

UPDATED

KFC

KELOWNA FLIGHTCRAFT GROUP

No. 1, 5655 Kelowna Airport, Kelowna, British Columbia VIV 1S1
Telephone: 1 (604) 765 1481
Fax: 1 (604) 765 1489
PRESIDENT: Barry Lapointe
VICE-PRESIDENT: Jim Rogers
OPERATORS MANAGER: Greg Carter
CV5800 PROJECT MANAGER: Bill De Meester

KFC Group comprises four companies, each specialising in one area of aviation but complementing the others:

Kelowna Flightcraft Air Charter Ltd responds to all kinds of charter, from fire patrols to time-sensitive cargo or personnel flights. Fleet comprises two Boeing 727s, 12 Convair 580s, one Gulfstream I, two Cessna 402Bs and a DC-3.

Kelowna Flightcraft Ltd totally maintains KFC Air Charter fleet and undertakes work for outside customers; overhaul, manufacturing and engineering capabilites have Canadian MoT approval and FAA (FAR Pt 43.17) recognition.

Kelowna Flight-Comm Avionics Inc intensively involved in CV5800 rework programme, and others involving civilian, military and government aircraft.

Kelowna Flightcraft R & D Ltd involved in research and development of major aircraft modifications including STCs and STAs such as CV5800, long-range fuel system for Convair 580, cargo door modifications and AirResearch GTC 85-90 APU modification.

VERIFIED

KFC CV5800

TYPE: Stretched conversion of Convair 580 (applicable to any Convair 340/440/580 airframe).

PROGRAMME: Launched 1984; structural work started January 1990; first flight (N5800) 11 February 1992. The second production aircraft was expected to fly in September 1994. The first production aircraft began commercial service 2 January 1994. The CV5800 received FAA certification 11 December 1993 and DoT certification 17 December 1993.

CURRENT VERSIONS: **Cargo:** As first prototype. *Detailed description applies to this version.*

Passenger: Under development; to seat 76 people at 81 cm (32 in) pitch.

COSTS: Development programme $16.5 million (1989); standard aircraft $6 million.

DESIGN FEATURES: Fuselage stretched by 4.34 m (14 ft 3 in) and strengthened for 2,268 kg (5,000 lb) increase in gross weight; wing also strengthened internally; standard Convair 340/580 tail unit; Honeywell four-tube EFIS and other modern avionics; Allison 501 turboprops and Hamilton Standard propellers; configured to cargo class E standard, with cargo conveyance system available optionally. Will meet Stage 3 noise and Cat. II landing standards; modification claimed to extend aircraft life by 100,000 hours.

FLYING CONTROLS: Conventional mechanical (cables), all with trim and servo tabs; vortex generators added to tailplane; control column bob-weight and increased spring tension system; Honeywell autopilot.

STRUCTURE: Conventional aluminium alloy stressed skin.

LANDING GEAR: Twin-wheel nose and main units retract forward hydraulically. Menasco oleo-pneumatic shock-absorbers and Goodyear wheels and tyres on each unit; mainwheel tyres size 39 × 13-16, pressure 5.86 bars (85 lb/sq in); nosewheel tyres size 29 × 7.5-14, pressure 4.34 bars (63 lb/sq in). Goodyear disc brakes and Hytrol anti-skid units; nose unit steerable ±62°. Minimum ground turning radius about 19.81 m (65 ft 0 in).

POWER PLANT: Two Allison 501-D22G turboprops (each 3.430 kW; 4,600 shp at 13,820 rpm); Hamilton Standard 54H60-77/117 four-blade constant-speed propellers (reversible-pitch optional). Standard fuel capacity 6,549 litres (1,730 US gallons; 1,440 Imp gallons) in 3,274 litre (865 US gallon; 720 Imp gallon) tank in each wing; increase to 7,874 litres (2,080 US gallons; 1,732 Imp gallons) total available optionally. Single pressure fuelling point in each lower wing skin. Oil capacity 32 litres (8.4 US gallons; 7 Imp gallons).

ACCOMMODATION: Flight crew of two. Crew/passenger door at front and cargo door at rear of cabin, both on port side.

Overwing emergency exit each side. Entire accommodation pressurised and air conditioned.

SYSTEMS: Convair/AiResearch pressurisation/air conditioning system (maximum differential 0.29 bar; 4.16 lb/ sq in). Dual primary hydraulic systems (pressure 207 bars; 3,000 lb/sq in), plus emergency standby. Electrical power at 400 Hz supplied by two 24 V DC generators, two 240 V AC alternators and two 12 V batteries. Gaseous oxygen system. Hot-air de-icing and anti-icing of wings, tail unit, propellers and engine airtakes. AiResearch GTC 85-90 APU for electric (DC) engine starting.

AVIONICS: Honeywell SPZ-4500 digital autopilot, dual FZ-450 flight management computers. Primus II nav/com/ ident radio system, EDZ-803 four-tube EFIS. Primus 650 weather radar, dual VG-14A/C-14A AHRS, and A-A300 radio altimeter.

DIMENSIONS, EXTERNAL:

Wing span	32.10 m (105 ft 4 in)
Length overall	29.18 m (95 ft 9 in)
Height overall	8.58 m (28 ft 1⅔ in)
Tailplane span	11.15 m (36 ft 7 in)
Wheel track	7.62 m (25 ft 0 in)
Wheelbase	7.97 m (26 ft 1¾ in)
Propeller diameter	4.11 m (13 ft 6 in)
Propeller ground clearance	0.33 m (13 in)
Cabin door (fwd, port): Height	2.11 m (6 ft 11 in)
Width	0.91 m (3 ft 0 in)
Cargo door (rear, port): Height	1.83 m (6 ft 0 in)
Width	3.05 m (10 ft 0 in)
Emergency exits (each): Height	0.66 m (2 ft 2 in)
Width	0.48 m (1 ft 7 in)

DIMENSIONS, INTERNAL:

Cabin excl flight deck: Length	20.80 m (68 ft 3 in)
Max width	2.97 m (9 ft 9 in)
Max height	1.96 m (6 ft 5 in)
Floor area	48.12 m² (518 sq ft)
Volume (cargo)	87.67 m³ (3,096 cu ft)

AREAS:

Wings, gross	85.47 m² (920.0 sq ft)

WEIGHTS AND LOADINGS:

Operating weight empty (cargo verison)	15,043 kg (33,166 lb)
Max payload (cargo version)	9,903 kg (21,834 lb)
Max standard fuel	5,257 kg (11,591 lb)
Max T-O weight	28,576 kg (63,000 lb)
Max landing weight	26,308 kg (58,000 lb)
Max zero-fuel weight	24,947 kg (55,000 lb)
Max wing loading	334.3 kg/m² (68.48 lb/sq ft)
Max power loading	4.17 kg/kW (6.85 lb/shp)

PERFORMANCE: Still being determined

VERIFIED

KFC CV5800 stretched conversion of the Convair 580/C-131F

1994

CHILE

ENAER

EMPRESA NACIONAL DE AERONÁUTICA DE CHILE

Avenida José Miguel Carrera 11087, P.36½, Santiago
Telephone: 56 (2) 528 2735 and 528 2823
Fax: 56 (2) 528 2699
Telex: 645115 ENAER CT
PRESIDENT: Caupolicán Boisset
DIRECTOR OF MARKETING: Alejandro Vargas

ENAER is a state owned company formed in 1984 from the IndAer industrial organisation, set up by the Chilean Air Force in 1980. Aircraft manufacturing started in 1980 with the assembly of 27 Piper PA-28 Dakota light aircraft for Chilean Air Force and flying club use. With a 1989 workforce of about 1,900 people, ENAER's current activities are the design and production of aircraft and electronic warfare equipment.

ENAER's current major programmes are the T-35 Pillán trainer and T-36/A-36 Halcón (CASA C-101) jet trainer/ attack aircraft, plus development of Chile's first lightplane of indigenous design, the Ñamcu (Eaglet). It also undertakes upgrade programmes for the Chilean Air Force including conversion of Beechcraft 99s for maritime surveillance and retrofitting FACh Hawker Hunters with a Caiquen II radar warning receiver system. Latest such programme is the Pantera, an airframe/avionics upgrade for the service's Mirage 50s.

UPDATED

ENAER Pantera (Dassault) Mirage 50C with foreplanes and upgraded nav/attack system *(Kenneth Munson)*

1994

ENAER PANTERA 50C (PANTHER)

TYPE: Combat aircraft upgrade (Dassault Mirage 50).

PROGRAMME: Pantera 50C is a programme to upgrade the airframe and avionics of eight Mirage 50FCs, six 50CHs and two 50DCH's with technical assistance from Israel Aircraft Industries (IAI), launched in mid-1980s. Flight testing (with foreplanes only) began 1986; first flight of fully upgraded aircraft at end of 1988; second aircraft completed by March 1992; three more then under way, but programme slowed by lack of funding. Aircraft serve with 4° Esquadrão of Brigada Aérea IV at Carlos Ibánez AB, southern Chile.

DESIGN FEATURES: Forward fuselage untapered plug of approximately 1.00 m (3 ft 3¼ in) inserted immediately ahead of windscreen; non-movable canard surfaces shoulder-mounted on engine intake trunks; in-flight refuelling probe added.

AVIONICS: New suite includes INS, Elbit radar, computerised HUD, modified electrical/hydraulic/weapon control systems ENAER Caiquen III RWR and Eclipse chaff/flare dispensing system.

VERIFIED

ENAER/IAI (Northrop) F-5E upgraded aircraft of the Chilean Air Force *(Charles Bickers)*

1994

ENAER/IAI (NORTHROP) F-5 UPGRADE

TYPE: Combat aircraft upgrade (Northrop F-5E/F).

PROGRAMME: In March 1994 the first built upgraded ENAER F-5E was exhibited. This aircraft originated from a contract signed in 1990 for Israeli Aircraft Industries (IAI) to upgrade 12 F-5Es and two F-5Fs of the Chilean Air Force and for ENAER to install the modification kits.

DESIGN FEATURES: Modification kits consists of an Elta EL/ M-2032B multimode fire control radar, an improved WDNS, El-Op HUD, HOTAS, Astronautics (Israel) modular mission and display processor and a multiple databus. Nose modifications to accommodate larger radar antenna include deletion of the port M39 20 mm gun and relocation of some subsystems.

VERIFIED

ENAER T-35 PILLÁN (DEVIL)

Spanish Air Force designation: E.26 Tamiz

TYPE: Two-seat fully aerobatic basic (T-35A/C) and instrument (T-35B/D) military trainer.

PROGRAMME: First two prototypes (first flight 6 March 1981) developed by Piper; followed by three Piper kits for ENAER assembly (first flight, by FACh s/n 101, 30 January 1982); then slight redesign, replacing all-moving tailplane by electrically trimmable tailplane with conventional elevator, increasing rudder mass balance and deepening canopy; ENAER series production started September 1984; first flight of production T-35A, 28 December 1984; deliveries to FACh began 31 July 1985. Supply of ENAER T-35C kits to CASA started 27 December 1985 (first flight 12 May 1986), completed September 1987; T-35D deliveries to Panama began 1988; first flight of single-seat T-35S made 5 March 1988; T-35A/B deliveries to FACh completed by Spring 1990; export deliveries completed 1991.

VERSIONS: **T-35A:** Primary trainer version for Chilean Air Force. In service with Escuela de Aviación Capitan Avalos at El Bosque AB, Santiago. *Detailed description applies to this version except where indicated.*

T-35B: Chilean Air Force instrument trainer, with more comprehensive instrumentation.

T-35C: Primary trainer for Spanish Air Force (designation **E.26 Tamiz**), assembled by CASA from ENAER kits.

T-35D: Instrument trainer for Panama and Paraguay.

T-35S: Single-seat prototype (CC-PZB); evaluation continuing in 1990; 313 kW (420 shp) Allison 250-B17 turboprop planned for any production version. Illustrated in *Jane's All the World's Aircraft 1991-92*.

T-35DT Turbo Pillán: Turboprop version; see Soloy in USA section.

UPGRADES: **Soloy:** T-35DT. See separate entry.

OPERATORS: Versions of the T-35 Pillan are in service with the armed forces of the following countries: Chile (80), Panama (6), Paraguay (15) and Spain (35+).

DESIGN FEATURES: Based on Piper Cherokee series (utilising many components of PA-28 Dakota and PA-32 Saratoga); cleared to FAR Pt 23 (Aerobatic category) and military standards for basic, intermediate and instrument flying training. Low-wing, tandem-seat design; sweptback vertical tail, non-swept horizontal surfaces.

Wing section NACA 65$_2$-415 on constant chord inboard panels, NACA 65$_2$-415 (modified) at tips; incidence 2° at root, –0° 30′ at tip; dihedral 7° from roots.

FLYING CONTROLS: Manual/mechanical actuation of primary three-axis surfaces (ailerons slotted, elevators/rudder mass balanced); single-slotted wing flaps, aileron trim tab (port) and tailplane/elevator trim all electrically actuated; variable incidence tailplane.

STRUCTURE: Main structure of aluminium alloy and steel, with riveted skins, except for glassfibre engine cowling, wingtips and tailplane tips. Single-spar fail-safe wings, with components mainly from PA-28-236 Dakota (leading-edges) and PA-32R-301 Saratoga (trailing-edges), modified for shorter span; vertical tail virtually identical with Dakota; tailplane uses some standard components from Dakota and PA-31 (Navajo/Cheyenne); tailcone from Cherokee components, modified for narrower fuselage.

LANDING GEAR: Hydraulically retractable tricycle type, with single wheel on each unit. Main units retract inward, steerable nosewheel rearward. Mininum ground turning radius 6.20 m (20 ft 4 in). Piper oleo-pneumatic shock-absorber in each unit. Emergency free-fall extension. Cleveland mainwheels and McCreary tyres size 6.00-6 (8 ply), nosewheel and tyre size 5.00-5 (6 ply). Tyre pressures: 2.62 bars (38 lb/sq in) on mainwheels, 2.41 bars (35 lb/sq in) on nosewheel. Single-disc air-cooled hydraulic brake on each mainwheel. Parking brake.

POWER PLANT: One 224 kW (300 hp) Textron Lycoming IO-540-K1K5 flat-six engine, driving a Hartzell HC-C3YR-4BF/FC7663R three-blade constant-speed metal propeller. Fuel in two integral aluminium tanks in wing leading-edges, total capacity 291.5 litres (77 US gallons; 64.1 Imp gallons), of which 272.5 litres (72 US gallons; 60 Imp gallons) are usable. Overwing gravity refuelling point on each wing. Oil capacity 11.4 litres (3 US gallons; 2.5 Imp gallons). Fuel and oil systems permit unlimited inverted flight.

ACCOMMODATION: Two vertically adjustable seats, with seat belts and shoulder harnesses, in tandem beneath one-piece transparent jettisonable canopy which opens sideways to starboard. One-piece acrylic windscreen, and one-piece window in glassfibre fairing aft of canopy. Rear (instructor's) seat 22 cm (8.7 in) higher. Dual controls standard. Baggage compartment aft of rear cockpit, with external access on port side. Cockpits ventilated; cockpit heating and canopy demisting by engine bleed air.

SYSTEMS: Electrically operated hydraulic system, at 124 bars (1,800 lb/sq in) pressure for landing gear retraction and 44.8 bars (650 lb/sq in) for gear extension; separate system at 20.7 bars (300 lb/sq in) for wheel brakes. Electrical system is 24 V DC, powered by 28 V 70 A engine-driven Prestolite alternator and 24 V 10 Ah battery, with inverter for AC power at 400 Hz to operate RMIs and attitude indicators. External power socket. No oxygen or de-icing provisions.

AVIONICS: *T-35A:* Two Collins VHF-251 com transceivers, two Collins AMR-350 audio panels, one Clark Isocom, and one each Collins VIR-351 VOR, ADF-650A and TOR-950 IFF. *T-35C:* One Bendix/King KTR 908, two KFS 598A control units, two KMA 244 audio panels and two KR 87 ADF. Blind-flying instrumentation and full IFR capability in T-35B and D.

DIMENSIONS, EXTERNAL:

Wing span	8.84 m (29 ft 0 in)
Wing aspect ratio	5.7
Length overall	8.00 m (26 ft 3 in)
Height overall	2.64 m (8 ft 8 in)
Tailplane span	3.05 m (10 ft 0 in)
Wheel track	3.02 m (9 ft 11 in)
Wheelbase	2.09 m (6 ft 10¼ in)
Propeller diameter	1.93 m (6 ft 4 in)

DIMENSIONS, INTERNAL:

Cockpit: Length	3.24 m (10 ft 7½ in)
Max width	1.04 m (3 ft 5 in)
Max height	1.48 m (4 ft 10¼ in)

AREAS:

Wings, gross	13.69 m² (147.34 sq ft)

WEIGHTS AND LOADINGS:

Weight empty, equipped	930 kg (2,050 lb)
Fuel weight	210 kg (462 lb)
Max aerobatic T-O weight	1,315 kg (2,900 lb)

ENAER T-35B Pillán trainer aircraft of the Chilean Air Force

1995

ENAER T-35D Pillán trainer aircraft of the Panamanian Air Force

1995

ENAER T-35C (E.26 Tamiz) of the Spanish Air Force. Constructed by CASA from kits ***(Peter March)***

1995

Max T-O and landing weight 1,338 kg (2,950 lb)
Max wing loading 97.73 kg/m² (20.03 lb/sq ft)
Max power loading 5.98 kg/kW (9.83 lb/hp)
PERFORMANCE (at max T-O and landing weight, ISA):
Never-exceed speed (VNE) 241 kts (446 km/h; 277 mph)
Max level speed at S/L 168 kts (311 km/h; 193 mph)
Cruising speed:
75% power at 2,680 m (8,800 ft) 144 kts (266 km/h; 166 mph) IAS
55% power at 5,120 m (16,800 ft) 138 kts (255 km/h; 159 mph) IAS
Stalling speed: flaps up 67 kts (125 km/h; 78 mph)
flaps down 62 kts (115 km/h; 72 mph)
Max rate of climb at S/L 465 m (1,525 ft)/min
Time to: 1,830 m (6,000 ft) 4 min 42 s
3,050 m (10,000 ft) 8 min 48 s
Service ceiling 5,840 m (19,160 ft)
Absolute ceiling 6,250 m (20,500 ft)
T-O run 287 m (940 ft)
T-O to 15 m (50 ft) 494 m (1,620 ft)
Landing from 15 m (50 ft) 509 m (1,670 ft)
Landing run 238 m (780 ft)
Range with 45 min reserves:
75% power at 2,440 m (8,000 ft) 590 n miles (1,093 km; 679 miles)
55% power at 3,660 m (12,000 ft) 650 n miles (1,204 km; 748 miles)
Range, no reserves:
75% power at 2,440 m (8,000 ft) 680 n miles (1,260 km; 783 miles)
55% power at 3,660 m (12,000 ft) 735 n miles (1,362 km; 846 miles)
Endurance at S/L: 75% power 4 h 24 min
55% power 5 h 36 min
g limits +6/−3

NEW ENTRY

CHINA, PEOPLE'S REPUBLIC

CAC

CHENGDU AIRCRAFT CORPORATION

PO Box 800, Chengdu, Sichuan
Telephone: 86 (28) 669629
Fax: 86 (28) 669816
Telex: 60132 CCDAC CN
GENERAL MANAGER: Hou Jianwu
DEPUTY GENERAL MANAGER: Li Shaoming

Founded in 1958, Chengdu Aircraft Corporation is a major centre for the development and production of fighter aircraft. Current activity includes several models of the J-7/F-7 fighter series, and continuing limited batch production of the JJ-5/FT-5 fighter/trainer developed from the Soviet MiG-17. Chengdu is subcontracted to McDonnell Douglas of the USA to manufacture 100 nosecones for MD-80 series airliners, both for the Shanghai MD-82 programme and for the US production line. First nosecone delivered 13 December 1991. The CAC facility occupies a site area of 510 ha (1,260 acres), and had a 1992 workforce of about 22,000. Output includes a number of non-aerospace products.

VERIFIED

CAC FT-5 two-seat trainer aircraft of the Sri Lankan Air Force *(Peter Steinemann)* *1995*

CAC (MIKOYAN) JJ-5

Chinese name: Jianjiji Jiaolianji-5 (Fighter training aircraft 5) or Jianjiao-5
Westernised designation: FT-5

TYPE: Two-seat trainer aircraft.

PROGRAMME: This tandem two-seat version of the J-5 (Chinese built MiG-17) was developed at Chengdu in 1965 and flew for the first time on 8 May 1966. In essence, it combines the tandem cockpits and forward fuselage of the MiG-15UTI with the rest of the airframe of the J-5A (Chinese MiG-17PF), though retaining the latter's lipped intake, the small fairing indicating provision for a radar ranging gunsight in the front cockpit. Other changes include use of a non-afterburning Xian (XAE) WP5D turbojet, rated at 26.48 kN (5,952 lb st); and reduction of the armament to a single Type 23-1 (23 mm) gun, carried in a removable belly pack, with the barrel to the starboard side of the nosewheel doors.

Certificated for mass production at the end of 1966, the JJ-5 is still the standard advanced trainer of the Chinese air forces, to which pupil pilots graduate after basic training on the NAMC CJ-6 (which see). A total of 1,061 had been built by the end of 1986, and limited batch production was continuing in 1989-90.

OPERATORS: Versions of the JJ-5 (FT-5) are in service with the armed services of the following countries: Albania (35), China (500), Pakistan (20) Sri Lanka (2), Sudan, Tanzania and Zimbabwe (2).

WEIGHTS AND LOADINGS:
Weight empty, equipped 4,080 kg (8,995 lb)
Normal T-O weight 5,401 kg (11,907 lb)
Max T-O weight 6,215 kg (13,700 lb)

PERFORMANCE:
Normal operating speed 418 kts (775 km/h; 482 mph)
Max rate of climb at S/L 1,620 m (5,315 ft)/min
Service ceiling 14,300 m (46,900 ft)
T-O run 760 m (2,493 ft)
Landing run 780-830 m (2,559-2,723 ft)
Range with max fuel at 12,000 m (39,370 ft) 664 n miles (1,230 km; 764 miles)
Max endurance at 13,700 m (45,000 ft) with two 400 l (105.5 US gallon; 88 Imp gallon) drop tanks 2 h 38 min

VERIFIED

CAC FT-5 two-seat trainer aircraft of the Pakistan Air Force *(Peter Steinemann)* *1995*

CAC (MIKOYAN) J-7

Chinese name: Jianjiji-7 (Fighter aircraft 7) or Jian-7
Westernised designation: F-7

TYPE: Single-seat fighter and close support aircraft.

PROGRAMME: Soviet licence to manufacture MiG-21F-13 and its R-11F-300 engine granted to Chinese government 1961, when some pattern aircraft and component knock down (CKD) kits delivered, but necessary technical documentation not completed; assembly of first J-7 using Chinese made components began early 1964; original plan in 1964-65 was for Chengdu and Guizhou factories to become main airframe/engine production centres for J-7, backed up by Shenyang until these were fully productive, but plans affected by onset of cultural revolution. Static testing completed November 1965; first flight of Shenyang built J-7, 17 January 1966; Chengdu production of J-7 I began June 1967; development of J-7 II began 1975, followed by first flight 30 December 1978 and production approval September 1979. Development of F-7M and J-7 III started 1981; J-7 III first flight 26 April 1984; F-7M revealed publicly October 1984, production go-ahead December 1984, named Airguard early 1986; first F-7P deliveries to Pakistan 1988; first F-7MPs to Pakistan mid-1989.

VERSIONS: **J-7**: Initial licence version using Chinese made components, built at Shenyang; few only.

J-7 I: Initial Chengdu version for PLA Air Force (1967), with variable intake shock cone and second 30 mm gun; not accepted in large numbers, due mainly to unsatisfactory escape system (front-hinged canopy, to which ejection seat was attached).

F-7A: Export counterpart of J-7 I, supplied to Albania and Tanzania.

J-7 II: Modified and improved development of J-7 I, with WP7B turbojet of increased thrust (43.15 kN; 9,700 lb st dry, 59.82 kN; 13,448 lb st with afterburning); 720 litre (190.2 US gallon; 158.4 Imp gallon) centreline drop tank for increased range; brake-chute relocated at base of rudder to improve landing performance and shorten

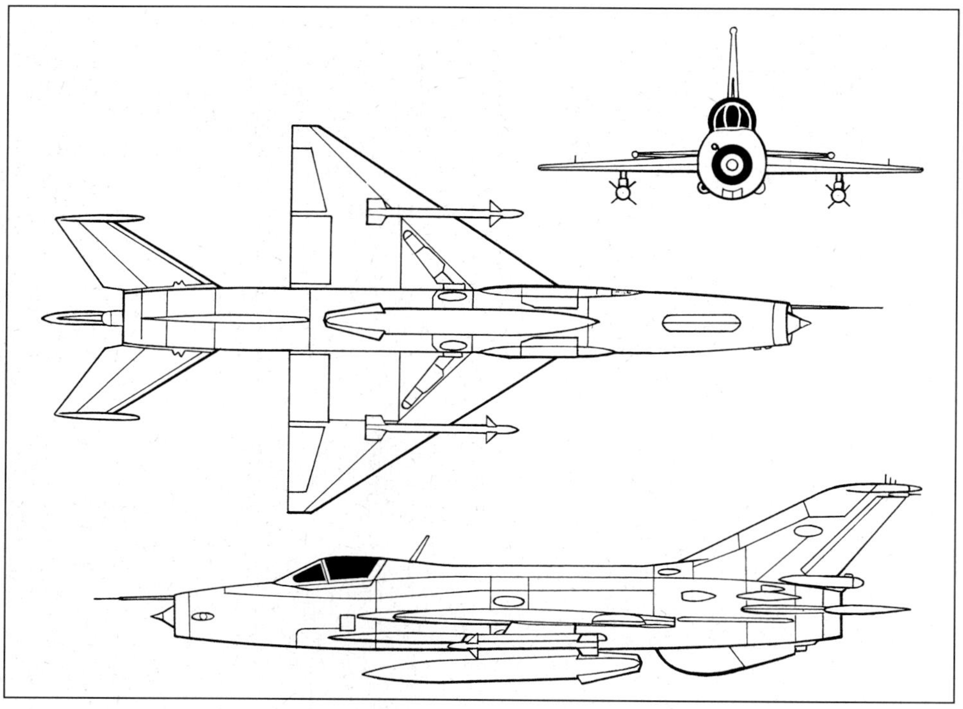

CAC (Mikoyan) F-7 *(Jane's/Mike Keep)*

1994

run; rear-hinged canopy, jettisoned before ejection seat deploys; new Chengdu Type II seat offering ejection at zero height and speeds down to 135 knots (250 km/h; 155 mph); and new Lanzhou compass system. Small batch production (for example 14 in 1989) continuing.

F-7B: Export version of J-7 II, with R550 Magic missile capability; supplied to Egypt and Iraq in 1982-83.

F-7BS: Hybrid version supplied to Sri Lanka 1991; has F-7B fuselage/tail and Chinese avionics (no HUD and so on), combined with four-pylon wings of F-7M. Equips No. 5 Squadron.

F-7M Airguard: Upgraded export version, developed from J-7 II; new avionics imported from May 1979 included GEC-Marconi Avionics HUDWAC (head-up display and weapon aiming computer); new ranging radar, air data computer, radar altimeter and IFF; more secure com radio; improved electrical power generation system for the new avionics; two additional underwing stores points; improved WP7B(BM) engine; birdproof windscreen; strengthened landing gear; ability to carry PL-7 air-to-air missiles; nose probe relocated from beneath intake to top lip of intake, offset to starboard. Exported to Bangladesh, Iran and Zimbabwe. In production. *Description applies to this version except where indicated.*

F-7P Airguard: Variant of F-7M (briefly called Skybolt), embodying 24 modifications to meet specific requirements of Pakistan Air Force, including ability to carry four air-to-air missiles (Sidewinders) instead of two and fitment of Martin-Baker Mk 10L ejection seat. Delivered 1988-91.

F-7MP: Further modified variant of F-7P; improved cockpit layout and navigation system incorporating Collins AN/ARN-147 VOR/ILS receiver, AN/ARN-149 ADF and Pro Line II digital DME-42. Avionics (contract for up to 100 sets) delivered to China from early 1989.

J-7 III: Chinese equivalent of MiG-21MF, much redesigned from J-7 II with blown flaps and all-weather, day/night capability. Main improvements are more powerful WP13 engine; additional fuel in deeper dorsal spine; JL-7 (J-band) interception radar, with correspondingly larger nose intake and centrebody radome; sideways opening (to starboard) canopy, with centrally located rearview mirror; improved HTY-4 low-speed/zero height ejection seat; more advanced fire control system; twin-barrel 23 mm gun under fuselage (with HK-03D optical gunsight); broader-chord vertical tail surfaces, incorporating antennae for LJ-2 omnidirectional RWR in hemispherical fairing each side at base of rudder; increased weapon/stores capability (four underwing stations), similar to that of F-7M; and new or additional avionics (which see). Joint development by Chengdu and Guizhou (GAIC); in production and service.

F-7 III: Improved version of J-7 III; was due to enter PLA Air Force service 1992.

J-7E: Upgraded version of J-7 II with modified wing, retaining existing leading-edge sweep angle of 57° inboard but reduced sweep of only 42° outboard; span increased by 1.17 m (3 ft 10 in) and area by 1.88 m² (20.2 sq ft), giving 8.17 per cent more wing area; four underwing stations instead of two, outer pair each plumbed for 480 litre (127 US gallon; 105.6 Imp gallon) drop tank; new WP7F version of WP7 engine, rated at 44.13 kN (9,921 lb st) dry and 63.74 kN (14,330 lb st) with afterburning; armament generally as listed for F-7M, but capability extended to include PL-8 air-to-air missiles; *g* limits: 8 (up to Mach 0.8) and 6.5 (above Mach 0.8); avionics include head-up display and air data computer. Reportedly due to have made first flight April 1990.

Super-7: Proposed development of F-7M; described separately.

JJ-7/FT-7: Tandem two-seat operational trainer, based on J-7 II and MiG-21US; developed at Guizhou.

UPGRADES: **CAC** F-7M. See versions.

CAC J-7E. See versions.

CAC Super-7. See following entry.

OPERATORS: Versions of the J-7 (F-7) are in service with the armed forces of the following countries: Albania (15), Bangladesh (16), China (500), Egypt (90), Myanmar, Sri Lanka (4), Sudan, Tanzania (10) and Zimbabwe (20+).

DESIGN FEATURES: Diminutive tailed delta with clipped tips to mid-mounted wings; circular-section fuselage with dorsal spine; nose intake with conical centrebody; swept tail, with large vertical surfaces and ventral fin.

Wing anhedral 2° from roots; incidence 0°; thickness/chord ratio approximately 5 per cent at root, 4.2 per cent at tip; quarter-chord sweepback 49° 6′ 36″; no wing leading-edge camber.

FLYING CONTROLS: Manual operation, with autostabilisation in pitch and roll; hydraulically boosted inset ailerons; plain trailing-edge flaps, actuated hydraulically; forward-hinged door type airbrake each side of underfuselage below wing leading-edge; third forward-hinged airbrake under fuselage forward of ventral fin; airbrakes actuated hydraulically; hydraulically boosted rudder and all-moving, trimmable tailplane.

STRUCTURE: All-metal; wings have two primary spars and auxiliary spar; semi-monocoque fuselage, with spine housing control pushrods, avionics, single-point refuelling cap and fuel tank; blister fairings on fuselage above and below each wing to accommodate retracted mainwheels.

LANDING GEAR: Inward retracting mainwheels, with 600 × 200 mm tyres (pressure 11.5 bars; 166.8 lb/sq in) and LS-16 disc brakes; forward retracting nosewheel, with 500 × 180 mm tyre (pressure 7.0 bars; 101.5 lb/sq in) and LS-15 double-acting brake. Nosewheel steerable ±47°. Minimum ground turning radius 7.04 m (23 ft 1¼ in). Tail braking parachute at base of vertical tail.

POWER PLANT: One Chengdu WP7B(BM) turbojet (43.15 kN; 9,700 lb st dry, 59.82 kN; 13,448 lb st with afterburning) in F-7M; WP13 turbojet (40.21 kN; 9,039 lb st dry, 64.72 kN; 14,550 lb st with afterburning) in J-7 III/F-7 III. Total internal fuel capacity 2,385 litres (630 US gallons; 524.5 Imp gallons), contained in six flexible tanks in fuselage and two integral tanks in each wing. Provision for carrying a 500 or 800 litre (132 or 211.3 US gallon; 110 or 176 Imp gallon) centreline drop tank, and/or a 500 litre drop tank on each outboard underwing pylon. Maximum internal/external fuel capacity 4,185 litres (1,105.3 US gallons; 920.5 Imp gallons).

ACCOMMODATION: Pilot only, on Chengdu Aircraft Industrial Corporation zero-height/low-speed ejection seat operable between 70 and 459 knots (130-850 km/h; 81-528 mph) IAS. Martin-Baker Mk 10L seat in Pakistani F-7P/MP. One-piece canopy, hinged at rear to open upward. J-7 III/F-7 III canopy opens sideways to starboard.

SYSTEMS: Improved electrical system in F-7M, using three static inverters, to cater for additional avionics. Jianghuai YX-3 oxygen system.

AVIONICS (F-7M): GEC-Marconi Avionics suite includes Type 956 HUDWAC, AD 3400 two-band UHF/VHF multifunction com system, Type 226 Skyranger ranging radar with ECCM, and an air data computer. Other avionics include Chinese Type 602 IFF transponder, Type 0101 HR A/2 radar altimeter, WL-7 radio compass, and XS-6A marker beacon receiver. HUDWAC (head-up display and weapon aiming computer) provides pilot with displays for instrument flying, with air-to-air and air-to-ground weapon aiming symbols integrated with flight-instrument symbology. It can store 32 weapon parameter functions, allowing for both current and future weapon variants. In air-to-air combat its four modes (missiles, conventional gunnery, snapshoot gunnery, dogfight and standby aiming reticle) allow for all eventualities. Navigation function includes approach mode.

AVIONICS (J-7 III): New avionics include GT-4 ECM jammer, FJ-1 flight data recorder, Type 605A ('Odd Rods' type) IFF; angle of attack vane and air data probe similar to those of F-7M; Beijing Aeronautical Instruments Factory KJ-11 twin-channel autopilot.

ARMAMENT (F-7M): Two 30 mm Type 30-1 belt-fed cannon, with 60 rds/gun, in fairings under front fuselage just forward of wingroot leading-edges. Two hardpoints under each wing, of which outer ones are wet for carriage of drop tanks. Centreline pylon used for drop tank only. Each inboard pylon capable of carrying a PL-2, -2A, -5B or -7 missile or, at customer's option, a Matra R.550 Magic; one 18-tube pod of Type 57-2 (57 mm) air-to-air and air-to-ground rockets; one Type 90-1 (90 mm) seven-tube pod of air-to-ground rockets; or a 50, 150, 250 or 500 kg bomb. Each outboard pylon can carry one of above rocket pods, a 50 or 150 kg bomb, or a 500 litre drop tank.

ARMAMENT (J-7 III): One 23 mm Type 23-3 twin-barrel gun in ventral pack. Five external stores stations can carry two to four PL-5B air-launched missiles; four Qingan HF-16B 12-round launchers for Type 57-2 or seven-round pods of Type 90-1 rockets; or two 500 kg, four 250 kg or 10 100 kg

CAC (Mikoyan) F-7 of the Bangladesh Air Force *(Peter Steinemann)*

1994

bombs, in various combinations with 500 litre (one centreline and/or one under each wing) or 800 litre (underfuselage station only) drop tanks.

DIMENSIONS, EXTERNAL:
Wing span: except J-7E 7.154 m (23 ft 5⅝ in)
J-7E 8.32 m (27 ft 3½ in)
Wing chord: at root 5.508 m (18 ft 0¾ in)
at tip 0.462 m (1 ft 6¼ in)
Wing aspect ratio: except J-7E 2.22
J-7E 2.78
Length overall: excl nose probe 13.945 m (45 ft 9 in)
incl nose probe 14.885 m (48 ft 10 in)
Fuselage: Length 12.177 m (39 ft 11½ in)
Max diameter 1.341 m (4 ft 4¾ in)
Height overall 4.103 m (13 ft 5½ in)
Tailplane span 3.14 m (10 ft 3½ in)
Wheel track 2.692 m (8 ft 10 in)
Wheelbase 4.806 m (15 ft 9¼ in)

AREAS:
Wings, gross: except J-7E 23.00 m² (247.6 sq ft)
J-7E 24.88 m² (267.8 sq ft)
Ailerons (total): except J-7E 1.18 m² (12.70 sq ft)
Trailing-edge flaps (total) 1.87 m² (20.13 sq ft)
Fin 3.48 m² (37.46 sq ft)
Rudder 0.97 m² (10.44 sq ft)
Tailplane 3.94 m² (42.41 sq ft)

WEIGHTS AND LOADINGS:
Weight empty 5,275 kg (11,629 lb)
Normal max T-O weight with two PL-2 or PL-7 air-to-air missiles: F-7M 7,531 kg (16,603 lb)
J-7 III 8,150 kg (17,967 lb)
Wing loading at normal max T-O weight:
F-7M 327.43 kg/m² (67.10 lb/sq ft)
J-7 III 354.35 kg/m² (72.58 lb/sq ft)
Power loading at normal max T-O weight:
F-7M 125.5 kg/kN (1.230 lb/lb st)
J-7 III 125.9 kg/kN (1.234 lb/lb st)

PERFORMANCE (F-7M at normal max T-O weight with two PL-2 or PL-7 air-to-air missiles, except where indicated):
Never-exceed speed (VNE) above 12,500 m (41,010 ft)
Mach 2.35 (1,346 kts; 2,495 km/h; 1,550 mph)
Max level speed between 12,500 and 18,500 m (41,010-60,700 ft)
Mach 2.05 (1,175 kts; 2,175 km/h; 1,350 mph)
Unstick speed
167-178 kts (310-330 km/h; 193-205 mph)
Touchdown speed
162-173 kts (300-320 km/h; 186-199 mph)
Max rate of climb at S/L 10,800 m (35,435 ft)/min
Acceleration from Mach 0.9 to 1.2 at 5,000 m (16,400 ft)
35 s
Max sustained turn rate: Mach 0.7 at S/L 14.7°/s
Mach 0.8 at 5,000 m (16,400 ft) 9.5°/s
Service ceiling 18,200 m (59,710 ft)
Absolute ceiling 18,700 m (61,350 ft)
T-O run 700-950 m (2,297-3,117 ft)
Landing run with brake-chute
600-900 m (1,969-2,953 ft)
Typical mission profiles:
combat air patrol at 11,000 m (36,000 ft) with two air-to-air missiles and three 500 l drop tanks, incl 5 min combat 45 min
long-range interception at 11,000 m (36,000 ft) at 351 n miles (650 km; 404 miles) from base, incl Mach 1.5 dash and 5 min combat, stores as above
hi-lo-hi interdiction radius, out and back at 11,000 m (36,000 ft), with three 500 l drop tanks and two 150 kg bombs 324 n miles (600 km; 373 miles)
lo-lo-lo close air support radius with four rocket pods, no external tanks 200 n miles (370 km; 230 miles)
Range: two PL-7 missiles and three 500 l drop tanks
939 n miles (1,740 km; 1,081 miles)
self-ferry with one 800 l and two 500 l drop tanks, no missiles 1,203 n miles (2,230 km; 1,385 miles)
g limit +8

PERFORMANCE (J-7 III at normal max T-O weight):
Max operating Mach number 2.1
Unstick speed (with afterburning)
173 kts (320 km/h; 199 mph)
Touchdown speed (with flap blowing)
135-146 kts (250-270 km/h; 155-168 mph)
Min level flight speed 140 kts (260 km/h; 162 mph)
Max rate of climb at S/L 9,000 m (29,525 ft)/min
Service ceiling 18,000 m (59,050 ft)
Acceleration from Mach 1.2 to 1.9 at 13,000 m (42,650 ft)
3 min 27 s
Air turning radius at 5,000 m (16,400 ft) at Mach 1.2
5,093 m (16,710 ft)
T-O run (with afterburning) 800 m (2,625 ft)
Landing run (with flap blowing, drag-chute and brakes deployed) 550 m (1,805 ft)
Range: internal fuel only 518 n miles (960 km; 596 miles)
with 800 l belly tank
701 n miles (1,300 km; 807 miles)
with 800 l belly tank and two 500 l underwing tanks 898 n miles (1,664 km; 1,034 miles)
g limits: up to Mach 0.8 +8.5
above Mach 0.8 +7

VERIFIED

CAC (Mikoyan) F-7 *(Peter Steinemann)*

1994

CAC (Mikoyan) F-7M of the Bangladesh Air Force *(Peter Steinemann)*

1995

CAC (Mikoyan) FT-7 two-seat trainer aircraft of the Sri Lankan Air Force *(Peter Steinemann)*

1995

CAC SUPER-7

TYPE: Proposed export development of F-7M Airguard.

PROGRAMME: Agreement between CATIC and Grumman (USA) for joint preliminary design signed 21 October 1988; Grumman participation suspended by US government mid-1989. Preliminary design and wind tunnel testing completed; CATIC seeking alternative partner to continue programme.

DESIGN FEATURES: Lateral air intakes for more powerful engine; 'solid' ogival nosecone for modern fire control radar; wings of enlarged span and area, with leading-edge slats and additional pair of hardpoints inboard for air-to-air missiles; enlarged dorsal spine housing additional fuel; single-point pressure refuelling; easier access engine compartment; arrester hook; modified ventral fin; strengthened main landing gear with larger tyres; new straight-leg, steerable nosewheel unit; belly mounted twin-barrel 23 mm gun instead of two internal 30 mm; new cockpit, incorporating HUD and new ejection seat; revised ECS for avionics cooling.

CUSTOMERS: Pakistan regarded as main market.

DIMENSIONS, EXTERNAL:

Wing span	8.98 m (29 ft 5½ in)
Length overall	15.30 m (50 ft 2½ in)
Height overall	4.13 m (13 ft 6½ in)
Wheel track	2.79 m (9 ft 1¾ in)
Wheelbase	5.59 m (18 ft 4 in)

AREAS:

Wings, gross	approx 24.62 m² (265.0 sq ft)

WEIGHTS AND LOADINGS (estimated):

Internal fuel	2,327 kg (5,130 lb)
Design gross weight	9,100 kg (20,062 lb)
Max T-O weight	11,295 kg (24,900 lb)

PERFORMANCE (estimated):

Max level speed	above Mach 1.8
Service ceiling	16,765 m (55,000 ft)
T-O distance	555 m (1,821 ft)
Landing distance	860 m (2,822 ft)
Mission radius:	
air-to-air (hi-hi-hi)	475 n miles (880 km; 547 miles)
air-to-ground (hi-lo-hi)	329 n miles (610 km; 379 miles)
g limit (max)	+8.5

VERIFIED

Model of the proposed Super-7 advanced development of the CAC F-7M

1994

HAMC

HARBIN AIRCRAFT MANUFACTURING COMPANY

PO Box 201, Harbin, Heilongjian 150066
Telephone: 86 (451) 8602122
Fax: 86 (451) 8602061
Telex: 87082 HAF CN

GENERAL MANAGER: Yang Shouwen

Harbin had its origin in the plant of the Manshu Aeroplane Manufacturing Company, one of several aircraft and aero-engine factories established in Manchukuo (Manchuria) by the Japanese in 1938. After the Communist regime came to power in mainland China in 1949 it was re-established in 1952 and re-equipped with Soviet assistance, and since then has been responsible for production of the H-5 twin-jet light bomber, a reverse engineered version of the Soviet Ilyushin Il-28, and the nationally designed SH-5 amphibian and Y-11 aircraft. Landing gear doors for the British Aerospace 146 are produced under a 1981 agreement with BAe, and doors and wing components for the Shorts 3-60.

Harbin is also the chief centre for helicopter production, which began with the Mil Mi-4 (Chinese Z-5, first flown on 14 December 1958; total of 545 built. It is currently responsible for the Aerospatiale Dauphin 2 (Z-9A) manufacturing programme. The workforce numbers about 17,000.

UPDATED

HARBIN (ILYUSHIN) H-5

Chinese name: Hongzhaji-5 (bomber aircraft 5) or Hong-5

NATO reporting names: Beagle (H-5) and Mascot (HJ-5)

TYPE: Three-seat tactical light bomber.

PROGRAMME: After receiving large numbers of Ilyushin Il-28 three-seat tactical light bombers from the Soviet Union, China began building its own equivalent, the H-5, in 1966. Production at Harbin, ended in 1982. About 500 H-5s are believed to equip the air force of the People's Liberation Army, with about 130 more in service with the PLA Navy. Some may be configured for nuclear weapon delivery.

VERSIONS: **H-5 (Hongzhaji-5 or Hong-5):** Standard three-seat tactical light bomber, similar to basic Il-28. Some early examples were exported to Albania. Production also included torpedo-bomber version similar to Soviet Il-28T.

HJ-5 (Hongzhaji Jiaolianji-5 or Hongjiao-5): Two-seat operational and pilot training version, similar to Soviet Il-28U (NATO reporting name 'Mascot'). Armament and ventral ground mapping radar fairing deleted; 'solid' nose; second 'stepped' cockpit (with dual controls) ahead of and below pilot's cockpit. Two or three supplied to each operational H-5 unit.

HZ-5 (Hongzhaji Zenchaji-5 or Hongzhen-5): Three-seat tactical reconnaissance version, similar to Soviet Il-28R. Wingtip auxiliary fuel tanks standard; weapons bay occupied by alternative packs containing cameras or electronic sensors.

In 1985 China ordered modified Rushton low level towed targets from the FR Group in the UK. These are for use with H-5s of the PLA Navy, to simulate sea-skimming anti-ship missiles.

OPERATORS: Versions of the H-5 are in service with the armed forces of China (450).

The description which follows applies primarily to the standard Il-28/H-5 bomber:

DESIGN FEATURES: Cantilever shoulder-wing monoplane with non-swept leading-edges and tapered trailing-edges. TsAGI SR5S wing section with maximum thickness/chord ratio of 12 per cent. Incidence 0° 38'. Dihedral 3° from roots. 'Solid' fairing aft of pilot's cockpit incorporates a dielectric panel. Single ventral radome standard, forward of weapons bay; some aircraft have two such radomes. All-swept cruciform structure. Fin on root platform built integrally with fuselage, has leading-edge sweep of 45°. Fixed-incidence tailplane has 33° sweepback on leading-edges and 7° dihedral.

FLYING CONTROLS: Hydraulically actuated trailing-edge slotted flaps inboard and outboard of each engine nacelle, with settings of 0, 20 and 50°. Plain ailerons which deflect 15° up and 20° down. Trim tab in each aileron. Trim tabs in rudder and each elevator.

STRUCTURE: Conventional semi-monocoque structure of circular cross-section. Glazed nose with optically flat bomb aiming panel.

POWER PLANT: Two Harbin Wopen-5 WP-5 (Klimov VK-1A) non-afterburning turbojet engines, each rated at 26.5 kN (5,952 lb st) and mounted in underwing pods. Fuel in five flexible fuselage tanks (three forward and two aft of weapons bay), integral wing tanks, and (standard on HZ-5, optional on other models) wingtip auxiliary tanks. Total capacity, including tip tanks, 7,908 litres (2,089 US gallons; 1,740 Imp gallons). Refuelling points in fuselage (four), wings and each tip tank. Provision for assisted take-off JATO rocket under fuselage on each side.

ACCOMMODATION: Flight crew of three (instructor and pupil only in HJ-5), all in pressurised and air conditioned accommodation. Pilot on ejection seat in single 'fighter' type cockpit, under jettisonable canopy which opens sideways to starboard. Navigator/bomb aimer, also on ejection seat, occupies a position forward, below and to starboard of pilot, access to which is via an upward-opening jettisonable hatch above the nose and offset to starboard (in HJ-5, roof hatch of forward cockpit hinges sideways to starboard). Access to radio operator/rear gunner's position is via a power operated downward-opening hatch in underside of rear fuselage, which also serves as an escape hatch for this member of the crew. Dual controls in HJ-5.

ARMAMENT AND OPERATIONAL EQUIPMENT: Two fixed, forward-firing 23 mm cannon (each with 100 rds) in lower forward fuselage, one each side of nosewheel bay; associated gyro gunsight in pilot's cockpit. Two similar guns, each with 225 rds, on movable ball mounting in tail turret. Internal weapons bay in mid-fuselage, with normal and maximum capacities of 1,000 kg (2,205 lb) and 3,000 kg (6,614 lb) respectively. Typical loads may include four 500 kg or eight 250 kg bombs or one large or two smaller torpedoes, mines or depth charges. Some H-5s may be configured for nuclear weapon delivery. Provision in standard H-5 for single vertical camera, installed beneath rearmost forward fuselage fuel tank. HZ-5 can carry from three to five cameras in the weapons bay, plus 12 to 18 flares or photoflash bombs.

DIMENSIONS, EXTERNAL (standard H-5):

Wing span (excl tip tanks)	21.45 m (70 ft 4½ in)
Wing chord, mean	2.955 m (9 ft 8½ in)
Wing aspect ratio	7.55
Length of fuselage (excl tail guns)	17.65 m (57 ft 11 in)
Fuselage: Max diameter	1.8 m (5 ft 10¾ in)
Distance betwen c/l of engine nacelles	6.80 m (22 ft 3¾ in)
Height overall	6.70 m (21 ft 11¾ in)
Tailplane span	7.10 m (23 ft 3½ in)
Wheel track	7.40 m (24 ft 3½ in)
Wheelbase	approx 8.10 m (26 ft 7 in)

AREAS:

Wings, gross	60.80 m² (654.45 sq ft)

WEIGHTS AND LOADINGS (standard H-5):

Weight empty, equipped	12,890 kg (28,417 lb)
Fuel load: normal	3,800 kg (8,377 lb)
max (incl 200 kg; 441 lb in tip tanks)	6,600 kg (14,550 lb)
Internal weapon load: normal	1,000 kg (2,205 lb)
max	3,000 kg (6,614 lb)
Normal T-O weight	18,400 kg (40,565 lb)
Max T-O weight	21,200 kg (46,738 lb)
Wing loading:	
at normal T-O weight	approx 303 kg/m² (62 lb/sq in)
at max T-O weight	approx 349 kg/m² (71.5 lb/sq in)
Power loading: at normal T-O weight	approx 347.5 kg/kN (3.4 lb/lb st)
at max T-O weight	approx 400 kg/kN (3.9 lb/lb st)

PERFORMANCE (standard H-5 at normal T-O weight except where indicated):

Max level speed: at S/L	432 kts (800 km/h; 497 mph)
at 4,500 m (14,760 ft)	487 kts (902 km/h; 560 mph)
at 12,000 m (39,370 ft)	434 kts (805 km/h; 500 mph)
Typical cruising speed	415 kts (770 km/h; 478 mph)
Unstick speed:	
at normal T-O weight	119 kts (220 km/h; 137 mph)
at max T-O weight	126 kts (234 km/h; 145 mph)
Touchdown speed	100 kts (185 km/h; 115 mph)
Rate of climb: max, at S/L	900 m (2,952 ft)/min
at 5,000 m (16,400 ft)	630 m (2,067 ft)/min
at 8,000 m (26,250 ft)	420 m (1,378 ft)/min
at 12,000 m (39,370 ft)	72 m (236 ft)/min
Time to 5,000 m (16,400 ft)	6 min 30 s
Time to 10,000 m (32,800 ft)	18 min
Service ceiling	12,300 m (40,350 ft)
T-O run: at normal T-O weight	875 m (2,870 ft)
at max T-O weight	1,159 m (3,773 ft)
Landing run at landing weight of 14,690 kg (32,385 lb)	1,170 m (3,838 ft)
Range with max fuel, at T-O weight:	
at 410 kts (760 km/h; 472 mph) at 1,000 m (3,280 ft)	612 n miles (1,135 km; 705 miles)
at 415 kts (770 km/h; 478 mph) at 10,000 m (32,800 ft)	1,176 n miles (2,180 km; 1,355 miles)
at 232 kts (430 km/h; 267 mph) at 10,000 m (32,800 ft)	1,295 n miles (2,400 km; 1,490 miles)

VERIFIED

HARBIN Y-11

Chinese name: Yunshuji-11 (Transport aircraft 11) or Yun-11

TYPE: Twin-engined agricultural and general purpose aircraft.

PROGRAMME: This twin-engined utility aircraft was designed and developed in China as a potential replacement for the Y-5 (Chinese built An-2). First flight took place on 30 December 1975.

Construction of a small preproduction series of about 15 aircraft began in 1977, and these were used in top-dressing and pest control operations in 1977-78. The Y-11 is now used primarily in agricultural, forestry and geophysical survey applications. Other possible applications include short-haul and aeromedical transportation, fishery protection, firefighting and flying training.

OPERATORS: Versions of the Y-11 are in service with the armed forces of China (15).

DESIGN FEATURES: Braced high-wing monoplane, with constant chord from root to tip. Wing section NACA 4412. No dihedral.

FLYING CONTROLS: All-metal drooping ailerons and electrically actuated fabric covered two-section double-slotted flaps along full-span of trailing-edges. All-metal leading-edge automatic slats from nacelle to tip of each wing, with smaller inboard slat on each side between nacelle and fuselage. Trim tab in each aileron. Inset tab in rudder and port elevator.

STRUCTURE: Two-spar structure with aluminium alloy skin, bonded between spars, riveted elsewhere. Small stub wings at cabin floor level support the main landing gear units; bracing strut from each stub wing out to approximately mid-span. The fuselage is a conventional semi-monocoque all-metal structure of basically rectangular cross-section, swept upward at rear. The tail unit is a cantilever non-swept metal structure, with low-set tailplane and small dorsal fin. Fabric covered horn balanced rudder and elevators.

LANDING GEAR: Non-retractable tricycle type, with oleo-pneumatic shock-absorber in each tail unit. Twin-wheel main units, attached to underside of stub wings. Single steerable nosewheel. Mainwheel tyres size 500 × 150 mm, pressure 2.90-3.45 bars (42-50 lb/sq in). Nosewheel tyre size 400 × 150 mm, pressure 2.90 bars (42 lb/sq in). Small bumper under tailcone. Pneumatic brakes.

POWER PLANT: Two 213 kW (285 hp) Quzhou (Chuchow) Huosai-6A (Chinese development of Ivchenko/Vedeneev AI-14RF) nine-cylinder radial air-cooled engines, each driving a two-blade variable-pitch propeller, underslung from wings and fitted with louvred intakes in front of cylinders to control cooling. Two metal fuel tanks between spars of each outer wing, with smaller tank in each engine nacelle. Total fuel capacity 530 litres (140 US gallons; 116 Imp gallons). Normal fuel load carried on agricultural missions is 285 litres (75 US gallons; 62.5 Imp gallons).

ACCOMMODATION: Crew of two on flight deck, with separate forward-opening door on port side for access. Dual controls. Cabin accommodates seven passengers normally (with removable folding jump seat for an eighth passenger), or equivalent cargo. Cargo/passenger double door on port side of fuselage, in line with wing trailing-edge.

SYSTEMS: Pneumatic system for engine starting and wheel brakes, supplied by pump driven by each engine and back-up air cylinder. Electrical system includes 1.5 kW generator on each engine and 30 Ah storage battery.

AVIONICS AND EQUIPMENT: Radio; operational equipment according to mission. Agricultural version has hopper with capacity of 900 kg (1,984 lb) or 900 litres (238 US gallons; 198 Imp gallons), and six rotary atomisers for spraying; it can cover an area of 200 km² (77.2 sq miles) in one mission. Geophysical survey version has magnetometer boom projecting 1.30 m (4 ft 3¼ in) beyond the tail, other equipment in wingtip containers and a camera installation in the cabin floor.

DIMENSIONS, EXTERNAL:

Wing span	17.0 m (55 ft 9¼ in)
Wing aspect ratio	8.5
Length overall	12.017 m (39 ft 5¼ in)
Height overall	4.64 m (15 ft 2¾ in)
Width of stub wings	3.612 m (11 ft 10¼ in)
Elevator span	5.10 m (16 ft 8¾ in)
Wheel track (c/l of shock-absorbers)	3.45 m (11 ft 3¾ in)
Wheelbase	3.642 m (11 ft 11½ in)
Propeller diameter	2.40 m (7 ft 10½ in)
Distance between propeller centres	4.27 m (14 ft 0 in)
Cargo door: Width	0.988 m (3 ft 3 in)
Height	1.22 m (4 ft 0 in)

DIMENSIONS, INTERNAL:

Cabin: Length	3.58 m (11 ft 9 in)
Max width	1.27 m (4 ft 2 in)
Max height	1.48 m (4 ft 10¼ in)

AREA:

Wings, gross	34.0 m² (365.97 sq ft)

WEIGHTS:

Weight empty	2,050 kg (4,519 lb)
Max fuel load	390 kg (860 lb)
Max payload	870 kg (1,918 lb)
Normal T-O and landing weight	3,250 kg (7,165 lb)
Max T-O weight	3,500 kg (7,715 lb)

PERFORMANCE (at normal T-O weight):

Max level speed	119 kts (220 km/h; 137 mph)
Cruising speed: 75% power	102 kts (190 km/h; 118 mph)
65% power	94 kts (175 km/h; 109 mph)
57% power	89 kts (165 km/h; 102 mph)
Speed for agricultural operation	86 kts (160 km/h; 99 mph)
Stalling speed, flaps up	57 kts (105 km/h; 65 mph)
Max rate of climb at S/L	246 m (807 ft)/min
Service ceiling	4,000 m (13,125 ft)
STOL T-O landing run	140 m (459 ft)
Range at 3,000 m (9,845 ft) with max fuel, no reserves	537 n miles (995 km; 618 miles)

UPDATED

NAMC

NANCHANG AIRCRAFT MANUFACTURING COMPANY

PO Box 5001-506, Nanchang, Jiangxi
Telephone: 86/Nanchang 41112/41512
Fax: 86/Nanchang 41112 ext 2272
Telex: 95068 NAMC CN
GENERAL MANAGER: Wu Mingwang
INFORMATION: Feng Jinghua

Created in 1951, Nanchang was responsible for licence production of the Soviet Yak-18 trainer, of which it built 379 (as the CJ-5) betwen 1954 and 1958, and continues to manufacture its own CJ-6A development of that aircraft. In the 1960s it shared in the large production programme for the J-6 fighter (Chinese development of the MiG-19), from which it subsequently developed the Q-5/A-5 attack aircraft series. Between 1957-68, NAMC produced 727 examples of the Y-5 (Chinese An-2) biplane, including 114 as local service transports and 229 for agricultural use. Its latest design, the N-5A, is a dedicated agricultural aircraft which made its first flight at the end of 1989. An upgrade programme for the A-5 is under way, and the K-8 jet trainer is being developed jointly with Pakistan. NAMC occupies a 500 ha (1,235 acre) site, with 10,000 m² (107,639 sq ft) of covered space, and had a workforce of more than 20,000 in the late 1980s. About 80 per cent of its activities are non-aerospace.

VERIFIED

NAMC CJ-6 predecessor of the CJ-5 on final approach

1994

NAMC BT-6 basic trainer aircraft of the Bangladesh Air Force *(Peter Steinemann)*

1995

NAMC CJ-6

Chinese name: Chuji Jiaolianji-6 (Basic training aircraft 6) or Chujiao-6
Westernised designation: PT-6
Bangladeshi military designation: BT-6

TYPE: Single-engine basic trainer and agricultural aircraft.

PROGRAMME: The CJ-6's predecessor, the CJ-5, one of the first aircraft to be mass produced in post-1949 China, was a licence-built version of the Soviet Yak-18 basic trainer. It first flew on 3 July 1954, and 379 were built at Nanchang from 1954-58, for air force, navy and CAAC use.

Design of the CJ-6, as a Chinese engineered successor to the CJ-5, was initiated at Shenyang in the Autumn of 1956, a prototype powered by a 108 kW (145 hp) Mikulin M-11ER radial engine flying for the first time on 27 August 1958. Flight trials with this power plant proving disappointing, a modified version was tested (first flight 18 July 1960) with a 194 kW (260 hp) Ivchenko (AI-14R) engine. Responsibility for the CJ-6 was subsequently transferred to Nanchang, where further redesign preceded the first flight of a production-standard prototype on 15 October 1961. Production go-ahead was given in 1962 for the aircraft (in January) and the Chinese HS6 version of the AI-14R engine (in June), and a total of 1,796 CJ-6s (all versions) had been built by the end of 1986.

VERSIONS: **CJ-6A:** Standard version since 1965 signifying the introduction of an uprated HS6A engine.

CJ-6B: Armed version of the CJ-6A. There were 10 built from 1964-66.

Haiyan: Civil agricultural version (see following entry) has been developed and a six-seat utility version has been proposed.

OPERATORS: Versions of the CJ-6 are in service with the armed forces of Bangladesh (35), China (1,500), North Korea (100+) and Zambia (10+).

The details which follow apply to the standard CJ-6A basic trainer:

STRUCTURE: All-metal cantilever low-wing monoplane. Two-spar wings, with detachable, tapered and dihedralled outer panels. Retractable tricycle landing gear, with low-pressure mainwheel tyres, suitable for operation from grass strips.

POWER PLANT: One 213 kW (285 hp) Zhuzhou (SMPMC) HS6A nine-cylinder air-cooled radial engine, driving a Baoding J9-G1 two-blade constant-speed propeller. Fuel capacity (two tanks) 100 litres (26.4 US gallons; 22 Imp gallons).

DIMENSIONS, EXTERNAL:

Wing span	10.18 m (33 ft 4¾ in)
Length overall	8.46 m (27 ft 9 in)
Height overall	3.25 m (10 ft 8 in)

WEIGHTS AND LOADINGS:

Weight empty	1,172 kg (2,584 lb)
Max fuel	110 kg (243 lb)
Max T-O weight	1,419 kg (3,128 lb)

PERFORMANCE:

Max level speed	155 kts (286 km/h; 178 mph)
Landing speed	62 kts (115 km/h; 72 mph)
Max rate of climb at S/L	380 m (1,248 ft)/min
Service ceiling	5,080 m (16,665 ft)
T-O run	280 m (920 ft)
Landing run	350 m (1,150 ft)
Endurance	3 h 36 min

UPDATED

NAMC HAIYAN (PETREL)

TYPE: Single-engine agricultural and patrol aircraft.

PROGRAMME: To meet a national requirement for a multi-purpose agricultural and forestry aircraft, NAMC began work on a conversion of its CJ-6 basic trainer (see preceding entry) in April 1985.

VERSIONS: **Haiyan A:** Prototype flying for the first time on 17 August 1985. Little change to the basic, proven airframe of the CJ-6 was necessary. To cater for the higher operating rates involved, the Haiyan A was fitted with a more powerful (257 kW; 345 hp) version of the HS-6A engine and a new design propeller. Removal of the rear seat allowed a 400 kg (882 lb) insecticide tank to be installed, and another 200 kg (441 lb) of chemical was accommodated in the leading-edge of the wing centre-section. Volumetric capacity in the rear cockpit, without removing the instrument panel, is sufficient to allow 800 kg (1,764 lb) to be carried in this location if desired. The dispersal system consists of four Type 751 underwing sprinkler heads, fed by a modified LB-4 fuel pump, and it can be used for both low and ultra-low spraying.

Initial flight tests have proved the practicality of the conversion, work was continuing in late 1985/early 1986 mainly to perfect the dispersal system and improve the aircraft's take-off and landing performance.

Haiyan B (proposed): Specialised agricultural and forestry version for crop spraying (dry or liquid pesticide or fertiliser), seed sowing and forest firefighting.

Haiyan C (proposed): Patrol and observation version, with normal CJ-6 rear seat accommodation but having increased fuel capacity to extend endurance to over six hours. Suitable for forestry and fishery patrol, cartography, aerial photography, geological survey, coastal and border patrol, with appropriate equipment according to customer requirements.

The following details apply to the Haiyan A prototype:

DIMENSIONS, EXTERNAL: As for CJ-6

WEIGHTS:

Weight empty	1,214 kg (2,676 lb)
Max T-O weight	2,035 kg (4,486 lb)

PERFORMANCE (at max T-O weight):

Max level speed	160 kts (297 km/h; 185 mph)
Normal operating speed	86 kts (160 km/h; 99 mph)
Operating height: max	6,250 m (20,500 ft)
min	1 m (3 ft)
T-O run	280 m (919 ft)
Landing run	350 m (1,148 ft)
Range	421 n miles (780 km; 484 miles)
Endurance	4 h 11 min
Swath width	30 m (98 ft)

VERIFIED

NAMC Q-5 (A-5)

Chinese name: Qiangjiji-5 (Attack aircraft 5) or Qiang-5
Westernised designation: A-5
NATO reporting name: Fantan

TYPE: Single-seat close air support, ground attack and air-to-air combat aircraft.

PROGRAMME: Development of this twin-jet attack aircraft, derived from the J-6/MiG-19 produced in China, began with a design proposal submitted by Shenyang in August 1958. Responsibility for the programme was assigned to Nanchang, but the prototype programme was cancelled in 1961. It was, however, kept alive by a small team of enthusiasts, and was officially resumed two years later, the first prototype eventually making its initial flight on 4 June 1965. On receipt of a preliminary design certificate, a preproduction batch was authorised at the end of that year, but further modifications, to the fuel, armament, hydraulic and other systems, were found necessary, leading to the flight testing of two much modified prototypes beginning in October 1969. Series production of the Q-5 was approved at the end of that year, and deliveries began in 1970. Several of these initial production aircraft were adapted for nuclear weapon delivery tests in the early 1970s.

VERSIONS: **Q-5:** Initial production version, with internal fuselage bay approximately 4.00 m (13 ft 1½ in) long for two 250 kg or 500 kg bombs, two underfuselage attachments in adjacent bay for two similar bombs, and two stores pylons beneath each wing; Series 6 WP6 turbojets; brake-chute in tailcone, between upper and lower 'pen-nib' fairings.

Q-5 I: Extended payload/range version, first proposed in 1976; flight tested in late 1980 and certificated for production on 20 October 1981. Internal bomb bay blanked off; space used to enlarge main fuselage fuel tank and add a flexible tank, and underfuselage stores points increased to four; improved series WP6 engines; modified landing gear; brake-chute relocated under base of rudder; improved Type I rocket ejection seat; HF/SSB transceiver added. Some Q-5 Is adapted for PLA Naval Air Force to carry two underfuselage torpedoes; these reportedly have Doppler type nose radar and 20 m (66 ft) sea-skimming capability with C-801 anti-shipping missiles.

Q-5 IA: Improved Q-5 I, with additional underwing hardpoint each side (increasing stores load by 500 kg; 1,102 lb), new gun/bomb sighting systems, pressure refuelling, and added warning/countermeasures systems.

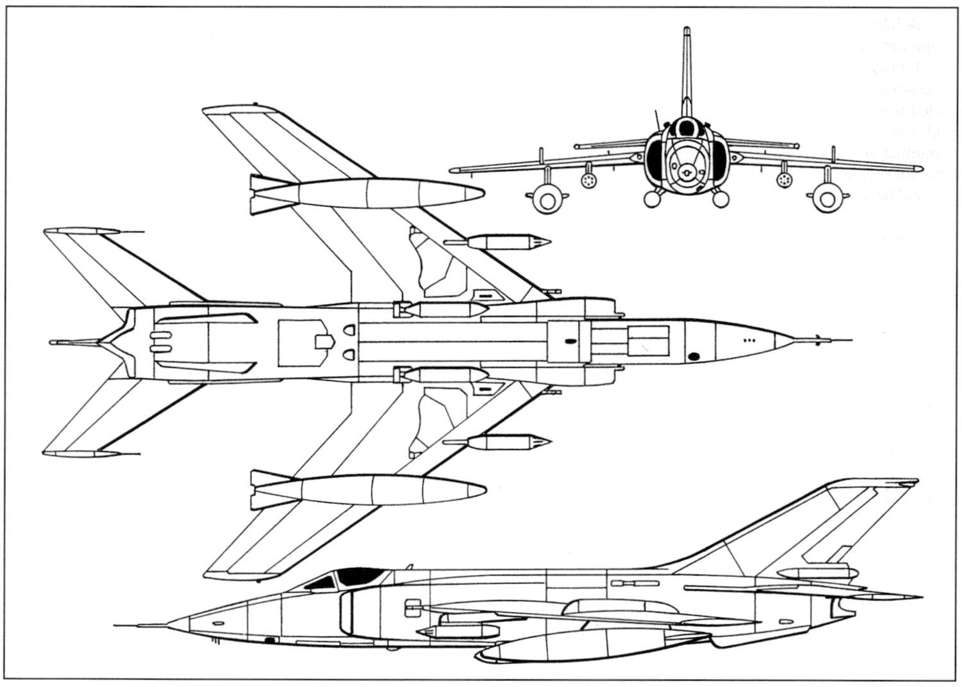

NAMC A-5 Fantan *(Jane's/Mike Keep)*

1994

NAMC A-5 of the Pakistan Air Force *(Peter Steinemann)*

1995

NAMC A-5 of the Pakistan Air Force *(Peter Steinemann)*

1994

Q-5 II: As Q-5 IA, but fitted (or retrofitted) with radar warning receiver.

Q-5K Kong Yun: Upgraded version proposed for Chinese armed forces. Development abandoned in 1990.

A-5C: Export version for Pakistan Air Force, involving 32 modifications from Q-5 I, notably upgraded avionics, Martin-Baker PKD10 zero/zero ejection seats and adaptation of hardpoints for 356 mm (14 in) lugs compatible with PAF weapons, including Sidewinder missiles. Three prototypes built. Original order for 52, placed in April 1981, delivered between January 1983 and January 1984, equipping Nos. 7, 16 and 26 Squadrons of Pakistan Air Force. Further 98, ordered in 1984, also now delivered. Bangladesh ordered 20.

A-5M: Upgraded export version of Q-5 II under development. Described separately.

Including some 200 for export, nearly 1,000 Q-5s (all versions) have been built to date. A design study was carried out in the mid-1980s by FRL in the UK to equip the Q-5 as a receiver for in-flight refuelling, with a Xian H-6 bomber adapted to act as the tanker aircraft.

UPGRADES: **NAMC**: Q-5 II. See Versions.

NAMC/Alenia: A-5M. See separate entry in this section.

NAMC/Thompson-CSF: A-5K Kong Yun (Cloud). See separate entry in this section.

OPERATORS: Versions of the Q-5/A-5 are in service with the armed forces of the following countries: Bangladesh (20), China (550), North Korea (40), Myanmar (20+) and Pakistan (50+).

The following description applies to the Q-5 II and A-5C except where otherwise indicated:

DESIGN FEATURES: Cantilever all-metal mid-wing monoplane, of low aspect ratio, with 4° anhedral from roots. Sweepback at quarter-chord 52° 30′. Deep, full chord boundary layer fence on each upper surface at mid-span. Air intake on each side of fuselage, abreast of cockpit; twin jetpipes side by side at rear. Top and bottom 'pen-nib' fairings aft of nozzles. Centre-fuselage is 'waisted' in accordance with area rule. Dorsal spine fairing between rear of cockpit and leading-edge of fin. Shallow ventral strake under each jetpipe. Tailplane has 6° 30′ anhedral and anti-flutter weight projecting forward from each tip. Tail warning antenna in tip of fin.

FLYING CONTROLS: Inboard of each fence is a hydraulically actuated Gouge flap, the inner end of which is angled to give a trailing-edge at right angles to side of fuselage. Hydraulically actuated (by irreversible servo) internally balanced aileron outboard of each fence. Forward-hinged, hydraulically actuated door type airbrake under centre of fuselage, forward of bomb attachment points. Electrically operated inset trim tab at inboard end of port aileron. Mechanically actuated mass balanced rudder, with electrically operated inset trim tab. One-piece hydraulically actuated (by irreversible servo) all-moving tailplane.

STRUCTURE: The wing is a multispar basic structure with ribs and stressed skin, with three-point attachment to fuselage. The fuselage is a conventional all-metal structure of longerons, stringers and stressed skin, built in forward and rear portions which are detachable aft of wing trailing-edge to provide access to engines. The tail unit is a cantilever all-metal stressed skin structure, with sweepback on all surfaces.

LANDING GEAR: Hydraulically retractable wide-track tricycle type, with single wheel and oleo-pneumatic shock-absorber on each unit. Main units retract inward into wings, non-steerable nosewheel forward into fuselage, rotating through 87° to lie flat in gear bay. Mainwheels have size 830 × 205 mm tubeless tyres, and disc brakes. Tail braking parachute, deployed when aircraft is 1 m (3.3 ft) above the ground, in bullet fairing at root of vertical tail trailing-edge beneath rudder.

POWER PLANT: Two Shenyang WP6 turbojets, each rated at 25.50 kN (5,732 lb st) dry and 31.87 kN (7,165 lb st) with afterburning, mounted side by side in rear of fuselage. Improved WP6A engines (see A-5M entry for details) available optionally. Lateral air intake with small splitter plate, for each engine. Hydraulically actuated nozzles. Internal fuel in three forward and two rear fuselage tanks with combined capacity of 3,648 litres (964 US gallons; 802.5 Imp gallons). Provision for carrying a 760 litre (201 US gallon; 167 Imp gallon) drop tank on each centre underwing pylon, to give maximum internal/external fuel capacity of 5,168 litres (1,366 US gallons; 1,136.5 Imp gallons). When centre wing stations are occupied by bombs, a 400 litre (105.7 US gallon; 88 Imp gallon) drop tank can be carried instead on each outboard underwing pylon.

ACCOMMODATION: Pilot only, under one-piece jettisonable canopy which is hinged at rear and opens upward. Downward view over nose, in level flight, is 13° 30′. Low-speed seat allows for safe ejection within speed range of 135-458 knots (250-850 km/h; 155-528 mph) at zero height or above. Aircraft in Pakistan service have been refitted with Martin-Baker PKD10 zero/zero seats. Armour plating in some areas of cockpit to protect pilot from anti-aircraft gunfire. Cockpit pressurised and air conditioned.

SYSTEMS: Dual air conditioning systems, one for cockpit environment and one for avionics cooling. Two independent hydraulic systems, each operating at pressure of 207 bars (3,000 lb/sq in). Primary system actuates landing gear extension and retraction, flaps, airbrake and afterburner nozzles; auxiliary system supplies power for aileron and all-moving tailplane boosters. Emergency system, operating pressure 108 bars (1,570 lb/sq in), for actuation of main landing gear. Electrical system (28 V DC) powered by two 6 kW engine-driven starter/generators, with two inverters for 115 V single-phase and 36 V three-phase AC power at 400 Hz.

AVIONICS: Include CT-3 VHF com transceiver, WL-7 radio compass, WG-4 low altitude radio altimeter, LTC-2 horizon gyro, YD-3 IFF, Type 930 radar warning receiver and XS-6 marker beacon receiver. Combat camera in small 'teardrop' fairing on starboard side of nose (not on export models). 'Odd Rods' type IFF aerials under nose on Q-5 variants, replaced on A-5C by a single blade antenna. Space provision in nose and centre-fuselage for additional or updated avionics, including an attack radar.

NAMC A-5 of the Pakistan Air Force *(Peter Steinemann)*

1994

EQUIPMENT: Landing light under fuselage, forward of nosewheel bay and offset to port; taxying light on nosewheel leg.

ARMAMENT: Internal armament consists of one 23 mm cannon (Norinco Type 23-2K), with 100 rds, in each wingroot. Ten attachment points normally for external stores: two pairs in tandem under centre-fuselage, and three under each wing (one inboard and two outboard of mainwheel leg). Fuselage stations can each carry a 250 kg bomb (Chinese 250-2 or 250-3, US Mk 82 or Snakeye, French Durandal, or similar). Inboard wing stations can carry 6 kg or 25 lb practice bombs, a pod containing eight Chinese 57-2 (57 mm), seven 68 mm, seven Norinco 90-1 (90 mm) or four 130-1 (130 mm) rockets. Centre wing stations can carry a 500 kg or 750 lb bomb, a BL755 600 lb cluster bomb, a Chinese 250-2 or -3 bomb, US Mk 82 or Snakeye, French Durandal or similar, or a C-801 anti-shipping missile. Normal bomb-carrying capacity is 1,000 kg (2,205 lb), maximum capacity 2,000 kg (4,410 lb). Instead of bombs, the centre wing stations can each carry a 760 litre drop tank (see Power Plant paragraph) or an ECM pod. The outboard wing stations can each be occupied by a 400 litre drop tank (when the larger tank is not carried on the centre wing station) or by air-to-air missiles such as the Chinese PL-2, PL-2B, PL-7, AIM-9 Sidewinder and Matra R.550 Magic. Within the overall maximum T-O weight, all stores mentioned can be carried provided that CG shift remains within the allowable operating range of 31 to 39 per cent of mean aerodynamic chord, and more than 22 external stores configurations are possible. The aircraft carries an SH-1J or ABS1A optical sight for level and dive bombing, or for air-to-ground rocket launching. Aircraft in Chinese service can carry a single 5-20 kT nuclear bomb.

DIMENSIONS, EXTERNAL (Q-5 II):

Wing span	9.70 m (31 ft 10 in)
Wing chord (mean aerodynamic)	3.097 m (10 ft 2 in)
Wing aspect ratio	3.37
Length overall: incl nose probe	16.255 (53 ft 4 in)
excl nose probe	15.415 m (50 ft 7 in)
Height overall	4.516 m (14 ft 9¾ in)
Wheel track	4.40 m (14 ft 5¼ in)
Wheelbase	4.01 m (13 ft 2 in)

AREAS:

Wings, gross	27.95 m² (300.85 sq ft)
Vertical tail surfaces (total)	4.64 m² (49.94 sq ft)
Horizontal tail surfaces:	
movable	5.00 m² (53.82 sq ft)
total, incl projected fuselage area	8.62 m² (92.78 sq ft)

WEIGHTS AND LOADINGS:

Weight empty	6,494 kg (14,317 lb)
Fuel: max internal	2,827 kg (6,232 lb)
two 400 l drop tanks	620 kg (1,367 lb)
two 760 l drop tanks	1,178 kg (2,597 lb)
max internal/external	4,005 kg (8,829 lb)
Max external stores load	2,000 kg (4,410 lb)
Max T-O weight: clean	9,530 kg (21,010 lb)
with max external stores	12,000 kg (26,455 lb)
Max wing loading: clean	341 kg/m² (69.9 lb/sq ft)
with max external stores	429 kg/m² (87.9 lb/sq ft)
Max power loading: clean	149.5 kg/kN (1.47 lb/lb st)
with max external stores	188.3 kg/kN (1.85 lb/lb st)

PERFORMANCE (at max clean T-O weight, with afterburning, except where indicated):

Max limiting Mach number (VNE)	Mach 1.5
Max level speed: at 11,000 m (36,000 ft)	Mach 1.12 (643 kts; 1,190 km/h; 740 mph)
at S/L	653 kts (1,210 km/h; 752 mph)
T-O speed: clean, 15° flap	162 kts (300 km/h; 186 mph)
with max external stores, 25° flap	178 kts (330 km/h; 205 mph)
* Landing speed: 25° flap, brake-chute deployed	150-165 kts (278-307 km/h; 172-191 mph)
* Max rate of climb at 5,000 m (16,400 ft)	4,980-6,180 m (16,340-20,275 ft)/min
Service ceiling	15,850 m (52,000 ft)
T-O run:	
* clean, 15° flap	700-750 m (2,300-2,460 ft)
with max external stores, 25° flap	1,250 m (4,100 ft)
Landing run:	
25° flap, brake-chute deployed	1,060 m (3,480 ft)
Combat radius with max external stores, afterburners off:	
lo-lo-lo (500 m; 1,640 ft)	216 n miles (400 km; 248 miles)
hi-lo-hi (8,000/500/8,000 m; 26,250/1,640/26,250 ft)	324 n miles (600 km; 373 miles)
Range at 11,000 m (36,000 ft) with max internal and external fuel, afterburners off	nearly 1,080 n miles (2,000 km; 1,243 miles)
g limits:	
with full load of bombs and/or drop tanks	5
with drop tanks empty	6.5
clean	7.5

* *depending upon airfield elevation and temperature*

VERIFIED

NAMC A-5K KONG YUN (CLOUD) UPGRADE

TYPE: Combat aircraft upgrade (Q-5/A-5).

PROGRAMME: In a programme similar to that of the A-5M (see next entry), initiated in June 1987, a French team led by Thomson-CSF carried out prototype upgrading of the Q-5 with a new avionics suite. This programme was intended to

NAMC A-5K Fantan upgraded with Thomson-CSF TMV 630. This programme was stopped in 1990

1994

Thomson-CSF TMV 630 airborne rangefinder
1994

fulfil a PLA Air Force requirement, but was abandoned in 1990.

AVIONICS: Thomson-CSF TMV 630 airborne rangefinder.

VERIFIED

NAMC A-5M UPGRADE

TYPE: Combat aircraft upgrade (Q-5 II).

PROGRAMME: This improved version of the Q-5 II, intended for export, is the subject of a programme started on 1 August 1986 between CATIC and Alenia to upgrade the aircraft's avionics by incorporating a new all-weather nav/attack system similar to that used in the AMX aircraft. The M (for Modified) version also has improved WP6A turbojets with dry and afterburning ratings of 29.42 kN (6,614 lb st) and 36.78 kN (8,267 lb st) respectively.

Two A-5M prototypes were ordered, and the first of these (converted from Q-5 IIs) made its initial flight on 30 August 1988, but was lost in a crash on 17 October 1988. The second prototype flew on 8 March 1989, and a replacement for the first was subsequently completed. On 19 February 1991 it was announced that development and flight testing had been successfully completed, and that an agreement was being negotiated to initiate production to meet potential orders. No order has been announced to date of publication.

DESIGN FEATURES: The nav/attack system is designed around two Singer central digital computers and a dual-redundant MIL-STD-1553B databus with plenty of growth potential. Other new sensors and equipment include a Pointer 2500 ranging radar, Litton LN-39A inertial navigation system, Alenia HUD-25 head-up display, air data computer, three-axis gyro package, RW-30 radar warning receiver, chaff/flare dispenser, HSI, AG-5 attitude indicator, static inverters, mode controls, and an interface unit linking these with the aircraft's existing AR-3201 VHF com radio, radio altimeter, radio compass, marker beacon receiver, IFF and armament system.

ARMAMENT: The number of external stores stations is increased to 12 by adding a fourth pylon beneath each outer wing, with some redistribution of the weapons carried on each wing station, and the PL-5B is added to the range of air-to-air missiles. External stores configurations are otherwise essentially the same as for the Q-5 II/A-5C, except that the underwing drop tanks can be of 1,140 litre (301 US gallon; 251 Imp gallon) capacity. The maximum external stores load of 2,000 kg (4,410 lb) remains unchanged.

DIMENSIONS, EXTERNAL: As for Q-5 II/A-5C except:

Length overall 15.366 m (50 ft 5 in)
Height overall 4.53 m (14 ft 10¼ in)

WEIGHTS AND LOADINGS:

Weight empty 6,728 kg (14,833 lb)
Max T-O weight: clean 9,769 kg (21,537 lb)
with max external stores 12,200 kg (26,869 lb)
Max wing loading: clean 349.5 kg/m² (71.58 lb/sq ft)
with max external stores 436.5 kg/m² (89.40 lb/sq ft)
Max power loading: clean 132.8 kg/kN (1.30 lb/lb st)
with max external stores 165.8 kg/kN (1.63 lb/lb st)

PERFORMANCE (estimated):

Max level speed at S/L at clean T-O weight 658 kts (1,220 km/h; 758 mph)
Max level flight Mach number at 11,000 m (36,000 ft) at clean T-O weight 1.205
Unstick speed, with afterburning:
no external stores, 15° flap 162 kts (300 km/h; 187 mph)
full external stores, 25° flap 174 kts (322 km/h; 200 mph)
Landing speed, 25° flap, brakes on and brake-chute deployed (depending upon AUW) 150-166 kts (278-307 km/h; 173-191 mph)
Max vertical rate of climb at 5,000 m (16,400 ft) at clean T-O weight, with afterburning 6,900 m (22,638 ft)/min
Service ceiling at clean T-O weight 16,000 m (52,500 ft)
T-O run, with afterburning:
no external stores, 15° flap 911 m (2,989 ft)
full external stores, 25° flap 1,250 m (4,101 ft)
Landing run, 25° flap, brakes on and brake-chute deployed 1,060 m (3,478 ft)
Combat radius with full external stores:
out at 8,000 m (26,250 ft), combat at 500 m (1,640 ft) and back at 11,000 m (36,000 ft) 280 n miles (518 km; 322 miles)
out, combat and back all at 500 m (1,640 ft) 174 n miles (322 km; 200 miles)
Range at 11,000 m (36,000 ft) with two 760 l (200 US gallon; 167 Imp gallon) drop tanks 1,080 n miles (2,000 km; 1,243 miles)

VERIFIED

NAMC/Alenia prototype A-5M improved capability attack aircraft, with new nav/attack system and digital avionics
1994

SAC

SHENYANG AIRCRAFT CORPORATION

PO Box 328, Shenyang, Liaoning
Telephone: 86 (24) 462680
Telex: 80018 SAMC CN
GENERAL MANAGER: Tang Qiansan
DEPUTY MANAGER: Xu Guosheng

One of the pioneer centres of fighter design in China, Shenyang now occupies a site area of more than 800 ha (1,976 acres) and has a workforce of over 20,000 people. Between 1956 and 1959 it produced 767 examples of the J-5 (licence MiG-17F), and from 1963 was the major producer (others were produced at Nanchang) of the J-6 series (Chinese versions of the MiG-19), including 634 tandem two-seat JJ-6 fighter/trainers. Initial development and production of the J-7 series (see Chengdu Aircraft Corporation entry) began at Shenyang, and SAC's principal current programme concerns the J-8 air superiority fighter. Aerospace products account for about 30 per cent of SAC's present activities, and include subcontract manufacture of cargo doors for the Boeing 757 and Boeing Canada Dash 8; rudders for the British Aerospace ATP; wing ribs and emergency exit hatches for the Airbus A320; tailcone, landing gear door and pylon components for the Lockheed C-130 Hercules; and other machined parts for BAe, Boeing, Deutsche Airbus and Saab-Scania.

VERIFIED

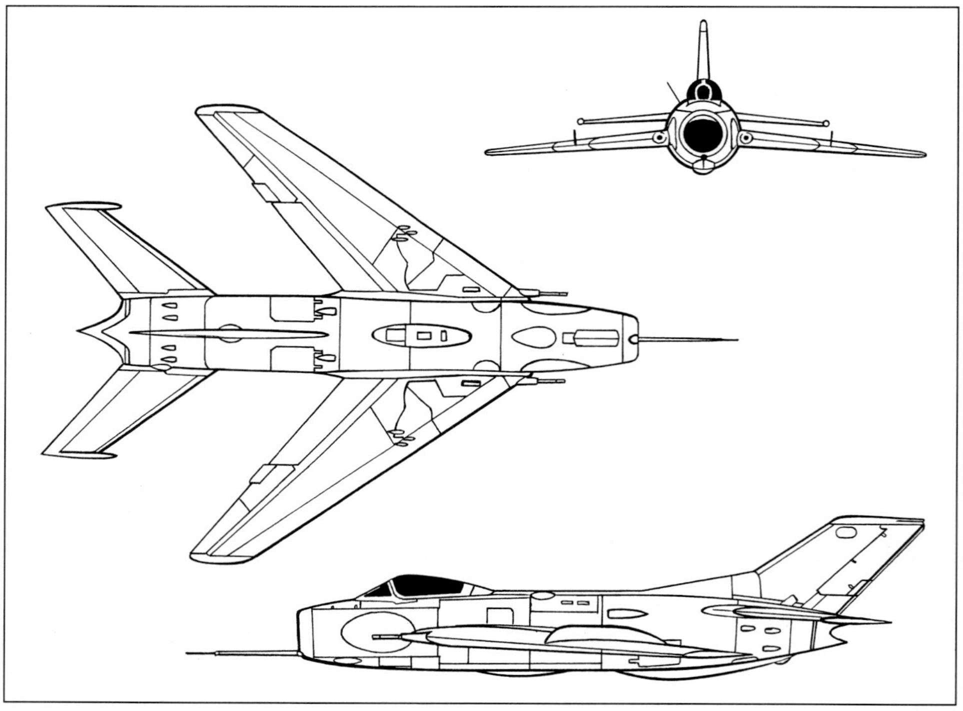

SAC F-6 single-seat day fighter *(Jane's/Mike Keep)*
1994

SHENYANG (MIKOYAN) J-6 (F-6)

Chinese name: Type 6 Jianjiji (Type 6 Fighter) and JJ-6 Jianjiji Jiaolinji-6
Export versions: F-6
NATO reporting names: Farmer-C (MiG-19SF) and Farmer-D (MiG-19PF)

TYPE: Single-seat day fighter, attack and tactical reconnaissance aircraft.

PROGRAMME: The J-6 (F-6) is basically a MiG-19 fighter built in China. Its original design was initiated by the Mikoyan bureau in Russia, where the I-350 (M) or I-360 prototype, with non-afterburning Mikulin AM-5 engines, flew for the first time on or about 18 September 1953.

In the Soviet Union the MiG-19 was phased out of production by the end of the 1950s, but a licence agreement for its manufacture in China was signed in January 1958. Many MiG-19s had been delivered to China in knocked down form before the deterioration of Moscow-Beijing relations. The designation **J-6** was given to the Chinese version of the MiG-19S fighter, which first flew in December 1961 and from mid-1962 it became standard equipment in the Chinese Air Force of the People's Liberation Army.

Production of the J-6 was stepped up from about 1966, and several thousand have since been built, including counterparts of the Soviet MiG-19PF/PM and MiG-19SF versions. China has also developed a number of variants. One of these is a tactical reconnaissance model, similar to the MiG-19R, with a camera pack in the lower forward fuselage. The J-6 from Xinjiang (Sinkiang) air base in Fujian (Fukien) Province, whose pilot defected to Taiwan in July 1977, was one of the original Soviet supplied MiG-19s converted to this configuration. It was equipped with a vertical/oblique camera installation in the lower forward fuselage, from which the underfuselage 30 mm cannon had been deleted. This aircraft was reported to belong to the 1st Reconnaissance Air Unit; the two wingroot cannon showed no traces of having been fired. Some of the limited all-weather models have a different centrebody

SAC F-6 single-seat day fighter *(Peter Steinemann)*

1994

SAC F-6 single-seat day fighters of Pakistan Air Force *(Peter Steinemann)*

1994

SAC F-6 day fighter showing bifurcated nose intake *(Peter Steinemann)*

1994

radar housing of longer, slim-conical form. Another variant in service is the **FT-6**, a trainer version with a tandem two-seat cockpit installation generally similar to that of the Soviet built MiG-19UTI. The Q-5 'Fantan-A' strike fighter, derived from the J-6, is described separately.

Immediately after the Indo-Pakistan war of September 1965, China offered J-6s to Pakistan. There were 40 supplied initially and the first Pakistan Air Force squadron was operational within a year. Subsequent deliveries brought the total of J-6s acquired by Pakistan to 135, and by 1981 these equipped nine PAF air defence and ground attack squadrons. PAF J-6s have been fitted with rails for a Sidewinder missile under each wing, and are currently being equipped to carry an underbelly auxiliary fuel tank. By Spring 1974 the Air Wing of the Tanzanian People's Defence Force had received sufficient J-6s for a single squadron; 40 have been delivered to Egypt, including some two-seat FT-6s, and well over 100 others to Albania, Bangladesh, Cambodia and Vietnam.

In early 1980, US aerospace industry visitors to China were told that Shenyang had the capacity to build 30 J-6s per month but it appeared to be involved primarily in overhaul programmes at that time. It is believed that production now continues at a rate sufficient only to provide aircraft for domestic replacements and to fulfil possible export orders.

VERSIONS: **F-6**: Export designation.

J-6: Corresponds to MiG-19S.

J-6A/B: Corresponds to MiG-19PM all-weather fighter.

J-6C: Day fighter version.

JJ-6: Two-seat fighter trainer.

JZ-6: Day/night reconnaissance version.

The initial production **MiG-19** day fighter, known to NATO as 'Farmer-A', was powered by two AM-5F (Forsirovanny: increased power) turbojets each rated at 22.06 kN (4,960 lb st) dry and 29.81 kN (6,702 lb st) with afterburning. This version, which had a maximum level speed of Mach 1.1 at altitude, began to enter service with the Soviet air defence force in early 1955. It was later redesignated **MiG-19F** after being refitted with R-9BF engines and other improvements developed for subsequent versions.

The latter included a ventral airbrake, dorsal spine fairing and enlarged dorsal fin, but the first major airframe improvement was the introduction of an all-moving tailplane to replace the original tailplane/elevators configuration, the latter having proved ineffective. This brought a change of designation to **MiG-19S** (Stabiliser: tailplane), and in this form the day fighter received the NATO reporting name 'Farmer-C'. Other changes introduced in the MiG-19S included Tumansky (Mikulin) R-9B engines (rated at 25.50 kN (5,732 lb st dry and 32.36 kN; 7,275 lb st with afterburning); built-in armament of three 30 mm NR-30 cannon (instead of the single 37 mm N-37 and two wing-mounted 23 mm guns in the original MiG-19); and an attachment under each wing for a bomb or air-to-surface rocket. Normal T-O weight of this version was 7,400 kg (16,314 lb). When fitted with R-9BF engines, with a modified afterburner, it has a 200 kg (441 lb) greater T-O weight and is designated **MiG-19SF**.

Meanwhile, in about 1957, a version with limited all-weather capability was put into production as the **MiG-19P** (Perekhvatchik: interceptor), and was allocated the NATO reporting name 'Farmer-B'. Powered by R-9BF engines, it had a small Izumrud (Emerald) radar scanner inside its engine air intake and a ranging unit in the intake top lip. An armament of two wing-mounted NR-23 cannon was carried. The later **MiG-19PF** differed in having NR-30 wing guns and two 212 mm ARS-212 underwing rockets; the **MiG-19PM** (Modifikatsirovanny: modified) differed from the PF in having four first generation radar-homing missiles (NATO 'Alkali') instead of guns. The PF and PM were both given the NATO reporting name 'Farmer-D'.

Other versions produced in the USSR included the **MiG-19R** (Razvedchik: reconnaissance), with cameras in the bottom of the front fuselage, and two wing-mounted NR-30 cannon; and the **MiG-19UTI** tandem two-seat operational trainer.

OPERATORS: Versions of the J-6/F-6 are in service with the armed forces of the following countries: Albania (25), Bangladesh (25+), China (3,000), Egypt (80+), North Korea (100), Pakistan (90+), Sudan (10), Tanzania (10), and Zambia (10+).

The following description is based on known details of the basic MiG-19SF, modified where possible to apply specifically to the Chinese J-6:

DESIGN FEATURES: Cantilever all-metal mid-wing monoplane. Wing section TsAGI S-12S at root, SR-7S at tip. Thickness/chord ratio 8.74 per cent (root), 8 per cent (tip). Anhedral 4° 30′. Sweepback at quarter-chord 55°. Large full chord boundary layer fence, 32 mm (1¼ in deep), above each wing at mid-span to enhance aileron effectiveness. Fuselage has top and bottom pen-nib fairings aft of nozzles. Entire rear fuselage detaches at wing trailing-edge for engine servicing. Shallow ventral strake under rear of fuselage. Upward-hinged pitot boom mounted on lower lip of nose intake. Anti-flutter weight projecting forward from each tailplane tip. Sweepback on tailplane leading-edges 57° 37′. Fin tip incorporates antenna for tail warning radar.

FLYING CONTROLS: Entire wing trailing-edge formed by aerodynamically balanced aileron (outboard) and large Fowler-type TsAGI flap, both hydraulically powered. Compressed air emergency extension system for flaps. Trim tab in port aileron. Plate-type spoiler beneath each wing, forward of aileron, to improve lateral control. Spoiler actuation is coupled with that of aileron, and takes place only when aileron is deflected downwards. Forward-hinged door-type airbrake, operated hydraulically, one each side of fuselage aft of wing trailing-edge. Forward-hinged perforated door-type airbrake under centre-fuselage. The tail unit has hydraulically actuated one-piece horizontal surfaces, with electrical emergency actuation. Stick-to-tailplane gearing via electromechanical linkage, reduces required stick forces during high-*g* manoeuvres. Mass balanced rudder, with electrically actuated trim tab. Large dorsal fin and dorsal spine enclosing actuating rods for tail control surfaces.

STRUCTURE: The wing is a three-spar structure, with auxiliary spar, ribs, and stressed skin of 1.5-2.0 mm sheet duralumin. Main spar is of 30 HGNSA; auxiliary spar provides mountings for wing guns and main landing gear members. The fuselage is a conventional semi-monocoque structure of circular section, with divided (bifurcated) air intake in nose and side by side twin orifices at rear. The tail unit is a conventional all-metal structure, of similar construction to wings with stressed skin of 1.2 mm duralumin.

SAC F-6 day fighter of the Bangladesh Air Force *(Peter Steinemann)*

1995

LANDING GEAR: Wide track retractable tricycle type, with single wheel on each unit. Hydraulic actuation, nosewheel retracting forward, main units inward into wing roots. Pneumatic emergency extension system. All units of levered-suspension type, with oleo-pneumatic shock-absorbers. Nose unit is steerable, self-centring, and fitted with hydraulic shimmy damper. Main units have KT-37 wheels, and 660 × 220 mm tyres, pressure 8.83 bars (128 lb/sq in). Nosewheel tyre size 500 × 180; pressure 5.88 bars (85.3 lb/sq in). Pneumatically operated brakes on main wheels, with pneumatic emergency back-up. Pneumatically deployed brake parachute housed in bottom of rear fuselage above ventral strake. Small tail bumper.

POWER PLANT: Two Shenyang built WP-6 developments of Tumansky (Mikulin bureau) R-9BF turbojet, each rated at 25.50 kN (5,732 lb st) dry and 31.88 kN (7,167 lb st) with afterburning. Hydraulically actuated nozzles. Two main fuel tanks in tandem between cockpit and engines, and two smaller tanks under forward end of engine tailpipes, total capacity 2,155 litres (569 US gallons; 474 Imp gallons). Provision for two 880 litre (211 US gallon; 176 Imp gallon) or 1,520 litre (400 US gallon; 334 Imp gallon) underwing drop tanks, raising maximum total fuel capacity to 3,755 litres (991 US gallons; 826 Imp gallons) and 5,195 litres (1,371 US gallons; 1,143 Imp gallons) respectively; provision on Pakistan Air Force J-6s for underfuselage tank.

ACCOMMODATION: Pilot only, on Martin-Baker PKD10 zero/zero ejection seat, under rearward-sliding blister canopy. In emergency, canopy is jettisoned by an explosive charge at the lock, after which it is carried away by the slipstream. Fluid anti-icing system for windscreen. Cockpit pressurised, heated and air conditioned.

SYSTEMS: Cockpit pressurised by air conditioning system mounted on top of fuselage aft of cockpit, using compressor bleed air. Constant temperature maintained by adjustable electric thermostat. Two independent hydraulic systems. Main system, powered by pump on starboard engine, actuates landing gear retraction and extension, flaps, spoilers, airbrakes and afterburner nozzle mechanism. System for tailplane and aileron boosters powered by a pump on the port engine, and can also be supplied by the main system should the booster system fail. Electrical system powered by two DC starter/generators, supplemented by a battery, providing 27 V DC, and 115 V 400 Hz and 36 V 400 Hz AC.

AVIONICS AND EQUIPMENT: Standard avionics include VHF transceiver, blind-flying equipment, radio compass, radio altimeter, tail-warning system, navigation lights, taxying light on nosewheel leg, and landing light in bottom of front fuselage.

ARMAMENT: Installed armament of two or three 30 mm NR-30 belt-fed cannon, one in each wingroot (deleted on FT-6) and (not on MiG-19PF) one under starboard side of nose. Aircraft supplied to Pakistan have an attachment under each wing for a Harbin built AIM-9 Sidewinder air-to-air missile, outboard of drop tank. More usual is the provision of one or two attachments inboard of each tank. Packs of eight air-to-air rockets can be carried on these inboard points, or on the drop tank attachments. Alternative underwing loads can include four air-to-air guided weapons, two 250 kg (or 500 lb) bombs, or single rockets of up to 212 mm calibre. Optical gunsight. Gun camera in top lip of air intake of MiG-19SF; Izumrud airborne interception radar in centre of nose intake of MiG-19PF, with ranging unit in top lip of intake.

DIMENSIONS, EXTERNAL:

Wing span	9.20 m (30 ft 2¼ in)
Wing chord at root	3.73 m (12 ft 3¾ in)
Wing chord at tip	1.278 m (4 ft 2¼ in)
Wing chord, mean	3.02 m (9 ft 10¾ in)
Wing aspect ratio	3.24
Length overall (MiG-19SF):	
incl nose probe	14.90 m (48 ft 10½ in)
excl nose probe	12.60 m (41 ft 4 in)
Length overall (FT-6):	
excl nose probe	approx 13.44 m (44 ft 1 in)
Length of fuselage	11.82 m (38 ft 9½ in)
Fuselage: Max diameter	1.45 m (4 ft 9 in)
Height overall	3.88 m (12 ft 8¾ in)
Tailplane span	5.00 m (16 ft 4¾ in)
Wheel track	4.15 m (13 ft 7½ in)

AREAS:

Wings, gross	25.00 m² (269 sq ft)
Ailerons (total)	1.56 m² (16.79 sq ft)
Trailing-edge flaps (total)	3.44 m² (37.03 sq ft)
Airbrakes (three, total)	1.50 m² (16.15 sq ft)
Ventral strake	0.614 m² (6.61 sq ft)
Fin, incl dorsal fin	4.62 m² (49.73 sq ft)
Rudder, incl tab	1.90 m² (20.45 sq ft)
Tailplane	4.62 m² (49.73 sq ft)

WEIGHTS AND LOADINGS:

Weight empty, nominal	5,760 kg (12,700 lb)
Normal T-O weight	7,600 kg (16,755 lb)
Max T-O weight	8,700 kg (19,180 lb)
Max wing loading	348 kg/m² (71.28 lb/sq ft)
Max power loading	136.4 kg/kN (1.34 lb/lb st)

PERFORMANCE:

Max level speed at 10,000 m (32,800 ft)	783 kts (1,452 km/h; 902 mph)
Cruising speed	512 kts (950 km/h; 590 mph)
Stalling speed, flaps up	189 kts (350 km/h; 218 mph)
Landing speed	127 kts (235 km/h; 146 mph)
Max rate of climb at S/L	6,900 m (22,635 ft)/min
Time to service ceiling	8 min 12 s
Service ceiling	17,900 m (58,725 ft)
Absolute ceiling	19,870 m (65,190 ft)
T-O run, with afterburning	515 m (1,690 ft)
T-O run, with underwing tanks, no afterburning	900 m (2,953 ft)
T-O to 25 m (82 ft), with afterburning	1,525 m (5,000 ft)
T-O to 25 m (82 ft), with underwing tanks, no afterburning	1,880 m (6,170 ft)
Landing from 25 m (82 ft), with brake-chute	1,700 m (5,580 ft)
Landing from 25 m (82 ft), without brake-chute	1,980 m (6,495 ft)
Landing run, with brake-chute	600 m (1,970 ft)
Landing run, without brake-chute	890 m (2,920 ft)
Combat radius with 800 l external tanks	370 n miles (685 km; 426 miles)
Normal range at 14,000 m (46,000 ft)	750 n miles (1,390 km; 863 miles)
Max range with 1,520 l external tanks	1,187 n miles (2,200 km; 1,366 miles)
Max endurance at 14,000 m (46,000 ft)	2 h 38 min

VERIFIED

SAC J-8

Chinese name: Jianjiji-8 (Fighter aircraft 8) or Jian-8
Westernised designation: F-8
NATO reporting name: Finback

TYPE: Twin-engine air superiority and secondary ground attack aircraft.

PROGRAMME: Development of the J-8 began at Shenyang in 1964, the first of two original prototypes making its first flight on 5 July 1969. Although all other J-8 activity was suspended during the cultural revolution, flight trials were allowed to continue, and these prototypes accumulated 663 hours of flying, in 1,025 flights, before production was eventually authorised in 1979.

VERSIONS: **J-8 ('Finback-A'):** Initial clear weather day fighter, powered by two Liyang (LMC) WP7B turbojets

SAC J-8 I Finback-A powered by two Liyang (LMC) WP7B turbojets

1994

(each 43.15 kN; 9,700 lb st dry and 59.82 kN; 13,448 lb st with afterburning) and armed with single twin-barrel 23 mm cannon and four wing-mounted PL-2B air-to-air missiles. Single intake in nose with conical centrebody. Approved for production in 1979, but built in small numbers only. Retrofitted from 1984 with Sichuan SR-4 fire control radar in intake centrebody. In service.

J-8 I ('Finback-A'): Improved version of J-8, with same power plant and armament but fitted from outset with Sichuan SR-4 radar in intake centrebody. Three prototypes built, one being lost before flight testing; first flight, by second aircraft, made on 24 April 1981. Initial batch production authorised in July 1985. More than 100 now in service, including upgraded J-8s, by early 1990, but gradually being supplanted by J-8 II.

J-8 II ('Finback-B'): All-weather dual-role version (high-altitude interceptor and ground attack), embodying some 70 per cent redesign compared with J-8 I. Main configuration change is to 'solid' nose and twin lateral air intakes, to provide nose space for fire control radar and other avionics, and increased air flow for more powerful WP13A II turbojets. Four prototypes built (first flight 12 June 1984), plus two for static/fatigue testing. In production and service; 'several dozen' built by early 1990, but being manufactured in small economic batches rather than at a steady monthly rate.

UPGRADES: **SAC/Grumman:** Two J-8 IIs were delivered to Grumman Aerospace Corporation for prototype avionics upgrading under a joint Chinese/US programme known as Peace Pearl, but this programme was embargoed by the US government in mid-1989 and cancelled by the Chinese government in 1990.

The following details should be regarded as provisional estimates only.

DIMENSIONS, EXTERNAL:

Wing span	10.0 m (32 ft 9¾ in)
Length overall	19.0 m (62 ft 4 in)
Height overall	5.2 m (17 ft 0¾ in)

AREA:

Wings, gross	40.0 m² (430 sq ft)

WEIGHTS:

Weight empty	12,000 kg (26,455 lb)
Normal T-O weight	17,000 kg (37,480 lb)
Max T-O weight	19,000 kg (41,890 lb)

PERFORMANCE:

Max level speed at 12,000 m (39,370 ft)	Mach 2.3 (1,320 kts; 2,450 km/h; 1,520 mph)
Max rate of climb at S/L	12,000 m (39,370 ft)/min
Service ceiling	18,000 m (59,050 ft)
Range	500-100 n miles (925-1,850 km; (575-1,150 miles)

VERIFIED

SAC J-8 II Finback-A improved version of the J-8 fitted with Sichuan SR-4 radar in intake centrebody
1994

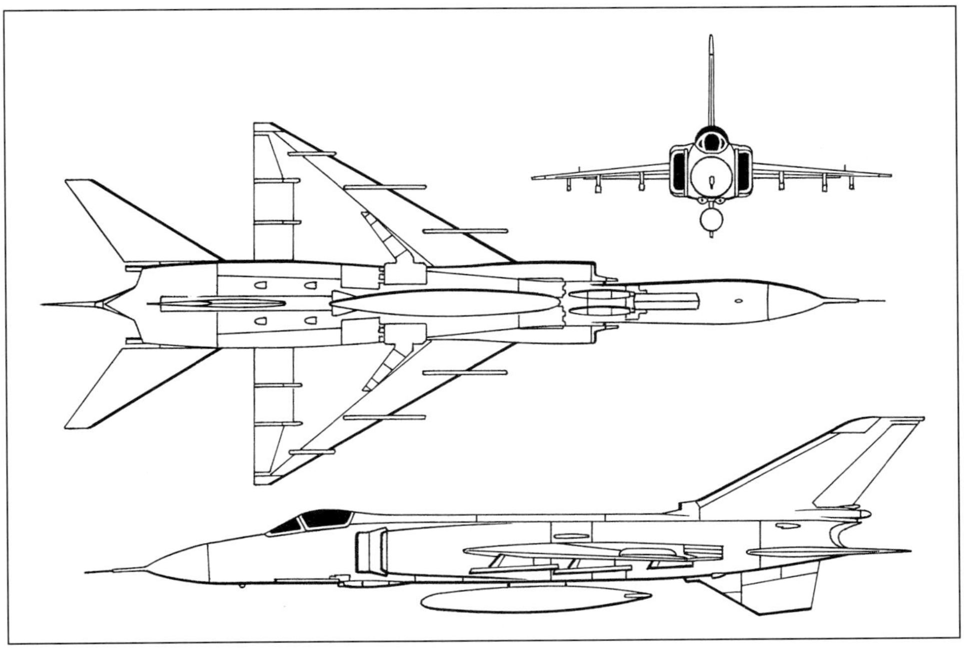

SAC J-8 II Finback-A air superiority and secondary ground attack aircraft ***(Jane's/Mike Keep)***
1994

XAC

XIAN AIRCRAFT COMPANY

PO Box 140, Xian, Shaanxi 710000
Telephone: 86 (29) 7214959, 7214960 and 7216929
Fax: 86 (29) 6202263 and 6203707
Telex: 70101 XAC CN

PRESIDENT: Wang Qinping
SALES MANAGER: Wang Zhigang

Aircraft factory established at Xian 1958; current XAC has covered area of some 300 ha (741.3 acres) and 1992 workforce of 19,730, of whom about 90 per cent are engaged in aircraft production. Earlier programmes have included licence production of Tu-16 twin-jet bomber (as H-6); major current programmes concern JH-7 fighter-bomber and Y-7 transport series of An-24/26 derivatives. Subcontract work includes glassfibre header tanks, water float pylons, ailerons and various doors for Canadair CL-215/415 amphibians and various components for Airbus A300, Boeing 737/747 and ATR 42.

UPDATED

XAC (ANTONOV) Y-7

Chinese name: Yunshuji-7 (Transport aircraft 7) or Yun-7

NATO reporting name: Coke

TYPE: Twin-turboprop short/medium-range transport aircraft.

PROGRAMME: Civil and military examples of the Antonov An-24 twin-turboprop transport aircraft (40 of which were purchased from the former USSR) have been in service with CAAC and the PLA Air Force since about 1970. The Y-7, a reverse engineered version of this 48/52-passenger aircraft, received its Chinese certificate of airworthiness in 1980, following the completion of three prototypes (first flight 25 December 1970) plus two additional airframes for static and fatigue testing. Public debut by a preproduction Y-7 took place on 17 April 1982, and production started later that year.

VERSIONS: **Y-7:** Initial production version: first flight announced 1 February 1984. Deliveries to CAAC began shortly afterwards; scheduled passenger services began 29 April 1986. There were 20 built, including at least two for PLA Naval Air Force.

Y7-100: Improved version, developed in 1985 by conversion of one Y-7 (B-3499) by Hong Kong Aircraft Engineering Company (HAECO). Winglets added; new three-person flight deck layout, all-new cabin interior with 52 reclining seats, windscreen de-icing, new HF/VHF communications, new navigation equipment, and installation of oxygen, air data and environmental control systems. Meets BCAR standards. First production Y7-100 flown in late 1985; certification awarded 23 January 1986. Approximate total of 45 delivered December 1990; production continuing.

Y7-200A: Improved Y7-100, with Pratt & Whitney Canada turboprops and Collins EFIS 85/86 avionics. First flight 26 December 1993.

Y7-200B: Improved Y7-100 for domestic market, with Dongan WJ5E turboprops and new four-blade propellers, more advanced avionics (Collins EFIS 85/86), higher maximum lift coefficient, improved stall characteristics, lower fuel consumption. Overall length increased by 0.74 m (2 ft 5¼ in); empty weight reduced by 500 kg (1,102 lb). First flight 26 November 1990; deliveries scheduled to begin in 1992. Expected performance improvements include reduction of stalling speed from 92 knots (170 km/h; 106 mph) to 76 knots (140 km/h; 87 mph); a 300 m (985 ft) reduction in field length; a range increase (with maximum payload) of 162 n miles (300 km; 186 miles); and Cat. II landing capability.

Y7H-500: Military and civil cargo version, derived from Antonov An-26.

OPERATORS: Versions of the Y-7 are in service with the armed forces of China (2).

The following description is based on the An-24RV, modified where possible to refer to the Y-7 and Y7-100.

DESIGN FEATURES: Cantilever high-wing monoplane, with 2° 12′ 2″ anhedral on outer panels. Incidence 3°. Sweepback at quarter-chord on outer panels 6° 50′. Winglet at each tip (being retrofitted also on Y-7). Tailplane dihedral 9°.

FLYING CONTROLS: Mass balanced servo-compensated ailerons, with large glassfibre trim tabs. Hydraulically operated Fowler flaps along entire wing trailing-edges inboard of unpowered ailerons; single-slotted flaps on centre-section, double-slotted outboard of nacelles. Servo tab and trim tab in each aileron. All tailplane controls operated manually. Balance tab in each elevator, trim tab and spring tab in rudder.

STRUCTURE: The wing is an all-metal two-spar structure, built in five sections: constant-chord centre-section, two tapered inner wings and two tapered outer panels. The fuselage is an all-metal semi-monocoque structure in front, centre and rear portions of bonded/welded construction. The tail unit is a cantilever all-metal structure, with a single ventral fin.

LANDING GEAR (An-24RV): Retractable tricycle type with twin wheels on all units. Hydraulic actuation, with emergency gravity extension. All units retract forward. Mainwheels are size 900 × 300 mm, tyre pressure 5.39-5.88 bars (78.2-85.3 lb/sq in); nosewheel size 700 × 250 mm, tyre pressure 3.92 bars (56.8 lb/sq in). (Mainwheel tyre pressures variable to cater for different types of runway.) Disc brakes on mainwheels; steerable and castoring nosewheel unit.

POWER PLANT: Two Harbin WJ-5A-1 turboprops, each rated at 2,080 kW (2,790 shp) for T-O and 1,976 kW (2,650 shp) at ISA + 23°C; four-blade constant-speed fully feathering propellers with elongated spinners. Fuel in integral wing tanks immediately outboard of nacelles, and four bag-type tanks in centre-section, total capacity 5,550 litres (1,466 US gallons; 1,220 Imp gallons). Provision for four additional tanks in centre-section. Pressure refuelling point in starboard engine nacelle; gravity refuelling point above each tank. One 8.83 kN (1,985 lb st) Type PY 19A-300 auxiliary turbojet (or Chinese equivalent) in starboard engine nacelle for engine starting, to improve take-off and in-flight performance, and to reduce stability and handling problems if one turboprop engine fails in flight.

ACCOMMODATION: Crew of three (Y7-100) or five (Y-7) on flight deck plus cabin attendant. Standard layout has four-abreast seating, with centre aisle, for 48 (Y-7) or 52 (Y7-100) in air conditioned soundproofed (by Tracor) and pressurised cabin. Galley (by Lermer) and toilet at rear on starboard side. Baggage compartments forward and aft of passenger cabin, plus overhead stowage bins in cabin. Passenger door on port side, at rear of cabin, is of airstair type. Doors to forward and rear baggage compartments on starboard side. All doors open inward. Electric windscreen de-icing in (Y7-100).

SYSTEMS: Hamilton Standard environmental control system in Y7-100 (cabin pressure differential in An-24RV is 0.29 bars (4.27 lb/sq in). Main and emergency hydraulic systems, pressure 152 bars (2,200 lb/sq in), for landing gear actuation, nosewheel steering, flaps, brakes, windscreen wipers and propeller feathering. Electrical system in An-24RV includes two 27 V DC starter/generators, two alternators to provide 115 V 400 Hz AC supply, and two inverters for 36 V 400 Hz three-phase AC. Puritan-Bennet passenger oxygen system optional in Y7-100.

AVIONICS AND EQUIPMENT: Standard communications equipment comprises Collins 618M-3 dual VHF, Collins 628T-3 single HF, Becker audio selection and intercom, and Sundstrand AC-557C cockpit voice recorder. Standard navigation equipment comprises dual ADI-84A, dual EHSI-74 electronics HSI, dual RMI-36, FGS-65 flight guidance system, dual 51RV-4B VOR/ILS, dual DME-4, dual DF-206, ADF, 860F-4 radio altimeter, 621A-6A ATC transponder, 51Z-4 marker beacon receiver and CWC-80 instrument warning system, all by Collins; Litton LTN-211 VLF/Omega navigation system; Sperry MHRS dual compass system, dual altitude reference and Primus 90 colour weather radar; IDC air data system; Sundstrand UFDR flight data recorder; and KJ-6A autopilot. Gables control units. Other instrumentation by Gould, IDC, Sextant Avionique and Smiths.

DIMENSIONS, EXTERNAL:

Wing span: Y-7	29.20 m (95 ft 9½ in)
Y7-100 over winglets	29.637 m (97 ft 2¾ in)
Wing chord: at root	3.50 m (11 ft 5¾ in)
at tip	1.095 m (3 ft 7 in)
Wing aspect ratio	11.7
Length overall	24.218 m (78 ft 9 in)
Height overall	8.553 m (28 ft 0¾ in)
Fuselage: Max width	2.90 m (9 ft 6¼ in)
Max depth	2.50 m (8 ft 2½ in)
Tailplane span	9.08 m (29 ft 9½ in)
Wheel track (c/l of shock struts)	7.90 m (25 ft 11 in)
Wheelbase	7.90 m (25 ft 11 in)
Passenger door (port, rear): Height	1.40 m (4 ft 7 in)
Width	0.75 m (2 ft 5½ in)
Height to sill	1.40 m (4 ft 7 in)
Baggage compartment door (stbd, fwd):	
Height	1.10 m (3 ft 7¼ in)
Width	1.20 m (3 ft 11¼ in)
Height to sill	1.30 m (4 ft 3 in)
Baggage compartment door (stbd, rear):	
Height	1.41 m (4 ft 7½ in)
Width	0.75 m (2 ft 5½ in)

DIMENSIONS, INTERNAL:

Cabin:	
Length, incl flight deck: Y-7	9.90 m (32 ft 5¾ in)
Y7-100	10.50 m (34 ft 5½ in)
Max width	2.80 m (9 ft 2¼ in)
Max height	1.90 m (6 ft 2¾ in)
Volume: Y7-100	56.0 m^3 (1,987 cu ft)
Baggage compartment volume (Y7-100):	
fwd	4.50 m^3 (159 cu ft)
rear	6.70 m^3 (237 cu ft)

AREAS:

Wings, gross	74.98 m^2 (807.1 sq ft)
Vertical tail surfaces (total)	13.38 m^2 (144 sq ft)
Horizontal tail surfaces (total)	17.23 m^2 (185.5 sq ft)

WEIGHTS AND LOADINGS:

Operating weight empty: Y-7	14,235 kg (31,383 lb)
Y7-100	14,900 kg (32,849 lb)
Max fuel (both)	4,790 kg (10,560 lb)
Max payload (both)	4,700 kg (10,362 lb)
Max T-O and landing weight (both)	21,800 kg (48,060 lb)
Max wing loading (both)	290.7 kg/m^2 (59.6 lb/sq ft)
Max power loading (both)	5.24 kg/kW (8.61 lb/shp)

PERFORMANCE (Y7-100 except where indicated):

Max level speed	279 kts (518 km/h; 322 mph)
Max cruising speed at 4,000 m (13,125 ft)	
Y-7	258 kts (478 km/h; 297 mph)
Y7-100	261 kts (484 km/h; 301 mph)
Econ cruising speed at 6,000 m (19,685 ft)	228 kts (423 km/h; 263 mph)
Max rate of climb at S/L	458 m (1,504 ft)/min
Service ceiling	8,750 m (28,700 ft)
Service ceiling, one engine out	3,900 m (12,800 ft)
T-O run at S/L, FAR Pt 25: ISA	1,248 m (4,095 ft)
ISA + 20°C	1,398 m (4,590 ft)
Landing run	620 m (2,035 ft)
Range: max (52-passenger) payload	491 n miles (910 km; 565 miles)
max standard fuel	1,025 n miles (1,900 km; 1,180 miles)
standard and auxiliary fuel	1,306 n miles (2,420 km; 1,504 miles)

UPDATED

XAC Y7-100, distinguishable by the winglets, is the improved version of the Y-7

1993

XAC Y7-100 twin-turboprop transport aircraft

1994

CZECH REPUBLIC

AERO

AERO VODOCHODY AKCIOVÁ SPOLEČNOST (Aero Vodochody Company Ltd)

25070 Odolena Voda
Telephone: 42 (2) 664 10041
Fax: 42 (2) 685 8041
Telex: 121169 AERO C

MANAGING DIRECTOR: Ing Zděnek Chalupník
CHIEF DESIGNERS:
Ing Dobroslav Rak (training aircraft)
Ing Jan Mikula (passenger aircraft)

This factory was established on 1 July 1953. Aero produced together with LET Kunovice about 3,600 L-29 Delfin jet trainers from 1963-74 and has produced more than 2,800 L-39s since 1972.

VERIFIED

AERO L-29 DELFIN

TYPE: Two-seat jet basic and advanced trainer.

PROGRAMME: The L-29 was evolved by a team led by K. Tomas and Z. Rublic. The first prototype, the XL-29 flew for the first time on 5 April 1959, powered by a Bristol Siddeley Viper turbojet engine. The second prototype, equipped with a Czech M 701 turbojet engine began its flight trials in July 1960. The preproduction version (third prototype) completed its tests in 1961, and the L-29 was approved for quantity production.

Manufacture was centred at the Vodochody (Aero) and Kunovice plants which were devoted almost entirely to producing the L-29 to meet large orders from the Czechoslovak Air Force, the former Soviet Union, Bulgaria, Hungary, the German Democratic Republic, Syria, Egypt, Romania, Iraq, Indonesia, Nigeria and Uganda.

The first production aircraft was completed in April 1963, a month ahead of schedule, and by the beginning of 1974 more than 2,000 were supplied to the USSR for use by the Soviet and other Warsaw Pact air forces, the remainder being for the Czechoslovak Air Force or for export.

Production of the L-29 Delfin ended during 1974.

VERSIONS: **L-29R:** Counter-insurgency version, with nose cameras and underwing stores.

L-29A: Single-seat aerobatic version.

OPERATORS: Versions of the L-29 are in service with the armed forces of the following countries: Afghanistan (24), Azerbaijan (60), Czech Republic (13), Egypt (50), Ghana (8), Mali (6), Romania (30), Slovakia (16), Syria (60), Uganda (4 in storage) and Ukraine (20).

DESIGN FEATURES: Cantilever mid-wing monoplane. Wing section NACA 64_2A217 at root, NACA 64_2A212 at tip. Dihedral 0° on centre section, 3° on outer wings. Incidence 1° 30′. Variable incidence tailplane and elevator mounted at tip of fin.

FLYING CONTROLS: All-metal ailerons and hydraulically operated Fowler flaps. Trim tab in each aileron. Hydraulically operated sideways opening perforated airbrakes on each side of rear fuselage. Tailplane incidence is linked with landing flaps. Trim tabs in elevator. Adjustable tab in rudder.

STRUCTURE: All-metal stressed skin wing structure with single main-spar at 40 per cent chord. The fuselage is an all-metal monocoque structure of circular section. Front portion forms pressurised compartment for the crew, radio equipment and electronics. Centre-fuselage is integral with wing centre-section and contains main fuel tanks. Engine mounting is attached to rear bulkhead of centre-fuselage. Rear fuselage is connected to centre portion through eight attachment points to permit quick removal for engine servicing. Zones of high stress have steel alloy reinforcement.

LANDING GEAR: Retractable tricycle type. Hydraulic retraction, mainwheels inward into wingroots, nosewheel rearward into fuselage. Oleo-pneumatic shock-absorbers. Single wheel on each unit. Barum low-pressure tyres, size 600 × 180 on main wheels, 420 × 150 on nosewheel. Pneumatic brakes.

POWER PLANT: One M 701c 500 turbojet engine rated at 890 kg/kN (1,960 lb) at 15,400 rpm, for take-off. Two fuel tanks in fuselage, capacity 690 litres (182.5 US gallons; 152 Imp gallons) and 360 litres (94.9 US gallons; 79 Imp gallons) respectively. Total internal fuel capacity 1,050 litres (277.4 US gallons; 231 Imp gallons). Total usable capacity 962 litres (254 US gallons; 211 Imp gallons). Provision for two 150 litre (39.6 US gallon; 33 Imp gallon) underwing auxiliary tanks. Inverted-flying tank in rear main tank permits 15 seconds of inverted flight. All tanks of aluminium alloy.

ACCOMMODATION: Crew of two in tandem on synchronised ejection seats in air conditioned and pressurised cabin. Rear seat raised 15 cm (6 in) higher than front seats. Fittings for *g* suits. Canopy over front cockpit opens sideways to starboard. Rearward-sliding canopy over rear cockpit.

ARMAMENT: Provision for gun camera and gunsight, and either two bombs of up to 100 kg, eight air-to-ground rockets or two 7.62 machine gun pods under the wings.

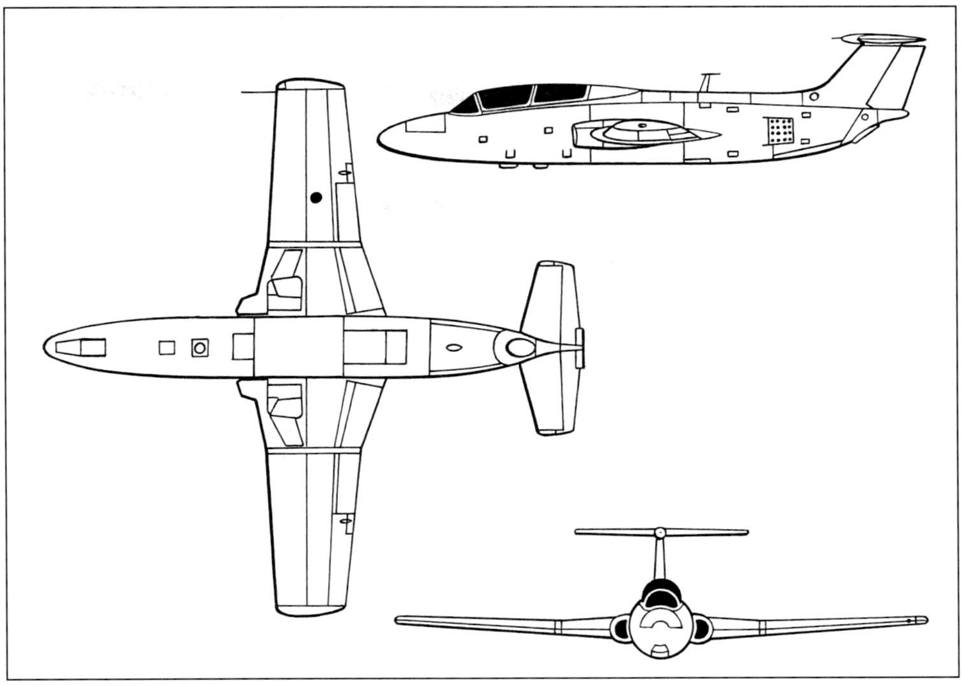

Aero Vodochody L-29 Delfin *(Jane's/Mike Keep)*

1994

Aero Vodochody L-29 Delfin *(Peter March)*

1994

Aero Vodochody L-29 Delfin two-seat jet trainer

1994

DIMENSIONS, EXTERNAL:
Wing span 10.29 m (33 ft 9 in)
Wing chord: at root 2.70 m (8 ft 10 in)
at tip 1.40 m (4 ft 7 in)
Wing aspect ratio 5.36
Length overall 10.81 m (35 ft 5½ in)
Height over tail 3.13 m (10 ft 3 in)
Tailplane span 3.34 m (10 ft 11½ in)
Wheel track 3.44 m (11 ft 3½ in)
Wheelbase 3.90 m (12 ft 9½ in)

WEIGHTS AND LOADINGS:
Weight empty 2,280 kg (5,027 lb)
Normal T-O weight 3,280 kg (7,231 lb)
Max permissible loaded weight with external tanks 3,540 kg (7,804 lb)
Max wing loading 179 kg/m² (36.7 lb/sq ft)
Max power loading 3.95 kg/kN (3.95 lb/lb st)

PERFORMANCE (at AUW of 3,250 kg; 7,165 lb):
Max never-exceed speed 442 kts (820 km/h; 510 mph)
Limiting Mach number 0.75
Max level speed at 5,000 m (16,400 ft) 353 kts (655 km/h; 407 mph)
Max level speed at S/L 332 kts (615 km/h; 382 mph)
Landing speed 78 kts (145 km/h; 90 mph)
Stalling speed: flaps up 87 kts (160 km/h; 100 mph)
flaps down 71 kts (130 km/h; 81 mph)
Max rate of climb at S/L 840 m (2,755 ft)/min
Service ceiling 11,000 m (36,000 ft)
T-O run 550 m (1,805 ft)
Landing run 440 m (1,444 ft)
Max range on internal fuel at 5,000 m (16,400 ft) 344 n miles (640 km; 397 miles)
Max range with external tanks at 5,000 m (16,400 ft) 480 n miles (894 km; 555 miles)
Endurance on internal fuel at 247 kts (460 km/h; 285 mph) 1 h 47 min
Endurance with external tanks 2 h 30 min

VERIFIED

LET

LET AERONAUTICAL WORKS

68604 Kunovice
Telephone: 42 (632) 411111
Fax: 42 (632) 61352

DIRECTOR AND GENERAL MANAGER: Zdeněk Pernica
MANAGING DIRECTOR: Josef Skrášek
TECHNICAL DIRECTOR: Jan Fridrich
CHIEF DESIGNER: Miroslav Pešák

Established at Kunovice in 1950, producing Yak-11 trainer as C-11; now producing L-410UVP-E and developing L-610; also produces equipment for radar and computer technology. LET Aeronautical Works is now a joint stock company fully owned by Aero Holding AS. A process of asset and debt restructuring is to be completed in 1995 resulting in the revitalisation of the company.

UPDATED

LET L-410 twin-turboprop light transport aircraft

1994

LET L-410 TURBOLET

TYPE: Twin-turboprop general purpose light transport aircraft.

PROGRAMME: Design of the L-410 twin-turboprop light transport aircraft was started in 1966, by a team led by Ing Ladislav Smrcek. The first prototype (OK-YKE), powered by Pratt & Whitney Aircraft of Canada PT6A-27 turboprop engines, flew for the first time on 16 April 1969. Three additional PT6A-engined prototypes were completed subsequently; the second of these was later test-flown with Hartzell four-blade propellers in a successful demonstration of reduced aircraft vibration and cabin noise levels.

VERSIONS: **L-410A:** Initial passenger/cargo version, powered by 533 kW (715 ehp) Pratt & Whitney Aircraft of Canada PT6A-27 engines. Entered service with Czechoslovak domestic operator Slov-Air in late 1971.

L-410AF: Aerial photography/survey version, generally similar to L-410A but with larger, wider and extensively glazed nose compartment. One built, which was exported to Hungary in 1974.

L-410M/T: Similar to L-410A, but with Motorlet M 601 A engines and seats for up to 17 passengers. First flown 1973; first production example delivered 1976. Total of 110 built, including prototype. Superseded in 1979 by L-410UVP.

L-410UVP: Standard production version from beginning of 1979; first of three prototypes flown on 1 November 1977. Changes include increased wing span and area; fuselage lengthened by 0.47 m (1 ft 6½ in) compared with L-410M; enlarged vertical tail surfaces; dihedral tailplane; improved cockpit systems and additions to standard instrumentation; introduction of spoilers, automatic bank control flaps, automatic propeller feathering, and anti-skid system for main landing gear units; fabric covered elevators and rudder; and lateral model M 601 engines and V 508 propellers. Basic version is for passenger transport duties, but cabin can be converted easily to all-cargo, aeromedical, parachutist or firefighting configurations; aircraft can also be equipped for aerial photography or calibration of ground navigation aids. The L-410UVP can operate from grass, sand and gravel strips as well as from paved runways, and in snow and ice conditions. In service with many of the former Warsaw Pact civilian airlines. Total 508 built.

OPERATORS: Versions of the L-410 (not including L-410UVP-E) are in service with the armed forces of the following countries: Czech Republic (17+), Estonia, El Salvador, Hungary (7), Latvia (2), Libya (20), Lithuania (2) and Slovakia (9).

The following description applies to the L-410UVP:

DESIGN FEATURES: Cantilever high-wing monoplane. Wing section NACA 63A418 at root, NACA 63A412 at tip. Dihedral 1° 45′. Incidence 2° at root, −0° 30′ at tip. No sweepback at front spar. Vertical tail surfaces swept back 35°. One-piece tailplane, with 7° dihedral from roots, mounted part-way up fin.

FLYING CONTROLS: Hydraulically actuated double-slotted metal flaps, with both slots variable. Spoiler forward of each flap. All-metal ailerons, forward of which are pop-up bank control surfaces that come into operation automatically during single-engine operation and decrease the lift on the side of the running engine. Balance tab in rudder and each elevator.

STRUCTURE: The wing is a conventional all-metal two-spar torsion-box structure, attached to fuselage by four-point mountings. Chemically machined skin with longitudinal reinforcement. The fuselage is a conventional all-metal semi-monocoque spot welded and riveted structure, built in three main portions. The tail unit is a conventional cantilever structure, of all-metal construction except for elevators and rudder, which are fabric-covered.

SYSTEMS: Gumotex Breclav pneumatic de-icing of leading-edges.

LANDING GEAR: Retractable tricycle type, with single wheel on each unit. Hydraulic retraction, nosewheel forward, mainwheels inward to lie flat in fairing on each side of fuselage. Technometra Radotin oleo-pneumatic shock-absorbers. Non-braking nosewheel, with servo-assisted steering, fitted with 548 × 221 mm (9.00-6) tubeless tyre, pressure 2.74 bars (39.8 lb/sq in). Nosewheel is also steerable by rudder pedals. Mainwheels fitted with 718 × 306 m (12.50-10) tubeless tyres, pressure 3.14 bars (45.5 lb/sq in). All wheels manufactured by Moravan Otrokovice, tyres by Baramtech Zlin. Moravan Otrokovice hydraulic disc brakes, parking brake and anti-skid units on mainwheels. Metal ski landing gear, with plastics under-surface, optional.

POWER PLANT: Two 544 kW (730 ehp) Walter M 601 B turboprop engines, each driving an Avia V 508 B three-blade reversible-pitch fully feathering metal propeller. At higher ambient temperatures, engine power can be increased to 590 kW (790 ehp) for short periods by water injection into compressor. De-icing for propeller blades (electrical) and water lower intakes; anti-icing flaps inside each nacelle. Eight bag-type fuel tanks in wings, total capacity 1,290 litres (341 US gallons; 284 Imp gallons). Total oil capacity (including oil in cooler) 22 litres (5.75 US gallons; 4.8 Imp gallons). Water tank capacity (for injection into compressor) 11 litres (2.8 US gallons; 2.4 Imp gallons).

ACCOMMODATION: Crew of one or two on flight deck. Dual controls standard. Standard accommodation in main cabin for 15 passengers, with pairs of adjustable seats on starboard side of aisle and single seats opposite, all 76 cm (30 in) pitch. Baggage compartment (at rear, accessible from cabin), toilet and wardrobe standard in this version. Cabin heated by engine bleed air. Alternative layouts include all-cargo; ambulance, accommodating six stretchers, five sitting patients and a medical attendant; accommodation for 14 parachutists and a despatcher/instructor; firefighting configuration, carrying 12 firefighters and a pilot/observer. All-cargo version has protective floor covering, crash nets on each side of cabin, and tiedown provisions; floor is at truckbed height. Aircraft can also be equipped for aerial photography or for calibration or ground navigation aids. Double upward-opening doors aft on port side, with stowable steps; right hand door serves as passenger entrance and exit. Both doors open for cargo loading, and can be removed for paratroop training missions. Rearward-opening door, at front on starboard side, serves as emergency exit.

SYSTEMS: No APU, air conditioning or pressurisation systems. Duplicated hydraulic systems, No. 1 system actuating landing gear, flaps, spoilers, automatic pitch trim surfaces, mainwheel brakes, nosewheel steering and windscreen wipers. No. 2 system for emergency landing gear extension, flap actuation and parking brake. Electrical system includes AC power from three three-phase 115 V 400 Hz inverters, guaranteeing against a loss of power for essential instruments; DC power from two 5.6 kW generators and two 25 Ah batteries.

AVIONICS: Standard instrumentation provides for flight in IMC conditions, with all basic instruments duplicated and three artificial horizons. Communications include two VHF with a range of 65 n miles (120 km; 75 miles) at 1,000 m (3,280 ft) altitude and crew intercom. Standard navigation instruments include artifical horizons (three); barometric altimeters, airspeed indicators, rate of climb indicators, turn indicators, RMIs, gyro compasses, ILS, and ARK-15M ADF with range of 97 n miles (180 km; 112 miles) at 1,000 m (3,280 ft) altitude (two of each); and radio altimeter with ground proximity warning, ASI with stall warning, magnetic compass, GMK-1GE VOR, and ILS with marker beacon receiver (one of each). Cockpit, instrument and passenger cabin lights, navigation lights, three landing lights in nose (each with two levels of light intensity), crew and cabin fire extinguishers, windscreen wipers, and alcohol spray for windscreen and wiper de-icing, are also standard. Flight data recorder, cockpit voice recorder, SSR repeater and encoding altimeter, transponder, and electrically heated flight deck windows, are all optional.

DIMENSIONS, EXTERNAL:

Wing span	19.478 m (63 ft 10¾ in)
Wing chord at root	2.534 m (8 ft 3¾ in)
Length overall	14.467 m (47 ft 5½ in)
Fuselage: Max width	2.08 m (6 ft 10 in)
Max depth	2.10 m (6 ft 10¾ in)
Height overall	5.829 m (19 ft 1½ in)
Tailplane span	6.736 m (22 ft 1¼ in)
Wheel track	3.65 m (11 ft 11½ in)
Wheelbase	3.666 m (12 ft 0¼ in)
Propeller diameter	2.50 m (8 ft 2½ in)
Distance between propeller centres	4.816 m (15 ft 9½ in)
Passenger/cargo door (port, aft):	
Height	1.46 m (4 ft 9½ in)
Width overall	1.25 m (4 ft 1 in)
Width (passenger door only)	0.80 m (2 ft 7½ in)
Height to sill	0.79 m (2 ft 7 in)

LET L-410T training version of the LET L-410UVP *(LET Aeronautical Works)*

1995

Emergency exit door (stbd, fwd):
Height 0.97 m (3 ft 2¼ in)
Width 0.66 m (2 ft 2 in)

DIMENSIONS, INTERNAL:
Cabin, excl flight deck: Length 6.34 m (20 ft 9½ in)
Max width 1.95 m (6 ft 4¾ in)
Max height 1.658 m (5 ft 5¼ in)
Aisle width at 0.4 m (1 ft 3¾ in) above cabin floor 0.34 m (1 ft 1½ in)
Floor area 9.69 m² (104.3 sq ft)
Volume 17.86 m³ (630.7 cu ft)
Baggage compartment volume (rear) 0.77 m³ (27.2 cu ft)

AREAS:
Wings, gross 35.18 m² (378.67 sq ft)
Ailerons (total) 2.89 m² (31.11 sq ft)
Automatic bank control flaps (total) 0.49 m² (5.27 sq ft)
Trailing-edge flaps (total) 5.92 m² (63.72 sq ft)
Spoilers (total) 0.87 m² (9.36 sq ft)
Fin 4.49 m² (48.33 sq ft)
Rudder, incl tab 2.81 m² (30.25 sq ft)
Tailplane 6.41 m² (69.00 sq ft)
Elevators, incl tabs 3.15 m² (33.91 sq ft)

WEIGHTS AND LOADINGS:
Basic empty weight 3,725 kg (8,212 lb)
Max fuel 1,000 kg (2,205 lb)
Max payload 1,310 kg (2,888 lb)
Max T-O weight 5,700 kg (12,566 lb)
Max landing weight 5,500 kg (12,125 lb)
Max zero-fuel weight 5,170 kg (11,398 lb)

PERFORMANCE (at max T-O weight, ISA, except where indicated):
Never-exceed speed 194 kts (360 km/h; 224 mph) EAS
Max cruising speed 197 kts (365 km/h; 227 mph)
Econ cruising speed 162 kts (300 km/h; 186 mph)
Stalling speed:
flaps up 78 kts (144 km/h; 90 mph) EAS
flaps down, at max landing weight 61 kts (112 km/h; 70 mph) EAS
Max rate of climb at S/L 468 m (1,535 ft)/min
Rate of climb at S/L, one engine out 90 m (295 ft)/min
Max operating altitude 6,000 m (19,700 ft)
Service ceiling, one engine out 2,850 m (9,350 ft)
T-O run 400 m (1,312 ft)
T-O to 10.5 m (35 ft) 520 m (1,706 ft)
Landing from 9 m (30 ft) at max landing weight 810 m (2,657 ft)
Landing run at max landing weight 328 m (1,000 ft)
Range at 3,000 m (9,850 ft) with max fuel and 850 kg (1,874 lb) payload, 30 min reserves 561 n miles (1,040 km; 646 miles)
Range with max payload and 505 kg (1,113 lb) fuel, 30 min reserves 248 n miles (460 km; 285 miles)

UPDATED

LET L-410M with Motorlet M 601 A turboprop engines *(LET Aeronautical Works)*

1995

ZLIN

MORAVAN INC

CR-765 81 Otrokovice
Telephone: 42 (67) 922041 and 924351
Fax: 42 (67) 922518 and 922479
MANAGING DIRECTOR: Milan Charvát
COMMERCIAL DIRECTOR: Ing Karel Kubena
TECHNICAL DIRECTOR: Ing Antonin Partika
MARKETING MANAGER: Jiri Aron

Formed 18 September 1934 as Zlinská Letecká Akciová Společnost (Zlin Aviation Company Ltd) in Zlin; manufacture of Zlin aircraft started 1933 by Masarykova Letecká Liga (Masaryk League of Aviation); company renamed Moravan after Second World War; also now manufactures aircraft equipment. Current workforce approximately 1,500.

Tenders for majority shareholding invited during 1993; main bidder by mid-1994 was Bata shoe company of Canada.

NEW ENTRY

ZLIN 42 M

TYPE: Two-seat light training and touring aircraft.

PROGRAMME: The prototype Zlin 42 was first flown on 17 October 1967, and in 1969 began undergoing flight trials prior to certification. Standard power plant of the initial version, which entered production in 1971, was the 134 kW (180 hp) M 137 A engine, with which the Z 42 conforms to FAR Pt 23 airworthiness specifications in the Aerobatic category and is suitable for service in climates with temperatures between +40° and −20°C. The principal customer for this version was the German Democratic Republic, which acquired several dozen for military liaison duties.

Prototype was first flown in November 1972, certification under FAR Pt 23 in the Aerobatic (+6*g* to −3.5*g*) and Normal categories obtained in 1973. This differs from the initial version in having a constant-speed propeller and a tail fin identical to that of the Z 43. Production of the Zlin 42 M began in 1974. Total Zlin 42 production (all versions) was in excess of 235.

VERSIONS: **42 M:** Initial two-seat version powered by one 134 kW (180 hp) Avia M 137 AZ inverted engine.

Zlin 43: Four-seat version with one 157 kW (210 hp) Avia M 337 A inverted six-cylinder air-cooled in-line engine.

UPGRADES: **Zlin:** Zlin 42 MU. At the request of a number of operators of the initial production version, their aircraft have, during overhaul, been fitted retrospectively with Avia V 503 A two-blade metal propellers of the type fitted to later production Z 42 M. These aircraft were redesignated Zlin 42 MU.

OPERATORS:: Versions of the 42 are in service with the armed forces of the following countries: Bulgaria, Czech Republic (33).

The following description applies to the subsequent Zlin 42 M production version.

DESIGN FEATURES: Cantilever low-wing monoplane. Wing section NACA $63_2416.5$. Dihedral 6° from roots. Sweepforward 4° 20′ at quarter-chord. Cantilever tail unit.

FLYING CONTROLS: All-metal slotted ailerons and flaps all have

Zlin 42 trainer aircraft of the Czech Republic Air Force *(Peter March)*

1995

same dimensions. Mass balanced flaps and ailerons operated mechanically by control rods. Ground adjustable tab in each aileron. Tail unit control surfaces have partial mass and aerodynamic balance. Trim tab in elevator and rudder. Rudder actuated by control cables, elevator by control rods.

STRUCTURE: The wing is an all-metal structure with a single main spar. Engine cowlings of sheet metal. Centre-fuselage of welded steel tube truss construction, covered with laminated glassfibre panels. The rear fuselage is an all-metal semi-monocoque structure. Tail unit of all-metal structure.

LANDING GEAR: Non-retractable tricycle type, with nosewheel offset to port. Oleo-pneumatic nosewheel shock-absorber. Mainwheels carried on flat spring steel legs. Nosewheel steering by means of rudder pedals. Single wheel on each unit. Mainwheels and Barum tyres size 420 × 150, pressure 1.86 bars (27 lb/sq in); nosewheel and Barum tyre size 350 × 135, pressure 2.45 bars (35.6 lb/sq in). Hydraulic brakes on mainwheels can be operated from either seat. Parking brake standard. Wheel fairings and skis optional.

POWER PLANT: One 134 kW (180 hp) Avia M 137 AZ inverted six-cylinder air-cooled in-line engine, with low-pressure injection pump, driving a two-blade Avia V 503 A fully automatic constant-speed propeller. Fuel tanks in each wing leading-edge, with total capacity of 130 litres (34.3 US gallons; 28.6 Imp gallons). Fuel and oil systems permit aerobatic and inverted flying. Oil capacity 12 litres (3.2 US gallons; 2.6 Imp gallons).

ACCOMMODATION: Individual side by side seats for two persons, the pilot's seat being to port. Both are adjustable for height and permit the use of back-type parachutes. Baggage space aft of seats. Cabin and windscreen heating and ventilation. Forward-opening door on each side of cabin. Dual controls standard.

SYSTEMS: Electrical system includes a 600 W 27 V engine-driven generator and Teledyne battery. External power source can be used for starting the engine.

AVIONICS AND EQUIPMENT: VHF radio and IFR instrumentation optional.

DIMENSIONS, EXTERNAL:

Wing span	9.11 m (29 ft 10¾ in)
Wing chord (constant)	1.42 m (4 ft 8 in)
Length overall	7.07 m (23 ft 2½ in)
Height overall	2.69 m (8 ft 10 in)
Elevator span	2.90 m (9 ft 6 in)
Wheel track	2.33 m (7 ft 7¾ in)
Wheelbase	1.66 m (5 ft 5¼ in)
Propeller diameter	2.00 m (6 ft 6¾ in)

DIMENSIONS, INTERNAL:

Cabin: Length	1.80 m (5 ft 10¾ in)
Width	1.12 m (3 ft 8 in)
Height	1.20 m (3 ft 11¼ in)
Baggage space	0.2 m³ (7.1 cu ft)

AREAS:

Wings, gross	13.15 m² (141.5 sq ft)
Ailerons (total)	1.408 m² (15.16 sq ft)
Trailing-edge flaps (total)	1.408 m² (15.16 sq ft)
Fin	0.54 m² (5.81 sq ft)
Rudder, incl tab	0.81 m² (8.72 sq ft)
Tailplane	1.23 m² (13.24 sq ft)
Elevator, incl tab	1.36 m² (14.64 sq ft)

WEIGHTS AND LOADINGS:

Basic weight, empty	645 kg (1,422 lb)
Max T-O weight: Aerobatic	920 kg (2,028 lb)
Normal	970 kg (2,138 lb)
Max wing loading: Aerobatic	70 kg/m² (14.3 lb/sq ft)
Normal	74 kg/m² (15.2 lb/sq ft)
Max power loading:	
Aerobatic	6.87 kg/kW (11.27 lb/hp)
Normal	7.24 kg/kW (11.88 lb/hp)

PERFORMANCE (at max Aerobatic T-O weight):

Max level speed at S/L, ISA	122 kts (226 km/h; 140 mph)
Cruising speed at 600 m (1,975 ft), ISA	116 kts (215 km/h; 134 mph)
Stalling speed, power off:	
flaps up	54 kts (100 km/h; 62.5 mph) CAS
flaps down	49 kts (89 km/h; 56 mph) CAS
Max rate of climb at S/L	312 m (1,025 ft)/min
Service ceiling	4,250 m (13,950 ft)
T-O run	185 m (607 ft)
T-O to 15 m (50 ft)	380 m (1,245 ft)
Landing from 15 m (50 ft)	410 m (1,345 ft)
Landing run	135 m (443 ft)
Range with max standard fuel	286 n miles (530 km; 329 miles)
g limits: Aerobatic	+6.0/–3.5
Normal	+3.8/–1.5

NEW ENTRY

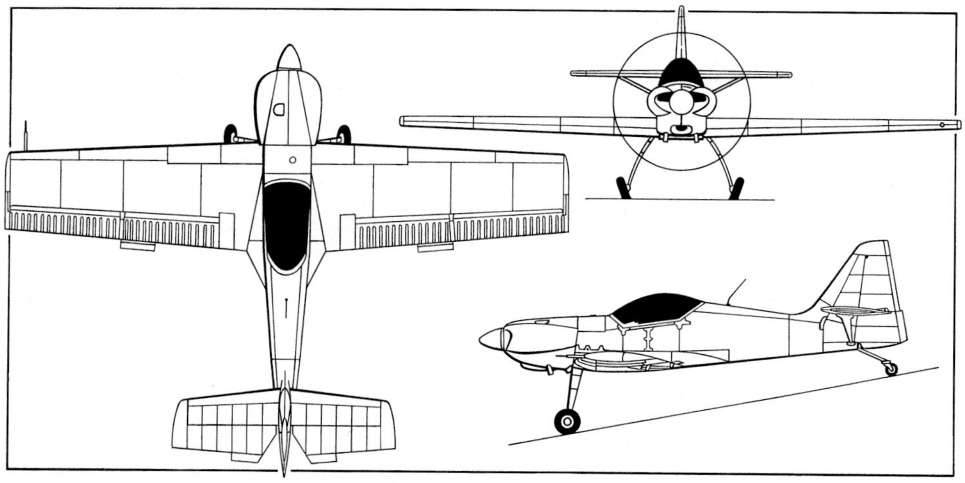

Zlin 50 LS aerobatic aircraft with 224 kW (300 hp) Lycoming engine ***(Jane's/Mike Keep)***

1995

Zlin 50 LS single-seat aerobatic aircraft ***(Peter March)***

1995

ZLIN 50

TYPE: Single-seat aerobatic aircraft.

PROGRAMME: First flight (Z 50 LS) 29 June 1981; FAR Pt 23 certification 1982; won European Aerobatic Championships 1983, World Aerobatic Championships 1984 and 1986.

VERSIONS: **Z 50 L:** Initial version, with 194 kW (260 hp) Textron Lycoming AEIO-540-D4B5; superseded.

Z 50 LA: Modified L with propeller pitch control and propeller speed governor.

Z 50 LE: Short-span version with T-O weight reduced by 50 kg (110 lb); first flight 1990; only two built.

Z 50 LS: Principal version; more powerful engine than earlier models; total 65 built by mid-1991. *Detailed description applies to this version.*

Z 50 M: Longer and slimmer nose than L series, with 134 kW (180 hp) Avia M 137 AZ in-line engine and constant-speed propeller; first flight (OK-080) 25 April 1988; Aerobatic and Normal certification February 1989; prototype and two production aircraft built by January 1989; two more completed for 1992 delivery to South Africa. Overall length 6.96 m (22 ft 10 in), tailplane area 2.24 m² (24.1 sq ft), otherwise dimensionally as Z 50 LS.

CUSTOMERS: Total of 75 (72 L/LA/LE/LS and three M) built by 1 January 1992, with two more M then under construction. Exported to Bulgaria, Germany, Hungary, Italy, Poland, Romania, South Africa, Spain and UK.

DESIGN FEATURES: Provision for fitting range-extending tip tanks. Wing sections NACA 0018 (root), 0012 (tip); dihedral 1° 7′ 24″.

FLYING CONTROLS: Long-span, two-segment mass balanced ailerons on each wing (no flaps); inner ailerons have geared tabs, port outer has ground adjustable tab; trim tab and geared tab on elevators, geared tab on rudder; ailerons and elevators operated by rods, rudder by cables; fixed incidence tailplane.

STRUCTURE: All-metal except for fabric covered elevators and rudder; single tip-to-tip main spar with auxiliary spar; duralumin skins; stressed skin fuselage; braced tailplane.

LANDING GEAR: Non-retractable tailwheel type. Mainwheels carried on flat-spring titanium cantilever legs. Mechanical mainwheel brakes actuated by rudder pedals. Fully castoring tailwheel, with flat-spring shock-absorption, has automatic locking device to maintain aircraft on a straight track during taxying, take-off and landing. Mainwheel tyres size 350 × 135 mm, pressure 2.5 bars (36 lb/sq in); tailwheel tyre size 200 × 80 mm, pressure 1.0 bar (14.5 lb/sq in). Mainwheel fairings optional.

POWER PLANT: One 224 kW (300 hp) Textron Lycoming AEIO-540-L1B5D flat-six engine, driving a Hoffmann HO-V123K-V/200AH three-blade constant-speed wooden propeller. Single main fuel tank in fuselage, aft of firewall, capacity 60 litres (15.9 US gallons; 13.2 Imp gallons). Auxiliary 50 litre (13.2 US gallon; 11 Imp gallon) tank can be attached to each wingtip for cross-country flights only. Fuel and oil systems designed for full aerobatic manoeuvres, including inverted flight. Oil capacity 12 litres (3.2 US gallons; 2.6 Imp gallons).

ACCOMMODATION: Single seat under fully transparent sideways opening (to starboard) bubble canopy, which can be jettisoned in emergency. Seat and backrest adjustable, permit use of back type parachute. Cockpit ventilated by sliding panel in canopy.

SYSTEMS: Electrical system includes 24 V LUN 2111 alternator as main power source and two 12 Ah batteries. External power socket in fuselage side for engine starting.

AVIONICS: VHF radio optional.

DIMENSIONS, EXTERNAL:

Wing span	8.58 m (28 ft 1¾ in)
Wing span over tip tanks	9.03 m (29 ft 7½ in)
Wing chord: at root	1.73 m (5 ft 8¼ in)
at tip	1.21 m (3 ft 11¾ in)
Wing aspect ratio	5.9
Length overall (tail up)	6.62 m (21 ft 8¾ in)
Height over tail (static)	2.075 m (6 ft 9¾ in)
Elevator span	3.44 m (11 ft 3½ in)
Wheel track	1.90 m (6 ft 2¾ in)
Wheelbase	5.05 m (16 ft 7 in)
Propeller diameter	2.00 m (6 ft 6¾ in)
Propeller ground clearance (tail up)	0.31 m (1 ft 0¼ in)

AREAS:

Wings, gross	12.50 m² (134.55 sq ft)
Ailerons (total)	2.80 m² (30.14 sq ft)
Fin	0.59 m² (6.35 sq ft)
Rudder, incl tab	0.81 m² (8.72 sq ft)
Tailplane	1.66 m² (17.87 sq ft)
Elevators (total, incl tabs)	1.20 m² (12.92 sq ft)

WEIGHTS AND LOADINGS (A: Aerobatic, N: Normal category):

Weight empty: A	600 kg (1,322 lb)
N	610 kg (1,345 lb)
Max T-O weight: A	760 kg (1,675 lb)
N	840 kg (1,852 lb)
Max wing loading: A	60.8 kg/m² (12.45 lb/sq ft)
N	67.2 kg/m² (13.76 lb/sq ft)
Max power loading: A	3.40 kg/kW (5.58 lb/hp)
N	3.75 kg/kW (6.17 lb/hp)

PERFORMANCE (at max Aerobatic T-O weight):

Never-exceed speed (VNE)	181 kts (337 km/h; 209 mph) CAS
Max level speed at 500 m (1,640 ft), ISA	166 kts (308 km/h; 191 mph)

Max cruising speed at 500 m (1,640 ft), ISA	148 kts (275 km/h; 171 mph)
Stalling speed, engine idling	56 kts (103 km/h; 81 mph) CAS
Max rate of climb at S/L, ISA	840 m (2,755 ft)/min
Service ceiling, ISA	8,175 m (26,820 ft)
T-O run	150 m (492 ft)
T-O to 15 m (50 ft)	300 m (985 ft)
Landing from 15 m (50 ft)	530 m (1,740 ft)
Landing run	300 m (985 ft)
g limits: A	+8/–6
N	+3.8/–1.5

NEW ENTRY

Zlin 50 LS single-seat aerobatic aircraft *(Peter March)* *1995*

FINLAND

VALMET

VALMET AVIATION INDUSTRIES INC

SF-35600 Halli
Telephone: 358 (42) 8291
Fax: 358 (42) 829600
Telex: 28269
PRESIDENT AND CEO: Markku Valtonen
DIRECTOR, TRAINING AIRCRAFT: Jorma Lemmetyinen

Valmet and predecessors have built 30 types of aircraft, including 19 of Finnish design, since 1922; current activities include manufacturing Valmet L-90 TP RediGO turboprop trainer; production of aircraft and aero-engine parts; and maintenance, overhaul, repair and modification of aircraft and accessories, aero-engines and accessories, marine diesel engines, accessories and gears, and foundry products. Contract signed 1989 for Valmet to produce all tail surfaces for Saab 2000; will assemble 57 F/A-18Cs for Finnish Air Force from CKD kits during period 1995-2000, and will also co-produce fuselage dorsal and side skin panels.

Facilities in Kuorevesi and Linnavuori; workforce about 950 at 1 January 1994.

NEW ENTRY

Valmet L-70 (Vinka) Miltrainer aircraft of the Finnish Air Force *1995*

Valmet L-70 (Vinka) Miltrainer aircraft of the Finnish Air Force *1995*

VALMET L-70 MILTRAINER

Finnish Air Force name: Vinka

TYPE: Two-seat training or two/four-seat touring aircraft.

PROGRAMME: A development contract for the L-70 was placed with Valmet by the Finnish Air Force on 23 March 1973. The aircraft, which was originally designated Leko-70, an abbreviation of 'Lentokone', the Finnish word for 'aeroplane', first flew on 1 July 1975. It is named Vinka (a cold Arctic wind) by the Finnish Air Force, to whom 30 Vinkas were delivered during 1980-82.

The Miltrainer is designed for Aerobatic or Utility use as a two-seater. In civil use, in the Normal category, it is capable of seating up to four persons, depending upon the amount of baggage carried, and fulfils the requirements of FAR Pt 23 in all three categories. In addition to these requirements, the Finnish Air Force specified some special military strength and other requirements for the aircraft. For instance, the Vinka has a fatigue life of more than 8,000 flying hours in heavy military use.

Normal roles include primary flying training, aerobatic training, night and instrument flying training, observation and liaison, tactical training, and ambulance duties. Secondary roles can include search and rescue, supply dropping, weapon training, photo reconnaissance, television monitoring/transmission, glider or target towing, and agricultural operations. The design permits the use of a ski landing gear.

OPERATORS: Versions of the L-70 are in service with the armed forces of Finland (28).

DESIGN FEATURES: Cantilever low-wing monoplane. Wing section NACA 63_2A615 (modified). Dihedral 6° from roots. Incidence 2°. Fail-safe structure, attached to fuselage by steel fittings, comprising main spar, auxiliary spar, ribs and stringers of constant chord except for forward-swept wingroot leading-edges. Tail unit has slight sweepback on vertical surfaces and shallow dorsal fin.

FLYING CONTROLS: Electrically operated slotted flaps and mass balanced ailerons on trailing-edges all of aluminium alloy riveted construction. Ailerons actuated by stainless steel control cables. Flaps and ailerons have fluted skins. Spring tab in each aileron. Elevators and rudder aerodynamically and mass balanced and actuated by stainless steel control cables. Geared trim tabs in rudder and each elevator.

STRUCTURE: The fuselage is an aluminium alloy semi-monocoque fail-safe structure of frames and longerons with riveted skin. Welded steel tube engine mount and wing carry-through structure. Stainless steel firewall. Cockpit floor of panels of bonded sandwich. Riveted aluminium alloy skin (fluted on flaps and ailerons). The tail unit is an aluminium alloy structure with riveted skin (fluted on fin, rudder and elevators).

LANDING GEAR: Non-retractable tricycle type. Cantilever main legs. Automotive Products oleo-pneumatic shock-absorber in each unit. Cleveland 40-75S mainwheels, with Goodyear 6.00-6 6 ply tyres, pressure 1.86 bars (27 lb/sq in); Goodyear 395-32926 castoring and self-centring nosewheel, with Goodyear 5.00-5 4 ply tyre, pressure 2.07 bars (30 lb/sq in). Nosewheel steering optional. Cleveland 30-52K hydraulic disc brakes and parking brake. Provision for fitting Finncraft skis.

POWER PLANT: One 149 kW (200 hp) Avco Lycoming AEIO-360-A1B6 flat-four engine, driving a Hartzell HC-C2YK-4F/FC 7666A-2 two-blade constant-speed propeller with spinner. Christen-801 fuel and oil systems permit up to 1½ minutes of continuous inverted flight in Aerobatic category. Semi-integral bonded sandwich fuel tank in each wingroot ahead of main spar; total capacity 170 litres (45 US gallons; 37.4 Imp gallons). Gravity fuelling point in top of each tank. Safom reticulated polyurethane foam filling for fuel tanks is optional. Oil capacity 7.5 litres (2 US gallons; 1.65 Imp gallons).

ACCOMMODATION: Side by side seats for instructor and pupil in trainer version, with integral longitudinal central console which serves also to reinforce fuselage floor. Dual controls standard, but instructor's or pupil's control column can be removed if desired. Windscreen and one-piece rearward-sliding fully transparent jettisonable canopy, with steel tube turnover frame. Canopy can be locked in partially open position if required. Provision for two more seats at rear, which can be removed to make room for additional baggage. Up to 280 kg (617 lb) of baggage or freight can

Valmet L-70 (Vinka) Miltrainer aircraft flying in formation

1995

be carried internally if aircraft is flown as a single-seater. As ambulance, can accommodate one stretcher patient and medical attendant in addition to pilot. Cockpit heated and ventilated.

SYSTEMS: 28 V DC electrical system, with Prestolite 24 V 70 A alternator and 25 Ah Ni/Cd battery. Ground power receptacle. No hydraulic or pneumatic systems.

AVIONICS AND EQUIPMENT: Standard avionics include two VHF transceivers, one ADF, one VOR/ILS with indicator, two RMI, one gyrosyn compass system, and intercom. Standard equipment includes accelerometer, dual airspeed indicators, dual artificial horizons, clock, magnetic compass, dual rate of climb indicators, dual turn and slip indicators, outside air temperature gauge, and tachometer; electrically heated pitot static head; inertia reel shoulder harnesses for front seats; first aid kit; internal and external corrosion proofing; instrument lighting; cockpit utility lights for map reading and as standby instrument lighting; warning lights for alternator, battery temperature and directional gyro; anti-collision beacon; landing and taxying lights in starboard wing leading-edge; navigation lights; and fire extinguisher. Equipment for secondary roles may include reflex gunsight and external load control panel; glider or target towing hook; one long-focus or four short-focus vertical cameras (provision for 35 × 40 cm; 13.8 × 15.75 in aperture, with quick-release panel, in floor of rear cockpit); four underwing pylons (total capacity 300 kg: see following paragraph) for stores; and dispersal equipment for agricultural missions.

ARMAMENT AND OPERATIONAL EQUIPMENT: Four underwing attachments, the inner pair each stressed for 150 kg (330.5 lb) and the outer pair for 100 kg (220 lb) each; max external load 300 kg (661 lb). As single-seater, can carry four 50 kg bombs; two 100 kg bombs plus two flare pods; four pods each with eighteen 37 mm or six 68 mm rockets; two pods each with twin 7.62 or 5.56 mm machine guns and 1,000 rds/pod; two such gun pods and two flare pods; two pods each with single 12.7 mm machine gun and 150 rds/pod; or two reconnaissance or photographic pods. As two-seater, typical loads can include four or eight anti-tank missiles, depending upon type and size; one TV pod (with transmitter) and one searchlight pod; three 10-person liferafts and a searchlight pod; or three 6-person emergency rescue packs and a searchlight pod.

DIMENSIONS, EXTERNAL:
- Wing span: 9.63 m (31 ft 7¼ in)
- Wing chord (constant over most of span): 1.53 m (5 ft 0¼ in)
- Wing aspect ratio: 6.93
- Length overall: 7.50 m (24 ft 7¼ in)
- Height overall: 3.31 m (10 ft 10¼ in)
- Tailplane span: 3.60 m (11 ft 9¾ in)
- Wheel track: 2.30 m (7 ft 6½ in)
- Wheelbase: 1.61 m (5 ft 3½ in)
- Propeller diameter: 1.88 m (6 ft 2 in)
- Propeller ground clearance: 0.25 m (9¾ in)

AREAS:
- Wings, gross: 14.00 m^2 (150.70 sq ft)
- Ailerons (total): 1.412 m^2 (15.20 sq ft)
- Trailing-edge flaps (total): 1.90 m^2 (20.45 sq ft)
- Fin: 0.87 m^2 (9.36 sq ft)
- Rudder, incl tab: 0.79 m^2 (8.50 sq ft)
- Tailplane: 2.01 m^2 (21.64 sq ft)
- Elevators, incl tabs: 1.01 m^2 (10.87 sq ft)

WEIGHTS AND LOADINGS:
- Operating weight empty, equipped: 767 kg (1,691 lb)
- Max payload with full fuel: 380 kg (838 lb)
- Max T-O weight: Aerobatic: 1,040 kg (2,293 lb)
  - Utility: 1,050 kg (2,315 lb)
  - Normal: 1,250 kg (2,756 lb)
- Max wing loading:
  - Aerobatic: 74.3 kg/m^2 (15.22 lb/sq ft)
  - Utility: 75.0 kg/m^2 (15.36 lb/sq ft)
  - Normal: 89.3 kg/m^2 (18.53 lb/sq ft)
- Max power loading:
  - Aerobatic: 6.97 kg/kW (11.47 lb/hp)
  - Utility: 7.04 kg/kW (11.58 lb/hp)
  - Normal: 8.38 kg/kW (13.78 lb/hp)

PERFORMANCE (at AUW of 1,000 kg; 2,205 lb):
- Never-exceed speed: 193 kts (360 km/h; 223 mph)
- Max level speed at S/L: 127 kts (235 km/h; 146 mph)
- Cruising speed (75% power) at 1,525 m (5,000 ft): 120 kts (222 km/h; 138 mph)
- Stalling speed, power off:
  - flaps up: 53 kts (98 km/h; 61 mph)
  - flaps down: 46 kts (85 km/h; 53 mph)
- Max rate of climb at S/L: 342 m (1,120 ft)/min
- Service ceiling: 5,000 m (16,400 ft)
- T-O run: 230 m (755 ft)
- Landing run: 175 m (575 ft)
- Min ground turning radius: 8.00 m (26 ft 3 in)
- Range with max fuel, no reserves: 513 n miles (950 km; 590 miles)
- Endurance at S/L (65% power): 4 h 48 min
- Max endurance, no reserves: 6 h 12 min
- *g* limits: Aerobatic: +6/−3
  - Utility: +4.4/−2.02
  - Normal: +3.3/−1.8

NEW ENTRY

FRANCE

AEROSPATIALE

AEROSPATIALE SNI

37 boulevard Montmorency, F-75781 Paris Cedex 16
Telephone: 33 (1) 42 24 24 24
Fax: 33 (1) 42 24 21 32
Telex: AISPA 640025 F

PRESIDENT AND CEO: Louis Gallois
DIRECTOR OF INFORMATION AND COMMUNICATIONS: Patrice Kreis

AIRCRAFT GROUP

PRESIDENT AND COO: Claude Terrazzoni
AIRBUS PROGRAMME DIRECTOR: Gerard Blanc
REGIONAL TRANSPORT PROGRAMME DIRECTOR: Antoine Bouvier
DIRECTOR OF COMMUNICATIONS: André Bloch

WORKS AND FACILITIES:
- Toulouse. PLANT MANAGER: Christian Beugnet
- Nantes Bouguenais. PLANT MANAGER: Jean Noël Guillou
- Saint-Nazaire. PLANT MANAGER: Jacques Crusson
- Méaulte. PLANT MANAGER: Claude Berlan

AVIATION SUBSIDIARIES OF AEROSPATIALE GROUP

Maroc Aviation (part holding)
Société de Construction d'Avions de Tourisme et d'Affaires (SOCATA) (listed separately in this section)
SOGERMA-SOCEA
Société d'Exploitation et de Constructions Aéronautiques (SECA)
Sextant Avionique (50:50 with Thomson-CSF)

Aerospatiale formed 1 January 1970 by French government-directed merger of Sud-Aviation, Nord-Aviation and SEREB.

Besides programmes listed below, Aerospatiale has 37 per cent share in Airbus Industrie (see International section); main Airbus assembly plant is at Aerospatiale base at Toulouse. Group helicopter activities now vested in Eurocopter SA, which see in International section. Part holding in Aerospatiale (20 per cent) held by SOGEPA. As of 1 January 1995 Aerospatiale adopted a new structure based on two entities: operating units and groups.

UPDATED

AEROSPATIALE HELICOPTERS

These now listed under Eurocopter in International section. International programmes in which Aerospatiale was a partner.

UPDATED

AEROSPATIALE LIGHT AIRCRAFT (SOCATA)

Products of Aerospatiale's light aircraft manufacturing subsidiary, SOCATA, are listed separately in this section.

VERIFIED

AEROSPATIALE/ALENIA ATR 42/72

Equally owned Groupement d'Intéret Economique formed with Aeritalia (now Alenia) 5 February 1982 to manufacture 42- and 72-passenger turboprop regional transports.

VERIFIED

AEROSPATIALE ATSF

Aerospatiale's own studies of advanced supersonic airliner (Avion de Transport Supersonique Futur) continue. Company has joined the other international SST studies with Germany, Japan, UK and USA.

VERIFIED

VLCT/UHCA

Aerospatiale participating in German, Japanese, UK and USA study outside Airbus Industrie of feasibility of a very large commercial transport/ultra-high capacity airliner seating between 500 and 800 passengers.

VERIFIED

EUROFLAG

Aerospatiale is one of the companies co-operating in development of this future military tactical transport.

VERIFIED

EUROPATROL

Aerospatiale is one of a group of European companies investigating development of a new international maritime patrol aircraft.

VERIFIED

AEROSPATIALE (FOUGA) CM 170 MAGISTER and SUPER MAGISTER

Israeli Defence Force name: Tzukit

TYPE: Light twin-jet two-seat trainer.

PROGRAMME: The Magister is a light two-seat twin-jet trainer, which was developed originally for the French Air Force. The contract for three prototypes was signed on 27 June 1951, and the first of these flew for the first time on 23 July 1952. A preproduction order for 10 aircraft was placed in June 1953 and the first aircraft of this order made its maiden flight on 7 July 1954.

The initial series production order for 95 aircraft was signed on 13 January 1954, and the first of these aircraft flew at Toulouse on 29 February 1956. Subsequent contracts have increased the total number of Magister-series aircraft built to 921.

Of the German Magisters, 188 were built under licence by Flugzeug-Union-Sud. Similarly, 59 of the Finnish aircraft were built by Valmet OY and 36 of the Israeli aircraft by Israel Aircraft Industries.

VERSIONS: **CM 170 Magister:** Basic version with two 880 lb (400 kg) st Marboré IIA turbojets. Total of 761 ordered.

CM 170 Super Magister: Similar to Magister, but with 1,058 lb (480 kg) st Marboré VI turbojets. First flown on 28 August 1962. Latest orders for 130 aircraft for French Air Force, included in totals above, and seven for Brazil, were for this model which continued in production until mid-1969. 112 had been delivered by 31 March 1968.

CM 175 Zéphyr: Similar to Magister but developed for carrier operations. See next entry.

UPGRADES: **IAI:** AMIT upgrade. See separate entry.

OPERATORS: Versions of the CM 170 are in service with the armed forces of the following countries: Algeria (20), Bangladesh (15), Belgium (18), Cameroon (10), El Salvador (5), France (70), Gabon (4), Ireland (5+), Israel (80), Lebanon (5), Libya (12), Morocco (20+), Senegambia (5) and Togo (4).

The following description applies to both the Magister and Super Magister:

DESIGN FEATURES: Cantilever mid-wing monoplane. NACA 64 Series wing section, thickness/chord ratio varying from 19 per cent at root to 12 per cent at tip. Aspect ratio 7.42. Chord 1.98 m (6 ft 6 in) at root, 0.87 m (2 ft 10 in) at tip. No dihedral. Incidence 2°. Leading-edge sweepback 13°. Butterfly-type tail unit with included angle of 110°.

FLYING CONTROLS: Hydraulically operated all-metal slotted flaps. Automatic tabs in ailerons. All but early versions have servo-control ailerons. Retractable airbrakes in upper and lower surfaces. Aerodynamically balanced elevators.

STRUCTURE: The wing is a single spar aluminium alloy stressed skin structure. The fuselage is an all-metal semi-monocoque stressed skin structure. The tail unit is an all-metal single-spar structure.

LANDING GEAR: Messier retractable tricycle type. Hydraulic actuation. Nosewheel has anti-shimmy device. Messier wheels and hydraulic brakes. Mainwheels size 480 × 180 mm. Nosewheel size 350 × 130 mm.

POWER PLANT: Two Turbomeca Marboré turbojet engines (details under model descriptions above). Main fuel in two fuselage tanks of 255 litres (67 US gallons; 56 Imp gallons) and 475 litres (125 US gallons; 104 Imp gallons) capacity respectively. Total internal fuel capacity 730 litres (193 US gallons; 160 Imp gallons). Two non-jettisonable wingtip tanks (125 litres; 33 US gallons; 27.5 Imp gallons each) may be fitted. Oil capacity 24.4 litres (6.4 US gallons; 5.4 Imp gallons).

ACCOMMODATION: Tandem cockpits under transparent canopy. Hinged sections over cockpits open upward and rearward. Pupil in front seat, instructor in rear. All flying, brake and engine controls duplicated. Emergency landing gear and airbrake controls in front cockpit only.

SYSTEMS: Cockpits pressurised and air conditioned by SEMCA turbo-refrigeration equipment driven by air bleeds from engine compressors. Individual oxygen supply with regulator in each cockpit.

ELECTRONICS AND EQUIPMENT: UHF and VHF radio, VOR, IFF, radio compass, interphone, and blind-flying equipment standard. Provision for Tacan navigation equipment in place of radio compass.

ARMAMENT: Two 7.5 or 7.62 mm machine guns (200 rpg) in fuselage nose. Hyro gunsight in each cockpit, that in rear cockpit fitted with additional periscopic sight. Racks for two 25 kg (55 lb) air-to-ground rockets, one type 181 Matra launcher with 18 × 37 mm rockets, one launcher with 7 × 68 mm rockets, one 50 kg (100 lb) bomb or one Nord AS 11 guided missile may be fitted under each wing.

DIMENSIONS, EXTERNAL:

Wing span: without tip tanks	11.4 m (37 ft 5 in)
with tip tanks	12.15 m (39 ft 10 in)
Length overall	10.06 m (33 ft 0 in)
Height over tail	2.80 m (9 ft 2 in)
Tail span	4.38 m (14 ft 4½ in)
Wheel track	3.80 m (12 ft 6¾ in)
Wheelbase	4.49 m (14 ft 9 in)

AREAS:

Wings, gross	17.30 m² (186.1 sq ft)
Ailerons (total)	1.10 m² (11.84 sq ft)
Flaps (total)	2.10 m² (22.6 sq ft)
Horizontal tail area (projected)	3.71 m² (40 sq ft)
Vertical tail area (projected)	2.60 m² (28 sq ft)

WEIGHTS AND LOADINGS (M: Magister, SM: Super Magister):

Weight empty with fixed equipment:	
M	2,150 kg (4,740 lb)
SM	2,310 kg (5,093 lb)
Average T-O weight, without external tanks:	
M and SM	2,850 kg (6,280 lb)
Average T-O weight with external tanks:	
M and SM	3,100 kg (6,835 lb)
Max T-O weight: M	3,200 kg (7,055 lb)

PERFORMANCE (M: Magister, SM: Super Magister):

Max level speed at S/L: M	350 kts (650 km/h; 403 mph)
SM	378 kts (700 km/h; 435 mph)
Max level speed at 9,000 m (30,000 ft):	
M	386 kts (715 km/h; 444 mph)
SM	402 kts (745 km/h; 463 mph)
Max permissible Mach number.	0.82
Max permissible diving speed	464 kts (860 km/h; 535 mph)
Max rate of climb at S/L:	
at 2,850 kg (6,280 lb) AUW:	
M	1,020 m (3,345 ft)/min
SM	1,200 m (3,935 ft)/min
at 3,100 kg (6,835 lb) AUW:	
M	900 m (2,950 ft)/min
SM	1,080 m (3,540 ft)/min
Service ceiling at 3,100 kg (6,835 lb) AUW:	
M	11,000 m (36,000 ft)
SM	13,500 m (44,300 ft)
T-O run at 3,100 kg (6,835 lb) AUW:	
M	655 m (2,150 ft)
SM	580 m (1,900 ft)
T-O to 15 m (50 ft) at 3,100 kg (6,835 lb) AUW:	
M	930 m (3,050 ft)
SM	670 m (2,200 ft)
Range at 9,000 m (30,000 ft) with 120 l (31 US gallons; 26 Imp gallons) fuel reserve:	
at 2,850 kg (6,280 lb) AUW:	
M	499 n miles (925 km; 575 miles)
SM	507 n miles (940 km; 585 miles)
at 3,100 kg (6,835 lb) AUW:	
M	648 n miles (1,200 km; 735 miles)
SM	675 n miles (1,250 km; 775 miles)
Endurance at 9,000 m (30,000 ft) with reserves:	
at 2,850 kg (6,280 lb) AUW: M	1 h 55 min
SM	2 h 0 min
at 3,100 kg (6,835 lb) AUW: M	2 h 40 min
SM	2 h 50 min

Aerospatiale (Fouga) CM 170 Magister two-seat trainer aircraft *(Daniel March)*

1995

Aerospatiale (Fouga) CM 170 Magister two-seat trainer aircraft *(Peter March)*

1995

NEW ENTRY

AEROSPATIALE (FOUGA) CM 175 ZEPHYR

TYPE: Light twin-jet, carrier-based trainer.

PROGRAMME: Generally similar to the Magister, but incorporates certain modifications to meet naval requirements. These include catapult equipment, arrester gear and two separate rearward sliding hoods for the crew. First of two prototypes flew 30 May 1959. French Navy ordered 30 aircraft.

OPERATORS: Versions of the CM 175 are in service with the armed forces of France (16).

DESIGN FEATURES: Generally as for CM 170 Magister.

FLYING CONTROLS: Generally as for CM 170 Magister.

STRUCTURE: Generally as for CM 170 Magister.

LANDING GEAR: Basically same as for Magister, but long-stroke Messier oleo-pneumatic shock-absorbers compress during retraction. Kléber-Colombes 10HP wheels and Dunlop 04 tyres. Messier hydraulic brakes.

Aerospatiale (Fouga) CM 175 Zéphyr carrier-based trainer version of the CM 170 Magister *(Daniel March)*

1995

Wheel track 3.80 m (12 ft 6¾ in). Wheelbase 4.70 m (15 ft 5 in).

POWER PLANT: Two Turbomeca Marboré IIB turbojet engines (400 kg; 880 lb st each). Otherwise as for Magister, except smaller fuselage fuel tanks has capacity of only 235 litres (62 US gallons; 51.7 Imp gallons) compared with 255 litres (67 US gallons; 56 Imp gallons) on Magister.

ACCOMMODATION: Basically same as for Magister, but with separate rearward-sliding canopy over each seat. Controls for emergency landing gear release, flaps, airbrakes, brakes and arrester hook in front cockpit only.

ARMAMENT: Same as for Magister.

DIMENSIONS, EXTERNAL: Same as for Magister, except:

Height	2.95 m (9 ft 8 in)

WEIGHTS AND LOADINGS:

Weight empty, equipped	2,350 kg (5,180 lb)
Normal loaded weight	3,200 kg (7,055 lb)
Max permissible loaded weight	3,400 kg (7,495 lb)
Max landing weight	3,100 kg (6,835 lb)

PERFORMANCE:

Max speed at S/L	340 kts (630 km/h; 392 mph)
Max speed at 7,000 m (22,950 ft)	350 kts (650 km/h; 403 mph)
Landing speed	90 kts (167 km/h; 104 mph)
Min catapult speed at max AUW	95 kts (176 km/h; 109 mph)
Rate of climb at S/L	900 m (2,950 ft)/min

NEW ENTRY

Aerospatiale (Nord 262) Frégate twin-engined light transport aircraft of the French Navy *(Andrew March)* *1995*

AEROSPATIALE N 262 and FRÉGATE

TYPE: Twin-engined light transport aircraft.

PROGRAMME: Design of the N 262 began in the Spring of 1961, and the prototype (F-WKVR) flew for the first time on 24 December 1962. It was followed by three preproduction aircraft, which were built at Châtillon-sous-Bagneux and assembled at Melun-Villaroche flight test centre. Final assembly of production models was undertaken at Bourges. 110 built.

VERSIONS: **N 262 Series A:** Preceded by Series B Standard early production version, with 805 kW (1,080 ehp Bastan) VIC turboprop engines. Received FAA Type Approval on 15 March 1965. First production Series A was airframe number 9 (F-WLHX), delivered to Lake Central Airlines (now Allegheny) on 17 August 1965.

N 262 Series B: Designation of first four production aircraft only, built for Air Inter. Same power plant as Series A. First Series B (F-BLHS, airframe number 4) flown for first time on 8 June 1964. Received SGAC certification on 16 July 1964. Entered service 24 July 1964.

Frégate (formerly N262 Series C and D): Version for both civil and military use, with more powerful Bastan VII turboprop engines, having improved single-engine ceiling, cruising speed and T-O performance at 'hot and high' airfields. Power plant dispenses with water-methanol system of Series A and B and has higher initial TBO. An N 262 (airframe number 36) began flying experimentally with Bastan VIIA engines in July 1968, and was also test-flown with different wingtips which bestow improved low-speed handling. The Frégate entered production in 1970, alongside the Series A, from the 74th aircraft. Certification granted 24 December 1970.

N 262 E: Converted N 262 As for navigation and flight engineer training. Aircraft also used for inshore surveillance.

OPERATORS: Versions of the N 262 are in service with the armed forces of the following countries: Angola (4), Burkina Faso (2) and France (45+).

The following description applies specifically to the Frégate, but is generally applicable to all versions:

DESIGN FEATURES: Cantilever high-wing monoplane. Wing section NACA 23016 (modified) at root, NACA 23012 (modified) at tip. Dihedral 3° from root. Incidence 3°. No sweepback. Cantilever tail unit. Fixed incidence tailplane.

FLYING CONTROLS: Sealed all-metal ailerons. Balance tab in starboard aileron. Electrically controlled hydraulically actuated all-metal three-position flaps in inner and outer sections on each trailing-edge. Tail unit control surfaces fabric-covered. One controllable tab and one balance tab in rudder and each elevator.

STRUCTURE: The wing is an all-metal two-spar fail-safe structure in conventional light alloys. The fuselage is a light alloy fail-safe semi-monocoque structure built up from 39 circular main and secondary frames covered with skin panels arranged circumferentially in sets of four. The tail unit is a metal structure built as a separate unit and bolted to rear fuselage frame.

LANDING GEAR: Retractable tricycle type, designed and manufactured by ERAM, with single wheel on each unit. Electrohydraulic retraction, nosewheel forward, mainwheels rearward into fairings on sides of fuselage. ERAM oleo-pneumatic nitrogen-filled shock-absorbers. Mainwheels have Dunlop or Kléber-Colombes tyres size 12.50-16, pressure 4.00 bars (58 lb/sq in). Nosewheel has Dunlop or Kléber-Colombes Type 06 tyre, size 9.00-6, pressure 3.50 bars (51 lb/sq in). Goodyear hydraulic disc brakes, with anti-skid units. Self-centring nosewheel is fitted with shimmy damper and is steerable hydraulically.

POWER PLANT (Frégate): Two 854 kW (1,145 ehp) Turbomeca Bastan VII turboprop engines, each driving a Ratier Forest FH 206-1 four-blade constant-speed fully feathering metal propeller. Six bag-type flexible fuel tanks between wing spars, forming two groups of three tanks with provision for cross-feed and having a total usable capacity of 2,000 litres (528 US gallons; 440 Imp gallons). Provision for two additional optional bag tanks in wing centre-section, each of 285 litres (75 US gallons; 62.5 Imp gallons) usable capacity, giving a maximum usable capacity of 2,570 litres (678 US gallons; 565 Imp gallons). Refuelling point above outer wing tank on each side. Pressure refuelling point at front of starboard side main landing gear fairing. No fuel dump system. Oil capacity 23 litres (6 US gallons; 5 Imp gallons). Electrical anti-icing of engine intakes, spinners and propellers, with additional anti-icing of intakes by engine bleed air.

ACCOMMODATION: Crew of two on flight deck, with central jump-seat at rear for a third crew member if carried. Standard airline version has seating for 26 passengers at 810 mm (32 in) pitch; maximum seating for 29 at 710 mm (28 in) pitch, in three-abreast rows, with two seats on starboard side of aisle and single seat on port side. Movable forward bulkhead, to cater for variable mixed cargo (in front)/passenger (at rear) layouts. Bulkhead can be located in two intermediate positions, to provide 20 or 14 seats at 810 mm (32 in) pitch in rear of cabin, with 9.7 m³ (342 cu ft) or 13.2 m³ (467 cu ft) of cargo space respectively in front part of cabin. Galley, toilet and (on 26-seat version) separate coat space at rear of cabin. For quick-change passenger/cargo operation, foldaway seats can be installed which, when folded, give an available width for cargo of 1.68 m (5 ft 6 in) throughout entire cabin length. Alternative layouts include a six-person executive suite forward with 10 passengers aft; ambulance version with accommodation for 12 stretchers and two medical attendants; or aerial survey version with wide range of cameras and survey equipment and fully equipped darkroom. Military versions can be fitted out to carry 18 paratroops or 29 troops, or as 22-seat utility transports. Naval versions (Series A) can be equipped for target towing, artillery and missile observation, radar calibration or crew training duties. Standard transport versions have two-section passenger door at rear on port side, the lower half of which has built-in airstairs, and a large cargo door at front on the port side. Emergency exits at front of cabin on each side, at rear on starboard side, and on port side of flight deck. Standard baggage compartments between flight deck and cabin on each side. All accommodation is pressurised, soundproofed and air conditioned. Windscreen has electrical anti-icing.

SYSTEMS: SEMCA air conditioning system using bleed air from engine. Maximum pressure differential 0.29 bars (4.20 lb/sq in). Auxiliary ventilation via ram-air inlet at front of port main landing gear fairing. Hydraulic system, operated by two engine-driven pumps at pressure of 207 bars (3,000 lb/sq in), actuates landing gear, nosewheel steering, flaps, brakes and gust locks. Electrically driven (27 V DC) back-up pump and 100 bars (1,450 lb/sq in) surge accumulator. Hand pump for emergency operation of flaps, landing gear and gust locks. Kléber-Colombes (Goodrich licence) pneumatic de-icing boots on wing outer

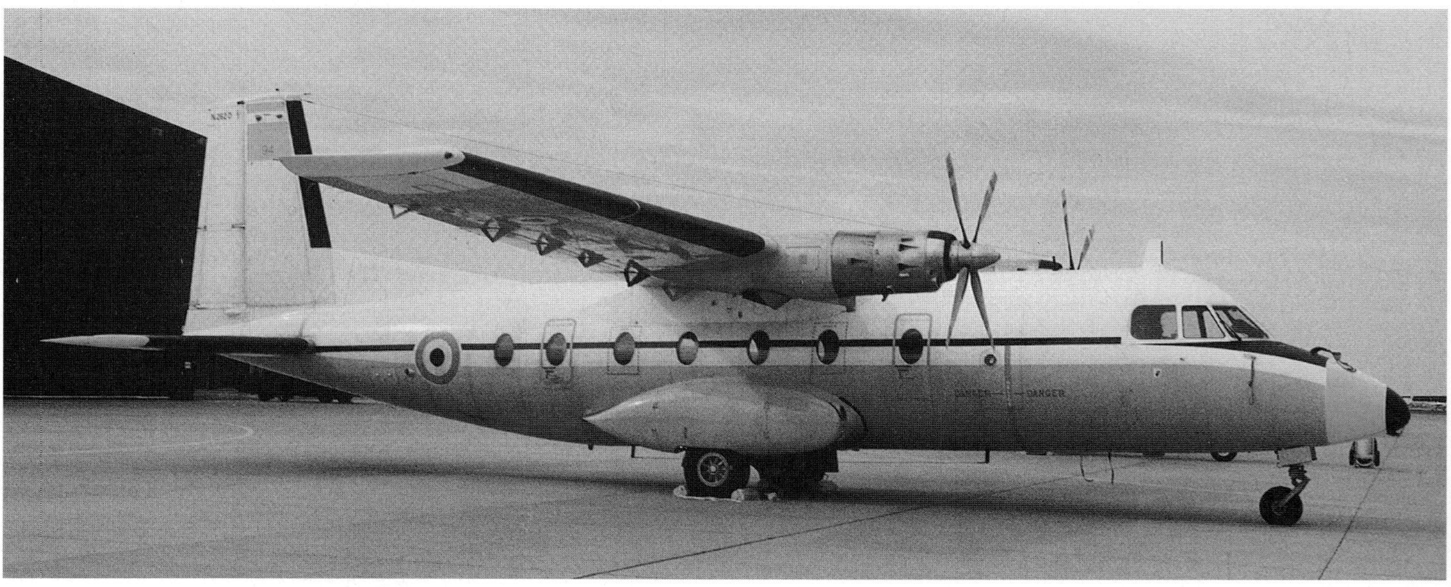

Aerospatiale (Nord 262) Frégate twin-engined light transport aircraft of the French Air Force *(Peter March)*

1995

leading-edges and tail unit leading-edges. Two 24/27 V 40 Ah Ni/Cd batteries, in rear fuselage, and two 9 kVA engine-driven starter/generators provide 28 V DC electrical supply for engine starting, feathering pumps and rotary inverters. External 28 V DC power receptacle. AC system includes two engine-driven 12 kVA three-phase alternators providing 115/200 V 400 Hz power for engine anti-icing, windscreen heating and anti-icing, and heating for galley. Two single-phase 750 VA rotary inverters provide continuous 115 V 400 Hz AC supply for flight deck instruments. System also includes four 115/26 V 400 Hz auto-transformers. Optional APU, in port landing gear fairing, provides power for electrical services, engine starting and cabin air conditioning.

ELECTRONICS AND EQUIPMENT: Standard equipment includes two Collins 618 M 1 VHF, two Collins 51 RV 1 VOR/ILS, Collins 51 Z 4 marker beacon receiver, Collins DF 203 ADF, Collins 331 A6A course indicator, SFIM A 213 flight recorder, Sperry C 14 gyro compass, two Allen RMI, one Bendix OMI, interphone and public address systems. Emergency equipment includes oxygen masks and cylinders, fire extinguishers, liferafts and emergency radio. Optional equipment includes HF radio, autopilot, second gyro compass, second ADF, weather radar, ATC transponder, radio altimeter and DME; and choice of flight director/recorder, VHF, VOR/ILS and marker beacon receiver.

DIMENSIONS, EXTERNAL (Frégate):
Wing span 22.60 m (74 ft 1¾ in)
Wing chord: at root 3.10 m (10 ft 2 in)
at tip 1.80 m (5 ft 10¾ in)
Wing aspect ratio 9.10
Length overall 19.28 m (63 ft 3 in)
Height overall 6.21 m (20 ft 4½ in)
Fuselage: Max diameter 2.45 m (8 ft 0½ in)
Tailplane span 8.80 m (28 ft 10½ in)
Wheel track 3.13 m (10 ft 3 in)
Wheelbase 7.23 m (23 ft 9 in)
Propeller diameter 3.20 m (10 ft 6 in)
Distance between propeller centres 5.91 m (19 ft 4¾ in)
Passenger door (rear, port): Height 1.66 m (5 ft 5¼ in)
Width 0.68 m (2 ft 3 in)
Height to sill 1.08 m (3 ft 6½ in)
Cargo door (fwd, port): Height 1.53 m (5 ft 0¼ in)
Width 1.28 m (4 ft 2½ in)
Height to sill 1.08 m (3 ft 6½ in)
Emergency exit doors (fwd, port and stbd):
Height 1.38 m (4 ft 6¼ in)
Width 0.51 m (1 ft 8 in)
Emergency exit door (aft, stbd):
Height 0.92 m (3 ft 0¼ in)
Width 0.51 m (1 ft 8 in)

DIMENSIONS, INTERNAL:
Cabin, incl baggage space and toilet:
Length 10.61 m (34 ft 10 in)
Max width 2.15 m (7 ft 1 in)
Width at floor 1.66 m (5 ft 5¼ in)
Max height 1.80 m (5 ft 10¾ in)
Floor area 17.0 m^2 (183 sq ft)
Volume 32.5 m^3 (1,146 cu ft)
Baggage hold: port 1.9 m^3 (67 cu ft)
stbd 2.6 m^3 (92 cu ft)

AREAS (Frégate):
Wings, gross 55.79 m^2 (601 sq ft)
Ailerons (total) 4.07 m^2 (43.8 sq ft)
Trailing-edge flaps (total) 8.98 m^2 (96.6 sq ft)
Fin 10.1 m^2 (108.7 sq ft)
Rudder, incl tabs 3.75 m^2 (40.4 sq ft)
Tailplane 18.0 m^2 (193.7 sq ft)
Elevators, incl tabs 4.54 m^2 (48.8 sq ft)

WEIGHTS AND LOADINGS (Frégate):
Basic weight empty 6,200 kg (13,668 lb)
Manufacturer's weight empty, equipped 6,959 kg (15,342 lb)
Basic operating weight 7,225 kg (15,928 lb)
Max payload 3,075 kg (6,779 lb)
Max T-O weight 10,800 kg (23,810 lb)
Max ramp weight 10,850 kg (23,920 lb)
Max landing weight 10,450 kg (23,040 lb)
Max zero-fuel weight 10,300 kg (22,710 lb)
Max wing loading 193 kg/m^2 (39.5 lb/sq ft)
Max power loading 6.32 kg/kW (10.2 lb/ehp)

PERFORMANCE (Frégate, at max T-O weight except where indicated):
Max level speed 225 kts (418 km/h; 260 mph)
Max and econ cruising speed 220 kts (408 km/h; 254 mph)
Normal operating limit speed 214 kts (397 km/h; 247 mph)
Max speed: with landing gear extended 154 kts (285 km/h; 177 mph)
with 15° flap 143 kts (265 km/h; 165 mph)
with 35° flap 127 kts (235 km/h; 146 mph)
Final approach speed 90 kts (167 km/h; 104 mph)
Stalling speed: flaps up, at max landing weight 86 kts (159 km/h; 99 mph)
wheels and flaps down, at max landing weight 74 kts (136 km/h; 85 mph)
Max rate of climb at S/L 420 m (1,380 ft)/min
Service ceiling 8,690 m (28,500 ft)
Service ceiling, one engine out, at AUW of 9,525 kg (21,000 lb) 4,920 m (15,000 ft)
Min ground turning radius 8.0 m (26 ft 3 in)
Runway LCN at max weight 8
T-O run 570 m (1,870 ft)
T-O to 10.7 m (35 ft) 1,070 m (3,510 ft)
Landing from 15 m (50 ft) 530 m (1,740 ft)
Range: with max fuel, no reserves 1,295 n miles (2,400 km; 1,490 miles)
with max fuel, FAA reserves 985 n miles (1,825 km; 1,135 miles)
with 26 passengers and baggage, no reserves 780 n miles (1,450 km; 900 miles)
with 26 passengers and baggage, FAA reserves 550 n miles (1,020 km; 633 miles)

NEW ENTRY

AEROSPATIALE SE 210 CARAVELLE

TYPE: Twin-jet medium-range airliner.

PROGRAMME: The Caravelle twin-jet short/medium-range airliner was designed by the former SNCA du Sud-Est (later Sud-Aviation), and was ordered in prototype form by the Sécrétariat d'Etat à l'Air in January 1953. The first of two prototypes flew for the first time on 27 May 1955, and the second on 6 May 1956. There were 280 built of which approximately 45 still in service.

VERSIONS: **Caravelle I and IA:** Initial production series, Rolls-Royce Avon RA.29 Mk 522 (Series I) or Mk 522A (Series IA) turbojet engines. First production model flew on 18 May 1958. Entered service in mid-1959. FAA Type Approval 8 April 1959. Nineteen Srs I and 13 Srs IA built; all except one were later converted to Caravelle III standard.

Caravelle III: Second production version, with 50.71 kN (11,400 lb st) Avon RA.29 Mk 527 turbojets and standard accommodation for 64-80 passengers. First flew on 30 December 1959. First delivery in April 1960 to Alitalia, which later converted its four Series III to VI-N standard. FAA Type Approval 12 July 1960. One Caravelle III, airframe number 42, was fitted with General Electric CJ805-23C turbofan engines. In this form it was designated Caravelle VII, and flew for the first time on 29 December 1960, but was later restored to Srs III standard.

Caravelle VI-N: Two Avon RA.29 Mk 531 engines each 54.26 kN (12,200 lb st). Accommodation for 16-20 first class and 55-60 economy class passengers, or 80 economy class passengers. First flew on 10 September 1960. First delivery (to Sabena) in January 1961.

Caravelle VI-R: Similar to Series VI-N, but with modified windscreen for improved visibility, Rolls-Royce Avon 533R turbojet engines each 56.0 kN (12,600 lb st) fitted with thrust reversers, and spoilers in three sections on the trailing-edge of each wing. Prototype flew on 6 February 1961. FAA Type Approval 5 June 1961. Entered service 14 July 1961.

Caravelle 10 R: Similar to Series VI-R, but with Pratt & Whitney JT8D-7 turbofan engines each 62.27 kN (14,000 lb st) fitted with thrust reversers. Change in fuselage structure gives considerable increase in capacity of lower holds. Prototype flew for the first time on 18 January 1965.

Caravelle II R: Mixed passenger/freight version, derived from Series 10 R and also powered by two Pratt & Whitney JT8D-7 turbofan engines.

Increase of 0.93 m (36.6 in) in length of front fuselage. Floor strength increased from 600 kg/m^2 (123 lb/sq ft) to 1,000 kg/m^2 (205 lb sq ft) over a length of 9.0 m (29 ft 6 in). Number of cargo attachment rails increased from four to seven, and a cargo door, size 3.32 m × 1.825 m (10 ft 10¾ in × 6 ft 0¼ in), provided on port side of front fuselage. In addition, various parts of the fuselage have been strengthened. Typical mixed traffic payloads are: 12 first class passengers and 50 tourist passengers, with 46.5 m^3 (1,642 cu ft) of cargo space; 50 tourist class passengers with 66.0 m^3 (2,331 cu ft) of cargo space. In all-freight configuration, the Caravelle 11 R has 115 m^3 (4,061 cu ft) of cargo space. In all-passenger form, it can be fitted with 89 or 99 seats. The first Caravelle 11 R flew for the first time 21 April 1967.

Caravelle 12: Essentially, this is the airframe of the Caravelle Super B (see below), with an even longer fuselage, and powered by 64.50 kN (14,500 lb st) Pratt & Whitney JT8D-9 turbofan engines. Additional 2.00 m (6 ft 6¾ in) section inserted in the fuselage ahead of the wing leading-edge, and a second section 1.21 m (3 ft 11½ in) long added aft of the trailing-edge. Landing gear is in the strengthened form developed for the Super B at its highest operating weight of 56,000 kg (123,460 lb). Wings, engine pods, nose and tail unit, systems and equipment as for the Super B; but there is some strengthening of the central fuselage above the wings, enlargement of two of the emergency exits, and revision of the interior layout. Accommodation provides five-abreast seating for 128 tourist-class passengers, arranged as 12 rows at front and nine rows in rear of cabin with seats at 81 cm (32 in) pitch; three rows in centre, in line with emergency exits; and two four-abreast rows at extreme rear. An alternative tourist layout seating 118 passengers at 86 cm (34 in) pitch is available. A mixed class layout can be provided for 88 tourist class passengers at the latter seat pitch and 16 first-class passengers (four-abreast) at 96 cm (38 in) pitch. A passenger/cargo layout is also available, for 100 tourist passengers at 34 in pitch and 7.10 m^3 (250 cu ft) of cargo space.

Caravelle Super B: The first Caravelle Super B (airframe No. 169) flew for the first time 3 March 1964. It is normally powered by two 62.27 kN (14,000 lb st) Pratt & Whitney JT8D-7 turbofan engines, although the first examples were fitted with JT8D-1 engines. It differs from other Caravelles (except the Srs 11 R and Srs 12) in having a 1.0 m (3 ft 3½ in) longer fuselage, providing accommodation for 68 first class passengers (four-abreast), 86 mixed class passengers, or 104 tourist passengers.

Aerodynamic refinements include an extension forward of the wing leading-edge near the root, an increase from 35 to 45° in the operating travel of the flaps, which are of the double-slotted type, the addition of a bullet fairing at the intersection of the rudder and elevators, and a greater tailplane span than any other version except the Srs 12. First delivery of a Super B, to Finnair, was made on 25 July 1964. The Super B is also sometimes referred to as the Series 10 B.

UPGRADES: All Series I and II upgraded to Caravelle III standard. Alitalia converted its four Caravelle Series III to Caravelle VI-N standard.

OPERATORS: Versions of the Caravella are in service with the armed forces of the following countries France: (1) and Sweden (2).

DESIGN FEATURES: Cantilever low-wing monoplane. NACA 651212 wing section with cambered leading-edge. Sweepback at quarter-chord 20°. Dihedral 3°. Incidence at root 2°.

FLYING CONTROLS: Two-piece all-metal ailerons on each wing, operated hydraulically by duplicated actuators, with electric standby power. Hydraulically actuated Fowler flaps. Airbrakes on upper and lower surfaces ahead of flaps. Three-section spoilers on trailing-edge of each wing of Series VI-R. Hydraulically powered rudder and elevators, using duplicated actuators, with electric standby power.

STRUCTURE: Wing in two sections joined on fuselage centreline. All-metal three-spar structure, with spanwise stringers riveted to skin. The fuselage is a circular-section all-metal semi-monocoque structure. Maximum diameter 3.20 m (10 ft 6 in). The tail unit is a cantilever all-metal structure. Sweepback on tailplane 30° at quarter-chord.

LANDING GEAR: Retractable tricycle type. Hispano

Aerospatiale SE 210 Caravelle Super B at Gatwick Airport *(Peter J. Cooper)*

shock-absorbers. Twin nosewheel unit retracts forward. Each main unit has a four-wheel bogie and retracts inward. Goodyear, Firestone, Fléber-Colombes, Dunlop or Goodrich nosewheel tyres, size 26 × 7.75-13. Mainwheels and tubeless tyres of Goodyear, Dunlop, Firestone or Kléber-Colombes manufacture, size 35 × 9.00-17. Depending on type of tyre and AUW, nosewheel tyre pressures vary from 6.0-7.6 kg/cm² (85-109 lb/sq in). Similarly, pressure in front main bogies varies from 7.0-8.2 kg/cm² (100-117 lb/sq in) and in rear main bogies from 10.9-12.8 kg/cm² (155-182 lb/sq in). Hydraulic retraction. Maxaret anti-skid brakes on mainwheels.

POWER PLANT: Two turbojet or turbofan engines mounted in nacelles on each side of the rear fuselage just ahead of unit (details under Versions). Fuel in four integral tanks in wings, with total capacity of 19,000 litres (5,019 US gallons; 4,180 Imp gallons). Super B has provision for additional centre tank, increasing total capacity to 22,000 litres (5,812 US gallons; 4,840 Imp gallons).

ACCOMMODATION: Crew compartment for two or three persons. Details of passenger accommodation under Versions above. Entire accommodation pressurised. Main access to cabin aft through door under rear fuselage with hydraulically operated integral steps. Steps serve as tail support when lowered. Further door on port side at front of cabin. Two toilets, coat rooms and light baggage racks aft of cabin. Two galleys, one forward and one aft of cabin.

SYSTEMS: Air conditioning system utilises two turbo-compressors, driven by engine-bleed air, and includes a cold air unit. Pressure differential 0.57 bars (8.25 lb/sq in). Hydraulic system, pressure 172.37 bars (2,500 lb/sq in), for landing gear actuation, nosewheel steering, brakes, flying controls and airbrakes. Electrical system includes two 30 V DC engine-driven generators and inverters for 115 V 400 c/s AC. Series 12 and Super B have APU for engine starting, air conditioning of flight deck and cabin on ground and in flight up to maximum cruising altitude, and operation of a third 40 kVA alternator both on the ground and in the air. Provision for Aérospatiale/Lear automatic landing system for operation in Cat. IIIA weather conditions. Thermal de-icing on wings and tailplane.

DIMENSIONS, EXTERNAL:
- Wing span — 34.30 m (112 ft 6 in)
- Wing aspect ratio — 8.02
- Wing chord:
  - at root (except Srs 12 and Super B) — 6.33 m (20 ft 9 in)
  - at tip — 2.23 m (7 ft 4 in)
- Length overall: Srs III, VI 10 R — 32.01 m (105 ft 0 in)
  - Srs 11 R — 32.71 m (107 ft 4 in)
  - Srs 12 — 36.24 m (118 ft 10½ in)
  - Super B — 33.01 m (108 ft 3½ in)
- Height overall:
  - except Srs 12 and Super B — 8.72 m (28 ft 7 in)
  - Srs 12 and Super B — 9.01 m (29 ft 7 in)
- Tailplane span:
  - except Srs 12 and Super B — 10.60 m (34 ft 9 in)
  - Srs 12 and Super B — 12.00 m (39 ft 4 in)
- Wheel track (c/l of shock struts) — 5.21 m (17 ft 0 in)
- Wheelbase: Srs III, VI, 10 R — 11.79 m (38 ft 7 in)
  - Srs 11 R — 12.72 m (41 ft 8¾ in)
  - Srs 12 — 14.80 m (48 ft 6½ in)
  - Super B — 12.50 m (41 ft 0 in)
- Passenger door (fwd, port): Height — 1.69 m (5 ft 6½ in)
  - Width — 0.91 m (3 ft 0 in)
  - Height to sill — 2.35 m (7 ft 8½ in)
- Crew door (fwd, stbd): Height — 1.22 m (4 ft 0 in)
  - Width — 0.61 m (2 ft 0 in)
- Cargo compartment doors (underfloor, stbd):
  - Height — 0.91 m (3 ft 0 in)
  - Width — 0.76 m (2 ft 6 in)

DIMENSIONS, INTERNAL:
- Cabin, excl flight deck:
  - Length: Srs III, VI, 10 R — 22.50 m (73 ft 8½ in)
    - Srs 11 R — 23.37 m (76 ft 8½ in)
    - Srs 12 (incl toilet and rear compartments) — 26.40 m (86 ft 7 in)
    - Srs 12 (excl toilet and rear compartments) — 22.24 m (72 ft 11 in)
    - Super B — 23.45 m (76 ft 11½ in)
- Width at floor: except Srs 12 — 2.72 m (8 ft 11 in)
  - Srs 12 — 2.69 m (8 ft 10 in)
- Width at armrest — 3.00 m (9 ft 9½ in)
- Max height — 2.00 m (6 ft 7 in)
- Floor area: Srs III, VI, 10 R — 60.00 m² (646 sq ft)
  - Srs 11 R — 62.25 m² (670 sq ft)
  - Srs 12 — 71.28 m² (767 sq ft)
  - Super B — 62.70 m² (675 sq ft)
- Volume: Srs III, VI, 10 R — 110 m³ (3,885 cu ft)
  - Srs 11 R — 124 m³ (4,379 cu ft)
  - Srs 12 — 142 m³ (5,015 cu ft)
  - Super B — 125 m³ (4,414 cu ft)
- Freight holds (main cabin):
  - except Super B — 5.70 m³ (201 cu ft)
  - Super B — 4.33 m³ (153 cu ft)
- Freight holds (underfloor): Srs III — 9.60 m³ (339 cu ft)
  - Srs VI, 10R — 10.60 m³ (374 cu ft)
  - Srs 12 (rear) — 5.00 m³ (176 cu ft)
  - Srs 12 (fwd) — 11.50 m³ (406 cu ft)
  - Super B — 12.00 m³ (423 cu ft)

AREAS:
- Wings, gross — 146.7 m² (1,579 sq ft)
- Ailerons (total) — 7.84 m² (84.4 sq ft)
- Trailing-edge flaps (total) — 24.70 m² (265.8 sq ft)
- Fin — 10.00 m² (107.6 sq ft)
- Rudder — 5.50 m² (59.2 sq ft)
- Tailplane — 21.55 m² (232.0 sq ft)
- Elevators — 6.45 m² (69.4 sq ft)

WEIGHTS AND LOADINGS:
- Manufacturer's weight empty:
  - Srs III — 24,185 kg (53,320 lb)
  - Srs VI-N — 24,915 kg (54,922 lb)
  - Srs VI-R — 26,280 kg (57,935 lb)
  - Srs 10 R — 26,725 kg (58,920 lb)
  - Srs 11 R — 28,841 kg (63,585 lb)
  - Srs 12 — 29,500 kg (65,050 lb)
  - Super B — 27,623 kg (60,897 lb)
- Basic operating weight: Srs III — 27,210 kg (59,985 lb)
  - Srs VI-N — 27,330 kg (60,250 lb)
  - Srs VI-R — 28,655 kg (63,175 lb)
  - Srs 10 R — 29,075 kg (64,100 lb)
  - Srs 12 — 31,800 kg (70,100 lb)
  - Super B — 30,055 kg (66,260 lb)
- Max payload: Srs III — 8,400 kg (18,520 lb)
  - Srs VI-N — 7,900 kg (17,415 lb)
  - Srs VI-R — 8,200 kg (18,080 lb)
  - Srs 10 R — 9,400 kg (20,720 lb)
  - Srs 11 R — 9,095 kg (20,050 lb)
  - Srs 12 — 13,200 kg (29,100 lb)
  - Super B — 9,100 kg (20,060 lb)
- Max T-O weight: Srs III — 46,000 kg (101,413 lb)
  - Srs VI-N — 48,000 kg (105,822 lb)
  - Srs VI-R — 50,000 kg (110,230 lb)
  - Srs 10 R, 11 R, Super B (standard u/c) — 52,000 kg (114,640 lb)
  - Srs 10 R, 11 R, Super B (strengthened u/c) — 54,000 kg (119,050 lb)
  - Srs 12 — 56,000 kg (123,460 lb)
  - Super B — 56,000 kg (123,460 lb)
- Max landing weight: Srs III — 43,800 kg (96,560 lb)
  - Srs VI-N — 45,700 kg (100,750 lb)
  - Srs VI-R — 47,620 kg (104,990 lb)
  - Srs 10 R, 11 R, 12, Super B — 49,500 kg (109,130 lb)
- Max zero-fuel weight:
  - Srs III, VI-N — 35,500 kg (78,260 lb)
  - Srs VI-R — 37,000 kg (81,570 lb)
  - Srs 10 R — 38,500 kg (84,880 lb)
  - Srs 11 R — 40,000 kg (88,185 lb)
  - Srs 12 — 45,000 kg (99,200 lb)
  - Super B — 39,500 kg (87,080 lb)
- Max wing loading: Srs III — 313 kg/m² (64.2 lb/sq ft)
  - Srs VI-N — 326 kg/m² (66.8 lb/sq ft)
  - Srs VI-R — 341 kg/m² (69.8 lb/sq ft)
  - Srs 10 R, 11 R — 354 kg/m² (72.6 lb/sq ft)
  - Srs 12 — 381 kg/m² (78.0 lb/sq ft)
  - Super B — 368 kg/m² (75.4 lb/sq ft)
- Max power loading: Srs III — 4.44 kg/kg st (4.44 lb/lb st)
  - Srs VI-N — 4.33 kg/kg st (4.33 lb/lb st)
  - Srs VI-R — 4.37 kg/kg st (4.37 lb/lb st)
  - Srs 10 R, 11 R — 4.10 kg/kg st (4.10 lb/lb st)
  - Srs 12, Super B — 4.25 kg/kg st (4.25 lb/lb st)

PERFORMANCE:
- Max cruising speed at 7,620 m (25,000 ft):
  - Srs III at AUW 40,820 kg (90,000 lb) — 434 kts (805 km/h; 500 mph)
  - Srs VI-N at AUW of 41,730 kg (92,000 lb) — 456 kts (845 km/h; 525 mph)
  - Srs VI-R at AUW of 43,000 kg (94,800 lb) — 456 kts (845 km/h; 525 mph)
  - Srs 10 R at AUW of 47,250 kg (104,170 lb) — 432 kts (800 km/h; 497 mph)
  - Srs 11 R at AUW of 52,000 kg (114,640 lb) — 432 kts (800 km/h; 497 mph)
  - Srs 12 at AUW of 50,000 kg (110,230 lb) — 437 kts (810 km/h; 503 mph)
  - Super B at AUW of 52,000 kg (114,640 lb) — 445 kts (825 km/h; 512 mph)
- Best-cost cruising speed at 10,670 m (35,000 ft):
  - Srs III at AUW of 42,750 kg (94,250 lb) — 391 kts (725 km/h; 450 mph)
  - Srs VI-N at AUW of 43,750 kg (96,450 lb) — 426 kts (790 km/h; 490 mph)
  - Srs VI-R at AUW of 34,500 kg (100,310 lb) — 424 kts (785 km/h; 488 mph)
  - Srs 10 R at AUW of 47,250 kg (104,170 lb) — 405 kts (750 km/h; 466 mph)
  - Super B at AUW of 47,750 kg (105,270 lb) — 432 kts (800 km/h; 497 mph)
- T-O distance at max T-O weight:
  - Srs III — 1,830 m (6,000 ft)
  - Srs VI-N — 1,950 m (6,400 ft)
  - Srs VI-R — 2,073 m (6,800 ft)
  - Srs 10 R (AUW of 52,000 kg; 114,640 lb) — 1,950 m (6,400 ft)
- T-O balanced field length at max T-O weight:
  - Srs 11 R (CAR) — 1,950 m (6,400 ft)
  - Srs 12 (ISA at S/L) — 2,240 m (7,345 ft)
  - Srs 12 (ISA + 15°C) — 2,380 m (7,805 ft)
  - Super B (SR 422 B) — 2,090 m (6,850 ft)
- Landing distance at max landing weight:
  - Srs III — 1,800 m (5,900 ft)
  - Srs VI-N — 1,965 m (6,450 ft)
  - Srs VI-R — 1,720 m (5,650 ft)
  - Srs 10 R — 1,620 m (5,315 ft)
  - Srs 11 R — 1,550 m (5,085 ft)
  - Super B — 1,580 m (5,180 ft)
- Landing distance at AUW of 49,000 kg (108,025 lb), ISA at S/L: Srs 12 — 1,570 m (5,150 ft)
- Range with 7,620 kg (16,800 lb) payload, 260 n miles (480 km; 300 mile) diversion and SR 427 reserves:
  - Srs III — 995 n miles (1,845 km; 1,146 miles)
  - Srs VI-N — 1,350 n miles (2,500 km; 1,553 miles)
  - Srs VI-R — 1,380 n miles (2,560 km; 1,590 miles)
  - Srs 10 R — 1,780 n miles (3,295 km; 2,050 miles)
  - Super B — 1,758 n miles (3,260 km; 2,025 miles)
- Range with 8,550 kg (18,850 lb) payload, max fuel, no reserves:
  - Srs 12 — 2,055 n miles (3,810 km; 2,367 miles)
- Range with 8,550 kg (18,850 lb) payload, reserves for 199 n miles (370 km; 230 mile) diversion:
  - Srs 12 — 1,478 n miles (2,740 km; 1,702 miles)
- Range with max payload, 260 n miles (480 km; 300 mile) diversion and SR 427 reserves:
  - Srs III — 915 n miles (1,700 km; 1,056 miles)
  - Srs VI-N — 1,270 n miles (2,350 km; 1,460 miles)
  - Srs VI-R — 1,240 n miles (2,300 km; 1,430 miles)
  - Srs 10 R (at 52,000 kg; 114,640 lb AUW) — 1,565 n miles (2,900 km; 1,800 miles)
  - Srs 10 R (at 54,000 kg; 119,050 lb AUW) — 1,865 n miles (3,455 km; 2,145 miles)
  - Super B — 1,435 n miles (2,655 km; 1,650 miles)
- Range with max payload, with reserves:
  - Srs 11 R (at 52,000 kg; 114,640 lb AUW) — 1,241 n miles (2,300 km; 1,430 miles)
  - Srs 11 R (at 54,000 kg; 19,050 lb AUW) — 1,511 n miles (2,800 km; 1,740 miles)
- Range with max payload, reserves for 199 n miles (370 km; 230 miles) diversion:
  - Srs 12 — 872 n miles (1,620 km; 1,005 miles)
- Range with max payload, no reserves:
  - Srs 12 — 1,428 n miles (2,650 km; 1,645 miles)

UPDATED

AEROSPATIALE SN 601 CORVETTE

TYPE: Twin-engine transport aircraft.

PROGRAMME: The Corvette was designed to fulfil a variety of roles, including executive transport, air taxi, ambulance, freighter or training aircraft. It can be equipped for radio aids calibration or aerial photography.

VERSIONS: **SN 600**: Prototype only (F-WRSN), with two rear-mounted Pratt & Whitney Aircraft of Canada JT15D-1 turbofan engines, each rated at 9.81 kN (2,200 lb st). First flew on 16 July 1970, and completed more than 270 flying hours before being lost in a crash on 23 March 1971.

SN 601: Initial production Corvette, with two JT15D-4 turbofan engines and a longer fuselage than the prototype. The first SN 601 (F-WUAS) was completed in 1972, together with two airframes for static and fatigue testing. It flew for the first time on 20 December 1972, and was followed by the second SN 601 (F-WRNZ) on 7 March 1973, the third (F-WUQP) on 12 January 1974. Type certification of this version was received from the SGAC on 28 May 1974 and from the FAA on 24 September 1974. Deliveries began in September 1974. Total of 40 built.

DESIGN FEATURES: Cantilever low-wing monoplane of all-metal construction. Thickness/chord ratio 13.65 per cent at root, 11.5 per cent at tip. Dihedral 3° on outer panels. Sweepback 22° 32′ on leading-edge. Cantilever tail unit with sweepback on all surfaces.

FLYING CONTROLS: Manually operated aluminium alloy ailerons and electrically operated double-slotted long travel trailing-edge flaps of aluminium alloy and honeycomb construction. Three-section spoiler forward of each outer flap. Hydraulically actuated airbrakes inboard of spoilers, above and below each wing. Electrically actuated trim tab in port aileron. Electrically actuated variable-incidence tailplane. Manually operated elevators and rudder. Electrically actuated trim tab in rudder.

STRUCTURE: Conventional two-spar fail-safe structure, of aluminium alloy. The fuselage is an aluminium alloy semi-monocoque fail-safe structure of circular cross-section. The tail unit is a cantilever aluminium alloy structure, with tailplane mounted on fin. Sweepback on all surfaces.

LANDING GEAR: Hydraulically retractable tricycle type, with hydraulic shock-absorbers and single wheel on each unit. Mainwheels retract inward, nosewheel forward into fuselage. Low-pressure tyres of 245 mm (10 in) diameter on mainwheels and 152 mm (6 in) diameter on nosewheel. Mainwheel tyre pressure 5.45 bars (79 lb/sq in); nosewheel tyre pressure 4.48 bars (65 lb/sq in). Hydraulic brakes and anti-skid units. Nosewheel steerable.

POWER PLANT: Two 11.12 kN (2,500 lb st) Pratt & Whitney Aircraft of Canada JT15D-4 turbofan engines, mounted in pod on each side of rear fuselage. Two integral wing fuel tanks, with total capacity of 1,660 litres (439 US gallons; 365 Imp gallons). Provision for tip-tanks of approximately 700 litres (185 US gallons; 154 Imp gallons) total capacity.

ACCOMMODATION: Crew of one or two on flight deck. Normal seating for 6 to 14 passengers in single seats on each side of centre aisle. Galley, toilet and baggage compartments available to customer's requirements. Two-part door, with built-in airstairs, at front on port side; upper part of door is hinged at top, lower part is hinged at bottom of doorway.

SYSTEMS: Cabin air conditioning and pressurisation by engine bleed air; maximum differential 0.59 bars (8.6 lb/sq in). Hydraulic system for actuating landing gear, nosewheel steering, wheel brakes and airbrakes. Main electrical system includes two 10.5 kW 28.5 V DC starter/generators, one 36 Ah battery and two inverters for 400 Hz AC supply. TKS type de-icing of leading-edges.

ELECTRONICS AND EQUIPMENT: Blind-flying instrumentation standard. Radio, radar or other special equipment to customer's requirement.

DIMENSIONS, EXTERNAL:

Wing span	12.87 m (42 ft 2½ in)
Wing aspect ratio	7.45
Length overall	13.83 m (45 ft 4½ in)
Length of fuselage	12.90 m (42 ft 4 in)
Fuselage diameter	1.70 m (5 ft 7 in)
Height overall	4.23 m (13 ft 10½ in)
Tailplane span	5.00 m (16 ft 4¾ in)
Wheel track	2.75 m (8 ft 5¼ in)
Wheelbase	5.22 m (17 ft 1½ in)
Passenger door: Height	1.31 m (4 ft 3½ in)
Width	0.71 m (2 ft 4 in)
Mean height to sill	0.85 m (2 ft 9½ in)

DIMENSIONS, INTERNAL:

Cabin, excl flight deck: Max length	5.73 m (18 ft 9½ in)
Max width	1.56 m (5 ft 1½ in)
Max height	1.52 m (5 ft 0 in)
Floor area	6.60 m² (71 sq ft)
Volume	9.93 m³ (351 cu ft)
Baggage compartment volume (10-passenger) layout	1.08 m³ (38.1 cu ft)

AREAS:

Wings, gross	22.0 m² (236.8 sq ft)
Vertical tail surfaces	4.22 m² (45.4 sq ft)
Horizontal tail surfaces	5.47 m² (58.9 sq ft)

WEIGHTS:

Manufacturer's weight empty	3,510 kg (7,738 lb)
Max T-O weight	6,600 kg (14,550 lb)
Max ramp weight	6,650 kg (14,660 lb)
Max landing weight	5,700 kg (12,550 lb)
Max zero-fuel weight	5,600 kg (12,345 lb)

PERFORMANCE (at max T-O weight, except where indicated):

Max cruising speed at 9,000 m (30,000 ft)	410 kts (760 km/h; 472 mph)
Econ cruising speed at 11,900 m (39,000 ft)	306 kts (566 km/h; 352 mph)
Stalling speed, flaps and landing gear down at max landing weight	91 kts (168 km/h; 105 mph)
Max rate of climb at S/L	823 m (2,700 ft)/min
Service ceiling	12,500 m (41,000 ft)
T-O balanced field length (FAR 25)	1,390 m (4,560 ft)
Landing distance at max landing weight	755 m (2,480 ft)
Max range with tip tanks, 45 min reserves:	
Max cruise power	1,290 n miles (2,390 km; 1,485 miles)
Econ cruise power	1,380 n miles (2,555 km; 1,588 miles)
Range with 12 passengers, 45 min reserves:	
Max cruise power	800 n miles (1,480 km; 920 miles)
Econ cruise power	840 n miles (1,555 km; 967 miles)

OPERATIONAL NOISE CHARACTERISTICS:

T-O noise level at 3.5 n miles (6.5 km; 4 miles) from start of T-O roll: without power reduction	81.2 EPNdB
with power reduction	74 EPNdB
Approach noise level at 1.08 n miles (2 km; 1.24 miles) from landing threshold on 3° glide slope	90 EPNdB

VERIFIED

Aerospatiale SN 601 Corvette executive transport aircraft

1994

Aerospatiale SN 601 Corvette with wingtip fuel tanks for extended range

1994

AEROSPATIALE/COTAM TRANSALL C-160 RENOVE AVIONICS AND SYSTEMS UPGRADE

TYPE: Military and civil cargo aircraft avionics and systems upgrade.

PROGRAMME: Research and development to upgrade the French Air Force C-160 Transall aircraft began 1988. First flight 6 February 1992. Prototype delivery 30 April 1993. First series aircraft May 1994. Final delivery 1999.

DESIGN FEATURES: Programme consists of five system architectures:

(1) Aircraft position and attitude. Three types of new sensor: one inertial reference unit (IRU), one attitude and heading reference unit (AHRU) and one GPS receiver based on the acquisition of signals from a constellation of satellites.

(2) Flight management system (FMS), built around two Gemini 10 computers to which are associated three control and display units.

(3) Radio management system (RMS), enabling the control of radio-communications transceivers and radio-navigation receivers.

(4) Electronic flight instrument system (EFIS)-854 TF, consisting of one electronic attitude director indicator (EADI) and one electronic horizontal situation indicator (EHSI) for the pilot and co-pilot.

(5) Head-up display system consisting of optical head unit (OHU) and head-up display computer (HUDC), used for all types of mission, providing essential flight information for the tactical final phases of assault landing and air dropping.

The programme also includes a self-protection system consisting of decoys and radar warning.

UPDATED

Transall C-160 view of modernised pilot's instruments panels *(Aerospatiale)*

1994

Transall C-160 view of modernised navigator's instrument panel *(Aerospatiale)*

1994

DASSAULT

DASSAULT AVIATION

9 Rond-Point Champs Elysées-Marcel Dassault, F-75008 Paris
Telephone: 33 (1) 43 59 14 70
Fax: 33 (1) 42 56 22 49
PRESS INFORMATION OFFICE: 27 rue du Professeur Pauchet, F-92420 Vaucresson
Telephone: 33 (1) 47 95 86 87
Fax: 33 (1) 47 95 86 80/47 95 87 40
WORKS: F-92214 St Cloud, F-95101 Argenteuil, F-78141 Vélizy-Villacoublay, F-33127 Martignas, F-33701 Bordeaux-Mérignac, F-33630 Cazaux, F-64205 Biarritz-Parme, F-13801 Istres, F-74371 Argonay, F-59472 Lille-Seclin, F-86000 Poitiers
CHAIRMAN AND CEO: Serge Dassault
VICE-PRESIDENT, ENGINEERING, RESEARCH, CO-OPERATION: Bruno Revellin-Falcoz
VICE-PRESIDENT, ECONOMICS AND FINANCE: Charles Edelstenne
VICE-PRESIDENT, INDUSTRIAL AND SOCIAL AFFAIRS: Michel Herchin
SENIOR VICE-PRESIDENT, INTERNATIONAL AFFAIRS: Pierre Chouzenoux
DIRECTOR OF COMMUNICATIONS: Jean-Pierre Robillard

Former Avions Marcel Dassault-Breguet Aviation formed from merger of Dassault and Breguet aircraft companies in December 1971; French government acquired 20 per cent of stock in January 1979, raised to 46 per cent in November 1981; present company name adopted April 1990. Some shares hold double voting rights so that French government has majority control, with 55 per cent. On 17 September 1992, Dassault joined with Aerospatiale in state-owned joint holding company, SOGEPA (which see), which has 35 per cent holding in Dassault Aviation; Dassault and Aerospatiale now pool resources and co-ordinate R&D strategy, although each remains a separate entity.

Employees in 1993 totalled 9,800. Business divided 65 per cent military, 25 per cent civil and 10 per cent space. Dassault assembles and tests its civil and military aircraft in its own factories, but operates wide network of subcontractors. Other products include flight control system components, maintenance and support equipment and CAD/CAM software (CATIA). Dassault Aviation is component of Dassault Industries, which also includes Dassault Falcon Service, Dassault Belgique Aviation and Dassault Electronique.

Dassault Aviation has shared in Atlantique programme with Belgium, Germany and Italy; SEPECAT (Jaguar) with UK; and Alpha Jet with Dornier. Offset manufacture of Dassault aircraft components arranged with Belgium, Egypt, Greece and Spain. Dassault makes fuselages for Fokker at Biarritz-Parme.

Dassault has produced 1,100 executive jets and received orders for almost 2,700 Mirages of all types; over 6,500 aircraft built since 1945.

UPDATED

DASSAULT/DORNIER ALPHA JET

Details of the Alpha Jet programme can be found in the International section of this edition.

VERIFIED

DASSAULT/BAe JAGUAR

Details of the Jaguar programme can be found under SEPECAT in the International section of this edition.

VERIFIED

DASSAULT MIRAGE III

TYPE: Single-seat fighter-bomber intruder aircraft.

PROGRAMME: Mirage III was designed initially as a Mach 2 high-altitude all-weather interceptor, capable of performing ground support missions and requiring only small airstrips. Developed versions include a two-seat trainer, long-range fighter-bomber and reconnaissance aircraft. A total of more than 1,400 Mirage III/5/50s of all types have been ordered and delivered for service in 20 countries, including licensed production abroad.

The experimental prototype flew for the first time on 17 November 1956, powered by a SNECMA Atar 101G turbojet with afterburner (44.1 kN; 9,900 lb st).

VERSIONS: **Mirage III-B:** Two-seat version of the III-A preseries production version.

Mirage III-BE: Two-seat version of the III-E.

Mirage III-C: All-weather interceptor and day ground attack fighter.

Mirage III-D: Two-seat version built initially in Australia for the RAAF. Similar French built models ordered by 12 countries, including six more for Australia. Atar 9C afterburning turbojet engine. Tandem seating under one-piece canopy; radar deleted, but fitted with radio beacon equipment. Intended primarily as a trainer, but suitable for strike sorties, carrying air-to-surface armament. Total of 186 Mirage III-B/III-D/5 two-seaters sold to 20 countries.

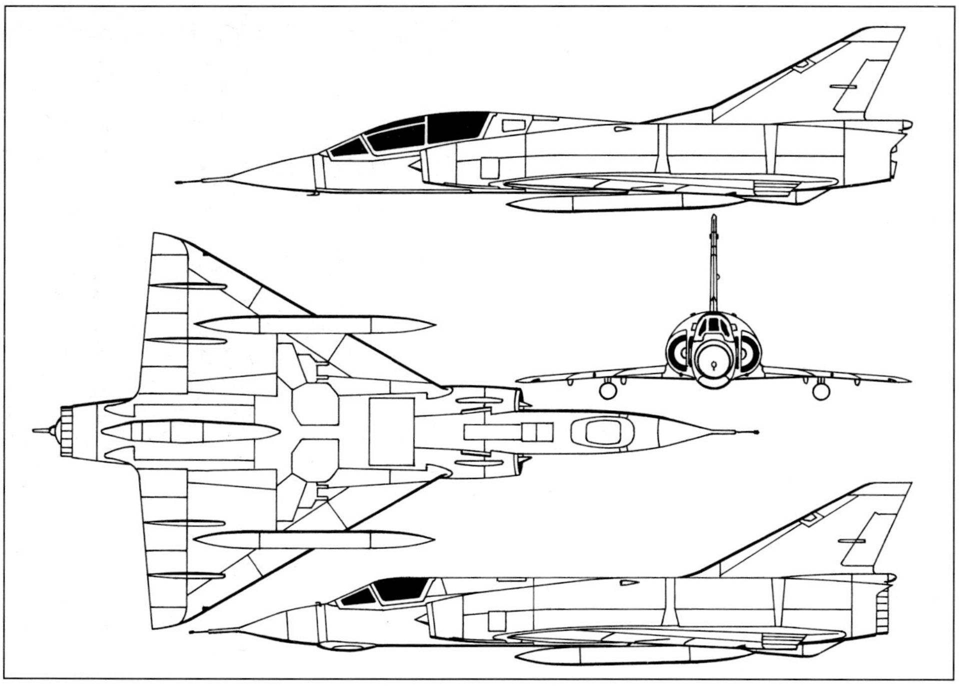

Dassault Mirage III. Additional side-view (top) of two-seat version *(Jane's/Mike Keep)* *1994*

Mirage IIID2Z/EZ/R2Z: Built for South Africa with Atar 9K-50 turbojet engines. Delivered in 1974-75. Rebuilt by Altas Aviation and renamed Cheetah D/E/R.

Mirage III-E: Long-range fighter-bomber/intruder version with Atar 9C afterburning turbojet engine, 532 have been built for 13 air forces. First of three prototypes flew on 5 April 1961, and the first delivery of a production III-E was made in January 1964. Thirty III-Es of the 4e Escadre of the French Air Force, equipping two squadrons at Luxeuil were carriers of the 15 kT AN 52 tactical nuclear weapon. They have since been replaced by the Mirage 2000.

Mirage III-R: Reconnaissance version of III-E. Set of five Omera type 31 cameras, in place of radar in nose, can be focused in four different arrangements for very low altitude, medium altitude, high altitude and night reconnaissance missions. Self-contained navigation system. Provision for air-to-surface armament. Two prototypes, of which the first flew in November 1961. Total of 159 production models ordered, including Mirage 5-Rs for nine air forces.

Mirage III-RD: Similar to III-R but with improved Doppler navigation system in fairing under front fuselage, gyrostabilised gunsight and nose pack containing Omera 40 and 33 cameras. Twenty built for French Air Force; other with avionics changes, for Pakistan. An unspecified version related to the French Air Force version has provision for carrying SAT Cyclope infra-red tracking equipment in a modified nosecone.

Mirage III-S: Developed from Mirage III-E. There were 36 supplied to Swiss Air Force.

UPGRADES: **Altas Aviation:** Cheetah see separate entry.

Dassault: For full details see Advanced Technology Upgrade Programmes in this section.

SAGEM: Mirage III upgrade for Pakistan Air Force. See separate entry.

Swiss Federal Aircraft Factory (F+W): See Mirage III Upgrade Programme.

OPERATORS: Versions of the Mirage III are in service with the armed forces of the following countries: Argentina (39), Brazil (20), Chile (1), France (12+), Pakistan (70+), South Africa (29) and Switzerland (50+).

The following description refers to the Mirage III-E, but is generally applicable to all versions.

DESIGN FEATURES: Cantilever low-wing monoplane of delta planform, with conical camber. Thickness/chord ratio 4 per cent to 3.4 per cent. Anhedral 1°. No incidence. Sweepback on leading-edge 60° 34'. Tail unit has sweptback fin.

FLYING CONTROLS: Trailing-edge of each wing comprises two elevons for pitch and roll control, and an inboard flap which also has an elevator function. All control surfaces hydraulically powered by Dassault twin-cylinder actuators with artificial feel. Airbrakes, comprising small panels hinged to upper and lower wing surfaces, near leading-edge. Hydraulically actuated rudder only. Dassault twin-cylinder actuators with artificial feel.

STRUCTURE: All-metal torsion box structure; stressed skin of machined panels with integral stiffeners. The fuselage is an all-metal structure, 'waisted' in accordance with the area rule. Tail unit consists of fin and rudder only.

LANDING GEAR: Retractable tricycle type, with single wheel on each unit. Hydraulic retraction, nosewheel rearward main units inward. Messier-Hispano-Bugatti shock-absorbers and disc brakes. Mainwheel tyres size 750 × 230-15/1, pressure 5.9-9.8 bars (85.5-142 lb/sq in). Nosewheel tyre size 450 × 390-05. Braking parachute.

POWER PLANT: One SNECMA Atar 9C turbojet engine (60.8 kN; 13,670 lb st with afterburning), fitted with an

Dassault Mirage III-DBR of the Brazilian Air Force *(Aviaplans/F. Robineau)* *1994*

overspeed system which is engaged automatically from Mach 1.4 and permits a thrust increase of approximately 8 per cent in the high supersonic speed range. Movable half-cone centrebody in each air intake. Optional and jettisonable SEPR 844 single-chamber rocket motor (14.7 kN; 3,300 lb st) under engine bay. Space for structural fuel tank, capacity 550 litres (145 US gallons; 121 Imp gallons), when rocket motor is not fitted. Four flexible fuel tanks around engine air inlet ducts, combined capacity 1,020 litres (269 US gallons; 224 Imp gallons). Two integral fuel tanks in each wing combined capacity 1,370 litres (361 US gallons; 301 Imp gallons). Total available internal fuel (without rocket motor) 2,940 litres (775.8 US gallons; 646 Imp gallons). Provision for this to be augmented by two 625, 1,100, 1,300 or 1,700 litre (165, 290, 343, 449 US gallon; 137, 242, 285 or 374 Imp gallon) underwing drop tanks; 500 litre (132 US gallon; 110 Imp gallon) non-jettisonable supersonic tanks; JL-100 jettisonable tanks each housing both 250 litres (66 US gallons; 55 Imp gallons) fuel and air-to-surface rockets; Bidon Cyclope jettisonable tanks each housing 1,100 litres (290 US gallons; 242 Imp gallons) fuel and electronic equipment, or Bidon Homing jettisonable tanks housing 850 litres (224 US gallons; 187 Imp gallons) fuel and electronic equipment.

ACCOMMODATION: Single seat under rearward-hinged canopy. Hispano built Martin-Baker RM4 zero altitude/90 knots (267 km/h; 104 mph) ejection seat.

SYSTEMS: Two separate air conditioning systems for cockpit and avionics. Two independent hydraulic systems, pressure 207 bars (3,000 lb/sq in), for flying controls, landing gear and brakes. Power for DC electrical system for 24 V 40 Ah batteries and a 26.5 V 9 kW generator. AC electrical system power provided by one 200 V 400 Hz transformer and one 200 V 10 kVA alternator.

AVIONICS AND EQUIPMENT: Duplicated UHF, Tacan, Doppler, CSF Cyrano II fire control radar in nose, navigation computer and automatic gunsight. Central gyro and other avionics provide accurate and stabilised heading information. CSF 97 sighting system gives air-to-air facility for cannon and missiles, air-to-ground facility for dive bombing or LABS, and navigation facility for horizon and heading.

ARMAMENT: Ground attack armament consists normally of two 30 mm DEFA 552A guns in fuselage, each with 125 rounds of incendiary, high-explosive or armour-piercing ammunition, and two 454 kg (1,000 lb) bombs, or an AS 30 air-to-surface missile under the fuselage and 454 kg (1,000 lb) bombs under the wings. Total external load, on five hardpoints 4,000 kg (8,818 lb). Alternative underwing stores include combined tank/bomb carriers, each with 500 litres (132 US gallons; 110 Imp gallons) of fuel and 907 kg (2,000 lb) of bombs; JL-100 pods, each with 250 litres (66 US gallons; 55 Imp gallons) of fuel and 18 rockets; jettisonable underwing fuel tanks. For interception duties, one Matra R.530 air-to-air missile can be carried under fuselage, with optional guns and two Matra Magic missiles.

DIMENSIONS, EXTERNAL:

Wing span	8.22 m (26 ft 11½ in)
Wing ratio	1.94
Length overall: III-E	15.03 m (49 ft 3½ in)
III-R	15.50 m (50 ft 10¼ in)
Height overall	4.50 m (14 ft 9 in)
Wheel track	3.15 m (10 ft 4 in)
Wheelbase: III-E	4.87 m (15 ft 11¾ in)

AREAS:

Wings, gross	35.00 m² (376.7 sq ft)
Vertical tail surfaces (total)	4.5 m² (48.4 sq ft)

WEIGHTS AND LOADINGS:

Weight empty: III-E	7,050 kg (15,540 lb)
III-R	6,600 kg (14,550 lb)
T-O weight 'clean': III-E	9,600 kg (21,165 lb)
Max T-O weight: III-E, R	13,700 kg (30,200 lb)
Max wing loading: III-E, R	393.1 kg/m² (80.53 lb/sq in)

PERFORMANCE (Mirage III-E, in 'clean' condition with guns installed, except where indicated):

Max level speed at 12,000 m (39,375 ft)	Mach 2.2 (1,268 kts; 2,350 km/h; 1,460 mph)
Max level speed at S/L	750 kts (1,390 km/h; 863 mph)
Cruising speed at 11,000 m (36,000 ft)	Mach 0.9
Approach speed	183 kts (340 km/h; 211 mph)
Landing speed	157 kts (290 km/h; 180 mph)
Time to 11,000 m (36,000 ft), Mach 0.9	3 min
Time to 15,000 m (49,200 ft), Mach 0.9	6 min 50 s
Service ceiling at Mach 1.8	17,000 m (55,775 ft)
Ceiling, using rocket motor	23,000 m (75,450 ft)
T-O run according to mission (up to max T-O weight)	700-1,600 m (2,295-5,250 ft)
Landing run, using brake parachute	700 m (2,295 ft)
Combat radius, ground attack	647 n miles (1,200 km; 745 miles)

UPDATED

Dassault Mirage III-E of the French Air Force

1994

Dassault Mirage III-BE of the French Air Force *(Sirpa Air)*

1994

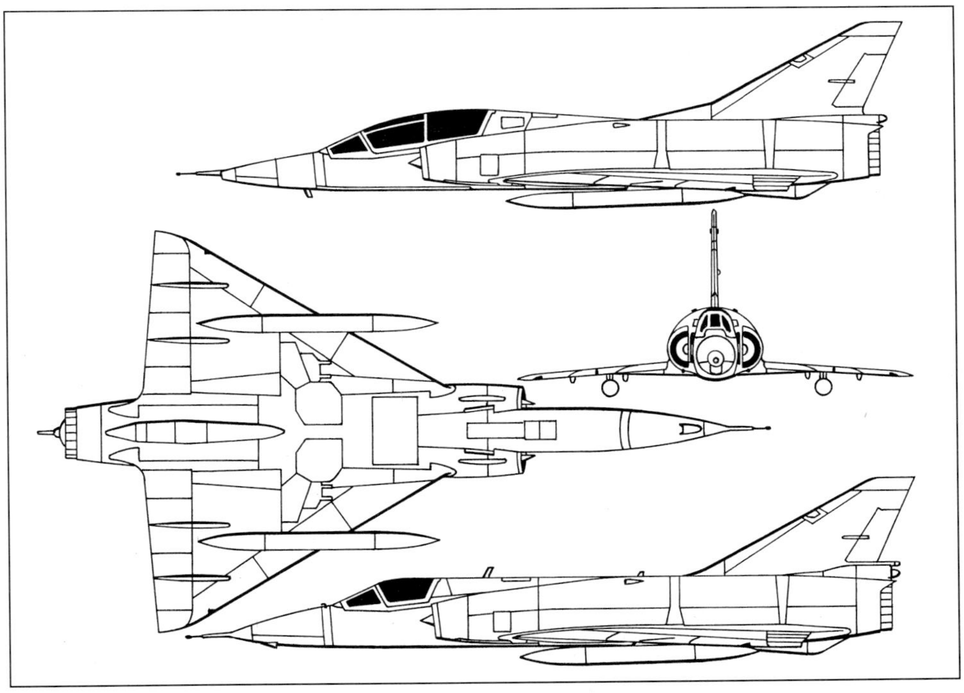

Dassault Mirage 5 ground attack aircraft. Additional side-view (top) of two-seat version *(Jane's/Mike Keep)*

1994

DASSAULT MIRAGE 5

TYPE: Single- and two-seat ground attack aircraft.

PROGRAMME: The Mirage 5 is a ground attack aircraft using the same airframe and engine as the Mirage III-E. The basic VFR version has simplified avionics, 470 litres (123 US gallons; 103 Imp gallons) greater fuel capacity than the III-E, in a tank between the engine air intakes, and considerably extended stores carrying capability. It combines the full Mach 2+ capability of the Mirage III, and its capability to operate from semi-prepared airfields, with simpler maintenance. In ground attack configuration, up to 4,000 kg (8,818 lb) of weapons and 1,000 litres (264 US gallons; 220 Imp gallons) of fuel can be carried externally on seven wing and fuselage attachment points. The Mirage 5 can also be flown as an interceptor, with two Magic or Sidewinder air-to-air missiles and 4,700 litres (1,241 US gallons; 1,034 Imp gallons) of external fuel. At customer's option any degree of IFR/all-weather operation was provided for, with reduced fuel or weapons load. The Mirage 5 was flown for the first time on 19 May 1967.

VERSIONS: **Mirage 5:** Initial version developed for Israel with Atar 9C-3 turbojet engines as a clear weather fighter bomber.

Mirage 5D: Export version based on Mirage 5 airframe. Radar ommitted.

Mirage 5R: Reconnaissance version with cameras fitted in modified nosecone.

Nesher/Dagger: Israeli built Mirage 5 named Nesher.

Subsequently transferred to Argentina and renamed Dagger.

UPGRADES: **Dassault:** For details see Advanced Technology Upgrade Programmes in this section.

SABCA: Mirsip/Elkan programme see separate entry in Belgium section.

SAGEM: Participation in the Mirsip/Elkan programme see separate entry in this section.

SNECMA: Contract to re-engine Colombian Air Force Mirage 5s with Atar 9K50 turbojets.

OPERATORS: Versions of the Mirage 5 are in service with the armed forces of the following countries: Abu Dhabi (29), Argentina (20+), Colombia (14), Egypt (75+), Gabon (9), Libya (80+), Pakistan (60+), Peru (16) and Zaire (8).

The structural description of the Mirage III-E is generally applicable to the Mirage 5, with the following exceptions:

ARMAMENT: Seven attachment points for external loads with multiple launchers permitting a maximum load of more than 4 tonnes. Ground attack weapons are similar to those carried by the III-E. For interception, two Magic or Sidewinder missiles can be carried under the wings.

EQUIPMENT: Optional equipment on the latest versions includes an inertial nav system and nav/attack system, with head-up display and either Agave multipurpose radar or an air-to-surface laser rangefinder and Aïda II radar.

DIMENSIONS, EXTERNAL: As Mirage III-E except:
- Length overall — 15.55 m (51 ft 0¼ in)

WEIGHTS AND LOADINGS: As Mirage III-E except:
- Weight empty — 6,600 kg (14,550 lb)

PERFORMANCE (in 'clean' condition, with guns installed, except where indicated): As III-E, plus:
- Combat radius with 907 kg (2,000 lb) bomb load:
  - hi-lo-hi — 700 n miles (1,300 km; 808 miles)
  - lo-lo-lo — 350 n miles (650 km; 404 miles)
- Ferry range with three external tanks — 2,158 n miles (4,000 km; 2,485 miles)

UPDATED

Dassault Mirage 5 of the Egyptian Air Force *(AMD-BA/Aviaplans)*

1994

DASSAULT MIRAGE 50

TYPE: Single-seat multimission fighter aircraft.

PROGRAMME: The Mirage 50 retains the basic airframe of the Mirage III/5 series, but is powered by a higher rated SNECMA Atar 9K-50 turbojet, as fitted in the Mirage F1s of the French Air Force and 10 other air forces. This gives 70.6 kN (15,873 lb st) with afterburning, representing a thrust increase of between 17 and 23 per cent compared with Mirage III/5s.

The prototype Mirage 50 flew for the first time on 15 April 1979. First customer was the Chilean Air Force which ordered 16.

The Mirage 50 is a multimission fighter, suitable for air superiority duties with guns and dogfight missiles, air patrol and supersonic interception, and ground attack combined with self-defence capability. It can carry the full range of operational stores, armament and equipment developed for the Mirage III/5 series, plus Agave or Cyrano IVM multifunction radar, with Matra Magic or 530 air-to-air missiles respectively, an inertial nav/attack system and head-up display. It is available in reconnaissance configuration. A two-seat training version is also available. Improvements compared with other delta-wing Mirages include a 15-20 per cent shorter take-off run, improved armament/fuel load, higher rate of climb, faster acceleration and better manoeuvrability. Maximum internal fuel capacity is 3,410 litres (900.7 US gallons; 750 Imp gallons). Underwing tanks can increase total capacity to 6,810 litres (1,799 US gallons; 1,498 Imp gallons).

UPGRADES: **Dassault:** For full details see Advanced Technology Upgrade Programmes in this section.

Cardoen: Pantera 50C. See separate entry.

OPERATORS: Versions of the Mirage 50 are in service with the armed forces of the following countries: Chile (16) and Venezuela (15).

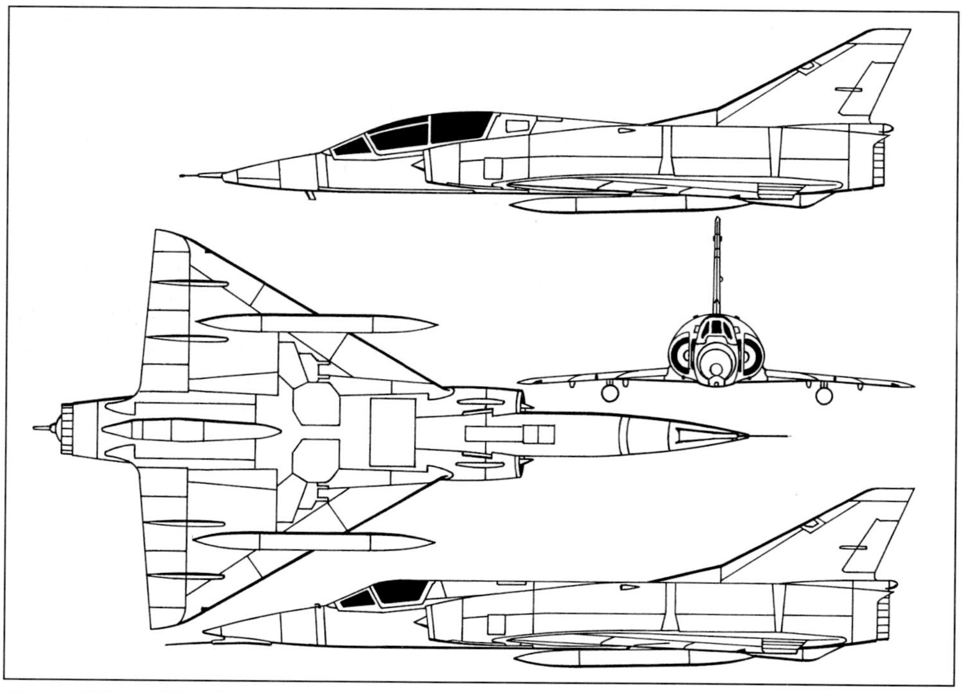

Dassault Mirage 50 multimission fighter. Additional side-view (top) of two-seat version *(Jane's/Mike Keep)*

1994

DIMENSIONS, EXTERNAL: As Mirage III-E except:
- Length overall — 15.56 m (51 ft 0½ in)

WEIGHTS AND LOADINGS:
- Weight empty, equipped — 7,150 kg (15,765 lb)
- T-O weight 'clean' — 9,900 kg (21,825 lb)
- Max T-O weight — 13,700 kg (30,200 lb)

PERFORMANCE:
- Max level speed at altitude — Mach 2.2 (750 kts; 1,390 km/h; 863 mph IAS)
- Max rate of climb at S/L — 11,160 m (36,600 ft)/min
- Time to 17,700 m (58,000 ft) at Mach 2, with two Magic missiles — 6 min 48 s
- Service ceiling at Mach 2 — 18,000 m (59,055 ft)
- T-O run with two magic missiles — 915 m (3,000 ft)

Dassault Mirage 50 multimission fighter with Aerospatiale AM 39 Exocet

1994

Dassault Mirage 50 with SNECMA Atar 9K-50 turbojet

1994

T-O run, at max T-O weight 1,830 m (6,000 ft)
Combat radius with two 400 kg bombs and max external fuel with reserves:
lo-lo-lo at Mach 0.6 370 n miles (685 km; 425 miles)
hi-lo-hi at Mach 0.85/0.9
675 n miles (1,205 km; 775 miles)

VERIFIED

DASSAULT MIRAGE IV

TYPE: Two-seat supersonic bomber aircraft.

PROGRAMME: The Mirage IV tandem two-seat supersonic delta-wing bomber was designed specifically to deliver a nuclear weapon. Its development and production were undertaken in association with many other companies. In particular, Sud Aviation took responsibility for the wing and rear fuselage, and Breguet for the tail surfaces.

The original prototype Mirage IV, which flew for the first time on 17 June 1959, was a scaled-up derivative of the Mirage III fighter, powered by two SNECMA Atar 09 turbojets (each 58.8 kN; 13,225 lb st with afterburner) and with a take-off weight of approximately 25,000 kg (55,115 lb). It reached a speed of Mach 1.9 during its 14th test flight in July 1959, and exceeded Mach 2 during its 33rd test flight.

The prototype was followed by three preproduction Mirage IVs the first of which flew on 12 October 1961. Powered by two 62.76 kN (14,110 lb st) Atar 09Cs, this aircraft was slightly larger than the first and was more representative of the production Mirage IV-A with a large circular radome under its centre-fuselage, ahead of the semi-submerged nuclear free-fall bomb. The first preproduction aircraft was used for bombing trials and development at Colomb-Béchar. The second preproduction Mirage IV was similar, and was used to develop the navigation system and for flight refuelling trials with a Boeing KC-135F Stratotanker.

The last of the three preproduction aircraft flew on 23 January 1963, and was a completely operational model with Atar 09Ks, full equipment, including flight refuelling noseprobe, and armament.

The French Air Force ordered a total of 50 production Mirage IV-As for delivery during the period 1964-65. A follow-on order for 12 more was placed subsequently, and the Mirage IV-A became operational with the French Air Force's 91e Escadre de Bombardement at Mont-de-Marsan, the 93e Escadre at Istres, and with a third unit at St Dizier east of Paris.

The production Mirage IV-A was given advanced electronic navigation and bombing equipment. However, a two-year study of the problems of low-altitude penetration showed that the Mirage IV-A could be adapted for a low level role with minor modifications. These were introduced on the line and incorporated retrospectively into aircraft already built.

VERSIONS: **Mirage IV-A:** Initial version.

Mirage IV-P: Eighteen of the Mirage IV strategic bombers operated by the Commandement des Forces Aériennes Stratégiques (CFAS) of the French Air Force have been modified to carry the ASMP medium-range air-to-surface nuclear missile. Navigation and targeting capabilities are improved by installation of a Thomson-CSF Arcana pulse-Doppler radar and dual inertial systems. Uprated EW equipment includes, typically, a Thomson-CSF jamming pod and a chaff dispensing pod on the two outboard underwing hardpoints.

The modified aircraft have been redesignated Mirage IV-P (for pénétration), attained operational capability with Escadron de Bombardement 1/91 'Gascogne' at Mont-de-Marsan on 1 May 1986, followed by Escadron de Bombardement 2/91 'Marne' at St Dizier on 1 December 1986. A few aircraft are allocated to the OCU: CIFAS 'Aquitaire' at Bordeaux. The following description refers to the Mirage IV-A, although the major differences between this model and the IV-P are detailed above.

OPERATORS: Versions of the Mirage IV are in service with the armed forces of France (18).

DESIGN FEATURES: Cantilever mid-wing monoplane. Thickness/chord ratio varies from 3.8 to 3.2 per cent. Sweepback on leading-edge 60 per cent. Single notch in leading-edge of each wing at about 60 per cent span. Tail unit consists of sweptback fin and rudder only. Housing for Aerazur brake parachute, with clamshell doors, at base of rudder.

FLYING CONTROLS: Entire trailing-edge of each wing made up to two elevons operated by Dassault electrohydraulic twin-cylinder servo control units, with signals from auto-stabiliser fed into the system. Hydraulically actuated airbrakes above and below each wing near wingroot

Dassault Mirage 50EV of the Venezuelan Air Force *(Aviaplans/F. Robineau)*

1995

Dassault Mirage IV-P strategic bomber *(AMD-BA/Aviaplans)*

1994

Dassault Mirage IV-P strategic bomber of the French Air Force *(Sirpa Air)*

1995

Dassault Mirage IV strategic bomber *(Armée de l'Air)*

1994

leading-edge. No high-lift devices, conical camber spoiler or tabs. Rudder actuated in same way as elevons. No tabs.

STRUCTURE: The fuselage is an all-metal semi-monocoque structure. Rear portion of fuselage comprises two circular ducts housing turbojets. Underside of rear fuselage recessed to accommodate weapon. The tail unit is a swept-back fin and rudder only, of multispar construction.

LANDING GEAR: Hydraulically retractable tricycle type. Twin-wheel steerable nose unit retracts rearward. Four-wheel bogie main units retract inward into fuselage. Messier oleo-pneumatic shock-absorbers, wheels and brakes.

POWER PLANT: Two SNECMA Atar 9K-50 turbojet engines, each rated at 70.61 kN (15,873 lb st) with afterburning. Movable half-cone centrebodies in air intakes. Integral fuel tanks between main and rear spars in each wing and in leading-edge forward of landing gear bay. Fuel tanks occupy most of centre-fuselage between rear cockpit and engine bay. Further tanks between double skins outboard of each engine air-duct, under air-ducts and engines, and in leading-edge of tail fin. Provision for carrying a large jettisonable fuel tank of 2,000 litres (528 US gallons; 440 Imp gallons) capacity under each wing. Whilst ferrying, an external tank can be carried in the underfuselage weapon recess. Flight refuelling nose probe. Two groups of six JATO units can be attached under the wings for short-field take-off.

ACCOMMODATION: Crew of two in tandem on Martin-Baker (Hispano built) Mk BM.4 ejection seats. Rearward-hinged canopy over pilot's seat. Rearward hinged cover over navigator's position.

SYSTEMS: No de-icing systems.

AVIONICS AND EQUIPMENT: Includes Thomson-CSF Cyrano IV radar under centre-fuselage. Marconi Doppler, Dassault computer and countermeasures equipment, SFENA (now Sextant Avionique) autopilot and Omera-Robot cameras. Mirage IV-P uses Arcana pulse-Doppler radar and dual inertial systems, Thomson-CSF jamming pod, Philips BOZ-100 chaff/flare pod, and radar warning receivers.

ARMAMENT: One 60 kT nuclear weapon semi-recessed under fuselage, or 16 × 1,000 lb conventional bombs, or four Martel air-to-surface missiles under the wings and fuselage. Mirage IV-P carries the ASMP (Air-Sol Moyenne Portée) supersonic nuclear air-to-surface missile, with a yield and range of 150 kT and 40-54 n miles (75-100 km; 46-62 miles) respectively. ASMP is intended for stand-off attacks against heavily defended targets, such as airfields and command and communications centres.

DIMENSIONS, EXTERNAL:

Wing span	11.85 m (38 ft 10½ in)
Length overall	23.50 m (77 ft 1in)
Height overall	5.65 m (18 ft 6½ in)

AREAS:

Wings, gross	78.00 m² (840.0 sq ft)

WEIGHTS AND LOADINGS:

Weight empty	14,500 kg (31,965 lb)
Average T-O weight	31,600 kg (69,665 lb)
Max T-O weight	33,475 kg (73,800 lb)

PERFORMANCE:

Max level speed at 11,000 m (36,000 ft)	Mach 2.2
Combat speed at high altitude	Mach 1.8
Service ceiling	20,000 m (65,615 ft)
Tactical radius	668 n miles (1,250 km; 771 miles)
Ferry range	2,158 n miles (4,000 km; 2,485 miles)

VERIFIED

DASSAULT MIRAGE F1

Spanish Air Force designation: C.14

TYPE: Single-seat multimission fighter and attack aircraft.

PROGRAMME: The Mirage F1 prototype flew for the first time on 23 December 1966 and was followed by three preseries aircraft.

The primary role of the single-seat Mirage **F1-C** production version, to which the detailed description applies, is that of all-weather interception at any altitude. It is equally suitable for visual ground attack missions, carrying a variety of external loads beneath the wings and fuselage.

VERSIONS: **F1-B:** Two-seat version of F1-C, the first of which made its first flight on 26 May 1976.

F1-D: Two-seat version of F1-E.

F1-E: Single-seat multirole air superiority/ground attack/reconnaissance version for export customers, with an inertial navigation system, nav/attack central computer, CRT head-up display, and a large inventory of external stores.

F1-R: Single-seat (French Air Force **F1-CR**) day and night reconnaissance variant.

Production of the F1-A ground attack version, with reduced equipment, a retractable flight refuelling probe and increased fuel, has been completed.

Many F1-Cs of the French Air Force were delivered or modified to **F1-C-200** standard by installation of an 8 cm (3.15 in) fuselage plug for a removable flight refuelling probe. Export customers who have F1s equipped with refuelling probes include Iraq, Morocco, South Africa and Spain.

By January 1990, a total of 731 Mirage F1s had been ordered and deliveries made to: France (251 including five prototypes), Ecuador (two F1-JE and 16 JA, equivalent to F1-B and E), Greece (40 F1-CG), Iraq (98 F1-EQs and 15 BQs), Jordan (two F1-BJ, 17 CJ - to be upgraded to EJ— and 17 EJ), Kuwait (six F1-BK and 27 CK), Libya (six F1-BD, 16 AD and 16 ED), Morocco (30 F1-CH and 20 EH), Qatar (two F1-DDA and 13 EDA), South Africa (32 F1-AZ and 16 CZ) and Spain (six F1-BE, 45 CE and 22 EE). Production ended in 1990 with the last of a follow-on batch of 15 for Iraq, but these, plus five from an earlier order, have been embargoed. Deliveries by January 1991 were 721, reportedly including some (embargoed Iraqi) aircraft to the Kuwait Air Force in exile.

The first production F1 flew on 15 February 1973 and was delivered officially to the French Air Force on 14 March 1973. The first unit to receive the F1 was the 30e Escadre at Reims, which became operational in early 1974. This now has two and a half squadrons of F1-Cs (the half detached to Djibouti) and an OCU squadron of F1-Bs, whilst the 12e Escadre de Chasse at Cambrai has three squadrons of F1-Cs. F1-C-200s made surplus by conversion of units to Mirage 2000Cs are to be issued to 13e Escadre at Colmar in the tactical attack role. Designated **Mirage F1-CT**, they will receive INS and navigation computer, as described for the Mirage F1-CR, plus Martin-Baker F10M ejection seats, radar warning receivers, chaff/flare dispensers, secure radio, laser rangefinder and air-to-ground ordnance. The 1990 defence budget included funding for the first 19 conversions, plus authorisation for a further 22, deliveries being due from late 1991. A prototype conversion was begun at Biarritz-Parme in 1989 and was to begin trials at Istres in 1991. Remaining aircraft will be re-worked at the Atelier Industriel de l'Air at Clermont-Ferrand. Authorisation for a further 14 conversions was authorised in 1992. A first squadron of the 13th Fighter Wing is operational.

Deliveries of the F1-C series to the French Air Force totalled 162, made up of 83 F-1Cs and 79 F1-C-200s. Twenty tandem-seat F-1Bs were delivered to an OCU squadron, initially at Orange, from June 1980; each aircraft is equipped with the same radar, weapon system and air-to-air missiles as the F1-C, but has no internal guns. Fuel capacity is reduced by 450 litres (119 US gallons; 99 Imp gallons); empty weight increased by 200 kg (441 lb); and the forward fuselage extended by 30 cm (11¾ in).

The French Air Force also purchased F1-CRs to re-equip the three squadrons of the 33e Escadre de Reconnaissance, at Strasbourg. These aircraft are designated **F1-CR-200** (having a fixed in-flight refuelling probe) and differ from the F1-C in being fitted with the IVMR model of Cyrano radar (with additional ground mapping, contour mapping, air-ground ranging and blind let-down modes), a SAGEM Uliss 47 inertial platform and Dassault Electronique 182 navigation computer. A SAT SCM2400 Super Cyclope infra-red linescan reconnaissance system replaces the starboard gun, and an undernose bay houses either a 75 mm Thomson-TRT 40 panoramic camera or a 150 mm Thomson-TRT 33 vertical camera. French aircraft have a secondary ground attack role and may also carry a centreline podded sensor in the form of a Thomson-CSF Raphaël TH SLAR or a Thomson-CSF Astac electronic reconnaissance system for detection of ground radars. Ongoing modification is making French F1-CRs compatible with a FLIR pod and a TV reconnaissance pod. Several types of sensor pod are available for fitment to export Mirage F1-Es. The first of two F1-CR-200 prototypes flew on 20 November 1981. There were 64 (including the prototypes) ordered for the French Air Force. The first production F1-CR-200 flew on 10 November 1982, and the first squadron (2/33) became operational in July 1983. Deliveries were completed in 1987 and the third and last squadron of 33e Escadre converted from Mirage IIIRs in 1988.

Export F1-Cs have a radar similar to Cyrano IV or IVM. Export F1-Es have radar similar to Cyrano IVMR but repackaged to save space. Mirage F1-EQ5s and EQ6s of the Iraqi Air Force are equipped to carry Exocet anti-ship missiles and laser-guided weapons such as the AS 30L missile and Matra 400 kg laser-guided bomb. They have Thomson-CSF Agave radar.

The Mirage F1 was produced by Dassault Aviation in co-operation with the Belgian company SABCA, in which Dassault Aviation has a parity interest, and CASA of

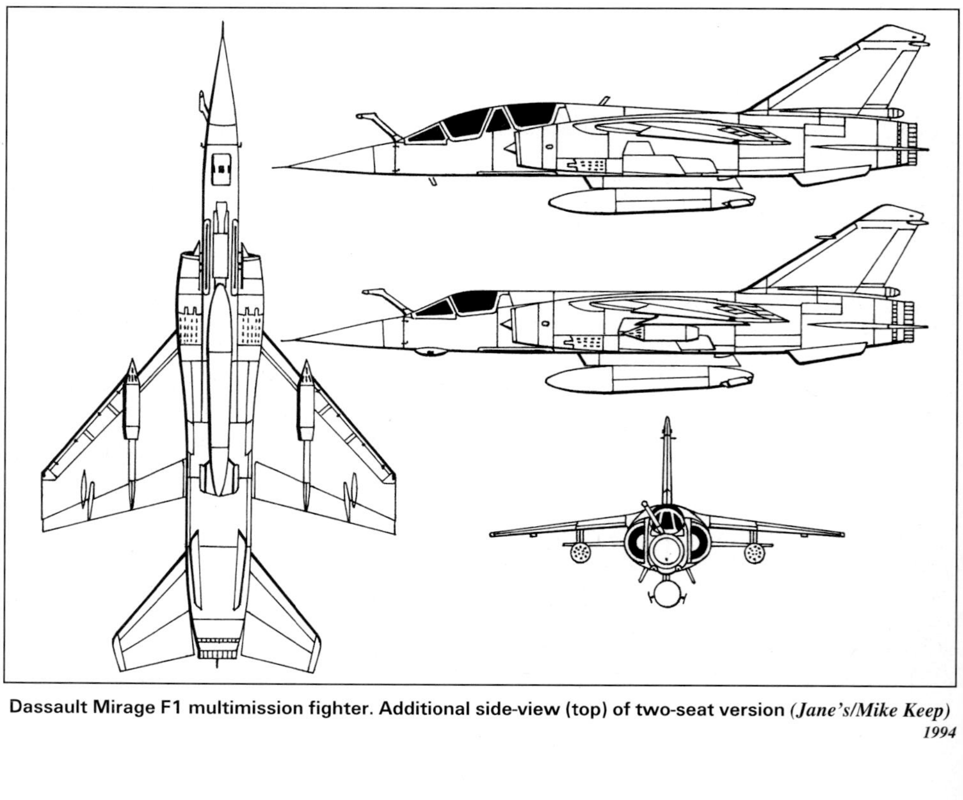

Dassault Mirage F1 multimission fighter. Additional side-view (top) of two-seat version *(Jane's/Mike Keep)*

1994

Spain, which built fuselage sections for all Mirage F1s ordered. Dassault Aviation also has a technical and industrial co-operation agreement with the Armaments Development and Production Corporation of South Africa Ltd (Armscor), whereby the latter company has rights to build the Mirage F1 under licence. Mirage F1 flying hours passed 1 million in June 1989.

UPGRADES: **F1-CR-200:** In 1991 an upgrade programme was initiated to install a new INS, improved main computer and a HUD the same as in the Mirage 2000D allowing a full night and day FLIR capability, with other improvements in the ECM and electronic reconnaissance systems. See Versions.

Grinaker Avitronics: Grinaker Avitronics of South Africa has developed an EW upgrade package for the Mirage F1. The system provides radar warning, a chaff dispensing system and an active ECM system for self-protection. System does not restrict aircraft weapon load and has negligible aerodynamic effects. System includes: RWS-130 radar warning receiver and an SPJ-200 CW jammer.

Spanish Air Force: Requirement to upgrade the Spanish Air Force Mirage F1-CE, F1-EE and F1-C aircraft between January 1995 and 1999 to extend their operational lives beyond 2010. No prime contractor announced as of going to press. Upgrade to include installation of: radar altimeter and GPS, onboard computer, stores management system, colour VHF camera and an ECM suite (ENSA-AL-300 radar warning receiver with ALE-40 decoys. All aircraft will be fitted to use the Barax detector/jammer, the original communications fit is to be replaced with Have Quick II units and IFF/SIF Mode 4 Crypto will be added.

OPERATORS: Versions of the Mirage F1 are in service with the armed forces of the following countries: Ecuador (14), France (135), Greece (30+), Iraq, Jordan (31), Kuwait (19), Libya (50+), Morocco (39), South Africa (30) and Spain (50+).

The following abbreviated description applies to the F1-C-200 production version for the French Air Force, except where indicated:

DESIGN FEATURES: High-mounted, swept, tapered anhedral wings with extended chord on outer two-thirds of wing. Optional refuelling probe on upper surface of nosecone. All-moving, mid-mounted, swept tapered tailplane. Large, swept single-spar fin.

STRUCTURE: The wing is an all-metal two-spar torsion box structure. Trailing-edge control surfaces of honeycomb sandwich construction, with carbonfibre aileron skin. The fuselage conventional all-metal semi-monocoque structure. Titanium alloy is used for landing gear trunnions, engine firewall and certain other major structures. High tensile steel wing attachment points.

FLYING CONTROLS: Entire leading-edge can be drooped hydraulically (manually for T-O and landing, automatic in combat). Two differentially operating double-slotted flaps and one aileron on each trailing-edge. Two spoilers on each wing, ahead of flaps. Large hydraulically actuated door type airbrake in forward underside of each intake trunk. All-moving tailplane mid-set on fuselage, and actuated hydraulically by electric or manual control.

LANDING GEAR: Retractable tricycle type, by Messier-Bugatti. Hydraulic retraction, nose unit rearward, main units upward into rear of intake trunk fairings. Twin wheels on each unit. Messier-Bugatti brakes and anti-skid units. Brake parachute in bullet fairing at base of rudder.

POWER PLANT: One SNECMA Atar 9K-50 turbojet, rated at 70.6 kN (15,873 lb st) with afterburning. Fuel in integral tanks in wings (combined capacity 375 litres; 99 US gallons; 82.5 Imp gallons), and three main tanks and one inverted-flight supply tank (combined capacity 3,925 litres; 1,037 US gallons; 863.5 Imp gallons) in fuselage. Total internal fuel capacity 4,300 litres (1,136 US gallons; 946 Imp gallons). Provision of two jettisonable auxiliary fuel tanks (each 1,200 litres, 317 US gallons; 264 Imp gallons) to be carried on inboard wing pylons, plus a single tank of 2,200 litres (581 US gallons; 484 Imp gallons) capacity on the underfuselage station. Non-retractable, but removable, optional flight refuelling probe on starboard side of nose.

ACCOMMODATION: SEM Martin-Baker F1RM4 ejection seat for pilot, under rearward canopy (SEM Martin-Baker F10M rocket seat in latest F1-Cs and in F1-E and F1-CR. Two Mk 10 seats with inter-seat sequence system in F1-B). No delay through-the-canopy escape system.

SYSTEMS: Two independent hydraulic systems, pressure 207 bars (3,000 lb/sq in) each, for landing gear actuation, flaps and flying controls. Electrical system includes two Auxilec 15 kVA variable speed alternators. DC power provided by two transformer-rectifiers operating in conjunction with battery.

AVIONICS: Thomson-CSF Cyrano IV fire control radar in nose. LMT Tacan, LMT NR-AI-4-A IFF, remote setting interception system, three-axis generator, central air data computer SFIM spherical indicator with ILS pointers, Sextant Avionique Type 63 navigation indicator, SFENA 505 autopilot and CSF head-up display, with wide field of view double converter. (Standard equipment on F-1E includes SAGEM Uliss 47 INS, Dassault Electronique 182 central digital computer for nav/attack computations, TH C8F VE-120C CRT head-up display, Sextant Avionique air data computer and digital armament/nav control panels).

Dassault Mirage F1CR of the French Air Force *(Sirpa Air)*

1995

Dassault Mirage F1C of the French Air Force *(Jamie Hunter)*

1994

Dassault Mirage F1C of the French Air Force *(Jamie Hunter)*

1994

ARMAMENT: Standard installed armament of two 30 mm DEFA 553 cannon, with 135 rds/gun, mounted in lower central fuselage. Maximum theoretical external combat load 6,300 kg (13,890 lb), comprising 2,100 kg (4,630 lb) on centreline pylon; 1,300 kg (2,866 lb) on each inner wing pylon; 550 kg (1,213 lb) on each outer wing pylon; 150 kg (331 lb) on each wingtip AAM rail (Matra 550 Magic or AIM-9 Sidewinder) and 100 kg (220 lb) of chaff/flare dispensers on each fuselage shoulder position. Practical maximum operational load is 4,000 kg (8,818 lb). Externally mounted weapons for interception role include Matra Super 530 air-to-air missiles under inboard wing pylons. For ground attack, typical loads may include one Armat anti-radar missile, one AM 39 Exocet anti-ship missile, or up to 14 250 kg bombs, 30 anti-runway bombs or 144 Thomson-Brandt rockets. Other possible external loads include auxiliary fuel tanks, a Thomson-CSF Atlis laser designator pod with AS 30L missiles or 400 kg laser-guided bombs, air-to-surface missiles, and sensor pods comprising a Thomson-CSF Raphaël SLAR, a Thomson-TRT/Dassault Harold long-range oblique photographic unit (1,700 mm Thomson-TRT 38), Thomson-TRT/Dassault COR2 multipurpose pod (visual spectrum and IR linescan), a Thomson-CSF TMV 018 Syrel real-time electro-optical reconnaissance pod, a Dassault Nora real-time video pod and a Thomson-CSF Astac ground radar detector unit. A Thomson-CSF BF radar warning receiver is standard, and a range of jamming (Barracuda, Barem, Barrax) and chaff/flare (Phimat, Sycomor) pods may be fitted. Some Moroccan F1-EHs and Spanish F1-CEs have chaff/flare dispensers scabbed to each side of rear fuselage.

DIMENSIONS, EXTERNAL (F1-C):
Wing span: without missiles 8.40 m (27 ft 6¾ in)
over Magic missiles approx 9.32 m (30 ft 6¾ in)
Length overall: F1-C 15.23 m (49 ft 11¾ in)
F1-C-200 15.30 m (50 ft 2½ in)
F1-B 15.53 m (50 ft 11½ in)
Height overall 4.50 m (14 ft 9 in)

AREAS:
Wings, gross 25.00 m² (269.1 sq ft)

WEIGHTS AND LOADINGS (F1-C):
Weight empty 7,400 kg (16,314 lb)
T-O weight, clean 10,900 kg (24,030 lb)
Max T-O weight 16,200 kg (35,715 lb)

PERFORMANCE (F1-C):
Max level speed: high altitude Mach 2.2
low altitude Mach 1.2 (800 kts; 1,480 km/h; 920 mph EAS)
Max rate of climb at S/L (with afterburning) 12,780 m (41,930 ft)/min
Service ceiling 20,000 m (65,600 ft)
Stabilised supersonic ceiling 16,000 m (52,500 ft)
T-O run (AUW of 11,500 kg; 25,355 lb) 600 m (1,970 ft)
Landing run (AUW of 8,500 kg; 18,740 lb) 670 m (2,200 ft)
Combat radius:
hi-lo-hi at Mach 0.75/0.88, with 14 250 kg bombs and max internal fuel, with reserves 230 n miles (425 km; 265 miles)
lo-lo with one Exocet and two external tanks, with reserves and incl missile flight path 378 n miles (700 km; 435 miles)
Combat air patrol endurance, with two Super 530 missiles and underbelly tank, with reserves, incl one attack at ceiling 2 h 15 min

UPDATED

MIRAGE ADVANCED TECHNOLOGY UPGRADE PROGRAMMES

TYPE: Combat aircraft upgrades (Mirage III/5/50 and Mirage F1).

PROGRAMME: Although the first generation Mirage (types III, 5 and 50) remains available to special order, series production has now ended. At the beginning of 1990, orders totalled 1,422, of which 1,415 had been delivered, including 949 exported. Many of these remain in service, or are in storage awaiting resale.

Since 1977, Dassault has been involved in programmes to update the navigation and attack systems, flight aids, radio com/nav, power plant and other features of in-service Mirage III/5/50 aircraft. In particular, several air forces have awarded Dassault contracts to install an inertial platform, digital computer, CRT head-up display, air-to-ground laser rangefinder and other equipment for improved navigational accuracy, easier target acquisition, and high bombing precision in the various CCIP (continuous computation of the impact point) or CCRP (continuous computation of the release point) modes, including standoff capability through the introduction of CCRP with initial point. Combat efficiency in the air-to-air gunnery mode is improved considerably by display of a highly accurate hot-line on the HUD.

Another major improvement available for the Mirage III/5/50 series is a flight refuelling kit offering an increase of 30 to more than 100 per cent in radius of action. Already ordered by several air forces for their Mirage 5s, this system was demonstrated in flight before becoming generally available to Mirage operators in 1986. It involves lengthening the nose of the aircraft by 90 mm (3½ in) to accommodate system changes associated with a non-retractable probe on the starboard side, forward of the windscreen, and a single-point pressure refuelling port for both internal and external tanks. With the addition of a pressure refuelling system, time for refuelling on the ground is reduced from 15 to 3 minutes.

Over half the air forces operating Mirage III/5/50s have now opted for update programmes, some of which are undertaken at least partly by local organisations and are of sufficient complexity to warrant mention in other national sections in this book. Brief details are listed below.

UPGRADES: **Argentina:** From 1982 onwards, up to 26 IAI Daggers (Israeli built Mirage 5s) underwent three-phase 'Finger Ia', 'IIa' and 'IIIa' upgrade programme, eventually receiving a laser rangefinder, INS, HUD, ECM equipment and refuelling probe. Nine ex-Peruvian Mirage 5Ps, similarly modified to 'Mara' standard.

Belgium: A Mirage Systems Improvement Programme (MIRSIP) is being applied by SABCA to 15 Mirage 5BAs and five Mirage 5BDs. The aircraft were originally to be operated by the Belgian Air force but restructuring has now made the aircraft surplus to requirements. The first upgraded Mirage flew in December 1992. In July 1994 the Chilean government agreed to purchase the upgraded aircraft as well as an additional five aircraft which will not be upgraded. Additions, incorporated by SABCA as main contractor, include a GEC Ferranti HUD, Thomson-CSF TMV 630 laser-range, SAGEM nav/attack avionics (Uliss 92 INS and UTR 90 computer), fixed canard foreplanes, single-point pressure refuelling, pilot's liquid oxygen system, strobe anti-collision lights and complete re-wiring. Martin-Baker Mk 10L ejection seats are being installed under a separate contract. An unspecified reconnaissance pod will be acquired for some aircraft.

Brazil: First of seven ex-French Mirage IIIEs (including two converted to tandem-seat trainers) handed over by Dassault on 30 September 1988 following a complete overhaul and addition of foreplanes; 10 existing Mirage IIIEBRs and two IIIDBRs being modified in Brazil from December 1989.

Chile: Mirage 50C update with foreplanes and Israeli avionics in progress.

Colombia: In 1988, two Mirage 5COD trainers received Kfir C7-type avionics and half-size Kfir canards. Base Arsenal at Madrid converting 10 Mirage 5COAs and two CORs to similar standards (but with 75 per cent canards) as Mirage 5M (Modification). First aircraft flown in January 1989.

Egypt: Improvement programme for some Mirages has been completed.

France: In February 1993 the French government approved a plan to upgrade 37 French Air Force Mirage 2000RDM to Mirage 2000-5 standard. Upgrade consists of fitting the Thompson-CSF RDY radar, new cockpit displays. Upgraded M53 turbofans and Matra MICA air-to-air missiles.

Pakistan: Update programme completed on original aircraft. Fifty ex-Australian Mirage IIIOs acquired in 1990 for re-work at the Mirage Rebuild Factory, Kamra, with some export sales in prospect.

Peru: Some 12 Mirage 5P/P3s and three 5DP/DP3s fitted with refuelling probe, laser-ranger and other improvements, as 5P4 and 5DP4.

South Africa: Mirage IIIs being fitted with foreplanes and Israeli avionics as Atlas Cheetah (which see). Mirage IIICZ withdrawn from service October 1990. Mirage F1 upgrade programme currently under way. See Mirage F1 entry.

Spain: CESELSA, with CASA as subcontractor, will install Emerson Electric AN/APQ-159 radar, Honeywell AN/AYK-14 mission computer, a HUD, radar warning receiver, HOTAS controls, radar altimeter, Tacan, AN/ALE-40 chaff/flare dispenser and multifunction displays, plus pressure refuelling and in-flight refuelling probes. Rework involves 18 Mirage IIIEEs and five two-seat IIIBEs—the latter with provision for buddy refuelling pods.

Switzerland: Foreplanes and avionics improvements being installed locally in Mirage IIIS/RS fleet by F+W.

Venezuela: Six single-seat and two two-seat aircraft to receive Atar 9K-50 power plants and improved avionics, under designations Mirage 50EV and 50DV respectively, deliveries beginning on 22 October 1990. Venezuelan aircraft have canards, a Uliss 81 INS, Cyrano IVM3 radar and an in-flight refuelling probe. Armament options will include Magic 2 AAMs and AM39 Exocet anti-ship missiles. These aircraft are being supplemented by six new Mirage 50EVs, one new Mirage 50DV and three refurbished second-hand Mirage 50EVs. These Mirages are the last to be produced after 35 years' continuous production.

No modification programmes have been announced for Abu Dhabi, Gabon, Libya or Zaïre. The type has been withdrawn from use in Australia, Israel and Lebanon.

Several standards of equipment for modified or new airframes have been offered by Dassault in recent years, including the Mirage 3NG and Mirage 50M. Most recent of these, the **Mirage IIIEX**, was revealed in 1989 as a modified IIIE (being replaced in French service) with a longer, Mirage F1-type nose; underfuselage strakes (as on Mirage 5D series); an in-flight refuelling probe, offset to port, ahead of the cockpit; and fixed canards. The Doppler fairing below the forward fuselage is deleted.

Israel Aircraft Industries (which see), seeking to export surplus Kfirs to nations not authorised to receive the GE J79 turbojet, is undertaking trial installation of a SNECMA Atar 9K-50, thereby 're-inventing' the Mirage 50.

VERIFIED

DASSAULT MIRAGE 2000 UPGRADES

TYPE: Combat aircraft upgrade.

PROGRAMME: The following programmes have been announced. For full description of Mirage 2000, see *Jane's All the World's Aircraft 1995-96*.

UPGRADES: **Mirage 2000C-5:** FFr4,600 million ($830 million) conversion programme announced November 1992 for upgrading 37 late production 2000Cs to **2000C-5** standard (see 2000-5 entry, below) for continued French service; contract awarded 25 November 1993; conversion work by Dassault on initial 15 began 1994, with 15 to follow in 1995; all to be completed by 1997, but only one reworked aircraft funded in 1994 defence budget. RDI radars from these upgraded aircraft to be retrofitted in early production (RDM) Mirage 2000Cs.

2000-3: Private venture upgrade, begun 1987, with Rafale-type multifunction (five-CRT) cockpit displays known as APSI (advanced pilot system interface); prototype BX1 (ex-No. 501) flew 10 March 1988.

2000-4: Private venture integration of Matra MICA AAM. First guided flight against target drone, 9 January 1992.

2000-5: Multirole upgrade incorporating -3 and -4 improvements, plus Thomson-CSF RDY radar and new central processing unit. Thomson-CSF VEH 3020 HUD and ICMS Mk 2 countermeasures; Matra Super 530D and BAe Sky Flash as alternatives to MICA; laser-guided bombs and ASMs, or Apache stand-off dispenser in air-to-ground role. First Mirage flight of RDY radar in BY1 (ex-No. B01, later BY2) 10 March 1988; first flight of full 2000-5 (two-seat) 24 October 1990 (same aircraft; no serial number); first single-seater (conversion of trials aircraft CY1) flown 27 April 1991. Options include SNECMA M88-P20 uprated by 4 per cent to 98.06 kN (22,046 lb st); available for export from 1996.

NEW ENTRY

DASSAULT (BREGUET) 1150 ATLANTIC

TYPE: Twin-engine Maritime Patrol Aircraft (MPA).

PROGRAMME: On 9 July 1974, Dassault-Breguet delivered the 18th Breguet 1150 Atlantic maritime patrol aircraft ordered by the Italian government. This completed the production programme for 87 operational Atlantics of the basic type, made up of 40 aircraft for the French Navy, three of which were passed on to Pakistan. Manufacture was shared by companies in France, Germany, Italy and Pakistan, with additional airframe components supplied by the Belgian ABAP group and some equipment from the USA and UK.

UPGRADES: **Alenia:** See separate entry.

Dornier: See separate entry.

OPERATORS: Versions of the Atlantic are in service with the armed forces of the following countries: France (31), Germany (19), Italy (18) and Pakistan (4).

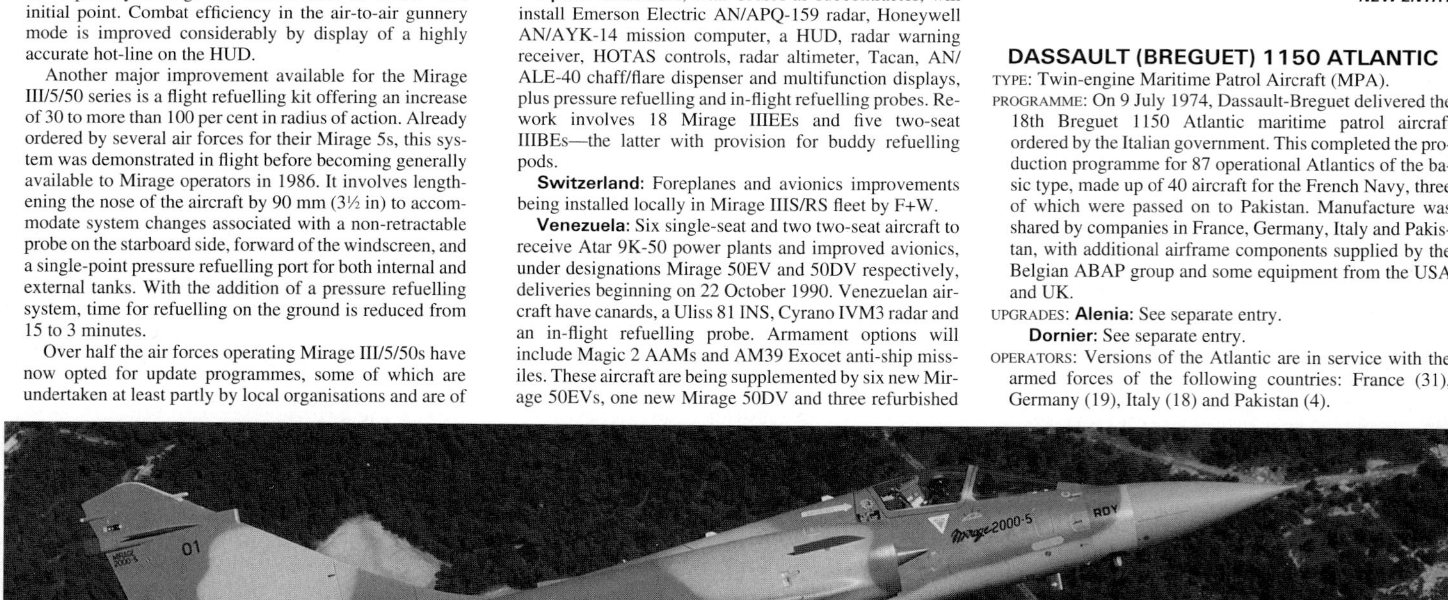

Dassault Mirage 2000-5 with Thomson-CSF RDY radar *(Aviaplans/F. Robineau)*

The following description applies to the basic version.

DESIGN FEATURES: Cantilever mid-wing monoplane. Wing section NACA 64 series. Dihedral on outer wings only. Fixed-incidence tailplane.

FLYING CONTROLS: Conventional all-metal ailerons actuated by SAMM twin-cylinder jacks. All-metal slotted flaps, with bonded light alloy honeycomb filling, over 75 per cent of span. Three hinged spoilers on upper surface of each outer wing, forward of flaps. Metal airbrake above and below each wing. No trim tabs.

Tail unit control surfaces operated through SAMM twin-cylinder jacks. No trim tabs.

STRUCTURE: All-metal three-spar fail-safe structure, with bonded light alloy honeycomb skin panels on torsion box and on main landing gear doors. The fuselage is an all-metal 'double-bubble' fail-safe structure, with bonded honeycomb sandwich skin on pressurised central section of upper fuselage, weapons bay doors and nosewheel door. The tail unit is a cantilever all-metal structure with bonded honeycomb sandwich skin panels on torsion boxes.

SYSTEMS: Kléber-Colombes pneumatic de-icing boots on wing leading-edges. Kléber-Colombes pneumatic de-icing boots on tail unit leading-edges.

LANDING GEAR: Retractable tricycle type, supplied by Messier-Hispano, with twin wheels on each unit. Hydraulic retraction, nosewheels rearward, main units forward into engines nacelles. Kléber-Colombes dimpled tyres, size 39 × 13-16 on mainwheels, 26 × 7.75-13 on nosewheels. Tyre pressures: main 9.52 bars (138 lb/sq in), nose 6.07 bars (88 lb/sq in). Messier-Hispano disc brakes with Maxaret anti-skid units.

POWER PLANT: Two 4,553 kW (6,106 ehp) SNECMA-built Rolls-Royce Tyne RTy.20 Mk 21 turboprop engines, each driving a Ratier-built HSD four-blade constant-speed propeller. Six integral fuel tanks with total capacity of 21,000 litres (5,547 US gallons; 4,619 Imp gallons). Provision for wingtip tanks to be fitted.

ACCOMMODATION: Normal flight crew of 12 comprising observer in nose; pilot and co-pilot on flight deck; a tactical co-ordinator, navigator, two sonobuoy operators, and radio, radar and ECM/MAD/Autoycus operators in tactical compartment; and two observers in beam positions. On long-range patrol missions a further 12 men can be carried as relief crew. The upper, pressurised section of the fuselage, from front to rear, comprises the nose observer's compartment, flight deck, tactical operations compartment, rest compartment for crew, and beam observers' compartment.

SYSTEMS: SEMCA air conditioning and pressurisation system. Hydraulic system pressure 207 bars (3,000 lb/sq in). Electrical system provides 28.5 V DC, 115/200 V variable-frequency AC and 115/200 V stabilised-frequency AC. AirResearch GTCP 85-100 APU in starboard side of front fuselage, adjacent radar compartment, for engine starting and ground air conditioning, can also power one 20 kVA AC alternator and one 4 kW DC generator for emergency electrical power supply.

ARMAMENT AND OPERATIONAL EQUIPMENT: Main weapons carried in bay in unpressurised lower fuselage. Weapons include all NATO standard bombs, 175 kg (385 lb) US or French depth charges. HVAR rockets, homing torpedoes, including types such as the Mk 44 Brush or LX.4 with acoustic heads, or four underwing air-to-surface missiles with nuclear or high explosive warheads. Electronic equipment includes a retractable Thomson-CSF radar installation, a MAD tailboom and an electrical countermeasures pod at the top of the tailfin. Sonobuoys are carried in a compartment aft of the main weapons bay, while the whole of the upper and lower rear fuselage acts as a storage compartment for sonobuoys and marker flares. Compartment for retractable Thomson-CSF radar 'dustbin' forward of the main weapons bay. Forward of this, the lower nose section acts as additional storage for military equipment and the APU. Weapons system includes Plotac optical tactical display, 80 × 80 cm (31.5 × 31.5 in) in size, consisting of separate tables for search display and localisation and attack display. At 1:30,000 scale, this gives coverage of any area 21,950 × 21,950 m (24,000 × 24,000 yd) to an accuracy of 1 mm (that is less than 30.5 m; 100 ft at that scale). Heading references provided by duplicated gyrosopic platforms of the 3-gyro (1° of freedom) 4-gimbals type, with magnetic compasses as back-up system. Janus-type Doppler has stabilised antenna and works in the Ke band to provide direct indication of ground speed and drift. In case of failure an automatic switch is made to the air data system. The analogue type navigation computer is accurate to 0.25 per cent. The MAD is of the atomic resonance type and uses light simulation techniques. Plotac system has provision to accept additional detectors. Radar has 'sea-return' circuits and stabilised antenna enabling it to detect a submarine snorkel at up to 40 n miles (75 km; 46 miles) even in rough seas.

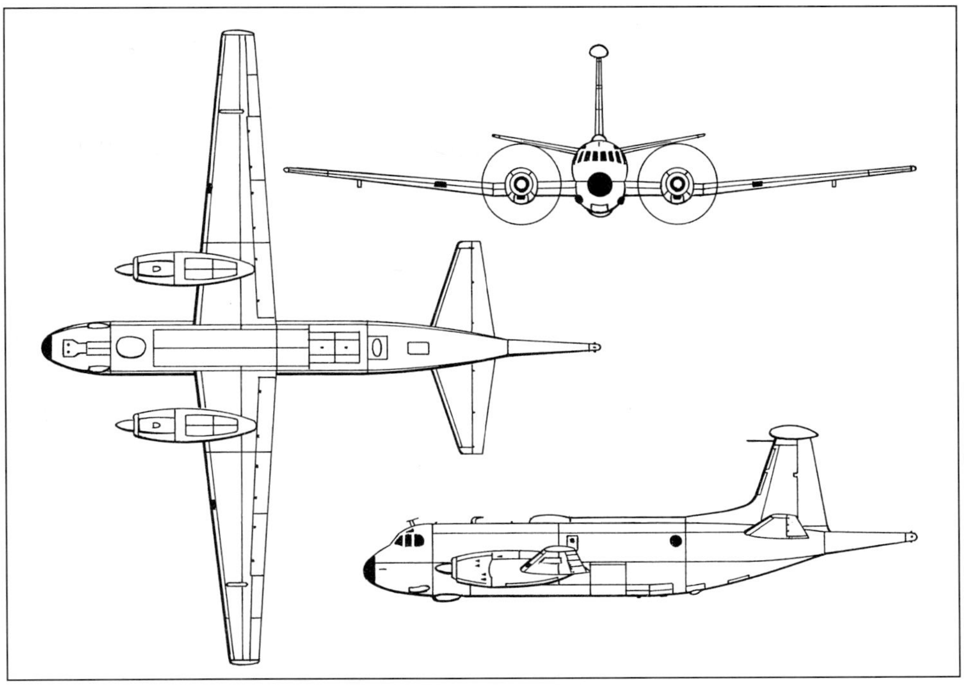

Dassault (Breguet) 1150 Atlantic *(Jane's/Mike Keep)*

1994

Dassault (Breguet) 1150 Atlantic of the French Navy *(Aviaplans)*

1994

Dassault (Breguet) 1150 Atlantic of the French Navy *(Peter J. Cooper)*

1994

DIMENSIONS, EXTERNAL:	
Wing span	36.30 m (119 ft 1 in)
Wing aspect ratio	10.94
Length overall	31.75 m (104 ft 2 in)
Height overall	11.33 m (37 ft 2 in)
Fuselage: Max width	2.90 m (9 ft 6 in)
Max depth	4.00 m (13 ft 1½ in)
Tailplane span	12.31 m (40 ft 4½ in)
Wheel track	9.00 m (29 ft 6¼ in)
Wheelbase	9.44 m (31 ft 0 in)
Propeller diameter	4.88 m (16 ft 0 in)
DIMENSIONS, INTERNAL:	
Tactical compartment: Length	8.60 m (28 ft 2½ in)
Height	1.93 m (6 ft 4 in)
Max width	2.70 m (8 ft 10½ in)
Rest compartment: Length	5.10 m (16 ft 8¾ in)
Height	1.93 m (6 ft 4 in)
Max width	2.70 m (8 ft 10½ in)
Beam observer's compartment:	
Length	1.00 m (3 ft 3¼ in)
Main weapons bay: Length	9.00 m (29 ft 6¼ in)
Height	1.55 m (5 ft 1 in)
Height under wing	1.00 m (3 ft 3¼ in)
Max width	2.20 m (7 ft 2½ in)
AREAS:	
Wings, gross	120.34 m² (1,295 sq ft)
Ailerons	5.40 m² (58.0 sq ft)
Trailing-edge flaps (total)	26.80 m² (288.4 sq ft)
Spoilers (total)	1.66 m² (17.8 sq ft)
Fin	16.64 m² (179.1 sq ft)

Rudder	5.96 m² (64.1 sq ft)
Tailplane	32.5 m² (349.7 sq ft)
Elevators	8.28 m² (89.1 sq ft)

WEIGHTS AND LOADINGS:

Useful load	18,551 kg (40,900 lb)
Max zero-fuel	34,473 kg (76,000 lb)
Max T-O weight	43,500 kg (95,900 lb)

PERFORMANCE (at max T-O weight):

Max level speed at high altitudes	355 kts (658 km/h; 409 mph)
Cruising speed	300 kts (556 km/h; 345 mph)
Service ceiling	10,000 m (32,800 ft)
T-O to 10.7 m (35 ft), ISA	1,500 m (4,925 ft)
T-O to 10.7 m (35 ft), ISA + 17°C, 15° flap	1,700 m (5,575 ft)
Max range	4,854 n miles (9,000 km; 5,590 miles
Max endurance at patrol speeds of 169 kts (320 km/h; 195 mph)	18 h

VERIFIED

Dassault (Breguet) Mystère-Falcon 20 modified by Flight Refuelling Ltd with underwing ECM training pod *1994*

DASSAULT MYSTÈRE-FALCON 20 AND 200

US Coast Guard designation: HU-25 Guardian

TYPE: Twin-turbofan executive transport and multirole aircraft.

PROGRAMME: The Mystère-Falcon 200 twin-turbofan light transport was based on the Mystère 20 design, first flown in prototype form on 4 May 1963. Manufacture of the Mystère-Falcon 200 began with aircraft c/n 401, first flown on 30 April 1980, concurrent with the production rundown of the earlier Mystère-Falcon 20F series, the last of which (c/n 486) came off the assembly line in late 1983. The Model 200 had been introduced, originally as the Mystère-Falcon 20H, at the 1981 Paris Air Show, with AlliedSignal turbofans in place of the F's General Electric CF700s, larger integral fuel tankage in the rear fuselage, redesigned wingroot fairings, automatic slat extension, and many important systems changes. Certification was achieved on 21 June 1981.

One aircraft was built in 1988, completing production of 473 Mystère-Falcon 20s and 35 Mystère-Falcon 200s. Several have been, and are being, converted for specific duties, as listed below, while Mystère-Falcon 20s are eligible for re-engining to 20-5 standard (described separately under the Garrett entry in USA section).

VERSIONS: **Calibration:** There have been 10 Mystère-Falcons, in several different variants, delivered to the French DGAC, French Air Force, and authorities in Spain (designation TM.11), Indonesia and Iran, for navaid calibration. Most are equipped with Dassault-designed high/low level navigation facility calibration systems, some in the form of a removable console.

Airline crew training: Mystère-Falcon 20s have been used by Air France to train pilots for its jet airliners, with up to five aircraft being used simultaneously. Japan Air Lines also used three of this version.

Quick-change and cargo: A quick-change kit, consisting of an assembly of nets and supports keeps the centre aisle free and allows direct access to nine freight compartments. Total usable volume of these compartments is 6.65 m³ (235 cu ft), and transformation from executive configuration to cargo configuration, or vice versa, takes less than one hour. A different specific cargo conversion was performed on 33 aircraft in the USA. For both versions the maximum zero-fuel weight of 9,980 kg (22,000 lb) allows a payload of up to 3,000 kg (6,615 lb).

Target towing: A Mystère-Falcon 20 is used by the French Air Force for target towing. It carries a Secapem target on an inboard hardpoint under each wing and a pod containing a winch and cable on each of two outboard hardpoints. Missions of up to 2 hours duration can be flown, cruising at up to 300 knots (555 km/h; 345 mph) at 450 m (1,500 ft) or 270 knots (500 km/h; 310 mph) at 4,500 m (15,000 ft). The hardpoints (650 kg; 1,433 lb inboard, 750 kg; 1,650 lb outboard) can be used to carry alternative stores if required. Sixteen former Federal Express cargo aircraft operated by FR Aviation of the UK provide Royal Navy target facilities with equipment including an RM30A target winch, AN/ALQ-167 radar jammer, BOZ-3 chaff dispenser, AN/ALE-43 chaff/flare dispenser and ATRS-5 radar simulator, all mounted under the wings. Late in 1990 FR acquired five Falcon 20s from the Canadian Forces for conversion and subseqent operation under anticipated French and NATO contracts.

Aerial photography: This version has two ventral camera bays fitted with optical glass windows. It is operated for high-altitude photography, survey and scientific research in several countries. The camera installation can be supplemented by a multispectral scanner and other scientific loads.

Systems trainer: Five aircraft fitted with the combat radar and navigation systems of various Mirage types are in service with the French Air Force for training pilots. These comprise two Mystère **20NA**s (Système de Navigation-Attaque — equivalent to the Mirage IIIE); and single examples of the **20SNR** (Système de Navigation-Reconnaissance — Mirage E1-CR) and versions equivalent to the Mirage 2000N and Mirage IV-P.

Ambulance: Up to three stretchers can be accommodated, together with a large supply of oxygen and equipment for intensive care and monitoring of patients. Cabinets near the door are removed to facilitate the loading of stretchers.

Electronic warfare: Norway, Canada and Morocco have been followed by several other nations, including Pakistan and Spain, in operating Mystère-Falcon 20 aircraft modified for ECM duties such as radar and communications intelligence and jamming.

Remote sensing: In 1988, a Falcon 20 owned by Innotech was fitted with infra-red mapping equipment supplied by the Canada Centre for Remote Sensing and used for fire-spotting by the forestry authorities in Idaho, USA.

Versions of the Falcon 20 supplied to the US Coast Guard have the following designations:

HU-25A Guardian: Basic version delivered in 1982-83 for search and rescue and offshore surveillance. Total of 41 includes modified HU-25Bs and Cs.

HU-25B: Retrospective designation for variant tasked with location of sea pollution and identification of vessels responsible. Equipment comprises one of the six Aerojet Aireye detection systems ordered for the Guardian in the form of a Motorola AN/APS-131 SLAR pod under the forward fuselage, offset, starboard; a Texas Instruments RS-18C linescan unit in a starboard underwing pod; and a laser illuminated TV under the port wing.

HU-25C: Designation of eight Guardians converted to identify and track air or seaborne drug smugglers by means of a fighter type Westinghouse AN/APG-66 radar in the nose, and turret-mounted Texas Instruments WF-360 FLIR. Also fitted with secure HF/UHF/VHF-FM radio communications. Entered service 30 May 1988.

Two used Falcon 200s acquired by the Chilean Navy are being fitted with Thomson-CSF TRES (Tactical Radar and ESM System), including Varan radar, and will be capable of firing AM 39 Exocet.

UPGRADES: See Versions.

FR Aviation: See separate entry.

OPERATORS: Versions of the Falcon 20/200 are in service with the armed forces of the following countries: Belgium (2), Central African Republic (1), Chile (2), Dubai (1), Egypt (2), France (20), Iran (6), Iraq, Libya (2), Morocco (2), Norway (3), Pakistan (1), Peru (1), Portugal (3), Spain (5), Sudan (1), Syria (2), Tunisia (1) and Venezuela (4).

An abbreviated specification follows:

DESIGN FEATURES: Cantilever low-wing monoplane. Thickness/chord ratio varies from 10.5 to 8 per cent. Dihedral 2°. Incidence 1° 30′. Sweepback at quarter-chord 30°. Variable-incidence tailplane.

FLYING CONTROLS: Hydraulically actuated airbrakes forward of the hydraulically actuated two-section, single-slotted flaps. No trim tabs on tail.

STRUCTURE: All-metal (copper bearing alloys) fail-safe torsion box structure with machined stressed skin. The fuselage is an all-metal semi-monocoque structure of circular cross-section, built on fail-safe principles. The tailplane is a cantilever all-metal structure, mounted halfway up fin.

LANDING GEAR: Retractable tricycle type, by Messier-Bugatti, with twin wheels on all three units. Hydraulic retraction. Maximum steering angle of nosewheel ±50° for taxying, ±180° for towing.

POWER PLANT: Two Garrett ATF 3-6A-4C turbofans (each rated at 23.13 kN; 5,200 lb st). Optional thrust reversers are produced by Hurel-Dubois. Fuel in two integral tanks in wings and large integral tank in rear fuselage, with total capacity of 6,000 litres (1,585 US gallons; 1,320 Imp gallons).

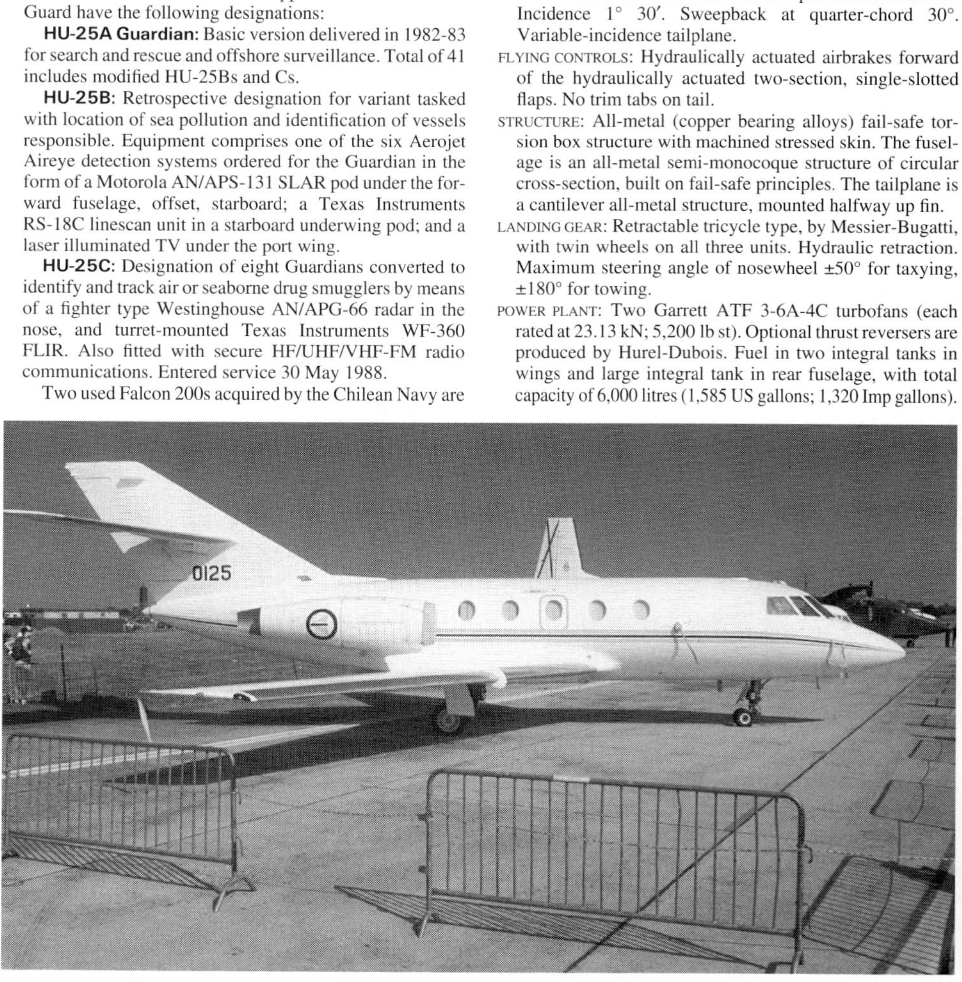

Dassault (Breguet) Mystère-Falcon 20 of the Norwegian Air Force ***(Paul Jackson)*** *1994*

ACCOMMODATION: Flight deck for crew of two, with airline type instrumentation. Airstair door, with handrail, on port side. Main cabin normally seats nine passengers. Alternative arrangement provides 12 compact seats at a pitch of 76 cm (30 in).

SYSTEMS: Duplicated air conditioning and pressurisation system. Two independent hydraulic systems. Electrical system includes a 9 kW 28 V DC starter/generator on each engine, three 750 kVA inverters and two 36 Ah batteries. Solar T40 APU optional. Wing leading-edges and engine air inlets anti-iced with LP compressor bleed air.

AVIONICS: Collins FCS-80 flight control system standard, with dual Collins EFIS-86C electronic flight instrument system using colour CRTs. Standard avionics include duplicated VHF, VOR, DME and ATC transponder, one weather radar and one radio altimeter.

DIMENSIONS, EXTERNAL:

Wing span	16.32 m (53 ft 6½ in)
Wing aspect ratio	6.5
Length: overall	17.15 m (56 ft 3 in)
fuselage	15.55 m (51 ft 0 in)
Height overall	5.32 m (17 ft 5 in)
Wheel track	3.34 m (10 ft 11½ in)
Wheelbase	5.74 m (18 ft 10 in)
Passenger door: Height	1.52 m (5 ft 0 in)
Width	0.80 m (2 ft 7½ in)
Height to sill	1.09 m (3 ft 7 in)

DIMENSIONS, INTERNAL:

Cabin, incl fwd baggage space and rear toilet:	
Length	7.26 m (23 ft 10 in)
Max width	1.79 m (5 ft 10½ in)
Max height	1.70 m (5 ft 7 in)
Volume	20.0 m³ (706 cu ft)
Baggage space (cabin)	0.65 m³ (23 cu ft)
Baggage compartment (rear fuselage)	0.80 m³ (28.2 cu ft)

AREAS:

Wings, gross	41.00 m² (441.3 sq ft)

WEIGHTS AND LOADINGS:

Weight empty, equipped	8,250 kg (18,190 lb)
Payload with max fuel	1,265 kg (2,790 lb)
Max fuel	4,845 kg (10,680 lb)
Max T-O and ramp weight	14,515 kg (32,000 lb)
Max zero-fuel weight	10,200 kg (22,500 lb)
Max landing weight	13,100 kg (28,800 lb)
Max wing loading	354.0 kg/m² (72.5 lb/sq ft)
Max power loading	313.8 kg/kN (3.08 lb/lb st)

PERFORMANCE:

Max operating Mach No. 0.865

Max operating speed:

at S/L 350 kts (648 km/h; 402 mph) IAS

at 6,100 m (20,000 ft) 380 kts (704 km/h; 438 mph) IAS

Max cruising speed at 9,150 m (30,000 ft) at AUW of 11,340 kg (25,000 lb) 470 kts (870 km/h; 541 mph)

Econ cruising speed at 12,500 m (41,000 ft) 420 kts (780 km/h; 485 mph)

Stalling speed 84 kts (156 km/h; 97 mph)

Service ceiling 13,715 m (45,000 ft)

Min ground turning radius about nosewheel 12.80 m (42 ft 0 in)

FAR 25 balanced T-O field length with eight passengers and full fuel 1,420 m (4,660 ft)

FAR 121 landing distance with eight passengers, FAR 121 reserves 1,130 m (3,710 ft)

Range with max fuel and eight passengers at long-range cruising speed, 45 min reserves 2,510 n miles (4,650 km; 2,890 miles)

UPDATED

DASSAULT GARDIAN 2

TYPE: Twin-turbofan surface vessel attack, electronic surveillance and countermeasures aircraft.

PROGRAMME: The Gardian 2 is a Falcon 200 fitted with a Thomson-CSF Varan radar for maritime detection, a Sextant Avionique Omega navigation system and four underwing hardpoints. With additional equipment, it can perform the following missions:

VERSIONS: **Target designation:** This includes over-the-horizon targeting for maritime forces or coastal missile batteries; missile mid-course retargeting; control of surface operations; and strike guidance against surface ships or land objectives. Equipment includes a navigation table, UHF modem to transmit data, V/UHF DF, and interrogator. Options include ESM, search windows, inertial platform, VHF/FM, HF and track-while-scan radar system.

AM 39 Exocet attack: As well as two Exocet sea-skimming air-to-surface missiles, this requires an inertial platform, Omega/INS interface, AM 39 interface and controls, and IFF interrogator. Options are track-while-scan, navigation table and ESM.

Electronic surveillance and countermeasures: This requires either Thomson-CSF DR 2000 ESM and navigation table, or an integrated system including a Thomson-CSF DR 4000 ESM, a computer, the Varan radar, an inertial platform and tactical visualisation elements from the Atlantique 2 system. Options include an IFF interrogator, AM 39 installation, track-while-scan, countermeasures or decoy pods, elint equipment, HF/VHF/UHF comint equipment and V/UHF DF.

Target towing: As for Falcon 200.

In all cases, the cabin can be arranged to permit secondary transport missions.

OPERATORS: Versions of the Gardian 2 are in service with the armed forces of France (2).

VERIFIED

DASSAULT MYSTÈRE-FALCON 100

TYPE: Twin-turbofan executive transport aircraft.

PROGRAMME: The 226th and final Mystère-Falcon 100, was delivered to Aeropersonal in September 1990. Like its predecessors, it is a small executive jet for five to eight passengers, with compound swept wings fitted with high-lift devices, and powered by Garrett TFR731-2 turbofans.

Fuselages were provided by Potez at Aire-sur Adour, which assembles components built by SOGERMA, SOCEA and SOCATA. Wings came from CASA of Spain; tail units and nose assemblies from IAM of Italy; and many other components such as tailfins, doors and emergency exits from Latécoère at Toulouse.

By comparison with the Mystère-Falcon 10, the 100 has an increase of 225 kg (496 lb) in maximum T-O weight and higher maximum ramp weight; a fourth cabin window on the starboard side, opposite the door; a larger heated, unpressurised rear baggage compartment; and a Collins five CRT EFIS-85 instrument package.

In December 1986, the Falcon 100 was certificated by the DGAC for Cat. II approaches in commercial operation.

Under a state sponsored research and development programme, Aerospatiale and Dassault Aviation manufactured a set of resin-impregnated carbonfibre wings for a Falcon 10 designated **V10F**. Dassault Aviation made the port wing, Aerospatiale the starboard wing, retaining the aerodynamic form of the standard metal wings. The V10F (F-WVPR, c/n 5) flew for the first time on 21 May 1985 and received DGAC certification on 16 December 1985. It made a total of 40 test flights and built up flying hours subsequently as one of the aircraft in the charter fleet of Europe Falcon Service.

The following description applies to the standard production Mystère-Falcon 100:

DESIGN FEATURES: Cantilever low-wing monoplane with increased sweepback on inboard leading-edges.

FLYING CONTROLS: Wing has leading-edge slats and double-slotted trailing-edge flaps and plain ailerons. Two-section spoilers above each wing, forward of flaps.

STRUCTURE: The wing is an all-metal torsion box structure. The fuselage is an all-metal semi-monocoque structure, designed to fail-safe principles. The tail unit is a cantilever all-metal structure, similar to that of Falcon 200.

LANDING GEAR: Retractable tricycle type, manufactured by Messier-Hispano-Bugatti, with twin wheels on each main gear unit, single wheel on nose gear. Hydraulic retraction, main units inward, nosewheel forward. Oleo-pneumatic shock-absorbers. Mainwheel tyres size 22 × 5.75 in, pressure 9.31 bars (135 lb/sq in). Nosewheel tyre size 18 × 5.75 in, pressure 6.55 bars (95 lb/sq in).

POWER PLANT: Two Garrett TFE731-2 turbofans (each 14.4 kN; 3,230 lb st), pod-mounted on sides of rear fuselage. Thrust reversers optional. Fuel in two integral tanks in wings and two integral feeder tanks in rear fuselage, with total capacity of 3,340 litres (882 US gallons; 735 Imp gallons). Separate fuel system for each engine, with provision for cross-feeding. Pressure refuelling system.

ACCOMMODATION: Crew of two on flight deck, with dual controls and airline type instrumentation. Provision for third crew member on a jump seat. Seating arrangements differ from aircraft to aircraft in accordance with customer preference. All have a two/three-place sofa in the rear of the cabin, with further seats for a total of up to eight passengers. There is an internal baggage compartment behind the sofa, and a small galley and toilet forward of the passenger accommodation. Clamshell door at the front, on the port side, with built-in steps.

SYSTEMS: Duplicated air conditioning and pressurisation systems supplied with air bled from both engines. Pressure differential 0.61 bars (8.8 lb/sq in). Two independent hydraulic systems, each of 207 bars (3,000 lb/sq in) pressure and with twin engine-driven pumps and emergency electric pump, to actuate primary flight controls, flaps, landing gear, wheel brakes, spoilers, yaw damper and nosewheel steering. Plain hydraulic reservoir pressurised at 1.47 bars (21 lb/sq in). 28 V DC electrical system with a 9 kW DC starter/generator on each engine, three 750 VA 400 Hz 115 V inverters and two 23 Ah batteries. Automatic emergency oxygen system.

AVIONICS AND EQUIPMENT: Standard avionics include Collins APS 80 autopilot and EFIS-85, ADC 80 air data computer and electrical instruments, dual VHF, VOR, DME, transponders, ADF and intercom systems, Collins weather radar, and radio altimeter. Optional avionics are a 718 U5M HF transceiver, Collins APS 85 autopilot and EFIS-86C, and Global GNS 500 flight management system and long-range navigation system.

Dassault (Breguet) Mystère-Falcon 100 *1994*

Dassault (Breguet) Gardian 2 with Thomson-CSF Varan radar *1994*

Dassault (Breguet) Mystère-Falcon 100 of Lyon-Air, equipped for operations to Cat. II landing minima

1994

DIMENSIONS, EXTERNAL:

Wing span	13.08 m (42 ft 11 in)
Wing chord (mean)	2.046 m (6 ft 8½ in)
Wing aspect ratio	7.1
Length overall	13.86 m (45 ft 5¾ in)
Length of fuselage	12.47 m (40 ft 11 in)
Height overall	4.61 m (15 ft 1½ in)
Tailplane span	5.82 m (19 ft 1 in)
Wheel track	2.86 m (9 ft 5 in)
Wheelbase	5.30 m (17 ft 4¾ in)
Passenger door: Height	1.47 m (4 ft 10 in)
Width	0.80 m (2 ft 7 in)
Height to sill	0.884 m (2 ft 10¾ in)
Emergency exit (stbd side, over wing):	
Height	0.914 m (3 ft 0 in)
Width	0.508 m (1 ft 8 in)

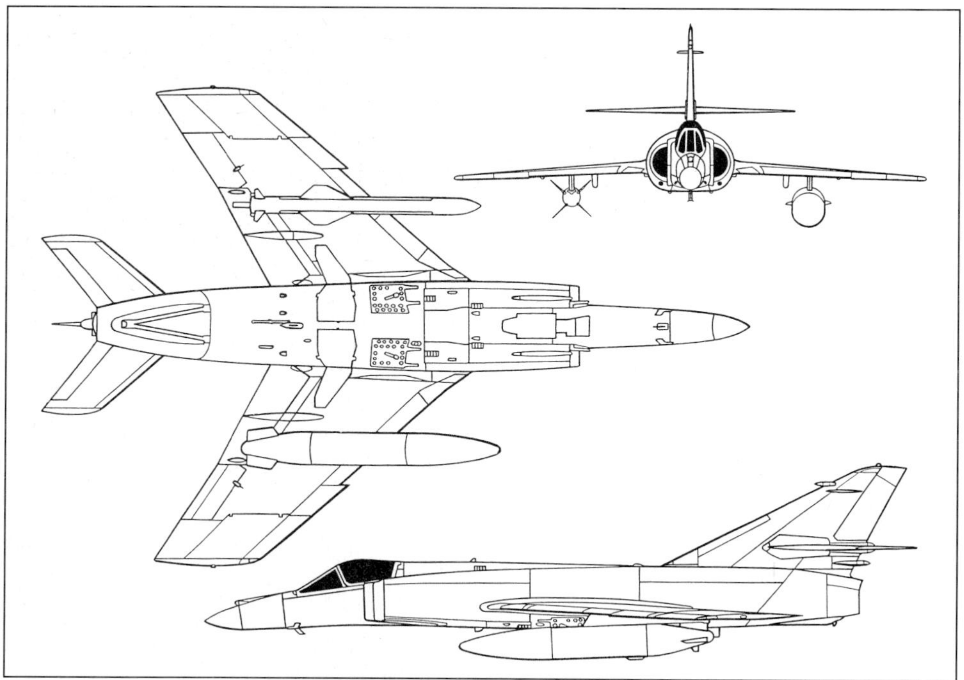

Dassault Super Etendard *(Jane's/Mike Keep)*

1994

Dassault Super Etendard Modernisé *(Aviaplans/F. Robineau)*

1994

DIMENSIONS, INTERNAL:

Cabin, excl flight deck: Length	4.70 m (15 ft 5 in)
Max width	1.55 m (5 ft 1 in)
Max height	1.45 m (4 ft 9 in)
Volume	7.11 m^3 (251 cu ft)
Baggage compartment volume:	
Cabin	0.72 m^3 (25.4 cu ft)
Rear	0.81 m^3 (28.6 cu ft)

AREAS:

Wings, gross	24.1 m^2 (259.4 sq ft)
Vertical tail surfaces (total)	4.54 m^2 (48.87 sq ft)
Horizontal tail surfaces	6.75 m^2 (72.65 sq ft)

WEIGHTS AND LOADINGS:

Weight empty, equipped	5,055 kg (11,145 lb)
Max payload	1,305 kg (2,875 lb)
Payload with max fuel	840 kg (1,852 lb)
Max fuel	2,680 kg (5,910 lb)
Max T-O weight	8,755 kg (19,300 lb)
Max ramp weight	8,800 kg (19,400 lb)
Max zero-fuel weight	6,540 kg (14,420 lb)
Max landing weight	8,000 kg (17,640 lb)

PERFORMANCE (at AUW OF 8,280 kg; 18,254 lb):

Never-exceed speed at S/L 350 kts (648 km/h; 402 mph)
Max operating Mach number 0.87
Max cruise Mach number at 10,670 m (35,000 ft) 0.84
Max cruising speed at 7,620 m (25,000 ft) 492 kts (912 km/h; 566 mph)
Approach speed 100 kts (185 km/h; 115 mph)
Operational ceiling (four passengers, full fuel) 13,715 m (45,000 ft)
FAR 25 balanced T-O field length with four passengers and fuel for a 1,000 n miles (1,850 km; 1,150 mile) stage, 45 min reserves 960 m (3,150 ft)
FAR 25 balanced T-O field length, with four passengers and max fuel 1,325 m (4,350 ft)
FAR 121 landing field length, with four passengers and 45 min reserves 1,065 m (3,495 ft)
Range with four passengers and 45 min reserves 1,565 n miles (2,900 km; 1,802 miles)

VERIFIED

DASSAULT SUPER ETENDARD

TYPE: Single-seat transonic carrier-based strike fighter.

PROGRAMME: The Super Etendard was developed for the French Navy as an updated version of its Etendard IV-M carrier-based fighter, which served with operational squadrons since 1962. The airframe and equipment of the Super Etendard were expected to be 90 per cent common with those of the IV-M, except for the nav/attack system. In fact, the installation of a more powerful turbojet engine and equipment of enhanced capability, together with the adoption of improved aerodynamic features and modern manufacturing techniques, made the Super Etendard 90 per cent new. It is basically a transonic single-seat strike fighter, for low and medium altitude operations from ships in the class of the French Navy's *Clemenceau* and *Foch*. Its equipment includes a highly sophisticated and accurate nav/attack integrated avionics system. Inherent long range is increased by flight refuelling capability, and it is able to operate as a tanker for other aircraft.

The Atar 8K-50 turbojet engine is a non-afterburning version of the Atar 9K-50 used in the Mirage F1 multimission fighter and attack aircraft. It has a lower specific fuel consumption than the Atar 8 fitted in the Etendard IV-M. The thrust increase of about 10 per cent, combined with a new leading-edge and redesigned flaps, allows a significant increase in gross weight for catapulting and, hence, permits increased fuel load and armament.

There were two prototypes produced by conversion of standard IV-M airframes. The first of these flew for the first time on 28 October 1974. Its programme included engine development, followed in 1978 by tests of the Super Etendard's external load-carrying capability and firing trials of the Exocet AM39 air-to-surface anti-ship missile.

The second prototype, which flew for the first time on 25 March 1975, was used for tests of the Super Etendard's navigation system and bombing capabilities. Its subsequent tasks included shipboard operation under open-sea conditions in waters other than the Mediterranean, where all early trials took place.

It was intended originally to build 100 production aircraft, but the number was reduced to 71 in order to conform with budget limitations. The first aircraft flew on 24 November 1977. Deliveries began on 28 June 1978, when the third production aircraft was accepted officially by the French Navy, and Dassault-Breguet (now Dassault Aviation) had delivered 61 Super Etendards to this service by 1 May 1982. The first export order, for 14, was placed by the Argentine Navy in 1979, and eight of these had been delivered by 1 May 1982. Their first known operational deployment took place on 4 May 1982, when an Exocet missile launched from a Super Etendard destroyed the Royal Navy's Type 42 destroyer HMS *Sheffield* off the Falkland Islands.

Production of the original, carrier-based, version of this single-seat transonic strike fighter for the navies of France and Argentina ended in 1983.

Offers by Dassault to re-launch Super Etendard production for land-based operations, and to transfer production to IPTN of Indonesia, have not been taken up.

UPGRADES: **Dassault/Atelier Aéronautique:** Super Etendard Modernisé. In late 1986 the French Navy signed a contract to initiate study of a weapon system modernisation, including replacement of the existing Agave radar with the Dassault Electronique Anémone, having twice the detection range against surface targets; a new SAGEM INS and UAT 90 computer; a Thomson-CSF wide-angle HUD with TV or IR overlay; and cockpit modernisation, including an electronic warfare display screen, HOTAS controls and provision for pilot's night vision goggles. An additional fuselage stores pylon was added. Minimal structural changes are required for the aircraft to achieve an extended life of 6,500 hours by 2005, when replacement by Rafale Ms is to begin.

A prototype conversion flew at Dassault's Istres plant on 5 October 1990. Initial conversions will temporarily retain Agave radar, as Anémone will not be available until mid-1994. Thomson-CSF TMV 011 Sherloc radar warning receivers will also be a later retrofit. Funding for the first 10 conversions was included in the 1991 defence budget. First unit (Flotille 17F based at Laudivisiau is now operational. In the near future, new air-to-surface capabilities will be added (that is capability of laser-guided weapons like AS 30L.

OPERATORS: Versions of the Super Etendard are in service with the armed forces of the following countries: Argentina (12) and France (71).

DESIGN FEATURES: Cantilever mid-wing monoplane. Thickness/chord ratio varies from 6 per cent at root to 5 per cent at tip. Anhedral 3° 30′. Sweepback at quarter-chord 45°. All-moving tailplane. All tail surfaces swept.

FLYING CONTROLS: Inset ailerons, hydraulically powered by Dassault irreversible dual circuits with artificial feel. Spoiler on top surface of each wing, ahead of special double-slotted flap with second slot in form of an integral 'gutter'. Flap travel increased by comparison with Etendard VI-M. Hydraulically powered drooping leading-edges, with extended chord on outer panels. Perforated airbrake under each side of centre-fuselage. All-moving tailplane (with electrically powered pitch trim) and rudder are powered in same way as ailerons.

STRUCTURE: The wing is an all-metal two-spar torsion-box structure; stressed skin of machined panels with integral stiffeners. Tips fold upward for carrier stowage. The fuselage is a semi-monocoque structure, 'waisted' in accordance with area rule. The tail unit is a cantilever all-metal structure, with tailplane mid-set on fin.

LANDING GEAR: Retractable tricycle type, with single wheel on each unit, manufactured by Messier-Hispano-Bugatti. Nosewheel retracts rearward, main units inward into wings and fuselage. Messier-Hispano-Bugatti oleo-pneumatic shock-absorbers and disc brakes. Mainwheel tyres size 30 × 7.7-16; nosewheel tyre size 490 × 155-9. Brake-chute in fairing at junction of fin tailplane trailing-edges.

POWER PLANT: One SNECMA Atar 8K-50 non-afterburning turbojet, rated at 49 kN (11,025 lb st). Fuel in integral tanks in wings and rubber tanks in fuselage, with total capacity of 3,270 litres (845 gallons; 719 Imp gallons). Provision for an external tank of 1,100 litres (290 US gallons; 242 Imp gallons) under each wing, and a 600 litre (158 US gallon; 132 Imp gallon) centreline tank or flight refuelling 'buddy' pack under the fuselage. Retractable flight refuelling probe in fairing in front of windscreen.

ACCOMMODATION: Pilot only, on Hispano-built Martin-Baker SEMMB CM4A lightweight ejection seat in pressurised and air conditioned cockpit. Extensively armoured.

SYSTEMS: Duplicated hydraulic circuits for flying controls, landing gear, brakes and airbrakes, and wing leading-edge droop.

AVIONICS AND EQUIPMENT: SAGEM-Kearfott ETNA inertial navigation and attack system; Thomson-CSF/ESD (now Dassault Electronique) Agave lightweight searchtrack/designation/telemetry/navigation radar; Thomson-CSF VE-120 head-up display; Crouzet Type 97 navigation display, armament control panel and selector box, and Type 66 air data computer; TRT radio altimeter, SFIM three-axis attitude indicator; LMT micro-Tacan and IFF, and Socrat VOR.

ARMAMENT: Two DEFA 30 mm guns, each with 125 rds, in bottom of engine air intake trunks. Underfuselage attachments for two 250 kg bombs. Four underwing attachments for 400 kg bombs, Magic air-to-air missiles or rocket pods. Optionally, one Exocet AM 39 air-to-surface missile under starboard wing, and one external fuel tank under port wing. French Navy Super Etendards are equipped to carry tactical nuclear weapons.

DIMENSIONS, EXTERNAL:

Wing span	9.60 m (31 ft 6 in)
Width, wings folded	7.80 m (25 ft 7 in)
Wing aspect ratio	3.23
Length overall	14.31 m (46 ft 11½ in)
Height overall	3.86 m (12 ft 8 in)
Wheel track	3.50 m (11 ft 6 in)
Wheelbase	4.80 m (15 ft 9 in)

AREA:

Wings, gross	28.4 m² (305.7 sq ft)

WEIGHTS AND LOADINGS:

Weight empty	6,500 kg (14,330 lb)
Max fuel, incl two 1,100 l underwing tanks	4,800 kg 10,580 lb)
Max weapon load, internal fuel only	2,100 kg (4,630 lb)
Mission T-O weight	9,450-12,000 kg (20,833-26,455 lb)

PERFORMANCE:

Max level speed at height	approx Mach 1
Max level speed at low altitude	637 kts (1,180 km/h; 733 mph)
Approach speed for shipboard landing at AUW of 7,800 kg (17,200 lb)	135 kts (250 km/h; 155 mph)
Service ceiling	13,700 m (45,000 ft)
Radius of action with AM 39 missile and two external tanks, hi-lo-hi	460 n miles (850 km; 530 miles)

VERIFIED

Dassault Super Etendard of the French Navy *(Peter J. Cooper)*

1994

DASSAULT (BREGUET 1050) ALIZÉ

TYPE: Three-seat carrier-based anti-submarine aircraft.

PROGRAMME: The Breguet Type 1050 Alizé is a three-seat carrier-based anti-submarine 'hunter-killer'. The aircraft is a derivative of the Type 960 Vultur. The first prototype Alizé flew for the first time on 6 October 1956 and was followed by five preproduction aircraft. By the end of 1958 these logged 2,300 flying hours. This includes catapulting and deck landing trials on the carrier *Arromanches*. Orders for 75 Alizés were placed by the French Navy for use from both light and fleet-type aircraft carriers. The first was delivered on 20 May 1959. A further 12, ordered by the Indian Navy, were delivered from 1961. These were supplemented by a further five surplus French aircraft.

UPGRADES: **Dassault/Atelier Aéronautique:** In the early 1980s, 28 Aéronavale Alizé shipboard maritime patrol aircraft were modernised. A further, less extensive rework for 24 of these was begun at the Cuers naval workshops in 1990, involving addition of a SAGEM MTP16 microcomputer integrated with a system of tactical data processing and transmission, improved decoy systems and INS datalinks. The Alizé will be retained in land- and sea-based service until early in the next century.

Thomson-TRT: In early 1993 the French Navy ordered the Thomson-TRT Chlio forward-looking infra-red (FLIR) system for its Dassault (Breguet) Alizé aircraft.

OPERATORS: Versions of the Alizé are in service with the armed forces of the following countries: France (27) and India (7).

DESIGN FEATURES: Cantilever low-wing monoplane, with hydraulically folding outer panels. Thickness/chord ratio 15 per cent at root, 10 per cent at tip. The tail unit has tailplane and tailfin.

FLYING CONTROLS: Double-slotted flaps in inboard and outboard sections in each wing. Tail unit has balanced elevators and horn-balanced rudder.

STRUCTURE: The fuselage is a semi-monocoque structure. Conventional cantilever-type tailplane.

LANDING GEAR: Retractable tricycle type, the mainwheels retracting forward into combined landing gear/armament nacelles on the wing leading-edges. Twin wheels on each main unit, size 6.5 × 10. Steerable nosewheel. Hydraulic brakes and landing gear retraction. Retractable arrester hook under rear fuselage.

POWER PLANT: One 2,100 ehp Rolls-Royce Dart RDa 7 Mk 21 turboprop engine, driving a Rotol four-blade variable-pitch metal propeller with spinner. Five fuel tanks, comprising one 114 litre (30 US gallon; 25 Imp gallon) tank in each inner wing section and one 1,409 litre (372 US gallon; 310 Imp gallon) fuselage tank.

ACCOMMODATION: Enclosed cabin with side by side seating for the pilot and forward radar operator at front and sideways-facing seat for rear operator behind on starboard side.

WEAPONS: Weapons bay in fuselage accommodation, one torpedo or three 160 kg depth charges. Racks for two 160 kg or 175 kg depth charges under inner sections of wings, and for six 5 in rockets or two AS 12 missiles under outer sections. Sonobuoys in front of wheel fairing.

DIMENSIONS, EXTERNAL:

Wing span	15.60 m (51 ft 2 in)

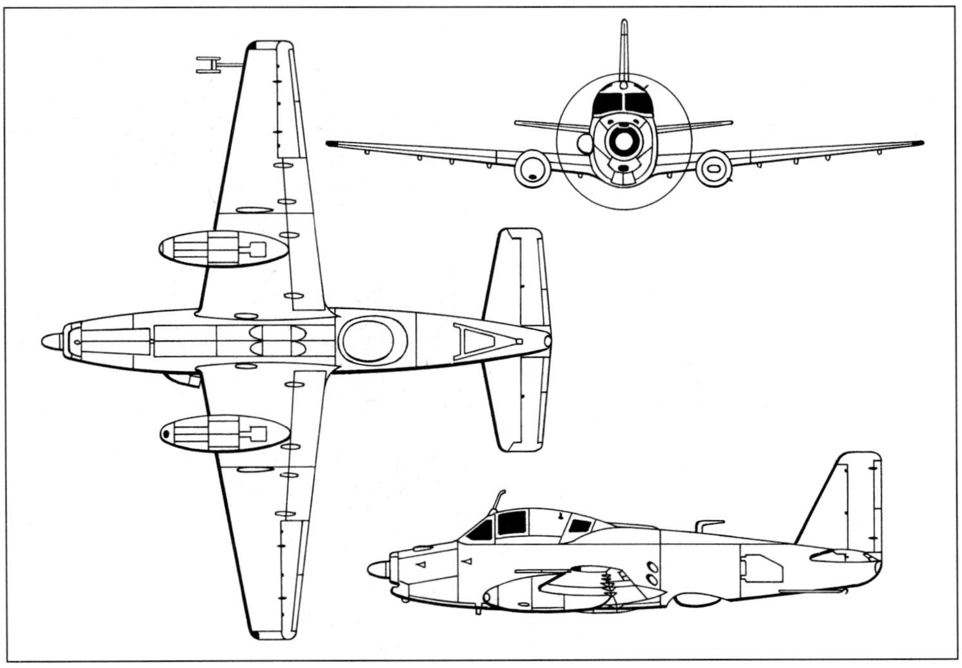

Dassault (Breguet) Alizé anti-submarine aircraft *(Jane's/Mike Keep)*

1994

Dassault (Breguet) Alizé anti-submarine aircraft with Rolls-Royce Dart RDa7 Mk 21 turboprop *(Aviaplans/F. Robineau)* *1994*

Dassault (Breguet) Alizé anti-submarine aircraft with wings folded for storage *(MAP)* *1994*

Length overall	13.86 m (45 ft 5¾ in)
Width folded	7.00 m (22 ft 11½ in)
Height overall	5.00 m (16 ft 5 in)
Wheel track	4.87 m (15 ft 11¾ in)
Wheelbase	5.67 m (18 ft 7¼ in)
AREA:	
Wings, gross	36.00 m² (387.5 sq ft)
WEIGHTS AND LOADINGS:	
Weight empty, equipped	5,700 kg (12,566 lb)
Max T-O weight	8,200 kg (18,078 lb)
PERFORMANCE:	
Max level speed at 3,050 m (10,000 ft)	280 kts (518 km/h; 322 mph)
Max level speed at S/L	248 kts (460 km/h; 286 mph)
Patrol speed	130-200 kts (240-370 km/h; 149-230 mph)
Max rate of climb at S/L	420 m (1,380 ft)/min
Service ceiling	8,000 m (26,250 ft)
T-O run from land	575 m (1,886 ft)
Landing run from land	470 m (1,542 ft)
Range	1,349 n miles (2,500 km; 1,553 miles)
Max endurance with 475 l auxiliary tank	7 h 40 min
Normal endurance	5 h 10 min

VERIFIED

DASSAULT (VOUGHT) F-8P CRUSADER UPGRADE

TYPE: Carrier-based combat aircraft upgrade.

PROGRAMME: Originally due to have been retired on 30 June 1993, the French Navy's Vought F-8E(FN) Crusader carrier-based interceptors will be updated for up to five more years of service before their operating squadron, 12 Flottille at Landivisiau, has completed conversion to Rafale Ms. In 1989, Dassault was contracted to determine the required changes, although work will be performed by the navy's own workshops at Atelier d'Aviation de Cuers, on 17 of the 20 Crusaders remaining from 42 delivered.

DESIGN FEATURES: Apart from a radar warning receiver, modifications are expected to concentrate on alleviation of airframe fatigue, replacement of wiring and ejection seats, and a navigation equipment update. Upgrade also includes engine refurbishment as well as communications and self-protection systems improvements. Redeliveries began in June 1992. The last of the 17 was undergoing work in April 1995.

VERIFIED

MUDRY

AVIONS MUDRY et CIE

Aérodrome de Bernay, BP 214, F-27302 Bernay
Telephone: 33 32 43 47 34
Fax: 33 32 43 47 90
Telex: MUDRY 180 587 F

DIRECTORS GENERAL:
Auguste Mudry (President)
Danielle Baron (Assistant Director General)

CHIEF DESIGNER: Jean Marie Klinka

Mudry Aviation Ltd
Flager County Airport, Box 18T 7, Florida 32110, USA
Telephone: 1 (904) 437 9700
Fax: 1 (904) 437 1170

PRESIDENT: Daniel Heligoin

Company established at Bernay 1958 and operated alongside CAARP of Beynes; two companies consolidated at Bernay; design office at Courcelles; total workforce 50. Plan to certify Sukhoi Su-26MX in France lapsed for administrative reasons; CAP X no longer to be produced in Russia; Mudry has manufactured Aéronautic Baroudeur microlight in small numbers. Production of CAP 21 (1993-94 and earlier *Jane's All the World's Aircraft*) has ended.

NEW ENTRY

Mudry CAP 21 single-seat aerobatic aircraft *(Peter March)* *1995*

MUDRY CAP 21

TYPE: Single-seat competition aerobatic aircraft.

PROGRAMME: First flight 23 June 1980; first delivery May 1982; 39 built when production ceased in January 1990; aircraft now offered as alternative to CAP 231.

VERSIONS: **CAP 21:** Standard production version to which description applies.

CAP 21-260: A Mudry CAP 21 registered in Italy (I-SIVM) was retrofitted with a 194 kW (260 hp) engine by its Italian owner, Sergio Dallan. The aircraft embodies several other modifications including a main spar which permits load factors of ±10*g*; larger wingroot fairings on both the leading- and trailing-edges; longer exhaust pipes which reduce noise; repositioned main landing gear legs with root fairings to reduce airflow disturbance at the wingroot leading-edge; a larger rudder; improved aileron control linkage; substitution of a one-piece canopy; movement of the reclined pilot's seat further aft to offset the effect on the CG of a heavier power plant; replacement of the standard 40 litre (10.5 US gallon; 8.8 Imp gallon) fuel tank with a 60 litre (15.8 US gallon; 13.2 Imp gallon) tank; and installation of a ferry tank under the pilot's seat.

DESIGN FEATURES: Symmetrical V16F wing section; thickness/chord ratio 16 per cent; dihedral 1° 30′; no twist.

FLYING CONTROLS: Mechanical controls; trim tab in each elevator and servo tab in each aileron; roll rate 180°/s at 135 knots (250 km/h; 155 mph).

STRUCTURE: All-wood airframe with plywood skinning on wings and fuselage; single-spar wing; cantilever tail surfaces.

LANDING GEAR: Non-retractable tailwheel type. Cantilever glassfibre main legs, with streamline fairings over wheels. Disc brakes.

POWER PLANT: One 149 kW (200 hp) Textron Lycoming AEIO-360-A1B flat-four engine, driving a two-blade Hartzell variable-pitch propeller. Fixed-pitch propeller optional. Normal fuel tank capacity 40 litres (10.5 US gallons; 8.8 Imp gallons). Maximum fuel capacity 75 litres (20 US gallons; 16.5 Imp gallons), with 15 litre (4 US gallon; 3.3 Imp gallon) gravity tank for inverted flying. Christen inverted oil system.

ACCOMMODATION: Single glassfibre seat under rearward sliding transparent canopy. Special aerobatic shoulder harness.

DIMENSIONS, EXTERNAL:	
Wing span	8.08 m (26 ft 6 in)
Wing aspect ratio	7.0
Length overall	6.46 m (21 ft 2½ in)
Height overall	1.52 m (5 ft 0 in)
AREAS:	
Wings, gross	9.40 m² (101.18 sq ft)
Ailerons (total)	1.42 m² (15.28 sq ft)
Fin	0.780 m² (8.40 sq ft)
Rudder	0.725 m² (7.80 sq ft)
Elevators	1.18 m² (12.70 sq ft)

Mudry CAP 21 single-seat aerobatic aircraft with Textron Lycoming AEIO-360-A1B flat-four engine *(Peter March)*

1995

WEIGHTS AND LOADINGS:
- Weight empty: 500 kg (1,103 lb)
- Max T-O weight: Aerobatic: 620 kg (1,367 lb)
  - Normal: 750 kg (1,653 lb)
- Max wing loading:
  - Aerobatic: 65.95 kg/m² (13.5 lb/sq ft)
- Max power loading:
  - Aerobatic: 4.16 kg/kW (6.84 lb/hp)

PERFORMANCE:
- Never-exceed speed (VNE): 200 kts (372 km/h; 231 mph)
- Max level speed at S/L: 151 kts (280 km/h; 174 mph)
- Max cruising speed (75% power): 135 kts (250 km/h; 155 mph)
- Stalling speed: 49 kts (90 km/h; 56 mph)
- Max rate of climb at S/L: 780 m (2,560 ft)/min
- Rate of roll: 200°/s
- Range (with supplementary tank): 431 n miles (800 km; 497 miles)
- *g* limits: +8/−6

NEW ENTRY

SAGEM

SOCIÉTÉ D'APPLICATIONS GÉNÉRALES D'ELECTRICITÉ ET DE MECANIQUE

27 rue Leblanc, F-75512 Paris Cedex 15
Telephone: 33 (1) 40 70 63 63
Fax: 33 (1) 40 70 67 24
Telex: 205255 F

CHAIRMAN AND CEO: Pierre Faurre (PDG)
SENIOR VICE-PRESIDENT, MANAGING DIRECTOR, NAVIGATION AND DEFENCE DIVISION: Daniel Dupuy
MANAGING DIRECTOR, MISSILE GUIDANCE AND SPACE DEPARTMENT: M. Arnold
MANAGING DIRECTOR, AERONAUTIC DEPARTMENT: J. J. Floch
MANAGING DIRECTOR, NAVY DEPARTMENT: P. Dufour
MANAGING DIRECTOR, LAND SYSTEMS DEPARTMENT: J. Paccard
MANAGING DIRECTOR, INTERNATIONAL AFFAIRS (NAVIGATION AND DEFENSE DIVISION): J. Paccard
MANAGING DIRECTOR, BUSINESS DEVELOPMENT (NAVIGATION AND DEFENSE DIVISION): M. Argouse
PUBLIC RELATIONS EXECUTIVE: Christine de Cambray

SAGEM was formed in 1925. The company is a major French engineering concern specialising in the manufacture of high-precision mechanical, optical and electronic devices. Equipment is supplied for navigation and defence, telecommunications and industrial applications. Inertial navigation systems are produced for installation on aircraft, ground vehicles, satellite launchers, missiles and naval vessels. Other activities are in the fields of electro-optical equipment, aircraft retrofits, digital computers and image processing.

VERIFIED

MAESTRO UPGRADE PACKAGE

TYPE: Combat aircraft upgrade package.

DESIGN FEATURES: Modular avionics enhancement system targeted for retrofit operations (MAESTRO). This system designed to be tailored to customer's specific requirements consists basically of the following: full inertial and GPS systems (ULISS or SIGMA), TERCOR terrain contour matching for stealth navigation and blind attack (EBS 1501 bubble memory), wide field of view HUD, FLIR compatible (IRIS FLIR), HOTAS pilot interface, air-to-ground and air-to-air fire control including radar or laser rangefinder, self-protection system and an all-digital third generation mission-planning system (CIRCE 20001).

VERIFIED

Dassault Mirage 5 upgraded with SAGEM MAESTRO system *(SAGEM)*

1995

MIRAGE III UPGRADE (PAKISTAN)

TYPE: Upgraded delta-wing multirole combat aircraft.

PROGRAMME: SAGEM holds full responsibility for a programme to upgrade the Mirage III fleet of the Pakistan Air Force. First production order issued in June 1993. Two versions are being retrofitted, trainer/strike and multirole fighter. The first version has been fully qualified and production is in progress.

DESIGN FEATURES: Upgrade plans include air-to-ground capability as well as installation of a pulse Doppler radar.

UPDATED

MIRAGE 5 UPGRADE (BELGIUM)

TYPE: Upgraded delta-wing air-to-ground/recce combat aircraft.

DESIGN FEATURES: SAGEM is responsible for design, production and integration of the weapon delivery, navigation and reconnaissance system, part of the Mirage 5 upgrade programme (MIRSIP) for the Belgian Air Force, including supply of nav/attack unit (UNA 92B), the TERCOR terrain contour matching system with 3D digital map database and the UTR 90 computer.

PROGRAMME: The MIRSIP programme was awarded full operational clearance by customer in May 1993. The production phase is to be completed in early 1995 for the upgrade of 15 single-seat Mirage 5 BA and five two-seat Mirage 5 BD for trainer/fighter and attack/recce versions. See also SABCA. Restructuring within the Belgian Air Force has made these aircraft surplus to requirements. In July 1994 the Chilean government agreed to purchase the upgraded aircraft as well as an additional five which will not be upgraded.

UPDATED

ELKAN UPGRADE PROGRAMME

TYPE: Delta-wing air-to-ground/recce combat aircraft upgrade.

PROGRAMME: Programme to supply the Mirage 5 aircraft recently bought from Belgium with a weapons system upgrade. Qualification is in progress. Production to be completed by the end of 1995.

DESIGN FEATURES: Package to include an integrated weapon delivery and navigation system organised around the multifunction nav/attack unit (UNA 92B), integrating autonomous inertial navigation, multichannel GPS receiver, MIL-STD-1553 databus control and mission computation in the same unit. A fully digital mission planning system (CIRCE 2001) is also included to prepare the missions at squadron level and load the data transfer cartridges with all relevant navigation and tactical data.

NEW ENTRY

PZL MIELEC IRYDA M-93 (UPGRADED IRYDA I-22)

TYPE: Upgraded swept, high-wing light attack/advanced trainer aircraft.

PROGRAMME: SAGEM has been selected to upgrade the avionics system on the PZL Mielec Iryda I-22 upgrade in co-operation with the Polish Aviation Institute and the PZL Mielec manufacturer. Flight qualification has been completed. Demonstrated weapon accuracy at par with guided weapons.

DESIGN FEATURES: Systems for the upgraded version, the Iryda M-93, include an inertial navigation system (ULISS with embedded GPS), wide field of view HUD, full colour EFIS, and extended attack capability.

UPDATED

PZL Iryda advanced trainer with SAGEM upgraded avionics suite *(SAGEM)* *1995*

SECA

SOCIÉTÉ D'EXPLOITATION ET DE CONSTRUCTIONS AÉRONAUTIQUES (Subsidiary of Aerospatiale)

PO Box 132, F-93352 Le Bourget
Telephone: 33 (1) 49 34 54 44
Fax: 33 (1) 48 35 96 27
Telex: SECAVIA BRGET 235710 F
MANAGER, PUBLIC RELATIONS: M Laroche

UPDATED

SECA/(FOKKER) F27 FRIENDSHIP ARAT SCIENTIFIC RESEARCH UPGRADE

TYPE: Twin-turboprop scientific research aircraft upgrade.

PROGRAMME: SECA, the aircraft overhaul and modification centre at Paris Le Bourget, has converted a Fokker F27 Friendship Mk 100 to a flying laboratory for the Institute Géographique National. Known as the F27 ARAT (Avion de Recherche Atmosphérique et de Télédétection), it differs substantially from standard in the following ways: addition of Thomson-CSF Varan radar in a ventral radome; a removable nose-probe; two survey camera windows in the lower fuselage; a laser window in the upper starboard

SECA (Fokker) F27 Friendship ARAT flying laboratory *1994*

rear fuselage; 30kVA APU for the operation of scientific equipment during flight; four attachments for various small sensor pods around the forward fuselage; underwing pylons; and three multipurpose equipment panels two above the fuselage and one below.

VERIFIED

SECA (Fokker) F27 Friendship ARAT flying laboratory nose on, showing removable nose-probe *1994*

SOCATA

SOCIÉTÉ DE CONSTRUCTION D'AVIONS DE TOURISME ET D'AFFAIRES (Subsidiary of Aerospatiale)

Le Terminal, Bâtiment 413, Zone d'Aviation d'Affaires, F-93352 Le Bourget Cedex
Telephone: 33 (1) 49 34 69 70
Fax: 33 (1) 49 34 69 71
PRESIDENT AND DIRECTOR GENERAL: Jean-Marc de Raffin Dourny
TECHNICAL DIRECTOR: Jean-Louis Rabilloud
CUSTOMER SERVICES AND BUSINESS DEVELOPMENT DIRECTOR: Emile Escalé
COMMERCIAL DIRECTOR: C. Duplay
PROMOTION AND COMMUNICATION: Mme C. Vallette

This company, formed in 1966, is a subsidiary of Aerospatiale, responsible for producing all of the group's piston-engined light aircraft, as well as the Epsilon primary/basic trainer. By 31 January 1993 sales of the TB series of light aircraft, including the international TBM 700, totalled more than 1,500.

SOCATA also produces components for the Airbus A300, A320, A330/340, Lockheed C-130, ATR 42, Mystère-Falcon 100, 200 and 50 business aircraft, and Super Puma, Dauphin and Ecureuil helicopters. It is responsible for overhaul and repair of MS 760 Paris light jet aircraft.

SOCATA's works cover an area of 56,000 m² (602,775 sq ft).

UPDATED

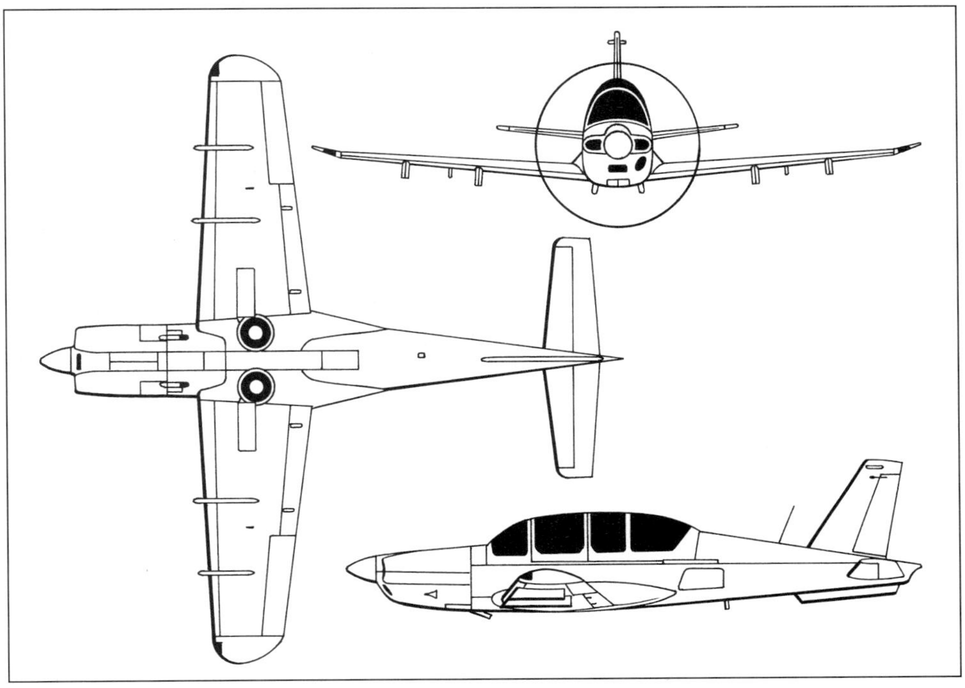

SOCATA TB 30 Epsilon ***(Jane's/Mike Keep)*** *1994*

SOCATA TB 30 EPSILON

TYPE: Two-seat military primary/basic trainer and counter-insurgency aircraft.

PROGRAMME: First details of the **Epsilon** tandem two-seat primary/basic trainer were released in September 1978. Purpose of the project was to meet French Air Force requirements for a propeller driven aircraft for use in the initial stages of a more cost-effective pilot training scheme than that which was currently in operation.

A development contract from the French Air Force for two ground test airframes, was announced by Aerospatiale in June 1979. The first prototype flew on 22 December that year, followed by the second prototype on 12 July 1980. On 6 January 1982, a manufacturing programme was approved, covering delivery of 150 Epsilons at the rate of 30 a year. A contract for the first production batch of 30 aircraft was received on 5 March 1982. The first production Epsilon flew on 29 June 1983; deliveries to the Centre d'Expériences Aériennes Militaires (CEAM) at Mont-de-Marsan began on 29 July 1983, and to Groupement École 315 at Cognac/Chateaubriand in June 1984. Following abandonment of plans to equip other flying schools, GE 315 had been issued with 150 Epsilons by late 1989. Of these, 148 aircraft are still in service.

Until 1989 the Epsilon programme was handled by the Aircraft Division of Aerospatiale, as prime contractor responsible for the entire programme. Design and manufacture were subcontracted to SOCATA, the company's light aircraft subsidiary at Tarbes, which also took full marketing responsibility. In October 1987, an order was announced for 18 Epsilons for Esquadra 104 of the Portuguese Air Force at Sintra. The first of these was handed over at Tarbes on 18 January 1989; the remainder have been assembled by OGMA in Portugal to provide the first 120 hours of pilot training.

An armed version is available to export customers, with four underwing hardpoints for a total 300 kg (661 lb) of external stores with pilot only, or 200 kg (441 lb) with crew of two. Empty weight of this version is 929 kg (2,048 lb), maximum T-O weight 1,400 kg (3,086 lb), and *g* limits +6/−3. An Epsilon armed with two twin 7.62 mm machine gun pods could loiter for 30 minutes at low altitude over a combat area 170 n miles (315 km; 195 miles) from its base. First export order, for three armed Epsilons, was placed by the Togolese Air Force in Autumn of 1984. These aircraft (c/n 51, 55 and 59) were delivered in August 1986 and were followed by a single attrition replacement ordered in 1987. Sales of Epsilons by Spring 1989 totalled 172, plus two demonstrator/trials aircraft.

The first prototype Epsilon was used as testbed for the 335 kW (450 shp) TP 319 turboprop developed by Turbomeca from the TM 319 turboshaft, and was subsequently further modified under the new name *Oméga*.

SOCATA TB 30 Epsilon military basic/primary trainer and counter-insurgency aircraft *1994*

SOCATA TB 30 Epsilon first prototype

1994

OPERATORS: Versions of the Epsilon are in service with the armed forces of the following countries: France (140), Portugal (16) and Togo (3).

The following description applies mainly to the basic version of the Epsilon, as operated by the French Air Force:

DESIGN FEATURES: Cantilever low-wing monoplane. Wing section RA 1643 at root, RA 1243 at tip. Thickness/chord ratio 16 per cent at root, 12 per cent at tip. Dihedral 5°. Incidence 2°. Slightly upturned wingtips. Fixed incidence tailplane, with dihedral. Shallow ventral fin.

FLYING CONTROLS: Electrically actuated single-slotted flaps. Light alloy ailerons with spring tabs. Balanced elevators and rudder, with spring tab in each elevator and ground adjustable tab on rudder.

STRUCTURE: The wing is an all-metal light alloy structure, with single main spar and rear auxiliary spar, built in two panels attached directly to sides of the fuselage. Press-formed ribs and heavy gauge skin without stringers. The fuselage is a light alloy semi-monocoque structure of four longerons, frames and heavy gauge skin, without stringers. The tail unit is a cantilever single-spar light alloy structure. Fixed surfaces metal covered; elevators and rudder covered with polyester fabric.

LANDING GEAR: Electrohydraulically retractable tricycle type, with single wheel on each unit. Inward retracting main units and rearward retracting castoring nosewheel. Mainwheel tyres size 380 × 150; nosewheel tyre size 330 × 130. Independent hydraulic single-disc brake on each mainwheel. Parking brake.

POWER PLANT: One 224 kW (330 hp) Textron Lycoming AEIO-540L1B5D flat-six engine, driving a Hartzell HC-C2YR-4()F/FC 8475-6R two-blade constant-speed metal propeller, with spinner. Fuel in two wing leading-edge tanks, with fuel capacity of 210 litres (55.5 US gallons; 46 Imp gallons). Refuelling points on wing upper surface. Christen system to permit up to 2 minutes inverted flight.

ACCOMMODATION: Two seats in tandem, with rear seat raised by 70 mm (2.7 in). Rudder pedals are mechanically adjustable fore and aft. Two-component sliding Plexiglas canopy, with emergency jettison system, plus sideways hinged (to port) windscreen, providing access to instruments for servicing. Full dual controls. Baggage compartment aft of cabin, with door on port side.

SYSTEMS: Hydraulic systems for actuating landing gear and brakes; 28 V electrical system includes engine-driven alternator; battery for engine starting and emergency use; ground power socket in port side of rear fuselage. Cabin heated and ventilated. Windscreen demister.

AVIONICS: Standard installation includes blind-flying instrumentation, standby artificial horizon, VHF, UHF, automatic and manual VOR, transponder, ILS capability and Tacan. Optional equipment includes a Rockwell-Collins ETC-500F electronic tactical control unit providing centralised control and display of all radio communications and radio navigation equipment.

ARMAMENT (not on French Air Force Epsilons): Four underwing hardpoints; outboard points each able to carry 80 kg (176 lb), inboard points 160 kg (352 lb). Alternative loads include two Matra CM pods each containing two 7.62 mm machine guns, four Matra F2D launchers for six XF1 68 mm rockets, six Bavard F4B practice bombs, two 125 kg bombs, two Alkan 500 grenade launchers with 10 Lacroix rounds each, and four land or sea survival kit containers. Associated equipment includes Alkan 663 stores racks, SFOM 83A3 sight and Alkan E105C firing control box.

DIMENSIONS, EXTERNAL:

Wing span	7.92 m (25 ft 11¾ in)
Wing chord: at root	1.46 m (4 ft 9½ in)
at tip	0.92 m (3 ft 0¼ in)
Wing aspect ratio	7.0
Length overall	7.59 m (24 ft 10¾ in)
Height overall	2.66 m (8 ft 8¾ in)
Tailplane span	3.20 m (10 ft 6 in)
Wheel track	2.30 m (7 ft 6½ in)
Wheelbase	1.80 m (5 ft 10¾ in)
Propeller diameter	1.98 m (6 ft 6 in)
Propeller ground clearance	0.25 m (10 in)

AREAS:

Wings, gross	9.00 m² (96.9 sq ft)
Fin	1.02 m² (10.98 sq ft)
Tailplane	2.00 m² (21.53 sq ft)

WEIGHTS AND LOADINGS:

Weight empty, equipped	932 kg (2,055 lb)
Fuel weight	150 kg (330 lb)
Max T-O and landing weight	1,250 kg (2,755 lb)
Max wing loading	139 kg/m² (28.4 lb/sq ft)
Max power loading	5.58 kg/kW (9.18 lb/hp)

PERFORMANCE (at max weight):

Never-exceed speed	281 kts (530 km/h; 323 mph)
Max level speed at S/L	204 kts (378 km/h; 236 mph)
Max cruising speed (75% power) at 1,830 m (6,000 ft)	193 kts (358 km/h; 222 mph)
Approach speed	80 kts (148 km/h; 92 mph)
Stalling speed, flaps and landing gear down, power off	62 kts (115 km/h; 72 mph)
Max rate of climb at S/L	564 m (1,850 ft)/min
Service ceiling	7,010 m (23,000 ft)
T-O run	410 m (1,345 ft)
T-O to 15 m (50 ft)	440 m (1,444 ft)
Landing run	250 m (820 ft)
Endurance (65% power)	3 h 45 min
g limits	+6.7/−3.35

VERIFIED

GERMANY

DASA

DAIMLER-BENZ AEROSPACE AG

PO Box 801109, D-81663 Munich
Telephone: 49 (89) 607-0
Fax: 49 (89) 607-26481
Telex: 5287-0 DASAM D
Teletext: 89702+DASAM D

CHAIRMAN, DEUTSCHE AEROSPACE SUPERVISORY BOARD:
Edzard Reuter

PRESIDENT OF BOARD OF MANAGEMENT AND CEO:
Jürgen E. Schrempp

MEMBERS OF THE BOARD OF MANAGEMENT:
Dr Manfred Bischoff (Finance and Control)
John Tucker (President Propulsion Systems Group and Chairman Board of Management MTU München GmbH)
Werner Heinzmann (President Space Systems Group, President Defence and Civil Systems Group, Chairman Board of Management Dornier GmbH)
Dr Hartwig Knitter (Personnel)
Hartmut Mehdorn (President Aircraft Group)

Daimler-Benz, formed 19 May 1989, is the aircraft, defence, space and propulsion systems arm of the Daimler-Benz group; integrating Dornier, Messerschmitt-Bölkow-Blohm (MBB), MTU Motoren- und Turbinen-Union München and Telefunken SystemTechnik (TST); major reorganisation completed on 30 September 1992 included merger of TST (Ulm) with MBB (Ottobrunn) and transfer of Daimler-Benz business to former MBB, now trading as Daimler-Benz; simultaneously German government transferred its 20 per cent share in Deutsche Airbus (Hamburg) to Daimler-Benz with retrospective effect from 1 January 1992; Hamburg company operating as Daimler-Benz Aerospace Airbus, a subsidiary of Daimler-Benz Aerospace, since 1 October 1992.

Workforce at 31 December 1994 was 77,000. Company organised into groups covering Aircraft, Space Systems, Propulsion Systems and Defence and Civil Systems; these groups have 12 divisions; all groups aim at international strategic co-operation.

Efforts to become an international competitor and equal partner in international programmes have included MTU general co-operation agreement with Pratt & Whitney, formation of Eurocopter with Aerospatiale, and senior membership of Airbus consortium through Daimler-Benz Aerospace Airbus.

Daimler-Benz Aerospace negotiated throughout 1992 to acquire 51 per cent of Netherlands company Fokker; on 27 April 1993 DASA acquired 78 per cent of a Fokker holding company shared with the Netherlands government; new capital also injected; DASA will acquire remaining government holding in three years; Fokker aircraft line to continue and new Fokker 70 (which see) continues to be developed.

Former East German Elbe Flugzeugwerke incorporated into Daimler-Benz Aerospace Airbus and Flugzeugwerke Ludwig Felde into DASA Propulsion Group during 1991.

Aircraft Group consists of the Military Aircraft, Regional Aircraft, Daimler-Benz Aerospace Airbus, Eurocopter and Fokker divisions. Activities include Airbus family, Tornado, Alpha Jet, Dornier regional airliners and helicopters. Participation in Airbus consortium is 37.9 per cent.

Space Systems Group consists of the Satellite Systems Division and the Orbital Infrastructure Division. Group produces satellites for environmental and weather observation, reconnaissance and verification; international activities include Ariane 4 and 5 and the Columbus space station (APM) group is developing concepts for new applications and potential commercialisation.

Defence and Civil Systems Group consists of Dynamic Systems, Sensor Systems, Information and Communication Systems; products include anti-tank, anti-ship and anti-aircraft missiles, dispenser systems, drones and training systems, radar systems, systems for ammunition disposal, environmental protection systems and a variety of other high technology civil systems.

Propulsion Systems Group consists of Aircraft and Land/Marine Propulsion, Applications Divisions. Aircraft propulsion projects include repair and overhaul of large civil engines and military engines, and joint development with Rolls-Royce, IAE, Pratt & Whitney and Turbomeca of such engines as RB199, EJ200, V2500, PW300, RTM 322 and MTR 390.

UPDATED

DAIMLER-BENZ AEROSPACE AIRBUS GmbH

PO Box 950109, Kreetslag 10, D-21111 Hamburg
Telephone: 49 (40) 7437-0
Fax: 49 (40) 743 4422
Telex: 21950-0 DA D

WORKS: Hamburg, Bremen, Einswarden, Varel, Stade, Lemwerder, Munich, Laupheim Speyer, Dresden

PRESIDENT AND CEO: Dr. Gustav Humbert

CHAIRMAN, SUPERVISORY BOARD: Jürgen E. Schrempp

DIRECTOR, MARKETING AND SUPPORT: Horst Emker

DIRECTOR, ENGINEERING DESIGN AND TECHNOLOGY:
Jürgen Thomas

DIRECTOR, PERSONNEL: Hans-Joachim Gante

DIRECTOR, TECHNICAL ENGINEERING: Gerhard Eisen

DIRECTOR, FINANCES AND CONTROLLING:
Dr Pieter Deisch

Daimler-Benz Aerospace Airbus develops and manufactures about one-third of Airbus airliner family and about one-quarter of Fokker 100. Through the organisation Airspares and acting on behalf of Airbus Industrie, Daimler-Benz Aerospace Airbus is responsible for provision of spare parts worldwide for Airbus fleet. In Aircraft Service Centre (ASC) in Lemwerder, Transall is just one of the many aircraft maintained and overhauled for operational service until 2010.

Daimler-Benz Aerospace Airbus (16,500 employees in 8 locations) produces fuselage sections and vertical tails of A300-600, A310, A319, A320, A321, A330 and A340. Company is main cabin furnishing centre for Airbus family (except for A330/A340) and fits all movable wing parts to wing torsion boxes produced by British Aerospace. A321, a stretched version of A320 and A319 is assembled by Daimler-Benz Aerospace Airbus. Aft fuselage for Fokker 100 and Fokker 70 is manufactured by Elbe Flugzeugwerke GmbH in Dresden.

UPDATED

AIRBUS SUPER TRANSPORTER

Together with Aerospatiale, Daimler-Benz Aerospace Airbus has established Special Aircraft Transport International Company (SATIC). The German-French joint venture is responsible for manufacture of Super Guppy successor Airbus Super Transporter (AST), based on Airbus A300-600.

VERIFIED

VERY LARGE COMMERCIAL TRANSPORT

Daimler-Benz Aerospace Airbus is working intensively in overall framework of Airbus Industrie consortium on study for a Very Large Commercial Transport (VLCT) for more than 600 passengers. Study currently performed by company under designation A2000 is German Airbus partner's contribution.

VERIFIED

FLA

In the Euroflag group, company is co-operating with Aerospatiale, British Aerospace, CASA, Alenia, Flabel, OGMA and Tusas on a successor for Transall and Hercules aircraft.

SUPERSONIC COMMERICAL TRANSPORT (SCT)

An international study group is working on a new supersonic airliner. Participants in study group on Supersonic Commercial Transport (SCT) include Aerospatiale (France), Alenia (Italy), Boeing Commercial Airplane Group (USA), British Aerospace (United Kingdom), Daimler-Benz Aerospace Airbus (Germany), Japanese Aircraft Industries (Japan), McDonnell Douglas (USA) and Tupolev (Russia).

VERIFIED

CRYOPLANE

Daimler-Benz Aerospace Airbus is leading a group of companies that includes Tupolev and Kuznetsov in Russia to examine development of airliner fuelled with liquid hydrogen. Joint feasibility study in Germany and Russia was completed in early 1993.

VERIFIED

Transall C-160D cockpit upgraded by Daimler-Benz Aerospace Airbus

1995

DAIMLER-BENZ AEROSPACE (TRANSALL) C-160D STRUCTURAL/AVIONICS UPGRADE

TYPE: Military transport aircraft avionics and structural upgrades.

PROGRAMME: Daimler-Benz Aerospace Airbus is responsible for the maintenance and modification of the German Air Force C-160 Transall aircraft. As part of a continuing programme, the company has increased the aircraft life up to the year 2010 as well as initiating an autonomous navigation and flight control ANA/FRA upgrade.

DESIGN FEATURES: ANA/FRA system consists of the installation of autonomous navigation unit high integration air data computers (HIADC) (GMAv), central display unit (CDU), electronic horizontal situation indicator (EHSI), navigation display unit, GPS system, flight management system (FMS), (Rockwell Collins FMS 800) inertial navigation system (INS), attitude director indicator (ADI), autopilot and attitude heading reference system (AHRS).

NEW ENTRY

DASA

DORNIER LUFTFAHRT GmbH (Subsidiary of Dornier GmbH — part of Daimler-Benz Aerospace AG)

HEADQUARTERS: Dornier Airfield, PO Box 1103, D-82230 Wessling
Telephone: +49 (8153) 30-0
Fax: +49 (8153) 30-2055
Telex: 526 450 doas d

Dornier GmbH, formerly Dornier-Metallbauten, formed 1922 by late Professor Claude Dornier, has operated as a GmbH since 22 December 1972. Daimler-Benz AG acquired majority holding (65.5 per cent) in Dornier GmbH in 1985, but had reduced this to 57.55 per cent by 1 January 1989, when new three-group Dornier company structure came into being with Silvius Dornier (21.22 per cent) and Claudius Dornier heirs (21.22 per cent) as other shareholders. Daimler-Benz shareholding since assumed by Daimler-Benz Aerospace AG.

All of Dornier's aviation activities now undertaken by Dornier Luftfahrt GmbH at Oberpfaffenhofen, which is wholly owned subsidiary of Dornier GmbH. Dornier Luftfahrt is now the basis of Daimler-Benz Aerospace Regional Aircraft Division.

Dornier Luftfahrt GmbH manufactures the Dornier 328 and Dornier 228 regional airliners; subcontractor to Daimler-Benz Aerospace Airbus; supports the 18 NATO Boeing E-3A Sentry AWACS and NATO trainer/cargo aircraft (TCA); technical and logistic servicing of German Navy Breguet 1150 Atlantic 1 and contributes to Atlantique 2; life extension modifications to 168 Bell UH-1D utility helicopters; service centre for business jets and regional turboprops.

VERIFIED

Dornier (Daimler-Benz Aerospace) Do 28D Skyservant of the German Navy *(P. Tompkins)*

1994

DORNIER Do 28 D-2 SKYSERVANT

TYPE: Twin-engined STOL transport and utility aircraft.

PROGRAMME: Despite its designation, the Skyservant inherited only the basic configuration of the Do 28. The prototype (D-INTL) first flew 23 February 1966. Type approval for the basic Do 28 D was granted 24 February 1967, and for the development Do 28 D-1 on 6 November 1967. FAA certification of the Do 28 D-1 was granted 19 April 1968. Military type approval of the Do 28 D-1 was granted January 1970, and the Do 28 D-2 late 1971, in accordance with Milspec standards. Initial deliveries of the Skyservant were made Summer 1967; by mid-1979 sales totalled 260 aircraft, to more than 25 countries. Of these 245 were delivered by June 1979 including 101 Do 28 D-2s to the Federal German Luftwaffe for general duties and 20 to the Navy Air Arm for support duties. Others went into service with the air forces of Cameroon, Ethiopia, Israel, Nigeria, Somalia, Thailand, Turkey and Zambia. Production of a further batch of 20 aircraft was then under way.

On 15 March 1972 M. F. Tuytjens set six international records in a Do 28 D-1 Skyservant in Class C1e for piston-engined business aircraft in the 3,000-6,000 kg weight category.

OPERATORS: Versions of the Do 28 D-2 are in service with the armed forces of the following countries: El Salvador (5), Germany (7), Greece (10), Kenya (7), Malawi (4), Morocco (2), Niger (2), Nigeria (15), Thailand (3) and Turkey (5).

The following details apply to the Do 28 D-2, which introduced a number of aerodynamic and detail refinements, including a 192 kg (423 lb) increase in AUW and more extensive standard equipment.

DESIGN FEATURES: Cantilever high-wing monoplane. Wing section NACA 23018 (modified), with nose slot in the outer half of each wing. Dihedral 1° 30′. Incidence 4°.

FLYING CONTROLS: Double-slotted ailerons and flaps have metal structure, partly Eonnex-covered. Balance tabs on ailerons. Tail unit has all-moving horizontal surface, with combined anti-balance and trim tab. Trim tab in rudder.

STRUCTURE: The wing is an all-metal box-spar structure. The fuselage is a conventional all-metal stressed-skin structure. The tail unit is a cantilever all-metal structure, with rudder and horizontal surfaces partly Eonnex-covered.

LANDING GEAR: Non-retractable tailwheel type. Dornier oleo-pneumatic shock-absorbers on main units. Glassfibre sprung tailwheel unit, mainwheel tyres size 8.50-10, pressure 3.38 bars (49 lb/sq in). Twin-contact tailwheel tyre size 5.50-4, pressure 2.76 bars (40 lb/sq in). Double-disc hydraulic brakes. Fairings on main legs and wheels standard.

POWER PLANT: Two 283 kW (380 hp) Lycoming IGSO-540-A1E flat-six engines, mounted on stub wings and each

Dornier (Daimler-Benz Aerospace) Do 28D Skyservant twin-engine STOL aircraft *(P. Tompkins)*

1994

driving a Hartzell Type HC-B3W30-2B/W10151B-8R three-blade constant-speed and fully feathering propeller. Fuel tanks in engine nacelles, with total usable capacity of 893 litres (236 US gallons; 196.5 Imp gallons). Refuelling points above nacelles. Provision for two underwing auxiliary fuel tanks with combined capacity of 474 litres (124.9 US gallons; 104 Imp gallons). Total capacity of separate oil tanks; 33 litres (8.7 US gallons; 7.25 Imp gallons).

ACCOMMODATION: Pilot and either co-pilot or passenger side by side on flight deck. Dual controls standard. Main cabin fitted with up to 12 seats, with aisle, or 13 seats inward-facing folding seats, or five stretchers and five folding seats, all layouts including toilet and/or baggage compartment and/or darkroom for aerial survey missions aft of cabin. Alternatively, cabin can be stripped for cargo-carrying. Door on each side of flight deck. Emergency exit on starboard side of cabin. Combined two-section passenger and freight door on port side of cabin, at rear.

SYSTEMS: Pneumatic de-icing optional.

EQUIPMENT: Standard equipment includes dual controls, dual brake system, directional slaved gyro, cabin heating, 100A alternators and provisions for de-icing system and IFR com/nav antennae installation; other equipment to customer's specifications.

DIMENSIONS, EXTERNAL:

Wing span	15.55 m (51 ft 0¼ in)
Wing chord (constant)	1.90 m (6 ft 2¾ in)
Wing aspect ratio	8.3
Length overall	11.41 m (37 ft 5¼ in)
Height overall	3.90 m (12 ft 9½ in)
Tailplane span	6.61 m (21 ft 8¼ in)
Wheel track	3.52 m (11 ft 6 in)
Wheelbase	8.63 m (28 ft 3¾ in)
Propeller diameter	2.36 m (7 ft 9 in)
Passenger door (port, rear):	
Height	1.34 m (4 ft 4¾ in)
Width	0.65 m (2 ft 1½ in)
Height to sill	0.60 m (1 ft 11½ in)
Freight door (port, rear):	
Height	1.34 m (4 ft 4¾ in)
Width	1.28 m (4 ft 2 ½ in)

DIMENSIONS, INTERNAL:

Cabin, excl flight deck and rear baggage compartment:	
Max length	3.97 m (13 ft 0½ in)
Max width	1.37 m (4 ft 6 in)
Max height	1.52 m (4 ft 11⅞ in)
Floor area	5.30 m² (57.05 sq ft)
Volume	8.10 m³ (286 cu ft)

AREAS:

Wings, gross	29.00 m² (312 sq ft)
Ailerons (total)	2.64 m² (28.4 sq ft)
Trailing-edge flaps (total)	4.80 m² (51.6 sq ft)
Fin, incl dorsal fin	3.65 m² (39.3 sq ft)
Rudder, incl tab	1.40 m² (15.1 sq ft)
Tailplane, incl tab	7.65 m² (82.3 sq ft)

WEIGHTS AND LOADINGS:

Weight empty, standard	2,328 kg (5,132 lb)
Max T-O weight	3,842 kg (8,470 lb)
Max T-O weight with optional external fuel tanks	4,015 kg (8,852 lb)
Max ramp weight	3,862 kg (8,514 lb)
Max landing weight	3,650 kg (8,050 lb)
Max wing loading	132 kg/m² (27.1 lb/sq ft)
Max power loading	6.79 kg/kW (11.14 lb/hp)

PERFORMANCE (at max T-O weight):

Max level speed at 3,050 m (10,000 ft)	175 kts (325 km/h; 202 mph)
Max cruising speed at 3,050 m (10,000 ft)	165 kts (306 km/h; 190 mph)
Max cruising speed at S/L	147 kts (273 km/h 170 mph)
Econ cruising speed, 50% power at 3,050 m (10,000 ft)	130 kts (241 km/h; 150 mph)
Stalling speed, power off, flaps down	56.5 kts (104 km/h; 65 mph)
Min control speed, power on, flaps down	35 kts (65 km/h; 40 mph)
Max rate of climb at S/L	354 m (1,160 ft)/min
Rate of climb at S/L, one engine out	55 m (180 ft)/min
Service ceiling	7,680 m (25,200 ft)
Service ceiling, one engine out	2,805 m (9,200 ft)
T-O run	280 m (920 ft)
T-O to 15 m (50 ft)	415 m (1,360 ft)
Landing from 15 m (50 ft)	390 m (1,280 ft)
Landing run	201 m (660 ft)
Range with max payload	566 n miles (1,050 km; 652 miles)

VERIFIED

DORNIER 128-6

TYPE: Twin-turboprop STOL transport and utility aircraft.

PROGRAMME: The first turboprop development of the Skyservant was the Do 28 D-5X prototype, which flew for the first time on 9 April 1978 with two 447 kW (600 shp) Avco Lycoming LTP 101-600-1As derated to 298 kW (400 shp). It was redesignated Dornier 128-6X when retrofitted with Pratt & Whitney Canada PT6A-110 turboprops, and made its first flight with this power plant on 4 March 1980.

Except for reinforced landing gear and strengthening of the engine support structure, the airframe of the Dornier 128-6 is generally similar to that of the piston-engined 128-2, and the aircraft was available also in maritime patrol aircraft (MPA) form with a large undernose radome.

Certification by the LBA was granted in March 1981; the first production Dornier 128-6, an aircraft for Lesotho Airways, was delivered in July 1981. A small number of 128-6MPAs, equipped with MEL Marec radar, was ordered by the Cameroon Air Force for Maritime patrol, and in 1982 a contract for 18 Dornier 128-6 utility aircraft was placed by the Nigerian Air Force. The Nigerian Air Force has 21 Dornier 128-6 in service in early 1992. Production was phased by early 1986.

One Dornier 128-6 named *Polar I* was specially equipped to support a German expedition during the Winter (South Polar Summer) of 1983-84. It has wheel-ski landing gear, weather radar, de-icing system and a survival kit for 20 days. It was used primarily to transport field teams to unprepared sites, and to ferry equipment such as sleds and skidoos.

OPERATORS: Versions of the Do 128-6 are in service with the armed forces of the following countries: Cameroon (2) and Nigeria (21).

DESIGN FEATURES: Cantilever high-wing monoplane. The wing-section NACA 23018 (modified), with nose slot in the outer half of each wing. Dihedral 1° 30′. Incidence 4°.

FLYING CONTROLS: Double-slotted ailerons and flaps have metal structure, partly Eonnex-covered. Balance tabs in ailerons. The tail unit has all-moving horizontal surface, with combined anti-balance and trim tab. Trim tab in rudder.

STRUCTURE: The wing is an all-metal box spar structure. The fuselage is a conventional all-metal skin structure. The tail unit is a cantilever all-metal structure, with rudder and horizontal surfaces partly Eonnex-covered.

LANDING GEAR: Non-retractable tailwheel type. Dornier oleo-pneumatic shock-absorbers in main units, glassfibre sprung tailwheel unit. Mainwheel tyre size 8.50-10, pressure 3.38 bars (49 lb/sq in). Twin contact tailwheel tyre size 5.50-4, pressure 2.7 bars (40 lb/sq in). Double disc hydraulic brakes. Fairings on main legs and wheels standard.

POWER PLANT: Two Pratt & Whitney Canada PT6A-110 turboprop engines, each derated to 298 kW (400 shp), mounted on stub wings and each driving a Hartzell B3TN-3D/T102B-9.5 three-blade constant-speed and fully feathering metal propeller with spinner. Fuel tanks in rear of engine nacelles, with total usable capacity of 893 litres (235.4 US gallons; 196 Imp gallons). Refuelling points above nacelles. Provision for two underwing auxiliary fuel tanks with combined capacity of 474 litres (124.9 US gallons; 104 Imp gallons). Total capacity of separate oil tanks, 33 litres (8.71 US gallons; 7.25 Imp gallons).

ACCOMMODATION: Pilot and either co-pilot or passenger side by side on flight deck. Dual controls standard. Main cabin normally equipped to carry 10 passengers in pairs, with centre aisle, or five stretchers and five folding seats, all layouts including toilets and/or baggage compartment and/or darkroom for aerial survey missions aft of cabin. Alternatively, cabin can be stripped for cargo carrying. Door on each side of flight deck. Emergency exit on starboard side of cabin. Combined two-section passenger and freight door on port side of cabin, at rear.

SYSTEMS: Pneumatic de-icing boots on leading-edges optional.

EQUIPMENT: Provision for com/nav antennae installation, and avionics, to customer's requirements. Standard equipment includes directional slaved gyro, cabin heating, dual brake system, 100A alternators and provisions for optional de-icing system.

DIMENSIONS, EXTERNAL:

Wing span	15.85 m (52 ft 0 in)
Wing chord (constant)	1.90 m (6 ft 2¾ in)
Wing aspect ratio	8.33
Length overall (in flying attitude)	11.41 m (37 ft 5¼ in)
Height overall (static)	3.90 m (12 ft 9½ in)
Tailplane span	6.61 m (21 ft 8¼ in)
Wheel track	3.50 m (11 ft 6 in)
Wheelbase	8.63 m (28 ft 3¾ in)
Distance between propeller centres	4.20 m (13 ft 9½ in)
Passenger door (port, rear):	
Height	1.34 m (4 ft 4¾ in)
Width	0.64 m (2 ft 1¼ in)
Height to sill	0.60 m (1 ft 11½ in)
Freight door (port, rear):	
Height	1.34 m (4 ft 4¾ in)
Width, incl passenger door	1.28 m (4 ft 2½ in)

DIMENSIONS, INTERNAL:

Cabin, excl flight deck and rear baggage compartment:	
Max length	3.97 m (13 ft 0½ in)
Max width	1.37 m (4 ft 6 in)
Max height	1.52 m (4 ft 11¾ in)
Floor area	5.30 m² (57.05 sq ft)
Volume	8.00 m³ (282.5 cu ft)
Baggage compartment volume	0.90 m³ (31.8 cu ft)

Dornier (Daimler-Benz Aerospace) Do 128-6 Turbo-Skyservant

1994

AREAS:

Wings, gross	29.00 m² (312.2 sq ft)
Ailerons (total)	2.64 m² (28.4 sq ft)
Trailing-edge flaps (total)	4.80 m² (51.6 sq ft)
Fin, incl dorsal fin	3.65 m² (39.3 sq ft)
Rudder, incl tab	1.40 m² (15.1 sq ft)
Tailplane, incl tab	7.65 m² (82.3 sq ft)

WEIGHTS AND LOADINGS:

Weight empty	2,540 kg (5,600 lb)
Max payload	1,273 kg (2,806 lb)
Max T-O weight	4,350 kg (9,590 lb)
Max landing weight	4,140 kg (9,127 lb)
Max wing loading	150 kg/m² (30.72 lb/sq ft)
Max power loading	7.30 kg/kW (11.99 lb/shp)

PERFORMANCE (at max T-O weight except where indicated):

Max level speed	183 kts (339 km/h; 211 mph)
Max cruising speed at 3,050 m (10,000 ft)	178 kts (330 km/h; 205 mph)
Cruising speed, 75% power at 3,050 m (10,000 ft)	165 kts (305 km/h; 190 mph)
Cruising speed, max range power at 3,050 m (10,000 ft)	140 kts (259 km/h; 161 mph)
Cruising speed, 50% power at 3,050 m (10,000 ft)	138 kts (256 km/h; 159 mph)
Landing speed	63 kts (116 km/h; 72 mph)
Max rate of climb at S/L	384 m (1,260 ft)/min
Rate of climb at S/L, one engine out	54 m (177 ft)/min
Service ceiling	9,935 m (32,600 ft)
T-O to 15 m (50 ft)	554 m (1,820 ft)
Landing from 15 m (50 ft) at max landing weight	503 m (1,650 ft)
Range with 805 kg (1,774 lb) payload, no reserves	788 n miles (1,460 km; 907 miles)
Range with max fuel at cruising speed	858 n miles (1,590 km; 988 miles)
Range with max fuel at max range power	985 n miles (1,825 km; 1,134 miles)

VERIFIED

Dornier (Daimler-Benz Aerospace) 228-201 of the German Navy

1994

DORNIER 228

TYPE: Twin-turboprop STOL light transport.

PROGRAMME: The design of the Dornier 228 complies with US FAR Pt 23 requirements, including Amendment 23, and Appendix A of FAR Pt 135. One prototype of each initial version was built; the first of these, the Dornier 228-100 (D-IFNS) made its first flight on 28 March 1981. The 228-200 (D-ICDO) flew for the first time 9 May 1981. A static test airframe of the 228-200 was also completed. Service life is for 62,500 flights without major structural repair.

British CAA and American FAA certification were granted on 17 April and 11 May 1984 respectively, followed by Australian certification 11 October 1985. In addition, LBA certification has been accepted by the licensing authorities of Bhutan, Canada, India, Japan, Malaysia, Nigeria, Norway, Sweden and Taiwan.

Excluding Indian licence production, firm orders for the Dornier 228 (all versions) totalled 231 by December 1994 from 82 customers in 35 countries; deliveries at the time had reached 170.

VERSIONS **228-100:** Basic version, with standard accommodation for 15 passengers in airline seats at 76 cm (30 in) pitch. German (LBA) certification awarded 18 December 1981. Deliveries began in February 1982 and entered service with Norving Flyservice A/S in Norway Summer 1982. Suitable for a wide range of other duties, including freight or mixed cargo/passenger transport, executive travel, air taxi service, photogrammetry, airways calibration, training ambulance or search and rescue (SAR) operations, and paramilitary missions. Two ordered for SAR and other special duties by Australia's National Safety Council in 1987 are fitted with bubble windows each side of the cabin, a wide roller cabin door for SAR or firefighting operations, lightweight 360° Bendix radar, an Omega Loran system, 20-man liferaft, a rear fuselage launch chute for smoke or colour markers and flares, and provision for two stretchers.

228-101: Identical to 228-100 except for reinforced fuselage and different mainwheel tyres, to permit higher operating weights, and installation of engine fire extinguishing system to conform to SFAR Pt 41b. Introduced in 1984. In service with the German Navy (1).

228-200: Lengthened fuselage, providing standard accommodation for 19 passengers at 76 cm (30 in) seat pitch and a larger rear baggage compartment, but otherwise generally similar to 228-100. Certificated by German LBA 6 September 1982. One delivered to Japan's National Aerospace Laboratory in 1988 was specially equipped by AAR Oklahoma Inc, USA, for use as an in-flight simulator to evaluate aerodynamic and flight control system characteristics for future aircraft, including an automatic landing system for a future spaceplane.

228-201: Introduced in 1984. Identical to 228-200 except for changes noted under 228-101. Accommodation for 19 passengers with two pilots, 20 passengers with one pilot, 15-17 passengers with two pilots plus galley and lavatory, or nine passengers as a corporate transport. One (98+78) evaluated by MFG 5 of the German Navy and by German Air Force during 1986-87. Two specially equipped -201s, delivered in 1985-86 to DLR for meterological (remote sensing) and environmental forestry and pollution control work, have ventral camera/scanner apertures and numerous attachments for special sensors and other equipment under the wings and fuselage.

228-202: Designed to offer increases in payload/range performance, compared with 228-201, for only a slight difference in empty weight. Certificated by LBA and FAA in August and September 1986 respectively; available since Autumn 1987.

228-203F Freight Liner: Freighter version, with higher payload and operating weights. Additional crew door on starboard side. Wing, propeller and windscreen deicing. Colour weather radar for all-weather operation. Primary structural life expectancy of at least 56,000 flights. Five dome lights in main cabin and six partitioning cargo nets. Floor rails. In addition to main cabin, forward and aft baggage compartments have capacities of 120 kg (265 lb) and 210 kg (463 lb) respectively.

228-212: Generally as -201, but with 6,400 kg (14,110 lb) and 6,100 kg (13,448 lb) take-off and landing weights respectively, allowing increased payload on short route segments. Reduced empty weight. Maximum fuel weight 1,885 kg (4,155 lb). Increased engine T-O rating (to 579 kW; 776 shp at S/L, ISA). Strengthened landing gear (with modified anti-skid system and carbon brakes), wing boxes and fuselage; two underfuselage fins to improve STOL and low-speed flying; increase in maximum flaps extended speed from 150 to 160 knots IAS; modification of elevator and rudder control (electrically actuated rudder trim); and relocating the battery from the nose to the landing gear bay. Modernised avionics and equipment. Certificated by the LBA April 1989.

228 Troop, Paratroop and Ambulance: Based on the 228-200. Troop accommodates 17, 20 or 22 fully equipped troops, and can be adapted quickly for paratroop operations. Fold-up seats along sides. Lightweight toilet, roller door, military nav/com, and loadmaster intercom. Paratroop accommodates 16, 19 or 21 persons, plus jumpmaster. Similar equipment to Troop, but no toilet. Ambulance also has roller door and accommodates six stretchers in pairs and nine sitting casualties/attendants. Optional small galley, toilet, refrigerator, oxygen system and cabin intercom. Can carry full casevac load at cruising speed over a distance of 850 n miles (1,575 km; 978 miles).

228 Maritime Patrol: Described separately.

On 29 November 1983 contracts were signed covering the transfer of technology in a progressive programme to manufacture versions of the Dornier 228 under licence in India, by Hindustan Aeronautics Ltd. A production run of about 150 aircraft is envisaged, for various Indian organisations and customers, and was prefaced by the delivery of five 228-201s to Vayudoot, the Indian regional airline in 1984-85, and three 228-101s to the Indian Coast Guard in 1986-87. Meanwhile, Dornier began delivery of complete sets of aircraft assemblies to India in early 1985, and the first of an HAL assembled 228 was made 31 January 1986.

UPGRADES: **FR Aviation:** Maritime patrol version. See separate entry.

OPERATORS: Versions of the Do 228 are in service with the armed forces of the following countries: Germany (1), India (26), Italy (8), Malawi (4), Mauritius (2), Niger (2) and Nigeria (3).

DESIGN FEATURES: Cantilever high-wing monoplane. Dornier A-5 supercritical wing section. No dihedral or anhedral. Sweepback on leading-edge of outer panels 8°. Cantilever tailplane.

FLYING CONTROLS: Fowler single-slotted trailing-edge flaps. Ailerons can be drooped symmetrically to augment trailing-edge flaps and are operated differentially to serve as conventional ailerons. Trim coupling of flaps and tailplane optional. All-moving tailplane, with horn balance elevators. Trim tab in rudder.

STRUCTURE: The wing is a two-spar rectangular centre-section and two tapered outer panels ending in raked tips. Wingtips of CFRP; Kevlar used in construction of wing rib webs, wingroot fairings and ailerons. Remainder of wing is of light alloy construction. The fuselage is a conventional stressed skin unpressurised structure of light alloy, built in five sections. GFRP nosecone and Kevlar landing gear fairings. The tail unit is a cantilever structure, with rudder and horizontal surfaces Eonnex covered. GFRP is used for tips of tailplane and elevators and also for tips of rudder and fin. Hybrid composites used for fin leading-edges.

LANDING GEAR: Retractable tricycle type, with single mainwheels and twin-wheel nose unit. Main units retract forward and inward into fairings built on to the lower fuselage. Hydraulically steerable nosewheels retract forward. Goodyear wheels and tyres size 8.50-10 on mainwheels (12 ply rating on 228-100, 10 ply rating on 228-200); size 6.00-6, 6 ply rating, on nosewheels. Low pressure tyres optional. Goodyear brakes on mainwheels.

POWER PLANT: Two 533 kW (715 shp) Garrett TPE331-5-252D turboprops in -100/-101, each driving Hartzell HC-B4TN-5ML/LT10574 four-blade constant-speed fully feathering reversible-pitch metal propeller; -200 series have 578.7 kW (776 shp) version of same engine. Primary box forms an integral fuel tank with a total usable capacity of 2,386 litres (630 US gallons; 525 Imp gallons). Oil capacity per engine 5.9 litres (1.56 US gallons; 1.30 Imp gallons). Optional APU.

ACCOMMODATION: Crew of one or two, and accommodation for 15 to 20 passengers as described under model listings, or 16, 19 or 21 paratroops plus jumpmaster, or 17, 20 or 22 fully equipped troops in military transport configuration. Pilots' seats adjustable fore and aft. Individual seats down each side of the cabin with a central aisle. Flight deck door on port side (both sides on -203F). Combined two-section passenger and freight door with integral steps, on port side of cabin at rear. One emergency exit on port side of cabin, two on starboard side (deleted on -203F). Baggage compartment at rear of cabin, accessible externally and from cabin; capacity 210 kg (463 lb). Enlarged baggage door optional. Additional baggage space in fuselage nose, with separate access; capacity 120 kg (265 lb). Modular units using seat rails for rapid changes of role. In air ambulance configuration, six stretchers carried in three double units plus nine seated patients/medical attendants.

SYSTEMS: Entire accommodation heated and ventilated. Air conditioning system optional. Heating by engine bleed air. Hydraulic system, pressure 207 bars (3,000 lb/sq in), for landing gear, brakes and nosewheel steering. Handpump for emergency landing gear extension. Primary 28 V DC electrical system, supplied by two 28 V 250 A engine-driven starter/generators and two 24 V 25 Ah Ni/Cd batteries. Two 350 VA inverters supply 115/26 V 400 Hz AC system. APU optional. Air intake anti-icing standard. De-icing system optional for wing and tail unit leading-edges, windscreen and propellers.

AVIONICS: Instrumentation for IFR flight standard. Autopilot optional, to permit single-pilot IFR operation. Standard avionics include dual Bendix/King KY 196 VHF com, dual KN 53 VOR/ILS and KN 72 VOR/LOC converters; KMR 675 marker beacon receiver, dual or single KR 87 ADF and KT 76A transponder; dual or single Aeronetics 7137 RMI; dual or single DME; two Honeywell GH14B gyro horizons; two King KPI 552 HSIs; dual ASIs; dual altimeters; dual ADIs; dual VSIs; Becker audio selector and intercom.

EQUIPMENT: Standard equipment includes complete internal and external lighting, hand fire extinguisher, first aid kit, gust control locks and tiedown kit. Wide range of optional avionics and equipment available, including weather radar. For geophysical role, equipment includes VLF magnetometer in nose probe, VLF or protomagnetometer in port wing fairing, gamma ray detector, camera in rear of cabin with operator's position in forward section of cabin, navigation telescope, emergency equipment, and magnetometer in tail 'sting'; operator's position for magnetometers on port side of cabin.

DIMENSION, EXTERNAL:

Wing span	16.97 m (55 ft 8 in)
Wing aspect ratio	9.0
Length overall: 100 series	15.04 m (49 ft 4⅛ in)
200 series	16.56 m (54 ft 4 in)
Height overall	4.86 m (15 ft 11½ in)
Tailplane span	6.45 m (21 ft 2 in)
Wheel track	3.30 m (10 ft 10 in)

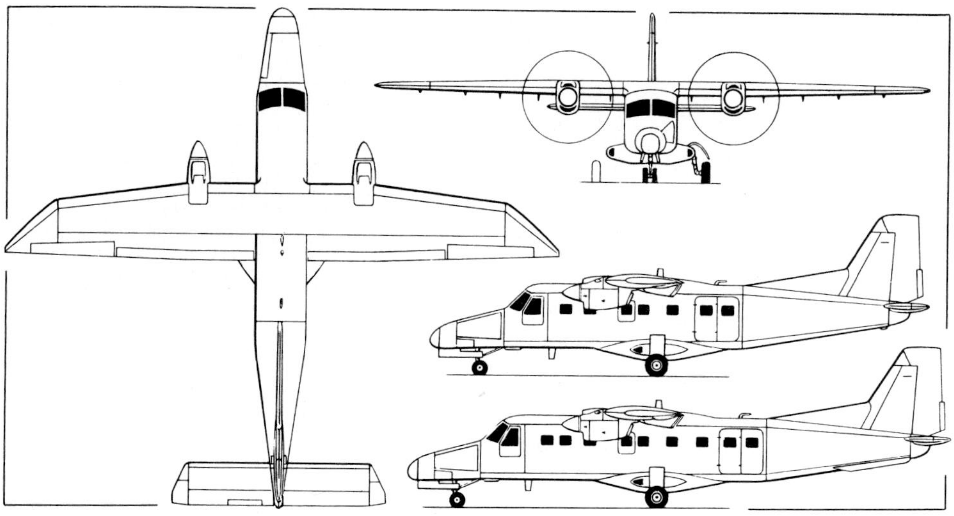

Dornier (Daimler-Benz Aerospace) 228-100 light transport, with additional side view (bottom) of 228-212
1994

Wheelbase: 100 series 5.53 m (18 ft 1¾ in)
200 series 6.29 m (20 ft 7½ in)
Propeller diameter 2.73 m (8 ft 11½ in)
Propeller ground clearance 1.08 m (3 ft 6½ in)
Passenger door (port rear):
Height 1.34 m (4 ft 4¾ in)
Width 0.64 m (2 ft 1¼ in)
Height to sill 0.60 m (1 ft 11½ in)
Freight door (port, rear):
Height 1.34 m (4 ft 4¾ in)
Width, incl passenger door 1.28 m (4 ft 2½ in)
Emergency exits (each): Height 0.66 m (2 ft 2 in)
Width 0.48 m (1 ft 7 in)
Baggage door (nose): Height 0.50 m (1 ft 7½ in)
Width 1.20 m (3 ft 11¼ in)
Standard baggage door (rear):
Height 0.90 m (2 ft 11½ in)
Width 0.53 m (1 ft 9 in)

DIMENSIONS, INTERNAL:
Cabin, excl flight deck and rear baggage compartment:
Length: 100 series 6.33 m (20 ft 9 in)
200 series 7.08 m (23 ft 2¾ in)
Max width 1.346 m (4 ft 5 in)
Max height 1.55 m (5 ft 1 in)
Floor area: 100 series 8.50 m² (91.49 sq ft)
200 series 9.56 m² (102.9 sq ft)
Volume: 100 series 13.00 m³ (459.1 cu ft)
200 series 14.70 m³ (519.1 cu ft)
Rear baggage compartment volume:
100 series, standard 1.20 m³ (42.4 cu ft)
100 series, optional; 200 series standard 2.60 m³ (91.8 cu ft)
Nose baggage compartment volume 0.89 m³ (31.4 cu ft)
Cargo volume (203F):
main cabin 14.70 m³ (519.1 cu ft)
nose 0.89 m³ (31.4 cu ft)
rear fuselage 2.60 m³ (91.8 cu ft)

AREAS:
Wings, gross 32.00 m² (344.3 sq ft)
Ailerons (total) 2.708 m² (29.15 sq ft)
Trailing-edge flaps (total) 5.872 m² (63.21 sq ft)
Fin, incl dorsal fin 4.50 m² (48.44 sq ft)
Rudder, incl tab 1.50 m² (16.15 sq ft)
Horizontal tail surfaces (total) 8.33 m² (89.66 sq ft)

WEIGHTS AND LOADINGS:
Weight empty, standard: 100 2,980 kg (6,570 lb)
101 2,990 kg (6,592 lb)
200 3,086 kg (6,803 lb)
201 3,096 kg (6,825 lb)
202 3,221 kg (7,101 lb)
Operating weight empty:100 3,413 kg (7,524 lb)
101 3,423 kg (7,546 lb)
200 3,547 kg (7,820 lb)
201 3,737 kg (8,238 lb)
202 3,698 kg (8,153 lb)
212 3,742 kg (8,249 lb)
Max payload: 100 2,127 kg (4,689 lb)
101 2,117 kg (4,667 lb)
200 1,853 kg (4,085 lb)
201 1,903 kg (4,195 lb)
202 1,895 kg (4,178 lb)
203F 2,300 kg (5,070 lb)
212 1,848 kg (4,074 lb)
Max ramp weight: 100, 200 5,730 kg (12,632 lb)
101, 201 6,010 kg (13,250 lb)
202 6,230 kg (13,735 lb)
212 6,430 kg (14,175 lb)
Max T-O weight: 100, 200 5,700 kg (12,566 lb)
101, 201 5,980 kg (13,183 lb)
202 6,200 kg (13,668 lb)
203F 6,500 kg (14,330 lb)
*212 6,400 kg (14,110 lb)
Max landing weight: 100, 101, 200 5,700 kg (12,566 lb)
201, 202 5,900 kg (13,007 lb)
212 6,100 kg (13,448 lb)
Max wing loading:
100, 200 178.1 kg/m² (36.48 lb/sq ft)
101, 201 186.9 kg/m² (38.28 lb/sq ft)
202 193.7 kg/m² (39.67 lb/sq ft)
203F 203.1 kg/m² (41.60 lb/sq ft)
212 200.0 kg/m² (40.96 lb/sq ft)
Max power loading:
100, 200 5.35 kg/kW (8.79 lb/shp)
101, 201 5.17 kg/kW (8.49 lb/shp)
202 5.36 kg/kW (8.81 lb/shp)
203F 4.19 kg/kW (9.23 lb/shp)
212 5.53 kg/kW (9.09 lb/shp)
**Increasable to 6,600 kg (14,550 lb) in special cases*

PERFORMANCE (at max T-O weight, S/L, ISA, except where indicated):
Never-exceed speed (all) 255 kts (472 km/h; 293 mph) IAS
Max operating speed (V_{MO}):
212 223 kts (413 km/h; 256 mph) IAS
Max cruising speed at 3,050 m (10,000 ft) (all) 231 kts (428 km/h; 266 mph)
Cruising speed at 4,575 m (15,000 ft), average cruise weight of 5,300 kg (11,684 lb):
212 220 kts (408 km/h; 253 mph)
Max cruising speed at S/L (all) 199 kts (370 km/h; 230 mph)
Econ cruising speed (all) 180 kts (333 km/h; 207 mph)
Stalling speed, flaps up:
100 79 kts (146 km/h; 91 mph) IAS
200 81 kts (150 km/h; 93 mph) IAS
Stalling speed, flaps down:
100 63 kts (117 km/h; 73 mph) IAS
200 67 kts (124 km/h; 77 mph) IAS
Max rate of climb at S/L:
100, 200 618 m (2,025 ft)/min
101, 201 582 m (1,910 ft)/min
202 546 m (1,790 ft)/min
203F 516 m (1,695 ft)/min
212 570 m (1,870 ft)/min
Rate of climb at S/L, one engine out:
100, 200 162 m (531 ft)/min
101, 201 138 m (450 ft)/min
202 126 m (413 ft)/min
212 134 m (440 ft)/min
Service ceiling, 30.5 m (100 ft)/min rate of climb:
100, 200 9,020 m (29,600 ft)
101, 201 8,535 m (28,000 ft)
Service ceiling, one engine out, 30.5 m (100 ft)/min rate of climb: 100, 200 4,265 m (14,000 ft)
202 4,210 m (13,810 ft)
203F 4,116 m (13,500 ft)
Service ceiling, one engine out, 15 m (50 ft)/min rate of climb 3,962 m (13,000 ft)
T-O run: 100 411 m (1,350 ft)
101 442 m (1,450 ft)
202 686 m (2,250 ft)
T-O to 10.7 m (35 ft): 203F 732 m (2,400 ft)
T-O to 15 m (50 ft): 100 564 m (1,850 ft)
101 592 m (1,945 ft)
200 750 m (2,461 ft)
201 625 m (2,050 ft)
202 686 m (2,250 ft)
Landing from 15 m (50 ft) at max landing weight:
100 600 m (1,968 ft)
200 620 m (2,034 ft)
202, max braking 536 m (1,760 ft)
203F 558 m (1,831 ft)
212 512 m (1,680 ft)
Range at 3,050 m (10,000 ft) with max passenger payload, max cruising speed:
100 724 n miles (1,343 km; 834 miles)
101 939 n miles (1,740 km; 1,081 miles)
200 323 n miles (600 km; 372 miles)
202 540 n miles (1,000 km; 621 miles)
Max range at 3,050 m (10,000 ft) at econ cruising speed, reserves for 85 n miles (157 km; 98 mile) diversion, 45 min hold and 15% fuel remaining:
201 164 n miles (305 km; 189 miles)
203F (max payload) over 400 n miles (741 km; 460 miles)
203F (1,360 kg; 3,000 lb) payload over 1,200 n miles (2,224 km; 1,380 miles)
Max range at 3,050 m (10,000 ft) at econ cruising speed, max passenger load and 45 min reserves:
202 610 n miles (1,130 km; 702 miles)
Range at 3,050 m (10,000 ft) at econ cruising speed, max fuel load and 45 min reserves:
202 1,460 n miles (2,704 km; 1,680 miles)
Range at 3,050 m (10,000 ft) with 19 passengers, reserves for 50 nm (93 km; 57 mile) diversion, 45 min hold and 5% fuel remaining:
212 at max cruising speed 560 n miles (1,038 km; 645 miles)
212 at max range speed630 n miles (1,167 km; 725 miles)
Range with 775 kg (1,708 lb) payload, conditions as above:
212 at max cruising speed 1,160 n miles (2,150 km; 1,335 miles)
212 at max range speed 1,320 n miles (2,446 km; 1,520 miles)

UPDATED

DORNIER 228 MARITIME PATROL

TYPE: Twin-turboprop STOL Maritime Patrol Aircraft (MPA).

PROGRAMME: Developed for Indian Coast Guard, Royal Thai Navy and others for maritime and fisheries patrol and border patrol.

DESIGN FEATURES: Modifications to standard 228 include major corrosion protection, radome beneath fuselage, four wing hardpoints for searchlight, Micronair spraypod and other equipment, roller door for dropping survival equipment and chute in rear cabin for dropping smoke markers and flares.

POWER PLANT: As for standard 228, with optional auxiliary

Dornier (Daimler-Benz Aerospace) 228 Maritime Patrol Aircraft of the Royal Thai Navy
1994

fuel tanks to increase fuel capacity to 2,886 litres (762 US gallons; 635 Imp gallons).

ACCOMMODATION: Pilot and co-pilot with full dual controls and instruments as standard; co-pilot operates optional searchlight; two bubble observation windows in front of cabin (180° view) and photography window on port side which can be opened in flight. Console for radar operator on port side of cabin incorporating radar display, digital navigation display and intercom controls. Rest area on starboard side of rear cabin with optional folding table, galley or refrigeration and toilet.

AVIONICS: Standard items as detailed for 228-212. Optional exchange of standard avionics to Collins Pro Line II. Com/nav equipment comprises Collins AN/ARC-182 VHF/UHF transceiver, Collins DF-301E VHF/UHF direction finder, Collins radio altimeter ALT-55B, low altitude warning system, weather radar, Global/Wulfsberg GNS-500-5 nav system with search pattern mode, and optional GPS and Loran C. Bendix/King RDR-1500B maritime surveillance radar with 360° scan in underfuselage radome. Optional Litton AN/APS-504(V)5, THORN EMI Super Searcher and Eaton AN/APS-128. Day/night mission equipment includes Honeywell forward-looking infra-red system (FLIR), stabilised long-range observation system (SLOS), night vision goggles, Spectrolab 80 Mcd searchlight, markers and flares, loud-hailer and Nikon hand-held camera with data annotation.

VERIFIED

DORNIER 228 MARITIME POLLUTION SURVEILLANCE

TYPE: Twin-turboprop STOL Maritime Pollution Surveillance aircraft.

PROGRAMME: Developed for Netherlands Coast Guard (Kustwacht) and Finnish Frontier Guard.

DESIGN FEATURES: Modifications for pollution control as for Maritime Patrol version, but with large floor cutouts to carry an IR/UV scanner plus photographic, television and IR cameras.

POWER PLANT: As for Maritime Patrol version.

ACCOMMODATION: Crew as for Maritime Patrol version. Main operator workstation on port side front of cabin, with all necessary controls and displays for remote sensing equipment; equipment rack for all the different computers, video recorders, power supplies and hard copy units on port side in mid-cabin; observer workstation with similar controls and displays as in operator console, is at rear of cabin, in front of toilet compartment. Two bubble observation window port and starboard in front of cabin, each bulged to give 180° view; two optical photo windows port and starboard in rear cabin can be opened in flight; rest area on starboard side of mid-cabin has seats and folding table.

AVIONICS: As for Maritime Patrol version, but also Racal R-Nav II navigation management system with search pattern mode, Decca, GPS, and Bendix/King RDR 1400C weather and search radar with 18 in antenna.

EQUIPMENT: Primary surveillance sensors from Terma Elektronik AS, Denmark, include Terma side-looking airborne radar (SLAR) with underfuselage antenna, and Daedalus ABS 3500 bi-spectral IR/UV scanner. To secure evidence of pollution, 228 equipped with downward-looking colour television, Nikon photographic and IR cameras and, for photographic documentation, Nikon hand-held camera interfacing with navigation system.

VERIFIED

DORNIER 228 PHOTOGRAMMETRY/ GEO-SURVEY

TYPE: Twin-turboprop photogrammetry/geo-survey aircraft.

PROGRAMME: Besides being used in a purely photogrammetric version, 228 can serve wide variety of users as working platform in earth sciences field. The 228 allows good access to sensors and other equipment both in flight and for maintenance.

DESIGN FEATURES: Modifications to standard 228 as for Maritime Patrol version, but production model includes large sliding hatch in cabin floor for sensor installation. Wing hardpoints support different sensors, and various antennae are mounted on fuselage and tail.

POWER PLANT: As for Maritime Patrol version.

ACCOMMODATION: Crew dependent on mission, but basically as described for Maritime Patrol version.

AVIONICS: As for Maritime Patrol version.

OPTIONAL EQUIPMENT: Photogrammetry version has aerial survey cameras installed in floor cutout, Wild or Zeiss navigation telescope, operator station, flight track camera, intercom system, toilet modified as darkroom, rest area with folding table, and small galley or refrigerator.

Geo-survey version has VLF electromagnetometer mounted in nose thimble, magnetometer in tail sting, VLF or proton magnetometer installed in wingtips, electromagnetic reflection system mounted on wing hardpoints, and gamma-ray detector in lower fuselage. Aerial survey camera installed in floor cutout in rear fuselage.

VERIFIED

ATLANTIC 1 MODERNISATION PROGRAMME

TYPE: Anti-Submarine Warfare (ASW) aircraft upgrade (Atlantic 1).

PROGRAMME: The Federal German Navy fleet of 20 Atlantic 1s entered service in 1965. Of these, 15 are allocated to ASW and long-range maritime reconnaissance, and these aircraft have been modernised by Dornier under a DM200 million programme to extend their capability into the mid-1990s.

DESIGN FEATURES: The original search radar is replaced by a new Texas Instruments system with digital cockpit display. A new Loral ESM system, installed in the wingtip pods, has increased frequency range, improved direction finding accuracy and automatic analysis. Passive underwater sonar detection is improved by means of an Emerson Electric eight-channel modification kit with increased frequency range and digital signal processing. A new Dornier sonobuoy launcher replaces the original rotary launcher; and all these improvements are backed up by replacing the former tape recorder with a new IRIG standard model, and adding a Litton LN-33 second navigation system with Decca updated inertial platform, primarily for tactical navigation.

VERIFIED

MILITARY AIRCRAFT

MILITARY AIRCRAFT DIVISION OF DAIMLER-BENZ AEROSPACE AG

PO Box 801160, D-81663 Munich
Telephone: +49 (89) 607-0
HEAD OF STRATEGIC BUSINESS UNIT: Hans-Juergen Wolfert

Major military aircraft activities of the Group include aircraft armament and airborne reconnaissance systems, disarmament verification systems, simulation and training systems, and research into advanced aircraft systems, materials and manufacturing technologies. The division also makes Airbus subassemblies.

VERIFIED

Daimler-Benz Aerospace F-4F ICE upgraded Phantoms flying at 'Top Gun' Point Mugu Test Range, California

1994

DASA F-4F ICE UPGRADE PROGRAMME

TYPE: Combat aircraft upgrade.

PROGRAMME: Under German Defence Ministry programme known as ICE (improved combat effectiveness), 110 Luftwaffe F-4F Phantom IIs, primarily those of fighter wings JG 71, JG 72 and JG 74, to be upgraded for lookdown/shootdown capability against multiple targets. Programme, for which DASA Military Aircraft Division is prime contractor, initiated late 1983 and reached end of definition phase two years later; entered full-scale development phase December 1986; two development aircraft flying since 1990. First flight of fully modified F-4F ICE 2 May 1990. AMRAAM firing trials from October 1991 to September 1992 at US Navy Pacific Weapon Test Range at Point Mugu. Production ICE conversion started July 1991; first delivery April 1992; by end of 1994, approximately 70 aircraft had been delivered from Daimler-Benz Aerospace's Manching plant and 12 more from Luftwaffe Depot Maintenance Facility (LW Werft) 62. ICE development contract completed 31 December 1993. The ICE programme is scheduled to be completed in late 1996.

DESIGN FEATURES: Features include replacement of existing Westinghouse AN/APQ-120 radar with all-digital multimode Hughes AN/APG-65, built under licence in Germany by Telefunken SystemTechnik (now DASA Defence and Civil Systems Group), fitting up to four Hughes AIM-120 (AMRAAM) air-to-air missiles, new TST radar control console, optimisation (by Hughes) of cockpit display installation of new Litef digital fire control computer, Honeywell H-423 laser inertial platform, GEC-Marconi Avionics CPU-143/A digital air data computer, new IFF system, Frazer-Nash AMRAAM launchers, MIL-STD-1553B digital databus with Advanced Operational Flight Program software, and improved resistance to electronic jamming and other countermeasures.

Further 40 Luftwaffe F-4Fs, serving in fighter-bomber role with JBGs 35 and 36, were upgraded in early 1993 (databus, INS and ADC only).

UPDATED

IABG

INDUSTRIEANLAGEN-BETRIEBSGESELLSCHAFT mbH (IABG)

Einsteinstrasse 20, D-85521 Ottobrunn
Telephone: +49 (89) 6088-0
Fax: +49 (89) 60 882417
EXECUTIVE DIRECTOR: Dr William E. Sweeney Jr
EXECUTIVE DIRECTOR: Dr Ing Rudolf F. Schwarz
MANAGER, PUBLIC RELATIONS: Dipl Wirtsch Ing Volker Huber

Test and Analysis Center

Einsteinstrasse 20, D-85521 Ottobrunn
Telephone: +49 (89) 6088 2610
Fax: +49 (89) 6088 4066
HEAD OF DIVISION: Dipl Ing Wolfgang Raasch
DEPUTY: Wilhelm E. Pfrang
MARKETING MANAGER: Dipl Ing Wolfgang Mohr

Company active in scientific and technical services including systems engineering, operations research studies, logistics support; environmental studies, realistic field tests, defence weapons systems analysis, simulation and evaluation as well as aircraft and spacecraft technology, qualification and acceptance testing, fatigue and static strength tests.

The Test and Analysis Center is active or has been involved in structure tests and life extension programmes on the following aircraft: Airbus A300, A310, A320, A330/340,

Dassault Breguet Atlantic I, Dassault/Dornier Alpha Jet, Dornier Do 31, Dornier Seastar, Egrett G-500, Eurocopter BK 117, Eurofighter 2000, Fiat G91, Grob G 115, Lockheed F-104 Starfighter, Panavia Tornado, Pilatus PC-7 Turbo-Trainer, Transall C-160.

VERIFIED

IABG structural fatigue test facilities undertaking analysis of Airbus passenger airliner
1994

GREECE

HAI

HELLENIC AEROSPACE INDUSTRY LTD

Tanagra, PO Box 23, GR-32009 Schimatari
Telephone: 30 (1) 8836 711, (262) 52000, 52126 and 52161
Fax: 30 (1) 8838 714, (262) 52170 and 52102
Telex: 299306 HAIGR
CHAIRMAN/MANAGING DIRECTOR: Athanasios Nezis
GENERAL MANAGER: Evangelos Ladis
SALES/MARKETING DIRECTOR: Nick Vassilopoulos

Established in 1975, state-owned Hellenic Aerospace Industry (HAI) operates an Aircraft Facility, Engine Facility, Electronics/Avionics Division and Manufacturing Facility based in Tanagra works occupying 180 ha (445 acre) site with 150,000 m^2 (1,614,585 sq ft) covered floor space and 1994 workforce of nearly 3,000. The company has extensive capabilities in the fields of civil and military aircraft manufacture, repair, maintenance, overhaul and modification/upgrade kits (including crash damage). Customers include countries in Middle East, Africa, Europe and North America, including for full support Mirage F1, Atar 09K50, F-4 Phantom, A-7 Corsair, Lockheed C-130 depot level maintenance, J79 for USAFE and RAF. Major manufacturing includes F-16 rear fuselages and intakes, Airbus A300 and A310 door frames, assembly and test of SNECMA M53. In October 1994 HAI signed an agreement with Dassault Aviation of France to produce airframe parts for the Dassault Falcon 900 EX.

NEW ENTRY

Hellenic Aerospace engineers undertaking overhaul work on Greek Air Force Dassault Mirage F1-C *(HAI)*
1995

INDIA

HAL

HINDUSTAN AERONAUTICS LIMITED

CORPORATE OFFICE: PO Box 5150, 15/1 Cubbon Road, Bangalore 560 001
Telephone: 91 (80) 526 5201
Fax: 91 (812) 577533 and 226 8758
Telex: 845 2266 HAL IN
CHAIRMAN: R. N. Sharma

Hindustan Aeronautics Limited (HAL) was formed on 1 October 1964, and has 12 manufacturing divisions at seven locations (Bangalore, Nasik, Koraput, Hyderabad, Kanpur, Lucknow and Korwa), plus a Design Complex. The total workforce is over 37,000.

Kanpur Division is producing the HPT-32 *ab initio* trainer and assembles, under licence, the Dornier 228. Nasik and Koraput Divisions are manufacturing airframes and engines of the MiG-27 in collaboration with the former Soviet Union. Hyderabad Division manufactures avionics for all aircraft produced by HAL, as well as airport radars. Lucknow Division is producing aircraft accessories under licence from manufacturers in France, former USSR and the UK. Korwa manufactures inertial navigation systems.

In addition to HAL's manufacturing programmes, major design and development activities include the Advanced Light Helicopter (ALH) programme.

UPDATED

BANGALORE COMPLEX

Post Bag 1785, Bangalore 560 017
Telephone: 91 (812) 565201 and 561020
Telex: 91 2234
MANAGING DIRECTOR: Dr C. G. Krishnadas Nair

Aircraft Division
Post Bag 1788, Bangalore 560 017
Telephone: 91 (812) 561020 and 565201
Fax: 91 (812) 5265188
Telex: 91 2234
GENERAL MANAGER: B. Haridass

Helicopter Division
Post Bag 1790, Bangalore 560 017
Telephone: 91 (812) 565201 and 561020
Fax: 91 (812) 5584717
Telex: 91 2764
GENERAL MANAGER: U. D. Paradkar
Telephone: 91 (812) 558 2924

The Bangalore Complex is subdivided into an Aircraft Division, Helicopter Division, Aerospace Division, Engine Division, Overhaul Division, Services Division, Foundry and Forge Division, Flight Operations, and Design Complex. It is engaged in the manufacture of the SEPECAT Jaguar International combat aircraft and its Adour engine. The Complex also undertakes repair and overhaul of airframes, engines, and allied instruments and accessories. A contract to supply up to 150 sets of tailplanes for the British Aerospace ATP transport aircraft was signed in May 1987.

VERIFIED

HAL (SEPECAT) JAGUAR INTERNATIONAL

Indian Air Force name: Shamsher (Assault Sword)

TYPE: Single-seat tactical attack aircraft and two-seat operational trainer.

PROGRAMME: Forty Jaguar Internationals with Adour Mk 804 engines delivered from UK; 45 more with Adour Mk 811s assembled in India; further 31 being manufactured under licence in India (first delivery early 1988); further 15 (ordered 1988) were cancelled 1989. Delivery was completed by 1991. Production has been launched for an additional 15 aircraft.

VERSIONS: **Jaguar B:** Two-seat trainer version.

Jaguar S: Standard single-seat attack version; *description applies to this model, except where indicated.*

Anti-shipping Jaguar: Jaguars of IAF No. 6 Squadron, assigned to anti-shipping role, have Thomson-CSF Agave radar, Smiths Industries DARIN nav/attack system and Sea Eagle air-to-surface missiles; first modified aircraft delivered January 1986; eight delivered by end of 1992; two more to follow.

OPERATORS: Versions of the HAL (Sepecat) Jaguar are in service with the armed forces of India (115).

DESIGN FEATURES: Cantilever shoulder-wing monoplane. Anhedral 3°. Sweepback 40° at quarter-chord. Cantilever tailplane. Ventral fins beneath rear fuselage. Tailplane sweepback at quarter-chord 40° on horizontal, 43° on vertical surfaces. All-moving slab type tailplane with 10 anhedral.

FLYING CONTROLS: No ailerons. Lateral control by two-section spoilers, forward of outer flap on each wing in association, at low speeds, with differential tailplane. Hydraulically actuated (by screwjack) full-span double-slotted trailing-edge flaps. Fairey Hydraulics powered flying controls. Two halves of tailplane can operate differentially to supplement the spoilers. No separate elevators. Controls powered by hydraulics.

STRUCTURE: The wing is an all-metal two-spar torsion box structure; skin machined from solid aluminium alloy, with integral stiffeners. Wing main portion built as single unit with three-point attachment to each side of fuselage. Outer panels fitted with slat which also gives effect of extended chord leading-edge. The fuselage is an all-metal (mainly aluminium) structure built in three main units, making use of sandwich panels and, around the cockpit, honeycomb panels. Local use of titanium alloy in engine bay area. The tail unit is a cantilever all-metal, two-spar structure with aluminium alloy sandwich panels. Rudder and outer panels and trailing-edge of tailplane have honeycomb core.

POWER PLANT: Two HAL built Rolls-Royce Turbomeca Adour Mk 811 turbofans, each rated at 24.6 kN (5,520 lb st) dry and 37.4 kN (8,400 lb st) with afterburning. Fixed geometry air intake on each side of fuselage aft of cockpit. Fuel in six tanks, one in each wing and four in fuselage. Total internal fuel capacity 4,200 litres (1,110 US gallons; 924 Imp gallons). Armour protection for critical fuel system components. Provision for carrying three auxiliary drop tanks, each of 1,200 litres (317 US gallons; 264 Imp gallons) capacity, on fuselage and inboard wing pylons. Provision for in-flight refuelling, with retractable probe forward of cockpit on starboard side.

ACCOMMODATION (single-seater): Enclosed cockpit for pilot, with rearward hinged canopy and Martin-Baker IN9B zero/zero ejection seat. Windscreen bulletproof against 7.5 mm rifle fire.

ACCOMMODATION (trainer): Crew of two in tandem on Martin-Baker IN9B Mk II zero/zero ejection seats. Individual rearward hinged canopies. Rear seat 38 cm (15 in) higher than front seat. Bulletproof windscreen, as in single-seat version.

SYSTEMS: Air conditioning and pressurisation systems maintain automatically, throughout the flight envelope, comfortable operating conditions for crew. The systems also control temperature in certain equipment bays. Two independent hydraulic systems, powered by two Vickers engine-driven pumps. Hydraulic pressure 207 bars (3,000 lb/sq in). First system (port engine) supplies one channel of each actuator for flying controls, hydraulic motors which actuate flaps and slats, landing gear retraction and extension, brakes and anti-skid units. Second system supplies other half of each flying control actuator, two further hydraulic motors actuating slats and flaps, airbrake and landing gear emergency extension jacks, nosewheel steering and wheel brakes. In addition, there is an emergency hydraulic power transfer unit. Electrical power provided by two 15 kVA AC generators, either of which can sustain functional and operational equipment without load shedding. DC power provided by two 4 kW transformer-rectifiers. Emergency power for essential instruments provided by 15 Ah battery and static inverter. De-icing, rain clearance and demisting standard. Liquid oxygen system, which also pressurises pilot's anti-*g* suit. Jaguar is fully power controlled in all three axes and is automatically stabilised as a weapons platform by gyros which sense disturbances and feed appropriate correcting data through a computer to the power control assemblies, in addition to human pilot manoeuvre demands. Power controls are all of duplex tandem arrangement, with mechanical and electrical servo-valves of Fairey platen design. Air-to-air combat capability can be enhanced by inclusion of roll/yaw dampers, to increase lateral stability, and by increasing slat and flap angles.

HAL (SEPECAT) Jaguar International of the Indian Air Force *1994*

AVIONICS: Include Smiths Industries DARIN (display attack and ranging inertial navigation) nav/attack system, incorporating SAGEM Uliss 82 INS, GEC-Marconi HUDWAC (head-up display and weapon aiming computer) and GEC-Marconi COMED 2045 (combined map and electronic display). HAL IFF-400 AM.

ARMAMENT: Two 30 mm Aden cannon in lower fuselage aft of cockpit in single-seater; single Aden gun on port side in two-seater. One stores attachment on fuselage centreline and two under each wing. Centreline and inboard wing points can each carry up to 1,134 kg (2,500 lb) of weapons, outboard underwing points up to 567 kg (1,250 lb) each. Typical alternative loads include one air-to-surface missile and two 1,200 litre (317 US gallon; 264 Imp gallon) drop tanks; eight 1,000 lb bombs; various combinations of free-fall, laser-guided, retarded or cluster bombs; overwing Matra R.550 Magic missiles; air-to-surface rockets; or a reconnaissance camera pack.

DIMENSIONS, EXTERNAL:
- Wing span: 8.69 m (28 ft 6 in)
- Length overall, incl probe:
  - single-seat: 16.83 m (55 ft 2½ in)
  - two-seat: 17.53 m (57 ft 6¼ in)
- Height overall: 4.89 m (16 ft 0½ in)
- Wheel track: 2.41 m (7 ft 11 in)
- Wheelbase: 5.69 m (18 ft 8 in)

AREAS:
- Wings, gross: 24.18 m² (260.27 sq ft)

WEIGHTS AND LOADINGS:
- Typical weight empty: 7,000 kg (15,432 lb)
- Max external stores load (incl overwing): 4,763 kg (10,500 lb)
- Normal T-O weight (single-seater, with full internal fuel and ammunition for built-in cannon): 10,954 kg (24,149 lb)
- Max T-O weight with external stores: 15,700 kg (34,612 lb)
- Max wing loading: 649.3 kg/m² (133 lb/sq ft)
- Max power loading: 209.9 kg/kN (2.06 lb/lb st)

PERFORMANCE:
- Max level speed:
  - at S/L: Mach 1.1 (729 kts; 1,350 km/h; 840 mph)
  - at 11,000 m (36,000 ft): Mach 1.6 (917 kts; 1,699 km/h; 1,056 mph)
- Landing speed: 115 kts (213 km/h; 132 mph)
- T-O run: clean: 565 m (1,855 ft)
  - with four 1,000 lb bombs: 880 m (2,890 ft)
  - with eight 1,000 lb bombs: 1,250 m (4,100 ft)
- T-O to 15 m (50 ft) with typical tactical load: 940 m (3,085 ft)
- Landing from 15 m (50 ft) with typical tactical load: 785 m (2,575 ft)
- Landing run:
  - normal weight, with brake-chute: 470 m (1,540 ft)
  - normal weight, without brake-chute: 680 m (2,230 ft)
  - overload weight, with brake-chute: 670 m (2,200 ft)
- Typical attack radius, internal fuel only:
  - hi-lo-hi: 460 n miles (852 km; 530 miles)
  - lo-lo-lo: 290 n miles (537 km; 334 miles)
- Typical attack radius with external fuel:
  - hi-lo-hi: 760 n miles (1,408 km; 875 miles)
  - lo-lo-lo: 495 n miles (917 km; 570 miles)
- Ferry range with external fuel: 1,902 n miles (3,524 km; 2,190 miles)
- *g* limits: +8.6 (+12 ultimate)

VERIFIED

HAL HJT-16 Mk I/IA KIRAN (RAY OF LIGHT)

TYPE: Two-seat jet basic trainer.

PROGRAMME: The HJT-16 was designed by an Indian team led by Dr V. M. Ghatage to meet the needs of the Indian Air Force for a basic trainer. It was powered by a Rolls-Royce Viper 11 turbojet engine, and flew for the first time on 4 September 1964. A second prototype flew in August 1965, and the first of 24 preproduction **Kiran Mk Is** was delivered to the IAF in March 1968.

From the 119th aircraft, a hardpoint was fitted beneath each wing, enabling the Kiran to be used also for armament training and light attack duties. This version is designated **Mk IA**.

Production of the Kiran I/IA totalled 190 by September 1981 to meet a requirement of both the Indian Air Force and Navy.

OPERATORS: Versions of the HAL HJT-16 MkI/IA are in service with the armed forces of India (220).

DESIGN FEATURES: Cantilever low-wing monoplane. Wing section NACA 23015 at root, NACA 23012 at tip. Dihedral 4° from roots. Incidence 0° 30′ at root. Two full-chord boundary layer fences on upper surface of each wing.

HAL HJT-16 Kiran Mk IA two-seat jet basic trainer *1994*

HAL HJT-16 Kiran Mk I two-seat jet basic trainer of the Indian Air Force

1994

FLYING CONTROLS: Frise-type differential ailerons, with balance tab in each aileron and ground adjustable tab on port aileron. Hydraulically actuated trailing-edge split flaps. Hydraulically actuated door-type airbrake under centre of fuselage. Electrically operated variable-incidence tailplane. Ground-adjustable tab on rudder.

STRUCTURE: The wing is a conventional all-metal three-spar structure. The fuselage is an all-metal semi-monocoque structure of light alloy. The tail unit is a cantilever all-metal structure.

LANDING GEAR: Retractable tricycle type, of HAL manufacture. Hydraulic actuation. Main units retract inward into fuselage; self-centring twin-contact non-steerable nosewheel retracts forward. Oleo-pneumatic shock-absorbers. Mainwheel tyres size 19 × 6.25-9, pressure 6.21 bars (90 lb/sq in). Nosewheel tyre size 15.4 × 4-6, pressure 4.83 bars (70 lb/sq in). Hydraulic brakes, without cooling.

POWER PLANT: One 11.12 kN (2,500 lb st) Rolls-Royce Viper 11 turbojet engine. Internal fuel in two main saddle tanks in fuselage 209 litres (55.2 US gallons; 46 Imp gallons), wing centre-section collector tank 282 litres (74.5 US gallons; 62 Imp gallons) and two outboard wing tanks 218 litres (57.7 US gallons; 48 Imp gallons), giving total capacity of 1,137 litres (300 US gallons; 250 Imp gallons). Provision for two underwing tanks with total capacity of 454 litres (120 US gallons; 100 Imp gallons). System permits 30 seconds of inverted flying.

ACCOMMODATION: Crew of two side by side in air conditioned and pressurised cockpit, on Martin-Baker H4HA zero-altitude fully automatic ejection seats. Clamshell type canopy. Dual controls and duplicated blind-flying instrumentation.

Air conditioning system has maximum pressure differential of 0.12 bars (1.75 lb/sq in). Dowty Hydraulic system for landing gear, flaps and airbrake, pressure 207 bars (3,000 lb/sq in). Accumulator for manual emergency system. Electrical system is of 28V DC single-wire earth return type, with two 24V 25Ah batteries. Normalair pressure-demand oxygen system.

AVIONICS: STR 9X/M 10-channel VHF transceiver, AX-3 single-channel VHF standby set and Marconi AD 722 ADF (initially: BEL built Bendix DFA in later aircraft). Landing light in nose.

ARMAMENT: Hardpoint beneath each wing, each capable of carrying a 500 lb bomb, an HAL pod containing two 7.62 mm machine guns, a pod containing seven 68 mm SNEB rockets, or a 227 litre (60.1 US gallon; 50 Imp gallon) drop tank. Vinten G-90 Mk 1 gun camera and two 8B/2465 (Mk IIIN) reflector gunsights.

DIMENSIONS, EXTERNAL:

Wing span	10.70 m (35 ft 1¼ in)
Wing chord: at root	2.35 m (7 ft 8½ in)
at tip	1.02 m (3 ft 4 in)
Wing aspect ratio	6.03
Length overall	10.60 m (34 ft 9 in)
Height overall	3.635 m (11 ft 11 in)
Tailplane span	3.90 m (12 ft 9½ in)
Wheel track	2.42 m (7 ft 11 in)
Wheelbase	3.50 m (11 ft 6 in)

AREAS:

Wings, gross	19.00 m² (204.5 sq ft)
Ailerons (total, incl tabs)	1.55 m² (16.68 sq ft)
Flaps total	2.34 m² (25.19 sq ft)
Fin, incl dorsal fin	1.576 m² (16.96 sq ft)
Rudder, incl tab	0.714 m² (7.69 sq ft)
Tailplane	2.58 m² (27.77 sq ft)
Elevators (total)	1.14 m² (12.27 sq ft)

WEIGHTS AND LOADINGS:

Weight empty	2,560 kg (5,644 lb)
Max T-O weight: 'clean'	3,650 kg (8,047 lb)
with two 50 Imp gallon drop tanks	4,235 kg (9,336 lb)
Max wing loading 'clean'	192 kg/m² (39.3 lb/sq ft)

PERFORMANCE:

Max level speed:	
at S/L	375 kts (695 km/h; 432 mph)
at 9,150 m (30,000 ft)	371 kts (688 km/h; 427 mph)
Max cruising speed	175 kts (324 km/h; 201 mph)
Stalling speed: flaps and landing gear up	85-92 kts (158-171 km/h; 98-106 mph)
flaps and landing gear down	74-78 kts (137-145 km/h; 85.5-90 mph)
Ceiling	9,150 m (30,000 ft)
Time to 9,150 m (30,000 ft)	20 min
Min ground turning radius	5.50 m (18 ft 0½ in)
T-O run	442 m (1,450 ft)
Endurance on internal fuel at 230 kts (426 km/h; 265 mph) at 9,150 m (30,000 ft)	1 h 45 min

VERIFIED

HAL HJT-16 KIRAN Mk II

TYPE: Two-seat jet trainer and light attack aircraft.

PROGRAMME: This version of the Kiran, for armament training and counter-insurgency duties was developed from the Mks I/IA by the HAL design bureau at Bangalore. The first prototype made its initial flight on 30 July 1976. A second prototype was flown in February 1979. Principal differences include improved weapon carrying capability, a more powerful engine, updated avionics and an improved hydraulic system. The Rolls-Royce Orpheus 701-01 turbojet, replacing the Viper engine of the Mks I/IA, gives the Kiran Mk II improved maximum speed, climb and manoeuvrability.

The design and development phase was completed by March 1983. An order for 40 Kiran Mk IIs was placed by the Indian Air Force, deliveries beginning in 1984. Deliveries ended in March 1989.

OPERATORS: Versions of the HJT-16 Kiran Mk II are in service with the armed forces of India (60).

DESIGN FEATURES: Cantilever low-wing monoplane. Wing section NACA 23015 at root, NACA 23012 at tip. Dihedral 4° from roots. Incidence 0° at root. Sweepback 7° 18′ at quarter-chord. Two full-chord boundary layer fences on upper surface of each wing. Hydraulically actuated door type airbrake under centre-fuselage, in line with flaps. Variable incidence tailplane.

FLYING CONTROLS: Frise differential ailerons, with balance tab in each aileron and ground adjustable tab on port aileron. Hydraulically actuated trailing-edge split flaps. Electrically operated variable incidence tailplane, with elevators.

STRUCTURE: The wing is a conventional all-metal three-spar structure of aluminium/copper alloy. The fuselage is an all-metal semi-monocoque fail-safe structure of light alloy. The tail unit is a cantilever all-metal structure of aluminium/copper alloy.

LANDING GEAR: Retractable tricycle type, of HAL manufacture. Hydraulic actuation. Main units retract inward into fuselage; self-centring twin-contact non-steerable nosewheel retracts forward. HAL oleo-pneumatic shock-absorbers. Dunlop mainwheel tyres size 19 × 6.25-9, pressure 9.24 bars (134 lb/sq in). Dunlop nosewheel tyre size 6.00-4, pressure 6.90 bars (100 lb/sq in). Dunlop hydraulic brake.

POWER PLANT: One Rolls-Royce Orpheus 701-01 turbojet engine, rated at 18.4 kN (4,130 lb st). Internal fuel in two flexible saddle tanks in fuselage (each 209 litres; 55.2 US gallons; 46 Imp gallons), one 282 litre (74.46 US gallon; 62 Imp gallon) collector tank in wing centre-section, and two outboard integral wing tanks (each 218 litres; 57.7 US gallons; 48 Imp gallons), giving total internal capacity of 1,340 litres (354 US gallons; 295 Imp gallons). Provision for four underwing tanks with total capacity of 908 litres (240 US gallons; 200 Imp gallons). One refuelling point in each outer wing and two in fuselage. Oil capacity 11.4 litres (3 US gallons; 2.5 Imp gallons).

ACCOMMODATION: Crew of two side by side in air conditioned and pressurised cockpit, on Martin-Baker H4HA zero-altitude fully automatic ejection seats. Clamshell type canopy, hinged at rear and opening upward. Dual controls and duplicated blind-flying instruments.

SYSTEMS: Bootstrap type air conditioning system, maximum pressure differential of 0.12 bars (1.75 lb/sq in). Hydraulic system, pressure 207 bars (3,000 lb/sq in), actuates landing gear, flaps, airbrake and wheel brakes, and canopy emergency release. Accumulator for manual emergency. Electrical system is of 28 V DC single-wire earth return type, with 5.1 kW generator, single 40 Ah Ni/Cd battery for standby power, and 250 V inverter for AC power to supply avionics, instruments and other ancillary systems. High pressure demand type gaseous oxygen system.

AVIONICS: HAL VUC 201 V/UHF multichannel com transceiver; HAL COM 150A four-channel standby UHF; HAL ARC-610A ADF; HAL IFF Mk 10 (BAT). Blind-flying instrumentation standard. Landing light in nose.

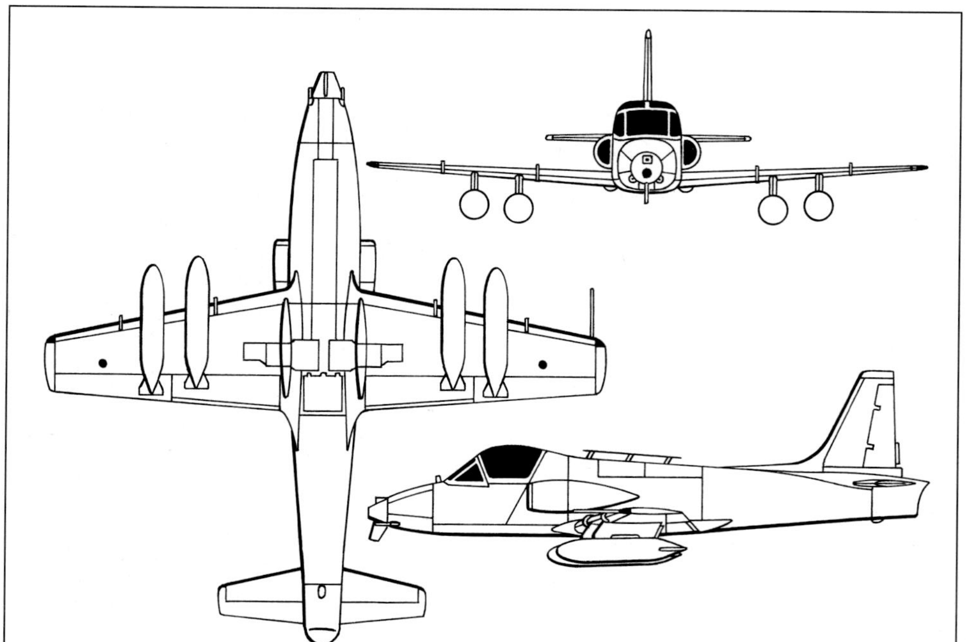

HAL HJT-16 Kiran Mk II *(Jane's/Mike Keep)*

1994

ARMAMENT: Two 7.62 mm Aden machine guns in nose, with 150 rds/gun; G90 gun camera, and Ferranti ISIS gunsight with Teledyne camera. Two pylons under each outer wing, each with an ejector release unit capable of carrying a 227 litre (60 US gallon; 50 Imp gallon) drop tank, a 250 kg bomb, a reusable pod containing 18 68 mm SNEB rockets, or a CBLS-200 carrier with four 25 lb practice bombs.

DIMENSIONS, EXTERNAL:

Wing span	10.70 m (35 ft 1¼ in)
Wing chord: at root	2.35 m (7 ft 8½ in)
at tip	1.02 m (3 ft 4 in)
Length overall	10.25 m (33 ft 7½ in)
Fuselage: Max width	1.36 m (4 ft 5½ in)
Height overall	3.635 m (11 ft 11 in)
Tailplane span	3.90 m (12 ft 9½ in)
Wheel track	2.42 m (7 ft 11 in)
Wheelbase	3.50 m (11 ft 6 in)

DIMENSIONS, INTERNAL:

Cockpit: Length	1.94 m (6 ft 4¼ in)
Max width	1.06 m (3 ft 5¾ in)
Max height	1.41 m (4 ft 7½ in)

AREAS:

Wings, gross	19.00 m² (204.5 sq ft)
Ailerons (total, incl tabs)	1.17 m² (12.59 sq ft)
Flaps (total)	2.34 m² (25.19 sq ft)
Fin, incl dorsal fin	2.29 m² (24.65 sq ft)
Rudder, incl tab	0.536 m² (5.77 sq ft)
Tailplane	3.72 m² (40.04 sq ft)
Elevators (total)	0.87 m² (9.36 sq ft)

WEIGHTS AND LOADINGS:

Weight empty, equipped	2,966 kg (6,539 lb)
Max fuel load	1,679 kg (3,701 lb)
Normal T-O weight	4,122 kg (9,087 lb)
Max T-O weight	4,950 kg (10,913 lb)
Max landing weight	4,300 kg (9,480 lb)
Max wing load	260 kg/m² (53.25 lb/sq ft)
Thrust weight ratio	0.454

PERFORMANCE (at max T-O weight, ISA):

Never-exceed speed	463 kts (858 km/h; 533 mph)
Max level speed at S/L	380 kts (704 km/h; 437 mph)
Max cruising speed at 4,575 m (15,000 ft)	335 kts (621 km/h; 386 mph) IAS
Econ cruising speed at 4,575 m (15,000 ft)	225 kts (417 km/h; 259 mph) IAS
Stalling speed: flaps and landing gear up	100 kts (185 km/h; 115 mph) IAS
flaps and landing gear down	85 kts (158 km/h; 98 mph) IAS
Max rate of climb at S/L	1,600 m (5,250 ft)/min
Service ceiling	12,000 m (39,375 ft)
Min ground turning radius	6.55 m (21 ft 6 in)
Runway LCN	6
T-O run	470 m (1,540 ft)
T-O to and landing from 15 m (50 ft)	700 m (2,300 ft)
Landing run	465 m (1,525 ft)
Range at 6,000 m (19,680 ft) with max internal fuel	332 n miles (615 km; 382 miles)

VERIFIED

HAL HJT-16 Kiran Mk II in HAL markings

1994

HAL HJT-16 Kiran Mk II two-seat jet trainer and light attack aircraft

1994

KANPUR DIVISION

Post Bag 225, Kanpur 208 008
Telephone: 91 (512) 43071 and 43074
Fax: 91 (512) 350505
Telex: 91 325 243 HALK IN
GENERAL MANAGER: Sri A. P. Arya

VERIFIED

HAL (HAWKER SIDDELEY) HS 748

TYPE: Twin-turboprop transport aircraft.

PROGRAMME: The first Indian built 748 flew on 1 November 1961, followed by the second one on 13 March 1963. The first four Indian 748s were Srs 1 aircraft, utilising components imported from the UK.

The first Indian built Srs 2 flew for the first time on 28 January 1964; by January 1973, 39 Srs 2 aircraft had been built and delivered. Most of the airframe components were manufactured by HAL from locally produced raw materials. The aircraft's 1,570 kW (2,105 ehp) Dart 531 turboprop engines were built by the Bangalore Complex of HAL. Some HF, VHF and other radio equipment was manufactured by Bharat Electronics Ltd of Bangalore.

Indian Airplanes placed an initial order for 14 Srs 2 aircraft; the first of these was delivered on 28 June 1967 and the last in March 1970. Since then a further 10 have been built of which three were delivered to Indian Airlines and three were modified for photographic survey role; others were delivered to the Indian Directorate General of Civil Aviation (for calibration role and the National Remote Sensing Agency).

HAL has also produced the aircraft for the Indian Air Force. These include 10 VIP executive transports, 10 as navigation trainers and 18 as trainers for pilots of multi-engined aircraft. In early 1992 the Indian Air Force had a total fleet of 60 748s.

A prototype HS 748(M) military freighter, developed by Kanpur Division, flew for the first time on 16 February 1972 and successfully completed flight trials with the Indian Air Force. An order to Kanpur Division for 10 more HS 748s was announced in June 1975. These aircraft were built to the military freighter standard. See also British Aerospace entry.

OPERATORS: Versions of the HAL (Hawker Siddeley) are in service with the armed forces of India (60+).

VERIFIED

HAL (Hawker Siddeley) HS 748 twin-turboprop transport aircraft of the Indian Air Force

1994

HAL (HAWKER SIDDELEY) HS 748 ASWAC UPGRADE

TYPE: ASWAC upgrade.

PROGRAMME: Under a programme started in 1985, a number of Kanpur built HS 748s are to be modified to fulfil an ASWAC (airborne surveillance, warning and control) requirement with the Indian Air Force, and an aerodynamic testbed fitted with a 4.80 m (15 ft 9 in) rotodome was flown for the first time on 5 November 1990. The (empty) rotodome was designed and built by Daimler-Benz Aerospace under subcontract to HAL, although production units are expected to be manufactured by the latter with technical assistance from the German company. Airframe modification and mounting of the radome was undertaken by Kanpur Division.

The ASWAC programme is managed by India's Defence Research and Development Organisation. Much of the onboard avionics are being designed by the Electronics and Radar Development Establishment, and will be manufactured by BEL (Bharat Electronics Ltd). Foreign participation is expected in developing the main radar.

VERIFIED

HAL (Hawker Siddeley) HS 748 Airborne Surveillance, Warning and Control (ASWAC) aircraft
1994

MiG COMPLEX

Ojhar Township Post Office, Nasik 422 207, Maharashtra
Telephone: 91 (22) 77901
Fax: 91 (253) 375825
Telex: 91 752241
MANAGING DIRECTOR: H. K. Anand
GENERAL MANAGER, NASIK DIVISION: N. R. Mohanty
GENERAL MANAGER, KORAPUT DIVISION: V. M. Akolkar

The MiG Complex was originally formed with the Nasik, Koraput and Hyderabad Divisions of HAL, which, under an agreement concluded in 1962, built the airframes, power plants and avionics of MiG-21 series fighters under licence from the former USSR. Indian production of MiG-21s was phased out in 1986-87, as production of the MiG-27M increased. The Hyderabad Division is now a part of the Accessories Complex.

VERIFIED

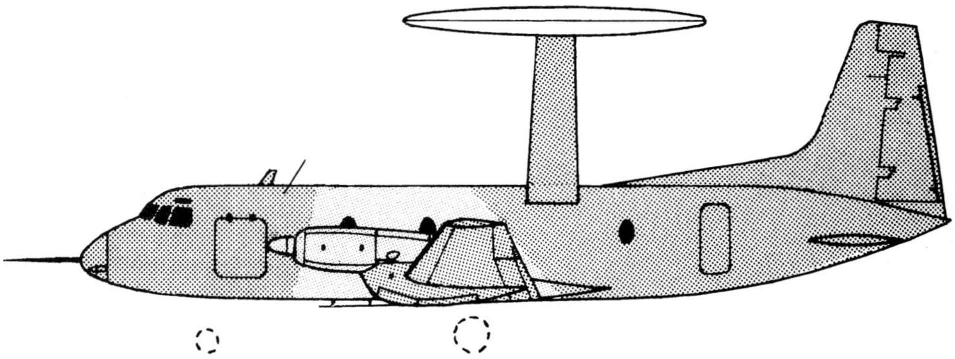

HAL prototype ASWAC conversion of the BAe (HS) 748 *(Jane's/Derek Ballington)*
1995

HAL (MIKOYAN) MiG-21

Indian Air Force designations: Type 66-400, 66-600, 74, 76, 77 and 96
NATO reporting name: Fishbed

TYPE: Single-seat multirole fighter and two-seat operational trainer.

PROGRAMME: Several versions of the MiG-21 were supplied to or manufactured in India, including the MiG-21F (IAF designation Type 74) and MiG-21PF (Type 76).

VERSIONS: **MiG-21FL:** Soviet export designation of the late production model MiG-21PF. First version to be manufactured in India, following the supply of some aircraft from Soviet production and an initial order for 60 in 1964. Production by HAL, initially from imported components and engines, began in late 1966; about 100 were assembled in this way, with first deliveries to IAF in 1967. First examples built from raw materials with 60 per cent indigenous content, handed over to IAF on 19 October 1970, and by December 1971 all MiG-21 versions were in service with Nos. 1, 4, 8, 28, 29, 30, 45 and 47 squadrons. Indian production was completed in 1973 and is believed to have totalled 196, IAF designation Type 77.

MiG-21M: Production version with R-11F2S-300 engine. First aircraft handed over to IAF on 14 February 1973, and about 15 delivered by Spring 1974. IAF designation Type 96. Fifty Soviet built MiG-21PFMAs, to supplement Indian production, placed in service with Nos. 7 and 108 squadrons.

MiG-21MF: Improved production version with more powerful Tumansky R-13-300 turbojet engine and increased fuel. Total of 150 ordered at delivery rate of 30 a year.

MiG-21U: Tandem two-seat training versions (NATO reporting name *Mongol*). Small batch of MiG-21U Mongol A (type 66-400 Series) delivered from USSR in 1965. Main version in service is the Mongol B.

OPERATORS: Versions of the HAL (Mikoyan) MiG-21 are in service with the armed forces of India (410+).

DIMENSIONS, EXTERNAL (MiG-21MF):

Wing span	7.15 m (23 ft 5½ in)
Length, incl pitot boom	15.76 m (51 ft 8½ in)
Fuselage length, intake lip to jetpipe nozzle	12.30 m (40 ft 4¼ in)
Height overall	4.10 m (13 ft 5½ in)
Tailplane span	3.70 m (12 ft 8 in)
Wheel track	2.69 m (8 ft 10 in)
Wheelbase	4.81 m (15 ft 9½ in)

AREAS:

Wings, gross	23.0 m² (247.0 sq ft)

WEIGHTS AND LOADINGS (MiG-21MF):

Weight empty	5,843 kg (12,882 lb)
T-O weight:	
with four K-13A missiles	8,200 kg (18,078 lb)
with two K-13A missiles and two 490 l (130 US gallon; 108 Imp gallon) drop tanks	8,950 kg (19,730 lb)
with two K-13As and three drop tanks	9,400 kg (20,725 lb)
Max T-O weight	9,800 kg (21,605 lb)
Max wing loading	426.0 kg/m² (87.5 lb/sq ft)
Max power loading	151.4 kg/kN (1.48 lb/lb st)

PERFORMANCE (MiG-21MF):
Max level speed: above 11,000 m (36,000 ft)
Mach 2.05 (1,175 kts; 2,175 km/h; 1,353 mph)

HAL MiG-21M of the Indian Air Force designated Type 96
1994

at low altitude
Mach 1.06 (701 kts; 1,300 km/h; 807 mph)
Landing speed 146 kts (270 km/h; 168 mph)
Design ceiling 18,000 m (59,050 ft)
Practical ceiling about 15,250 m (50,000 ft)
T-O run at normal AUW 800 m (2,625 ft)
Landing run 550 m (1,805 ft)
Combat radius (hi-lo-hi):
with four 250 kg bombs, internal fuel
200 n miles (370 km; 230 miles)
with two 250 kg bombs and drop tanks
400 n miles (740 km; 460 miles)
Range, internal fuel only
593 n miles (1,100 km; 683 miles)
Ferry range, with three external tanks
971 n miles (1,800 km; 1,118 miles)
PERFORMANCE (MiG-21US, clean):
Max level speed: above 12,200 m (40,000 ft)
Mach 2.02 (1,159 kts; 2,150 km/h; 1,335 mph)
at S/L Mach 1.06 (701 kts; 1,300 km/h; 807 mph)
Max rate of climb at S/L 6,400 m (21,000 ft)/min
Rate of climb at 11,000 m (36,000 ft)
3,050 m (10,000 ft)/min
Time to 1,500 m (4,920 ft) 20 s
Turn rate at 4,575 m (15,000 ft):
instantaneous (Mach 0.5) 11.1°/s
instantaneous (Mach 0.9) 13.4°/s
sustained (Mach 0.9) 7.5°/s
T-O run 700 m (2,297 ft)

VERIFIED

HAL (MIKOYAN) MiG-21 UPGRADE

TYPE: Combat aircraft upgrade.
PROGRAMME: It was reported in October 1992 that the Indian Air Force was considering upgrading approximately 100 MiG-21bis Type 75 to offset delays in the LCA programme. Probable improvements would include a more powerful engine (GE F404 preferred), improved avionics (including AN/APG-66 radar), greater firepower and strengthened airframe. Estimated programme cost Rs400 million ($1.33 million). On 3 May 1994 India's Deputy Defence Minister announced Mikoyan had been awarded this contract to upgrade 100 MiG-21bis Fishbeds with an option for 70 more. See also Mikoyan MiG-21-93 Upgrade.

VERIFIED

HAL (MIKOYAN) MiG-27M MLU UPGRADE

TYPE: Combat aircraft upgrade.
PROGRAMME: HAL intends to carry out a mid-life update on its MiG-27M aircraft. Programme is to include Smiths/SAGEM/DARIN nav/attack system as in its IAF Jaguar Internationals.

VERIFIED

INDONESIA

IPTN

INDUSTRI PESAWAT TERBANG NUSANTARA (Nusantara Aircraft Industries Ltd)

Bandung (Head Office & Plant)
PO Box 1562 BD, GPM 1st Floor, Jl. Pajajaran No. 154 Bandung - 40174
Telephone: 62 (22) 632145, 632043
Fax: 62 (22) 631696, 632145
Telex: 28295 IPTNBD IA - 28108 IPTNBD IA
Jakarta (Branch & Sales Office)
PO Box 3752 JKT, BBD Plaza 14th & 15th Floor, Jl. Imam Bonjol No. 61 Jakarta - 10310
Telephone: 62 (21) 3103833, 322247, 322395, 325911
Fax: 62 (21) 3100081, 325319, 333357
Telex: 61246 IPTNJK IA - 61392 IPTNJK IA
PRESIDENT: Prof Dr Ing B. J. Habibie
DIRECTOR OF COMMERCE: Ir Hari Laksono
VICE-DIRECTOR OF COMMERCE: Ir Heru Santoso
CHIEF ENGINEER: Ir Bambang Pamungkas
PRESS RELATIONS MANAGER: Drs Soleh Affandi

This company was officially inaugurated as PT Industri Pesawat Terbang Nurtanio (Nurtanio Aircraft Industry Ltd) on 23 August 1976, when the government of Indonesia implemented a decision to centralise all existing facilities in the establishment of a single new aircraft industry. The original capital was provided by combining the assets of Pertamina's Advanced Technology and Aeronautical Division with those of the former Nurtanio Aircraft Industry. The present name of the company was adopted in late 1985. A weapons system division, located in Menang Tasikmalaya, West Java, develops and produces the weaponry fitted to aircraft built by the company for military customers. By the end of 1990 the company had contracts for the sale of 270 aircraft, with negotiations then under way for a further 50.

IPTN is jointly responsible with CASA of Spain for development and production of the Airtech CN-235, as well as continuing licence manufacture of the NC-212 Aviocar, NBO-105, NAS-332 Super Puma and NBell-412. Studies for the N-442 small helicopter, with Daimler-Benz Aerospace (MBB) of Germany, are continuing. Major subcontracting includes the manufacture of components for the Boeing 737 and 767, Fokker 100 and General Dynamics F-16, and for Pratt & Whitney engines. A 1,393 m² (15,000 sq ft) power plant maintenance centre can maintain, overhaul and repair Allison 250, P&WC PT6A, Textron Lycoming LTS 101 and Turbomeca Makila turboshafts and Garrett TPE 331 and General Electric CT7 turboprop engines. The company had a workforce of approximately 16,000 employees in 1995, and occupies a 70 ha (173 acres) site with 600,000 m² (6,458,340 sq ft) of covered accommodation (latest figures received).

UPDATED

IPTN (Lockheed) L-100-30 Hercules converted from military transport to civil passenger aircraft *(IPTN)*
1995

IPTN (LOCKHEED) HERCULES CONVERSION

TYPE: Modified military cargo transport (C-130) to civil passenger configuration.
PROGRAMME: Conversion of two L-100-30 Hercules; first (PK-MLT) undertaken by LACI with IPTN assistance, second (PK-MLS) by IPTN alone; STC obtained April 1990; redelivered August 1991 and April 1992.
CUSTOMERS: Merpati Nusantara Airlines (two).
DESIGN FEATURES: Airframe strengthened (incl thicker wing ribs) for 12,700 kg (28,000 lb) increase in max T-O weight; seating for 97 passengers installed; cabin headroom increased by 17.5 cm (6.9 in); intercom, interphone, smoke detectors and emergency lighting installed; ice warning system reinstated; hydraulic and environmental control systems modified.

VERIFIED

IPTN/Boeing Surveiller aircraft of the Indonesian Air Force *(Boeing)*
1995

IPTN/BOEING (BOEING) B737-200 SURVEILLER UPGRADE

TYPE: Maritime surveillance aircraft conversion.
PROGRAMME: IPTN is upgrading the remaining two of three Indonesian Air Force Surveiller aircraft as part of a $117 million four-year contract awarded to Boeing Defense & Space Group. Work on the first began in Spring 1993. This aircraft was redelivered in August 1993.
DESIGN FEATURES: Upgrade consists of the installation of a new mission avionics system based on Boeing's digital processing and display system (DPDS), a real-time display for the side-looking airborne modular multimission radar (SLAMMR), nose-mounted search radar and IFF equipment.

NEW ENTRY

INTERNATIONAL PROGRAMMES

AIRBUS

AIRBUS INDUSTRIE

1 Rond Point Maurice Bellonte, F-31707 Blagnac Cédex, France
Telephone: 33 61 93 33 33
Fax: 33 61 93 37 92
Telex: AIRBU 530526 F
PARIS OFFICE: 12 bis avenue Bosquet, F-75007 Paris, France
Telephone: 33 (1) 45 51 40 95
MANAGING DIRECTOR: Jean Pierson
CHIEF OPERATING OFFICER: Heribert Flosdorff
SENIOR VICE-PRESIDENT, COMMERCIAL: Stuart Iddles
MANAGER, TECHNICAL PRESS: David Velupillai
AIRFRAME PRIME CONTRACTORS:
Aerospatiale: see under France
British Aerospace Airbus: see under UK
CASA: see under Spain
Daimler-Benz Aerospace Airbus: see under Germany

Airbus Industrie set up December 1970 as Groupement d'Intérêt Economique to manage development, manufacture, marketing and support of A300; this management now extends to A300-600, A310, A320, A321 and A340/A330. Airbus Industrie is responsible for all work (total workforce about 30,000) on these programmes by partner companies; Aerospatiale has 37.9 per cent interest in Airbus Industrie, Daimler-Benz Aerospace Airbus 37.9 per cent, British Aerospace 20 per cent, CASA 4.2 per cent. Fokker is an associate in A300 and A310 and Belairbus in A310, A320 and A330/A340. Alenia of Italy manufactures front fuselage plug for A321.

Airbus Industrie deliveries declined from 163 in 1991 to 157 in 1992 and 138 in 1993. Cancellations by end of 1993 appeared to total 28. New orders totalled 38 in 1993 compared with 136 in 1992. First A340 delivered to Air France in March 1993 was 1,000th Airbus airliner. Total order backlog at end 1993, not including subsequent changes, was 667. Planned output is 138 aircraft in 1995 and 154 in 1996.

Subsidiaries include Aeroformation and Airbus Industrie of North America. Airbus Industrie China set up in Beijing early 1994 with training centre and spares store; Xian Aircraft Company and Shenyang Aircraft Corporation make parts for Airbus aircraft.

In March 1994, Airbus signed an agreement with Indonesia's IPTN to assist in flight testing of N-250-100 turboprop transport (see under Indonesia) between end 1994 and certification in 1997.

Airbus Industrie joined its four member companies in March 1993 in studying Very Large Commercial Transport (see UHCA/VLCT/NLA in this section) with Boeing.

UPDATED

AIRBUS A300

TYPE: Large-capacity widebodied short/medium-range transport aircraft.

PROGRAMME: The A300 is basically a widebodied aircraft with underwing pods for two turbofan engines. The underwing location of the power plant enables the A300 to use any advanced technology turbofan engine in the 22,700 kg (50,000 lb) thrust class. The engines have considerable development potential and are installed in pods interchangeable with those of the McDonnell Douglas DC-10 Series 30. The CF6-50K of 24,040 kg (53,000 lb st) and CF6-50L of 24,493 kg (54,000 lb st) became available in the late 1970s.

Six airframes were involved in the certification programme, including one static test specimen and separate components, to cover the complete structure, for fatigue tests.

Construction of the first A300, a B1, began in September 1969. This aircraft made its first flight on 28 September 1972, and was followed by the second B1 on 5 February 1973. Together with the first two A300B2 aircraft, the B1s flew more than 1,580 hours by the time French and German certification was granted on 15 March 1974. This was followed by FAA certification on 30 May 1974 and covers automatic approach and landing in Cat. II weather conditions. Certification for Cat. IIIA was granted on 30 September 1974. The first automatic landing was made by the second B1 development aircraft, on 2 May 1973.

VERSIONS: **A300B1**: Initial version with 22,226 kg (49,000 lb st) CF6-50A engines, overall length of 50.97 m (167 ft 2½ in) and maximum seating for 302 passengers. First and second development aircraft are built to this standard.

A300B2: Basic production version. Third and fourth aircraft are to this configuration, and flew for the first time on 28 June and 20 November 1973 respectively. Next three (that is first three production) aircraft for Air France; the first of these flew for the first time on 15 April 1974 and was delivered on 11 May to Air France, it entered service with Air France on 23 May 1974 on the Paris-London route. Air France aircraft are equipped to accommodate 26 first class and 225 economy class passengers.

Airbus Industrie A300B4 at Heathrow ***(Peter J. Cooper)***

1994

A300B2-100: Initial production version. Type certificated by DGAC and LBA on 15 March 1974; entered service with Air France on 23 May 1974. Thirty delivered.

A300B2-200: Basically as B2-100 but with leading-edge Krueger flaps, and same wheels and brakes as B4. First flown on 30 July 1976. First delivery, to South African Airways, 23 November 1976. Twenty-four ordered including three B2-200FF for VASP with two-man forward-facing cockpit.

A300B2-300: Basically as B2-200 but increased zero-fuel and landing weights for increased payload and multi-stop flexibility. Four delivered to SAS, later converted to B4-100s.

A300B2-600: *For full details see Jane's All the World's Aircraft 1995-96.*

A300B4-100: Basic longer range version. Developed from original A300B2, with Krueger flaps; first flown on 26 December 1974. French and West German certification granted 26 March 1975, FAA Type Approval 30 June 1976. Entered service on 1 June 1975. Sixty-six ordered originally, plus converted B2-300s.

A300B4-200: Compared with -100 series has reinforced wings and fuselage, improved landing gear optional, additional fuel tank in rear cargo hold, and higher T-O weight. By 1 July 1984 100 had been ordered, including -200 convertible and 200FF (forward-facing cockpit) versions.

A300B4-600: Advanced version of the B4-200. *For full details see Jane's All the World's Aircraft 1995-96.*

A300C4: Freighter version.

Aerospatiale was responsible for the manufacture of the entire nose section (including the flight deck), lower centre fuselage and engine pylons, and for final assembly. MBB (now part of Daimler-Benz Aerospace) was responsible for the manufacture of the forward fuselage, between the flight deck and wing box, the upper centre fuselage, the rear fuselage and the vertical tail surfaces. British Aerospace had design responsibility for the wings, the building of the wing fixed structures, and collaborated with Fokker, which built the wingtips and wing moving surfaces. Wing assembly was carried out by MBB. CASA manufactured the horizontal tail surfaces, the port and starboard forward passenger doors and the landing gear doors.

Large A300 sections were flown from their places of manufacture in Europe to the final assembly line in Toulouse on board Super Guppy outsize cargo aircraft. After assembly, painting in customer's colour scheme was carried out at Toulouse. Aircraft were then flown to Hamburg for installation of interior furnishings and equipment before returning to Toulouse for final customer acceptance.

General Electric engines assembled under licence by SNECMA; some components were also licence-built by SNECMA (France) and MTU (Germany). The whole power plant assembly below the engine/pylon interface is virtually identical to that of the McDonnell Douglas DC-10-30, and nacelles are supplied by McDonnell Douglas. Nacelles for Pratt & Whitney engines manufactured by Rohr (Germany) and are interchangeable with those of JT9D-59-powered DC-10-40 aircraft and JT9D-70-powered Boeing 747s.

DESIGN FEATURES: Cantilever mid-wing monoplane. Thickness/chord ratio 10.5 per cent. Sweepback 28° at quarter-chord. Variable incidence tailplane.

FLYING CONTROLS: Each wing has three-section leading-edge slats (no slat cutout over the engine) and three double-slotted Fowler flaps on trailing-edge; Krueger flap on leading-edge wingroot (B2-200/B4); an all-speed aileron between inboard flap and outer pair; and low-speed aileron outboard of the outer pair of flaps. Lift dump facility by combination of three spoilers (outboard) and two airbrakes (inboard) on each wing, forward of outer pair of flaps, plus two additional airbrakes forward of inboard flap. CFRP spoilers, developed by MBB/VFW, evaluated on 12 aircraft in scheduled service. The flaps extend over 84 per cent of each half span, and increase the wing chord by 25 per cent when fully extended. The datum of the all-speed aileron is deflected downward by up to 1° with flap operation to maintain trailing-edge continuity with deflected flaps. Drive mechanism for flaps and slats are similar to one another, each powered by twin motors driving ball screwjacks on each surface with built-in protection against asymmetric operation. Two slat positions for take-off and landing. Preselection of the airbrake/lift dump lever allows automatic extension of the lift dumpers on touchdown. All flight controls are powered by triplex hydraulic servo-jacks, with no manual reversion. Tailplane powered by two motors driving a fail-safe ball screwjack.

STRUCTURE: The wing is a primary two-spar box structure, integral with fuselage and incorporating fail-safe principles, built of high strength aluminium alloy. Third spar across inboard sections. Machined skin with open-sectioned stringers. The fuselage is a conventional semi-monocoque structure of circular cross-section, with frames and open Z-section stringers. Built mainly of high strength aluminium alloy, with steel or titanium for some major components. Skin panels integrally machined in areas of high stress. Honeycomb panels or restricted glassfibre laminates for secondary structures. The tail unit is a cantilever all-metal structure, with sweepback on all surfaces, variable incidence tailplane and separately controlled elevators.

LANDING GEAR: Hydraulically retractable tricycle type, of Messier-Hispano-Bugatti design, with Messier-Hispano-Bugatti/Liebherr/Dowty shock-absorbers and wheels standard. Twin-wheel nose unit retracts forward, main units inward into fuselage. Free-fall extension. Each four-wheel main unit comprises two tandem-mounted bogies, interchangeable left with right. Standard bogie size is 927 × 1,397 mm (36½ × 55 in); wider bogie of 978 × 1,397 mm (38½ × 55 in) is available on B4-100/200. Mainwheel tyre sizes and pressures as follows: B2, 46 × 16-20 (standard bogie) 12.4 bars (180 lb/sq in); B2-200, 49 × 17-20 (standard bogie) at 11.4 bars (165 lb/sq in); B4-100, 46 × 16-20 (standard), 49 × 17-20 (standard) or 49 × 17-20 (wide bogie), with respective pressures of 14.2, 11.9 and 10.6 bars (206, 172 and 154 lb/sq in); B4-200, 49 × 17-20 (standard) or 49 × 17-20 (wide bogie), with respective pressures of 12.4 and 11.1 bars (180 and 161 lb/sq in). Nosewheel tyres size 40 × 14-16 on all these B2/B4 models, with pressures of 8.6 bars (125 lb/sq in) on B2 variants and 9.0 bars (131 lb/sq in) on B4 variants. Steering angles 65°/95°. Messier-Hispano/Liebherr/Dowty hydraulic disc brakes standard on all mainwheels. Duplex anti-skid units fitted, with a third standby hydraulic supply for wheel brakes. Bendix or Goodrich wheels and brakes available optionally.

POWER PLANT: Underwing location of the power plant enables the A300 to use any advanced technology turbofan engine in the 222.5 kN (50,000 lb) thrust class. It has been offered with the following engines: two 227 kN (51,000 lb st) General Electric CF6-50C turbofans (B20101/201 and B4-101/201); two 233.5 kN (52,500 lb st) General Electric CF6-50C2 (B4-103/203); or two 236 kN (53,000 lb st) Pratt & Whitney JT9D-59A (B2-220 and B4-120/220) or -59B (B4-221).

These engines are installed in pods interchangeable with those of the McDonnell Douglas DC-10 Series 30 and 40.

Fuel in two integral tanks in each wing, with total usable capacity of 44,000 litres (11,624 US gallons; 9,679 Imp gallons) in B2. Fifth integral tank in wing centre section of B4, increasing total usable capacity to 58,100 litres (15,348 US gallons; 12,780 Imp gallons). For B4-200 series an optional self-contained fuselage fuel tank is available, capacity 6,000 litres (1,585 US gallons; 1,320 Imp gallons). This unit fits into the rear cargo hold, where it takes the place of two standard LD3 containers. Two

Airbus Industrie A300B4-203 of Air Jamaica leaves British Aerospace Aviation Services at Filton with new livery ***(British Aerospace Aviation Services)***

1995

standard refuelling points beneath each wing, outboard of engine pylons.

ACCOMMODATION (B2 and B4): Crew of three on flight deck (two on aircraft for Garuda, Tunis Air and VASP), plus two observers' seats. Seating for between 220 and 320 passengers in main cabin in six-, seven-, eight- or nine-abreast layout with 79/86 cm (31/34 in) seat pitch. Typical economy class layout has 269 seats, eight-abreast with two aisles, at 86 cm (34 in) seat pitch. This layout includes one galley aft and one forward, two galleys in mid-cabin, and one galley and four toilets aft, with provision for a second toilet forward and an additional galley aft. Up to 336 passengers can be carried at 76 cm (30 in) seat pitch in single-class high density layout. Closed hat racks on each side, forming baggage lockers (maximum individual capacity 0.062 m² (2.19 cu ft)). Provision for central double-sided rack. Two outward parallel-opening plug-type passenger doors ahead of wing leading-edge on each side, and one each side at rear. Underfloor baggage/cargo holds fore and aft of wings, with doors on starboard side. The forward hold will accommodate four 2.24 × 3.17 m (88 × 125 in) pallets or 12 LD3 or IATA A1 containers on aircraft with the original door with a width of 2.44 m (8 ft 0 in). When the wider (2.69 m; 8 ft 10 in) door is fitted pallet size can be increased to 2.43 × 3.17 m (96 × 125 in). The rear hold will accommodate eight LD3 containers each of 4.25 m³ (150 cu ft) capacity. Additional bulk loading of freight provided for in an extreme rear compartment with usable volume of 16.0 m³ (565 cu ft). The latter compartment can be used for the transport of livestock. Entire accommodation is pressurised, including freight, baggage and avionics compartments.

SYSTEMS: Air for air conditioning system can be provided from engines, the APU or a high pressure ground source. Supply is controlled by separate and parallel bootstrap units, each of which includes a flow limiting unit, cooler unit, water separator and temperature control unit. In addition, air from each engine passes through a pressure precooler unit. Distribution in flight deck and three cabin areas, with independent regulation. Two independent automatic systems, with manual override, control the cabin altitude, its rate of change and the differential pressure. Cabin pressure differential for normal operations is 0.57 bars (8.25 lb/sq in). Hydraulic system comprises three fully independent circuits, operating simultaneously. Fluid used is a fibre resistant phosphate-ester type, working at a pressure of 207 bars (3,000 lb/sq in). The three circuits provide triplex power for primary flying controls; if any circuit fails, full control of the aircraft is retained without any necessity for action by crew. All three circuits supply the all-speed and low-speed ailerons, rudder and elevator; blue circuit additionally supplies tail trim, spoilers, slats and rudder variable-gear unit; green circuit additionally supplies airbrakes, spoilers, slats, elevator artificial feel units, flaps, steering, wheel brakes and normal landing gear requirements; yellow circuit additionally supplies tail trim, airbrakes, lift dumpers, rudder variable gear unit, elevator artificial feel units, flaps, wheel-brakes and steering. Each circuit normally powered by engine-driven self-regulating pumps, one on each for the green circuit and one each for the blue and yellow circuits. Dowty Rotol ram-air turbine-driven pump provides standby hydraulic power should both engines become inoperative. Main electrical power is supplied by two Westinghouse three-phase constant-frequency AC generators mounted on the engines. A third identical generator, driven by the APU, can supply power in flight, to replace a failed engine-driven generator, and on the ground. Supply frequency is 400 Hz and voltage is 115/200 V. Any one generator can supply sufficient power to operate all equipment and systems necessary for take-off and landing. A conventional generator CSD system is installed, the two units being mounted on opposite sides of the engine gearbox with the CSD driving an air-cooled generator at a constant 8,000 rpm. Each generator is rated at 90 kVA, with overload ratings of 135 kVA for 5 minutes and 180 kVA for 5 seconds. The APU generator is driven at constant speed through a gearbox. Three unregulated transformer-rectifier units (TRUs) supply 28 V DC power. Three 24 V 25 Ah Ni/Cd batteries are used for the APU starting and fuel control, engine starter control, standby lights and, by selection, emergency busbar. This busbar and a 115 V 400 Hz static inverter provide standby power in flight if normal power is unavailable. This system is separated completely from the main system. Hot air protection for engine intakes and slat sections on the wings outboard of the engine. Garrett TSCP 700-5 or (from c/n 246) GTCP 331-250 APU in tailcone, exhausting upward. The installation incorporates APU noise attenuation. Fire protection system is self-contained, and firewall panels protect main structure from an APU fire. The APU can be operated on the ground, in flight up to 10,675 m (35,000 ft), and in icing conditions. Relights are possible up to 7,620 m (25,000 ft). Aircraft is completely independent of ground power sources, since all major services can be operated by the APU.

AVIONICS: Standard communications avionics include two VHF sets and one Selcal system, plus interphone and passenger address systems. An accident and voice recorder are also installed. Standard navigation avionics include two VOR, two ILS, two radio altimeters, marker beacon receiver, two ADF, two DME, two ATC transponders and a weather radar. Sperry digital air data computer standard. Most other avionics are to the customer's requirements, only those related to the blind landing system (ILS and radio altimeter) being selected and supplied by the manufacturer. Additional optional avionics include one or two HF, third VHF, second marker beacon receiver, second radar, navigation computer and pictorial display. Pilot and co-pilot each have an integrated instrument system combining heading and attitude (three SAGEM MGC 10/ARINC 569 are standard in B2, but they can be replaced by MGC 30/ARINC 571 Mk 1 inertial sensors, which are modular with the MGC 10) SFENA (now Sextant Avionique) autopilot/flight director system; and radio information. The SFENA/Smiths/Bodenseewerk automatic flight control system includes a comprehensive range of en route facilities such as VOR coupling, heading select, height acquire, turbulence, rate of descent (if required) and control wheel steering, in addition to the normal height, speed, pitch-and-roll attitude and heading locks. An optional speed reference system with built-in windshear protection is available.

EQUIPMENT: Dual automatic landing system provides coupled approach and automatic landing facilities suitable for Cat. 111A operation. The system is designed to allow extension to Cat. 111B automatic landing capability. An automatic braking system is available.

DIMENSIONS, EXTERNAL (B2-100-200, B4-100/200):

Wing span	44.84 m (147 ft 1 in)
Wing aspect ratio	7.73
Length overall	53.62 m (175 ft 1 in)
Length of fuselage	52.03 m (170 ft 8½ in)
Fuselage: Max diameter	5.64 m (18 ft 6 in)
Height overall	16.53 m (54 ft 3 in)
Tailplane span	16.94 m (55 ft 7 in)
Wheel track	9.60 m (31 ft 6 in)
Wheelbase (c/l of shock-absorbers)	18.60 m (61 ft 0 in)
Passenger doors (each): Height	1.93 m (6 ft 4 in)
Width	1.07 m (3 ft 6 in)
Height to sill: fwd	4.60 m (15 ft 1 in)
centre	4.80 m (15 ft 9 in)
rear	5.50 m (18 ft 0½ in)
Emergency exits (each): Height	1.60 m (5 ft 3 in)
Width	0.61 m (2 ft 0 in)
Height to sill	4.87 m (15 ft 10 in)
Underfloor cargo door (fwd):	
Height	1.71 m (5 ft 7½ in)
Width	2.44 m (8 ft 0 in) or 2.69 m (8 ft 10 in)
Height to sill	2.56 m (8 ft 4¾ in)
Underfloor cargo door (rear):	
Height	1.71 m (5 ft 7½ in)
Width	1.81 m (5 ft 11¼ in)
Height to sill	2.96 m (9 ft 8½ in)
Underfloor cargo door (extreme rear):	
Height	0.95 m (3 ft 1 in)
Width	0.95 m (3 ft 1 in)
Height to sill	3.30 m (10 ft 10 in)

DIMENSIONS, INTERNAL:

Cabin, excl flight deck: Length	39.15 m (128 ft 6 in)
Max width	5.28 m (17 ft 4 in)
Max height	2.54 m (8 ft 4 in)
Underfloor cargo hold:	
Length: fwd	10.60 m (34 ft 9¼ in)
rear	6.89 m (22 ft 7¼ in)
extreme rear	3.10 m (10 ft 2 in)

Airbus Industrie A300B2 equipped with 278 seats

1994

Airbus Industrie A300 fly-by-wire widebodied short/medium-range transport aircraft

1994

Max height 1.76 m (5 ft 9 in)
Max width 4.20 m (13 ft 9¼ in)
Underfloor cargo hold volume:
fwd 75.1 m^3 (2,652 cu ft)
rear 46.8 m^3 (1,652 cu ft)
extreme rear 16.0 m^3 (565 cu ft)
Max total volume for bulk loading 140.0 m^3 (4,944 cu ft)

AREAS (B2, B4):
Wings, gross 260.0 m^2 (2,798.6 sq ft)
Leading-edge slats (total) 30.513 m^2 (328.44 sq ft)
Krueger flaps (total) 1.115 m^2 (12.00 sq ft)
Trailing-edge flaps (total) 46.599 m^2 (501.59 sq ft)
All-speed ailerons (total) 7.217 m^2 (77.68 sq ft)
Low-speed ailerons (total) 5.568 m^2 (59.93 sq ft)
Spoilers (total) 5.403 m^2 (58.16 sq ft)
Airbrakes (total) 8.105 m^2 (87.24 sq ft)
Fin 45.235 m^2 (486.90 sq ft)
Rudder 13.570 m^2 (146.07 sq ft)
Tailplane 69.454 m^2 (747.60 sq ft)
Elevators (total) 17.850 m^2 (192.14 sq ft)

WEIGHTS AND LOADINGS:
Manufacturer's weight empty:
B2-100/120 77,062 kg (169,892 lb)
B2-200/220 77,427 kg (170,697 lb)
B4-100/120 79,070 kg (174,319 lb)
B4-200/220 79,833 kg (176,000 lb)
Typical operating weight empty:
B2-100/120, B2-200 85,910 kg (189,400 lb)
B2-220 87,500 kg (192,905 lb)
B4-100 (basic) 88,100 kg (194,227 lb)
B4-120 (basic) 89,300 kg (196,873 lb)
B4-200 (optional) 88,500 kg (195,109 lb)
B4-220 (optional) 89,700 kg (197,755 lb)
Max payload (structural):
B2-100/120 34,590 kg (76,258 lb)
B2-200 34,600 kg (76,280 lb)
B2-220 33,000 kg (72,752 lb)
B4-100 (basic) 35,900 kg (79,146 lb)
B4-120 (basic) 34,700 kg (76,500 lb)
B4-200 (optional) 35,200 kg (77,602 lb)
B4-220 (optional) 34,300 kg (75,618 lb)
Max usable fuel:
B2-200/220 (basic) 34,000 kg (74,957 lb)
B4-100/120 (basic) 47,500 kg (104,720 lb)
Max T-O weight: B2-100 137,000 kg (302,030 lb)
or 142,000 kg (313,055 lb)
B2-200 142,000 kg (313,055 lb)
B4-100 150,000 kg (330,690 lb)
or 153,000 kg (337,305 lb)
or 157,500 kg (347,230 lb)
B4-200 165,000 kg (363,760 lb)
Max ramp weight: B2-200 142,900 kg (315,040 lb)
B4-200 165,900 kg (365,745 lb)
Max landing weight:
B2-100 127,500-134,000 kg (281,090-295,420 lb)
B2-200 130,000-134,000 kg (286,600-295,420 lb)
B4-100 133,000-136,000 kg (293,215-299,825 lb)
B4-200 134,000 kg (295,420 lb)
Max zero-fuel weight:
B2-100 116,500-124,000 kg (256,835-273,375 lb)
B2-200 120,500 kg (265,655 lb)
or 124,000 kg (273,375 lb)
B4-100 122,000-126,000 kg (268,960-277,780 lb)
B4-200 124,000 kg (273,375 lb)
Max wing loading: B2-200 546 kg/m^2 (111.8 lb/sq ft)
B4-100 606 kg/m^2 (124.1 lb/sq ft)
B4-200 635 kg/m^2 (130.0 lb/sq ft)
Max power loading:
B2-201 (CF6-50C) 312.8 kg/kN (3.07 lb/lb st)
B2-220 (JT9D-59A) 300.8 kg/kN (2.95 lb/lb st)
B4-101 (CF6-50C) 346.9 kg/kN (3.40 lb/lb st)
B4-103 (CF6-50C2) 337.3 kg/kN (3.31 lb/lb st)
B4-203 (CF6-50C2) 353.3 kg/kN (3.46 lb/lb st)
B4-220 (JT9D-59A) 349.6 kg/kN (3.43 lb/lb st)
B4-221 (JT9D-59B) 340.2 kg/kN (3.34 lb/lb st)

PERFORMANCE (at max T-O weight except where indicated):
Max operating speed (VMO): B2 from S/L to 8,475 m (27,800 ft), B4 from S/L to 7,740 m (25,400 ft)
345 kts (639 km/h; 397 mph) CAS
Max operating speed Mach number (MMO):
B2 above 8,475 m (27,800 ft) 0.86
B4 above 7,740 m (25,400 ft) 0.82
Max cruising speed at 7,620 m (25,000 ft):
B2, B4 492 kts (911 km/h; 567 mph)
Typical high-speed cruise at 9,145 m (30,000 ft):
B2, B4 495 kts (917 km/h; 570 mph)
Typical long-range speed at 9,450 m (31,000 ft):
B2, B4 457 kts (847 km/h; 526 mph)
Approach speed at typical weight:
B2 131 kts (243 km/h; 151 mph)
B4 132 kts (245 km/h; 152 mph)
Max operating altitude: B2 10,675 m (35,000 ft)
Min ground turning radius (wingtips)
33.51 m (109 ft 11¼ in)
Runway LCN at max T-O weight (A: flexible pavement of 51 cm (20 in) thickness, B: rigid pavement of 76 cm (30 in) radius of relative stiffness):
B2-100: A 75
B 68
B2-200: A 73
B 67
B4-1 (46 × 16-20 tyres): A 86
B 79
B4-100 (49 × 17-20 tyres): A 78
B 70
B4-100 (49 × 17-20 tyres, wide bogie): A 70
B 62
B4-200 (49 × 17-20 tyres): A 85
B 74
B4-200 (49 × 17-20 tyres): A 77
B 67
T-O field length (S/L, ISA + 15°C):
B2-200 1,753 m (5,750 ft)
B2-220 1,692 m (5,550 ft)
B4-100 2,332 m (7,650 ft)
B4-120 2,286 m (7,500 ft)
B4-200 2,972 m (9,750 ft)
B4-220 2,774 m (9,100 ft)
Landing field length at typical weight:
B2-200/220 1,463 m (4,800 ft)
B2-100/120 1,494 m (4,900 ft)
Range with 269 passengers and baggage, FAR international reserves):
B2-200 1,850 n miles (3,430 km; 2,130 miles)
B2-220 1,750 n miles (3,240 km; 2,015 miles)
B4-100 2,650 n miles (4,910 km; 3,050 miles)
B4-120 2,600 n miles (4,820 km; 2,990 miles)
B4-200/220 2,900 n miles (5,375 km; 3,340 miles)
Range with max fuel, FAR international reserves, zero payload:
B2-200 2,500 n miles (4,630 km; 2,875 miles)
B2-220 2,450 n miles (4,540 km; 2,820 miles)
B4-100/200 3,400 n miles (6,300 km; 3,915 miles)
B4-120/220 3,500 n miles (6,485 km; 4,030 miles)

OPERATIONAL NOISE LEVELS (FAR Pt 36):
T-O: B2 90 EPNdB
B4-100/120 92 EPNdB
B4-200/220 94 EPNdB
Approach: B2, B4 101 EPNdB
Sideline: B2, B4 95 EPNdB

VERIFIED

AIRBUS A300C4/F4

TYPE: Large-capacity widebodied short/medium-range transport aircraft.

PROGRAMME: The A300C4 is a convertible freighter version of the A300B4, available with the same range of power plant options. Main differences from the B4 are a large upper deck cargo door, a reinforced cabin floor, a smoke detection system in the main cabin, and an interior trim adaptable to the freighter role. The upper deck cargo door is on the opposite side to that of the underfloor holds, enabling loading or unloading to be carried out simultaneously at all positions.

In the freight mode the loading system, consisting of ball mats, roller tracks and electrical drive units, is fitted to the existing seat rails, and permits the carriage of up to 13 2.24 × 3.17 m (88 × 125 in) pallets, 12 2.44 × 3.17 m (96 × 125 in) pallets, or eight of the former plus five of the latter, in basic configurations. Total upper deck volume thus varies between 173 m^3 (6,100 cu ft) and 179 m^3 (6,315 cu ft). A 9*g* barrier net is installed in the front of the cabin. Total cargo-carrying capability of the A300C4 is approximately 41,000 kg (90,390 lb).

For customers requiring it, the A300C4 has the capability for conversion to passenger or mixed passenger/cargo configuration. Typical options include accommodation (in mainly eight-abreast seating) for up to 315 passengers on the upper deck; or 145 passengers (seven/eight-abreast) plus six 2.24 × 3.17 m (88 × 125 in) pallets; or 75 passengers plus nine 2.24 × 3.17 m (88 × 125 in) pallets; or 14 2.44 × 3.17 m (88 × 125 in) or 2.19 × 3.17 m (96 × 125 in) pallets; or 20 2.44 × 3.17 m (86 × 125 in) pallets.

The first A300C4 (c/n 83), for Hapag Lloyd of West Germany, made its initial flight in mid-1979. After fitting of the upper deck cargo door, as a retrospective conversion by MBB, it was delivered in January 1980. Subsequent A300C4s were built as C4s, the forward fuselage being delivered with the cargo door already fitted.

Airbus Industrie has also defined an **A300F4** freighter version, with a maximum payload of 46,000 kg (101,410 lb). This is similar to the C-4, but with all passenger provisions removed completely and the cabin windows replaced by metal blanking plates. It can carry up to 20 2.24 × 3.17 m (88 × 125 in) pallets on the upper deck, or eight 2.44 × 2.44 × 3.04 m (96 × 96 × 120 in) ISO containers plus five 2.44 × 3.17 m (96 × 125 in) pallets. Loadable volume of the upper deck of the A300F4 is 294.5 m^3 (10,400 cu ft).

The description of the A300B4 applies generally also to the convertible C4, except as outlined above and detailed below:

DIMENSIONS, EXTERNAL: As A300B4, plus:
Upper deck cargo door (fwd, port):
Height (projected) 2.57 m (8 ft 5¼ in)
Width 3.58 m (11 ft 9 in)

DIMENSIONS, INTERNAL: As A300B4, except:
Cabin upper deck usable for cargo:
Length 31.78 m (104 ft 3 in)
Min height 2.16 m (7 ft 1 in)

Max height 2.44 m (8 ft 0 in)
Volume 173-179 m^3 (6,100-6,315 cu ft)
Max total volume for bulk loading (upper and lower decks) 286 m^3 (10,087 cu ft)

WEIGHTS AND LOADINGS:
Manufacturer's weight empty (basic):
passenger mode 81,000 kg (178,575 lb)
freight mode 81,900 kg (180,560 lb)
pure freighter 78,300 kg (172,620 lb)
Manufacturer's weight empty (optional):
passenger mode 81,300 kg (179,235 lb)
freight mode 82,200 kg (181,220 lb)
pure freighter 78,600 kg (173,285 lb)
Operating weight empty (basic):
* passenger mode 89,000 kg (196,210 lb)
freight mode 82,900 kg (182,765 lb)
pure freighter 79,000 kg (174,165 lb)
Operating weight empty (optional):
* passenger mode 89,200 kg (196,650 lb)
freight mode 83,200 kg (183,425 lb)
pure freighter 79,400 kg (175,045 lb)
Max payload (structural) (basic):
passenger mode 35,000 kg (77,160 lb)
freight mode 41,100 kg (90,610 lb)
pure freighter 45,000 kg (99,210 lb)
Max payload (structural) (optional):
passenger mode 36,800 kg (81,130 lb)
freight mode 42,800 kg (94,360 lb)
pure freighter 46,600 kg (102,735 lb)
Max T-O weight: basic 157,500 kg (347,230 lb)
optional 165,000 kg (363,765 lb)
Max landing weight: basic 134,000 kg (295,420 lb)
optional 136,000 kg (299,830 lb)
Max zero-fuel weight: basic 124,000 kg (273,375 lb)
optional 126,000 kg (277,780 lb)
* *incl weight of underfloor cargo hold containers and pallets*

PERFORMANCE (estimated):
Range with max (structural) payload, allowances for ground manoeuvring, 30 min hold at 460 m (1,500 ft), no diversion:
basic 2,100 n miles (3,890 km; 1,825 miles)
optional 2,500 n miles (4,635 km; 2,170 miles)
Range with max fuel, no payload, allowances as above:
passenger mode 3,900 n miles (7,230 km; 3,390 miles)
freight mode 4,100 n miles (7,600 km; 3,560 miles)
pure freighter 4,200 n miles (7,785 km; 3,650 miles)

VERIFIED

BRITISH AEROSPACE/ AEROSPATIALE CONCORDE

AIRFRAME PRIME CONTRACTORS:
British Aerospace (see separate entry).
Aerospatiale (see separate entry).

POWER PLANT PRIME CONTRACTORS:
Rolls-Royce.
Société Nationale d'Etude et de Construction de Moteurs d'Aviation (SNECMA).

VERIFIED

CONCORDE

TYPE: Four-jet supersonic transport aircraft.

PROGRAMME: Anglo-French negotiations concerning the development of a supersonic transport aircraft culminated on 29 November 1962 in the signing of two agreements, one between the French and British governments, the other between the manufacturers to whom the project was entrusted. The agreements provided for the manufacture of two Concorde prototypes, followed by two preproduction aircraft and two airframes for static and fatigue testing. The static test programme was completed in September 1973, and this airframe was tested to destruction in June 1974. Fatigue testing was programmed to continue until two aircraft 'lives' (about 48,000 flights) had been attained.

The planned flight test programme, involving the two prototype, two preproduction and first four production Concordes, achieved its target of 5,335 hours flying at the time when the full passenger-carrying certificate of airworthiness was granted by the SGAC and CAA in late 1975.

By the beginning of 1979 a total of 19 Concordes had flown, including 15 production models. The 16th and last production aircraft first flew on 20 April 1979. In development and test flying with the manufacturers 2,930 flights had then been made, of which 1,800 involved flight at supersonic speeds. A total of 6,560 hours block time had been amassed, including 2,450 hours supersonic. In addition, up to the Summer of 1979, airline service had involved 7,600 flights amounting to 26,000 hours.

Airframe development and production of the Concorde were undertaken jointly by Aerospatiale and BAe, with two final assembly lines, at Toulouse and Filton respectively. There was no duplication of main production jigs.

Aerospatiale was responsible for development and production of the rear cabin section, wings and wing control surfaces, hydraulic systems, flying controls, navigation systems, radio and air conditioning. The automatic flight control system was designed by Marconi (now GEC-Marconi) in the UK and SFENA (now Sextant Avionique) in France, under contract to Aerospatiale. BAe was responsible for the three forward sections of the fuselage, the rear fuselage and vertical tail surfaces, the engine nacelles and ducting, the electrical system, sound and thermal insulation, oxygen system, fuel system, engine installation, and fire warning and extinguishing systems.

The first two production aircraft were flown on 6 December 1973 at Toulouse and 13 February 1974 at Filton, each attaining a speed of approximately 868 knots (1,610 km/h; 1,000 mph) on its first flight. Third and fourth production aircraft flew for the first time at Toulouse and Filton on 31 January and 27 February 1975 respectively. Special category certificates of airworthiness were granted in May and June 1975 by the SGAC and CAA, anticipating the full airworthiness certificates which were granted on 13 October and 5 December 1975. The fifth and sixth production aircraft made their first flights on 25 October and 5 November 1975 respectively, and the next four production aircraft were delivered to Air France and British Airways (two each) in 1976. Nos. 11 and 12 first flew 10 February and 17 March 1977, followed by No. 14 on 21 April 1978, No. 13 on 26 June 1978, and No. 15 on 26 December 1978.

British Airways and Air France each received seven aircraft. Air France has since removed one aircraft from service permanently.

In 1977-78 BAe flight tested modifications to the wing and tail controls surfaces designed to reduce drag and fuel consumption. These modifications which consisted of an approximate 0.61 m (2 ft) increase in the fin chord and an approximate 0.05 m (2 in) increase in the chord of the elevons and rudder were designed to reduce fuel consumption by approximately 680 kg (1,500 lb) on a flight of 3,500 n miles (6,485 km; 4,030 miles). In 1978 also, BAe flight tested a new thinner air intake lip. This modification reduced fuel consumption by approximately 1,360 kg (3,000 lb) on a supersonic flight of 3,500 n miles (6,485 km; 4,030 miles).

UPGRADES: **British Airways:** See separate entry below.

British Aerospace: Collins TCAS II/ATC models equipment will be fully operational by end of 1995.

The following description applies to the production Concorde:

DESIGN FEATURES: Cantilever low-wing of ogival delta planform. Thickness/chord ratio 3 per cent at root 2.15 from nacelle outboard. Slight anhedral.

FLYING CONTROLS: Six elevons in three pairs on trailing-edge of each wing, of aluminium alloy honeycomb construction. Each elevon pair is independently operated by a tandem jack, each half supplied from an independent hydraulic source and controlled by a separate electrical system. Dowty Boulton Paul power control units. Hydraulic artificial feel units protect the aircraft against excessive aerodynamic loads induced by pilot through over-control. Autostabilisation is provided. Autopilot control is by signals fed into normal control circuit. No high-lift devices. Two independent two-section aluminium honeycomb rudder controlled in same way as elevons.

STRUCTURE: The wing is a continuous camber structure with multispar torsion box, manufactured mainly from RR.58 (AU2GN) aluminium alloy. Integrally machined components used for highly loaded members and skin panels. In centre wing, spars are continuous across fuselage, the spars and associated frames being built as single assemblies extending between the engine nacelles. Forward wing sections built as separate components attached to each side of fuselage, spar loads being transferred to cross-members in lower part of main fuselage frames. The fuselage is a mainly conventional pressurised aluminium alloy semi-monocoque structure of constant cross-section, with unpressurised nose and tail cones. Hoop frame at approximately 0.55 m (21.5 in) pitch support integrally machined panels having closely pitched longitudinal stringers. Window surround in passenger cabin formed of integral skin-stringer panels machined from aluminium alloy planks. Nose is drooped hydraulically to improve forward view during take-off, initial climb, approach and landing. Retractable visor is raised hydraulically to fair in windscreen in cruising flight. The tail unit is a vertical fin and rudder only. Fin is multispar torsion box of similar construction to wings. Each engine nacelle consists of hydraulically controlled variable-area (by ramp) air intake, engine bay and nozzle support structure. Intakes are of RR.58 or AU2GN aluminium alloy with steel leading-edges. The engine bay has an Inconel centre wall with aluminium alloy forward doors and titanium rear doors. The nozzle bay, aft of the rear spar, is of welded Stresskin sandwich panels and heat-resistant nickel alloys. Reverser buckets, which are also used as secondary nozzle, are actuated by ball-screw jacks driven by compressed air through flexible shafts. Leading-edges of intakes, rear ramp sections and intake auxiliary door are electrically de-iced.

BAe/Aerospatiale Concorde four-jet supersonic passenger aircraft *(Paul Tompkins)*

BAe/Aerospatiale Concorde with four Rolls-Royce/SNECMA Olympus 593 Mk 610 turbojets *(Paul Tompkins)*

1995

Engine nose bullet and inlet guide vanes are de-iced by hot engine bleed air.

LANDING GEAR: Hydraulically retractable tricycle type. Messier-Hispano nose and main units, with either Kléber, Dunlop or Goodyear wheels and tyres. Twin-wheel steerable nose unit retracts forward. Four-wheel bogie main units retract inward. Oleo-pneumatic shock-absorbers. Mainwheels and tyres size 47 × 15.75-22, pressure 12.9 bars (187 lb/sq in). Nosewheels and tyres size 31 × 10.75-14, pressure 12 bars (174 lb/sq in). Dunlop carbon disc brakes. SNECMA (Hispano) SPAD anti-skid units. Retractable tail bumper. Strengthened wheels of 15.75 × 22.1 diameter, fitted with reinforced tyres at 16 bars (232 lb/sq in) were fitted in September 1982 to all aircraft.

POWER PLANT: Four Rolls-Royce/SNECMA Olympus 593 Mk 610 turbojet engines, each rated at 169.3 kN (38,050 lb st) with 17 per cent afterburning, and Type 28 thrust reversers. Fuel system is used also as heat sink and to maintain aircraft trim. All tanks are of integral construction and are in two groups, with total usable capacity of 119,786 litres (31,645 US gallons; 26,350 Imp gallons). Main group comprises five tanks in each wing and four tanks in fuselage and maintains CG automatically in cruising flight. Trim tank group (three tanks) comprises two tanks at the front and a tank of 13,150 litres (3,473 US gallons; 2,892 Imp gallons) capacity in fuselage beneath tail fin. This group maintains correct relationship between CG and aerodynamic centre of pressure by transferring fuel rearward during acceleration and forward during return to subsonic flight. Four pressure refuelling points in bottom fairing, two forward of each main landing gear unit. Oil capacity 22.75 litres (6 US gallons; 5 Imp gallons) per engine.

ACCOMMODATION: Pilot and co-pilot side by side on flight deck, with third crew member behind on starboard side. Provision for supernumary seat behind pilot. Wide variety of four-abreast layouts to suit individual requirements of airlines. With all normal toilet and galley service facilities, up to 128 economy class passengers can be carried with 86 cm (34 in) seat pitch. A version with 144 passenger seats at 81 cm (31 in) pitch is available. Normal configuration for both British Airways and Air France is now 100 passengers. Toilets at front and centre of cabin. Baggage space under forward cabin and aft of cabin. Passenger doors forward of cabin and amidships on port side, with service doors opposite. Baggage door aft of cabin on starboard side. Emergency exits in rear half of cabin on each side. Two galley areas.

SYSTEMS: BAe air conditioning system, comprising four independent subsystems, with Hamilton Standard heat exchangers. Pressure differential 0.74 bars (10.7 lb/sq in). In each subsystem the air passes through a primary ram-air heat exchanger to an air cycle cold-air unit, and then through secondary air/air and air/fuel heat exchangers. The air is then mixed with hot air and fed to cabins, flight deck, baggage holds, landing gear, equipment and radar bays. Hydraulic services utilise two primary systems and one standby, pressure 276 bars (4,000 lb/sq in), each actuated by two engine-driven pumps. Temperature of the oronite M.2V fluid is limited by heat exchangers, main systems actuate flying control surfaces, artificial feel units, landing gear, wheel brakes, nosewheel steering, windscreen visor, nosecone droop, engine intake ramps and fuel pumps in rear transfer tank. Electrical system powered by four 60 kVA engine-driven constant-speed brushless alternators giving 200/115 V AC at 400 Hz. Four 150 A transformer-rectifiers and two 25 Ah batteries provide 28 V DC supply. Wing leading-edges ahead of air intakes are de-iced electrically. No tail unit de-icing.

AVIONICS: SFENA/Marconi Avionics (now Sextant Avionique and GEC-Marconi respectively) automatic flight control system (AFCS). AC Delco Carousel IVA primary navigation system comprises three identical inertial platforms, each coupled to a digital computer to form three self-contained units, two VOR/ILS systems, one ADF (GEC-Marconi AD-380 in British Airways aircraft), two DME systems, one marker, two receiver RCA AVQ-X weather radars and two TRT (now Thomson TRT-Défense) AHV-5 radio altimeters. Plessey (now Siemens Plessey) flight data recording system in British Airways aircraft. Provision for supplementary system including a long distance radio fixing system of the Loran C type. Optional equipment includes a second ADF. Basic communications equipment consists of two VHF and two HF transceivers, one Selcal decoder and two ATC transponders (Cossor SSR 2700 in British Airways aircraft). Nose radome by Reinforced Microwave Plastics. Provision for a third VHF transceiver and datalink equipment.

BAe/Aerospatiale Concorde of British Airways *(Paul Tompkins)*

1995

DIMENSIONS, EXTERNAL:

Wing span	25.56 m (83 ft 10 in)
Wing aerodynamic reference chord at root	27.66 m (90 ft 9 in)
Wing aspect ratio	1.7
Length overall	62.10 m (203 ft 9 in)
Height overall	11.40 m (37 ft 5 in)
Fin aerodynamic reference chord at base	10.59 m (34 ft 9 in)
Wheel track	7.72 m (25 ft 4 in)
Wheelbase	18.19 m (59 ft 8¼ in)
Passenger doors (each): Height	1.67 m (5 ft 5¾ in)
Width	0.76 m (2 ft 6 in)
Height to sill: fwd	4.88 m (16 ft 0 in)
amidships	4.74 m (15 ft 7 in)
Service doors (each): Height	1.22 m (4 ft 0 in)
Width	0.61 m (2 ft 0 in)
Height to sill: fwd	4.88 m (16 ft 0 in)
amidships	4.75 m (15 ft 7½ in)
Baggage hold door (underfloor):	
Length	0.99 m (3 ft 3 in)
Width	0.84 m (2 ft 9⅕ in)
Height to sill	3.54 m (11 ft 7 in)
Baggage hold door (rear, stbd):	
Height	1.52 m (5 ft 0 in)
Width	0.76 m (2 ft 6 in)
Height to sill	4.04 m (13 ft 3 in)

DIMENSIONS, INTERNAL:

Cabin: Length, flight deck door to rear pressure bulkhead, incl galley and toilets	39.32 m (129 ft 0 in)
Width	2.63 m (8 ft 7½ in)
Height	1.96 m (6 ft 5 in)
Volume	238.5 m^3 (8,440 cu ft)
Baggage/freight compartments:	
underfloor	6.43 m^3 (227 cu ft)
rear fuselage (total)	13.31 m^3 (470 cu ft)

AREAS:

Wings, gross	358.25 m^2 (3,856 sq ft)
Elevons (total)	32.00 m^2 (344.44 sq ft)
Fin (excl dorsal fin)	33.91 m (365 sq ft)
Rudder	10.40 m^2 (112 sq ft)

WEIGHTS AND LOADINGS:

Operating weight empty	78,700 kg (173,500 lb)
Typical payload	11,340 kg (25,000 lb)
Max payload	12,700 kg (28,000 lb)
Max T-O weight	185,065 kg (408,000 lb)
Max zero-fuel weight	92,080 kg (203,000 lb)
Max landing weight	111,130 kg (245,000 lb)
Max wing loading	approx 488 kg/m^2 (1,000 lb/sq ft)
Max power loading	approx 268 kg/kN (2.5 lb/lb st)

PERFORMANCE:

Max cruising speed at 15,635 m (51,000 ft)	Mach 2.04 or 530 kts CAS, whichever is the lesser, equivalent to TAS of 1,176 kts (2,179 km/h; 1,345 mph)
T-O speed	214 kts (397 km/h; 246 mph)
Landing speed	162 kts (300 km/h; 187 mph)
Rate of climb at S/L	1,525 m (5,000 ft)/min
Service ceiling	approx 18,290 m (60,000 ft)
Min ground turning radius	21.7 m (71 ft 0 in)
Runway LCN at max T-O weight	90
T-O weight to 10.7 m (35 ft)	3,410 m (11,200 ft)
Landing from 10.7 m (35 ft)	2,220 m (7,300 ft)
Range with max fuel, FAR reserves and 8,845 kg (19,500 lb)	3,550 n miles (6,580 km; 4,090 miles)
Range with max payload, FAR reserves: at Mach 0.95 at 9,100 m (30,000 ft)	2,760 n miles (5,110 km; 3,180 miles)

OPERATIONAL NOISE CHARACTERISTICS (FAR Pt 36):

T-O noise level	119.5 EPNdB
* Approach noise level	116.7 EPNdB
Sideline noise level	112.2 EPNdB

* *Approach noise level can be reduced by at least 7 EPNdB by using decelerated approach procedure*

UPDATED

BRITISH AEROSPACE/AEROSPATIALE CONCORDE INTERIOR/EXTERIOR UPGRADE

TYPE: Supersonic passenger jet Interior/exterior upgrade.

PROGRAMME: Programme carried out by British Airways to restyle the interior and exterior of its fleet of Concordes. First refurbished aircraft went back into service June 1993.

DESIGN FEATURES: Programme consisted of stripping down exterior paintwork and repainting, installing new seats with back supports, increasing size of overhead bins, upgrading the galley facilities, music system (with CD player) and the cabin lighting system. This work coincided with full engineering works to the aircraft which consisted of undercarriage replacement as well as X-rays of the hull and rewiring work.

VERIFIED

DASSAULT/DORNIER

PARTICIPATING COMPANIES:
Dassault Aviation: see under France
Dornier Luftfahrt: see under Germany

VERIFIED

DASSAULT/DORNIER ALPHA JET

TYPE: Tandem-seat jet basic, low-altitude and advanced trainer and close support/battlefield reconnaissance aircraft.

PROGRAMME: Started 1970, initially for air forces of France and Germany; four prototypes (first flight 26 October 1973). Production authorised March 1975; has now ended, but any version remains available. Two new versions have been proposed to meet specifications for the Italian Air Force (IAF) and the Royal Australian Air Force (RAAF).

VERSIONS: **Advanced trainer/light attack** (formerly Alpha Jet E): Ordered for the air forces of France (176), Belgium (33), Egypt (30) designated **MS1**, Ivory Coast (7), Morocco (24), Nigeria (24), Qatar (6) and Togo (6).

Close support version (formerly Alpha Jet A): Ordered for Federal German Luftwaffe (175). First flown (A1) 12 April 1978. Delivered in 1979-83 for JaboGs 41, 43 and 49 (51 aircraft each) and the Luftwaffe base at Beja in Portugal (18) for weapons training. These were earmarked to form an additional combat unit (JaboG 44) in the event of an emergency.

Alpha Jet 2: Ordered by Egypt (15, including 11 built by AOI) and Cameroon (7). (The Egyptian Air Force version is designated **MS2**.) Improved attack version, incorporating nav/attack system; uprated Larzac 04-C20 engines.

Alpha Jet ATS: Advanced training system version fitted with state of the art cockpit controls, and displays such as CRT raster HUD combined with rear cockpit HUD and lateral multifunction displays and multifunction keyboards in each cockpit. Proposed for training in the use of nav/attack systems of future combat aircraft, including training in the operation of such sensors as radar, FLIR, laser and ECM systems. Under development. None yet ordered.

UPGRADES: **Dassault/Dornier:** Update programme for armament and avionics, to be developed in 1991-93, would include improved instruments, navigation and air data sensors as well as a stall warning indicator; but early 1991 report indicated that Luftwaffe has proposed phasing out of fleet between 1991 and 1993. German Air Force Alpha Jets were retrofitted with Larzac 04-C20 turbofan engines. The Portuguese received 50 ex-German Air Force Alpha Jets in early 1994.

OPERATORS: Versions of the Alpha Jet are in service with the armed forces of the following countries: Belgium (30), Cameroon (5), Egypt (34), France (130), Germany (35), Ivory Coast (4), Morocco (24), Portugal (40), Qatar (6) and Togo (5).

FLYING CONTROLS: Hydraulically actuated ailerons/rudder/all-moving tailplane, all with trimmable artificial feel; Fowler slotted flaps and twin rear fuselage airbrakes, also actuated hydraulically.

POWER PLANT: Two SNECMA/Turbomeca Larzac 04-C6 turbofans, each rated at 13.24 kN (2,976 lb st), mounted on sides of fuselage; 14.12 kN (3,175 lb st) Larzac 04-C20 turbofans retrofitted to Luftwaffe aircraft, standard on Alpha Jet 2 and optional for other versions. Internal fuel capacity 1,900 litres (502 US gallons; 418 Imp gallons) or 2,040 litres (539 US gallons; 449 Imp gallons). Provision for 310 or 450 litre (82 or 119 US gallon; 68 or 99 Imp gallon) capacity drop tank on each outer wing pylon, plus (on Alpha Jet 2) a 450 or 625 litre (119 or 165 US gallon; 99 or 137.5 Imp gallon) tank on each inboard wing pylon. Pressure refuelling (point near starboard intake) standard for all tanks, including drop tanks. Gravity system for fuselage tanks and drop tanks. Fuel system permits inverted flying.

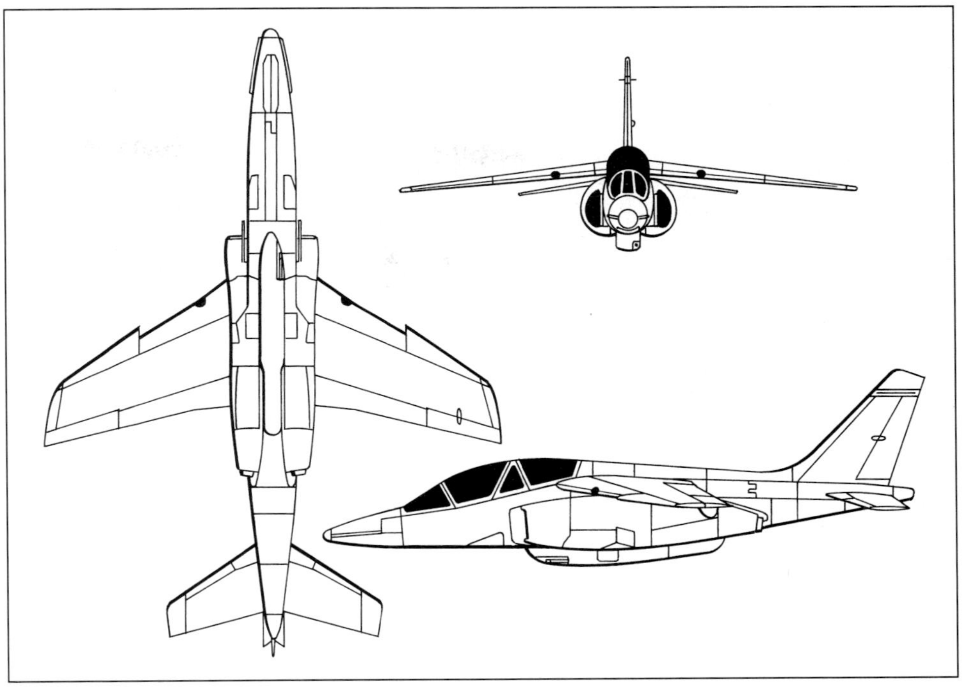

Dassault/Dornier Alpha Jet *(Jane's/Mike Keep)*

1994

ACCOMMODATION: Two persons in tandem, in pressurised cockpit under individual upward opening canopies. Dual controls standard. Rear seat (for instructor in trainer versions) is elevated. French trainer versions fitted with Martin-Baker AJRM4 ejection seats, operable (including ejection through canopy) at zero height and speeds down to 90 knots (167 km/h; 104 mph). Martin-Baker B10N zero/zero seats in aircraft for Belgian Air Force, E10N for Egypt and Q10N in those for Qatar. Aircraft for Germany fitted with licence-built (by MBB, now part of Daimler-Benz Aerospace) Stencel S-III-S3AJ zero/zero ejection seats. Baggage compartment in tailcone, with access door on starboard side.

SYSTEMS: Cockpit air conditioning and demisting system. Cabin pressure differential 0.30 bars (4.3 lb/sq in). Two independent and redundant hydraulic systems, each 207 bars (3,000 lb/sq in), with engine-driven pumps (emergency electric pump on one circuit), for actuating control surfaces, landing gear, brakes, flaps, airbrakes, and (when fitted) nosewheel steering. Pneumatic system, for cockpit pressurisation and air conditioning, occupants' pressure suits and fuel tank pressurisation, is supplied by compressed air from engines. Main electrical power supplied by two 28 V 9 kW starter/generators, one on each engine. Circuit includes a 36 Ah Ni/Cd battery for self-starting and two static inverters for supplying 115 V AC current at 400 Hz to auxiliary systems. External ground DC power receptacle in port engine air intake. Hydraulic and electrical systems can be sustained by either engine in the event of the other becoming inoperative. Liquid-film anti-icing system; de-icing by electrical heater mats. Oxygen mask for each occupant, supplied by liquid oxygen converter of 10 litres (2.64 US gallons; 2.2 Imp gallons) capacity. Emergency gaseous oxygen bottle for each occupant.

AVIONICS: Standard avionics, according to version, include V/UHF and VHF or UHF transceivers, IFF/SIF, VOR/ILS/marker beacon receiver, Tacan, radio compass, gyro platform and intercom. Landing light on starboard mainwheel leg, taxying light on port leg. Alpha Jet 2 has Sagem Uliss 81 inertial platform (replacing SFIM 550) and Una 81 nav/attack unit, Thomson-CSF VE 110C head-up display with film or video camera (VEM 130 in Lancier), Thomson-CSF TMV 630 laser rangefinder, TRT AHV 9 radar altimeter and Dassault Electronique Digibus digital multiplexed avionics databus.

ARMAMENT: For armament training and close support, can be equipped with an underfuselage jettisonable pod containing a 30 mm DEFA or 27 mm Mauser cannon with 150 rounds; or an underfuselage pylon for one 250 kg bomb, one 400 kg modular bomb, or a target towing system. Provision also for two hardpoints under each wing, with non-jettisonable adaptor pylons. On these can be carried M155 launchers for 18 68 mm rockets; HE or retarded bombs of 125, 250 or 400 kg; 625 lb cluster dispensers; 690 or 825 lb special purpose tanks; practice launchers for bombs or rockets; Dassault CC-420 underwing 30 mm gun pods, each with 180 rounds; or two drop tanks (see Power Plant). Provision for air-to-air or air-to-surface missiles such as Sidewinder, Magic or Maverick, or reconnaissance pod. Total load for all five stations more than 2,500 kg (5,510 lb). Dassault CEM-1 (combined external

Dassault/Dornier Alpha Jet tandem two-seat advanced trainer, close support/battlefield reconnaissance aircraft of the French Air Force *(Dassault Aviation/Aviaplans)*

1995

Dassault/Dornier Alpha Jet MS2 of the Egyptian Air Force *(Dassault Aviation/Aviaplans)*

1995

multistore) carriers can be attached to inboard underwing pylons, permitting simultaneous carriage of mixed fuel/bomb/rocket loads, including six rockets and four practice bombs, or 18 rockets with one 500 lb bomb, or six penetration bombs, or grenades or other stores. A special version of the CEM-1 allows carriage of a reconnaissance pod containing four cameras (three Omera 61 cameras and an Omera 40 panoramic camera) and a decoy launcher. Luftwaffe aircraft equipped with ML Aviation twin stores carriers, CBLS 200 practice bomb and rocket launcher carriers, and ejector release units. Fire control system for air-to-air or air-to-ground firing, dive bombing and low-level bombing. Firing by pupil pilot (in front seat) is governed by a safety interlock system controlled by the instructor, which energises the forward station trigger circuit and illuminates a fire clearance indicator in the pupil's cockpit. Thomson-CSF 902 sight and film or video gun camera in French version; Kaiser/VDO KM 808 sight and gun camera in German aircraft.

DIMENSIONS, EXTERNAL:
Wing span 9.11 m (29 ft 10¾ in)
Length overall: trainer 11.75 m (38 ft 6½ in)
close support version, incl probe 13.23 m (43 ft 5 in)
Height overall (at normal T-O weight) 4.19 m (13 ft 9 in)
Tailplane span 4.33 m (14 ft 2½ in)
Wheel track 2.71 m (8 ft 10¾ in)
Wheelbase 4.72 m (15 ft 5¾ in)

AREAS:
Wings, gross 17.50 m² (188.4 sq ft)

WEIGHTS AND LOADINGS:
Weight empty, equipped: trainer 3,345 kg (7,374 lb)
close support version 3,515 kg (7,749 lb)
Fuel (internal) 1,520 kg (3,351 lb)
or 1,630 kg (3,593 lb)
Fuel (external) 500 kg (1,102 lb)
or 720 kg (1,587 lb)
or 1,440 kg (3,174 lb)
Max external load more than 2,500 kg (5,510 lb)
Normal T-O weight:
trainer, clean 5,000 kg (11,023 lb)
Max T-O weight:
with external stores 8,000 kg (17,637 lb)
Max wing loading: clean 285.7 kg/m² (58.52 lb/sq ft)
with external stores 457.1 kg/m² (93.62 lb/sq ft)
Max power loading:
04-C6 engines: clean 188.82 kg/kN (1.85 lb/lb st)
with external stores 302.11 kg/kN (2.96 lb/lb st)
04-C20 engines: clean 177.05 kg/kN (1.74 lb/lb st)
with external stores 283.29 kg/kN (2.78 lb/lb st)

PERFORMANCE (at normal clean T-O weight, except where indicated):
Max level speed at 10,000 m (32,800 ft):
Larzac 04-C6 Mach 0.85
Larzac 04-C20 Mach 0.86
Max level speed at S/L:
Larzac 04-C6 540 kts (1,000 km/h; 621 mph)
Larzac 04-C20 560 kts (1,038 km/h; 645 mph)
Max speed for flap and landing gear extension
200 kts (370 km/h; 230 mph)
Approach speed 110 kts (204 km/h; 127 mph)
Landing speed at normal landing weight
92 kts (170 km/h; 106 mph)
Stalling speed: flaps and landing gear up
116 kts (216 km/h; 134 mph)
flaps and landing gear down
90 kts (167 km/h; 104 mph)
Max rate of climb at S/L 3,660 m (12,000 ft)/min
Rate of climb at S/L, one engine out, at 4,782 kg (10,542 lb) AUW, in landing configuration
330 m (1,085 ft)/min
Time to 9,150 m (30,000 ft) less than 7 min
Service ceiling 14,630 m (48,000 ft)
T-O run: Larzac 04-C6 370 m (1,215 ft)
Larzac 04-C20 320 m (1,050 ft)
Landing run at usual landing weight
approx 500 m (1,640 ft)
Low altitude radius of action (trainer): clean, max internal fuel 291 n miles (540 km; 335 miles)
with external tanks 361 n miles (670 km; 416 miles)
High altitude radius of action (trainer), reserves of 15% internal fuel: clean, max internal fuel
664 n miles (1,230 km; 764 miles)
with external tanks 782 n miles (1,450 km; 901 miles)
Lo-lo-lo mission radius (close support version), incl combat at max continuous thrust and 54 n miles (100 km; 62 mile) dash: with belly gun pod and underwing weapons
210 n miles (390 km; 242 miles)
with belly gun pod, underwing weapons and external tanks 340 n miles (630 km; 391 miles)
Hi-lo-hi mission radius (close support version), incl combat at max continuous thrust and 54 n miles (100 km; 62 mile) dash: with belly gun pod and underwing weapons
315 n miles (583 km; 363 miles)
with belly gun pod, underwing weapons and external tanks 580 n miles (1,075 km; 668 miles)
Ferry range (internal fuel and four 450 l external tanks)
more than 2,160 n miles (4,000 km; 2,485 miles)
Endurance (internal fuel only):
low altitude more than 2 h 30 min
high altitude more than 3 h 30 min
g limits +12/–6.4 ultimate

UPDATED

EHI

EH INDUSTRIES LIMITED

PARTICIPATING COMPANIES:
Agusta: see under Italy
Westland: see under UK

CO-CHAIRMEN:
Alan Jones
Amedeo Caporaletti

MANAGING DIRECTOR: E. Striano

EH Industries formed June 1980 by Westland Helicopters and Agusta (50 per cent each) to undertake joint development of new anti-submarine warfare helicopter for Royal Navy and Italian Navy within an integrated programme under which naval, army and civil variants now being produced. Programme handled on behalf of both governments by UK Ministry of Defence; Westland has design leadership for commercial version, Agusta for rear loading military/utility version; naval version being developed jointly for UK and Italian navies and export. IBM/Westland selected to manage Royal Navy programme in 1991; Canadian programme launched October 1992 and terminated 12 months later.

In March 1994 EHI teamed with McDonnell Douglas Helicopter Systems to offer minimally modified, US built EH 101 to US armed forces for Medium Lift Replacement Helicopter programme, complementing Bell/Boeing V-22 Osprey.

NEW ENTRY

EH INDUSTRIES EH 101 UPGRADE

TYPE: Multirole, military helicopter upgrade.

PROGRAMME: During 1993 a study was initiated for improved systems for the EH 101 helicopter which was also to be available as a retrofit.

DESIGN FEATURES: Package to include improved sensors and processing equipment as well as uprated versions of the RTM 322 turboshaft engines. For full description of EH 101 see *Jane's All the World's Aircraft 1995-96.*

NEW ENTRY

EUROCOPTER

EUROCOPTER SA

72 boulevard de Courcelles, F-75017 Paris, France
Telephone: 33 (1) 49 34 44 44
Fax: 33 (1) 49 34 44 47
Telex: 212836 F AISPA X

CHAIRMEN:
Jean-François Bigay
Dr Siegfried Sobotta

DIRECTOR OF COMMUNICATIONS: X. Poupardin

Formed 16 January 1992 by merger of Aerospatiale and MBB (Daimler-Benz Aerospace) helicopter divisions. Share capital held on two levels; Eurocopter Holding SA owned 60 per cent by Aerospatiale and 40 per cent by Daimler-Benz Aerospace; capital of Eurocopter SA held 75 per cent by Eurocopter Holding and 25 per cent by Aerospatiale. Eurocopter is ready for other companies to join.

During its first year, Eurocopter took orders for 159 helicopters (10 Super Puma/Cougars, 12 Lamas, 102 Ecureuil/Fennecs, 17 Dauphin/Panthers, 10 BO 105s and eight BK 117s), 57 of them military and over 92 per cent for export. Total sales during 1992, including NH 90 programme, worth FFr15.2 billion (DM4.4 billion). Eurocopter accounted for 44 per cent of world civil helicopters registered in 1992 and 79 per cent of world public service helicopters, producing 51 per cent of whole non-military market.

Extremely difficult 1993 for Eurocopter, as for other manufacturers, but company held its market share, especially with larger Super Puma/Cougar. On a turnover of FFr10.21 billion (10.5 per cent down on 1992) Eurocopter made a loss of FFr300 million (about $50 million), but FFr250 million of this is provision against risks. New orders in 1993 covered 166 helicopters worth FFr9.3 billion; company raised its market share in world civil turbine helicopters from 51.5 per cent in 1991 to 56 per cent in 1993, when only 260 civil turbine helicopters were sold worldwide. World military market share, excluding USA and Russia, climbed back up to 24 per cent. Company needs to sell 20 more Cougar/Super Pumas in 1994. Eurocopter products cover 75 to 80 per cent of the range of helicopters in terms of size and capacity, but company intends to expand this to 95 per cent, particularly with hoped for co-operative development of Russian Mil Mi-38.

Workforce was reduced by 1,060 to 10,500 during 1993 and some 2,000 more redundancies were avoided by the sale of 43 Cougar/Super Pumas.

All national and international subsidiaries and partially owned companies concerned with helicopters formerly belonging to Aerospatiale or Deutsche Aerospace are now controlled by Eurocopter SA through Eurocopter Participations, which see.

Works are at La Courneuve (Paris) and Marignane (Marseille) in France and Ottobrunn (Munich) and Donauwörth in Germany.

VERIFIED

EUROCOPTER INTERNATIONAL

72 boulevard de Courcelles, F-75017 Paris, France
Telephone: 33 (1) 49 34 44 44
Fax: 33 (1) 49 34 44 47
Telex: 212836 F AISPA X

CHAIRMAN: Jean P. Bernadet
DEPUTY CHAIRMAN: Werner Reinl

Wholly owned subsidiary of Eurocopter SA, formed 6 May 1991; responsible for marketing all group helicopters.

VERIFIED

EUROCOPTER FRANCE

PO Box 13, F-13725 Marignane, France
Telephone: 33 42 85 85 85
Fax: 33 42 85 85 00
Telex: 410975 F

PRESIDENT DIRECTOR GENERAL: Jean-François Bigay

Industrial concern in charge of development, production and product support of current and future products originating in French part of group.

VERIFIED

EUROCOPTER DEUTSCHLAND

PO Box 9801140, D-8000 Munich 80, Germany
Telephone: 49 (89) 607-0
Fax: 49 (89) 607 24915
Telex: 5287-027 MBB D

CHAIRMAN: Dr Siegfried Sobotta

PUBLIC RELATIONS EXECUTIVE: Christina Gotzhein

Industrial concern in charge of development, production and product support of current and future products originating in German part of group.

VERIFIED

EUROCOPTER PARTICIPATION

PO Box 13, F-13725 Marignane, France
DIRECTOR: Philippe Harache

Company formed January 1993 as subsidiary of Eurocopter SA to manage 16 Eurocopter subsidiaries outside France and Germany, including the principal ones listed below. Others are Jamaero (Singapore), Helibras (Brazil), Eurocopter de Mexico, Lansav (South Africa), MBB Helicopter Systems (UK), Helicopteros Espana. Partly owned are MBB Kutlutas Helikopterleri (Turkey), Euroaircraft Services (Malaysia), MBB Helicopter and Transport (Nigeria), Philippine Helicopter Services, Eurocopter International Belgium, and Eurocopter do Brasil.

VERIFIED

AMERICAN EUROCOPTER CORPORATION

2701 Forum Drive, Grand Prairie, Texas 75053-4005, USA
Telephone: 1 (214) 641 1000
Fax: 1 (214) 641 3550
CHAIRMAN: Guy Essautier
CEO: David Smith

Combines former MBB Helicopter of West Chester, Pennsylvania, and Aerospatiale Helicopters of Grand Prairie, Texas, both of which have modification and assembly facilities as well as sales activities. MBB site now concentrating on support. Aerospatiale Helicopters had agreement with LTV (now Vought) to propose AS 565 Panther to US Army as utility transport. Two AS 350BAs modified at Grand Prairie began demonstrations at US Army Fort Rucker helicopter training base on 14 September 1992 in connection with New Training Helicopter competition, subsequently won by Bell TH-67 Creek.

VERIFIED

EUROCOPTER CANADA LTD

PO Box 250, 1100 Gilmore Road, Fort Erie, Ontario L2A 5M9, Canada
Telephone: 1 (416) 871 7772
Fax: 1 (416) 871 3320
VICE-PRESIDENT AND GENERAL MANAGER:
Richard W. Harwood

Formerly MBB Helicopter Canada; producer and design authority for Eurocopter BO 105 LS; completes and sells BO 105 CBS and BK 117 in Canada; partner in EC 135 (BO 108), development of which was partly financed by Canadian government.

VERIFIED

EUROCOPTER SERVICE JAPAN

Owned by Eurocopter SA (51 per cent), Nozaki (24.5 per cent) and Sony Trading (24.5 per cent); 410 Eurocopter helicopters of various types, a large proportion of Japanese civil helicopter fleet, and 75 Eurocopter/Kawasaki BK 117s, in service; company supports, modifies and sells them.

VERIFIED

EUROCOPTER INTERNATIONAL PACIFIC

PO Box 51, Bankstown, NSW 2200, Australia
Telephone: 61 (2) 794 9900
Fax: 61 (2) 791 0195

Sales and support company for Eurocopter products and programmes.

VERIFIED

EUROCOPTER TIGER GmbH

Gustav Heinemann Ring 135 (PO Box 830356), D-8000 Munich 83, Germany
Telephone: 49 (89) 638250-0
Fax: 49 (89) 638250-50
CEOs: Bernard Darrieus, Ingo Jaschke

Company formed 18 September 1985 to manage development and manufacture of Tiger/Tigre/Gerfaut battlefield helicopter for French and German armies; because it is working on a single government contract, Eurocopter Tiger is not a full member of Eurocopter; executive authority for programme is DFHB (Deutsch-Französisches Hubschrauberbüro) in Koblenz; procurement agency is German government BWB (Bundesamt für Wehrtechnik und Beschaffung). Tiger is being offered for British Army and Netherlands battlefield competitions. More Aerospatiale helicopter designations were amended in January 1990, the changes being applied retrospectively. The prefix AS (for Aerospatiale) now replaces SA (referring to the former Sud-Aviation) in most cases. Additionally, military versions add 200 to their former numerical designation and adopt a standardised role suffix (except for the Gazelle): A—Armed, land-based; C—anti-tank; M—maritime, non-combatant; S—maritime, armed; and U—utility. Some helicopters were renamed, military versions of Super Puma, Ecureuil and Dauphin becoming Cougar, Fennec and Panther respectively. By December 1990 Aerospatiale had sold approximately 8,500 helicopters in 115 countries.

VERIFIED

AEROSPATIALE SA 315B

Indian Army name: Cheetah

TYPE: Turbine-driven general purpose helicopter.

PROGRAMME: Design of the SA 315B Lama began in late 1968, initially to meet a requirement of the Indian armed forces, and a prototype was flown for the first time on 17 March 1969. French certification was granted on 30 September 1970 and FAA Type Approval on 25 February 1972.

The Lama combines features of the Alouette II and III, having the airframe (with some reinforcement) of the former and the dynamic components, including the Artouste power plant and rotor system, of the AS 316 Alouette III.

During demonstration flights in the Himalayas in 1969 a Lama, carrying a crew of two and 140 kg (308 lb) of fuel, made the highest landings and take-offs ever recorded, at a height of 7,500 m (24,600 ft).

On 21 June 1972, a Lama set a helicopter absolute height record of 12,442 m (40,820 ft). The pilot was Jean Boulet, holder of the previous record in an SE 3150 Alouette.

The production Lama is capable of transporting an external load of 1,135 kg (2,500 lb) at an altitude of more than 2,500 m (8,200 ft). In an agricultural role, it can be fitted with spraybars and an underbelly tank of 1,135 litres (300 US gallons; 250 Imp gallons) capacity, developed jointly by Aerospatiale Helicopter Corporation and Simplex Manufacturing Company. The tank is equipped with an electrical emergency dump system.

An alternative agricultural installation, using two side-mounted glassfibre tanks of Simplex manufacture, was shown for the first time at the 1976 US National Agriculture Aviation Association convention. Up to 1,000 kg (2,200 lb) of liquid chemicals can be carried in these tanks, which weigh 132 kg (290 lb) empty. A high performance electric pump dispenses up to 455 litres (120 US gallons; 100 Imp gallons)/min at 2.06 bars (30 lb/sq in) boom pressure.

A total of 407 Lamas was manufactured by Aerospatiale and production is now only carried out by Hindustan Aeronautics Limited (HAL) of India. Helibras of Brazil produced seven SA 315B helicopters. HAL had produced a total of 222 by 31 March 1993.

OPERATORS: Versions of the SA 315B are in service with the armed forces of the following countries: Angola (2), Argentina (4), Cameroon (1), Chile (6), Ecuador (3), India (HAL Cheetah - 140), Morocco (2), Pakistan (5), Peru (8) and Togo (2).

DESIGN FEATURES: Glazed cabin has light metal frame. Centre and rear of fuselage have a triangulated steel tube framework. Rotor system and drive as for Alouette III.

LANDING GEAR: Skid type, with removable wheels for ground manoeuvring. Pneumatic floats for normal operation from water, and emergency flotation gear, inflatable in the air, are available.

POWER PLANT: One 649 kW (870 shp) Turbomeca Artouste IIIB turboshaft engine, derated to 410 kW (550 shp). Fuel tank in fuselage centre-section, with capacity of 575 litres (152 US gallons; 126 Imp gallons), of which 573 litres (151 US gallons; 126 Imp gallons) are usable.

ACCOMMODATION: Glazed cabin seats pilot and passenger side by side in front and three passengers behind. Provision for external sling loads of up to 1,135 kg (2,500 lb). Can be equipped for rescue (hoist capacity 160 kg; 352 lb). Liaison, observation, training, agricultural, photographic and other duties. As an ambulance, can accommodate two stretchers and a medical attendant.

Aerospatiale SA 315B turbine-driven general purpose helicopter ***(Eurocopter)***

1995

Aerospatiale SA 315B Lama on lift duties

1994

DIMENSIONS, EXTERNAL:

Main rotor diameter	11.02 m (36 ft 1¾ in)
Tail rotor diameter	1.91 m (6 ft 3¼ in)
Main rotor blade chord (constant)	0.35 m (13⅘ in)
Length overall, both rotors turning	12.92 m (42 ft 4¾ in)
Length of fuselage	10.26 m (33 ft 8 in)
Height overall	3.09 m (10 ft 1¾ in)
Skid track	2.38 m (7 ft 9¾ in)

WEIGHTS AND LOADINGS:

Weight empty	1,021 kg (2,251 lb)
Normal max T-O weight	1,950 kg (4,300 lb)
Max T-O weight with externally slung cargo	2,300 kg (5,070 lb)

PERFORMANCE (A: at AUW of 1,950 kg; 4,300 lb, B: at AUW of 2,300 kg; 5,070 lb, with slung load):

Never-exceed speed at S/L: A	133 kts (210 km/h; 130 mph)
Max cruising speed: A	103 kts (192 km/h; 119 mph)
B	65 kts (120 km/h; 75 mph)
Max rate of climb at S/L: A	330 m (1,080 ft)/min
B	234 m (758 ft)/min
Service ceiling: A	5,400 m (17,715 ft)
B	3,000 m (9,840 ft)
Hovering ceiling IGE: A	5,050 m (16,565 ft)
B	2,950 m (9,675 ft)
Hovering ceiling OGE: A	4,600 m (15,090 ft)
B	1,550 m (5,085 ft)
Range with max fuel: A	278 n miles (515 km; 320 miles)

VERIFIED

AEROSPATIALE SA 316B ALOUETTE III

TYPE: Turbine-driven general purpose helicopter.

PROGRAMME: Alouette III helicopter was developed from the Alouette II, with larger cabin, greater power, improved equipment and higher performance. The prototype flew for the first time on 28 February 1959, and a total of 1,455 Alouette IIIs had been delivered for civil and military operation in 74 countries by 1 May 1985 which was the final year of production by Aerospatiale although Brasov of Romania continued manufacturing the SA 316B for a short period. All Romanian production has now ceased.

Those aircraft delivered before 1969 were designated SE 3160. The subsequent Artouste-engined SA 316B has strengthened main and rear rotor transmissions, higher AUW and increased payload. It flew for the first time on 27 June 1968; first deliveries were made in 1970, and this version received FAA Type Approval on 25 March 1971. The SA 319B, with Astazou engine, is described separately, but is included in the total sales figure above.

The sale of Alouette IIIs to India, Romania and Switzerland included licence agreements for manufacture of the aircraft in those countries. Quantities involved were 250 in India, 180 in Romania and 60 in Switzerland.

In 1977, an SA 316B operated by Trans North Turbo Air for the Canadian Park Service evacuated a mountain climber, suffering from pulmonary edema and frostbite, from a point at the 4,235 m (13,900 ft) level on the east ridge of Mount Logan. This is thought to be the greatest height at which a hoist rescue has been effected.

OPERATORS: Versions of the SA 316B are in service with the armed forces of the following countries: Argentina (4), Belgium (33), Burundi (4), Congo (2), Ecuador (3), Equatorial Guinea (2), France (5), Guinea Bissau (1), Ireland (8), Lebanon (2), Libya (4), Malawi (1), Malta (3), Mexico (3), Mozambique (3), Myanmar (10), Pakistan (12), Peru (9), Portugal (18), South Africa (60), Tunisia (7), United Arab Emirates (7), Zaïre (7) and Zimbabwe (35).

DESIGN FEATURES: Three-blade main and anti-torque rotors. All-metal main rotor blades, of constant chord, on articulated hinges, with hydraulic drag-hinge dampers. Rotor brake standard. Welded-steel tube centre-section, carrying the cabin at the front and a semi-monocoque tailboom. The tail unit is a cantilever all-metal fixed tailplane, with twin endplate fins, mounted on tailboom. Main rotor driven through planetary gearbox, with freewheel for autorotation.

Take-off drive for tail rotor at lower end of main gearbox, from where a torque shaft runs to a small gearbox which supports the tail rotor and houses the pitch-change mechanism. Cyclic and collective pitch controls are powered.

LANDING GEAR: Non-retractable tricycle type, manufactured by Messier-Hispano-Bugatti. Nosewheel is fully castoring. Provision for pontoon landing gear.

POWER PLANT: One 649 kW (870 shp) Turbomeca Artouste IIIB turboshaft engine derated to 425 kW (570 shp). Fuel in single tank in fuselage centre-section, with capacity of 575 litres (152 US gallons; 126.5 Imp gallons), of which 573 litres (151 US gallons; 126 Imp gallons) are usable.

ACCOMMODATION: Normal accommodation for pilot and six persons, with three seats in front and a four-person folding seat at rear of cabin. Two baggage holds in centre-section, on each side of the welded structure and enclosed by the centre-section fairings. Provision for carrying two stretchers athwartships at rear of cabin, and two other persons, in addition to pilot. All passenger seats removable to enable aircraft to be used for freight-carrying. Provision for external sling for loads of up to 750 kg (1,650 lb). One forward-opening door on each side, immediately in front of two rearward-sliding doors. Dual controls and cabin heating optional.

OPERATIONAL EQUIPMENT (military version): In the assault role, the Alouette III can be equipped with a wide range of weapons. A 7.62 mm AA52 machine gun (with 1,000 rds) can be mounted athwartships on a tripod behind the pilot's seat, firing to starboard, either through a small window in the sliding door or through the open doorway with the door locked open. The rear seat is removed to allow the gun mounting to be installed. In this configuration, maximum accommodation is for pilot, co-pilot, gunner and one passenger, although normally only the pilot and gunner would be carried. Alternatively, a 20 mm MG 151/20 cannon (with 480 rds) can be carried on an open turret-type mounting on the port side of the cabin. For this installation all seats except that of the pilot are removed, as is the port side cabin door, and the crew consists of pilot and gunner. Instead of these guns, the Alouette III can be equipped with four AS 11 or two AS 12 wire-guided missiles on external jettisonable launching rails, with an APX-Bézu 260 gyro-stabilised sight, or 68 mm rocket pods.

DIMENSIONS, EXTERNAL:

Main rotor diameter	11.02 m (36 ft 1¾ in)
Tail rotor diameter	1.91 m (6 ft 3¼ in)
Main rotor blade chord (each)	0.35 m (13⅘ in)
Length overall, rotor turning	12.84 m (42 ft 1½ in)
Length of fuselage, tail rotor turning	10.17 m (33 ft 4⅓ in)
Width overall, blades folded	2.60 m (8 ft 6¼ in)
Height to top of rotor head	3.00 m (9 ft 10 in)
Wheel track	2.60 m (8 ft 6¼ in)

WEIGHTS AND LOADINGS:

Weight empty	1,143 kg (2,520 lb)
Max T-O weight	2,200 kg (4,850 lb)

PERFORMANCE (standard version, at max T-O weight):

Never-exceed speed at S/L	113 kts (210 km/h; 130 mph)
Max cruising speed at S/L	100 kts (185 km/h; 115 mph)
Max rate of climb at S/L	260 m (850 ft)/min
Service ceiling	3,200 m (10,500 ft)
Hovering ceiling IGE	2,850 m (9,350 ft)
Hovering ceiling OGE	1,500 m (4,920 ft)
Range with max fuel at S/L	267 n miles (495 km; 307 miles)
Range at optimum altitude	290 n miles (540 km; 335 miles)

VERIFIED

AEROSPATIALE SA 318C ALOUETTE ASTAZOU

Aerospatiale SA 318C on crop-spraying duties

1994

TYPE: Turbine-driven general purpose helicopter.

PROGRAMME: Developed from the SE 313B Alouette II, the prototype Alouette II Astazou flew for the first time on 31 January 1961. Extension of the Alouette II airworthiness certificate to the Alouette II Astazou was subsequently granted in France (18 February 1964) and the USA (25 November 1964). A total number of 1,305 Alouette II had been sold to 126 operators in 46 countries by the Spring of 1975, the final year of manufacture.

OPERATORS: Versions of the SA 318C are in service with the armed forces of Cameroon (3) and Peru (3).

DESIGN FEATURES: Three-blade main rotor, two-blade anti-torque rotor. All-metal main rotor blades on articulated hinges, with hydraulic drag-hinge dampers. Blades may be folded towards rear. Main rotor driven through planetary gearbox, with freewheel or autorotation. Take-off drive for tail rotor at lower end of main gearbox, from where a torque shaft runs to a small gearbox which supports the tail rotor and houses the pitch-change mechanism. Cyclic and collective pitch controls are powered.

STRUCTURE: Glazed cabin has light metal frame. Centre and rear fuselage have a triangulated steel tube framework.

LANDING GEAR: Skid type, with removable wheels for ground manoeuvring. Raised skid gear available for flying crane operation. Pneumatic floats for normal operation from water and emergency flotation gear, inflatable in the air are available.

POWER PLANT: One 530 shp Turbomeca engine derated to 360 shp and fitted with a centrifugal clutch. Fuel tank, capacity 580 litres (153 US gallons; 127.5 Imp gallons), in centre-fuselage.

ACCOMMODATION: Glazed cabin seats pilot and passenger side by side in front and three passengers behind. Can be adapted for flying crane (payload 600 kg; 1,322 lb), rescue (hoist capacity 120 kg; 265 lb), liaison, observation, training, agricultural, photographic, ambulance and other duties. As an ambulance can accommodate two stretchers and a medical attendant internally.

DIMENSIONS, EXTERNAL:

Main rotor diameter	10.20 m (33 ft 5⅝ in)
Tail rotor diameter	1.91 m (6 ft 3¼ in)
Length overall, rotors turning	12.10 m (39 ft 8½ in)
Fuselage length, tail rotor turning	9.75 m (31 ft 11¾ in)
Width overall, blades folded	2.38 m (7 ft 9¾ in)
Height overall	2.75 m (9 ft 0 in)
Skid track	2.38 m (7 ft 9¾ in)

WEIGHTS AND LOADINGS:

Weight empty	890 kg (1,961 lb)
Max T-O weight	1,650 kg (3,630 lb)

PERFORMANCE (at max T-O weight):

Max level speed at S/L	110 kts (205 km/h; 127 mph)
Max cruising speed at S/L	97 kts (180 km/h; 112 mph)
Max rate of climb at S/L	396 m (1,300 ft)/min
Service ceiling	3,300 m (10,800 ft)
Hovering ceiling in ground effect	1,550 m (5,085 ft)
Hovering ceiling out of ground effect	900 m (2,950 ft)
Range: with max fuel at S/L	388 n miles (720 km; 447 miles)
with 600 kg (1,322 lb) payload	53 n miles (100 km; 62 miles)
with 480 kg (1,058 lb) payload	161 n miles (300 km; 186 miles)
Max endurance at S/L	5 h 18 min

VERIFIED

Aerospatiale Alouette III of the French Navy

1994

AEROSPATIALE SA 319B ALOUETTE II ASTAZOU

TYPE: Turbine-driven general purpose helicopter.

PROGRAMME: The SA 319B Alouette III Astazou is a direct development of the SA 316B, from which it differs principally in having an Astazou XIV turboshaft engine (649 kW; 870 shp), derated to 447 kW (600 shp) with increased thermal efficiency and a 25 per cent reduction in fuel consumption. A prototype SA 319 was completed in 1967. The production total is included in the figures given under the 316B entry. In addition to versions comparable with those of the SA 316B, a specially equipped naval version was produced, as follows:

OPERATORS: Versions of the SA 319B are in service with the armed forces of the following countries: Argentina (3), Cameroon (3), France (40), Greece (4), Mexico (3), Pakistan (2), Peru (2), Spain (6) and Surinam (2).

OPERATIONAL EQUIPMENT (naval version): The Alouette III can fulfil a variety of shipborne roles; features common to all naval configurations include a quick-mooring harpoon to ensure instant and automatic mooring on landing and before take-off, a nosewheel locking device, and folding main rotor blades. For detecting and destroying small surface craft such as torpedo boats, it was equipped with a SFENA (now Sextant Avionique) three-axis stabilisation system, OMERA ORB 31 radar, APX-Bézu 260 gyro-stabilised sight and two AS 12 wire-guided missiles. For the ASW role, it can carry two Mk 44 homing torpedoes beneath the fuselage, or one torpedo and magnetic anomaly detection (MAD) gear in a streamlined container which is towed behind the helicopter on a 50 m (150 ft) cable. The aircraft can be used for search and rescue when the cabin floor is protected by an anti-corrosion covering to prevent sea water from reaching vital components. Rescue hoist (capacity 225 kg; 500 lb) mounted on port side of fuselage.

WEIGHTS AND LOADINGS (standard SA 319B):

Weight empty	1,146 kg (2,527 lb)
Max T-O weight	2,250 kg (4,960 lb)

PERFORMANCE:

Max level speed at S/L	118 kts (220 km/h; 136 mph)
Max rate of climb at S/L	270 m (885 ft)/min
Hovering ceiling IGE	3,100 m (10,170 ft)
Hovering ceiling OGE	1,700 m (5,575 ft)
Range with six passengers (80 kg; 176 lb each) T-O weight at S/L	325 n miles (605 km; 375 miles)

VERIFIED

AEROSPATIALE SA 321 SUPER FRELON

TYPE: Three-engine heavy-duty helicopter.

PROGRAMME: The Super Frelon is a three-engined multi-purpose helicopter derived from the smaller SA 3200 Frelon. Under a technical co-operation contract, Sikorsky Aircraft, USA, provided assistance in the development of the Super Frelon, in particular with the detail specifications, design, construction and testing of the main and tail rotor system. Under a further agreement, the main gearcase and transmission box were produced in Italy by Fiat.

The first prototype of the Super Frelon (originally designated SA (AS) 3210-01) flew on 7 December 1962, powered by three 985 kW (1,320 shp) Turmo $IIIG_2$ engines, and represented the troop transport version. The second prototype, flown on 28 May 1963, was representative of the naval version, with stabilising floats on the main landing gear supports. Four preproduction aircraft followed, and the French government ordered an initial production series of 17, designated SA (AS) 321G, in October 1965. Approximately 100 Super Frelons have been delivered for civil and military duties in eight countries.

VERSIONS: **SA 321F**: Commercial airliner, designed to carry 34-37 passengers in a standard of comfort comparable to that of fixed-wing airliners, over 94 n miles (175 km; 108 miles) stage lengths at a cruising speed of 124 knots (230 km/h; 143 mph), with 20 minutes reserve fuel. The prototype was designed in accordance with US FAR Pt 29 regulations and flew for the first time on 7 April 1967. Type certification was granted by the SGAC on 27 June 1968 and by the FAA on 29 August 1968.

SA 321G: Anti-submarine helicopter. First version of the AS 321 to enter production. The first SA 321G flew on 30 November 1965 and deliveries began in early 1966. Twenty-four built. In service with the French Navy (12). The SA 321G can also be operated from the French helicopter carriers.

SA 321H: Version for air force and army service, without stabilising floats or external fairings on each side of lower fuselage. Turmo $IIIE_6$ engines instead of Turmo $IIIC_6$ in other versions. No de-icing equipment fitted.

SA 321Ja: Utility and public transport version, intended to fulfil the main roles of personnel and cargo transport. Designed to carry a maximum of 27 passengers. External loads of up to 5,000 kg (11,023 lb) can be suspended from the cargo sling and carried 27 n miles (50 km; 31 miles), the aircraft returning to base without load. An internal payload of 4,000 kg (8,818 lb) can be carried over 100 n miles (185 km; 115 miles) at 124 knots (230 km/h; 143 mph) with 20 minutes fuel reserves. The SA 321Ja prototype flew for the first time on 6 July 1967. A French certificate of airworthiness was granted in December 1971.

OPERATORS: Versions of the SA 321 are in service with the

Aerospatiale Alouette III of the Sécurité Civile

1994

Aerospatiale Alouette III of the Swiss Air Force *(Peter Cooper)*

1995

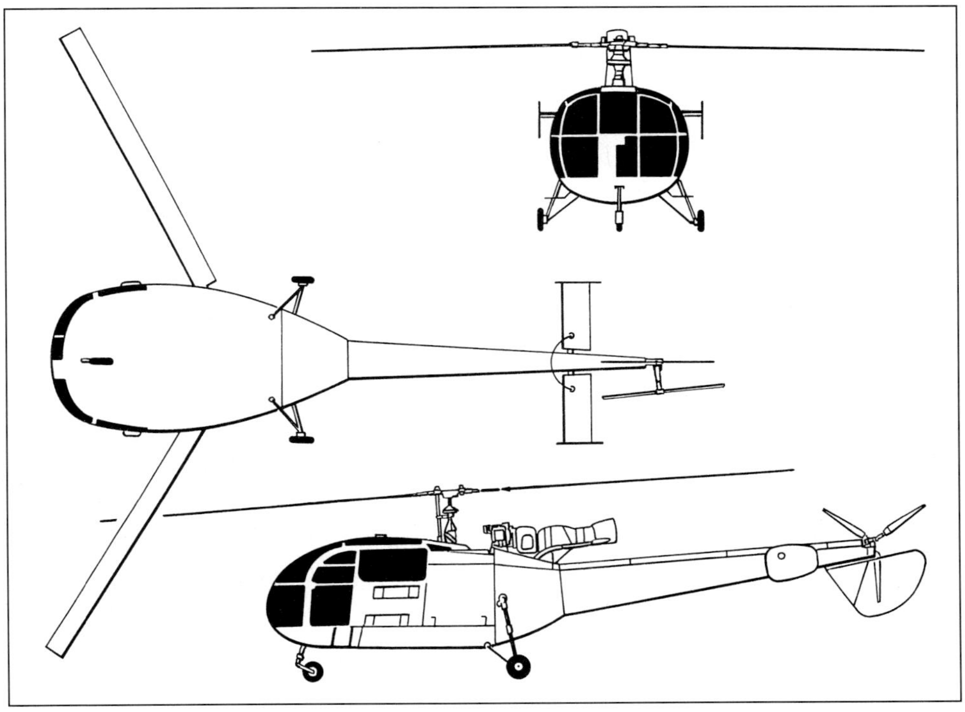

Aerospatiale Alouette III *(Jane's/Mike Keep)*

1994

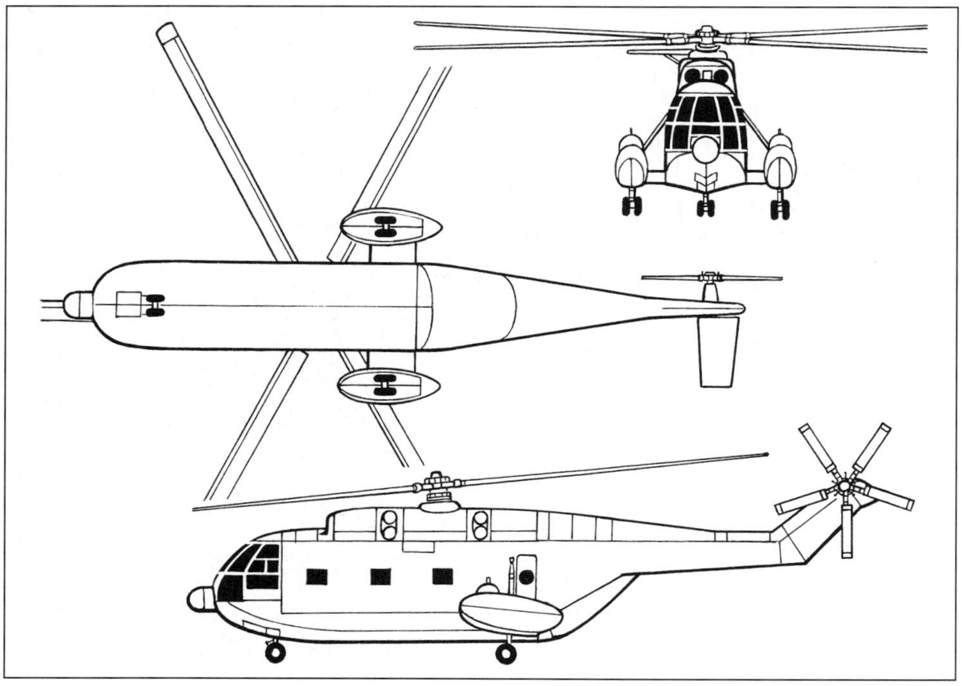

Aerospatiale SA 321 Super Frelon *(Jane's/Mike Keep)*

1994

armed forces of the following countries: China, France (16), Iraq (10) and Libya (12).

DESIGN FEATURES: Six-blade main rotor and five-blade anti-torque tail rotor. Main rotor head consists basically of two six-armed star-plates carrying the drag and flapping hinges for each blade. Each main blade is 8.60 m (28 ft 2½ in) long, with constant-chord and NACA 00012 section. Rearward folding of all six main rotor blades of SA 321G is accomplished automatically by hydraulic jacks, simultaneously with automatic folding of the tail rotor pylon. The rotor can be stopped within 40 seconds by a boosted disc-type rotor brake fitted to this shaft. Main rotor 207 and 212 rpm. Tail rotor 990 rpm. The driveshaft from the rear engine is geared directly to the shaft from the port forward engine. The two forward engines have a common reduction gear from which an output shaft drives the main rotor shaft through helical gearing. There are two reduction gear stages on the main rotor shaft. The tail rotor shaft is driven by gearing from the shaft linking the rear and port forward engines and incorporates two-stage reduction.

FLYING CONTROLS: The root of each blade carries a fitting for pitch control and each blade has an individual hydraulic damper to govern movement in the drag plane.

STRUCTURE: Main rotor blades of all-metal construction, with D-section main spar forming leading-edge. Tail rotor of similar construction to main rotor, with blades 1.60 m (5 ft 3 in) long. Boat-hull fuselage of conventional metal semi-monocoque construction, with watertight compartments inside planing bottom. On the SA 321G, there is a small stabilising float attached to the rear landing gear support structure on each side. The tail section of the SA 321G folds for stowage. Small fixed stabilisers on starboard side of tail rotor pylon on all versions. The SA 321F does not have stabilising floats, but large external fairings on each side of the centre-fuselage serve a similar purpose and also act as baggage containers.

LANDING GEAR: Non-retractable tricycle type, by Messier-Hispano-Bugatti. Twin wheels on each unit. Oleo-pneumatic shock-absorbers can be shortened on the SA 321G to reduce height of aircraft for stowage. Magnesium alloy wheels, all of same size. Tyre pressure 6.9 bars (100 lb/sq in). Optionally, low pressure (3.45 bars; 50 lb/sq in) tyres may be fitted. Hydraulic disc brakes on mainwheels. Nosewheel unit is steerable and self-centring.

POWER PLANT: Three 1,170 kW (1,570 shp) Turbomeca Turmo $IIIC_6$ turboshaft engines ($IIIE_6$ engines in SA 321H); two mounted side by side forward of main rotor shaft and one aft of rotor shaft. Fuel in flexible tanks under floor of centre-fuselage, with total standard capacity of 3,975 litres (1,050 US gallons; 874 Imp gallons) in SA 321G/H and 3,900 litres (1,030 US gallons; 858 Imp gallons) in SA 321Ja. Optional auxiliary fuel tankage comprises two 500 litre (132 US gallon; 110 Imp gallon) external tanks on all models, two 500 litre (132 US gallon; 110 Imp gallon) internal tanks in the SA 321G, and three 666 litre (175.9 US gallon; 146.5 Imp gallon) tanks in the 321H/Ja.

ACCOMMODATION (military version): Crew of two on flight deck, with dual controls and advanced all-weather equipment. SA 321G carries three other flight crew, and has provision for 27 passengers. SA 321H transport accommodates 27-30 troops, 5,000 kg (11,023 lb) of internal or external cargo, or 15 stretchers and two medical attendants. Main cabin is ventilated and soundproofed. Sliding door on starboard side of front fuselage. Rear-loading ramp is actuated hydraulically and can be opened in flight.

ACCOMMODATION (SA 321F): Airliner seats for up to 37 passengers (34 if toilets are installed) in three-abreast rows with centre aisle. Alternative layouts for 8, 14 or 23 passengers, with toilets, or 11, 17 or 26 passengers without toilets, the remainder of the cabin space being blanked off by movable partitions and used for carriage of freight; with these configurations, unused seats are folded against the cabin wall. All seats and interior furnishings are designed for quick removal when helicopter is to be used for all-freight services. To cater for operations over marshland or water, the hull and lateral cargo compartments are sealed sufficiently to permit an occasional landing on water.

ACCOMMODATION (SA 321Ja): Seating for up to 27 passengers in the personnel transport role. As a cargo transport, external loads of up to 5,000 kg (11,023 lb) can be suspended from the cargo sling. Loading of internal cargo (up to 5,000 kg; 11,023 lb) is effected via the rear ramp doors, with the assistance of a Tirefor hand winch.

ARMAMENT AND OPERATIONAL EQUIPMENT: The ASW SA 321G operates normally in tactical formations of three or four

Aerospatiale SA 321G three-engine heavy-duty helicopter *(Peter Cooper)*

1995

aircraft, each helicopter carrying the full range of detection, tracking and attack equipment, including a self-contained navigation system associated with a Doppler radar, a 360° radar with transponder and display console, and dipping sonar. Four homing torpedoes can be carried in pairs on each side of the main cabin. Both the SA 321G and H can be fitted with an anti-surface vessel weapon system, consisting of two Exocet missiles and launch installation associated with an Omera-Segid Héraclès ORB 31D or ORB 32 radar for target designation. Other equipment is provided for secondary duties such as towing and minesweeping. Rescue hoist of 275 kg (606 lb) capacity standard.

DIMENSIONS, EXTERNAL:
Main rotor diameter 18.90 m (62 ft 0 in)
Tail rotor diameter 4.00 m (13 ft 1½ in)
Main rotor blade chord (each) 0.54 m (1 ft 9¼ in)
Tail rotor blade chord (each) 0.30 m (11⅘ in)
Length overall, rotors turning 23.03 m (75 ft 6⅝ in)
Length of fuselage, tail rotor turning 20.08 m (65 ft 10¾ in)
Length of fuselage 19.40 m (63 ft 7¾ in)
Length overall:
SA 321G, blades and tail folded 17.07 m (56 ft 0 in)
Width overall: SA 321F, incl baggage containers 5.04 m (16 ft 6⅜ in)
SA 321G, blades and tail folded 5.20 m (17 ft 0¾ in)
Width of fuselage 2.24 m (7 ft 4¼ in)
Height over tail rotor (normal) 6.76 m (22 ft 2¼ in)
Height overall:
SA 321G, blades and tail folded 4.94 m (16 ft 2½ in)
Wheel track 4.30 m (14 ft 1 in)
Wheelbase 6.56 m (21 ft 6¼ in)
Cabin door: Height 1.55 m (5 ft 1 in)
Width 1.20 m (3 ft 11¼ in)
Rear-loading ramp: Length 1.90 m (6 ft 2¾ in)
Width 1.90 m (6 ft 2¾ in)

DIMENSIONS, INTERNAL:
Cabin length: SA 321F 9.67 m (31 ft 9 in)
SA 321G and Ja 7.00 m (22 ft 11½ in)
Width: SA 321F 1.96 m (6 ft 5 in)
SA 321G and Ja, at floor 1.90 m (6 ft 2¾ in)
Height: SA 321F 1.80 m (5 ft 10¾ in)
SA 321G and Ja 1.83 m (6 ft 0 in)
Usable volume: SA 321G and Ja 25.3 m³ (893 cu ft)

WEIGHTS AND LOADINGS:
Weight empty, standard aircraft:
SA 321G 6,863 kg (15,130 lb)
SA 321H 6,702 kg (14,775 lb)
SA 321Ja 6,868 kg (15,141 lb)
Max T-O weight 13,000 kg (28,660 lb)

PERFORMANCE (at max T-O weight):
Never-exceed speed at S/L 148 kts (275 km/h; 171 mph)
Cruising speed at S/L 134 kts (248 km/h; 154 mph)
Cruising speed at S/L, one engine out 113 kts (210 km/h; 130 mph)
Max rate of climb at S/L 300 m (985 ft)/min
Rate of climb at S/L, one engine out 146 m (479 ft)/min
Service ceiling 3,100 m (10,170 ft)
Service ceiling, one engine out 1,200 m (3,940 ft)
Hovering ceiling IGE 1,950 m (6,400 ft)
Normal range at S/L 440 n miles (815 km; 506 miles)
Normal range at S/L, one engine out 496 n miles (920 km; 572 miles)
Range at S/L with 3,500 kg (7,716 lb) payload 549 n miles (1,020 km; 633 miles)
Endurance in ASW role 4 h

VERIFIED

Aerospatiale SA 321G Super Frelon of the French Navy *(R. A. Cooper)*

1994

Aerospatiale SA 321 Super Frelon. Civilian version

1994

Aerospatiale AS 330 Puma HC.1 of the Royal Air Force *(Paul Jackson)*

1994

AEROSPATIALE SA 330 PUMA

TYPE: Medium-sized transport helicopter.

PROGRAMME: The twin-engined SA 330 Puma was developed initially to meet a French Army requirement for a medium-sized *hélicoptère de manoeuvre*, capable of operating by day and night in all weathers and climates. In 1967, the SA 330 was selected for the RAF Tactical Transport Programme, and was included in the joint production agreement between Aerospatiale and Westland in the UK.

The first of two SA 330 prototypes flew on 15 April 1965, and the last of six preproduction models on 30 July 1968, followed in September by the first production aircraft.

The final French production versions were as follows:

VERSIONS: **SA 330J/L:** Civil (J) and military (L) versions introduced in 1976 with main rotor blades of composite materials. Increased maximum T-O weight.

On 25 April 1978, the SA 330J became the first helicopter outside the Soviet Union to be certificated for all-weather operations, including flight in icing conditions. Equipment for this comprises thermal de-icing of the main rotor blades; thermal anti-icing of the tail rotor blades; special lengthened air intakes to ensure normal air supply into engines regardless of ambient conditions, including protection against sand and sea spray; and installation of weather radar.

A total of 697 SA 330 Pumas had been sold by 1987 when Aerospatiale ceased production to concentrate on the AS 320 Super Puma. Sole source.

UPGRADES: **Atlas Aviation:** AS 330 Puma conversions. See separate entry.

IAR: IAR-330 Puma 2000. See separate entry.

Westland/Racal: RAF avionics upgrade. See separate entry.

OPERATORS: Versions of the SA 330 are in service with the armed forces of the following countries: Algeria (5), Brazil (6), Cameroon (2), Chile (10), Ecuador (1), Ethiopia (1), France (30), Gabon (4), Iraq (22), Ivory Coast (3), Kenya (3), Kuwait (6), Lebanon (9), Malawi (1), Mexico (3), Nepal (2), Nigeria (2), Pakistan (1), Portugal (10), Senegambia (2), South Africa (80), Spain (5), Sudan (12), Togo (1), United Arab Emirates (11), UK (41) and Zaïre (9).

DESIGN FEATURES: Four-blade main rotor, with a fully articulated hub and integral rotor brake. The blade cuffs, equipped with horns, are connected by link-rods to the swashplate, which is actuated by three hydraulic twin-cylinder servo-control units. Attachment of each blade to its sleeve by two quick-disconnect pins enables blades to be folded back quickly by manual methods. The five-blade tail rotor has flapping hinges only, and is located on the starboard side of the tailboom. Mechanical shaft and gear drive. Main gearbox on top of cabin behind engines, has two separate inputs from the engines and five reduction stages. The first stage drives, from each engine, an intermediate shaft directly driving the alternator and the ventilation fan, and indirectly driving the two hydraulic pumps. At the second stage the action of the two units becomes synchronised on a single main driveshaft by means of free-wheeling spur gears. If one or both engines are stopped, this enables the drive gears to be rotated by the remaining turbine or the autorotating rotor, thus maintaining drive to the ancillary systems when the engines are stopped. Drive

Aerospatiale AS 330 Puma with flotation equipment

1994

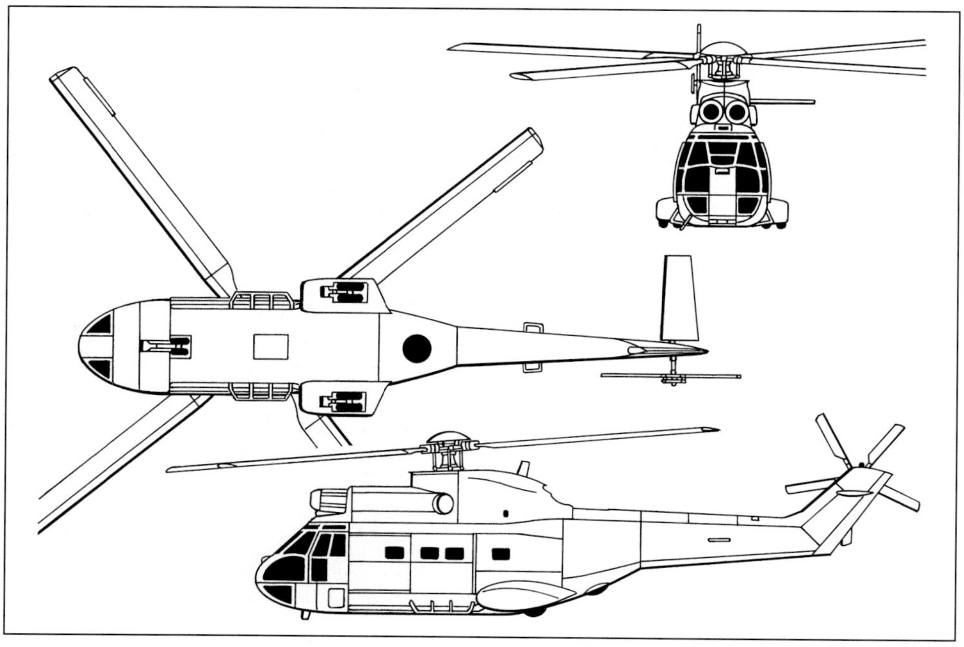

Aerospatiale AS 330 Puma *(Jane's/Mike Keep)*

1994

Aerospatiale SA 330 (RAF HC. Mk 1) Puma medium-sized helicopter

1995

to the tail rotor is via shafting and an intermediate angle gearbox, terminating at a right-angle tail rotor gearbox. Turbine output 23,000 rpm, main rotor shaft 265 rpm. Tail rotor shaft 1,278 rpm. The hydraulically controlled rotor brake, installed on the main gearbox, permits stopping of the rotor 15 seconds after engine shutdown.

STRUCTURE: Each of the moulded blades is made up of a glass-fibre roving spar, a composite glassfibre and carbonfibre fabric skin, with Moltoprene/honeycomb filler. The leading-edge is covered with a stainless steel protective section. The fuselage is a conventional all-metal semi-monocoque structure. Local use of titanium alloy under engine installation, which is outside the main fuselage shell. Monocoque tailboom supports the tail rotor on the starboard side and a horizontal stabiliser on the port side.

SYSTEMS: Optional blade de-icing system, with heating mat protected by titanium shielding on leading-edge of each main and tail rotor blade.

LANDING GEAR: Messier-Hispano-Bugatti semi-retractable tricycle type, with twin wheels on each unit. Main units retract upward hydraulically into fairings on sides of fuselage; self-centring nose unit retracts rearward. When landing gear is down, the nosewheel jack is extended and the mainwheel jacks are telescoped. Dual-chamber oleo-pneumatic shock-absorbers. All tyres same size (7.00-6), of Dunlop or Kléber-Colombes tubeless type, pressure 6.0 bars (85 lb/sq in) on all units. Hydraulic differential disc brakes, controlled by foot pedals. Lever-operated parking brake. Emergency pop-out flotation units can be mounted on rear landing gear fairings and forward fuselage.

POWER PLANT: Two Turbomeca Turmo IVC turboshaft engines, each with maximum rating of 1,175 kW (1,575 shp) and fitted with intake anti-icing. Engines are mounted side by side above cabin forward of the main rotor assembly and separated by a firewall. They are coupled to the main rotor transmission box, with shaft drive to the tail rotor, and form a completely independent system from the fuel tanks up to the main gearbox inputs. Fuel in four flexible tanks and one auxiliary tank beneath cargo compartment floor, with total capacity of 1,544 litres (407.7 US gallons; 339.5 Imp gallons). Provision for additional 1,900 litres (502 US gallons; 418 Imp gallons) in four auxiliary ferry tanks installed in cabin. External auxiliary tanks (two each 350 litres; 92.3 US gallons; 77 Imp gallons capacity) are available. For long-range missions (mainly offshore) one or two special internal tanks each 215 litres (56.7 US gallons; 47.25 Imp gallons) can be fitted in the cabin. Each engine is supplied normally by a pair of interconnected primary tanks, the lower halves of which have self-sealing walls for protection against small calibre projectiles. Refuelling point on starboard side of main cabin. Oil capacity 22 litres (5.76 US gallons; 4.8 Imp gallons) for engines, 25.5 litres (6.73 US gallons; 5.6 Imp gallons) for transmission.

ACCOMMODATION: Crew of one or two side by side on anti-crash seats on flight deck, with jump-seat for third crew member if required. Door on each side of flight deck on later versions. Internal doorway connects flight deck with folding seat in doorway for an extra crew member or cargo supervisor. Dual controls standard. Accommodation in main cabin for 16 individually equipped troops, six stretchers and six seated patients, or equivalent freight. The number of troops can be increased to 20 in the high-density version. Strengthened floor for cargo-carrying, with lashing points. Jettisonable sliding door on each side of main cabin; or port-side door with built-in steps and starboard-side door in VIP or airline configurations. Removable panel on underside of fuselage, at rear of main cabin, permits loads to be accommodated and also serves as an exit on SA 330L version. Removable door with integral steps for access to baggage racks on SA 330J version. A hatch in the floor below the centreline of the main rotor is provided for carrying loads of up to 3,200 kg (7,055 lb) on an internally mounted cargo sling. A fixed or retractable rescue hoist (capacity 275 kg; 606 lb) can be mounted externally on the starboard side of the fuselage. The cabin can be equipped in 8/9/12-seat VIP, 17-seat commuter or 20-seat high-density layouts, with baggage compartment and/or toilet facilities in the rear of cabin. Cabin and flight deck are heated, ventilated and soundproofed. Demisting, de-icing, washers and wipers for pilot's windscreens.

SYSTEMS: Two independent hydraulic systems, each 172 bars (2,500 lb/sq in), supplied by self-regulating pumps driven by the main gearbox. Each system supplies one set of servo unit chambers, the left-hand system supplying in addition, the autopilot, landing gear, rotor brake and wheel brakes. Freewheels in main gearbox ensure that both systems remain in operation, for supplying the servo controls, if the engines are stopped in flight. Other hydraulically actuated systems can be operated on the ground from the main gearbox, or by external power through the ground power receptacle. There is also an independent auxiliary system, fed through a handpump, which can be used in an emergency to lower the landing gear and pressurise the accumulator for the parking brake on the ground. Three-phase 200 V AC electrical power supplied by two 15 kVA 400 Hz alternators driven by the port side intermediate shaft from the main gearbox and available on the ground under the same conditions as the hydraulic ancillary systems. 28.5 V 10 kW DC power provided from the AC system by two transformer-rectifiers. Main aircraft battery used for self-starting and emergency power in flight. For the latter purpose, an emergency 400 VA inverter can supply the essential navigation equipment from the battery, permitting at least 20 minutes continued flight in the event of a main power failure. De-icing of engines and engine air intakes by warm air bled from compressor. Anti-snow shield for Winter operations.

AVIONICS AND EQUIPMENT: Optional communications equipment includes VHF, UHF, tactical HF and HF/SSB radio installations and intercom system. Navigational equipment includes radio compass, radio altimeter, VLF Omega, Decca navigator and flight log, Doppler, and VOR/ILS with glide path. Autopilot, with provision for coupling to self-contained navigation and microwave landing systems. Full IFR instrumentation available optionally. The search and rescue version has nose-mounted Bendix RDR 1400 or RCA Primus 40 or 50 search radar, Doppler, and Decca self-contained navigation system, including navigation computer, polar indicator, roller-map display, hover indicator, route mileage indicator and ground speed and drift indicator.

ARMAMENT (optional): A wide range of armament can be carried, including side-firing 20 mm cannon, axial-firing 7.62 mm machine guns, missiles and rockets.

DIMENSIONS, EXTERNAL:

Main rotor diameter	15.00 m (49 ft 2½ in)
Tail rotor diameter	3.04 m (9 ft 11½ in)
Distance between rotor centres	9.20 m (30 ft 2¼ in)
Blade chord, main rotor	0.60 m (1 ft 11½ in)
Ground clearance of tail rotor	2.00 m (6 ft 6¾ in)
Length overall	18.5 m (59 ft 6½ in)
Length of fuselage	14.06 m (46 ft 1½ in)
Length, blades folded	14.80 m (46 ft 6¾ in)
Width, blades folded	3.50 m (11 ft 5¾ in)
Height overall	5.14 m (16 ft 10½ in)
Height to top of rotor head	4.38 m (14 ft 4½ in)
Width over wheel fairings	3.00 m (9 ft 10 in)
Wheel track	2.38 m (7 ft 10¾ in)
Wheelbase	4.045 m (13 ft 3 in)
Passenger cabin doors, each: Height	1.35 m (4 ft 5 in)
Width	1.35 m (4 ft 5 in)
Height to sill	1.00 m (3 ft 3¼ in)
Floor hatch, rear of cabin: Length	0.98 m (3 ft 2¾ in)
Width	0.70 m (2 ft 3½ in)

DIMENSIONS, INTERNAL:

Cabin: Length	6.05 m (19 ft 10 in)
Max width	1.80 m (5 ft 10¾ in)
Max height	1.55 m (5 ft 1 in)
Floor area	7.80 m² (84 sq ft)
Usable volume	11.40 m³ (403 cu ft)

AREAS:

Main rotor blades (each)	4.00 m² (43 sq ft)
Tail rotor blades (each)	0.28 m² (3.01 sq ft)
Main rotor disc	177.0 m² (1,905 sq ft)
Tail rotor disc	7.30 m² (78.6 sq ft)
Horizontal stabiliser	1.34 m² (14.4 sq ft)

WEIGHTS AND LOADINGS:

Weight empty, standard aircraft:	
SA 330J	3,766 kg (8,303 lb)
SA 330L	3,615 kg (7,970 lb)
Max T-O and landing weight	7,400 kg (16,315 lb)

Certification for T-O weight of 7,500 kg (16,535 lb) (for cargo-sling mission only)

PERFORMANCE (SA 330J/L. A: at 6,000 kg; 13,230 lb AUW, B: at 7,400 kg; 16,315 lb AUW):

Never-exceed speed: A	158 kts (294 km/h; 182 mph)
B	142 kts (263 km/h; 163 mph)
Max cruising speed: A	146 kts (271 km/h; 168 mph)
B	139 kts (258 km/h; 160 mph)
Max rate of climb at S/L: A	552 m (1,810 ft)/min
B	366 m (1,200 ft)/min
Service ceiling (30 m; 100 ft/min rate of climb):	
A	6,000 m (19,680 ft)
B	4,800 m (15,750 ft)
Hovering ceiling IGE: A, ISA	4,400 m (14,435 ft)
A, ISA + 20°C	3,700 m (12,135 ft)
B, ISA	2,300 m (7,545 ft)
B, ISA + 20°C	1,600 m (5,250 ft)
Hovering ceiling OGE: A, ISA	4,250 m (13,940 ft)
A, ISA + 20°C	3,600 m (11,810 ft)
B, ISA	1,700 m (5,575 ft)
B, ISA + 20°C	1,050 m (3,445 ft)
Max range at normal cruising speed, no reserves:	
A	309 n miles (572 km; 355 miles)
B	297 n miles (550 km; 341 miles)

UPDATED

AEROSPATIALE SA 341/342 GAZELLE

TYPE: Five-seat light utility helicopter.

PROGRAMME: The first prototype of this five-seat lightweight helicopter (designated SA 340) made its first flight on 7 April 1967, powered by an Astazou III engine. It was followed by a second prototype on 12 April 1968 and then by four preproduction SA 341 Gazelles.

The first production SA 341 Gazelle flew for the first time on 6 April 1971, with a larger cabin than its predecessors, enlarged tail unit, additional door on the starboard side at rear (optional on production aircraft) and uprated Astazou IIIA engine.

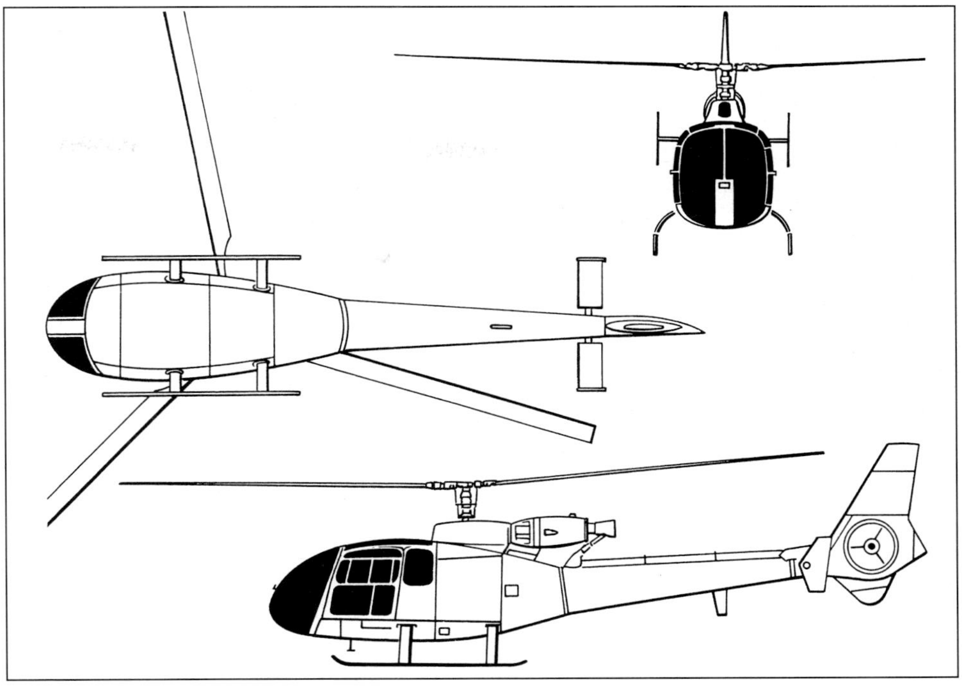

Aerospatiale SA 342 Gazelle *(Jane's/Mike Keep)* ***1994***

Under an Anglo-French agreement signed in 1967, the SA 341 Gazelle was produced jointly with Westland Helicopters Ltd, and was also built under licence in Yugoslavia. Over 800 have been sold to 130 operators in 32 countries.

VERSIONS: **SA 341B:** British Army version, with Astazou IIIN engine designated Gazelle AH Mk 1.

SA 341C: British Navy version. Designated Gazelle HT. Mk 2.

SA 341D: Royal Air Force training version. Designated Gazelle HT. Mk 3. Operated by No. 2 FTS, based at Shawbury.

SA 341E: Royal Air Force communications version. Designated Gazelle HCC. Mk 4.

SA 341F: Original French Army version, with Astazou IIIC engine; 166 procured.

SA 341G: Civil version, with Astazou IIIA engine. Certificated by SGAC on 7 June 1972 and by the FAA on 18 September 1972. In January 1975, it was announced that the SA 341G had become the first helicopter in the world authorised to be flown by a single pilot under IFR Cat. I conditions. It has since been certificated for IFR Cat. II operation, with ceiling of 30 m (100 ft) and 365 m (1,200 ft) forward visibility. Equipment fitted to the aircraft which qualified for this FAA certification comprised a Sperry flight director coupled to SFENA (now Sextant Avionique) servo-dampers. A version known as the **Stretched Gazelle**, has the rear portion of the cabin modified to provide an additional 20 cm (8 in) of leg room for the rear seat passengers.

SA 341H: Military export version, with Astazou IIIB engine.

Three Class E1c records were set up by the SA 341-01 at Istres on 13 and 14 May 1971. These were: 167.28 knots (310.00 km/h; 192.62 mph) in a straight line over a 3 km course; 168.36 knots (312.00 km/h; 193.87 mph) in a straight line over a 15/25 km course; and 159.72 knots (296.00 km/h; 183 mph) over a 100 km closed circuit.

SA 342J: Similar to SA 342L, for commercial operators. Higher maximum T-O weight. Improved fenestron tail rotor. Certificated by DGAC 27 April 1976. Deliveries began 1977.

SA 342K: Military version first flown 11 May 1973 and supplied to Kuwait. 650 kW (870 shp) Astazou VIXH turboshaft engine, with momentum-separation shrouds over intakes.

SA 342L$_1$: Military counterpart of SA 342J, with improved fenestron tail rotor.

SA 342M: Version for ALAT (French Army Light Aviation Corps). Differs from SA 342L$_1$ in having an ALAT instrument panel. Optional equipment specified as standard by ALAT includes SFIM PA 85G autopilot, Sextant Avionique Nadir self-contained navigation system, Decca 80 Doppler and night-flying equipment. An exhaust deflector remains optional. Maximum T-O weight initially 1,900 kg (4,188 lb). French orders totalled 188 SA 342Ms each armed with four HOT missiles and gyrostabilised sight for anti-tank duties.

UPGRADES: **Eurocopter:** SA 342M ATAM/Viviane. See Eurocopter SA 342 Gazelle weapons upgrades.

RN Air Yard Fleetlands: Commenced redelivery in June 1995 of initial batch of Army Westland AH Mk 1 Gazelle Helicopters refitted with GEC-Marconi Defence Systems Rewarder radar warning receiver (RWR). Rewarder is the service name for the AWARE-3 RWR developed by Ferranti before its acquisition by GEC.

OPERATORS: Versions of the SA 341/342 are in service with the armed forces of the following countries: Angola (10), Burundi (2), Cameroon (4), China (8), Croatia, Cyprus (3), Ecuador (13), Egypt (80), France (340), Gabon (5), Guinea (1), Iraq (50), Ireland (2), Kenya (1), Kuwait (20), Lebanon (7), Libya (40), Malawi (1), Morocco (30), Qatar (16),

Westland built SA 342 Gazelle with Mi-8 Hip shadowing along former West Berlin border *(Roy Michell)* ***1994***

Senegambia (1), Syria (55), Trinidad and Tobago (2), Tunisia (6), United Arab Emirates (11) and UK (30).

The following details apply to all SA 341 versions:

DESIGN FEATURES: Three-blade semi-articulated main rotor and 13-blade shrouded-fan anti-torque tail rotor, known as 'fenestron' or 'fan-in-fin'. Rotor head and rotor mast form a single unit. The main rotor blades are of NACA 0012 section, attached to NAT hub by flapping hinges. There are no drag hinges. Main rotor blades can be folded manually for stowage. Main reduction gearbox forward of engine, which is mounted above rear part of cabin. Intermediate gearbox beneath engine, rear gearbox supporting the tail rotor. Main rotor/engine rpm ratio 378.3:6,179. Tail rotor/engine rpm ratio 5,774:6,179. Small horizontal stabiliser on tailboom, ahead of tail rotor fin.

Aerospatiale SA 342 Gazelle of the French ALAT forces

1995

Aerospatiale SA 342M Gazelle armed with a Giat M621 20 mm cannon on starboard mount and 2.75 in folding fin rocket pod on port side

1995

FLYING CONTROLS: Rotor brake standard.

STRUCTURE: Each blade has a single leading-edge spar of plastics material reinforced with glassfibre, a laminated glass-fabric skin and honeycomb filler. Tail rotor blades are of die-forged light alloy, with articulation for pitch change only. Cockpit structure is based on a welded light alloy frame which carries the windows and doors. This is mounted on a conventional semi-monocoque lower structure consisting of two longitudinal box sections connected by frames and bulkheads. Central section, which encloses the baggage hold and fuel tank and supports the main reduction gearbox, is constructed of light alloy honeycomb sandwich panels. Rear section, which supports the engine and tailboom is of similar construction. Honeycomb sandwich panels are also used for the cabin floors and transmission platform. Tailboom is of conventional sheet metal construction, as are the horizontal tail surfaces and the tail fin.

LANDING GEAR: Steel tube skid type. Wheel can be fitted at rear end of each skid for ground handling. Provision for alternative float or ski landing gear.

POWER PLANT: One Turbomeca Astazou IIIA turboshaft, installed above fuselage aft of cabin and delivering 440 kW (590 shp) for take-off (maximum continuous rating also 440 kW; 590 shp). Main fuel tank in fuselage, usable capacity 445 litres (117.7 US gallons; 98 Imp gallons). Provision for 90 litre (23.8 US gallon; 19.8 Imp gallon) auxiliary tank beneath baggage compartment and/or 200 litre (52 US gallon; 44 Imp gallon) ferry tank inside rear cabin. Total possible usable fuel capacity 735 litres (193.4 US gallons; 161 Imp gallons). Refuelling point on starboard side of cabin. Oil capacity 13 litres (3.36 US gallons; 2.8 Imp gallons) for engine, 3.5 litres (0.92 US gallons; 0.77 Imp gallons) for gearbox.

ACCOMMODATION: Crew of one or two on side by side seats in front of cabin, with bench seat to the rear for a further three persons. The bench seat can be folded into floor wells to leave a completely flat cargo floor. Access to baggage compartment via rear cabin bulkhead, or via optional door on starboard side. Cargo tie-down points in cabin floor. Forward-opening car-type door on each side of cabin, immediately behind which are rearward-opening auxiliary cargo loading doors. Baggage compartment at rear of cabin. Ventilation standard. Dual controls optional.

SYSTEMS: Hydraulic system, pressure 40 bars (570 lb/sq in), serves three pitch change jacks for main rotor head and one for tail rotor. 28 V DC electrical system supplied by 4 kW engine-driven generator and 40 Ah battery. Optional 26 V AC system, supplied by 0.5 kVA alternator at 115/200 V 400 Hz.

ELECTRONICS AND EQUIPMENT: Optional communications equipment includes UHF, VHF, HF, intercom systems and homing aids. Optional navigation equipment includes radio compass, radio altimeter and VOR. Blind-flying instrumentation standard on 341B and F, optional on other versions. A variety of operational equipment can be fitted, according to role, including a 700 kg (1,540 lb) cargo sling, 135 kg (300 lb) rescue hoist, one or two stretchers (internally), or photographic and survey equipment.

ARMAMENT: Military loads can include two pods of Matra or Brandt 2.75 in or 68 mm rockets, four AS 11 or two AS 12 wire-guided missiles with APX-Bézu 334 gyrostabilised sight, four or six HOT wire-guided missiles with APX 397 gyrostabilised sight, two forward-firing 7.62 mm machine guns, reconnaissance flares or smoke markers.

DIMENSIONS, EXTERNAL:	
Main rotor diameter	10.50 m (34 ft 5½ in)
Tail rotor diameter	0.695 m (2 ft 3⅜ in)
Main rotor blade chord (constant)	0.30 m (11¾ in)
Distance between rotor centres	5.85 m (19 ft 2¼ in)
Length overall	11.97 m (39 ft 3⁵⁄₁₆ in)
Length of fuselage	9.53 m (31 ft 3³⁄₁₆ in)
Width, rotors folded	2.015 m (6 ft 7⁵⁄₁₆ in)
Height to top of rotor head	2.72 m (8 ft 11⅛ in)
Height overall	3.15 m (10 ft 2⅝ in)
Skid track	2.015 m (6 ft 7⁵⁄₁₆ in)
Main cabin doors, each: Height	1.05 m (3 ft 4⁵⁄₁₆ in)
Width	1.00 m (3 ft 3¼ in)
Height to sill	0.63 m (2 ft 0¾ in)
Auxiliary cabin doors, each: Height	1.05 m (3 ft 4⁵⁄₁₆ in)
Width	0.48 m (1 ft 6¾ in)
Height to sill	0.63 m (2 ft 0¾ in)
DIMENSIONS, INTERNAL:	
Cabin: Length	2.20 m (7 ft 2⁵⁄₁₆ in)
Max width	1.32 m (4 ft 4 in)
Max height	1.21 m (3 ft 11⅝ in)
Floor area	1.50 m² (16.1 sq ft)
Volume	1.80 m³ (63.7 cu ft)
Baggage hold volume	0.45 m³ (15.9 cu ft)
AREAS:	
Main rotor blades, each	1.57 m² (16.9 sq ft)
Tail rotor blades, each	0.007 m² (0.075 sq ft)
Main rotor disc	86.5 m² (931 sq ft)
Tail rotor disc	0.37 m² (3.98 sq ft)
Fin	0.45 m² (4.84 sq ft)
Tailplane	1.80 m² (19.4 sq ft)

WEIGHTS AND LOADINGS:
- Weight empty: 341G — 917 kg (2,022 lb)
- 341H — 908 kg (2,002 lb)
- Max T-O and landing weight: 341G/H — 1,800 kg (3,970 lb)
- Max disc loading: 341G/H — 19.5 kg/m^2 (4 lb/sq ft)

PERFORMANCE (at max T-O weight):
- Never-exceed speed at S/L: SA 341 — 167 kts (310 km/h; 193 mph)
- Max cruising speed at S/L: SA 341 — 142 kts (264 km/h; 164 mph)
- Econ cruising speed at S/L: SA 341 — 126 kts (233 km/h; 144 mph)
- Max rate of climb at S/L: SA 341 — 540 m (1,770 ft)/min
- Service ceiling: SA 341 — 5,000 m (16,400 ft)
- Hovering ceiling IGE: SA 341 — 2,850 m (9,350 ft)
- Hovering ceiling OGE: SA 341 — 2,000 m (6,560 ft)
- Range at S/L with max fuel: SA 341 — 361 n miles (670 km; 416 miles)
- Range with pilot and 500 kg (1,102 lb) payload: SA 341 — 193.5 n miles (360 km; 223 miles)

VERIFIED

AEROSPATIALE SA 342 GAZELLE WEAPONS UPGRADES

TYPE: Combat helicopter upgrade.

PROGRAMME: First flight 7 April 1967, powered by Astazou III fixed-shaft turbine.

VERSIONS: **SA 342M ATAM:** 30 French Army machines each retrofitted with four Matra air-to-air Mistrals (ATAM) and T2000 sight.

SA 342M Viviane: Up to 70 French Army SA 342 normally fitted with daytime HOT sight being retrofitted with Viviane stabilised direct view/IR/laser roof-mounted sight to allow night firing of HOT missiles.

CUSTOMERS: French Army ALAT operating 170 SA 341L1 and 188 SA 342M in early 1992; total (all variants and customers) 1,254 delivered and 1,116 operating 30 June 1991, excluding those produced by Westland in UK and Soko in former Yugoslavia; total 628 SA 341 delivered and 542 in service on 30 June 1991; total 626 SA 342 delivered and 574 in service on 30 June 1991.

POWER PLANT: One 640 kW (858 shp) Turbomeca Astazou fixed-shaft turboshaft with optional momentum separator on intake and upturned exhaust pipe; engine and rotor directly coupled and turn at constant speed, power controlled by fuel input. Usable fuel 545 litres (144 US gallons; 120 Imp gallons).

ACCOMMODATION: Pilot with gunsight on right; weapon aimer/observer on left; three seats in rear if equipment allows.

AVIONICS: ALAT instrument panel; VHF/UHF/FM communications radio and homing; SFIM PA 85G autopilot; Sextant Avionique Nadir/Decca Doppler 80 self-contained navigation system; night-flying equipment.

DIMENSIONS, EXTERNAL:
- Main rotor diameter — 10.50 m (34 ft 5½ in)
- Length overall, rotors turning — 11.97 m (39 ft 3$\frac{5}{16}$ in)
- Height overall — 3.19 m (10 ft 5½ in)

WEIGHTS AND LOADINGS:
- Max T-O weight: 342M — 2,100 kg (4,630 lb)
- Max disc loading — 21.94 kg/m^2 (4.49 lb/sq ft)

PERFORMANCE:
- Max cruising speed at S/L — 140 kts (260 km/h; 161 mph)

VERIFIED

Eurocopter BO 105 LS especially designed for 'hot-and-high' operations

1994

EUROCOPTER BO 105

TYPE: Five/six-seat light helicopter.

PROGRAMME: The first prototype of the BO 105 with hingeless titanium rotor hub and composite rotor blades first flew on 16 February 1967.

Production of 100 BO 105 M (VBH) and 212 BO 105 P (PAH-1) military versions for the Federal German Army ended in 1984.

BO 105s have been used by MBB to flight test a TST (formerly AEG) mast-mounted radar system, and a Lucas Aerospace bolt-on undernose turret to permit off-axis firing of Stinger air-to-air missiles, controlled by a GEC Ferranti helmet pointing system. This latter helicopter has joined one German Army BO 105 M and two Army BO 105 Ps (PAH-1s) being flown at test and operational training centres to evaluate the weapon and sighting systems needed for the BSH-1 escort helicopter. A BO 105 has been flown to test the OPST-1 DLR-MBB-Liebherr full authority, intelligent triplex, optically signalled yaw control that relieves the pilot of the task of yaw/collective axis co-ordination.

By January 1991 more than 1,300 BO 105s of all models had been delivered to 37 countries in five continents. All BO 105 helicopters for the world market are manufactured and assembled at the company's Donauwörth facility, with the exceptions of those for the Spanish market that are assembled in Spain by CASA, those for the Indonesian market which are manufactured and assembled in that country by IPTN, and BO 105 LS helicopters produced at MBB's Fort Erie facility in Ontario, Canada.

VERSIONS: **BO 105 CB:** Standard production version since 1975, with two Allison 250-C20B engines, operable in air temperatures ranging from –45 to +54°C. LBA certification received in November 1976. Specially equipped versions were delivered to the Mexican Navy (12) and Swedish Army (20) by September 1988. In Spain, CASA assembled 57 of an initial 60 for the Spanish Army for armed reconnaissance (18), observation (14) and anti-tank missions (28), and has continued assembly.

BO 105 CBS: Version with increased rear seat leg room in a cabin extended by a 0.25 m (10 in) plug. Available in five-seat executive or six-seat high density configurations. Identified by small additional window aft of rear door on each side. Marketed in the USA by MBB Helicopter Corporation under the name **Twin Jet II**. Certificated in early 1983 by FAA for IFR operation in accordance with SFAR Pt 29-4, requiring two pilots, radar, Loran C and a separate battery, but not a stability augmentation system, although SAS is available as an option. The Swedish Air Force has four BO 105 CBSs, equipped for IFR search and rescue.

BO 105 LS: Re-engined high-altitude version of CBS produced exclusively in Canada. Five preproduction models produced in Germany during 1984.

BO 105 M (VBH): Liaison and light observation helicopter for the Federal German Army, with strengthened transmission gearing, reinforced rotor components, a tail rotor with improved thrust and performance, a rupture-proof fuel system, and a landing gear able to absorb higher energy levels. Production of 100 approved by the Federal government, to replace Alouette IIs. Deliveries completed in 1984. In early 1992 there were 96 in service with the German Army.

BO 105 P (PAH-1): Anti-tank version of the BO 105 M, with same airframe improvements as BO 105 M, outriggers to carry six Euromissile HOT missiles, a stabilised sight above the co-pilot's, and a Singer AN/ASN-128 Doppler navigation system. The Federal German government gave its approval for the procurement of 212 PAH-1s for the Federal German Army. Deliveries began on 4 December 1980 and were completed in mid-1984. There were 108 in service in early 1992. First PAH-1 unit was Heeresfliegerregiment 16 at Celle. Empty equipped weight of the PAH-1, including pilot and weapons operator, is 1,913 kg (4,217 lb). Maximum T-O weight is 2,400 kg (5,291 lb). At maximum continuous power and ISA the PAH-1 has a forward rate of climb at S/L of 540 m (1,770 ft)/min and can hover OGE with T-O power at 1,580 m (5,180 ft). At maximum continuous power it has a maximum cruising speed at S/L of 119 knots (220 km/h; 137 mph) and a certificated service ceiling of 4,200 m (13,780 ft).

UPGRADES: **Eurocopter BO 105/PAH-1/VBH/BSH-1:** See separate entry.

OPERATORS: Versions of the BO 105 are in service with the armed forces of the following countries: Bahrain (3), Brunei (6), Chile (6), Colombia (4), Germany (300), Iraq (75), Kenya (1), Mexico (11), Netherlands (25), Nigeria (24), Peru (24), Philippines (10), Spain (68), Sweden (20), Trinidad and Tobago (1) and United Arab Emirates (6).

The following description applies to the BO 105 CBS except where indicated:

DESIGN FEATURES: Four-blade main rotor, comprising rigid titanium head and GFRP blades, with titanium anti-erosion strip and pendulous vibration damper on each blade. NACA 23012 lifting aerofoil with drooped leading-edge and reflexed trailing-edge. Two-blade semi-rigid tail rotor. Tail rotor gearbox on fin. Main rotor 424 rpm. Tail rotor 2,220 rpm. Main transmission utilises two bevel gear input stages with freewheeling clutches and a spur collector gear stage. Planetary reduction gear; three auxiliary drives for accessories. Main transmission rated for twin-engine input of 257 kW (345 shp) per engine, or a single-engine input of 283 kW (380 shp).

FLYING CONTROLS: Main rotor has roller bearings for pitch change. Main rotor brake standard.

STRUCTURE: Folding of two main rotor blades optional. Tail rotor blades of GFRP, with stainless steel anti-erosion

Aerospatiale SA 342 Gazelle with Viviane upgrade

1994

strip. The fuselage is a conventional light alloy semi-monocoque structure of pod and boom type. Glassfibre reinforced cowling over power plant. Titanium sheet engine deck. Horizontal stabiliser of conventional light alloy construction with small endplate fins.

LANDING GEAR: Skid type, with cross-tubes designed for energy absorption by plastic deformation in the event of a heavy landing. Inflatable emergency floats can be attached to skids.

POWER PLANT: Two 313 kW (420 shp) Allison 250-C20B turboshafts, each with a maximum continuous rating of 298 kW (400 shp). Bladder fuel tanks under cabin floor, capacity 580 litres (153.2 US gallons; 127.5 Imp gallons), of which 570 litres (150.6 US gallons; 125.3 Imp gallons) are usable. Fuelling point on port side of cabin. Auxiliary tanks in freight compartment available optionally. Oil capacity: engine 12 litres (3.2 US gallons; 2.6 Imp gallons), gearbox 11.6 litres (3.06 US gallons; 2.55 Imp gallons).

ACCOMMODATION: Pilot and co-pilot or passenger on individual longitudinally adjustable front seats with safety belts and automatic locking shoulder harnesses. Optional dual controls. Bench seat at rear for three persons, removable for cargo and stretcher carrying. A full EMS version is available. Both cabin and cargo compartment have panelling, sound insulation and floor covering. Entire rear fuselage aft of seats and under power plant available as freight and baggage space, with access through two clamshell doors at rear. Two standard stretchers can be accommodated side by side in ambulance role. One forward-opening hinged and jettisonable door and one sliding door on each side of cabin. Ram air and electrical ventilation system. Heating system optional.

SYSTEMS: Tandem fully redundant hydraulic system, pressure 103.5 bars (1,500 lb/sq in), for powered main rotor controls. System flow rate 6.2 litres (1.64 US gallons; 1.36 Imp gallons)/min. Bootstrap/oil reservoir, pressurised at 1.7 bars (25 lb/sq in). Electrical system powered by two 150 A 28 V DC starter/generators and a 24 V 25 Ah Ni/Cd battery; external power socket.

AVIONICS: Wide variety of avionics available including weather radar, Doppler navigation, SAS and autopilot.

EQUIPMENT: Standard equipment includes basic flight instruments, engine instruments, heated pitot, tiedown rings in cargo compartment, cabin and cargo compartment dome lights, position lights and collision warning lights. Options include dual controls, heating system, windscreen wiper, rescue winch, landing light, searchlight, externally mounted loudspeaker, fuel dump valve, external load hook, settling protectors, snow skids, and manual main rotor blade folding.

Eurocopter BO 105 CB twin-engined helicopter operated by the Bavarian Police Helicopter Squadron *(Eurocopter)* *1995*

Eurocopter BO 105 CB of the Federal German Police *1994*

Eurocopter BO 105 PAH-1A1 upgraded PAH-1 helicopter *1994*

DIMENSIONS, EXTERNAL:

Main rotor diameter	9.84 m (32 ft 3½ in)
Tail rotor diameter	1.90 m (6 ft 2¾ in)
Main rotor blade chord	0.27 m (10⅝ in)
Tail rotor blade chord	0.18 m (7 in)
Distance between rotor centres	5.95 m (19 ft 6¼ in)
Length: incl main and tail rotors	11.86 m (38 ft 11 in)
excl rotors: CB	8.56 m (28 ft 1 in)
CBS	8.81 m (28 ft 11 in)
fuselage pod: CB	4.30 m (14 ft 1 in)
CBS	4.55 m (14 ft 11 in)
Height to top of main rotor head	3.02 m (9 ft 11 in)
Width over skids: unladen	2.53 m (8 ft 3½ in)
laden	2.58 m (8 ft 5½ in)
Rear-loading doors: Height	0.64 m (2 ft 1 in)
Width	1.40 m (4 ft 7 in)

DIMENSIONS, INTERNAL:

Cabin, incl cargo compartment:	
Max width	1.40 m (4 ft 7 in)
Max height	1.25 m (4 ft 1 in)
Volume	4.80 m^3 (169 cu ft)
Cargo compartment: Length	1.85 m (6 ft 0¾ in)
Max width	1.20 m (3 ft 11¼ in)
Max height	0.57 m (1 ft 10½ in)
Floor area	2.25 m^2 (24.2 sq ft)
Volume	1.30 m^3 (45.9 cu ft)

AREAS:

Main rotor disc	76.05 m^2 (818.6 sq ft)
Tail rotor disc	2.835 m^2 (30.5 sq ft)

WEIGHTS AND LOADINGS:

Weight empty, basic: CB	1,277 kg (2,815 lb)
CBS	1,301 kg (2,868 lb)
Standard fuel (usable)	456 kg (1,005 lb)
Max fuel, incl auxiliary tanks	776 kg (1,710 lb)
Max T-O weight	2,500 kg (5,511 lb)
Max disc loading	32.9 kg/m^2 (6.74 lb/sq ft)

PERFORMANCE (at max T-O weight):

Never-exceed speed (VNE) at S/L	131 kts (242 km/h; 150 mph)
Max cruising speed at S/L	131 kts (242 km/h; 150 mph)
Best range speed at S/L	110 kts (204 km/h; 127 mph)
Max rate of climb at S/L, max continuous power	444 m (1,457 ft)/min
Vertical rate of climb at S/L, T-O power	90 m (295 ft)/min
Max operating altitude	3,050 m (10,000 ft)
Hovering ceiling T-O power: IGE	1,525 m (5,000 ft)
OGE	457 m (1,500 ft)
Range with standard fuel and max payload, no reserves:	
at S/L	300 n miles (555 km; 345 miles)
at 1,525 m (5,000 ft)	321 n miles (596 km; 370 miles)
Ferry range with auxiliary tanks, no reserves:	
at S/L	519 n miles (961 km; 597 miles)
at 1,525 m (5,000 ft)	550 n miles (1,020 km; 634 miles)
Endurance with standard fuel and max payload, no reserves: at S/L	3 h 24 min

VERIFIED

EUROCOPTER BO 105/PAH-1/VBH/BSH-1

TYPE: Combat helicopter upgrade.

PROGRAMME: Production of 100 BO 105 M/VBH (Verbindungs und Beobachtungs Hubschrauber) scouts and 212 BO 105 P/PAH-1 (Panzer Abwehr Hubschrauber) anti-tank helicopters completed 1984; fitting of new rotor blades, improved oil cooling and intakes to 209 remaining PAH-1s approved September 1990; of these, 155 to have HOT 2 missile system with lightweight launchers and digital avionics and 54 were planned to become BSH (Begleitschutz Hubschrauber).

VERSIONS: **PAH-1 Phase I (PAH-1A1):** Current retrofit from original PAH-1 for improved day fighting capability; first in service Summer 1991; programme to be completed mid-1994.

BSH: Begleitschutz Hubschrauber escort helicopter; conversion of 54 from PAH-1s cancelled early 1993.

PAH-1 Phase II: Added night firing capability with roof-mounted sight cancelled in budget adjustments January 1993.

CUSTOMERS: German Army, as above; Swedish Army adopted ESCO HeliTOW system.

COSTS: PAH-1 Phase II cost reported as DM630 million in January 1992.

DESIGN FEATURES: New rotor blades and lighter HOT system increase useful load by 180 kg (397 lb).

Data substantially as for BO 105 CB except:

WEIGHTS AND LOADINGS:

Empty weight, without fuel, missiles or crew 1,688 kg (3,721 lb)

German anti-tank mission weight 2,380 kg (5,247 lb)

PERFORMANCE:

Hovering ceiling, OGE, at mission weight 2,100 m (6,890 ft)

VERIFIED

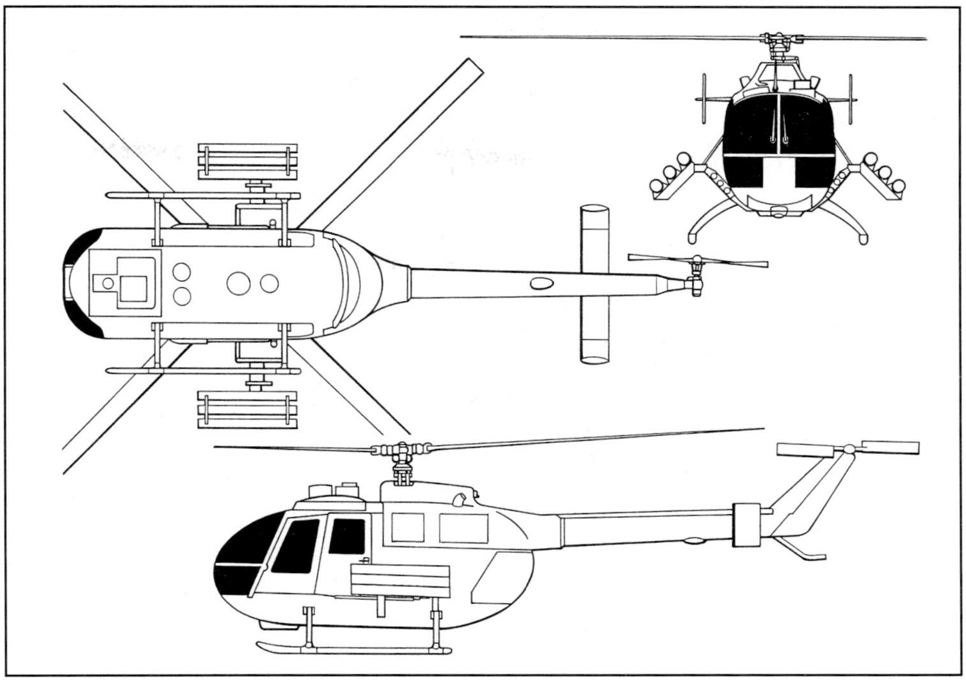

Eurocopter BO 105 PAH-1 Anti-tank helicopter of the German Army *(Jane's/Mike Keep)* *1995*

MCDONNELL DOUGLAS/BAe

PARTICIPATING COMPANIES:

McDonnell Douglas: see under USA

British Aerospace: see under UK

GENERAL MANAGER, AV-8: Patrick J. Finneran Jr

VICE-PRESIDENT AND GENERAL MANAGER, T45TS: Robert H. Soucy Jr

VERIFIED

MCDONNELL DOUGLAS/BRITISH AEROSPACE HARRIER II

US Marine Corps designations: AV-8B and TAV-8B

RAF designations: Harrier GR. Mk 5, 5A and 7, and T. Mk 10

Spanish Navy designation: VA.2 Matador II

TYPE: Single-seat V/STOL close support, battlefield interdiction night attack and reconnaissance aircraft.

PROGRAMME: Present collaborative programme began with two YAV-8B (converted AV-8A) aerodynamic prototypes (first flights 9 November 1978 and 19 February 1979); followed by four FSD aircraft (first flight 5 November 1981); first 12 pilot production AV-8Bs ordered FY82 (first flight 29 August 1983), deliveries to USMC beginning 12 January 1984; development programme for night attack version announced November 1984; first flights of RAF GR. Mk 5 development aircraft 30 April (ZD318) and 31 July 1985 (ZD319); first USMC operational AV-8B squadron (VMA-331) achieved IOC August 1985; first flight of two-seat TAV-8B (No. 162747) 21 October 1986; first flight of night attack AV-8B prototype (162966) 26 June 1987; first GR. Mk 5 for RAF (ZD324) handed over 1 July 1987; TAV-8B deliveries (to VMAT-203) began August 1987; EAV-8B deliveries to Spain 1987-88; production contract for new-build GR. Mk 7s placed April 1988; first flight of Pegasus 11-61 power plant (ZD402) 10 June 1989; first production night attack AV-8B (163853) delivered to VMA-214 on 15 September 1989; first flight of RAF GR. Mk 7 (development aircraft, converted from GR. Mk 5) 29 November 1989; 27 GR. Mk 7s ordered April 1988 (later increased to 34); first flight of production Mk 7 (ZG471) May 1990; production contract for T. Mk 10 placed February 1990; 24 FY91 AV-8Bs for USMC will be to Harrier II Plus standard (announced 4 December 1990). First flight of T. Mk 10, 7 April 1994.

VERSIONS: **AV-8B Harrier II:** US Marine Corps single-seat close support version. Night attack avionics (including FLIR bulge ahead of windscreen) from 167th AV-8B (163853) onwards (see Avionics paragraph), plus (from No. 182, 163874 and TAV-8B No. 16, 164120, in December 1990) uprated F402-RR-408 (Pegasus 11-61) engine.

AV-8B Harrier II Plus: Radar equipped night attack version; described separately.

TAV-8B Harrier II: US Marine Corps two-seat operational trainer, with longer forward fuselage and 0.43 m (1 ft 5 in) taller vertical tail than AV-8B; two cockpits in tandem; two underwing stores stations only; BAe major subcontractor for this version.

EAV-8B: Manufacturer's designation for Spanish Navy **VA.2 Matador II** single-seat export version.

Harrier GR. Mk 5: Royal Air Force single-seat battlefield air interdiction/close air support version. Two additional underwing stations, for Sidewinder missile carriage.

McDonnell Douglas/BAe GR. 7 of the RAF showing Zeus EW system on the nosecone *(BAe)* *1995*

McDonnell Douglas/BAe Harrier II GR. 7 V/STOL aircraft of the Royal Air Force *(BAe)* *1994*

Harrier GR. Mk 5A: Interim designation for 19 GR. Mk 5s pending eventual upgrade to full GR. Mk 7 standard.

Harrier GR. Mk 7: Royal Air Force single-seat night attack version, based on GR. Mk 5. A programme to develop a night attack version of the Harrier II was announced by McDonnell Douglas in November 1984, and a USMC prototype of this version flew in 1987. The new nav/attack equipment includes two FD 4512 HUDs and a HUDWAC, an upgraded digital INS unit, projected map display, high-resolution head-down TV-type raster display in each cockpit, and video recording facilities, and is compatible with the aircraft's nose-mounted FLIR sensor. 'Production' Nightbird Harrier IIs to be undertaken as a retrofit programme, requiring an enhanced (ACCS 2500) Computing Devices mission computer, and a multipurpose colour display with a separate video map generator.

Harrier T. Mk 10: Royal Air Force operational trainer for GR. Mk 7, based on TAV-8B airframe with eight underwing pylons, FLIR and night vision equipment of GR. Mk 7, but no ARBS.

UPGRADES: **BAe Military Aircraft Division:** Completed project study in February 1995 for feasibility of integrating GEC-Marconi Avionics TIALD thermal imaging and laser designator pod onto Harrier GR. 7 aircraft of the RAF.

McDonnell Douglas/BAe Harrier T. Mk 10 two-seat trainer on its maiden flight (7 April 1994) *(BAe)*

1995

McDonnell Douglas/BAe GR. Mk 7: See Versions.

McDonnell Douglas/BAe Harrier II Plus: See separate entry.

McDonnell Douglas/BAe T. Mk 10: See Versions.

McDonnell Douglas: Contract awarded by Italian Navy to fit its two Harrier TAV-8B trainer aircraft with Rolls-Royce Pegasus F402-RR-408 engines. BAe and Rolls-Royce acted as principal subcontractors. Programme completed November 1994.

McDonnell Douglas: Technology Demonstrator Harrier II Plus being used as a testbed to prove concept of wingtip missile rails for AIM-9 Sidewinder missiles.

OPERATORS: Versions of the AV-8B are in service with the armed forces of the following countries: India (27), Italy (18), Spain (19), UK (150) and USA (190).

CUSTOMERS: US Marine Corps ordered total of 280 by FY91, including four FSDs (ordered FY79), 24 TAV-8Bs and 24 Harrier II Plus; target procurement of 300 AV-8Bs and 28 TAV-8Bs unlikely to be achieved. VMAT-203 received first AV-8B (161573) 12 January 1984 and first TAV-8B (162963) March 1987; also operated by VMA-211 (re-equipped 1990, second Night Attack squadron), VMA-214 (first with Night Attack version; initial aircraft, 163853, delivered 15 September 1989), VMA-223 (1 October 1987), VMA-231 (September 1985), VMA-311 (1989), VMA-331 (commissioned 30 January 1985 as first operational unit), VMA-513 (January 1987) and VMA-542 (1986); one squadron disbands in FY93. Requirement to modify 114 earlier aircraft (built 1983-89) to Plus standard with new fuselage shells and power plant upgraded to -408 series, between FY94 and FY2000.

Royal Air Force ordered total of 110 by 1990, comprising two FSDs, 41 GR. Mk 5s, 19 GR. Mk 5As, 34 GR. Mk 7s and 14 T. Mk 10s (last mentioned ordered March 1990). First service flight of Mk 5 (ZD324) with No. 233 OCU, 30 March 1988; deliveries to No. 1 Squadron from 23 November 1988 and unit redeclared to NATO on 2 November 1989; to No. 3 Squadron from 17 March 1989 (ZD401); Mk 5A delivered mainly to storage from 21 August 1990 (ZD432); first Mk 7 delivery to A&AEE, Boscombe Down, 5 June 1990 (ZG472); to Strike/Attack Operational Evaluation Unit from 17 August 1990 (ZG473)—unit flew RAF's first NVG Harrier mission, 11 December 1990; to No. 4 Squadron from 12 September 1990 (ZG473); re-equipped No. 3 Squadron from 30 November 1990 (ZG479). Single-seat deliveries completed early 1992. Contract for BAe upgrade of 58 (less any attrition) Mks 5/5A to Mk 7 awarded 2 November 1990; first ex-Mk 5 (ZD380) redelivered to RAF 21 December 1990; first ex-Mk 5A (ZD430) on 9 April 1991.

Spanish Navy received 12 EAV-8Bs for Novena Escuadrilla (Eslla 009) between 6 October 1987 and September 1988; operational aboard *Principe de Asturias*; single TAV-8B ordered early 1992; eight new-build Harrier II Plus required; remaining 11 EAV-8Bs to be converted to Plus standard; procurement target 21, including attrition replacement.

McDonnell Douglas/BAe AV-8B Harrier II V/STOL aircraft of the USMC

1994

Italian Navy ordered a total of 18; two AV-8Bs ordered May 1989 and 16 Harrier II Plus, of which first three ordered late 1991 (to come from USMC batch); option on eight; first delivery (two TAV-8Bs, MM55032-55033, to *Giuseppe Garibaldi* 23 August 1991.

Total firm orders 405; commitments 24.

COSTS: £200 million (1990 contract) for 14 T. Mk 10 aircraft. $3,000 million (estimated) for rebuild of 114 AV-8Bs to Plus configuration.

DESIGN FEATURES: Differences compared with Harrier GR. Mk 3/AV-8A include bigger wing and longer fuselage; use of graphite epoxy (carbonfibre) composite materials for wings and parts of fuselage and tail unit; adoption of supercritical wing section; addition of LIDS (lift improvement devices: strakes to replace gun/ammunition pods when armament not carried, plus retractable fence panel forward of pods) to augment lift for vertical take-off; larger wing trailing-edge flaps and drooped ailerons; redesigned forward fuselage and cockpit; redesigned engine air intakes to provide more VTO/STO thrust and more efficient cruise; two additional wing stores stations; wing outriggers relocated at mid-span to provide better ground manoeuvring capability; leading-edge root extensions (LERX) to enhance instantaneous turn rate and air combat capability; landing gear strengthened to cater for higher operating weights and greater external stores loads. Wing span and area increased by approximately 20 and 14.5 per cent respectively compared with GR. Mk 3/AV-8A; leading-edge sweep reduced by 10°; thickness/chord ratios 11.5 per cent (root)/7.5 per cent (tip); marked anhedral on wings and variable incidence tailplane.

FLYING CONTROLS: Hydraulic actuation (by Fairey irreversible jacks) of drooping ailerons and slab tailplane; rudder actuated mechanically; single-slotted trailing-edge flaps with slot closure doors; manoeuvring at airspeeds below wing-borne flight by jet reaction control valves in nosecone and tailcone and at each wingtip and by thrust vectoring; LIDS 'box' traps air cushion bounced off ground by engine exhaust in VTOL modes, providing enough extra lift to enable aircraft to take off vertically at a gross weight equal to its maximum hovering gross weight; large forward hinged airbrake beneath fuselage aft of rear main landing gear bay.

STRUCTURE: One-piece wing (including main multispar torsion box, ribs and skins), ailerons, flaps, LERX, outrigger pods and fairings, forward part of fuselage, LIDS, tailplane and rudder, are manufactured mainly from graphite epoxy (carbonfibre) and other composites; centre and rear fuselage, wing leading-edges (reinforced against bird strikes on RAF aircraft), wingtips, tailplane leading-edges and tips, and fin, are of aluminium alloy; titanium used for front and rear underfuselage heatshields and small area forward of windscreen. McDonnell Douglas/BAe work split is 60/40 for AV-8B and EAV-8B, 50/50 for RAF aircraft. McDonnell Douglas builds entire wing, front and forward centre-fuselage (including nosecone, air intakes, heatshields, engine access doors and forward fuel tanks) and underfuselage fences/strakes, for all aircraft, plus tailplanes for USMC and Spanish aircraft, and assembles all USMC/Spanish fuselages; BAe builds rear centre and rear fuselage (including blast and heatshields, centre and rear fuel tanks, dorsal air intakes and tail bullets), fins and rudders, and the complete jet reaction control system, for all aircraft, plus tailplanes for RAF aircraft, and assembles all RAF fuselages; final assembly is by McDonnell Douglas for USMC/Italy/Spain, BAe for RAF.

LANDING GEAR: Retractable bicycle type of Dowty design, permitting operation from rough unprepared surfaces of very low CBR (California Bearing Ratio). Hydraulic actuation, with nitrogen bottle for emergency extension. Single steerable nosewheel retracts forward, twin coupled mainwheels rearward, into fuselage. Small outrigger units, at approximate mid-span between flaps and ailerons, retract rearward into streamline pods. Telescopic oleo-pneumatic main and outrigger gear; levered suspension nosewheel leg. Dunlop wheels, tyres, multidisc carbon brakes and anti-skid system. Mainwheel tyres (size 26.0 × 7.75-13.00) and nosewheel tyre (size 26.0 × 8.75-11) all have pressure of 8.62 bars (125 lb/sq in). Outrigger tyres are size 13.5 × 6.00-4.00, pressure 10.34 bars (150 lb/sq in). McDonnell Douglas responsible for entire landing gear system.

POWER PLANT: One 105.87 kN (23,800 lb st) Rolls-Royce F402-RR-408 (Pegasus 11-61) vectored thrust turbofan in AV-8B (95.42 kN; 21,450 lb st F402-RR-406A/Pegasus 11-21 in aircraft delivered before December 1990); one 96.75 kN (21,750 lb st) Pegasus Mk 105 in Harrier GR. Mk 5; Mk 152-42 in EAV-8B. Redundant digital engine control system (DECS), with mechanical back-up, standard from March 1987. Zero-scarf front nozzles. Air intakes have an elliptical lip shape, leading-edges reinforced against bird strikes, and a single row of auxiliary intake doors. Access to engine accessories through top of fuselage, immediately ahead of wing. Integral fuel tanks in wings; total internal fuel capacity (fuselage and wing tanks) 4,319 litres (1,141 US gallons; 950 Imp gallons). Water injection tank with capacity of approximately 227 kg (500 lb). Retractable bolt-on in-flight refuelling probe optional. Each of the four inner underwing stations capable of carrying a 1,135 litre (300 US gallon; 250 Imp gallon) auxiliary fuel tank.

ACCOMMODATION: Pilot only, on zero/zero ejection seat (UPC/Stencel for USMC, Martin-Baker for RAF), in pressurised, heated and air conditioned cockpit. AV-8B cockpit raised approximately 30.5 cm (12 in) in comparison to AV-8A/YAV-8B, with redesigned one-piece wraparound windscreen (thicker on RAF aircraft than on those for USMC) and rearward sliding bubble canopy, to improve all-round field of view. Windscreen de-icing. Windscreens and canopies for all aircraft manufactured by McDonnell Douglas.

SYSTEMS: No. 1 hydraulic system has flow rate of 43 litres (11.4 US gallons; 9.5 Imp gallons)/min; flow rate of No. 2 system is 26.5 litres (7.0 US gallons; 5.8 Imp gallons)/min. Reservoirs nitrogen pressurised at 2.76-5.52 bars (40-80 lb/sq in). Other systems include Westinghouse variable speed constant frequency (VSCF) solid-state electrical system, Lucas Mk 4 gas-turbine starter/APU, Clifton Precision onboard oxygen generating system (OBOGS), and Graviner Firewire fire detection system. Dorsal airscoop at base of fin for avionics bay cooling system.

AVIONICS: Include dual Collins RT-1250A/ARC U/VHF com (GEC Avionics AD3500 ECM-resistant U/VHF-AM/FM in RAF aircraft), R-1379B/ARA-63 all-weather landing receiver (AV-8B only), RT-1159A/ARN-118 Tacan, RT-1015A/APN-194(V) radar altimeter, Honeywell CV-3736/A com/nav/identification data converter, Bendix/King RT-1157/APX-100 IFF (Cossor IFF 4760 transponder for RAF), Litton AN/ASN-130A inertial navigation system (replaced by GEC Ferranti FIN 1075 with RAF), AiResearch CP-1471/A digital air data computer, Smiths Industries SU-128/A dual combining glass HUD and CP-1450/A display computer, IP-1318/A CRT Kaiser digital display indicator, and (RAF only) GEC Ferranti moving map display. Litton AN/ALR-67(V)2 fore/aft looking RWR (AV-8B only), UK MoD AN/ARR-51 FLIR receiver, Goodyear AN/ALE-39 flare/chaff dispenser (upper and lower rear fuselage) (Tracor AN/ALE-40 in RAF aircraft). Primary weapon delivery sensor system for AV-8B and GR. Mks 5/7 is Hughes Aircraft AN/ASB-19(V)2 or (V)3 Angle Rate Bombing Set, mounted in nose and comprising a dual-mode (TV and laser) target seeker/tracker. ARBS functions in conjunction with Control Data Corporation CP-1429/AYK-14(V) mission computer (Computing Devices ACCS 2000 for RAF), Smiths Industries AN/AYQ-13 stores management system, display computer, HUD and digital display indicator. Flight controls that interface with reaction control system provided by Honeywell AN/ASW-46(V)2 stability augmentation and attitude hold system, currently being updated to high AOA capable configuration. RAF aircraft have an accident data recorder. Night attack versions equipped with GEC Sensors nose-mounted FLIR, Smiths Industries wide-angle HUD/HDD, digital colour moving map display (Honeywell for USMC, GEC Avionics for RAF) and pilot's NVGs (variant of GEC Ferranti Nite-Op) with compatible cockpit lighting. Provision for Sanders AN/ALQ-164 defensive ECM pod on centreline pylon. GR. Mks 5/7 have Marconi Defence Systems Zeus internal ECM system comprising advanced RWR and multimode jammer with Northrop RF transmitter; and Siemens Plessey missile approach warning (MAW) equipment, mounted in tailboom, which automatically activates appropriate countermeasures upon detecting approach of enemy missiles. Vinten VICON 18 Srs 403 reconnaissance pod evaluated for RAF 1991. RAF defensive aids include Bofors BOL chaff dispenser in rear of Sidewinder launch rails (Phimat pod on port outer wing pylon pending BOL availability).

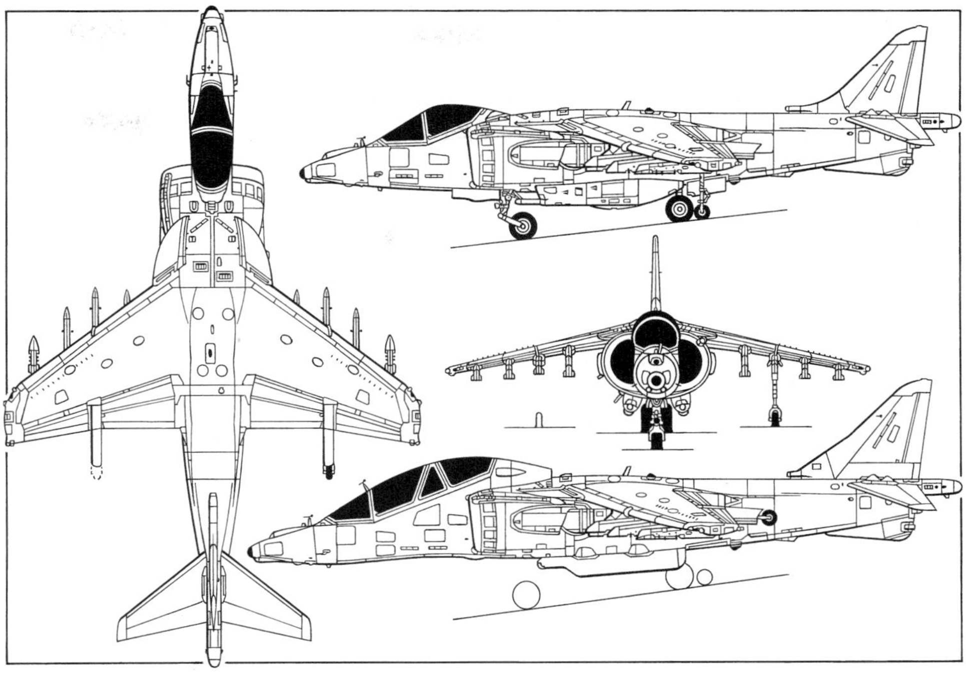

McDonnell Douglas/BAe Harrier GR. Mk 7 V/STOL close support aircraft, with additional side view (bottom) of T. Mk 10 *(Jane's/Mike Keep)*

1994

EQUIPMENT: Back-up mechanical instrumentation includes ASI, altimeter, AOA indicator, attitude indicator, cabin pressure altitude indicator, clock, flap position indicator, HSI, standby compass, turn and slip indicator, and vertical speed indicator. Anti-collision, approach, formation, in-flight refuelling, landing gear position, auxiliary exterior lights, and console, instrument panel and other internal lighting.

ARMAMENT: Two underfuselage packs, mounting on port side a five-barrel 25 mm cannon based on General Electric GAU-12/U, and 300-round container on starboard side, in AV-8B; or (RAF) two 25 mm Royal Ordnance Factories cannon with 100 rds/gun (derived from 30 mm Aden). Single 454 kg (1,000 lb) stores mount on fuselage centreline, between gun packs. Three stores stations under each wing on AV-8B, stressed for loads of up to 907 kg (2,000 lb) inboard, 454 kg (1,000 lb) on intermediate stations, and 286 kg (630 lb) outboard. Four inner wing stations are wet, permitting carriage of auxiliary fuel tanks; reduced manoeuvring limits apply when tanks mounted on intermediate stations. RAF aircraft and new production Harrier II Plus have additional underwing station, for Sidewinder air-to-air missile, ahead of each outrigger wheel fairing. Typical weapons include two or four AIM-9L Sidewinder, Magic or AGM-65E Maverick missiles, or up to six Sidewinders; up to 16 540 lb free-fall or retarded general purpose bombs, 12 BL 755 or similar cluster bombs, 1,000 lb free-fall or retarded bombs, 10 Paveway laser-guided bombs, eight fire bombs, 10 Matra 155 rocket pods (each with 18 68 mm SNEB rockets), or (in addition to underfuselage gun packs) two underwing gun pods. ML Aviation BRU-36/A bomb release units standard on all versions. TAV-8B can carry six Mk 76 practice bombs or two LAU-68 rocket launchers for weapons training.

DIMENSIONS, EXTERNAL:

Wing span	9.25 m (30 ft 4 in)
Wing aspect ratio	4.0
Length overall (flying attitude):	
AV-8B	14.12 m (46 ft 4 in)
TAV-8B	15.32 m (50 ft 3 in)
GR. Mks 5/7	14.36 m (47 ft 1½ in)
T. Mk 10	15.79 m (51 ft 9½ in)

McDonnell Douglas/BAe Technology Demonstrator aircraft with wingtip missile rails for AIM-9 Sidewinder air-to-air missiles *(McDonnell Douglas)*

1995

Height overall 3.55 m (11 ft 7¾ in)
Tailplane span 4.24 m (13 ft 11 in)
Outrigger wheel track 5.18 m (17 ft 0 in)

AREAS:
Wings, excl LERX, gross 21.37 m² (230.0 sq ft)
LERX (total): Pegasus 11-21 0.81 m² (8.7 sq ft)
Pegasus 11-61 1.24 m² (13.4 sq ft)
Ailerons (total) 1.15 m² (12.4 sq ft)
Trailing-edge flaps (total) 2.88 m² (31.0 sq ft)
Ventral fixed strakes (total) 0.51 m² (5.5 sq ft)
Ventral retractable fence (LIDs) 0.24 m² (2.6 sq ft)
Ventral airbrake 0.42 m² (4.5 sq ft)
Fin 2.47 m² (26.6 sq ft)
Rudder, excl tab 0.49 m² (5.3 sq ft)
Tailplane 4.51 m² (48.5 sq ft)

WEIGHTS AND LOADINGS (single-seaters, except where indicated):
Operating weight empty (incl pilot and unused fuel):
AV-8B 6,336 kg (13,968 lb)
GR. Mk 7 7,050 kg (15,542 lb)
TAV-8B 6,451 kg (14,223 lb)
Max fuel: internal only* 3,519 kg (7,759 lb)
internal and external* 7,180 kg (15,829 lb)
Max external stores: Pegasus 11-61 6,003 kg (13,235 lb)
Pegasus 11-21/Mk 105† 4,899 kg (10,800 lb)
Max useful load (incl fuel, stores, weapons, ammunition, and water injection for engine):
VTO approx 3,062 kg (6,750 lb)
STO more than 7,710 kg (17,000 lb)
Basic flight design gross weight for 7*g* operation 10,410 kg (22,950 lb)
Max T-O weight:
435 m (1,427 ft) STO 14,061 kg (31,000 lb)
S/L VTO, ISA:
AV-8B/Pegasus 11-61 9,342 kg (20,595 lb)
GR. Mk 7 8,700 kg (19,180 lb)
S/L VTO, 32°C 8,142 kg (17,950 lb)
Design max landing weight 11,340 kg (25,000 lb)
Max vertical landing weight 9,043 kg (19,937 lb)

* *205 kg (453 lb) less in TAV-8B*
† *throughout full manoeuvring envelope*

PERFORMANCE:
Max Mach number in level flight:
at S/L 0.87 (575 kts; 1,065 km/h; 661 mph)
at altitude 0.98
STOL T-O run at max T-O weight:
ISA 435 m (1,427 ft)
32°C 518 m (1,700 ft)
Operational radius with external loads shown:
short T-O (366 m; 1,200 ft), 12 Mk 82 Snakeye bombs, internal fuel, 1 h loiter 90 n miles (167 km; 103 miles)
hi-lo-hi, short T-O (366 m; 1,200 ft), seven Mk 82 Snakeye bombs, two 300 US gallon external fuel tanks, no loiter 594 n miles (1,101 km; 684 miles)
deck launch intercept mission, two AIM-9 missiles and two external fuel tanks 627 n miles (1,162 km; 722 miles)
Unrefuelled ferry range, with four 300 US gallon external tanks:
tanks retained 1,638 n miles (3,035 km; 1,886 miles)
tanks dropped 1,965 n miles (3,641 km; 2,263 miles)
Combat air patrol endurance at 100 n miles (185 km; 115 miles) from base 3 h
g limits +8/−3

UPDATED

MCDONNELL DOUGLAS/BRITISH AEROSPACE HARRIER II PLUS UPGRADE

TYPE: Enhanced capability derivative of AV-8B.

PROGRAMME: Intention to develop radar-equipped version of AV-8B announced as McDonnell Douglas/BAe private venture June 1987; radar integration efforts (with Hughes Aircraft) started 1988; tri-national MoU (USA/Italy/Spain) of 28 September 1990 approved joint funding to develop and integrate AN/APG-65 radar; US Navy contract of 3 December 1990 authorised development of prototype and completion to Harrier II Plus standard of 24 AV-8Bs ordered in FY90-91; production MoU signed by USA/Italy/Spain March 1992; prototype (AV-8B No. 205, 164129) first flight scheduled for October 1992; deliveries to USMC due to start in Spring 1993 and finish November 1994 (beginning No. 233, 164542). Harrier consortium adds Hughes (for radar), Alenia (Italy) and CASA (Spain); Italy and Spain each have 15 per cent share.

CUSTOMERS: US Marine Corps (24); orders expected from Italian Navy (16, with option for eight more) and Spanish Navy (eight, plus 11 conversions); USMC considering remanufacture of 114 earlier AV-8Bs to Harrier II Plus standard. Refer also to AV-8B entry.

COSTS: $181.5 million (December 1990 contract) for development and production of 24 for USMC.

DESIGN FEATURES: Generally as for night attack AV-8B with F402-RR-408 engine, plus Hughes Aircraft AN/APG-65 multimode pulse Doppler radar (extending nose by 0.43 m; 1 ft 5 in), FLIR, and weapon capability extended to include AMRAAM, Sparrow, Sea Eagle and Harpoon; enlarged LERX. RAF type (eight pylon) wing. Fatigue life of 6,000 hours; improved ECM.

DIMENSIONS, EXTERNAL: As AV-8B except:
Length overall 14.55 m (47 ft 9 in)

WEIGHTS AND LOADINGS:
Operating weight empty 6,740 kg (14,860 lb)
Max T-O weight 14,061 kg (31,000 lb)

PERFORMANCE (estimated, with 137 m; 450 ft short T-O deck run, 6.5° ski-jump, 20 kts (37 km/h; 23 mph) wind over deck, air temperature 32°C, optimum cruise conditions, incl reserves for landing):
Anti-shipping combat radius with two Harpoons, two Sidewinders and two 1,136 l (300 US gallon; 250 Imp gallon) drop tanks 609 n miles (1,128 km; 701 miles)
Combat air patrol (incl 2 min combat) with four AMRAAM and two 300 US gallon tanks: time on station:
at 100 n miles (185 km; 115 mile) radius 2 h 42 min
at 200 n miles (370 km; 230 mile) radius 2 h 6 min
Sea surveillance combat radius (incl 50 n miles; 92 km; 57 mile dash at S/L) with two Sidewinders and two 300 US gallon tanks 608 n miles (1,127 km; 700 miles)

VERIFIED

McDonnell Douglas/BAe AV-8B Harrier II Plus with the Hughes APG-65 radar

1994

PANAVIA

PANAVIA AIRCRAFT GmbH

im Airport Business Centre, Am Söldnermoos 17, D-85399 Hallbergmoos, Germany
Telephone: 49 (811) 801238/9
Fax: 49 (811) 801386

MANAGING DIRECTOR: Oskar Friedrich
PUBLIC RELATIONS MANAGER: W. D. Gnamm

PARTICIPATING COMPANIES:
Alenia: see under Italy
British Aerospace: see under UK
Daimler-Benz Aerospace: see under Germany

Panavia formed 26 March 1969 as industrial prime contractor to design, develop and produce an all-weather MRCA (multirole combat aircraft) for air forces of Germany (including Navy), Italy, Netherlands and UK; Netherlands withdrew July 1969, shareholdings then being readjusted to UK and Germany 42.5 per cent each, Italy 15 per cent. Tornado programme, one of largest European industrial ventures yet undertaken, is guided and monitored on behalf of the three governments by NAMMO (NATO MRCA Management Organisation), whose executive agency NAMMA (formed 15 December 1968) is collocated with Panavia; Tornado production involves three major versions: IDS (interdictor/strike), ECR (electronic combat and reconnaissance) and ADV (air defence variant).

UPDATED

PANAVIA TORNADO IDS

RAF designations: GR. Mks 1, 1A, 1B and 4

TYPE: All-weather close air support/battlefield interdiction, interdiction/counter-air strike, naval strike and reconnaissance aircraft.

PROGRAMME: Six government feasibility study (originally involving Belgium and Canada) initiated 17 July 1968; project definition began 1 May 1969; development phase started 22 July 1970; structural design completed August 1972; first flight 14 August 1974 by first of nine prototypes; Tornado name adopted September 1974; German procurement approved 19 May 1976; production programme initiated 29 July 1976 by three government MoU for 809 aircraft in six batches (640 IDS, 165 ADV, plus four preseries aircraft brought up to IDS production standard); first flight 5 February 1977 by first of six preseries Tornados.

Italian production approved 8 March 1977; first flights by production IDS in UK and Germany 10 and 27 July 1979 respectively; deliveries for operational conversion training (to Tri-national Tornado Training Establishment at RAF Cottesmore) began 1 July 1980; first flight by Italian production aircraft 25 September 1981; operational squadron deliveries began 1982 to RAF (6 January), Germany (Navy, 2 July) and Italy (27 August); first export delivery (to Saudi Arabia) March 1986; batch 7 contract (57 IDS, 35 ECR and 32 ADV) awarded 10 June 1986; contract to develop mid-life update for RAF GR. Mk 1 awarded 16 March 1989; 26 GR. Mk 4 (part of proposed batch 8) cancelled 18 June 1990; completion of IDS/ECR production to batch 7 orders 1992.

VERSIONS: **ECR:** Electronic combat and reconnaissance version, utilising IDS airframe; described separately.

German Air Force IDS: Equips 10 squadrons; two each with JBGs 31 (Nörvenich), 33 (Büchel) and 34 (Memmingen), and two in reconnaissance role with AG 51 (Schleswig/Jagel) all NATO-assigned, plus two (one until mid-1994) with weapons training unit JBG 38 (Jever); others at TTTE in UK and WTD 61 at Manching; all German Tornados cleared for in-flight refuelling by USAF KC-10A and KC-135 tankers; mid-life improvement programme under development for German and Italian Tornados, to confer more accurate navigation for blind attacks, improved capability for sortie generation, increased range and target acquisition capability, reduced penetration altitude, improved ESM, better threat suppression and greater reliability/maintainability. Principal weapons carried are MW-1, AIM-9L, AGM-65, AGM-88.

German Navy IDS: Equips two operational squadrons: (NATO-assigned) with Marinefliegergeschwader 2 (Eggebeck) for strike missions against sea and coastal targets, plus one (1/MFG 2) for reconnaissance using MBB/Alenia multisensor pod; mid-life update under development (see preceding paragraph). Principal weapons carried are BL 755, AIM-9L, AGM-88, Kormoran 1 and 2.

Italian Air Force IDS: In service with four operational squadrons: Gruppi 154 (6° Stormo at Brescia-Ghedi), 155 (50° Stormo, Piacenza) Kormoran anti-shipping Gruppo

156 (36° Stormo, Gioia del Colle) and from September 1993, 102° Gruppo of 6° Stormo: additionally with Reparto Sperimentale di Volo at Pratica di Mare and TTTE in UK; mid-life update under development (see German Air Force paragraph for details). Principal weapons carried are MW-1, AIM-9L, AGM-65D, AGM-88, Kormoran 1.

Royal Air Force GR. Mk 1: UK IDS version, equipping four NATO-assigned squadrons with RAF Germany (Nos. IX, 14, 17 and 31 at Brüggen), plus UK training units (TTTE at Cottesmore, TWCU/No. XV (Reserve) Squadron at Lossiemouth), Strike/Attack Operational Evaluation Unit (Boscombe Down), A&AEE, and DRA Bedford. Squadron deliveries (to No. IX) began 6 January 1982; modification for tactical nuclear weapon carriage began 1984; first combat use (Gulf War) 17 January 1991; to be redesignated GR. Mk 4 (which see) after receiving mid-life update. Principal weapons carried are JP 233, BL 755, AIM-9L, ALARM, CPU-123. Interim Phase 2 modifications introduced 1990 include RAM on forward-facing surfaces. Have-Quick 2 secure radios, filters for NVG use, GPS and provision for 2,250 litre (594 US gallon; 495 Imp gallon) drop tanks. Mk XII Mode 4 IFF for 1991 Gulf War deployment. Principal weapons carried are WE177B, 1,000 lb bomb, JP233, BL 755, AIM-9L, ALARM, Paveway II. No. IX Squadron operational with ALARM, 1 January 1993 (but weapon first used two years previously); most TIALD designator-capable Tornados grouped in No. 14 Squadron from Autumn 1993. Two former SACEUR strike squadrons transferred to maritime strike/attack 1993-94 (see GR. Mk 1B).

Royal Air Force GR. Mk 1A: UK day/night all-weather tactical reconnaissance version, equipping No. 2 Squadron (Laarbruch) and No. 13 (Honington); one development aircraft (ZA402, first flight 11 July 1985) followed by 15 others also converted from GR. Mk 1 (delivered from 3 April 1987) and 14 new-production Mk 1As (delivered from 13 October 1989). Retains air-to-surface role except for deletion of guns; identifiable by small underbelly blister fairing (immediately behind laser rangefinder pod) and transparent side panels for BAe SLIR (sideways looking infra-red) system and Vinten Linescan 4000 surveillance system; has Computing Devices Company signal processing and video recording system (first video-based tac/recce system with replay facility) offering capability for future real-time reconnaissance data relay; first operational mission in Gulf War, night of 18/19 January 1991.

Royal Air Force GR. Mk 1B: Retrofit by RAF at St Athan of 24 GR. Mk 1s for maritime attack; IOC 1994 with four BAe Sea Eagle (or two, plus two drop tanks); second-phase modification to permit up to five Sea Eagles (or four, plus one tank). First aircraft is ZA407 which entered rework at BAe, Warton, on 20 January 1993. No. 27 Squadron reassigned from SACEUR strategic strike to maritime role and simultaneously re-numbered No. 12 Squadron 1 October 1993, to Lossiemouth 7 January 1994. No. 617 Squadron similarly retasked and transferred Marham to Lossiemouth, 27 April 1994. First delivery was ZA456 to No. 617 Squadron on 14 February 1994.

Royal Air Force GR. Mk 4 MLU: Designation to be applied to between 80 and 140 GR. Mk 1s after receiving MLU (mid-life update); modifications include new GEC-Marconi Defence Systems EW; updated weapon control system; FLIR; GPS; and interface for TIALD pod. Several other systems abandoned in 1993 as cost-saving measure, including terrain-referenced navigation and terrain-following 3D display. Also deleted was redesigned engine air intake of lower radar cross-section. Prototype Mk 4 (ZD708) first flight 29 May 1993; followed by ZG773; some MLU items also tested in P15/XZ631.

Royal Saudi Air Force IDS: Equips two squadrons (Nos. 7 and 66), each including some configured for reconnaissance; deliveries March 1986 to October 1987 (first 20), further 28 (completing initial order) between May 1989 and 25 November 1991. Second batch of 48 confirmed 28 January 1993 and contract signed June 1993. Principal weapons carried are JP 233, AIM-9L, ALARM; Sea Eagle on order.

Tornado 2000: Proposed successor to RAF GR. Mk 4; would be single-seat, optimised for low-level/high-speed/long-range penetration and able to carry standoff weapons; longer fuselage than current IDS, containing increased fuel tankage; faceted nose section and pitot intakes to minimise radar signature.

UPGRADES: **British Aerospace GR. Mk 1:** Upgrade for tactical nuclear weapon. See Versions.

British Aerospace GR. Mk 1A: UK day/night all-weather tactical reconnaissance conversion. See Versions.

British Aerospace: GR. Mk 1B retrofit of GR. Mk 1s for maritime attack. See Versions.

British Aerospace GR. Mk 4: See Versions. Go ahead given May 1994.

Daimler-Benz Aerospace MLI: Proposed mid-life improvement (MLI) programme. Initially covering new main computer, integration of Ada software language and new weapons data interface comprising FLIR, GPS and EW systems. Programme for German and Italian Air Force Tornados.

Daimler-Benz Aerospace: Proposed programme to fit two German Air Force Tornado IDSs with a reconnaissance pod.

OPERATORS: Versions of the Tornado IDS are in service with the armed forces of the following countries: Germany (290+), Italy (76), Saudi Arabia (48) and UK (220).

CUSTOMERS: Italy (100: Air Force 88 strike including one preseries, plus 12 dual), Saudi Arabia (48: Air Force 28 attack, six recce, 14 dual; 48 more to follow) and UK (228: Air Force 164 strike, 14 recce, and 51 dual).

DESIGN FEATURES: Continuously variable geometry shoulder wings, with leading-edge sweep angles of 25° (minimum) and 67° (maximum) on movable portions, 60° on fixed inboard portions; modest overall dimensions; high wing loading to minimise low-altitude gust response; swivelling wing pylons to retain stores alignment with fuselage; sweep limited to 63° if 2,250 litre drop tanks carried.

FLYING CONTROLS: Full span double-slotted fixed-vane flaperons (four segments per side), all-moving tailplane (tailerons) and inset rudder, all actuated by electrically controlled tandem hydraulic jacks; full span wing leading-edge slats (three segments each side); two upper-surface spoilers/lift dumpers forward of each central pair of flaperons; tailerons operate together for pitch control and differentially for roll control; spoilers provide augmented roll control at unswept and intermediate wing positions at low speed; Krueger flap on leading-edge of each wing glove box; door type airbrake each side on top of rear fuselage; wing sweep hydraulically powered via ballscrew actuators (aircraft can land safely with wings fully swept if sweep mechanism fails); triple-redundant CSAS (command stability augmentation system), APFD (autopilot/flight director) and TFE (terrain-following E-scope), as detailed under Avionics paragraph.

STRUCTURE: Basically all-metal (mostly aluminium alloy with integrally stiffened skins, titanium alloy for wing carry-through box and pivot attachments); FRP for nosecone, dielectric panels and interface between fixed and movable portions of wings; Teflon plated wing pivot bearings; elastic seal between outer wings and fuselage sides; nosecone hinges sideways to starboard for access to radar antennae; slice of fuselage immediately aft of nosecone also hinges to starboard for access to forward avionics bay and rear of radars; passive ECM antenna fairing near top of fin; ram air intake for heat exchanger at base of fin. Alenia builds entire outer wings (including moving surfaces), with Microtecnica as prime contractor for sweep system; BAe (Warton) builds front and rear fuselage portions (including engine installation) and entire tail unit; MBB is prime contractor for centre-fuselage (including engine intake ducts, wing centre-section box, pivot mechanism, and interface with outer wings); radar-transparent nosecone by Telefunken SystemTechnik.

The following details apply to the basic IDS production version; subsystem details are listed by team leader only, for the sake of clarity.

LANDING GEAR: Hydraulically retractable tricycle type, with forward retracting twin-wheel steerable nose unit. Single-wheel main units retract forward and upward into centre section of fuselage. Emergency extension system, using nitrogen gas pressure. Development and manufacture of the complete landing gear and associated hydraulics is headed by Dowty (UK). Dunlop aluminium alloy wheels, hydraulic multidisc brakes and low-pressure tyres (to permit operation from soft, semi-prepared surfaces) and Goodyear anti-skid units. Mainwheel tyres size 30 × 11.50-14.5, Type VIII (24 or 26 ply); nosewheel tyres size 18 × 5.5, Type VIII (12 ply). Runway arrester hook beneath rear of fuselage.

POWER PLANT: Two Turbo-Union RB199-34R turbofans, fitted with bucket type thrust reversers and installed in rear fuselage with downward opening doors for servicing and engine change. Mk 101 engines of early production aircraft nominally rated at 38.7 kN (8,700 lb st) dry and 66.0 kN (14,840 lb st) with afterburning (uninstalled); RAF aircraft have their engines downrated to 37.7 kN (8,475 lb st) in squadron service (37.0 kN; 8,320 lb st dry for TTTE) to extend service life. Mk 103 engines, introduced in May 1983 (engine number 761), are dry rated nominally at 40.5 kN (9,100 lb st) uninstalled (38.5 kN; 8,650 lb st for RAF) and provide 71.5 kN (16,075 lb st) with afterburning. RAF ordered 100 modification kits in 1983 to upgrade Mk 101 engined aircraft to Mk 103 standard. All internal fuel in multicell Uniroyal self-sealing integral fuselage tanks and/or wing box tanks, all fitted with press-in fuel sampling and water drain plugs, and all refuelled from a single-point NATO connector. Capacity of these tanks totals approximately 6,090 litres (1,610 US gallons; 1,340

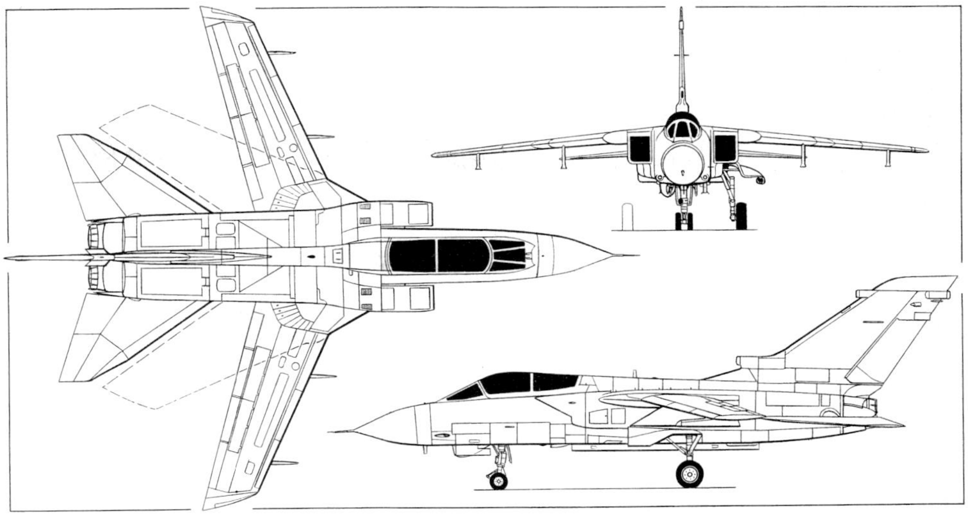

Panavia Tornado IDS multirole combat aircraft *(Jane's/Dennis Punnett)*

1994

Panavia Tornado GR. 1A of the Royal Air Force *(BAe/Geoff Lee)*

1994

Panavia Tornado GR. 1 of the Royal Air Force *(BAe/Geoff Lee)*

1994

Imp gallons). Additional 551 litre (145.5 US gallon; 121 Imp gallon) tank in fin (on RAF aircraft only). Detachable and retractable in-flight refuelling probe can be mounted on starboard side of fuselage, adjacent cockpit. Provision for one or two drop tanks to be carried beneath fuselage (1,500 litres; 396 US gallons; 330 Imp gallons) and single tanks on the shoulder pylons and inboard underwing pylons (1,500 or 2,250 litres; 396 or 594 US gallons; 330 or 495 Imp gallons). Some German Navy and Italian Air Force aircraft adapted to carry a Sargent-Fletcher Type 28-300 1,135 litre (300 US gallon; 250 Imp gallon) buddy type hose/drogue refuelling pod. Dowty afterburning fuel control system. Telefunken Systemtechnik intake de-icing system.

ACCOMMODATION: Crew of two on tandem Martin-Baker Mk 10A zero/zero ejection seats under Kopperschmidt one-piece canopy, which is hinged at rear and opens upward. Flat centre armoured windscreen panel and curved side panels, built by Lucas Aerospace, incorporate Sierracote electrically conductive heating film for windscreen anti-icing and demisting. Canopy (and windscreen in emergency) demisted by engine bleed air. Windscreen is hinged at front and can be opened forward and upward, allowing access to back of pilot's instrument panel. Seats provide safe escape at zero altitude and at speeds from zero up to 630 knots (1,166 km/h; 725 mph) IAS.

SYSTEMS: Cockpit air conditioned and pressurised (maximum differential 0.36 bars; 5.25 lb/sq in) by Normalair-Garrett conventional air cycle system (with bootstrap cold air unit) using engine bleed air with ram air precooler, Marston intercooler, and Teddington temperature control system. Nordmicro air intake control system, and Dowty engine intake ramp control actuators. Two independent hydraulic systems, each of 276 bars (4,000 lb/sq in pressure), are supplied from two separate, independently driven Vickers pumps, each mounted on an engine accessory gearbox. Each system is supplied from a separate bootstrap type reservoir. Systems provide fully duplicated power for primary flight control system, tailerons, rudder, flaps, slats, wing sweep, pitch Q-feel system, and refuelling probe. Port system also supplies power for Krueger flaps, inboard spoilers, port air intake ramps, canopy, and wheel brakes; starboard system for airbrakes, outboard spoilers, starboard air intake ramps, landing gear, nosewheel steering, and radar stabilisation and scanning. Main system includes Dowty accumulators and Teves power pack. Fairey Hydraulics system for actuation of spoilers, rudder and taileron control. Provision for reversion to single-engine drive of both systems, via a mechanical cross-connection between the two engine auxiliary gearboxes, in the event of a single engine failure. In the event of a double engine flameout, an emergency pump in No. 1 system has sufficient duration for re-entry into the engine cold relight boundary. Flying control circuits are protected from loss of fluid due to leaks in other circuits by isolating valves which shut off the utility circuits if the reservoir contents drop below a predetermined safety limit level. Electrical system consists of a 115/200 V AC three-phase 400 Hz constant frequency subsystem and a 28 V DC subsystem. Power is generated by two Rotax automatically controlled oil-cooled brushless AC generators integrated with a constant-speed drive unit and driven by the engines via a KHD accessory gearbox. Normally, each engine drives its own accessory gearbox, but provision is also made for either engine to drive the opposite gearbox through a cross-drive system. In the event of a generator failure, the remaining unit can supply the total aircraft load. Both gearboxes and generators can be driven by APU when aircraft is on ground. The generators supply two main AC busbars and an AC essential busbar. DC power is provided from two fan-cooled transformer/rectifier units (power being derived from the main AC system), these feeding power to two main DC busbars, one essential DC busbar and a battery busbar. Either TRU can supply total aircraft DC load. A fifth DC busbar is provided for maintenance purposes only. Battery is a rechargeable Ni/Cd type, and provides power for basic flightline servicing and for starting APU. In the event of main electrical system or double TRU failure, it is connected automatically to the essential services busbar to supply essential electrical loads. Normalair-Garrett demand type oxygen system, using a 10 litre (2.6 US gallon; 2.2 Imp gallon) lox converter. Emergency oxygen system installed on each seat. GEC Avionics flow metering system. Eichweber fuel gauging system and Flight Refuelling flexible couplings. Graviner fire detection and extinguishing systems. Rotax contractors. Smiths engine speed and temperature indicators.

AVIONICS: Communications equipment includes GEC-Plessey PTR 1721 (UK and Italy) or Rohde und Schwarz (Germany) UHF/VHF transceiver; Telefunken Systemtechnik UHF/ADF (UK and Germany only); SIT emergency UHF with Rohde und Schwarz switch; BAe HF/SSB aerial tuning unit; Rohde und Schwarz (UK and Germany) or Montedel (Italy) HF/SSB radio; Ultra communications control system; GEC Avionics central suppression unit (CSU); Leigh voice recorder; Chelton UHF communications and landing system aerials.

Primary self-contained nav/attack system includes a European built Texas Instruments multimode forward-looking, terrain-following ground mapping radar; GEC Ferranti FIN 1010 three-axis digital inertial navigation system (DINS) and combined radar and map display; Decca Type 72 Doppler radar system, with Kalman filtering of the Doppler and inertial inputs for extreme navigational accuracy; Microtecnica air data computer; Litef Spirit 3 central digital computer (64K initially, 224K on current production aircraft); Alenia radio/radar altimeter (to be replaced on RAF aircraft by GEC Sensors AD1990 covert radar altimeter); Smiths electronic head-up display with Davall camera; GEC Ferranti nose-mounted laser rangefinder and marked target seeker; GEC Avionics TV tabular display; Astronautics (USA) bearing distance and heading indicator and contour map display. Defensive equipment includes Siemens (Germany) or Cossor SSR-3100 (UK and Saudi Arabia) IFF transponder; and Elettronica ARI 23284 radar warning receiver (being replaced in GR. Mk 1 from 1987 by Marconi Defence Systems Hermes RHWR). Production batches 6 and 7 (556th IDS onwards) incorporate an MIL-STD-1553B databus, upgraded radar warning equipment and active ECM, an improved missile control unit, and integration of HARM anti-radar missile.

Flight control system includes a GEC Avionics triplex command stability augmentation system (CSAS), incorporating fly-by-wire and autostabilisation; GEC Avionics autopilot and flight director (APFD), using two self-monitoring digital computers; GEC Avionics triplex transducer unit (TTU), with analog computing and sensor channels; GEC Avionics terrain-following E-scope (TFE), Fairey quadruplex electrohydraulic actuator; and Microtecnica air data set. The APFD provides preselected attitude, heading or barometric height hold, heading and track acquisition, and Mach number or airspeed hold with autothrottle. Flight director operates in parallel with, and can be used as back-up for the autopilot, as a duplex digital system with an extensive range of modes. Automatic approach, terrain-following and radio height-holding modes are also available. Other instrumentation includes Smiths HSI, VSI and standby altimeter; Lital standby AHRS; SEL (with Setac) or (in UK aircraft) GEC Avionics AD2770 (without Setac) Tacan; Cossor CILS 75/76 ILS; Bodenseewerk attitude director indicator; Dornier flight data recorder. Marconi Sky Shadow (jamming/deception) and BOZ 101 (Germany), 102 (Italy) or 107 (UK) chaff/flare ECM pods. Telefunken Systemtechnik Cerberus II or III jammer pods on German and Italian aircraft. GEC Ferranti TIALD (thermal imaging airborne laser designator) night/adverse visibility pods for RAF No. IX Squadron Tornados. (Can also carry similar Thomson-CSF CLDP pod.) Various terrain reference navigation systems have also been developed, including BAe Terprom, GEC Spartan, and Ferranti Penetrate, as have night vision systems incorporating FLIR and NVGs.

EQUIPMENT: German Navy and Italian Air Force Tornados can carry an MBB/Alenia multisensor reconnaissance pod on the centreline pylon. RAF GR. Mk 1A fitted with infra-red cameras in ammunition bay.

ARMAMENT: Fixed armament comprises two 27 mm IWKA-Mauser cannon, one in each side of the lower forward fuselage, with 180 rds/gun. Other armament varies according to version, with emphasis on the ability to carry a wide range of advanced weapons. A GEC Avionics stores management system is fitted; Sandall Mace 355 and 762 mm (14 and 30 in) ejector release units standard on UK Tornados; German and Italian aircraft use multiple weapon carriage system (MWCS) ejector release units. ML Aviation CBLS 200 practice bomb carriers are also standard. The battlefield interdiction version is capable of carrying weapons for hard or soft targets. Weapons are carried on seven fuselage and wing hardpoints: one centreline pylon fitted with a single ejection release unit (ERU), two fuselage shoulder pylons each with three ERUs, and, under each wing, one inboard and one outboard pylon each with a single ERU. Among the weapons carried by the IDS Tornado are the Sidewinder air-to-air, and ALARM or HARM anti-radiation missiles; JP 233 low-altitude airfield attack munition dispenser; CPU-123B Paveway laser-guided bomb; Maverick, Sea Eagle and Kormoran air-to-surface missiles; napalm; BL 755 cluster bombs (277 kg; 611 lb

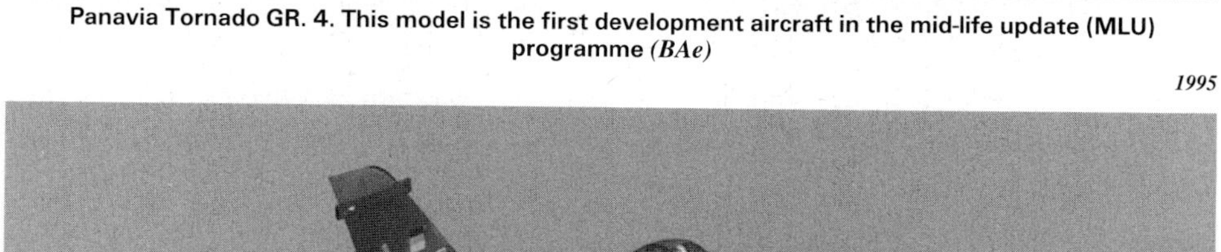

Panavia Tornado GR. 4. This model is the first development aircraft in the mid-life update (MLU) programme *(BAe)*

1995

Panavia Tornado IDS of the German Air Force *(MTU)*

1995

Mk 1 or 264 kg; 582 lb Mk 2); MW-1 munitions dispenser; 1,000 lb bombs; smart or retarded bombs; BLU-1B 750 lb fire bombs; Matra 250 kg ballistic and retarded bombs; Lepus flare bombs; LAU-51A and LR-25 rocket launchers.

DIMENSIONS, EXTERNAL:
- Wing span: fully spread 13.91 m (45 ft 7½ in)
  - fully swept 8.60 m (28 ft 2½ in)
- Length overall 16.72 m (54 ft 10¼ in)
- Height overall 5.95 m (19 ft 6¼ in)
- Tailplane span 6.80 m (22 ft 3½ in)
- Wheel track 3.10 m (10 ft 2 in)
- Wheelbase 6.20 m (20 ft 4 in)

AREAS:
- Wings, gross (to fuselage c/l, 25°sweepback) 26.60 m² (286.3 sq ft)

WEIGHTS AND LOADINGS:
- Basic weight empty approx 13,890 kg (30,620 lb)
- Weight empty, equipped 14,091 kg (31,065 lb)
- Fuel (approx):
  - internal: wing/fuselage tanks 4,600 kg (10,140 lb)
  - fin tank (RAF only) 440 kg (970 lb)
  - drop tanks (each): 1,500 l 1,197 kg (2,640 lb)
  - 2,250 l 1,796 kg (3,960 lb)
- Nominal max external stores load more than 9,000 kg (19,840 lb)
- Max T-O weight:
  - clean, full internal fuel 20,411 kg (45,000 lb)
  - with external stores approx 27,950 kg (61,620 lb)

PERFORMANCE:
- Max Mach number in level flight at altitude, clean 2.2
- Max level speed:
  - clean above 800 kts (1,480 km/h; 920 mph) IAS
  - with external stores Mach 0.92 (600 kts; 1,112 km/h; 691 mph)
- Landing speed approx 115 kts (213 km/h; 132 mph)
- Time to 9,150 m (30,000 ft) from brake release less than 2 min
- Automatic terrain-following down to 61 m (200 ft)
- Required runway length less than 900 m (2,950 ft)
- Landing run 370 m (1,215 ft)
- Max 360° rapid roll clearance with full lateral control 4*g*
- Radius of action with heavy weapons load, hi-lo-hi 750 n miles (1,390 km; 863 miles)
- Ferry range approx 2,100 n miles (3,890 km; 2,420 miles)
- *g* limit +7.5

UPDATED

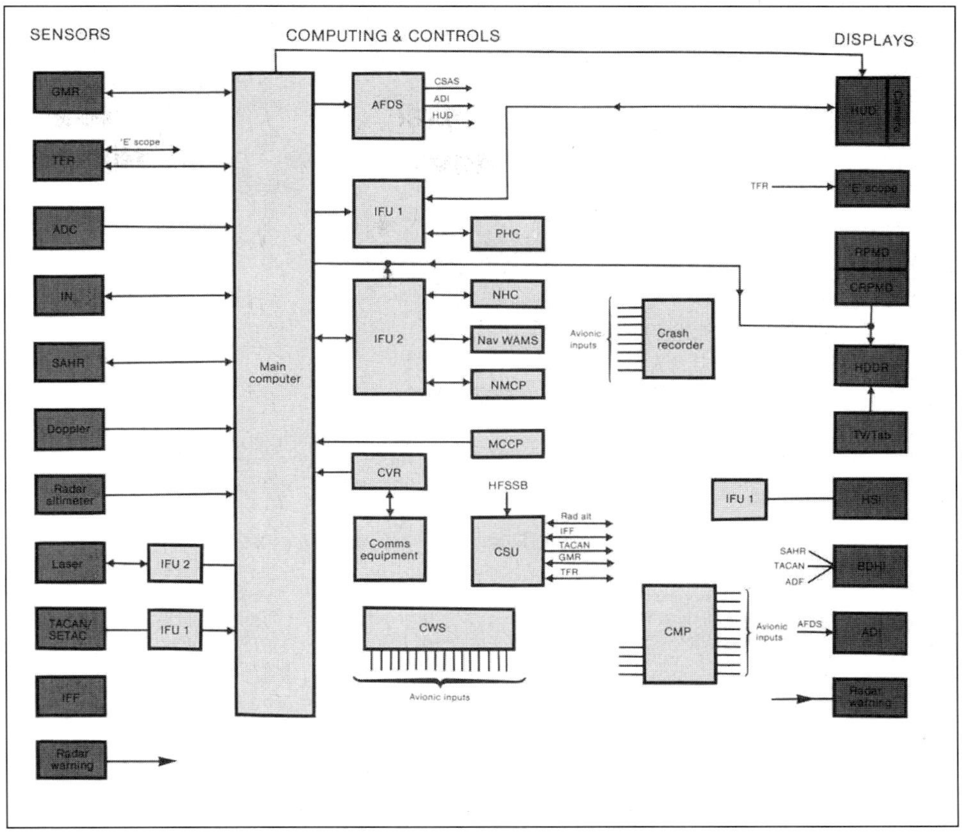

Panavia Tornado IDS avionics system

1995

PANAVIA TORNADO ECR

TYPE: Electronic combat and reconnaissance version of Tornado IDS.

PROGRAMME: Selected by German Luftwaffe to supplement existing in-service tactical reconnaissance aircraft such as Wild Weasel F-4G Phantom; 35 included in batch 7 production contract signed 10 June 1986; two development aircraft (s/n 9803 and 9878) converted from IDS (first flight 18 August 1988); first production aircraft (s/n 4623) made first flight 26 October 1989; deliveries (to 2/JBG 38 initially, later to 3/JBG 32) began 21 May 1990, ended with s/n 4657, 28 January 1992, but not operational until April 1993 when emitter location system became available. First Tornado ECR with ELS redelivered to Luftwaffe on 8 February 1993. All aircraft concentrated in JBG 32 from mid-1994. Italian prototype (conversion of IDS MM7079) rolled out 19 March 1992; first flight 20 July 1992; to re-equip current IDS squadron. 155° Gruppo of 50° Stormo at Piacenza from 1996.

VERSIONS: **ECR:** Intended for standoff reconnaissance and border control, reconnaissance via image-forming and electronic means, electronic support, and employment of anti-radar guided missiles.

IT ECR: Italian equivalent to German ECR.

OPERATORS: Versions of the Tornado ECR are in service with the armed forces of the following countries: Germany (35) and Italy (8).

The description of the IDS Tornado applies generally also to the ECR version except as follows:

POWER PLANT: Mk 105 version of RB199 engine, providing approximately 10 per cent more thrust than Mk 103.

AVIONICS: Include Texas Instruments ELS (emitter location system); Honeywell/Sondertechnik infra-red linescan; Zeiss FLIR; onboard processing/storing/transmission systems for reconnaissance data; advanced tactical displays for pilot and weapons officer.

ARMAMENT: Both internal cannon deleted; external load stations can be used for ECR or fighter-bomber missions, or a combination of both; in ECR role will normally be configured to carry two HARM missiles, two AIM-9L Sidewinders, an active ECM pod, chaff/flare dispenser pod, and two 1,500 litre (396 US gallon; 330 Imp gallon) underwing fuel tanks.

VERIFIED

PANAVIA TORNADO ADV

RAF designations: Tornado F. Mks 2, 2A and 3

TYPE: All-weather air defence interceptor, air superiority fighter and combat patrol aircraft.

PROGRAMME: Feasibility studies for ADV (air defence variant) for UK, begun in 1968, given impetus by MoD Air Staff Target 395 of 1971 for interceptor with advanced radar and Sky Flash air-to-air missiles; full-scale development authorised 4 March 1976; three prototypes (first flight 27 October 1979) included in production batch 1; first flight by F. Mk 2 production aircraft 5 March 1984; last F. Mk 2 delivered 9 October 1985; first flight by F. Mk 3 made 20 November 1985; first export order (by Saudi Arabia) placed 26 September 1985.

VERSIONS: **Royal Air Force F. Mk 2:** Designation of first 18 (batch 4) production ADVs, with RB199 Mk 103 engines; currently in store or ground training use except for one each with BAe at Warton, A&AEE at Boscombe Down, DRA at Boscombe Down, (previously Farnborough). DRA aircraft (ZD902; dual control) first flown 18 August 1992 as TIARA—Tornado Integrated Avionics Research Aircraft; to receive Blue Vixen radar (Sea Harrier FRS. Mk 2 programme), GEC-Marconi IRST system, GEC-Marconi Defence Systems RHWR and JTIDS Class 2. Front cockpit configured as single-seat fighter, with three multifunction displays, helmet-mounted sight and, possibly, voice activated functions.

Panavia Tornado ECR of the German Air Force

1994

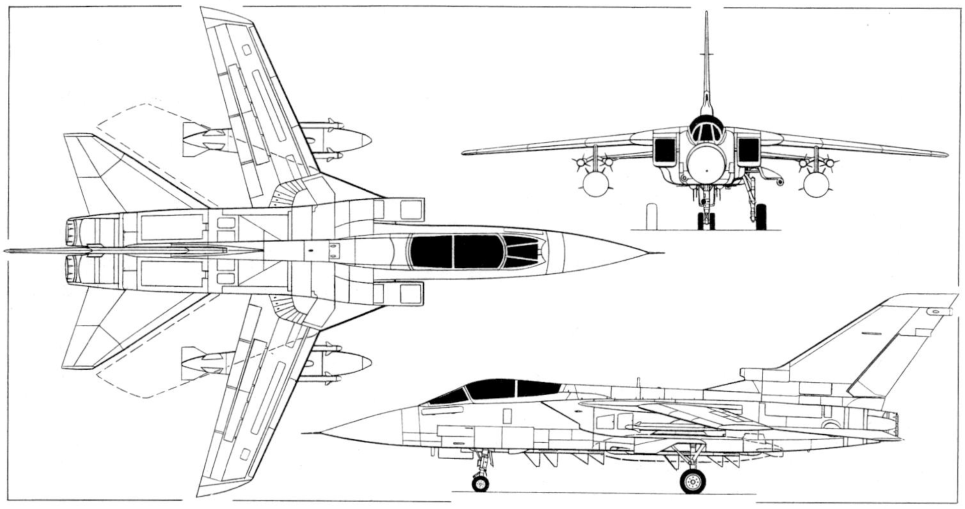

Panavia Tornado F. Mk 3 all-weather air defence interceptor *(Jane's/Dennis Punnett)*

1994

Royal Air Force F. Mk 2A: Designation to be applied to F. Mk 2s after being upgraded largely to F. Mk 3 standard except for retention of Mk 103 engines. Programme abandoned.

Royal Air Force F. Mk 3: Definitive production version (batches 5-7), delivered from 28 July 1986 (to No. 229 OCU/65 Squadron at Coningsby, which became No. 56 (Reserve) Squadron on 1 July 1992); now equips six UK air defence squadrons (Nos. 5 and 29 at Coningsby, Nos. 11 and 25 at Leeming and Nos. 43 and 111 at Leuchars); No. 23 Squadron disbanded, 28 February 1994 at Leeming. In Falkland Islands, four aircraft issued to No. 1435 Flight from July 1992. Primary missions are air defence of UK, protection of NATO's northern and western approaches, and long-range air defence of UK maritime forces; main differences from F. Mk 2 are uprated (Mk 104) engines, automatic wing sweep and manoeuvring systems, and improved avionics (see Flying Controls, Structure, Power Plant and Avionics paragraphs). Stage 1 update (new-build from Block 13, plus retrofits), from 1989, introduced HOTAS-type 'combat stick' for pilot, type 'AA' radar upgrade, improvements to Hermes RHWR, 5 per cent combat boost switch for power plants, and chaff/flare dispensers beneath rear fuselage. All except chaff/flare systems implemented by early 1992, although auto wing sweep disconnected. Stage 2G radar upgrade under trial in 1992 for installation in mid-1990s. Nos. 5 and 29 Squadrons to be reissued with Block 15/16 aircraft in 1994, after retrofit with JTIDS datalink (NATO Link 16). Aircraft of Tornado F3 Operational Evaluation Unit (formed, Coningsby, April 1987) conducted service trials, including first major test, 27 October 1993, with data transfer between two RAF E-3D AWACS. French E-3F and two Tornados.

Royal Saudi Air Force ADV: Equips No. 29 Squadron at Dhahran; deliveries began 20 March 1989; ended 8 October 1990. Formation of No. 34 Squadron abandoned; all assigned to No. 29 from early 1993.

Italian Air Force F. Mk 3: Leased from RAF. Agreed by UK and Italian defence ministers, 17 November 1993, IAF to lease 24 ex-RAF aircraft for 10 years. Crew training by No. 56 (Reserve) Squadron at RAF Coningsby; first 12 aircraft to 12° Gruppo of 36° Stormo at Gioia del Colle, Autumn 1995; second 12 to 18° Gruppo of 37° Stormo at Trapani/Birgi. Minor modifications required for compatibility with Alenia Aspide AAM.

UPGRADES: **Panavia Tornado F. Mk 3:** Stage 1 update and Stage 2G radar upgrade. See Royal Air Force F. Mk 3 in Versions.

OPERATORS: Versions of the Tornado ADV are in service with the armed forces of the following countries: Italy (leased from RAF 24), Saudi Arabia (24), UK (170+).

DESIGN FEATURES: Structural changes reduce drag, especially at supersonic speed, compared with IDS version, and longer fuselage provides more space for avionics and additional 10 per cent internal fuel.

FLYING CONTROLS: Similar to IDS, but with AWS (automatic wing sweep), AMDS (automatic manoeuvre device system) and SPILS (spin prevention and incidence limiting system); AWS allows scheduling of four different sweep angles (25° at speeds up to Mach 0.73, 45° from there up to Mach 0.88, 58° up to Mach 0.95 and 67° above Mach 0.95), enabling specific excess power at transonic speeds and turning capability at subsonic speeds to be maximised; buffet-free handling can be maintained, to limits defined by SPILS, by using AMDS, which schedules with wing incidence to deploy either flaperons and slats at 25° sweep angle or slats only at 45° (beyond 45°, both flaperons and slats are scheduled in); fly-by-wire CSAS/APFD system modified for increased roll rate and reduced pitch stick forces.

STRUCTURE: Generally as IDS version except: fuselage lengthened forward of front cockpit to accommodate longer radome, and aft of rear cockpit to allow Sky Flash missiles to be carried in two tandem pairs; CG shift compensated by extending fixed inboard portions of wings to increase chord and give 67° leading-edge sweep angle; Krueger flaps deleted; afterburner nozzles extended by 360 mm (14 in) on F. Mk 3, requiring modification to adjacent contours of rudder and tailerons; one internal gun deleted; wing/tailplane/fin leading-edges of 20 F. Mk 3s coated with radar absorbent material (RAM) for early 1991 Gulf operations.

LANDING GEAR: As IDS version, but nosewheel steering augmented to minimise 'wander' on landing.

POWER PLANT: Two Turbo-Union RB199-34R Mk 104 turbofans, each with uninstalled rating of 40.5 kN (9,100 lb st) dry and 73.5 kN (16,520 lb st) with afterburning. Lucas digital engine control. Fuselage fuel capacity (including fin tank) approximately 7,250 litres (1,915 US gallons; 1,595 Imp gallons). Internally mounted, fully retractable in-flight refuelling probe in port side of nose, adjacent cockpit. Provision for drop tanks of 1,500 or 2,250 litres (396 or 594 US gallons; 330 or 495 Imp gallons) capacity to be carried on the shoulder pylons and underwing pylons.

ACCOMMODATION: As for IDS version.

SYSTEMS: Generally as described for IDS version, with the addition of a radar-dedicated cold air unit to cool the Foxhunter radar, and a pop-up ram air turbine to assist recovery in the event of engine flameout at high altitude in a zoom climb.

AVIONICS: Among those in the IDS Tornado which are retained in the ADV are the communications equipment (GEC-Plessey VHF/UHF transceiver, SIT emergency UHF, Rohde und Schwarz HF/SSB, Ultra communications control system and Epsylon cockpit voice recorder); GEC Avionics triplex fly-by-wire CSAS and APFD system; Litef Spirit 3 central digital computer (with capacity increased from 64K to 224K) and data transmission system; Smiths electronic head-up and navigator's head-down display; GEC Ferranti FIN 1010 inertial navigation system (to which is added a second 1010 to monitor the head-up display); GEC Sensors Tacan; Cossor ILS; and Cossor IFF transponder. Those deleted include the Texas Instruments nose radar, Decca 72 Doppler radar with terrain-following, GEC Ferranti laser rangefinder and marked target seeker, and Lital standby AHRS.

Nose-mounted Marconi Defence Systems AI Mk 24 Foxhunter multimode track-while-scan pulse Doppler radar with FMICW (frequency modulated interrupted continuous wave), with which is integrated a new Cossor IFF-3500 interrogator and a radar signal processor to suppress ground clutter. This system is intended to enable the aircraft to detect targets more than 100 n miles (185 km; 115 miles) away, and to track several targets simultaneously. A ground mapping mode for navigation back-up is available. GEC Ferranti is subcontractor for the Foxhunter transmitter and aerial scanning mechanism. New data processor, being introduced during early 1990s, offers final Foxhunter standard considerably more capable

Panavia Tornado F. Mk 3 all-weather air defence interceptor of the Royal Air Force

1994

Panavia Tornado F. Mk 3 of the Royal Air Force, carrying four Skyflash and four Sidewinder missiles
1994

TORNADO PRODUCTION

		IDS		IDS		IDS		IDS		IDS		ADV		ADV			
		RAF		GAF		GN		IAF		RSAF		RAF		RSAF		Total	Cum
Batch	**Block**	**S**	**T**	**S**	**T**	**S**	**T**	**S**	**T**	**S**	**T**	**S**	**T**	**S**	**T**		**total**
1	1	8	8	3	11							2	1			33	33
1	2	3	4	—	3											10	43
2	3	8	7	7	3	1	—	2	2							30	73
2	4	18	4	8	5			3	2							40	113
2	5	13	5			11	5	5	1							40	153
3	6	25	7			32	—	9	4							77	230
3	7	35	1	24	12			14	1							87	317
4	8	23	6	27	4			13	—			—	6			79	396
4	9	22	2	29	4			14	—			10	2			83	479
5	10	2	—	29	—	—	5	11	2	11	3	12	6			81	560
5	11			2	—	32	—	16	—	3	3	22	12			90	650
5	12					2	—									2	652
6	12			21	5	9	2					39	7			83	735
6	13			5	8	13	—					12	10	18	6	72	807
7	14	14[1]	—	18[2]	—					10	4	7	—			53	860
7	15	7	6	17[2]	—					10[3]	4	17	—			61	921
7	16											—	8			8	929
		178	**50**	**190**	**55**	**100**	**12**	**87**	**12**	**34**	**14**	**121**	**52**	**18**	**6**		
		228		**245**		**112**		**99**		**48**		**173**		**24**			
Subtotal							**732 IDS**						**197 ADV**				
Prototype		3	1	3	—			2	—							9	
Preseries		1	1[4]	2[4]	1			1[4]	—							6	
		234		**251**		**112**		**102**		**48**		**173**		**24**			
							747						**197**				**944**
In hand								48 (RSAF)									**48**
Totals							**795 IDS**						**197 ADV**				**992**

Notes: [1]New-build GR. Mk 1A
[2]Tornado ECR
[3]Including six reconnaissance (equivalent of GR. Mk 1A)
[4]Refurbished to production standard
RAF-Royal Air Force
GAF-German Air Force
GN-German Navy
IAF-Italian Air Force
RSAF-Royal Saudi Air Force
S-single control
T-dual control

NEW ENTRY

than earlier versions of this radar, in particular more automation to improve close combat capability. Modification kits will bring radars already in service up to the new standard. A pilot's head-down display is added, a GEC Ferranti displayed data video recorder (DDVR) replaces the navigator's wet-film display recorder, and a Marconi Defence Systems Hermes modular RHWR is added. Head-up/head-down displays are on front instrument panel only, radar control and datalink presentations on rear panel only; both panels have weapon control and RHWR displays. A GEC Ferranti FH 31A AC driven 3 in horizon gyro in the rear cockpit, in addition to providing an attitude display for the navigator, feeds pitch and roll signals to other avionics systems in the aircraft in certain modes. Lucas digital electronic engine control unit (DECU 500). ESM (electronic surveillance measures) and ECCM are standard; a Singer-Kearfott ECM-resistant datalink system, interoperable with other NATO systems, is under development for installation later. Because of its comprehensive avionics the Tornado ADV can contribute significantly to the transfer of vital information over the entire tactical area and can, if necessary, partially fulfil the roles of both AEW and ground-based radar. Smiths Industries/Computing Devices Company missile management system (MMS), which also controls tank jettison, has provision for pilot override, optimised for visual attack. Studies being undertaken for 1553B multiplex digital databus associated with AMRAAM and Sidewinder replacement.

ARMAMENT: Fixed armament of one 27 mm IWKA-Mauser cannon in starboard side of lower forward fuselage. Four BAe Sky Flash semi-active radar homing medium-range air-to-air missiles semi-recessed under the centre-fuselage, carried on internally mounted Frazer-Nash launchers; two European built NWC AIM-9L Sidewinder infra-red homing short-range air-to-air missiles on each inboard underwing station (outboard stations not used on RAF ADVs). The Sky Flash missiles, each fitted with an MSDS monopulse seeker head, can engage targets at high altitude or down to 75 m (250 ft), in the face of heavy ECM, and at standoff ranges of more than 25 n miles (46 km; 29 miles). Release system permits the missile to be fired over the Tornado's full flight envelope. For the future, the ADV will be able to carry, instead of Sky Flash and Sidewinder, up to six Hughes AIM-20 AMRAAM or BAe Active Sky Flash medium-range and four new-generation short-range air-to-air missiles.

DIMENSIONS, EXTERNAL: As for IDS version, except:
Length overall 18.68 m (61 ft 3½ in)

WEIGHTS AND LOADINGS (approx):
Operational weight empty 14,500 kg (31,970 lb)
Fuel (approx):
internal: wing/fuselage tanks 5,216 kg (11,500 lb)
fin tank as for IDS version
drop tanks as for IDS version
Max external fuel 5,806 kg (12,800 lb)
Nominal max external stores load 8,500 kg (18,740 lb)
Max T-O weight 27,986 kg (61,700 lb)

PERFORMANCE:
Max Mach number in level flight at altitude, clean 2.2
Max level speed, clean
800 kts (1,480 km/h; 920 mph) IAS
Rotation speed, depending on AUW
145-160 kts (269-297 km/h; 167-184 mph)
Normal touchdown speed 115 kts (213 km/h; 132 mph)
Demonstrated roll rate at 750 kts (1,390 km/h; 864 mph) and up to 4g 180°/s
Operational ceiling approx 21,335 m (70,000 ft)
T-O run:
with normal weapon and fuel load 760 m (2,500 ft)
ferry configuration (four 1,500 l drop tanks and full weapon load) approx 1,525 m (5,000 ft)
T-O to 15 m (50 ft) under 915 m (3,000 ft)
Landing from 15 m (50 ft) approx 610 m (2,000 ft)
Landing run with thrust reversal 370 m (1,215 ft)
Intercept radius: supersonic
more than 300 n miles (556 km; 345 miles)
subsonic
more than 1,000 n miles (1,853 km; 1,151 miles)
Endurance 2 h combat air patrol at 300-400 n miles (555-740 km; 345-460 miles) from base, incl time for interception and 10 min combat

UPDATED

SEPECAT

SOCIÉTÉ EUROPÉENNE DE PRODUCTION DE L'AVION E.C.A.T.

PARTICIPATING COMPANIES:
British Aerospace: see under UK
Dassault Aviation: see under France

PRESIDENT: J. P. Weston (BAe)
VICE-PRESIDENT: R. Dubost (Dassault)

This Anglo-French company formed May 1966 by Breguet Aviation and British Aircraft Corporation to design and produce Jaguar strike fighter/trainer; production now in India only (see HAL entry).

VERIFIED

SEPECAT JAGUAR

TYPE: Single-seat tactical support aircraft and two-seat operational advanced trainer.

PROGRAMME: The Jaguar was designed by Breguet (now Dassault Aviation) and BAe to meet a common requirement of the French and British Air Forces, laid down in early 1965, which called for a dual role aircraft to be used as an advanced and operational trainer and a tactical support aircraft.

All 402 aircraft from the original orders for the Royal Air Force (202) and Armée de l'Air (200) had been delivered by the end of 1981. One additional two-seat Jaguar B for the RAF was delivered in late 1982. These aircraft were delivered with 22.75/32.5 kN (5,115/7,305 lb st) Adour Mk 102 turbofan engines.

VERSIONS: **Jaguar A:** French single-seat tactical support version. 160 ordered, final 30 fitted with Martin Marietta/

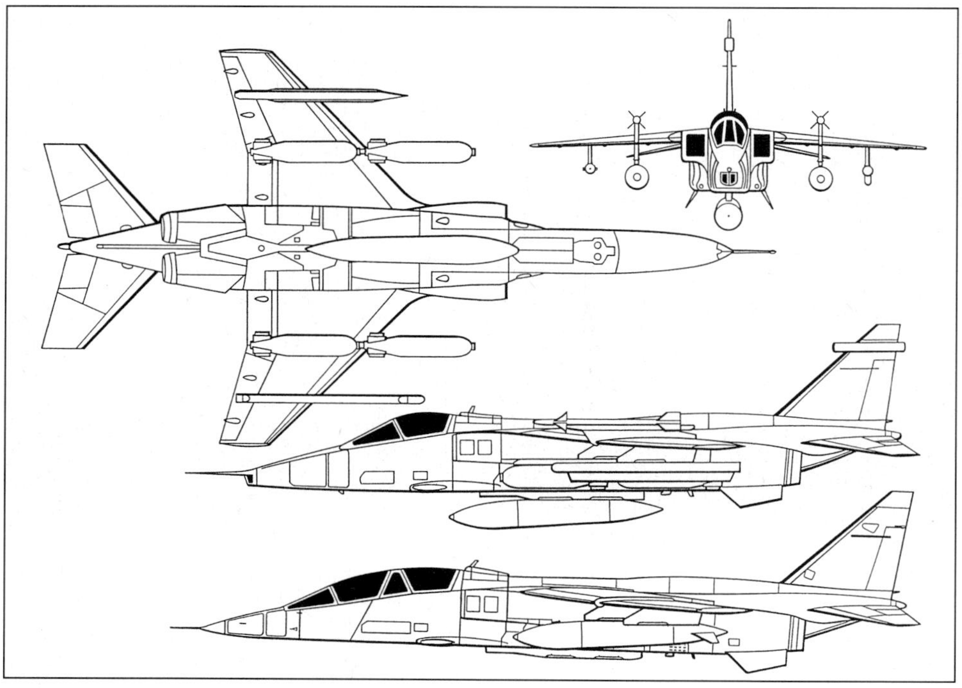

Sepecat Jaguar single-seat tactical support aircraft and two-seat operational trainer *(Jane's/Mike Keep)*

1994

Thomson-CSF ATLIS II target TV acquisition and laser designation pod.

Jaguar B (RAF designation Jaguar T. Mk 2): British two-seat operational training version. Total of 38 built.

Jaguar E: French two-seat advanced training version. Total of 40 built.

Jaguar S (RAF designation Jaguar GR. Mk 1): British single-seat version basically similar to A but with Ferranti FIN 1064 (originally GEC-Marconi navigation and weapon-aiming subsystem - NAVWASS). Total 165 built. Jaguar GR. 1 upgraded to Jaguar GR. 1A.

Jaguar GR. Mk 1A: Upgraded GR. Mk 1. See Upgrades in this entry.

Jaguar International: Export version. See also Hindustan Aeronautics Ltd.

Jaguar International is the export version of the aircraft, the first example of which (G27-266) made its initial flight on 19 August 1976. This version has Ardour Mk 804 or more powerful Mk 811 engines, which give improved combat performance with substantially enhanced manoeuvrability and acceleration in the low-level speed range. Other customer options include overwing pylons compatible with Matra R.550 Magic or similar dogfight missiles; a multipurpose radar such as the Thomson-CSF Agave: up to four anti-shipping weapons such as Sea Eagle, Harpoon, Exocet and Kormoran on the underwing and underfuselage hardpoints: and night sensors such as low light level TV.

Initial orders for Jaguar International were placed by the Sultan of Oman's Air Force (12) and Ecuadorean Air Force (12), each order including two two-seaters. These aircraft were powered by Ardour Mk 804 engines. Deliveries to Ecuador were made in January-November 1977, and to Oman between March 1977 and July 1978 SOAF aircraft are fitted with a GEC Avionics 920ATC NAVWASS computer and carry AIM-9P Sidewinder air-to-air missiles on the outboard underwing pylons. Deliveries to Oman of a further 12, powered by Ardour Mk 811 engines, were completed by November 1983.

Under a 1979 agreement, an initial batch of 40 Jaguar Internationals with Ardour Mk 804 engines was purchased from Britain by the Indian government; the agreement provided for a further 45 (with Ardour Mk 811 engines) to be assembled in India from European built components, leading to the eventual manufacture of 31 additional aircraft under licence by Hindustan Aeronautics Ltd (HAL) of Bangalore. The first Jaguar assembled at Bangalore (JS136) made its initial flight 31 March 1982. As an interim measure, the Indian Air Force borrowed 18 RAF Jaguars, the first two of which were handed over 19 July 1979. By early 1984 15 of these (including one two-seater) had been returned. The Indian Air Force's first Jaguar squadron (No. 14) became operational by the Summer of 1980, and the second (No. 5) in August 1981. All 40 European built Jaguars for India had been delivered by the end of 1982. Those assigned to anti-shipping duty have had nose-mounted Agave radar and air-to-surface missiles. (See India section.)

UPGRADES: **DRA:** See separate entry.

Sepecat GR. 1A: Between 1978 and 1984 RAF Jaguars were refitted with uprated Adour Mk 104 engines, equivalent to the Mk 804 which powered early Jaguar Internationals. Modifications to RAF Jaguars taking part in early 1991 Gulf War included RWR upgraded to Sky Guardian standard; Vinten LOROP pod fitted; second U/VHF radio (with modified antenna) installed; engine control amplifier modified for increased thrust. French Jaguar units in the Gulf War included aircraft from the 7e and 11e Escadres.

OPERATORS: Versions of the Jaguar are in service with the armed forces of the following countries: Ecuador (9), India (115), Nigeria (15), Oman (22) and UK (50).

DESIGN FEATURES: Cantilever shoulder-wing monoplane. Anhedral 3°. Sweepback 40° at quarter-chord. Tail unit has sweepback at quarter-chord 40° on horizontal, 43° on vertical surfaces. All-moving slab-type tailplane with 10° anhedral. Ventral fins beneath rear fuselage.

FLYING CONTROLS: Outer panels fitted with slat which also gives effect of extended chord leading-edge. No ailerons: control by two-section spoilers, forward of outer flap on each wing, in association (at low speeds) with differential tailplane. Hydraulically operated (by screwjack) full span double-slotted trailing-edge flaps. Fairey hydraulics powered flying controls. Leading-edge slats can be used in combat. In 1983 tests began on a carbonfibre wing at BAe Military Aircraft at Warton. All-moving slab type tailplane, with 10° anhedral, the two halves of which can operate differentially to supplement the spoilers. No separate elevators. Fairey Hydraulic powered flying controls.

STRUCTURE: Wing is an all-metal two-spar torsion box structure; skin machined from solid aluminium alloy, with integral stiffeners. Entire wing British built. Main portion built as single unit, with three-point attachment to each side of fuselage. The fuselage is an all-metal structure, mainly aluminium, built in three main units and making use of panels and, around the cockpit(s), honeycomb panels. Local use of titanium alloy in engine bay area. Forward and centre fuselage, up to and including main undercarriage bays, and including cockpit(s), main system installations, forward fuel tanks and landing gear, are of French construction. Air intakes, and entire fuselage aft of mainwheel bays, including engine installation, rear fuel tanks and complete tail assembly, are British built. Two door type airbrakes under rear fuselage, immediately aft of each mainwheel well. Structure and systems aft of cockpit(s), identical for single-seat and two-seat versions. The tail unit is a cantilever all-metal structure, covered with aluminium alloy sandwich panels. Rudder and outer panels and trailing-edge of tailplane have honeycomb core. Entire tail unit British built.

LANDING GEAR: Messier-Hispano-Bugatti retractable tricycle type, all units having Dunlop wheels and low-pressure tyres for rough field operation. Hydraulic retraction, with oleo-pneumatic shock-absorbers. Forward retracting main units each have twin wheels, tyre size 615 × 225-10, pressure 5.8 bars (84 lb/sq in). Wheels pivot during retraction to stow horizontally in bottom of fuselage. Single rearward retracting nosewheel, with tyre size 550 × 250-6 and pressure 3.9 bars (57 lb/sq in). Twin landing/taxying lights in nosewheel door. Dunlop hydraulic brakes. Anti-skid units and arrester hook standard. Irvin brake parachute of 5.5 m (18 ft 0½ in) diameter in fuselage tailcone.

POWER PLANT: Two Rolls-Royce Turbomeca Adour Mk 804 turbofan engines, rated at 23.7 kN (5,320 lb st) dry and 35.75 kN (8,040 lb st) with afterburning, in aircraft for Ecuador, India (first 40) and Oman (first 12). Ardour Mk 811, rated at 24.6 kN (5,520 lb st) dry and 41.23 kN (9,270 lb st) with afterburning, in remaining aircraft for India, second 12 for Oman and those for Nigeria. Fixed geometry air intake on each side of fuselage aft of cockpit. Fuel in six tanks, one in each wing and four in fuselage.

Sepecat Jaguar GR. 1A of the RAF with overwing AIM-9 Sidewinder missiles *(BAe/Geoff Lee)*

1994

Total internal fuel capacity 4,200 litres (1,109.7 US gallons; 924 Imp gallons). Armour protection for critical fuel system components. In basic tactical sortie the loss of fuel from one tank at halfway point would not prevent the aircraft from regaining its base. Provision for carrying three auxiliary drop tanks, each of 1,200 litres (317 US gallons; 264 Imp gallons) capacity, on fuselage and inboard wing pylons. Provision for in-flight refuelling, with retractable probe forward of cockpit on starboard side.

ACCOMMODATION (trainer): Crew of two in tandem on Martin-Baker 9B Mk II ejection seats. Individual rearward hinged canopies. Rear seat 38 cm (15 in) higher than front seat. Windscreen bulletproof against 7.5 mm rifle fire.

ACCOMMODATION (single-seater): Enclosed cockpit for pilot, with rearward hinged canopy and Martin-Baker E9B (Ecuador), O9B (Oman) or IN9B (India) ejection seat as in two-seaters. Bulletproof windscreen, as in two-seat version.

SYSTEMS: Air conditioning and pressurisation systems maintain automatically, throughout the flight envelope, comfortable operating conditions for crew, and also control temperature in certain equipment bays. Two independent hydraulic systems, powered by two Vickers engine-driven pumps. Hydraulic pressure 207 bars (3,000 lb/sq in). First system (port engine) supplies one channel of each actuator for flying controls, hydraulic motors which actuate flaps and slats, landing gear retraction and extension, brakes and anti-skid units. Second system supplies other half of each flying control actuator, two further hydraulic motors actuating slats and flaps, airbrake and landing gear emergency extension jacks, nosewheel steering and wheel brakes. In addition, there is an emergency hydraulic power transfer unit. Electrical power provided by two 15 kVA AC generators, either of which can sustain functional and operational equipment without load shedding. DC power provided by two 4 kW transformer-rectifiers. Emergency power for essential instruments provided by 15 Ah battery and static inverter. De-icing, rain clearance and demisting standard. Liquid oxygen system, which also pressurises pilot's anti-*g* suit. Jaguar is fully power controlled in all three axis and is automatically stabilised as a weapons platform by gyros which sense disturbances and feed appropriate correcting data through a computer to the power control assemblies, in addition to human pilot manoeuvre demands. Power controls are all of duplex tandem arrangement, with mechanical and electrical servo-valves of Fairey platen design. Air-to-air combat capability can be enhanced by inclusion of roll/yaw dampers, to increase lateral stability, and by increasing slat and flap angles.

AVIONICS AND OPERATIONAL EQUIPMENT: Differ according to individual customer requirements; details have not all been made public, but first 40 for India have a Smiths head-up display similar to that in RAF Jaguars. Indian assembled Jaguars will have a raster cursive head-up display, Sagem inertial navigation and weapon aiming system, and Ferranti COMED 2045 combined map and electronic display.

Sepecat Jaguar A of the French Air Force

1994

ARMAMENT: Two 30 mm Aden or DEFA 553 cannon in lower fuselage aft of cockpit in single-seater; single Aden gun on port side in two-seater. One stores attachment on fuselage centreline and two under each wing. Centreline and inboard wing points can each carry up to 1,134 kg (2,500 lb) of weapons, outboard underwing points up to 567 kg (1,250 lb) each. Maximum external stores load, including overwing loads, 4,763 kg (10,500 lb). Typical alternative loads include one Martel AS.37 anti-radar missile and two 1,200 litre (317 US gallon; 264 Imp gallon) drop tanks; eight 1,000 lb bombs; various combinations of free-fall and retarded bombs, Hunting BL755 or Beluga cluster bombs, Matra R.550 Magic missiles and air-to-surface rockets, including the 68 mm SNEB rocket; a reconnaissance camera pack; or two drop tanks. Jaguar International can also carry two Matra Magic air-to-air missiles on overwing pylons; aircraft for Oman carry two AIM-9P Sidewinders on outboard underwing pylons.

DIMENSIONS, EXTERNAL:

Wing span	8.69 m (28 ft 6 in)
Wing chord: at root	3.58 m (11 ft 9 in)
at tip	1.13 m (3 ft 8½ in)
Wing aspect ratio	3.12
Length overall:	
incl probe: single-seater	16.83 m (55 ft 2½ in)
two-seater	17.53 m (57 ft 6¼ in)
excl probe: single-seater	15.52 m (50 ft 11 in)
two-seater	16.42 m (53 ft 10½ in)
Height overall	4.89 m (16 ft 0½ in)
Tailplane span	4.53 m (14 ft 10¼ in)
Wheel track	2.41 m (7 ft 11 in)
Wheelbase	5.69 m (18 ft 8 in)

AREAS:

Wings, gross	24.18 m² (260.27 sq ft)
Leading-edge slats (total)	1.05 m² (11.30 sq ft)
Trailing-edge flaps (total)	4.12 m² (44.35 sq ft)
Spoilers (total)	0.90 m² (9.67 sq ft)
Vertical tail surfaces (total)	3.90 m² (42.00 sq ft)
Horizontal tail surfaces (total)	7.80 m² (83.96 sq ft)

WEIGHTS AND LOADINGS:

Typical weight empty	7,000 kg (15,432 lb)
Normal T-O weight (single-seater, with full internal fuel and ammunition for built-in cannon)	10,954 kg (24,149 lb)
Max T-O weight with external stores	15,700 kg (34,612 lb)
Max wing loading	649.3 kg/m² (133 lb/sq ft)
Max power loading:	
Ardour Mk 804	219.6 kg/kN (2.15 lb/lb st)
Ardour Mk 811	209.9 kg/kN (2.06 lb/lb st)

PERFORMANCE:

Max level speed:	
at S/L	Mach 1.1 (729 kts; 1,350 km/h; 840 mph)
at 11,000 m (36,000 ft)	Mach 1.6 (917 kts; 1,699 km/h; 1,056 mph)
Landing speed	115 kts (213 km/h; 132 mph)
T-O run: 'clean'	565 m (1,855 ft)
with four 1,000 lb bombs	880 m (2,890 ft)
with eight 1,000 lb bombs	1,250 m (4,100 ft)
T-O to 15 m (50 ft) with typical tactical load	940 m (3,085 ft)
Landing from 15 m (50 ft) with typical tactical load	785 m (2,575 ft)
Landing run:	
normal weight, with brake-chute	680 m (2,230 ft)
overload weight, with brake-chute	670 m (2,200 ft)
Typical attack radius, internal fuel only:	
hi-lo-hi	460 n miles (825 km; 530 miles)
lo-lo-lo	290 n miles (537 km; 334 miles)
Typical attack radius with external fuel:	
hi-lo-hi	760 n miles (1,408 km; 875 miles)
lo-lo-lo	495 n miles (917 km; 570 miles)
Ferry range with external fuel	1,902 n miles (3,524 km; 2,190 miles)
g limits	+8.6/+12 (ultimate)

Sepecat Jaguar GR. 1 of the Royal Air Force ***(BAe)***

1995

UPDATED

SOKO/AVIOANE

PARTICIPATING COMPANIES:
Soko: see under Bosnia-Hercegovina
Utva: see under Yugoslavia
Avioane: see under Romania

VERIFIED

SOKO J-22 ORAO (EAGLE) AND AVIOANE IAR-93

TYPE: Single-seat close support, ground attack and tactical reconnaissance aircraft, with secondary capability as low level interceptor. Combat capable two-seat versions used also for advanced flying and weapon training.

PROGRAMME: Joint design by Yugoslav and Romanian engineers, started 1970 under original project name Yurom, to meet requirements of both air forces; two single-seat prototypes started in each country 1972, making simultaneous first flights 31 October 1974; first flight in each country of a two-seat prototype 29 January 1977; each manufacturer then built 15 preproduction aircraft (first flights 1978); series production began in Romania (IAv Craiova, now Avioane) 1979, in Yugoslavia (Soko) 1980; Yugoslav production stopped by damage and dismantling of Mostar factory in Bosnia-Hercegovina in 1992.

VERSIONS: **IAR-93A:** Romanian single- and two-seat versions with non-afterburning Viper Mk 632-41 turbojets; first flight 1981; production completed.

IAR-93B: Romanian single- and two-seat versions with afterburning Viper Mk 633-47 turbojets; first flight 1985. Avionics of two-seaters being upgraded, started early 1993.

IJ-22 and INJ-22 Orao 1: Yugoslav non-afterburning preseries aircraft for tactical reconnaissance (INJ-22 also a conversion trainer); total 15 single-seat IJ-22 and two-seat INJ-22 built.

NJ-22 Orao: Production two-seat tactical reconnaissance version, some with afterburning engines; first flight 18 July 1986; total 35 ordered (all delivered). Improved version planned.

J-22 Orao 2: Production single-seat attack version, some with afterburning engines; first flight 20 October 1983; in production.

UPGRADES: **Avioane:** IAR-93B. See Versions.

Soko: J-22 Orao. See separate entry.

OPERATORS: Versions of the J-22/IAR-93 are in service with the armed forces of the following countries: Romania (200), Bosnia-Hercegovina (Serb forces 6) and Yugoslavia (40).

DESIGN FEATURES: Wings of NACA 65A-008 (modified) section, shoulder-mounted with 0° incidence and 3° 30′ anhedral from roots; sweepback 35° at quarter-chord and approximately 43° on outer leading-edges; incidence 0°; inboard leading-edges extended forward (except on prototypes/preproduction) at 65° sweepback; small boundary

Soko J-22 Orao of the former Yugoslav Air Force *(Peter Steinemann)*

1995

layer fence on upper surface of each outer panel; fuselage has hydraulically actuated door type perforated airbrake underneath on each side forward of mainwheel bay, dorsal spine fairing, pen-nib fairing above exhaust nozzles, and detachable rear portion for access to engine bays; all-sweptback tail surfaces, plus ventral strake beneath rear fuselage each side.

FLYING CONTROLS: Internally balanced plain ailerons; single-slotted trailing-edge flaps, two-segment leading-edge slats (three-segment on prototype/preproduction aircraft), low-set all-moving tailplane and inset rudder, all hydraulically actuated (except leading-edge slats: electrohydraulic), with PPT servo-actuators for primary control surfaces; no aileron or rudder tabs.

STRUCTURE: Conventionally built, almost entirely of aluminium alloy except for honeycomb rudder and tailplane on later production aircraft; two-spar wings with ribs, stringers and partially machined skin; wing spar box forms integral fuel tanks on production aircraft; preseries aircraft have rubber fuel cells, forward of which are sandwich panels.

LANDING GEAR: Hydraulically retractable tricycle type of Messier-Bugatti design, with single-wheel hydraulically steerable nose unit and twin-wheel main units. All units retract forward into fuselage. Two-stage PPT hydro/nitrogen shock-absorber in each unit. Mainwheels and tubeless tyres size 615 × 225 mm; pressure 4.5 bars (65.3 lb/sq in). Nosewheel and tubeless tyre size 450 × 190 mm, pressure 4.0 bars (58.0 lb/sq in), on all afterburning versions. Hydraulic disc brakes on each mainwheel unit, and electronic anti-skid system. Minimum ground turning radius 7.00 m (22 ft 11½ in). Bullet fairing at base of rudder contains hydraulically deployed 4.2 m (13 ft 9½ in) diameter braking parachute.

POWER PLANT (non-afterburning versions): Two 17.79 kN (4,000 lb st) Turbomecanica/Orao (licence-built Rolls-Royce) Viper Mk 632-41 turbojets, mounted side by side in rear fuselage; air intake on each side of fuselage, below cockpit canopy. Preseries aircraft have seven fuselage tanks and two collector tanks, with combined capacity of 2,480 litres (655 US gallons; 545.5 Imp gallons), and two 235 litre (62 US gallon; 51.75 Imp gallon) rubber fuel cells in wings, giving total internal fuel capacity of 2,950 litres (779 US gallons; 649 Imp gallons). J-22 and NJ-22 have five fuselage tanks, two collectors and two integral wing tanks for total internal capacity of 3,120 litres (824 US gallons; 686 Imp gallons). Provision for carrying three 500 litre (132 US gallon; 110 Imp gallon) auxiliary fuel tanks, one on underfuselage stores attachment and one inboard under each wing. Pressure refuelling point in fuselage below starboard air intake; gravity refuelling points in fuselage near starboard wing trailing-edge and in each external tank. Oil capacity 6.25 litres (1.65 US gallons; 1.37 Imp gallons).

POWER PLANT (afterburning versions): Two Turbomecanica/Orao (licence-built Rolls-Royce) Viper Mk 633-47 turbojets, each rated at 17.79 kN (4,000 lb st) dry and 22.24 kN (5,000 lb st) with afterburning. Production aircraft have five fuselage and two collector tanks, plus two integral wing tanks, giving total internal capacity of 3,120 litres (824 US gallons; 686 Imp gallons). Drop tanks, refuelling points and oil capacity as for non-afterburning versions.

ACCOMMODATION: Single-seat or tandem two-seat cockpit(s), with Martin-Baker zero/zero seat for each occupant (RU10J in IAR-93, Y-10LB in Orao), capable of ejection through canopy. Canopy of single-seat IAR-93A and J-22 hinged at rear and actuated electrically to open upward; single-seat IAR-93B, and all two-seaters, have manually operated canopies opening sideways to starboard. All accommodation pressurised, heated and air conditioned. Dual controls in two-seat versions.

SYSTEMS: Bootstrap type environmental control system for cockpit pressurisation (maximum differential 0.214 bar; 3.1 lb/sq in), air conditioning, and windscreen de-icing/demisting. Two Prva Petoletka independent hydraulic systems, each of 207 bars (3,000 lb/sq in) pressure, flow rate 48 litres (12.7 US gallons; 10.6 Imp gallons)/s, for actuation of leading-edge slats, trailing-edge flaps, ailerons, tailplane, rudder, airbrakes, landing gear extension/retraction, mainwheel brakes, nosewheel steering, brake-chute and afterburner nozzles. Back-up system for landing gear, wheel brakes and primary flight controls. No pneumatic system. Main electrical system is 28 V DC, supplied by two Lucas BC-0107 9 kW engine-driven starter/generators through two voltage regulators and a switching system, and a 24 V 36 Ah Ni/Cd battery; four 300 VA static inverters for 115 V AC power at 400 Hz. High pressure (150 bars; 2,175 lb/sq in) gaseous oxygen system for crew.

AVIONICS: Standard avionics include VHF/UHF air-to-air and air-to-ground com radio (20 W transmission power); gyro unit (Honeywell SGP500 twin-gyro platform in Orao), radio altimeter, ADF, radio compass and marker beacon receiver; IFF (Romanian aircraft only); and GEC-Marconi three-axis stability augmentation system, incorporating a basic bank/attitude hold autopilot and emergency wings-level facility. Oraos also have Collins VIR-30 VOR/ILS and DME-40; and Iskra SO-1 RWR. Reconnaissance pod (1-18 GHz VRT10 radio surveillance pod, photographic/jamming pod or photographic/IR pod on IJ/INJ-22), or P10-65-13 passive jamming pod (Orao) on underfuselage station. Chaff and IR decoy launch pods (up to three per aircraft) can also be carried.

EQUIPMENT: Landing light under nose, forward of nosewheel bay; taxying light on nosewheel shock strut.

ARMAMENT (IAR-93A/B): Two 23 mm GSh-23L twin-barrel cannon in lower front fuselage, below engine air intakes, with 200 rds/gun. Gun camera and GEC-Marconi D282 gyro gunsight. Five external stores stations, of which inboard underwing pair and fuselage centreline station are each stressed for loads up to 500 kg (1,102 lb); outboard underwing stations stressed for up to 300 kg (661 lb) each, giving maximum external stores load of 1,500 kg (3,307 lb). Typical weapon loads can include two or three 500 kg bombs; four or five 250 kg bombs; four multiple carriers each with three 100 kg or 50 kg bombs; two such multiple carriers plus two L-57-16MD launchers each with 16 57 mm rockets; four L-57-16MD launchers; four launchers each with two 122 mm, one 128 mm or one 240 mm rocket (122 and 240 mm not used on Orao); a GSh-23L cannon pod with four L-57-16MD rocket launchers; four 160 kg KPT-150 or similar munition dispensers; or four L-57-32 launchers each with 32 57 mm rockets. Some IAR-93Bs equipped to carry up to eight air-to-air missiles, on twin launch rails, on four underwing stations. Centreline and inboard underwing points can carry drop tanks.

ARMAMENT (Orao): Guns, gun camera, drop tanks and centreline reconnaissance pod as for IAR-93. All four wing stations stressed for 500 kg (1,102 lb), and fuselage station for 800 kg (1,763 lb), giving maximum external stores capacity of 2,800 kg (6,173 lb). Typical weapon loads include five 50 kg, 100 kg, 250 kg or 500 kg bombs; four multiple carriers for total of 12 50 kg or 100 kg or eight 250 kg bombs; four FLAB-350 napalm bombs (each 360 kg; 794 lb); five BL755 bomblet dispensers, or eight on four multiple carriers; 16 BRZ-127 5 in HVAR rockets; four pods of L-57-16MD or L-128-04 (4 × 128 mm) rockets, or eight pods on multiple carriers; five 500 kg AM-500 sea mines; or two launch rails for AGM-65B Maverick or Yugoslav developed Grom air-to-surface missiles. The 100 kg and 250 kg bombs can be parachute retarded.

DIMENSIONS, EXTERNAL:	
Wing span	9.30 m (30 ft 6¼ in)
Wing chord: at root	4.20 m (13 ft 9⅜ in)
at tip	1.40 m (4 ft 7⅛ in)
Wing aspect ratio	3.33
Length overall, incl probe:	
single-seater	14.90 m (48 ft 10⅝ in)
two-seater	15.38 m (50 ft 5½ in)
Length of fuselage: single-seater	13.02 m (42 ft 8½ in)
two-seater	14.44 m (47 ft 4½ in)
Fuselage: Max width	1.62 m (5 ft 3¾ in)
Height overall	4.52 m (14 ft 10 in)
Tailplane span	4.59 m (15 ft 0¾ in)
Wheel track (c/l of shock struts)	2.50 m (8 ft 2½ in)
Wheelbase: single-seater	5.40 m (17 ft 8½ in)
two-seater	5.88 m (19 ft 3½ in)
AREAS:	
Wings, gross	26.00 m² (279.86 sq ft)
Ailerons (total)	2.39 m² (25.73 sq ft)
Trailing-edge flaps (total)	3.33 m² (35.84 sq ft)
Leading-edge slats (total)	2.22 m² (23.90 sq ft)
Fin	3.55 m² (38.21 sq ft)
Rudder	0.88 m² (9.47 sq ft)
Tailplane	7.31 m² (78.68 sq ft)
WEIGHTS AND LOADINGS:	
Weight empty, equipped: IAR-93A	6,150 kg (13,558 lb)
IAR-93B	5,750 kg (12,676 lb)
J-22	5,500 kg (12,125 lb)
IJ-22	5,755 kg (12,687 lb)
Max internal fuel: IAR-93A	2,457 kg (5,416 lb)
IAR-93B	2,400 kg (5,291 lb)
J-22	2,430 kg (5,357 lb)
IJ-22	2,360 kg (5,203 lb)
Max external stores load: IAR-93A	1,500 kg (3,307 lb)
IAR-93B	2,500 kg (5,511 lb)
J-22, IJ-22	2,800 kg (6,173 lb)
Normal T-O weight clean:	
IAR-93A	8,826 kg (19,458 lb)
IAR-93B	8,400 kg (18,519 lb)
J-22	8,170 kg (18,012 lb)
IJ-22 (with recce pod)	8,500 kg (18,739 lb)
Max T-O weight: IAR-93A	10,326 kg (22,765 lb)
IAR-93B	10,900 kg (24,030 lb)
J-22	11,080 kg (24,427 lb)
IJ-22	9,500 kg (20,944 lb)
Max landing weight: IAR-93A	8,826 kg (19,458 lb)
IAR-93B	9,360 kg (20,635 lb)
J-22	9,500 kg (20,944 lb)
IJ-22	8,600 kg (18,960 lb)
Max wing loading: IAR-93A	397.1 kg/m² (81.3 lb/sq ft)
IAR-93B	419.2 kg/m² (85.9 lb/sq ft)
J-22	426.1 kg/m² (87.3 lb/sq ft)
IJ-22	365.4 kg/m² (74.8 lb/sq ft)
Max power loading:	
IAR-93A	289.8 kg/kN (2.84 lb/lb st)
IAR-93B	245.2 kg/kN (2.40 lb/lb st)

Soko NJ-22 Orao tandem-seat operational trainer of the former Yugoslav Air Force

1994

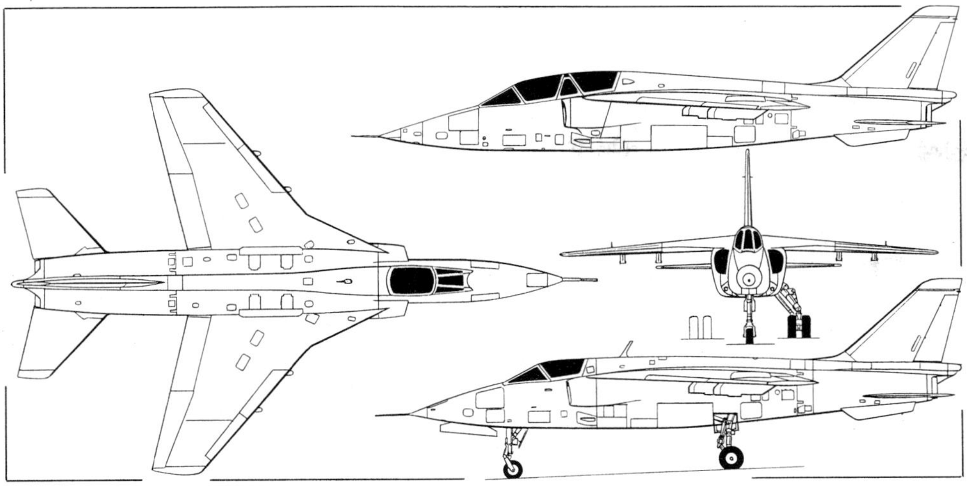

Soko single-seat J-22 Orao with additional side view (top) of two-seat NJ-22 *(Jane's/Dennis Punnett)*
1994

Soko J-22 Oraos of the former Yugoslav Air Force
1994

J-22 249.2 kg/kN (2.44 lb/lb st)
IJ-22 213.7 kg/kN (2.09 lb/lb st)

PERFORMANCE (IAR-93A: at max T-O weight, IAR-93B: at 8,400 kg; 18,519 lb T-O weight, J-22: at clean T-O weight with 50% internal fuel, IJ-22: at T-O weight of 8,500 kg; 18,739 lb, except where indicated):

Max level speed at S/L:
IAR-93A 577 kts (1,070 km/h; 665 mph)
IAR-93B 586 kts (1,086 km/h; 675 mph)
J-22 610 kts (1,130 km/h; 702 mph)
IJ-22 566 kts (1,050 km/h; 652 mph)

Max level speed at altitude:
J-22 at 11,000 m (36,100 ft) Mach 0.96
IJ-22 at 8,000 m (26,250 ft) Mach 0.93

Max cruising speed: IAR-93A at 7,000 m (22,965 ft) 394 kts (730 km/h; 453 mph)
IAR-93B at 5,000 m (15,240 ft) 587 kts (1,089 km/h; 676 mph)
J-22, IJ-22 at 11,000 m (36,100 ft) Mach 0.7

Stalling speed at S/L:
IAR-93A 130 kts (241 km/h; 150 mph)
IAR-93B 148 kts (274 km/h; 171 mph)
J-22, IJ-22, gear and flaps down 100 kts (185 km/h; 115 mph)

Max rate of climb at S/L:
IAR-93A 2,040 m (6,693 ft)/min
IAR-93B 3,900 m (12,800 ft)/min
J-22 5,340 m (17,520 ft)/min
IJ-22 2,280 m (7,480 ft)/min

Service ceiling: IAR-93A 10,500 m (34,450 ft)
IAR-93B 13,600 m (44,625 ft)
J-22 15,000 m (49,210 ft)
IJ-22 13,500 m (44,290 ft)

Time to 6,000 m (19,685 ft): J-22 1 min 20 s
IJ-22 3 min 12 s

T-O run: IAR-93A 1,500 m (4,921 ft)
IAR-93B 800 m (2,625 ft)
J-22 at 9,443 kg (20,818 lb) with four BL755s 880 m (2,888 ft)
IJ-22 1,200 m (3,937 ft)

T-O to 15 m (50 ft): IAR-93A 1,600 m (5,249 ft)
IAR-93B 1,150 m (3,775 ft)
J-22 1,255 m (4,118 ft)
IJ-22 1,800 m (5,906 ft)

Landing from 15 m (50 ft): IAR-93A 1,650 m (5,413 ft)
IAR-93B 1,520 m (4,987 ft)
IAR-93B with brake-chute 990 m (3,250 ft)
J-22 1,295 m (4,249 ft)
IJ-22 1,400 m (4,594 ft)

Landing run: IAR-93A 720 m (2,362 ft)
IAR-93B 1,050 m (3,445 ft)
IAR-93B with brake-chute 690 m (2,265 ft)
J-22 755 m (2,477 ft)
IJ-22 850 m (2,789 ft)

Landing run with brake-chute:
IAR-93A, IAR-93B 670 m (2,200 ft)
J-22 530 m (1,739 ft)
IJ-22 600 m (1,969 ft)

Tactical radius, IAR-93A, IAR-93B:
lo-lo-lo-with four rocket launchers, 5 min over target 140 n miles (260 km; 161 miles)
hi-hi-hi patrol with three 500 l drop tanks, 45 min over target 205 n miles (380 km; 236 miles)
lo-lo-hi with two rocket launchers, six 100 kg bombs and one 500 l drop tank, 10 min over target 243 n miles (450 km; 280 miles)
hi-hi-hi with four 250 kg bombs and one 500 l drop tank, 5 min over target 286 n miles (530 km; 329 miles)

Tactical radius, J-22:
hi-lo-hi with four BL755s and 1,500 l centreline drop tank, 2 min reheat 282 n miles (522 km; 324 miles)
hi-lo-hi with four 500 kg air mines and 1,500 l centreline drop tank, 1.2 min reheat 248 n miles (460 km; 286 miles)
hi-lo-hi with eight 250 kg bombs and one 500 l drop tank, 2 min reheat 200 n miles (370 km; 300 miles)

Ferry range:
IAR-93A, IAR-93B, with three 500 l drop tanks 1,025 n miles (1,900 km; 1,180 miles)
J-22, with two 500 l drop tanks, at 6,000 m (19,685 ft) 712 n miles (1,320 km; 820 miles)

g limits: IAR-93A, IAR-93B, J-22, IJ-22 +8/−4.2

UPDATED

SOKO/AVIOANE J-22 (EAGLE) UPGRADE

TYPE: Combat aircraft upgrade.

PROGRAMME: Proposal to upgrade with modern avionics including integration of radar and inertial nav/attack system via new databus. Plans also exist to retrofit air intake and wing leading-edge de-icing system.

VERIFIED

TRANSALL

ARBEITSGEMEINSCHAFT TRANSALL

AIRFRAME COMPANIES:
Aerospatiale: see under France
Daimler-Benz Aerospace: see under Germany

UPDATED

TRANSALL C-160 (Second Series)

TYPE: Twin-turboprop cargo and para-drop aircraft.

PROGRAMME: The Transall (Transporter Allianz) group was formed in January 1959 by MBB, Nord Aviation and VFW, and undertook development and production of the C-160 twin-turboprop transport for the air forces of France (50), Germany (90), South Africa (9) and Turkey (20). This initial production was shared between the three participating companies and ended in 1972. Production of a second series was authorised in 1977 to meet an additional French order and requests from other countries. Under a new industrial agreement, signed on 29 October 1976, production was shared Aerospatiale (50 per cent) and MBB (50 per cent), with a single final assembly line at Toulouse. Aerospatiale was responsible for the manufacture of the

Transall C-160 transport operating with the International Red Cross in Uganda *(Gerald C. Cauderay)*
1994

Transall C-160 transport of the German Air Force *(Daimler-Benz Aerospace Airbus)*

1994

wings, wing/fuselage fairings, fuselage doors, emergency exits and engine nacelles. Manufacture of the fuselage, main landing gear fairings and all tail surfaces were the responsibility of MBB. The engine was manufactured by Rolls-Royce, SNECMA, MTU and FN Herstal. Components were airlifted to Toulouse by Super Guppy transport for final assembly and flight testing. The main improvements in this second series production aircraft are updated avionics, and extended range resulting from a reinforced wing with an optional additional fuel tank in the centre-section.

The French Air Force placed an initial order for 25 (increased to 29 in 1982). There were 10 fitted at the outset with in-flight refuelling equipment (hose reel and drogue type) in the port main landing gear fairing to permit their operation as tankers; five others incorporated provisions for this equipment and are capable of rapid adaptation to the tanker role if needed. All have a 4.00 m (13 ft 2 in) receiver boom mounted above and behind the flight deck. They are capable of refuelling carrier-based aircraft of the French Navy, as well as French Air Force combat aircraft.

First flight of three aircraft of the new series took place at Toulouse on 9 April 1981. Deliveries to the French Air Force began in 1981 and went into service with Escadrons 1/64 'Béarn' and 2/64 'Anjou' of the 64e Escadre de Transport at Evreux. The four additional aircraft were equipped as communications relay aircraft on behalf of the nation's nuclear deterrent forces from 1987. To ensure maximum survivability and effectiveness in a nuclear combat environment, the aircraft are equipped as flight refuelling/receivers. They are designated **Astarté** *(Avion-Station-relais de Transmission exceptionelles)* and operated under the overall Ramses *(réseau amont maillé stratégique et de survie)* programme. Equipment includes a US built VLF system, installed by Thomson-CSF, of the kind fitted to Tacamo EC-130s of the US Navy.

In addition to the French order, six second series Transalls have been delivered to the Indonesian government, and are operated by Pelita Air Service to assist in the country's transmigration of inhabitants from Java to less heavily populated islands. Production of the second series ended in 1985.

VERSIONS: **C-160D:** German Air Force designation.

C-160F: French Air Force designation.

C-160G: Gabriel elint version.

C-160H: Astarté. Relay aircraft of French Air Force.

UPGRADES: **Aerospatiale:** C-160H Astarté. See following.

Aerospatiale: C-160G Gabriel. See following.

Aerospatiale/COTAM: French Air Force C-160 Transall upgrade. See separate entry in Aerospatiale section.

Daimler-Benz Aerospace Airbus: German Air Force ANA/FRA upgrade. See separate entry.

GEC-Marconi: Awarded contract in January 1994 to supply its high integration air data computer (HIADC) to the German Air Force C-160D ANA/FRA upgrade programme.

Rockwell: Awarded contract in March 1994 to supply FMS-800 flight management system (FMS) as well as GPS receivers and to automate routine tasks as part of the German Air Force ANA/FRA upgrade programme.

OPERATORS: Versions of the Transall are in service with the armed forces of the following countries: France (76), Germany (85) and Turkey (20).

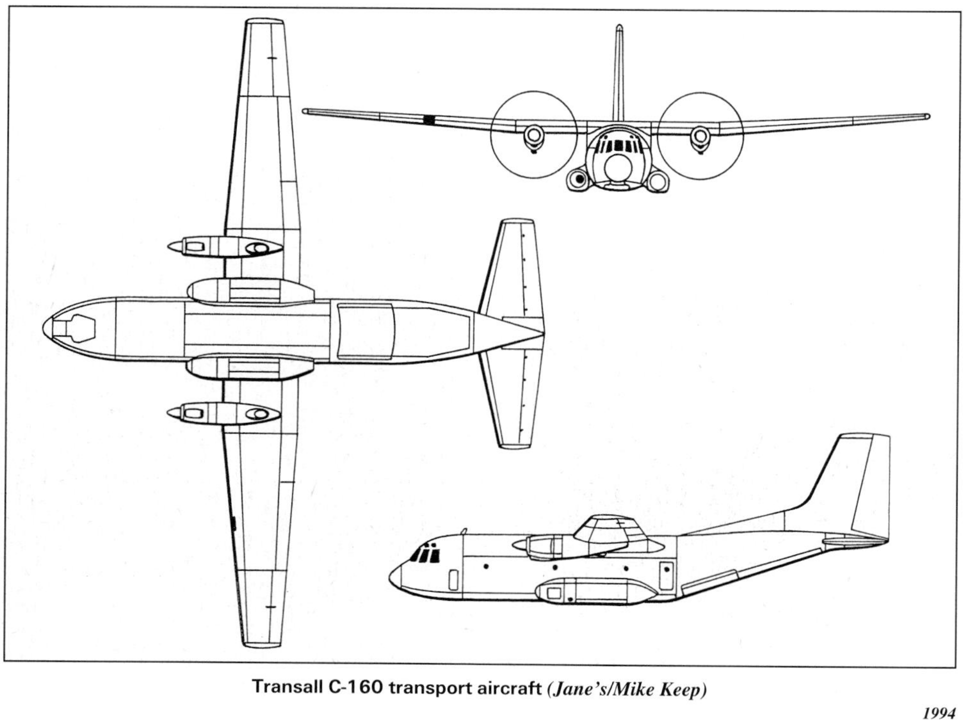

Transall C-160 transport aircraft *(Jane's/Mike Keep)*

1994

DESIGN FEATURES: Cantilever high-wing monoplane. Dihedral on outer wings 3° 26′. Cantilever tailplane.

FLYING CONTROLS: All-metal ailerons and hydraulically operated double-slotted flaps. Hydraulically operated airbrakes (inboard, above and below wings) and spoilers (outboard) forward of flaps on each wing.

STRUCTURE: All-metal two-spar structure designed on fail-safe principles. Wing in three sections; a centre-section, which carries the engines, and two outer panels. The fuselage is of aluminium alloy (2024-T3) semi-monocoque structure of circular basic section, flattened at the bottom, and designed on fail-safe principles. Underside of upswept rear fuselage lowers to form loading ramp for vehicles. The tail unit is a cantilever aluminium alloy (2024-T3) structure with a large dorsal fin.

LANDING GEAR: Retractable tricycle type, built by Messier-Hispano-Bugatti and Liebherr Aerotechnik. Hydraulic retraction and hydraulic/pneumatic shock-absorption. Each main unit comprises two pairs of wheels in tandem and is mounted inside a fairing on the side of the fuselage. Wheels can be raised to lower the fuselage for loading. Steerable twin-wheel nose unit. Mainwheel tyres size 15.00 × 16; nosewheel tyres size 12.5 × 16. Tyre pressure 3.79 bars (55 lb/sq in) on main units, 3.14 bars (45.5 lb/sq in) on nose unit. Messier-Hispano-Bugatti brakes.

POWER PLANT: Two 4,549 kW (6,100 ehp) Rolls-Royce Tyne RTy.20 Mk 22 turboprop engines, each driving a Ratier Forest built BAe 4/8000/6 four-blade, constant-speed, fully feathering, reversible-pitch propeller. Single-point pressure refuelling: gravity refuelling available optionally. Fuel in four integral wing tanks with total capacity of 19,050 litres (5,032 US gallons; 4,190 Imp gallons). Additional wing centre-section tank optional, capacity 9,000 litres (2,379 US gallons; 1,980 Imp gallons). Probe in-flight refuelling. Hose reel and drogue type in-flight refuelling tanker equipment optional. Water-methanol usable capacity 318.5 litres (84 US gallons; 70 Imp gallons). Oil capacity (total) 68.4 litres (18 US gallons; 15 Imp gallons).

ACCOMMODATION: Pressurised accommodation for crew of four, comprising pilot, co-pilot, flight engineer and navigator. Typical payloads include 93 troops or 61-68 fully equipped paratroops; 62 stretchers and four attendants; armoured vehicles, tanks and tractors not exceeding maximum permissible payload weight. Flight deck and cargo compartment air conditioned and pressurised in flight and on the ground. Power-assisted controls. Paratroop door on each side immediately aft of the landing gear fairings; hydraulically operated rear-loading ramp. The floor and all doors are of truckbed height. The floor is provided with lashing points of 5,000 kg (11,023 lb) capacity, arranged in a 51 cm (20 in) grid, and 12,000 kg (26,455 lb) capacity on the sidewalls, and is stressed to carry large military vehicles. Loads which cannot be driven in can be taken onboard rapidly by an automatic translation and stowing system. Individual loads of up to 8,000 kg (17,637 lb) can be airdropped, including drops at low level altitude (3-9 m; 10-30 ft or during touch-and-go).

Normalair-Garrett pressurisation and air conditioning system, differential 0.302-0.322 bars (4.38-4.67 lb/sq in). Two separate primary hydraulic systems, pressure 175 bars (2,500 lb/sq in), for flying controls, loading ramp, landing gear, wheel brakes, flaps, spoilers, airbrakes, nosewheel steering and other auxiliaries. Two more systems, pressure 175 bars (2,500 lb/sq in), for emergency and ground services, as well as a handpump-driven emergency system.

AC electrical system includes two 60 VA 380-580 Hz generators, one 60 VA 400 Hz generator and two 9 kVA 400 Hz generators. 28 V DC system and 40 Ah batteries. Garrett GTCP-85-160 APU in forward part of port main undercarriage fairing.

AVIONICS AND EQUIPMENT: Socrat TRAP138 VHF; TRT TRAP139 UHF; C.I.T. TRAP33 HF; TEAM AS1227B PA system; TEAM TF.AP14 intercom; EAS RNA720 VOR/ILS; Socrat NRAN19 ADF; LMT NRAP1A Tacan; LMT NRAI2A IFF; Omera ORB37 weather radar; TRT NR.AS11 radio altimeter; CSF MR250 marker beacon receiver; Jaeger 6411 encoding altimeter; SFIM CADV automatic flight control system incorporating two verical gyros, two compasses, PA51 autopilot, DV86 flight director, RDN72 Doppler navigation radar.

Export version available with Collins 618M-3A VHF com; Collins 628T1 HF; Collins 51RV4 VOR/ILL; Collins DF206 ADF; Collins 860E-4 DME; Collins 621A6A ATC; Bendix RDR1E weather radar; VDO HM510-51 encoding altimeter; Collins 51Z4 marker beacon receiver; Collins FPC75 GPWS.

Transall C-160 aircraft of the French Air Force *(Aerospatiale)*

1995

DIMENSIONS, EXTERNAL:

Wing span	40.00 m (131 ft 3 in)
Wing chord: at root	4.84 m (15 ft 10½ in)
at tip	2.428 m (7 ft 11½ in)
mean	4.176 m (13 ft 8½ in)
Wing aspect ratio	10
Length overall, excl probe	32.40 m (106 ft 3½ in)
Fuselage: Max diameter	4.30 m (14 ft 1¼ in)
Height overall	11.65 m (38 ft 2¾ in)
Tailplane span	14.50 m (47 ft 7 in)
Wheel track	5.10 m (16 ft 9 in)
Wheelbase	10.48 m (34 ft 4½ in)
Propeller diameter	5.49 m (18 ft 0 in)
Distance between propeller centres	10.90 m (35 ft 9¼ in)
Crew door (fwd, port): Height	1.22 m (4 ft 0 in)
Width	0.62 m (2 ft 0½ in)
Paratroop door (each side): Height	1.90 m (6 ft 2½ in)
Width	0.90 m (3 ft 0 in)
Rear-loading ramp: Length	3.70 m (12 ft 1½ in)
Width	3.15 m (10 ft 3½ in)
Emergency exits:	
Main hold, fwd, stbd side (one):	
Height	0.88 m (2 ft 10½ in)
Width	0.54 m (1 ft 9¼ in)
Flight deck roof (one); roof of main hold, fwd (one); and two in roof of main hold at rear (one each side of dorsal fin): Height	0.54 m (1 ft 9¼ in)
Width	0.64 m (2 ft 1¼ in)

DIMENSIONS, INTERNAL:

Cabin, excl flight deck and ramp:	
Length	13.51 m (44 ft 4 in)
Max width	3.15 m (10 ft 3½ in)
Max height	2.98 m (9 ft 8½ in)
Floor area	42.6 m² (458.5 sq ft)
Volume	115.0 m³ (4,061 cu ft)
Cabin, incl ramp: Length	17.21 m (56 ft 6 in)
Floor area	54.25 m² (584 sq ft)
Volume	139.9 m³ (4,940 cu ft)

AREAS:

Wings, gross	160 m² (1,722 sq ft)
Ailerons (total)	6.88 m² (74.06 sq ft)
Trailing-edge flaps (total, extended)	34.54 m² (371.8 sq ft)
Spoilers (total)	0.80 m² (8.61 sq ft)
Fin: excl dorsal fin	29.50 m² (317.5 sq ft)
incl dorsal fin	36.00 m² (387.5 sq ft)
Rudder	10.20 m² (109.8 sq ft)
Tailplane	33.50 m² (360.6 sq ft)
Elevators	10.30 m² (110.9 sq ft)

WEIGHTS AND LOADINGS:

Min operating weight empty	28,000 kg (61,730 lb)
Typical operating weight empty	29,000 kg (63,935 lb)
Max fuel load: standard	15,295 kg (33,720 lb)
optional	22,520 kg (49,648 lb)
Max payload	16,000 kg (35,275 lb)
Max T-O weight	51,000 kg (112,435 lb)
Max zero-fuel weight	45,000 kg (99,210 lb)
Max landing weight	47,000 kg (103,615 lb)
Max wing loading	319 kg/m² (65.34 lb/sq ft)
Max power loading	5.61 kg/kW (9.22 lb/ehp)

PERFORMANCE (at max T-O weight except where indicated):

Never-exceed speed:	
at 4,875-9,145 m (16,000-30,000 ft)	Mach 0.64
below 4,875 m (16,000 ft)	320 kts (593 km/h; 368 mph)
Max level speed at 4,875 m (16,000 ft)	277 kts (513 km/h; 319 mph)
Stalling speed, flaps down	95 kts (177 km/h; 110 mph)
Max rate of climb at S/L	396 m (1,300 ft)/min
Rate of climb at S/L, one engine out	91 m (300 ft)/min
Service ceiling at 45,000 kg (99,210 lb) AUW	8,230 m (27,000 ft)
Service ceiling, one engine out at 45,000 kg (99,210 lb) AUW	3,050 m (10,000 ft)
T-O run, 20° flap	715 m (2,346 ft)
T-O to 10.5 m (35 ft), 20° flap	990 m (3,248 ft)
Landing from 15 m (50 ft), 40° flap, at max landing weight without propeller reversal	869 m (2,850 ft)
Landing run, normal	550 m (1,800 ft)
Min ground turning radius	28.60 m (93 ft 10 in)
Range, reserve of 5% initial fuel, allowance for 30 min hold at S/L, OWE of 29,000 kg (63,935 lb):	
with 8,000 kg (17,640 lb) payload	2,750 n miles (5,095 km; 3,166 miles)
with 16,000 kg (35,275 lb) payload	1,000 n miles (1,853 km; 1,151 miles)
Max ferry range with centre-section wing tank	4,780 n miles (8,858 km; 5,504 miles)

UPDATED

TRANSALL C-160 GABRIEL

TYPE: Electronic surveillance upgrade.

PROGRAMME: Aerospatiale responsible for integrating a complete sigint electronic intelligence system which was developed by Thomson-CSF. The Gabriel aircraft converted were two second series models which are now in service with the French Air Force.

DESIGN FEATURES: The package includes a sigint system, designed to undertake: detection, analysis, identification and localisation of radar emissions and a comint subsystem designed for detection, interception, classification, listen-in, analysis and localisation of radio communications stations. Upgrade work included structural modifications for antenna installation and a new layout for 10 operators.

AVIONICS: System includes: Thomson UHF TRAP 139, Thomson HF 200 W TRAP, Collins HF 1 kW HF121 with antenna trailing wire, TEAM 150 CCS 55A, TEAM ICS 47A, two Sagem IMS Ulyss 45C, Thomson DME820, two Sextant HSI (4 × 5 in)/ADI (5 × 5 in), Thomson IFF MRAI 7.

NEW ENTRY

TRANSALL C-160H ASTARTE

TYPE: Airborne communications relay upgrade.

PROGRAMME: Aerospatiale modified four second series C-160 Transall aircraft to carry a Thomson-CSF designed communication centre and a Rockwell International designed VLF transmitter. Aircraft primarily designated for communications with submerged nuclear ballistic missile submarines of the French Navy. Aircraft can also be used for communications with other French weapon systems. The four aircraft are a major element in the RAMSES communication network.

DESIGN FEATURES: Main elements of upgrade consist of: new operational layout including communications centre for five operators, VLF transmitter and Dual trailing wire antennae as well as protection from nuclear electronic pulse.

NEW ENTRY

IRAQ

IAF

IRAQI AIR FORCE

Ministry of Defence, Bab Al-Muadam, Baghdad

BAGHDAD 1

TYPE: AEW aircraft upgrade.

PROGRAMME: AEW version of Iraqi Il-76MD ('Candid-B') converted by Iraqi Air Force; first shown at Iraqi defence exhibition 1989; locally built Thomson-CSF Tigre surveillance radar, normally trailer-mounted, installed inverted under tail of Il-76; signal processing modified to reduce ground clutter; radio and radar ESM added; navigation system modified; claimed to have well over 180° scan; transmission to ground by voice or real-time datalink; Iraqi-designed generator; can track and identify targets at 189 n miles (350 km; 217 miles); four radar operators; reported used during Iran-Iraq war. Status of aircraft currently unknown.

VERIFIED

ADNAN 1

TYPE: AEW aircraft upgrade.

PROGRAMME: Improved AEW aircraft based on Il-76MD ('Candid-B') with back-to-back radar antennae in rotating

Ilyushin Il-76 Adnan 1 AEW aircraft

1994

aerial housing (rotodome); two large strakes under rear fuselage maintain directional stability and nose-down pitching moment at low airspeed; rotodome diameter estimated at 9 m (29 ft 6 in) and about 4 m (13 ft 1 in) above fuselage; detection range reported as 'few hundred kilometres'.

Baghdad television reported start of testing in December 1990; at least three Adnans reported operational at start of Gulf War Desert Storm on 17 January 1991; one put out of action at Al Taqaddum 23 January 1991; two others were flown to Iran late January.

Tanker version of Il-76 shown on Iraqi television 20 January 1991 with single hose/drogue pod at base of rear-loading ramp. Status of aircraft currently unknown.

VERIFIED

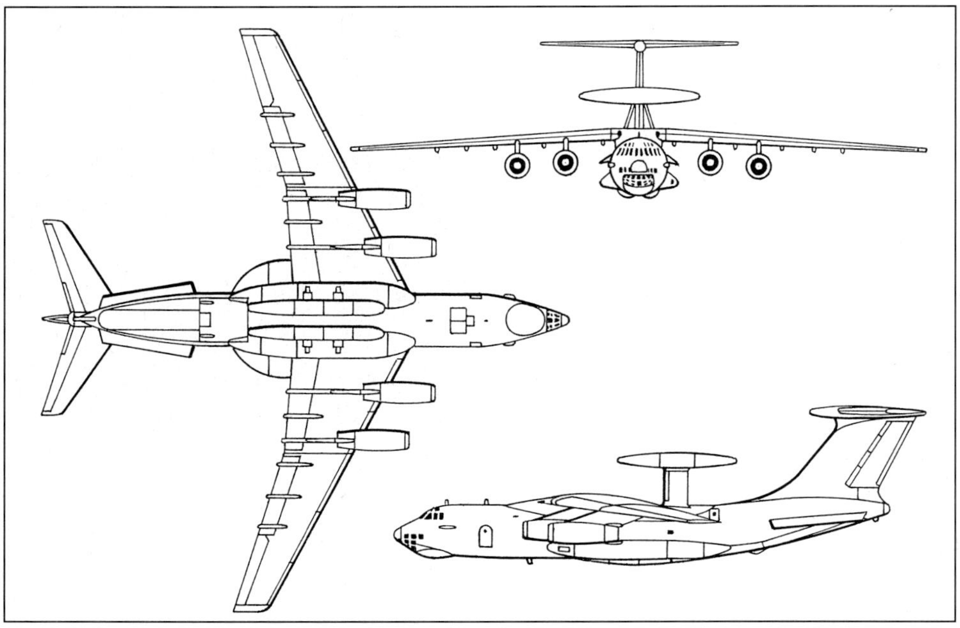

Ilyushin Il-76 Adnan 1 AEW aircraft *(Jane's/Mike Keep)*
1994

ISRAEL

ELBIT

ELBIT LTD

Advanced Technology Centre, PO Box 539, Haifa 31053
Telephone: 972 (4) 315315
Fax: 972 (4) 550002/551623
MARKETING PROMOTION MANAGER: Ilana Gelfer

Company origins can be traced back to 1971 when it began to design, develop and manufacture airborne systems in the fields of weapons delivery, navigation, stores management, display systems and mission computers. Elbit has been involved in a number of upgrade programmes undertaken by IAI including A-4, CH-53 (as systems integrator), F-4, F-5, F-15, F-16, Kfir and Mirage. The company announced 24 May 1993 that it had signed an agreement worth $300 million to upgrade 100 MiG-21 aircraft of the Romanian Air Force over a six-year period. This programme has suffered some delay over negotiations with Mikoyan to achieve design approval of Romanian requirement for airframe changes. In June 1995 the company announced that it was now offering complete modular avionics and structural upgrade packages for the CH-53, F-4, F-5, F-16 and MiG-21.

NEW ENTRY

ELBIT (MIKOYAN) MiG-21 UPGRADE

TYPE: Combat aircraft upgrade.
PROGRAMME: Signed agreement 24 May 1993 to upgrade 100 MiG-21 aircraft of the Romanian Air Force. Contract commenced in 1994 and is to last six years. Elbit is collaborating with the Romanian company, Aerostar.
CUSTOMERS: Romanian Air Force (100).
COSTS: $300 million.
DESIGN FEATURES: No details released yet but thought to include: radar (Elta), IFF, RWR, radio, radar altimeter, weapons system, HUD/UFCP, DASH, navigation system (RLG, GPS, ILS, radio/nav), multifunction displays, multirole computer CADC.

NEW ENTRY

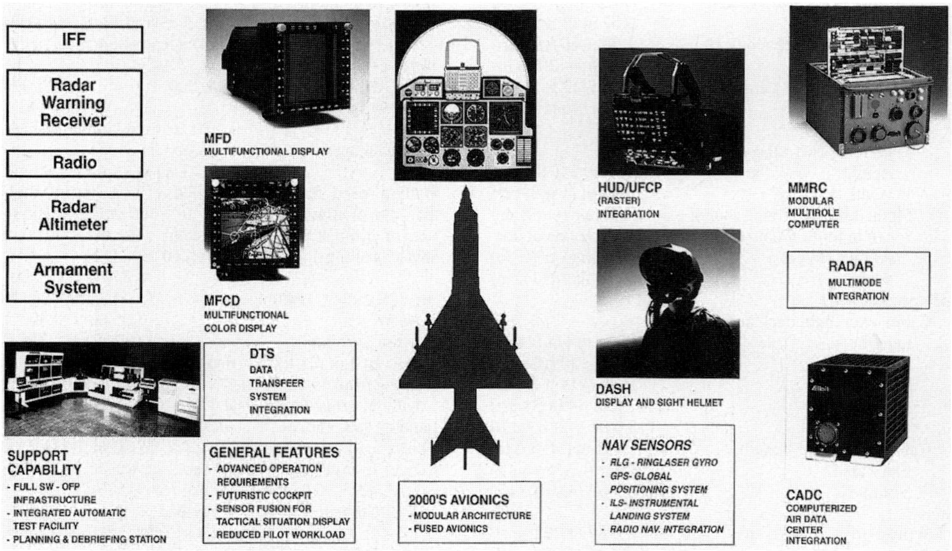

Elbit (Mikoyan) MiG-21 upgrade package ordered by Romanian Air Force *(Elbit)*
1995

ELBIT (NORTHROP) F-5 UPGRADE

TYPE: Combat aircraft upgrade.
PROGRAMME: Currently engaged on upgrade programmes with three undislosed customers.
DESIGN FEATURES: Offers complete package including advanced man/machine interface (MMI) and complete cockpit design, structural improvements, weapon delivery, navigation, day and night vision tactical systems and colour displays.
AVIONICS: Installation includes digital mission computer, air data computer, ring laser gyro (RLG), wide-angle HUD with upfront controller, HOTAS controls, colour video HUD camera and recorder, multifunction display and/or multifunction colour display, data transfer system, multimode radar, VOR/ILS and GPS. System architecture is based on MIL-STD-1553 database which can be expanded to include specialised mission equipment such as FLIR, ECM, advanced RWR and advanced weapons. Further systems include integrated comm/nav IFF integrated autonomous air combat manoeuvring (ACMI) and display and sight helmet (DASH).

NEW ENTRY

Elbit (Northrop) F-5 upgrade package, ordered by three countries *(Elbit)*
1995

ELBIT (AERO VODOCHODY) L-39 UPGRADE

TYPE: Trainer aircraft upgrade.

PROGRAMME: Engaged in programme with undisclosed customer to upgrade L-89 with western avionics.

AVIONICS: Core upgrade includes: weapon delivery and navigation system (WDNS), head-up display (HUD), upfront control panel (UFCP), central mission computer, ring laser gyro/inertial navigation system (RLG/INS), video camera and VTR and multifunction display (MFD) in back seat. Optional systems include stand-alone autonomous air combat manoeuvring instrumentation (A^2CMI), display and sight helmet (DASH), radar and radar warning receiver (RWR).

NEW ENTRY

(MCDONNELL DOUGLAS) F-4 PHANTOM 2000 UPGRADE

TYPE: Combat aircraft upgrade.

PROGRAMME: Acting as avionics systems integrator for the Israeli Air Force F-4 2000 upgrade. First redelivery of aircraft took place 9 April 1989.

DESIGN FEATURES: Upgrade is designed to integrate new air-to-air and air-to-ground weapons and sensors and the interface with aircraft systems.

AVIONICS: Upgrade includes: MIL-STD-1553 databus interfaced Westinghouse Norden Systems APG-76 multimode high-resolution radar, El-Op (Kalser licence) wide-angle diffractive optics HUD, Elbit multifunction electronic displays for both crewmen, digital WDNS and HOTAS (hands-on throttle and stick), orbit integrated com and com/nav systems, improved ECM and self-protection systems. Core of system is derivative of ACE-3 currently fitted in IAF F-16C/Ds.

NEW ENTRY

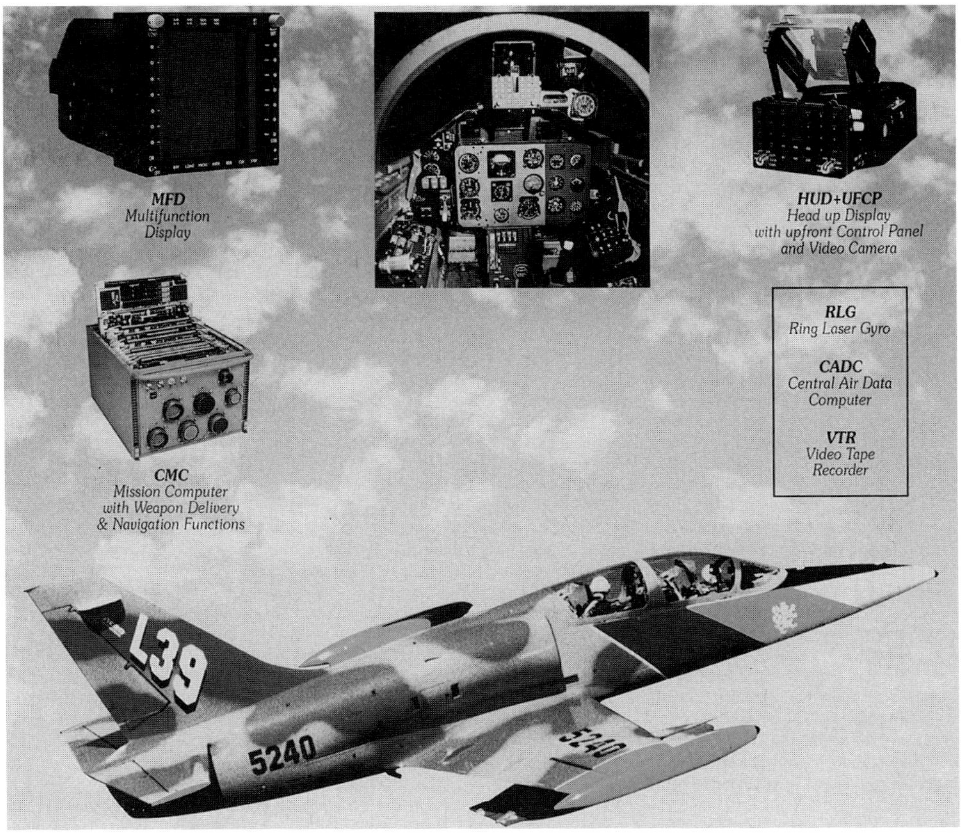

Elbit (Aero Vodochody) L-39 western avionics upgrade package *(Elbit)*

1995

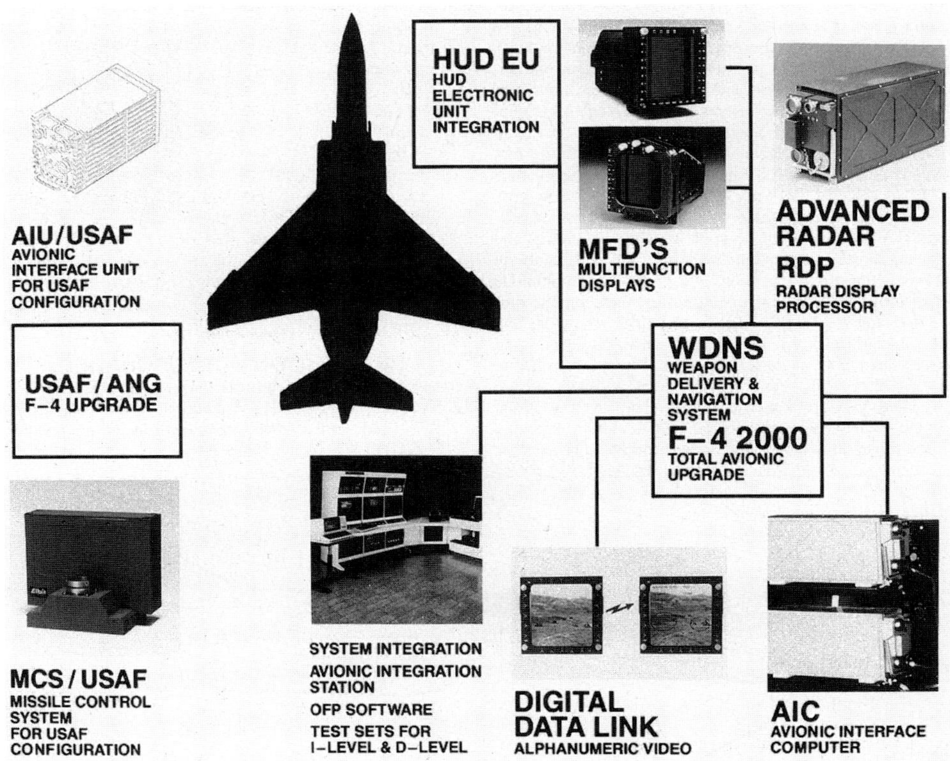

Elbit (McDonnell Douglas) F-4 Phantom 2000 Upgrade *(Elbit)*

1995

IAI

ISRAEL AIRCRAFT INDUSTRIES LTD

Ben-Gurion International Airport, Israel 70100
Telephone: 972 (3) 9353111, 9358111
Fax: 972 (3) 9353131, 9712290
Telex: ISRAVIA 381014, 381033, 381002

CHAIRMAN: Z. Zur (Gen Retd)

PRESIDENT AND CEO: M. Keret

VICE-PRESIDENTS:
- D. Arzi (General Manager, Bedek Aviation Group)
- O Harai (General Manager, Military Aircraft Group)
- M. Boness (General Manager, Commercial Aircraft Group)
- S. Alkon (General Manager, Electronics Group)
- S. Peretz (Finance)
- M. Broder (Marketing)
- D. Suslik (Director of Corporate Communications)

Israel Aircraft Industries is a multinational aerospace family of divisions and subsidiaries. As Israel's largest high-tech concern, IAI provides science-based solutions to a broad spectrum of needs in the air, land, sea and space environments.

During 1994, IAI reorganised its structure into four groups: Commercial Aircraft Group, Military Aircraft Group, Electronics Group and Bedek Aviation Group. Among other reasons, the restructuring reflects IAI's goals of attracting further foreign co-operation and investment in the commercial enterprises.

IAI's 1994 sales of $1.4 billion includes 80 per cent exports to five continents. These sales are the result of a rapid and growing expansion into the civil aerospace sector, and IAI's continued position as a prime source for customer tailored defence and civil systems. All these activities are based on increasing performance as a prime contractor, subcontractor, and joint ventures partner, which include: B747 conversions to combi and full cargo aircraft; total upgrades of military aircraft; the serial production of the Astra family of business jets with the addition currently under development of the Astra Galaxy long-range business aircraft; military and civil structures, components, simulators and training aids.

New areas include medical systems, environmental systems, and civil space products and services. The Amos communication satellite scheduled for launch in 1995. IAI's defence and support products include: air and ground unmanned vehicles. In 1994, IAI flew a new UAV, the Heron, designed for long endurance missions. Secure communications, offensive and defensive missile systems, the continuation of developing the Arrow anti-tactical ballistic missile (ATBM) which intercepted successfully a surrogate tactical ballistic missile in a flight test in June 1994. Radars, precision guided munitions, stabilised optronic systems, mine and obstacle breaching systems, EW, elint/ESM, sigint and comint/comjam. IAI had a workforce of 13,500 in 1994.

UPDATED

IAI LAVI (YOUNG LION) TECHNOLOGY DEMONSTRATOR

TYPE: Technology demonstrator aircraft.

PROGRAMME: The Lavi multirole combat aircraft received programme go-ahead in February 1980, and full-scale development started in October 1982. First flight of prototype B-1 took place on 31 December 1986, the second (B-2) making its initial flight on 30 March 1987. The programme envisaged production of at least 300 aircraft, including about 60 combat-capable two-seat operational trainers, but was terminated by the Israeli government on 30 August 1987 after 82 flights due to severe budgetary constraints. IAI decided to continue to validate the main tasks of the programme, utilising the B-3 (third prototype) as a technology demonstrator (TD). This airframe utilises some components from one of the Lavi prototypes, but is of two-seat configuration, with elevons approximately 15 per cent larger than on the Lavi prototypes. Its prime objective is to serve as a demonstrator of advanced systems and technologies developed and produced by the Israeli industry, and to act as a testbed for future developments. The first flight of the TD was made on 25 September 1989.

Some of the avionics upgrading packages now offered by IAI for various aircraft derive from the TD systems and will have been tested and validated on the TD. The Lavi programme was terminated by the Israeli government 30 August 1987, but IAI continues to fly the B-3 two-seat prototype as a technology demonstrator (TD).

POWER PLANT: One Pratt & Whitney PW1120 afterburning turbojet (military rating approximately 55.6 kN; 12,500 lb st dry and 82.7 kN; 18,600 lb st with afterburning). Ventral single-shock intake based on that of General Dynamics F-16. Maximum fuel capacity 3,330 litres (880 US gallons; 732 Imp gallons) in integral wing tanks, plus 5,095 litres (1,346 US gallons; 1,121 Imp gallons) externally.

ACCOMMODATION: Two Martin-Baker Mk 10 lightweight zero/zero ejection seats, in tandem, under teardrop cockpit canopy.

AVIONICS: Elta electronic warfare self-protection system for rapid threat identification (IFF) and flexible response (ECM). This computer-based, fully automatic system uses active and passive countermeasures, including internally and externally podded power-managed noise and deception jammers. Elbit integrated display system includes a Hughes wide angle holographic HUD (in front cockpit), three multifunction display (two monochrome and one colour), display computers, and communications controller. Pilot can operate most systems through a single El-Op upfront control. Lear Siegler/MBT quadruple-redundant digital fly-by-wire flight control system, with stability augmentation, MBT control unit and Moog servo-actuators. No mechanical back-up. Sundstrand actuation system, with geared rotary actuators, for leading-edge flaps. Cockpit designed to minimise pilot workload in high *g* and dense threat environment, with full HOTAS (hands on throttle and stick) operation. Elta EL/M-2032 multimode pulse Doppler radar, incorporating automatic target acquisition and track-while-scan in the air-to-air mode, and beam-sharpened ground mapping/terrain-avoidance and sea search in the air-to-surface mode. The radar's coherent transmitter and stable multichannel receiver ensure reliable look-up/look-down performance over a broad band of frequencies, as well as high resolution mapping. Elta programmable signal processor, backed by a network of distributed, embedded computers, provides optimum allocation of computer power and considerable flexibility for algorithm updating and system growth. Advanced versions of Elbit ACE-4 mission computer (128 K memory) and SMS-86 stores management systems, both compatible with dual MIL-STD-1553B databuses; SMS-86 capable of managing both conventional and smart weapons and sensors. Elta ARC-740 fully computerised UHF com radio system, Elistra radar warning receiver and Astronautics air data computer. Taman TINS 1700 advanced inertial navigation system and GPS planned for installation.

ARMAMENT: Internally mounted 30 mm cannon, with helmet sight. Four underwing hardpoints for air-to-surface missiles, bombs, rockets and other stores; inboard pair wet for carriage of auxiliary fuel tanks. Seven underfuselage stores attachments (three tandem pairs plus one on centreline). Infra-red air-to-air missile at each wingtip.

DIMENSIONS, EXTERNAL:

Wing span	8.78 m (28 ft 9⅔ in)
Length overall	14.57 m (47 ft 9⅔ in)
Height overall	4.78 m (15 ft 8¼ in)
Wheel track	2.31 m (7 ft 7 in)
Wheelbase	3.86 m (12 ft 8 in)

AREAS:

Wings, gross	33.05 m² (355.75 sq ft)

WEIGHTS AND LOADINGS:

Max fuel: internal (usable)	2,624 kg (5,785 lb)
external	4,164 kg (9,180 lb)
Max ordnance (excl air-to-air missiles)	2,721 kg (6,000 lb)
Max external load	7,257 kg (16,000 lb)
T-O weight: basic	9,990 kg (22,024 lb)
max	18,370 kg (40,500 lb)
Max wing loading	555.6 kg/m² (113.8 lb/sq ft)
Combat thrust/weight ratio	1.07

PERFORMANCE (estimated):

Max level speed above 11,000 m (36,000 ft)	Mach 1.85 or 800 kts (1,482 km/h; 921 mph) CAS
Rotation speed	140 kts (259 km/h; 161 mph)
Low-altitude penetration speed: two infra-red missiles and eight 750 lb M117 bombs	538 kts (997 km/h; 619 mph)
two infra-red missiles and two 2,000 lb Mk 84 bombs	597 kts (1,106 km/h; 687 mph)
Air turning rate at Mach 0.8 at 4,575 m (15,000 ft): sustained	13.2°/s
max	24.3°/s
Max rate of roll	300°/s
T-O run	approx 305 m (1,000 ft)
Combat radius: air-to-ground, lo-lo-lo	600 n miles (1,112 km; 691 miles)
air-to-ground, hi-lo-hi with two Mk 84 or six Mk 82 bombs	1,150 n miles (2,131 km; 1,324 miles)
air-to-air, combat air patrol	1,000 n miles (1,853 km; 1,151 miles)
g limit	+7.2 cleared for TD

VERIFIED

IAI KFIR (LION CUB) UPGRADES

TYPE: Single-seat strike, ground attack and fighter aircraft.

PROGRAMME: A total of 212 was built (27 C1s and 185 C2s/TC2s, most of the latter subsequently being upgraded to C7/TC7).

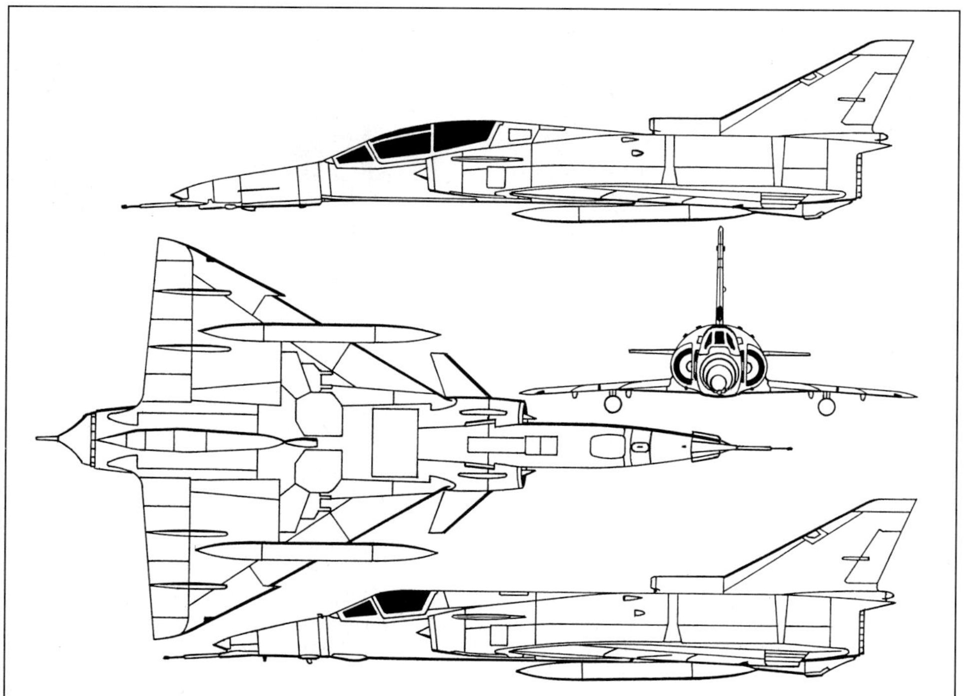

IAI Kfir strike, ground attack and fighter aircraft with additional side-view (top) of two-seat version *(Jane's/Mike Keep)* ***1994***

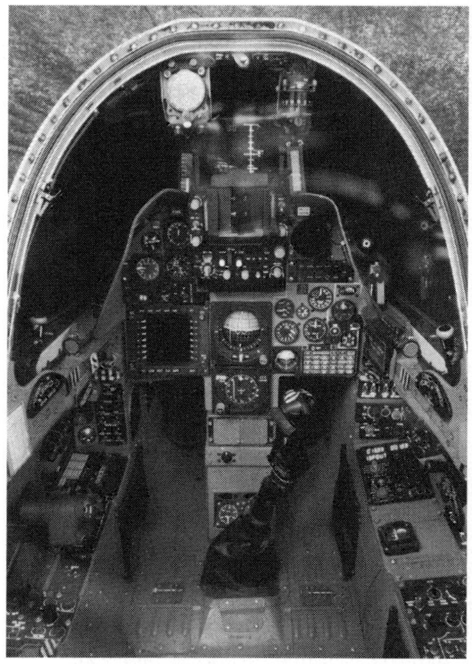

IAI Kfir C-7 cockpit with Elbit weapons delivery and navigation system (WDNS) ***1994***

VERSIONS: **C2:** Modified and major production version first flown in 1974. Changes from C1 include detachable, sweptback foreplanes on engine intake trunks, small strakes each side of nose and extended chord outer wings to improve airfield and combat performance. Late production versions had lengthened nose to accommodate larger EL/M-2001B radar. No longer in Israeli service. Twelve C2 Kfirs supplied to Ecuador.

C7: Upgraded conversion of C2, with 'combat plus' optional increase in engine thrust and improved avionics, delivered to Israeli Air Force from mid-1983. Other features include new HOTAS (hands on throttle and stick) cockpit installation and two additional external stores stations. Colombia ordered 13 in 1988, delivery of which was made in 1989.

TC2: Tandem two-seat operational trainer version of C2 with drooped and lengthened nose. First flown in 1981. One in service in Ecuador.

TC7: Designation of TC2 when upgraded to C7 standard. Two exported to Colombia with C7s mentioned above.

French government approval was given in Autumn 1989 for IAI to purchase five SNECMA Atar 9K-50 turbojet engines for test installation in Kfirs as part of an upgrade and renewed export marketing programme that would be unhampered by US restrictions involving the aircraft's present J79 engines. First flight by an Atar engined Kfir was planned for the second half of 1990.

Kfir-2000: Further upgraded version designed by IAI Lahav plant and offered to potential customers in 1993. Avionics and equipment, based on those in Lavi TD, include Elta EL/2032 look-down/shoot-down, track-while-scan radar in enlarged nose radome; Astronautics mission, armament and display computer, HOTAS cockpit with wide-angle HUD, upfront control panel and two MFDs (one coloured) and ring laser gyro INS and GPS, all linked through a MIL-STD-1553B digital databus.

OPERATORS: Versions of the Kfir are in service with the armed forces of the following countries: Colombia (13), Ecuador (10) and Israel (150).

The following abbreviated description applies to the C7 and TC7.

POWER PLANT: One General Electric J79-J1E turbojet (modified GE-17), with variable area nozzle, rated at 52.89 kN (11,890 lb st) dry and 79.45 kN (17,860 lb st) with afterburning. Adjustable half-cone centrebody in each air intake. Internal fuel in five fuselage and four integral wing tanks. Total internal capacity 3,243 litres (857 US gallons 713.4 Imp gallons). Refuelling point on top of fuselage, above forward upper tank. Wet points for the carriage of one drop tank beneath each wing (inboard), and one under fuselage; these tanks may be of 500, 825, 1,300 or 1,700 litres (132, 218, 343.5 or 449 US gallons); 110, 181.5, 286 or 374 Imp gallons) capacity; maximum external fuel capacity 4,700 litres (1,242 US gallons); 1,034 Imp gallons). Provision for boom/receptacle or probe/drogue in-flight refuelling system, and for single-point pressure refuelling.

ACCOMMODATION: Pilot only, on Martin-Baker IL10P zero/zero ejection seat (two tandem seats in TC7), under rearward hinged upward opening canopy. Cockpit pressurised, heated and air conditioned.

AVIONICS: Compared with C2, C7 differs in having an improved HOTAS cockpit installation, WDNS-341 weapons delivery and navigation system, Elbit System 82 computerised stores management and release system, video subsystems, smart weapons delivery capability and updated electronic warfare systems. The ranging radar is an Elta EL/M-2001B.

ARMAMENT: Fixed armament of one 30 mm DEFA 552 cannon in underside of each engine air intake (140 rds/gun). Nine hardpoints (five under fuselage and two under each wing) for external weapons, ECM pods or drop tanks. For interception duties, Sidewinder and/or Python 3/4 infra-red homing air-to-air missile can be carried under each outer wing. The Kfir 2000 is configured to carry two BVR active radar homing missiles. Ground attack version can carry up to six 500 lb bombs, two Shrike, Maverick or GBU-15 air-to-surface weapon under the fuselage, and two 1,000 lb or eight adapted 500 lb bombs (conventional, smart or concrete dibber type) under the wings. Alternative weapons can include Mk 82/83/84 and M117/118 bombs; CBU-52/58 and TAL-1/2 cluster bombs; LAU-3A/10A/32A rocket launchers; flares, chaff, Elta EL/L-8202 ECM and other podded system.

DIMENSIONS, EXTERNAL:

Wing span	8.22 m (26 ft 11½ in)
Foreplane span	3.73 m (12 ft 3 in)
Length overall, incl probe	15.65 m (51 ft 4¼ in)
Height overall	4.55 m (14 ft 11¼ in)
Wheel track	3.20 m (10 ft 6 in)
Wheelbase	4.87 m (15 ft 11¾ in)

AREAS:

Wings, gross	34.8 m² (374.6 sq ft)
Foreplanes (total)	1.66 m² (17.87 sq ft)

WEIGHTS AND LOADINGS:

Weight empty (interceptor, estimated)	7,285 kg (16,060 lb)
Max usable fuel: internal	2,572 kg (5,670 lb)
external	3,727 kg (8,217 lb)
Max external stores	6,085 kg (13,415 lb)
Typical combat weight:	
interceptor, 50% internal fuel, two Python missiles	9,390 kg (20,700 lb)
interceptor, two 500 l drop tanks, two Python missiles	11,603 kg (25,580 lb)
combat air patrol, three 1,300 l drop tanks, two Python missiles	14,270 kg (31,460 lb)
ground attack, two 1,300 l drop tanks, seven 500 lb bombs, two Python missiles	14,670 kg (32,340 lb)
Max clean T-O weight	10,415 kg (22,961 lb)
Max T-O weight	16,500 kg (36,376 lb)
Wing/foreplay loading at 9,390 kg (20,700 lb) combat weight	257.5 kg/m² (52.8 lb/sq ft)

PERFORMANCE:

Max level speed above 11,000 m (36,000 ft)	Mach 2.2
Max sustained level speed at height, clean	Mach 2.0
Max level speed at S/L, clean	750 kts (1,389 km/h; 863 mph)
Max rate of climb at S/L	14,000 m (45,930 ft)/min
Time to 15,240 m (50,000 ft), full internal fuel, two Shafrir missiles	5 min 10 s
Stabilised ceiling (combat configuration)	17,680 m (58,000 ft)
Turn performance at 4,575 m (15,000 ft), combat weight of 9,390 kg (20,700 lb):	
turn rate: sustained	9.6°/s
instantaneous	18.9°/s
turn radius: sustained	1,326 m (4,350 ft)
instantaneous	671 m (2,200 ft)
T-O run at max T-O weight	1,450 m (4,750 ft)
Landing from 15 m (50 ft) at 11,566 kg (25,500 lb) landing weight	1,555 m (5,100 ft)
Landing run at 11,566 kg (25,500 lb) landing weight	1,280 m (4,200 ft)
* Combat radius, 20 min fuel reserves:	
high-altitude interception, one 825 l and two 1,300 l drop tanks, two Python missiles	419 n miles (776 km; 482 miles)
combat air patrol, one 1,300 l and two 1,700 l drop tanks, two Python missiles, incl 60 min loiter	476 n miles (882 km; 548 miles)
ground attack, hi-lo-hi, two 800 lb and two 500 lb bombs, two Python missiles, one 1,300 l and two 1,700 l drop tanks	640 n miles (1,186 km; 737 miles)
Ferry range:	
three 1,300 l drop tanks	1,614 n miles (2,991 km; 1,858 miles)
one 1,300 l and two 1,700 l drop tanks	1,744 n miles (3,232 km; 2,008 miles)
g limit	+7.5

* *Can be increased by 30 per cent with one in-flight refuelling*

UPDATED

IAI ARAVA

TYPE: Twin-turboprop STOL light transport aircraft.

PROGRAMME: Design of the Arava light STOL transport started 1966; the first of two flying prototypes made its initial flight 27 November 1969 and the second 8 May 1971.

The Arava was first certificated as a civil aircraft, by the FAA in April 1972. This version, designated IAI 101, did not go into production, but formed the basis for the initial

IAI Kfir C-7 multimission fighter aircraft of the Israeli Air Force *(IAI)*

1995

IAI Kfir C-7 multimission fighter showing air-to-air missiles *(IAI)*

1994

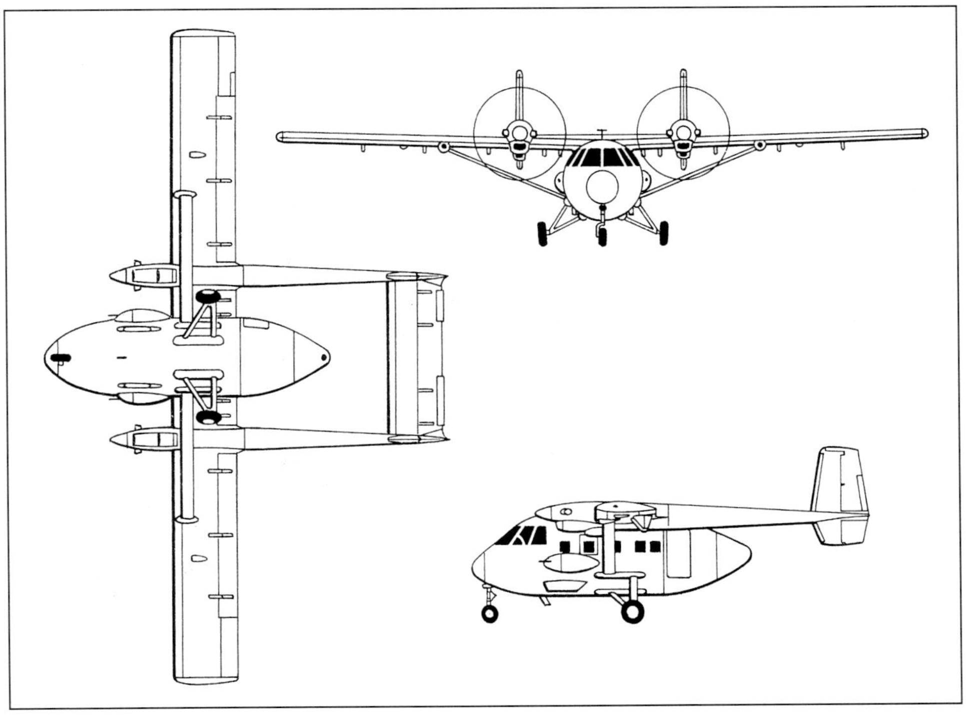

IAI Arava 201 STOL light transport *(Jane's/Mike Keep)*

1994

IAI Arava 201 STOL aircraft loading jeep

1994

production Arava 102 (civil) and 201 (military) transport versions.

More than 90 Aravas had been delivered by the end of 1987, most of these being military 201s with the majority of military and civil sales being to Latin America. In 1984 the Israeli Air Force began to replace its Douglas C-47s in the main transport and trainer roles with a mixture of Arava 201s and 202s. Duties include that of multi-engined trainer for transport pilots.

VERSIONS: **Arava 101B.** Modified civil version with PT6A-36 engines and accommodation for 19 passengers or 2,393 kg (5,275 lb) of cargo. Improved cabin interior, and enhanced performance at higher ambient temperatures. Certificated by Israeli CAA and FAA under SFAR Pt 41C. Only known customer was Airspur of Los Angeles, which received four of a cargo version (marketed in USA as **Cargo Commuter-liner**) in 1982-83.

Arava 102. Initial production civil version, based on 101; certificated by Israel CAA to FAR Pt 23 (Normal category) and FAR Pt 135 (annex A). Accommodation for 20 passengers in airline-standard four-abreast configuration, with toilet. Available also in a VIP configuration for up to 12 passengers, as an all-cargo transport, as a medical clinic for flying doctor services, and in versions for mapping, mining research, rainmaking and bridge construction, as flying laboratories for agriculture and health ministries, and for supplying oil prospecting units.

Arava 201. Initial military transport version. Prototype first flew on 7 March 1972. Three lease-operated by Israeli Air Force in October 1973, others delivered 1984-85. Standard equipment enables a wide variety of missions to be undertaken. Total of more than 70 sold, including some equipped for maritime surveillance duties, fitted with either an AD-9 modification to extend the range and detection capability of the standard search/weather radar, or a more advanced detection system. Available also in several electronic warfare configurations, with various pallet-mounted elint and ESM packages, ventral or fuselage-side 'dustbin' radome, rearward-facing scanners mounted on the fuselage tailcone, a 60 kVA APU for additional electrical power generation, and numerous blade and whip type antennae above and below fuselage, on top of tailbooms, and elsewhere.

Arava 202. Modified military version, flight tested in 1976-77. Differs principally in being longer, and having a fully 'wet' wing containing approximately 726 kg (1,600 lb) more fuel, wingtip winglets, and a boundary layer fence just inboard of each tip. Powered by 559 kW (750 shp) PT6A-36 engines; single-point pressure refuelling system. The winglet modification (but not the increased fuel capacity) is available as a retrofit modification of existing Aravas. Several 202s operate in South America, and deliveries to the Israeli Air Force have included some of this version.

UPGRADES: **IAI**: Arava 202 wingtip retrofit. See Versions.

OPERATORS: Versions of the Arava are in service with the armed forces of the following countries: Bolivia (1), Cameroon (2), Colombia (2), Ecuador (4), Guatemala (6), Honduras (2), Israel (10), Mexico (12), Swaziland (2), Thailand (3) and Venezuela (4).

The following description applies to the Arava 201, except where otherwise indicated:

DESIGN FEATURES: Braced high-wing monoplane with single streamline section strut each side. Wing section NACA 63(215)A 417. Dihedral 1°30′. Incidence 0°27′. No sweepback. Endplate winglets of NASA (Whitcomb) profile standard on 202, optional for other models. Fixed incidence tailplane.

FLYING CONTROLS: Frise light alloy ailerons. Electrically operated double-slotted light alloy flaps. Scoop type alloy spoilers, for lateral control, above wing at 71 per cent chord. Electrically actuated trim tab in port aileron. Geared tab and electrically actuated trim tab in elevator and geared trim tab in each rudder.

STRUCTURE: Wing has light alloy two-spar torsion box structure. The fuselage is a conventional semi-monocoque light alloy structure of stringers, frames and single-skin panels, of increased length on 202. The tail unit is a cantilever light alloy structure, with twin fins and rudders, carried on twin booms extending rearward from engine nacelles. Tailbooms were built by IAI Technologies Division.

LANDING GEAR: Non-retractable tricycle type, of Servo-Hydraulics Lod manufacture, with single mainwheels and steerable nosewheel. Mainwheels carried on twin struts, incorporating oleo-pneumatic shock-absorbers. Mainwheel size 11.00-12, tyre pressure 3.31 bars (48 lb/sq in). Disc brakes on main units.

POWER PLANT: Two 559 kW (750 shp) Pratt & Whitney Canada PT6A-34 turboprops, each driving a Hartzell HC-B3TN three-blade hydraulically actuated fully feathering reversible-pitch metal propeller, (PT6A-36 engines of same T-O rating in 101B and 202). Electric de-icing of propellers optional. Two integral fuel tanks in each wing, with total usable capacity (except 202) of 1,665 litres (440 US gallons; 366 Imp gallons). Four overwing refuelling points. Optional pressure refuelling point (standard on 202) behind fuselage/strut fairing. Two cabin-mounted tanks, each of 1,022 litres (270 US gallons; 225 Imp gallons), are available optionally for self-ferry flights.

ACCOMMODATION: Crew of one or two on flight deck, with door on starboard side. Airline type seating for up to 20 passengers in Arava 102 (19 in 101B, 24 in 202), plus toilet. Arava 201 can accommodate 24 fully equipped troops (30 in 202), or 16 paratroops and two dispatchers (20 + 3 in 202). Outward door at rear of cabin, opposite which, at floor level, is an emergency exit/baggage door on the starboard side. Rear doors were built by IAI Combined Technologies Division. Fuselage tailcone is hinged to swing sideways through more than 90° to provide unrestricted access to main cabin. Alternative interior configurations available for ambulance role (10 stretchers and two medical attendants in Arava 102 and 201; 12 stretchers and five medical attendants/sitting patients in 202); as all-freight transport carrying (typically) a jeep-mounted recoilless rifle and its four-man crew; or as a maritime patrol aircraft fitted with search radar and other special equipment (see Arava 201 model listing paragraph) Emergency exit on each side, forward of wing leading-edge.

SYSTEMS: Hydraulic system (pressure 172 bars; 2,500 lb/sq in) for brakes and nosewheel steering only. Electrical system includes two 28 V 170 A DC engine-driven starter/generators, a 28 V 40 Ah Ni/Cd battery, and two 250 115/26 V 400 Hz static inverters.

AVIONICS: Blind-flying instrumentation standard. Optional avionics include VHF, VOR/ILS, ADF, marker beacon receiver, weather radar and PA system.

AVIONICS (electronic warfare version): Elta EL/L-8310 manually operated elint/ESM (electronic intelligence/surveillance) system (L-8311 or L-8312 systems optional); Elta EL/K-7010 jamming system; plus 60 kVA auxiliary generator to provide necessary additional electrical power.

ARMAMENT: Optional 0.50 mm Browning machine gun pack on each side of fuselage, above a pylon for a pod containing six 82 mm rockets. Rearward firing machine gun optional. Librascope gunsight.

DIMENSIONS, EXTERNAL:

Wing span: 201	20.96 m (68 ft 9 in)
202	21.63 m (70 ft 11½ in)
Wing chord (constant)	2.09 m (6 ft 10½ in)
Wing aspect ratio	10.1
Length overall: 201	13.03 m (42 ft 9 in)
202	13.47 m (44 ft 2¼ in)
Length of fuselage pod: 201	9.33 m (30 ft 7 in)
202	10.23 m (33 ft 6¾ in)
Diameter of fuselage	2.50 m (8 ft 2 in)
Height overall: 201	5.21 m (17 ft 1 in)
202	5.22 m (17 ft 1½ in)
Tailplane span (c/l of tailbooms)	5.21 m (17 ft 1 in)
Wheel track	4.01 m (13 ft 2 in)
Wheelbase: 201	4.62 m (15 ft 2 in)
202	5.12 m (16 ft 9½ in)
Propeller diameter	2.59 m (8 ft 6 in)
Propeller ground clearance	1.75 m (5 ft 9 in)
Crew door (fwd, stbd): Height	0.93 m (3 ft 0½ in)
Width	0.48 m (1 ft 7 in)
Passenger door (rear, port):	
Height	1.57 m (5 ft 2 in)
Width	0.62 m (2 ft 0½ in)
Airdrop opening, tailcone removed:	
Height	1.75 m (5 ft 9 in)
Width	2.33 m (7 ft 8 in)
Emergency/baggage door (rear, stbd):	
Height	1.12 m (3 ft 8 in)
Width	0.61 m (2 ft 0 in)
Emergency window exits (each):	
Height	0.66 m (2 ft 2 in)
Width	0.48 m (1 ft 7 in)

DIMENSIONS, INTERNAL:

Cabin, excl flight deck and hinged tailcone:	
Length: 201	3.87 m (12 ft 8 in)
202	4.77 m (15 ft 7¾ in)
Max width	2.33 m (7 ft 8 in)
Max height	1.75 m (5 ft 9 in)
Floor area: 201	7.16 m^2 (77.07 sq ft)
202	8.83 m^2 (95.05 sq ft)
Volume: 201	12.7 m^3 (449.2 cu ft)
202	16.3 m^3 (575.6 cu ft)
Baggage compartment volume	2.60 m^3 (91.8 cu ft)
Tailcone volume	3.20 m^3 (113 cu ft)

AREAS:

Wings, gross	43.68 m^2 (470.2 sq ft)

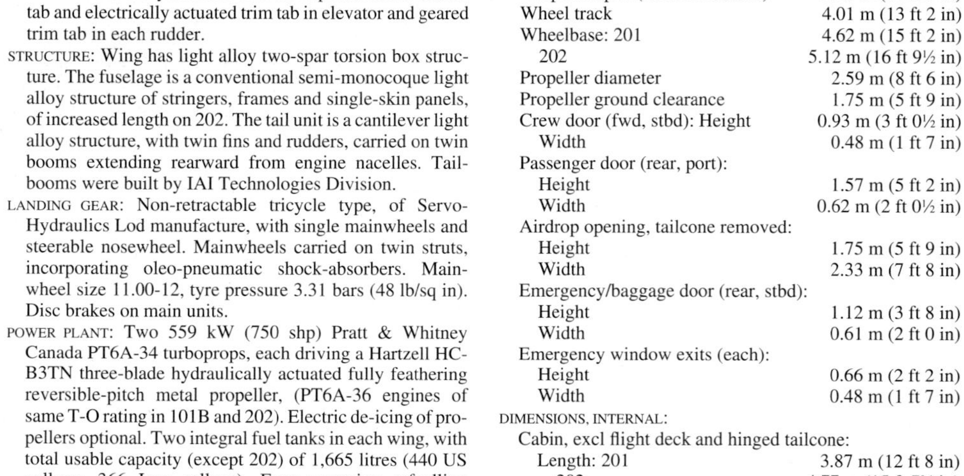

IAI Arava 202 STOL aircraft. with wingtip winglets ***(IAI)***

1994

IAI Arava 'Cargo Commuter Liner' STOL aircraft *(IAI)*

1994

Ailerons (total)	1.75 m² (18.84 sq ft)
Trailing-edge flaps (total)	8.80 m² (94.72 sq ft)
Spoilers (total)	0.85 m² (9.2 sq ft)
Fins (total)	4.86 m² (52.31 sq ft)
Rudder (total, incl tabs)	3.44 m² (37.03 sq ft)
Tailplane	9.36 m² (100.75 sq ft)
Elevator, (incl tabs)	2.79 m² (30.03 sq ft)

WEIGHTS AND LOADINGS:

Basic operating weight empty:	
201	3,999 kg (8,816 lb)
202	4,111 kg (9,063 lb)
Max payload: 101B	2,393 kg (5,275 lb)
102, 201	2,313 kg (5,100 lb)
202	2,495 kg (5,500 lb)
Max T-O weight: 101B	6,867 kg (15,140 lb)
101, 201	6,804 kg (15,000 lb)
202	7,711 kg (17,000 lb)
Max landing weight: 201	6,804 kg (15,000 lb)
202	7,416 kg (16,349 lb)
Max zero-fuel weight: 201	6,350 kg (14,000 lb)
Max wing loading: 101B	157.2 kg/m² (32.20 lb/sq ft)
102, 201	155.8 kg/m² (31.90 lb/sq ft)
202	176.5 kg/m² (36.16 lb/sq ft)
Max power loading: 101B	6.14 kg/kW (10.09 lb/shp)
102, 201	6.09 kg/kW (10.00 lb/shp)
202	6.90 kg/kW (11.33 lb/shp)

PERFORMANCE (at max T-O weight; 201 except where indicated):

Never-exceed speed	176 kCAS (325 km/h; 202 mph)
Econ cruising speed at 3,050 m (10,000 ft)	
201	168 kts (311 km/h; 193 mph)
202	150 kts (278 km/h; 173 mph)
Stalling speed, flaps up	75 kts (140 km/h; 87 mph)
flaps down	62 kts (115 km/h; 72 mph)
Max rate of climb at S/L	393 m (1,290 ft)/min
Rate of climb at S/L, one engine out	55 m (180 ft)/min
Service ceiling	7,620 m (25,000 ft)
Service ceiling, one engine out	2,375 m (7,800 ft)
STOL T-O run: 201	293 m (960 ft)
202	400 m (1,315 ft)
STOL T-O run to 15 m (50 ft)	463 m (1,520 ft)
STOL landing from 15 m (50 ft)	469 m (1,540 ft)
STOL landing run	250 m (820 ft)
Range with max payload, 45 min reserves:	
except 202	140 n miles (259 km; 161 miles)
202	340 n miles (630 km; 392 miles)
Max range with 1,587 kg (3,500 lb) payload 45 min reserves:	
except 202	540 n miles (1,000 km; 622 miles)
202	860 n miles (1,594 km; 990 miles)

UPDATED

IAI 1124 WESTWIND

TYPE: Twin-turbofan civil transport and military maritime patrol aircraft.

PROGRAMME: The Westwind had its origins in the Jet Commander designed in the USA and flown for the first time 27 January 1963. Production was transferred in 1968 to Israel Aircraft Industries, which continued to develop and market successively improved versions. In 1975, from aircraft c/n 187, the Garrett TFE731 turbofan became the standard power plant. Initial version with the engine was the 1124 Westwind (53 built). This was followed in 1978 (from c/n 240) by the improved 1124 Westwind I: a (1124N) variant.

VERSIONS: **1124 Westwind.** Initial turbofan-powered production version, announced in September 1974 and introduced in the following year. Prototypes were two converted 1123 Westwinds, the first of which made its first flight on 21 July 1975. FAA certification received in Spring 1976, with deliveries beginning shortly afterwards. Total of 53 built (c/n 187-239), of which c/n 239 later became the prototype for 1124A Westwind 2.

1124 Westwind I. Basic turbofan-powered production version, announced in September 1978 and introduced from c/n 240 onwards. Improved version of 1124 Westwind, differing chiefly in having an optional 317 kg (700 lb) increase in fuel load, installed in a removable tank in the forward baggage compartment; an increase of approximately 5 per cent in cabin useful volume, achieved by relocation of some avionics and by lowering the floor in the toilet compartment; RCA Primus 400 colour weather radar as standard; and improved fuel and environmental control systems. Orders received from Rhein-Flugzeugbau for four aircraft, equipped for target simulation on behalf of the Federal German Armed Forces.

1124N Sea Scan. Maritime version, announced in 1976. Prototype was converted from an aircraft (4X-CJA, c/n 154) which had served previously as prototype for the 1124 Westwind. Three specially equipped 1123N Sea Scans, delivered to the Israeli Navy for coastal patrol, tactical support and anti-terrorist duties, have since been brought up to 1124N standard and equipped with thrust reversers, single-point pressure refuelling, anti-corrosion protection, fuselage side stores pylons, bubble windows, Litton APS-504(V)2 360° search radar, Global GNS-500A VLF/Omega navigation system, operators' consoles, galley and toilet. Three additional 1124Ns are in Australian Customs service liveries. These aircraft are owned and operated by Skywest Aviation Pty of Perth, Western Australia. A low-altitude search range of 1,379 n miles (2,555 km; 1,588 miles), and search endurance of more than 6 hours 30 minutes enables the Sea Scan to cover a search area of 82,740 n miles² (268,056 km²; 103,496 sq miles) along a 60 n miles (111 km; 69 mile) search band at a height of 915 m (3,000 ft). Increased search range and endurance to 2,500 n miles (4,633 km; 2,878 miles) and over 8 hours can be attained at altitudes of up to 13,720 m (45,000 ft). Available for specific operational requirements, with equipment to customer's specification.

Preliminary design and evaluation studies of a second-generation Sea Scan have been completed by IAI. In addition to its routine anti-terrorist low-level maritime patrol functions, the new Sea Scan multimission maritime patrol aircraft (MPA) can be deployed for ASW, signal intelligence (sigint) and anti-shipping air-to-surface missile attack operations. In the ASW role, search, detection, tracking, identification and attack are carried out using high-performance maritime search radar, ESM, sonobuoys, onboard signal analysis, colour multipurpose displays (MPDs), trailing MAD, long-range gyrostabilised sighting system (GSSS) and torpedoes. Search, localisation and attack at 100 n miles (185 km; 115 miles) from base can be performed for approximately 5 hours, enabling a landing back at base with 45 minutes reserve fuel. Replacing torpedoes with Gabriel Mk III air-to-surface missiles, and removing some specific ASW equipment (sonobuoys and MAD), allows anti-shipping missile attacks to be made from a standoff range of 32 n miles (60 km; 37 miles) at a distance greater than 1000 n miles (1,853 km; 1,151 miles) from base. Comint, elint and IDF equipment installed in the aircraft permits long-range high-altitude sigint operations with an endurance of more than 8 hours.

1124A Westwind 2. Developed version of the Westwind I for improved 'hot and high' field performance, range and economy of operation. Prototype (4X-CMK c/n 239, converted from early production 1124 Westwind) flown for the first time 24 April 1979. Certificated by Israeli CAA 11 December 1979 and by FAA 17 April 1980. New modified 'Sigma' wing of IAI section, NASA type winglets above tip tanks, flat (instead of 'trenched') cabin floor, increased seated headroom, airline type flushing toilet, relocated overhead passenger service units, and other improvements. First delivery (of the prototype to Helicol of Colombia) made 16 May 1980.

1125 Astra. Improved version of the Westwind 2. In production. Deliveries of turbofan-powered Westwind totalled approximately 250.

OPERATORS: Versions of the Westwind are in service with the armed forces of the following countries: Honduras (1) and Israel (3).

DESIGN FEATURES: Cantilever mid-wing monoplane. Wing section NACA 64A212 on Westwind I, IAI modified 'Sigma' 1 on Westwind 2. Dihedral 2°. Incidence 1° at root, −1° at tip. Sweepback 4°37′ at quarter-chord. Permanently attached wingtip fuel tanks, with (Westwind 2 only) NASA type winglet on upper surface. The tail unit is a cantilever all-metal structure. Tail unit has 28° sweepback at tailplane quarter-chord and 35° sweepback at fin quarter-chord. Variable incidence tailplane.

FLYING CONTROLS: Manually operated all-metal ailerons. Electrically operated all-metal double-slotted Fowler trailing-edge flaps. Dropped and cambered glassfibre covered leading-edges (Westwind I only). Electrically operated trim tab in port aileron. Hydraulically actuated speed brake and two 1 ft dumpers above each wing, forward of flap. All skins chemically milled and fully sealed. All primary surfaces, including ailerons tab, are fully mass balanced. Variable incidence tailplane, actuated electrically. Manually operated statically balanced elevators and rudder. Electrically operated trim tab in rudder.

STRUCTURE: The wing is an aluminium alloy flush riveted two-spar fail-safe structure. The fuselage is an all-metal

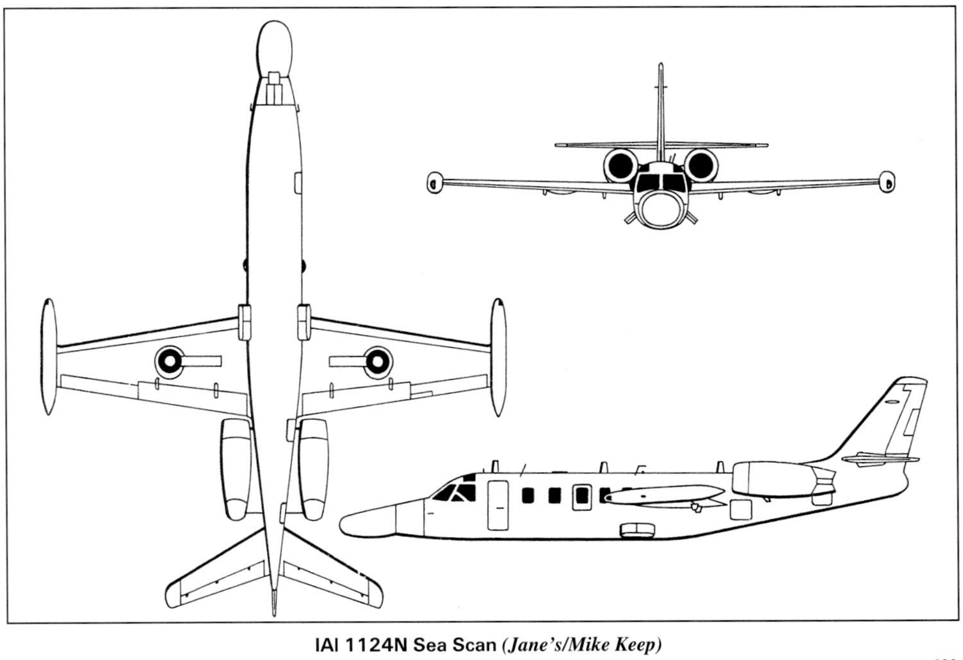

IAI 1124N Sea Scan *(Jane's/Mike Keep)*

1994

IAI 1124A Westwind 2. Note wingtip winglets *(IAI)*

1994

IAI 1124N Sea Scan of the Israeli Air Force *(R. A. Cooper)*

1994

semi-monocoque flush riveted structure of aluminium alloy and steel sheet, with chemically milled skins. Built in two main sections and joined at rear pressure bulkhead. Forward section, except for nosecone, is fully pressurised and fail-safe.

LANDING GEAR: Hydraulically retractable tricycle type, mainwheels retracting outward into wings, twin nosewheels rearward. No doors over mainwheel when retracted. Oleo-pneumatic shock-absorbers. Single wheels on main units, pressure (Westwind I) 10.69 bars (155 lb/sq in). Nose unit steerable and self-centring. Nosewheel tyre pressure (Westwind I) 3.45 bars (50 lb/sq in). Westwind 2 has Goodyear wheels, with size 16 × 4.4 (main) and 24 × 9.50-10.5 tyres (nose), pressures 9.86 bars and 3.79 bars (143 and 55 lb/sq in) respectively. Goodyear multiple-disc brakes, with Hydro-Aire fully modulated anti-skid system having automatic computer/sensor to prevent wheel lock and maintain brake effectiveness. Parking brake.

POWER PLANT: Two 16.46 kN (3,700 lb st) Garrett TFE731-3-100G turbofan engines, with Grumman thrust reversers, pod-mounted on side of rear fuselage. Eighty-five per cent of wing area forms an integral fuel tank, and additional fuel is carried separately in wingtip tanks and single rear fuselage tank. Total usable capacity (Westwind I) of 4,920 litres (1,299 US gallons; 1,082 Imp gallons) including wingtip tanks. Increased weight option permits additional 317 kg (700 lb) of fuel (397 litres; 105 US gallons; 87 Imp gallons) to be carried in a removable tank in forward baggage compartment. Capacity increased on Westwind 2, with 2,089 litres (552 US gallons; 460 Imp gallons) in each main wing tank 428 litres (113 US gallons; 94 Imp gallons) in each wingtip tank, and 379 litres (100 US gallons; 83 Imp gallons) in rear fuselage auxiliary tank, giving total usable capacity of 5,413 litres (1,430 US gallons; 1,191 Imp gallons). Single-point pressure refuelling on starboard side of fuselage; gravity points in each wing upper surface, each tip tank, and for fuselage auxiliary tank. Oil capacity 5.7 litres (1.5 US gallons; 1.25 Imp gallons) per engine.

ACCOMMODATION: Standard seating for pilot, co-pilot and seven passengers, or up to a maximum of 10 passengers, in pressurised and air conditioned cabin. Elliptical cabin section in Westwind 2 increases seated headroom and allows a flat rather than 'trenched' cabin floor, an airline type flushing toilet, and improved placing of the overhead passenger service units. Standard passenger layout comprises six individual tracked and swivelling seats, with two tables, plus a one-person divan. Fully enclosed toilet compartment at rear of cabin on starboard side. Plug type door at front on port side, provides access to both cabin and flight deck. Emergency exit on each side, forward of wing. Pressurised baggage compartment in rear of cabin, adjacent toilet; two heated but unpressuried compartments for up to 476 kg (1,050 lb) of baggage in rear of fuselage, each with separate external access on port side.

SYSTEMS: Garrett three-spool freewheeling turbine air conditioning system; pressurisation differential 0.61 bars (8.8 lb/sq in) normal, 0.62 bars (9.0 lb/sq in) maximum. Primary hydraulic system, pressure 138 bars (2,000 lb/sq in), operates through two engine-driven pumps to actuate landing gear, wheel brakes, nosewheel steering, speed brakes, lift dumpers and thrust reversers. Electrically operated emergency system, pressure 69 bars (1,000 lb/sq in), for brakes only. Pneumatic system, using engine bleed air, for wing and tailplane de-icing boots only. DC electrical system with two 350 A 28 V engine-driven starter/generators and two 28 V 37 Ah long-life Ni/Cd batteries. One main bus for each generator, connected to the central battery bus. Two 1 kVA solid state inverters provide 115 V AC power at 400 Hz, each being independently capable of supplying the entire AC load if required. Engine air intakes anti-iced by engine bleed air. Oxygen system supplied by pressurised cylinder of 1.36 m³ (48 cu ft) capacity. Electrically heated windscreen pitot system angle of attack sensor. Engine fire extinguishing system. No APU.

AVIONICS (Westwind I): Full dual IFR instrumentation standard, including Collins dual VHF-20A com, dual VIR-30A nav, dual DME-40 and ADF-60A. Other avionics include Collins NCS-31A radar navigation and control system (Global) Navigation NS-500A VLF in Sea Scan, RCA Primus 400 weather radar, and dual Sperry C-14 compass system. Collins FCS-105 flight control system, FD-109Z flight director and AP-105 autopilot. Canadian Marconi CMA-734 Omega navigation system approved for use in US and North Atlantic airspace.

AVIONICS (Westwind 2): Standard avionics and equipment (all Collins except where indicated) include dual VHF-20A VHF com, dual VIR-30A VHF nav, IAI nv switching system, FCS-80 flight control system, FDS-85 flight director, APS-80 autopilot, ADS-80 air data system, FMS-90 navigation system, DME-40 DME, ADF-60A ADF, dual RMI-36 RMIs, dual TDR-90 transponders, ALT-50A radio altimeter, ALI-80A encoding altimeter (pilot), Kollsman B4420 digital altimeter (co-pilot), MSI-80C Mach/airspeed indicator (pilot), IDC Mach/airspeed indicator (co-pilot), VNI-80 vertical nav indicator (pilot), Teledyne SLZ-9706-DGLE vertical nav indicator (co-pilot), PRE-80A preselector/alerter, dual 346B-3 audio systems, RNS-300 radar navigation system, WXR-300 weather radar, Teledyne SLZ-9618-5 angle of attack system, dual Sperry C-14 compasses, HSI-84 co-pilot's HSI, Sperry GH-14B co-pilot's attitude gyro, J.E.T. A1-804 standby attitude gyro, and Davco 811-B digital clock. Landing light in nose of each wingtip tank. Optional avionics include dual VHF-20B (instead of VHF-20A) and single VHF-251 VHF com; HF718U-5 and HF-220 HF com; Litton LTN-211, Collins LRN-85 or Global GNS-500A-2 VLF/Omega nav; Collins EFIS-85A; Global GNS-1000 airborne flight information system (AFIS); second FDS-85, second ADC-80J for co-pilot's FDA-85, and comparator warning annunciator system; second ADF-60A, second DME-40 FPA-80 Flight Profile ADV, Collins TAI-80A SAT/TAS indicator, DRI-55 digital radio altimeter, ALT-55B (instead of ALT-50A) radio altimeter; Fairchild 5424-501 flight data recorder; Fairchild A-100 cockpit voice recorder; Kollsman ALT B4515 co-pilot's encoding altimeter (instead of B44420); Davco 811-B co-pilot's digital clock, Hobbs hour meter, Dorne and Margolin ELT-6 Emergency locator transmitter, Devore Tel-Tail lights, Wulfsberg Flitefone III system, and ICD cabin display.

DIMENSIONS, EXTERNAL:

Wing span, incl tip tanks	13.65 m (44 ft 9½ in)
excl tip tank	13.16 m (43 ft 2 in)
Wing chord; at root	3.20 m (10 ft 6 in)
at tip	1.07 m (3 ft 6 in)
Westwind 2 at tip	1.17 m (3 ft 10¼ in)
Wing aspect ratio	6.51
Length overall	15.93 m (52 ft 3 in)
Fuselage: Max width	1.57 m (5 ft 2 in)
Max depth	1.83 m (6 ft 0 in)
Height overall	4.81 m (15 ft 9½ in)
Tailplane span	6.40 m (21 ft 0 in)
Wheel track	3.35 m (11 ft 0 in)
Wheelbase	7.79 m (25 ft 6¾ in)
Passenger door: Height	1.32 m (4 ft 4 in)
Width	0.61 m (2 ft 0 in)
Height to sill	0.51 m (1 ft 8 in)
Baggage compartment door (main):	
Height	0.61 m (2 ft 0 in)
Width	0.56 m (1 ft 10 in)
Height to sill	0.91 m (3 ft 0 in)
Baggage compartment door (rear):	
Height	0.38 m (1 ft 3 in)
Width	0.51 m (1 ft 8 in)
Height to sill	1.27 m (4 ft 2 in)
Emergency exits (each):	
Height	0.66 m (2 ft 2 in)
Width	0.51 m (1 ft 8 in)

DIMENSIONS, INTERNAL:

Cabin, incl flight deck and toilet:	
Length Westwind 2	6.08 m (19 ft 11¼ in)
Cabin, excl flight deck:	
Length: Westwind I	4.72 m (15 ft 6 in)
Westwind 2	4.74 m (15 ft 6½ in)
Max width	1.45 m (4 ft 9 in)
Max height	1.50 m (4 ft 11 in)
Floor area	6.52 m² (70.2 sq ft)
Volume	9.83 m³ (347 cu ft)
Baggage compartment: fwd (main)	1.13 m³ (40 cu ft)
rear	0.40 m³ (14 cu ft)
cabin	0.25 m³ (9 cu ft)

AREAS (A: Westwind I, B: Westwind 2):

Wings, gross	28.64 m² (308.26 sq ft)
Ailerons (total)	1.43 m² (15.40 sq ft)
Trailing-edge flaps (total): A	3.86 m² (41.58 sq ft)
B	3.85 m² (41.40 sq ft)
Speed brakes/lift dumpers (total)	1.37 m² (14.80 sq ft)
Fin: A	3.52 m² (37.94 sq ft)
B	3.02 m² (32.52 sq ft)
Rudder, incl tab: A	0.99 m² (10.66 sq ft)
B	1.02 m² (11.00 sq ft)
Tailplane: A	4.87 m² (52.42 sq ft)
B	4.86 m² (52.28 sq ft)
Elevators (total)	1.64 m² (17.66 sq ft)

WEIGHTS AND LOADINGS (A: Westwind I, B: Westwind 2):

Weight empty, equipped: A	5,578 kg (12,300 lb)
Basic operating weight empty:	
A (typical)	5,760 kg (12,700 lb)
B	6,010 kg (13,250 lb)
Max fuel: B	4,345 kg (9,580 lb)
Max payload: A	1,496 kg (3,300 lb)
A (optional)	1,542 kg (3,400 lb)
B	1,474 kg (3,250 lb)
Max T-O weight: A	10,365 kg (22,850 lb)
A (optional), B	10,660 kg (23,500 lb)
Max ramp weight: A	10,430 kg (23,000 lb)
A (optional), B	10,725 kg (23,650 lb)
Max landing weight: A, B	8,620 kg (19,000 lb)
Max zero-fuel weight: A, B	7,485 kg (16,500 lb)
Max cabin floor loading	976 kg/m² (200 lb/sq ft)
Max wing loading: A	361.73 kg/m² (74.13 lb/sq ft)
A (optional), B	372.02 kg/m² (76.23 lb/sq ft)
Max power loading: A	314 kg/kN (3.09 lb/lb st)
A (optional), B	324 kg/kN (3.18 lb/lb st)

PERFORMANCE (Westwind I, at max T-O weight of 10,639 kg; 23,450 lb, except where indicated):

Max level speed, S/L to 5,900 m (19,400 ft)
471 kts (872 km/h; 542 mph)

Max operating speed, S/L to 5,900 m (19,400 ft)
360 kts (666 km/h; 414 mph) IAS
Max operating speed Mach No. from 5,900 m (19,400 ft) to 13,725 m (45,000 ft) Mach 0.765
Econ cruising speed at 12,500 m (41,000 ft)
400 kts (741 km/h; 460 mph)
Stalling speed, flaps and landing gear down, at max landing weight 99 kts (183 km/h; 114 mph) CAS
Max rate of climb at S/L 1,524 m (5,000 ft)/min
Max operating altitude 13,715 m (45,000 ft)
FAA T-O balanced field length 1,495 m (4,900 ft)
T-O balanced field length at 8,165 kg (18,000 lb) AUW
945 m (3,100 ft)
Landing distance from 15 m (50 ft) at max landing weight, with thrust reversal 625 m (2,050 ft)
Landing distance from 15 m (50 ft) at 6,350 kg (14,000 lb) AUW, with thrust reversal 518 m (1,700 ft)
Range with seven passengers and baggage, IFR reserves
more than 2,150 n miles (3,983; 2,475 miles)
Max range with two passengers and baggage, 45 min reserves
more than 2,600 n miles (4,815 km; 2,993 miles)
Range with long-range fuel tank, five passengers and baggage, IFR reserves, at T-O weight of 10,660 kg (23,500 lb)
2,400 n miles (4,446 km; 2,763 miles)
Range with long-range fuel tank, two passengers and baggage, 45 min reserves, at T-O weight of 10,660 kg (23,500 lb)
2,900 n miles (5,375 km; 3,339 miles)

PERFORMANCE (Westwind 2, at max T-O weight except where indicated):
Never exceed, max level and max cruising speed at 8,840 m (29,000 ft) 469 kts (868 km/h; 539 mph)
Econ cruising speed between 11,890 and 12,500 m (39,000-41,000 ft) 390 kts (723 km/h; 449 mph)
Stalling speed at max landing weight, flaps down, engines idling 99 kts (184 km/h; 114 mph) CAS
Max rate of climb at S/L 1,524 m (5,000 ft)/min
Rate of climb at S/L, one engine out 250 m (820 ft)/min
Max certificated ceiling 13,715 m (45,000 ft)
Service ceiling, one engine out:
at 9,072 kg (20,000 lb) gross weight
6,400 m (21,000 ft)
at 7,030 kg (15,500 lb) gross weight
9,450 m (31,000 ft)
Min ground turning radius 14.50 m (47 ft 7 in)
T-O run 1,218 m (3,995 ft)
T-O balanced field length 1,600 m (5,250 ft)
Landing run from 15 m (50 ft) at max landing weight
747 m (2,450 ft)
Landing run at max landing weight 534 m (1,750 ft)
Range, NBAA VFR reserves:
with max payload (10 passengers)
2,390 n miles (4,430 km; 2,750 miles)
with max fuel and four passengers
2,905 n miles (5,385 km; 3,345 miles)

OPERATING NOISE LEVELS FAR Pt 36 (at max T-O weight):
T-O: Westwind I, normal MTOW 84.2 EPNdB
Westwind 2 85.1 EPNdB
Approach: Westwind I, normal MTOW 93.0 EPNdB
Westwind 2 92.8 EPNdB
Sideline: Westwind I, normal MTOW 88.4 EPNdB
Westwind 2 88.5 EPNdB

UPDATED

LAHAV DIVISION

IAI's Lahav Division offers a number of upgrade possibilities for existing combat aircraft. Modular modernisation packages include new avionics, airframe modifications and refits.

NEW ENTRY

IAI (McDONNELL DOUGLAS) F-15 UPGRADE

TYPE: Combat aircraft upgrade.

PROGRAMME: Lahav Division involved in upgrading of F-15 structures, maintainability and various utility and mission systems.

UPDATED

IAI (MIKOYAN) MiG-21-2000 UPGRADE

TYPE: Combat aircraft upgrade.

PROGRAMME: Lahav Division offers MiG-21 upgrade and associated training and ground support facilities on international market.

DESIGN FEATURES: MiG-21bis displayed at 1993 Paris Air Show had Elta EL/M-2032 look-down/shoot-down radar with Doppler beam sharpening (replacing former Soviet 'Jaybird'); El-Op HUD, monochrome HDD and colour tactical display; new one-piece windscreen; Martin-Baker ejection seat; and single mission computer compatible with a MIL-STD-1553 databus. Possible weapons include Rafael Python 3 infra-red AAMs and Mk 82 laser-guided

IAI 1124N Sea Scan customs maritime patrol aircraft *(IAI)*

1994

IAI Lahav Division undertaking USAF F-15 overhaul and modification *(IAI)*

1995

IAI (Mikoyan) MiG-21-2000 upgrade

1994

bombs. Other possible options in upgrade package include DASH (display and sight helmet), increased internal/external fuel capacity and airframe life extension.

Lahav Division is currently building its MiG-21-2000 demonstrator. First flight took place on 24 May 1995.

UPDATED

IAI (MIKOYAN) CAMBODIAN AIR FORCE MiG-21BIS UPGRADE

TYPE: Combat aircraft upgrade.

PROGRAMME: Commenced on first four aircraft in early 1995. Programme to be completed by end of 1995 and to include between 10 and 20 aircraft.

DESIGN FEATURES: Package to include structural modifications, maintenance improvements, flight safety improvements, improved external stores as well as advanced navigation and communications equipment. Not thought to include radar.

NEW ENTRY

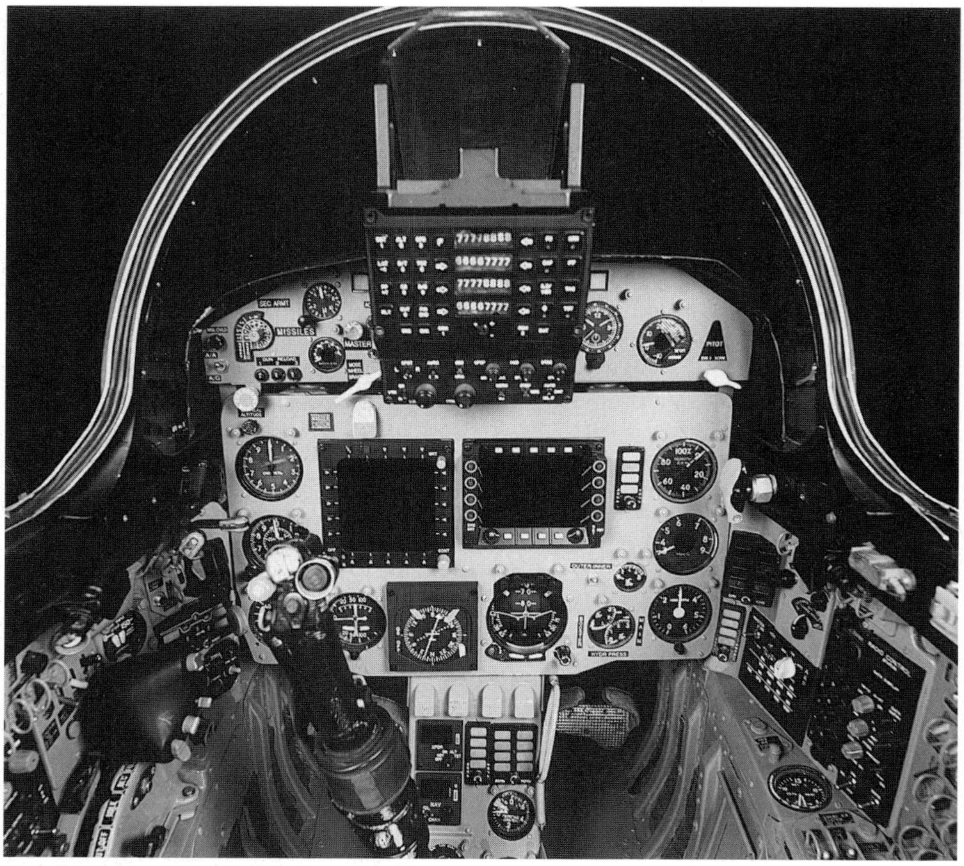

IAI Lahav Division's MiG-21-2000 advanced cockpit layout *(IAI)*
1995

BEDEK AVIATION GROUP

Ben-Gurion International Airport, Israel 70100
Telephone: 972 (3) 935 3743
Fax: 972 (3) 935 7066

CORPORATE VICE-PRESIDENT AND GENERAL MANAGER: A. Arzi

Bedek Aviation Group is internationally approved as a single-site civil and military airframe, power plant, systems and accessory service and upgrading centre. It has three operating divisions: aircraft division - maintenance, upgrading and conversion; engine division - engine maintenance; components division - components maintenance, modification, which with the group headquarters, have a workforce of 3,200 housed in 110,000 m² (1,184,030 sq ft) of covered space. Bedek performs most IAI civil and military upgrading and retrofit programmes, current examples including the programmes described in this entry and conversion of a large number of transport aircraft to passenger/cargo, tanker, EW and reconnaissance mission configurations. Additional ongoing work includes the turnaround inspection, overhaul, upgrading, repair, retrofit, outfitting and testing of more than 25 types of aircraft, among these are various models of Boeing: 707, 727, 737, 747, 767, McDonnell Douglas: DC-8, DC-9, DC-10, MD-11, MD-80, Lockheed L-100, C-130; Airbus A-300, A-310; YAK-40, YAK-42, TU-134, TU-154; AN-12, AN-72P; IL-86, IL-96; S-2E (Tracker).

Power plants processed encompass 30 types of civil and military piston, turboprop, turbojet and turbofan engine and their components, including the JT9D, JT3D, JT8D, Series PT6, F-100, F-110, J79, T-56250C, CFM56-2-3. More than 6,000 types of accessories and instruments are serviced including: APU and propellers, transmission, CSD and IDG, oxygen and safety equipment, communication and navigation, electrical, hydraulic, pneumatic system, flight control and radar systems, transducers instruments. The Group provides total technical support to several international operators, and holds warranty and/or approved service centre appointments from domestic and foreign air regulatory agencies, air arms and a large number of leading aerospace manufacturers. Bedek Aviation is approved as a repair and overhaul agency by most major civil aviation authorities such as the Israeli CAA, Israeli Air Force, US Military, FAA (USA), CAA (UK) and LBA (Germany); these approvals cover aircraft, engine and accessory service.

UPDATED

IAI (Douglas) A-4 Skyhawk modernised version
1994

IAI (McDonnell Douglas) F-4 Phantom 2000 upgrade *(IAI)*
1994

IAI (DOUGLAS) A-4 SKYHAWK UPGRADE

TYPE: Combat aircraft upgrade.

DESIGN FEATURES: Modifications (already applied to Israeli Air Force A-4s) include life extension, complete rewiring, dual-disc mainwheel brakes, nosewheel steering, wing spoilers, two additional underwing hardpoints, engine tailpipe extension, brake parachute in fairing under tail. Wingroot cannon 30 mm; modern WDNS; extended nose compartment and saddleback hump for more and lighter avionics; flare/chaff dispensers under tail. Dual control two-seat modification also offered.

VERIFIED

IAI (McDONNELL DOUGLAS) F-4 PHANTOM 2000 UPGRADE

TYPE: Combat aircraft upgrade.

PROGRAMME: Israeli Air Force (IAF) programme to extend life beyond 2000, also available for export; first flight Phantom 2000 prototype 11 August 1987; first redelivery to IAF 9 April 1989; first operational use 5 February 1991. Prototype completed by IAF; airframe changes to production aircraft by IAI; redeliveries exceeded 20 by mid-1991 and then continuing at average two per month.

CUSTOMERS: Israeli Air Force. Reported to be bidding for contract to upgrade Greek and South Korean F-4 aircraft. Reported to be on verge of concluding contract with Turkish Air Force for 50 F-4Es.

DESIGN FEATURES: Modifications include reinforced skins and fuel cells in fuselage and wings; complete rewiring; dual

MIL-STD-1553B databuses; replaced and rerouted hydraulic lines; number of avionics boxes reduced; built-in test features added; small strakes added above intake flanks to improve stability and manoeuvrability; cockpit comfort and instrument layout modernised. New IAF avionics installation includes Westinghouse Norden Systems APG-76 multimode high resolution radar, El-Op (Kaiser licence) wide angle diffractive-optics HUD, Elbit multi-function electronic displays for both crewmen, digital WDNS with HOTAS (hands on throttle and stick), Orbit integrated com and com/nav systems, improved ECM and self-protection systems; core of system is derivative of ACE-3 currently fitted in IAF F-16C/Ds; overall avionics integrated is Elbit Computers Ltd.

VERIFIED

IAI (NORTHROP) F-5 Plus: SHAHAM UPGRADE

TYPE: Combat aircraft upgrade.

CUSTOMERS: Chilean Air Force.

DESIGN FEATURES: Avionics improvements made by Bedek Aviation Group include HUD, two multifunction displays and HOTAS, modern pulse Doppler radar, video camera and recorder, optional helmet-mounted display, improved weapon delivery and ECM, new weapons and maintainability improvements. Other options include modern ejection seat, secure communications, flight refuelling and podded or nose-mounted reconnaissance equipment.

UPDATED

IAI (NORTHROP) F-5 CHILEAN AIR FORCE UPGRADE

TYPE: Combat aircraft upgrade.

PROGRAMME: Upgrade of 12 F-5Es and two F-5Fs made by Bedek Aviation Group, ordered early 1990.

DESIGN FEATURES: Programme includes Elta EL/M-2032B multimode pulse Doppler fire control radar, improved WDNS, INS, El-Op HUD, two head-down CRT displays, HOTAS, Aeronautic (Israel) air data computer, modular mission and display processor and multiple databus, and full EW suite (passive RWR, jammer and chaff/flare dispenser). Nose modifications to accommodate larger radar antenna include removal of port M39 20 mm cannon and relocation of some subsystems. Refit reportedly includes provision of Python air-to-air missiles. First two (one F-5E, one F-5F) under flight test early 1993, redelivered September 1993. Remainder to be converted by ENAER Chile under IAI supervision. First Chilean built aircraft delivered 1994.

UPDATED

IAI (MARSH/GRUMMAN) S-2UP TRACKER UPGRADE

TYPE: Combat aircraft upgrade.

PROGRAMME: Customer revealed as Argentine Navy ($30 million contract to upgrade one S-2E and provide kits for customer to upgrade five more). Bedek Aviation Group's Shaham Plant teamed with Marsh Aviation Company (see entry in main US section); upgrade includes refit with Garrett TPE331-15 turboprops and five-blade propellers; improved air conditioning and oxygen systems; wing-mounted high candlepower searchlight; new autopilot; modernisation of other flight systems; payload and take-off weights, and performance, increased without loss of carrier launch and retrieval capability.

Bedek has also developed a higher level S-2E upgrade package incorporating specified radar plus ASW, EW and/or a day/night reconnaissance system, all controlled through one multifunction display. Argentine Navy upgrade apparently did not include these features.

VERIFIED

IAI (AEROSPATIALE) AMIT FOUGA

Israeli Air Force name: Tzukit

TYPE: Combat aircraft upgrade.

PROGRAMME: The AMIT Fouga (Advanced Multimission Improved Trainer) was engineered by the Bedek Aviation Division of IAI to Israeli Air Force requirements, to enable its Fouga Magisters to remain as standard IAF trainers during the 1980s. It is, in effect, completely rebuilt and modernised, and is a dedicated trainer with all armament removed, although its retains capability for patrol and aerial photographic missions. The upgrade programme is available for other operators of this aircraft. The Israeli Air Force was reported in 1990 to be seeking a Tzukit replacement.

VERIFIED

IAI (Northrop) F-5 Plus upgrade *(IAI)*

1994

IAI (Marsh/Grumman) S-2 Tracker upgrade *(IAI)*

1994

IAI (Aerospatiale) AMIT Fouga upgrade *(IAI)*

1994

IAI (BOEING) 707/720 UPGRADES

TYPE: Passenger/transport aircraft upgrades.

DESIGN FEATURES: Bedek Aviation Group has refurbished and resold numerous Boeing 707s and 720s, often after conversion from passenger to cargo. Probe/drogue or flying boom air refuelling tanker and other configurations, including a sigint/tanker conversion with wingtip refuelling pods, aft hose reel and Elta EL/L-8300 sigint system. AEW Phalcon version mounting an Elta Electronics Phalcon solid-state L-band radar with six conformal phased-array antennae: two on each side of the fuselage, one in an enlarged nose and one under the tail delivered to Chilean air force in March 1995. In addition to the radar, the Phalcon system incorporates a sophisticated monopulse IFF, wide range ESM system and a comint data processing system for tactical situation display. Modifications involved in the tanker conversion include local reinforcement of the outer wings, supports for additional fuel tanks where applicable, and fuselage reinforcement for the boom support point of aft hose reel exit; an additional hydraulic system to power the fuel pumps and boom or aft reel; adaptation of the fuel supply system to the tanker role; electrical system changes to add external illumination, refuelling system controls, boom operator's station with 3D electro-optical viewing system, director lights for pilots of receiver aircraft; and avionics upgrade to individual customer requirements.

Tanker mixed configuration, with a centreline boom and two probe/drogue underwing refuelling pods, have been delivered to the Israeli Air Force, and four similar conversions were ordered in 1989 for the Royal Australian Air Force.

WEIGHTS AND LOADINGS (707-320C tanker, approx):

Operational weight empty	65,770 kg (145,000 lb)
*Internal fuel weight	72,575 kg (160,000 lb)
**Additional fuel weight	up to 13,605 kg (30,000 lb)
Tanker T-O weight	151,950 kg (335,000 lb)

**90,300 l (23,855 US gallons; 19,863 Imp gallons)*
***17,034 l (4,500 US gallons; 3,747 Imp gallons)*

UPDATED

IAI (BOEING) 747-100 and -200 FREIGHTER UPGRADES

TYPE: Passenger/transport aircraft upgrades.

DESIGN FEATURES: Bedek Aviation Group converted a Boeing 747-100 freighter configuration prototype, for certification in 1990. Changes include installing a 3.05 × 3.40 m (10 ft 9 in × 11 ft 2 in) upward opening main deck cargo door aft of the wing on the port side, with local reinforcement of the fuselage; reinforcing the cabin floor to increase load-carrying capacity; installing a fully powered ball mat/roller cargo handling system and restraint system, and a bulkhead between the passenger and cargo compartments; and interior modifications to select passenger/cargo combinations. Basic configuration options to be offered are all-cargo, with up to 29 main deck standard pallets or containers; Combi, with passengers at front and 7 to 13 pallets aft; and all-passenger, with interior layout to customer's specification. Versions to accommodate non-standard containers, and similar conversions of the Model 747-200 can be produced.

Several conversion programmes of Boeing 747-100s to all-cargo configuration were carried out during 1990-94 including Lufthansa (6), SAA (1), Polaris (7), Air France (1).

WEIGHTS AND LOADINGS (747-100 Combi, estimated):

Operating weight empty	148,325 kg (327,000 lb)
Max payload	98,883 kg (218,000 lb)
Max T-O weight	334,750 kg (738,000 lb)
Max landing weight	265,350 kg (585,000 lb)
Max zero fuel weight	247,435 kg (545,500 lb)

UPDATED

IAI (LOCKHEED) C-130/L-100 HERCULES UPGRADES

TYPE: Transport aircraft upgrade.

DESIGN FEATURES: Bedek Aviation Group has already accomplished several successful conversions of C-130 series aircraft to such configurations as air refuelling tanker and sigint platform, with appropriate airframe modifications and avionics upgrade. Operational configurations currently being offered for any C-130B to C-130H variant, or their L-100 commercial counterparts, include: probe and drogue air refuelling tanker, with an additional 11,356 litre (3,000 US gallon; 2,498 Imp gallon) cargo compartment fuel tank plus two underwing fuel pods; maritime surface patrol and ASW, with appropriate surveillance, acoustic, MAD, armament or stores management systems, and operator stations; C^3I and electronic warfare platform, with comint, elint, communications and EW systems to customer's requirements; search and rescue, with a rescue kit, flare storage/launcher and operator station on a logistic pallet installed on the rear-loading ramp; emergency assistance, with an insulated cabin mounted on a logistic pallet for ambulance or flying hospital missions, firefighting configuration with up to 11,356 litres (3,000 US gallons; 2,498 Imp gallons) of water or retardant in pallet-mounted tanks in the cargo hold; and VIP, 65-seat passenger/cargo combi transport, with full airliner type seating, galley and toilet

IAI (Boeing) 707 conversion to a mixed probe/drogue-flying boom air refueller

1994

IAI (Boeing) 707 AEW Phalcon conversion

1994

IAI Bedek Group (Boeing) B747 converted passenger/cargo aircraft *(IAI)*

1995

facilities, pallet-mounted in an air conditioned environment.

WEIGHTS AND LOADINGS (C-130H tanker, approx):

Operational weight empty	35,380 kg (78,000 lb)
*Internal fuel weight	29,030 kg (64,000 lb)
**Additional fuel weight	10,885 kg (24,000 lb)
Tanker T-O weight	75,295 kg (166,000 lb)
Max overload T-O weight	79,380 kg (175,000 lb)

**36,643 l (9,860 US gallons; 8,060 Imp gallons)*
***13,627 l (3,600 US gallons; 2,997 Imp gallons)*

UPDATED

IAI COCKPIT UPGRADES

TYPE: Transport aircraft cockpit upgrades.

PROGRAMME: As future development programmes, Bedek Aviation intends to convert existing three-man flight decks of transport aircraft for two-man operation. Conversion will include advanced monitoring and control systems, including engine indicating and crew alerting system (EICAS). Candidate aircraft include the Boeing 727 and McDonnell Douglas DC-10.

VERIFIED

IAI Bedek Group (Sikorsky) CH-53D Yasur 2000 upgrade *(IAI)*

1995

Technologies Division

PO Box 190, Lod Industrial Zone 71101
Telephone: 972 (8) 239111
Fax: 972 (8) 222792
Telex: 381520 SHLD IL
GENERAL MANAGER: David Arzi

This division is the parent facility to four separate plants: SHL (Servo Hydraulics Lod), Ramta Structures and Systems, MATA Helicopters, and Golan Industries. SHL designs, develops and manufactures hydraulic system components, hydraulic flight control servo-systems, landing gears and brake systems; and produces air actuated chucks, miniature gears, clutches and brakes. Among others, its products equip the Kfir, Arava, Westwind and Astra, and IDF Sikorsky Black Hawk helicopters; manufacturing approvals are held from Boeing, Dornier, General Dynamics and General Electric, among others. Ramta undertakes metal and advanced composites fabrication for the F-4 Phantom, F-16 Fighting Falcon, E-2C Hawkeye, Kfir, Westwind and Astra, as well as manufacturing ground vehicles and patrol boats. MATA repairs, reconfigures and remanufactures helicopter structures and components, and produces equipment and systems for rotating-wing aircraft. Golan designs and manufactures aircraft crew and passenger seats (including designing crashworthy troop seats for the Bell/Boeing V-22 Osprey), aircraft wheels and cockpit controls.

VERIFIED

IAI YASUR 2000 (SIKORSKY) CH-53D UPGRADE

TYPE: Combat helicopter upgrade.

DESIGN FEATURES: Headed by MATA Helicopters, this programme is aimed at upgrading the Israeli Air Force fleet of some 30 Sikorsky CH-53D heavy lift helicopters.

DESIGN FEATURES: Programme go-ahead given June 1990; first flight of first Yasur 2000, 4 June 1992; redelivery began 2 February 1993. The Yasur 2000 embodies structural changes to extend service life well into the next century, and improved avionics such as a new mission computer, moving map display, two multifunction displays and a new autopilot system. Other modifications include internal auxiliary fuel tanks, flight refuelling boom, rescue hoist, crashworthy seats, cockpit armour, improved launching lights, internal batteries, electric pump for reloading APP accumulator and new APP clutch. Elbit is avionics systems integrator for the Yasur 2000.

VERIFIED

RADA ELECTRONIC INDUSTRIES LTD

HEADQUARTERS: PO Box 54 Beit Shean, Israel 10900
Telephone: 972 (6) 585811
Fax: 972 (52) 555176
CHAIRMAN: Abraham Perelman
PRESIDENT AND CEO: Haim Nissenson
DIRECTOR: Yair Grinberg
VICE-PRESIDENT, COMMERCIAL OPERATIONS, AND DIRECTOR: Nathan Metuki
DIRECTOR: Eles Dobronsly
SECRETARY: Mordechai Perera
VICE-PRESIDENT, MARKETING: Meir Shariv
VICE-PRESIDENT, US OPERATIONS: Itzhak Almog
DIRECTOR OF FINANCE AND CONTROLS: Yossi Kedem

Rada Electronic Industries Ltd is an Israel based company with subsidiaries in Israel as well as Belgium, Netherlands and the USA. The group operates in two business sectors: the development, production and marketing of aerospace and defence-related equipment and the distribution of electronic and computer components. RADA was incorporated in Israel in 1970. Main plants are in the industrial zone at Beit-Shean.

The company is involved in the upgrading of aircraft and is responsible for the Autonomous Combat-manoeuvres. Evaluation (ACE) System (which see). The ACE system eliminates the necessity for ACMI type pods, ground tracking and telemetry.

VERIFIED

ITALY

AEREA

AEREA SpA

HEAD OFFICE AND PLANT: Via Celafonia 18, I-20156 Milano
Telephone: 39 (2) 38000218
Fax: 39 (2) 33402676
Telex: 333494 AEREA I
CHAIRMAN: Dott Ing Marco Bognetti
MANAGING DIRECTOR: Dott Ing Silvano Mantovani
MARKETING DIRECTOR: Dott Ing Vincenzo Riviello

Aerea commenced operations in 1927. The company is responsible for the design, development, production, qualification, installation and maintenance of aircraft weapons carriage and release systems. Main activities are in the fields of: bomb racks, ejector release units, missiles rail launchers, rocket launchers, suspension systems, sound underwater signalling bombs and training dispensers for bombs and rockets. The company is also active in the production of door gun posts and chaff/flare dispensers.

NEW ENTRY

AEREA (MIKOYAN) MiG-21 UNIVERSAL STORES CARRIAGE SYSTEMS UPGRADE

TYPE: Combat aircraft mission capabilities upgrade.

PROGRAMME: In early 1993 Aerea and ANPK MiG initiated preliminary discussion on a collaborative programme to develop a modular stores carriage system for future MiG-21 upgrade programmes. The two companies signed a Memorandum of Understanding (MoU) on 10 August 1993.

DESIGN FEATURES: Concept of package is to offer universal system which will be capable of operating both Warsaw Treaty and NATO standard unguided weapons. Package to include systems for 250 kg and 500 kg bombs, 2.75 in, 50 mm, 68 mm and 81 mm rocket launchers, bomb/rocket dispensers as well as flare/chaff dispensers.

NEW ENTRY

AERITALIA — *see Alenia*

AERONAUTICA MACCHI

AERONAUTICA MACCHI SpA

Via P. Foresio, I-21040 Venegogno Superiore
Telephone: 39 (331) 865912
Fax: 39 (331) 865910
CHAIRMAN: Dott Fabrizio Foresio

The original Macchi company was founded in 1913 in Varese, and produced a famous line of high-speed flying boats and seaplanes. On 1 January 1981 the Aeronautica Macchi group reorganised its structure, transforming itself into a holding company and transferring all of its operating activities to a newly formed, wholly owned company known as Aermacchi SpA. The group includes, besides Aermacchi SpA, the subsidiary companies Aero Engineering (aeronautical design), SICAMB (airframe and equipment manufacturing, including licence production of Martin-Baker ejection seats), OMG (precision machining), and Logic (electronics equipment). A 25 per cent holding in Aeronautica Macchi was acquired by Aeritalia (now Alenia) in 1983. Transfer of all company activities to the Venegogno airfield site completed 31 August 1993.

VERIFIED

AERMACCHI

AERMACCHI SpA (Subsidiary of Aeronautica Macchi SpA)

Via P. Foresio, PO Box 101, I-21040 Venegogno Superiore
Telephone: 39 (331) 813111
Fax: 39 (331) 827595
CHAIRMAN: Dott Fabrizio Foresio
MANAGING DIRECTOR: Dott Ing Giorgio Brazzelli
GENERAL MANAGERS:
Dott Ing Bruno Cussigh
Dott Ing Romano Antichi
TECHNICAL DIRECTOR: Dott Ing Massimo Lucchesini
COMMERCIAL MANAGER: Dott Cesare Cozzi

Aermacchi, which celebrated its 75th anniversary in 1988, is the aircraft manufacturing company of the Aeronautica Macchi group. Its plants at Venegogno airfield occupy a total area of 274,000 m² (2,949,310 sq ft), including 52,000 m² (559,720 sq ft) of covered space; the flight test centre has covered space of 5,100 m² (54,900 sq ft) in a total area of 28,000 m² (301,390 sq ft). Total workforce at the end of 1994 was 1,800.

In addition to its aircraft programmes, Aermacchi is active in the field of aerospace ground equipment, with a complete line of hydraulic, electric and pneumatic ground carts for servicing civil and military aircraft, and also has important roles in the Ariane and EFA programmes.

UPDATED

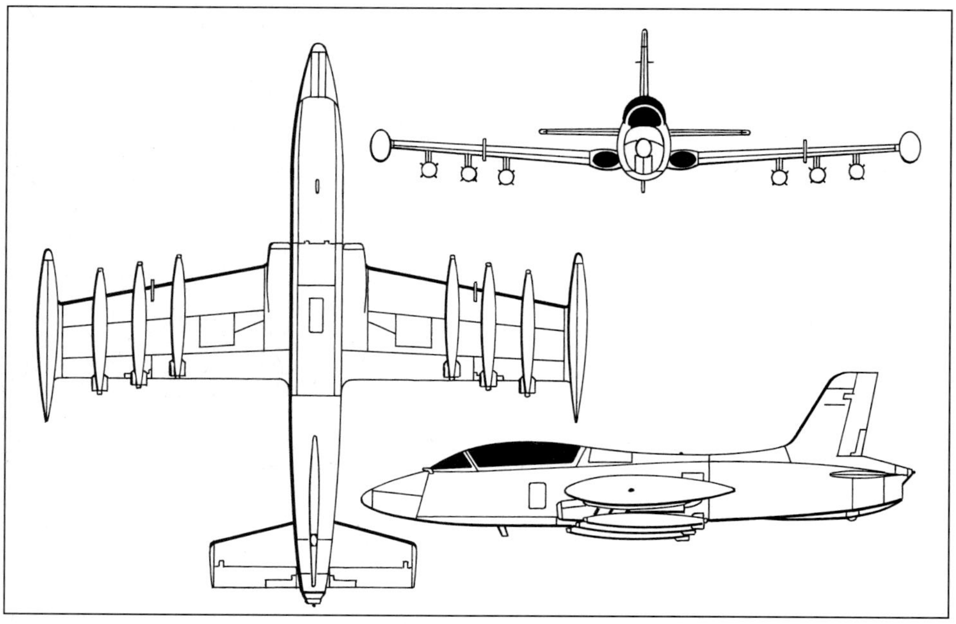

Aermacchi M.B. 326 two-seat training version ***(Jane's/Mike Keep)***

1994

Atlas Impala I (M.B. 326K) single-seat operational trainer with light ground attack capabilities

1994

Atlas Impala II (M.B. 326M) two-seat training aircraft

1994

AERMACCHI M.B. 326

TYPE: Two-seat basic and advanced trainer or single-seat operational trainer.

PROGRAMME: The first prototype of the original Aermacchi M.B. 326 jet trainer flew for the first time on 10 December 1957, powered by a Rolls-Royce Viper 8 turbojet engine. The more powerful Viper 11 engine powers six production versions of the aircraft built for the air forces of Italy (M.B. 326 and 326E), Tunisia (M.B. 326B), Ghana (M.B. 326F), Australia (M.B. 326H) and South Africa (M.B. 326M), and one version built for Alitalia (M.B. 326D).

The initial South African version, built by Atlas, was the Impala Mk 1.

VERSIONS: **AT-26 Xavante:** Brazilian built (by Embraer) version of the M.B. 326GC.

Atlas Impala I: South African built (by Atlas Aviation) version of the M.B. 326M.

Atlas Impala II: South African built (by Atlas Aviation) version of the M.B. 326K with Viper 540 turbojet.

M.B. 326E: Two-seat advanced trainer and weapons trainer approximately 12 built (including six conversions) for the Italian Air Force's *Scuola di Volo Bsaica Iniziale* at Lecce. Viper 11 engine (11.12 kN; 2,500 lb st), plus strengthened wings and six underwing hardpoints of the M.B. 326GB. New avionics and equipment included UHF radio, miniaturised Tacan, gyroscopic weapons sight and gun camera.

M.B. 326GB: Two-seat dual-control advanced training and attack version, with airframe modifications and Viper 20 Mk 540 engine. M.B. 326G prototype flew for the first time in Spring 1967; similar to M.B. 326B. Customers include Argentine Navy (8), air forces of Zaïre (17) and Zambia (23). In addition 167 similar **M.B. 326GCs** were built in Brazil by Embraer for the Brazilian Air Force, as the **AT-26 Xavante**, and for the air force of Togo (3).

M.B. 326GC: Brazilian built M.B. 326 designated AT-26 Xavante. See above.

M.B. 326K: Single-seat operational trainer and light ground attack version. Retains most of the structure and systems of M.B. 326GB, but has more powerful Viper 632 engine, no second cockpit, additional fuselage fuel tanks, increased weapon-carrying capacity. Prototype with Viper 540 engine first flown 22 August 1970; second prototype, with Viper 632, first flew 1971. Delivered to Dubai Defence Force (6), Ghana Air Force (9), South African Air Force (reported), Tunisian Air Force (8) and Zaïre Air Force (6). In South Africa, Atlas Impala Mk 2 is a Viper 540 powered version based on the M.B. 326K.

M.B. 326L: Two-seat advanced trainer, combining airframe of single-seat M.B. 326K with standard two-seat dual control cockpit installation. Two delivered to Dubai and four to Tunisian Air Force.

UPGRADES: **RAAF M.B. 326H:** 19 M.B. 326 of the Royal Australian Air Force being rewinged and 35 RAAF M.B. 326s being refurbished.

OPERATORS: Versions of the M.B. 326 are in service with the armed forces of the following countries: Argentina (4), Australia (30+), Brazil (180), Ghana (9), Italy (60), Paraguay (9), South Africa (200), Togo (4), Tunisia (19), United Arab Emirates (5), Zaïre (15) and Zambia (18).

DESIGN FEATURES: Cantilever mid-wing monoplane. Wing section NACA 6A series (modified). Thickness chord ratio 13.7 per cent at root, 12 per cent at tip. Dihedral 2° 55′. Incidence 2° 30′.

FLYING CONTROLS: Manually operated all-metal ailerons and hydraulically operated slotted flaps. Electrically actuated balance and trim tab in port aileron. Geared balance tab in starboard aileron. Hydraulically operated dive-brake under centre-fuselage. Electrically operated trim tab in rudder and each elevator.

STRUCTURE: The wing is an all-metal two-spar structure in three sections, of which the centre-section is integral with the fuselage. The fuselage is an all-metal structure. The tail unit is an all-metal structure.

POWER PLANT (GB): One Rolls-Royce Viper 20 Mk 540 turbojet engine, rated at 15.17 kN (3,410 lb st). Fuel in flexible rubber main tank in fuselage, capacity 782 litres (206 US gallons; 172 Imp gallons) and two 305 litre (80 US gallon; 67 Imp gallon) non-jettisonable wingtip tanks. Total standard fuel capacity 1,392 litres (367 US gallons; 306 Imp gallons). Provision for two 332 litre (88 US gallon; 73 Imp gallon) jettisonable underwing tanks, to give a total capacity of 2,056 litres (543 US gallons; 452 Imp gallons). Single-point pressure refuelling receptacle under fuselage. Fuel dump valves permit quick emptying of tanks.

POWER PLANT (K and L): One Rolls-Royce Viper Mk 632-43 turbojet engine, rated at 18.79 kN (4,000 lb st). M.B. 326L fuel capacity same as for GB. Fuel in M.B. 326K contained in three rubber fuselage tanks and two permanent wingtip tanks, total usable capacity 1,660 litres (440 US gallons;

366 Imp gallons). Provision to install self-sealing fuselage tanks and reticulated foam anti-explosive filling in all tanks, including those at wingtips. Two underwing stations equipped normally to carry jettisonable auxiliary tanks of up to 334 litres (90 US gallons; 75 Imp gallons) each. Single-point. pressure refuelling receptacle and auxiliary gravity refuelling points.

ACCOMMODATION (GB and L): Crew of two in tandem under a one-piece moulded perspex canopy which hinges sideways to starboard. Pressurised cockpit, differential 0.21 bars (3.0 lb/sq in) in GB, 0.24 bars (3.5 lb/sq in) in L. Dual controls and instruments. Blind-flying screens for pupil. Martin-Baker Mk 04A lightweight ejection seats in GB; Mk 06A zero/zero seats in L.

ACCOMMODATION (K): Pilot only, on Martin-Baker Mk 6 zero/zero ejection seat in pressurised and air-conditioned cockpit (differential 0.24 bars (3.5 lb/sq in). Separately controlled canopy jettison system provided, but seat is fitted with breakers to permit ejection through canopy in extreme emergency. Canopy hinges sideways to starboard. Provision for armour protection for pilot and other vital areas.

ARMAMENT (GB and L): Provision for up to 1,814 kg (4,000 lb) of armament on six underwing attachments. Typical weapon loads include following alternatives; two LAU-3/A packs each containing 19 2.75 in FFAR rockets and two packs each containing eight Hispano-Suiza SURA 80 mm rockets; two 12.7 mm gun pods and four packs each containing six SURA 80 mm rockets; one 7.62 mm Minigun, one 12.7 mm gun pod, two Matra 122 rocket packs and two packs each containing six SURA 80 mm rockets; two 500 lb bombs and eight 5 in HVAR rockets; two AS 12 missiles; one 12.7 mm gun pod, one reconnaissance pack containing four Vinten cameras and two 272 kg (600 lb) drop-tanks, or two pods each containing 30 mm DEFA gun. SFOM Type 83 fixed gunsight or Ferranti LFS 5/102A gyrosight. Gun camera in nose.

ARMAMENT (K): Standard fixed armament of two 30 mm DEFA electrically operated cannon in lower front fuselage, with 125 rds/gun. Six underwing pylons, the inboard four stressed to carry up to 454 kg (1,000 lb) each and the outboard pair up to 340 kg (750 lb) each. Maximum external military load (with reduced fuel) is 1,814 kg (4,000 lb). Each pylon fitted with standard NATO 355 mm (14 in) MA-4A stores rack. Typical loads may include two 750 lb and four 500 lb bombs, four napalm containers, two AS 11 or AS 12 air-to-surface missiles, two machine gun pods, two Matra 550 air-to-surface missiles, six SUU-11A/A 7.62 mm Minigun pods, and various Matra or other launchers for 37, 68, 100 mm, 2.75 or 5 in rockets. A four-camera tactical reconnaissance pod can be carried on the port inner pylon without affecting the weapon capability of the other five stations.

DIMENSIONS, EXTERNAL:

Wing span over tip tanks: GB	10.854 m (35 ft 7¼ in)
K	10.850 m (35 ft 7 in)
Wing area, gross	19.30 m² (207.7 sq ft)
Wing aspect ratio	6.08
Length overall	10.673 m (35 ft 0¼ in)
Height overall	3.72 m (12 ft 2 in)
Wheel track	2.485 m (8 ft 2 in)
Wheelbase	5.157 m (13 ft 7½ in)

WEIGHTS AND LOADINGS (M.B. 326GB. A: Trainer, B: Attack):

Basic operating weight, excl crew:	
A	2,685 kg (5,920 lb)
B*	2,558 kg (5,640 lb)
Max zero-fuel weight: A	2,849 kg (6,280 lb)
B*	2,640 kg (5,820 lb)
Max T-O weight (full internal fuel, wingtip and underwing tanks): A	4,577 kg (10,090 lb)
B, no armament	4,447 kg (9,805 lb)
B, with 769 kg (1,695 lb) armament)	5,216 kg (11,500 lb)
Max T-O weight (max armament):	
B*, with fuel in fuselage tank only and 1,962 kg (4,325 lb) armament)	5,216 kg (11,500 lb)
Max wing loading	269.5 kg/m² (55.2 lb/sq ft)
Max power loading	343 kg/kN (3.37 lb/lb st)

* *Without tip tanks and aft ejection seat*

WEIGHTS AND LOADINGS (M.B. 326K):

Weight empty, equipped	3,123 kg (6,885 lb)
T-O weight 'clean'	4,645 kg (10,240 lb)
Typical operational T-O weights:	
patrol and reconnaissance	5,048 kg (11,130 lb)
photographic reconnaissance	5,111 kg (11,270 lb)
Max T-O weight	5,897 kg (13,000 lb)
Max landing weight	5,443 kg (12,000 lb)
Normal design landing weight	4,535 kg (10,000 lb)
Max wing loading	350 kg/m² (62.4 lb/sq ft)

PERFORMANCE: (M.B. 326GB. A: Trainer at typical weight of 3,937 kg; 8,680 lb representing max T-O weight without underwing tanks, B: Attack version at combat weight of 4,763 kg; 10,500 lb, C: Attack version at max T-O weight):

Brazilian AT-26 Xavante (M.B. 326GC) operational trainers

1994

Aermacchi 326B two-seat operational trainer of the Tunisian Air Force *(Aermacchi)*

1995

Never-exceed speed:	
A	Mach 0.82 (469 kts; 871 km/h; 541 mph EAS)
B	Mach 0.75 (419 kts; 778 km/h; 483 mph EAS)
Max level speed: A	468 kts (867 km/h; 539 mph)
Max cruising speed: A	430 kts (797 km/h; 495 mph)
Max rate of climb at S/L: A	1,844 m (6,050 ft)/min
B	1,082 m (3,550 ft)/min
C	845 m (3,100 ft)/min
Time to 3,050 m (10,000 ft): B	3 min 10 s
C	4 min 0 s
Time to 6,100 m (20,000 ft): A	4 min 10 s
B	8 min 0 s
C	9 min 20 s
Time to 9,150 m (30,000 ft): A	7 min 40 s
B	15 min 0 s
C	18 min 40 s
Time to 12,200 m (40,000 ft): A	13 min 5 s
Service ceiling: A	14,325 m (47,000 ft)
B	11,900 m (39,000 ft)
T-O run, ISA: A	412 m (1,350 ft)
B	640 m (2,100 ft)
C	845 m (2,770 ft)
T-O to 15 m (50 ft) ISA: A	555 m (1,820 ft)
B	866 m (2,840 ft)
C	1,140 m (3,740 ft)
Landing from 15 m (50 ft), ISA:	
A at landing weight of 3,175 kg (7,000 lb)	631 m (2,070 ft)
B at landing weight of 4,195 kg (9,250 ft)	802 m (2,630 ft)
Range (A, with 113 litres; 30 US gallons; 25 Imp gallons reserve): fuselage and tip tanks	998 n miles (1,850 km; 1,150 miles)
fuselage, tip and underwing tanks	1,320 n miles (2,445 km; 1,520 miles)
Combat radius (c):	
max fuel, 769 kg (1,695 lb) armament, 90 kg (200 lb) fuel reserve, out at 6,100 m (20,000 ft), return at 7,620 m (25,000 ft)	350 n miles (648 km; 403 miles)
fuselage tank only, 1,814 kg (4,000 lb) armament, 90 kg (200 lb) fuel reserve, cruise at 3,050 m (10,000 ft) 5 min over target	69 n miles (130 km; 80 miles)
max fuel, 771 kg (1,700 lb) armament, 90 kg (200 lb) fuel reserve, cruise at 3,050 m (10,000 ft), 1 h 50 min patrol at 150 m (500 ft) over target	49.5 n miles (92 km; 57 miles)

PERFORMANCE (M.B. 326K. A: aircraft 'clean' at AUW of 4,390 kg; 9,680 lb, B: armed aircraft at 5,443 kg; 12,000 lb AUW):

Max design limit speed at S/L	500 kts (927 km/h; 576 mph) EAS
Max limiting Mach number	0.82
Max level speed at 1,525 m (5,000 ft):	
A	480 kts (890 km/h; 553 mph)
Max level speed at 9,150 m (30,000 ft):	
B	370 kts (686 km/h; 426 mph)
Stalling speed, flaps up:	
A	102 kts (190 km/h; 118 mph) CAS
B	113 kts (211 km/h; 131 mph) CAS
Stalling speed, flaps down:	
A	91 kts (169 km/h; 105 mph) CAS
B	102 kts (190 km/h; 118 mph) CAS
Max rate of climb at S/L: A	1,980 m (6,500 ft)/min
B	1,143 m (3,750 ft)/min
Time to 10,670 m (35,000 ft): A	9 min 30 s
B	23 min
Runway LCN at max T-O weight	5
T-O run, ISA: A	411 m (1,350 ft)
B	670 m (2,200 ft)
T-O run, ISA +20°C: A	518 m (1,700 ft)
B	815 m (2,675 ft)
T-O to 15 m (50 ft), ISA: A	572 m (1,875 ft)
B	914 m (3,000 ft)
T-O to 15 m (50 ft), ISA +20°C: A	709 m (2,325 ft)
B	1,158 m (3,800 ft)
Typical combat radius:	
B (internal fuel and 1,280 kg; 2,822 lb external weapons), lo-lo-lo	145 n miles (268 km; 167 miles)
B (reduced fuel and 1,814 kg; 4,000 lb external weapons), lo-lo-lo	70 n miles (130, km; 81 miles)
visual reconnaissance with two external fuel tanks	400 n miles (740 km; 460 miles)
photo-reconnaissance with two auxiliary tanks and camera pod, hi-lo-hi	560 n miles (1,036 km; 644 miles)
Max ferry range (two underwing tanks)	more than 1,149 n miles (2,130 km; 1,323 miles)
g limits	+7.33/−3.5

UPDATED

AGUSTA

AGUSTA SpA

A Finmeccanica company

Via Giovanni Agusta 520, I-21017 Cascina Costa di Samarate (VA)
Telephone: 39 (331) 229111
Fax: 39 (331) 222595
Telex: 332569 AGUCA I
OFFICES:
Via Sicilia, I-00187 Rome
Telephone: 39 (6) 49801
Fax: 39 (6) 6799944
Telex: 614398 AGURO I
CHAIRMAN: Gen Basilio Cottone
CEO: Amedeo Caporaletti
MARKETING AND SALES: Dott Enrico Guerra

Formed in 1977; the Agusta group completely reorganised from 1 January 1981 under new holding company Agusta SpA; became part of Italian public holding company EFIM, employing nearly 10,000 people in 12 factories in various parts of Italy. Workforce in Italy and abroad was about 6,000 in 1993 and turnover was LIt1,000 billion. The order book was LIt3,500 billion.

As part of recovery of liquidated state-owned EFIM, Agusta group, including OMI, OTO Melara, Breda Meccanica Bresciana (now OTO BREDA), Galileo and SMA, was transferred to Finmeccanica from 1 January 1993; Agusta finally integrated into Finmeccanica on 12 February 1994.

Various domestic activities of Agusta are grouped under location of works. Familiar names of Costruzioni Aeronautiche Giovanni Agusta, Caproni Vizzola Costruzioni Aeronautiche and BredaNardi Costruzioni Aeronautiche no longer used; SIAI-Marchetti initially became Sesto Calende works, but reverted to SIAI-Marchetti in early 1994; in addition, Elicotteri Meridionali is domestic affiliate and Agusta is parent of Agusta Aerospace Corporation of Philadelphia, USA, and no longer has international affiliates in South Korea and Turkey.

UPDATED

AGUSTA A 109A

TYPE: Twin-engine general purpose helicopter.

PERFORMANCE: The basic version of the Agusta 109A high-speed, high-performance twin-engined helicopter accommodates a pilot and seven passengers, and has a large baggage compartment in the rear of the fuselage. Alternatively, the A 109 can be adapted for freight-carrying, as an ambulance, for search and rescue and other duties.

The first three A 109 flying prototypes flew for the first time on 4 August 1971. RAI and FAA certification for VFR operation was announced 1 June 1975, and certification for IFR single-pilot operation was obtained on 20 January 1977. Certification has since been granted in Canada, France, Germany, Philippines, Sweden, Switzerland and the UK. Deliveries started in 1976 and totalled 250 of this pre Mk II version.

Agusta successfully carried out trials of a four unit flotation gear for the A 109A, manufactured by Garrett Air Cruisers of New Jersey.

OPERATORS: Versions of the A 109 are in service with the armed forces of the following countries: Argentina (10), Belgium (45+), Italy (85+), Libya (2), Slovenia (1), UK (4) and Venezuela (10).

The following description applies to the standard A 109A.

DESIGN FEATURES: Fully articulated four-blade single main rotor and port-side two-blade semi-rigid delta-hinged tail rotor. Sweptback vertical fins (above and below fuselage). Motor rotor/engine rpm ratio 1:15.62; tail rotor/engine rpm ratio 1:2.80. take-off drive from coupling gearbox drives tail rotor via an output shaft and tail rotor gearbox. Main transmission assembly housed in fairing above passenger cabin, driving main rotor through a coupling gearbox and 90° main reduction gearbox. Transmission ratings 516 kW (692 shp) for take-off and maximum continuous twin-engined operation, with maximum contingency rating of 598 kW (802 shp) for 6 seconds. Ratings for single-engined operation are 298 kW (400 shp) for take-off, 287 kW (385 shp) maximum continuous, and 334 kW (448 shp) maximum contingency for 10 seconds.

FLYING CONTROLS: Main rotor blades have a droop-snoot aerofoil section. Non-swept elevator, mounted on rear of tailboom. Elevator linked to collective pitch control.

STRUCTURE: Main rotor blades have thickness/chord ratio of 11.3 per cent at root and 6 per cent at tip, and are attached to hub by tension/torsion straps. Main rotor blades are of aluminium alloy bonded construction, with a honeycomb core, have swept tips, stainless steel tip caps and leading-edge strips, and are protected against corrosion. A manual blade-folding capability and rotor brake are optional. Tail rotor blades are of aluminium alloy, bonded at the trailing-edge, with a honeycomb core and stainless steel leading-edge strip. The fuselage is a pod and boom type, of aluminium alloy construction, built in four main sections: nose, cockpit, passenger cabin and tailboom.

LANDING GEAR: Retractable tricycle type, with oleo-pneumatic shock-absorbers on each unit. Single mainwheels and castoring (45° each side of centre) and self-centring nosewheel. Hydraulic retraction, nosewheel forward, mainwheels upward into fuselage. Hydraulic emergency extension and locking. Disc brakes on mainwheels. All tyres are of tubeless type, and of same size (360 × 135.6) and pressure (5.9 bars; 85 lb/sq in). Tailskid under ventral fin. Emergency pop-out flotation gear and fixed snow skis optional.

POWER PLANT: Two Allison 250-C20B turboshaft engines (each 313 kW; 420 shp for T-O, 298 kW; 400 shp maximum continuous power, 276 kW; 370 shp maximum cruise power, derated to 258 kW; 346 shp for twin-engined operation), mounted side by side in upper rear fuselage and separated from passenger cabin and from each other by firewalls. Two fuel tanks in lower rear fuselage, combined capacity 560 litres (148 US gallons; 123 Imp gallons), of which 550 litres (145 US gallons; 121 Imp gallons) are usable. Refuelling point in each side of fuselage, near top of each tank. Oil capacity 7.7 litres (2 US gallons; 1.7 Imp gallons) for each engine and 12 litres (3.1 US gallons; 2.6 Imp gallons) for transmission. Provision for internal auxiliary tanks containing 138 or 198 kg (304 or 436 lb) of fuel.

ACCOMMODATION: Crew of one or two on flight deck, with pilot seated on right. Dual controls optional. Main cabin seats up to six passengers on three forward- or rearward-facing seats in centre, plus three forward-facing seats at rear. A seventh passenger can be carried in lieu of second crew member. Four/five-seat VIP layout available, with refreshment and music centre. Forward-opening crew door and passenger door on each side. Large space at rear of cabin for up to 150 kg (331 lb) of baggage, with access via forward-opening door on port side. Centre row of seats removable to permit use as freight transport. Ambulance version can accommodate two stretchers, one above the other, and two medical attendants, in addition to the pilot. External freight can be transported on a centre of gravity hook. Sliding doors can be installed for rescue missions.

SYSTEMS: Two identical independent Magnaghi hydraulic systems, pressure 103.5 bars (1,500 lb/sq in), supply dual flight servo-controls and provide emergency power in the event of engine failure. Magnaghi utility hydraulic system, pressure 69 bars (1,000 lb/sq in), for activation of landing gear, wheel and rotor braking, nosewheel locking, and emergency back-up. 28 V DC electrical system, using two 30 V 15 0A engine-driven starter/generators, and one 24 V 13 Ah Ni/Cd battery (22 Ah heavy-duty battery on IFR version). Single phase AC power at 400 Hz supplied by two 115/26 V 250 VA solid-state static inverters. Third inverter as emergency back-up on IFR version. External power receptacle. Engine anti-icing system, using engine bleed air.

AVIONICS: Standard instrumentation, plus Collins avionics for VFR or IFR operation, to customers' requirements, including VHF-20A VHF/AM com (dual in IFR version), AG-06 intercom, VIR-31A VOR/ILS with VOR/LOC, glide slope and marker beacon receiver, TDR-90 transponder, ADF-60A ADF and DME. Optional avionics include AA-215 radio altimeter with LGCS, Helics II flight director and autotrim, AFCS, pilot's navigation instruments, co-pilot's flight navigation instruments, standby attitude indicator, two- or three-axis autopilot, weather radar and area nav.

EQUIPMENT: Depending upon mission, may include internal cargo platform, external cargo sling, externally mounted rescue hoist, first aid kit, stretchers, container for up to 9,800 litres (2,588 US gallons; 2,155 Imp gallons) of fire retardant, or equipment for exploration, thermal mapping, survey, or powerline control duties.

DIMENSIONS, EXTERNAL:

Diameter of main rotor	11.00 m (36 ft 1 in)
Diameter of tail rotor	2.03 m (6 ft 8 in)
Length overall, rotors turning	13.05 m (42 ft 9¾ in)
Length of fuselage	10.706 m (35 ft 1½ in)
Fuselage: Max width	1.42 m (4 ft 8 in)
Height over tail fin	3.30 m (10 ft 10 in)
Elevator span	2.88 m (9 ft 5½ in)
Width over mainwheels	2.45 m (8 ft 0½ in)
Wheelbase	3.535 m (11 ft 7¼ in)
Passenger doors (each): Height	1.06 m (3 ft 5¾ in)
Width	1.15 m (3 ft 9¼ in)
Height to sill	0.65 m (2 ft 1½ in)
Baggage door (port, rear): Height	0.51 m (1 ft 8 in)
Width	1.00 m (3 ft 3¼ in)

DIMENSIONS, INTERNAL:

Cabin, excl flight deck: Length	1.63 m (5 ft 4¼ in)
Max width	1.32 m (4 ft 4 in)
Max height	1.28 m (4 ft 2½ in)
Volume	2.82 m^3 (100 cu ft)
Baggage compartment volume	0.52 m^3 (18.4 cu ft)

AREAS:

Main rotor blades (each)	1.84 m^2 (19.8 sq ft)
Tail rotor blades (each)	0.203 m^2 (2.185 sq ft)
Main rotor disc	3.24 m^2 (34.87 sq ft)

WEIGHTS AND LOADINGS:

Basic weight empty:	
standard (pilot and seven passengers)	1,415 kg (3,120 lb)
offshore oil support (IFR)	1,604 kg (3,536 lb)
ambulance	1,657 kg (3,653 lb)
firefighting	1,596 kg (3,518 lb)
Max external slung load	907 kg (2,000 lb)
Max baggage	150 kg (331 lb)
Typical T-O weight:	
offshore oil support (IFR)	2,596 kg (5,723 lb)
ambulance (IFR)	2,409 kg (5,311 lb)
Max normal T-O weight	2,450 kg (5,400 lb)
Max certificated weight	2,600 kg (5,732 lb)
Max disc loading	27.4 kg/m^2 (5.60 lb/sq ft)
Max power loading	4.15 kg/kW (6.82 lb/shp)

PERFORMANCE (S/L, ISA except where indicated. A: AUW of 2,250 kg; 4,960 lb, B: AUW of 2,450 kg; 5,400 lb, C: AUW of 2,600 kg; 5,732 lb):

Never-exceed speed:	
A, B	168 kts (311 km/h; 193 mph)
C	160 kts (269 km/h; 184 mph)
Max cruising speed: A	151 kts (280 km/h; 174 mph)
B	144 kts (267 km/h; 166 mph)
C	142 kts (263 km/h; 163 mph)
Econ cruising speed: A	126 kts (233 km/h; 145 mph)
B	125 kts (232 km/h; 144 mph)
C	124 kts (230 km/h; 143 mph)
Max rate of climb at S/L: A	567 m (1,860 ft)/min
B	494 m (1,620 ft)/min
C	451 m (1,480 ft)/min
Rate of climb at S/L, one engine out:	
A	152 m (500 ft)/min
B	104 m (340 ft)/min
C	67 m (220 ft)/min
Service ceiling, 30.5 m (100 ft)/min rate of climb, at max continuous power: A	5,485 m (18,000 ft)
B	4,970 m (16,300 ft)
C	4,600 m (15,100 ft)
Service ceiling, one engine out, 30.5 m (100 ft)/min rate of climb, at max continuous power:	
A	2,440 m (8,000 ft)
B	1,615 m (5,300 ft)
C	945 m (3,100 ft)
Hovering ceiling IGE: A	3,750 m (12,300 ft)
B	2,985 m (9,800 ft)
C	2,410 m (7,900 ft)

Agusta A 109A high-speed twin-engine helicopter ***1994***

Hovering ceiling IGE, ISA +20°C:
A 2,955 m (9,700 ft)
B 2,135 m (7,000 ft)
C 1,400 m (4,600 ft)
Hovering ceiling OGE: A 2,835 m (9,300 ft)
B 2,040 m (6,700 ft)
Hovering ceiling OGE, ISA +20°C:
A 2,040 m (6,700 ft)
B 1,220 m (4,000 ft)
Range with max fuel, no reserves:
A 315 n miles (583 km; 363 miles)
B 305 n miles (565 km; 351 miles)
C 296 n miles (548 km; 341 miles)
Endurance with max fuel, no reserves: A 3 h 43 min
B 3 h 30 min
C 3 h 15 min

VERIFIED

Agusta A 109A of the Italian Army carrying a Meteor Mirach Drone

1994

AGUSTA A 109A (MILITARY, NAVAL and POLICE VERSIONS)

TYPE: Twin-engine combat helicopter.

PROGRAMME: Several non-commercial versions of the A 109A were developed by Agusta. In general, their configuration, structure and power plant are similar to those of the standard civil production versions, although certain versions were available with non-retractable landing gear. Features of some or all military and naval versions include, as standard, dual controls and instrumentation; rotor brake; sliding doors; environmental control system; emergency flotation gear; armoured seats; crashworthy fuel tanks; heavy-duty battery; particle separator; external cargo hook; multipurpose universal supports for external stores; rescue hoist; high-load cargo floor; and infra-red suppression system. The naval versions, specially configured for shipboard compatibility, can be equipped with four-axis ASE, radar altimeter, internal auxiliary fuel tanks, non-retractable landing gear, search radar, anchorage points for deck lashings, and an automatic navigation system.

VERSIONS: **Aerial scout:** For forward area combat reconnaissance, and command and direction of attack helicopter team. Secondary capability for support of covert operations; artillery observation and adjustment; radio relay, and emergency rescue of combat aircrew. Can be armed with a 7.62 mm flexible machine gun, with stabilised sight, plus two XM157 launchers each for seven 2.75 in rockets. Normal crew of three.

Light attack against tanks and other hardpoint targets such as air defence weapons, vehicles and bunkers for which it can be armed with TOW or HOT missile systems.

Light attack against softpoint targets such as automatic weapons and/or troop formations. Various combinations of armament include pintle-mounted MG3 7.62 mm machine gun in each doorway, with 600 rds/gun; a flexible, remotely controlled, externally mounted 7.62 mm gun with 1,000 rds; twin-trainable, remotely controlled, externally mounted MG3s, with a total of 2,500 rounds; two external pods each containing one or two 7.62 mm machine guns or one 12.7 mm gun, with varying ammunition capacities; or two gun pods and two launchers each for six 68 mm, seven 70, 75 or 81 mm, or 14, 18 or 28 50 mm rockets. Normal crew of two.

Command and control: For target designation and direction of helicopter attack force. Can be armed with combination of rockets and flexible machine guns, as described in previous paragraph.

Utility: With accommodation or equipment for up to seven troops (transport role); two stretcher patients and two medical attendants in addition to pilot (as ambulance); externally mounted electrically operated 150 kg (331 lb) capacity rescue hoist above rear door on starboard side; or mounted underfuselage hook for 970 kg (2,000 lb) slung load. In ambulance version, cabin sidewalls are extended outwards, to enable stretchers to be installed across width of cabin, and modified port side loading doors fitted.

ESM/ECM: Electronic warfare version, for military and naval use. Available in two basic forms: with passive ESM equipment only in cabin, plus weapon systems if required, and with passive ESM plus modularised active ECM (jamming), plus any required weapons. Passive ECM include radar warning and locating equipment, interferometer, and an electromagnetic emission analyser. Provision for chaff dispenser to be mounted on tailboom.

Naval: Primary naval missions are anti-submarine classification and attack, anti-surface vessel, electronic warfare, stand-off missile guidance and reconnaissance. Secondary capabilities for search and rescue, troop transportation, ambulance, flying crane, coastguard patrol, and intership liaison duties. Configurations for electronic warfare and utility roles are generally similar to those described in preceding 'Utility' and 'ESM/ECM' paragraphs. There is standard accommodation for a two- or three-man crew, and complete instrumentation for day and night sea operation in all weathers; ASE is supplied by the standard duplicated hydraulic systems, and MAD by the self-contained third system. Electrical system capability is increased to cater for high power demand; a four-axis cross-country autopilot system and emergency flotation gear are optional.

For the ASW role, specialised equipment includes one or two homing torpedoes and six marine markers. Detection of the submarine can be carried out either by the parent ship (in which case the A 109A is acting as a weapon carrier system) or by onboard retractable classification and localisation equipment (MAD). For the ASV role the naval A 109A carries a high-performance long-range search radar with high discrimination in rough sea conditions. The surface attack is performed with AS 12 or AM 10 air-to-surface wire-guided missiles. For the TG-2 (standoff missile guidance) mission, the helicopter is equipped with a special system to control and guide a ship-launched Otomat missile. For armed patrol, the naval A 109A is equipped with a search radar and armament to customer's requirements. The coastguard patrol configuration includes a low-light level TV camera and a special installation for external high-efficiency loudspeakers.

Police and other patrol duties: For police and patrol (including armed patrol) and surveillance, coastal patrol, pollution patrol, overland and oversea search and rescue, forestry patrol and firefighting, and similar utility missions. Principal equipment for SAR versions includes search radar, rescue hoist, stretcher/first aid kits, radar altimeter, skis or emergency flotation gear, ASE and flare/smoke grenades. For aerial patrol it can include external loudspeakers, low-light level TV, pollution monitoring equipment system for spraying chemical retardants, and other items depending upon requirements of mission.

VERIFIED

Agusta-Sikorsky ASH-3D of the Peruvian Navy

1994

AGUSTA-SIKORSKY AS-61

TYPE: Twin-engine amphibious all-weather anti-submarine helicopter.

PROGRAMME: Agusta licence manufacture of the Sikorsky S-61S and S-61R in various civil and military forms started in 1967, and deliveries of anti-submarine ASH-3Ds to the Italian Navy began in 1969. Additional orders were placed, both for the Italian armed forces and for other navies, in various configurations including ASW, VIP transport and rescue. Most recent customers include the navies of Brazil (4) and Argentina (2).

VERSIONS: **ASH-3D/TS:** (Transporto Speciale) VIP transport helicopter serving with the Italian and other air forces. Agusta is exclusively responsible for repair and overhaul. Production could be restarted within 36 months.

ASH-3H: Multirole naval version.

Apart from some local strengthening, uprated engines and an improved tail surface, the Agusta built airframe remains essentially similar to that of the Sikorsky built SH-3D/H, of which production has ended. The Agusta ASH-3H is capable of operation in the roles of anti-submarine search, classification and strike; anti-surface vessel (ASV); anti-surface missile defence (ASMD); electronic warfare (EW); tactical troop lift; search and rescue (SAR); vertical replenishment, and casualty evacuation.

Starting in mid-1987, an ASH-3 was used as a testbed for the BAe/Bendix HELRAS long-range high-resolution dipping sonar intended for naval versions of the Anglo-Italian EH 101 Sea King replacement helicopter.

OPERATORS: Versions of the AS-61 are in service with the armed forces of the following countries: Egypt (2), Iran (2), Italy (38), Libya (1), Peru (8) and Venezuela (4).

Agusta-Sikorsky AS-61A VIP helicopter

1994

The following description applies to the ASH-3H:

DESIGN FEATURES: Five-blade main and tail rotors. Both engines drive through freewheel units and rotor brake to main gearbox. Steel driveshafts. Tail rotor shaft driven through intermediate and tail gearboxes. Main rotor/engine rpm ratio 1:93.43. Tail rotor engine rpm ratio 1:16.7.

FLYING CONTROLS: Rotor brake standard.

STRUCTURE: All-metal fully articulated oil lubricated main rotor. Flanged cuffs on blades bolted to matching flanges on all-steel rotor head. Main rotor blades are interchangeable and are provided with an automatic folding system. All-metal tail rotor. Single-step boat hull of all-metal semi-monocoque construction. Tail section folds to reduce stowage requirements. Fixed strut-braced stabiliser on starboard side of tail section.

TAIL SURFACE: Fixed strut-braced stabiliser on starboard side of tail section.

LANDING GEAR: Amphibious. Land gear consists of two twin-wheel main units, which are retracted rearward hydraulically into stabilising floats, and non-retractable tailwheel. Oleo-pneumatic shock-absorbers. Mainwheels and tubeless tyres size 6.50-10 Type III, pressure 4.83 bars (70 lb/sq in). Tailwheel and tyre size 6.00-6. Hydraulic disc brakes. Boat hull and pop-out flotation bags in stabilising floats permit emergency operation from water.

POWER PLANT: Two 1,118 kW (1,500 shp) General Electric T58-GE-100 turboshaft engines, mounted side by side above cabin. An optional anti-ice/sand shield can be provided. Fuel in underfloor bag tanks with a total capacity of 3,180 litres (840 US gallons; 700 Imp gallons). Internal auxiliary fuel tanks may be fitted for long-range ferry purposes. Pressure and gravity refuelling points.

ACCOMMODATION: Crew of four in ASW role (pilot, co-pilot and two sonar operators); accommodation for up to 31 paratroops in troop lift role, 15 stretchers and a medical attendant in casualty evacuation configuration, and up to 25 survivors in SAR role. Dual controls. Crew door at rear of flight deck on port side. Large loading door at rear of cabin on starboard side.

SYSTEMS: Three main hydraulic systems. Primary and auxiliary systems operate main rotor control. Utility system for landing gear, winches and blade folding, pressure 207 bars (3,00 lb/sq in). Electrical system includes two 20 kVA 200 V three-phase 400 Hz engine-driven generators, a 260 V single-phase AC supply fed from the aircraft's 22 Ah Ni/Cd battery through an inverter, and DC power provided as a secondary system from two 200 A transformer-rectifier units. Accessories driven by power take-off on tail rotor shaft.

ARMAMENT AND OPERATIONAL EQUIPMENT (ASW/ASV roles): As equipped for these roles the ASH-3H is a fully integrated all-weather weapons system, capable of operating independently of surface vessels, and has the following equipment and weapons to achieve this task: low-frequency 360° depth AQS-18/AQS-13F sonar; Doppler radar and ASW automatic navigation system; SMA/APS-707 radar with one or two transceivers, with 'chin' radome for 360° coverage; radio altimeter; AFCS; marine markers and smoke floats; two or four homing torpedoes (A 244 AS, Mk 44 or Mk 46); or four depth charges. The AFCS provides three-axis stabilisation in pilot-controlled manoeuvres, attitude hold heading, heading hold and height hold on cruising flight; controlled transition manoeuvres to and from hover; automatic height control and plan position control in the hover; and trim facility. According to the threat, the Agusta SH-3H can be equipped with medium-range (four AS 12 air-to-surface wire-guided) missiles or long-range (two Marte Mk 2 or Exocet AM39/Harpoon type) missiles. The OTO Melara Marte Mk 2 is an all-weather day and night 'fire and forget' anti-ship missile with a range of 13.5 n miles (25 km; 15.5 miles); guidance: sea-skimming in elevation, terminal radar active homing in azimuth. The SMA/APS-707 radar has been specially designed to operate in a dense electronic emission environment and has a special interface to draw out target data to feed the computer for the long-range missiles. Provisions are also incorporated for the installation of MAD and advanced EW systems.

OPERATIONAL EQUIPMENT (Search and rescue and transport roles): Search radar, and variable speed hydraulic rescue hoist of 272 kg (600 lb) capacity mounted above starboard side cargo door.

DIMENSIONS, EXTERNAL:

Main rotor diameter	18.90 m (62 ft 0 in)
Main rotor blade chord	0.46 m (1 ft 6¼ in)
Tail rotor diameter	3.23 m (10 ft 7 in)
Distance between rotor centres	11.10 m (36 ft 5 in)
Length overall, both rotors turning	21.91 m (71 ft 10¾ in)
Length of fuselage	16.69 m (54 ft 9 in)
Length, main rotor and tail pylon folded	14.40 m (47 ft 3 in)
Width (over sponsons), rotors folded	4.98 m (16 ft 4 in)
Height to top of rotor head	4.74 m (15 ft 6½ in)
Height overall, main rotor and tail pylon folded	4.93 m (16 ft 2 in)
Height overall, tail rotor turning	5.23 m (17 ft 2 in)
Wheel track	3.96 m (13 ft 0 in)
Wheelbase	7.18 m (23 ft 6½ in)
Crew door (fwd, port): Height	1.68 m (5 ft 6 in)
Width	0.91 m (3 ft 0 in)
Height to sill	1.14 m (3 ft 9 in)
Main cabin door (stbd): Height	1.52 m (5 ft 0 in)
Width	1.73 m (5 ft 8 in)
Height to sill	1.14 m (3 ft 9 in)

AREAS:

Main rotor blades (each)	4.14 m^2 (44.54 sq ft)
Tail rotor blades (each)	0.22 m^2 (2.38 sq ft)
Main rotor disc	280.5 m^2 (3,019 sq ft)
Tail rotor disc	8.20 m^2 (88.30 sq ft)
Stabiliser	1.86 m^2 (20.00 sq ft)

WEIGHTS AND LOADINGS:

Internal load capacity (cargo)	2,720 kg (6,000 lb)
Max external load capacity (with low response sling)	3,630 kg (8,000 lb)
Max T-O weight	9,525 kg (21,000 lb)

PERFORMANCE:

Never-exceed speed	144 kts (267 km/h; 165 mph)
Typical cruising speed	120 kts (222 km/h; 138 mph)
Max rate of climb at S/L	670 m (2,200 ft)/min
Service ceiling	3,720 m (12,200 ft)
Hovering ceiling: IGE	2,500 m (8,200 ft)
OGE	1,130 m (3,700 ft)
Range with 31 troops	314 n miles (582 km; 362 miles)
Range with max standard fuel	630 n miles (1,166 km; 725 miles)

VERIFIED

AGUSTA-SIKORSKY AS-61R (HH-3F) PELICAN

TYPE: Twin-engined amphibious helicopter.

PROGRAMME: Agusta began production of this multipurpose search and rescue helicopter in 1974, and deliveries began in 1976. The production line was reopened to make two for the National Civil Protection Service (SNPC) and another 13 rescue helicopters for the Italian Air Force. These had a new radar, Loran, FLIR and navigation computer, all of which were retrofitted in the remaining 19 of the original production batch.

OPERATORS: Versions of the AS-61R are in service with the armed forces of Italy.

DESIGN FEATURES: Five-blade fully articulated all-metal main rotor. Conventional tail rotor with five aluminium blades. Steel driveshafts. Tail rotor shaft driven through intermediate gearbox and tail gearbox. Main rotor/engine rpm ratio 1:93.43. Tail rotor/engine rpm ratio 1:16/7. Twin turbines drive through freewheeling units and rotor brake to main gearbox.

FLYING CONTROLS: Rotor brake standard.

STRUCTURE: Flanged cuffs on blades bolted to matching flanges on rotor head. Control by rotating and stationary swashplates. Blades do not fold. The fuselage is an all-metal semi-monocoque structure of pod and boom type. Cabin of basic square section. Rear-loading ramp. Strut-braced horizontal stabiliser on starboard side of tail rotor pylon.

LANDING GEAR: Hydraulically retractable tricycle type, with twin wheels on each unit. Mainwheels retract forward into sponsons, each of which provides 2,176 kg (4,797 lb) of buoyancy and, with boat hull, permits amphibious operation. Oleo-pneumatic shock-absorbers. All wheels and tyres tubeless Type III rib, size 22.1 × 6.50-10, pressure 6.55 bars (95 lb/sq in). Hydraulic disc brakes.

POWER PLANT: Two 1,118 kW (1,500 shp) General Electric T58-GE-100 turboshafts, mounted side by side above cabin, immediately forward of main transmission. Fuel in four bladder tanks beneath cabin floor, with total capacity of 4,225 litres (1,116 US gallons; 929 Imp gallons), of which 4,183 litres (1,105 US gallons; 920 Imp gallons) are usable. Provisions for removable internal auxiliary fuel tanks. Refuelling point on port side of fuselage. Total oil capacity 26.5 litres (7 US gallons; 5.8 Imp gallons).

ACCOMMODATION: Crew of two side by side on flight deck, with dual controls. Provision for flight engineer or attendant. Accommodation in SAR configuration for 10 passenger seats and six stretchers; utility version can accommodate up to 26 troops on foldable, safety belt equipped seats, 15 stretchers plus two medical attendants, or cargo. Jettisonable sliding door on starboard side at front of cabin. Internal door between cabin and flight deck. Hydraulically operated rear-loading ramp, in two hinged sections, giving opening with minimum width of 1.73 m (5 ft 8 in) and headroom of up to 2.21 m (7 ft 3 in). Ramp can be operated when helicopter is on the water. Reinforced (41 kg/m^2; 200 lb/sq ft loading) cargo floor in utility version.

SYSTEMS: Primary and auxiliary hydraulic systems, pressure 103.5 bars (1,500 lb/sq in), for flying control servos. Utility hydraulic system, pressure 207 bars (3,000 lb/sq in), for landing gear, rear ramp and winches. Pneumatic system, pressure 207 bars (3,000 lb/sq in), for emergency blow-down landing gear extension. Electrical system includes 24

V 22 Ah battery, two 20 kVA 115 V AC generators and one 300A DC generator. APU standard.

AVIONICS: SAR version has comprehensive suite for that role (including search/nav radar, Loran, FLIR and nav computer).

EQUIPMENT: SAR version has 272 kg (600 lb) capacity rescue hoist. Nightsun searchlight, detachable rescue platform for use when afloat, auxiliary flotation system, loudhailer set and sea anchor. Equipment for utility missions can include low response external cargo sling.

Agusta-Sikorsky ASH-3D of the Italian Navy

1994

DIMENSIONS, EXTERNAL:

Main rotor diameter	18.90 m (62 ft 0 in)
Main rotor blade chord	0.46 m (1 ft 6¼ in)
Tail rotor diameter	3.15 m (10 ft 4 in)
Distance between rotor centres	11.22 m (36 ft 10 in)
Length: overall, excl radome	22.25 m (73 ft 0 in)
fuselage	17.45 m (57 ft 3 in)
Width over landing gear	4.82 m (15 ft 10 in)
Height: to top of rotor head	4.90 m (16 ft 1 in)
overall	5.51 m (18 ft 1 in)
Wheel track	4.06 m (13 ft 4 in)
Wheelbase	5.21 m (17 ft 1 in)
Cabin door (fwd, stbd): Height	1.65 m (5 ft 4¾ in)
Width	1.22 m (4 ft 0 in)
Height to sill	1.27 m (4 ft 2 in)
Rear ramp: Length	4.29 m (14 ft 1 in)
Width	1.85 m (6 ft 1 in)

DIMENSIONS, INTERNAL:

Cabin (excl flight deck): Length	7.89 m (25 ft 10½ in)
Max width	1.98 m (6 ft 6 in)
Max height	1.91 m (6 ft 3 in)
Floor area	approx 15.16 m^2 (168 sq ft)
Volume	approx 29.73 m^3 (1,050 cu ft)

AREAS:

Main rotor blades (each)	3.71 m^2 (39.9 sq ft)
Tail rotor blades (each)	0.22 m^2 (2.35 sq ft)
Main rotor disc	280.5 m^2 (3,019 sq ft)
Tail rotor disc	7.80 m^2 (83.9 sq ft)
Stabiliser	2.51 m^2 (27.0 sq ft)

WEIGHTS AND LOADINGS:

Weight empty	6,010 kg (13,255 lb)
Max cargo payload: internal	2,270 kg (5,000 lb)
external	3,628 kg (8,000 lb)
Normal T-O weight	9,635 kg (21,247 lb)
Max T-O weight	10,000 kg (22,050 lb)
Max disc loading	35.65 kg/m^2 (7.30 lb/sq ft)

PERFORMANCE (at normal T-O weight except where indicated):

Max level speed at S/L:

normal T-O weight	141 kts (261 km/h; 162 mph)
max T-O weight	138 kts (255 km/h; 159 mph)

Cruising speed at S/L, AUW of 9,072 kg (20,000 lb), ISA +20°C: for best range 130 kts (241 km/h; 150 mph)

for best endurance 75 kts (139 km/h; 86 mph)

Max rate of climb at S/L	408 m (1,340 ft)/min
Service ceiling: normal T-O weight	3,385 m (11,100 ft)
Hovering ceiling IGE	2,195 m (7,200 ft)
Min ground turning radius	11.29 m (37 ft 0½ in)
Runway LCN at max T-O weight	approx 4.75

Typical mission profiles (ISA + 20°C):

SAR: Loiter for 5 h in search area 50 n miles (92 km; 57 miles) from base, hover for 30 min to rescue survivors, return to base and land, 10% fuel remaining

Utility: Fly 240 n miles (445 km; 276 miles) from base, pick up 24 fully equipped troops and return to base, landing with 272 kg (600 lb) of fuel remaining

Range with max standard fuel, no reserves 770 n miles (1,427 km; 886 miles)

VERIFIED

SIAI-MARCHETTI

MAIN WORKS: Sesto Calende (Varese)

AIRFIELD AND WORKS: Vergiate and Malpensa

Founded in 1915, SIAI-Marchetti produced a wide range of military and civil landplanes and flying-boats up to the end of the Second World War. Now known simply as the Sesto Calende works of Agusta, current products include piston, turboprop and turbofan-powered trainers. Since the 1970s it has been engaged in the co-production of licence-built Boeing CH-47C, Bell 204/205/212/412, and Sikorsky S-61A, SH-3D/H and H-3F helicopters.

On 6 October 1988 a memorandum of intent was signed with Grumman Aircraft Systems (a division of Grumman Corporation), to offer a version of the S.211 for the USAF/USN's Joint Primary Aircraft Training System (JPATS) requirement.

SIAI undertakes the overhaul and repair of various types of aircraft (notably the C-130 Hercules, DHC-5 Buffalo and Cessna Citation II). It participates in national or multinational programmes, producing parts for the Alenia G222, Panavia Tornado, AMX, Airbus A310 and Atlantique 2.

The works at Sesto Calende, Vergiate and Malpensa total 1,370,267 m^2 (14,749,416 sq ft) in area, of which 119,494 m^2 (1,286,221 sq ft) are covered.

VERIFIED

SIAI-MARCHETTI SF.260

TYPE: Two/three-seat fully acrobatic military/civil light aircraft.

PROGRAMME: The prototype for the SF.260 series, known as the F.250, was designed by Dott Ing Stelio Frati and built by Aviamilano. Flown for the first time on 15 July 1964, it was powered by a 186.5 kW (250 hp) Textron Lycoming engine and was certificated for aerobatic flying.

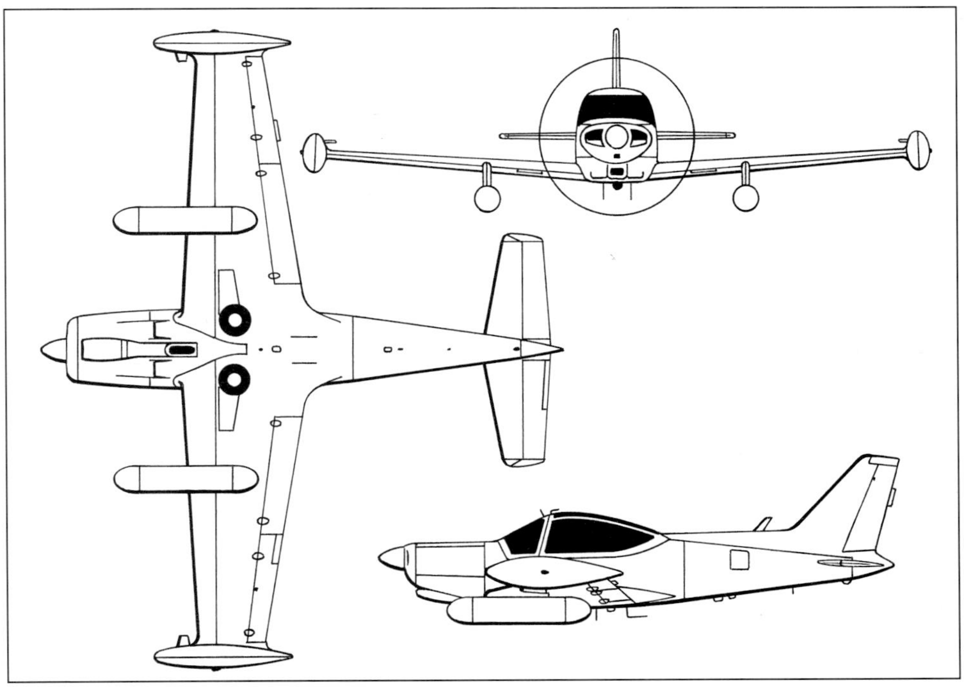

SIAI-Marchetti SF.260W Warrior *(Jane's/Mike Keep)*

1994

SIAI-Marchetti SF.260M military light aircraft

1994

The version developed initially for civil production was manufactured, at first under licence from Aviamilano, by SIAI-Marchetti, and is designated SF.260. It received FAA type approval on 1 April 1966. Subsequently SIAI-Marchetti became the official holder of the type certificate and of all manufacturing rights in the SF.260.

VERSIONS: **SF.260D:** Improved version and updated civil version.

SF.260M: Two/three-seat military trainer, developed from civil SF.260A and first flown on 10 October 1970. Introduced a number of important structural and aerodynamic improvements, many of which were subsequently applied to later models. Meets requirements for basic flying training; instrument flying; aerobatics, including deliberate spinning; night flying; navigation flying; and formation flying.

SF.260W Warrior: Trainer/tactical support version of SF.260M, first flown (I-SJAV) in May 1972. Two or four underwing pylons, for up to 300 kg (661 lb) of external stores, and cockpit stores selection panel. Able to undertake a wide variety of roles, including low-level strike; forward air control; forward air support; armed reconnaissance; and laison. Also meets same requirements as SF.260M for use as a trainer.

SF.260SW Sea Warrior surveillance/SAR/supply version.

SF.260TP: Turboprop version (see *Jane's All The World's Aircraft*).

UPGRADES: **EVS:** Completed upgrading nine Belgian Air Force SF 260M aircraft with IFR equipment. Modifications were carried out by the Belgian Elementary Flying School (EVS) at Goetsenhoeven. Programme utilised off-the-shelf COTS instruments.

OPERATORS: Versions of the SF 260 are in service with the armed forces of the following countries: Belgium (35+), Bolivia (3), Brunei (2), Burkina Faso (5), Burundi (7), Chad (3), Ethiopia (12), Haiti (5), Ireland (7), Italy (38), Libya (200+), Philippines (35), Singapore (26), Somalia (4), Sri Lanka (15+), Thailand (10), Tunisia (18), Turkey (40), Uganda (2), United Arab Emirates (5), Zaïre (9) and Zimbabwe (30).

The following description is generally applicable to all piston-engined models unless otherwise stated.

DESIGN FEATURES: Low-mounted straight wings with wingtip pods. Wing section 64_1-212 (modified) at root, NACA 64_1-210 (modified) at tip. Dihedral 6° 20′ from roots (5° on SF.260D). Incidence 2° 45′ at root, 0° at tip. No sweepback.

FLYING CONTROLS: Differentially operating Frise ailerons, and electrically actuated single-slotted flaps. Flaps operated by torque tube and mechanical linkage, ailerons by pushrods and cables. Servo tab in each aileron.

STRUCTURE: The wing is an all-metal light alloy safe-life structure, with single main spar and auxiliary rear spar, built in two portions bolted together at centreline and attached to fuselage by six bolts. Press-formed ribs. Skin butt joined and flush riveted.

LANDING GEAR: Electrically retractable tricycle type, with manual emergency actuation. Inward retracting main gear, of trailing arm type, and rearward retracting nose unit, each embodying Magnaghi oleo-pneumatic shock-absorber (type 2/22028 in main units). Each welded steel tube main leg is hinged to the main and rear spars. Nose unit is of leg and fork type, with coaxial shock-absorber and torque strut. Cleveland P/N 3080A mainwheels, with size 6.00-6 tube and tyre (6 ply rating), pressure 2.45 bars (35.5 lb/sq in). Cleveland P/N 40-77A nosewheel, with size 5.00-5 tube and tyre (6 ply rating), pressure 1.96 bars (28.4 lb/sq in). Cleveland P/N 3000-500 independent hydraulic single-disc brake and parking brake on each mainwheel. Nosewheel steering (±20°) operated directly by rudder pedals to which it is linked by pushrods.

POWER PLANT: One 194 kW (260 hp) Textron Lycoming O-540-E4A5 flat-six engine, driving a Hartzell HC-C2YK-1BF/8477-8R two-blade constant-speed metal propeller. AEIO-540-D4A5 engine available optionally. Fuel in two light alloy tanks in wings, capacity of each 49.5 litres (13.1 US gallons; 10.9 Imp gallons); and two permanent wingtip tanks, capacity of each 72 litres (19 US gallons; 15.85 Imp gallons). Total internal fuel capacity 243 litres (64.2 US gallons; 53.5 Imp gallons), of which 235 litres (62.1 US gallons; 51.7 Imp gallons) are usable. Individual refuelling point on top of each tank. In addition, SF.260W may be fitted with two 80 litre (21.1 US gallon; 17.5 Imp gallon) auxiliary tanks on underwing pylons. Oil capacity (all models) 11.4 litres (3 US gallons; 2.5 Imp gallons).

ACCOMMODATION (SF.260M; W similar): Side by side front seats (for instructor and pupil in SF.260M), with third seat centrally at rear. Front seats individually adjustable fore and aft, with forward-folding backs and provision for back type parachute packs. Dual controls standard. All three seats equipped with lap belts and shoulder harnesses. Baggage compartment aft of rear seat. Upper portion of canopy tinted. Emergency canopy release handle for each front seat occupant. Steel tube windscreen frame for protection in the event of an overturn.

SYSTEMS (SF.260M; other models generally similar): Hydraulic system for mainwheel brakes only. No pneumatic system. 24 V DC electrical system of single-conductor negative earth type, including 70 A Prestolite engine-mounted alternator/rectifier and 24 V 24 Ah Varley battery, for engine starting, flap and landing gear actuation, fuel booster pumps, electronics and lighting. Sealed battery compartment in rear of fuselage on port side. Connection of an external power source automatically disconnects the battery. Heating system for carburettor air intake. Emergency electrical system for extending landing gear if normal electrical actuation fails; provision for mechanical extension in the event of total electrical failure. Cabin heating, and windscreen de-icing and demisting, by heat exchanger using engine exhaust air. Additional manually controlled warm air outlets for general cabin heating. Oxygen system optional.

AVIONICS (SF.260M; W generally similar): Basic instrumentation to customer's requirements. Blind-flying instrumentation and communications equipment optional: typical selection includes dual Collins 20B VHF com; Collins VIR-31A VHF nav; Collins ADF-60A; Collins TDR-90 ATC transponder; Collins PN-101 compass; ID-90-000 RMI; and Gemelli AG04-1 intercom. Instrument panel can be slid rearward to provide access to rear of instruments.

EQUIPMENT: Military equipment to customer's requirements. External stores can include one or two reconnaissance pods with two 70 mm automatic cameras, or two supply containers. Landing light in nose, below spinner.

ARMAMENT (SF.260W): Two or four underwing hardpoints, able to carry external stores on NATO standard pylons up to a maximum of 300 kg (661 lb) when flown as a single-seater. Typical alternative loads can include one or two SIAI gun pods, each with one or two 7.62 mm FN machine guns and 500 rounds; two Aerea AL-8-70 launchers each with eight 2.75 in rockets; two LAU-32 launchers each with seven 2.75 in rockets; two Aerea AL-18-50 launchers each with 18 2 in rockets; two Aerea AL-8-68 launchers each with eight 68 mm rockets; two Aerea AL-6-80 launchers each with six 81 mm rockets; two LUU-2/B parachute flares; two SAMP EU 32 125 kg general purpose bombs or EU 13 120 kg fragmentation bombs; two SAMP EU 70 50 kg general purpose bombs; Mk 76 11 kg practice bombs; two cartridge throwers for 70 mm multipurpose cartridges, F 725 flares or F 130 smoke cartridges. One or two photo-reconnaissance pods with two 70 mm automatic cameras; two supply containers.

DIMENSIONS, EXTERNAL:
Wing span over tip tanks 8.35 m (27 ft 4¾ in)
Wing chord: at root 1.60 m (5 ft 3 in)
mean aerodynamic 1.325 m (4 ft 4¼ in)
at tip 0.784 m (2 ft 6⅞ in)
Wing aspect ratio (excl tip tanks) 6.3
Wing taper ratio 2.2
Length overall 7.10 m (23 ft 3½ in)
Fuselage: Max width 1.10 m (3 ft 7¼ in)
Max depth 1.042 m (3 ft 5 in)
Height overall 2.41 m (7 ft 11 in)
Elevator span 3.01 m (9 ft 10½ in)
Wheel track 2.274 m (7 ft 5½ in)
Wheelbase 1.66 m (5 ft 5¼ in)
Propeller diameter 1.93 m (6 ft 4 in)
Propeller ground clearance 0.32 m (1 ft 0½ in)

DIMENSIONS, INTERNAL:
Cabin: Length 1.66 m (5 ft 5¼ in)
Max width 1.00 m (3 ft 3¼ in)
Height (seat cushion to canopy) 0.98 m (3 ft 2½ in)
Volume 1.50 m³ (53 cu ft)
Baggage compartment volume 0.18 m³ (6.36 cu ft)

AREAS:
Wings, gross 10.10 m² (108.70 sq ft)
Ailerons (total, incl tabs) 0.762 m² (8.20 sq ft)
Trailing-edge flaps (total) 1.18 m² (12.70 sq ft)
Fin 0.76 m² (8.18 sq ft)
Dorsal fin 0.16 m² (1.72 sq ft)
Rudder, incl tab 0.60 m² (6.46 sq ft)
Tailplane 1.46 m² (15.70 sq ft)
Elevator, incl tab 0.96 m² (10.30 sq ft)

WEIGHTS AND LOADINGS:
Manufacturer's basic weight empty:
M 755 kg (1,664 lb)
W 770 kg (1,697 lb)
Weight empty, equipped: D 755 kg (1,664 lb)

SIAI-Marchetti SF.260W Warrior two/three-seat aircraft

1994

M 815 kg (1,797 lb)
W 830 kg (1,830 lb)
Fuel: in-wing and wingtip tanks (all versions) 169 kg (372.5 lb)
underwing tanks (W only) 114 kg (251.5 lb)
Typical mission weights:
M, trainer (clean) 1,140 kg (2,513 lb)
W, two 47 kg (103.5 lb) machine gun pods and full internal fuel 1,163 kg (2,564 lb)
W, one Alkan 500B cartridge thrower, one two-camera reconnaissance pod and full internal fuel 1,182 kg (2,605 lb)
W, trainer with 94 kg (207 lb) external stores 1,249 kg (2,753 lb)
W, self-ferry with two 80 litre (21.1 US gallon; 17.5 Imp gallon) underwing tanks 1,285 kg (2,833 lb)
W, two 125 kg bombs and 150 kg (331 lb) internal fuel 1,300 kg (2,866 lb)
W, two AL-8-70 rocket launchers and 160 kg (353 lb) internal fuel 1,300 kg (2,866 lb)
Max T-O weight: D, M, Aerobatic 1,100 kg (2,425 lb)
D, Utility 1,100 kg (2,425 lb)
M, Utility 1,200 kg (2,645 lb)
W, max permitted 1,300 kg (2,866 lb)
Max wing loading: D 109 kg/m² (22.4 lb/sq ft)
M 119 kg/m² (24.4 lb/sq ft)
W 129 kg/m² (26.4 lb/sq ft)
Max power loading: D 5.68 kg/kW (9.33 lb/hp)
M 6.19 kg/kW (10.17 lb/hp)
W 6.70 kg/kW (11.01 lb/hp)

PERFORMANCE (D: at AUW of 1,102 kg; 2,430 lb, M: at AUW of 1,200 kg; 2,645 lb, W: at 1,300 kg; 2,866 lb, except where indicated):
Never-exceed speed (VNE):
D, M 235 kts (436 km/h; 271 mph)
Max level speed at S/L:
D 187 kts (347 km/h; 215 mph)
M 180 kts (333 km/h; 207 mph)
W 165 kts (305 km/h; 190 mph)
Max cruising speed (75% power):
D at 3,050 m (10,000 ft) 178 kts (330 km/h; 205 mph)
M at 1,500 m (4,925 ft) 162 kts (300 km/h; 186 mph)
W at 1,500 m (4,925 ft) 152 kts (281 km/h; 175 mph)
Stalling speed, flaps and landing gear up:
M 74 kts (137 km/h; 86 mph)
W 88 kts (163 km/h; 102 mph)
Stalling speed, flaps and landing gear down:
D 60 kts (111 km/h; 70 mph)
M 68 kts (126 km/h; 79 mph)
W 72 kts (134 km/h; 83 mph)
Max rate of climb at S/L: D 546 m (1,791 ft)/min
M 457 m (1,500 ft)/min
W 381 m (1,250 ft)/min
Time to 1,500 m (4,925 ft): M 4 min
W 6 min 20 s
Time to 2,300 m (7,550 ft): M 6 min 50 s
W 10 min 20 s
Time to 3,000 m (9,850 ft): M 10 min
W 18 min 40 s
Service ceiling: D 5,790 m (19,000 ft)
M 4,665 m (15,300 ft)
W 4,480 m (14,700 ft)
T-O run at S/L: D 480 m (1,575 ft)
M 384 m (1,260 ft)
T-O to 15 m (50 ft) at S/L: M 606 m (1,988 ft)
W 825 m (2,707 ft)
Landing from 15 m (50 ft) at S/L: D 445 m (1,460 ft)
M 539 m (1,768 ft)
W 645 m (2,116 ft)
Landing run at S/L: D, M 345 m (1,132 ft)
Operational radius:
W, 6 h 25 min single-seat armed patrol mission at 1,163 kg (2,564 lb) AUW, incl 5 h 35 min over operating area, 20 kg (44 lb) fuel reserves 50 n miles (92 km; 57 miles)
W, 3 h 38 min single-seat strike mission, incl two 5 min loiters over separate en route target areas, 20 kg (44 lb) fuel reserves 250 n miles (463 km; 287 miles)
W, 4 h 54 min single-seat strike mission, incl 5 min over target area, 20 kg (44 lb) fuel reserves 300 n miles (556 km; 345 miles)
W, 4 h 30 min single-seat photo-reconnaissance mission at 1,182 kg (2,605 lb) AUW, incl three 1 h loiters over separate en route operating areas, 20 kg (44 lb) fuel reserves 150 n miles (278 km; 172 miles)
W, 6 h 3 min two-seat self-ferry mission with two 80 litre (21.1 US gallon; 17.5 Imp gallon) underwing tanks, at 1,285 kg (2,833 lb) AUW, 30 kg (66 lb) fuel reserves 926 n miles (1,716 km; 1,066 miles)
Range with max fuel:
D (two-seat) 805 n miles (1,490 km; 925 miles)
M (two-seat) 890 n miles (1,650 km; 1,025 miles)
g limits (M):
at max Aerobatic T-O weight +6/−3
at max Utility T-O weight without external load +4.4/−2.2

UPDATED

SIAI-Marchetti SF.260W Warrior. This example was the first trainer/tactical support version
1994

ELICOTTERI MERIDIONALI SpA

Formed with assistance from Agusta and began to operate in October 1967. It remains a separate commercial entity affiliated to Agusta. In 1968 EM (Elicotteri Meridionali SpA) acquired rights to the co-production, marketing and servicing of the Boeing CH-47C Chinook transport helicopter for customers in Italy and certain foreign countries. Italian production of the CH-47C airframe is undertaken by the Sesto Calende works.

EM, whose works occupy a total area of more than 300,000 m² (3,229,170 sq ft), participates in the manufacturing programmes for the Agusta A 109 and A 129, Agusta-Bell 212/412, and Agusta-Sikorsky S-61 variants. It has complete facilities for overhaul, repair and field assistance. EM is the designated overhaul organisation for all types of Italian Army helicopter, and is also distributor in Italy for Allison 250 turboshaft engines.

VERIFIED

Elicotteri Meridionali (Boeing) CH-47C Plus with T55-L-712E engines on relief duty in Mogadishu *(Paul Jackson)*
1994

EM (BOEING) CH-47C CHINOOK AND CH-47C PLUS UPGRADE

TYPE: Twin-engine transport helicopter.

PROGRAMME: Italian manufacture of CH-47C began in Spring 1970 for Italian Army Aviation; later customers included Egypt (15), Greece (10), Iran (68 of 95 originally ordered), Libya (20), Morocco (nine), and US Army (11).

UPRGRADES: **EM. CH-47C PLUS:** Produced for operation by Italian Army on behalf of Civil Protection Agency; upgrading programme began to fit earlier aircraft with new Textron Lycoming T-55-L-712E engines, composite rotor blades and more advanced transmission system; maximum T-O weight increased to 2,680 kg (50,000 lb).

EM. ESFC: Agusta developed, jointly with Hosp Ital SpA (a division of Cogefar) of Milan, an **ESFC** (emergency surgery flying centre) version of Chinook for use as mobile hospital; one delivered to Italian Army in 1987; six more CH-47Cs ordered for Italian Army Light Aviation unit at Castelnuovo di Porto specialising in disaster relief and firefighting.

Italian Army's fleet of Chinooks being overhauled at rate of three a year by EM; first of 23 then-operational aircraft redelivered March 1986; roles include firefighting using 5,000 litre (1,321 US gallon; 1,100 Imp gallon) metal tank.

VERIFIED

ALENIA

(A Finmeccanica Company)
Via E Petrolini 2, I-00197 Rome
Telephone: 39 (6) 807781
Fax: 39 (6) 8072215/8075184
Telex: 611395 Alenia I

PRESIDENTS:
Fausto Cereti
Enrico Gimelli

GENERAL MANAGER, FINANCE: Paolo Micheletta
GENERAL MANAGER, OPERATIONS: Giorgio Zappa
SECRETARY GENERAL: Massimo Rizzo
VICE-PRESIDENT, PRESS AND PR: Fabio Dani
MARKETING AND SALES: Nicolas Zalonis

Alenia was founded in 1990 following the merger of Aeritalia and Selenia. In February 1993 Alenia was merged into Finmeccanica. Main responsibilities include participation in the following programmes: AMX with Aeronautica Macchi and Embraer (Brazil), Eurofighter 2000, Boeing 767/777, McDonnell Douglas MD-90/MD-80/MD-11, ATR 42/72 Regional Transport Aircraft, Panavia Tornado wings as well as G222 structural parts. Further activities include the design, development and production of the following systems: missile, radar command and control, naval and underwater, military communications and Unmanned Airborne Vehicles (UAV).

The company is also active in the fields of: Air/Maritime Traffic Control, radio communications, information systems for telecommunications networks, pollution control systems as well as equipment for space systems. Upgrade specific activity includes weapons modernisation and integration on the Lockheed F-104 Starfighters for the Italian Air Force (qv), McDonnell Douglas DC-8/DC-10/MD-11 freighter conversions (qv), Atlantic 1 modernisation (qv), Boeing 727 re-engining (qv) and Boeing 707 tanker conversions (qv).

VERIFIED

ALENIA (FIAT) G91

TYPE: Lightweight single-seat tactical fighter-bomber, reconnaissance aircraft and two-seat trainer.

PROGRAMME: The G91Y is a twin-engined development of the earlier single-engined Fiat G91, based upon the airframe of the G91T version.

Two G91Y prototypes were built, the first of which flew for the first time on 27 December 1966. They were followed by 20 preseries G91Ys for the Italian Air Force, the first of which was flown in July 1968.

Delivery of the initial series of 35 production G91Ys to the Italian Air Force began in September 1971, and was completed by mid-1973. Delivery of an additional 10 was completed by mid-1976. The G91Y is in the process of being decommissioned from the air forces of Italy and Portugal.

VERSIONS: **G91T:** Twin-seat advanced trainer. In service with the Italian Air Force.

G91Y: Twin-engine, single-seat light tactical bomber and reconnaissance aircraft.

OPERATORS: Versions of the G91Y/T are in service with the armed forces of Italy (70).

DESIGN FEATURES: Cantilever low-wing monoplane. Laminar-flow section. Sweepback at quarter-chord 37° 40′ 38″. Variable incidence tailplane. Auxiliary fin beneath each side of rear fuselage.

FLYING CONTROLS: Ailerons with hydraulic servo control. Electrically actuated slotted trailing-edge flaps. Automatic full-span leading-edge slats. Electrically actuated variable-incidence tailplane.

STRUCTURE: The wing is a two-spar structure, with milled skin panels and detachable leading-edges. The fuselage is a semi-monocoque structure. Rear fuselage detachable for engine replacement. Two-door type airbrakes under the centre-fuselage. The tail unit is a cantilever structure.

LANDING GEAR: Retractable tricycle type of Messier-Hispano design. Hydraulic actuation. Mainwheel tyre pressure 3.93 bars (57 lb/sq in). Hydraulic brakes. Brake-chute housed at base of rudder. Arrester hook under rear fuselage.

POWER PLANT: Two General Electric J85-GE-13A turbojet engines (each 12.1 kN; 2,720 lb st dry, 18.15 kN; 4,080 lb st with afterburning), mounted side by side in rear fuselage. Provision for JATO units for assisted take-off. Fuel in main tanks in fuselage and inner wing panels with total capacity of 3,200 litres (844 US gallons; 703 Imp gallons). Provision for underwing auxiliary tanks.

ACCOMMODATION: Pilot only, on fully automatic zero/zero ejection seat, under electrically actuated rearward-hinged jettisonable canopy. Cockpit armoured, pressurised and air-conditioned.

ARMAMENT: Two 30 mm DEFA cannon and cameras in nose. Four underwing attachments for 1,000 lb bombs, 340 kg (750 lb) napalm tanks, four 7 × 2 in rocket packs, four 28 × 2 in rocket packs or four 5 in rocket containers.

AVIONICS: Nav/attack system includes Computing Devices of Canada 5C-15 position and homing indicator, Sperry SYP-820 twin-axis gyro platform, Bendix King RDA-12 Doppler radar and Air Research air data computer, Ferranti ISIS B gyro gunsight. Smiths electronic head-up display, Honeywell AN/APN-171 radar altimeter and Marconi-Elliot AD 370 ADF.

Alenia (Fiat) G91Y of the Italian Air Force. This aircraft is in the process of being decommissioned from both the Italian and Portuguese air forces

1994

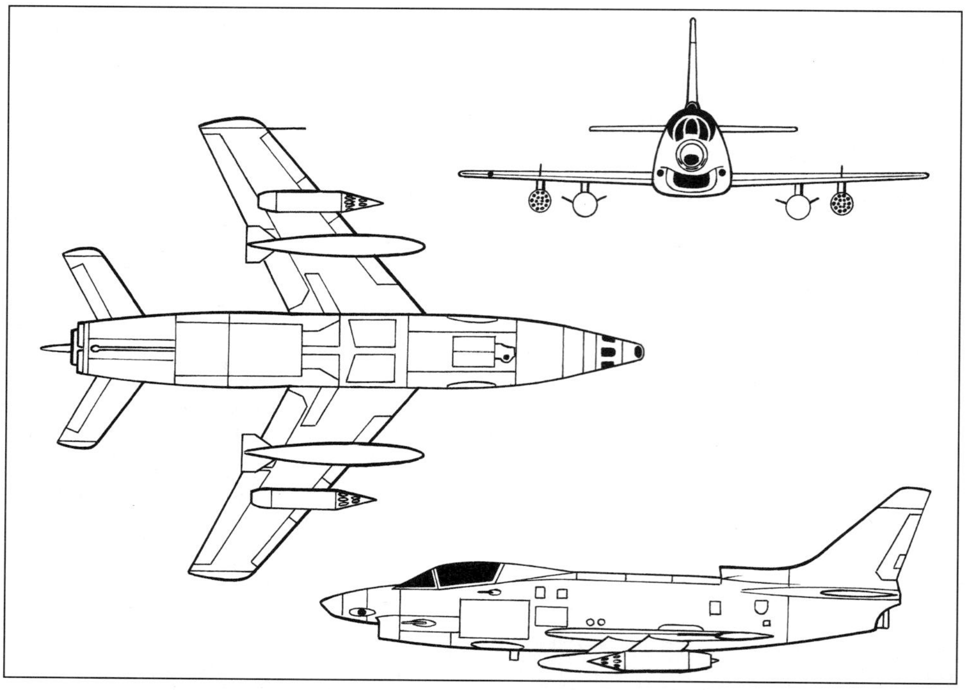

Alenia (Fiat) G91Y single-seat bomber/reconnaissance aircraft *(Jane's/Mike Keep)*

1994

Alenia (Fiat) G91T of 32 Stormo of the Italian Air Force *(P. Tompkins)*

1994

DIMENSIONS, EXTERNAL:

Wing span	9.01 m (29 ft 6½ in)
Wing chord at root	2.526 m (8 ft 3½ in)
Wing chord at tip	1.274 m (4 ft 2¼ in)
Wing aspect ratio	4.475
Length overall	11.67 m (38 ft 3½ in)
Height overall	4.43 m (14 ft 6 in)
Tailplane span	4.00 m (13 ft 1½ in)
Wheel track	2.94 m (9 ft 8 in)
Wheelbase	3.56 m (11 ft 8 in)

AREAS:

Wings, gross	18.13 m^2 (195.15 sq ft)
Ailerons (total)	1.742 m^2 (18.75 sq ft)
Trailing-edge flaps (total)	1.736 m^2 (18.69 sq ft)
Fin (excl ventral fins)	1.753 m^2 (18.87 sq ft)
Rudder	0.398 m^2 (4.28 sq ft)
Horizontal tail surfaces (total)	2.810 m^2 (30.25 sq ft)

WEIGHTS AND LOADINGS:

Weight empty	3,900 kg (8,598 lb)
Normal T-O weight	7,800 kg (17,196 lb)
Max T-O weight (semi-prepared surface)	7,000 kg (15,432 lb)
Max T-O weight (hard runway)	8,700 kg (19,180 lb)
Max wing loading	480 kg/m^2 (98.3 lb/sq ft)
Max power loading	239.7 kg/kN (2.35 lb/lb st)

PERFORMANCE (at max T-O weight, except where indicated):

Max level speed at 9,145 m (30,000 ft)	Mach 0.95
Max level speed at S/L	600 kts (1,100 km/h; 690 mph)
Stalling speed, flaps down	125 kts (230 km/h; 143 mph)
Max rate of climb at S/L:	
with afterburning	5,180 m (17,000 ft)/min
without afterburning	2,134 m (7,000 ft)/min
Time to 12,200 m (40,000 ft):	
with afterburning	4 min 30 s
without afterburning	11 min
Service ceiling	12,500 m (41,000 ft)
Service ceiling, one engine out (with afterburning)	6,000 m (19,685 ft)
* T-O run: hard runway	1,220 m (4,000 ft)
semi-prepared runway	915 m (3,000 ft)
semi-prepared surface, with JATO	457 m (1,500 ft)
* T-O run to 15 m (50 ft):	
hard runway	1,830 m (6,000 ft)
semi-prepared surface	1,372 m (4,500 ft)
semi-prepared surface, with JATO	762 m (2,500 ft)
Landing from 15 m (50 ft)	600 m (1,970 ft)
Combat radius (A: with 1,814 kg; 4,000 lb external armament, B: with two 400 l; 88 Imp gallon; 105 US gallon auxiliary fuel tanks):	
A, lo-lo-lo	200 n miles (370 km; 230 miles)
A, lo-lo-hi	250 n miles (463 km; 288 miles)
A, hi-lo-hi	305 n miles (565 km; 351 miles)
B, lo-lo-lo	360 n miles (666 km; 414 miles)
B, lo-lo-hi	450 n miles (833 km; 518 miles)
B, hi-lo-hi	550 n miles (1,018 km; 633 miles)
Ferry range with max fuel	1,890 n miles (3,500 km; 2,175 miles)

VERIFIED

ALENIA (LOCKHEED) F-104S ASA STARFIGHTER UPGRADE

TYPE: Lightweight single-seat tactical fighter-bomber and reconnaissance aircraft weapons upgrade.

PROGRAMME: Aeritalia production of the F-104S ended in March 1979 after the manufacture of 246 aircraft, including 40 for the Turkish Air Force. Development was initiated in 1982 of a weapons system updating programme for 147 of the Italian Air Force's F-104Ss. Known as ASA (Aggiornamento Sistema d'Arma).

DESIGN FEATURES: Programme included fitting the FIAR R21G/M1 Setter look-down/shoot-down radar; advanced ECM; improved IFF and altitude reporting system; improved electrical system; improved weapons delivery (armament computer and time delay unit); and a new automatic pitch control computer. Weapons include Aspide 1A medium-range and AIM-9L Sidewinder short-range air-to-air missiles. The 147th and final ASA was redelivered in 1993.

VERIFIED

Alenia (Lockheed) F-104S ASA Starfighter upgrade

1994

ALENIA (AERITALIA) G222

US Air Force designation: C-27A Spartan

TYPE: Pressurised twin-turboprop tactical transport.

PROGRAMME: Designed by Ing Giuseppe Gabrielli, the Aeritalia (originally Fiat) now Alenia G222 was conceived in four separate configurations, three of which were halted at the research project stage. Two unpressurised prototypes were built of the military transport version of which the first (MM582) made its initial flight on 18 July 1970 and the second (MM583) on 22 July 1971. The first prototype was handed over to the Italian Air Force on 21 December 1971 for operational evaluation. One airframe was completed for static and one for fatigue testing.

Several major Italian airframe companies share in the construction programme, including Aermacchi (outer wings); Piaggio (wing centre-section); SIAI-Marchetti (tail unit); CIRSEA (landing gear); and IAM (miscellaneous airframe components). Wing flaps are contributed by Hellenic Aerospace Industries. Fuselages are built by Alenia's Transport Aircraft Group, in the Pomigliano d'Arco works near Naples; final assembly takes place at the Capodichino Works, Naples. Still in production.

VERSIONS: **G222:** Standard military transport, to which the detailed description mainly applies. First delivery of a production G222 was made in November 1976 to the air force of Dubai, which ordered one. This was followed in 1977 by the first of three for the Argentine Army, and in early 1981 by two for the Somali Air Force. Two others originally ordered by Somalia were later cancelled. This version also ordered by Venezuelan Army (2) and Air Force (6), and Nigerian Air Force (5); deliveries to these countries were completed in 1985. Principal customer is the Italian Air Force, which has ordered 44 (30 standard transports, eight G222SAA, four G222RM and two G222GE), of which 43 had been delivered by mid-1986. The first G222 for the Italian Air Force (MM62101) flew on 23 December 1975, and deliveries began on 21 April 1978. These aircraft are in service with the 46a Aerobrigata at Pisa-San Giusto, and are operated primarily in the roles of troop, paratroop and cargo transport, or for aeromedical duties. Six quick-change kits, produced by Alenia, are held by the Italian Air Force for in the field conversions to the aeromedical configuration. This latter version has been used in support of Red Cross relief operations in Kampuchea, Peru and elsewhere. There were five 222s ordered by the Italian Ministry for Civil Defence, to create a rapid-intervention squadron for firefighting, aeromedical evacuation, and airlift of supplies to earthquake and other disaster areas. First of these aircraft had been delivered by Spring 1986.

G222RM: Flight inspection version (radiomisure), specially equipped for in-flight calibration of ground radio navigation and communication facilities (VOR, ILS, Tacan, NDB, AVHF and UHF). The automatic flight inspection system, based on an inspection programme recorded in a central computer, determines the exact aircraft position, mixing the information from an inertial reference unit with data from a DME or with visual fixes performed with a vertical TV, and processes signals from a set of dedicated receivers to provide the parameters of facilities under inspection. These parameters are available on CRT displays and on a paper recorder and printer. Only one flight inspector is necessary, in addition to the two-man flight crew and ample space remains in the rear of the hold.

G222SAA: Firefighting version (Sistema Aeronautico Antincendio), with specially designed modular dispersal system for water or retardant (see 'Equipment'). Eight delivered to Italian Air Force, which has used them extensively and successfully in many parts of Italy.

G222T: Version with Rolls-Royce Tyne turboprops, larger diameter propellers and higher operating weights. Twenty built, including two in VIP transport configuration, for Libyan Arab Air Force.

G222VS (Versione Speciale): Electronic warfare version, first flown 9 March 1978. Carrying a pilot, co-pilot and up to 10 systems operators, it has a modified cabin fitted with racks and consoles for detection, signal processing and data recording equipment, and an electrical system providing up to 40 kW of power for its operation. Externally distinguishable by small 'thimble' radome beneath the nose and a larger 'doughnut' radome on top of the tail fin. Two ordered by Italian Air Force, the first of which was delivered in 1983; in service with the 71° Gruppo Guerra Elettronica at Pratica di Mare. IAF designation is **G222GE** (Guerra Elettronica). Dimensions, weights and

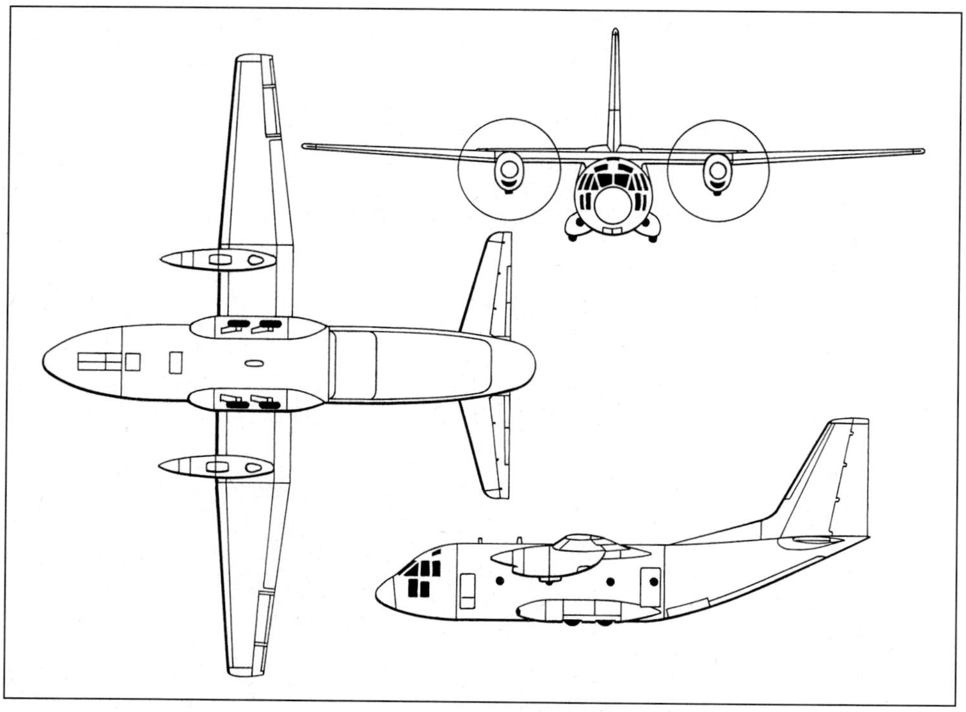

Alenia (Aeritalia) G222 twin-turboprop tactical transport aircraft *(Jane's/Mike Keep)*

1994

performance similar to those of the standard troop transport.

C-27A Spartan (G222-710): US Air Force transport for Panama and South America; fitted with mission equipment by Chrysler Technologies Airborne Systems (see CTAS in US section); US Air Force ordered total of 10 in August 1990 and February 1991; eight options. See also different weights and performance of USAF C-27A in Chrysler Technologies entry.

UPGRADES: **Chrysler Technologies C-27A Spartan:** See separate entry.

Italian Air Force: Completed testing an upgraded avionics suite proposed for its fleet of G222 transport aircraft. Programme includes; SMA-710 weather radar, Litton Italia LTN-92 and a five-channel GPS system. Upgrade allows operations in areas without radio navigation aids.

OPERATORS: Versions of the G222 are in service with the armed forces of the following countries: Argentina (3), Congo, Italy (37), Libya (20), Nigeria (5), Thailand (6), United Arab Emirates (1), Venezuela (8) and Yemen.

DESIGN FEATURES: Cantilever high-wing monoplane with maximum thickness/chord ratio of 15 per cent. Dihedral 2° 3′ on outer panels. Cantilever tailplane unit is with swept fin and variable incidence, swept tailplane.

FLYING CONTROLS: Double-slotted flaps extend over 60 per cent of trailing-edge. Two-section hydraulically actuated spoilers ahead of each outboard flap segment, used also as lift dumpers on landing. Spoilers and flaps fully powered by tandem hydraulic actuators. Manually operated ailerons, each with inset servo tab. Two tabs in each elevator; no rudder tabs. Rudder fully powered by tandem hydraulic actuators; elevators operated manually.

STRUCTURE: The wing is an aluminium alloy three-spar fail-safe box structure, built in three portions. One-piece constant-chord centre-section fits into recess in top of fuselage and is secured by bolts at six main points. Outer panels tapered on leading- and trailing-edges. Upper surface skins are of 7075-T6 alloy, lower surfaces have bonded metal skins with metal honeycomb core. The fuselage is a pressurised fail-safe structure of aluminium alloy stressed skin construction and circular cross-section. Easily removable stiffened floor panels. The tail unit is a cantilever safe-life structure of aluminium alloy, with sweptback three-spar fin and slightly swept two-spar variable incidence tailplane. Rudder and elevator of metal honeycomb construction.

POWER PLANT (except G222T): Two Fiat built General Electric T64-GE-P4D turboprops, each flat rated at 2,535 kW (3,400 shp) at ISA +25°C and driving a Hamilton Standard 63E60-27 three-blade variable- and reversible-pitch propeller with spinner. Fuel in integral tanks: two in the outer wings, combined capacity 6,800 litres (1,796 US gallons; 1,495 Imp gallons), and two centre-section tanks, combined capacity 5,200 litres (1,374 US gallons; 1,143 Imp gallons), with cross-feed provision to either engine. Total overall fuel capacity 12,000 litres (3,170 US gallons; 2,638 Imp gallons).

ACCOMMODATION: Normal crew of three (two pilots and radio operator/flight engineer) on flight deck. Provision for loadmaster or jumpmaster when required. Standard troop transport version has 32 foldaway sidewall seats and 21 stowable seats for 53 fully equipped troops, and also carries two 20-man liferafts stowed in the wing/fuselage and a single nine-man liferaft in the cargo compartment. Paratroop transport version can carry up to 40 fully equipped paratroops, and is fitted with the 32 sidewall seats and liferafts as in the troop transport version, plus eight stowable seats, door jump platforms and static lines. Five-person VIP lounge plus seats for 16 other passengers in VIP transport version. Cargo transport version can accept standard pallets of up to 2.24 m (88 in) wide, and can carry up to 9,000 kg (19,840 lb) of freight. Provision is made for 135 cargo tiedown points, on a 51 cm (20 in) square NATO standard grid, and a 1,500 kg (3,306 lb) capacity cargo hoist. Typical Italian military equipment loads can include two CL-52 light trucks; one CL-52 with a 105 mm L4 howitzer or 1 ton trailer; Fiat AR-59 Campagnola reconnaissance vehicle with 106 mm recoilless gun or 250 kg (550 lb) trailer; or five standard A-22 freight pallets. In the aeromedical role the G222 can accommodate 36 stretchers and four medical attendants. A second toilet can be installed, and provision can be made to increase the water supply and to install electrical points and hooks for medical treatment bottles. In this version, the cabin oxygen system is available to all stretcher positions. Crew door is forward of cabin on port side. Passenger doors, at front and rear of main cabin on starboard side and at rear on port side, can be used also as emergency exits. Two emergency hatches in cabin roof, forward and aft of wing carry-through structure. Hydraulically operated rear-loading ramp and upward-opening door in underside of upswept rear fuselage, which can be opened in flight for airdrop operations. In cargo version, five pallets of up to 1,000 kg (2,205 lb) each can be airdropped from rear-opening, or a single pallet of up to 5,000 kg (11,023 lb). Paratroop jumps can be made either from this opening or from the rear side doors. Windscreens and quarter-light panels are de-iced and demisted electrically. Wipers and screen wash for both windscreens. Entire accommodation pressurised and air-conditioned.

SYSTEMS: Pressurisation system maintains a cabin differential of 0.41 bars (5.97 lb/sq in), giving a 1,200 m (3,940 ft) environment at altitudes up to 6,000 m (19,680 ft). Air conditioning system uses engine bleed air during flight; on ground, it is fed by compressor bleed air from APU to provide cabin heating to a minimum of 18°C. Garrett 113.3 kW (152 hp) APU, installed in starboard main landing gear fairing, provides power for engine starting, hydraulic pump and alternator actuation, air conditioning on ground, and all hydraulic and electrical systems necessary for loading and unloading on ground. Two independent hydraulic systems, each of 207 bars (3,000 lb/sq in) pressure. No.1 system actuates flaps, spoilers, rudder, wheel brakes and (in emergency only) landing gear extension; No. 2 system actuates flaps, spoilers, rudder, wheel brakes, nosewheel steering, landing gear extension and retraction, rear ramp/door and windscreen wipers. Auxiliary hydraulic system, fed by APU powered pump, can take over from No. 2 system in flight, if both main systems fail, to operate essential services. In addition, a standby handpump is provided for emergency use to lower the landing gear and, on the ground, to operate the ramp/door and parking brakes. Three 45 kVA alternators, one driven by each engine through constant-speed drive units and one by the APU, provide 115/200 V three phase AC electrical power at 400Hz. 28 V DC power is supplied from the main AC buses via two transformer-rectifiers, with 24 V 34 Ah Ni/Cd battery and static inverter for standby and emergency power. External AC power socket. Electric de-icing of spinners and propeller leading-edges. Engine intakes anti-iced by electrical/hot air system. Liquid oxygen system for crew and passengers (with cabin wall outlets); this system can be replaced by a gaseous oxygen system if required. Emergency oxygen system available for all occupants in the event of a pressurisation failure. Pneumatically inflated de-icing boots on outer leading-edges, using engine bleed air.

AVIONICS: Standard communications equipment includes 3,500-channel UHF, two 1,360-channel VHF-AM, 920-channel VHF-FM, 28,000-channel HF/SSB, crew intercom and PA system. Navigation equipment includes Omega system, with TAS computer, autopilot, flight director, two compasses, and two vertical gyros; and an integrated ground-based system incorporating two VOR, marker beacon receiver, two ILS ADF, two Tacan or DME, and horizontal situation indicator. Other avionics include Meteo weather radar, with secondary terrain-mapping mode; radar altimeter; and IFF/ATC transponder including altitude reporting. Provision for head-up display. Landing light on nosewheel leg.

EQUIPMENT (G222SAA): Modular palletised firefighting pack can be installed in under two hours without any modification to the basic transport aircraft. The module consists of a 6,000 litre (1,320 Imp gallon; 1,588 US gallon) tank

Alenia (Aeritalia) G222 twin-turboprop tactical aircraft *(Peter J. Cooper)*

1994

Alenia (Aeritalia) G222 twin-turboprop tactical transport aircraft of the Italian Air Force *(Alenia)*

1995

and four pressurised air containers to activate the pneumatic actuators and discharge the retardant through the rear ramp/door opening via two nozzles. Length of area covered averages 300 m (985 ft).

DIMENSIONS, EXTERNAL:

Wing span	28.70 m (94 ft 2 in)
Wing chord: at root	3.40 m (11 ft 1¾ in)
at tip	1.685 m (5 ft 6¼ in)
Wing aspect ratio	10.0
Length overall	22.70 m (74 ft 5½ in)
Height overall	9.80 m (32 ft 1¾ in)
Fuselage: Max diameter	3.55 m (11 ft 7¾ in)
Tailplane span	12.40 m (40 ft 8¼ in)
Wheel track	3.668 m (12 ft 0½ in)
Wheelbase (to c/l of main units)	6.23 m (20 ft 5¼ in)
Propeller diameter	4.42 m (14 ft 6 in)
Distance between propeller centres	9.50 m (31 ft 2 in)
Rear-loading ramp/door: Width	2.45 m (8 ft 0½ in)
Height	2.25 m (7 ft 4½ in)

DIMENSIONS, INTERNAL:

Main cabin: Length	8.58 m (28 ft 1¾ in)
Width	2.45 m (8 ft 0½ in)
Height	2.25 m (7 ft 4½ in)
Floor area: excl ramp	21.00 m^2 (226.0 sq ft)
incl ramp	25.68 m^2 (276.4 sq ft)
Volume	58.0 m^3 (2,048 cu ft)

AREAS:

Wings, gross	82.00 m^2 (882.6 sq ft)
Ailerons (total)	3.65 m^2 (39.29 sq ft)
Trailing-edge flaps (total)	18.40 m^2 (198.06 sq ft)
Spoilers (total)	1.65 m^2 (17.76 sq ft)
Fin (incl dorsal fin)	12.19 m^2 (131.21 sq ft)
Rudder	7.02 m^2 (75.56 sq ft)
Tailplane	19.09 m^2 (205.48 sq ft)
Elevators (total)	4.61 m^2 (49.62 sq ft)

WEIGHTS AND LOADINGS (standard version except where indicated):

Weight empty	14,590 kg (32,165 lb)
Weight empty, equipped	15,400 kg (33,950 lb)
Operating weight empty (standard and SAA)	15,700 kg (34,610 lb)
Max payload (cargo)	9,000 kg (19,840 lb)
Equipment module (SAA)	2,200 kg (4,850 lb)
Retardant (SAA)	6,800 kg (14,990 lb)
Max fuel load	9,400 kg (20,725 lb)
Fuel (SAA)	3,330 kg (7,340 lb)
Max T-O weight (standard and SAA)	28,000 kg (61,730 lb)
Max landing weight (standard and SAA)	26,500 kg (58,420 lb)
Max zero-fuel weight	24,400 kg (53,790 lb)
Max cargo floor loading	750 kg/m^2 (155 lb/sq ft)
Max wing loading	341.5 kg/m^2 (69.9 lb/sq ft)
Max power loading	5.52 kg/kW (9.08 lb/shp)

PERFORMANCE (standard G222 transport, at max T-O weight except where indicated):

Max level speed a 4,575 m (15,000 ft)	291 kts (540 km/h; 336 mph)
Long-range cruising speed at 6,000 m (19,680 ft)	237 kts (439 km/h; 273 mph)
Airdrop speed (paratroops or cargo)	110-140 kts (204-259 km/h; 127-161 mph) IAS
Drop speed (G222SAA, T-O configuration)	120 kts (222 km/h; 138 mph)
Stalling speed, flaps and landing gear down	84 kts (155 km/h; 97 mph)
Time to 4,500 m (14,760 ft)	8 min 35 s
Max rate of climb at S/L	520 m (1,705 ft)/min
Rate of climb at S/L, one engine out	125 m (410 ft)/min
Service ceiling	7,620 m (25,000 ft)
Service ceiling, one engine out	5,000 m (16,400 ft)
Optimum height above ground during drop (G222SAA)	50-100 m (165-300 ft)
T-O run	662 m (2,172 ft)
T-O run to 15 m (50 ft)	1,000 m (3,280 ft)
Landing from 15 m (50 ft)	775 m (2,543 ft)
Landing run at max landing weight	545 m (1,788 ft)
Accelerate/stop distance	1,200 m (3,937 ft)
Min ground turning radius	20.80 m (68 ft 3 in)
Range with max payload, optimum cruising speed and height	740 n miles (1,371 km; 852 miles)
Range with 36 stretchers and 4 medical attendants	1,349 n miles (2,500 km; 1,553 miles)
Range with max retardant load (SAA)	540 n miles (1,000 km; 621 miles)
Ferry range with max fuel	2,500 n miles (4,633 km; 2,879 miles)
g limit	+2.5

VERIFIED

Alenia (Aeritalia) G222 in C-27A configuration for the USAF *(Alenia)*

1995

ALENIA (BREGUET) 1150 ATLANTIC 1 UPGRADES

TYPE: Maritime patrol aircraft upgrade.

PROGRAMME: The Alenia subsidiary Officine Aeronavali Venezia has been working on two upgrade programmes for the 18 Breguet 1150 Atlantic 1 aircraft of the Italian Air Force. These are the ALCO and ESM programmes.

DESIGN FEATURES: ALCO. Undertaken in collaboration with Dassault Aviation. Programme commenced in 1986 and all 18 aircraft were completed in 1993. Package included installation of advanced: search radar, navigation system and acoustic system.

ESM. Programme started in 1988 and consisted of the installation of a new ESM system to enhance the aircraft's sensitivity in tracing and analysing the source of signals emissions. To date nine aircraft have been upgraded and it is hoped that the remaining nine will be upgraded within two years.

VERIFIED

ALENIA (McDONNELL DOUGLAS) DC-8 FREIGHTER CONVERSION

TYPE: Four-engine jet transport aircraft freighter conversion.

PROGRAMME: Under a US programme launched in 1976, McDonnell Douglas converted 20 DC-8 passenger transports into specialised freighters. In 1982 responsibility for the DC-8 freighter conversion programme was taken over by Officine Aeronavali Venezia SpA (a wholly owned subidiary of Finmeccanica), which delivered its first conversion in February 1986. A total of 47 conversions had been performed by April 1994, mainly for the DC-8-70 series.

VERIFIED

ALENIA (BOEING) B707 TANKER CONVERSION

TYPE: Four-engine transport aircraft tanker conversion.

PROGRAMME: In 1988 Alenia was awarded the contract for the conversion of four commercial 707 aircraft into tanker/transport for the Italian Air Force (AMI). Extensive flight testing activity was planned and successfully carried out, in order to qualify the whole system and verify the final performance against the Italian Air Force requirement. Alenia is now involved in the 707 tanker fleet support on the base of a turnkey contract received from the Italian MoD.

DESIGN FEATURES: The final configuration of Air Refueller and Airborne Command Post with cargo and passenger capability is equipped with a three-refuelling point developed by Boeing Military Aircraft for the Italian Air Force. Alenia designed and installed the mission avionics (CCTV included), the cargo and VIP interiors, the APU and airstair.

VERIFIED

ALENIA/DEE HOWARD (BOEING) B727QF RE-ENGINING PROGRAMME

TYPE: Three-engine transport aircraft re-engine programme.

PROGRAMME: Contract signed with UPS in 1990. First prototype, designated 727QF, flew on 18 April 1992 and FAA certification was obtained at the end of the same year. The re-engining programme was developed in the USA by the Alenia subsidiary, Dee Howard (which see). Flight deck modernisation and airframe ageing are coupled to the programme. See also Dee Howard entry.

DESIGN FEATURES: Replacement of existing engines with Rolls-Royce Tay 651-54 engines. The new engines provide greater thrust on take-off whilst reducing the fuel consumption by 18 per cent. Noise and emission reductions to meet FAA Cat. III requirements are matched with improved performance and reliability.

CUSTOMERS: UPS has committed $400 million to re-engine 40 of its 727-100 aircraft with options for an additional 40 aircraft.

VERIFIED

ALENIA (McDONNELL DOUGLAS) DC-10 FREIGHTER CONVERSION

TYPE: Four-engine transport aircraft freighter conversion.

PROGRAMME: The DC-10 freighter conversion programme was launched in 1990 to be a successor to the DC-8 programme. The conversion is performed under licence from McDonnell Douglas and provides for the modification of any DC-10 series to carry freight with payload capability of up to 180,000 lb. Four aircraft had been converted by April 1994.

VERIFIED

ALENIA (MCDONNELL DOUGLAS) MD-11 FREIGHTER CONVERSION

TYPE: Three-engine transport aircraft freighter conversion.

PROGRAMME: Alenia, through Aeronavali, intends to launch the MD-11 freighter conversion programme to increase performance and integrate freight carrying capabilities. It is hoped the aircraft will fill the gap between the long-range small aircraft and long-range large aircraft. One MD-11 had been converted by April 1994.

VERIFIED

Alenia (McDonnell Douglas) DC-8 freighter converison *1994*

Alenia (Boeing) converted B707 tanker *1994*

Alenia (McDonnell Douglas) DC-10 freighter conversion *1994*

PIAGGIO

INDUSTRIE AERONAUTICHE E MECANICHE RINALDO PIAGGIO SpA

Via Cibrario 4, I-16154 Genova Sestri, Genoa
Telephone: 39 (10) 6481
Fax: 39 (10) 6520160
Telex: 270695 AERPIA I
WORKS: Genova Sestri, Finale Ligure (SV), and Wichita, Kansas, USA
BRANCH OFFICE: Via A. Gramsci 34, I-00197 Rome
PRESIDENT: Dott Rinaldo Piaggio
DIRECTOR GENERAL AND CEO: Ing Roberto Mannu
DIRECTOR OF OPERATIONS: Ing Aldo Gianni
MARKETING AND SALES: Enzo Traini

The original Piaggio company began the construction of aeroplanes in its Genova Sestri plant in 1916, and later in the Finale Ligure works. The present company was formed on 29 February 1964, and has since operated as an independent concern. It employs about 1,500 people in three production divisions and in Italy has a total covered works area (Genova Sestri and Finale Ligure) of approximately 120,000 m² (1,291,670 sq ft). In the USA, Piaggio Aviation Inc at Wichita, Kansas, was founded by Piaggio on 9 September 1987 with head office in Dover, Delaware. The Wichita factory of approximately 10,000 m² (107,640 sq ft) is producing Avanti fuselages. In addition to aircraft of its own design, Piaggio is producing components for the Alenia G222, Panavia Tornado and AMX. Aeritalia, now Alenia (which see), acquired a 31 per cent holding in Piaggio in 1988, which was adjusted to 24.5 per cent in 1992 and then raised to 30.9 per cent in early 1993 following further restructuring.

VERIFIED

Piaggio PD.808 of the 14 Stormo Italian Air Force ***(Peter J. Cooper)***

1994

R. PIAGGIO PD.808

TYPE: Twin-jet light utility aircraft.

PROGRAMME: The PD.808 is a 6/10-seat light jet utility aircraft suitable for both civil and military use. The Italian Defence Ministry assisted Piaggio by purchasing the two prototypes (each powered by 1,360 kg; 3,000 lb st Viper Mk 525 engines) and by providing test facilities. The first of these prototypes made its first flight on 29 August 1964.

In mid-1965, the Italian Defence Ministry ordered 25 production PD.808s. These aircraft have more powerful Viper Mk 526 engines and differ from the original prototype in having larger tip tanks, longer dorsal fin and forward-sliding nose fairing. In early 1994 the Italian Air Force operated 20 PD.808s for calibration and ECM duties.

VERSIONS: **PD.808 VIP**: Six-seat version for the Italian Air Force, used for government and military VIP transport duties.

PD.808 TA: Nine-seat version for military transport duties and navigation training. RAI and FAA certification 19 October 1968.

PD.808 ECM: Electronic countermeasures military version, with accommodation for two pilots and three equipment operators.

PD.808 RM: Radio calibration version, equipment for medium and high-altitude calibration of navigation aids.

OPERATORS: Versions of the PD.808 are in service with the armed forces of Italy (18).

The following description applies to the standard production aircraft with Viper Mk 526 engines:

DESIGN FEATURES: Cantilever low-wing monoplane. Wing section DES 0010-1.1-40/11° (modified) at root, DES 0008-1.1-40/9° (modified at tip). Dihedral 3°. Incidence 1°. Sweepback 1° 50′ at quarter-chord. Fixed incidence tailplane.

FLYING CONTROLS: All-metal ailerons, each with geared and trim tab. Hydraulically operated single-slotted trailing-edge flaps. Hydraulically operated spoilers on top of each wing operate on nosewheel contact during the landing. Trim and geared tabs in elevator, trim tab in rudder.

STRUCTURE: The wing is an aluminium alloy fail-safe three-spar structure. The fuselage is a circular-section all-metal fail-safe structure with machined frames. Hydraulically operated airbrakes under fuselage. The tail unit is a cantilever all-metal structure, with fixed-incidence tailplane.

LANDING GEAR: Retractable tricycle type with single wheel on each unit. Hydraulic retraction, all units retracting forward. Oleo-pneumatic shock-absorbers. Goodyear mainwheels and tyres size 24.15 × 7.65-10, pressure 9.49 kg/cm² (135 lb/sq in). Goodyear nosewheel and tyre, size 17.9 × 5.70-8, pressure 9.49 kg/cm² (135 lb/sq in). Aerostatica R.F.91 braking parachute, diameter 2.40 m (7 ft 1½ in). Messier hydraulic brakes. Hydraulic nosewheel steering.

POWER PLANT: Two Rolls-Royce Bristol Viper Mk 526 turbojet engines (each rated at 1,524 kg; 3,360 lb st) mounted on sides of rear fuselage. Fuel in two integral tanks in wings, with total capacity 1,935 litres (511 US gallons; 425.5 Imp gallons) and two integral wingtip tanks with total capacity 1,792 litres (473 US gallons; 394 Imp gallons). Total fuel capacity 3,727 litres (983 US gallons; 819.5 Imp gallons). Refuelling points on tip tanks. Oil capacity 13 kg (28 lb).

ACCOMMODATION: Crew of one or two on flight deck, with dual controls. Downward-opening plug type door, with built-in steps, at front of cabin on port side. Cabin pressurised and air-conditioned. Standard seating in PD.808 VIP for six persons in individual seats, with bar and toilet at front of cabin and baggage space aft of rear seats. Further baggage space in nose, with access by sliding nose forward. Up to nine seats in cabin in PD.808 TA military transport version. Trainer version of PD.808 TA has one main student station in the co-pilot's seat, with two or three more student stations in cabin. Two pilots and three equipment operators in PD.808 ECM version. With cabin stripped, there is space for 5.24 m² (185 cu ft) of freight, survey camera stations, navigation training or radio calibration equipment. Cabin can be adapted to ambulance configuration, with space for two stretchers, two seated casualties and a medical attendant.

SYSTEMS: AiResearch simple air cycle air conditioning and pressurisation system, using engine bleed air. Pressure differential 0.62 kg/cm² (8.8 lb/sq in). Hydraulic system, pressure 210 kg/cm² (3,000 lb/sq in), operates landing gear, flaps, spoilers, airbrakes, nosewheel steering and wheel brakes. Pneumatic de-icing and anti-icing. Electrical system includes two DC generators, three inverters, two AC generators and two batteries.

AVIONICS: Standard equipment includes VHF radio, ADF, glide slope and marker beacon receivers. DME and weather radar optional.

DIMENSIONS, EXTERNAL:

Wing span over tip tanks	13.20 m (43 ft 3½ in)
Wing chord (mean)	2.438 m (8 ft 0 in)
Wing aspect ratio	6.25
Length overall	12.85 m (42 ft 2 in)
Height overall	4.80 m (15 ft 9 in)
Tailplane span	5.43 m (17 ft 9½ in)
Wheel track	3.68 m (12 ft 0¾ in)
Wheelbase	4.50 m (14 ft 9 in)
Passenger door (forward, port):	
Height	1.26 m (4 ft 1½ in)
Weight	0.65 m (2 ft 1½ in)

DIMENSIONS, INTERNAL:

Cabin: Length	4.47 m (14 ft 8 in)
Max width	1.64 m (5 ft 4½ in)
Max height	1.45 m (4 ft 9 in)
Floor area	4.66 m² (50.16 sq ft)
Volume	8.24 m³ (291 cu ft)

AREAS:

Wings, gross	20.9 m² (225 sq ft)
Ailerons (total)	1.12 m² (12.10 sq ft)
Trailing-edge flaps (total)	2.32 m² (25.00 sq ft)
Spoilers (total)	1.25 m² (13.45 sq ft)
Airbrakes	0.56 m² (6.00 sq ft)
Fin	3.34 m² (36.0 sq ft)
Rudder, incl tab	0.54 m² (5.81 sq ft)
Tailplane	6.50 m² (70.0 sq ft)
Elevator, incl tabs	1.75 m² (18.9 sq ft)

WEIGHTS AND LOADINGS:

Weight empty, equipped	4,830 kg (10,650 lb)
Max payload	862 kg (1,900 lb)
Max T-O weight	8,165 kg (18,000 lb)
Max ramp weight	8,300 kg (18,300 lb)
Max zero-fuel weight	5,902 kg (13,000 lb)
Max landing weight	7,257 kg (16,000 lb)
Max wing loading	390.6 kg/m² (80 lb/sq ft)
Max power loading	2.70 kg/kg st (2.70 lb/lb st)

PERFORMANCE (at max T-O weight, except where indicated):

Max never-exceed speed
425 kts (788 km/h; 489 mph) EAS between S/L and 4,260 m (14,000 ft) Mach 0.85 above 4,260 m (14,000 ft)
Max level speed at 5,945 m (19,500 ft)
460 kts (852 km/h; 529 mph)
Max cruising speed above 11,000 m (36,000 ft)
432 kts (800 km/h; 497 mph)
Econ cruising speed at 12,500 m (41,000 ft)
390 kts (722 km/h; 449 mph)
Stalling speed, landing configuration, at landing weight of 5,902 kg (13,000 lb) 90 kts (167 km/h; 104 mph)
Max rate of climb at S/L at AUW of 7,176 kg (15,821 lb)
1,650 m (5,400 ft)/min

Service ceiling	13,715 m (45,000 ft)
Service ceiling, one engine out	7,925 m (26,000 ft)
Min ground turning radius	7.00 m (22 ft 11¾ in)
Runway LCN	16
T-O run	841 m (2,760 ft)
T-O to 10.7 m (35 ft)	970 m (3,180 ft)
FAR 25 T-O field length	1,695 m (5,560 ft)

Landing from 15 m (50 ft), with four passengers and baggage, 45 min reserve 912 m (2,990 ft)
Range with max fuel and 381 kg (840 lb) payload, 45 min reserve 1,148 n miles (2,128 km; 1,322 miles)

VERIFIED

Piaggio PD.808RM radio calibration version of the 14 Stormo Italian Air Force ***(Peter J. Cooper)***

1994

JAPAN

FUJI

FUJI HEAVY INDUSTRIES LTD (Fuji Jukogyo Kabushiki Kaisha)

Subaru Building, 7-2 1-chome, Shinjuku-ku, Tokyo 160
Telephone: 81 (3) 3347 2525
Fax: 81 (3) 3347 2588
Telex: 0 232 2268 FUJI J
PRESIDENT: Isamu Kawai
UTSUNOMIYA MANUFACTURING DIVISION: 1-11 Yonan 1-chome, Utsunomiya, Tochigi 320
Telephone: 81 (286) 581111

Aerospace Division

MANAGING DIRECTOR, GENERAL MANAGER AEROSPACE DIVISION: Yasuyaki Kogure
GENERAL MANAGER, COMMERCIAL BUSINESS DEPARTMENT: Kisaburo Wani
GENERAL MANAGER, MARKETING AND SALES DEPARTMENT: Yujiro Yoshida
DIRECTOR AND GENERAL MANAGER, AEROSPACE MANUFACTURING DIVISION: Tsutou Ono
DEPUTY GENERAL MANAGER OF AEROSPACE MANUFACTURING DIVISION: Ryuji Matsuda

Fuji Heavy Industries, established on 15 July 1953, is a successor to the Nakajima aircraft company, which was established in 1917 and built 25,935 aircraft up to the end of the Second World War.

The Utsunomiya Manufacturing Division (Aircraft and Rolling Stock Plants) occupies a site of 599,816 m² (6,456,359 sq ft) including a floor area of 188,490 m² (2,028,887 sq ft) and in April 1994 employed 2,940 people.

In addition to the T-5, Fuji is currently producing the Bell Model 205B and AH-1S Huey Cobra helicopters. It is building wing main assemblies for JMSDF Lockheed P-3C Orions; main landing gear doors and some titanium airframe parts for Japanese built McDonnell Douglas F-15J fighters; and wings, tailplanes and canopies for the Kawasaki T-4 jet trainer. Commercial aircraft components are produced for the Boeing 747 (spoilers, inboard and outboard ailerons), Boeing 757 (outboard trailing-edge flaps), Boeing 767 (wing/body fairings and main landing gear doors), McDonnell Douglas MD-11 (outboard ailerons) and Fokker 50 (rudder and elevators), Boeing 777 (centre wing, wing-to-body fairings, and main landing gear doors).

Fuji and Kawasaki are investigating the configuration of the National Aeronautical Laboratory medium-range fan-lift VTOL airliner. Fuji is also involved in research on the Japanese SST and hypersonic transport and has an important composites department.

VERIFIED

FUJI KM-2B

JASDF designation: T-3

TYPE: Two-seat primary trainer.

PROGRAMME: The KM-2B is a modification of the original KM-2 development of the Beechcraft T-34A Mentor, combining the airframe and power plant of the KM-2 with the two-seat cockpit installation of the T-34A. The first KM-2B (JA3725) was flown for the first time on 26 September 1974, and received JCAB category A certification on 28 November 1974.

In 1975 the JASDF selected the KM-2B to replace the T-34A in the primary trainer role, and 50 were ordered with the designation T-3. The first of these made its first flight on 17 January 1978, and 44 had been delivered to the JASDF by February 1981.

UPGRADES: **Fuji T-5:** See separate entry.

OPERATORS: Versions of the T-3 (KM-2B) are in service with the armed forces of Japan (50).

DESIGN FEATURES: Cantilever low-wing monoplane. Wing section NACA 23016.5 at root, NACA 23012 at tip. Dihedral 6° from roots. Incidence 4° at root geometric twist. No sweep at quarter-chord. Cantilever tail unit. Fixed-incidence tailplane.

FLYING CONTROLS: Aluminium alloy single-slotted flaps and plain ailerons. Servo tab in each aileron; port tab controllable for trim. Controllable tab in each elevator and anti-servo tab in rudder.

STRUCTURE: The wing is an all-metal tapered two-spar structure with stressed skin. The fuselage is an all-metal conventional semi-monocoque structure. The tail unit is an all-metal structure.

LANDING GEAR: Electrically retractable tricycle type, with single wheel and oleo-pneumatic shock-absorber on each unit. Nosewheel retracts rearward, main wheels inward into wings. Goodyear wheels and Type III tyres on all units: size 6.50-8 (6 ply), pressure 2.34 bars (34 lb/sq in) on main units; size 5.00-5 (4 ply), pressure 2.76 bars (40 lb/sq in) on nose unit. Goodyear automatically adjustable single-disc brakes on main units.

POWER PLANT: One 254 kW (340 hp) Avco Lycoming IGSO-480-A1F6 flat-six engine, driving a Hartzell HC-A3V20-1F/V9333N-3 three-blade constant-speed propeller with spinner. One metal and one bladder-type fuel tank in each wing, total capacity 265 litres (70 US gallons; 58.3 Imp gallons). Refuelling points on top of wings. Oil capacity 11.35 litres (3 US gallons; 2.5 Imp gallons).

ACCOMMODATION: Crew of two in tandem, on adjustable seats in heated and ventilated cabin. Dual controls standard. Framed canopy, with rearward-sliding section over each seat. Space for 13.6 kg (30 lb) of baggage aft of rear seat.

SYSTEMS AND AVIONICS: Blind-flying instrumentation standard. Electrical system includes 28 V 100 A DC generator and two 160 VA static inverters. Standard avionics include ICS, VHF, Tacan, and ATC transponder with Mode C.

DIMENSIONS, EXTERNAL:

Wing span	10.004 m (32 ft 10 in)
Wing chord: at root	2.13 m (7 ft 0 in)
at tip	1.07 m (3 ft 6 in)
Wing aspect ratio	6.1
Length overall	8.036 m (26 ft 4¼ in)
Height overall	3.023 m (9 ft 11 in)
Tailplane span	3.712 m (12 ft 2¼ in)
Wheel track	2.924 m (9 ft 7 in)
Wheelbase	2.266 m (7 ft 5¼ in)
Propeller diameter	2.286 m (7 ft 6 in)
Propeller ground clearance	0.25 m (9¾ in)

DIMENSIONS, INTERNAL:

Cabin: Max width	approx 0.90 m (2 ft 11½ in)
Max height	approx 1.30 m (4 ft 3¼ in)

AREAS:

Wings, gross	16.50 m² (177.6 sq ft)
Ailerons (total)	1.07 m² (11.52 sq ft)
Trailing-edge flaps (total)	1.98 m² (21.31 sq ft)
Fin	0.97 m² (10.44 sq ft)
Rudder, incl tab	0.61 m² (6.57 sq ft)
Tailplane	3.46 m² (37.24 sq ft)
Elevators, incl tab	1.39 m² (14.96 sq ft)

WEIGHTS AND LOADINGS:

Weight empty	1,120 kg (2,469 lb)
Normal operational T-O weight	1,510 kg (3,329 lb)
Max T-O weight	1,542 kg (3,400 lb)
Max wing loading	93.5 kg/m² (19.15 lb/sq ft)
Max power loading	6.07 kg/kW (10.00 lb/hp)

PERFORMANCE (at normal operating T-O weight):

Never-exceed speed	223 kts (413 km/h; 257 mph) EAS
Max level speed at 2,440 m (8,000 ft)	198 kts (367 km/h; 228 mph)
Max cruising speed at 2,440 m (8,000 ft)	177 kts (328 km/h; 204 mph)
Econ cruising speed at 2,440 m (8,000 ft)	137 kts (254 km/h; 158 mph)
Stalling speed: flaps up	66 kts (123 km/h; 76 mph)
flaps down	54 kts (100 km/h; 62 mph)
Max rate of climb at S/L	463 m (1,520 ft)/min
Service ceiling	8,170 m (26,800 ft)
T-O run	265 m (870 ft)
T-O to 15 m (50 ft)	503 m (1,650 ft)
Landing from 15 m (50 ft)	436 m (1,430 ft)
Landing run	238 m (780 ft)
Range with max fuel	520 n miles (965 km; 600 miles)

UPDATED

FUJI T-5

TYPE: Two/four-seat primary trainer; turboprop version (KM-2Kai) of KM-2.

PROGRAMME: Fuji refitted KM-2 with Allison 250-B17D turboprop in 1984; first flown 28 June 1984 as KM-2D; JCAB certification (Aerobatic and Utility categories) gained 14 February 1985; ordered March 1987 as replacement for 31 JMSDF KM-2s; first flight of production KM-2Kai 27 April 1988; deliveries began 30 August 1988.

CURRENT VERSIONS: **T-5:** For JMSDF; Aerobatic and Utility applications; *description applies to this version.*

OPERATORS: Versions of the T-5 are in service with the armed forces of Japan (34).

CUSTOMERS: See table. Total of 32 T-5s delivered to JMSDF by December 1994 for 201 Squadron of Ozuki Air Training Group.

FUJI T-5 FUNDING (JMSDF)

FY	Qty	Cum Total
1986	1	1
1987	2	3
1988	3	6
1989	2	8
1990	7	15
1991	9	24
1992	5	29
1993	3	32
1994	—	32
1995	2	34

DESIGN FEATURES: Turboprop engine; sliding canopy; modernised cockpit; visibility, payload and cockpit volume increased.

Wing section NACA 23016.5 at root, NACA 23012 at tip; no sweep; dihedral 6°; incidence 4° at root, 1° at tip.

FLYING CONTROLS: Mechanically actuated plain ailerons, balanced elevators and rudder; single-slotted wing flaps; aileron anti-servo tabs, port tab controllable for trim; controllable tabs in elevators; rudder anti-servo tab.

STRUCTURE: All-metal light alloy.

LANDING GEAR: Electrically retractable tricycle type, with emergency manual control. Oleo-pneumatic shock-absorber in each unit. Main units retract inward into wings, nose unit rearward into fuselage. Single Parker wheel and Goodyear tyre on each main unit, size 6.50-8 (6 ply); Goodyear nosewheel and tyre, size 5.00-5 (4 ply). Nose unit steerable ±16°. Parker single-disc hydraulic brakes. Minimum ground turning radius 7.38 m (24 ft 2½ in).

POWER PLANT: One Allison 250-B17D turboprop, flat rated at 261 kW (350 shp), driving a Hartzell HC-B3TP-7A/T10173-18 three-blade constant-speed fully feathering propeller. Two bladder type fuel tanks in each wing, one of 94.6 litres (25 US gallons; 20.8 Imp gallons) capacity and one of 87 litres (23 US gallons; 19.2 Imp gallons); total capacity 363 litres (96 US gallons; 80 Imp gallons). Gravity refuelling point in top surface of each wing. Oil capacity 9.5 litres (2.5 US gallons; 2.1 Imp gallons).

ACCOMMODATION: Enclosed cabin seating two persons side by side, with dual controls (Aerobatic version), or four persons in pairs in Utility version. Rearward sliding canopy; optional IFR training hood. Accommodation heated and ventilated.

SYSTEMS: Hydraulic system for brakes only. Electrical system includes 30 V 150 A starter/generator, two 160 VA static inverters, and 24 Ah battery for engine starting and emergency.

AVIONICS: *Comms:* UHF and VHF radio; intercom; SIF.

Flight: Include ADF and Tacan.

DIMENSIONS, EXTERNAL:

Wing span	10.04 m (32 ft 11¼ in)
Wing chord: at root	2.13 m (6 ft 11¾ in)
at tip	1.07 m (3 ft 6¼ in)
Wing aspect ratio	6.11
Length overall	8.44 m (27 ft 8¼ in)
Height overall	2.96 m (9 ft 8½ in)
Elevator span	3.71 m (12 ft 2 in)
Wheel track	2.92 m (9 ft 7 in)
Wheelbase	2.27 m (7 ft 5½ in)
Propeller diameter	2.12 m (6 ft 11½ in)
Propeller ground clearance	0.37 m (1 ft 2½ in)

DIMENSIONS, INTERNAL:

Cabin: Length	2.90 m (9 ft 6¼ in)
Max width	1.27 m (4 ft 2 in)
Max height	1.33 m (4 ft 4½ in)

AREAS:

Wings, gross	16.50 m² (177.6 sq ft)
Ailerons (total, incl tabs)	1.09 m² (11.73 sq ft)
Trailing-edge flaps (total)	1.98 m² (21.31 sq ft)
Fin, incl dorsal fin	1.28 m² (13.78 sq ft)

Fuji T-5 two/four-seat primary trainer of the Japanese Air Self-Defence Force ***(Fuji)***

1995

Fuji T-5 primary trainer with one Allison 250-B17D turboprop engine *(Fuji)*

1995

Rudder, incl tab 0.66 m² (7.10 sq ft)
Tailplane 3.46 m² (37.24 sq ft)
Elevators (total, incl tabs) 1.39 m² (14.96 sq ft)

WEIGHTS AND LOADINGS (A: Aerobatic, U: Utility):
Weight empty: A, U 1,082 kg (2,385 lb)
Max fuel weight: A, U 644 kg (1,420 lb)
Max T-O weight: A 1,585 kg (3.494 lb)
U 1,805 kg (3,979 lb)
Max wing loading: A 96.06 kg/m² (19.67 lb/sq ft)
U 109.39 kg/m² (22.40 lb/sq ft)
Max power loading: A 6.07 kg/kW (9.98 lb/shp)
U 6.92 kg/kW (11.37 lb/shp)

PERFORMANCE (at max Aerobatic T-O weight except where indicated):
Never-exceed speed (VNE) 223 kts (413 km/h; 256 mph) EAS
Max level speed at 2,440 m (8,000 ft) 193 kts (357 km/h; 222 mph)
Econ cruising speed at 2,440 m (8,000 ft) 155 kts (287 km/h; 178 mph)
Stalling speed, flaps and landing gear down, power off 56 kts (104 km/h; 65 mph)
Max rate of climb at S/L 518 m (1,700 ft)/min
Service ceiling 7,620 m (25,000 ft)
T-O run 302 m (990 ft)
T-O to 15 m (50 ft) 430 m (1,410 ft)
Landing from 15 m (50 ft) 515 m (1,690 ft)
Landing run 174 m (570 ft)
Range with max payload (Utility version), MIL-C-5011A reserves 510 n miles (945 km; 587 miles)

NEW ENTRY

FUJI-BELL UH-1H and ADVANCED 205B

TYPE: Single-engined general purpose helicopters.

PROGRAMME: Fuji (now only production source) manufactures Bell UH-1H under sub-licence from Mitsui and Company, Bell's Japanese licensee; earlier production programme involved 34 Bell 204Bs and 22 Bell 204B-2s. First flight of Fuji built UH-1H, 17 July 1973. Joint Fuji-Bell upgrade of 205 produced Advanced 205B (formerly 205A-1); first flight of prototype (N19AL) in Texas 23 April 1988; demonstrated to US Army that year and in Japan, Far East and Southeast Asia 1989.

VERSIONS: **UH-1H:** Standard production model for JGSDF; *described in detail below*.

Advanced 205B (UH-1J): Upgrade of Bell 205, with UH-1N/212 type tapered rotor blades, Textron Lycoming T53-L-703 engine, 212 type transmission rated at 962 kW (1,290 shp); none yet ordered.

UH-1J: Upgraded UH-1H.

UPGRADES: **Fuji:** UH-1J. See separate entry.

OPERATORS: Versions of the UH-1H/J are in service with the armed forces of Japan (144).

CUSTOMERS: Total of 133 UH-1Hs ordered for JGSDF by March 1992. Further 16 approved for FY93.

DESIGN FEATURES: Same airframe and dynamic components as Bell UH-1H but has tractor tail rotor and Kawasaki built engine; two-blade teetering main rotor; interchangeable metal blades being replaced by new composite blades; stabilising bar above and at right angles to main rotor blades; underslung feathering axis head; two-blade tail rotor; main rotor rpm 294-324. See Versions for Advanced 205B.

FLYING CONTROLS: Duplicated hydraulic fully powered controls with adjustable centring spring on cyclic stick for trimming; small synchronised elevator on rear fuselage, connected to cyclic control to increase allowable CG travel.

STRUCTURE: All-metal, except for new main rotor blades (see Design Features); honeycomb tail rotor blades.

LANDING GEAR: Tubular skid type. Lock-on ground handling wheels and inflatable nylon float bags available.

Fuji (Bell) UH-1H of the JGSDF

1994

Fuji (Bell) UH-1H single-engine general purpose helicopter

1994

POWER PLANT: One 1,044 kW (1,400 shp) Kawasaki built Textron Lycoming T53-K-13B turboshaft, mounted aft of transmission on top of fuselage and enclosed in cowlings. Transmission rating 820 kW (1,100 shp). Five interconnected rubber fuel cells, total capacity 844 litres (223 US gallons; 186 Imp gallons), of which 799 litres (211 US gallons; 176 Imp gallons) are usable. Overload fuel capacity of 1,935 litres (511 US gallons; 425 Imp gallons) usable, obtained by installation of kit comprising two 568 litre (150 US gallon; 125 Imp gallon) internal auxiliary fuel tanks interconnected with basic fuel system.

ACCOMMODATION: Pilot and 11-14 troops, or six stretchers and medical attendant, or 1,759 kg (3,880 lb) of freight. Crew doors open forward and are jettisonable. Two doors each side of cargo compartment, front door hinged to open forward and is removable, rear door slides aft. Forced air ventilation system.

AVIONICS: FM, UHF, VHF radios, IFF transponder, Gyromatic compass system, DF set, VOR receiver and intercom standard.

EQUIPMENT: Standard equipment includes bleed air heater and defroster, comprehensive range of engine and flight instruments, power plant fire detection system, 30V 300A DC starter/generator, navigation, landing and anti-collision lights, controllable searchlight, hydraulically boosted controls. Optional equipment includes external cargo hook, auxiliary fuel tanks, rescue hoist, 150,000 BTU muff heater.

DIMENSIONS, EXTERNAL:
Main rotor diameter 14.63 m (48 ft 0 in)
Tail rotor diameter 2.59 m (8 ft 6 in)
Main rotor blade chord 0.53 m (1 ft 9 in)
Tail rotor blade chord 2.56 m (8 ft 4¾ in)
Length:
overall (main rotor fore and aft) 17.62 m (57 ft 9⅝ in)
fuselage 12.77 m (41 ft 10¼ in)
Height: overall, tail rotor turning (excl fin tip antenna) 4.41 m (14 ft 5½ in)
to top of main rotor head 3.60 m (11 ft 9¾ in)
Tailplane span 2.84 m (9 ft 4 in)
Width over skids 2.91 m (9 ft 6½ in)

DIMENSIONS, INTERNAL:
Cabin: Max width 2.34 m (7 ft 8 in)
Max height 1.25 m (4 ft 1¼ in)
Volume (excl flight deck) approx 6.23 m³ (220 cu ft)

AREAS:
Main rotor disc 168.11 m² (1,809.56 sq ft)
Tail rotor disc 5.27 m² (56.7 sq ft)

WEIGHTS AND LOADINGS:

Weight empty, equipped	2,390 kg (5,270 lb)
Max T-O and landing weight	4,309 kg (9,500 lb)
Max disc loading	25.63 kg/m^2 (5.25 lb/sq ft)
Max power loading	5.26 kg/kW (8.64 lb/shp)

PERFORMANCE (at max T-O weight):

Max level and max cruising speed	110 kts (204 km/h; 127 mph)
Max rate of climb at S/L	488 m (1,600 ft)/min
Service ceiling	3,840 m (12,600 ft)
Hovering ceiling: IGE	4,145 m (13,600 ft)
OGE	335 m (1,100 ft)
Range at S/L	252 n miles (467 km; 290 miles)

VERIFIED

Fuji (Bell) UH-1J upgraded UH-1H helicopter of the Japanese Ground Self-Defence Force *1995*

FUJI (BELL) UH-1J UPGRADE

TYPE: Combat helicopter upgrade (UH-1).

PROGRAMME: First flight of UH-1J 23 April in Texas 1988. Demonstrated in USA in 1988 and in Southeast Asia in 1989. Fuji delivered first UH-1J in September 1993.

DESIGN FEATURES: UH-1J is an upgraded UH-1H, with increased payload/agility with UH-1N/212 type tapered rotor blades. Installation of IR jammer, wire strike protection system, MIL-STD-1553B databus and improved avionics and night operation capability by NVG compatible cockpit.

CUSTOMERS: JGSDF to replace UH-1H with UH-1J, 12 delivered in FY93, 13 in FY94 and expected 13 in FY95.

POWER PLANT: One 1,500 shp Kawasaki built Textron Lycoming T53-K-70 turboshaft. Transmission rated at 962 kW (1,290 shp).

ACCOMMODATION: Crew of two and 13 troops.

DIMENSIONS, EXTERNAL:

Length, overall	17.44 m (57 ft 2½ in)
Width, overall	2.86 m (9 ft 4½ in)
Height, overall	4.39 m (14 ft 4¾ in)

WEIGHTS AND LOADINGS:

Internal	4,763 kg (10,500 lb)
External	5,080 kg (11,200 lb)

PERFORMANCE:

Max speed	130 kts (240 km/h; 150 mph)

VERIFIED

KANEMATSU

KANEMATSU CORPORATION

2-1 Shibaura 1-chome, Minato-ku, Tokyo 105-05
Telephone: 81 (3) 5440 8710
Fax: 81 (3) 5440 6525
Telex: 22333/4
PRESIDENT: Masao Yosomiya
SALES DIVISION MANAGER: Yoshihiro Miwa

Kanematsu is prime contractor for outfitting Raytheon Hawker 800s (formerly BAe 125 Corporate 800) to JASDF specifications. Full description of standard aircraft under Raytheon Corporate Jets; Fuji is system integrator and responsible for U-125/U-125A maintenance.

UPDATED

Kanematsu (Raytheon) Hawker 800 (U-125) Navaid upgrade to replace the JASDF MU-2J aircraft *1994*

KANEMATSU (RAYTHEON) HAWKER 800 NAVAID UPGRADE

TYPE: Twin-turbofan aircraft for navaid calibration and SAR conversion.

PROGRAMME: Raytheon Hawker 800 (see UK section) selected under JASDF FC-X programme to replace Mitsubishi MU-2J in navaid calibration, and also selected under JASDF HS-X programme to replace MU-2S in SAR roles respectively. Kanematsu Corporation acting as prime contractor, with Fuji as systems integrator, outfitting aircraft to JASDF specification; first U-125 delivered to JASDF 18 December 1992. The third U-125 aircraft was delivered 1 November 1994. The first U-125A aircraft was delivered to JASDF at the end of February 1995.

VERSIONS: **U-125**: For navaid flight check role, to replace MU-2J.

U-125A: Search and rescue version, to replace MU-2S; 360° search radar, FLIR, airdroppable marker flares and rescue equipment.

CUSTOMERS: JASDF (three U-125 and three U-125A).

UPDATED

Kanematsu (Raytheon) Hawker 800 (U-125A) SAR upgrade to replace the JASDF MU-2S aircraft ***(Kanematsu)*** *1995*

KAWASAKI

KAWASAKI JUKOGYO KABUSHIKI KAISHA (Kawasaki Heavy Industries Ltd)

Kobe Crystal Tower, 1-3 Higashi-Kawasaki-Cho 1-chome, Chuo-ku, Kobe
TOKYO AND AEROSPACE GROUP OFFICE: World Trade Center Building, 4-1 Hamamatsu-Cho 2-chome, Minato-ku, Tokyo 105
Telephone: 81 (3) 3435 1595
Fax: 81 (3) 3436 3037
Telex: 242-4371 KAWAJU J
PRESIDENT: Hiroshi Ohba

Aerospace Group

EXECUTIVE MANAGING DIRECTOR AND GROUP SENIOR GENERAL MANAGER: Ryozo Tsutsui
WORKS: Gifu, Nagoya 1 and 2, Akashi, Seishin and Harima

Kawasaki Aircraft Company built many US aircraft under licence from 1955; amalgamated with Kawasaki Dockyard Company and Kawasaki Rolling Stock Manufacturing Company to form Kawasaki Heavy Industries Ltd 1 April 1969; Aerospace Group employs some 4,500 people; Kawasaki has 25 per cent holding in Nippi (which see).

Kawasaki is currently prime contractor on T-4 programme and for OH-X new small observation helicopter; co-developer and co-producer, with Eurocopter, of BK 117 helicopter; manufactures MD 500 helicopters under licence (344 delivered by March 1994); is prime contractor for Japanese licence production of P-3C/Update II and III for JMSDF and CH-47 Chinooks for JGSDF and JASDF. Subcontract work includes rear fuselages, wings and tail units for Mitsubishi built F-15 Eagles; and forward and centre-fuselage panels and wing ribs for Boeing 767. Nominated as prime contractor for maintenance and support of JASDF E-2C Hawkeyes and C-130 Hercules. Undertaking feasibility study for JDA for future C-1/C-130 transport replacement.

Kawasaki also extensively involved in satellites and launch vehicles, and HOPE orbiting spaceplane to operate with NASA Space Station; is member of International Aero Engines consortium and produces Lycoming T53 and T55 engines under licence; overhauls engines; and builds hangars, docks, passenger bridges and similar airport equipment.

UPDATED

KAWASAKI C-1

TYPE: Medium-range twin-turbofan transport aircraft.

PROGRAMME: The C-1 medium-sized troop and freight transport was designed to meet the JASDF's requirement for a replacement for its former fleet of Curtiss C-46s. Preliminary design, and the construction of two XC-1 flying prototypes (plus one airframe for static tests), was

Kawasaki C-1 medium-range twin-turbofan transport aircraft

1995

undertaken by NAMC. The first flying prototype, assembled at Kawasaki's Gifu factory, made its first flight on 12 November 1970, followed by the second on 16 January 1971. These were handed over to the JDA on 24 February and 20 March 1971 respectively. Further development and evaluation tests by the JDA were completed in March 1973.

Two preproduction aircraft had been delivered by the end of February 1974. Airframe fatigue testing by the JDA was completed in November 1974.

Eleven production C-1s were delivered between December 1974 and the end of March 1976. An additional 13 C-1s were ordered during FY75; these had all been delivered by the end of February 1978, including two long-range aircraft with an additional 4,732 litre (1,250 US gallon; 1,042 Imp gallon) wing centre-section fuel tank. Two more long-range C-1s were ordered during FY77, and one in FY79; production ended in 1981 with the delivery of the 31st aircraft. All C-1s were camouflaged by 1982.

VERSIONS: **C-1**: Standard version to which the following description applies.

EC-1: Under a March 1983 Defence Agency contract, Kawasaki modified the 21st production C-1 (JASDF 78-1021) for use as an ECM training aircraft. The aircraft first flew on 3 December 1984 and was handed over to the JASDF's Air Proving Wing in late January 1985 for evaluation. Transferred to the Electronic Warfare Training Unit in June 1986. Equipped with a domestically built TRDI/Mitsubishi Electric XJ/ALQ-5 ECM system. The EC-1 has a large bulbous nose and tail radomes, a blister on each side of the forward fuselage aft of the flight deck side windows, and blisters on each side of the rear fuselage and two antennae beneath the fuselage.

UPGRADES: **Kawasaki**: EC-1. See Versions.

OPERATORS: Versions of the C-1 are in service with the armed forces of Japan.

DESIGN FEATURES: Cantilever high-wing monoplane. Wings have 20° sweepback at quarter-chord, with slightly increased leading-edge sweep inboard of engine pylons. Thickness/chord ratio 12 per cent at root, 11 per cent at tip. Anhedral 5° 30′ from centre-section. The tail unit is an alloy cantilever T tail with sweepback on all surfaces (30° at fin quarter-chord, 25° at tailplane quarter-chord). Tailplane has 5° anhedral. Variable incidence tailplane.

FLYING CONTROLS: Two quadruple-slotted flaps on each trailing-edge. Forward of these, on each wing, are three flight spoilers and a ground spoiler. Drooping leading-edge slats in four sections on each wing. Manually operated aileron outboard of each outer flap. Trim tab in port aileron. Tailplane with elevators. Balance tab in each elevator and anti-balance tab in rudder. Elevators and rudder are each operated by two independent hydraulic actuator systems; the elevators can be operated manually in an emergency.

STRUCTURE: Conventional two-spar fail-safe structure of aluminium alloy, including control surfaces. The fuselage is a conventional semi-monocoque fail-safe circular cross-section structure of aluminium alloy. Aluminium alloy tail unit.

LANDING GEAR: Hydraulically retractable tricycle type, of Sumitomo design. Each main unit has two pairs of wheels in tandem, retracting forward into fairings built on to the sides of the fuselage. Forward-retracting nose unit has twin wheels. Mainwheel tyres size 35 × 10.7-16, nosewheel tyre size 28 × 9-12; tyre pressure (all units) 6.21 bars (90 lb/sq in). Oleo shock-absorbers. Kayaba wheels with Dunlop tyres, which on main units have pressure of 5.17 bars (75 lb/sq in). Kayaba hydraulic brakes (two-rotor on first 10 production aircraft, three-rotor from 11th aircraft onward); Sumisei anti-skid units.

POWER PLANT: Two 64.5 kN (14,500 lb st) Mitsubishi (Pratt & Whitney) JT8D-M-9 turbofan engines, installed in pylon-mounted underwing pods and fitted with thrust reversers. Four integral wing fuel tanks with total capacity of 15,200 litres (4,012 US gallons; 3,344 Imp gallons). Single pressure refuelling point for all tanks, plus overwing gravity refuelling point for each tank.

ACCOMMODATION: Crew of five, comprising pilot, co-pilot, navigator, flight engineer and load supervisor. Escape hatch in flight deck on starboard side. Flight deck and main cabin pressurised and air conditioned. Standard complements are as follows: troops (maximum) 60, paratroops (maximum) 45, stretchers 36 plus attendants. As a cargo carrier, loads can include a 2½ ton truck, a 105 mm howitzer, two ¾ ton trucks or three Jeeps. Up to three preloaded freight pallets, 2.24 m (7 ft 4 in) wide and 2.74 m (9 ft 0 in) long, can be carried. Floor is stressed for loads of up to 7 kg/cm² (100 lb/sq in). Access to flight deck via downward-opening door, with built-in stairs, on port side of forward fuselage. Paratroop door on each side of fuselage, aft of wing trailing-edge. For airdropping, the rear-loading ramp/door can be opened in flight to the full cabin cross-section.

SYSTEMS: Pressurisation and air conditioning systems utilise engine bleed air. APU in front section of starboard landing gear fairing supplies electrical power on ground and in the air, and bleed air on ground. Three independent hydraulic systems, each 207 bars (3,000 lb/sq in). No. 1 system actuates flight controls; No. 2 system actuates flight controls, high-lift devices, landing gear, nosewheel steering and brakes; No. 3 system is used for cargo door operation, and provides emergency back-up for brakes and high-lift devices. Electrical power supplied by three 40 kVA AC generators, two engine-driven and air-cooled and one driven by the APU. 28 V DC power is obtained from AC source through a transformer-rectifier. One 24 V 30 Ah Ni/Cd battery for emergency DC power.

AVIONICS AND EQUIPMENT: Standard avionics include autopilot, Doppler radar, radio altimeter, HF, VHF and UHF radio, ADF, UHF/DF, marker beacon, VOR/ILS, Tacan, SIF, dual compass system and flight director system. Optional avionics include Loran and weather radar.

DIMENSIONS, EXTERNAL:

Wing span	30.60 m (100 ft 4¾ in)
Wing chord: at root	6.30 m (20 ft 8 in)
at tip	2.00 m (6 ft 6¾ in)
Wing aspect ratio	7.8
Length overall	29.00 m (95 ft 1¾ in)
Length of fuselage	26.50 m (86 ft 11¼ in)
Height overall	9.99 m (32 ft 9¼ in)
Tailplane span	11.30 m (37 ft 1 in)
Wheel track	4.40 m (14 ft 5¼ in)
Wheelbase	9.33 m (30 ft 7¼ in)
Rear-loading ramp/door: Length	2.67 m (8 ft 9¼ in)
Width	2.70 m (8 ft 10¼ in)
Height to sill	1.25 m (4 ft 1¼ in)

DIMENSIONS, INTERNAL:

Cabin: Max length	10.80 m (35 ft 5¼ in)
Max width	3.60 m (11 ft 9¾ in)
Max height	2.55 m (8 ft 4½ in)

Kawasaki EC-1 ECM training aircraft of the JASDF ***(Peter Steinemann)***

1995

Kawasaki C-1 transport aircraft with two Mitsubishi (Pratt & Whitney) JT8D-M-9 turbofan engines *(Peter Steinemann)*

1995

Floor area	28.6 m² (308 sq ft)
Volume (excl ramp area)	73.8 m³ (2,606 cu ft)

AREAS:

Wings, gross	120.5 m² (1,297 sq ft)
Ailerons (total)	3.4 m² (36.6 sq ft)
Trailing-edge flaps (total)	22.9 m² (246.5 sq ft)
Spoilers (total)	8.9 m² (95.8 sq ft)
Fin	15.8 m² (170.1 sq ft)
Rudder, incl tabs	6.4 m² (68.9 sq ft)
Tailplane	18.3 m² (197.0 sq ft)
Elevators, incl tabs	6.5 m² (70.0 sq ft)

WEIGHTS AND LOADINGS:

Weight empty	23,320 kg (51,410 lb)
Weight empty, equipped	24,300 kg (53,572 lb)
Normal payload	7,900 kg (17,416 lb)
Max payload	11,900 kg (26,235 lb)
Normal T-O weight	38,700 kg (85,320 lb)
Max T-O weight	45,000 kg (99,210 lb)
Max wing loading	373.4 kg/m² (76.48 lb/sq ft)
Max power loading	350 kg/kN (3.42 lb/lb st)

PERFORMANCE (at normal T-O weight except where indicated):

Max level speed at 7,620 m (25,000 ft) at 35,450 kg (78,150 lb) AUW	435 kts (806 km/h; 501 mph)
Econ cruising speed at 10,670 m (35,000 ft) at 35,450 kg (78,150 lb) AUW	354 kts (657 km/h; 408 mph)
Max rate of climb at S/L	1,065 m (3,500 ft)/min
Service ceiling	11,580 m (38,000 ft)
Service ceiling, one engine out	5,485 m (18,000 ft)
T-O run	640 m (2,100 ft)
T-O to 15 m (50 ft)	910 m (3,000 ft)
Landing from 15 m (50 ft) at 36,860 kg (81,260 lb) weight	823 m (2,700 ft)
Landing run at 36,860 kg (81,260 lb) weight	455 m (1,500 ft)
Range with max fuel and 2,200 kg (4,850 lb) payload	1,810 n miles (3,353 km; 2,084 miles)
Range with 7,900 kg (17,416 lb) payload	700 n miles (1,300 km; 807 miles)

NEW ENTRY

KAWASAKI P-2J

JMSDF designation: P-2J

TYPE: Four-engined anti-submarine and maritime patrol aircraft.

PROGRAMME: The P-2J was developed by Kawasaki, originally under the designation GK-210, to meet a JMSDF requirement for a new anti-submarine aircraft to replace its P2V-7 Neptunes in service during the 1970s. Design, based very closely upon that of the P2V-7 (P-2H), began in October 1961. Work on the conversion of a standard P2V-7 as the P-2J prototype began in June 1965, and this aircraft flew for the first time on 21 July 1966.

The first production P-2J was flown on 8 August 1969, and was delivered to the JMSDF on 7 October. A further 81 P-2Js were built, and the last was delivered during March 1979. There are currently two P-2Js still in service with the JMSDF.

VERSIONS: **P-2J:** Initial version for ASW and maritime patrol duties. *Description refers to this version.* Withdrawn from service.

EP-2J: Converted P-2J with equipment for electronic intelligence (elint) duties. Equipment includes: HLR-105 and HLR-106 systems. Withdrawn from service.

UP-2J: Converted P-2J with equipment for target-towing, ECM training and drone launch. Withdrawn from service.

UPGRADES: **Kawasaki:** Electronics modernisation programme undertaken in the mid-1970s. See Electronics.

OPERATORS: Versions of the P-2J are no longer in service with the armed forces of Japan.

DESIGN FEATURES: Cantilever all-metal mid-wing monoplane with taper on outer panels. Wing section NACA 2419 (modified) at root, NACA 4410.5 at tip. Dihedral 5° on outer panels. Incidence 3° 30′ at root. No sweepback. Wing designed to give temporary flotation in event of ditching. Tail unit is virtually unchanged from P2V-7 except for an increase in rudder area by extending the chord by 0.3 m (1 ft) at the top.

FLYING CONTROLS: All-metal ailerons, each incorporating a spring and trim tab. Fowler type all-metal inboard and outboard trailing-edge flaps. All-metal two-section spoilers in upper surface of outer wing panels, inboard of ailerons.

STRUCTURE: The wing has centre-section box beam continuous through fuselage. The fuselage is a conventional all-metal unpressurised semi-monocoque structure, basically as P2V-7 (P-2H) but with extra 1.27 m (4 ft 2 in) section inserted between wing leading-edge and cockpit to house improved electronic equipment. The tail unit is an all-metal structure. Tail unit incorporates 'Varicam' (variable camber), a movable trimming surface between the fixed tailplane and each elevator which is operated by hydraulically driven screwjack. Spring tab and trim tab in rudder, balance tab and spring tab in each elevator.

LANDING GEAR: Retractable tricycle type, with single steerable nosewheel and twin-wheel main units. Hydraulic retraction, nosewheel rearward, mainwheels forward into inboard engine nacelles. Sumimoto Precision oleo-pneumatic shock-absorbers. Goodyear Type VII tubeless tyres on all units, size 9.9 × 34-14 on nosewheel and 13 × 39-16 on mainwheels. Tyre pressures 6.3 kg/cm² (90 lb/sq in) on nosewheel, 7.0 kg/m² (100 lb/sq in) on mainwheels. Goodyear disc brakes and on/off-type anti-skid units on main units.

POWER PLANT: Two 2,850 ehp Japanese built General Electric T64-IHI-10 turboprop engines, mounted on wing centre-section and each driving a Sumimoto Precision 63E60-19 three-blade variable-pitch metal propeller. Outboard of these, on underwing pylons, are two pod-mounted Ishikawajima-Harima J3-IHI-7C turbojets, each rated at 1,400 kg st (3,085 lb). Fuel in inboard and outboard wing tanks with total capacity of 11,433 litres (3,020 US gallons; 2,515 Imp gallons), plus 1,514 litres (400 US gallons; 333 Imp gallons) in port wingtip tank. For Ferry purposes a 2,650 litre (700 US gallon; 583 Imp gallon) auxiliary tank can be installed in the weapons bay. Oil capacity 23.6 litres (6.2 US gallons; 5.2 Imp gallons) for each turboprop and 11 litres (2.9 US gallons; 2.4 Imp gallons) for each turbojet engine.

ACCOMMODATION: Crew of 12, including two pilots on flight deck, seven men in tactical compartment in forward fuselage and three aft of centre-section wing box beam. Aft of tactical compartment, in centre-fuselage are an ordnance room, galley and toilet. Crew escape hatches in flight deck, tactical and ordnance compartments. All accommodation heated, ventilated and air conditioned.

SYSTEMS: Primary hydraulic system, pressure 210 kg/m² (3,000 lb/sq in), for Varicam (tail unit), landing gear and nosewheel steering. Secondary system, pressure 105 kg/m² (1,500 lb/sq in), for flaps spoilers, jet pod doors, mainwheel brakes and propeller braking. Two 40kVA generators provide 115/200 V AC power at 400 Hz. DC power from three 28 V 200 A transformer-rectifiers. Thermal de-icing of wing leading-edges. Tail unit has thermal de-icing.

ELECTRONICS (originally fitted): Communications and navigation equipment comprised HIC-3 intercom system, AN/ARC-552 UHF transceiver, HRC-6 VHF transceiver, two HRC-7 HF transceivers with N-CU-58/HRC couplers, HGC-1 teletypewriter with N-CV-55/HRC converter, RRC-15 HF emergency radio set, AN/APN-153(V) Doppler navigation system, AN/AYK-2 navigation computer, N-PT-3 plotter, N-OA-35/HSA tactical plotter, HRN-4 Loran, two HRN-2 ADF, AN/ARA-50 UHF/DF, AN/ARN-52(V) Tacan, AN/APN-171(V) radar altimeter, HRN-3 marker beacon receiver, N-O-41 altitude warning oscillator, and PB-60J autopilot. ASW equipment included AN/APS-80(J) search radar, AN/APA-125A radar display, HLR-1 ECM, AN/ASQ-10A MAD, AN/ASR-3 trail detector, two AN/ARR-52A(V) sonobuoy receivers, AQA-1 sonobuoy indicator, AQA-5 Jezebel recorder, AN/ASA-20B Julie recorder, N-R-86/HRA OTPI, AN/ASA-16 tactical display system, AN/APX-68 SIF transponder, HPX-1 IFF interrogator, N-KY-22/HPX IFF decoder, AN/ASA-50 ground speed and bearing computer, N-RO-5/HMH BT recorder, HQH-101 ionisation data recorder, and HSA-1 sonobuoy data display system. Search light in starboard wingtip pod.

ELECTRONICS (modernised aircraft): Communications and navigation equipment comprises HIC-3 interphone, HRC-110 UHF transceiver, HRC-106 VHF transceiver, two HRC-107 HF transceivers, two N-CU-58/HRC HF antenna couplers, HGC-102 teletypewriter, HGA-101 TTY converter, RRC-15 emergency radio, AN/APN-187-N1 Doppler radar, AN/AYK-2 navigation computer, N-PT-3 navigation plotter, N-OA-35/HSA tactical plotter, HRN-104 Loran, HRN-101 ADF, AN/ARA-50 UHF/DF, HRN-105 Tacan, AN/APN-171-N1 radar altimeter, HRN-106 VOR/ILS receiver, and PB-60J autopilot. ASW equipment includes AN/APS-80-N search radar, AN/APA-125-N radar indicator, HLR-101 ECM, HSQ-101 MAD, HSA MAD compensator, HSA-103 SAD, AN/ASA-20B Julie recorder, AN/AQA-5-N Jezebel recorder, HQA-101 active sonobuoy indicator, AN/ARR-52A(V) sonobuoy receiver, N-R-86/HRA OTPI, AN/ASA-16 integrated tactical display, HQH-101 tape recorder, AN/APX-68-N SIF transponder, HPX-101 IFF interrogator, N-KY-22/HPX IFF decoder, AN/ASA-50 ground speed tactical bearing computer and HSA-1 sonobuoy data display system. Searchlight in starboard wingtip pod.

Kawasaki P-2J of the Japanese Maritime Self-Defence Force (JMSDF) *(Peter Steinemann)*

1994

DIMENSIONS, EXTERNAL:

Wing span	29.78 m (97 ft 8½ in)
Wing span over tip tanks	30.87 m (101 ft 3½ in)
Wing chord: at root	4.45 m (14 ft 7½ in)
at tip	2.22 m (7 ft 3½ in)
Wing aspect ratio	10
Length overall	29.23 m (95 ft 10¾ in)
Height overall	8.93 m (29 ft 3½ in)
Tailplane span	10.36 m (34 ft 0 in)
Wheel track (c/l of shock struts)	7.62 m (25 ft 0 in)
Wheelbase	8.84 m (29 ft 0 in)
Propeller diameter	4.43 m (14 ft 6½ in)
Distance between propeller centres	7.62 m (25 ft 0 in)

DIMENSIONS, INTERNAL (tactical compartment):

Cabin: Length	5.49 m (18 ft 0 in)
Max width	2.03 m (6 ft 8 in)
Max height (at c/l)	1.55 m (5 ft 1 in)
Floor area	11.15 m^2 (120 sq ft)
Volume	14.16 m^3 (500 cu ft)

AREAS:

Wings, gross	92.9 m^2 (1,000 sq ft)
Ailerons (total)	5.87 m^2 (63.2 sq ft)
Trailing-edge flaps (total)	16.70 m^2 (179.6 sq ft)
Spoilers (total)	1.41 m^2 (15.2 sq ft)
Fin	17.65 m^2 (190.0 sq ft)
Rudder, incl tabs	4.04 m^2 (43.5 sq ft)
Tailplane	21.50 m^2 (231.0 sq ft)
Elevators, incl tabs and Varicam	8.45 m^2 (91.0 sq ft)

WEIGHTS AND LOADINGS:

Weight empty	19,277 kg (42,500 lb)
Max T-O weight	34,019 kg (75,000 lb)
Max zero-fuel weight	23,087 kg (50,900 lb)
Max landing weight	28,122 kg (62,000 lb)
Max wing loading	366 kg/m^2 (75.00 lb/sq ft)
Max power loading	5.97 kg/ehp (13.16 lb/ehp)

PERFORMANCE (at max T-O weight):

Max never-exceed speed	350 kts (649 km/h; 403 mph)
Max cruising speed	217 kts (402 km/h; 250 mph)
Econ cruising speed at 3,050 m (10,000 ft)	200 kts (370 km/h; 230 mph)
Stalling speed, flaps down	90 kts (166 km/h; 103 mph)
Max rate of climb at S/L	550 m (1,800 ft)/min
Service ceiling	9,150 m (30,000 ft)
T-O to 15 m (50 ft)	1,100 m (3,600 ft)
Landing run from 15 m (50 ft)	880 m (2,880 ft)
Range with max fuel	2,400 n miles (4,450 km; 2,765 miles)

VERIFIED

KAWASAKI (LOCKHEED) P-3C ORION

TYPE: Land-based maritime patrol and ASW aircraft.

PROGRAMME: Kawasaki is prime contractor for JMSDF Orions; first of four P-3Cs assembled by Kawasaki from US-supplied knocked-down components flown 17 March 1982 and delivered 26 May to Fleet Squadron 51 at Atsugi Air Base; production continues.

VERSIONS: **P-3C/Update II and III:** Versions built under licence in Japan; Japan Defence Agency plans to modernise current Update II configuration to Update III (see Lockheed) from 1996.

EP-3: Two Orions, ordered FY87 and FY88, equipped for electronic surveillance; first flight (9171) October 1990; delivered March and November 1991; third ordered in FY92; fourth ordered in FY93 and fifth planned in FY94 or FY95; NEC and Mitsubishi Electric low- and high-frequency detector systems.

UP-3C: One flying testbed for JMSDF; ordered in FY91. First flight November 1994; delivery in February 1995.

UP-3D: Two ECM trainers for JMSDF; to be ordered during FY94-95.

Proposed Variants: Ocean surveillance, military transport and systems testbed variants proposed to Japan Defence Agency.

UPGRADES: **Lockheed Update II and III:** See above and Lockheed entry.

OPERATORS: Versions of the Kawasaki (Lockheed) P-3C Orion are in service with the armed forces of Japan.

CUSTOMERS: 109 to be purchased by JMSDF, of which 105 ordered by FY93, one more P-3C and one more UP-3D approved for FY94; first three (US built) P-3Cs handed over to JMSDF in April 1981; 98 Kawasaki built P-3Cs delivered by 31 March 1995; two squadrons of 10 aircraft each based at Atsugi, two at Hachinohe, two at Kanoya, one at Shimofusa, one at Iwakuni and two at Naha.

STRUCTURE: Kawasaki builds centre-fuselages and is responsible for final assembly and flight testing; Fuji, Mitsubishi, Nippi and ShinMaywa participate in production of airframe; IHI manufactures Allison T56-IHI-14 engines.

VERIFIED

Kawasaki P-2J anti-submarine and maritime patrol aircraft *(Peter Steinemann)*

1994

Kawasaki (Lockheed) P-3C Orion maritime patrol aircraft of the JMSDF

1994

Kawasaki (Lockheed) EP-3 electronic surveillance aircraft

1994

MITSUBISHI

MITSUBISHI JUKOGYO KABUSHIKI KAISHA (Mitsubishi Heavy Industries Ltd)

5-1 Marunouchi 2-chome, Chiyoda-ku, Tokyo 100
Telephone: 81 (3) 3212 3111
Fax: 81 (3) 3212 9865
Telex: J22443

NAGOYA AEROSPACE SYSTEMS WORKS: 10 Oye-cho, Minato-ku, Nagoya 455

PRESIDENT: Kentaro Aikawa

EXECUTIVE VICE-PRESIDENTS:
Yoshitake Makise
Hideo Hirotsu

MANAGING DIRECTOR AND GENERAL MANAGER (Aircraft and Special Vehicle Headquarters): Yutaka Hineno

GENERAL MANAGER, AIRCRAFT DEPARTMENT: Isao Fukuchi

Mitsubishi began the production of aircraft in 1921, and manufactured 18,000 aircraft of approximately 100 different types prior to 1945, as well as 52,000 engines in the 1,000-2,500 hp range. The present Komaki South plant was built in 1952 and, together with the Oye and Komaki North plants, was later consolidated as Nagoya Aircraft Works, with a combined floor area of 552,463 m^2 (5,946,666 sq ft).

Mitsubishi developed the YS-11, MU-2, MU-300 and the CCV T-2 and is prime contractor for the Japanese F-4E and F-15, and for the FS-X. It was prime contractor for the T-2 supersonic trainer and F-1 close air support aircraft for the JASDF, with Fuji, Nippi and ShinMaywa as principal subcontractors; and is currently producing forward and rear fuselages for JMSDF Lockheed P-3C Orions, under subcontract to Kawasaki (which see). Other subcontract work includes manufacture of rear passenger cabin sections of the Boeing 767 jet transport. Part of this work is, in turn, subcontracted by Mitsubishi to ShinMaywa. Mitsubishi also manufactures tailcones for the McDonnell Douglas MD-11.

The company also produces aero-engines and participates in the H-I and H-II launchers and the Japanese Experimental Module for the US Space Station and for the HOPE orbiting spaceplane.

The Japan Defence Agency has authorised service introduction of the AAM-3 short-range air-to-air missile developed by Mitsubishi with NEC seeker and proximity fuze and Komatsu warhead. AAM-3 will replace the AIM-9 on F-15Js and F-4EJs.

VERIFIED

MITSUBISHI MU-2

TYPE: Twin-turboprop utility transport.

PROGRAMME: The MU-2 is a twin-turboprop STOL utility transport aircraft, the basic design of which was begun in 1960. Prototype construction began in 1962 and the first aircraft was flown on 14 September 1963.

VERSIONS: **MU-2A:** Three built. Powered by two Turbomeca Astazou IIK turboprop engines.

MU-2B: 34 built. Powered by AiResearch TPE331-25AB turboprop engines. Otherwise generally similar to MU-2A. First MU-2B flew on 11 March 1965. Japanese Type Approval received on 15 September 1965. FAA Type Approval on 4 November 1965.

MU-2C: Four built. Unpressurised liaison and reconnaissance/support version for JGSDF. First flown on 11 May 1967, and first production aircraft delivered on 30 June 1967. Wingtip fuel tanks replaced by fuselage tank aft of cabin. One vertical and one swing-type oblique camera in photographic version. Optional equipment included two 13 mm nose guns, bombs and rockets.

MU-2D: 18 built. Superseded MU-2B as standard commercial model. First flown on 5 March 1966. Differs from MU-2B only in having integral fuel tanks instead of bladder type, and additional metal tanks in outer wings.

MU-2E: 16 built. Search and rescue version for JASDF. Doppler radar in nose 'thimble', bulged observation window on each side of rear fuselage, and port side sliding door to allow lifeboat to be dropped. Unpressurised cabin. Extensive nav/com equipment, increased fuel capacity of 1,389 litres (367 US gallons; 305.6 Imp gallons) and maximum T-O weight increased to 4,560 kg (10,053 lb). First flight on 15 August 1967 (designation **MU-2S**).

MU-2F: 95 built. Commercial version developed from MU-2D, with uprated AiResearch TPE331-1-151A turboprop engines of 705 ehp and extra fuel in enlarged wingtip tanks. First flown on 6 October 1967, MU-2F received its JCAB Type Certificate on 3 April 1968 and FAA Type Certificate on 6 August 1968.

MU-2G: 46 built. 'Stretched' version, developed from MU-2F and retaining the same power plant. Interior cabin length increased by 2.78 m (9 ft 1½ in), and overall length by 1.90 m (6 ft 2¾ in), with added external fairings on fuselage sides into which the main landing gear units retract. Nosewheel unit repositioned further aft and retracts forward instead of rearward as on other MU-2 models. Other modifications include increase in fin height and area. Prototype (JA 8737) first flown on 10 January 1969; JCAB Type Certificate received on 2 June 1969 and FAA Type Certificate on 14 July 1969.

MU-2J: 108 built. Basically similar to MU-2G, but with more powerful AiResearch TPE331-6-251M turboprop engines and maximum cruising speed of 324 knots (600 km/h; 373 mph). JCAB certification received early 1971. First flown August 1970.

MU-2K: 83 built. Development of MU-2F, with same power plant as MU-2J. First flown in August 1971. JCAB certification February 1972. FAA Type Approval May 1972.

MU-2L: 36 built. Basically similar to MU-2J, but with maximum T-O weight increased to 5,250 kg (11,575 lb), representing a net increase of more than 272 kg (600 lb) in useful load. JCAB certification June 1974, FAA Type Approval July 1974.

MU-2M: 27 built. Basically similar to MU-2K, but with maximum T-O weight increased to 4,750 kg (10,470 lb). JCAB and FAA certification dates as for MU-2L.

MU-2N: 36 built. Basically similar to MU-2L, but with AiResearch TPE331-5-252M turboprop engines and four-blade propellers. Received FAA Type Approval 1977.

MU-2P: 31 built. Basically similar to MU-2M, but with AiResearch TPE331-5-252M turboprop engines and four-blade propellers. Received FAA Type Approval January 1977.

Marquise: Over 130 built by MAI in the USA. Basically similar to the MU-2N, but with Garrett TPE331-10-501M turboprop engines, four-blade propellers and increased fuel capacity. Manufacturer's designation MU-2B-60. First flown on 13 September 1977. Certificated 2 March 1978.

Solitaire: Over 60 built by MAI in the USA. Basically similar to the MU-2P, but with Garrett TPE331-10-501 turboprop engines, rated in this installation at 495.5 kW (665 shp) each. Propeller and fuel capacity details as for Marquise. Manufacturer's designation MU-2B-40. First flown 28 October 1977. Certificated 2 March 1978.

OPERATORS: Versions of the MU-2 are in service with the armed forces of Japan (48).

The following description applies generally to the MU-2J, K, L, M, N and P, except where a specific version is indicated.

DESIGN FEATURES: Cantilever high-wing monoplane. Wing section NACA 64A415 at root, NACA 63A212 (modified) at tip. No dihedral. Incidence 2°. Washout 3°. Sweepback 0° 21′ at quarter-chord. Small auxiliary fin beneath each side of rear fuselage on MU-2J, L and N. Cantilever tail unit.

FLYING CONTROLS: Spoilers for lateral control, between rear spar and flaps. Electrically actuated full-span double-slotted Fowler type flaps of aluminium alloy and plastics construction. Outboard flap section each side incorporates trim aileron. All primary controls manually operated. Trim tab in rudder and each elevator.

STRUCTURE: The wing is a one-piece two-spar all-metal structure with chemically milled skins of 2024 and 7075 aluminium alloy. The fuselage is a circular-section aluminium alloy semi-monocoque structure. The tail unit is a cantilever structure of aluminium alloy, except for top of fin, which is of reinforced plastics.

LANDING GEAR (MU-2J, L and N): Retractable tricycle type, with single-wheel on each main unit and twin-wheel steerable nose unit. All wheels retract electrically, nosewheel forward, mainwheels upward into fairings on fuselage sides. Manual back-up system provided. Oleo-pneumatic shock-absorbers. Mainwheel tyres Type III, size 8.50-10 (10 ply). Nosewheel tyres Type III, size 5.00-5 (6 ply). Tyre pressure 2.76-4.76 bars (40-69 lb/sq in) on main units, 3.79 bars (55 lb/sq in) on nose unit. Goodrich single-disc nine-spot hydraulic brakes.

LANDING GEAR (MU-2K, M and P): Retractable tricycle type, with single-wheel on each main unit and twin-wheel steerable nose unit. All wheels retract electrically into fuselage. Oleo-pneumatic shock-absorbers. Mainwheel tyres Type III 8.50-10 (10 ply). Nosewheel tyres Type III 5.00-5 (6 ply). Mainwheel tyre pressure 2.76-4.21 bars (40-61 lb/sq in), nosewheel tyre pressure 3.79 bars (55 lb/sq in). Goodrich single-disc nine-spot hydraulic brakes.

POWER PLANT (MU-2J, K, L and M): Two AiResearch TPE331-6-251M turboprop engines, each rated at 540 kW (724 ehp) in MU-2J, K and M; and 579 kW (776 ehp) in MU-2L. Hartzell HC-B3TN-5/T10178HB-11 or 11R fully feathering three-blade reversible-pitch constant-speed propellers. Fuel in wing tanks, total usable capacity 705 litres (186 US gallons; 155 Imp gallons) and two fixed wingtip tanks with total usable capacity of 682 litres (180 US gallons; 150 Imp gallons). Maximum total fuel capacity of 1,387 litres (366 US gallons; 305 Imp gallons). Oil capacity 11.8 litres (3.1 US gallons; 2.6 Imp gallons).

POWER PLANT (MU-2N and P): Two AiResearch TPE331-5-252M turboprop engines, each rated at 579 kW (776 ehp) in MU-2N and 540 kW (724 ehp) in MU-2P. Hartzell HC-B4TN-5DL/LT10282B-5.3R four-blade fully feathering constant-speed reversible-pitch propellers. Fuel in wing tanks, total usable capacity 697 litres (184 US gallons; 153.5 Imp gallons) and two fixed wingtip tanks with total usable capacity of 682 litres (180 US gallons; 150 Imp gallons). Maximum total fuel capacity 1,379 litres (364 US gallons; 303.5 Imp gallons). Oil capacity 11.8 litres (3.1 US gallons; 2.6 Imp gallons).

ACCOMMODATION (MU-2J, L and N): Seats for pilot and co-pilot or passenger on flight deck. Seating in main cabin, on rearward- and forward-facing seats, for four to 11 persons. Separate compartment at rear of cabin provides coat locker, toilet and baggage compartment. Door at rear of cabin on port side with built-in steps. Emergency exit door under wing on starboard side.

ACCOMMODATION (MU-2K, M and P): Seats for pilot and co-pilot or passenger on flight deck. Typical seating for five passengers in main cabin, two on individual rearward-facing seats, three on forward-facing rear bench seat. Optional arrangements for up to nine persons, including pilot. Pressurised baggage compartment over mainwheel bays, capacity 100 kg (220 lb). Non-pressurised baggage compartment aft of mainwheel bays, capacity 70 kg (154 lb). Space for coats and small baggage at rear of cabin. Door under wing on port side. Emergency exit door opposite main door.

SYSTEMS: Air cycle pressurisation and air conditioning system. Differential 0.34 bars (5.0 lb/sq in) in MU-2J and K; 0.41 bars (6.0 lb/sq in) in MU-2L, M, N and P. Hydraulic braking system. 28 V DC primary electrical system, supplemented by 115 V AC system for instruments and electronics. DC power supplied by two 30 V 200 A generators and two 24 V 40 Ah Ni/Cd batteries. Oxygen system standard. Pneumatic de-icing boots.

ELECTRONICS AND EQUIPMENT: Blind-flying instrumentation standard. Standard equipment includes VOR/LOC, glide slope, ADF and marker beacon receivers; ATC transponder; DME; VHF or other communications systems; compass systems; autopilot; and weather radar.

DIMENSIONS, EXTERNAL:

Wing span over tip tanks	11.94 m (39 ft 2 in)
Wing chord (mean)	1.54 m (5 ft 1 in)
Wing aspect ratio	7.71
Length overall: J, L, N	12.02 m (39 ft 5 in)
K, M, P	10.13 m (33 ft 3 in)
Length of fuselage: J, L, N	11.84 m (38 ft 10 in)
K, M, P	9.97 m (32 ft 8 7/12 in)
Height overall: J, L, N	4.17 m (13 ft 8 in)
K, M, P	3.94 m (12 ft 11 in)
Tailplane span	4.80 m (15 ft 9 in)
Wheel track: J, L, N	2.40 m (7 ft 11 in)
K, M, P	2.36 m (7 ft 9 in)
Wheelbase: J, L, N	4.40 m (14 ft 5 in)
K, M, P	4.52 m (14 ft 10 in)

Mitsubishi MU-2 twin-turboprop utility aircraft

1994

Propeller diameter: J, K, L, M	2.29 m (7 ft 6 in)
N, P	2.48 m (8 ft 2 in)
Distance between propeller centres	4.50 m (14 ft 9 in)
Propeller ground clearance: J	0.77 m (2 ft 6⅓ in)
K	0.65 m (2 ft 1⅗ in)
L	0.77 m (2 ft 6⅓ in)
M	0.63 m (2 ft 0⅔ in)
N	0.67 m (2 ft 2⅓ in)
P	0.53 m (1 ft 8⅔ in)
Cabin door: Height	1.22 m (4 ft 0 in)
Width	0.75 m (2 ft 5½ in)
Emergency exit door: Height	0.72 m (2 ft 4½ in)
Width	0.70 m (2 ft 3⅗ in)
DIMENSIONS, INTERNAL:	
Cabin length: J, L, N	5.99 m (19 ft 8 in)
K, M, P	3.35 m (11 ft 0 in)
Max width: J, K, L, M, N, P	1.50 m (4 ft 11 in)
Max height: J, K, L, M, N, P	1.30 m (4 ft 3⅕ in)
Baggage compartment: J, L, N	1.08 m^3 (38.00 cu ft)
K	0.89 m^3 (31.30 cu ft)
M, P	1.22 m^3 (43.10 cu ft)
AREAS (all versions):	
Wings, gross	16.55 m^2 (178 sq ft)
Flaps (total)	3.90 m^2 (42.0 sq ft)
Spoilers (total)	0.54 m^2 (5.82 sq ft)
Fin: J, L, N	2.85 m^2 (30.70 sq ft)
K, M, P	2.77 m^2 (29.80 sq ft)
Rudder, incl tab	1.17 m^2 (12.60 sq ft)
Tailplane	4.01 m^2 (43.30 sq ft)
Elevators, incl tabs	1.40 m^2 (15.07 sq ft)
WEIGHTS AND LOADINGS:	
Weights empty, equipped: J	3,085 kg (6,800 lb)
K	2,685 kg (5,920 lb)
L	3,165 kg (6,975 lb)
M	2,760 kg (6,085 lb)
N	3,205 kg (7,065 lb)
P	2,800 kg (6,175 lb)
Max ramp weight: J	4,922 kg (10,850 lb)
K	4,522 kg (9,970 lb)
L, N	5,272 kg (11,625 lb)
M, P	4,770 kg (10,520 lb)
Max T-O weight: J	4,900 kg (10,800 lb)
K	4,500 kg (9,920 lb)
L, N	5,250 kg (11,575 lb)
M, P	4,750 kg (10,470 lb)
Max landing weight: J	4,655 kg (10,260 lb)
K	4,280 kg (9,435 lb)
L, N	5,000 kg (11,025 lb)
M, P	4,515 kg (9,955 lb)
Max wing loading: J	296 kg/m^2 (60.6 lb/sq ft)
K	272 kg/m^2 (55.7 lb/sq ft)
L, N	317.4 kg/m^2 (65.0 lb/sq ft)
M, P	287.1 kg/m^2 (58.8 lb/sq ft)
Max power loading: J, L, N	4.54 kg/kW (7.46 lb/ehp)
K	4.16 kg/kW (6.85 lb/ehp)
M, P	4.40 kg/kW (7.23 lb/ehp)

PERFORMANCE (at respective AUWs, except where indicated, of 4,175 kg; 9,200 lb (J), 3,960 kg; 8,730 lb (K), 4,695 kg; 10,350 lb (L and N), 4,195 kg; 9,250 lb (M and P)):

Max operating speed: J, K, L, M, N, P (all weights)	250 kts (465 km/h; 290 mph) CAS
Max cruising speed at 4,575 m (15,000 ft):	
J	300 kts (555 km/h; 345 mph)
K	320 kts (590 km/h; 365 mph)
L and N at 4,175 kg (9,200 lb) AUW	295 kts (550 km/h; 340 mph)
M and P at 3,960 kg (8,730 lb) AUW	320 kts (590 km/h; 365 mph)
Econ cruising speed at 7,620 m (25,000 ft):	
J	265 kts (490 km/h; 305 mph)
K	270 kts (500 km/h; 310 mph)
L and N at 4,175 kg (9,200 lb) AUW	260 kts (480 km/h; 300 mph)
Econ cruising speed at 8,535 m (28,000 ft):	
M and P at 3,960 kg (8,730 lb) AUW	270 kts (500 km/h; 310 mph)
Stalling speed, flaps up:	
J	96 kts (177 km/h; 110 mph) CAS
K	95 kts (175 km/h; 109 mph) CAS
L, N	100 kts (185 km/h; 115 mph) CAS
M, P	97 kts (180 km/h; 112 mph) CAS
Stalling speed, flaps down:	
J, M, P	73 kts (135.5 km/h; 84 mph) CAS
K	71 kts (132 km/h; 82 mph) CAS
L, N	76.5 kts (142 km/h; 88 mph) CAS
Max rate of climb at S/L: J	820 m (2,690 ft)/min
K	945 m (3,100 ft)/min
L, N	800 m (2,630 ft)/min
M, P	865 m (2,840 ft)/min
Rate of climb at S/L, one engine out:	
J	260 m (845 ft)/min
K	280 m (920 ft)/min
L, N	205 m (675 ft)/min
M, P	230 m (760 ft)/min
Service ceiling: J	9,380 m (30,800 ft)
K	10,110 m (33,200 ft)
L, N	9,020 m (29,600 ft)
M, P	9,810 m (32,200 ft)
Service ceiling, one engine out: J	5,700 m (18,700 ft)
K	6,030 m (19,800 ft)
L, N	4,710 m (15,450 ft)
M, P	5,490 m (18,000 ft)
T-O to 15 m (50 ft) at max T-O weight:	
J	570 m (1,870 ft)
K	520 m (1,700 ft)
L, N	660 m (2,170 ft)
M, P	550 m (1,800 ft)
Landing run from 15 m (50 ft):	
J at 3,190 kg (8,620 lb) AUW	510 m (1,670 ft)
K at 3,510 kg (7,740 lb) AUW	455 m (1,490 ft)
L and N at 4,300 kg (9,470 lb) AUW	575 m (1,880 ft)
M and P at 3,780 kg (8,340 lb) AUW	485 m (1,600 ft)
Max range with wing tanks and wingtip tanks, full, 30 min reserves: J at 7,620 m (25,000 ft)	1,350 n miles (2,500 km; 1,550 miles)
K at 7,620 m (25,000 ft)	1,460 n miles (2,700 km; 1,680 miles)
L and N at 7,620 m (25,000 ft)	1,260 n miles (2,330 km; 1,450 miles)
M and P at 8,535 m (28,000 ft)	1,460 n miles (2,700 km; 1,680 miles)

VERIFIED

Mitsubishi MU-2 aircraft ***(Peter March)***

1994

MITSUBISHI T-2

TYPE: Two-seat supersonic jet trainer.

PROGRAMME: The T-2, the first supersonic aircraft developed by the Japanese aircraft industry, is a twin-engined two-seat jet trainer designed to meet the requirements of the JASDF.

Mitsubishi was selected as prime contractor for the development programme in September 1967. Design, under the leadership of Dr Kenji Ikeda, was followed by the completion of four XT-2 flying prototypes plus static and fatigue test airframes. The first XT-2 (19-5101) flew for the first time on 20 July 1971, the second (29-5102) on 2 December 1971, and the third and fourth on 28 April and 20 July 1972.

Mitsubishi, as prime contractor was responsible for fuselage construction, final assembly and flight testing of production aircraft. Major programme subcontractors were Fuji (wings and tail unit), Nippi (pylons and launchers) and Shin Meiwa (drop tanks).

Under contract to the Technical Research and Development Institute, Mitsubishi converted one T-2 as a CCV (control configured vehicle). *The following description applies to the standard production T-2/T-2A:*

VERSIONS: **T-2:** Advanced trainer.

T-2A: Combat trainer.

T-2CCV: Testbed aircraft for the development of the F-1 single-seat close support fighter aircraft.

OPERATORS: Versions of the T-2 are in service with the armed forces of Japan (80+).

DESIGN FEATURES: Cantilever all-metal shoulder-wing monoplane. Wing section NACA 65 series (modified). Thickness/chord ratio 4.66 per cent. Anhedral 9° from roots. Sweepback on leading-edge 68° at root, 42° 29′ inboard of extended chord outer panels and 36° on outer panels; basic sweepback at quarter-chord 35° 47′. One-piece all-moving swept tailplane with 15° anhedral. Small ventral fin under each side of fuselage at rear.

FLYING CONTROLS: Electrically actuated aluminium honeycomb leading-edge flaps, the outer portions of which have extended chord. Electrically actuated all-metal single-slotted flaps, with aluminium honeycomb trailing-edges over 70 per cent of each half-span. No conventional ailerons. Lateral control by hydraulically actuated all-metal two-section slotted spoilers ahead of flaps. Two hydraulically actuated door type airbrakes under centre of fuselage, aft of mainwheel bays. Hydraulically actuated all-moving swept tailplane. Hydraulically actuated rudder.

STRUCTURE: The wing is a multispar torsion box machined from tapered thick panels and constructed mainly of 7075 aluminium alloy. The fuselage is a conventional all-metal semi-monocoque structure, mainly of 7075 aluminium alloy. Approximately 10 per cent of structure, by weight, is

Mitsubishi T-2 two-seat supersonic trainer

1994

Mitsubishi T-2 of the JASDF *(Peter Steinemann)*

1994

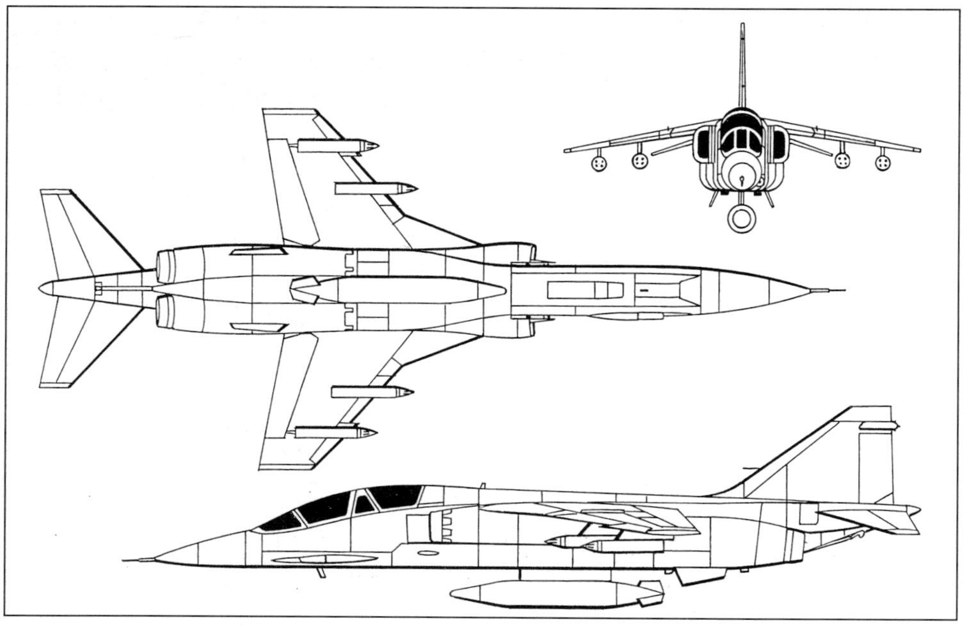

Mitsubishi T-2 two-seat supersonic trainer of the JASDF *(Jane's/Mike Keep)*

1994

Mitsubishi T-2 two-seat trainer of the JASDF *(Peter Steinemann)*

1995

Mitsubishi F-1 single-seat close support aircraft *(Peter Steinemann)*

1994

of titanium, mostly around engine bays. The tail unit is a cantilever all-metal structure. Tail unit inner leading-edges of titanium; outer leading-edges of aluminium. Trailing-edges of aluminium honeycomb construction.

LANDING GEAR: Hydraulically retractable tricycle type, with pneumatic back-up for emergency extension. Main units retract forward into fuselage, nose unit rearward. Single wheel on each unit. Nosewheel steerable through 72°. Oleo-pneumatic shock-absorbers. Nosewheel tyre size 18 × 5.5 Type VII (14 ply rating), maximum pressure 14.82 bars (215 lb/sq in); mainwheel tyres size 25 × 6.75 Type VII (18 ply rating), maximum pressure 20.69 bars (300 lb/sq in). Hydraulic brakes and Hydro-Aire anti-skid units. Runway arrester hook beneath rear fuselage. Brake parachute in tailcone.

POWER PLANT: Two Rolls-Royce Turbomeca Adour Mk 801A turbofan engines, each rated at 20.95 kN (4,710 lb st) dry and 31.45 kN (7,070 lb st) with afterburning, mounted side by side in centre of fuselage. Engines licence-built by Ishikawajima-Harima, under designation TF40-IHI-801A. Fixed geometry air intake, with auxiliary 'blow-in' intake doors, on each side of fuselage aft of rear cockpit. Fuel in seven fuselage tanks with total capacity of 3,823 litres (1,010 US gallons; 841 Imp gallons). Pressure refuelling point in starboard side of fuselage, forward of mainwheel bay. Three 821 litre (217 US gallon; 180 Imp gallon) auxiliary fuel tanks can be carried beneath the wings and fuselage.

ACCOMMODATION: Crew of two in tandem on Daiseru built Weber ES-7J zero/zero ejection seats in pressurised and air conditioned cockpits, separated by windscreen. Rear seat elevated 28 cm (11 in) above front seat. Individual manually operated rearward-hinged jettisonable canopies.

SYSTEMS: Cockpit air conditioning system. Two independent hydraulic systems, each 207 bars (3,000 lb/sq in), for flight controls, landing gear and utilities. Pneumatic bottle for landing gear extension. Primary electrical power from two 12/15 kVA AC generators.

AVIONICS: Mitsubishi Electric J/ARC-51 UHF, Nippon Electric J/ARN-53 Tacan and Toyo Communications J/APX-101 SIF/IFF. Mitsubishi Electric J/AWG-11 search and ranging radar in nose, with Mitsubishi Electric (Thomson-CSF) head-up display in cockpit. Lear Siegler 5010BL attitude and heading reference system. Liquid oxygen equipment.

ARMAMENT (combat trainer version): One Vulcan JM61 multibarrel 20 mm cannon in lower fuselage, aft of cockpit on port side. Attachment point on underfuselage centreline and two under each wing for drop tanks or other stores. Wingtip attachments for air-to-air missiles.

DIMENSIONS, EXTERNAL:	
Wing span	7.88 m (25 ft 10¼ in)
Wing chord: at root	4.172 m (13 ft 8¼ in)
at tip	1.133 m (3 ft 8½ in)
Wing aspect ratio	3
Wing taper ratio	3.7
Length overall, incl probe	17.85 m (58 ft 6¾ in)
Length of fuselage	17.31 m (56 ft 9½ in)
Height overall	4.39 m (14 ft 5 in)
Tailplane span	4.33 m (14 ft 2½ in)
Wheel track	2.82 m (9 ft 3 in)
Wheelbase	5.72 m (18 ft 9 in)
AREAS:	
Wings, gross	21.17 m² (227.9 sq ft)
Airbrakes (total)	0.96 m² (10.33 sq ft)
Vertical tail surfaces (total, excl ventral fins)	5.00 m² (53.82 sq ft)
Horizontal tail surfaces (total)	6.70 m² (72.12 sq ft)
WEIGHTS AND LOADINGS:	
Operational weight empty	6,309 kg (13,909 lb)
Max T-O weight	12,800 kg (28,219 lb)
PERFORMANCE (in 'clean' configuration except where indicated):	
Max level speed	Mach 1.6
Max rate of climb at S/L	10,670 m (35,000 ft)/min
Service ceiling	15,240 m (50,000 ft)
T-O run	910 m (3,000 ft)

VERIFIED

MITSUBISHI F-1

TYPE: Single-seat close support fighter.

PROGRAMME: Following the JASDF's decision to develop a single-seat close air support fighter from the T-2 supersonic trainer, design of this aircraft began in 1972. The second and third production T-2 trainers (59-5106 and 59-5107) were converted as prototypes, in which form they made their first flights on 3 and 7 June 1975 respectively. They were delivered to the JASDF Air Proving Wing at Gifu in July and August 1975, and after a year of flight test and evaluation the aircraft was type approved in November 1976 and officially designated F-1.

The F-1 is still in service with the JASDF (74).

DESIGN FEATURES: Airframe, power plant and systems, generally similar to T-2, but with the rear cockpit area modified as avionics compartment for bombing computer, inertial navigation system and radar warning system.

ACCOMMODATION: Generally similar to T-2, but without rear seat and with 'solid' fairing in place of second canopy.

AVIONICS AND EQUIPMENT: UHF; Tacan; IFF/SIF; Mitsubishi

Electric J/AWG-12 nose-mounted air-to-air and air-to-ground radar, with Mitsubishi Electric (Thomson-CSF) head-up display; Japan Aviation Electronics J/ASN-1 (Ferranti 6TNJ-F) inertial navigation system; radio altimeter; air data computer; Automatic Flight Control System; Mitsubishi Electric J/ASQ-1 fire control system and bombing computer (replacing original fire control and bombing computer from January 1982, for compatibility with ASM-1 missile); strike camera system; radar homing and warning system; attitude and heading reference system.

ARMAMENT: Single JM61 multibarrel 20 mm cannon. One underfuselage and four underwing hardpoints, as in T-2, with detachable multiple ejector racks. Primary weapon is the Mitsubishi ASM-1 air-to-surface missile, of which two can be carried on underwing stations. Bombs of 500 or 750 lb can be carried on all five external stations, up to a maximum weight of 2,721 kg (12,500 lb). CBU-87/B Cluster Bombs can be carried on all five external stations. The four underwing stations can each be used for rocket pods such as the JLAU-3A (with 19 70 mm), RL-7 (seven 70 mm) and RL-4 (four 125 mm). For air-to-air combat the F-1 can carry up to four AIM-9 Sidewinder missiles, one at each wingtip and one on each of the outboard underwing hardpoints. For long-range missions, the F-1 can carry up to three auxiliary fuel tanks (see Power Plant paragraph for Mitsubishi T-2).

DIMENSIONS, EXTERNAL:	
Wing span	7.88 m (25 ft 10¼ in)
Wing chord: at root	4.172 m (13 ft 8¼ in)
at tip	1.133 m (3 ft 8½ in)
Wing aspect ratio	3
Wing taper ratio	3.7
Length overall, incl probe	17.85 m (58 ft 6¾ in)
Length of fuselage	17.31 m (56 ft 9½ in)
Height overall	4.48 m (14 ft 8¼ in)
Tailplane span	4.33 m (14 ft 2½ in)
Wheel track	2.82 m (9 ft 3 in)
Wheelbase	5.72 m (18 ft 9 in)
AREAS:	
Wings, gross	21.17 m² (227.9 sq ft)
Airbrakes (total)	0.96 m² (10.33 sq ft)
Vertical tail surfaces (total, excl ventral fins)	5.00 m² (53.82 sq ft)
Horizontal tail surfaces (total)	6.70 m² (72.12 sq ft)
WEIGHTS AND LOADINGS:	
Operational weight empty	6,418 kg (14,149 lb)
Max T-O weight	13,700 kg (30,203 lb)
PERFORMANCE: Generally similar to T-2 clean, except for:	
T-O run at max T-O weight	1,280 m (4,200 ft)

VERIFIED

MITSUBISHI F-4EJKai and RF-4EJ UPGRADES

TYPE: Tactical fighter and interceptor (F-4EJKai) and reconnaissance fighter (RF-4EJ) upgrades.

PROGRAMME: Mitsubishi prime contractor in major equipment and weapon system update for JASDF F-4s; prototype F-4EJKai first flight 17 July 1984, delivered 13 December 1984.

VERSIONS: **F-4EJKai:** Avionics and weapons upgrade of F-4EJ; some expected to be allocated to close air support instead of interceptor duties.

RF-4EJ: 17 F-4EJs being converted to this model, joining 12 existing aircraft. Upgraded avionics. Mitsubishi in charge of programme.

CUSTOMERS: Current plans to convert 100 of JASDF's 124 remaining F-4EJs to F-4EJKai configuration and 17 to RF-4EJ; total of 86 F-4EJKai conversions under way, of which 83 completed by January 1995.

DESIGN FEATURES: F-4EJKai has Westinghouse J/APG-66J radar; other advanced avionics include Litton J/ASN-4 INS, HUD and J/APR-4Kai RWR; look-down/shoot-down capability with AIM-7E/F Sparrows or AIM-9P/L Sidewinders; can carry two ASM-1/-2 anti-shipping missiles. RF-4E upgrades receiving Texas Instruments AN/APQ-172 forward-looking radars and Melco-developed variant of Thomson-CSF Astac elint pod.

UPDATED

MITSUBISHI F-15J EAGLE

TYPE: Combat aircraft upgrade.

PROGRAMME: Plans now emerging for potential upgrade of JASDF F-15J fleet include possible adoption of stealth materials and advanced equipment currently being developed for FS-X, at estimated unit cost of about $19.2 million; service would like to modify a demonstrator by 1996 prior to fleet retrofit. Some F-15Js also planned to be adapted as escorts for FS-X anti-shipping missions, carrying a dedicated ECM pod containing an active phased-array radar; TRDI has reqested approximately $6 million to begin pod development in FY94 for completion in FY96.

VERIFIED

Mitsubishi F-1 single-seat combat aircraft of the JASDF ***(Mitsubishi)***

1995

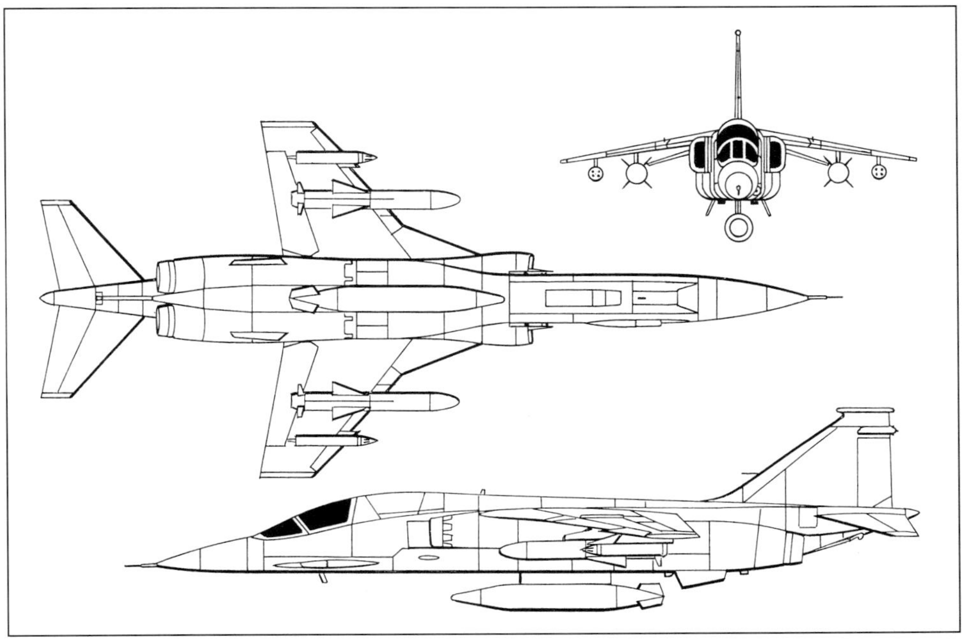

Mitsubishi F-1 single-seat close support aircraft of the JASDF ***(Jane's/Mike Keep)***

1994

Mitsubishi F-4EJKai upgraded tactical fighter and interceptor

1994

NIPPI

NIHON HIKOKI KABUSHIKI KAISHA (Japan Aircraft Manufacturing Co Ltd)

HEAD OFFICE AND YOKOHAMA PLANT: 3175 Showa-machi, Kanazawa-ku, Yokohama 236
Telephone: 81 (45) 773 5100
Fax: 81 (45) 773 5102
CHAIRMAN: Teruaki Yamada
PRESIDENT: Kanji Sonoda
ATSUGI PLANT: 28 Soyagi 2-chome, Yamato 242
Telephone: 81 (462) 652012
Fax: 81 (462) 652048
Telex: 3872 308 NIPPI A J

Yokohama plant, including head office, opened in 1935. It now has 74,975 m² (809,730 sq ft) floor area and employs about 1,100 people; manufactures wing in-spar ribs for Boeing 767/777, elevators for Boeing 757, underwing 'Y-barrels' for McDonnell Douglas MD-11, components and assemblies for Kawasaki built P-3C Orion (engine nacelles) and T-4 (pylons) and Mitsubishi built F-15J (pylons and launchers), forebody for Kawasaki CH-47J, V2500 core cowl, shelters for Mitsubishi built Patriot ground-to-air missiles, body structures and extendable masts for Japanese satellites, tail units for Japanese built rocket vehicles, and targets for Japan Defence Agency.

Atsugi plant has a 42,707 m² (461,236 sq ft) floor area and employs about 600 people. It is chiefly engaged in overhaul, repair and maintenance of various types of aeroplanes and helicopters, including those of JDA and Maritime Safety Agency, and carrier based aircraft of US Navy. Kawasaki has 25 per cent holding in Nippi.

UPDATED

Nippi (NAMC) YS-11EA electronic intelligence aircraft of the JASDF

1995

NIPPI (NAMC) YS-11EA

TYPE: Electronic support measures aircraft, converted from NAMC YS-11C transport.

PROGRAMME: Development started September 1987. Construction of first prototype started February 1988. First flight 12 September 1991; delivered to JASDF 20 December 1991.

OPERATORS: Versions of the YS-11EA are in service with the armed forces of Japan (1).

CUSTOMERS: Japan Air Self-Defence Force (one). Aircraft delivered 20 December 1991 and commenced service in July 1992.

DESIGN FEATURES: Main differences from standard YS-11 are replacement of Rolls-Royce Dart RDa.10 turboprops with 2,605 kW (3,493 ehp) GE T64-IHI-10Js and Sumitomo built Hamilton Standard 63E60-27 propellers, plus installation of NEC J/ALQ-7 ECM system.

WEIGHTS AND LOADINGS:

Max design T-O weight	25,700 kg (56,659 lb)
Max landing weight	24,500 kg (53,998 lb)
Max zero-fuel weight	22,500 kg (49,590 lb)
Max wing loading	271.1 kg/m² (55.53 lb/sq ft)

PERFORMANCE:

Max level speed at 4,270 m (14,000 ft) 287 kts (531 km/h; 330 mph)
Max rate of climb at S/L 494 m (1,620 ft)/min
Service ceiling at AUW of 25,000 kg (55,115 lb) 8,230 m (27,000 ft)
Range with max fuel, cruising at 7,315 m (24,000 ft) 1,440 n miles (2,667 km; 1,657 miles)
Range with max payload 700 n miles (1,296 km; 805 miles)

UPDATED

SHINMAYWA

SHINMAYWA INDUSTRIES LTD

Nippon Building, 6-2 Otemachi 2-chome, Chiyoda-ku, Tokyo 100
Telephone: 81 (3) 3245 6611
Fax: 81 (3) 3245 6616
Telex: 222 2431 SMIC T J
CHAIRMAN OF THE BOARD: Shinji Tamagawa
PRESIDENT: Shiko Saikawa
HEAD OFFICE: 5-25 Kosone 1-chome, Nishinomiya 663
Telephone: 81 (798) 47 0331
Telex: 5644493
KONAN PLANT: Ogi Higashinada-ku, Kobe City

Aircraft Division

EXECUTIVE MANAGING DIRECTOR AND GENERAL MANAGER: Junpei Matsuo
GENERAL MANAGER, KONAN PLANT: Takao Masuda
GENERAL MANAGER, BUSINESS PLANNING: Junsei Nagai (Tokyo Office)
GENERAL MANAGER, SALES: Yushi Tanaka (Tokyo Office)
WORKS: Konan and Tokushima

The former Kawanishi Aircraft Company became Shin Meiwa Industry Company in 1949 and established itself as a major overhaul centre for Japanese and US military and commercial aircraft. The company was renamed ShinMaywa in 1992.

ShinMaywa's principal activities are production of the US-1A long-range STOL search and rescue amphibian for the JMSDF, and overhaul work on amphibians.

ShinMaywa also produces components for other aircraft, including drop tanks for the Kawasaki T-4 trainer; external drop tanks for Mitsubishi built McDonnell Douglas F-15J Eagles; nosecones and tailcones, ailerons and trailing-edge flaps for the Kawasaki built Lockheed P-3C; tailplanes for the Mitsubishi built Sikorsky SH-60J; internal cargo handling system for the Kawasaki built Boeing CH-47J; wing and tail engine pylons for the McDonnell Douglas MD-11; thrust reverser doors for the McDonnell Douglas MD-80, under subcontract to Rohr Industries Inc; tailplane trailing-edges for the Boeing 757/767, under subcontract to Vought Aircraft Company; and other components for the Boeing 767, under subcontract to Mitsubishi. The company also modified five Learjet 36As into U-36A naval fleet training support aircraft for the JMSDF, with one attrition aircraft being modified for delivery March 1995.

ShinMaywa has been studying and looking for partners to develop its Amphibious Air Transport System, which is a 30/50-passenger airliner powered by two wing-mounted turbofans with upper surface blowing. Range would vary from 500 n miles (926 km; 575 miles) with full payload to 1,200 n miles (2,224 km; 1,380 miles) with full fuel. Take-off distance would be 1,000 m (3,280 ft) on water and 800 m (2,624 ft) on soft ground. Cruising speed would be between 300 and 360 knots (556 and 667 km/h; 345 and 414 mph).

VERIFIED

SHINMAYWA US-1A

TYPE: Four-turboprop STOL search and rescue amphibian.

PROGRAMME: First flown 16 October 1974; first delivery (as US-1) 5 March 1975; all now have T64-IHI-10J engines as US-1As.

VERSIONS: **US-1A:** SAR amphibian, developed from PS-1 ASW flying-boat; manufacturer's designation SS-2A. Data apply to this version.

UPGRADES: **ShinMaywa:** Firefighting amphibian: PS-1 modified in 1976 to firefighting configuration; 7,348 kg (16,200 lb) capacity water tank in centre-fuselage aft of step. Since then, US-1A modified experimentally, with more than 13,608 kg (30,000 lb) tank capacity; tank system developed by Conair of Canada.

ShinMaywa: Company reported to be discussing upgrade and possible wider use of US-1A with JMSDF; additional missions could include passenger and logistic transport. Main upgrade likely to be uprated engines, more modern AFCS and greater use of composites.

OPERATORS: Versions of the US-1A are in service with the armed forces of Japan (15).

CUSTOMERS: There were 15 US-1/1As ordered, of which 14 delivered by March 1994. No. 71 SAR Squadron of the JMSDF maintains fleet structure of seven aircraft at Iwakuni and Atsugi bases. Recent additional orders are for attrition because of phase out of older aircraft.

DESIGN FEATURES: Boundary layer control system and extensive flaps for propeller slipstream deflection for very low landing and take-off speeds; low-speed control and stability enhanced by blowing rudder, flaps and elevators, and by use of automatic flight control system (see Flying Controls). Fuselage high length/beam ratio; V shaped single-step planing bottom, with curved spray suppression strakes along sides of nose and spray suppressor slots in lower fuselage sides aft of inboard propeller line; double-deck interior. Large dorsal fin.

FLYING CONTROLS: Automatic flight control system controlling elevators, rudder and outboard flaps. Hydraulically powered ailerons, elevators with tabs and rudder, all with 'feel' trim. High-lift devices include outboard leading-edge slats over 17 per cent of wing span and large outer and inner blown trailing-edge flaps deflecting 60° and 80° respectively; outboard flaps can be linked with ailerons; inboard flaps, elevators and rudder blown by BLC system. Two spoilers in front of outer flaps on each wing. Inverted slats on tailplane leading-edge.

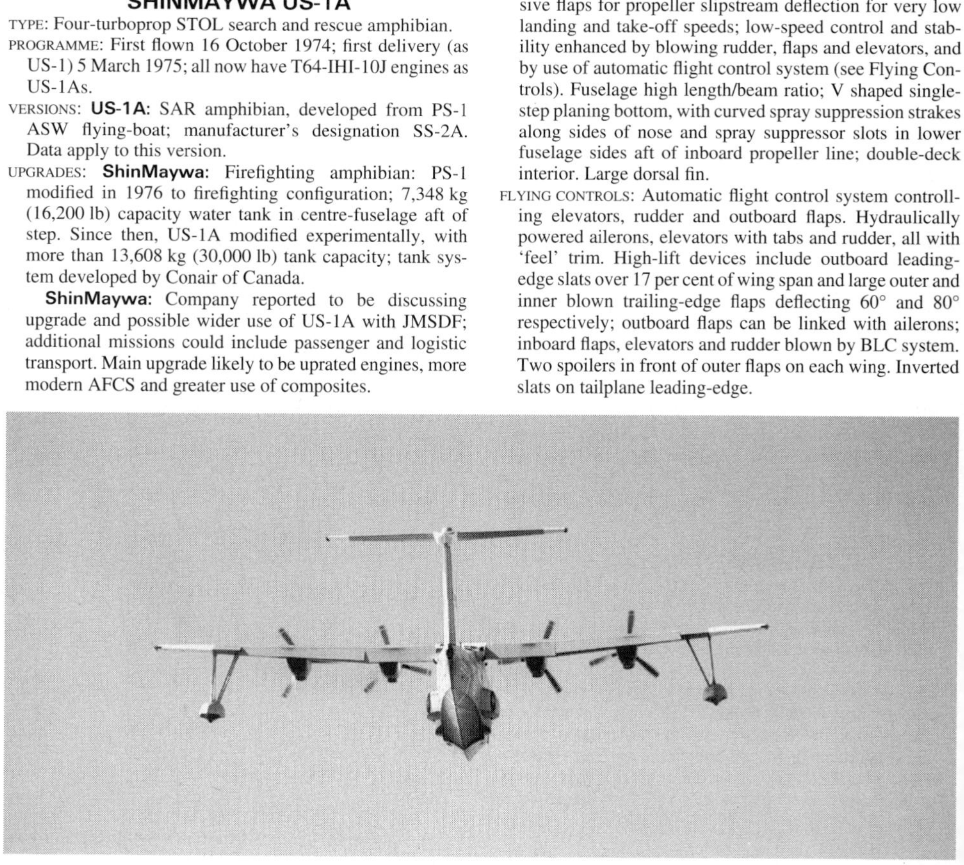

ShinMaywa US-1A four-turboprop STOL aircraft

1994

ShinMaywa US-1A search and rescue (SAR) amphibian aircraft of the JMSDF *(ShinMaywa)*

1995

STRUCTURE: All-metal; two-spar wing box.

LANDING GEAR: Flying-boat hull, plus hydraulically retractable Sumitomo tricycle landing gear with twin-wheels on all units. Steerable nose unit. Oleo-pneumatic shock-absorbers. Main units, which retract rearward into fairings on hull sides, have size 40 × 14-22 (Type VII) tyres, pressure 7.79 bars (113 lb/sq in). Nosewheel tyres size 25 × 6.75-18 (Type VII), pressure 20.69 bars (300 lb/sq in). Three-rotor hydraulic disc brakes. No anti-skid units. Minimum ground turning radius 18.80 m (61 ft 8¼ in) towed, 21.20 m (69 ft 6¾ in) self-powered.

POWER PLANT: Four 2,605 kW (3,493 ehp) Ishikawajima built General Electric T64-IHI-10J turboprops, each driving a Sumitomo built Hamilton Standard 63E60-27 three-blade constant-speed reversible-pitch propeller. Fuel in five wing tanks, with total usable capacity of 11,640 litres (3,075 US gallons; 2,560.5 Imp gallons) and two fuselage tanks (10,849 litres; 2,866 US gallons; 2,386.5 Imp gallons); total usable capacity 22,489 litres (5,941 US gallons; 4,947 Imp gallons). Pressure refuelling point on port side, near bow hatch. Oil capacity 152 litres (40.2 US gallons; 33.4 Imp gallons). Aircraft can be refuelled on open sea, either from surface vessel or from another US-1A with detachable at sea refuelling equipment.

ACCOMMODATION: Crew of three on flight deck (pilot, co-pilot and flight engineer), plus navigator/radio operator's seat in main cabin. Latter can accommodate up to 20 seated survivors or 12 stretchers, one auxiliary seat and two observers' seats. Sliding rescue door on port side of fuselage, aft of wing.

SYSTEMS: Cabin air conditioning system. Two independent hydraulic systems, each 207 bars (3,000 lb/sq in). No. 1 system actuates ailerons, outboard flaps, spoilers, elevators, rudder and control surface 'feel'; No. 2 system actuates ailerons, inboard and outboard flaps, wing leading-edge slats, elevators, rudder, landing gear extension/retraction and lock/unlock, nosewheel steering, mainwheel brakes and windscreen wipers. Emergency system, also of 207 bars (3,000 lb/sq in), driven by 24 V DC motor, for actuation of inboard flaps, landing gear extension/retraction and lock/unlock, and mainwheel brakes. Oxygen system for all crew and stretcher stations. Garrett GTCP85-131J APU provides power for starting main engines and shaft power for 40 kVA emergency AC generator. BLC system includes a C-2 compressor, driven by a 1,014 kW (1,360 shp) Ishikawajima built General Electric T58-IHI-10-M2 gas turbine, housed in upper centre portion of fuselage, which delivers compressed air at 14 kg (30.9 lb)/s and pressure of 1.86 bars (27 lb/sq in) for ducting to inner and outer flaps, rudder and elevators. Electrical system includes 115/200 V three-phase 400 Hz constant frequency AC and three transformer-rectifiers to provide 28 V DC. Two 40 kVA AC generators, driven by Nos. 2 and 3 main engines. Emergency 40 kVA AC generator driven by APU. 24 V emergency DC power from two 34 Ah Ni/Cd batteries. De-icing of wing and tailplane leading-edges. Fire detection and extinguishing systems standard.

AVIONICS: HIC-3 interphone, HRC-107 HF, N-CU-58/HRC antenna coupler, HGC-102 teletypewriter, HRC-106 radio, HRC-113 radio, HRN-101 ADF, AN/ARA-50 UHF/DF, HRN-105B Tacan, HRN-115-1 GPS nav system, HRN-107B-1 VOR/ILS receiver, AN/APN-171 (N2) radio altimeter, HPN-101B wave height meter, AN/APC-187C Doppler radar, AN/AYK-2 navigation computer, A/A24G-9 TAS transmitter, N-PT-3 dead reckoning plotting board, HRA-5 nav display, AN/APS-115-1 search radar, AN/APX-68-NB IFF transponder, RRC-22 emergency transmitter and N-ID-66/HRN BDHI.

ShinMaywa modification of a PS-1 to firefighting configuration

1994

EQUIPMENT: Marker launcher, 10 marine markers, six green markers, two droppable message cylinders, 10 float lights, pyrotechnic pistol, parachute flares, two flare storage boxes, binoculars, two rescue equipment kits, two droppable liferaft containers, rescue equipment launcher, lifeline pistol, lifeline, three lifebuoys, loudspeaker, hoist unit, rescue platform, lifeboat with outboard motor, camera and 12 stretchers. Sea anchor in nose compartment. Stretchers can be replaced by troop seats.

DIMENSIONS, EXTERNAL:

Wing span	33.15 m (108 ft 9 in)
Wing chord: at root	5.00 m (16 ft 4¾ in)
at tip	2.39 m (7 ft 10 in)
Wing aspect ratio	8.1
Length overall	33.46 m (109 ft 9¼ in)
Height overall	9.95 m (32 ft 7¾ in)
Tailplane span	12.36 m (40 ft 8½ in)
Wheel track	3.56 m (11 ft 8¼ in)
Wheelbase	8.33 m (27 ft 4 in)
Propeller diameter	4.42 m (14 ft 6 in)
Rescue hatch (port side, rear fuselage):	
Height	1.58 m (5 ft 2¼ in)
Width	1.46 m (4 ft 9½ in)

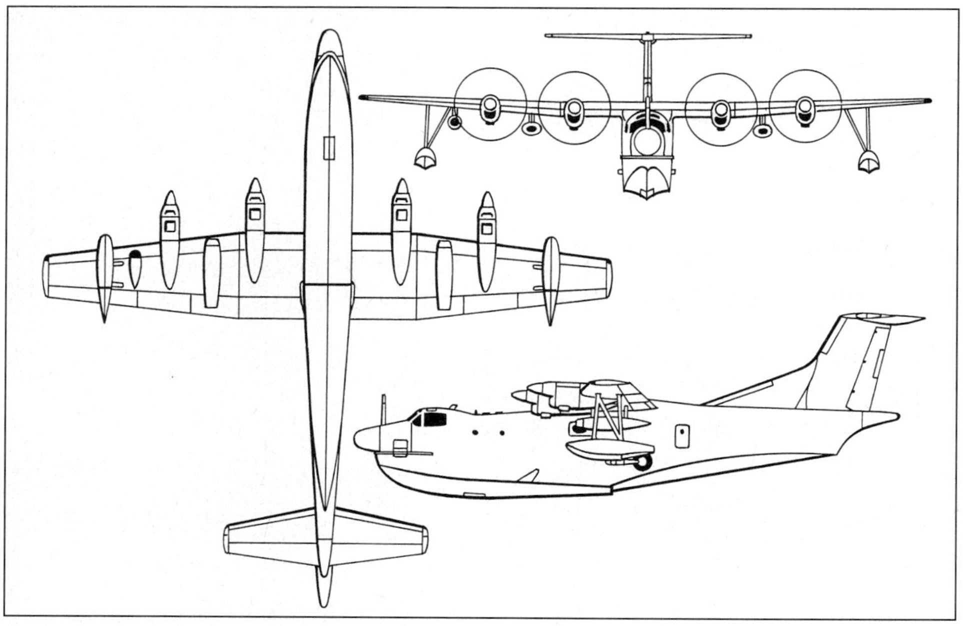

ShinMaywa US-1A of the JMSDF *(Jane's/Mike Keep)*

1994

AREAS:

Wings, gross	135.82 m² (1,462.0 sq ft)
Ailerons (total)	6.40 m² (68.90 sq ft)
Inner flaps (total)	9.40 m² (101.18 sq ft)
Outer flaps (total)	14.20 m² (152.85 sq ft)
Leading-edge slats (total)	2.64 m² (28.42 sq ft)
Spoilers (total)	2.10 m² (22.60 sq ft)
Fin	17.56 m² (189.0 sq ft)
Dorsal fin	6.32 m² (68.03 sq ft)
Rudder	7.01 m² (75.50 sq ft)
Tailplane	23.04 m² (248.0 sq ft)
Elevators, incl tab	8.78 m² (94.50 sq ft)

WEIGHTS AND LOADINGS (search and rescue):

Manufacturer's weight empty	23,300 kg (51,367 lb)
Weight empty, equipped	25,500 kg (56,218 lb)
Usable fuel: JP-4	17,518 kg (38,620 lb)
JP-5	18,397 kg (40,560 lb)
Max oversea operating weight	36,000 kg (79,365 lb)
Max T-O weight: from water	43,000 kg (94,800 lb)
from land	45,000 kg (99,200 lb)
Max wing loading	331.4 kg/m² (67.9 lb/sq ft)
Max power loading	4.32 kg/kW (7.10 lb/ehp)

PERFORMANCE (search and rescue, land T-O. A: at 36,000 kg; 79,365 lb weight, B: at 43,000 kg; 94,800 lb, C: at max T-O weight):

Max level speed: C	276 kts (511 km/h; 318 mph)
Max level speed at 3,050 m (10,000 ft): A	282 kts (522 km/h; 325 mph)
Cruising speed at 3,050 m (10,000 ft): C	230 kts (426 km/h; 265 mph)
Max rate of climb at S/L: A	713 m (2,340 ft)/min
C	488 m (1,600 ft)/min
Service ceiling: A	8,655 m (28,400 ft)
C	7,195 m (23,600 ft)
T-O to 15 m (50 ft) on land, 30° flap, BLC on (ISA): C	655 m (2,150 ft)
T-O distance on water, 40° flap, BLC on (ISA): B	555 m (1,820 ft)
Landing from 15 m (50 ft) on land, AUW of 36,000 kg (79,365 lb), 40° flap, BLC on, with reverse pitch (ISA): A	810 m (2,655 ft)
Landing distance on water, AUW of 36,000 kg (79,365 lb), 60° flap, BLC on (ISA): A	220 m (722 ft)
Runway LCN requirement: B	42

Max range at 230 kts (426 km/h; 265 mph) at 3,050 m (10,000 ft) 2,060 n miles (3,817 km; 2,372 miles)

UPDATED

KOREA, SOUTH

KA

KOREAN AIR

KAL Building, CPO Box 864, 41-3 Seosomun-Dong, Chung-Ku, Seoul
Telephone: 82 (2) 751 7114
Telex: KALHO K 27526
CHAIRMAN AND CEO: C. H. Cho
PRESIDENT: Y. H. Cho

Aerospace Division

Marine Center Building 22FL, 118-2ka, Namdaemunro, Chungku, Seoul
Telephone: 82 (2) 726 6240
Fax: 82 (2) 756 7929
Telex: KALHO K 27526 (SELDBKE)
EXECUTIVE VICE-PRESIDENT: Y. T. Shim
MANAGING VICE-PRESIDENT, PLANT OPERATION: B. S. Lee

Korean Institute of Aeronautical Technology (KIAT)

VICE-PRESIDENT: S. Y. Yoo
Address/telephone/telex/fax details as for Aerospace Division

Aerospace Division, one of several divisions of Korean Air, was established in 1976 to manufacture and develop aircraft.

By late 1993, the Aerospace Division occupied a 705,810 m² (7.6 million sq ft) site at Kim Hae, including a floor area of 239,000 m² (2.6 million sq ft) and had a workforce of 2,130 people.

Korean Air is supported by Daewoo and Samsung in producing between 80 and 90 Sikorsky UH-60Ps worth over $500 million under licence in a five-year programme up to 1995. Korean Air will first assemble, paint and flight test seven UH-60Ps, and manufacture the entire airframe by the end of the programme. Dynamic components and rotors will come from the USA. The UH-60P is essentially the current US Army UH-60L with T700-GE-701C engines and 2,535 kW (3,400 shp) transmission, with added avionics. Korean Air will also make parts for and assemble the T700 engines. Eventually, Korean Air might market and manufacture entire UH-60s for other countries in the region and become a supplier to Sikorsky. The first UH-60P was delivered to Korean Air late in 1990.

Since 1978, Korean Air has overhauled Republic of Korea Air Force aircraft. Programmed depot maintenance of US military aircraft in the Pacific area began in 1979, including structural repair of F-4 Phantoms, systems modifications for the F-16 Fighting Falcon, MSIP upgrading of the F-15 Eagle and overhaul of C-130 Hercules transports. KA began deliveries in April 1988 of wingtip extensions and flap track fairings for the Boeing 747 and fuselage components for the McDonnell Douglas MD-11.

As a part of the Korean aircraft industry development programme from 1988, Korean Air developed microlight aircraft, the five-seat Chang-Gong 91, and a modified Model 520MK light helicopter, derived from the Model 500, in association with McDonnell Douglas. Also projected is a primary trainer for the Korean Air Force, designated KTX-1.

VERIFIED

MALAYSIA

AIROD

AIROD SENDIRIAN BERHAD (Subsidiary of Aerospace Industries (Malaysia)

Locked Bag 4004, Subang International Airport, Pejabat Pos Kampung Tunku, 47309 Petaling Jaya
Telephone: 60 (3) 746 5112
Fax: 60 (3) 746 1395
Telephone: MA 37910
MANAGING DIRECTOR AND CEO: Dato Ariff Awang
SENIOR MARKETING DIRECTOR: Dato Wan Majid
PUBLIC RELATIONS DIRECTOR: Col (Ret'd) M. A. Theseira

Airod formed 1985 as joint venture of Aerospace Industries Malaysia and Lockheed Services International, USA. In late 1994 Lockheed Aircraft Services reduced its share in Airod to 30 per cent from 40 per cent.

UPDATED

SIKORSKY S-61A NURI UPGRADE

TYPE: Multipurpose military helicopters.
PROGRAMME: March 1991 contract to upgrade 34 aircraft between 1991 and end of 1993 with modernised navigation and other avionics; first example redelivered late 1991; 21st handed over 16 February 1993.
CUSTOMERS: Royal Malaysian Air Force (34).
COSTS: Contract value $50 million.

Sikorsky S-61A upgrade showing nose being prepared for installation of WX radar

1994

DESIGN FEATURES: To enhance operational capability and reduce crew workload.

AVIONICS: Include new nose-mounted WX weather radar, Doppler navigation system and radar altimeter.

VERIFIED

Sikorsky S-61A Nuri avionics upgrade, showing overview of the cockpit layout including new instrument panel
1994

NETHERLANDS

FOKKER

NV KONINKLIJKE NEDERLANDSE VLIEGTUIGENFABRIEK FOKKER

CORPORATE CENTRE: PO Box 12222, NL-1100 AE Amsterdam-Zuidoost
Telephone: 31 (20) 605 6666
Fax: 31 (20) 605 7015
Telex: 11526 FMHS NL
CHAIRMAN: B. J. A. van Schaik
OPERATING COMPANIES:
Fokker Aircraft BV, PO Box 12222, NL-1100 AE Amsterdam-Zuidoost
Telephone: 31 (20) 605 6666
Fax: 31 (20) 605 7015
Telex: 11526 FMHS NL
Fokker Aircraft Services BV, PO Box 3, NL-4630 AA Hoogerheide
Fokker Space and Systems BV, PO Box 32070, NL-2303 DB Leiden
Fokker Special Products BV, PO Box 59, NL-7900 AB Hoogeveen
Aircraft Financing and Trading BV, PO Box 12222, NL-1100 AE Amsterdam-Zuidoost
Avio-Diepen BV, PO Box 5952, NL-2280 HZ Rijswijk

Royal Netherlands Aircraft Factory Fokker founded by Anthony Fokker 21 July 1919. Since 1 January 1987, Fokker has had six operating companies. Main products are Fokker 50, 60, 70 and 100 airliners and their derivatives. Fokker Defence Marketing handles all defence activities.

Agreement reached 27 April 1993 for Deutsche (now Daimler-Benz) Aerospace (DASA) to acquire 51 per cent holding in Fokker; DASA to acquire Dutch government's 31.8 per cent holding in two stages, immediately and in 1996; DASA established new company, Fokker Holding, owning 51 per cent of NV Fokker, with remaining 49 per cent held by private shareholders; other companies could join Fokker Holding.

Total Fokker workforce reduced from 12,000 to 8,500 by late 1994; Dordrecht plant closed; output to be increased from 40 aircraft a year in 1994 to 46 in 1995; this compares with output of 45 Fokker 100s and 22 50s in 1993; management being streamlined and production facilities made more flexible; supplier prices being renegotiated. Despite this, Fokker is increasing its market share.

Schiphol plant, Amsterdam, active in Fokker 50, 60, 70 and 100 assembly and test flying facilities, design offices, spare parts stores, R&D department, numerically controlled milling department, electronics division, space integration and test facilities, and computer facilities.

Drechtsteden plant active on detail production and component assembly for Fokker 50, 60, 70 and 100 and F-16.

Ypenburg active in installation of F-16 centre fuselages and construction of composites components for Fokker 50, 60, 70 and 100 and Westland Lynx (radomes and fairings); new 30,000 m^2 (322,917 sq ft) composites and bonding plant began operating May 1990.

Fokker Aircraft Services BV situated at Woensdrecht, specialises in maintenance, overhaul, repair and modification of civil and military aircraft. ELMO plant, also at Woensdrecht, produces electrical and electronic systems and cable harnesses.

Hoogeveen (Fokker Special Products BV) engaged in industrial products activities, such as licence programmes, shelters, missile launchers, pylons, fuel tanks and thermoplastics components.

During 1993, Fokker undertook design and manufacture of empennage of Gulfstream G V; first delivery expected mid-1995.

As part of European F-16 programme, Fokker was responsible for component manufacture and assembly of F-16s for Netherlands (213, delivered between June 1979 and 27 February 1992), Norway (72) and Denmark (12); it continues to produce F-16 centre-fuselages, wing moving surfaces, main landing gear doors and legs, tailplanes, rudders and fin leading-edges for USAF and export assembly lines of Lockheed Fort Worth.

UPDATED

FOKKER F27 FRIENDSHIP

TYPE: Twin-turboprop short-range transport.

PROGRAMME: The first of two F27 prototypes made its first flight 24 November 1955, and the second 29 January 1957. Deliveries by Fokker began in November 1958, and were continuous through almost 30 years. Production of the F27 came to an end in 1986. An order for two by the Royal Thai Navy, announced 23 June 1986, brought the overall sales total to 786 to 168 customers in 63 countries.

VERSIONS: **Mk 100:** 83 built, including two corporate orders.

Mk 200: 115 built including one corporate order. Basic airliner or executive model. First flight 20 September 1959.

Mk 300: 13 built.

Mk 400M: Military version, with accommodation for 46 parachute troops, 6,025 kg (13,283 lb) of freight or 24 stretchers and nine attendants. Large cargo door and enlarged parachuting door on each side. First flight 24 April 1965. A cartographic version with two super wide-angle cameras, remotely controlled from central navigation station, and navigation sight; inertial navigation system, with digital readout at navigation station and recorded on each picture; photography through optical glass window panes, and electrically operated window doors. First flew 24 August 1973.

Mk 500: Similar to Mk 200 but with lengthened fuselage and large cargo door. First flight 15 November 1967. The French Ministère des Postes de Télécommunications ordered 15 with special large doors on both sides.

Mk 600: Similar to Mk 200, but with a large cargo door. Does not have the reinforced and watertight flooring of the Combiplane. Can be fitted with quick-change interior, featuring roller tracks and palletised seats. First flight 1968.

F27 Maritime, Maritime Enforcer: Surveillance versions, based on the Mk 200; described separately.

Any of the above models can be fitted, at customer's option with a Dowty Rotol very-rough-field landing gear giving increased overall height and propeller ground clearance. Rough-field gear versions went into service with Air Tanzania, Air Zaïre, Aramco, Burma Airways, Oman Aviation Services and Somali Airways. An F27 was modified by CAC of Australia, for use by the Royal Australian Navy for mapping water depths around the continent by means of laser technology. The equipment chosen included a THORN EMI Electronics (Australia) Pty laser airborne depth sounder (LADS); structural changes to the F27 included a window (for the laser beam) in the lower fuselage, and the provision of underwing auxiliary fuel tanks.

UPGRADES: **CAC:** An F27 was modified by CAC of Australia, for use by the Royal Australian Navy for mapping water depths around the continent by means of laser technology. The equipment chosen included a THORN EMI Electronics (Australia) Pty laser airborne depth sounder (LADS); structural changes to the F27 included a window (for the laser beam) in the lower fuselage, and the provision of underwing auxiliary fuel tanks.

Fokker F27-500 of TAM Transportes Aereos Regionais SA ***(Fokker)***
1995

Fokker F27 in the colours of Lauda Air of Austria. Since withdrawn from service with this airline *(Fokker)*

1995

Fokker: Any of the above models can be fitted, at customer's option with a Dowty Rotol very-rough-field landing gear giving increased overall height and propeller ground clearance. Rough-field gear versions went into service with Air Tanzania, Air Zaïre, Aramco, Burma Airways, Oman Aviation Services and Somali Airways.

Seca: Arat. See separate entry.

OPERATORS: Versions of the F27 are in service with the armed forces of the following countries: Algeria (3), Argentina (13), Bolivia (5), Ghana (4), Guatemala (3), Iceland (1), Indonesia (7), Iran (18), Ivory Coast (1), Myanmar (4), Netherlands (14), Nigeria (2), Pakistan (3), Philippines (11), Senegambia (6), Spain (3), Sudan (1), Thailand (5), USA (2) and Uruguay (2).

DESIGN FEATURES: Cantilever high-wing monoplane. Wing section NACA 64-421 at root, 64-415 at tip. Dihedral 2° 30′. Incidence 3° 30′. Cantilever tail unit.

FLYING CONTROLS: Mechanically operated single-slotted flaps. Electrically operated trim tab in each aileron. Trim tab in each elevator.

STRUCTURE: The wing is an all-metal riveted and metal-bonded two-spar stressed skin structure, consisting of centre-section and two detachable outer sections. Detachable honeycomb core sandwich leading-edges. GRP trailing-edges. The fuselage is an all-metal stressed skin structure, built to fail-safe principles, with cylindrical portions metal bonded and conical parts riveted. Fuselage is pressurised between rear bulkhead of nosewheel compartment and circular pressure bulkhead aft of the baggage compartment. Length of pressurised section 16.16 m (53 ft 0 in), except for Mk 500 in which the pressurised section is 17.66 m (57 ft 11 in) long. The slightly flattened fuselage bottom is reinforced by underfloor members. The tail unit is a cantilever all-metal skin structure. Fin and tailplane, as well as leading-edges of surfaces, are detachable.

LANDING GEAR: Retractable tricycle type. Pneumatic retraction. Dowty oleo-pneumatic shock-absorbers. Dunlop wheels, tyres and brakes. Twin-wheel main units retract rearward into engine nacelles. Single-wheel steerable nose unit retracts forward into non-pressurised nosecone. Mainwheel tyre pressure 5.62 bars (81.5 lb/sq in), nosewheel tyre pressure 3.87 bars (56 lb/sq in). Pneumatic brakes on mainwheels, with Dunlop Maxaret automatic anti-skid system. Provision on all above versions for Dowty Rotol very-rough-field landing gear in which, at 19,730 kg (43,500 lb) AUW, the total stroke in the main gear is lengthened from 305 mm (12 in) to 406 mm (16 in), increasing the aircraft's static height and propeller ground clearance by 76 mm (3 in). Low-pressure mainwheel tyres are fitted, pressure 4.2 bars (61 lb/sq in) below 18,143 kg (40,000 lb) AUW and 4.57 bars (66 lb/sq in) at higher operating weights. Nose unit is of levered suspension type, with tyre pressure of 3.87 bars (56 lb/sq in).

POWER PLANT (from Autumn 1984): Two Rolls-Royce Dart Mk 552 turboprop engines, each developing 1,648 kW (2,210 shp) plus 2.34 kN (525 lb st) for take-off. Four-blade Dowty Rotol constant-speed propellers. All F27s from early 1983 fitted with intake 'hush kits', decreasing engine noise level during approach and taxying by approximately 6 EPNdB. Integral fuel tanks in outer wings, capacity 5,136 litres (1,357 US gallons; 1,130 Imp gallons). Optionally, wing bag tanks for an additional 2,289 litres (605 US gallons; 503.5 Imp gallons) may be fitted. Overwing fuelling, but pressure refuelling optional. Provision for carrying two 950 litre (251 US gallon; 209 Imp gallon) external fuel tanks under wings. Methylbromide fire extinguishing system with flame detectors.

ACCOMMODATION (Mk 400M): Folding canvas seats, with safety harnesses alongside cabin side for up to 46 paratroops. Toilet and provision for medical supply box or pantry unit at rear. Ambulance version can accommodate 24 USAF type stretchers, in eight tiers of three, with seats at front and rear for up to nine medical attendants or sitting casualties. All-cargo version fitted with skid strips, tie-down fittings, protection plates and hinged hat racks. Dispatch door on each side of fuselage at rear for dropping supplies and personnel.

ACCOMMODATION (Mk 500): Main cabin has standard seating for 52 passengers four-abreast at 76 cm (30 in) pitch; alternative layouts enable up to 60 passengers to be carried at 72 cm (28.5 in) pitch.

ACCOMMODATION (executive and VIP versions): Can be furnished to customer's specification, but in the basic layout the cabin is divided into three sections: a conference room with six seats, a rest room with settee and divan, and a lounge with four seats. Toilet, galley, wardrobe, baggage space and seat for attendant in forward fuselage. Second toilet and baggage space at rear.

SYSTEMS: Kleber-Colombes rubber boot de-icers on wing leading-edges. Garrett pressurisation and air conditioning system utilises two Rootes type engine-driven blowers. Choke heating and air-to-air heat exchanger; optional bootstrap cooling system. Pressure differential 0.29 bars (4.16 lb/sq in) in Mk 500. No hydraulic system. Dunlop pneumatic system, pressure 235 bars (3,400 lb/sq in), for landing gear retraction, nosewheel steering and brakes. Emergency pneumatic circuits for landing gear extension and brakes. Bendix 28 V electrical system supplied by two 375 A 28 V DC engine-driven generators. Secondary system supplied via two 115 V 400 Hz AC constant frequency inverters. Variable frequency AC power supply, from 120/208 V 15 kVA engine-driven alternators, for anti-icing and heating. Two 24 V 40 Ah Ni/Cd batteries. 1.12 m³ (39.4 cu ft) oxygen system for pilots.

AVIONICS AND EQUIPMENT: VHF and HF transceivers, VHF navigation system (including glide slope) ADF, ILS marker beacon receiver, dual gyrosyn compass system, Fairchild flight data recorder, intercom system, Bendix weather radar and Smiths autopilot. Marquette windscreen wipers. Sperry SPZ-600 automatic flight control system standard from 1983, comprising fail-passive autopilot and a flight director with Cat II option.

DIMENSIONS, EXTERNAL:

Wing span	29.00 m (95 ft 1¾ in)
Wing chord: at root	3.45 m (11 ft 4 in)
at tip	1.40 m (4 ft 7 in)
Wing aspect ratio	12
Length overall: except Mk 500	23.56 m (77 ft 3½ in)
Mk 500	25.06 m (82 ft 2½ in)
Fuselage: Max width	2.70 m (8 ft 10¼ in)
Max height	2.79 m (9 ft 1¾ in)
Max height overall, standard landing gear:	
except Mk 500	8.50 m (27 ft 11 in)
Mk 500	8.71 m (28 ft 7¼ in)
Height overall, rough-field landing gear:	
except Mk 500	8.59 m (28 ft 2 in)
Tailplane span	9.75 m (32 ft 0 in)
Wheel track (c/l shock-absorbers)	7.20 m (23 ft 7½ in)
Wheelbase: except Mk 500	8.74 m (28 ft 8 in)
Mk 500	9.74 m (31 ft 11¼ in)
Propeller diameter	3.50 m (11 ft 6 in)
Propeller ground clearance: standard landing gear:	
except Mk 500	0.94 m (3 ft 1 in)
Mk 500	0.99 m (3 ft 3 in)
rough-field landing gear:	
except Mk 500	1.02 m (3 ft 4¼ in)
Passenger door (rear, port): Height	1.65 m (5 ft 5 in)
Width	0.73 m (2 ft 4¾ in)
Height to sill: except Mk 500	1.22 m (4 ft 0 in)
Mk 500	1.39 m (4 ft 6¾ in)
Service/emergency door (rear, stbd):	
Height	1.12 m (3 ft 8 in)
Width	0.74 m (2 ft 5 in)
Height to sill	0.99 m (3 ft 3 in)
Large cargo door (Mk 500): Height	1.78 m (5 ft 10 in)
Width	2.32 m (7 ft 7½ in)
Height to sill	1.03 m (3 ft 4½ in)
Dispatch doors (Mk 400M only, rear port and stbd each):	
Height	1.65 m (5 ft 5 in)
Width	1.19 m (3 ft 11 in)
Height to sill	1.22 m (4 ft 0 in)

DIMENSIONS, INTERNAL:

Cabin, excl flight deck:	
Length: except Mk 500	14.46 m (47 ft 5 in)
Mk 500	15.96 m (52 ft 4 in)
Max width	2.49 m (8 ft 2 in)
Max height	1.93 m (6 ft 4 in)
Floor area (excl toilet):	
except Mk 500	26.0 m² (280 sq ft)
Mk 500	30.2 m² (325 sq ft)
Volume (excl toilet):	
except Mk 500	56.0 m³ (1,978 cu ft)
Mk 500	65.5 m³ (2,313 cu ft)

Fokker F27 Mk 400M of the Argentine Air Force

1994

Freight hold (fwd), max: Mk 500	5.58 m³ (197 cu ft)
Freight hold (rear), max: all versions	2.83 m³ (100 cu ft)
AREAS:	
Wings, gross	70.00 m² (753.5 sq ft)
Ailerons (total)	3.51 m² (37.80 sq ft)
Trailing-edge flaps (total)	12.72 m² (136.90 sq ft)
Vertical tail surfaces (total)	14.20 m² (153 sq ft)
Horizontal tail surfaces (total)	16.00 m² (172 sq ft)
WEIGHTS AND LOADINGS:	
Manufacturer's weight empty:	
Mk 400M	11,213 kg (24,720 lb)
Mk 500, 52 seats	12,243 kg (26,992 lb)
Operating weight empty:	
Mk 400M, all-cargo	11,479 kg (25,307 lb)
Mk 400M, medical evacuation	11,902 kg (26,240 lb)
Mk 400M, paratrooper	11,655 kg (25,696 lb)
Mk 500, 52 seats	12,701 kg (28,000 lb)
Max payload (weight limited):	
Mk 400M, all-cargo	6,438 kg (14,193 lb)
Mk 400M, medical evacuation	6,015 kg (13,260 lb)
Mk 400M, paratrooper	6,261 kg (13,804 lb)
Mk 500, 52 seats	5,896 kg (13,000 lb)
Fuel load: standard (all versions)	4,123 kg (9,090 lb)
with optional wing bag tanks	5,978 kg (13,180 lb)
Max T-O weight (all versions)	20,820 kg (45,900 lb)
Max landing weight:	
Mks 400M and 500	19,050 kg (42,000 lb)
optional all versions	19,731 kg (43,500 lb)
Max zero-fuel weight (all versions)	18,234 kg (40,200 lb)
Max wing loading (all versions)	291.6 kg/m² (59.75 lb/sq ft)
Max power loading (all versions)	6.39 kg/kW (10.5 lb/shp)
PERFORMANCE (at weight indicated):	
Normal cruising speed at 6,100 m (20,000 ft) and AUW of 17,237 kg (38,000 lb): all versions	259 kts (480 km/h; 298 mph)
Rate of climb at S/L, AUW of 18,143 kg (40,000 lb):	
all civil versions	451 m (1,480 ft)/min
military versions	494 m (1,620 ft)/min
Service ceiling at AUW of 17,237 kg (38,000 lb):	
all civil versions	8,990 m (29,500 ft)
military versions	9,145 m (30,000 ft)
Service ceiling, one engine out, at 17,237 kg (38,000 lb):	
all versions	3,565 m (11,700 ft)
military versions	4,055 m (13,300 ft)
Runway LCN at max T-O weight, standard landing gear:	
rigid pavement	18
flexible pavement	15
Required T-O field length (ICAO-PAMC) at AUW of 18,143 kg (40,000 lb), all civil versions:	
S/L, ISA	988 m (3,240 ft)
S/L, ISA +15°C	1,088 m (3,570 ft)
915 m (3,000 ft), ISA	1,210 m (3,970 ft)
Required T-O field length (military) at AUW of 18,143 kg (40,000 lb), military versions:	
S/L, ISA	704 m (2,310 ft)
S/L, ISA +15°C	765 m (2,510 ft)
915 m (3,000 ft), ISA	838 m (2,750 ft)
Required landing field length (ICAO-PAMC) at AUW of 16,329 kg (36,000 lb), all civil versions:	
S/L	1,003 m (3,290 ft)
1,525 m (5,000 ft)	1,076 m (3,530 ft)
Required landing field length (military) at AUW of 17,010 kg (37,500 lb), military versions:	
S/L	579 m (1,900 ft)
915 m (3,000 ft)	622 m (2,040 ft)
Range (ISA, zero wind conditions) with FAR 121.645 reserves for alternate, 30 min hold at 3,050 m (10,000 ft) and 10% flight fuel: Mk 500, 52 passengers	935 n miles (1,741 km; 1,082 miles)
Military transport range (ISA, zero wind conditions) at max T-O weight, reserves for 30 min hold at S/L and 5% initial fuel: Mk 400M, all-cargo, max standard fuel	1,195 n miles (2,213 km; 1,375 miles)
Mk 400M, all-cargo, max possible fuel	2,370 n miles (4,389 km; 2,727 miles)
Military combat radius, conditions as above:	
400M, all-cargo, max standard fuel	625 n miles (1,158 km; 719 miles)
Mk 400M, all-cargo, max possible fuel	1,230 n miles (2,278 km; 1,416 miles)
Max endurance at 6,100 m (20,000 ft):	
Mk 400M, max standard fuel	7 h 25 min
Mk 400M, max possible fuel	12 h 47 min
OPERATIONAL NOISE LEVELS (FAR Pt 36; without hush kit):	
T-O	90.6 EPNdB
Approach: Mk 500	98.9 EPNdB
Sideline	92.2 EPNdB

VERIFIED

FOKKER F27 (SURVEILLANCE VERSIONS)

TYPE: Twin-turboprop maritime patrol aircraft.

PROGRAMME: A maritime patrol version of the Friendship, designed to meet the requirements of various coastal agencies throughout the world which require a cost-effective surveillance aircraft for coastal patrol, fishery protection, search and rescue and similar offshore duties was defined in July 1975, and shortly afterwards Fokker began converting an ex-airline F27 to serve as a prototype/demonstration aircraft (PH-FCX). This prototype made its first flight in February 1976.

VERSIONS: **F27 Maritime:** Basic unarmed maritime patrol version, for duties which include coastal surveillance, search and rescue and environmental control. It is operated by a crew of up to six persons, and its standard fuel capacity gives it an endurance of 10-12 hours, or a range of up to 2,700 n miles (5,000 km; 3,107 miles), depending on the mission. In service with Indian Navy (2), New Zealand Navy (3 in storage), Pakistani Navy, Peruvian Navy (2), and air forces of Angola (1), Netherlands (2), Nigeria (4), Philippines (3) and Spain (3). Three others, for the Royal Thai Navy are equipped to carry armament, but are not to full Maritime Enforcer standard.

Maritime Enforcer: Version of F27 for armed surveillance, anti-submarine and anti-shipping warfare, with enhanced avionics and provisions for carrying external stores (armament chosen and installed by operator).

UPGRADES: **Royal Netherlands Air Force:** Reported to be fitting its F27s involved with UN support operations with a defensive aids system (DAS), consisting of an Elta pulse Doppler missile advance warning system (MAWS) and a dispenser subsystem using a Terma ALE-40.

Thomson-CSF: Contract signed in 1993 to install an Ocean Master radar (T-CST/DASA) and a T-CSF airborne electronic warfare ESM DR 3000 A system on an F27 of the Pakistan Navy.

DESIGN FEATURES: Same as F27 except that the airframe is heavily treated with anti-corrosive measures; in tail unit, only the port elevator has a trim tab; and 'teardrop' windows are fitted to the flight deck.

LANDING GEAR: As described in the F27, but with tyre pressures of 5.52 bars (80 lb/sq in) on main units and 3.80 bars (55 lb/sq in) on nose unit. With long-stroke main gear fitted, pressure in the low-pressure mainwheel tyres is 4.50 bars (65 lb/sq in), and in tyre on the levered suspension nose unit 3.80 bars (55 lb/sq in).

POWER PLANT: Two Rolls-Royce Dart Mk 552 turboprop engines as described in main F27 entry. Four-blade Dowty Rotol propellers. Integral fuel tanks in outer wings, total capacity 5,140 litres (1,357 US gallons; 1,130 Imp gallons). Overwing gravity and pressure refuelling. Additional centre-wing tank of 2,310 litres (610 US gallons; 508 Imp gallons) capacity, and two 938 litre (248 US gallon; 206.5 Imp gallon) tanks on underwing pylons, giving a total fuel capacity of 9,326 litres (2,463 US gallons; 2,051 Imp gallons). Methyl bromide fire extinguishing system, with flame detectors. Water-methanol tank in each nacelle, combined capacity 303 litres (80 US gallons; 67 Imp gallons).

ACCOMMODATION: Flight compartment seats two pilots side by side, with folding seat for third crew member if required. Main cabin of F27 Maritime fitted out as tactical compartment (for two to four operators), containing advanced avionics, galley, toilet and crew rest area. Maritime Enforcer accommodates crew of seven including two pilots: tactical co-ordinator (Taco) responsible for off-airways navigation and overall efforts of mission crew; acoustic sensor operator (ASO) to handle active and passive sonobuoys, acoustic receivers and processor display system; non-acoustic sensor operator (NASO) controlling search radar and electronic surveillance subsystem; and two observers. Bubble windows for observers are provided at rear of main cabin. Rear cabin door is openable in flight. Standard cargo door at front on port side, with sill at truckbed height. Cargo holds forward and aft of main cabin.

SYSTEMS: Generally as for F27, except that bootstrap cooling system is standard; cabin pressure differential is 0.38 bars (5.5 lb/sq in); secondary electrical system has a third 115 V 400 Hz AC constant frequency inverter, and oxygen system includes individual supply for each tactical crew member.

AVIONICS AND EQUIPMENT: Com/nav equipment comprises Collins 618T-3 HF transceiver, two Collins 618M-3 VHF transceivers (three in Enforcer), Collins AN/ARC-159 UHF transceiver (VHF/UHF in Enforcer), interphone, crew address system (Enforcer). Litton LTN-72 inertial navigation system, IDC air data computer, dual Sperry C-9 gyrocompasses (not in Enforcer), Collins DF-206 radio compass, Collins 51Z-4 marker beacon receiver, Honeywell AN/APN-198 radio altimeter (two in Enforcer), Collins DF-301E VHF/UHF direction finder (UHF/VHF-AM/FM in Enforcer), two Collins 51RV-4 VOR/ILS receivers, two Collins HSIs (F27 Maritime only), Smiths SEP-2E/M autopilot (single in F27 maritime, none in Enforcer), Collins 621A-6A ATC transponder (F27 Maritime only).

OPERATIONAL EQUIPMENT: Both F27 and Enforcer versions fitted with Litton 360° search radar in ventral dome; AN/APS-504(V)2 in Maritime, AN/APS-504(V)5 in Enforcer. Additional mission equipment in Enforcer include GEC Avionics central tactical computer and display system, radar detection and display system, on-top position indicator/receiver, dual sonobuoy signal receivers, GEC Avionics AQS 902 (LAPADS) sonar, and sonobuoy processing system. Both passive and active sonobuoys are carried (up to 40 of SSQ-36, SSQ-41B or SSQ-47B type, up to 120 smaller buoys, or a mixture of both sizes), and launched from the internal stores area in the rear of the cabin. Marconi Defence Systems electronic surveillance and monitoring equipment to detect radar transmissions, which can be classified and recorded and their bearings transferred to the tactical display. An infra-red detection system (IRDS), Teledyne Electronics MAD. A datalink with available ground or shipborne systems can be provided. Searchlight pod optional, on central starboard wing pylon.

ARMAMENT (Maritime Enforcer): Fokker only installed provisions for armament; weapon mix and purchase left up to customer. Alkan stores management system. Two 907 kg (2,000 lb) stores attachments on the fuselage and three under each wing (capacities 295 kg; 650 lb inboard, 680 kg; 1,500 lb in centre, and 113 kg; 250 lb outboard). Typical ASW armament can include two or four Mk 44, Mk 46, Sting Ray or A244/S torpedoes and/or depth bombs. For anti-shipping warfare, two AM39 Exocet, AGM-65F Maverick, AGM-84A Harpoon, Sea Skua, Sea Eagle or similar air-to-surface missiles can be carried. Auxiliary fuel tanks can be carried on the central underwing pylons.

DIMENSIONS, EXTERNAL: As for F27 Mks 200/400/600, except:	
Height overall	8.70 m (28 ft 6½ in)
Cabin volume (excl flight deck)	60.5 m³ (2,136 cu ft)
WEIGHTS AND LOADINGS: (A: Maritime, B: Maritime Enforcer):	
Manufacturer's weight empty:	
A	12,519 kg (27,600 lb)
B	13,725 kg (30,260 lb)
Operating weight empty: A	13,314 kg (29,352 lb)
B (typical)	14,568 kg (32,117 lb)
Max fuel (incl pylon tanks): A	7,257 kg (16,000 lb)
B	7,511 kg (16,500 lb)

Fokker F27 Mk 200 Maritime of Netherlands Air Force

1994

Normal max T-O weight: both 20,410 kg (45,000 lb)
Operational necessity weight:
A 21,545 kg (47,500 lb)
B 21,320 kg (47,000 lb)
Emergency overload T-O weight:
B 22,680 kg (50,000 lb)
Max landing weight: both 18,600 kg (41,000 lb)
Max zero-fuel weight: both 17,900 kg (39,500 lb)
Max wing loading 291.6 kg/m² (59.75 lb/sq in)
Max power loading 6.39 kg/kW (10.5 lb/shp)

PERFORMANCE (at normal T-O weight except where indicated, A: Maritime, B: Maritime Enforcer):
Never-exceed speed, AUW of 17,237 kg (38,000 lb), S/L to 6,100 m (20,000 ft):
A 259 kts (480 km/h; 298 mph) CAS
B 256 kts (474 km/h; 294 mph) CAS
Normal cruising speed of 6,100 m (20,000 ft), AUW of 17,237 kg (38,000 lb):
both 250 kts (463 km/h; 287 mph)
Normal operating speed at 6,100 m (20,000 ft), AUW of 17,237 kg (38,000 lb):
A 227 kts (420 km/h; 261 mph) CAS
Patrol speed at 457 m (1,500 ft):
A 150-180 kts (227-333 km/h; 172-207 mph)
Stalling speed, flaps up:
A 96 kts (178 km/h; 111 mph) CAS
Max rate of climb at S/L, ISA, AUW of 18,143 kg (40,000 lb): A 442 m (1,450 ft)/min
Time to 6,100 m (20,000 ft): B 27 min
Service ceiling: A 8,990 m (29,500 ft)
Service ceiling, one engine out: A 3,565 m (11,700 ft)
Runway LCN (42% deflection) at 15,875 kg (35,000 lb) AUW: A
rigid pavement, l = 76.2 cm (30 in) 10.4
flexible pavement, h = 25.4 cm (10 in) 11.4
flexible pavement, h = 12.7 cm (5 in) 9.0
Runway CBR, unpaved soil, h = 25.4 cm (10 in), 3,000 passes: A
AUW of 15,875 kg (35,000 lb) 6.25
AUW of 20,410 kg (45,000 lb) 7.85
T-O run at S/L, ISA:
A at 20,410 kg (45,000 lb) T-O weight 975 m (3,200 ft)
B at 21,320 lb (47,000 lb) T-O weight 1,525 m (5,000 ft)
T-O run at S/L, ISA +20°C:
A at 20,410 kg (45,000 lb) T-O weight 1,080 m (3,545 ft)
B at 21,320 kg (47,000 lb) T-O weight 1,700 m (5,575 ft)
Landing distance (unfactored, ISA at S/L):
A at landing weight of 19,731 kg (43,500 lb): 610 m (2,000 ft)
A at landing weight of 13,607 kg (30,000 lb) 530 m (1,740 ft)
B at landing weight of 15,422 (34,000 lb) 560 m (1,837 ft)
Ferry range at 6,100 m (20,000 ft) with 4,536 kg (10,000 lb) payload, 30 min loiter and 5% reserves:
A 1,000 n miles (1,850 km; 1,150 miles)
Max range at 7,010-7,620 m (23,000-25,000 ft), reserves for 30 min hold, 5% flight fuel remaining:
B 2,698 n miles (5,000 km; 3,107 miles)
Max endurance: B 10 h 24 min

UPDATED

Fokker F28-2000 aircraft of Air Gabon ***(Fokker)***

1995

FOKKER F28 FELLOWSHIP

TYPE: Twin-turbofan short/medium-range airliner.

PROGRAMME: The F28 Fellowship twin-turbofan short/medium-haul transport was developed in collaboration with other European aircraft manufacturers, with pre-financing from the Netherlands Agency for Aerospace Programmes and through a loan guaranteed by the Dutch government.

Production was undertaken by Fokker in association with MBB in Germany and Short Brothers in the UK. Fokker was responsible for the front fuselage, to a point just aft of the flight deck, the centre-fuselage and wingroot fairings. MBB built the cylindrical fuselage section between the wing leading-edge and flight deck, the fuselage from the wing trailing-edge to the rear pressure bulkhead, the rear fuselage and tail unit, and the engine nacelles and support stubs. Shorts was responsible for wings and other components, including mainwheel and nosewheel doors.

First flight of the prototype F28 (PH-JHG) was made on 9 May 1967, and the second prototype, PH-WEV, flew on 3 August 1967. The third F28 (PH-MOL) flew for the first time on 20 October 1967 and was brought up to production standard in early Summer 1968.

The Dutch RLD granted a C of A to the F28 on 24 February 1969, and the first delivery (of the fourth aircraft, to LTU) was made on the same day. The aircraft received FAA type approval on 24 March 1969, German certification 30 March 1969 and British CAA type approval June 1979. RLD certification for operation from unpaved runways was granted in mid-1972. The 200th Fellowship, a Mk 400 for Garuda, was delivered on 30 January 1984 along with 17 other F28s.

A total of 236 had been ordered by the end of 1985.

VERSIONS: **Mk 1000:** Initial short fuselage version with seating for up to 65 passengers.

Mk 1000C: Basically the same as Mk 1000 but capable of all-cargo or mixed passenger/cargo operations with large freight door at front on port side aft of passenger door.

Mk 2000: Similar to Mk 1000 except for lengthened fuselage, permitting an increase in accommodation for up to 79 passengers in all-tourist configuration. First flight 28 April 1971.

Mk 3000: Similar to Mk 4000 but with short fuselage seating up to 65 passengers. Available also in 15-passenger VIP or executive layout, with range of up to 2,200 n miles (4,074 km; 2,533 miles).

Mk 4000: High-density long fuselage version, first flown 20 October 1976, to seat up to 85 passengers at 74 cm (29 in) pitch. Two additional overwing emergency exits (making a total of four).

Aircraft for East-West Airlines differ from the standard Mk 4000 in having centre-wing bag tanks adding 3,293 litres (870 US gallons; 724 Imp gallons) to the standard fuel capacity, giving a total of 13,041 litres (3,445 US gallons; 2,868 Imp gallons). They also have a Global VLF/Omega navigation system, and are configured for 72 passengers only, at a seat pitch of 84 cm (33 in).

UPGRADE: **FAS F28 Life Extension:** See separate entry.

The following details apply to both the standard Mk 3000 and Mk 4000, except where a specific model is indicated.

DESIGN FEATURES: Cantilever low/mid-wing monoplane. Wing section NACA 0000-X 40Y series camber varying along span. Thickness/chord ratio up to 14 per cent on inner panels, 10 per cent at tip. Dihedral 2° 30′. Sweepback at quarter-chord 16°. Variable incidence T tail.

FLYING CONTROLS: Irreversible hydraulically operated ailerons. Emergency manual operation of ailerons through tabs. Hydraulically operated Fowler double-slotted flaps over 70 per cent of each half-span with electric emergency extension. Five-section hydraulically operated lift dumpers in front of flap on each wing. Trim tab in each aileron. Hydraulically actuated variable incidence T tailplane. Electric emergency actuation of tailplane. Hydraulically boosted elevators. Hydraulically operated rudder with duplicated actuators and emergency manual operation.

STRUCTURE: The wing is a single-cell two-spar light alloy torsion box structure, comprising centre-section, integral with fuselage, and two outer panels. Fail-safe construction. Lower skin made of three planks, taper rolled top skin. Forged ribs in centre-section, built-up ribs in outer panels. Double skin leading-edge with ducts for hot air de-icing. The fuselage is a circular-section semi-monocoque light alloy fail-safe structure made up of skin panels with Redux-bonded Z stringers. Bonded doubler plates at door and window cutouts. Quickly detachable sandwich (metal/end-grain balsa) floor panels. Hydraulically operated petal brakes form aft end fuselage. The tail unit is a cantilever light alloy structure. Tail unit has honeycomb sandwich skin panels used extensively, in conjunction with multiple spars.

LANDING GEAR: Retractable tricycle type of Dowty Rotol manufacture, with twin-wheels on each unit. Hydraulic retraction, nosewheels forward, main units inward into fuselage. Oleo-pneumatic shock-absorbers. Goodyear wheels, tyres and electronically controlled braking system. Steerable nosewheel. Mainwheel tyres size 40 × 14, 16 ply rating, pressure 5.27 bars (76.5 lb/sq in).

POWER PLANT: Two Rolls-Royce RB183-2 Mk 555-15P turbofan engines with blade cooling (each 44 kN; 9,900 lb st), flat rated to 29.7°C, pod-mounted on sides of rear fuselage and fitted with acoustic intake liner and internal mixer. No water injection or thrust reversers. Thermal anti-icing for air intakes. Integral fuel tanks in each outer wing panel with total usable capacity of 9,740 litres (2,574 US gallons; 2,143 Imp gallons). Optional seven bladder tank units in wing centre-section with total usable capacity of 3,300 litres (872 US gallons; 726 Imp gallons). Single refuelling point under starboard wing, near root.

ACCOMMODATION: Crew of two side by side on flight deck, with jump seat for third crew member. Electrically heated windscreen. Pantry/baggage space immediately aft of flight deck on starboard side, followed by entrance lobby with hydraulically operated airstair door on port side,

Fokker F28 Mk 3000 Fellowship of Garuda Indonesian Airlines

1994

Fokker F28 Fellowship passenger aircraft in the colours of Air Mauritanie

1995

service and emergency door on starboard side, and seat for cabin attendant. Additional emergency door on each side of main cabin, over wing (two each side on Mk 4000). Main cabin layout of Mk 3000 can be varied to accommodate 55, 60 or 65 passengers five-abreast at 94, 81/84 or 79 cm (37, 32/33 or 31 in) seat pitch respectively. In Mk 4000, layout can accommodate up to 85 passengers at 74 cm (29 in) pitch. Aft of cabin are a wardrobe (port), baggage compartment (port) and toilet compartment (starboard). Underfloor cargo compartments fore and aft of wing, with single door on starboard side of forward hold, with one door on rear hold of each version.

SYSTEMS: Garrett air conditioning system, using engine bleed air. Maximum pressure differential 0.51 bars (7.45 lb/sq in). Two independent hydraulic systems, pressure 207 bars (3,000 lb/sq in). Primary system for flight controls, landing gear, nosewheel steering and brakes; secondary system for duplication of certain essential flight controls. Westinghouse all-AC electrical system utilises two 20 kVA engine-driven generators to supply three-phase constant frequency 115/200 V 400 Hz power. One 20 Ah battery for starting APU and for emergency power. Garrett GTCP 36-4A APU, mounted aft of rear pressure bulkhead, for engine starting, ground air conditioning and ground electrical power to drive a third AC generator for standby use on essential services in flight. Hot air de-icing on ducted wing leading-edges. Double skin leading-edges for hot air de-icing.

AVIONICS AND EQUIPMENT: Standard avionics include Collins VHF navigation system (with glide slope), DME marker beacon receiver, RCA weather radar, ADF, ATC transponder, dual compass system, interphone and public address systems, Smiths SEP6 autopilot, Collins FD 108 flight director, flight guidance caution system, Fairchild flight data recorder. Menasco powered flight controls. Thermal bleed air system for wing leading-edges, tailplane leading-edge and engine air intakes. Optional equipment to customer's requirements, including equipment for operation in Cat. II weather minima.

DIMENSIONS, EXTERNAL:
Wing span 25.07 m (82 ft 3 in)
Wing chord at root 4.80 m (15 ft 9 in)
Length overall: 3000 27.40 m (89 ft 10¾ in)
4000 29.61 m (97 ft 1¾ in)
Length of fuselage: 3000 24.55 m (80 ft 6½ in)
4000 26.76 m (87 ft 9½ in)
Fuselage: Max width 3.30 m (10 ft 10 in)
Height overall 8.47 m (27 ft 9½ in)
Tailplane span 8.64 m (28 ft 4¼ in)
Wheel track (c/l of shock-absorbers) 5.04 m (16 ft 6½ in)
Wheelbase: 3000 8.90 m (29 ft 2½ in)
4000 10.35 m (33 ft 11½ in)
Passenger door (fwd, port): Height 1.93 m (6 ft 4 in)
Width 0.86 m (2 ft 10 in)
Cargo door (fwd, port, optional):
Height 1.87 m (6 ft 1¾ in)
Width 2.49 m (8 ft 2 in)
Height to sill 2.24 m (7 ft 4¼ in)
Service/Emergency door (fwd, stbd):
Height 1.27 m (4 ft 2 in)
Width 0.61 m (2 ft 0 in)
Emergency exits (centre each):
Height 0.91 m (3 ft 0 in)
Width 0.51 m (1 ft 8 in)
Freight hold doors (each):
Height (fwd, each) 0.90 m (2 ft 11½ in)
Height (rear) 0.80 m (2 ft 7½ in)
Width (fwd, each) 0.95 m (3 ft 1½ in)
Width (rear) 0.89 m (2 ft 11 in)
Height to sill (fwd, each) 1.47 m (4 ft 10 in)
Height to sill (rear) 1.59 m (5 ft 2½ in)
Baggage door (rear, port, optional):
Height 0.60 m (1 ft 11½ in)
Width 0.51 m (1 ft 8 in)

DIMENSIONS, INTERNAL:
Cabin, excl flight deck:
Length: 3000 13.10 m (43 ft 0 in)
4000 15.31 m (50 ft 3 in)
Max length of seating area:
3000 10.74 m (35 ft 2¾ in)
4000 12.95 m (42 ft 6¾ in)
Max width 3.10 m (10 ft 2 in)
Max height 2.02 m (6 ft 7¼ in)
Floor area: 3000 38.4 m^2 (413.3 sq ft)
4000 44.8 m^2 (482.2 sq ft)
Volume: 3000 71.5 m^3 (2,525 cu ft)
4000 83.0 m^3 (2,931 cu ft)
Freight hold (underfloor, fwd):
3000 6.90 m^3 (245 cu ft)
4000 8.70 m^3 (308 cu ft)
Freight hold (underfloor, rear):
3000 3.80 m^3 (135 cu ft)
4000 4.84 m^3 (171 cu ft)
Baggage hold (aft of cabin), max:
both 2.30 m^3 (81.22 cu ft)

AREAS:
Wings, gross 79.00 m^2 (850 sq ft)
Ailerons (total) 2.67 m^2 (28.74 sq ft)
Trailing-edge flaps (total) 14.00 m^2 (150.7 sq ft)
Fuselage airbrakes (total) 3.62 m^2 (38.97 sq ft)
Fin (incl dorsal fin) 12.30 m^2 (132.4 sq in)
Rudder 2.30 m^2 (24.6 sq ft)
Tailplane 19.50 m^2 (209.9 sq ft)
Elevators (total) 3.84 m^2 (41.33 sq ft)

WEIGHTS AND LOADINGS:
Operating weight empty:
3000, 65 seats 16,965 kg (37,400 lb)
4000, 85 seats 17,645 kg (38,900 lb)
Max weight limited payload:
3000 8,437 kg (18,600 lb)
4000 10,478 kg (23,100 lb)
Fuel load: standard 7,820 kg (17,240 lb)
with optional centre-section tanks 10,469 kg (23,080 lb)
Max T-O weight: both 33,110 kg (73,000 lb)
Max zero-fuel weight: 3000 25,400 kg (56,000 lb)
4000 28,122 kg (62,000 lb)
Max landing weight: 3000 29,030 kg (64,000 lb)
4000 31,524 kg (69,500 lb)
Max cabin floor loading:
all passenger versions 366 kg/m^2 (75 lb/sq ft)
Max wing loading: both 407 kg/m^2 (83.4 lb/sq ft)
Max power loading: both 367.5 kg/kN (3.6 lb/lb st)

PERFORMANCE (Mks 3000 and 4000 at AUW of 29,000 kg; 63,934 lb, ISA, except where indicated):
Never-exceed speed
Mach 0.83 (390 kts; 723 km/h; 449 mph) EAS
Max permissible operating speed
Mach 0.75 (330 kts; 611 km/h; 380 mph) EAS
Max cruising speed at 7,000 m (23,000 ft)
455 kts (843 km/h; 523 mph)
Econ cruising speed at 9,150 m (30,000 ft)
366 kts (678 km/h; 421 mph)
Max cruising altitude 10,675 m (35,000 ft)
Min ground turning radius: 3000 9.60 m (31 ft 6 in)
4000 10.90 m (35 ft 9 in)
Runway LCN at max T-O weight:
rigid pavement 25
flexible pavement 20
FAR T-O field length at max T-O weight:
S/L 1,585 m (5,200 ft)
S/L, ISA + 10°C 1,635 m (5,364 ft)
S/L, ISA + 15°C 1,710 m (5,610 ft)
610 m (2,000 ft), ISA 1,710 m (5,610 ft)
915 m (3,000 ft), ISA 1,820 m (5,970 ft)
FAR landing field length at max landing weight:
S/L 1,065 m (3,495 ft)
1,525 m (5,000 ft) 1,276 m (4,185 ft)
Range, high-speed schedule, FAR 121.645 reserves:
* 3000, 65 passengers
1,480 n miles (2,743 km; 1,704 miles)
4000, 85 passengers
1,025 n miles (1,900 km; 1,180 miles)
Range, long-range schedule, FAR 121.645 reserves:
* 3000, 65 passengers
1,710 n miles (3,169 km; 1,969 miles)
4000, 85 passengers
1,125 n miles (2,085 km; 1,295 miles)

* *With wing centre-section tanks*

OPERATIONAL NOISE LEVELS (ICAO ANNEX 16):
T-O 86.3 EPNdB
Approach 94.0 EPNdB
Sideline 99.9 EPNdB

VERIFIED

FOKKER

FOKKER AIRCRAFT SERVICES

PO Box 3, NL-4630 AA Hoogerheide
Telephone: 31 (1646) 18000
Fax: 31 (1646) 14073
Telex: 78063 FOW NL
PRESIDENT: G. L. M. Hamers
VICE-PRESIDENT MARKETING SALES: Erik Goedhart MBA

Maintenance, modification and repair services for F27, F28, Fokker 50, 60, 70 and 100 as well as Boeing 737 series, Lockheed F-16, Northrop F-5 and Westland Lynx.

Capabilities include line, hangar and heavy maintenance of airframes, major overhaul of components and systems, aircraft conversions (cargo, special missions), interior refurbishment, VIP and corporate refitting, installation of moving belt loading systems, corrosion control and ageing aircraft programmes, avionics upgrading, APU installation, painting/stripping, non-destructive testing, calibration and contract engineering.

Fokker Aircraft Services has hangar facilities totalling 16,000 m^2 (172,160 sq ft) and handles some 260 aircraft per year. Approvals include RLD, FAA, CAA, LBA, JAR-145, ISO 9001 and AQAP-1.

UPDATED

FAS (WESTLAND) LYNX AVIONICS UPGRADE

TYPE: Combat helicopter upgrade.

PROGRAMME: Contract starting April 1993 to convert Royal Netherlands Navy Lynx helicopter to meet new operational requirements. Contract to be completed in 1994. FAS is co-operating with Westland Helicopters in this reconfiguration programme.

DESIGN FEATURES: Programme includes installation of customised avionics suite consisting of engineering and installation of the following systems: VHF-UHF-Mariphone Communication System, VUMCS, VOR/ILS, distance measuring equipment (DME), electronic horizontal situation indicator (EHSI), identification friend or foe (IFF), direction finding (DF) and a radar altimeter. Programme also includes installation of wide band secure voice system, remote-controlled UHF AN/ARC (V), modified audio selector boxes as well as a radio selector box.

UPDATED

FAS F-16 PACER SLIP

TYPE: Combat aircraft structural upgrade.

PROGRAMME: Work being carried out in conjunction with the Royal Netherlands Air Force to implement the PACER SLIP (Structural Life Improvement Programme) on 180 aircraft.

DESIGN FEATURES: FAS in co-operation with Lockheed Fort Worth (formerly General Dynamics) will apply the ECP1910 (Engineering Change Proposal) established for this programme. Work includes the design and production of doublers for bulkhead reinforcement as well as cold working of pipe hole connections through the bulkheads.

VERIFIED

FOKKER F28 FELLOWSHIP LIFE EXTENSION PROGRAMME

TYPE: Short/medium-range transport life extension programme.

PROGRAMME: In 1982 a Structural Integrity Programme (SIP) was introduced by Fokker to increase the design life of the F28 from 60,000 to 90,000 flight cycles (FCs). In late 1992 Fokker introduced a further programme to increase FCs from 90,000 to 100,000.

DESIGN FEATURES: Fokker has adopted an aircraft-to-aircraft approach whereby each aircraft is inspected when it has reached between 85,000 and 90,000 FCs, all structural modifications to be made before 90,000 FCs followed by repetitive inspections and retirement life replacements between 90,000 and 100,000 FCs.

VERIFIED

FAS (BOEING) 737 MODIFICATION CAPABILITIES

TYPE: Twin-turbofan passenger aircraft modification.

PROGRAMME: Company is responsible for a fast turnaround overhaul/repair/modification package for all types of Boeing 737.

DESIGN FEATURES: Programme consists of a choice of: base checks (up to and including D-checks), component overhaul, avionics upgrading, interior refurbishment, structural repairs (including engineering support), corrosion prevention and control programmes, aircraft conversion, VIP and corporate refitting, stripping, painting as well as non-destructive testing.

NEW ENTRY

FAS (WESTLAND) LYNX AVIONICS UPGRADE EVALUATION

TYPE: Combat helicopter upgrade.

PROGRAMME: Contract to design and evaluate new avionics for Royal Dutch Navy Lynx helicopters to be completed on one prototype by October 1995.

DESIGN FEATURES: Programme consists of the installation of the following systems for evaluation: forward-looking infra-red (FLIR), global positioning system (GPS) and a radar warning receiver (RWR).

NEW ENTRY

NEW ZEALAND

RNZAF

ROYAL NEW ZEALAND AIR FORCE

Air Staff, HQ NZ Defence Force, Stout Street, Wellington
Telephone: 64 (4) 496 0515
Telex: 3513 DEF COM NZ
DIRECTOR OF OPERATIONS: Wg Cdr Vaughan Paul

VERIFIED

RNZAF (Douglas) A-4K upgraded Skyhawks

1995

RNZAF A-4K SKYHAWK UPGRADE

TYPE: Combat aircraft upgrade.

PROGRAMME: Under the project name Kahu (Maori for 'Hawk'), the remaining 20 of the RNZAF's fleet of A-4 Skyhawks (15 out of 17 A-4s and five TA-4s) have been upgraded and refurbished. Installation kits were made by Pacific Aerospace Corporation and fitted by Safe Air. The first Kahu A-4K was handed over on 6 July 1989 and the last in December 1990. Smith Industries acted as prime contractor.

DESIGN FEATURES: The upgrade included new wing spars and systems for launching; AGM-65 Maverick, AIM-9L Sidewinder and GBU-16 laser-guided bomb capability, plus Westinghouse AN/APG-66 (NZ) radar with maritime target tracking; HOTAS controls and provision for FLIR and night vision goggles; MIL-STD-1553B databus; Collins AN/ARC-182 VHF/UHF and AN/ARC-159 UHF standby radios, VIR-130 VOR/ILS with glide slope and AN/ARN-118 Tacan; Smiths AN/APN-194 radar altimeter; Hazeltine AN/APX-72 IFF; General Instrument AN/ALR-66 (VE) radar warning receiver; Garrett digital air data computer; and Goodyear AN/ALE-39 chaff/flare dispenser. Weapons also include CRV-7 rockets and GP bombs (Mk 82/83). Upgrade incorporates up-load/download capability via data transfer module.

ARMAMENTS: AGM-65 Maverick air-to-ground missile, AIM-9L Sidewinder air-to-air missile, GBU-16 laser-guided bomb, CRV-7 rockets, GP bombs (Mk 82/83) as well as two internally mounted 20 mm cannon.

UPDATED

RNZAF (Douglas) A-4K upgraded Skyhawk cockpit
1994

RNZAF (LOCKHEED) P-3K ORION PROJECT RIGEL UPGRADE

TYPE: Maritime patrol aircraft upgrade.

PROGRAMME: Under a project named Rigel 1 six Royal New Zealand Air Force P-3B Orion aircraft were updated in 1981. In 1987 the aircraft were fitted with the LTN-211 inertial navigation systems. In 1995 the aircraft was fitted with GPS. The aircraft are also scheduled to undergo two

RNZAF (Lockheed) P-3K Orion showing upgraded surveillance suite *(RNZAF)*

1995

more upgrade programmes. Project Sirius will upgrade the ESM and acoustic processing system. Project Sirius is scheduled for circa 2000. The second project, 'Project Kestrel', involves rewinging the aircraft. This will take place from mid-1997. The aircraft nacelles will also be refurbished in Project Kestrel.

DESIGN FEATURES: Project Rigel upgrade package consisted of the installation of: LTN-72 INS, LTN-211 Omega, APS-134 radar, AAS-36 IRDS, new data handling system comprising an MIL-STD-1553A databus, three AYK-14 digital computing units (DCU), four display generator units (DGU), three multipurpose displays, pilot display and controller, ASH-33(V) digital tape system (DTS), video recorder reproducer, high-speed printer and five multifunction keysets.

NEW ENTRY

PAKISTAN

PAC

PAKISTAN AERONAUTICAL COMPLEX

Kamra, District Attock

WORKS: F-6 Rebuild Factory; Mirage Rebuild Factory; Kamra Avionics and Radar Factory; Aircraft Manufacturing Factory (all at Kamra)

Telephone: 92 (51) 580260/5

Fax: 92 (51) 584162

Telex: 5601 PAC KAMRA PK

DIRECTOR GENERAL: Air Marshal Dilawar Hussain

MANAGING DIRECTORS:

Air Cdre Tariq Hamid (AMF)

Air Cdre Muhammad Idreas Malik (F-6RF)

Air Cdre Niaz Husain (MRF)

Air Cdre Azfar Ali Khan (KARF)

Pakistan Aeronautical Complex is organ of Pakistan Ministry of Defence; consists of four factories, as follows:

Aircraft Manufacturing Factory (AMF) came into operation mid-1981, as licence production centre for Saab Safari/Supporter (Pakistani name Mushshak); major facilities include equipment for all Mushshak GFRP component manufacture; 1992 workforce approximately 1,000. Collaborating with NAMC in China in developing Karakorum 8 jet trainer.

F-6 Rebuild Factory, or F-6RF, established 1980 primarily for overhauling Pakistan's Shenyang F-6s and their accessories; has expanded its role by undertaking overhaul of FT-5, F-6, FT-6 and A-5C aircraft and their accessories. Rebuild and overhaul of Chinese F/FT-7s and accessories began May 1992.

F-6RF authorised to manufacture 7,000 spare parts for FT-5, F/FT-6 and A-5C aircraft and 1,140 litre (301 US gallon; 250 Imp gallon) F-6 auxiliary fuel tanks; production of 500 and 800 litre (132 and 211 US gallon; 110 and 176 Imp gallon) supersonic drop tanks for F-7P began mid-1991; 1993 workforce approx 2,500.

F-6RF possesses modern technical facilities for various engineering processes such as surface treatment, heat treatment, forging, casting, non-destructive testing, and machine tools required to manufacture items from raw materials. As well as conventional machines, modern machines such as CNC lathe, EDM die-sinking and EDM wire-cutting are utilised for various engineering processes. Precision casting facility and advanced metrological centre being established.

Mirage Rebuild Factory (MRF) began operating 1978; has site area of over 81 ha (200.15 acres) and nearly 2,000 engineers and technicians; can accomplish complete overhaul of Mirage III/5, Atar 9C engine, and all associated aircraft components and engine accessories; current overhaul capacity 8-10 aircraft and over 50 engines each year; can overhaul/rebuild third country Mirage III/5s, engines, components and accessories; overhauling United Arab Emirates Air Force Mirages since 1988; overhaul of 42 Dassault/Commonwealth Mirage IIIOAs and eight IIIDs, bought from Australia; first IIID received January 1991. Structural repair of Mirage fuselages has been established and now being upgraded to undertake wings as well.

Facility being upgraded to undertake increased life core (ILC) modification, overhaul and upgrade of Pratt & Whitney F100-PW-220E turbofans and F-16 engine jet fuel starters, and will soon have limited capability to service and overhaul F100 engine accessories.

Kamra Avionics and Radar Factory (KARF) began operating 1987; employs over 250 engineers and technicians. At present rebuilding Siemens MPDR-45E radars, complex components and electronics modules, and caterpillar/Siemens power generators. Has modern, environmentally controlled and ASD equipped electronics workshops, environmental test chambers, sophisticated test measurement, and Pagamat ATE diagnostic equipment. Shortly to undertake rebuilding of Siemens control and reporting centre and MPDR-60/90 radars. Involved in co-production of airborne radar for fighters and upgrading of Mirage III avionics suite.

VERIFIED

POLAND

PZL MIELEC

WYTWÓRNIA SPRZETU KOMUNIKACYJNEGO 'PZL-MIELEC' SA (Transport Equipment Corporation 'PZL-Mielec' Joint Stock Company)

ul. Wojska Polskiego 3, PL-39-300 Mielec

Telephone: 48 (196) 7010

Fax: 48 (14) 214785

Telex: 632293 pl

GENERAL MANAGER: Wiesław Pastuła

Largest and best equipped aircraft factory in Poland; founded 1938 and had produced over 15,000 aircraft by 1 January 1993. In addition to production of aircraft of its own design and ex-Soviet types, manufactures components for Ilyushin Il-86 (including fins, tailplanes, engine pylons, wing slats and flaps) and Il-96; began manufacture and subassembly of components for Socata TB series light aircraft in March 1991.

UPDATED

PZL MIELEC TS-11 ISKRA (SPARK)

TYPE: Fully aerobatic two-seat jet primary and advanced trainer (A, B and D) and single-seat reconnaissance aircraft (C and DF).

PROGRAMME: Designed in 1957 under the supervision of Docent Ing T. Soltyk, the TS-11 Iskra two-seat jet trainer was produced as a replacement for the piston-engined TS-8 Bies. Four prototypes were built during 1958-59, the first of these being used for static testing. First flight, on 5 February 1960, was made by the second aircraft, followed later in the same year by the third and fourth prototypes. Type approval was received in mid-1961, and quantity production began at Mielec in 1963. The formal handing over of the first Iskra to the Polish Air Force took place in March 1963, and the aircraft entered service in 1964. Iskras of the Polish Air Force logged a total of 100,000 flying hours by 1978. Another 30,000 flying hours had been accumulated in India.

Early production aircraft were powered by a 7.66 kN (1,720 lb st) HO-10 Polish-designed axial flow turbojet engine. In April 1964, flight testing began using the intended power plant, the more powerful SO-1; from the latter half of the 1960s, production Iskras were powered either by the SO-1 or by the modified but similarly rated SO-3.

About 500 Iskras had been built by mid-1979, in the following versions:

VERSIONS: **Iskra-Bis A:** Basic two-seat version, for primary and advanced training. Two underwing hardpoints for external weapons. Withdrawn.

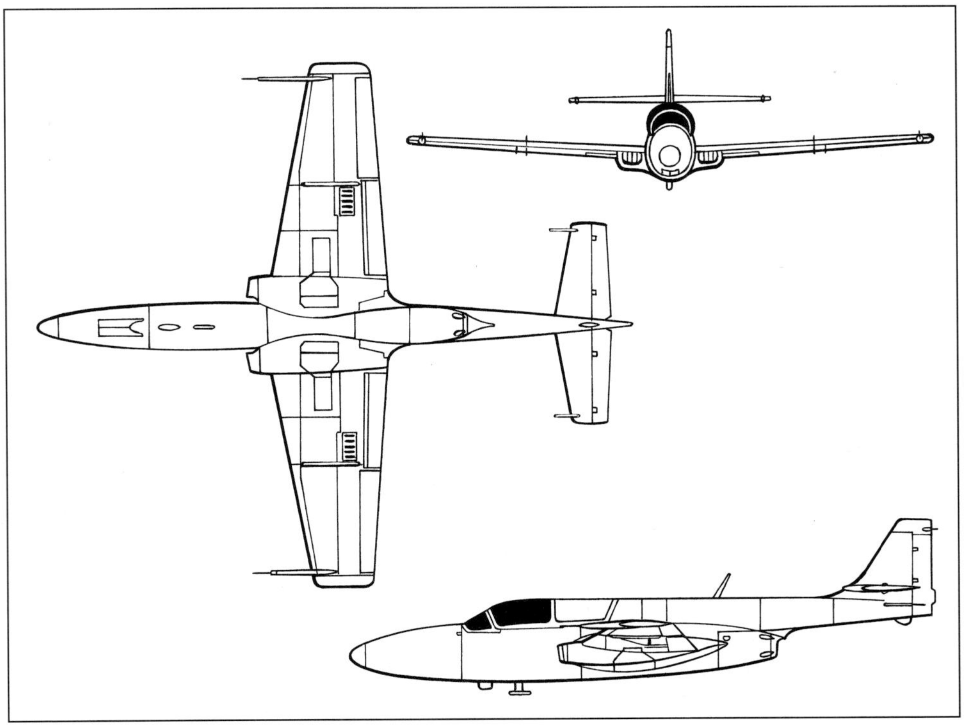

PZL Mielec TS-11 Iskra two-seat trainer *(Jane's/Mike Keep)*

1994

Iskra-Bis B: Two-seat primary and advanced trainer, with four underwing attachments for missiles or other weapons. Prototype designated Iskra 100. Withdrawn.

Iskra-Bis C: Single-seat reconnaissance version. Prototype, designated Iskra 200, first flew in June 1972. Increased fuel capacity for greater range; camera in fuselage floor, aft of cockpit. Withdrawn.

Iskra-Bis D: Similar to Bis B, but able to carry a wider selection of external weapons. Prototype also designated Iskra 200. Export version supplied to Indian Air Force (50) in 1976. In production. Still in service.

Iskra-Bis DF: Two-seat combat and reconnaissance trainer, with increased armament capability of Bis D, plus provision for three cameras, one in each air intake fairing and one in fuselage floor beneath rear cockpit. In production. Still in service.

UPGRADE: **PZL Mielec:** TS-11R Iskra. See separate entry.

OPERATORS: Versions of the TS-11 Iskra are in service with the armed forces of the following countries: India (38) and Poland (90+).

The following description applies to the Iskra-Bis B, D and DF, except where indicated:

DESIGN FEATURES: Cantilever mid-wing monoplane. Wing section NACA 64209 at root, NACA 64009 at tip. Sweepback at quarter-chord 7°. Marked dihedral. One boundary layer fence on each wing. Anti-flutter weight fairing projecting from each wing near tip. The tail unit is a cantilever structure with variable incidence. Fin integral with fuselage. Anti-flutter weight fairing projecting from each half of tailplane at tip.

FLYING CONTROLS: Hydraulically servo-assisted, aerodynamically balanced ailerons. Hydraulically actuated two-section double-slotted flaps and airbrakes (maximum deflection 87°). Tailplane actuated electrically. Mass and

PZL Mielec TS-11 Iskra of the Polish Air Force *(Piotr Butowski)*

1994

aerodynamically balanced elevators and rudder. Ground adjustable tab on rudder, fixed balance tab in port elevator.

STRUCTURE: The wing is an all-metal torsion box structure with two steel main spars and duralumin stressed skin. The fuselage is an all-metal, semi-monocoque structure of pod and boom type. Two-spar tailplane.

LANDING GEAR: Retractable tricycle type with single wheel on each unit. Nosewheel retracts forward, mainwheels inward into wingroot air intake trunks. Hydraulic actuation, with pneumatic emergency extension. Mainwheels size 600 × 180, tyre pressure 5.38 bars (78 lb/sq in). Nosewheel size 400 × 150, tyre pressure 3.45 bars (50 lb/sq in). Oleo-pneumatic shock-absorbers. Disc brakes on main wheels. Castoring and self-centring nosewheel, with shimmy damper.

POWER PLANT: One SO-3 turbojet, rated at 9.81 kN (2,205 lb st), or (Iskra-Bis DF) higher powered SO-3W, rated at 10.79 kN (2,426 lb st), mounted in fuselage aft of cockpit section, with nozzle under tailboom. Fuel in two 315 litre (82 US gallon; 69 Imp gallon) integral wing tanks, one rubber 500 litre (132 US gallon; 110 Imp gallon) fuselage main tank (700 litre; 184 US gallon; 154 Imp gallon in single-seaters) and one rubber 70 litre (18.6 US gallon; 15.5 Imp gallon) fuselage collector tank. Total fuel capacity 1,200 litres (263.5 Imp gallons) in two-seaters, 1,400 litres (370 US gallons; 308 Imp gallons) in single-seaters. Fuel system permits up to 40 seconds of inverted flight.

ACCOMMODATION: Crew of one, or two in tandem, on lightweight ejection seat(s), under a one-piece hydraulically actuated rearward-hinged upward-opening jettisonable canopy. Cockpit pressurised and air conditioned. Rear seat of trainers slightly raised.

SYSTEMS: Hydraulic system, pressure 138 bars (2,000 lb/sq in), for actuation of ailerons, flaps, airbrakes, landing gear, canopy and mainwheel brakes. Pneumatic system, pressure 118 bars (1,710 lb/sq in), for cockpit pressurisation, anti-icing and gun charging. Emergency pneumatic system for landing gear extension, flaps and emergency braking. Electrical power provided by 28.5 V AW-30 generator and 24 V 28 Ah battery, for engine starting, instruments, lights and armament control system; 115 V converter for AC power. Air conditioning, oxygen, ethyl alcohol anti-icing and CO_2 fire extinguishing systems standard.

AVIONICS: Trainers have complete dual controls and instrumentation, including blind-flying panels. Standard avionics also include R-800, R-802G or R-802W VHF com; ARK-9 or ARL-1601 radio compass; RW-UM radio altimeter; MRP-56P marker beacon receiver; SPU-2P crew intercom in two-seaters; and IFF.

ARMAMENT: Forward-firing 23 mm cannon in nose on starboard side, with S-13 gun camera. Two or four attachments for a variety of underwing stores, including bombs of up to 100 kg (220 lb), eight-barrel rocket pods and 7.62 mm gun pods.

DIMENSIONS, EXTERNAL:

Wing span	10.06 m (33 ft 0 in)
Wing chord: at root	2.254 m (7 ft 4¾ in)
at tip	1.162 m (3 ft 9¾ in)
Wing aspect ratio	5.71
Length overall	11.15 m (36 ft 7 in)
Height overall	3.50 m (11 ft 5½ in)
Tailplane span	3.84 m (12 ft 7¼ in)
Wheel track	3.47 m (11 ft 4½ in)
Wheelbase	3.44 m (11 ft 3½ in)

AREAS:

Wings, gross	17.50 m² (188.37 sq ft)
Ailerons (total)	1.48 m² (15.93 sq ft)
Trailing-edge flaps (total)	1.74 m² (18.73 sq ft)
Fin	1.55 m² (16.68 sq ft)
Rudder (incl tab)	0.70 m² (7.53 sq ft)
Tailplane	2.38 m² (25.62 sq ft)
Elevators (incl tab)	1.16 m² (12.48 sq ft)

WEIGHTS AND LOADINGS:

Weight empty (trainer versions):	
B	2,494 kg (5,498 lb)
D/DF	2,560 kg (5,644 lb)
Normal T-O weight for 570 l (150.6 US gallons; 125.5 Imp gallons) internal fuel: B	3,184 kg (7,019 lb)
D/DF	3,243 kg (7,150 lb)
Normal T-O weight with 1,200 l (317 US gallons; 263.5 Imp gallons) internal fuel: B	3,704 kg (8,166 lb)
D/DF	3,734 kg (8,232 lb)
T-O weight (reconnaissance version):	
DF	3,787 kg (8,349 lb)
Max T-O weight with full external armament:	
B	3,810 kg (8,400 lb)
D/DF	3,840 kg (8,465 lb)
Max wing loading: D/DF	219 kg/m² (44.85 lb/sq ft)
Max power loading: D (SO-3)	387.4 kg/kN (3.8 lb/lb st)
DF (SO-3W)	355.9 kg/kN (3.5 lb/lb st)

PERFORMANCE (at normal T-O weight with full internal fuel, except where indicated):

Never-exceed speed	Mach 0.8 (404 kts; 750 km/h; 446 mph)
Max level speed at 5,000 m (16,400 ft):	
B, D	388 kts (720 km/h; 447 mph)
DF	415 kts (770 km/h; 478 mph)
Normal cruising speed:	
B, D/DF	324 kts (600 km/h; 373 mph)
Unstick speed: B, D/DF	102 kts (190 km/h; 118 mph)
Landing speed: B, D/DF	92 kts (170 km/h; 106 mph)
Stalling speed, power off, flaps down:	
B, D/DF	81 kts (150 km/h; 93.5 mph)
Max rate of climb at S/L: B, D	840 m (2,755 ft)/min
DF	1,164 m (3,820 ft)/min
Time to 5,000 m (16,400 ft): DF	5 min 18 s
Time to 6,000 m (19,685 ft): B, D	9 min 36 s
Time to 7,000 m (22,975 ft): B, D	13 min 36 s
Time to 11,000 m (36,000 ft): DF	36 min 0 s
Service ceiling: B	11,140 m (36,550 ft)
D/DF	11,000 m (36,000 ft)
T-O run: B	725 m (2,380 ft)
D	750 m (2,460 ft)
DF	650 m (2,135 ft)
T-O to 15 m (50 ft), flaps down: B	1,100 m (3,609 ft)
D	1,190 m (3,904 ft)
DF	1,090 m (3,575 ft)
Landing from 15 m (50 ft), flaps down:	
B	1,000 m (3,280 ft)
D/DF	1,110 m (3,642 ft)
Landing run: B	660 m (2,165 ft)
D/DF	700 m (2,300 ft)
Range at 7,000 m (22,975 ft):	
B with 570 l fuel	235 n miles (435 km; 270 miles)
D/DF with 570 l fuel	243 n miles (450 km; 280 miles)
B with 1,200 l fuel	626 n miles (1,160 km; 720 miles)
g limits (ultimate): B, D/DF	+8.0/−4.0

UPDATED

PZL MIELEC TS-11R ISKRA UPGRADE

TYPE: Coastal reconnaissance modification of Iskra-Bis DF.

PROGRAMME: Original Iskra production totalled four prototypes and 423 production aircraft (31 Iskra 100, 45 Iskra 100-Bis A, 134 Iskra 100-Bis B, five Iskra 200 ART-Bis C and 208 Iskra 200 SB-Bis DF). TS-11R is replacement for SBLiM-2As (modified MiG-15UTIs) in Polish Naval Air Force, which were retired in early 1991; prototype (c/n 3H1917) converted at Mielec June 1991, delivered at end of year to 7th Special Air Regiment of PNAF at Siemirowice for evaluation; five more delivered subsequently.

DESIGN FEATURES: Modified by installing weather radar in reshaped nose behind dielectric nosecone and replacing dual controls in rear cockpit with radar VDU and artificial horizon.

AVIONICS: Bendix/King RDS-81 weather radar in nose; GPS navigation system; three (Russian) AFA-39 optical cameras.

VERIFIED

PZL Mielec TS-11R Iskra prototype conversion *(Piotr Butowski)*

1994

PZL Mielec TS-11 Iskra two seat advanced jet trainer *(Piotr Butowski)*

1994

PZL SWIDNIK SA

ZYGMUNTA PULAWSKIEGO-PZL SWIDNIK (Zygmunt Pulawski Transport Equipment Manufacturing Centre, Swidnik)

Al. Lotników Polskich 1, PL-21-045 Swidnik k/Lublina
Telephone: 48 (81) 12061, 12071, 13061, 13071 and 13081
Fax: 48 (81) 13505, 12173 and 13358
Telex: 0642301 WSK PL
GENERAL MANAGER: Mieczyslaw Majewski, MScEng
DIRECTOR OF RESEARCH AND DEVELOPMENT:
Ryszard Kochanowski, MScEng
MARKETING MANAGER: Andrzej Stachyra

Swidnik factory established 1951; engaged initially in manufacturing components for LiM-1 (MiG-15) jet fighter; began licence production of Soviet Mi-1 helicopter in 1955, building some 1,700 as SM-1s, followed by 450 Swidnik developed SM-2s; design office formed at factory to work on variants/developments of SM-1 and original projects such as SM-4 Latka.

Swidnik works named after famous pre-War PZL designer, Zygmunt Pulawski, September 1957; currently employs about 5,000 persons; production concentrates on W-3 and W-3A Sokól, Kania, and Soviet designed Mi-2; also manufactures PW-5 sailplane (World Class competition winner) and components for Aerospatiale/Alenia ATR 72. Polish government announced intention to privatise Swidnik factory 22 February 1991 (effective January 1992), initially as state-owned limited company. This process is still taking place.

Plans announced early 1993 to set up subassembly factory for Sokól in Halifax, Nova Scotia, to be known as PZL Canada, to assemble W-3 for North/South American and Pacific Rim markets, possibly with Canadian or US engines and avionics. This has not yet been achieved.

UPDATED

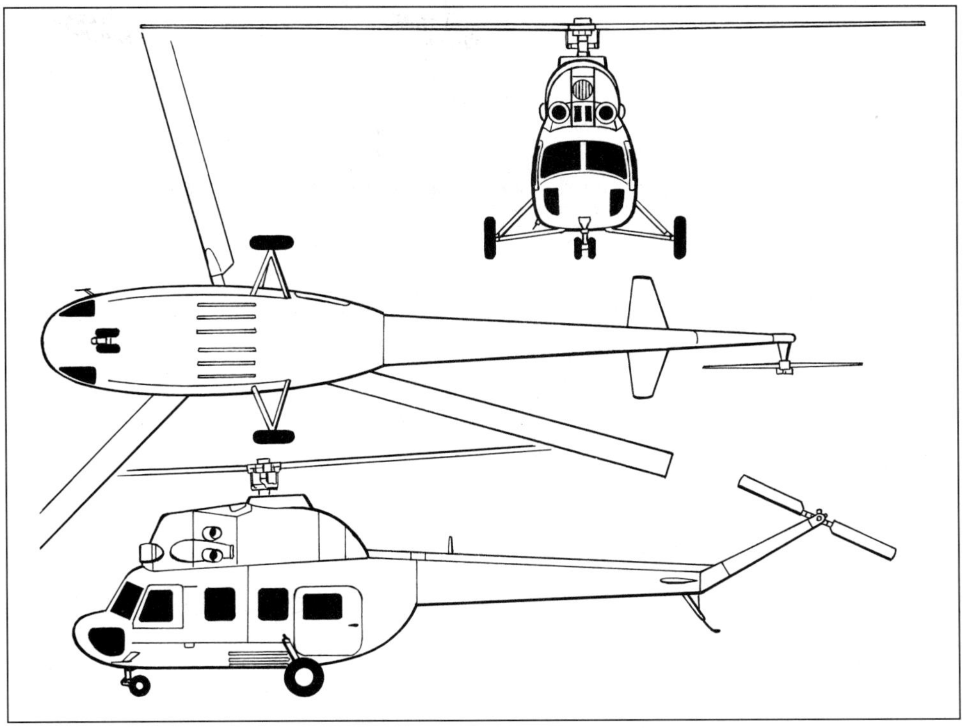

PZL Swidnik (Mil) Mi-2 *(Jane's/Mike Keep)* *1994*

PZL Swidnik (Mil) Mi-2 of the former Czechoslovakian Air Force *(Peter March)* *1994*

PZL SWIDNIK (MIL) Mi-2

NATO reporting name: Hoplite

TYPE: Twin-turbine general purpose light helicopter.

PROGRAMME: Designed in USSR by Mikhail L. Mil and first flown September 1961; January 1964 agreement assigned further development, production and marketing exclusively to Polish industry; first flight of Polish example 4 November 1965; series production began 1965. Has undergone continuous development and upgrading, with versions for new applications developed to meet specific customers' requirements. Production at standstill early 1992, but reported continuing on limited basis 1994.

VERSIONS: **Civil Mi-2**: Recent versions include examples for convertible passenger/cargo transport; ambulance and rescue (**Mi-2R**); agricultural dusting and spraying (conventional or ULV); freighter with external sling and electric hoist; training; and aerial photography (able to carry photographic, photogrammetric, thermal imaging or TV cameras for oblique or vertical pictures. *Details below refer to basic civil Mi-2, except where indicated.*

Mi-2B: Different electrical system and more modern navigational aids; manufactured in same versions (except agricultural) as basic Mi-2, and has same flight performance; empty equipped weights 2,300 kg (5,070 lb) for passenger version, 2,293 kg (5,055 lb) for cargo version; T-O weight unchanged; no rotor blade de-icing. Production total not large.

Mi-2RM: Naval version.

Mi-2T: Military transport version.

Mi-2URN: Combat support/armed reconnaissance version; as Mi-2US but with two Mars 2 launchers (each 16 S-5 57 mm unguided rockets) instead of pylon-mounted gun pods; PKV gunsight in cockpit for aiming all weapons; in service from 1973.

Mi-2URP: Anti-tank version; cabin side outriggers for four 9M14M Malyutka (AT-3 'Sagger') wire-guided missiles; four additional missiles in cargo compartment; in service from 1976; later version can carry four 9M32 Strela 2 missiles.

Mi-2US: Gunship version; 23 mm NS-23KM cannon on port side of fuselage, two 7.62 mm gun pods on each side pylon, two other 7.62 mm PK type pintle-mounted machine guns in rear of cabin.

OPERATORS: Versions of the Mi-2 are in service with the armed forces of the following countries: Belarus (10), Bulgaria (14), Cuba (5), Czech Republic (65), Djibouti (2), Hungary (40), Latvia (6), Nicaragua (4), Poland (100), Russia (500), Slovakia (15+), Syria (20) and Ukraine (80).

CUSTOMERS: Over 5,250 built for civil and military operators, majority exported; military operators include air forces of Bulgaria, Czech Republic, Germany, Hungary, Iraq, North Korea, Libya, Poland (approximately 130 Mi-2CH, T, URN, URP, US and other variants, mostly with No. 49 and No. 56 Combat Helicopter Regiments and No. 47 Helicopter Training Regiment), Russia, Slovakia and Syria; civil operators in European and various developing countries, including agricultural models in Bulgaria, Czech Republic, Egypt, Iraq, Poland and Russia.

DESIGN FEATURES: Three-blade main rotor with hydraulic blade vibration dampers; flapping, drag and pitch hinges on each blade; anti-flutter weights on leading-edges, balancing plates on trailing-edges. Coil spring counterbalance in main and tail rotor systems; pitch change centrifugal loads on tail rotor carried by ribbon type steel torsion elements. Blades do not fold; rotor brake fitted. Main rotor blade section NACA 230-12M. Main rotor shaft driven via gearbox on each engine; three-stage WR-2 main gearbox, intermediate gearbox and tail rotor gearbox; main rotor/engine rpm ratio 1:24.6, tail rotor/engine rpm ratio 1:4.16; main gear box provides drive for auxiliary systems and take-off for rotor brake; freewheel units permit disengagement of failed engine and autorotation.

FLYING CONTROLS: Hydraulic system for cyclic and collective pitch control boosters; variable incidence horizontal stabiliser, controlled by collective pitch lever.

STRUCTURE: Main and tail rotor blades have extruded duralumin spar with bonded honeycomb trailing-edge pockets; pod and boom fuselage of sheet duralumin, bonded and spot welded or riveted or longerons and frames, in three main assemblies (nose incuding cockpit, central section and tailboom); steel alloy main load bearing joints.

LANDING GEAR: Non-retractable tricycle type, plus tailskid. Twin-wheel nose unit. Single wheel on each main unit. Oleo-pneumatic shock-absorbers in all units, incuding tailskid. Main shock-absorbers designed to cope with both normal operating loads and possible ground resonance. Mainwheel tyres size 600 × 180, pressure 4.41 bars (64 lb/sq in). Nosewheel tyres size 400 × 125, pressure 3.45 bars (50 lb/sq in). Pneumatic brakes on mainwheels. Metal ski landing gear optional.

POWER PLANT: Two 298 kW (400 shp) Polish built Isotov GTD-350 turboshafts, mounted side by side above cabin. Fuel in single rubber tank, capacity 600 litres (158.5 US gallons; 131 Imp gallons), under cabin floor. Provision for carrying 238 litre (63 US gallon; 52.4 Imp gallon) external tank on each side of cabin. Refuelling point in starboard side of fuselage. Oil capacity 25 litres (6.6 US gallons; 5.4 Imp gallons).

ACCOMMODATION: Normal accommodation for one pilot on flight deck (port side). Seats for up to eight passengers in air conditioned cabin, comprising back to back bench seats for three persons each, with two optional extra starboard side seats at rear, one behind the other. All passenger seats removable for carrying up to 700 kg (1,543 lb) of internal freight. Access to cabin via forward hinged doors on each side at front of cabin and aft on port side. Pilot's sliding window jettisonable in emergency. Ambulance version has accommodation for four stretchers and medical attendant, or two stretchers and two sitting casualties. Side by side seats and dual controls in pilot training version. Cabin heating, ventilation and air conditioning standard.

SYSTEMS: Cabin heating, by engine bleed air, and ventilation; heat exchangers warm atmospheric air for ventilation system during cold weather. Hydraulic system, pressure 65 bars (940 lb/sq in), for cyclic and collective pitch control boosters. Hydraulic fluid flow rate 7.5 litres (1.98 US gallons; 1.65 Imp gallons)/min. Vented reservoir, with gravity feed. Pneumatic system, pressure 49 bars (710 lb/sq in), for mainwheel brakes. AC electrical system, with two STG-3 3 kW engine-driven starter/generators and 208 V 16 kVA three-phase alternator. 24 V DC system, with two 28 Ah lead-acid batteries. Main and tail rotor blades de-iced electrically; engine air intake de-icing by engine bleed air. Electric de-icing of windscreen.

PZL Swidnik (Mil) Mi-2 Polish ambulance version

1994

AVIONICS: Standard items include two transceivers (MF/HF), gyro compass, radio compass, radio altimeter, intercom system and blind-flying panel. Nose and tail warning radar fitted to some military versions.

EQUIPMENT: Agricultural version carries hopper on each side of fuselage (total capacity 1,000 litres; 264 US gallons; 220 Imp gallons of liquid or 750 kg; 1,650 lb of dry chemical) and either a spraybar to rear of cabin on each side or distributor for dry chemicals under each hopper. Swath width covered by spraying version is 40-45 m (130-150 ft). For search and rescue, electric hoist, capacity 120 kg (264 lb), is fitted. In freight role an underfuselage hook can be fitted for suspended loads of up to 800 kg (1,763 lb). Polish press has illustrated version equipped for laying smokescreens. Electrically operated wiper for pilot's windscreen. Freon fire extinguishing system, for engine bays and main gearbox compartment, can be actuated automatically or manually.

ARMAMENT: See Versions.

DIMENSIONS, EXTERNAL:
Main rotor diameter 14.50 m (47 ft 6⅞ in)
Main rotor blade chord (constant, each) 0.40 m (1 ft 3¾ in)
Tail rotor diameter 2.70 m (8 ft 10¼ in)
Length overall, rotors turning 17.42 m (57 ft 2 in)
Length of fuselage 11.40 m (37 ft 4¾ in)
Height to top of rotor head 3.75 m (12 ft 3½ in)
Stabiliser span 1.85 m (6 ft 0¾ in)
Wheel track 3.05 m (10 ft 0 in)
Wheelbase 2.71 m (8 ft 10¾ in)
Tail rotor ground clearance 1.59 m (5 ft 2¾ in)
Cabin door (port, rear): Height 1.065 m (3 ft 5¾ in)
Width 1.115 m (3 ft 8 in)
Cabin door (stbd, front): Height 1.11 m (3 ft 7¾ in)
Width 0.75 m (2 ft 5½ in)
Cabin door (port, front): Height 1.11 m (3 ft 7¾ in)
Width 0.78 m (2 ft 6¾ in)

DIMENSIONS, INTERNAL:
Cabin:
Length: incl flight deck 4.07 m (13 ft 4¼ in)
excl flight deck 2.27 m (7 ft 5½ in)
Mean width 1.20 m (3 ft 11¼ in)
Mean height 1.40 m (4 ft 7 in)

AREAS:
Main rotor blades (each) 2.40 m^2 (25.83 sq ft)
Tail rotor blades (each) 0.22 m (2.37 sq ft)
Main rotor disc 166.4 m^2 (1,791.11 sq ft)
Tail rotor disc 5.73 m^2 (61.68 sq ft)
Horizontal stabiliser 0.70 m^2 (7.53 sq ft)

WEIGHTS AND LOADINGS:
Weight empty, equipped:
passenger version 2,402 kg (5,295 lb)
cargo version 2,372 kg (5,229 lb)
ambulance version 2,410 kg (5,313 lb)
agricultural version 2,372 kg (5,229 lb)
Basic operating weight empty:
single-pilot versions 2,365 kg (5,213 lb)
dual control version 2,424 kg (5,344 lb)
Max payload, excl pilot, oil and fuel 800 kg (1,763 lb)
Normal T-O weight (and max T-O weight of agricultural version) 3,550 kg (7,826 lb)
Max T-O weight (special versions) 3,700 kg (8,157 lb)
Max disc loading 22.4 kg/m^2 (4.6 lb/sq ft)

PERFORMANCE (at 3,550 kg; 7,826 lb T-O weight):
Never-exceed speed (VNE) at 500 m (1,640 ft):
agricultural version 84 kts (155 km/h; 96 mph)
other versions 113 kts (210 km/h; 130 mph)
Max cruising speed at 500 m (1,640 ft):
agricultural version (without agricultural equipment) 102 kts (190 km/h; 118 mph)
other versions 108 kts (200 km/h; 124 mph)
Max level speed with agricultural equipment 84 kts (155 km/h; 96 mph)
Econ cruising speed at 500 m (1,640 ft):
for max range 102 kts (190 km/h; 118 mph)
for max endurance 54 kts (100 km/h; 62 mph)
Max rate of climb at S/L 270 m (885 ft)/min
Time to 1,000 m (3,280 ft) 5 min 30 s
Time to 4,000 m (13,125 ft) 26 min
Service ceiling 4,000 m (13,125 ft)
Hovering ceiling: IGE approx 2,000 m (6,560 ft)
OGE approx 1,000 m (3,280 ft)
Range at 500 m (1,640 ft):
max payload, 5% fuel reserves 91 n miles (170 km; 105 miles)
max internal fuel, no reserves 237 n miles (440 km; 273 miles)
max internal and auxiliary fuel, 30 min reserves 313 n miles (580 km; 360 miles)
max internal and auxiliary fuel, no reserves 430 n miles (797 km; 495 miles)
Endurance at 500 m (1,640 ft), no reserves:
max internal fuel 2 h 45 min
max internal and auxiliary fuel 5 h
Endurance (agricultural version), 5% reserves:
spraying 40 min
dusting 50 min

VERIFIED

PZL SWIDNIK KANIA/KITTY HAWK

TYPE: Twin-turboshaft multipurpose light helicopter.

PROGRAMME: Developed in collaboration with Allison in USA; two prototypes produced by converting Mi-2 airframes; first flight of first prototype (SP-PSA) 3 June 1979. Polish certification of the Kania was carried out in two stages.

The first took place in 1979-81 and resulted, on 1 October 1981, in a supplementary type certificate to that of the Mi-2. The second stage, concerning a considerably improved **Kania Model 1** version was carried out during 1982-86 under the leadership of Stanisław I. Markisz. Improvements included among others, redesigned cockpit and cabin layout, engine and flight controls as well as engine and transmission cowlings. In February 1986 this version of the Kania was granted a separate type certificate as an FAR Pt 29 (Transport Cat. B) day and night SVFR multipurpose utility helicopter with Cat. A engine isolation. Three built by early 1991.

VERSIONS: **Kania Model 1:** Intended for passenger transport (with standard, executive or customised interiors), cargo transport (internal or slung load), agricultural (LV and ULV spraying/spreading/dusting), medevac, training, rescue, and aerial surveillance configurations.

CUSTOMERS: Four prototypes and six or seven production aircraft built.

DESIGN FEATURES: Three-blade fully articulated main rotor and two-blade seesaw tail rotor. Main gearbox equipped with freewheel units, oil cooling system, oil temperature and pressure gauges and switches, tacho-generator with low and high rpm warning, air compressor, and a spare power pad of 19.1 kW (25.6 shp) at 8000 rpm. Steel engine driveshafts, each with two crowned tooth couplings. Tail rotor driveshaft of duralumin tube, with similar crowned tooth couplings and anti-friction bearings. Hoist and cargo sling attachments points standard. Transmission includes main rotor, intermediate and tail rotor gearboxes, each with individual lubrication system.

FLYING CONTROLS: Three hydraulic boosters for longitudinal lateral and collective pitch control augmentation.

STRUCTURE: Glassfibre/epoxy blades on both rotors. Conventional semi-monocoque fuselage and circular-section tailboom. Glassfibre/epoxy horizontal stabiliser at end of tailboom.

LANDING GEAR: Non-retractable tricycle type, plus tailskid. Twin-wheel castoring and self-centring nose unit; single wheel on each main unit. Pneumatic brakes on mainwheels.

POWER PLANT: Two Allison 250-C20B turboshafts, mounted side by side above cabin; each rated at 313 kW (240 shp) for T-O, 30 minutes twin-engine emergency power and one engine out maximum continuous power, and 276 kW (370 shp) for normal cruise. Automatic and manual torque sharing control systems standard. Two separate fuel boost systems, each with fuel filter bypass switch, fuel pressure gauge and switch, connected by crossfeed. Standard usable fuel capacity of 600 litres (158.5 US gallons; 131 Imp gallons), with provision for additional 423 litres usable (111.75 US gallons; 93 Imp gallons) in optional auxiliary tanks. Fuel quantity gauge and fuel reserve warning. Two separate oil systems, each with oil cooling, temperature and pressure gauges, oil filter bypass pop-up and chip warning. Each engine equipped with starter/generator, engine fuel pump effective for cruise after both boost pumps out, N1 and N2 tacho-generators, TOT gauge and switch, start counter, and 'engine out' warning. Dual engine inlet anti-icing standard, each engine compartment equipped with fire detection system and with automatic and manual fire extinguishing systems.

ACCOMMODATION: Pilot (port side), and co-pilot or passenger, on adjustable and removable front seats, each fitted with safety belt. Dual controls optional. Accommodation for up to eight more persons, on two three-person bench seats and a single or double seat at rear of cabin, all with safety belts. Seats removable for carriage of cargo (up to 1,200 kg; 2,645 lb), two or four stretchers plus medical attendants, agricultural or other specialised equipment. Access to cabin via jettisonable door on each side at front (port side of sliding type) and larger passenger/cargo door at rear on port side. Pilot's windscreen wiper standard, co-pilot's optional. Cargo and stretcher tiedown points in cabin floor. Cabin soundproofing and ventilation standard; heating, carpets, double pane windows, pilot's heated windscreen, all optional. Baggage compartment at rear of cabin. Cockpit and cabin lighting standard.

SYSTEMS: Hydraulic system, with pressure gauge and switch, standard. Compressed air system, with accumulator and system gauges, standard. Ventilation standard, with individually controllable fresh air outlets; Casey cabin heaters optional, with individual control of hot air flow and central control of overall cabin temperature. DC electrical system based on two 28 V 150 A starter/generators and a 25 Ah Ni/Cd battery, with ground power receptacle. Ground/battery power, battery overtemperature and 'generator out' warnings standard. A 16 kVA AC generator and/or 115 V 250 A static inverter are optional; this AC system is equipped with AC generator and AC 115 V warnings. Dual fire detection and extinguishing systems for engines standard. Electric de-icing of main and tail rotor blades (including icing and 'system out' warnings) optional.

AVIONICS: Primary instrumentation includes attitude altitude airspeed, turn and slip, and rate of climb indicators; magnetic compass and gyro compass; HSI; clock; VHF com transceiver; and full range of power plant and systems control, monitoring and warning instruments. Optional radio-navigation avionics include digital ADF, R/Nav (VOR 1), audio panel, VOR/LOC/glide slope converter, transponder with or without altitude encoder, marker beacon receiver, DME, second VHF com transceiver, VOR 2 receiver, HF com transceiver, and radar altimeter.

EQUIPMENT: Standard equipment includes dual anti-collision lights, navigation lights, portable fire extinguisher, tool kit and first aid kit. Fluorescent tube cabin lighting and/or individual lights optional. According to mission, the Kania can be equipped with an 800 kg (1,763 lb) capacity stabilised cargo sling; 120 kg (265 lb) capacity hoist (275 kg; 606 lb hoist under test); stretchers and casualty care equipment; or equipment for a variety of agricultural duties.

DIMENSIONS, EXTERNAL:
Main rotor diameter 14.558 m (47 ft 9¼ in)
Tail rotor diameter 2.70 m (8 ft 10¼ in)
Length overall, rotors turning 17.47 m (57 ft 3¾ in)
Length of fuselage 12.03 m (39 ft 5½ in)
Height to top of rotor head 3.75 m (12 ft 3½ in)
Stabiliser span 1.84 m (6 ft 0½ in)
Wheel track 3.05 m (10 ft 0 in)
Wheelbase 2.71 m (8 ft 10¾ in)

DIMENSIONS, INTERNAL:
Cabin: Length, incl flight deck 4.07 m (13 ft 4¼ in)
Max width 1.50 m (4 ft 11 in)
Max height 1.62 m (5 ft 3¾ in)
Floor area 5.68 m^2 (61.1 sq ft)
Volume 7.76 m^3 (274.0 cu ft)
Baggage compartment volume 0.45 m^3 (15.89 cu ft)

AREAS:
- Main rotor disc 166.50 m^2 (1,792.2 sq ft)
- Tail rotor disc 5.725 m^2 (61.6 sq ft)

WEIGHTS:
- Basic weight empty 2,000 kg (4,409 lb)
- Max load in cabin 1,200 kg (2,645 lb)
- Max cargo sling load 800 kg (1,763 lb)
- Max agricultural chemical load 1,000 kg (2,205 lb)
- Max load in baggage compartment 100 kg (220 lb)
- Normal T-O weight 3,350 kg (7,385 lb)
- Max T-O weight 3,550 kg (7,826 lb)

PERFORMANCE (clean aircraft at S/L, ISA, zero wind, at normal T-O weight):
- Max cruising speed 116 kts (215 km/h; 135 mph)
- Econ cruising speed 102 kts (190 km/h; 118 mph)
- Max rate of climb (T-O power) 525 m (1,725 ft)/min
- Rate of climb, one engine out 61 m (200 ft)/min
- Service ceiling 4,000 m (13,125 ft)
- Hovering ceiling: IGE 2,500 m (8,200 ft)
- OGE 1,375 m (4,510 ft)
- Range at econ cruising speed:
  - standard fuel, 30 min reserves 232 n miles (430 km; 267 miles)
  - standard fuel, no reserves 266 n miles (493 km; 306 miles)
  - max fuel, 30 min reserves 432 n miles (800 km; 497 miles)
  - max fuel, no reserves 466 n miles (863 km; 536 miles)

VERIFIED

PZL Swidnik Kania twin-turboshaft light helicopter

1994

PORTUGAL

OGMA

OFICINAS GERAIS DE MATERIAL AERONÁUTICO (General Aeronautical Material Workshops)

P-2615 Alverca
Telephone: 351 (1) 9581000
Fax: 351 (1) 9581288 and 9580401
Telex: 14479 OGMA P
DIRECTOR: Maj-Gen Adriano de Aldeia Portela
DEPUTY DIRECTOR: Col A. A. Nogueira Pinto
COMMERCIAL MANAGER: Lt Col M. Renato Oliveira
PUBLIC RELATIONS AND ADVERTISING MANAGER: Jorge M. F. Pires

OGMA, founded in 1918, is the department of the Portuguese Air Force responsible for maintenance and repair, at depot level, of its aircraft, avionics, engines, ground communications and radar equipment, and can undertake similar work for civil or military national or foreign customers. OGMA has a total covered area of 116,000 m^2 (1,248,612 sq ft), and a workforce of approximately 2,700 people.

Under a contract signed in 1959, OGMA undertakes IRAN, refurbishing and rehabilitation, periodic inspection and emergency maintenance and crash repair of US Air Force and Navy aircraft. For Aerospatiale of France, OGMA has manufactured main and tail rotor structures for the SA 315B Lama and some components for other helicopters.

OGMA's engine repair and maintenance facility, with a covered area of 28,000 m^2 (301,390 sq ft), overhauls military and commercial turbojets and turbofans (up to 146.8 kN; 33,000 lb st), and turboprop and turboshaft engines of up to 5,667 kW (7,600 shp). In addition to two fully computerised test cells, this facility is equipped with plasma spray, two vacuum furnaces, complete cleaning and electroplating facilities, non-destructive testing, shot-peening and other specific equipment. Besides work for the Portuguese Air Force, OGMA also overhauls, under contract, Artouste III and Turmo IV turboshaft engines for Turbomeca of France; and is a maintenance/overhaul centre for Allison T56 engines and gearboxes for the USAF and other customers.

OGMA performs major maintenance on C-130/L-100 Hercules transport aircraft as a Lockheed Service Center, and on Alouette III, Puma and Ecureuil helicopters as an Aerospatiale Service Station. It has recently been appointed as a line service centre for Dassault Falcon 20 and Falcon 50 aircraft and Garrett TFE731 engines.

The Avionics Division has premises covering an area of approximately 6,400 m^2 (68,900 sq ft), fully equipped to the latest demands in the field of maintenance for new generation avionics, communications systems, test equipment and calibration laboratories. OGMA is licensed by Litton Systems of Canada to carry out level 2 and 2A maintenance on LTN-72 INS equipment.

VERIFIED

OGMA (Aerospatiale) AS 330 Puma S1 re-engined with Makila turboshaft engines
1994

OGMA (AEROSPATIALE) AS 330 PUMA RE-ENGINE PROGRAMME

TYPE: Combat helicopter re-engine programme.

PROGRAMME: During early 1991, OGMA was engaged in a programme to re-engine 10 Portuguese Air Force AS 330 Puma S1 helicopters with Makila turboshaft engines. These aircraft were also retrofitted with composite blades.

VERIFIED

ROMANIA

AEROSTAR

BACAU Grup Industrial Aeronautic (Aeronautical Industrial Group)

9 Condorilor Street, R-5500 Bacău
Telephone: 40 (93) 130070
Fax: 40 (93) 170513-161113
Telex: 21339 AERO R
DIRECTOR GENERAL: Dipl Eng Ioan Florin Vornicelu
PRODUCTION DIRECTOR: Dipl Eng Constantin Prescură
TECHNICAL DIRECTOR: Dipl Eng Ioan Prescură
COMMERCIAL DIRECTOR: Dipl Eng Dan Onică

Factory founded 1953 as URA (later IRAv, then IAv Bacau), originally as repair centre for Romanian Air Force Yak-17/23, MiG-15/17/19/21 and Il-28/H-5 front-line military aircraft and Aero L-29/L-39 jet trainers; this work still continuing for later of these types; built first Romanian prototype of IAR-93 between 1972 and 1974. Five other specialised work sections: landing gears hydraulic and pneumatic equipment; special production; light aircraft; engines and reduction gears; and avionics. Factories have production buildings floor area totalling 50,396 m² (542,457 sq ft) plus further 31,618 m² (340,333 sq ft) of auxiliary buildings; workforce in early 1992 totalled 8,385, including 1974 technical personnel. Landing gears and/or hydraulic/pneumatic equipment have been produced for IAR-316B (Alouette), IAR-330 (Puma) and Ka-126 helicopters; IAR-93, IAR-99 and Rombac 1-11 jet aircraft; and Iak-52 and IAR-822/823/827 and AG-6 light aircraft. Engines include former Soviet M-14P for Iak-52, M-14V26 for Ka-26 and RU-19A-300 APU for An-24; reduction gears include R-26 and VR-126. Avionics factory produces radio altimeters, radio compasses, marker beacon receivers, IFF and other radio/radar items.

VERIFIED

AEROSTAR (MIKOYAN) MiG-21 UPGRADE

TYPE: Combat aircraft upgrade.
PROGRAMME: Collaborating with Elbit on upgrade programme for 100 MiG-21 aircraft of Romanian Air Force. For full details of programme see separate entry in Elbit section.

VERIFIED

IAR

IAR SA (formerly ICA)

1 Aeroportului Street, PO Box 198, R-2200 Brasov
Telephone: 40 (921) 50015
Fax: 40 (921) 51304
Telex: 61266 ICAER R
MANAGING DIRECTOR: Dipl Eng N. Banea

Factory, created 1968, continues work begun in 1926 by IAR-Brasov and undertaken 1950-59 as URMV-3 Brasov; approximately 3,000 workforce in 1994. Currently manufactures Romanian designed light aircraft, Puma helicopters under licence from Eurocopter France (as IAR-330) and Russian Ka-126; IS-28/29 series sailplanes/motor gliders; aircraft components and equipment.

VERIFIED

IAR IAR-823

TYPE: Two/five-seat cabin monoplane.

PROGRAMME: Design of the IAR-823 training and touring light aircraft was started at IMFCA in May 1970, by a team led by Dipl Eng Radu Manicatide. Construction of a prototype began at ICA-Brasov Autumn 1971, and this aircraft made its first flight in July 1973. Construction and testing were in compliance with FAR Pt 23, and the aircraft is certificated for aerobatic, utility and normal category operation. The first production aircraft flew in 1974 and 87 had been delivered to the Romanian Air Force and Romanian flying clubs by Summer 1982. The IAR IAR-823 is still in service with the Romanian Air Force (35+).

As a two-seater, the IAR IAR-823 is fully aerobatic and is intended for training duties. With a rear bench seat for up to three more persons it is suitable as an executive, taxi or touring aircraft. Provision is made for two underwing pylons for the carriage of drop tanks or practice weapons.

DESIGN FEATURES: Cantilever low-wing monoplane. Wing section NACA 23012 (modified). Dihedral 7° from roots. Incidence 3° at root, 1° at tip. Cantilever tail unit.

FLYING CONTROLS: Electrically actuated fabric covered metal single-slotted flaps and fabric covered Frise-slotted metal ailerons. Ground adjustable tab on each aileron. Electrically actuated automatic trim tab in each elevator; controllable tab in rudder.

STRUCTURE: The wing is a conventional all-metal structure with single main spar and rear auxiliary spar; three-point attachment to fuselage. Riveted spars, ribs and skin of corrosion-proof aluminium alloy. Leading-edges riveted, and sealed to ribs and main spar to form main torsion box and integral fuel tanks. The fuselage is an all-metal semi-monocoque structure. Glassfibre engine cowling. The tail unit is a cantilever metal structure. Two-spar duralumin covered fin and tailplane; fabric covered duralumin horn balanced rudder and elevators.

LANDING GEAR: Retractable tricycle type, with steerable nosewheel. Electrical retraction, main units inward, nose unit rearward. Emergency manual actuation. Oleo-pneumatic shock-absorbers. Mainwheel tyres size 6.00-6, pressure 2.93 bars (42.5 lb/sq in). Nosewheel tyre size 355 × 150 mm. Independent hydraulic mainwheel brakes, pedal controlled from left front seat. Shimmy damper on nose unit. No wheel doors.

POWER PLANT: One 216 kW (290 hp) Avco Lycoming IO-540-GID5 flat-six engine, driving a Hartzell HC-92WK-1D/W 9350-4.6 two-blade constant-speed metal propeller. Fuel in four integral wing tanks, total capacity 360 litres (95 US gallons; 79 Imp gallons). Provision for two 70 litre (18.5 US gallon; 15.4 Imp gallon) drop tanks on underwing pylons.

ACCOMMODATION: Fully enclosed cabin, seating two persons side by side on individual adjustable front seats, with removable bench seat at rear for up to three more people. Dual controls standard in training versions, optional in other versions. Upward hinged window/door (optionally jettisonable) on each side of cabin, which is soundproofed, heated and ventilated. Compartment at rear of cabin for up to 40 kg (88 lb) of baggage. Equipment and layout can be varied for use as air taxi, executive or freight transport, ambulance, liaison or photographic aircraft.

IAR-823 two/five-seat training and touring aircraft

1994

SYSTEMS AND AVIONICS: Electrical system, including 50A alternator and 24 V 30 Ah battery, for engine starting, elevator tab and landing gear actuation, radio communications, landing and navigation lights, and cabin and instrument lighting. Standard avionics include VFR instrumentation and TR 800 transceiver. Optional equipment according to mission, includes blind-flying instrumentation and, in civil transport version, marker beacon receiver, nav/com radio, VOR/ILS, ADF and autopilot.

DIMENSIONS, EXTERNAL:

Wing span	10.00 m (32 ft 9¾ in)
Wing chord: at c/l	2.00 m (6 ft 6¾ in)
at tip	1.00 m (3 ft 3¼ in)
Wing aspect ratio	6.7
Length overall	8.315 m (27 ft 3½ in)
Height overall	2.86 m (9 ft 4¾ in)
Wheel track	2.48 m (8 ft 1¾ in)
Wheelbase	1.86 m (6 ft 1¼ in)
Propeller diameter	2.23 m (7 ft 4 in)

AREAS:

Wings, gross	15.00 m² (161.5 sq ft)
Ailerons (total)	1.20 m² (12.92 sq ft)
Trailing-edge flaps (total)	1.78 m² (19.16 sq ft)
Horizontal tail surfaces (total)	3.30 m² (35.52 sq ft)
Vertical tail surfaces (total)	1.50 m² (16.15 sq ft)

WEIGHTS AND LOADINGS (A: Aerobatic, U: Utility, N: Normal category):

Weight empty: A	910 kg (2,006 lb)
U	930 kg (2,050 lb)
N	950 kg (2,094 lb)
Max T-O weight: A	1,190 kg (2,623 lb)
U	1,380 kg (3,042 lb)
N	1,500 kg (3,307 lb)
Max wing loading: A	79.3 kg/m² (16.24 lb/sq ft)
U	92.0 kg/m² (18.85 lb/sq ft)
Max power loading: A	5.51 kg/kW (9.05 lb/hp)
U	6.39 kg/kW (10.49 lb/hp)
N	6.94 kg/kW (11.40 lb/hp)

PERFORMANCE (A: at Aerobatic max T-O weight, B: at AUW of 1,400 kg; 3,086 lb):

Never-exceed speed: B	215 kts (400 km/h; 248 mph)
Max level speed at S/L: A, B	162 kts (300 km/h; 186 mph)
Max cruising speed (75% power):	
A at S/L	154 kts (285 km/h; 177 mph)
B at 1,750 m (5,750 ft)	162 kts (300 km/h; 186 mph)
Econ cruising speed (60% power):	
A at S/L	140 kts (260 km/h; 162 mph)
B at 3,050 m (10,000 ft)	156 kts (290 km/h; 180 mph)
Stalling speed:	
flaps up: A	61 kts (112 km/h; 70 mph)
B	63 kts (115 km/h; 72 mph)
flaps down, power off:	
A	55 kts (102 km/h; 64 mph)
B	49 kts (90 km/h; 56 mph)
Max rate of climb at S/L: A, B	420 m (1,380 ft)/min
Service ceiling: A, B	5,600 m (18,375 ft)
T-O run: A, B	160 m (525 ft)
T-O run to 15 m (50 ft): A, B	300 m (984 ft)
Landing run from 15 m (50 ft): A, B	250 m (820 ft)
Landing run: A, B	200 m (656 ft)

Range with max fuel: A701 n miles (1,300 km; 807 miles)
B, according to mission and payload, 1 h reserves
431-970 n miles (800-1,800 km; 497-1,118 miles)

Endurance, according to mission and payload	3-6 h
g limits	+6/−3

VERIFIED

IAR IAR-330 PUMA 2000

TYPE: Combat helicopter upgrade.

PROGRAMME: Design being finalised late 1992; prototype roll-out was due 1993.

VERSIONS: Upgrade to be offered on **IAR-330J** and **-330L**.

DESIGN FEATURES: More modern and higher powered engines (type not specified); EFIS flight deck and GPS navigation; obstacle avoidance database; advanced night vision system. Elbit of Israel prime contractor for new avionics.

ACCOMMODATION: 23 seats or high density and greater cargo loads.

AVIONICS: Standard aircraft will include HOCAS (hands off cyclic and stick), helmet-mounted HUD and GPS navigation. Options include El-Op multisensor stabilised integrated system (MSIS) with nose-mounted TV/FLIR cameras, digital image processing and communications system, engine performance indicator and radar/laser warning systems. New EFIS flight deck will use 1553B databus technology and versatile central onboard computer, plus comprehensive mission management software including laser rangefinder/target designator and weapon delivery system. Night vision system comprises helmet-mounted NVGs for both pilots, supplemented by HUD. Multifunction displays (MFDs) to include five different vector or raster moving maps, ILS landing information, and video images of battlefield from TV/FLIR cameras or prerecorded tapes. Downlink option for MFD imagery.

ARMAMENT: 20 mm guns and Hellfire or Swingfire missiles, or other ordnance of customer's choice.

VERIFIED

RUSSIA

BERIEV (TANTK)

TAGANROG AVIATSIONNYI NAUCHNO-TEKHNICHESKIY KOMPLEKS IMIENI G. M. BERIEVA (TANTK) (Taganrog Aviation Scientific-Technical Complex named after G. M. Beriev)

Instrumentalniy Toupic 347927, Taganrog
Telephone: 7 (86344) 49839, 49901, 49964
Fax: 7 (86344) 41454

PRESIDENT AND GENERAL DESIGNER: Gennady S. Panatov

This OKB founded by Georgy Mikhailovich Beriev (1902-1979) in 1932; except during the Second World War, 1942-45, it has been based at Taganrog, in the northeast corner of Sea of Azov; since 1948 has been primary centre for Russian seaplane development. In 1990 was redesignated as shown.

TANTK now includes the experimental design bureau, experimental production facilities, a flight test complex, economic, financial and logistics support services, with test bases and proving grounds at the Black Sea and Sea of Azov. Its products are experimental prototypes of amphibious aircraft and wing-in-ground-effect (WIG) vehicles, together with test reports and technical documentation for their series production. It undertakes design and development of unconventional aircraft in response to requests for proposals from other companies, testing of aircraft and assemblies in maritime conditions, and training of aircrew and ground personnel for seaplane operation.

VERIFIED

Beriev Be-12 'Mail' over the Sea of Japan *(G. Jacobs)*

1994

Beriev Be-12 'Mail' twin-turboprop anti-submarine aircraft

1994

BERIEV Be-12 (M-12) TCHAIKA (SEAGULL)

NATO reporting name: Mail

TYPE: Twin-turboprop anti-submarine and maritime patrol amphibian.

PROGRAMME: First flew 1960; displayed publicly at 1961 Soviet Aviation Day Display, Tushino Airport, Moscow; production of estimated 100 began 1964 to replace Be-6s on anti-submarine and surveillance duties out to some 200 n miles (370 km; 230 miles) from shore bases.

UPGRADES: **Beriev:** Be-12P. See separate entry.

OPERATORS: Versions of the Be-12 are in service with the armed forces of the following countries: Russia (90) and Ukraine (20+).

CUSTOMERS: Soviet Naval Aviation had an estimated 75 in 1991, with Northern and Black Sea Fleets (95 in service in total); no exports, although Be-12 seen in temporary Egyptian insignia in early 1970s.

DESIGN FEATURES: General configuration inherited from piston-engined Be-6, with sharply cranked wing to raise propellers clear of water, single-step hull of high length/beam ratio, with spray dams on sides of nose.

FLYING CONTROLS: Mechanically actuated; hydraulically boosted ailerons, each with two electrically operated tabs; elevators and horn-balanced rudders hydraulically boosted; electrically operated trim tab in each elevator and rudder; hydraulically actuated trailing-edge flaps in two sections on each wing, passing under engine nacelle to wingroot.

STRUCTURE: Conventional all-metal; two-spar wings with considerable centre-section dihedral, slight anhedral on outer panels; semi-monocoque hull with conventional planing bottom; dihedral tailplane with endplate fins and rudders.

LANDING GEAR: Hydraulically retractable tailwheel type, comprising single wheel main units which retract upward through 180° to lie flush within sides of hull, and a rearward-retracting steerable tailwheel. Oleo-pneumatic mainwheel shock-absorbers. Except for top of each mainwheel, all units are fully enclosed by doors when retracted. Non-retractable wingtip floats.

POWER PLANT: Two Ivchenko AI-20D turboprops, each rated at 3,862 kW (5,180 ehp) and driving an AV-68I four-blade variable-pitch propeller. Metal cowlings open downward in halves, permitting their use as servicing platforms. Fuel tanks, between spars in wings and in fuselage, with total capacity of approximately 11,000 litres (2,905 US gallons, 2,420 Imp gallons).

ACCOMMODATION: Not pressurised. Crew of five on flight deck. Glazed navigation and observation station in nose. Astrodome observation station in top of rear fuselage. Side hatches in rear fuselage permit loading while afloat.

SYSTEMS: Hydraulic system actuates flaps and landing gear. Two engine-driven generators power 28 V DC electrical system.

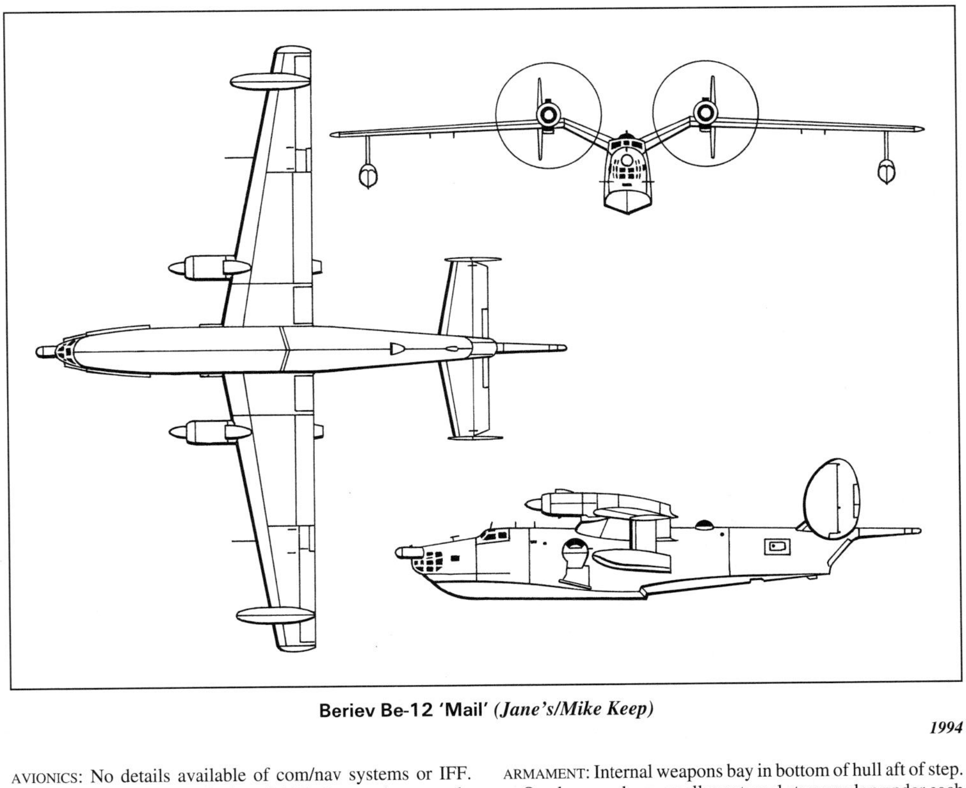

Beriev Be-12 'Mail' *(Jane's/Mike Keep)*

1994

AVIONICS: No details available of com/nav systems or IFF. Radome above nose glazing. MAD (magnetic anomaly detection) sting extends rearward from tail.

EQUIPMENT: APU exhausts through aperture in port side of rear fuselage.

ARMAMENT: Internal weapons bay in bottom of hull aft of step. One large and one smaller external stores pylon under each outer wing panel, for torpedoes, depth charges, mines and other stores.

DIMENSIONS, EXTERNAL:

Wing span	29.84 m (97 ft 8¾ in)
Wing aspect ratio	8.99
Length overall	30.17 m (99 ft 0 in)
Height overall	7.00 m (22 ft 11½ in)
Propeller diameter	4.85 m (16 ft 0 in)
AREAS:	
Wings, gross	99 m² (1,065.2 sq ft)
WEIGHTS AND LOADINGS:	
Max operational load	10,000 kg (22,045 lb)
Max T-O weight	36,000 kg (79,344 lb)
Max wing loading	363.6 kg/m² (74.5 lb/sq ft)
Max power loading	4.39 kg/kW (7.22 lb/ehp)
PERFORMANCE:	
Max level speed	296.8 kts (550 km/h; 342 mph)
Normal operating speed	172 kts (320 km/h; 199 mph)
Rate of climb at S/L	912 m (2,990 ft)/min
Service ceiling	11,280 m (37,000 ft)
Range with max fuel	4,050 n miles (7,500 km; 4,660 miles)

VERIFIED

Beriev Be-12P firefighter conversion

1994

Be-12P FIREFIGHTER UPGRADE

TYPE: Twin-turboprop firefighter upgrade.

PROGRAMME: Beriev, in accordance with the TANTK's conversion policy has upgraded several Be-12s for firefighting duties. The Be-12P has already taken part in firefighting operations within the Russian Federation.

DESIGN FEATURES: Airframe installed with water tanks with a total capacity of 6 m³ water intake systems for scooping water during aquaplaning, systems for water intake from airfield sources as well as systems for dropping water on fires.

UPDATED

BERIEV A-50

NATO reporting name: Mainstay

TYPE: Four-turbofan airborne early warning and control aircraft.

PROGRAMME: Development from Ilyushin Il-76 transport began in 1970s to replace Tu-126s of APVO; production began early 1980s; maintained at rate of five a year until 1990 when only two delivered, causing problems for fighter force; service entry 1984; two operated round-the-clock over Black Sea during 1991 Gulf War, monitoring USAF flights from Turkey to Iraq and watching for possible stray US cruise missiles heading for CIS territory. Continued development by Beriev OKB.

OPERATORS: Versions of the A-50 are in service with the armed forces of Russia (15).

DESIGN FEATURES: Derivative of Il-76, with conventionally located rotating 'saucer' radome (diameter 9 m; 29 ft 6 in), lengthened fuselage forward of wings, flight refuelling noseprobe, satellite nav/com antenna above fuselage forward of wing, new IFF and comprehensive ECM; normal nose glazing around navigator's station replaced by non-transparent fairings; intake for avionics cooling at front of dorsal fin; no rear gun turret; small aerodynamic surface on each outer landing gear fairing, 'shadowing' circular form of rotating radome aft of wing trailing-edge.

FLYING CONTROLS: Hydraulically boosted; manual operation possible in emergency; mass balanced ailerons, with

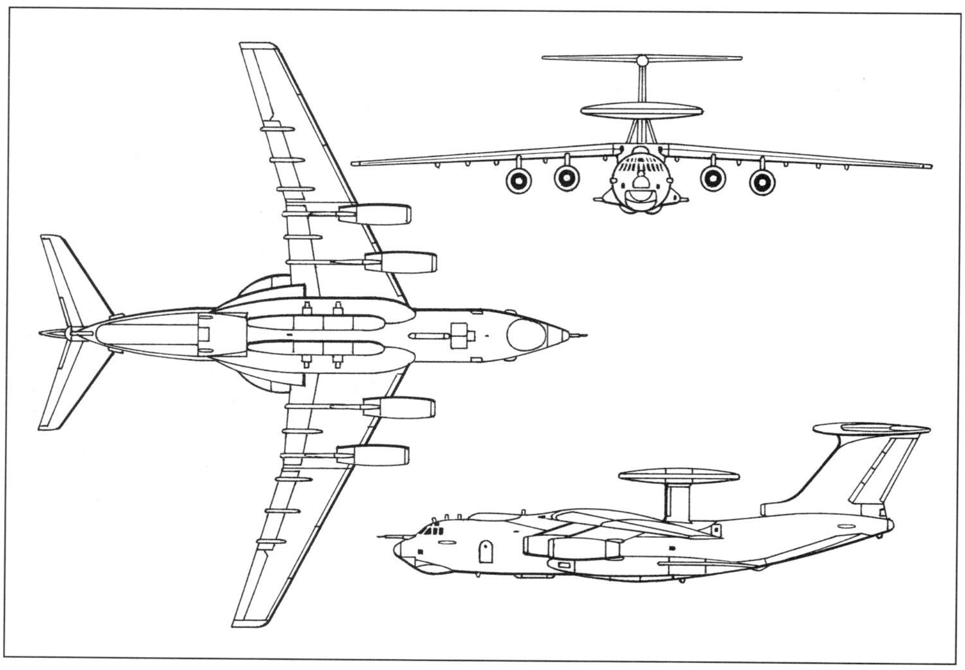

Beriev (Ilyushin) A-50 'Mainstay' AEW&C aircraft *(Jane's/Mike Keep)*

1994

Beriev (Ilyushin) A-50 'Mainstay' AEW&C aircraft *(Swedish Air Force)*

1994

balance/trim tabs; two-section triple-slotted trailing-edge flaps over approximately 75 per cent of each semi-span; eight upper-surface spoilers forward of flaps on each wing, four on each inner and outer panel; leading-edge slats over almost entire span, two on each inner panel, three on each outer panel; variable incidence T tailplane; elevators and rudder aerodynamically balanced, each with tab.

STRUCTURE: All-metal; five-piece wing of multispar fail-safe construction, centre-section integral with fuselage; basically circular-section semi-monocoque fail-safe fuselage; underside of upswept rear fuselage made up of two outward hinged clamshell doors, upward hinged panel between doors, and downward hinged ramp; all tail surfaces sweptback.

LANDING GEAR: Hydraulically retractable tricycle type. Steerable nose unit has two pairs of wheels, side by side, with central oleo. Main gear on each side has two units in tandem, each unit with four wheels on single axle. Low-pressure tyres size 1.300 × 480 mm on mainwheels. 1.100 × 330 mm on nosewheels. Nosewheels retract forward. Main units retract inward into two large ventral fairings under fuselage, with additional large fairing on each side of lower fuselage over actuating gear. During retraction mainwheel axles rotate around leg, so that wheels stow with axles parallel to fuselage axis (that is wheels remain vertical but at 90° to direction of flight). All doors on wheel wells close when gear is down, to prevent fouling of legs by snow, ice and mud. Oleo-pneumatic shock-absorbers. Tyre pressure can be varied in flight from 2.5 to 5 bars (36-73 lb/sq in) to suit different landing strip conditions. Hydraulic brakes on mainwheels.

POWER PLANT: Four Aviadvigatel D-30KP turbofans, each 117.7 kN (26,455 lb st), in individual underwing pods. Each pod on large forward-inclined pylon and fitted with clamshell thrust reverser. Integral fuel tanks between spars of inner and outer wing panels. Total fuel capacity 109,480 litres (28,922 US gallons; 24,083 Imp gallons). In-flight refuelling difficult because of severe buffeting induced by rotating rotodome in tanker's slipstream.

ACCOMMODATION: Normal crew of 15. No onboard toilet.

AVIONICS: Considered capable of detecting and tracking aircraft and cruise missiles flying at low-altitude over land and water, and of helping to direct fighter operations over combat areas as well as enhancing air surveillance and defence of CIS. Colour CRT displays for radar observers; RWR; satellite datalink to ground stations.

EQUIPMENT: Flare pack on each side of rear fuselage; wingtip countermeasures pods under development.

PERFORMANCE: Normally operates on figure-of-eight course at 10,000 m (33,000 ft), with 54 n miles (100 km; 62 miles) between centres of the two orbits.

VERIFIED

ILYUSHIN

AVIATSIONNYI KOMPLEKS IMIENI S. V. ILYUSHINA (Aviation Complex named after S. V. Ilyushin)

45g Leningradsky Prospekt, 125190 Moscow
Telephone: 7 (095) 157 33 12, 943 85 09
Fax: 7 (095) 212 21 32, 212 02 75
Telex: 411956 Sokol
E-mail: ilyushin@sovam.com
GENERAL DESIGNER: Genrikh V. Novozhilov
CHIEF DESIGNER: I. Ya. Katyrev
FOREIGN ECONOMIC RELATIONS DEPARTMENT:
V. A. Belyakov

Ilyushin OKB is named after Sergei Vladimirovich Ilyushin, who died 9 February 1977, aged 82. OKB was founded 1933. About 60,000 aircraft of Ilyushin design have been built.

VERIFIED

ILYUSHIN Il-18

NATO reporting name: Coot

TYPE: Four-engined passenger transport.

PROGRAMME: The Il-18 prototype, named Moskva (Moscow), flew for the first time on 4 July 1957 and production models entered service with Aeroflot in 1959. Production is believed to have exceeded 700 aircraft, of which more than 100 were exported for use by commercial airlines. Military operators in early 1994 include Afghanistan (1), China (approximately 9) and Syria (approximately 4).

An anti-submarine version, the Il-38 is in service and described separately. Another version the Il-20 (NATO reporting name **'Coot-A'**) is also described separately.

VERSIONS: **Il-18A:** Produced from 1957-1959 with NK-4 and AE-20 engines. Accommodation for 89 passengers and take-off weight of 58 tons. Max payload 12 tons.

Il-18B: Modified 14-18A with AE-20 engine. T-O weight 61.2 tons and max payload 14 tons.

Il-18V: Standard version for Aeroflot, with four 2,983 kW (4,000 ehp) AI-20K or AI-20M turboprop engines and fuel capacity of 23,700 litres (6,261 US gallons; 5,213 Imp gallons). Accommodation for 89-100 passengers in two different layouts.

Il-18E: Developed version with 3,169 kW (4,250 ehp) AI-20M engines. Same fuel capacity as Il-18V. Accommodation can be increased to 110 mixed class or 122 tourist class in Summer, by deleting coat storage space essential in Winter. Production from 1965-1966. Main version for 100 passengers. Improved interior.

Il-18D: Generally similar to Il-18E, but with additional centre-section fuel tankage, increasing total capacity to 30,000 litres (7,926 US gallons; 6,600 Imp gallons). Increased all-up weight. Production from 1965-1969. Engine AE-20K T-O power 4,250 hp. T-O weight 64 ton. With increased range.

Il-18D 'Cyclone': Multipurpose atmosphere research lab.

Il-18D ORR: Ocean fish reconnaissance aircraft.

Il-18GR: Cargo version.

OPERATORS: Versions of the IL-18 are in service with the armed forces of the following countries: Afghanistan (1), Belarus, China (7), Kazakhstan, North Korea (4), Romania (2), Russia (20), Syria (4) and Vietnam (2).

DESIGN FEATURES: Cantilever low-wing monoplane. Cantilever tailplane.

FLYING CONTROLS: All-metal ailerons are mass balanced and aerodynamically compensated, and fitted with spring tabs. Manually operated flying controls. Electrically actuated double-slotted flaps. Trim tabs in rudder and elevators. Additional spring tab in rudder. Manually operated flying controls. Electrothermal de-icing.

STRUCTURE: The wings have a mean thickness/chord ratio of 14 per cent and are of an all-metal structure. Three spars in centre-section, two in outer wings. The fuselage is a circular-section all-metal monocoque structure. The structure is of fail-safe type, and appears to employ rip top doublers around window cutouts, door frames and more heavily loaded skin panels. The tail unit is a cantilever all-metal structure.

LANDING GEAR: Retractable tricycle type. Hydraulic actuation. Four-wheel bogie main units, with 930 mm × 305 mm tyres and hydraulic brakes. Steerable (45° each way) twin nosewheel unit, with 700 mm × 250 mm tyres. Tyre pressures: main 7.86 bars (114 lb/sq in), nose 5.86 bars (85 lb/sq in). Hydraulic brakes and nosewheel steering. Pneumatic emergency braking.

POWER PLANT: Four Ivchenko AI-20 turboprops, driving AV-681 four-blade reversible-pitch propellers. Ten flexible bag-type fuel tanks in inboard panel of each wing and integral tank in outboard panel with capacity of 23,700 litres (6,261 US gallons; 5,213 Imp gallons). The Il-18D has additional bag tanks in centre-section, giving total capacity of 30,000 litres (7,926 gallons; 6,600 Imp gallons). Pressure fuelling through four international standard connections in inner nacelles. Provision for overwing fuelling. Oil capacity 58.5 litres (15 US gallons; 12.85 Imp gallons) per engine.

ACCOMMODATION: Crew of five, comprising two pilots, navigator, wireless operator and flight engineer. Flight deck is separated from remainder of fuselage by a pressure bulkhead to reduce the hazards following a sudden decompression of either. Standard 110-seat high-density version has a forward cabin containing 24 seats six-abreast, successively, an entrance lobby with two toilets on the starboard side, two large wardrobes in line with the propellers, the main cabin containing 71 seats in six-abreast rows, a galley/pantry opposite the rear door, a rear cabin containing 15 seats five-abreast, and a rear toilet compartment. Deletion of the wardrobes enables two or more rows of seats to be installed in the main cabin in Summer, increasing maximum capacity to 122 seats. In 90-seat configuration, all seating is five-abreast, with 20 passengers in the front cabin, 55 in centre cabin and 15 in rear cabin. Again, two more rows of seats can replace the wardrobes in Summer, increasing the capacity to 100 seats. The 65-seat layout in the Il-18D has 14 seats (5-5-4) in front cabin, 43 seats (4-5-5-5-5-5-5-5-4) in centre cabin and eight seats (4-4) in rear cabin. Pressurised cargo holds under floor forward and aft of the wing, and a further, unpressurised hold aft of the rear pressure bulkhead.

SYSTEMS: Cabin pressurised to maximum differential of 0.49 bars (7.1 lb/sq in). Electrical system includes eight 12 kW DC generators and 28.5 V single-phase AC inverters. Hydraulic system, pressure 207 bars (3,000 lb/sq in), for landing gear retraction, nosewheel steering brakes and flaps. Electrothermal de-icing on wings and tail unit.

EQUIPMENT: Equipment includes dual controls and blind-flying panels, weather radar and ILS indicators, automatic navigation equipment, two automatic radio compasses, radio altimeter.

DIMENSIONS, EXTERNAL:

Wing span	37.4 m (122 ft 8½ in)
Wing chord: at root	5.61 m (18 ft 5 in)
at tip	1.87 m (6 ft 2 in)
Wing aspect ratio	10
Length overall	35.9 m (117 ft 9 in)
Height overall	10.17 m (33 ft 4 in)
Tailplane span	11.8 m (38 ft 8½ in)
Wheel track	9.0 m (29 ft 6 in)

Ilyushin Il-18 four-engine passenger transport aircraft

1994

Ilyushin Il-18 used by the Ministry of Fisheries based at Murmansk

1994

Ilyushin Il-18 four-engine passenger transport aircraft

1995

Ilyushin Il-20 'Coot-A' elint/reconnaissance aircraft shadowed by BAe Sea Harrier FRS Mk. 1 from the aircraft carrier HMS *Invincible* during exercise 'Cold Winter'

1994

Wheelbase 12.78 m (41 ft 10 in)
Propeller diameter 4.50 m (14 ft 9 in)
Passenger doors (each):
Height 1.40 m (4 ft 7 in)
Width 0.76 m (2 ft 6 in)
Height to sill 2.90 m (9 ft 6 in)
Freight hold doors (underfloor, each):
Height 0.90 m (2 ft 11 in)
Width 1.20 m (3 ft 11 in)

DIMENSIONS, INTERNAL:
Flight deck:
Volume 9.36 m (330 cu ft)
Cabin, excl fight deck:
Length approx 24.0 m (79 ft 0 in)
Max width 3.23 m (10 ft 7 in)
Max height 2.00 m (6 ft 6 in)
Volume 238 m^3 (8,405 cu ft)
Baggage and freight holds (underfloor and aft of cabin):
Total 29.3 m^3 (1,035 cu ft)

AREAS:
Wings, gross 140 m^2 (1,507 sq ft)
Ailerons (total) 9.11 m^2 (98.05 sq ft)
Trailing-edge flaps (total) 27.15 m^2 (292.2 sq ft)
Vertical tail surfaces (total) 17.93 m^2 (193.0 sq ft)
Rudder 6.83 m^2 (73.52 sq ft)
Horizontal tail surfaces (total) 27.79 m^2 (299.13 sq ft)
Elevators (total) 11.80 m^2 (127.0 sq ft)

WEIGHTS AND LOADINGS:
Weight empty, equipped (90-seater):
Il-18E 34,630 kg (76,350 lb)
Il-18D 35,000 kg (77,160 lb)
Max payload 13,500 kg (29,750 lb)
Max T-O weight:
Il-18V, E 61,200 kg (134,925 lb)
Il-18D 64,000 kg (141,100 lb)
Max wing loading (Il-18D) 457 kg/m^2 (93.6 lb/sq ft)
Max power loading (Il-18D) 5.05 kg/kW (8.30 lb/ehp)

PERFORMANCE (at max T-O weight):
Max cruising speed:
Il-18V 351 kts (650 km/h; 404 mph)
Il-18E, D 364 kts (675 km/h; 419 mph)
Econ cruising speed:
Il-18V 324 kts (600 km/h; 373 mph)
Il-18E, D 337 kts (625 km/h; 388 mph)
Operating height:
Il-18D 8,000-10,000 m (26,250-32,800 ft)
T-O run:
Il-18E 1,100 m (3,610 ft)
Il-18D 1,300 m (4,265 ft)
Landing run:
Il-18E, D 850 m (2,790 ft)
Range with max fuel, 1 h reserves:
Il-18E 2,805 n miles (5,200 km; 3,230 miles)
Il-18D 3,508 n miles (6,500 km; 4,040 miles)
Range with max payload, 1 h reserves:
Il-18E 1,728 n miles (3,200 km; 1,990 miles)
Il-18D 1,997 n miles (3,700 km; 2,300 miles)

UPDATED

ILYUSHIN Il-20

NATO reporting name: Coot-A

TYPE: Military elint/reconnaissance variant of Il-18 four-turboprop transport.

PROGRAMME: First observed 1978; small number operational.

OPERATORS: Versions of the IL-20 are in service with the armed forces of the following countries: (5+) and Ukraine (5+).

DESIGN FEATURES: Il-18 airframe basically unchanged; under-fuselage container, approximately 10.25 m (33 ft 7½ in) long and 1.15 m (3 ft 9 in) deep, assumed to house side-looking radar; container, approximately 4.4 m (14 ft 5 in) long and 0.88 m (2 ft 10½ in) deep, on each side of forward fuselage contains door over camera or other sensor; antennae and blisters include eight on under-surface of centre and rear fuselage, with two large plates projecting above forward fuselage.

The following abbreviated details of Il-18D indicate likely features retained by Il-20.

POWER PLANT: Four 3,169 kW (4,250 ehp) Ivchenko AI-20M turboprops, each driving an AV-68I four-blade reversible-pitch propeller. There are 10 flexible fuel tanks in inboard panel of each wing and integral tank in outboard panel, with a total capacity of 23,700 litres (6,261 US gallons; 5,213 Imp gallons). Some Il-18 airliners have additional bag tanks in centre-section, giving a total capacity of 30,000 litres (7,925 US gallons; 6,600 Imp gallons).

DIMENSIONS, EXTERNAL:
Wing span 37.42 m (122 ft 9¼ in)
Wing chord: at root 5.61 m (18 ft 5 in)
at tip 1.87 m (6 ft 2 in)
Wing aspect ratio 10
Length overall 35.9 m (117 ft 9 in)
Height overall 10.17 m (33 ft 4 in)
Tailplane span 11.80 m (38 ft 8½ in)
Wheel track 9.00 m (29 ft 6 in)
Wheelbase 12.78 m (41 ft 10 in)

Propeller diameter 4.50 m (14 ft 9 in)
Cabin doors (each): Height 1.40 m (4 ft 7 in)
Width 0.76 m (2 ft 6 in)
Height to sill 2.90 m (9 ft 6 in)

DIMENSIONS, INTERNAL:
Flight deck: Volume 9.36 m³ (330 cu ft)
Cabin, excl flight deck:
Length (approx) 24.0 m (79 ft 0 in)
Max width 3.23 m (10 ft 7 in)
Max height 2.00 m (6 ft 6¾ in)
Volume 238 m³ (8,405 cu ft)

AREAS:
Wings, gross 140 m² (1,507 sq ft)

WEIGHTS AND LOADINGS (Il-18D airliner):
Max payload 13,500 kg (29,750 lb)
Max T-O weight 64,000 kg (141,100 lb)
Max wing loading 457.1 kg/m² (93.6 lb/sq ft)
Max power loading 5.05 kg/kW (8.30 lb/ehp)

PERFORMANCE (Il-18D airliner, at max T-O weight):
Max cruising speed 364 kts (675 km/h; 419 mph)
Econ cruising speed 337 kts (625 km/h; 388 mph)
Operating height 8,000-10,000 m (26,250-32,800 ft)
T-O run 1,300 m (4,265 ft)
Landing run 850 m (2,790 ft)
Range, 1 h reserves:
with max fuel 3,508 n miles (6,500 km; 4,040 miles)
with max payload1,997 n miles (3,700 km; 2,300 miles)

VERIFIED

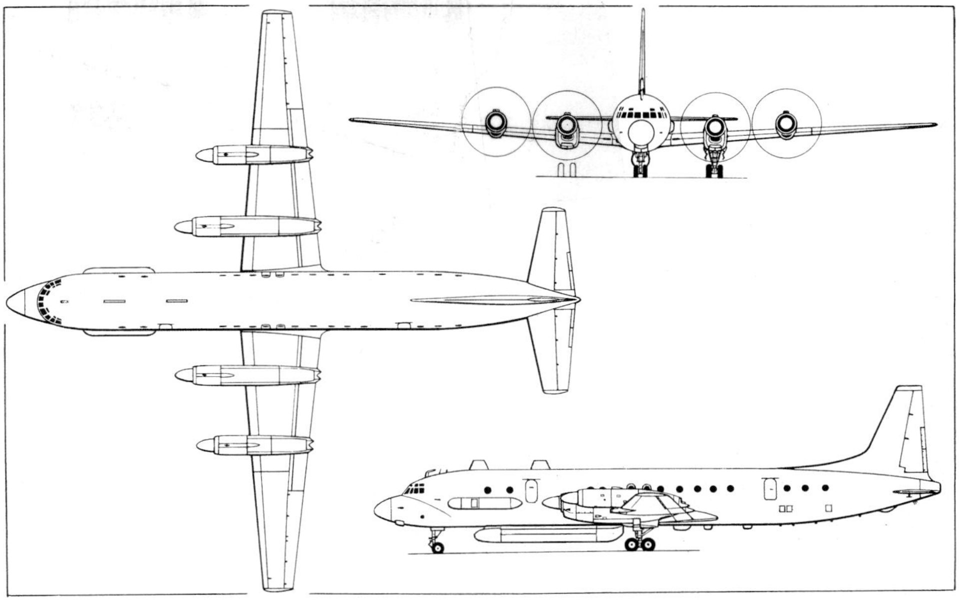

Ilyushin Il-20 'Coot-A' elint/reconnaissance development of the Il-18 airliner *(Jane's/Dennis Punnett)*
1994

ILYUSHIN Il-22

NATO reporting name: Coot-B

TYPE: Four turboprop command post.

DESIGN FEATURES: Many Il-22 airborne command post adaptations of the Il-18 transport are operational with Russian Air Force. No details are available, but it would be logical to expect a variety of external fairings and antennae differing from one aircraft to another depending on its specific duties.

VERIFIED

Ilyushin Il-38 'May' four-turboprop maritime patrol craft
1994

ILYUSHIN Il-38

NATO reporting name: May

TYPE: Intermediate-range shore-based four-turboprop maritime patrol aricraft.

PROGRAMME: Development of Il-18 airliner, first reported 1970; about 59 serve currently with former Soviet Naval Aviation; deployed periodically to Libya and Syria, and to Yemen for patrols over Red Sea, Gulf of Aden, Arabian Sea and Indian Ocean. India placed only export order 1975.

OPERATORS: Versions of the Il-38 are in service with the armed forces of the following countries: India (4) and Russia (50).

DESIGN FEATURES: Basic Il-18 airframe, with lengthened fuselage, and wings moved forward to cater for effect of role equipment and stores in CG position; few cabin windows; large undernose radome; MAD tail sting; wing dihedral 3° from roots; mean thickness/chord ratio 14 per cent.

FLYING CONTROLS: Flying controls cable actuated; mass and aerodynamically balanced ailerons with electric trim tabs; hydraulically assisted elevators and rudder, each with electric trim tab; additional rudder spring tab; hydraulically actuated double-slotted wing trailing-edge flaps.

STRUCTURE: All-metal; three-spar wing centre-section, two spars in outer wings; circular-section fail-safe semi-monocoque fuselage, with rip-stop doublers around window cutouts, door frames and more heavily loaded skin panels.

LANDING GEAR: Retractable tricycle type, strengthened in comparison to Il-18. Hydraulic actuation. Four-wheel bogie main units, with 930 × 305 mm tyres and hydraulic brakes. Steerable (45° each way) twin nosewheel unit, with 700 × 250 mm tyres. Hydraulic brakes and nosewheel steering. Pneumatic emergency braking.

POWER PLANT: Four Ivchenko AI-20M turboprops, each rated at 3,169 kW (4,250 shp), driving AV-68I four-blade reversible-pitch metal propellers. Multiple bag type fuel tanks in centre-section and in inboard panel of each wing, and integral tank in outboard panel, with a total capacity of 30,000 litres (7,925 US gallons; 6,600 Imp gallons). Pressure fuelling through four international standard connections in inner nacelles. Provision for overwing fuelling. Oil capacity 58.5 litres (15.45 US gallons; 12.85 Imp gallons) per engine. Engines started electrically.

ACCOMMODATION: Pilot and co-pilot side by side on flight deck, with dual controls; flight engineer to rear. Number of operational crew believed to be nine, but unconfirmed. Flight deck is separated from main cabin by a pressure bulkhead to reduce hazards following sudden decompression of either. Main cabin has few windows and contains search equipment, electronic equipment and crew stations appropriate to role. Door is on starboard side at rear of cabin (location of Il-18 service door).

SYSTEMS: Cabin pressurised to maximum differential of 0.49 bar (7.1 lb/sq in). Electrical system includes eight engine-driven generators for 28 V DC and 115 V 400 Hz AC supply. Hydraulic system, pressure 207 bars (3,000 lb/sq in), for landing gear retraction, nosewheel steering, brakes, elevator and rudder actuators, flaps, weapon bay doors and radar antennae. Electrothermal de-icing system for wings and tail unit.

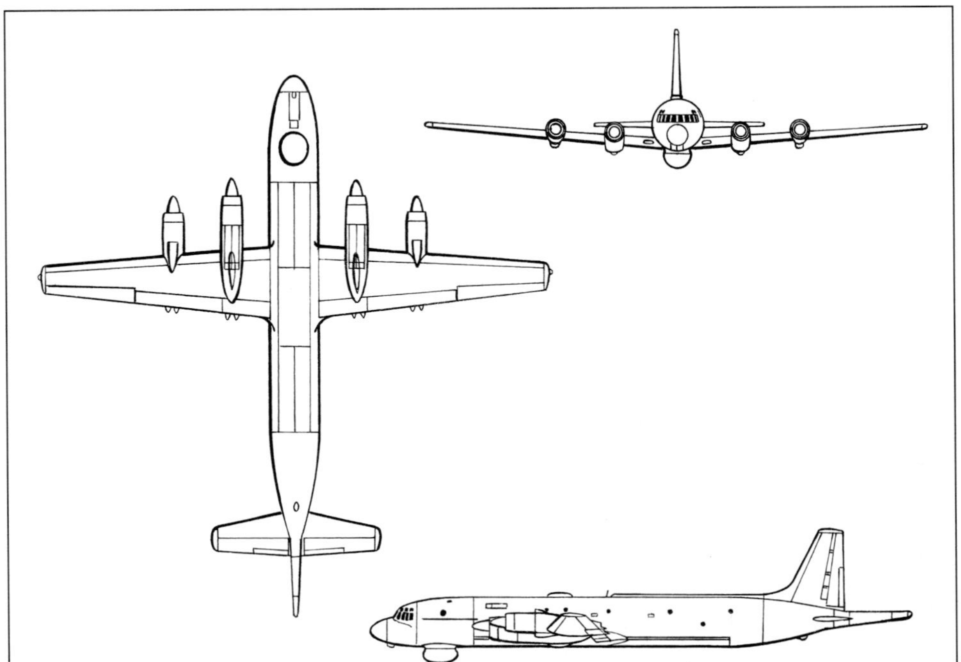

Ilyushin Il-38 'May' intermediate-range shore-based maritime patrol aircraft *(Jane's/Mike Keep)*
1994

AVIONICS: Navigation/weather radar in nose. Search radar (NATO 'Wet Eye') in undernose radome. MAD tail 'sting'. Automatic navigation equipment, radio compasses and radio altimeter probably similar to those of Il-18.

ARMAMENT: Two weapons/stores bays forward and aft of wing carry-through structure on most aircraft, to accommodate a variety of attack weapons and sonobuoys.

DIMENSIONS, EXTERNAL:
As listed under Il-20 entry, except:
Length overall 39.60 m (129 ft 10 in)

WEIGHTS AND LOADINGS:
Weight empty 36,000 kg (79,367 lb)

Max T-O weight	63,500 kg (140,000 lb)
Max wing loading	453.6 kg/m² (92.9 lb/sq ft)
Max power loading	5.01 kg/kW (8.24 lb/ehp)

PERFORMANCE:

Max level speed at 6,400 m (21,000 ft)	390 kts (722 km/h; 448 mph)
Max cruising speed at 8,230 m (27,000 ft)	330 kts (611 km/h; 380 mph)
Patrol speed at 600 m (2,000 ft)	216 kts (400 km/h; 248 mph)
Min flying speed	103 kts (190 km/h; 118 mph)
T-O run	1,300 m (4,265 ft)
Landing run with propeller reversal	850 m (2,790 ft)
Range with max fuel	3,887 n miles (7,200 km; 4,473 miles)
Patrol endurance with max fuel	12 h

VERFIFIED

Ilyushin Il-38 'May' over the Pacific Ocean ***(G. Jacobs)***

1994

ILYUSHIN Il-62

NATO reporting name: Classic

TYPE: Four-turbofan long-range transport.

PROGRAMME: Announced 24 September 1962, when the first prototype (CCCP-06156) was inspected by the late Premier Krushchev, the standard Il-62 is a long-range airliner, with four Kuznetsov turbofan engines mounted in horizontal pairs on each side of the rear fuselage. It accommodates up to 186 passengers and was designed to fly on ranges equivalent to Moscow-New York (approximately 4,155 n miles; 7,700 km; 4,800 miles) with more than 150 passengers and reserve fuel.

The Kuznetsov engines were not ready in time for the first flight of the first prototype, which took place January 1963, with four 73.55 kN (16,535 lb st) Lyulka AL-7 engines installed. This aircraft was followed by a second prototype and three preproduction aircraft. Series production then started at Kazan, and Aeroflot introduced the Il-62 on to its Moscow-Montreal service on 15 September 1967, as a replacement for the Tu-114.

The Il-62 inaugurated Aeroflot's Moscow-New York service in July 1968, and has been used subsequently on many other routes, including Moscow-Paris and Moscow-Tokyo. Production is reported to have exceeded 250 including developed Il-62M/Mks. Current operators include Aeroflot (about 150), CAAC of China, Choson Minhang of North Korea, CSA Czechoslovak Airlines, Cubana, the now defunct Interflug of the former East Germany, Polish Airlines LOT, Tarom of Romania and the Czechoslovak government for VIP operation.

DESIGN FEATURES: Cantilever low-wing monoplane. Sweepback 32° 30′ at quarter-chord. Extended chord leading-edge on outer two-thirds of each wing. Cantilever tail unit.

FLYING CONTROLS: Each wing fitted with three-section manually operated ailerons, electrically actuated slotted flaps and two hydraulically operated spoiler sections forward of flaps. Trim tab and spring loaded servo tab in each centre aileron, spring loaded servo tab in each inner aileron. Tail unit has electrically actuated variable incidence T tailplane. Manually operated rudder fitted with yaw damper, trim tab and servo tab. Manually operated elevators have two automatic trim tabs and two manual trim tabs.

STRUCTURE: The wing is an all-metal structure with four spars inboard, two at tip. Removable leading-edge. The fuselage is a conventional all-metal semi-monocoque structure. Frames are duralumin stampings and pressings. Integrally pressed skin panels at highly stressed areas. Floors are sandwich panels with foam plastics filler. Nosecone hinges upward for access to radar. The tail unit is a cantilever all-metal structure.

LANDING GEAR: Hydraulically retractable tricycle type. Forward-retracting twin-wheel steerable nose unit. Emergency extensions by gravity. Oleo-nitrogen shock-absorber in each unit. Each main unit carries a four-wheel bogie and retracts inward into wingroots. Mainwheel tyres size 1,450 × 450, pressure 9.31 bars (135 lb/sq in). Nosewheel tyres size 930 × 305, pressure 7.86 bars (114 lb/sq in). Hydraulic disc brake and inertia type electric anti-skid unit on each mainwheel, supplemented by large tail parachute. Parking brakes. Hydraulic twin-wheel strut is extended downward to support rear fuselage during loading and unloading.

POWER PLANT: Four Kuznetsov NK-8-4 turbofan engines, each rated at 103 kN (23,150 lb st), mounted in horizontal pairs on each side of rear fuselage. Thrust reverser on each outboard engine. Hot air anti-icing system for engine intakes. Automatically controlled fuel system, with seven integral tanks, three in wing centre-section, two in each outer panel. Each engine has its own independent fuel system with cross-feed. Total fuel capacity 100,000 litres (26,418 US gallons; 21,998 Imp gallons). Four standard international underwing pressure refuelling points. Eight gravity refuelling sockets. Total oil capacity 204 litres (54 US gallons; 45 Imp gallons).

ACCOMMODATION: Crew of five (two pilots, navigator, radio operator and flight engineer) on flight deck. Provision for two supernumerary pilot/navigators. Basic two-cabin layout, and galley, toilet and wardrobe facilities, are unchanged in the three main versions, only the width and pitch of the seats can be varied. In the 186-passenger version, there are 72 seats in the forward cabin and 114 in the rear cabin, all six-abreast and all at a seat pitch of 86 cm (34 in). In the 168-seat configuration, increased pitch reduces capacity to 66 in the forward cabin and 102 in the rear cabin. The 114-passenger version has 45 seats in the forward cabin and 69 in the rear cabin, all five-abreast, except for four-abreast rear row by door. A first class/deluxe version for 85 passengers is available, with 45 seats in forward cabin and 40 four-abreast sleeperette chairs with footrests in rear cabin. Passenger doors forward of front cabin and between cabins on port side. Total of five toilets, opposite forward door, between cabins starboard and aft of rear cabin (both sides). Electrically powered galley/pantry amidships and wardrobes in each version. Pressurised baggage and freight compartments under cabin floor, forward and aft of wing. Unpressurised baggage/cargo compartment at extreme rear of fuselage. All compartments have tiedown fittings and rails in floor, and removable nets to restrain cargo.

SYSTEMS: Air conditioning and pressurisation system maintains sea level conditions up to 7,000 m (23,000 ft) and gives equivalent of 2,100 m (6,900 ft) at 13,000 m (42,600 ft). Pressure differential 0.62 bars (9.0 lb/sq in). Hydraulic system, pressure 207 bars (3,000 lb/sq in), for landing gear retraction, nosewheel steering, brakes, spoilers and windscreen wipers. Emergency hydraulic system, powered by electric motor, for nosewheel steering, mainwheel extension and spoiler control. Three-phase 200/115 V AC electrical supply from four 40kVA engine-driven generators (optional 27 V DC system with either 18 kW engine-driven generators). Four transformer-rectifiers and four batteries for DC supply. Electric windscreen de-icing. TA-6 APU in tailcone.

AVIONICS: Standard avionics include two-channel autopilot, navigation computer, air data system, HF and UHF radio, VOR/ILS, RMI, Doppler, radio altimeter and weather radar. Polyot automatic flight control system optional.

DIMENSIONS, EXTERNAL:

Wing span	43.20 m (141 ft 9 in)
Length overall	53.12 m (174 ft 3½ in)
Length of fuselage	49.00 m (160 ft 9 in)
Height overall	12.35 m (40 ft 6¼ in)
Tailplane span	12.23 m (40 ft 1½ in)
Fuselage height	4.10 m (13 ft 5½ in)
Fuselage width	3.75 m (12 ft 3½ in)
Wheel track	6.80 m (22 ft 3½ in)
Wheelbase	24.49 m (80 ft 4½ in)
Passenger doors (each):	
Height	1.83 m (6 ft 0 in)
Width	0.86 m (2 ft 9¾ in)
Height to sill	3.55 m (11 ft 8 in)
Emergency exit (galley service) door:	
Height	1.38 m (4 ft 6¼ in)
Width	0.61 m (2 ft 0 in)
Emergency exits (overwing):	
Height	0.91 m (2 ft 11¾ in)
Width	0.51 m (1 ft 8 in)
Front cargo hold door:	
Height	1.31 m (4 ft 3½ in)
Width	1.26 m (4 ft 1½ in)
Height to sill	1.90 m (6 ft 3 in)
Second cargo hold door:	
Height	1.00 m (3 ft 3¼ in)
Width	1.26 m (4 ft 1½ in)
Height to sill	1.90 m (6 ft 3 in)
Third cargo hold door:	
Height	0.70 m (2 ft 3½ in)
Width	0.70 m (2 ft 3½ in)
Height to sill	2.26 m (7 ft 5 in)
Rear cargo hold door:	
Height	1.15 m (3 ft 9 in)

Ilyushin Il-62 'Classic' four-turbofan long-range transport aircraft

1994

Width 1.07 m (3 ft 6 in)
Height to sill 3.68 m (12 ft 0¾ in)

DIMENSIONS, INTERNAL:
Cabin:
Max height 2.12 m (6 ft 11½ in)
Max width 3.49 m (11 ft 5¼ in)
Volume 163 m^3 (5,756 cu ft)
Total volume of pressure cell 396 m^3 (13,985 cu ft)
Cargo hold volume:
Front 22.7 m^3 (801 cu ft)
Second 12.6 m^3 (445 cu ft)
Third 6.9 m^3 (243 cu ft)
Rear 5.8 m^3 (205 cu ft)

AREAS:
Wings, gross 279.55 m^2 (3,009 sq ft)
Ailerons (total) 16.25 m^2 (174.9 sq ft)
Spoilers (total) 9.54 m^2 (102.7 sq ft)
Flaps (total) 43.48 m^2 (468.0 sq ft)
Horizontal tail surfaces (total) 40.00 m^2 (430.5 sq ft)
Vertical tail surfaces (total) 35.60 m^2 (383.2 sq ft)

WEIGHTS AND LOADINGS:
Weight empty 66,400 kg (146,390 lb)
Operating weight empty 69,400 kg (153,000 lb)
Max payload 23,000 kg (50,700 lb)
Max fuel 83,325 kg (183,700 lb)
Max ramp weight 167,000 kg (368,000 lb)
Max T-O weight 162,000 kg (355,150 lb)
Max landing weight 105,000 kg (231,500 lb)
Max zero-fuel weight 93,500 kg (206,130 lb)
Max wing loading 572 kg/m^2 (117.2 lb/sq ft)

PERFORMANCE:
Normal cruising speed
442-486 kts (820-900 km/h; 509-560 mph)
Normal cruising height
10,000-12,000 m (33,000-39,400 ft)
Landing speed
119-129 kts (220-240 km/h; 137-149 mph)
Max rate of climb at S/L 1,080 m (3,540 ft)/min
FAR T-O field length:
ISA at S/L 3,250 m (10,660 ft)
ISA + 20°C at S/L 3,915 m (12,840 ft)
FAR landing field length:
ISA at S/L 2,800 m (9,185 ft)
ISA + 20°C at S/L 2,950 m (9,680 ft)
Range with max payload, 66,700 kg (147,050 lb) fuel, 1 h fuel reserves 3,612 n miles (6,700 km; 4,160 miles)
Range with 80,000 kg (176,370 lb) fuel and 100,000 kg (22,045 lb) payload, 1 h fuel reserves
4,963 n miles (9,200 km; 5,715 miles)

VERIFIED

ILYUSHIN Il-62M/MK

NATO reporting name: Classic

TYPE: Four-turbofan long-range airliner.

PROGRAMME: First displayed publicly at the 1971 Paris Air Show, the **Il-62M** is a developed version of the Il-62, with no dimensional changes to the airframe. It is fitted with more powerful turbofans, of a different type, with clamshell thrust reversers on the outboard engine of each pair, offering a lower approach speed and improved air flow over the rear of the nacelles. An additional fuel tank is installed in the tail fin, contributing (with the improved specific fuel consumption of the engines) to the longer range of this version.

Revised layout of the flight deck equipment, and improved navigation and radio communications equipment, are features of the Il-62M. Control wheels of different design allow the pilots a better field of view, and the aircraft's automatic flight control system permits automatic landings in ICAO Cat. II conditions, with planned extension to Cat. III. The wing spoilers of this version can be utilised differentially to enhance roll control.

Additional emergency and rescue equipment is installed on the Il-62M. Unlike the Il-62, it has a containerised baggage and freight system, with mechanised loading and unloading.

The Il-62M exhibited in Paris in 1971 and 1973 was the prototype (SSSR-86673). Production models entered service on Aeroflot's Moscow-Havana route in 1974 and took over progressively all of the airline's very long distance services.

Since 1985, the Il-62M and Il-62MK have been updated by improvements to the navigation system, based on a new triplex INS, and by changes to the turbofans and pods to reduce noise and air pollution.

The basic structural description of the Il-62 applies also to the Il-62M. The main innovations are as follows:

POWER PLANT: Four Soloviev D-30KU turbofans, each rated at 107.9 kN (24,250 lb st), mounted in horizontal pairs on each side of rear fuselage. Clamshell thrust reverser on each outboard engine. Remainder of power plant basically as for Il-62, but additional fuel tank in tail fin, giving total capacity of 105,300 litres (27,817 US gallons; 23,162 Imp gallons).

ACCOMMODATION: Alternative configurations for up to 174 economy class, 168 tourist class or 140 mixed class passengers. In the basic tourist class version there are two toilets opposite the forward door, on the starboard side, aft of the flight deck. The forward cabin contains 66 seats, all six-abreast in threes with centre aisle. Galley/pantry, coat stowage and toilet amidships. Rear cabin contains 102 seats, six-abreast in threes with centre aisle. Two toilets and wardrobe to rear of this cabin. Doors as on Il-62. Two emergency exits on each side, over wing. Forward underfloor baggage and freight hold accommodates nine containers, each weighing approximately 45 kg (100 lb) empty and with a capacity of 600 kg (1,322 lb) and 1.6 m^3 (56.5 cu ft). Rear hold accommodates five similar containers. Two compartments for non-containerised cargo. Total baggage and freight capacity 48 m^3 (1,695 cu ft).

AVIONICS: Duplicated SAU-1T automatic flight control system provides for automatic control from a height of 200 m (660 ft) after take-off to a height of 30 m (100 ft) on the approach to land; DISS-013 Doppler indicator and NV-PB-1 navigation computer; TKS-P course sensing system; TsGV-10P vertical master gyros; SVS-PN-15 air data system; Kurs-MP-2 radio navigation system, utilising VOR, ILS or SP-50 beacons; GROZA radar; SD-67 DME; ARK-15 ADF; RV-5 radio altimeter; SO-70 IFF transponder; MIKRON 2-24 MHz HF radio; LANDASH 118-135 MHz VHF radio; VESHANIE public address and in-flight entertainment system.

DIMENSIONS AND AREAS:
Same as for Il-62

WEIGHTS (Il-62M):
Max payload 23,000 kg (50,700 lb)
Max T-O weight 165,000 kg (363,760 lb)
Max landing weight 105,000 kg (231,500 lb)
Max zero-fuel weight 94,600 kg (208,550 lb)

PERFORMANCE (Il-62M, at max T-O weight):
Normal cruising speed
442-486 kts (820-900 km/h; 509-560 mph)
Normal cruising height
10,000-12,000 m (33,000-39,400 ft)
Balanced T-O distance (ISA, S/L) 3,300 m (10,830 ft)
Landing run (ISA, S/L) 2,500 m (8,200 ft)
Range: with max payload, with 5,100 kg (11,240 lb) fuel reserves 4,210 n miles (7,800 km; 4,846 miles)
with 10,000 kg (22,045 lb) payload, with reserves
5,400 n miles (10,000 km; 6,215 miles)

VERIFIED

Ilyushin Il-62M long-range transport aircraft *(Peter Cooper)*

1995

Ilyushin Il-62M four-turbofan long-range airliner

1994

ILYUSHIN Il-86

NATO reporting name: Camber

TYPE: Four-turbofan medium-range widebodied passenger transport.

PROGRAMME: Construction of two prototypes started 1974; first flight 22 December 1976 from Ilyushin OKB headquarters at old Moscow Central Airport, Khodinka, from 1,820 m (5,970 ft) runway; to official flight test centre, by prototype SSSR-86000: first production Il-86 (SSSR-86002) flew at Voronezh assembly plant 24 October 1977; first delivery (SSSR-86004) to Aeroflot 24 September 1979; scheduled services began 26 December 1980; first international service. Moscow-East Berlin, 3 July 1981; four delivered 1993; production terminated with final Il-86 for Aeroflot 1994; total of 95 built; including four command posts; remainder all commercial.

UPGRADES: **Ilyushin:** Il-86 Command post. See following entry.

CUSTOMERS: Aeroflot and its successors (80), China Northern Airlines (3).

DESIGN FEATURES: Conventional low/mid swept-wing; two-deck fuselage, intended to be entered via lower deck doors, into stowage compartments for coats and hand baggage, and up stairways to passenger deck (deletion of lower deck airstairs and internal stairways optional); additional centreline bogie between main landing gear units; dihedral from roots on wings and tailplane; wing sweepback 35° at quarter-chord; all tail surfaces swept.

FLYING CONTROLS: Hydraulic actuation, without manual reversion for primary surfaces; aileron and two-section double-slotted flaps occupy entire trailing-edge of each wing; multisection upper-surface spoilers and airbrakes forward of each flap section; full-span leading-edge slats; variable incidence tailplane; rudder and elevators each two-section.

STRUCTURE: All-metal; inner wings three-spar, outer panels two-spar; shallow fence above wing in line with each engine pylon; circular-section semi-monocoque fuselage; floors of both decks of honeycomb and carbonfibre reinforced plastics.

LANDING GEAR: Retractable four-unit type. Forward retracting steerable twin-wheel nose unit; three four-wheel bogie main units. Two of latter retract inward into wingroot fairings; third is mounted centrally under fuselage, slightly forward of the others, and retracts forward. Mainwheel tyre size 1,300 × 480 mm; nosewheel tyres size 1,120 × 450 mm.

POWER PLANT: Four KKBM (Kuznetsov) NK-86 turbofans, each 127.5 kN (28,660 lb st), on pylons forward of wing leading-edges. Combined thrust reverse/noise attenuators. Integral fuel tanks in wings, capacity 114,000 litres (30,116 US gallons; 25,077 Imp gallons).

ACCOMMODATION: Two pilots and flight engineer, with provision for navigator. Flight engineer's seat normally faces to starboard, aft of co-pilot, but can pivot to central forward-facing position to enable engineer to operate throttles. Upper deck, on which all seats are located, divided into three separate cabins by wardrobes, a serving area connected by elevator to lower deck galley, and cabin staff accommodation with eight toilets at front (two) and rear (six) of aircraft. Unusually large windows, indirect lighting in walls and in ceiling panels, and enclosed baggage lockers at top of side walls. Preponderance of metal and natural fibre materials rather than plastics throughout cabins to enhance safety in an emergency. Up to 350 passengers in basic nine-abreast seating throughout, with two aisles, each 55 cm (21.6 in) wide. Mixed class layout for 28 passengers six-abreast in front cabin, and 206 passengers eight-abreast in other two cabins. Three airstair doors (made in Kharkov) hinge down front port side of lower deck; one is forward of wing, others aft. Four further doors at upper deck level on each side, for emergency use (using dual inflatable escape slides) and for use at airports where utilisation of high level boarding steps or bridges preferred. Coats and hand baggage stowed on lower deck before passengers climb one of three fixed stairways to main deck. (Optional deletion of lower deck airstair doors and stairways reduces operating weight empty by 3,000 kg; 6,610 lb and permits installation of 25 more seats on upper deck). Cargo holds on lower deck accommodate heavy or registered baggage and freight in eight standard LD3 containers, or 16 LD3 containers if some carry-on baggage racks omitted. Access via upward hinged doors forward of starboard wingroot leading-edge and at side of rear hold. Containers can be loaded and unloaded by self-propelled truck with built-in roller conveyor. Films can be shown in flight, and there is choice of 12 tape recorded audio programmes. A bar-buffet can be provided on lower deck in place of baggage and freight accommodation in forward vestibule.

SYSTEMS: Four self-contained hydraulic systems, each operated by one of engines, for actuation of flying control surfaces, tailplane variable incidence, spoilers airbrakes, slats, flaps, landing gear, nosewheel steering, wheel brakes, anti-skid system, and upper level doors when passenger gangways used. All hot pipelines of air conditioning system, and all fuel supply lines, outside pressure cell. Primary 200/155 V 400 Hz AC electrical system powered by four 40 kVA engine-driven generators. Secondary 36 V three-phase AC and 27 V DC systems. Five accumulators and static transformer. Smoke detection sensors in baggage, freight and equipment stowage areas. Pulse generating de-icing system consuming 500 times less energy than conventional hot air or electrical system. APU in tailcone.

AVIONICS: All avionics equipment within pressurised part of fuselage. Flight control and nav systems provide for automatic climb to selected height, control of rate of climb and automatic descent, and automatic landing in ICAO Cat. IIIA conditions. Preprogrammable Doppler nav system with readout display screen, on which microfilmed maps, can be projected. Position of aircraft indicated by cursor, driven by computer. Nav system updated automatically by inputs from VOR or VOR/DME radio beacons. Collins TCAS II flight tested and available from early 1994.

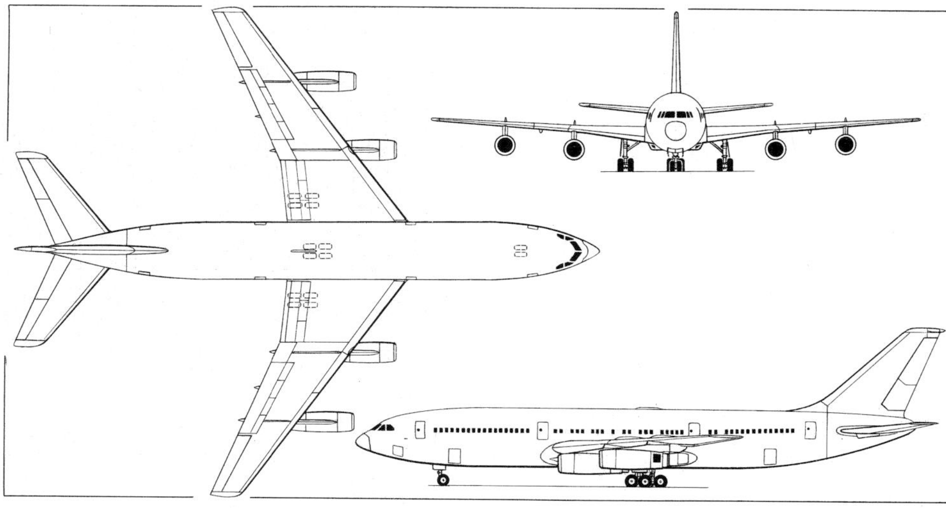

Ilyushin Il-86 four-turbofans widebodied passenger transport aircraft

1995

Ilyushin Il-86 medium-range passenger transport *(Peter Cooper)*

1995

DIMENSIONS, EXTERNAL:

Wing span	48,06 m (157 ft 8¼ in)
Length overall	59.54 m (195 ft 4 in)
Fuselage: Length	56.10 m (184 ft 0¾ in)
Max diameter	6.08 m (19 ft 11½ in)
Height overall	15.81 ft (51 ft 10½ in)
Tailplane span	20.57 m (67 ft 6 in)
Wheel track (c/l of outer shock struts)	11.15 m (36 ft 7 in)
Wheelbase	21.34 m (70 ft 0 in)

DIMENSIONS, INTERNAL:

Main cabins: Height	2.61 m (8 ft 7 in)
Max width	approx 5.70 m (18 ft 8½ in)

AREAS:

Wings, gross	320.0 m² (3,444 sq ft)

WEIGHTS AND LOADINGS:

Max payload	42,000 kg (92,600 lb)
Max fuel	88,350 kg (194,775 lb)
Max T-O weight (dependent on size and type of runway)	190,000-208,000 kg (418,875-458,560 lb)
Max landing weight	175,000 kg (385,800 lb)
Max wing loading	650.0 kg/m² (133.1 lb/sq ft)
Max power loading	407.9 kg/kN (4.0 lb/lb st)

PERFORMANCE (designed):

Normal cruising speed at 9,000-11,000 m (30,000-36,000 ft) 486-512 kts (900-950 km/h; 559-590 mph)

Approach speed 130-141 kts (240-260 km/h; 149-162 mph)

Field length for T-O and landing 2,300-2,600 m (7,550-8,530 ft)

* Range: with 40,000 kg (88,185 lb) payload 1,944 n miles (3,600 km; 2,235 miles)

* with max fuel 2,480 n miles (4,600 km; 2,858 miles)

* *Reports suggest that these design ranges are not being achieved. The former East German airline Interflug quoted a max range of 1,350 n miles (2,500 km; 1,550 miles) in its sale literature*

Ilyushin Il-86 with four KKBM (Kuznetsov) NK-86 turbofan engines

1995

NEW ENTRY

ILYUSHIN Il-86 COMMAND POST

NATO reporting name: Maxdome

TYPE: Airborne command post version of Il-86 transport.

PROGRAMME: First observed at Zhukovsky Flight Research Centre 1992; four seen to be completed at that time (SSSR-86146-49).

DESIGN FEATURES: Large boat-shaped fairing above fuselage forward of wings; large pod with ram air intake under each inner wing; long and shallow dished fairing forward of fin root; strake antenna under rear fuselage; small fin-like component on port side lower fuselage, carrying what appears to be drogue for VLF training wire aerial. SSSR-86146 and -86147 have large blade aerials above centre and rear fuselage and under forward fuselage.

NEW ENTRY

Ilyushin Il-86 airborne command post *(Sebastian Zacharias)* *1995*

KAMOV

VERTOLETNYI NAUCHNO-TEKHNICHESKIY KOMPLEKS IMENI N. I. KAMOVA (VNTK) (Helicopter Scientific and Technology Complex named after N. I. Kamov)

March 8th Street, Lubertsy, 140007 Moscow Region
Telephone: 7 (095) 700 3204, 171 3743
Fax: 7 (095) 700 3071
Telex: 206112 Kamov

GENERAL DESIGNER: Sergei Victorovich Mikheyev, PhD

DEPUTY GENERAL DESIGNER: Veniamin Kasjanikov

CHIEF DESIGNERS:
Juri Sokovikov
Vyacheslav Krygin
Evgeny Pak

Formed in 1947, this OKB continues work of Prof Dr Ing Nikolai Ilyich Kamov, a leading designer of rotating wing aircraft from late 1920s, who died on 24 November 1973, aged 71; all Kamov helicopters in current service have coaxial contrarotating rotors; Ka-62, under development, has single main rotor.

VERIFIED

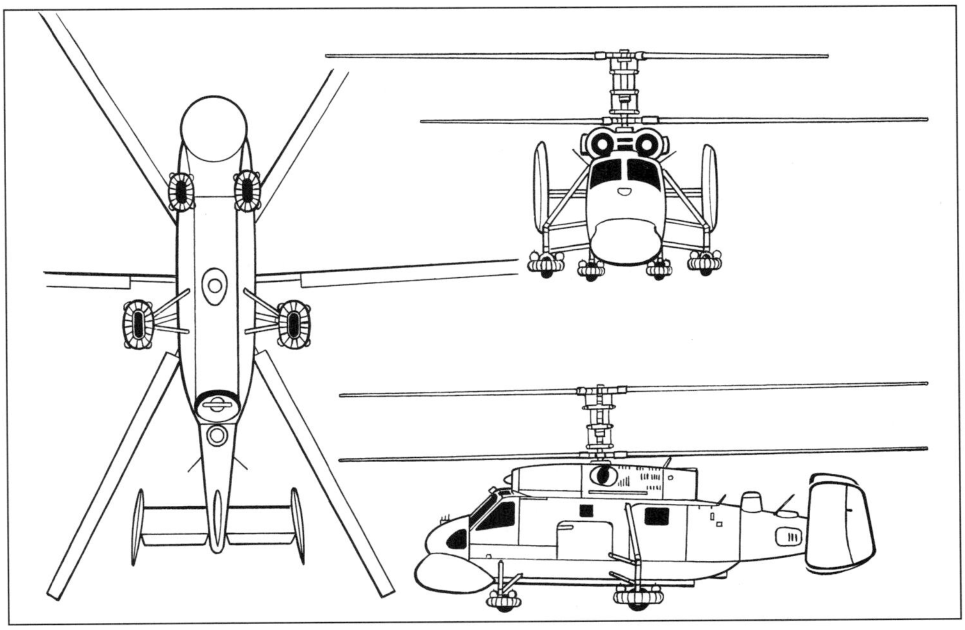

Kamov Ka-25 'Hormone' multipurpose military helicopter *(Jane's/Mike Keep)* *1994*

KAMOV Ka-25

NATO reporting name: Hormone

TYPE: Twin-turbine multipurpose military helicopter.

PROGRAMME: Prototype flew 1961; shown in Soviet Aviation Day flypast, Tushino Airport, Moscow, July 1961, carrying two dummy air-to-surface missiles (ASMs not fitted to production aircraft); about 460 built 1966-75, of which 88 remain operational with Russian Navy.

VERSIONS: Reportedly more that 25 versions, three major variants in service:

Ka-25BSh ('Hormone A'): Ship-based anti-submarine helicopter, operated from Russian missile frigates, cruisers, helicopter carriers and carrier/cruisers. Major shortcoming is lack of automatic hover capability, preventing night and adverse weather use of dipping sonar. Being replaced by Ka-27PL ('Helix-A').

Ka-25PL ('Hormone-A'): Ship-based anti-submarine helicopter, operated from former Soviet Navy missile frigates, cruisers, helicopter carriers and carrier/cruisers of 'Kiev' class; major shortcoming is lack of automatic hover capability, preventing night and adverse weather use of dipping sonar. Being replaced progressively by Ka-27PL ('Helix-A').

Ka-25T ('Hormone-B'): Special electronics version, providing over-the-horizon target acquisition for ship-launched cruise missiles including SS-N-3B (NATO 'Shaddock') from 'Kresta I' cruisers, SS-N-12 ('Sandbox') from 'Kiev' and 'Slava' class cruisers, SS-N-19 ('Shipwreck') from battle cruisers *Kirov* and *Frunze*, and SS-N-22 ('Sunburn') from 'Sovremenny' class destroyers. 'Kiev' and 'Kirov' class ships each carry three 'Hormone-Bs', other classes one; larger undernose radome (NATO 'Big Bulge') than Ka-25BSh, with spherical undersurface; cylindrical radome under rear cabin for datalink; when radar operates, all landing gear wheels can retract upward to minimise interference to emissions; cylindrical fuel container each side of lower fuselage.

Kamov Ka-25PL 'Hormone-A' anti-submarine helicopter of the CIS Naval Aviation *1994*

Kamov Ka-25 'Hormone' coaxial rotor helicopter

1994

Kamov Ka-26 agricultural chemical spray version. Note air filter in nose *(Dr Fabian Zoltan)*

1994

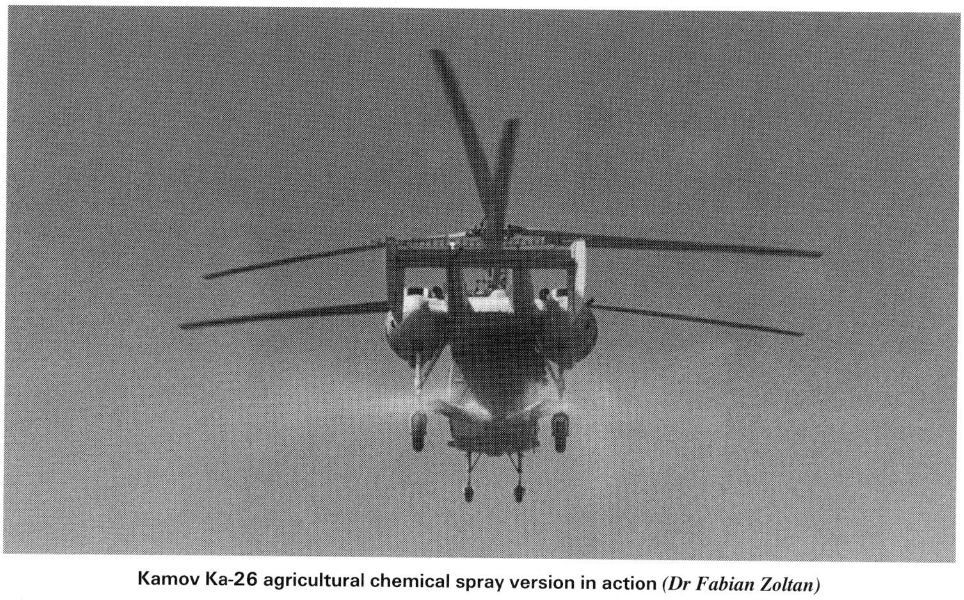

Kamov Ka-26 agricultural chemical spray version in action *(Dr Fabian Zoltan)*

1994

Ka-25BShZ: Equipped to tow minesweeping gear.

Ka-25PS ('Hormone-C'): Search and rescue version with special role equipment, including hoist.

OPERATORS: Versions of the Ka-25 are in service with the armed forces of the following countries: Bulgaria (2), India (5), Russia (130+), Syria (5), Ukraine (5), Vietnam (15) and Yugoslavia (4).

Following details apply to Ka-25BSh:

DESIGN FEATURES: Use of folding three-blade coaxial rotors, requiring no tail rotor, and triple tail fins ensures compact stowed overall dimensions on board ship; engines above cabin and external mounting of operational equipment and auxiliary fuel leaves interior uncluttered.

POWER PLANT: Two 671 kW (900 shp) Glushenkov GTD-3F turboshafts, mounted side by side above cabin, forward of rotor driveshaft, on early aircraft. Later aircraft have 738 kW (990 shp) GTD-3BM turboshafts. Independent fuel supply to each engine. Provision for carrying external fuel tank on each side of cabin.

ACCOMMODATION: Pilot and co-pilot side by side on flight deck, with rearward sliding door on each side. Entry to main cabin is via a rearward sliding door to rear of main landing gear on port side. Cabin large enough to contain 12 folding seats for passengers.

AVIONICS: Equipment available for all versions includes autopilot, navigational system, radio compass, radio communications, lighting system for all-weather operation by day or night, and hoist mounted above cabin door. IFF antennae (NATO 'Odd Rods') above nose and alongside central tail fin. Dipping sonar housed in compartment at rear of main cabin, immediately forward of tailboom, and search radar (NATO 'Short Horn') in flat-bottom undernose radome (diameter 1.25 m; 4 ft 1 in) on anti-submarine version, which can have a canister of sonobuoys mounted externally aft of the starboard main landing gear. Most aircraft have a cylindrical housing for ESM above the tailboom, with a shallow blister fairing to the rear of the cylindrical housing. Similar housing under rear of cabin for datalink equipment.

EQUIPMENT: Cylindrical container on each side of lower fuselage for markers, smoke generators or beacons. The port side container has been seen housing reconnaissance cameras.

ARMAMENT: Doors under the fuselage of some aircraft enclose a weapons bay for two 450 mm (18 in) ASW torpedoes, nuclear or conventional depth charges and other stores.

DIMENSIONS, EXTERNAL:	
Rotor diameter (each)	15.74 m (51 ft 7¾ in)
Length of fuselage	9.75 m (32 ft 0 in)
Height to top of rotor head	5.37 m (17 ft 7½ in)
Width over tail fins	3.76 m (12 ft 4 in)
Wheel track: front	1.41 m (4 ft 7½ in)
rear	3.52 m (11 ft 6½ in)
Cabin door: Height	1.10 m (3 ft 7¼ in)
Width	1.20 m (3 ft 11¼ in)
DIMENSIONS, INTERNAL:	
Cabin, excl flight deck:	
Length	3.95 m (12 ft 11½ in)
Max width	1.50 m (4 ft 11 in)
Max height	1.25 m (4 ft 1¼ in)
AREAS:	
Main rotor disc (each)	194.6 m^2 (2,095 sq ft)
WEIGHTS AND LOADINGS:	
Weight empty	4,765 kg (10,505 lb)
Max T-O weight	7,500 kg (16,535 lb)
PERFORMANCE:	
Max level speed	113 kts (209 km/h; 130 mph)
Normal cruising speed	104 kts (193 km/h; 120 mph)
Service ceiling	3,350 m (11,000 ft)
Range, with reserves:	
with standard fuel	217 n miles (400 km; 250 miles)
with external tanks	351 n miles (650 km; 405 miles)

VERIFIED

KAMOV Ka-26

NATO reporting name: Hoodlum-A

TYPE: Light twin-engined multipurpose helicopter.

PROGRAMME: Prototype first flew 1965; production aircraft entered agricultural service in Soviet Union 1970; 850 built for many civil and military roles.

OPERATORS: Versions of the Ka-26 are in service with the armed forces of Hungary (18).

CUSTOMERS: Delivered for civilian use in 15 countries; military operators include the Hungarian Air Force.

DESIGN FEATURES: Airframe comprises backbone structure carrying flight deck, coaxial contrarotating rotors, landing gear, engine pods and twin tailbooms; space aft of flight deck, between main landing gear units and under rotor transmission, can accommodate interchangeable modules for passenger/freight transport, air ambulance, aerial survey, forest firefighting, mineral prospecting, pipeline and power transmission line construction, search and rescue, and (of primary importance) agricultural equipment.

Abbreviated details of Ka-26 follow:

POWER PLANT: Two 242.5 kW (325 hp) Vedeneyev M-14V-26 air-cooled radial piston engines, mounted in pods on short stub wings at top of fuselage.

ACCOMMODATION: Fully enclosed cabin, with door on each side, fitted out normally for operation by single pilot; second seat and dual controls optional. Cabin warmed and demisted by air from combustion heater, which also heats passenger compartment when fitted. Air filter on nose of agricultural version. For agricultural work, a chemical hopper (capacity 900 kg; 1,985 lb) and dust spreader or spraybars are fitted in module space, on the aircraft's centre of gravity. This equipment is quickly removable and can be replaced by a cargo/passenger pod accommodating four or six persons, with provision for a seventh passenger beside the pilot; or two stretcher patients, two seated casualties and a medical attendant in ambulance role. Alternatively, the Ka-26 can be operated with either an open platform for hauling freight or a hook for slinging bulky loads at the end of a cable or in a cargo net.

DIMENSIONS, EXTERNAL:

Rotor diameter (each)	13.00 m (42 ft 7¾ in)
Vertical separation between rotors	1.17 m (3 ft 10 in)
Length of fuselage	7.75 m (25 ft 5 in)
Height overall	4.05 m (13 ft 3½ in)
Width: over engine pods	3.64 m (11 ft 11½ in)
over agricultural spraybars	11.20 m (36 ft 9 in)
Tailplane span	4.60 m (15 ft 1 in)
Wheel track: mainwheels	2.42 m (7 ft 11½ in)
nosewheels	0.90 m (2 ft 11½ in)
Wheelbase	3.48 m (11 ft 5 in)
Passenger pod door: Height	1.40 m (4 ft 7 in)
Width	1.25 m (4 ft 1¼ in)

DIMENSIONS, INTERNAL:

Passenger pod:	
Length, floor level	1.83 m (6 ft 0 in)
Width, floor level	1.25 m (4 ft 1¼ in)
Headroom	1.40 m (4 ft 7 in)

AREAS:

Main rotor disc (each)	132.7 m² (1,430 sq ft)

WEIGHTS AND LOADINGS:

Operating weight, empty: stripped	1,950 kg (4,300 lb)
cargo/platform	2,085 kg (4,597 lb)
cargo/hook	2,050 kg (4,519 lb)
passenger	2,100 kg (4,630 lb)
agricultural	2,216 kg (4,885 lb)
Fuel weight: transport	360 kg (794 lb)
other versions	100 kg (220 lb)
Payload: transport	900 kg (1,985 lb)
agricultural duster	1,065 kg (2,348 lb)
agricultural sprayer	900 kg (1,985 lb)
with cargo platform	1,065 kg (2,348 lb)
flying crane	1,100 kg (2,425 lb)
Normal T-O weight: transport	3,076 kg (6,780 lb)
agricultural	2,980 kg (6,570 lb)
Max T-O weight: all versions	3,250 kg (7,165 lb)

PERFORMANCE (at max T-O weight):

Max level speed	91 kts (170 km/h; 105 mph)
Max cruising speed	81 kts (150 km/h; 93 mph)
Econ cruising speed	49-59 kts (90-110 km/h; 56-68 mph)
Agricultural operating speed range	16-62 kts (30-115 km/h; 19-71 mph)
Service ceiling	3,000 m (9,840 ft)
Service ceiling, one engine out	500 m (1,640 ft)
Hovering ceiling at AUW of 3,000 kg (6,615 lb):	
IGE	1,300 m (4,265 ft)
OGE	800 m (2,625 ft)
Range with seven passengers, 30 min fuel reserves	215 n miles (400 km; 248 miles)
Max range with auxiliary tanks	647 n miles (1,200 km; 745 miles)
Endurance at econ cruising speed	3 h 42 min

VERIFIED

Kamov Ka-26 with coaxial contrarotating rotors *(Dr Fabian Zoltan)* *1995*

MiG

AVIATSIONNYI NAUCHNO-PROMSHLENNYI KOMPLEKS - ANPK MiG IMIENI A. I. MIKOYANA (A. I. Mikoyan Aviation Scientific-Industrial Complex 'MiG' - ANPK MiG)

6 Leningradskoye Shosse, 125299 Moscow
Telephone: 7 (095) 158 2321
Fax: 7 (095) 943 0027
Teletape: 111338

GENERAL DESIGNER: Rostislav Apollosovich Belyakov

In 1971 Artem I. Mikoyan, the original head of this OKB, was succeeded as General Designer by Academician Rostislav A. Belyakov. He supervised completion of production of the MiG-23 and 25, designed under A.I. Mikoyan, while later aircraft such as MiG-27 and 29 (both in many versions) and MiG-31 were developed and put into service.

In addition to new combat aircraft, ANPK MiG is developing training and light transport aircraft, and others with air cushion landing gear. It is also manufacturing horizontal tail surfaces for Dassault Falcon 900 business jets.

VERIFIED

Mikoyan MiG-15 'Fagot'. Large number of former Warsaw Treaty 'Fagots' in private hands in the USA, such as this one, believed to be ex-Polish Air Force *(Peter March)* *1994*

MIKOYAN MiG-15

NATO reporting name: Fagot
Chinese designation: J-2 (F-2)
Polish designation: LiM-2A/M

TYPE: Single-seat jet fighter.

PROGRAMME: The first MiG-15 Fagot flew as the I-310 development single-seat fighter on 30 December 1947. It was powered by one of 25 British Rolls-Royce Nene turbojets supplied to the former Soviet Union, now Commonwealth of Independent States (CIS) with 30 Derwent engines. Production of the MiG-15 began with Soviet RD-45 engine (a Nene development), later superseded by the Klimov VK-1 on the improved MiG-15bis. MiG-15s were also built in China, Czechoslovakia and Poland. The single-seat MiG-15bis and tandem two-seat MiG-15UTI trainer were the first jet aircraft to be built in China and were identical in all important respects to their Soviet counterparts.

VERSIONS: **MiG-15:** Standard single-seat fighter.

MiG-15bis: Improved fighter with Klimov VK-1 engine. See separate entry.

MiG-15UTI: Two-seat combat/trainer. See separate entry.

OPERATORS: Versions of the MiG-15 are in service with the armed forces of the following countries: Algeria (30), Angola (9), Congo (1), Cuba (25), Guinea Republic (1), Guinea-Bissau (2), Mali (1), Mongolia (3), Romania (12) and Yemen (4).

Mikoyan MiG-15 'Fagot' in Vietnamese Air Force colours *(Peter March)* *1994*

DESIGN FEATURES: Mid-wing cantilever monoplane. Sweepback at the leading-edge 35° at 25 per cent chord. Dihedral −3°. Thickness/chord ratio 11 per cent. Two fences on each wing. Variable incidence high-mounted tailplane.

FLYING CONTROLS: Hydraulically operated Fowler split flaps. Ailerons have both mass and sealed aerodynamic balances. Airbrakes on rear fuselage.

STRUCTURE: The wing is an all-metal, light alloy, stressed-skin structure with two main I section spars. The fuselage is a semi-monocoque light alloy stressed-skin structure in two main units: a forward assembly of nose and mid-sections and a rear section. Rear section joined by quick-release bolts at the rear wing spar attachment. The tail unit is a monoplane type with high-mounted tailplane.

LANDING GEAR: Retractable tricycle. Air-oil shock-absorbers. Mainwheels, with levered suspension, raised inwards, nosewheel forward. Hydraulic retraction, with emergency pneumatic system. Pneumatic brakes.

POWER PLANT: One RD-45 centrifugal-flow turbojet engine (2,270 kg; 4,515 lb st). Bifurcated nose air inlet. Fuel tanks in mid-fuselage section between the divided air intake ducts. Total internal fuel capacity 1,250 litres (330 US gallons; 275 Imp gallons). Drop tanks 600 litres (160 US gallons; 133 Imp gallons) each may be carried, one under each wing.

ACCOMMODATION: Pressurised cockpit forward of wing leading-edge with sliding canopy. Maximum pressure 4.2 lb/sq in from 7,930 m (26,000 ft). Oxygen system is provided for emergency use. Ejection seat with automatic release is standard.

ARMAMENT: Standard armament consists of two NS (Nudelmann-Suranov) 23 mm cannon below port side of nose and one 37 mm N (Nudelmann) cannon below starboard side. Guns mounted in a carriage which can be lowered for easy maintenance. Gyro gunsight. Rockets or two 1,000 lb bombs may be carried under wing.

DIMENSIONS, EXTERNAL:

Wing span	10.10 m (33 ft 1½ in)
Length	11.10 m (36 ft 4 in)
Height	3.40 m (11 ft 2 in)

WEIGHTS AND LOADINGS:

Weight empty	3,780 kg (8,320 lb)
Weight loaded (combat)	5,120 kg (11,270 lb)
Max loaded weight (with external fuel or bombs)	6,465 kg (14,240 lb)

PERFORMANCE:

Max speed	approx 518 kts (1,072 km/h; 670 mph)
Stalling speed	175 km/h (109 mph)
Initial rate of climb	3,170 m (10,400 ft)/min
Service ceiling	15,550 m (51,000 ft)

UPDATED

MIKOYAN MiG-15bis

TYPE: Single-seat jet fighter.

STRUCTURE: As for standard MiG-15.

POWER PLANT: One 'VK-1' centrifugal-flow turbojet engine (2,270 kg; 3,766 lb st).

FUEL: T-77 kerosene. Internal capacity 1,410 litres (372 US gallons; 310.2 Imp gallons); external fuel 490 litres (129 US gallons; 108 Imp gallons).

ACCOMMODATION: Pilot's seat below rear sliding canopy.

ARMAMENT AND EQUIPMENT: One 37 mm cannon with 40 rounds on starboard side of lower front fuselage; two 23 mm NS cannon each with 80 rounds on port side. Gyro gunsight. Attachments for auxiliary fuel tanks or 550 kg (1,100 lb) of bombs.

DIMENSIONS, EXTERNAL:

Wing span	10.10 m (33 ft 1½ in)
Length	11.10 m (36 ft 4 in)
Wing area	23.70 m² (255 sq ft)

WEIGHTS AND LOADINGS:

Weight loaded	5,028 kg (11,085 lb)

PERFORMANCE:

Max speed	1,100-1,200 km/h (684-746 mph)
Stalling speed (clean)	210 km/h (131 mph)
Stalling speed (flap and L/G down)	190 km/h (118 mph)
Ceiling	above 15,000 m (49,200 ft)
Endurance	about 2 h

UPDATED

MIKOYAN MiG-15UTI

NATO reporting name: Midget

TYPE: Two-seat conversion trainer developed from MiG-15.

STRUCTURE: Similar to fighter type.

POWER PLANT: One RD-45 turbojet engine (270 kg; 3,766 lb st).

ACCOMMODATION: Pupil and instructor sit in tandem under two separate canopies.

ARMAMENT: Guns not carried but fairings are retained.

DIMENSIONS, EXTERNAL:

Wing span	10.10 m (33 ft 1½ in)
Length	11.10 m (36 ft 4 in)
Height	3.40 m (11 ft 2 in)

PERFORMANCE:

Similar to that of MiG-15.

UPDATED

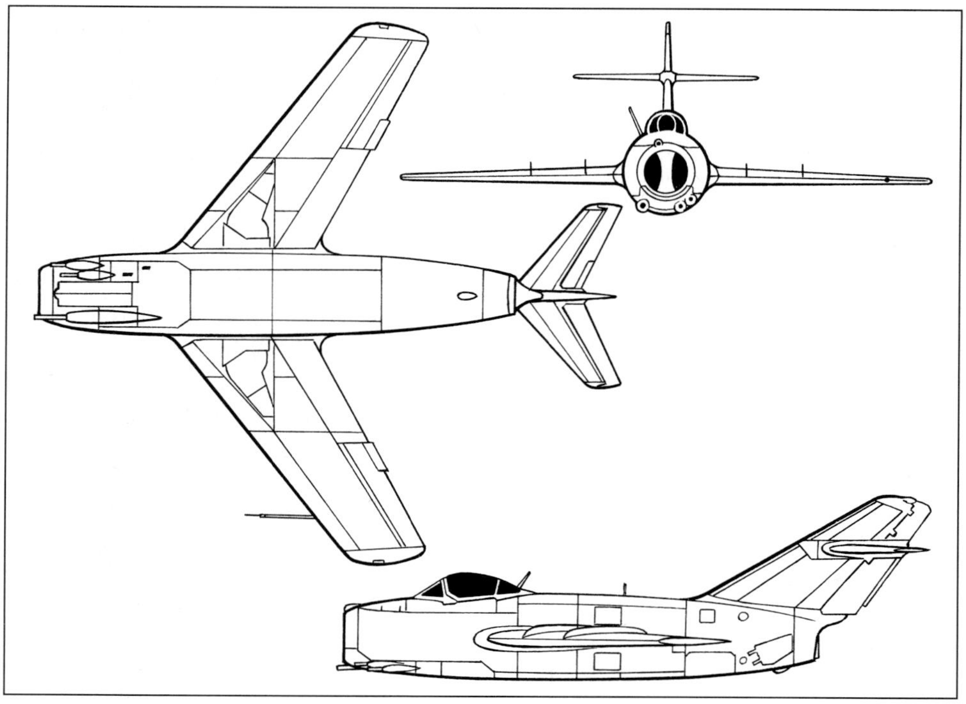

Mikoyan MiG-15 'Fagot' *(Jane's/Mike Keep)*

1994

MIKOYAN MiG-17

NATO reporting name: Fresco

TYPE: Single-seat fighter-bomber.

PROGRAMME: Less than a year after the MiG-15 had been sanctioned for production (1948), this design bureau initiated work on a follow-on fighter that would approach the speed of sound mainly through refinement of the basic airframe configuration. The first prototype received the designation I-330, and flew for the first time in January 1950. Claims that the I-330 managed to better Mach 1 during test flights are believed to be unsubstantiated, but the overall improvements in performance were important.

Following the loss of the first prototype, a second and further improved prototype took over, allowing testing to be completed in 1951, and production of the MiG-17 was given the go-ahead. Compared to the MiG-15, the MiG-17 had a lengthened fuselage with softer taper, larger area tail surfaces to benefit handling characteristics, and thinner section wings with rounded tips. Indeed, the wings were designed from scratch, with the inner leading-edges extended forward; this resulted in greater root chord and varying leading-edge sweepback (45° along inner portions, 42° on outer panels). A mark of identification was the MiG-17's three boundary layer fences on each wing.

Mikoyan MiG-17 'Fresco' single-seat fighter-bomber in Polish Air Force colours, but no longer in service with them

1994

Mikoyan MiG-15UTI 'Midget' two-seat trainer *(Piotr Butowski)*

1994

Production began with a day fighter model (NATO 'Fresco-A'), which retained the VK-1 engine. The later MiG-17PF introduced all-weather capability, housing Izumrud S-band radar in a 'bullet' radome at the centre of the nose air intake and in an extension on the upper lip of the intake. Subsequently, this S-band radar was superseded by an E/F-band version of 'Scan Fix', which still gave neither a large antenna nor a wide angle of scan and is now thought obsolescent.

In addition to the specialised two-seat trainer derivative of the MiG-17, known as the JJ-5 and exclusive to Chinese production (described in the China section), the MiG-17 was also built (apart from in the USSR) in China, Czechoslovakia and Poland, with whom it was known as the J-5, S-104 and LiM-5 respectively. Soviet production alone is believed to have totalled some 6,000 aircraft, with the other countries adding at least half that number again.

VERSIONS: **MiG-17F** ('Fresco-C'): Day fighter, superseding 'Fresco-A' in production in 1953. VK-1F engine gave way to more powerful VK-1A. MIG-17F was the most produced version.

MiG-17PF ('Fresco-D'): Limited all-weather version with radar as noted above. Initially (at least) fitted with afterburning Klimov VK-1F engine, rated at 33.14 kN (7,450 lb) thrust with afterburning.

J-5: Chinese built version of the MiG-17F, constructed at Shenyang. Production is thought to have begun in late 1956; by 1959 these were almost entirely of Chinese local manufacture. VK-1 type engines were built at Harbin.

J-5A or J-5Jia: Chinese version of the MiG-17PF, constructed at Shenyang.

F-5/-5A: Chinese export models of the MiG-17, including those to Albania (30).

OPERATORS: Versions of the MiG-17 are in service with the armed forces of the following countries: Algeria (50), Angola (8), Congo (8), Cuba (18), Ethiopia (15), Guinea Republic (4), Guinea-Bissau (3), Madagascar (4), Mali (5), Mongolia (12), Romania (6) and Syria (30).

The following details refer specifically to the MiG-17F 'Fresco-C':

DESIGN FEATURES: Cantilever mid-wing monoplane. Sweepback 45° at roots, 42° on outer panels. Anhedral 3°. Three boundary layer fences on each wing.

FLYING CONTROLS: Split Fowler type flaps. Bulged rear airbrakes.

STRUCTURE: The fuselage is a semi-monocoque structure. The tail unit has sharply swept surfaces. Fin tip dielectric aerial.

LANDING GEAR: Retractable tricycle type. Mainwheel tyres diameter 66 cm.

POWER PLANT: One Klimov VK-1A turbojet engine, developing 33.83 kN (7,605 lb st) with afterburning. Normal fuel load in internal tanks 1,410 litres (372 US gallons; 310 Imp gallons). A 400 litre (106 US gallon; 88 Imp gallon) external tank may be fitted at half-span on each wing.

ACCOMMODATION: Pilot only in pressurised cockpit with ejection seat. Rearward sliding canopy.

ARMAMENT: One 37 mm Nudelmann-Suranov NS-37 cannon and two 23 mm Nudelmann-Rikhter NR-23 cannon, or three 23 mm cannon. Provision for four underwing packs of 8 × 55 mm air-to-air rockets or a total of 500 kg (1,102 lb) of bombs under the wings.

AVIONICS: See Programme.

DIMENSIONS, EXTERNAL:

Wing span	9.63 m (31 ft 7 in)
Length overall (without radar)	11.26 m (36 ft 9½ in)
Length overall (with radar)	12.33 m (40 ft 4 in)
Height overall	3.80 m (12 ft 5½ in)
Wheel track	3.85 m (12 ft 7½ in)

WEIGHTS:

Weight empty	3,930 kg (8,664 lb)
Max T-O weight	6,069 kg (13,379 lb)

PERFORMANCE:

Max level speed at 3,000 m (9,845 ft)	617 kts (1,145 km/h; 711 mph)
Rate of climb at S/L	3,900 m (12,795 ft)/min
Service ceiling	16,600 m (54,460 ft)
Max range, with external tanks and bombs	755 n miles (1,400 km; 870 miles)

UPDATED

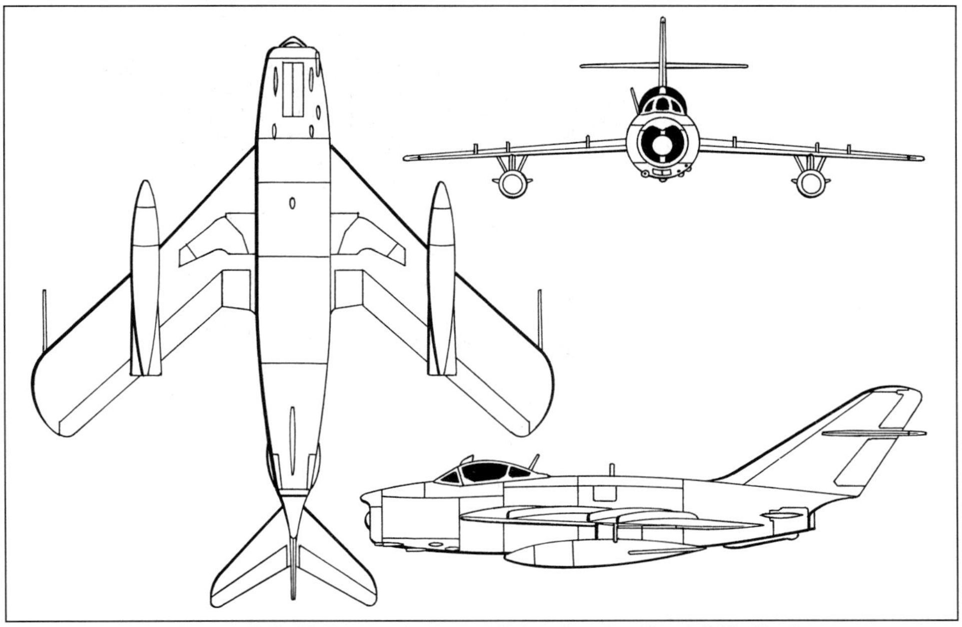

Mikoyan MiG-17 'Fresco' single-seat fighter-bomber *(Jane's/Mike Keep)*

1994

Mikoyan MiG-17 'Fresco'. This aircraft in Polish Air Force colours but no longer in service with them

1994

MIKOYAN MiG-19

NATO reporting name: Farmer

TYPE: Single-seat interceptor fighter.

PROGRAMME: Soon after test flying began of the MiG-17 prototypes, work started on a supersonic fighter that entered service in 1955 as the MiG-19. It was the last of the early style MiG jet fighters, but the first in the MiG-15/17/19 series to possess true supersonic performance in level flight. However, its traditional configuration belied its real strengths. Retaining a proven layout may well have been a prudent move because, in 1951, more radical Yakovlev and Lavochkin prototype fighters totally failed to meet a slightly earlier specification for a transonic fighter. The barrel fuselage with its wide circular nose air intake housed twin engines in all MiG-19s with the exception of the very first prototype (which made its maiden flight in 1952), while the wing design was wonderfully suited to a wide speed range. The third prototype, the 1350/M, which first flew in September 1953, was the first to have the full MiG-19 configuration as we know it.

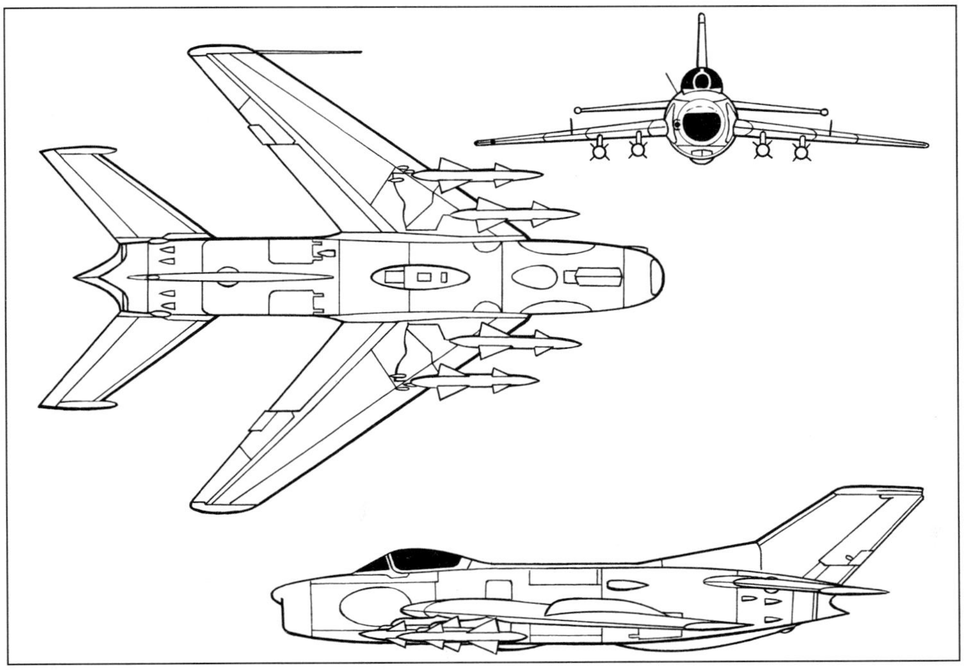

Mikoyan MiG-19PM 'Farmer' *(Jane's/Mike Keep)*

1994

Unfortunately for the MiG-19, the experience of MiG-15s in the Korean War led the MiG bureau to pass on quickly from swept-wing fighters to a very different aircraft with a much improved power/weight ratio, the MiG-21. Indeed, it was the success of the MiG-21, which joined more air forces in greater number than any other jet fighter in history, that appeared to date the MiG-19 prematurely, and Soviet production up to 1959 covered fewer than half the number of MiG-17s built.

Perhaps the greatest compliment to the MiG-19 has been its amazingly successful production run in China. As the Shenyang J-6 (which see), it entered production in the early 1960s and first joined the Air Force of the People's Liberation Army in mid-1962, following genuine MiG-19s

Mikoyan 'Farmer' in service with the air forces of Afghanistan (F-6), Cuba and North Korea (F-6)

1994

Mikoyan MiG-19 'Farmer' single-seat interceptor and fighter aircraft

1994

previously assembled in China from Soviet knock-down components. Chinese production of the J-6 greatly exceeded Soviet MiG-19 production, and only in the mid-1980s was full-scale building run down. Ironically, the Chinese initially built only small numbers of the J-7, a MiG-21 derivative.

VERSIONS: **MiG-19SF** ('Farmer-C'): Controlled stabiliser day fighter-bomber, powered by two 31.9 kN (7,165 lb st) RD-9B turbojet engines. Three 30 mm NR-30 cannon. Underwing attachments for two air-to-air missiles, two rockets of up to 212 calibre, two packs of eight air-to-air rockets, two 250 kg bombs, drop tanks or other stores. Has the all-moving tailplane and underfuselage as well as fuselage side airbrakes introduced on the MiG-19S.

MiG-19PF ('Farmer-D'): (*Perekhvatchik*: interceptor). Limited all-weather model with Izumrud (Emerald) radar scanner inside its engine air intake in a 'bullet' radome and ranging unit in the intake top lip. Armed with two wingroot cannon. Lower cockpit canopy.

MiG-19PM ('Farmer-D'): (*Modifikatsirovanny*; modified). As for MiG-19PF, but able to carry four first generation 'Alkali' missiles. No cannon.

MiG-19R: Cannon armed reconnaissance model, with two nose cameras.

J-6: Chinese built version of the MiG-19 for domestic use. See separate entry.

F-6: Chinese built version of the MiG-19 for export. See separate entry.

OPERATORS: Versions of the MiG-19 are in service with the armed forces of Cuba (30).

DESIGN FEATURES: Cantilever monoplane. Wings have slight anhedral and are swept at 55° at 25 per cent of chord. Single fence on each wing at mid-span. Air intake in nose and twin jet exits under rudder. Swept horizontal and vertical surfaces on tail unit.

FLYING CONTROLS: All-moving tailplane.

STRUCTURE: The fuselage is a circular section with straight top line and curved ventral line.

LANDING GEAR: Tricycle type with inward-retracting mainwheels.

POWER PLANT: Two axial flow turbojets with reported thrust of 3,300 kg (7,273 lb st) each. Later versions have central shock-cone in intake, possible housing radar. Two or four 400 litre (105 US gallon; 88 Imp gallon) external tanks can be carried on underwing pylons.

ACCOMMODATION: Single cockpit well forward above nose.

ARMAMENT: One 30 mm cannon in bottom of fuselage nose. Two 30 mm cannon in each wingroot. Provision for underwing pods, each containing 8 × 50 mm rockets, or two air-to-air missiles.

DIMENSIONS:

Wing span	9.75m (32 ft)
Length	11.43 m (37 ft 6 in)

WEIGHTS AND LOADINGS:

Weight loaded	9,000 kg (19,840 lb)

PERFORMANCE:

Max level speed reported as	Mach 1.54

UPDATED

MIKOYAN MiG-21

NATO reporting names: Fishbed and Mongol

TYPE: Single-seat multirole fighter and two-seat operational trainer.

PROGRAMME: Development began to meet Autumn 1953 official requirements for short-range interceptor; tailed delta configuration selected for production late 1956 after flight testing of prototypes with swept and delta wings; Ye-6 preproduction prototype flew late 1957; MiG-21 production authorised 1958; deliveries began late that year; production completed except in China, where developed versions are manufactured by Chengdu Aircraft Corporation (CAC) and Guizhou Aviation Industry (GAIC).

VERSIONS: **MiG-21F** ('Fishbed-C'): First major production version, also built in the former Czechoslovakia. Still in production in improved forms in China, with Chinese designation **J-7**. Short-range clear-weather fighter, with radar ranging equipment and a Tumansky R-11 turbojet rated at 42.25 kN (9,500 lb st) dry and 56.4 kN (12,676 lb st) with afterburning (designation of engine given in Soviet press statements as TDR Mk R37F). Two underwing pylons for UV-16-57 pods, each containing 16 57 mm rockets, or K-13 air-to-air missiles, and one NR-30 cannon in starboard side of fuselage (one each side on early aircraft and on the 10 supplied to India). Internal fuel capacity of 2,340 litres (618 US gallons; 515 Imp gallons), plus underfuselage pylon for external fuel tank of 490 litres (129.4 US gallons; 108 Imp gallons) capacity. Small nose air intake of approximately 69 cm (27 in) diameter, with movable three-shock centrebody housing the radar ranging equipment. Undernose pitot boom, which folds upward on ground to reduce risk of ground personnel walking into it. Transparent blister cockpit canopy which hinges upward about base of integral flat bulletproof windscreen. Transparent rearview panel (not on aircraft built in Czechoslovakia) aft of canopy at front of shallow dorsal spine fairing. Large blade antenna at rear of this panel, with small secondary antenna midway along spine. Fowler flap between fuselage and aileron on each trailing-edge, with fairing plate under wing at outer extremity. Small forward-hinged airbrake under fuselage, forward of ventral fin; two further forward-hinged airbrakes, on each side of underfuselage in line with wingroot leading-edges, integral with part of cannon fairings. Brake parachute housed inside small door on port underside of rear fuselage, with cable attachment under rear part of ventral fin. Semi-encapsulated escape system, in which canopy is ejected with seat, forming shield to protect pilot from slipstream, until the seat has been slowed by its drogue chute. Leading-edge of fin extended forward on all but early aircraft, to increase chord.

MiG-21PF ('Fishbed-D'): Basic model of second series of operational versions with mid-fuselage waisted in accordance with area rule and forward fuselage of less tapered form. Intake enlarged to diameter of approximately 91 cm (36 in) and housing larger centrebody for R1L search/track radar (NATO 'Spin Scan A') to enhance all-weather capability (designation suffix letter 'P', standing for *Perekhvatchik*, is applied to aircraft adapted for all-weather interception from an earlier designed role). Remainder of airframe generally similar to that of MiG-21F, but pitot boom repositioned above air intake; cannon armament and fairings deleted, permitting simplified design for forward airbrakes; larger mainwheels and tyres, requiring enlarged blister fairing on each side of fuselage, over wing, to accommodate wheel in retracted position; dorsal spine fairing widened and deepened aft of canopy, to reduce drag and house additional fuel tankage, and rearview transparency deleted; primary blade antenna repositioned to mid-spine and secondary antenna deleted. Uprated R-11 turbojet, giving 58.4 kN (13,120 lb st) with afterburning. Internal fuel capacity increased to 2,850 litres (753 US gallons, 627 Imp gallons) in seven fuselage tanks. Late production aircraft have attachments for a rocket-assisted take-off unit (RATOG) aft of each main landing gear bay, and provision for a flap blowing system known as *Sduva Pogranichnovo Sloya* (SPS), which reduces the normal landing speed by some 22 knots (40 km/h; 25 mph). Flaps are larger than original Fowler type, do not move aft, and lack outboard fairing plates.

Fishbed-E: Basically similar to 'Fishbed-C' but with broad chord vertical tail surfaces. Brake parachute repositioned into acorn fairing, made up of clamshell doors, at

Mikoyan MiG-21PFM 'Fishbed-F' of the Polish Air Force ***(Piotr Butowski)***

1994

base of rudder, above jet nozzle. Provision for GP-9 underbelly pack, housing GSh-23 twin-barrel 23 mm gun, in place of centreline pylon, with associated predictor sight and electrical ranging system. Identified in 1964.

MiG-21FL: Export version of late model MiG-21PF series, with broad chord vertical tail surfaces and brake parachute housing at base of rudder but no provision for SPS or RATOG. About 200 were initially assembled and later built under licence in India by Hindustan Aeronautics Ltd (which see), with the IAF designation Type 77. R-11-300 turbojet rated at 38.25 kN (8,598 lb st) dry and 60.8 kN (13,668 lb st) with afterburning. Suffix letter 'L' (*Lokator*) indicates the installation of Type R2L ('Spin Scan B') search/track radar, reported to have lock-on range of 10 n miles (19 km; 12 miles) but to be ineffective at heights below about 915 m (3,000 ft) because of ground 'clutter'. Can be fitted with GP-9 underbelly gun pack. Identified in 1966.

MiG-21PFS or MiG-21PF(SPS): Similar to 'Fishbed-D', but with SPS as standard production installation.

MiG-21PFM ('Fishbed-F'): Successor to interim MiG-21PFS, embodying all the improvements introduced progressively of the PF and PFS, the suffix letter 'M' indicating an exportable version of an existing design. Leading-edge of fin extended forward a further 45 cm (18 in). Small dorsal fin fillet eliminated. Additional refinements, including sideways hinged (to starboard) canopy and conventional windscreen quarter-lights; simple ejection seat instead of semi-encapsulated type; and large dielectric portion at tip of tail fin. R2L radar. Maximum permissible speed at low-altitude reported to be 593 knots (1,100 km/h; 683 mph). Built also in the former Czechoslovakia.

MiG-21PFMA ('Fishbed-J'): Multirole version, with four underwing pylons instead of former two; deepened dorsal spine fairing above fuselage; improved radar (NATO 'Jay Bird'); Tumansky R-11F2S-300 turbojet, rated at 38.25 kN (8,598 lb st) dry and 60.8 kN (13,668 lb st) with afterburning; 0-130KM/200 ejection seat; armament one GSh-23 twin-barrel 23 mm gun, two AA-2/2D ('Atoll') infra-red air-to-air missiles on inboard pylons, two radar-homing AA-2Cs ('Atolls') or drop fuel tanks on outboard pylons.

MiG-21R ('Fishbed-H'): Tactical reconnaissance version, basically as MiG-21PFMA; external pod for forward facing or oblique cameras, or elint sensors, on centreline pylon; suppressed ECM antenna at mid-point on dorsal spine; optional radar warning receivers in wingtip fairings.

MiG-21MF ('Fishbed-J'): As MiG-21PFMA, but with higher rated, lighter weight, Tumansky R-13-300 turbojet; gun gas deflector beneath suction relief door forward of each wingroot; entered Soviet service 1969.

MiG-21M: Export variant of MiG-21MF; R-11F2S-300 engine; built as Indian Air Force Type 96 by Hindustan Aeronautics, with deliveries from 1973-81.

MiG-21RF ('Fishbed-H'): Tactical reconnaissance version of MiG-21MF; equipment as MiG-21R.

MiG-21SMB ('Fishbed-K'): As MiG-21MF, but deep dorsal spine extends rearward as far as brake-chute housing, for maximum fuel tankage and optimum aerodynamic form; deliveries began 1971.

MiG-21bis-A ('Fishbed-L'): Third generation multirole air combat/ground attack version; Tumansky R-25-300 turbojet, rated at 73.6 kN (16,535 lb st) with afterburning; updated avionics; generally improved construction standards; wider and deeper dorsal fairing than MiG-21MF; capacity of seven internal self-sealing fuel tanks increased to 2,900 litres (766 US gallons; 638 Imp gallons); weight empty 5,353 kg (11,800 lb); maximum T-O weight 9,500 kg (20,940 lb).

MiG-21bis-B ('Fishbed-N'): Advanced version of 'Fishbed-L'; further improved avionics indicated by ILS antennae (NATO 'Swift Rod') under nose and on fin tip; two radar-homing AA-2C ('Atoll') outboard and two AA-8 ('Aphid') inboard, or four AA-8s; one version can carry nuclear weapons; rate of climb at AUW of 6,800 kg (15,000 lb), with 50 per cent fuel and two 'Atolls', is 17,700 m (58,000 ft)/min; weight empty 6,000 kg (13,225 lb); maximum T-O weight clean 8,500 kg (18,740 lb); produced also by HAL, India, 1980-87.

MiG-21SMT ('Fishbed-K'): Version of MiG-21 with R-13F-300 turbojet and improved VHF-UHF com.

MiG-21U ('Mongol'): Two-seat trainer; initial 'Mongol-A' generally similar to MiG-21F except for two cockpits in tandem with sideways hinged (to starboard) double canopy, larger mainwheels and tyres of MiG-21PF, one-piece forward airbrake, above intake pitot and no gun. 'Mongol-B' has broader chord vertical tail surfaces and under-rudder brake chute housing of later operational variants, deeper dorsal spine and no dorsal fin.

MiG-21US ('Mongol-B'): As later MiG-21U, but provision for SPS flap blowing; retractable periscope for instructor in rear seat; internal fuel capacity 2,400 litres (634 US gallons; 528 Imp gallons).

MiG-21UM ('Mongol-B'): Two-seat trainer counterpart of MiG-21MF; R-13 turbojet; four underwing stores pylons.

UPGRADES: **Aerea:** MiG-21 Modular weapons storage upgrade. See separate entry.

Aerostar: MiG-21 upgrade programme. See separate entry.

Elbit: Collaborating with IAI on Romanian Air Force MiG-21 upgrade of approximately 100 aircraft.

Chengdu Aircraft Corporation (CAC): F-7M, F-7MP, Super 7. See separate entry for F-7.

Czech Republic: Requirement to upgrade fleet of 36 aircraft.

Egypt: Fleet of aircraft upgraded in early 1980s. Equipment fitted included: Teledyne navigation and IFF units, Head-Up Displays manufactured by GEC Avionics and Tracor ALE-40 Chaff/flare dispensers. The weapon system was also adapted for AIM-9P Sidewinder air-to-air missiles.

Hindustan Aeronautics Ltd (HAL): See separate entry.

Hungary: Twenty aircraft being retrofitted with western IFF units and radio sets.

IAI: MiG-21-2000. See separate entry.

Israel Aircraft Industries (IAI): Contract to upgrade Cambodia's fleet of 19 grounded MiG-21 aircraft. Work commenced in January 1995 and is expected to be completed by the end of 1995.

Mikoyan: MiG-21-93 upgrade. See separate entry.

SOKOL: MiG-21-93 upgrade. See separate entry.

TEREM: Bulgarian state company licensed in 1994 to upgrade all models of the MiG-21. The company plans to upgrade six MiG-21s for use as demonstration models.

OPERATORS: Versions of the MiG-21 are in service with the armed forces of the following countries: Afghanistan (60+), Algeria (80), Angola (70), Azerbaijan (2), Belarus, Bulgaria (40+), Congo (14), Croatia (15), Cuba (200), Czech Republic (220), Egypt (100), Ethiopia (65), Finland (24), Guinea Republic (7), Hungary (76), India (400+), Iraq (80+), Laos (40+), Libya (50), Madagascar (15), Mali (10), Mongolia, Mozambique, Nigeria (30), North Korea (150), Poland (130+), Romania (112), Russia (300), Slovakia (45), Sudan (9), Syria (240+), Uganda (5), Vietnam (170+), Yemen (70), Yugoslavia (125) and Zambia (13).

Following details refer to MiG-21MF:

DESIGN FEATURES: Diminutive tailed delta with clipped tips to mid-mounted wings; circular-section fuselage with prominent dorsal spine; nose intake with large three-position centrebody; swept tail, with large vertical surfaces and ventral fin; 2° wing anhedral from roots; TsAGI section,

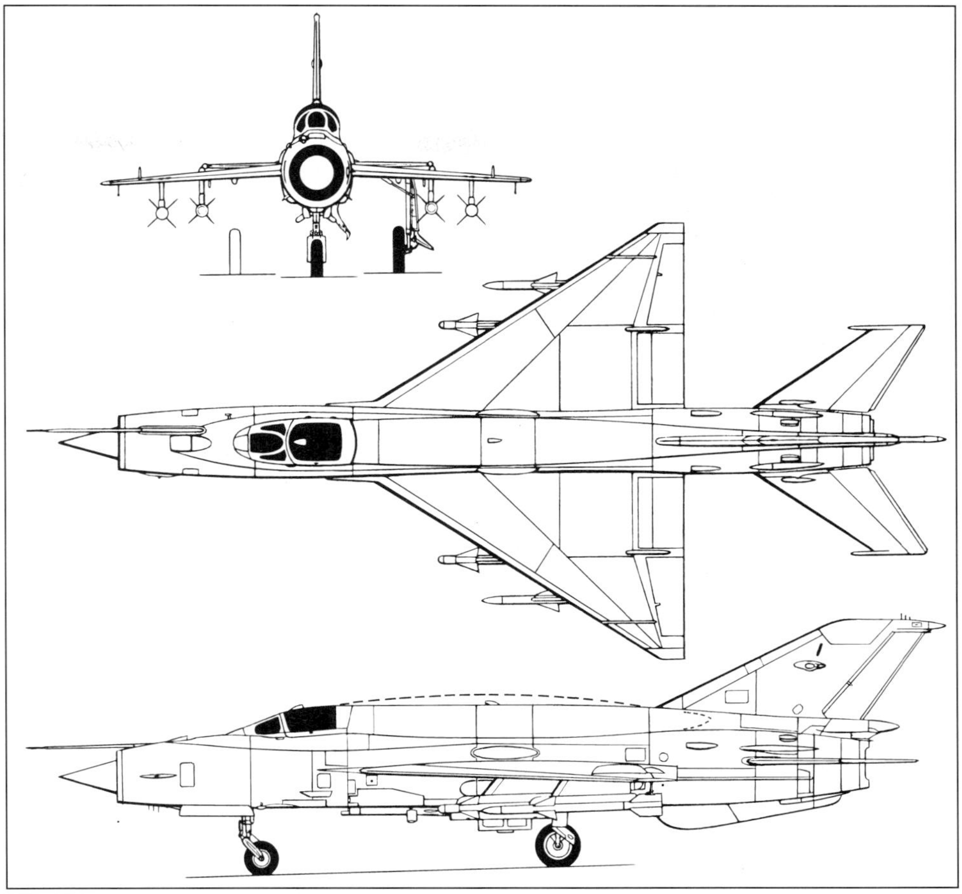

Mikoyan MiG-21SMT 'Fishbed-K' *(Jane's/Dennis Punnett)*

1994

Mikoyan MiG-21PF 'Fishbed-D' of the Hungarian Air Force *(Peter J. Cooper)*

1994

Mikoyan MiG-21bis of the Hungarian Air Force *(Peter J. Cooper)*

1995

thickness/chord ratio 5 per cent at root, 4.2 per cent at tip; leading-edge sweep 57°; no wing leading-edge camber.

FLYING CONTROLS: Manual operation, with autostabilisation in pitch and roll; hydraulically boosted inset ailerons; blown plain trailing-edge flaps, actuated hydraulically; forward-hinged door type airbrake each side of underfuselage below wing leading-edge; third forward-hinged airbrake under fuselage forward of ventral fin; airbrake actuated hydraulically; hydraulically boosted rudder and all-moving horizontal surface with two gearing ratios for varying combinations of altitude and airspeed; tailplane trim switch on control column; no tabs.

STRUCTURE: All-metal; wings have two primary spars and auxiliary spar; semi-monocoque fuselage, with spine housing control pushrods, avionics, single-point refuelling cap and fuel tank; blister fairings on fuselage above and below each wing to accommodate retracted mainwheels.

LANDING GEAR: Hydraulically retractable tricycle type, with single wheel on each unit; all units housed in fuselage when retracted. Forward-retracting non-steerable nosewheel unit, tyre size 500 × 180 mm; inward retracting mainwheels which turn to stow vertically inside fuselage. Size 800 × 200 mm tyres on mainwheels, inflated to approximately 7.93 bars (115 lb/sq in), ruling out normal operation from grass runways. Pneumatic disc brakes on all three wheels, supplied from compressed air bottles. Steering by differential mainwheel braking. Wheel doors remain open when legs are extended. Brake parachute housed inside acorn fairing at base of rudder.

POWER PLANT: One Tumansky R-13-300 turbojet, rated at 41.55 kN (9,340 lb st) dry and 64.73 kN (14,550 lb st) with afterburning. Fuel tanks in fuselage, and two integral tanks in each wing, with total capacity of 2,600 litres (687 US gallons; 572 Imp gallons), of which approximately 1,800 litres (475 US gallons; 396 Imp gallons) are usable within CG limits at low speed. Provision for carrying one finned external fuel tank, capacity 490 litres (130 US gallons; 108 Imp gallons) or 800 litres (211 US gallons; 176 Imp gallons), on underfuselage pylon and two 490 litre drop tanks on outboard underwing pylons. Two jettisonable solid propellant JATO rockets can be fitted under rear fuselage, aft of wheel doors.

ACCOMMODATION: Pilot only, on zero/130 ejection seat with spring loaded arm at top which ensures that seat cannot be operated unless hood is closed. Canopy is sideways hinged, to starboard, and is surmounted by a small rear-view mirror. Flat bulletproof windscreen. Cabin air conditioned. Armour plating forward and aft of cockpit.

SYSTEMS: Duplicated hydraulic system, supplied by engine-driven pump, with back-up by flow powered electric pump, and emergency electric tailplane trim.

AVIONICS: Search and track radar (NATO 'Jay Bird') in intake centrebody, with search range of 10.8 n miles (20 km; 12.5 miles). Other standard avionics include VOR, ARK automatic radio compass, IFF and Sirena 3 radar warning system with an indicator marked in 45° sectors in front of and behind the aircraft. Gyro gunsight maintains precision up to 2.75*g*. Automatic ranging can be fed into gunsight. Full blind-flying instrumentation, with attitude and heading indicators driven by remote central gyro platform.

ARMAMENT: One twin-barrel 23 mm GSh-23 gun, with 200 rounds, in belly pack. Four underwing pylons for weapons or drop tanks. Typical loads for interceptor role include two AA-2/2D (K-13A) 'Atoll' air-to-air missiles on inner pylons and two radar homing AA-2C 'Atolls' or two UV-16-57 rocket packs (16 57 mm rockets) on outer pylons; or two drop tanks and two AA-2/2D or AA-2C 'Atoll'. Typical loads for ground attack role are four UV-16-57 rocket packs, two 500 kg and two 250 kg bombs; or four 240 mm S-24 air-to-surface rockets.

DIMENSIONS, EXTERNAL (MiG-21MF):
Wing span 7.15 m (23 ft 5½ in)
Length, incl pitot boom 15.76 m (51 ft 8½ in)
Fuselage length, intake lip to jetpipe nozzle
12.30 m (40 ft 4¼ in)
Height overall 4.10 m (13 ft 5½ in)
Tailplane span 3.70 m (12 ft 8 in)
Wheel track 2.69 m (8 ft 10 in)
Wheelbase 4.81 m (15 ft 9½ in)

AREAS:
Wings, gross 23.0 m² (247.0 sq ft)

WEIGHTS AND LOADINGS (MiG-21MF):
Weight empty 5,843 kg (12,882 lb)
T-O weight:
with four K-13A missiles 8,200 kg (18,078 lb)
with two K-13A missiles and two 490 litre (130 US gallon; 108 Imp gallon) drop tanks
8,950 kg (19,730 lb)
with two K-13As and three drop tanks
9,400 kg (20,725 lb)
Max T-O weight 9,800 kg (21,605 lb)
Max wing loading 426.0 kg/m² (87.5 lb/sq ft)
Max power loading 151.4 kg/kN (1.48 lb/lb st)

PERFORMANCE (MiG-21MF):
Max level speed above 11,000 m (36,000 ft)
Mach 2.05 (1,175 kts; 2,175 km/h; 1,353 mph)
Max level speed at low-altitude
Mach 1.06 (701 kts; 1,300 km/h; 807 mph)
Landing speed 146 kts (270 km/h; 168 mph)
Design ceiling 18,000 m (59,050 ft)
Practical ceiling about 15,250 m (50,000 ft)
T-O run at normal AUW 800 m (2,625 ft)
Landing run 550 m (1,805 ft)
Combat radius (hi-lo-hi):
with four 250 kg bombs, internal fuel
200 n miles (370 km; 230 miles)
with two 250 kg bombs and drop tanks
400 n miles (740 km; 460 miles)
Range, internal fuel only
593 n miles (1,100 km; 683 miles)
Ferry range, with three external tanks
971 n miles (1,800 km; 1,118 miles)

PERFORMANCE (MiG-21US, clean):
Max level speed above 12,200 m (40,000 ft)
Mach 2.02 (1,159 kts; 2,150 km/h; 1,335 mph)
Max level speed at S/L
Mach 1.06 (701 kts; 1,300 km/h; 807 mph)
Max rate of climb at S/L 6,400 m (21,000 ft)/min
Rate of climb at 11,000 m (36,000 ft)
3,050 m (10,000 ft)/min
Time to 1,500 m (4,920 ft) 20 s
Turn rate at 4,575 m (15,000 ft):
instantaneous (Mach 0.5) 11.1°/s
instantaneous (Mach 0.9) 13.4°/s
sustained (Mach 0.9) 7.5°/s
T-O run 700 m (2,297 ft)

UPDATED

MIKOYAN/SOKOL (MIKOYAN) MiG-21-93 UPGRADE

TYPE: Combat aircraft upgrade.

PROGRAMME: Modernisation package to enhance the combat capability of MiG-21bis and MiG-21M/MF aircraft and to increase service life up to the years 2010-2015. On 3 May 1994 India's Deputy Defence Minister announced that a contract had been awarded to Mikoyan and Sokol to upgrade 100 MiG-21bis aircraft with an option for 70 more.

CUSTOMERS: Indian Air Force (100 aircraft with option for further 70 aircraft).

COSTS: Estimated costs for Indian Air Force upgrade are $1.33 million per aircraft.

DESIGN FEATURES: Programme includes: total replacement of onboard avionics, new slim-line canopy as well as improved power supply and air conditioning. Systems installation consists of Kopyo multifunctional coherent pulse Doppler radar with slot antenna, redesigned cockpit with head-up display (HUD), multifunction display (MFD) and hands-on throttle and stick (HOTAS) systems as well as a helmet-mounted target designator, inertial reference system (IRS), navigation computer, ADU, radio navigation/approach system, communications station, IFF system and ground air datalink. The radar warning receiver (RWR) provides electronic countermeasures (ECM) using onboard chaff/flare dispensers and anti-radar/radiation missiles. The design of the avionics system provides for integration of non-Russian system.

ARMAMENTS: Weapons include GSh-23 gun, 100/250 and 500 kg bombs, S-5 and S-8 unguided rocket projectiles, R-60 guided missiles, R-73 air-to-air missiles with IR homing warhead, R-27 medium-range guided missile with semi-active or IR homing warhead, RVV-AE medium-range active radar-guided AA missiles. Further air-to-air surface missiles include: S-13 and S-24 unguided rockets, KAB-500 correctable air bombs with TV homing warhead, X-25MP and X-31P anti-radiation missiles, X-31A anti-ship supersonic missile.

UPDATED

MIKOYAN MiG-23

NATO reporting names: Flogger-A, B, C, E, F, G, H and K

TYPE: Single-seat variable geometry air combat fighter and two-seat operational trainer.

PROGRAMME: Development began 1964; 23-11/1 prototype first flew 10 June 1967 and was displayed during Aviation Day flypast, Domodyedovo Airport, Moscow, 9 July 1967; preseries aircraft delivered to Soviet air forces 1970; initial series production interceptors delivered 1973; with MiG-27, superseded MiG-21 as primary equipment Soviet tactical air forces and APVO home defence interceptor force; production in USSR ended mid-1980s but continues in India; replacement of early variants with MiG-29s and Su-27s continues.

VERSIONS: **MiG-23** ('Flogger-A'): Prototype shown at Domodyedovo 9 July 1967. One Lyulka AL-7F-1 afterburning turbojet, rated at 98.1 kN (22,046 lb st).

MiG-23S ('Flogger-A'): Preproduction version, with AL-7F-1 engine. Issued to complete fighter regiment in 1971 for development.

MiG-23SM ('Flogger-A'): As MIG-23S, but with four APU-13 pylons for external stores added under engine air intake ducts and fixed inboard wing panels.

MiG-23U: Tandem two-seat training counterpart of MiG-23S.

MiG-23M ('Flogger-B'): Most produced production version; first flown June 1972; single-seat air combat fighter; first aircraft of former Soviet Union with demonstrated ability to track and engage targets flying below its own altitude; Soyuz/Khachaturov R-29-300 turbojet, rated at 122.5 kN (27,540 lb st) with afterburning; no wing leading-edge flaps initially (retrofitted later); Sapfir-23D-Sh J-band radar (NATO 'High Lark'); Sirena-3 radar warning system; Doppler; TP-23 infra-red search/track pod under cockpit; standard in Soviet air forces from about 1975.

MiG-23MF ('Flogger-B'): Export version of MiG-23M, in service with non-Soviet Warsaw Pact air forces from 1978.

MiG-23UB ('Flogger-C'): Tandem two-seat operational training/combat version; Tumansky R-27F2M-300 turbojet, rated at 98 kN (22,045 lb st) with afterburning; individual canopy over each seat; rear seat raised, with retractable periscopic sight; deepened dorsal spine fairing aft of rear canopy. First flown May 1969; in production 1970-78.

MiG-23MS ('Flogger-E'): Export version of MiG-23M with R-27F2M-300 engine; equipped to lower standard; smaller radar ('Jay Bird', search range 15 n miles; 29 km; 18 miles, tracking range 10 n miles; 19 km; 12 miles) in shorter nose radome; no infra-red sensor or Doppler; armed with R-3S (K-13T; NATO AA-2 'Atoll') or R-60 (K-60; NATO AA-8 'Aphid') air-to-air missiles and GSh-23 gun.

Mikoyan MiG-23 'Flogger' single-seat fighter aircraft

1994

Mikoyan MiG-23ML 'Flogger-G' of the former Czechoslovakian Air Force *(Peter J. Cooper)*

1994

MiG-23B ('Flogger-F'): Single-seat light attack aircraft based on MiG-23S interceptor airframe; forward fuselage redesigned; instead of ogival radome, nose sharply tapered in side elevation, housing PrNK Sokol-23S nav/attack system; twin-barrel 23 mm GSh-23L gun retained in bottom of centre-fuselage; armour on sides of cockpit; wider, low-pressure tyres; Lyulka AL-21F-300 turbojet, rated at 11.27 kN (25,350 lb st) with afterburning; fuel tanks designed to fill with neutral gas as fuel level drops, to prevent explosion after impact; active and passive ECM; six attachments under fuselage and wings for wide range of weapons; project started 1969; first flight 20 August 1970; 24 built; developed as MiG-23BN/BM/BK and MiG-27 series.

MiG-23BN ('Flogger-F'): As MiG-23B except for Soyuz/Khachaturov R-29B-300 turbojet, rated at 11.27 kN (25,350 lb st) with afterburning, and Sokol-23N nav/attack system.

MiG-23BM ('Flogger-F'): As MiG-23BN except for PrNK-23 nav/attack system slaved to a computer.

MiG-23BK: Further equipment changes, NATO reporting name 'Flogger-H' identifies aircraft with small fairing for radar warning receiver each side of bottom fuselage, forward of nosewheel doors. Iraqi aircraft have Dassault type fixed flight refuelling probe forward of windscreen.

MiG-23ML ('Flogger-G'): Much redesigned and lightened version (L of designation for logkiy: light) built in series 1976-81; basically as MiG-23M, but Soyuz/Khachaturov R-35-300 turbojet; rear fuselage fuel tank deleted; much smaller dorsal fin; modified nosewheel leg; Sapfir-23ML lighter weight radar; new undernose pod for TP-23M IRST; new missles. *Detailed description applies to MiG-23ML.*

MiG-23P ('Flogger-G'): Modified version of MiG-23ML; digital nav system computer guides aircraft under automatic control from the ground and informs pilot when to engage afterburning and to fire his missiles and gun.

MiG-23MLD ('Flogger-K'): Mid-life update of MiG-23ML (D of designation stands for dorabotannyy: modified); identified by dogtooth notch at junction of each wing glove leading-edge and intake trunk; system introduced to extend and retract leading-edge flaps automatically when wing sweep passes 33° (system disengaged and flaps retracted when speed exceeds 485 knots; 900 km/h; 560 mph and wings at 72° sweep); new IFF antenna forward of windscreen; R-73A (NATO AA-11 'Archer') close-range air-to-air missiles on fuselage pylons; pivoting pylons under outer wings; radar warning receivers and chaff/flare dispensers added; built-in simulation system enables pilot to train for weapon firing and air-to-surface missile guidance without use of gun or missiles.

UPGRADES: **Mikoyan:** MiG-23MLD 'Flogger-K' mid-life update of MiG-23ML (see Versions).

OPERATORS: Versions of the MiG-23 are in service with the armed forces of the following countries: Afghanistan, Algeria (60), Angola (80), Belarus (70), Bulgaria (50), Cuba (50+), Czech Republic (75+), Ethiopia (50), Hungary (20), India (130), Iraq (80+), Kazakhstan (30), Libya (135+), North Korea (60), Poland (32), Romania (30), Russia (1,000), Sudan (6), Syria (150+), Ukraine (120), Vietnam (5) and Yemen (25).

DESIGN FEATURES: Shoulder-wing variable geometry configuration; sweep variable manually in flight or on ground to 16°, 45° or 72° (values given in manuals and on pilot's panel; true values 18° 40′, 47° 40′ and 74° 40′ respectively); two hydraulic wingspeed motors driven separately by main and control booster systems; if one system fails, wing sweep system remains effective at 50 per cent normal angular velocity; rear fuselage detachable between wing and tailplane for engine servicing; lower portion of large ventral fin hinged to fold to starboard when landing gear extended, for ground clearance; leading-edge sweepback 72° on fixed-wing panels, 57° on horizontal tail surfaces, 65° on fin.

FLYING CONTROLS: Hydraulically actuated; full-span single-slotted trailing-edge flaps, each in three sections; outboard sections operable independently when wings fully swept; no ailerons; two-section upper surface spoilers/lift dumpers, forward of mid and inner flap sections each side, operate differentially in conjunction with horizontal tail surfaces (except when disengaged at 72° sweep), and collectively for improved runway adherence and braking after touchdown; leading-edge flap on outboard two-thirds of each main (variable geometry) panel, coupled to trailing-edge flaps; all-moving horizontal tail surfaces operated differentially and symmetrically for aileron and elevator function respectively; ground adjustable tab on each horizontal surface; rudder actuated by hydraulic booster with spring artificial feel; four door type airbrakes, two on each side of rear fuselage, above and below horizontal tail surface.

STRUCTURE: All-metal; two main spars and auxiliary centre spar in each wing; extended chord (dogtooth) on outer panels visible when wings swept; fixed triangular inboard wing panels; welded steel pivot box carry-through structure; basically circular section semi-monocoque fuselage, flattened each side of cockpit; lateral air intake trunks blend into circular rear fuselage; splitter plate, with boundary layer bleeds, forms inboard face of each intake; two rectangular auxiliary intake doors in each trunk, under inboard wing leading-edge, are sucked open to increase intake area at take-off and low airspeeds; pressure relief vents under rear fuselage; fin and forward portion of horizontal surfaces conventional light alloy structures; rudder and rear of horizontal surfaces have honeycomb core.

LANDING GEAR: Hydraulically retractable tricycle type; single wheel on each main unit and steerable twin-wheel nose unit; mainwheel tyres size 830 × 300; nosewheel tyres size 520 × 125; main units retract inward into rear of air intake trunks; main fairings to enclose these units attached to legs; small inboard fairing for each wheel bay hinged to fuselage belly. Nose unit, with mudguard over each wheel, retracts rearward. Mainwheel disc brakes and anti-skid units. Brake parachute, area 21 m² (226 sq ft), in cylindrical fairing at base of rudder with split conic doors.

POWER PLANT: One Soyuz/Khachaturov R-35-300 turbojet, rated at up to 127.5 kN (28,660 lb st) with maximum afterburning. Three fuel tanks in fuselage, aft of cockpit, and six in wings; internal fuel capacity 4,250 litres (1,122 US gallons; 935 Imp gallons). Variable geometry air intakes and variable nozzle. Provision for jettisonable external fuel tank, capacity 800 litres (211 US gallons; 176 Imp gallons), on underfuselage centreline; two more under fixed-wing panels.Two additional external tanks of same capacity may be carried on non-swivelling pylons under outer wings for ferry flights, with wings in fully forward position. Attachment for assisted take-off rocket each side of fuselage aft of landing gear.

ACCOMMODATION: Pilot only, on zero/130 ejection seat in air conditioned and pressurised cockpit, under small actuated rearward hinged canopy. Bulletproof windscreen.

AVIONICS: Modernised SAU-23AM automatic flight control system coupled to Polyot short-range navigation and flight system. Sapfir-23ML J-band multimode radar (NATO 'High Lark 2': search range 38 n miles; 70 km; 43 miles, tracking range 29 n miles; 55 km; 34 miles) behind dielectric nosecone; no radar scope; instead, picture is projected

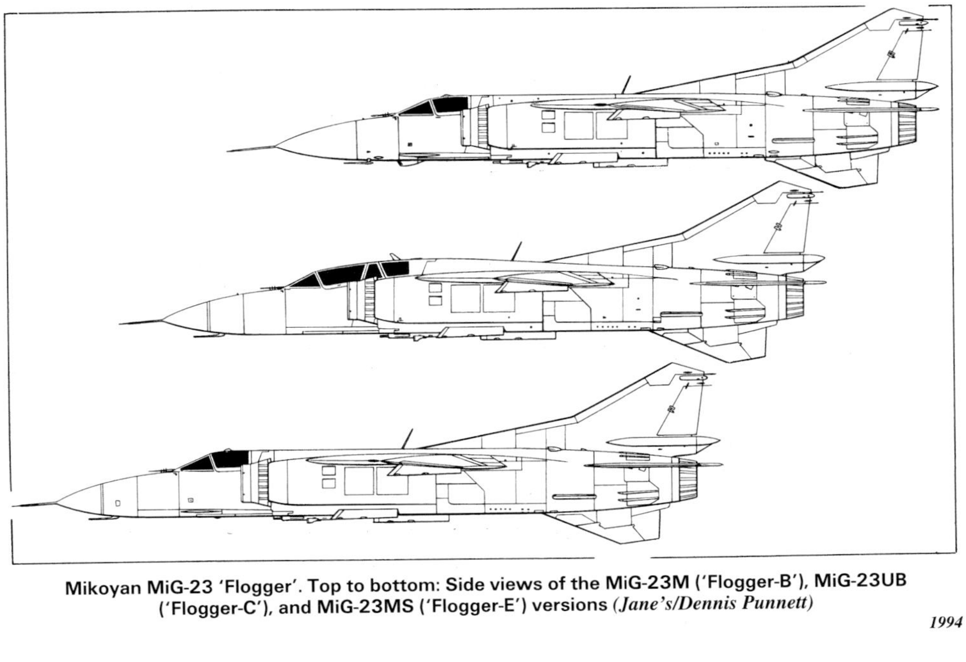

Mikoyan MiG-23 'Flogger'. Top to bottom: Side views of the MiG-23M ('Flogger-B'), MiG-23UB ('Flogger-C'), and MiG-23MS ('Flogger-E') versions *(Jane's/Dennis Punnett)*

1994

onto head-up display. RSBN-6S short-range radio nav system; ILS, with antennae (NATO 'Swift Rod') under radome and at tip of fin trailing-edge; suppressed UHF antennae form tip of fin and forward fixed portion of ventral fin; yaw vane above fuselage aft of radome; angle of attack sensor on port side. SRO-2 (NATO 'Odd Rods') IFF antenna immediately forward of windscreen. TP-23M undernose infra-red sensor rod, Sirena-3 radar warning system, and Doppler equipment standard on CIS version. Sirena-3 antennae in horns at inboard leading-edge of each outer wing and below ILS antenna on fin.

EQUIPMENT: ASP-17ML gunsight; small electrically heated rearview mirror on top of canopy; retractable landing/taxying light under each engine air intake.

ARMAMENT: One 23 mm GSh-23L twin-barrel gun in fuselage belly pack; large flash eliminator around muzzles; 200 rounds. Two pylons in tandem under centre-fuselage, one under each engine air intake duct, and one under each fixed inboard wing panel, for radar guided R-23R (K-23R; NATO AA-7 'Apex'), infra-red R-23T (K-23T; AA-7 'Apex') and/or infra-red R-60T (AA-8 'Aphid') air-to-air missiles, B-8 pack of 20 80 mm S-8 air-to-surface rockets, UB-32-57 packs of 32 57 mm S-5 rockets, S-24 240 mm rockets, bombs, container weapons, UPK-23-250 pods containing a GSh-23L gun, various sensor and equipment pods or other external stores. Use of twin launchers under air intake ducts permits carriage of four R-60 missiles, plus two R-23 on underwing pylons.

DIMENSIONS, EXTERNAL:

Wing span: fully spread	13.965 m (45 ft 10 in)
fully swept	7.779 m (25 ft 6¼ in)
Length overall: incl nose probe	16.71 m (54 ft 10 in)
Fuselage length: nosecone tip to jetpipe nozzle	15.65 m (51 ft 4¼ in)
Height overall	4.82 m (15 ft 9¾ in)
Wheel track	2.658 m (8 ft 8¾ in)
Wheelbase	5.772 m (18 ft 11¼ in)

AREAS:

Wings, gross: spread	37.25 m² (402.0 sq ft)
swept	34.16 m² (367.7 sq ft)

WEIGHTS AND LOADINGS:

Weight empty	10,200 kg (22,485 lb)
Max external weapon load	3,000 kg (6,615 lb)
T-O weight	14,700-17,800 kg (32,405-39,250 lb)
Max wing loading: spread	476.6 kg/m² (97.6 lb/sq ft)
swept	521.0 kg/m² (106.7 lb/sq ft)
Max power loading	139.6 kg/kN (1.37 lb/lb st)

PERFORMANCE:

Max level speed: at height, 72° sweep	Mach 2.35 (1,350 kts; 2,500 km/h; 1,553 mph)
at height, 16° sweep	Mach 0.88 (507 kts; 940 km/h; 584 mph)
at S/L, 72° sweep	Mach 1.10 (728 kts; 1,350 km/h; 838 mph)
Max rate of climb at S/L	14,400 m (47,250 ft)/min
Service ceiling	18,500 m (60,700 ft)
T-O run	500 m (1,460 ft)
Landing run	750 m (2,460 ft)
Combat radius: with six air-to-air missiles	620 n miles (1,150 km; 715 miles)
with 2,000 kg (4,410 lb) of bombs	378 n miles (700 km; 435 miles)
Range: with max internal fuel	1,050 n miles (1,950 km; 1,210 miles)
with three external tanks	1,520 n miles (2,820 km; 1,750 miles)
g limit: below Mach 0.85	+8.5
above Mach 0.85	+7.5

UPDATED

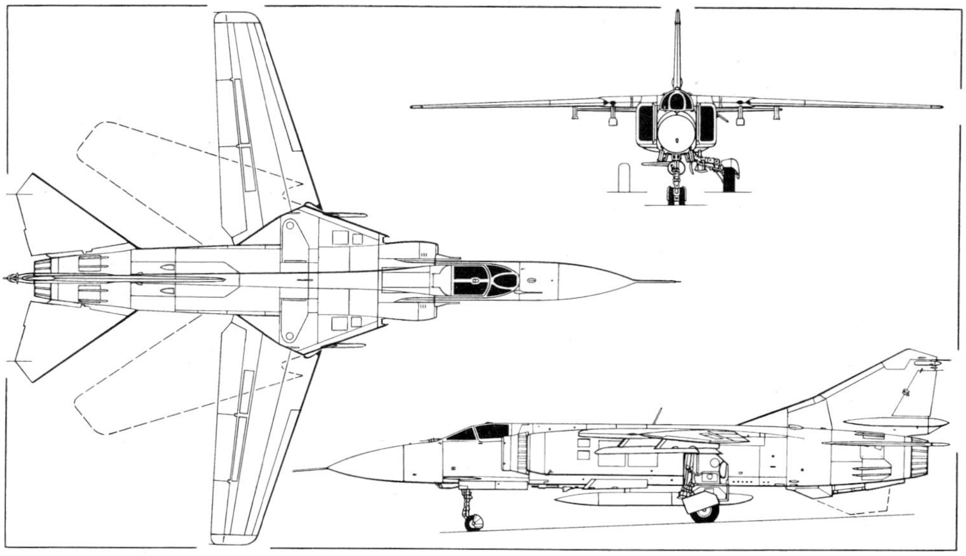

MiG-23MLD ('Flogger-K') single-seat interceptor post mid-life update *Jane's/Dennis Punnett)*

1994

Mikoyan MiG-23ML with R-35-300 turbojet engine *(Peter J. Cooper)*

1995

Mikoyan MiG-25 'Foxbat' single-seat interceptor

1994

MIKOYAN MiG-25

NATO reporting name: Foxbat

TYPE: Single-seat interceptor, reconnaissance aircraft and two-seat conversion trainer.

PROGRAMME: Design started 1959 as Ye-155P supersonic high-altitude interceptor to counter all potential threats, from low-flying cruise missiles to A-11 (SR-71A reconnaissance aircraft) under US development; programme launched officially February 1962; Ye-155R reconnaissance version designed and built 1961-62; Ye-155R-1 first to fly 6 March 1964; Ye-155P-1 interceptor prototype flew 9 September 1964; production as MiG-25/25R series, completed mid-1980s.

VERSIONS: **MiG-25P** ('Foxbat-A'): Single-seat interceptor derived from Ye-155P-1 prototype; two R-15B-300 turbojets, each 100.1 kN (22,500 lb st) with afterburning and with 150 hour service life; Smertch-A (NATO 'Fox Fire') radar, search range 54 n miles (100 km; 62 miles), tracking range 27 n miles (50 km; 31 miles); CIS aircraft all converted to MiG-25PDS.

MiG-25RB ('Foxbat-B'): Single-seat high-altitude reconnaissance bomber, derived from Ye-155R-1 prototype; production began as MiG-25R, for reconnaissance only, in 1969; bombing capability added to redesignated RB in 1970; no guns or air-to-air missiles; R-15BD-300 turbojets; any one of three interchangeable photographic/elint modules, with five camera windows and flush dielectric panels, carried aft of small dielectric nosecap, instead of interceptor's Smertch radar; slightly reduced wing span; wing leading-edge sweep constant 41° from root to tip; first aircraft produced in former USSR with INS updated by Doppler; specially developed automatic bombing system makes possible all-weather day/night precision attacks at supersonic speed from heights above 20,000 m (65,600 ft) against targets with known geographic co-ordinates, carrying four 500 kg bombs under wings, two under fuselage; SRS-4A/B elint equipment; fuel tank in each fin, providing additional 700 litres (185 US gallons; 154 Imp gallons) capacity; provision for 5,300 litre (1,400 US gallon; 1,165 Imp gallon) underbelly tank; able to fly long distances at cruising speed of Mach 2.35, maximum speed of Mach 2.83 with full bomb load.

MiG-25RBV and RBT ('Foxbat-B'): As MiG-25RB, with different equipment, including SRS-9 elint on RBV. Produced 1978-82.

MiG-25PU ('Foxbat-C'): Training version of MiG-25P; redesigned nose section containing separate cockpit with individual canopy for instructor, forward of standard cockpit and at lower level; gun firing and weapon release simulation standard; some systems modified and updated, permitting simulation of failures; no radar in nose; no combat capability; first roll-out 1968; maximum speed limited to Mach 2.65.

MiG-25RU ('Foxbat-C'): Training version of MiG-25R; identical to MiG-25PU except for absence of combat simulation system; no reconnaissance sensors; first roll-out 1972.

MiG-25RBK ('Foxbat-D'): Produced simultaneously with RB series from 1972; reconnaissance modules contain different elint and other avionics and no cameras; bombing capability retained.

MiG-25RBS ('Foxbat-D'): As MiG-25RBK but with different sensors; followed RBK in production 1975-82; all RBs upgraded to **MIG-25RBSh** during servicing from 1981. Other equipment changes produced **MiG-25RBF** in 1981.

MiG-25PD ('Foxbat-E'): Development of MiG-25P, produced 1978-82; uprated R-15BD-300 turbojets with life of 1,000 hours; Sapfir-25 radar and IRST, giving look-down/shoot-down capability comparable with MiG-23M; two R-40 (K-40; NATO AA-6 'Acrid') and four R-60 (K-60; NATO AA-8 'Aphid') missiles; provision for same underbelly tank as on MiG-25R series.

MiG-25PDS ('Foxbat-E'): As MiG-25PD but converted from MiG-25P, from 1979. Front fuselage lengthened by 250 mm (10 in) to accommodate flight refuelling equipment on some aircraft. *Detailed description applies to MiG-25PDS, except where indicated.*

MiG-25BM ('Foxbat-F'): Defence suppression aircraft derived from MiG-25RB; development started 1972; produced 1982-85; ECM in place of reconnaissance module in lengthened nose, with dielectric panel each side; small blister each side at rear of radome; dielectric panel at front of each outboard weapon pylon; underbelly auxiliary fuel tank as MiG-25R series; four Kh-58 (NATO AS-11 'Kilter') anti-radiation missiles underwing to attack surface-to-air missile radars over standoff ranges.

UPGRADES: **Mikoyan:** MiG-25PDS. See Versions.
Mikoyan: MiG-25RB conversions. See Versions.

OPERATORS: Versions of the MiG-25 are in service with the armed forces of the following countries: Algeria (35+), Azerbaijan (5), Belarus (60), Bulgaria (3), India (8), Iraq (20), Kazakhstan (13), Libya (70), Russia (300), Syria (40) and Ukraine (90).

DESIGN FEATURES: With MiG-31 derivative, is fastest combat aircraft yet identified in squadron service; original role demanded high-speed high-altitude capability and weapon system for attack over considerable range; high swept wing with anti-flutter body (maximum diameter 30 cm; 11.8 in) at each tip; slim front fuselage, with ogival nosecone, blended into rectangular air intake trunks with wedge intakes; inner wall of intakes curved at top and not parallel with outer wall; hinged panel forms lower intake lip, enabling area to be varied electronically; fuselage undersurface dished between engines; all-swept tail surfaces; twin 11° outward canted fins and twin outward canted ventral fins, all with large flush antennae; wing anhedral 5° from roots; leading-edge sweepback 42° 30′ inboard, 41° outboard of each outer missile pylon; sweepback at quarter-chord 32°; sweepback on tailplane 50°, fins 60°. Two shallow upper surface fences on each wing, in line with weapon pylons.

FLYING CONTROLS: Aileron at centre of each wing trailing-edge; plain flap on inboard 37 per cent; all-moving horizontal tail surfaces able to operate differentially at high-speeds; inset rudders; no tabs; airbrakes above and below jetpipes at rear of fuselage.

STRUCTURE: Airframe 80 per cent tempered steel, 8 per cent titanium in areas subject to extreme heating, such as wing and tail unit leading-edges, 11 per cent D19 heat-resistant aluminium alloy, by weight; two main wing spars forming torsion box, auxiliary front spar and two auxiliary rear spars; 14 primary fuselage frames, many intermediate frames and stringers.

LANDING GEAR: Retractable tricycle type. Single wheel, with high pressure tyre of 1.30 m (51.2 in) diameter, on each forward-retracting main unit; wheel stows vertically between air intake duct and outer skin of each trunk. Twin-wheel forward-retracting nose unit. Twin circular (60 m^2; 645 sq ft) or cruciform (50 m^2; 538 sq ft) brake-chutes in fairing above and between jet nozzles.

POWER PLANT: Two Soyuz/Tumansky R-15BD-300 single-shaft turbojets, each rated at 110 kN (24,700 lb st) with afterburning, in compartment of silver-coated steel. Fuel in two welded structural tanks occupying 70 per cent of volume of fuselage, between cockpit and engine bay, in saddle tanks around intake ducts, and in integral tank in each wing, filling almost entire volume inboard of outer fence; total capacity 17,660 litres (4,665 US gallons; 3,885 Imp gallons); provision for 5,300 litre (1,400 US gallon; 1,165 Imp gallon) underbelly tank.

ACCOMMODATION: Pilot only, on KM-1 zero-height/70 knot (130 km/h; 81 mph) ejection seat similar to that fitted to some versions of MiG-21. Canopy hinged to open sideways, to starboard.

SYSTEMS: Electronic fuel control system.

AVIONICS: Sapfir-25 fire control radar in nose, search range 54 n miles (100 km; 62 miles), tracking range 40 n miles (75 km; 46 miles), forward of avionics compartment housing navigation radar; K-10T radar scope; infra-red search/track sensor pod under front fuselage. SO-63 transponder; SRZO-2 (NATO 'Odd Rods') IFF, and SOD-57M ATC/SIF, with antennae in starboard fin tip; Sirena-3 360° radar warning system with receivers in centre of each wingtip anti-flutter body and starboard fin tip; SAU-155 automatic flight control system; unidentified ECCM, decoys and jammers. RSB-70/RPS HF, R-832M VHF-UHF; R-831 UHF communications equipment; RSBN-6S short-range nav; SP-50 (NATO 'Swift Rod') ILS; MRP-56P marker beacon receiver, RV-UM or RV-4 radio altimeter, and ARK-10 radio compass.

EQUIPMENT: Retractable landing light under front of each intake trunk. Back-up optical weapon sight.

ARMAMENT: No gun. Air-to-air missiles on four underwing attachments; originally one radar-guided R-40R (K-40R; NATO AA-6 'Acrid') and one infra-red R-40T (K-40T; NATO AA-6 'Acrid') under each wing; alternatively, one R-40 and two R-60s under each wing and, later, one R-23 (K-23; NATO AA-7 'Apex') and two R-73A (NATO AA-11 'Archer') or R-60T (NATO AA-8 'Aphid') under each wing.

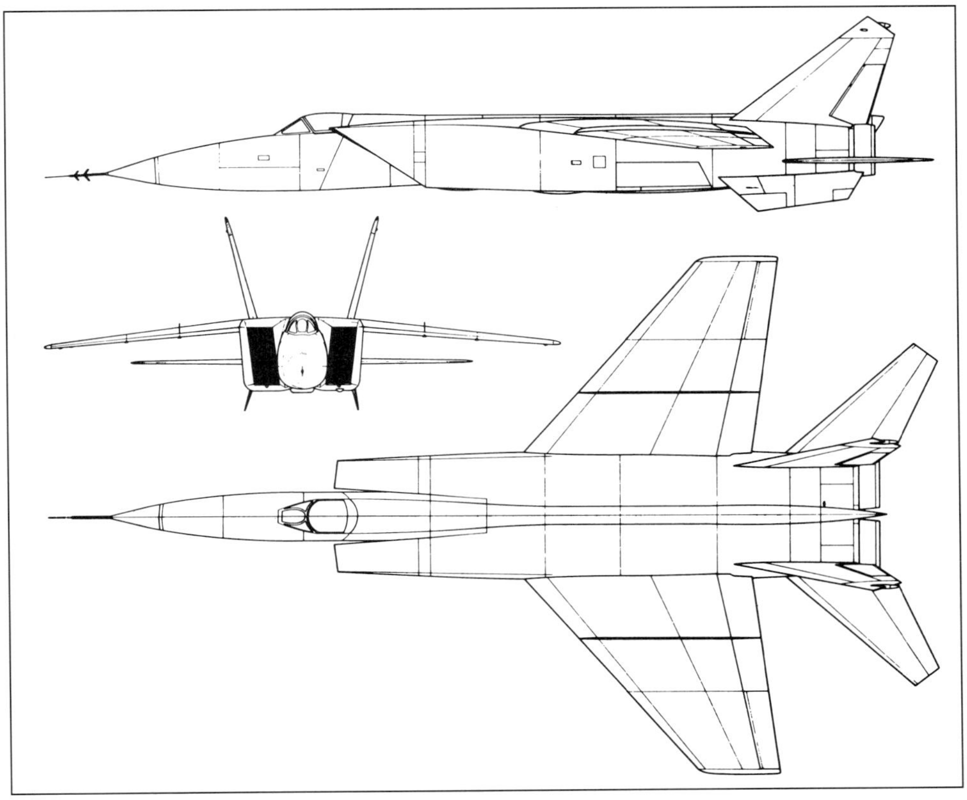

Mikoyan MiG-25P 'Foxbat-A'

1994

Mikoyan MiG-25 'Foxbat' carrying four AA-6 'Acrid' missiles

1994

DIMENSIONS, EXTERNAL:

Wing span: MiG-25P	14.015 m (45 ft 11¾ in)
MiG-25RB	13.418 m (44 ft 0¼ in)
Wing aspect ratio: MiG-25P	3.4
Length overall	23.82 m (78 ft 1¾ in)
Length of fuselage	19.40 m (63 ft 7¾ in)
Height overall	6.10 m (20 ft 0¼ in)
Wheel track	3.85 m (12 ft 7½ in)
Wheelbase	5.14 m (16 ft 10½ in)

AREAS:

Wings, gross: MiG-25P	61.40 m^2 (660.9 sq ft)
Vertical tail surfaces (total)	16.00 m^2 (172.2 sq ft)
Horizontal tail surfaces (total)	9.81 m^2 (105.6 sq ft)

WEIGHTS AND LOADINGS:

Max internal fuel: P	14,570 kg (32,120 lb)
R series	15,245 kg (33,609 lb)
Max fuel with underbelly Tank: P	18,940 kg (41,755 lb)
Take-off weight:	
P, clean, max internal fuel	34,920 kg (76,985 lb)
P, four R-40 missiles, max internal fuel	36,720 kg (80,950 lb)
R series, normal	37,000 kg (81,570 lb)
R series, max	41,200 kg (90,830 lb)
Max wing loading: P	598 kg/m^2 (122.5 lb/sq ft)
R series	671 kg/m^2 (137.4 lb/sq ft)
Max power loading: P	166.9 kg/kN (1.64 lb/lb st)
R series	187.3 kg/kN (1.84 lb/lb st)

PERFORMANCE:

Max permitted Mach number at height: P, R series 2.83
Max level speed: at 13,000 m (42,650 ft):
P, R series 1,620 kts (3,000 km/h; 1,865 mph)
at S/L: P, R series
Mach 0.98 (647 kts; 1,200 km/h; 745 mph)
T-O speed: P 195 kts (360 km/h; 224 mph)
Landing speed: P 157 kts (290 km/h; 180 mph)
Time to 20,000 m (65,000 ft) at Mach 2.35: P 8.9 min
Time to 19,000 m (62,335 ft): R series, clean 6.6 min
R series, with 2,000 kg (4,410 lb) of bombs 8.2 min
Service ceiling: P 20,700 m (67,900 ft)
R series, clean 21,000 m (68,900 ft)
T-O run: P 1,250 m (4,100 ft)
Landing run, with brake-chute: P 800 m (2,625 ft)
Range with max internal fuel:
P, supersonic 675 n miles (1,250 km; 776 miles)
P, subsonic 933 n miles (1,730 km; 1,075 miles)
R series, supersonic
882 n miles (1,635 km; 1,015 miles)
R series, subsonic
1,006 n miles (1,865 km; 1,158 miles)
Range with 5,300 litre external tank:
R series, supersonic
1,150 n miles (2,130 km; 1,323 miles)
R series, subsonic
1,295 n miles (2,400 km; 1,491 miles)
Endurance: P 2 h 5 min
g limit: P, supersonic +4.5

UPDATED

MIKOYAN MiG-29 EXPORT UPGRADE OPTIONS

TYPE: Combat aircraft upgrade.

PROGRAMME: The MiG Complex now offers a choice of modifications to its export MiG-29 aircraft. For full details of the MiG-29, see *Jane's All The World's Aircraft* 1995-96.

DESIGN FEATURES: Foreign customers of the aircraft can opt for the following upgrades: installation of two drop wing tanks with total capacity of 2,300 litres (607 US gallons; 506 Imp gallons) to increase range up to 1,618 n miles (3,000 km 1,864 miles), radar upgrade for use with PBB-AE active homing air-to-air missiles targeting two aircraft simultaneously, aircraft flight control system with improved handling and stability performance at high angles of attack (max permissible angle of attack increased to 30°), combat load increased from 2 to 4 tons, in-flight refuelling system, ground mapping mode for use with anti-submarine radar active homing missiles, modes for operation of air-to-ground radar passive homing missiles, modes for operation of air-to-ground pin-point television and laser homing missiles. Upgrade also includes the following western navigational systems: TACAN AN/ARN-118(V) navigational system, ILS-71 landing system, P-800П1 radio communication station with emergency frequency of 243 MHz, GPS TNL-1000 satellite navigation system and CO-69M RBS responder. All the instruments are also available with Imperial measurement system (feet and knots and so on).

NEW ENTRY

MIL

MOSKOVSKY VERTOLYOTNY ZAVOD (MVZ) IMENI M. L. MILYA (Moscow Helicopter Plant named after M. L. Mil)

2 Sokolnichyesky Val, 107113 Moscow
Telephone: 7 (095) 264 9083
Fax: 7 (095) 264 5571
Telex: 412144 MIL SU

CHIEF DIRECTOR, GENERAL DESIGNER: Mark V. Vineberg

OKB founded 1947 by Mikhail Leontyevich Mil, who was involved with Soviet gyroplane and helicopter development from 1929 until his death on 31 January 1970, aged 60. His original Mi-l, first flown September 1948 and introduced into service 1951, was first series production helicopter built in USSR. More than 25,000 helicopters of Mil design built, representing 95 per cent of all helicopters in CIS.

Mil design and production facilities to be integrated into group comprising Mil Moscow, Kazan and Rostov plants, a helicopter operating company, financial and insurance interests. Associates will include Ulan-Ude production centre, Arsenyev and Viatka factories.

VERIFIED

MIL Mi-4

NATO reporting name: Hound

TYPE: General purpose helicopter.

PROGRAMME: The Mi-4 was standard equipment in the Soviet armed forces, Aeroflot and with many other civil and military operators throughout the world. Several thousand were produced but only about 20 remain in service with the armed forces of the CIS. These are fulfilling liaison duties.

VERSIONS: **Mi-4**: Basic military version with underfuselage gondola for navigator. Production said to have started in 1952. Civil freight version is generally similar, with double clamshell rear-loading doors.

Soviet films of military exercises, released in 1968 showed a close support version of the Mi-4, armed with a gun in the front of the underfuselage nacelle and air-to-surface rockets.

Mi-4P: Passenger-carrying version, with seats for 8-11 passengers, which entered service on Aeroflot's Simferopol-Yalta route in the Crimea in November 1958. As an ambulance it can carry eight stretchers and medical attendant. This version has square windows instead of the circular windows of the military version, and has no underfuselage gondola. The wheels are often fitted with spats.

Mi-4S: Agricultural version. Large chemical container in main cabin, capacity 1,000 kg (2,200 lb) of dust or 1,600 litres (422 Imp gallons; 352 US gallons) of spray. Container is fitted with mechanical distributor for dry chemicals, which are spread through bifurcated ducts by hydraulically actuated fan in duct which replaces the military ventral gondola. Liquids are sprayed from bars mounted aft of mainwheels. Rate of spread is up to 18 litres (5 US gallons; 4 Imp gallons) or 20 kg (44 lb) per second, with swath width of 40-80 m (130-260 ft), at forward speed of 32 knots (60 km/h; 37 mph).

All versions are able to be fitted with two large inflatable pontoons, mounted so that the wheels of the landing gear protrude slightly beneath them, for amphibious operation.

In Spring 1965, details were given of a series of high-altitude tests made with an Mi-4 fitted with two-speed supercharger and all-metal main rotor. After engaging the second speed at 4,650 m (15,240 ft), the aircraft climbed to 8,000 m (26,240 ft). It was also operated at an airfield height of 5,000 m (16,400 ft).

OPERATORS: Versions of the Mi-4 are in service with the armed forces of the following countries: Afghanistan (12), Albania (18), Algeria (30), China (355+), Cuba (20) and North Korea (20).

The following specifications apply to the standard Mi-4.

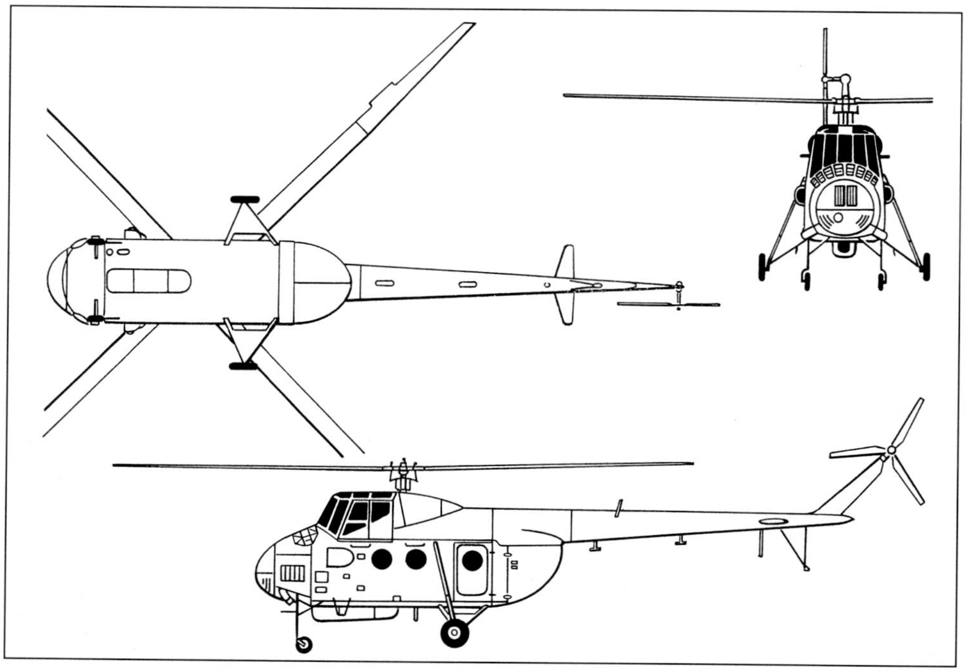

Mil Mi-4 'Hound' general purpose helicopter ***(Jane's/Mike Keep)***

1994

Mil Mi-4 'Hound' helicopters engaged in troop deployment ***(Tass)***

1994

DESIGN FEATURES: Four-bladed main rotor with hydraulic servo-control, and three-bladed anti-torque rotor at starboard side of tailboom.

STRUCTURE: Main rotor blades were originally tapered, with steel spars and plywood covering; since 1961 they have been of constant-chord all-metal construction. All-metal semi-monocoque structure of pod and boom type, with provision for clamshell doors under the tailboom attachment point in freight-carrying version.

LANDING GEAR: Non-retractable four-wheel type. All units fitted with shock-absorbers. Nosewheels are fully castoring. Spatts optional. Provision for fitting pontoons.

POWER PLANT: One 1,700 hp ASh-82 V 18-cylinder air-cooled radial engine mounted in fuselage nose.

ACCOMMODATION: Crew of two on flight deck, with underfuselage gondola for observer in military version. Commercial version carries 8-16 passengers in heated, ventilated and soundproofed cabin, with door at rear on port side. Aft of cabin are a toilet, wardrobe and compartment for 100 kg (220 lb) of baggage. Ambulance version carries eight stretchers and attendant. Freight version has clamshell rear doors. Military version carries up to 14 troops, 1,600 kg (3,525 lb) of freight or vehicles such as a GAZ-69 'Jeep', 76 mm anti-tank gun or two motorcycle/sidecar combinations.

ELECTRONICS AND EQUIPMENT: Radio and instrumentation for night and bad weather flying are standard equipment.

SYSTEMS: Liquid leading-edge de-icing system.

DIMENSIONS, EXTERNAL:	
Main rotor diameter	21.0 m (68 ft 11 in)
Tail rotor diameter	3.60 m (11 ft 10 in)
Length of fuselage	16.80 m (55 ft 1 in)
Height overall	5.18 m (17 ft 0 in)
Wheel track (front)	1.53 m (5 ft 0 in)
Wheel track (rear)	3.82 m (12 ft 6 in)
Wheelbase	3.79 m (12 ft 5 in)
Rear-loading door (freighter):	
Height	1.60 m (5 ft 3 in)
Width	1.86 m (6 ft 1½ in)
DIMENSIONS, INTERNAL:	
Cabin volume	16 m^3 (565 cu ft)
AREA:	
Main rotor disc	346 m^2 (3,724 sq ft)
WEIGHTS AND LOADINGS:	
Max payload	1,740 kg (3,835 lb)
Normal T-O weight	7,500 kg (16,535 lb)
Max T-O weight	7,800 kg (17,200 lb)
PERFORMANCE (at max T-O weight):	
Max level speed at 1,500 m (4,920 ft)	113 kts (210 km/h; 130 mph)
Econ cruising speed	86 kts (160 km/h; 99 mph)
Service ceiling	5,500 m (18,000 ft)
Range with 11 passengers and 100 kg (220 lb) baggage	134 n miles (250 km; 155 miles)
Range with eight passengers and 100 kg (220 lb) baggage	217 n miles (400 km; 250 miles)

VERIFIED

MIL Mi-6 and Mi-22

NATO reporting name: Hook

TYPE: Twin-turbine heavy transport helicopter.

PROGRAMME: Joint military/civil requirement issued 1954; prototype flew 5 June 1957 as by far the world's largest helicopter of that time; five built for development testing; initial preseries of 30; more than 800 built for civil/military use, ending 1981; developments included Mi-10 and Mi-10K flying cranes; Mi-6 dynamic components used in duplicated form on V-12 (Mi-12) of 1967, which remains the largest helicopter yet flown.

VERSIONS: **Mi-6** ('Hook-A'): Basic transport, as described in detail.

Mi-6VKP ('Hook-B'): Command support version with dorsal 'clothesline' antenna.

Mi-22 ('Hook-C'): Developed command support version with single large dorsal blade antenna on forward part of tailboom.

OPERATORS: Versions of the Mi-6 and Mi-22 are in service with the armed forces of the following countries: Algeria (4), Belarus (30), Egypt (6), Ethiopia (10), Iraq (15), Kazakhstan, Laos (1), Peru (6), Russia (265+), Syria (10), Ukraine (40) and Vietnam (10).

Abbreviated details follow.

DESIGN FEATURES: Five-blade main rotor and four-blade tail rotor. Main rotor blades each have tapered steel tube spar, to which are bonded built-up metal aerofoil sections. Conventional transmission. Main reduction gearbox drives tail rotor, fan AC generators and hydraulic pumps. Intermediate reduction gearbox fitted with special fan. Two small cantilever removable shoulder wings, mounted above main landing gear struts, offload rotor by providing some 20 per cent of total lift in cruising flight. Tail rotor support acts as vertical stabiliser.

FLYING CONTROLS: Blades have coincident flapping and drag hinges and fixed tabs. Main rotor shaft inclined forward at 5° to vertical. Control via large welded swashplate. Hydraulically actuated powered controls. Main rotor collective-pitch control interlocked with throttle controls. Variable incidence horizontal stabiliser near end of tailboom for trim purposes.

STRUCTURE: The fuselage is a conventional all-metal riveted semi-monocoque structure of pod and boom type.

POWER PLANT: Two 4,101 kW (5,500 shp) Soloviev D-25 V (TV-2BM) turboshafts, mounted side by side above cabin, forward of main rotor shaft. Eleven internal fuel tanks, with total capacity of 6,315 kg (13,922 lb), and two external tanks, on each side of cabin, with total capacity of 3,490 kg (7,695 lb). Provision for two additional ferry tanks inside cabin, with total capacity of 3,490 kg (7,695 lb).

ACCOMMODATION: Crew of five, consisting of two pilots, navigator, flight engineer and radio operator. Four jettisonable doors and overhead hatch on flight deck. Electrothermal anti-icing system for glazing of flight deck and navigator's compartment. Equipped normally for cargo operation, with easily removable tip-up seats along sidewalls. When these seats are supplemented by additional seats installed in centre of cabin, 65-90 passengers can be carried, with cargo or baggage in the aisles. Normal military seating is

MIL Mi-4 general purpose helicopter

1995

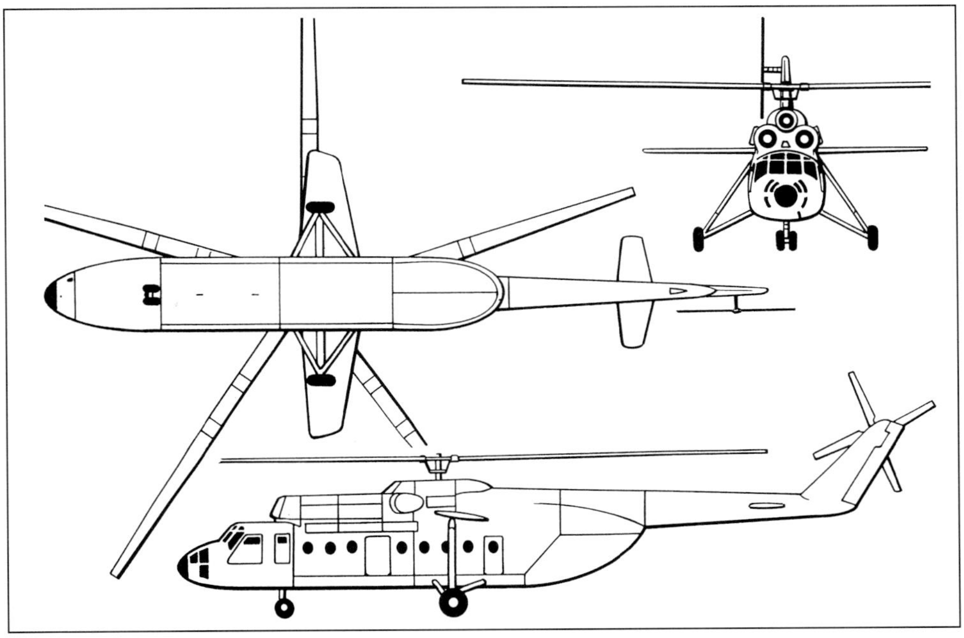

Mil Mi-6 'Hook' heavy transport helicopter *(Jane's/Mike Keep)*

1994

Mil Mi-6 twin-turbine transport and heavy lift helicopter

1995

for 70 combat equipped troops. As an air ambulance, 41 stretcher cases and two medical attendants on tip-up seats can be carried. One of attendant's stations is provided with intercom to flight deck, and provision is made for portable oxygen installations for the patients. Cabin floor is stressed for loadings of 2,000 kg/m² (410 lb/sq ft), with provision for cargo tiedown rings. Rear clamshell doors and ramps are operated hydraulically. Standard equipment includes an electric winch of 800 kg (1,765 lb) capacity and pulley block system. Central hatch in cabin floor for cargo sling system for bulky loads. Three jettisonable doors, fore and aft of main landing gear on port side and aft of landing gear on starboard side.

AVIONICS: VHF and HF communications radio, intercom, radio altimeter, radio compass, three-channel autopilot, marker beacon receiver, directional gyro and full all-weather instrumentation.

ARMAMENT: Some military Mi-6s have a 12.7 mm machine gun in the nose.

DIMENSIONS, EXTERNAL:

Main rotor diameter	35.00 m (114 ft 10 in)
Tail rotor diameter	6.30 m (20 ft 8 in)
Length: overall, rotors turning	41.74 m (136 ft 11½ in)
fuselage, excl nose gun and tail rotor	33.18 m (108 ft 10½ in)
Height (overall	9.86 m (32 ft 4 in)
Wing span	15.30 m (50 ft 2½ in)
Wheel track	7.50 m (24 ft 7¼ in)
Wheelbase	9.09 m (29 ft 9¾ in)
Rear-loading doors: Height	2.70 m (8 ft 10¼ in)
Width	2.65 m (8 ft 8¼ in)
Passenger doors:	
Height: front door	1.70 m (5 ft 7 in)
rear doors	1.61 m (5 ft 3½ in)
Width	0.80 m (2 ft 7½ in)
Sill height: front door	1.40 m (4 ft 7¼ in)
rear doors	1.30 m (4 ft 3¼ in)
Central hatch in floor	1.44 m (4 ft 9 in) × 1.93 m (6 ft 4 in)

DIMENSIONS, INTERNAL:

Cabin: Length	12.00 m (39 ft 4½ in)
Max width	2.65 m (8 ft 8¼ in)
Max height: at front	2.01 m (6 ft 7 in)
at rear	2.50 m (8 ft 2½ in)
Cabin volume	80 m³ (2,825 cu ft)

AREAS:

Main rotor disc	961.1 m² (10,356 sq ft)

WEIGHTS AND LOADINGS:

Weight empty	27,240 kg (60,055 lb)
Max internal payload	12,000 kg (26,450 lb)
Max slung cargo	8,000 kg (17,637 lb)
Full load: internal	6,315 kg (13,922 lb)
with external tanks	9,805 kg (21,617 lb)
Max T-O weight with slung cargo at altitudes under 1,000 m (3,280 ft)	38,400 kg (84,657 lb)
Normal T-O weight	40,500 kg (89,285 lb)
Max T-O weight for VTO	42,500 kg (93,700 lb)
Max disc loading	44.17 kg/m² (9.05 lb/sq ft)

PERFORMANCE (at max T-O weight for VTO):

Max level speed	162 kts (300 km/h; 186 mph)
Max cruising speed	135 kts (250 km/h; 155 mph)
Service ceiling	4,500 m (14,750 ft)
Range with 8,000 kg (17,637 lb) payload	338 n miles (620 km; 385 miles)
Range with external tanks and 4,500 kg (9,920 lb) payload	540 n miles (1,000 km; 621 miles)
Max ferry range (tanks in cabin)	781 n miles (1,450 km; 900 miles)

VERIFIED

MIL Mi-9

NATO reporting name: Hip-G

Designation Mi-9 applies to airborne command post variant of Mi-8; 'hockey stick' antennae projecting from rear of cabin and from undersurface of tailboom, aft of Doppler radar box; rearward inclined short whip antenna above forward end of tailboom; strakes on fuselage undersurface.

VERIFIED

MIL Mi-14

NATO reporting name: Haze

TYPE: Twin-turbine shore-based amphibious helicopter.

PROGRAMME: Development of Mi-8; first flew September 1969, under designation **V-14** and with Mi-8 power plant; changed to Mi-17 engines for production, which continues.

VERSIONS: **Mi-14PL** ('Haze-A'): Basic ASW version; four crew; large undernose radome; OKA-2 retractable sonar in starboard rear of planing bottom, forward of two probable sonobuoy or signal flare chutes; APM-60 towed magnetic anomaly detection (MAD) bird stowed against rear of fuselage pod (moved to lower position on some aircraft); weapons include torpedoes, bombs and depth charges in enclosed bay in bottom of hull; WAS-5M-3 liferaft (in all versions).

Mi-14W: Polish designation of Mi-14PL.

Mi-14BT ('Haze-B'): Mine countermeasures version; fuselage strake, for hydraulic tubing, and air conditioning

Mil Mi-6 'Hook' with stub wings removed for crane duties

1994

Mil Mi-6 'Hook' unloading caterpillar truck in the oil fields of the Tiumen region *(Tass)*

1994

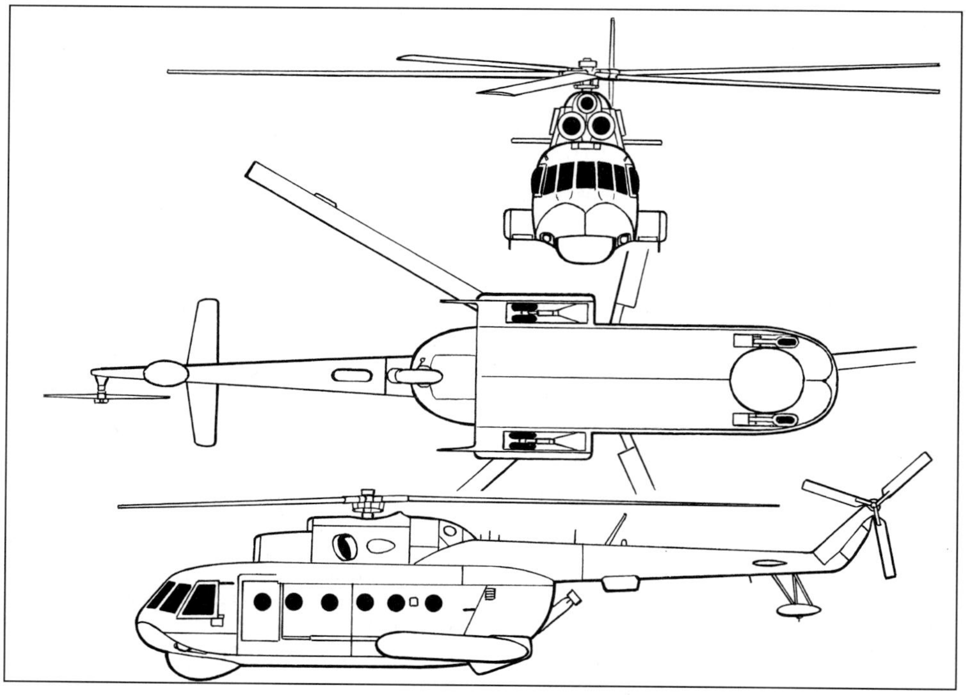

Mil Mi-14 'Haze' amphibious helicopter *(Jane's/Mike Keep)*

1994

pod on starboard side of cabin; no MAD; container for searchlight to observe MCM gear during deployment and retrieval under tailboom, forward of Doppler box.

Mi-14PS ('Haze-C'): Search and rescue version, carrying 10 20-place liferafts; room for 10 survivors in cabin; provision for towing many more survivors in liferafts; fuselage strake and air conditioning pod as Mi-14BT; double width sliding door at front of cabin on port side, with retractable rescue hoist able to lift up to three persons in basket, searchlight each side of nose and under tailboom.

UPGRADES: **Isolair:** Firefighter conversion. See separate entry.

OPERATORS: Versions of the Mi-14 are in service with the armed forces of the following countries: Bulgaria (6), Cuba (10), Ethiopia (2), Libya (12), Poland (15), Romania (6), Russia (90), Ukraine (40) and Yugoslavia (4).

DESIGN FEATURES: Developed from Mi-8; overall dimensions, power plant and dynamic components as Mi-17; new features include boat hull, sponson carrying inflatable flotation bag each side at rear and small float under tailboom; fully retractable landing gear with two forward-retracting single wheel nose units and two rearward retracting twin-wheel main units.

POWER PLANT: Two Isotov TV3-117 turboshaft engines.

AVIONICS (Mi-14PL): Type 12-M undernose radar, R-842-M HF transceiver, R-860 VHF transceiver, SBU-7 intercom, RW3 radio altimeter, ARK-9 and ARK-U2 ADFs, DISS-15 Doppler. Chrom Nikiel IFF, AP34-B autopilot/autohover system and SAU-14 autocontrol system.

DIMENSIONS, EXTERNAL:

Main rotor diameter	21.29 m (69 ft 10¼ in)
Length overall, rotors turning	25.30 m (83 ft 0 in)
Height overall	6.93 m (22 ft 9 in)

AREAS:

Main rotor disc	356 m² (3,832 sq ft)

WEIGHTS AND LOADINGS:

Max T-O weight	14,000 kg (30,865 lb)
Max disc loading	39.3 kg/m² (8.05 lb/sq ft)

PERFORMANCE:

Max level speed	124 kts (230 km/h; 143 mph)
Max cruising speed	116 kts (215 km/h; 133 mph)
Normal cruising speed	110 kts (205 km/h; 127 mph)
Service ceiling	3,500 m (11,500 ft)
Range with max fuel	612 n miles (1,135 km; 705 miles)
Endurance with max fuel	5 h 56 min

UPDATED

Mil Mi-14PL 'Haze-A' of the Polish Navy *(Piotr Butowski)*
1994

Mil Mi-14PS 'Haze-C' of the Polish Navy *(Piotr Butowski)*
1994

SOKOL

SOKOL GOVERNMENT-OWNED AIRCRAFT PLANT

2 Chaadaeva Street, Nizhniy Novgorod 603045
Telephone: 7 (8312) 467 127
Fax: 7 (8312) 247 966
Telex: 151101 ANGAR

NEW ENTRY

SOKOL/MIKOYAN (MIKOYAN) MiG-21-93 UPGRADE

TYPE: Combat aircraft upgrade.

PROGRAMME: Modernisation package to enhance the combat capability of MiG-21bis and MiG-21M/MF aircraft and to increase service life up to the years 2010-2015. On 3 May 1994 India's Deputy Defence Minister announced that a contract had been awarded to Sokol and Mikoyan to upgrade 100 MiG-21bis aircraft with an option for 70 more.

CUSTOMERS: Indian Air Force (100 with option for 70 more).

COSTS: Estimated to be $1.33 million per aircraft.

DESIGN FEATURES: Programme includes: total replacement of onboard avionics, new slimline canopy as well as improved power supply and air conditioning. Systems installation consists of Kopyo multifunctional coherent pulse Doppler radar with slot antenna, redesigned cockpit with head-up display (HUD), multifunction display (MFD) and hands-on throttle and stick (HOTAS) systems as well as a helmet-mounted target designator, inertial reference system (IRS), navigation computer, ADU, radio navigation/approach system, communications station, IFF system and ground air datalink. The radar warning receiver (RWR) provides electronic countermeasures (ECM) using onboard chaff/flare dispensers and anti-radar/radiation missiles. The design of the avionics system provides for integration of non-Russian systems.

ARMAMENT: Weapons include GSh-23 gun, 100/250 and 500 kg bombs, S-5 and S-8 unguided rocket projectiles, R-60 guided missiles, R-73 air-to-air missiles with IR homing warhead, R-27 medium-range guided missile with semi-active or IR homing warhead, RVV-AE medium-range active radar-guided AA missiles. Further air-to-air surface missiles include: S-13 and S-24 unguided rockets, KAB-500 correctable air bombs with TV homing warhead, X-25MP and X-31P anti-radiation missiles, X-31A anti-ship supersonic missile.

NEW ENTRY

SUKHOI

SUKHOI DESIGN BUREAU AVIATION SCIENTIFIC-INDUSTRIAL COMPLEX

23A Polikarpov Street, 125284 Moscow
Telephone: 7 (095) 945 6525
Fax: 7 (095) 200 4243
Telex: 414716 SUHOI SU

GENERAL DESIGNER AND CEO: Mikhail Petrovich Seemonov
FIRST DEPUTY GENERAL DESIGNER: Aleksei I. Knishev
DIRECTOR, MANUFACTURER: Vladimir N. Avramenko
DEPUTY GENERAL DESIGNERS:
Alexander F. Barkovsky
Mikhail A. Pogosian
Boris V. Rakitin (Sport Aviation Projects)

OKB named for Pavel Osipovich Sukhoi, who headed it from 1939 until his death in September 1975. It remains one of two primary Russian centres for development of fighter and attack aircraft, and is widening its activities to include civilian aircraft, under konversiya programme, in some cases with Western partners.

Sukhoi also active in design and construction of large ground effect vehicles, including 100/150-passenger A.90.150 Ekranoplan described in *Jane's High-Speed Marine Craft 1993-94*.

UPDATED

SUKHOI Su-7B

NATO reporting names: Fitter-A and Moujik

TYPE: Single-seat ground attack fighter.

PROGRAMME: The first prototype of this single-seat fighter, designated S-1, was flown for the first time by test pilot A. G. Kochetkov on 8 September 1955, and was displayed in prototype form in the flypast over Moscow in the 1956 Soviet Aviation Day. It was the first Soviet aircraft to have all-moving horizontal tail surfaces and a fore and aft translating air intake centrebody to adjust supersonic air flow. Power plant was, successively, an AL-7 turbojet and AL-7F; armament comprised three NR-30 guns.

The S-1 prototypes were followed by a number of S-2s, embodying certain aerodynamic refinements, and by a

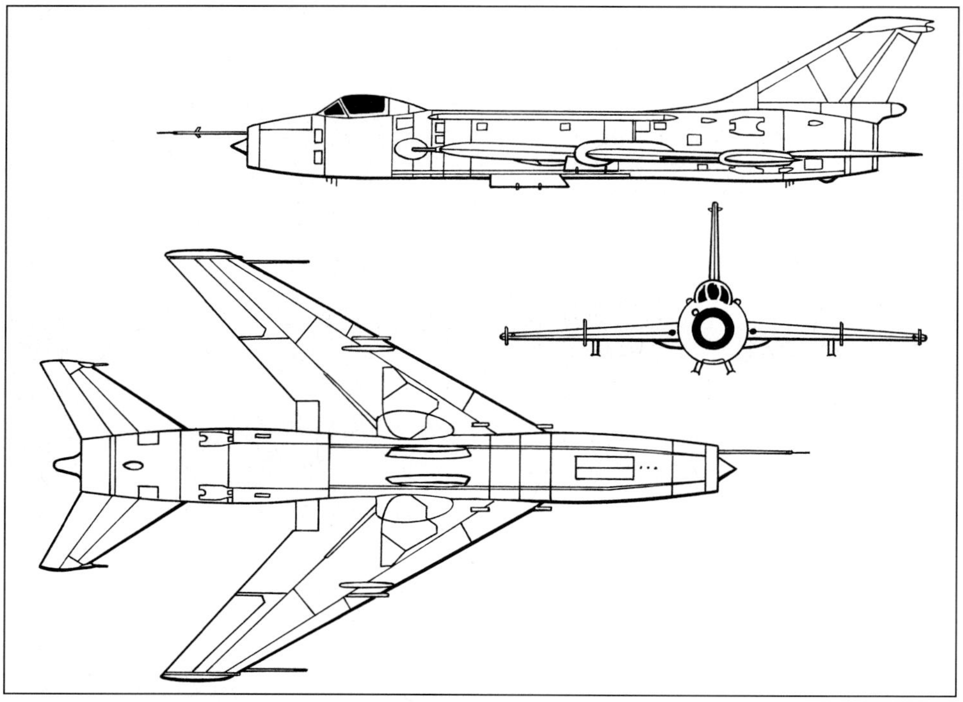

Sukhoi Su-7B 'Fitter-A' *(Jane's/Mike Keep)*

1994

Sukhoi Su-7BKL 'Fitter-A' of the Polish Air Force. 'Fitter-As' now withdrawn *(Piotr Butowski)*

1994

Sukhoi Su-7BKL 'Fitter-A' of the former Soviet Air Force. Not included in the CIS inventory *(Piotr Butowski)*

1994

small number of preseries aircraft designated **Su-7**. After evaluation of these, a new prototype, known as the S-22, was built in fighter-bomber form. It was flown for the first time in April 1959, by E. Soloviev; and the S-22 was ordered into series production as the **Su-7B** fighter-bomber, air combat and reconnaissance aircraft, to which NATO gave the reporting name **'Fitter-A'**.

VERSIONS: **Su-7:** Preseries aircraft.

Su-7B: Fighter-bomber air combat and reconnaissance aircraft to which NATO gave the reporting name 'Fitter-A'. Not in service.

Su-7BM (S-22): Version with improved flight control and navigation equipment as well as two slim duct fairings along the top of the centre-fuselage. Other new features were introduced progressively including: zero-altitude ejection seat, Sirena tail warning radar, uprated engine, a second pair of underwing stores pylons, large blast panels on the sides of the front fuselage by the muzzles of the wingroot guns, JATO attachments under the rear fuselage. In service with Algeria and possibly Iraq.

Su-7BKL (S-22KL = koleso-lyzhny, wheel-ski): Version with an ALF1-1-200 turbojet, a low pressure nosewheel, small extensible skid outboard of each mainwheel.

Su-7BMK: Last single-seat production version strengthened to permit heavier load. In service with North Korea.

Su-7U: Two-seat operational trainer. Still in inventories of Algerian and possibly Iraqi air forces.

Su-7UM/UMK (U-22): Two-seat operational trainer version. NATO reporting name Moujik. Presumed withdrawn from service.

OPERATORS: Versions of the Su-7 are in service with the armed forces of the following countries: Afghanistan (25+), Algeria (12), Iraq and North Korea (30).

The following description applies to the Su-7BMK:

DESIGN FEATURES: Cantilever mid-wing monoplane. Wing thickness/chord ratio 8 per cent. No dihedral or anhedral. Sweepback 60° on leading-edges. Wing chord is extended giving a straight trailing-edge on inboard section of each wing. Two boundary layer fences on each wing at approximately mid-span and immediately inboard of tip. Two intake suction relief doors on each side of nose. Two slim duct fairings along top of centre-fuselage.

FLYING CONTROLS: Hydraulically powered spring loaded ailerons. Large chord flaps over entire trailing-edge from root to inboard end of aileron on each wing. No slats or tabs. Two hydraulically actuated door type airbrakes at top and bottom on each side of rear fuselage. The tail unit has hydraulically powered control surfaces, all-moving horizontal surfaces, with anti-flutter bodies projecting forward from tips. Conventional rudder with yaw damper. No tabs.

STRUCTURE: Conventional all-metal two-spar structure. The fuselage is a conventional all-metal semi-monocoque structure of circular section. Break point at wing trailing-edge permits removal of rear fuselage for engine servicing. The tail unit is a cantilever all-metal structure with 55° sweepback at quarter-chord on all surfaces.

LANDING GEAR: Retractable tricycle type, with single wheel on each unit. Steerable nosewheel retracts forward, main units inward into wings. Differential brakes on mainwheels. Twin brake-chutes in large container with clamshell doors, at base of rudder.

POWER PLANT: One Lyulka AL-7F-1-100 (TRD-31) turbojet engine, rated at 66.64 kN (14,980 lb st) dry and 94.08 kN (21,150 lb st) with afterburning. Time taken for afterburner light up 6-7 seconds. Variable area afterburner nozzle. Saddle fuel tanks in centre-fuselage and integral tanks between spars of inner wings. Total internal fuel capacity 2,940 litres (776 US gallons; 647 Imp gallons). Gravity fuelling points above fuselage tanks and each wing tank. Provision for two drop tanks side by side under fuselage, with total capacity of 1,200 litres (317 US gallons; 264 Imp gallons); and two ferry tanks, total capacity 1,800 litres (475 US gallons; 396 Imp gallons) on inner wing pylons. Two SPRD-100 solid propellant rocket units, each 29.4 kN (6,610 lb st), can be attached under rear fuselage to shorten T-O run.

ACCOMMODATION: Pilot only in pressurised cockpit, on KS-4 zero-altitude rocket-powered ejection seat, under rearward sliding blister canopy. Flat windscreen of armoured glass. Rearview mirror on top of canopy,

SYSTEMS: Main and standby hydraulic systems, with emergency pump, for actuating flying controls, flaps, airbrakes, landing gear, nosewheel steering and afterburner nozzle. Cockpit heating system. KKO-2 oxygen system. Pneumatic system adequate for engine starting and three afterburner engagements per sortie, with reserves. Electrical system includes navigation lights and retractable taxying light under nose.

AVIONICS AND EQUIPMENT: Standard avionics include VHF/UHF radio, ILS, RSIU very short-wave fighter radio, ADF, transponder, SRO-2M (NATO 'Odd Rods') IFF, Sirena 3 tail warning radar, ranging radar in air intake centrebody, autopilot. Launcher for Very cartridges or chaff under starboard wingroot leading-edge. Provision for vertical and oblique cameras in belly aft of nosewheel bay.

ARMAMENT: Two 30 mm NR-30 guns, each with 70 rounds, in wingroot leading-edges. ASP-5ND gyro gunsight. Six external stores pylons. Two underbelly pylons and inner underwing pylons each capable of carrying 750 kg (1,650 lb); outer underwing pylons each stressed for

500 kg (1,100 lb). Stores include UB-16-57U rocket pods (each 16 57 mm S-5, S-5M or S-5K rockets), S-24 250 kg concrete-piercing guided rockets, S-3K unguided rockets, and free-fall bombs (usually two 750 kg and two 500 kg), including nuclear weapons. When underbelly fuel tanks are fitted, maximum external weapons load is 1,000 kg (2,205 lb).

DIMENSIONS, EXTERNAL:

Wing span	8.77 m (28 ft 9¼ in)
Length overall, incl probe	16.80 m (55 ft 1½ in)
Height overall	4.80 m (15 ft 9 in)

AREAS:

Wings, gross	23.0 m² (247 sq ft)

WEIGHTS AND LOADINGS:

Operating weight empty	8,328 kg (18,360 lb)
Normal T-O weight	12,000 kg (26,450 lb)
Max T-O weight	13,440 kg (29,630 lb)

PERFORMANCE:

Max level speed at 12,200 m (40,000 ft):
'clean' Mach 1.6 (917 kts; 1,700 km/h; 1,055 mph)
with external stores Mach 1.2 (685 kts; 1,270 km/h; 788 mph)

Max level speed at S/L:
without afterburning 460 kts (850 km/h; 530 mph)
with afterburning 625 kts (1,158 km/h; 720 mph)

Rotation speed for T-O	195 kts (360 km/h; 224 mph)
Approach speed	195 kts (360 km/h; 224 mph)
Max rate of climb at S/L	9,000 m (29,525 ft)/min
Service ceiling	18,000 m (59,050 ft)
T-O run	2,400 m (7,875 ft)
Combat radius	135-187 n miles (250-345 km; 155-215 miles)
Max range	780 n miles (1,450 km; 900 miles)
Fuel consumption with afterburning at S/L	360 kg (794 lb)/min

UPDATED

SUKHOI Su-15

NATO reporting name: Flagon

TYPE: Single-seat twin-jet all-weather interceptor and two-seat combat trainer.

PROGRAMME: Developed to 1962 requirement; T-5 prototype flew approximately 1964; 10 development aircraft displayed on 1967 Soviet Aviation Day, Moscow; fully operational early 1970s. Su-15 now being replaced.

VERSIONS: **Su-15TM** ('Flagon-E'): Single-seat interceptor; longer span wings than earlier versions, with compound sweep; Tumansky R-13F-300 turbojets; additional fuel; uprated avionics, including Typhoon (NATO 'Twin Scan') radar; major production version operational second half 1973.

Su-15bis ('Flagon-F'): Final production version; ogival nose radome instead of earlier conical type; generally as 'Flagon-E' but Tumansky R-25-300 engines.

'Flagon-G': Tandem two-seat training version of 'Flagon-F'; probable combat capability; individual rearward hinged canopy over each seat; periscope above rear canopy to enhance forward view; overall length unchanged.

OPERATORS: Versions of the Su-15 are in service with the armed forces of the following countries: Russia (200+) and Ukraine (50+).

DESIGN FEATURES: Designed as Mach 2.5 interceptor to succeed Su-11; original mid-mounted wings of simple delta form replaced in current versions by wings with extended outer panels, giving compound sweep of 60° inboard, 47° outboard; no dihedral or anhedral; boundary layer fence above weapon pylon on each wing at approximately 70 per cent span; basically circular cockpit section with large ogival dielectric nosecone; centre-fuselage faired into toed-in rectangular section air intake ducts; 60° leading-edge sweepback on all tail surfaces; anhedral tailplane with anti-flutter bodies near tips.

FLYING CONTROLS: Each wing trailing-edge comprises aileron and large-chord plan flap; all-moving tailplane; conventional rudder; no trim tabs; top and bottom door type airbrakes each side of rear fuselage, forward of tailplane.

STRUCTURE: Conventional all-metal.

LANDING GEAR: Tricycle type, with single wheel on each main unit and twin nosewheels, all on levered-suspension legs. Mainwheels retract inward into wings and intake ducts; nosewheels retract forward, with blistered doors. Nosewheels steerable. Brake-chute at base of fin.

POWER PLANT: Two turbojets, with variable area nozzles, mounted side by side in rear fuselage. These are Tumansky R-13F2-300s, each rated at 64.73 kN (14,500 lb st) with afterburning. Ram air intakes, with variable ramps on splitter plates, embodying vertical slots for boundary layer control. Blow-in auxiliary inlets between main intake and wing leading-edge in side of each duct. Fuel tanks in centre-fuselage and wings.

ACCOMMODATION: Single zero/zero ejection seat in enclosed cockpit, with rearward sliding blister canopy. Rearview mirror above canopy of some aircraft.

AVIONICS: Large I-band radar (NATO 'Twin Scan') in nose, SOD-57M ATC/SIF nav system SRO-2 (NATO 'Odd Rods') IFF, Sirena-3 radar warning system.

ARMAMENT: Two pylons for external stores under each wing. Normal armament comprises one radar homing and one infra-red homing AA-3 air-to-air missile (NATO 'Anab') on outboard pylons, and an infra-red homing AA-8 close-range missile ('Aphid') on each inboard pylon. Side by side pylons under centre-fuselage for weapons, including GSh-23L 23 mm gun pods, or external fuel tanks.

DIMENSIONS, EXTERNAL (estimated):

Wing span	9.15 m (30 ft 0 in)
Length overall	21.33 m (70 ft 0 in)
Height overall	5.10 m (16 ft 8½ in)

Sukhoi Su-7U two-seat operational trainer

1995

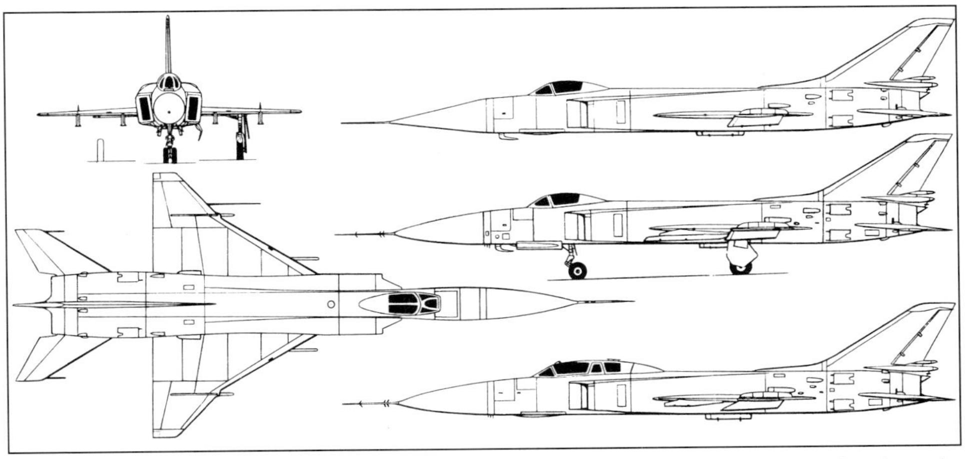

Sukhoi Su-15 ('Flagon-F') single-seat twin-jet all-weather interceptor, with additional side elevations of 'Flagon-E' (top) and 'Flagon-G' (bottom)

1994

Sukhoi Su-15bis 'Flagon-F' twin-jet interceptor armed with AA-3 'Anab' missiles and gun pods *(Swedish Coast Guard/Air Patrol)*

1994

Sukhoi Su-15 'Flagon' single-seat all-weather interceptor

1994

WEIGHTS AND LOADINGS (estimated):

Weight empty	11,000 kg (24,250 lb)
Max T-O weight	18,000 kg (39,680 lb)
Max power loading	139.0 kg/kN (1.36 lb/lb st)

PERFORMANCE (estimated):

Max level speed above 11,000 m (36,000 ft) with external stores	Mach 2.1
Time to 11,000 m (36,000 ft)	2 min 30 s
Service ceiling	20,000 m (65,600 ft)
Combat radius	538 n miles (998 km; 620 miles)

VERIFIED

SUKHOI Su-17, Su-20 and Su-22

NATO reporting names: Fitter-C, -D, -E, -F, -G, -H, -J and -K

TYPE: Single-seat variable geometry ground attack fighter, reconnaissance aircraft and two-seat combat trainer.

PROGRAMME: Prototype S-22I or Su-7IG (Izmenyaemaya Geometriya; variable geometry) was minimal conversion of fixed-wing Su-7 (NATO 'Fitter-A'); only 4.2 m (13 ft 9 in) of each wing pivoted, outboard of large fence and deepened inboard glove panel; first flew 2 August 1966; shown at Aviation Day display July 1967; given NATO reporting name 'Fitter-B'; two squadrons of **Su-17** 'improved Fitter-Bs' in Soviet air forces 1972; AL-21F-3 engine then replaced AL-7 in major Soviet air force production versions, beginning with 'Fitter-C'. Production ended 1991.

VERSIONS: **Su-17M** (S-32M, 'Fitter-C'): Single-seat attack aircraft; AL-21F-3 engine; eight stores pylons; additional wing fence on each glove panel; curved dorsal fin; operational with CIS air forces and Naval Aviation since 1971 in relatively small numbers. *Detailed description applies to basic Su-17, except where indicated otherwise.*

Su-17R: Reconnaissance version of Su-17M.

Su-17M-2/M-2D (S-32M2, 'Fitter-D'): Generally as Su-17M, but forward fuselage lengthened by 0.38 m (15 in) and drooped 3° to improve pilot's view while keeping intake face vertical; added undernose Doppler navigation radar pod; Klen laser rangefinder in intake centrebody.

Su-17UM-3 ('Fitter-G'): Two-seat trainer version of Su-17M-3 with combat capability; drooped forward fuselage and deepened spine like Su-17UM-2D; taller vertical tail surfaces; removable ventral fin; starboard wingroot gun only; laser rangefinder in intake centrebody.

Su-17M-3 (S-52, 'Fitter-H'): Improved single-seater; same deepened spine and tail modifications as Su-17UM-3; Doppler navigation radar internally in deepened undersurface of nose; gun in each wingroot; launcher for R-60 (AA-8 'Aphid') air-to-air missile between each pair of underwing pylons; approximately 165 'Fitter-H/Ks' equipped for tactical reconnaissance carry, typically, centreline sensor pod, active ECM pod under port wing glove, two underwing fuel tanks.

Su-17M-4 (S-54, 'Fitter-K'): Single-seat version, identified 1984; cooling air intake at front of dorsal fin; otherwise as Su-17M-3. Maximum weapon load 4,250 kg (9,370 lb), including nuclear weapons, bombs, rocket pods, S-25 tube-launched rockets with 325 mm head, 23 mm SPPU-22 gun pods, two R-3 or R-13M (AA-2 'Atoll'), R-60 (AA-8 'Aphid') or R-73A (AA-11 'Archer') air-to-air missiles, Kh-23 (AS-7 'Kerry') or Kh-25ML (AS-10 'Karen') air-to-surface missiles, or a reconnaissance pod. When four SPPU-22 gun pods are fitted, with downward attack capability, the two underfuselage pods can be arranged to fire rearward. Chaff/flare and decoy dispensers standard.

Su-20 (S-32MK, 'Fitter-C'): Export version of Su-17M.

Su-20R: Reconnaissance version of Su-20.

All Su-17s and Su-20s have AL-21F-3 engines; some Su-22 export aircraft have Tumansky R-29BS-300 (112.8 kN; 25,350 lb st with afterburning) in more bulged rear fuselage, with rearranged small external air intakes on rear fuselage and shorter plain metal shroud terminating fuselage, as follows:

Su-22U ('Fitter-E'): Tandem two-seat trainer developed from Su-17M-2, with Tumansky engine; no Doppler pod; deepened dorsal spine fairing for additional fuel tankage; port wingroot gun deleted.

Su-22 ('Fitter-F'): Export Su-17M-2; modified undernose electronics pod, R-29 engine; gun in each wingroot; weapons include R-3 (AA-2 'Atoll') air-to-air missiles; aircraft supplied to Peru had Sirena-2 limited coverage radar warning system and virtually no navigation aids; some basic US supplied avionics retrofitted.

Su-22UM-3K ('Fitter-G'): Export Su-17UM-3; AL-21F-3 or R-29B engine.

Su-22M-3 ('Fitter-J'): As Su-17M-3 but R-29 engine; internal fuel tankage 6,270 litres (1,656 US gallons; 1,379 Imp gallons); more angular dorsal fin; AA-2 ('Atoll') air-to-air missiles.

Su-22M-4 ('Fitter-K'): As Su-17M-4; AL-21F-3 engine.

Su-22M-5: Franco-Russian upgrade proposal. See following entry.

UPGRADES: **Sukhoi:** Su-22 ('Fitter-F'). See Versions.

Sukhoi Sextant Avionique: Su-22M-5. See following entry.

OPERATORS: Versions of the Su-17/20/22 are in service with the armed forces of the following countries: Afghanistan (50), Algeria (30+), Angola (35+), Bulgaria (22), Czech Republic (55), Hungary (20), Iraq, Libya (90), Peru (50+), Poland (120), Russia (450+), Slovakia (20), Syria (60), Ukraine (40), Vietnam (100) and Yemen (50).

Description applies to basic Su-17, except where indicated.

DESIGN FEATURES: Modest amount of variable geometry added to original fixed-wing Su-7 permitted doubled external load from strips little more than half as long, and 30 per cent greater combat radius; progressive refinements led to very effective final versions. Conventional mid-wing all-swept monoplane, except for variable geometry outer wings with manually selected positions of 28°, 45°, 63°; wide span fixed centre-section glove panels; basically circular fuselage with dorsal spine; ram intake with variable shock-cone centrebody; pitot on port side of nose, transducer to provide pitch and yaw data for fire control computer starboard; anti-flutter bodies near tailplane tips.

FLYING CONTROLS: Slotted ailerons operable at all times; slotted trailing-edge flap on each variable geometry wing panel operable only when wings spread; area-increasing flap on each centre-section glove panel; full-span leading-edge slats on variable geometry wing panels; top and bottom door type airbrakes each side of rear fuselage, forward of tailplane; all-moving horizontal tail surfaces; conventional rudder; no tabs.

STRUCTURE: All-metal; semi-monocoque fuselage; large main wing fence on each side, at junction of fixed and movable panels, square-cut at front, with attachment for external store; shorter fence above glove panel each side.

LANDING GEAR: Retractable tricycle type, with single wheel on each unit. Nosewheel retracts forward, requiring blistered door to enclose it. Main units retract inward into centre-section. Container for single cruciform brake-chute between base of rudder and tailpipe.

POWER PLANT: One Saturn/Lyulka AL-21F-3 turbojet, rated at 76.5 kN (17,200 lb st) dry and 110 kN (24,700 lb st) with afterburning. Fuel capacity increased to 4,550 litres (1,202 US gallons; 1,000 Imp gallons) by added tankage in dorsal spine fairing. Provision for carrying up to four 800 litre (211 US gallon; 176 Imp gallon) drop tanks on outboard wing pylons and under fuselage. When underfuselage tanks are carried only the two inboard wing pylons may be used for ordnance, to a total weight of 1,000 kg (2,204 lb). Two solid propellant rocket units can be attached to rear fuselage to shorten T-O run.

ACCOMMODATION: Pilot only, on ejection seat, under rearward hinged transparent canopy. Rearview mirror above canopy.

AVIONICS: SRD-5M (NATO 'High Fix') I-band ranging radar in intake centrebody; ASP-5ND fire control system; HUD standard; Sirena-3 radar warning system providing 360° coverage, with antennae in slim cylindrical housing above brake-chute container and in each centre-section leading-edge, between fences; SRO-2M IFF; SOD-57M ATC/SIF, with transponder housing beneath brake-chute container; SP-50 ILS, RSB-70 HF and RSIU-5/R-831 UHF/VHF.

ARMAMENT: Two 30 mm NR-30 guns, each with 80 rounds, in wingroot leading-edges. Total of nine weapon pylons (one on centreline, two tandem pairs under fuselage, one under each centre-section leading-edge, one under each main wing fence) for more than 3,175 kg (7,000 lb) of bombs, including nuclear weapons, rocket pods, 23 mm gun pods and guided missiles such as the air-to-surface AS-7 (NATO 'Kerry'), AS-9 ('Kyle') and AS-10 ('Karen').

DIMENSIONS, EXTERNAL:

Wing span, fully spread	13.80 m (45 ft 3 in)
fully swept	10.00 m (32 ft 10 in)
Wing aspect ratio: fully spread	4.8
fully swept	2.7
Length overall, incl probes	18.75 m (61 ft 6¼ in)
Fuselage length	15.40 m (50 ft 6¼ in)
Height overall	5.00 m (16 ft 5 in)

AREAS (estimated):

Wings, gross: fully spread	40.0 m² (430.0 sq ft)
fully swept	37.0 m² (398.0 sq ft)

WEIGHTS AND LOADINGS (Su-17M-4):

Max external stores	4,250 kg (9,370 lb)
Normal T-O weight	16,400 kg (36,155 lb)

Sukhoi Su-22M-4 'Fitter-K' of the Czech Republic Air Force ***(Peter J. Cooper)***

1994

Sukhoi Su-22 of the Czech Republic Air Force ***(P. Tompkins)***

1994

Max T-O weight 19,500 kg (42,990 lb)
Max wing loading: spread 487.5 kg/m² (100 lb/sq ft)
swept 527 km/m² (108 lb/sq ft)
Max power loading 177.3 kg/kN (1.74 lb/lb st)

PERFORMANCE (Su-17M-4):
Max level speed: at height Mach 2.09
at S/L Mach 1.14 (755 kts; 1,400 km/h; 870 mph)
Service ceiling 15,200 m (49,865 ft)
T-O run 900 m (2,955 ft)
Landing run 950 m (3,120 ft)
Range with max fuel:
at high-altitude 1,240 n miles (2,300 km; 1,430 miles)
at low-altitude 755 n miles (1,400 km; 870 miles)

UPDATED

SUKHOI Su-22M-5 UPGRADE

TYPE: Combat aircraft upgrade.

PROGRAMME: Sukhoi announced in early 1995 that it was offering an avionics upgrade based on a Franco-Russian-designed suite to all current operators of the aircraft.

DESIGN FEATURES: Upgrade consists of the installation of the following equipment: MIL-STD-1553 databus, PrNK-55 Nav attack system; Phazatron/Thomson-CSF Phathom radar probably based on the Super Kopyo; Sextant Avionique mission computer, Head-Up Display (HUD), two MFS55 cockpit displays, Thomson-CSF TMV 630 laser rangefinder, convertible laser designation pod (CLDP) laser designator, Rubis FLIR nav/attack pod, Barem ECM pod, Sherloc radar warning receiver (RWR) NSS100-P GPS receiver, NC12 TACAN, Thomson-CSF AHV6 radio altimeter, ERA 2000 VHF/UHF communications radio and NR 810A VOR/ILS/Marker receiver. Further proposals include changes to the electrical system, in-flight refuelling and hands-on-throttle-and stick (HOTAS).

NEW ENTRY

SUKHOI Su-24

NATO reporting name: Fencer

TYPE: Two-seat variable geometry 'frontal bomber' reconnaissance and EW aircraft.

PROGRAMME: Design started 1964 under Yevgeniy S. Felsner, Paval Sukhoi's successor, to replace Il-28 and Yak-28 attack aircraft; T-6-1 prototype, first flown June 1967 and now at Monino, had fixed delta wings with downswept tips, and four Koliesov auxiliary booster motors mounted vertically in fuselage for improved take-off performance; T-6-2IG variable geometry prototype chosen for production; first flight January 1970; by 1981, delivery rate 60-70 a year; production of Su-24M/MR/MP continues.

VERSIONS **Su-24** ('Fencer-A'): Had rectangular rear fuselage box enclosing jet nozzles; few early aircraft only; deployed with trials unit 1974.

Su-24 ('Fencer-B'): First operational version, 1976; deeply dished bottom skin to rear fuselage box between jet nozzles; larger brake-chute housing at base of rudder.

Su-24 ('Fencer-C'): Introduced 1981; important avionics changes; multiple nose fitting instead of former simple probe; triangular fairing for RWR on side of each engine air intake, forward of fixed wingroot, and each side of fin tip; chord of fin leading-edge extended forward, except at tip, giving kinked profile.

Su-24M ('Fencer-D'): Major attack version; first flew 1977; entered service 1983; believed to have terrain-following radar instead of earlier terrain-avoidance system; added flight refuelling capability, with centrally mounted retractable probe forward of windscreen; nose approximately 0.75 m (2 ft 6 in) longer to accommodate new avionics bay; large overwing fences with integral extended wingroot glove pylons when carrying Kh-29 (NATO AS-14 'Kedge') missiles; undernose antennae deleted; laser ranger/designator housing aft of nosewheel bay; single long noseprobe; export version is **Su-24MK**.

Su-24MR ('Fencer-E'): Reconnaissance version of Su-24M used by tactical and naval air forces; internal equipment includes Shtik side-looking airborne multi-mission radar in nose, Zima IR reconnaissance system, Aist-M TV reconnaissance system, panoramic and oblique cameras in ventral fairing. A Shpil-2M laser pod can be carried on centreline, with a Tangazh elint pod or Efir-1M radiation detector pod on starboard underwing swivelling pylon and two R-60 air-to-air missiles under port wing. Data can be transmitted to ground by data link. No overwing fences; shorter nose radome, with flush dielectric side panels on nose; domed centre-fuselage air intake for heat exchanger; 'hockey stick' antenna at bottom of fuselage under each engine air intake nose section; provision for two 3,000 litre (792 US gallon; 660 Imp gallon) underwing auxiliary fuel tanks; flight refuelling and air-to-surface missile capabilities retained; tactical air force units include two squadrons on Chinese border; deliveries to Baltic fleet, replacing Tu-16s, began Summer 1985.

Su-24MP ('Fencer-F'): Electronic warfare/jamming/sigint version to replace Yak-28 'Brewer-E'; dielectric nose panels differ from those of Su-24MR; added small fairing under nose; no underside electro-optics; 'hockey stick' antennae as on Su-24MR; centreline EW pod.

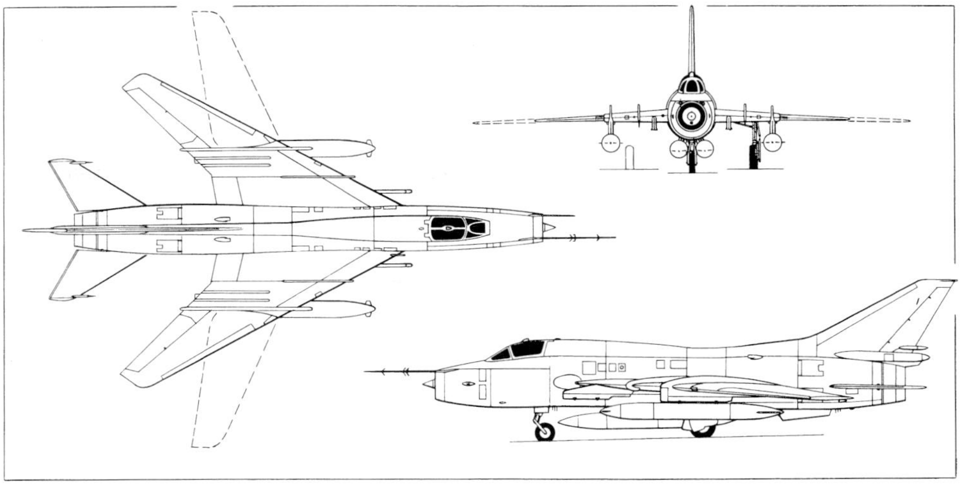

Sukhoi Su-17M-4 'Fitter-K', final version of the variable geometry 'Fitter' series *(Jane's/Dennis Punnett)*
1994

OPERATORS: Versions of the Su-24 are in service with the armed forces of the following countries: Iran (24), Kazakhstan (35+), Libya (15), Russia (600+), Syria (20) and Ukraine (250+).

DESIGN FEATURES: Variable geometry shoulder wing; slight anhedral from roots; triangular fixed glove box; three-position (16°, 45°, 68°) pivoted out panels; slab-sided rectangular section fuselage; integral engine air intake trunks, each with splitter plate and outer lip inclined slightly downward; chord of lower part of tail fin extended forward, giving kinked leading-edge; leading-edge sweepback 30° on inset rudder, 50° on horizontal tail surfaces; basic operational task, as Soviet designated 'frontal bomber', to deliver wide range of air-to-surface missiles for defence suppression, with some hard target kill potential; specially developed long-range navigation system and electro-optical weapons systems make possible penetration of hostile airspace at night or in adverse weather with great precision, to deliver ordnance within 55 m (180 ft) of target.

FLYING CONTROLS: Full-span leading-edge slats, aileron and two-section double-slotted trailing-edge flaps on each outer wing panel; differential spoilers forward of flaps for roll control at low speeds and use as lift dumpers on landing; airbrake under each side of centre-fuselage; inset rudder; all-moving horizontal tail surfaces operate collectively for pitch control, differentially for roll control, assisted by wing spoilers except when wings fully swept.

STRUCTURE: All-metal; semi-monocoque fuselage; two slightly splayed ventral fins.

LANDING GEAR: Retractable tricycle type, with twin-wheels on each unit. Main units retract forward and inward into air intake duct fairings; nose unit retracts rearward. Trailing link type of shock absorbers in main units and low-pressure

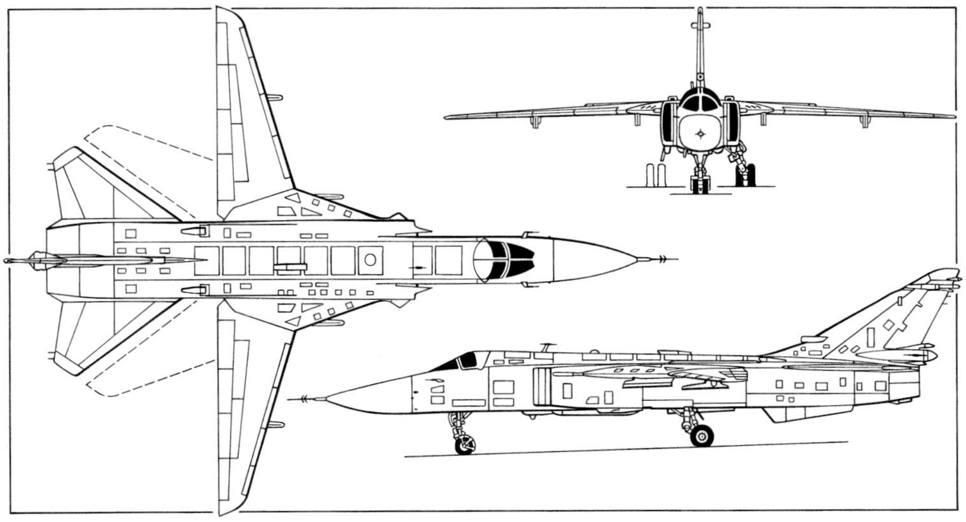

Sukhoi Su-24M 'Fencer-D' variable geometry attack aircraft *(Jane's/Mike Keep)*
1994

Sukhoi Su-24M 'Fencer-D' variable geometry attack aircraft of the Russian Air Force *(Jamie Hunter)*
1994

Sukhoi Su-24M 'Fencer-D' of the Russian Air Force believed to have terrain-following radar *(Jamie Hunter)* *1994*

tyres for operation from semi-prepared fields. Mudguard on nosewheels.

POWER PLANT: Two Saturn/Lyulka AL-21F-3A turbojets, each rated at 110 kN (24,700 lb st) with afterburning. Variable intake ramps. Internal fuel capacity, estimated at 13,000 litres (3,434 US gallons; 2,860 Imp gallons), can be supplemented by four 1,250 litre (330 US gallon; 275 Imp gallon) external tanks on underbelly and glove pylons. Probe-and-drogue flight refuelling capability, including operation as buddy tanker.

ACCOMMODATION: Crew of two (pilot and weapon systems officer) side by side on ejection seats. Cockpit width 1.65 m (5 ft 5 in). Jettisonable canopy, hinged to open upward and rearward in two panels, split on centreline.

AVIONICS: Large radar in nose. Laser ranger/designator under front fuselage. Radar warning receivers on sides of engine air intakes and tail fin. Active anti-radar suppression equipment.

ARMAMENT: Nine pylons under fuselage, each wingroot glove and outer wings for guided and unguided air-to-surface weapons, including TN-1000 and TN-1200 nuclear weapons, missiles such as AS-7 (NATO 'Kerry'), AS-10 ('Karen'), AS-11 ('Kilter'), AS-12 ('Kegler'), AS-13 ('Kingbolt') and AS-14 ('Kedge'), rockets of 57 mm to 370 mm calibre, bombs (typically 36 × 100 kg FAB-100), 23 mm gun pods or external fuel tanks. Two pivoting underwing pylons were the first of their kind observed on a Soviet aircraft. No internal weapons bay. One six-barrel 23 mm Gatling type gun inside fairing on starboard side of fuselage undersurface. Unidentified fairing on other side.

DIMENSIONS, EXTERNAL:
- Wing span: spread — 17.63 m (57 ft 10 in)
- swept — 10.36 m (34 ft 0 in)
- Length overall, incl probe — 24.53 m (80 ft 5¾ in)
- Height overall — 4.97 m (16 ft 3¾ in)
- Wheel track — 3.70 m (12 ft 1½ in)

AREAS:
- Wings, gross — 42.00 m² (452.1 sq ft)

WEIGHTS AND LOADINGS:
- Weight empty, equipped — 19,000 kg (41,885 lb)
- Max external stores — 8,000 kg (17,635 lb)
- Normal T-O weight — 36,000 kg (79,365 lb)
- Max T-O weight — 39,700 kg (87,520 lb)
- Max wing loading — 945.2 kg/m² (193.6 lb/sq ft)
- Max power loading — 180.5 kg/kN (1.77 lb/lb st)

PERFORMANCE:
- Max level speed, clean: at height — Mach 2.18
- at S/L — Mach 1.15
- Service ceiling — 17,500 m (57,400 ft)
- T-O run — 1,300 m (4,265 ft)
- Landing run — 950 m (3,120 ft)
- Combat radius:
  - lo-lo-lo — over 174 n miles (322 km; 200 miles)
  - lo-lo-hi 2,500 kg (5,500 lb) of weapons — 515 n miles (950 km; 590 miles)
  - hi-lo-hi with 3,000 kg (6,615 lb) of weapons and two external tanks — 565 n miles (1,050 km; 650 miles)

VERIFIED

SUKHOI Su-25 and Su-28

NATO reporting name: Frogfoot

TYPE: Single-seat close support aircraft and two-seat trainer.

PROGRAMME: Development began 1968; prototype, known as T-8-1, flew 22 February 1975, with two 25.5 kN (5,732 lb st) non-afterburning versions of Tumansky RD-9 turbojet and underbelly twin-barrel AO-17A 30 mm gun in fairing; in eventual developed form, second prototype, T-8-2, had more powerful non-afterburning versions of R-13, designated R-95Sh, wingtip avionics/speed brake pods, underwing weapon pylons, and internal gun; observed by satellite at Ramenskoye flight test centre 1977, given provisional US designation 'Ram-J; entered production 1978 with R-95 turbojets; trials unit, followed by squadron of 12, sent to Afghanistan for co-ordinated low-level close support of Soviet ground forces in mountain terrain, with Mi-24 helicopter gunships; fully operational 1984; attack versions built initially at Tbilisi, Georgia; production at Tbilisi ended 1989 (approximately 330 built); production now at Ulan-Ude; all production for CIS completed 1991-92; see separate entry on Su-25T/Su-34.

VERSIONS: **Su-25** ('Frogfoot-A'): Single-seat close support aircraft. Export version **Su-25K** (kommercheskiy; commercial). *Detailed description applies basically to late production Su-25K.*

Su-25UB ('Frogfoot-B'): Tandem two-seat operational conversion and weapons trainer; first photographs Spring 1989; rear seat raised considerably, giving hump-back appearance: separate hinged portion of continuous framed canopy over each cockpit; taller tail fin, increasing overall height to 5.20 m (17 ft 0¾ in); new IFF blade antenna forward of windscreen instead of SRO-2 (NATO 'Odd Rods'); weapons pylons and gun retained. Export version **Su-25UBK**.

Su-25UT ('Frogfoot-B'): As Su-25UB, but without weapons; prototype first flew 6 August 1985; demonstrated 1989 Paris Air Show as **Su-28**; overall length 15.36 m (50 ft 4¾ in); few only.

Su-25UTG (G for gak; hook) ('Frogfoot-B'): As Su-25UT, with added arrestor hook under tail; used initially for deck landing training on dummy flight deck marked on runway at Saki naval airfield, Ukraine; on 1 November 1989 was third aircraft to land for trials on carrier *Admiral of the Fleet Kuznetsov*, after Su-27K and MiG-29K; 10 built in 1989-90; five in Ukrainian use at Saki; one lost; four at Severomorsk, Kola Peninsula, for service individually on *Kuznetsov*; to be supplemented by Su-25UBPs.

Su-25UBP: Ten standard Su-25UBs were to be converted to Su-25UBP (Palybnyi: shipborne) in 1993 for service on *Admiral of the Fleet Kuznetsov*. Reportedly cancelled.

Su-25BM (BM for buksir misheney; target towing aircraft): As Su-25 attack aircraft, with added underwing pylons for rocket propelled targets released for missile training by fighter pilots.

Su-25T: For full details see *Jane's All the World's Aircraft* 1995-96 Su-39 entry.

UPGRADES: **Sukhoi Su-25UBP**: See Versions.

OPERATORS: Versions of the Su-25/28 are in service with the armed forces of the following countries: Azerbaijan, Belarus (90), Bulgaria (35+), Czech Republic (25), Georgia, Hungary (20), Iraq (20+), North Korea (20), Russia (190+), Slovakia (13) and Ukraine. (35).

DESIGN FEATURES: Shoulder-mounted wings; approximately 22° sweepback; anhedral from roots; extended chord leading-edge dogtooth on outer 50 per cent each wing; wingtip pods each split at rear to form airbrakes that project above and below pod when extended; retractable landing light in base of each pod, outboard of small glareshield and

Sukhoi Su-25UB 'Frogfoot-B' tandem two-seat operational conversion *(L. Freundt)* *1994*

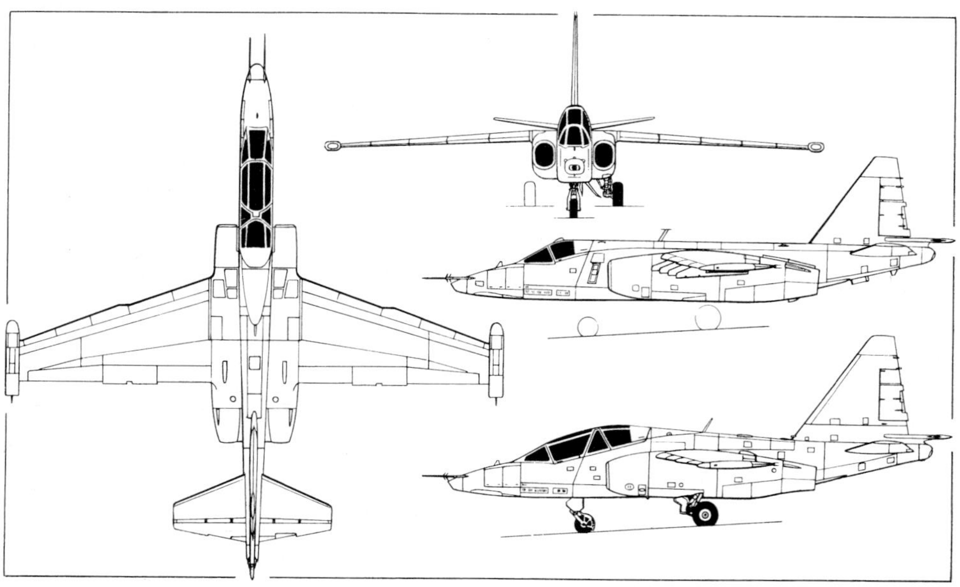

Sukhoi Su-25UT (Su-28) 'Frogfoot-B', with added side elevation (centre) of Su-25K *(Jane's/Dennis Punnett)* *1994*

aft of dielectric nosecap for ECM; semi-monocoque fuselage, with 24 mm (0.94 in) welded titanium armoured cockpit; pitot on port side of nose, transducer to provide data for fire control computer on starboard side; conventional tail unit; variable incidence tailplane, with slight dihedral.

Emphasis on survivability led to features accounting for 7.5 per cent of normal T-O weight, including armoured cockpit; pushrods instead of cables to actuate flying control surfaces (duplicated for elevators); damage-resistant main load-bearing members; widely separated engines in stainless steel bays; fuel tanks filled with reticulated foam for protection against explosion.

Maintenance system packaged into four pods for carriage on underwing pylons; covers onboard systems checks, environmental protection, ground electrical power supply for engine starting and other needs, and pressure refuelling from all likely sources of supply in front-line areas; engines can operate on any fuel likely to be found in combat area, including MT petrol and diesel oil.

FLYING CONTROLS: Hydraulically actuated ailerons, with manual back-up; multiple tabs in each aileron; double-slotted two-section wing trailing-edge flaps; full-span leading-edge slats, two segments per wing; manually operated elevators and two-section inset rudder; upper rudder section operated through sensor vanes and transducers on nose probe and automatic electromechanical yaw damping system; tabs in lower rudder segment and each elevator.

STRUCTURE: All-metal; three-spar wings; semi-monocoque slab-sided fuselage.

LANDING GEAR: Hydraulically retractable tricycle type, mainwheels retract to lie horizontally in bottom of engine air intake trunks. Single wheel with low-pressure tyre on each levered suspension unit, with oleo-pneumatic shock-absorber, mudguard on forward-retracting steerable nosewheel, which is offset to port; mainwheel tyres size 840 × 360 mm, pressure 9.3 bars (135 lb/sq in); nosewheel tyre size 660 × 200 mm, pressure 7.35 bars (106 lb/sq in); brakes on mainwheels. Twin cruciform brake-chutes housed in tailcone.

POWER PLANT: Two Soyuz/Tumansky R-195 turbojets in long nacelles at wingroots, each 44.18 kN (9,921 lb st); 5 mm thick armour firewall between engines; current upgraded R-195 turbojets have pipe-like fitment at end of tailcone, from which air is expelled to lower exhaust temperature and so reduce infra-red signature; non-waisted under surface to rear cowlings, which have additional small airscoops (as three-view). Fuel tanks in fuselage between cockpit and wing front spar, and between rear spar and fin leading-edge, and in wing centre-section; provision for four PTB-1500 external fuel tanks on underwing pylons.

ACCOMMODATION: Single K-36L zero/zero ejection seat under sideways hinged (to starboard) canopy, with small rear-view mirror on top; flat bulletproof windscreen. Folding ladder for access to cockpit built into port side of fuselage.

SYSTEMS: 28 V DC electrical system, supplied by two engine-driven generators.

AVIONICS: Laser rangefinder and target designator under flat sloping window in nose. SRO-1P (NATO 'Odd Rods') or (later) SRO-2 IFF transponder, with antennae forward of windscreen and under tail. Sirena-3 radar warning system antenna above fuselage tailcone.

EQUIPMENT: ASO-2 V chaff/flare dispensers (total of 256 flares) can be carried above root of tailplane and above rear of engine ducts. Strike camera in top of nosecone.

ARMAMENT: One twin-barrel AO-17A 30 mm gun with rate of fire of 3,000 rds/min in bottom of front fuselage on port side, with 250 rounds (sufficient for a one second burst during each of five attacks). Eight large pylons under wings for 4,400 kg (9,700 lb) of air-to-ground weapons, including UB-32A rocket pods (each 32 × 57 mm S-5), B-8M1 rocket pods (each 20 × 80 mm S-8) 240 mm S-24 and 330 mm S-25 guided rockets, Kh-23 (NATO AS-7 'Kerry'), Kh-25 (AS-10 'Karen') and Kh-29 (AS-14 'Kedge') air-to-surface missiles, laser-guided rocket-boosted 350 kg, 490 kg and 670 kg bombs, 500 kg incendiary anti-personnel and chemical cluster bombs, and SPPU-22 pods each containing a 23 mm GSh-23 gun with twin barrels that can pivot downward for attacking ground targets, and 260 rounds. Two small outboard pylons for R-3S (K-13T; NATO AA-2D 'Atoll') or R-60 (AA-8 'Aphid') air-to-air self-defence missiles.

Sukhoi Su-25K 'Frogfoot-A' of the Czech Republic Air Force *(Peter J. Cooper)* *1994*

Sukhoi Su-25UB tandem two-seat operational conversion and weapons training aircraft *1995*

DIMENSIONS, EXTERNAL (Su-25K):

Wing span	14.36 m (47 ft 1½ in)
Wing aspect ratio	6.12
Length overall	15.53 m (50 ft 11½ in)
Height overall	4.80 m (15 ft 9 in)

AREAS:

Wings, gross	33.7 m² (362.75 sq ft)

WEIGHTS AND LOADINGS:

Weight empty	9,500 kg (20,950 lb)
Max T-O weight	14,600-17,600 kg (32,187-38,800 lb)
Max landing weight	13,300 kg (29,320 lb)
Max wing loading	522.2 kg/m² (107.0 lb/sq ft)
Max power loading	199.2 kg/kW (1.96 lb/lb st)

PERFORMANCE:

Max level speed at S/L	Mach 0.8 (526 kts; 975 km/h; 606 mph)
Max attack speed, airbrakes open	372 kts (690 km/h; 428 mph)
Landing speed (typical)	108 kts (200 km/h; 124 mph)
Service ceiling: clean	7,000 m (22,965 ft)
with max weapons	5,000 m (16,400 ft)
T-O run: typical	600 m (1,970 ft)
with max weapon load from unpaved surface	under 1,200 m (3,935 ft)
Landing run: normal	600 m (1,970 ft)
with brake-chutes	400 m (1,312 ft)
Range with 4,400 kg (9,700 lb) weapon load and two external tanks:	
at S/L	405 n miles (750 km; 466 miles)
at height	675 n miles (1,250 km; 776 miles)
g limits: with 1,500 kg (3,306 lb) of weapons	+6.5
with 4,400 kg (9,700 lb) of weapons	+5.2

UPDATED

TUPOLEV

AVIATSIONNYI NAUCHNO-TEKHNICHESKIY KOMPLEKS IMIENI A. N. TUPOLEVA (Aviation Scientific-Technical Complex named after A. N. Tupolev) (ANTK)

15 Naberejnaia Akademika Tupoleva, 111250 Moscow
Telephone: 7 (095) 267 2508
Fax: 7 (095) 261 0868, 261 7141

GENERAL DIRECTOR: Valentin Klimov
ASSISTANT TO GENERAL DIRECTOR: Andrei I. Kandalov
CHIEF DESIGNERS:
Lev Aronovich Lanovski (Commercial Aircraft)
Dmitry S. Markov
L. L. Selyakov
Alexander S. Shengardt

Andrei Nikolayevich Tupolev was leading figure in Central Aero-Hydrodynamic Institute (TsAGI) in Moscow from when it was founded, in 1918, until his death 23 December 1972, aged 84. The Bureau that bears his name concentrates primarily on large military and civil aircraft.

UPDATED

TUPOLEV Tu-16

NATO reporting name: Badger

TYPE: Twin-jet medium bomber, maritime reconnaissance/attack and electronic warfare aircraft.

PROGRAMME: Prototype flown by N. S. Rybka under OKB designation Tu-88 27 April 1952 was overweight; Andrei Tupolev delayed production until second prototype flew in 1953 with uprated AM-3A turbojets and 5,500 kg (12,125 lb) weight reduction; maximum speed increased to 535 knots (992 km/h; 616 mph) at cost of maximum IAS of only 378 knots (700 km/h; 435 mph) at low-altitude instead of originally required Mach 0.9; deliveries began 1954; nine Tu-16s took part in May Day 1954 flypast over Moscow, 54 in Aviation Day flypast 1955; approximately 2,000 produced in many versions; manufacture ended late 1950s; late versions all converted bombers; all but earliest have uprated AM-3M (RD-3M) engines.

VERSIONS: **Tu-16A** ('Badger A'): 1954; original strategic bomber version, with conventional or nuclear free-fall weapons; glazed nose, with small undernose radome; defensive armament of seven 23 mm AM-23 guns; licence production, as **H-6**, by Xian Aircraft Company of China ended in late 1980s.

Tu-16KS ('Badger-B'): 1954; first version with air-to-surface missiles (two KS-1; AS-1 'Kennel' turbojet-powered ASMs underwing); now retired.

Tu-16Z: 1955; experimental flight refuelling tanker, using unique wingtip-to-wingtip transfer technique (see production Tu-16N).

Tu-16T: 1959; torpedo bomber.

Tu-16S Korvet: 1965; Tu-16T adapted for search and rescue duties; large radio-controlled rescue boat under fuselage; continues in use.

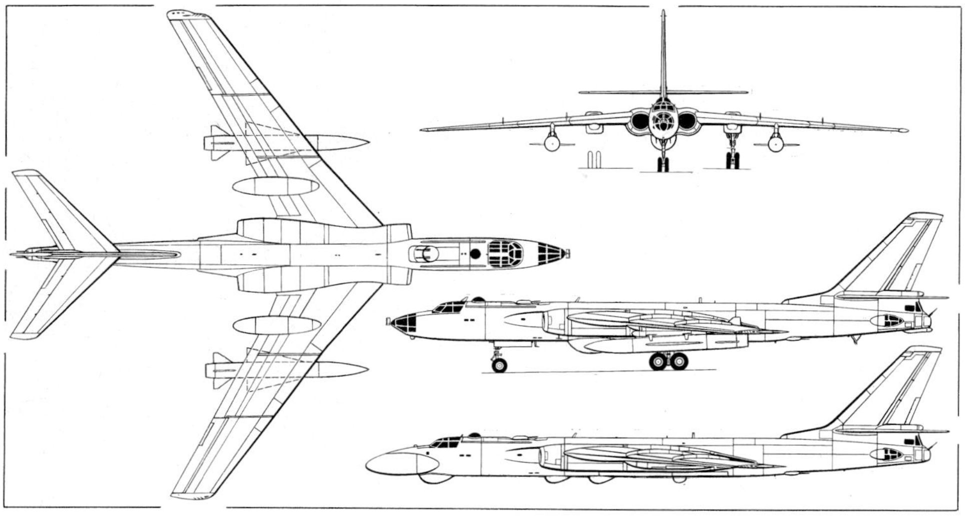

Tupolev Tu-16K-26 'Badger-G Mod', with additional side view (bottom) of Tu-16R 'Badger-D' *(Jane's/Dennis Punnett)*

1994

Tu-16K-10 ('Badger-C'): 1958; anti-shipping version; first seen 1961 Soviet Aviation Day display; Mikoyan K-10S (NATO AS-2 'Kipper') turbojet-powered missile in underbelly recess; wide nose radome (NATO 'Puff Ball') instead of glazing and nose gun of Tu-16A; no provision for free-fall bombs; about 100 supplied to Northern, Baltic, Black Sea and Pacific Fleet shore bases.

Tu-16R ('Badger-D'): Maritime/electronic reconnaissance version, introduced early 1960s; nose as Tu-16K-10 but larger undernose radome; three elint radomes in tandem under weapons bay; large number of cameras in weapons bay. Tu-16Rs carry crew of eight to ten.

Tu-16 ('Badger-E'): Photographic/electronic reconnaissance version; as Tu-16A but cameras in weapons bay; two additional radomes under fuselage, larger one aft.

Tu-16R ('Badger-F'): Similar to 'Badger-E' but elint pod on pylon under each wing; late versions have various small radomes under centre-fuselage.

Tu-16K-11/16 ('Badger-G'): 1959; as Tu-16KS but with two KSR-11 (K-11) or KSR-2 (K-16; AS-5 'Kelt') underwing rocket-powered missiles,which can be carried over range greater than 1,735 n miles (3,220 km; 2,000 miles). Free-fall bombing capability retained; delivered mainly to Naval anti-shipping squadrons.

Tu-16K-26 ('Badger-G Mod'): Modified to carry KSR-5 (K-26; AS-6 'Kingfish') Mach 3 missile, with nuclear or conventional warhead, under each wing; large radome, presumably associated with missile operation, under centre-fuselage, replacing chin radome; external device on glazed nose may help to ensure correct attitude of Tu-16K-26 during missile launch; delivered to Northern, Black Sea and Pacific Fleets.

Tu-16K-10-26 ('Badger-C Mod'): 1962; modified to carry two KSR-5 (K-26; AS-6 'Kingfish') missiles underwing, in addition to K-10S capability.

Tu-16N: 1963; as Tu-16A but equipped as flight refuelling tanker, using wingtip-to-wingtip technique to refuel other Tu-16s, or probe-and-drogue system to refuel Tu-22s; added tankage in bomb bay.

Tu-16PP (Postanovchik Pomiekh: jammer) ('Badger-H'): stand-off or escort ECM aircraft; primary function chaff dispensing to protect missile carrying strike force; two teardrop radomes, fore and aft of weapons bay, house passive receivers to identify enemy radar signals and establish length of chaff strips to be dispensed; dispensers with total capacity of up to 9,075 kg (20,000 lb) of chaff located in weapons bay, with three chutes in doors; hatch aft of weapons bay; two blade antennae aft of bay (that on port side with V twin blades); glazed nose and chin radome.

Tu-16PP ('Badger-J'): ECM version specialised for active jamming in all frequencies; modified from standard bomber; some equipment in canoe shape radome protruding from weapons bay, surrounded by heat exchangers and exhaust ports; anti-radar noise jammers operated in A- to I-bands inclusive; glazed nose as Tu-16A; some aircraft (as illustrated) have large flat-plate antennae at wingtips.

Tu-16R ('Badger-K'): Electronic reconnaissance variant; nose as Tu-16A; two teardrop radomes, inside and forward of weapons bay (closer together than on 'Badger-H'); four small pods on centreline in front of rear radome; chaff dispenser aft of weapons bay.

Tu-16 ('Badger-L'): Naval electronic warfare variant, like 'Badger-G' but with equipment of the kind fitted to the Tu-95 'Bear-G', including an ECM nose thimble, small pods on centre-fuselage forward of engine air intakes, and 'solid' extended tailcone containing special equipment instead of tail gun position; sometimes has pod on pylon under each wing.

Tupolev Tu-16R 'Badger-D' maritime/electronic reconnaissance aircraft

1994

Tu-16LL: Testbed for AL-7F-1, VD-7, D-36 and other engines, and equipment.

Tu-16 Tsiklon: 1979; meteorological laboratory.

Tu-104G: Redesignation of Tu-16s used by Aeroflot for urgent mail transportation.

B-6D: Chinese export version of H-6D. See below.

H-6A: Initial Chinese-built (Xian) version. First flown 24 December 1968.

H-6D: Chinese-built (Xian) maritime strike version.

UPGRADES: **Tupolev:** Tu-16K-10-26 (Badger-C-Mod). See Versions.

OPERATORS: Versions of the Tu-16 are in service with the armed forces of the following countries: China (H-6A/D 120), Iraq, Russia (65) and Ukraine (75).

DESIGN FEATURES: All-swept high mid-wing configuration; heavy engine nacelles from root fairings; wing section PR-1-10S-9, with 15.7 per cent thickness/chord ratio, on inboard panels; SR-11-12 section, of 12 per cent thickness/chord, outboard; anhedral 3° from roots; incidence 1°; sweepback 41° inboard, 37° outboard at leading-edges, 35° at quarter-chord; 42° sweepback at quarter-chord on tail fin and tailplane; tailplane incidence −1° 30′.

FLYING CONTROLS: Conventional, with control wheel and rudder pedals; hydraulically boosted, with large trim tabs in each aileron, elevators and rudder; except in region of landing gear nacelles, entire wing trailing-edges comprise aerodynamically balanced Frise/TsAGI ailerons and electrically operated outboard and inboard sections of TsAGI (modified Fowler) flaps.

STRUCTURE: All-metal; two spar wings made in centre-section (integral with fuselage), inner and outer sections; nacelle for landing gear at junction of inner and outer wing panels each side; circular semi-monocoque fuselage in five sections; nose houses navigator's pressure cabin with double-glazed nose panels in magnesium alloy frame, pilot's pressure cabin, forward gunner's cabin and radar equipment; second and fourth sections house fuel tanks, with weapon compartment between them; tail section contains pressure cabin for radio operator/gunner and rear gunner.

LANDING GEAR: Hydraulically retractable tricycle type. Twin-wheel nose unit retracts rearward. Main four-wheel bogies retract into housings projecting beyond the wing trailing-edge. Mainwheel tyres size 1,100 × 330 mm; nose-wheel tyres size 900 × 275 mm. Anti-skid brakes on mainwheels.

POWER PLANT: Early Tu-16s have two Mikulin AM-3A turbojets, each rated at 85.21 kN (19,155 lb st) at sea level. Later aircraft fitted with RD-3M-500 (AM-3M) turbojets, each rated at 93.05 kN (20,920 lb st). Engines semi-recessed into sides of fuselage, giving unplanned area ruling. Divided air intake ducts: main duct passes through wing torque box between spars; secondary duct passes under wing to feed into primary air flow in front of engine. Engines separated from wings and fuselage by firewalls. Jetpipes inclined outward 3° to shield fuselage from effects of exhaust gases. Fuel in 27 wing and fuselage tanks, with total capacity of 43,800 litres (11,570 US gallons; 9,635 Imp gallons). Provision for underwing auxiliary fuel tanks and for flight refuelling. Tu-16 tankers trail hose from starboard wingtip; receiving equipment is in port wingtip extension.

ACCOMMODATION Normal crew of six on ejection seats, with two pilots side by side on flight deck. Navigator/bombardier, on seat with armoured sides and base, in glazed nose of all versions except 'Badger-C and -D'. Manned tail position plus lateral observation blisters in rear fuselage under tailplane. Entry via two hatches in bottom of fuselage, in front and rear structural sections.

AVIONICS: PBR-4 Rubin 1 mapping radar; SP-50M for IFR to ICAO Cat. I standard; RSDN Chaika Loran; NavSat receiver; AP-6E autopilot, NAS-1 nav including DISS Trassa Doppler, R-807 and R-808 HF; RSIU-3M UHF; IFF, two ARK-15 ADFs, SPU-10 intercom; RV-5 and RV-18 radar altimeters; Sirena-2 radar warning receivers.

EQUIPMENT: Differs according to role; all Tu-16s can carry cameras in small section of fuselage, ahead of weapons bay, for reconnaissance.

ARMAMENT: PV-23 integrated armament firing control system. Forward dorsal and rear ventral barbettes each containing two 23 mm AM-23 guns. Two similar guns in tail position controlled by an automatic gun ranging radar set. Seventh, fixed, gun on starboard side of nose of versions with nose glazing. Bomb load of up to 9,000 kg (19,800 lb) delivered from weapons bay 6.5 m (21 ft) long in standard bomber, under control of navigator. Normal bomb load 3,000 kg (6,600 lb). Naval versions can carry air-to-surface winged standoff missiles.

DIMENSIONS, EXTERNAL ('Badger-C'):

Wing span	32.99 m (108 ft 3 in)
Mean aerodynamic chord	5.021 m (16 ft 5¾ in)
Length overall	34.80 m (114 ft 2 in)
Height overall	10.36 m (34 ft 0 in)
Basic diameter of fuselage	2.50 m (8 ft 2½ in)
Tailplane span	11.75 m (38 ft 6½ in)
Wheel track	9.775 m (32 ft 0¾ in)
Wheelbase	10.91 m (35 ft 9½ in)

AREAS:

Wings, gross	164.65 m² (1,772.3 sq ft)
Ailerons (total)	14.77 m² (159.0 sq ft)
Flaps (total)	25.17 m² (270.9 sq ft)

Vertical tail surfaces	23.30 m² (250.8 sq ft)
Horizontal tail surfaces	34.45 m² (370.8 sq ft)
WEIGHTS AND LOADINGS ('Badger-G'):	
Weight empty, equipped	37,200 kg (82,000 lb)
Weight of max fuel	34,360 kg (75,750 lb)
Normal T-O weight	75,000 kg (165,350 lb)
Max landing weight	50,000 kg (110,230 lb)
Max wing loading	455.5 kg/m² (93.3 lb/sq ft)
Max power loading	403.0 kg/kN (3.95 lb/lb st)
PERFORMANCE ('Badger-G', at max T-O weight):	
Max level speed at 6,000 m (19,700 ft)	566 kts (1,050 km/h; 652 mph)
Service ceiling	15,000 m (49,200 ft)
Range with 3,000 kg (6,600) bomb load	3,885 n miles (7,200 km; 4,475 miles)

VERIFIED

Tupolev Tu-16K-26 'Badger-G Mod' with AS-6 'Kingfish' Mach 3 nuclear/conventional missiles

1994

TUPOLEV Tu-22

NATO reporting name: Blinder

TYPE: Twin-jet supersonic bomber and maritime reconnaissance aircraft.

PROGRAMME: Development started 1956. First shown publicly 1961 when 10 took part in Soviet Aviation Day flypast, Moscow; one had AS-4 (NATO 'Kitchen') missile semi-recessed in weapons bay; 22 shown in 1967 display at Domodedovo, all with refuelling probes, most with AS-4s; total of about 250 manufactured.

VERSIONS: **Tu-22** ('Blinder-A'): Basic bomber with supersonic dash capability; fuselage weapons bay for free-fall nuclear and conventional bombs (from FAB-250 to FAB-900); entered limited service.

Tu-22K ('Blinder-B'): As 'Blinder-A' but weapons bay doors redesigned to carry Kh-22 (AS-4 'Kitchen') air-to-surface missile semi-recessed; larger radar and partially retractable flight refuelling probe on nose. Free-fall bombing capability retained.

Tu-22R ('Blinder-C'): Daylight reconnaissance version; with six windows in weapons bay doors for three pairs of long focal length cameras; chaff dispensing chute aft of weapons bay; flight refuelling probe. On some aircraft, tail gun is deleted and a 2.4 m (7 ft 10½ in) tail extension, housing ECM equipment, is fitted.

Tu-22R ('Blinder-C Mod'): Capability extended to night reconnaissance by addition of centreline conformal pack, approximately 5.50 m (18 ft) long and 1.0 m (3 ft 3 in) wide, possibly containing IR and other sigint systems together with photo-flares. 'Solid' tailcone extension, housing ECM. Possible ESM installation just forward of wing on each side.

Tu-22U ('Blinder-D'): Training version: raised cockpit for instructor aft of standard flight deck, with stepped-up canopy.

Tu-22P ('Blinder-E'): Electronic warfare conversion, with avionics and cooling systems in weapons bay. Ventral fairing, approximately 6.3 m (21 ft) long and 0.3 m (1 ft) deep, under centreline, heat exchanger blister at rear; four sweptback jamming antennae at corners of weapons bay doors; modifications to nosecone; additional dielectric panels.

UPGRADES: **Tupolev:** Tu-22R 'Blinder-C-Mod'. See Versions.

OPERATORS: Versions of the Tu-22 are in service with the armed forces of the following countries: Libya (6), Russia (90) and Ukraine (70+).

DESIGN FEATURES: All-swept mid-wing configuration; engine nacelles mounted above rear fuselage, each side of fin, with translating intake lips; main landing gear units retract into pods on wing trailing-edge; small wingtip pods; wing section modified TsAGI SR-5S; constant slight anhedral from roots; leading-edge sweepback approximately 50° inboard of fence on each wing, 45° outboard, with small acutely swept segment at root; fuselage basically circular section, with area ruling at wingroots.

FLYING CONTROLS: Hydraulically powered control surfaces; two-section ailerons, with tab in each inboard section; tracked plain flaps inboard and outboard of each landing gear pod; all-moving horizontal tail surfaces at bottom of fuselage; aerodynamically balanced rudder, with inset tab.

STRUCTURE: All-metal, semi-monocoque fuselage.

LANDING GEAR: Retractable tricycle type. Wide track four-wheel bogie main units retract rearward into pods built on to wing trailing-edges. Oleo-pneumatic shock-absorbers. Main legs designed to swing rearward for additional cushioning during taxying and landing on rough runways. Twin-wheel nose unit retracts rearward. Small retractable skid to protect rear fuselage in tail-down landing or take-off. Twin brake-chutes standard.

POWER PLANT: Two Koliesov VD-7 turbojets, each rated at 137.5 kN (30,900 lb st) with afterburning, mounted in nacelles above rear fuselage, on each side of tail fin. Lip of each intake is in the form of a ring which can be translated forward by jacks for take-off. Air entering ram intake is then supplemented by air ingested through annular slot between ring and main body of nacelle. Jetpipes have convergent-divergent nozzle inside outer fairing. Semi-retractable flight refuelling probe on nose of

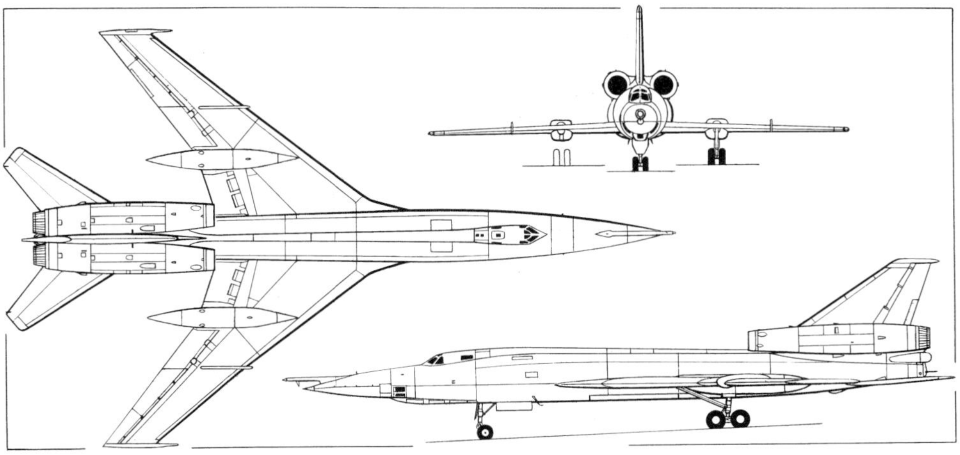

Tupolev Tu-22 'Blinder-A' twin-jet supersonic reconnaissance bomber aircraft *(Jane's/Dennis Punnett)*

1994

Tupolev Tu-22P 'Blinder-E' electronic warfare (EW) version *(Piotr Butowski)*

1994

Tupolev Tu-22P 'Blinder-E' *(Piotr Butowski)*

1994

'Blinder-B/C', with triangular guard underneath to prevent drogue damaging nosecone.

ACCOMMODATION: Crew of three in tandem. Row of windows in bottom of fuselage, aft of nose radome, at navigator/systems operator's station. Pilot has upward ejection seat; other crew members have downward ejection seats.

AVIONICS: Nav/attack radar in nose; tail warning and gun fire control radar ('Bee Hind') at base of rudder.

EQUIPMENT: Chaff/flare countermeasures dispensers and bombing assessment cameras carried in rear of wheel pods of some aircraft.

ARMAMENT: Weapons bay in centre-fuselage, with double-fold doors on 'Blinder-A'. Special doors with panels shaped to accommodate recessed AS-4 ('Kitchen') missile on 'Blinder-B'. Single 23 mm NR-23 gun in radar directed tail turret.

DIMENSIONS, EXTERNAL (estimated):

Wing span	23.50 m (77 ft 1¼ in)
Length overall	42.60 m (139 ft 9¾ in)
Height overall	10.00 m (32 ft 9¾ in)
Weapons bay doors:	
Length	7.0 m (22 ft 11½ in)
Width	1.65 m (5 ft 5 in)

AREAS:

Wings, gross	162.0 m² (1,744 sq ft)

WEIGHTS AND LOADINGS:

Max fuel load	42,500 kg (93,695 lb)
Max weapon load	12,000 kg (26,455 lb)
Normal T-O weight	85,000 kg (187,390 lb)
Max T-O weight	92,000 kg (202,820 lb)
Max T-O weight with four JATO rockets	94,000 kg (207,230 lb)
Normal landing weight	60,000 kg (132,275 lb)
Max wing loading (without JATO)	568.0 kg/m² (116.3 lb/sq ft)
Max power loading (without JATO)	284.3 kg/kN (2.79 lb/lb st)

PERFORMANCE (RD-7M-2 engines):

Max level speed at 12,200 m (40,000 ft)	Mach 1.52 (870 kts; 1,610 km/h; 1,000 mph)
Landing speed	168 kts (310 km/h; 193 mph)
Service ceiling, supersonic	13,300 m (43,635 ft)
T-O run	2,250 m (7,385 ft)
Landing run	2,170 m (7,120 ft)
Landing run with brake-chutes	1,650 m (5,415 ft)
Radius of action	700-1,188 n miles (1,300-2,200 km; 807-1,365 miles)
Range with max fuel	2,645 n miles (4,900 km; 3,045 miles)
Ferry range	3,050 n miles (5,650 km; 3,510 miles)

VERIFIED

TUPOLEV Tu-22M (Tu-26)

NATO reporting name: Backfire

TYPE: Twin-engined variable geometry medium bomber and maritime reconnaissance/attack aircraft.

PROGRAMME: NATO revealed the existence of a Soviet variable geometry bomber programme Autumn 1969; prototype observed July 1970 on the ground near Kazan manufacturing plant, western Russia; confirmed subsequently as twin-engined design by Tupolev OKB; at least two prototypes built, with first flight estimated 1969, up to 12 preproduction models by early 1973, for development testing, weapons trials and evaluation; production has been 30 a year.

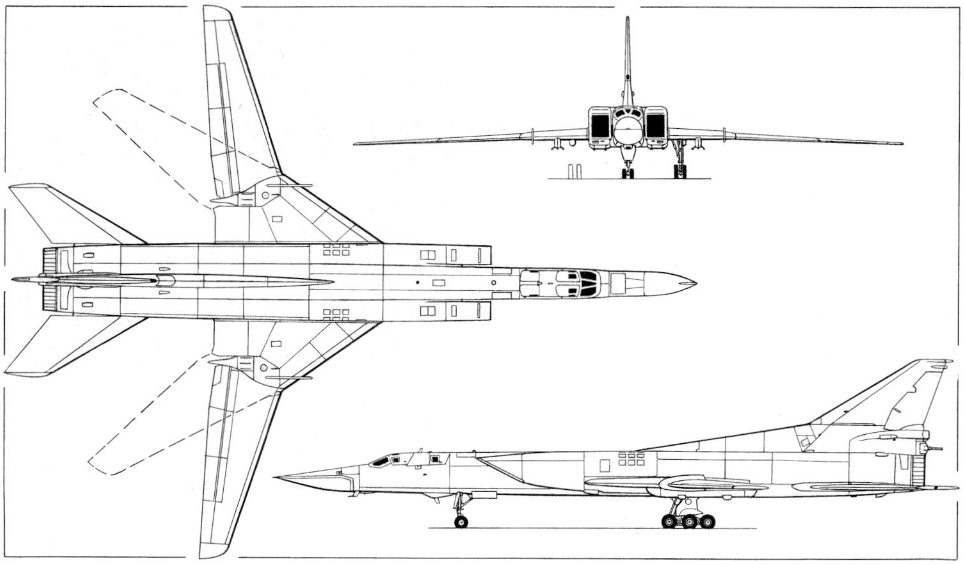

Tupolev Tu-22M-3 'Backfire-C' long-range bomber and maritime reconnaissance aircraft *(Jane's/Dennis Punnett)* ***1994***

VERSIONS: **Tu-22M-1** ('Backfire-A'): Initial version, thought to have equipped one squadron; large pods on wing trailing-edges; these did not, as originally believed, house retracted main landing gear, which always retracted into fuselage; slightly inclined lateral air intakes, with large splitter plates.

Tu-22M-2 ('Backfire-B'): First series production version; increased wing span; wing trailing-edge pods eliminated except for shallow underwing fairings, no longer protruding beyond trailing-edge; seen usually with optional flight refuelling nose probe removed and its housing replaced by long fairing. Initial armament was normally one air-to-surface missile (NATO AS-4 'Kitchen') semi-recessed under fuselage; current aircraft have rack for a 'Kitchen' under each fixed-wing centre-section panel, although fuselage mount retained; external stores racks seen frequently under engine air intake trunks; two GSh-23 twin-barrel 23 mm guns in tail mounting, initially beneath ogival radome, now with drum-shape radome of larger diameter. *Detailed description applies specifically to Tu-22M-2.*

Tu-22M-3 ('Backfire-C'): Advanced long-range bomber and maritime version; first known deployment with Black Sea fleet air force 1985; new engines of approximately 25 per cent higher rating; wedge type engine air intakes; upturned nosecone; no visible flight refuelling probe; rotary launcher in weapons bay for six RKV-500B (AS-16 'Kickback') short-range attack missiles; provision for four more underwing as alternative to standard two AS-4 'Kitchens'; single GSh-23 twin-barrel 23 mm gun, with barrels superimposed, in aerodynamically improved tail mounting, beneath large drum-shape radome.

OPERATORS: Versions of the Tu-22M are in service with the armed forces of the following countries: Russia (160) and Ukraine (30).

DESIGN FEATURES: Capable of performing nuclear strike, conventional attack and anti-ship missions; low-level penetration features ensure better survivability than for earlier Tupolev bombers; not expected to become ALCM carriers, although used for development launches, deployment of RKV-500B (AS-16 'Kickback') short-range attack missiles in Tu-22Ms has increased significantly their weapon carrying capability. Low/mid-wing configuration; large-span fixed centre-section and two outer steering sleeves variable from 20 to 65° sweepback; no anhedral or dihedral, but wing section so thin that outer panels flex considerably in flight; leading-edge fence towards tip of centre-section each side; basically circular fuselage forward of wings, with ogival dielectric nosecone; centre-fuselage faired into rectangular section air intake trunks, each with large splitter plate and assumed to embody complex variable geometry ramps; no external area ruling of trunks; all-swept tail surfaces, with large dorsal fin.

FLYING CONTROLS: Full-span leading-edge slat, aileron and three-section slotted trailing-edge flaps aft of spoilers/lift dumpers on each steering sleeve integral rudder.

LANDING GEAR: Retractable tricycle type; each mainwheel bogie comprises three pairs of wheels in tandem, with two forward pairs farther apart than rear pairs; bogies pivot inwards from vestigial fairing under centre-section on each side into bottom of fuselage.

POWER PLANT: Two unidentified turbofans, side by side in rear fuselage, each more than 2,500 kg with afterburning. Fuel is in integral tanks in wing central section and steering sleeves and in fuselage tanks.

ACCOMMODATION: Pilot and co-pilot side by side, under upward opening gull-wing doors hinged on centreline; two crew members further aft, as indicated by position of windows between flight deck and air intakes.

AVIONICS: Large missile targeting and navigation radar (NATO 'Down Beat') inside dielectric nosecone; radar ('Box Tail') for tail turret, above guns. Fairing with flat glazed front panel under front fuselage, for video camera to provide visual assistance for weapon aiming from high-altitude. Very advanced ECM and ECCM; infra-red missile approach warning sensor above fuselage aft of cockpit; eight chaff/flare multiple dispensers in bottom of each engine duct between wingroot and tailplane, another in each tailplane root fairing.

ARMAMENT: Maximum offensive weapon load three Kh-22 (NATO AS-4 'Kitchen') air-to-surface missiles, one semi-recessed under centre-fuselage, one under fixed centre-section panel of each wing; or 24,000 kg (52,910 lb) of conventional bombs or mines, half carried internally and half on racks under wings and engine air intake trunks. Internal bombs can be replaced by rotary launcher for six Kh-15P (AS-16 'Kickback') short-range attack missiles, with four more underwing as alternative to Kh-22s. Normal weapon load is single Kh-22 or 12,000 kg (26,455 lb) of bombs. Typical loads two FAB-3000, eight FAB-1500, 42 FAB-500 or 69 FAB-250 or -100 bombs (figures indicated weight in kg), or eight 1,500 kg or 18 500 kg mines. One GSh-23 twin-barrel 23 mm gun, with barrels superimposed, in radar directed tail mounting.

DIMENSIONS, EXTERNAL (Tu-22M-3):

Wing span: fully spread	34.28 m (112 ft 5¾ in)
fully swept	23.30 m (76 ft 5½ in)
Wing aspect ratio: fully spread	6.40
fully swept	3.09
Length overall	42.46 m (139 ft 3¾ in)
Height overall	11.05 m (36 ft 3 in)
Weapons bay: Length	approx 7.00 m (22 ft 11½ in)
Width	approx 1.80 m (5 ft 10¾ in)

AREAS:

Wings, gross: 20° sweep	183.58 m² (1,976.1 sq ft)
65° sweep	175.8 m² (1,892.4 sq ft)

WEIGHTS AND LOADINGS (Tu-22M-3):

Max weapon load	24,000 kg (52,910 lb)

Tupolev Tu-22M-3 'Backfire-C' of the Russian Air Force *(Peter J. Cooper)* ***1994***

Fuel load approx 50,000 kg (110,230 lb)
Max T-O weight 124,000 kg (273,370 lb)
Max T-O weight with JATO 126,400 kg (278,660 lb)
Normal landing weight 78,000 kg (171,955 lb)
Max landing weight 88,000 kg (194,000 lb)
Max wing loading (without JATO):
20° sweep 675.45 kg/m² (138.34 lb/sq ft)
65° sweep 705.35 kg/m² (144.45 lb/sq ft)
Max power loading (without JATO)
253 kg/kN (2.48 lb/lb st)

PERFORMANCE (Tu-22M-3):
Max level speed: at high-altitude
Mach 1.88 (1,080 kts; 2,000 km/h; 1,242 mph)
at low altitude
Mach 0.86 (567 kts; 1,050 km/h; 652 mph)
Nominal cruising speed at height
485 kts (900 km/h; 560 mph)
T-O speed 200 kts (370 km/h; 230 mph)
Normal landing speed 154 kts (285 km/h; 177 mph)
Service ceiling 13,300 m (43,635 ft)
T-O run 2,000-2,100 m (6,560-6,890 ft)
Normal landing run 1,200-1,300 m (3,940-4,265 ft)
Unrefuelled combat radius:
supersonic, hi-hi-hi, 12,000 kg (26,455 lb) weapons
810-1,000 n miles (1,500-1,850 km; 930-1,150 miles)
subsonic, lo-lo-lo, 12,000 kg (26,455 lb) weapons
810-900 n miles (1,500-1,665 km; 930-1,035 miles)
subsonic, hi-lo-hi, 12,000 kg (26,455 lb) weapons
1,300 n miles (2,410 km; 1,495 miles)
subsonic, hi-hi-hi, max weapons
1,188 n miles (2,200 km; 1,365 miles)
g limit +2.5

VERIFIED

Tuplolev Tu-22M-2 'Backfire-B' long-range bomber

1994

TUPOLEV Tu-95 and Tu-142

NATO reporting name: Bear

TYPE: Four-turboprop long-range bomber and maritime reconnaissance aircraft.

PROGRAMME: **Tu-95/1** prototype, with four 8,950 kW (12,000 ehp) Kuznetsov 2TV-2F turboprops, flew 1952, was destroyed during testing. **Tu-95/2** with 8,950 kW (12,000 ehp) TV-12 turboprops, flew 1955. Seven **Tu-95**s (NATO 'Bear-A') took part in 1955 Aviation Day flypast; operational with strategic attack force 1956; **Tu-95M** ('Bear-A') was modernised production version; experimental **Tu-95K** of 1956 airdropped the MiG-19 SM-20 aircraft equipped to test features of the Kh-20 missile system; production **Tu-95K-20** ('Bear-B') of 1959 was armed with a Kh-20 (AS-3 'Kangaroo') air-to-surface missile; **Tu-95KD** of 1961 was similar to Tu-95K-20, with added flight refuelling nose probe. 'Bear' series remained in almost continuous, latterly small scale, production for 38 years, ending 1992. Variants included **Tu-96** of 1956, a high-altitude high-speed bomber development with NK-16 engines, built but not flown; **Tu-116/114D**, with Tu-95 airframe adapted as civil aircraft; **Tu-119**, a Tu-95M converted but not flown as testbed for a nuclear engine; and production **Tu-126** (NATO 'Moss') AEW&C aircraft (see 1990-91 *Jane's All the World's Aircraft*).

VERSIONS: **Tu-95RTs** ('Bear-D'): Maritime reconnaissance aircraft, first identified 1967; glazed nose; undernose radar (NATO 'Short Horn'); large underbelly radome for I-band surface search radar ('Big Bulge'); elint blister fairing each side of rear fuselage; nose refuelling probe; variety of blisters and antennae, including streamlined fairing on each tailplane tip. Defensive armament comprises three pairs of 23 mm NR-23 guns in remotely controlled rear dorsal and ventral barbettes and manned tail turret; two glazed blisters on rear fuselage, under tailplane, used for sighting by gunner controlling all these guns; dorsal and ventral barbettes can also be controlled from station aft of flight deck. Housing for I-band tail warning radar ('Box Tail') above tail turret is larger than on previous variants; no offensive weapons; tasks include pinpointing of maritime targets for missile launch crews on ships and aircraft that are themselves too distant for precise missile aiming and guidance; about 15 operational, probably converted from 'Bear-A' strategic bombers. A 'Bear-D' was the first Tu-95 seen, 1978, with faired tailcone housing special equipment instead of normal tail turret and radome.

Tu-95MR ('Bear-E'): Strategic reconnaissance conversion of Tu-95M ('Bear'A); armament; refuelling probe and rear fuselage elint fairings as Tu-95RTs; six camera windows in weapons bay, in pairs.

Tu-142 ('Bear-F'): Anti-submarine version; prototype flew 1968; first of extensively redesigned Tu-142 series; double-slotted flaps; longer fuselage forward of wings; rudder of increased chord. Deployed initially by Naval Aviation 1970; re-entered production mid-1980s. Initial 'Bear-Fs' had 12-wheel main landing gear bogies, retracting into enlarged and lengthened fairings aft of inboard engine nacelles, and undernose radar; main underfuselage J-band radar housing considerably farther forward than on 'Bear-B' and smaller; no large blister fairings under and on sides of rear fuselage; two cargo compartments for sonobuoys, torpedoes, nuclear or conventional depth charges in rear fuselage, one replacing usual rear ventral gun turret and leaving tail turret as sole defensive gun position. Later 'Bear-Fs' identified as follows:

Mod 1: Reverted to standard size nacelles and standard four-wheel main landing gear bogies; chin-mounted J-band radar deleted; fewer protrusions.

Mod 2 (**Tu-142M**): First flown 1975; nose lengthened by 2 m; roof of flight deck raised; angle of flight refuelling probe lowered by 4°; inertial navigation system standard.

Mod 3: MAD boom added to fin tip; fairings at tailplane tips deleted.

Mod 4: Chin radar reinstated; self-protection ECM thimble radome on nose; other fairings added; observation blister each side of rear fuselage deleted; entered service 1985; further deliveries 1991.

All versions of Tu-142M can carry eight Kh-35 active radar homing anti-ship missiles in underwing pairs; IOC is scheduled for 1994.

Tu-95K-22 ('Bear-G'): Bomber and elint conversion of 'Bear-B/C'); two Kh-22 (AS-4 'Kitchen') air-to-surface missiles, on pylon under each wingroot; new large undernose radome ('Down Beat'); ECM thimble is located in fuselage nose; 'solid' tailcone containing special equipment, as on some 'Bear-Ds'; ventral gun turret sole defensive armament.

Tu-95M-5: Missile carrier, with two KSR-5 (AS-6 'Kingfish') missiles, flown 1972.

Tu-95M-55: Flown 1976; carrier for unidentified missile.

Tu-95MS ('Bear-H'): Late production version; crew of seven; based on Tu-142 airframe but fuselage same length as Tu-95. Initial **Tu-95MS6** version carries six RK-55 (AS-15A 'Kent') long-range cruise missiles on an internal rotary launcher; the **Tu-95MS16** carries two more under each wingroot, and a cluster of three between each pair of engines, for a total of 16. Built at Kuybyshev; achieved IOC 1984; larger and deeper radome built into nose; small fin tip fairing; no elint blister fairings on sides of rear fuselage; ventral gun turret deleted; some aircraft have single twin-barrel 23 mm gun instead of usual pair in tail turret, active electronic jammer; RWR and missile warning receivers; chaff/flare dispensers.

Tu-142MR ('Bear-J'): Identified 1986; Soviet counterpart of US Navy E-6A and EC-130Q Tacamo, with VLF communications avionics to maintain on-station/all-ocean link between national command authorities and nuclear missile armed submarines under most operating conditions; large ventral pod for VLF trailing wire antenna, several kilometres long, under centre-fuselage in weapons

Tupolev Tu-95MS 'Bear-H' strategic bomber ***(Peter J. Cooper)***

1994

bay area; undernose fairing as 'Bear-F' Mod 4; fin tip pod with trailing-edge of kind on some 'Bear-Hs'; satcom dome aft of flight deck canopy; modified 'Bear-F' airframe. With Northern and Pacific Fleets.

Tu-95U: Conversion of Tu-95M for training.

OPERATORS: Versions of the Tu-95 and Tu-142 are in service with the armed forces of the following countries: Russia (250+) and Ukraine (20).

DESIGN FEATURES: Unique large, high-performance, four-turboprop combat aircraft, able to carry largest air-launched missiles and outsize radars; all-swept mid-wing configuration; fuselage same diameter as US Boeing B-29/Soviet Tu-4, with similar crawlway linking crew compartments fore and aft of weapons bay; main landing gear retracts into wing trailing-edge nacelles; contraprops with high tip speeds; wing section SR-5S, thickness/chord ratio 12.5 per cent at root; slight anhedral; sweepback at quarter-chord 37° on inner wings, 35° outer panels.

FLYING CONTROLS: All flying control surfaces hydraulically boosted; three-segment aileron and two-segment area-increasing flap each wing; trim tab in each inboard aileron segment; upper-surface spoiler forward of each inboard aileron; adjustable tailplane incidence; trim tab in rudder and each elevator.

STRUCTURE: All-metal; four spars in each inner wing, three outboard; three boundary layer fences above each wing; circular section semi-monocoque fuselage containing three pressurised compartments; tail gunner's compartment not accessible from others.

LANDING GEAR: Hydraulically retractable tricycle type. Main units consist of four-wheel bogies, with tyres approximately 1.50 m (5 ft) diameter and hydraulic internal expanding brakes. Twin-wheels on steerable nose unit. All units retract rearward, main units into nacelles built on to wing trailing-edge. Retractable tail bumper consisting of two small wheels. Braking parachute may be used to reduce landing run.

POWER PLANT: Four KKBM Kuznetsov NK-12M V turboprops, each 11,033 kW (14,795 ehp); eight-blade contra-rotating reversible-pitch Type AV-60N propellers. Fuel in wing tanks, normal capacity 95,000 litres (25,100 US gallons; 20,900 Imp gallons). Flight refuelling probe above nose extends forward further 0.5 m (1 ft 7 in) as it enters probe; fuel flows through duct on starboard side of front fuselage to main tanks; flush light each side of probe in upper part of nose aids night refuelling.

ACCOMMODATION: Crew of seven; pilots side by side on pressurised flight deck; other four crew in forward compartment face rearward; com operator behind pilot, nav/defensive systems operator further aft on port side; flight engineer behind co-pilot, with spare seat for an observer, sixth flight crew member equivalent to US bombardier/navigator, aft on central seat; gunner in rear turret compartment, with entry via ventral hatch. Remainder of crew enter through hatch in top of nosewheel bay. No ejection seats. Conveyor in flight deck floor carries crew members to hatch in nose-wheel bay, with landing gear lowered, in emergency. Astrodome in roof over sixth crew member.

SYSTEMS: Flight crew accommodation pressurised. Thermal anti-icing of wing and tailplane leading-edges. Gas-turbine APU in dorsal fin, with exhaust above tailplane leading-edge.

AVIONICS: ('Bear-D'): Large I-band radar (NATO 'Big Bulge') in blister fairing under centre-fuselage, for reconnaissance and to provide data on potential targets for anti-shipping aircraft or surface vessels. In latter mode, PPI presentation is datalinked to missile launch station. Four-PRF range J-band circular and sector scan navigation radar (NATO 'Short Horn'). I-band tail warning radar (originally NATO 'Bee Hind'; later 'Box Tail') in housing at base of rudder. SRO-2 IFF (NATO 'Odd Rods'), A-321 ADF, A-322Z Doppler radar, A-325Z/321B Tacan/DME and ILS.

EQUIPMENT ('Bear-D'): Two remotely controlled chaff/flare dispensers.

DIMENSIONS, EXTERNAL ('Bear-F'):

Wing span	51.10 m (167 ft 8 in)
Wing aspect ratio	8.39
Length overall	49.50 m (162 ft 5 in)
Height overall	12.12 m (39 ft 9 in)
Propeller diameter	5.60 m (18 ft 4½ in)

AREAS:

Wings, gross	311.1 m² (3,349 sq ft)

WEIGHTS AND LOADINGS (A: 'Bear-F' Mod 3, B: 'Bear-H'):

Weight empty: B	90,000 kg (198,415 lb)
Max fuel: A	87,000 kg (191,800 lb)
Max T-O weight: A	185,000 kg (407,850 lb)
B	188,000 kg (414,470 lb)
Max landing weight:	
B	135,000 kg (297,620 lb)
Max wing loading: A	594.7 kg/m² (121.8 lb/sq ft)
B	604.3 kg/m² (123.8 lb/sq ft)
Max power loading: A	4.19 kg/kW (6.89 lb/shp)

PERFORMANCE ('Bear-H'):

Max level speed at S/L	350 kts (650 km/h; 404 mph)
at 7,620 m (25,000 ft)	Mach 0.83 (499 kts; 925 km/h; 575 mph)
at 11,600 m (38,000 ft)	Mach 0.78 (447 kts; 828 km/h; 515 mph)
Nominal cruising speed	384 kts (711 km/h; 442 mph)
T-O speed	162 kts (300 km/h; 187 mph)
Landing speed, approx	146 kts (270 km/h; 168 mph)
Service ceiling: normal	12,000 m (39,370 ft)
with max weapons	9,100 m (29,850 ft)
Combat radius with 11,340 kg (25,000 lb) payload:	
unrefuelled	3,455 n miles (6,400 km; 3,975 miles)
with one in-flight refuelling	4,480 n miles (8,300 km; 5,155 miles)

VERIFIED

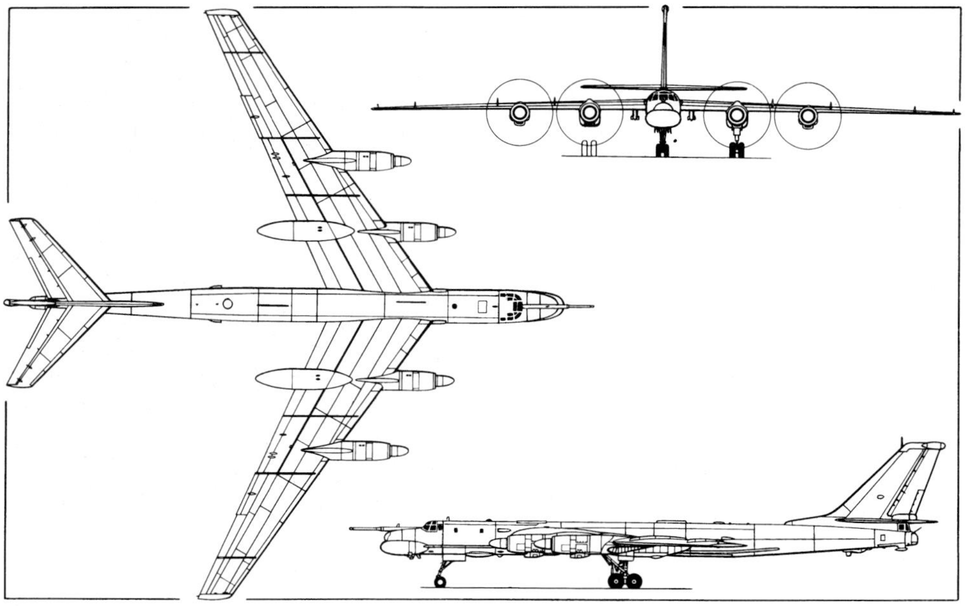

Tupolev Tu-95MS 'Bear-H' strategic bomber *(Jane's/Dennis Punnett)*

1994

Tupolev Tu-95RTS 'Bear-D' over the South China Seas *(G. Jacobs)*

1994

TUPOLEV Tu-134UBL and Tu-134BSh

TYPE: Conversions of Tu-134B airliners (NATO 'Crusty') for training pilots and navigators of CIS strategic aviation.

PROGRAMME: Conversions produced by Kharkov factory; first deliveries of Tu-134UBL to 184th Regiment, flying Tu-160, Priluki air base, early 1991.

VERSIONS: **Tu-134UBL** (uchebno-boevoi dla lotchikov: trainer for pilots): New nose, containing Tu-160 radar, increases overall length nearly 5 m to 41.92 m (137 ft 6½ in); aerodynamic characteristics almost unchanged at angles of attack up to 30° (20° at high-speed); deterioration of longitudinal and lateral stability at greater angles of attack requires reduced maximum T-O weight of 44,250 kg (97,550 lb) from Tu-134B's 47,600 kg (104,935 lb); maximum speed at 10,000 m (32,800 ft) reduced slightly to 464 knots (860 km/h; 534 mph).

Tu-134BSh (bombardirovochno-shturmanskiy: bomber navigational): As Tu-134UBL but Tu-22M radar in nose for fourth year training of student navigators at military school, Tambov; consoles in cabin with bombsight displays for 12 pupils; underwing racks for training bombs.

OPERATORS: Versions of the Tu-134UBL are in service with the armed forces of Russia and Ukraine.

VERIFIED

TUPOLEV Tu-154

NATO reporting name: Careless

TYPE: Three-engined medium-range transport aircraft.

PROGRAMME: The Tu-154, announced Spring 1966, was intended to replace the Tu-104, Il-18 and An-10 on medium/long stage lengths of up to 3,240 n miles (6,000 km; 3,725 miles). It is able to operate from airfields with a class B surface, including packed earth and gravel. Normal flight can be maintained after shutdown of any one engine. Single-engine flight is possible at a lower altitude.

The first of six prototype and preproduction models flew for the first time on 4 October 1968. The seventh Tu-154 was delivered to Aeroflot for initial route proving and crew training early 1971. Mail and cargo flights began in May. Initial passenger carrying services were flown for a few days early Summer 1971 between Moscow and Tbilisi. Regular services began 9 February 1972, over the 700 n miles (1,300 km; 800 mile) route between Moscow and Mineralnye Vody, in the North Caucasus. International services began with a proving flight between Moscow and Prague 1 August 1972.

VERSIONS: **Tu-154:** Initial version announced in 1966.

Tu-154A: Development of the Tu-154. First entered service in 1974. Described separately.

Tu-154B/B-2: Development of the Tu-154A. Entered production in 1977. Described separately.

Tu-154M: For full details see *Jane's All The World's Aircraft* 1995-96.

Tu-154S: Cargo version, announced in 1981 initially as a conversion of the Tu-154B. Described separately.

UPGRADES: **Tupolev:** Tu-154S cargo version. See separate entry.

OPERATORS: Versions of the Tu-154 are in service with the armed forces of the following countries: Belarus (2), Czech Republic (1), Germany (2), Kazakhstan (1), North Korea (3), Poland (1), Russia (20+), Slovakia (1) and Ukraine (5).

The following details apply to the basic Tu-154.

DESIGN FEATURES: Cantilever low-wing monoplane. Sweepback 35° at quarter-chord. Variable incidence tailplane. Sweepback of 40° at quarter-chord on horizontal surfaces, 45° on leading-edge of vertical surfaces.

FLYING CONTROLS: Five-section slat on outer 80 per cent of each wing leading-edge. Triple-slotted flaps. Four-section spoilers forward of flaps on each wing. Outboard sections supplement ailerons for roll control. Section inboard of landing gear housing serves as airbrake and lift dumper; two middle sections can be used as airbrakes in flight. All control surfaces actuated hydraulically. Tail unit control surfaces hydraulically actuated by irreversible servo controls.

STRUCTURE: The wing is a conventional; all-metal three-spar fail-safe structure; centre spar extending to just outboard of inner edge of aileron on each wing. The fuselage is a conventional all-metal semi-monocoque fail-safe structure of circular section. The tail unit is a cantilever all-metal structure. Rudder of honeycomb construction.

SYSTEMS: Hot air de-icing of wing leading-edge. Slats are electrically heated. Leading-edges of fin and tailplane and engine intake de-iced by hot air.

LANDING GEAR: Retractable tricycle type. Hydraulic actuation. Main units retract rearward into fairings on wing trailing-edge. Each consists of a bogie made up of three pairs of wheels, size 930 × 305, in tandem; tyre pressure 7.86 bars (114 lb/sq in). Steerable anti-shimmy twin-wheel nose unit has wheels size 800 × 225 and retracts rearward. Disc brakes and anti-skid units on mainwheels.

POWER PLANT: Three Kuznetsov NK-8-2 turbofan engines, each rated at 93.2 kN (20,950 lb st), one on each side of rear fuselage and one inside extreme rear of fuselage. Two lateral engines fitted with upper and lower thrust reversal grilles. Integral fuel tanks in wings; standard capacity 41,140 litres (10,860 US gallons; 9,050 Imp gallons). Maximum fuel capacity 46,825 litres (12,360 US gallons; 10,300 Imp gallons). Single point refuelling standard.

ACCOMMODATION: Flight crew of two pilots and flight engineer; provision for navigator aft of pilot and folding seats for additional pilots or instructors. There are basic passenger versions for a total of 167, 158, 152, 146 and 128 passengers. Each has a toilet at the front (starboard), removable galley amidships and three toilets aft. Coat storage, folding seat and inflatable evacuation chute in each entrance lobby. Standard economy class version has 54 seats in six-abreast rows, with two tables between front rows, in forward cabin; and 104 seats in six-abreast rows (rear two rows four-abreast) in rear cabin at seat pitch of 75 cm (29.5 in). The 167-seat high density version differs in having one further row of six seats in the forward cabin and reduced galley facilities. The tourist class versions carry 146 passengers at a seat pitch of 81 cm (32 in) or 152 at a pitch of 87 cm (34.25 in) with reduced galley facilities. The 128-seat version has only 24 first class seats, four-abreast at a pitch of 102 cm (40 in), in the forward cabin. There is also an all-cargo version. Passenger doors are forward of front cabin and between cabins on the port side, with emergency and service doors opposite. All four doors open outwards. Four emergency exits, two over wing on each side. Two pressurised baggage holds under main cabin floor, with two inward opening doors. Normal provision for mechanised loading and unloading of baggage and freight in containers. Smaller unpressurised hold under rear cabin for carrying spare parts or special cargo such as radioactive isotopes.

SYSTEMS: Air conditioning system pressure differential 0.62 bars (9.0 lb/sq in). Three independent hydraulic systems; working pressure 207 bars (3,000 lb/sq in). No. 1 system, powered by two pumps driven by centre engine and port engine, operates landing gear, brakes and all control surfaces. No. 2 system, powered by a pump driven by centre engine, actuates nosewheel steering, the second flying controls circuit and landing gear emergency extension. No. 3 system, powered by a pump on starboard engine, actuates the third flying controls circuit and second landing gear emergency extension circuit. Three-phase 200/115 V AC electrical system, supplied by three 40kVA alternators. 28 V DC system. APU standard, driving 40kVA alternator and 12 kW starter/generator.

AVIONICS AND EQUIPMENT: Automatic flight control system standard, including automatic navigation on preprogrammed route under control of navigational computer with en route checks by ground radio beacons (including VOR, VOR/DME) or radar, and automatic approach by ILS to ICAO Cat. II standards (development to Cat. III standard in hand). Moving-map ground position indicator, HF and VHF radio and radar standard. Safety equipment includes inflatable liferafts.

DIMENSIONS, EXTERNAL:

Wing span	37.55 m (123 ft 2½ in)
Length overall	47.90 m (157 ft 1¾ in)
Height overall	11.40 m (37 ft 4¾ in)
Diameter of fuselage	3.80 m (12 ft 5½ in)
Tailplane span	13.40 m (43 ft 11½ in)
Wheel track	11.50 m (37 ft 9 in)
Wheelbase	18.92 m (62 ft 1 in)
Passenger doors (each): Height	1.73 m (5 ft 7 in)
Width	0.80 m (2 ft 7½ in)
Height to sill	3.10 m (10 ft 2 in)
Servicing door: Height	1.28 m (4 ft 2½ in)
Width	0.61 m (2 ft 0 in)
Emergency door: Height	1.28 m (4 ft 2½ in)
Width	0.64 m (2 ft 1¼ in)
Emergency exits (each): Height	0.90 m (2 ft 11½ in)
Width	0.48 m (1 ft 7 in)
Main baggage hold doors (each):	
Height	1.20 m (3 ft 11¼ in)
Width	1.35 m (4 ft 5 in)
Height to sill	1.80 m (5 ft 10¾ in)
Rear (unpressurised) hold:	
Height	0.90 m (2 ft 11½ in)
Width	1.10 m (3 ft 7¼ in)
Height to sill	2.20 m (7 ft 2½ in)

DIMENSIONS, INTERNAL:

Cabin: Width	3.58 m (11 ft 9 in)
Height	2.02 m (6 ft 7½ in)
Volume	163.20 m^3 (5,763 cu ft)
Main baggage holds: front	21.50 m^3 (759 cu ft)
rear	16.50 m^3 (582 cu ft)
Rear underfloor hold	5.00 m^3 (176 cu ft)

AREAS:

Wings, gross	201.45 m^2 (2,169 sq ft)
Horizontal tail surfaces	40.55 m^2 (436.48 sq ft)
Vertical tail surfaces	31.72 m^2 (341.43 sq ft)

WEIGHTS AND LOADINGS:

Operating weight empty	43,500 kg (95,900 lb)
Normal payload	16,000 kg (35,275 lb)
Max payload	20,000 kg (44,090 lb)
Max fuel	33,150 kg (73,085 lb)
Max ramp weight	90,300 kg (199,077 lb)
Normal T-O weight	84,000 kg (185,188 lb)
Max T-O weight	90,000 kg (198,416 lb)
Normal landing weight	68,000 kg (149,915 lb)
Max landing weight	80,000 kg (176,370 lb)
Max zero-fuel weight	63,500 kg (139,994 lb)

PERFORMANCE (at max T-O weight, except where indicated):

Max level speed:
above 11,000 m (36,000 ft) Mach 0.90
at low altitudes 283 kts (525 km/h; 326 mph) IAS

Max cruising speed at 9,500 m (31,150 ft)
526 kts (975 km/h; 605 mph)

Best-cost cruising speed at 11,000-12,000 m (36,000-39,350 ft)
Mach 0.85 (486 kts; 900 km/h; 560 mph)

Long-range cruising speed at 11,000-12,000 m (36,000-39,350 ft)
Mach 0.80 (459 kts; 850 km/h; 528 mph)

Approach speed 127 kts (235 km/h; 146 mph)
Min ground turning radius 24.60 m (80 ft 8½ in)
T-O run at normal T-O weight, ISA 1,140 m (3,740 ft)

Balanced runway length at max T-O weight, FAR standard: ISA, S/L 2,100 m (6,890 ft)
ISA + 20°C, S/L 2,420 m (7,940 ft)

Landing field length at max landing weight, FAR standard:
ISA, S/L 2,060 m (6,758 ft)
ISA + 20°C, S/L 2,217 m (7,273 ft)

Range at 11,000 m (36,000 ft) with standard fuel, reserves for 1 h and 6% of total fuel:
at 486 kts (900 km/h; 560 mph), with T-O weight of 84,000 kg and max payload (158 passengers, baggage, and 5 tonnes of cargo and mail)
1,360 n miles (2,520 km; 1,565 miles)
as above, T-O weight of 90,000 kg
1,867 n miles (3,460 km; 2,150 miles)
as 459 kts (850 km/h; 528 mph), with T-O weight of 84,000 kg and max payload as above
1,510 n miles (2,800 kg; 1,740 miles)
as above, T-O weight of 90,000 kg
2,050 n miles (3,800 km; 2,360 miles)
max range with 13,650 kg (30,100 lb) payload
2,850 n miles (5,280 km; 3,280 miles)

Range at 11,000 m (36,000 ft) with optional centre-wing tanks, reserves as above:
with 9,000 kg (19,840 lb) payload (95 passengers)
3,453 n miles (6,400 km; 3,977 miles)
with 6,700 kg (14,770 lb) payload (70 passengers)
3,723 n miles (6,900 km; 4,287 miles)

UPDATED

TUPOLEV Tu-154A and Tu-154B

NATO reporting name: Careless

TYPE: Three-engined, medium-range transport aircraft.

PROGRAMME: A developed version of the Tu-154, with the designation **Tu-154A**, was reported in early 1973, with the first flight scheduled for later that year. An article in the April 1975 issue of the Soviet magazine *Grazhdanskaya Aviatsiya* recorded that this aircraft had entered service with Aeroflot in April 1974 and that production Tu-154s were to be put into scheduled operation during 1975.

The Tu-154A is dimensionally unchanged in comparison to the original model, and is able to carry a normal payload of 152 passengers in Summer and 144 in Winter. Alternative configurations provide seats for 168 passengers on high density routes, or 12 first class and 128 tourist class. Changes have centred mainly on the power plant, equipment and systems, to permit an increased gross weight, improve performance and reliability, and reduce servicing requirements.

The power plant consists of three Kuznetsov NK-8-2U turbofan engines, each uprated to 103 kN (23,150 lb st). Increased maximum take-off and landing weights allow extra fuel to be carried, raising the maximum capacity to 39,750 kg (87,630 lb). An additional tank, capacity 6,600 kg (14,550 lb), is mounted between the front and centre spars in the centre-section. It is intended primarily as a ballast tank for ferrying, and the fuel it contains can be pumped into the main system only on the ground. When the aircraft carries less than a full payload, this tank can be filled and its contents can be transferred to the main tanks at a destination airport, so reducing purchases of fuel outside the operator's home country. Other fuel system improvements have been made to the anti-icing fluid additive system; the centre-section tanks can be purged with CO_2 in the event of a forced landing with the wheels retracted.

The controls for the flaps, leading-edge slats and tailplane are interconnected, so that when the flaps are operated the tailplane is trimmed 3° down. An override switch caters for CG conditions which require a movement of more than 3°.

Additional emergency exits in the rear fuselage meet international requirements. The floor of the baggage holds has been strengthened to prevent damage by sharp edged packages and baggage; and a smoke warning system has been introduced in the holds.

The electrical system has been modified in comparison with the Tu-154 and employs three alternators, on separate supply circuits, to provide 200/115 V AC power. Two circuits supply all electrical services; the third supplies the electrical anti-icing system for the leading-edge slats. If one alternator fails, the remaining primary alternator can provide for all essential services, supplemented by the alternator on the APU. The duplicated DC electrical system embodies three rectifiers, of which one is for emergency use in the event of a failure of either of the others.

An ABSU automatic approach and landing system is fitted. This met ICAO Cat. I requirements initially, but was to be uprated to Cat. II later. Other equipment changes include the provision of duplicated radio compass, radio altimeter and DME; and the introduction of two-speed windscreen wipers and a system to indicate angle of bank limitations. An MSRP-64 flight recorder covers some 80 parameters, and a Mars-B voice recorder with open microphone is standard.

Tupolev Tu-154B with Kuznetsov NK-8-2U turbofans and modified spoilers for improved lateral control at low speeds *(Peter J. Cooper)*

1994

Servicing requirements and costs were reduced considerably on the Tu-154A, for which the servicing cycle is 300, 900 and 1,800 hours.

In 1977, production was switched to a further improved version, designated **Tu-154B**, since refined as the **Tu-154B-2**. This retains the NK-8-2U turbofans of the Tu-154A, but has modified spoilers for improved lateral control at low speeds, and is fitted with Thomson-CSF/SFIM automatic flight control and navigation equipment approved for ICAO Cat. II automatic landings. Maximum take-off and zero-fuel weights were increased; and rearward extension of the usable cabin space enables 154-180 passengers to be carried despite the introduction of two more emergency exit doors, immediately forward of the engine nacelle on each side, making a total of six.

In the typical 169-seat high density configuration, the passengers are seated mostly six-abreast with centre aisle at a pitch of 75 cm (29.5 in), with 68 in the forward cabin and 101 in the rear cabin. The forward door, on the port side between the flight deck and passenger accommodation, leads into a vestibule with a galley/pantry on the port side and a toilet opposite. There are three more toilets at the rear of the cabin, and each of the two doorways contain three seats for cabin staff. The basic all-tourist version differs in having 62 seats forward and 98 aft, with added coat stowage opposite the forward door and a further galley amidships. In the 154-seat mixed class layout, the forward cabin is divided in two, with a separate lounge for up to 24 first class passengers at 96 cm (38 in) seat pitch.

Improvements were made to the avionics, notably to simplify take-off and landing procedures. A different radar is fitted, and the fuel tank used as ballast on the Tu-154A can be used normally, as part of the standard fuel system of the Tu-154B/B-2. The main landing gear units are fitted with small hydraulic jacks which swivel the front axles to compensate for yaw on the runway on landing, and so reduce tyre wear.

WEIGHTS AND LOADINGS (A, Tu-154A; B, Tu-154B):
Basic operating weight: B 50,775 kg (111,940 lb)
Normal payload: A 16,000 kg (35,275 lb)
Max payload: A 18,000 kg (39,680 lb)
B 19,000 kg (41,887 lb)
Max fuel: B 39,750 kg (87,633 lb)
Max T-O weight: A 94,000 kg (207,235 lb)
B 98,000 kg (216,050 lb)
Max zero-fuel weight: B 71,000 kg (156,525 lb)
Max landing weight (normal):
A, B 78,000 kg (171,960 lb)
Max landing weight (emergency):
A 92,000-94,000 kg (202,825-207,235 lb)

PERFORMANCE (A, Tu-154A; B, Tu-154B, at max T-O weight except where indicated):
Max level speed:
A 310 kts (575 km/h; 357 mph) IAS
except with less than 7,150 kg (15,763 lb) fuel at heights above 7,000 m (23,000 ft)
Normal cruising speed at up to 12,000 m (39,370 ft):
A Mach 0.85 (486 kts; 900 km/h; 560 mph)
B 486-513 kts (900-950 km/h; 560-590 mph)
Required runway length: B 2,200 m (7,218 ft)
Range with payload of 16,000 kg (35,275 lb):
A 1,725-1,780 n miles (3,200-3,300 km; 1,985-2,050 miles)
Range with 120 passengers and baggage, with reserves:
B 2,160 n miles (4,000 km; 2,485 miles)
Range with max payload, with reserves:
B 1,485 n miles (2,750 km; 1,705 miles)

VERIFIED

Tupolev Tu-154B-1 three-engined medium-range transport aircraft *(Peter Cooper)*
1995

Tupolev Tu-154B-2 with modified spoilers for improved lateral control at low speeds *(Peter Cooper)*
1995

TUPOLEV Tu-154S

TYPE: Three-engine, medium-range cargo aircraft.

PROGRAMME: This freight carrying version of the Tu-154 was announced Autumn 1982. It is being offered initially as a conversion of the Tu-154B, with an unobstructed cargo volume of 72 m³ (2,542 cu ft) in the main hold. A freight door 2.80 m (9 ft 2¼ in) wide and 1.87 m (6 ft 1½ in) high is installed in the port side of the hold, forward of the wing, with a ball mat inside and roller tracks the full length of the floor of the hold. Typical loads include nine standard international pallets measuring 2.24 × 2.74 m (88 × 108 in), plus additional freight in the standard underfloor baggage holds which have a volume of 38 m³ (1,341 cu ft). Nominal range of the Tu-154C, with 20,000 kg (44,100 lb) of cargo, is 1,565 n miles (2,900 km; 1,800 miles).

VERIFIED

TUPOLEV Tu-160

NATO reporting name: Blackjack

TYPE: Four-engined variable geometry long-range strategic bomber.

PROGRAMME: Designed as Aircraft 70 under leadership of V. I. Bliznuk; prototype observed by satellite at Ramenskoye flight test centre 25 November 1981 (photograph in 1982-83 *Jane's All the World's Aircraft*), first flew 19 December 1981; US Defense Secretary Frank Carlucci invited to inspect 12th aircraft built, at Kubinka air base, near Moscow, 2 August 1988; deliveries to 184th Regiment, Priluki air base, Ukraine, began May 1987; production at Kazan airframe plant ended 1992.

UPGRADES: **Tupolev.** Tu-160SC. Space launch vehicle. See following entry.

OPERATORS: Versions of the Tu-160 are in service with the armed forces of the following countries: Russia (21) and Ukraine (20).

DESIGN FEATURES: Intended for high-altitude standoff role carrying ALCMs and for defence suppression, using short-range attack missiles similar to US Air Force SRAMs, along path of bomber making low-altitude penetration to attack primary targets with free-fall nuclear bombs or missiles; this implies capability of subsonic cruise/supersonic dash at almost Mach 2 at 18,300 m (60,000 ft) and transonic flight at low-altitude. About 20 per cent longer than USAF B-1B, with greater unrefuelled combat radius and much higher maximum speed; low-mounted variable geometry wings, with very long and sharply swept fixed root panel; small diameter circular fuselage; horizontal tail surfaces mounted high on fin, upper portion of which is pivoted one-piece all-moving surface; large dorsal fin; engines mounted as widely separated pairs in underwing ducts, each with central horizontal V wedge intakes and jetpipes extending well beyond wing centre-section trailing-edge; manually selected outer wing sweepback 20°, 35° and 65°; when wings fully swept, inboard portion of each trailing-edge flap hinges upward and extends above wing as large fence; unswept tail fin; sweptback horizontal surfaces, with conical fairing for brake-chute aft of intersection.

FLYING CONTROLS: Fly-by-wire. Full-span leading-edge flaps, long-span double-slotted trailing-edge flap and inset drooping aileron on each wing; all-moving vertical and horizontal one-piece tail surfaces.

STRUCTURE: Slim and shallow fuselage blended with wingroots and shaped for maximum hostile radar signal deflection.

LANDING GEAR: Twin nosewheels retract rearward; main gear comprises two bogies, each with three pairs of wheels; retraction very like that on Tu-154 airliner; as each leg pivots rearward bogie rotates through 90° around axis of centre pair of wheels, to lie parallel with retracted leg; gear retracts into thickest part of wing, between fuselage and inboard engine on each side; so track relatively small. Nosewheel tyres size 1,080 × 400 mm; mainwheel tyres size 1,260 × 425 mm.

POWER PLANT: Four Samara/Trud NK-321 turbofans, each 245 kN (55,115 lb st) with afterburning. In-flight refuelling probe retracts into top of nose.

ACCOMMODATION: Four crew members in pairs, on individual K-36 ejection seats; one window each side of flight deck can be moved inward and rearward for ventilation on ground; flying controls use fighter type sticks rather than yokes or wheels; crew enter via nosewheel bay.

AVIONICS: Radar in slightly upturned dielectric nosecone claimed to provide terrain-following capability; fairing with flat glazed front panel, under forward fuselage, for video camera to provide visual assistance for weapon aiming; astro-inertial nav with map display; no head-up display or CRTs; active jamming self-defence system.

ARMAMENT: No guns. Internal stowage for up to 16,330 kg (36,000 lb) of free-fall bombs, short-range attack missiles or ALCMs; a rotary launcher can be installed in each of two 12.80 m (42 ft) long weapon bays, carrying 12 Kh-15P

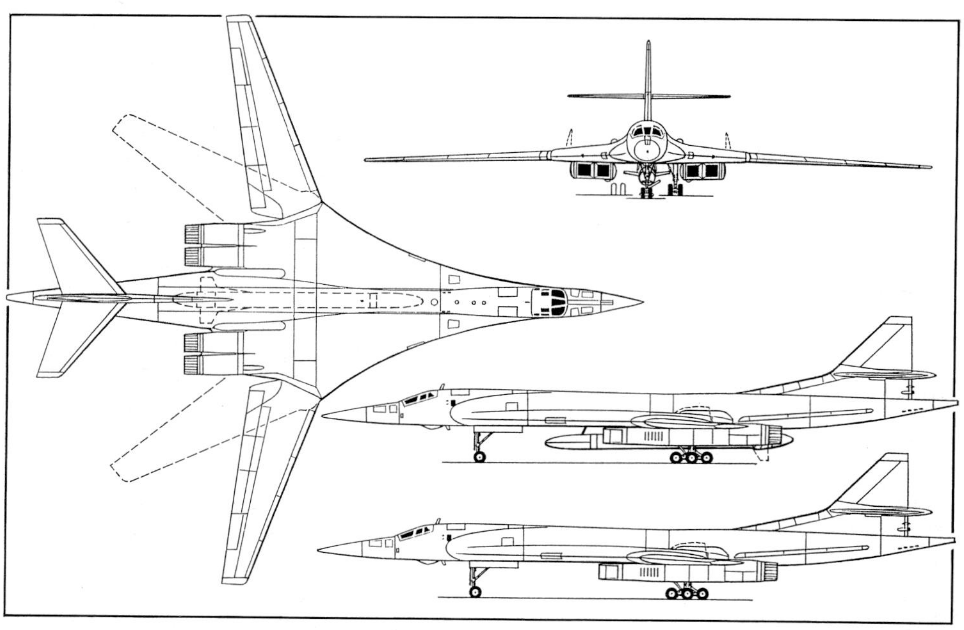

Tupolev Tu-160 strategic bomber with additional side view of Tu-160SC *(Jane's/Dennis Punnett)*
1995

(AS-16 'Kickback') SRAMs or six ALCMs, currently RK-55s (AS-15 'Kent'), intended to be superseded by supersonic AS-19s ('Koala').

DIMENSIONS, EXTERNAL:

Wing span: fully spread (20°)	55.70 m (182 ft 9 in)
35° sweep	50.70 m (166 ft 4 in)
fully swept (65°)	35.60 m (116 ft 9¾ in)
Length overall	54.10 m (177 ft 6 in)
Height overall	13.10 m (43 ft 0 in)
Tailplane span	13.25 m (43 ft 5¾ in)
Wheel track	5.40 m (17 ft 8½ in)
Wheelbase	17.88 m (58 ft 8 in)

AREA:

Wings gross: fully spread	232.0 m² (2,497 sq ft)

WEIGHTS AND LOADINGS:

Weight empty	110,000 kg (242,500 lb)
Max fuel	160,000 kg (352,735 lb)
Max weapon load	16,330 kg (36,000 lb)
Max T-O weight	275,000 kg (606,260 lb)
Max power loading	280 kg/kN (2.75 lb/lb st)

PERFORMANCE:

Max level speed at 12,200 m (40,000 ft)	Mach 2.05 (1,200 kts; 2,220 km/h; 1,380 mph)
Cruising speed at 13,700 m (45,000 ft)	Mach 0.9 (518 kts; 960 km/h; 596 mph)
Max rate of climb at S/L	4,200 m (13,780 ft)/min
Service ceiling	15,000 m (49,200 ft)
T-O run at max AUW	2,200 m (7,220 ft)
Landing run at max landing weight	1,600 m (5,250 ft)
Radius of action at Mach 1.5	1,080 n miles (2,000 km; 1,240 miles)
Max unrefuelled range	6,640 n miles (12,300 km; 7,640 miles)
g limit	+2

VERIFIED

Tupolev Tu-160 'Blackjack' variable geometry strategic bomber *(Piotr Butowski)*

1994

TUPOLEV Tu-160SC

TYPE: Conversion of Tu-160 bomber as space launch vehicle.

PROGRAMME: Announced at Asian Aerospace '94, Singapore; no satellite launch contracts by eary 1994.

COSTS: Payload launch cost $6,000-8,000/kg.

DESIGN FEATURES: Commercialised verson of Tu-160 bomber (which see), as carrier component of Burlak aviation space launch complex; Burlak two-stage rocket, carrying payload, under fuselage on centreline mount. Design payload injection into orbit at height of 200-1,000 km (124-620 miles): 800-600 kg (1,765-1,323 lb) into polar orbit; 1,100-840 kg (2,425-1,850 lb) into equatorial orbit.

ACCOMMODATION: Crew of four.

DIMENSIONS, EXTERNAL: As Tu-160 bomber, except:

Burlak rocket: Length overall	21.40 m (70 ft 2½ in)
Diameter	1.50 m (4 ft 11 in)
Burlak payload module: Length	3.80 m (12 ft 5½ in)
Diameter	1.40 m (4 ft 7¼ in)

WEIGHTS AND LOADINGS:

Burlak launch weight	30,000 kg (66,135 lb)
Max T-O weight	275,000 kg (606,260 lb)

PERFORMANCE:

Speed for Burlak launch: at 9,000-11,000 m (29,500-36,000 ft)	Mach 0.8
at 12,000-13,000 m (39,350-42,650 ft)	Mach 1.7
Nominal cruising speed	Mach 0.77 (459 kts; 850 km/h; 528 mph)
Required runway length	3,500 m (11,500 ft)

NEW ENTRY

Tupolev Tu-160 'Blackjack' strategic bomber on approach

1994

YAKOVLEV

JOINT-STOCK COMPANY A.S. YAKOVLEV DESIGN BUREAU

68 Leningradsky Prospekt, 125315 Moscow
Telephone: 7 (095) 157 5737
Fax: 7 (095) 157 4726

CHAIRMAN AND GENERAL DESIGNER: Alexander N. Dondukov
PRESIDENT AND GENERAL DIRECTOR: Oleg F. Demchenko
FIRST VICE-PRESIDENT AND FIRST DEPUTY GENERAL DESIGNER: Vladimir G. Dmitriev
DEPUTY GENERAL DIRECTOR (EXTERNAL ECONOMIC RELATIONS), VICE-PRESIDENT: Arkady I. Gurtovoy
DEPUTY GENERAL DIRECTOR (FLIGHT TESTS): Andrei A. Sinitsin
VICE-PRESIDENTS AND CHIEF DESIGNERS:
Vladimir G. Dimitriev (Yak-242, Yak-46)
Nicolas N. Dolzhenkov (Yak-130)
Vladimir A. Mitkin (Yak-48/IAI Galaxy)
Alexei G. Rakhimbaev (Yak-42, Yak-142)
CHIEF DESIGNERS:
Sergoi A. Yakovlev (Yak-3, Yak-40 TL, Yak-77)
Dmitry K. Drach (Yak-54, Yak-55M, Yak-112)
Yuri I. Yankevich (Yak-18T, Yak-52, Yak-58, UAV)
CHIEF OF DEPARTMENTS:
MARKETING AND SALES: Anatoly S. Ivanov
EXTERIOR: Evgeny M. Tarasov
INFORMATION/PRESS SERVICE: Yuri V. Zasypkin

Founder of this OKB, Alexander Sergeyevich Yakovlev, died on 22 August 1989, aged 83. He was one of most versatile Soviet designers. Products of his OKB ranged from transonic long-range fighters to Yak-24 tandem-rotor helicopter, an operational V/STOL carrier-based fighter and a variety of training, competition aerobatic and transport aircraft.

In 1992 Yak Corporation was established to manage aircraft marketing, design, development, production, sales and aftersales support. The Corporation includes A. S. Yakovlev Design Bureau as well as the Saratov and Smolensk airframe plants. In 1991 Industrial Association Skorost created to cooperate in production of the Yak-242. In 1994 Hyundai-Yak Aerospace Co Ltd was created. This is a joint venture corporation with South Korean Hyundai Group which is aimed at developing, marketing and selling a wide range of small and medium-sized aircraft. Collaborative agreements have been concluded with Israel Aircraft Industries on Yak-48/Astra Galaxy design and production, and with Aermacchi (Italy) on Yak-130.

UPDATED

YAKOVLEV Yak-38

NATO reporting name: Forger

TYPE: Single-seat ship-based V/STOL combat aircraft and two-seat trainer.

PROGRAMME: Prototype **Yak-36M** flew 1971; production began 1975 as world's second operational V/STOL jet combat aircraft after UK Harrier; deployed with Navy development squadron on *Kiev*, first of class of four 40,000 ton carrier/cruisers to put to sea 1976; subsequently 12 production Yak-38s on this ship and on each of sister ships, *Minsk, Novorossiysk* and *Admiral of the Fleet Gorshkov* (then *Baku*), later, on 65,000 ton aircraft carrier *Admiral of the Fleet Kuznetsov*; experimental operation from specially configured ro-ro ships reported. In process of retirement from service.

VERSIONS: **Yak-38** ('Forger-A'): Basic single-seat combat aircraft; primary roles reconnaissance, strikes against small ships and fleet defence against shadowing maritime reconnaissance aircraft.

Yak-38M ('Forger-A'): Developed single-seater; Soyuz/Tumansky R-28V-300 turbojet (65.7 kN; 14,770 lb st); each liftjet uprated to 31.9 kN (7,165 lb st); areas of engine air intakes and liftjet compartment correspondingly increased. Operational since 1984; one of first Yak-38Ms instrumented to check new features shown at 1992 Farnborough Air Show.

Yak-38U ('Forger-B'): Two-seat trainer; second cockpit forward of normal cockpit, with ejection seat at lower level under continuous transparent canopy; to compensate for longer nose, plug inserted in fuselage aft of wing, lengthening constant-section portion without modification of tapering rear fuselage; no ranging radar or weapon pylons; two on each 'Kiev' class ship.

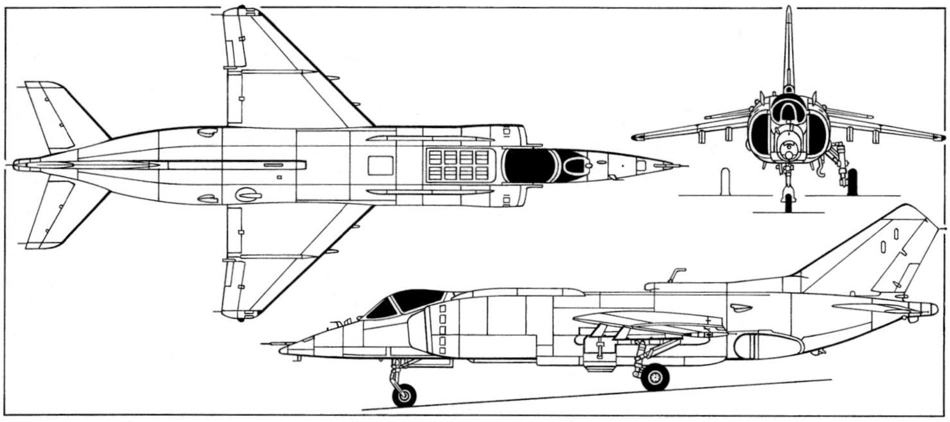

Yakovlev Yak-38M 'Forger-A' single-seat V/STOL carrier-based combat aircraft *(Jane's/Mike Keep)* *1994*

Yakovlev Yak-38M 'Forger-A' single-seat V/STOL aircraft *(Peter J. Cooper)* *1994*

Yakovlev Yak-38 'Forger-A' single-seat V/STOL aircraft *(Peter J. Cooper)* *1994*

OPERATORS: Versions of the Yak-38 are in service with the armed forces of Russia (30).

DESIGN FEATURES: Very small mid-mounted folding wings; thickness/chord ratio estimated maximum 6 per cent; constant anhedral from roots; leading-edge sweepback approximately 45°; jet reaction control valve with upper and lower slots in each wingtip; basically oval section fuselage; integral engine air intake ducts, with boundary layer splitter plates and downward inclined lips; row of small blow-in auxiliary intake doors short distance aft of each intake; rearward hinged door over liftjets, immediately aft of canopy, with 16 spring loaded louvres; location of corresponding side-hinged underfuselage doors conforms with forward tilt of lift engines; positions of these doors controlled automatically during take-off and landing as part of control system; fence on each side of door above liftjets, presumably to prevent ingestion of reflected exhaust efflux; yaw reaction control nozzle to each side of small tailcone; no reaction control system in nose; all-swept tail unit, with tailplane anhedral; air intake at front of long duct extending forward from base of fin, to cool avionics bay in rear fuselage.

FLYING CONTROLS: Aircraft appears to be extremely stable during take-off and landing; initially, take-off was always vertical, with rear vectored thrust nozzles up to 10° forward of vertical; this was followed by smooth conversion approximately 5-6 m (15-20 ft) above deck, achieved by lowering aircraft's nose about 5° below horizon and maintaining this attitude until aircraft accelerated to 30-40 knots (55-75 km/h; 35-46 mph); at this speed, 5° nose-up attitude was assumed, and accelerating transition continued by vectoring aft the nozzles of the propulsion engine.

This VTO technique superseded by STOL take-off, with short forward run, made possible by automatic control system which ensures lift engines are brought into use, and thrust vectoring rear nozzles rotated, at optimum point in take-off run; STOL take-off can be assumed to offer improved payload/range.

Landing procedure begins with gradual descent from far astern, with final 400 m (1,300 ft) flown essentially level, about 30 m (100 ft) above water; aircraft crosses ship's stern with about 5 knot (10 km/h; 6 mph) closure rate, 10-14 m (35-45 ft) above flight deck, then flares gently to hover and descends vertically; precise landings are ensured by automatic control system, perhaps in association with laser devices lining each side of rear deck. Fully automatic control system ensures synchronisation of engine functioning, aerodynamic control operation, jet reaction nozzle operation, stabilisation and guidance.

Conventional aileron on each outer wing panel, with setback hinges and inset trim tab; large single-slotted Fowler flap on each inboard panel; no leading-edge flaps or slats; rudder and each elevator have setback hinges and trim tab.

STRUCTURE: All-metal; each wing comprises two light alloy panels of approximately equal span; outer panel folds vertically upward for shipboard stowage; semi-monocoque fuselage and tail unit of light alloy.

LANDING GEAR: Retractable tricycle type. Single wheel on each unit, with legs of trailing link type with oleo-pneumatic shock absorption. Nose unit retracts rearward, main units forward into fuselage. Small bumper under upward curving rear fuselage.

POWER PLANT: Primary power plant is a Tumansky R-27V-300 turbojet (68 kN; 15,300 lb st), mounted in the centre-fuselage and exhausting through a single pair of hydraulically actuated vectoring side nozzles aft of the wings. No afterburner is fitted. Two RKBM RD-36-35FVR liftjets (each approximately 30 kN; 6,725 lb st) in tandem immediately aft of cockpit, inclined forward at 13° from vertical, exhausting downward, and used also to adjust pitch and trim. Fuel tanks in fuselage, forward and aft of main engine. Drop tanks, each estimated to have capacity of 600 litres (158 US gallons; 132 Imp gallons), can be carried on under-wing pylons.

ACCOMMODATION: Pilot only, on zero/zero ejection seat under sideways hinged (to starboard) transparent canopy. Electronic system ejects pilot automatically if aircraft pitch/roll angles and pitch/roll rates indicate emergency.

AVIONICS: Ranging radar in nose. IFF (NATO 'Odd Rods') antennae forward of windscreen. Other avionics in rear fuselage.

ARMAMENT: No installed armament. Two pylons under fixed panel of each wing for 2,600-3,600 kg (5,730-7,935 lb) of external stores, including gun pods each containing a 23 mm twin-barrel GSh-23 cannon, rocket packs, bombs weighing up to 500 kg each, AS-7 short-range air-to-surface missiles (NATO 'Kerry'), armour-piercing anti-ship missiles, AA-8 air-to-air missiles ('Aphid') and auxiliary fuel tanks.

DIMENSIONS, EXTERNAL (estimated):

Wing span	7.32 m (24 ft 0 in)
Width, wings folded	4.88 m (16 ft 0 in)
Length overall: 'Forger-A'	15.50 m (50 ft 10¼ in)
'Forger-B'	17.68 m (58 ft 0 in)
Height overall	4.37 m (14 ft 4 in)
Tailplane span	3.81 m (12 ft 6 in)
Wheel track	2.90 m (9 ft 6 in)
Wheelbase	5.50 m (18 ft 0 in)

AREAS (estimated):

Wings, gross	18.5 m² (199 sq ft)

WEIGHTS AND LOADINGS (estimated):

Basic operating weight, incl pilot(s):	
'Forger-A'	7,485 kg (16,500 lb)
'Forger-B'	8,390 kg (18,500 lb)
Max T-O weight	11,700 kg (25,795 lb)
Max wing loading: 'Forger-A'	632.4 kg/m² (129.6 lb/sq ft)
Max power loading: 'Forger-A'	91.4 kg/kN (0.90 lb/lb st)

PERFORMANCE ('Forger-A', estimated, at max T-O weight):

Max level speed at height	Mach 0.95 (545 kts; 1,009 km/h; 627 mph)
Max level speed at S/L	Mach 0.8 (528 kts; 978 km/h; 608 mph)
Max rate of climb as S/L	4,500 m (14,750 ft)/min
Service ceiling	12,000 m (39,375 ft)
Combat radius:	
with air-to-air missiles and external tanks, 75 min on station	100 n miles (185 km; 115 miles)
with max weapons, lo-lo-lo	130 n miles (240 km; 150 miles)
with max weapons, hi-lo-hi	200 n miles (370 km; 230 miles)

UPDATED

SINGAPORE

SA

SINGAPORE AEROSPACE LTD

540 Airport Road, Paya Lebar, Singapore 1953
Telephone: 65 287 1111
Fax: 65 280 9713 and 280 8213
Telex: RS 43255 SAMKG
CHAIRMAN AND CEO: Lim Hok San
PRESIDENT: Quek Poh Huat
VICE-PRESIDENT MARKETING: Michael Ng
CORPORATE COMMUNICATIONS: Shirley Tan

Formed early 1982 as government owned Singapore Aircraft Industries Ltd, controlled by Ministry of Defence Singapore Technology Holding Company Pte Ltd; renamed Singapore Aerospace April 1989; total 1994 workforce nearly 4,000.

Major programmes include refurbishing and A-4 to TA-4 conversion for RSAF and other air forces; overhaul, repair and refurbishing of many types including C-130 Hercules, F-5E/F Tiger II, Hunter, Strikemaster and several models of Bell and Eurocopter helicopters. Work started at Seletar; new 15,000 m² (161,460 sq ft) factory at Paya Lebar opened October 1983.

Assembly of 30 SIAI-Marchetti S.211s (see under Agusta in Italian section) begun for RSAF 1985; 17 of 22 Super Pumas for RSAF assembled by SA; company is partner in EC 120 helicopter programme with Eurocopter and CATIC (see International section); to manufacture 100 shipsets of nosewheel doors worth $12 million for Boeing 777 (option for another 100), and 150 sets of passenger doors for Fokker 100.

Six subsidiaries are as follows:

Singapore Aerospace Engineering Pte Ltd,
Seletar West Camp, Singapore 2879
Telephone: 65 481 5955
Fax: 65 482 0245
Telex: RS 25507 SAMAIR
VICE-PRESIDENT/GENERAL MANAGER: Bob Tan

Maintenance, modification and repair of civil and military aircraft and helicopters; upgrade programmes for A-4 Skyhawk and F-5E Tiger II; assembly of S.211 jet trainers.

Singapore Aerospace Systems Pte Ltd,
505A Airport Road, Paya Lebar, Singapore 1953
Telephone: 65 287 2222
Fax: 65 284 4414
Telex: RS 55851 SAERO
VICE-PRESIDENT/GENERAL MANAGER: Foo Hee Liat

Maintenance, overhaul and repair of civil and military aircraft components and equipment; authorised service centre for Aerospatiale, Bell Helicopter Textron, Grimes Aerospace, IMI Marston, Lucas Aerospace and Rockwell Collins.

Singapore Aerospace Engines Pte Ltd,
501 Airport Road, Paya Lebar, Singapore 1953
Telephone: 65 285 1111
Fax: 65 282 3010
Telex: RS 33268 SAENG
VICE-PRESIDENT/GENERAL MANAGER: Chong Kok Pan

Overhaul and repair of Pratt & Whitney JT8D and JT15D, Rolls-Royce Avon Mk 207, Wright J65, Allison T56/501, Turbomeca Makila 1A, General Electric J85 and F404 and Lycoming T53 gas-turbine engines.

Singapore Aerospace Manufacturing Pte Ltd,
503 Airport Road, Paya Lebar, Singapore 1953
Telephone: 65 284 6255
Fax: 65 288 0965 and 284 2704
Telex: RS 38216 SAMPL
VICE-PRESIDENT/GENERAL MANAGER: Goh Chin Khee

Manufacture of aircraft structure and aero-engine components, external stores and composites structures. Contracts include Airbus A320 rear passenger doors, Boeing 777 nosewheel doors and Fokker 100 passenger doors.

Singapore Aerospace Supplies Pte Ltd,
540 Airport Road, Paya Lebar, Singapore 1953
Telephone: 65 287 1111
Fax: 65 284 1167 and 280 6179
Telex: RS 51158 SASUPP
VICE-PRESIDENT/GENERAL MANAGER: Donald Seow

Supplies wide range of parts and components for civil and military aircraft; material support specialist for Singapore Aerospace.

Singapore Aviation Services Company Pte Ltd,
11 Changi North Crescent, Singapore 1750
Telephone: 65 545 0988
Fax: 65 545 6757
VICE-PRESIDENT/GENERAL MANAGER: Chiang Woon Seng

Specialises in commercial aircraft engineering, especially Section 41 modification of Boeing 747, heavy maintenance and structural modification, widebody interior conversions and refurbishing, ageing aircraft modification, avionics upgrade and modification, painting, finishing and corrosion control programme.

VERIFIED

SA (MCDONNELL DOUGLAS) A-4S SUPER SKYHAWK

TYPE: Combat aircraft upgrade.

PROGRAMME: First two McDonnell Douglas A-4S Skyhawks fitted with General Electric non-afterburning F404-GE-100D in 1986; Phase I (engine) conversion of 52 launched 1987; Phase II (avionics) upgrade started in 1991.

VERSIONS: **A-4S-1**: First upgrade, re-engined only: see Power Plant.

A-4SU: Second upgrade stage: see Avionics.

CUSTOMERS: RSAF received 52 A-4S-1 and TA-4S Phase I aircraft (first squadron declared operational March 1992); Phase II being introduced; further airframes and engines reported being acquired.

DESIGN FEATURES: F404 engine increases dash speed by 15 per cent, initial climb by 35 per cent and acceleration by 40 per cent; turn rate and take-off also improved. Wing section NACA 0008 to 0005; quarter-chord sweep 33° 12′.

FLYING CONTROLS: Powered controls; electrically variable incidence tailplane; automatic slats; airbrakes in fuselage sides; lift-dumping spoilers above split flaps.

POWER PLANT: One 48.04 kN (10,800 lb st) General Electric F404-GE-100D non-afterburning turbofan with gas-turbine starting. Internal fuel in integral wing tanks (combined capacity 2,142.5 litres; 566 US gallons; 471.3 Imp gallons) and self-sealing centre-fuselage tank (870.5 litres; 230 US gallons; 191.5 Imp gallons); total capacity 3,013 litres (796 US gallons; 662.8 Imp gallons). One 1,136 or 1,514 litre (300 or 400 US gallon; 250 or 333 Imp gallon) centreline drop tank; 568 or 1,136 litre (150 or 300 US gallon; 125 or 250 Imp gallon) drop tank on each inboard wing pylon. Single-point fuelling receptacle in rear fuselage; all tanks allow pressure or gravity fuelling; optional centreline buddy refuelling store contains 1,136 litres (300 US gallons; 250 Imp gallons), hydraulic hose/drogue unit with 18.3 m (60 ft) hose and ram-air driven hydraulic pump (only in-wing and underwing fuel is transferable; transfer rate about 681 litres; 180 US gallons; 150 Imp gallons/min).

ACCOMMODATION: McDonnell Douglas Escapac ejection seat(s); individual rear hinged upward opening canopy(ies); bullet-resistant windscreen; pressurised and air conditioned.

AVIONICS: Phase II avionics comprise GEC-Marconi (formerly GEC Ferranti) 4150 HUD, new HDD, Litton LN-93 laser inertial navigation system and Maverick missile interface.

ARMAMENT: Two 20 mm Mk 12 guns in wingroots; centreline and four underwing pylons for bombs, rockets, air-to-surface missiles and gun pods; missiles and tanks on inner pylons only.

DIMENSIONS, EXTERNAL:

Wing span	8.38 m (27 ft 6 in)
Wing aspect ratio	2.91
Length overall	12.72 m (41 ft 8⅜ in)
Height overall	4.57 m (14 ft 11⅞ in)
Wheel track	2.37 m (7 ft 9½ in)
Wheelbase	3.64 m (11 ft 11⅓ in)

AREAS:

Wings, gross	24.14 m² (259.82 sq ft)

WEIGHTS AND LOADINGS:

Operating weight empty	4,649 kg (10,250 lb)
Max fuel weight: internal	2,364 kg (5,213 lb)
external (one 400 and two 300 US gallon tanks)	2,961 kg (6,529 lb)
Max T-O weight	10,206 kg (22,500 lb)
Max landing weight	7,257 kg (16,000 lb)
Max zero-fuel weight	7,841 kg (17,287 lb)
Max wing loading	422.8 kg/m² (86.6 lb/sq ft)
Max power loading	212.6 kg/kN (2.08 lb/lb st)
Thrust/weight ratio	0.48

PERFORMANCE (at max T-O weight except where indicated):

Never-exceed speed (VNE) at S/L	628 kts (1,163 km/h; 723 mph)
Max level speed at S/L	609 kts (1,128 km/h; 701 mph)
Max cruising speed at 9,150 m (30,000 ft)	445 kts (825 km/h; 512 mph)
Econ cruising speed at 10,670 m (35,000 ft)	424 kts (786 km/h; 488 mph)
Stalling speed at S/L	133 kts (247 km/h; 154 mph)
Max rate climb at S/L	3,326 m (10,913 ft)/min
Combat ceiling	12,200 m (40,000 ft)
T-O run	1,220 m (4,000 ft)
T-O to 15 m (50 ft)	1,768 m (5,800 ft)
Landing from 15 m (50 ft) at max landing weight	1,590 m (5,215 ft)
Landing run at max landing weight	1,372 m (4,500 ft)
Range, 113 kg (250 lb) fuel reserves:	
with max payload	625 n miles (1,158 km; 720 miles)
with max internal/external fuel	2,046 n miles (3,791 km; 2,356 miles)

VERIFIED

SA (McDonnell Douglas) A-4S Super Skyhawk

1994

SA (NORTHROP) F-5 UPGRADE

TYPE: Combat aircraft upgrade.

DESIGN FEATURES: SA currently offering three-level avionics upgrade of F-5 aircraft. Start-up installation comprises digital mission computer, air data computer, ring laser gyro INS, wide-angle HUD with up-front controller, HOTAS controls, colour video HUD camera and recorder. Second level adds multifunction display and/or multifunction colour display, data transfer system, multimode radar, VOR/ILS, Tacan and radar altimeter. System architecture, based on MIL-STD-1553B databus, can be expanded to include specialised mission equipment such as FLIR, ECM, advanced RWR and new advanced weapons.

VERIFIED

SA (NORTHROP) F-5E/RF-5E CONVERSION

TYPE: Combat aircraft reconnaissance conversion.

PROGRAMME: Under a three-year programme started in 1990, eight of the RSAF's 28 F-5E Tiger II fighters are being converted to RF-5E TigerEye reconnaissance configuration. The modifications, which are undertaken by Singapore Aerospace with engineering and on-site technical support from Northrop, include changes to the forward part of the fuselage and installation of pallet-mounted cameras and other sensors, and updated com/nav equipment.

VERIFIED

SOUTH AFRICA

AERO SERVICES

AERO SERVICES (PTY) LTD

Corner of Bravo and Charlie Avenue, PO Box 30043, Jet Park 1459, Boksburg
Telephone: 27 (11) 397 5346
Fax: 27 (11) 397 5003
MANAGING DIRECTOR: Chris Scott

Responsible for thermo-forming of plastics, injection moulding of rigid and self-skinning foam structures, cutting and sewing of fabric and leather furnishings as well as composite and sandwich panel structures for aircraft interiors. Sidewall panels and bin systems for B707, 727, 737 and Fokker F27. Interior refurbishment of Sikorsky S-61 helicopters and Douglas DC-3 aircraft.

UPDATED

Aero Services (Douglas) DC-3 interior refurbishment *1995*

AERO SERVICES (DOUGLAS) DC-3 INTERIOR CONVERSION KITS

TYPE: Aircraft interior upgrade.
PROGRAMME: As of July 1995 six DC-3 interiors delivered. Completed conversion requires 200 to 300 man hours.
DESIGN FEATURES: Upgrade includes thermo-formed sidewall and ceiling panels, overhead stowage bins. Optional features include forward and aft bulkheads, galley and toilet installations, refurbished or new seating and carpets or soft furnishings.

VERIFIED

AERO SERVICES (SIKORSKY) S-61 INTERIOR UPGRADE

TYPE: Aircraft interior upgrade.
PROGRAMME: As of July 1995 two S-61 interior delivered to Court Helicopters with one more set in progress.
DESIGN FEATURES: Programme includes an all-thermo-formed rigid plastic system with centreline lighting and air gaspers at each passenger position.

UPDATED

Aero Services (Sikorsky) S-61 interior refurbishment *1995*

AEROSUD

AEROSUD (PTY) LTD

PO Box 2262, Halfway House 1685
Telephone: 27 (11) 315 4390
Fax: 27 (11) 315 3924
DIRECTOR, MARKETING DESIGN: Kobus de Villiers

Responsible for design and upgrading of airframe structures and systems including pneumatic, hydraulic, fuel, environmental control, electrical and servo systems as well as the design and production of engine intake and exhaust modifications. Company also capable of performing systems integration and ground and flight testing. Aerosud has been involved in a number of major combat aircraft upgrade programmes, including the installation of chaff/flare and avionic systems as well as aircraft engine and system modifications. Other programmes include aircraft interior refurbishment and spray wash systems for power lines.

UPDATED

Sikorsky S-61 fitted with Aerosud long-range fuel tank *1994*

AEROSUD (SIKORSKY) S-61 LONG-RANGE TANK UPGRADE

TYPE: Sikorsky S-61 long-range fuel tank upgrade.
DESIGN FEATURES: Removable fuel tank (1,250 litre capacity) mounted under the centre-fuselage, using the four existing cargo sling mountings. Fuel transferred to aircraft by two

pumps. External sight glass fitted for gaugeing. The tank can be jettisoned in emergencies. Minor modifications required to electrical and fuel systems.

VERIFIED

AEROSUD (SIKORSKY) S-61 DUST FILTER UPGRADE

TYPE: Sikorsky S-61 dust filter upgrade.

DESIGN FEATURES: Installation of a self-cleaning vortex tube dust filter over the existing engine bell mouth air intakes. Creation of centrifugal force field in the air flow forces dust particles to outer radial edge. Efficiency of dust filter system is 93 per cent using SAE course. Pressure loss is 1,004 kPa.

UPDATED

AEROSUD/RUSJET FIGHTER AIRCRAFT RE-ENGINE PROGRAMME

TYPE: Combat aircraft re-engine programme.

PROGRAMME: Collaboration between Aerosud and Rusjet, a joint venture company consisting of Marvol, Klimov, Mikoyan as well as other Russian companies. Programme consists of development of turbofan engine for use in re-engining MiG-21, Mirage F1 and Mirage III aircraft. A Mirage F1 has conducted 50 hours of flight tests in South Africa with the new engine. Complete flight envelope has been examined including weapons flights. The Cheetah version has completed ground tests and is due to fly in May 1995.

DESIGN FEATURES: The new engine is the Klimov SMR-95 turbofan developed using technology from the RD-35 programme.

UPDATED

AEROSUD/SAAF GENERIC AVIONICS SYSTEM

TYPE: Combat aircraft avionics upgrade.

PROGRAMME: Collaboration between South African Air Force (SAAF) and aerospace companies including: ATE, AMS, Teklogic, Xcel, Kentron, Megkon, General Computing, Barcom and Avitronics to design and develop a generic avionics upgrade suite for use in a variety of military aircraft. Tests currently being carried out on SAAF Dassault Mirage F1 aircraft.

DESIGN FEATURES: The programme is to be carried out in two phases. Phase one is the 'kernel' avionics suite and air-to-ground weapons capability. The second phase will add air-to-air systems and advanced functions. Aerosud is responsible for the physical integration of the subsystems into cockpit and fuselage as well as design of all mechanical and structural components, stress analysis, mass and balance analysis, environmental control system research, process control and planning, manufacture of all mechanical and structural items, installation and final inspection.

AVIONICS: Avionics suite to be based around an MIL-STD-1553 databus. Phase one equipment to consist of an advanced mission computer with multiplexed man/machine interface, flat panel cockpit control and display unit (CDU), multifunction head-up display (HUD) displaying both stroke and raster images and inertial navigation system (INS). Phase two includes follow-on development of subsystems for air-to-air weapons delivery, hybridised inertial and satellite navigation system (GPS), tactical situation display, helmet display, full day and night capability, fully integrated communication system, electronic countermeasures (ECM) unit and forward-looking infra-red (FLIR) system.

NEW ENTRY

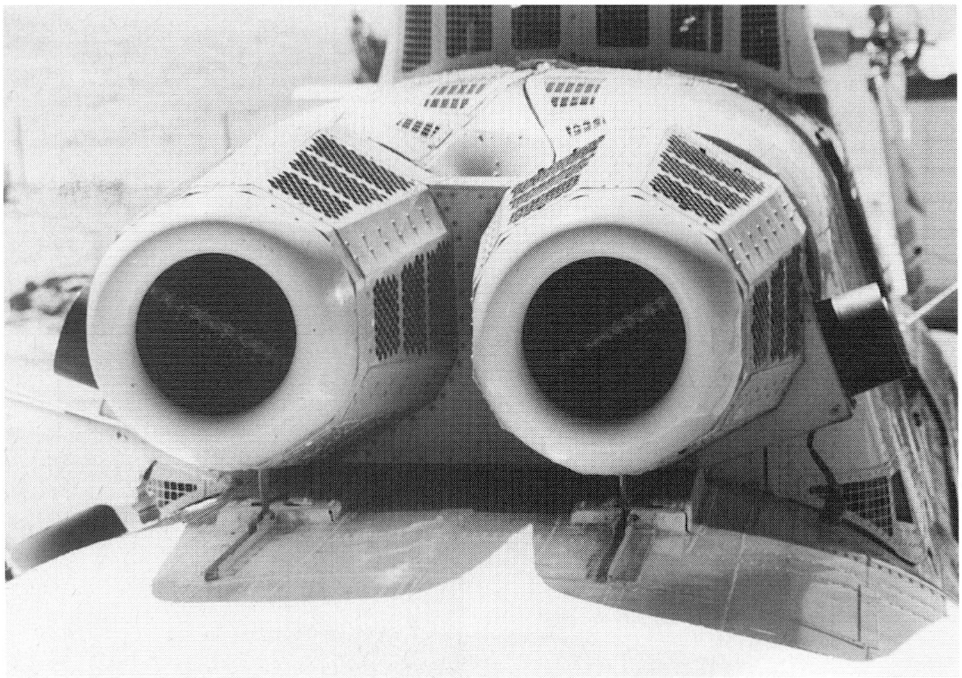

Sikorsky S-61 fitted with Aerosud vortex tube dust filter

1994

Aerosud/Rusjet (Dassault) Mirage F1 re-engined with an SMR-95 turbofan engine

1995

Aerosud/SAAF (Dassault) Mirage F1 cockpit demonstrator for the Generic Avionics System Upgrade

1995

ATLAS

ATLAS AVIATION (PTY) LIMITED

(A Division of Denel (Pty) Ltd)
PO Box 11, Kempton Park 1620, Transvaal
Telephone: 27 (11) 927 4117
Fax: 27 (11) 973 5353
Telex: 742403 BONAERO
GENERAL MANAGER: J. J. Eksteen
EXECUTIVE MARKETING MANAGER: C. J. Maree
SIMERA
(A Division of Denel (Pty) Ltd)
PO Box 117, Kempton Park 1620, Transvaal
Telephone: 27 (11) 927 4443
Fax: 27 (11) 395 1280
DIVISIONAL GENERAL MANAGER: J. J. Eksteen
EXECUTIVE MARKETING MANAGER: R. J. Watson
PUBLIC RELATIONS MANAGER: S. J. Basch

Atlas Aviation, which was founded in 1963, changed its name from Atlas Aircraft Corporation of South Africa with effect from 1 April 1992. The company has extensive manufacturing, design and development facilities for airframes, engines, missiles and avionics. It developed the Cheetah version of the Mirage III and the Rooivalk gunship helicopter as well as the V3B and V3C missiles and many weapons installations.

Restructuring of Armscor on 1 April 1992 created Denel as self-sufficient commercial industrial group, in which Atlas is military aircraft manufacturing and maintenance branch of Simera, which is starting up civil aircraft work. Simera/Altas workforce in 1994 was about 4,000.

ATLAS IMPALA Mk 2

TYPE: Single-seat operational trainer.

PROGRAMME: Based on the MB-326K, South African manufacture began with the assembly of seven aircraft from Italian built components, the first of which was handed over to the SAAF on 22 April 1974.

POWER PLANT: One Rolls-Royce Viper Mk 540 turbojet engine.

VERIFIED

ATLAS CHEETAH

TYPE: Single-seat fighter and reconnaissance aircraft and two-seat operational trainer.

PROGRAMME: Revealed July 1985; first Cheetah DZ, converted from SAAF two-seat Mirage III-D2Z serial 845, rolled out July 1986; operational Summer 1987.

VERSIONS: **Cheetah DZ**: First conversions were two-seaters, originally III-BZ, DZ and D2Z. Two-seater more elaborately equipped, probably to act as pathfinder for single-seaters.

Cheetah EZ and R2: Conversions of up to 15 Mirage III-EZ, four III-RZ and four III-R2Z single-seaters; no R2 conversions yet confirmed.

Cheetah ACW: Upgraded version, revealed April 1992 (ACW: advanced combat wing); fixed and drooped wing leading-edges, kinked outboard of dog-tooth, reduce supersonic drag, increase sustained turn rate by 14 per cent and allow increase in-wing fuel tankage; maximum T-O weight 600 kg (1,323 lb) higher than Cheetah EZ. Prototype converted from Mirage III-R2Z Serial 855; has reportedly been flown down to 80 knots (148 km/h; 92 mph) and at 33° angle of attack.

Project 855: Reported February 1994 that 12 Cheetah EZs to be further upgraded in two-year, $1.8 billion programme, main feature of which to be new Elta EL/M-2035 fire control radar; most of work to be undertaken by Atlas. Use of '855' as project number suggests that ACW demonstrator may serve as prototype for new installation.

OPERATORS: Versions of the Cheetah are in service with the armed forces of South Africa (28).

DESIGN FEATURES: Two-seater has longer nose containing multimode radar, RWR antennae, avionics cooling system and external cable ducts from nose to centre-fuselage equipment bay; like Israeli Kfirs, Cheetah has dog-tooth outboard leading-edges and fixed foreplanes; ordinary fences replace slot fences; single-seater has only ranging radar in nose. Nose probe higher and offset to port on R2. About 50 per cent of airframe renewed.

FLYING CONTROLS: As Mirage III, but with small strakes beside nose to direct yaw-correcting vortices against fin at high angles of attack.

POWER PLANT: One Snecma Atar 9K-50 turbojet (70.6 kN; 15,873 lb st with afterburning). Bolt-on flight refuelling probe on starboard side of cockpit.

AVIONICS: EZ has ranging radar in nose, RWR antennae in nose and fin trailing-edge, apparently chaff/flare dispenser in former rocket motor fairing, radar altimeter, Elbit HUD and nav/attack system (perhaps including inertial system) and possibly South African made helmet-mounted sight. DZ two-seater has larger look-down search and track radar in nose with large cooling system at base of pitot head.

ARMAMENT: Normal Mirage III 30 mm DEFA gun in fuselage underside; Armscor V3B Kukri or V3C Darter dogfight

Atlas Impala Mk 2 (foreground) Impala Mk 1 (background) of the SAAF

1994

Atlas Cheetah EZ single-seat development of the Dassault Mirage III

1994

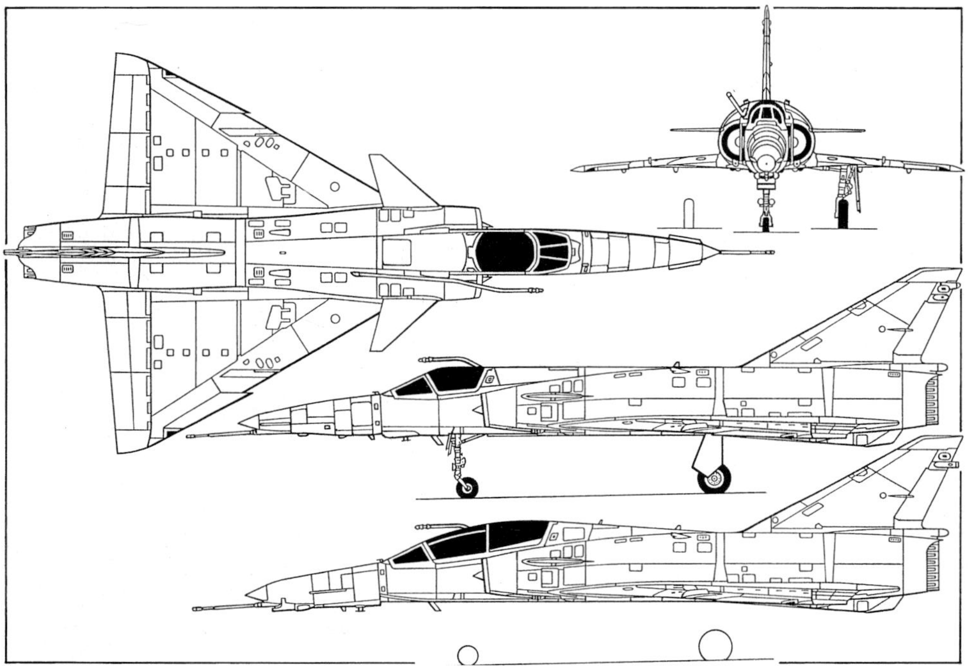

Altas Cheetah EZ single-seat fighter, with additional side view (bottom) of Altas Cheetah DZ *(Jane's/Mike Keep)*

1994

Altas Cheetah DZ two-seat development of the Dassault Mirage III *(Herman Potgieter)*

1994

missiles; possible Matra R.550 semi-active air-to-air missiles; AS.30 air-to-surface missiles, possibly with designator pod. Cheetah seen loaded with eight 500 lb bombs, two underwing tanks and two V3B missiles.

VERIFIED

ATLAS (EUROCOPTER/AEROSPATIALE) AS 330 PUMA UPGRADES

TYPE: Combat helicopter upgrade.

VERSIONS: **Oryx** (previously known as Gemsbok): AS 330 airframe refitted with Makila power plant, uprated transmission similar to that of Rooivalk, cockpit displays redesigned for single pilot operation, ventral fin and tailplane of AS 532 Cougar, plus nose radome; programme started 1984; entered service from 1988 as replacement for SAAF SA 321 Super Frelons.

Gunship: Four Atlas proposals offered as possible quick-change, lower cost alternative to Rooivalk, some of which already evaluated by SAAF. Strap-on beams aft of cabin sliding doors, each with two pylons and wingtip missile rail; Atlas armoured seats; still able to transport 12 troops; Grinaker Avitronics integrated airborne communications system. **Option 1**: With Kentron TC-20 ventral gun turret and helmet sight. **Option 2:** With four launchers for 68 mm rockets for area suppression. **Option 3**: As Option 2, but Kentron nose-mounted HSOS (helicopter stabilised optronic sight) added; integrated on development aircraft scheduled for late 1994. **Option 4**: Full anti-tank version with ZT-3 Swift laser-guided missiles.

UPDATED

Atlas (Eurocopter/Aerospatiale) Oryx upgraded AS 330 Puma helicopter aboard SAS *Drakensberg* in Abu Dhabi *(Paul Beaver)*

1995

Altas (Eurocopter/Aerospatiale) AS 330 Puma gunship conversion with display of full weapon carrying potential

1995

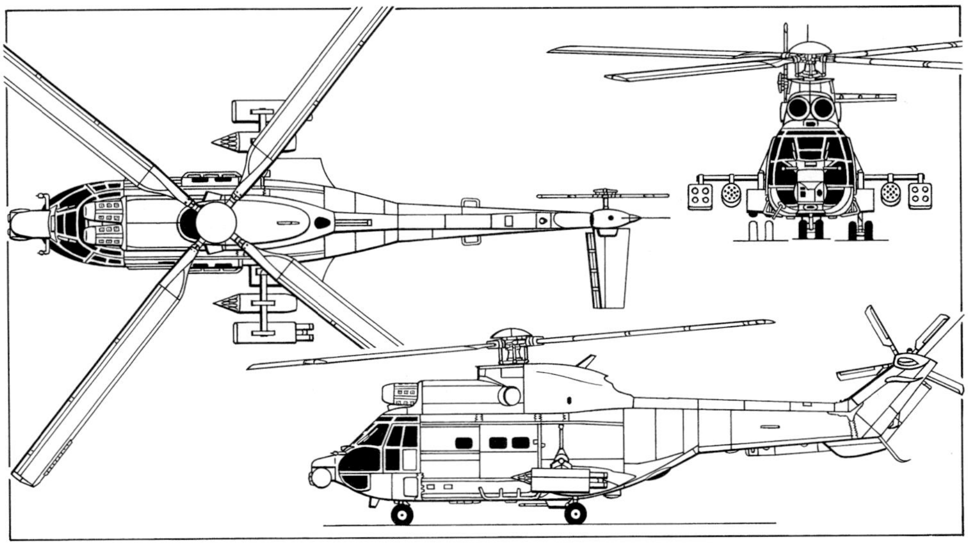

Atlas gunship upgrade of the Eurocopter (Aerospatiale) AS 330 Puma *(Jane's/Mike Keep)*

1994

HUNTING

HUNTING AVIATION (PTY) LIMITED

Private Bag 1010, Germiston 1400
Telephone: 27 (11) 827 8940
Fax: 27 (11) 824 2217
Telex: 74-6830
CHAIRMAN AND MANAGING DIRECTOR: Richard Charter
FINANCIAL DIRECTOR: Len Busschau
OPERATIONS DIRECTOR: Steve Wybenga
MARKETING AND NEW BUSINESS DEVELOPMENT DIRECTOR: Steve Duley
AIRMOTIVE AND COMPONENTS DIVISION MANAGER: Jan Viljoen
FIXED-WING DIVISION MANAGER: Carlos Barata
HELICOPTER DIVISION MANAGER: Meyer Potgieter

The origins of Hunting Aviation (Pty) Ltd can be traced back to 1935 when Airservices Pty Ltd was founded to maintain AOC's aircraft. In 1952 the company adopted the name Field Aviation Services (Africa) (Pty). In 1992 the company was renamed Hunting Aviation (Pty) Ltd. The company is organised into three main divisions: Airmotive & Components, Fixed-Wing and Helicopter. Facilities cover a total area of 1,858 m² (20,000 sq ft).

Responsibilities are in the fields of fixed- and rotary-wing aircraft overhaul, modification, repair, upgrade conversion and refurbishment as well as support, repair and overhaul of engines (turbine and piston), propellers and components. Further responsibilities include the manufacture of aviation parts.

NEW ENTRY

PROFESSIONAL AVIATION

PROFESSIONAL AVIATION SERVICES (PTY) LTD

Terminal Building, Lanseria Airport, PO Box 3171, Randburg 2125
Telephone 27 (11) 701 3320
Fax: 27 (11) 659 1336/1331
Telex: 4-21865 SA
DIRECTORS:
R. C. H. Garbett
C. T. Garbett

Formed 1979; buys, sells, charters, leases and manages many types of fixed-wing aircraft and helicopters; maintains and supports aircraft; undertakes turboprop conversions of DC-3/C-47s.

UPDATED

Professional Aviation (Douglas) DC-3 Jet Prop ***(Peter Cooper)***

1995

PROFESSIONAL AVIATION (DOUGLAS) JET PROP DC-3 AMI

TYPE: Stretched C-47 with turboprop retrofit.

CUSTOMERS: Four civil conversions.

COSTS: Approximately SAR5 million ($3.0 million) per aircraft (1992).

DESIGN FEATURES: All lifed items zero-lifed; fuselage lengthened by 1.02 m (3 ft 4 in) forward of wing centre-section; fully overhauled main and tail landing gear (with new disc brakes and tyres), wiring and hydraulic lines; one-piece windscreens; new instrument panel; new electrical system; turboprop engines and five-blade propellers; upgraded avionics; optional auxiliary fuel system; redesigned interior for cargo and/or passengers. Substantial payload/range gains. Conversion takes approximately six months.

POWER PLANT: Two 1,062 kW (1,424 shp) Pratt & Whitney Canada PT6A-65AR turboprops; Hartzell five-blade metal propellers with autofeathering, reverse pitch and Beta control. Option for outer-wing auxiliary fuel tanks increasing capacity by 1,514 or 3,028 litres (400 or 800 US gallons; 333 or 666 Imp gallons).

ACCOMMODATION: Interior volume increased by 35 per cent, permitting layouts for 34-40 passengers, passenger/cargo mix or all-cargo; quick-change from cargo to passenger interior in 60 minutes; galley optional. Cargo door at rear on port side; cargo liners and floor tracks installed.

SYSTEMS: New 24 V electrical system with two 250 A starter/generators. Full fire detection and extinguishing systems. Propeller and engine inlet de-icing.

AVIONICS: Bendix/King package including VLF/Omega, HF, GPS, GPWS, flight data recorder and cockpit voice recorder; coupled three-axis autopilots, certificated for Cat. III operation.

DIMENSIONS, EXTERNAL: As C-47 except:
Length overall 20.68 m (67 ft 10¼ in)
Propeller diameter 2.82 m (9 ft 3 in)
Cargo door: Height 1.50 m (4 ft 11 in)
Width 1.50 m (4 ft 11 in)

WEIGHTS AND LOADINGS:
Basic weight empty 7,257 kg (16,000 lb)
Max payload: standard 4,000 kg (8,818 lb)
optional 4,500 kg (9,921 lb)
Max T-O weight 12,202 kg (26,900 lb)

PERFORMANCE:
Max cruising speed 185 kts (343 km/h; 213 mph)
Max rate of climb at S/L 305 m (1,000 ft)/min
Service ceiling, OEI 4,265 m (14,000 ft)
Min runway requirement 1,000 m (3,280 ft)
Min landing distance 300 m (984 ft)
Range with max payload 350 n miles (648 km; 403 miles)
Endurance with max payload 3 h
Max IFR range at 1,220 m (4000 ft) with 2,000 kg (4,409 lb) payload and max auxiliary fuel, reserves for 1 h diversion and 45 min hold 2,000 n miles (3,706 km; 2,303 miles)
Max endurance, conditions as above 14 h

UPDATED

Professional Aviation (Douglas) DC-3 passenger conversion

1994

Professional Aviation (Douglas) DC-3 all-cargo conversion

1994

SAAF

SOUTH AFRICAN AIR FORCE

South African Air Logistic Command, Chief of Staff Logistics, Private Bag X199, Pretoria 0001
Telephone: 27 (12) 312 2716
CHIEF OF AIR STAFF, LOGISTICS: Major General G. C. Nell

UPDATED

SAAF (DOUGLAS) DC-3 TURBO-DAK UPGRADE

TYPE: Transport and maritime patrol aircraft upgrade.

PROGRAMME: The SAAF launched this upgrade programme in late 1987 to extend operational lives and enhance capabilities of its 39 DC-3 Dakotas. Due for completion in late 1994. A maritime patrol Turbo-Dak is also being developed with specialised avionics and long-range tanks to extend range to 4,800 km.

DESIGN FEATURES: Programme includes re-engining original radial engines with PW PT-6 turboprop engines, updating the avionics with South African equipment, updating the electrical and hydraulic systems and adapting the control surfaces to account for improved performance. The new engines require a fuselage stretch of 1 m in order to maintain the original CG.

PERFORMANCE:

Cruise speed	378 km/h (235 mph)
Range	2,830 km (1,759 miles)
Max payload	4,400 kg (9,697 lb)

VERIFIED

South African Air Force DC-3 Turbo-Dak re-engined DC-3 with Pratt & Whitney PT-6 turboprops
1994

SPAIN

AISA

AERONAUTICA INDUSTRIAL SA

Carretera del Aeroclub, PO Box 27094, E-28044 Madrid
Telephone: 34 (1) 396 3247
Fax: 34 (1) 2083958
Telex: E 49872 MADRID
PRESIDENT: J. A. Pérez-Nievas
GENERAL MANAGER: Carlos Herraiz
DESIGN MANAGER: Rafael Moreno

This company was founded in 1923, assuming its present title in 1934. Its design office has, since the Second World War, been responsible for several liaison, training and sporting aircraft for the Spanish Air Force and aeroclub flying schools. The Cuatros Vientos factory has a covered area of 10,270 m² (110,545 sq ft) and employs about 200 people.

AISA repairs and overhauls US aircraft, in particular the Beechcraft B55 Barons and F33 Bonanzas operated by the Spanish Air Force and the National School of Aeronautics. It also repairs and overhauls Bell 47, 204, 205, 206 and 212, and Boeing CH-47, helicopters for the Spanish Army, Navy, Air Force and civilian operators. As a subcontractor to Messier-Bugatti, it is producing landing gear shock-absorbers and hydraulic actuators for the Dassault Mirage F1, 2000 and Falcon series, Dassault/Dornier Alpha Jet, and other European aviation programmes. Under subcontract to CASA, it produces structural components for the C-212 Aviocar and Airbus transports; and, in offset programmes, helicopter structures and hydraulic components for Aerospatiale and Agusta. Involved in Eurofighter 2000 programme.

VERIFIED

AISA UPGRADE ACTIVITY

TYPE: Fixed- and rotary-wing upgrades.

PROGRAMME: AISA took part in modernising Spanish Air Force Mirage IIIs. Company currently participating in the upgrading of Spanish Boeing CH-47Cs to CH-47Ds. Company generally dedicated to aircraft upgrading and modification.

VERIFIED

CASA

CONSTRUCCIONES AERONAUTICAS SA

Avenida de Aragón 404, E-28022 Madrid
Telephone: 34 (1) 585 7268
Fax: 34 (1) 585 7666/7
Telex: 27418 CASA E
WORKS: Getafe, Illescas, Tablada, San Pablo, San Fernando, Puerto Real and Cádiz
PRESIDENT AND CHAIRMAN OF THE BOARD: Paul Herranz
PUBLIC RELATIONS AND PRESS MANAGER: José de Sanmillán

CASA was founded in 1923 and is now owned 99.285 per cent by the Spanish state holding company TENEO, 0.7098 per cent by DASA and 0.005 per cent by other minority shareholders.

In 1989, CASA formed Industria de Turbopropulsores (ITP) to make parts for and assemble turbojet engines, including the EJ200 for the European Fighter Aircraft, the F404 for the F/A-18, possibly the R-R Pegasus for Spanish Matadors (Harriers) and the eventual engine for the CASA AX. A second company, Compania Espanola de Sistemas Aeronauticas (CESA) was formed with a 40 per cent holding by Lucas Aerospace. The CASA space division has been greatly expanded and a largely automated composites manufacturing plant has been set up at Illescas, near Toledo.

The largest of CASA's present programmes is its 4.2 per cent share in Airbus (see International section). It makes horizontal tail surfaces, landing gear doors, wing ribs and skins, leading- and trailing-edges and passenger doors for the A300/310/320, many of them in composites. Similar work is being carried out for the A330/340.

Second largest programme is CASA's share in the European Fighter Aircraft (EFA). CASA is active in all four of EFA's joint teams for avionics, control systems, flight management system and structure. The last named includes sharing the starboard wing with BAe and the rear fuselage with Alenia. CASA is responsible for integration and software creation of the EFA communication subsystem.

CASA's own aircraft programmes include the C-212 Aviocar, the C-101 Aviojet and the CASA 3000. The CN-235 is shared with IPTN under Airtech and CASA is looking for a partner to produce the new AX operational trainer.

In October 1989, CASA won the competition to design, stress, test and manufacture the entire wing of the Saab 2000 turboprop commuter. CASA is using the same advanced metal bonding technique as used for the Saab 340, but more composites are included.

CASA has designed and is manufacturing the tailplane of the McDonnell Douglas MD-11 and makes outer flaps for the Boeing 757.

In past years, CASA has made centre-fuselages for Dassault Mirage F1s and components for McDonnell Douglas F/A-18s, both for the Spanish Air Force. It has assembled 57 MBB BO 105s for the Spanish Army and 19 more for Spanish government agencies, and two BK 117s for ICONA. It assembled 12 of the 18 AS 532B$_1$ Cougars for the Spanish Army. CASA manufactures tail components for the Sikorsky S-70 and would assemble and test S-70s purchased for the Spanish armed forces. CASA has also assembled 40 ENAER T-35C Pilláns for the Spanish Air Force under the designation E.26 Tamiz.

CASA has seven factories covering 335,000 m² (3,605,900 sq ft) and employed 9,200 personnel in 1993.

UPDATED

CASA UPGRADE AND MAINTENANCE ACTIVITY

TYPE: Fixed- and rotary-wing upgrades.

PROGRAMME: For more than 35 years, the Maintenance Division of CASA has been maintaining and modernising aircraft and helicopters for various customers, but mainly for the USAF and the Spanish Air Force (FAE) including F-4

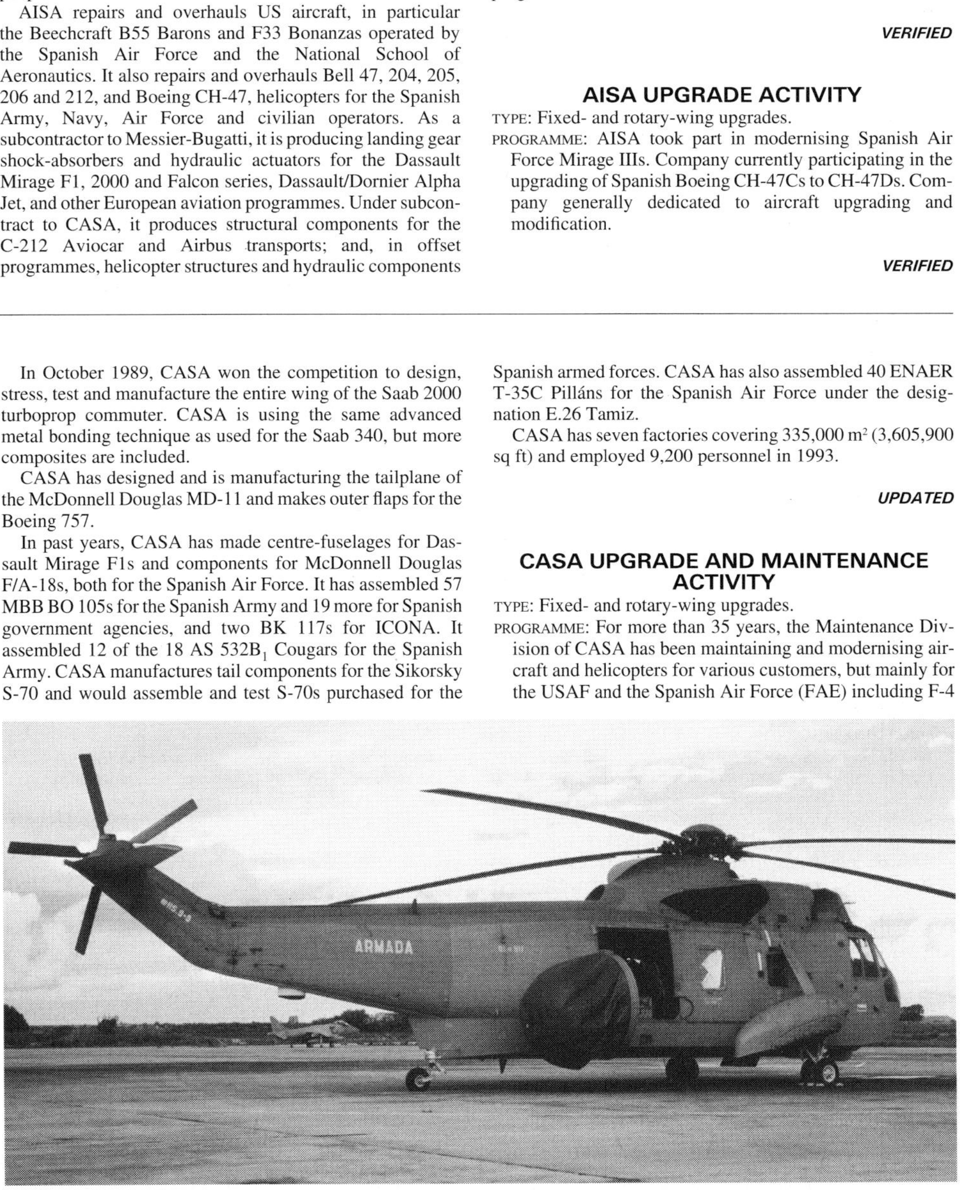

CASA upgrade of Spanish Navy SH-3D with THORN EMI Searchwater AEW radar
(Press Office Sturzenegger)
1994

Phantom IIs, F-15 Eagles and F-18 Hornets. The Maintenance Division is currently performing large modernisation programmes on the F-5 Tiger. CASA also installed THORN EMI Searchwater AEW radars in three Spanish Navy SH-3Ds.

VERIFIED

CASA C-212-100 AVIOCAR

Spanish military designation: T.12B/C and TR.12A

TYPE: Twin-turboprop STOL utility transport aircraft.

PROGRAMME: The C-212 Aviocar twin-turboprop light utility STOL transport was evolved by CASA to fulfil a variety of military and civil roles, but primarily to replace the mixed fleet of Junkers Ju 52/3m (T.2), Douglas DC-3 (T.3) and CASA-207 Azor (T.7) transport aircraft formerly in service with the Spanish Air Force. The first prototype flew for the first time 26 March 1971, the second 23 October 1971.

The C-212 is able to fulfil six main roles, including: 16-seat paratroop transport, military freighter, ambulance, photographic aircraft, crew trainer or 19-seat passenger transport. The C-212 has been certificated to joint military and civilian standards by Instituto Nacional de Técnica Aerospacial (INTA), which was also responsible for the flight test programme. It has STOL capability that enables it to use unprepared landing strips about 400 m (1,310 ft) in length, and has been optimised for operation in remote areas with a poor infrastructure.

Eight preproduction Aviocars were ordered initially by the Spanish Air Ministry; the first of these made its first flight on 17 November 1972, and all had flown by February 1974.

VERSIONS: **C-212A (Spanish Air Force designation T.12B):** Military utility transport. In service with the air forces of the following countries: Indonesia (10), Jordan (2), Portugal (4) and Spain (50+).

C-212AV: VIP transport aircraft operated by the air forces of Jordan (1) and Spain.

C-212B (Spanish Air Force designation TR.12A): Photographic survey version. Six of the eight preproduction Aviocars were completed in this configuration for the Spanish Air Force, and were each equipped with two Wild RC-10 aerial survey cameras. Four others have been delivered to the Portuguese Air Force.

C-212C: Commercial transport version. In use by the Royal Thailand Rain Industry in rainmaking configuration.

C-212E: Navigation trainer. Two of the eight preproduction Aviocars were completed to this configuration for the Spanish Air Force.

C-212 Series 200: Improved version with more powerful Garrett TPE 331-10 engines. Described separately.

OPERATORS: Versions of the C-212-100 are in service with the armed forces of the following countries: Chile (9), Indonesia, Jordan (17), Portugal (10), Spain (70) and Uruguay (2).

DESIGN FEATURES: Cantilever high-wing monoplane. Wing section NACA 65_3-218. Incidence 2° 30′. No dihedral or sweepback. Cantilever tail unit with dorsal fin fairing forward of fin. Tailplane mid-mounted on rear of fuselage.

FLYING CONTROLS: All-metal ailerons and double-slotted trailing-edge flaps. Trim tab in port aileron. Trim tab in rudder and each elevator.

STRUCTURE: The wing is an all-metal light alloy fail-safe structure. The fuselage is a semi-monocoque fail-safe structure of light alloy construction. The tail unit is a two-spar all-metal structure.

LANDING GEAR: Non-retractable tricycle type, with single mainwheels and single steerable nosewheel. CASA oleo-pneumatic shock-absorbers. Dunlop wheels and tyres, main units size 11.00-12 (8 ply) Type III, nose unit size 8.00-7 Type III. Tyre pressure (all units) 3.10 bars (45 lb/sq in). Dunlop hydraulic disc brakes on mainwheels.

POWER PLANT: Two 559 kW (750 shp) Garrett-AiResearch TPE 331-5-251C turboprop engines, each driving a Hartzell HC-B4TN-5CL-LT10282HB+4 four-blade constant-speed (Beta mode on ground) metal propeller (three-blade HC-B3TN-5E in prototype and early preseries aircraft). Fuel in four outer-wing tanks, with total capacity of 2,100 litres (554.8 US gallons; 462 Imp gallons). Oil capacity 6 litres (1.59 US gallons; 1.32 Imp gallons) per engine.

ACCOMMODATION: Crew of two on flight deck. For the troop transport role, the main cabin can be fitted with 16 inward-facing seats along the cabin walls, to accommodate paratroops and an instructor/jumpmaster; or seats for 18 fully equipped troops. As an ambulance, the cabin would normally be equipped to carry 12 stretcher patients and two medical attendants. As a freighter, the Aviocar can carry up to 2,100 kg (4,630 lb) of cargo in the main cabin, including light vehicles. Photographic version is equipped with two Wild RC-10 vertical cameras and a darkroom. Navigation training version accommodation consists of individual desks for an instructor and five pupils, in two rows with appropriate instrument installations. The civil passenger transport version has standard seating for 19 persons in five rows of three (one to port and two to starboard of centre aisle) at 79 cm (31 in) pitch, plus two rows of two seats. VIP transport version, furnished to customer's requirements, those of the Spanish and Jordanian air forces were furnished with seats for 12 passengers (four forward and eight aft, with cabin divider), with two foldaway tables in forward section and toilet forward and baggage compartment aft of main passenger areas. Access to main cabin via two doors on the port side, one aft of (and providing access to) the flight deck and one aft of the wing trailing-edge. In addition, there is a two-section underfuselage loading ramp/door aft of the main cabin; this door is openable in flight for the discharge of paratroops or cargo, and is fitted with external wheels, to allow the door to remain open during ground manoeuvring. There is an emergency exit door aft of the wing trailing-edge on the starboard side. All versions have a toilet at the forward end of the main cabin on the starboard side, with a baggage compartment opposite on the port side. In the civil transport version, the interior of the rear-loading door can be used for additional baggage stowage.

SYSTEMS: Unpressurised cabin. Hydraulic system, pressure 138 bars (2,000 lb/sq in), operates mainwheel brakes, flaps, nosewheel steering and ventral cargo door. Electrical system is supplied by two 9 kW starter/generators. Rubber de-icing boots of leading-edges.

AVIONICS AND EQUIPMENT: Radio and radar equipment includes VHF, UHF, VOR ILS and one ADF. Blind-flying instrumentation standard. Optional avionics include Tacan, SIF/IFF and a second ADF.

DIMENSIONS, EXTERNAL:

Wing span	19.00 m (62 ft 4 in)
Wing chord: at root	2.50 m (8 ft 2½ in)
at tip	1.50 m (4 ft 11 in)
Wing aspect ratio	9
Length overall	15.16 m (44 ft 9 in)
Height overall	6.68 m (21 ft 11 in)
Tailplane span	7.40 m (24 ft 3¼ in)
Wheel track	3.10 m (10 ft 2 in)
Wheelbase	5.55 m (18 ft 2½ in)
Propeller diameter	2.69 m (8 ft 10 in)
Distance between propeller centres	5.30 m (17 ft 4¾ in)
Passenger door (port, aft):	
Max height	1.58 m (5 ft 2¼ in)
Max width	0.70 m (2 ft 3½ in)
Crew and servicing door (port, fwd):	
Max height	1.10 m (3 ft 7¼ in)
Max width	0.60 m (1 ft 11⅝ in)
Rear-loading door: Max length	3.66 m (12 ft 0 in)
Max width	1.70 m (5 ft 7 in)
Max height	1.80 m (5 ft 10¾ in)

DIMENSIONS, INTERNAL:

Cabin (between flight deck and rear-loading door):	
Length	6.50 m (21 ft 4 in)
Max width	2.10 m (6 ft 10¾ in)
Max height	1.80 m (5 ft 10¾ in)
Floor area	10.50 m^2 (113.0 sq ft)
Volume	22.00 m^3 (776.9 cu ft)

AREAS:

Wings, gross	40.0 m^2 (430.56 sq ft)
Ailerons (total)	2.45 m^2 (26.37 sq ft)
Trailing-edge flaps (total)	7.38 m^2 (79.44 sq ft)
Fin	4.25 m^2 (45.75 sq ft)
Rudder, incl tab	2.02 m^2 (31.74 sq ft)
Tailplane	7.36 m^2 (79.22 sq ft)
Elevators, incl tabs	4.36 m^2 (46.93 sq ft)

WEIGHTS AND LOADINGS:

Manufacturer's weight empty	3,700 kg (8,157 lb)
Weight empty, equipped	3,905 kg (8,609 lb)
Max payload	2,100 kg (4,630 lb)
Fuel load	1,600 kg (3,527 lb)
Max T-O weight	6,500 kg (14,330 lb)
Max zero-fuel weight	6,000 kg (13,227 lb)
Max landing weight	6,250 kg (13,780 lb)

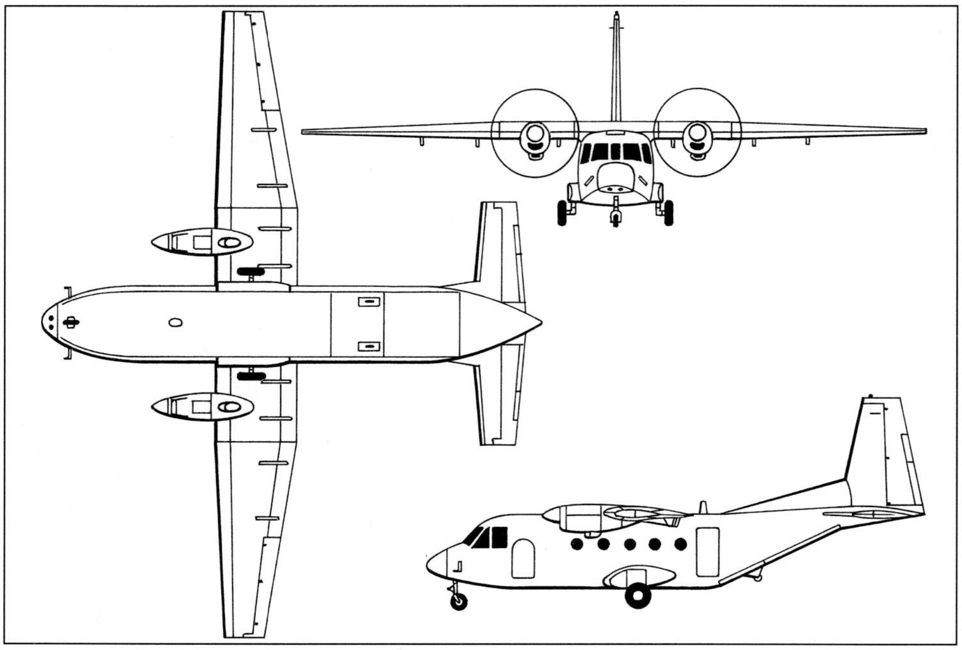

CASA C-212-100 Aviocar twin-turboprop STOL aircraft *(Jane's/Mike Keep)* ***1994***

CASA C-212-100 Aviocar navigation trainer of the Spanish Air Force (FAE) *(Peter J. Cooper)* ***1994***

Max wing loading 162.5 kg/m² (33.3 lb/sq ft)
Max power loading 5.81 kg/kW (9.55 lb/shp)

PERFORMANCE (at max T-O weight except where indicated):
- Never-exceed speed 253 kts (471 km/h; 292 mph) EAS
- Max level speed at 3,660 m (12,000 ft) 194 kts (359 km/h; 223 mph)
- Econ cruising speed at 3,660 m (12,000 ft) 148 kts (275 km/h; 171 mph)
- Stalling speed: flaps up 74 kts (137 km/h; 85.5 mph)
  - flaps down 61.5 kts (113 km/h; 70.5 mph)
- Max rate of climb at S/L 518 m (1,700 ft)/min
- Max rate of climb at S/L, one engine out 93 m (305 ft)/min
- Service ceiling 8,140 m (26,700 ft)
- Service ceiling, one engine out 4,115 m (13,500 ft)
- T-O run 350 m (1,148 ft)
- T-O run to 15 m (50 ft) 484 m (1,588 ft)
- Landing run from 15 m (50 ft) 518 m (1,700 ft)
- Landing run 331 m (1,085 ft)
- Range at 3,050 m (10,000 ft):
  - with max fuel and 1,045 kg (2,303 lb) payload 949 n miles (1,760 km; 1,093 miles)
  - with max payload 258 n miles (480 km; 298 miles)

UPDATED

CASA C-212-100 Aviocar STOL aircraft of the Spanish Air Force (FAE) *(Peter March)*

1994

CASA C-212 AVIOCAR (ASW and MARITIME PATROL VERSIONS)

Swedish Navy designation: Tp89

TYPE: Twin-turboprop ASW and maritime patrol aircraft.

PROGRAMME: CASA has developed specialised versions of the C-212 equipped for anti-submarine and maritime patrol duties. They are generally similar to the standard C-212 except for addition of a nose radome and various external antennae.

OPERATORS: Deliveries had been made by early 1991 to the Spanish Air Force (nine Series 100/200 for SAR duties), Spanish Ministry of Finance (five), Swedish Navy (one for ASW), Swedish Coast Guard (four, equipped with SLAR and IR/UV search gear), Venezuelan Navy (eight for maritime patrol), Mexican Navy (10 for maritime patrol), Portuguese Air Force (five for maritime surveillance and anti-pollution control).

POWER PLANT: As for standard C-212. Auxiliary underwing fuel tanks, total capacity 1,000 litres (269 US gallons; 220 Imp gallons).

ACCOMMODATION (ASW version): Pilot and co-pilot on flight deck, with OTPI and additional central console for radar repeater; controls for radio navigation, Doppler, DME, ADF, UHF/DF, Omega and VOR/ILS; weapons delivery controls; and intervalometer for rockets. Avionics rack on port side, aft of pilot, for com/nav equipment; second rack on starboard side, aft of co-pilot, contains avionics for mission equipment (radar, sonobuoys, MAD and ESM). Immediately aft of the latter rack, along the starboard side of the cabin, are three control consoles for the mission crew members. The first console has the radar control and display, ESM control and display, and intercom switch control. The second has the tactical display and control, MAD recorder and control, and intercom switch (ICS). The rearmost of the three incorporates intercom switch, sonobuoy receiver control unit, acoustic control panel, and acoustic control and display units.

ACCOMMODATION (maritime patrol versions): Pilot and co-pilot on flight deck, with central console for radar repeater; controls for radio navigation, Doppler, DME, ADF, UHF/DF, Omega, VOR/ILS and searchlight. Avionics rack on port side, aft of pilot, for com/nav and radar equipment. On starboard side of cabin is a console for the radar operator that incorporates radar display, long-range system CDU repeater and ASW and ICS controls. Posts for two observers at rear of cabin.

AVIONICS: Communications equipment includes one HF and two VHF transceivers, single UHF, and interphone. Navigation equipment includes automatic flight control system, flight director, VOR/ILS (including VOR/LOC), glide slope and marker beacon receiver, DME, ADF, UHF/DF, radar altimeter, VLF/Omega, autopilot and compass. ASW version has underfuselage search radar with 360° scan, electronic support measures (ESM), sonobuoy processing system (SPS), OTPI, MAD, tactical processing system (TPS), IFF/SIF transponder; maritime patrol version has underfuselage AN/APS-128 100 kW search radar with optional FLIR.

EQUIPMENT: Sonobuoy and smoke marker launcher, searchlight, smoke markers and camera in maritime patrol version.

ARMAMENT: Includes option to carry torpedoes such as Mk 46 and StingRay; air-to-surface missiles such as Sea Skua and AS 15TT; and air-to-surface rockets.

UPDATED

CASA 212 Aviocar maritime patrol aircraft of the Swedish Navy equipped with SLAR and IR/UV search equipment

1994

CASA 212 Aviocar search and rescue (SAR) aircraft of the Spanish Air Force

1994

CASA C-212 AVIOCAR (ELINT/ECM VERSION)

TYPE: Twin-turboprop elint/ECM aircraft.

PROGRAMME: A version of the Aviocar for electronic intelligence and electronic countermeasures duties entered development in 1981. At least six (all Series 200s) have been ordered by undisclosed customers, and two C-212s previously delivered to the Portuguese Air Force were modified retrospectively for elint/ECM duties. This version is also available for the Series 300.

The elint/ECM version carries equipment for automatic signal interception, classification and identification in dense signal environments, data from which enable a map to be drawn plotting the position and characteristics of hostile radars. Emitters for the jamming part of the mission are also carried.

VERIFIED

CASA C-212 SERIES 200 AVIOCAR

Spanish military designations: TR.12D and D.3B
Swedish military designation: Tp 89

TYPE: Twin-turboprop STOL utility aircraft.

CASA C-212-200 Aviocar STOL utility transport aircraft

1994

PROGRAMME: The C-212 200 Series Aviocar is an improved version of the C-212-5 Series 100. The Series 200 has more powerful TPE331-10 engines and increased maximum T-O weight. Aircraft c/n 138 and 139 served as prototypes for this version, making their first flights on 30 April and 20 June 1978 respectively.

Certificated in March 1979 under FAR Pt 25, the Series 200 can be operated under FAR Pt 121 and Pt 135 conditions, and is well within the noise requirements of FAR Pt 36.

By February 1995 sales of the Aviocar (all versions: approximately 135 being Series 100) had reached 447 (207 civil and 240 military). Production of the Series 200 ended in 1987. There are still approximately 78 C-212s in service with the Spanish Air Force although not all of these are Series 200. ASW/maritime patrol and elint/ECM versions of the Aviocar are described separately.

OPERATORS: Versions of the C-212-200 are in service with the armed forces of the following countries: Angola (9), Chad (1), Colombia (5), Costa Rica (2), Equatorial Guinea (3), Indonesia, Mexico (8), Paraguay (4), Spain (9), Sweden (3), Thailand (8), Transkei (2), United Arab Emirates (4), Venezuela (5) and Zimbabwe.

The following description applies to the Series 200 transport aircraft:

DESIGN FEATURES: Cantilever high-wing monoplane. Wing section NACA 65_3-218. Incidence 2° 30′. No dihedral or sweepback. Cantilever tail unit with dorsal fin fairing forward of fin. Tailplane mid-mounted on rear of fuselage.

FLYING CONTROLS: All-metal ailerons and double-slotted trailing-edge flaps. Trim tab in port aileron. Trim tab in rudder and each elevator.

STRUCTURE: The wing is an all-metal light alloy fail-safe structure. The fuselage is a semi-monocoque fail-safe structure of light alloy construction. The tail unit is a two-spar all-metal structure.

LANDING GEAR: Non-retractable tricycle type, with single mainwheels and single steerable nosewheel. CASA oleo-pneumatic shock-absorbers. Goodyear wheels and tyres, main units size 11.00-12 Type III (10 ply rating), nose unit size 24-7-7 Type VII (8 ply rating). Tyre pressure 3.86 bars (56 lb/sq in) on main units, 3.72 bars (54 lb/sq in) on nose unit. Goodyear hydraulic disc brakes on mainwheels. No brake cooling. Anti-skid system optional.

POWER PLANT: Two Garrett TPE331-10R-513C turboprops, each flat rated at 671 kW (900 shp) and driving a Dowty Rotol R-313 four-blade constant-speed fully feathering reversible-pitch propeller (Hartzell propellers on aircraft built before July 1983). Fuel in four integral wing tanks, with total capacity of 2,040 litres (539 US gallons; 449 Imp gallons), of which 2,000 litres (528 US gallons; 440 Imp gallons) are usable. Gravity refuelling point above each tank. Additional fuel can be carried in one 1,000 litre (264 US gallon; 220 Imp gallon) or two 750 litre (198 US gallon; 165 Imp gallon) optional ferry tanks inside cabin, and/or two 500 litre (132 US gallon; 110 Imp gallon) auxiliary underwing tanks. Single pressure refuelling point in starboard wing leading-edge; gravity point for each wing tank. Oil capacity 6.5 litres (1.7 US gallons; 1.4 Imp gallons) per engine.

ACCOMMODATION: Crew of two on flight deck; cabin attendant in civil version. For troop transport role, main cabin can be fitted with 26 inward-facing seats along cabin walls, plus three forward-facing seats, to accommodate 23 paratroops with an instructor/jumpmaster; or seats for 26 fully equipped troops. As an ambulance, cabin is normally equipped to carry 12 stretcher patients and up to four medical attendants. As a freighter, up to 2,700 kg (5,952 lb) of cargo can be carried in main cabin, including two LD1, LD727/DC-8 or three LD3 pallets, or light vehicles. Cargo system which is certificated to FAR Pt 25, includes roller loading/unloading system and 9*g* barrier net. Photographic version is equipped with two Wild RC-10A vertical cameras and a darkroom. Navigation training version has individual desks for instructor and five pupils, in two rows, with appropriate instrument installations. Civil passenger transport version has standard seating for up to 26 persons in mainly four-abreast layout at 72 cm (28.5 in) pitch, with provision for quick change to all-cargo or mixed passenger/cargo interior. Toilet, galley and 400 kg (882 lb) capacity baggage compartment standard. VIP transport version can be furnished to customer's requirements. Forward- and outward-opening door on port side immediately aft of flight deck: passenger door on port side aft of wing; inward-opening emergency exit opposite each door on starboard side. Additional emergency exits in roof and floor of flight deck. A two-section underfuselage loading ramp/door aft of main cabin is openable in flight for discharge of paratroops or cargo, and is fitted with external wheels to allow door to remain open during ground manoeuvring. Interior of rear door can be used for additional baggage stowage. Final production examples of Series 200 have the new toilet/galley kit, and rear baggage compartment (without ramps), as introduced for the Series 300. Entire accommodation heated and ventilated; air conditioning optional.

SYSTEMS: Freon cycle or engine bleed air conditioning system optional. Hydraulic system, pressure 138 bars (2,000 lb/sq in), provides duplicated circuit via electric pump for mainwheel brakes, flaps, nosewheel steering and ventral cargo ramp/door. Handpump for emergency use in case of electric failure. Electrical system is supplied by two 9 kW starter/generators, three batteries and three static converters. Pneumatic boot de-icing of wing and tail unit leading-edges, electric anti-icing of propellers and windscreens. Oxygen system for crew (including cabin attendant); two portable oxygen cylinders for passenger supply. Engine and cabin fire protection systems. Rubber boots and engine bleed air for leading-edge de-icing.

AVIONICS AND EQUIPMENT: Standard avionics include Bendix/King or Collins VHF com, VOR/ILS, ADF, DME, ATC transponder, radio altimeter, intercom (with Gables control) and PA system; King directional gyro; Sperry flight director; Bendix/King weather radar. Blind-flying instrumentation standard. Optional avionics include second King or Collins ADF and transponder; Collins HF com; Global Omega nav; Sperry autopilot; Bendix/King weather radar; Martech emergency radio beacon; and Fairchild flight data and cockpit voice recorders.

ARMAMENT (military versions, optional): Two machine gun pods, two rocket launchers, or one launcher and one gun pod, on hardpoints on fuselage sides (capacity 250 kg; 551 lb each).

DIMENSIONS, EXTERNAL:

Wing span (standard, without optional glassfibre extended tips)	19.00 m (62 ft 4 in)
Wing chord: at root	2.49 m (8 ft 2 in)
at tip	1.50 m (4 ft 11 in)
Wing aspect ratio	9.0
Length overall	15.15 m (49 ft 8½ in)
Fuselage: Max width	2.30 m (7 ft 6½ in)
Height overall	6.30 m (20 ft 8 in)
Tailplane span	8.40 m (27 ft 6¾ in)
Wheel track	3.10 m (10 ft 2 in)
Wheelbase	5.55 m (18 ft 2½ in)
Propeller diameter: Hartzell	2.79 m (9 ft 2 in)

CASA C-212-200 Aviocar commuter aircraft

1995

Dowty 2.74 m (9 ft 0 in)
Propeller ground clearance (min) 1.32 m (4 ft 4 in)
Distance between propeller centres 5.30 m (17 ft 4¾ in)
Passenger door (port, rear):
Max height 1.58 m (5 ft 2¼ in)
Max width 0.70 m (2 ft 3½ in)
Crew and servicing door (port, fwd):
Max height 1.10 m (3 ft 7¼ in)
Max width 0.58 m (1 ft 10¾ in)
Rear-loading door: Max length 3.66 m (12 ft 0 in)
Max width 1.70 m (5 ft 7 in)
Max height 1.80 m (5 ft 10¾ in)
Emergency exit (stbd, fwd): Height 1.10 m (3 ft 7¼ in)
Width 0.58 m (1 ft 10¾ in)
Emergency exit (stbd, rear): Height 0.94 m (3 ft 1 in)
Width 0.55 m (1 ft 9¾ in)
DIMENSIONS, INTERNAL:
Cabin (excl flight deck and rear-loading door):
Length 6.50 m (21 ft 4 in)
Max width 2.10 m (6 ft 10¾ in)
Max height 1.80 m (5 ft 10¾ in)
Floor area 11.33 m^2 (121.1 sq ft)
Volume 23.0 m^3 (812.2 cu ft)
Cabin: volume incl flight deck and rear-loading door 27.0 m^3 (953 cu ft)
Baggage compartment volume 2.90 m^3 (102.4 cu ft)
AREAS:
Wings, gross (standard) 40.0 m^2 (430 sq ft)
Ailerons (total, incl tab) 2.44 m^2 (26.26 sq ft)
Trailing-edge flaps (total) 7.47 m^2 (80.41 sq ft)
Fin, incl dorsal fin 6.27 m^2 (67.49 sq ft)
Rudder, incl tab 2.05 m^2 (22.07 sq ft)
Tailplane 11.90 m^2 (128.09 sq ft)
Elevators, incl tabs 4.36 m^2 (46.93 sq ft)
WEIGHTS AND LOADINGS:
Manufacturer's weight empty 3,780 kg (8,333 lb)
Weight empty, equipped (cargo) 4,400 kg (9,700 lb)
Max payload (cargo) 2,700 kg (5,952 lb)
Max fuel: standard 1,600 kg (3,527 lb)
with underwing auxiliary tanks 2,400 kg (5,291 lb)
Max T-O weight 7,700 kg (16,975 lb)
Max ramp weight 7,750 kg (17,085 lb)
Max landing weight 7,450 kg (16,424 lb)
Max zero-fuel weight 7,100 kg (15,653 lb)
Max cabin floor loading 732 kg/m^2 (150 lb/sq ft)
Max wing loading 192.5 kg/m^2 (39.45 lb/sq ft)
Max power loading 5.74 kg/kW (9.43 lb/shp)
PERFORMANCE (at max T-O weight, ISA):
Max operating speed (VMO) 202 kts (374 km/h; 232 mph) IAS
Max cruising speed at 3,050 m (10,000 ft) 197 kts (365 km/h; 227 mph)
Normal cruising speed at 3,050 m (10,000 ft) 187 kts (346 km/h; 215 mph)
Stalling speed, T-O configuration 78 kts (145 km/h; 90 mph)
Max rate of climb at S/L 474 m (1,555 ft)/min
Rate of climb at S/L, one engine out 108 m (355 ft)/min
Service ceiling 8,535 m (28,000 ft)
Service ceiling, one engine out 3,505 m (11,500 ft)
T-O run 440 m (1,445 ft)
T-O distance (FAR Pt 25, unfactored) 610 m (2,000 ft)
T-O to 15 m (50 ft) 630 m (2,065 ft)
Landing from 15 m (50 ft) 505 m (1,655 ft)
Landing distance (FAR Pt 25, unfactored) without propeller reversal 550 m (1,805 ft)
Landing run 200 m (656 ft)
Min ground turning radius 14.99 m (49 ft 3¼ in)
Range at max cruising speed, no reserves:
with max payload 220 n miles (408 km; 253 miles)
with max fuel 950 n miles (1,760 km; 1,094 miles)
g limits +3/−1.2

UPDATED

SWEDEN

SAAB-SCANIA

SAAB-SCANIA AKTIEBOLAG

S-581 88 Linköping
Telephone: 46 (13) 18 00 00
Fax: 46 (13) 18 18 02
Telex: 50040 SAABLG S
PRESIDENT AND CEO: Lars V. Kylberg
Saab Aircraft AB
S-581 88 Linköping
Telephone: 46 (13) 18 20 00
Fax: 46 (13) 18 27 22
PRESIDENT: Hans Krüger
PUBLIC RELATIONS DIRECTOR: Henry Stenson
Saab Military Aircraft
S-581 88 Linköping
Telephone: 46 (13) 18 00 00
Fax: 46 (13) 18 18 02
GENERAL MANAGER: Hans Ahlinder
PUBLIC RELATIONS DIRECTOR: Jan Ahlgren
SAAB 340 MARKETING:
Saab Aircraft International Ltd, Leworth House, 14-16 Sheet Street, Windsor, Berkshire SL4 1BG, UK
Telephone: 44 (1753) 817520
Fax: 44 (1753) 858884
Telex: 847 815 SFIWIN G
PRESIDENT: Martin Craigs
PUBLIC RELATIONS AND PROMOTIONAL SERVICES:
Mike Savage
Saab Aircraft of America Inc, Loudoun Technical Center, 21300 Ridgetop Circle, Sterling, Virginia 22170, USA
PRESIDENT: John J. Faherty
PUBLIC RELATIONS: Ron Sherman
Telephone: 1 (703) 406 7200
Fax: 1 (703) 406 7224

The original Svenska Aeroplan AB was founded at Trollhättan in 1937 for the production of military aircraft. In 1939 this company was amalgamated with the Aircraft Division (ASJA) of the Svenska Järnvägsverkstäderna rolling stock factory in Linköping, where the main aerospace factory is now located. The company's name was changed to Saab Aktiebolag in May 1965. During 1968 Saab merged with Scania-Vabis, to strengthen the two companies' position in automotive products. Malmö Flygindustri (MFI) was acquired in the same year.

Saab-Scania has more than 26,000 employees. As from 1994 Saab-Scania has combined its defence operations in Saab Defense, which comprises Saab Military Aircraft and the former members of the Combitech Group; Saab Missiles, Saab Instruments and Saab Training Systems. Saab Defense and Saab Combitech together with Saab Aircraft AB, form Business Area Saab Aircraft and Defense. Saab Missiles and Saab Instruments were merged and renamed Saab Dynamics in Spring 1995.

Production of the JA 37 Viggen ended in 1990. Saab-Scania's current aerospace activities include production of the Saab 340B and development of the JAS 39 Gripen. Since 1939 the company has delivered more than 2,000 military jet aircraft and more than 1,500 piston-engined aircraft. Since 1962, it has had a dealership for Schweizer (Hughes) helicopters in Scandinavia and Finland. Since 1978, it has manufactured inboard wing flaps and vanes for the McDonnell Douglas MD-80 series, and now also produces composite spoilers for the MD-82/83. A 25,000 m^2 (269,100 sq ft) factory at Linköping, for final assembly of the Saab 340, was completed in July 1982; an extension to this facility, to accommodate wing and tail unit production, was inaugurated 26 June 1986. A further 15,000 m^2 (161,445 sq ft) expansion has taken place over the past year to accommodate increased production of the Saab 340B and assembly of the Saab 2000.

In the electronics field, current production items include computer systems, autopilots, fire control and bombing systems for piloted aircraft, and electronics for guided missiles. Spaceborne computers, electro-optical fire control systems and field artillery computer systems are also under development and in production. Announced joint marketing agreement with BAe of Saab JAS 39 at Paris Air Show 1995.

UPDATED

SAAB 32 LANSEN

Swedish Air Force designations: A 32A, S 32C, J 32B/D/E

TYPE: Two-seat all-weather attack fighter.

PROGRAMME: The Saab-32 was designed and built for the Swedish Air Force in accordance with a specification laid down by the Swedish Air Board for an aircraft primarily intended for all-weather attack operations against ground and sea targets.

The first prototype of the Saab-32 made its first flight on 3 November 1952, powered by an early version of the British built Rolls-Royce Avon axial flow turbojet engine. The production A 32A, however, is fitted with the Avon RA.7R with afterburner (Swedish Air Force designation RM 5) built under licence in Sweden by Svenska Flygmotor AB of Trollhättan.

During 1953 a Lansen prototype exceeded the speed of sound under complete control during dive tests, one of the first aircraft of its class in the world to do so.

VERSIONS: **A 32A:** First production version for all-weather attack duties. Deliveries began late in 1955. Withdrawn from service.

J 32B: All-weather and night fighter version first flown 7 January 1957. Differs from the A 32A mainly in having a considerably more powerful Svenska Flygmotor built Avon engine (RM 6 in Sweden) fitted with an enlarged Swedish developed afterburner, giving considerably increased rate of climb and speed. The J 32B also features a new and powerful armament, and a new navigation and fire control system. In service with Royal Swedish Air Force.

J 32D: Version used for target towing. In service with Royal Swedish Air Force.

J 32E: ECM version. See separate entry.

S 32C: Reconnaissance version with camera-nose. 44 aircraft delivered between 1958-59. Withdrawn from service.

UPGRADE: **Saab:** J 32E Lansen. See separate entry.

OPERATORS: Versions of the Saab 32 are in service with the armed forces of Sweden (24).

The description below refers specifically to the A 32A all-weather attack version, but is also generally applicable to the J 32B:

DESIGN FEATURES: Swept low-wing cantilever monoplane with 35° sweepback at 25 per cent wing chord. Thin laminar flow wing section. Aspect ratio 4.5. Wing fence on each wing.

FLYING CONTROLS: Ailerons operated by hydraulic booster system of Saab design. Fowler flaps. Four airbrakes in rear fuselage sides. Elevator operated by hydraulic booster system.

STRUCTURE: The wing is an all-metal flush riveted stressed skin structure. The fuselage is an all-metal structure with flush riveted stressed skin. Entire rear fuselage is quickly removable for access to engine. The tail unit is a cantilever monoplane type. Movable tailplane mounted on fairings used to smooth the air flow around the tailplane.

LANDING GEAR: Retractable nosewheel type. Hydraulically (Saab) retractable, mainwheels into the fuselage, the nosewheel forwards. Goodyear wheels and brakes, with Dunlop Maxaret anti-skid device.

Saab J 32B all-weather and night fighter aircraft with Swedish-built RR Avon Series 200 engine

1995

POWER PLANT: One Svenska Flygmotor built Rolls-Royce Avon RA.7R (Swedish designation RM 5) axial flow turbojet engine, with re-heat installation. Air intakes in fuselage sides.

ACCOMMODATION: Pressurised cockpit for pilot and observer. Ejector seats of Saab design. The aircraft may be equipped with dual controls to facilitate crew training.

ARMAMENT: Four fuselage-mounted 20 mm cannon. Various combinations of underwing stores can be carried including bombs (50, 100, 250 or 500 kg), rockets (13.5, 15 or 18 cm calibre) and so on. The aircraft can also carry the Type 304 air-to-surface guided missile.

DIMENSIONS, EXTERNAL:

Wing span	13.00 m (42 ft 8 in)
Length	14.94 m (49 ft 1 in)
Height	4.65 m (15 ft 3 in)

WEIGHTS AND LOADINGS:

Weight empty	approx 7,000 kg (15,400 lb)
Max T-O weight	13,600 kg (30,000 lb)

PERFORMANCE:

Max speed	over 594.6 kts (1,100 km/h; 700 mph)
Landing speed	108 kts (200 km/h; 125 mph)
Service ceiling	approx 15,000 m (49,200 ft)

UPDATED

Saab J 32B Lansen (background) Saab J 32E Lansen (foreground) *(Peter Liander)*

1994

Saab J 35OE of the Austrian Air Force *(Peter Cooper)*

1995

Saab J 35OE Draken of the Austrian Air Force *(Saab)*

1994

SAAB J 32E LANSEN ELECTRONIC WARFARE AND COUNTERMEASURES UPGRADE

TYPE: Two-seat ECM and EW aircraft upgrade.

PROGRAMME: Details of this extensively modified electronic warfare and countermeasures version of the Lansen were released in 1985. It equips F13M target flying squadron of the Swedish Air Force, which is based at Malmslätt, near Linköping, but operates throughout Sweden and provides target flying services for other countries. Currently, F13M consists of two flying units. Flying unit 85 performs most of the Air Force's signal reconnaissance missions, using two specially equipped Tp85 Caravelles, supported by a Fairchild Metro for flight training. Flying unit 32 has 14 J 32E (for electronic) Lansens, six J 32D Lansen target tugs and three dual-control J 32B Lansens for training and radioactive sampling. The J 32Bs are equipped with FFV Aerotech gamma pods (three under each wing), for collecting gamma-emitting radionuclides in the tropopause. The J 32Es can be used for a variety of missions, including jamming and countermeasures training, and 'aggressor' flying for SwAF combat squadrons. The J 32E will remain in service up to 2005.

DESIGN FEATURES: The J 32E is a conversion of the standard J 32B, fitted with new flight instrumentation, a civil and military transponder, modernised autopilot and a range of specialised mission equipment, including:

Ingeborg, a microcomputer signal homing receiver for the S-, C- and part of L-band, which works in parallel with the Adrian and G 24 systems to make possible optimum jamming. Its three antennae are inside the aircraft's dielectric nosecone. The control unit replaces the original radar scope in the navigator's cockpit.

G 24, a nose-mounted jamming transmitter which exists in three versions, covering the L-, S- and C-bands, and is for use against ground and ship radars.

Boz 3 chaff dispenser pods, usually carried on each of the two outboard underwing pylons. Operator can select different dispensing programmes covering various radar bands.

Petrus jamming pod is intended mainly for use against X-band fighter, attack aircraft and anti-aircraft radars. Microcomputer control equipment can transmit camouflage (roar) and disguise jamming, and generate radar signatures within wide ranges, ahead and rearward, and also warns of attacks from the rear. Petrus is carried on the J 32E's inboard underwing pylons.

Adrian jamming pod. Externally similar to, and interchangeable with, Petrus. Used mainly against ground and ship S- and C-band radars. Has forward and rearward pointing antennae.

Mera is a computerised radio jamming equipment and homing receiver for the VHF and UHF bands. Jamming can take place at several frequencies, such as FM/AM (roar), pulse and oscillating roar. Using a tape-recorded disguise, replay of commands and music can be transmitted. The jamming operator can also transmit misleading commands on a selected frequency.

UPDATED

SAAB 35 DRAKEN (DRAGON)

TYPE: Single-seat supersonic all-weather fighter, reconnaissance and attack aircraft.

PROGRAMME: The Saab 35 Draken single-seat fighter was originally designed to intercept bombers in the transonic speed range at high altitude, and carries radar equipment to accomplish this under all weather conditions. It is also able to carry substantial weapon loads for attack duties or cameras for photographic reconnaissance .

The first of three prototypes made its first flight on 25 October 1955, and the first version, the J 35A, entered service with the Swedish Air Force at the beginning of 1960. Subsequently the Draken went through several stages of development, and continuously improved versions for the Swedish Air Force included the J 35B, D and F fighter versions, the SK 35C trainer version and the S 35E reconnaissance version. Production of these models has been completed. Versions of the J 35 are still in service with the air forces of Austria (24), Finland (45) and Sweden (approximately 40). Four Danish aircraft (two-seaters) have been exported to the US National Test Pilot School in California for training pilots and flight test engineers.

VERSIONS: **Saab 35X:** Long-range fighter/attack/reconnaissance version developed for the export market. Externally similar to the J 35F but has greatly increased attack capabilities (maximum external load 4,500 kg; 9,920 lb) and range. For reconnaissance duties, a nose similar to that of the J 35E is fitted. T-O run with nine 1,000 lb bombs is 1,210 m (4,030 ft).

Saab 35A: Initial fighter version.

Saab 35B: Improved fighter version.

Saab 35BS: Finnish Air Force air defence aircraft.

Saab 35C: Trainer version.

Saab 35CS: Finnish Air Force operational trainer aircraft.

Saab 35D: Improved fighter version.

Saab 35E: Reconnaissance version.

Saab 35F: Fighter version.

Saab 35FS: Finnish Air Force fighter version.

Saab 35J: Upgraded J 35F. See separate entry.

Saab 35OE: Austrian Air Force fighter version.

Saab 35XD: Danish Air Force fighter version.

UPGRADES: **Saab Military Aircraft:** Saab J 35J see separate entry.

Saab Military Aircraft: Plan to upgrade Austrian J 35OE with underwing weapons pylons for air-to-air missiles (AIM-9 Sidewinders).

Valmet: Plan to upgrade Austrian J 350E with radar warning receiver and chaff/flare dispenser. First aircraft completed late 1994.

OPERATORS: Versions of the Saab 35 are in service with the armed forces of the following countries: Austria (24), Finland (45) and Sweden (74).

The following description applies to the basic Saab 35X:

DESIGN FEATURES: Cantilever 'double delta' mid-wing monoplane. Sweepback on centre wing leading-edge 80°, on outer wing leading-edge 57°. Thickness/chord ratio 5 per cent. Central wing integral with fuselage. Conventional delta shape fin and rudder.

FLYING CONTROLS: Two pairs of airbrakes, above and below rear fuselage. Elevons on wing trailing-edge comprise two inboard and two outboard surfaces, the latter being mass-balanced. Each control surface servo-operated by two hydraulic tandem jacks fed by two separate hydraulic systems. No aerodynamic loading on control surfaces is fed into stick and rudder pedals. Stick forces are generated artificially. Three-axis stabilisation system.

STRUCTURE: The wing is an all-metal stressed skin structure with some bonding. Outer wing panels, attached to centre wing with a bolt joint, have relatively thick skin on a framework of spars and ribs and can be detached for transportation by road. The fuselage is an all-metal structure, in front and rear sections, connected to each other by a bolt joint. Fuselage front section integral with front of centre wing structure.

LANDING GEAR: Retractable tricycle type. Hydraulic actuation. Main units retract outward, the legs shortening during retraction to reduce the space required inside wing. Nosewheel retracts forward and is steerable. Tyre pressures 10-13 kg/cm² (142-185 lb/sq in) on nose unit, 12-17 kg/cm² (171-242 lb/sq in) on main units. Goodyear double-disc brakes and Dunlop anti-skid brake units. Dual retractable tailwheels. Brake parachute in fairing above rear fuselage. Arrester hook optional.

POWER PLANT: One Volvo Flygmotor (Rolls-Royce licence) Avon 300 series engine (Swedish Air Force designation RM6C) with Swedish developed afterburner. Static thrust approximately 5,800 kg (12,790 lb) dry and 8,000 kg (17,650 lb) with afterburner. Internal fuel in inner wing tanks. Total internal fuel capacity 2,825 litres (747 US gallons; 565 Imp gallons). Provision for external tanks under fuselage and wings, increasing total capacity to 4,925 litres (1,303 US gallons; 985 Imp gallons). Additional internal tanks can be fitted in place of guns for ferry purposes. Single-point pressure fuelling system, capacity 840 litres (222 US gallons; 185 Imp gallons)/min.

ACCOMMODATION: Pressurised and air conditioned cockpit, with fully automatic Saab 73SE-F rocket-assisted ejection seat and GQ parachute system permitting ejection within the normal flight envelope and down to 54 knots (100 km/h; 62 mph) on the ground. Rearward hinged canopy.

SYSTEMS: Duplicated hydraulic system, with two independent pumps for control surface and landing gear actuation. Third pump, for emergency use, is driven by ram air in case of engine failure. Three-phase AC electrical system supplies 200/115 V 400 Hz power via a 20 kVA engine-driven generator or, in emergency, via a 3.5 kVA generator in emergency power unit. Equipment requiring DC power is fed from these AC systems via two rectifiers giving 2.2 kW at 29 V. One 6 Ah accumulator acts as a buffer. The power permits three engine starting attempts.

ELECTRONICS AND EQUIPMENT: Complete radar equipment with nose scanner and pilot's scope, as well as Saab S7 collision-course fire control equipment. Honeywell autopilot, with air data system, stick-steering and various following modes. Vertical type instruments. Celsius FR28 VHF/UHF. DME.

ARMAMENT: Six attachment points (each 454 kg; 1,000 lb) for external stores; two under each wing and two under fuselage. Stores can consist of air-to-air missiles and unguided air-to-surface rocket pods (19 × 7.5 cm), 12 × 13.5 cm Bofors air-to-ground rockets, nine 454 kg (1,000 lb) bombs. Two or four RB24 Sidewinder air-to-air missiles can be carried under wings and fuselage. One 30 mm Aden cannon.

Saab J 35F single-seat fighter aircraft *(Saab)*

1994

DIMENSIONS, EXTERNAL:

Wing span	9.40 m (30 ft 10 in)
Wing aspect ratio	1.77
Width, outer wing panels removed	4.40 m (14 ft 5 in)
Length overall	15.35 m (50 ft 4 in)
Height overall	3.89 m (12 ft 9 in)
Wheel track	2.70 m (8 ft 10½ in)
Wheelbase	4.00 m (13 ft 1 in)

AREAS:

Wings, gross	49.20 m² (529.6 sq ft)
Fin	5.25 m² (56.51 sq ft)
Rudder	0.95 m² (10.23 sq ft)

WEIGHTS AND LOADINGS:

T-O weight clean	10,354 kg (23,856 lb)

PERFORMANCE (A: AUW of 25,130 lb, B: AUW of 32,165 lb):

Max level speed with afterburning: A	Mach 2
B	Mach 1.4
Max rate of climb at S/L with afterburning:	
A	10,500 m (34,450 ft)/min
B	6,900 m (22,650 ft)/min
Time to 11,000 m (36,000 ft) with afterburning:	
A	2 min 36 s
Time to 15,000 m (49,200 ft) with afterburning:	
A	5 min
T-O run with afterburning: A	650 m (2,130 ft)
B	1,170 m (3,840 ft)
T-O to 15 m (50 ft) with afterburning:	
A	960 m (3,150 ft)
B	1,550 m (5,080 ft)
Landing run at normal landing weight:	
A and B	530 m (1,740 ft)
Radius of action (hi-lo-hi, internal fuel only):	
A	343 n miles (635 km; 395 miles)
Radius of action (hi-lo-hi) with two 1,000 lb bombs and two drop tanks: B	541 n miles (1,003 km; 623 miles)
Ferry range with max internal and external fuel	1,754 n miles (3,250 km; 2,020 miles)

UPDATED

SAAB J 35J DRAKEN UPGRADE

TYPE: Combat aircraft upgrade.

PROGRAMME: Under the revised designation **J 35J** (previously J 35F Mod or J 35F-Ny: new), 66 J 35F Draken fighters of F10 Wing of the Swedish Air Force, based near Ängelholm in southern Sweden, have been updated to extend their service life until the end of the 1990s when they will be replaced by the JAS 39 Gripen. There are about 50 J 35 still flying. Saab Military Aircraft was responsible for modification and redelivery, with FFV Aerotech as subcontractor.

DESIGN FEATURES: Two additional inboard underwing pylons are fitted, allowing four drop tanks and two air-to-air missiles, or four missiles (two RB24 Sidewinders and two RB27 Falcons) and two 550 litre (145.3 US gallon; 121 Imp gallon) tanks. Operation of the automatic gun and weapons electronics are changed, radar and IFF improved, infra-red missile target seeker upgraded, altitude warning system added, and instruments improved. F10 Wing will be the only unit operating this version of the Draken. Redeliveries were completed in 1991.

VERIFIED

Saab J 35J Draken upgraded aircraft with additional underwing hardpoints ***(Saab)***

1994

SAAB JA 37 VIGGEN (THUNDERBOLT)

TYPE: Single-seat all-weather multipurpose combat aircraft.

PROGRAMME: The Saab 37 Viggen multimission combat aircraft was produced to fulfil the primary roles of attack, interception, reconnaissance and training. In 1994 eight fighter squadrons, six combined fighter/attack and reconnaissance squadrons were in service with the Swedish Air Force.

The first of seven prototypes flew for the first time on 8 February 1967; the seventh was the prototype for the two-seat SK 37 operational trainer. First squadron delivery (of AJ 37s to F7) was made in June 1971.

Production deliveries of the AJ 37, SF 37, SH 37 and SK 37 versions totalled 180 (110, 26, 26 and 18). The final version (149 built) was the JA 37 interceptor, the last of which was delivered on 29 June 1990, bringing overall Viggen production to 329.

VERSIONS: **AJ 37**: Single-seat all-weather attack version with secondary interceptor capability.

JA 37: Single-seat interceptor with more powerful Volvo Flygmotor RM8B engine. Improved performance and capability for attack missions. Four elevon hydraulic actuators under each wing instead of three as on other versions. Modified taller tail fin similar to that of the SK 37.

SF 37: Single-seat all-weather armed photographic reconnaissance version to replace the S 35E Draken. Modified nose containing cameras and other equipment.

SH 37: Single-seat all-weather maritime reconnaissance version to replace the S 32C Lansen. Primarily to survey, register and report activities in the neighbourhood of Swedish territory.

SK 37: Tandem two-seat dual-control training version in which the rear cockpit takes the place of some electronics and forward fuselage fuel tank and is fitted with bulged hood and periscopes. Modified taller fin. Capable of secondary attack role.

AJS 37: Planned modification of 115 Viggens with broader range of weapons and additional avionics to allow multirole operational to be built up prior to the introduction of the JAS 39 Gripen. See following entry.

UPGRADES: **Saab Military Aircraft**: AJS 37. See Versions and following entry.

OPERATORS: Versions of the JA 37 are in service with the armed forces of Sweden (129).

DESIGN FEATURES: Tandem arrangement of delta foreplane, with trailing-edge flaps, and a rear-mounted delta main wing with two-section hydraulically actuated powered elevons on each trailing-edge. Main wing has compound sweep on leading-edge plus dog-tooth extensions on outer panels. Extensive use of metal bonded honeycomb panels for wing control surfaces, foreplane flaps and main landing gear doors.

The fuselage is a conventional structure, using light metal forgings, heat resistant plastics bonding, titanium and metal bonded honeycomb. Four plate type airbrakes round rear fuselage.

Pitch and lateral controls in the trailing-edge of the wing with conventional fin and rudder.

LANDING GEAR: Retractable tricycle type of Saab origin, built by Motala Verkstad and designed for a maximum rate of sink of 5 m (16.4 ft)/s. Power steerable twin-wheel nose unit retracts forward. Each main unit has two wheels in tandem and retracts inward into main wing and fuselage. Main oleos shorten during retraction. Nosewheel tyres size 18 × 5.5, pressure 10.7 bars (155 lb/sq in). Mainwheel tyres size 26 × 6.6, pressure 14.8 bars (215 lb/sq in). Goodyear wheels and brakes. Dunlop anti-skid system.

POWER PLANT: One Volvo Flygmotor RM8B turbofan, rated at 72.1 kN (16,203 lb st) dry and 125 kN (28,108 lb st) with afterburning. Thrust reverser doors actuated automatically by compression of oleo as nose gear strikes runway. Fuel in one tank in each wing, saddle tank over engine, one tank in each side of fuselage, and one aft of cockpit. Pressure refuelling point beneath starboard wing. Provision for jettisonable external tank on centreline pylon.

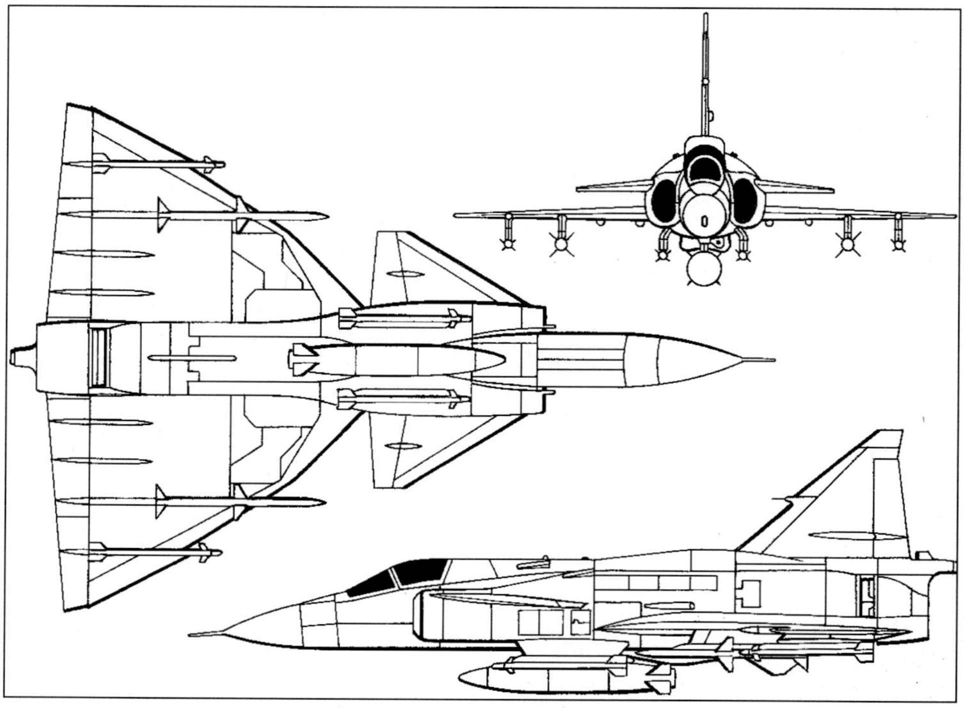

Saab JA 37 Viggen single-seat all-weather multipurpose combat aircraft ***(Jane's/Mike Keep)***

1994

Saab SH 37 Viggen maritime reconnaissance aircraft ***(Saab)***

1994

ACCOMMODATION: Pilot only, on Saab-Scania fully adjustable rocket-assisted zero/zero ejection seat beneath rearward hinged clamshell canopy. Cockpit pressurisation, heating and air conditioning by engine bleed air. Birdproof windscreen.

AVIONICS: Advanced target search and acquisition system, based on high-performance long-range Ericsson PS-46/A pulse Doppler radar resistant to variations of weather, altitude, clutter and ECM. Automatic speed control system, Smiths electronic head-up display, Bofors Aerotronics aircraft attitude instruments, radio and fighter link equipment, Plessey Electronic SKC-2037 central digital computer, Garrett LD-5 digital air data computer, Plessey Electronic KT-70L inertial measuring equipment, Honeywell/Saab-Scania SA07 digital automatic flight control system, Honeywell radar altimeter, SATT radar warning system, Ericsson radar display system and electronic countermeasures, and AIL Tactical Instrument Landing System (TILS). Most avionics connected to central digital computer, which is programmed to check out and monitor these systems both on ground and during flight. Ram air cooling for avionics compartment.

ARMAMENT: Permanent underbelly pack, offset to port side of centreline, containing one 30 mm Oerlikon KCA long-range cannon with 150 rounds. Three underfuselage and four underwing hardpoints. Armament can include two BAe Sky Flash (Swedish designation RB71) and six AIM-9L Sidewinder (RB74) air-to-air missiles. For air-to-surface attack, a total of 24 135 mm rockets can be carried in four pods.

DIMENSIONS, EXTERNAL:

Main wing span	10.60 m (34 ft 9¼ in)
Main wing aspect ratio	2.4
Foreplane span	5.45 m (17 ft 10½ in)
Length: overall (incl probe)	16.40 m (53 ft 9¾ in)
fuselage	15.58 m (51 ft 1½ in)
Height: overall	5.90 m (19 ft 4¼ in)
main fin folded	4.00 m (13 ft 1½ in)
Wheel track	4.76 m (15 ft 7½ in)
Wheelbase (c/l of shock-absorbers)	5.69 m (18 ft 8 in)

AREAS:

Main wings, gross	46.00 m² (495.1 sq ft)
Foreplanes, outside fuselage	6.20 m² (66.74 sq ft)

WEIGHTS AND LOADINGS (approx):

T-O weight: clean	16,400 kg (36,146 lb)
with normal armament	17,000 kg (37,478 lb)

PERFORMANCE:

Max level speed: at high altitude	above Mach 2
at 100 m (330 ft)	Mach 1.2
Approach speed	approx 119 kts (220 km/h; 137 mph)
Time to 10,000 m (32,800 ft) from brakes off, with afterburning	less than 1 min 40 s
T-O run	approx 400 m (1,310 ft)
Landing run (with reverser)	approx 500 m (1,640 ft)
Required landing field length: conventional landing	1,000 m (3,280 ft)
no-flare landing	500 m (1,640 ft)
Tactical radius with external armament: hi-lo-hi	over 540 n miles (1,000 km; 620 miles)
lo-lo-lo	over 270 n miles (500 km; 310 miles)

UPDATED

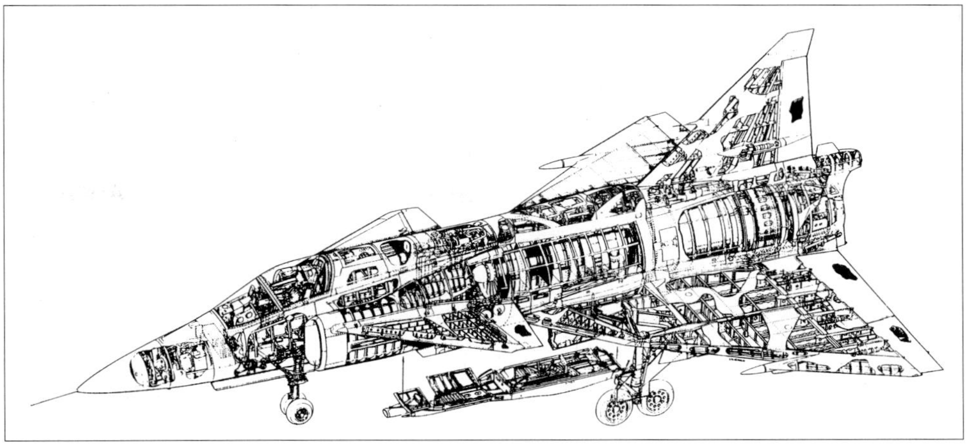

Saab JA 37 single-seat all-weather combat aircraft

1995

Saab JA 37 Viggen interceptor of the Swedish Air Force

1995

SAAB AJS 37 VIGGEN (THUNDERBOLT)

TYPE: Multirole attack interceptor and reconnaissance fighter upgrade.

PROGRAMME: Last production Viggen (JA 37) delivered 29 June 1990; AJS 37 conversion announced May 1991; converting 75 AJ, SF and SH 37s into interchangeable defence, attack and reconnaissance fighters; redeliveries began mid-1993 and are due for completion in 1996.

CUSTOMERS: Total 329 Viggens produced for Swedish Air Force, comprising 110 AJ 37, 26 SF 37, 26 SH 37, 18 SK 37 and 149 JA 37.

COSTS: AJS 37 programme SEK300 million (1991) ($48.3 million).

DESIGN FEATURES: Combines all three missions in one aircraft.

AVIONICS: New computer-based threat analysis/mission planning system; multiprocessors and databus to increase computing power and improved communications; updated ECM.

ARMAMENT: Wider range of weapons to include DASA DWS 39 and six types of missile including RB05, RB15F, RB24/24J and RB74; same reconnaissance pod as JAS 39 Gripen.

NEW ENTRY

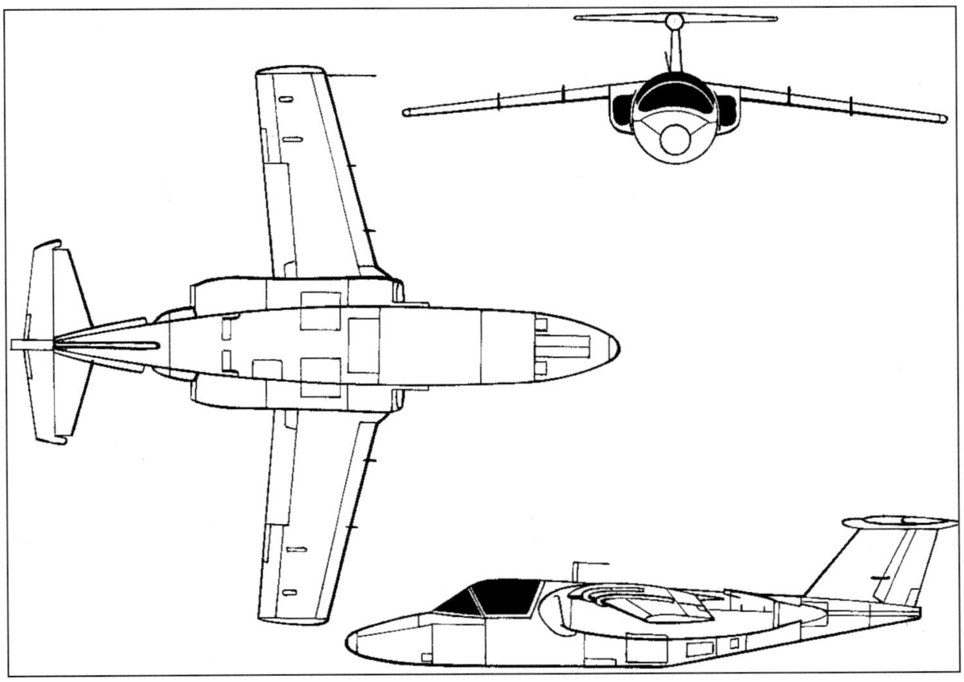

Saab 105 twin-jet multipurpose aircraft *(Jane's/Mike Keep)*

1994

SAAB 105

Swedish Air Force designation: SK60

Austrian Air Force designation: Saab 105OE

TYPE: Twin-jet multipurpose aircraft.

PROGRAMME: The Saab 105 first flew in 1963. Between 1966 and 1969, a total of 150 production aircraft were delivered to the Swedish Air Force, under the overall designation SK60. Of these 140 remain operational in five versions. The SK60 is unique in that it is the only aircraft in the Swedish Air Force used for everything from primary to tactical training.

VERSIONS: **SK60A:** Two-seat primary, basic and advanced trainer.

SK60B: Two-seat light attack and advanced trainer.

SK60C: Two-seat light attack, reconnaissance and advanced trainer aircraft.

SK60D: Four-seat liaison aircraft.

SK60E: Four-seat liaison aircraft with civil avionics.

Saab 105OE: Austrian version of the Saab 105XT.

Saab 105XT: Export version with more powerful General Electric J85-17 engines.

UPGRADES: **Saab Military Aircraft:** Swedish Air Force SK60 underwent a life extension programme in 1987 to continue operations of 140 aircraft until 2010. Wings strengthened, new parachutes and more comfortable harness introduced. Further programme to update avionics is envisaged.

Saab Military Aircraft: Austrian Air Force Saab 105OE to undergo life extension programme to continue operations until 2010.

Saab 105OEs of the Austrian Air Force

1994

Saab SK60B two-seat twin-jet multipurpose aircraft

1995

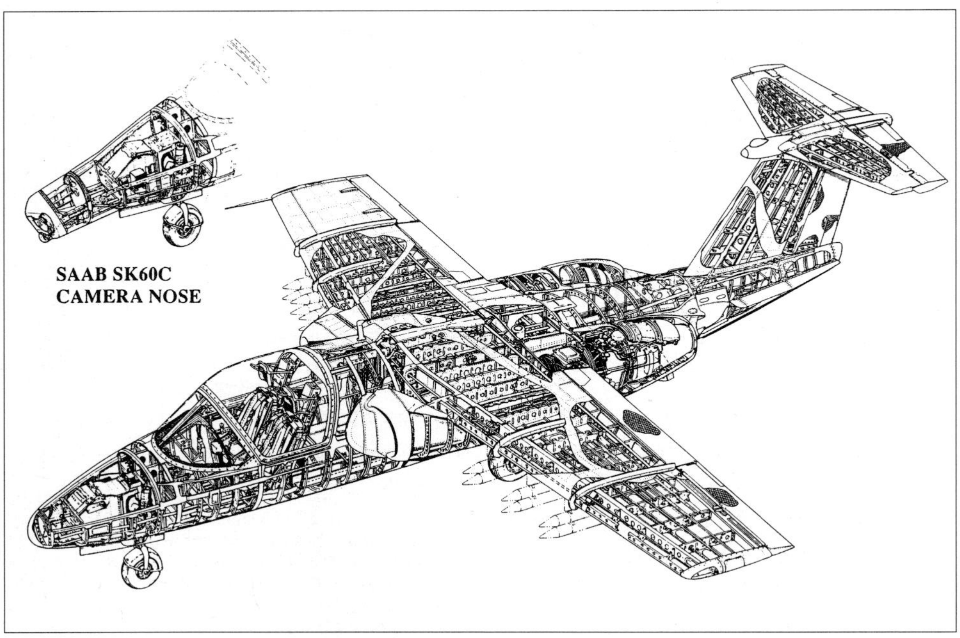

Saab SK60A/B with SK60C camera nose above

1995

Saab Military Aircraft/Williams-Rolls: Programme signed in November 1993 to re-engine 115 SK60 (Swedish Air Force) twin-engine trainers (with an option for 20 more) with Williams/Rolls-Royce FJ44 turbofan engines. Programme to take place 1994-1998. See following entry.

The following details refer to the Saab 105OE unless otherwise stated:

DESIGN FEATURES: Cantilever shoulder-wing monoplane. Sweepback 12° 48′ at quarter-chord. Anhedral 6°. Thickness/chord ratio 10.3 per cent at root, 12 per cent at tip. Two small fences on upper surfaces of each wing. The tailplane is a cantilever structure with tailplane mounted at tip of fin. Small ventral fin mounted on aft fuselage.

FLYING CONTROLS: Aerodynamically balanced ailerons with boosted control. Geared servo tab in each aileron; starboard tab adjustable mechanically for trimming. Hydraulically operated single-slotted flaps. Hydraulically operated perforated airbrakes pivoted in transverse slots in lower fuselage aft of landing gear. Tail unit control surfaces statically and aerodynamically balanced. Electrically operated trim tab in rudder.

STRUCTURE: The fuselage is a semi-monocoque structure.

LANDING GEAR: Retractable tricycle type with hydraulic actuation. Main units retract into fuselage. Forward-retracting hydraulically steerable nosewheel. Hydraulic disc brakes with anti-skid.

POWER PLANT: The standard engine at present in all SK60 aircraft (A-E) if the Turbomeca-SNECMA Aubisque (RM9B) turbofan. During 1995-98 115 aircraft will be equipped with Williams-Rolls FJ44 (RM15) turbofan engines. The Saab 105OE retains General Electric J85-17B turbojet engines.

ACCOMMODATION: Two side by side ejection seats. Alternative provision for four fixed seats.

AVIONICS: Wide range of navigational and communication equipment. Standard installation includes one VHF and one UHF unit with audio control, one VOR/ILS with marker beacon, one ADF, one transponder and one DME.

ARMAMENT: Three attachment points under each wing. Weapons include; bombs, including flare bombs and napalm, rockets, air-to-surface and air-to-air missiles as well as miniguns or camera pod.

DIMENSIONS, EXTERNAL (SK60A, Saab 105OE):
Wing span 9.50 m (31 ft 2 in)
Length 10.80 m (35 ft 4 in)
Height 2.70 m (8 ft 9 in)
Wing area 16.30 m² (175.5 sq ft)

WEIGHTS AND LOADINGS (trainer):
Empty weight, equipped: SK60A 2,770 kg (6,106 lb)
Saab 105OE 2,960 kg (6,526 lb)
Fuel (max internal): SK60A 1,050 kg (2,315 lb)
Saab 105OE 1,620 kg (3,571 lb)
T-O weight: SK60A, normal/max fuel 3,680/3980 kg (8,113/8,774 lb)
Saab 105OE 4,750 kg (10,472 lb)

PERFORMANCE:
T-O run: SK60A 680 m (2,231 ft)
Saab 105OE 405 m (1,329 ft)
Landing run: SK60A 490 m (1,608 ft)
Saab 105OE 620 m (2,034 ft)
Max speed:
at sea level: SK60A 402 kts (745 km/h; 463 mph)
Saab 105OE 513 kts (950 km/h; 591 mph)
at 6,000 m (20,000 ft):
SK60A 424 kts (785 km/h; 488 mph)
Saab 105OE 502 kts (930 km/h; 578 mph)
Max initial rate of climb:
SK60A 25.0 m/s (4,921 ft/min)
Saab 105OE 59.0 m/s (11,614 ft/min)
Service ceiling: SK60A 12,000 m (39,370 ft)
Saab 105OE 13,700 m (44,950 ft)
Ferry range (no fuel reserves):
SK60A 1,377 n miles (2,550 km; 1,585 miles)
Saab 105OE 1,296 n miles (2,400 km; 1,490 miles)

UPDATED

SAAB SK60 (SAAB 105) RE-ENGINE PROGRAMME

TYPE: All-aspect jet trainer.

PROGRAMME: Of 150 Saab 105s delivered to Swedish Air Force 1966-69, about 140 remain in service and have, since 1987, been its only dedicated trainer, covering all stages from primary through basic and advanced to tactical training. Upgrade programme then began to ensure continued service until at least 2010; have already been refitted with strengthened wings, new seat harnesses and new parachutes. Latest instalment of upgrade programme is 4 November 1993 contract to refit with Williams-Rolls FJ44 turbofans and introduce more modern flight instruments; 115 aircraft to receive new engines (with option for further 20); Saab Military Aircraft undertaking initial batch of 10, for first flight Spring 1995 and redelivery during 1996; thereafter, most of work will be carried out at Swedish Air Force Flight Academy (F5 Wing at Ljungbyhed), with 115th redelivery scheduled for Summer 1998.

CUSTOMERS: Swedish Air Force (115 plus option for 20).

COSTS: Re-engineering programme approximately SEK900 million.

DESIGN FEATURES: Engine intakes modified as necessary for new power plant. Lighter engines decrease maximum T-O weight by 500 kg (1,102 lb); extra power significantly increases climb rate and T-O performance; other gains include greater fuel efficiency, reduced maintenance time and cost, and lower noise and exhaust emission levels.

POWER PLANT: Twin 7.28 kN (1,636 lb st) Turbomeca/SNECMA Aubisque turbofans replaced by 8.45 kN (1,900 lb st) Williams-Rolls FJ44s turbofan engines.

NEW ENTRY

SAAB 340A

TYPE: Twin-turboprop regional transport aircraft.

PROGRAMME: First details of the 340, then called Saab-Fairchild 340, were announced January 1980 and the go-ahead on joint design, development, manufacture and marketing was given in September 1980. Saab took complete control of the programme in November 1985 and Fairchild continued as a subcontractor until 1987, when the designation was changed to Saab 340.

The first prototype (SE-ISF) made the first flight on 25 January 1983. The fourth (first production) 340A (SE-E04) flew 5 March 1984. Ten European nations and the USA took part in the joint certification process to FAR/JAR 25, resulting in Swedish certification May 1984 and virtually simultaneous certification by the other countries in June.

The 340A first went into service, with Crossair, in June 1984 and the first in the USA entered service in August that year. The first corporate 340A was delivered in November 1985. From mid-1985, engine power was increased from 1,215 kW (1,630 shp) to 1,294 kW (1,735 shp) and propeller diameter was increased. Earlier aircraft were retrofitted. An improved cabin with new lining and larger overhead bins, plus improved fittings, was introduced in mid-1988. It was designed and fitted by Metair Aircraft of the UK, meeting the FAA fire resistance standards that became mandatory August 1990.

The 340B hot-and-high version was announced late 1987 and replaced the 340A on the production line from c/n 160. It was certificated on 3 July 1989 and the first aircraft was delivered to Crossair in September 1989. By 1 February 1991, firm orders for the 340A and B totalled 354, of which more than half were for US customers and the rest for Europe, Australia, Southeast Asia and South America. The 200th 340 was handed over 14 August 1990, and a total of 229 340As and 340Bs had been delivered by 31 March 1991.

UPGRADES: **Ultra Electronics:** Installation of active noise control (ACN) systems as retrofit programme. See separate entry.

DESIGN FEATURES: Cantilever low-wing monoplane. Basic wing section NASA MS(1)-0313 with thickness/chord ratios of 16 and 12 per cent at root and tip respectively. Dihedral 7° from roots. Incidence 2° at root. Sweepback 3° 36′ at quarter-chord. The tail unit is a cantilever structure, with sweptback vertical and non-swept horizontal surfaces, the latter having marked dihedral.

FLYING CONTROLS: Hydraulically actuated single-slotted trailing-edge flaps with aluminium alloy spars, honeycomb panels faced with aluminium sheet, and leading/trailing-edges of Kevlar. Ailerons have Kevlar skins and glassfibre leading-edges. Electrically operated geared/trim tab in each aileron. Geared trim tab in each elevator; spring trim tab in rudder.

STRUCTURE: Tapered two-spar wings embodying fail-safe principles. Stringers and skins of 2024/7075 aluminium alloy. Wingroot/fuselage fairings of Kevlar sandwich. The fuselage is a conventional fail-safe/safe-life semi-monocoque pressurised metal structure, of circular cross-section. Built in three portions: nose (including flight deck), passenger compartment, and tail section (incorporating baggage compartment). All doors of aluminium honeycomb. Nosecone of Kevlar; cabin floor of carbonfibre sandwich. Fin integral with fuselage. Construction similar to that of wings, with tailplane and fin of aluminium honeycomb. Rudder and elevators have Kevlar skins and glassfibre leading-edges.

LANDING GEAR: Retractable tricycle type, of AP Precision Hydraulics design and manufacture, with twin Goodyear wheels and oleo-pneumatic shock-absorber on each unit. Hydraulic actuation. All units retract forward, main units into engine nacelles. Mainwheel doors of Kevlar sandwich. Hydraulically steerable nose unit (60° to both left and right), with shimmy damper. Mainwheel tyres size 24 × 7.7-10, pressure 6.89 bars (100 lb/sq in); nosewheel tyres size 18 × 6.0-6, pressure 3.79 bars (55 lb/sq in). Independent Goodyear carbon hydraulic disc brakes on main units, with anti-skid control.

POWER PLANT: Two General Electric CT7-5A2 turboprops, each rated at 1,294 kW (1,735 shp). Dowty Rotol four-blade slow-turning constant-speed propellers with full autofeathering and reverse pitch capability, each with spinner and glassfibre/polyurethane foam/carbonfibre moulded blades. Fuel in each tank in each outer wing; total capacity 3,220 litres (850.5 US gallons; 708 Imp gallons). Single-point pressure refuelling inlet in starboard outer wing panel. Overwing gravity refuelling point in each wing.

ACCOMMODATION: Two pilots and provision for observer on flight deck; attendant's seat (forward, port) in passenger cabin. Main cabin accommodates up to 35 passengers, in 11 rows of three, with aisle, and rearward-facing seat(s) on starboard side at front. One or both rearward-facing seats can be replaced by an optional galley module and/or baggage/wardrobe module. Seat pitch 76 cm (30 in). Standard provision for galley, wardrobe or storage module on port side at front of cabin, regardless of installations on starboard side. Toilet at front or rear of cabin. In former case, QC operation (conversion from passenger to freight interior or vice versa) is possible. Also available is a VIP to airliner convertible, as well as a fixed-installation combi with 19 passengers and 1,500 kg (3,307 lb) of cargo. Passenger door (plug type) at front of cabin on port side, with separate airstair. Type II emergency exit opposite this on starboard side, and Type III over wing on each side. Overhead crew escape hatch in flight deck roof. Baggage space under each passenger seat; overhead storage bins. Main baggage/cargo compartment aft of passenger cabin, with large plug type door on port side. Entire accommodation pressurised, including baggage compartment.

SYSTEMS: Hamilton Standard environmental control system (maximum pressure differential 0.48 bar; 7.0 lb/sq in) maintains a S/L cabin environment up to an altitude of 3,660 m (12,000 ft) and a 1,525 m (5,000 ft) environment up to the maximum cruising altitude of 7,620 m (25,000 ft). Single on-demand hydraulic system, operating between 138 and 207 bars (2,000-3,000 lb/sq in), for actuation of landing gear, wheel and propeller braking, nosewheel steering and wing flaps. System is powered by single 28 V DC electric motor driven pump, rated delivery 9.5 litres (2.5 US gallons; 2.1 Imp gallons)/min. Self-pressurising main reservoir with 5.08 litres (0.18 cu ft) capacity, operating at pressure of 1.79-2.69 bars (26-39 lb/sq in). Hydraulic back-up via four accumulators and pilot operated handpump, working via an emergency reservoir of 2.5 litres (0.09 cu ft) capacity. Electrical power supplied by two 28 V 400 A DC engine-driven starter/generators, each connected to a separate busbar. Variable frequency 115/200 V for heating circuits provided by two 26 kVA AC generators; single-phase 115 V and 26 V AC at 400 Hz for avionics provided by static inverters. Two 40 Ah Ni/Cd batteries for ground power and engine starting; standby 5 Ah lead-acid battery for emergency use. External power receptacle. Pneumatic boot de-icing of wing and tail unit leading-edges, using engine bleed air. Flight deck windows have electric anti-icing and electrically driven windscreen wipers. Electric anti-icing is provided also for engine air intakes, propellers and pitot heads. Demisting by means of air conditioning system. Plug-in connections for oxygen masks. Kidde engine fire detection system. Duncan/Garrett GTCP 36-150W APU kit certificated for installation as optional extra, to provide standby and emergency electrical power, main engine starting assistance, ground preheating and precooling, and other power support functions. Pneumatic boot de-icing of wing leading-edges. Pneumatic boot de-icing of fin and tailplane leading-edges.

AVIONICS: Standard avionics include all equipment required for FAR 121 operations. Aircraft is equipped with Bendix/King Gold Crown III or Collins Pro Line II com/nav radios, and a Collins integrated digital flight guidance and autopilot system (FGAS) consisting of attitude and heading reference units, electronic (CRT) flight display units, fail-passive autopilot/flight director system, colour weather radar, air data system with servo instruments, and radio altimeter. Lucas Aerospace electroluminescent flight deck instrument panel array. Dowty Aerospace microprocessor-based flight deck central warning system. Rosemount pitot static tubes, total temperature sensors and stall warning system. Provision for additional avionics to customer's requirements.

DIMENSIONS, EXTERNAL:

Wing span	21.44 m (70 ft 4 in)
Wing chord: at root	2.837 m (9 ft 3¾ in)
at tip	1.0645 m (3 ft 5⅝ in)
Wing aspect ratio	11.0
Length overall	19.72 m (64 ft 8½ in)
Fuselage: Max diameter	2.31 m (7 ft 7 in)
Height overall	6.86 m (22 ft 6 in)
Tailplane span	8.67 m (28 ft 5¼ in)
Wheel track	6.71 m (22 ft 0 in)
Wheelbase	7.14 m (23 ft 5 in)
Propeller diameter: initially	3.20 m (10 ft 6 in)
current	3.35 m (11 ft 0 in)
Propeller ground clearance	0.58 m (1 ft 11 in)
Distance between propeller centres	6.71 m (22 ft 0 in)
Passenger door: Height	1.60 m (5 ft 3 in)
Width	0.66 m (2 ft 2 in)
Height to sill	1.63 m (5 ft 4 in)
Cargo door: Height	1.30 m (4 ft 3 in)
Width	1.35 m (4 ft 5 in)
Height to sill	1.68 m (5 ft 6 in)
Emergency exit (fwd, stbd): Height	1.32 m (4 ft 4 in)
Width	0.51 m (1 ft 8 in)
Emergency exits (overwing, each):	
Height	0.91 m (3 ft 0 in)
Width	0.51 m (1 ft 8 in)

DIMENSIONS, INTERNAL:

Cabin, excl flight deck, incl toilet and galley:	
Length	10.39 m (34 ft 1 in)
Max width	2.16 m (7 ft 1 in)
Width at floor	1.70 m (5 ft 7 in)
Max height	1.83 m (6 ft 0 in)
Volume	33.5 m³ (1,183.0 cu ft)
Baggage/cargo compartment volume	6.8 m³ (240.0 cu ft)

AREAS:

Wings, gross	41.81 m² (450.0 sq ft)
Ailerons (total)	2.12 m² (22.84 sq ft)
Trailing-edge flaps (total)	8.07 m² (86.84 sq ft)
Fin (incl dorsal fin)	10.53 m² (113.38 sq ft)
Rudder (incl tab)	2.76 m² (29.71 sq ft)
Tailplane	13.30 m² (143.16 sq ft)
Elevators (total, incl tabs)	3.29 m² (35.40 sq ft)

WEIGHTS AND LOADINGS:

Typical operating weight empty	7,899 kg (17,415 lb)
Max payload (weight limited)	3,668 kg (8,087 lb)
Max fuel load	2,581 kg (5,690 lb)
Max ramp weight	12,383 kg (27,300 lb)
Max T-O weight	12,700 kg (28,000 lb)
Max landing weight	12,020 kg (26,500 lb)
Max zero-fuel weight	11,567 kg (25,500 lb)
Max wing loading	303.75 kg/m² (62.22 lb/sq ft)
Max power loading	4.91 kg/kW (8.07 lb/shp)

PERFORMANCE (at MTOW, ISA, except where indicated):

Max operating speed (VMO)	250 kts (463 km/h; 288 mph)
Max operating Mach number (MMO)	0.5

Saab 340A twin-turboprop regional transport aircraft

1994

Max cruising speed:
at 4,575 m (15,000 ft) 272 kts (504 km/h; 313 mph)
Best range cruising speed at 7,620 m (25,000 ft)
250 kts (463 km/h; 288 mph)
Stalling speed: flaps up 104 kts (193 km/h; 120 mph)
T-O flap 93 kts (173 km/h; 107 mph)
approach flap 87 kts (162 km/h; 101 mph)
landing flap 82 kts (152 km/h; 95 mph)
Max rate of climb at S/L 548 m (1,800 ft)/min
Rate of climb at S/L, one engine out 167 m (550 ft)/min
Service ceiling: standard 7,620 m (25,000 ft)
Service ceiling, one engine out (net) 3,960 m (13,000 ft)
FAR Pt 25 required T-O field length:
at S/L 1,212 m (3,975 ft)
at S/L, ISA + 15°C 1,295 m (4,250 ft)
at 1,525 m (5,000 ft) 1,509 m (4,950 ft)
at 1,525 m (5,000 ft), ISA + 15°C 1,935 m (6,350 ft)
FAR Pt 25 required landing field length (at MLW):
at S/L 1,180 m (3,870 ft)
at 1,525 m (5,000 ft) 1,335 m (4,380 ft)
Min ground turning radius 15.85 m (52 ft 0 in)
Runway LCN: flexible pavement 8
rigid pavement 10
Range with 35 passengers and baggage, reserves for 45 min hold at 1,525 m (5,000 ft) and 100 n mile (185 km; 115 mile) diversion: at max cruising speed
570 n miles (1,056 km; 656 miles)
at long-range cruising speed
630 n miles (1,167 km; 725 miles)
Range with 30 passengers, reserves as above:
at max cruising speed
845 n miles (1,566 km; 973 miles)
at long-range cruising speed
940 n miles (1,742 km; 1,082 miles)
OPERATIONAL NOISE LEVELS (FAR Pt 36, and ICAO 16):
T-O (with cutback) 85.7 EPNdB
Sideline 87.6 EPNdB
Approach 89.6 EPNdB

UPDATED

SAAB 340B AEW & C

TYPE: Airborne early warning and control aircraft.
PROGRAMME: Prototype aircraft ordered by the Swedish Air Force in February 1993. Five more ordered 23 December 1993. First flight (SE-C42) 17 January 1994 (minus antenna). Deliveries began in early 1995. For full description of Saab 340B see *Jane's All the World's Aircraft 1995-96*.
CUSTOMERS: Swedish Air Force (5).
COSTS: Total costs approximately SEK400 million.
DESIGN FEATURES: Basic airframe is a Saab 340B with an Ericsson PS-890 Erieye S-band side-looking airborne reconnaissance radar. APU fitted for electrical power and mission system cooling.
DIMENSIONS, EXTERNAL:
Length (antenna) 9 m (29.5 ft)
WEIGHTS AND LOADINGS:
Weight (antenna) 900 kg (1,984 lb)

NEW ENTRY

Saab 340B AEW & C aircraft of the Swedish Air Force *(Saab)*

1995

SWITZERLAND

PILATUS

PILATUS FLUGZEUGWERKE AG
PILATUS AIRCRAFT LTD
PILATUS CONSTRUCTIONS AERONAUTIQUES SA
(A division of Oerlikon-Bührle Group)

CH-6370 Stans
Telephone: 41 (41) 63 61 11
Fax: 41 (41) 613351
Telex: 866202 PIL CH
CEO AND PRESIDENT: O. J. Schweuk
SENIOR VICE-PRESIDENTS:
Oscar Bründler (Finance and Controlling)
Andrea N. Muggli (Logistics)
O. Masefield (Research and Development)
HEAD OF PRODUCTION: Erich Bodenmann
HEAD OF GENERAL AVIATION: Ignaz Gretener

Pilatus Flugzeugwerke AG was formed in December 1939. It is a subsidiary of the Oerlikon-Bührle Group. Current products are the PC-6 B2-H4 Turbo-Porter, PC-7 Turbo Trainer, the PC-7 Mk II Turbo-Trainer, PC-9 and PC-12.

On 24 January 1979 Pilatus purchased the assets of Britten-Norman (Bembridge) Ltd of the UK, which has operated since then under the name Pilatus Britten-Norman Ltd (which see) as a subsidiary of Pilatus Aircraft Ltd. The company is also involved in overhaul and maintenance of civil and military aircraft (both fixed- and rotary-wing) and has entered modified (Mk II) PC-9 with Beech Aircraft for US Air Force/Navy JPATS trainer programme. Also active in manufacture of components for F/A-18 aircraft. It has been reported that the company are planning to set up PC-7/PC-9 assembly lines at its Pilatus Britten-Norman subsidiary in the UK.

UPDATED

Pilatus PC-6/B2-H2 Turbo-Porter STOL utility aircraft *(Pilatus)*

1995

Pilatus PC-6/B2-H4 Turbo-Porter fitted with Wipline amphibious floats *(Pilatus)*

1995

PILATUS PC-6 TURBO-PORTER

US Army designation: UV-20A Chiricahua
Thailand armed version: AU-23 Peacemaker

TYPE: Single-engine STOL utility aircraft.
PROGRAMME: The PC-6 is a single-engined multipurpose utility aircraft with STOL characteristics, permitting operation from unprepared strips under harsh environmental and terrain conditions. It can be converted rapidly from a freighter to a passenger transport, and adapted for a great number of missions, including supply dropping, search and rescue, ambulance, aerial survey and photography, parachuting, cropspraying, water bombing, rainmaking and glider or target towing. The PC-6 can operate from soft ground, snow, glacier or water.

The first piston-engined prototype made its first flight 4 May 1959. The B2 can be fitted with an air inlet filter for operation in desert conditions and for agricultural applications.
VERSIONS: **PC-6/A Turbo-Porter:** Version with 523 shp Turboméca Astazou IIE or IIG turboprop engine and Ratier Figeac FH 76-1-07/FH 76.207 electrically controllable fully feathering reversible-pitch propeller, diameter 2.50 m (8 ft 2½ in). Oil capacity 8 litres (2.2 US gallons; 1.8 Imp gallons). Original certification 26 November 1962; later models, with increased AUW, are designated PC-6/A-H1 and PC-6/A-H2, certificated on 10 December 1963 and 26 August 1964 respectively.

PC-6/A1 Turbo-Porter: Version with 573 shp Turboméca Astazou XII turboprop engine and Hamilton Standard 23LF-351/1017A hydraulically controllable fully feathering reversible-pitch propeller, diameter 2.59 m (8 ft 6 in). Oil capacity 12 litres (3.1 US gallons; 2.6 Imp gallons). Certificated as PC-6/A1-H2 on 29 February 1968.

PC-6/A2 Turbo-Porter: Version with 573 shp

Pilatus PC-6/B1-H2 of Zimex Aviation *(Pilatus)*

1995

Turboméca Astazou XIVE turboprop engine. Other details as for PC-6/A1. Certificated as PC-6/A2-H2 on 22 January 1971.

PC-6/B Turbo-Porter: Version with 550 shp Pratt & Whitney (UACL) PT6A-6A turboprop engine and Hartzell HC-B3TN-3C/T-10173CH hydraulically controllable fully feathering reversible-pitch, constant-speed propeller, diameter 2.56 m (8 ft 5 in). Oil capacity 13 litres (3.5 US gallons; 2.9 Imp gallons). Certificated as PC-6/B-H2 on 8 June 1965.

PC-6/B1 Turbo-Porter: Similar to PC-6/B but with 550 shp PT6A-20 turboprop engine and HC-B3TN-3D/T-10173 or /T-10178 propeller. First flight May 1966. Certificated as PC-6/B1-H2 on 6 August 1966. Total of 112 built by December 1970 (73 by Pilatus and 39 by Fairchild Hiller).

PC-6/B2-H2 Turbo-Porter: Certificated on 30 June 1970 and powered by a PT6A-27 turboprop engine.

PC-6/B2-H4: For CAR.3 operations (commercial operations with fare paying passengers) the maximum take-off weight is increased by 600 kg (1,323 lb), resulting in a payload increase of 570 kg (1,257 lb). This was achieved by improving the aerodynamic efficiency of the wings with new tip fairings, enlarging the dorsal fin, installing uprated mainwheel shock-absorbers and a new tailwheel assembly, and a slight strengthening of the airframe. While the H4 modification can be retrofitted to all existing PC-6/B1-H2 and B2-H2 models equipped with electrically operated longitudinal trim, all new-production Porters from mid-1985 are of the H4 version only. Still in production. For further details see *Jane's All the World's Aircraft 1995-96.*

PC-6/C Turbo-Porter: Version with 575 shp AiResearch TPE 331-25D turboprop engine and Hartzell HC-B3TN-3D/T-10178 hydraulically controllable fully feathering reversible-pitch constant-speed propeller, diameter 2.56 m (8 ft 5 in). Oil capacity 11 litres (2.9 US gallons; 2.4 Imp gallons). Prototype, built by Fairchild Hiller, first flown in October 1965; first Pilatus built example flown 4 March 1966. Certificated as PC-6/C-H2 on 1 December 1966.

PC-6/C1 Turbo-Porter: Similar to PC-6/C but with 576 shp TPE 331-1-100 turboprop engine. Certificated as PC-6/C1-H2 on 15 July 1970.

PC-6/D-H3 Porter: Latest piston-engined version, first flown (HB-FFW) 2 April 1970. Fitted with 500 hp Lycoming TIO-720 turbocharged engine and distinguishable by its larger, sweptback vertical tail surfaces and modified wingtips. Increased T-O weight and range compared with earlier Porter models. Further development and type certification is dependent upon availability of production engine.

UPGRADES: **Micronair:** The PC-6 can be fitted with two Micronair underwing spray pods with a capacity of 50 US gallons each. This system is ideally suited for locust control as the aircraft can be converted quickly by the operator and saves having an aircraft dedicated to this role.

Pilatus: The H4 modification can be retrofitted to all existing PC-6/B2-H2 models equipped with electrically operated longitudinal trim.

Pilatus: The PC-6/B2-H4 was certificated for use with Wipline amphibious floats in Summer 1994.

OPERATORS: Versions of the PC-6 are in service with the armed forces of the following countries: Argentina (2), Austria (12), Bolivia (16), Chad (2), Ecuador (3), France (5), Iran (15), Myanmar (7), Peru (12), Switzerland (18), Thailand (25) and Yugoslavia (14).

The following description applies to the B2-H2 model.

DESIGN FEATURES: Braced high-wing monoplane, with single streamline-section bracing strut each side. Wing section NACA 64-514 (constant). Dihedral 1°. Incidence 2°. Cantilever tail unit.

FLYING CONTROLS: Entire wing trailing-edge hinged, inner sections consisting of electrically operated all-metal double-slotted flaps and outer sections of all-metal single-slotted ailerons. No airbrakes. Trim tabs and/or Flettner tabs on ailerons optional. Fixed tabs are mandatory if these are not fitted. Variable incidence tailplane. Flettner tabs on elevator.

STRUCTURE: Single-spar all-metal structure. The fuselage is an all-metal semi-monocoque structure. All-metal tail unit.

LANDING GEAR: Non-retractable tailwheel type. Oleo shock-absorbers of Pilatus design on all units. Steerable/lockable tailwheel. Goodyear Type II mainwheels on GA 284 tyres size 24 × 7 or 7.5 × 10; pressure 2.21 bars (32 lb/sq in); oversize Goodyear Type III wheels and tyres optional, size 11.0 × 12; pressure 0.88 bars (12.8 lb/sq in). Goodyear tailwheel with size 5.00-4 tyre. Goodyear disc brakes. Alternative Pilatus wheel/ski gear or Edo 58-4580 or 679-4930 floats may be fitted.

POWER PLANT: One 410 kW (550 shp) Pratt & Whitney Aircraft of Canada PT6A-27 turboprop engine, driving a Hartzell HC-B3TN-3D/T-10178 propeller. Standard fuel in integral wing tanks, capacity 480 litres (170 US gallons; 142 Imp gallons) maximum. Two underwing auxiliary tanks, each of 190 litres (50 US gallons; 42 Imp gallons), available optionally.

ACCOMMODATION: Cabin has pilot's seat forward on port side, with one passenger seat alongside, and is normally fitted with six quickly removable seats, in pairs, to the rear of these additional passengers. Up to 10 persons can be carried in high-density layout. Floor is level, flush with door sill, and is provided with seat rails. Forward-opening door beside each front seat. Large rearward-sliding door on starboard side of main cabin or large rearward-sliding doors on both sides of main cabin. Hatch (optional equipment) in floor 0.58 × 0.90 m (1 ft 10¾ in × 2 ft 11½ in), openable from inside cabin, for installation of aerial camera or for supply dropping. Hatch in cabin rear wall 0.50 × 0.80 m (1 ft 7 in × 2 ft 7 in) permits stowage of six passenger seats or accommodation of freight items up to 5.0 m (16 ft 5 in) in length. Walls lined with lightweight soundproofing and heat insulation material. Adjustable heating and ventilation systems provided. Dual controls optional.

SYSTEMS: Cabin heated by bleed air from engine compressor. Scott 8500 oxygen system optional. 200 A 30 V starter/generator and 24 V 34 Ah Ni/Cd battery.

DIMENSIONS, EXTERNAL:

Wing span	15.13 m (49 ft 8 in)
Wing span over navigation lights	15.20 m (49 ft 10½ in)
Wing chord (constant)	1.90 m (6 ft 3 in)
Wing aspect ratio	7.96
Length overall	10.90 m (35 ft 9 in)
Height overall (tail down)	3.20 m (10 ft 6 in)
Elevator span	5.12 m (16 ft 9½ in)
Wheel track	3.00 m (9 ft 10 in)
Wheelbase	7.87 m (25 ft 10 in)
Propeller diameter	2.56 m (8 ft 5 in)
Cabin double door (port) and sliding door (starboard):	
Height	1.04 m (3 ft 5 in)
Width	1.58 m (5 ft 2¼ in)

DIMENSIONS, INTERNAL:

Cabin, from back of pilot's seat to rear wall:	
Length	2.30 m (7 ft 6½ in)
Max width	1.16 m (3 ft 9½ in)
Max height (at front)	1.28 m (4 ft 2½ in)
Height at rear wall	1.18 m (3 ft 10½ in)
Floor area	2.67 m^2 (28.6 sq ft)
Volume	3.28 m^3 (107 cu ft)

AREAS:

Wings, gross	28.80 m^2 (310 sq ft)
Ailerons (total)	3.83 m^2 (41.2 sq ft)
Flaps (total)	3.76 m^2 (40.5 sq ft)
Fin	1.70 m^2 (18.3 sq ft)
Rudder, incl tab	0.96 m^2 (10.3 sq ft)
Tailplane	4.03 m^2 (43.4 sq ft)
Elevator, incl tab	4.22 m^2 (45.4 sq ft)

WEIGHTS AND LOADINGS:

Weight empty, equipped	1,215 kg (2,678 lb)
Max T-O and landing weight:	
Normal (CAR.3)	2,200 kg (4,850 lb)
Restricted (CAR.8)	2,700 kg (6,100 lb)
Max cabin floor loading	488 kg/m^2 (100 lb/sq ft)
Max wing loading (Normal)	76.4 kg/m^2 (15.65 lb/sq ft)
Max power loading (Normal)	5.37 kg/kW (8.82 lb/shp)

PERFORMANCE (at max T-O weight, Normal category):

Never-exceed speed	151 kts (280 km/h; 174 mph) IAS
Max cruising speed at 3,050 m (10,000 ft)	140 kts (259 km/h; 161 mph) TAS
Econ cruising speed at 3,050 m (10,000 ft)	129 kts (240 km/h; 150 mph) TAS
Stalling speed, power off:	
flaps up	50 kts (93.5 km/h; 58 mph)
flaps down	44 kts (82 km/h; 51 mph)
Max rate of climb at S/L	482 m (1,850 ft)/min
Service ceiling	9,150 m (30,025 ft)
T-O run	110 m (360 ft)
T-O run to 15 m (50 ft)	235 m (771 ft)
Landing from 15 m (50 ft)	220 m (722 ft)
Landing run	73 m (240 ft)
Max range, no reserves:	
internal fuel only	560 n miles (1,036 km; 644 miles)
with external fuel	875 n miles (1,620 km; 1,007 miles)
Endurance: internal fuel only	4 h 20 min
with external fuel	6 h 45 min
g limits	+3.72/−1.50

UPDATED

PILATUS PC-6 TURBO-PORTER (AGRICULTURAL VERSIONS)

TYPE: Agricultural aircraft conversion.

PROGRAMME: The Turbo-Porter can, if required, be equipped for agricultural duties, the necessary equipment being easily removable when not required, to permit the use of the aircraft for other work. Approximately 40 Turbo-Porters have been completed in agricultural configuration: these are in service in Indonesia, Sudan, Switzerland, Thailand, Venezuela and Zaïre.

DESIGN FEATURES: For liquid spraying, a stainless steel tank (capacity 1,330 litres; 292.5 Imp gallons; 351.5 US gallons) is installed behind the two front seats, and 46- or 62-nozzle spraybooms are fitted beneath the wings. In this configuration the aircraft can cover a swath width of 45 m (148 ft). An ultra-low-volume system, using four to six atomisers or two to six Micronairs, is also available, permitting increase in swath width up to 400 m (1,310 ft).

For dusting with granulated materials, the lower part of the standard tank can be replaced by a discharge and dispersal door permitting coverage of a swath width of up to 20 m (66 ft). A Transland spreader can be fitted for dust application (swath up to 30 m; 100 ft). Effective swath width of these versions is 13-40 m (43-131 ft), the optimum being approximately 20 m (66 ft).

Both versions are fitted with small doors in the fuselage sides, giving access to the tank/hopper for servicing, removal or replenishment, and two single seats or a bench seat for three persons can be installed aft of the tank. Optional items include an engine air intake screen and a loading door for chemical in the top of the fuselage.

AVIONICS AND EQUIPMENT: Optional avionics include Decca Mk 8A navigator, Decca Hi-Fix radio, Decca Doppler 72 radar, gyrosyn CL-11 compass and SR 54A radio altimeter.

WEIGHTS AND LOADINGS (liquid spray system):
- Weight empty, incl spray system, oil and pilot 1,440 kg (3,170 lb)
- Fuel 380 kg (837 lb)
- Chemical 950 kg (2,093 lb)
- Max T-O weight 2,770 kg (6,100 lb)
- Max landing weight 2,200 kg (4,850 lb)

PERFORMANCE (liquid spray version, PT6A-27 engine, at max T-O weight):
- Never-exceed speed 120 kts (222 km/h; 138 mph)
- Operating speed approx 90 kts (167 km/h; 104 mph)
- Operating height 6-8 m (20-26 ft)
- Stalling speed, power off, flaps down 49 kts (91 km/h; 57 mph)
- T-O run 180 m (590 ft)
- T-O to 15 m (50 ft) 390 m (1,280 ft)
- Landing from 15 m (50 ft) 345 m (1,132 ft)
- Landing run 130 m (427 ft)
- Spraying duration with full spray tank 6 min

NEW ENTRY

SWISS FEDERAL AIRCRAFT FACTORY (F+W)

EIDGENÖSSISCHES FLUGZEUGWERK—FABRIQUE FEDERALE D'AVIONS—FABBRICA FEDERALE D'AEROPLANI

CH-6032 Emmen
Telephone: 41 (41) 59 41 11
Fax: 41 (41) 55 25 88
MANAGING DIRECTOR: Hansjürg Kobelt

F+W is the Swiss government's industrial basis for interdisciplinary services in the field of civil and military aviation, space technologies and guided weapon systems. Activities include research, development, engineering, manufacture, assembly and testing. Major areas are: combat aircraft final assembly, systems integration and improvements; aircraft, helicopter and RPV structural framework; avionic and test equipment; wind tunnel testing; environmental testing; guided missiles production and improvements. F+W is main contractor in licence production or Swiss industry participation programmes in the military field.

Current programmes for the Swiss government include subassembly manufacturing for and final assembly of the F/A-18 combat aircraft; development and production of the Ranger unmanned air vehicle for the ADS reconnaissance and surveillance system; licence production of the Stinger anti-aircraft missile; improvement of the anti-tank Dragon and TOW missiles.

Recent or current structure work includes production of wing slats and tracks for the McDonnell Douglas MD-80/90 series and wingtips for the Airbus A320 series; development and production of external fuel tanks for the Dassault Rafale and of the horizontal stabiliser for the Pilatus PC-12.

In the Ariane space programme, F+W, as a subcontractor to Contraves, has been responsible for the final assembly and instrumentation of the payload fairing of Ariane 1, 3 and 4 and is involved in the development of Ariane 5. In 1995 F+W had covered facilities of 140,000 m² (1,506,400 sq ft) and a workforce of 650 personnel.

UPDATED

Swiss Federal Aircraft Factory (Dassault) Mirage III-DS two-seat aircraft *(F+W)*

1994

Swiss Federal Aircraft Factory (Dassault) Mirage III-S upgraded single-seat combat aircraft of the Swiss Air Force *(F+W)*

1995

F+W MIRAGE IMPROVEMENT PROGRAMME

TYPE: Combat aircraft upgrade.

PROGRAMME: The Swiss government approved funding in 1985 to retrofit Swiss Air Force Mirage IIIs. There are 30 III-S, 18 III-RS, two III-BS and two III-DS. Main items are a fixed canard and strake on each side of the nose. These improve manoeuvrability and increase stability in yaw near the upper limit of the flight envelope.

DESIGN FEATURES: Other improvements include new audible warning and visual angle of attack monitoring systems, to alert the pilot when approaching limits of the flight envelope; Martin-Baker Mk 6 zero/zero ejection seats in place of the present Mk 4 seats; addition of infra-red and passive/active ECM; provision of more powerful VHF radios; wing refurbishing; ability to carry two underwing 500 litre (132 US gallon; 110 Imp gallons) drop tanks and a 730 litre (193 US gallon; 160.5 Imp gallon) centreline tank; mounting of improved blast deflectors for the two internal guns, to allow firing at high angles of attack; and a new camouflage paint scheme. The retrofit programme was finished in 1993.

VERIFIED

F+W (NORTHROP) F-5E/F TIGER UPGRADES

TYPE: Combat aircraft upgrade.

PROGRAMME: F+W has carried out structural and avionics improvements for the Swiss Air Force F-5E/F Tiger aircraft.

DESIGN FEATURES: Structural improvements include redesigned horizontal stabiliser and upper longeron with Swiss Fix. Avionics modifications include integration of radar warning system on F-5E/F and integration of a training jammer on F-5F.

VERIFIED

TAIWAN

AIDC

AERO INDUSTRY DEVELOPMENT CENTER

PO Box 90008-10, Taichung 40772
Telephone: 886 (4) 252 3051/2
Fax: 886 (4) 256 2282
OTHER WORKS: Kangshan
DIRECTOR GENERAL: Gen Wen-Li Lin
VICE-DIRECTOR GENERAL: Dr Shih-sen Wang

Established 1969 to succeed Bureau of Aircraft Industry (BAI), formed 1946, which moved to Taiwan in 1948; now subsidiary of government Chung Shan Institute of Science and Technology (CSIST).

Produced 118 Bell UH-1H (Model 205) helicopters under licence 1969-76 for Chinese Nationalist Army; built PL-1A prototype (based on Pazmany PL-1) and 55 PL-1B Chien-Shou primary trainers for Republic of China Air Force between 1968 and 1974; built under licence 248 Northrop F-5E Tiger IIs and 36 two-seat F-5Fs between 1974 and 1986.

Designed and produced T-CH-1 Chung-Hsing turboprop basic trainer; developed and produced AT-3A/B Tzu-Chung twin-turbofan advanced trainer; has begun production of IDF Ching-Kuo fighter.

Taiwan Ministry of Defence announced in 1993 intention to privatise AIDC; new business development unit set up early 1994; reorganisation with commercial company structure planned to start July 1994. Workforce in 1994 of 6,000 personnel.

VERIFIED

AIDC AT-3 TSU-CHIANG

TYPE: Tandem two-seat twin-turbofan military trainer and close support aircraft.

PROGRAMME: Development contract placed July 1975; first flights of two prototypes (0801 and 0802) 16 September 1980 and 30 October 1981; production of order for 60 started in March 1982; first flight of production aircraft (73-6003) 6 February 1984; deliveries began March 1984, completed by early 1990, but wingroot leading-edge improvements flight tested on prototypes in early 1992.

VERSIONS: **AT-3:** Advanced two-seat military jet trainer and close air support airplane. *Description applies to this version, except where indicated.*

AT-3A: Single-seat version of the AT-3, with the navigation/attack system (NAS). It is designed for air-to-surface attack including outstanding close air support and anti-ship capability. One prototype conversion flown 1989.

AT-3B: Two-seat flight trainer and close air support airplane, with the navigation/attack system (NAS).

UPGRADES: **AIDC:** 20 AT-3As converted to AT-3B close air support aircraft.

OPERATORS: Versions of the AT-3 are in service with the armed forces of Taiwan (60).

CUSTOMERS: RoCAF (60 AT-3). AIDC plans to produce AT-3A/B in the future, and still retaining the operation of prototype at present.

DESIGN FEATURES: Supercritical wing section; quarter-chord sweepback 7° 20′; thickness/chord ratio 10 per cent; dihedral 0° 46′; incidence 1° 30′.

FLYING CONTROLS: Hydraulic fully powered slab tailplane, rudder and sealed-gap ailerons; yaw damper; electro-hydraulically controlled single-slotted flaps; two airbrake panels forward of mainwheel wells.

STRUCTURE: One-piece multispar metal wing with thick machined skins forming torsion box; fin integral with rear fuselage; three-part fuselage; steel, magnesium, and graphite/epoxy components in fuselage; airbrake panels of laminated graphite composites; ailerons have metal honeycomb core.

LANDING GEAR: Hydraulically retractable tricycle type, with single wheel on each unit. Main units retract inward into fuselage, nosewheel forward. Oleo-pneumatic shock-absorber in each unit. Two-position extending nose leg increases static angle of attack by 3° 30′, to reduce T-O run, and is shortened automatically during retraction. Emergency extension by gravity. Mainwheels and tyres size 24 × 8.00-13, presure 8.96 bars (130 lb/sq in). Hydraulically steerable nose unit, with wheel and tyre size 18 × 6.50-8, pressure 5.51 bars (80 lb/sq in). All-metal multidisc brakes.

POWER PLANT: Two Garrett TFE731-2-2L non-afterburning turbofans (each 15.57 kN; 3,500 lb st), installed in nacelle on each side of fuselage. Inclined ram air intakes, each with splitter plate abreast of rear cockpit. Engine starting by onboard battery or ground power. All fuel carried in fuselage, in two equal-size rubber impregnated nylon bladder tanks, with combined capacity of 1,630 litres (430.6 US gallons; 385.6 Imp gallons). Two independent fuel systems, one for each engine, with crossfeed to allow fuel from either or both systems to be fed to either or both engines. Pressure fuelling point forward of, and below, port air intake for internal and external tanks. A 568 litre (150 US gallon; 125 Imp gallon) auxiliary drop tank can be carried on each inboard underwing pylon. Oil capacity 5.7 litres (1.5 US gallons; 1.25 Imp gallons) total, 1.9 litres (0.5 US gallons; 0.42 Imp gallons) usable. Fire warning and extinguishing systems for each engine bay.

ACCOMMODATION: Crew of two in tandem on zero-zero ejection (through canopy) seats, under individual manually operated canopies which open sideways to starboard. Crew separated by internal windscreen. Independent miniature detonation chord (MDC) system to break each canopy for ground and in-flight emergency egress. MDC can be operated from outside cockpit on ground. Rear seat elevated 30 cm (12 in). Dual controls standard.

SYSTEMS: AiResearch bootstrap air cycle ECS, for cockpit air conditioning and pressurisation (maximum differential 0.34 bar; 5 lb/sq in), canopy seal, demisting, and pressurisation of *g* suits, hydraulic reservoirs and external fuel tanks. Two independent hydraulic systems, pressure 207 bars (3,000 lb/sq in), with engine-driven pumps (flow rate 34.4 litres; 9.09 US gallons; 7.57 Imp gallons/min). Air type reservoir, pressurised at 2.41 bars (35 lb/sq in). Flight control hydraulic system provides power only for operation of primary flying control surfaces. Utility system serves primary flying control surfaces, landing gear, landing gear doors, airbrakes, wheel brakes, nosewheel steering and stability augmentation system. Primary electrical power supplied by two 28 V 12 kW DC starter/generators, one on each engine. One 40 Ah Ni/Cd battery for engine starting. Two static inverters supply AC power at 400 Hz. External DC power socket on starboard side of centre fuselage. Hydraulic and electrical systems can be sustained by either engine. Liquid oxygen system, capacity 5 litres (1.3 US gallons; 1.1 Imp gallons), for crew.

AVIONICS: *AT-3:* Most radio and nav equipment located in large avionics bays in forward fuselage. Standard avionics include UHF com, intercom, IFF/SIF, Tacan, panel-mounted VOR/ILS/marker beacon indicator, AHRS and angle of attack system, plus full blind-flying instrumentation. Wide range of optional avionics available.

AT-3A/B: The avionic system developed for the AT-3A/B is the navigation/attack system (NAS) which integrates the data from the laser gyro inertial navigation system (LINS), Tacan, radar altimeter, radar, anti-ship missile control unit and Sidewinder control system in the computer/interface unit, and the navigation, A/A, A/G functions were displayed on HUD and dual head-down display units (DU).

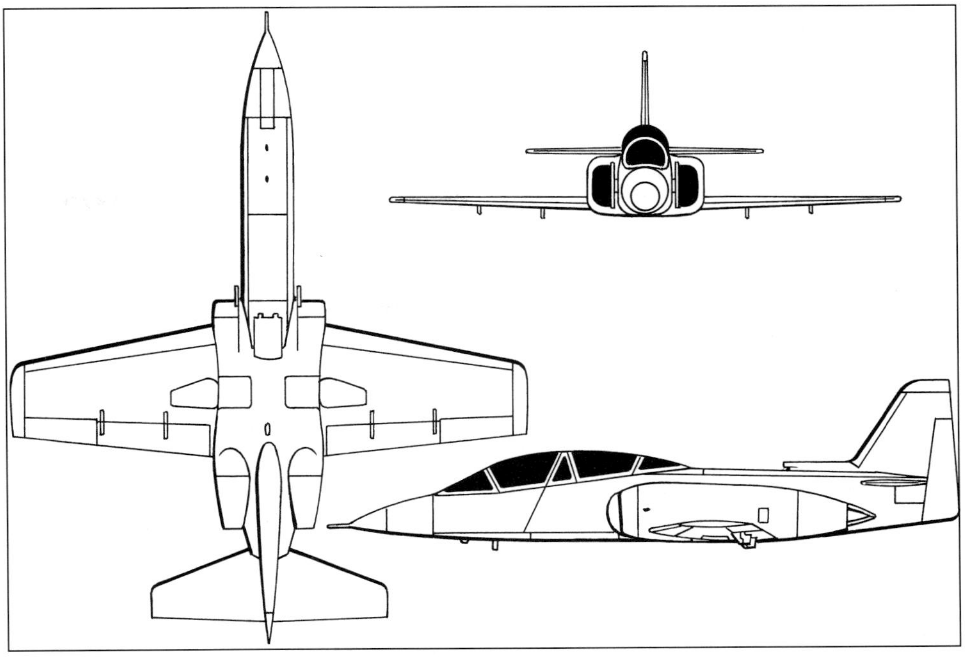

AIDC AT-3 Tsu-Chiang two-seat trainer aircraft ***(Jane's/ Mike Keep)***

1994

AIDC AT-3A, this the first production version ***(Ivan Chou)***

1994

AIDC AT-3A single-seat attack aircraft of the Taiwanese Air Force ***(RoCAF)***

1995

EQUIPMENT: Can be equipped with A/A 37U-15TTS aerial target system, carried on centreline and outboard pylons; aerial gunnery/missile target system (AGTS/AMTS), carried on inboard pylon.

ARMAMENT: Manually adjustable gunsight and camera in forward cockpit. Large weapons bay beneath rear cockpit can house variety of stores, including quick-change semi-recessed machine gun packs. Disposable weapons can be carried on centreline pylon (stressed for 907 kg; 2,000 lb load), two inboard underwing pylons (each 635 kg; 1,400 lb and capable of accepting triple ejector racks), two outboard underwing pylons (each 272 kg; 600 lb), and wingtip launch rails (each of 91 kg; 200 lb capacity), subject to maximum external stores load of 2,721 kg (6,000 lb). Weapons that can be carried include GP, SE, cluster and fire bombs; SUU-25A/A, -25C/A and -25E/A flare dispensers; LAU-3/A, -3A/A, -3B/A, -10/A, -/10A/A, -60/A, -68A/A and -68B/A rocket launchers; wingtip infrared air-to-air missiles; and rocket pods, practice bombs and bomb or rocket training dispensers.

DIMENSIONS, EXTERNAL:

Wing span	10.46 m (34 ft 3¾ in)
Wing chord: at root	2.80 m (9 ft 2¼ in)
at tip	1.40 m (4 ft 7 in)
Wing aspect ratio	5.0
Length overall, incl nose probe	12.90 m (42 ft 4 in)
Height overall	4.36 m (14 ft 3¾ in)
Tailplane span	4.83 m (15 ft 10¼ in)
Wheel track	3.96 m (13 ft 0 in)
Wheelbase	5.49 m (18 ft 0 in)

AREAS:

Wings, gross	21.93 m² (236.05 sq ft)
Ailerons (total)	1.33 m² (14.32 sq ft)
Trailing-edge flaps (total)	2.53 m² (27.23 sq ft)
Fin	3.45 m² (37.14 sq ft)
Rudder	1.15 m² (12.38 sq ft)
Tailplane	5.02 m² (54.04 sq ft)

WEIGHTS AND LOADINGS:

Weight empty, equipped	3,855 kg (8,500 lb)
Max fuel: internal	1,270 kg (2,800 lb)
external	884 kg (1,950 lb)
Max external stores load	2,721 kg (6,000 lb)
Normal T-O weight: trainer, clean	5,216 kg (11,500 lb)
Max T-O weight with external stores	7,938 kg (17,500 lb)
Max landing weight	7,360 kg (16,225 lb)
Max wing loading	362 kg/m² (74.14 lb/sq ft)
Max power loading	254.9 kg/kN (2.5 lb/lb st)

PERFORMANCE (at max T-O weight):

Max limiting Mach number	1.05
Max level speed: at S/L	485 kts (898 km/h; 558 mph)
at 11,000 m (36,000 ft)	Mach 0.85 (488 kts; 904 km/h; 562 mph)
Max cruising speed at 11,000 m (36,000 ft)	Mach 0.83 (476 kts; 882 km/h; 548 mph)
Stalling speed: flaps and landing gear up	100 kts (185 km/h; 115 mph)
flaps and landing gear down	90 kts (167 km/h; 104 mph)
Max rate of climb at S/L	3,078 m (10,100 ft)/min
Service ceiling	14,625 m (48,000 ft)
T-O run	458 m (1,500 ft)
T-O to 15 m (50 ft)	671 m (2,200 ft)
Landing from 15 m (50 ft)	945 m (3,100 ft)
Landing run	671 m (2,200 ft)
Range with max internal fuel	1,230 n miles (2,279 km; 1,416 miles)
Endurance with max internal fuel	3 h 12 min

UPDATED

AIDC AT-3B trainer (in foreground) and prototype AT-3A single-seat attack version

1995

TURKEY

TAI

TURKISH AEROSPACE INDUSTRIES INC (TUSAS HAVACILIK VE UZAY SANAYII A.S.)

PO Box 18, TR-06692 Kavaklidere, Ankara
Telephone: 90 (312) 811 18 00
Fax: 90 (312) 811 14 25
MANAGING DIRECTOR: Jerry R. Jones
DEPUTY MANAGING DIRECTOR: Dr Birol Altan
DIRECTOR OF PROGRAMMES: Kaya Ergenç

TAI is a majority owned Turkish company made up of Turkish (51 per cent) and US (49 per cent) partners and was formed 15 May 1984. Shareholders (by percentages) are Turkish Aircraft Industries (49), Turkish Armed Forces Foundation (1.9), Turkish Air League (0.1), Lockheed of Turkey Inc (42) and General Electric (7).

The objectives of TAI are to construct an aircraft facility capable of manufacturing modern weapon systems, helicopters and aerospace vehicles, modernise aircraft, perform research and development work in the field of aerospace, and to design, develop and manufacture total weapon systems for the defence of Turkey and commercial aircraft for national requirements as well as for international sales.

TAI facilities cover an area of 2.3 million m² (24,754,900 sq ft) with a factory floor area of over 130,000 m² (1,076,390 sq ft). The production facilities include fabrication and assembly, chemical processing, paint shops, fuel calibration, chemical processing and composite and metal bonding buildings, which are equipped with high technology machinery and equipment capable of producing modern aircraft. The manufacturing and assembly building covers 62,500 m² (672,687 sq ft) under one roof. The fabrication area includes numerical control machining centres, complete forming capability, detail fabrication shops, harness manufacture and hydraulic tubing. TAI also has an advanced computer capability.

The company has a workforce of 2,331 employees. TAI has a modern training centre, which provides highest quality training on the aircraft related skills and techniques since 1986. In addition to TAI employees, special training has also been given to the Turkish Air Force (TUAF) trainees. The centre also provides training on F-16 assembly to 64 Koreans.

Research and development activities are carried out in collaboration with Lockheed, NASA and NATO-AGARD. The company is engaged in design and development of UAV and a dual purpose passenger/cargo aircraft for 19 passengers and also participates in the development of the Future Large Aircraft.

TAI is looking to expand modernisation business from F-16s into F-4 and F-5s. The company also carries out studies to participate in projects that would respond other potential needs of TUAF such as developing and co-producing aircraft to meet the requirements for 'open skies' as well as acquiring air refuelling, airborne warning and control capabilities. To this end TAI is in contact with TUAF and potential producers of such aircraft.

TAI is ready to perform depot level maintenance, Falcon-up, and other modification and modernisation of all aircraft in the inventory of the Turkish armed forces as well as of other countries in the region.

VERIFIED

TAI (LOCKHEED FORT WORTH) F-16C/D

TYPE: Single/two-seat multirole fighter.

PROGRAMME: TAI is under contract to Lockheed Forth Worth to co-produce 152 of 160 F-16C/D for the Turkish Air Force under Peace Onyx-I programme. Of these aircraft 127 were delivered to TUAF by the end of 1993. TAI is also under contract to produce aft and centre fuselages and wings of F-16 aircraft for USAF. Within the framework of the follow-on programme Peace Onyx-II TAI will produce 80 additional aircraft of block 50 configuration for TUAF.

VERIFIED

TAI (CASA) CN-235M

TYPE: Twin-turboprop civil and military transport.

PROGRAMME: TAI produces 50 CASA CN-235M light transport aircraft for TUAF within the framework of the subcontract signed with CASA of Spain. Light Transport Aircraft programme, which covers seven stages will end in late 1997. First flight by Turkish assembled CN-235 made 24 September 1992. 14 aircraft were delivered by December 1994.

UPDATED

TAI (AGUSTA) SF-260D

TYPE: Two/three-seat fully aerobatic military light aircraft.

PROGRAMME: TAI assembled 34 SF-260D primary trainers

TAI (SIAI-Marchetti) SF-260D fully aerobatic military light aircraft

1995

under the subcontract signed with Agusta of Italy. First aircraft was delivered 24 July 1991. The programme was completed with the aircraft delivered to TUAF in December 1993. For full specifications see Agusta entry.

VERIFIED

TAI F-16 UPGRADE

TYPE: Single/two-seat multirole fighter upgrade.

PROGRAMME: TAI has started modification work with the installation of electronic countermeasures equipment (LORAL ALQ-178) on the Turkish Air Force F-16s. Thirteen F-16s were redelivered to TUAF by December 1993.

TAI will also start in 1994 'Falcon-up' modification which aims structural development of F-16s under Peace Onyx-I programme.

VERIFIED

UKRAINE

ANTONOV

ANTONOV DESIGN BUREAU

1 Tupolev Street, Kiev 252062
Telephone: 7 (044) 442 6415
Fax: 7 (044) 449 9996
Telex: 131309 OZON SU

GENERAL DESIGNER: Pyotr Vasilyevich Balabuyev
CHIEF DESIGNER: O. Bogdanov

Antonov OKB was founded in 1946 by Oleg Konstantinovich Antonov, who died 4 April 1984, aged 78. In current production are An-32 and An-124 at Kiev, An-72/74 at Kharkov. An-124 wings are manufactured in Tashkent, and airlifted to Kiev and Ulyanovsk for final assembly. Any production of An-225 is likely to be at Kiev. Small An-2 and An-28 are built by PZL Mielec, Poland.

UPDATED

Antonov An-2 single-engine general purpose biplane *(Paul Beaver)*

1995

ANTONOV An-2

NATO reporting name: Colt

TYPE: Single-engined general purpose biplane.

PROGRAMME: Prototype flew as SKh-1 31 August 1947. More than 5,000 An-2s were built at Kiev, ending in mid-1960s after limited manufacture of specialised agricultural An-2M. Production transferred to PZL Mielec, Poland, from where more than 11,730 delivered since 1960. China acquired licence and has built Yunshuji-5 (Y-5) versions 1957 to date.

VERSIONS: **An-2D:** Polish-built paratroop trainer.

An-2L: Water-bombing version. Entered service in 1969.

An-2M Russia (USSR): Agricultural version for crop dusting/spraying.

An-2M Poland: Polish designation of An-2V floatplane.

An-2P: Basic general-purpose version.

An-2R: Polish designation for Russian (USSR) An-2S.

An-2S (Poland): Designation of Polish-built An-2P ambulance version.

An-2S Russia (USSR): Agricultural version of An-2P for crop dusting/spraying.

An-2T: Polish-built version mixed cargo version equivalent to An-2P.

An-2TP: Polish-built luxury passenger/cargo version.

An-2V: Floatplane version.

An-2ZA: Meteorological version for high-altitude research.

Y-5B: Chinese-built agricultural and forestry version.

Y-5N: Chinese-built civil transport version.

UPGRADES: **Antonov:** An-3 turboprop conversion. See following entry.

OPERATORS: Versions of the An-2 are in service with the armed forces of the following countries: Afghanistan (12), Albania (10), Belarus (50), Bulgaria (2), China (300), Croatia (3), Cuba (20), Egypt (10), Laos (10), Lithuania (24), Mali (2), Mongolia (25), Nicaragua (5), North Korea (200), Poland (24), Romania (10), Russia (300), Ukraine (150) and Vietnam (20).

DESIGN FEATURES: Unequal-span single-bay biplane. Dihedral both wings approximately 2° 48′. I-type interplane struts. The fuselage is an all-metal stressed skin semi-monocoque structure of circular section forward of cabin, rectangular in the cabin section and oval in the tail section. The tail unit is a braced structure with fin integral with rear fuselage.

FLYING CONTROLS: Drooping ailerons and electrically actuated automatic slots on upper wings only. Trim tab in port aileron. Electrically actuated slotted flaps on all four wings. Trim tabs in rudder and port elevator.

STRUCTURE: The wing is an all-metal two-spar structure, fabric covered aft of front spar. All-metal fuselage.

LANDING GEAR: Non-retractable split-axle type with long-stroke oleo shock-absorbers. Mainwheel tyres size 800 × 260 mm, pressure 43 lb/sq in (3 kg/cm²). Pneumatic brakes. Self-centring tailwheel, tyre size 470 × 210 mm. Wheels interchangeable with floats or skis.

POWER PLANT: One 1,000 hp Shvetsov ASh-62 driving a Type V509-D9 four-blade constant-speed metal propeller. Six fuel tanks in upper wings with total capacity of 1,200 litres (317 US gallons; 264 Imp gallons). Oil capacity 120 litres (31.7 US gallons; 26.5 Imp gallons).

ACCOMMODATION: Equipped normally for operation by pilot only, in air conditioned cockpit, separated from main cabin by pressure bulkhead. Side windows of flight deck bulged to give good downward field of vision. Entry via port side of canopy, which hinges upward on centreline. Dual controls and second seat can be fitted for training or when aircraft is used for other duties. Easily convertible to carry passengers in tip-up wall seating, or for freight transport. Chemical hopper normally installed in cabin. Upward-opening freight door on port side of cabin is removable for parachuting or supply dropping. Smaller rearward-opening passenger door inset in freight door.

SYSTEMS: 28 V electrical system includes a 3,000 W engine-driven generator and 30 Ah battery.

ELECTRONICS AND EQUIPMENT: Standard equipment includes blind-flying instrumentation, radio and agricultural dusting and spraying gear. Equipment available for special duties includes two fixed semi-automatic AFA-33/50, or one AFA-33/75 and one AFA-18/21, cameras for photographic survey works; Geiger-Müller apparatus, magnetometer, and so on, for geological surveying.

DIMENSIONS, EXTERNAL:

Wing span (upper)	18.18 m (59 ft 8½ in)
Wing span (lower)	14.24 m (46 ft 8½ in)
Length overall	12.95 m (42 ft 6 in)
Height overall	4.20 m (13 ft 9½ in)

WEIGHTS AND LOADINGS:

Max payload	1,500 kg (3,300 lb)
Max T-O weight	5,500 kg (12,125 lb)

PERFORMANCE (at max T-O weight):

Max level speed at 1,750 m (5,750 ft)	253 km/h (157 mph)
Service ceiling	4,350 m (14,270 ft)
Max T-O run on grass	200 m (655 ft)
Landing run	100 m (330 ft)
Range with max fuel	488 n miles (905 km; 562 miles)

ANTONOV An-3

TYPE: Turboprop development of An-2 agricultural variant.

PROGRAMME: Reported Spring 1972 as expected competitor to Polish turbofan WSK-PZL Mielec M-15 for standardised agricultural use in Soviet Union and eastern Europe; confirmed 1979 that prototype produced by retrofitting piston-engined An-2 with 706 kW (946 shp) Glushenkov TVD-10 turboprop; state trials 1982-83, following rejection of M-15, and announcement that production An-3 would have 1,081 kW (1,450 shp) Glushenkov TVD-20 turboprop; plans announced for large scale conversion of An-2s to An-3s, but no subsequent confirmation of programme launch; announcement in 1988 that new 1,140 kW (1,528 shp) Glushenkov TVD-1500 turboprop developed for An-3.

DESIGN FEATURES: Longer and slimmer nose than An-2, housing turboprop, with small plugs fore and aft of wings to lengthen fuselage; cockpit farther forward, sealed and air conditioned; multipanelled starboard cockpit windows replaced by single large blister window; instrumentation, electrical and fuel systems new; access to cockpit of agricultural version via small door on port side; this door omitted on transport An-3, with access via main cabin door and airtight door between cabin and flight deck.

POWER PLANT: One 1,140 kW (1,528 shp) Glushenkov TVD-1500 turboprop. Total of 1,200 litres (317 US gallons; 264 Imp gallons) of fuel carried in six tanks in upper wings.

Antonov An-3 agricultural biplane with Glushenkov TVD-20 turboprop engine

1994

EQUIPMENT: Chemical spraytank for agricultural duty has capacity of 2,200 litres (581 US gallons; 484 Imp gallons), more than 50 per cent increase over that of An-2. Dusting equipment available.

DIMENSIONS, EXTERNAL:

Length overall	14.33 m (47 ft 0 in)

WEIGHTS AND LOADINGS:

Max T-O weight	5,800 kg (12,787 lb)
Max wing loading	81.10 kg/m² (16.61 lb/sq ft)
Max power loading	5.09 kg/kW (8.37 lb/shp)

PERFORMANCE:

Normal cruising speed	97 kts (180 km/h; 112 mph)
Rate of climb at S/L with max payload	240 m (785 ft)/min

ANTONOV An-12

NATO reporting name: Cub

TYPE: Four-turboprop transport and electronic warfare aircraft.

PROGRAMME: Prototype flew 1958, with Kuznetsov NK-4 turboprops, as rear-loading development of An-10 airliner; more than 900 built with AI-20K engines for military and civil use, ending in USSR in 1973; standard medium-range paratroop and cargo transport of former Soviet Military Transport Aviation (VTA) from 1959; replacement with Il-76 began 1974; Shaanxi Aircraft Company, China, manufactures redesigned Yunshuji-8 (Y-8) transport version and derivatives.

VERSIONS: **Cub (An-12BP):** Basic transport, fewer than 150 now with VTA, plus 200 with Soviet air armies and air forces of military districts and groups of forces; 120 Aeroflot An-12s form military reserve; An-12BP has tail gun turret.

Cub-A: Elint version; as 'Cub' but blade aerials on front fuselage, aft of flight deck, and other changes.

Cub-B: Elint conversion of basic 'Cub' for Soviet Naval Air Force; two additional radomes under forward and centre fuselage, plus other antennae; about 10 produced.

Cub-C: ECM variant with several tons of electric generation, distribution and control gear in cabin, palletised jammers for at least five wavebands faired into belly, and chaff/flare dispensers; glazed nose and undernose radar of transport retained. Ogival 'solid' fuselage tailcone, housing electronic equipment, replaces usual gun position.

Cub-D: ECM variant for active countermeasures, with pods each side of front fuselage and tail fin. Soviet Navy has about 29 'Cub-Cs and Ds'.

Y-8: Chinese built prototype.

Y-8A: Chinese built helicopter carrier. Main cabin height increased by 120 mm (4.72 in) by deleting internal gantry; downward-opening rear ramp/door.

Y-8B: Chinese licence-built civil transport. Military equipment deleted; weight reduced by 1,720 kg (3,792 lb); some avionics differ.

Y-8C: Chinese licence-built fully pressurised version for civil and military applications, developed with Lockheed collaboration.

Y-8D: Chinese licence-built export version, with main avionics by Collins and Litton.

Y-8E: Chinese licence-built drone carrier version. Forward pressure cabin accommodates drone controller's console; carrier/launch trapeze for one drone under each outer wing panel.

Y-8F: Chinese licence-built livestock carrier, with cages to hold up to 500 sheep or goats.

Y-8H: Chinese licence-built aerial survey version.

Y-8X: Chinese licence-built maritime patrol version, with Western com/nav, radar, surveillance and search equipment; larger chin radome.

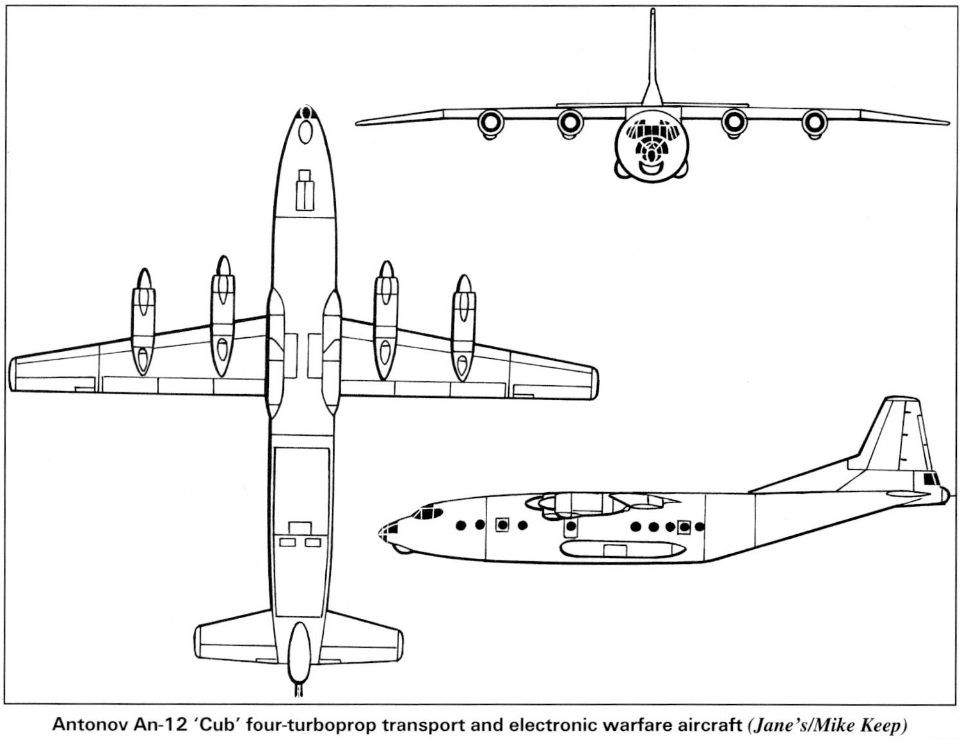

Antonov An-12 'Cub' four-turboprop transport and electronic warfare aircraft *(Jane's/Mike Keep)* *1994*

OPERATORS: Verions of the An-12 are in service with the armed forces of the following countries: Afghanistan (12), Belarus, China (25), Czech Republic (1), Ethiopia (4), Iraq (5), Russia (10), Slovak Republic (1), Sri Lanka (2), Ukraine (40), Yemen (1) and Yugoslavia (12).

The following abbreviated details apply to the standard Soviet built military An-12BP transport:

POWER PLANT: Four 2,942 kW (3,945 ehp) Ivchenko AI-20K turboprops, driving AV-68 four-blade reversible-pitch propellers. All fuel in 22 bag tanks in wings, total normal capacity 13,900 litres (3,672 US gallons; 3,058 Imp gallons). Maximum capacity 18,100 litres (4,781 US gallons; 3,981 Imp gallons).

ACCOMMODATION: Pilot and co-pilot side by side on flight deck. Engineer's station on starboard side, behind co-pilot. Radio operator in well behind pilot, facing outward. Navigator in glazed nose compartment. Rear gunner in tail turret. Crew door on port side forward of wing. No integral rear-loading ramp. Access to freight hold via large door under upswept rear fuselage, comprising two longitudinal halves which can be hinged upward inside cabin to provide access for direct loading of freight from trucks. Undersurface of fuselage aft of this door is formed by a further, rear-hinged, door which retracts upward into fuselage to facilitate loading and unloading. Equipped to carry 90 troops or 60 paratroops, all of whom can be despatched in under one minute, with rear door panels folded upward.

ARMAMENT: Two 23 mm NR-23 guns in tail turret.

DIMENSIONS, EXTERNAL:

Wing span	38.00 m (124 ft 8 in)
Wing chord (mean)	3.452 m (11 ft 4 in)
Wing aspect ratio	11.85
Length overall	33.10 m (108 ft 7¼ in)
Height overall	10.53 m (34 ft 6½ in)
Tailplane span	12.20 m (40 ft 0¼ in)
Wheel track	5.42 m (17 ft 9½ in)
Wheelbase	10.82 m (35 ft 6 in)
Propeller diameter	4.50 m (14 ft 9 in)
Rear-loading hatch: Length	7.70 m (25 ft 3 in)
Width	2.95 m (9 ft 8 in)

DIMENSIONS, INTERNAL:

Cargo hold: Length	13.50 m (44 ft 3½ in)
Max width	3.50 m (11 ft 5¾ in)
Max height	2.60 m (8 ft 6¼ in)
Volume	97.2 m³ (3,432.6 cu ft)

AREAS:

Wings, gross	121.70 m² (1,310 sq ft)

WEIGHTS AND LOADINGS:

Weight empty	28,000 kg (61,730 lb)
Max payload	20,000 kg (44,090 lb)
Normal T-O weight	55,100 kg (121,475 lb)
Max T-O weight	61,000 kg (134,480 lb)
Max wing loading	501.2 kg/m² (102.6 lb/sq ft)
Max power loading	5.18 kg/kW (8.52 lb/shp)

PERFORMANCE:

Max level speed	419 kts (777 km/h; 482 mph)
Max cruising speed	361 kts (670 km/h; 416 mph)
Min flying speed	88 kts (163 km/h; 101 mph)
Landing speed	108 kts (200 km/h; 124 mph)
Max rate of climb at S/L	600 m (1,970 ft)/min
Service ceiling	10,200 m (33,500 ft)
T-O run	700 m (2,300 ft)
Landing run	500 m (1,640 ft)
Range: with max payload	1,942 n miles (3,600 km; 2,236 miles)
with max fuel	3,075 n miles (5,700 km; 3,540 miles)

UPDATED

Antonov An-12 'Cub' of the Czech Republic Air Force *(P. Tompkins)* *1994*

ANTONOV An-22 ANTHEUS

NATO reporting name: Cock

TYPE: Long-range turboprop heavy transport.

PROGRAMME: Prototype first flew 27 February 1965; production terminated early, in 1974; about 45 used by VTA and Aeroflot.

VERSIONS: Early An-22s had small nose glazing, radar under starboard landing gear fairing; for production, weather/collision avoidance radar above new nose glazing, and undernose navigation/mapping radome. One An-22 (SSSR-644460), with original nose glazing and undernose radome, adapted to deliver An-124 wings non-stop 1,693 n miles (3,136 km; 1,949 miles) from Valery Chkalov plant in Tashkent, where built, to Kiev assembly works; each 23 tonne wing carried over fuselage on mounts that prevent flexing of An-22 in flight being transmitted to An-124 wing; added central tail fin preserves An-22 handling qualities in flight.

OPERATORS: Versions of the An-22 are in service with the armed forces of Russia (50).

The following abbreviated details apply to the standard production An-22:

POWER PLANT: Four 11,185 kW (15,000 shp) Kuznetsov NK-12MA turboprops, each driving a pair of four-blade contrarotating propellers.

ACCOMMODATION: Crew of five or six on pressurised flight deck, with navigator's station in nose. Pressurised cabin for 28-29 passengers aft of flight deck, separated from main cabin by bulkhead containing two doors. Unpressurised main cabin, with reinforced titanium floor, tiedown fittings and rear-loading ramp. When ramp lowers, a large door which forms the underside of the rear fuselage retracts upward inside fuselage to permit easy loading of tall vehicles. Rails in roof of cabin for four travelling gantries continue rearward on underside of this door. Two winches, used in conjunction with gantries, each have a capacity of 2,500 kg (5,500 lb). Door in each landing gear fairing, forward of wheels, for crew and passengers.

DIMENSIONS, EXTERNAL:

Wing span	64.40 m (211 ft 4 in)
Length overall	approx 57.92 m (190 ft 0 in)
Height overall	12.53 m (41 ft 1½ in)
Propeller diameter	6.20 m (20 ft 4 in)

DIMENSIONS, INTERNAL:

Main cabin: Length	33.00 m (108 ft 3 in)
Max width	4.40 m (14 ft 5 in)
Max height	4.40 m (14 ft 5 in)
Volume	640 m^3 (22,600 cu ft)

AREAS:

Wings, gross	345 m^2 (3,713 sq ft)

WEIGHTS AND LOADINGS:

Weight empty, equipped	114,000 kg (251,325 lb)
Max payload	80,000 kg (176,350 lb)
Max fuel	43,000 kg (94,800 lb)
Max T-O weight	250,000 kg (551,160 lb)
Max wing loading	724.6 kg/m^2 (148.4 lb/sq ft)
Max power loading	5.59 kg/kW (9.19 lb/shp)

PERFORMANCE:

Max level speed	399 kts (740 km/h; 460 mph)
T-O run	1,300 m (4,260 ft)
Landing run	800 m (2,620 ft)

Range: with max fuel and 45,000 kg (99,200 lb) payload
5,905 n miles (10,950 km; 6,800 miles)
with max payload
2,692 n miles (5,000 km; 3,100 miles)

UPDATED

ANTONOV An-24

NATO reporting name: Coke

TYPE: Twin-turboprop short-range transport.

PROGRAMME: Development of this twin-turboprop transport was started in 1958, to replace piston-engined types on Aeroflot's internal feederline routes. The An-24 was intended originally to carry 32-40 passengers, but when the prototype flew in April 1960 it had been developed into a 44-seater. It was followed by a second prototype and five preproduction An-24s. Flight trials were stated to be complete in September 1962 and the An-24 entered service on Aeroflot's routes from Moscow to Voronezh and Saratov in September 1963. More than 50 million passengers and 500,000 tonnes of cargo had been carried by Aeroflot An-24Vs by 1971.

The An-24 is designed to operate from airfields of limited size, with paved or natural runways, and can be fitted with rocket-assisted take-off units to permit operation with a full load of cargo at ambient temperatures above 30°C. Two were taken to the Antarctic in late 1969, to replace pistion-engined Il-14s used previously for flights between Antarctic stations.

By 1976, a total of about 1,100 An-24s had been produced, several hundred of them for Aeroflot.

On the prototype, the engine nacelles extended only a little past the wing trailing-edges: production An-24s have lengthened nacelles with conical rear fairings. A ventral tail fin was also added on production models of the passenger-carrying version, which are supplemented by the An-24T and An-26 specialised freight-carrying versions of the same basic design.

Production versions of the An-24 are designed for a service life of 30,000 hours and 15,000 landings, and are available in a variety of forms, as follows:

VERSIONS: **An-24V Srs I:** Standard version in production in 1967. Equipped normally to carry a crew of three (pilot, copilot/radio operator/navigator and cabin attendant) and up to 50 passengers. Powered by two 2,500 ehp Ivchenko AI-24 turboprop engines. TG-16 self-contained starter-generator in rear of starboard engine nacelle. Mixed passenger-freight, convertible cargo/passenger, all-freight and executive versions available.

An-24V Srs II: Standard version, seating up to 50 passengers. Superseded Srs I (with 2,550 ehp AI-24 engines) in 1968, and described in detail below. Basically as Srs I, powered by two Ivchenko AI-24A turboprop engines, with water injection. Can have crew of up to five (two pilots, navigator, radio operator and, on jump seat, an engineer or cargo handler) on flight deck. TG-16 self-contained starter/generator in rear of starboard engine nacelle. Mixed passenger/freight, convertible cargo/passenger, all-freight and executive versions available.

An-24P: Firefighting (*Pozharny*) version, which underwent evaluation in the USSR in 1971. Special provisions for enabling firefighters to be parachuted from a height of 800-1,200 m (2,625-3,940 ft) to deal with forest fires.

An-24RT: Generally similar to An-24T but with Type RU auxiliary turbojet in starboard nacelle, as on An-24RV.

An-24RV: Generally similar to Srs II version of An-24V, but with an 8.83 kN (1,985 lb st) Type RU 19-300 auxiliary turbojet engine in starboard nacelle instead of starter/generator. This turbojet is used for engine starting, to improve take-off performance and to improve performance in the air. It permits take-off with a full payload from airfields up to 3,000 m (9,840 ft) above S/L and at temperatures up to ISA + 30°C. It also ensures considerably improved stability and handling characteristics after a failure of one of the turboprop engines in flight. Maximum T-O weight is increased by 800 kg (1,760 lb) at S/L ISA and by 2,000 kg (4,410 lb) at S/L ISA + 30°C by use of the auxiliary turbojet. An An-24RV was demonstrated at the 1967 Paris Air Show.

An-24T: Generally similar to An-24V Srs II but equipped as specialised freighter. Crew of five, consisting of pilot, co-pilot, navigator, radio operator and flight engineer. Normal passenger door at rear of cabin is deleted and replaced by a belly freight door at the rear of the cabin. This hinges upward and to the rear, providing a hatchway for cargo loading. An electrically powered winch, capacity 1,500 kg (3,300 lb), is used to hoist crates through the hatch and runs on a rail in the cabin ceiling to position the payload inside the cabin. Electrically or manually powered conveyor, capacity 4,500 kg (9,920 lb), flush with cabin floor. Fewer windows. Folding seats along walls of cabin. Emergency exit hatches in side and in floor at front of cabin. Rear cargo door permits airdropping of payload or parachutists. Provision for stretcher carrying in air ambulance role. Single ventral fin replaced by twin ventral fins,

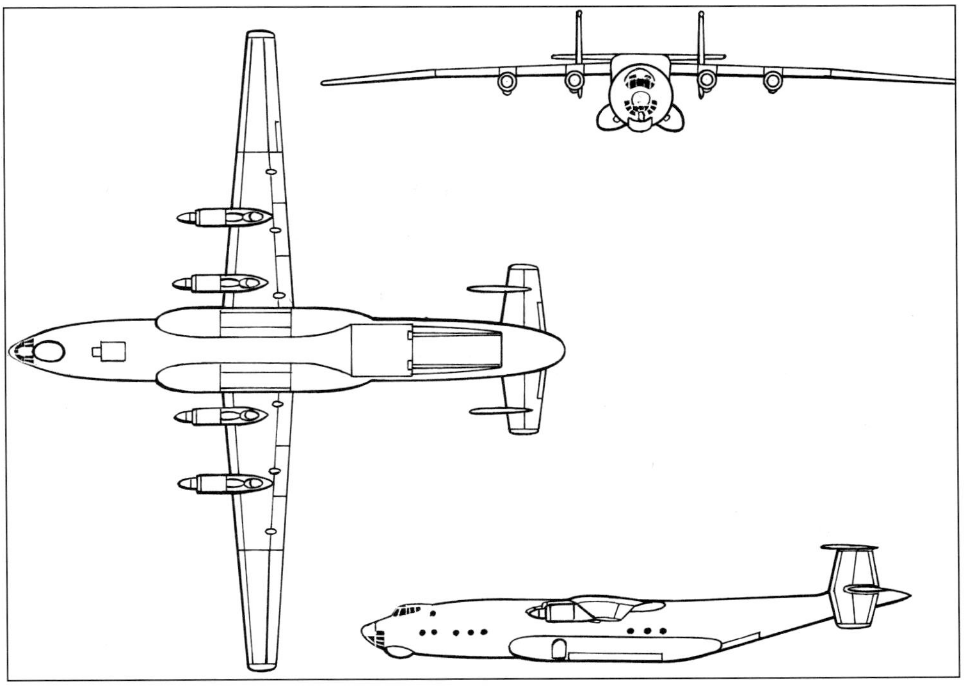

Antonov An-22 'Cock' four-turboprop transport aircraft *(Jane's/Mike Keep)*

1994

Antonov An-22 'Cock' four-turboprop heavy transport aircraft

1994

Antonov An-22 'Cock' carrying the wing of an An-124 *(Tass)*

1994

Antonov An-24 twin-turboprop short-range transport aircraft

1994

forming Vee, aft of cargo door. An An-24T was displayed at the 1967 Paris Air Show, and this version serves with several airlines.

Y-7: Chinese licence-built version see *Jane's All The World's Aircraft 1994-95.*

Y7-100: Improved Chinese version see *Jane's All The World's Aircraft 1994-95.*

Y7-200A: Improved Chinese version see *Jane's All The World's Aircraft 1994-95.*

Y7-200B: Improved Chinese version see *Jane's All The World's Aircraft 1994-95.*

Y7H-500: Improved Chinese version see *Jane's All The World's Aircraft 1994-95.*

OPERATORS: Versions of the An-24 are in service with the armed forces of the following coutries: Bulgaria (5), Congo (5), Cuba (4), Czech Republic (3), Guinea Republic (1), Guinea-Bissau (1), Hungary (2), Kazakhstan (15), Laos (6), Mali (2), Mongolia (6), Romania (10), Russia (100), Slovak Republic (2), Somalia (2), Sudan (5), Syria (2), Vietnam (9) and Yemen (7).

The following description refers to the basic An-24V Srs II, unless otherwise noted, but is generally applicable to all versions except for the detailed differences noted above:

DESIGN FEATURES: Cantilever high-wing monoplane. Wing has 2° anhedral on outer panels. Incidence 3°. Sweepback at quarter-chord on outer panels 6° 50′. The tail unit is a cantilever all-metal structure with ventral fin (two ventral fins on An-24T/RT versions). 9° dihedral on tailplane.

FLYING CONTROLS: Mass-balanced servo-compensated ailerons with large trim tabs. Hydraulically operated Fowler flaps along entire wing trailing-edges inboard of unpowered ailerons; single-slotted flaps on centre-section, double-slotted outboard of nacelles. Servo and trim tabs in each aileron. All tail unit controls manually operated. Trim tabs in elevators. Trim tab and spring tab in rudder.

STRUCTURE: The wing is an all-metal two-spar built in five sections: centre-section, two inner wings and two outer wings. Trim tabs of glassfibre. Wing skin is attached by electrical spot-welding. The fuselage is an all-metal semi-monocoque structure in front, centre and rear portions of bonded/welded construction.

LANDING GEAR: Retractable tricycle type with twin-wheels on all units. Hydraulic retraction. Emergency extension by gravity. All units retract forward. Mainwheels size 900 × 300-700, tyre pressure 3.45-4.90 bars (50-71 lb/sq in). Nosewheels size 700 × 250, tyre pressure 2.45-3.45 bars (35.5-50 lb/sq in). Tyre pressures variable to cater for different types of runway. Disc brakes on mainwheels. Steerable and castoring nosewheel unit.

POWER PLANT (all versions): Two 1,902 kW (2,550 ehp) Ivchenko AI-24A turboprop engines (with provision for water injection; weight of water 68 kg; 150 lb), each driving an AV-72 four-blade constant-speed fully feathering propeller. Electrical de-icing system for propeller blades and hubs; hot air system for engine air intakes. Fuel in integral tanks immediately outboard of nacelles, and four bag type tanks in centre-section, total capacity 5,550 litres (1,220 Imp gallons). Provision for four additional tanks in centre-section. Pressure refuelling socket in starboard engine nacelle. Gravity fuelling point above each tank. Carbon dioxide inert gas system to create fireproof condition inside fuel tanks. Oil capacity 53 litres (11.5 Imp gallons). One 8.83 kN (1,985 lb st) Type RU 19-300 auxiliary turbojet in starboard nacelle of An-24RV and An-24RT. Provision for fitting rocket-assisted take-off units on cargo versions.

ACCOMMODATION (An-24V/RV): Crew of three (pilot, co-pilot/radio operator/navigator and one stewardess). Provision for carrying navigator, radio operator and engineer. Normal accommodation for 44-52 passengers in air conditioned and pressurised cabin. Standard layout has baggage and freight compartments on each side aft of flight deck; then the main cabin with 52 forward-facing reclining seats, in pairs at a pitch of 72 cm (28.3 in), on each side of centre aisle (optionally two small sofas for babies at rear, instead of two of the seats); buffet and stewardess' seat, and toilet, opposite door to rear of cabin; and wardrobes at rear. Passenger door on port side, aft of cabin, is of airstair type. Door on starboard side for freight hold (front). All doors open inward. The 46-seat version has a removable partition aft of the fifth row of seats, instead of one row of seats. The mixed passenger/cargo version is laid out normally for 36 passengers, with 14 m^3 (495 cu ft) forward hold for baggage, freight and mail, and rear wardrobe and baggage hold (2.8 m^3; 99 cu ft). A typical de luxe or executive layout retains the forward and aft baggage and freight holds of the airliner version but has the main cabin divided into three compartments. The forward compartment contains four pairs of seats, in aft-facing and forward-facing pairs with tables between, and a buffet. Next comes a similar cabin without the buffet, followed by a sleeping compartment, with a sofa, two seats and table. At the rear is the standard toilet compartment opposite the airstair door, and large wardrobe space.

ACCOMMODATION (An-24T/RT): Provision for crew of up to five, with optional cargo handler. Door at front of cabin on starboard side. Upward-opening cargo door in belly at rear of cabin. Maximum overall dimensions of cargo packages that can be handled are 1.1 × 1.5 × 2.6 m (43.3 × 59 × 102 in) or 1.3 × 1.5 × 2.1 m (51.2 × 59 × 82.7 in). Toilet (port side) and emergency exit door in belly, immediately aft of flight deck. Folding seats, in two-, three- and four-place units, for 30 paratroops or 38 equipped soldiers along walls of main cabin. Ambulance configuration is equipped to carry 24 stretcher cases and one medical attendant. Cargo loading system includes rails in floor, electric winch, overhead gantry, tie-down fittings, nets and harness. Electrical de-icing system for windscreens.

SYSTEMS: Air conditioning system uses hot air tapped from the 10th compressor stage of each engine, with a heat exchanger and turbocooler in each nacelle. Cabin pressure differential 0.29 bars (4.27 lb/sq in). Main and emergency hydraulic systems, pressure 151.7 bars (2,200 lb/sq in), for landing gear retraction, nosewheel steering, flaps, brakes, windscreen wipers, propeller feathering and, on An-24T, operation of cargo and emergency escape doors. Hand-pump to operate doors only and build up pressure in main system. Electrical system includes two 27 V DC starter/generators, two alternators to provide 115 V 400 Hz AC supply and two inverters for 36 V 400 Hz three-phase AC supply. An-24T has permanent oxygen system for pilot, installed equipment for other crew members and three portable bottles for personnel in cargo hold. Thermal de-icing system on wing. All tail unit leading-edges incorporate thermal de-icing.

ELECTRONICS AND EQUIPMENT (An-24T/RT): Standard radio equipment includes two R-802V VHF transceivers, R-836 HF transmitter and US-8 receiver, SPU-7 intercom, two ARK-11 ADF, RV-2 radio altimeter, SP-50 ILS and KRP-F glide path receiver, GRP-2 glide slope receiver and MRP-56 marker receiver, and RPSN-2AN weather, obstruction and navigation radar. Flight and navigational equipment includes an AP-28L1 autopilot, TsGV-4 master vertical gyro, GPK-52AP directional gyro, GIK-1 gyro compass, two ZK-2 course setting devices, two AGD-1 artificial horizons, AK-59P astro-compass, NI-50BM-K ground position indicator and other standard blind-flying instruments, plus three clocks. Optional OPB-1R sight for pinpoint dropping of cargo and determination of navigational data.

DIMENSIONS, EXTERNAL:

Wing span	29.20 m (95 ft 9½ in)
Wing aspect ratio	11.7
Length overall	23.53 m (77 ft 2½ in)
Height overall	8.32 m (27 ft 3½ in)
Width of fuselage	2.90 m (9 ft 6 in)
Depth of fuselage	2.50 m (8 ft 2½ in)
Tailplane span	9.08 m (29 ft 9½ in)
Wheel track (c/l shock struts)	7.90 m (25 ft 11 in)
Wheelbase	7.89 m (25 ft 10½ in)
Propeller diameter	3.90 m (12 ft 9½ in)
Propeller ground clearance	1.145 m (3 ft 9 in)
Passenger door (port, aft, except on An-24T):	
Height	1.40 m (4 ft 7 in)
Width	0.75 m (2 ft 5½ in)
Height to sill	1.40 m (4 ft 7 in)
Freight compartment door (stbd, fwd):	
Height	1.10 m (3 ft 7¼ in)
Width	1.20 m (3 ft 11¼ in)
Height to sill	1.30 m (4 ft 3 in)
Baggage compartment door (stbd, aft, except on An-24T):	
Height	1.41 m (4 ft 7½ in)
Width	0.75 m (2 ft 5½ in)
Cargo door (belly, rear, An-24T only):	
Length	2.85 m (9 ft 4 in)
Width: max	1.40 m (4 ft 7 in)
min	1.10 m (3 ft 7¼ in)
Height above ground	1.25-1.62 m (4 ft 1 in-5 ft 4 in)
Emergency exit (An-24T, side):	
Height	0.60 m (1 ft 11½ in)
Width	0.50 m (1 ft 7½ in)
Emergency exit (An-24T, underfuselage):	
Length	1.155 m (3 ft 9½ in)
Width	0.70 m (2 ft 3½ in)

DIMENSIONS, INTERNAL:

Main passenger cabin (52-seater):	
Length	9.69 m (31 ft 9½ in)
Max width	2.76 m (9 ft 1 in)
Max height	1.91 m (6 ft 3 in)
Floor area	39.95 m^2 (430 sq ft)
Cargo hold (An-24T): Length	15.68 m (51 ft 5½ in)
Width	2.17 m (7 ft 1½ in)
Height	1.765 m (5 ft 9½ in)
Volume	50 m^3 (1,765 cu ft)

AREAS:

Wings, gross	74.93 m^2 (807.1 sq ft)
Horizontal tail surfaces (total)	17.23 m^2 (185.5 sq ft)
Vertical tail surfaces (total, excl dorsal fin)	13.38 m^2 (144.0 sq ft)

WEIGHTS AND LOADINGS:

Weight empty: An-24V	13,300 kg (29,320 lb)
An-24T	14,060 kg (30,997 lb)
Basic operating weight: An-24T	14,698 kg (32,404 lb)
Fuel weight:	
An-24T with max payload	1,800 kg (3,968 lb)
An-24T for max range	4,760 kg (10,494 lb)
Max payload (ISA, S/L):	
An-24V, An-24RV	5,500 kg (12,125 lb)
An-24T	4,612 kg (10,168 lb)
An-24RT	5,700 kg (12,566 lb)
Max ramp weight: An-24T	21,100 kg (46,540 lb)
Max T-O and landing weight:	
An-24V, An-24T, S/L, ISA	21,000 kg (46,300 lb)
An-24V, An-24T, S/L, ISA + 30°C	19,800 kg (43,650 lb)
An-24RV, An-24RT, S/L, ISA or ISA + 30°C	21,800 kg (48,060 lb)
Max wing loading: An-24V	276 kg/m^2 (56.53 lb/sq ft)

PERFORMANCE (at max T-O weight):

Normal cruising speed at 6,000 m (19,700 ft)	243 kts (450 km/h; 280 mph)
Max range cruising speed at 7,000 m (23,000 ft)	243 kts (450 km/h; 280 mph)
T-O speed:	
An-24T	97-100 kts (180-185 km/h; 112-115 mph)
Landing speed:	
An-24V	89 kts (165 km/h; 103 mph) CAS
An-24T	87-95 kts (160-175 km/h; 100-109 mph)
Max rate of climb at S/L: An-24V	114 m (375 ft)/min
An-24RV	204 m (670 ft)/min
Rate of climb at S/L, one engine out:	
An-24V, ISA	84 m (275 ft)/min
An-24V, ISA + 30°C, with water injection	84 m (275 ft)/min
An-24RV, ISA	174 m (570 ft)/min
An-24RV, ISA + 30°	90 m (295 ft)/min
Service ceiling: An-24V, An-24T	8,400 m (27,560 ft)
An-24RV, An-24RT	9,000 m (29,525 ft)
Service ceiling, one engine out:	
An-24T	2,750 m (9,020 ft)
T-O run: An-24V	600 m (1,970 ft)
An-24T	640 m (2,100 ft)
Balance T-O runway: An-24T, ISA	1,720 m (5,645 ft)
An-24T, ISA + 15°C	1,750 m (5,745 ft)
Landing run at AUW of 20,000 kg (44,100 lb):	
An-24T	880 m (1,903 ft)
Landing from 15 m (50 ft) at AUW of 20,000 kg (44,100 lb): An-24T	1,590 m (5,217 ft)
Range with max payload, with reserves:	
An-24V, An-24RV	296 n miles (550 km; 341 miles)
An-24T, An-24RT	344 n miles (640 km; 397 miles)

Range, with max fuel: An-24V, 45 min fuel reserves
1,293 n miles (2,400 km; 1,490 miles)
An-24T, with 1,612 kg (3,554 lb) payload, no reserves 1,618 n miles (3,000 km; 1,864 miles)

UPDATED

ANTONOV An-26

NATO reporting name: Curl

TYPE: Twin-turboprop pressurised short-haul transport.

PROGRAMME: First exhibited 1969 Paris Air Show; more than 1,000 built before superseded in production by An-32; derivative Y-7H-500 built by Xian Aircraft Company (see XAC, China).

VERSIONS: **An-26** ('Curl-A'): Original version; electrically/manually operated conveyor flush with cabin floor for freight handling.

An-26B ('Curl-A'): Improved version, announced 1981, to carry three standard freight pallets, each 2.44 m (8 ft) long, 1.46 m (4 ft 9½ in) wide and 1.60 m (5 ft 3 in) high, with total weight of 5,500 kg (12,125 lb). Improved freight handling equipment.

An-26 ('Curl-B'): Signals intelligence (sigint) version, many short blade antennae mounted on fuselage.

OPERATORS: Versions of the An-26 are in service with the armed forces of the following countries: Afghanistan (20), Angola (10), Belarus (20), Benin (2), Bulgaria (3), Cape Verde, China, Czech Republic (4), Ethiopia (1), Germany (1), Hungary (7), Iraq (6), Kazakhstan (15), Laos (3), Lithuania (3), Mali (2), Mongolia (1), Mozambique (5), Nicaragua (4), Poland (26), Romania (6), Russia (250), Slovak Republic (2), Somalia (1), Syria (4), Ukraine (30), Vietnam (40), Yemen (6), Yugoslavia (6) and Zambia (4).

CUSTOMERS: Civil customers include Aero Caribbean (Cuba), Aeronica (Nicaragua), Air Mongol, Alyemda (Yemen), Ariana Afghan Airlines, CAAC (China), Cubana, Syrianair and Tarom (Romania).

DESIGN FEATURES: Generally similar to earlier An-24RT specialised freighter, with auxiliary turbojet; more powerful turboprops and redesigned 'beaver-tail' rear fuselage. Oleg Antonov's special loading ramp forms underside of rear fuselage when retracted, slides forward under rear of cabin to facilitate direct loading and when airdropping cargo. Wing anhedral 2° on outer panels; incidence 3°; sweepback on outer panels 6° 50′ at quarter-chord, 9° 41′ on leading-edge; swept vertical and horizontal tail; tailplane dihedral 9°.

FLYING CONTROLS: Mechanical controls; mass balanced servo compensated ailerons with electrical glassfibre trim tabs; manual tab in each elevator; electrical trim/servo tab in rudder; hydraulically actuated tracked and slotted TsAGI flaps, single-slotted on centre-section, double-slotted outboard of nacelles.

STRUCTURE: Conventional light alloy; two-spar wing, built in centre, two inner and two detachable outer sections, with skin attached by electrical spot welding; bonded/welded semi-monocoque fuselage in front, centre and rear portions, with 'bimetal' (duralumin-titanium) bottom skin for protection during operation from unpaved airfields; blister on each side of fuselage forward of rear ramp carriers track to enable ramp to slide forward; large dorsal fin; ventral strake on each side of ramp.

LANDING GEAR: Hydraulically retractable tricycle type, with twin-wheels on each unit. Emergency extension by gravity. All units retract forward. Shock-absorbers of oleo-nitrogen type on main units; nitrogen-pneumatic type on nose unit. Mainwheel tyres size 1,050 × 400 mm, pressure 5.9 bars (85 lb/sq in). Nosewheel tyres size 700 × 250 mm, pressure 3.9 bars (57 lb/sq in). Mainwheels fitted with hydraulic disc brakes and anti-skid units. Nosewheels can be steered hydraulically through 45° each side while taxying and are controllable through ±10° during take-off and landing.

POWER PLANT: Two 2,103 kW (2,820 ehp) Ivchenko AI-24VT turboprops, each driving a four-blade constant-speed fully feathering propeller. Electric de-icing system for propeller blades and hubs; hot air system for engine air intakes. One 7.85 kN (1,765 lb st) RU 19A-300 auxiliary turbojet in starboard nacelles for use, as required, at take-off, during climb and in level flight, and for self-contained starting of main engines. Two independent but interconnected fuel systems, with 5,500 kg (12,125 lb) of fuel, contained in integral tanks in inner wings and 10 bag tanks in centre-section. Pressure refuelling socket in starboard engine nacelle. Gravity fuelling point above each tank area. Carbon dioxide inert gas system to create fireproof condition inside fuel tanks.

ACCOMMODATION: Basic crew of five (pilot, co-pilot, radio operator, flight engineer and navigator), with station at rear of cabin on starboard side for loading supervisor or load dispatcher. Optional domed observation window for navigator on port side of flight deck. Toilet on port side aft of flight deck; crew door, small galley and oxygen bottle stowage on starboard side. Emergency escape hatch in door immediately aft of flight deck. Large downward hinged rear ramp/door, hinged to an anchorage mounted on tracks running forward under the blister fairings. This enables ramp/door to slide forward under fuselage for direct loading on to cabin floor or for airdropping of freight. When doing so, its rear is supported by the pivoted swinging arm on each side which also raises and lowers door in the alternative fixed-hinge mode. Door can be locked in any intermediate position. Electrically powered mobile winch, capacity 2,000 kg (4,409 lb), hoists crates through rear entrance and runs on a rail in the cabin ceiling to position payload in cabin. Electrically and manually operated conveyor, capacity 4,500 kg (9,920 lb), built-in flush with cabin floor of original An-26, facilitates loading and airdropping of freight. An-26B has removable rollgangs, mechanism for moving pallets inside hold, and moorings, enabling two men to load and unload three pallets in 30 minutes. Rollgangs can be stowed against sides of cabin. Both versions can accommodate a variety of motor vehicles, including GAZ-69 and UAZ-469 military vehicles, or cargo items up to 1.50 m (59 in) high by 2.10 m (82.6 in) wide. Height of rear edge of cargo door surround above the cabin floor is 1.50 m (4 ft 11 in). Cabin is pressurised and air conditioned, and can be fitted with a row of tip-up seats along each wall to accommodate a total of 38 to 40 persons. Conversion to troop transport role, or to an ambulance for 24 stretcher patients and a medical attendant, takes 20 to 30 minutes in the field.

SYSTEMS: Air conditioning system uses hot air tapped from the 10th compressor stage of each engine, with a heat exchanger and turbocooler in each nacelle. Cabin pressure differential 0.29 bars (4.27 lb/sq in). Main and emergency hydraulic systems, pressure 151.7 bars (2,200 lb/sq in), for landing gear retraction, nosewheel steering, flaps, brakes, windscreen wipers, propeller feathering and operation of cargo ramp and emergency escape doors. Handpump to operate doors only and build up pressure in main system. Electrical system includes two 27 V DC starter/generators on engines, a standby generator on the auxiliary turbojet, and three storage batteries for emergency use. Two engine-driven alternators provide 115 V 400 Hz single-phase AC supply, with standby inverter. Basic source of 36 V 400 Hz three-phase AC supply is two inverters, with standby transformer. Permanent oxygen system for pilot, installed equipment for other crew members and three portable bottles for personnel in cargo hold. Bleed air thermal de-icing system for wing and tail unit leading-edges. Electric windscreen de-icing.

Antonov An-26 'Curl' twin-turboprop pressurised short-haul transport aircraft ***(P. Tompkins)***
1994

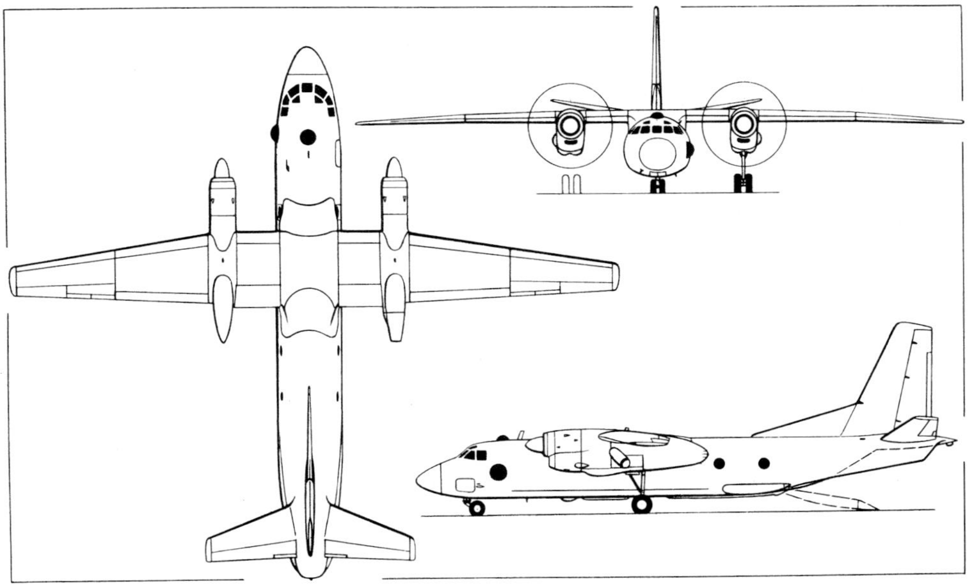

Antonov An-26 twin-turboprop short-haul transport aircraft ***(Jane's/Dennis Punnett)***
1994

AVIONICS: Standard com/nav avionics comprise two VHF transceivers, HF, intercom, two ADF, radio altimeter, glide path receiver, glide slope receiver, marker beacon receiver, weather/navigation radar, directional gyro and flight recorder. Optional avionics include a flight director system, astrocompass and autopilot.

EQUIPMENT: Standard equipment includes parachute static line attachments and retraction devices, tiedowns, jack to support ramp sill, flight deck curtains, sun visors and windscreen wipers. Optional items include OPB-1R sight for pinpoint dropping of freight, medical equipment, and liquid heating system.

ARMAMENT: Provision for bomb rack on fuselage below each wingroot trailing-edge.

DIMENSIONS, EXTERNAL:

Wing span	29.20 m (95 ft 9½ in)
Wing aspect ratio	11.7
Length overall	23.80 m (78 ft 1 in)
Height overall	8.575 m (28 ft 1½ in)
Width of fuselage	2.90 m (9 ft 6 in)
Depth of fuselage	2.50 m (8 ft 2½ in)
Tailplane span	9.973 m (32 ft 8¾ in)
Wheel track (c/l shock struts)	7.90 m (25 ft 11 in)
Wheelbase	7.651 m (25 ft 1¼ in)
Propeller diameter	3.90 m (12 ft 9½ in)
Propeller ground clearance	1.227 m (4 ft 0¼ in)
Crew door (stbd, front): Height	1.40 m (4 ft 7 in)
Width	0.60 m (1 ft 11½ in)
Height to sill	1.47 m (4 ft 9¾ in)
Loading hatch (rear): Length	3.40 m (11 ft 1¾ in)
Width at front	2.40 m (7 ft 10½ in)
Width at rear	2.00 m (6 ft 6¾ in)
Height to sill	1.47 m (4 ft 9¾ in)
Height to top edge of hatchway	3.014 m (9 ft 10¾ in)
Emergency exit (in floor at front):	
Length	1.02 m (3 ft 4¼ in)
Width	0.70 m (2 ft 3½ in)
Emergency exit (top): Diameter	0.65 m (2 ft 1½ in)
Emergency exits (one each side of hold):	
Height	0.60 m (1 ft 11½ in)
Width	0.50 m (1 ft 7½ in)

DIMENSIONS, INTERNAL:
Cargo hold: Length of floor 11.50 m (37 ft 8¾ in)
Width of floor 2.40 m (7 ft 10½ in)
Max height 1.91 m (6 ft 3 in)
AREAS:
Wings, gross 74.98 m² (807.1 sq ft)
Vertical tail surfaces (total, incl dorsal fin) 15.85 m² (170.61 sq ft)
Horizontal tail surfaces (total) 19.83 m² (213.45 sq ft)
WEIGHTS AND LOADINGS:
Weight empty 15,020 kg (33,113 lb)
Normal payload 4,500 kg (9,920 lb)
Max payload 5,500 kg (12,125 lb)
Normal T-O and landing weight 23,000 kg (50,706 lb)
Max T-O and landing weight 24,000 kg (52,911 lb)
Max wing loading 320.1 kg/m² (65.6 lb/sq ft)
Max power loading 5.71 kg/kW (9.38 lb/ehp)
PERFORMANCE (at normal T-O weight):
Cruising speed at 6,000 m (19,685 ft) 237 kts (440 km/h; 273 mph)
T-O speed 108 kts (200 km/h; 124 mph) CAS
Landing speed 102 kts (190 km/h; 118 mph) CAS
Max rate of climb at S/L 480 m (1,575 ft)/min
Service ceiling 7,500 m (24,600 ft)
T-O run, on concrete 780 m (2,559 ft)
T-O to 15 m (50 ft) 1,240 m (4,068 ft)
Landing from 15 m (50 ft) 1,740 m (5,709 ft)
Landing run, on concrete 730 m (2,395 ft)
Min ground turning radius 22.30 m (73 ft 2 in)
Range, no reserves:
with max payload 594 n miles (1,100 km; 683 miles)
with max fuel 1,376 n miles (2,550 km; 1,584 miles)

Antonov An-26 twin-turboprop short-haul aircraft of the Slovak Air Force *(P. Tompkins)*

1994

ANTONOV An-30

NATO reporting name: Clank

TYPE: Twin-turboprop aerial survey aircraft.

PROGRAMME: First flown 1974; built in small numbers.

VERSIONS: **An-30:** Basic photographic aircraft, as described in detail.

An-30M 'Sky Cleaner': Carries eight modular containers of granular carbon dioxide, instead of photographic equipment, in main cabin, this is seeded into clouds to induce precipitation over arable land, increase snow cover or fight forest fires; it can also protect specific areas, such as a large city, against excessive precipitation by causing it to be distributed in surrounding regions. External pod on each side of centre-fuselage houses six multiple dispensers (similar to military IRCM flare dispensers) for total of 384 meteorological cartridges that are fired into clouds. Capacity of each chemical modular container 130 kg (286 lb). Rate of discharge of carbon dioxide granules 0.8-6 kg (1.75-13.25 lb)/min. Chin radome standard.

OPERATORS: Versions of the An-30 are in service with the armed forces of the following countries: Bulgaria (1), China (8), Cuba (1), Romania (3), Russia, Ukraine and Vietnam (2).

DESIGN FEATURES: Developed from An-24RT and An-26; nose extensively glazed for navigator; flight deck raised to improve access to navigator's compartment; fewer windows in main cabin; standard port side cabin door at rear, forward freight door on starboard side, and load hoisting/conveyor system retained.

FLYING CONTROLS: As An-26.

STRUCTURE: As An-26.

LANDING GEAR: As An-26.

POWER PLANT: Turboprops and auxiliary turbojet as An-26; water injection system for turboprops; AV-72T four-blade constant-speed fully feathering and reversible-pitch propellers; maximum fuel capacity 6,200 litres (1,638 US gallons; 1,364 Imp gallons).

ACCOMMODATION: Flight crew of five (pilot, co-pilot, flight engineer, radio operator and navigator); two photographer/surveyors; provision for conversion to transport by placing cover plates over camera apertures; toilet, buffet and crew rest area with armchairs and couches standard.

SYSTEMS: All accommodation pressurised and air conditioned.

EQUIPMENT: Equipment for map-making and air survey. Provision for automatic or semi-automatic photography; cameras, control desk, darkroom and film storage in main cabin.

AVIONICS: Radar optional in chin fairing.

DIMENSIONS, EXTERNAL:
Wing span 29.20 m (95 ft 9½ in)
Wing aspect ratio 11.4
Length overall 24.26 m (79 ft 7 in)
Height overall 8.32 m (27 ft 3½ in)
Tailplane span 9.09 m (29 ft 10 in)
Fuselage, nominal diameter 2.90 m (9 ft 6¼ in)
Wheel track (c/l of oleos) 7.90 m (25 ft 11 in)
Wheelbase 7.65 m (25 ft 1¼ in)
Propeller diameter 3.90 m (12 ft 9½ in)
Propeller ground clearance 1.20 m (3 ft 11¼ in)
AREAS:
Wings, gross 74.98 m² (807.1 sq ft)
Vertical tail surfaces (total, incl dorsal fin) 15.85 m² (170.61 sq ft)
Horizontal tail surfaces (total) 17.23 m² (185.46 sq ft)
WEIGHTS AND LOADINGS:
Basic operating weight 15,590 kg (34,370 lb)
Weight of aerial photography equipment 650 kg (1,433 lb)
Max T-O and landing weight 23,000 kg (50,706 lb)
Max wing loading 306.75 kg/m² (62.8 lb/sq ft)
Max power loading 5.47 kg/kW (8.99 lb/ehp)
PERFORMANCE:
Max level speed 291 kts (540 km/h; 335 mph)
Cruising speed at 6,000 m (19,685 ft) 232 kts (430 km/h; 267 mph)
Landing speed 95 kts (175 km/h; 109 mph)
Service ceiling:
with APU operating 8,300 m (27,230 ft)
without APU 7,300 m (23,950 ft)
T-O run on concrete 710 m (2,330 ft)
Landing run on concrete 670 m (2,198 ft)
Range with max fuel, no reserves 1,420 n miles (2,630 km; 1,634 miles)

UPDATED

Antonov An-30 'Clank' twin-turboprop aerial survey aircraft

1994

UNITED KINGDOM

AIR UK ENGINEERING LTD

AIR UK ENGINEERING LTD

Liberator Road, Norwich Airport, Norwich, Norfolk NR6 6ER
Telephone: 44 (1603) 254400
Fax: 44 (1603) 254402
Telex: 975153/4
Sita: NW1 ENUK

MANAGING DIRECTOR: Robert Nunn
COMMERCIAL DIRECTOR: Steve Kaye
FINANCIAL DIRECTOR: Steve Whitby
PLANNING/LOGISTICS DIRECTOR: Andy Beale
PRODUCTION DIRECTOR: Andy Marshall
QUALITY ASSURANCE DIRECTOR: Tony Oliver
TECHNICAL COLLEGE DIRECTOR: Adam Livingstone

Air UK Engineering Ltd is the maintenance facility of Air UK Ltd, Britain's third largest scheduled airline. Capabilities cover the Fokker F100, Fokker F50, Fokker F27, BAe 146. Support facilities include painting, galley manufacture, luxury interiors and spares trading. Workshop back-up incorporates airframe, undercarriage, pneumatic, hydraulic, electrical instrumentation, radio and avionic component overhaul and rectification. Over 30 per cent of work is carried out for third party customers from across the world.

Air UK Engineering's training courses are recognised by the British Civil Aviation Authority, ICAO and have worldwide DCA approval.

The Technical College provides engineering and pilots ground school type courses for the above aircraft, as well as Basic, Apprentice, Management and specialised training, under their UK CAA *ab initio* approvals and recognitions.

UPDATED

AIR UK (FOKKER) F27 FRIENDSHIP REFURBISHMENT

TYPE: Transport aircraft interior refurbishment.

PROGRAMME: Interior designed by Air UK Engineering for Air UK Ltd's fleet of Fokker F27 Friendship aircraft. Package available to other Fokker F27 users as well as operators of the Fokker F50.

DESIGN FEATURES: Programme includes: renewal of cabin trim, carpeting and seat covers (incorporating motif), improved seating without loss of leg room, improved overhead bins and redesigned passenger service units as well as sculptured ceiling, side panels and wash lighting.

VERIFIED

Air UK (Fokker) F50 undergoing maintenance and refurbishment work *(Air UK)*
1995

Air UK (Fokker) F27 refurbished interior *(Air UK)*
1994

BAE

BRITISH AEROSPACE plc

HEADQUARTERS: Warwick House, Farnborough Aerospace Centre, Farnborough, Hampshire GU14 6YU
Telephone: 44 (1252) 373232
Fax: 44 (1252) 383000
CHAIRMAN: Robert P. Bauman
CEO: Richard Evans, CBE
DIRECTOR, COMMUNICATIONS: Locksley Ryan
SUBSIDIARIES:
British Aerospace Defence Ltd
  Dynamics Division
  Military Aircraft Division
  Royal Ordnance Division
  Systems and Services Division
  BAe Systems and Equipment (BASE) *See following entry*
  British Aerospace Flying College Ltd
  BAe Australia
  BAe Ansett Flying College (50 per cent)
  BAe SEMA (50 per cent)
British Aerospace Airbus
Regional Aircraft
  Avro International Aerospace
  Asset Management Organisation
Jetstream Aircraft *See following entry*
BAe Inc
Arlington Securities
BAe International
Aerostructures Company
Engineering

In 1977 the former companies of British Aircraft Corporation (Holdings) Ltd, Hawker Siddeley Aviation Ltd, Hawker Siddeley Dynamics Ltd and Scottish Aviation Ltd, were brought together through nationalisation. In 1981 the structure of British Aerospace was changed from a corporation in public ownership to a public limited company in the private sector, and in 1985 HM Government sold its remaining shareholding. Since then the company has been restructured several times.

British Aerospace PLC is one of the largest private companies in the UK. Main areas of responsibility are in the fields of military and civil aircraft, guided weapons, ammunition, explosives, guns, electronic warfare systems and armoured tracked vehicles.

UPDATED

BAC ONE-ELEVEN

TYPE: Twin-engine short/medium-range turbofan transport.

PROGRAMME: Details of the One-Eleven were announced 9 May 1961, simultaneously with the news that British United Airways had ordered 10. Design and manufacture was shared between three BAC factories, at Weybridge, Filton and Hurn. There were 230 built.

VERSIONS: **Series 200:** Initial production model for short- and medium-range duties.

Series 300: Similar to Srs 200 but developed for high payload.

Series 400: Similar to Srs 300 but developed for US markets.

Series 475: Combines standard fuselage and accommodation of Series 400 with wings and power plant of Series 500 and a modified landing gear system, using low pressure tyres, to permit operation from secondary low-strength runways with poorer grade surfaces. The Srs 400/500 development aircraft (G-ASYD) was converted to serve as prototype and flew for the first time on 27 August 1970. First production Series 475 flew for the first time 5 April 1971. Certification and first production delivery (to Faucett of Peru) took place in July 1971. The three Srs 475s ordered by the Sultan of Oman's Air Force were fitted with a quick-change passenger/cargo interior layout and a 3.05 × 1.85 m (10 ft 0 in × 6 ft 1 in) forward freight door.

Series 500: Derived from the 300/400, this version incorporates a lengthened fuselage which accommodates 97-119 passengers, with a flight crew of two. Wingtip extensions increase span by 1.52 m (5 ft 0 in). Take-off performance improved by increased wing area and by installation of two Rolls-Royce Spey Mk 512 DW turbofans, each rated at 5,692 kg (12,550 lb st). Main landing gear strengthened and heavier wing plank stringers used to cater for increased AUW.

Prototype converted from Srs 400 development aircraft, flew for the first time 30 June 1967. First Srs 500 production aircraft flew on 7 February 1968. ARB certification 15 August 1968. Deliveries to BEA began 29 August 1968, and regular services 17 November 1968.

Series 525: Seven aircraft supplied to Tarom.

Romaero Rombac One-Eleven Series 495: Romaero-built version manufactured in Romania. Based on UK-built Srs 400 fuselage and Series 560 power plant. For full details see *Jane's All the World's Aircraft 1995-96.*

Romaero Rombac One-Eleven Series 497: Convertible passenger/cargo version being developed by Romaero of Romania. For full details see *Jane's All the World's Aircraft 1995-96.*

Romaero Rombac One-Eleven Series 560: Romaero-built version manufactured on Romania based on UK-built Series 300/400. For full details see *Jane's All the World's Aircraft 1995-96*.

Romaero Airstar 2000: Basically as Series 560 but re-engined with Rolls-Royce Tay 650 turbofans. For full details see *Jane's All the World's Aircraft 1995-96*.

UPGRADES: **Dee Howard:** BAC One-Eleven 2400 2500: Programme no longer current.

Rolls-Royce: A One-Eleven hush kit, comprising an intake duct lining, a bypass duct lining, an acoustically lined jetpipe and six-chute exhaust silencer, was flown for the first time, on the Srs 475 development aircraft, on 14 June 1974. It was designed to reduce the area within the 90 EPNdB noise contour by approximately 50 per cent, giving a noise footprint equivalent to that of a twin-turboprop aircraft. The effectiveness of the hush kit has been proven and demonstrated by comprehensive testing, performance penalties proving to be lower than estimated, with a 0.75 per cent thrust loss on take-off and a 2 per cent increase in fuel consumption. The installed weight of production versions is less than 181 kg (400 lb).

OPERATORS: Versions of the BAC One-Eleven are in service with the armed forces of the following countries: Oman (3), Romania (1) and UK (4).

DESIGN FEATURES: Cantilever low-wing monoplane. Modified NACA cambered wing section. Thickness/chord ratio 12.5 per cent at root, 11 per cent at tip. Dihedral 2°. Incidence 2° 30′. Sweepback 20° at quarter-chord. Variable incidence T tail.

FLYING CONTROLS: Ailerons of Redux-bonded light alloy honeycomb, manually operated through servo tabs. Port servo tab used for trimming. Light alloy Fowler flaps hydraulically operated through Lucas actuators. Light alloy spoiler/airbrakes on upper surface of wing, operated hydraulically through Dowty Boulton Paul actuators. Hydraulically actuated lift dumpers, inboard of spoilers, are standard on Srs 400, 475 and 500; structural provision for them on Srs 300. Flaps on Series 475 have a glassfibre coating. Variable-incidence T tailplane, controlled through duplicated Lucas hydraulic system. Elevators and rudder actuated hydraulically through Dowty Boulton Paul tandem jacks.

STRUCTURE: All-metal wing structure of copper-based aluminium alloy, built on fail-safe principles. Three-shear-web torsion box with integrally machined skin/stringer panels. The fuselage is a conventional circular-section all-metal fail-safe structure with continuous frames and stringers. Skin made from copper-based aluminium alloy. The tail unit is a cantilever all-metal fail-safe structure. Fin integral with rear fuselage.

LANDING GEAR: Retractable tricycle type, with twin wheels on each unit. Hydraulic retraction, nose unit forward, main units inward. Oleo-pneumatic shock-absorbers manufactured by BAC. Hydraulic nosewheel steering. Dunlop wheels, tubeless tyres and four-plate heavy-duty hydraulic disc brakes on Srs 200, 300 and 400; five-plate heavy-duty hydraulic disc brakes on Srs 475 and 500. Maxaret anti-skid units on Srs 200 and 300. Hytrol Mk III anti-skid units on Srs 400, 475 and 500. Mainwheel tyres size 40 × 12, pressure (Srs 200) 8.8 bars (128 lb/sq in); (Srs 300, 400) 9.7 bars (141 lb/sq in); (Srs 500) 11.01 bars (160 lb/sq in). Dunlop 44 × 16 tyres on Srs 475, pressure 5.7 bars (83 lb/sq in). Nosewheel tyres size 24 × 7.25, pressure (Srs 200) 7.6 bars (110 lb/sq in); (Srs 300, 450, 500) 7.6 bars (110 lb/sq in). Dunlop 24 × 7.7 tyres on Srs 475, pressure 7.2 bars (105 lb/sq in). All tyre pressures are given for aircraft at mid-CG position and operating at maximum taxi weight.

POWER PLANT: Two turbofan engines, mounted in pod on each side of rear fuselage (details under 'Series' descriptions). Fuel in integral wing tanks of 10,160 litres (2,684 US gallons; 2,235 Imp gallons) and centre-section tank (optional on Series 200) of 3,864 litres (1,021 US gallons; 850 Imp gallons) capacity; total fuel capacity 14,024 litres (3,704 US gallons; 3,085 Imp gallons). Optional 1,591 litres (420 US gallons; 350 Imp gallons) and 3,182 litre (840 US gallon; 700 Imp gallon) fuel tanks are available to increase total capacity. Pressure refuelling point in fuselage forward of wing on starboard side. Provision for gravity refuelling. Oil capacity (total engine oil) 13.66 litres (3.6 US gallons; 3 Imp gallons) per engine.

BAe (BAC) One-Eleven Series 525 at Heathrow Airport *(Peter J. Cooper)*

1994

ACCOMMODATION (all versions except Srs 500): Crew of two on flight deck and up to 89 passengers in main cabin. Single class or mixed class layout, with movable divider bulkhead to permit any first/tourist ratio. Typical mixed class layout has 16 first class (four-abreast) and 49 tourist (five-abreast) seats. Galley units normally at front on starboard side. Coat space available on port side aft of flight deck and on Srs 200 and 300 at rear vestibule. One toilet at rear on starboard side in Srs 200. Two toilets in Series 300, 400 and 500, in front and rear combinations (Srs 300 has one front port, one rear starboard; Srs 400 and 500 have one each side at rear). Ventral entrance with hydraulically operated airstair. Forward passenger door on port side incorporates optional power-operated airstair. Galley service door forward on starboard side. Two baggage and freight holds under floor, fore and aft of wings, with doors on starboard side. Forward freight door on Srs 475s for the Royal Air Force of Oman. Entire accommodation air conditioned.

ACCOMMODATION (Srs 500): Crew of two on flight deck and up to 119 passengers in main cabin. Two additional overwing emergency exits, making two on each side. Otherwise generally similar to other versions.

SYSTEMS: Fully duplicated air conditioning and pressurisation systems with main components by Normalair-Garrett. Air bled from engine compressors through heat exchangers. Maximum pressure differential .51 bars (7.5 lb/sq in). Hydraulic system, pressure 206.7 bars (3,000 lb/sq in), operates flaps, spoilers, rudder, elevators, tailplane, landing gear, brakes, nosewheel steering, ventral and forward airstairs and windscreen wipers. No pneumatic system. Electrical system utilises two 30 kVA Plessey/Westinghouse AC generators, driven by Plessey constant-speed drive and starter units, plus a similar generator mounted on the APU and shaft driven. AiResearch gas-turbine APU in tailcone to provide ground electrical power, air conditioning and engine starting, also some system checkout capability. APU is run during take-off to eliminate performance penalty of bleeding engine air for cabin air conditioning. Thermal de-icing of wing leading-edges with engine bleed air. Leading-edges of fin and tailplane de-iced by engine bleed air.

ELECTRONICS AND EQUIPMENT: Communications and navigation equipment generally to customer's individual requirements. Typical installation includes dual VHF communications equipment to ARINC 546, dual VHF navigation equipment to ARINC 547A, including glide slope receivers, marker receiver, flight/service interphone system, Marconi AD 370, Bendix DFA 73 or Collins DF 203 ADF, ATC transponder to ARINC 532D, Collins 860 E2 DME, Ekco E 190 or Bendix RDR 1E weather radar. Sperry C9 or CL11 compass systems and Collins FD 108 flight director system (dual) are also installed. The autopilot is the Elliot 2000 Series system and provision is made on the Srs 500 for additional equipment, including automatic throttle control, for low weather minimal operation.

DIMENSIONS, EXTERNAL (Srs 475, 500):

Wing span	28.50 m (93 ft 6 in)
Wing chord: at root	5.01 m (16 ft 5 in)
at tip	1.61 m (5 ft 3½ in)
Wing aspect ratio	8.5
Length overall: Srs 475	28.50 m (93 ft 6 in)
Srs 500	32.61 m (107 ft 0 in)
Length of fuselage: Srs 475	25.55 m (83 ft 10 in)
Srs 500	29.67 m (97 ft 4 in)
Height overall	7.47 m (24 ft 6 in)
Tailplane span	8.99 m (29 ft 6 in)
Wheel track	4.34 m (14 ft 3 in)
Wheelbase: Srs 475	10.08 m (33 ft 1 in)
Srs 500	12.62 m (41 ft 5 in)
Passenger door (fwd, port): Height	1.73 m (5 ft 8 in)
Width	0.82 m (2 ft 8 in)
Height to sill	2.13 m (7 ft 0 in)

BAe (BAC) One-Eleven Series 475 with Rolls-Royce hush-kit fitted

1994

BAe (BAC) One-Eleven Series 500 of British Airways

1994

Ventral entrance: Height 1.83 m (6 ft 0 in)
Width 0.66 m (2 ft 2 in)
Height to sill 2.13 m (7 ft 0 in)
Freight door (fwd, stbd):
Height (projected) 0.79 m (2 ft 7 in)
Width 0.91 m (3 ft 0 in)
Height to sill 1.09 m (3 ft 7 in)
Freight door (rear, stbd):
Height (projected) 0.66 m (2 ft 2 in)
Width 0.91 m (3 ft 0 in)
Height to sill 1.30 m (4 ft 3 in)
Freight door (fwd, Srs 475 SOAF):
Height 1.85 m (6 ft 1 in)
Width 3.05 m (10 ft 0 in)
Galley service door (fwd, stbd):
Height (projected) 1.22 m (4 ft 0 in)
Width 0.69 m (2 ft 3 in)
Height to sill 2.13 m (7 ft 0 in)

DIMENSIONS, INTERNAL (Srs 475):
Cabin, excl flight deck: Length 17.31 m (56 ft 10 in)
Max width 3.16 m (10 ft 4 in)
Max height 1.98 m (6 ft 6 in)
Floor area approx 47.00 m² (506 sq ft)
Freight hold, fwd 10.02 m³ (354 cu ft)
Freight hold, rear 4.42 m³ (156 cu ft)

DIMENSIONS, INTERNAL (Srs 500):
Cabin, excl flight deck: Length 21.44 m (70 ft 4 in)
Total floor area approx 61.78 m² (665 sq ft)
Freight hold (total volume) 19.45 m³ (687 cu ft)

AREAS (Srs 475, 500):
Wings, gross 95.78 m² (1,031 sq ft)
Ailerons (total) 2.86 m² (30.8 sq ft)
Flaps (total) 16.30 m² (175.6 sq ft)
Spoilers (total) 2.30 m² (24.8 sq ft)
Vertical tail surfaces (total) 10.90 m² (117.4 sq ft)
Rudder, incl tab 3.05 m² (32.8 sq ft)
Horizontal tail surfaces (total) 23.90 m² (257.0 sq ft)
Elevators, incl tab 6.55 m² (70.4 sq ft)

WEIGHTS AND LOADINGS:
Operating weight empty: Srs 475 23,464 kg (51,731 lb)
Srs 500 24,758 kg (54,582 lb)
Max payload: Srs 475 9,647 kg (21,269 lb)
Srs 500 11,983 kg (26,418 lb)
Max T-O weight:
Srs 475 41,730-44,678 kg (92,000-98,500 lb)
Srs 500 45,200-47,400 kg (99,650-104,500 lb)
Max ramp weight: Srs 475 44,905 kg (99,000 lb)
Srs 500 47,625 kg (105,000 lb)
Max landing weight:
Srs 475 38,100-39,462 kg (84,000-87,000 lb)
Srs 500 39,462 kg (87,000 lb)
Max zero-fuel weight: Srs 475 33,112 kg (73,000 lb)
Srs 500 36,741 kg (81,000 lb)
Max wing loading: Srs 475 435.5 kg/m² (89.2 lb/sq ft)
Srs 500 472 kg/m² (96.7 lb/sq ft)
Max power loading: Srs 475 400.3 kg/kN (3.92 lb/lb st)
Srs 500 424.7 kg/kN (4.16 lb/lb st)

PERFORMANCE (at max T-O weight):
Never-exceed speed (structural) 410 kts (760 km/h; 472 mph) EAS
Max level cruising speed at 6,400 m (21,000 ft) 470 kts (871 km/h; 541 mph) TAS
Fuel econ cruising speed at 7,620 m (25,000 ft) 400 kts (742 km/h; 461 mph) TAS
Stalling speed (T-O flap setting):
Srs 475 99 kts (184 km/h; 114 mph) EAS
Srs 500 105 kts (195 km/h; 121 mph)
Rate of climb at S/L at 300 kts (555 km/h; 345 mph) EAS:
Srs 475 756 m (2,480 ft)/min
Srs 500 695 m (2,280 ft)/min
Max cruising height 1,670 m (35,000 ft)
Min ground turning radius: Srs 475 17.07 m (56 ft 0 in)
Srs 500 17.98 m (59 ft 0 in)
Runway LCN at max weight, rigid pavement (1:40):
Srs 475 32
Srs 500 53
T-O run at S/L, ISA: Srs 475 1,676 m (5,500 ft)
Srs 500 1,981 m (6,500 ft)
Balance T-O to 10.7 m (35 ft) at S/L, ISA:
Srs 475 1,798 m (5,900 ft)
Srs 500 2,225 m (7,300 ft)
Landing distance (BCAR) at S/L, ISA, at max landing weight: Srs 475 1,439 m (4,720 ft)
Landing run at S/L, ISA at max landing weight:
Srs 475 826 m (2,710 ft)
Still-air range with max fuel, ISA with reserves for 200 n miles (370 km; 230 mile) diversion and 45 min hold:
Srs 475 1,997 n miles (3,700 km; 2,300 miles)
Srs 500 1,880 n miles (3,484 km; 2,165 miles)
Still-air range with typical capacity payload, ISA reserves as above: Srs 475 at 44,678 kg (98,500 lb) 1,619 n miles (3,000 km; 1,865 miles)
Srs 500 at 47,400 kg (104,500 lb) 1,480 n miles (2,744 km; 1,705 miles)
Srs 475 executive aircraft with additional 3,182 litres (840 US gallons; 700 Imp gallons) fuel has equivalent range of 2,549 n miles (4,725 km; 2,936 miles)

UPDATED

BAC 167 STRIKEMASTER

TYPE: Two-seat, basic and advanced trainer.

PROGRAMME: The BAC 167 Strikemaster was developed from the BAC 145 Series. It has the same airframe but is powered by a Rolls-Royce Viper Mk 535 turbojet engine (15.2 kN; 3,410 lb st) and has eight underwing hardpoints, enabling it to carry up to 1,360 kg (3,000 lb) of stores. This makes it particularly suitable for counter-insurgency combat operations.

The first BAC 167 was flown for the first time on 26 October 1967. The Strikemaster is still in service with the air forces of the following countries: Ecuador, Oman and Saudi Arabia.

VERSIONS: **Mk 80/A**: In service with the Royal Saudi Air Force.

Mk 82: In service with the Sultan of Oman's Air Force.

Mk 83: In service with the Botswana Defence Force.

Mk 89: In service with the Ecuadorean Air Force (FAE).

OPERATORS: Versions of the BAC 167 Strikemaster are in service with the armed forces of the following countries: Botswana (7), Ecuador (16), Kenya (1), Oman (13) and Sudan (3).

The following description applies to the later versions of the Strikemaster, such as the Mk 88 but is also substantially applicable to earlier versions:

DESIGN FEATURES: Cantilever low-wing monoplane. Wing section NACA 23015 (modified) at root, NACA 4412 (modified) at tip. Dihedral 6°. Incidence 3° at root, 0° at tip.

FLYING CONTROLS: Metal-covered ailerons with balance tabs. Hydraulically operated slotted flaps. Hydraulically operated airbrakes and lift spoilers on wings at rear spar

BAe (BAC) 167 Strikemaster two-seat basic and advanced trainer

1994

BAe (BAC) 167 Strikemaster of the Ecuadorean Air Force

1994

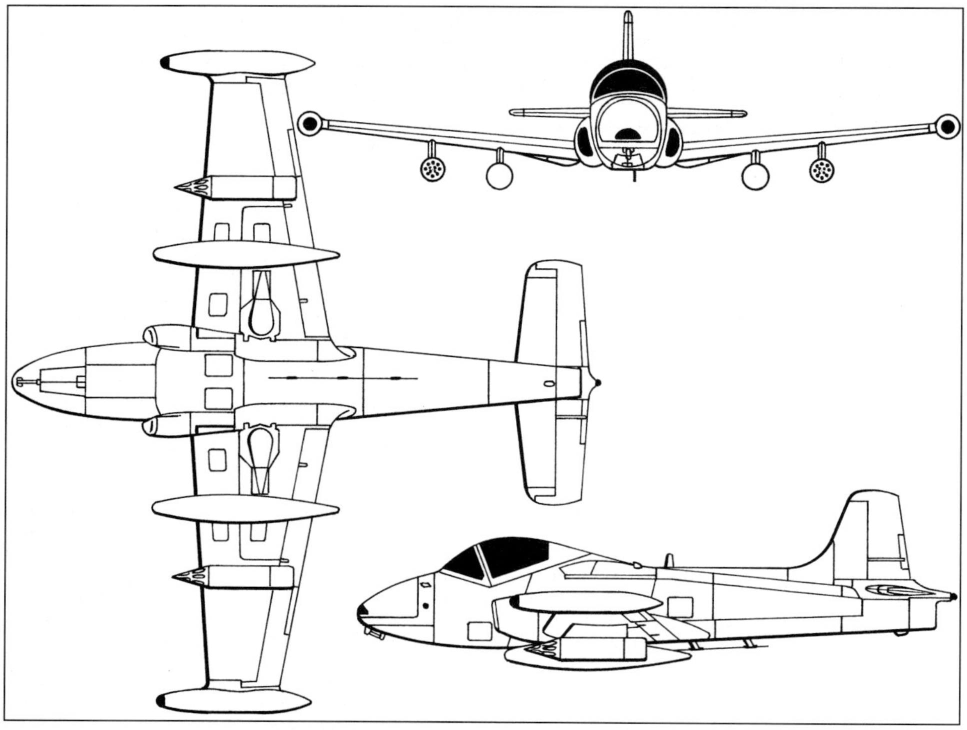

BAe (BAC) 167 Strikemaster two-seat trainer ***(Jane's/Mike Keep)***

1994

BAe (BAC) 167 Strikemaster two-seat trainer of the Sultan of Oman's Air Force ***(BAe)***

1995

position ahead of flaps. Combined trim and balance tab in starboard elevator; balance tabs in port elevator and rudder.

STRUCTURE: All-metal structure, with main and subsidiary spars having three-point attachment to fuselage. The fuselage is an all-metal semi-monocoque stressed-skin structure, built in three parts, comprising bulkheads, built-up frames and longerons covered with light alloy panels. Hinged nose cap provides access to pressurisation, oxygen, radio and electrical equipment. The tail unit is a cantilever all-metal structure. One-piece tailplane, interchangeable elevators, fin and rudder. Fixed surfaces covered with smooth and movable surfaces with fluted alloy skin.

LANDING GEAR: Hydraulically retractable tricycle type. Mainwheels retract inward into wings, nosewheel forward. Dowty oleo-pneumatic shock-absorbers. Dunlop mainwheels with tubeless tyres size 21 × 6.75-9, pressure 6.90 bars (100 lb/sq in). Dunlop nosewheel and tubeless tyre size 6.00-4, pressure 6.21 bars (90 lb/sq in). Dunlop hydraulic disc brakes.

POWER PLANT: One Rolls-Royce Bristol Viper Mk 535 turbojet engine (15.2 kN; 3,140 lb st) in fuselage aft of cockpit. Lateral intake on each side of forward fuselage. Internal fuel capacity (one integral tank outboard and three bag tanks inboard in each wing) is 1,227 litres (324 US gallons; 270 Imp gallons). Refuelling point near each wingtip. Two wingtip fuel tanks, total capacity 436 litres (115 US gallons; 96 Imp gallons), are a standard fit at all times. All tanks in wings are connected. System designed to permit 18 seconds of inverted flight. Oil capacity 8 litres (2.1 US gallons; 1.75 Imp gallons).

ACCOMMODATION: Two persons side by side in pressurised cabin, on Martin-Baker ejection seats (Mk PB4/1 and PB4/2), suitable for use down to ground level and 90 knots (167 km/h; 104 mph). Power-operated rearward sliding canopy. Dual controls standard.

SYSTEMS: Pressurisation and air conditioning system by Normalair and Tiltman Langley, differential 0.21 bars (3 lb/sq in), using engine bleed air. Hydraulic system, pressure 145 bars (2,100 lb/sq in), for landing gear, flaps, airbrakes, lift spoilers and wheel brakes. Engine-driven generator provides 28 V DC supply. Three 25 Ah batteries. Two inverters supply phased AC to flight instruments and fire warning system. Automatically controlled gaseous oxygen system for each crew member.

EQUIPMENT: Varies in different Mks to meet individual customer's requirements. The following radio equipment has been installed in various combinations: ARC 51 BX and ARC 52 UHF; PV 141 UHF homer; D 403 UHF standby; PTR 175 UHF/VHF; Collins 618M VHF; Collins 618 FIA VHF standby; ARI 18129/2 Violet picture; Sunair ASB 100 and SA 14-RaHF; and PTR 446, SSR 1600 and SSRT 2100 IFF. The following navigation equipment has been installed in various combinations: Bendix 220B UHF; Bendix CNS 240 B VHF; RCA ACQ-75 DME; AD 370B and ADF 722; Bendix 221 VOR/ILS; and ARN 84, ARN 52 and ARN 65 Tacan.

ARMAMENT: Two 7.62 mm FN machine guns, with 550 rds/gun; one in the lower lip of each air intake duct. Later versions have SFOM gunsights; GM2L reflector gunsights fitted to some earlier models. Provision for a G90 gun camera and a Smiths camera sight recorder. Four underwing strongpoints for the carriage of external stores. Typical underwing loads include two 341 and two 227 litre (90 and 60 US gallon; 75 and 50 Imp gallon) drop tanks; four Matra launchers each containing 18 68 mm SNEB rockets; four 540 LAU 68 rocket launchers, each with seven rockets; four 540 lb ballistic or retarded bombs, four 250 or 500 kg bombs; four PMBR carriers, each with six practice bombs; light-series bomb carriers to carry 8.5, 19 or 25 lb practice bombs; BAC/Vinten five-camera reconnaissance pod; or banks of SURA 80 mm rockets, with four rockets per bank. Other armament, to specific customer requirements, can include napalm tanks, 65 or 125 kg bombs, 2.75 or 3 in rockets, and 7.62 or 20 mm gun packs. Maximum T-O weight of 5,215 kg (11,500 lb) includes one pilot only, full usable (internal and wingtip tanks) and 1,200 kg (2,650 lb) of external stores. Maximum possible external stores load 1,360 kg (3,000 lb).

DIMENSIONS, EXTERNAL:
- Wing span over tip tanks: 11.23 m (36 ft 10 in)
- Wing chord: at root: 2.33 m (7 ft 8 in)
  - at tip: 1.27 m (4 ft 2 in)
- Wing aspect ratio: 5.84
- Length overall: 10.27 m (33 ft 8½ in)
- Height overall: 3.34 m (10 ft 11½ in)
- Tailplane span: 3.27 m (10 ft 8⅞ in)
- Wheel track: 3.27 m (10 ft 8⅞ in)
- Wheelbase: 2.93 m (9 ft 7⅔ in)

AREAS:
- Wings, gross: 19.85 m² (213.7 sq ft)
- Ailerons (total): 1.77 m² (19.06 sq ft)
- Trailing-edge flaps (total): 2.30 m² (24.80 sq ft)
- Fin: 0.86 m² (9.30 sq ft)
- Rudder: 1.00 m² (10.74 sq ft)
- Tailplane: 2.18 m² (23.51 sq ft)
- Elevators, incl tabs: 1.93 m² (20.80 sq ft)

WEIGHTS AND LOADINGS:
- Operating weight empty, equipped, incl crew: 2,810 kg (6,195 lb)
- Typical T-O weights:
  - pilot conversion training, two crew, full internal fuel: 4,219 kg (9,303 lb)
  - armament training, two crew, full internal fuel, practice armament (bombs and racks): 4,808 kg (10,600 lb)
  - ferry role, two crew, full internal fuel plus inboard and outboard drop tanks: 5,213 kg (11,493 lb)
- Max T-O weight: 5,215 kg (11,500 lb)

PERFORMANCE (at max T-O weight except where indicated):
- Never-exceed speed: 450 kts (834 km/h; 518 mph)
- Max level speed, with 50% fuel 'clean'
  - at S/L: 391 kts (724 km/h; 450 mph)
  - at 5,485 m (18,000 ft): 418 kts (774 km/h; 481 mph)
  - at 6,100 m (20,000 ft): 410 kts (760 km/h; 472 mph)
- Stalling speed at 4,309 kg (9,500 lb) AUW:
  - flaps up: 98.5 kts (182 km/h; 113 mph)
  - flaps down: 85.5 kts (158 km/h; 98 mph)
- Max rate of climb at S/L (training, full internal fuel): 1,600 m (5,250 ft)/min
- Time to height (training, full internal fuel):
  - to 9,150 m (30,000 ft): 8 min 45 s
  - to 12,200 m (40,000 ft): 15 min 30 s
- Service ceiling: 12,200 m (40,000 ft)
- T-O to 15 m (50 ft):
  - at 3,579 kg (7,930 lb) AUW (training): 579 m (1,900 ft)
  - at 5,215 kg (1,500 lb) AUW (combat): 1,067 m (3,500 ft)
- Landing from 15 m (50 ft):
  - at 2,948 kg (6,500 lb) AUW (training): 732 m (2,400 ft)
  - at 5,103 kg (11,250 lb) AUW (aborted armed sortie): 1,295 m (4,250 ft)
- Combat radius (hi-lo-hi), 5 min over target, 10% reserves:
  - with 1,360 kg (3,000 lb) weapons load: 215 n miles (397 km; 247 miles)

with 907 kg (2,000 lb) weapons load
355 n miles (656 km; 408 miles)
with 454 kg (1,000 lb) weapons load
500 n miles (925 km; 575 miles)
Combat radius (lo-lo-lo), at S/L, 5 min over target, 10% reserves: with 1,360 kg (3,000 lb) weapons load
126 n miles (233 km; 145 miles)
with 907 kg (2,000 lb) weapons load
175 n miles (323 km; 201 miles)
with 454 kg (1,000 lb) weapons load
240 n miles (444 km; 276 miles)
reconnaissance mission
300 n miles (555 km; 345 miles)
Range with 91 kg (200 lb) fuel reserves:
at 3,789 kg (8,355 lb) AUW (training)
629 n miles (1,166 km; 725 miles)
at 4,558 kg (10,500 lb) AUW (combat)
1,075 n miles (1,992 km; 1,238 miles)
at 5,125 kg (11,500 lb) AUW (max T-O)
1,200 n miles (2,224 km; 1,382 miles)

UPDATED

BAe (BAC) SA-3-120 Bulldog during test flight from Prestwick for the Malaysian Air Force

1994

BAe SA-3-120 BULLDOG SERIES 120

Swedish military designation: Sk 61

United Kingdom military designation: T. Mk 1

TYPE: Two/three-seat primary trainer.

PROGRAMME: The first flight of the Beagle-built Bulldog prototype was made 19 May 1969. A second prototype, completed by Scottish Aviation, was flown on 14 February 1971; it was later refurbished, issued with a Normal category C of A, and delivered to a private owner under the designation Model 104. The first production Bulldog flew 22 June 1971 and received full ARB certification 30 June 1971. The first 98 production Bulldogs were of the Series 100 version, built for Kenya, Malaysia and Sweden.

Production then continued with the Series 120, which was awarded full CAA certification on 12 February 1973. The Srs 120 has been built for the RAF 130 T. Mk 1/Model (121), and the air forces of Botswana (6), Ghana (13), Hong Kong (2), Jordan (13), Kenya (9), Lebanon (6) and Nigeria (32).

VERSIONS: **Model 121:** For Royal Air Force, designated **T. Mk 1**. First flight 30 January 1973.

Model 122: For Ghanaian Air Force.

Model 123: For Nigerian Air Force (20+ in service).

Model 124: One aircraft used as a company demonstration aircraft.

Model 125: Built for the Jordanian Royal Academy of Aeronautics. Transferred to the Royal Jordanian Air Force in 1978.

Model 126: Built for the Lebanese Air Force (five still in service).

Model 127: Built for the Kenyan Air Force (12 still in service).

Model 130 T. Mk 1: In service with the Royal Air Force.

OPERATORS: Versions of the SA-3-120 Bulldog are in service with the armed forces of the following countries: Ghana (10), Jordan (10), Kenya (7), Lebanon (5), Malaysia (10), Nigeria (18), Sweden (60) and UK (116).

DESIGN FEATURES: Cantilever low-wing monoplane. Wing-section NACA 63_2615. Dihedral 6° 30′. Incidence 1° 9′ at root.

FLYING CONTROLS: Electrically operated slotted trailing-edge flaps and slotted ailerons of similar construction. Ground adjustable tab on starboard aileron.

STRUCTURE: Conventional single-spar riveted stressed-skin structure of light alloy. The fuselage is a conventional light alloy stressed-skin semi-monocoque structure. The tail unit is a cantilever two-spar light alloy structure. Fixed-incidence tailplane. Full-span trim tab in starboard elevator. Manually operated trim tab in rudder. Ventral fin.

LANDING GEAR: Non-retractable tricycle type, with single wheel on each unit. Steerable nosewheel with Automotive Products oleo-pneumatic shock-absorber and Goodyear wheel and tyre, size 5.00-5, pressure 2.76 bars (40 lb/sq in). Main units have Automotive Products oleo-pneumatic shock-absorbers and Goodyear wheels and tyres, size 6.00-6, pressure 2.07 bars (30 lb/sq in). Goodyear hydraulic disc brakes on mainwheels. Optional ski landing gear.

POWER PLANT: One 149 kW (200 hp) Avco Lycoming IO-360-A1B6 flat-four engine, driving a Hartzell HC-C2YK-4F/FC7666A-2 two-blade constant-speed metal propeller with spinner. Avco Lycoming AEIO-360-A1B6 engine available optionally, permitting up to 20 seconds of inverted flight. Four removable metal fuel tanks, two in each wing, with total usable capacity of 145.5 litres (38 US gallons; 32 Imp gallons). Refuelling on top of each wing. Oil capacity 7.6 litres (2 US gallons; 1.67 Imp gallons).

ACCOMMODATION: Enclosed cabin seating pilot and co-pilot or trainee side by side with dual controls, with space at rear for observer's seat or up to 100 kg (220 lb) of baggage. Rearward sliding jettisonable transparent canopy. Cabin heated and ventilated.

SYSTEMS: Heat exchanger for cabin heating. Hydraulic system, pressure 40 bars (580 lb/sq in), for mainwheel brakes only. Vacuum-type pneumatic system available optionally. 24 V DC power from engine-driven alternator and 24 V 25 Ah storage battery. No oxygen or de-icing systems.

AVIONICS AND EQUIPMENT: VHF, UHF or HF com radio to individual customer's requirements; panel can accommodate dual VHF nav, ATC transducer, ADF and other navaids. Blind-flying instrumentation standard.

ARMAMENT AND OPERATIONAL EQUIPMENT: Standard aircraft is unarmed, but has provision for installation of four underwing hardpoints to which can be attached various loads including unguided or wire-guided air-to-surface weapons; 7.62 mm machine gun pods, grenade launchers; practice or active bombs up to 50 kg; markers; supply containers; leaflet dispensers; and rescue and survival equipment. Maximum underwing load 290 kg (640 lb).

DIMENSIONS, EXTERNAL:
Wing span 10.06 m (33 ft 0 in)
Wing chord: at root 1.51 m (4 ft 11¾ in)
at tip 0.86 m (2 ft 9¾ in)
Wing aspect ratio 8.4
Length overall 7.09 m (23 ft 3 in)
Height overall 2.28 m (7 ft 5¾ in)
Tailplane span 3.35 m (11 ft 0 in)
Wheel track 2.03 m (6 ft 8 in)
Wheelbase 1.40 m (4 ft 7 in)
Propeller diameter 1.88 m (6 ft 2 in)
Propeller ground clearance 0.26 m (10¼ in)

DIMENSIONS, INTERNAL:
Cabin: Length 2.11 m (6 ft 11 in)
Max width 1.14 m (3 ft 9 in)
Max height 1.02 m (3 ft 4 in)

AREAS:
Wings, gross 12.02 m² (129.4 sq ft)
Ailerons (total) 0.87 m² (9.4 sq ft)
Trailing-edge flaps (total) 1.30 m² (13.95 sq ft)
Vertical tail surfaces (total) 2.11 m² (22.72 sq ft)
Horizontal tail surfaces (total) 2.55 m² (27.50 sq ft)

WEIGHTS AND LOADINGS:
Weight empty, equipped 649 kg (1,430 lb)
Basic operating weight empty 669 kg (1,475 lb)
Max T-O weight:
normal and semi-aerobatic 1,066 kg (2,350 lb)
fully aerobatic 1,015 kg (2,238 lb)
Max wing loading 88.6 kg/m² (18.15 lb/sq ft)
Max power loading 7.15 kg/kW (11.75 lb/hp)

PERFORMANCE (at max T-O weight):
Never-exceed speed (structural)
210 kts (389 km/h; 241 mph)
Max level speed at S/L 130 kts (241 km/h; 150 mph)
Max cruising speed at 1,220 m (4,000 ft)
120 kts (222 km/h; 138 mph)
Econ cruising speed at 1,220 m (4,000 ft)
105 kts (194 km/h; 121 mph)
Stalling speed, flaps down, power off
53 kts (98 km/h; 61 mph) EAS
Max rate of climb at S/L 315 m (1,034 ft)/min
Service ceiling 4,875 m (16,000 ft)
Min ground turning radius 9.75 m (32 ft 0 in)
T-O run 274 m (900 ft)
T-O to 15 m (50 ft) 427 m (1,400 ft)
Landing run from 15 m (50 ft) 363 m (1,190 ft)
Landing run 153 m (500 ft)
Range with max fuel, 55% power, no reserves
540 n miles (1,000 km; 621 miles)
Endurance with max fuel, conditions as above 5 h
g limits: semi-aerobatic +4.4/−1.8
fully aerobatic +6/−3

UPDATED

BAe (BAC) SA-3-120 Bulldog two/three-seat primary trainer ***(P. Tompkins)***

1994

BAe (HS) TRIDENT

TYPE: Short/medium-range jet airliner.

PROGRAMME: The BAe (HS) (formerly de Havilland D.H. 121) Trident was ordered into production initially to meet BEA's requirements for a short-haul 600 mph airliner for service from 1963-64 onwards. Design was started in 1957 and construction of the first airframe began 29 July 1959. The first Trident (G-ARPA), a production aircraft for BEA, flew for the first time 9 January 1962.

VERSIONS: **Trident 1:** Initial version. Withdrawn from service.

Trident 1E: Improved version. Withdrawn from service.

Trident 2E: Developed version; 15 ordered by BEA in August 1965, with accommodation for up to 115 passengers. Overall length unchanged. Fuel capacity increased, enabling the aircraft to operate non-stop over BEA's longest routes, between London and the Middle East. Maximum T-O weight increased considerably, but take-off performance improved by use of more powerful (5,411 kg; 11,930 lb st) Rolls-Royce Spey RB163-25 Mk 512-5W turbofan engines. Landing distances reduced by use of leading-edge slats, as in Trident 1E, and by increased wing span. Low-drag (Küchemann) wingtips. Some strengthening of undercarriage, and of wing and fuselage by use of thicker panels, is offset by other weight savings including greater use of titanium. BEA machines were furnished to carry 97 tourist class passengers, compared with 88 in other operators' Trident 1s, and are fitted with automatic landing equipment. The first Trident 2E (G-AVFA) flew for the first time 27 July 1967, and the first aircraft for BEA (G-AVFC) was delivered on 15 February 1968. BEA referred to this variant as the Trident 2, and scheduled services began 18 April 1968.

A high-density version is available, having a maximum of 149 seats.

The following details apply generally to the Trident 2E.

DESIGN FEATURES: Cantilever low-wing monoplane. Wing sections designed with high critical drag rise Mach number for economical operation at ultimate subsonic cruising speeds. Mean thickness/chord ratio approximately 9.8 per cent. Dihedral 3°. Incidence 6° 30′ at root, 1° 30′ at tip. Sweepback at quarter-chord 35°. Cantilever tail unit with tailplane mounted at tip.

FLYING CONTROLS: Conventional all-metal ailerons actuated by triplexed power control system without manual reversion. Three independent hydraulic systems work continuously in parallel and power three separate jacks of Fairey manufacture at each primary flying control surface. Two all-metal double-slotted trailing-edge flaps on each wing. Krueger leading-edge flap at each wingroot. All flaps operated by screw jacks and hydraulic motors of Hobson manufacture. One all-metal spoiler on 2E forward of outer flap on each wing, acts also as airbrake/lift dumpers. Lift dumpers forward of inner flaps. No trim tabs. Full-span leading-edge slats in four sections per wing, operated by screw jacks and extending on curved titanium tracks. All-moving tailplane with geared slotted flap on trailing-edge to assist negative life coefficient for take-off and landing. No trim tabs. Power control systems as for ailerons.

STRUCTURE: Main wing is continuous from wingtip to wingtip, and comprises a six-cell centre-section box extending across the fuselage, a two-cell box from the wingroot out to 40 per cent of the semi-span, and from there a single-cell box to the wingtip. The entire wing box is subdivided to form integral fuel tanks. Skins and stringers are of aluminium alloy as are the leading-edge and trailing-edge flaps. Extensive use of Reduxing between skins and stringers. Structure is fail-safe, except for slat and flap tracks which are safe-life components tested to at least six times the aircraft life. The fuselage consists of a pressure shell extending back to the engines and a rear fuselage carrying the engines and the tail unit. Fuselage is a semi-monocoque fail-safe structure of aluminium-copper alloys. No structural bulkheads in pressure cell. Unpressurised cut-outs for nose and main landing gear and wing centre-section. The tail unit is a cantilever all-metal structure.

LANDING GEAR: Retractable tricycle type. Hydraulic retraction. Hawker Siddeley (main units) and Lockheed (nose) oleo-pneumatic shock absorbers. Each main unit consists of two twin-tyred wheels mounted on a common axle: during retraction the leg twists through nearly 90° and lengthens by 15 cm (6 in) enabling wheels to stow within the circular cross-section of the fuselage. Nose unit has twin-wheels and is offset 61 cm (2 ft 0 in) to port, retracting transversely. Dunlop wheels, tyres and multi-plate disc brakes, with Maxaret anti-skid units. Trident 3B has mainwheel tyres size 36 × 10, pressure 11.4 bars (165 lb/sq in), and nosewheel tyres size 29 × 8, pressure 8.55 bars (124 lb/sq in).

POWER PLANT: Three Rolls-Royce Spey turbofan engines (details under 'Series' descriptions). Two in pods, one on each side of rear fuselage; one inside rear fuselage. Additionally, Trident 3B has a 2,381 kg (5,250 lb st) Rolls-Royce RB162-86 turbojet installed in tail, below the rudder, to boost T-O and climb-out. Five integral fuel tanks, four in wings and one in centre-section. Total usable fuel capacity: Trident 2E, 29,094 litres (7,680 US gallons; 6,400 Imp gallons); Trident 3B, 25,548 litres (6,744 US gallons; 5,620 Imp gallons). One pressure refuelling point under each wing. Oil capacity 13.5 litres (3.6 US gallons; 3 Imp gallons) per engine.

ACCOMMODATION: Crew of three on flight deck. Mixed class version has galley (starboard) and toilet (port) at front, then a 12-seat first class compartment, with seats in pairs on each side of central aisle, two galleys (port), 79-seat tourist class cabin with three-seat units on each side of aisle, and two toilets at rear. All-tourist version in BEA service has 97 seats, six-abreast, with galley and toilet at front and two toilets at rear. Provision can be made for 132 passengers in high-density seating arrangement, or 149 passengers with seven-abreast. Two inward-opening plug-type passenger doors, at front and centre of cabin on port side, with provision for built-in airstairs. Doors for crew and servicing at front and amidships on starboard side. Large underfloor baggage holds forward and aft of wing. All crew and passenger accommodation air conditioned. Provision for air conditioning forward part of forward baggage hold for animals.

SYSTEMS: Hawker Siddeley Dynamics air conditioning and pressurisation system, differential 0.57 bars (8.25 lb/sq in). Two independent supplies, each capable of maintaining full cabin pressurisation; with emergency ram air system for use below 2,440 m (8,000 ft). Three independent hydraulic systems operating all flying controls, landing gear, nosewheel steering, brakes and windscreen wipers. Each system powered by separate engine-driven pump, operating continuously in parallel at 206.7 bars (3,000 lb/sq in), using Skydrol fluid. Back-up hydraulic power supplied by two electrically driven pumps, and emergency power from drop out air turbine, capable of feeding any one system. Pneumatic system for toilet flushing, forward water system, stall recovery system and for pressurising hydraulic reservoirs. Electrical system comprises three separate channels, supplied by three 27.5 kVA brushless generators. Emergency 30 minute AC and DC supply available from 24 V battery. AiResearch GTCP 85C APU for engine starting and cabin air conditioning, driving generator to provide 40 kVA of electrical power from which hydraulic systems can also be actuated through standby pumps.

ELECTRONICS AND EQUIPMENT: To customer's specification. Provision for duplicated VOR/ILS, including a third localiser for three-channel automatic landing guidance; integration of navigational aids with flight system, providing coupling facilities for all flight modes except take-off; duplicated ADF, VHF and HF with selective calling; C- or X-band weather radar; triplicated radio altimeters for automatic landing; Doppler; DME and transponder.

BAe (HS) Trident 2E short/medium-range jet airliner of the Air Force of the People's Liberation Army of China *(Grahame Griffiths)*

1994

DIMENSIONS, EXTERNAL:

Wing span	29.87 m (98 ft 0 in)
Length overall	34.97 m (114 ft 9 in)
Height overall	8.23 m (27 ft 0 in)
Tailplane span	10.44 m (34 ft 3 in)
Wheel track (centres of shock-struts)	5.83 m (19 ft 1¼ in)
Wheelbase	13.41 m (44 ft 0 in)
Passenger doors: Height	1.78 m (5 ft 10 in)
Width	0.71 m (2 ft 4 in)
Min height to sill	2.87 m (9 ft 5 in)
Max height to sill	3.12 m (10 ft 3 in)
Crew and service doors: Height	1.22 m (4 ft 0¼ in)
Width	0.61 m (2 ft 0 in)
Height to sill	approx 2.74 m (9 ft 0 in)
Emergency exits (above centre-section port and stbd):	
Height	1.03 m (3 ft 4 in)
Width	0.51 m (1 ft 8 in)
Baggage hold doors (fwd, stbd):	
Height (vertical)	0.89 m (2 ft 11 in)
Width	1.22 m (4 ft 0 in)
Height to sill	1.37 m (4 ft 6 in)
Baggage hold door (rear, port):	
Mean height (vertical)	0.81 m (2 ft 8 in)
Width	0.89 m (2 ft 11 in)
Height to sill	1.37 m (4 ft 6 in)

DIMENSIONS, INTERNAL:

Cabin, excl flight deck: Length	20.46 m (67 ft 1½ in)
Max width	3.44 m (11 ft 3½ in)
Max height	2.02 m (6 ft 7½ in)
Floor area	65.77 m² (708 sq ft)
Volume	125.7 m³ (4,440 cu ft)
Freight hold (fwd)	13.88 m³ (490 cu ft)
Freight hold (rear)	7.65 m³ (270 cu ft)

AREAS:

Wings, gross	135.73 m² (1,461 sq ft)
Ailerons (total)	4.89 m² (52.5 sq ft)
Spoilers	1.42 m² (15.3 sq ft)
Fin	18.76 m² (202 sq ft)
Rudder	4.84 m² (52.1 sq ft)
Tailplane	28.80 m² (310 sq ft)

WEIGHTS AND LOADINGS:

Operating weight, empty	33,203 kg (73,200 lb)
Max payload	12,156 kg (26,800 lb)
Max T-O weight	65,090 kg (143,500 lb)
Max zero-fuel weight	45,359 kg (100,000 lb)
Max landing weight	51,261 kg (113,000 lb)

PERFORMANCE (at max T-O weight):

Max diving speed (design limit) Mach 0.95

Typical high-speed cruise Mach 0.88 at 8,230 m (27,000 ft) 525 kts (972 km/h; 605 mph)

Econ cruising speed Mach 0.88 at 9,150 m (30,000 ft) 518 kts (959 km/h; 596 mph)

T-O field length for 1,610 km (1,000 miles) stage, with 9,697 kg (21,378 lb) payload 1,950 m (6,400 ft)

Range with max fuel* and 7,266 kg (16,020 lb) payload 2,171 n miles (4,025 km; 2,500 miles)

Range with typical space-limited payload* of 9,679 kg (21,378 lb) 2,110 n miles (3,910 km; 2,430 miles)

* *Reserves for 217 n miles (450 km; 250 miles) diversion, 45 min hold at 4,570 m (15,000 ft), final reserve, 4.5% en route allowance and allowances for taxi prior to take-off, circuit approach and land at destination and taxi after landing.*

VERIFIED

BAe (HS) SUPER 748

TYPE: Twin-engined passenger or freight transport.

PROGRAMME: Design of this short/medium turboprop airliner started in January 1959, initially as the Avro 748. The first prototype flew 24 June 1960, followed by a second on 10 April 1961. The Super 748 embodies all the improvements of the earlier Series 2B, together with significant new developments, including an advanced flight deck, new style galley, and engine hush kits. A new mark of engine, the Dart RDa.7 Mk 552, offers up to 12 per cent reduction in fuel consumption over previous marks. A system for dynamic balancing of the propellers to reduce cabin vibration and noise levels is optional. On flight deck, push button select/indicators replace old technology switches, and not only select the services required, but indicate the system status by a colour coded system of illumination. Baggage capacity is increased by 0.28 m³ (10 cu ft). Options include a large rear freight door with an opening of 2.67 × 1.72 m (8 ft 9 in × 5 ft 7¾ in), together with a strengthened cabin floor capable of supporting an overall floor loading of 976 kg/m² (200 lb/sq ft)

VERSIONS: **748 Series 1:** Initial production version with 1,800 ehp Dart RDa.6 Mk 514 engines. First production 748 Series 1 (G-ARMV) flew on 30 August 1961, and received its C of A in January 1962. Production completed.

748 Series 2: Developed version with 2,105 ehp Dart RDa.7 Mk 531 engines but otherwise similar to Series 1. The second prototype Series 1 was re-engined with Dart 7s and first flew as a Series 2 on 6 November 1961. It was subsequently fitted out as a demonstration aircraft and was sold in August 1967. First production Series 2 flew in August 1962. C of A received in October 1962.

BAe (HS) Super 748-200 Series 2 of Makung Airlines ***(BAe)***

1995

748 Series 2A: Superseded Series 2 in production from mid-1967. Differs only in having 2,280 ehp Dart RDa.7 Mk 532-2L turboprop engines, giving improved performance. In service with the Royal Australian Air Force and Navy.

BAe 748 Military Transport: The military transport version has the large rear freight door and strengthened floor available for the civil transport, it also has fixed fittings to undertake a wide range of military roles. Optional military overload take-off and landing weights give improved payload/range capabilities. A total of 52 military 748s were exported, of which 18 were fitted with the rear freight door and strengthened floor.

The BAe 748 is the subject of a manufacturing agreement with the Indian government, and 89 aircraft were assembled from British built components by Hindustan Aeronautics Ltd. Of these 17 were for Indian Airlines and 72 for the Indian Air Force (more than 60 of these are still in service).

UPGRADES: **RAAF:** Project AIR 5232. Programme to update the systems on six of the eight Royal Australian Air Force (RAAF) navigation training aircraft. Improved systems to include: ground-mapping radar (similar to the one in use with the RAF Dominie T.1), off-the-shelf navigation computer and simulated radar warning receiver (RWR). Programme also includes installation of GPS as part of the RAAF Joint Project 5195 to install GPS in all RAAF, RAN and Army tactical aircraft. The GPS system will allow fusion of radar, GPS and inertial navigation system data from the existing Litton LTN-92 Inertial Navigation System (INS). First aircraft planned for redelivery in 1997.

OPERATORS: Versions of the BAe (HS) 748 are in service with the armed forces of the following countries: Australia (10), Brazil (12), Burkina Faso (2), Ecuador (4), India (60), South Korea (2), Nepal (1), New Zealand (9), Sri Lanka (3), Tanzania (3), Thailand (8) and Zambia (1).

The following description applies to the basic Super 748, except where indicated:

DESIGN FEATURES: Cantilever low-wing monoplane. Wing section NACA 23018 at root, NACA 4412 at tip. Dihedral 7°. Incidence 3°. Sweepback 2° 54′ at quarter-chord. Fixed incidence tailplane.

FLYING CONTROLS: All-metal setback hinge, shielded horn balance, manually operated ailerons and electrically actuated Fowler flaps. Geared tab in each aileron. Trim tab in starboard aileron. Manually operated tail unit controls. Trim tabs in elevators and rudder. Spring tab in rudder.

STRUCTURE: All-metal two-spar fail-safe structure. No cutouts in spars for engines or landing gear. The fuselage is an all-metal semi-monocoque riveted fail-safe structure, of circular section. The tail unit is a cantilever all-metal structure.

LANDING GEAR: Retractable tricycle type, with hydraulically steerable nose unit. All wheels retract forward hydraulically. Mainwheels retract into bottom of engine nacelles forward of front wing spar. Dowty Rotol shock-absorbers. Twin-wheels, with Dunlop tyres, on all units. Mainwheels size 32 × 10.75-14. Nosewheel size 25.65 × 8.5-10. Standard tyre pressure: mainwheels 5.03 bars (73 lb/sq in); nosewheels 3.79 bars (55 lb/sq in). Minimum tyre pressures: mainwheels 4.48 bars (65 lb/sq in); nosewheels 3.45 bars (50 lb/sq in). Dunlop disc brakes with Maxaret anti-skid units. No brake cooling.

POWER PLANT: Two 1,700 kW (2,280 ehp) Rolls-Royce Dart RDa.7 Mk 552-2 turboprop engines, each driving a Dowty Rotol four-blade constant-speed fully feathering propeller. Engine hush kits, externally evident by forward extensions of nose cowl by approximately 7.5 cm (3 in). Provision for automatic injection of water methanol into live engine in the event of an engine failure on take-off. Fuel in two integral wing tanks, with total capacity of 6,550 litres (1,729 US gallons; 1,440 Imp gallons). Tanks modified to provide increased wing bending relief. Underwing pressure refuelling and overwing gravity refuelling. Oil capacity 14.2 litres (3.75 US gallons; 3.12 Imp gallons) per engine.

ACCOMMODATION (commercial): Crew of two on flight deck, and cabin attendant. Accommodation for 40-58 passengers in paired seats on each side of central gangway. Baggage compartment forward of cabin, with provision for steward's seat. Galley, toilet and baggage compartment and steward's seat can be replaced by freight hold with moving partition between hold and passenger cabin. Main passenger door, on port side at rear, with smaller door on starboard side to serve as baggage door and emergency exit. Crew and freight door on port side at front. Hydraulically operated stairs.

ACCOMMODATION (military transport): Up to 58 troops in airline type seats. Provision for forward and aft baggage compartments and hydraulically operated stairs. In paratroop role up to 48 paratroops and dispatchers can be accommodated on sidewall folding seats with safety harness. Dropping by static line or free-fall. For casualty evacuation up to 24 stretchers and nine nursing staff can be carried, with provision for medical supplies and equipment. For supply dropping a guided roller conveyer system allows twelve 340 kg (750 lb) or six 680 kg (1,500 lb) loads to be dropped within 6 seconds. Large cargo door will accept items up to 1.42 × 1.42 × 3.66 m (4 ft 8 in × 4 ft 8 in × 12 ft) or small diameter pipes over 12 m (39 ft 4 in) in length. Onboard freight hoist and palletised freight system available. Quickly removable VIP cabin available, and a variety of VIP layouts, with separate toilet, telephone and wide range of options.

SYSTEMS: Normalair automatic pressurisation and air conditioning system, giving equivalent altitude of 2,440 m (8,000 ft) at 7,620 m (25,000 ft). Pressure differential 0.38 bars (5.5 lb/sq in). Hydraulic system, pressure 172 bars (2,500 lb/sq in), for landing gear retraction, nosewheel steering, brakes and propeller brakes. No pneumatic system. One 9 kW 28 V DC generator and one 22 kVA alternator on each engine. Two 1,800 VA static inverters.

AVIONICS AND EQUIPMENT: Collins or Bendix solid-state avionics. Blind-flying instrumentation and Bendix RDR 1300 colour weather radar. Standard equipment by Sperry includes SPZ-500 multimode autopilot/flight director system, air data computer, SPI-402 10 cm (4 in) flight director instruments, GH-14 vertical gyro and C-14 Gyrosyn compass systems. Provision for flight data recorder, flight deck voice recorder and GPWS.

DIMENSIONS, EXTERNAL (Super 748):

Wing span	31.23 m (102 ft 5½ in)
Wing chord at root	3.49 m (11 ft 5¾ in)
Wing aspect ratio	12.668
Length overall	20.42 m (67 ft 0 in)
Fuselage: Max diameter	2.67 m (8 ft 9 in)
Height overall	7.57 m (24 ft 10 in)
Tailplane span	10.97 m (36 ft 0 in)
Wheel track	7.54 m (24 ft 9 in)
Wheelbase	6.30 m (20 ft 8 in)
Propeller diameter	3.66 m (12 ft 0 in)
Propeller ground clearance	0.61 m (2 ft 0 in)
Passenger door (port, rear): Height	1.57 m (5 ft 2 in)
Width	0.76 m (2 ft 6 in)
Height to sill	1.84 m (6 ft 0½ in)
Freight and baggage door (fwd):	
Height	1.37 m (4 ft 6 in)
Width	1.22 m (4 ft 0 in)
Height to sill	1.84 m (6 ft 0½ in)
Baggage door (rear, stbd): Height	1.24 m (4 ft 1 in)
Width	0.64 m (2 ft 1 in)
Height to sill	1.84 m (6 ft 0½ in)
Optional freight door (rear, port):	
Height	1.72 m (5 ft 7¾ in)
Width	2.67 m (8 ft 9 in)

DIMENSIONS, INTERNAL:

Cabin, excl flight deck: Length	14.17 m (46 ft 6 in)
Max width	2.46 m (8 ft 1 in)
Max height	1.92 m (6 ft 3½ in)
Floor area	27.50 m² (296 sq ft)
Volume	56.35 m³ (1,990 cu ft)
Max total freight holds	9.54 m³ (337 cu ft)

AREAS:

Wings, gross	77.00 m² (828.87 sq ft)
Ailerons (total)	3.98 m² (42.90 sq ft)
Trailing-edge flaps (total)	14.81 m² (159.40 sq ft)
Fin	9.81 m² (105.64 sq ft)
Rudder, incl tabs	3.66 m² (39.36 sq ft)
Tailplane	17.55 m² (188.9 sq ft)
Elevators, incl tab	5.03 m² (54.10 sq ft)

WEIGHTS AND LOADINGS (A: Super 748; B: military transport):

Basic operating weight, incl crew:	
A	12,327 kg (27,176 lb)
B	11,656 kg (25,697 lb)
Max payload: A	5,136 kg (11,323 lb)
B	5,807 kg (12,802 lb)
B, optional overload	7,848 kg (17,302 lb)
Max T-O weight: A, B	21,092 kg (46,500 lb)
B, optional overload	23,133 kg (51,000 lb)
Max zero-fuel weight: A, B	17,463 kg (38,500 lb)
B, optional overload	19,504 kg (43,000 lb)
Max landing weight: A, B	19,504 kg (43,000 lb)
B, optional overload	21,546 kg (47,500 lb)
Max wing loading: A	273.9 kg/m² (56.1 lb/sq ft)
Max power loading: A	6.20 kg/kW (10.2 lb/ehp)

BAe (HS) 748 Super 2B of LIAT (1974) Ltd based in Antigua ***(BAe)***

1995

PERFORMANCE (Super 748 at Max T-O weight unless otherwise stated):

Cruising speed at AUW of 17,236 kg (38,000 lb)	244 kts (452 km/h; 281 mph)
Max rate of climb at S/L at AUW of 17,236 kg (38,000 lb)	433 m (1,420 ft)/min
Service ceiling	7,620 m (25,000 ft)
Min ground turning radius	11.82 m (39 ft 0 in)
Runway LCN	9 to 18
T-O run (BCAR)	1,134 m (3,720 ft)
Balanced field length: BCAR	1,393 m (4,570 ft)
BCAR, 840 n miles (1,557 km; 967 mile) sector, 44 passengers and reserves for 200 n miles (370 km; 230 miles) plus 45 min hold	963 m (3,160 ft)
T-O to 15 m (50 ft)	1,158 m (3,800 ft)
Landing field length (BCAR)	1,036 m (3,400 ft)
Landing from 15 m (50 ft)	625 m (2,050 ft)
Landing run	387 m (1,270 ft)
Range with max payload, reserves for 200 n miles (370 km; 230 miles) plus 45 min hold	926 n miles (1,715 km; 1,066 miles)
Range with max fuel, 3,359 kg (7,803 lb) payload and reserves for 200 n miles (370 km; 230 miles) plus 45 min hold	1,560 n miles (2,892 km; 1,797 miles)

OPERATIONAL NOISE LEVELS (FAR Pt 36):

T-O	88.7 EPNdB
Approach	92.8 EPNdB
Sideline	93.3 EPNdB

UPDATED

BAe (HS) 748-2A of the Burkina Faso Air Force

1994

BAe (HS) 748MF Andover C.1 of the Empire Test Pilots School *(Paul Tompkins)*

1994

BAe (HS) 748MF

Royal New Zealand Air Force designation: Andover C. Mk 1

TYPE: Twin-turboprop military transport.

PROGRAMME: The Hawker Siddeley (formerly Avro) 748MF is a rear-loading military transport aircraft which was ordered by the RAF for service with RAF Transport Command as the Andover C. Mk 1. It utilites, with minor modifications, the same front fuselage, wing, landing gear and tail unit as the commercial 748 Srs 2. The majority of the aircraft systems are also similar. Its performance includes the ability to operate into and out of unprepared strips 460 m (1,500 ft) long.

Design work began in April 1962. Construction of a prototype was started two months later and this aircraft (a conversion of the civil 748 G-ARRV) flew for the first time 21 December 1963.

More powerful Dart turboprops have been fitted and the 748MF has a unique variable landing gear supplied by Dowty, which will enable the position of the floor of the cabin to be adjusted easily, both vertically and horizontally, to mate precisely with the tailboard of the loading or unloading vehicle.

During trials in January 1962, the prototype HS 748 Srs 2 demonstrated the potential of the 748MF by manoeuvring, taking off and landing in deep mud, carrying heavy payloads. On one occasion, it took off and landed with a payload of nearly 5½ tons, with one of its turboprop engines stopped.

OPERATORS: The Andover is in service with the Royal New Zealand Air Force (9).

The general description of the 748 Srs 2 applies equally to the 748MF, with the following differences.

DESIGN FEATURES: Wing aspect ratio 1.55. Mean chord 2.59 m (8 ft 6 in). Upswept rear fuselage containing integral loading ramp. The tailplane has been moved up to base of fin.

Twin mainwheels on each unit, size 34 × 11.75-14, pressure 5.27 kg/cm² (75 lb/sq in). Twin nosewheels size 8.5-10, pressure 3.87 kg/cm² (55 lb/sq in).

POWER PLANT: Two Rolls-Royce Dart RDa.12 turboprop engines, each rated at 3,245 ehp, and driving a Dowty Rotol four-blade constant-speed reversible-pitch propeller, diameter 4.42 m (14 ft 6 in). Integral fuel tanks in each wing with total capacity of 10,310 litres (2,723 US gallons; 2,268 Imp gallons). Underwing pressure refuelling and overwing gravity fuelling. Oil capacity 14.2 litres (3.75 US gallons; 3.12 Imp gallons) per engine.

ACCOMMODATION: Pilot and co-pilot side by side on flight deck, with optional seat for navigator/radio operator. Accommodation in main cabin for 52 troops, 40 paratroops, 24 stretcher patients and attendants or up to 6,849 kg (15,099 lb) of freight or vehicles. Typical military loads for airdropping would include a Ferret Mk 2/3 scout car and a ¼ ton Land Rover Mk 5 or three Land Rovers. Toilets in fore and aft cabin on starboard side. Rear toilet is removed when transporting freight or vehicles. Outward and upward-opening door for crew and freight at front on port side. There is an inward-opening door for passengers and for paratroop dropping on port side and an outward-opening baggage door on starboard side. Rear-loading ramp can be lowered in flight. Height of ramp sill when in line with cargo floor is 2.31 m (7 ft 2 in) in its normal position, but aircraft can be lowered on its main undercarriage, reducing this height to 1.14 m (3 ft 9 in) above ground. Entire cabin is pressurised and air conditioned. Introduction of a new rear fuselage with integral loading ramp.

BAe (HS) 748MF Andover C.1. Note the different propeller markings *(Paul Tompkins)*

1994

SYSTEMS: Normalair pressurisation and air conditioning system, differential 0.38 bars (5.5 lb/sq in). Hydraulic system, pressure 197 kg/cm² (2,800 lb/sq in), actuates landing gear nosewheel steering, propeller brakes and rear-loading ramp. No pneumatic system or APU. One 9 kW 28 V DC generator and one 30 kVA alternator on each engine. Three 115 V AC three-phase 400 c/s inverters.

ELECTRONICS AND EQUIPMENT: Bendix or Collins solid-state radio and radar. Blind-flying instrumentation standard. Weather radar and autopilot to customer's requirements.

DIMENSIONS, EXTERNAL:

Wing span	29.87 m (98 ft 0 in)
Length overall	23.75 m (77 ft 11 in)
Height over tail	8.97 m (29 ft 5 in)
Tailplane span	9.14 m (30 ft 0 in)
Wheel track	8.45 m (27 ft 9 in)
Wheelbase	7.04 m (23 ft 1 in)
Passenger door (port, rear): Height	1.75 m (5 ft 9 in)
Width	0.91 m (3 ft 0 in)
Height to sill	variable
Rear-loading ramp: Height	1.68 m (5 ft 6 in)
Width	2.08 m (6 ft 10 in)
Height to sill	0.99-2.18 m (3 ft 3 in-7 ft 2 in)
Forward freight	1.37 m (4 ft 6 in)
Width	1.22 m (4 ft 0 in)
Height to sill	variable
Baggage door (stbd, rear): Height	1.22 m (4 ft 0 in)
Width	0.76 m (2 ft 6 in)
Height to sill	variable

DIMENSIONS, INTERNAL:

Cabin, excluding flight deck, incl canopy:	
Length	16.76 m (55 ft 0 in)
Max width	2.44 m (8 ft 0 in)
Max height	1.91 m (6 ft 3 in)
Floor area (usable)	34.83 m² (375 sq ft)
Volume (usable)	62.30 m³ (2,200 cu ft)

AREAS:

Wings, gross	77.20 m² (831.4 sq ft)
Ailerons (total)	3.94 m² (42.4 sq ft)
Trailing-edge flaps (total)	14.84 m² (159.8 sq ft)
Fin	12.65 m² (136.1 sq ft)
Rudder, incl tabs	4.27 m² (46.0 sq ft)
Tailplane	17.84 m² (192.0 sq ft)
Elevators, incl tabs	3.82 m² (41.1 sq ft)

WEIGHTS AND LOADINGS:

Basic operating weight, incl crew	12,550 kg (27,666 lb)
Max payload	6,849 kg (15,099 lb)
Max T-O weight	22,680 kg (50,000 lb)
Max zero-fuel weight	19,504 kg (43,000 lb)
Max landing weight	21,590 kg (47,600 lb)
Max wing loading	293.4 kg/m² (60.1 lb/sq ft)
Max power loading	3.49 kg/eshp (7.70 lb/eshp)

PERFORMANCE:

Max level speed 265 kts (491 km/h; 305 mph)
Max permissible diving speed Mach 0.65 (300 kts; 555 km/h; 345 mph)
Normal cruising speed at 18,145 kg (40,000 lb) AUW 243 kts (450 km/h; 280 mph)
Stalling speed, clean, at max landing weight 108 kts (200 km/h; 124 mph)
Stalling speed, flaps and wheels down 75 kts (145 km/h; 90 mph)

Rate of climb at S/L	365 m (1,200 ft)/min
Service ceiling	7,300 m (24,000 ft)
T-O run, STOL	314 m (1,030 ft)
T-O to 15 m (50 ft), STOL	595 m (1,950 ft)
Landing run 15 m (50 ft), STOL	460 m (1,510 ft)
Landing run, STOL	308 m (1,010 ft)

Range with max fuel, allowances for 370 km (230 mile) diversion, 30 min standoff at 305 m (1,000 ft) and 5% stage fuel reserve 1,955 n miles (3,620 km; 2,250 miles)
Range with max payload, reserves as above 429 n miles (795 km; 495 miles)

UPDATED

BAe Matador AV-8S single-seat close-support aircraft of the Spanish Navy ***(BAe)***

1995

BAe Harrier T. Mk 2 close-support and reconnaissance trainer aircraft of the Royal Air Force ***(BAe)***

1995

BAe HARRIER

RAF designation: Harrier GR. Mk 3 and T. Mks 4/4A
RN designations: T. Mks 4N/8N
Indian Navy designation: T. Mk 60
Spanish Navy designation: VA.1 Matador (AV-8S and TAV-8S)

TYPE: V/STOL close support and reconnaissance aircraft.

PROGRAMME: World's first operational fixed-wing V/STOL attack fighter; supplanted in production by AV-8B Harrier II/Harrier GR. Mks 5/7 except limited manufacture as two-seat compatible trainer for Sea Harrier, as detailed below; following February 1990 order for two-seat Harrier IIs, plans abandoned to convert some RAF T. Mk 4s to T. Mk 6 standard with night vision equipment for training pilots of GR. Mk 7s fitted with FLIR.

Two dual-control Harriers employed in test programmes, comprising XW175 at RAE Bedford as Vectored-thrust Advanced Aircraft Control (VAAC) Harrier; and XW267 with 'Nightbird' night vision equipment.

VERSIONS: **Harrier GR. Mk 1, 1A and 3:** Single-seat close support and tactical reconnaissance versions for the RAF. First production GR. Mk 1 flew 28 December 1967. Deliveries made initially to Harrier Conversion Team at RAF Wittering, beginning with SV744 on 9 April 1969, and to No. 1 Squadron, formed at the same base 18 July 1969. Delivered subsequently to Nos. 3, 4 and (temporarily) 20 Squadrons in West Germany. These aircraft were designated Harrier GR. Mk 1 when fitted initially with Pegasus 101 engines. GR. Mk 1/1A no longer in service. GR. Mk 3 no longer in front-line service. AV-8S still in service with the Spanish Navy.

Harrier T. Mk 2, 2A, 4, 4A and 6 and Harrier T. Mk 4RN: Two-seat versions retaining the full combat capability of the single-seater in terms of equipment fit and weapon carriage. There is a large degree of commonality in structure and system components, ground support equipment and flight and ground crew training. Differences includes a longer nose section forward of the wing leading-edge, with two cockpits in tandem; tailcone approximately 1.83 m (6 ft) longer than that of the single-seat model; and enlarged fin surfaces. The two-seat Harrier can be used operationally with the rear seat and compensating tail ballast removed, thus minimising the weight penalty over its single-seat counterpart. First of two development aircraft flew 24 April 1969, and the first of 25 production aircraft for the RAF on 3 October 1969. The two-seater entered RAF service July 1970, and the last aircraft (also the final first-generation Harrier for the RAF) was delivered on 2 October 1987. The RAF Harrier T. Mk 2, like the GR. Mk 1, was powered originally by the Pegasus 101 engine.

Harrier T. Mk 8: Upgraded Harrier T. Mk 4/4N for the Royal Navy.

Harrier T. Mk 60: Two-seat operational trainer version for Indian Navy. T. Mk 4A configuration, but with complete Sea Harrier avionics except for Blue Fox radar. Four ordered. Deliveries to India began in 1984.

Matador AV-8S/TAV-8S: Ordered through the USA for the Spanish Navy. Nine in service.

Sea Harrier FRS. Mks 1, 51 and F/A. Mk 2: Described separately.

UPGRADES: **BAe Harrier GR. 3:** When retrofitted subsequently with the Pegasus 102 they were redesignated GR. Mk 1A. Aircraft now in service have the Pegasus 103 engines and are designated GR. Mk 3. A total of 118 production GR. 1/3s was built for the Royal Air Force up to 1987, excluding prototypes. In total, 14 of them took part in the Falklands campaign in 1982.

BAe Harrier T. Mk 2A/4A: The designations T. Mk 2A and T. Mk 4 apply to aircraft retrofitted with, respectively, the Pegasus 102 and 103. The T. Mk 4A reverts to the original pointed nose, without LRMTS, to make it lighter for training duties with No. 233 OCU at RAF Wittering. One further T. Mk 4A was funded by the Royal Navy, but operated by the RAF. Genuine Royal Navy two-seaters are designated Harrier T. Mk 4RN; three were delivered, of which one was lost in 1985. Two-seat Harrier production for the UK thus totalled 31, including prototypes. It is expected that some RAF T. Mk 4s will be converted to T. Mk 6 standard, with night vision equipment for training pilots of GR. Mk 5s fitted eventually with FLIR sensors and redesignated GR. Mk 7.

BAe V-8A T. Mk 8: Authorisation February 1992 for five naval Harrier trainers to receive cockpits representative of Sea Harrier F/A Mk 2 (including FIN 1031B INS, but without radar), being redesignated T. Mk 8.

OPERATORS: Versions of the Harrier are in service with the armed forces of the following countries: India (4), Italy (18), Spain (19), UK (133) and USA (166).

DESIGN FEATURES:
Technical details as for Sea Harrier, except:

LANDING GEAR: T. Mk 4 tyre pressures 6.90 bars (100 lb/sq in) on nose unit, 6.55 bars (95 lb/sq in) on main and outrigger units.

POWER PLANT: One Rolls-Royce Pegasus Mk 103 (or navalised Mk 104) vectored thrust turbofan (95.6 kN; 21,500 lb st).

ACCOMMODATION: Crew of two (Mk 4) on Martin-Baker Mk 9D zero/zero rocket ejection seats.

AVIONICS: GEC-Plessey U/VHF, Ultra standby UHF, GEC Avionics AD 2770 Tacan and Cossor IFF, GEC Ferranti FE 541 inertial navigation and attack system (INAS), with Honeywell C2G compass, Smiths electronic head-up display of flight information, and air data computer. Marconi ARI.18223 radar warning receiver. GEC Ferranti Type 106 laser ranger and marked target seeker (LRMTS) in most RAF Harriers.

DIMENSIONS, EXTERNAL: As for Sea Harrier except:

Wing span	7.70 m (25 ft 3 in)
Length overall: two-seat (laser nose)	17.50 m (57 ft 5 in)
Height overall: two-seat	4.17 m (13 ft 8 in)

WEIGHTS AND LOADINGS:

Weight empty (pilot/s plus four pylons; no guns): T. Mk 4	6,693 kg (14,755 lb)
T. Mk 4A	6,568 kg (14,480 lb)
Internal fuel	2,295 kg (5,060 lb)

BAe Harrier GR. Mk 3 close-support and tactical reconnaissance aircraft of the RAF ***(BAe)***

1995

Max T-O weight: two-seat 11,880 kg (26,200 lb)
Max wing loading 636.0 kg/m² (130.3 lb/sq ft)
Max power loading 124.27 kg/kN (1.22 lb/lb st)

PERFORMANCE (single-seat):
Max level speed at S/L 635 kts (1,176 km/h; 730 mph)
Max Mach number in a dive at height 1.3
Time to 12,200 m (40,000 ft) from vertical T-O 2 min 23 s
Service ceiling 15,600 m (51,200 ft)
T-O run: with 2,270 kg (5,000 lb) payload at max T-O weight approx 305 m (1,000 ft)
Range: hi-lo-hi with 1,995 kg (4,400 lb) payload 360 n miles (666 km; 414 miles)
lo-lo with 1,995 kg (4,400 lb) payload 200 n miles (370 km; 230 miles)
Ferry range 1,850 n miles (3,425 km; 2,129 miles)
Range with one in-flight refuelling more than 3,000 n miles (5,560 km; 3,455 miles)
Endurance:
combat air patrol 100 n miles (185 km; 115 miles) from base 1 h 30 min
with one in-flight refuelling more than 7 h
g limits +7.8/–4.2

UPDATED

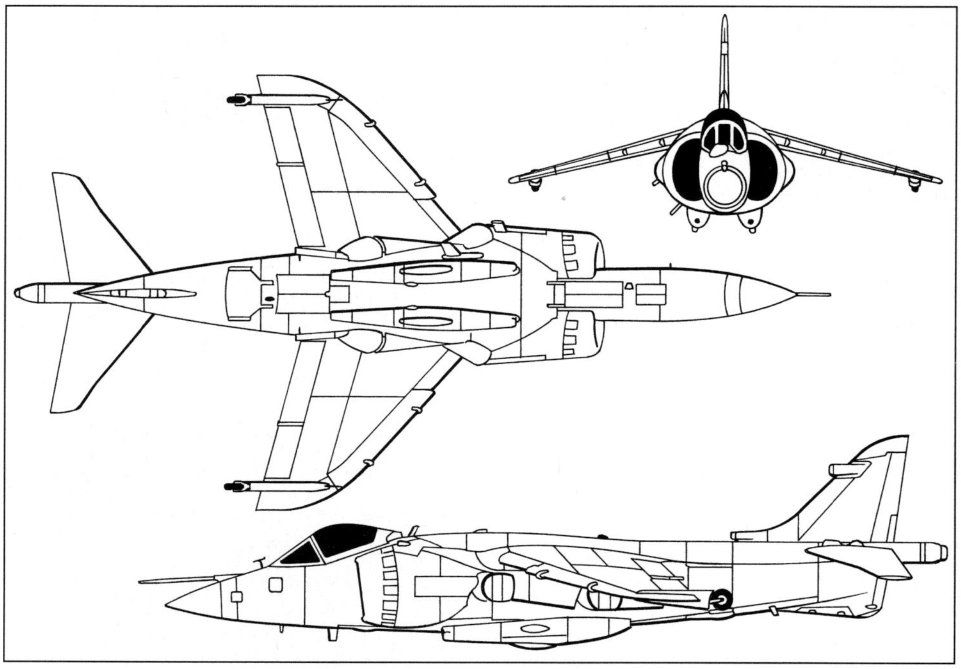

BAe Sea Harrier FRS. Mk 1 ***(Jane's/Mike Keep)***

1994

BAe SEA HARRIER

RN designations: FRS. Mk 1 and F/A. Mk 2
Indian Navy designation: FRS. Mk 51

TYPE: V/STOL fighter, reconnaissance and strike aircraft.

PROGRAMME: Development of P1184 Sea Harrier announced by British government 15 May 1975; first flight (XZ450) 20 August 1978; first delivery to Royal Navy (XZ451) 18 June 1979; first ship trials (HMS *Hermes*) November 1979.

Ski jump launching ramp (proposed by Lt Cdr D. R. Taylor, RN) take-off trials ashore 1977, and at sea from 30 October 1980; HMS *Invincible* and *Illustrious* first fitted with 7° ramps, HMS *Ark Royal* 12°; latter allows 1,135 kg (2,500 lb) increased load for same take-off run or 50 to 60 per cent shorter run at same weight; HMS *Invincible* recommissioned with 13° ramp 18 May 1989; HMS *Illustrious* began similar 2½ year re-work May 1991.

VERSIONS: **FRS. Mk 1:** Initial Royal Navy version, Pegasus 104 engine; first used operationally during Falkland Islands campaign 1982, from HMS *Hermes* and *Invincible* (29 flew 2,376 sorties, destroying 22 enemy aircraft in air-to-air combat without loss; four lost in accidents and two to ground fire). Total 37 remained, June 1993, including Mk 2 conversions.

FRS. Mk 51: Similar to Mk 1, for Indian Navy.

F/A. Mk 2: Differs externally from Mk 1 by less pointed nose radome; longer rear fuselage, resulting from 35 cm (1 ft 1¾ in) plug aft of wing trailing-edge; revisions of antennae and external stores. Internal changes include GEC-Marconi Blue Vixen pulse Doppler radar, offering all-weather lookdown/shootdown capability, with inherent track-while-scan, multiple target engagement, greatly increased missile launch range, enhanced surface target acquisition, and improved ECCM performance. Current weapons plus AIM-20 AMRAAM on Dowty/Frazer-Nash launch rails compatible with AIM-9L Sidewinder.

Improved systems built around MIL-STD-1553B databus, with dual redundant data highway, allowing computerised time sharing of information processed in databus control and interface unit.

UK MoD gave BAe project definition contract January 1985 for mid-life update of RN Sea Harrier FRS. Mk 1s; upgraded aircraft redesignated F/A. Mk 2; aerodynamic development F/A. Mk 2 converted at Dunsfold from Mk 1 ZA195; first flight 19 September 1988; first flight of second development aircraft (XZ439) 8 March 1989; contract for conversion of further 29 Mk 1s to F/A. Mk 2s signed by UK MoD 7 December 1988; modifications begun at Kingston October 1990, continuing at Dunsfold and Brough; redelivery from 2 April 1993, for later augmentation by newly built F/A. Mk 2s; additional five conversions under consideration.

New F/A. Mk 2 AAM launch rail first tested by live AIM-9L (from Mk 1) 2 November 1988; AMRAAM trials began in USA (using XZ439) with first launch 29 March 1993, flying from Eglin AFB; airborne testing of Blue Vixen radar began in RAE One-Eleven (ZF433), completing 114 hour/121 sortie programme November 1987; development work transferred to RAE BAe (XW930), first flown with A-version radar 26 August 1988; second 125-600B (ZF130) given full F/A. Mk 2 weapon system, including representative cockpit in co-pilot's position and Sidewinder acquisition round on underwing pylon (first flight at Woodford 20 May 1988; began development flying at Dunsfold December 1988; not fitted with B-version Blue Vixen radar until September 1989); first flight of B-version radar in Sea Harrier XZ439 24 May 1990. First F/A. Mk 2 deck landing, by ZA195 on HMS *Ark Royal*, 7 November 1990. Operational Evaluation Unit for Mk 2 formed at Boscombe Down, June 1993.

Redesign of cockpit includes new HUD and dual multipurpose HDDs; proposed JTIDS terminal delayed following cancellation by US Navy of originally proposed equipment; JTIDS/Sea Harrier integration study (for retrofit) undertaken by BAe 1992-93. All time-critical weapon systems controls positioned on up-front control panel or on throttle and stick (HOTAS).

Wingtip extensions of 20 cm (8 in) and 30 cm (1 ft 0 in) test-flown to enhance stability carrying AMRAAM, but proved unnecessary by 1990 trials.

UPGRADES: **BAe Sea Harrier F/A. Mk 2:** See Versions.

OPERATORS: Versions of the Sea Harrier are in service with the armed forces of the following countries: India (23) and UK (20).

CUSTOMERS: Royal Navy ordered three development aircraft plus batches of 21, 10, 14 and nine by September 1984; all built as Mk 1s, last completed June 1988; intent to order at least 10 F/A. Mk 2s revealed March 1990, converted to firm order for 18 in January 1994. Naval Intensive Flying Trials Unit (No. 700A Squadron) commissioned at RNAS Yeovilton 18 September 1979 and became normally shore based No. 899 HQ Squadron, April 1980; front line units Nos. 800 and 801 Squadrons, with eight aircraft each (previously five); non-radar T. Mk 4N two-seat trainers three received.

First 'production' conversions to Mk 2 (XZ497) retained by BAe; second aircraft (ZE695) was first delivered 2 April 1993; nine delivered by end of 1993. Operational Evaluation Unit formed at Boscombe Down June 1993; last Mk 1 departed 899 Squadron, December 1993, in preparation for Mk 2 pilot training; initial loss sustained 5 January 1994.

Six similar FRS. Mk 51s handed over to Indian Navy from January 1983; delivered (after pilot training in UK) between December 1983 and July 1984; used by No. 300 (White Tiger) Squadron from INS *Vikrant*; two T. Mk 60 two-seat trainers received (see Harrier entry); 10 more FRS. Mk 51s and one T. Mk 60 ordered by Indian government November 1985; delivered December 1989 to September 1991; letter of intent for seven more FRS. Mk 51s and one T. Mk 60 issued September 1986, to equip INS *Viraat* (former HMS *Hermes*); third batch deliveries began September 1991 and completed April 1992. Total orders 98.

The following description applies to Sea Harrier FRS. Mk 1 except where indicated otherwise:

DESIGN FEATURES: Single-engined V/STOL system with four rotatable exhaust nozzles that can be set through 98.5° from fully aft position; short take-off made with nozzles initially full aft, then turned partially downward for lift-off and continued forward acceleration; nozzles can be vectored at high-speed to tighten turn radius or decelerate suddenly; control at less than wingborne airspeed automatically transferred to puffer jets at wingtips, nose and tail, also enhancing combat manoeuvres.

Main differences from land-based Harriers include elimination of magnesium components, introduction of raised cockpit, revised operational avionics, and installation of multimode GEC-Marconi radar with air-to-air intercept and air-to-surface modes in redesigned nose that folds to port; Pegasus 104 turbofan of Mk 1 incorporates additional anti-corrosion features and generates more electrical power than land-based Pegasus 103. See Versions for F/A. Mk 2 features.

Wing section BAe (HS) design; thickness/chord ratio 10 per cent at root, 5 per cent at tip; anhedral 12°; incidence 1° 45′; and sweepback at quarter-chord 34°.

FLYING CONTROLS: Plain ailerons irreversibly operated by tandem hydraulic jacks; one-piece variable incidence tailplane, with 15° anhedral, irreversibly operated by tandem hydraulic jacks; manually operated rudder with trim tab; flaps; jet reaction control valve built into front of each outrigger wheel fairing and in nose and tailcone; large airbrake under fuselage; ventral fin under rear fuselage.

STRUCTURE: One-piece aluminium alloy three-spar safe-life wing with integrally machined skins (Brough built); entire wing unit removable to provide access to engine; revised inboard one-third of F/A. Mk 2 wing incorporates additional fence, kinked leading-edge, re-positioning of dog-tooth fillet closer to fuselage, and reduction of overwing vortex generators from 12 to 11; 67.3 cm (2 ft 2½ in) wing extensions available for ferrying; ailerons, flaps, rudder and tailplane trailing-edge of bonded aluminium alloy honeycomb construction; safe-life fuselage of frames and stringers, mainly aluminium alloy but with titanium skins

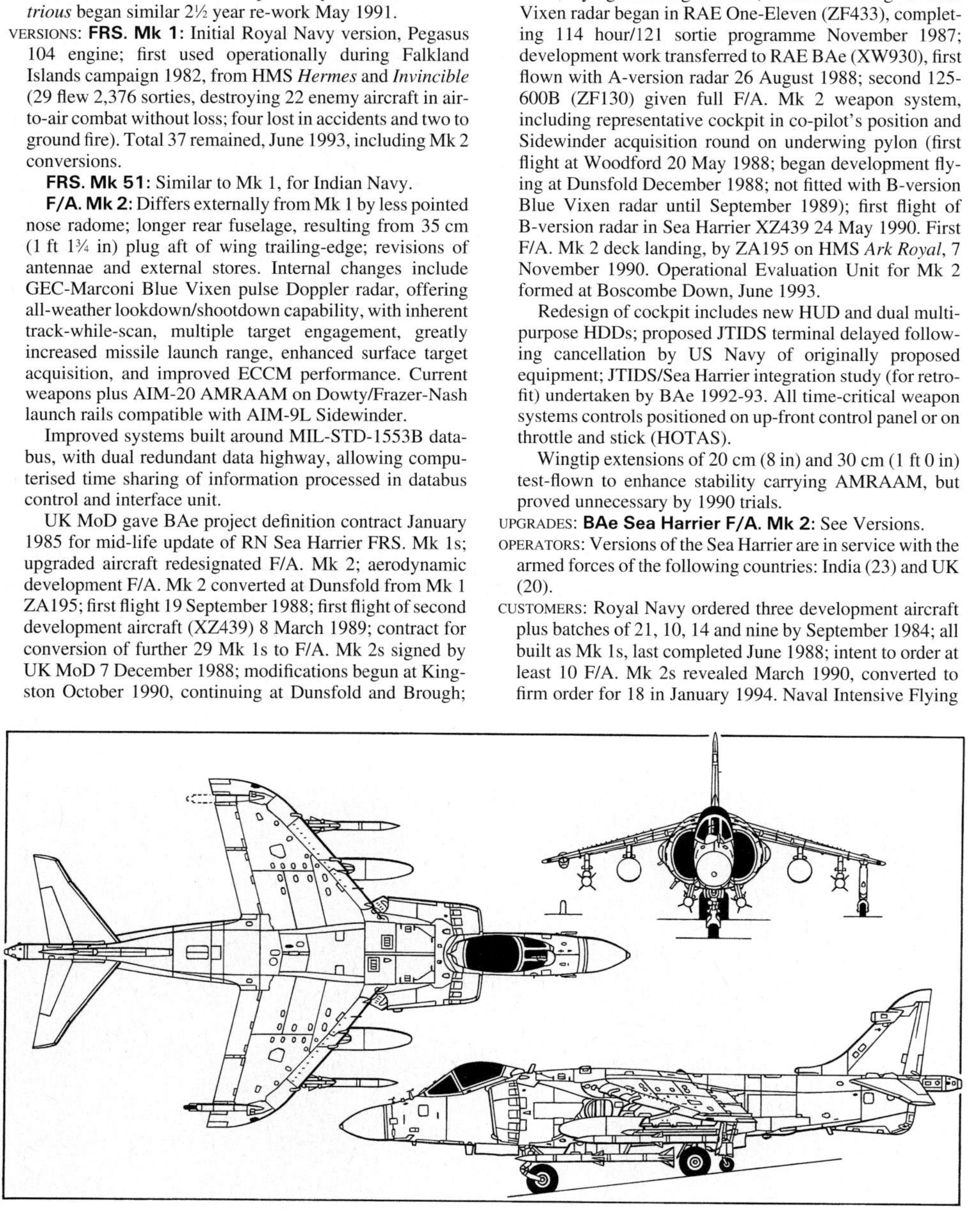

BAe Sea Harrier F/A. Mk 2 ***(Jane's/Mike Keep)***

1994

at rear and some titanium adjacent engine and other special areas; access to power plant through top of fuselage, ahead of wings; F/A. Mk 2 has deepened and stiffened nose structure, plus new rear fuselage, lengthened by 35 cm (1 ft 1¾ in); fin tip carries suppressed VHF aerial.

LANDING GEAR: Retractable bicycle type of Dowty Aerospace manufacture, permitting operation from rough unprepared surfaces of CBR as low as 3 to 5 per cent. Hydraulic actuation, with nitrogen bottle for emergency extension of landing gear. Single steerable nosewheel retracts forward, twin coupled mainwheels rearward, into fuselage. Small outrigger units retract rearward into fairings slightly inboard of wingtips. Nosewheel leg of levered suspension liquid spring type. Dowty telescopic oleo-pneumatic main and outrigger gear. Dunlop wheels and tyres, size 26.00 × 8.75-11 (nose unit), 27.00 × 7.74-13 (main units) and 13.50 × 6.4 (outriggers). Dunlop multidisc brakes and Dunlop-Hytrol adaptive anti-skid system.

POWER PLANT: One Rolls-Royce Pegasus Mk 104 or (FRS. Mk 2) Mk 106 vectored thrust turbofan (95.6 kN; 21,500 lb st), with four exhaust nozzles of the two-vane cascade type, rotatable through 98.5° from fully aft position. Engine bleed air from HP compressor used for jet reaction control system and to power duplicated air motor for nozzle actuation. Low drag intake cowls each have eight automatic suction relief doors aft of leading-edge to improve intake efficiency by providing extra engine air at low forward or zero speeds. A 227 litre (60 US gallon; 50 Imp gallon) tank supplies demineralised water for thrust restoration in high ambient temperatures for STO, VTO and vertical landings. Fuel in five integral tanks in fuselage and two in wings, with total capacity of approximately 2,865 litres (757 US gallons; 630 Imp gallons). This can be supplemented by two 455 litre (120 US gallon; 100 Imp gallon) jettisonable combat tanks, or two 864 litre (228 US gallon; 190 gallon) tanks, or two 1,500 litre (396 US gallon; 330 Imp gallon) ferry tanks on the inboard wing pylons. Ground refuelling point in port rear nozzle fairing. Provision for in-flight refuelling probe above the port intake cowl.

ACCOMMODATION: Pilot only, on Martin-Baker Mk 10H zero/zero rocket ejection seat which operates through the miniature detonating cord equipped canopy of the pressurised, heated and air conditioned cockpit. Seat raised 28 cm (11 in) compared with Harrier. Manually operated rearward sliding canopy. Birdproof windscreen, with hydraulically actuated wiper. Windscreen washing system.

SYSTEMS: Three-axis limited authority autostabiliser for V/STOL flight. Pressurisation system of BAe design, with Normalair-Garrett and Delaney Gallay major components; maximum pressure differential 0.24 bar (3.5 lb/sq in). Two hydraulic systems; flow rate: System 1, 36 litres (9.6 US gallons; 8 Imp gallons)/min; System 2, 23 litres (6 US gallons; 5 Imp gallons)/min. Systems, pressure 207 bars (3,000 lb/sq in), actuate Fairey flying control and general services and a retractable ram air turbine inside top of rear fuselage, driving a small hydraulic pump for emergency power. Turbine deleted from FRS. Mk 2. Hydraulic reservoirs nitrogen pressurised at 2.75 to 5.5 bars (40 to 80 lb/sq in). AC electrical system with transformer-rectifiers to provide required DC supply. Two 15 kVA generators. Two 28 V 25 Ah batteries, one of which energises a 24 V motor to start Lucas Mk 2 gas-turbine starter/APU. This unit drives a 6 kVA auxiliary alternator for ground readiness servicing and standby. Bootstrap cooling unit for equipment bay, with intake at base of dorsal fin. Autopilot function on Fairey Hydraulics, giving throughput to aileron and tailplane power controls as well as to three-axis autostabiliser. British Oxygen liquid oxygen system of 4.5 litres (1.2 US gallons; 1 Imp gallon) capacity in Royal Navy aircraft; Indian Navy has gaseous oxygen system.

AVIONICS: Nose-mounted GEC-Marconi Blue Fox (Blue Vixen in F/A. Mk 2) multimode radar, with TV raster daylight viewing tube which conveys flight information, as well as radar data, to pilot. New and larger Smiths electronic HUD and 20,000 word digital weapon aiming computer. Autopilot, radar altimeter and Decca Doppler 72 radar. GEC-Marconi self-aligning attitude and heading reference platform and digital navigation computer. Radio navaids include UHF homing, GEC-Marconi AD 2770 Tacan with offset facility and I-band transponder. Radio com by multichannel GEC-Marconi PTR 377 U/VHF, with VHF standby via D 403M transceiver. Passive electronic surveillance and warning of external radar illumination by receiver with forward and rear hemisphere antennae in fin and tailcone respectively. Intended 1994 retrofit of Mk XII IFF; GPS in 1997; JTIDS in 1997-99.

EQUIPMENT: Optically flat panel in nose, on port side, for F.95 oblique camera, which is carried as standard. A cockpit voice recorder with in-flight playback facility supplements the reconnaissance cameras, and facilitates rapid debriefing and mission evaluation.

ARMAMENT: No built-in armament. Combat load carried on four underwing and one underfuselage pylons, all with ML ejector release units. Inboard wing points and fuselage point stressed for loads up to 907 kg (2,000 lb) each, and outboard underwing pair for loads up to 295 kg (650 lb) each; two strake fairings under the fuselage can each be replaced by a 30 mm Aden gun pod and ammunition or, on FRS. Mk 2, by two AIM-120 AMRAAMs. Aircraft cleared for operations with maximum external load exceeding 2,270 kg (5,000 lb), and has flown with weapon load of 3,630 kg (8,000 lb). F/A. Mk 2 outboard pylons re-stressed to 454 kg (1,000 lb). Able to carry 30 mm guns, bombs, rockets and flares of UK and US designs. Alternative stores loads of RN Sea Harriers include a WE177 nuclear bomb; free-fall (1,030 lb) and parachute-retarded (1,120 lb) bombs; Lepus flares; and ML CBLS 100 carriers for Ports-

BAe Sea Harrier FRS. Mk 1 of the Royal Navy

1994

BAe Harrier F/A. Mk 2 upgraded FRS.1 V/STOL fighter aircraft of the Royal Navy ***(BAe)***

1995

BAe Harrier F/A. Mk 2 fighter, reconnaissance and strike aircraft *(BAe)*

1995

mouth Aviation 3 kg and 14 kg practice bombs. Four AIM-9 Sidewinder missiles carried on the outboard underwing pylons (Matra Magic instead of Sidewinder on Indian Navy aircraft); provision for two air-to-surface missiles of Sea Eagle or Harpoon type. FRS. Mk 2 accommodates up to four AIM-120 AMRAAMs, or two AIM-120s and four AIM-9L Sidewinders, on Frazer-Nash common rail launchers. BAe ALARM anti-radiation missile may replace AIM-120.

DIMENSIONS, EXTERNAL:

Wing span: normal	7.70 m (25 ft 3 in)
ferry	9.04 m (29 ft 8 in)
Length overall: FRS. Mk 1	14.50 m (47 ft 7 in)
F/A. Mk 2	14.17 m (46 ft 6 in)
Length overall, nose folded:	
FRS. Mk 1	12.73 m (41 ft 9 in)
F/A. Mk 2	13.16 m (43 ft 2 in)
Height overall	3.71 m (12 ft 2 in)
Tailplane span	4.24 m (13 ft 11 in)
Outrigger wheel track	6.76 m (22 ft 2 in)
Wheelbase, nosewheel to mainwheels	approx 3.45 m (11 ft 4 in)

AREAS:

Wings, gross	18.68 m² (201.1 sq ft)
Fin (excl ventral fin): two-seat	3.57 m² (38.4 sq ft)
Rudder, incl tab	0.49 m² (5.3 sq ft)
Tailplane	4.41 m² (47.5 sq ft)

WEIGHTS AND LOADINGS (FRS. Mk 1):

Operating weight empty	6,374 kg (14,052 lb)
Max fuel: internal	2,295 kg (5,060 lb)
external	2,404 kg (5,300 lb)
Max weapon load: STO	3,630 kg (8,000 lb)
VTO	2,270 kg (5,000 lb)
Max T-O weight	11,880 kg (26,200 lb)
Max wing loading	636.0 kg/m² (130.3 lb/sq ft)
Max power loading	124.27 kg/kN (1.22 lb/lb st)

PERFORMANCE (FRS. Mk 1):

Max Mach number at high-altitude	1.25
Max level speed at low altitude	above 640 kts (1,185 km/h; 736 mph) EAS
Typical cruising speed:	
high altitude, for well over 1 h on internal fuel	above Mach 0.8
low altitude	350-450 kts (650-833 km/h; 404-518 mph)
with rapid acceleration to	600 kts (1,100 km/h; 690 mph)
STO run at max T-O weight, without ski jump	approx 305 m (1,000 ft)
Time from alarm to 30 n miles (55 km; 35 miles) combat area	under 6 min
High-altitude intercept radius, with 3 min combat and reserves for VL	400 n miles (750 km; 460 miles)
Attack radius	250 n miles (463 km; 288 miles)
g limits	+7.8/−4.2

COMBAT PROFILES (F/A. Mk 2, from carrier fitted with a 12° ski jump ramp, at ISA + 15°C and with a 20 knot; 37 km/h; 23 mph wind over the deck):

Combat air patrol: Up to 1½ hours on station at a radius of 100 n miles (185 km; 115 miles), carrying four AMRAAMs, or two AMRAAMs and two 30 mm guns, plus two 864 litre (228 US gallon; 190 Imp gallon) combat drop tanks.

Reconnaissance: Low level cover of 130,000 nm² (446,465 km²; 172,380 sq miles) at a radius of 525 n miles (970 km; 600 miles) from the carrier, with outward and return flights at medium/high level, carrying two 30 mm guns and two 864 litre (228 US gallon; 190 Imp gallon) combat drop tanks. Overall flight time 1 hour 45 minutes.

Surface attack (hi-lo-hi): Radius of action to missile launch 200 n miles (370 km; 230 miles), carrying two Sea Eagle missiles and two 30 mm guns.

Take-off deck run for the above missions is 137 m, 107 m and 92 m (450 ft, 350 ft and 300 ft) respectively, with vertical landing.

Interception: A typical deck-launched interception could be performed against a Mach 0.9 target at a radius of 116 n miles (215 km; 133 miles), or a Mach 1.3 target at 95 n miles (175 km; 109 miles), after initial radar detection of the approaching target at a range of 230 n miles (425 km; 265 miles), with the Sea Harrier at 2 minutes alert status, carrying two AMRAAM missiles.

UPDATED

BAe (HAWKER) HUNTER

TYPE: Single-seat ground attack aircraft fighter (Mk 9) and two-seat operational and advanced trainer (Mk 66).

PROGRAMME: A total of 1,985 Hunters were built, including 460 manufactured by Fokker and Aviolanda in the Netherlands and by Avions Fairey and SABCA in Belgium.

VERSIONS: **F (GA) Mk 9**: Development of Mk 6 single-seat fighter for ground support duties with the RAF in the Middle East, with Avon Mk 207 turbojet, brake parachute and ability to carry 1,045 litre (230 Imp gallon; 276 US gallon) drop tanks in addition to ground attack weapons. In service with the air forces of Chile and Zimbabwe.

F Mk 58: Fighter/ground attack aircraft in service with the Swiss Air Force. Withdrawn from service.

F Mk 70: Fighter aircraft in service with the Lebanese Air Force.

FGA Mk 71/FR 71A: Fighter-bomber and fighter-reconnaissance aircraft operated by the Chilean Air Force.

T Mk 66: Trainer operated by the Lebanese Air Force.

T Mk 67/72: Trainer aircraft operated by the Chilean Air Force.

T Mk 68: Trainer operated by the Swiss Air Force. Withdrawn from service.

T Mk 80: Trainer operated by the Zimbabwean Air Force.

UPGRADES: **Enaer:** In the late 1980s Enaer fitted Chilean Air Force single-seat Hunters with domestically developed Caiquen II radar warning receivers (RWR) and Eclipse chaff/flare dispensers. Aircraft redesignated Aguila (eagle).

OPERATORS: Versions of the Hunter are in service with the armed forces of the following countries: Chile (20), Lebanon (6), UK (2) and Zimbabwe (11).

The following description applies specifically to the Hunter F (GA) Mk 9 and T. Mk 66, but is generally applicable to all versions:

DESIGN FEATURES: Cantilever mid-wing monoplane. Hawker high-speed symmetrical wing section. Aspect ratio 3.33 (Mk 9) or 3.25 (Mk 66). Mean wing chord 3.11 m (10 ft 2½ in). Thickness/chord ratio 8.5 per cent. Anhedral 1°. Incidence 1° 30′ at root. Sweepback at quarter-chord 39° 54′. Variable incidence tailplane.

FLYING CONTROLS: All-metal ailerons, powered hydraulically by Fairey Hydro-boosters. Electrically operated trim tab in port aileron. Hydraulically operated all-metal split flaps. Airbrake in form of hinged flap conforming to curvature of underside of rear fuselage. Electrically operated (by Rotax actuator) variable incidence tailplane. Elevators actuated hydraulically through Fairey Hydro-boosters. Interconnection between elevators and tailplane actuator makes provision for operation of the units as an electrically operated 'flying tail', which can be cut out by operation of a switch in the cockpit if desired. Electrically operated trim tab in rudder.

STRUCTURE: Conventional all-metal stressed-skin structure, with extended leading-edge chord on outer wings. The fuselage is an all-metal semi-monocoque stressed-skin structure built in three sections; nose section containing cockpit, armament pack and nosewheel unit; centre-section with integral wingroot stubs, engine mounting attachments and intake ducts; and detachable, rear fuselage with integral fin base and removable jet-pipe and tailcone unit. The tail unit is a cantilever all-metal structure with sweepback on all surfaces.

LANDING GEAR: Retractable tricycle type, with single wheel on each unit. Hydraulic actuation, with emergency pneumatic system. Mainwheels retract inward into wingroots, nosewheel forward. Dowty liquid-spring shock-absorbers. Dunlop mainwheels type AH51338 (alternatively type AH50701 or AH50207 on Mk 9) and tyres size 29 × 6.25-16, pressure 14.0 kg/cm² (200 lb/sq in). Dunlop nosewheel type AH9336 and tyre size 19 × 6.25-9, pressure 8.10 kg/cm² (115 lb/sq in). Dunlop hydraulic brakes with Maxaret anti-skid units. One 3.05 m (10 ft) diameter ring-slot brake parachute in fairing over jet-pipe nozzle.

POWER PLANT: One 4,540 kg (100,000 lb) st Rolls-Royce Avon Mk 203 (Hunter Mk 66) or Mk 207 (Hunter Mk 9) turbojet engine. Four flexible bag-type tanks in fuselage and four in each wing. Capacities: front fuselage 909 litres (240 US gallons; 200 Imp gallons), rear fuselage 236 litres (62 US gallons; 52 Imp gallons). Total fuel capacity 1,781 litres (470 US gallons; 392 Imp gallons). Underwing attachments two 454 litre (132 US gallon; 100 Imp gallon) or two 1,040 litres (276 US gallon; 230 Imp gallon) jettisonable tanks. Pressure refuelling through coupling in port wheel bay.

ACCOMMODATION (Mk 9): Pressurised and air conditioned cockpit with sliding jettisonable canopy. Martin-Baker Type 2H or 3H fully automatic ejection seat.

ACCOMMODATION (Mk 66): Pressurised cockpit for two side by side with dual controls. Two Martin-Baker 4H automatic lightweight ejector seats.

ARMAMENT (Mk 9): Four 30 mm Aden guns (150 rds/gun) in self-contained removable package in underside of fuselage nose. Gun pack can be winched down for re-arming and servicing etc. Automatic gun-ranging radar in front fuselage with scanner in nose radome, and gyro gunsight. The following external stores may be carried: on the inboard wing pylons: two 454 kg (1,000 lb) bombs, two 227 kg (500 lb) bombs, two carriers each with two 11.4 kg (25 lb) practice bombs, two clusters of six 3 in rockets, two containers, each with either 24 or 37 2 in folding fin rockets, two 454 litre (120 US gallon; 100 Imp gallon) phenolic asbestos drop tanks or two 1,040 litre (276 US gallon; 230 Imp gallon) steel drop tanks; on the outboard wing pylons: two 454 litre (120 US gallon; 100 Imp gallon) drop tanks, or up to 24 3 in No. 1 Mk 5 RPs with 12 lb warheads. The 454 litre (120 US gallon; 100 Imp gallon) drop tanks may

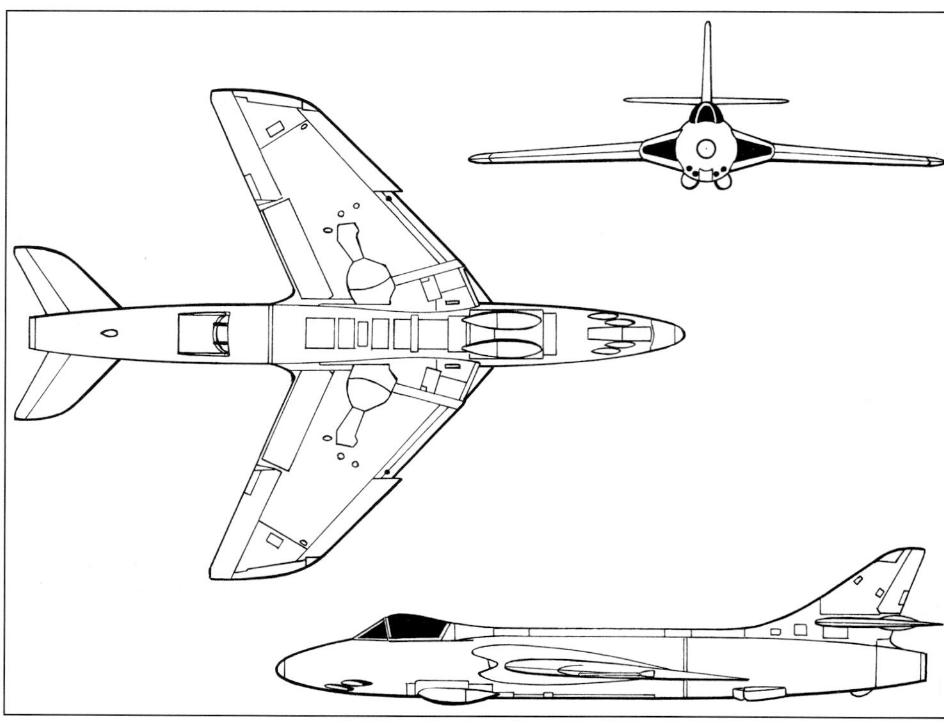

BAe (Hawker) Hunter single-seat ground attack fighter aircraft *(Jane's/Mike Keep)*

199

be used as napalm bombs. Combinations of the above loads may be carried.

ARMAMENT (Mk 66): Two 30 mm Aden guns in fairings under the nose. Two gunsights, one for each pilot. All the above-mentioned external stores may be carried.

SYSTEMS: Marston-Excelsior air conditioning and pressurisation system, supplied alternatively by engine bleed air or from ram air intake in nose. Pressure differential 0.25 kg/cm² (3.5 lb/sq in). Hydraulic system, pressure 210 kg/m² (3,000 lb/sq in), for actuating landing gear, wheel brakes, airbrake, flaps, ailerons and elevators, supplied by Dowty pump. 28 V electrical system includes two Type 517 engine-driven generators and two type J 24 V 25 Ah batteries, plus two 12 V 4 Ah batteries in series for emergency supply.

ELECTRONICS AND EQUIPMENT: Radio installation by Plessey consists of a UHF communications system and UHF standby system. A system to give the pilot audio warning of loss of hydraulic pressure is linked with UHF system. Murphy DME. Cossor IFF. Ekco radar ranging equipment.

DIMENSIONS, EXTERNAL:

Wing span	10.26 m (33 ft 8 in)
Length overall: Mk 9	13.98 m (45 ft 10½ in)
Mk 66	14.90 m (48 ft 10½ in)
Height overall	4.01 m (13 ft 2 in)
Tailplane span	3.60 m (11 ft 10 in)
Wheel track	4.50 m (14 ft 9 in)
Wheelbase	4.80 m (15 ft 9 in)

AREAS:

Wings, gross	32.43 m² (439 sq ft)
Ailerons (total)	2.46 m² (26.52 sq ft)
Flaps (total): Mk 9	2.50 m² (26.93 sq ft)
Mk 66	2.90 m² (31.20 sq ft)
Fin	2.69 m² (28.90 sq ft)
Rudder, with tabs	0.57 m² (6.10 sq ft)
Tailplane	3.49 m² (37.60 sq ft)
Elevators	1.51 m² (16.30 sq ft)

WEIGHTS AND LOADINGS:

Weight empty: Mk 9	6,020 kg (13,270 lb)
Mk 66	5,965 kg (13,150 lb)
Max military load: Mk 9	3,357 kg (7,400 lb)
Mk 66	3,220 kg (7,100 lb)
Max T-O weight: Mk 9	10,885 kg (24,000 lb)
Mk 66	10,660 kg (23,500 lb)
Max landing weight	7,600 kg (16,750 lb)

PERFORMANCE (Mks 9 and 66, with 2,988 litres (793 US gallons, 660 Imp gallons) of fuel in drop tanks):

Max level speed	Mach 0.92
Max diving speed (no Mach limit)	620 kts (1,149 km/h; 714 mph)
Max cruising speed	Mach 0.84
Econ cruising speed	399 kts (740 km/h; 460 mph)
Stalling speed	125 kts (232 km/h; 114 mph)
Rate of climb at S/L	approx 2,440 m (8,000 ft)/min
Service ceiling	15,250 m (50,000 ft)
T-O run	640 m (2,100 ft)
T-O to 15 m (50 ft)	1,050 m (3,450 ft)
Landing from 15 m (50 ft)	1,417 m (4,650 ft)
Landing run	960 m (3,150 ft)
Range, no reserves	1,600 n miles (2,965 km; 1,840 miles)

UPDATED

BAe (Hawker) Hunter FGA. Mk 58 of the Swiss Air Force equipped with AGM-65B Maverick missiles
1994

BAe (Hawker) Hunter F. Mk 58 of the Swiss Air Force *(Peter J. Cooper)*
1994

BAe (Hawker) Hunter GA. Mk 11 of the Royal Navy *(Paul Tompkins)*
1994

BAe NIMROD

TYPE: Four-turbofan maritime patrol aircraft.

PROGRAMME: The Nimrod was developed to replace the Shackleton maritime reconnaissance aircraft of RAF Strike Command, with which it is scheduled to serve until well into the 1990s. Design of the Nimrod, as the Hawker Siddeley 801, began in June 1964, and government authority to proceed was announced in June 1965.

Based substantially upon the airframe of the Hawker Siddeley (de Havilland) Comet 4C, the Nimrod was a new production aircraft with 1.98 m (6 ft 6 in) shorter, modified pressurised fuselage, underslung pannier for operational equipment and weapons; Rolls-Royce Spey engines (instead of Avon turbojets of the Comet), with wider air intakes to allow for the greater mass flow. Other external changes include enlarged flight deck main windows and 'eyebrow' windows; ESM and MAD equipment, in glassfibre fairings on top of the fin and in the tailboom respectively; and a searchlight in the starboard wing external fuel tank. The search radar is housed in a glassfibre fairing which forms the nose of the unpressurised lower fuselage.

The Nimrod was designed to combine the advantages of high altitude, fast transit speed with low wing loading and good low speed manoeuvring capabilities when operating in its primary roles of anti-submarine warfare, surveillance and anti-shipping strike. When required, two of the four Spey engines can be shut down to extend endurance, and the aircraft can cruise and climb on only one engine. A wide range of weapons can be carried in the 14.78 m (48 ft 6 in) long bomb bay, and large numbers of sonobuoys and markers can be carried and released from the pressurised rear fuselage area.

In addition to its surveillance and ASW roles, the Nimrod can be used for day and night photography. As supplied originally to the RAF, the aircraft had a standoff surface missile capability. This was subsequently deleted but was reactivated on some aircraft during the Falklands campaign in 1982. The Nimrod MR. Mk 1 could carry 16 additional personnel in the self-support role; the MR. Mk 2 can carry only 10 without the removal of equipment.

Two prototypes were built, utilising existing Comet 4C airframes. The first of these, fitted with Spey engines, flew for the first time 23 May 1967 and was used for aerodynamic testing. The second retained its original Avon engines, was flown for the first time 31 July 1967, and was used for development of the nav/tac system and special maritime equipment. Both are now in storage in interim MR. Mk 2 condition.

VERSIONS: **Nimrod MR. Mk 1**: Initial production version. First flown 28 June 1968. In total 46 delivered, with 11 being allocated for conversion to AEW. Mk 3. Remainder upgraded to Mk 2.

Nimrod R. Mk 1: Three aircraft (additional) to the 46 MR. Mk 1s ordered for RAF Strike Command were delivered to No 51 Sqn at RAF Wyton. These aircraft are employed for electronic reconnaissance and can be identified by the absence of a MAD tailboom. They are all fitted with flight refuelling probes. An upgrade programme called STARWINDOW is under way and the first modified aircraft was declared operational in early 1995. XW 666 was lost on 16 May 95 when it ditched in the Moray Firth with a double engine fire. The aircraft was on a post-major servicing airtest, so had no role equipment on board. The crew escaped with minor injuries.

Nimrod MR. Mk 2: The RAF refitted 35 MR. Mk 1s with new communications equipment and advanced tactical sensors, ESM and navigation systems, under a

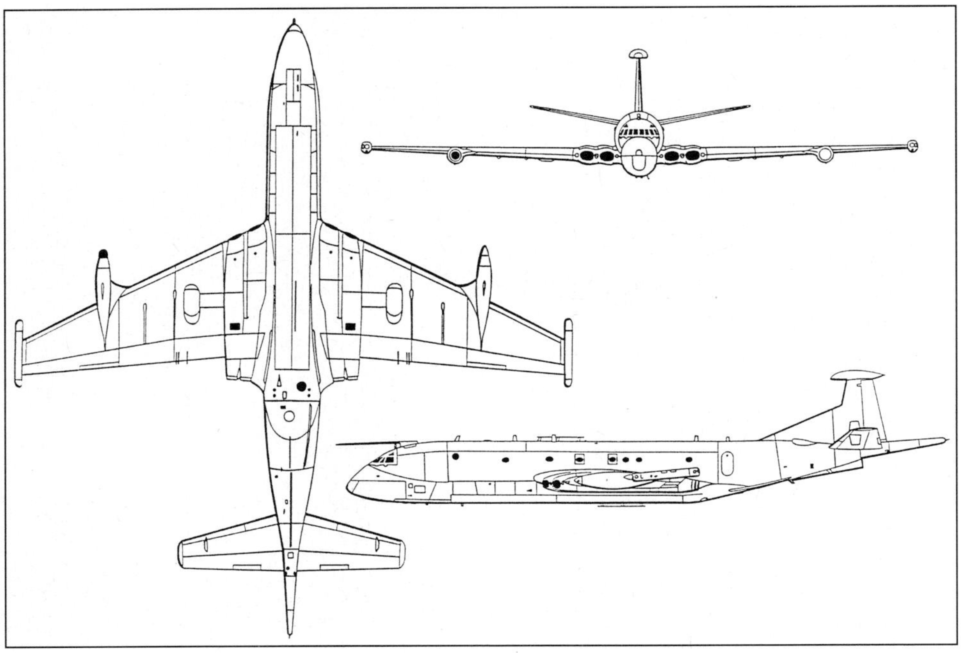

BAe Nimrod MR. Mk 2 maritime patrol aircraft ***(Jane's/Mike Keep)***

1994

BAe Nimrod R. Mk 1 reconnaissance aircraft ***(Paul Jackson)***

1994

programme which began in 1975 (one was lost due to bird strike). Redelivery started 23 August 1979. These aircraft were redesignated MR. Mk 2, and repainted in a NATO approved camouflage scheme. Equipment includes an advanced search radar, offering greater range and sensitivity coupled with a higher data processing rate; a new acoustic processing system, developed by GEC Avionics, which is compatible with a wide range of existing and projected sonobuoys, and a Loral early warning support measures (EWSM) equipment in a pod at each wingtip. Aircraft deployed to Ascension Island during Spring 1982 Falklands campaign were fitted with Sidewinder air-to-air missiles for self-defence, and were given an attack capability with bombs, Sting Ray torpedoes and, later, Harpoon missiles. Air-to-air refuelling probes were fitted at that time to 16 aircraft (redesignated MR. Mk 2P) making possible flights of up to 19 hours with one additional pilot and navigator. Provision for such probes, Sidewinder and Harpoon missile installations, have now been made on all MR. Mk 2 aircraft. Associated with these changes are an added ventral fin, small finlets above and below the tailplane on each side, and 11 vortex generators on the leading-edge of each outer wing.

Ample space and power is available in the basic Nimrod design to accept additional or alternative sensors such as sideways-looking infra-red linescan, low light level TV and digital processing of intercepted signals.

UPGRADES: **BAe:** Nimrod R. 1, MR. 2. See Versions.

BAe/Boeing: Teamed for a bid to re-life and upgrade Nimrod, as part of RAF replacement maritime patrol aircraft (RMPA) competition. Refurbished Nimrods to be equipped with new sensors, a new mission systems avionics package and upgraded weapon systems including a new defensive aids subsystems.

E-Systems: Nimrod R. 1. Awarded fixed-price contract in August 1992. Completed first of three conversions in early 1995. Upgrade package, code-named STARWINDOW, includes improved systems for digital processing, direction finding and real-time downloading.

OPERATORS: Versions of the Nimrod are in service with the UK armed forces (31).

The following description applies to the Nimrod MR. Mk 1 and 2:

DESIGN FEATURES: Cantilever low/mid-wing monoplane, of metal construction. Sweepback 20° at quarter-chord. Small finlets near leading-edge of tailplane on each side.

FLYING CONTROLS: All-metal ailerons, operated through duplicated hydraulic and mechanical units. Trim tab in each aileron. Plain flaps outboard of engines, operated hydraulically. Rudder and elevators operated through duplicated hydraulic and mechanical units. A glassfibre pod on top of the fin houses ESM equipment. Trim tab in each elevator.

STRUCTURE: The wings are an all-metal structure, comprising centre-section, two stub wings and two outer panels. Extensive use of Redux metal-to-metal bonding. The fuselage is an all-metal semi-monocoque structure. The circular section cabin space is fully pressurised. Below this is an unpressurised pannier housing the bomb bay, radome and additional space for operational equipment. Segments of this pannier are free to move relative to each other, so that structural loads in the weapons bay are not transmitted to the pressure cell. Glassfibre nose radome and tailboom. The tail unit is a cantilever all-metal structure, with large dorsal and small ventral fin.

LANDING GEAR: Retractable tricycle type. Four-wheel tandem bogie main units, with size 36 × 10-18 Dunlop tyres, pressure 12.76 bars (185 lb/sq in). Twin-wheel nose unit, with size 30 × 9-15 Dunlop tyres, pressure 6.21 bars (90 lb/sq in).

POWER PLANT: Four Rolls-Royce RB168-20 Spey Mk 250 turbofans, each rated at 54 kN (12,140 lb st). Reverse thrust fitted on two outer engines. Fuel in fuselage keel tanks, and permanent external tank on each wing leading-edge, with total capacity of 48,780 litres (12,886 US gallons; 10,730 Imp gallons). Provision for up to six removable tanks in weapons bay. Flight refuelling probe over flight deck.

ACCOMMODATION: Normal crew of 21, comprising pilot, co-pilot, and flight engineer on flight deck; routine navigator, tactical navigator, radio operator, two sonic systems operators, ESM/MAD operator, and two observers/stores loaders in main (pressurised) cabin, which is fitted out as a tactical compartment. In this compartment, from front to rear, are a toilet on the port side; stations for the two navigators (stbd), radio and radar operators (port), and sonic systems operators (stbd) in the forward section; ESM/MAD operator's station, galley, four-seat dining area, rest quarters and sonobuoy stowage in the middle section; and buoy and marker launch area in the rear section. Three hemispherical observation windows forward of wings (one port, two stbd), giving 180° field of view. Two normal doors, emergency door, and four overwing emergency exits. Weapons bay can be utilised for additional fuel tanks (see under Power Plant) or for the carriage of cargo. Provision is made for a trooping role, in which configuration 45 passengers can be accommodated if some rear fuselage equipment is removed.

SYSTEMS: Air conditioning by engine bleed air; Smith-Kollsman pressurisation system, with additional Normalair-Garrett conditioning pack on Mk 2 aircraft, maximum differential 0.603 bars (8.75 lb/sq in). Anti-icing and bomb bay heating by engine bleed air. Lockheed hydraulic system, pressure 172 bars (2,500 lb/sq in), for duplicated flying control power units, landing gear shock-absorbers, steering and door jacks, weapons bay door jacks, camera aperture door jacks, and self-sealing couplings for water charging, ground test, engine bay and ancillary services. Lucas APU provides high pressure air for engine starting. Electrical system utilises four 60 kVA engine-driven alternators, with English Electric constant-speed drives to provide 200 V 400 Hz three-phase AC supply. Secondary AC comes from two 115 V three-phase static transformers, with duplicate 115/26 V two-phase static transformers which also feed a 1 kVA frequency changer providing a 115 V 1,600 Hz single-phase supply for radar equipment. Emergency supplies for flight instruments are provided by a 115 V single-phase static inverter. DC supply is by four 28 V transformer-rectifier units, backed up by two Ni/Cd batteries. Hot air anti-icing system for wings. Hot air anti-icing system for tail unit.

AVIONICS AND EQUIPMENT (MR. Mk 1): Routine navigation by Decca Doppler Type 67M/GEC Avionics E3 heading reference system, with reversionary heading from a Sperry gyro compass system, operating in conjunction with a Ferranti routine dynamic display. Tactical navigation, and stores selection and release, by GEC Avionics nav/attack system utilising an 8K GEC Avionics 920B digital computer. Tactical display station provided by updated information about aircraft position with present and past track, sonobuoy positions, range circles from sonobuoys, ESM bearings, MAD marks, radar contacts and visual bearings. Course information can be displayed automatically to the pilots on the flight director system; alternatively, the computer can be coupled to the autopilot to allow the tactical navigator to direct the aircraft to predicted target interception, weapon release point, or any other point on the tactical display. ASW equipment includes Sonic 1 CASV-21D surface vessel detection radar in nose; Thomson-CSF ESM (electronic support measures) equipment in pod on top of fin; and Emerson Electronics ASQ-10A MAD (magnetic anomaly detector) in extended tailboom. Strong Electric 70 million candlepower searchlight at front of starboard external wing fuel tank. Aeronautical and General Instruments F.126 and F.135 cameras for day and night photography respectively, the latter having Chicago Aero Industries electronic flash equipment. Smith SFS.6 automatic flight control system, embodying SEP.6 three-axis autopilot, integrated with the navigation and tactical system. Twin Plessey PTR 175 UHF/VHF, and GEC Avionics AD 470 HF, communications transceivers; twin GEC Avionics AD 260 VOR/ILS; GEC Avionics AD 2770 Tacan; Decca Loran C/A; GEC Avionics AD 360 ADF; Honeywell AN/APN-17(V) radar altimeter. Yaw damper and Mach trim standard.

AVIONICS AND EQUIPMENT (MR. Mk 2): New and more flexible operational system, using three-spar separate processors for tactical system, based on a 920 ATC computer with a greater storage capacity than that of MR. Mk 1, to provide improved computing and display facilities and, in conjunction with a Ferranti inertial navigation system, improved navigation capabilities. THORN EMI Searchwater long-range surface vessel detection radar, with its own data processing subsystem incorporating a Ferranti FM 1600D digital computer. This system presents a clutter-free picture, can detect and classify surface vessels, submarine snorts and periscopes at extreme ranges, can track several targets simultaneously and is designed to operate in spite of countermeasures. AQS 901 acoustics processing and display system, based on twin GEC Avionics 920 computers, is compatible with a wide range of passive and active sonobuoys, either in existence or under development, including the Australian BARRA passive directional sonobuoy, the Canadian TANDEM, the US SSQ-41 and SSQ-53, and the Ultra A size X17255 command active multibeam sonobuoy (CAMBS), with a performance similar to that of helicopter dipping sonars. Communication improved by the installation of twin GEC Avionics AD 470 HF transceivers (instead of the original single AD 470), and a radio teletype and encryption system. Loral EWSM equipment in two wingtip pods. Onboard crew training system developed by the Maritime Aircraft Systems Division of GEC Avionics Ltd. Known as ACT-1 (Airborne Crew Trainer Mk 1), it consists of a single exercise control unit comprising a control and display panel with push-buttons and a reel of magnetic tape containing the software programme, by means of which the AQS 901 processing and display system can operate in a training mode. Using the ACT-1, which physically resembles a TV game, one crew member can play the part of a submarine, trying to outwit his colleague operating the AQS 901 detection system. Although not a replacement for ground-based simulator training, the

ACT-1 onboard system enables a Nimrod captain to train his crew in authentic operational conditions, without the expenditure of sonobuoys.

ARMAMENT (MR. Mk 2): 14.78 m (48 ft 6 in) long weapons bay, with two pairs of doors, in unpressurised lower fuselage pannier, able to carry up to six lateral rows of ASW weapons, including up to nine torpedoes as well as bombs. Alternatively, up to six auxiliary fuel tanks can be fitted in the weapons bay, or a combination of fuel tanks and weapons can be carried. To ensure weapon serviceability, the weapons bay is heated when the ambient temperature falls below +5°C. Bay approximately 9.14 m (30 ft 0 in) long in rear pressurised part of the fuselage for storing and launching of active and passive sonobuoys and marine markers. Two rotary launchers, each capable of holding six size A sonobuoys, are used when the cabin is unpressurised; two single-barrel launchers are used when the aircraft is pressurised. A hardpoint is provided beneath each wing, just outboard of the mainwheel doors, on which can be carried two AIM-9 Sidewinder air-to-air missiles, a Harpoon air-to-surface missile, rocket or cannon pod, or mine, according to mission requirements.

DIMENSIONS, EXTERNAL:

Wing span	35.00 m (114 ft 10 in)
Wing chord: at root	9.00 m (29 ft 6 in)
at tip	2.06 m (6 ft 9 in)
Wing aspect ratio	6.2
Length overall: MR. Mk 2	39.35 m (129 ft 1 in)
R. Mk 1, incl probe	36.60 m (120 ft 1 in)
Height overall	9.08 m (29 ft 8½ in)
Tailplane span	14.51 m (47 ft 7¼ in)
Wheel track	8.60 m (28 ft 2½ in)
Wheelbase	14.24 m (46 ft 8½ in)

DIMENSIONS, INTERNAL:

Cabin (incl flight deck, navigation and ordnance areas, galley and toilet): Length	26.82 m (88 ft 0 in)
Max width	2.95 m (9 ft 8 in)
Max height	2.08 m (6 ft 10 in)
Volume	124.14 m³ (4,384 cu ft)

AREAS:

Wings, gross	197.0 m² (2,121 sq ft)
Ailerons (total)	5.63 m² (60.6 sq ft)
Trailing-edge flaps (total)	23.37 m² (251.6 sq ft)
Fin and rudder (above tailplane centreline)	10.96 m² (118 sq ft)
Dorsal fin	5.67 m² (61 sq ft)
Tailplane	40.41 m² (435 sq ft)
Elevators (incl tabs)	12.57 m² (135.3 sq ft)

WEIGHTS AND LOADINGS (MR. Mk 2):

Typical weight empty	39,000 kg (86,000 lb)
Max disposable load	6,120 kg (13,500 lb)
Fuel load: standard tanks	38,940 kg (85,840 lb)
max with six auxiliary tanks in weapons bay	45,785 kg (100,940 lb)
Normal max T-O weight	80,510 kg (177,500 lb)
Max overload T-O weight	87,090 kg (192,000 lb)
Typical landing weight	54,430 kg (120,000 lb)

PERFORMANCE (MR. Mk 1):

Max operational necessity speed, ISA + 20°C	500 kts (926 km/h; 575 mph)
Max transit speed, ISA + 20°C	475 kts (880 km/h; 547 mph)
Econ transit speed, ISA + 20°C	425 kts (787 km/h; 490 mph)
Typical low-level patrol speed (two engines)	200 kts (370 km/h; 230 mph)
Operating height range	S/L to 12,800 m (42,000 ft)
Min ground turning radius	27.1 m (89 ft 0 in)
Runway LCN at T-O weight of 82,550 kg (182,000 lb)	50
T-O run at 80,510 kg (177,500 lb) AUW, ISA at S/L	1,463 m (4,800 ft)
Unfactored landing distance at 54,430 kg (120,000 lb): Landing weight, ISA at S/L	1,615 m (5,300 ft)
Typical ferry range	4,500-5,000 n miles (8,340-9,265 km; 5,180-5,755 miles)
Typical endurance	12 h

UPDATED

BAe Nimrod MR. Mk 2 upgraded maritime patrol aircraft *(Paul Tompkins)*

1994

BAe Nimrod MR. Mk 2 four-turbofan maritime patrol aircraft *(BAe)*

1995

BAe (VICKERS) VC2 VISCOUNT

TYPE: Four-engined turboprop airliner.

PROGRAMME: Design of the Viscount was started in 1945 and construction of the prototype, known as the Vickers Type 630 began in March 1946. The prototype flew for the first time 16 July 1948, and was followed by the larger Type 700 prototype on 19 April 1950. The initial production model in the 700 Series was the Viscount 701 with which BEA opened the World's first regular turboprop airline services in April 1953.

When production of the Viscount ended in March 1964, a total of 444 had been built, made up of the 630 prototype, a similar airframe used as a flying testbed for two Rolls-Royce Tay turbojets, 288 Series 700 Viscounts, 68 of the stretched Series 800 and 86 Series 810 Viscounts with more powerful engines.

UPGRADES: **BWA**: Life extension. See separate entry.

DESIGN FEATURES: Cantilever low-wing monoplane. Wing section NACA 63 Series (modified). Aspect ratio 9.17. Wing chord 4.54 m (14 ft 10 in) at root, 1.35 m (4 ft 5 in) at tip. Dihedral 3°. Incidence 2° 30′. Cantilever tail unit with 14° dihedral on tailplane.

FLYING CONTROLS: All-metal Westland-Irving ailerons. Electrically operated all-metal double-slotted flaps in three sections on each wing. Spring tabs on all tail unit control surfaces.

STRUCTURE: Single-spar all-metal structure, with two subsidiary spanwise members and stressed alclad skin. The fuselage is an all-metal stressed-skin semi-monocoque structure of circular cross-section. Entire fuselage pressurised, except for nosewheel retraction bay and tailcone. The tail unit is a cantilever all-metal structure.

LANDING GEAR: Retractable tricycle type with twin-wheels on each unit. All units retract forward hydraulically. Steerable nosewheel. Vickers oleo-pneumatic shock-absorbers on all three units. Dunlop or Goodyear wheels, tyres and hydraulic disc brakes, with Maxaret anti-skid units. Mainwheel tyres size 36 × 10.75-16.5, pressure 9.21 kg/cm² (131 lb/sq in). Nosewheel tyres size 24 × 7.25-12, pressure 6.20 bars (90 lb/sq in).

POWER PLANT: Four 1,990 ehp Rolls-Royce Dart RDa. 7/1 Mk 525 turboprop engines, each driving a four-blade Rotol or DH fully feathering propeller, diameter 3.05 m (10 ft 0 in). Fuel in four interconnected bag-type tanks in wings with total capacity of 8,637 litres (2,282 US gallons; 1,900 Imp gallons). Provision for one slipper tank on leading-edge of each wing outboard of outer engines, each with capacity of 659 litres (174 US gallons; 145 Imp gallons). Two-point pressure refuelling. Provision for gravity refuelling. Oil capacity 56.8 litres (15 US gallons; 12.5 Imp gallons).

ACCOMMODATION: Two pilots on flight deck, plus cabin attendants. Up to 52 first class or 70 tourist passengers in a variety of layouts, all with galleys and toilets. Passenger doors at front and rear of cabin on port side, forward door with optional airstairs. Servicing door at rear on starboard side. Two freight holds under floor forward of wing, with doors on starboard side, and one at rear of cabin. Entire accommodation is air conditioned.

SYSTEMS: Air conditioning and pressurisation by three Roots engine-driven superchargers. Pressure differential 0.45 kg/cm² (6.5 lb/sq in). Freon refrigerating unit optional. Hydraulic system, pressure 172.3 bars (2,500 lb/sq in), for landing gear, nosewheel steering and brakes. Four 9 kW engine-driven generators supply 27.5 V electrical power. Thermal de-icing of wing leading-edges. Thermal de-icing of fin and tailplane leading-edges.

ELECTRONICS AND EQUIPMENT: VHF and HF communications and navigation equipment by Bendix, Collins or Remler, to customer's requirements. Blind-flying instrumentation standard. Weather radar optional.

DIMENSIONS, EXTERNAL:

Wing span	28.56 m (93 ft 8½ in)
Length overall	26.11 m (85 ft 8 in)

BAe (Vickers) VC2 Viscount turboprop airliner

1994

Height over tail	8.16 m (26 ft 9 in)
Tailplane span	10.92 m (35 ft 9¾ in)
Wheel track (centreline oleos)	7.26 m (23 ft 10 in)
Wheelbase	9.00 m (29 ft 6 in)
Passenger door (fwd, port): Height	1.63 m (5 ft 4 in)
Width	0.91 m (3 ft 0 in)
Height to sill	2.29 m (7 ft 6 in)
Baggage door (fwd, stbd): Height	0.76 m (2 ft 6 in)
Width	1.07 m (3 ft 6 in)
Height to sill	1.40 m (4 ft 7 in)
Baggage door (rear, stbd): Height	0.76 m (2 ft 6 in)
Width	1.07 m (3 ft 6 in)
Height to sill	1.32 m (4 ft 4 in)

DIMENSIONS, INTERNAL:

Cabin, excl flight deck: Length	17.37 m (57 ft 0 in)
Width at arms-rest height	3.05 m (10 ft 0 in)
Max height	1.96 m (6 ft 5 in)
Volume (excl rear freight compartment)	75.89 m³ (2,680 cu ft)
Freight holds (underfloor) total	7.08 m³ (250 cu ft)
Freight hold (aft of cabin)	3.40 m³ (120 cu ft)

AREAS:

Wings, gross	89.47 m² (963 sq ft)
Ailerons (total)	4.40 m² (47.4 sq ft)
Trailing-edge flaps (total)	14.26 m² (153.5 sq ft)
Fin	8.92 m² (96.0 sq ft)
Rudder, incl tab	5.76 m² (62.0 sq ft)
Tailplane	12.49 m² (134.5 sq ft)
Elevators, incl tab	9.62 m² (103 sq ft)

WEIGHTS AND LOADINGS:

Weight empty, equipped	18,753 kg (41,565 lb)
Typical max payload	6,577 kg (14,500 lb)
Max T-O weight	32,886 kg (72,500 lb)
Max zero-fuel weight	26,082 kg (57,500 lb)
Max landing weight	29,030 kg (64,000 lb)
Max wing loading	368.1 kg/m² (75.4 lb/sq ft)
Max power loading	4.13 kg/ehp (9.11 lb/ehp)

PERFORMANCE (at max T-O weight):

Max recommended cruising speed at 6,100 m (20,000 ft)	310 kts (575 km/h; 357 mph)
Econ cruising speed at 5,485 m (18,000 ft)	305 kts (566 km/h; 352 mph)
Approach speed at max landing weight	137 kts (225 km/h; 140 mph)
Service ceiling	7,620 m (25,000 ft)
Balanced T-O field length	1,807 m (5,930 ft)
Landing from 15 m (50 ft) at max landing weight	1,355 m (4,450 ft)
Range with max fuel and 6,485 kg (14,300 lb) payload, no reserves	1,528 n miles (2,830 km; 1,760 miles)
Range with max payload, no reserves	1,498 n miles (2,775 km; 1,725 miles)

VERIFIED

BAe (BAC) VC10 SERIES 1100

RAF designations: VC10 C. Mk 1/1K and VC10 K. Mk 2/3/4

TYPE: Four-engined jet airliner.

PROGRAMME: Design work for the VC10 began March 1958 and construction of the prototype was started January 1959. The prototype flew for the first time on 29 June 1962. ARB Certification was given on 23 April 1964, and the VC10 entered service on BOAC's route to West Africa 29 April 1964.

VERSIONS: **C. Mk 1:** Transport version in service with the RAF. All being converted to C. Mk 1(K) tanker transport aircraft.

C. Mk 1(K): New designation for the 13 aircraft undergoing conversion by FR Aviation. See following entry and separate entry for FR Aviation.

K. Mk 2: Tanker conversion. Five in service with the RAF. See following entry.

K. Mk 3: Tanker conversion. Four in service with the RAF. See following entry.

K. Mk 4: Tanker conversion. Four in service with the RAF. See following entry.

UPGRADES: **BAe.** VC10 K. Mk 2/3/4. See separate entry.

FR Aviation: C. Mk 1 to C. Mk 1(K) conversion. See separate entry.

RAF: Six aircraft fitted with GEC-Marconi Defence Systems E- through J-band ARI 18223/17 Sky Guardian 200 RWRs during Operation Granby. Three additional aircraft being fitted with the system.

OPERATORS: Versions of the VC-10 are in service with the UK armed forces (27).

The following description applies generally to all airline models of the 1100 Series:

DESIGN FEATURES: Cantilever low-wing monoplane. High-lift wing section. Aspect ratio 7.29. Chord 10.23 m (33 ft 6⅔ in) at root. Wing thickness/chord ratio: Model 1101 13 per cent at root to 10 per cent at station 179: 9.75 per cent contant from 179 to tip: other versions have 4 per cent leading-edge chord extension from root to station 476. Mean dihedral 3°. Incidence 4° at root. Sweepback at quarter-chord 32° 30′. One boundary layer fence on each wing. Tailplane mounted on top of fin.

FLYING CONTROLS: Aluminium alloy ailerons, with honeycomb core trailing-edge. Each aileron divided in two sections, with individual Boulton Paul electrohydraulic power control unit. No trim tabs. Three Redux-bonded laminated light alloy spoilers in top surface of each wing, with individual hydraulic actuation, work in conjunction with ailerons to provide differential rolling power. Spoilers also used as airbrakes. Hydraulically actuated light alloy Fowler flaps, in five sections on each wing. Slats, in four sections on each wing, extending over major part of leading-edge and slats. Hydraulically actuated variable incidence tailplane. Electrohydraulic power-operated elevators, each in two sections with individual Boulton Paul power control units. Rudder in three sections, each with individual Boulton Paul power control unit. No trim tabs.

STRUCTURE: All-metal structure, with four spanwise members in inner wings, reducing to two outboard, and wing skins integrally machined from light alloy billets. Stress levels low throughout structure. Extensive use of fail-safe techniques. The fuselage is an oval light alloy semi-monocoque fail-safe structure, employing conventional frames and stringers, and making extensive use of integrally machined panels. The tail unit is a cantilever all-metal structure.

LANDING GEAR: Hydraulically retractable tricycle type. Main units are inward-retracting four-wheel bogies. Twin-wheels on forward-retracting steerable nose unit. Vickers oleo-pneumatic shock absorbers. Dunlop mainwheels and tubeless tyres size 50 × 18. Dunlop nosewheels and tubeless tyres size 39 × 13. Dunlop Hydraulic disc brakes and Maxaret anti-skid units. Provision for direct braking.

POWER PLANT: Four Rolls-Royce Conway RCo. 42 turbofan engines (each 9,525 kg; 21,000 lb st) in lateral pairs on each side of rear fuselage. No silencers. Reversers on outboard engines only. Fuel in integral tanks in wings and transfer tanks in centre-section and outer wings, total capacity 81,486 litres (21,527 US gallons; 17,925 Imp gallons). Refuelling point in rear inboard edge of each wing. Oil capacity 22 litres (4.8 Imp gallons; 5.7 US gallons).

ACCOMMODATION: Basic flight crew of three: captain, co-pilot and flight engineer. Provision for navigator and supernumerary. Standard six-abreast economy class seating for 135 passengers. Provision for up to 151 passengers. Four galleys in 135-seat layout, two forward and two aft. Three toilets at rear of cabin and two forward. Two passenger doors forward of wing on port side. Two underfloor freight holds, with doors at front of fuselage on starboard side, at rear of fuselage on port side and under rear fuselage. Two service doors forward and aft of wing on starboard side. Entire accommodation pressurised and air conditioned.

SYSTEMS: Two independent Normalair air conditioning and pressurisation systems with vapour-cycle cooling packs and four Godfrey engine-driven compressors. Pressure differential 0.62 bars (9 lb/sq in). Two independent hydraulic systems, pressure 206 bars (3,000 lb/sq in). No pneumatic system. 200/115 V electrical system supplied by four Westinghouse 40 kVA alternators. Emergency electric ram-air turbine. No APU in commercial models.

ELECTRONICS AND EQUIPMENT: Fully duplicated HF and VHF communications and navigation equipment by Marconi and Bendix. Ekco radar. Elliot-Bendix auto-flare with space provision for ultimate incorporation of automatic landing equipment.

DIMENSIONS, EXTERNAL:

Wing span	44.55 m (146 ft 2 in)
Length overall	48.36 m (158 ft 8 in)
Length of fuselage	40.74 m (133 ft 8 in)
Height over tail	12.04 m (39 ft 6 in)
Tailplane span	13.36 m (43 ft 10 in)
Wheel track	6.53 m (21 ft 5 in)
Wheelbase	20.08 m (65 ft 10½ in)
Passenger doors (each): Height	1.83 m (6 ft 0 in)
Width	0.86 m (2 ft 10 in)
Height to sill: fwd	3.05 m (10 ft 0 in)
rear	3.25 m (10 ft 8 in)
Cargo door (Models 1102, 1103): Height	2.13 m (7 ft 0 in)
Width	3.55 m (11 ft 8 in)
Freight hold door (fwd, stbd): Height	1.22 m (4 ft 0 in)
Width	1.52 m (5 ft 0 in)
Height to sill	1.75 m (5 ft 9 in)
Freight hold door (rear, port): Height	1.12 m (3 ft 8 in)
Width	1.37 m (4 ft 6 in)
Height to sill	2.29 m (7 ft 6 in)
Freight hold door (ventral, rear): Length	0.76 m (2 ft 6 in)
Width	0.76 m (2 ft 6 in)
Service doors (each): Height	1.52 m (5 ft 0 in)
Width	0.61 m (2 ft 0 in)
Height to sill: fwd	3.05 m (10 ft 0 in)
rear	3.50 m (11 ft 6 in)

DIMENSIONS, INTERNAL:

Cabin, excl flight deck: Length	28.14 m (92 ft 4 in)
Max width	3.50 m (11 ft 6 in)
Max height	2.26 m (7 ft 5 in)
Floor area	92.9 m² (1,000 sq ft)
Volume	189.7 m³ (6,700 cu ft)
Freight hold (fwd)	18.55 m³ (655 cu ft)
Freight hold (rear)	23.65 m³ (835 cu ft)

AREAS:

Wings, gross: Except Model 1101	272.4 m² (2,932 sq ft)
Model 1101	264.9 m² (2,851 sq ft)
Ailerons (total)	13.57 m² (146 sq ft)
Trailing-edge flaps (total)	47.20 m² (508 sq ft)
Leading-edge slats (total)	24.15 m² (260 sq ft)
Spoilers (total)	9.75 m² (105 sq ft)
Vertical tail surfaces (gross)	44.22 m² (476 sq ft)
Rudder	9.29 m² (100 sq ft)

BAe (BAC) VC10 C.1 of the Royal Air Force, being refuelled by a VC10 Tanker *(SAC Paul Bolland)*

1994

Horizontal tail surfaces (gross) 59.27 m² (638 sq ft)
Elevators 13.57 m² (146 sq ft)

WEIGHTS AND LOADINGS (airline models):
Basic operating weight 86,668 kg (146,979 lb)
Max payload 18,039 kg (39,769 lb)
Max T-O weight 141,520 kg (312,000 lb)
Max ramp weight 142,425 kg (314,000 lb)
Max zero-fuel weight 85,000 kg (187,400 lb)
Max landing weight 97,975 kg (216,000 lb)
Max wing loading 537 kg/m² (110 lb/sq ft)
Max power loading 3.7 kg/kg st (3.7 lb/lb st)

PERFORMANCE (airline models, at max T-O weight):
Max level speed Mach 0.86
Max permissible diving speed Mach 0.94
Max cruising speed 493 kts (914 km/h; 568 mph)
Econ cruising speed at 11,600 m (38,000 ft) 478 kts (886 km/h; 550 mph)
Rate of climb at S/L 585 m (1,920 ft)/min
Service ceiling 12,800 m (42,000 ft)
T-O to 10.7 m (35 ft) 2,525 m (8,280 ft)
Landing from 15 m (50 ft) 1,945 m (6,380 ft)
Range with max fuel, no allowances 5,273 n miles (9,765 km; 6,070 miles)
Range with max payload, no allowances 4,381 n miles (8,115 km; 5,040 miles)

PERFORMANCE (military model 1106):
As for Super VC10 Srs 1150, except:
Stalling speed (max T-O weight):
Flaps down, slats out 120 kts (222 km/h; 138 mph) EAS
Flaps up, slats in 166 kts (308 km/h; 191 mph) EAS
Stalling speed (max landing weight):
Flaps down, slats out 98 kts (182 km/h; 113 mph) EAS
Flaps up, slats in 138 kts (256 km/h; 159 mph) EAS
Rate of climb at S/L 930 m (3,050 ft)/min

UPDATED

BAe (BAC) VC10 C.1 of the Royal Air Force *(SAC Paul Bolland)*

1994

BAe (BAC/VICKERS) VC10 C. Mk 1(K), K. Mk 2, 3 and 4 TANKERS

TYPE: Tanker upgrade.

PROGRAMME: In addition to original VC10 C. Mk 1 transports (delivered 1966-68 and used by No. 10 Squadron), RAF's No. 101 Squadron operates tanker conversions of former airliners as five **VC10 K. Mk 2s** (delivered from July 1983) and four **Super VC10 K. Mk 3s** (delivered from February 1985).

Modification at BAe Filton involved installation of internal Flight Refuelling Mk 17B hose and drum unit (HDU), plus wing-mounted FR. Mk 32/2800 Pods.

BAe (Regional Aircraft) Ltd at Manchester awarded two-part contract 1990 to modify 13 aircraft to tankers; first part (Air Staff Requirement 415) covers five unconverted Super VC10s stored at Abingdon to become **K. Mk 4s** with Mk 17B HDU in fuselage and two underwing Mk 32s; also included are general refurbishment, and installation of military avionics (including air-to-air Tacan) and closed-circuit TV; JTIDS terminals will be installed at later stage for relay of information between ground stations and AEW and interceptor aircraft. Work performed at Filton after aircraft returned to airworthy condition at Abingdon; ZD242 to Filton 27 July 1990, to launch conversion programme; service in 1994. Second part of programme (ASR 416), eight No. 10 Squadron transports to be converted into **C. Mk 1(K)s** by FR Aviation at Bournemouth, each having secondary tanker capability with two Mk 32 HDUs underwing (plus Tacan and TV); first aircraft redelivered December 1992; option held on similar rework of five remaining Mk 1s; neither Mk 1(K) nor Mk 4 conversions involve additional fuel tanks. Rework of option aircraft in progress.

Details of modifications include:

WEIGHTS AND LOADINGS:
Max T-O weight: K. Mk 2 142,000 kg (313,056 lb)
K. Mk 3 151,900 kg (334,882 lb)
K. Mk 4 151,900 kg (334,882 lb)
Theoretical max fuel weight*:
K. Mk 2 78,170 kg (172,335 lb)
K. Mk 3 83,420 kg (183,909 lb)
Practical max fuel weight:
K. Mk 2 74,000 kg (163,142 lb)
K. Mk 3 80,000 kg (176,370 lb)
K. Mk 4 70,200 kg (154,780 lb)

* *Max T-O weight exceeded when all tanks full*

UPDATED

BAe (BAC) VC10 K. Mk 4 tanker upgrade

1994

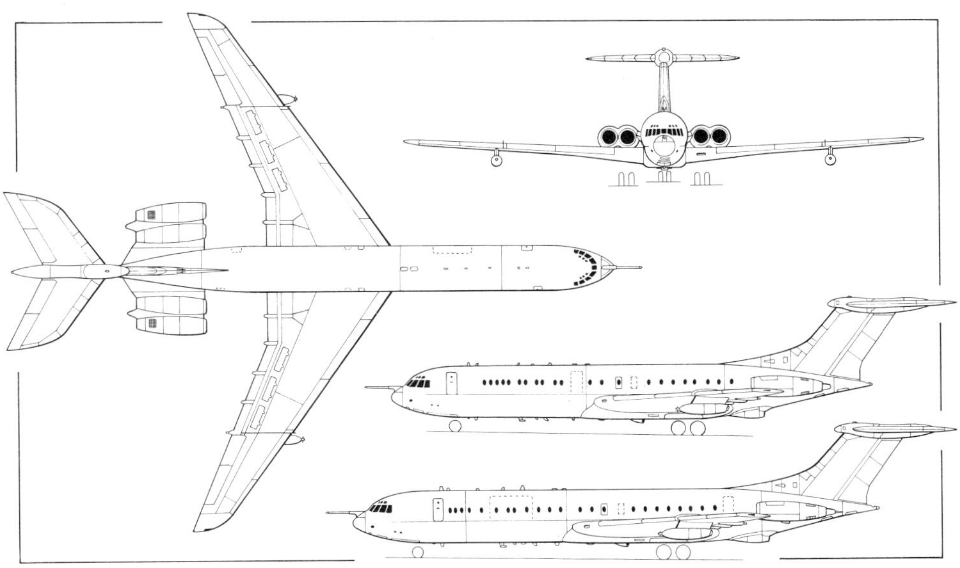

BAe (BAC) VC10 K. Mk 3 and additional view of K. Mk 2 (dotted lines of sealed 'inoperative doors' and escapes) *(Jane's/Mike Keep)*

1994

BAe (ENGLISH ELECTRIC) CANBERRA

TYPE: Long-range multipurpose aircraft.

PROGRAMME: Designed by the former English Electric Company to specification B.3/45, the Canberra remained in production for over 10 years. The first prototype (VN799) flew for the first time on 13 May 1949. Deliveries of the production Canberras to the Royal Air Force began in January 1951 and versions were ordered subsequently by many Commonwealth and foreign air forces. The Canberra is still in service with the air forces of Argentina (5+), India (40+), and Peru, as well as the Royal Air Force and Navy.

Excluding an unknown number of PR. Mk 9 aircraft, in total 1,352 Canberras were built; 901 in the UK by English Electric, Short Brothers & Harland, A. V. Roe and Handley Page; 48 were built under licence in Australia by the Department of Defence Production; and 403 under licence in the USA (under the US designation B-57) by the Martin company.

In all 27 Mks of Canberra, including seven USAF B-57 versions were produced, some of them specifically for overseas customers. UK production included four B. Mk 1 prototypes, 415 B. Mk 2 (two prototypes, 407 MoA, six overseas), 37 PR. Mk 3 (two prototypes, 35 MoA), 74 T. Mk 4 (one prototype, 63 MoA, 10 overseas), one prototype M. Mk 5, 103 B. Mk 6 (one prototype, 90 MoA, 12 overseas), 22 B(I) Mk 6 and 75 PR Mk 7 (one prototype, 74 MoA). The principal UK Mks were as follows:

VERSIONS: **B(I) Mk 8:** Two-seat long-range night intruder in service with the air force of Peru.

B(I) Mk 56: In service with the Air Force of Peru.

B(I) Mk 58: Modified version of the B(I). In service with the Indian Air Force.

B(2): ECM version in service with the Royal Air Force.

E. Mk 15: Radar and calibration version in service with the Royal Air Force.

PR. Mk 9: Photo-reconnaissance version in service with the Royal Air Force.

PR. Mk 57: Modified version of the PR. Mk 7 photo-

BAe (English Electric) Canberra PR. 9 photographic reconnaissance aircraft of the RAF *(BAe)*

1995

reconnaissance version in service with the Indian Air Force.

T. Mk 4: One in service with the Royal Air Force.

T. Mk 54: Trainer version. In service with the air forces of India and Peru.

UPGRADES: **FR Aviation:** Relife programme. See separate entry.

OPERATORS: Versions of the Canberra are in service with the armed forces of the following countries: Argentina (6), Chile (2), India (46), Peru (22) and UK (9).

DESIGN FEATURES: Cantilever mid-wing monoplane. RAE/D symmetrical high-speed wing section. Wing aspect ratio 5.79 m (19 ft 0 in) at root, 2.34 m (7 ft 8 in) at tip. Wing thickness/chord ratio 12 per cent at root, 9 per cent at tip. Dihedral 2° on centre-section, 4° 21′ on outer wings. Incidence 2°. Variable incidence dihedral (7° 57′) tailplane.

FLYING CONTROLS: Aluminium alloy Irving-Westland pressure-balanced ailerons with spring tab in each. Four hydraulically operated split trailing-edge flaps. Airbrakes consist of drag channels which can be extended from top and bottom wing surfaces aft of spar. Tailplane hinged at leading-edge and operated by English Electric actuator in rear fuselage. Spring tab in port elevator. Mass-balanced rudder with spring tab.

STRUCTURE: Aluminium alloy single-spar structure comprising a single-cell torsion box structure inboard of engines and two-cell torsion box outboard. The fuselage is an all-metal semi-monocoque stressed structure of circular section, built in nose, centre and rear portions. The tail unit is a cantilever all-metal structure, except forward portion of fin which is of wood construction, with plywood covering.

LANDING GEAR: Hydraulically retractable tricycle type. Main units retract inward, nosewheel rearward. Each main unit, of English Electric design, is an oleo-pneumatic shock-absorber strut carrying a single wheel inboard. Dowty levered-suspension liquid-spring fully castoring nose unit with twin-wheels. Dunlop AH 51337 21 in mainwheels and tyres, pressure 5.51-11.01 bars (80-160 lb/sq in) depending on AUW. Dunlop AH 9590 nosewheels and tyres, pressure 6.89 bars (100 lb/sq in). Dunlop hydraulic disc brakes with Maxaret anti-skid units.

POWER PLANT: Two Rolls-Royce Avon 109 turbojet engines (each 3,357 kg; 7,400 lb st). Fuel in three main tanks in upper part of fuselage above bomb-bay and two integral tanks in wings. Forward and centre-fuselage tanks are of internally braced self-sealing type, capacity 2,364 litres (624 US gallons; 520 Imp gallons) and 1,441 litres (380 US gallons; 317 Imp gallons) respectively. Rear lace-supported bag tank capacity 2,455 litres (648 US gallons; 540 Imp gallons). Integral wing tanks each 2,046 litres (540 US gallons; 450 Imp gallons). Total internal fuel capacity 10,351 litres (2,734 US gallons; 2,277 Imp gallons). Oil capacity (total) 15.5 litres (4.1 US gallons; 3.4 Imp gallons). Provision for auxiliary wingtip tanks, each with capacity of 1,109 litres (293 US gallons; 244 Imp gallons). Refuelling points above centre-fuselage on port side and on top of each wing.

ACCOMMODATION: Pilot on Martin-Baker Type 2CB ejection seat under fighter-type canopy offset to port side of fuselage. Navigator's seat totally enclosed in nose. Door on starboard side of nose. Cabin air conditioned.

SYSTEMS: English Electric pressurisation and air conditioning system, using engine-bleed air. Hydraulic system, pressure 190 kg/cm² (2,700 lb/sq in), for landing gear retraction, brakes, airbrakes, flaps and bomb-bay doors. No pneumatic system. Two 28 V 9 kW Type 519 engine driven generators and four 12 V 40 Ah batteries.

ARMAMENT: In bomber role can carry six 1,000 lb bombs or one 4,000 lb and two 1,000 lb bombs, or eight 500 lb bombs internally, plus up to 2,000 lb of stores on underwing pylons. In interdictor role, a pack of four Hispano cannon is installed in rear of weapons bay, leaving room in forward part for 16 4.5 in flares or three 1,000 lb bombs. Equipped to carry Nord AS.30 air-to-surface missiles during 1963. Nuclear weapons can be carried.

DIMENSIONS, EXTERNAL (Mks 8 and 9):

Wing span: Mk 8	19.51 m (64 ft 0 in)
Mk 9	20.67 m (67 ft 10 in)
Span over tip tanks: Mk 8	19.96 m (65 ft 6 in)
Length overall: Mk 8	19.96 m (65 ft 6 in)
Mk 9	20.32 m (66 ft 8 in)
Height overall	4.77 m (15 ft 8 in)
Tailplane span	8.36 m (27 ft 5 in)
Wheel track: Mk 8	4.80 m (15 ft 9 in)
Mk 9	4.82 m (15 ft 9¾ in)
Wheelbase	4.64 m (15 ft 2¾ in)

AREAS:

Wings, gross: Mk 8	89.19 m² (960 sq ft)
Mk 9	97.08 m² (1,045 sq ft)
Ailerons (total)	6.69 m² (72 sq ft)
Trailing-edge flaps (total): Mk 8	5.96 m² (64.20 sq ft)
Mk 9	6.06 m² (65.20 sq ft)
Fin	3.34 m² (35.91 sq ft)
Rudder, incl tab	2.85 m² (30.62 sq ft)
Tailplane	11.95 m² (128.60 sq ft)
Elevators, incl tab	5.78 m² (62.20 sq ft)

WEIGHTS AND LOADINGS (Mk 8):

Basic operating weight (interdictor)	12,678 kg (27,950 lb)
Max T-O weight	24,925 kg (54,950 lb)
Max zero-fuel weight	15,050 kg (33,180 lb)
Max landing weight	18,145 kg (40,000 lb)

PERFORMANCE:

Max speed at S/L	Mach 0.68 (446 kts; 827 km/h; 517 mph)
Max speed at 12,200 m (40,000 ft) at 19,760 kg (44,000 lb) AUW	470 kts (871 km/h; 541 mph)
Rate of climb at S/L	1,035 m (3,400 ft)/min
Service ceiling	14,630 m (48,000 ft)
T-O to 15 m (50 ft)	1,830 m (6,000 ft)
Landing from 15 m (50 ft)	1,190 m (3,900 ft)
Range with max fuel, no reserves	3,153 n miles (5,840 km; 3,630 miles)
Range with max load, no reserves at 600 m (2,000 ft) with 10 min over target at full power	699 n miles (1,295 km; 805 miles)

UPDATED

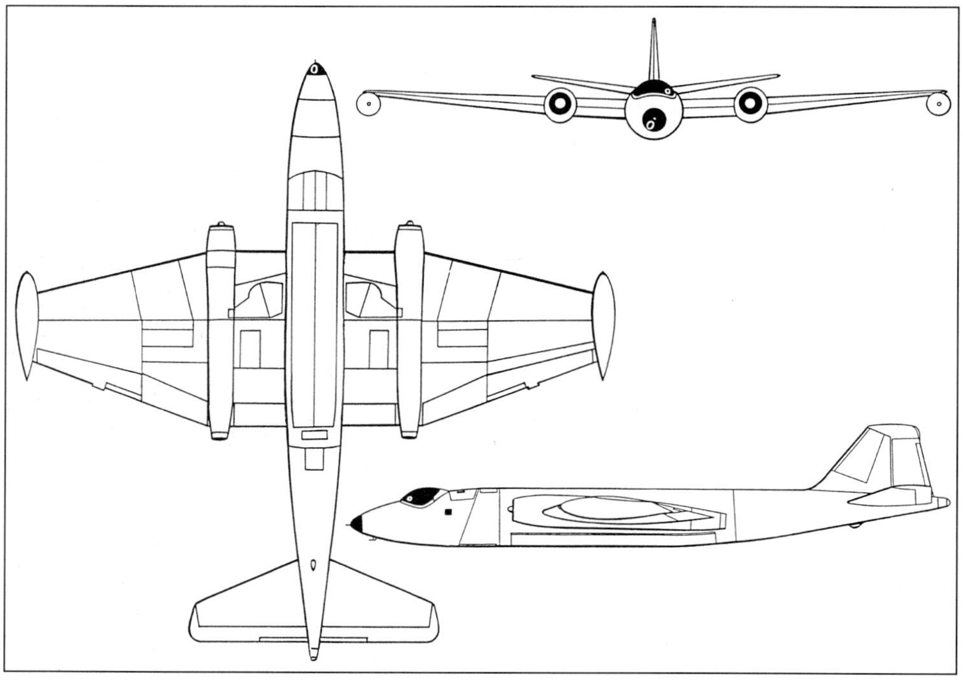

BAe (English Electric) Canberra, long-range multipurpose aircraft *(Jane's/Mike Keep)*

1994

BAe (English Electric) Canberra PR. 9 of the RAF powered by two Rolls-Royce Avon turbojet engines *(BAe)*

1995

British Aerospace F/A. Mk 2 Sea Harrier retrofitted with BASE radar radome ***(BASE)***

1994

BASE

British Aerospace (Systems and Equipment Limited)

Clittaford Road, Southway, Plymouth PL6 6DE
Telephone: 44 (1752) 695695
Fax: 44 (1752) 695500
Telex: 45708
MANAGING DIRECTOR: Nigel Randall
DIRECTOR OF SYSTEMS BUSINESS: Terry Bleakman
HEAD OF EQUIPMENT BUSINESS: Mike Baker
MANAGER, REINFORCED AND MICROWAVE PLASTICS BUSINESS: Trevor Cook
HEAD OF PUBLIC AFFAIRS: Chris Tear

Stand-alone company within the British Aerospace Defence Company. BASE is active in the field of high technology aerospace and defence engineering as well as motion sensing, navigation, tracking and communications. Headquarters at Plymouth has a 15 acre site with 25,000 m^2 of covered space. Stevenage site houses reinforced and microwave plastics activities. BASE is extensively equipped to design, develop and manufacture radomes and equipment for the testing and support of radomes as well as electronic, electromechanical and software products. Products include LINS 300 laser inertial navigation unit, TERPROM digital terrain system, SCR500 flight data recorder, SCR500-030 and SCR500-120 cockpit voice recorders, SCR500-630 and SCR500-660 cockpit voice/digital flight data recorders, SCR500-1620 combined flight data recorder and the VITS Video Image Tracking System.

VERIFIED

BASE RADOME UPGRADE PROGRAMME

TYPE: Combat aircraft radome upgrade.
PROGRAMME: Design and development of radomes for possible upgrading of Mikoyan MiG-21, Northrop F-5 and Mirage III aircraft. Programme follows successful design, development and manufacture of radome chosen for the British Aerospace F/A. Mk 2 Sea Harrier upgrade.

VERIFIED

BASE (LOCKHEED) F-16 MLU TERPROM UPGRADE

TYPE: Combat aircraft digital terrain system upgrade.
PROGRAMME: Selected to supply the TERPROM system for the US Air Force F-16 Mid-Life Update programme.
DESIGN FEATURES: TERPROM uses digital terrain database together with radar altimeter to correct and refine aircraft navigation outputs. System uses Ada software. For full details see *Jane's Avionics 1994-95*.

VERIFIED

(NORTHROP) F-5E TIGER II INERTIAL NAVIGATION UPGRADE

TYPE: Combat aircraft navigation system upgrade.
PROGRAMME: Subcontracted by Smiths Industries in 1987 to supply the LINS 300 Inertial Navigation System for a Northrop F-5E Tiger II upgrade programme awarded by an undisclosed air force.
DESIGN FEATURES: The LINS 300 system outputs in MIL-STD-1553B, ARINC 429 and synchro formats. For full details see *Jane's Avionics 1994-95*.

VERIFIED

BASE/HONEYWELL (BRITISH AEROSPACE) F/A 2 SEA HARRIER NAVIGATION UPGRADE

TYPE: Combat aircraft navigation upgrade.
PROGRAMME: Base announced 12 June 1995 that it had teamed with Honeywell to offer the LINS-764G Ring Laser Gyro for the UK F/A 2 Sea Harrier Navigation upgrade.
DESIGN FEATURES: Base will adapt the LINS-764G ring laser gyro (RLG) inertial navigation system (INS) with embedded global positioning system (GPS) to the F/A 2 application, supply system integration experience, implement an unaided at-sea align capability as well as provide the lead technical, programme management and logistics support for the system.

NEW ENTRY

BWA

BRITISH WORLD AIRLINES

Viscount House, Southend Airport, Essex SS2 6YL
Telephone: 44 (1702) 354435
Fax: 44 (1702) 331914
Telex: 995687 and 995576
MANAGING DIRECTOR, AIRLINE: Mr R. Sturman
MANAGING DIRECTOR, ENGINEERING: B. Stone
SALES DIRECTOR: M. T. Sessions
FLIGHT OPERATIONS DIRECTOR: Capt I. Vanderbeek

BWA's (lease and charter airline) engineering division is licensed by CAA and BAe to perform Viscount life extension modifications; eight Viscounts operated under contract by BWA on passenger and night freight services.

UPDATED

BAe (Vickers) VC2 Viscount turboprop re-lifed and operated by BWA ***(BWA)***

1995

BAC VISCOUNT LIFE EXTENSION

TYPE: Freighter aircraft re-life.
PROGRAMME: By the end of 1990, 12 74-seat Viscount Series 800s comprehensively overhauled, including rib and fuselage reconditioning, to requirements formulated by BWA/BAe; now CAA certificated for further 15 years' service or 75,000 flights. In total, eight are in service with BWA; approximately 50 other airworthy Viscounts eligible for rework worldwide.

UPDATED

DRA/RAF

ROYAL AIR FORCE
DEFENCE RESEARCH AGENCY

SEPECAT JAGUAR TIALD UPGRADE

RAF designation: GR.1B.

TYPE: Combat aircraft upgrade.

PROGRAMME: In June 1994 the RAF gave the go-ahead to upgrade 12 Sepecat Jaguar aircraft with the GEC-Marconi Avionics Thermal Imaging Airborne Laser Designator (TIALD) pod. Initial aircraft to be upgraded by the Defence Reseach Agency at Boscombe Down. Remainder to be upgraded by the RAF at Boscombe Down and St Athan.

DESIGN FEATURES: System includes installation of TIALD pod, MIL-STD-1553 databus interface card, GMAv Type A4 wide-angle HUD, GMAv computer symbol generator (as used on the Panavia Tornado GR. 4 Mid-Life Update) as well as replacing the aircrafts' projected-map display with a multipurpose colour display (MPCD) and a digital map generator. Further features include providing control of TIALD through true HOTAS capability as well as installing an additional video recorder for TIALD imagery.

UPDATED

DRA/RAF (Sepecat) Jaguar GR.1B upgraded with TIALD pod on inner hardpoint of starboard wing
1995

E-2G RESEARCH GROUP

E-2G RESEARCH GROUP

HEADQUARTERS: 2nd Floor, 1 Northumberland Avenue, Trafalgar Square, London WC2
Telephone: 44 (171) 487 3626
Fax: 44 (171) 487 3817
DIRECTOR: W. Owen

The E-2G Research Group, is primarily a specialist support aviation consultancy concerned with the use of fixed- and rotary-wing aircraft in counter-insurgency and anti-narcotics operations.

The company is responsible for the upgrading of existing airframes to suit whatever operational requirements may exist with the prime concern being to create flexible, rather than role specific upgrades.

VERIFIED

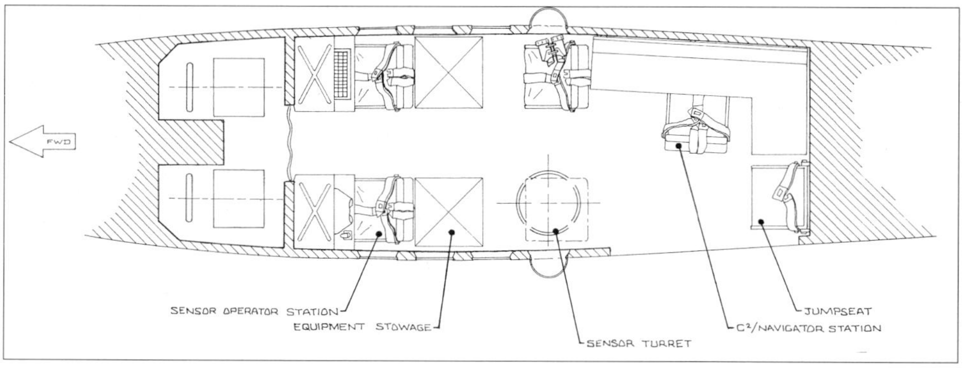

E-2G (Antonov) An-2 interior C^3I upgrade
1994

An-2 (C^3I) UPGRADE

TYPE: Single-engined general purpose biplane upgrade.

PROGRAMME: The following is a generic example of the type of upgrade possible to configure the An-2 for duties including: airborne command and control, Elint/EW and reconnaissance/observation.

AVIONICS: NVG-compatible cockpit with dual GPS and flight instrument upgrade as required.

SYSTEMS (EW): VERSATRIX multifrequency scanner/transmitter (100-1000k Hz) and DIRECTRIX system. Audio tape recording system for Elint analysis. Anti-IR flair kit.

SYSTEMS (Observation): Gyrostabilised sensor turret utilising either existing vertical camera port or mounted internally using a door or window position according to customer's requirements. Video format recorded material is positional coded from the cockpit GPS. Port and starboard observation blisters. Anti-IR flare kit.

VERIFIED

FAIREY HYDRAULICS LTD

FAIREY HYDRAULICS LTD (Part of the Fairey Group)

Claverham, Bristol BS19 4NF
Telephone: 44 (1934) 835224
Fax: 44 (1934) 835337
Telex: 444518 FAIRCL G
MANAGING DIRECTOR: Simon Frost
MANAGER, PUBLICITY: R J Harris
AEROSPACE MARKETING MANAGER: Stephen Page

The company is active in the design, development and manufacture of actuation systems for military and civil fixed- and rotary-wing aircraft, guided weapons, combat vehicles and industrial applications. Main fields of activity are: computer-operated servo-systems, high power density devices, complex electromechanical, hydromechanical and critically lifted mechanisms as well as aircraft landing gear, interface units and database management, critical software and subsystems management. The company has carried out work on the following types of aircraft: BAe Hawk, EH Industries, EH-101, Eurocopter HH-65 Dolphin, Eurofighter 2000, Lockheed L-1011 TriStar, McDonnell Douglas/BAe AV-8B Harrier II, Panavia Tornado, Sepecat Jaguar and the Spearfish Torpedo.

NEW ENTRY

FAIREY (LOCKHEED) L-1011 TRISTAR CARGO DOOR ACTUATOR

TYPE: Commerical cargo door conversion.

PROGRAMME: Contract awarded by Marshall Aerospace in late 1994 to design and supply the actuation system for operation of the cargo doors being installed by Marshalls on Lockheed L-1011 TriStar aircraft. For more details see separate entry in Marshall Aerospace section.

DESIGN FEATURES: Complete system comprises a hydraulic actuator and a hydraulic control unit. The hydraulic actuator, which is a simplex unit with two sub-actuators, provides the primary force to open doors. The sub-actuators operate the lock mechanism.

CUSTOMERS: Marshall Aerospace.

NEW ENTRY

FLS AEROSPACE

FLS AEROSPACE

Long Border Road, Stansted Airport, Stansted, Essex CM24 1RE
Telephone: 44 (1279) 680068
Fax: 44 (1279) 680047
Telex: 81422
Sita: STNLF7X
CHAIRMAN CEO: Steffen Harpøth
COO: Peter Latham
QUALITY DIRECTOR: Phil Hosey
GENERAL MANAGER AIRLINE SERVICES: Ian Ludlow
SALES EXECUTIVE: William McNaught
MAINTENANCE DIRECTOR: Ron Corfield
FINANCE DIRECTOR: Mike James
GENERAL MANAGER COMPONENT SERVICES: Vilhelm Hahn-Petersen
GENERAL MANAGER MATERIAL SERVICES: Phillip Baldwin
HEAD OF PUBLIC RELATIONS: Bryan Southgate

FLS Aerospace, a wholly owned subsidiary of FLS Industries A/S of Denmark, is a major independent aerospace organisation with capabilities in the fields of aircraft heavy maintenance and modification, engineering, interior/exterior repainting and coating, overhaul and repair, as well as interior refurbishment and component support services. Further capabilities are in the field of in-flight entertainment systems.

The company has 12,355 m² (133,000 sq ft) of covered hangar space at its Stansted facilities and 12,542 m² (135,000 sq ft) at its Manchester facilities.

FLS Aerospace has undertaken work on the following aircraft types: Boeing B707, B727, B737, B747, B757, B767, Douglas DC-8, DC-9, MD-80, DC-10, Airbus A300, A310,

A320, British Aerospace BAC-111, BAe 146, BAe 748, Jetstream, Fokker F27, F28, F50, F100, Lockheed L-1011 and Shorts 360. Approvals include: JAR 145, UK CAA, ISO 9002, EN 29002, PCA, MoD US FAA.

FLS was awarded a repeat contract in mid-1994 for depot level maintenance of a further range of RAF E-3D AWACS components. This contract is supplementary to a five year contract signed in January 1992.

UPDATED

FLS INTERIOR REFURBISHMENT

TYPE: Passenger aircraft refurbishment.

PROGRAMME: Complete interior refit programme for the following aircraft types: Boeing B707, B727, B737, B747, B757, British Aerospace BAC-111, BAe 146, BAe 748, Douglas DC-8, DC-10, MD-80, Airbus A300, A310, A320, Lockheed L-1011 TriStar, Fokker F27 Friendship and Shorts 360.

DESIGN FEATURES: Package includes: seating reconfigurations, escape path lighting, toilet conversions, safety equipment relocation, galley installation, floor coverings, interior repainting. Package also includes installation of in-flight entertainment systems as well as overhead bin refinishing and ceiling and wall panelling.

NEW ENTRY

FLS Stansted covered facilities of 12,355 m² (133,000 sq ft) *1995*

FLS EXTERIOR RECOATING

TYPE: Exterior paint stripping and repainting.

PROGRAMME: Recent programmes include: Executive and corporate operators, Ryanair, British Airways, Continental Airlines, United Airlines, European Aviation, Brymon European, and UK MoD.

DESIGN FEATURES: New livery, application, latest technology High Solids Paint used when specified, interior specialist paints used, chemical and plastic media stripping expertise.

COSTS: Vary.

NEW ENTRY

FLS (BAe/HAWKER) 125 DOMINIE OVERHAUL AND CONVERSION

TYPE: Military navigation trainer aircraft upgrade.

PROGRAMME: Awarded contract by the UK MoD in early 1995 for the overhaul and conversion of Royal Air Force Dominie T.1 navigation trainer aircraft. Contract to run from 1 April 1995 until 31 March 1998.

DESIGN FEATURES: Programme includes overhaul and conversion as required of 11 major airframe assemblies including: ailerons, main plane leading-edges, under carriages and tail planes. Contract also covers spare parts.

CUSTOMERS: UK MoD.

NEW ENTRY

FLS (BOEING 747) PYLON RETROFIT

TYPE: Large passenger aircraft pylon retrofit.

PROGRAMME: FLS announced in June 1995 that it had received a contract to carry out its second Boeing 747 engine pylon (strut) modification. This is a result of the Boeing Service Bulletin (SB) issued in November 1994 which called for a series of strengthening measures to be carried out on B747 engine pylons.

DESIGN FEATURES: Strengthening measures depend upon the individual aircraft's history and engine mod standard. Package can include improvements to strut, the wing nacelle rib and underwing fittings.

NEW ENTRY

FRA

FR AVIATION LTD

Bournemouth International Airport, Christchurch, Dorset BH23 6NE
Telephone: 44 (1202) 409000
Fax: 44 (1202) 487899

Established in 1985 from the Airfield Divison of Flight Refuelling Limited, the company is a subsidiary within Cobham Plc. FR Aviation owns the largest fleet (20) of Dassault 20 Aircraft in the world. These aircraft are primarily engaged in the provision of contractorised air defence training on behalf of the Royal Navy, Royal Air Force and NATO. Duties include: threat simulation, electronic warfare training and target towing. Additional aircraft owned by the company include: three Dornier 228s and one Pilatus Britten-Norman BN-2-T Islander engaged in economic zone surveillance around the shores of the UK (on behalf of the UK Government) as well as a Cessna Conquest for airfield and navigation aid flight inspection. A further Cessna Conquest and two Beech Barons are owned/operated by FR Aviation on other military tasks (target towing). FR Aviation has developed extensive engineering facilities to support its aircraft and holds ISO 9000/ISO 9000-3, JAR 145 and FAA Foreign Repair Station Approvals.

Capabilities include avionics and component overhaul through to engine overhaul and full-scale aircraft repairs and conversion. Flight test and certification are performed on site.

The company is also responsible for the selection, modification and operation of aircraft for maritime surveillance and has capabilities in the selection, purchase, installation and operation of military specification 360° radars, infra-red/colour TV cameras, navigation, communications and data/video link equipment. Training for pilots, equipment operators and observers can be given, tailored to the operational requirements of the aircraft operator or government agency.

NEW ENTRY

FR Aviation (BAe/BAC) VC10 passenger/tanker conversion *(FR Aviation)* *1995*

FR AVIATION (BAE/BAC) VC10 C. MK 1(K) TANKER CONVERSION

TYPE: Large passenger/tanker conversion.

PROGRAMME: Contract awarded 1990 for initial batch of eight aircraft, later increased to 13 aircraft. Programme due for completion in 1996. Work is being carried out in collaboration with sister company Flight Refuelling Ltd which is also part of Cobham Plc.

DESIGN FEATURES: Conversion consists of installation of two-point air-to-air refuelling equipment which will give the aircraft dual passenger transport and in-flight refuelling capabilities. Programme also includes upgrading the communications and navigation equipment.

NEW ENTRY

FR AVIATION (DASSAULT) FALCON 20 STORES CARRIAGE UPGRADE

TYPE: Twin-turbofan passenger aircraft conversion.

PROGRAMME: First undertaken in 1991 for an undisclosed Canadian customer. Company has also carried out this conversion on its own fleet of aircraft.

DESIGN FEATURES: Package includes design, development, manufacture and installation of up to four underwing hardpoints per aircraft. Significant flutter analysis and validation is also undertaken by the company. Hardpoints are suitable for the carriage of pods for air defence training including Electronic Warfare (EW), threat simulation and target towing.

NEW ENTRY

FR AVIATION (BAE/ENGLISH ELECTRIC) CANBERRA LIFE-EXTENSION

TYPE: Reconnaissance aircraft life-extension programme.

PROGRAMME: Awarded contract in 1988 by the RAF to extend the lives of its 25 Canberra PR.9 aircraft. Programme due for completion in mid/late 1995.

DESIGN FEATURES: Package includes: production and embodiment of major airframe repair schemes. Total support for this work is available on site.

NEW ENTRY

FR AVIATION (NORTH AMERICAN) F-100D SUPER SABRE LIFE-EXTENSION

TYPE: Equipment test aircraft life-extension.

PROGRAMME: Contract for first aircraft awarded by an American company in mid-1994. Contract for second aircraft is expected in 1995.

DESIGN FEATURES: Complete refurbishment entailed demarrying the wing and fuselage, comprehensive overhaul and a large number of structural modifications including an extensive fuselage longeron repair.

NEW ENTRY

FR Aviation (Dassault) Falcon 20 fitted with underwing hardponts

1995

JETSTREAM AIRCRAFT LTD

JETSTREAM AIRCRAFT LTD (Wholly owned subsidiary of British Aerospace plc)

Prestwick International Airport, Ayrshire KA9 2RW
Telephone: 44 (1292) 79888
Fax: 44 (1292) 79703
Telex: 77432

CHAIRMAN: M. J. (Mike) Turner
MANAGING DIRECTOR: Allan MacDonald
MARKETING DIRECTOR: Nick Godwin
SALES DIRECTOR: Jonathan Walton
DIRECTOR, PRODUCT ENGINEERING: William B. Black
DIRECTOR, TECHNICAL ENGINEERING: Gordon McConnell
PRODUCT PROMOTIONS MANAGER: Robert Gillies
PUBLIC AFFAIRS: Peter Black

Jetstream Aircraft Limited is responsible for design, manufacture, marketing, sales and customer support for Jetstream family of twin-turboprop aircraft. Production, marketing and customer support of the ATP was moved to Prestwick (from Woodford) following reorganisation of BAe's commercial aircraft business. Jetstream 61, a development of ATP with more powerful PW127D engines (for improved hot and high performance) and 70 seats at 78 cm (31 in) pitch, is added to the range and available from third quarter 1994. A new generation of fast turboprops, the 50-seat Jetstream 51 and 70-seat Jetstream 71, has also been proposed. Signed marketing agreement on 6 June 1995 with Avro and ATR to market aircraft in collaboration. Agreement comes into force in January 1996.

NEW ENTRY

JETSTREAM 31

TYPE: Light commuter executive transport.

PROGRAMME: Development of this current version of the Jetstream was launched by British Aerospace on 5 December 1978. A development aircraft (G-JSSD), converted from a Jetstream 1 built by Handley Page, flew for the first time on 28 March 1980. Full production go-ahead was given in January 1982, and the first production Jetstream 31 (G-TALL) made its first flight on 18 March 1982. On 29 June 1982 the Jetstream 31 was certificated to BCAR Section D in the UK. US certification under SFAR 41C followed on 30 November 1982. German (LBA) certification was gained in July 1983, Australian (DOA) certification in February 1984, Swedish (BCA) in December 1984, Dutch (RLD) in January 1985 and Canadian (TC) in May 1987. First deliveries, to customers in Germany and the UK, took place in December 1982, and the 200th delivery was made on 3 October 1988.

VERSIONS: **Airliner:** Designed to carry 18/19 passengers. Able to operate up to 680 n miles (1,260 km; 783 miles) stage length, without refuelling, with 18 passengers, baggage and full IFR reserves.

Corporate: Executive version, designed for eight to ten passengers, and able to carry nine passengers and baggage for 1,050 n miles (1,945 km; 1,208 miles) with full IFR reserves. Typical interior has six fully reclining and swivelling chairs, a three-place divan, galley for hot and cold meal service, cocktail cabinet, wardrobe and washroom/toilet.

Executive Shuttle: Intended for the large company, shuttling its personnel between factories, or for the business charter market. With typical layout for 12 passengers, this version has a range of 1,050 n miles (1,945 km; 1,208 miles) with full IFR reseves.

Special Role: Intended for various specialist applications such as military communications, casualty evacuation, multi-engine training, cargo operations, airfield calibration, resources survey, and protection. A patrol version designated **Jetstream 31EZ** was developed for operation in exclusive economic zones (for example, offshore patrol and surveillance), with underbelly 360° scan radar, increased fuel, observation windows and searchlight. Two aircraft, specially equipped with Tornado IDS avionics, were delivered to the Royal Saudi Air Force in November 1987 for navigator training.

Jetstream Super 31: See *Jane's All the World's Aircraft*.

Jetstream T. Mk 3: Training version, in service with the Royal Navy. Eyebrow windows above flight deck windscreen to improve all-round view. Interior fitted with two observer training consoles with radar indicator, TANS computer and Doppler. Racal ASR 360 search radar mounted under the fuselage. Four ordered in April 1984, for operation by No. 750 Naval Air Squadron based at Culdrose, Cornwall, for helicopter observer training. All delivered by October 1986.

DESIGN FEATURES: Cantilever low-wing monoplane. Wing section NACA 63A418 at root. NACA 63A412 at tip. Dihedral 7° from roots. Incidence 2° at root, 0° at tip. Sweepback 0° 34′ at quarter-chord. Cantilever tail unit with fixed incidence tailplane.

FLYING CONTROLS: Manually operated alloy Frise ailerons. Hydraulically operated aluminium alloy double-slotted flaps. No slats or leading-edge flaps. Trim tab in each aileron. The tail unit has manually operated control surfaces. Trim tabs in rudder and each elevator.

STRUCTURE: Aluminium alloy fail-safe structure of front, main and rear spars with chordwise ribs. Wing skins chemically etched and reinforced with bonded spanwise stringers. The fuselage is a conventional aluminium alloy semi-monocoque fail-safe structure with chemically milled skin panels. Nose and tail sections unpressurised.

The tail unit is a cantilever two-spar aluminium alloy structure.

LANDING GEAR: Retractable tricycle type, with nosewheel steering (±45°). Hydraulic retraction, mainwheels inward into wings, twin nosewheels forward. British Aerospace oleo-pneumatic shock-absorbers in all units. Dunlop wheels and tyres; mainwheel tyre size 28 × 9.00-12, pressure 3.93 bars (57 lb/sq in); nosewheel tyres size 6.00-6, pressure 2.34 bars (34 lb/sq in). Anti-skid units.

POWER PLANT: Two 701 kW (940 shp) Garret TPE331-10UG turboprops, each driving a Dowty Rotol four-blade variable- and reversible-pitch fully-feathering metal propeller. Fuel in integral tank in each wing, total capacity 1,718 litres (454 US gallons; 378 Imp gallons). Refuelling point on top of each outer wing. Water methanol injection optional.

ACCOMMODATION: Two seats side by side on flight deck, with provision for dual controls, though aircraft can be approved (subject to local regulations) for single pilot operation. Main cabin can be furnished in commuter layout for up to 19 passengers at 76/79 cm (30/31 in) pitch, or with executive interior for 8/10 passengers, but optional layouts are available, including a QC (quick change) option enabling an operator to change from an 18-seat layout to 12-seat executive configuration in around 1¼ hours. Downward-opening passenger door, with integral airstairs, at rear of cabin on port side. Emergency exit over wing on starboard side. Baggage compartment in rear of cabin, aft of main door. Entire accommodation pressurised, heated, ventilated and air conditioned. Toilet standard; galley and bar optional. Jetstream completion is by Field Aircraft at East Midlands Airport, Castle Donington, UK.

SYSTEMS: Air conditioning system with cabin pressurisation at maximum differential of 0.38 bars (5.5 lb/sq in), providing a 2,440 m (8,000 ft) cabin altitude at 7,620 m (25,000 ft). Single hydraulic system, pressure 138 bars (2,000 lb/sq in), with two engine-driven pumps, each capable of supplying 20.7 litres (5.46 US gallons; 4.55 Imp gallons)/min. One pump is capable of supplying all hydraulic systems. Combined air/oil reservoir, pressurised to 1.24

Jetstream 31 T. Mk 3 training aircraft of the Royal Navy

1995

bars (18 lb/sq in), for main and emergency supply, for actuation of flaps, landing gear, brakes and nosewheel steering. APU optional. Goodrich rubber de-icing system for leading-edges.

DIMENSIONS, EXTERNAL:

Wing span	15.85 m (52 ft 0 in)
Wing chord: at root	2.19 m (7 ft 2½ in)
at tip	0.80 m (2 ft 7¼ in)
Wing aspect ratio	9.95
Length overall	14.37 m (47 ft 1½ in)
Length of fuselage	13.40 m (43 ft 11½ in)
Height overall	5.38 m (17 ft 8 in)
Fuselage: Max diameter	1.98 m (6 ft 6 in)
Tailplane span	6.60 m (21 ft 8 in)
Wheel track	5.94 m (19 ft 6 in)
Wheelbase	4.60 m (15 ft 1 in)
Propeller diameter	2.69 m (8 ft 10 in)
Passenger door: Height	1.42 m (4 ft 8 in)
Width	0.86 m (2 ft 10 in)
Emergency exit: Height	0.91 m (3 ft 0 in)
Width	0.56 m (1 ft 10 in)

DIMENSIONS, INTERNAL:

Cabin, excl flight deck: Length	7.39 m (24 ft 3 in)
Max width	1.85 m (6 ft 1 in)
Max height	1.80 m (5 ft 11 in)
Floor area	8.35 m² (90 sq ft)
Volume (trimmed aircraft)	16.99 m³ (600 cu ft)
Baggage compartment volume:	
Airliner	2.13-2.74 m³ (75.2-96.7 cu ft)
Corporate	1.34-1.48 m³ (47.2-52.2 cu ft)
Baggage pod (optional)	1.39 m³ (49 cu ft)

AREAS:

Wings, gross	25.20 m² (271.3 sq ft)
Ailerons, aft of hinge line (total)	1.52 m² (16.4 sq ft)
Trailing-edge flaps (total)	3.25 m² (35.0 sq ft)
Vertical tail surfaces (total)	7.72 m² (83.1 sq ft)
Horizontal tail surfaces (total)	7.80 m² (84.0 sq ft)

WEIGHTS AND LOADINGS:

Operating weight empty	4,488 kg (9,894 lb)
Baggage pod weight	59 kg (130 lb)
Max fuel	1,372 kg (3,024 lb)
Max payload	1,805 kg (3,980 lb)
Max capacity of baggage pod	197 kg (435 lb)
Max T-O weight: standard	6,950 kg (15,322 lb)
US domestic	6,900 kg (15,212 lb)
Max ramp weight	7,000 kg (15,432 lb)
Max landing weight	6,600 kg (14,550 lb)
Max zero-fuel weight	6,300 kg (13,889 lb)
Max wing loading	275.8 kg/m² (56.5 lb/sq ft)
Max power loading	4.96 kg/kW (8.15 lb/shp)

PERFORMANCE (at max T-O weight except where stated):

Max cruising speed at max cruise power at 4,570 m (15,000 ft)	263 kts (488 km/h; 303 mph)
Econ cruising speed at 7,620 m (25,000 ft)	230 kts (426 km/h; 264 mph)
Stalling speed, flaps down	86 kts (159 km/h; 99 mph)
Max rate of climb at S/L	635 m (2,080 ft)/min
Rate of climb at S/L, one engine out	119 m (390 ft)/min
Certificated ceiling	7,620 m (25,000 ft)
Service ceiling, one engine out	3,660 m (12,000 ft)
Min ground turning radius about nosewheel	6.50 m (21 ft 4 in)
T-O field length: BCAR Section D	1,440 m (4,724 ft)
T-O to 15 m (50 ft): SFAR 41C	975 m (3,200 ft)
Landing field length, at max landing weight:	
BCAR Section D	1,235 m (4,052 ft)
SFAR 41C/FAR 135	1,165 m (3,820 ft)
Accelerate/stop distance:	
SFAR 41C	1,362 m (4,470 ft)
Range	see individual model listings

NEW ENTRY

Jetstream 31 navigation training aircraft with Panavia Tornado rear cockpit installed

1995

Jetstream 31 passenger transport aircraft with Garrett TPE331 turboprop engines

1995

MANN

MANN AVIATION FLIGHT SYSTEMS LTD

Fairoaks Airport, Chobham, Woking, Surrey GU24 8HX
Telephone: 44 (1276) 857444
Fax: 44 (1276) 857510
DIRECTOR: Brian Chipping

Company active in the field of design, development, manufacture and installation of airborne instruments including navigation, communication and night vision systems.

NEW ENTRY

MANN (WESTLAND) SEA KING UPGRADE

TYPE: Anti-submarine warfare (ASW) helicopter upgrade.

PROGRAMME: Contract awarded 29 June 1993 to install dipping sonar in six Sea King aircraft of Pakistan Navy. First flight took place May 1995. First aircraft redelivered July 1995. Work now started on second aircraft. Contract due for completion in December 1996.

DESIGN FEATURES: Programme consists of design and manufacture of installation kits for GEC Avionics AQS 928G/206 dipping sonar system. Kit design and manufacture at Fairoaks airport (England). Installation taking place at PN Base Mehran, Karachi.

CUSTOMERS: Pakistan Navy (six aircraft).

COSTS: £3.4 million.

NEW ENTRY

MARSHALL AEROSPACE

MARSHALL OF CAMBRIDGE AEROSPACE LTD

The Airport, Cambridge CB5 8RX
Telephone: 44 (1223) 373737
Fax: 44 (1223) 321032
Telex: 81208
MANAGING DIRECTOR: M. A. Bloomfield
ENGINEERING DIRECTOR: R. E. Ward
MARKETING DIRECTOR: M. G. Milne
PUBLIC RELATIONS: G. McA. Bacon

Marshall Aerospace (previously the aircraft division of Marshall of Cambridge Engineering Limited) specialises in the modification, overhaul and repair of aircraft both civil and military, including the design and installation of avionic modifications and outfitting. Several space projects have also been completed. The company has the full upgrade capability of a main manufacturer from design, through component manufacture to installation, test flying, certification and product support. Appointed Sister Design Authority for RAF C-130 fleet, also Lockheed Service Centre in 1966, design for RAF TriStar fleet in 1985; Gulfstream approved Service Centre for Europe from 1960 and Citation Service Station since 1974. FAA approval as extension of Douglas production facility in 1991. Sister design company for E-3 AWACs 1995.

UPDATED

MARSHALL (LOCKHEED) HERCULES C. MK 3

TYPE: Logistics aircraft fuselage stretch.

PROGRAMME: It was announced in 1978 by the Ministry of Defence that a contract had been signed with Lockheed Corporation to modify 30 Hercules C. Mk 1s (C-130Ks) of the Royal Air Force to the **Hercules C. Mk 3** configuration.

DESIGN FEATURES: The modification involved stretching the fuselage by the insertion of a 2.54 m (8 ft 4 in) plug

Marshall Aerospace (Lockheed) C. Mk 3 of the RAF in new grey camouflage *(Marshall)*

1995

forward of the wings and a 2.03 m (6 ft 8 in) plug aft of the wings, thus producing the same fuselage dimensions and capacity as those of the L-100-30 commercial Hercules. The first aircraft (XV223) was modified by Lockheed in Marietta, Georgia, in 1979; the remaining 29 were modified by Marshall of Cambridge. The programme was completed in 1985.

Examples of volumetric load differences include:

Cabin volume, incl ramp:	
C. Mk 1	127.4 m³ (4,500 cu ft)
C. Mk 3	171.5 m³ (6,057 cu ft)
Palletised loads: C. Mk 1	5
C. Mk 3	7
Land Rovers plus trailers: C. Mk 1	3 + 2
C. Mk 3	4 + 3

VERIFIED

MARSHALL (LOCKHEED) HERCULES C. MK 1K

TYPE: Logistics aircraft tanker/receiver conversion.

PROGRAMME: Six Royal Air Force Lockheed Hercules C. Mk 1 aircraft were converted by Marshall into flight refuelling tanker/receivers by fitting a Mk 17 hose drum unit (HDU) on the cargo compartment ramp, a drogue deployment box on the outside of the cargo bay door, four long-range fuel tanks in the cabin, and an in-flight refuelling probe over the flight deck. Modified in this way, each aircraft retains the availability of a pressurised cargo compartment for normal flight, but depressurises when operating with the drogue deployed. The first converted aircraft, designated Hercules **C. Mk 1K**, first flew 8 June 1982 and became operational with No. 1312 Flight from Ascension Island early August 1982. Total of five Mk 1s have been further modified by Marshall with pod under each wingtip, containing 'Orange Blossom' ESM (equivalent to Racal MIR-2 'Orange Crop') equipment.

In 1982 16 RAF Hercules C. Mk 1 aircraft were fitted with a probe to provide an in-flight refuelling capability and are designated **C. Mk 1P**. During the Falklands campaign, one of them remained in the air for 28 hours on an operational mission, creating what was then a record for duration in a Hercules. The RAF subsequently decided that all C. Mk 1 and C. Mk 3 Hercules would be equipped with probes, and the first **C. Mk 3P** (XV214) was returned to RAF Lyneham in February 1986, after conversion. Modification of the RAF's 31 Mk 1 and 30 Mk 3 Hercules was completed in 1989.

In 1987, some RAF Hercules began receiving Loral AN/ALQ-157 IR jamming equipment and chaff/flare dispensers on each side of the rear fuselage, and a contract was placed with General Instrument for AN/ALR-66 radar warning receivers. At the same time, 20 Hercules C. Mk 3s were equipped with shallow underfuselage radome for Sierra Research AN/APN-169B station-keeping equipment to permit formation parachute-dropping in all weathers.

VERIFIED

MARSHALL (LOCKHEED) C-130K REFURBISHMENT STUDY

TYPE: Logistics aircraft refurbishment study.

PROGRAMME: The company has undertaken an extensive study for the UK Ministry of Defence to determine what refurbishment would be necessary to keep the C-130K in service with an acceptable availability rate.

NEW ENTRY

MARSHALL (LOCKHEED) L-1011 TRISTAR TANKER CONVERSION

TYPE: Military tanker conversion.

PROGRAMME: Six Lockheed L-1011-500 TriStar purchased from British Airways 1982; converted to in-flight refuelling tankers or tanker/transports for RAF. Work started 1983.

VERSIONS: **TriStar K. Mk 1**: Two interim (ZD950 and 953) and two full (ZD951 and 949) conversions to tanker/passenger standard; each has two Flight Refuelling Mk 17T hose drum units (HDUs) and seven fuel tanks in fore and aft baggage compartments, totalling additional 45,386 kg (100,060 lb) of fuel; overall total fuel capacity 142,111 kg (313,300 lb); HDUs deliver fuel at 1,814 kg (4,000 lb)/min at 3.45 bars (50 lb/sq in); flight refuelling receiver probe above forward fuselage; closed circuit TV for monitoring refuelling; first K. Mk 1 (ZD950) flight 9 July 1985; maximum T-O weight 244,940 kg (540,000 lb); received CA Release to Service 1 December 1988; ZD953 redelivered to RAF 25 March 1986, followed by ZD951 and 949 mid-July 1986; interim models since converted to KC. Mk 1s.

TriStar KC. Mk 1: TriStars ZD948 and 952 converted to tanker/freighter role; ZD950 and 953 (see K. Mk 1) now this standard; additional modifications include 2.64 × 3.56 m (104 × 140 in) cargo door in port side, forward of wing leading-edge, and cargo handling system; passenger floor strengthened for high-density loads; all items on cabin floor pallet-mounted; in passenger role, pallets carry

Marshall Aerospace (Lockheed) C-130 Hercules C. Mk 1K, the first tanker. Note the four strakes on cargo door as well as the drogue and indicator lights

1994

Marshall Aerospace (Lockheed) L-1011 TriStar converted to military tanker *(P. Tompkins)*

1994

Marshall Aerospace (Lockheed) L-1011 TriStar converted to KC. Mk 1 tanker/freighter for the RAF

1994

approximately 190 seats plus galleys and baggage; ZD948 redelivered to RAF 9 December 1988; ZD952 and 953 redelivered 1989 and ZD950 December 1990.

Marshall completed design study to fit FRL Mk 32B pods to K. Mk 1s, KC. Mk 1s and C. Mk 2(K)s; Mk 32 pods can deliver fuel simultaneously at 1,134 kg (2,500 lb)/min at 3.45 bars (50 lb/sq in).

General Instrument AN/ALR-66 radar warning receivers to be fitted to RAF TriStars; all assigned to No. 216 Squadron, RAF Brize Norton.

UPDATED

MARSHALL (LOCKHEED) L-1011 TRISTAR CIVILIAN FREIGHTER CONVERSION

TYPE: Civil freighter conversion.

PROGRAMME: First civilian L-1011 Tristar freighter handed over to American International Airways on 16 August 1995.

DESIGN FEATURES: Based on Marshall experience in converting the TriStar to a military freighter, a civil conversion is now available for L-1011 TriStar variants. This conversion has an enlarged door sized 2.92 × 3.94 m (115 × 155 in), a single scissor jack instead of two vertical jacks and a bulkhead instead of the cargo restraint net fitted to the military freighter version. Depending on the TriStar version and its particular modification state, the Marshall TriStar freighter conversion will carry up to 64,411 kg (142,000 lb) for 3,000 n miles (5,556 km; 3,456 miles), using 23 main deck pallets and up to 24 LD3 containers in the underfloor cargo holds.

CUSTOMERS: American International Airways (8). Four more on order.

VERIFIED

MARSHALL (LOCKHEED) L-1011 TRISTAR SUPPLEMENTAL FUEL TANKS

TYPE: Civil fuel tank supplement.

DESIGN FEATURES: Marshall has produced a supplemental fuel tank for Delta Air Lines to extend the range of its TriStar-500s. The tank fits in the forward cargo hold and contains 5,897 kg (13,000 lb) of fuel and can be removed when not required. Tank and equipment weigh approximately 907 kg (2,000 lb) of which 680 kg (1,500 lb) removable.

VERIFIED

MARSHALL (LOCKHEED) L-1011 TRISTAR PEGASUS SPACE BOOSTER

TYPE: L-1011 TriStar satellite launch upgrade.

PROGRAMME: Marshall has modified a TriStar-1 to carry a Pegasus XL air-launched wing rocket which can put a satellite of up to 450 kg (1,000 lb) in orbit. The TriStar arrived at Cambridge in June 1992 and was handed over to Orbital Sciences Corporation in November 1993. The Pegasus XL weighs 23,000 kg (52,000 lb). The first live drop was completed on 27 June 1994. First succesful launch of Orbital Sciences Pegasus satellite carrier took place on 3 April 1995.

UPDATED

MARSHALL (MCDONNELL DOUGLAS) DC-10 WEIGHT UPGRADES

TYPE: Commercial passenger aircraft weight increase conversion kit.

PROGRAMME: Marshall has designed and manufactured for Douglas a weight upgrade kit to increase the maximum take-off weight of the DC-10-10 from 195,048 to 202,305 kg (430,000 to 446,000 lb). There have been 10 kits produced.

VERIFIED

MARSHALL (MCDONNELL DOUGLAS) MD-11 SUPPLEMENTARY FUEL TANKS

TYPE: Commercial aircraft fuel increase conversion.

PROGRAMME: Based on the TriStar supplementary fuel tanks mentioned earlier, Marshall has produced similar tanks for the MD-11 which are in service with Delta Air Lines. The MD-11 installation uses two supplemental tanks, each similar to the TriStar supplemental tank, in the forward cargo bay hold, providing 11,800 kg (26,000 lb) fuel. As with the TriStar, installation tanks can easily be removed and the fixed installation is 590 kg (1,300 lb). Marshall is now designing, in collaboration with McDonnell Douglas, an improved and lighter supplemental tank for the MD-11ER.

UPDATED

MARSHALL AVIONIC UPGRADES

TYPE: Military aircraft avionics upgrades.

PROGRAMME: Marshall has upgraded several RAF Hercules with specialist avionic improvements and has completely revised the flight deck ergonomics of eight C-130s for an overseas air force, including integrating a wide range of new sensors into an upgraded autopilot. In collaboration with Smiths Industries the company has been awarded a contract to update the navigation systems on 30 RAF Hercules. The first installation is currently being undertaken. In addition Marshall has undertaken a wide range of avionic upgrades on other aircraft. These have included: improving the nav attack system for F-5E aircraft for an overseas air force; installation of a helicopter gun turret controlled by helmet-mounted sight; SATCOM installations on TriStar, Gulfstream and B707; fitment of a FLIR installation on a Citation; revision of the avionic suite fitted to the RAF Bulldog fleet. Currently 11 RAF Dominies (BAe 125) are being upgraded with new avionics in association with Racal Radar Defence Systems (formerly THORN EMI) using their Super Searcher radar.

UPDATED

MARSHALL INTERIOR UPGRADES

TYPE: Commercial passenger aircraft interior refurbishment.

PROGRAMME: In addition to making structural improvements to aircraft interiors, Marshall has also undertaken a wide range of furnishing upgrades. Projects have included upgrade of 10 BA TriStar-200s (Sunset III), complete revision of three MD-11 interiors, outfitting BAe 125, Gulfstream and Citation interiors for executive customers. Marshall has also produced a VIP conversion of an RJ 85 for Pelita of Indonesia in co-operation with Design Research Associates.

VERIFIED

Marshall Aerospace (BAe) RJ 85 interior conversion for Pelita of Indonesia *(Marshall)*

1995

Marshall Aerospace (Lockheed) L-1011 TriStar converted to Pegasus XL satellite launcher *(Marshall)*

1995

PILATUS BRITTEN-NORMAN

PILATUS BRITTEN-NORMAN LTD (Subsidiary of Oerlikon-Bührle Holding Ltd)

The Airport, Bembridge, Isle of Wight PO35 5PR
Telephone: 44 (1983) 872511
Fax: 44 (1983) 873246
Telex: 86277 PBNBEM G
MANAGING DIRECTOR: Anthony Stansfeld
TECHNICAL DIRECTOR: Robert Wilson
HEAD OF MARKETING: Walter Stark

Pilatus Aircraft Ltd of Switzerland acquired Britten-Norman (Bembridge) Ltd in 1979, including Isle of Wight facilities and former Fairey SA Islander/Trislander production hardware ot Gosselies, Belgium.

Britten-Norman became a member of the Fairey Group on 31 August 1972, when the assets of Britten-Norman Ltd were acquired by the Fairey Group. The announcement assured the future of the business after some doubt following the original company's difficulties which had arisen in 1971.

Final Trislander delivered to the Botswana Defence Force in September 1984 but Audrey Promotions of New South Wales, Australia have purchased 10 kits of components produced by Fairey at Gosselies.

In February 1995, Pilatus Britten-Norman were audited to BS EN ISO 9001 1994 quality management standard and to Aerospace Sector Certification Scheme 157 Approval. The company was also awarded JAR 145, the joint aviation authority standard required for carrying out repair, overhaul and modification work to BN-2 series aircraft operated by JAA member countries.

UPDATED

PILATUS BRITTEN-NORMAN BN-2A ISLANDER

TYPE: Twin-engine feeder line transport aircraft.

PROGRAMME: The Islander BN-2A is a utility 10-seat aircraft and the predecessor of the Islander BN-2B which is still in production. Design work began in April 1964 and construction of the prototype (G-ATCT) was started in September of the same year. It flew for the first time on 13 June 1965, powered by two 157 kW (210 hp) Rolls-Royce Continental IO-360-B engines and with wings of 13.72 m (45 ft) span. Subsequently, the prototype was re-engined with more powerful Lycoming O-540 engines, with which it flew for the first time on 17 December 1965. The wing span was also increased by 1.22 m (4 ft) to bring the prototype to production standard.

Deliveries of Islanders began in August 1967, and by June 1977 more than 750 aircraft of the various models had been delivered to operators in 117 countries. There are 560 BN-2A Islanders currently in operation with commercial, military and government operators in over 100 countries worldwide. Largest military fleet operators include the Belgian Army (10), Indian Navy (13) and Philippine Air Force (18). However, the BN-2A is still used predominently by the traditional Islander customer: the small air taxi and commuter operator.

The basic BN-2A Islander was available with a choice of two alternative power plants and either standard 14.94 m (49 ft 0 in) span wings or wingtip extensions having raked tips and containing auxiliary fuel tanks.

The version of the Islander with 224 kW (300 hp) fuel-injection engines was first introduced in 1970, deliveries beginning in November of that year.

An amphibious version of the Islander was announced early Summer 1975.

VERSIONS: **Pilatus Britten-Norman Defender:** Military version of the BN-2A for a wide variety of roles, including SAR, internal security, long-range patrol, troop transport, forward air control, logistic support and casualty evacuation.

Pilatus Britten-Norman Maritime Defender: Generally similar to the Defender version. The maritime version differs by having a modified nose section to accommodate a large search radar.

UPGRADES: **Jonas Aircraft:** A Rajay turbocharging installation was developed in the USA by Jonas Aircraft, the New York based distributors for Britten-Norman in the western hemisphere. The Rajay installation is a bolt-on unit, for manual operation, which can be fitted on to standard 194 kW (260 hp) engines. The superchargers have the effect of increasing the single-engined ceiling to 3,810 m (12,500 ft) and twin-engined ceiling to 7,925 m (26,000 ft). Cruising speed is also increased from 139 kts (275 km/h; 160 mph) at 2,135 m (7,000 ft) to 146 kts (270 km/h; 168 mph) at 3,050 m (10,000 ft).

Pilatus Britten-Norman: A series of modification kits have been introduced by the company. These kits were made available on production aircraft and as retrofit packages.

OPERATORS: Versions of the BN-2A are in service with the armed forces of the following countries: Angola (13), Botswana (7), Ghana (8), Haiti (1), Jamaica (2), Madagascar (3), Mauritania (5), Mexico (6), Panama (1), Philippines (6), Rwanda (2), Seychelles (1), United Arab Emirates (1), Venezuela (1), Zaïre (1) and Zimbabwe (6).

The following description applies to the standard BN-2A, unless otherwise stated:

DESIGN FEATURES: Cantilever high-wing monoplane. NACA 23012 constant wing section. No dihedral. Incidence 2°. No sweepback. Flared wingtips of Britten-Norman design. Raked-back extended wingtips optional. Fixed incidence tailplane.

FLYING CONTROLS: Slotted ailerons and single-slotted flaps of metal construction. Flaps operated electrically, ailerons by pushrods and cables. Ground adjustable tab on starboard aileron. Fixed-incidence tailplane with mass-balanced elevator. Rudder and elevator are actuated by pushrods and cables. Trim tabs in rudder and elevator.

STRUCTURE: Conventional riveted two-spar torsion-box structure in one piece, using L72 aluminium-clad aluminium alloys. The fuselage is a conventional riveted four-longeron semi-monocoque structure of pressed frames and stringers and metal skin, using L72 aluminium alloys. Optional 1.15 m (3 ft 9¼ in) nose extension for baggage stowage. The tail unit is a cantilever two-spar structure, with pressed ribs and metal skin, using L72 aluminium-clad aluminium alloys.

LANDING GEAR: Non-retractable tricycle type, with twin-wheels on each main unit and single steerable nosewheel. Cantilever main legs mounted aft of rear spar. All three legs fitted with Lockheed oleo-pneumatic shock-absorbers. All five wheels and tyres size 16 × 7-7, supplied by Goodyear. Tyre pressure: main 2.41 bars (35 lb/sq in); nose 2.00 bars (29 lb/sq in). Foot-operated air-cooled Cleveland hydraulic brakes on main units. Parking brake. Wheel/ski gear available optionally.

POWER PLANT: Two Lycoming flat six engines, each driving a Hartzell HC-C2YK-2B or -2C two-blade metal constant-speed feathering propeller. Standard power plant is the 194 kW (260 hp) O-540-E4CS, but the 224 kW (300 hp) IO-540-K1B5 was available at customer's request. Optional Rajay turbocharging installation on 194 kW (260 hp) engine, to improve high-altitude performance. Integral fuel tank between spars on each wing, outboard of engine. Total fuel capacity (standard) 518 litres (137 US gallons; 114 Imp gallons). With auxiliary tanks in wingtip extensions, total capacity is increased to 741 litres (196 US gallons; 163 Imp gallons). Additional pylon-mounted underwing auxiliary tanks, each 227 litres (60 US gallons; 50 Imp gallons) capacity, available optionally. Refuelling point in upper surface of wing above each internal tank. Total oil capacity 22.7 litres (6 US gallons; 5 Imp gallons).

ACCOMMODATION: Up to 10 persons, including pilot, on side by side front seats and four bench seats. No aisle. Seat backs fold forward. Access to all seats via three forward-opening doors, forward of wing and at rear of cabin on port side and forward of wing on starboard side. Baggage compartment at rear of cabin, with port side loading door in standard versions. Exit in emergency by removing door windows. Special executive layouts available. Can be operated as a freighter, carrying more than a ton of cargo; in this configuration the passenger seats can be stored in the rear baggage bay. In ambulance role, up to three stretchers and two attendants can be accommodated. Other layouts possible, including photographic and geophysical survey, parachutist transport to trainer (with accommodation for up to eight parachutists and a dispatcher), public health spraying and crop spraying. A firefighting version is available.

SYSTEMS: Southwind cabin heater standard, 45,000 BTU Stewart Warner combustion unit, with circulating fan, provides hot air for distribution at floor level outlets and at windscreen demisting slots. Fresh air, boosted by propeller slipstream, is ducted to each seating position for on-ground ventilation. Electrical DC power, for instruments, lighting and radio, from one or two engine-driven 24 V 50A self-rectifying alternators and a controller to main busbar and circuit-breaker assembly in nose bay. Emergency busbar with automatic changeover provides a secondary route for essential services. A 24 V 17 Ah heavy-duty lead-acid battery for independent operation, ground power receptacle provided. Optional electrical de-icing of propellers and windscreen, and pneumatic de-icing of wing and tail unit leading-edges. Intercom system, including second headset, and passenger address system are standard. Oxygen system available optionally for all versions. BTR-Goodrich pneumatic de-icing boots optional for wings. Pneumatic de-icing of tailplane and fin optional.

DIMENSIONS, EXTERNAL:

Wing span	14.94 m (49 ft 0 in)
Wing chord, constant	2.03 m (6 ft 8 in)
Wing chord at root	2.19 m (7 ft 2½ in)
Wing aspect ratio	7.4
Length overall	10.86 m (35 ft 7¾ in)
Fuselage: Max width	1.21 m (3 ft 11½ in)
Max depth	1.46 m (4 ft 9¾ in)
Height overall	4.18 m (13 ft 8¾ in)
Tailplane span	4.67 m (15 ft 4 in)
Wheel track (c/l of shock-absorbers)	3.61 m (11 ft 10 in)
Wheelbase	3.99 m (13 ft 1¼ in)
Propeller diameter	2.59 m (8 ft 6 in)
Cabin door (front, port):	
Height	1.10 m (3 ft 7½ in)
Width: top	0.64 m (2 ft 1¼ in)
Height to sill	0.59 m (1 ft 11¼ in)
Cabin door (front, starboard):	
Height	1.10 m (3 ft 7½ in)
Max width	0.86 m (2 ft 10 in)
Height to sill	0.57 m (1 ft 10½ in)
Cabin door (rear, port): Height	1.09 m (3 ft 7 in)
Width: top	0.635 m (2 ft 1 in)
bottom	1.19 m (3 ft 11 in)
Height to sill	0.52 m (1 ft 8½ in)
Baggage door (rear, port): Height	0.69 m (2 ft 3 in)

DIMENSIONS, INTERNAL:

Passenger cabin, aft of pilot's seat:	
Length	3.05 m (10 ft 0 in)
Max width	1.09 m (3 ft 7 in)
Max height	1.27 m (4 ft 2 in)
Floor area	2.97 m² (32 sq ft)
Volume	3.68 m³ (130 cu ft)
Baggage space aft of passenger cabin	1.39 m³ (49 cu ft)
Freight capacity:	
aft of pilot's seat, incl rear cabin baggage space	4.70 m³ (166 cu ft)
with four bench seats folded into rear cabin baggage space	3.68 m³ (130 cu ft)

AREAS:

Wings, gross	30.19 m² (325.0 sq ft)
with extended tips	31.31 m² (337.0 sq ft)
Ailerons (total)	2.38 m² (25.6 sq ft)
Flaps (total)	3.62 m² (39.0 sq ft)
Fin	3.41 m² (36.64 sq ft)
Rudder, incl tab	1.60 m² (17.2 sq ft)
Tailplane	6.78 m² (73.0 sq ft)
Elevator, incl tabs	3.08 m² (33.16 sq ft)

WEIGHTS AND LOADINGS (A: standard wings, B: extended wings, C: 194 kW; 260 hp and D: 224 kW; 300 hp engines):

Weight empty, equipped (without electronics):	
C	1,627 kg (3,588 lb)
D	1,695 kg (3,738 lb)
Max T-O weight (A, B)	2,993 kg (6,600 lb)
Max zero-fuel weight (BCAR):	
A, C, D	2,855 kg (6,300 lb)
B, C, D	2,810 kg (6,200 lb)
Max landing weight (A, B)	2,855 kg (6,300 lb)
Max wing loading: A	99.1 kg/m² (20.3 lb/sq ft)
B	95.7 kg/m² (19.6 lb/sq ft)
Max floor loading, without cargo panels	586 kg/m² (120 lb/sq ft)
Max power loading: C	7.71 kg/kW (12.7 lb/hp)
D	6.68 kg/kW (11.0 lb/hp)

PERFORMANCE (at max T-O weight, ISA. C: 194 kW; 260 hp and D: 224 kW; 300 hp engines):

Never-exceed speed: C, D (standard wings)	177 kts (327 km/h; 203 mph) IAS
C, D (extended wings)	184 kts (340 km/h; 211 mph) IAS

Pilatus Britten-Norman BN-2A Islander twin-engine transport aircraft *(Pilatus)*

1995

Max level speed at S/L:
C 147 kts (273 km/h; 170 mph)
D 156 kts (290 km/h; 180 mph)
Max cruising speed (75% power) at 2,135 m (7,000 ft):
C 139 kts (257 km/h; 160 mph)
D 147 kts (273 km/h; 170 mph)
Cruising speed (67% power) at 2,750 m (9,000 ft):
C 137 kts (254 km/h; 158 mph)
D 146 kts (270 km/h; 168 mph)
Cruising speed (59% power) at 3,960 m (13,000 ft)
C 133 kts (246 km/h; 153 mph)
D 143 kts (264 km/h; 164 mph)
Stalling speed, flaps up:
C 50 kts (92 km/h; 57 mph) IAS
Stalling speed, flaps down:
C 40 kts (74 km/h; 46 mph)
Max rate of climb at S/L: C 296 m (970 ft)/min
D 347 m (1,140 ft)/min
Rate of climb at S/L, one engine out:
C 58 m (190 ft)/min
D 61 m (200 ft)/min
Absolute ceiling: C 4,635 m (15,200 ft)
D 6,020 m (19,750 ft)
Service ceiling: C 4,025 m (13,200 ft)
Service ceiling, one engine out:
C, standard wings 1,770 m (5,800 ft)
C, extended wings 2,040 m (6,700 ft)
D, standard wings 1,890 m (6,200 ft)
D, extended wings 2,180 m (7,150 ft)
Min ground turning radius 9.45 m (31 ft 0 in)
T-O run at S/L, zero wind, hard runway:
C 169 m (555 ft)
D 201 m (660 ft)
T-O run at 1,525 m (5,000 ft): D 285 m (936 ft)
T-O run at 15 m (50 ft) at S/L, zero wind, hard runway:
C 332 m (1,090 ft)
D 335 m (1,100 ft)
T-O to 15 m (50 ft) at 1,525 m (5,000 ft):
D 475 m (1,560 ft)
Landing run from 15 m (50 ft) at S/L, zero wind, hard runway: C, D 292 m (960 ft)
Landing distance at 1,525 m (5,000 ft):
D 350 m (1,150 ft)
Landing run at 1,525 m (5,000 ft):
D, ISA + 20°C 169 m (555 ft)
Landing run at S/L, zero wind, hard runway:
C, D 137 m (450 ft)
Range at 75% power at 2,135 m (7,000 ft):
C, standard wings 622 n miles (1,153 km; 717 miles)
C, extended wings 903 n miles (1,673 km; 1,040 miles)
Range at 67% power at 2,750 m (9,000 ft):
C, standard wings 713 n miles (1,322 km; 822 miles)
C, extended wings 1,036 n miles (1,920 km; 1,193 miles)
Range at 59% power at 3,960 m (13,000 ft):
C, standard wings 755 n miles (1,400 km; 870 miles)
C, extended wings 1,096 n miles (2,032 km; 1,263 miles)

UPDATED

PILATUS BRITTEN-NORMAN BN-2B (AL.1) UPGRADE

TYPE: Communications/trainer aircraft upgrade.

PROGRAMME: Awarded contract in early 1995 to upgrade the British Army Air Corp's fleet of seven AL.1 Islanders to give an improved pilot's field of view. Work to be completed during 1996.

NEW ENTRY

PILATUS BRITTEN-NORMAN AGRICULTURAL ISLANDER

TYPE: Agricultural aircraft.

PROGRAMME: The agricultural version of the basic BN-2A Islander has specially designed pods which are fitted beneath each wing. Each pod consists of a chemical tank, pump and a rotary atomiser for spraying liquid chemicals. The combined volume of the tanks is 380 litres (100.396 US gallons; 83.589 Imp gallons) and the total weight of the system when charged is 496 kg (1,094 lb). The system has been evolved so that the cabin area is kept clean and uncontaminated by chemical deposit or odour, and can be converted rapidly for conventional passenger or military use.

VERIFIED

PILATUS BRITTEN-NORMAN FIREFIGHTER

TYPE: Firefighting aircraft.

PROGRAMME: This firefighting version of the basic BN-2A Islander, first announced at the 1976 Farnborough Air Show, is equipped with four specially designed interconnected liquid tanks which are mounted on a 9 *g* restraint structure attached to the cabin floor pickups for the standard cabin seating. Rapid filling is accomplished by means of an external standard fire equipment pump, via a short connecting hose which is withdrawn through the aircraft's baggage bay door. Content gauges are not required as an overflow system ensures that filling beyond the payload capability of the aircraft is impossible. Large dual vents prevent airlocks or flow-back during fillings and lightly pressurise the system in flight to ensure rapid discharge of the liquid fire retardant. Each of the four containers has an outlet positioned approximately on the aircraft's centreline, and the contents of the tanks can be dumped simultaneously or on a two-shot basis. During either method of discharge, changes to the aircraft's CG and trim are negligible.

Total capacity of the system is 800 litres (211 US gallons; 176 Imp gallons), subject to the availability of suitable ground-based pumping equipment, can be charged with water or fire retardant in approximately two minutes. Optimum speed for discharge is 65 knots (120 km/h; 75 mph) at an altitude of 60 m (200 ft), the load being distributed over an area of 100 × 15 m (330 × 50 ft) in 2.5 seconds. Test flying of a demonstration aircraft (G-BDHU) began in 1976.

VERIFIED

PILATUS BRITTEN-NORMAN BN-2A Mk III TRISLANDER

TYPE: Three-engine feederline transport aircraft.

PROGRAMME: Autumn 1970 Britten-Norman introduced an enlarged development of the twin-engined Islander, with a third engine mounted at the rear and a lengthened fuselage seating up to 17 passengers.

The prototype Trislander was produced by converting the second prototype of the twin-engined Islander (G-ATWU), adding a 2.29 m (7 ft 6 in) length of parallel-section fuselage forward of the wing, reinforcing the rear fuselage and fitting a new main landing gear with larger wheels and tyres. The tail unit was modified to act as a mount for the third engine. This aircraft made its first flight 11 September 1970, appearing at the SBAC display at Farnborough later the same day. Production aircraft have additional fin area above the rear engine.

The prototype was later dismantled and its fuselage used for structural testing. By the end of 1970 construction had begun of three production aircraft by converting standard Islander airframes from the production line, and this system was adopted for all production aircraft, thus maintaining maximum flexibility on what became a totally integrated Islander/Trislander production line. The first production Trislander (G-AYTU) was flown 6 March 1971, and the first delivery (to Aurigny Air Services; see list at back of this edition) in the Channel Islands was made 29 June 1971.

ARB certification of the Trislanders, awarded 14 May 1971, approved the aircraft for both VFR and IFR operation and full public transport with one pilot and up to 17 passengers. FAA certification followed 4 August 1971, to FAR Pt 23 and to the latest air taxi requirements of SFAR Pt 23 and Appendix A of FAR Pt 135. The Appendix A standard is higher than that met by most other commuter aircraft at that time, and was achieved primarily because of continued take-off capability and fatigue-free structure.

By mid-1979 orders had been received for more than 80 aircraft. Of these more than 60 were delivered to customers in Africa, Australasia, Canada, Indonesia, South America, UK and USA. There were 49 Trislanders still in service as of June 1994. The UK Channel Islands operator Aurigny Air Services is the largest fleet operator with nine aircraft.

VERSIONS: **BN-2A Mk III-2:** Standard version with extended nose containing baggage compartment.

BN-2A Mk III-3: As Mk III-2, with an autofeather system which feathers the propeller automatically should an engine fail on take-off.

BN-2A Mk III-4: As Mk III-3, plus a standby rocket engine to provide additional thrust should an engine fail on take-off.

DESIGN FEATURES: Cantilever high-wing monoplane. NACA 23012 constant wing section. No dihedral. Incidence 2°. No sweepback. Flared-up wingtips of Britten-Norman design, with raked tips. Fixed incidence tailplane.

FLYING CONTROLS: Slotted ailerons and electrically operated single-slotted permanently drooped flaps of metal construction. Ground adjustable tab in starboard aileron. Fixed incidence tailplane (with raked tips) and elevators are similar in construction to those of the Islander. Trim tab in rudder.

STRUCTURE: Conventional riveted two-spar torsion-box wing structure in one piece using aluminium-clad aluminium alloys. Increases in skin gauges and spar laminates compared with twin-engined versions. Structure is strictly safe-life, but with several fail-safe features and principles. The fuselage is a conventional riveted four-longeron semi-monocoque structure of pressed frames and stringers and metal skin, using L72 aluminium-clad aluminium alloys. Some reinforcement of fuselage aft of wing to support weight of rear engine. The tail unit is a cantilever structure, using L72 aluminium-clad aluminium alloys, with low aspect ratio main fin which also acts as mount for the third engine.

LANDING GEAR: Non-retractable tricycle type, with twin-wheel main units and single steerable nosewheel. Cantilever main legs mounted aft of rear spar. All five wheels and tyres are Cleveland size 7.00-6. Tyre pressure 3.10 bars (45 lb/sq in) on main units, 2.00 bars (29 lb/sq in) on nose unit. Cleveland foot-operated disc brakes on main units. Parking brake. No anti-skid units. Fairings fitted to main gear extension tubes below the engine nacelle and above the shock-absorber attachment bolts.

POWER PLANT: Three 194 kW (260 hp) Lycoming O-540-E4C5 flat-six engines (two mounted on wings and one on vertical tail), each driving a Hartzell HC-C2YK-2G/C8477-4 two-blade constant-speed fully feathering metal propeller. Automatic feathering device available as an option. Standby rocket engine, mounted just below rear tail-engine nacelle, is available as an option; weighing 21 kg (46.2 lb), this provides 1.56 kN (350 lb st) for 12 seconds. Fuel in two integral tanks between front and rear wing spars, outboard of the engine nacelles, and two tanks in wingtips. Total fuel capacity 746 litres (197 US gallons; 164 Imp gallons). Overwing refuelling point above each tank. Oil capacity 34 litres (9 US gallons; 7.5 Imp gallons).

ACCOMMODATION: Up to 18 persons, including pilot, in pairs on bench seats at approximately 79 cm (31 in) pitch. Access to all seats provided by five broad-hinged rearward-opening car type doors, two on port side and three on starboard side. Baggage compartment at rear of cabin, with external baggage door on port side. Exit in emergency by removing window panels in front four passenger doors. Heating, ventilation and sound insulation standard. Ambulance or VIP interior layouts at customer's option. Dual controls optional.

SYSTEMS: One Southwind cabin heater fitted as standard. DC electrical system includes two 24 V 50A self-rectifying alternators, supplying the instruments, lighting and radio, and a 24 V 17 Ah battery. No hydraulic or pneumatic systems, except for self-contained hydraulic brakes. BTR-Goodrich pneumatic de-icing boots optional for wings. BTR-Goodrich pneumatic de-icing boots for tailplane optional.

AVIONICS AND EQUIPMENT: Optional avionics include Bendix M4C or Mitchell Century III autopilot, a wide range of Bendix, King or Narco VHF or HF nav/com equipment, including ADF and DME. Optional equipment includes windscreen de-icing, second cabin heater, cargo tiedowns, anti-collision strobe beacons, emergency exit beta lights, electric propeller de-icing and pneumatic de-icing systems.

DIMENSIONS, EXTERNAL:

Wing span	16.15 m (53 ft 0 in)
Wing chord (constant)	2.03 m (6 ft 8 in)
Wing aspect ratio	7.95
Length overall	15.01 m (49 ft 3 in)

Pilatus Britten-Norman BN-2A Mk III Trislander three-engine transport aircraft

1994

Fuselage: Max width 1.21 m (3 ft 11½ in)
Max depth 1.46 m (4 ft 9¾ in)
Height overall 4.32 m (14 ft 2 in)
Tailplane span 6.48 m (21 ft 3 in)
Wheel track (c/l of shock-absorbers) 3.35 m (11 ft 0 in)
Wheelbase 7.12 m (23 ft 4¼ in)
Propeller diameter 2.03 m (6 ft 8 in)
Propeller ground clearance 0.69 m (2 ft 3 in)
Distance between propeller centres (wing engines) 3.61 m (11 ft 10 in)
Passenger doors (stbd, fwd and centre):
Height 1.10 m (3 ft 7½ in)
Max width 0.89 m (2 ft 10⅞ in)
Height to sill 0.57 m (1 ft 10½ in)
Passenger doors (port, fwd and rear):
Height 1.09 m (3 ft 7 in)
Max width 1.21 m (3 ft 11⅞ in)
Height to sill 0.57 m (1 ft 10½ in)
Passenger door (stbd, rear): Height 1.09 m (3 ft 7 in)
Width 0.75 m (2 ft 5½ in)
Baggage compartment door (rear, port):
Height 0.66 m (2 ft 2 in)
Width 0.44 m (1 ft 5⅓ in)
Nose baggage compartment door (port, optional):
Width 0.79 m (2 ft 7 in)

DIMENSIONS, INTERNAL:
Cabin: Length excl flight deck but incl rear baggage compartment 8.24 m (27 ft 0½ in)
Max width 1.09 m (3 ft 7 in)
Max height 1.27 m (4 ft 2 in)
Floor area 7.85 m² (84.45 sq ft)
Volume 9.54 m³ (337 cu ft)
Rear baggage compartment volume 0.71 m³ (25.0 cu ft)
Nose baggage compartment volume (optional) 0.62 m³ (22.0 cu ft)

AREAS:
Wings, gross 31.31 m² (337.0 sq ft)
Ailerons (total) 2.38 m² (25.6 sq ft)
Trailing-edge flaps (total) 3.62 m² (39.0 sq ft)
Fin 5.83 m² (62.7 sq ft)
Rudder, incl tab 1.13 m² (12.2 sq ft)
Tailplane 8.36 m² (90.0 sq ft)
Elevators 2.42 m² (26.0 sq ft)

WEIGHTS AND LOADINGS:
Weight empty, equipped (without electronics) 2,650 kg (5,843 lb)
Max T-O and landing weight 4,536 kg (10,000 lb)
Max wing loading 144.8 kg/m² (29.67 lb/sq ft)
Max power loading 7.79 kg/kW (12.8 lb/hp)

PERFORMANCE (at max T-O weight, ISA):
Max level speed at S/L 156 kts (290 km/h; 180 mph)
Cruising speed (75% power) at 1,980 m (6,500 ft) 144 kts (267 km/h; 166 mph)
Cruising speed (67% power) at 2,470 m (9,000 ft) 138 kts (256 km/h; 159 mph)
Cruising speed (50% power) at 3,960 m (13,000 ft) 130 kts (241 km/h; 150 mph)
Max rate of climb at S/L 298 m (980 ft)/min
Rate of climb at S/L, one engine out 86 m (283 ft)/min
Absolute ceiling 4,450 m (14,600 ft)
Service ceiling 4,010 m (13,150 ft)
Service ceiling, one engine out 2,105 m (6,900 ft)
T-O run at S/L, zero wind, hard runway 393 m (1,290 ft)
T-O to 15 m (50 ft) at S/L, zero wind, hard runway 594 m (1,950 ft)
Landing from 15 m (50 ft) at S/L, zero wind, hard runway 440 m (1,445 ft)
Landing run at S/L, zero wind, hard runway 259 m (850 ft)
Max still-air range at 59% cruising power 868 n miles (1,610 km; 1,000 miles)

VERIFIED

RACAL

RACAL AVIONICS LTD (Part of the Racal Electronics Group)

88 Bushey Road, London SW20 0JH
Telephone: 44 (181) 946 8011
Fax: 44 (181) 946 7430
MANAGING DIRECTOR: S O'Malley
MARKETING DIRECTOR: P. L. Gallagher
SENIOR PUBLICITY MANAGER: T. J. Rodgers

The origins of the company can be traced back to April 1980 when Decca Ltd was acquired by Racal Electronics. Decca's former activities in the fields of airborne navigation, flight management, ground-based aviation equipment and other systems such as airport surveillance radar were combined into one business to create Racal Avionics. Current responsibilities are in the fields of design development and manufacture of systems for: satellite communications, airborne navigation, navigation management, avionics management and ground-based navigation. The company also offers turnkey facilities, full aircraft design services and worldwide product support.

NEW ENTRY

RACAL AVIONICS (AEROSPATIALE) SA 330 PUMA NAVIGATION UPDATE (PNU)

TYPE: Military transport helicopter navigation upgrade.
PROGRAMME: Contract awarded in January 1993 to Racal Avionics (prime) by the RAF to upgrade the navigation systems on its fleet of 42 Puma HC Mk 1 helicopters. Contract extension signed in May 1994 (value £2 million) for additional avionics. All helicopters to be redelivered in 1996.
COSTS: $13.6 million at outset.
DESIGN FEATURES: Programme consists of the installation of the Supertans system and a new GPS receiver to offer P(Y) Code rather than C/A (coarse acquisition) Code together with TACAN and VOR/ILS/MKR. Elements of the instrumentation are being upgraded to enhance instrument flight rules (IFR) and cleared flight operations along airways by installing dual EHSIs that use LCD technology. The programme also consists of new cabling for digital systems, the provision of a MIL-STD-1388 integrated logistics support package which includes a five-year contractor logistics support programme.
AVIONICS: The Racal computer is linked to a Racal Doppler and Rockwell MAGR Global Positioning System (GPS) sensor as well as omnidirectional radio (VOR) and distance measuring equipment (DME). System accepts, correlates and integrates inputs from both sensors. The display is Night Vision Goggle (NVG) compatible.

RAF Aerospatiale Puma HC Mk 1 upgraded cockpit suite installed by Racal Avionics *(Racal Avionics)* *1995*

NEW ENTRY

RACAL

RACAL RADAR DEFENCE SYSTEMS (Part of the Racal Electronics Group)

Manor Royal, Crawley, West Sussex RH10 2PZ
Telephone: 44 (1293) 644125
Fax: 44 (1293) 611770

The Racal Radar Defence Systems business in Crawley (Southern England) formed part of the THORN EMI Sensors Group before its acquisition by Racal Electronics in 1995. The company forms part of Racal Electronic's Defence Radar and Avionics Group.

NEW ENTRY

RACAL (BAe/HAWKER DOMINIE T. Mk 2 NAVIGATION UPGRADE

TYPE: Navigation trainer avionics upgrade.
PROGRAMME: Contract awarded in 1992 to THORN EMI to upgrade 11 Royal Air Force Dominie T. Mk 1 trainers with Super Searcher radar. Flight trials began in 1995 at RAF Boscombe Down and Marshall Aerospace in Cambridge. Programme to be completed in 1996.
DESIGN FEATURES: Responsible as prime contractor for modification of the aircraft as well as installation of Super Searcher radar, associated radio equipment, avionics, navigation mission computing and systems integration. Contract includes complete logistic support package.
CUSTOMERS: Royal Air Force (11 aircraft).
COSTS: Contract valued at £20 million.

RAF BAe 125 Dominie T. Mk 2 navigation trainer during flight trials undergoing flight trials for the Super Searcher radar installed by Racal Radar Defence Systems *(Racal)* *1995*

NEW ENTRY

RAYTHEON CORPORATE JETS INC — *see US section*

SHORTS

SHORT BROTHERS PLC (Subsidiary of Bombardier Inc)

PO Box 241, Airport Road, Belfast BT3 9DZ
Telephone: 44 (1232) 458444
Fax: 44 (1232) 732974
Telex: 74688
OTHER WORKS: Newtownards, Castlereagh, Belfast (3), Dunmurry, Newtownabbey and Bournemouth
LONDON OFFICE: 14 Queen Anne's Gate, London SW1H 9AA
Telephone: 44 (171) 222 4555
Fax: 44 (171) 976 8505
PRESIDENT: R. W. R. McNulty, CBE
DIRECTOR, COMMUNICATION AND PUBLIC AFFAIRS: A. McRitchie

After more than 40 years in UK government ownership, Shorts, oldest established aircraft manufacturer in the world, was acquired in October 1989 by Bombardier of Canada. Shorts is European Group of Bombardier Aerospace, which also comprises North American Group. Shorts operates internationally from headquarters in Belfast, Northern Ireland, with subsidiary offices in London, Washington, Kuala Lumpur and Bahrain; currently employs approximately 9,500 personnel worldwide on broad-based work programme. Since 1989 Shorts has been transformed through a £200 million focused investment programme in plant, machinery and facilities, which has extended significantly its capabilities and profitability.

Production of Shorts 330 and 330-UTT (total 136 built) ended in September 1992; manufacturer of S312 (Embraer licence) Tucano completed early 1993.

The two manufacturing centres, Newtownabbey and Dunmurry, cater for advanced composites programme; over 27,871 m² (300,000 sq ft) production floor area allocated to manufacture of components for Boeing, Rolls-Royce, International Aero Engines, Fokker and British Aerospace. Shorts has major partnership role in Canadair Regional Jet, responsible for designing and manufacturing 9.75 m (32 ft) long centre section of fuselage, including forward and aft fuselage extension plugs, as well as wing flaps, ailerons, spoilers and spoilerons, and design of vanes. Shorts designed and manufactured the complete fuselage and tailplane for the Learjet 45 Business Jet. The first fuselage was delivered to Learjet in Wichita in October 1994.

Shorts designed, developed and now produces Fokker 100/70 wings as risk-sharing partner with Fokker of Netherlands and Daimler-Benz Aerospace Airbus of Germany. Shorts also currently supplies components for Boeing 737, 747, 757 and 777, and a large range of metal bonded assemblies are supplied to all of Boeing's current commercial aircraft programmes. Announced 28 March 1994 that Shorts will undertake 10 per cent of BAe work-share in European Future Large Military Aircraft (see under Euroflag in International section).

Shorts is source of high technology aerostructures: its Nacelle Systems Division specialises in design and manufacture of engine nacelle and turbofan components. The division makes a significant contribution to many Rolls-Royce power plant programmes, and makes innovative use of new materials and technologies to improve product lines. Company is sole source supplier of nacelle components for RB211 engines used on Boeing 747, 757 and 767, and for design, development and manufacture of nacelles for Trent 775 for new generation of large widebody aircraft such as A330. Shorts builds complete nacelles for Lycoming LF 507 on all RJ Avroliner variants and is responsible for design, development and manufacture of some 40 per cent of nacelles for IAE V2500 D series engine. Division also supplies complete nacelle for GE CF34 on Canadair Regional Jet and Challenger. In December 1992 Shorts and Hurel-Dubois announced formation of International Nacelle Systems (INS), to offer complete nacelle capability to aircraft and engine manufacturers worldwide. In partnership with Hurel-Dubois, Shorts will supply complete nacelles for the BMW Rolls-Royce BR710 engines selected for the Bombardier Global Express and Gulfstream V. It also designs and builds the complete nacelle for the Allison 300T engine for the Embraer 145 regional. Through a separate entity, Société International de Nacelles (Toulouse) (SINT), the two companies have established an aero-engine podding facility at Toulouse, France.

In March 1993 Shorts and Thomson-CSF announced the creation of a joint venture to exploit the synergies between Shorts close air defence systems and Thomson-CSF's electronics systems expertise. Shorts Missile System Limited (SMS) will offer fully integrated close air defence systems to customers worldwide. Starburst was in service with British Forces during Gulf War, when its reliability was stated by UK MoD to be 'unparalleled'. Defence Systems Division markets Shorland range of armoured vehicles, Skeet aerial target, and a weapon simulation and training product known as S1 Multi Arms Trainer.

UPDATED

SHORTS SC.7 SKYVAN Srs 3 and 3M

TYPE: Civil or military STOL utility light transport aircraft.

PROGRAMME: Design of the SC.7 Skyvan was started as a private venture in 1959, and the first prototype (G-ASCN) flew for the first time on 17 January 1963, with continental GTSIO-520 piston engines. It was re-engined with 388 kW (520 shp) Astazou II turboprops and first flew in its new form 2 October 1963. The change to Garrett TPE331 turboprops was made on the Skyvan Series 3 in 1967.

Orders for all versions of the Skyvan totalled 150. Production ceased in 1987.

VERSIONS: **Skyvan Srs 3**: Civil version, which superseded Srs 2 in 1968. First Srs 3 to fly was the second development aircraft, G-ASZI, which had been equipped with Astazous. The first flight with Garrett engines was made 15 December 1967, and a second aircraft (G-ASZJ) re-engined with TPE331s flew on 20 January 1968.

Skyvan Srs 3M: Military version of Srs 3, modified internally to accept optional equipment for typical military missions. Prototype (G-AXPT) flew for the first time in early 1970. Suitable for paratrooping and supply dropping, assault landing, troop transport, casualty evacuation, staff transport and vehicle ordnance transport.

Skyvan Srs 3M-200: Following an engineering product review programme in early 1982, the Skyvan was cleared in non-civil applications at maximum T-O weight of 6,804 kg (15,000 lb).

In February 1970 the Skyvan became the first aircraft to be certificated under the British Air Registration Board's new Civil Airworthiness Requirements for STOL operations.

UPGRADES: **THORN EMI**: THORN EMI (now Racal Radar Defence Systems) produced an airborne early warning version with a large nose radome for its Skymaster I-band radar. Two additional tail fins, aligned with the sides of the fuselage, maintained directional stability. Two or three radar operators could be housed in the cabin. Radar horizon is about 100 n miles (185 km; 115 miles) from the Skyvan's cruising height, and patrol time about 2 hours.

OPERATORS: Versions of the Skyvan are in service with the armed forces of the following countries: Austria (2), Botswana (2), Ghana (6), Guyana (1), Mauritania (1), Mexico (2), Nepal (3), Oman (15), United Arab Emirates (1), Singapore (6) and Thailand (3).

The following description applies to the final production version of the Srs 3 and 3M:

DESIGN FEATURES: Braced high-wing monoplane. Wing section NACA 63A series (modified). Thickness/chord ratio 14 per cent. Dihedral 2° 2′. Incidence 2° 30′. Cantilever two-spar tail unit with twin fins and rudders.

FLYING CONTROLS: All-metal single-slotted ailerons. Geared tabs in port and starboard ailerons, with manual trim in starboard aileron. All-metal single-slotted flaps. Fixed incidence tailplane. Geared trim tabs in outer elevators and rudders.

STRUCTURE: Wing is a light alloy structure consisting of a two-cell box with wing skins made up of a uniform outer sheet bonded to a corrugated inner sheet. The fuselage is a light alloy structure. Nose and crew cabins of conventional skin/stringer design. Elsewhere, the fuselage structure consists of double skin panels (flat outer sheets bonded to inner corrugated sheets), stabilised by frames. The tail unit is a cantilever all-metal two-spar structure.

LANDING GEAR: Non-retractable tricycle type. Single wheel on each unit. Steerable nosewheel. Main units carried on short sponsons. Electrohydraulics oleo-pneumatic shock-absorbers. Mainwheel tyres size 11.00-12, nosewheel tyre size 7.50-10. Tyre pressure (all units standard) 2.76 bars (40 lb/sq in). Hydraulically operated disc brakes with differential braking for steering. Low pressure tyres 2.07 bars (30 lb/sq in) available optionally.

POWER PLANT: Two 533 kW (715 shp) Garrett TPE331-2-201A turboprop engines, each driving a Hartzell HC-B3TN-5/T10282H three-blade variable-pitch propeller. Fuel in four tanks in pairs on top of fuselage between wingroots, each pair consisting of one tank of 182 litres (48 US gallons; 40 Imp gallons) capacity and one of 484 litres (127 US gallons; 106 Imp gallons) capacity. Total fuel capacity of 1,332 litres (352 US gallons; 293 Imp gallons). Provision for increase in total fuel capacity to 1,773 litres (468.4 US gallons; 390 Imp gallons) by installing four specially designed tanks in spaces between fuselage frames on each side, beneath main fuel tank. Oil capacity 7.73 litres (2.0 US gallons; 1.7 Imp gallons).

ACCOMMODATION: Crew of one, with provision for two. Accommodation (Srs 3) for up to 19 passengers, or 12 stretcher patients and attendants, or 2,085 kg (4,600 lb) of freight, vehicles or agricultural equipment. Srs 3M can accommodate 22 equipped troops; 16 paratroops and a dispatcher; 12 stretcher cases and two medical attendants; or 2,358 kg (5,200 lb) of freight. It carries its own lightweight vehicle loading ramps and has a one-piece door which leaves the fuselage threshold entirely clear of obstructions. Executive version provides luxury accommodation and equipment for nine passengers. Full width rear-loading door, and forward door on each side of crew compartment. Rear door can be opened in flight to permit the parachuting of loads up to 1.37 m (4 ft 6 in) in height. Cockpit and cabin heated by engine bleed air mixed with fresh air from intake in nose. Cabin unpressurised. Some aircraft fitted with Rolamat cargo-loading equipment.

SYSTEMS: Hydraulic system, pressure 172 bars (2,500 lb/sq in), operates flaps, wheel brakes and nosewheel steering. Bootstrap hydraulic reservoir. No pneumatic system. Electrical system utilises two busbars, operating independently, each connected to a 28 V 125A DC static inverter. General services are 28 V DC, some radio and instruments 115V AC.

AVIONICS AND EQUIPMENT: Radio optional. Typical installation for operations in Europe and USA consists of duplicated VHF, duplicated VOR/ILS, marker beacon receiver and ADF. Provision for HF, DME, transponder, Bendix M4D autopilot and weather radar. Blind-flying instrumentation standard.

EQUIPMENT (Srs 3M): Port side blister window for an air dispatcher; two anchor cables for parachute static lines; a guard rail beneath the tail to prevent control surface fouling by the static lines; inward-facing paratroop seats with safety nets; parachute signal light; mounts for NATO stretchers; and roller conveyors for easy loading and paradropping of pallet-mounted supplies.

DIMENSIONS, EXTERNAL:

Wing span	19.79 m (64 ft 11 in)
Wing chord (constant)	1.78 m (5 ft 10 in)
Length overall: without radome	12.21 m (40 ft 1 in)
with radome	12.60 m (41 ft 4 in)
Height overall	4.60 m (15 ft 1 in)
Tailplane span	5.28 m (17 ft 4 in)
Wheel track	4.21 m (13 ft 10 in)
Wheelbase	4.52 m (14 ft 10 in)
Propeller diameter	2.59 m (8 ft 6 in)
Propeller ground clearance	1.52 m (5 ft 0 in)
Crew and passenger doors (fwd, port and stbd):	
Height	1.52 m (5 ft 0 in)
Width	0.51 m (1 ft 8 in)
Height to sill	1.14 m (3 ft 9 in)
Rear-loading door: Height	1.78 m (5 ft 10 in)
Width	1.98 m (6 ft 6 in)
Height to sill	0.74 m (2 ft 5 in)

DIMENSIONS, INTERNAL:

Cabin, excl flight deck: Length	5.67 m (18 ft 7 in)
Max width	1.98 m (6 ft 6 in)
Max height	1.98 m (6 ft 6 in)
Floor area	11.15 m² (120 sq ft)
Volume	22.09 m³ (780 cu ft)

AREAS:

Wings, gross	35.12 m² (378 sq ft)
Ailerons (total)	3.00 m² (32.3 sq ft)
Trailing-edge flaps (total)	5.86 m² (63.1 sq ft)
Fins	7.62 m² (82.0 sq ft)
Rudders, incl tabs	2.41 m² (25.9 sq ft)
Tailplane	7.53 m² (81.0 sq ft)
Elevators, incl tabs	3.62 m² (39.0 sq ft)

WEIGHTS AND LOADINGS (with 1,332 litres; 351 US gallons; 293 Imp gallons of fuel):

Basic operating weight: Srs 3	3,331 kg (7,344 lb)
Srs 3M	3,356 kg (7,400 lb)
Srs 3M-200, equipped	3,768 kg (8,307 lb)
Typical operating weight as freighter:	
Srs 3	3,447 kg (7,600 lb)
Srs 3M	3,456 kg (7,620 lb)
Typical operating weight:	
Srs 3, with passengers	3,674 kg (8,100 lb)
Srs 3M, with troops	3,778 kg (8,330 lb)
Srs 3M-200 paratroop transport	3,943 kg (8,692 lb)
Max payload for normal T-O weight:	
Srs 3	2,086 kg (4,600 lb)
Srs 3M, 3M-200	2,358 kg (5,200 lb)
Max payload for overload T-O weight:	
Srs 3M	2,721 kg (6,000 lb)
Max fuel weights: standard tanks	1,052 kg (2,320 lb)
with extra tanks	1,415 kg (3,120 lb)
Max T-O weight: Srs 3, normal	5,670 kg (12,500 lb)
Srs 3M, normal	6,214 kg (13,700 lb)
Srs 3M, overload	6,577 kg (14,500 lb)
Srs 3M-200	6,804 kg (15,000 lb)
Max landing weight: Srs 3	5,670 kg (12,500 lb)
Srs 3M	6,123 kg (13,500 lb)
Srs 3M-200	6,577 kg (14,500 lb)
Max wing loading: Srs 3	163.6 kg/m² (33.5 lb/sq ft)
Srs 3M	179.1 kg/m² (36.7 lb/sq ft)
Srs 3M-200	196.3 kg/m² (40.2 lb/sq ft)
Max power loading: Srs 3	5.32 kg/kW (8.74 lb/shp)
Srs 3M	6.17 kg/kW (9.58 lb/shp)
Srs 3M-200	6.38 kg/kW (10.49 lb/shp)

PERFORMANCE (at max T-O weight, except where indicated with 1,332 litres; 351 US gallons; 293 Imp gallons of fuel):

Shorts SC.7 Skyvan Srs 3M of the Ghana Air Force *(Peter J. Cooper)*

1994

Never-exceed speed:
  Srs 3, 3M 217 kts (402 km/h; 250 mph) EAS
Max cruising speed at 3,050 m (10,000 ft):
  Srs 3, 3M at max continuous power 175 kts (324 km/h; 202 mph)
  Srs 3M-200 at max continuous power at AUW of 6,577 kg (14,500 lb) 174 kts (322 km/h; 200 mph)
  Srs 3, 3M at cruise power 168 kts (311 km/h; 193 mph)
  Srs 3M-200 at cruise power at AUW of 6,577 kg (14,500 lb) 166 kts (308 km/h; 191 mph)
Econ cruising speed at 3,050 m (10,000 ft):
  Srs 3, 3M 150 kts (278 km/h; 173 mph)
  Srs 3M-200 at AUW of 6,577 kg (14,500 lb) 150 kts (278 km/h; 173 mph)
Stalling speed, flaps down:
  Srs 3 60 kts (111 km/h; 69 mph) EAS
  Srs 3M 62 kts (115 km/h; 71 mph) EAS
Max rate of climb at S/L: Srs 3 500 m (1,640 ft)/min
  Srs 3M 466 m (1,530 ft)/min
Service ceiling at 30.5 m (100 ft)/min climb:
  Srs 3 6,858 m (22,500 ft)
  Srs 3M 6,705 m (22,000 ft)
Service ceiling, one engine out (15 m; 50 ft/min climb):
  Srs 3 3,810 m (12,500 ft)
  Srs 3M 2,895 m (9,500 ft)
Min ground turning radius 3.76 m (12 ft 4 in)
Runway LCN at AUW of 5,670 kg (12,500 lb):
  standard tyres 3.5
  low pressure tyres 3.0
T-O run, STOL, unfactored: Srs 3 213 m (700 ft)
  Srs 3M 238 m (780 ft)
  Srs 3M-200 290 m (950 ft)
T-O run (normal): Srs 3 (BCAR) 512 m (1,680 ft)
T-O to 10.7 m (35 ft), Transport Group A, ISA at S/L:
  Srs 3 1,020 m (3,350 ft)
T-O to 15 m (50 ft), STOL, unfactored:
  Srs 3 320 m (1,050 ft)
  Srs 3M 384 m (1,260 ft)
  Srs 3M-200 488 m (1,600 ft)
T-O to 15 m (50 ft):
  Srs 3 (BCAR, normal) 610 m (2,000 ft)
  Srs 3 (BCAR, STOL) 482 m (1,580 ft)
  Srs (FAR Pt 23) 488 m (1,600 ft)
Landing run from 15 m (50 ft) at max landing weight:
  Srs 3 (BCAR, normal) 622 m (2,040 ft)
  Srs (BCAR, STOL) 567 m (1,860 ft)
  Srs 3 (FAR Pt 23) 451 m (1,480 ft)
  Srs 3M (STOL, unfactored) 425 m (1,395 ft)
  Srs 3M-200 (STOL, unfactored) 457 m (1,500 ft)
Landing run from 15 m (50 ft), Transport Group A, ISA at S/L, at max landing weight 1,010 m (3,320 ft)
Landing run from 9 m (30 ft), at max landing weight:
  Srs 3 (STOL, unfactored) 351 m (1,150 ft)
  Srs 3 (BCAR, STOL) 500 m (1,640 ft)
Landing run, at max landing weight:
  Srs 3M (STOL, unfactored) 212 m (695 ft)
Range at long-range cruising speed, 45 min reserves:
  Srs 3 600 n mile (1,115 km; 694 miles)
  Srs 3M 580 n miles (1,075 km; 670 miles)
Range at long-range cruising speed, 45 min reserves:
  Srs 3 (typical freighter) with 1,815 kg (4,000 lb) payload, ISA 162 n miles (300 km; 187 miles)
  Srs 3M (typical freighter with 2,268 kg (5,000 lb) payload, ISA 208 n mile (386 km; 240 miles)
  Srs 3M-200 with 1,815 kg (4,000 lb) payload, ISA + 10°C 540 n miles (1,000 km; 621 miles)

UPDATED

SHORTS 330

TYPE: Twin-turboprop transport.

PROGRAMME: Originally known as SD3-30; first flight of first prototype (G-BSBH) 22 August 1974; CAA certification to full Transport Category 18 February 1976; US FAR Pt 25 and Pt 36 approval 18 June 1976; subsequent approvals from Canadian Department of Transport, German LBA and Australian Department of Transport. First order (three), by Command Airways of Poughkeepsie, New York, 14 August 1974; deliveries began June 1976; first entered service with Time Air, 24 August 1976.

VERSIONS: **330-200:** Standard passenger version, as detailed.

330-UTT: Military utility tactical transport version. Described separately.

Sherpa: Freighter version of 330, with ramp type full width rear-loading door. Described separately.

UPGRADES: **Field Aircraft Services of Calgary:** Contract from the US Army to modify four ex-airline 330s to military standard for operations in Kwajalein area of pacific.

OPERATORS: Versions of the Shorts 330 are in service with the armed forces of the following countries: Thailand (2), United Arab Emirates (1) and USA (23).

CUSTOMERS: Orders and options for 330, 330-UTT and Sherpa totalled 170 by start of 1991; Sherpa used by US Air Force as **C-23A** and US Army National Guard as **C-23B**; US Army also leases six ex-airline 330s for operation in Kwajalein area of Pacific.

DESIGN FEATURES: Derived from smaller STOL Skyvan, retaining latter's proven characteristics, including large square-section unpressurised cabin with low floor level, braced high-mounted wings, twin tail unit, and safe-life concept and design philosophy for structural components; conforms with CAB Pt 298 (US); meets FAR Pt 36 noise requirements by substantial margin.

First 26 Shorts 330s powered by 875 kW (1,173 shp) PT6A-45As; next 40 given PT6A-45Bs; subsequent aircraft fitted with more powerful PT6A-45Rs; and higher equipment standard; optional Goodrich pneumatic boot de-icing on wing and tail leading-edges.

NACA 63A series (modified) wing sections; thickness/chord ratio 18 per cent at root, 14 per cent on outer panels; dihedral 3° on outer panels.

The following description applies to standard 330-200 passenger version:

FLYING CONTROLS: Single-slotted ailerons with geared trim tabs; twin unshielded horn aerodynamic balance rudders; full-span elevator, aerodynamically balanced by setback hinges; geared trim tabs in elevator and starboard rudder (port rudder, trim only); single-slotted three-section flaps; fixed incidence tailplane.

STRUCTURE: All-metal safe-life construction; wing centre-section (integral with top of centre-fuselage) tapered on leading- and trailing-edges, and is two-spar single-cell box with conventional skin and stringers; strut-braced wing outer panels are reinforced Skyvan constant-chord units, each with two-cell box and having smooth outer skin bonded to corrugated inner skin; fuselage in two main portions; nose portion (including flight deck, nosewheel bay and forward baggage compartment), centre (including main wing spar attachment frames and lower transverse beams carrying main landing gear and associated fairings), and rear portion (including aft baggage compartment and tail unit attachment frames); nose and rear fuselage of skin/stringer design, remainder smooth outer skin bonded to corrugated inner skin and stabilised by frames; two-spar twin-fin tail unit, with reinforced tailplane leading-edge.

LANDING GEAR: Menasco retractable tricycle type, with single wheel on each unit. Main units carried on short sponsons, into which the wheels retract hydraulically. Oleo-pneumatic shock-absorbers. Nosewheel is steerable ±50°. Mainwheel tyre size 34 × 10.75-16; nosewheel tyre size 9.6. Normal tyre pressures: main units 5.45 bars (79 lb/sq in), nose unit 3.79 bars (55 lb/sq in).

POWER PLANT: Two 893 kW (1,198 shp) Pratt & Whitney Canada PT6A-45R turboprops, each driving a Hartzell

Shorts 330 30-seat regional airliner in Thai Airways markings

1994

five-blade constant-speed fully feathering metal low-speed propeller. Fuel tanks in wing centre-section/fuselage fairing; total usable capacity increased from original 2,182 litres (576 US gallons; 480 Imp gallons) to 2,546 litres (672.5 US gallons; 560 Imp gallons) in January 1985. Normal cross-feed provisions to allow for pump failure. Single pressure refuelling point in starboard landing gear fairing, backed by three gravity refuelling points in fuselage spine.

ACCOMMODATION: Crew of two on flight deck, plus cabin attendant. Dual controls standard. Standard seating for 30 passengers, in 10 rows of three at 76 cm (30 in) pitch, with wide aisle. Seat rails fitted to facilitate changes in configuration. Galley, toilet and cabin attendant's seat at rear. Large overhead baggage lockers. Entire accommodation soundproofed and air conditioned. Baggage compartments in nose and to rear of cabin, each with external access and capable of holding a combined total of 500 kg (1,100 lb) of baggage. Passenger door is at rear of cabin on port side. Passenger version has two emergency exits on the starboard side, two on the port side (including passenger door) and one in the flight deck roof. Mixed traffic version has full access to these emergency exits. For mixed passenger/freight operation a partition divides the cabin into a rear passenger area (typically for 18 persons) and a forward cargo compartment, the latter being loaded through a large port side door capable of admitting ATA 'D' type containers. In all-cargo configuration the cabin can accommodate up to seven 'D' type containers, with ample space around them for additional freight. Cabin floor is flat throughout its length, and is designed to support loadings of 181 kg (400 lb)/ft run at 610.3 kg/m^2 (125 lb/sq ft). Locally reinforced areas of higher strength are also provided. Seat rails can be used as cargo lashing points. Freight loading is facilitated by the low-level cabin floor.

SYSTEMS: Hamilton Standard air conditioning system, using engine bleed air. Hydraulic system of 207 bars (3,000 lb/sq in), supplied by engine-driven pumps, operates landing gear, nosewheel steering, flaps and brakes (at half pressure) and includes emergency accumulators. Air/oil reservoir pressurised to 1.72 bars (25 lb/sq in) at 20°C. Main electrical system, for general services, is 28 V DC and is of the split busbar type with cross-coupling for essential services. Lucas 28 V DC starter/generator for engine starting and aircraft services, with separate 1.5 kW 200 V AC output for windscreen anti-icing and demisting. Special AC sources of 115 V and 26 V available at 400 Hz for certain instruments, avionics and fuel booster pumps. Anti-icing standard for engine intake ducts, inlet lips and propellers. Optional de-icing of wing and tailplane leading-edges.

AVIONICS: Wide range of radio and navigation equipment available to customer's requirements. Typical standard avionics comprise duplicated VHF communications and navigation systems, two glide slope/marker beacon receivers, two ILS repeaters, two radio magnetic indicators, one ADF, one transponder, one DME, PA system and weather radar. Flight data recorder and voice recorder available as standard options.

EQUIPMENT: Passenger safety equipment standard.

DIMENSIONS, EXTERNAL:

Wing span	22.76 m (74 ft 8 in)
Wing chord (standard mean)	1.85 m (6 ft 0¾ in)
Length overall	17.69 m (58 ft 0½ in)
Width of fuselage	2.24 m (7 ft 4 in)
Height overall	4.95 m (16 ft 3 in)
Tailplane span	5.68 m (18 ft 7¾ in)
Wheel track	4.24 m (13 ft 10⅞ in)
Wheelbase	6.15 m (20 ft 2 in)
Propeller diameter	2.82 m (9 ft 3 in)
Propeller ground clearance	1.83 m (6 ft 0 in)
Cabin floor: Height above ground	0.94 m (3 ft 1 in)
Passenger door (port, rear): Height	1.57 m (5 ft 2 in)
Width	0.71 m (2 ft 4 in)
Height to sill	0.94 m (3 ft 1 in)
Cargo door (port, fwd): Height	1.68 m (5 ft 6 in)
Width	1.42 m (4 ft 8 in)
Height to sill	0.94 m (3 ft 1 in)

DIMENSIONS, INTERNAL:

Cabin: Max length, incl toilet	9.47 m (31 ft 1 in)
Max width	1.93 m (6 ft 4 in)
Max height	1.93 m (6 ft 4 in)
Floor area	18.77 m^2 (202 sq ft)
Volume (all-cargo)	34.83 m^3 (1,230 cu ft)
Baggage compartment volume:	
nose	1.27 m^3 (45 cu ft)
rear of cabin	2.83 m^3 (100 cu ft)
Cabin overhead lockers (total)	1.13 m^3 (40 cu ft)

AREAS:

Wings, gross	42.1 m^2 (453.0 sq ft)
Ailerons (total, aft of hinges)	2.55 m^2 (27.5 sq ft)
Trailing-edge flaps (total)	7.74 m^2 (83.3 sq ft)
Fins (total)	8.65 m^2 (93.1 sq ft)
Rudders (total, aft of hinges)	2.24 m^2 (24.1 sq ft)
Tailplane (total)	7.77 m^2 (83.6 sq ft)
Elevator (total, aft of hinges)	2.55 m^2 (27.5 sq ft)

WEIGHTS AND LOADINGS:

Weight empty, equipped (incl crew of three):	
330-200 for 30 passengers	6,680 kg (14,727 lb)
Fuel	2,032 kg (4,480 lb)
Max payload for normal max T-O weight:	
30 passengers and baggage	2,653 kg (5,850 lb)
cargo	3,400 kg (7,500 lb)
Max T-O weight	10,387 kg (22,900 lb)
Max landing weight	10,251 kg (22,600 lb)
Max wing loading	246.8 kg/m^2 (50.55 lb/sq ft)
Max power loading	5.81 kg/kW (9.56 lb/shp)

PERFORMANCE (at max T-O weight, ISA at S/L, except where indicated):

Max cruising speed at 3,050 m (10,000 ft), AUW of 9,525 kg (21,000 lb) 190 kts (352 km/h; 218 mph)

Econ cruising speed at 3,050 m (10,000 ft), AUW of 9,525 kg (21,000 lb) 160 kts (296 km/h; 184 mph)

Stalling speed, flaps and landing gear up 90 kts (167 km/h; 104 mph) EAS

Stalling speed at max landing weight, flaps and landing gear down 73 kts (136 km/h; 85 mph) EAS

Max rate of climb at S/L 360 m (1,180 ft)/min

Service ceiling, one engine out, AUW of 9,072 kg (20,000 lb) 3,500 m (11,500 ft)

Min ground turning radius about nosewheel 7.77 m (25 ft 6 in)

T-O distance (FAR Pt 25 and BCAR Gp A):

ISA	1,042 m (3,420 ft)
ISA + 15°C	1,295 m (4,250 ft)

Landing distance, AUW of 9,072 kg (20,000 lb):

BCAR	1,143 m (3,750 ft)
FAR	1,030 m (3,380 ft)

Runway LCN at max T-O weight 10.7

Range with max passenger payload, cruising at 3,050 m (10,000 ft), no reserves 473 n miles (876 km; 544 miles)

Range with max fuel, cruising at 3,050 m (10,000 ft), no reserves:

passenger version, 1,966 kg (4,335 lb) payload 915 n miles (1,695 km; 1,053 miles)

cargo version, 2,306 kg (5,085 lb) payload 758 n miles (1,403 km; 872 miles)

OPERATIONAL NOISE LEVELS (FAR Pt 36):

T-O	88.9 EPNdB
Sideline	84.7 EPNdB
Approach	92.9 EPNdB

UPDATED

SHORTS 330-UTT

TYPE: Military utility tactical transport.

PROGRAMME: Military version of 330; production began 7 September 1982.

CUSTOMERS: Royal Thai Army received two and Royal Thai Border Police two 1984-85; latter now also operates one army aircraft.

DESIGN FEATURES: Basic airframe and power plant of 330 unchanged; maximum payload increased to 3,630 kg (8,000 lb); maximum operational necessity T-O weight 11,158 kg (24,600 lb); strengthened cabin floor, reconfigured avionics panel.

ACCOMMODATION: Up to 33 troops, 30 paratroops plus jumpmaster (exit via inward-opening rear door each side), or 15 stretchers plus four seated personnel.

PERFORMANCE:

Cruising speed at 3,050 m (10,000 ft), AUW of 9,979 kg (22,000 lb):

high-speed cruise, max continuous power	201 kts (372 km/h; 231 mph)
long-range cruise	160 kts (296 km/h; 184 mph)

Max rate of climb at S/L at normal max T-O weight of 10,387 kg (22,900 lb):

two engines	381 m (1,250 ft)/min
one engine	89 m (290 ft)/min
STOL T-O run at S/L, 15° flap	415 m (1,360 ft)
STOL T-O to 15 m (50 ft), 15° flap	644 m (2,110 ft)

STOL landing from 15 m (50 ft) at AUW of 9,525 kg (21,000 lb), flaps down, propeller reversal 488 m (1,600 ft)

STOL landing run, conditions as above 235 m (770 ft)

Range with 30 fully armed assault troops 600 n miles (1,112 km; 691 miles)

VERIFIED

SHORTS SHERPA

US Air Force designation: C-23A

US Army National Guard designations: C-23B and C-23B+

TYPE: Twin-turboprop freight/utility version of 330-200.

PROGRAMME: First flight of prototype 23 December 1982; first EDSA C-23A flown 6 August 1984; all 18 delivered by 6 December 1985; C-23A fleet was based at Zweibrücken, Germany, for transporting high-priority spares between over 20 peacetime USAF bases in Europe; redeployed in USA 1991; eight now with US Forestry Service, six with US Army, four with USAF. Deliveries of 10 enhanced C-23Bs began September 1990, to US Army National Guard; one each to State Area Commands in Alabama, California, Connecticut, Mississippi, Oregon, Puerto Rico and Utah; three to Aviation Classification and Repair Depot, Missouri, six more C-23Bs ordered for US Army National Guard January 1992 for delivery by end September 1992.

UPGRADES: **Shorts:** C-23B+ conversion programme for the US Army National Guard. See separate entry.

CUSTOMERS: Eighteen C-23As ordered by US Air Force March 1984, for 10th Military Airlift Squadron (MAC) in EDSA (European Distribution System Aircraft) role; US Department of Army ordered 10 C-23Bs October 1988 to replace DHC-4 Caribous with Army National Guard.

Civil registered Sherpa delivered to Venezuelan government to support hydro-electric power development schemes.

DESIGN FEATURES: Retains features of all passenger 330-200, allowing utility passenger role; forward freight door and wide-body hold of 330-200 unchanged; power-assisted full-width rear cargo ramp/door for through loading (operated inside/outside aircraft, and lowered to variety of positions to simplify loading from range of ground equipment); forward baggage compartment of 330-200 retained; standard airline containers can be accommodated in main cabin, up to LD3 size, making aircraft particularly suited to short haul cargo feeder services; loads can include two ½ ton vehicles or bulky cargo; cabin also suitable for specialist role equipment; roller conveyor systems, including optional pallet locks picking up on aircraft's standard seat rails. No passenger cabin windows in C-23A.

ArNG aircraft embody strengthened wings, more powerful 1,061 kW (1,424 shp) PT6A-65AR turboprop and five-blade propellers of Shorts 360, plus uprated landing gear, more advanced flight deck instrumentation and air-openable rear freight doors for supply dropping; ramp upper section retracts inward and upward, while bottom section lowers to provide drop platform; maximum payload increased to 3,302 kg (7,280 lb).

The following details (except Avionics) apply to C-23B:

ACCOMMODATION: Crew of two on flight deck, plus optional jump seat facility. Dual controls standard. Aircraft air conditioned throughout. Baggage compartment in nose with external access. Passenger door at rear of cabin on port

Shorts C-23A Sherpa operated by the US Air Force

1994

Shorts C-23B Sherpa operated by the US Army National Guard

1994

side. Cargo door at front of cabin on port side. Power-assisted full-width rear-loading ramp/door. In an all-cargo configuration the cabin can accommodate up to seven CO8 or four LD3 containers. Cabin floor is flat throughout its length and is designed to support 181 kg (400 lb)/ft run at 610.3 kg/m² (125 lb/sq ft). The locally reinforced centre cabin area is able to carry 272 kg (600 lb) per foot run at 732.4 kg/m² (150 lb/sq ft). A further 272 kg (600 lb) total load can be stowed on ramp/door. Seat rails can be used as cargo lashing points. Freight loading facilitated by level cabin floor. Alternative loads: seating capacity for 30 personnel in airline standard seats; 27 paratroops plus jumpmaster on side-facing seats; 15 stretchers plus three attendants; two Land Rover/Jeep class vehicles; various aircraft engines and associated transport dollies.

AVIONICS: Avionics in C-23As for US Air Force include single UHF and HF radios, dual VHF-AM/FM, two flight directors, dual VOR/ILS, a Litton LTN-96 ring laser gyro inertial navigation system, Tacan, dual ADF, flight data recorder, cockpit voice recorder, IFF transponder, GPWS, radar altimeter, Collins RNS-300 colour weather radar with terrain-mapping.

DIMENSIONS, EXTERNAL: As for 330-200 except:

Wing span	22.81 m (74 ft 10 in)
Height overall	5.00 m (16 ft 5 in)
Wheel track	4.26 m (14 ft 0 in)
Propeller ground clearance	1.82 m (5 ft 11½ in)
Main cabin door (port, rear):	
Height	1.59 m (5 ft 2½ in)
Width	0.69 m (2 ft 3 in)
Cargo door (port, fwd): Height	1.66 m (5 ft 5½ in)
Width	1.41 m (4 ft 7½ in)
Rear ramp/door: opening inward, clear opening	1.56 × 1.79 m (5 ft 1½ in × 5 ft 10½ in)
opening outward, clear opening	1.92 × 1.79 m (6 ft 3½ in × 5 ft 10½ in)

DIMENSIONS, INTERNAL:

Cabin: Max length	9.09 m (29 ft 10 in)
Width at side seat rails	1.89 m (6 ft 2½ in)
Height	1.97 m (6 ft 5½ in)
Floor area	17.18 m² (185.0 sq ft)
Volume	35.28 m³ (1,246 cu ft)
Baggage compartment volume (nose)	1.27 m³ (45 cu ft)

AREAS:

Ailerons (gross)	3.27 m² (35.2 sq ft)
Rudders (total)	2.86 m² (30.8 sq ft)
Elevator (total)	3.60 m² (38.8 sq ft)

WEIGHTS AND LOADINGS:

Operational weight empty (freight role)	7,276 kg (16,040 lb)
Max payload	3,302 kg (7,280 lb)
Max T-O weight	11,612 kg (25,600 lb)
Max landing weight	11,385 kg (25,100 lb)
Max wing loading	275.91 kg/m² (56.51 lb/sq ft)
Max power loading	6.50 kg/kW (10.68 lb/shp)

PERFORMANCE (at max T-O weight, ISA at S/L except where indicated):

Max cruising speed at 3,050 m (10,000 ft) and AUW of 9,980 kg (22,000 lb)	194 kts (359 km/h; 223 mph)
Normal cruising speed	180 kts (333 km/h; 207 mph)
Stalling speed, flaps and landing gear up	97 kts (179 km/h; 111 mph)
Stalling speed at max landing weight, flaps and landing gear down	78 kts (145 km/h; 90 mph)
Max rate of climb at S/L	445 m (1,460 ft)/min
Service ceiling, OEI, AUW of 9,979 kg (20,000 lb)	3,660 m (12,000 ft)
T-O run at max T-O weight	564 m (1,850 ft)
T-O run at 15 m (50 ft)	802 m (2,630 ft)
Landing from 15 m (50 ft) at max landing weight	586 m (1,920 ft)
Landing run	345 m (1,130 ft)
Range: with max payload and no reserves	446 n miles (827 km; 514 miles)
with 2,318 kg (5,110 lb) payload	1,031 n miles (1,912 km; 1,188 miles)

VERIFIED

Shorts C-23A Sherpa twin-turboprop freight/utility aircraft unloading vehicles

1994

SHORTS C-23B+ SHERPA CONVERSION

TYPE: Military freighter/utility conversion.

PROGRAMME: Contract to supply the US Army National Guard with a further 20 (option for 10 more) C-23B Sherpa aircraft. Programme involves converting Shorts 330-200 civilian models to C-23B standard. Programme started in 1994 and is due to last three years. US Army National Guard exercised 8 of their 10 options in November 1994.

COSTS: $100 million for supply of the 20 C-23B aircraft and $44 million for the optional 10 aircraft.

DESIGN FEATURES: Programme to include removal of a 91 cm (36 in) plug from the forward fuselage and the rear fuselage to be replaced aft of the passenger door by a C-23B aft fuselage (with rear-loading ramp). Other structural modifications to include a new C-23B tailplane and twin fins, adjustment of the engine and engine cowls and increase in fuel tank capacity.

AVIONICS: New avionics to include updated Collins communication, navigation, autopilot and colour weather radar systems, a 3M Stormscope, and Bendix IFF transponder as well as a flight management system (FMS).

UPDATED

SHORTS 360-300 and 360-300F

TYPE: Twin-turboprop commuter transport (300) and freighter (300F).

PROGRAMME: Details released 10 July 1980; first flight of prototype (G-ROOM) 1 June 1981 (PT6A-45 engines); PT6A-65Rs subsequently fitted; CAA certification 3 September 1982; FAA certification to FAR Pt 25 and Pt 36 early November 1982; first flight of production aircraft 19 August 1982; launch customer of 360-300F Rheinland Air Service of Dusseldorf, Germany (two delivered from March 1989).

VERIONS: **360:** Initial production version with PT6A-65Rs.

360 Advanced: Production version from November 1985: 1,062 kW (1,424 shp) PT6A-65ARs.

360-300: Six-blade synchrophasing propellers; cambered wing lift struts; low-drag engine nacelle exhaust stubs; lightweight seats; PT6A-67R engines giving increased maximum cruising speed and improved weight/altitude/temperature limits, offering better hot and high performance, improved en route climb and allowing higher maximum T-O weight. Optional Cat. II autopilot; wet sink toilet facility; supplementary ground air conditioning system; protective liners for freighting.

360-300F: Freighter adaptation of 360-300; up to 4,536 kg (10,000 lb) load; optional enlarged port forward cargo door enabling loading of LD3-size containers; optional roller conveyor with pallet locks, directional transfer mat, side guidance rails, and forward cargo restraint barrier.

CUSTOMERS: 36 entered commercial service with Suburban Airlines of Pennsylvania 1 December 1982; first 360-300 to Philippine Airlines 18 March 1987; first 360-300 certificated for 39 passengers operated by Capital Airlines in UK

Shorts 360-300 twin-turboprop commuter transport aircraft

1994

from October 1987; by January 1991 over 160 Shorts 360s were in service worldwide.

DESIGN FEATURES: Stretched (0.91 m; 3 ft fuselage plug ahead of wings) development of 330, specifically for short haul airline operations over about 120 n mile (222 km; 138 mile) average stage lengths; six extra passengers; strengthened outer wing panels and bracing struts; tapered rear fuselage with increased baggage capacity; drag-reducing single fin/rudder tail unit; more powerful, fuel efficient engines. 360-300F particularly suited to overnight parcel services.

FLYING CONTROLS: Single fin/rudder; constant-chord tailplane; elevators and rudder with trim tabs.

STRUCTURE: Similar to 330-200 but lengthened fuselage, tapering for new all-metal two-spar tail unit; strengthened outer wing panels/struts; heavy-duty floor panels for freighting.

LANDING GEAR: Similar to Shorts 330-200, but of Dowty design with Dunlop tyres. Mainwheel tyres size 37 × 11.75-16, pressure 5.38 bars (78 lb/sq in). Maxaret anti-skid units standard. Nosewheel steerable ±55°.

POWER PLANT: Two 1,062 kW (1,424 shp) Pratt & Whitney Canada PT6A-67R turboprops, each driving a Hartzell advanced technology six-blade constant-speed fully feathering propeller. New engine naccelles with low-drag exhaust stubs on Series 300. Fuel capacity 2,182 litres (576 US gallons; 480 Imp gallons).

ACCOMMODATION: Crew of two on flight deck, plus cabin attendant. Dual controls standard. Main cabin accommodation similar to Shorts 330-200, but seating 36 passengers in 12 rows of three (optionally, 39 passengers). Standard ground and in-flight air conditioning. Large overhead baggage lockers. Baggage compartments in nose and to rear of cabin, each with external access, giving equivalent of almost 0.17 m³ (6 cu ft) of baggage space per passenger (0.20 m³; 7.2 cu ft per passenger if locker space is included). Self-contained passenger stairs.

SYSTEMS: Generally as for Shorts 330-200 except for electrical system, which has Lear Siegler 28 V 300 A DC starter/generators and three 400 VA single-phase static inverters for AC power. Full de-icing and anti-icing systems standard.

AVIONICS: From Collins Pro Line II range, including dual FDS-65 flight director systems, dual VHF-21A com, dual VIR-32 VHF nav, dual DME-42, dual TDR-90 transponders, dual RMI-36, ADF-60A, dual MCS-65 magnetic compasses, and WXR-220 colour weather radar, plus Sundstrand Mk II GPWS, Honeywell YG7500 radar altimeter, Fairchild A100A voice recorder and GEC Plessey PV1584G data recorder. Options include Collins HF-230 HF com, APS-65 Cat. II autopilot and second ADF.

DIMENSIONS, EXTERNAL: As for Shorts 330-200 except:

Wing span	22.80 m (74 ft 9½ in)
Length overall	21.58 m (70 ft 9⅝ in)
Height overall	7.27 m (23 ft 10¼ in)
Tailplane span	7.19 m (23 ft 7 in)
Wheelbase	7.06 m (23 ft 2 in)
Propeller diameter	2.74 m (9 ft 0 in)
Propeller ground clearance	1.78 m (5 ft 10 in)
Rear door sill height	0.98 m (3 ft 2¼ in)

DIMENSIONS, INTERNAL:

Cabin: Length	11.02 m (36 ft 2 in)
Max width	1.93 m (6 ft 4 in)
Max height	1.93 m (6 ft 4 in)
Passenger compartment volume	41.06 m³ (1,450 cu ft)
Baggage compartment volume:	
forward	1.27 m³ (45 cu ft)
rear	4.81 m³ (170 cu ft)
lockers	1.47 m³ (52 cu ft)

AREA: As for Shorts 330-200 except:

Wings, gross	42.18 m² (454.0 sq ft)
Vertical tail surfaces (total)	8.49 m² (91.4 sq ft)
Horizontal tail surfaces (total)	9.85 m² (106.0 sq ft)

WEIGHTS AND LOADINGS:

Typical operating weight empty	7,870 kg (17,350 lb)
Max payload:	
36 passengers and baggage	3,184 kg (7,020 lb)
cargo	3,765 kg (8,300 lb)
Max fuel load	1,741 kg (3,840 lb)
Max T-O weight	12,292 kg (27,100 lb)
Max ramp weight	12,337 kg (27,200 lb)
Max landing weight	12,020 kg (26,500 lb)
Max wing loading	291.4 kg/m² (59.7 lb/sq ft)
Max power loading	5.79 kg/kW (9.52 lb/shp)

PERFORMANCE (to FAR Pt 25: at max T-O weight except where indicated:

Cruising speed at 3,050 m (10,000 ft) and 11,340 kg (25,000 lb) aircraft weight
216 kts (400 km/h; 249 mph)

Max rate of climb at S/L: ISA	290 m (952 ft)/min
ISA + 15°C	282 m (925 ft)/min
Service ceiling, one engine out	2,665 m (8,750 ft)
Min ground turning radius about nosewheel	8.21 m (26 ft 11 in)
Balanced T-O field length: ISA	1,305 m (4,280 ft)
ISA + 15°C	1,402 m (4,600 ft)
Landing distance at max landing weight: ISA	1,220 m (4,000 ft)
Runway LCN	14.1

Range at 3,050 m (10,000 ft), cruising at 216 kts (400 km/h; 249 mph), 50 n mile (93 km; 47 mile) diversion, 45 min hold, 54 kg (119 lb) fuel allowance, 36 passengers with baggage at 86 kg (190 lb) each
402 n miles (745 km; 463 miles)

Range as above at 182 kts (337 km/h; 210 mph) with 31 passengers and baggage
636 n miles (1,178 km; 732 miles)

VERIFIED

SHORTS S312 TUCANO

RAF designation: Tucano T. Mk 1

TYPE: Two-seat turboprop basic trainer and (exports) counter-insurgency.

PROGRAMME: Co-operation agreement between Shorts and Embraer of Brazil to develop new version of EMB-312 Tucano, to meet or exceed all UK MoD Air Staff Target 412 requirements as Jet Provost replacement for RAF, announced May 1984; selected by UK government 21 March 1985, partly because least expensive of competitors; first flight of EMB-312 with required Garrett engine (PP-ZTC) in Brazil 14 February 1986; air-freighted to UK after 14.35 flight test hours, reassembled as G-14-007 (later G-BTUC); first flight of G-14-007 in UK 11 April 1986; first flight of production T. Mk 1 (ZF135) 30 December 1986; formal roll-out 20 January 1987; ZF135 delivered to Aeroplane and Armament Experimental Establishment, Boscombe Down, 26 June 1987 for provisional type certificate trials; deliveries to Central Flying School, Scampton, began (with ZF138) 16 June 1988; flying by Tucano Course Design Team (CFS) started early August 1988; formal handover of first aircraft 1 September 1988. Shorts also responsible for logistic supplies; Airwork has maintenance contract at Church Fenton, Hunting Aircraft at Cranwell and (from 1 April 1993) Scampton, and Shorts at Finningley; five GEC-Marconi simulators delivered to bases from 1990. Weapon trials completed April 1991.

VERSIONS: **T. Mk 1:** Standard version to which following description applies.

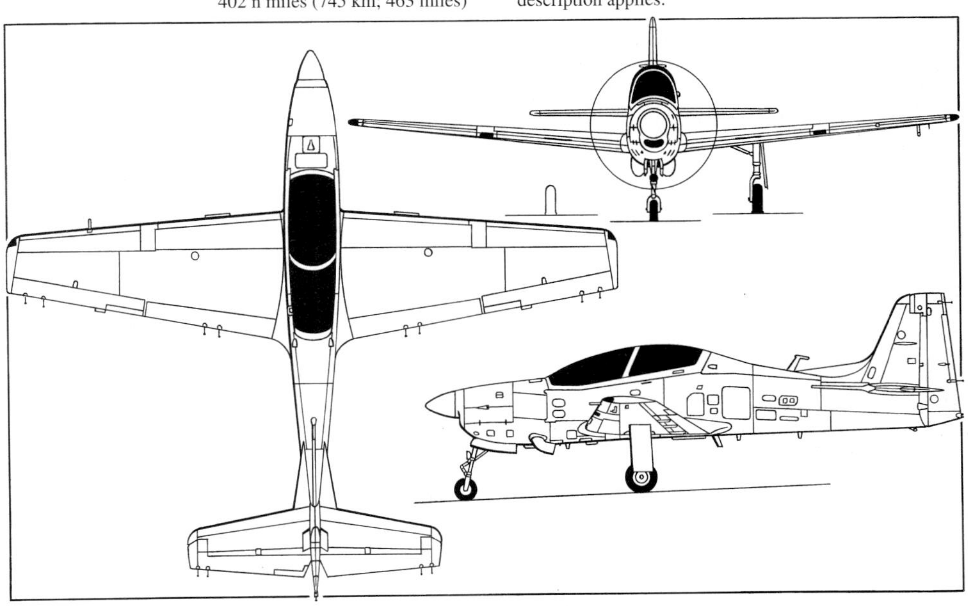

Shorts S312 Tucano basic trainer with Garrett TPE331-12B turboprop ***(Jane's/Mike Keep)***

1995

Shorts S312 Tucano two-seat basic trainer aircraft *(Shorts)*

1995

T. Mk 51: Version exported to Kenya for weapons training.

T. Mk 52: Version exported to Kuwait.

UPGRADES: **Shorts:** In January 1990 Shorts was contracted by the UK Ministry of Defence to modify the first 50 completed aircraft with strengthened flying controls, revised com/nav equipment and structural changes to extend fatigue life as introduced on production line.

OPERATORS: Versions of the Shorts Tucano are in service with the UK armed forces (130).

DESIGN FEATURES: Based on EMB-312, but only 20 per cent commonality. Modifications include TPE331-12B engine offering higher speed (particularly at low level) and increased rate of climb; ventral airbrake to control descent speed; structural strengthening for increased manoeuvre loads and fatigue life (12,000 hours); new cockpit layout; use of UK equipment; wing incidence 1° 13′; four wing strongpoints for armament training/light attack (for exports).

FLYING CONTROLS: Strengthened; hydraulically actuated ventral airbrake.

STRUCTURE: 7075-T73511 and 7075-T76, and 2024-T3 sheet, aluminium alloy two-spar wing torsion box; wing leading-edges strengthened against birdstrikes.

LANDING GEAR: Stressed for maximum sink rate of 4.0 m (13 ft)/s. Nosewheel unit supplied by Fairey Hydraulics; steerable ±20°. Dunlop wheels and tyres. Dunlop hydraulic single-disc brakes on mainwheels. Parking brake.

POWER PLANT: One 820 kW (1,100 shp) Garrett TPE331-12B-701A turboprop, driving a Hartzell HC-D4N-5C/D9327K four-blade constant-speed fully feathering reversible-pitch propeller. Additional 38 kW (51 ehp) derived from exhaust. Provision for two external tanks with total capacity of 640 litres (169 US gallons; 142 Imp gallons). Oil capacity 9.0 litres (2.4 US gallons; 2.0 Imp gallons).

ACCOMMODATION: Instructor and pupil in tandem, on Martin-Baker 8LC Mk 1 lightweight ejection seats effective down to zero altitude and 70 knots (130 km/h; 81 mph) and with through-canopy capability. Lucas two-piece canopy strengthened to withstand 1 kg (2.2 lb) birdstrike at 240 knots (444 km/h; 276 mph) and incorporating miniature detonating cord. No pressurisation.

SYSTEMS: Cockpit air conditioning by Normalair-Garrett with engine bleed air plus recirculated cockpit air through a regenerative turbofan system. Single hydraulic system, pressure 217.5 bars (3,150 lb/sq in), for landing gear extension and retraction, landing gear doors and airbrake. Accumulator to lower landing gear in emergency; pressure 125 bars (1,810 lb/sq in). DC electrical power provided by a 28 V 250 A Lucas starter/generator and two 24 Ah Ni/Cd batteries. Static inverter (260 VA) for 115 V single-phase AC power at 400 Hz. Normalair-Garrett oxygen system from a single bottle, capacity 2,250 litres (80 cu ft) at 124 bars (1,800 lb/sq in). Emergency oxygen bottle, capacity 70 litres (2.5 cu ft), mounted on each ejection seat. Engine air intake de-iced by engine bleed air; propeller, pitot head, static vents, and stall warning vane de-iced electrically.

AVIONICS: Standard avionics package includes Collins avionics control system. Bendix/King IFF, Collins DME, Collins VOR/ILS/MB.J.E.T. gyromagnetic compass. Collins VHF/UHF Collins VHF and Racal CCS; navigation data presented on EHSI and RMI in each cockpit; avionics have dual control; aircraft integrated monitoring system fitted as standard. Optional ADF and Tacan; provision for optional stores management system and optical gunsight; optional armament package includes Avimo optical gunsight, Saab/Mekel gunsight camera, ML Aviation stores management system. Base Ten weapon control electronics and other equipment from Dowty and Guardian Electronics.

ARMAMENT: Optional provision on export variants for four underwing hardpoints; inboard points stressed to 290 kg (640 lb) and outboard points to 225 kg (496 lb). Loads can include SAMP Type 25,250 kg bombs, SAMP Type BL61 125 kg bombs. LAU-32 rocket launcher, Matra F2b rocket launcher, FN Herstal twin 12.7 mm machine gun pod and FN Herstal heavy machine gun pods. ML Aviation CBLS200 practice bomb carriers; ERU-119 ejector release units.

DIMENSIONS, EXTERNAL:

Wing span	11.28 m (37 ft 0 in)
Propeller diameter	2.39 m (7 ft 10 in)
Propeller ground clearance	0.32 m (12.6 in)

AREAS:

Fin, excl dorsal fin	2.08 m² (22.40 sq ft)
Rudder, incl tab	1.57 m² (16.9 sq ft)
Tailplane, incl fillets	4.57 m² (49.20 sq ft)

WEIGHTS AND LOADINGS: A: aerobatic configuration, B: full weapons configuration):

Basic weight empty (max): A	2,210 kg (4,872 lb)
B	2,232 kg (4,920 lb)
Max internal fuel: A, B	545 kg (1,202 lb)
Max T-O weight: A	2,935 kg (6,470 lb)
B	3,600 kg (7,937 lb)
Max landing weight: A	2,935 kg (6,470 lb)
B	3,600 kg (7,937 lb)
Max zero-fuel weight: A	2,318 kg (5,110 lb)
Max wing loading: A	151.3 kg/m² (31.0 lb/sq ft)
B	185.6 kg/m² (38.02 lb/sq ft)
Max power loading: A	3.58 kg/kW (5.88 lb/shp)
B	4.39 kg/kW (7.21 lb/shp)

PERFORMANCE (typical weight of 2,900 kg; 6,393 lb, except where indicated):

Never-exceed speed (VNE)	330 kts (611 km/h; 380 mph)
Max level and max continuous speeds:	
ISA at S/L	269 kts (498 km/h; 310 mph)
ISA at 3,050 m (10,000 ft)	277 kts (513 km/h; 319 mph)
Max speed, flaps and/or landing gear down	145 kts (269 km/h; 167 mph)
Stalling speed, power off at 2,700 kg (5,952 lb) AUW:	
flaps and landing gear up	77 kts (141 km/h; 88 mph) EAS
flaps and landing gear down	70 kts (130 km/h; 81 mph) EAS
Max rate of climb at S/L	997 m (3,270 ft)/min
Service ceiling	10,365 m (34,000 ft)
T-O run	363 m (1,190 ft)
T-O to 15 m (50 ft)	590 m (1,930 ft)
Landing from 15 m (50 ft)	625 m (2,050 ft)
Landing run	360 m (1,180 ft)
Radius of action, 500 kg (1,102 lb) weapon load, 5 min over target	390 n miles (723 km; 499 miles)
Range at 7,620 m (25,000 ft) with max fuel, 10% reserves:	
internal fuel only	954 n miles (1,767 km; 1,099 mil
with external fuel	1,790 n miles (3,317 km; 2,061 mil
Endurance at econ cruising speed at 7,620 m (25,000 ft) 10% reserves	5 h 12 min
g limits:	
aerobatic at 2,935 kg (6,470 lb) AUW	+6.5/−3.3
with weapons, at 3,600 kg (7,937 lb) AUW	+4.4/−2.2
with weapons, at 3,100 kg (6,834 lb) AUW, stores less than 185 kg (408 lb) each	+6.0/−3.0

NEW ENTRY

Shorts S312 Tucano T. Mk 1 basic trainer of the Royal Air Force *(Shorts)*

1995

SMITHS INDUSTRIES AEROSPACE

SMITHS INDUSTRIES AEROSPACE (part of Smiths Industries Plc)

65 Finchley Road, London NW11 8DS
Telephone: 44 (181) 458 3232
Fax: 44 (181) 209 0526
CHAIRMAN: N. V. Barber
DIRECTOR, PRESS RELATIONS: R. Plumley

DEFENCE SYSTEMS DIVISION:
PO Box 188, Bishops Cleeve, Cheltenham, Gloucester GL52 4ZA
Telephone: 44 (1242) 673333
Fax: 44 (1242) 661798
MANAGING DIRECTOR: N. J. Wilton

Smiths Industries Aerospace is organised into the Civil Systems, Defence Systems, Engine & Fuel Systems, Naval & Marine Systems and the Product Support divisons. The company is active in the fields of design, development and manufacture of systems for civil and military aircraft (both fixed- and rotary-wing), naval and merchant vessels, fighting vehicles, space platforms and air traffic control (ATC) facilities.

The divisions contribute extensively to upgrade programmes and the Defense Systems division in North America led the project Kahu contract to upgrade the RNZAF fleet of A-4 Skyhawks.

NEW ENTRY

SMITHS INDUSTRIES (LOCKHEED) C-130 NAVIGATION UPGRADE

TYPE: Military transport aircraft upgrade.

PROGRAMME: In August 1994 the Royal Air Force awarded a contract to Smiths Industries Aerospace Defence Systems Division to upgrade half of its fleet of 59 Hercules C-130 aricraft. The programme is scheduled to be completed by September 1996. Smiths Industries Aerospace Product Support Division will be responsible for a five-year in-service support programme. The RAF has been granted five years to exercise an option for the upgrade of the remaining half of its Hercules fleet.

DESIGN FEATURES: The programme consists of the installation of a derivative of the Self-Contained Navigation System (SCNS) currently installed in over 550 USAF C-130 aircraft.

NEW ENTRY

ULTRA

ULTRA ELECTRONICS

Bridport Road, Greenford, Middlesex UB6 8UA
Telephone: 44 (181) 813 4321
Fax: 44 (181) 813 4322
Telex: 934512
CEO: Dr Julian Blogh
MARKETING DIRECTOR: Richard Lane

Ultra Electronics was formed from the Dowty Electronics Systems Division in October 1993 as a result of a management buy-in. The original seven businesses include Sonar and Communication Systems, Command and Control Systems, Controls Division, Electrics Division, Weapons Systems, Ocean Systems and Magnetics Division. During 1995 the company acquired the two new businesses; Noise and Vibration Systems and the Helitune Division.

Ultra Electronics is active in the design, development and manufacture of active noise control systems for aircraft cabins, active and passive sonobuoys, and compact High-Pressure Pure Air Generators (HPPAGs) for missile cooling and pneumatic ejection of weapons from aircraft. Ultra also designs and manufactures command and control equipment for ships and submarines, communication systems, aircraft flight controls and engine accessories, cockpit systems, underwater measurement ranges and sidescan sonars. In 1995 the group employed 1,260 people.

NEW ENTRY

NOISE AND VIBRATION DIVISION

1 Cambridge Business Park Cowley Road, Cambridge CB4 4WZ
Telephone: 44 (1223) 426699
Fax: 44 (1223) 426696
MANAGING DIRECTOR: Dr Colin Ross
MANAGER, SALES AND MARKETING: David Brown

Origins of Noise and Vibration division can be traced back to April 1995 when Ultra Electronics acquired the aircraft noise cancellations business of Noise Cancellation Technologies Inc. Main responsibilities include the design, development, manufacture and installation of active noise control system for twin-prop and twin-jet regional and commercial/business aircraft. The company is also developing an active tuned vibration attenuator system which acts directly on the structure and therefore does not require loudspeakers. The following aircraft have been fitted with Ultra noise control systems: ATR 42-500, Saab 340B Plus, and Saab 2000. In January 1996 Ultra systems will be fitted on de Havilland DHC-8 Dash 8 series 100/200/300.

NEW ENTRY

ULTRA (SAAB) 340 ACTIVE NOISE CONTROL RETROFIT SYSTEM

TYPE: Twin-prop quiet cabin upgrade.

PROGRAMME: Development of retrofit system commenced in 1991 and has been tested on the following aircraft: ATR 42, Beech King Air, Beech Starship, Canadair Challenger 601, de Havilland DHC-8 Dash 8, Jetstream 41, MD 80 Saab 340 and Saab 2000. System first went into service on a production Saab 2000 in August 1994. Offered as a retrofit package on Saab 340 since early 1995.

DESIGN FEATURES: System consists of about 24 loudspeakers and 48 microphones positioned in the aircraft trim. The microphones provide a central controller with information about the residual noise in cabin in order to maintain the system's performance as the propeller speed and sound levels change. System can reduce the average sound level in the cabin by 10 dB.

NEW ENTRY

WESTLAND

WESTLAND GROUP PLC

Yeovil, Somerset BA20 2YB
Telephone: 44 (1935) 75222
Fax: 44 (1935) 702131
Telex: 46277 WHL YEO G
CHAIRMAN: Alan Jones
CEO: David Wright
PUBLIC RELATIONS DIRECTOR: Christopher Loney

Westland Aircraft Ltd (now Westland Group plc) formed July 1935, taking over aircraft branch of Petters Ltd (known previously as Westland Aircraft Works) that had designed/built aircraft since 1915; entered helicopter industry having acquired licence to build US Sikorsky S-51 as Dragonfly 1947; developed own Widgeon from Dragonfly; technical association with Sikorsky Division of United Technologies continued after decision to concentrate on helicopter design, development and construction.

Acquisition of Saunders-Roe Ltd 1959, Helicopter Division of Bristol Aircraft Ltd and Fairey Aviation Ltd 1960, and British Hovercraft Corporation's Aerospace Division 1983, plus subsequent restructuring into divisions, which later consolidated into current limited liability companies, as Westland Helicopters Ltd, Westland Aerospace Ltd and Westland Technologies Ltd. GKN acquired Westland in 1994.

Current programmes include construction of Lockheed C-130J, Saab 2000 and Jetstream 41 as well as composite engine cowlings for Boeing Canada DHC-8 Dash 8 and Dornier 328 transmissions gears and gearboxes for other companies; composite structures for Airbus, Boeing and McDonnell Douglas aircraft. ECS for Dornier 328, ATR 42/72, Jetstream 41. Westland also involved in work on the B-1, B-2, F-22, Tornado, Hawk and F/A-18 Hornet programmes.

EH Industries Ltd is joint Westland/Agusta (Italy) management company supporting EH 101 helicopter, collaboration with Agusta extended to include design, manufacture and marketing across joint product range; EHI Inc (USA) is a subsidiary of EHI Ltd; Westland Group activities in USA and Central America represented by wholly owned subsidiary, Westland Inc.

UPDATED

WESTLAND HELICOPTERS LIMITED

Yeovil, Somerset BA20 2YB
Telephone: 40 (1935) 75222
Fax: 44 (1935) 704201
Telex: 46277 WHL YEO G
MANAGING DIRECTOR: Richard I. Case
MARKETING DIRECTOR: P. F. J. Burton
HEAD OF HM GOVERNMENT BUSINESS: G. N. Cole
PUBLIC RELATIONS MANAGER: Vivien Davis

Sea King and Lynx in production; Westland and Agusta of Italy collaborate on EH 101 development and manufacture. An agreement with United Technologies permits Sikorsky Black Hawk production as WS 70. Other activities include construction of carbonfibre/glassfibre main rotor blades to replace metal blades on S-61s, SH-3s and Westland Sea Kings; advanced design composite main rotor blades successfully tested on Lynx, TT300 and EH 101.

Under June 1989 agreement, Westland obtained co-production rights for McDonnell Douglas AH-64 Apache. UK MoD ordered 67 WAH-64D Apache attack helicopters in July 1995..

UPDATED

WESTLAND SEA KING

TYPE: Anti-submarine, search and rescue airborne early warning helicopter.

PROGRAMME: Licence to develop/manufacture Sikorsky S-61 obtained 1959; developed initially for Royal Navy as advanced ASW helicopter with prolonged endurance; SAR, tactical troop transport, casualty evacuation, cargo-carrying, long-range self-ferry secondary roles; ordered for RN 1967; first flight of production HAS. Mk 1 (XV642) 7 May 1969.

VERSIONS: **Sea King AEW. Mk 2A:** Developed mid-1982 to give Royal Navy airborne early warning capability; 10 (plus trials vehicle) converted from HAS. Mk 2As; THORN EMI Searchwater radar in air pressurised container on swivel mounting. Plans exist to upgrade AEW. Mk 2A with new IFF, an enhanced communications fit and the incorporation of a radar video recorder. See also Upgrade section for UK MoD MSU.

Sea King HAR. Mk 3: Uprated version for SAR with RAF; first flight 6 September 1977; 16 HAR. Mk 3s delivered by 1979, plus three in 1985. Operated by No. 202 Squadron at Finningly (HQ) and detachments at Boulmer, Brawdy, Manston, Lossiemouth and Leconfield, plus No. 78 Squadron on Falkland Islands; two flight crew, air electronics/winch operator and loadmaster/winchman; up to six stretchers, or two stretchers and 11 seated survivors, or 19 persons; nav system includes Decca TANS F computer, accepting Mk 19 Decca nav receiver and Type 71 Doppler inputs; MEL radar; No. 78 Squadron helicopters fitted with RWR and chaff/flare dispensers. The Mk 3 production line is to be re-opened to fulfill an order of six aircraft for the RAF. These aircraft will have digital radar, FLIR and other improvements, which are to be retrofitted into earlier aircraft.

Sea King HAR. Mk 3A: Order for six aircraft officially announced 19 February 1992 and effected in October 1992 (deliveries: two in 1995 and four in 1996). Designated HAR. Mk 3A, these to have Racal RNAV2 computer, Racal Doppler 91, BAe GM9 compass, Smiths-Newmark SN500 flight control system, Collins HF9000 HF radio, Cossor STR2000 series GPS, Motorola MX-1000(R) mountain rescue radio, THORN EMI ARI5955/2 search radar, Rockwell Collins AN/ARC-182 VHF/UHF, Bendix/King VOR/ILS and Bendix/King DME/ADF. Operated by No. 202 Squadron at Boulmer (HQ) and detachments at Boulmer, Brawdy, Manston, Lossiemouth

Westland Sea King Mk 50 of the Australian Navy *(Westland)*

1995

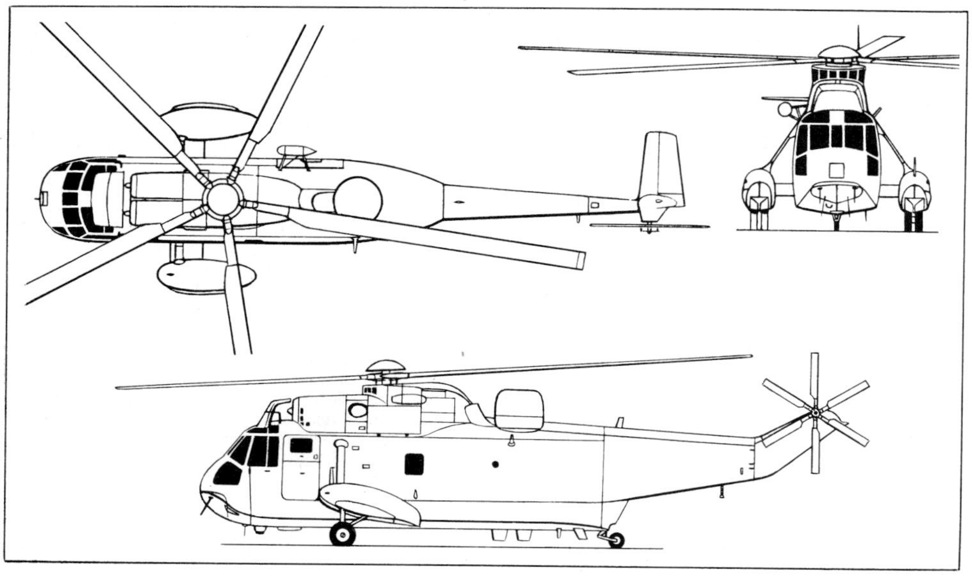

Westland Sea King HAS. Mk 5 anti-submarine helicopter *(Jane's/Dennis Punnett)*

1994

and Leconfield, plus No. 78 Squadron on Falkland Islands. Revised UK organisation, by 1996, to be Nos. 22/202 Squadrons with detachments at Boulmer, Chivenor (from April 1994), Leconfield, Lossiemouth, Valley (from mid-1996) and Wattisham (from mid-1994); plus Sea King OCU at St Mawgan (from April 1993). Accommodation comprises two flight crew, air electronics/winch operator and loadmaster/winchman; up to six stretchers, or two stretchers and 11 seated survivors, or 19 persons; nav system of initial production Mk 3 includes Decca TANS F computer, accepting Mk 19 Decca nav receiver and Type 71 Doppler inputs; THORN EMI ARI5955 radar; No. 78 Squadron helicopters fitted with RWR and chaff/flare dispensers.

Sea King HC. Mk 4: Utility version of Commando Mk 2 for Royal Navy.

Sea King HAR. Mk 5: Four HAS. Mk 5s with ASW avionics stripped 1987-88 for SAR; operational from 1 April 1988 with No. 771 Squadron at Culdrose.

Sea King HAS. Mk 5: Updated ASW/SAR version for Royal Navy; 30 new aircraft handed over 2 October 1980 to July 1986; one HAS. Mk 1, 20 HAS. Mk 2s and 35 HAS. Mk 2As brought to same standard by 1987 at Fleet Air Arm workshops; four became HAR. Mk 5s and others HAS. Mk 6s; nav/attack system utilises TANS G coupled to Decca 71 Doppler and MEL Sea Searcher radar (in larger radome); Racal MIR-2 Orange Crop ESM, passive sonobuoy dropping equipment, and associated GEC Avionics LAPADS acoustics processing and display equipment; four crew, with sonar operator also monitoring LAPADS as additional crew station; cabin enlarged by moving rear bulkhead 1.72 m (5 ft 7¾ in) aft; maximum T-O weight 9,525 kg (21,000 lb).

New equipment allows pinpoint of enemy submarine at greater range and attack with torpedoes; can monitor signals from own sonobuoys and those dropped by RAF Nimrod in joint search; can remain on station for long periods up to 87 n miles (160 km; 100 miles) from ship.

Sea King HAS. Mk 6: Uprated RN ASW version; large blade aerial under starboard side of nose; five new aircraft ordered from October 1987 and delivered January to August 1990; 25 HAS. Mk 5s being retrofitted to standard at RN Fleetlands workshop using Westland supplied kits; 44 more kits followed; first flight of conversion (Mk 5 XZ581) 15 December 1987; first flight of new Mk 6 (ZG816) 7 December 1989; entered service (ZA136) with Intensive Flight Trials Unit within No. 824 Squadron (detached from Culdrose to Prestwick) 15 April 1988, squadron later disbanding; issued to 819 Squadron, Prestwick, April 1989; and Culdrose squadrons 810 (November 1989), 820 (January 1990), 826 (February 1990, disbanded 30 July 1993); 814 (October 1990); and 706 (June 1991). Mk 6 Operational Evaluation Unit is 'E' Flight of 810 Squadron. By early 1994, 76 Squadron reverted to Mk 5; 819 Squadron operating Mks 5/6; 810, 814 and 820 Squadrons wholly Mk 6.

AQS-920G-DS enhanced sonar system (31 ordered from Maritime Aircraft Systems Division of GEC Avionics under 1987 contract, plus upgrade to standard of 112 previous AQS-902C sonobuoy processing systems), replacing Mk 5's analog computing element of Plessey 195 dipping sonar with digital processor (changing designation to GEC Ferranti 2069: 44 ordered initially in June 1989 and delivered from August 1991), and presenting integrated information from sonobuoys and dipping sonar on single CRT display; sonar dunking depth increased from 75 m (245 ft) to about 213 m (700 ft); GEC-Plessey PTR 446 improved IFF; upgraded ESM to Orange Reaper standard; two GEC Sensors AD3400 VHF/UHF secure speech radios; CAE Electronics internal MAD retrofit awarded to Westland, December 1991 for 73 conversion kits (clearance trials completed 1989); 227 to 363 kg (500 to 800 lb) weight-saving offers improved performance (equivalent to 30 minutes extra fuel).

Advanced Sea King: Rolls-Royce Gnome H.1400-IT engines, 1,092 kW (1,465 shp); uprated main gearbox with emergency lubrication and strengthened main lift frames; composite main and tail rotor blades; improved search radar; maximum AUW 9,752 kg (21,500 lb) for improved payload/range; through-life costs reduced.

Sea King Mk 41: SAR version in service with the German Navy.

Sea King Mk 42/42A: ASW/SAR version in service with the Indian Navy.

Sea King Mk 42C: Assault transportation version in use with the Indian Navy.

Sea King Mk 43/43A/43B: Search and rescue version in service with the Norwegian Air Force.

Sea King Mk 42B: Advanced Sea King ASW/ASV version for Indian Navy; 20 ordered; GEC Avionics AQS-902 sonobuoy processor and tactical processing system; MEL Super Searcher radar; integrated Alcatel HS-12 dipping sonar; Chelton 700 series homing; Marconi Hermes ESM; BAe Sea Eagle missile capability; first flight (IN513) 17 May 1985; two handed over (IN515/516) 16 January 1989; last delivered 12 December 1990.

Sea King Mk 43B: Norwegian Air Force SAR helicopter; one (temporarily ZH566) delivered early 1992, supplementing 11 Mk 43s received 1972-78 (nine survivors to be upgraded to Mk 43B standard with additional Bendix/King nose radar, Racal Doppler 91, R-NAV2 and Mk 32 Decca, plus 2000F FLIR; redelivery began July 1992 with 071).

Sea King Mk 45: ASW/ASV version in service with the Pakistan Navy.

Sea King Mk 47: ASW/anti-ship version in service with the Egyptian Air Force.

Sea King Mk 48: SAR version in service with the Belgian Air Force.

Sea King Mk 50/50A: ASW/utility version in service with the Australian Navy.

UPGRADES: **Mann Aviation Flight Systems**: Contract to upgrade six Sea King helicopters of Pakistan Navy. See separate entry.

Westland Sea King HAS. Mk 3A: Proposal to retrofit HAS. Mk 3 to HAS. Mk 3As. For the Royal Air Force. See Versions.

Sikorsky/Airod (Nuri) S-61 Upgrade: See Airod entry (Malaysia).

UK MoD AEW7 MSU: Issued an invitation to tender in early 1995 for the mission systems update (MSU) and associated support of the Royal Navy's 10 Westland Sea King AEW2A aircraft to bring them up to AEW7 standard. Central to the MSU is the radar system update (RSU). Other features include improved datalink and central tactical systems. Upgraded aircraft are due to enter service in the year 2000.

Westland Sea King HAS. Mk 5: Conversion of one HAS. Mk 1, 20 HAS. Mk 2s and 35 HAS. Mk 2As to HAS. Mk 5 standard for the Royal Navy. See Versions.

Westland HAS. Mk 6: Conversion of 25 HAS. Mk 5 to HAS. Mk 6 standard. For the Royal Navy. See Versions.

Westland Mk 43B: Conversion of nine Mk 43 to Mk 43B standard for the Norwegian Air Force. See Versions.

Westland/SABCA Mk 48: Contract placed January 1994 by Belgian Air Force to upgrade their fleet of five Sea King Mk 48 with Bendix RDR 1500B search radar, FLIR Systems FLIR 2000F, Racal RNS252 navigation system and Canadian Marconi CMA 3012 global positioning system (GPS). Kits to be installed by SABCA.

Westland Helicopters Ltd: Contract awarded July 1994 to extend the lives (until 2008) of the Royal Australian Navy's Sea King helicopters at the HMAS Naval air

Westland Sea King AEW. Mk 2A standard operated by the Royal Navy

1994

station at Nowra, South Sydney. Cost of contract is $40 million. Contract to be completed mid-1996 includes radio navigation, avionics and structural upgrades.

OPERATORS: Versions of the Westland Sea King are in service with the armed forces of the following countries: Australia (7), Belgium (5), Egypt (5), Germany (22), India (39), Norway (16), Pakistan (6) and UK (141).

CUSTOMERS: Total Sea King/Commando planned production: 328, of which 320 built by January 1995. Royal Navy also received four Sikorsky S-61Ds for development 1966-67; (one remains. Recent developments to these include three German **Mk 41s** converted for anti-ship role (anti-ship Sea Skua capability subsequently removed); upgrade of Norwegian **Mk 43s** (see above); five Belgian Air Force **Mk 48s** (delivered 1976) given composite main rotor blades and new Doppler and TANS to upgrade nav systems (by Westland under 1989 contract; work completed early 1991); most other operators retrofitted composite blades.

Following details apply to the Advanced Sea King:

DESIGN FEATURES: Based on SH-3 airframe and rotor system; Rolls-Royce Gnome turboshaft engines; transmission rating 2,200 kW (2,950 shp); specialised equipment to UK requirements; composite rotor blades; new five-blade tail rotor for increased capability in side wind; unbraced tail stabiliser; increased fuel capacity. Automatic main rotor blade folding and spreading is standard; for shipboard operation the tail pylon can also be folded.

FLYING CONTROLS: Mk 31 AFCS provides radio altitude displays for both pilots; artificial horizon displays; three-axis stabilisation in pilot-controlled manoeuvres; attitude hold, heading hold and height hold in cruising flight; controlled transition manoeuvres to and from the hover; automatic height control and plan position control in the hover; and an auxiliary trim facility.

POWER PLANT: Two 1,238 kW (1,660 shp) (maximum contingency rating) Rolls-Royce Gnome H.1400-1T turboshafts, mounted side by side above cabin. Fuel in six underfloor bag tanks, total capacity 3,714 litres (981 US gallons; 817 Imp gallons). Internal auxiliary tank, capacity 863 litres (228 US gallons; 190 Imp gallons), may be fitted for long-range ferry purposes. Pressure refuelling point on starboard side, two gravity points on port side. Flat plate debris guard for engine air intakes. Optional Centrisep air cleaner unit.

ACCOMMODATION: Crew of four in ASW role; accommodation for up to 22 survivors or 18 if radar fitted in SAR role; and up to 28 troops in utility role. Alternative layouts for nine stretchers and two attendants; or 15 VIPs. Two-section airstair door at front on port side, cargo door at rear on starboard side. Entire accommodation heated and ventilated. Cockpit doors and windows, and two windows each side of cabin, can be jettisoned in an emergency.

SYSTEMS: Three main hydraulic systems. Primary and auxiliary systems operate main rotor control. System pressure 103.5 bars (1,500 lb/sq in); flow rate 22.7 litres/min at 87.9 bars (6 US gallons; 5 Imp gallons/min at 1,275 lb/sq in). Unpressurised reservoir. Utility system for main landing gear, sonar and rescue winches, blade folding and rotor brake. System pressure 207 bars (3,000 lb/sq in); flow rate 41 litres/min at 186.2 bars (10.8 US gallons; 9 Imp gallons/min at 2,700 lb/sq in). Unpressurised reservoir. Electrical system includes two 20 kVA 200 V three-phase 400 Hz engine-driven generators, a 26 V single-phase AC supply fed from the aircraft's 40 Ah Ni/Cd battery through an inverter, and DC power provided as a secondary system from two 200 A transformer-rectifier units.

AVIONICS (ASW models): As equipped for this role, the Sea King is a fully integrated all-weather hunter/killer weapon system, capable of operating independently of surface vessels, and the following equipment and weapons can be fitted to achieve this task: GEC Ferranti 2069, GEC-Plessey Type 195, Bendix/King AN/AQS-13B or Alcatel HS-312 dipping sonar, GEC Avionics Doppler navigation system, MEL Super Searcher radar in dorsal radome, transponder beneath rear fuselage, Honeywell AN/APN-171 radar altimeter, BAe GM9B Gyrosyn compass system, Louis Newmark Mk 31 automatic flight control system. Observer/navigator has tactical display on which sonar contacts are integrated with search radar and navigational information. Radio equipment comprises Collins AN/ARC-182 UHF/VHF and homer, Ultra D 403M standby UHF, Collins 718U-5 HF radio, Racal B693 intercom, Telebrief system and IFF provisions. CAE Electronics AN/ASQ-504(V) internal MAD ordered for RN Sea Kings in 1987 and fitted from 1988 onwards. Whittaker Electronic Systems AN/ALQ-167 Yellow Veil modular jamming equipment installed internally in Mk 5 from about 1986.

Westland Sea King Mk 43B search and rescue (SAR) of the Royal Norwegian Air Force

1994

AVIONICS (non-ASW models): A wide range of radio and navigation equipment may be installed, including VHF/UHF communications, VHF/UHF homing, radio compass, Doppler navigation system, radio altimeter, VOR/ILS, radar and transponder, of Collins, GEC-Plessey, Honeywell and GEC Avionics manufacture. A Honeywell compass system and a Louis Newmark automatic flight control system are also installed.

EQUIPMENT: Two No. 4 marine markers, four No. 2 Mk 2 smoke floats, Ultra Electronics mini-sonobuoys, in ASW versions. Sea Kings equipped for search and rescue have a Breeze BL 10300 variable speed hydraulic rescue hoist of 272 kg (600 lb) capacity mounted above the starboard side cargo door. Second electric hoist optional.

ARMAMENT: Up to four Mk 46, Whitehead A244S or Sting Ray homing torpedoes, or four Mk 11 depth charges or one Clevite simulator. For secondary role a mounting is provided on the rear frame of the starboard door for a general purpose machine gun. Sea King is also in service with Exocet in Pakistan and Qatar and with Sea Eagle in India.

DIMENSIONS, EXTERNAL:

Main rotor diameter	18.90 m (62 ft 0 in)
Tail rotor diameter	3.16 m (10 ft 4 in)
Length: overall, rotors turning	22.15 m (72 ft 8 in)
main rotor folded	17.42 m (57 ft 2 in)
rotors and tail folded	14.40 m (47 ft 3 in)
Height: overall, rotors turning	5.13 m (16 ft 10 in)
rotors spread and stationary	4.85 m (15 ft 11 in)
to top of rotor head	4.72 m (15 ft 6 in)
Fuselage: Length	17.02 m (55 ft 10 in)
Max width	2.16 m (7 ft 1 in)
Width: overall, rotors folded	
with flotation bags	4.98 m (16 ft 4 in)
without flotation bags	4.77 m (15 ft 8 in)
Wheel track (c/l of shock-absorbers)	3.96 m (13 ft 0 in)
Wheelbase	7.14 m (23 ft 5 in)
Cabin door (port): Height	1.68 m (5 ft 6 in)
Width	0.91 m (3 ft 0 in)
Cargo door (stbd): Height	1.52 m (5 ft 0 in)
Width	1.73 m (5 ft 8 in)
Height to sill	1.14 m (3 ft 9 in)

DIMENSIONS, INTERNAL:

Cabin: Length	7.59 m (24 ft 11 in)
Max width	1.98 m (6 ft 6 in)
Max height	1.92 m (6 ft 3½ in)
Floor area (incl area occupied by radar, sonar)	13.94 m² (150 sq ft)
Volume	28.03 m³ (990 cu ft)

AREAS:

Main rotor disc	280.6 m² (3,020.3 sq ft)
Tail rotor disc	7.8 m² (83.9 sq ft)

WEIGHTS AND LOADINGS (A: anti-submarine, B: anti-surface vessel; C: airborne early warning, D: SAR, E: troop transport, F: external cargo, G: VIP):

Basic weight: with sponsons	5,447 kg (12,009 lb)
without sponsons	5,426 kg (11,963 lb)
Weight empty, equipped (typical):	
A	7,331 kg (16,163 lb)
B	7,625 kg (16,810 lb)
C	7,417 kg (16,353 lb)
D	6,087 kg (13,409 lb)
E	5,572 kg (12,308 lb)
F	5,572 kg (12,308 lb)
G	6,998 kg (15,429 lb)
Max T-O weight	9,752 kg (21,500 lb)
Max underslung or internal load	3,628 kg (8,000 lb)
Max disc loading	34.75 kg/m² (7.12 lb/sq ft)
Max power loading	4.44 kg/kW (7.29 lb/shp)

PERFORMANCE (at max T-O weight, ISA):

Never-exceed speed (VNE, British practice) at S/L	122 kts (226 km/h; 140 mph)
Cruising speed at S/L	110 kts (204 km/h; 126 mph)
Max rate of climb at S/L	564 m (1,850 ft)/min
Max vertical rate of climb at S/L	246 m (808 ft)/min
Service ceiling, one engine out	1,220 m (4,000 ft)
Max contingency ceiling (1 h rating)	1,067 m (3,500 ft)
Hovering ceiling: IGE	1,707 m (5,600 ft)
OGE	1,067 m (3,500 ft)
Radius of action: A (2 h on station, incl three torpedoes)	125 n miles (231 km; 144 miles)
B (2 h on station, incl two Sea Eagles)	110 n miles (204 km; 126 miles)
C (2 h 24 min on station)	100 n miles (185 km; 115 miles)
D (picking up 20 survivors)	220 n miles (407 km; 253 miles)
E (28 troops) range	300 n miles (556 km; 345 miles)
F (1,814 kg; 4,000 lb external load)	225 n miles (417 km; 259 miles)
G	580 n miles (1,075 km; 668 miles)
Range with max standard fuel, at 1,830 m (6,000 ft)	800 n miles (1,482 km; 921 miles)
Ferry range with max standard and auxiliary fuel, at 1,830 m (6,000 ft)	940 n miles (1,742 km; 1,082 miles)

PERFORMANCE (at typical mid-mission weight):

Never-exceed speed (VNE, British practice) at S/L	146 kts (272 km/h; 169 mph)
Cruising speed at S/L	132 kts (245 km/h; 152 mph)

UPDATED

Westland Sea King Mk 48 of the Belgian Navy. Belgian Mk 48 currently being upgraded by Westland and SABCA

1994

WESTLAND COMMANDO

TYPE: Twin-turboshaft tactical military helicopter.

PROGRAMME: First flight 12 September 1973; first flight HC. Mk 4 (ZA290) 26 September 1979; HC. Mk 4s delivered November 1979 to October 1990; HC. Mk 4 ZF115 (28th delivery) first Commando type completed with composites main rotor blades (first flight 14 November 1985); composite blades retrofitted to HC. Mk 4s.

VERSIONS: **Mk 1:** Designation of first five Commandos, ordered on behalf of the Egyptian Air Force by the Saudi Arabian government. Minimally modified version; essentially a standard Sea King Mk 41 aircraft, able to transport up to 21 troops. First two delivered to Egypt January/February 1974.

Commando Mk 2: Major production version, to which detailed description applies. Flew for the first time (G-17-12) on 16 January 1975. Saudi Arabian order included 17 Mk 2s and two VIP Mk 2Bs for the Egyptian Air Force. Four Mk 2s (three Mk 2As and one VIP Mk 2C) delivered to Qatar Emiri Air Force. The Egyptian Air Force also received in 1979-80 four Commando Mk 2Es equipped with Elettronica IHS-6 ECM/ESM for an electronic warfare role.

Commando Mk 3: Eight ordered by Qatar. First flown (QA30) 14 June 1982; deliveries between December 1982 and January 1984. Provision for Armament, including

Westland HC. Mk 4 Commando with underslung load of 4¾ ton Land Rover FFR hard top

1994

Westland HC. Mk 4 with underslung load of Royal Ordnance 105 mm light gun

1994

Exocet missiles and anti-tank missiles. Based on Advanced Sea King.

Sea King HC. Mk 4: Royal Navy utility Commando Mk 2; folding main rotor blades; folding tail pylon; non-retractable landing gear; 28 equipped troops or 2,720 kg (6,000 lb) cargo internally; 3,628 kg (8,000 lb) maximum slung load; parachuting/abseiling equipment; Decca TANS with chart display and Decca 71 Doppler nav system; 7.62 mm cabin machine gun; can operate in Arctic or tropics; serves with Nos. 707, 845 and 846 (Naval Air Commando) and 772 (SAR) Squadrons. Additionally, No. 848 Squadron formed 16 November 1990 for Gulf War service. Two **Mk 4Xs** with less operational equipment delivered to RAE (including for Blue Kestrel radar trials) 1982-83; one **Mk 4** to Empire Test Pilots' School 3 May 1989.

OPERATORS: Versions of the Westland Commando are in service with the armed forces of the following countries: Egypt (five Mk 1s, 17 Mk 2s, two Mk 2Bs, four Mk 2Es) and Qatar (three Mk 2As, one Mk 2C, eight Mk 3s, conversion began 1990 of Qatari Mk 2s to **Mk 3** standard with new nav/com equipment). Six Indian Sea King Mk 42Cs and 40 RN Sea King HC. Mk 4s essentially Commandos (three of latter lost, all in Falklands conflict); three UK MoD Mk 4s. Total production 89; last delivery (HC. Mk 4 ZG822) 15 October 1990.

Following data applies to final production aircraft:

DESIGN FEATURES: Based on Sea King; optimised payload/range and endurance for tactical troop transport, logistic support and cargo transport, and casualty evacuation primary roles, or air-to-surface strike and SAR secondary roles; five-blade composite main rotor, attached to hub by multiple bolted joint; NACA 0012 blade section; rotor brake; optional automatic main rotor blade folding; five-blade composite tail rotor; twin input four-stage reduction main gearbox, with single bevel intermediate and tail gearboxes; main rotor/engine rpm ratio 93.43; tail rotor/engine rpm ratio 15.26; unpressurised; stub wings instead of Sea King sponsons; non-retractable landing gear.

STRUCTURE: Light alloy stressed skin; optional tail pylon folding.

LANDING GEAR: Non-retractable tailwheel type, with twin-wheel main units. Oleo-pneumatic shock-absorbers. Main-wheel tyres size 6.50-10, tailwheel tyre size 6.00-6.

POWER PLANT: Two 1,238 kW (1,660 shp) (maximum contingency rating) Rolls-Royce Gnome H.1400-1T turboshafts, mounted side by side above cabin. Transmission rating 2,013 kW (2,700 shp). Fuel in six underfloor bag tanks, total capacity 3,714 litres (981 US gallons; 817 Imp gallons). Internal auxiliary tank, capacity 863 litres (228 US gallons; 190 Imp gallons), may be fitted for long-range ferry purposes. Pressure refuelling point on starboard side, two gravity points on port side. Flat plate debris guard for engine air intakes. Optional Centrisep air cleaner unit.

ACCOMMODATION: Crew of two on flight deck. Seats along cabin sides, and single jump seat, for up to 28 troops. Overload capacity 45 troops. Two-piece airstair door at front on port side, cargo door at rear on starboard side. Entire accommodation heated and ventilated. Cockpit doors and windows, and two windows each side of main cabin, are jettisonable in an emergency.

SYSTEMS: Primary and secondary hydraulic systems for flight controls. No pneumatic system. Electrical system includes two 20 kVA alternators.

AVIONICS: Wide range of radio, radar and navigation equipment available to customer's requirements.

EQUIPMENT: Cargo sling and rescue hoist optional.

ARMAMENT: Wide range of guns, missiles, and so on may be carried, to customer's requirements: typically, pintle-mounted gun (7.62 or 20 mm) in cabin doorway and machine gun pod (0.50 in or 7.62 mm) on each side of forward fuselage with reflector sight for pilot. If fitted, sponsons may mount one rocket pod each (FZ M159C 2.75 in; Matra F4 68 mm; Thomson Brandt 68-22 68 mm; Medusa 81 mm; or SNIA HL-12 80 mm).

DIMENSIONS, EXTERNAL:

Main rotor diameter	18.90 m (62 ft 0 in)
Tail rotor diameter	3.16 m (10 ft 4 in)
Distance between rotor centres	11.10 m (36 ft 5 in)
Main rotor blade chord	0.46 m (1 ft 6¼ in)
Length overall, rotors turning	22.15 m (72 ft 8 in)
fuselage	17.02 m (55 ft 10 in)
Height: overall, rotors turning	5.13 m (16 ft 10 in)
to top of rotor head	4.72 m (15 ft 6 in)
Wheel track (c/l of shock-absorbers)	3.96 m (13 ft 0 in)
Wheelbase	7.21 m (23 ft 8 in)
Passenger door (fwd, port):	
Height	1.68 m (5 ft 6 in)
Width	0.91 m (3 ft 0 in)
Cargo door (rear, stbd): Height	1.52 m (5 ft 0 in)
Width	1.73 m (5 ft 8 in)

DIMENSIONS, INTERNAL: As Advanced Sea King (SAR version)

AREAS: As for Advanced Sea King, plus:

Main rotor blades (each)	4.14 m^2 (44.54 sq ft)
Tail rotor blades (each)	0.23 m^2 (2.46 sq ft)
Tailplane	1.80 m^2 (19.40 sq ft)

WEIGHTS AND LOADINGS:

Operating weight empty (troop transport, two crew, typical)	5,620 kg (12,390 lb)
Max T-O weight	9,752 kg (21,500 lb)
Max underslung load	3,628 kg (8,000 lb)

PERFORMANCE (at max T-O weight): As given for Advanced Sea King, plus:

Range with max payload (28 troops), reserves for 30 min standoff 214 n miles (396 km; 246 miles)

VERIFIED

WESTLAND WASP

TYPE: Five/six-seat general purpose helicopter.

VERSIONS: **WASP HAS. Mk 1:** Version for Royal Navy, developed from Scout with folding tail and special landing gear for deck operations. Intended primarily for operation from small platforms on frigates and destroyers in anti-submarine weapon-carrying role, the normal load being two Mk 44 torpedoes. Also employed for search and rescue, training and other subsidiary duties. First Wasp HAS. Mk 1 for Royal Navy flew 28 October 1962, and deliveries began in second half of 1963.

OPERATORS: Versions of the Westland Wasp are in service with the armed forces of the following countries: Indonesia (9), Malaysia (6) and New Zealand (5).

DESIGN FEATURES: Four-blade main rotor, with all-metal blades carried on fully articulated hub. Torsion blade suspension system. Two-blade tail rotor with metal blades. Rotors driven through steel shafting. Primary gearbox at rear of engine, secondary gearbox at base of pylon, angle

Westland HC. Mk 4 Commando landing Royal Marines from HMS *Sir Tristram* during Exercise Dynamic Impact which took place in Sardinia 1994

1994

gearbox at base of fin, tail rotor gearbox at top of fin. Main rotor/engine rpm 1:71. Tail rotor/engine rpm ratio 1:15.

FLYING CONTROLS: Main rotor hub has drag and flapping hinges. Rotor brake standard. Tail rotor has flapping hinge.

STRUCTURE: The fuselage is a conventional aluminium alloy stressed skin structure. Front section forms the cabin, fuel tank bays and aft compartment. Rear section is a tapered boom terminating in a fin which carries the tail rotor. Horizontal stabiliser of light alloy construction mounted on starboard side of fin opposite tail rotor.

LANDING GEAR: Non-retractable four-wheel type. All four wheels castor and are carried on Lockheed shock-absorber struts. All wheels and tubeless tyres are Dunlop, size 15 × 4.75-6.5, pressure 60 lb/sq in (4.22 kg/cm²). Dunlop dog clutch brakes. Flotation gear standard.

POWER PLANT: One 710 shp (derated) Rolls-Royce Bristol Nimbus 503 turboshaft engine, mounted above fuselage to rear of cabin. Fuel in three interconnected flexible tanks in fuselage below main rotor, with total capacity of 705 litres (186 US gallons; 155 Imp gallons). Refuelling point on starboard side of decking. Oil capacity 7 litres (1.8 US gallons; 1.5 Imp gallons).

ACCOMMODATION: Two seats side by side at front of cabin, with bench seat for three persons at rear. Four doors, by front and rear seats on each side of cabin. Rear seats removable for cargo carrying. Heater standard.

SYSTEMS: Delaney Galley/Westland 1 kW cabin heating and windscreen demisting system. Hydraulic system, pressure 73.9 bars (1,050 lb/sq in), operating servo jacks for rotor head controls and rotor brake. No pneumatic system. 28 V DC electrical supply from engine-driven generator. Limited supply by 15 or 23 Ah battery. Three-phase 115 V 400 Hz AC provided by inverter.

ELECTRONICS AND EQUIPMENT: PTR.170 and PV.141 UHF and UHF homing radio, and standby UHF. Intercom taken from side tone of UHF T/R. Blind-flying instrumentation standard. Equipment includes autostabilisation/autopilot system, with radio altimeter.

DIMENSIONS, EXTERNAL:

Main rotor diameter	9.83 m (32 ft 3 in)
Tail rotor diameter	2.29 m (7 ft 6 in)
Distance between rotor centres	6.24 m (20 ft 5½ in)
Length overall, rotors turning	12.29 m (40 ft 4 in)
Length of fuselage	9.24 m (30 ft 4 in)
Width, rotors folded	2.64 m (8 ft 8 in)
Height to top of rotor hub	2.72 m (8 ft 11 in)
Overall height, tail rotor turning	3.56 m (11 ft 8 in)
Wheel track	2.44 m (8 ft 0 in)
Wheelbase	2.44 mm (8 ft 0 in)
Cabin doors (fwd, each): Height	1.13 m (3 ft 8½ in)
Width	0.94 m (3 ft 1 in)
Height to sill	0.74 m (2 ft 5 in)
Cabin doors (rear, each): Height	1.13 m (3 ft 8½ in)
Width	1.07 m (3 ft 6 in)
Height to sill	0.74 m (2 ft 5 in)

DIMENSIONS, INTERNAL:

Cabin: Length	1.84 m (6 ft 0½ in)
Max width	1.55 m (5 ft 1 in)
Max height	1.35 m (4 ft 5 in)

AREAS:

Main rotor blades (each)	1.44 m² (15.5 sq ft)
Tail rotor blades (each)	0.17 m² (1.82 sq ft)
Main rotor disc	75.90 m² (816.86 sq ft)
Tail rotor disc	4.10 m² (44.16 sq ft)
Tailplane (semi-span)	0.31 m² (3.32 sq ft)

WEIGHTS AND LOADINGS:

Manufacturer's weight empty	1,566 kg (3,452 lb)
Max payload, external cargo	680 kg (1,500 lb)
Max fuel load	562 kg (1,240 lb)
Max T-O and landing weight	2,495 kg (5,500 lb)
Max disc loading	32.71 kg/m² (6.70 lb/sq ft)

PERFORMANCE (at max T-O weight):

Max level speed at S/L	104 kts (193 km/h; 120 mph)
Max permissible driving speed	109 kts (203 km/h; 126 mph)
Max and econ cruising speed	96 kts (177 km/h; 110 mph)
Max rate of climb at S/L	439 m (1,440 ft)/min
Vertical rate of climb at S/L	183 m (600 ft)/min
Practical manoeuvring ceiling	3,720 m (12,200 ft)
Hovering ceiling in ground effect	3,810 m (12,500 ft)
Hovering ceiling out of ground effect	2,682 m (8,800 ft)
Max range with standard fuel	263 n miles (488 km; 303 miles)
Range with max fuel, including allowances of 5 min for T-O and landing, and 15 min cruising at best cruising height, with four passengers	234 n miles (435 km; 270 miles)

VERIFIED

Westland HAS. Mk 1 Wasp of the Brazilian Navy, since withdrawn from service

1994

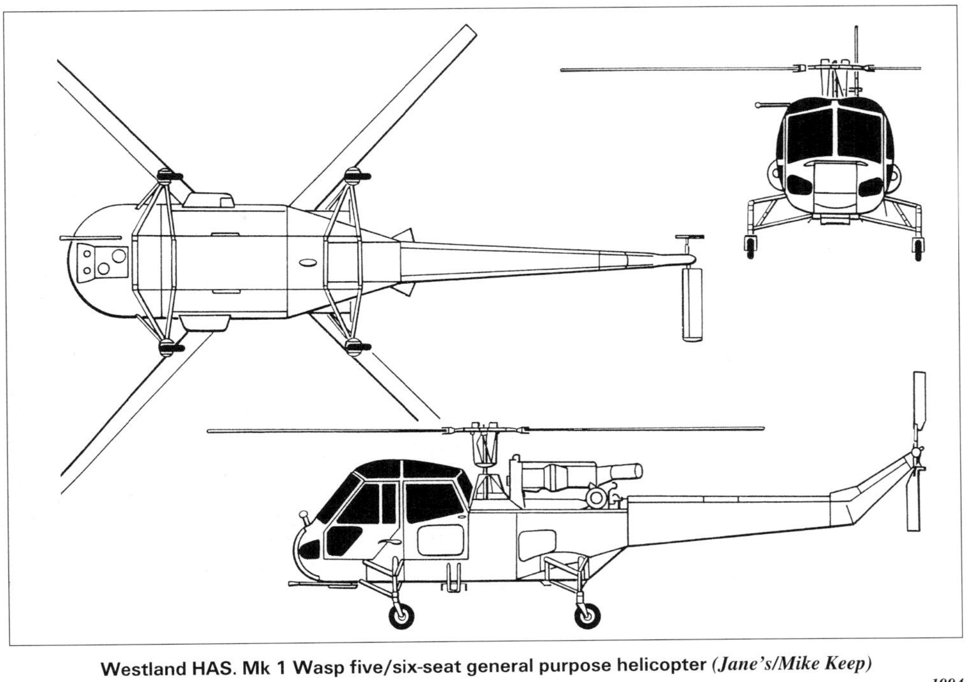

Westland HAS. Mk 1 Wasp five/six-seat general purpose helicopter ***(Jane's/Mike Keep)***

1994

WESTLAND WESSEX

TYPE: Single-rotor medium duty helicopter.

PROGRAMME: The Westland Wessex is a turbine-powered development of the Sikorsky S-58.

VERSIONS: **Wessex HAS. Mk 1:** Initial production version, developed for the Royal Navy, with one 1,450 shp Napier Gazelle 161 shaft-turbine. Re-engined with a 1,100 shp Gazelle NGa.11, flew for the first time 17 May 1957. Withdrawn from service.

Wessex HC. Mk 2: High-performance development of the Mk 1 with two coupled 1,350 shp Bristol Siddeley Gnome Mk 110/111 shaft-turbine engines. Power limitation of 1,550 shp at rotor head. Prototype converted from Wessex 1, flew for the first time 18 January 1962, and the first production model (XR588) 5 October 1962. Still in service with the DRA Bedford (1), 2 FTS (11), 72 Squadron Aldergrove (25), 28 Squadron Sek Kong (8) and 22 Squadron (17) SAR.

Wessex Mk 3: Similar to Mk 1, but with 1,850 shp Gazelle NGa.18 165 shaft-turbine engine. Not in service.

Wessex HCC. Mk 4: Queen's lift helicopter.

Wessex HC. Mk 5: SAR helicopter of the Royal Air Force based in Cyprus.

Wessex HU. Mk 5: Similar to Mk 2, for Commando assault duties from carriers of the Royal Navy. Design work began in April 1962 and construction of the prototype was started in May 1962. In service with A&EE (1) and 84 Sqn Akrotiri Cyprus (5).

Wessex HAS. Mk 31: Generally similar to Mk 1, but with a 1,540 shp Gazelle Mk 162 engine. Ordered for the Royal Australian Navy for anti-submarine duties from HMAS *Melbourne*. Withdrawn from service.

Wessex Mk 52: Similar to Mk 2, for Iraqi Air Force. Withdrawn from service.

Wessex Mk 53: Similar to Mk 2, for Ghanaian Air Force. Withdrawn from service.

Wessex Mk 60: Civil version in service with Uruguayan Navy.

OPERATORS: Versions of the Westland Wessex are in service with the armed forces of UK (62) and Uruguay (2).

DESIGN FEATURES: Main and tail rotor each have four blades. Blades attached to hub by taper bolts. Main rotor blades fold manually. Rotor brake fitted. Shaft drive to main rotor through double epicyclic gear. Shaft drive to tail rotor through intermediate and tail gearboxes. Tail end folds to port and forward for stowage. Tail rotor carried at tip of vertical stabilising fin. Small horizontal stabiliser inset in leading-edge of fin.

STRUCTURE: All blades of light alloy extruded spar and light alloy bonded trailing-edge structure. The fuselage is a light alloy semi-monocoque structure, with steel tube support structure for main rotor gearbox.

LANDING GEAR: Non-retractable tailwheel type. All three units fitted with Westland oleo-pneumatic shock-absorber. Dunlop wheels, tyres and hydraulic disc brakes. Tubeless treaded mainwheel tyres, size 6.00 × 11. Tailwheel tyre size 6.00 × 6.

POWER PLANT (Mk 2): One Bristol Siddeley Gnome Mk 110 and one Gnome Mk 111 shaft-turbine engines, with Type 10 coupling gearbox. Rated at 1,350 shp per engine; 1,550 shp at rotor head. Two flexible fuel tanks under cabin floor, total capacity 1,409 litres (372.8 US gallons; 310 Imp gallons). Provision for carrying two 500 litre (132 US gallon; 110 Imp gallon) auxiliary tanks in cabin for ferry purposes. Refuelling point in starboard side of fuselage. Oil capacity 9 litres (2.4 US gallons; 2 Imp gallons) per engine, 19 litres (5.1 US gallons; 4.25 Imp gallons) in main gearbox.

POWER PLANT (Mk 5): One Bristol Siddeley Gnome Mk 112 and one Gnome Mk 113 shaft-turbines, with Type 11 coupling gearbox. Otherwise as for Mk 2.

ACCOMMODATION (Mk 2): Crew of one to three according to role. Up to 16 passengers in main cabin, on folding troop seats, or up to eight stretchers in banks of four. Doors on each side of flight deck and on starboard side of cabin.

ACCOMMODATION (Mk 5): Crew of one to three according to role. Three fixed troop seats, and either 13 removable folding seats or eight stretchers, or 1,814 kg (4,000 lb) of freight.

SYSTEMS: Compressor bleed air for heating. Ambient air circulation by fan. High-pressure hydraulic system for

Westland Wessex HC. Mk 2 search and rescue (SAR) helicopter of the Royal Air Force *(P. Tompkins)*
1994

Westland Wessex HC. Mk 2 turbine-powered development of Sikorsky S-58 *(P. Tompkins)*
1994

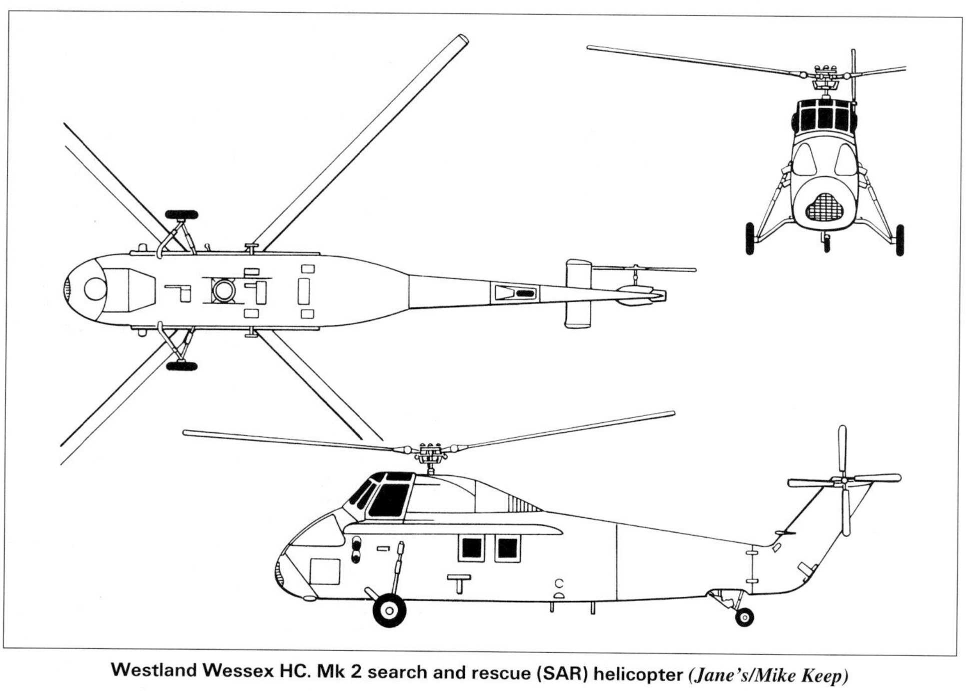

Westland Wessex HC. Mk 2 search and rescue (SAR) helicopter *(Jane's/Mike Keep)*
1994

powered flying controls and 272 kg (600 lb) capacity hoist. 24 V DC electrical system, with two 6 kW generators.

DIMENSIONS, EXTERNAL:	
Main rotor diameter	17.07 m (56 ft 0 in)
Tail rotor diameter	2.90 m (9 ft 6 in)
Length overall	20.03 m (65 ft 9 in)
Length of fuselage	14.74 m (48 ft 4½ in)
Length, blades and tail folded	11.73 m (38 ft 6 in)
Width, rotors folded	4.06 m (13 ft 4 in)
Height of top of rotor hub	4.39 m (14 ft 5 in)
Overall height	4.93 m (16 ft 2 in)
Wheel track	3.66 m (12 ft 0 in)
Cabin door: Height	1.22 m (4 ft 0 in)
Width	1.22 m (4 ft 0 in)
Height to sill	approx 0.91 m (3 ft 0 in)
DIMENSIONS, INTERNAL:	
Main cabin: Length	4.16 m (13 ft 8 in)
Max width	1.68 m (5 ft 6 in)
Max height	1.83 m (6 ft 0 in)
AREAS:	
Main rotor blade (each)	3.55 m² (38.25 sq ft)
Tail rotor blade (each)	0.27 m² (2.90 sq ft)
Main rotor disc	228.8 m² (2,463 sq ft)
Tail rotor disc	6.60 m² (71 sq ft)
WEIGHTS AND LOADINGS:	
Basic operating weight: Mk 2	3,767 kg (8,304 lb)
Mk 5	3,927 kg (8,657 lb)
Payload, with max fuel: Mk 2	1,117 kg (2,462 lb)
Mk 5	957 kg (2,109 lb)
Max T-O weight: Mk 2, Mk 5	6,120 kg (13,500 lb)
Max landing weight: Mk 2, Mk 5	6,120 kg (13,500 lb)
Max disc loading: Mk 2, Mk 5	26.85 kg/m² (5.5 lb/sq ft)
PERFORMANCE (at max T-O weight):	
Max level speed at S/L:	
Mk 2, Mk 5	114 kts (212 km/h; 132 mph)
Max cruising speed: Mk 2, Mk 5	204 km/h (127 mph)
Max rate of climb at S/L:	
Mk 2, Mk 5	503 m (1,650 ft)/min
Vertical rate of climb at S/L:	
Mk 2, Mk 5	168 m (550 ft)/min
Range with max fuel, 10% reserves:	
Mk 2, Mk 5	416 n miles (770 km; 478 miles)

VERIFIED

WESTLAND LYNX, UPGRADES

TYPE: Combat helicopter upgrade.

PROGRAMME: For full details of Westland Lynx, see *Jane's All the World's Aircraft 1994-95*.

UPGRADES: **Lynx AH. Mk 7**: Uprated British Army version, meeting GSR 3947; with improved systems, reversed-direction tail rotor with improved composite blades to reduce noise and enhance extended period hover at high weights; 4,876 kg (10,750 lb) AUW; 13 ordered, eight from Mk 5 contract (two cancelled); first flight (ZE376) 7 November 1985; 11th delivered July 1987.

RN workshops at Fleetlands converting Mk 1s to Mk 7s; first conversion (XZ641) redelivered 30 March 1988; box-type exhaust diffusers added from early 1989; programme continues; approximately 75 conversions by late 1992. Interim conversion is Lynx **AH. Mk 1GT** with uprated engines and rotors, but lacking Mk 7's improved electronic systems; first conversions (XZ195) 1991. GEC Ferranti (now GEC-Marconi) AWARE-3 radar warning receiver selected 1989 for retrofit, designated ARI23491; Mk 1 XZ668 to Westland for trial installation 22 November 1991.

Lynx Mk 8 HMA (formerly known as HAS. Mk 8): For Royal Navy; equivalent to export **Super Lynx**; passive identification system; 5,125 kg (11,300 lb) maximum T-O weight; improved (reversed-direction) tail rotor control; BERP composite main rotor blades; Racal RAMS 4000 central tactical system (CTS eases crew's workload by centrally processing sensor data and presents mission information on multifunction CRT display; 15 systems ordered 1987, 106 September 1989); original Seaspray Mk 1 radar repositioned in new chin radome; GEC-Marconi Sea Owl thermal imager (×5 or ×30 magnifying system on gimballed mount, with elevation +20 to −30° and azimuth +120 to −120°; ordered October 1989) in former radar position; MIR-2 ESM updated; three Mk 3s used in development programme as tactical system (XZ236), dummy Sea Owl/chin radome (ZD267) and avionics (ZD266) testbeds; see Jane's All the World's Aircraft 1995-96 for Phases 1 and 2 of Lynx Mk 8 programme.

Definitive Mk 8 (Phase 3) conversions begun 1992 with addition of Sea Owl, further radar and navigation upgrades, (including RACAL RNS252 'Super TANS', composites BERP main rotor blades and reversed-direction tail rotor. Conversion programme covers 49 aircraft in two phases. Phase one covers 25 aircraft. Westland is to convert 12 aircraft and supply 13 conversion kits to the Royal Navy at Fleetlands. Westlands had delivered eight converted aircraft and four conversion kits by August 1995. The 12th aircraft is due for delivery in February 1996 and the remaining kits are due for delivery in November 1995. Phase two of the contract covers 24 conversion kits which are to be delivered to Fleetlands between January and December 1997. Fleetlands is due to complete all conversion by the year 2000.

Lynx AH. Mk 9: UK Army Air Corps equivalent of export **Battlefield Lynx**; tricycle wheel landing gear; maximum T-O weight 5,125 kg (11,300 lb); advanced technology composites main rotor blades; exhaust diffusers; no TOW capability; first flight of prototype (converted company demonstrator XZ170) 29 November 1989; 16 new aircraft (beginning ZG884, flown 20 July 1990) ordered for delivery from 1991, plus eight Mk 7 conversions (contract awarded November 1991); for Nos. 672 and 664 Squadrons of 9 Regiment, Dishforth, to support 24th Airmobile Brigade; some outfitted as advanced command posts, remainder for tactical transport role. Deliveries to A&AEE, Boscombe Down from 22 May 1991 (ZG884); to No. 672 Squadron from 19 December 1991 (ZG889); No. 664 Squadron from June 1972; final aircraft (ZG923) flown 30 June 1992; first 'production' conversion (ZG538) to Westland for modifications 3 February 1992.

Lynx Mk 21A: Five remaining Brazilian Navy Lynx Mk 21 to be upgraded to Super Lynx Mk 21A standard. Contract placed in February 1994 includes nine new-built aircraft.

Other versions and operators where orders completed, Royal Netherlands Navy upgraded five UH-14As and eight SH-14Cs to **SH-14D** standard, with Alcatel dipping sonar, UHF radios, RWR, FLIR Systems Inc 2000HP FLIR, Trimble Type 2200 GPS, new radar altimeter, composites rotor blades and Mk 42 Gem power plants. Nine SH-14Bs, already with sonar, raised to SH-14D standards, but in interim SH-14Cs upgraded to SH-14B through deletion of MAD and addition of sonar. UH-14As are first full SH-14D conversions, from 1990; programme designated STAMOL (Standaardisatie en Modernisering Lynx); standard fleet comprising 16 with sonar and six with provisions for sonar installation. Completed early 1993.

Denmark upgrading Mk 80s to Mk 90 with Gem Mk 42 power plant, Racal tactical data system, Racal MIR-2 passive radar detection system, composites blades and four-bag emergency flotation system; maximum T-O weight increased to 4,876 kg (10,750 lb).

UPDATED

WESTLAND RACAL PUMA UPGRADES

TYPE: Combat helicopter upgrade.

PROGRAMME: Contract announced September 1994 for a navigation upgrade for 40 RAF Puma helicopters. Racal avionics is acting as prime contractor and systems integrator. Westland Industries, part of the Westland Group, is responsible for the design, installation and test work of new navigation equipment fit. Programme to last one year. See also Racal Avionics Ltd.

DESIGN FEATURES: New avionics suite consists of global positioning system (GPS), tactical air navigation system (TACAN), high-frequency omnidirectional radio (VOR), distance measuring equipment (DME) and integrated support packages.

COSTS: Contract award value of £3.5 million.

UPDATED

Westland Lynx AH. Mk 9 of which first production conversion was February 1992 *(Westland)*
1995

Westland Lynx Mk 3 used as testbed for Lynx HAS Mk 8 upgrade programme *(Westland)*
1995

Westland Lynx HMA. Mk 8 equivalent to export Super Lynx *(Westland)*
1994

UNITED STATES OF AMERICA

AAC

AEROSTAR AIRCRAFT CORPORATION

3608 South Davison Boulevard, Spokane, Washington 99204-5799
Telephone: 1 (509) 455 8872
Fax: 1 (509) 838 0831
PRESIDENT: Steve Speer
VICE-PRESIDENT, MARKETING: James S. Christy

Subsidiary of Machen formed to develop jet-powered version of former Piper Aerostar. Company also extensively upgrades piston-engined Aerostars.

VERIFIED

AAC AEROSTAR 3000

TYPE: Light aircraft engine conversion.

PROGRAMME: Pressurised Piper 601P, 602P and 700P re-engined with 8.45 kN (1,900 lb st) Williams FJ44 turbofans in underslung pods; cruising speed 400 knots (740 km/h; 460 mph); range 1,500 n miles (2,780 km; 1,725 miles). Conversion stopped; new-build Aerostar being developed with improved pressurisation and range, retaining published cruising speed. Programme cancelled.

VERIFIED

AAC Aerostar Super 700 conversion with Lycoming TIO-540-U2A turbocharged engines *(AAC)*

1994

AEROSTAR SUPER 700

TYPE: Light aircraft performance enhancement conversion.

PROGRAMME: Conversion package for previously manufactured Piper 601-P/602-P.

DESIGN FEATURES: Re-engine with Lycoming 350 HP TIO-540-U2A turbocharged piston engine. Package consists of a series of modifications: required modifications include turbo induction air intercooling system, overboost protection, low noise signature Hartzell propeller blades, low speed flight control enhancement system and fireproof engine hoses. Optional modifications include Iconel front exhaust system, Iconel exhaust tail pipes, stainless steel turbo mounting, crankcase breather oil recovery system, complete known icing certification kit, auxiliary fuel tank, 700-6 wheel brake system and auxiliary electric seal pump system.

WEIGHTS AND LOADINGS:

Ramp weight	2,883 kg (6,356 lb)
Gross T-O weight	2,864 kg (6,315 lb)
Gross landing weight	2,722 kg (6,000 lb)
Standard empty weight	1,915 kg (4,221 lb)

PERFORMANCE:

Max speed (TAS at mid-cruise weight)	274 kts (507 km/h; 315 mph)
Rate of climb (at S/L and gross weight):	
Two engine	572 m (1,875 ft)/min
One engine	117 m (383 ft)/min
Cruising speed (TAS at mid-cruise weight):	
75% power	261 kts (483 km/h; 300 mph)
65% power	245 kts (453 km/h; 282 mph)
55% power	225 kts (416 km/h; 259 mph)
T-O distance (at S/L, zero wind)	595 m (1,950 ft)
Landing distance (at S/L, zero wind)	439 m (1,440 ft)

VERIFIED

ADVANCED AIRCRAFT

ADVANCED AIRCRAFT CORPORATION

2106 Palomar Airport Road, Carlsbad, California 92008
Telephone: 1 (619) 435 3647
Telex: 249075 ATSD UR
PRESIDENT: Neil F. Martin
GENERAL MANAGER: Leland L. Dimon III

Company formed 1 July 1983; acquired production facilities of Riley Aircraft Manufacturing Inc; two Cessna turbine conversions offered in 1987, since when no further details.

UPDATED

ADVANCED AIRCRAFT TURBINE P-210

TYPE: Cessna P-210 Pressurised Centurion powered by P&WC PT6A-135 turboprop.

PROGRAMME: Previously known as Spirit 750; conversion developed by Riley Aircraft Manufacturing; on offer 1987.

DESIGN FEATURES: Powered by 559 kW (750 shp) PT6A-135 turboprop flat rated to 335.5 kW (450 shp) driving Hartzell three-blade constant-speed fully feathering reversible-pitch Q-tip propeller. Conversion includes Flint wingtip fuel tanks with combined capacity of 62.5 litres (16.5 US gallons; 13.7 Imp gallons); modified control surfaces; ventral fin; glassfibre aerodynamic cowling with inertial separator.

SYSTEMS: Conversion includes electric induction lip de-icing; electric propeller de-icing; FAA approved 24 V electrical system; Smiths 28 V 200 A starter/generator; 28 V 50 A standby generator; Gill 639T heavy duty battery; fuel computer; new engine instrument panel; 3M Stormscope weather radar.

DIMENSIONS, EXTERNAL: As for pressurised Centurion, except:

Length overall	9.17 m (30 ft 1 in)
Propeller diameter	1.98 m (6 ft 6 in)

WEIGHTS AND LOADINGS:

Weight empty	1,199 kg (2,621 lb)
Max ramp weight	1,822 kg (4,016 lb)
Max T-O weight	1,814 kg (4,000 lb)
Max landing weight	1,723 kg (3,800 lb)
Max wing loading	102.53 kg/m² (21.0 lb/sq ft)
Max power loading	5.41 kg/kW (8.88 lb/shp)

PERFORMANCE (at max T-O weight except where indicated):

Max cruising speed at 7,010 m (23,000 ft)	253 kts (470 km/h; 292 mph)
Econ cruising speed	213 kts (394 km/h; 245 mph)
Stalling speed: flaps up	67 kts (125 km/h; 78 mph) IAS
30% flap	63 kts (117 km/h; 73 mph) IAS
flaps down	58 kts (108 km/h; 67 mph) IAS
Max rate of climb at S/L	549 m (1,800 ft)/min
Service ceiling	above, 7,010 m (23,000 ft)
T-O to 15 m (50 ft) at AUW of 1,542 kg (3,400 lb)	334 m (1,095 ft)
Landing from 15 m (50 ft) at AUW of 1,542 kg (3,400 lb)	394 m (1,291 ft)
Range with max fuel, ISA, max cruise power at 7,010 m (23,000 ft), allowances for start, taxi, take-off, climb and descent, 45 min reserves	1,008 n miles (1,868 km; 1,160 miles)

VERIFIED

Advanced Aircraft Turbine P-210 turboprop conversion of Cessna Pressurised Centurion before addition of ventral fin

1994

ADVANCED AIRCRAFT REGENT 1500

TYPE: Cessna 421C Golden Eagle with P&WC PT6A turboprops.

PROGRAMME: Modification developed by Riley Aircraft Manufacturing Inc; first flight November 1979 at Palomar Airport; certificated 1980.

DESIGN FEATURES: Powered by two P&WC PT6A-135 turboprops, each flat rated at 559 kW (750 shp) and driving Hartzell three-blade, constant-speed autofeathering reversible-pitch propellers with Q-tips; fuel augmented by nacelle locker tanks with combined capacity of 568 litres (150 US gallons; 125 Imp gallons); optional all-glassfibre nacelle-end tanks increasing combined capacity to 605 litres (160 US gallons; 133 Imp gallons); advanced aerodynamic cowlings incorporating inertia separators; electric de-icing of intakes and propellers.

SYSTEMS: Generally as for Cessna 421C, except bleed air unit for pressurisation system; electrical system with 200 A starter/generators and heavy duty battery; engine fire detection system; and pneumatic leading-edge de-icing boots for wings and tail unit.

WEIGHTS AND LOADINGS:

Weight empty	2,404 kg (5,300 lb)
Max T-O weight	3,447 kg (7,600 lb)
Max zero-fuel weight	2,993 kg (6,600 lb)
Max landing weight	3,266 kg (7,200 lb)
Max power loading	3.08 kg/kW (5.07 lb/shp)

PERFORMANCE (at max T-O weight except where indicated):
Max cruising speed, at max cruise power, AUW of 3,220 kg (7,100 lb):
at 8,230 m (27,000 ft) 315 kts (584 km/h; 363 mph)
at 6,705 m (22,000 ft) 278 kts (515 km/h; 320 mph)
at 4,875 m (16,000 ft) 251 kts (465 km/h; 289 mph)
Max rate of climb at S/L:
at max T-O weight 945 m (3,100 ft)/min
at 3,175 kg (7,000 lb) AUW 1,310 m (4,300 ft)/min
Service ceiling 9,145 m (30,000 ft)
Service ceiling, one engine out, at 3,175 kg (7,000 lb) AUW 7,620 m (25,000 ft)
T-O run 462 m (1,517 ft)
T-O to 15 m (50 ft) 694 m (2,277 ft)
Landing from 15 m (50 ft) with propeller reversal 606 m (1,988 ft)
Landing run with propeller reversal 324 m (1,064 ft)
Range with max fuel, ISA, allowances for start, taxi, take-off, climb and descent, 45 min reserves:
max cruise power: at 4,875 m (16,000 ft) 832 n miles (1,542 km; 958 miles)
at 6,705 m (22,000 ft) 987 n miles (1,829 km; 1,136 miles)
at 8,230 m (27,000 ft) 1,169 n miles (2,166 km; 1,346 miles)
max range power: at 5,485 m (18,000 ft) 908 n miles (1,682 km; 1,045 miles)
at 6,075 m (22,000 ft) 1,112 n miles (2,060 km; 1,280 miles)
at 8,230 m (27,000 ft) 1,276 n miles (2,364 km; 1,469 miles)

Advanced Aircraft Regent 1500, with Pratt & Whitney Canada PT6A-135 turboprops
1994

VERIFIED

AERO UNION

AERO UNION CORPORATION

100 Lockheed Avenue, Chico, California 95926
Telephone: 1 (916) 896 3000
Fax: 1 (916) 893 8585
PRESIDENT: Dale P. Newton
SECRETARY-TREASURER/GENERAL MANAGER: Victor E. Alvistur
SENIOR VICE-PRESIDENT, MARKETING: John M. Oswald

Company established 1959 for aerial firefighting; manufactures tank systems for Douglas DC-3, DC-4, DC-6 and DC-7, Fairchild C-119, Grumman S-2, Lockheed C-130/L-100 Hercules, Lockheed P-3 Orion and Electra; also produces aircraft fuel tanks, airstairs, environmental control systems, cargo pallet roller systems, retardant delivery and aerial spraying systems, dorsal fins; sole manufacturer of Model 1080 Air Refuelling Store developed by Beechcraft and acquired by Aero Union in 1985.

UPDATED

Aero Union C-130 Aerial Spray System
1994

AERO UNION MODULAR AERIAL FIREFIGHTING SYSTEM (MAFFS)

TYPE: Firefighting tank, control and dispensing modules to fit USAF 463L pallet system.
CUSTOMERS: 10 sets of MAFFS delivered to US Air Force reserve, two sets to Italy, two to Portugal and three to Greece.
DESIGN FEATURES: Designed for aircraft such as Lockheed C-130 equipped for USAF 463L pallet system; MAFFS consists of five retardant tank modules, totalling 11,356 litres (3,000 US gallons; 2,498 Imp gallons), control module, dispensing module with two retractable nozzles making a swath 13-61 m (43-200 ft) wide and 180-610 m (590-2,000 ft) long according to height and speed. System empty weight is 2,177 kg (4,800 lb); powered by compressed air reservoir at each tank and battery power for control module; installed in C-130 in less than one hour; typical filling time 15 minutes; can be operated at 30-150 m (98-492 ft) at 130-140 knots (241-259 km/h; 150-161 mph). Conversion completed on Lockheed L-188 Electra in August 1994 with 11,356 litre (3,000 US gallon; 2,498 Imp gallon) constant flow tank.

UPDATED

AERO UNION C-130 AUXILIARY FUEL SYSTEM

TYPE: Transferable or usable extra fuel for C-130/L-100.
PROGRAMME: Development complete.
DESIGN FEATURES: Usable for in-flight refuelling, range extension or fuel transport; two cylindrical tanks mounted side by side of 6.1 m (20 ft) long platform loaded by 463L cargo handling system; total capacity 13,627 litres (3,600 US gallons; 2,998 Imp gallons); length 5.26 m (17 ft 3 in); width 3.0 m (9 ft 10¼ in); height, excluding plumbing, 1.68 m (5 ft 6 in); weight empty about 1,542 kg (3,400 lb); plumbing complies with MIL-F-17874; flow rate 1,136 litres (300 US gallons; 250 Imp gallons)/min. Unit interfaces with Lockheed air refuelling manifold; passage provided for crew access fore and aft of tanks.

VERIFIED

Aero Union Retardant Aerial Delivery System (RADS) C-130 Fireliner
1994

AERO UNION C-130 FIRELINER RETARDANT AERIAL DELIVERY SYSTEM (RADS)

TYPE: Version of SP-2H Firestar conversion (which see) for C-130.
PROGRAMME: First flight early 1990; service entry Spring 1990, following FAA certification 29 March 1990. Six sets delivered by October 1994.
DESIGN FEATURES: First conversion on C-130A, but adaptable to later marks. The retardant tank of 11,356 litres (3,000 US gallons; 2,498 Imp gallons), is a two part installation breaking at the floor line so aircraft can be converted back to a cargo configuration quickly. The drop system maintains a constant rate of flow regardless of the level of retardant remaining in the tank, this is achieved by an onboard computer that adjusts the tank door openings to compensate for changing fluid levels. Coverage levels are selected by the pilot. Five aircraft have been converted and flew during the 1991 fire season.

VERIFIED

AERO UNION C-130 AERIAL SPRAY SYSTEM

TYPE: Modular spray kit for unmodified Lockheed C-130.
DESIGN FEATURES: Self-contained modular system with two Aero Union bulk fuel transfer tanks, each of 6,813 litres (1,800 US gallons; 1,498 Imp gallons), mounted on 463L pallet; tanks modified to supply liquid to a retractable spray module attached to cargo ramp; boom hydraulically extended into airstream; when boom retracted, ramp can be fully closed; interchangeable spray nozzles vary application rate; ultra-low-rate nozzles optional; pressure and flow rate to boom monitored and controlled by loadmaster using portable electronic control panel.

VERIFIED

AERO UNION HELIBORNE AERIAL FIREFIGHTING SYSTEM (HAFFS)

TYPE: Designed for installation in large helicopters like Boeing CH-47/Model 234 and M.

VERSIONS: Smaller system, capacity 2,650-3,785 litres (700-1,000 US gallons; 583-833 Imp gallons) being developed for CH-46/Kawasaki KV107. Also evaluating derivative designs for Aerospatiale AS 332, EHI EH 101, Mil Mi-8, Mi-26 and Sikorsky UH-60.

DESIGN FEATURES: Capacity of 7,571 litres (2,000 US gallons; 1,665 Imp gallons) delivered through remotely operated, variable pattern foam cannon capable of delivering 2,271 litres (600 US gallons; 500 Imp gallons)/min at stand-off distance of more than 61 m (200 ft). All or part of payload can also be dropped in 10 seconds. HAFFS will saturate half-acre area at density of 40.3 l/m² (0.99 US gallons; 0.82 Imp gallons/sq ft). System empty weight 1,749 kg (3,857 lb).

VERIFIED

AERO UNION GRUMMAN HU-16B ALBATROSS TANKER CONVERSION

TYPE: Firefighting kit for HU-16B amphibian.

DESIGN FEATURES: Two tanks totalling 3,785 litres (1,000 US gallons; 833 Imp gallons) mounted side by side in fuselage and filled through two retractable probes extended while the aircraft is planing on the step; refilling takes 10 seconds at 70 knots (130 km/h; 81 mph); filling on land through single nozzle in fuselage side; tanks can be emptied individually or together.

WEIGHTS AND LOADINGS:

Weight empty	9,956 kg (21,950 lb)
Max T-O weight	15,036 kg (33,150 lb)

PERFORMANCE:

Max level speed	220 kts (408 km/h; 253 mph)
Cruising speed	162 kts (300 km/h; 186 mph)

VERIFIED

AERO UNION SP-2H FIRESTAR

TYPE: Civilian air tanker version of Lockheed P2V-7/SP-2.

PROGRAMME: Four currently flying.

DESIGN FEATURES: Westinghouse J34 booster turbojets and other military equipment removed; 7,571 litre (2,000 US gallon; 1,665 Imp gallon) capacity tank weighting 318 kg (700 lb) and two computer-controlled, electrohydraulically actuated dispersal doors that can deliver precisely metered flow rates of 189-2,650 litres (50-700 US gallons; 41.6-583 Imp gallons) per second. One-eighth, one-quarter, one-half or whole tankful can be selected; computerised logic allows selection of coverage density and quantity; drops are precise and uniform without overlaps or gaps; flow is constant regardless of quantity remaining in tank; drop patterns 50 per cent longer than typical multi-door system claimed for same volume of retardant.

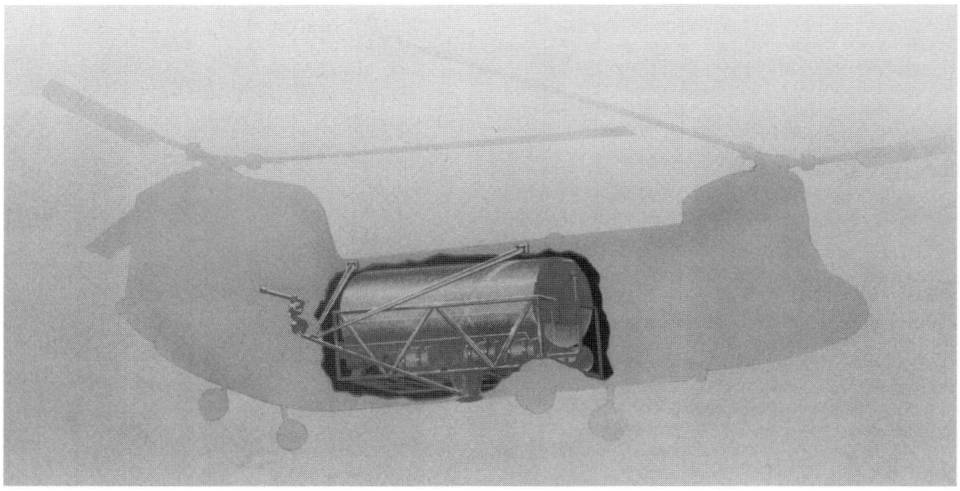

Aero Union Heliborne Aerial FireFighting System (HAFFS)

1994

Aero Union SP-2H Firestar air tanker version of Lockheed P2V-7

1994

WEIGHTS AND LOADINGS:

Weight empty	16,692 kg (36,800 lb)
Max payload	8,164 kg (18,000 lb)
Max T-O and landing weight	30,617 kg (67,500 lb)

PERFORMANCE (at max T-O weight):

Never-exceed speed (VNE)	270 kts (500 km/h; 311 mph)
Econ cruising speed	196 kts (361 km/h; 225 mph)
Max airdrop speed	145 kts (268 km/h; 167 mph)
Stalling speed, landing gear and flaps down	93 kts (172 km/h; 107 mph)
Max rate of climb at S/L	488 m (1,600 ft)/min
Rate of climb at S/L, one engine out	122 m (400 ft)/min
Certificated ceiling	5,485 m (18,000 ft)
Service ceiling, one engine out	5,365 m (17,600 ft)
Max ferry range, 30 min reserves	1,300 n miles (2,409 km; 1,497 miles)

VERIFIED

Aero Union P-3A Aerostar

1994

AERO UNION P-3A AEROSTAR

TYPE: Civil firefighting conversion of P-3A Orion.

PROGRAMME: First conversion early 1990 with temporary tank; installation of computer-controlled tank to follow. Six installations completed by October 1994.

DESIGN FEATURES: Initially 11,356 litre (3,000 US gallon; 2,498 Imp gallon) ventral tank with eight drop doors. Development planned of computer-controlled tank similar to those in SP-2H and C-130.

VERIFIED

AERO UNION CESSNA 208/208B CARAVAN FIREFIGHTER

TYPE: Cessna 208/208B Caravan I conversion.

DESIGN FEATURES: Ventrally mounted 1,893 litre (500 US gallon; 416 Imp gallon) aluminium tank attaches to standard cargo pod mountings; tank weighs 163 kg (360 lb) and has two drop doors over almost entire length of tank; preselected drop rates computer-modulated in eight increments to maintain flows of 3.8-30.3 litres (108 US gallons; 0.8-6.7 Imp gallons) per 9.3 m² (100 sq ft); integral injection system holds enough retardant for 10 drops; optional spray manifold kit for aerial insecticide spraying.

VERIFIED

AG-CAT

AG-CAT CORPORATION

Malden Industrial Park, Building 167, Malden, MO 63863
Telephone: 1 (314) 276 5770
Fax: 1 (314) 276 5776
PRESIDENT: Jim Krepps
VICE-PRESIDENT, SALES & MARKETING: Jim Kosier
MANAGER, OPERATIONS: Jack Krepps

On 28 March 1995, Schweizer Aircraft announced the sale of the Ag-Cat programme to Ag-Cat Corp.

NEW ENTRY

AG-CAT SERIES

TYPE: Agricultural biplane.

PROGRAMME: First flight of original Grumman Ag-Cat 27 May 1957; first deliveries 1959; production resumed in October 1981 with improved G-164B designated Ag-Cat Super-B, later joined by Ag-Cat Super-B Turbine.

VERSIONS: **Ag-Cat Super-B/600 (G-164B):** Basic model powered by 447.5 kW (600 hp) P&W R-1340 radial engine, driving Pacific Propeller Type 12D40/AG100 two-blade constant-speed metal propeller; improvements in Super-B include 40 per cent more hopper capacity and 0.97 m (3 ft 2 in) wider stainless steel gatebox and bottom loader as standard equipment; upper wing raised 20 cm (8 in) to improve pilot's view and increase load carrying capability, operating speed and climb rate. *Detailed description applies to this version except where indicated.*

Ag-Cat 450B (G-164B): Generally similar alternative version, available to special order, powered by 335.5 kW (450 hp) P&W R-985 radial engine; usable fuel 242 litres (64 US gallons; 53.3 Imp gallons); hopper volume 1.23 m³ (43.5 cu ft) and capacity 1,230 litres (325 US gallons; 271 Imp gallons).

Ag-Cat Super-B Turbine (G-164B): Similar to basic Super-B, but powered by choice of 373 kW (500 shp)

P&WC PT6A-11AG, 507 kW (680 shp) PT6A-15AG or 559 kW (750 shp) PT6A-34AG turboprop engines. Airframe now being offered to customers to install own engine; PT6A-15, PT6A-20, PT6A-27, PT6A-28 or PT6A-34 acceptable; new air induction filter; improved oil cooling system; fuel capacity increased to 454 litres (120 US gallons; 100 Imp gallons).

UPGRADES: **Marsh:** G-164 C-T Turbo Cat. See separate entry.

CUSTOMERS: Total 2,628 Ag-Cats built by Schweizer under contract, including 1,730 G-164As, 832 G-164Bs, 44 G-164Cs and 22 G-164Ds. Ag-Cat Super-B Turbine also manufactured under licence by Ethiopian Airlines Technical Services Division in Addis Ababa since December 1986.

DESIGN FEATURES: Wing section NACA 4412; dihedral 3°; incidence 6°; uncowled piston engines.

FLYING CONTROLS: Mechanical, with trim tab on port elevator; ground adjustable tabs on one aileron, rudder and starboard elevator; no flaps.

STRUCTURE: Each wing has two aluminium spars and sheet skin wrapped over whole upper surface and back to front spar on under surface; remainder of underside fabric covered; leading-edges in five panels to simplify repair; glassfibre wingtips; light alloy ailerons; wire-braced fabric covered tail surfaces on steel tube frame; welded steel tube fuselage with removable side panels.

LANDING GEAR: Non-retractable tailwheel type. Cantilever spring steel legs. Cleveland mainwheels with tyres size 8.50-10, 8 ply, pressure 2.42 bars (35 lb/sq in). Steerable tailwheel with tyre size 12.4-4.5, pressure 3.45 bars (50 lb/sq in). Cleveland heavy duty air-cooled hydraulic disc brakes. Parking brake.

POWER PLANT: One Pratt & Whitney nine-cylinder air-cooled radial engines with Pacific Propeller constant-speed propeller, or Pratt & Whitney Canada PT6A turboprop, as detailed under Versions. Fuel tanks in upper wing with combined usable capacity of 302 litres (80 US gallons; 66.6 Imp gallons). Single-point refuelling on upper surface of upper wing centre-section. Oil capacity 32.2 litres (8.5 US gallons; 7.1 Imp gallons).

ACCOMMODATION: Single seat in enclosed cockpit. Reinforced fairing aft of canopy for turnover protection. Canopy side panels open outward and down, canopy top upward and to starboard, to provide access. Baggage compartment. Cockpit pressurised against dust ingress and ventilated by ram air. Safety padded instrument panel. Air conditioning by J.B. Systems optional.

Schweizer Ag-Cat 450B (G-164B) agricultural aircraft

1994

SYSTEMS: Hydraulic system for brakes only. Optional electrical system with 24 V alternator, navigation and/or strobe lights, external power socket, and electric engine starter.

EQUIPMENT: Radio optional. Standard equipment includes control column lock, instrument glareshield, seat belt and shoulder harness, tinted windscreen, stall warning light, refuelling steps and assist handles, tiedown rings, and urethane paint external yellow finish.

Forward of cockpit, over CG, is a glassfibre hopper, capacity 1,514 litres (400 US gallons; 333 Imp gallons), for agricultural chemicals (dry or liquid) with distributor beneath fuselage. Low-volume, ULV or high-volume spray system, with leading- or trailing-edge booms. Emergency dump system for hopper load; can be used also for water bomber operations.

DIMENSIONS, EXTERNAL (A: Super-B/600, B: Super-B Turbine, C: 450B):

Wing span: upper	12.92 m (42 ft 4½ in)
lower	12.36 m (40 ft 6¾ in)
Wing chord, constant	1.47 m (4 ft 10 in)
Wing aspect ratio: upper wing	8.7
biplane, effective mean	5.5
Length overall: A	7.44 m (24 ft 5 in)
B	8.41 m (27 ft 7¼ in)
C	7.54 m (24 ft 9 in)
Height overall: A, C	3.51 m (11 ft 6 in)
B	3.68 m (12 ft 1 in)
Tailplane span	3.96 m (13 ft 0 in)
Wheel track	2.44 m (8 ft 0 in)
Wheelbase	5.59 m (18 ft 4 in)
Propeller diameter (max)	2.74 m (9 ft 0 in)
Propeller ground clearance: A	0.27 m (10⅘ in)
B	0.37 m (1 ft 2¾ in)
Cockpit door: Height	0.53 m (1 ft 9 in)
Width	0.64 m (2 ft 1 in)
Height to sill	0.71 m (2 ft 4 in)

DIMENSIONS, INTERNAL:

Cockpit: Length	1.27 m (4 ft 2 in)
Max width	0.76 m (2 ft 6 in)
Max height	1.14 m (3 ft 9 in)
Hopper volume	1.51 m^3 (53.5 cu ft)

AREAS:

Wings, gross	36.48 m^2 (392.7 sq ft)
Ailerons (total)	2.92 m^2 (31.4 sq ft)
Fin	1.67 m^2 (17.97 sq ft)
Rudder	1.12 m^2 (12.0 sq ft)
Tailplane	2.12 m^2 (22.8 sq ft)
Elevators	2.06 m^2 (22.2 sq ft)

WEIGHTS AND LOADINGS:

Weight empty equipped, spray and duster versions:	
A	1,656 kg (3,650 lb)
B	1,429 kg (3,150 lb)
C	1,508 kg (3,325 lb)
Certificated gross weight	2,358 kg (5,200 lb)
Max T-O weight (CAM.8)	3,184 kg (7,020 lb)
Max wing loading	87.42 kg/m^2 (17.91 sq ft)
Max power loading: A	7.12 kg/kW (11.71 lb/hp)
B	5.70-8.54 kg/kW (9.36-14.04 lb/shp)
C	9.49 kg/kW (15.6 lb/hp)

PERFORMANCE (A: Super-B/600, B: Super-B Turbine with PT6A-15AG engine, C: 450B):

Never-exceed speed (VNE)	136 kts (252 km/h; 157 mph)
Working speed: A, C	100 kts (185 km/h; 115 mph)
B	113 kts (209 km/h; 130 mph)
Stalling speed, power off	56 kts (103 km/h; 64 mph)
T-O run	120 m (394 ft)
T-O to 15 m (50 ft) at 2,358 kg (5,200 lb) certificated gross weight: A	320 m (1,050 ft)
B	274 m (900 ft)
C	396 m (1,300 ft)
Landing from 15 m (50 ft)	407 m (1,333 ft)
Landing run	157 m (513 ft)
Range with max fuel	172 n miles (318 km; 198 miles)
Design *g* limits, all versions	+4.2/−1

Schweizer Ag-Cat Super-B Turbine agricultural aircraft

1994

UPDATED

AIR TRACTOR

AIR TRACTOR INC

PO Box 485, Municipal Airport, Olney, Texas 76374
Telephone: 1 (817) 564 5616
Fax: 1 (817) 564 2348
Telex: 910 890 4792

PRESIDENT: Leland Snow

Air Tractor agricultural aircraft based on 30-year experience of Leland Snow, who produced Snow S-2 series, which later became Rockwell S-2R.

VERIFIED

AIR TRACTOR MODEL AT-301/ 301A/301B AIR TRACTOR

TYPE: Single-seat agricultural aircraft.

PROGRAMME: Design of the Model **AT-301** Air Tractor was initiated January 1971, and construction of the first prototype/preproduction aircraft started August 1972. This aircraft flew for the first time in September 1973, and received FAA certification under FAR Pt 23 in November that year. At that time the aircraft was also flight tested to meet FAR Pt 8 requirements.

VERSIONS: **AT-301A**, introduced in 1980, is identical to the AT-301 except for the installation of a larger hopper, with a 0.97 m (3 ft 2 in) wide gatebox for high application rates of dry chemicals.

By November 1991, a total of 1000 AT-301/301A/301Bs had been ordered, of which over 700 had been completed. **AT-301B** introduced in 1987 to replace the AT-301A, is identical to the AT-301 except for the installation of a larger hopper, with a 0.97 m (3 ft 2 in) wide gatebox for high application of dry chemicals.

DESIGN FEATURES: Cantilever low-wing monoplane. Wing section NACA 4415. Dihedral 3° 30′. Incidence 2°. No sweepback.

STRUCTURE: Conventional two-spar structure of 2024-T3 light alloy. Wing ribs and skins zinc chromated before assembly. Wingroots and skin overlaps sealed chemical entry. The fuselage is a welded structure of 4130N steel tube, oven stress relieved and oiled internally. Quickly detachable skins of 2024-T3 light alloy, with Camloc fasteners. Rear fuselage lightly pressurised to prevent chemical ingress. The tail unit is a light alloy structure, with cantilever fin and strut-braced tailplane.

FLYING CONTROLS: Ailerons of light alloy construction, interconnected with trailing-edge flap deflection of 30°. Electrically operated Fowler trailing-edge flaps of light alloy construction. No trim tabs. Fabric covered rudder and elevators. Trim tab in each rudder elevator.

Air Tractor AT-301 agricultural aircraft with Pratt & Whitney R-1340 air-cooled radial engine *(Air Tractor)*

1995

LANDING GEAR: Non-retractable tailwheel type. Cantilever spring steel main gear; flat spring suspension for castoring and lockable tailwheel. Cleveland mainwheels with tyres size 8,50-10 (8 ply), pressure 2.83 bars (41 lb/sq in). Tailwheel tyre size 12.50-4, pressure 2.42 bars (35 lb/sq in). Cleveland type 30-89 hydraulic disc brakes.

POWER PLANT: One 447 kW (600 hp) Pratt & Whitney R-1340 air-cooled radial engine, driving a Hamilton Standard 12D40/6101A-12 two-blade constant-speed metal propeller without spinner, or a Pacific Propeller 22D40 constant-speed Hydromatic propeller. Fuel in two integral wing tanks with combined capacity of 447 litres (126 US gallons; 105 Imp gallons). Refuelling points on upper surface of wings at root. Oil capacity 30.3 litres (8 US gallons; 6.7 Imp gallons).

ACCOMMODATION: Single seat in enclosed cabin which is sealed to prevent chemical ingress. Downward hinged window/door on each side. Baggage compartment in bottom of fuselage, aft of cabin, with door on port side. Cabin ventilation by 0.10 m (4 in) diameter airscoop.

SYSTEMS: Agricultural dispersal system comprises a 1,211 litre (320 US gallon; 266 Imp gallon) glassfibre hopper, mounted in the forward fuselage for the AT-301; 1,325 litre (350 US gallon; 291 Imp gallon) hopper for the AT-301A and AT-301B. Transland valve and strainer; 5 cm (2 in) Agrinautics pump; 5 cm (2 in) stainless steel plumbing; and up to 69 nozzles in spraybars mounted below and just aft of wing trailing-edges. 24 V electrical system, supplied by 60 A engine-driven alternator.

AVIONICS AND EQUIPMENT: Optional avionics include Narco Com 811 radio, King KX nav/com, King KR 87 ADF, Narco AT-150 transponder and Narco ELT-10 emergency locator transmitter. Optional equipment includes 600W retractable landing light in port wingtip; night flying package of strobe/navigation, instrument, post and dome lights; night working directional and altitude gyro instrument package; glassfibre engine cowling ring; Hydromatic propeller; Cleveland 29-11 wheels and high-flotation tyres; external power socket; and ferry fuel system. Alternative agricultural equipment includes Transland high volume and extra high volume spreader systems, and six- or eight-unit Micronair installations, and windscreen washer.

DIMENSIONS, EXTERNAL:

Wing span	13.75 m (45 ft 1¼ in)
Wing chord, constant	1.83 m (6 ft 0 in)
Wing aspect ratio	7.5
Length overall	8.23 m (27 ft 0 in)
Height overall	2.59 m (8 ft 6 in)
Propeller diameter	2.77 m (9 ft 1 in)

AREAS:

Wings, gross	25.08 m^2 (270 sq ft)
Ailerons, (total)	3.55 m^2 (38.2 sq ft)
Trailing-edge flaps (total)	3.34 m^2 (36.0 sq ft)
Fin	0.90 m^2 (9.7 sq ft)
Rudder	1.30 m^2 (14.0 sq ft)
Tailplane	2.42 m^2 (26.0 sq ft)
Elevators (incl, tabs)	2.36 m^2 (25.4 sq ft)

WEIGHTS AND LOADINGS (A: AT-301, B: 301B):

Weight empty, spray equipped: A	1,723 kg (3,800 lb)
B	1,746 kg (3,850 lb)
Certificated gross weight (FAR Pt 23):	
A, B	2,268 kg (5,000 lb)
Typical operating weight (CAM 8):	
A	3,356 kg (7,400 lb)
B	3,492 kg (7,700 lb)
Max wing loading: A	133.8 kg/m^2 (27.4 lb/sq ft)
B	139.1 kg/m^2 (28.5 lb/sq ft)
Max power loading: A	7.51 kg/kW (12.33 lb/hp)
B	7.81 kg/kW (12.83 lb/hp)

PERFORMANCE (at max T-O weight except where indicated):

Max level speed at S/L:	
A	146 kts (270 km/h; 168 mph)
B	142 kts (264 km/h; 164 mph)
Max level speed at S/L, with dispersal equipment:	
A	139 kts (257 km/h; 160 mph)
B	135 kts (251 km/h; 156 mph)
Cruising speed at 1,825 m (6,000 ft), spraypump removed for ferrying: A	135 kts (249 km/h; 155 mph)
B	132 kts (245 km/h; 152 mph)
Typical working speed:	
A, B	104-121 kts (193-225 km/h; 120-140 mph)
Stalling speed at 2,268 kg (5,000 lb) AUW:	
flaps up: A, B	64 kts (118 km/h; 73 mph)
flaps down: A, B	53 kts (98 km/h; 61 mph)
Stalling speed as usually landed:	
A, B	49 kts (90 km/h; 56 mph)
Max rate of climb at S/L, dispersal equipment installed, AUW of 2,268 kg (5,000 lb):	
A	488 m (1,600 ft)/min
B	457 m (1,500 ft)/min
T-O run at AUW of 2,268 lb (5,000 lb):	
A, B	244 m (800 ft)
T-O run at AUW of 3,175 kg (7,000 lb):	
A, B	396 m (1,300 ft)
Landing run at normal landing weight:	
A	91 m (300 ft)
Range with max fuel, no allowances:	
A	469 n miles (869 km; 540 miles)

VERIFIED

Air Tractor AT-301 single-seat agricultural aircraft

1994

ALLIEDSIGNAL

ALLIEDSIGNAL AEROSPACE

Los Angeles International Airport, 6201 West Imperial Highway, Los Angeles California 90045
Telephone: 1 (310) 568 3729
Fax: 1 (310) 568 3715
Telex: 181827 A/B AIRE AVI LSA
DIRECTOR, RETROFIT PROGRAMMES: Conrad Daigle

UPDATED

ALLIEDSIGNAL (DASSAULT) 731 FALCON 20B RETROFIT PROGRAMME

TYPE: Engine retrofit for in-service Falcon 20s.

PROGRAMME: Announced May 1987; new engine pylons developed by Dassault Aviation in France (which see); first flight development; Falcon 20F (F-WTFE) from Bordeaux-Mérignac 7 October 1988; first flight second aircraft (Falcon 20C F-WTFF) 26 January 1989; certificated USA and France March 1989.

CUSTOMERS: Total 80 Falcons of all models retrofitted to mid-1995, including first European conversion delivered by Europe Falcon Service on 6 July 1989.

COSTS: Standard aircraft $3.6 million.

DESIGN FEATURES: TFE731-5BR engine currently used in retrofits, with resulting Falcon 20B designation. Retrofit

AlliedSignal Aerospace 731 Falcon 20B

1994

undertaken by AlliedSignal Aerospace at Los Angeles, California, or Springfield, Illinois; Falcon Jet Service Center at Little Rock, Arkansas; Europe Falcon Service, Le Bourget Airport, Paris, France; and TransAirCo, Geneva, Switzerland.

UPDATED

ALLIEDSIGNAL (CESSNA) CITATION 500 UPGRADE PROGRAMME

TYPE: Light aircraft avionic upgrade.

PROGRAMME: Reported that AlliedSignal is offering upgrade to Citation 500, including Bendix/King EFS-50 EFIS and digital AFCS and state-of-the-art communications and navigation equipment. Certification tests began in March 1993, allowing initial redeliveries of upgraded aircraft in Autumn that year.

UPDATED

AMERICAN AVIATION

AMERICAN AVIATION INC

HEADQUARTERS: South 3608 Davison Boulevard, Spokane, Washington 99204
Telephone: 1 (509) 838 5354
Fax: 1 (509) 838 0831

American Aviation Inc is an aircraft engineering and development centre involved in the design, development and production of intercooling systems for general aviation applications. Systems are available for the Piper Navajo, CR and Chieftain as well as the Cessna 340, 340A, 414, 414A, 337, and the Beechcraft Duke and 56TC Baron. The company also designs and develops Ram Air Recovery pitot cowlings for the Piper Cheyenne I, II and IIXL.

VERIFIED

AVACELLE

AVACELLE INC

2203 Perimeter Road, Mobile, Alabama 36615
Telephone: 1 (334) 433 7707
Fax: 1 (334) 433 5002
PRESIDENT: Leo Marquez
COO: A. B. Stewart
EXECUTIVE VICE-PRESIDENT: Tom Stewart

Avacelle Inc was established to produce noise reduction nacelles for the Boeing 707-320 B/C type aircraft. The company has a manufacturing plant in Oklahoma City covering 47,000 sq ft.

UPDATED

AVACELLE (BOEING) B707 NOISE REDUCTION NACELLE ASSEMBLY

TYPE: Large passenger aircraft engine nacelle assembly.

PROGRAMME: Development of nacelle began Spring 1990. First nacelle was completed Spring 1992 and sent to P&W for testing. Production expected to begin in 1995.

DESIGN FEATURES: Assembly designed to bring Boeing B707-320B/C aircraft in line with FAR Pt 36 Stage three noise requirements. Assembly offers minimum change in T-O and climb thrust, minimum change in fuel mileage (depending on altitude, weight, performance and operational characteristics). Aircraft empty weight increase estimated at 2,300 lb.

COST: $3.85 million per set of four nacelles.

VERIFIED

Avacelle (Boeing) 707 noise reduction nacelle *(Avacelle)* *1995*

AYRES

AYRES CORPORATION

PO Box 3090, 1 Rockwell Avenue, Albany, Georgia 31706-3090
Telephone: 1 (912) 883 1440
Fax: 1 (912) 439 9790
Telex: 547629 AYRESPORT ABN
SALES MANAGERS:
Marvin H. Wilson (International)
Pete Lewis (Domestic)

Ayres Corporation bought manufacturing and world marketing rights to Thrush Commander-600 and -800 from Rockwell International General Aviation Division in November 1977.

UPDATED

AYRES THRUSH-600

TYPE: Single-seat agricultural aircraft.

PROGRAMME: The Thrush-600 has a 1.5 m^3 (53 cu ft) hopper able to contain up to 1,514 litres (400 US gallons; 333 Imp gallons) of liquid or 1,487 kg (3,280 lb) of dry chemicals. Corrosion-proofing is of activated Copon, and it is certificated to both CAR 3 and Normal category as well as CAM 8 Restricted category requirements.

The following description applies to the single-seat Thrush-600.

DESIGN FEATURES: Cantilever low-wing monoplane. Dihedral 3° 30′.

STRUCTURE: Two-spar structure of light alloy throughout wings, except for main spar caps of heat-treated SAE 4000 Series steel. Leading-edge formed by heavy main spar and

Ayres Thrush-600 two-seat agricultural aircraft *(A. March)*
1994

the nose skin. Wingroots sealed against chemical entry. The fuselage is a welded chrome-molybdenum steel tube structure covered with quickly removable light alloy panels. Underfuselage skin of stainless steel. The tail unit is a wire-braced welded chrome-molybdenum steel tube structure, fabric-covered. Streamline-section heavy duty stainless steel attachment fittings.

FLYING CONTROLS: Light alloy plain ailerons. Electrically operated flaps. Light alloy controllable trim tab in each elevator. Deflector cable from cockpit to fin-tip.

LANDING GEAR: Non-retractable tailwheel type units have rubber-in-compression shock-absorbtion and 29 × 11.00-10 wheels with 10 ply tyres. Hydraulically operated disc brakes. Parking brakes. Wire cutters on main gear. Steerable, locking tailwheel, size 12.5 × 4.5 in.

POWER PLANT: One 448 kW (600 hp) Pratt & Whitney R-1340 Wasp nine-cylinder air-cooled radial engine, driving a Hamilton Standard type 12D40/EAC AG-100-2 two-blade constant-speed metal propeller, is standard. Available optionally is a 448 kW (600 hp) Pezetel PZL-3S seven-cylinder air-cooled radial engine, driving a Dowty-Rotol type CR289-3 three-blade constant-speed metal propeller. One 200.5 litre (53 US gallon; 44 Imp gallon) integral tank in each wing, giving total fuel capacity of 401 litres (106 US gallons; 88 Imp gallons), of which 378.5 litres (100 US gallons; 83 Imp gallons) are usable.

ACCOMMODATION: Single adjustable seat in 'safety pod' sealed cockpit enclosure is standard, with steel tube overturn structure. Second, tandem seat, with or without dual controls, is optional, to provide accommodation for additional crew member, passenger, or flying instructor. Two overhead windows for improved view in turns (provisional). Downward hinged door on each side. Tempered safety glass windscreen. Dual inertia-reel safety harness standard. Baggage compartment. Windscreen wiper and washer.

SYSTEM: Electrical system powered by a 24 V 50 A engine-driven alternator; 70 A alternator optional. Lightweight 24 V 35 Ah battery.

AVIONICS AND EQUIPMENT: Hopper forward of cockpit with capacity of 1.50 m³ 53 cu ft) or 1,514 litres (400 US gallons; 333 Imp gallons). Hopper has a 0.33 m² (3.56 sq ft) lid, opened by two handles. Standard equipment includes Universal spray system with external 50 mm (2 in) stainless steel plumbing, 50 mm Root Model 67 pump with wooden fan, Transland gate, 50 mm valve, quick-disconnect pump mount and strainer. Streamlined spraybooms with outlets for 68 nozzles, 36 nozzles installed. Micro-adjust valve control (spray) and calibrator (dry). A 63 mm (2½ in) side-loading system is installed on the port side. Navigation lights, instrument lights and two rotating beacons. Optional equipment includes a rear cockpit to accommodate aft-facing crew member, or forward-facing seat for passenger, or flying instructor if optional dual controls installed; alternatively space can be used for cargo. Other optional items are a Transland high-volume spreader, agitator installation, extra high density spray configuration with 70 nozzles installed; Agrinautics electrically operated three-way valve, emergency shut-off valve, pump in lieu of Root pump and strainer in lieu of Transland strainer; Agvenco 6520 pump in lieu of Root pump. Six- or eight-unit AU3000 Micronair installation in lieu of standard booms and nozzles; Transland S-2 Boommaster with Q-D flange instead of standard gate and Root pump; night lights including wingtip turn lights, cockpit fire extinguisher, and water bomber configuration. Optional avionics include Bendix T-12C ADF or T-12D digital ADF; King KX 170B, KX 170BE, KX 175B or KX 175BE nav/com; KI 210C converter indicator; Narco Com-11A or Com-11B com transceiver; and Nav-11 nav receiver.

DIMENSIONS, EXTERNAL:
Wing span 13.54 m (44 ft 5 in)
Length overall (tail up) 8.34 m (27 ft 4½ in)
Height overall 2.79 m (9 ft 2 in)
Tailplane span 4.86 m (15 ft 11½ in)
Wheel track 2.72 m (8 ft 11 in)
Propeller diameter 2.74 m (9 ft 0 in)

AREA:
Wings, gross 30.34 m² (326.6 sq ft)

WEIGHTS AND LOADINGS:
Weight empty, equipped 1,678 kg (3,700 lb)
Max T-O weight (CAR 3) 2,721 kg (6,000 lb)
Max T-O weight (CAM 8) 3,130 kg (6,900 lb)
Max wing loading 103.0 kg/m² (21.1 lb/sq ft)
Max power loading 6.99 kg/kW (11.5 lb/hp)

PERFORMANCE (with spray equipment installed and at CAR 3 max T-O weight, unless otherwise indicated):
Max level speed 122 kts (225 km/h; 140 mph)
Max cruising speed, 70% power 108 kts (200 km/h; 124 mph)
Working speed, 70% power 91-100 kts (169-185 km/h; 105-115 mph)
Stalling speed: flaps up 61 kts (113 km/h; 70 mph)
flaps down 57.5 kts (107 km/h; 66 mph)
at normal landing weight, flaps up 50 kts (92 km/h; 57 mph)
at normal landing weight, flaps down 48 kts (89 km/h; 55 mph)
Max rate of climb at S/L 274 m (900 ft)/min
Service ceiling 4,574 m (15,000 ft)
T-O run 236 m (775 ft)
Landing run 152 m (500 ft)
Ferry range with max fuel at 70% power 350 n miles (648 km; 403 miles)

VERIFIED

Ayres Thrush Commander-600 dropping fertiliser over Compton Abbas Airfield, Dorset, UK *(Peter March)*
1994

BRS

BALLISTIC RECOVERY SYSTEMS INC

1845 Henry Avenue, South St Paul, Minnesota 55075
Telephone: 1 (612) 457 7491
Fax: 1 (612) 457 8651

CFO: Mark Thomas
CHAIRMAN: Boris Popov
PUBLIC RELATIONS MANAGER: Dan Johnson

VERIFIED

BRS GARD-150

TYPE: Whole-aircraft parachute recovery system.

PROGRAMME: Retrofit installation of parachute system designed for recovery of Cessna 150 aircraft. Gard-150 project commenced 1986 and received FAA approval for

BRS Gard-150 whole-aircraft recovery system installed and deployed on a Cessna Model 150
1994

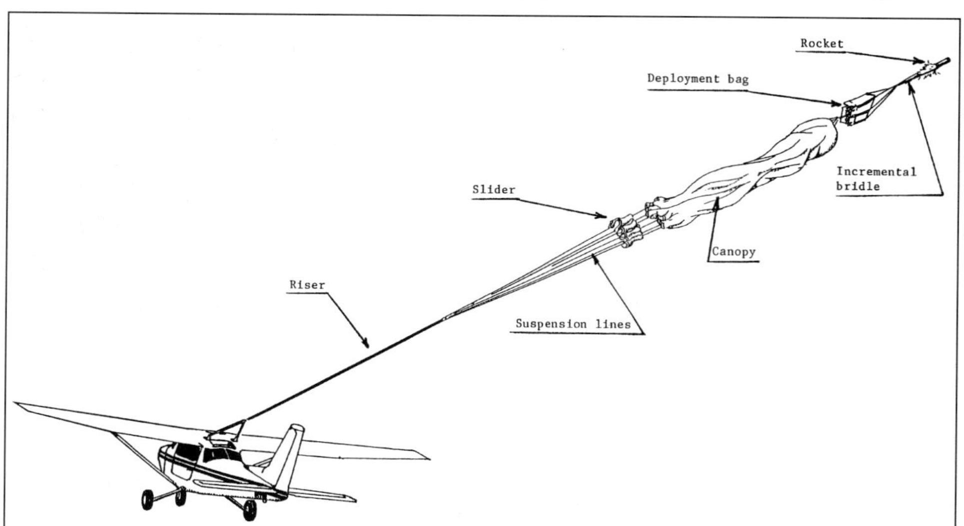

BRS Gard-150 whole-aircraft recovery system
1995

most Cessna aircraft in the 150-152 Model series May 1993. FAA approval is being sought for other aircraft types including: Piper Cub, Super Cub, Pacer as well as Taylorcraft aircraft. Development began in 1994 of larger systems for aircraft exceeding 3,175 kg (7,000 lb) and 643 km/h (400 mph).

DESIGN FEATURES: System consists of a rocket deployed parachute contained inside a box which is installed on the top of the canopy to the rear and protrudes into the cabin area by approximately 5 in. System deploys by pulling a handle within the cabin. System deployment time is under 1 second.

VERIFIED

BASLER

BASLER TURBO CONVERSIONS INC

255 West 35th Avenue, PO Box 2305, Oshkosh, Wisconsin 54903-2305
Telephone: 1 (414) 236 7820
Fax: 1 (414) 235 0381
PRESIDENT: Thomas R. Weigt
VICE-PRESIDENT: Bob Clark

Company associated with Basler Flight Service Inc at same address. New 6,968 m² (75,000 sq ft) factory occupied 21 January 1990. Employees 142. Additional centres being set up in Taiwan and Poland.

VERIFIED

Basler Turbo Conversions (Douglas) Turbo-67 DC-3/C-47 converted aircraft of the El Salvadorean Air Force *(FAES)*

1995

BASLER TURBO-67 DC-3 CONVERSION

TYPE: Turboprop conversion and stretch of DC-3.

PROGRAMME: Multiple STC received 27 February 1990; full certification 11 December 1990. 25 conversions completed by April 1995. Capacity at Oshkosh for 14 conversions simultaneously.

CUSTOMERS: Air Colombia (three), El Salvador Air Force, via USAF (four) Bolivian Air Force (one), Colombian Air Force (one), US Forest Service (two), United Technologies (one).

DESIGN FEATURES: Pratt & Whitney Double Wasps replaced by Pratt & Whitney Canada PT6A-67R turboprops. Fuselage reinforced and stretched by 1.02 m (3 ft 4 in) plug forward of wing to retain centre of gravity; front cabin bulkhead also moved 1.52 m (5 ft 0 in) forward; wing reinforced to accept extra load; rear door can be enlarged by upward-opening flap to take LD3 containers, of which five can be carried. New electrical system to FAR Pt 25 and upgraded hydraulics to give faster landing gear retraction; fuel system modified for jet fuel; optional additional tanks. Gross take-off and landing weight raised.

FLYING CONTROLS: All-metal control surfaces; lateral control improved; bob weight and down springs in elevator circuit; spring servo tab in rudder; dropped outboard leading-edge and new wingtips. Coupled autopilot to Cat. II limits.

LANDING GEAR: As for DC-3.

POWER PLANT: Two 1,062 kW (1,424 shp) P&WC PT6A-67R turboprops rated at 955 kW (1,281 shp) at 32°C (91°F), driving 2.92 m (9 ft 7 in) diameter Hartzell metal five-blade feathering and reversing propellers. Standard usable fuel 2,918 litres (771 US gallons; 642 Imp gallons); optional 5,837 litres (1,542 US gallons; 1,284 Imp gallons).

ACCOMMODATION: Two pilots in modernised cockpit. Normal 38 passengers for five LD3 containers with floor guides and rollers and upward-opening door extension.

SYSTEMS: Variable delivery hydraulic pumps; two 300 A starter/generators, four buses (main and emergency each side), two 12 V 88 Ah batteries, bleed air cable heating; propeller and engine intake de-icing standard, pneumatic boots for wing and tail optional; fire detection and extinguisher system.

AVIONICS: Full Bendix/King com/nav and remote area system.

WEIGHTS AND LOADINGS:

Weight empty	6,804 kg (15,000 lb)
Basic operating weight (equipped)	7,121 kg (15,700 lb)
Max internal fuel	2,346 kg (5,172 lb)
Max T-O and landing weight	13,041 kg (28,750 lb)
Max zero-fuel weight	11,884 kg (26,200 lb)
Max power loading	6.83 kg/kW (11.22 lb/shp)

PERFORMANCE:

Max cruising speed at 7,620 m (25,000 ft) 80% torque 210 kts (389 km/h; 241 mph)
Service ceiling, one engine out 3,960 m (13,000 ft)
Max range, 45 min reserves:
with long-range tanks at 7,620 m (25,000 ft), 80% torque, or 4,575 m (15,000 ft) 90% torque 2,260 n miles (4,188 km; 2,602 miles)
with standard tanks at 7,620 m (25,000 ft), 80% torque 1,000 n miles (1,853 km; 1,150 miles)

UPDATED

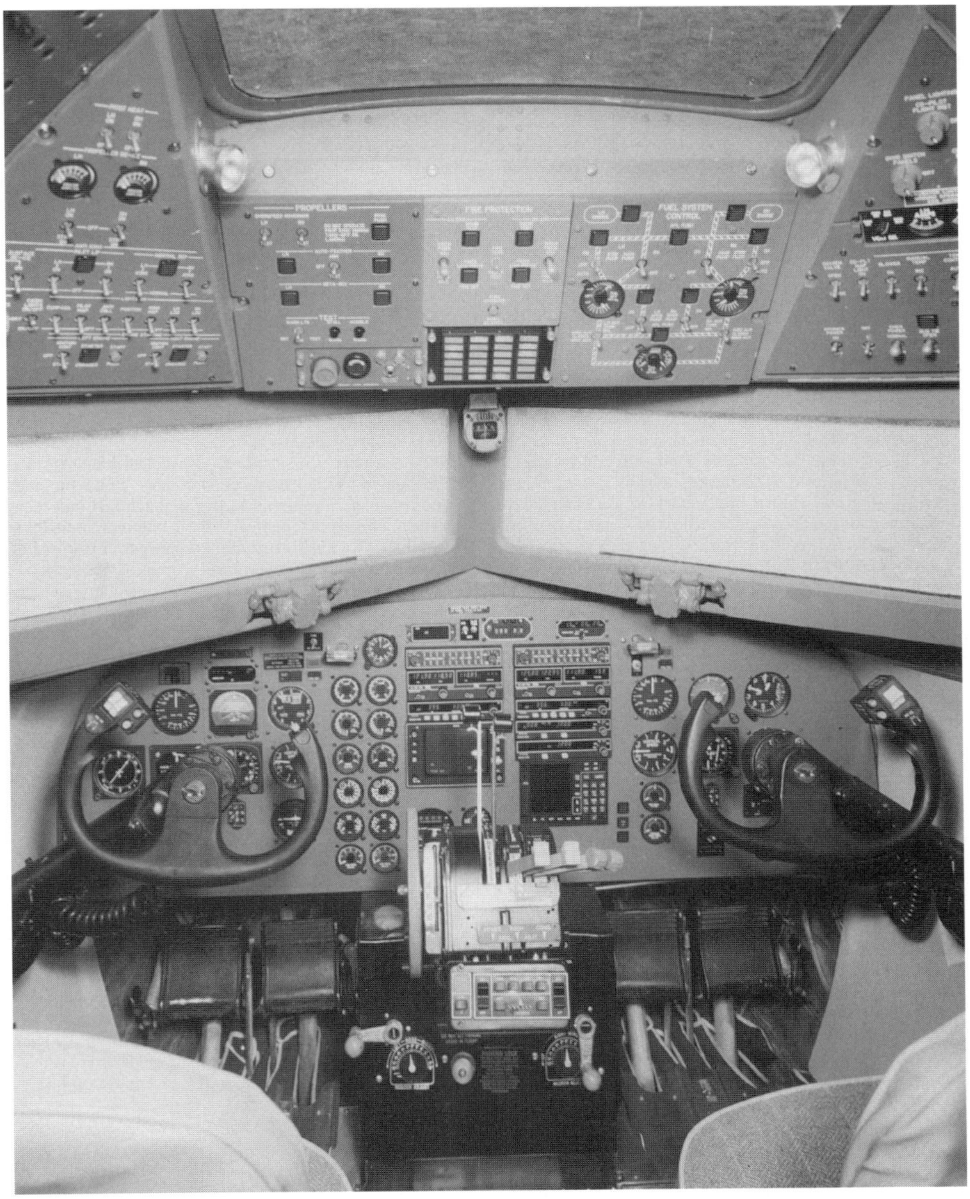

Basler Turbo Conversions Turbo-67 DC-3/C-47 avionics suites

1994

BEECHCRAFT — *aircraft now listed under Raytheon Aircraft Company in this section*

BELL

BELL HELICOPTER TEXTRON INC
(Subsidiary of Textron Inc)

PO Box 482, Fort Worth, Texas 76101
Telephone: 1 (817) 280 8415
Fax: 1 (817) 280 8221
PRESIDENT: Webb F. Joiner
SENIOR EXECUTIVE VICE-PRESIDENT: Lloyd Shoppa
SENIOR VICE-PRESIDENT, MARKETING: Peter H. Parsinen
SENIOR VICE-PRESIDENT, PRODUCT SUPPORT AND US GOVERNMENT BUSINESS: Fred Hubbard
VICE-PRESIDENT, RESEARCH AND ENGINEERING: Troy M. Gaffey
DIRECTOR, PUBLIC AFFAIRS AND ADVERTISING: Carl L. Harris

From 1970-81, Bell Helicopter Textron was an unincorporated division of Textron Inc; wholly owned subsidiary of Textron Inc from 3 January 1982. Bell Helicopter Canada formed at Montreal/Mirabel under contract with Canadian government October 1983; transfer to Mirabel of Model 206B JetRanger and Model 206L LongRanger production completed January 1987 to make room for V-22 Osprey. Production of Models 212/412 transferred mid-1988 and early 1989 respectively. Other programmes transferred to Canada include: Bell 230/430 and 407 and 407T.

More than 32,000 Bell helicopters manufactured worldwide, including over 9,000 civilian.

Bell helicopters built in the USA detailed here. Those built in Canada listed under Canada; several models built under licence by Agusta in Italy and Fuji in Japan; Korea Bell Helicopter Company (KBHC) will co-produce helicopters with Bell Helicopter Textron in Republic of Korea; Bell Helicopter de Venezuela CA, joint venture with Maquinarias Mendoza CA and Aerotecnica SA, established early 1984 in Caracas for marketing and support; Bell Helicopter Asia (Pte) Ltd is wholly owned Singapore-based company for marketing and support in Southeast Asia.

UPDATED

Bell 47G-4A three-seat general utility helicopter *(Peter Cooper)*

1995

BELL 47G

Military designations: OH-13 and TH-13

TYPE: Three-seat general utility helicopter.

PROGRAMME: First flight of prototype took place in December 1945. First helicopter to gain Approved Type Certificate. Over 5,000 manufactured. The Bell 47 has been licence-built by Agusta, Kawasaki and Westland.

VERSIONS: **Model 47G-3B-1:** Three-seater with outstanding high-altitude performance over a wide range of temperatures. Basically similar to earlier Model 47G-3, but Lycoming TVO-435-B1A supercharged engine uprated to 200 kW (270 hp), fuel capacity increased, cabin widened by 20 cm (8 in), and AUW increased by 45 kg (100 lb). A 4.5 kg (10 lb) weight has been added to each blade-tip to improve auto-rotation characteristics and manoeuvrability. Received FAA Type Approval 25 January 1963.

Model 47G-3B-2: Three-seater with outstanding high-altitude performance over a wide range of temperatures. Basically similar to earlier Model 47G-3, but Lycoming TVO-435-G1A supercharged engine uprated to 207 kW (280 hp), fuel capacity increased to 216 litres (57 US gallons; 48 Imp gallons), cabin widened by 20 cm (8 in) to 1.52 m (5 ft 0 in), and AUW increased by 45 kg (100 lb). A 4.5 kg (10 lb) weight has been added to each blade-tip to improve auto-rotation characteristics and manoeuvrability. FAA Type certificate issued 17 January 1968. The Royal New Zealand Air Force ordered seven 47G-3B-2s for delivery during 1970.

Model 47G-4A: Basic three-seat utility helicopter with 226 kW (305 hp) (derated to 207 kW; 280 hp) Lycoming VO-540-B1B3 engine. Cyclic and collective flying controls are hydraulically boosted. Airframe approved for 1,200 hours between overhauls. Dynamic components, cabin and centre-frame interchangeable with 47G-3B-2. Deliveries began in December 1965.

Model 47G-5: Low-cost model in which non-essential structures and components have been eliminated to reduce initial price and increase maximum useful load to 540 kg (1,191 lb). Powered by 196 kW (265 hp) Lycoming VO-435-B1A engine. Standard items include tinted Plexiglas canopy, doors, heavy-duty battery, 70 A 12 V alternator, synchronised elevator and twin 108 litre (28.5 US gallon; 24 Imp gallon) fuel tanks. A rotating beacon, landing light and full range of Model 47 optional extras are available. The standard Model 47 rotor system, with stabilising bar, and hydraulically boosted controls, retained.

AB-47G: Agusta-built version manufactured in Italy. In service with armed forces of Italy, Libya, Malta and Zambia.

OH-13: Three-seat observation and medical evacuation version in service with Greek, Pakistani and Turkish armed forces.

TH-13: Two-seat ot three-seat training version in service with Turkish armed forces.

UPGRADES: **Soloy:** Develops and manufactures turbine kits for the Bell 47.

OPERATORS: Versions of the Bell 47G are in service with the armed forces of the following countries: Colombia (8), Greece (4), Indonesia (8), Italy (21), Lesotho (1), Libya (8), Malaysia (7), Malta (4), New Zealand (5), Pakistan (12), Peru (18), Taiwan (10), Tanzania, Turkey (15), Uruguay (1) and Zambia (12).

The description which follows refers specifically to the Model 47G-4A.

DESIGN FEATURES: Two-blade semi-rigid main rotor with interchangeable blades or all-metal bonded construction and stabilising bar below at right-angles to blades. Two-blade all-metal tail rotor.

FLYING CONTROLS: Conventional swashplate assembly for cyclic-pitch and collective pitch control. Brake on main rotor optional. Through centrifugal clutch and two-stage planetary transmission. Shaft drive to tail rotor. Main rotor rpm 333-370. Small synchronised elevator at rear end of fuselage responds to the fore and aft motion of the cyclic-pitch control, to provide better stability and allow a greatly increased CG travel.

STRUCTURE: Fuselage built in three sections: centre, tail and cabin. Centre section has a welded steel-tube structure to carry the engine and cabin. Rear section is a triangular in cross-section steel tube structure which serves as tail rotor drive shaft support.

LANDING GEAR: Tubular skid type with small ground handling wheels and tie-down and towing attachments. Cross-tubes serve as supports for external loads such as litters or cargo bins. For amphibious operation, two air-inflated nylon floats are easily attached.

POWER PLANT: One vertically mounted 226 kW (305 hp) (derated to 207 kW; 280 hp) Lycoming VO-540-B1B3 six-cylinder horizontally opposed fan-cooled engine with clutch, drive shaft and rotor assembly in an integral unit. Two inter-connected saddle-mounted fuel tanks (216 litres; 57 US gallons; 48 Imp gallons total capacity) on CG and with gravity feed. See also Versions.

ACCOMMODATION: Side-by-side seats for three in cabin enclosed by a blue-tinted free-blown Plexiglas canopy. Door on each side, with sliding windows. For fair weather or specialised operations the doors are quickly removable. Can carry 455 kg (1,000 lb) cargo externally on under-fuselage hook.

Bell 47G-4A with Lycoming VO-540-B1B3 engine *(Peter Cooper)*

1995

ELECTRONICS AND EQUIPMENT: Standard equipment includes complete VFR flight and engine instruments, hydraulic boost controls, 28 V 50 A generator, electric starter, ground handling wheels, heavy-duty battery, and so on. Additional accessories available in FAA-approved kits include pontoons, night flying equipment, dusting and spraying equipment, cargo carriers, litters, dual controls, heater-defroster, rotor brake, inertial reel and shoulder harness, ARC, King or Narco 360-channel VHF transceivers, fire extinguisher, first aid kit, and so on.

DIMENSIONS, EXTERNAL:
- Main rotor diameter: 11.31 m (37 ft 1⅛ in)
- Main rotor blade chord: 28 cm (11 in)
- Tail rotor diameter: 1.78 m (5 ft 10 in)
- Length overall (main rotor fore and aft): 13.17 m (43 ft 2½ in)
- Length of fuselage: 9.90 m (32 ft 6 in)
- Width, rotors fore and aft: 2.59 m (8 ft 6 in)
- Width over skids: 2.29 m (7 ft 6 in)
- Height to top of rotor hub: 2.82 m (9 ft 3 in)

DIMENSIONS, INTERNAL:
- Cabin: Length: 1.50 m (4 ft 11 in)
- Max width: 1.52 m (5 ft 0 in)
- Max height: 1.37 m (4 ft 6 in)

AREAS:
- Main rotor blade (each): 1.59 m² (17.14 sq ft)
- Tail rotor blade (each): 0.11 m² (1.20 sq ft)
- Main rotor disc: 100.61 m² (1,083 sq ft)
- Tail rotor disc: 2.49 m² (26.8 sq ft)

WEIGHTS AND LOADINGS:
- Weight empty, with oil and equipped:
  - 47G-3B-2: 878 kg (1,937 lb)
  - 47G-4A: 852 kg (1,880 lb)
  - 47G-5: 776 kg (1,712 lb)
  - OH-13S: 999 kg (2,203 lb)
  - TH-13T: 1,133 kg (2,499 lb)
- Max T-O weight:
  - OH-13S, Ag-5, 47G-5: 1,293 kg (2,850 lb)
  - 47G-3B-2, 47G-4A, TH-13T: 1,338 kg (2,950 lb)

PERFORMANCE (at max T-O weight, except where indicated. 47G-5 figures for aircraft with synchronised elevator and second fuel tank):
- Max level speed at S/L:
  - all except Ag-5: 91 kts (169 km/h; 105 mph)
  - Ag-5: 78 kts (145 km/h; 90 mph)
- Recommended cruising speed at 1,525 m (5,000 ft):
  - OH-13S, TH-13T, 47G-3B-2: 72 kts (133 km/h; 83 mph)
  - 47G-4A: 77 kts (143 km/h; 89 mph)
  - 47G-5, Ag-5: 74 kts (137 km/h; 85 mph)
- Max rate of climb at S/L:
  - 47G-3B-2, TH-13T: 302 m (990 ft)/min
  - 47G-4A: 244 m (800 ft)/min
  - OH-13S: 168 m (550 ft)/min
- Service ceiling: 47G-3B-2: 5,608 m (18,400 ft)
  - 47G-4A: 3,415 m (11,200 ft)
  - Ag-5, 47G-5: 3,200 m (10,500 ft)
  - OH-13S: 5,640 m (18,500 ft)
  - TH-13T: 5,120 m (16,800 ft)
- Hovering ceiling in ground effect:
  - 47G-3B-2: 5,060 m (16,600 ft)
  - 47G-4A: 2,347 m (7,700 ft)
  - Ag-5, 47G-5: 1,800 m (5,900 ft)
  - OH-13S: 5,500 m (18,000 ft)
  - TH-13T: 4,875 m (16,000 ft)
- Hovering ceiling out of ground effect:
  - 47G-3B-2: 3,750 m (12,300 ft)
  - 47G-4A: 1,190 m (3,900 ft)
  - 47G-5, Ag-5: 412 m (1,350 ft)
  - OH-13S: 4,510 m (14,800 ft)
  - TH-13T: 3,200 m (10,500 ft)
- Range with max fuel at 1,525 m (5,000 ft), no reserves:
  - 47G-3B-2, OH-13S, TH-13T: 217 n miles (402 km; 250 miles)
  - 47G-4A: 225 n miles (416 km; 259 miles)
  - 47G-5, Ag-5: 222 n miles (411 km; 256 miles)

NEW ENTRY

BELL MODEL 204

US military designations: UH-1A/B/C/E/F/L, HH-1K and TH-1F/L Iroquois

Swedish military designation: Hkp 3B

TYPE: Single-rotor general purpose helicopter.

PROGRAMME: In 1955 the Bell Model 204 won a US Army design competition for a utility helicopter suitable for front-line casualty evacuation, general production version was originally designated HU-1, giving rise to the nickname 'Hueycopter', which survived the change of designation to UH-1. Official US Army name for the UH-1 series is Iroquois.

VERSIONS: **XH-40**: Three prototypes, the first of which flew in 1956.

YH-40: Six service test models.

UH-1: Nine preproduction models.

UH-1A: Initial production version, incorporating changes requested as a result of service testing. Six-seater, powered by an 641 kW (860 shp) Lycoming T53-L-1A turboshaft engine, derated to 574 kW (770 shp). Deliveries to the US Army began 30 June 1959 and were completed March 1961. Thirteen operated by utility Tactical Transport Helicopter Company in Vietnam were modified to carry 16 × 2.75 in air-to-surface rockets and two 0.30 in machine guns. A total of 14 were delivered for use as helicopter instrument trainers with dual controls and a device for simulated instrument instruction.

UH-1B: Development of UH-1A, initially with 716 kW (960 shp) T53-L-5 turboshaft. Subsequent deliveries with 820 kW (1,100 shp) T53-L-11. Crew of two and seven troops, or three stretchers, two sitting casualties and medical attendant, or 1,360 kg (3,000 lb) of freight. Rotor diameter 13.41 m (44 ft). Normal fuel capacity 625 litres (165 US gallons; 137 Imp gallons); overload capacity 1,250 litres (330 US gallons; 274 Imp gallons). For armed support duties, a rocket pack and electrically controlled machine gun can be mounted on each side of cabin. Other armament installations tested on UH-1B included General Electric M-5 nose-mounted 40 mm grenade launcher and XM-30 armament system, consisting of two side-mounted XM-140 30 mm cannon with central ammunition reservoir and fire control system. Deliveries began March 1961. This version was superseded by the UH-1C on the Bell assembly line, but production continued by Fuji in Japan in order to fulfil an order of 89 UH-1Bs for the JGSDF.

UH-1C: In September 1965, Bell introduced its Model 540 'door hinge' rotor, with blades of increased 69 cm (27 in) chord, on this developed version of the UH-1B, offering some increase in speed and a substantial increase in manoeuvrability through resistance to blade stall. Through reduced vibration and stress levels, the 540 rotor eliminates previous limitations on maximum level flight speed. T53-L-11 turboshaft, accommodation and armament as for UH-1B. Normal fuel capacity 916 litres (242 US gallons; 201 Imp gallons); overload 2,241 litres (592 US gallons; 493 Imp gallons). Superseded UH-1B in production for US Army, but itself superseded by AH-1G.

UH-1E: In March 1962, Bell won a design competition for an assault support helicopter for the US Marine Corps, to replace Cessna O-1B/C fixed-wing aircraft and Kaman OH-43D helicopters. Designated UH-1E, this version is generally similar to the UH-1B/C, but has a personnel hoist, rotor brake and Marine electronics. The 540 rotor and increased fuel capacity (as UH-1C) were introduced in 1965.

UH-1F: Following a design competition, it was announced in June 1963 that an initial batch of 25 UH-1F helicopters, based on the UH-1B were to be built for the USAF in 1963-64, and many more later, for missile site support duties. Each has a 948 kW (1,272 shp) General Electric T58-GE-3 turboshaft (derated to 820 kW; 1,100 shp), a 14.63 m (48 ft) rotor, normal fuel capacity of 945 litres (250 US gallons; 208 Imp gallons) and overload capacity of 1,552 litres (410 US gallons; 341 Imp gallons). This version can handle up to 1,815 kg (4,000 lb) of cargo at missile site silos, or carry a pilot and 10 passengers. The first UH-1F flew 20 February 1964. Subsequent contracts for a further 121 aircraft were completed in 1967. First delivery to an operational unit was made to the 4486th Test Squadron at Eglin AFB in September 1964. This model was used for classified psychological warfare missions in Vietnam.

TH-1F: Training version of UH-1F for USAF.

HH-1K: Sea-air rescue version for US Navy, which placed letter contract for 27 late in 1968, for delivery in 1970. The aircraft has the UH-1E airframe, T53-L-13 turboshaft engine (derated to 820 kW; 1,100 shp) and revised avionics.

TH-1L: Training version for the US Navy. Similar to UH-1E but with 1,044 kW (1,400 shp) Lycoming T53-L-13 turboshaft (derated to 820 kW; 1,100 shp) and improved electronics. Contract for 45 received 16 May 1968; the first of these was delivered to the US Navy at Pensacola, Florida, 26 November 1969.

UH-1L: Utility version of TH-1L for US Navy. Eight ordered, and delivered during 1969.

UH-1M: US Army version fitted with Hughes Aircraft Iroquois night fighter and night tracker (Infant) system to detect and acquire ground targets under low ambient lighting conditions, two sensors mounted on nose of cabin serve a low light level TV system with three cockpit displays and a direct-view system using an image intensifier at cockpit/gunner's station. Three UH-1Ms deployed with hunter-killer helicopter groups in Vietnam in early 1970 to evaluate system.

Bell 204 (UH-1B) of the Uruguayan Air Force *(Dennis Hughes)*

1994

RH-2 (Research Helicopter 2): One UH-1A was used as a flying laboratory for new instrument and control systems. Installations included an electronic control system and high-resolution radar in a large fairing above the flight deck, enabling the pilot to detect obstacles ahead of the aircraft in bad visibility.

HueyTug: It was announced on 3 September 1968 that a UH-1C had been retrofitted with a 2,125 kW (2,850 shp) Lycoming T55-L-7C turboshaft and 15.24 m (50 ft) 'door-hinge' rotor as the prototype of a new flying crane version able to lift a 3 ton external payload. Associated modifications, all of which can be applied retrospectively to existing UH-1s, include substitution of a 1,491 kW (2,000 hp) transmission and larger tail rotor, reinforcement of the airframe and fitment of a larger tailboom, and use of a stability control and augmentation system instead of the normal stabiliser bar. The HueyTug is designed to hover out of ground effect at 1,220 m (4,000 ft), 95°F, at 6,350 kg (14,000 lb) maximum T-O weight. Maximum level speed clean is 140 knots (259 km/h; 161 mph).

Model 204B: Commercial and military export version of UH-1B, with 10 seats, 820 kW (1,100 shp) T5311A turboshaft and 14.63 m (48 ft) rotor. Tailboom incorporates a 0.99 m³ (35 cu ft) baggage compartment, cabin doors with jettisonable emergency exits, passenger steps on each side of cabin, improved outside lights, commercial radio equipment, fire detection and extinguishing systems. Received FAA certification 4 April 1963.

AB 204B: Agusta-built version of Bell 204B manufactured in Italy. In service with armed forces of Sweden, Turkey and Yemen.

OPERATORS: Versions of the Bell 204 are in service with the armed forces of the following countries: Austria (21), Colombia (2), El Salvador (10), Honduras (2), Japan (25), Sweden (6), Thailand, Turkey (23), Uruguay (4) and Yemen (2).

The following description applies to the commercial Model 204B except where indicated:

DESIGN FEATURES: Two-blade all-metal semi-rigid main rotor with interchangeable blades. Usual Bell stabilising bar above and at right angles to main rotor blades. Underslung feathering axis hub. Two-blade all-metal tail rotor of honeycomb construction. Blades do not fold. Shaft-drive to both main and tail rotors. Main rotor rpm 295-324 (294-317 in UH-1F only).

FLYING CONTROLS: Small synchronised elevator on rear fuselage is connected to the cyclic control to increase allowable CG travel.

STRUCTURE: Main blades built up of extruded aluminium spars and laminates. Blade chord 53.3 cm (21 in). All-metal tail rotor blades. The fuselage is a conventional all-metal semi-monocoque structure.

TAIL SURFACE: Small synchronised elevator on rear fuselage is connected to the cyclic control to increase allowable CG travel.

LANDING GEAR: Tubular skid type. Lock-on ground handling wheels available.

POWER PLANT: One 820 kW (1,100 shp) Lycoming T5309A shaft-turbine mounted above fuselage aft of cabin. Two fuel tanks on CG, immediately aft of cabin, total capacity 916 litres (242 US gallons; 201 Imp gallons).

ACCOMMODATION: Crew of two side by side, with dual controls. Standard model has bench seats for eight passengers, three-abreast in centre row and five-abreast in rear row. Optional layouts include individual chairs with tip-up seats or special interiors to customer's requirements, with optional settee, cabinet, writing table, glass-panelled dividing wall between crew and passenger compartments. Two doors on each side, front one hinged to open forward, rear one sliding aft. Compartment for 182 kg (400 lb) of baggage. Passenger seats removable to provide 3.96 m³ (140 cu ft) of accessible cargo space for loads of up to 1,360 kg (3,000 lb) weight. Forced air ventilation system.

ELECTRONICS AND EQUIPMENT: Standard equipment includes hydraulic power boost on cyclic, collective and tail rotor flying controls. Optional equipment includes full all-weather flight instrumentation, multichannel select VHF transceivers, visual omni-range and ILS course indicator with heading and glide slope presentation, ADF, VHF marker beacon receivers, Sperry C-4 navigation compass, Lear VGI 5 in all-attitude flight indicator, and external sling for 1,820 kg (4,000 lb) of freight.

DIMENSIONS, EXTERNAL:
- Main rotor diameter: UH-1B, E — 13.41 m (44 ft 0 in)
  - 204B, UH-1D, F — 14.63 m (48 ft 0 in)
- Tail rotor diameter — 2.59 m (8 ft 6 in)
- Length overall (main rotor fore and aft):
  - 204B — 17.37 m (57 ft 0 in)
  - UH-1B, E — 16.15 m (53 ft 0 in)
  - UH-1D — 16.42 m (53 ft 10¾ in)
  - UH-1F — 17.40 m (57 ft 1 in)
- Length of fuselage: UH-1B, E — 12.98 m (42 ft 7 in)
  - 204B, UH-1D, F — 13.59 m (44 ft 7 in)
- Overall height: UH-1B, E — 3.87 m (12 ft 8½ in)
  - UH-1D — 4.08 m (13 ft 5 in)
  - UH-1F — 3.81 m (12 ft 6 in)
- Skid track — 2.39 m (7 ft 10 in)

DIMENSIONS, INTERNAL (204B):
- Cabin: Length — 2.59 m (8 ft 6 in)
  - Max width — 2.39 m (7 ft 10 in)
  - Max height — 1.47 m (4 ft 10 in)
- Baggage compartment — 0.85 m³ (30 cu ft)

AREAS:
- Main rotor blade (each):
  - 204B, UH-1D, F — 3.58 m² (38.5 sq ft)
- Main rotor disc: UH-1B, E — 141.2 m² (1,520 sq ft)
  - UH-1B, and E (540 rotor), 204B, UH-1D, F — 168.0 m² (1,808 sq ft)
- Tail rotor disc — 5.27 m² (56.8 sq ft)

WEIGHTS AND LOADINGS:
- Weight empty: UH-1B — 2,050 kg (4,519 lb)
  - UH-1B (540 rotor) — 2,196 kg (4,842 lb)
  - 204B — 2,086 kg (4,600 lb)
  - UH-1C (incl armoured crew seats) — 2,300 kg (5,071 lb)
  - UH-1D — 2,140 kg (4,717 lb)
  - UH-1E — 2,155 kg (4,750 lb)
  - UH-1E (540 rotor) — 2,293 kg (5,055 lb)
  - UH-1F — 2,010 kg (4,430 lb)
- Max payload: UH-1C — 1,205 kg (2,656 lb)
- Max T-O and landing weight:
  - UH-1B, 204B (normal), UH-1E — 3,856 kg (8,500 lb)
  - UH-1B and E (540 rotor) — 4,309 kg (9,500 lb)
  - UH-1C, D and H, 204B (external load) — 4,309 kg (9,500 lb)
  - UH-1F — 4,082 kg (9,000 lb)
- Max zero-fuel weight: UH-1C — 3,596 kg (7,927 lb)
- Max disc loading: UH-1C — 30.5 kg/m² (6.25 lb/sq ft)
- Max power loading: UH-1C — 3.91 kg/hp (8.63 lb/hp)

PERFORMANCE (at max T-O weight):
- Max level speed:
  - UH-1C at S/L — 128 kts (238 km/h; 148 mph)
- Max permissible diving speed:
  - UH-1C — 157 kts (290 km/h; 180 mph)
- Max cruising speed: 204B — 117 kts (217 km/h; 135 mph)
  - UH-1C at S/L — 128 kts (238 km/h; 148 mph)
- Econ cruising speed: 204B — 104 kts (193 km/h; 120 mph)
  - UH-1C at 1,525 m (5,000 ft) — 124 kts (230 km/h; 143 mph)

Bell 205 (UH-1H) general-purpose helicopter of the US Army *(Bell)*

1995

Rate of climb at S/L: UH-1C 425 m (1,400 ft)/min
UH-1H, 204B 488 m (1,600 ft)/min
Vertical rate of climb at S/L: UH-1C at 3,855 kg (8,500 lb) AUW 165 m (540 ft)/min
Service ceiling:
204B (3,400 kg; 7,500 lb AUW) 4,815 m (15,800 ft)
UH-1C 3,500 m (11,500 ft)
Hovering ceiling in ground effect:
204B (at 3,400 m; 7,500 ft AUW) 4,175 m (13,700 ft)
UH-1C 3,230 m (10,600 ft)
Hovering ceiling out of ground effect:
204B 732 m (2,400 ft)
UH-1C at 3,855 kg (8,500 lb) AUW 3,050 m (10,000 ft)
Range with max fuel, no allowances:
204B 286 n miles (530 km; 330 miles)
UH-1C 332 n miles (615 km; 382 miles)

UPDATED

BELL 205

US military designations: UH-1D/H/V, EH-1H and HH-1H
Canadian military designations: CH-118 Iroquois
Spanish military designation: HU.10B

TYPE: Single-rotor general-purpose helicopter.

PROGRAMME: Although basically similar to the earlier Model 204, the Model 205 introduced a longer fuselage, increased cabin space to accommodate more passengers as well as other changes.

VERSIONS: **CH-118:** Similar to UH-1H, for Mobile Command, Canadian Armed Forces. First of 10 delivered on 6 March 1968. Originally designated CUH-1H.

EH-1H: Electronic countermeasures configuration, with the Quick Fix I airborne communications interception, emitter locating and jamming system, including an AN/APR-39(V)2 radar warning receiver, XM130 chaff/flare dispenser and AN/ALQ-144 infra-red jammer. The FY81 budget added $5.1 million to convert initial Quick Fix IA systems in the EH-1H to Phase IB configuration, plus survivability equipment to protect the aircraft against known and postulated threats, including hot metal/plume suppression. By April 1981, three of the EH-1Hs had been delivered, with seven more to follow. However, the Quick Fix mission has been taken over by the much larger Sikorsky EH-60A version of the Black Hawk utility transport helicopter (which see).

HH-1H: It was announced on 4 November 1970 that a fixed price contract worth more than $9.5 million had been received from the USAF for 30 HH-1H aircraft (generally similar to the UH-1H) for use as local base rescue helicopters. Deliveries were completed during 1973.

UH-1D: This US Army version of the Model 205 Iroquois has a 820 kW (1,100 shp) Lycoming T53-L-11 turboshaft, 14.63 m (48 ft) rotor, normal fuel capacity of 832 litres (220 US gallons; 183 Imp gallons) and overload capacity of 1,968 litres (520 US gallons; 433 Imp gallons). Relocation of the fuel cells increases cabin space to 6.23 m^3 (220 cu ft), providing sufficient room for a pilot and 12 troops, or six litters and a medical attendant, or 1,815 kg (4,000 lb) of freight. A contract for a service test batch of seven YUH-1Ds was announced in July 1960 and was followed by further very large production orders from the US Army and from many other nations of the non-Communist world. First YUH-1D flew on 16 August 1961 and delivery to US Army field units began on 9 August 1963, when the second and third production UH-1Ds went to the 11th Air Assault Division at Fort Benning, Georgia. The UH-1D was superseded in production for the US Army by the UH-1H, but 352 UH-1Ds were built subsequently under licence in Germany for the German Army and Air Force. Prime contractor was Dornier.

UH-1H: Following replacement of the original T53-L-11 turboshaft by the 1,044 kW (1,400 shp) T53-L-13, the version of the Model 205 for the US Army was designated UH-1H. Deliveries of an initial series of 319 aircraft for the US Army began in September 1967. Subsequent orders included 300 more for the Army in January 1971, and nine for the RNZAF.

Additional orders for a total of 560 UH-1Hs were placed in 1971-73. An add-on contract for 54 more UH-1Hs, valued at $11.9 million, was awarded in September 1974, with delivery extending into 1976.

Under a licensing agreement concluded in 1969, the Republic of China is producing UH-1Hs for the Nationalist Chinese Army, with much of the manufacturing and assembly process being carried out at Taichung, Taiwan. The initial production programme was for 50 helicopters, already delivered. Subsequent orders have increased the total procurement to 118.

UH-1V: Approximately 220 UH-1Hs converted by US Army as a medevac version. Avionics and equipment in this version include a radio altimeter, AEL AN/ARN-124 DME, glide slope and rescue hoist.

The 7,000th Model 205/205A helicopter was completed in 1973.

AB 205A: Agusta-built version of the Bell 205 UH-1D. Also built in Germany by Dornier.

Model 205A-1: Civil and export designation of UH-1H. Also built in Italy by Agusta, Japan by Fuji and Taiwan by AIDC.

AB 205A-1: Agusta-built Bell Model 205A-1 (UH-1H) manufactured in Italy.

UPGRADES: **Bell:** Bell 205 UH-1HP Huey II upgrade. See separate entry.

Bogan Aerotech: Huey II. See separate entry.

Fuji-Bell: UH-1J and Advanced 205B. See separate entries.

Heli-Conversions: Super 205. See separate entry.

US Army: The US Army plans to retain at least 2,700 improved UH-1Hs in service beyond the year 2000 to perform such operations as resupply, troop transport, command and control, electronic warfare, medical evacuation and minefield emplacement. To make such a plan realistic, a product improvement programme for the Army's UH-1H fleet introduced improved or new avionics and equipment including an AN/ALQ-144 infra-red jammer, AN/APN-209 radar altimeter, AN/APR-39 radar warning receiver, AN/ARC-164 UHF/AM radio, AN/ARN-124 DME, XM130 chaff/flare dispenser, NOE communications (FM/HF), communications security, infra-red suppressor (hot metal and plume), altimeter lighting (5 V), crashworthy auxiliary fuel system, closed circuit refuelling, fuel tank vent, improved main input driveshaft, and main rotor mast plug. In addition, it is planned to introduce, as a minimum, new composite main rotor blades; improved stabiliser bar, tail rotor hub and servo cylinders; a split engine deck, and improved oil filtration; a night vision compatible cockpit; built-in Vibrex connections; an improved AN/ASN-43 gyro magnetic compass; and Doppler navigation.

A US Army Request for Proposal for composite main rotor blades for the UH-1H was issued on 16 November 1981. The Army's schedule called for a qualified blade to be ready for production after 32 months. Procurement of 6,000 blades was anticipated in 1985-89, at a cost of $20,000 or less per blade in FY81 dollars. Bell tendered a joint proposal with Boeing, and this team was awarded a $19 million development contract during 1982 by the US Army Aviation Research and Development Command. Bell designed the composite blade for the UH-1H, but both companies fabricated test blades and supported laboratory and flight testing to ensure compliance with Army requirements. The composite rotor blades provide a 6 per cent improvement in the UH-1H's hovering capability and a 5-8 per cent reduction in fuel flow in forward flight. Bell provided manufacturing tools and fixtures and transferred specific manufacturing knowledge to Boeing, so that both companies are equally capable and qualified to manufacture production blades, for which contracts are expected to exceed $100 million. The first flight of the composite rotor blades on a UH-1H took place in early 1985. Production deliveries began in January 1988.

US Army: UH-1V. See Versions.

Westland Engineering: $18 million contract for design and supply of speed decreasing gearbox initially for UH-1H re-engine programme.

OPERATORS: Versions of the Bell 205 are in service with the armed forces of the following countries: Argentina (4), Australia (25), Bolivia (19), Brazil (30), Canada (27), Chile (11), Colombia (30), Dominican Republic (9), El Salvador (60), Ethiopia (2), Germany (100), Greece (110), Guatemala (15), Honduras (7), Indonesia (16), Iran (75), Italy (90), Japan (140), Mexico (1), Morocco (40), Myanmar (12), New Zealand (14), Oman (20), Pakistan (15), Panama (8), Papua New Guinea (4), Peru (18), Philippines (65), Saudi Arabia (20), Singapore (20), South Korea (25), Spain (60), Surinam (1), Taiwan (120), Tunisia (2), Turkey (150), Uganda (3), United Arab Emirates (6), USA (2,780), Uruguay (3), Venezuela (12), Zambia (4) and Zimbabwe (2).

The following details refer specifically to the military UH-1H:

DESIGN FEATURES: Two-blade semi-rigid main rotor. Stabilising bar above and at right angles to main rotor blades. Underslung feathering axis head. Two-blade all-metal tail rotor. Shaft drive to both main and tail rotor. Transmission rating 820 kW (1,100 shp). Main rotor rpm 294-324.

FLYING CONTROLS: Small synchronised elevator on rear fuselage is connected to the cyclic control to increase allowable CG travel.

STRUCTURE: Interchangeable main blades built up of extruded aluminium spars and laminates. Tail rotor blades of honeycomb construction. Blades do not fold. The fuselage is a conventional all-metal semi-monocoque structure.

LANDING GEAR: Tubular skid type. Lock-on ground handling wheels and inflated nylon float bags available.

POWER PLANT: One 1,044 kW (1,400 shp) Textron Lycoming T53-L-13 turboshaft, mounted aft of the transmission on top of the fuselage and enclosed in cowlings. Five interconnected rubber fuel cells, total capacity 844 litres (223 US gallons; 186 Imp gallons), of which 799 litres (211 US gallons; 176 Imp gallons) are usable. Overload fuel capacity of 1,935 litres (511 US gallons; 425 Imp gallons) usable, obtained by installation of kit comprising two 568 litre (150 US gallon; 125 Imp gallon) internal auxiliary fuel tanks interconnected with the basic fuel system.

ACCOMMODATION: Pilot and 11-14 troops, or six litters and a medical attendant, or 1,759 kg (3,880 lb) of freight. Crew doors open forward and are jettisonable. Two doors on each side of cargo compartment; front door is hinged to open forward and is removable, rear door slides aft. Forced air ventilation system.

AVIONICS AND EQUIPMENT: FM, UHF, VHF radio sets, IFF transponder, Gyromatic compass system, direction finder set, VOR receiver and intercom standard. Optional nav/com systems. Standard equipment includes bleed air heater and defroster, comprehensive range of engine and flight instruments, power plant fire detection system, 30 V 300 A DC starter/generator, navigation, landing and

Bell Huey II upgrade of Bell 205 with Lycoming T53-L-703 turboshaft engine *(Bell)*

1995

anti-collision lights, controllable searchlight, hydraulically boosted controls. Optional equipment includes external cargo hook, auxiliary fuel tanks, rescue hoist, 150,000 BTU muff heater.

DIMENSIONS, EXTERNAL:

Main rotor diameter	14.63 m (48 ft 0 in)
Tail rotor diameter	2.59 m (8 ft 6 in)
Main rotor blade chord	0.53 m (1 ft 9 in)
Tail rotor blade chord	0.213 m (8.4 in)
Length: overall (main rotor fore and aft)	17.62 m (57 ft 9⅝ in)
fuselage	12.77 m (41 ft 10¾ in)
Height: overall, tail rotor turning (excl fin tip antenna)	4.41 m (14 ft 5½ in)
to top of main rotor head	3.60 m (11 ft 9¾ in)
Stabiliser span	2.84 m (9 ft 4 in)
Width over skids	2.91 m (9 ft 6½ in)

DIMENSIONS, INTERNAL:

Cabin: Max width	2.34 m (7 ft 8 in)
Max height	1.25 m (4 ft 1¼ in)
Volume (excl flight deck)	approx 6.23 m³ (220 cu ft)

AREAS:

Main rotor disc	168.11 m² (1,809.56 sq ft)
Tail rotor disc	5.27 m² (56.7 sq ft)

WEIGHTS AND LOADINGS:

Weight empty, equipped	2,363 kg (5,210 lb)
Basic operating weight (troop carrier mission)	2,520 kg (5,557 lb)
Mission weight	4,100 kg (9,039 lb)
Max T-O and landing weight	4,309 kg (9,500 lb)
Max zero-fuel weight	3,660 kg (8,070 lb)
Max disc loading	25.6 kg/m² (5.25 lb/sq ft)
Max power loading	4.13 kg/kW (8.63 lb/shp)

PERFORMANCE (at max T-O weight):

Never-exceed speed	110 kts (204 km/h; 127 mph)
Max level and cruising speed	110 kts (204 km/h; 127 mph)
Econ cruising speed at 1,735 m (5,700 ft)	110 kts (204 km/h; 127 mph)
Max rate of climb at S/L	488 m (1,600 ft)/min
Service ceiling	3,840 m (12,600 ft)
Hovering ceiling: IGE	4,145 m (13,600 ft)
OGE	1,220 m (4,000 ft)
Range with max fuel, no allowances, no reserves, at S/L	276 n miles (511 km; 318 miles)

NEW ENTRY

BELL 205 UH-1HP HUEY II UPGRADE

TYPE: Proposed upgrade of UH-1H.

DESIGN FEATURES: More powerful T53 engine; Bell 212 main rotor system; uprated transmission; Bell 212 tailboom with tractor tail rotor. Upgrade offers 28 per cent increase in hover ceiling, increased component life, reduced maintenance man-hours per flight hours and a 41 per cent reduction in direct cost of operation. Gross weight increase to 4,763 kg (10,500 lb) internal and 5,091 kg (11,200 lb) external; internal payload increases by 291 kg (642 lb).

Manufacture of upgraded Bell 205A/UH-1J continues under licence in Japan by Fuji (which see).

POWER PLANT: One 1,342 kW (1,800 shp) Textron Lycoming T53-L-703 turboshaft. Transmission rating increased to 962 kW (1,290 shp).

PERFORMANCE:

Max level speed	110 kts (204 km/h; 127 mph)
Max rate of climb at S/L	515 m (1,690 ft)/min
Service ceiling	4,905 m (16,100 ft)

UPDATED

BELL MODEL 209 HUEYCOBRA, SEACOBRA and SUPERCOBRA

US Army designations: AH-1G, AH-1Q and AH-1R

US Navy/Marine Corps designations: AH-1J, AH-1T and AH-1W

TYPE: Single-engine AH-1G/S and twin-engine (AH-1J/T/W) close support and attack helicopters.

PROGRAMME: Bell Helicopter Textron initiated the Model 209 in March 1965 as a company funded development of the UH-1B/C Iroquois intended specifically for armed helicopter missions. The original design combined the basic transmission and rotor system and (in its standard form) the power plant of the UH-1C with a new, streamlined fuselage designed for maximum speed, armament load and crew efficiency. Relatively small, its low silhouette and narrow profile make it easy to conceal with small camouflage nets or to move under cover of trees. Tandem seating provides the best possible field of view for the crew of two.

The Model 209 prototype made its first flight 7 September 1965, and the US Army's intention to order the aircraft was announced 11 March 1966, the initial model being known as the AH-1G HueyCobra. Total orders for all versions of the HueyCobra SeaCobra exceed 1,400.

VERSIONS: **AH-1G HueyCobra:** Original version for the US Army, powered by a single 1,044 kW (1,400 shp) Avco Lycoming T53-L-13 turboshaft, derated to 820 kW (1,100 shp) for T-O and maximum continuous rating. Development contract for two preproduction aircraft placed 4 April 1966, followed on 13 April by an initial order for 110 aircraft plus long-lead-time spares. Subsequent contracts raised the total US order to 1,075, deliveries of which began in June 1967. The US Marine Corps acquired 38 AH-1Gs during 1969, for transition training and initial deployment pending deliveries of the AH-1J. Six were supplied to Israel in 1974, and the Spanish Navy received eight (designated **Z.14**) for anti-shipping strike duties, giving an overall production total of 1,127 AH-1Gs. A number of AH-1Gs has been converted to **TH-1G** dual control trainers. Following the decision in 1977 to equip the HueyCobra with TOW missiles, 92 were converted to interim AH-1Q standard; all of these, and 286 other AH-1Gs, were converted to Mod AH-1S configuration by February 1984, completing contracts existing at that time. An order for 29 more conversions was announced in September 1984, followed by a contract for a further 29 in September 1985. One AH-1G was converted to **JAH-1G** as an armament testbed aircraft. Two US Army AH-1Gs have been acquired for night interception missions by the US Customs Service.

AH-1J SeaCobra: Initial twin turboshaft version for US Marine Corps, powered by a 1,342 kW (1,800 shp) Pratt & Whitney Canada T400-CP-400 coupled free-turbine turboshaft, a military version of the PT6T-3 Turbo Twin Pac. Engine and transmission flat rated at 820 kW (1,100 shp) continuous output, with increase to 932 kW

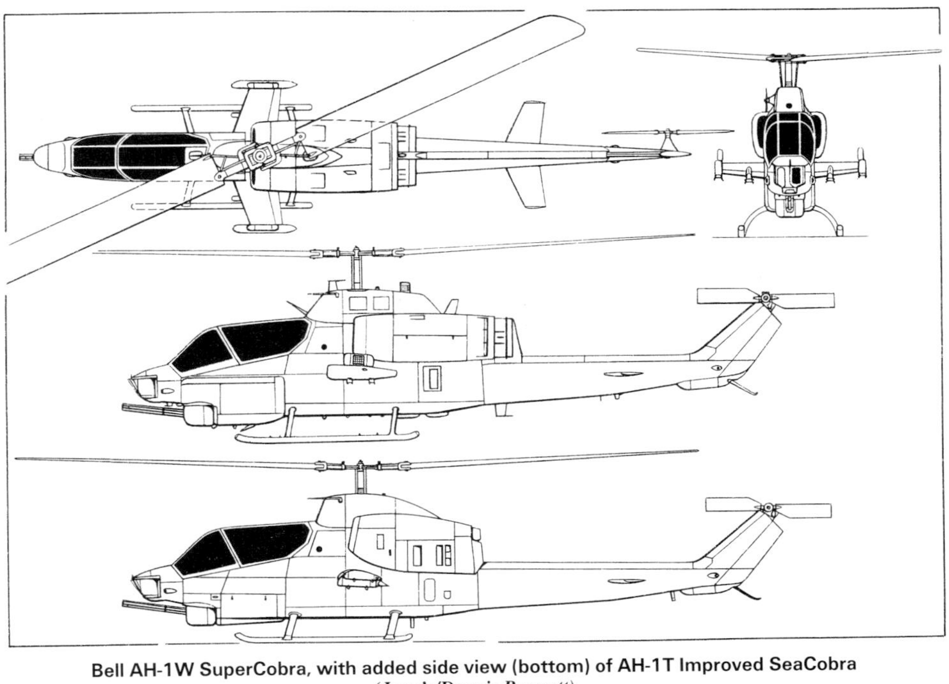

Bell AH-1W SuperCobra, with added side view (bottom) of AH-1T Improved SeaCobra *(Jane's/Dennis Punnett)*

1994

Bell AH-1W SuperCobra twin-engine close-support helicopter *(Bell)*

1995

Bell AH-1J SeaCobra of the US Marine Corps

1994

(1,250 shp) available for T-O or 5 minutes emergency power. Total of 69 delivered to US Marine Corps between mid-1970 and February 1975, the last two being converted later as prototypes for the AH-1T. About 58 USMC AH-1Js remained operational in May 1982, when US Naval Air Systems Command awarded Bell a $4.7 million contract for phase 1 of a two phase programme which called for the company to integrate a Hellfire missile system and night vision cockpit in these helicopters. A further 202 TOW-capable AH-1Js were supplied to the Imperial Iranian Army Aviation from 1974, the US Army acting as purchasing agent.

AH-1Q HueyCobra: Interim anti-armour version for US Army, converted from AH-1G to fire Hughes TOW anti-tank missiles. Total of 92 converted; subsequently upgraded to Mod AH-1S standard.

AH-1S HueyCobra: Advanced and modernised TOW-capable version for US Army; described separately.

AH-1T Improved SeaCobra: Improved version of twin-engined AH-1J for US Marine Corps. Last two AH-1Js modified as prototypes under a US Army Aviation Systems Command contract, with uprated components for significantly increased payload and performance. Incorporates features of AH-1J airframe, but embodies dynamic system of Bell Model 214, some technology developed for Bell Model 309 KingCobra, an upgraded power plant (1,469 kW; 1,970 shp T400-WV-402) and transmission capable of transmitting the full rated engine power. Initial contract for 10 announced 23 June 1975; total of 57 built, of which 51 were subsequently modified to TOW configuration. First AH-1T (USN serial number 59228) flew 20 May 1976, and was delivered to US Marine Corps 15 October 1977. All have been converted to AH-1W standard.

AH-1W SuperCobra: During 1980, Bell flight tested successfully an AH-1T powered by two General Electric T700-GE-700 turboshafts with a combined output in excess of 2,386 kW (3,200 shp). This installation was made in an AH-1T loaned by the US Marine Corps, as part of an R&D programme to establish the specification of a helicopter with enhanced capability for future procurement. Improvements were proposed for a qualification configuration, suitable for retrofit to existing AH-1Ts, included installation of General Electric T700-GE-401 turboshafts with a combined output of 2,423 kW (3,250 shp); a new combining gearbox; and a number of detail improvements. The T700-GE-401 has intermediate and contingency ratings of 1,260 kW (1,690 shp) and 1,285 kW (1,723 shp) respectively. The fuel system is designed to survive 23 mm damage.

A T700-GE-401 testbed helicopter, then designated AH-1T+, made its first flight 16 November 1983 and was evaluated by the US Marine Corps, beginning December 1983. Early 1984 Congressional approval was given for the procurement of 44 production AH-1W SuperCobras, 22 each in FY85 and FY86. The first AH-1W delivered on 27 March 1986 was scheduled to undergo a seven month test programme with Naval Air Systems Command. A second AH-1W began a three month electromagnetic interference test programme Spring 1986.

The first AH-1T uprated to AH-1W standard for the USMC was later fitted with a larger main rotor based on Bell's Model 680 bearingless research rotor. This test aircraft/rotor was flown approximately 100 hours but production funding was not available at the time.

Missions assigned to the AH-1W include anti-armour, troop carrying helicopter escort, multiple weapon fire support, armed reconnaissance, and search and target acquisition. A night targeting system is now being installed on all AH-1W SuperCobra aircraft by Israeli Aircraft Industries for USMC AH-1Ws and Israeli operated AH-1S HueyCobras. Maximum cruising speed is to be increased by 25 knots (46 km/h; 29 mph).

UPGRADES: **Bell:** AH-1Q. See Versions.

Bell: AH-1W. See Versions and separate entry.

Bell: AH-1W MLU. Three-phase upgrade programme for the USMC AH-1W comprising:

Phase 1: Currently underway, includes; installation of the Israeli designed Night Targeting System (NTS), plus canopy and cockpit modifications, embedded INS-GPS and new ARC-210 secure radios.

Phase 2: Installation of Integrated Weapon System (IWS) which entails retrofitting the entire USMC AH-1W fleet with 'glass' cockpit systems, digital communications, automatic target hand-off and Hellfire II integration. Given (ACAT) III category in June 1994.

Phase 3: Installation of a four-blade rotor system and new drive train, navigation FLIR, extra ordnance wing stations and augmented flight controls.

Bell: See following entry for full description of AH-1E/F/P/S and TH-1S.

OPERATORS: Versions of the Bell 209 are in service with the armed forces of the following countries: Iran (80), Israel (40), Japan (80), Jordan (24), Pakistan (20), South Korea (49), Taiwan (8), Turkey (12) and USA (1,100).

DESIGN FEATURES: AH-1G/J Model 540 two-blade wide-chord 'door hinge' main rotor, similar to that of UH-1C. Interchangeable blades. Blades do not fold. Two-blade flex beam tractor tail rotor. Main rotor blades have increased chord, and swept tips which reduce noise and improve high-speed performance. Tail rotor also similar to that of Model 214, with increased diameter and blade chord. Main rotor/engine rpm ratio 1:21.288. Tail rotor/engine rpm ratio 1:4.52. Shaft drive to both main and tail rotors. Main rotor rpm 294-324.

AH-1T/W rotor system and transmission similar to that of Bell Model 214, with strengthened main rotor head incorporating Lord Kinematics elastomeric and Teflon faced bearings.

Small mid-mounted stub-wings, to carry armament and offload rotor in flight.

FLYING CONTROLS: Rotor brake fitted. Tail rotor blade chord increased on AH-1J, which also has push/pull tail rotor controls. Elevator, of inverted aerofoil section, mid-mounted on tailboom forward of fin.

Bell AH-1F HueyCobra of the US Army *(Peter J. Cooper)*

1994

Bell AH-1W SuperCobra of the US Marine Corps ***(Bell)***

1995

STRUCTURE: Blades built up of extruded aluminium spars and laminates. Two-blade all-metal flex-beam tractor tail rotor on starboard side of honeycomb construction. The fuselage is a conventional all-metal semi-monocoque structure, with low silhouette and narrow profile. AH-1T/W have forward fuselage lengthened by insertion of a 0.305 m (1 ft 0 in) plug, to accommodate tankage for additional 181.5 kg (400 lb) of fuel, and tailboom lengthened by 0.79 m (2 ft 7 in). Sweptback vertical fin/tail rotor pylon, strengthened on twin-engined models to cater for increased power.

LANDING GEAR: Non-retractable tubular skid type. Ground handling wheels optional.

POWER PLANT: Single or twin turboshafts, as detailed under model listings. Fuel capacity: G and J, 1,014 litres (268 US gallons; 233 Imp gallons); T, two fuselage tanks, total capacity 1,158 litres (306 US gallons; 255 Imp gallons). Fuel loads, where known, are given under Weights and Loadings heading. Refuelling point in port side of fuselage, aft of cockpits. Oil capacity, 19 litres (5 US gallons; 4.2 Imp gallons).

ACCOMMODATION: Crew of two in tandem, with co-pilot/gunner in front seat and pilot at rear. Crew are protected by seats and side panels of Norton Company Noroc armour; other panels protect vital areas of aircraft.

SYSTEMS: Hydraulic system, pressure 207 bars (3,000 lb/sq in), with Abex pumps, for flight controls and other services. Maximum flow rate (primary system) 30.3 litres (8 US gallons; 6.7 Imp gallons)/min, (utility system) 19 litres (5 US gallons; 4.2 Imp gallons)/min. Closed bootstrap reservoir. Battery powered 28 V DC electrical system. Environmental control and fire detection systems.

AVIONICS (AH-1G): Communications equipment includes AN/ARC-54/131 FM radio; AN/ARC-51 and AN/ARC-134 voice com; KY-28 secure voice system.

AVIONICS (AH-1T): AN/ARC-159(V)1 UHF command set, AN/ARC-114A FM tactical set, AN/AIC-18 intercom, AN/ARN-84(V) Tacan, AN/ARA-50 UHF DF, AN/ASN-75B gyrosyn compass, AN/ARN-83 DF, AN/APN-171(V) radar altimeter, AN/APX-72 IFF transponder and AN/APN-154(V) radar beacon. Provision for TSEC/KY-28 com security unit and KIT-1A/TSEC Mk XII computer.

ARMAMENT AND OPERATIONAL EQUIPMENT (AH-1G): Initial production AH-1Gs were fitted with GAU-2B/A 7.62 mm Minigun in Emerson Electric TAT-102A undernose turret. This was superseded by an M28 turret, able to mount either two Miniguns (each with 4,000 rounds), or two M129 40 mm grenade launchers (each with 300 rounds), or one Minigun and one M129. The Miniguns in these turrets have two rates of fire, controlled by the gunner's trigger: 1,600 rds/min for searching or registry fire, or 4,000 rds/min attack. The M129 fires at a single rate of 400 rds/min. Four external stores attachments under stub-wings can accommodate 76 2.75 in rockets in four M159 launchers, 28 similar rockets in four M157 launchers, or two M18E1 Minigun pods. An initial batch of six AH-1Gs was delivered to the US Army in December 1969 equipped with a Bell/General Electric M 35 armament subsystem. This unit consists of an M61 six-barrel 20 mm automatic cannon on the port inboard wing station, having a firing rate of 750 rds/min. Two ammunition boxes faired flush to the fuselage below the stub-wings each accommodate 500 rds, and total installed weight of the system is 531 kg (1,172 lb). A total of 350 M35 kits was ordered subsequently by the US Army. All wing stores are symmetrically or totally jettisonable. In normal operation, the co-pilot/gunner controls and fires the turret armament, and pilot (aided by an M73 adjustable reflex rocket sight) normally fires the wing stores. The pilot can fire the turreted weapons only in the stowed (that is, dead ahead) position; the turret returns to the stowed position automatically when the gunner releases his grip on the slewing switch. The gunner also has the capability to fire the wing stores if required. Other operational equipment on the AH-1G includes an M130 chaff dispenser.

ARMAMENT (AH-1J): Electrically operated General Electric undernose turret, housing M197 three-barrel 20 mm weapon (a lightweight version of the M61 cannon). A 750 round ammunition container is located in the fuselage directly aft of the turret; firing rate is 750 rds/min, but a 16-burst limiter incorporated in the firing switch. Barrel length of 1.25 m (5 ft) makes it imperative that the M197 is centralised before wing stores are fired. Gun can be tracked 110° to each side, 18° upward and 50° downward. Four attachments under stub-wings for various loads, including LAU-68A/A (seven-tube) or LAU-61A/A (19-tube) 2.75 in rocket launchers, or M18E1 Minigun pods. Total possible armament load 245 kg (542 lb) internal, 998 kg (2,200 lb) external.

ARMAMENT (AH-1T): Chin turret as AH-1J. Underwing attachments for four LAU-61A, LAU-68A, LAU-68A/A, LAU-68B/A or LAU-69A 2.75 in rocket pods; or two CBU-55B fuel-air explosive weapons; four SU-44 flare dispensers; two M118 grenade launchers; Mk 45 parachute flares; or two Minigun pods. Alternative TOW or Hellfire air-to-surface missile installations.

DIMENSIONS, EXTERNAL:	
Main rotor diameter: G, J	13.41 m (44 ft 0 in)
T, W	14.63 m (48 ft 0 in)
Main rotor blade chord: G, J	0.69 m (2 ft 3 in)
T	0.84 m (2 ft 9 in)
Tail rotor diameter: G, J	2.59 m (8 ft 6 in)
T	2.96 m (9 ft 8½ in)
Tail rotor blade chord: G	0.21 m (8⅔ in)
J	0.29 m (11½ in)
T	0.305 m (1 ft 0 in)
Wing span (all)	3.23 m (10 ft 7 in)
Length overall, main rotor fore and aft:	
G	16.14 m (52 ft 11½ in)
J	16.26 m (53 ft 4 in)
T, W	17.68 m (58 ft 0 in)
Length of fuselage: G, J	13.59 m (44 ft 7 in)
T	14.68 m (48 ft 2 in)
Width of fuselage: G	0.965 m (3 ft 2 in)
J, T	0.98 m (3 ft 2½ in)
Height overall: G	4.12 m (13 ft 6¼ in)
J	4.15 m (13 ft 8 in)
T, W	4.33 m (14 ft 3 in)
Elevator span (all)	2.11 m (6 ft 11 in)
Width over skids (all)	2.13 m (7 ft 0 in)
Width over TOW missile pods: G	3.26 m (10 ft 8¾ in)
AREAS:	
Main rotor disc: G, J	141.26 m^2 (1,520.53 sq ft)
T	168.11 m^2 (1,809 sq ft)
Tail rotor disc: G, J	5.27 m^2 (56.75 sq ft)
T	6.88 m^2 (74.03 sq ft)
WEIGHTS AND LOADINGS:	
Operating weight empty, incl amounts shown for crew, fluids, avionics and armour:	
G (404 kg; 891 lb)	2,754 kg (6,073 lb)
J (398 kg; 877 lb)	3,294 kg (7,261 lb)
Weight empty: T	3,642 kg (8,030 lb)
W	4,672 kg (10,300 lb)
Operating weight empty: T	3,904 kg (8,608 lb)
Mission fuel load:	
G (871 l; 230 US gallons; 192 Imp gallons)	680 kg (1,500 lb)
J	725 kg (1,600 lb)
T	944 kg (2,081 lb)
Max useful load (fuel and disposable ordnance):	
J	1,144 kg (2,523 lb)
T	2,445 kg (5,392 lb)
Mission weight: G	4,266 kg (9,407 lb)
J	4,523 kg (9,972 lb)
Max T-O and landing weight: G	4,309 kg (9,500 lb)
J	4,535 kg (10,000 lb)
T	6,350 kg (14,000 lb)
W	6,690 kg (14,750 lb)
PERFORMANCE (at max T-O weight, ISA):	
Never-exceed speed: G	190 kts (352 km/h; 219 mph)
J	180 kts (333 km/h; 207 mph)
Max level speed at S/L:	
G, T	149 kts (277 km/h; 172 mph)
J	180 kts (333 km/h; 207 mph)
W	189 kts (350 km/h; 218 mph)
Max crosswind speed for hovering:	
J	40 kts (74 km/h; 46 mph)
Vertical rate of climb at S/L: T	92 m (301 ft)/min

Bell AH-1P HueyCobra of the US Army

1994

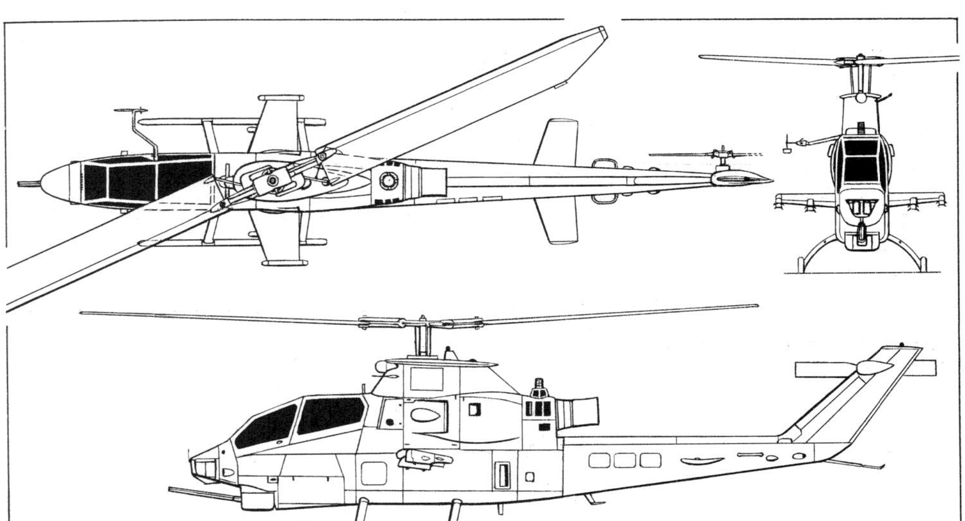

Bell AH-1F HueyCobra (Textron Lycoming T53-L-703 turboshaft) ***(Jane's/Dennis Punnett)***

1994

Bell AH-1F upgraded helicopter armed with TOW missile system *(Bell)*

1995

Bell AH-1S HueyCobra of the US Army equipped with TOW missile system

1994

Max rate of climb at S/L:
- G, normal rated power: 375 m (1,230 ft)/min
- J, normal rated power: 332 m (1,090 ft)/min
- T: 544 m (1,785 ft)/min

Rate of climb at S/L, one engine out:
- W: over 244 m (800 ft)/min

Service ceiling:
- G, normal rated power: 3,475 m (11,400 ft)
- J, normal rated power: 3,215 m (10,550 ft)
- T, max continuous power: 2,255 m (7,400 ft)

Hovering ceiling IGE: G: 3,015 m (9,900 ft)
- J: 3,794 m (12,450 ft)
- W: 4,495 m (14,750 ft)

Hovering ceiling OGE: T: 365 m (1,200 ft)
- W: 914 m (3,000 ft)

Combat radius at 138 kts (255 km/h; 158 mph) at S/L:
- T: 108 n miles (200 km; 124 miles)

Range at S/L with max fuel no reserves:
- G: 325 n miles (602 km; 374 miles)
- J: 335 n miles (620 km; 385 miles)
- T: 310 n miles (574 km; 356 miles)
- W: 343 n miles (635 km; 395 miles)

UPDATED

BELL MODEL 209 HUEYCOBRA (MODERNISED VERSIONS/UPGRADES)

US Army designations: AH-1E, AH-1F, AH-1P, AH-1S and TH-1S

TYPE: Two-seat close support and attack helicopter.

PROGRAMME: AH-1S first ordered as TOW-capable version of AH-1G in 1975; programme included conversion of earlier AH-1Gs and three-stage production of new aircraft with various degrees of upgrading; all versions designated AH-1S until March 1987, when new-build AH-1s allotted dormant UH-1 Iroquois suffixes AH-1P, AH-1E and AH-1F.

VERSIONS: **AH-1S:** Formerly AH-1S(MOD); 92 AH-1Qs (early TOW-capable AH-1G) upgraded by 1979; 87 AH-1Qs upgraded in 1986-88 with Textron Lycoming T53-L-703 engines, Kaman rotor blades (see AH-1P) and TOW system, but retaining original curved canopies; total includes 15 in **TH-1S** Night Stalker configuration for training AH-64 crews to operate night vision system and integrated helmet and display sighting system (IHADSS).

AH-1P: First batch of 100 new-production TOW Cobras (formerly called Production AH-1S), beginning with 76-22567, delivered 1977-78, two becoming AH-1F prototypes; improvements include flat-plate canopy, upturned exhaust, improved nap-of-the-earth (NOE) instrument panel, continental US (CONUS) navigation equipment, radar altimeter, improved communication radios, uprated engine and transmission, push/pull anti-torque control and, from 67th aircraft onwards, Kaman composite rotor blades with tapered tips.

AH-1E: Formerly Enhanced Cobra Armament System or Up-gun AH-1S; next 98 new-build aircraft, from 77-22673, with AH-1P improvements plus universal 20 mm gun turret (invariably fitted with long barrel 20 mm cannon); improved wing stores management system for 2.75 in rockets; automatic compensation for off-axis gun firing; 10kVA alternator for increased power. Delivered 1978-79.

AH-1F: Fully upgraded TOW version, previously designated Modernised AH-1S; 149 manufactured for US Army, beginning 78-23095, in 1979-86, including 50 transferred to Army National Guard; also 378 AH-1Gs converted to full AH-1F standard between November 1979 and June 1982, including 41 **TAH-1F** trainers; improvements of AH-1P and AH-1E added, plus new fire control system having laser rangefinder and tracker, ballistics computer, low airspeed sensor probe, Kaiser pilot's head-up display, Doppler navigation system, IFF transponder, infra-red jammer above engine, hot metal and plume infra-red suppressor, closed-circuit refuelling, new secure voice communications, Kaman composite rotor blades.

Retrofits: Later modifications have included C-Nite equipment fitted to 50 US Army AH-1Fs (reduced from planned 500), Air-to-Air Stinger (ATAS) and Cobra Fleet Life Extension (C-Flex), engine air filter, redesigned swashplate, M43 nuclear/biological/chemical mask, AN/AVR-2 laser warning and improved SCAS roll modifications. C-Nite FLIR for TOW sight delivered 1990 to US Army's Aviation Battalion in South Korea. C-Flex items already completed include Nite Fix lighting, AH-1G-to-AH-1S upgrade and K-Flex driveshaft; remaining C-Flex work includes rotor improvements, improved TOW test set and radio upgrade.

Proposed retrofits: With sufficient orders, Bell is marketing upgrade package including kit to uprate engine to T53-L-703-70X offering 1,491 kW (2,000 shp); retrofit 14.02 m (46 ft 0 in) diameter Bell 412 four-blade main rotor with composite yoke instead of titanium hub; stretch tailboom by 51 cm (20 in); needle bearing tail rotor, as on Model 412 and Israeli AH-1F; three-axis digital stability augmentation; reliability and maintainability modifications. Such an AH-1F would carry 454 kg (1,000 lb) more weapons, manoeuvre down to $-0.5g$ and achieve OGE hover ceiling of 1,067 m (3,500 ft) in ISA+20°C at weight of 4,990 kg (11,000 lb).

UPGRADES: **General Electric:** Announced in early 1994 that it was to re-engine the Turkish Army's 12 AH-1W Super Cobra helicopters with T700-GE-401 turboshaft engines.

Taman: Reported to have won contract in early 1995 to upgrade Turkish Cobra Helicopters with night targeting systems (NTS). Contract value thought to be $12 million.

OPERATORS: Japan received two AH-1Es and converted to AH-1F; AH-1F manufactured under licence by Fuji with 79 funded to end of FY91. Israel (six AH-1E, 30 AH-1F), Jordan (24 AH-1F), Pakistan (20), South Korea (70 AH-1F/J with C-Nite), Thailand (four AH-1F) and Turkey (25 AH-1E/P).

COSTS: $9.85 million per unit (South Korea, including spares and support), 1990.

DESIGN FEATURES: Transmission rated at 962 kW (1,290 shp) for take-off and 845 kW (1,134 shp) continuous; Kaman composite blades, fitted from 67th AH-1P onwards, tolerate hits by 23 mm shells, have tungsten carbide bearing sleeves and outer 15 per cent of blade is tapered in chord and thickness; tailboom strengthened against 23 mm hits; airframe has infra-red suppressant paint finish.

POWER PLANT: One 1,342 kW (1,800 shp) Textron Lycoming T53-L-703 turboshaft. Closed circuit refuelling on AH-1F. Fuel capacity 980 litres (259 US gallons; 216 Imp gallons). Upward-facing exhaust on AH-1E; IR suppression nozzle on AH-1F.

ACCOMMODATION: Flat-plate canopy has seven planes of viewing surfaces, designed to minimise glint and reduce possibility of visual detection during nap-of-the-earth (NOE) flying; it also provides increased headroom for pilot. Improved instrument layout and lighting, compatible with use of night vision goggles. Improved, independently operating window/door ballistic jettison system to facilitate crew escape in emergency.

SYSTEMS: 10 kVA 400 Hz AC alternator with emergency bus added to electrical system. Hydraulic system pressure 103.5 bars (1,500 lb/sq in), maximum flow rate 22.7 litres (6 US gallons; 5 Imp gallons)/min. Open reservoir. Battery driven Abex standby pump, for use in event of main hydraulic system failure, can be used for collective pitch control and for boresighting turret and TOW missile system. Improved environmental control and fire detection systems.

AVIONICS: Standard lightweight avionics equipment (SLAE) includes AN/ARC-114 FM, AN/ARC-164 UHF/AM voice com, and E-Systems (Memcor Division) AN/ARC-115 VHF/AM voice com (compatible with KY-58 single-channel secure voice system). Other avionics include AN/ASN-128 Doppler nav system in AH-1F; HSI; VSI; radar altimeter; push/pull anti-torque controls for tail rotor; co-pilot's standby magnetic compass. C-Flex upgrade includes introduction of Magnavox AN/ARC-164(V) UHF/AM, Collins AN/ARC-186 VHF/AM-FM, ITT AN/ARC-201 (SINCGARS) VHF/FM, and LaBarge AN/ARN-89B D/F.

ARMAMENT: M65 system with eight Hughes TOW missiles, disposed as two two-round clusters on each outboard

Bell AH-1(4B)W helicopter close support and attack helicopter

1994

Bell AH-1(4B)W helicopter of the US Marine Corps

1994

underwing station. Inboard wing stations remain available for other stores. Beginning with first AH-1E, M28 (7.62/40 mm) turret in earlier HueyCobras replaced by new electrically powered General Electric universal turret, designed to accommodate 20 mm weapon and improve standoff capability, although only 20 mm M197 three-barrel cannon (with 750 rounds) mounted in this turret. Rate of fire 675 rds/min. Turret position is controlled by pilot or co-pilot/gunner through helmet sights, or by co-pilot using M65 TOW missile system's telescopic sight unit. Field of fire up to 110° to each side of aircraft, 20.5° upward and 50° downward. Also from first AH-1E, helicopter equipped with Baldwin Electronics M138 wing stores management subsystem, providing means to select and fire, singly or in groups, any one of five types of external 2.75 in rocket store. These mounted in launchers each containing 7 or 19 tubes, additional to TOW missile capability.

In addition to these installations, first AH-1F introduced fire control subsystem which includes Kaiser head-up display for pilot, Teledyne Systems digital fire control computer for turreted weapon and underwing rockets, omnidirectional airspeed system to improve cannon and rocket accuracy, Hughes laser rangefinder (accurate over 10,000 m; 32,800 ft), and Rockwell AN/AAS-32 automatic airborne laser tracker. Other operational equipment includes Hughes LAAT stabilised sight, GEC Avionics M-143 air data subsystem, Bendix/King AN/APX-100 solid state IFF transponder, Sanders AN/ALQ-144 infra-red jammer (above engine), suppressor for infra-red signature from engine hot metal and exhaust plume, and AN/APR-39 radar warning receiver.

DIMENSIONS, EXTERNAL:
Main rotor diameter 13.41 m (44 ft 0 in)
Main rotor blade chord (from 67th AH-1P onward) 0.76 m (2 ft 6 in)
Tail rotor diameter 2.59 m (8 ft 6 in)
Tail rotor blade chord 0.305 m (1 ft 0 in)
Wing span 3.28 m (10 ft 9 in)
Length overall, rotors turning 16.18 m (53 ft 1 in)
Width of fuselage 0.99 m (3 ft 3 in)
Height to top of rotor head 4.09 m (13 ft 5 in)
Width over TOW pods 3.56 m (11 ft 8 in)
Elevator span 2.11 m (6 ft 11 in)
Width over skids 2.13 m (7 ft 0 in)

AREAS:
Main rotor disc 141.26 m² (1,520.23 sq ft)
Tail rotor disc 5.27 m² (56.75 sq ft)

WEIGHTS AND LOADINGS (AH-1S):
Operating weight empty 2,993 kg (6,598 lb)
Mission weight 4,524 kg (9,975 lb)
Max T-O and landing weight 4,535 kg (10,000 lb)
Max disc loading 32.10 kg/m² (6.58 lb/sq ft)
Max power loading 4.72 kg/kW (7.75 lb/shp)

PERFORMANCE (AH-1S at max T-O weight, ISA):
Never-exceed speed (VNE) (TOW configuration) 170 kts (315 km/h; 195 mph)
Max level speed (TOW configuration) 123 kts (227 km/h; 141 mph)
Max rate of climb at S/L, normal rated power 494 m (1,620 ft)/min
Service ceiling, normal rated power 3,720 m (12,200 ft)
Hovering ceiling IGE 3,720 m (12,200 ft)
Range at S/L with max fuel, 8% reserves 274 n miles (507 km; 315 miles)
g limits +2.5/–0.5

UPDATED

BELL AH-1(4B)W

TYPE: Close support and attack helicopter with four-blade main rotor.

PROGRAMME: Prototype AH-1W (converted from AH-1T 161022) subsequently modified to AH-1(4B)W; first flight 24 January 1989; evaluated by US Marine Corps in 1990. Rotor retrofit planned for USMC AH-1Ws. Programme still under consideration in July 1995. A plan to upgrade the AH-1W with integrated weapons system (IWS) has been under consideration since the Gulf War but was shelved after the UK MoD chose the AH-64D instead of the AH-1 Cobra Venom.

DESIGN FEATURES: Four-blade bearingless composite main rotor based on Bell 680 rotor. Additional improvements include expanded *g* envelope; uprated transmission. Maximum weight increased by 930 kg (2,050 lb); speed increased by 20 knots (37 km/h; 23 mph); digital flight control system.

FLYING CONTROLS: Blade roots of controlled flexibility to allow flapping, lead-lag and pitch-change movements without bearings; roots surrounded by rigid cuffs which transmit pitch-change commands from control rods to blade roots.

STRUCTURE: Two blades folded manually for compact storage.

Data generally as for AH-1W except as follows:

ARMAMENT: Cranked, six-station stub-wings, including overwing position for two air-to-air missiles at each tip. Total ordnance capacity 1,775 kg (3,914 lb), including 750 rounds for nose cannon and two chaff/flare dispensers.

DIMENSIONS, EXTERNAL:
Main rotor blade chord 0.63 m (2 ft 1 in)

WEIGHTS AND LOADINGS:
Max T-O and landing weight 7,620 kg (16,800 lb)

PERFORMANCE (at 6,120 kg; 13,494 lb mid-mission weight):
Never-exceed speed (VNE) over 200 kts (371 km/h; 230 mph)
Max level speed 170 kts (315 km/h; 196 mph)
Max cruising speed 160 kts (297 km/h; 184 mph)
g limit +3.39

UPDATED

BELL KIOWA

US Army designation: OH-58

Canadian military designation: CH-136

TYPE: Turbine-powered light observation helicopter.

PROGRAMME: On 8 March 1968 the US Army named Bell as the winner of its reopened light observation helicopter competition, and awarded the company the first increment of a total order for 2,200 OH-58 helicopters.

VERSIONS: **OH-58A:** Generally similar to the Model 206A and each powered by a 236.5 kW (317 shp) Allison T63-A-700 turboshaft. Major difference concerns the main rotor, that of the Kiowa having an increased diameter. There are also differences in the internal layout and avionics. The first OH-58A was delivered to the US Army 23 May 1969 and deployment in Vietnam began early Autumn 1969.

A total of 74 COH-58As, generally similar to the OH-58A, were delivered to the Canadian Forces, and were designated CH-136. Of these, 14 of them, used for basic helicopter training at Portage la Prairie, Manitoba, were replaced by the same number of new JetRanger IIIs from 1981, and reissued to other units.

Under a co-production agreement with the Australian government, 56 Model 206B-1 Kiowa military light observation helicopters (similar to the OH-58A) were delivered over an eight year period. The initial 12 206B-1s were built by Bell. Commonwealth Aircraft Corporation was prime Australian licensee, with responsibility for final assembly of the remainder. Only the engines and avionics were supplied from US sources.

OH-58B: Delivery of 12 helicopters to the Austrian Air Force was completed in 1976.

OH-58C: Under a US Army development qualification contract placed on 30 June 1976, Bell converted an OH-58A to an improved standard. This involved installation of a flat glass canopy to reduce glint, an uprated Allison T63-A-720 turboshaft, and an IR reduction package. Two additional OH-58As were modified to OH-58C configuration, for preproduction flight testing by Bell and the US Army, and production modification of 435 OH-58As to OH-58C standard was completed in March 1985 at Bell Helicopters Amarillo plant. An additional 150 modifications were carried out by Israel Aircraft Industries (IAI) for the US Army in West Germany. The final configuration includes a new instrument panel, modifications to reduce vulnerability in combat, CONUS (Continental US) navigation equipment, day optics, improved avionics and improved maintenance features. The additional power significantly improves high-altitude, hot-weather performance.

UPGRADES: **Bell:** OH-58C. See Versions.

Bell: In September 1985 Bell began to deliver to the US Army improved tail rotor configuration kits for installation in the OH-58A and OH-58C. The kit increases the amount of tail rotor thrust available, giving the pilot more capability to correct an uncommanded yaw at low speeds, as experienced during nap-of-the-earth flying.

Bell: OH-58D AHIP programme. See separate entry.

IAI: OH-58C. See Versions.

The following description applies to OH-58C aircraft:

DESIGN FEATURES: Two-blade semi-rigid seesaw type main rotor, employing preconing underslinging to ensure smooth operation. Blades are of standard Bell 'droop snoot' section. Transmission improvements include a

Bell OH-58A light observation helicopter of the US Army *(Bell)*

1995

Bell OH-58C light observation helicopter of the US Army *(Bell)*

1995

Bell OH-58C light observation helicopter with flat-glass canopy *(Bell)*

1995

four-pinion upper planetary, with new thrust bearing and 'fly dry' capability. Shaft to tail rotor single-stage bevel gearbox protected by cover. Main rotor/engine rpm ratio 1:17.44; main rotor 354 rpm. Tail rotor/engine rpm ratio 1:2.353.

FLYING CONTROLS: Each blade is connected to the head by means of a grip, pitch change bearings and a tension-torsion strap assembly. Rotor brake available as optional kit.

STRUCTURE: Main rotor blades have a D-shape aluminium spar, bonded aluminium alloy skin, honeycomb core and a trailing-edge extension. Two tail rotor blades have bonded aluminium skin but no core, main rotor blades do not fold, but modification to permit manual folding is possible. Forward cabin section is made up of two aluminium alloy beams and 25 mm (1 in) thick aluminium honeycomb sandwich. Rotor, transmission and engine are supported by upper longitudinal beams. Upper and lower structures are interconnected by three fuselage bulkheads and a centre-post to form an integrated structure. Intermediate section is of aluminium alloy semi-monocoque construction. Aluminium monocoque tailboom. A low-glare canopy design reduces the solar glint signature. The windscreens are slightly convex to assist rain removal and increase their strength. The tail unit has a fixed stabiliser of aluminium monocoque construction, with inverted aerofoil section. Fixed vertical tail fin in sweptback upper and ventral sections, made of aluminium honeycomb with aluminium alloy skin.

LANDING GEAR: Aluminium alloy tubular skids bolted to extruded cross-tubes. Tubular steel skid ventral fin to protect tail rotor in tail down landing. Special high skid gear (0.25 m; 10 in greater ground clearance) available for use in areas with high brush. Pontoons or stowed floats, capable of in-flight inflation, available as optional kits.

POWER PLANT: One 313 kW (420 shp) Allison T63-A-720 turboshaft. 'Black Hole' exhaust stacks and hot metal shroud for infra-red suppression. Fuel tank below and behind aft passenger seat, total usable 276 litres (73 US gallons; 61 Imp gallons). Refuelling point on starboard side of fuselage, aft of cabin. Oil capacity 5.6 litres (1.5 US gallons; 1.25 Imp gallons).

ACCOMMODATION: Forward crew compartment seats pilot and co-pilot/observer side by side. Entrance to this compartment is provided by single door on each side of fuselage. The cargo/passenger compartment, which has its own doors, one on each side, provides approximately 1.13 m^3 (40 cu ft) of cargo space, or provision for two passengers by installation of two seat cushions, seat belts and shoulder harnesses. A redesigned instrument panel houses new avionics, and all flight instruments have been modified for night operations using night vision goggles. An improved defrost/defog air circulation system increases the aircraft's mission readiness.

SYSTEMS: Hydraulic system, pressure 41.5 bars (600 lb/sq in), for cyclic, collective and directional controls. Maximum flow rate 7.57 litres (2 US gallons; 1.65 Imp gallons)/min. Open reservoir. Electrical supply from 150 A starter/generator. One 24 V 13 Ah Ni/Cd battery.

AVIONICS: C-6533/ARC intercommunication subsystem, AN/ARC-114 VHF-FM, AN/ARC-115 VHF-AM, AN/ARC-116 UHF-AM, AN/ARN-89 ADF, AN/ASN-43 gyro magnetic compass, AN/APX-100 IFF transponder TSEC/KY-28 communications security set, C-8157/ARC control indication, MT-380/ARC mounting, TS-1843/APX transponder test set and mounting, KIT-1A/TSEC computer and mounting, duplicate AN/ARC-114, AN/APR-39 radar warning, ID-1351 C/A HBI, ID-1347 C/ARN CDI; and provisions for AN/ARN-123(V)1 CONUS nav, AN/APN-209 radar altimeter and YG-1054 proximity warning.

ARMAMENT: Standard equipment is the M27 armament kit, utilising the 7.62 mm Minigun.

DIMENSIONS, EXTERNAL:

Main rotor diameter	10.77 m (35 ft 4 in)
Tail rotor diameter	1.65 m (5 ft 5 in)
Main rotor blade chord	0.33 m (1 ft 1 in)
Distance between rotor centres	5.96 m (19 ft 6½ in)
Length: overall rotors turning	12.49 m (40 ft 11¾ in)
fuselage	9.93 m (32 ft 7 in)
Height: over tail fin	2.54 m (8 ft 4 in)
overall	2.91 m (9 ft 6½ in)
Stabiliser span	1.97 m (6 ft 5¾ in)
Width over skids	1.92 m (6 ft 3½ in)

AREAS:

Main rotor blades (total)	3.55 m^2 (38.26 sq ft)
Main rotor disc	90.93 m^2 (978.8 sq ft)
Tail rotor blades (total)	0.22 m^2 (2.37 sq ft)
Tail rotor disc	2.14 m^2 (23.04 sq ft)
Stabiliser	0.90 m^2 (9.65 sq ft)

WEIGHTS AND LOADINGS (A: OH-58A, B: OH-58C):

Weight empty: A	664 kg (1,464 lb)
B	825 kg (1,818 lb)
Operating weight: A	1,049 kg (2,313 lb)
B	1,104 kg (2,434 lb)
Max T-O and landing weight: A	1,360 kg (3,000 lb)
B	1,451 kg (3,200 lb)
Max zero-fuel weight: A	1,145 kg (2,252 lb)
B	1,200 kg (2,646 lb)
Max disc loading: A	14.9 kg/m^2 (3.07 lb/sq ft)
B	15.9 kg/m^2 (3.27 lb/sq ft)

PERFORMANCE (A: OH-58A at observation mission gross weight of 1,255 kg; 2,768 lb, ISA except where indicated, B: OH-58C at mission gross weight of 1,311 kg; 2,890 lb):

Never-exceed speed at S/L:	
A, B	120 kts (222 km/h; 138 mph)
Cruising speed for max range:	
A, B	102 kts (188 km/h; 117 mph)
Loiter speed for max endurance:	
A, B	49 kts (91 km/h; 56 mph)
Max rate of climb at S/L: A	543 m (1,780 ft)/min
B	549 m (1,800 ft)/min
Service ceiling: A	5,760 m (18,900 ft)
B	5,640 m (18,500 ft)
Hovering ceiling: IGE: A, B	4,025 m (13,200 ft)
OGE: A	2,680 m (8,800 ft)
B	2,955 m (9,700 ft)
OGE (armed scout mission):	
A at 1,360 kg (3,000 lb)	1,830 m (6,000 ft)
B at 2,500 kg (3,200 lb)	2,500 m (8,200 ft)
Max range at S/L, 10% reserves:	
A, B	259 n miles (481 km; 299 miles)
Max range at S/L, armed scout mission at 1,451 kg (3,200 lb), no reserves:	
B	264 n miles (490 km; 305 miles)
Endurance at S/L, no reserves: A, B	3 h 30 min

UPDATED

BELL MODEL 406 (AHIP) UPGRADE

US Army designations: OH-58D Kiowa and Kiowa Warrior

TYPE: Two-seat scout and attack helicopter.

PROGRAMME: Bell won US Army Helicopter Improvement Program (AHIP) 21 September 1981; first flight of OH-58D 6 October 1983; deliveries started December 1985; first based in Europe June 1987. Production running at minimum economic rate of three a month, compared with capacity for 12.

VERSIONS: **Prime Chance:** Fifteen special armed OH-58Ds modified from September 1987 under Operation Prime Chance for use against Iranian high-speed boats in Gulf; delivery started after 98 days, in December 1987; firing clearance for Stinger, Hellfire, 0.50 in gun and seven-tube rocket pods completed in seven days; aircraft remained at sea in Gulf 1991, operated by A and B Troops of 4th Squadron/17th Aviation Regiment (known as task Force 118 until January 1991).

Kiowa Warrior: Armed version, to which all planned OH-58Ds are being modified; integrated weapons pylons, uprated transmission and engine, lateral C of G limits increased, raised gross weight, EMV protection of avionics bays, localised strengthening, radar warning receiver, IR jammer, video recorder, SINCGARS radios, laser warning receiver and tilting vertical fin, armament same as Prime Chance.

MultiPurpose Light Helicopter (MPLH): Further modification of Kiowa Warrior features: squatting landing gear, quick-folding rotor blades horizontal stabiliser and tilting fin to allow helicopter to be transported in cargo aircraft and flown to cover 10 minutes after unloading from C-130. Later additions include cargo hook for up to 907 kg (2,000 lb) slung load and fittings for external carriage of six outward-facing troop seats or two stretchers.

UPGRADES: **Bell:** Stage 1 of Multi-Stage Improvement Program (MSIP) includes fitting Global Positioning System receiver, improved Doppler, digital data loader and MIL-STD-1750 processors.

Bell: Stealth Kiowa Warrior. $200,000 (1992) stealth kit includes reshaped nose, engine and gearbox cowling of radar absorbent material, cuffs over blade roots and tail rotor hub; sleeve over rotor shaft and control rods, windowless rear door and radar absorbent coatings on nose, windows, mast-mounted sight and landing gear; 18 modified for Gulf War in 1991, but were not deployed.

Other upgrades include mast-mounted sight processor, night pilotage system, engine to 19.5 per cent high/hot performance improvement, extended range and Honeywell computer processor improvement.

Under an R&D effort, Bell has reconfigured insertion prototype. The nose has been enlarged to accommodate systems boxes moved from the cabin area to provide space to transport two to three passengers. An INS/GPS navigation system, digital map, night pilotage FLIR, mast-mounted sight and two processor upgrades are included in the prototype.

Honeywell: Awarded contract 13 December 1994 to develop software to incorporate the Honeywell embedded-global-positioning-system-aided inertial navigation system (EGI) into the OH-58D integrated cockpit control display system as part of the Multi-Stage Improvement Program (MSIP) for the US Army. Contract value: $1.2 million.

CUSTOMERS: US Army: initial plan to modify 592 OH-58A/B to OH-58D reduced to 477; again reduced to 207, but currently stands at 383 (16, 44, and 39 in FY83/84/85, plus 36 per year in FY87-93, 15 in FY94 and 14 in FY95). Stated army requirement is 507. New-build Kiowa Warrior from 207th aircraft, May 1991; initially to C and D Troop of 4-17 Aviation, Fort Bragg. First retrofit contract (28

Bell OH-58D Kiowa Warrior firing Hellfire missile *(Bell)*

1995

Bell OH-58D Kiowa Warrior firing 2.75 in rockets

1995

helicopters) to Bell January 1992. Main effort now is provision for armament and accompanying upgrades in new aircraft and retrofitting existing OH-58D to armed Kiowa Warrior configuration; all Kiowa Warriors to be raised to ultimate MultiPurpose Light Helicopter (MPLH) standard. Total of 340 delivered by February 1995.

COSTS: $7.46 million (1993) programme unit cost.

DESIGN FEATURES: Four-blade Bell soft in plane rotor with carbon composites yoke, elastomeric bearings and composites blades. Transmission rating: Kiowa 339 kW (455 shp) continuous; Kiowa Warrior 410 kW (550 shp) continuous. Main rotor 395 rpm; tail rotor 2,381 rpm. McDonnell Douglas/Northrop mast-mounted sight containing TV and IR optics and laser designator/ranger; Honeywell integrated control of mission functions, navigation, communications, systems and maintenance functions based on large electronic primary displays for pilot and observer/gunner; hands-on cyclic and collective controls for all combat functions; airborne target handover system in all OH-58Ds operates air-to-air as well as air-to-ground using digital frequency-hopping; system indicates location and armament state of other helicopters; some OH-58Ds have real-time video downlink capable of relaying of US Army and Air Force aircraft to headquarters 22 n miles (40 km; 25 miles) away or, via satellite, to remote locations.

FLYING CONTROLS: Full-powered controls, including tail rotor, with four-way trim and trim release; stability and control augmentation system (SCAS) using AHRS gyro signals; automatic bob-up and return to hover mode; Doppler blind hover guidance mode; co-pilot/observer's cyclic stick can be disconnected from controls and locked centrally.

STRUCTURE: Basic OH-58 structure reinforced; armament cross-tube fixed above rear cabin floor; avionics occupy rear cabin area, baggage area and nose compartment.

LANDING GEAR: Light alloy tubular skids bolted to extruded cross-tubes.

POWER PLANT: One Allison 250-C30R (T703-Ad-700) turboshaft, C30X with improved diffuser in Kiowa Warrior with an intermediate power rating of 485 kW (650 shp) at S/L, ISA. Transmission rating: Kiowa 339 kW (455 shp) continuous; Kiowa Warrior 410 kW (550 shp) continuous. One self-sealing crash resistant fuel cell, capacity 399 litres (105.4 US gallons; 87.8 Imp gallons) located aft of the cabin area. Refuelling point on starboard side of fuselage. Oil capacity 5.7 litres (1.5 US gallons; 1.2 Imp gallons).

ACCOMMODATION: Pilot and co-pilot/observer seated side by side. Door on each side of fuselage. Accommodation is heated and ventilated.

SYSTEMS: Single hydraulic system, pressure 69 bars (1,000 lb/sq in), for main and tail rotor controls and SCAS system. Maximum flow rate 11.36 litres (3 US gallons; 2.5 Imp gallons)/min. Open-type reservoir. Primary electrical power provided by 10 kVA 400 Hz three-phase 120/208 V AC alternator with 200 A 28 V DC transformer-rectifier unit for secondary DC power. Back-up power provided by 500 VA 400 Hz single-phase 115 V AC solid state inverter and 200 A 28 V DC starter/generator.

AVIONICS: Multifunction displays for vertical and horizontal situation indication, mast-mounted sight day/night viewing and communications control, with selection via control column handgrip switches. Five com transceivers, datalink and secure voice equipment. Plessey (PESC) AN/ASN-157 Doppler strapdown INS. Equipped for day/night VFR. Mast-mounted sight houses 12× magnification TV camera, auto-focusing IR thermal imaging sensor and laser rangefinder/designator, with automatic target tracking and in-flight automatic boresighting. Night vision goggles; AHRS; and airborne target handover subsystem (ATHS). Phase 1 additions, introduced on production line in 1990, include doubled computer capacity to 88 k, added weapons selection/aiming, IR jammer, second RWR and laser warning, video recorder, data transfer system, SINCGARS and Have-Quick II radios, ANVIS display and symbology system and EMV hardening.

EQUIPMENT: NBC mask in Phase 1 and Warrior aircraft.

ARMAMENT: Four Stinger air-to-air or Hellfire air-to-surface missiles, or two seven-round 2.75 in rocket pods, or one 0.50 in machine gun, mounted on pylons on cabin sides. IR jammer standard on armed version.

DIMENSIONS, EXTERNAL:

Main rotor diameter	10.67 m (35 ft 0 in)
Main rotor blade chord (mean)	0.24 m (9½ in)
Tail rotor diameter	1.65 m (5 ft 5 in)
Length: overall, rotors turning	12.85 m (42 ft 2 in)
fuselage, excl rotors	10.31 m (33 ft 10 in)
Width, rotors folded	1.97 m (6 ft 5½ in)
Height: to top of rotor head	2.59 m (8 ft 6 in)
overall	3.90 m (12 ft 9½ in)
Skid track	1.88 m (6 ft 2 in)
Cabin doors (port and stbd, each):	
Height	1.04 m (3 ft 5 in)
Width	0.91 m (3 ft 0 in)
Height to sill	0.66 m (2 ft 2 in)

AREAS:

Main rotor blades (each)	1.30 m² (13.95 sq ft)
Tail rotor blades (each)	0.13 m² (1.43 sq ft)
Main rotor disc	89.37 m² (962.0 sq ft)
Tail rotor disc	2.14 m² (23.04 sq ft)
Fin	0.85 m² (9.1 sq ft)

WEIGHTS AND LOADINGS:

Weight empty	1,281 kg (2,825 lb)
Max fuel weight	321 kg (707 lb)
Max T-O and landing weight: Kiowa	2,041 kg (4,500 lb)
Kiowa Warrior	2,495 kg (5,500 lb)
Max zero-fuel weight	1,711 kg (3,773 lb)
Max disc loading	22.95 kg/m² (4.7 lb/sq ft)
Max power loading: Kiowa	6.02 kg/kW (9.89 lb/shp)
Kiowa Warrior	5.82 kg/kW (9.56 lb/shp)

PERFORMANCE (at max T-O weight, clean):

Never-exceed speed (V_{NE})	130 kts (241 km/h; 149 mph)
Max level speed at 1,220 m (4,000 ft)	128 kts (237 km/h; 147 mph)
Max cruising speed at 610 m (2,000 ft)	120 kts (222 km/h; 138 mph)
Econ cruising speed at 1,220 m (4,000 ft)	110 kts (204 km/h; 127 mph)
Max rate of climb: at S/L, ISA	469 m (1,540 ft)/min
at 1,220 m (4,000 ft), 35°C (95°F)	over 366 m (1,200 ft)/min
Vertical rate of climb: at S/L, ISA	232 m (760 ft)/min
at 1,220 m (4,000 ft), 35°C (95°F)	over 152 m (500 ft)/min
Service ceiling	over 3,660 m (12,000 ft)
Hovering ceiling: IGE, ISA	over 3,660 m (12,000 ft)
OGE, ISA	3,415 m (11,200 ft)
OGE, 35°C (95°F)	1,735 m (5,700 ft)
Range with max fuel, no reserves	300 n miles (556 km; 345 miles)
Endurance	2 h 30 min

UPDATED

Bell OH-58D Kiowa Warrior with 2.75 in rocket launcher on starboard pylon and 0.50 in machine gun on port side ***(Bell)***

1995

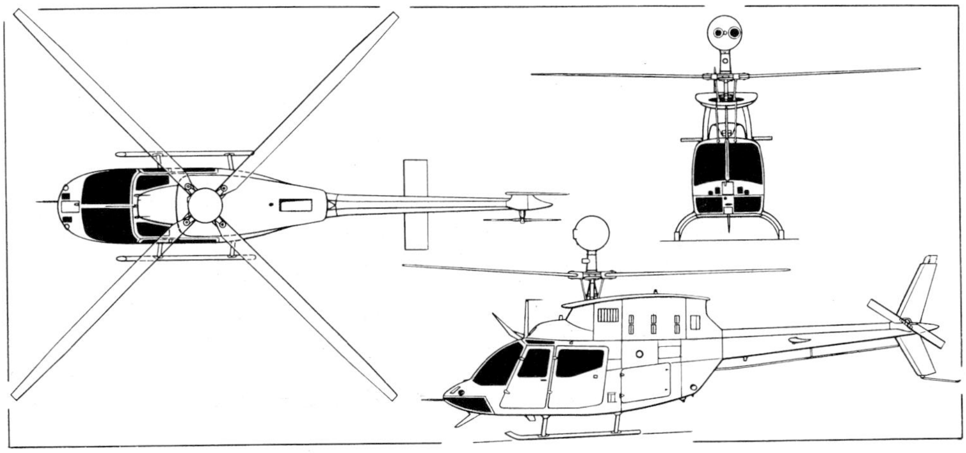

Bell OH-58D Kiowa scout and attack helicopter ***(Jane's/Dennis Punnett)***

1995

BOEING

THE BOEING COMPANY

PO Box 3707, Seattle, Washington 98124
Telephone: 1 (206) 655 2121
Fax: 1 (206) 655 1171
Telex: 329430
CHAIRMAN AND CEO: Frank A. Shrontz
PRESIDENT: Philip M. Condit

Company founded July 1916. Boeing Defense & Space Group formed in January 1990 to co-ordinate the Aerospace & Electronics, Helicopters, Military Airplanes and Advanced Systems divisions of The Boeing Company. Simultaneously, the former Military Airplanes divisions at Wichita reduced in size by the transfer of some activities to Boeing Commercial Airplane Group.

Operating components of The Boeing Company include:

BOEING DEFENSE & SPACE GROUP
Electronic Systems Division
Military Airplanes Division
See next entry
Helicopters Division
Follows Military Airplanes Division
Missiles & Space Division
Product Support Division
BOEING COMMERCIAL AIRPLANE GROUP
Follows Helicopters Division

VERIFIED

BOEING DEFENSE & SPACE GROUP

PO Box 3999, Seattle, Washington 98124-2499
Telephone: 1 (206) 773 2816
Telex: 329430
PRESIDENT: C. G. 'Jerry' King
PUBLIC RELATIONS OFFICER: P. B. Dakan

In addition to Military Airplanes and Helicopters Divisions, the Defense & Space Group controls Electronic Systems Division, Missiles & Space Division and Product Support Division. Subsidiaries include Argosystems, Boeing Aerospace Operations, Boeing Defense and Space Group-Irving, Boeing Defense and Space Group-Corinth.

VERIFIED

BOEING MILITARY AIRPLANES (Division of Boeing Defense & Space Group)

PO Box 3999, Seattle, Washington 98124
Telephone: 1 (206) 655 1198
Fax: 1 (206) 655 7012
Telex: 329430
VICE-PRESIDENT AND GENERAL MANAGER: Richard Hardy
PUBLIC RELATIONS MANAGER: Randy Harrison

On 2 January 1990 Boeing Military Airplanes, Wichita, was designated part of new Military Airplanes division based at Seattle and absorbing Boeing Advanced Systems. Civil airliner activities at Wichita subordinated to Boeing Commercial Airplane Group, Seattle.

NASA ordered conversion of a second 747-100 to carry Space Shuttle Orbiter vehicle early 1988; delivered 20 November 1990; Boeing Military Airplanes offers to upgrade McDonnell Douglas F-4 Phantoms with digital avionics and conformal fuel tanks or pylons; contracted to design and manufacture new navigation and weapon delivery system for 600 Air National Guard, Air Force Reserve and US Air Force F-4s.

UPDATED

BOEING 707 MILITARY CONVERSIONS

TYPE: Military conversions of the Boeing 707 transport aircraft.

PROGRAMME: For programme history and general description of the Boeing 707 see separate entry in the Boeing Commercial Aircraft section.

VERSIONS: **E-3 Sentry:** Airborne early warning aircraft. See E-3 entry in this section.

KE-3: Saudi Arabian Air Force tanker aircraft. See separate entry in this section.

E-6A Mercury: Take charge and move out (TACAMO) communications aircraft for ballistic missile submarines. See separate entry in this section.

E-8B J-STARS: Joint surveillance and target attack radar aircraft. See separate entry in Northrop Grumman section.

C-18A: General training aircraft.

EC-18B: Advanced range instrumentation aircraft (ARIA) with very large nose radome.

EC-18D: Cruise missile control aircraft.

C-135A/B: Special transport versions. Few remain in service.

C-135C: Transport reconversions of WC-135B weather reconnaissance aircraft.

C-135E: Upgraded version of C-135A with Pratt & Whitney JT3D (military designation TF33) turbofan engines.

C-135F: Designation of French tanker model.

C-135FR: Upgraded C-135F aircraft.

EC-135A/G/L: Airborne command post relay aircraft.

EC-135C/H/J/K/P/Y: Airborne command post aircraft.

EC-135E: Range support test aircraft.

KC-135D/E: Re-engined KC-135A tanker aircraft with TF33 (Pratt & Whitney JT3D) turbofan engines. See separate entry in this section for KC-135 aircraft.

KC-135Q: Converted KC-135 versions with nav/com equipment.

KC-135R: Re-engined aircraft with CFM56 turbofan engines.

KC-135T: Re-engined KC-135Q aircraft with CFM56 (military designation F108).

NC-135: Development and trials aircraft.

NKC-135A/E: Fleet electronic warfare aircraft.

OC-135B: Reconnaissance aircraft. See separate entry in USAF section.

RC-135S Cobra Ball: Missile test reconnaissance aircraft with large windows in upper fuselage.

RC-135U: Strategic reconnaissance aircraft with cheek fairings.

RC-135V Rivet Joint: Reconnaissance aircraft.

RC-135W: Reconnaissance aircraft with thimble nose radome.

TC-135B: Trainer aircraft for OC-135B aircraft.

TC-135S: Trainer aircraft for RC-135S 'Cobra Ball' aircraft.

TC-135W: Trainer aircraft for RC-135U/V/W aircraft.

WC-135B: Meteorological aircraft.

C-137B/C: General transport aircraft.

EC-137D: Crew conversion trainer aircraft.

NEW ENTRY

BOEING E-3 SENTRY

French designation: E-3F Système de Détection Aéroporté
UK designation: E-3D Sentry AEW. Mk 1
US designation: E-3B/C AWACS

TYPE: Mobile, flexible, survivable, jamming resistant, high capacity radar station and command, control and communications centre; airborne warning and control system (AWACS).

PROGRAMME: Two prototype EC-137Ds used to test competing radars; Westinghouse selected; full-scale development completed 1976. USAF received 34 E-3s by June 1984, including two prototypes.

First production E-3A delivered to 552nd Airborne Warning and Control Wing, Tactical Air Command, at Tinker AFB, Oklahoma, on 24 March 1977; initial operational capability (IOC) April 1978. At various times, E-3s deployed to Iceland, Germany, Saudi Arabia, Sudan, the Mediterranean area, South West Asia and the Pacific and in support of drug enforcement programme. E-3As began to work with NORAD continental air defence 1 January 1979.

The 552nd Wing has three AWACS squadrons and supporting units. Overseas units include the 960th, 961st and 962nd AWAC squadrons based respectively at NAS Keflavik, Iceland, Kadena AB, Okinawa, Japan, and Elmendorf, Alaska, providing command and control capability to CINCLANT (through Commander, Iceland Defence Force) and CINCPAC.

Production completed 1991; 1,010th and final 707 airframe, an E-3, retained by Boeing for tests and trials; upgrades continue; Boeing to transfer avionics to 767 airframe as successor AWACS (which see).

VERSIONS: **Core E-3A:** Initial standard of 23 production USAF Sentries.

E-3B: Block 20 modification updated two EC-137Ds and 22 USAF Core E-3As to E-3B standard by adding ECM-resistant voice communications, a third HF and five more UHF radios (making 12), faster IBM CC-2 computer with larger memory, five more SDCs (making 14), Westinghouse austere maritime surveillance capability to main radar, provision for Have-Quick anti-jamming in UHF radios, self-defence, and radio teletype. First E-3B redelivered to USAF 18 July 1984; remaining 23 modified with Boeing kits at Tinker AFB. Under Project Snappy, unspecified sensor installed in seven USAF E-3B/Cs by January 1991 for participation in Gulf War; eight further aircraft modified.

US/NATO Standard E-3A: Original standard for USAF aircraft Nos. 26 to 34 of which deliveries began December 1981, and of updated aircraft No. 3. Additions include full maritime surveillance capability, CC-2 computer, additional HF radios, ECM-resistant voice communications, radio teletype, provision for self-defence, and ECM. There are 18 NATO E-3As to this standard.

E-3C: USAF Block 25 modification of 10 USAF E-3As began 1984. Adds five more SDCs, five more UHF radios and (first use 1 February 1991) Have-Quick A-Nets secure communications system.

E-3D Sentry AEW. Mk 1: British government announced order for six, 18 December 1986 and exercised option for seventh in October 1987. CFM56-2A-3 power plants. First flight E-3D (ZH101) 11 September 1989; first flight fully equipped 5 January 1990; flown to UK on 3 November 1990 after combined British-French airworthiness certification and in-flight refuelling trials. Second aircraft (ZH102) flown to Waddington 4 July 1990 for fitting out by a consortium led by British Aerospace (which see) and handed over to RAF as first receipt, 24 March 1991. Final UK E-3D (ZH107), the 1,009th airframe of the Boeing 707 family, first flown 14 June 1991 and delivered to RAF 12 May 1992. Sentry Training Squadron formed at Waddington 1 June 1990. No. 8 Squadron, RAF, formed with E-3Ds at Waddington 1 July 1991; declared to NATO 1 July 1992; RAF aircraft are E-3D component, NAEWF. Boeing gave UK 130 per cent industrial offset.

E-3F SDA: France ordered three with CFM56-2A-3 power plants, February 1987; fourth aircraft ordered 1987, but options for two more dropped August 1988. First (No. 201) flown 27 June 1990; delivered UTA Industries at Le Bourget for fitting-out 10 October 1990; to Avord for integration trials and handover 19 December 1990; official delivery to CEAM for military acceptance, 22 May 1991. Delivery of remaining E-3Fs followed 27 July and 11 September 1991 and 15 February 1992. 36e Escadre de Détection Aéroportée formed at Avord 1 March 1990 as

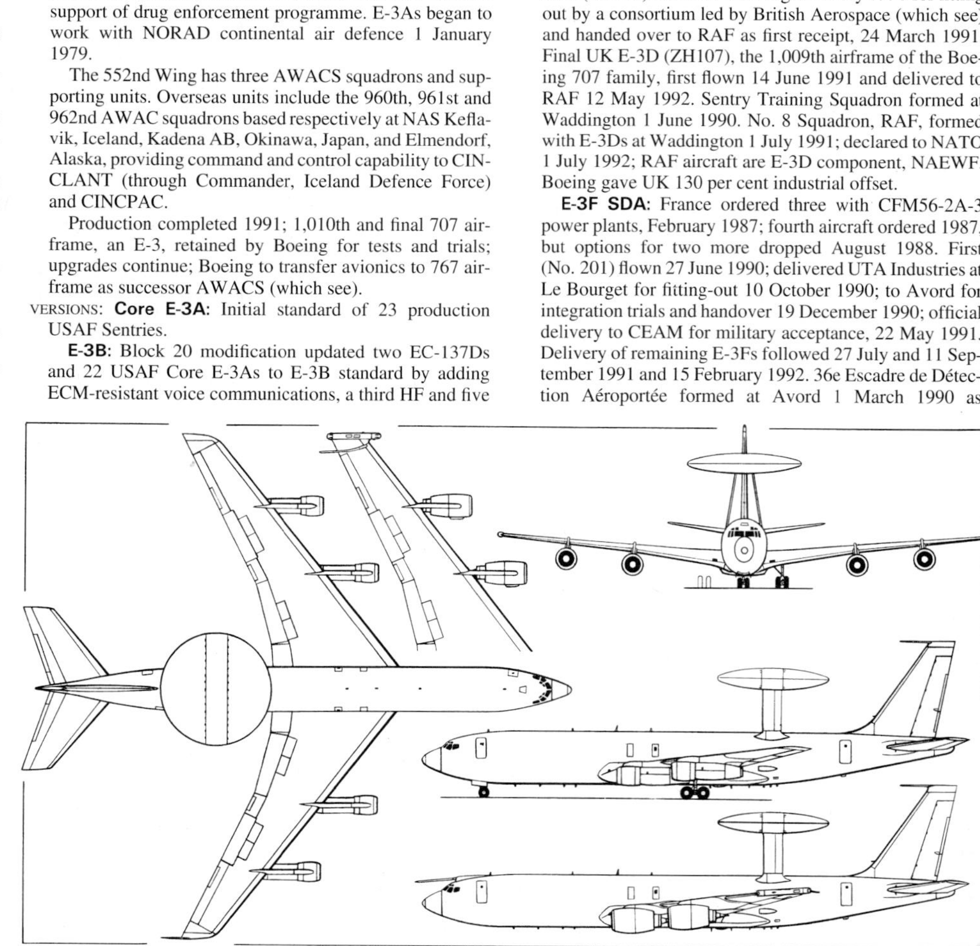

Boeing E-3A Sentry, with lower side view and wing scrap view of Royal Air Force E-3D Sentry AEW. Mk 1
(Jane's/Dennis Punnett)

1994

operating wing of two squadrons (Escadrons 1/36 'Berry' and 2/36 'Nivernais'). Boeing gave France 130 per cent industrial offset.

JE-3C: In May 1987, Boeing received $241.5 million USAF contract for full-scale development and integration into USAF and NATO E-3s of AN/AYR-1 ESM system to detect signals from hostile and friendly targets. First installation in JE-3C temporary test aircraft 73-1674; tested over 1,000 hours in 155 sorties, beginning September 1990; second in NATO E-3A LX-N-90442, redelivered October 1991; flew 15 trial sorties, early 1992.

KE-3A: Boeing designation for Saudi Arabian tanker based on E-3 airframe; CFM56 engines; no rotodome or other surveillance equipment; eight built.

UPGRADES: **Boeing:** E-3B. See Versions.

Boeing: E-3C. See Versions.

Boeing: In 1987 the USAF awarded Boeing Defense & Space Group an ICON (Integration Contract) contract to upgrade US and NATO E-3s with an AN/AYR-1 electronic support measures (ESM) system, passive surveillance capability and offer other block enhancements to the US E-3 fleet. The latter improvements include upgrading of the Joint Tactical Distribution Systems (JTIDS) to TADIL-J (Tactical Digital Information Link-J); increased computer capacity; and ability to use the Global Positioning System (GPS) to pinpoint AWACS location anywhere in the world. In March 1989 Boeing was authorised to begin production of Have-Quick A-NETS, an improved communication system that provides secure, anti-jam radio contact with other AWACS, friendly aircraft and ground stations.

In 1993, NATO's AWACS modernisation programme intensified. Boeing delivered upgraded computer systems to NATO as part of the Memory Upgrade Programme which included upgrading the existing IBM CC2 computer to the CC2E model. Trials of NATO E-3 AN/AYR-1 began in February 1995.

Daimler-Benz Aerospace: Contract for NATO E-3 upgrade placed January 1993; designated Mod Block 1; includes new colour displays, Have-Quick secure radios and Link 16 (JTIDS) datalink; initial two aircraft under 1993 contract; follow-on order for remaining 16 modifications placed mid-1993; all retrofit work by DASA in Germany.

USAF: Extend Sentry programme. Plan to upgrade most, if not all aircraft subsystems to extend the aircraft's life to 2025.

Westinghouse: In 1989 Westinghouse was awarded a development contract for upgrading AN/APY-1/2 radars of USAF E-3s with new processors, displays and pulse compression to double performance against small targets like cruise missiles. If successful, this will be applied to all 34 E-3B/Cs from 1994 onwards.

CUSTOMERS: Total of 68 ordered (USAF 34, NATO 18, Saudi Arabia five, UK seven, France four); last delivered March 1992. First NATO production E-3A flew at Renton 18 December 1980 and delivered to system integrator Dornier at Oberpfaffenhofen 19 March 1981; all 18 delivered to NATO between 22 January 1982 and 25 April 1985. NATO E-3As assigned to Nos. 1, 2 and 3 Squadrons and Training Centre of E-3A component, NATO Airborne Early Warning Force, based at Geilenkirchen, Germany. Up to six NATO E-3As detached to forward operating locations (FOL) at Konya (Turkey), Previza (Greece), Trapani/Birgi (Italy) and Ørland (Norway). NAEWF has NATO Command status and is staffed by personnel from Belgium, Canada, Denmark, Greece, Italy, Netherlands, Norway, Portugal, Turkey and USA, plus non-aircrew personnel from Luxembourg. ESM systems upgrade under way for 1995 completion. Seven-year E-3A modernisation plan agreed June 1990, subject to funding.

COSTS: $129 million, flyaway, NATO 1982; $222 million programme unit cost, UK 1987; France $254 million, 1987. NATO ESM upgrade $230 million (1990); NATO E-3A modernisation $700 million (1990). E-3D programme cost $1.3 billion (1987); E-3F programme cost $550 million (1987).

DESIGN FEATURES: Cantilever low-wing monoplane. Dihedral 7°. Incidence 2°. Sweepback at quarter-chord 35°. All-metal two-spar fail-safe structure. Centre-section continuous through fuselage. Normal outboard aileron, and small inboard aileron on each wing, built of aluminium honeycomb panels. Two tracked and slotted flaps and one fillet flap of aluminium alloy on each wing. Full span leading-edge flaps. Four hydraulically operated aluminium alloy spoilers on each wing, forward of flaps. Primary controls are aerodynamically balanced and manually operated through spring tabs. Lateral control at low speeds by all four ailerons, supplemented by spoilers which are interconnected with the ailerons. Lateral control at high speeds by inboard aileron and spoilers only. Operation of flaps adjusts linkage between inboard and outboard ailerons to permit outboard operation with flaps extended. Spoilers may also be used symmetrically as speed brakes. Thermal anti-icing of wing leading-edges.

The fuselage is an all-metal semi-monocoque fail-safe structure with cross-section made up of two circular arcs of different radii, the larger above, faired into smooth-contoured oval. Structure strengthened by comparison with that of the commercial Model 707-320.

The tail unit is a cantilever all-metal structure. Electrically and manually operated variable incidence tailplane. Powered rudder. Anti-balance tab and trim tab in rudder. Trim and control tabs in each elevator.

Details below refer to USAF E-3s except where indicated:

FLYING CONTROLS: Essentially as for Boeing 707 family.

STRUCTURE: RAF aircraft have additional wing stringers outboard of outer engines because of wingtip pods and trailing-edge HF antennae. Otherwise, as Boeing 707.

POWER PLANT: Four Pratt & Whitney TF33-PW-100/100A turbofans, each rated at 93.4 kN (21,000 lb st), mounted in pods beneath the wings. Fuel contained in integral wing tanks. Usable fuel 90,528 litres (23,915 US gallons; 19,913 Imp gallons). Provision for in-flight refuelling, with receptacle for boom over flight deck. Four CFM56-2A-2/3 turbofans on French, Saudi and UK aircraft. SOGERMA in-flight refuelling probe in addition to receptacle on E-3D and E-3F.

ACCOMMODATION: Basic E-3A operational crew of 17 includes a flight crew complement of four plus 13 AWACS specialists, though this latter number can vary for tactical and defence missions. Full crew complement is two pilots, navigator, flight engineer, tactical director, fighter allocator, two weapons controllers, surveillance controller, link manager, three surveillance operators, communications operator, radar technician, communications technician and computer display technician. E-3B/C have extra five AWACS specialist positions increasing crew to 18, plus four on flight deck. Aft of flight deck, from front to rear of fuselage, are communications, data processing and other equipment bays; multipurpose consoles; communications, navigation and identification equipment; and crew rest area, galley and parachute storage rack.

SYSTEMS: A liquid cooling system provides protection for the radar transmitter. An air cycle pack system, a draw-through system, and two closed loop ram-cooled environmental control systems ensure a suitable environment for crew and avionics equipment. Electrical power generation has a 600 kVA capability. Distribution centre for mission equipment power and remote avionics in lower forward cargo compartment. Rear cargo compartment houses radar transmitter and an APU. External sockets allow intake of power when aircraft is on ground. Two separate and independent hydraulic systems power essential flight and mission equipment, but either system can satisfy requirements of both equipment groups in an emergency.

AVIONICS: Elliptical cross-section rotodome of 9.14 m (30 ft) diameter and 1.83 m (6 ft) maximum depth, mounted 3.35 m (11 ft) above fuselage, comprises four essential elements: a turntable, strut-mounted above rear fuselage, supporting rotary joint assembly to which are attached sliprings for electrical and waveguide continuity between rotodome and fuselage; structural centre-section of aluminium skin and stiffener supporting the Westinghouse AN/APY-1 surveillance radar (AN/APY-2 from No. 25 onwards, and in all export E-3s) and IFF/TADIL-C antennae, radomes, auxiliary equipment for radar operation and environmental control of the rotodome interior; liquid cooling of the radar antennae; and two radomes of multi-layer glassfibre sandwich material, one for surveillance radar and one for IFF/TADIL-C array. For surveillance operations rotodome is hydraulically driven at 6 rpm, but during non-operational flights it is rotated at only ¼ rpm, to keep bearings lubricated. Radar operates in E/F-band and can function as both a pulse and/or a pulse Doppler radar for detection of aircraft targets. A similar pulse radar mode with additional pulse compression and sea clutter adaptive processing is used to detect maritime/ship traffic. Radar is operable in six modes: PDNES (pulse Doppler non-elevation scan), when range is paramount to elevation data; PDES (pulse Doppler elevation scan), providing elevation data with some loss of range; BTH (beyond the horizon), giving long-range detection with no elevation data; Maritime, for detection of surface vessels in various sea states; Interleaved, combining available modes for all-altitude longer range aircraft detection, or for both aircraft and ship detection; and Passive, which tracks enemy ECM sources

Boeing JE-3C temporary test aircraft of the US Air Force ***(Boeing)***

1995

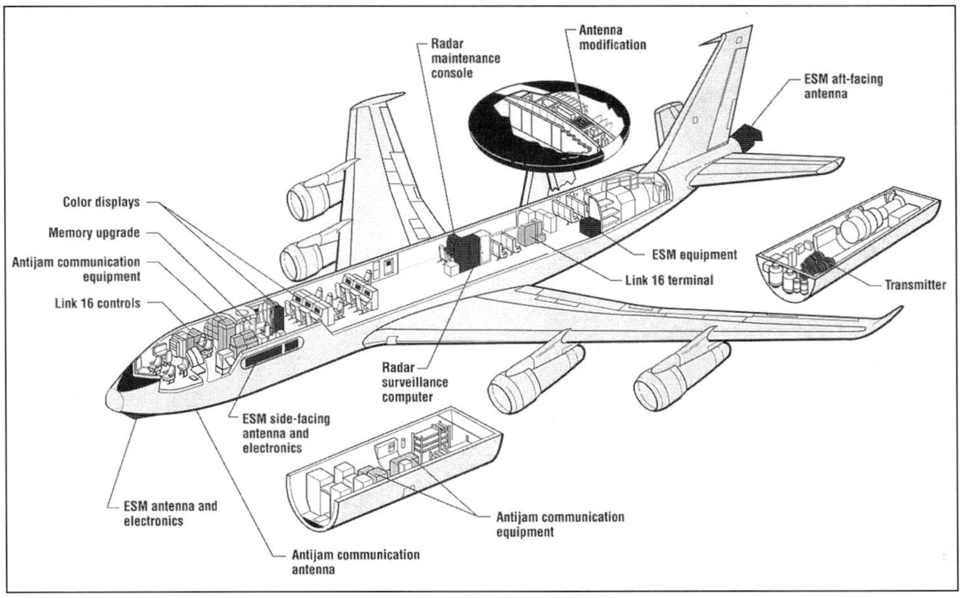

Boeing E-3A NATO Sentry diagram showing major systems modernisation ***(Boeing)***

1995

Boeing E-3D Sentry AEW. Mk 1, powered by CFM56-2A-3 engines ***(P. Tompkins)***

1994

Boeing E-3A Sentry in service with NATO

1994

without transmission-induced vulnerability. Radar antennae, spanning about 7.32 m (24 ft), and 1.52 m (5 ft) deep, scan mechanically in azimuth, and electronically from ground level up into the stratosphere. Heart of the data processing capability of the first 24 aircraft in their original core E-3A form is an IBM 4 Pi CC-1 high-speed computer (see 1987-88 *Jane's All the World's Aircraft* for details). From 25th aircraft, the new and improved IBM CC-2 computer was installed from the start, with a main storage capacity of 665,360 words. Data display and control are provided by Hazeltine high resolution colour situation display consoles (SDC) and auxiliary display units (ADU). The E-3B carries 14 SDCs and two ADUs. Navigation/guidance relies upon two Delco AN/ASN-119 Carousel IV inertial navigation platforms, a Northrop AN/ARN-120 Omega set which continuously updates the inertial platforms, and a Teledyne Ryan AN/APN-213 Doppler velocity sensor to provide airspeed and drift information. Communications equipment provides HF, VHF and UHF channels through which information can be transmitted or received in clear or secure mode, in voice or digital form. A Bendix/King weather radar is carried in the nose. Identification is based on an Eaton (AIL) AN/APX-103 interrogator set which is the first airborne IFF interrogator to offer complete AIMS Mk X SIF air traffic control and Mk XII military identification friend or foe (IFF) in a single integrated system. Simultaneous Mk X and Mk XII multitarget and multimode operations allow the operator to instantaneously obtain the range, azimuth and elevation, code identification and IFF status of all targets within radar range. NATO E-3As carry, and USAF aircraft have provisions for, a radio teletype. All aircraft from No. 25 have an inboard underwing hardpoint on each side. There is no current requirement for either USAF or NATO AWACS to carry weapons; but on NATO E-3As these hardpoints can be used to mount additional podded items of ECM equipment. E-3Ds carry Loral 1017 'Yellow Gate' ESM pods at the wingtips.

DIMENSIONS, EXTERNAL:

Wing span: normal	44.42 m (145 ft 9 in)
E-3D	44.98 m (147 ft 7 in)
Length overall	46.61 m (152 ft 11 in)
Height overall	12.73 m (41 ft 9 in)

WEIGHTS AND LOADINGS (E-3D):

Fuel weight (JP4)	70,510 kg (155,448 lb)
Normal T-O weight	147,417 kg (325,000 lb)
Max T-O weight	150,820 kg (332,500 lb)
Max ramp weight	151,953 kg (335,000 lb)

PERFORMANCE:

Max level speed	460 kts (853 km/h; 530 mph)
Service ceiling: TF33	over 8,850 m (29,000 ft)
CFM56	over 9,145 m (30,000 ft)
Endurance on station, 870 n miles (1,610 km; 1,000 miles) from base	6 h
Max unrefuelled endurance: TF33	more than 11 h
CFM56	more than 10 h

UPDATED

BOEING ADVANCED AIRBORNE COMMAND POST

USAF designation: E-4B

TYPE: Advanced airborne command aircraft.

PROGRAMME: On 28 February 1973 the US Air Force's Electronic Systems Division announced from its headquarters at Hanscom Field, Bedford, Massachusetts, that it had awarded the Boeing Company a $59 million fixed price contract for the supply of two Model 747-200Bs to be adapted as **E-4A** airborne command posts under the 481B Advanced Airborne Command Post (AABNCP) programme. A further contract valued at more than $27.7 million was awarded in July 1973 for a third aircraft; in December 1973 a fourth aircraft was contracted at $39 million. This was to be fitted with more advanced equipment (see below) and designated E-4B.

The E-4s were intended to replace EC-135 Airborne Command Posts of the National Military Command System and Strategic Air Command which are military versions of the Model 707. AABNCPs provide the critical communications link between US National Command Authority and the nation's strategic retaliatory forces during and following a nuclear or conventional attack on the USA. They were also equipped with wiring that would be needed to add an ICBM launching ('Looking Glass') capability if ground control centres became inoperative, but the associated 'black boxes' were not fitted and there is no longer any intention to do so.

E-Systems won the contract to install interim equipment in the three E-4As. This involved transfer and integration of equipment removed from EC-135s, providing integration aircraft with increased endurance and the ability to carry an expanded battle staff. The E-4A's floor space can accommodate almost three times the payload of the EC-135.

The first E-4A flew for the first time on 13 June 1973, and was delivered to Andrews AFB, Maryland, December 1974. The second and third also consigned to Andrews AFB, were received in May and September 1975. In their initial form, they were operated as National Emergency Airborne Command Posts (NEACPs), and provided operational experience invaluable in finalising the design equipment installed in the E-4B.

The third and fourth aircraft differed initially from the first two in having General Electric CF6-50E turbofan engines, each rated at 233.5 kN (52,000 lb st), instead of the JT9Ds normally fitted to aircraft of the 747 series; CF6-50Es were fitted retrospectively to the first two aircraft during 1976, and have since been upgraded to CF6-50E2 standard, to improve fuel economy and T-O thrust under high ambient temperature conditions. The total planned was four E-4Bs, comprising the fourth aircraft, and the three E-4As brought up to the same standard retrospectively. Contracts covering modification of one E-4A to E-4B configuration, with options to modify the other two, were announced 26 June 1980. The two options were duly exercised during December 1980 and October 1981, and the first converted E-4B was redelivered to the USAF 15 July 1983. The second was redelivered 18 May 1984, and the third 30 January 1985.

Boeing, E-Systems and a team comprising Electrospace Systems Inc of Richardson, Texas; Collins Radio Division of Rockwell International Corporation, Dallas, Texas; RCA Corporation of Morristown, New Jersey; and Burroughs Corporation, Federal and Special Systems Group, of Paoli, Pennsylvania, were responsible for designing and installing the advanced command post equipment in the E-4B, under a programme managed by Oklahoma City Air Logistics Centre. The first E-4B was delivered to the US Air Force in August 1975 in testbed configuration, with flight refuelling equipment but without the planned command, control and communications equipment. Next stage involved the installation of the 1,200kVA electrical system (two 150kVA generators on each engine) designed to support the advanced avionics. Finally the operational systems were added, and the first flight of the fully equipped E-4B took place 10 June 1978. US Air Force tests of operational capability began later that year.

The first E-4B (75-0125) was redelivered to the US Air Force 21 December 1979, and entered service January 1980. It has accommodation for a larger battle staff than that carried by the E-4A; an air conditioning system of 226.5 m^3 (8,000 cu ft)/min capacity to cool avionics components; nuclear thermal shielding; acoustic controls; an improved technical control facility; and new super high frequency (SHF) and dual Collins VLF/LF communications systems, the latter employing trailing short-wire and long-wire antennae, of which the long-wire system has an antenna 4.3 n miles (8 km; 5 miles) in length. The SHF antennae are housed in a dorsal fairing which is a recognition characteristic of the E-4B.

Strategic Air Command (SAC) is the sole operational manager of the AABNCP force. The main operating base for the E-4 fleet is at Offut AFB Nebraska.

UPGRADES: **Rockwell:** Contract awarded by ARINC Research Corp to supply FMS-800 flight management system for installation in E-4 aircraft of the USAF. FMS-800 is a fully integrated system including a control display unit with embedded GPS (CDU-900). For more details of FMS-800, see *Jane's Avionics 1995-96.*

ACCOMMODATION (E-4B): Up to 94 crew members on three decks. Upper deck contains flight deck and flight crew rest area. Access to main deck compartments is by aisle on starboard side; these compartments include the NCA (National Command Authority) area, conference room, briefing room, battle staff work area, communications control centre, technical control centre (where operators monitor and maintain quality of communications links), and crew rest area. Forward and rear lower lobes house electronic equipment, an onboard maintenance area and a winch operator's station for the long-wire VLF antenna. The NCA's senior advisor conference room is equipped with a screen, and secure telephones at the conference table. Briefing room for second-level advisory staff contains table, podium and a viewing screen served also by the projection room. Battle staff area accommodates up to 30 crew members responsible for information flow into and out of aircraft; their two-position consoles contain work surfaces and facilities for communications and data storage.

AVIONICS (E-4B): Command and control avionics, powered by 1,200kVA electrical power generation system, include

Boeing E-4B National Emergency Airborne Command Post (NEACP)

1994

Boeing E-6A TACAMO II prototype long endurance communications aircraft

1995

13 external communication systems operating through 46 antennae with configurations ranging from a small dish for SHF satellite links (in larger fairing aft of upper deck) to an 8 km (5 mile) trailing wire for VLF and LF communication. Ability to use satellite systems reduces dependence on ground stations and protects against jamming and direct track attempts. A long-range link, established with the high power VLF system, resists atmospheric nuclear effects and is very difficult to jam. The HF, MF, VHF and UHF bands provide additional two-way radio channels. Secure voice and teletype links are achieved through HF, UHF, and SHF bands, and the E-4B's high-speed secure-record communications equipment interfaces to the automatic digital network. The E-4B system is capable of tying in to commercial telephone and radio networks, and could potentially be used for radio broadcasts to the general population. When it is on the ground it can also be connected to a ground communications network, which can be disconnected quickly. Other E-4B avionics and instrumentation include search radar in nosecone, Tacan VHF Omni navigation, dual radio altimeters, glide slope and marker beacon receiver.

DIMENSIONS, EXTERNAL:

Wing span	59.64 m (195 ft 8 in)
Length overall	70.51 m (231 ft 4 in)
Height overall	19.33 m (63 ft 5 in)

DIMENSIONS, INTERNAL:

Floor area (three decks, total)	511 m² (5,500 sq ft)

WEIGHTS AND LOADINGS:

Max fuel weight	150,395 kg (331,565 lb)
Max T-O weight	362,875 kg (800,000 lb)
Max ramp weight	364,235 kg (803,000 lb)

PERFORMANCE:

T-O run for 8 h endurance	1,525 m (5,000 ft)
Mission endurance	72 h
Unrefuelled endurance	more than 12 h

UPDATED

BOEING E-6A TACAMO II

TYPE: Long endurance communications relay aircraft carrying US Navy airborne very low frequency (AVLF) system.

PROGRAMME: US Navy contract placed with Boeing Aerospace 29 April 1983 to replace EC-130Q Hercules TACAMO (take charge and move out); first full test flight of prototype (162782) 1 June 1987; two production E-6As ordered FY86; three, three and seven production E-6As ordered in FY87-89, making 16; first two delivered to VQ-3 squadron, Barber's Point, Hawaii, 2 August 1989; VQ-3 withdrew last C-130 in August 1990, having eight E-6As for Pacific area; VQ-4 assigned seven at Patuxent River, Maryland, for Atlantic area; first of these delivered January 1991. By early 1991, continuous TACAMO airborne alert terminated; both squadrons transferred to Tinker AFB, Oklahoma, in 1992.

UPGRADES: **Chrysler Technologies:** E-6B upgrade programme. See separate entry.

CUSTOMERS: US Navy 16.

DESIGN FEATURES: Substantially as for USAF E-3, but powered by four CFM International turbofans.

FLYING CONTROLS: Substantially as for E-3. Boeing contract 1991 to Bendix/King to develop digital AFCS.

STRUCTURE: 75 per cent common with E-3, including EMP nuclear hardening; radome support structure deleted. Additions include wingtip ESM/Satcom pods and HF antenna fairings, increased corrosion protection, forward freight door of 707-320C. Local strengthening to overcome stresses caused by banked orbit with antenna deployed: flap guide vanes modified; three minor changes to be made to cure tail fin flutter.

POWER PLANT: Four 106.76 kN (24,000 lb st) CFM International F108-CF-100 (CFM56-2A-2) turbofans in individual underwing pods, as on some export E-3s. Fuel contained in integral tanks in wings, with single-point refuelling. In-flight refuelling via boom receptacle above flight deck.

ACCOMMODATION: Basic militarised interior sidewalls, ceilings and lighting are similar to those of the E-3A. Interior divided into three main functional areas: forward of wings (flight deck and crew rest area), overwing (five-man mission crew), and aft of wings (equipment). Forward crew area, 50 per cent common with that of E-3A, accommodates a four-man crew on flight deck. Compartment immediately aft contains dining, toilet, and eight-bunk rest area for spare crew carried on extended or remote deployment missions. Crew enter by ladder and hatch in floor of this compartment. Overwing compartment with communications and other consoles, their operators, and an airborne communications officer (ACO). To the rear is compartment containing R/T racks, transmitters, trailing wire antennae and their reels. Bale-out door at rear on starboard side.

SYSTEMS: Some 75 per cent of the E-6A's systems are the same as the E-3. Among those retained are the liquid cooling system for the transmitters, 'draw-through' cooling system for other avionics, the 600 kVA electrical power generation system, APU, liquid oxygen system, and Mil-spec hydraulic oil.

AVIONICS: Three Collins AN/ARC-182 VHF/UHF com transceivers, all with secure voice capability; five Collins AN/ARC-190 HF com (one transceiver, one receive only); and Hughes Aircraft AN/AIC-32 crew intercom with secure

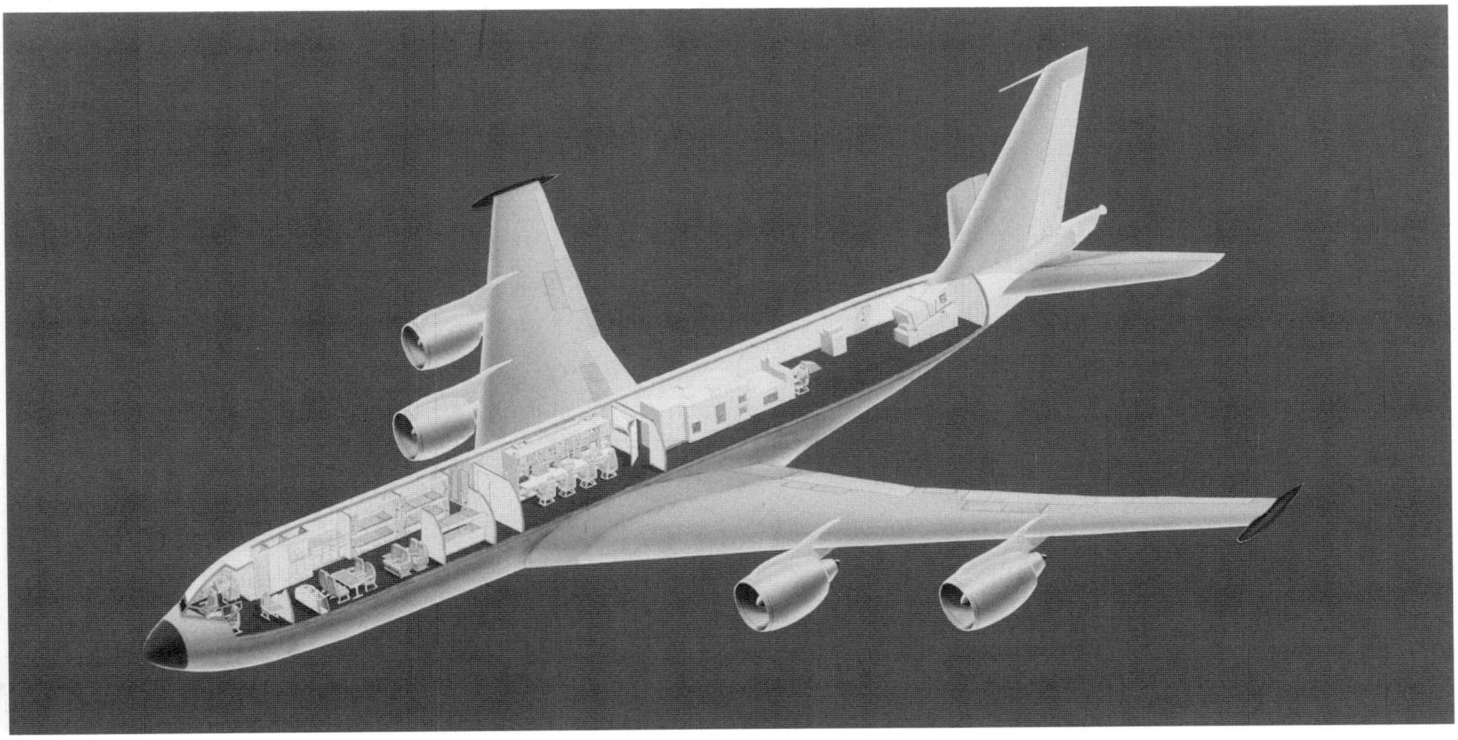

Boeing E-6A TACAMO II cutaway drawing showing crew stations

1994

voice capability. External aerials for Satcom UHF reception in each wingtip pod; fairings beneath each pod house antenna for standard HF reception. Navigation by triplex Litton LTN-90 ring laser gyro-based inertial reference system integrated with a Litton LTN-211 VLF/Omega system and duplex Smiths Industries SFM 102 digital/analog flight management computer system (FMCS). Bendix/King AN/APS-133 colour weather radar, in nosecone, with capability for short-range terrain-mapping, tanker beacon homing and waypoint display. Honeywell AN/APN-222 high/low-range (0-15,240 m; 0-50,000 ft) radio altimeter, and Collins low-range (0-762 m; 0-2,500 ft) radar altimeter, with ILS and GPWS. General Instrument AN/ALR-66(V)4 electronic support measures (ESM), in each wingtip pod, provide information on threat detection, identification, bearing and approximate range. In overwing compartment, overseen by ACO, is a new communications central console, which incorporates ERCS (emergency rocket communications system) receivers, cryptographic equipment, new teletypes, tape recorders, and other communications equipment, all hardened against electromagnetic interference. In each operational area the E-6A links upward with airborne command posts and the Presidential E-4, to satellites, and to the ERCS; and downward to VLF ground stations and the SSBN fleet. Mean time between failures of complete mission avionics is less than 20 hours, but the E-6 is able to carry spares, and a spare crew, to permit extended missions of up to 72 hours with in-flight refuelling, and/or deployment to remote bases, where it is capable of autonomous operation.

EQUIPMENT: Main VLF antenna is a 7,925 m (26,000 ft) long trailing wire aerial (LTWA), with a 41 kg (90 lb) drogue at the end, which is reeled out from the middle part of the rear cabin compartment through an opening in the cabin floor. The LTWA, with its drogue, weighs about 495 kg (1,090 lb) and creates some 907 kg (2,000 lb) of drag when fully deployed. Acting as a dipole is a much shorter (1,220 m; 4,000 ft) trailing wire (STWA), winched out from beneath the tailcone. At patrol altitude, with the LTWA deployed, the aircraft enters a tight orbit and the wire stalls, causing it to be almost vertical (70 per cent verticality is required for effective sub-sea communications). Signals transmitted through the trailing wire antennae use 200 kW of power, and can be received by submerged SSBNs via a towed buoyant wire antenna.

ARMAMENT: None.

DIMENSIONS, EXTERNAL:

Wing span	45.16 m (148 ft 2 in)
Length overall	46.61 m (152 ft 11 in)
Height overall	12.93 m (42 ft 5 in)
Wheel track	6.73 m (22 ft 1 in)
Wheelbase	17.98 m (59 ft 0 in)
Forward cargo door: Height	2.34 m (7 ft 8 in)
Width	3.40 m (11 ft 2 in)
Height to sill	3.20 m (10 ft 6 in)

AREAS:

Wings, gross	283.4 m² (3,050.0 sq ft)

WEIGHTS AND LOADINGS:

Operating weight empty	78,378 kg (172,795 lb)
Max fuel	70,305 kg (155,000 lb)
Max T-O weight	155,128 kg (342,000 lb)
Max wing loading	547.4 kg/m² (112.13 lb/sq ft)
Max power loading	363.26 kg/kN (3.56 lb/lb st)

PERFORMANCE (S/L, ISA, estimated):

Dash speed	530 kts (981 km/h; 610 mph)
Cruising speed at 12,200 m (40,000 ft)	455 kts (842 km/h; 523 mph)
Patrol altitude	7,620-9,150 m (25,000-30,000 ft)
Service ceiling	12,800 m (42,000 ft)
Critical field length	2,042 m (6,700 ft)
Max effort T-O run	1,646 m (5,400 ft)
Max effort T-O run with fuel for 2,500 n miles (4,630 km; 2,875 miles)	732 m (2,400 ft)
Landing run at max landing weight	793 m (2,600 ft)
Mission range, unrefuelled	6,350 n miles (11,760 km; 7,307 miles)
Endurance: unrefuelled	15 h 24 min
on-station, 1,000 n miles (1,850 km; 1,150 miles) from T-O	10 h 30 min
with one refuelling	28 h 54 min
with multiple refuelling	72 h

UPDATED

Boeing B-52H eight-jet heavy bomber *(P. Tompkins)*

1994

Boeing B-52H heavy bomber. B-52H to be upgraded to dual nuclear/conventional role *(P. Tompkins)*

1994

BOEING B-52 STRATOFORTRESS

TYPE: Eight-jet heavy bomber.

PROGRAMME: Full-scale development of system to carry AGM-86B air-launched cruise missile started early 1978; 98 B-52Gs and 95 B-52Hs modified to each carry 12 AGM-86B on inboard external pylons plus SRAM or other armament internally; first unit to reach IOC with AGM-86B with 416th Bomb Wing of SAC at Griffiss AFB December 1982; all aircraft fitted by late 1989; programme completed FY90.

VERSIONS: **B-52G:** Withdrawn from service.

B-52H: In 1994, 95 still in service; does not need strakelet fairings, as all carry cruise missiles. Only the B-52H remains in service with the USAF. All to be modified for conventional as well as nuclear role.

UPGRADES: **Boeing:** Modification of B-52Hs to carry eight AGM-86B, SRAM, Advanced Cruise Missile or free-fall nuclear bombs on internal common strategic rotary launcher (CSRL) started 1982; first flight September 1985; 98 CSRLs to be produced; Boeing Military Airplanes manufactured 95 modification kits, plus support for US Air Force programme managed by Oklahoma City Air Logistics Center at Tinker AFB and conducted at Kelly AFB, Texas; first fully modified aircraft delivered to Carswell AFB April 1988; first full alert with internal and external stowage September 1989.

B-52H also being equipped for General Dynamics AGM-129A advanced cruise missile to augment AGM-86B; captive carry tests of 12 on underwing pylons began early 1989.

Boeing: The B-52H 'Conventional Mission Upgrade Programme' involves transferring the conventional weapons capability from the B-52Gs which have been retired from service. Upgrade includes: capability to accommodate HAVE NAP and Harpoon missiles and the universal bomb bay adaptor as well as integrated conventional stores and Global Positioning System (GPS). First aircraft to be modified arrived at Boeing Space & Defense in June 1993. A total of 21 conversions to be completed by end 1995.

The following details apply to B-52H:

DESIGN FEATURES: Cantilever all-metal semi-monocoque with anhedral and 35° sweepback. Hydraulically actuated variable incidence tailplane.

Boeing B-52H heavy bomber being modified to carry AGM-86B SRAM, advanced cruise missiles and free-fall nuclear bombs *(P. Tompkins)*

1995

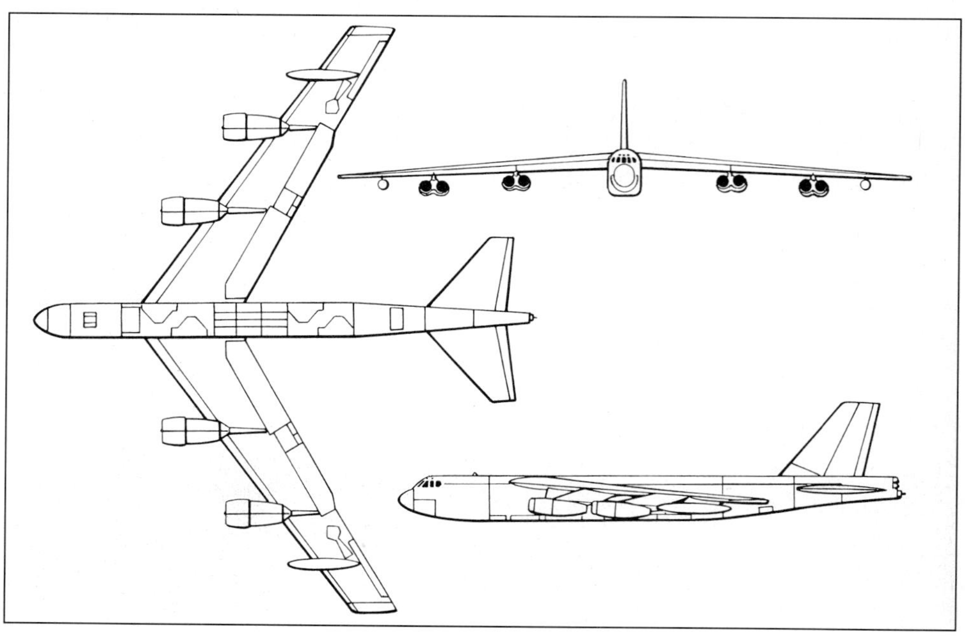

Boeing B-52H eight-jet heavy bomber *(Jane's/Mike Keep)*

1994

FLYING CONTROLS: Lateral control by serrated spoilers on upper surface of wings. Spoilers can be used as airbrakes. Area-increasing flaps. All control surfaces operated manually through servo tabs.

STRUCTURE: All-metal stressed skin structure. The fuselage is an all-metal semi-monocoque structure. The tail unit is a cantilever all-metal structure.

LANDING GEAR: Four individually retracted twin-wheel main units in tandem pairs, retracting into fuselage fore and aft of bomb-bay. Two forward units have hydraulic steering. All four units can be set at a selected degree of castor in flight or on the ground for 'crosswind' landings. Small outrigger units retract inward into wings outboard of outer engine positions. Ribbon braking parachute. 13.4 m (44 ft) diameter, in compartment in top of fuselage aft of rudder.

POWER PLANT (B-52H): Eight 75.6 kN (17,000 lb st) Pratt & Whitney TF33-P-3 turbofans. Fuel capacity 174,130 litres (46,000 US gallons; 38,303 Imp gallons) internally, plus two 2,650 litre (700 US gallon; 583 Imp gallon) underwing drop tanks.

ACCOMMODATION (B-52G/H): Crew of six (pilot and co-pilot, side by side on flight deck, navigator, radar navigator, ECM operator and gunner).

AVIONICS: Boeing OAS (offensive avionics system), introduced from 1980, is a digital solid state system, and includes Tercom (terrain comparison) guidance, a Teledyne Ryan Doppler radar, Honeywell AN/ASN-131 gimballed electrostatic airborne inertial navigation system (GEANS), IBM/Raytheon AN/ASQ-38 analog bombing/navigation system with IBM digital processing, Smiths attitude heading and reference system, Honeywell radar altimeter, Honeywell controls and displays, and Norden Systems modernised strategic radar. Under Phase II of the programme, completed by FY89, all 152 B-52Gs and 95 Hs were equipped with OAS. All currently operational B-52Hs have an AN/ASQ-151 electro-optical viewing system (EVS) to improve low level penetration capability. The EVS sensors are housed in two steerable, side by side chin turrets. The starboard turret houses a Hughes Aircraft AN/AAQ-6 forward-looking infra-red (FLIR) scanner, while the port turret contains a Westinghouse AN/AVQ-22 low light level TV camera. Phase VI avionics include Motorola AN/ALQ-122 SNOE (Smart Noise Operation Equipment) and Northrop AN/ALQ-155(V) advanced ECM; an AFSATCOM kit which permits worldwide communication via satellite; a Dalmo Victor AN/ALR-46 digital radar warning receiver; Westinghouse AN/ALQ-153 pulse Doppler tail warning radar and ITT Avionics AN/ALQ-172(V) ECM.

ARMAMENT (B-52H): Single 20 mm Vulcan multibarrel cannon in tail turret instead of four machine guns. All aircraft being equipped to carry 12 AGM-86 cruise missiles externally and eight internally on CSRL. See also Upgrades, above.

DIMENSIONS, EXTERNAL:
Wing span 56.39 m (185 ft 0 in)
Length overall 49.05 m (160 ft 10⁹⁄₁₀ in)
Height overall 12.40 m (40 ft 8 in)
Wheel track (c/l of shock struts) 2.51 m (8 ft 3 in)
Wheelbase 15.48 m (50 ft 3 in)

DIMENSIONS, INTERNAL:
Weapons bay volume 29.53 m³ (1,043 cu ft)

AREAS:
Wings, gross 371.6 m² (4,000 sq ft)

WEIGHTS AND LOADINGS:
Max T-O weight more than 221,350 kg (488,000 lb)
Max wing loading approx 595.7 kg/m² (122.0 lb/sq ft)
Max power loading (approx):
G 452.1 kg/kN (4.44 lb/lb st)
H 366.0 kg/kN (3.59 lb/lb st)

PERFORMANCE:
Max level speed at high altitude Mach 0.90 (516 kts; 957 km/h; 595 mph)
Cruising speed at high altitude Mach 0.77 (442 kts; 819 km/h; 509 mph)
Penetration speed at low altitude Mach 0.53 to 0.55 (352-365 kts; 652-676 km/h; 405-420 mph)
Service ceiling 16,765 m (55,000 ft)
T-O run: H 2,900 m (9,500 ft)
Range with max fuel, without in-flight refuelling:
H more than 8,685 n miles (16,093 km; 10,000 miles)

VERIFIED

PRODUCT SUPPORT DIVISION (Division of Boeing Defense & Space Group)

PO Box 7730, Wichita, Kansas 67277
Telephone: 1 (316) 526 3902
Fax: 1 (316) 523 5369
VICE-PRESIDENT AND GENERAL MANAGER: Lee Gantt
PUBLIC RELATIONS MANAGER: Carolyn Russell

Product Support Division formed 1991 for modification and post-production support of Boeing military products. Programmes include support of B-52 Stratofortress, KC-135 tanker/transport, VC-137, TC-43, C-18, C-22 and E-4B; offensive avionics system of Rockwell B-1B; mission data preparation systems; and all-composites replacement wing for US Navy A-6 Intruder. Contracts for 178 new wing sets for A-6E placed by March 1988; final set delivered May 1992. Additional wings required following cancellation of A-12 programme; Boeing to build further 120 wings for final assembly at Seattle; first delivery January 1995.

VERIFIED

Boeing KC-135R strategic flight refuelling tanker/transport aircraft *1994*

Boeing KC-135R refuelling a Rockwell International B-1B Lancer *1994*

BOEING KC-135 STRATOTANKER

TYPE: Strategic flight refuelling tanker/transport with numerous C-135 special-mission variants (not all with tanker capability).

PROGRAMME: Between 1957 and 1965, the US Air Force received 729 KC-135A tankers, 18 C-135A and 30 C-135B transports, 14 EC-135C and three EC-135J command posts, four RC-135A and 10 RC-135C survey aircraft (B and C versions have turbofans). Current US Air Force fleet is 730 aircraft of all versions, including 411 KC-135A/R, two C-135As, one NC-135A, 10 NKC-135As (excluding two for the US Navy), four EC-135As, four C-135Bs, seven WC-135Bs, three C-135Cs, 13 EC-135Cs, four KC-135Ds, three C-135Es, one NKC-135E, four EC-135Es, 163 KC-135Es, four EC-135Gs, four EC-135Hs, four EC-135Js, two EC-135Ks, five EC-135Ls, four EC-135Ps, 54 KC-135Qs, two RC-135Ss, one TC-135S, two RC-135Us, eight RC-135Vs, six RC-135Ws, one TC-135W, one RC-135X and two EC-135Ys.

From 1975 to 7 November 1988 Boeing extended life of every KC-135 beyond year 2020 by replacing sections of underwing skins and other modifications; selection of 97.9 kN (22,000 lb st) CFM56-2B-1 turbofan for evaluation on a KC-135A announced 1980 (see KC-135R below); Air National Guard and AFRes KC-135As and 23 special missions aircraft re-engined with used airline JT3D-3B turbofans between 1981 and 1988 (see KC-135E below).

VERSIONS: **KC-135A:** Initial tanker-transport built by Boeing for the USAF. With four Pratt & Whitney J57-P-59W turbojet engines.

KC-135R: CFM56-powered USAF tanker; first flight 4 August 1982; additional modifications included upgraded electrical and hydraulic systems, performance and fuel management system, upgraded flight control system, strengthened main landing gear, dual APUs for quick engine starting. First nine production conversions funded FY82; 363 modification kits (54,000 parts each) and 320 conversions funded by October 1992; 300th conversion redelivered Summer 1992; power plants on order for further 163 conversions. USAF Systems Command reported to have received funding to test Flight Refuelling Mk 32B hose-reel refuelling pod on one KC-135R. USAF considering Boeing offer of two-man EFIS cockpit modification costing under $1 million.

Compared with KC-135A, KC-135R can offload 65 per cent more fuel at 1,500 n mile (2,775 km; 1,725 mile) radius and 150 per cent more fuel at 2,500 n mile (4,630 km; 2,875 mile) radius departing at average gross weight; take-off run 762 m (2,500 ft) shorter; 90 EPNdB footprint 98 per cent smaller; maximum T-O weight increased from 136,800 kg (301,600 lb) to 146,285 kg (322,500 lb); maximum fuel load increased from 86,047 kg (189,702 lb) to 92,210 kg (203,288 lb).

KC-135E: Urgent re-engining of reserve refuelling squadrons based near built-up areas. USAF bought retired Boeing 707-100B/720B/320B/320C airliners and spare JT3D-3B engines; between 1981 and January 1988, all 138 Air National Guard and Air Force Reserve KC-135As and 23 special missions aircraft re-engined; additional 14 conversions ordered September 1988 and completed by Boeing Louisiana Inc by September 1990, shared between six squadrons; further nine completed by mid-1991; total 184; five-rotor Mark II/III wheel brakes also fitted. Four **KC-135D**s received JT3Ds without change of designation.

C-135FR: See Upgrades.

Boeing C-135FR of the French Air Force with wing-mounted hose and drogue pods installed by Boeing Defense & Space Group *1994*

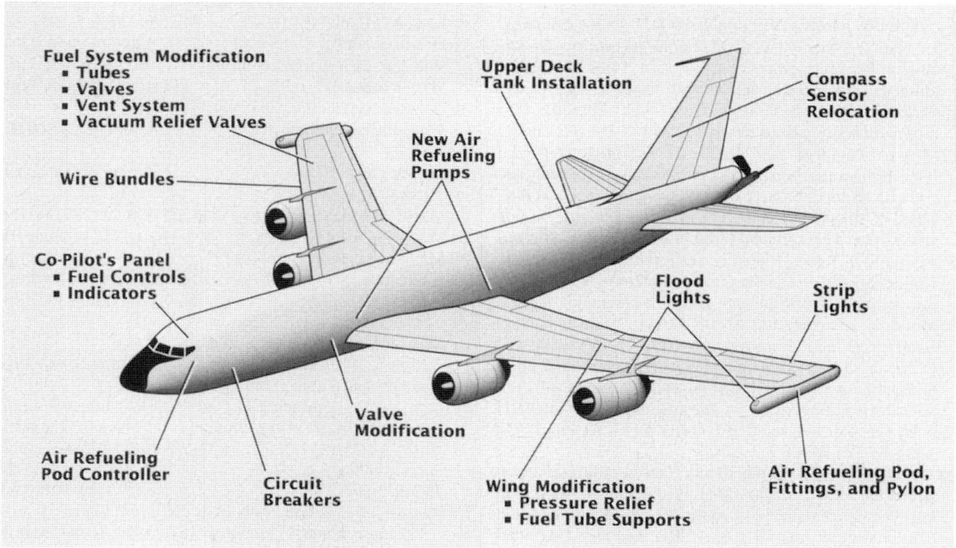

Boeing refuelling pod modification for the C-135FR

1994

Boeing 707-320C converted to tanker/transport

1994

Other versions: See under Programme.

UPGRADES: **Boeing**: Flight Refuelling Ltd Mk 32B wing pods installed under each wing by Boeing from September 1991.

Boeing: In June 1994, Turkey requested the re-work, overhaul and modification of seven of its KC-135A tankers to KC-135R configuration. Turkey received the aircraft from the USAF under the terms of the Southern Region Amendment whereby the aircraft were transferred free of charge but Turkey finances all modifications. In 1995 Turkey reduced the contract to seven aircraft. Turkey received two KC-135A on lease, until the KC-135R aircraft are delivered in 1997. Cost of upgrade is $315 million.

Boeing: 12 **C-135F** delivered to France in 1964; one lost; structural modifications applied to remaining 11; converted to **C-135FR**s and redelivered between 26 August 1985 and 13 April 1988 with CFM56 power plants; Dassault Electronique Adèle radar warning receivers installed from 1990.

DIMENSIONS, EXTERNAL:

Wing span	39.88 m (130 ft 10 in)
Length overall	41.53 m (136 ft 3 in)
Height overall	11.68 m (38 ft 4 in)
Wheel track	6.73 m (22 ft 1 in)
Wheelbase	13.92 m (45 ft 8 in)

WEIGHTS AND LOADINGS (basic tanker version):

Operating weight empty	44,663 kg (98,466 lb)
Design T-O weight	111,130 kg (245,000 lb)
Max T-O weight	134,715 kg (297,000 lb)
Design landing weight	83,915 kg (185,000 lb)

PERFORMANCE (refuelling missions at max T-O weight):

Average cruising speed at 9,300-13,700 m (30,500-45,000 ft)	462 kts (856 km/h; 532 mph)
T-O speed	167 kts (310 km/h; 192 mph)
Rate of climb at S/L	393 m (1,290 ft)/min
Rate of climb at S/L, one engine out	177 m (580 ft)/min
Time to 9,300 m (30,500 ft)	27 min
T-O run	2,760 m (9,050 ft)
T-O field length	3,260 m (10,690 ft)
Landing run at 47,720 kg (105,200 lb) AUW	580 m (1,900 ft)
Transfer radius with 3,055 kg (6,734 lb) fuel reserve	1,000 n miles (1,850 km; 1,150 miles)
Total mission time	5½ h

UPDATED

BOEING 707 TANKER/TRANSPORT CONVERSIONS

Royal Saudi Air Force designation: KE-3A
Spanish Air Force designation: T.17

TYPE: Export tanker modification of 707-320 airliner.

PROGRAMME: Launched 1982; first flight of ex-TWA 707-320C converted demonstrator early 1983; earlier limited conversions completed before 1982.

VERSIONS: See Design Features for choice of alternative quick-change roles.

CUSTOMERS: Only customers for new-build 707 tanker/transports were Saudi Arabia (eight illogically designated **KE-3A**s powered by CFM56-2A-2 engines, delivered September 1987) and Iran (four **707-3J9C**s with Beech 1800 wingtip pods installed by Boeing). Refurbished and converted 707 airliners supplied to Australia (four 707-338Cs with FRL Mk 32 wing pods by ASTA using IAI conversion kits); Brazil (four **KC-137**s with Beech pods, converted by Boeing); Canada (two **CC-137**s with provision for Beech 1800 wingtip pods, converted by Boeing); Israel (six 707s converted by IAI with Sargent-Fletcher Company 34-000 pods); Italy (four 707-382Bs with SFC 34-000 pods and FR 480C fuselage HDU, being converted by Alenia Officine Aeronavali; first flight, by MM62148, 31 October 1990); Morocco (one 707-138B with Beech pods converted by AMIN); Peru (one 707-323C, IAI conversion with FRL pods); Spain (two **T.17**s with SFC 34-000 pods, converted by Boeing); Venezuela (one converted by IAI, 1990).

COSTS: $9.6 million per aircraft (Australia 1990) excluding value of airframe.

DESIGN FEATURES: When not refuelling, track-mounted interiors can be fitted for all-cargo, all-passenger or combi, aeromedical or VIP. Options include nose-mounted receiver probe; pods by Beech, Sargent-Fletcher Company or Flight Refuelling Ltd; rear fuselage hose and drogue unit; KC-135-type centreline boom. Systems can be mixed to give multiple capability.

Optional extra 19,040 litre (5,030 US gallon; 4,188 Imp gallon) underfloor fuel tank in rear cargo hold; with this and standard 90,301 litre (23,855 US gallon; 19,863 Imp gallon) wing tanks and triple refuelling points, 707 tanker/transport can transfer 55,878 kg (123,190 lb) of fuel at 1,000 n mile (1,850 km; 1,150 mile) radius. Other basic modifications include refuelling control panel for flight engineer, improved hydraulic system and fuel pumps, strengthened outer wing, new wingtips, military avionics, TV monitor and boom operator's or pod observer's station. Quick-change cabin allows additional roles, such as coastal patrol, ECM, maritime missions, tactical command and control. Pylons or bays can be provided to dispense sonobuoys or other sensors, mines, bombs, air-to-air or stand-off missiles, and chaff or flares.

VERIFIED

BOEING 767 AWACS

TYPE: Airborne warning and control system successor to E-3.

PROGRAMME: Boeing announced, December 1991, definition studies for modified 767-200ER airliner with Westinghouse AN/APY-2 radar; project sustained by Japanese interest; engine selection October 1992; J-STARS (radar reconnaissance), military transport and tanker versions also possible.

CUSTOMERS: Japanese government purchase decision, December 1992; parliamentary approval given early 1993; total requirement for four. First two ordered November 1993; modification by Boeing Product Support Division at Wichita from January 1995; avionics installation at Seattle from October 1995; seven-month flight-test programme at Seattle from April 1996; mission system equipment installation from October 1996 (No. 2 aircraft first); delivery of first two aircraft in January 1998.

DESIGN FEATURES: For full description see Boeing 767 entry in *Jane's All the World's Aircraft 1994-95*. Additionally, substantial structural modifications (eight frames and nine floor beams) to accommodate rotodome; ventral fins may be added on rear fuselage.

POWER PLANT: Two 273.6 kN (61,500 lb st) General Electric CF6-80C2 turbofans modified for additional electrical generation. In-flight refuelling receptacle.

ACCOMMODATION: Two-man flight crew; up to 18 mission crew.

SYSTEMS: Two 150 kVA generators on each engine replace single 90 kVA units; total 600 kVA.

AVIONICS: Generally as E-3C; capable of accepting latest AWACS upgrades such as cruise missile and UAV detection.

DIMENSIONS, EXTERNAL:

Wing span	47.57 m (156 ft 1 in)
Length overall	48.51 m (159 ft 2 in)
Height overall	15.85 m (52 ft 0 in)
Radome: Diameter	9.14 m (30 ft 0 in)
Thickness	1.83 m (6 ft 0 in)

WEIGHTS AND LOADINGS:

Max T-O weight	171,004 kg (377,000 lb)

PERFORMANCE:

Service ceiling	10,360-13,100 m (34,000-43,000 ft)
Range, unrefuelled	4,500-5,000 n miles (8,340-9,266 km; 5,182-5,758 miles)

Boeing 767 AWACS showing mission equipment and crew station configuration ***(Boeing)***

1995

Endurance:
at 1,000 n mile (1,853 km; 1,150 mile) radius 7 h
at 300 n mile (556 km; 345 mile) radius 10 h
with in-flight refuelling 22 h

VERIFIED

F-4 PHANTOM NWDS UPGRADE

TYPE: Combat aircraft upgrade.

PROGRAMME: Boeing Military Airplanes won a contract in 1986 to develop and install a modern digital avionics suite and weapons delivery system in more than 300 Tactical Air Command F-4Es and RF-4Cs.

DESIGN FEATURES: The NWDS (navigation and weapons delivery system) upgrade includes a databus link to a ring laser gyro inertial navigation unit, a new mission computer, avionics interface unit, data transfer system and new cockpit controls and displays. Using a technology employed in B-1B, F-16 and A-6 aircraft, the NWDS enables F-4E aircrews to deliver weapons with accuracy equal to newer front-line fighter-bombers. The system's ring laser gyro computes the aircraft's position and allows for precision navigation reconnaissance and accurate weapons delivery.

Work was undertaken at the aircraft's home base, the first return to service being an F-4E of the 163rd TFS, Indiana ANG, at Fort Wayne on 8 September 1989. Withdrawn from service.

VERIFIED

BOEING HELICOPTERS (Division of Boeing Defense & Space Group)

Boeing Center, PO Box 16858, Philadelphia, Pennsylvania 19142
Telephone: 1 (215) 591 2700
Fax: 1 (215) 591 2701
Telex: 510 669 2217 BOEMORA MOR
PRESIDENT AND GENERAL MANAGER: Denton Hanford
VICE-PRESIDENT, RESEARCH AND ENGINEERING: William Walls
MANAGER, PUBLIC RELATIONS: Madelyn Bush

Vertol Aircraft Corporation (formerly Piasecki Helicopter Corporation) purchased in 1960, becoming Vertol Division of Boeing; now Helicopter Division of Boeing Defense & Space Group. Produced more than 2,500 tandem-rotor helicopters for US military services and export. Main production programme, modernisation of early CH-47s to CH-47D for US Army. Production of H-46 Sea Knight SR&M improvement kits for US Navy and Marine Corps completed January 1989.

Boeing Helicopters teamed with Bell Helicopter Textron in V-22 Osprey programme. Joined First Team with Sikorsky Aircraft 3 June 1985 to compete for US Army LH (RAH-66 Comanche) light helicopter, with Boeing integrating avionics; declared competition winner April 1991. All-composite rotor blade for Bell UH-1H developed for US Army; first blade delivered June 1989 to Army Engineering Flight Agency for interchangeability check with Bell-produced blade.

Boeing Helicopters produces fixed leading-edge components for Boeing 737/747/757/767 and metal leading-edge slats for 757.

Workforce 7,000 in early 1994. Manufacturing plant at Ridley Township, Pennsylvania, has 325,150 m² (3,500,000 sq ft) of covered floor space; flight test centre at Greater Wilmington, Delaware, has 8,565 m² (92,200 sq ft). An 11,055 m² (119,000 sq ft) development facility and 17,466 m² (188,000 sq ft) office/computer centre added at Ridley Township early 1987.

VERIFIED

Boeing CH-46E Sea Knight of the USMC with (SR&M) safety, reliability and maintenance retrofits *1994*

BOEING MODEL 107

USN and USMC designations: CH-46/UH-46 Sea Knight
Swedish military designation: Hkp 4B/C
Other military designations: KV-107

TYPE: Twin-engine transport helicopter.

PROGRAMME: In 1956, Vertol began preliminary design and engineering of a twin-turbine transport helicopter for commercial and military applications. The main objective was to take full advantage of the high power, small size and light weight of the shaft-turbine engines, then becoming available. To achieve the best possible hovering performance, the traditional Vertol tandem-rotor layout was retained, and the turbines were mounted above the rear of the cabin, on each side of the aft rotor pylon. This resulted in maximum unobstructed cabin area and permits the use of a large rear ramp for straight-in loading of vehicles and bulky freight.

Construction of a prototype, designated Model 107, was started in May 1957, and this aircraft flew for the first time on 22 April 1958, powered by two 641 kW (860 shp) Lycoming T53 turboshaft engines. It was designed for water landing capability, without the addition of special flotation gear or boat hull design, and was intended to carry 23-25 passengers in normal airline standard accommodation.

As a result of experience with this prototype, including extensive demonstration tours in North America, Europe and the Far East, several advanced versions were developed. The current versions in service with the US armed forces are the UH/CH-46D (USN 40+) and UH/CH-46-E (USMC 244+). Versions in service with the Canadian armed forces are CH-113/A (13). Versions in service with the Swedish Navy are 107 HPK-4 (10).

VERSIONS: **107 Model II:** Standard commercial version, with two 932 kW (1,250 shp) General Electric CT58 turboshaft. Available as an airliner with roll-out rear baggage container or utility model with rear-loading ramp. A prototype, modified at company expense from one of the three YHC-1A (CH-46A) military helicopters built for evaluation by the US Army, flew for the first time 25 October 1960, followed by the first production model on 19 May 1961. FAA certification was received 26 January 1962. Eight, with 25-seat interior were delivered subsequently to New York Airways, who introduced the 107 Model II to their scheduled services 1 July 1962. Of the eight, three were purchased by Pan American World Airways, who leased them to New York Airways. Kawasaki received permission to licence build the CH-46 in December 1965.

107 Model IIA: In Spring 1968, Kawasaki obtained Japanese and US Type certification of the 107 Model IIA airliner, with 1,044 kW (1,400 shp) General Electric CT58-140-1 turboshaft engines, to increase performance at high operating temperatures.

CH-46A (formerly HRB-1) **Sea Knight:** This US Marine Corps assault transport version of the 107 Model II has the specified military mission of carrying a crew of three and 17-25 fully equipped troops or 1,814 kg (4,000 lb) of cargo over a combat radius of 100 n miles (185 km; 115 miles) at 129 knots (240 km/h; 150 mph). An initial batch of 14 was ordered in February 1961, followed by annual repeat orders which brought the total number ordered to 600 by the end of 1970, including CH-46Ds and CH-46Fs. An integrated loading system permits rapid loading, by one man, under field conditions. A powered blade-folding system enables the rotor blades to be folded quickly by a pilot-operated control, to simplify handling on board aircraft carriers. The CH-46A is powered by two 932 kW (1,250 shp) General Electric T58-GE-8B turboshafts, and has all weather capabilities. In a rescue role, it can retrieve 20 persons up to 91 n miles (168 km; 105 miles) from its base and can carry 15 stretcher patients and two attendants. First CH-46A flew on 16 October 1962. The US Navy Board of Inspection and Survey tests required for fleet release were completed in November 1964, and four Marine Squadrons were operating CH-46As by June 1965. In September 1965, the US Department of Defense ordered Boeing to increase production of the CH-46/D by 100 per cent. The CH-46 commenced service in Vietnam in March 1966. Withdrawn from service.

CH-46D Sea Knight: Generally similar to CH-46A, but with 1,044 kW (1,400 shp) General Electric T58-GE-10 turboshaft and cambered rotor blades. CH-46s delivered from August 1966 to June 1968 were of this version. A CH-46D Sea Knight handed over to the US Marine Corps in August 1968 was the 1,000th twin turbine tandem-rotor helicopter to be completed by Boeing. Withdrawn from service.

CH-46E Sea Knight: Upgraded CH-46A with 1,394 kW (1,870 shp) General Electric T58-GE-16 turboshafts and other modifications including provision of crash attenuating seats for pilot and co-pilot, a crash and combat resistant fuel system, and improved rescue system. Initial fleet modifications began in 1977, and the first CH-46E modified at Cherry Point NV, Naval Rework Facility was rolled out 3 August 1977.

CH-46F Sea Knight: Generally similar to CH-46D, with the same engines and rotor blades. Contains additional electronics equipment. All CH-46s delivered since July 1968 were of this version. Withdrawn from service.

UH-46A Sea Knight: Similar to CH-46A. Ordered by US Navy for operation from AFS or AOE combat supply ships to transport supplies, ammunition and missiles as

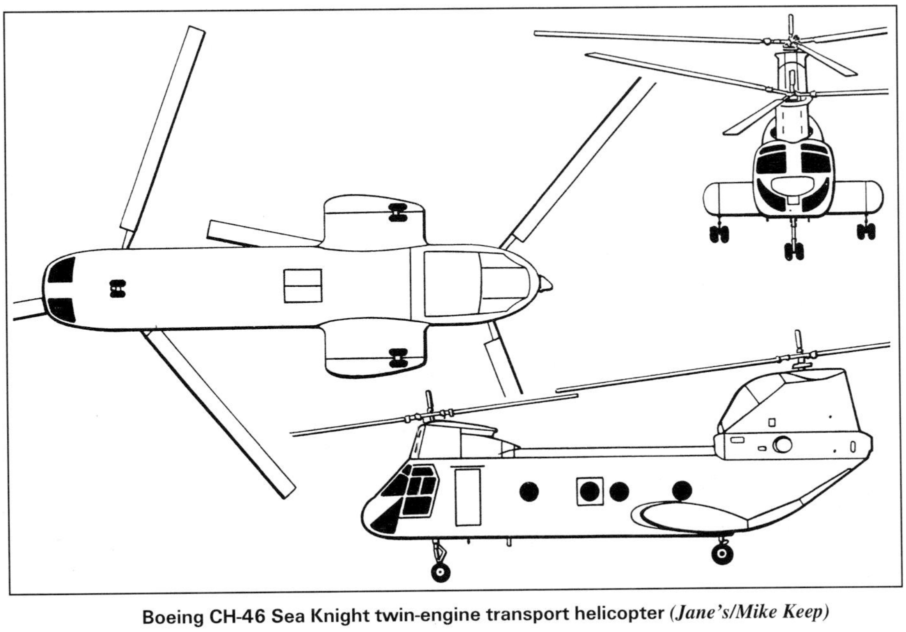

Boeing CH-46 Sea Knight twin-engine transport helicopter *(Jane's/Mike Keep)* *1994*

Boeing CH-46E Sea Knight of the USS *Milwaukee*

1994

well as aviation spares to combatant vessels underway at sea. Secondary tasks include transfer of personnel, and SAR. First deliveries to Utility Helicopter Squadron One, Ream Field, California in July 1964, followed by deliveries to Utility Helicopter Squadron Four at Norfolk in December 1964. Withdrawn from service.

UH-46D Sea Knight: Generally similar to UH-46A, but with 1,044 kW (1,400 shp) General Electric T58-GE-10 shaft turbine engines and cambered rotor blades. UH-46s delivered since September 1966 were of this version.

CH-113 Labrador: Six utility models delivered to RCAF in 1963-64 for SAR duties. Generally similar to CH-46A. Two 932 kW (1,250 shp) General Electric T58-GE-8B turboshafts. Larger capacity fuel tanks (total 3,408 litres; 900 US gallons; 750 Imp gallons) giving a range of over 567 n miles (1,050 km; 650 miles).

CH-113A Voyageur: Twelve aircraft in a similar configuration to that of CH-46A, were delivered to Canadian Army in 1964-65. They are used as troop and cargo carriers in logistical and tactical missions.

Hkp 4C: Built for Royal Swedish Navy (45) and Air Force (10) in 1962-63, with Bristol Siddeley Gnome H.1200 turboshaft engines and fuel tanks of 3,786 litres (1,000 US gallons; 832 Imp gallons) capacity. Naval version has equipment for anti-submarine and mine countermeasures operations. Since upgraded with Gnome H.1400 turboshafts and new avionics.

HH-46D: Rescue version in service with the US Navy.

UH-46D: Base utility and rescue helicopter. In service with the US Navy.

KV-107II/IIA: Kawasaki-built versions of the Model 107 manufactured in Japan.

UPGRADES: **Boeing**: CH-46E. See Versions.

Boeing Defense & Space: Awarded contract Feb 1995 by US Navy to supply 120 Dynamic Components Upgrade (DCU) kits, Lot III of a contract to supply 400 DCU kits.

Boeing of Canada: Under a Canadian government contract Boeing of Canada Ltd has upgraded six CH-113 and five CH-113A helicopters to a single, improved maritime search and rescue standard under a three-phase Canadian Forces programme known as SARCUP (Search And Rescue Capability Upgrading Programme). An improved rescue hoist is also fitted.

The third and final phase of SARCUP began concurrently with the second (Speedline) phase, and the first two aircraft (one CH-113 Labrador and one CH-113A Voyageur) were redelivered to the CAF in the Autumn of 1982. Following their acceptance the nine remaining aircraft which underwent the earlier interim avionics programme and/or Speedline phases were cycled through Boeing of Canada. This phase required each helicopter to remain in plant for about seven months, with the final delivery scheduled for June 1984.

In October 1983 Boeing of Canada was awarded a Canadian government contract to upgrade three more CH-113As to the SARCUP configuration. This programme, known as VOFUP (Voyageur Follow-on Update Programme).

The SARCUP improvements extend the aircraft's range, and enable them to fly search and rescue missions in adverse weather conditions by day and by night. Production of the improvement kits, aircraft modification and flight testing are undertaken by Boeing of Canada Ltd's Arnprior Division.

OPERATORS: Versions of the Model 107 are in service with the armed forces of the following countries: Canada (22), Japan (100), Saudi Arabia (17), Sweden (14) and USA (297).

DESIGN FEATURES: Two three-blade rotors in tandem, rotating in opposite directions. The CH/UH-46 has power-operated blade folding. Power is transmitted from each engine through individually overrunning clutches into the aft transmission, which combines the engine outputs, thereby providing a single power output to the interconnecting shaft which enables both rotors to be driven by either engine.

STRUCTURE: Square-section stressed-skin semi-monocoque structure built primarily of high strength bare and alclad aluminium alloy. Transverse bulkheads and built-up frames support transmission, power plant and landing gear. Loading ramp forms undersurface of upswept rear fuselage on utility and military models. Baggage container replaces ramp on airliner version. Fuselage is sealed to permit operation from water.

LANDING GEAR: Non-retractable tricycle type, with twin-wheels on all three units. Oleo-pneumatic shock-absorbers manufactured by Loud (main gear) and Jarry (nose gear). Goodyear tubeless tyres size 8 × 5.5, pressure 10.55 kg/cm² (150 lb/sq in), on all wheels. Goodyear disc brakes.

POWER PLANT: See Versions.

ACCOMMODATION (107 Model II): Standard accommodation for two pilots, stewardess and 25 passengers in airliner. Seats in eight rows, in pairs on port side and single seats on starboard side (two pairs at rear of cabin) with central aisle. Airliner fitted with parcel rack and a roll-out baggage container, with capacity of approximately 680 kg (1,500 lb), located in underside of rear fuselage. Ramp of utility model is power-operated on the ground or in flight and can be removed or left open to permit carriage of extra long cargo.

ACCOMMODATION (CH/UH-46): Crew of three, 25 troops and troop commander. Door at front of troop compartment on starboard side. Door is split type; upper half rolls on tracks to stowed position in fuselage crown, lower half is hinged at the bottom and opens outward, with built-in steps. Loading ramp and hatch at rear of fuselage can be opened in flight or on the water. Floor has centre panel stressed for 1,464 kg/m² (300 lb/sq ft). A row of rollers on each side for handling standard military pallets or wire baskets. Outer portion of floor is vehicle treadway stressed for 454 kg (1,000 lb) rubber-tyred wheel loads. Cargo and personnel hoist system includes a variable-speed winch capable of 907 kg (2,000 lb) cable pull at 9 m (30 ft)/min for cargo loading or 272 kg (600 lb) cable pull at 30 m (100 ft)/min for personnel hoisting, it can be operated by one man. A 4,535 kg (10,000 lb) capacity hook for external loads is installed in a cargo hatch in the floor.

SYSTEMS (CH/UH-46): Cabin heated by Janitrol combustion heater. Hydraulic system provides 105 kg/cm² (1,500 lb/sq in) pressure for flying control boost, 210 kg/cm² (3,000 lb/sq in) for other services. Electrical system includes two 40 kVA AC generators and a Leland 200 A DC generator. Solar APU provides power for starting and systems check-out.

ELECTRONICS AND EQUIPMENT: Blind-flying instrumentation standard. CH-46 has dual stability augmentation systems and automatic trim system.

DIMENSIONS, EXTERNAL:

Main rotor diameters (each):

107-II, CH-113, CH/UH-46A	15.24 m (50 ft 0 in)
CH/UH-46D, CH-46F	15.54 m (51 ft 0 in)
Distance between rotor centres	10.16 m (33 ft 4 in)
Length overall, blades turning:	
107-II, CH-113, CH/UH-46A	25.40 m (83 ft 4 in)
CH/UH-46D, CH-46F	25.70 m (84 ft 4 in)
Length of fuselage: 107-II, CH-113	13.59 m (44 ft 7 in)
CH/UH-46A, CH-46F	13.66 m (44 ft 10 in)
Width, rotors folded	4.42 m (14 ft 6½ in)
Height to top of rear rotor hub	5.09 m (16 ft 8½ in)
Wheel track	3.92 m (12 ft 10½ in)
Wheelbase	7.57 m (24 ft 10 in)
Passenger door (fwd): Height	1.60 m (5 ft 3 in)
Width	0.91 m (3 ft 0 in)

DIMENSIONS, INTERNAL:

Cabin, excl flight deck: Length	7.37 m (24 ft 2 in)
Normal width	1.83 m (6 ft 0 in)
Max height	1.83 m (6 ft 0 in)
Floor area: 107-II	13.47 m² (145 sq ft)
CH/UH-46A and D, CH-46F and CH-113 (incl ramp)	16.72 m² (180 sq ft)
Volume (usable): 107-II, CH-113, CH/UH-46A and D and CH-46F	24.5 m³ (865 cu ft)

AREAS:

Main rotor blade (each):	
107-II, CH-113, CH/UH-46A	3.48 m² (37.50 sq ft)
CH/UH-46D, CH-46F	3.70 m² (39.85 sq ft)
Main rotor disc (total):	
107-II, CH-113, CH/UH-46A	364.6 m² (3,925 sq ft)
CH/UH-46D, CH-46F	379.6 m² (4,086 sq ft)

WEIGHTS AND LOADINGS:

Weight empty, equipped: 107-II	4,868 kg (10,732 lb)
CH-113	5,104 kg (11,251 lb)
CH/UH-46A	5,627 kg (12,406 lb)
CH/UH-46D	5,927 kg (13,067 lb)
CH-46F	6,051 kg (13,342 lb)
Mission T-O weight: CH-113	8,797 kg (19,394 lb)
Max T-O and landing weight:	
107-II	8,618 kg (19,000 lb)
CH-113, CH/UH-46A	9,706 kg (21,400 lb)
CH/UH-46D, CH-46F	10,433 kg (23,000 lb)
Max disc loading: 107-II	23.60 kg/m² (4.84 lb/sq ft)
CH-113, CH/UH-46A	26.61 kg/m² (5.45 lb/sq ft)
CH/UH-46D, CH-46F	27.48 kg/m² (5.63 lb/sq ft)
Max power loading: 107-II	3.54 kg/shp (7.8 lb/shp)
CH-113, CH/UH-46A	3.98 kg/shp (8.77 lb/shp)
CH/UH-46D, CH-46F	4.00 kg/shp (8.84 lb/shp)

PERFORMANCE (107-II and CH/UH-46A at AUW of 8,618 kg; 19,000 lb, CH-113 at AUW of 8,482 kg; 18,700 lb, CH/UH-46D and CH-46F at AUW of 9,435 kg; 20,800 lb):

Max permissible speed:	
107-II, CH-113	145 kts (270 km/h; 168 mph)
CH/UH-46A	138 kts (256 km/h; 159 mph)
CH/UH-46D, CH-46F	144 kts (267 km/h; 166 mph)
Max cruising speed:	
107-II, CH-113	136 kts (253 km/h; 158 mph)
CH/UH-46A	135 kts (249 km/h; 155 mph)
CH/UH-46D, CH-46F	143 kts (266 km/h; 165 mph)
Econ cruising speed:	
107-II, CH-113	125 kts (232 km/h; 144 mph)
CH/UH-46A	131 kts (243 km/h; 151 mph)
CH/UH-46D, CH-46F	134 kts (248 km/h; 154 mph)
Max rate of climb at S/L (normal rated power):	
107-II	439 m (1,440 ft)/min
CH-113	465 m (1,525 ft)/min
CH/UH-46A	438.7 m (1,439 ft)/min
CH/UH-46D, CH-46F	523 m (1,715 ft)/min
Service ceiling (normal rated power):	
107-II	3,960 m (13,000 ft)
CH-113	4,265 m (14,000 ft)
CH/UH-46A	3,960 m (13,000 ft)
CH/UH-46D, CH-46F	4,265 m (14,000 ft)
Service ceiling (military power), one engine out, yaw, 248 rpm: 107-II	107 m (350 ft)
CH-113	366 m (1,200 ft)
CH/UH-46A	S/L
CH/UH-46D, CH-46F	320 m (1,050 ft)
Hovering ceiling in ground effect:	
107-II	2,560 m (8,400 ft)
CH-113	2,985 m (9,800 ft)
CH/UH-46A	2,765 m (9,070 ft)
CH/UH-46D, CH-46F	2,895 m (9,500 ft)
Hovering ceiling out of ground effect:	
107-II, CH-113	2,012 m (6,600 ft)
CH/UH-46A	1,707 m (5,600 ft)
CH/UH-46D, CH-46F	1,753 m (5,750 ft)

Ranges: 107-II utility with 3,000 kg (6,600 lb) payload, 10% fuel reserve 94 n miles (175 km; 109 miles)

CH-113 with 907 kg (2,000 lb) payload to hover IGE at 3,090 m (10,150 ft), 10% fuel reserve 577 n miles (1,070 km; 665 miles)

CH-113 with 2,270 kg (5,000 lb) payload to hover IGE at 4,070 m (13,350 ft), 10% fuel reserve 405 n miles (751 km; 467 miles)

CH/UH-46A at AUW of 8,722 kg (19,229 lb) with 1,815 kg (4,000 lb) payload, 10% fuel reserve 199 n miles (370 km; 230 miles)

CH/UH-46A at AUW of 9,706 kg (21,400 lb) with 2,753 kg (6,070 lb), 10% fuel reserve 199 n miles (370 km; 230 miles)

CH/UH-46D at AUW of 9,435 kg (20,800 lb) with 2,064 kg (4,550 lb) payload, 10% fuel reserve 206 n miles (383 km; 238 miles)

CH/UH-46D at AUW of 10,433 kg (23,000 lb) with 3,062 kg (6,750 lb) payload, 10% fuel reserve 198 n miles (367 km; 228 miles)

CH-46F at AUW of 9,435 kg (20,800 lb) with 1,939 kg (4,275 lb) payload, 10% fuel reserve 206 n miles (383 km; 238 miles)

CH-46F at AUW of 10,433 kg (23,000 lb) with 2,937 kg (6,475 lb) payload, 10% fuel reserve 198 n miles (367 km; 228 miles)

VERIFIED

BOEING MODELS 114 and 414

US Army designation: CH-47 Chinook
Canadian Forces designation: CH-147
Royal Air Force designation: Chinook HC. Mk 1/2
Spanish Army designation: HT.17

TYPE: Tandem-rotor, twin-turbine transport helicopter.

PROGRAMME: Design of all-weather medium transport helicopter for US Army began 1956; first flight of first of five YCH-47As 21 September 1961.

VERSIONS: **CH-47A:** Initial production version, powered by two 1,640 kW (2,200 shp) Lycoming T55-L-5 or 1,976 kW (2,650 shp) T55-L-7 turboshaft engines. Operation of the CH-47A by the Vietnamese Air Force (VNAF) began in 1971, four delivered to Royal Thai Air Force. Withdrawn from service.

CH-47B: Developed version with 2,125 kW (2,850 shp) T55-L-7C turboshaft engines, redesigned rotor blades with cambered leading-edge, blunted rear rotor pylon, and strakes along ramp and fuselage for improved flying qualities. First of two prototypes flew for the first time early October 1966. Deliveries began 10 May 1967. Withdrawn from service.

CH-47C: Developed version with uprated transmissions and 2,796 kW (3,750 shp) T55-L-11A; integral fuel capacity increased to 3,944 litres (1,042 US gallons; 867.6 Imp gallons); first flight 14 October 1967; 270 delivered to US Army from Spring 1968; 182 US Army CH-47Cs retrofitted with composite rotor blades; integral spar inspection system (ISIS) introduced 1973 together with crashworthy fuel system retrofit kit. Transmissions of some As and Bs upgraded to CH-47C standard.

CH-47D: US Army contract to modify one each of CH-47A, B and C to prototype Ds placed 1976; first flight 11 May 1979; first production contract October 1980; first flight 26 February 1982; initial operational capability (IOC) achieved 28 February 1984 with 101st Airborne Division; second multiyear production contract for 144 CH-47Ds awarded 13 January 1989, bringing total CH-47D (and MH-47E) ordered to 472; deliveries reached 447 by December 1993; production rate four per month; programme completion was due October 1993; US regular Army deliveries completed October 1990 with 17th operating unit (C Company, 228th Aviation Regiment, Fort Wainright, Alaska); deliveries to National Guard (Texas) began 1988.

CH-47D update includes strip down to bare airframe, repair and refurbish, fit Textron Lycoming T55-L-712 turboshafts, uprated transmissions with integral lubrication and cooling, composite rotor blades, new flight deck compatible with night vision goggles (NVG), new redundant electrical system, modular hydraulic system, advanced automatic flight control system, improved avionics and survivability equipment, Solar T62-T-2B APU operating hydraulic and electrical systems through accessory gear drive, single-point pressure refuelling, and triple external cargo hooks. Composites account for 10 to 15 per cent of structure. About 300 suppliers involved.

Boeing CH-47C Chinook of the Italian Army *(Peter J. Cooper)*

1994

At maximum gross weight of 22,680 kg (50,000 lb), CH-47D has more than double useful load of CH-47A. Sample loads include M198 towed 155 mm howitzer, 32 rounds of ammunition and 11-man crew, making internal/external load of 9,980 kg (22,000 lb); D5 caterpillar bulldozer weighing 11,225 kg (24,750 lb) on centre cargo hook; US Army Milvan supply containers carried at up to 130 knots (256 km/h; 159 mph); up to seven 1,893 litre (500 US gallon; 416 Imp gallon), 1,587 kg (3,500 lb) rubber fuel blivets carried on three hooks. Four CH-47Ds converted for in-flight refuelling from C-130 at up to 120 knots (222 km/h; 138 mph) by day and night and in moderate turbulence, approved July 1988; graphite fuel boom at lower starboard side contains telescoping aluminium tube and can accept flow rate of 568 litres (150 US gallons; 125 Imp gallons)/min to refuel completely in six minutes; first delivery to US Army July 1988. The final remanufactured CH-47D rolled off the assembly line at Boeing Helicopters 15 April 1994. For full details, specifications and dimensions see *Jane's All the World's Aircraft 1995-96.*

CH-47D Special Operations Aircraft: Two battalions of 160th Special Operations Aviation Regiment (at Fort Campbell, Kentucky, and Hunter AAF, Georgia) equipped pending availability of MH-47E with CH-47D SOA fitted with refuelling probes, thermal imagers, Bendix/King RDR-1300 weather radar, improved communications and navigation equipment, and two pintle-mounted 7.62 mm machine guns. Navigator/commander's station fitted in some SOAs. At least two (83-24110 and 83-24118) converted to **MH-47D**.

GCH-47D: At least 12 Chinooks grounded for engineer training.

MH-47E: Special Forces variant. Prototype development contract 2 December 1987; 11 of intended 51 MH-47Es funded in FY88 and FY89 and taken from CH-47D remanufacture; prototype (88-0267) flew 1 June 1990; first 11 (of 24 intended) delivered from November 1992 to 2 Battalion of 160th Special Operations Aviation Regiment at Fort Campbell, Kentucky. Later helicopters earmarked for 3 Battalion/160 SOAR at Hunter AAF, Georgia, and 1/245th Aviation Battalion (SOA), Oklahoma National Guard, Lexington (eight and 16 MH-47Es respectively).

Mission profile 5½ hour covert deep penetration over 300 n mile (560 km; 345 mile) radius in adverse weather, day or night, all terrain with 90 per cent success probability. Requirements include self-deployment to Europe in stages up to 1,200 n miles (1,677 km; 1,042 miles), 44-troop capacity, powerful defensive weapons and ECM. Equipment includes IBM-Allied Bendix integrated avionics with four-screen NVG-compatible EFIS; dual MIL-STD-1553 digital databuses; AN/ASN-145 AHRS; jamming-resistant radios; Rockwell-Collins CP1516-ASQ automatic target handoff system; inertial AN/ASN-137 Doppler, Rockwell-Collins AN/ASN-149(V)2 GPS/Navstar receiver and terrain referenced positioning navigation systems; Rockwell-Collins ADF-149; laser (Perkin-Elmer AN/AVR-2), radar (E-Systems AN/APR-39A) and missile (Honeywell AN/AAR-47) warning systems; ITT AN/ALQ-136(V) pulse jammer and Northrop AN/ALQ-162 CW jammer; Tracor M-130 chaff/flare dispensers; TI AN/APQ-174 radar with modes for terrain-following down to 30 m (100 ft), terrain-avoidance, air-to-ground ranging and ground mapping; Hughes AN/AAQ-16 FLIR in chin turret; digital moving map display; uprated T55-L-714 turboshafts with FADEC; increased fuel capacity; additional troop seating (44 maximum); OBOGS; rotor brake; 272 kg (600 lb) rescue hoist with 61 m (200 ft) usable cable; two M-2 0.50 in window-mounted machine guns (port forward: starboard aft); provisions for Stinger AAMs using FLIR for sighting. This system largely common with equivalent Sikorsky MH-60K.

Chinook HC. Mk 1/2: RAF versions (35 of 41 remaining); designation CH-47-352; all HC. Mk 1s upgraded to HC. Mk 1B; UK MoD authorised Boeing to update 33 Mk 1Bs to Mk 2, equivalent to CH-47D, October 1989; changes include new automatic flight control system, updated modular hydraulics, stronger transmission, improved T62-T-2B APU, airframe reinforcements, low IR paint scheme, long-range fuel system and standardisation of defensive aids package (IR jammers, chaff/flare dispensers, missile approach warning and machine gun mountings). Conversion continues from 1991 to July 1995. Chinook HC. Mk 1B ZA718 began flight testing Chandler Evans/Hawker Siddeley dual-channel, FADEC system for Mk 2 in October 1989. Same helicopter to Boeing, March 1991; rolled out as first Mk 2 19 January 1993; to UK for A&AEE trials, March 1993.

HT.17 Chinook: Spanish Army version.

Boeing 234: Commercial version, now out of production. See separate entry.

Boeing 414: Export military version. Now superseded by CH-47D International Chinook.

CH-47D International Chinook: Boeing 414-100 first sold to Japan; Japan Defence Agency ordered two for JGSDF and one for JASDF Spring 1984; first flight (N7425H) January 1986 and, with second machine, delivered to Kawasaki April 1986 for fitting out; co-production arrangement by Spring 1994, 15 in orders for JASDF and 36 for JGSDF; final total of at least 54

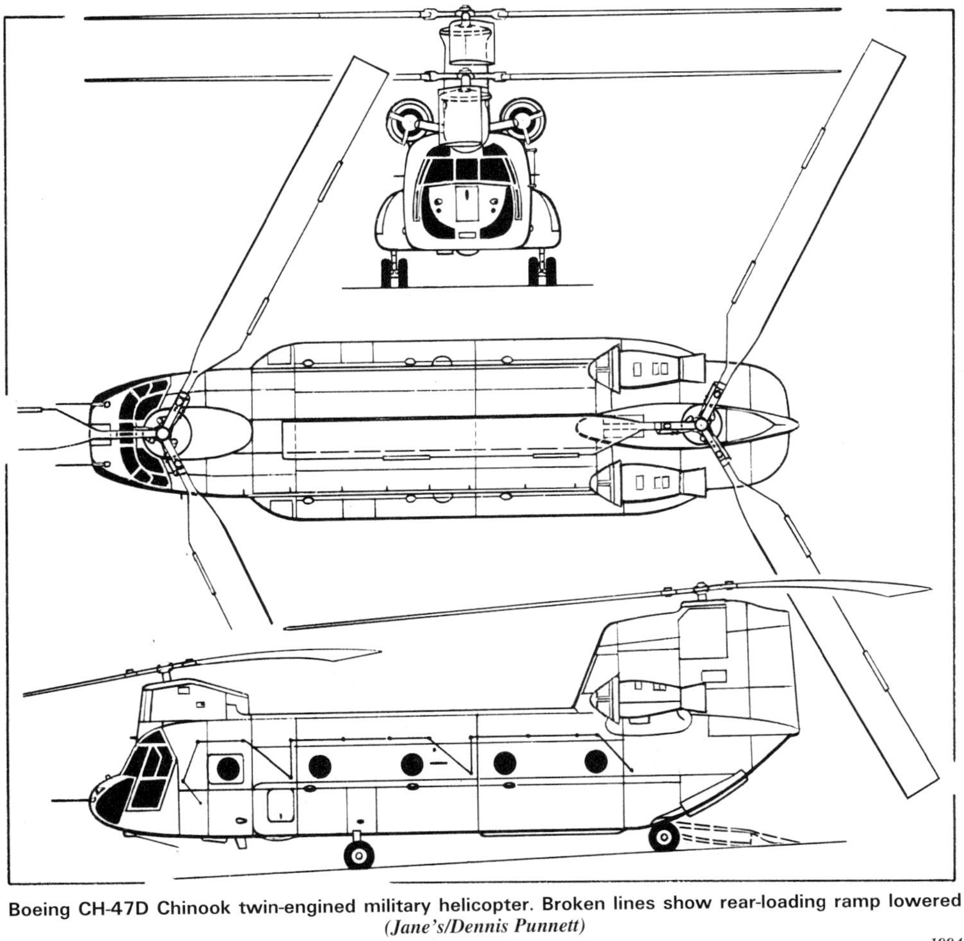

Boeing CH-47D Chinook twin-engined military helicopter. Broken lines show rear-loading ramp lowered *(Jane's/Dennis Punnett)*

1994

Boeing CH-47D Chinook carrying MLRS launch tubes

1994

Boeing MH-47E prototype showing extended nose and refuelling probe ***(Boeing)***

1995

expected. International Chinook available in four versions with combinations of standard or long-range (MH-47E type fuel tanks and T55-L-712 SSB or T55-L-714.

Advanced Chinook: Chinook upgrade study with 3,729 kW (5,000 shp) class engines, redesigned rotor blades and hubs, and additional (MH-47E) fuel.

UPGRADES: **Boeing:** CH-47C. See Versions.

Boeing: HC. Mk 2. See Versions.

Boeing: CH-47D. In June 1993 the Dutch MoD acquired seven former Canadian CH-47 and awarded a contract to Boeing to upgrade them to CH-47D standard. First aircraft redelivered July 1995.

Boeing: Advanced Chinook. Chinook upgrade study with 3,729 kW (5,000 shp) class engines, redesigned rotor blades and hubs, and additional (MH-47E) fuel.

Boeing: CH-47D. See Versions.

Boeing: MH-47E. Special Forces variant. See Versions.

The following description applies to the CH-47C:

DESIGN FEATURES: Two three-blade rotors, rotating in opposite directions and driven through interconnecting shaft which enable both rotors to be driven by either engine. Rotor blades, of a modified NACA 00123 section. Two blades of each rotor can be folded manually. All bearings are submerged completely in oil. Power is transmitted from each engine through individual overrunning clutches, into the combiner transmission, thereby providing a single power output to the interconnecting shafts. Rotor/engine rpm ratio 64:1.

FLYING CONTROLS: Rotor heads are fully articulated, with pitch, flapping and drag hinges.

STRUCTURE: Rotor blades have cambered leading-edge, a strengthened steel spar structure and honeycomb-filled trailing-edge boxes. The fuselage is a square-section all-metal semi-monocoque structure. Loading ramp forms undersurface of upswept rear fuselage. Fairing pods along bottom of each side are made of metal honeycomb sandwich and are sealed and compartmented, as is the underfloor section of the fuselage for buoyancy during operation from water.

LANDING GEAR: Menasco non-retractable quadricycle type, with twin-wheels on each forward unit and single wheels on each rear unit. Oleo-pneumatic shock-absorbers on all units. Rear units fully castoring and steerable; power steering installed on starboard rear unit. All wheels are government-furnished size 24 × 7.7-VII, with tyres size 8.50-10-III, pressure 4.62 bars (67 lb/sq in). Two single-disc hydraulic brakes. Provision for fitting detachable wheel-skis.

POWER PLANT: Two 2,796 kW (3,750 shp) Lycoming T55-L-11A turboshaft engines, mounted on each side of rear rotor pylon. Combined transmission rating 5,369 kW (7,200 shp); maximum single-engine transmission limit 3,430 kW (4,600 shp). Self-sealing fuel tanks in external pods on sides of fuselage. Total fuel capacity is 4,137 litres (1,093 US gallons; 910 Imp gallons), or 3,944 litres (1,042 US gallons; 867 Imp gallons) when equipped with Crashworthy Fuel System. Refuelling points above tanks. Total oil capacity 14 litres (3.7 US gallons; 3 Imp gallons).

ACCOMMODATION: Two pilots on flight deck, with dual controls. Jump seat is provided for crew chief or combat commander. Jettisonable door on each side of flight deck. Depending on seating arrangement, 33 to 44 troops can be accommodated in main cabin, or 24 stretchers plus two attendants, or vehicles and freight. Typical loads include a complete artillery section with crew and ammunition. All components of the Pershing missile system are transportable by Chinooks. Extruded magnesium floor designed for distributed load of 1,465 kg/m² (300 lb/sq ft) and concentrated load of 1,136 kg (2,500 lb) per wheel in tread position. Floor contains 83 × 2,270 kg (5,000 lb) tiedown fittings and eight × 4,540 kg (10,000 lb) fittings. Rear-loading ramp can be left completely or partially open, or can be removed to permit transport of extra long cargo and in-flight parachute or free drop delivery of cargo and equipment. Main cabin door, at front on starboard side, comprises upper hinged section which can be opened in flight and lower section with integral steps. Lower section is jettisonable. Up to 12,700 kg (28,000 lb) can be carried on external cargo hook.

SYSTEMS: Cabin heated by 200,000 BTU heater-blower. Hydraulic system provides pressures of 207 bars (3,000 lb/sq in) for flying controls, and 276 bars (4,000 lb/sq in) for engine starting. Electrical system includes two 20 kVA alternators driven by transmission drive system. Solar T62 APU runs accessory gear drive, thereby operating all hydraulic and electrical systems.

AVIONICS AND EQUIPMENT: All government furnished, including UHF communications and FM liaison sets, transponder, intercom, omni-receiver, ADF and marker beacon receiver. Blind-flying instrumentation standard. Special equipment includes dual electrohydraulic stability augmentation system, automatic/manual speed trim system, hydraulically powered winch for rescue and cargo handling purposes, cargo and rescue hatch in floor, integral work stands and steps for maintenance, rearview mirror, provisions for paratroops' static line and for maintenance davits for removal of major components.

DIMENSIONS, EXTERNAL:	
Diameter of rotors (each)	18.29 m (60 ft 0 in)
Main rotor blade chord	0.64 m (2 ft 1¼ in)
Distance between rotor centres	11.94 m (39 ft 2 in)
Length overall, rotors turning	30.18 m (99 ft 0 in)
Length of fuselage	15.54 m (51 ft 0 in)
Width, rotors folded	3.78 m (12 ft 5 in)
Height to top of rear rotor hub	5.68 m (18 ft 7⅝ in)
Wheel track (c/l of shock-absorbers)	3.20 m (10 ft 6 in)
Wheelbase	6.86 m (22 ft 6 in)
Passenger door (fwd, stbd): Height	1.68 m (5 ft 6 in)
Width	0.91 m (3 ft 0 in)
Height to sill	1.09 m (3 ft 7 in)
Rear-loading ramp entrance: Height	1.98 m (6 ft 6 in)
Width	2.31 m (7 ft 7 in)
Height to sill	0.79 m (2 ft 7 in)
DIMENSIONS, INTERNAL:	
Cabin, excl flight deck: Length	9.20 m (30 ft 2 in)
Width (mean)	2.29 m (7 ft 6 in)
Width at floor	2.52 m (8 ft 3 in)
Height	1.98 m (6 ft 6 in)
Floor area	21.0 m² (226 sq ft)
Usable volume	41.7 m³ (1,474 cu ft)
AREAS:	
Rotor blades (each)	5.86 m² (63.1 sq ft)
Main rotor discs (total)	525.3 m² (5,655 sq ft)

UPDATED

BOEING MODEL 234 COMMERCIAL CHINOOK

TYPE: Commercial tandem-rotor twin-turboshaft helicopter.

PROGRAMME: Launched by British Airways Helicopters order for three in 1978 (later increased to six) for North Sea offshore operations; first flight 19 August 1980; FAA and

Boeing Chinook HC. Mk 2 of the RAF

1994

CAA certification 19 and 26 June 1981 respectively; first service with BAH (now British International Helicopters) 1 July 1981; BAH deliveries completed 1 June 1982; FAA and CAA certificated 234 LR Combi Summer 1982. Commercial Chinooks now withdrawn from service over the North Sea.

VERSIONS: **234 LR Long Range:** About twice CH-47 fuel load in composites tanks attached to fuselage flanks with anti-vibration mounts; flight deck floor on shock-mounts; 44-passenger interior (on shock mounts), with toilet and galley, based on Boeing airliners; walk-on baggage bins on rear ramp; alternative mixed passenger/cargo or all-cargo layouts.

234 ER Extended Range: Typical configurations are 17 passengers and two tanks for additional 875 n miles (1,621 km; 1,008 miles) or 32 passengers and single cabin fuel tank; FAA certificated May 1983.

234 UT Utility: External fuel cells replaced by two cylindrical tanks in forward fuselage: FAA supplemental type certificate October 1981 at maximum gross weight 23,133 kg (51,000 lb) with external loads up to 12,700 kg (28,000 lb) on single hook; approved for 24 passengers and operation at up to 3,660 m (12,000 ft) at full gross weight. No known orders.

234 MLR Multipurpose Long Range: Similar to 234 LR but with utility interior; can be reconfigured in eight hours; four men can handle cabin cylindrical fuel tanks and ramp baggage bins.

CUSTOMERS: Six 234 LR delivered to British BAH/BIH (one lost) and three to Helikopter Service SA of Norway 1981 to 1985; five former BIH aircraft now operated by Columbia Helicopters in Oregon for logging and firefighting; two second-hand 234s delivered to Trump Air 1989 for New York-Atlantic City service; two second-hand 234 ERs leased to Arco Alaska in 1983 and one to ERA Helicopters in 1985.

LANDING GEAR: As described for CH-47D, but with tyre pressures of 8.55 bars (124 lb/sq in) on forward gear, 7.20 bars (104.4 lb/sq in) on rear gear. Wheel-ski gear optional.

POWER PLANT: Two Textron Lycoming AL 5512 turboshafts, pod-mounted on sides of rear rotor pylon. Each engine has maximum T-O rating of 3,039 kW (4,075 shp), maximum continuous rating of 2,218.5 kW (2,975 shp), and 30 minutes contingency rating of 3,247 kW (4,355 shp). Long-range model has two fuel tanks, one in each fuselage side fairing, with total capacity of 7,949 litres (2,100 US gallons; 1,749 Imp gallons). Utility model has two drum-shape internal tanks, with total capacity of 3,702 litres (978 US gallons; 814 Imp gallons). Extended-range model has both fuselage side and internal drum tanks. Single-point pressure refuelling.

ACCOMMODATION: Two pilots side by side on flight deck, with dual controls. Passenger cabin of long-range model seats up to 44 persons four-abreast, with centre aisle. Each seat has overhead bin and underseat stowage for carry-on baggage; larger items are stowed over the rear ramp in the main baggage compartment. Galley, with cabin attendant's seat, and toilet, are standard, between flight deck and cabin. Basic FAA/CAA approved combi versions offer 8-32 passenger seats, with cargo at rear of cabin, loaded via rear ramp; or 22-32 passenger seats, with cargo stowed on only one side of cabin's centre aisle. All passenger facilities can be removed, and heavy duty floor installed, for freight-only service. Passenger door at front of cabin on starboard side. Crew door on each side of flight deck. Cabin floor supported by dynamically tuned fittings to reduce vibration. Hydraulically powered cargo ramp can be stopped at any intermediate position to match the level of the loading vehicle being used. Single central cargo hook is standard on utility model for carrying external loads of up to 12,700 kg (28,000 lb). Optional dual tandem hooks for precision operations and for load stability in high-speed flight; or three tandem hooks for delivering multiple loads.

SYSTEMS: Heating and ventilation systems maintain comfortable flight deck/cabin temperature in ambient temperatures down to −32°C. Duplicated flying control, hydraulic and electrical systems, as described for CH-47D/Model 414. Solar T62T-2B APU, rated at 71 kW (95 shp), drive auxiliary gearbox on rear transmission to start engines and provide power for two flying control system hydraulic pumps and two alternators. All critical systems heated to inhibit ice build-up.

AVIONICS: Duplicated full blind-flying instrumentation, weather radar, and dual four-axis automatic flight control system with built-in test equipment, provide all-weather capability.

EQUIPMENT: Optional equipment includes passenger interior furnishings for the utility model, combi interior, downward-shining cargo load light, rescue hoist of 272 kg (600 lb) capacity, glassfibre wheel-skis, an ice detector probe, and ditching equipment that includes two liferafts, each with an overload capacity of 36 persons. Standard items include integral work platforms, and a maintenance panel that allows 26 separate checks to be made from a single ground level position.

DIMENSIONS, EXTERNAL: As CH-47D except:

Length of fuselage	15.87 m (52 ft 1 in)
Width over fuselage side fairings	4.78 m (15 ft 8 in)
Wheel track (fwd)	3.20 m (10 ft 6 in)
Wheel track (rear)	3.40 m (11 ft 2 in)
Wheelbase	7.87 m (25 ft 9⁹⁄₁₀ in)

DIMENSIONS, INTERNAL:

Passenger cabin	as for CH-47D
Baggage compartment volume	4.42 m^3 (156.0 cu ft)
Utility model, cargo hold volume	41.03 m^3 (1,449 cu ft)

WEIGHTS AND LOADINGS:

Manufacturer's weight empty: LR	11,748 kg (25,900 lb)
ER	12,020 kg (26,500 lb)
MLR	11,113 kg (24,500 lb)
UT	9,797 kg (21,600 lb)
Operating weight empty: LR	12,292 kg (27,100 lb)
ER	12,406 kg (27,350 lb)
MLR	11,400 kg (25,134 lb)
UT	10,002 kg (22,050 lb)
Fuel load: LR, MLR	6,391 kg (14,091 lb)
ER	9,368 kg (20,653 lb)
UT	2,976 kg (6,562 lb)
Max payload: LR, MLR, internal	9,072 kg (20,000 lb)
ER, UT, internal	8,731 kg (19,250 lb)
ER, MLR, UT, external	12,700 kg (28,000 lb)
Max T-O weight:	
ER, LR, MLR, internal load	22,000 kg (48,500 lb)
UT, internal load	19,051 kg (42,000 lb)
ER, LR, MLR, UT, external load	23,133 kg (51,000 lb)

PERFORMANCE:

Never-exceed speed (VNE):
- ER, LR, MLR: 150 kts (278 km/h; 173 mph)
- UT: 140 kts (259 km/h; 161 mph)

Max cruising speed at 610 m (2,000 ft):
- ER, LR, MLR, internal load, at 20,411 kg (45,000 lb) AUW: 145 kts (269 km/h; 167 mph)
- UT, internal load, at 19,051 kg (42,000 lb) AUW: 140 kts (259 km/h; 161 mph)

Cruising speed for optimum range, at 610 m (2,000 ft):
- ER, LR, MLR, UT, internal load, at all gross weights: 135 kts (250 km/h; 155 mph)

Max rate of climb at S/L at max T-O weight:
- ER, LR, MLR, internal load: 360 m (1,180 ft)/min
- UT, internal load: 457 m (1,500 ft)/min

Operational ceiling:
- ER, LR, MLR, UT: 4,575 m (15,000 ft)

Hovering ceiling IGE:
- ER, LR, MLR, internal load: 2,590 m (8,500 ft)
- UT, internal load: 4,085 m (13,400 ft)

Hovering ceiling OGE:
- ER, LR, MLR, internal load: 820 m (2,700 ft)
- UT, internal load: 3,505 m (11,500 ft)

Range with 45 min IFR reserves:
- LR, 44 passengers: 530 n miles (982 km; 610 miles)
- ER, 17 passengers: 830 n miles (1,538 km; 956 miles)
- LR, MLR, with max fuel: 620 n miles (1,149 km; 714 miles)
- ER with max fuel: 1,035 n miles (1,918 km; 1,192 miles)
- UT with max internal load: 229 n miles (424 km; 264 miles)
- UT with max external load: 145 n miles (269 km; 167 miles)

Max endurance: LR, MLR, internal load: 5 h 18 min
- ER, internal load: 8 h 25 min
- UT, external load: 2 h 18 min

Boeing 234 Commercial Chinook in Trump Air colours now Shuttle Inc (trading as USAir Shuttle)
1994

VERIFIED

BOEING COMMERCIAL AIRPLANE GROUP

PO Box 3707, Seattle, Washington 98124-2207
Telephone: 1 (206) 237 2121
Fax: 1 (206) 237 1706
Telex: 0650 329430 BOEING CO C

PRESIDENT: Ronald B. Woodard

EXECUTIVE VICE-PRESIDENTS:
Richard R. Albrecht (Business Acquisition)
Robert L. Dryden (Airplane Production)

SENIOR VICE-PRESIDENTS:
Fred R. Howard (Business Resources)
Alan R. Mulally (Airplane Development and Definition)
Thomas E. Schick (Airplane Support)

Boeing Commercial Airplane Group, headquartered at Renton, near Seattle, assembles jetliners at two major sites. The Renton site produces the 737 and 757 standard-body jetliners. The Everett site is where the wide-bodied 747, 767 and the new 777 jetliners are produced.

Deliveries in 1994 included 121 737s, 40 747s, 69 757s and 50 767s. First deliveries of the new 777 begin in May 1995. By 31 December 1994, total orders for Boeing jetliners of all types, including military derivatives, stood at 8,731 while deliveries totalled 7,772.

UPDATED

BOEING MODEL 707

TYPE: Four-engine transport aircraft.

PROGRAMME: First flight Model 367-80, original prototype of 707, 15 July 1954; developed version ordered in large numbers by US Air Force as KC-135 (Model 717); commercial developments of prototype were 707 and 720. Last commercial 707 was 707-320C for Moroccan government delivered March 1982.

Total 1,010 commercial and military 707/720s ordered and delivered. Production ended mid-1991. Boeing Military Airplane offers tanker/transport conversions of ex-airline 707s; 707/720 conversions also offered by Israel Aircraft Industries and Comtran Ltd, USA.

VERSIONS: **367-80:** The prototype tanker-transport, flown on 15 July 1954. Powered originally by Pratt & Whitney JT3P turbojets (each 4,300 kg; 9,500 lb st), with maximum T-O weight of 72,570 kg (160,000 lb). Wing span 39.60 m (130 ft 0 in).

707-020: Alternative designation for Model 720.

707-120: First production version. Intended primarily for continental use, but capable of full load over-ocean operation on many routes. Four Pratt & Whitney JT3C-6 turbojet engines (each 6,124 kg; 13,500 lb st). Longer fuselage than prototype. Fuselage width increased by 0.41 m (16 in) to 3.76 m (12 ft 4 in). Accommodation for up to 181 passengers. First production aircraft flew for the first time on 20 December 1957.

707-120B: Development of 707-120, with four 7,718 kg (17,000 lb st) Pratt & Whitney JT3D-1 or 8,165 kg (18,000 lb st) JT3D-3 turbofan engines and design improvements incorporated originally in the Boeing 720. These included a new inboard wing leading-edge and four additional segments of leading-edge flaps. First 707-120B flew for the first time on 22 June 1960. FAA approval was received on 1 March 1961.

707-220: Airframe and accommodation identical with those of the 707-120 but powered by four Pratt & Whitney JT4A-3 turbojet engines (each 7,167 kg; 15,800 lb st). Received FAA Type Approval on 5 November 1959.

707-320 Intercontinental: Long-range over-ocean version with increased wing span and longer fuselage. Four Pratt & Whitney JT4A-3 or -5 (each 7,167 kg; 15,800 lb st) or JT4A-9 turbojet engines. Accommodation for up to 189 passengers. The 16th production 707 was the first 707-320 and flew for the first time on 11 January 1959. Received FAA Type Approval on 15 July 1959.

707-320B Intercontinental: Development of the 707-320 with four Pratt & Whitney JT3D-3 or -3B turbofan engines (each 8,165 kg; 18,000 lb st), fitted with double thrust reversers. New leading- and trailing-edge flaps, low-drag wingtips and other refinements. First 707-320B flew 31 January 1962. FAA Type Approval received 31 May 1962, and the type entered service in June 1962 with Pan American.

707-320C Convertible: Cargo or mixed cargo-passenger version of 707-320B with 2.31 × 3.4 m (91 × 134 in) forward cargo door and Boeing developed cargo loading system, using pallets or containers. Cargo space comprises 210 m^3 (7,415 cu ft) on full main deck and 48 m^3 (1,700 cu ft) in two lower deck holds. Accommodation for up to 215 passengers. Received FAA Type Approval 30 April 1963, and first entered service in June 1963 with Pan American.

707-320C Freighter: All cargo version of the Convertible. Passenger facilities eliminated, increasing payload by 1,242 kg (2,736 lb). The full main deck

Boeing 707-320B with Pratt & Whitney JT3D-3 turbofans

1994

accommodates 215.5 m³ (7,612 cu ft) of cargo, and 50 m³ (1,770 cu ft) is accommodated in the two lower decks.

707-420 Intercontinental: A 707-320 but powered by four Rolls-Royce Conway Mk 508 turbofan engines (each 7,945 kg; 17,500 lb st). Received Type Approval 12 February 1960.

UPGRADES: **Alenia:** B707 tanker conversion. See separate entry.

Boeing Product Support Division: B707 tanker/transport conversion. See separate entry.

Boeing Product Support Division: KC-135. See separate entry.

Boeing Military Airplanes: E-3 Sentry. See separate entry.

Boeing Military Airplanes: E-6A TACAMO II. See separate entry.

Comtran: Head of State, hushkit and Super Q conversions. See separate entry.

Israeli Aircraft Industries: B707 tanker/transport conversion. See separate entry.

Northrop/Grumman: E-8 (J-STARS). See separate entry.

OPERATORS: Versions of the B707 are in service with the armed forces of the following countries: Argentina (5), Australia (5), Brazil (4), Canada (5), Chile (4), Colombia (1), Egypt (3), France (4), Germany (4), Indonesia (1), Iran (14), Israel (13), Italy (4), Libya (7), Morocco (1), NATO (21), Pakistan (2), Peru (1), Romania (2), Saudi Arabia (14), South Africa (4), Spain (3), Togo (1), United Kingdom (7), United States (669), Venezuela (1), Yugoslavia (1) and Zaïre (1).

The following description applies to the 707-320C Convertible:

DESIGN FEATURES: Cantilever low-wing monoplane. Dihedral 7°. Incidence 2°. Sweepback at quarter-chord 35°.

FLYING CONTROLS: Normal outboard aileron and small inboard aileron on each wing. Two Fowler flaps and one fillet flap of aluminium alloy on each wing. Full-span leading-edge flaps. Four hydraulically operated aluminium alloy spoilers on each wing, forward of flaps. Primary flying controls are aerodynamically balanced and manually operated through spring tabs. Lateral control at low speeds by all four ailerons, supplemented by spoilers which are interconnected with the ailerons. Lateral control at high speeds by inboard ailerons and spoilers only. Operation of flaps adjusts linkage between inboard and outboard ailerons to permit outboard operation with extended flaps. Spoilers may be used symmetrically as speed brakes. Anti-balance tab and trim tab in rudder. Trim and control tabs in each elevator. Electrically and manually operated variable incidence tailplane. Powered rudder.

STRUCTURE: All-metal two-spar fail-safe structure. Centre-section continuous through fuselage. Ailerons built of aluminium honeycomb panels. The fuselage is an all-metal semi-monocoque fail-safe structure with cross-section made up of two circular arcs of different radii, the larger above, faired into smooth-contoured ellipse. The tail unit is a cantilever all-metal structure.

LANDING GEAR: Hydraulically retractable tricycle type. Main units are four-wheel bogies which retract inward into underside of thickened wingroot and fuselage. Landing gear doors close when legs fully extended. Gear can be extended in flight to give maximum rate of descent of 4,570 m (15,000 ft)/min when used in conjunction with spoilers. Boeing oleo-pneumatic shock-absorbers. Mainwheels and tyres size 46 × 16. Nosewheel and tyres size 39 × 13. Tyre pressures: mainwheels 12.41 bars (180 lb/sq in), nosewheels 7.93 bars (115 lb/sq in). Multidisc brakes by Goodyear. Hydro-Aire flywheel detector type anti-skid units.

POWER PLANT: Four Pratt & Whitney JT3D-7 turbofan engines, each developing 84.5 kN (19,000 lb st), in pods underwing. Fuel in four main, two reserve and one centre main integral wing tanks with total capacity of 90,299 litres (23,855 US gallons; 19,863 Imp gallons). Provision for both pressure and gravity refuelling. Total oil capacity 114 litres (30 US gallons; 25 Imp gallons).

ACCOMMODATION: Maximum accommodation for up to 219 passengers. Typical arrangement has 14 first class seats, a four-seat lounge and 133 coach class seats, with four galleys and five toilets. There are two passenger doors, forward and aft on port side. Galley servicing doors forward and aft on starboard side. Cargo compartments fore and aft of wing in lower segment of fuselage below cabin floor. Entire accommodation, including cargo compartments, air conditioned and pressurised.

SYSTEMS: Air-cycle air conditioning and pressurisation system, using three AiResearch engine-driven turbocompressors. Pressure differential 0.59 bars (8.6 lb/sq in). Hydraulic system, pressure 207 bars (3,000 lb/sq in), for landing gear retraction, nosewheel steering, brakes, flaps, flying controls and spoilers. Electrical system includes four 30 kVA or 40 kVA 115/200 V 3-phase 400 Hz AC alternators and four 75 A transformer-rectifiers giving 28 V DC. APU optional. Thermal anti-icing of wing leading-edges.

AVIONICS AND EQUIPMENT: To customer's specification.

DIMENSIONS, EXTERNAL:

Wing span	44.42 m (145 ft 9 in)
Wing chord: at root	10.33 m (33 ft 10⁷⁄₁₀ in)
at tip	2.84 m (9 ft 4 in)
Wing aspect ratio	7.056
Length overall	46.61 m (152 ft 11 in)
Length of fuselage	44.35 m (145 ft 6 in)
Width of fuselage	3.76 m (12 ft 4 in)
Height overall	12.93 m (42 ft 5 in)
Tailplane span	13.95 m (45 ft 9 in)
Wheel track	6.73 m (22 ft 1 in)
Wheelbase	17.98 m (59 ft 0 in)
Passenger doors (each): Height	1.83 m (6 ft 0 in)
Width	0.86 m (2 ft 10 in)
Height to sill (fwd)	3.25 m (10 ft 8 in)
Height to sill (aft)	3.20 m (10 ft 6 in)
Cargo door: Height	2.34 m (7 ft 8 in)
Width	3.40 m (11 ft 2 in)
Height to sill	3.20 m (10 ft 6 in)
Forward baggage compartment door:	
Height	1.27 m (4 ft 2 in)
Width	1.22 m (4 ft 0 in)
Height to sill	1.55 m (5 ft 1 in)
Rear baggage compartment door (fwd):	
Height	1.24 m (4 ft 1 in)
Width	1.22 m (4 ft 0 in)
Height to sill	1.47 m (4 ft 10 in)
Rear baggage compartment door (aft):	
Height	0.89 m (2 ft 11 in)
Width	0.76 m (2 ft 6 in)
Height to sill	1.93 m (6 ft 4 in)

DIMENSIONS, INTERNAL:

Cabin, excl flight deck: Length	33.93 m (111 ft 4 in)
Max width	3.55 m (11 ft 8 in)
Max height	2.34 m (7 ft 8 in)
Floor area	106.18 m² (1,143 sq ft)
Volume	228.6 m³ (8,074 cu ft)
Baggage compartment (fwd)	23.65 m³ (835 cu ft)
Baggage compartment (rear)	24.50 m³ (865 cu ft)

AREAS:

Wings, gross	283.4 m² (3,050 sq ft)
Ailerons (total)	11.24 m² (121 sq ft)
Trailing-edge flaps (total)	44.22 m² (476 sq ft)
Leading-edge flaps	14.31 m² (154 sq ft)
Fin	30.47 m² (328 sq ft)
Rudder, incl tabs	9.48 m² (102 sq ft)
Tailplane	58.06 m² (625 sq ft)
Elevator, incl tabs	14.03 m² (151 sq ft)

WEIGHTS AND LOADINGS:

Basic operating weight, empty:	
Passenger	66,406 kg (146,400 lb)
Cargo	64,000 kg (141,100 lb)
Max payload: Cargo	40,324 kg (88,900 lb)
Max T-O weight	151,315 kg (333,600 lb)
Max ramp weight	152,405 kg (336,000 lb)
Max zero-fuel weight	104,330 kg (230,000 lb)
Max landing weight	112,037 kg (247,000 lb)
Max wing loading	537.1 kg/m² (110 lb/sq ft)
Max power loading	448 kg/kN (4.39 lb/lb st)

PERFORMANCE (at average cruising weight unless indicated otherwise):

Never-exceed speed	Mach 0.95
Max level speed	545 kts (1,010 km/h; 627 mph)
Max cruising speed at 7,620 m (25,000 ft)	525 kts (973 km/h; 605 mph)
Econ cruising speed	478 kts (886 km/h; 550 mph)
Stalling speed (flaps down, at max landing weight)	105 kts (195 km/h; 121 mph)
Max rate of climb at S/L	1,219 m (4,000 ft)/min
Service ceiling	11,885 m (39,000 ft)
CAR T-O to 10.7 m (35 ft)	3,054 m (10,020 ft)
CAR landing from 15 m (50 ft)	1,095 m (6,250 ft)
Landing run	785 m (2,575 ft)
Range with max fuel, 14 first class and 133 tourist passengers, long-range step cruise, international reserves	5,000 n miles (9,265 km; 5,755 miles)
Range with 36,287 kg (80,000 lb) cargo, long-range step cruise, international reserves	3,150 n miles (5,835 km; 3,625 miles)

UPDATED

BOEING MODEL 720

TYPE: Four turbojet passenger.

DESIGN FEATURES: The Boeing 720 is the intermediate-range member of the Boeing jet transport family. Although a completely different design from the weight and structural strength standpoints, the 720 is almost identical to the 707-120 in external outline and main dimensions, aerodynamic design and control systems. This has made possible the use of 707 passenger cabin interiors, flight deck, systems components and most 707 interchangeable and replaceable parts and spares.

The most important aerodynamic change compared with the 707-120 is a refinement to the wing leading-edge, which increases the angle of sweepback and decreases the thickness/chord ratio, with consequent improvement in take-off performance and cruising speed. These modifications were incorporated subsequently on the Boeing 707-120B.

Major weight saving has been achieved by lightening the structure to the extent made possible by reducing the standard fuel load.

VERSIONS: **720:** Basic model, powered by four Pratt & Whitney JT3C-7 (each 5,670 kg; 12,500 lb st) or JT3C-12 (each 5,902 kg; 13,000 lb st) turbojets. The first 720 flew 23 November 1959. FAA Type Approval received 30 June 1960, and the type entered service with United Air Lines 5 July 1960. Production completed.

720B: Developed version, powered by four Pratt & Whitney JT3D-1 (each 7,710 kg; 17,000 lb st) or JT3D-3 (each 8,165 kg; 18,000 lb st) turbofan engines. First 720B flew 6 October 1960. FAA Type Approval received 3 March 1961.

The description of the Boeing 707-120B applies equally to the Boeing 720 and 720B, with the following exceptions:

LANDING GEAR: Mainwheels and tyres size 40 × 14, pressure 10.2 kg/cm² (145 lb/sq in). Nosewheels and tyres size 34 × 9.9, pressure 7.0 kg/cm² (100 lb/sq in). Goodrich multidisc brakes with Hydro-Air flywheel detector type anti-skid units.

POWER PLANT: Four turbojet (720) or turbofan (720B) engines mounted in pods underwings (details under Versions). Fuel in four main, two reserve and one centre main integral wing tanks. Capacity varies with individual model, but Boeing give maximum standard fuel capacities as 56,136 litres (14,830 US gallons; 12,348 Imp gallons) for 720 and 56,325 litres (14,880 US gallons; 12,390 Imp gallons) for 720B. Provision for both pressure and gravity refuelling. Oil capacity 91 litres (24 US gallons; 20 Imp gallons).

ACCOMMODATION: Crew of three or four on flight deck. Typical seating arrangement provides for 38 first class passengers and 74 tourist passengers, with three galleys and three toilets. Doors and cargo compartments as for 707-120B.

DIMENSIONS, EXTERNAL:

Wing span	39.87 m (130 ft 10 in)
Length overall	41.68 m (136 ft 9 in)
Length of fuselage	39.77 m (130 ft 6 in)
Width of fuselage	3.76 m (12 ft 4 in)
Height overall	12.67 m (41 ft 6½ in)
Tailplane span: 720	12.09 m (39 ft 8 in)
720B	13.21 m (43 ft 4 in)
Wheel track	6.68 m (21 ft 11 in)
Wheelbase	15.44 m (50 ft 8 in)
Doors as for 707-120/120B	

DIMENSIONS, INTERNAL:

Cabin, excl flight deck: Length	29.41 m (96 ft 6 in)
Max width	3.55 m (11 ft 8 in)
Max height	2.31 m (7 ft 7 in)
Floor area	93.18 m² (1,003 sq ft)
Volume	194.3 m³ (6,860 cu ft)
Baggage compartment (fwd)	19.48 m³ (688 cu ft)
Baggage compartment (aft)	19.54 m³ (690 cu ft)

AREAS: 720 as for 707-120, 720B as for 707-120B.

WEIGHTS AND LOADINGS:

Basic operating weight, empty:	
720	50,260 kg (110,800 lb)
720B	51,204 kg (112,883 lb)
Max payload: 720	12,790 kg (28,200 lb)
720B	19,692 kg (43,117 lb)
Max T-O weight: 720	103,870 kg (229,000 lb)
720B	106,140 kg (234,000 lb)
Max ramp weight: 720	104,330 kg (230,000 lb)
720B	106,590 kg (235,000 lb)
Max zero-fuel weight: 720	63,050 kg (139,000 lb)
720B	70,760 kg (156,000 lb)
Max landing weight	79,380 kg (175,000 lb)
Max wing loading: 720	459.4 kg/m² (94.1 lb/sq ft)
720B	451.4 kg/m² (92.8 lb/sq ft)
Max power loading: 720	4.77 kg/kg st (4.77 lb/lb st)
720B	3.25 kg/kg st (3.25 lb/lb st)

PERFORMANCE (at average cruising weight):

Max level speed	545 kts (1,010 km/h; 627 mph)
Max permissible diving speed	Mach 0.95
Max cruising speed at 7,620 m (25,000 ft):	
720	510 kts (945 km/h; 587 mph)
720B	530 kts (983 km/h; 611 mph)
Econ cruising speed at 12,190 m (40,000 ft)	484 kts (897 km/h; 557 mph)
Stalling speed, flaps down, at max landing weight:	
720	99 kts (184 km/h; 114 mph)
720B	101 kts (187 km/h; 116 mph)
Rate of climb at S/L: 720	731 m (2,400 ft)/min
720B	1,243 m (4,080 ft)/min
Service ceiling: 720	12,200 m (40,000 ft)
720B	12,800 m (42,000 ft)
CAR T-O to 10.7 m (35 ft): 720	2,865 m (9,400 ft)
720B	1,966 m (6,450 ft)
CAR landing from 15 m (50 ft): 720	1,890 m (6,200 ft)
720B	1,935 m (6,350 ft)
Landing run: 720	790 m (2,590 ft)
720B	825 m (2,710 ft)
Range with max fuel, allowances for climb and descent, no reserves:	
720	4,549 n miles (8,430 km; 5,240 miles)
720B	4,967 n miles (9,205 km; 5,720 miles)
Range with max payload, allowances for climb and descent, no reserves:	
720	3,680 n miles (6,820 km; 4,235 miles)
720B	3,610 n miles (6,690 km; 4,155 miles)

VERIFIED

BOEING MODEL 727

Military designation: C-22B/C

TYPE: Three-turbofan transport aircraft.

PROGRAMME: First 727-100 rolled out 27 November 1962; first flight 9 February 1963; FAA certification 24 December 1963; into service with Eastern Air Lines 1 February 1964. 727-200 flew 27 July 1967; FAA certification 30 November 1967; into service with Northeast Airlines 14 December 1967. Last of 572 727-100s delivered October 1972 and of 1,260 727-200s September 1984. Current re-engining programmes by Valsan Partners and Dee Howard (which see). Being announced its intention to supplement the four-jet 707/720 series with the three-engined short/medium-range 727 on 5 December 1960. The 727 switched to a rear-engined configuration, but has an upper fuselage section identical with that of the 707/720 and many parts and systems are interchangeable between the three types.

VERSIONS: **727-100:** Standard for up to 131 passengers. Three JT8D-7 turbofans each rated at 6,350 kg (14,000 lb st) to 29°C (84°F) are standard, with JT8D-9s flat-rated at 6,577 kg (14,500 lb st) optional. AUW of basic model is 64,410 kg (142,000 lb), but 727-100 is available with AUW of 68,947 kg (152,000 lb), 72,570 kg (160,000 lb) and 76,655 kg (169,000 lb).

727-100C: Convertible cargo-passenger version, with standard AUW of 72,570 kg (160,000 lb) and optional AUW of 76,655 kg (169,000 lb). Announced on 22 July 1964. Identical with 727-100 except for installation of heavier flooring and floor beams and same large cargo door as on the 707-320C. Effective opening is 2.18 m (7 ft 2 in) high by 3.40 m (11 ft 2 in) wide, and 727-100C is able to utilise same cargo pallets and handling system as 707-320C. Galleys and seats quickly removable and hatracks stowable, permitting conversion for mixed passenger/cargo or all-cargo services in under two hours. Payloads are 94 mixed-class passengers; 52 passengers and baggage, plus 10,295 kg (22,700 lb) of cargo on four pallets; or 17,236 kg (38,000 lb) of cargo on eight pallets. The full eight-pallet payload can be carried more than 1,477 n miles (2,737 km; 1,700 miles). A 13,605 kg (30,000 lb) payload can be carried 1,997 n miles (3,700 km; 2,300 miles).

727-100QC: Same as 727-100C, except that, by using palletised galleys and seats, and advanced cargo loading techniques, complete conversion from all-passenger to all-cargo configuration can be made in less than 30 minutes. All-cargo, high-AUW version has ramp weight of 77,110 kg (170,000 lb), maximum T-O weight of 76,655 kg (169,000 lb) and maximum landing weight of 64,640 kg (142,500 lb).

727-200: Passenger version of the lengthened 727-200 announced by Boeing 12 May 1971, at an initial ramp weight of 86,635 kg (191,000 lb), and first delivered in June 1972. The interior features the 'Superjet look', and a large overhead stowage bin compartment available at no extra cost. Availability of successively more powerful engines made weight increases up to a maximum ramp weight of 95,254 kg (210,000 lb) possible for aircraft with JT8D-15/17/17R turbofans. A performance data computer system is standard equipment to provide onboard information for optimising flight profile and fuel consumption.

727-200F: A freighter version with strengthened fuselage structure became available in 1981. Powered by Pratt & Whitney JT8D-17A engines, it can accommodate 11 pallets, each 2.23 × 3.17 m (7 ft 4 in × 10 ft 5 in), in a windowless cabin with a large door at the front on the port side. Federal Express placed an order for 15 aircraft in September 1981.

C-22B/C: Military VIP, general transport and cargo aircraft.

UPGRADES: **Boeing:** Retrofit kits for the 'Superjet look' are offered, as are kits for larger overhead stowage bins compartments. Entry is via hydraulically operated integral aft stairway under centre engine and door at front on port side with optional Weber Aircraft electrically operated airstairs.

Boeing/Federal Express: 727RE 'Quiet 727'. See following entry.

Dee Howard: B727 Update. See separate entry.

Hamilton Aviation: Clamshell and sliding door conversions. See separate entry.

Pemco Aeroplex: Combi conversion. See separate entry.

Snow Aviation: Re-engine programme. See separate entry.

OPERATORS: Versions of the B727 are in service with the armed forces of the following countries: Burkino Faso (1), Cameroon (1), Congo (1), Iran (1), Mexico (6), New Zealand (2), Nigeria (1), Panama (1), Qatar (1), Senegambia (1), Taiwan (4), Yemen (1) and Zaïre (1).

The following details apply to the 727-200 passenger aircraft:

DESIGN FEATURES: Cantilever low-wing monoplane. Special Boeing aerofoil sections. Thickness/chord ratio from 13 per cent to 9 per cent. Dihedral 3°. Incidence 2°. Sweepback at quarter-chord 32°.

FLYING CONTROLS: Hydraulically powered aluminium ailerons, inboard (high speed) and outboard (low speed) units, operate in conjunction with flight spoilers. Triple-slotted trailing-edge flaps constructed primarily of aluminium and aluminium honeycomb. Four aluminium leading-edge slats on outer two-thirds of wing. Three Krueger leading-edge flaps on inboard third of wing, made from magnesium or aluminium castings. Seven spoilers on each wing, consisting of five flight spoilers outboard and two ground spoilers inboard. Spoilers function also as airbrakes. Balance tab in each outboard aileron; control tab in each aileron. Controls are hydraulically powered dual systems with automatic reversion to manual control. Dual-powered variable incidence tailplane, with direct manual reversion. Hydraulically powered dual elevator control system with control tab manual reversion. Hydraulically powered rudders, utilising two main systems with back-up third system for lower rudder. Anti-balance tabs; rudder trim by displacing system neutral.

STRUCTURE: Primary structure is a two-spar aluminium box with conventional ribs. Upper and lower surfaces are of riveted skin/stringer construction. There are no chordwise splices in the primary structure from the fuselage to the wingtip. Advanced 727-200s at gross weight options have modified stringers and in-spar webs, as well as upper and lower surface wing skins of increased gauge. Structure is fail-safe. The fuselage is a semi-monocoque fail-safe structure, with aluminium alloy skin reinforced by circumferential frames and longitudinal stringers. The tail unit is a Cantilever T tail, built primarily of aluminium alloys. Actuators manufactured primarily by Weston, National Water Lift and Bertea.

LANDING GEAR: Hydraulically retractable tricycle type, with twin-wheels on all three units. Nosewheels retract forward, main gear inward into fuselage. Boeing oleo-pneumatic shock-absorbers. Goodrich nose-gear wheels, tyres and brakes are standard on all models. Goodrich and Bendix are both approved suppliers of main gear wheels for all Model 727s. Nosewheels and tyres size 32 × 11.5 Type VIII. Main gear wheels size 49 × 17, with tyres size 50 × 21 Type VII are standard.

POWER PLANT: Three Pratt & Whitney JT8D-9A turbofan engines with thrust reversers and full sound attenuation, each flat rated at 64.5 kN (14,500 lb st) to 29°C (84°F), are standard. Optionally, JT8D-15s rated at 68.9 kN (15,500 lb st), JT8D-17s rated at 71.2 kN (16,000 lb st) or JT8D-17Rs with automatic performance reserve (APR) and rated at 77.4 kN (17,400 lb st), can be fitted. The JT8D-17R engines are normally operated at an alternative rating of 72.9 kN (16,400 lb st); APR senses any significant loss in thrust by an engine during take-off and initial climb, automatically increasing thrust on the other engines to 77.4 kN (17,400 lb st). This feature significantly improves performance from hot/high airports. Each engine has individual fuel system fed from integral tanks in wings, but all three tanks are interconnected. Optional fuselage fuel tanks can be installed, displacing forward and/or aft cargo compartment volume. Standard total capacity 30,623 litres (8,090 US gallons; 6,736 Imp gallons). Modular design bladder cell tanks with dual fuel barrier can be installed to contain up to approximately 9,387 litres (2,480 US gallons; 2,065 Imp gallons). Single pressure fuelling point, rated at 2,271 litres (600 US gallons; 500 Imp gallons)/min, near leading-edge underside of starboard wing at mid-span. Total usable oil capacity 45.5 litres (12 US gallons; 10 Imp gallons).

Boeing 720B four-engine transport aircraft

1994

Boeing 727-200 passenger aircraft

1994

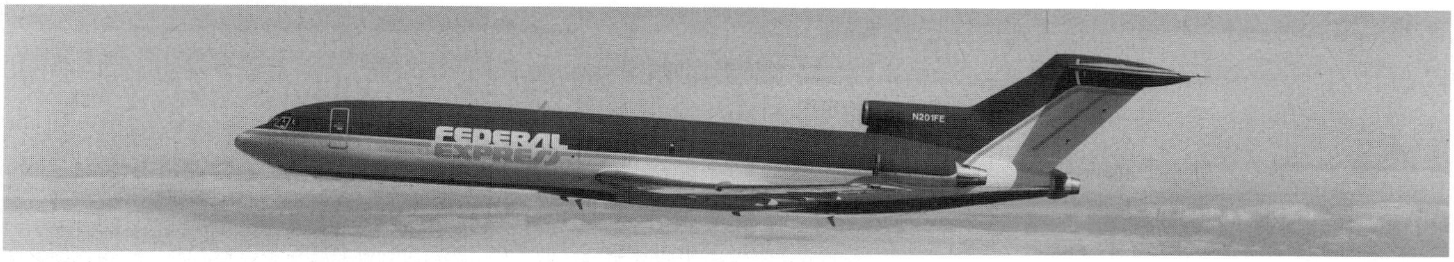

Boeing 727-200 freighter of Federal Express

1994

ACCOMMODATION: Crew of three on flight deck. Basic accommodation for 145 passengers (14 first class, 131 tourist class, four- to six-abreast). Maximum capacity 189 passengers. Two galleys forward and two aft. One toilet forward and two aft. Other layouts to customer's requirements. A 'Superjet look' passenger interior design is standard. The widebody effect is achieved (without any changes in cross-section dimensions) by lighting and architectural redesign. Two Type III emergency exits in mid-cabin on each side and aft service door on each side. The starboard forward service door is opposite the port forward passenger door. Two heated and pressurised cargo compartments under floor, forward and aft of main landing gear bay. Each compartment has one outward-opening cargo door; a second door is optional for the aft compartment.

SYSTEMS: Garrett air conditioning and pressurisation system, using engine bleed air combined with air cycle refrigeration. Pressure differential 0.59 bars (8.6 lb/sq in). Three independent 207 bar (3,000 lb/sq in) hydraulic systems, utilising Boeing Material Specification BMS 3-11 hydraulic fluid, provide power for flying controls, landing gear and aft airstairs. Electrical system includes three 40 kVA 400 Hz constant frequency AC generators, three 50 A transformer-rectifier units, one 22 Ah battery. Garrett APU provides electrical power and compressed air for engine starting and air conditioning on ground. Thermal anti-icing of wing leading-edges by engine bleed air.

AVIONICS AND EQUIPMENT: Standard avionics, available from alternative sources, include flight and service attendant's intercom, passenger address system, ARINC 531 Selcal, dual VHF com, ARINC 594 ground proximity warning system, ARINC 557 voice recorder, ARINC 542 flight recorder and remote encoder, marker beacon receiver, ARINC 552 radio altimeter, ARINC 572 ATC transponder, dual DME, ARINC 570 ADF, dual ARINC 547 VHF nav systems, dual radio direction/distance measuring indicators with dual directional gyros and compass systems, dual FD110 flight director systems, single-channel Mod Blk V autopilot, dual vertical gyros, dual yaw dampers, performance data computer system, dual ARINC 545 digital air data computer systems, push-button audio selector panels, and VHF nav/com frequency preselect. Optional avionics, available from several alternative sources, include a third VHF com, single or dual HF com, third nav, Omega, single or dual ARINC 564 X-band weather radar, second ADF, Z-15 flight director systems, dual autopilot channels, roll monitor and flare coupler, third vertical gyro, and full range autothrottle/speed control system.

DIMENSIONS, EXTERNAL:
- Wing span: 32.92 m (108 ft 0 in)
- Wing chord: at root: 7.70 m (25 ft 3 in)
  - at tip: 2.34 m (7 ft 8 in)
- Wing aspect ratio: 7.07
- Length overall: 46.69 m (153 ft 2 in)
- Length of fuselage: 41.51 m (136 ft 2 in)
- Height overall: 10.36 m (34 ft 0 in)
- Tailplane span: 10.90 m (35 ft 9 in)
- Wheel track: 5.72 m (18 ft 9 in)
- Wheelbase: 19.28 m (63 ft 3 in)
- Passenger door (ventral): Length: 1.93 m (6 ft 4 in)
  - Width: 0.81 m (2 ft 8 in)
- Passenger door (fwd): Height: 1.83 m (6 ft 0 in)
  - Width: 0.86 m (2 ft 10 in)
  - Height to sill: 2.67 m (8 ft 9 in)
- Service door (fwd): Height: 1.65 m (5 ft 5 in)
  - Width: 0.84 m (2 ft 9 in)
- Service doors (rear, each): Height: 1.52 m (5 ft 0 in)
  - Width: 0.76 m (2 ft 6 in)
- Baggage hold door (fwd): Height: 1.07 m (3 ft 6 in)
  - Width: 1.37 m (4 ft 6 in)
- Baggage hold door (rear): Height: 1.12 m (3 ft 8 in)
  - Width: 1.37 m (4 ft 6 in)

DIMENSIONS, INTERNAL:
- Cabin (aft of flight deck to rear pressure bulkhead):
  - Length: 28.24 m (92 ft 8 in)
  - Max width: 3.55 m (11 ft 8 in)
  - Max height: 2.11 m (6 ft 11 in)
  - Floor area: 91.05 m² (980 sq ft)
  - Volume: 188.4 m³ (6,652 cu ft)
- Baggage hold (fwd): 20.1 m³ (710 cu ft)
- Baggage hold (rear): standard: 23.1 m³ (815 cu ft)
  - with optional second door: 21.1 m³ (745 cu ft)

AREAS:
- Wings, gross: 157.9 m² (1,700 sq ft)
- Ailerons (total): 5.30 m² (57 sq ft)
- Trailing-edge flaps (total): retracted: 26.10 m² (281 sq ft)
  - extended: 36.04 m² (388 sq ft)
- Flight spoilers (total): 7.41 m² (79.8 sq ft)
- Fin: 33.07 m² (356 sq ft)
- Rudder, incl tabs: 6.31 m² (66 sq ft)
- Tailplane: 34.93 m² (376 sq ft)
- Elevators, incl tabs: 8.83 m² (95 sq ft)

WEIGHTS AND LOADINGS (A: 727-200 at brake release weight of 83,820 kg; 184,800 lb, B: 727-200 at brake release weight of 86,405 kg; 190,500 lb, C: 727-200 at brake release weight of 95,027 kg; 209,500 lb, D: 727-200F freighter at brake release weight of 92,124 kg; 203,100 lb):
- Operating weight empty (basic specification):
  - A: 44,633 kg (98,400 lb)
  - B: 45,178 kg (99,600 lb)
  - C: 46,675 kg (102,900 lb)
- Operating weight empty (typical airline):
  - A: 45,360 kg (100,000 lb)
- Operating weight empty (not incl BFE cargo system):
  - D: 41,684 kg (91,898 lb)
- Max payload (structural, based on airline operating weight empty): A: 18,144 kg (40,000 lb)
- Max payload, incl cargo system:
  - D: 28,622 kg (63,102 lb)
- Max T-O weight: A: 83,820 kg (184,800 lb)
  - B: 86,405 kg (190,500 lb)
  - C: 95,027 kg (209,500 lb)
  - D: 92,124 kg (203,100 lb)
- Max ramp weight: A: 84,275 kg (185,800 lb)
  - B: 86,635 kg (191,000 lb)
  - C: 95,254 kg (210,000 lb)
  - D: 92,533 kg (204,000 lb)
- Max zero-fuel weight: A: 62,595 kg (138,000 lb)
  - B: 63,500 kg (140,000 lb)
  - C: 65,315 kg (144,000 lb)
  - D: 70,306 kg (155,000 lb)
- Max landing weight: A, B: 70,080 kg (154,500 lb)
  - C: 73,028 kg (161,000 lb)
  - D: 75,296 kg (166,000 lb)
- Max wing loading: A: 530.7 kg/m² (108.7 lb/sq ft)
  - B: 546.8 kg/m² (112 lb/sq ft)
  - C: 600.5 kg/m² (123 lb/sq ft)
- Max power loading: A: 433.2 kg/kN (4.2 lb/lb st)
  - B: 418.0 kg/kN (4.1 lb/lb st)
  - C: 434.5 kg/kN (4.3 lb/lb st)

PERFORMANCE (A: at brake release weight of 83,820 kg; 184,800 lb, B: at brake release weight of 86,405 kg; 190,500 lb, C: at brake release weight of 95,027 kg; 209,500 lb, except where indicated):
- Max operating speed: Mach 0.90
- Max level speed: A at 6,585 m (21,600 ft): 549 kts (1,017 km/h; 632 mph)
  - B, C at 6,250 m (20,500 ft): 539 kts (999 km/h; 621 mph)
- Max cruising speed: A at 6,705 m (22,000 ft): 514 kts (953 km/h; 592 mph)
  - B at 7,530 m (24,700 ft): 520 kts (964 km/h; 599 mph)
- Econ cruising speed at 9,145 m (30,000 ft): 471 kts (872 km/h; 542 mph)
- Stalling speed at S/L, flaps down: at 72,575 kg (160,000 lb): 104 kts (193 km/h; 120 mph)
- Initial cruise altitude: 10,060 m (33,000 ft)
- Min ground turning radius: 24.49 m (80 ft 4 in)
- Runway LCN at max weight of 86,635 kg (191,000 lb), optimum tyre pressure and 0.51 m (20 in) flexible pavement: 50 × 21 tyres: 70 m
- CAR T-O distance to 10.7 m (35 ft):
  - A: 2,880 m (9,450 ft)
  - B: 2,554 m (8,380 ft)
  - C: 2,804 m (9,200 ft)
- CAR landing distance from 15 m (50 ft):
  - at 71,668 kg (158,000 lb): 1,430 m (4,690 ft)
- Range at long-range cruising speed, with fuel load as specified and payload of 12,474 kg (27,500 lb), ATA domestic reserves:
  - A with 30,623 l (8,090 US gallons; 6,736 Imp gallons): 2,000 n miles (3,706 km; 2,303 miles)
  - B with 33,878 l (8,950 US gallons; 7,452 Imp gallons): 2,180 n miles (4,040 km; 2,510 miles)
  - C with 36,831 l (9,730 US gallons; 8,102 Imp gallons): 2,370 n miles (4,392 km; 2,729 miles)
- Range with 18,144 kg (40,000 lb) payload, at long-range cruising speed, ATA domestic reserves:
  - A: 1,530 n miles (2,835 km; 1,762 miles)
  - B: 1,615 n miles (2,993 km; 1,860 miles)
  - C: 2,160 n miles (4,003 km; 2,487 miles)

OPERATIONAL NOISE LEVELS (with JT8-15 engines FAR Pt 36):
- T-O at brake release weight of 86,405 kg (190,500 lb): 100 EPNdB
- Approach at 70,080 kg (154,500 lb) landing weight and 30° flap: 100.4 EPNdB
- Sideline: 102.2 EPNdB

UPDATED

BOEING (FEDERAL EXPRESS) 727RE 'QUIET 727'

TYPE: Re-engined/quietened Boeing 727-100 and -200.

PROGRAMME: Prototype conversion of 727 (Sterling Airways 727-200 OY-SAS), to meet FAR Pt 36 Stage 3 and ICAO Annex 16 Chapter 3 noise limits, completed June 1988; first flight 12 July 1988; Supplemental Type Certificate awarded October 1988; certification of first customer 727-100 conversion July 1990; additional Winglet System conversions certificated in 1991.

DESIGN FEATURES: Modification work by Rockwell International, Palmdale and Sterling Airways; work included acoustic treatment of centre engine and removal of thrust reverser, and replacement of outer engines by JT8D-217Cs; empty weight increased by 1,542 kg (3,400 lb), zero-fuel weight increased by 1,361 kg (3,000 lb), and maximum landing weight reduced by 907 kg (2,000 lb); performance improvements include 24 per cent higher maximum S/L climb rate, 27 per cent shorter T-O distances and 300 n miles (556 km; 345 miles) longer range; maximum range of executive configuration expected to be greater than 4,000 n miles (7,412 km; 4,606 miles).

Flight testing of 727 with winglets completed November 1990; demonstrated specific fuel consumption reduction of at least five per cent and could be further improved when wing and flap testing completed. Combined with re-engining, drag reduction modifications claimed to reduce fuel burn by 16-18 per cent. 727-100 winglets cerificated January 1991 and -200 winglets March 1991.

OPERATIONAL NOISE LEVELS (727-200RE):
- T-O: 95.6 EPNdB
- Approach: 98.3 EPNdB
- Sideline: 97.4 EPNdB

NEW ENTRY

Boeing 727-27C (Combi) aircraft of First Air *(Peter J. Cooper)*

1995

Boeing 737-200 twin-turbofan on approach

1994

BOEING MODEL 737-200

US Air Force designations: CT-43A and T-43A

TYPE: Twin-turbofan short-range transport aircraft.

PROGRAMME: Design of original 737 began 11 May 1964; first flight of 737-100, 9 April 1967; FAA certification 15 December 1967; 30 built. Superseded by 737-200; first flight 8 August 1967; added to 737-100 type certificate 21 December 1967; first delivery to United Air Lines 29 December 1967. Last of 1,114 737-200s delivered August 1988; total includes 19 **T-43A** navigation trainers for US Air Force and three **Surveillers** (now being upgraded) for Indonesian Air Force.

The original Model 737 was designed to utilise many components and assemblies already in production for the Boeing 727. Design began on 11 May 1964, and the first Model 737 flew on 9 April 1967. The Boeing 737 was the third commercial transport to reach sales of over 1,000, and in June 1987 became the world's best selling commercial airliner when orders for all models surpassed the previous record total of 1,831 sales set by the Boeing 727. They passed the 2,000 mark 3 March 1988.

VERSIONS: **Advanced 737-200:** Standard model, with maximum ramp weight of 52,615 kg (116,000 lb) and maximum T-O weight of 52,390 kg (115,500 lb). JT8D-15A engines (each rated 68.9 kN; 15,500 lb st) standard; JT8D-17A (71.2 kN; 16,000 lb st) optional; basic fuel capacity of 19,532 litres (5,160 US gallons; 4,296 Imp gallons). Accommodation for 120 passengers and cargo, with 81 cm (32 in) pitch seating, or up to 130 passengers in 76 cm (30 in) pitch seating with no reduction in cabin facilities. Use of graphite composites in place of former aluminium honeycomb in later production aircraft reduced the weight of the rudder, elevators and ailerons. Coupled with further use of composites in a new advanced technology interior, this resulted in a total weight reduction of more than 454 kg (1,000 lb).

Advanced 737-200C/QC: Convertible passenger/cargo model with strengthened fuselage and floor, and a large two-position upper deck cargo door with effective opening of 2.15 × 3.40 m (7 ft 0½ in × 11 ft 2 in). The quick-change feature allows more rapid conversion by using palletised passenger seating and other special interior furnishings. Typical mixed interior configuration provides accommodation for three cargo pallets and 65 passengers with 81 cm (32 in) pitch seating. Total of 104 built and included in overall total of 1,114 Model 737-200s.

Corporate version: Same as standard Advanced 737-200, except interiors are adapted to special business and executive requirements. Additional fuel capacity offered by installation of fuel cells in lower cargo compartments. With maximum fuel this model can carry a 1,134 kg (2,500 lb) payload up to 4,000 n miles (7,412 km; 4,606 miles).

Advanced 737-200 High Gross Weight Structure: Higher gross weight models of the Advanced 737-200/200C, for longer range use, in two versions one has a maximum ramp weight of 56,700 kg (125,000 lb) and a maximum T-O weight of 56,472 kg (124,500 lb) with JT8D-15A or JT8D-17A engines, and a fuel capacity of either 21,009 litres (5,550 US gallons; 4,621 Imp gallons) or 22,598 litres (5,970 US gallons; 4,971 Imp gallons). The additional capacity for increased range capability is provided by a 1,476 litre (390 US gallon; 325 Imp gallon) or a 3,066 litre (810 US gallon; 674 Imp gallon) fuel tank installed in the aft lower cargo compartment. The second version, with a maximum ramp weight of 58,332 kg (128,600 lb), maximum T-O weight of 58,105 kg (128,100 lb), design landing weight of 48,534 kg (107,000 lb), and maximum zero-fuel weight of 43,091 kg (95,000 lb), has approximately 650 n miles (1,204 km; 748 miles) greater range capability than the standard Advanced 737-200. Sectors of 2,300 n miles (4,262 km; 2,648 miles) can be served with a 130-passenger payload and typical fuel reserves. Aircraft is identical to the Advanced 737-200 except for the auxiliary fuel tank, new wheels, tyres and brakes.

Most 737-200 versions meet FAR Pt 36 and ICAO Annex 16 in respect of noise characteristics.

In January 1987 the Boeing 737-200 with JT8D-9/9A/15/15A/17/17A engines received FAA approval for extended range operations, permitting flights over water or undeveloped land areas by these twin-engined commercial airliners without being subject to restrictions requiring such aircraft to remain within one hour's flying time of a suitable airport at single-engined cruising speeds. Operators must, however, apply for regulatory permission to operate extended range services with the aircraft.

An FAA certificated kit is available which enables the Model 737 to operate from unpaved or gravel runways. The kit includes a vortex dissipator for each engine, consisting of a short boom that protrudes from under each engine's forward edge. The boom is capped by a plug with downward-facing orifices. Pressurised engine bleed air forced through these orifices destroys any ground level vortex and prevents small pieces of gravel being ingested by the engines. Other items include a gravel deflection 'ski' on the nosewheel, deflectors between the main landing gear wheels, protective shields over hydraulic tubing and speed brake cable on the main gear strut, glassfibre reinforcement of lower inboard flap surfaces application of Teflon-base paint to fuselage and wing undersurfaces and provision of more robust DME and VHF antennae.

British Airways Advanced 737-200s known as Super 737s are equipped with advanced flight deck avionics, including a Honeywell SP-177 digital automatic flight control system. Lufthansa took delivery of 38 Advanced 737-200s with similar equipment. Cat. IIIA certification of these AFCS versions was granted 2 December 1981.

Surveiller: Specially equipped maritime surveillance 737-200 for Indonesian Air Force (three built), fitted also with 14 first class and 88 tourist class seats so that they can be used for government transport purposes.

UPGRADES: **Boeing:** Surveiller. The Boeing Defense & Space Group received a four-year $117 million contract to upgrade the mission avionics in all three of Indonesia's Surveiller aircraft. Work on the first began Spring 1993 and this was redelivered August 1993. The upgrade includes a new mission avionics system based on Boeing's Digital Processing and Display System (DPDS), a real-time display for the SLAMMR (Side-Looking Airborne Modular Multimission Radar), a nose-mounted search radar and IFF equipment. The remaining two aircraft are being modified at the Bandung facility of BPPT/IPTN, the Indonesian owned state aerospace company. See also IPTN in the Indonesia section.

Fokker Aircraft Services: Boeing 737 modification capabilities. See separate entry.

OPERATORS: Versions of the B737 are in service with the armed forces of the following countries: Brazil (2), Egypt (1), India (2), Indonesia (3), Mexico (2), Niger (1), Pakistan (1), Saudi Arabia (1), South Korea (1), Thailand (1), United States (15) and Venezuela (1).

The following description applies to the commercial versions of the 737-200:

DESIGN FEATURES: Low-wing monoplane. Special Boeing wing sections. Average thickness/chord ratio 12.89 per cent. Dihedral 6° at root. Sweepback at quarter-chord 25°. Variable incidence tailplane.

FLYING CONTROLS: Ailerons of graphite composite construction. Triple-slotted trailing-edge flaps of aluminium, with trailing-edges of aluminium honeycomb. Aluminium alloy Krueger flaps on leading-edge, inboard of nacelles. Three leading-edge slats of aluminium alloy with aluminium honeycomb trailing-edge on each wing from engine to wingtip. Two aluminium honeycomb flight spoilers on each outer wing serve both as airbrakes in the air and for lateral control, in association with ailerons. Two aluminium honeycomb ground spoilers on each wing, one outboard and one inboard of engine, are used only during landing. Ailerons are powered by two hydraulic systems with manual reversion. Trailing-edge flaps are hydraulically powered, with electrical back-up. Leading-edge slats and Krueger flaps are symmetrically powered by one hydraulic system normally, and by a second hydraulic system for alternate extension. Flight spoilers are symmetrically powered by the two main individual hydraulic systems. Variable control surfaces. Elevator has dual hydraulic power, with manual reversion. Rudder is powered by a dual actuator from two main hydraulic systems, with a standby hydraulic actuator and system. Tailplane trim has dual electric drive motors, with manual back-up. Elevator control tabs for manual reversion are locked out during hydraulic actuation.

STRUCTURE: Aluminium alloy dual-path fail-safe two-spar structure. The fuselage is an aluminium alloy semi-monocoque fail-safe structure. The tail unit is a cantilever aluminium alloy multispar structure, with graphite composite control surfaces.

LANDING GEAR: Hydraulically retractable tricycle type, with free-fall emergency extension. Nosewheels retract forward, main units inward. No main gear doors: wheels form wheel well seal. Twin wheels on each main and nose unit. Boeing oleo-pneumatic shock-absorbers. Mainwheels and tyres size 40 × 14-16 (low pressure 40 × 18-17 tyres, or C40 × 14-21/H40 × 14.5-19 tyres with heavy duty wheel brakes, are available optionally). Nosewheels and tyres size 24 × 7.7 (low pressure 24.5 × 8.5 tyres available

Boeing 737-200 Surveiller of the Indonesian Air Force

1994

Boeing 737-200 twin-turbofan short-range aircraft

1994

optionally). Bendix or Goodrich multidisc brakes. Hydro-Aire Mk III anti-skid units and automatic brakes standard.

POWER PLANT: Two Pratt & Whitney JT8D turbofans (details in individual model listings), in underwing pods. High performance target type thrust reversers, with full sound attenuation quiet nacelles. All models have standard fuel capacity of up to 19,532 litres (5,160 US gallons; 4,296 Imp gallons), with integral fuel cells in wing centre-section as well as two integral wing tanks. Long-range version has auxiliary fuel tank in rear lower cargo compartment, giving maximum fuel capacity of 22,598 litres (5,970 US gallons; 4,971 Imp gallons). Single-point pressure refuelling through leading-edge of starboard wing. Fuelling rate 1,135 litres (300 US gallons; 250 Imp gallons)/min. Auxiliary overwing fuelling points. Total oil capacity 41.5 litres (11 US gallons; 9.1 Imp gallons).

ACCOMMODATION: Crew of two side by side on flight deck. Details of passenger accommodation given under individual model descriptions. Passenger versions are equipped with forward airstair and a rear airstair is optional. Convertible passenger/cargo versions have the rear airstair as standard and forward airstair is optional. One plug type door at each corner, with passenger doors on port side and service doors on starboard side. Overwing emergency exit on each side. Basic passenger cabin has one lavatory and one galley at each end. Large-volume hand baggage overhead bins. Provision for a wide variety of interior arrangements. Freight holds forward and aft of wing, under floor.

SYSTEMS: Air conditioning and pressurisation system utilises engine bleed air. Maximum differential 0.52 bars (7.5 lb/sq in). Two functionally independent hydraulic systems with a third standby system, using fire resistant hydraulic fuel, for flying controls, flaps, slats, landing gear, nosewheel steering and brakes: pressure 207 bars (3,000 lb/sq in). No pneumatic system. Electrical supply provided by engine-driven generators. Garrett APU for air supply and electrical power in flight and on the ground, as well as engine starting. Engine bleed air for anti-icing supplied to engine nose cowls and all wing leading-edge slats.

AVIONICS AND EQUIPMENT: Equipment to satisfy FAA Cat. II low weather minimum criteria is standard, as well as SIADS Inc performance data computer system. Autopilot, specially designed for ILS localiser and glide slope control wheel steering. Optional equipment will permit Cat. IIIA capability. Very low frequency (VLF-Omega) navigation systems and a range of flight management systems with various levels of automation, including autothrottle and automatic flight control, are available as options.

DIMENSIONS, EXTERNAL:
- Wing span — 28.35 m (93 ft 0 in)
- Wing chord: at root — 4.71 m (15 ft 3 in)
  - at tip — 1.60 m (5 ft 3 in)
- Wing aspect ratio — 8.8
- Length overall — 30.53 m (100 ft 2 in)
- Length of fuselage — 29.54 m (96 ft 11 in)
- Height overall — 11.28 m (37 ft 0 in)
- Tailplane span — 10.97 m (36 ft 0 in)
- Wheel track — 5.23 m (17 ft 2 in)
- Wheelbase — 11.38 m (37 ft 4 in)
- Main passenger door (port, fwd):
  - Height — 1.83 m (6 ft 0 in)
  - Width — 0.86 m (2 ft 10 in)
  - Height to sill — 2.62 m (8 ft 7 in)
- Passenger door (port, rear): Height — 1.83 m (6 ft 0 in)
  - Width — 0.76 m (2 ft 6 in)
  - Width with airstair — 0.86 m (2 ft 10 in)
  - Height to sill — 2.74 m (9 ft 0 in)
- Emergency exits (overwing, port and stbd, each):
  - Height — 0.97 m (3 ft 2 in)
  - Width — 0.51 m (1 ft 8 in)
- Galley service door (stbd, fwd):
  - Height — 1.65 m (5 ft 5 in)
  - Width — 0.76 m (2 ft 6 in)
  - Height to sill — 2.62 m (8 ft 7 in)
- Service door (stbd, rear): Height — 1.65 m (5 ft 5 in)
  - Width — 0.76 m (2 ft 6 in)
  - Height to sill — 2.74 m (9 ft 0 in)
- Freight hold door (stbd, fwd): Height — 1.22 m (4 ft 0 in)
  - Width — 1.30 m (4 ft 3 in)
  - Height to sill — 1.30 m (4 ft 3 in)
- Freight hold door (stbd, rear): Height — 1.22 m (4 ft 0 in)
  - Width — 1.22 m (4 ft 0 in)
  - Height to sill — 1.55 m (5 ft 1 in)

DIMENSIONS, INTERNAL:
- Cabin, incl galley and toilet:
  - Length — 20.88 m (68 ft 6 in)
  - Max width — 3.53 m (11 ft 7 in)
  - Max height — 2.13 m (7 ft 0 in)
  - Floor area — 63.8 m² (687 sq ft)
  - Volume — 131.28 m³ (4,636 cu ft)
- Freight hold (fwd): Volume — 10.48 m³ (370 cu ft)
- Freight hold (rear): Volume — 14.30 m³ (505 cu ft)

AREAS:
- Wings, gross — 102.00 m² (1,098.0 sq ft)
- Ailerons (total) — 2.49 m² (26.8 sq ft)
- Trailing-edge flaps (total) — 16.87 m² (181.6 sq ft)
- Slats (total) — 6.52 m² (70.2 sq ft)
- Ground spoilers (total) — 3.68 m² (39.6 sq ft)
- Flight spoilers (total) — 2.64 m² (28.4 sq ft)
- Fin — 20.81 m² (224.0 sq ft)
- Rudder — 5.22 m² (56.2 sq ft)
- Tailplane — 28.99 m² (312.0 sq ft)
- Elevators, incl tabs (total) — 6.55 m² (70.5 sq ft)

WEIGHTS AND LOADINGS (standard aircraft at brake release weight of 52,390 kg; 115,500 lb except where indicated):
- Operating weight empty (JT8D-17A engines):
  - 200 — 27,445 kg (60,507 lb)
  - 200C all-passenger — 28,828 kg (63,555 lb)
  - 200C all-cargo — 27,231 kg (60,034 lb)
  - 200QC all-passenger — 30,141 kg (66,450 lb)
  - 200QC all-cargo — 27,500 kg (60,629 lb)
- Max payload: 200 — 15,645 kg (34,493 lb)
  - 200C all-passenger — 14,263 kg (31,445 lb)
  - 200C all-cargo — 15,860 kg (34,966 lb)
  - 200QC all-passenger — 12,950 kg (28,550 lb)
  - 200QC all-cargo — 15,590 kg (34,371 lb)
- Max T-O weight (all models):
  - basic — 52,390 kg (115,500 lb)
  - optional — 53,070 kg (117,000 lb) or 56,472 kg (124,500 lb) or 58,105 kg (128,100 lb)
- Max ramp weight (all models):
  - basic — 52,615 kg (116,000 lb)
  - optional — 53,295 kg (117,500 lb) or 56,700 kg (125,000 lb) or 58,332 kg (128,600 lb)
- Max zero-fuel weight (all models):
  - basic — 43,091 kg (95,000 lb)
  - optional for 200C — 44,906 kg (99,000 lb)
- Max landing weight (all models):
  - basic — 46,720 kg (103,000 lb)
  - optional — 47,627 kg (105,000 lb) or 48,534 kg (107,000 lb)
- Wing loading (all models):
  - basic — 575.5 kg/m² (117.9 lb/sq ft)
  - max optional — 638.2 kg/m² (130.7 lb/sq ft)
- Power loading (JT8D-17A engine, all models):
  - basic — 368 kg/kN (3.61 lb/lb st)
  - max optional — 408 kg/kN (4.00 lb/lb st)

WEIGHTS AND LOADINGS (at brake release weight of 56,472 kg; 124,500 lb):
- Operating weight empty — 27,574 kg (60,790 lb)
- Max payload — 15,517 kg (34,210 lb)
- Max T-O weight — 56,472 kg (124,500 lb)
- Max ramp weight — 56,700 kg (125,000 lb)
- Max zero-fuel weight — 43,091 kg (95,000 lb)
- Max landing weight — 48,534 kg (107,000 lb)
- Max wing loading — 620.24 kg/m² (127.04 lb/sq ft)
- Max power loading (JT8-17A engines) — 397 kg/kN (3.9 lb/lb st)

PERFORMANCE (ISA, with JT8D-17A engines):
- Max operating speed, all models — Mach 0.84 (350 kts; 648 km/h; 402 mph EAS)
- Max cruising speed, at an average cruise weight of 45,359 kg (100,000 lb) at 10,060 m (33,000 ft) — 462 kts (856 km/h; 532 mph)
- Econ cruising speed at 10,060 m (33,000 ft) — Mach 0.73
- Stalling speed, flaps down, at 46,720 kg (103,000 lb) landing weight — 102 kts (189 km/h; 117 mph)
- Runway LCN (at max ramp weight of 52,615 kg; 116,000 lb, optimum tyre pressure and 20 in flexible pavement): 40 × 14-16 tyres — 53
  - C40 × 14-21 tyres — 53
  - C40 × 18-17 tyres — 38
- FAR T-O distance to 10.7 m (35 ft), 737-200 at 49,435 kg (109,000 lb) AUW and 28.9°C (84°F):
  - JT8D-9A engines — 2,027 m (6,650 ft)
  - JT8-17A engines — 1,615 m (5,300 ft)
- FAR landing distance from 15 m (50 ft), 737-200 at landing weight of 46,720 kg (103,000 lb) — 1,372 m (4,500 ft)
- Min ground turning radius — 17.58 m (57 ft 8 in)
- Range, JT8D-17A engines, FAR domestic reserves, cruising at 10,060 m (33,000 ft), at 52,615 kg (116,000 lb) ramp weight with 115 passengers and 19,533 l (5,160 US gallons; 4,296 Imp gallons) fuel — 1,855 n miles (3,437 km; 2,136 miles)
- Range, conditions as above, except 58,332 kg (128,600 lb) ramp weight and 22,599 l (5,970 US gallons; 4,971 Imp gallons) fuel — 2,530 n miles (4,688 km; 2,913 miles)
- Range, conditions as above, with 130 passengers — 2,255 n miles (4,179 km; 2,596 miles)

OPERATIONAL NOISE LEVELS (JT8D-9 engines and nacelle acoustic treatment, FAR Pt 36):
- T-O at 52,390 kg (115,500 lb) brake release weight — 95.3 EPNdB
- Sideline at 52,390 kg (115,500 lb) brake release weight — 100.6 EPNdB
- Approach at 46,720 kg (103,000 lb) max landing weight — 101.1 EPNdB

UPDATED

BOEING MODEL 747

US Air Force designations: VC-25A and E-4

TYPE: High-capacity, widebody, long-range airliner.

PROGRAMME: Programme announced 13 April 1966 (first ever widebody jet airliner), with Pan American order for 25; official programme launch 25 July 1966; first flight 9 February 1969; FAA certification 30 December 1969; first delivery (to Pan Am) 12 December 1969; first route service New York-London flown 22 January 1970. In May 1990 Boeing decided to market only the -400; by 31 March 1991, two 747-200F Freighters remained to be delivered.

Produced as 747-100 (167), 747-100B (9), 747SP (45), 747-100SR (29), 747-200B (226, including 2 USAF VC-25A 'Air Force One'), 747-200C Convertible (13), 747-200F Freighter (73), 747-200M (77), 747-300 (56), 747-300M (21) and 747-300SR (4); total 720. Four other 747-200Bs modified by Boeing Aerospace as **E-4** command post aircraft; 19 Pan American 747s modified as passenger/cargo **C-19A**s by Boeing Military Airplanes for Civil Reserve Air Fleet (currently being operated by Evergreen International).

VERSIONS: **747-100**: The original 747-100 was introduced into commercial service in January 1970, and 167 were sold. Two of these have since been converted to -200B Combis. The -100B incorporates strengthened wing, fuselage and landing gear structure. Initial order, with 213.5 kN (48,000 lb st) JT9D-7F engines, was placed by Iran Air in 1978, and this aircraft was delivered on 5 July 1979. Subsequent versions, with maximum taxi weights of 323,411 kg (713,000 lb), 334,751 kg (738,000 lb), and 341,555 kg (753,000 lb), allow for the installation of a variety of optional engines in addition to the basic 206.8 kN (46,500 lb st) General Electric CF6-45A2, including the 233.5 kN (52,500 lb st) CF6-50E2, and 236.25 kN (53,110 lb st) Rolls-Royce RB211-524D4. Nine were built.

747SP: Lighter weight, shorter bodied derivative of the 747-100B. Described separately.

747-100SR: This short-range version of the 747-100B embodies structural changes required for high take-off and landing cycles. It is available at maximum taxi weights up to 273,515 kg (603,000 lb) with the same engines as the 747-100B. The first 747-100SR flew 4 September 1973 and was delivered 26 September 1973. Total of 29 were built.

747-200B: Passenger version, with same accommodation as 747-100B. First flown 11 October 1970 and certificated 23 December 1970; deliveries began 15 January 1971. Available at taxi weights up to 379,200 kg (836,000 lb). Standard engines include 243.5 kN (54,750 lb st) JT9D-7R4G2; 233.5 kN (52,500 lb) CF6-50E2 and 236.25 kN (53,110 lb st) RB211-524D4. Optional engines include 252.2 kN (56,700 lb st) CF6-80C2B1 and 236.25 kN (53,110 lb st) RB211-524D4-B. Total of 227 ordered. Selected by US Air Force for Presidential transport to replace VC-137 aircraft.

747-200C Convertible: Version of 747-200 which can be converted from all-passenger to all-cargo, or five combinations of both. The first 747-200C flew 23 March 1973, was certificated on 17 April, and was delivered to World Airways on 27 April 1973. Maximum T-O weight of 377,840 kg (833,000 lb) with Pratt & Whitney JT9D-7R4G2, Rolls-Royce RB211-524D4, General Electric CF6-50E2 or CF6-80C2.

747-200F Freighter: Version of the 747-200 capable of delivering 90,720 kg (200,000 lb) of palletised cargo over a range of 4,480 n miles (8,300 km; 5,159 miles). Described separately.

747-200M: Combi version of the basic 747-200B, incorporating a cargo door in the port side of the fuselage,

Boeing 747 Combi passenger/freight transport

1994

aft of the wing. This permits main deck layouts for passengers only, or for passengers and up to 12 main deck pallets/containers, with passenger and cargo areas separated by removable bulkhead. The first modification to Combi configuration was carried out on a Sabena 747-100, and redelivery was made in February 1974. The first production Combi was delivered to Air Canada in March 1975.

747-300: Version with extended upper deck to increase passenger accommodation. Described separately.

747-300M: Combi version of 737-300. Described separately.

747-300SR: Short-range version of 747-300.

747-400 and 400M: Advanced and Combi versions of 747-300.

E-4: Advanced Airborne Command post version 747, developed for the US Air Force. Four built.

VC-25A: Presidential aircraft known as Air Force One.

UPGRADES: **Boeing:** C-19A. A major modification to enhance capability is available to operators of 747-100 and -200B aircraft. Performed at Boeing facilities, it includes installation of a side cargo door. The side cargo door modification enables an operator to convert 747-100 or -200B passenger aircraft into version with main deck cargo capability. Variations include an all-cargo Special Freighter, a 6- or 12-pallet Combi, and an all-passenger or all-cargo Convertible. Like the Combi, the Convertible configuration can also be operated in a mixed passenger/cargo mode. The major elements of the side door modification include installation of a 3.05 m (10 ft 0 in) high by 3.40 m (11 ft 2 in) wide side cargo door, strengthened main deck floor, a fully powered or manual cargo handling system, and an option to increase certificated design weights of the aircraft. Boeing Military Airplanes is modifying 19 747s into the passenger/cargo convertible configuration to support the US Air Force's CRAF (Civil Reserve Air Fleet) enhancement programme. The first converted aircraft was redelivered to Pan American (now defunct) on 31 May 1985. These aircraft are known by the military designation C-19A and are currently being operated by Evergreen International.

Boeing: The 747 can be fitted or retrofitted with a performance management system (PMS) developed by Boeing and Delco. The PMS computer memory is programmed with airline economic factors and the performance characteristics of the individual aircraft/engine combination, and receives continuous inputs of altitude, airspeed, air temperature, fuel flow, wind velocity and other data during flight. Using these data, coupled with existing autopilot, autothrottle and inertial navigation systems. The PMS calculates, displays and controls automatically the optimum or desired airspeed, engine power setting, attitude and flight path of the aircraft for minimum fuel burn and/or minimum operating cost. The first complete system was delivered for airline service in June 1982. Results of in-service airline evaluations show trip fuel burn reductions in excess of one per cent.

Boeing: E-4. See separate entry.

OPERATORS: Versions of the B747 are in service with the armed forces of the following countries: Brunei (1), Iran (11), Japan (2), Oman (1), Saudi Arabia (1) and the United States (6).

The following description applies to the 747-200B:

FLYING CONTROLS: Low-speed outboard ailerons, high-speed inboard ailerons; eight overwing flight spoiler panels assist ailerons and act as airbrakes; variable incidence tailplane; no trim tabs; all surfaces fully powered hydraulically. High lift devices include triple-slotted trailing-edge flaps; 10 leading-edge flaps outboard; three sections Krueger flaps inboard; four overwing ground spoiler panels inboard.

STRUCTURE: Wing and tail surfaces are aluminium alloy dual-path fail-safe structures; aluminium honeycomb spoiler panels. Frame/stringer/stressed skin fuselage containing some bonding.

LANDING GEAR: Hydraulically retractable tricycle type. Twin-wheel nose unit retracts forward. Main gear comprises four four-wheel bogies: two, mounted side by side under fuselage at wing trailing-edge, retract forward; two, mounted underwings, retract inward. Disc brakes on all mainwheels, with individually controlled anti-skid units.

POWER PLANT: Four Pratt & Whitney, General Electric or Rolls-Royce turbofans, in pods pylon-mounted on wing leading-edges. Fuel in seven integral tanks: centre wing tank, two inboard main tanks, two outboard main tanks and two inboard reserve tanks. Fuselage tank optional. Refuelling point on each wing between inboard and outboard engines.

ACCOMMODATION: Normal operating crew of three, on flight deck above level of main deck. Observer station and provision for second observer station are provided. Crew rest area available as option at rear of upper deck. Basic accommodation for 452 passengers, made up of 32 first class and 420 economy class, which includes a 32-passenger upper deck (extended on 747-400, which see). Alternative layouts accommodate 447 economy class passengers in nine-abreast seating or 516 10-abreast, with 32 passengers on upper deck. All versions have two aisles. Five passenger doors on each side, of which two forward of wing on each side are normally used. Freight holds under floor, forward and aft of wing, with doors on starboard side. One door on forward hold, two on rear hold. Aircraft is designed for fully mechanical loading of baggage and freight. An optional side cargo door is available for passenger, convertible and freighter versions of the Model 747. Installed aft of door 4 on the port side of the fuselage, it allows the carriage of main deck cargo on passenger versions. Addition of this door to the freighter allows loads up to 3.05 m (10 ft) in height to be accommodated aft of the flight deck, and also makes possible simultaneous nose and side cargo handling.

SYSTEMS: Air cycle air conditioning system. Pressure differential 0.61 bar (8.9 lb/sq in). Four independent hydraulic systems, pressure 207 bars (3,000 lb/sq in), maximum capacity 265 litres (70 US gallons; 58 Imp gallons)/min at 196.5 bars (2,850 lb/sq in), each with one engine-driven and one pneumatically driven pump. The latter pumps supplement or substitute for engine-driven pumps. Reservoir in each system, pressurised by engine bleed air via a pressure regulation module. Reservoir relief valve pressure is nominal 4.48 bars (65 lb/sq in). A small AC powered electric pump is installed to charge the brake accumulator during towing of the aircraft. Electrical supply from four air-cooled 60 kVA generators mounted one on each engine. Two 60 kVA generators (supplemental cooling allows 90 kVA each) mounted on APU for ground operation and to supply primary electrical power when engine-mounted generators are not operating. Three-phase 400 Hz constant frequency AC generators, 115/200 V output. 28 V DC power obtained from transformer-rectifier units. 24 V 36 Ah Ni/Cd battery for selected ground functions and as in-flight back-up. Gas-turbine APU for pneumatic and electrical supplies.

AVIONICS: Standard avionics include three ARINC 566 VHF communications systems, two ARINC 533A HF communications systems, one ARINC 531 Selcal, three ARINC 547 VOR/ILS navigation systems, two ARINC 570 ADF, marker beacon receiver, two ARINC 568 DME, two ARINC 572 ATC, three ARINC 552 low-range radio altimeters, two ARINC 564 weather radar units, three ARINC 561 inertial navigation systems, two heading reference systems, ARINC 573 flight recorder, ARINC 557 cockpit voice recorder, integrated electronic flight control system with autothrottle and roll-out guidance to provide automatic stabilisation, path control and pilot assist functions for Cat. II and III landing conditions, two ARINC 565 central air data systems, stall warning system, central instrument warning system, ground proximity warning system, attitude and navigation instrumentation, and standby attitude indication.

EQUIPMENT: ARINC 412 interphone, ARINC 560 passenger address system, multiple passenger service and entertainment system.

DIMENSIONS, EXTERNAL:

Wing span	59.64 m (195 ft 8 in)
Length: overall	70.66 m (231 ft 10 in)
fuselage	68.63 m (225 ft 2 in)
Height overall	19.33 m (63 ft 5 in)
Tailplane span	22.17 m (72 ft 9 in)
Wheel track	11.00 m (36 ft 1 in)
Wheelbase	25.60 m (84 ft 0 in)
Passenger doors (10, each): Height	1.93 m (6 ft 4 in)
Width	1.07 m (3 ft 6 in)
Height to sill	approx 4.88 m (16 ft 0 in)
Baggage door (front hold): Height	1.68 m (5 ft 6 in)
Width	2.64 m (8 ft 8 in)
Height to sill	approx 2.64 m (8 ft 8 in)
Baggage door (fwd door, rear hold):	
Height	1.68 m (5 ft 6 in)
Width	2.64 m (8 ft 8 in)
Height to sill	approx 2.69 m (8 ft 10 in)

Boeing 747-200 passenger aircraft of the Middle East Airlines *(Peter J. Cooper)*

1994

Boeing 747-100 passenger aircraft of United Airlines *(Peter J. Cooper)*

1994

Boeing 747-200B (VC-25A) Air Force One presidential aircraft *(Boeing)*

1995

Bulk loading door (rear door on rear hold):
Height 1.19 m (3 ft 11 in)
Width 1.12 m (3 ft 8 in)
Height to sill approx 2.90 m (9 ft 6 in)
Optional cargo door (port): Height 3.05 m (10 ft 0 in)
Width 3.40 m (11 ft 2 in)

DIMENSIONS, INTERNAL:
Cabin, incl toilets and galleys:
Length 57.00 m (187 ft 0 in)
Max width 6.13 m (20 ft 1½ in)
Max height 2.54 m (8 ft 4 in)
Floor area, passenger deck 327.9 m^2 (3,529 sq ft)
Volume, passenger deck 789 m^3 (27,860 cu ft)
Cargo hold (fwd, containerised) volume 78.4 m^3 (2,768 cu ft)
Cargo hold (rear, containerised) volume 68.6 m^3 (2,422 cu ft)
Bulk volume 28.3 m^3 (1,000 cu ft)

AREAS:
Wings, reference area 511 m^2 (5,500 sq ft)
Ailerons (total) 20.6 m^2 (222 sq ft)
Trailing-edge flaps (total) 78.7 m^2 (847 sq ft)
Leading-edge flaps (total) 48.1 m^2 (518 sq ft)
Spoilers (total) 30.8 m^2 (331 sq ft)
Fin 77.1 m^2 (830 sq ft)
Rudder 22.9 m^2 (247 sq ft)
Tailplane 136.6 m^2 (1,470 sq ft)
Elevators (total) 32.5 m^2 (350 sq ft)

WEIGHTS AND LOADINGS (Letters are used to denote engine installations as follows: A: JT9D-7R4G2, B: CF6-45A2, C: CF6-50E2, D: CF6-80C2, E: RB211-524D4):
Operating weight empty (approx) for max available gross weights:
747-100SR (550 passengers):
B 162,431 kg (358,100 lb)
747-100B (366 passengers):
B 169,417 kg (373,500 lb)
E 171,866 kg (378,900 lb)
747-200B (366 passengers):
A 169,961 kg (374,700 lb)
C 171,548 kg (378,200 lb)
D 172,728 kg (380,800 lb)
E 173,998 kg (383,600 lb)
747-200M (257 passengers and seven pallets):
A 171,821 kg (378,800 lb)
C 173,408 kg (382,300 lb)
D 174,587 kg (384,900 lb)
E 175,858 kg (387,700 lb)
747-200C (366 passengers):
A 175,313 kg (386,500 lb)
C 176,901 kg (390,000 lb)
D 178,080 kg (392,600 lb)
E 179,350 kg (395,400 lb)
747-200C (28 pallets): A 163,973 kg (361,500 lb)
C 165,561 kg (365,000 lb)
D 166,740 kg (367,600 lb)
E 168,010 kg (370,400 lb)
Max fuel weight:
747-100B, 747-100SR: B, C 147,181 kg (324,480 lb)
E 148,324 kg (327,000 lb)
747-200B, 747-200C: A, E 165,289 kg (364,400 lb)
C, D 164,141 kg (361,870 lb)
Max structural payload:
747-100SR (550 passengers):
B 76,385 kg (168,400 lb)
747-100B (366 passengers):
B, C 69,399 kg (153,000 lb)
E 66,950 kg (147,600 lb)
747-200B (366 passengers):
A 68,855 kg (151,800 lb)
C 67,268 kg (148,300 lb)
D 66,088 kg (145,700 lb)
E 64,818 kg (142,900 lb)
747-200M (257 passengers and seven pallets):
A 75,387 kg (166,200 lb)
C 73,799 kg (162,700 lb)
D 72,620 kg (160,100 lb)
E 71,350 kg (157,300 lb)
747-200C (366 passengers):
A 92,306 kg (203,500 lb)
C 90,718 kg (200,000 lb)
D 89,539 kg (197,400 lb)
E 88,269 kg (194,600 lb)
747-200C (28 pallets): A 103,646 kg (228,500 lb)
C 102,058 kg (225,000 lb)
D 100,879 kg (222,400 lb)
E 99,609 kg (219,600 lb)
Max T-O weight:
747-100SR: B 235,870 kg (520,000 lb)
or 272,155 kg (600,000 lb)
747-100B: B, E 322,050 kg (710,000 lb)
or 333,390 kg (735,000 lb)
or 340,195 kg (750,000 lb)
747-200B, -200M, -200C:
A, C, D, E 351,534 kg (775,000 lb)
or 356,070 kg (785,000 lb)
or 362,875 kg (800,000 lb)
or 371,945 kg (820,000 lb)
or 377,840 kg (833,000 lb)
Max ramp weight:
747-100B: B, C, E 341,555 kg (753,000 lb)
747-200B, -200M, -200C:
A, C, D, E 379,200 kg (836,000 lb)
Max zero-fuel weight:
747-100SR: B 238,815 kg (526,500 lb)
747-100B: B, E 238,815 kg (526,500 lb)
747-200B: A, C, D, E 238,815 kg (526,500 lb)
747-200M: A, C, D, E 247,205 kg (545,000 lb)
747-200C: A, C, D, E 267,620 kg (590,000 lb)
Max landing weight:
747-100SR: B 229,065 kg (505,000 lb)
or 238,135 kg (525,000 lb)
747-100B: B, E 255,825 kg (564,000 lb)
747-200B: A, C, D, E 255,825 kg (564,000 lb)
or 265,350 kg (585,000 lb)
or 285,765 kg (630,000 lb)
747-200M: A, C, D, E 285,765 kg (630,000 lb)
747-200C: A, C, D, E 285,765 kg (630,000 lb)

PERFORMANCE (747-100B at max T-O weight of 340,195 kg; 750,000 lb, 747-200B at max T-O weight of 377,840 kg; 833,000 lb, except where indicated. Engines as designated under Weights and Loadings):
Max level speed at 9,150 m (30,000 ft):
747-100B at AUW of 272,160 kg (600,000 lb):
B 525 kts (973 km/h; 604 mph)
747-200B at AUW of 317,515 kg (700,000 lb):
A 527 kts (977 km/h; 607 mph)
C 523 kts (968 km/h; 602 mph)
D 530 kts (981 km/h; 610 mph)
E 522 kts (967 km/h; 601 mph)
Cruise ceiling, all versions 13,715 m (45,000 ft)
Min ground turning radius 22.86 m (75 ft 0 in)
Runway LCN (W: 334,750 kg; 738,000 lb, X: 341,555 kg; 753,000 lb, Y: 366,500 kg; 808,000 lb, Z: 379,200 kg; 836,000 lb max taxi weight on h = 0.51 m; 20 in flexible pavement): W 81
X 83
Y 86
Z 88
Runway LCN (weights as above, on 11.02 m; 40 in rigid pavement): W 87
X 89
Y 93
Z 95
FAR T-O distance to 10.7 m (35 ft) at S/L, ISA:
747-100B: B 3,050 m (10,000 ft)
747-200B: A 3,170 m (10,400 ft)
C 3,292 m (10,800 ft)
D 3,079 m (10,100 ft)
E 3,155 m (10,350 ft)
FAR landing field length:
747-100B, -200B at 255,825 kg (564,000 lb) 1,881 m (6,170 ft)
747-200B at 265,350 kg (585,000 lb) 1,942 m (6,370 ft)
747-200B at 285,765 kg (630,000 lb) 2,112 m (6,930 ft)
Range, long-range cruise, typical international reserves of 5% trip fuel, 200 n mile (371 km; 230 mile) alternate, 30 min hold at 457 m (1,500 ft):
747-100B with 366 passengers and baggage:
B 5,500 n miles (10,193 km; 6,333 miles)
747-200B with 366 passengers and baggage:
A 6,550 n miles (12,138 km; 7,542 miles)
C 6,350 n miles (11,760 km; 7,307 miles)
D 6,900 n miles (12,778 km; 7,940 miles)
E 6,600 n miles (12,223 km; 7,595 miles)
Ferry range, long-range cruise, reserves as above:
747-100B: B 7,300 n miles (13,520 km; 8,400 miles)
747-200B: A 7,450 n miles (13,797 km; 8,573 miles)
C 7,200 n miles (13,334 km; 8,285 miles)
D 7,900 n miles (14,630 km; 9,091 miles)
E 7,500 n miles (13,890 km; 8,631 miles)

OPERATIONAL NOISE LEVELS (As per FAR Pt 36, A: RB211-524C engines at brake release weight (BRW) of 340,195 kg; 750,000 lb and landing weight of 265,350 kg; 585,000 lb, B: JT9D-7R4G2 at BRW of 377,840 kg; 833,000 lb and landing weight of 285,765 kg; 630,000 lb):
T-O: A 104 EPNdB
B 106 EPNdB
Approach: A 107 EPNdB
B 107 EPNdB
Sideline: A 97 EPNdB
B 99 EPNdB

Boeing 747-200B (VC-25A) Air Force One upper deck interior *(Boeing)*

1995

UPDATED

BOEING MODEL 747-200F FREIGHTER

TYPE: Four-engine cargo transport aircraft.

PROGRAMME: The 747-200F Freighter is a version of the standard Model 747-200, capable of delivering 90,720 kg

(200,000 lb) of containerised or palletised main deck cargo over a range of more than 4,500 n miles (8,340 km; 5,180 miles). The first 747-200F flew for the first time 30 November 1971. It was certificated 7 March 1972 and delivered to Lufthansa two days later.

To ensure maximum utilisation, the 747-200F has a special loading system that enables two men to handle and stow the maximum load of up to 112,400 kg (247,800 lb) in 30 minutes.

The 747-200F can carry up to 29 containers measuring 3.05 × 2.44 × 2.44 m (10 ft long, 8 ft high and 8 ft wide), plus 30 lower lobe containers, each of 4.90 m³ (173 cu ft) capacity, and 22.65 m³ (800 cu ft) of bulk cargo. The main deck can accommodate ANSI/ISO containers of up to 12.2 m (40 ft) in length, and many combinations of pallets and igloos. The lower hold can accommodate combinations of IATAI-A1 or -A2, and ATA LD-1 or -3 half-width containers, full-width or main-deck baggage containers, and many combinations of pallets and igloos.

The nose loading door, which is hinged just below the flight deck to allow it to swing forward and upward, gives clear access to the main deck to facilitate the handling of long or large loads. A side door is available as an option, allowing simultaneous nose and side loading. The side cargo door will accept palletised loads up to 3.05 m (10 ft 0 in) in height.

The description of the Model 747-200B applies also to the Model 747-200F except as follows:

DESIGN FEATURES: As for Model 747-200B, except for nose cargo loading door, which is hinged at the top and opens forward and upward. No windows in freight hold.

ACCOMMODATION: Normal operating crew of three on flight deck. Nose cargo loading door, hinged at top. Lower lobe cargo doors, on starboard side, one forward and one aft of wing. Bulk compartment cargo door, on starboard side aft of lower lobe cargo door. Two doors for crew on port side of aircraft. Aircraft is designed for fully mechanical loading of freight.

DIMENSIONS, EXTERNAL: As for 747-200B except:
- Crew doors (two, each): Height 1.93 m (6 ft 4 in)
  - Width 1.07 m (3 ft 6 in)
  - Height to sill approx 4.88 m (16 ft 0 in)
- Nose cargo loading door: Height 2.49 m (8 ft 2 in)
  - Width at top (min) 2.67 m (8 ft 9 in)
  - Max width 3.81 m (12 ft 6 in)
  - Height to sill approx 4.90 m (16 ft 1 in)

DIMENSIONS, INTERNAL:
- Main cargo deck: Height 2.54 m (8 ft 4 in)
  - Max width at floor level 5.92 m (19 ft 5 in)
- Lower lobe: Width at floor level 3.18 m (10 ft 5 in)
- Total cargo volume 687 m³ (24,260 cu ft)

AREAS:
- Wings, reference area 511 m² (5,500 sq ft)
- Ailerons (total) 20.6 m² (222 sq ft)
- Trailing-edge flaps (total) 78.7 m² (847 sq ft)
- Leading-edge flaps (total) 48.1 m² (518 sq ft)
- Spoilers (total) 30.8 m² (331 sq ft)
- Fin 77.1 m² (830 sq ft)
- Rudder 22.9 m² (247 sq ft)
- Tailplane 136.6 m² (1,470 sq ft)
- Elevators (total) 32.5 m² (350 sq ft)

WEIGHTS AND LOADINGS (Letters are used to denote engine installations as follows: A: JT9D-7R4G2, B: CF6-50E2, C: CF6-80C2, D: RB211-524D4):
- Operating weight empty (approx) for max available gross weights: A 155,219 kg (342,200 lb)
  - B 156,807 kg (345,700 lb)
  - C 157,986 kg (348,300 lb)
  - D 159,256 kg (351,100 lb)
- Max fuel weights as for 747-200B
- Max payload (29 pallets): A 112,400 kg (247,800 lb)
  - B 110,812 kg (244,300 lb)
  - C 109,633 kg (241,700 lb)
  - D 108,363 kg (238,900 lb)
- Max T-O weight: A, B, C, D 351,535 kg (775,000 lb) or 356,070 kg (785,000 lb) or 362,875 kg (800,000 lb) or 371,945 kg (820,000 lb) or 377,840 kg (833,000 lb)
- Max ramp weight: A, B, C, D 379,200 kg (836,000 lb)
- Max zero-fuel weight: A, B, C, D 267,620 kg (590,000 lb)
- Max landing weight: A, B, C, D 285,765 kg (630,000 lb)

PERFORMANCE (at max T-O weight of 377,840 kg; 833,000 lb, except where indicated. Engines as designated under Weights and Loadings):
- Max level speed at AUW of 317,515 kg (700,000 lb), at 9,150 m (30,000 ft):
  - A 527 kts (977 km/h; 607 mph)
  - B 523 kts (968 km/h; 602 mph)
  - C 530 kts (981 km/h; 610 mph)
  - D 522 kts (967 km/h; 601 mph)
- Cruise ceiling 13,715 m (45,000 ft)
- Min ground turning radius 22.86 m (75 ft 0 in)
- FAR T-O distance to 10.7 m (35 ft) at S/L, ISA:
  - A 3,170 m (10,400 ft)
  - B 3,292 m (10,800 ft)
  - C 3,079 m (10,100 ft)
  - D 3,155 m (10,350 ft)
- FAR landing field length:
  - at AUW of 255,825 kg (564,000 lb) 1,881 m (6,170 ft)
  - at AUW of 265,350 kg (585,000 lb) 1,942 m (6,370 ft)
  - at AUW of 285,765 kg (630,000 lb) 2,112 m (6,930 ft)
- Range, long-range step cruise, typical international reserves of 5% trip fuel, 200 n mile (371 km; 230 mile) alternate, 30 min hold at 457 m (1,500 ft) with 90,270 kg (200,000 lb) payload:
  - A 4,700 n miles (8,704 km; 5,408 miles)
  - B 4,550 n miles (8,426 km; 5,236 miles)
  - C 4,900 n miles (9,075 km; 5,639 miles)
  - D 4,650 n miles (8,612 km; 5,351 miles)
- Ferry range, long-range cruise, reserves as above:
  - A 7,900 n miles (14,630 km; 9,091 miles)
  - B 7,650 n miles (14,168 km; 8,803 miles)
  - C 8,300 n miles (15,371 km; 9,551 miles)
  - D 7,950 n miles (14,723 km; 9,148 miles)

VERIFIED

Boeing 747SP passenger aircraft for use on lower density routes *(Boeing)*

1995

BOEING MODEL 747SP

TYPE: Four-engine passenger aircraft.

PROGRAMME: A lower weight longer range version of the basic Model 747, the 747SP (special performance), was introduced for use on lower density routes. Retaining a 90 per cent commonality of components with the standard Model 747, the major change is a reduction in overall length of 14.35 m (47 ft 1 in). It made its first flight 4 July 1975, and FAA certification was received 4 February 1976. First delivery was made on 5 March that year. Over 40 have been produced.

The description of the basic Model 747-100B/-200B applies also to the 747SP, except for the following details:

DESIGN FEATURES: As for Model 747. Wing structural materials are of reduced gauge. Large flap track fairings replaced by small link fairings. New wing/body fairings and leading-edge fillets. The fuselage is as for Model 747, except length reduced.

FLYING CONTROLS: Generally as for Model 747 except that trailing-edge flaps are of single-slotted variable pivot type.

STRUCTURE: The tail unit is similar to 747, but tailplane span increased by 3.05 m (10 ft). Two segment elevators. Height of fin increased by 1.52 m (5 ft 0 in). Double-hinged rudder.

LANDING GEAR: As for Model 747, except structural weight reduced. Mainwheel tyres size 46 × 16, pressure 12.63 bars (183 lb/sq in). Nosewheel tyres size 49 × 17, pressure 13.8 bars (200 lb/sq in). Higher gross weight aircraft use 747-100 wheels and brakes. Modified 747-100 wheels and brakes. Modified 747-100 steel brakes by Bendix.

POWER PLANT: Four General Electric CF6-45A2 or CF6-50E2F turbofans, each of 206.8 kN (46,500 lb st); four Pratt & Whitney JT9D-7A turbofans, each of 205.7 kN (46,250 lb st); four Rolls-Royce RB211-52B2 turbofans, each of 222.8 kN (50,100 lb st), RB211-524C2 engines each of 229.5 kN (51,600 lb st), or RB211-524D4 engines each of 236.25 kN (53,110 lb st). Fuel system and oil capacity as for Model 747-100B, but with an additional 5,966 litres (1,576 US gallons; 1,312 Imp gallons) reserve fuel, providing a total capacity of 190,625 litres (50,359 US gallons; 41,932 Imp gallons).

ACCOMMODATION: Normal operating crew of three on flight deck above level of main deck. Observer station and provision for second observer station are provided. An overhead crew rest area located above the main deck in the aft section of the passenger cabin is optional, with accommodation for four bunks and four seats; or eight bunks and two seats; or two bunks; two seats and five sleeper seats. Accommodation for 299 passengers on main deck, with 28 first class seats in forward area and 10-abreast seating throughout the major part of the main cabin. Seating for 32 passengers on upper deck, giving total capacity of 331 passengers. Maximum high-density accommodation for 440 passengers. Four doors on each side, two forward and two aft of the wing. Crew door on starboard side giving access to upper deck. Freight holds under floor, forward and aft of wing box, each with one door on starboard side.

SYSTEMS: Air cycle air conditioning system. Pressure differential 0.61 bar (8.9 lb/sq in). Four independent hydraulic systems, pressure 207 bars (3,000 lb/sq in), maximum capacity 265 litres (70 US gallons; 58 Imp gallons)/min at 196.5 bars (2,850 lb/sq in), each with one engine-driven and one pneumatically driven pump. The latter pumps supplement or substitute for engine-driven pumps. Reservoir in each system, pressurised by engine bleed air via a pressure regulation module. Reservoir relief valve pressure is nominal 4.48 bars (65 lb/sq in). A small AC powered electric pump is installed to charge the brake accumulator during towing of the aircraft. Electrical supply from four air-cooled 60 kVA generators mounted one on each engine. Two 60 kVA generators (supplemental cooling allows 90 kVA each) mounted on APU for ground operation and to supply primary electrical power when engine-mounted generators are not operating. Three-phase 400 Hz constant frequency AC generators, 115/200 V output. 28 V DC power obtained from transformer-rectifier units. 24 V 36 Ah Ni/Cd battery for selected ground functions and as in-flight back-up. Gas-turbine APU for pneumatic and electrical supplies.

AVIONICS: Standard avionics include three ARINC 566 VHF communications systems, two ARINC 533A HF communications systems, one ARINC 531 Selcal, three ARINC 547 VOR/ILS navigation systems, two ARINC 570 ADF, marker beacon receiver, two ARINC 568 DME, two ARINC 572 ATC, three ARINC 552 low-range radio altimeters, two ARINC 564 weather radar units, three ARINC 561 inertial navigation systems, two heading reference systems, ARINC 573 flight recorder, ARINC 557 cockpit voice recorder, integrated electronic flight control system with autothrottle and roll-out guidance to provide automatic stabilisation, path control and pilot assist functions for Cat. II and III landing conditions, two ARINC 565 central air data systems, stall warning system, central instrument warning system, ground proximity warning system, attitude and navigation instrumentation, and standby attitude indication.

EQUIPMENT: ARINC 412 interphone, ARINC 560 passenger address system, multiple passenger service and entertainment system.

DIMENSIONS, EXTERNAL: As for 747-100B/200B except:
- Length overall 56.31 m (184 ft 9 in)
- Height overall 19.94 m (65 ft 5 in)
- Tailplane span 25.22 m (82 ft 9 in)
- Wheelbase 20.52 m (67 ft 4 in)

DIMENSIONS, INTERNAL:
- Cabin, incl toilets and galleys:
  - Length 42.27 m (138 ft 8 in)
  - Max width 6.13 m (20 ft 1½ in)
  - Max height 2.54 m (8 ft 4 in)
  - Floor area, passenger deck 253.2 m² (2,725 sq ft)
  - Volume, passenger deck 613.34 m³ (21,660 cu ft)
- Baggage hold volume (fwd) 48.99 m³ (1,730 cu ft)
- Baggage hold volume (rear, containerised) 48.99 m³ (1,730 cu ft)
- Bulk compartment volume (rear) 11.33 m³ (400 cu ft)

AREAS: As for 747-100B/200B except:
- Ailerons (total) 20.37 m² (219.3 sq ft)
- Trailing-edge flaps (total) 78.78 m² (848 sq ft)
- Fin 82.22 m² (885 sq ft)
- Tailplane 142.51 m² (1,534 sq ft)

WEIGHTS AND LOADINGS (RB211-524D4 engines):
- Operating weight empty (approx, with 276 passengers) 151,454 kg (333,900 lb)
- Max fuel weight 149,361 kg (329,285 lb)
- Max payload 34,518 kg (76,100 lb) or 41,322 kg (91,100 lb)

Boeing 747-300 long-range passenger aircraft

1994

Max T-O weight (dry engines) 285,765 kg (630,000 lb)
or 299,370 kg (660,000 lb)
or 303,905 kg (670,000 lb)
or 312,980 kg (690,000 lb)
or 315,700 kg (696,000 lb)
or 317,515 kg (700,000 lb)
Max ramp weight 284,485 kg (636,000 lb)
or 302,090 kg (666,000 lb)
or 306,630 kg (676,000 lb)
or 315,700 kg (696,000 lb)
or 318,875 kg (703,000 lb)
Max zero-fuel weight 185,975 kg (410,000 lb)
or 192,775 kg (425,000 lb)
Max landing weight 204,115 kg (450,000 lb)
or 210,920 kg (465,000 lb)

PERFORMANCE (with RB211-524D4 engines, at T-O weight of 317,515 kg; 700,000 lb, except where indicated):
Never-exceed speed Mach 0.92
Max level speed, AUW of 226,795 kg (500,000 lb) at 9,150 m (30,000 ft) 538 kts (996 km/h; 619 mph)
Service ceiling 13,745 m (45,100 ft)
Min ground turning radius over outer wingtip 22.25 m (73 ft 0 in)
Runway LCN (Y: 302,090 kg; 666,000 lb, Z: 317,515 kg; 700,000 lb taxi weight on h = 0.51 m; 20 in flexible pavement): Y 70
Z 75
Runway LCN (weights as above, on 11.02 m; 40 in rigid pavement): Y 76
Z 80
FAR T-O distance to 10.7 m (35 ft) at S/L, ISA 2,347 m (7,700 ft)
FAR landing field length:
at AUW of 204,115 kg (450,000 lb) 1,594 m (5,230 ft)
at AUW of 210,920 kg (465,000 lb) 1,646 m (5,400 ft)
Range, long-range step cruise, typical international reserves, with 276 passengers and baggage 6,650 n miles (12,324 km; 7,658 miles)
Ferry range, long-range step cruise, typical international reserves 8,000 n miles (14,826 km; 9,212 miles)

VERIFIED

BOEING MODEL 747-300

TYPE: Four-engine passenger aircraft.

PROGRAMME: On 12 June 1980, Boeing announced an option for the Model 747 which incorporates structural changes to the aircraft's upper deck area to increase passenger carrying capacity and provide an optional crew rest area, as described for the Model 747SP. The upper forward fuselage is extended aft by 7.11 m (23 ft 4 in) to increase upper deck accommodation from 32 to a maximum of 69 passengers in all-economy class configuration. Seating is six-abreast, with a single aisle, and panniers between the outer seats and cabin wall are provided for hand baggage. Alternative configurations include 26 first class sleeper seats on the extended upper deck. In addition, seven seats can be accommodated on the main deck as a result of deleting the standard circular stairway. It is replaced by a new straight stairway at the rear of the upper deck area. Two new doors 1.83 m (6 ft 0 in) high and 1.07 m (3 ft 6 in) wide replace the existing 1.22 × 0.61 m (4 ft 0 in × 2 ft 0 in) upper deck exits. Other structural changes include the provision of a new emergency exit and additional windows. The extended upper deck option was made available initially on existing aircraft of the 747-100 and -200 series, which then became known as **747-300**s; maximum T-O weights are unchanged, but operating weight empty is increased by about 4,220 kg (9,310 lb). Most efficient high-speed cruise is increased from Mach 0.84 to 0.85 by the revised upper contours. Flight testing with JT9D-7R4G2 engines began 5 October 1982, followed on 10 December by the first 747-300 with CF6-50E2 engines, and FAA certification of the -300 was announced by Boeing on 7 March 1983. Initial deliveries were to Swissair and UTA, with whom the aircraft entered service on 28 March and 1 April 1983 respectively.

VERSIONS: **747-300**: Basic model.

747-300SR: Short-range version of the 747-300.

747-300M: Combi version of the 747-300.

The detailed specification for the Boeing 747-100B/200B applies also to the 747-300 except as follows:

WEIGHTS AND LOADINGS (A: JT9D-7R4G2, B: CF6-50E2, C: CF6-80C2, D: RB211-524D4):
Operating weight empty (approx) for max available gross weights:
747-300 (400 passengers):
A 174,134 kg (383,900 lb)
B 175,721 kg (387,400 lb)
C 176,901 kg (390,000 lb)
D 178,171 kg (392,800 lb)
747-300 (289 passengers and seven pallets):
A 175,585 kg (387,100 lb)
B 177,173 kg (390,600 lb)
C 178,352 kg (393,200 lb)
D 179,622 kg (396,000 lb)
Max fuel weights as for 747-200B
Max payload:
747-300 (400 passengers):
A 68,538 kg (151,100 lb)
B 66,950 kg (147,600 lb)
C 65,771 kg (145,000 lb)
D 64,501 kg (142,200 lb)
747-300M (289 passengers and seven pallets):
A 80,694 kg (177,900 lb)
B 79,106 kg (174,400 lb)
C 77,927 kg (171,800 lb)
D 76,657 kg (169,000 lb)
Max T-O weight:
747-300, -300M: A, B, C, D 351,535 kg (775,000 lb)
or 356,070 kg (785,000 lb)
or 362,875 kg (800,000 lb)
or 371,945 kg (820,000 lb)
or 377,840 kg (833,000 lb)
Max ramp weight:
747-300, -300M: A, B, C, D 379,200 kg (836,000 lb)
Max zero-fuel weight:
747-300: A, B, C, D 242,670 kg (535,000 lb)
747-300M: A, B, C, D 256,280 kg (565,000 lb)
Max landing weight:
747-300: A, B, C, D 260,360 kg (574,000 lb)
or 265,350 kg (585,000 lb)
or 285,765 kg (630,000 lb)
747-300M: A, B, C, D 285,765 kg (630,000 lb)

PERFORMANCE (at max T-O weight of 377,840 kg; 833,000 lb, except where indicated. Engines as designated under Weights and Loadings):
Max level speed at AUW of 317,515 kg (700,000 lb) at 9,150 m (30,000 ft):
A 530 kts (982 km/h; 610 mph)
B 529 kts (979 km/h; 608 mph)
C 538 kts (996 km/h; 619 mph)
D 526 kts (974 km/h; 605 mph)
FAR T-O distance to 10.7 m (35 ft) at S/L, ISA:
A 3,170 m (10,400 ft)
B 3,292 m (10,800 ft)
C 3,079 m (10,100 ft)
D 3,155 m (10,350 ft)
FAR landing field length:
at AUW of 255,825 kg (564,000 lb) 1,881 m (6,170 ft)
at AUW of 265,350 kg (585,000 lb) 1,942 m (6,370 ft)
at AUW of 285,765 kg (630,000 lb) 2,112 m (6,930 ft)
Range, long-range cruise, typical international reserves of 5% trip fuel, 200 n mile (371 km; 230 mile) alternate, 30 min hold at 457 m (1,500 ft), with 400 passengers and baggage: A 6,300 n miles (11,675 km; 7,254 miles)
B 6,100 n miles (11,297 km; 7,020 miles)
C 6,700 n miles (12,408 km; 7,710 miles)
D 6,250 n miles (11,575 km; 7,192 miles)
Ferry range, long-range cruise, reserves as above:
A 7,250 n miles (13,436 km; 8,348 miles)
B 7,000 n miles (12,964 km; 8,055 miles)
C 7,750 n miles (14,353 km; 8,918 miles)
D 7,300 n miles (13,520 km; 8,400 miles)

VERIFIED

BOEING 747-400 PERFORMANCE IMPROVEMENT PACKAGE (PIP)

TYPE: High-capacity, long-range transport aircraft performance enhancement. For full details of 747-400 see *Jane's All the World's Aircraft 1995-96*.

PROGRAMME: Announced in April 1993. Implementation depends on customer demand. First stage PIP flight tested in leased United Airlines 747-400 in May 1993. Second stage potentially available by mid-1996.

DESIGN FEATURES: First stage includes gross weight increase of 907 kg (2,000 lb) or more, longer chord dorsal fin made of CFRP, slower transfer of fuel from tailplane trim tank to prolong trim drag alleviation, and wing spoilers held down more tightly to reduce drag and leakage. This stage to be applied to production aircraft as soon as ready and to be retrofittable. Second stage PIP would increase gross weight to 421,840 kg (930,000 lb) and range to 8,000 n miles (14,816 km; 9,206 miles).

NEW ENTRY

BAI

BOGAN AEROTECH INC

2601 Gravel Road, Fort Worth, Texas 76118
Telephone: 1 (817) 595 1701
Fax: 1 (817) 595 3108
PRESIDENT: Jay H. Golding
EXECUTIVE VICE-PRESIDENT, COO, CFO: James R. Falik
DIRECTOR, MARKETING: Stephen E. Ryan

VERIFIED

BOGAN (BELL) HUEY II UPGRADE

TYPE: Utility helicopter upgrade.

PROGRAMME: Bogan is teamed with: Bell Helicopter and Lycoming to offer the Huey II upgrade. Bell certified upgrade kit for Bell Model 204/205 (UH-1) helicopters to increase engine power from 1,044 kW (1,400 shp) to 1,342 kW (1,800 shp) and increase hover ceiling at 4,309 kg (9,500 lb) ISA + 20°C by 275 per cent. Model 205 kits available from mid-1995.

DESIGN FEATURES: Installation of Bell 212 drive train and tailboom. Optional installation of: cockpit management system (MIL-STD-1553) databus, global positioning system (GPS), Doppler and other navigation systems (VOR/ILS, APF, Tacan), stabilised FLIR, night vision goggles (NVG) instrumentation, VHF, UHF HF communications, voice privacy, hook, hoist, improved particle separator, engine wash kit, digital fuel gauging system as well as underseat auxiliary fuel tank and engine monitoring system.

VERIFIED

BRANSON

BRANSON AIRCRAFT CORPORATION

3790 Wheeling Street, Denver, Colorado 80239
Telephone: 1 (303) 371 9112
Fax: 1 (303) 371 1813
Telex: 45-4577 BRANSON DVR
PRESIDENT: Carl F. Branson
EXECUTIVE VICE-PRESIDENT: Roger P. Kirwan

Founded in 1966 for special design and custom manufacturing of auxiliary fuel tanks, special interiors and equipment for civil aircraft.

VERIFIED

BRANSON (CESSNA) CITATION EXTENDED RANGE FUEL SYSTEM

TYPE: Business jet extended range fuel system modification.

PROGRAMME: Branson holds FAA FAR Pt 25 supplemental type certificate for 454 litre (120 US gallon; 100 Imp gallon) fuel tank installation for Cessna Models 500, 501 and Citation I.

DESIGN FEATURES: Installation weighs 36 kg (80 lb) and allows 295 kg (650 lb) increase in gross weight. The tank is of double wall, aluminium construction and is located close to aircraft centre of gravity for minimal CG shift. A new pressure bulkhead is constructed in front of the tank, isolating tank from cabin. Outer wall of tank is separated from the inner wall by a sealed vapour space vented outside the aircraft. Baggage compartment is reduced by 38 cm (15 in). Space remains for toilet and baggage.

WEIGHTS AND LOADINGS:

New gross ramp weight	5,738 kg (12,650 lb)
Empty system weight	36 kg (80 lb)

PERFORMANCE:

Additional fuel	454 l (120 US gallons; 100 Imp gallons)
New range	951 n miles (1,763 km; 1,905 miles)
Additional flying time	approx 1 h

VERIFIED

BRANSON (CESSNA) CITATION II WEIGHT INCREASE

TYPE: Business jet weight increase modification.

PROGRAMME: Branson holds FAA Pt 25 supplemental type certificate for 544 kg (1,200 lb) weight increase for Cessna Model 550 Citation II.

DESIGN FEATURES: STC kit includes higher performance wheels, brakes and tyres. Aircraft can carry full fuel and up to seven passengers. Quick-change cargo system (glassfibre liner and net/tiedown restraints) also available increasing payload to 1,224 kg (2,700 lb).

WEIGHTS AND LOADINGS:

New ramp weight	6,668 kg (14,700 lb)
New T-O weight	6,577 kg (14,500 lb)
Max passenger/baggage weight	1,279 kg (2,820 lb)

VERIFIED

BRANSON (CESSNA) CITATION EXTENDED WIDTH CARGO DOOR

TYPE: Business jet cargo door extension.

PROGRAMME: Cabin door modification for Cessna I, II and S/II. Meets FAA FAR Pt 25 requirements. Installation takes six weeks.

DESIGN FEATURES: Provides 0.91 m (3 ft 0 in) wide opening for loading stretchers, palletised cargo, research equipment, large aerial cameras and so on. Installation does not affect seating arrangement or furnishings. Door weighs 28.6 kg (63 lb).

VERIFIED

BRANSON (CESSNA) CITATION AIR AMBULANCE EQUIPMENT

TYPE: Air ambulance modification.

PROGRAMME: Medical equipment installation for Cessna Citation I, II and S/II. First installation in Cessna Citation II of Air Express of Oslo, Norway.

DESIGN FEATURES: Installation includes wider cabin door, one or two stretchers, medical oxygen supply, compressed air, vacuum and electrical outlets.

VERIFIED

BRANSON (MITSUBISHI) DIAMOND IA LONG-RANGE TANK AND AFT BAGGAGE COMPARTMENT

TYPE: Auxiliary fuel tank and baggage compartment installations.

PROGRAMME: Auxiliary long-range fuel tank and aft baggage compartment for Mitsubishi Diamond IA. FAA approval gained for Branson gross weight increase kit raising maximum T-O weight to 7,031 kg (15,500 lb).

DESIGN FEATURES: Auxiliary tank mounted against rear pressure bulkhead and holds 363 litres (96 US gallons; 80 Imp gallons). Also for Diamond IA is a transfer system from wing tank to fuselage tank avoiding the need to fill tanks separately. Branson tailcone baggage compartment holds maximum 204 kg (450 lb).

VERIFIED

Branson extended range fuel tank with inspection hatch removed

1994

Branson extended width cargo door for Cessna Citation I, II and S/II

1994

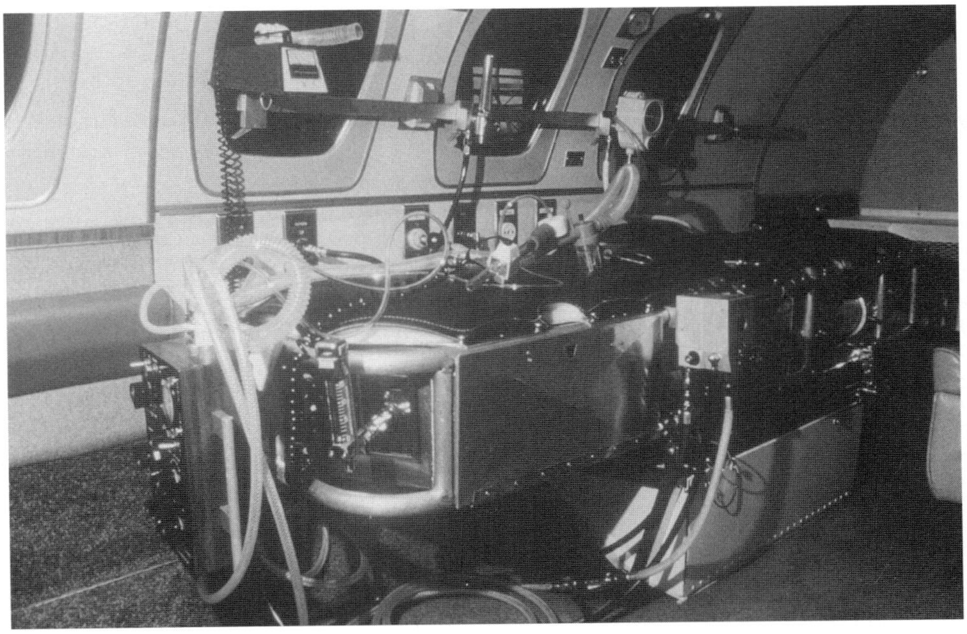

Branson Cessna Citation flying intensive care unit

1994

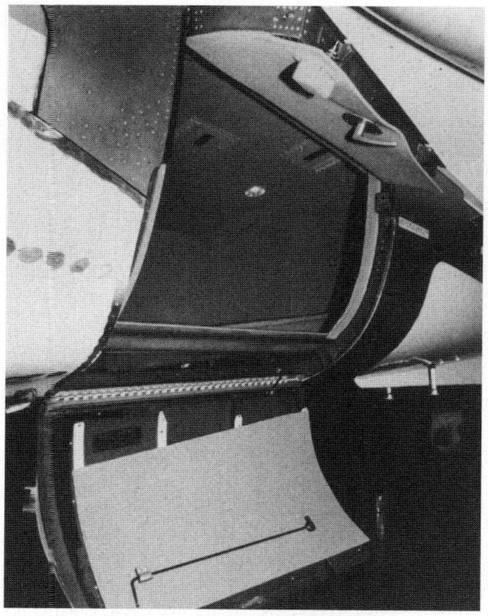

Branson Mitsubishi Diamond IA aft baggage compartment

1994

BRANSON LEARJET 55 LONG-RANGE TANKS

TYPE: Long-range fuel tank installation.

PROGRAMME: Installation of long-range fuel tanks for Learjet Model 55. Installation takes four weeks for smaller tanks and five weeks for the larger ones.

PERFORMANCE:

Additional fuel:

Large tank	757 l (200 US gallons; 166.5 Imp gallons)
Small tank	378.5 l (100 US gallons; 83 Imp gallons)

VERIFIED

BRANSON (FOKKER) F27 AND (FAIRCHILD) FH-227 LARGE CARGO DOOR

TYPE: Cargo door installation.

PROGRAMME: Large cargo door installation for Fokker F27 Friendships and Fairchild F227. FAA Pt 25 supplemental certificate acquired from Fairchild. Conversion possible in USA, Europe and Asia.

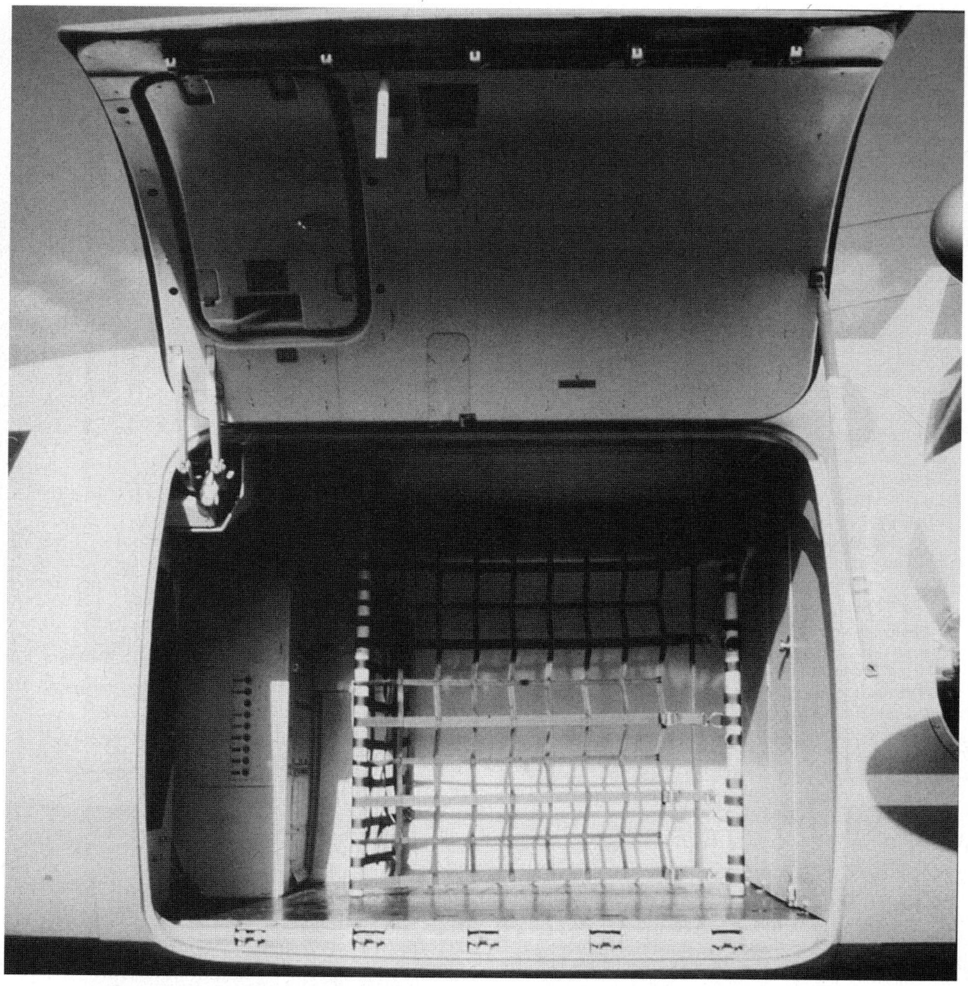

Branson large cargo door installed on a Fokker F27 Friendship

1994

DESIGN FEATURES: Door located in port side just aft of flight deck; electrically operated. Door opens 110° and 170°. Crew door incorporated. Conversions add approximately 238 kg (525 lb) to aircraft empty weight.

DIMENSIONS, EXTERNAL:

Door width	2.30 m (7 ft 6½ in)
Door height	1.83 m (6 ft 0 in)

VERIFIED

BUSH

BUSH CONVERSIONS INC

PO Box 431, Udall, Kansas 67146
Telephone: 1 (316) 782 3851
VICE-PRESIDENT: Barbara Williams

Company offers 'Taildragger' conversions for Cessna 150/152, 172/Skyhawk, R172K and 175, designed by former Ralph Bolen Inc.

VERIFIED

BOLEN 'TAILDRAGGER' CONVERSIONS

TYPE: Tailwheel conversion of tricycle Cessnas (as listed).

DESIGN FEATURES: Average increase of 8.5 to 10.5 knots (16 to 19.5 km/h; 10 to 12 mph) in speed; better performance from short and rough fields; tighter turning radius on ground; simpler operation on floats and skis.

FLYING CONTROLS: Unchanged.

STRUCTURE: Two bulkheads and new landing gear attachment box added to forward fuselage; Cessna 172 and 175 receive new cantilever landing gear legs; for Cessna 150 series F, G, H, J and K, existing legs are retained, but fitted to new box; earlier Cessna 150s have to receive legs of newer models with 15 × 6.00-6 tyre and brake. Tailwheel unit includes Scott 3200 wheel; leg attached by stress plates and stringers, without removal of skin. CG not displaced; STC includes operation on Fluidyne snow skis; no weight penalty with Cessna 150/152; new main legs and extra structure of Model 172/175 modification increase weight by 13.6 kg (30 lb).

Bolen 'Taildragger' conversion of a Cessna Aerobat 135 *(John Cook)*

1994

VERIFIED

CALIFORNIA HELICOPTER

CALIFORNIA HELICOPTER INTERNATIONAL

2935 Golf Course Drive, Ventura, California 93003-7604
Telephone: 1 (805) 644 5800
Fax: 1 (805) 644 5132
CHAIRMAN OF THE BOARD: Douglas F. Wax
PRESIDENT AND COO: Gary Podolny
VICE-PRESIDENT: William Dvorak

Rights to manufacture turbine conversion kits for Sikorsky S-58 and spare parts, plus support of worldwide S-58/S-58T fleet, bought from Sikorsky in 1981 after latter had converted or produced conversion kits for about 146 S-58s. California Helicopter offers dynamic component exchange service for S-58/S-58T.

UPDATED

CALIFORNIA HELICOPTER (SIKORSKY) S-58T

TYPE: Twin-turbine conversion of Sikorsky S-58.

CUSTOMERS: Civil customers include New York Airways (four) in 14-passenger layout for New York airports shuttle service.

DESIGN FEATURES: FAA and CAA approval for IFR operation.

POWER PLANT: Pratt & Whitney Canada PT6T-6 Twin-Pac rated at 1,398 kW (1,875 shp) for T-O and 1,249 kW (1,675 shp) maximum continuous.

WEIGHTS AND LOADINGS:

Weight empty	3,437 kg (7,577 lb)
Max T-O and landing weight	5,896 kg (13,000 lb)
Max disc loading	25.8 kg/m² (5.29 lb/sq ft)

PERFORMANCE (at max T-O weight. A: PT6T-3, B: PT6T-6):

Max level speed at S/L:	
A, B	120 kts (222 km/h; 138 mph)
Cruising speed: A, B	110 kts (204 km/h; 127 mph)
Hovering ceiling OGE: A	1,433 m (4,700 ft)
B	1,980 m (6,500 ft)
Single-engine absolute ceiling: A	640 m (2,100 ft)
B	1,280 m (4,200 ft)

Range with 1,071 l (283 US gallons; 236 Imp gallons) max usable fuel, incl 20 min reserves at cruising speed:

A	260 n miles (481 km; 299 miles)
B	242 n miles (447 km; 278 miles)

VERIFIED

California Helicopter conversion of a Sikorsky S-58 to S-58T twin-turbine
1994

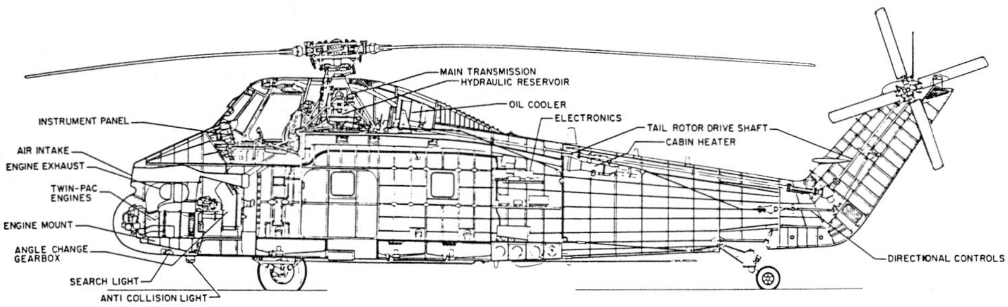

California Helicopter (Sikorsky) S-58T twin-turbine conversion *(California Helicopter)*
1995

CALIFORNIA MICROWAVE

CALIFORNIA MICROWAVE INC
(Government Electronics Division)

6022 Variel Avenue, PO Box 2800, Woodland Hills, California 91367
Telephone: 1 (818) 992 8000
Fax: 1 (818) 992 5079
Telex: 910 494 2794
PRESIDENT: J. Russell
GENERAL MANAGER: R. Medlin

VERIFIED

CALIFORNIA MICROWAVE (DE HAVILLAND) DHC-7 ARL and DHC-7 ARL-M UPGRADES

TYPE: US Army drug interdiction aircraft.

PROGRAMME: Prototype Grisly Hunter (now ARL) airborne reconnaissance low system, adapted from Developmental Sciences SkyEye UAV, installed in CASA C-212 (88-3210) by DSC; aircraft crashed 1 December 1989. Request for proposals issued February 1990 for 'production' prototype; options on further six; based at Howard AFB, Panama, for Caribbean patrols; range 1,200 n miles (2,224 km; 1,382 miles)/min. California Microwave chosen contractor August 1990; second-hand de Havilland Dash 7 as sensor platform; prototype was completed in early 1992.

The first three aircraft were delivered in May 1993. Two are SIGINT only aircraft integrated by ESL, now part of TRW Avionics and Surveillance, and one is IMINT only from California Microwave (CM). In mid-1995 the US Army lost funding for the final three of the nine aircraft. CM has since won the sole-source contract as prime contractor and systems integrator for the next and final three aircraft contract. These will combine the IMINT and SIGINT systems into one multifunction aircraft and will be known as ARL-M. The first two ARL-M are due for delivery in July 1996 and the third in November 1996. The first three aircraft will then be upgraded to the ARL-M standard.

DESIGN FEATURES: CM and Lockheed are supplying the SIGINT system which is common with the US Army's TACJAM-A electronic warfare (EW) system. IMINT mission equipment includes a ReconOptical infra-red linescanner, Westinghouse FLIR and Istec daylight imaging system. The airframe is to be modified by West Virginia Air International. Two further upgrades will be carried out by FY99: a joint tactical terminal and links to a theatre-level intelligence broadcast network.

COSTS: FY94-95 $50 million. FY96 request for $18.4 million. FY97 plans for $32.2 million.

UPDATED

CAT

COMMUTER AIR TECHNOLOGY

14700 North Airport Drive, Suite 206, Scottsdale, Arizona 85260
Telephone: 1 (602) 951 6288
Fax: 1 (602) 998 1239
CEO: Keith Nickels
PRESIDENT: Sharon Thuell

VERIFIED

CAT TRANSREGIONAL 250 and ST 17

TYPE: Commuter modifications of Beechcraft Super King Air 200.

PROGRAMME: Transregional 250 launched in mid/late 1980s (then as CAT 200) as 13-seat commuter modification of Super King Air; now marketed via Priority Aviation Leasing; ST 17 due for certification mid/late 1990s.

VERSIONS: **Transregional 250 CATPASS:** Basic 13-passenger version, formerly CAT 200. Known by certification name CAT Performance and Safety System (CATPASS); incorporates most Raisbeck King Air modifications (which see), maximum zero-fuel weight increased to 4,990 kg (11,000 lb); also cargo version.

ST 17: Fuselage stretched 1.22 m (4 ft 0 in) to hold 17 passengers and 1.13 m³ (40 cu ft) nose baggage compartment; Beechcraft King Air 1.22 m (4 ft 0 in) wide cargo door optional; 2.12 m³ (75 cu ft) cargo pod optional.

CUSTOMERS: CAT 200 in service with Mesa Airlines, New Mexico; first PAL lease of Transregional 250 (to Elgaz of Poland; two, with third on option) announced March 1991; other leases then under negotiation with operators in Africa (four), Brazil (four), Bulgaria (three) and Mexico (six); PAL has ordered five conversions with options for five from CAT. Unnamed French regional airline has three options for ST 17.

COSTS: $2.2 million (1991) for ST 17.

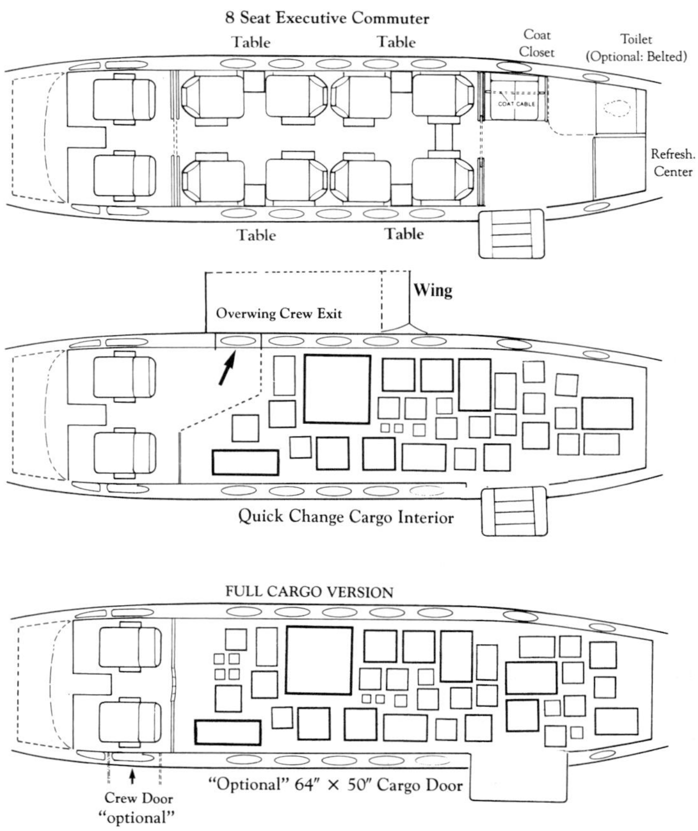

Commuter Air Technology modifications for Beechcraft Super King Air
1994

PERFORMANCE (ST 17):
- Max level speed 288 kts (534 km/h; 331 mph)
- Typical cruising speed 270 kts (500 km/h; 311 mph)
- Service ceiling 7,620 m (25,000 ft)
- Service ceiling, one engine out 6,035 m (19,800 ft)
- T-O field length, ISA + 15°C 887 m (2,910 ft)
- Landing field length, ISA + 15°C 805 m (2,640 ft)
- Max range, with reserves 1,200 n miles (2,224 km; 1,382 miles)

VERIFIED

Commuter Air Technology CATPASS 250 *1994*

CESSNA

CESSNA AIRCRAFT COMPANY (Subsidiary of Textron Inc)

PO Box 7704, Wichita, Kansas 67277-7706
Telephone: 1 (316) 941 6000
Fax: 1 (316) 941 7812
Telex: 417 400
CHAIRMAN AND CEO: Russell W. Meyer Jr
EXECUTIVE VICE-PRESIDENT, MARKETING PRODUCT SUPPORT: Gary W. Hay
PUBLIC RELATIONS: David Franson

Founded by late Clyde V. Cessna 1911; incorporated 7 September 1927; former Pawnee and Wallace aircraft divisions in Wichita consolidated in Aircraft Division mid-1984; acquired by General Dynamics as wholly owned subsidiary 1985. Acquired by Textron February 1992.

Owned subsidiaries include McCauley Accessory Division, Dayton, Ohio; Cessna Finance Corporation in Wichita. Sold 49 per cent interest in Reims Aviation of France to Compagnie Française Chaufour Investissement (CFCI) February 1989. CFCI continues manufacturing Cessna F 406 Caravan II and holds option to build Cessna single-engined aircraft when Cessna restarts production.

Total 178,242 aircraft produced by early 1995, 180 aircraft delivered in 1993, including: 13 Model 208 Caravan Is, 44 Model 208B Caravans, 14 Citation IIs, 44 Citation Vs, 34 Citationjets, 13 Citation VIs and 11 Citation VIIs. Total employees 5,700 on 1 January 1994. Range includes Caravan I and II and seven variants of Citation.

UPDATED

CESSNA SUSPENDED PRODUCTION

Production of the types listed below was suspended during the mid-1980s; there are no restoration plans at present time.

152: Total 7,500 Model 152 and Model 152 Aerobats produced by December 1986, including 640 by Reims Aviation in France.

Skyhawk: Total 35,773 civil Model 172/Skyhawks produced by 31 December 1987, including 2,144 F 172s by Reims Aviation in France. Additional 864 T-41A, T-41B, T-41C and T-41D Mescaleros produced as military basic trainers from 1966-1983.

Cutlass RG: Airframe of Model 172 Skyhawk with retractable landing gear of Skylane RG. Total 1,159 Cutlass RGs sold by 31 December 1987.

Skylane: Total 19,812 Model 182/Skylanes of various models built by 31 December 1987, including 169 F 182s produced by Reims Aviation.

Skylane RG and Turbo Skylane RG: Total 2,102 Skylane RGs produced by 31 December 1987, including 73 assembled in France as Reims 182 Skylane RGs.

185 Skywagon: Total 4,356 Model 185 Skywagons produced by 31 December 1987, including 497 military U-17A/B/Cs.

AG Truck and AG Husky: Total 1,949 AG Trucks and 386 AG Huskys produced; production suspended 1985.

Stationair 6 and Turbo Stationair 6 (US Air Force designation of Turbo Stationair: U-26A): Stationairs originated from U206 Skywagon and TU206 Turbo Skywagon; name changed to Stationair 6 and Turbo Stationair 6 in 1978, denoting six-seat capacity. Total 7,652 Model 206 Skywagons and Stationairs, including 643 de luxe Super Skylanes of similar design, produced by 31 December 1978.

Centurion, Turbo Centurion and Pressurised Centurion: Total 8,453 Model 210/Centurions, plus 51 Pressurised Centurions, delivered by 31 December 1987.

T303 Crusader: Total 297 T303 Crusaders delivered by 31 December 1987.

402C: Total 1,540 Model 402C Utililiners and 402C Businessliners delivered by 31 December 1987.

414A Chancellor: Total 1,067 Model 414/Chancellors delivered by 31 December 1987.

421 Golden Eagle: Total 1,909 early 421s, Model 421B Golden Eagle, 421B Executive Commuter and 421C Golden Eagle produced before production was suspended in 1985.

VERIFIED

Cessna 152 II of RS Pilot Training ***(Peter March)*** *1994*

CESSNA 152

TYPE: Two-seat cabin monoplane.

PROGRAMME: During 1977 Cessna introduced a new two-seat cabin monoplane to replace the Model 150, which had been in production for almost 20 years. Designated Model 152, it differed primarily in having a more powerful engine using low-lead 100 octane fuel, an improved 'gull wing' propeller, and power plant installation and cowling changes to reduce engine noise and vibration.

VERSIONS: **152:** Standard version, *as described.*

152 II: As Model 152, but including as standard a factory installed avionics and equipment package which includes a Cessna Series 300 nav/com, with variable intensity radio light, dual controls, vertical speed indicator, turn co-ordinator, outside air temperature gauge, attitude and directional gyros and vacuum system, sun visors and omniflash beacon; adding 15 kg (34 lb) to basic empty weight.

152 Trainer: Has the avionics and equipment package as detailed on the Model 152 II, plus intercom system, Cessna Series 300 transponder, economy mixture indicator, flight hour recorder, all-purpose control wheel, tinted cabin windows, quick drain oil valve, and refuelling steps and handles, adding a total of 20 kg (45 lb) to basic empty weight.

152 Aerobat: Described separately.

Versions of the Model 152 for 1983 introduce as standard on Avco Lycoming O-235-N2C engine with improved combustion chamber configuration, a vacuum system warning light, revised instrument panel for easier maintenance, quieter avionics cooling fan, and cabin interior refinements.

A total of 7,234 Model 152s had been built by 1 April 1984, including 565 F 152s built by Reims Aviation in France.

OPERATORS: Versions of the Cessna 152 are in service with the armed forces of the following countries: Bangladesh (4) and Bolivia (9).

DESIGN FEATURES: Braced high-wing monoplane. Wing section NACA 2412 (tips symmetrical). Dihedral 1°. Incidence 1° at root, 0°at tip. Cantilever tail unit with swept vertical surfaces. Tailplane abrasion boots optional.

FLYING CONTROLS: Modified Frise ailerons and electrically actuated NACA single-slotted trailing-edge flaps. No trim tabs. Trim tab in starboard elevator. Ground adjustable rudder tab.

STRUCTURE: The wing is an all-metal structure of light alloy. Conical camber glassfibre wingtips optional. Flaps of light alloy. The fuselage is a conventional semi-monocoque structure of light alloy. Tailplane of light alloy construction.

LANDING GEAR: Non-retractable tricycle type. Land-O-Matic cantilever main legs, each comprising a one-piece machined conically tapered spring steel tube. Steerable nosewheel on oleo-pneumatic shock-absorber strut. Mainwheel tubed tyres size 15 × 6.00-6, 4 ply rating, pressure 2.0 bars (29 lb/sq in); 6.00-6 4 ply rated tyres optional, pressure 1.45 bars (21 lb/sq in). Nosewheel tubed tyre size 5.00-5, 4 ply rating, pressure 2.07 bars (30 lb/sq in). Toe operated single-disc hydraulic brakes. Parking brake. Rudder pedal extensions and wheel fairings optional.

POWER PLANT: One 80.5 kW (108 hp) Avco Lycoming O-235-N2C flat-four engine, driving a McCauley Type 1A103/TCM6958 two-blade fixed-pitch metal propeller with spinner. Fuel tanks in wings, with total capacity of 98 litres (26 US gallons; 21.7 Imp gallons), of which 92.75 litres (24.5 US gallons; 20.4 Imp gallons) are usable. Optional long-range tanks have a total capacity of 147.5 litres (39 US gallons; 32.5 Imp gallons), of which 142 litres (37.5 US gallons; 31.3 Imp gallons) are usable. Refuelling points on upper surface of wing. Oil capacity 6.6 litres (1.75 US gallons; 1.5 Imp gallons).

ACCOMMODATION: Enclosed cabin seating two side by side. Vertically adjustable seats for pilot and co-pilot; inertia reel shoulder harness and dual controls optional on standard model. Baggage compartment behind seats, backs of which hinge forward. Baggage capacity 54 kg (120 lb). Optional 'family seat' can be fitted in baggage space, for two children not exceeding 54 kg (120 lb) in combined weight. Door, with opening window, on each side. Heating and ventilation standard. Dual windscreen defrosters standard. Cabin skylight windows optional.

SYSTEMS: Hydraulic system for brakes only. Electrical system includes a 28 V 60 A alternator and 24 V 12.75 Ah battery. Heavy-duty 24 V 15.5 Ah battery optional.

AVIONICS AND EQUIPMENT: Optional avionics include Series 300 nav/com with remote VOR/LOC indicator (standard on 152 II and Trainer), 300 ADF, 300 VOR/ILS indicator,

Cessna 152 II two-seat cabin monoplane ***(Peter March)*** *1994*

300 transponder (standard on Trainer), Series 400 glide slope receiver, 400 marker beacon receiver, 400 transponder, altitude encoder, and intercom system (standard on Trainer). Standard equipment includes safety belts, map compartment, control locks, stall warning device, baggage retaining net, cabin dome lights, variable intensity instrument panel red floodlights, wing-mounted (port side) landing and taxi lights, navigation lights and full-flow oil filter. Optional equipment (standard on 152 II and Trainer) includes dual controls, rate of climb indicator, turn coordinator, outside air temperature gauge, gyro installation and vacuum system, sun visors, omni-flash beacon and landing light. Other optional equipment includes artificial horizon, directional gyro, electric clock, quartz clock, economy mixture indicator, exhaust gas temperature indicator, flight hour recorder, true airspeed indicator, turn and bank indicator, cabin fire extinguisher, tinted windows, control wheel map light, instrument post lights, emergency locator transmitter, heated pitot, white strobe lights, anti-precipitation static kit, corrosion proofing, engine winterisation kit, external power socket, refuelling steps and handles, towbar, full-flow oil filter, and quick drain oil valve.

DIMENSIONS, EXTERNAL:
Wing span: standard 9.97 m (32 ft 8½ in)
with optional conical wingtips 10.11 m (33 ft 2 in)
Wing chord: at root 1.63 m (5 ft 4 in)
at tip 1.13 m (3 ft 8½ in)
Wing aspect ratio 6.7
Length overall 7.34 m (24 ft 1 in)
Height overall 2.59 m (8 ft 6 in)
Tailplane span 3.05 m (10 ft 0 in)
Wheel track 2.32 m (7 ft 7¼ in)
Wheelbase 1.47 m (4 ft 10 in)
Propeller diameter 1.75 m (5 ft 9 in)
Propeller ground clearance 0.305 m (1 ft 0 in)
Passenger doors (each): Width 0.84 m (2 ft 9¼ in)
Height 0.78 m (2 ft 6¾ in)

AREAS:
Wings, gross: standard 14.59 m² (157.0 sq ft)
with optional conical wingtips 14.82 m² (159.5 sq ft)

WEIGHTS AND LOADINGS (A: standard 152, B: 152 II, C: Trainer):
Weight empty: A 501 kg (1,104 lb)
B 516 kg (1,138 lb)
C 521 kg (1,149 lb)
Max T-O and landing weight:
A, B, C 757 kg (1,670 lb)
Max ramp weight: A, B, C 760 kg (1,675 lb)
Max wing loading 51.3 kg/m² (10.5 lb/sq ft)
Max power loading 9.40 kg/kW (15.5 lb/hp)

PERFORMANCE (at max T-O weight, ISA):
* Max level speed at S/L 109 kts (202 km/h; 125 mph)
* Max cruising speed, 75% power at 2,440 m (8,000 ft) 106 kts (196 km/h; 122 mph)
Stalling speed, power off:
flaps up 48 kts (89 km/h; 55 mph) CAS
flaps down 43 kts (80 km/h; 49 mph) CAS
Max rate of climb at S/L 218 m (715 ft)/min
Service ceiling 4,480 m (14,700 ft)
T-O run 221 m (725 ft)
T-O to 15 m (50 ft) 408 m (1,340 ft)
Landing from 15 m (50 ft) 366 m (1,200 ft)
Landing run 145 m (475 ft)
Range, recommended lean mixture, with allowances for start, taxi, T-O, climb and 45 min reserves at 75% power:
standard fuel, 75% power at 2,440 m (8,000 ft) 315 n miles (583 km; 362 miles)
max fuel, 75% power at 2,440 m (8,000 ft) 540 n miles (1,000 km; 621 miles)
Range, allowances as above, but with 45 min reserves at 45% power:
standard fuel, econ cruising power at 3,050 m (10,000 ft) 370 n miles (685 km; 425 miles)
max fuel, econ cruising power at 3,050 m (10,000 ft) 625 n miles (1,158 km; 719 miles)

* *With wheel speed fairings which increase speeds by approx 2 kts (3.7 km/h; 2.3 mph)*

UPDATED

CESSNA MODEL 152 AEROBAT

TYPE: Two-seat cabin aerobatic monoplane.

PROGRAMME: The Model 152 Aerobat combines the economy and versatility of the standard Model 152 with aerobatic capability. Structural changes allow the Aerobat to perform 'unusual attitude' manoeuvres and it is licensed in the Aerobatic category for load factors of +6 *g* and −3 *g* at full gross weight, permitting the performance of barrel and aileron rolls, snap rolls, Immelmann turns, Cuban eights, spins, vertical reversements, lazy eights and chandelles.

DESIGN FEATURES: Equipment of the aircraft differs only slightly from that of the standard 152. Series 300 nav/com with remote VOR/LOC indicator, quick-release cabin doors, removable seat cushions and reclining backs, quick-release lap-straps, and shoulder harnesses, are standard, as are two tinted skylights in the cabin roof which offer extra field of view. Distinct external styling provides immediate recognition of the Aerobat's role. The 152 II package detailed for the Model 152 is available also for the Aerobat.

DIMENSIONS: As for Model 152

Reims/Cessna FA 152 Aerobat aircraft ***(Peter March)***

1994

AREAS: As for Model 152

WEIGHTS AND LOADINGS: As for Model 152 except:
Weight empty 513 kg (1,131 lb)

PERFORMANCE (at max T-O weight, ISA): As for Model 152 except:
* Max level speed at S/L 108 kts (200 km/h; 124 mph)
* Max cruising speed, 75% power at 2,440 m (8,000 ft) 105 kts (194 km/h; 121 mph)
Range, recommended lean mixture, with allowances for engine start, taxi, T-O, climb and 45 min reserves at 75% power:
standard fuel, 75% power at 2,440 m (8,000 ft) 310 n miles (574 km; 357 miles)
max fuel, 75% power at 2,440 m (8,000 ft) 530 n miles (982 km; 610 miles)
Range, allowances as above, but with 45 min reserves at 45% power:
standard fuel, econ cruising power at 3,050 m (10,000 ft) 365 n miles (676 km; 420 miles)
max fuel, econ cruising power at 3,050 m (10,000 ft) 615 n miles (1,139 km; 708 miles)

* *With wheel speed fairings which increase speeds by approx 2 kts (3.7 km/h; 2.3 mph)*

VERIFIED

CESSNA 172 SKYHAWK

US Air Force designation: T-41 Mescalero

TYPE: Four-seat cabin monoplane.

PROGRAMME: The Skyhawk is certificated as a floatplane, and can be fitted with skis. A version designated F 172 was produced by Reims of France.

In 1985 a model of the Skyhawk was introduced which had new optional equipment including an electrical standby vacuum pump, a new DME with continuous LCD readouts of distance, groundspeed and time-to-station information, and a 12 month limited airframe warranty and 12 month unlimited hours engine warranty.

A total of 35,545 commercial aircraft in the Model 172/Skyhawk series had been built by the time production was suspended including 2,124 F 172s built in France by Reims. In addition, 864 were built between 1966-1983 as T-41A, T-41B, T-41C and T-41D Mescalero military basic trainers.

UPGRADES: **Bolen Conversions:** Taildragger modifications. See separate entry.

RAM Aircraft Corporation: RAM/Cessna 172. See separate entry.

OPERATORS: Versions of the Cessna 172 are in service with the armed forces of the following countries: Argentina (5), Bolivia (5), Burkina Fasco (1), Chile (14), Colombia (10), Dominican Republic (4), Ecuador (2), El Salvador (7), Greece (21), Guatemala (6), Haiti (1), Honduras (5), Ireland (7), Madagascar (4), Nicaragua (5), Pakistan (3), Peru (14), Philippines (12), Saudi Arabia (13), South Korea (15), Trinidad and Tobago (1), USA (50) and Uruguay (1).

The following description applies to the ***Model 172P***:

DESIGN FEATURES: Braced high-wing monoplane. NACA 2412 wing section. Dihedral 1° 44′. Incidence 1° 30′ at root, −1° 30′ at tip.

FLYING CONTROLS: Modified Frise all-metal ailerons. Electrically controlled NACA all-metal single-slotted flaps inboard of ailerons. Trim tab in starboard elevator. Ground adjustable tab in rudder, in-flight adjustable trim tab optional.

STRUCTURE: All-metal wing structure, except for conical camber glassfibre wingtips. Single bracing strut on each side. The fuselage is an all-metal semi-monocoque structure. The tail unit is a cantilever all-metal structure. Sweepback on fin 35° at quarter-chord.

LANDING GEAR: Non-retractable tricycle type. Cessna Land-O-Matic cantilever main legs, each comprising a one-piece machined conically tapered spring steel tube. Nosewheel is carried on an oleo-pneumatic shock-strut and is steerable with rudder up to 10° and controllable up to 30° on either side. Cessna mainwheels with tubed tyres size 6.00-6, 4 ply rating, pressure 1.93 bars (28 lb/sq in). Nosewheel tyre size 5.00-5, 6 ply rating, pressure 2.34 bars (34 lb/sq in). Hydraulic disc brakes. Parking brake. Wheel fairings optional. Edo Model 89-2000 floats optional.

POWER PLANT: One 119 kW (160 hp) Avco Lycoming O-320-D2J flat-four engine driving a McCauley Type 1C160/DTM7557 two-blade fixed-pitch metal propeller. Optional floatplane version has a McCauley Type 1A175/ETM8042 propeller. One fuel tank in each wing, total capacity 163 litres (43 US gallons; 36 Imp gallons). Usable fuel 151.4 litres (40 US gallons; 33 Imp gallons). Provision for long-range tanks, giving total capacity of 204 litres (54 US gallons; 45 Imp gallons), of which 189 litres (50 US gallons; 41 Imp gallons) are usable; or extra long-range system, using integral tanks in wings to provide total capacity of 257 litres (68 US gallons; 56 Imp gallons), of which 234 litres (62 US gallons; 51 Imp gallons) are usable. Oil capacity 7.5 litres (2 US gallons; 1.7 Imp gallons). Oil cooler and full-flow oil filter standard.

ACCOMMODATION: Cabin seats four in two pairs, with optional fully articulating front seats. Seat belts and shoulder restraints on all seats standard. Dual controls standard. Baggage aft of rear seats, capacity 54 kg (120 lb). An optional foldaway seat can be fitted in baggage space, for one or two children not exceeding 54 kg (120 lb) total weight. Door on each side of cabin, giving access to all seats, simplifies loading if rear seats are removed and cabin used for freight. Pilot's window opens; co-pilot's opening side window optional. Baggage door on port side. Combined heating and ventilation system; air conditioning optional. Dual windscreen defrosters. Glassfibre soundproofing. Overhead skylights and underwing courtesy lights optional.

SYSTEMS: Electrical system includes 28 V 60 A alternator, electric engine starter and 24 V 12.75 Ah battery. Heavy-duty 24 V 15.5 Ah battery optional. Air conditioning system of 14,000 BTU capacity optional. Vacuum system for blind-flying instruments. Standby electrically driven vacuum system optional.

AVIONICS AND EQUIPMENT: Optional avionics include Sperry Series 300 720-channel nav/com with remote VOR/LOC indicator, ADF and transponder, VOR/ILS indicator, Series 400 glide slope receiver, marker beacon receiver, transponder, altitude encoder, SDM-77A DME, intercom, 200A or 300A Nav-O-Matic single-axis autopilot. Stan-

Cessna 172E Skyhawk ***(Peter March)***

1994

Cessna 172 Skyhawk Floatplane ***(Peter March)***

1994

dard equipment includes artificial horizon, directional gyro, electric clock, outside air temperature gauge, rate of climb indicator, sensitive altimeter, turn co-ordinator, sun visors, single-cylinder engine priming system, full-flow oil filter, and towbar. Optional equipment includes true airspeed indicator, alternate static source, emergency locator transmitter, heated pitot, navigation light detectors, variable-intensity radio light, courtesy lights, and omni-flash beacon. Other operational equipment includes carburettor air temperature gauge, economy mixture indicator, flight hour recorder, quartz clock, all-purpose control wheel, reclining and vertically adjustable front seats, inertia reel shoulder harness, headrests, overhead skylights, starboard side storm window, glareshield, fire extinguisher, rear seats with individual reclining backs, internal corrosion proofing, rear seat ventilation system, tinted windows, ventilation fan, control wheel map light, map and instrument panel lights, wingtip strobe lights, anti-precipitation static kit, floatplane kit, glider tow hook, external socket, quick drain of oil valve, tailplane abrasion boots, wing strut and fuselage steps and handles for refuelling, and winterisation kit.

DIMENSIONS, EXTERNAL (L: landplane, F: floatplane):
- Wing span: 10.92 m (35 ft 10 in)
- Wing chord: at root: 1.63 m (5 ft 4 in)
- at tip: 1.12 m (3 ft 8½ in)
- Wing aspect ratio: 7.52
- Length overall: L: 8.20 m (26 ft 11 in)
- F: 8.13 m (26 ft 8 in)
- Height overall: L: 2.68 m (8 ft 9½ in)
- F: 3.63 m (11 ft 11 in)
- Tailplane span: 3.43 m (11 ft 3 in)
- Wheel track: L: 2.53 m (8 ft 3½ in)
- Wheelbase: L: 1.63 m (5 ft 4 in)
- Propeller diameter: L: 1.91 m (6 ft 3 in)
- F: 2.03 m (6 ft 8 in)
- Passenger doors (each): Height: 1.01 m (3 ft 3¾ in)
- Width: 0.89 m (2 ft 11 in)

AREAS:
- Wings, gross: 16.17 m² (174 sq ft)
- Ailerons (total): 1.70 m² (18.3 sq ft)
- Trailing-edge flaps (total): 1.97 m² (21.20 sq ft)
- Fin: 1.04 m² (11.24 sq ft)
- Rudder, incl tab: 0.69 m² (7.43 sq ft)
- Tailplane: 2.00 m² (21.56 sq ft)
- Elevators, incl tab: 1.35 m² (14.53 sq ft)

WEIGHTS AND LOADINGS (L: landplane, F: floatplane):
- Weight empty, equipped: L: 650 kg (1,433 lb)
- F: 733 kg (1,615 lb)
- Max T-O and landing weight: L: 1,089 kg (2,400 lb)
- F: 1,007 kg (2,220 lb)
- Max ramp weight: L: 1,092 kg (2,407 lb)
- F: 1,010 kg (2,227 lb)
- Max wing loading: L: 67.3 kg/m² (13.8 lb/sq ft)
- F: 62.0 kg/m² (12.7 lb/sq ft)
- Max power loading: L: 9.15 kg/kW (15.0 lb/hp)
- F: 8.44 kg/kW (13.9 lb/hp)

PERFORMANCE (L: landplane, F: floatplane, at max T-O weight, ISA):
- Never-exceed speed:
  - L, F: 158 kts (292 km/h; 181 mph) IAS
- Max level speed at S/L:
  - L: 123 kts (228 km/h; 141 mph)
  - F: 96 kts (178 km/h; 111 mph)
- Max cruising speed (75% power):
  - L at 2,440 m (8,000 ft): 120 kts (222 km/h; 138 mph)
  - F at 1,220 m (4,000 ft): 95 kts (176 km/h; 109 mph)
- Stalling speed, flaps up:
  - L: 51 kts (95 km/h; 59 mph) CAS
  - F: 48 kts (89 km/h; 55 mph) CAS
- Stalling speed, flaps down:
  - L: 46 kts (85 km/h; 53 mph)
  - F: 44 kts (82 km/h; 51 mph) CAS
- Max rate of climb at S/L: L: 213 m (700 ft)/min
  - F: 226 m (740 ft)/min
- Service ceiling: L: 3,960 m (13,000 ft)
  - F: 4,570 m (15,000 ft)
- T-O run: L: 272 m (890 ft)
  - F: 427 m (1,400 ft)
- T-O to 15 m (50 ft): L: 495 m (1,625 ft)
  - F: 658 m (2,160 ft)
- Landing run from 15 m (50 ft): L: 390 m (1,280 ft)
  - F: 410 m (1,345 ft)
- Landing run: L: 165 m (540 ft)
  - F: 180 m (590 ft)
- Range, at recommended lean mixture, with allowances for engine start, taxi, T-O, climb and 45 min reserves at 75% power:
  - max cruising speed:
    - L, standard fuel: 440 n miles (815 km; 506 miles)
    - F, standard fuel: 360 n miles (666 km; 414 miles)
    - L, 235 l (62 US gallons; 52 Imp gallons) fuel: 755 n miles (1,399 km; 869 miles)
    - F, 189 l (50 US gallons; 42 Imp gallons) fuel: 475 n miles (879 km; 546 miles)
- Range, allowances as above, but with 45 min reserves at 45% power:
  - econ cruising speed at 3,050 m (10,000 ft):
    - L, standard fuel: 520 n miles (963 km; 598 miles)
    - F, standard fuel: 435 n miles (806 km; 501 miles)
    - L, 235 l (62 US gallons; 52 Imp gallons) fuel: 875 n miles (1,620 km; 1,007 miles)
    - F, 189 l (50 US gallons; 42 Imp gallons) fuel: 565 n miles (1,046 km; 650 miles)

UPDATED

CESSNA 180 SKYWAGON

TYPE: One/six-seat cabin monoplane.

PROGRAMME: The Model 180 Skywagon has a typical Cessna braced high-wing monoplane layout, but with a tailwheel type of landing gear.

VERSIONS: **180 Skywagon:** Basic model, as described, which introduced as standard improvements for 1981 a new anti-precipitation static navigation antenna when factory installed radios are ordered, and an avionics cooling fan. New options include an intercom system, a lightweight headset, and DME and R/Nav equipment which together provide an IFR-certificated area navigation system.

180 Skywagon II: As Skywagon, plus factory installed avionics package which includes Series 300 nav/com with remote VOR indicator, ADF and an emergency locator transmitter. Optional Nav Pac provides a second nav/com, with Series 400 glide slope and marker beacon receiver. Standard equipment is as Skywagon, plus long-range fuel tanks with a total usable capacity of 284 litres (75 US gallons; 62.5 Imp gallons), artificial horizon, clock, directional gyro, exhaust gas temperature gauge, outside air temperature gauge, turn co-ordinator, vertical speed indicator, dual controls, sun visor, heated pitot, map light, quick drain oil plug, and external power socket.

A total of 6,161 Model 180s had been built by 1 January 1981. Production ended after delivery of the 1981 quota.

OPERATORS: Versions of the Cessna 180 are in service with the armed forces of the following countries: El Salvador (6), Honduras (2), Indonesia (5), Burma (Myanmar) (6) and Nicaragua (5).

DESIGN FEATURES: Wings generally similar to Skyhawk. Wing dihedral 1° 44′. Unswept cantilever tail unit.

FLYING CONTROLS: Generally similar to Skyhawk. Tail unit normally without trim tabs. Manually operated rudder trim available optionally.

STRUCTURE: The wing is generally similar to that of the Skyhawk. The fuselage is an all-metal semi-monocoque structure. Identical to Cessna 185 fuselage, except firewall and mounting brackets for dorsal fin.

LANDING GEAR: Non-retractable tailwheel type. Cessna cantilever spring steel main legs. Tailwheel has tapered tubular spring. Mainwheels and nylon tube-type tyres size 6.00-6 (optionally 8.00-6). Cessna tailwheel size 8.00 × 2.80. Tyre pressure, mainwheels 2.07 bars (30 lb/sq in), tailwheel 3.79 to 4.48 bars (55 to 65 lb/sq in) according to load. Hydraulic disc brakes. Parking brake. Wheel and brake fairings optional. Alternative Edo Model 628-296 floats, snow skis or amphibious gear.

POWER PLANT: One 171.5 kW (230 hp) Continental O-470-U flat-six engine, driving a McCauley 2A34C203/90DA-8 two-blade constant-speed metal propeller. Two fuel tanks in wings, with total standard capacity of 333 litres (88 US gallons; 73.3 Imp gallons), of which 318 litres (84 US gallons; 70 Imp gallons) are usable. Oil capacity 11.5 litres (3 US gallons; 2.5 Imp gallons).

ACCOMMODATION: Standard seating is for a pilot only, with a choice of three optional arrangements. Maximum seating is for six persons in three pairs, without baggage space. With fewer seats there is space at rear of cabin for up to 181 kg (400 lb) of baggage. Door on each side of cabin, plus optional cargo door and baggage compartment door on port side. Starboard door has quick-release hinge pins so that it can be removed when loading bulky cargo. Fifth and sixth passenger seats, attached to aft wall of cabin, can be folded when space is required for cargo. Hinged window each side. Instrument lighting controls are transistorised. Heating, ventilation, and windscreen defroster standard. Fully articulating seats for pilot and co-pilot, child's foldaway seat for the rear cabin and safety belts for rear-seat passengers are available optionally. Dual controls optional (standard on Skywagon II).

SYSTEMS: Hydraulic system for brakes only. Electrical system powered by 28 V 60 A alternator. 24 V 33 Ah battery. Oxygen system, 1.36 m³ (48 cu ft) capacity, optional.

AVIONICS AND EQUIPMENT: Optional avionics include Cessna Series 300 360-channel com transceiver, 360-channel nav/com with 160-channel nav and remote VOR indicator, 720-channel com and 200-channel nav with remote VOR/LOC indicator or VOR/ILS indicator, ADF with digital tuning, marker beacon receiver with three lights and aural signal, transponder with 4096 code capability, Collins R/Nav and DME-450C Nav-O-Matic single-axis autopilot with heading control and VOR intercept and track; or Series 400 720-channel com transceiver, 720-channel nav/com with remote VOR/LOC or VOR/ILS indicator, transponder with 4096 code capability, glide slope receiver and ADF with digital tuning. Standard avionics for Skywagon II include Cessna 300 Series nav/com with remote VOR indicator, ADF and an emergency locator transmitter. Standard equipment includes control locks, audible stall warning, baggage restraint net, cabin dome light, instrument panel red floodlights, and landing and taxi lights. Optional equipment (standard on Skywagon II) includes long-range fuel tanks, artificial horizon, clock, directional gyro, exhaust gas temperature gauge, outside air temperature gauge, turn co-ordinator, vertical speed indicator, dual controls, sun visor, heated pitot, map light, quick drain oil valve and external power socket. Other optional equipment includes carburettor air temperature gauge, economy mixture indicator, flight hour recorder, true airspeed indicator, turn and bank indicator, co-pilot's seat installation, pilot and co-pilot headrests, rudder pedal extensions, stowable starboard side rudder pedals, inertia reel shoulder harness, twin beverage pack, tailwheel lock, alternate static source, emergency locator transmitter, navigation light detectors, cargo tiedown fittings, bubble and tinted windows, de luxe interior, internal corrosion proofing, cabin fire extinguisher, photographic provisions, stretcher installation, courtesy lights, boom microphone, control wheel with map light and microphone switch,

Cessna Model 180 Skywagon one/six-seat cabin monoplane ***(Peter March)***

1994

Cessna Model 180F Skywagon braced high-wing monoplane *(Peter March)*

1994

auxiliary instrument lights, instrument panel post lights, omni-flash beacon, strobe light, amphibious kit, floatplane kit, ski kit, engine winterisation kit, agricultural sprayer system, non-congealing oil cooler, oil dilution system, overall paint scheme, and tailplane abrasion boots.

DIMENSIONS, EXTERNAL:

Wing span	10.92 m (35 ft 10 in)
Wing chord: at root	1.63 m (5 ft 4 in)
at tip	1.12 m (3 ft 8 in)
Wing aspect ratio	7.52
Length overall: Landplane	7.81 m (25 ft 7½ in)
Skiplane	8.47 m (27 ft 9½ in)
Floatplane	8.23 m (27 ft 0 in)
Amphibian	8.38 m (27 ft 6 in)
Height overall: Landplane, skiplane	2.36 m (7 ft 9 in)
Floatplane	3.71 m (12 ft 2 in)
Amphibian	3.86 m (12 ft 8 in)
Tailplane span	3.35 m (11 ft 0 in)
Wheel track: Landplane	2.26 m (7 ft 5 in)
Propeller diameter:	
Landplane, skiplane	2.08 m (6 ft 10 in)
Floatplane, amphibian	2.29 m (7 ft 6 in)
Passenger doors (each): Height	1.01 m (3 ft 3¾ in)
Width	0.89 m (2 ft 11 in)

AREAS:

Wings, gross	16.16 m² (174 sq ft)
Ailerons (total)	1.70 m² (18.3 sq ft)
Trailing-edge flaps (total)	1.97 m² (21.23 sq ft)
Fin	0.84 m² (9.01 sq ft)
Dorsal fin	0.19 m² (2.04 sq ft)
Rudder	0.68 m² (7.29 sq ft)
Tailplane	1.94 m² (20.94 sq ft)
Elevators	1.40 m² (15.13 sq ft)

WEIGHTS AND LOADINGS:

Weight empty, equipped:	
Skywagon landplane	749 kg (1,650 lb)
Skywagon II landplane	772 kg (1,701 lb)
Floatplane	888 kg (1,958 lb)
Skiplane	813 kg (1,792 lb)
Amphibian	1,004 kg (2,213 lb)
Max T-O weight:	
Landplane, skiplane	1,270 kg (2,800 lb)
Floatplane, amphibian	1,338 kg (2,950 lb)
Max ramp weight:	
Landplane, skiplane	1,275 kg (2,810 lb)
Floatplane, amphibian	1,343 kg (2,960 lb)
Max wing loading:	
Landplane, skiplane	78.6 kg/m² (16.1 lb/sq ft)
Floatplane, amphibian	83.0 kg/m² (17.0 lb/sq ft)
Max power loading:	
Landplane, skiplane	7.41 kg/kW (12.2 lb/hp)
Floatplane, amphibian	7.80 kg/kW (12.8 lb/hp)

PERFORMANCE (at max T-O weight, ISA):

Never-exceed speed:	
Landplane	167 kts (309 km/h; 192 mph)
Max level speed at S/L:	
* Landplane	148 kts (274 km/h; 170 mph)
Floatplane, amphibian, skiplane	129 kts (240 km/h; 149 mph)
Max cruising speed (75% power) at 2,440 m (8,000 ft):	
* Landplane	142 kts (264 km/h; 164 mph)

* *These speeds are 1 kt (1.9 km/h; 1.2 mph) higher with optional speed fairings installed*

Floatplane, amphibian	123 kts (228 km/h; 142 mph)
Skiplane	124 kts (230 km/h; 143 mph)
Econ cruising speed at 3,050 m (10,000 ft):	
Landplane	105 kts (195 km/h; 121 mph)
Floatplane, amphibian	99 kts (183 km/h; 114 mph)
Skiplane	88 kts (162 km/h; 101 mph)
Stalling speed, flaps up, power off:	
all versions	53 kts (98.5 km/h; 61 mph) CAS
Stalling speed, flaps down, power off:	
all versions	48 kts (88.5 km/h; 55 mph) CAS
Max rate of climb at S/L:	
Landplane	335 m (1,100 ft)/min
Floatplane, amphibian	296 m (970 ft)/min
Skiplane	277 m (910 ft)/min
Service ceiling: Landplane	5,395 m (17,700 ft)
Floatplane, amphibian	4,663 m (15,300 ft)
Skiplane	4,480 m (14,700 ft)
T-O run: Landplane	190 m (625 ft)
Floatplane	354 m (1,160 ft)
Amphibian: on land	213 m (700 ft)
on water	354 m (1,160 ft)
T-O to 15 m (50 ft): Landplane	367 m (1,205 ft)
Floatplane	579 m (1,900 ft)
Amphibian: on land	401 m (1,315 ft)
on water	579 m (1,900 ft)
Landing from 15 m (50 ft): Landplane	416 m (1,365 ft)
Floatplane	524 m (1,720 ft)
Amphibian: on land	442 m (1,450 ft)
on water	524 m (1,720 ft)
Landing run: Landplane	146 m (480 ft)
Floatplane	224 m (735 ft)
Amphibian: on land	226 m (740 ft)
on water	224 m (735 ft)

Range, at recommended lean mixture, with allowances for start, taxi, T-O, climb and 45 min reserves at 45% power:

max fuel, max cruising speed at 2,440 m (8,000 ft):	
Landplane	785 n miles (1,454 km; 903 miles)
Floatplane, amphibian	680 n miles (1,260 km; 783 miles)
Skiplane	685 n miles (1,270 km; 789 miles)
max fuel, econ cruising speed at 3,050 m (10,000 ft):	
Landplane	1,010 n miles (1,872 km; 1,163 miles)
Floatplane, amphibian, skiplane	815 n miles (1,510 km; 938 miles)

UPDATED

CESSNA CENTURION and TURBO CENTURION

TYPE: Six-seat cabin monoplane.

PROGRAMME: The original Model 210 (now Centurion), which flew in January 1957, followed the general formula of the Cessna series of all-metal high-wing monoplanes, but was the first to have a retractable tricycle landing gear.

Later versions of the Model 210/Centurion have a full cantilever wing, eliminating the bracing struts used on earlier models. The design started 24 October 1964 and construction of a prototype began 29 November 1964. The first T210 (now Turbo Centurion) with the new wing flew 18 June 1965.

In December 1970 Cessna announced the introduction of two new versions known as Centurion II and Turbo Centurion II. These differ from the Centurion and Turbo Centurion by having, as standard equipment a factory installed IFR avionics package offering a cost saving on avionics equipment, plus articulating front seats and all-purpose control wheel. A pressurised Centurion was introduced in late 1977.

VERSIONS: **Centurion:** Standard model, with 224 kW (300 hp) Continental IO-520-L flat-six engine, driving a McCauley D3A34C404/80VA-00 three-blade constant-speed metal propeller.

Centurion II: Identical to Centurion but with a factory installed package comprising a Series 300 nav/com with 720-channel com, 200-channel nav and VOR/LOC indicator, ADF, transponder, basic avionics kit, 200A Nav-O-Matic single-axis autopilot, emergency locator transmitter, true airspeed and economy mixture indicators, alternate static source, pilot's all-purpose control wheel, instrument post lights, radio stack lights, vertically adjustable and reclining co-pilot's seat, heated pitot and stall warning transmitter, underwing courtesy lights, navigation light detectors, omni-flash beacon and external power socket. An optional Nav Pac provides a second Series 300 nav/com, and Series 400 glide slope and marker beacon receivers, adding a further 7 kg (16 lb) to basic empty weight.

Turbo Centurion: Generally similar to Centurion, but powered by a 231 kW (310 hp) Continental TSIO-520-R turbocharged engine, driving a McCauley Type D3A34C402/90DFA-10 three-blade constant-speed metal propeller. Standard equipment of this version includes a pressure controller, turbocharger valve, overboost control, economy mixture indicator, alternate static source, map light, all-purpose control wheel and provisions for an oxygen system.

Turbo Centurion II: Identical to Turbo Centurion but with additional standard equipment as detailed for Centurion II, except that transponder is Series 400.

Pressurised Centurion/Pressurised Centurion II: Introduced in late 1977 (see separate entry).

The original versions received FAA Type Approval on 23 August 1966. A total of 8,397 plus 843 Pressurised Centurions were produced prior to the suspension of production in 1985.

UPGRADES: **Advanced Aircraft:** P-210. See separate entry.

RAM Aircraft Corporation: RAM/Cessna T210. See separate entry.

Riley International Corporation: Riley Rocket Cessna P-210. See separate entry.

Sierra Industries: Cessna landing gear door modification. See separate entry.

DESIGN FEATURES: Cantilever high-wing monoplane. Wing section NACA 64_2A215 at root, NACA 64_1A412 (A = 0.5) at tip. Dihedral 1° 30′. Incidence 1° 30′ at root, −1° 30′ at tip. Fixed incidence tailplane.

FLYING CONTROLS: All-metal Frise ailerons. Electrically actuated all-metal Fowler flaps. Ground adjustable tab in each aileron. Controllable trim tabs in rudder and starboard elevator. Electrical operation (standard with Series 400B autopilot).

STRUCTURE: All-metal structure, except for glassfibre conical camber tips. The fuselage is an all-metal semi-monocoque structure. The tail unit is a cantilever all-metal structure with 36° sweepback on fin.

LANDING GEAR: Hydraulically retractable tricycle type with single wheel on each unit. Nose unit retracts forward, main units aft and inward. Chrome vanadium tapered steel tube main legs. Steerable nosewheel with oleo-pneumatic shock-absorber. Cleveland or McCauley wheels with tubed tyres; mainwheel tyres size 6.00-6, 8 ply rating, pressure 3.79 bars (55 lb/sq in). Nosewheel tyre size 5.00-5, 6 ply rating, pressure 3.45 bars (50 lb/sq in). Turbo Centurion has a 5.00-5 nosewheel tyre of 10 ply rating, pressure 6.07 bars (88 lb/sq in). Cessna hydraulic disc brakes. Parking brake.

Cessna 210N Turbo Centurion *(A. March)*

1994

POWER PLANT: One flat-six engine, as described under model listings. Integral fuel tanks in wings, with maximum total usable capacity of 329 litres (87 US gallons; 72 Imp gallons). Optional long-range tanks increase total usable capacity to 435 litres (115 US gallons; 95.8 Imp gallons). Refuelling points above wing. Oil capacity 10.5 litres (2.75 US gallons; 2.29 Imp gallons). Electrically heated propeller de-icing boots optional.

ACCOMMODATION: Six persons in pairs in enclosed cabin. Fifth and sixth seats have folding backs to accommodate articles up to 2.01 m (6 ft 7 in) long. Openable window on port side standard; optional for starboard side. Dual controls standard. Forward hinged door on each side of cabin. Baggage space aft of rear seats, capacity 109 kg (240 lb), with outside door on port side. Combined heating and ventilation system. Windscreen defroster standard; electric anti-icing optional.

SYSTEMS: Integral hydraulic-electric unit for landing gear operation. Hydraulic system for brakes. Electrical power supplied by 28 V engine-driven alternator; 24 V 95 A alternator or dual 28 V 60 A alternators optional. 24 V 12.75 Ah battery standard, 24 V 15.5 Ah heavy-duty battery optional. Oxygen system of 2.1 m³ (74 cu ft) or 2.15 m³ (76 cu ft) capacity optional. Vacuum system standard. Back-up electrically driven vacuum system optional. Air conditioning system optional. Pneumatic de-icing system optional. Tailplane abrasion boots optional, but cannot be installed with optional pneumatic de-icing system.

AVIONICS AND EQUIPMENT: Optional 'Value Group A' avionics and equipment include Series 300 nav/com, ADF, transponder (400 series in Turbo Centurion), alternate static source, economy mixture indicator, all-purpose control wheel on pilot's side, instrument post lighting, omniflash beacon, underwing courtesy lights, map light, radio stack lights, emergency locator beacon, adjustable, reclining-back seats for front passengers, navigation light detectors, heated pitot and stall warning systems and ground service plug receptacle. 'Value Group B' adds antenna and coupler, second nav/com and Series 400 glide slope and marker beacon receivers. Optional avionics include 400B Nav-O-Matic two-axis autopilot, 400B IFCS system and Series 400 nav/com, VOR/LOC, VOR/ILS, ADF, DME, transponder, encoding altimeter, RMI, area navigation system; King-Pac avionics package (with 400B Nav-O-Matic or IFCS); Primus 100, Bendix RDR-160 or King KWX-56 weather radar, intercom system, radar altimeter and radio-telephone. Standard equipment includes electric clock, outside air temperature gauge, rate of climb indicator, sensitive altimeter, turn co-ordinator, vacuum system with altitude and directional gyros, cylinder head temperature gauge, recording tachometer, armrests, control locks, tinted windscreen and windows, glareshield, sun visors, audible stall and landing gear warning systems, baggage restraint net, adjustable fresh air vents, dome lights, control pedestal light, engine instrument lights, map lights, variable intensity instrument panel red floodlights, pilot's fore and aft/vertically adjustable and reclining seat, co-pilots fore and aft adjustable and reclining seat, seat belts, shoulder restraints, soundproofing, omni-vision rear window, baggage compartment tiedown rings and lights, landing and taxi light, navigation lights, quick fuel drains and sampler cup, overall paint scheme, cabin steps, jack pads, and towbar. Optional equipment for all versions includes an Alcor combustion analyser, clock/fuel computer, quartz clock, turn and bank indicator, flight hour recorder, electroluminescent instrument lights, electric elevator trim system, inertia reel shoulder harnesses, cabin fire extinguisher, headrests, writing desk, stereo entertainment centre, corrosion proofing, de-icing system, ice detector light, wingtip-mounted strobe lights, fin light, engine priming system, tailplane abrasion boots and retractable cabin entrance step (starboard). Engine winterisation kit available for Centurion only.

DIMENSIONS, EXTERNAL:

Wing span	12.41 m (38 ft 10 in)
Wing chord: at root	1.68 m (5 ft 6 in)
Wing aspect ratio	8.13
Length overall	8.59 m (28 ft 2 in)
Height overall	2.95 m (9 ft 8 in)
Tailplane span	4.88 m (16 ft 0 in)
Wheel track	2.64 m (8 ft 8 in)
Wheelbase	1.83 m (6 ft 0 in)
Propeller diameter	2.03 m (6 ft 8 in)
Passenger door (each): Max height	1.02 m (3 ft 4 in)
Max width	0.91 m (3 ft 0 in)
Height to sill	0.91 m (3 ft 0 in)
Baggage compartment door:	
Max height	0.38 m (1 ft 2¾ in)
Max width	0.72 m (2 ft 4½ in)

DIMENSIONS, INTERNAL:

Cabin: Length	3.50 m (11 ft 6 in)
Max width	1.07 m (3 ft 6 in)
Max height	1.22 m (4 ft 0 in)
Floor area	2.69 m² (29.0 sq ft)
Volume	3.96 m³ (139.9 cu ft)
Baggage space	0.46 m³ (16.25 cu ft)

AREAS:

Wings, gross	17.23 m² (185.5 sq ft)
Trailing-edge flaps (total)	2.74 m² (29.50 sq ft)
Fin, incl dorsal fin	0.95 m² (10.26 sq ft)
Rudder, incl tab	0.65 m² (6.95 sq ft)

Cessna 210 Centurion *(Peter March)*

1994

WEIGHTS AND LOADINGS:

Weight empty: Centurion	1,007 kg (2,220 lb)
Turbo Centurion	1,052 kg (2,320 lb)
Max ramp weight: Centurion	1,752 kg (3,862 lb)
Turbo Centurion	1,868 kg (4,118 lb)
Max T-O weight: Centurion	1,746 kg (3,850 lb)
Turbo Centurion	1,860 kg (4,100 lb)
Max landing weight: Centurion	1,746 kg (3,850 lb)
Turbo Centurion	1,769 kg (3,900 lb)
Max wing loading:	
Centurion	101.3 kg/m² (20.75 lb/sq ft)
Turbo Centurion	107.95 kg/m² (22.10 lb/sq ft)
Max power loading:	
Centurion	7.79 kg/kW (12.83 lb/hp)
Turbo Centurion	7.64 kg/kW (12.61 lb/hp)

PERFORMANCE (at mid-cruise weight, ISA):

Never-exceed speed (both models)
200 kts (371 km/h; 230 mph) IAS

Max level speed:
Centurion at S/L 175 kts (324 km/h; 202 mph)
Turbo Centurion at 6,100 m (20,000 ft)
225 kts (417 km/h; 259 mph)

Max cruising speed: Centurion at 2,135 m (7,000 ft)
169 kts (313 km/h; 195 mph)
Turbo Centurion at 6,100 m (20,000 ft)
207 kts (384 km/h; 238 mph)

Stalling speed, flaps up, power off:

Centurion	63 kts (117 km/h; 73 mph)
Turbo Centurion	65 kts (120 km/h; 75 mph)

Stalling speed, flaps down, power off:

Centurion	53 kts (98 km/h; 61 mph)
Turbo Centurion	55 kts (102 km/h; 63 mph)

Max rate of climb at S/L:	
Centurion	323 m (1,060 ft)/min
Turbo Centurion	351 m (1,150 ft)/min
Service ceiling: Centurion	4,875 m (16,000 ft)
Turbo Centurion	8,840 m (29,000 ft)
T-O run: Centurion	370 m (1,215 ft)
Turbo Centurion	387 m (1,270 ft)
T-O to 15 m (50 ft): Centurion	625 m (2,050 ft)
Turbo Centurion	643 m (2,110 ft)
Landing from 15 m (50 ft): Centurion	483 m (1,585 ft)
Turbo Centurion	488 m (1,600 ft)
Landing run: Centurion	248 m (815 ft)
Turbo Centurion	251 m (825 ft)

Range, standard fuel, recommended lean mixture, allowances for start, taxi, T-O, climb, descent and 45 min reserves, max cruise power:
Centurion at 2,135 m (7,000 ft)
770 n miles (1,427 km; 887 miles)
Turbo Centurion at 6,100 m (20,000 ft)
705 n miles (1,307 km; 812 miles)

Range, standard fuel, recommended lean mixture, allowances as above except max range power:
Centurion at 3,050 m (10,000 ft)
1,010 n miles (1,872 km; 1,163 miles)
Turbo Centurion at 3,050 m (10,000 ft)
855 n miles (1,584 km; 985 miles)

Range, optional long-range tanks, allowances as above except max cruise power:
Centurion at 2,135 m (7,000 ft)
1,070 n miles (1,983 km; 1,232 miles)
Turbo Centurion at 6,100 m (20,000 ft)
1,015 n miles (1,881 km; 1,169 miles)

Range, optional long-range tanks, allowances as above, except max range power:
Centurion at 3,050 m (10,000 ft)
1,390 n miles (2,576 km; 1,601 miles)
Turbo Centurion at 3,050 m (10,000 ft)
1,190 n miles (2,205 km; 1,370 miles)

VERIFIED

CESSNA PRESSURISED CENTURION and PRESSURISED CENTURION II

TYPE: Six-seat cabin monoplane.

PROGRAMME: On 10 November 1977, Cessna announced the introduction of a pressurised version of the Centurion. A new model designated **210R**, was introduced in late 1984, and is generally similar to the 1985 model Centurion and Turbo Centurion, except as detailed. A new optional six-place oxygen system using a lightweight, filament wound bottle of 2.15 m³ (76.0 cu ft) capacity, offered on the Pressurised Centurion.

In early 1985 Cessna announced that it was test flying an experimental version of the Pressurised Centurion powered by an Allison 250 turboprop engine, but said that the company had no current plans for a production version of this aircraft.

A total of 843 Pressurised Centurions were delivered prior to suspension of production.

The description of the Centurion applies also to the Pressurised Centurion, except as follows:

STRUCTURE: The fuselage is a conventional semi-monocoque structure of light alloy, with fail-safe structure in the pressurised section.

LANDING GEAR: Nosewheel tyre same as Turbo Centurion.

POWER PLANT: One 242.4 kW (325 hp) Teledyne Continental TSIO-520-CE flat-six turbocharged engine, driving a McCauley D3A36C410/80VMB-0 three-blade constant-speed metal propeller with spinner.

ACCOMMODATION: As for Centurion, except forward-hinged door on port side of cabin. Large emergency exit on starboard side. Baggage space aft of cabin area, capacity 91 kg (200 lb). Four windows each side of cabin, two overhead windows above the rear seats. Cabin pressurised, heated and ventilated.

SYSTEMS: As for Centurion, except cabin pressurisation system by engine bleed air, maximum differential 0.23 bars (3.35 lb/sq in) permitting a cabin altitude of 3,690 m (12,100 ft) at 7,010 m (23,000 ft). Cabin heated by double heat exchange system using exhaust system heat.

Cessna Pressurised Centurion II *(Peter March)*

1994

Cessna Pressurised Centurion six-seat monoplane

1994

AVIONICS AND EQUIPMENT: As for Centurion except that 'Value Group B' options are incorporated into 'Value Group A' package.

DIMENSIONS, EXTERNAL:

Wing span	12.41 m (38 ft 10 in)
Wing chord: at root	1.68 m (5 ft 6 in)
Wing aspect ratio	8.13
Length overall	8.59 m (28 ft 2 in)
Height overall	2.95 m (9 ft 8 in)
Tailplane span	4.88 m (16 ft 0 in)
Wheel track	2.64 m (8 ft 8 in)
Wheelbase	1.83 m (6 ft 0 in)
Propeller diameter	2.03 m (6 ft 8 in)
Passenger door (each): Max height	1.02 m (3 ft 4 in)
Max width	0.91 m (3 ft 0 in)
Height to sill	0.91 m (3 ft 0 in)
Baggage compartment door:	
Max height	0.38 m (1 ft 2¾ in)
Max width	0.72 m (2 ft 4½ in)

DIMENSIONS, INTERNAL: As for Centurion except:

Baggage space	0.52 m³ (18.3 cu ft)

WEIGHTS AND LOADINGS:

Weight empty	1,121 kg (2,471 lb)
Max ramp weight	1,868 kg (4,118 lb)
Max T-O weight	1,860 kg (4,100 lb)
Max landing weight	1,769 kg (3,900 lb)
Max wing loading	107.90 kg/m² (22.10 lb/sq ft)
Max power loading	7.67 kg/kW (12.61 lb/hp)

PERFORMANCE (at mid-cruise weight, ISA, except where indicated):

Max level speed at 6,100 m (20,000 ft)
201 kts (372 km/h; 231 mph)

Max cruising speed at 7,010 m (23,000 ft)
213 kts (394 km/h; 245 mph)

Cruising speed at 3,050 m (10,000 ft)
185 kts (343 km/h; 213 mph)

Stalling speed, power off:	
flaps up	65 kts (121 km/h; 75 mph) CAS
flaps down	55 kts (102 km/h; 63 mph) CAS
Max rate of climb at S/L	351 m (1,150 ft)/min
Max operating altitude	7,620 m (25,000 ft)
T-O run	387 m (1,270 ft)
T-O to 15 m (50 ft)	643 m (2,110 ft)
Landing from 15 m (50 ft)	488 m (1,600 ft)
Landing run	251 m (825 ft)

Range, with standard fuel, recommended lean mixture, with fuel allowance for engine start, taxi, T-O, climb and 45 min reserves:

at max cruise power: at 7,010 m (23,000 ft)
720 n miles (1,334 km; 829 miles)

at 3,050 m (10,000 ft)
665 n miles (1,232 km; 766 miles)

at max range power, allowances and reserves as above:
at 3,050 m (10,000 ft)
855 n miles (1,585 km; 985 miles)

Range, with optional fuel allowances and reserves as above:

at max cruise power: at 7,010 m (23,000 ft)
1,040 n miles (1,928 km; 1,198 miles)

at 3,050 m (10,000 ft)
935 n miles (1,733 km; 1,077 miles)

at max range power, allowances and reserves as above:
at 3,050 m (10,000 ft)
1,190 n miles (2,205 km; 1,370 miles)

VERIFIED

CESSNA T303 CRUSADER

TYPE: Twin-engined cabin monoplane.

PROGRAMME: Cessna announced 17 February 1978 the first flight, three days earlier, of a lightweight twin-engined aircraft known as the Model 303. At that time it was a four-seat aircraft, with 119 kW (160 hp) engines, and made use of bonded structures and a supercritical wing section. It was superseded by a new Model T303, with six seats, 186 kW (250 hp) turbocharged engines and conventional construction, except for use of bonding in the integral fuel tank area. Following certification to the latest FAR Pt 23 regulations on 27 August 1981, first deliveries of T303s were made in October 1981. A total of 297 Crusaders were sold.

The Crusader has counter-rotating propellers, and an extensive range of equipment. Cessna claimed it was the first aircraft in its class to have full IFR equipment as standard.

DESIGN FEATURES: Cantilever low-wing monoplane. Wing section NACA 23017 at root, NACA 23015.5 outboard of engines, and NACA 23012 at tip. Dihedral 7°. Incidence 3° at root, 0° at tip.

FLYING CONTROLS: Plain ailerons and wide span electrically actuated single-slotted Fowler trailing-edge flaps of light alloy construction. Trim tab in starboard aileron. Trim tab in rudder and starboard elevator.

STRUCTURE: Conventional two-spar structure of light alloy, with sheet metal ribs and stringers, and upper and lower skins. Flow energiser vanes are located at the wing leading-edge at each side of the fuselage and engine nacelles. The fuselage is an oval section semi-monocoque structure of light alloy construction. The tail unit is a cantilever structure, primarily of light alloy, with horizontal surfaces mounted partway up fin. Fin and rudder swept back. Long dorsal fin.

LANDING GEAR: Retractable tricycle type, with single wheel on each unit. Mainwheels retract inward, nosewheel forward. Free-fall emergency extension, with manually operated hydraulic pump as back-up. Main units of articulated (trailing link) type and steerable nosewheel all have oleo-pneumatic shock-absorbers. Cleveland mainwheels with tubed tyres size 6.00-6, 8 ply rating, pressure 3.52 to 3.93 bars (51 to 57 lb/sq in). Nosewheel with tubed tyre size 6.00-6, 6 ply rating, pressure 2.48 to 2.90 bars (36 to 42 lb/sq in). Hydraulically actuated single-disc brakes. Parking brake. Heavy-duty mainwheels with 6.50-8 tyres of 8 ply rating and heavy-duty optional (tyre pressure unchanged).

POWER PLANT: Two 186 kW (250 hp) Continental flat-six turbocharged engines comprising (port) a TSIO-520-AE and (stbd) an LTSIO-520-AE, driving respectively a McCauley Type 3AF32C506/82NEB-8 and 3AF32C507/L82NEB-8 counter-rotating three-blade constant-speed fully feathering metal propeller with spinner. Fuel in integral wing tanks with combined usable capacity of 579 litres (153 US gallons; 127 Imp gallons). Refuelling point on upper surface of each wing. Oil capacity 17 litres (4.5 US gallons; 3.75 Imp gallons). Electrically heated propeller de-icing boots optional.

ACCOMMODATION: Standard seating for pilot and five passengers; or pilot and four passengers. Six individual forward-facing seats can be replaced by optional club arrangement. Outboard armrests standard; inboard retractable armrests and adjustable headrests normally optional, but standard with club seating. Flight deck divider curtain optional. Wide range of optional cabin furniture and equipment. Clamshell type two-piece cabin door, with integral airstair, on port side at aft end. Emergency exit at front of cabin on starboard side. Optional top hinged cargo door immediately aft of airstair door, for freight, air ambulance or casket carrying operations. Baggage stowage in nose compartment (with door on port side), wing lockers, and at rear of cabin, with combined capacity of 267 kg (590 lb). Accommodation heated and ventilated. Air conditioning optional. Windscreen defrosting standard. Electrically heated windscreen anti-icing optional.

SYSTEMS: Electrical system powered by two 28 V 60 A engine-driven alternators and a 24 V 15.5 Ah battery; 28 V 95 A alternators optional. Hydraulic system for landing gear operation powered by electrohydraulic unit. Separate hydraulic system for brakes only. Heating system includes a 45,000 BTU combustion heater. Provisions for oxygen system of 2.17 m³ (76.6 cu ft) capacity. Optional air conditioning system includes 13,000 BTU heater and requires dual 95 A alternators and 28 Ah battery. Engine fire detection system standard. Wing leading-edge de-icing boots optional. Tailplane and fin leading-edge de-icing boots optional.

AVIONICS AND EQUIPMENT: Standard avionics include Sperry Series 400 (485B) nav/com with VOR/ILS, nav/com with VOR/LOC, ADF glide slope and marker beacon receivers, transponder, and 400B Nav-O-Matic two-axis autopilot with slaved directional gyro. Yaw damper system. Optional avionics include 400B IFCS. HSI system. Series 400 R/Nav, DME, RMI, second glide slope receiver, encoding altimeter, altitude encoder, Series 800 encoding/alert altimeter, intercom system, Bendix RDR-160 colour or monochromatic, King KWX-56 colour or Sperry Primus 100 monochromatic weather radar systems, and King KT-96 radiotelephone. Standard equipment includes sensitive altimeter, electric clock, horizon and slaved directional gyros with dual vacuum system, outside air temperature gauge; rate of climb, true airspeed and turn and bank indicators, alternate static source; cylinder head temperature and fuel flow gauges; low-fuel warning lights, economy mixture indicator; engine instrument, instrument panel post, and map lights; annunciator, control pedestal and radio lights; all-purpose control wheel, dual controls, electric elevator trim, control locks, padded glareshield, sun visors, audible stall warning system, emergency locator transmitter, six individual fore and aft adjustable and reclining seats, seat belts, shoulder restraints, armrests, map and storage pocket, soundproofing, adjustable ventilators, pilot's storm window, tinted windows, cabin courtesy lights, omni-flash beacons; landing, navigation and taxi lights; full-flow oil filters, fuel quick drain valves and sampler cup, external corrosion proofing, heated pitot and stall warning transmitter, anti-precipitation static kit, tie-down rings, jack pads, nose-gear viewing mirror, towbar and external power socket. Optional equipment includes alternative sensitive altimeters, quartz clock, cabin heater hour meter, turn co-ordinator electronic fuel flow indicating system, flight hour recorder, co-pilot's flight instruments, cabin fire extinguisher, rudder gust lock, microphone/headset combination, pilot and co-pilot vertically adjustable seats, inertia reel shoulder harnesses for pilot and co-pilot, headrests, stereo installation, table, refreshment centre. 'Fasten seat belts - no smoking' sign, flight deck divider curtain, window curtains, passengers' reading lights, baggage compartment courtesy lights, ice detection light, strobe lights, starboard landing light, propeller synchrophaser, propeller automatic unfeathering system, fan-driven ventilation system, and fuselage ice protection plates.

DIMENSIONS, EXTERNAL:

Wing span	11.90 m (39 ft 0½ in)
Wing chord: at root	1.75 m (5 ft 9 in)
at tip	1.24 m (4 ft 1 in)
Length overall	9.27 m (30 ft 5 in)
Height overall	4.06 m (13 ft 4 in)
Tailplane span	5.18 m (17 ft 0 in)
Wheel track	3.81 m (12 ft 6 in)

Cessna T303 Crusader twin-engine monoplane

1994

Wheelbase	2.29 m (7 ft 6 in)
Propeller diameter	1.88 m (6 ft 2 in)
Propeller ground clearance	0.305 m (1 ft 0 in)
Passenger door (port, rear):	
Height, mean	1.17 m (3 ft 0 in)
Width	0.61 m (2 ft 0 in)
Emergency exit (stbd, fwd): Height	0.69 m (2 ft 3 in)
Max width	0.58 m (1 ft 11 in)
Cargo door (port, optional): Height	0.88 m (2 ft 10¾ in)
Width	0.81 m (2 ft 8 in)
Baggage door (port, fwd):	
Height, mean	0.38 m (1 ft 3 in)
Width	0.86 m (2 ft 10 in)
Wing locker doors (port and stbd):	
Length	0.61 m (2 ft 0 in)
Width	0.65 m (2 ft 1½ in)

DIMENSIONS, INTERNAL:

Cabin: Length (fwd to rear bulkhead):	
standard	4.14 m (13 ft 7 in)
with cargo door	4.30 m (14 ft 1½ in)
Max height	1.20 m (3 ft 11½ in)
Max width	1.21 m (3 ft 11¾ in)

AREAS:

Wings, gross	17.58 m² (189.2 sq ft)

WEIGHTS AND LOADINGS:

Weight empty	1,526 kg (3,364 lb)
Max T-O weight	2,336 kg (5,150 lb)
Max ramp weight	2,347 kg (5,175 lb)
* Max landing weight	2,268 kg (5,000 lb)
Max zero-fuel weight	2,200 kg (4,850 lb)
Max wing loading	132.8 kg/m² (27.2 lb/sq ft)
Max power loading	6.28 kg/kW (10.3 lb/hp)

* *with optional heavy duty wheels and brakes max landing weight is 2,336 kg (5,150 lb)*

PERFORMANCE (at max T-O weight, S/L, ISA, except where indicated):

Max level speed at 5,485 m (18,000 ft)	216 kts (400 km/h; 249 mph)
Max cruising speed, 71% power at 6,100 m (20,000 ft)	196 kts (363 km/h; 226 mph)
Cruising speed, 72% power at 3,050 m (10,000 ft)	180 kts (333 km/h; 207 mph)
Stalling speed, power off:	
flaps up	68 kts (126 km/h; 79 mph) CAS
flaps down	62 kts (115 km/h; 72 mph) CAS
Max rate of climb at S/L	451 m (1,480 ft)/min
Rate of climb at S/L, one engine out	67 m (220 ft)/min
Max operating altitude	7,620 m (25,000 ft)
Service ceiling, one engine out	3,960 m (13,000 ft)
T-O run	389 m (1,275 ft)
T-O run to 15 m (50 ft)	533 m (1,750 ft)
Landing from 15 m (50 ft)	442 m (1,450 ft)
Landing run	250 m (820 ft)
Range with max fuel, recommended lean mixture, with allowances for engine start, taxi, T-O, climb, descent and 45 min reserves:	
71% power at 6,100 m (20,000 ft)	895 n miles (1,658 km; 1,030 miles)
72% power at 3,050 m (10,000 ft)	835 n miles (1,546 km; 961 miles)
econ cruising speed at 6,100 m (20,000 ft)	1,005 n miles (1,861 km; 1,157 miles)
econ cruising speed at 3,050 m (10,000 ft)	1,020 n miles (1,889 km; 1,174 miles)

VERIFIED

CESSNA 305A BIRD DOG

TYPE: Two-seat liaison, observation and forward air control aircraft.

PROGRAMME: Developed for the US Army Field Forces as a light reconnaissance aircraft which could also be used for observation, liaison and training. First flown in 1949. The aircraft was designated L-19 and named Bird Dog. Redesignated O-1 in 1962. The last production L-19 was completed on 7 October 1954 bringing the total built for the US Forces to 2,480. More than 2,580 built in total.

DESIGN FEATURES: Braced high-wing monoplane with dihedral. Square wingtips. Cantilever tail unit with integral fin. Tailplane mounted high on aft fuselage. Fixed tricycle-type landing gear.

Cessna 305 (O-1) Bird Dog in Royal Norwegian Air Force (RNoAF) colours. Since withdrawn *(Patrick Allen)*

1995

Cessna Model 310 twin-engine five/six-seat monoplane

1994

FLYING CONTROLS: Electrically operated flaps retrofitted to most models.

OPERATORS: Versions of the Cessna 305 Bird Dog are in service with the armed forces of the following countries: Austria (5), Chile (2), France (5), South Korea (20), Libya (10), Malta (5), Pakistan (1), Taiwan (12), Thailand (70) and Turkey (9).

POWER PLANT: One Continental O-470-11 flat-six air-cooled piston engine. Fuel carried in two wing tanks. Total capacity 160 litres (42 US gallons; 37 Imp gallons).

DIMENSIONS, EXTERNAL:

Wing span	10.9 m (36 ft)
Length	7.6 m (24 ft 11½ in)
Height	2.28 m (7 ft 6 in)

WEIGHTS AND LOADINGS:

Weight empty	680 kg (1,498 lb)
Weight loaded (primary mission)	1,000 kg (2,200 lb)
Max weight loaded (alternative missions)	1,103 kg (2,430 lb)
Wing loading	59 kg/m² (12.1 lb/sq ft)

PERFORMANCE:

Max speed	112 kts (208 km/h; 130 mph)
Cruising speed (29% power at 1,524 m; 5,000 ft)	89.8 kts (166.4 km/h; 104 mph)
Stalling speed	46.6 kts (86.4 km/h; 54 mph)
Initial rate of climb	453 m (1,485 ft)/min
Service ceiling	6,985 m (22,900 ft)
Absolute ceiling	7,565 m (24,800 ft)
Cruising range	691 n miles (1,280 km; 800 miles)
T-O to 15.25 m (50 ft)	171 m (560 ft)
Landing from 15.25 m (50 ft)	183 m (600 ft)

NEW ENTRY

CESSNA 310 and 310 II

TYPE: Twin-engine five/six-seat monoplane.

PROGRAMME: The Model 310 is a twin-engined five/six-seat cabin monoplane, the prototype of which flew on 3 January 1953. It went into production in 1954. The Turbo 310 was added in late 1968, and the first production model was delivered in December 1968. On 21 December 1973 Cessna announced two new versions of the Model 310, known as the 310 II and the Turbo 310 II, which have factory installed IFR avionics plus other comfort and convenience features as standard. Altogether 5,241 commercial examples of the 310 were built, plus 196 delivered to the US Air Force as U-3A/Bs. Since withdrawn.

VERSIONS: **310:** Standard model, as described in detail, powered by two 212.5 kW (285 hp) Continental IO-520-MB flat-six engines, driving McCauley three-blade fully feathering constant-speed metal propellers.

310 II: Identical to 310, but having as standard equipment dual 300 Series nav/com with 720-channel com, 200-channel nav, VOR/LOC and VOR/ILS indicators; 300 Series ADF with digital tuning; 400 Series glide slope receiver; marker beacon receiver; transponder; 400B Nav-O-Matic autopilot with approach coupler; associated antennae; avionics cooling kit and panel; economy mixture indicator, outside air temperature gauge, dual controls, six individual seats, emergency locator transmitter, large baggage door, starboard landing light, rotating beacon, taxi light, auxiliary fuel system of 238.5 litres (63 US gallons; 52 Imp gallons) capacity, nosewheel fender, static wicks, and external power socket.

Turbo T310: Similar to 310, but with two 212.5 kW (285 hp) Continental TSIO-520-BB turbocharged engines, with automatic synchronisation, full-flow oil filters, absolute and pressure ratio controllers, overboost control valves and engine cowl flaps as standard.

Turbo T310 II: Identical to T310, but with the additional standard equipment as detailed for 310 II.

UPGRADES: **Colemill Enterprises Inc:** Colemill Executive 600. See separate entry.

RAM Aircraft Corporation: RAM/Cessna T310. See separate entry.

OPERATORS: Versions of the Cessna 310 are in service with the armed forces of the following countries: Bolivia (1), Colombia (6), Guatemala (1), Madagascar (1), Mexico (1), Tanzania (4), Trinidad and Tobago (1), Venezuela (2) and Zaire (10).

DESIGN FEATURES: Cantilever low-wing monoplane. Wing section NACA 23018 at centreline, NACA 23009 at tip. Dihedral 5°. Incidence 2° 30′ at root, –0° 30′ at tip.

FLYING CONTROLS: Electrically operated split flaps. Trim tab in port aileron.

STRUCTURE: All-metal structure. The fuselage is an all-metal monocoque structure. The tail unit is a cantilever all-metal structure, with 40° sweepback on fin at quarter-chord. Small ventral fin. Trim tabs in rudder and starboard elevator. Electrically operated elevator trim optional.

LANDING GEAR: Retractable tricycle type. Electromechanical retraction. Cessna oleo-pneumatic shock-absorber struts. Nosewheel steerable to 15° and castoring from 15 to 55° each side. Mainwheels size 6.50-10, tyre pressure 4.14 bars (60 lb/sq in). Nosewheel size 6.00-6, tyre pressure 2.76 bars (40 lb/sq in). Goodyear single-disc hydraulic brakes. Hydraulic parking brake.

POWER PLANT: Two flat-six engines, as described under individual model listings, driving three-blade constant-speed fully feathering metal propellers. Automatic propeller unfeathering system and propeller de-icing optional; automatic propeller synchrophaser for T310 and T310 II, optional for 310 and 310 II. Standard fuel in two permanently attached canted wingtip tanks, each holding 193 litres (51 US gallons; 42.5 Imp gallons), of which 189 litres (50 US gallons; 41.63 Imp gallons) are usable. Cross-feed fuel system. Optional fuel in two 77.5 litre (20.5 US gallon; 17 Imp gallon) rubber fuel cells installed between the wing spars outboard of each engine nacelle, two 43.5 litre (11.5 US gallon; 9.6 Imp gallon) rubber fuel cells further outboard in each wing, and two 77.5 litre (20.5 US gallon; 17 Imp gallon) wing locker fuel tanks, providing a maximum fuel capacity of 783 litres (207 US gallons; 172.4 Imp gallons), of which 768 litres (203 US gallons; 169 Imp gallons) are usable. Oil capacity 24.6 litres (6.5 US gallons; 5.4 Imp gallons).

ACCOMMODATION: Cabin normally seats five, two in front and three on cross-bench behind. Four alternative seating arrangements are available, with up to six individual seats in pairs, all of which can tilt and have fore and aft adjustment, individual air vents, reading lights and magazine pockets. Dual controls optional. Inertia seatbelts for two front seats (optional for rear seats). Pilot's storm window, port side. Cabin windows are double-glazed to reduce noise level. Large door on starboard side giving access to all seats. Cargo door, 1.02 m (3 ft 4 in) wide, for loading of bulky items, standard on 310 II and T310 II, optional on 310 and T310. Baggage compartment at rear of cabin, capacity 163 kg (360 lb), with internal and external access; locker for a further 54.5 kg (120 lb) of baggage in the rear of each engine nacelle; and baggage compartment in extended nose with capacity of 158 kg (350 lb). Total baggage capacity 430 kg (950 lb). Optional cabin accessories include writing desk, window curtains, vertical adjustment of pilot and co-pilot seats, all-leather seats, oxygen system and photographic survey provisions. Windscreen defrosting standard; windscreen alcohol de-icing system, and electrically heated windscreen panel optional.

SYSTEMS: Electrical system powered by two 28 V 50 A engine-driven alternators and 24 V 25 Ah battery. 100 A alternators optional. Oxygen system of 2.17 m³ (76.6 cu ft) or 1.37 m³ (48.3 cu ft) capacity optional; an automatic altitude compensating regulator is standard with this installation. Janitrol 45,000 BTU thermostatically controlled blower-type heater for cabin heating and windscreen defrosting. Cabin air conditioning system rated at 12,000 BTU. Vacuum system supplied by two engine-driven pumps with adequate capacity to cater for the pneumatic de-icing boots and flight instruments. Hydraulic system for brakes only.

AVIONICS AND EQUIPMENT: Optional avionics include Series 300 nav/com transceiver with 720-channel com and 200-channel nav with remote VOR/LOC or VOR/ILS indicator, ADF with digital tuning, 10-channel HF and flight director system; Series 400 nav/com transceiver with 720-channel com and 200-channel nav with remote VOR/LOC or VOR/ILS indicator, 40-channel glide slope, ADF with digital tuning and BFO, transponder with 4096 code capability, encoding altimeter, Nav-O-Matic 400B two-axis autopilot and integrated flight control system with RMI or HSI; or Series 1000 com transceiver, nav receiver, ADF, and glide slope receiver, with Series 800 DME and R/Nav system. Additional avionics options include Bendix RDR-160 colour weather radar, KNC 610 area nav, AVQ-75 DME, KN 65 DME, radar altimeter, locator beacon, yaw damper, boom microphone and headset. Standard equipment includes blind-flying instrumentation, quartz crystal clock, sensitive altimeter, control locks, pilot and co-pilot safety belts, sun visors, alternator failure warning, heated pitot, heated stall warning transmitter, heater overheat warning, landing gear warning device, armrests, cabin radio speaker, baggage straps, adjustable air vents, emergency exit window, aft omni-vision window, tinted dual-pane windows, hat shelf, super soundproofing, map light, instrument post lights, variable-intensity emergency floodlight, reading lights, navigation light detectors, landing light, navigation lights, full-flow oil coolers, heated fuel vents, nosewheel fender, overall paint scheme, quick drain fuel valves, retractable cabin step, and towbar. Optional equipment for Model 310/Turbo T310, but standard for 310 II/Turbo T310 II, as detailed in model listings. Optional equipment for all versions includes angle of attack indicator, co-pilot's blind-flying instrumentation, digital clock, instantaneous rate of climb indicator, flight hour recorder, synchronous tachometer, true airspeed indicator, turn co-ordinator, heated static sources, vertically adjustable seats with inertia reel shoulder harness for pilot and co-pilot, pilot's relief tube, rudder pedal locks, cabin curtain, rear window curtains, cabin fire extinguisher, 'total flood' cabin fire extinguisher, eight-track stereo with cabin speakers and stereo headsets, internal corrosion proofing, electroluminescent instrument panel lighting, baggage compartment courtesy lights, courtesy light timer, ice detection light, anti-collision light, second landing light, taxi light, three-light strobe system, carpet for nose baggage area, fuselage ice protection plates, photographic provisions, anti-icing kit, engine fire detection and extinguishing system, and radome nose. Additional optional items for the Model 310 and 310 II include automatic propeller synchrophaser and partial oxygen system plumbing.

DIMENSIONS, EXTERNAL:
Wing span 11.25 m (36 ft 11 in)
Wing chord: at root 1.71 m (5 ft 7½ in)
at tip 1.16 m (3 ft 9½ in)
Wing aspect ratio 7.3
Length overall 9.74 m (31 ft 11½ in)
Height overall 3.25 m (10 ft 8 in)
Tailplane span 5.18 m (17 ft 0 in)
Wheel track 3.59 m (11 ft 9½ in)
Wheelbase 2.80 m (9 ft 2¼ in)
Propeller diameter: 310, 310 II 1.94 m (6 ft 4½ in)
T310, T310 II 1.98 m (6 ft 6 in)

DIMENSIONS, INTERNAL:
Baggage compartment: Cabin 1.26 m³ (44.6 cu ft)
Nacelles (total) 0.52 m³ (18.5 cu ft)
Nose 0.59 m³ (21 cu ft)

AREAS:
Wings, gross 16.63 m² (179 sq ft)
Ailerons (total) 1.32 m² (14.17 sq ft)
Trailing-edge flaps (total) 2.14 m² (23.06 sq ft)
Fin, incl ventral fin 1.74 m² (18.70 sq ft)
Rudder, incl tab 1.09 m² (11.76 sq ft)
Tailplane 2.99 m² (32.15 sq ft)
Elevators, incl tabs 2.05 m² (22.10 sq ft)

WEIGHTS AND LOADINGS:
Weight empty: 310 1,523 kg (3,358 lb)
310 II 1,635 kg (3,604 lb)
T310 1,573 kg (3,467 lb)
T310 II 1,685 kg (3,714 lb)
Max ramp weight: all versions 2,511 kg (5,535 lb)
Max T-O weight: all versions 2,495 kg (5,500 lb)
Max landing weight: all versions 2,449 kg (5,400 lb)
Max zero-fuel weight: 310, 310 II 2,223 kg (4,900 lb)
T310, T310 II 2,275 kg (5,015 lb)
Max wing loading:
all versions 150 kg/m² (30.73 lb/sq ft)
Max power loading:
all versions 5.87 kg/kW (9.65 lb/hp)

PERFORMANCE (at max T-O weight, ISA, except speeds at mid-cruise weight):
Never-exceed speed:
all versions 227 kts (420 km/h; 261 mph) CAS
Max level speed:
310 at S/L 207 kts (383 km/h; 238 mph)
T310 at 4,875 m (16,000 ft)
237 kts (439 km/h; 273 mph)
Max cruising speed: 310, 75% power at 2,285 m (7,500 ft)
195 kts (361 km/h; 225 mph)
T310, 73.6% power at 3,050 m (10,000 ft)
201 kts (372 km/h; 231 mph)
T310, 73.6% power at 6,100 m (20,000 ft)
223 kts (413 km/h; 257 mph)
Econ cruising speed with max fuel:
310 at 3,050 m (10,000 ft)
139 kts (258 km/h; 160 mph)
T310 at 6,100 m (20,000 ft)
171 kts (317 km/h; 197 mph)
Min control speed (VMC):
all versions 81 kts (150 km/h; 93 mph)
Stalling speed, flaps and landing gear up, power off:
all versions 78 kts (145 km/h; 90 mph) CAS
Stalling speed, flaps and landing gear down, power off:
all versions 70 kts (130 km/h; 81 mph) CAS
Max rate of climb at S/L: 310 507 m (1,662 ft)/min
T310 518 m (1,700 ft)/min
Rate of climb at S/L, one engine out:
310 113 m (370 ft)/min
T310 119 m (390 ft)/min
Service ceiling: 310 6,020 m (19,750 ft)
T310 8,350 m (27,400 ft)
Service ceiling, one engine out: 310 2,225 m (7,400 ft)
T310 5,245 m (17,200 ft)
T-O run: 310 407 m (1,335 ft)
T310 398 m (1,306 ft)
T-O to 15 m (50 ft): 310 518 m (1,700 ft)
T310 507 m (1,662 ft)
Landing from 15 m (50 ft):
all versions, at 2,449 kg (5,400 lb) 546 m (1,790 ft)
Landing run:
all versions, at 2,449 kg (5,400 lb) 195 m (640 ft)
Range, recommended lean mixture, allowances for start, taxi, T-O, climb and 45 min reserves at selected cruise power:
310, 310 II, max cruising speed at 2,285 m (7,500 ft) with 272 kg (600 lb) usable fuel
440 n miles (816 km; 507 miles)
310, 310 II, as above with 552 kg (1,218 lb) usable fuel
1,078 n miles (1,997 km; 1,241 miles)
310, 310 II, econ cruising speed at 3,050 m (10,000 ft) with 272 kg (600 lb) usable fuel
631 m (1,170 km; 727 miles)
310, 310 II, as immediately above with 552 kg (1,218 lb) usable fuel
1,534 n miles (2,842 km; 1,766 miles)
T310, T310 II, max cruising speed at 6,100 m (20,000 ft) with 272 kg (600 lb) usable fuel
469 n miles (869 km; 540 miles)
T310, T310 II, as immediately above with 552 kg (1,218 lb) usable fuel
1,197 n miles (2,218 km; 1,378 miles)
T310, T310 II, econ cruising speed at 6,100 m (20,000 ft) with 272 kg (600 lb) usable fuel
586 n miles (1,086 km; 675 miles)
T310, T310 II, as immediately above with 552 kg (1,218 lb) usable fuel
1,451 n miles (2,689 km; 1,671 miles)

UPDATED

Cessna 318 T-37B two-seat primary trainer
1994

CESSNA 318

USAF designation: T-37

TYPE: Two-seat primary trainer.

PROGRAMME: The T-37 was the first jet trainer designed as such from the start to be used by the USAF. The first of two prototype XT-37s made its first flight 12 October 1954, and the first of an evaluation batch of 11 T-37As flew 27 September 1955.

VERSIONS: **T-37A:** Initial production version with 4.1 kN (920 lb st) Continental J69-T-9 turbojets. There were 534 built. Converted to T-37B standard by retrospective modification.

T-37B: Two 4.56 kN (1,025 lb st) Continental J69-T-25 turbojets. New Omni navigational equipment, UHF radio and instrument panel. First T-37B was accepted into service with the USAF in November 1959. The T-37B has been supplied to the Cambodian, Chilean, Pakistani and Royal Thai air forces.

T-37C: Basically similar to T-37B, but with provision for both armament and wingtip fuel tanks. Initial order for 34 placed by USAF for supply to foreign countries under Military Assistance programme. Portugal received 30, 18 of which were supplied, under this programme Peru received 15 and others were supplied to Brazil, Chile, Colombia, Greece, South Korea, Pakistan, Thailand and Turkey.

UPGRADES: **Rockwell Sabreliner:** T-37B SLEP programme. See separate entry.

OPERATORS: Versions of the T-37 are in service with the armed forces of the following countries: Chile (20), Colombia (5), Germany (35), Greece (31), South Korea (25), Pakistan (40), Turkey (60) and USA (530).

The following details refer to the T-37B:

DESIGN FEATURES: Cantilever low-wing monoplane. Wing section NACA 2418 at root, NACA 2412 at tip. Dihedral 3°. Incidence at root 3° 30'.

FLYING CONTROLS: Hydraulically operated high lift slotted flaps inboard of ailerons. Hydraulically actuated speed brake below forward part of fuselage in region of cockpit. Movable surfaces all have electrically operated trim tabs.

STRUCTURE: Two-spar aluminium alloy structure. The fuselage is an all-metal semi-monocoque structure. The tail unit is a cantilever all-metal structure. Fin integral with fuselage. Tailplane mounted one-third of way up fin.

LANDING GEAR: Hydraulically retractable tricycle type. Bendix oleo-pneumatic shock-absorbers. Steerable nosewheel. Tyres by General Tire and Rubber Company. Mainwheel tyres size 20 × 4.4. Nosewheel tyre size 16 × 4.4. General Tire and Rubber Company multiple disc hydraulic brakes.

POWER PLANT: Two Continental J69-T-25 turbojet engines (each 4.56 kN; 1,025 lb st). Six rubber-cell interconnected fuel tanks in each wing, feeding main tank in fuselage aft of cockpit. Total usable fuel capacity 1,170 litres (309 US gallons; 257 Imp gallons). Automatic fuel transfer by engine-driven pumps and a submerged booster pump. Provision for two 245 litre (65 US gallon; 54 Imp gallon) wingtip fuel tanks on T-37C ply. Oil capacity 11.8 litres (3.12 US gallons; 2.6 Imp gallons).

ACCOMMODATION: Enclosed cockpit seating two side by side, with dual controls. Ejection seats and jettisonable clamshell type canopy. Standardised cockpit layout, with flaps, speed brakes, trim tab, radio controls and so on, positioned and operated as in a standard USAF combat aircraft.

ELECTRONICS: Standard USAF UHF radio; Collins VHF navigation equipment and IFF.

ARMAMENT AND EQUIPMENT (T-37C only): Provision for two 250 lb bombs or four AIM-9 Sidewinder missiles. Associated equipment includes K14C computing gunsight and AN-N6 16 mm gun camera. For reconnaissance duties, KA-20 or KB-10A cameras, or HC217 cartographic camera, can be mounted in fuselage.

DIMENSIONS, EXTERNAL:
Wing span 10.30 m (33 ft 9⅓ in)
Wing chord (mean) 1.70 m (5 ft 7 in)
Wing chord at tip 1.37 m (4 ft 6 in)
Wing aspect ratio 6.2
Length overall 8.92 m (29 ft 3 in)
Height overall 2.80 m (9 ft 2⅓ in)
Tailplane span 4.25 m (13 ft 11½ in)
Wheel track 4.28 m (14 ft 0½ in)
Wheelbase 2.36 m (7 ft 9 in)

Cessna 318 T-37B of the USAF

1994

AREAS:
Wings, gross 17.09 m² (183.9 sq ft)
Ailerons (total) 1.05 m² (11.30 sq ft)
Trailing-edge flaps (total) 1.40 m² (15.10 sq ft)
Fin 1.07 m² (11.54 sq ft)
Rudder, incl tab 0.58 m² (6.24 sq ft)
Tailplane 3.25 m² (34.93 sq ft)
Elevators, incl tabs 1.09 m² (11.76 sq ft)

WEIGHTS AND LOADINGS (A: T-37B, B: T-37C):
Max T-O weight: A 2,993 kg (6,600 lb)
B 3,402 kg (7,500 lb)
Max wing loading: A 175.3 kg/m² (35.9 lb/sq ft)
B 199.2 kg/m² (40.8 lb/sq ft)
Max power loading: A 328.2 kg/kN (3.21 lb/lb st)
B 373 kg/kN (3.65 lb/lb st)

PERFORMANCE (at max T-O weight except as noted. A: T-37B, B: T-37C):
Max level speed: A, at 7,620 m (25,000 ft) 370 kts (685 km/h; 426 mph)
B, at 7,620 m (25,000 ft) 349 kts (647 km/h; 402 mph)
Normal cruising speed: A, at 7,620 m (25,000 ft) 330 kts (612 km/h; 380 mph)
B, at 7,620 m (25,000 ft) 310 kts (574 km/h; 357 mph)
Stalling speed: A 72 kts (134 km/h; 83 mph)
B 77 kts (143 km/h; 89 mph)
Max rate of climb at S/L: A 920 m (3,020 ft)/min
B 728 m (2,390 ft)/min
Service ceiling: A 10,700 m (35,100 ft)
B 9,115 m (29,900 ft)
Service ceiling, one engine out: A 6,125 m (20,100 ft)
B 4,115 m (13,500 ft)
T-O to 15 m (50 ft): A 625 m (2,050 ft)
B 838 m (2,750 ft)
Landing from 15 m (50 ft): A 823 m (2,700 ft)
B 1,036 m (3,400 ft)
Range with MIL-C-5011A reserves at 7,620 m (25,000 ft):
A 472 n miles (874 km; 543 miles)
B 703 n miles (1,302 km; 809 miles)
Range at normal rated power, 5% reserves at 7,620 m (25,000 ft): A 525 n miles (972 km; 604 miles)
B 738 n miles (1,367 km; 850 miles)
Max range, 5% reserves at 7,620 m (25,000 ft):
A 576 n miles (1,067 km; 663 miles)

UPDATED

CESSNA 318E DRAGONFLY

USAF designation: A-37

TYPE: Two-seat light strike aircraft.

PROGRAMME: The A-37 is a development of the T-37 trainer, intended for armed counter-insurgency operations from short unimproved airstrips. Two YAT-37D prototypes were produced initially, for evaluation by the USAF, by modifying existing T-37 airframes. The first of these flew for the first time 22 October 1963, powered by two 10.68 kN (2,400 lb st) General Electric J85-GE-5 turbojets.

VERSIONS: **A-37A (318D):** First 39 aircraft converted from T-37B trainers. Withdrawn from service in 1974.

A-37B (318E): Production version, of which design began in January 1967. Construction of prototype started in following month and flew for the first time in September 1967. A-37B has two General Electric J85-GE-17A turbojets, giving more than double the take-off power available for the T-37, permitting an almost doubled take-off weight.

T-37C: Basically similar to the T-37B but with provision for both armament and wingtip fuel tanks.

OPERATORS: Versions of the A-37 are in service with the armed forces of the following countries: Chile (35), Colombia (25), Dominican Republic (8), Ecuador (9), El Salvador (8), Guatemala (8), Honduras (15), South Korea (20), Thailand (11), and Uruguay (12).

The following details apply to the A-37B:

DESIGN FEATURES: Cantilever low-wing monoplane. Wing section NACA 2418 (modified) at root, NACA 2412 (modified) at tip. Dihedral 3° 38′ at root, 1° at tip. No sweep at 22.5 per cent chord. Fixed-incidence tailplane mounted one-third of way up fin.

FLYING CONTROLS: Conventional all-metal ailerons, with forward skin of aluminium and aft skin of magnesium alloy. Electrically operated trim tab in port aileron with force sensitive boost tabs in both ailerons, plus hydraulically operated slot-lip ailerons forward of the flap on the outboard two-thirds of flap span. Hydraulically operated all-metal slotted flaps of NACA 2h type. Electrically operated trim tabs in port elevator and rudder.

STRUCTURE: Two-spar aluminium alloy structure. The fuselage is an all-metal semi-monocoque structure. Hydraulically operated speed brake, measuring 1.14 m (3 ft 9 in) by 0.30 m (1 ft 0 in), below forward fuselage immediately aft of nosewheel well. Mountings for removable probe for in-flight refuelling on upper fuselage in front of cockpit. The tail unit is a cantilever all-metal structure. Fin integral with fuselage.

LANDING GEAR: Retractable tricycle type. Cessna oleo-pneumatic shock-absorber struts on all three units. Hydraulic actuation, mainwheels retracting inward, nosewheel forward. Steerable nosewheel. Goodyear tyres and single-disc brakes. Mainwheel tyres size 7.00-8 (14PR). Nosewheel tyre size 6.00-6 (6PR). Type pressure: mainwheels 7.58 bars (110 lb/sq in), nosewheel 2.55 bars (37 lb/sq in).

POWER PLANT: Two General Electric J85-GE-17A turbojet engines, each rated at 12.7 kN (2,850 lb st). Fuel tank in each wing, each with capacity of 428 litres (113 US gallons; 95 Imp gallons); two non-jettisonable tip tanks, each with capacity of 360 litres (95 US gallons; 79 Imp gallons); sump tank in fuselage, aft of cockpit, capacity 344 litres (91 US gallons; 75 Imp gallons). Total standard usable fuel capacity 1,920 litres (507 US gallons; 422 Imp gallons). Single-point refuelling through in-flight refuelling probe, with adaptor. Alternative refuelling through gravity filler cap in each wing and each tip tank. Four 378 litre (100 US gallon; 83 Imp gallon) auxiliary tanks can be carried on underwing pylons. Provision for in-flight refuelling through nose probe. Total oil capacity 9 litres (2.25 US gallons; 1.87 Imp gallons).

ACCOMMODATION: Enclosed cockpit seating two side by side, with dual controls, dual throttles, full flight instrument panel on port side, partial panel on starboard side, engine instruments in between. Full blind-flying instrumentation. Jettisonable canopy hinged to open upward and rearward. Standardised cockpit layout as in standard USAF combat aircraft. Cockpit air conditioned but not pressurised. Flak-curtains of layered nylon are installed around the cockpit. Windscreen defrosted by engine bleed air. A polycarbonate bird-resistant windscreen is available optionally.

SYSTEMS: AiResearch air conditioning system of expansion turbine type, driven by engine bleed air. Hydraulic system, pressure 103.5 bars (1,500 lb/sq in), operates landing gear, main landing gear doors, flaps, thrust attenuator, nosewheel steering system, speedbrake, stall spoiler, inlet screen. Pneumatic system, pressure 138 bars (2,000 lb/sq in), utilises nitrogen-filled 819 cc (50 cu in) air bottle for emergency landing gear extension. Electrical system includes two 38 V DC 300 A starter/generators, two 24 V Ni/Cd batteries, and provisions for external power source. One main inverter (2,500 VA three-phase 115 V 400 Hz), and one standby inverter (750 VA three-phase 115 V 400 Hz), to provide AC power.

ELECTRONICS: Radio and radar installations include UHF communications (AN/ARC-109A, ARC-151 and ARC-164), FM communications (FM-622A), Tacan (AN/ARN-65), ADF (AN/ARN-83), IFF (AN/APX-72), direction-finder (AN/ARA-50), VHF communications (VHF-20B), VOR/LOC, glide slope, marker beacon (VIR-31A) and interphone (AIC-18).

ARMAMENT AND OPERATIONAL EQUIPMENT: GAU-2B/A 7.62 mm Minigun installed in forward fuselage. Each wing has four pylon stations, the two inner ones carrying 394 kg (870 lb) each, the intermediate one 272 kg (600 lb) and the outer one 227 kg (500 lb). The following weapons in various combinations, can be carried on these underwing pylons: SUU-20 bombs and rocket pod, MK-81 or MK-82 bomb, BLU-32/B fire bomb, SUU-11/A gun pod, CBU-24/B or CBU-25A dispenser and bomb, M-117 demolition bomb, LAU-3/A rocket pod, CBU-12/A,

Cessna Model 318E A/T-37 Dragonfly of the USAF *(Peter March)*

1994

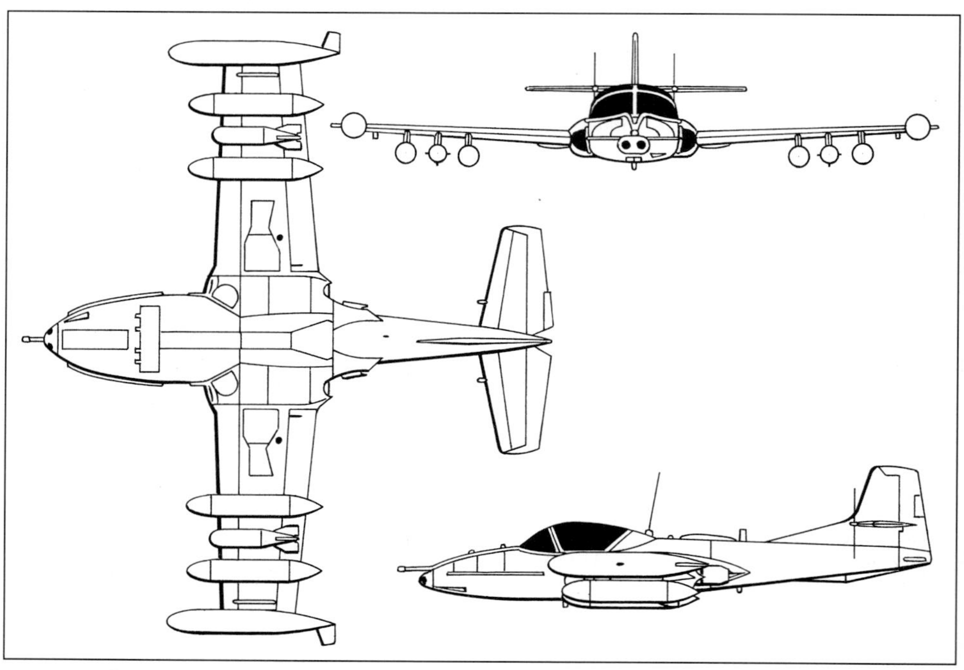

Cessna Model 318E A-37 Dragonfly two-seat, light strike aircraft *(Jane's/Mike Keep)*

1994

Cessna Model 318E (A-37B) two-seat light strike aircraft *(Cessna)*

1995

CBU-14/A or CBU-22/A dispenser and bomb, BLU-1C/B fire bomb, LAU-32/A or LAU-59/A rocket pod, CBU-19/A canister cluster and SUU-25/A flare launcher. Associated equipment includes an armament control panel, Chicago Aerial Industries CA-503 non-computing gunsight, KS-27 gun camera and KB-18A strike camera.

DIMENSIONS, EXTERNAL:

Wing span over tip tanks	10.93 m (35 ft 10½ in)
Wing chord at root	2.01 m (6 ft 7⅕ in)
Wing chord at tip	1.37 m (4 ft 6 in)
Wing aspect ratio	6.2
Length overall, excl refuelling probe	8.62 m (28 ft 3¼ in)
Height overall	2.70 m (8 ft 10½ in)
Tailplane span	4.25 m (13 ft 11½ in)
Wheel track	4.28 m (14 ft 0½ in)
Wheelbase	2.39 m (7 ft 10 in)

AREAS:

Wings, gross	17.09 m² (183.9 sq ft)
Ailerons (total)	1.05 m² (11.30 sq ft)
Trailing-edge flaps (total)	1.40 m² (15.10 sq ft)
Fin	1.07 m² (11.54 sq ft)
Rudder, incl tab	0.58 m² (6.24 sq ft)
Tailplane	3.25 m² (34.93 sq ft)
Elevators, incl tab	1.09 m² (11.76 sq ft)

WEIGHTS AND LOADINGS:

Weight empty, equipped	2,817 kg (6,211 lb)
Max T-O and landing weight	6,350 kg (14,000 lb)
Normal landing weight	3,175 kg (7,000 lb)
Max zero-fuel weight	4,858 kg (10,710 lb)
Max wing loading	319.3 kg/m² (65.4 lb/sq in)
Max power loading	250 kg/kN (2.1 lb/lb st)

PERFORMANCE (at max T-O weight, except as detailed otherwise):

Never-exceed speed (Mach limitation)	455 kts (843 km/h; 524 mph)
Max level speed at 4,875 m (16,000 ft)	440 kts (816 km/h; 507 mph)
Max cruising speed at 7,620 m (25,000 ft)	425 kts (787 km/h; 489 mph)
Stalling speed at max landing weight, wheels and flaps down	98.5 kts (182 km/h; 113 mph)
Stalling speed at normal landing weight, wheels and flaps down	75 kts (139 km/h; 86.5 mph)
Max rate of climb at S/L	2,130 m (6,990 ft)/min
Service ceiling	12,730 m (41,765 ft)
Service ceiling, one engine out	7,620 m (25,000 ft)
T-O run	531 m (1,740 ft)
T-O to 15 m (50 ft)	792 m (2,596 ft)
Landing from 15 m (50 ft) at max landing weight	2,012 m (6,600 ft)
Landing run at max landing weight	1,265 m (4,150 ft)
Landing run at normal landing weight	521 m (1,710 ft)
Range with max fuel, incl four 378 l (100 US gallon; 83 Imp gallon) drop tanks, at 7,620 m (25,000 ft) with reserves	878 n miles (1,628 km; 1,012 miles)
Range with max payload, incl 1,860 kg (4,100 lb) ordnance	399 n miles (740 km; 460 miles)

UPDATED

CESSNA 337 SKYMASTER and SKYMASTER II

Iranian Army designation: O-2

TYPE: Tandem-engined cabin monoplane.

PROGRAMME: This unorthodox all-metal four/six-seat business aircraft resulted from several years of study by Cessna aimed at producing a twin-engined aeroplane that would be simple to fly, low in cost, safe and comfortable, while offering all the traditional advantages of two engines. Construction of a full-scale mockup was started in February 1960 and completed two months later. The prototype flew for the first time 28 February 1961, followed by the first production model in August 1962. FAA Type Approval was received 22 May 1962 and deliveries of the original Model 336 Skymaster, with non-retractable landing gear, began in May 1963.

A total of 195 Model 336 Skymasters had been built by January 1965. In the following month, this version was superseded by the Model 337 Skymaster, with increased wing incidence, retractable landing gear, and other changes, making it a virtually new aeroplane. A total of 1,978 Model 336/337 Skymasters had been built by 1 January 1979, plus an additional 66 Reims Skymasters by Reims Aviation in France.

VERSIONS: **Skymaster**: Basic version, powered by two 156.5 kW (210 hp) Continental IO-360-GB flat-six engines, each driving a McCauley two-blade constant-speed and fully feathering metal propeller. *Detailed description applies to this version.*

Skymaster II: Generally similar to the Skymaster, but including the following equipment as standard: dual Series 300 nav/coms, Series 300 ADF, Series 400 transponder, glide slope, marker beacon, and Nav-O-Matic autopilot (400A Nav-O-Matic or Series 400 IFCS offered as alternative exchanges). Optional equipment includes directional and horizon gyros with associated vacuum system and suction gauge, true airspeed indicator, dual controls, extended range fuel system, ground power socket, heated pitot and stall warning, emergency locator transmitter, alternate static source.

Turbo-Skymaster: Generally similar to Skymaster, but powered by two 156.5 kW (210 hp) Continental TSIO-360-H flat-six turbocharged engines, each driving a two-blade constant-speed and fully feathering metal propeller.

Turbo-Skymaster II: Identical to Turbo-Skymaster, but with additional standard avionics and equipment as detailed for Skymaster II.

Pressurised Skymaster/Pressurised Skymaster II: Pressurised version with four windows on each side of cabin.

The 1979 versions of the Skymaster/Turbo-Skymaster have a number of standard and optional detail improvements to Series 400 avionics equipment.

OPERATORS: Versions of the Cessna 337 (including Reims-Cessna 337) are in service with the armed forces of the following countries: Bangladesh (2), Chad (5), Chile (2), Costa Rica (2), El Salvador (10), Equatorial Guinea (1), Guinea-Bissau (1), Haiti (6), Ivory Coast (2), Jamaica (1), South Korea (10), Madagascar (3), Mauritania (2), Mexico (2), Niger (2), Paraguay (1), Peru (1), Portugal (15), Senegambia (1), Sri Lanka (1), Thailand (9), Togo (2), Venezuala (2), Zaire (18), and Zimbabwe (16).

DESIGN FEATURES: Braced high-wing monoplane, with single streamlined bracing strut each side. Wing section NACA 2412 at root, NACA 2409 at tip. Dihedral 3°. Incidence 4° 30′ at root, 2° 30′ at tip. Cantilever tail unit with twin sweptback fins carried on two slim metal booms.

FLYING CONTROLS: Frise ailerons. Electrically operated all-metal single-slotted flaps. Ground adjustable tab in port aileron. Horn balanced rudder on both tail unit fins. Trim tab in starboard side of elevator, with optional electrical actuation.

STRUCTURE: Conventional all-metal two-spar wing structure with conical camber glassfibre wingtips. The fuselage is a conventional all-metal semi-monocoque structure.

LANDING GEAR: Hydraulically retractable tricycle type. Cantilever spring steel main legs. Steerable nosewheel with oleo-pneumatic shock-absorber. Mainwheels and tyres size 6.00-6. Nosewheel and tyre size 15 × 6.00-6. Mainwheel tyre pressure 3.80 bars (55 lb/sq in). Hydraulic disc brakes. Parking brake. Oversize wheels and heavy-duty brakes optional.

POWER PLANT: Two Continental flat-six engines, as detailed in model listings. Electrically operated cowl flaps. Propeller de-icing optional for forward propeller. Fuel in two main tanks in each wing, with total usable capacity of 333 litres (88 US gallons; 73.3 Imp gallons); two additional tanks in wings, with total usable capacity of 227 litres (60 US gallons; 50 Imp gallons), provide optional long-range system. Total usable capacity with optional tanks 560 litres (148 US gallons; 123.3 Imp gallons). Refuelling points above wings. Oil capacity (Skymaster/Skymaster II) 15 litres (4 US gallons; 3.3 Imp gallons); (Turbo-Skymaster/Turbo-Skymaster II) 17 litres (4.5 US gallons; 3.7 Imp gallons).

ACCOMMODATION: Standard accommodation for pilot and co-pilot on individual seats, with individual track-mounted seats for two passengers. Dual controls optional. Alternative arrangements include individual seats for fifth and sixth passengers. Optional cabin equipment includes fully articulating individual seats for passengers and matching headrests. Optional centre seats can be moved fore and aft to ease access to other seats. Space for 165 kg (365 lb) of baggage in four-seat version, with external loading door. Airstair door on starboard side. Cabin is heated, ventilated and soundproofed. Adjustable air vents and reading lights available to each passenger. Provision for carrying glass-fibre cargo pack, with capacity of 136 kg (300 lb), under fuselage; this reduces cruising speed by only 2.6 knots (5 km/h; 3 mph).

SYSTEMS: Electrical system supplied by two 38 A 28 V engine-driven alternators. 24 V battery. Hydraulic system for landing gear retraction and brakes.

AVIONICS AND EQUIPMENT: Optional avionics include Cessna Series 300 nav/com with 720-channel com and 200-channel nav with remote VOR/LOC or VOR/ILS indicator, ADF with digital tuning, marker beacon receiver with three lights and aural signal, DME, 10-channel HF transceiver, and Nav-O-Matic single-axis autopilot and integrated flight control system; or Series 400 nav/com with 720-channel com and 200-channel nav with VOR/LOC or VOR/ILS indicator, 40-channel glide slope receiver, ADF with digital tuning and BFO, transponder with 4096 code capability, Nav-O-Matic 400 or 400A two-axis autopilot and integrated flight control system, non-slaved HSI, aural/visual alert system with coupling to IFCS when installed, and Bendix RDR 160 weather radar. Standard equipment includes sensitive altimeter, airspeed indicator, rate of climb indicator, electric clock, outside air temperature gauge, audible stall warning device, engine synchronisation indicator, turn co-ordinator indicator, map light, sun visors, all-weather window, hinged window starboard side, navigation light detectors, elevator and aileron control locks, windscreen defroster, tinted windscreen and windows, dome light, reading lights, baggage net, omni-flash beacon, taxi light, anti-precipitation static kit, retractable tiedown rings, towbar, polyurethane external paint scheme and quick drain fuel tank valves. Optional equipment includes all-purpose control wheel with provision for map light, boom microphone switch, pitch trim switch, autopilot/electric trim disengage switch, blind-flying instrumentation, economy mixture indicator, true airspeed indicator, instrument post lights, approach plate holder, flight hour recorder, windscreen anti-icing panel, cabin fire extinguisher, baggage net, wall-mounted table, safety belts for third, fourth, fifth and sixth seats, internal corrosion proofing, emergency exit window port side, portable stretcher, cargo tiedown installation, full-flow oil filters, external power socket, propeller synchrophaser, winterisation kit, alternate static source, oxygen system, ice detection system, white strobe lights, photographic provisions, static wicks, pitot heating system, flush glide slope antenna, telescoping towbar and cargo rack.

Cessna 337 Skymaster (O-2A) of the USAF (since withdrawn) *(Peter March)*

1994

DIMENSIONS, EXTERNAL:
- Wing span 11.63 m (38 ft 2 in)
- Wing chord: at root 1.83 m (6 ft 0 in)
  - at tip 1.22 m (4 ft 0 in)
- Wing aspect ratio 7.18
- Length overall: Skymaster 9.07 m (29 ft 9 in)
  - Turbo-Skymaster 9.09 m (29 ft 10 in)
- Height overall 2.79 m (9 ft 2 in)
- Tailplane span 3.06 m (10 ft 0½ in)
- Wheel track 2.49 m (8 ft 2 in)
- Wheelbase 2.39 m (7 ft 10 in)
- Propeller diameter: Front 1.98 m (6 ft 6 in)
  - Rear 1.93 m (6 ft 4 in)
- Passenger door: Height 1.17 m (3 ft 10 in)
  - Width 0.91 m (3 ft 0 in)

DIMENSIONS, INTERNAL:
- Cabin: Length 3.02 m (9 ft 11 in)
  - Max width 1.12 m (3 ft 8¼ in)
  - Max height 1.30 m (4 ft 3¼ in)
  - Volume 3.62 m³ (128 cu ft)
- Baggage space 0.50 m³ (17 cu ft)

AREAS:
- Wings, gross 18.81 m² (202.5 sq ft)
- Ailerons (total) 1.43 m² (15.44 sq ft)
- Trailing-edge flaps (total) 3.43 m² (36.88 sq ft)
- Fins (total) 2.85 m² (30.68 sq ft)
- Rudders (total) 0.99 m² (10.70 sq ft)
- Tailplane 3.05 m² (32.82 sq ft)

WEIGHTS AND LOADINGS (A: Skymaster, B: Skymaster II, C: Turbo-Skymaster, D: Turbo-Skymaster II):
- Weight empty: A 1,270 kg (2,800 lb)
  - B 1,335 kg (2,943 lb)
  - C 1,306 kg (2,879 lb)
  - D 1,371 kg (3,022 lb)
- Max T-O weight: all versions 2,100 kg (4,630 lb)
- Max ramp weight: A, B 2,108 kg (4,648 lb)
  - C, D 2,110 kg (4,652 lb)
- Max landing weight: all versions 1,996 kg (4,400 lb)
- Max wing loading:
  - all versions 112 kg/m² (22.9 lb/sq ft)
- Max power loading:
  - all versions 6.71 kg/kW (11.0 lb/hp)

PERFORMANCE (at max T-O weight):
- Max level speed:
  - A, B at S/L 172 kts (319 km/h; 198 mph)
  - C, D at 6,100 m (20,000 ft) 207 kts (383 km/h; 238 mph)
- Max cruising speed:
  - A, B, 75% power at 1,675 m (5,500 ft) 169 kts (314 km/h; 195 mph)
  - C, D, 80% power at 6,100 m (20,000 ft) 200 kts (370 km/h; 230 mph)
  - C, D, 80% power at 3,050 m (10,000 ft) 182 kts (338 km/h; 210 mph)
- Stalling speed, flaps up, power off:
  - all versions 70 kts (130 km/h; 81 mph) CAS
- Stalling speed, flaps down, power off:
  - all versions 61 kts (113 km/h; 70 mph) CAS
- Max rate of climb at S/L: A, B 287 m (940 ft)/min
  - C, D 354 m (1,160 ft)/min
- Rate of climb at S/L, front engine only:
  - A, B 91 m (300 ft)/min
- Rate of climb at S/L, rear engine only:
  - A, B 98 m (320 ft)/min
- Single-engine rate of climb at S/L:
  - C, D 102 m (335 ft)/min
- Service ceiling: A, B 4,970 m (16,300 ft)
  - A, B, front engine only 2,105 m (6,900 ft)
  - A, B, rear engine only 2,165 m (7,100 ft)
  - C, D, single-engine 5,030 m (16,500 ft)
- Max certificated operating altitude:
  - C, D, single- or twin-engine 6,100 m (20,000 ft)
- T-O run: all versions 305 m (1,000 ft)
- T-O to 15 m (50 ft): all versions 511 m (1,675 ft)
- Landing from 15 m (50 ft):
  - all versions 503 m (1,650 ft)
- Landing run: all versions 213 m (700 ft)
- Range, recommended lean mixture, allowances for start, taxi, T-O, climb, and 45 min reserves at 45% power: A, B:
  - 75% power at 1,675 m (5,500 ft) with 239 kg (528 lb) usable fuel 545 n miles (1,011 km; 628 miles)
  - 75% power at 1,675 m (5,500 ft) with 403 kg (888 lb) usable fuel 990 n miles (1,835 km; 1,140 miles)
  - econ cruising power at 3,050 m (10,000 ft) with 239 kg (528 lb) usable fuel 670 n miles (1,242 km; 772 miles)
  - econ cruising power at 3,050 m (10,000 ft) with 403 kg (888 lb) usable fuel 1,235 n miles (2,288 km; 1,422 miles)
- C, D:
  - 80% power at 6,100 m (20,000 ft) with 239 kg (528 lb) usable fuel 520 n miles (964 km; 599 miles)
  - 80% power at 3,050 m (10,000 ft) with 239 kg (528 lb) usable fuel 490 n miles (908 km; 564 miles)
  - 80% power at 6,100 m (20,000 ft) with 403 kg (888 lb) usable fuel 975 n miles (1,807 km; 1,123 miles)
  - 80% power at 3,050 m (10,000 ft) with 403 kg (888 lb) usable fuel 905 n miles (1,677 km; 1,042 miles)
  - econ cruising power at 6,100 m (20,000 ft) with 239 kg (528 lb) usable fuel 590 n miles (1,093 km; 679 miles)
  - econ cruising power at 3,050 m (10,000 ft) with 239 kg (528 lb) usable fuel 580 n miles (1,075 km; 668 miles)
  - econ cruising power at 6,100 m (20,000 ft) with 403 kg (888 lb) usable fuel 1,125 n miles (2,084 km; 1,295 miles)
  - econ cruising power at 3,050 m (10,000 ft) with 403 kg (888 lb) usable fuel 1,080 n miles (2,002 km; 1,244 miles)

UPDATED

CESSNA 402C, 402C II and 402C III

TYPE: Ten-seat (optional nine-seat) convertible passenger/freight transport (Utiliner) or six/eight-seat business aircraft (Businessliner).

PROGRAMME: The original Model 402 was intended for the third-level airline market, with a convertible cabin and reinforced cabin floor of bonded honeycomb construction, enabling it to be changed quickly from a 10-seat commuter to a light cargo transport. On 8 December 1971 Cessna renamed the original 402 as the Model 402 Utiliner and introduced a version designated Model 402 Businessliner. Mk II versions of both aircraft were made available 29 October 1975, each including a package of factory installed equipment and avionics as standard; Model 402 III versions of both aircraft were introduced in 1978.

The 402C incorporates many of the improvements introduced in the Model 414A Chancellor. They include a new bonded 'wet' wing of increased span (without tip tanks), improved landing gear, more powerful engines, and many detail changes.

Cessna 402 Businessliner II

1994

Cessna 402 Businessliner

1994

VERSIONS: **402C Utiliner:** Basic version, as described in detail.

402C Businessliner: As basic version, except six/eight seats and optional side hinged door next to standard cabin door, to provide a total loading door width of 1.02 m (3 ft 4 in). Other options include folding business desks, stereo equipment, refreshment centre and cabin dividers.

402C II Utiliner: As basic version, plus the following factory installed equipment and avionics as standard: dual controls, flight recorder, economy mixture indicator, cabin hand fire extinguisher, starboard landing light, taxi light, locator beacon, external power socket, static dischargers, dual Cessna Series 300 nav/com with 720-channel nav and VOR/ILS and VOR/LOC indicators, Series 300 ADF, Series 400 marker beacon, glide slope and transponder, Series 400B Nav-O-Matic autopilot, and basic avionics kit comprising antennae, avionics cooling kit and audio system.

402C II Businessliner: As for 402C Businessliner, plus standard factory installed equipment and avionics detailed for Model 402C II Utiliner.

402C III Businessliner: As for Model 402C II Businessliner, except that standard equipment includes 100 A instead of 50 A engine-driven alternators, and the avionics package comprises Series 400 nav/com with HSI; second series 400 nav/com with VOR/ILS; dual Series 400 glide slope; Series 400 ADF, marker beacon, R/Nav, transponder; Series 800 encoding altimeter and altitude alerter; Series 1000 RMI; Bendix RDR-160 weather radar; Series 400B IFCS; yaw damper; and basic avionics kit, audio system, avionics cooling and antennae.

The same prototype served for the Models 401 and 402, and the FAA Type Certificate, awarded on 20 September 1966, also covered both types.

UPGRADES: **RAM Aircraft Corporation:** Re-engine package. See separate entry.

OPERATORS: Versions of the Cessna 402 are in service with the armed forces of the following countries: Bolivia (2), Comores (1), Haiti (1), Indonesia (2), Malaysia (10), Mexico (2), Paraguay (1) and Venezuela (1).

DESIGN FEATURES: Cantilever low-wing monoplane. Wing section NACA 23018 (modified) at aircraft centreline, NACA 23015 (modified) at centre-section/outerwing junction, NACA 23009 (modified) at tip. Fixed-incidence tailplane.

FLYING CONTROLS: All-metal ailerons and electrically actuated split flaps. Trim tab in port aileron. Trim tabs in rudder and starboard elevator. Electrical operation of trim tabs optional.

STRUCTURE: All-metal two-spar structure of light alloy with stamped ribs and surface skins reinforced with spanwise stringers. Outer wing panels of bonded construction. The fuselage is an all-metal semi-monocoque structure. The tail unit is a cantilever all-metal structure, with 40° sweepback on fin at quarter-chord.

LANDING GEAR: Hydraulically retractable tricycle type. Main units retract inward into wings, nosewheel unit rearward. No doors over mainwheels when retracted. Emergency extension system. Oleo-pneumatic shock-absorbers. Steerable nosewheel. Cleveland heavy-duty wheels. Mainwheel tyres 6.50-10, nosewheel tyre size 6.00-6. Cleveland heavy-duty hydraulic brakes. Parking brakes.

POWER PLANT: Two 242 kW (325 hp) Continental TSIO-520-VB flat-six turbocharged engines, each driving a three-blade constant-speed fully feathering metal propeller. Propeller synchronisation, automatic unfeathering and electrical de-icing optional. Integral wing fuel tanks with total capacity of 806 litres (213 US gallons; 177.4 Imp gallons), of which 772 litres (204 US gallons; 169.9 Imp gallons) are usable. Oil capacity 24.6 litres (6.5 US gallons; 5.4 Imp gallons).

ACCOMMODATION: Two seats side by side in the pilot's compartment. Dual controls standard on Model 402C II/III versions, optional for Model 402C versions. The Utiliner cabin has four individual seats in pairs and two double

Cessna 402 Businessliner II ten/nine-seat aircraft ***(Cessna)***

1995

seats. Passenger seats are 'Enviro-form' moulded honeycomb seats, glassfibre reinforced. Businessliner has four individual seats as standard, two additional seats optional, in the main cabin. Refreshment centre at aft of cabin. Passenger reading lights standard on Businessliner, optional on Utiliner. Door with built-in airstair on port side of cabin at centre. Storm windows for pilot and co-pilot. Tinted cabin windows. An emergency escape hatch is provided on the starboard side of the cabin. Optional cargo door and crew access door available. Baggage contained in area at rear of cabin, nose compartment, and wing lockers at rear of each engine nacelle, with combined capacity of 680 kg (1,500 lb). Cabin heated and ventilated. Windscreen defrosting standard. Electric anti-icing of pilot's and co-pilot's windows optional.

SYSTEMS: Electrical system powered by two 24 V 50 A alternators. 24 V 25 Ah battery. Battery can be sited optionally in nose baggage area. 100 A alternators optional, standard on 402C III version. Hydraulic system for brakes only. Vacuum system provided by two engine-driven pumps. Oxygen system of 1.25 m^3 (44 cu ft) or 3.25 m^3 (114.9 cu ft) capacity optional. Air conditioning system optional. Heating and ventilation system with 45,000 BTU gasoline heater standard.

AVIONICS AND EQUIPMENT: Optional avionics include Series 300 nav/com transceiver with 720-channel com and 200-channel nav with remote VOR/LOC or VOR/ILS indicator. ADF with digital tuning, 10-channel HF and flight director system; Series 400 nav/com transceiver with 720-channel com and 200-channel nav with remote VOR/LOC or VOR/ILS indicator, 40-channel glide slope, ADF with digital tuning and BFO, transponder with 4096 code capability, encoding altimeter, Nav-O-Matic 400A two-axis autopilot and integrated flight control system with optional RMI or HSI; or Series 1000 com transceiver, nav receiver, ADF, and glide slope, with Series 800 DME and R/Nav system. Additional equipment includes radiotelephone and CCC CIR-10 emergency locator transmitter. Standard equipment includes sensitive altimeter, quartz crystal clock, variable intensity floodlight, outside air temperature gauge, full blind-flying instrumentation, audible stall warning and landing gear indicators, cabin door 'Not locked' light, map light, heater overheat warning light, alternator failure lights, navigation light detectors, variable intensity instrument post lights, aileron and elevator control lock, sun visor, armrests, pilot and co-pilot safety belt system, super soundproofing, cabin radio speaker, adjustable cabin air ventilators, courtesy lights, retractable landing light, navigation lights, two rotating beacons, all-over paint scheme and towbar. Optional equipment, unless part of Model 402C II or III packages, includes digital clock, inertial shoulder restraint system for pilot and co-pilot, co-pilot's blind-flying instrumentation, economy mixture indicator, instantaneous rate of climb indicator, true airspeed indicator, rudder lock, flight hour recorder, turn co-ordinator, cabin hand fire extinguisher, 'total flood' cabin fire extinguisher, Utiliner or Businessliner interiors (including flight deck divider curtains, window curtains, headrests, reading lights, 'seat belt' and 'No smoking' signs and various arrangements of seats, tables, refreshment units and toilets), internal corrosion proofing, external power socket, ice detection light, second retractable landing light, three-light strobe system, propeller synchrophaser, photographic provisions, dual heated static source and static dischargers.

DIMENSIONS, EXTERNAL:
- Wing span: 13.45 m (44 ft 1½ in)
- Wing chord: at root: 1.77 m (5 ft 9¾ in)
  - at tip: 1.05 m (3 ft 5½ in)
- Length overall: 11.09 m (36 ft 4½ in)
- Height overall: 3.49 m (11 ft 5½ in)
- Tailplane span: 5.18 m (17 ft 0 in)
- Wheel track: 5.48 m (17 ft 11½ in)
- Wheelbase: 3.19 m (10 ft 5½ in)
- Propeller diameter: 1.94 m (6 ft 4½ in)
- Passenger door (standard): Height: 1.24 m (4 ft 1 in)
  - Width: 0.58 m (1 ft 11 in)
- Cargo door (optional): Height: 1.26 m (4 ft 1½ in)
  - Width: 1.00 m (3 ft 3½ in)
- Nose baggage doors (each): Height: 0.51 m (1 ft 8 in)
  - Width: 0.80 m (2 ft 7½ in)
- Nacelle baggage doors (each): Length: 0.61 m (2 ft 0 in)
  - Width: 0.62 m (2 ft 0½ in)

DIMENSIONS, INTERNAL:
- Cabin: Length: 4.83 m (15 ft 10 in)
  - Max width: 1.42 m (4 ft 8 in)
  - Max height: 1.30 m (4 ft 3 in)
  - Volume: 6.30 m^3 (222.4 cu ft)

AREAS:
- Wings, gross: 20.98 m^2 (225.8 sq ft)
- Fin: 3.52 m^2 (37.89 sq ft)
- Rudder, incl tab: 1.65 m^2 (17.77 sq ft)
- Tailplane: 5.64 m^2 (60.70 sq ft)
- Elevators, incl tab: 1.64 m^2 (17.63 sq ft)

WEIGHTS AND LOADINGS:
- Weight empty: Businessliner: 1,849 kg (4,077 lb)
  - Utiliner: 1,862 kg (4,105 lb)
  - Businessliner II: 1,914 kg (4,219 lb)
  - Utiliner II: 1,924 kg (4,241 lb)
  - Businessliner III: 1,967 kg (4,336 lb)
- Max T-O and landing weight: 3,107 kg (6,850 lb)
- Max ramp weight: 3,123 kg (6,885 lb)
- Max zero-fuel weight: 2,955 kg (6,515 lb)
- Max wing loading: 148.1 kg/m^2 (30.3 lb/sq ft)
- Max power loading: 6.42 kg/kW (10.5 lb/hp)

PERFORMANCE (at max T-O weight, except speeds are those at mid-cruise weight):
- Max level speed at 4,875 m (16,000 ft): 231 kts (428 km/h; 266 mph)
- Max cruising speed, 72% power:
  - at 6,100 m (20,000 ft): 213 kts (394 km/h; 245 mph)
  - at 3,050 m (10,000 ft): 194 kts (359 km/h; 223 mph)
- Econ cruising speed:
  - at 6,100 m (20,000 ft): 172 kts (318 km/h; 198 mph)
  - at 3,050 m (10,000 ft): 146 kts (270 km/h; 168 mph)
- Stalling speed, power off:
  - flaps up: 78 kts (145 km/h; 90 mph) CAS
  - flaps down: 68 kts (126 km/h; 78 mph) CAS
- Max rate of climb at S/L: 442 m (1,450 ft)/min
- Rate of climb at S/L, one engine out: 92 m (301 ft)/min
- Service ceiling: 8,200 m (26,900 ft)
- Service ceiling, one engine out: 450 m (1,480 ft)
- T-O run: 537 m (1,763 ft)
- T-O run to 15 m (50 ft): 669 m (2,195 ft)
- Landing from 15 m (50 ft): 757 m (2,485 ft)
- Landing run: 322 m (1,055 ft)
- Range, recommended lean mixture, allowances for start, taxi, T-O, climb, descent and 45 min reserves at 45% power:
  - 72% power at 6,100 m (20,000 ft) with 272 kg (600 lb) usable fuel: 419 n miles (776 km; 482 miles)
  - 72% power at 6,100 m (20,000 ft) with 555 kg (1,224 lb)usable fuel: 1,029 n miles (1,907 km; 1,185 miles)
  - 72% power at 3,050 m (10,000 ft) with 272 kg (600 lb) usable fuel: 401 n miles (744 km; 462 miles)
  - 72% power at 3,050 m (10,000 ft) with 555 kg (1,224 lb) usable fuel: 957 n miles (1,773 km; 1,102 miles)
  - econ cruising power at 6,100 m (20,000 ft) with 272 kg (600 lb) usable fuel: 461 n miles (855 km; 531 miles)
  - econ cruising power at 6,100 m (20,000 ft) with 555 kg (1,224 lb) usable fuel: 1,212 n miles (2,247 km; 1,396 miles)
  - econ cruising power at 3,050 m (10,000 ft) with 272 kg (600 lb) usable fuel: 484 n miles (896 km; 557 miles)
  - econ cruising power at 3,050 m (10,000 ft) with 555 kg (1,224 lb) usable fuel: 1,234 n miles (2,287 km; 1,421 miles)

UPDATED

CESSNA 414A CHANCELLOR

TYPE: Six/eight-seat pressurised light transport aircraft.

PROGRAMME: Cessna introduced the pressurised twin-engined Model 414 10 December 1969 as a 'step-up' aircraft for owners of Cessna or other light unpressurised twins. It combined the basic fuselage and tail unit of the Model 421 with the wing of the Model 402 and had 231 kW (310 hp) turbocharged Continental engines.

It was replaced in 1978 by the similar but much improved Model 414A Chancellor. Major changes included a new bonded 'wet' wing of increased span, extended nose and baggage area, and introduction of an external access door to the tailcone.

VERSIONS: **414A Chancellor:** Standard version. *Described in detail.*

414A Chancellor II: As Model 414A above, but with the following avionics and equipment as standard: dual Series 400 nav/com, with ARC, VOR/ILS and VOR/LOC; Series 400 ADF, DME, transponder, glide slope, marker beacon; 400B Nav-O-Matic autopilot; basic avionics kit, cooling kit, audio system, and all essential antennae; variable cabin pressure control system, economy mixture indicator, flight hour recorder, co-pilot's blind-flying instrumentation, cabin hand fire extinguisher, starboard landing light, taxi light, high-intensity strobe light, emergency locator beacon, external power socket, nosewheel fender and static dischargers.

414A Chancellor III: As standard Model 414A, with standard equipment of 414A II plus cabin air conditioning system and 100 A alternators, and an all-weather avionics package which includes dual Series 100 com, 1000 nav with HSI, 1000 nav with ARC, dual 1000 glide slope, 1000 ADF, RMI, Series 800 DME, R/Nav, transponder,

Cessna Model 414A Chancellor II six/eight-seat light transport aircraft

1995

Cessna 414 Chancellor six/eight-seat light pressurised aircraft

1994

encoding altimeter and altitude alerter; 800B IFCS; Series 400 marker beacon; AA-100 radio altimeter; Bendix RDR-160 weather radar; Series 1000 audio panel, basic avionics kit, cooling kit, and all essential antennae.

The Model 414A introduced the following improvements as standard; relocation of the tail position light, a redesigned cabin door latch, better accessibility to avionics, revised radio panel, repositioned outside air temperature gauge, easier access to landing gear hydraulic filters, and an improved fuel pickup valve.

UPGRADES: **RAM Aircraft Corporation:** Re-engine package. See separate entry.

OPERATORS: Versions of the Cessna 414 are in service with the armed forces of Bolivia (1).

DESIGN FEATURES: Cantilever low-wing monoplane. Wing section NACA 23018 (modified) at aircraft centreline, NACA 23015 (modified) at centre-section/outerwing junction, NACA 23009 (modified) at tip. Dihedral 5° on outer panels. Incidence 2° 30′ at root, –0° 30′ at tip. Fixed-incidence tailplane.

FLYING CONTROLS: All-metal ailerons and electrically actuated split flaps. Trim tab in port aileron. Trim tabs in rudder and starboard elevator.

STRUCTURE: All-metal two-spar structure of light alloy with stamped ribs and surface skins reinforced with spanwise stringers. Outer wing panels of bonded construction. The fuselage is a conventional all-metal semi-monocoque structure, with fail-safe structure in the pressurised section. The tail unit is a cantilever all-metal structure, with swept-back vertical surfaces.

LANDING GEAR: Hydraulically retractable tricycle type, main units retracting inward, nosewheel unit aft. Emergency extension by means of a 138 bar (2,000 lb/sq in) rechargeable nitrogen bottle. Oleo-pneumatic shock-absorbers. Steerable nosewheel. Mainwheel tyres size 6.50-10 (8 ply), nosewheel tyre size 6.00-6 (6 ply). Goodyear single-disc hydraulic brakes. Parking brakes.

POWER PLANT: Two 231 kW (310 hp) Continental TSIO-520-NB flat-six turbocharged engines, each driving a McCauley 3AF32C93M/82NC-5.5 three-blade constant-speed fully feathering metal propeller. Unfeathering pressure accumulator and electrical blade de-icing system optional. Fuel system with maximum usable capacity of 806 litres (213 US gallons; 177.4 Imp gallons). Oil capacity 24.6 litres (6.5 US gallons; 5.4 Imp gallons).

ACCOMMODATION: Two seats side by side in pilot's compartment. Optional curtain or solid divider with curtain, to separate pilot's compartment from main cabin. Standard seating arrangement for four forward-facing passenger seats. Optional arrangements provide for front passenger seats to face aft and forward-facing seventh and eighth seats. Individual consoles each include reading light and ventilator. Optional items include executive writing desk, tables, hat shelf, stereo equipment, electrically adjustable pilot's and co-pilot's seats, refreshment and thermos units, fore and aft cabin dividers, electric shaver converter, all-leather seats, passenger instrument console (clock, true airspeed indicator and altimeter) and intercom. Door is two-piece type with built-in airstairs in bottom portion, on port side of cabin at rear. Plug emergency escape hatch on starboard side of cabin. Double-pane cabin windows. Foul-weather windows for pilot and co-pilot, on each side of fuselage. Electrically de-iced windscreen optional. Baggage accommodated in nose compartment with external access doors, capacity 159 kg (350 lb), two wing lockers, capacity 54.5 kg (120 lb) each, and rear cabin area, capacity 226 kg (500 lb). Total baggage capacity 494 kg (1,090 lb). External access door to tailcone on starboard side.

SYSTEMS: Cabin pressurisation system, maximum differential 0.34 bars (5.00 lb/sq in), maintains sea level cabin conditions to an altitude of 3,350 m (11,000 ft), and a 3,050 m (10,000 ft) cabin altitude to a height of 8,075 m (26,500 ft). Electrical system powered by two engine-driven 28 V 50 A alternators. 24 V 25 Ah battery. 28 V 100 A alternators optional, standard on 414A III versions. Hydraulic system for brakes only. Vacuum system for blind-flying instrumentation and optional wing and tail unit de-icing system. Oxygen system of 3.25 m³ (114.9 cu ft) capacity, or emergency oxygen system of 0.31 m³ (11.0 cu ft) capacity optional. Air conditioning system optional, standard on 414A III.

AVIONICS AND EQUIPMENT: The various versions of the Model 414A have avionics as detailed in the model listings. Optional avionics for the basic 414A include those detailed for the 414 II/III, and alternative items from the Cessna Series 400, 800, 1000 range for all versions. Standard equipment includes sensitive altimeter, quartz crystal clock, dual controls, windscreen defroster, outside air temperature gauge, blind-flying instrumentation, audible stall warning device, instrument post lights, alternator failure lights, aileron and elevator control lock, aircraft systems monitoring device, heater overheat light, cabin door 'Not locked' light, sun visors, navigation light detectors aft cabin light, adjustable cabin air ventilators, window curtains, courtesy lights, reading lights, super soundproofing, non-congealing oil coolers, quick-drain fuel valves, heated stall warning transmitter, pitot and fuel vents, retractable landing light, overall paint scheme, propeller synchronisers, full-flow oil filters, navigation lights, rotating beacons and towbar. Optional equipment for Model 414A includes blind-flying instrumentation for co-pilot, economy mixture indicator, flight hour recorder, variable cabin pressure control system, cabin hand fire extinguisher, air conditioning system, 100 A alternators, locator beacon, starboard landing light, strobe lights, taxi light, static dischargers, external power socket, and nosewheel fender. Optional for the Model 414A II are 100 A alternators and cabin air conditioning system. Optional equipment for all versions includes electric elevator trim, angle of attack indicator, digital clock, true airspeed indicator, instantaneous rate of climb indicators, digital flow gauge with computer, rudder pedal lock, boom microphone, turn co-ordinator, electric or alcohol windscreen anti-icing, 'total flood' cabin fire extinguishing system, seventh and eighth seats, 'Fasten seat belt' and 'Oxygen' signs, toilet with privacy curtain, flight deck/cabin divider or curtain, executive table, refreshment centre, eight-track stereo installation, automatic timer to control courtesy lights, ventilating fan system, tinted windows, internal corrosion proofing, fuselage ice impact panels, ice detection lights, dual pitot system, radome nose, engine nacelle fire detection and extinguishing system and heavy-duty brakes.

DIMENSIONS, EXTERNAL:

Wing span	13.45 m (44 ft 1½ in)
Wing chord: at root	1.77 m (5 ft 9¾ in)
at tip	1.05 m (3 ft 5½ in)
Length overall	11.09 m (36 ft 4½ in)
Height overall	3.49 m (11 ft 5½ in)
Tailplane span	5.18 m (17 ft 0 in)
Wheel track	5.47 m (17 ft 11¾ in)
Wheelbase	3.18 m (10 ft 5¼ in)
Propeller diameter	1.94 m (6 ft 4½ in)
Passenger door: Height	1.21 m (3 ft 11½ in)
Width	0.58 m (1 ft 11 in)
Height to sill	1.21 m (3 ft 11½ in)

DIMENSIONS, INTERNAL:

Cabin: Length	4.42 m (14 ft 6 in)
Max width	1.40 m (4 ft 7 in)
Max height	1.29 m (4 ft 3 in)
Volume	6.1 m³ (215.6 cu ft)

AREAS:

Wings, gross	20.98 m² (225.8 sq ft)
Fin	3.52 m² (37.89 sq ft)
Rudder, incl tab	1.46 m² (15.72 sq ft)
Tailplane	4.15 m² (44.62 sq ft)
Elevators, incl tab	1.49 m² (16.08 sq ft)

WEIGHTS AND LOADINGS:

Weight empty: 414A	1,975 kg (4,354 lb)
414A II	2,052 kg (4,523 lb)
414A III	2,161 kg (4,764 lb)
Max T-O and landing weight	3,062 kg (6,750 lb)
Max ramp weight	3,078 kg (6,785 lb)
Max zero-fuel weight	2,955 kg (6,515 lb)
Max wing loading	145.94 kg/m² (29.89 lb/sq ft)
Max power loading	6.63 kg/kW (10.89 lb/hp)

PERFORMANCE (at max T-O weight, except speeds are those at mid-cruise weight):

Max level speed at 6,100 m (20,000 ft) — 239 kts (443 km/h; 275 mph)

Cruising speed, 74.8% power at 7,470 m (24,500 ft) — 224 kts (415 km/h; 258 mph)

Cruising speed, 74.8% power at 3,050 m (10,000 ft) — 193 kts (357 km/h; 222 mph)

Econ cruising speed at 7,620 m (25,000 ft) — 185 kts (342 km/h; 213 mph)

Econ cruising speed at 3,050 m (10,000 ft) — 144 kts (266 km/h; 166 mph)

Stalling speed, flaps up, power off: all versions — 82 kts (152 km/h; 95 mph) CAS

Stalling speed, flaps down, power off: all versions — 72 kts (133 km/h; 83 mph) CAS

Max rate of climb at S/L	482 m (1,580 ft)/min
Rate of climb at S/L, one engine out	88 m (290 ft)/min
Service ceiling	9,555 m (31,350 ft)
Service ceiling, one engine out	6,050 m (19,850 ft)
T-O run	666 m (2,185 ft)
T-O run to 15 m (50 ft)	791 m (2,595 ft)
Landing from 15 m (50 ft) at max landing weight	729 m (2,393 ft)
Landing run at max landing weight	309 m (1,013 ft)

Range, recommended lean mixture, with allowances for start, taxi, T-O, climb, descent and 45 min reserves at 45% power:

74.8% power at 7,470 m (24,500 ft) with 272 kg (600 lb) usable fuel — 465 n miles (861 km; 535 miles)

74.8% power at 7,470 m (24,500 ft) with 555 kg (1,224 lb) usable fuel — 1,147 n miles (2,126 km; 1,321 miles)

74.8% power at 3,050 m (10,000 ft) with 272 kg (600 lb) usable fuel — 434 n miles (805 km; 500 miles)

74.8% power at 3,050 m (10,000 ft) with 555 kg (1,224 lb) usable fuel — 1,026 n miles (1,901 km; 1,181 miles)

econ cruising power at 7,620 m (25,000 ft) with 272 kg (600 lb) usable fuel — 496 n miles (917 km; 571 miles)

econ cruising power at 7,620 m (25,000 ft) with 555 kg (1,224 lb) usable fuel — 1,286 n miles (2,383 km; 1,481 miles)

econ cruising power at 3,050 m (10,000 ft) with 272 kg (600 lb) usable fuel — 522 n miles (967 km; 601 miles)

econ cruising power at 3,050 m (10,000 ft) with 555 kg (1,224 lb) usable fuel — 1,294 n miles (2,398 km; 1,490 miles)

UPDATED

CESSNA 421

TYPE: Six/eight-seat pressurised light transport aircraft.

PROGRAMME: Cessna announced, 28 October 1965, a pressurised twin-engined business aircraft designated Model 421, the prototype of which had flown for the first time 14 October 1965. FAA Type Approval was received 1 May 1967 and deliveries commenced that same year.

Two developed versions were developed subsequently as the 421B Golden Eagle and 421B Executive Commuter,

Cessna 421 Golden Eagle six/eight-seat light pressurised transport aircraft

1994

remaining in production until replaced by the 421C in 1976.

VERSIONS: **421A:** Initial version.

421B Golden Eagle: First announced 10 December 1969, this is an improved version of the 421A. Principal changes from the earlier model comprise an increase of 0.71 m (2 ft 4 in) in overall length as a result of enlarging the nose section of the fuselage to provide more baggage and avionics capacity, an extension of 0.61 m (2 ft 0 in) in wing span to maintain take-off and cruise performance without the need to increase engine power, and strengthening of the landing gear to cater for a gross weight increased from 3,102 kg (6,840 lb) to 3,379 kg (7,450 lb). Other improvements include introduction of an Accru-Measure fuel monitoring system; flight instruments mounted in the basic T arrangement; new easy-to-read instruments with standard faces, larger lettering and numerals; instrument white post lighting; new lighting console providing precise control of all cockpit lighting; a systems annunciator panel; and more comfortable seating. New optional equipment included a high-intensity strobe light system; CAA conversion for UK operators; crew shoulder harnesses and dual heated pitot system.

421B Executive Commuter: First announced 16 February 1970, this is a 10-seat version of the above model, designed specifically for the commuter airline, commercial and corporate flying markets. It differs by having lightweight, easily removable seating to provide alternative passenger/cargo configuration; standard fuel capacity of 662 litres (175 US gallons; 145.7 Imp gallons) with optional tanks to allow a maximum fuel capacity of 852 litres (225 US gallons; 187.4 Imp gallons); and deleted rear cabin baggage area, resulting in a total avionics and baggage capacity of 453 kg (1,000 lb).

421C Golden Eagle: Standard versions, as described in detail.

421C Golden Eagle II: As Model 421C above but with the following avionics and equipment as standard: dual Series 400 nav/com, one with VOR/ILS, the other with VOR/LOC, ADF, transponder, DME, glide slope, marker beacon, 400B Nav-O-Matic, basic avionics and avionics cooling kits, associated antennae and slaved compass; variable cabin pressure control system, co-pilot's blind-flying instrumentation, economy mixture indicator, flight hour recorder, propeller synchroniser, cabin hand fire extinguisher, 100 A alternators, starboard landing light, high-intensity strobe lights, emergency locator beacon, external power socket, static dischargers and nosewheel fenders.

421C Golden Eagle III: As standard Model 421C, with standard equipment of 421C II plus cabin air conditioning system and 100A alternators, and an all-weather avionics package which includes dual Series 1000 com and glide slope, Series 1000 nav with HSI, nav with ARC, ADF, RMI and audio panel; Series 800 DME, R/Nav, transponder, encoding altimeter and altitude alerter; 800B IFCS; Series 400 marker beacon; AA-100 radio altimeter; RDR-160 weather radar; basic avionics and cooling kits, and all associated antennae.

UPGRADES: **Advanced Aircraft Corporation:** Advanced Aircraft Regent 1500. See separate entry.

RAM Aircraft Corporation: RAM/Cessna 421C and 421CW. See separate entry.

OPERATOR: Versions of the Cessna 421 are in service with the armed forces of the following countries: Bahamas (1), Ivory Coast (1), Nicaragua (1), Pakistan (1), Peru (1), Sri Lanka (1) and Turkey (4).

The following description applies to the standard Model 421C Golden Eagle:

DESIGN FEATURES: Generally similar to the Model 414A, which is as follows. Cantilever low-wing monoplane. Wing section NACA 23018 (modified) at aircraft centreline, NACA 23015 (modified) at centre-section/outerwing junction, NACA 23009 (modified) at tip. Dihedral 5° on outer panels. Incidence 2° 30′ at root, –0° 30′ at tip. Fixed-incidence tailplane.

FLYING CONTROLS: Generally similar to the Model 414A which has all-metal ailerons and electrically actuated split flaps. Trim tab in port aileron. Trim tabs in rudder and starboard elevator. The area of the fin and rudder has been increased from the Model 414A.

STRUCTURE: Generally similar to the Model 414A which has all-metal two-spar structure of light alloy with stamped ribs and surface skins reinforced with spanwise stringers. Outer wing panels of bonded construction. The fuselage is a conventional all-metal semi-monocoque structure, with failsafe structure in the pressurised section. The tail unit is a cantilever all-metal structure, with sweptback vertical surfaces.

LANDING GEAR: Hydraulically retractable tricycle type, main units retracting inward, nosewheel unit aft. Emergency extension by means of a 138 bar (2,000 lb/sq in) rechargeable nitrogen bottle. Oleo-pneumatic shock-absorbers. Steerable nosewheel. Mainwheel tyres size 6.50-10 (8 ply), nosewheel tyre size 6.00-6 (6 ply). Goodyear single-disc hydraulic brakes. Parking brake.

POWER PLANT: Two 280 kW (375 hp) Continental GTSIO-520-L flat-six geared and turbocharged engines, each driving a McCauley three-blade fully feathering constant-speed metal propeller. Standard total fuel capacity is 806 litres (213 US gallons; 177.4 Imp gallons), of which 780 litres (206 US gallons; 171.5 Imp gallons) are usable, contained in wet wing. Optional wing locker tanks provide a maximum usable capacity of 991 litres (262 US gallons; 218 Imp gallons). Oil capacity 24.6 litres (6.5 US gallons; 5.4 Imp gallons).

ACCOMMODATION: Generally similar to the Model 414A, which is as follows. Two seats side by side in pilot's compartment. Optional curtain or solid divider with curtain, to separate pilot's compartment from main cabin. Standard seating arrangement for four forward-facing passenger seats. Optional arrangements provide for front passenger seats to face aft and forward-facing seventh and eighth seats. Individual consoles each include reading light and ventilator. Optional items include executive writing desk, tables, hat shelf, stereo equipment, electrically adjustable pilot's and co-pilot's seats, refreshment and thermos units, fore and aft cabin dividers, electric shaver converter, all-leather seats, passenger instrument console (clock, true airspeed indicator and altimeter) and intercom. Door is two-piece type with built-in airstairs in bottom portion, on port side of cabin at rear. Plug emergency escape hatch on starboard side of cabin. Double-pane cabin windows. Foul-weather windows for pilot and co-pilot, on each side of fuselage. Electrically de-iced windscreen optional. Baggage accommodated in nose compartment with external access doors, capacity 159 kg (350 lb), two wing lockers, capacity 54.5 kg (120 lb) each, and rear cabin area, capacity 226 kg (500 lb). Total baggage capacity 494 kg (1,090 lb). External access door to tailcone on starboard side. In the Model 421C, the seats have tapered backs and headrests. The nose compartment can contain a total of 272 kg (600 lb) of baggage and avionics, and two wing lockers an additional 91 kg (200 lb) each, plus 226 kg (500 lb) in the rear cabin area, making a total capacity of 680 kg (1,500 lb).

SYSTEMS: Generally similar to the Model 414A, which is as follows. Cabin pressurisation system, maximum differential 0.34 bars (5.00 lb/sq in), maintains sea level cabin conditions to an altitude of 3,350 m (11,000 ft), and a 3,050 m (10,000 ft) cabin altitude to a height of 8,075 m (26,500 ft). Electrical system powered by two engine-driven 28 V 50 A alternators. 24 V 25 Ah battery. 28 V 100 A alternators optional. Hydraulic system for brakes only. Vacuum system for blind-flying instrumentation and optional wing and tail unit de-icing system. Oxygen system of 3.25 m³ (114.9 cu ft) capacity, or emergency oxygen system of 0.31 m³ (11.0 cu ft) capacity optional. Air conditioning system optional.

AVIONICS AND EQUIPMENT: Generally similar as for the Model 414A, which is as follows. The various versions of the Model 414A have avionics as detailed in the model listings. Optional avionics for the basic 414A include those detailed for the 414 II/III, and alternative items from the Cessna Series 400, 800, 1000 range for all versions. Standard equipment includes sensitive altimeter, quartz crystal clock, dual controls, windscreen defroster, outside air temperature gauge, blind-flying instrumentation, audible stall warning device, instrument post lights, alternator failure lights, aileron and elevator control lock, aircraft systems monitoring device, heater overheat light, cabin door 'Not locked' light, sun visors, navigation light detectors aft cabin light, adjustable cabin air ventilators, window curtains, courtesy lights, reading lights, super soundproofing, non-congealing oil coolers, quick-drain fuel valves, heated stall warning transmitter, pitot and fuel vents, retractable landing light, overall paint scheme, propeller synchronisers, full-flow oil filters, navigation lights, rotating beacons and towbar. Optional equipment for Model 414A includes blind-flying instrumentation for co-pilot, economy mixture indicator, flight hour recorder, variable cabin pressure control system, cabin hand fire extinguisher, air conditioning system, 100 A alternators, locator beacon, starboard landing light, strobe lights, taxi light, static dischargers, external power socket, and nosewheel fender. Optional for the Model 414A II are 100 A alternators and cabin air conditioning system. Optional equipment for all versions includes electric elevator trim, angle of attack indicator, digital clock, true airspeed indicator, instantaneous rate of climb indicators, digital flow gauge with computer, rudder pedal lock, boom microphone, turn coordinator, electric or alcohol windscreen anti-icing, 'total flood' cabin fire extinguishing system, seventh and eighth seats, 'Fasten seat belt' and 'Oxygen' signs, toilet with privacy curtain, flight deck/cabin divider or curtain, executive table, refreshment centre, eight-track stereo installation, automatic timer to control courtesy lights, ventilating fan system, tinted windows, internal corrosion proofing, fuselage ice impact panels, ice detection lights, dual pitot system, radome nose, engine nacelle fire detection and extinguishing system and heavy-duty brakes.

DIMENSIONS, EXTERNAL:	
Wing span	12.53 m (41 ft 1½ in)
Wing chord: at root	1.77 m (5 ft 9¾ in)
at tip	1.14 m (3 ft 8½ in)
Length overall	11.09 m (36 ft 4½ in)
Height overall	3.49 m (11 ft 5½ in)
Tailplane span	5.18 m (17 ft 0 in)
Wheel track	5.48 m (17 ft 11¾ in)
Wheelbase	3.18 m (10 ft 5¼ in)
Propeller diameter	2.29 m (7 ft 6 in)
AREAS:	
Wings, gross	19.97 m² (215 sq ft)
WEIGHTS AND LOADINGS:	
Weight empty: 421C Golden Eagle	2,077 kg (4,578 lb)
421C II Golden Eagle	2,160 kg (4,763 lb)
421C III Golden Eagle	2,258 kg (4,979 lb)
Max T-O weight	3,379 kg (7,450 lb)
Max ramp weight	3,402 kg (7,500 lb)
Max zero-fuel weight	3,054 kg (6,733 lb)
Max landing weight	3,266 kg (7,200 lb)
Max wing loading	169.4 kg/m² (34.7 lb/sq ft)
Max power loading	6.03 kg/kW (9.9 lb/hp)

PERFORMANCE (at max T-O weight, except speeds are those at mid-cruise weight):
Max level speed at 6,100 m (20,000 ft) 258 kts (478 km/h; 297 mph)
Max cruise speed, 73.5% power at 7,620 m (25,000 ft) 241 kts (447 km/h; 278 mph)
Max cruising speed, 73.5% power at 3,050 m (10,000 ft) 208 kts (386 km/h; 240 mph)
Econ cruising speed at 7,620 m (25,000 ft) 195 kts (361 km/h; 224 mph)
Econ cruising speed at 3,050 m (10,000 ft) 155 knoys (287 km/h; 178 mph)
Stalling speed, flaps up, power off:
all versions 83 kts (154 km/h; 96 mph) CAS
Stalling speed, flaps down, power off:
all versions 74 kts (137 km/h; 85 mph) CAS
Max rate of climb at S/L 591 m (1,940 ft)/min
Rate of climb at S/L, one engine out 107 m (350 ft)/min
Service ceiling 9,205 m (30,200 ft)
Service ceiling, one engine out 4,540 m (14,900 ft)
T-O run 544 m (1,786 ft)
T-O to 15 m (50 ft) 708 m (2,323 ft)
Landing run from 15 m (50 ft) 699 m (2,293 ft)
Landing run 219 m (720 ft)
Range, recommended lean mixture, with allowances for start, taxi, T-O, climb, descent and 45 min reserves at 45% power:
73.5% power at 7,620 m (25,000 ft) with 561 kg (1,236 lb) usable fuel 955 n miles (1,770 km; 1,100 miles)
73.5% power at 7,620 m (25,000 ft) with 713 kg (1,572 lb) usable fuel 1,271 n miles (2,356 km; 1,464 miles)
73.5% power at 3,050 m (10,000 ft) with 561 kg (1,236 lb) usable fuel 853 n miles (1,580 km; 982 miles)
73.5% power at 3,050 m (10,000 ft) with 713 kg (1,572 lb) usable fuel 1,123 n miles (2,081 km; 1,293 miles)
econ cruising power at 7,620 m (25,000 ft) with 561 kg (1,236 lb) usable fuel 1,092 n miles (2,023 km; 1,257 miles)
econ cruising power at 7,620 m (25,000 ft) with 713 kg (1,572 lb) usable fuel 1,487 n miles (2,755 km; 1,712 miles)
econ cruising power at 3,050 m (10,000 ft) with 561 kg (1,236 lb) usable fuel 1,088 n miles (2,017 km; 1,253 miles)
econ cruising power at 3,050 m (10,000 ft) with 713 kg (1,572 lb) usable fuel 1,464 n miles (2,713 km; 1,686 miles)

UPDATED

CESSNA 501 CITATION I

TYPE: Seven/nine-seat twin-turbofan executive transport aircraft.

PROGRAMME: Cessna announced, 7 October 1968, that it was developing an eight-seat pressurised executive turbofan aircraft named Fanjet 500, which would be able to operate from most airfields used by light and medium twin-engined aircraft. After the first flight of the prototype, on 15 September 1969, the aircraft's name was changed to Citation. Subsequently, the gross weight was increased for the first time and several other changes were made. These included a lengthened front fuselage, movement of the engine nacelles further aft, larger vertical tail, and resiting and introduction of dihedral on the tailplane.

In July 1971 Cessna announced that the first production Citation 0001 (N502CC) had made its first flight. Final FAA certification under FAR Pt 25 was awarded 9 September 1971, and has been followed by certification in many other countries.

The increase in take-off gross weight, to a maximum of 5,375 kg (11,850 lb), and the use of optional Rohr Industries thrust reversers, received FAA certification in February 1976. The improved Citation I became available in December 1976, with an increased wing span and JT15D-1A turbofan engines. It superseded the earlier model on the production line from c/n 350 onward and was certificated 15 December 1976. The first Citation I was delivered 21 December 1976.

VERSIONS: **Citation I:** Basic version. *As described in detail.*

501 Citation I/SP: Basically similar to the Citation I, but certificated to FAR Pt 23 for single-pilot operation. Type certification granted 7 January 1977. First aircraft delivered 25 January 1977.

UPGRADES: **Branson Aircraft Corporation:** Weight increase, extended range, cargo door and air ambulance modifications. See separate entries.

Sierra Industries: Sierra/Cessna Eagle series. See separate entry.

OPERATORS: Versions of the Citation I are in service with the armed forces of the following countries: Argentina (1), Ecuador (1), Mexico (1) and Venezuela (1).

DESIGN FEATURES: Cantilever low-wing monoplane without sweepback. Wing section at c/l NACA 23014 (modified), at wing station 247.95 NACA 23012. Incidence 2° 30′ at c/l, –0° 30′ at wing station 247.95. Dihedral 4°.

FLYING CONTROLS: Manually operated ailerons, with manual trim on port aileron. Electrically operated single-slotted trailing-edge flaps. Hydraulically operated aerodynamic speed brakes. Manually operated control surfaces. Electric elevator trim with manual override; manual rudder trim.

STRUCTURE: All-metal fail-safe structure with two primary spars, an auxiliary spar, three fuselage attachment points, and conventional ribs and stringers. The fuselage is an all-metal pressurised structure of circular section. Fail-safe design, providing multiple load paths. The tail unit is a cantilever all-metal structure. Horizontal surfaces have dihedral of 9°. Large dorsal fin and smaller ventral fin.

LANDING GEAR: Hydraulically retractable tricycle type with single wheel on each unit. Main units retract inward into wing, nose gear forward into fuselage nose. Free-fall and pneumatic emergency extension systems. Goodyear mainwheels with tyres size 22.0 × 8-10, 10 ply rating, pressure 6.90 bars (100 lb/sq in). Steerable nosewheel with Goodyear wheel and tyre size 18.0 × 4.4, 10 ply rating, pressure 8.27 bars (120 lb/sq in). Goodyear hydraulic brakes. Parking brake and pneumatic emergency brake system. Anti-skid system optional.

POWER PLANT: Two Pratt & Whitney Canada JT15D-1B turbofan engines, each rated at 9.77 kN (2,200 lb st) for take-off, mounted in pod on each side of rear fuselage. Rohr thrust reversers optional. Integral fuel tanks in wings, with combined usable capacity of 2,135 litres (564 US gallons; 469.6 Imp gallons).

ACCOMMODATION: Crew of two on separate flight deck, on fully adjustable seats, with seat belts and inertial reel shoulder harness, and sun visors. Fully carpeted main cabin equipped with two individual forward-facing seats aft, one single forward-facing seat centre port, one single aft-facing seat centre starboard and a fifth aft-facing corner lounge chair at front of cabin on starboard side, all with headrests. Refreshment unit at front of cabin. Toilet compartment and main baggage area at rear of cabin. Second baggage area in nose. Total baggage capacity 454 kg (1,000 lb). Cabin is pressurised, heated and air conditioned. Individual reading lights and air inlets for each passenger. Dropout constant flow oxygen system for emergency use. Plug type door with integral airstair at front on port side and one emergency exit on starboard side. Doors on each side of nose baggage compartment. Tinted windows, each with curtains. Optional layouts for crew of two and six or seven passengers, with executive table, flush toilet replace standard toilet, and choice of interior trims. Pilot's storm window, birdproof windscreen with de-fog system, anti-icing, standby alcohol anti-icing and bleed air rain removal system.

SYSTEMS: Pressurisation system supplied with engine bleed air, maximum pressure differential 0.59 bars (8.5 lb/sq in), maintaining an S/L cabin altitude to 6,720 m (22,040 ft), or a 2,440 m (8,000 ft) cabin altitude to 12,495 m (41,000 ft). Hydraulic system, pressure 103.5 bars (1,500 lb/sq in), with two pumps to operate landing gear and speed brakes. Separate hydraulic system for wheel brakes. Electrical system supplied by two 28 V 400 A DC starter/generators, with two 350 VA inverters and 24 V 40 Ah Ni/Cd battery. Oxygen system of 0.62 m³ (22 cu ft) capacity includes two crew demand masks and five dropout constant flow masks for passengers. High-capacity oxygen system optional. Engine fire detection and extinguishing systems. Wing leading-edge forward of each engine is electrically anti-iced. Pneumatic de-icing boots on outer leading-edges.

AVIONICS AND EQUIPMENT: Standard Cat. II avionics package on aircraft subsequent to c/n 275 comprises Sperry SPZ 500 flight control system with choice of single or double-cue command bars, including Sperry 500 autopilot, Sperry altimeter with altitude alerting and reporting functions, complete vertical navigation capability, air data computer, Sperry Model 600 (port)/Model 044 (starboard) horizontal situation indicator, Sperry ADI Model 600 command and control computer and autopilot servos, Collins WXR250C (Bendix RDR 1100 subsequent to c/n 590) continuous vision weather radar, two Collins VHF-20A com transceivers, two Collins VIR-30A nav receivers, Collins TDR-90 transponder, Collins DME-40, two Collins 332-C10 radio magnetic indicators, and Collins ADF-60. Provision for advanced instrumentation and avionics to

Cessna Citation I seven/nine-seat twin-turbofan executive transport aircraft

1994

customer's specification. Standard equipment includes encoding altimeter, instantaneous rate of climb, and turn and bank indicators, artificial horizon and directional gyros, cabin altitude and differential pressure gauge, outside air temperature gauge, flight hour recorder, control locks, storm window, battery temperature and low fuel level warning lights, high Mach/airspeed warning, stall warning device, baggage tiedown kit, cabin fire extinguisher, individual life vests, 'No smoking - fasten seat belt' sign, internally lit instruments, instrument standby lights, map light, entry light, cockpit dome light, individual reading lights, tailcone lights, tailcone compartment light, emergency exit lights, strobe lights, landing light, navigation lights, storm lights, taxi light, wing ice light, automatic engine start system, engine fire warning and extinguishing system, emergency battery pack, external power socket, static discharge wicks, towing and jack pad adaptors, inlet anti-icing, and surface de-icing system. Optional items include angle of attack indicator, radiotelephone, cockpit voice recorder, flight data recorder, ground proximity warning system, emergency transmitter, navigation chart case, flush toilets, refreshment cabinets, storage drawers, tables, engine fan synchroniser, gravel runway kit, and drag-chute.

DIMENSIONS, EXTERNAL:
Wing span 14.35 m (47 ft 1 in)
Wing aspect ratio 7.83
Length overall 13.26 m (43 ft 6 in)
Height overall 4.37 m (14 ft 4 in)
Tailplane span 5.74 m (18 ft 10 in)
Wheel track 3.84 m (12 ft 7 in)
Wheelbase 4.78 m (15 ft 8 in)
Cabin door (port): Height 1.30 m (4 ft 3 in)
Width 0.60 m (1 ft 11½ in)
Emergency exit (starboard):
Height 0.95 m (3 ft 1¼ in)
Width 0.56 m (1 ft 10 in)

DIMENSIONS, INTERNAL:
Cabin:
Length, front to rear bulkhead 5.33 m (17 ft 6 in)
Max width 1.50 m (4 ft 11 in)
Max height 1.32 m (4 ft 4 in)
Baggage space: cabin 1.10 m³ (39 cu ft)
nose 0.45 m³ (16 cu ft)

AREAS:
Wings, gross 25.9 m² (278.5 sq ft)
Horizontal tail surfaces (total) 6.56 m² (70.6 sq ft)
Vertical tail surfaces (total) 4.73 m² (50.9 sq ft)

WEIGHTS AND LOADINGS (from aircraft c/n 525 onward):
Weight empty, equipped 3,008 kg (6,631 lb)
Max fuel weight 1,727 kg (3,807 lb)
Max T-O weight 5,375 kg (11,850 lb)
Max ramp weight 5,443 kg (12,000 lb)
Max landing weight 5,148 kg (11,350 lb)
Max zero-fuel weight 3,810 kg (8,400 lb)
Optional max zero-fuel weight 4,309 kg (9,500 lb)

PERFORMANCE (at max T-O weight, ISA, except where indicated):
Max operating speed: S/L to 4,265 m (14,000 ft) 262 kts (486 km/h; 302 mph) IAS
4,265 m (14,000 ft) to 8,530 m (28,000 ft) 277 kts (513 km/h; 319 mph) IAS
above 8,530 m (28,000 ft) Mach 0.705
Cruising speed at mid-cruise weight of 4,536 kg (10,000 lb) at 10,670 m (35,000 ft) 357 kts (662 km/h; 411 mph)
Stalling speed at max landing weight 82 kts (152 km/h; 95 mph) CAS
Max rate of climb at S/L 829 m (2,719 ft)/min
Rate of climb at S/L, one engine out 252 m (826 ft)/min
Max certificated altitude 12,495 m (41,000 ft)
Service ceiling, one engine out 6,400 m (21,000 ft)
FAR 25 T-O balanced field length 893 m (2,930 ft)
T-O to 15 m (50 ft) 751 m (2,463 ft)
FAR 25 landing runway length at max landing weight 692 m (2,270 ft)
Balanced field length: at S/L, ISA 893 m (2,930 ft)
at 1,525 m (5,000 ft), 27° 1,800 m (5,900 ft)

Range with max fuel and 709 kg (1,562 lb) payload at 12,495 m (41,000 ft), allowances for T-O, climb, descent and 45 min reserves
1,328 n miles (2,459 km; 1,528 miles)

OPERATIONAL NOISE LEVELS (FAR Pt 36):

T-O	79 EPNdB
Approach	88 EPNdB
Sideline	87 EPNdB

UPDATED

CESSNA S550 CITATION S/II

US Navy designation: T-47A

TYPE: Six/eight-passenger improved Citation II.

PROGRAMME: Announced 4 October 1983; first flight 14 February 1984; certificated with exemption for single-pilot operation July 1984; first delivery late Summer 1984; first delivery Citation S/II ambulance late 1985 to Province of Manitoba, Canada. High-capacity brakes introduced as standard and offered as retrofit April 1987, reducing landing distance at maximum landing weight by 13 per cent.

VERSIONS: **S550:** Basic executive transport. *Detailed description applies to this version (c/n 0115 onwards) except where indicated.*

Ambulance: Carries single or double stretchers, up to four medical attendants and large quantities of medical oxygen.

T-47A (Cessna Model 552): Fifteen Citation S/IIs acquired by US Navy to replace T-39Ds in radar training role, part of five-year programme with three-year option, including provision of aircraft, simulators, maintenance and pilot services for training operators of air-to-air, intercept, air-to-ground and other radars. Differences from S/II include 12.89 kN (2,900 lb st) JT15D-5 turbofans, shorter wing to allow faster climb and Mach 0.733 at 12,200 m (40,000 ft), and Emerson nose-mounted AN/APQ-159 radar. Normal crew includes civilian pilot, Navy instructor, three students. First flight 15 February 1984; FAA certification 21 November 1984. Programme ended September 1991; aircraft returned to manufacturer.

UPGRADES: **Branson Aircraft Corporation:** Weight increase, extended range, cargo door and air ambulance modifications. See separate entries.

CUSTOMERS: US Navy (see above). Five specially equipped Citation S/IIs delivered to Flight Test Research Institute in Xian for Airborne Remote Sensing Centre of Chinese Academy of Sciences.

DESIGN FEATURES: Improvements, introduced on production line from c/n 506, include new wing aerofoil using Citation III supercritical technology, modified wing/fuselage fairing, extended inboard leading-edge, low drag engine pylon contours, sealed aileron/airbrake gaps, faired flap coves, hydraulically actuated flaps in two sections each side extending further inboard, P&WC JT15D-4B turbofans. New wing reduces cruise drag without sacrificing low speed and short-field capability.

Straight wing; incidence 2° 30′ at centreline, –0° 30′ at station 247.95; dihedral 4°; tailplane dihedral 9°.

Internal refinements include tailcone baggage volume increased to 0.79 m³ (28 cu ft), 12.7 cm (5 in) extra headroom in totally private toilet, soft-touch sound deadening headliners, Citation III-style seats with shoulder harness, lateral seat tracking for better head and elbow room, built-in lifejacket stowage, and redesigned sidewall air ducts giving greater insulation and heating and allowing more than 10 per cent extra aisle width.

Options include vanity unit for toilet, refreshment centres in composites, wide door for ambulance/cargo operations, Honeywell EFIS.

FLYING CONTROLS: Mechanically actuated ailerons assisted by geared trim tabs; elevator trim electrically actuated with mechanical standby; mechanical rudder trim; hydraulically actuated Fowler flaps; hydraulically operated airbrakes.

STRUCTURE: As described for Citation II, except ailerons and flaps have graphite composite structure.

LANDING GEAR: As for Citation II except mainwheel tyres have 12 ply rating, pressure 8.27 bars (120 lb/sq in). High-capacity brakes manufactured by Aircraft Wheel and Brake Division of Parker Hannifin Corporation.

POWER PLANT: As Citation II, but usable fuel capacity increased to 3,263 litres (862 US gallons; 718 Imp gallons).

ACCOMMODATION: Crew details as for Citation II. Seating for six to eight passengers in main cabin. Standard interior configuration provides for six passenger seats, two forward- and four aft-facing, each with headrest, seat belt and diagonal inertia reel harness; flushing toilet aft; tracked refreshment centre; forward cabin divider with privacy curtain, aft cabin divider with sliding doors. Passenger service units containing an oxygen mask, air vent and reading light for each passenger. Three separate baggage areas, one in nose section that is externally accessible, one in aft cabin area, and one in tailcone area, with a combined capacity of up to 658 kg (1,450 lb).

SYSTEMS: Pressurisation system as for Citation II but maintains sea level cabin altitude to 6,962 m (22,842 ft), or a 2,440 m (8,000 ft) cabin altitude to 13,105 m (43,000 ft). Hydraulic system as for Citation II. Electrical system as for Citation II, except starter/generators are 300 A. TKS Glycol anti-icing system on wing, tail surface de-icing eliminated.

AVIONICS: Standard avionics package comprises Honeywell SPZ-500 integrated flight director/autopilot system, with single-cue command bars, Honeywell C-14D compass system, Honeywell RD-450 (starboard) HSI, dual Collins VHF-22A com transceivers, dual Collins VIR-32 nav receivers with VOR/LOC, glide slope and marker beacon receivers, dual Collins RMI-30, Collins DME-42 with 339F-12 indicator, TDR-90 transponder, Collins ADF-60, and Honeywell Primus 300SL colour weather radar. Optional advanced avionics and instrumentation are available according to customer choice, and include Bendix/King Series III integrated EFIS, nav/com and radar systems.

DIMENSIONS, EXTERNAL:

Wing span over lights	15.90 m (52 ft 2½ in)
Wing chord (mean)	2.06 m (6 ft 9 in)
Wing aspect ratio	7.8
Length overall	14.39 m (47 ft 2½ in)
Height overall	4.57 m (15 ft 0 in)
Wheel track	5.36 m (17 ft 7 in)
Wheelbase	5.55 m (18 ft 2½ in)
Tailplane span	5.79 m (19 ft 0 in)
Cabin door (optional): Height	1.14 m (3 ft 9 in)
Width	0.89 m (2 ft 11 in)

DIMENSIONS, INTERNAL (S/II):

Cabin:	
Length, front to rear bulkhead	6.37 m (20 ft 10¾ in)
Max height	1.45 m (4 ft 9½ in)
Max width	1.49 m (4 ft 10¾ in)
Baggage capacity (total)	2.27 m³ (80.0 cu ft)

AREAS (S/II):

Wings, gross	31.83 m² (342.6 sq ft)
Horizontal tail surfaces (total)	6.48 m² (69.8 sq ft)
Vertical tail surfaces (total)	4.73 m² (50.9 sq ft)

WEIGHTS AND LOADINGS:

Weight empty, equipped	3,655 kg (8,059 lb)
Max baggage weight: cabin/tailcone	272 kg (600 lb)
nosecone	385 kg (850 lb)
Max ramp weight	6,940 kg (15,300 lb)
Max fuel weight	2,640 kg (5,820 lb)
Max T-O weight	6,849 kg (15,100 lb)
Max landing weight	6,350 kg (14,400 lb)
Max zero-fuel weight	4,990 kg (11,200 lb)
Max wing loading	215.17 kg/m² (44.07 lb/sq ft)
Max power loading	1.42 kg/kN (3.02 lb/lb st)

PERFORMANCE (at max T-O weight, except where indicated):

Max operating speed: S/L to 2,440 m (8,000 ft)
261 kts (483 km/h; 300 mph) IAS
2,440 m (8,000 ft) to 8,935 m (29,315 ft)
276 kts (511 km/h; 318 mph)
above 8,935 m (29,315 ft) Mach 0.721

Cruising speed at mid-cruise weight of 5,443 kg (12,000 lb) at 10,670 m (35,000 ft)
403 kts (746 km/h; 463 mph)

Stalling speed: clean, at max T-O weight
94 kts (174 km/h; 108 mph) CAS
at max landing weight
82 kts (152 km/h; 94 mph) CAS

Max rate of climb at S/L	926 m (3,040 ft)/min
Rate of climb at S/L, OEI	262 m (860 ft)/min
Max operating altitude	13,105 m (43,000 ft)
Service ceiling, OEI	7,315 m (24,000 ft)
T-O balanced field length (FAR Pt 25)	987 m (3,240 ft)

FAR 25 landing field length at max landing weight (high-capacity brakes) 805 m (2,640 ft)

Range with four passengers, two crew and baggage, zero wind, IFR reserves
1,739 n miles (3,223 km; 2,002 miles)

Range with max fuel
1,998 n miles (3,701 km; 2,300 miles)

OPERATIONAL NOISE LEVELS (FAR Pt 36):

T-O	78.0 EPNdB
Approach	91.0 EPNdB
Sideline	90.4 EPNdB

VERIFIED

CESSNA 650 CITATION III

TYPE: Twin-turbofan eight/11-seat long-range business jet.

PROGRAMME: First flight of prototype (N650CC) 30 May 1979; first flight of second prototype 2 May 1980; FAR Pt 25 Transport Category certification 30 April 1982; first internal delivery December 1982; first customer delivery Spring 1983; CAA certification April 1988; European certifications include Austria, Denmark, France, Germany, Italy, Spain, Sweden, Switzerland. New Masterpiece interior introduced at NBAA convention 1988.

CUSTOMERS: Total 189 delivered by 31 December 1990, including 15 during 1990. Production of Citation III, VI and VII sold to end of 1991.

DESIGN FEATURES: Wing has NASA supercritical aerofoil; dihedral 3°; sweepback at quarter-chord 25°. Tailplane anhedral 3°. Wing leading-edges de-iced by engine bleed air; tailplane electrically de-iced; fin unprotected.

FLYING CONTROLS: Pitch axis, variable incidence tailplane and elevator; rudder boosted to counteract asymmetric thrust; hydraulically powered ailerons with manual reversion after 3° movement by outboard spoiler panel; four hydraulically powered spoiler panels on each wing, of which outboard assists aileron, two centre panels act as airbrakes and all four panels used for emergency descent and lift dumping after touchdown. Electrically actuated trailing-edge flaps in three sections each side. Stall strips on inner and outer wings and small fence and turbulators ahead of outer flaps.

STRUCTURE: Conventional light alloy pressurised fuselage of circular section; fail-safe structure in pressurised area; light alloy tail surfaces; two-spar, fail-safe light alloy wing of bonded and riveted construction; wing built in three sections; flaps of Kevlar and graphite composites.

LANDING GEAR: Hydraulically retractable tricycle type. Main units retract inward into the undersurface of the wing centre-section, nosewheel forward and upward into the nose. Main units of trailing link type, each with twin wheels; steerable nose unit has a single wheel, maximum steering angle ±70-80°. Oleo-pneumatic shock-absorber in each unit. Hydraulically powered nosewheel steering, with an accumulator to provide steering after a loss of normal hydraulic power. Emergency landing gear extension by manual release and free-fall to locked position; pneumatic blowdown system for back-up. Mainwheel tyres size 22.0 × 5.75, 10 ply rating, pressure 10.20 bars (148 lb/sq in). Nosewheel tyre size 18.0 × 4.4, 10 ply rating, pressure 8.62 bars (125 lb/sq in). Fully modulated hydraulically powered anti-skid brake system. In the event of hydraulic system failure, an electrically driven standby pump provides pressure for the brakes. Emergency pneumatic brake system. Parking brake.

POWER PLANT: Two Garrett TFE731-3B-100S turbofans, each rated at 16.24 kN (3,650 lb st) for take-off, pod-mounted on sides of rear fuselage. Hydraulically operated Rohr target type thrust reversers standard. Two independent fuel systems, with integral tanks in each wing; usable capacity 4,183 litres (1,105 US gallons; 920 Imp gallons). Additional fuel cell behind rear fuselage bulkhead. Single-point pressure refuelling on starboard side of fuselage, to rear of wing trailing-edge. Gravity refuelling point on upper surface of each wing. A boost pump in the port wing fills the fuselage tank when pressure refuelling is not available. Engine intake anti-icing system.

ACCOMMODATION: Crew of two on separate flight deck, and up to nine passengers. Standard interior has six individual seats, with toilet at rear of cabin. The fuselage nose incorporates a radome, high-resolution radar, avionics bay and a storage compartment for crew baggage. Electrically heated baggage compartment in rear fuselage with external door on port side. Airstair door forward of wing on port side. Overwing emergency escape hatch on starboard side. Cabin is pressurised, heated and air conditioned. Windscreen anti-icing by engine bleed air, with alcohol spray back-up for port side of the windscreen. Windscreen defogging by warm air, and rain removal by engine bleed air and a mechanically actuated air flow deflector.

Cessna S550 Citation S/II improved Citation II

1994

Cessna 650 Citation III twin-turbofan long-range business jet

1994

SYSTEMS: Environmental control system, with separate controls of flight deck and cabin conditions. Direct engine bleed pressurisation system, with nominal pressure differential of 0.67 bar (9.7 lb/sq in), provides 2,440 m (8,000 ft) cabin environment to maximum certificated altitude and can maintain a sea level cabin environment to approximately 7,620 m (25,000 ft). Electrical system includes two 28 V 400 A DC starter/generators, two 200/115 V 5 kW three-phase engine-driven alternators, two 115 V 400 Hz solid-state static inverters, two 24 V 22 Ah Ni/Cd batteries and an external power socket in the tailcone. Hydraulic system of 207 bars (3,000 lb/sq in) powered by two engine-driven pressure compensated pumps for operation of spoilers, brakes, landing gear, nosewheel steering and thrust reversers. Hydraulic reservoir with integral reserve and an electrically driven hydraulic pump to provide emergency power. Oxygen system of 1.39 m³ (49 cu ft) capacity with automatic dropout constant-flow oxygen mask for each passenger and a quick-donning pressure demand mask for each crew member. Engine fire detection and extinguishing system.

AVIONICS: Standard avionics include a Honeywell SPZ-650 integrated flight director/autopilot system with AD650A ADI, RD650A HSI and C-14D compass system; Honeywell GH-14 ADI and RD450 HSI with C-14D compass system for co-pilot; AA-300 radio altimeter; dual Collins VHF-22A 720-channel com transceivers, dual VIR-32 nav receivers which include VOR, localiser, glide slope and marker beacon receivers, dual RMI-30, DME-42 DME, and TDR-90 transponder; Honeywell Primus 300SL colour weather radar; Collins ADF-60 ADF; J.E.T. standby attitude gyro; Teledyne angle of attack system; air data computer; dual Avtech audio amplifiers; and Telex microphones, headsets and speakers. A wide range of optional avionics is available including Bendix/King Series III integrated EFIS, nav/com and radar system.

EQUIPMENT: Standard equipment includes dual altimeters, Mach/airspeed indicators, angle of attack indicator, digital clock, instantaneous rate of climb indicators, outside air temperature gauge, crew seats with vertical, fore, aft and recline adjustments, seat belts, shoulder harnesses and inertia reels, six individual passenger seats, three forward- and three aft-facing with vertical, fore and aft adjustment, lateral tracking and recline adjustments, seat belts and shoulder harnesses, sun visors, flight deck divider with curtain, map case, openable storm windows, electroluminescent and edge-lit instrument panels, stall warning system, cockpit and cabin fire extinguishers, indirect cabin lighting, cabin aisle lights, door courtesy lights, 'Fasten seat belt—no smoking' signs, refreshment centre, cup holders, ashtrays, executive table, aft cabin divider with curtain, emergency exit signs, internal corrosion proofing, emergency battery pack, emergency portable cabin oxygen, navigation and recognition lights, dual landing and taxi lights, dual anti-collision strobe lights, red flashing beacon, dual wing ice lights, lightning protection, static discharge wicks and tiedown provisions.

DIMENSIONS, EXTERNAL:

Wing span	16.31 m (53 ft 6 in)
Wing mean aerodynamic chord	2.08 m (6 ft 9¾ in)
Wing aspect ratio	8.9
Length overall	16.90 m (55 ft 5½ in)
Height overall	5.12 m (16 ft 9½ in)
Tailplane span	5.60 m (18 ft 4½ in)
Wheel track	2.84 m (9 ft 4 in)
Wheelbase	6.50 m (21 ft 4 in)
Cabin door: Width	0.61 m (2 ft 0 in)
Height	1.37 m (4 ft 6 in)

DIMENSIONS, INTERNAL:

Cabin:	
Length, front to rear bulkhead	7.01 m (23 ft 0 in)
Length, aft of cockpit divider	5.66 m (18 ft 7 in)
Max width	1.73 m (5 ft 8 in)
Max height	1.78 m (5 ft 10 in)
Baggage capacity (aft)	1.88 m³ (66.4 cu ft)
Crew baggage compartment (nose)	0.17 m³ (6.0 cu ft)

AREAS:

Wings, gross	29.00 m² (312.0 sq ft)
Horizontal tail surfaces (total)	6.26 m² (67.4 sq ft)
Vertical tail surfaces (total)	6.04 m² (65.0 sq ft)

WEIGHTS AND LOADINGS:

Weight empty, standard	5,357 kg (11,811 lb)
Max fuel weight	3,349 kg (7,384 lb)
Max payload	1,583 kg (3,489 lb)
Max T-O weight	9,979 kg (22,000 lb)
Max ramp weight	10,070 kg (22,200 lb)
Max landing weight	9,072 kg (20,000 lb)
Max zero-fuel weight	6,940 kg (15,300 lb)
Max wing loading	344.2 kg/m² (70.51 lb/sq ft)
Max power loading	307.24 kg/kN (3.01 lb/lb st)

PERFORMANCE (at max T-O weight, ISA, except where indicated):

Max operating speed: S/L to 2,440 m (8,000 ft)
305 kts (565 km/h; 351 mph) IAS
at 11,130 m (36,525 ft)
278 kts (515 km/h; 320 mph) IAS
above 11,130 m (36,525 ft) Mach 0.851

Max cruising speed at 10,670 m (35,000 ft) and 7,257 kg (16,000 lb) cruise weight
472 kts (874 km/h; 543 mph)

Stalling speed: clean, at max T-O weight
125 kts (232 km/h; 144 mph) CAS
flaps and wheels down, at max landing weight
97 kts (180 km/h; 112 mph) CAS

Max rate of climb at S/L	1,127 m (3,700 ft)/min
Rate of climb at S/L, one engine out	245 m (805 ft)/min
Time to 13,100 m (43,000 ft)	33 min
Certificated ceiling	15,545 m (51,000 ft)
Ceiling, one engine out	7,165 m (23,500 ft)
FAR 25 T-O field length at S/L	1,579 m (5,180 ft)
FAR 25 landing field length at max landing weight	884 m (2,900 ft)
Turning circle based on nosewheel	6.63 m (21 ft 9 in)

Range, zero wind, with allowances for T-O, climb, descent and 45 min reserves (two crew, four passengers)
2,346 n miles (4,348 km; 2,701 miles)

g limits +3.2/−1

OPERATIONAL NOISE LEVELS (FAR Pt 36):

T-O	74.0 EPNdB
Approach	85.0 EPNdB
Sideline	81.0 EPNdB

VERIFIED

CHALK'S

CHALK'S INTERNATIONAL AIRLINES

750 SW 34th Street, Fort Lauderdale, Florida 33315-3566
Telephone: 1 (305) 359 0414
Fax: 1 (305) 359 5240
DIRECTOR, AIRCRAFT DEVELOPMENT: Lonny McClung

VERIFIED

CHALK'S (GRUMMAN) G-111 and G-111T CONVERSION PROGRAMMES

TYPE: Amphibious aircraft conversion.

PROGRAMME: Chalk's is the owner of the FAA type certificate for the G-111 Albatross and the oversight for the G-111T turboprop conversion programme. FAA approval for the G-111T programme was received in early 1994. Certification of first G-111T now expected in 1996 following joint venture agreement with Basler Turbo Conversions. First aircraft delivered to Basler in August 1995. This programme follows the G-111 structural modification programme which was managed by Resorts International and ran from 1975-83. Resorts International converted 13 aircraft, 12 of which are owned by Chalk's.

Chalk's (Grumman) G-111T computer generated image

1994

DESIGN FEATURES: G-111 structural modification programme included remanufacturing much of the airframe, addition of titanium wing spar caps and straps, addition of 28 or 30 passenger seats as well as addition of two emergency exit doors and improvements to the avionics and electrical systems. The G-111T turboprop programme consists of replacing the Wright 1820 radial engines with AlliedSignal Propulsion Engine Company (Garrett) TPE 331-14 HR/GR turboprop engines. This conversion is expected to be adaptable to both HU-16A and B model airframes.

WEIGHTS AND LOADINGS:

Max payload 3,402 kg (7,500 lb)

PERFORMANCE:

Cruise speed:

G-111T 220 kts (407 km/h; 252 mph) TAS

Range (internal fuel):

G-111T 700 n miles (1,296 km; 805 miles)

VERIFIED

Chalk's (Grumman) G-111 modified amphibious aircraft

1995

CHRYSLER

CHRYSLER TECHNOLOGIES AIRBORNE SYSTEMS INC (CTAS)

PO Box 154580, Waco, Texas 76715-4580
Telephone: 1 (817) 867 4202
Fax: 1 (817) 867 4230
Telex: 163346
PRESIDENT: R. H. Pacey
DIRECTOR, MARKETING: Keith Arnold

CTAS was established in 1989 after Chrysler transferred the airborne systems operations out of Electrospace Systems Inc. Services include development, manufacture, installation and integration of special purpose airborne systems and aircraft modification for military and commercial customers, including airborne EW IRCM and C^3 systems, wide-body heavy maintenance and VIP/head of state interior modifications.

UPDATED

CTAS (Alenia G222) C-27A STOL transport aircraft

1994

CTAS (ALENIA) C-27A SPARTAN UPGRADE

TYPE: Mid-range pressurised STOL transport upgrade.

PROGRAMME: US Air Force ordered five Alenia G222-710As (which see in Italian section) from Chrysler 20 August 1990; further five ordered 4 February 1991; aircraft procured from Alenia by Chrysler and modified for USAF with enhancements including autopilot, INS, new HF/VHF/UHF communications and specific mission equipment; first aircraft (I-RAIS) accepted 15 April 1991 and delivered to USA 17 April 1991; first flown after C-27A modifications, re-serialled 90-0170, 30 July 1991; to USAF 16 August 1991 for 45 days of operational testing at Pope AFB, North Carolina. CTAS provides aircrew academic and flight training, maintenance training and complete logistical support.

CUSTOMERS: US Air Force, 10; required for short-field operations by USSOUTHCOM in Panama; 90-0171 delivered 310th Airlift Squadron/61st Airlift Group, Howard AFB, Canal Zone, 26 September 1991. Tenth delivered 25 January 1993.

COSTS: First five bought for about $80 million; second five for $72.6 million.

Following description applies to the C-27A, except where indicated. Description of Alenia G222 basic airframe can be found under Alenia, Italy.

POWER PLANT: Two General Electric T64-GE-P4D turboprops, each flat rated at 2,535 kW (3,400 shp) at ISA + 25°C. New turbine inlet temperature indicators with analog pointer and digital readout added to C-27A engine instrument panel.

ACCOMMODATION: Normal crew of three (two pilots and loadmaster); loadmaster's seat by crew entrance door, facing aft in cargo compartment. Standard troop configuration has 34 foldaway sidewall seats with lifevest stowage compartment under each seat, and two 20-man liferafts stowed in wing/fuselage fairing. Paratroop version can carry up to 24 fully equipped paratroops and includes door jump platforms and static lines. Cargo version can accept three standard 463L 2.74 × 2.24 m (108 × 88 in) pallets, and carry up to 9,000 kg (19,840 lb) of freight. Provision for 135 cargo tiedown points, on 51 cm (20 in) square NATO standard grid. In combat operations, cargo compartment can accommodate one M1038 HMMWV, one M1009 CUCV, one 105 mm howitzer, or six A-22 CDS containers; in aeromedical role, 24 litters and four attendants. Crew and passenger door forward of cabin on left side. Paratroop doors at rear of cargo compartment on left and right sides; emergency exit in latrine, forward cargo compartment on right side. Three emergency exit hatches: one overhead in flight compartment, two in cargo compartment roof, forward and aft of wing. Hydraulically operated rear-loading ramp and upward-opening door on underside of upswept rear fuselage can be opened in flight for airdrop operations. Six CDS containers (up to 1,000 kg; 2,205 lb each), or a single pallet of up to 5,000 kg (11,023 lb), can be airdropped from rear opening; paratroops jump from either cargo ramp and door opening or from rear side paratroop doors.

SYSTEMS: Windscreen and quarter-light panels de-iced and demisted electrically; hydraulic wipers and liquid rain repellent system provided for both windscreens. Entire aircraft pressurised and air conditioned. Normal anti-skid braking provided by No. 2 hydraulic system with maximum braking pressure of 117 bars (1,700 lb/sq in). Emergency braking, without anti-skid, by No. 1 hydraulic system or hydraulic hand pump. Multiple disc brakes by BF Goodrich. Redesigned and reinforced flap tracks allow flaps to be lowered at higher airspeeds. Separate oxygen system for aeromedical missions; 12 internal power receptacles in cargo compartment for auxiliary and medical equipment, including eight power receptacles rated at 200 V AC, 400 Hz and two double fuse receptacles: one 115 V AC 60 Hz and one 220 V AC 60 Hz.

AVIONICS: Standard communications equipment includes 7,000-channel UHF radio with satcom and Have Quick I/II secure speech capability; two 4,080-channel VHF AM/FM (AN/ARC-186) radios with secure voice capability; VHF FM (AN/ARC-201A SINCGARS) radio; 28,000-channel (AN/ARC-190) HF radio with secure voice, selcal, anti-jam, and automatic link establishment capability; crew intercom and PA system. Navigation equipment includes LTN-92 INS, with TAS computer, autopilot, flight director, two compasses, two vertical gyros, dual Tacan-VOR/ILS-DME, marker beacon receiver, ADF, two radio magnetic indicators, magnetic compass, radar altimeter, and horizontal situation indicator. Other avionics include IFF/SIF, Bendix/King 1400C weather/search radar, ground proximity warning system, cockpit voice recorder, emergency locator transmitter, and flight data recorder. Davtron 880 digital readout clock on pilot's and co-pilot's yoke in the cockpit.

EQUIPMENT: Loadmaster provided with crew oxygen regulator, quick don mask/goggles, portable oxygen bottle, emergency escape breathing device and lifevest. Additional mission handling equipment includes loading winch, pry bar, anti-tilt jack, pallet extenders, and auxiliary ground loading ramps.

WEIGHTS AND LOADINGS:

Weight empty 16,103 kg (35,500 lb)

CTAS (Boeing) E-6B upgraded E-6A with dual TACAMO and ABNCP capabilities *(CTAS)*

1995

Max payload	6,740 kg (14,859 lb)
Max useful load	8,472 kg (18,678 lb)
Max fuel weight: JP4	9,348 kg (20,608 lb)
JP5	9,803 kg (21,612 lb)
Max T-O weight	25,800 kg (56,878 lb)
Max landing weight	24,200 kg (53,352 lb)

PERFORMANCE:

Max cruising speed	250 kts (463 km/h; 288 mph)
Airdrop speed	110-140 kts (204-259 km/h; 127-161 mph)
Service ceiling	6,706 m (22,000 ft)
T-O run (1,814 kg; 4,000 lb cargo, ISA + 15°C; 59°F)	457 m (1,500 ft)
Operating range	300 n miles (555 km; 345 miles)
Ferry range	1,500 n miles (2,779 km; 1,727 miles)

UPDATED

CTAS (BOEING) E-6B COMMAND POST MODIFICATION PROGRAMME

TYPE: Strategic communications relay aircraft upgrade.

PROGRAMME: Contract to provide performance improvements and avionics enhancements for the E-6A take charge and move out (TACAMO) aircraft. The E-6B programme has been established to upgrade TACAMO operational capabilities and cross-deck a subset of the strategic command's (STRATCOM's) EC-135 airborne command post (ABNCP) equipment to the E-6A aircraft. The modified aircraft (E-6B) will be capable of performing both the TACAMO and ABNCP missions.

DESIGN FEATURES: Major subsystems integrated into the E-6B aircraft are: airborne launch control system (ALCS), UHF C3 radio subsystem, digital airborne intercommunication switching system (DAISS), STU IIIR, secure facsimile, upgraded intercommunications system (ICS), MILSTAR airborne terminal system, mission computer system (MCS), UHF SATCOM receive system upgrade, time/frequency standards distribution system (T/FSDS), high-power transmitting set (HPTS) and an MIL-STD-1553B databus system.

NEW ENTRY

COLEMILL

COLEMILL ENTERPRISES INC

PO Box 60627, Cornelia Fort Air Park, Nashville, Tennessee 37206
Telephone: 1 (615) 226 4256
Fax: 1 (615) 226 4702
PRESIDENT: Ernest Colbert

Colemill specialises in performance improvements for single- and twin-engined aircraft.

VERIFIED

Colemill Panther Navajo re-engined Piper Navajo

1994

Colemill Executive 600 re-engined Cessna 310

1994

COLEMILL PANTHER NAVAJO

TYPE: Re-engined Piper Navajo and C/R Navajo.

VERSIONS: Optional Zip Tip winglets included in conversion or retrofitted under Supplemental Type Certificate awarded Summer 1982. Winglets improve stability at low airspeeds down to stall and add 4 to 9 knots (8 to 16 km/h; 5 to 10 mph) when cruising at between 45 and 65 per cent power at 3,960 to 7,620 m (13,000 to 25,000 ft), resulting in fuel savings.

DESIGN FEATURES: Basic conversion with more powerful engines, new cowlings, continuous running fuel pumps, digital fuel totaliser, wingtip-mounted landing lights, heavy-duty brakes and extended wingtips takes 10 to 14 days. Optional Zip Tip winglets increase wingspan to 13.15 m (43 ft 2 in) and wing area by 0.56 m² (6 sq ft).

Data below refer to Panther Navajo:

LANDING GEAR: Cleveland four-spot heavy-duty disc brakes.

POWER PLANT: Two 261 kW (350 hp) Textron Lycoming TIO-540-J2BD turbocharged engines, each driving a Hartzell four-blade constant-speed fully feathering metal propeller with Q-tips. Pressurised magnetos, Woodward propeller governors, synchrophasers and unfeathering accumulators standard. Fuel system as for basic Navajo, except for the addition of continuous running electrically operated fuel pumps.

EQUIPMENT: Generally as for standard Navajo, but existing fuel flow gauges are replaced by a Shadin Digiflow fuel management computer giving digital readout of fuel remaining/fuel consumed. Supplemental wingtip landing lights can be operated independently of the standard nosewheel-mounted landing light, prior to lowering of landing gear.

DIMENSIONS, EXTERNAL:

Wing span (excl winglets)	13.00 m (42 ft 8 in)

WEIGHTS AND LOADINGS:

Max power loading	5.65 kg/kW (9.3 lb/hp)

PERFORMANCE (at max T-O weight):

Max level speed	269 kts (498 km/h; 309 mph)
Max cruising speed, 75% power at optimum altitude	248 kts (459 km/h; 285 mph)
Cruising speed, 65% power: at 7,315 m (24,000 ft)	235 kts (435 km/h; 270 mph)
at 3,660 m (12,000 ft)	206 kts (381 km/h; 237 mph)
Max rate of climb at S/L	610 m (2,000 ft)/min
Rate of climb at S/L, one engine out	122 m (400 ft)/min
Short-field T-O run	229 m (750 ft)
T-O to 15 m (50 ft)	458 m (1,500 ft)
Landing from 15 m (50 ft)	427 m (1,400 ft)

VERIFIED

COLEMILL PANTHER II

TYPE: Re-engined Piper Chieftain.

DESIGN FEATURES: Conversion includes new 261 kW (350 hp) Textron Lycoming TIO-540-J2BD and LTIO-540-J2BD (opposite rotation) turbocharged engines, driving four-blade constant-speed fully feathering metal propellers with Q-tips; digital fuel management computer; Woodward propeller governors and synchrophase; and Cleveland four-spot heavy-duty brakes. Zip Tip winglets optional.

VERIFIED

COLEMILL EXECUTIVE 600

TYPE: Re-engined Cessna 310, Models F to Q.

DESIGN FEATURES: Conversion includes new 224 kW (300 hp) Teledyne Continental IO-520-E flat-six engines, driving McCauley three-blade propellers. Dimensions unchanged; empty weight increased by about 14 kg (30 lb).

WEIGHTS AND LOADINGS:

Max T-O weight	2,358 kg (5,200 lb)
Max power loading	5.26 kg/kW (8.67 lb/hp)

PERFORMANCE (at max T-O weight):

Max cruising speed, 75% power	205 kts (379 km/h; 236 mph)
Cruising speed, 65% power	195 kts (361 km/h; 224 mph)
Stalling speed, wheels and flaps down	64 kts (119 km/h; 74 mph)
Max rate of climb at S/L	762 m (2,500 ft)/min
Service ceiling	5,940 m (19,500 ft)
T-O to 15 m (50 ft)	518 m (1,700 ft)
Landing from 15 m (50 ft)	unchanged
Range with max fuel, 45 min reserves	1,050 n miles (1,944 km; 1,208 miles)

VERIFIED

Colemill Panther II re-engined Piper Chieftain

1994

COLEMILL PRESIDENT 600

TYPE: Re-engined Beechcraft B55 Baron.

CUSTOMERS: About 250 President 600 conversions delivered.

DESIGN FEATURES: Conversion includes 224 kW (300 hp) Teledyne Continental IO-520-E flat-six engines, driving three-blade propellers. Dimensions unchanged; empty weight increased by about 14 kg (30 lb).

WEIGHTS AND LOADINGS:

Max T-O weight	2,313 kg (5,100 lb)
Max power loading	5.16 kg/kW (8.5 lb/hp)

PERFORMANCE (at max T-O weight):

Max cruising speed, 75% power	203 kts (376 km/h; 233 mph)
Cruising speed, 65% power	193 kts (357 km/h; 222 mph)
Stalling speed, wheels and flaps down	66 kts (123 km/h; 76 mph)

Max rate of climb at S/L	823 m (2,700 ft)/min
Service ceiling	5,940 m (19,500 ft)
T-O to 15 m (50 ft)	497 m (1,631 ft)
Landing from 15 m (50 ft)	unchanged
Range with max fuel, 45 min reserves	1,050 n miles (1,944 km; 1,208 miles)

VERIFIED

Colemill President 600 re-engined Beechcraft B55 Baron

1994

COLEMILL FOXSTAR BARON

TYPE: Re-engined Beechcraft 55 or 58 Baron.

DESIGN FEATURES: Conversion includes new 224 kW (300 hp) Teledyne Continental IO-550-C engines, driving Hartzell Sabre Blade four-blade Q-tip propellers; Woodward propeller governors and synchrophasers; Shadin Digiflow fuel computer; Zip Tip winglets; 60 A alternators. Foxstar conversion FAA STC approved for all Model C55, D55, E55 and Model 58 Barons.

WEIGHTS AND LOADINGS:

Max T-O weight	2,449 kg (5,400 lb)
Max power loading	5.47 kg/kW (9.00 lb/hp)

PERFORMANCE (at max T-O weight):

Max cruising speed, 75% power	205 kts (380 km/h; 236 mph)
Cruising speed, 65% power	200 kts (371 km/h; 230 mph)
Stalling speed, landing gear and flaps down	74 kts (137 km/h; 85 mph)
Max rate of climb at S/L	561 m (1,840 ft)/min
Service ceiling	6,400 m (21,000 ft)
T-O to 15 m (50 ft)	610 m (2,000 ft)
Landing from 15 m (50 ft)	734 m (2,410 ft)
Range with max fuel, 45 min reserves	1,131 n miles (2,096 km; 1,302 miles)

VERIFIED

Colemill Foxstar Baron re-engined Beechcraft Baron

1994

COLEMILL STARFIRE BONANZA

TYPE: Re-engined Beechcraft Bonanza.

DESIGN FEATURES: FAA STC approved for all Beechcraft Model C33A, E33A, F33A, S35, V35A, V35B and A36 Bonanzas. Conversion includes 224 kW (300 hp) Teledyne Continental IO-550-B engine, driving Hartzell Sabre Blade four-blade Q-tip propeller; Woodward propeller governor; Shadin Digiflow fuel computer/totaliser; Zip Tip winglets; 60 A alternator.

PERFORMANCE:

Cruising speed	176 kts (326 km/h; 203 mph)
Max rate of climb at S/L	369 m (1,210 ft)/min
T-O run	296 m (971 ft)
T-O to 15 m (50 ft)	583 m (1,912 ft)

VERIFIED

Colemill Starfire Bonanza re-engined Beechcraft Bonanza

1994

COMTRAN

COMTRAN INTERNATIONAL INC

1770 Sky Place Boulevard, San Antonio, Texas 78216
Telephone: 1 (210) 821 6301
Fax: 1 (210) 822 7766
Telex: 767438 COMTRAN UD
PRESIDENT: Douglas Jaffe

Comtran International Inc is part of the Jaffe Group of companies. Activities are in the fields of design, development and installation of refurbishing and conversion programmes for Executive, VIP and Head of State aircraft. Work includes installation of improved interiors, avionics and communications systems as well as Stage II and III hush kits and maintenance work.

UPDATED

Comtran head of state conversion of a Boeing 707

1994

COMTRAN SUPER Q

TYPE: Passenger aircraft hush kit conversion.

DESIGN FEATURES: Includes engine noise reduction and cabin/flight deck updating. Comtran Q-707 nacelle hush kits include Rohr Industries DynaRohr liners with extended intake and fan exhaust ducts which meet FAR 36 Stage 2 and ICAO Annex 16 Chapter 2 requirements; 100 EPNdB footprint reduced from 5.6 n miles (10.4 km; 6.4 miles) for standard Boeing 707 to 2.8 n miles (5.2 km; 3.2 miles), without affecting anti-icing and thrust reversal.

Cabin has wide-body styling; lightweight seats; overhead bins; cabin entertainment system with video monitors and seat-mounted controls; cabin soundproofing; cabin headliner with 'wash' lighting; emergency escape lighting; overhauled forward and aft galleys and toilets; new liferafts, lifejackets and emergency escape chutes; new floor and side panel liners in baggage compartment. Full range of interiors available with up to 186 passengers in four-class layout.

Cabin meets ICAO, FAA and European S-2000 fire, safety and toxicity standards by means of fireblocking materials in all seats, non-toxic polycarbonate composite materials for overhead panels and seat trim, plus supplementary emergency floor lighting.

Flight deck consists of upgraded Collins avionics including FD109Y flight directors, long-range navigation systems, colour weather radar, new cockpit audio, and new wiring and switching as required.

PERFORMANCE: For typical mission with full passengers, 65,075 kg (145,450 lb) fuel including 9,072 kg (20,000 lb) reserve, maximum brake release weight 150,955 kg (332,800 lb), cruising at Mach 0.8 at 10,670 m (35,000 ft), estimated range is 4,365 n miles (8,089 km; 5,026 miles).

VERIFIED

COMTRAN (BOEING) 707 HEAD OF STATE CONVERSION

TYPE: Large passenger jet VIP conversion.

PROGRAMME: Conversion of Boeing 707 for VIP transport.

DESIGN FEATURES: Interior refurbishment consisting of: aft sitting room, wet bar/galley, main lounge/dining area, master state room, master bedroom and master bathroom. The conversion also includes installation of Stage II or Stage III hush kits.

POWER PLANT: Three Pratt & Whitney JT3D-1MC7 and one Pratt & Whitney JT3D-1MC6. Optional Stage II or Stage II hush kits.

AVIONICS: Installation includes the following avionics systems: dual Collins FD110 with inertial navigation systems interface (INSI), dual Collins 618M-2D VHF communications, dual Collins 51 RV-4 VHF navigation, dual Collins 621A-6A ATC transponders, dual Collins 860E-5 distance measuring equipment (DME), dual Collins 51Y-7 automatic direction-finders (ADF), Delco Carousel IV inertial navigation system (INS), Canadian Marconi 771 VLF/Omega, dual Collins 618T-2 HF communications, Color Sperry AVQ 90 weather radar, Bendix ALA-51A radio altimeter, Sundstrand Mk II ground proximity warning system, Motorola NA-126 long-range navigation, dual Sperry C-11B compass systems and Bendix PB autopilot.

EQUIPMENT: Installation includes: APU, two video entertainment systems with four colour monitors, stereo system with speakers and headsets, galley equipped with both microwave and convection ovens as well as self-contained boarding stairs.

DIMENSIONS, EXTERNAL:

Wing span	39.9 m (130 ft 10 in)
Length overall	41.2 m (135 ft 1 in)
Tail height	12.62 m (41 ft 5 in)

WEIGHTS AND LOADINGS:

Taxi weight	117,028 kg (258,000 lb)
T-O weight	117,028 kg (258,000 lb)
Landing weight	86,184 kg (190,000 lb)

PERFORMANCE:

Max cruise speed	515 kts (954 km/h; 593 mph)
Long-range cruise	480 kts (890 km/h; 553 mph)
Max operating altitude	12,802 m (42,000 ft)
Range	4,400 n miles (8,148 km; 5,068 miles)

VERIFIED

Comtran head of state conversion of Boeing 707 showing the main lounge/dining area
1994

DEE HOWARD

THE DEE HOWARD COMPANY (Subsidiary of Alenia)

9610 John Saunders Road, International Airport, PO Box 469001, San Antonio, Texas 78246-9001
Telephone: 1 (210) 821 4071
Fax: 1 (210) 821 4008
PRESIDENT AND CEO: Mr P. L. Ughi
VICE-PRESIDENT, ENGINEERING: David White
VICE-PRESIDENT, BUSINESS DEVELOPMENT: Armando Sassoli
DIRECTOR OF MARKETING AND SALES: Phil Gaudet

The Dee Howard Company is a subsidiary of the Italian aerospace company, Alenia. Main activities are in the fields of aircraft engineering and modifications as well as aircraft re-engineering, interior refurbishment (head of state aircraft), avionics upgrades and heavy maintenance. Further activities include the installation of thrust reversers on Falcon Jets.

UPDATED

Dee Howard (Boeing) 727-100 re-engined with Rolls-Royce Tay 651 turbofan engines
1995

DEE HOWARD (BOEING) 727 UPGRADE

TYPE: Large passenger aircraft avionics and engine upgrade.

PROGRAMME: Contract awarded by United Parcel Service to upgrade the avionics and re-engine 44 Boeing 727-100s. First aircraft flew in April 1992. FAA approval for November 1992. Redelivered 35 aircraft by December 1994. Contract completed August 1995.

CUSTOMERS: UPS (44) negotiating options. Dee Howard also completed the same upgrade for one aircraft belonging to a corporate customer and is currently negotiating contracts with new corporate customers.

DESIGN FEATURES: Installation of Collins EFIS and new air data system as part of a cockpit modernisation programme.

POWER PLANT: Retrofit of 67.17 kN (15,100 lb st) Rolls-Royce Tay 651 turbofans. Fuel capacity 24,252 kg (53,466 lb).

WEIGHTS AND LOADINGS:

Max T-O weight	76,885 kg (169,500 lb)
Max landing weight	64,638 kg (142,500 lb)
Max zero-fuel weight	59,875 kg (132,000 lb)

UPDATED

DEE HOWARD (DOUGLAS) DC-8 AVIONICS UPGRADE

TYPE: Large passenger aircraft avionics upgrade.

PROGRAMME: Contract to upgrade 49 Douglas DC-8-70 cockpits for United Parcel Services. First Douglas DC-8 flew in November 1990. Eight converted in 1991 and then 10 each year until contract completion. Contract completed in 1995.

CUSTOMERS: UPS (49) Dee Howard is negotiating with two other customers for this upgrade.

DESIGN FEATURES: Installation includes Collins EFIS and inertial navigation system (INS).

UPDATED

Dee Howard (Douglas) DC-8 upgraded cockpit
1994

E-SYSTEMS

E-SYSTEMS GREENVILLE DIVISION

PO Box 6056, Greenville, Texas 75403-6056
Telephone: 1 (903) 455 3450
Fax: 1 (903) 457 7674
Telex: 6829260 ESYSUW

The origins of the E-Systems Greenville Division can be traced back to 1945 when former North American Aviation employees led by Robert McCulloch and Bert Howard formed the Texas Engineering and Manufacturing Company. In 1952 the company changed its name to TEMCO Aircraft Corporation. In 1960 the corporation merged with Ling-Altec to form Ling-Temco Electronics Inc. This company acquired the Chance Vought Corporation and changed its name to Ling-Temco-Vought (LTV) Incorporated. In 1965 LTV reorganised and the former Aerosystems Division became LTV-Electrosystems. Later in 1965 LTV-Electrosystems acquired the Continental Electronics Company in Dallas and the LTV Military Electronics Division in Garland within this new organisation the Aerosystems Division became the Greenville Division. In 1972 LTV-Electrosystems became fully independent from LTV Inc and changed its name to E-Systems. Main activities of the Greenville Division include the design, development, manufacture, and installation of integrated airborne systems, as well as aircraft modification, repair and maintenance. The division is an FAA approved designated alteration station (DAS). The division has a 2,440 m (8,000 ft) runway and over 60,385 m² (650,000 sq ft) of covered facilities.

NEW ENTRY

E-SYSTEMS FLIGHT CALIBRATION CONVERSION

TYPE: Multimission turboprop and business jet calibration conversion.

PROGRAMME: Awarded contract by FAA for the supply of two additional flight inspection aircraft (one Learjet 60 and one Canadair Challenger 601-3A). This is part of a $400 million contract to supply the FAA with five Learjet 60 and two Challenger 601-3A flight inspection aircraft.

CUSTOMER: FAA (2).

DESIGN FEATURES: Systems consist of operator console and associated equipment racks. System interfaced with cockpit avionics and flight controls providing capability for either manual or automatic tracking of any desired flight plan.

AVIONICS: Console consists of: display, keyboard, AFIS controls, intercom controls, operator oxygen controls, radio magnetic indicator and marker beacon indicator. Equipment racks consist of Tacan bearing receiver, AFIS DME, printer, oscilloscope, controls, spectrum analyser, spare LRU storage, radio altimeter, data transfer unit. Flight data, time meter, TCAS receiver, HF communications radio transmitter, HF communications receiver, Loran receiver and signal acquisition unit. Cockpit avionics include: flight control system, radio beacon, frequency controller, cockpit radio magnetic indicator, VOR/ILSMKR receivers, distance measuring equipment (DME), MLS and Loran.

NEW ENTRY

EXCALIBUR

EXCALIBUR AVIATION COMPANY

8337 Mission Road, San Antonio, Texas 78214
Telephone: 1 (210) 927 6201
Fax: 1 (210) 927 6287
PRESIDENT: Michael M. Davis

Beechcraft Queen Air 65, A65 and 80 airframes are modified by Excalibur as Queenaire 800 and Queen Air A80s; B80 as Queenaire 8800. Conversion offers improved reliability, speed range and reduced operating costs.

VERIFIED

Excalibur Queenaire 800 re-engined Beechcraft Queen Air 80 *1992*

EXCALIBUR QUEENAIRE 800 and 8800

TYPE: Re-engined Beech Queen Air 65 and 80.

PROGRAMME: By August 1995 155 conversions had been completed.

CUSTOMERS: Civil operators, notably in South America; more than 50 US Army U-8Fs modified for US National Guard Bureau.

DESIGN FEATURES: Installation of two 298 kW (400 hp) Textron Lycoming IO-720-A1B eight-cylinder engines; Hartzell three-blade constant-speed fully feathering metal propeller; new engine mountings; new exhaust system; new low drag engine nacelles; new (or zero time overhauled and certificated) accessories; Excalibur fully enclosed wheel well doors.

WEIGHTS AND LOADINGS (A: Queenaire 800, B: Queenaire 8800):

Weight empty, equipped (average):	
A	2,449 kg (5,400 lb)
B	2,631 kg (5,800 lb)
Max T-O weight: A	3,628 kg (8,000 lb)
B	3,991 kg (8,800 lb)
Max landing weight: A	3,447 kg (7,600 lb)
B	3,792 kg (8,360 lb)
Max power loading: A	6.09 kg/kW (10.00 lb/hp)
B	6.70 kg/kW (11.00 lb/hp)

PERFORMANCE (at max T-O weight):

Cruising speed:	
75% power: A, B at 2,530 m (8,300 ft)	201 kts (372 km/h; 231 mph)
65% power: A, B at 3,050 m (10,000 ft)	195 kts (362 km/h; 225 mph)
45% power: A, B at 3,050 m (10,000 ft)	172 kts (319 km/h; 198 mph)
Stalling speed:	
gear and flaps up: A	80 kts (148 km/h; 92 mph)
B	86 kts (160 km/h; 99 mph)
gear and flaps down: A	68 kts (126 km/h; 78 mph)
B	70 kts (129 km/h; 80 mph)
Max rate of climb at S/L: A	468 m (1,535 ft)/min
B	454 m (1,490 ft)/min
Rate of climb at S/L, one engine out:	
A	110 m (360 ft)/min
B	76 m (250 ft)/min
Service ceiling: A	6,005 m (19,700 ft)
B	5,700 m (18,700 ft)
Service ceiling, one engine out: A	3,595 m (11,800 ft)
B	3,110 m (10,200 ft)
T-O to 15 m (50 ft): A	520 m (1,706 ft)
B	625 m (2,050 ft)
Landing from 15 m (50 ft): A	663 m (2,176 ft)
B	747 m (2,450 ft)

Range with max fuel at 3,050 m (10,000 ft), with 113.5 l (30 US gallons; 20.8 Imp gallons) reserves:

A	1,322 n miles (2,451 km; 1,523 miles)
B	1,547 n miles (2,867 km; 1,782 miles)

UPDATED

FAIRCHILD

FAIRCHILD INDUSTRIES INC

PO Box 10803, 300 West Service Road, Chantilly, Virginia 22021
Telephone: 1 (703) 478 5800
Fax: 1 (703) 478 5920

Fairchild Industries Inc closed/divested its remaining aeroplane manufacturing operations in 1987, and was subsequently acquired by and became a subsidiary of Banner Industries Inc (subsequently renamed The Fairchild Corporation), whose principal business is the manufacture of aerospace fasteners.

UPDATED

Fairchild Republic A-10A Thunderbolt II with AGM-65A Maverick missile *1994*

FAIRCHILD REPUBLIC THUNDERBOLT II

USAF designation: A-10A/OA-10

TYPE: Single-seat close support aircraft.

PROGRAMME: Fairchild Republic and Northrop each built two prototypes for evaluation under the US Air Force's A-X programme, initiated in 1967, for a close support aircraft. The first Fairchild Republic prototype (71-1369), designated YA-10A, flew for the first time 10 May 1972. It was announced 18 January 1973 that Fairchild was the winner of the competitive evaluation of the prototypes, and received a contract for six A-10A DT and E aircraft, the first of which flew 15 February 1975.

The first flight by a production A-10A Thunderbolt II (75-00258) was made 21 October 1975. Purchase of a total of 739 aircraft was planned (including the six DT and E aircraft); but funding was terminated in 1983 after a total of 713 production A-10s had been ordered. Delivery was completed 20 March 1984. There were still 327 aircraft in service with the USAF, USAF Reserve and ANG in early 1994. The Thunderbolt II was used during the 1991 Gulf War.

Export versions of the A-10 were available as single-seat night attack and two-seat combat-ready trainer aircraft. Night capability is provided by the addition of a Westinghouse WX-50 radar, Texas Instruments AAR-42 FLIR, Litton LN-39 inertial navigation system, Honeywell APN-194 radar altimeter, AiResearch digital air data computer, Ferranti 105 laser rangefinder and Kaiser head-up display. It is expected that night/adverse weather capability can be improved with the addition of a LANTIRN (low-altitude navigation targeting infra-red for night) fire control pod.

The first combat-ready A-10A wing was the 345th Tactical Fighter Wing, based at Myrtle Beach, South Carolina, to which deliveries began in March 1977.

VERSIONS: **A-10A:** Initial version. Single-seat close support aircraft.

OA-10A: Forward observation version.

UPGRADES: **GE Aerospace:** Low-Altitude Safety and Targeting Enhancements (LASTE). Programme consists of a new computer and software integrated with existing aircraft systems to give A-10A Thunderbolt II enhanced accuracy for both gunnery and bombing.

OPERATORS: Versions of the A-10 are in service with the US armed forces (380).

Fairchild Republic A-10A Thunderbolt II single-seat, close-support aircraft *(Paul Tompkins)*

1995

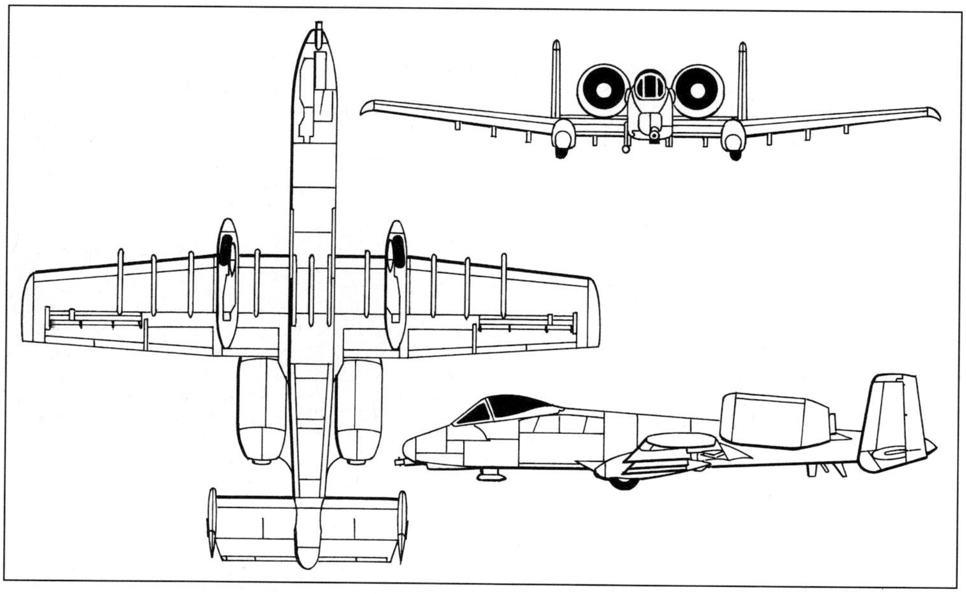

Fairchild Republic A-10A Thunderbolt II single-seat close support aircraft *(Jane's/Mike Keep)*

1994

DESIGN FEATURES: Cantilever low-wing monoplane, with wide chord, deep aerofoil section (NACA 6716 on centre-section and at start of outer panel, NACA 6713 at tip) to provide low wing loading. Incidence −1°. Dihedral 7° on outer panels.

FLYING CONTROLS: Wide span ailerons made up of dual upper and lower surfaces that separate to serve as airbrakes. Flaps, airbrakes and ailerons actuated hydraulically. Ailerons pilot-controlled by servo tab during manual reversion. Small leading-edge slat inboard of each mainwheel fairing. Redundant and armour-protected flight control system. Interchangeable elevators, each with an electrically operated trim tab. Rudders and elevators actuated hydraulically. Redundant and armour-protected flight control system.

STRUCTURE: Aluminium alloy three-spar structure, consisting of one-piece constant-chord centre-section and tapered outer panels with integrally stiffened skins and drooped (cambered) wingtips. Outer panel leading-edges and core of trailing-edges are of honeycomb sandwich. Four-point attachment of wings to fuselage, at front and rear spars. Two-segment, three-position trailing-edge slotted flaps, interchangeable right with left. The fuselage is a semi-monocoque structure of aluminium alloy (chiefly 2024 and 7075), with four main longerons, multiple frames, and lap-jointed and riveted skins. Built in front, centre and aft portions. Single-curvature components aft of nose portion, interchangeable right with left. Centre portion incorporates wing box carry-through structure. The tail unit is a cantilever aluminium alloy structure, with twin fins and interchangeable rudders mounted at the tips of constant-chord tailplane.

LANDING GEAR: Menasco retractable tricycle type with single wheel on each unit. All units retract forward, and have provision for emergency gravity extension. Interchangeable mainwheel units retract into non-structural pod fairings attached to the lower surface of the wings. When fully retracted approximately half of each wheel protrudes from the fairing. Steerable nosewheel is offset to starboard to clear firing barrel of gun. Mainwheels size 36 × 11, Type VII; nosewheel size 24 × 7.7-10, Type VII.

POWER PLANT: Two General Electric TF34-GE-100 high-bypass ratio turbofan engines, each rated at 40.3 kN (9,065 lb st), enclosed in separate pods, each pylon-mounted to the upper rear fuselage at a point approximately midway between the wing trailing-edges and the tailplane leading-edges. Fuel is contained in two tear-resistant and self-sealing cells in the fuselage, and two smaller, adjacent integral cells in the wing centre-section. Maximum internal fuel capacity 4,853 kg (10,700 lb). All fuel cells are internally filled with reticulated foam, and all fuel systems pipework is contained within the cells except for the feeds to the engines, which have self-sealing covers. Three 2,271 litre (600 US gallon; 499 Imp gallon) jettisonable auxiliary tanks can be carried on underwing and fuselage centreline pylons. Provision for in-flight refuelling using universal aerial refuelling receptacle slipway installation (UARRSI).

ACCOMMODATION: Single-seat enclosed cockpit, well forward of wings, with large transparent bubble canopy to provide all-round vision. Bulletproof windscreen. Canopy is hinged at rear and opens upward. Douglas ejection seat operable at speeds of 450 knots (834 km/h; 518 mph) down to zero speed at zero height. Entire cockpit structure is protected by an armoured 'bathtub' structure of titanium, capable of withstanding projectiles up to 23 mm calibre. Basic design work for a dual-control two-seat version was completed in the late 1970s.

SYSTEMS: Redundant control system incorporates two 207 bar (3,000 lb/sq in) primary hydraulic flight control systems, each powered by an engine-driven pump and a manual back-up. Hydraulic systems actuate flaps, flying control surfaces, landing gear, brakes and nosewheel steering. Two independent hydraulic motors, either of which is sufficient to sustain half-rate firing, supply drive for 30 mm gun barrel rotation. Electrical system includes two 30/40 kVA 115/200 V AC engine-driven generators and a standby battery and inverter. Auxiliary power unit. Environmental control system, using engine bleed air for cockpit pressurisation and air conditioning, pressurisation of pilot's *g* suit, windscreen anti-icing and rain clearance, fuel transfer, gun compartment purging, and other services.

ELECTRONICS AND EQUIPMENT: Head-up display giving airspeed, altitude, and dive angle; weapons delivery package with dual reticle optical sight for use in conjunction with underfuselage Pave Penny laser target seeker pod; target penetration aids; associated equipment for Maverick and other missile systems; IFF/SIF (AIMS); UHF/AM; VHF/FM; Tacan; UHF/DF; VOR/ILS; X-band transponder; all-altitude heading and attitude reference system (HARS); radar homing and warning (RHAW); secure voice communications; active or passive electronic countermeasures (ECM); armament control panel; and gun camera. Space provisions for HF/SSB, ILS/FDS, Loran C/D, ECM pod, chaff/flare dispenser and other 'growth' electronics and equipment.

ARMAMENT: General Electric GAU-8/A Avenger 30 mm seven-barrel cannon, mounted in nose with 2° depression and offset slightly to port so that as the barrels rotate the firing barrel is always on the aircraft's centreline. Gun and handling system for the linkless ammunitions are mechanically synchronised and driven by two motors fed from the aircraft's hydraulic system. The single drum magazine has a capacity of 1,350 rounds, and has a dual firing rate of either 2,100 or 4,200 rds/min. Four stores pylons under each wing (one inboard and three outboard of each mainwheel fairing), and three under fuselage, for maximum external load of 7.257 kg (16,000 lb). External load with full internal fuel is 5,482 kg (12,086 lb). The centreline pylon and the two flanking fuselage pylons cannot be occupied simultaneously. The centreline pylon has a capacity of 2,268 kg (5,000 lb); the two fuselage outer pylons and two centre-section underwing pylons 1,587 kg (3,500 lb) each; the two innermost outerwing pylons 1,134 kg (2,500 lb) each; and the four outermost wing pylons 453 kg (1,000 lb) each. These allow carriage of a wide range of stores, including twenty-eight 226 kg (500 lb) Mk 82 LDGP general purpose bombs; eight BLU-1 or BLU-27/B Rockeye II cluster bombs, sixteen CBU-52/71, ten AGM-65A Maverick missiles; Mk 82 and Mk 84 laser-guided bombs; Mk 84 electro-optically guided bombs; two SUU-23 pods; chaff or other jammer pods; or up to three drop tanks.

DIMENSIONS, EXTERNAL:

Wing span	17.53 m (57 ft 6 in)
Wing chord: at root	3.04 m (9 ft 11½ in)
mean	2.73 m (8 ft 11⅓ in)
at tip	1.99 m (6 ft 6⅔ in)
Wing aspect ratio	6.54
Length overall	16.26 m (53 ft 4 in)
Height overall	4.47 m (14 ft 8 in)
Tailplane span	5.74 m (18 ft 10 in)
Wheel track	5.25 m (17 ft 2½ in)
Wheelbase	5.40 m (17 ft 8¾ in)

AREAS:

Wings, gross	47.01 m² (506.0 sq ft)

Fairchild Republic A-10A Thunderbolt II of the USAF *(Peter J. Cooper)*

1994

Ailerons (total, incl tabs)	4.42 m² (47.54 sq ft)
Trailing-edge flaps (total)	7.99 m² (85.99 sq ft)
Leading-edge slats (total)	0.98 m² (10.56 sq ft)
Airbrakes (total)	8.06 m² (86.78 sq ft)
Fins (total)	7.80 m² (83.96 sq ft)
Rudders (total)	2.18 m² (23.50 sq ft)
Tailplane	8.31 m² (89.40 sq ft)
Elevators (total, incl tabs)	2.69 m² (29.00 sq ft)

WEIGHTS AND LOADINGS:

Manufacturer's empty weight	9,183 kg (20,246 lb)
Basic equipped weight, 'clean'	10,600 kg (23,370 lb)
Operating weight empty	10,710 kg (23,611 lb)
* Basic design weight, equipped	13,925 kg (30,700 lb)
**Forward airstrip weight	14,243 kg (31,400 lb)
Max T-O weight	21,500 kg (47,400 lb)
Max wing loading	449.88 kg/m² (92.14 lb/sq ft)
Max power loading	262.4 kg/kN (2.57 lb/lb st)
Thrust/weight ratio	0.6

* *incl six 500 lb bombs, 750 rds of ammunition and fuel for 300 n miles (55 km; 345 miles) with 20 min reserves*
***with four Mk 82 bombs*

PERFORMANCE (at max T-O weight except where indicated):

Never-exceed speed	450 kts (834 km/h; 518 mph)
Max combat speed at S/L, 'clean'	390 kts (722 km/h; 449 mph)
Combat speed at 1,525 m (5,000 ft), with six Mk 82 bombs	385 kts (713 km/h; 443 mph)
Cruising speed: at S/L	300 kts (555 km/h; 345 mph)
at 1,525 (5,000 ft)	342 kts (634 km/h; 394 mph)
Stabilised 45° dive speed below 2,440 m (8,000 ft), AUW of 15,932 kg (35,125 lb)	260 kts (481 km/h; 299 mph)
Max rate of climb at S/L at basic design weight	1,828 m (6,000 ft)/min
T-O distance: at max T-O weight	1,372 m (4,500 ft)
at forward airstrip weight	426 m (1,400 ft)
Landing distance: at max T-O weight	762 m (2,500 ft)
at forward airstrip weight	382 m (1,250 ft)
Operational radius:	
close air support and escort, 2 h loiter, 20 min reserves	250 n miles (463 km; 288 miles)
reconnaissance	400 n miles (740 km; 460 miles)
deep strike	540 n miles (1,000 km; 620 miles)
Ferry range, zero wind	2,300 n miles (4,200 km; 2,647 miles)

UPDATED

GENERAL DYNAMICS

GENERAL DYNAMICS CONVAIR DIVISION

PO Box 85377, San Diego, California 92186
Telephone: 1 (619) 542 6750
Fax: 1 (619) 542 7801
GENERAL MANAGER: Arthur J. Veitch
PUBLIC AFFAIRS REPRESENTATIVE: Patricia A. Gayton

Convair Division has approximately 1,800 employees, responsible for manufacturing the fuselages for McDonnell Douglas' MD-11 jetliner.

UPDATED

CONVAIR-LINER 340

TYPE: Twin-engined medium-range passenger aircraft.

PROGRAMME: Although based on the Model 240, the Convair-Liner 340 has a greater wing span and area, a longer fuselage, more powerful engines, greater all-up weight and many interior design improvements.

The fuselage of the 340 is the same diameter as the 240 but is 1.37 m (54 in) longer, achieved by inserting a 0.40 m (16 in) section forward and a 0.96 m (38 in) section aft of the wings. This extra length permits the introduction of four additional seats, or it can be used for cargo purpose.

The first Convair 340 flew 5 October 1951, and the first delivery to United Air Lines was made 31 March 1952. 209 commercial 340s plus 102 military versions built in total.

UPGRADES: **Convair:** 600/640. See separate entry.

Kelowna Aircraft Group: CV5800. See separate entry.

Super 580 Aircraft Company: Super 580. See separate entry.

DESIGN FEATURES: Low-wing cantilever monoplane. Wing aspect ratio 12. Mean aerodynamic chord 2.90 m (9 ft 6.3 in). The tail unit is a cantilever monoplane type.

FLYING CONTROLS: Rudder and elevators fitted with trim and servo control tabs.

STRUCTURE: The wing is an all-metal structure with stressed skin. The fuselage is an all-metal circular-section structure with stressed skin. The tail unit is an all-metal cantilever structure.

SYSTEMS: Thermal anti-icing on leading-edges.

LANDING GEAR: Retractable tricycle type. All three units fitted with twin wheels. Steerable nosewheel with the two wheels keyed to common axle so that wheel and axle rotate as a unit to eliminate shimmy and need for damper. Both struts and wheels designed for installation of Westinghouse Decelostat anti-wheel-skid units. Wheel track 7.62 m (25 ft).

POWER PLANT: Two Pratt & Whitney R-2800-CB16 18-cylinder radial air-cooled engines, each rated at 1,342 kW (1,800 hp) at 2,590 m (8,500 ft) in low blower and 1,268 kW (1,700 hp) at 4,240 m (14,500 ft) in high blower and with 1,790 kW (2,400 hp) available for take-off (with water injection). Engines mounted in aerodynamically clean nacelles incorporating what is known as ''aspirated cooling''. Air after passing around cylinder blocks enters a venturi section into which the exhaust gases are also ejected, the effect being to increase flow of cooling air. Air/exhaust mixture in a ratio of about 5/1 is ejected via twin tailpipes above rear end of nacelle. Exhaust gas augmentation is claimed to give an increase in speed of about 8 to 10.4 knots (16 to 19.2 km/h; 10 to 12 mph). Hamilton Standard Hydromatic constant-speed feathering and reversing airscrews, diameter 4.11 m (13 ft 6 in). Integral fuel tanks outboard of engine nacelles. Maximum fuel capacity 6,615 litres (1,750 US gallons; 1,458 Imp gallons). Aircraft delivered to Hawaiian Airlines were fitted with Pratt & Whitney R-2800-CB3 engines as well as three aircraft which were originally to be delivered to Northeast Airlines. Aircraft could also be equipped with CB-17 engines. Provision for 6,548 litres (1,730 US gallons; 1,442 Imp gallons).

ACCOMMODATION: Crew of three/four. Cabin seats 44 passengers in pairs on each side of central aisle. Integral self-contained stairway located forward of wings on port side. Cabin is pressurised and both sound and air conditioned. Radiant-wall heating and refrigeration maintains constant cabin temperature in the air and on the ground. Cargo compartment fore and aft of cabin, and below cabin floor forward of wings, with all access doors on starboard side.

DIMENSIONS, EXTERNAL:

Wing span	32.12 m (105 ft 4 in)
Length	24.14 m (79 ft 2 in)
Height (over tail)	8.59 m (28 ft 2 in)

AREAS:

Wing area	85.5 m² (920 sq ft)

WEIGHTS AND LOADINGS:

Weight empty	13,494 kg (29,750 lb)
Max gross T-O weight: standard	21,338 kg (47,000 lb)
with SB 340-144B	22,271 kg (49,200 lb)
with SB 340-239	22,543 kg (49,700 lb)
Max landing weight	21,092 kg (46,500 lb)
Max T-O wing loading	249.3 kg/m² (51.1 lb/sq ft)
Max T-O power loading	4.44 kg/hp (9.8 lb/hp)

PERFORMANCE:

Cruising speed (895 kW; 1,200 hp per engine) at 5,490 m (18,000 ft)	242 kts (448 km/h; 284 mph)
Range with 172 n miles (320 km; 200 miles) plus 45 min reserve; 895 kW (1,200 hp) per engine; 8.6 kts (16 km/h; 10 mph) head wind; 44 passengers and baggage; 4,373 l (1,157 US gallons; 964 Imp gallons) of fuel	1,740 n miles (3,225 km; 2,015 miles)
Max CAR operating height with one engine inoperative	2,745 m (9,000 ft)
Required CAR runway length for T-O at max T-O weight	1,425 m (4,675 ft)
Required CAR runway length for landing S/L at max landing weight	1,370 m (4,500 ft)

UPDATED

CONVAIR 440 METROPOLITAN

TYPE: Twin-engined medium-range Airliner.

PROGRAMME: The Model 440 is a development of the 340 with modifications to increase speed (by about 4.3 knots; 8 km/h; 5 mph) and to reduce the noise level in the cabin. To improve performance the engine cowling, baffles and the 'aspirated cooling' exhaust have been redesigned and aileron and flap seals have been fitted.

The cabin is provided with improved soundproofing and special inner window assemblies have been installed for the first eight rows of seats as an additional means to suppress engine, propeller and exhaust noise.

UPGRADES: **General Dynamics:** Convair 600/640. See separate entry.

Kelowna Flightcraft Group: CV5800. See separate entry.

Super 580 Aircraft Company: Super 580. See separate entry.

DESIGN FEATURES: Low-wing cantilever monoplane. NACA 63,4-120 at root, NACA 63,4-419 at wing STA 8 and NACA 63_2-415 at the tip. Wing section. Aspect ratio 12. Chord (mean) 2.90 m (9 ft 6.3 in). Dihedral 4° 50′ on centre wings; 6° 30′ on outer wings. Incidence 4°. The tail unit is a cantilever monoplane type.

Convair 340 twin-engined medium-range passenger aircraft

Convair 440 Metropolitan twin-engine medium-range passenger aircraft

1994

FLYING CONTROLS: All-metal aluminium alloy ailerons and Fowler flaps. Rudder and elevators fitted with trim and servo-control tabs.

STRUCTURE: All-metal stressed skin wing structure. The fuselage is an aluminium alloy circular-section structure with stressed skin covering.

SYSTEMS: Thermal anti-icing on leading-edges.

LANDING GEAR: Retractable tricycle type. All three units fitted with twin wheels. Steerable nosewheel. Oleo-pneumatic shock-absorber struts. Wheel track 7.62 m (25 ft).

POWER PLANT: Two Pratt & Whitney R-2800-CB-16 18-cylinder radial air-cooled engines, each rated at 1,342 kW (1,800 hp) at 2,590 m (8,500 ft) in low blower and 1,268 kW (1,700 hp) at 4,240 m (14,500 ft) in high blower, and with 1,790 kW (2,400 hp) available for take-off, with water injection, or two Pratt & Whitney R-2800-CB-17 engines, each rated at 1,417 kW (1,900 hp) at 2,135 m (7,000 ft) in low blower and 1,305 kW (1,750 hp) at 4,115 m (13,500 ft) in high blower, and with 1,864 kW (2,500 hp) available for take-off, with water injection. Three-blade Hamilton Standard Hydromatic automatic fully feathering and reversing airscrews, diameter 4.11 m (13 ft 6 in). Integral fuel tanks outboard of engine nacelles, with maximum capacity of 6,548 litres (1,730 US gallons; 1,442 Imp gallons). Oil capacity 227 litres (60 US gallons; 50 Imp gallons).

ACCOMMODATION: Crew of three/four. Cabin seats 44 or 52 passengers in pairs on each side of a central aisle. Integral self-contained stairway located forward of wings on port side. Cabin is pressurised, with radiant-wall heating and refrigeration to maintain a constant cabin temperature in the air and on the ground. Cargo compartments fore and aft of cabin and below cabin floor forward of wings.

DIMENSIONS, EXTERNAL:

Wing span	32.12 m (105 ft 4 in)
Length	24.14 m (79 ft 2 in)
Length with wx radar	24.84 m (81 ft 6 in)
Height	8.59 m (28 ft 2 in)

WEIGHTS AND LOADINGS (A: CB-16 engines, B: CB-17 engines):

Weight empty: A, B	14,199 kg (31,305 lb)
Max T-O weight: A	21,772 kg (48,000 lb)
B	22,271 kg (49,100 lb)
with SB 440-68	22,543 kg (49,700 lb)
Max landing weight: A, B	21,614 kg (47,650 lb)
Max wing loading: A	254.7 kg/m² (52.2 lb/sq ft)
B	260.6 kg/m² (53.4 lb/sq ft)
Max power loading: A	4.35 kg/hp (9.6 lb/hp)
B	4.45 kg/hp (9.82 lb/hp)

PERFORMANCE:

Max speed: A	268 kts (497 km/h; 309 mph)
B	269 kts (499 km/h; 310 mph)
Average cruising speed (895 kW; 1,200 hp per engine) at 6,100 m (20,000 ft) and AUW of 20,412 kg (45,000 lb): A, B	251 kts (465 km/h; 289 mph)
Stalling speed at 21,092 kg (46,500 lb) AUW: A, B	74 kts (137 km/h; 85 mph)
Range with 680 kg (1,500 lb) fuel reserve, 895 kW (1,200 hp) per engine, zero wind, 20,412 kg (45,000 lb) AUW at 6,100 m (20,000 ft): A	904 n miles (1,674 km; 1,040 miles)
B	1,130 n miles (2,092 km; 1,300 miles)
Rate of climb at S/L: A	363 m (1,192 ft)/min
B	384 m (1,260 ft)/min
Service ceiling: A	7,770 m (25,500 ft)
B	7,590 m (24,900 ft)
Max operating ceiling: A, B	7,010 m (23,000 ft)
Max CAR operating height with one engine inoperative at 20,412 kg (45,000 lb) AUW: A, B	2,895 m (9,500 ft)
Required CAR runway length for T-O at S/L at max T-O weight: A	1,503 m (4,930 ft)
B	1,525 m (5,000 ft)
Required CAR runway length for landing at S/L at 21,092 kg (46,500 lb) AUW: A	1,213 m (3,980 ft)
B	1,222 m (4,010 ft)

UPDATED

CONVAIR 600/640

TYPE: Twin-engined, medium-range transport turboprop upgrade.

PROGRAMME: The Convair 600 and 640 are, retrospectively, turboprop conversions of the twin-engined Convair-Liner 240 and 340/440 transports, developed as a joint undertaking by Rolls-Royce, UK and the Convair division of General Dynamics.

Conversion to Convair 600/640 standard involved replacing the original piston-engines with two 2,051 kW (2,750 shp) Rolls-Royce Dart RDa.10 or Mk 542-4/4K turboprops, driving four-blade Dowty Rotol type R-245/4-40-4.5/13 propellers, diameter 3.96 m (13 ft 0 in) for the Convair 600 and R.249/4-40-4.5/17 4.06 m (13 ft 4 in) for Convair 640, and the installation of new gearboxes, nacelles, combustion heaters, cabin air compressor and related cockpit instruments. Customers chose between buying the kits to carry out modifications themselves or have the work done by an approved modification centre. Interior modifications, to seat up to 56 persons in Convair 640 aircraft converted from Series 340/440 aircraft.

In passenger versions of the Convair 640 new seating is available which can be folded quickly against the cabin walls to accommodate cargo.

The Convair 600 has a standard fuel capacity of 3,785 litres (1,000 US gallons; 833 Imp gallons) and maximum capacity of 7,866 litres (1,550 US gallons; 1,292 Imp gallons) with long-range tanks. Capacity for 640 of 6,547 litres (1,730 US gallons; 1,442 Imp gallons).

Included in the modification kit was an AiResearch GTP30-95 APU, driving a 500A DC generator and cabin air compressor for ground air supply and both ground and in-flight electrical power. An enlarged cargo door can also be provided.

The first Convair 600 flew for the first time 20 May 1965 and received FAA supplemental Type Certificate 18 November 1965. The Convair 640 received its supplemental Type Certificate on 7 December 1965. The 600 entered service with Central Airlines 30 November 1965; the 640 with Caribair on 22 December 1965.

The descriptions of the Convair-Liner 240/340/440 series apply equally to the Convair 600 and 640, except for the details given above.

DIMENSIONS, EXTERNAL:*

Wing span: 600	27.98 m (91 ft 9 in)
640	32.12 m (105 ft 4 in)
Length overall: 600	22.77 m (74 ft 8 in)
640, standard	24.14 m (79 ft 2 in)
640, with nose radome	24.84 m (81 ft 6 in)
Length with wx radome	23.44 m (76 ft 11 in)
Height overall: 600	8.22 m (26 ft 11 in)
640	8.59 m (28 ft 2 in)

DIMENSIONS, INTERNAL:*

Cabin: Length: 600	10.19 m (33 ft 5 in)
640	11.15 m (36 ft 7 in)
Max width	2.62 m (8 ft 7 in)
Max height	2.00 m (6 ft 7 in)

* *Areas are unchanged by conversion*

WEIGHTS AND LOADINGS:

Weight empty: 600	12,872 kg (28,380 lb)
640 from 340	13,886 kg (30,615 lb)
640 from 440	14,118 kg (31,125 lb)
Max T-O weight: 600	20,955 kg (46,200 lb)
640	24,950 kg (55,000 lb)
Max zero-fuel weight: 600	17,915 kg (39,500 lb)
640	22,680 kg (50,000 lb)
Max landing weight: 600	19,958 kg (44,000 lb)
640	23,815 kg (52,500 lb)
Max wing loading: 600	276 kg/m² (56.5 lb/sq ft)
640	292 kg/m² (59.8 lb/sq ft)

PERFORMANCE (at max T-O weight):

Max level speed at 4,575 m (15,000 ft): 600	268 kts (497 km/h; 309 mph)
640	260 kts (482 km/h; 300 mph)
Max permissible diving speed: 600	234 kts (435 km/h; 270 mph) or Mach 0.550 CAS

Convair 600 twin-engine, medium-range passenger aircraft

1994

640 260 kts (482 km/h; 300 mph) or Mach 0.650 CAS
Max and econ cruising speeds: Same as max level speed at 4,575 m (15,000 ft)
Landing speed, power off, wheels and flaps down:
600 72 kts (134 km/h; 83 mph)
640 73 kts (136 km/h; 84 mph)
Rate of climb at S/L: 600 427 m (1,400 ft)/min
640 350 m (1,150 ft)/min
Rate of climb at S/L, one engine out:
600 134 m (440 ft)/min
640 99 m (325 ft)/min
Max operating ceiling 6,097 m (20,000 ft)
Service ceiling: 600 7,315 m (24,000 ft)
640 7,000 m (23,000 ft)
FAA T-O distance to 10.5 m (35 ft):
600 1,419 m (4,655 ft)
640 1,598 m (5,245 ft)
Range at 4,575 m (15,000 ft) with long-range tanks, 45 min reserve: 600 at 268 kts (497 km/h; 309 mph) 1,650 n miles (3,060 km; 1,900 miles)
Range at 4,575 m (15,000 ft) with standard fuel, 45 min reserve: 640 at 260 kts (482 km/h; 300 mph) 1,068 n miles (1,975 km; 1,230 miles)

UPDATED

GLOBAL

GLOBAL HELICOPTER TECHNOLOGY INC

5070 S. Collins St, Arlington, Texas 76018-1108
Telephone: 1 (817) 557 3391
Fax: 1 (817) 557 3392
PRESIDENT: Daniel W. Pettus
VICE-PRESIDENT: Dennis Halwes
COMMUNICATIONS: Charles Cole

Company specialises in helicopter design, testing and modification.

UPDATED

GLOBAL HELICOPTER HUEY 800

TYPE: Utility helicopter upgrade.

PROGRAMME: Conversion of Bell UH-1H with 986 kW (1,322 shp) LHTEC T800-LHT-800 engine; first flown 15 June 1992; FAA and military certification in progress. Cost of conversion $800,000. Frahm vibration absorber system optional; integral particle separator and FADEC control; MIL-STD-1553B databus; Hispano-Suiza reduction gearbox.

Claimed to have significantly lower fuel consumption than UH-1 with existing T53-L-13B or other engines currently proposed for refit. Prototype set new point-to-point distance record in Spring 1993 (Los Angeles to Atlanta), approximately 1,713 n miles (3,175 km; 1,973 miles), exceeding by more than 540 n miles (1,000 km; 621 miles) the previous record for helicopters in 3,000 to 4,500 kg (6,614 to 9,921 lb) class. Average fuel consumption was 140 kg (308.5 lb)/h, compared with more than 225 kg (496 lb)/h for standard UH-1; approximately 110 kg (242 lb) of fuel remained at end of flight.

Global Helicopter Huey 800 re-engined Bell UH-1H

1995

CUSTOMERS: The US Border Patrol took delivery of 5 Global Helicopter Huey 800 aircraft on 14 July 1995.

UPDATED

GRUMMAN — *see Northrop Grumman*

GULFSTREAM AEROSPACE

GULFSTREAM AEROSPACE CORPORATION

PO Box 2206, Savannah International Airport, Savannah, Georgia 31402-2206
Telephone: 1 (912) 965 3000
Fax: 1 (912) 965 3775/4171
Telex: 546470 GULF AERO
CHAIRMAN AND CEO: Theodore J. Forstmann
PRESIDENT AND COO: Fred A. Breidenbach
SENIOR VICE-PRESIDENT: W. W. Boisture
SENIOR VICE-PRESIDENT, ENGINEERING: Charles N. Coppi
SENIOR VICE-PRESIDENT, MARKETING: Robert H. Cooper
MANAGER, CORPORATE COMMUNICATIONS: Janie Rovolis

Original facilities purchased by Allen E. Paulson from Grumman Corporation in 1978; facilities expanded and two Gulfstream Models introduced when company purchased from Mr A. Paulson by Chrysler Corporation in 1985. Mr Paulson and Forstmann Little and Company completed repurchase from Chrysler Corporation 19 March 1990.

AiResearch facility at Long Beach International Airport, California, bought by Gulfstream in 1986 and expanded, can accommodate outfitting and completion of 17 Gulfstream IVs at a time. Gulfstream also operates subcontracting facility in Oklahoma City, Oklahoma. Corporations three facilities cover more than 185,806 m² (2 million sq ft) of manufacturing and servicing space.

VERIFIED

GULFSTREAM AEROSPACE GULFSTREAM II-B UPGRADE

TYPE: Business aircraft upgrade.

PROGRAMME: The first production Gulfstream II (no prototype was built) was flown by Grumman 2 October 1966. When production ended during 1979 a total of 256 had been manufactured.

Following resumption of the Gulfstream III programme by Gulfstream Aerospace in early 1978, the company announced Autumn that year that the Gulfstream III wing was to be offered as a retrofit for existing Gulfstream IIs. It was not until a year later, following certification of the Gulfstream III 22 September 1980, that work was started to adapt the Gulfstream III wing for this retrofit application. It also involved structural modifications to permit increases in T-O weight, operating speeds and cruising heights.

Gulfstream II c/n 70, the property of Southland Corporation, Dallas, Texas, was used for the prototype conversion, which completed the FAA certification programme 17 September 1981. The new wing has 27° 40′ of sweepback at quarter-chord, and incorporates winglets at the wingtips. Designated Gulfstream II-B in this new configuration, the prototype conversion (N711SC) was flown for the first time on 17 March 1981. This wing increases the range of the Gulfstream II and simultaneously provides a significant improvement in fuel efficiency.

UPGRADES: **Gulfstream:** Quiet Spey hush kit. See separate entry.

DIMENSIONS, EXTERNAL:
Wing span 23.72 m (77 ft 10 in)
Wing chord: at root 5.94 m (19 ft 6 in)
at tip 1.65 m (5 ft 5 in)
Wing aspect ratio 6.48
Length overall 24.97 m (81 ft 11 in)
Width of fuselage 2.39 m (7 ft 10 in)
Height overall 7.47 m (24 ft 6 in)
Tailplane span 8.23 m (27 ft 0 in)
Wheel track 4.22 m (13 ft 10 in)
Wheelbase 10.72 m (35 ft 2 in)
Passenger door (fwd, port): Height 1.57 m (5 ft 2 in)
Width 0.91 m (3 ft 0 in)
Baggage door (port, rear): Height 0.72 m (2 ft 4½ in)
Width 0.90 m (2 ft 11¾ in)
Emergency exits (four: two overwing each side):
Height 0.48 m (1 ft 7 in)
Width 0.66 m (2 ft 2 in)

DIMENSIONS, INTERNAL:
Cabin, incl galley and toilet:
Length 11.99 m (39 ft 4 in)
Max width 2.24 m (7 ft 4 in)
Max height 1.85 m (6 ft 1 in)
Floor area 19.16 m² (206.3 sq ft)
Volume 40.38 m² (1,427 cu ft)
Baggage hold volume 4.44 m³ (157 cu ft)

AREAS:
Wings, gross 86.83 m² (934.6 sq ft)

WEIGHTS AND LOADINGS:
Basic operating weight empty 17,304 kg (38,150 lb)
Max fuel weight 12,836 kg (28,300 lb)
Max payload 1,701 kg (3,750 lb)
Payload with max fuel 771 kg (1,700 lb)
Max T-O weight 31,615 kg (69,700 lb)
Max ramp weight 31,842 kg (70,200 lb)
Max landing weight 26,535 kg (58,500 lb)
Max zero-fuel weight 19,958 kg (44,000 lb)

PERFORMANCE:
Never-exceed speed Mach 0.85
Cruising Mach number at max power at 12,160 m (39,900 ft), AUW of 24,947 kg (55,000 lb) 0.85
Long-range cruising Mach number 0.77
Stalling speed at max landing weight, flaps down 103 kts (191 km/h; 119 mph)
Max rate of climb at S/L 1,300 m (4,270 ft)/min
Rate of climb at S/L, one engine out 448 m (1,470 ft)/min
Max operating altitude 13,710 m (45,000 ft)
FAA T-O distance 1,555 m (5,100 ft)
FAA landing distance 975 m (3,200 ft)
Range with eight passengers, NBAA IFR reserves 3,640 n miles (6,745 km; 4,190 miles)
Range with NBAA VFR reserves 4,000 n miles (7,524 km; 4,675 miles)

OPERATIONAL NOISE LEVELS (FAR Pt 36):
T-O 91 EPNdB
Approach 97 EPNdB
Sideline 103 EPNdB

VERIFIED

GULFSTREAM AEROSPACE GULFSTREAM III

USAF designation: C-20
US Navy designation: C-20D
US Army designation: C-20E

TYPE: Twin-turbofan executive and military communications aircraft.

PROGRAMME: The prototype Gulfstream III (N901G) made its first flight 2 December 1979. FAA certification was received 22 September 1980. A total of 206 had been built when production ended in 1986. The aircraft is still in service with the US Air Force (12+), Navy (3) and Army (2).

VERSIONS: **C-20A:** The US Air Force announced 7 June 1983 that it had awarded Gulfstream Aerospace a firm fixed price contract covering the lease of standard off-the-shelf Gulfstream III transports, with accommodation for five crew and 14 passengers, as replacements for Lockheed C-140 aircraft under the Air Force Special Air Missions task (C-SAM). The contract covered the lease of three aircraft during FY83-84. These aircraft, designated C-20A, have since been purchased outright and are based at Ramstein AB, Germany.

C-20B: In January 1986 the US Air Force awarded Gulfstream Aerospace a contract for the purchase of seven additional Gulfstream IIIs, delivery of which was completed by the end of 1987. These aircraft, outfitted with an advanced mission communications system and revised interior layout, are designated C-20B. These serve with the 89th Airlift Wing at Andrews AFB, Maryland.

Gulfstream III C-20A of the US Air Force

1994

C-20D/E: In 1986 the US Navy purchased two Gulfstream III aircraft designated C-20D, and in 1987 the US Army purchased two aircraft designated C-20E. The C-20D/E aircraft contained the same custom interior and outfitting configuration as the C-20B aircraft, but did not receive the Mission Communication System.

UPGRADES: **Gulfstream:** Quiet Spey hush kit. See separate entry.

OPERATORS: Versions of the Gulfstream III are in service with the armed forces of the following countries: Algeria (3), Angola (1), Bahrain (1), Brunei (1), Cameroon (1), Denmark (3), Egypt (2), Gabon (1), India (3), Ireland (1), Italy (2), Jordan (2), Mexico (2), Morocco (1), Oman (1), Saudi Arabia (1), USA (25) and Venezuela (1).

DESIGN FEATURES: Cantilever low-wing monoplane of light alloy construction. Wing section NACA 0012 modified at wing station 50; NACA 63A009.5 modified at wing station 145; NACA 64 series modified at wing station 385. Dihedral 3°. Incidence 3° 30′ at wing station 50, 1° 30′ at wing station 145, and −0° 30′ at wing station 414. Sweepback 27° 40′ at quarter-chord. By comparison with Gulfstream II, wing has extended chord leading-edges (0.76 m; 2 ft 6 in at root, 0.13 m; 5 in at tip). NASA (Whitcomb) wingtip winglets.

FLYING CONTROLS: One-piece single-slotted Fowler trailing-edge flaps. Spoilers forward of flaps assist in lateral control and can be extended for use as airbrakes. All control surfaces actuated hydraulically. Trim tab in port aileron. Trim tab in rudder and each elevator. Powered controls.

STRUCTURE: The fuselage is a conventional semi-monocoque structure of light alloy. Glassfibre nosecone hinged for access. The tail unit is a T tail structure of light alloy, except for composite rudder from c/n 435 effective mid-1984. Swept horizontal and vertical surfaces.

LANDING GEAR: Retractable tricycle type, with twin wheels on each unit. Inward retracting main units; steerable nose unit retracts forward. Mainwheel tyres size 34 × 9.25-16, pressure 12.0 bars (175 lb/sq in). Nosewheel tyres size 21 × 7.25-10, pressure 7.9 bars (115 lb/sq in). Goodyear air-cooled carbon brakes with Goodyear fully modulating anti-skid units.

POWER PLANT: Two Rolls-Royce Spey Mk 511-8 turbofan engines, each rated 50.7 kN (11,400 lb st), pod-mounted on sides of rear fuselage. Rohr target type reverser forms aft part of each nacelle when in stowed position. All fuel in integral tanks in wings, with total capacity of 15,868 litres (4,192 US gallons; 3,490 Imp gallons).

ACCOMMODATION: Crew of two or three. Standard seating for 19 passengers in pressurised and air conditioned cabin. Large baggage compartment at rear of cabin, capacity 907 kg (2,000 lb). Integral airstair door at front of cabin on port side. Electrically heated wraparound windscreen. C-20s have a crew of five and seating for 13 passengers.

SYSTEMS: Cabin pressurisation system with maximum differential of 0.65 bars (9.45 lb/sq in). Two independent hydraulic systems, one 103.5 bars (1,500 lb/sq in) and one 207 bars (3,000 lb/sq in). Maximum flow rate 83.3 litres (22 US gallons; 18.32 Imp gallons)/min. Two bootstrap type hydraulic reservoirs, 2.07 bars (30 lb/sq in) and 4.14 bars (60 lb/sq in) respectively. All flying controls hydraulically powered with manual reversion. APU in tail compartment. With effect from aircraft c/n 402, were scheduled for delivery in late 1983, the electrical systems are basically of 115/200 V 400 Hz, provided by three-phase variable speed alternators each rated at 36 kVA, two engine-driven, the third driven by the APU. Each engine-driven alternator is provided with a 30 kVA solid-state converter to give 25 kVA 115/200 V 400 Hz three-phase, but the alternator of the controlled speed APU does not need a converter. The capacity of a single converter is sufficient for the entire AC requirement of the aircraft. The system includes also a 300 A 28 V DC transformer-rectifier, an 800 VA 115 V 400 Hz single-phase solid-state battery-powered inverter for emergency AC power, two 24 V Ni/Cd batteries and an external power socket, giving a total weight saving of approximately 91 kg (200 lb) compared to the former system. Anti-icing by engine bleed air.

AVIONICS AND EQUIPMENT: Standard Gulfstream III avionics include three Collins VHF com, two Collins VIR-31B VHF nav; two Collins ADF-60A; two Collins DME-40; cockpit voice recorders, Sperry SPZ-800 automatic flight guidance and control system; and Collins WXR-250A, or WXR-30C; Bendix RDR-1200; or RCA Primus 40 WXD, or Primus 4000 weather radar. Collins and Sperry electronic flight instrument systems available optionally. Bendix Series III integrated avionics system also certificated. The C-20A and B aircraft have a Mission Communication System installed to provide additional worldwide communication capabilities for its passengers. The C-20D and E aircraft do not have the Mission Communication System.

DIMENSIONS, EXTERNAL:
Wing span 23.72 m (77 ft 10 in)
Wing aspect ratio 6.0
Wing chord at root (fuselage centreline) 5.94 m (19 ft 5⅞ in)
Winglet height 1.63 m (5 ft 4¼ in)
Length overall 25.32 m (83 ft 1 in)
Fuselage length 22.66 m (74 ft 4 in)
Height overall 7.43 m (24 ft 4½ in)
Tailplane span 8.23 m (27 ft 0 in)
Wheel track 4.17 m (13 ft 8 in)
Wheelbase 10.72 m (35 ft 2 in)
Passenger door (fwd, port): Height 1.57 m (5 ft 2 in)
Width 0.91 m (3 ft 0 in)
Baggage door (aft): Height 0.72 m (2 ft 4½ in)
Width 0.90 m (2 ft 11¾ in)

DIMENSIONS, INTERNAL:
Cabin: Length 12.60 m (41 ft 4 in)
Width 2.24 m (7 ft 4 in)
Height 1.85 m (6 ft 1 in)
Volume 42.53 m³ (1,502 cu ft)
Aft baggage compartment volume 4.44 m³ (157 cu ft)

AREAS:
Wings, gross 86.83 m² (934.6 sq ft)
Ailerons, incl tabs (total) 2.68 m² (28.86 sq ft)
Trailing-edge flaps (total) 11.97 m² (128.84 sq ft)
Flight spoilers (total) 2.87 m² (30.88 sq ft)
Ground spoilers (total) 4.59 m² (49.39 sq ft)
Winglets (total) 2.38 m² (25.60 sq ft)
Fin 10.92 m² (117.53 sq ft)
Rudder 4.16 m² (44.75 sq ft)
Horizontal tail surfaces (total) 12.70 m² (136.69 sq ft)
Elevators (total) 5.22 m² (56.22 sq ft)

WEIGHTS AND LOADINGS:
Manufacturer's weight empty 14,515 kg (32,000 lb)
Typical operating weight empty 17,236 kg (38,000 lb)
Max fuel load 12,836 kg (28,300 lb)
Typical payload 726 kg (1,600 lb)
Max T-O weight 31,615 kg (69,700 lb)
Max ramp weight 31,842 kg (70,200 lb)
Max zero-fuel weight 19,958 kg (44,000 lb)
Max landing weight 26,535 kg (58,500 lb)

PERFORMANCE:
Max cruising speed Mach 0.85 (501 kts; 928 km/h; 576 mph)
Long-range cruising speed Mach 0.77 (442 kts; 818 km/h; 508 mph)
Approach speed at max landing weight 136 kts (252 km/h; 157 mph)
Stalling speed at max landing weight 105 kts (195 km/h; 121 mph)
Max rate of climb at S/L 1,158 m (3,800 ft)/min
Rate of climb at S/L, one engine out 365 m (1,200 ft)/min
Max operating altitude 13,720 m (45,000 ft)
FAA balanced T-O field length 1,554 m (5,100 ft)
FAA landing distance 975 m (3,200 ft)
NBAA range, with eight passengers and baggage:
IFR reserves 3,650 n miles (6,760 km; 4,200 miles)
VFR reserves 4,100 n miles (7,598 km; 4,721 miles)

OPERATIONAL NOISE LEVELS (FAR Pt 36):
T-O 91 EPNdB
Approach 97 EPNdB
Sideline 103 EPNdB

UPDATED

GULFSTREAM AEROSPACE GULFSTREAM III SRA-1 (SPECIAL MISSIONS VERSION)

TYPE: Special missions aircraft.

PROGRAMME: Experience with specially equipped Gulfstream IIIs supplied to the Royal Danish Air Force for fishing patrol duties led Gulfstream Aerospace to announce in September 1983 development of a dedicated **SRA-1** (Surveillance and Reconnaissance Aircraft). The prototype SRA-1 (N47449) was first flown 14 August 1984, following a roll-out ceremony at Savannah, Georgia. As rolled out, the SRA-1 prototype featured wingtip-mounted antenna pods, but these were subsequently replaced by standard Gulfstream III winglets which were then fitted on all production Gulfstream IIIs unless customer specifications dictated otherwise. The SRA-1 had optional fully integrated or stand-alone systems, according to customer requirements, for electronic surveillance, command control, standoff high-altitude reconnaissance, maritime patrol surface surveillance, and anti-submarine warfare missions, or combinations of those roles.

The following other versions of the SRA-1 are available:

VERSIONS: **VIP:** Transportation, as a personnel/administrative transport with 18 passengers plus one attendant, for medical evacuation with 15 litter patients and two medical staff, or as a freight transport for up to 3,220 kg (7,100 lb) of priority cargo.

Electronic surveillance: The SRA-1 is equipped with 20 to 1200 MHz communictions intercept system; a 0.5 to 40 GHz electronic support measures (ESM) system; and VHF/UHF/HF communications for C³ functions; permitting accurate detection, location, analysis and classification of electronic signals. Operational capabilities include computer databases to provide automatic signal analysis. A typical electronic surveillance mission for the SRA-1 with a crew of 10 would have a gross weight of 31,743 kg (69,981 lb), providing a long-range endurance of 8.8 hours and a range of 3,565 n miles (6,607 km; 4,105 miles) with 1,361 kg (3,000 lb) fuel reserves (ISA, zero-wind). For a high-altitude loiter mission at the same gross weight and reserves, the SRA-1 can remain airborne for 9.6 hours, with a range of 3,527 n miles (6,036 km; 3,750 miles).

Maritime patrol: The SRA-1 is equipped with high-definition, all-weather, real-time search radar in a hinged nosecone; a FLIR system; ESM system; and stowage and manual launch systems for survival/rescue equipment, marine markers and flares. With a crew of five, mission gross weight of 30,810 kg (67,924 lb), and a fuel reserve of 1,361 kg (3,000 lb), ISA, zero-wind, the SRA-1 has a mission endurance of 9.2 hours and a range of 3,017 n miles (5,591 km; 3,474 miles) at a 9,140 m (30,000 ft) patrol altitude. For SAR missions, at a gross weight of 31,842 kg (70,200 lb) including a 1,032 kg (2,276 lb) retained payload, the SRA-1 can operate within a 1,000 n miles (1,853 km; 1,151 mile) radius of action and fly an 8 hour mission.

Reconnaissance: The SRA-1 is equipped with real-time, all-weather moving target indicator SLAR; long-range oblique photographic cameras, which may include electro-optical capability; and an ESM system. With a crew of seven the aircraft has a loiter endurance of 9.7 hours at altitudes above 7,620 m (25,000 ft).

ASW: The SRA-1 can be equipped with a high-definition maritime surveillance radar with periscope/snort detection; FLIR; a boom-mounted magnetic anomaly detector (MAD) extending from the rear fuselage beneath the fin/rudder; ESM; acoustic processing equipment; sonobuoys and automated data recording for post-flight analysis. Weapons on underwing hardpoints can include a wide

Gulfstream III C-20B of the US Air Force

1994

variety of air-to-surface and anti-shipping missiles and torpedoes.

WEIGHTS AND LOADINGS: As for Gulfstream III except:

Manufacturer's weight empty 14,834 kg (32,703 lb)
Basic weight empty (excl mission equipment) 16,408 kg (36,173 lb)
Max payload (cargo) 3,220 kg (7,100 lb)

PERFORMANCE: As for Gulfstream III except:

NBAA range at Mach 0.77 with crew of three and 725 kg (1,600 lb) payload:
IFR reserves 3,500 n miles (6,486 km; 4,030 miles)
VFR reserves 3,940 n miles (7,302 km; 4,537 miles)

VERIFIED

Gulfstream III SRA-1 special missions aircraft of the US Air Force

1994

GULFSTREAM STAGE III QUIET SPEY HUSH KIT

TYPE: Business aircraft hush kit.

PROGRAMME: Programme announced 1991 to assure GII, GIIB and GIII operators would be able to comply with FAR Stage III requirements. Preproduction hardware tests conducted 31 March to 1 April 1993 in Phoenix, Arizona.

DESIGN FEATURES: Hush kit fitted to the Rolls-Royce Spey turbofan engines.

VERIFIED

Gulfstream II fitted with Gulfstream 'Quiet Spey' hush kit

1994

HAMILTON AVIATION

HAMILTON AVIATION

PO Box 11746, Tucson, Arizona 85734-1746
Telephone: 1 (520) 294 3481
Fax: 1 (520) 741 1430

PRESIDENT: Gordon B. Hamilton
EXECUTIVE VICE-PRESIDENT: Gordon D. Hamilton
GENERAL ADMINISTRATOR: Alan R. Abate
VICE-PRESIDENT, MARKETING: Hugh M. Curley

Hamilton Aviation, established in 1947, operates from a 22 acre site on Tucson International airport. The company has 3,787 m² (40,000 sq ft) of covered facilities and a workforce of 300 personnel. Main responsibilities are in the fields of aircraft maintenance and modification of narrow body commercial jet aircraft including cargo conversions, interior refits, engine retrofits as well as hush kit installations. Work has been carried out in the following aircraft types: Boeing 727, 737, Convair 580, Fokker F27, Lockheed L-188 and C-130 as well as McDonnell Douglas DC-8, DC-9, and NAMC YS-11.

NEW ENTRY

HAMILTON AVIATION COMMERCIAL JET CARGO DOOR CONVERSION

TYPE: Clamshell type cargo door conversion for large commercial jets.

PROGRAMME: The company delivered four converted Boeing 727-200 and one McDonnell Douglas DC-9-30 aircraft during the period 6 July 1994 to 12 February 1995.

DESIGN FEATURES: Conversion includes: installation of cargo door, 9G bulkhead, freight handling system and class E compartment. Aircraft are also subject to: heavy 'C' level maintenance check, ageing aircraft structural modifications and CPCP as well as installation of GPS and Federal Express Stage III hush kits.

CUSTOMERS: Aeromexpress (two aircraft), Hunting Cargo Airlines, TNT Express Worldwide (Boeing) and National Aeronautics & Space Administation NASA (DC-9-30).

NEW ENTRY

HAMILTON AVIATION SLIDING DOOR INSTALLATION

TYPE: Large passenger aircraft cargo door conversion.

PROGRAMME: Announced late 1994, Hamilton Aviation's multiple STC will enable the sliding cargo door to be fitted to both narrow and wide-body commercial jet aircraft. The first installation of the door will be on a Boeing 727-200 followed by a McDonnell Douglas DC-10-30.

DESIGN FEATURES: Operational features of the door include: manual all-weather operation (inside and outside), door can be opened for loading in all weather conditions, door mechanism does not encroach into cargo loading area and is less susceptible to damage during loading. Door meets all FAR 25.783 requirements and maintains fuselage structural limits as well as having a fail-safe/damage tolerant design.

NEW ENTRY

Hamilton (Boeing) 727-200 cargo conversion with Stage III hush kits

1995

HELI-AIR

HELI-AIR (JAFFE HELICOPTER INCORPORATED)

119 Ida Road, Broussard, Louisiana 70518
Telephone: 1 (318) 837 2396
Fax: 1 (318) 837 2113
Telex: 510 1010 395

PRESIDENT: Gary J. Villiard
DIRECTOR OF OPERATIONS: David A. Brown

Member of Jaffe Group.

VERIFIED

HELI-AIR (BELL) 222SP

TYPE: Re-engined Bell 222.

PROGRAMME: First flight of Allison-engined Bell 222 test conversion (N5008Q) 10 November 1988; first flight of production conversion 12 January 1990; certificated 21 August 1991.

COSTS: $1.45 million (1992).

DESIGN FEATURES: Heli-Air developed modification to replace standard 510 kW (684 shp) LTS 101-750C-1 turboshafts in Bell 222A, 222B and 222UT with two 485 kW (650 shp) Allison 250-C30Gs. Performance unchanged except for improved engine-out performance and 36 kg (80 lb) greater useful load.

VERIFIED

HELI-CONVERSIONS

HELI-CONVERSIONS INCORPORATED

3830 Cross Street, Eugene, Oregon 97402-2497
Telephone: 1 (503) 688 4590
Fax: 1 (503) 688 0523
Telex: 364419
PRESIDENT: Birger Sterner
VICE-PRESIDENT: Vaden Francisco

Heli-Conversions Inc, based in Eugene, Oregon, specialises in the development and worldwide marketing of cost-reducing and performance-enhancing improvements for the Bell 205 and UH-1 series helicopters.

VERIFIED

Heli-Conversions Super 205 upgraded Bell 205 with Textron Lycoming T5317A turboshaft engine *1995*

HELI-CONVERSIONS' SUPER 205 UPGRADE

TYPE: Utility helicopter upgrade.
PROGRAMME: Upgrade package for Bell 205 and UH-1 series helicopters.
DESIGN FEATURES: Retrofit with the Textron Lycoming T5317A turboshaft engine and the Bell Model 212 main rotor system STCs. Additional features include: PLM Centrisep® air cleaner system, Howell Autotemp® digital EGT and engine usage monitoring system, T53-HC ''on-line'' engine wash kit, dual electric fuel boost pump system and a manual start fuel switch. Customers can select their own choice of the above installations to suit operational requirements.

UPDATED

ISOLAIR

ISOLAIR INC

20490 East Aschoff Road, Rhododendron, Oregon 97049
Telephone: 1 (503) 622 3010
Fax: 1 (503) 622 4274
DIRECTOR OF MARKETING: Gary Finlay

NEW ENTRY

ISOLAIR (MIL) TERMINATOR II

TYPE: Firefighting conversion of Mil Mi-14 (NATO 'Haze') helicopter (which see under Russia).
PROGRAMME: Announced September 1993 as ''world's only amphibious helicopterwater bomber''; low-time Mi-14s refurbished by Aerotec GmbH of Saarbrucken, Germany, before fitment with Isolair firefighting system and new avionics.
CUSTOMERS: Initial modification of six aircraft for wet-lease only; subsequent aircraft will be available for sale.
DESIGN FEATURES: Internal tank assembly housed in weapons bay ad cargo compartment; loading, mixing of foam or retardant, dump volume and drop length monitored and controlled from cockpit. System can perform 'running refill' on water at speeds up to 80 knots (148 km/h; 92 mph); with full water load, sufficient foam additive and fuel for 90 minutes' flying, Terminator II can fly up to 15 drop cycles per hour. System can be installed in approximately 56 man-hours and requires no major airframe modifications; principal change is fitment of multiple dump doors.

Two different self-loading methods used for fast drop cycles, both independent of ground support. From large, deep water areas, tank is scoop-filled in 60 seconds or less by high-speed taxying over surface; a snorkel probe is used to hover-refill from shallow water or portable tanks.

SYSTEMS: Electrical power requirement 200 A/28 V DC.
EQUIPMENT: Internal tank is 4.27 m (14 ft 0 in) long and can hold 4,600 litres (1,215 US gallons; 1,012 Imp gallons) of water plus 400 litres (106 US gallons; 88 Imp gallons) of foam or retardant additive.

WEIGHTS AND LOADINGS:

Weight of empty tank	601.5 kg (1,326 lb)
Max T-O weight	5,137 kg (11,326 lb)

PERFORMANCE:

Hover fill rate (max)	3,930 l (1,038 US gallons; 825 Imp gallons)/min
Hover fill time, probe submerged	70 s

NEW ENTRY

KAMAN

KAMAN AEROSPACE CORPORATION (Subsidiary of Kaman Corporation)

Old Windsor Road, PO Box No. 2, Bloomfield, Connecticut 06002
Telephone: 1 (203) 242 4461/243 7100
Telex: 710 425 3411
PRESIDENT AND CEO: Walter R. Kozlow
VICE-PRESIDENT, ENGINEERING: David J. White
MARKETING MANAGER: William H. G. Douglass
DIRECTOR, PUBLIC RELATIONS: J. Kenneth Nasshan
PUBLIC RELATIONS REPRESENTATIVE: Elizabeth K. Healy

Founded in 1945 by Charles H. Kaman, now Chairman of the Board of Kaman Corporation. Developed servo-flap control of helicopter main rotor, still used in H-2 Seasprite naval helicopter and K-Max® ''Aerial Truck'' commercial external lift helicopter. R&D programmes sponsored by US Army, Air Force, Navy and NASA include advanced design of helicopter rotor systems, blades and rotor control concepts, component fatigue life determination and structural dynamic analysis and testing. Kaman has undertaken helicopter drone programmes since 1953; is continuing advanced research in rotary wing unmanned aerial vehicles (UAVs).

Kaman is major subcontractor in many aircraft and space programmes, including design, tooling and fabrication of components in metal, metal honeycomb, bonded and composite construction, using techniques such as filament winding and braiding. Participates in programmes including Grumman A-6 and F-14, Bell/Boeing V-22, Boeing 737, 747, 757, 767 and 777, Sikorsky UH-60 and SH-60, and NASA Space shuttle Orbiter. Kaman also supplies acoustic engine ducts for P&W JT8D and thrust reversers for GE CF6-80C/E engines.

Kaman designed and, since 1977, has been producing all-composite rotor blades for Bell AH-1 Cobras for US and foreign armies; participated in AH-1 air-to-air Stinger (ATAS) programme.

UPDATED

KAMAN SEASPRITE and SUPER SEASPRITE

US Navy designation: SH-2

TYPE: Shipboard ASW helicopter with SAR, observation and utility capability.
PROGRAMME: First flight 2 July 1959; successive versions for US Navy SH-2F put back into production in 1981; from 1967 all single-engined SH-2A/B Seasprites progressively converted to twin-engined UH-2Cs with General Electric T58-GE-8Bs; later modified to Mk I Light Airborne Multi-Purpose System (LAMPS) standard to give small ships ASW, anti-ship surveillance and targeting (ASST), SAR and utility capability; all SH-2s subsequently upgraded to SH-2Fs, each with stronger landing gear, T58-GE-8F engines and improved rotor.

Operational deployment of LAMPS Mk I to HSL squadrons began 7 December 1971; by December 1991, more than 805,000 hours flown from ship classes FFG-7, DD-963, DDG-993, CG-47, FFG-1, FF-1052, FF-1040, CG-26, CGN-35, CGN-38 and BB-61; US Coast Guard WMEC and WHEC cutters being equipped to operate SH-2; eight active Navy LAMPS Mk I squadrons supplemented in 1984 by three Naval Air Reserve squadrons; 24 SH-2Fs transferred to HSL-74, 84 and 94 Reserve squadrons at South Weymouth (Massachusetts), North Island (California), and Willow Grove (Pennsylvania) as new aircraft delivered to HSLs; front-line squadrons are HSL-30, 32 and 34 at Norfolk, Virginia, and HSL-36 at Mayport, Florida, for Atlantic-based ships; and HSL-31, 33 and 35 at North Island, California, and Barber's Point, Hawaii, for the Pacific; active and Reserve LAMPS Mk I SH-2s will remain in service until 2010.

VERSIONS (in service 1991): **SH-2F:** LAMPS Mk I; 88 delivered between May 1973 and 1982, 75 of which operational February 1989; 16 SH-2Ds converted to SH-2F; first operational unit, HSL-33, deployed to Pacific 11 September 1973; 54 new SH-2Fs ordered FY82-86 (18, 18, six, six and six), delivered December 1989; another six ordered in FY87, being completed as SH-2G (see below). SH-2F fleet to be upgraded to SH-2G.

For operation in Arabian Gulf since 1987, 16 SH-2Fs augmented standard AN/ALR-66A(V)1 RWR and AN/ALE-39 chaff/flare dispensers by Sanders AN/ALQ-144 IR jammers on starboard side of tail rotor driveshaft, Honeywell AN/AAR-47 missile warning equipment, Collins AN/ARC-182 secure VHF/UHF radios, AN/DLQ-3 missile warning and jamming equipment, and Hughes AN/AAQ-16 FLIR under nose.

SH-2G Super Seasprite: SH-2F upgrade initiated FY87; airframe changes included replacing T58 with T700-GE-401 engines; fuel consumption improved by over 20 per cent. Avionics improvements include MIL-STD-1553B digital databus, onboard acoustic processor, multifunction raster display, AN/ASN-150 tactical navigation display, and 99-channel sonobuoys. SH-2G

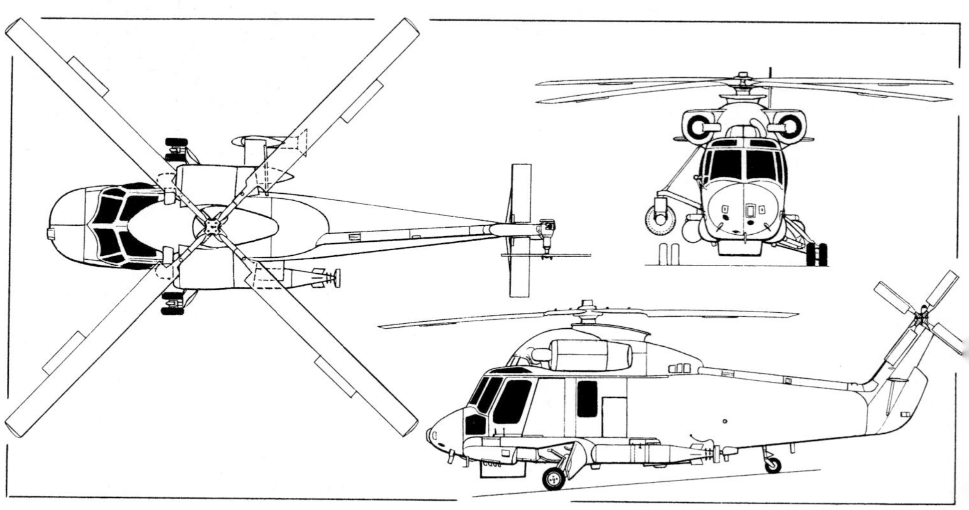

Kaman SH-2G Super Seasprite LAMPS Mk I helicopter *(Jane's/Dennis Punnett)* *1994*

qualified for dipping sonar, air-to-surface missiles, forward-looking infra-red sensors and various guns, rockets and countermeasures.

First flight of SH-2F 161653 as YSH-2G T700 engine testbed, April 1985; first flight with full avionics 28 December 1989; this helicopter delivered 1991, followed by six new-build SH-2Gs ordered FY87; no further new production envisaged. June 1987 contract launched conversion programme from SH-2F to SH-2G performed by Kaman; 18 conversion kits on order by January 1993, for completion by June 1994; rebuilds refurbished for further 10,000 flying hours. First production SH-2G (163541) flown March 1990; first delivery, to HSL-84, in December 1992; remaining five to follow in 1993; rebuilds thereafter. *Detailed description applies to SH-2G.*

SH-2G can be configured for Magic Lantern podded laser equipment for subsurface mine detection; ML-30 prototype equipment operationally tested in single SH-2F during Gulf War, 1991; two ML-90 prototypes ordered late 1991; delivered to US Navy, October 1992. Further unknown sensor reportedly under development.

UPGRADES: **Kaman:** SH-2G Super Seasprite. See Versions.

Kaman: In 1994 Egypt announced its intention to buy 10 SH-2F. These aircraft will be upgraded to SH-2G configuration. The Egyptian SH-2G(E) aircraft will be equipped with the AlliedSignal AQS-18A dipping sonar.

CUSTOMERS: US Navy is operating 16 SH-2G helicopters in two USN Reserve Squadrons HSL-94 in Willow Grove, Pennsylvania, and HSL-84 at Naval Air Station, North Island, in San Diego, California. Several of these aircraft have been converted from the SH-2F to SH-2G configuration. Taiwan received 12 SH-2Fs in 1994. These aircraft were ex-USN helicopters and will complement the 1052 class frigates currently leased from the US government.

COSTS: $12 million per conversion from SH-2F to SH-2G (1993 prices).

DESIGN FEATURES: Main rotor rpm 298; main and tail rotor blades folded manually; nose opens and folds back for shipboard stowage; lateral pylons for torpedoes or tanks; MAD bird in holder extending from starboard sides.

FLYING CONTROLS: Main rotor blades fixed on hub; pitch changed by trailing-edge tabs.

STRUCTURE: All-metal airframe with flotation hull; titanium main rotor hub and all-composite main rotor blades.

LANDING GEAR: Tailwheel type, with forward retracting twin mainwheels and non-retractable tailwheel. Liquid spring shock-absorbers in main gear legs; oleo-pneumatic shock-absorber in tailwheel unit, which is fully castoring for taxying but locked fore and aft for T-O and landing. Mainwheels have 8 ply tubeless tyres size 17.5 × 6.25-11, pressure 17.25 bars (250 lb/sq in); tailwheel 10 ply tube-type tyre size 5.00-5, pressure 11.04 bars (160 lb/sq in).

POWER PLANT: Two 1,285 kW (1,723 shp) General Electric T700-GE-401/401C turboshafts, one on each side of rotor pylon structure. Basic fuel capacity of 1,802 litres (476 US gallons; 396 Imp gallons), including up to two external auxiliary tanks with a combined capacity of 757 litres (200 US gallons; 166.5 Imp gallons). Ship-to-air helicopter in-flight refuelling (HIFR).

ACCOMMODATION: Crew of three, consisting of pilot, co-pilot/tactical co-ordinator, and sensor operator. One passenger with LAMPS equipment installed; four passengers or two litters with sonobuoy launcher removed. Provision for transportation of internal or external cargo. Space for additional troop seats.

SYSTEMS: Include dual 30 kVA electrical system and Turbomach T-62 gas-turbine APU.

AVIONICS: LAMPS Mk I mission equipment includes Canadian Marconi LN-66HP surveillance radar; General Instruments AN/ALR-66A(V)1 radar warning/ESM; Teledyne Systems AN/ASN-150 tactical management system; dual Collins AN/ARC-159(V)1 UHF radios; Texas Instruments AN/ASQ-81(V)2 magnetic anomaly detector; Computing Devices AN/UYS-503 acoustic processor; Flightline Electronics AN/ARR-84 sonobuoy receiver and AN/ARN-146 on-top position indicator; Tele-Dynamics AN/AKT-22(V)6 sonobuoy datalink; 15 DIFAR and DICASS sonobuoys; AN/ALE-39 chaff/flare dispensers; AN/ASQ-188 torpedo presetter. The US Navy plans to retrofit additional self-defence equipment in fleet SH-2Gs, consisting of Hughes AN/AAQ-16 FLIR, Sanders AN/ALQ-144 IR jammers, Loral AN/AAR-47 missile warning and Collins AN/ARC-182 VHF/UHF secure radio.

EQUIPMENT: Cargo hook for external loads, capacity 1,814 kg (4,000 lb); and folding rescue hoist, capacity 272 kg (600 lb).

ARMAMENT: One or two Mk 46 or Mk 50 torpedoes; eight Mk 25 marine smoke markers. Provision for pintle-mounted 7.62 mm machine gun in both cabin doorways.

DIMENSIONS, EXTERNAL:

Main rotor diameter	13.51 m (44 ft 4 in)
Main rotor blade chord	0.59 m (1 ft 11 in)
Tail rotor diameter	2.46 m (8 ft 1 in)
Tail rotor blade chord	0.236 m (9⅓ in)
Length: fuselage, excl tail rotor	12.19 m (40 ft 0 in)
overall (rotors turning)	16.08 m (52 ft 9 in)
nose and blades folded	11.68 m (38 ft 4 in)
Height: overall (rotors turning)	4.58 m (15 ft 0½ in)
blades folded	4.14 m (13 ft 7 in)
Width overall, incl MAD	3.74 m (12 ft 3 in)
Stabiliser span	2.97 m (9 ft 9 in)

Kaman SH-2G Seasprite upgraded SH-2F with T700-GE-401 engines

1995

Kaman SH-2F Seasprite shipborne ASW helicopter

1994

Kaman SH-2G Super Seasprite of the US Navy

1994

Kaman SH-2F Seasprite in service with the US Navy Reserve

1995

Wheel track (outer wheels) 3.30 m (10 ft 10 in)
Wheelbase 5.13 m (16 ft 10 in)
Tail rotor ground clearance 2.12 m (6 ft 11½ in)
AREAS
Main rotor blades (each) 3.96 m² (42.63 sq ft)
Tail rotor blades (each) 0.295 m² (3.175 sq ft)
Main rotor disc 143.41 m² (1,543.66 sq ft)
Tail rotor disc 4.77 m² (51.32 sq ft)
WEIGHTS AND LOADINGS:
Weight empty 3,447 kg (7,600 lb)
Max T-O weight 6,124 kg (13,500 lb)
Max disc loading 42.72 kg/m² (8.75 lb/sq ft)
PERFORMANCE (ISA):
Max level speed at S/L 138 kts (256 km/h; 159 mph)
Normal cruising speed 120 kts (222 km/h; 138 mph)
Max rate of climb at S/L 762 m (2,500 ft)/min
Rate of climb at S/L, one engine out 530 m (1,740 ft)/min
Service ceiling 7,285 m (23,900 ft)
Service ceiling, one engine out 4,816 m (15,800 ft)
Hovering ceiling: IGE 6,340 m (20,800 ft)
OGE 5,486 m (18,000 ft)
Max range, two external tanks 478 n miles (885 km; 500 miles)
Max endurance, two external tanks 5 h

UPDATED

LEARJET

LEARJET INC
(Subsidiary of Bombardier Inc)

PO Box 7707, Wichita, Kansas 67277-7707
Telephone: 1 (316) 946 2000
Fax: 1 (316) 946 2220
Telex: 417441
PRESIDENT AND CEO: Brian Barents
VICE-PRESIDENT, MARKETING AND SALES: Ted Farid
VICE-PRESIDENT, INTERNATIONAL MARKETING: Robert C. Williams
VICE-PRESIDENT, PRODUCT SUPPORT: Donald E. Grommesh
VICE-PRESIDENT, OPERATIONS: Ghislain Bourgue
VICE-PRESIDENT, CORPORATE AFFAIRS: William G. Robinson
VICE-PRESIDENT, ENGINEERING: William W. Greer
DIRECTOR, PUBLIC AFFAIRS: Jeff Miller

Acquisition of Learjet by Canada's Bombardier announced April 1990 and concluded 22 June 1990 for $75 million; name changed to Learjet Inc. Bombardier assumed responsibility for Learjet's line of credit.

Company originally founded 1960 by Bill Lear Snr as Swiss American Aviation Corporation (SAAC); transferred to Kansas 1962 and renamed Lear Jet Corporation; Gates Rubber Company bought about 60 per cent of company 1967; company renamed Gates Learjet Corporation; 64.8 per cent of company acquired by Integrated Acquisition Inc September 1987 and renamed Learjet Corporation; all manufacturing moved from Tucson, Arizona, to Wichita during 1988, leaving only customer service and modification centre. Wichita workforce 2,891 in 1993, total 3,751 in all locations.

Learjet is major subcontractor for Martin Marietta Manned Space Systems, Boeing, and US Air Force. Has subsidiary to maintain 83 Learjet 35As operating with US Air Force and Air National Guard (as C-21As). More than 1,700 Learjets produced to date.

Learjet bought manufacturing and marketing rights and tooling of Aeronca thrust reversers, for application to Learjet and other aircraft, March 1989.

Learjet delivered 38 aircraft in 1993, compared to 23 in 1992.

VERIFIED

Learjet 24 twin-jet light executive transport aircraft

1992

LEARJET 24F

TYPE: Twin-jet light executive transport aircraft.

PROGRAMME: The prototype Learjet twin-jet executive transport flew for the first time 7 October 1963 and deliveries of production Learjet 23 began 13 October 1964. After a total of 104 of this version had been delivered, it was superseded by the Learjet 24, which was certificated under Federal Air Regulations Part 25 (formerly CAR 4B). Deliveries of the Learjet 24 began in March 1966, and a total of 80 built. This was replaced by a developed version with more powerful engines, known as the Learjet 24B, which received FAA certification 17 December 1968, and then the 24D, 24C, 24E and 24F. The 24F, with Century III modifications, replaced the Learjet 24D and 24D/A in 1976. These latter aircraft were the first to feature a tail unit design in which the non-structural bullet at the junction of the tailplane and fin was deleted.

UPGRADES: **Dee Howard:** XR Learjet. See separate entry.

OPERATORS: Versions of the Learjet 24 are in service with the armed forces of the following countries: Ecuador (1), Slovenia (1) and Venezuela (1).

The following description applies to the Model 24F:

DESIGN FEATURES: Cantilever low-wing monoplane. Wing section NACA 64A109. Dihedral 2° 30′. Incidence 1°. Sweepback 13° at quarter-chord.

FLYING CONTROLS: Manually operated aerodynamically balanced. All-metal ailerons. Hydraulically operated all-metal single-slotted flaps. Hydraulically operated all-metal spoilers mounted on trailing-edge ahead of flaps. Electrically operated trim tab in port aileron. Balance tab in each aileron. Conventional manually operated control surfaces. Electrically operated trim tab in rudder.

STRUCTURE: All-metal eight spar structure with milled alloy skins. The fuselage is an all-metal flush-riveted semi-monocoque fail-safe structure. The tail unit is a cantilever all-metal sweptback structure, with electrically actuated variable-incidence T tailplane and small ventral fin.

LANDING GEAR: Retractable tricycle type, with twin wheels on each main unit and single steerable nosewheel. Hydraulic actuation, with back-up pneumatic extension. Oleo-pneumatic shock-absorbers. Mainwheels fitted with Goodyear 18 × 5.50 10 ply tyres, pressure 7.93 bars (115 lb/sq in). Nosewheel fitted with Goodyear Dual Chine tyre size 18 × 4.40 10 ply rating, pressure 7.24 bars (105 lb/sq in). Goodyear multiple-disc hydraulic brakes. Pneumatic emergency braking system. Parking brakes. Fully modulated anti-skid system.

POWER PLANT: Two General Electric CJ610-8A turbojet engines, each rated at 13.1 kN (2,950 lb st), mounted in pod on each side of fuselage aft of wings. Fuel in integral wing and wingtip tanks, bladder-type cell in the fuselage, with a total standard fuel capacity of 3,180 litres (840 US gallons; 700 Imp gallons). Oil capacity 3.75 litres (1 US gallon; 0.8 Imp gallon) per engine. Engine inlet anti-icing by bleed air.

ACCOMMODATION: Two seats side by side on flight deck, with dual controls. Up to six passengers in cabin, with one on inward-facing bench seat on starboard side at front, then two on forward or aft-facing armchairs with centre aisle, and three on forward-facing couch. Toilet and stowage space under front inward-facing seat, which can be screened from remainder of cabin by curtain. Refreshment cabinet opposite this seat. Baggage compartment aft of cabin. With back of rear bench seat folded down, baggage compartment and rear of cabin can be used to carry cargo or stretchers. Table at rear. In full cargo version, the rearward-facing armchair seats are also removed. Two-piece door, with upward-hinged portion and downward-hinged portion with integral steps, on port side of cabin at front. Emergency exit on starboard side. Cargo door optional. Windscreen anti-icing by engine bleed air with liquid methyl alcohol back-up.

SYSTEMS: Air conditioning by Freon R12 vapour cycle system, supplemented by ram-air heat exchanger during pressurised flight. Cabin pressurisation, by engine bleed air, has maximum differential of 0.65 bars (9.4 lb/sq in). Electrical system powered by dual 400 A 30 V DC starter/generators with AC power supplied from dual 1,000 VA solid-state inverters. Dual 24 V 41 Ah lead-acid battery; Ni/Cd batteries optional. Emergency battery pack optional. Dual engine-driven hydraulic pumps, each capable of maintaining full system pressure of 103.5 bars (1,500 lb/sq in). Auxiliary electrically driven hydraulic pump. Pneumatic system at pressure of 124 to 207 bars (1,800 to 3,000 lb/sq in) for emergency extension of landing gear and operation of wheel brakes. Engine fire detection and extinguishing system. Oxygen system for emergency use has crew demand masks and passenger dropout masks. Alcohol anti-icing system for radome. Anti-icing by engine bleed air ducted into wing leading-edges. Electrically heated de-icing of tailplane leading-edges.

AVIONICS AND EQUIPMENT: Complete nav/com systems, to full airframe airline standard, available to customer's requirements, comprising Collins, Bendix, Sperry or export equipment. All have Learjet autopilot as standard. A typical standard avionics package for the Learjet 24F includes dual Collins VHF-20A com, dual Collins VIR-30A VOR/ILS, dual marker lamps, dual Avtech audio heads, Collins DME-40, Collins TDS-90 transponder, Bendix RDR-1200 radar, Collins ADF-60 ADF, dual Allen 3137 RMI, Collins FD112V flight director, FC-110 flight control system, Collins/J.E.T. PN-101/5-4000 flight indicator (co-pilot), dual J.E.T. VG-206A vertical gyros, dual J.E.T. DN-101D directional gyros and IDC electric-encoding altimeter with altitude alerter and static defect correction module. Standard equipment includes birdproof windscreen; instrument panel flood, depressurisation warning and engine fire warning lights; Mach warning and Alpha Dot stall warning system; dual clocks; control locks; cabin dome reading and entry lights; cabin fire extinguisher; stereo system; ice detector, landing/taxi, navigation strobe and baggage compartment lights; anti-collision beacons; heated pitot tubes and static ports; lightning protection; and fire axe.

DIMENSIONS, EXTERNAL:
Wing span over tip tanks 10.84 m (35 ft 7 in)
Wing chord: at root 2.74 m (9 ft 0 in)
at tip 1.40 m (4 ft 7 in)
Wing aspect ratio 5.01
Length overall 13.18 m (43 ft 3 in)
Length of fuselage 12.50 m (41 ft 0 in)
Height overall 3.73 m (12 ft 3 in)
Tailplane span 4.47 m (14 ft 8 in)
Wheel track (c/l shock-absorbers) 2.51 m (8 ft 3 in)
Wheelbase 4.93 m (16 ft 2 in)
Cabin door: Height 1.57 m (5 ft 2 in)
Standard width 0.61 m (2 ft 0 in)
Optional width 0.91 m (3 ft 0 in)
Emergency exit: Height 0.71 m (2 ft 4 in)
Width 0.48 m (1 ft 7 in)
DIMENSIONS, INTERNAL:
Cabin, between pressure bulkheads:
Length 5.28 m (17 ft 4 in)
Max width 1.50 m (4 ft 11 in)
Max height 1.32 m (4 ft 4 in)
Volume, incl baggage compartment 6.97 m³ (246.0 cu ft)
Baggage compartment 1.13 m³ (40.0 cu ft)
AREAS:
Wings, gross 21.53 m² (231.77 sq ft)
Ailerons (total) 1.08 m² (11.70 sq ft)
Trailing-edge flaps (total) 3.42 m² (36.85 sq ft)
Spoilers 0.66 m² (7.05 sq ft)
Fin 3.47 m² (37.37 sq ft)
Rudder, incl tab 0.67 m² (7.18 sq ft)
Tailplane 5.02 m² (54.00 sq ft)
Elevators 1.31 m² (14.13 sq ft)
WEIGHTS AND LOADINGS:
Weight empty, equipped 3,234 kg (7,130 lb)
Operating weight empty 3,415 kg (7,530 lb)
Max payload 1,542 kg (3,400 lb)
Max T-O weight 6,123 kg (13,500 lb)
Max ramp weight 6,259 kg (13,800 lb)
Max wing bending weight 5,171 kg (11,400 lb)
Max landing weight 5,388 kg (11,800 lb)
Max wing loading 284.4 kg/m² (58.25 lb/sq ft)
Max power loading 233.7 kg/kN (2.29 lb/lb st)
PERFORMANCE (at max T-O weight, unless stated otherwise):
Never-exceed speed Mach 0.81
Max operating speed at 9,450 m (31,000 ft) 473 kts (877 km/h; 545 mph)
Max operating speed at 13,715 m (45,000 ft) 464 kts (859 km/h; 534 mph)

Econ cruising speed at 13,715 m (45,000 ft)
418 kts (774 km/h; 481 mph)
Stalling speed, 'clean'
111 kts (206 km/h; 128 mph) IAS
Stalling speed at max landing weight, wheels and flaps down 88 kts (162.5 km/h; 101 mph) IAS
Max rate of climb at S/L 2,164 m (7,100 ft)/min
Service ceiling 15,545 m (51,000 ft)
Service ceiling, one engine out 8,230 m (27,000 ft)
Min ground turning radius 10.46 m (34 ft 4 in)
T-O run 703 m (2,305 ft)
T-O to 10.7 m (35 ft), FAA balanced field length
1,005 m (3,297 ft)
Landing from 15 m (50 ft) at typical landing weight
700 m (2,295 ft)
Landing run at max landing weight 419 m (1,375 ft)
Range with four passengers, max fuel and 45 min reserves
1,355 n miles (2,512 km; 1,561 miles)

OPERATIONAL NOISE CHARACTERISTICS (FAR Pt 36):
T-O noise level 85.5 EPNdB
Approach noise level 95.3 EPNdB
Sideline noise level 103.7 EPNdB

UPDATED

LEARJET 25D

TYPE: Twin-jet light executive aircraft.

PROGRAMME: First flown 12 August 1966 as the Learjet 25, this aircraft is 1.27 m (4 ft 2 in) longer than the series 24 aircraft, and accommodates eight passengers and a crew of two. FAA certification in the air transport category (FAR Pt 25) was obtained 10 October 1967 and the initial delivery was made in November 1967. British CAA certification was received 26 June 1974.

The Learjet 25D introduced the same Century III improvements as the Learjet 24 series. The 'Softflite' handling package has also been standard on this model since 1 July 1979. In addition, 8° flap settings for take-off have been approved for the 25D, improving the high-altitude/hot day take-off performance. Factory-installed thrust reversers for the General Electric CJ610-8A turbojet engines optional.

Two Learjet 25Bs supplied to the Peruvian Air Force in 1974 were each fitted with an underbelly pack containing two Wild RC-10 aerial survey cameras.

UPGRADES: **Dee Howard:** XR Learjet. See separate entry.

OPERATORS: Versions of the Learjet 25 are in service with the armed forces of the following countries: Bolivia (1), Peru (2) and Yugoslavia (Serbia and Montenegro) (2).

The description of the Learjet 24F applies also to the Model 25D, except in the following details:

DIMENSIONS, EXTERNAL: As for Learjet 24F except:
Length overall 14.50 m (47 ft 7 in)
Wheelbase 5.84 m (19 ft 2 in)

DIMENSIONS, INTERNAL: As for Learjet 24F except:
Cabin, between pressure bulkheads:
Length 6.27 m (20 ft 7 in)
Volume, incl baggage compartment 8.47 m³ (299 cu ft)

WEIGHTS AND LOADINGS:
Weight empty, equipped 3,465 kg (7,640 lb)
Operating weight empty 3,647 kg (8,040 lb)
Max payload 1,524 kg (3,360 lb)
Max T-O weight 6,804 kg (15,000 lb)
Max ramp weight 7,030 kg (15,500 lb)
Max landing weight 6,033 kg (13,300 lb)
Max wing loading 315.9 kg/m² (64.7 lb/sq ft)
Max power loading 259.7 kg/kN (2.54 lb/lb st)

PERFORMANCE (at max T-O weight, unless stated otherwise):
As for Learjet 24F, except:
Max cruising speed at 12,500 m (41,000 ft)
Mach 0.81 (464 kts; 859 km/h; 534 mph)
Stalling speed, wheels and flaps down, at max landing weight 93 kts (172 km/h; 107 mph) IAS
Max rate of climb at S/L 1,920 m (6,300 ft)/min
Rate of climb at S/L, one engine out
526 m (1,725 ft)/min
Service ceiling, one engine out 7,165 m (23,500 ft)
Min ground turning radius 11.43 m (37 ft 6 in)
T-O run 843 m (2,765 ft)
T-O to 10.7 m (36 ft), FAA balanced field length
1,200 m (3,937 ft)
Landing run from 15 m (50 ft) at typical landing weight
686 m (2,250 ft)
Landing run at max landing weight 448 m (1,470 ft)
Range with four passengers, max fuel and 45 min reserves
1,437 n miles (2,663 km; 1,655 miles)

OPERATIONAL NOISE CHARACTERISTICS (FAR Pt 36):
T-O noise level 90.9 EPNdB
Approach noise level 103.7 EPNdB
Sideline noise level 95.2 EPNdB

UPDATED

LEARJET 25F

TYPE: Twin-jet light executive aircraft.

PROGRAMME: The first models of this longer range version of the basic Learjet 25 entered production in 1970. With the addition of a 772 litre (204 US gallon; 170 Imp gallon) fuselage fuel tank. The 25F has a non-stop range in excess of 1,826 n miles (3,383 km; 2,102 miles), plus fuel reserves. The cabin of the 24F is optionally convertible from a four/six-seat configuration to a two-bed sleeper compartment. It is otherwise the same as the Learjet 25D, and the description of that aircraft applies also to the 25F except as follows:

UPGRADES: **Dee Howard:** XR Learjet. See separate entry.

DIMENSIONS, INTERNAL:
Volume, incl baggage compartment 6.97 m³ (246 cu ft)

WEIGHTS AND LOADINGS:
Weight empty, equipped 3,436 kg (7,575 lb)
Operating weight empty 3,617 kg (7,975 lb)
Max payload 1,553 kg (3,425 lb)

PERFORMANCE (at max T-O weight):
Long-range cruising speed at 12,500 m (41,000 ft)
Mach 0.73 (418 kts; 774 km/h; 481 mph)
Max certificated operating altitude 15,545 m (51,000 ft)
Range with four passengers, max fuel, 45 min reserves
1,652 n miles (3,060 km; 1,902 miles)

VERIFIED

Learjet 25 twin-jet light executive transport aircraft

1994

LEARJET 25G

TYPE: Twin-jet light executive aircraft.

PROGRAMME: Announced 23 September 1980, this version of the Learjet 25 represents the first product of the co-operative agreement between the then Gates Learjet and the Dee Howard Company. It is similar in design to the Dee Howard XR Learjet (which see), and has a range more than 20 per cent better than other Model 25s. Features include an inboard-section glove on each wing, carrying additional fuel, a new nacelle pylon configuration that improves cruise performance; wingtip tank fin cuff of new design; pressure-tuned wing leading-edges; and a new span flow limiter. Range capability is further enhanced by resulting drag reduction.

UPGRADES: **Dee Howard:** XR Learjet. See separate entry.

Apart from the aerodynamic improvements, the description for the Learjet 25D applies also to the Learjet 25G, except as follows:

AVIONICS AND EQUIPMENT: Standard avionics as for the Learjet 25D, supplemented by J.E.T. FC-110 autopilot with dual yaw damper, IDC barometric altimeter (co-pilot), J.E.T. PS-835D and AI-804 emergency battery and attitude gyro, and dual Teledyne IVSI. Standard equipment includes Woodward engine synchroniser, LearAvia engine synchroscope, annunciator light package and flap preselect. RCA Primus 300 SL colour weather radar instead of Bendix radar. DN-104B directional gyros.

DIMENSIONS, INTERNAL: As for Learjet 25D except:
Cabin, between pressure bulkheads:
Length 6.38 m (20 ft 11 in)

AREAS:
Wings, gross 22.93 m² (246.8 sq ft)

WEIGHTS AND LOADINGS:
Weight empty, equipped 3,742 kg (8,250 lb)
Max fuel weight 2,991 kg (6,594 lb)
Max T-O weight 7,393 kg (16,300 lb)
Max ramp weight 7,620 kg (16,800 lb)
Max landing weight 6,214 kg (13,700 lb)
Max wing loading 322.5 kg/m² (66.05 lb/sq ft)
Max power loading 282.18 kg/kN (2.76 lb/lb st)

PERFORMANCE (estimated at S/L, ISA):
Cruising speed at 12,500 m (41,000 ft)
464 kts (860 km/h; 534 mph)
T-O balanced field length, FAR Pt 25 at max T-O weight 1,569 m (5,148 ft)
Landing distance (FAR Pt 91, at max landing weight)
820 m (2,690 ft)
Max range, crew of two and four passengers, allowances for taxi, T-O, climb, cruise at long-range power, descent and 45 min reserves
1,800 n miles (3,335 km; 2,073 miles)

VERIFIED

LEARJET 28 and 29 LONGHORN

TYPE: Twin-jet light executive aircraft.

PROGRAMME: Displayed for the first time at the US National Business Aircraft Association's annual convention on 27-29 September 1977, the prototype of the Learjet 28/29 series was generally similar to the 10-seat Learjet 25D, except that it introduced a wing of much increased span, fitted with supercritical winglets. Combined with a cambered leading-edge, this is claimed to improve climb performance, reduce drag, offer improved high-altitude cruise efficiency, reduce runway requirements and reduce approach speed. The 'Softflite' handling package became standard 1 July 1970. No wingtip fuel tanks are fitted.

Like the Learjet 25D the 28/29 is powered by two General Electric CJ610-8A turbojet engines, each rated at 13.1 kN (2,950 lb st). Total usable fuel capacity of the Model 28 is 2,646 litres (699 US gallons; 582 Imp gallons), and of the Model 29 is 3,035 litres (802 US gallons; 667.8 Imp gallons). The Model 28 accommodates a crew of two and eight passengers, the Model 29 a crew of two and six passengers. In other respects the 28/29 are similar to the 25D.

FAA certification of the Learjet 28/29 was received on 30 January 1979, and the first production deliveries were made shortly after.

OPERATORS: Versions of the Learjet 29 are in service with the Indian armed forces (2).

DIMENSIONS, EXTERNAL:
Wing span 13.35 m (43 ft 9½ in)
Length overall 14.51 m (47 ft 7 7/12 in)
Height overall 3.73 m (12 ft 3⅛ in)

DIMENSIONS, INTERNAL:
Cabin, excl flight deck:
Length: 28 4.37 m (14 ft 4 in)
29 3.86 m (12 ft 8 in)
Max width 1.49 m (4 ft 11 in)
Max height 1.32 m (4 ft 4 in)
Volume: 28 5.80 m³ (205 cu ft)
29 4.67 m³ (165 cu ft)
Baggage compartment: 28 0.85 m³ (30 cu ft)
29 0.76 m³ (27 cu ft)

AREAS:
Wings, gross 24.57 m² (264.5 sq ft)

WEIGHTS AND LOADINGS:
Weight empty: 28 3,750 kg (8,268 lb)
29 3,730 kg (8,224 lb)
Max payload: 28 1,058 kg (2,332 lb)
29 1,078 kg (2,376 lb)
Max fuel weight: 28 2,124 kg (4,684 lb)
29 2,437 kg (5,373 lb)
Max T-O weight 6,804 kg (15,000 lb)
Max ramp weight 7,030 kg (15,500 lb)

Learjet 28/29 twin-jet light executive aircraft

1992

Max landing weight 6,486 kg (14,300 lb)
Max wing loading 276.9 kg/m² (56.71 lb/sq ft)
Max power loading 246.52 kg/kN (2.42 lb/lb st)

PERFORMANCE (at max T-O weight, except where indicated):
- Never-exceed speed Mach 0.81
- Max level speed at 7,620 m (25,000 ft) 477 kts (833 km/h; 549 mph)
- Max cruising speed, mid-cruise weight, at 14,325 m (47,000 ft) 452 kts (837 km/h; 520 mph)
- Econ cruising speed, mid-cruise speed, at 15,545 m (51,000 ft) 408 kts (756 km/h; 470 mph)
- Stalling speed, flaps and wheels down, engines idling 89 kts (165 km/h; 102.5 mph) IAS
- Service ceiling 15,545 m (51,000 ft)
- Service ceiling, one engine out 8,840 m (29,000 ft)
- T-O to 9 m (30 ft) 927 m (3,040 ft)
- Landing distance from 15 m (50 ft) at max landing weight 833 m (2,734 ft)
- Range with four passengers, max fuel and 45 min reserves:
  - 28 1,137 n miles (2,107 km; 1,309 miles)
  - 29 1,376 n miles (2,550 km; 1,584 miles)

UPDATED

LEARJET 35A/36A SPECIAL MISSIONS VERSIONS

US Military designation: C-21A

TYPE: Special mission adaptations for Learjet 35A/36A.

PROGRAMME: For full details of standard Learjet 35A/36A aircraft see *Jane's All the World's Aircraft* 1995-96.

VERSIONS: **EC-35A:** Used for EW training simulation or as stand-off ECM/ESM platform. Still in limited production at customer demand.

PC-35A: Maritime patrol; equipment includes 360° sea surveillance digital radar, high-resolution television, forward-looking infra-red (FLIR), infra-red linescanner (IRLS), electronic support measures (ESM), magnetic anomaly detector, integrated tactical displays, and VLF Omega or other long-range navaids; hardpoint under each wing with Alkan 165B ejector for external stores up to 454 kg (1,000 lb); drop hatch for rescue gear, multitrack digital recorders, homing systems, ASW sonobuoy systems and data annotated hand-held cameras.

RC-35A and **RC-36A:** Reconnaissance versions; standard installations include long-range oblique photographic cameras (LOROP), side-looking synthetic aperture radars, and surveillance camera system in external pods. Geological versions delivered to China (see below).

UC-35A: Utility versions for transport, navaids calibration, medevac and target towing. Certificated tow systems include Hayes Universal Tow Target System (HUTTS) with or without MTR-101, Flight Refuelling LLHK, MRTT and EMT TGL targets, and Marquardt aerial target launch and recovery tow reel.

U-36A: Extensively modified for Japan Maritime Self-Defence Force (JMSDF). Delivered for target towing, antiship missile simulation and ECM; tip pods extended, in association with Shin Meiwa (now ShinMaywa), to house HWQ-1T missile seeker simulator, AN/ALQ-6 jammer and cameras. Additional equipment includes long-range ocean surveillance radar in underbelly fairing, AN/ALE-43 chaff dispenser, ARS-1-L high-speed tow sleeve with scoring, two-piece windscreen with electrical demisting for low-level missions, expanded underwing stores capability, greater maximum T-O and landing weights. Further deliveries to Japan expected during 1990s.

OPERATORS: Versions of the Learjet 35/36 are in service with the armed forces of the following countries: Argentina (6), Bolivia (1), Brazil (12), Chile (2), Finland (3), Iraq (5), Japan (5), Oman (1), Peru (2), Saudi Arabia (2), Switzerland (2), Thailand (2) and USA (84).

DESIGN FEATURES: Civilian and paramilitary missions include aerial survey, aeronautical research, airways calibration, ASW, atmospheric research, border patrol, ESM/ECM, geophysical survey, maritime patrol, pilot training, radar surveillance, reconnaissance, search and rescue, and weather modification.

PERFORMANCE (PC-35A):
- Operating speed:
  - at 11,275-12,500 m (37,000-41,000 ft) 415 kts (769 km/h; 478 mph)
  - at 4,575-7,620 m (15,000-25,000 ft) 319 kts (590 km/h; 367 mph)
  - S/L to 610 m (2,000 ft) 250 kts (463 km/h; 288 mph)
- Rate of climb at S/L 1,380 m (4,525 ft)/min
- Range:
  - at high altitude 2,249 n miles (4,168 km; 2,590 miles)
  - at medium altitude 1,617 n miles (2,996 km; 1,862 miles)
  - at low altitude 1,060 n miles (1,964 km; 1,220 miles)

UPDATED

Learjet 35A Special Missions aircraft

1994

LEARJET 55B and 55C

TYPE: Medium-range business jet.

PROGRAMME: First flight of first prototype Learjet 55 (N551GL) 19 April 1979; first flight of second prototype (N552GL) 15 November 1979; first flight of production aircraft 11 August 1980; FAR Pt 25 certification 18 March 1981; first delivery 30 April 1981. Phase I performance improvement package, certificated July 1983, included new triangular-wedge stall strips, automatic ground spoilers and automatic power reserve; Phase IA performance improvement package, introduced Spring 1984, includes high-energy, long-life brakes and modified landing gear; optional maximum T-O weight of 9,752 kg (21,500 lb) and maximum landing weight of 8,165 kg (18,000 lb) introduced from aircraft c/n 55-107; final Learjet 55C, No. 147, delivered December 1990 to UK; Learjet 55C now superseded by Learjet 60.

VERSIONS: **Learjet 55B:** Announced September 1986; optional maximum T-O weight became standard, wing improved, new interior, all-digital flight deck; better take-off performance, increased range and better mission flexibility.

Learjet 55C: Announced September 1987; incorporated Delta Fins which improved stability at all speeds, eliminated Dutch roll, softened stall, reduced approach speeds and ended need for stick puller/pusher and dual yaw dampers at departure.

Learjet 55C/ER: Extended range version with additional 163 kg (359 lb) fuel tank in tailcone baggage compartment; can be retrofitted.

Learjet 55C/LR: As 55C/ER, but with additional 386 litre (102 US gallon; 85 Imp gallon) tank between standard fuselage tank and ER optional tank; can be retrofitted; typical seating for seven and two crew.

UPGRADES: **Branson:** Long-range fuel tanks. See separate entry.

DESIGN FEATURES: Sweepback 13° at quarter-chord; advanced cambered leading-edge; Delta Fins, Softflite package and supercritical winglets. Anti-icing of wing leading-edge by engine bleed air and of tailplane leading-edge by electric heating.

FLYING CONTROLS: Manual flying controls with electric trim tabs on port aileron and rudder; electrically actuated variable incidence tailplane; hydraulically actuated slotted trailing-edge flaps; spoilers ahead of flaps.

Learjet U-36A in service with the Japanese Maritime Self-Defence Force (JMSDF)

1995

STRUCTURE: Multispar wing with cavity-milled skins; upper skin tapers in thickness towards tip; semi-monocoque fail-safe fuselage.

LANDING GEAR: Hydraulically retractable tricycle type, with twin wheels on each main unit and single steerable nosewheel (±55°). Mainwheel tyres size 17.5 × 5.75-8, 12 ply, pressure 12.4 bars (180 lb/sq in); nosewheel tyre size 18.0 × 4.4, 10 ply, pressure 7.24 bars (105 lb/sq in). High-pressure pneumatic system for emergency extension. Oleo-pneumatic shock-absorbers. Chined nosewheel tyre. Mainwheel doors are of composites construction. High-energy hydraulic braking system, with pneumatic back-up. Fully modulated anti-skid units.

POWER PLANT: Two Garrett TFE731-3A-2B turbofans, each rated at 16.46 kN (3,700 lb st), plus 0.80 kN (180 lb st) automatic power reserve, pod-mounted on sides of fuselage aft of wing. Thrust reversers standard. Fuel in integral wing tanks and a fuselage bladder tank, with a combined usable capacity of 3,782 litres (999 US gallons; 832 Imp gallons) in standard Model 55C; 3,982 litres (1,052 US gallons; 876 Imp gallons) in Model 55C/ER; 4,353 litres (1,150 US gallons; 957 Imp gallons) in Model 55C/LR. Single-point refuelling. Engine nacelle leading-edges and fan hubs anti-iced by engine bleed air.

ACCOMMODATION: Crew of two on flight deck, with dual controls. IPECO fully adjustable crew seats. Seating for four to eight passengers in a choice of six interior layouts. Two folding tables. Carpeted floor. Galley refreshment cabinet. Toilet. Baggage space at rear of cabin, and in fuselage nosecone and tailcone of basic Model 55C. Tailcone baggage space reduced on long-range versions, but 55C/LR has baggage area at front of cabin. Two-piece clamshell door at forward end of cabin on port side, with integral steps built into the lower section. Emergency exit/baggage door on starboard side of cabin.

SYSTEMS: Environmental control system comprises cabin pressurisation, ventilation, heating and cooling. Heating and pressurisation are provided by engine bleed air, with a maximum pressure differential of 0.65 bar (9.4 lb/sq in), maintaining a cabin altitude of 2,440 m (8,000 ft), to 15,545 m (51,000 ft). Freon vapour-cycle cooling system, supplemented by a ram air system. Anti-icing system includes distribution of engine bleed air to wing leading-edges, engine nacelle leading-edges and fan hubs, and pilot and co-pilot windscreens; electric anti-icing of tailplane leading-edge, pitot heads, stall warning vanes and static ports; and alcohol anti-icing of windscreens. Hydraulic system supplied by two engine-driven variable-volume constant-pressure pumps, one on each engine, each capable of maintaining alone the full system pressure of 103.5 bars (1,500 lb/sq in) for operation of landing gear, flaps and spoilers. Hydraulic system maximum flow rate 15 litres (4 US gallons; 3.33 Imp gallons)/min. Cylindrical reservoir pressurised to 1.38 bars (20 lb/sq in). Electrically driven hydraulic pump for emergency operation of all hydraulic services. Pneumatic system of 124 to 207 bars (1,800 to 3,000 lb/sq in) pressure for emergency extension of landing gear and operation of brakes. Electrical system powered by two 28 V 400 A engine-driven brushless generators, either of which is capable of maintaining adequate DC power to operate all electrical services; two 1 kVA solid-state inverters to provide AC power; and two 24 V 37 Ah lead-acid batteries. Oxygen system of 1.08 m³ (38 cu ft) capacity, with crew demand masks; dropout mask for each passenger, which is presented automatically if cabin altitude exceeds 4,265 m (14,000 ft). Sundstrand Turbomach APU optional (not on 55C/ER or 55C/LR).

AVIONICS: From aircraft c/n 134 onwards standard avionics include a Collins five-tube EFIS-85L-12 electronic flight instrumentation system with dual AHS-85 AHRS; Collins APS-85 digital autopilot with glareshield controller and dual flight data computers; dual ADC-82 air data systems; dual VHF-22A nav/coms with CTL-22 controls; dual VIR-32 nav receivers with CTL-32 controls; dual DME-42 with IND-42C indicators; ADF-60 with CTL-62 control; dual RMI-30; dual TDR-90 transponders with CTL-92 controls; ALT-55B radio altimeter, Collins WXR-350 Sensor weather radar, Universal UNS-1A long-range nav system; Bendix/King KHF 950 HF; Frederickson Jetcal-5 secal; Collins flight profile advisory system; and Wulfsberg Flitefone VI. Other standard flight deck equipment includes dual Collins encoding altimeters; MSI-80 Mach/airspeed indicators; VSI-80 instantaneous vertical speed indicators; Collins altitude preselector/alerter; TAS/SAT/TAT temperature indicator; IDC 2 in standby Mach/IAS indicator and altimeter; J.E.T. PS-835D and AI-804 emergency battery and attitude gyro; dual marker beacon systems; dual Avtech audio systems; dual Davtron 877 clocks; yaw damper; dual stall warning indicators; engine synchroniser and synchroscope; annunciator panels; avionics master switch; N_1 reminder; chip detector and flap preselect.

EQUIPMENT: Standard equipment includes sun visors; map lights; lifejackets; fire extinguisher; writing tables; dome lights; oxygen masks; manual storage compartments; and dividing panel between flight deck and cabin. Cabin equipment includes two folding tables; four individual seats with tracking, swivel, recline and sideways motions; three-seat forward-facing divan with storage and tracking and recline motions, single side-facing seat, galley refreshment cabinet with decanters, bottle cooler, ice chest, oven, tray carrier, hot and cold liquid containers and storage area; forward coat closet; stereo and tape storage; magazine rack; convenience panel with reading lights, air vents, cabin speakers and oxygen masks; lighting control panel; opaque and tinted window shades; indirect lighting; aisle lighting; entrance step light; flushing toilet with vanity mirror and light; baggage compartment light; passenger lifejackets; auxiliary cabin heater; fire extinguisher and axe; nose and tail baggage compartment lights; wing ice light and maintenance interphone jack plug sockets.

DIMENSIONS, EXTERNAL:

Wing span	13.34 m (43 ft 9 in)
Wing chord: at root	2.74 m (9 ft 0 in)
at tip	1.07 m (3 ft 6 in)
Wing aspect ratio	6.7
Length: overall	16.79 m (55 ft 1 in)
fuselage	15.93 m (52 ft 3 in)
Height overall	4.47 m (14 ft 8 in)
Tailplane span	4.47 m (14 ft 8 in)
Wheel track	2.51 m (8 ft 3 in)
Wheelbase	7.01 m (23 ft 0 in)
Cabin door: Height	1.70 m (5 ft 7 in)
Width	0.61 m (2 ft 0 in)

DIMENSIONS, INTERNAL:

Cabin: Length between pressure bulkheads:	
55C, 55C/ER	6.71 m (22 ft 0 in)
55C/LR	6.30 m (20 ft 8 in)
Length, cockpit/cabin divider to rear pressure bulkhead:	
55C, 55C/ER	5.08 m (16 ft 8 in)
55C/LR	4.67 m (15 ft 4 in)
Max width	1.80 m (5 ft 10¾ in)
Max height	1.74 m (5 ft 8½ in)
Volume, incl flight deck:	
55C, 55C/ER	13.37 m³ (472 cu ft)
55C/LR	12.88 m³ (455 cu ft)

Learjet 55C medium-range business jet

1994

Baggage capacity: 55C, rear cabin	0.93 m³ (33.0 cu ft)
55C, nose	0.17 m³ (6.0 cu ft)
55C, tail	0.52 m³ (18.5 cu ft)
55C/ER, total	1.36 m³ (48.0 cu ft)
55C/LR, total	1.81 m³ (64.0 cu ft)

AREAS:

Wings, gross	24.57 m² (264.5 sq ft)
Ailerons (total)	1.09 m² (11.70 sq ft)
Trailing-edge flaps (total)	3.42 m² (36.85 sq ft)
Winglets (total)	1.11 m² (12.00 sq ft)
Spoilers (total)	0.65 m² (7.05 sq ft)
Fin	4.67 m² (50.29 sq ft)
Rudder	0.99 m² (10.65 sq ft)
Tailplane	5.02 m² (54.00 sq ft)
Elevators (total)	1.31 m² (14.13 sq ft)

WEIGHTS AND LOADINGS:

Weight empty: 55C	5,832 kg (12,858 lb)
55C/ER	5,861 kg (12,922 lb)
55C/LR	5,920 kg (13,052 lb)
Typical operating weight: 55C	6,013 kg (13,258 lb)
55C/ER	6,042 kg (13,322 lb)
55C/LR	6,101 kg (13,452 lb)
Payload with max fuel: 55C	590 kg (1,302 lb)
55C/ER	625 kg (1,379 lb)
55C/LR	268 kg (519 lb)
Max payload: 55C	790 kg (1,742 lb)
55C/ER	761 kg (1,678 lb)
55C/LR	702 kg (1,548 lb)
Max fuel weight: 55C	3,035 kg (6,690 lb)
55C/ER	3,197 kg (7,049 lb)
55C/LR	3,496 kg (7,707 lb)
Max T-O weight: 55C	9,525 kg (21,000 lb)
55/ER and LR	9,752 kg (21,500 lb)
Max ramp weight: 55C	9,639 kg (21,250 lb)
55C/ER and LR	9,865 kg (21,750 lb)
Max zero-fuel weight: all versions	6,804 kg (15,000 lb)
Max landing weight: all versions	8,165 kg (18,000 lb)
Max wing loading: 55C	387.6 kg/m² (79.4 lb/sq ft)
55C/ER and LR	396.9 kg/m² (81.3 lb/sq ft)
Max power loading: 55C	289.3 kg/kN (2.84 lb/lb st)
55C/ER & LR	296.2 kg/kN (2.90 lb/lb st)

PERFORMANCE (at max T-O weight except where indicated):

Never-exceed speed (VNE):
below 2,440 m (8,000 ft) — 300 kts (555 km/h; 345 mph) IAS
2,440 m (8,000 ft) to 11,275 m (37,000 ft) — 350 kts (648 km/h; 403 mph) IAS
11,275 m (37,000 ft) to 13,715 m (45,000 ft) — Mach 0.81 to Mach 0.79
above 13,715 m (45,000 ft) — Mach 0.79

Max level speed at 9,150 m (30,000 ft) — 477 kts (884 km/h; 549 mph)
Max cruising speed at 12,500 m (41,000 ft) — 455 kts (843 km/h; 524 mph)
Econ cruising speed at 14,325 m (47,000 ft) — 419 kts (776 km/h; 482 mph)
Stalling speed — 106 kts (197 km/h; 122 mph)

Max rate of climb at S/L: 55C	1,273 m (4,176 ft)/min
55C/ER & LR	1,237 m (4,059 ft)/min
Rate of climb at S/L; one engine out:	
55C	378 m (1,240 ft)/min
55C/ER and LR	305 m (1,000 ft)/min
Max certificated ceiling	15,545 m (51,000 ft)
Min ground turning radius, about nosewheel	11.58 m (38 ft 0 in)
T-O balanced field length, FAR Pt 25:	
55C	1,536 m (5,039 ft)
55C/ER and LR	1,615 m (5,299 ft)
Landing distance, FAR Pt 91 at max landing weight	991 m (3,250 ft)

Range with crew of two, four passengers, allowances for taxi, T-O, climb, cruise at long-range power at 13,100 m (43,000 ft), descent and 45 min reserves:

55C	2,189 n miles (4,056 km; 2,520 miles)
55C/ER	2,305 n miles (4,271 km; 2,654 miles)
55C/LR	2,397 n miles (4,442 km; 2,760 miles)

OPERATIONAL NOISE LEVELS (FAR Pt 36):

T-O	86.3 EPNdB
Approach	90.7 EPNdB
Sideline	91.0 EPNdB

VERIFIED

LOCKHEED MARTIN

LOCKHEED MARTIN CORPORATION

6801 Rockledge Drive, Bethesda, Maryland 20817
Telephone: 1 (301) 897 6000
Fax: 1 (301) 897 6028
CHAIRMAN, AND CEO: Daniel M. Tellep
PRESIDENT: Norman Augustine

Former Lockheed Aircraft Corporation renamed Lockheed Corporation in September 1977. Merger with Martin Marietta announced 30 August 1994 and completed 15 March 1995. Combined corporation is largest defence contractor in USA, with annual sales worth $23 billion. Activities include design and production of aircraft, electronics, satellites, space systems, missiles, ocean systems, information systems, and systems for strategic defence and for command, control, communications and intelligence.

Lockheed Martin Corporation is largest US DoD contractor, largest NASA contractor and largest US DoE contractor, 1995 merger having resulted in emergence of company with 170,000 employees and organised into four major business groups: **Aeronautics Sector** (see next entry); **Information and Technology Services Sector** (comprising various elements including Lockheed Engineering & Science Company, Lockheed Information Management Services Company, Lockheed Space Operations Company, Martin Marietta Management & Data Systems Company, Martin Marietta Services Group, Martin Marietta Manned Space Systems Company, CalComp and Formtek); **Electronics Sector** (comprising Lockheed Canada, Lockheed Sanders Inc, Martin Marietta Armament Systems Company, Martin Marietta Communications Systems Company, Martin Marietta Control Systems Company, Martin Marietta Defense Systems Company, Martin Marietta Electronics & Missiles Company, Martin Marietta Government Electronic Systems Company and Martin Marietta Ocean Radar & Sensor Systems Company); and **Space and Strategic Missiles Sector** (comprising Lockheed Missiles & Space Company, Martin Marietta Astronautics Company and Martin Marietta Astrospace Company). Subsidiary concerns include Lockheed Martin Finance Inc, Sandia Corporation, Idaho Technologies Company and Martin Marietta Materials Inc. Joint ventures include Lockheed Air Terminal and Space Imaging Inc.

UPDATED

LOCKHEED MARTIN AERONAUTICS SECTOR (LMAS)

6801 Rockledge Drive, Bethesda, Maryland 20817
Telephone: 1 (301) 897 6121
Fax: 1 (301) 897 6252
PRESIDENT: James A. Blackwell, Jr
VICE-PRESIDENT, BUSINESS DEVELOPMENT:
James F. Madewell

LMAS includes six operating companies:

Lockheed Aeronautical Systems Company
Next entry
Lockheed Advanced Development Company
Follows Aeronautical Systems Company
Lockheed Fort Worth Company
Follows Advanced Development Company
Lockheed Aircraft Services Company
Follows Fort Worth Company
Lockheed Support Systems Inc
Martin Marietta Aero & Naval Systems Company

Prior to merger that culminated in creation of Lockheed Martin Aeronautics Sector, Lockheed aircraft manufacturing activity consolidated at Marietta in 1990, with move of P-3 Orion assembly line from Palmdale; established work on C-130 and P-3 programmes, plus F-22 development, continues at Marietta under management of Lockheed Aeronautical Systems Company. Lockheed Advanced Development Company retains autonomy at Palmdale within LMAS grouping, which has workforce in excess of 40,000.

UPDATED

LOCKHEED GALAXY

USAF designation: C-5

TYPE: Heavy logistics transport aircraft.

PROGRAMME: In early 1978 Lockheed received a USAF contract to manufacture two new sets of wings for the C-5A, of a design intended to increase service life of 30,000 hours. Apart from the moving surfaces, these wings are of virtually new design, using 7175-T73511 aluminium alloy for greater strength and increased resistance to corrosion and were built under subcontract by Avco Aerostructures Textron. The first re-winged Galaxy was redelivered to the USAF in early 1981, and the remaining 76 in operational service were fitted with the new wings in a modification programme which was completed in July 1987. Under a $10 million 1986 contract from the USAF, the Georgia division commenced modification of C-5As to enable them to carry oversized Space Shuttle cargo that was beyond existing airlift capability.

The 433rd MAW at Kelly AFB, Texas, became the first unit to receive the C-5A when the first of eight aircraft was delivered in December 1984. The Air National Guard's 105th MAG, based at Stewart Airport, Newburgh, New York, received its first C-5A in July 1985.

In the Summer of 1982, Congress approved a Lockheed proposal to manufacture a C-5N (N: new) version of the

Lockheed C-5B Galaxy aircraft of the US Air Force Air Mobility Command *(Paul Tompkins)*

1995

Galaxy, to meet an urgent USAF requirement for additional heavy lift capacity.

VERSIONS: **C-5A:** Initial version.

C-5B: Improved version. With similar internal arrangements and external aerodynamic configuration but with all the improvements developed for the C-5A.

The aircraft internal arrangements and external aerodynamic configuration are the same as those of the C-5A, but the new version includes all the changes, improvements and modifications incorporated in its predecessor during its years of service with the USAF.

The first C-5B was rolled out 12 July 1985, and made its first flight from Dobbins AFB 10 September 1985. First delivery to the USAF was made 8 January 1986, and deliveries were completed 17 April 1989.

C-5C: Modified aircraft with strengthened interior fuselage for the transport of satellites and other space related cargo. The front visor has also been sealed. Two aircraft converted.

C-5D: Proposed new-build aircraft with glass cockpit, improved engines (upgraded General Electric TF39, P&W F117 or Rolls-Royce RB211-535E4

UPGRADES: **Lockheed:** Re-winging and other modifications. See following entry.

Lockheed: Under a $6 million contract, Lockheed Aircraft Systems installed an ECM protection package to protect against infra-red guided missiles in 1993-94. The ECM package is designated Pacer Snow - Systems include: Tracor AN/ALE-40 flare dispensers and Loral AN/AAR-47 missile warning systems.

OPERATORS: Versions of the C-5 are in service with the US armed forces (94).

DESIGN FEATURES: Cantilever high-wing monoplane. Wing section NACA 0012 (modified) at 20 per cent span. NACA 0011 (modified) at 43.7 per cent span. Anhedral 5° 30′ at quarter-chord. Incidence 3° 30′ at root. Sweepback at quarter-chord 25°. The tail unit is a cantilever all-metal T tail. All surfaces swept; anhedral on tailplane.

FLYING CONTROLS: Statically balanced aluminium alloy ailerons. Modified Fowler aluminium trailing-edge flaps. Simple hinged aluminium alloy spoilers forward of flaps. No trim tabs. Sealed inboard slats and slotted outboard slats on leading-edges. Ailerons and spoilers operated by hydraulic servo actuators. Trailing-edge flaps and leading-edges actuated by ball screw jack and torque tube system. Variable incidence tailplane. Elevators and rudder operated through hydraulic servo actuators. Tailplane actuated through hydraulically powered screw jack.

STRUCTURE: Conventional fail-safe box structure of built-up spars and machined alloy extruded skin panels. Major wing assemblies built by Avco Aerostructures Textron and joined by Lockheed. Leading-edge slats and slat tracks, leading-edge ribs and ailerons built under subcontract by Canadair Ltd. The fuselage is a conventional semi-monocoque fail-safe structure of 7049-T73, 7050-T736, 7075-T73 and 7475 aluminium alloys. The tail unit components are single-cell box structures with integrally stiffened aluminium alloy skin panels. Elevators in four sections; rudder in two sections.

LANDING GEAR: Retractable tricycle type. Nose unit retracted rearward by hydraulically driven ballscrews. Main units rotated through 90° and retracted inward via hydraulically driven gearbox. Single nose shock-absorber and four main gear shock-absorbers are of Bendix oleo-pneumatic dual-chamber type. Four wheels on nose unit (steerable ±60°). Four main units (two in tandem on each side) each comprise a 'triangular footprint' six-wheel bogie made up of a pair of wheels forward of the shock-absorber and two pairs aft. All 28 tyres size 49 × 17-20 Type VII 26 ply. Tyre pressure: nosewheels 9.45 bars (137 lb/sq in), mainwheels 7.65 bars (111 lb/sq in) with in-flight deflation capability. Goodrich wheels, tyres and carbon disc brakes. Hydro-Aire fully modulating anti-skid units. Ground manoeuvrability enhanced by castoring rear main units.

POWER PLANT: Four General Electric TF39-GE-1C turbofans, each rated at 191.2 kN (43,000 lb st). Twelve integral fuel tanks in wings, between front and rear spars, comprising two outboard main tanks (each 13,874 litres; 3,665 US gallons; 3,052 Imp gallons), two inboard main tanks (each 14,755 litres; 3,898 US gallons; 3,246 Imp gallons), two outboard auxiliary tanks (each 18,034 litres; 4,764 US gallons; 3,967 Imp gallons), two inboard auxiliary tanks (each 18,401 litres; 4,861 US gallons; 4,078 Imp gallons); two outboard extended range tanks (each 15,865 litres; 4,191 US gallons; 3,490 Imp gallons), and two inboard extended range tanks (each 15,883 litres; 4,196 US gallons; 3,494 Imp gallons). Total capacity 193,624 litres (51,150 US gallons; 42,591 Imp gallons). Two refuelling points each side, in forward part of main landing gear pods. Flight refuelling capability, via inlet in upper forward fuselage, over flight engineer's station (compatible with KC-135 and KC-10 tankers). Oil capacity 138 litres (36.4 US gallons; 30.3 Imp gallons).

ACCOMMODATION: Standard crew of five, consisting of pilot, co-pilot, flight engineer and two loadmasters, with rest area for 15 people (relief crew, couriers and so on) at front of upper deck. Basic version has seats for 75 troops on rear part of upper deck, aft of wing box. Provision for carrying 270 troops on lower deck, but aircraft is primarily employed as freighter. Typical loads include two M1 Abrams tanks or 16¾ ton lorries; or one M1 and two Bradley armoured fighting vehicles; or 10 combat-ready Marine Corps LAV light armoured vehicles; or six AH-64 Apache attack helicopters; or 10 Pershing missiles with tow and launch vehicles; or 36 standard 463L load pallets. 'Visor' type upward-hinged nose, and loading ramp, permit straight-in loading into front of hold, under flight deck. Rear straight-in loading via ramp which forms undersurface of rear fuselage. Side panels of rear fuselage, by ramp, hinge outward to improve access on ground but do not need to open for airdrop operations in view of width of ramp. Ramp and associated side panels built under subcontract by Canadair Ltd. Provision of aerial delivery systems (ADS) kits for paratroops or cargo. Two passenger doors, one each on port and starboard sides at rear end of lower deck. One crew door on port side at forward end of lower deck. Five evacuation slides and four 25-person liferafts, all supplied by Garrett Air Cruisers. Entire accommodation pressurised and air conditioned.

Electronically controlled air conditioning and pressurisation systems: pressure differential 0.57 bars (8.2 lb/sq in). Four hydraulic systems, pressure 207 bars (3,000 lb/sq in) each, supply flying control and utility systems, with power supplied by two identical variable volume, constant pressure pumps on engine, each rated at 227 litres (60 US gallons; 50 Imp gallons)/min. Each system contains an unpressurised hydraulic reservoir. Electrical system includes four 60/80 kVA AC engine-driven generators. Two APUs provide auxiliary pneumatic and electrical power. Ground hydraulic power is supplied by two air turbine motors.

AVIONICS AND EQUIPMENT: Communications and navigation equipment to military requirements. Bendix colour weather radar. Three Delco inertial navigation units with triple-mix capabilities. Special equipment includes updated electronic malfunction detection, analysis and recording subsystem (MADAR II) which scans and analyses more than 800 test points.

DIMENSIONS, EXTERNAL:

Wing span	67.88 m (222 ft 8½ in)
Wing chord: at root	13.85 m (45 ft 5¼ in)
at tip	4.67 m (15 ft 4 in)
Wing aspect ratio	7.75
Length overall	75.54 m (247 ft 10 in)
Length of fuselage	70.29 m (230 ft 7¼ in)
Height overall	19.85 m (65 ft 1½ in)
Tailplane span	20.94 m (68 ft 8½ in)
Wheel track (between outer wheels)	11.42 m (37 ft 5½ in)

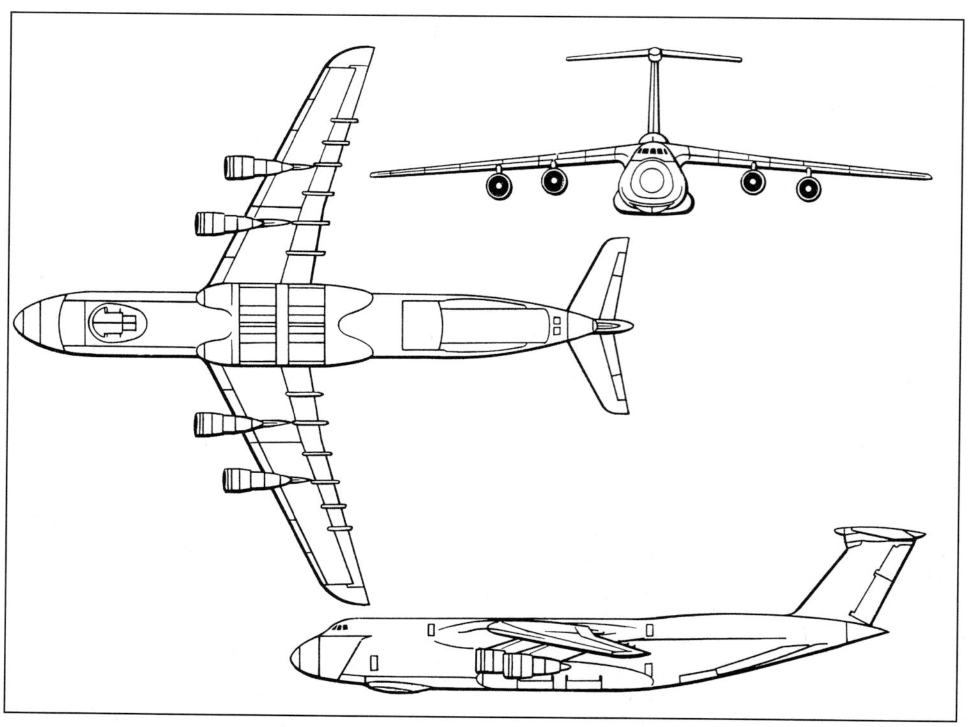

Lockheed C-5 Galaxy heavy transport aircraft *(Jane's/Mike Keep)*

1994

Wheelbase (c/l main gear to c/l nose gear) 22.22 m (72 ft 11 in)
Crew door (lower deck): Height 1.80 m (5 ft 10¾ in)
Width 1.02 m (3 ft 4 in)
Height to sill 3.94 m (12 ft 11 in)
Passenger door (lower deck): Height 1.83 m (6 ft 0 in)
Width 0.91 m (3 ft 0 in)
Height to sill 3.56 m (11 ft 8 in)
After loading opening ramp (ramp lowered):
Max height 3.93 m (12 ft 10¾ in)
Max width 5.79 m (19 ft 0 in)
Aft straight-in loading: Max height 2.90 m (9 ft 6 in)
Max width 5.79 m (19 ft 0 in)
Forward loading opening (ramp lowered or straight-in):
Max height 4.11 m (13 ft 6 in)
Max width 5.79 m (19 ft 0 in)
Height to floor (kneeled): forward 1.34 m (4 ft 4¾ in)
rear 1.45 m (4 ft 9 in)

DIMENSIONS, INTERNAL:
Cabin, excl flight deck:
Length: upper deck, forward 11.99 m (39 ft 4 in)
upper deck, rear 18.20 m (59 ft 8½ in)
lower deck, without ramps 36.00 m (121 ft 1 in)
lower deck, with ramps 44.09 m (144 ft 8 in)
Max width:
upper deck, forward 4.20 m (13 ft 9½ in)
upper deck, rear 3.96 m (13 ft 0 in)
lower deck 5.79 m (19 ft 0 in)
Max height: upper 2.29 m (7 ft 6 in)
lower deck 4.09 m (13 ft 5 in)
Floor area:
upper deck, forward 50.17 m^2 (540.0 sq ft)
upper deck, rear 72.10 m^2 (776.1 sq ft)
lower deck, without ramp 213.76 m^2 (2,300.9 sq ft)
Volume:
upper deck, forward 56.91 m^3 (2,010 cu ft)
upper deck, rear 170.46 m^3 (6,020 cu ft)
lower deck 985.29 m^3 (34,795 cu ft)

AREAS:
Wings, gross 576.0 m^2 (6,200 sq ft)
Ailerons (total) 23.49 m^2 (252.8 sq ft)
Trailing-edge flaps (total) 92.13 m^2 (991.7 sq ft)
Leading-edge slats (total) 60.25 m^2 (648.5 sq ft)
Spoilers (total) 40.01 m^2 (430.7 sq ft)
Fin 89.29 m^2 (961.1 sq ft)
Rudder 21.06 m^2 (226.7 sq ft)
Tailplane 89.73 m^2 (965.8 sq ft)
Elevators 24.03 m^2 (258.7 sq ft)

WEIGHTS AND LOADINGS (for 2.25*g*):
Operating weight empty, equipped 169,643 kg (374,000 lb)
Max payload 118,387 kg (261,000 lb)
Max fuel weight 150,815 kg (332,500 lb)
Max T-O weight 379,657 kg (837,000 lb)
Max zero-fuel weight 288,030 kg (635,000 lb)
* Max landing weight 288,415 kg (635,850 lb)
Max wing loading 659 kg/m^2 (135.48 lb/sq ft)
Max power loading 496.4 kg/kN (4.88 lb/lb st)
* *at 2.7 m (9 ft)/s descent rate*

PERFORMANCE (estimated at max T-O weight, except where indicated):
Never-exceed speed 402 kts (745 km/h; 463 mph) CAS
Max level speed at 7,620 m (25,000 ft) 496 kts (919 km/h; 571 mph)
Max cruising speed at 7,620 m (25,000 ft) 480-490 kts (888-908 km/h; 552-564 mph)
Econ cruising speed at 7,620 m (25,000 ft) 450 kts (833 km/h; 518 mph)
Stalling speed at max landing weight, 40° flap, power off 104 kts (193 km/h; 120 mph)
Max rate of climb at S/L 525 m (1,725 ft)/min
Service ceiling at AUW of 278,960 kg (615,000 lb) 10,895 m (35,750 ft)
Min ground turning radius:
about nosewheel 22.10 m (72 ft 6 in)
about wingtip 52.12 m (171 ft 0 in)
Runway LCN: asphalt 69
concrete 44
T-O run at S/L, ISA 2,530 m (8,300 ft)
T-O to 15 m (50 ft) at S/L, ISA 2,987 m (9,800 ft)
Landing from 15 m (50 ft), max landing weight at S/L, ISA 1,164 m (3,820 ft)
Landing run, max landing weight at S/L, ISA 725 m (2,380 ft)
Range with max payload, ISA, fuel reserves 5% of initial fuel plus 30 min loiter at S/L 2,982 n miles (5,526 km; 3,434 miles)
Range with max fuel, ISA reserves as above 5,618 n miles (10,411 km; 6,469 miles)

UPDATED

Lockheed C-5B Galaxy aircraft with General Electric TF39-GE-1C turbofan engines *(Paul Tompkins)* *1995*

Lockheed C-5A Galaxy heavy logistics transport aircraft *(Peter J. Cooper)* *1994*

LOCKHEED GALAXY UPGRADE

TYPE: Heavy transport aircraft upgrade.

PROGRAMME: Re-winging of remaining 77 of original 81 C-5As completed 1987; last of 50 C-5Bs delivered 17 April 1989. In November 1988 and October 1989, two C-5As redelivered to AFRes (433rd MAW, Kelly AFB, Texas) following Space Cargo Modification involving removal of upper troop deck aft of wing and redesign of rear ramp doors to provide same cargo volume as Space Shuttle. First 345-passenger conversion of C-5A for bulk trooping (68-0225) flew initial service between McGuire AFB and Spangdahlem, Germany, 27 April 1990. $2.36 million contract to Lockheed, June 1990, for two special operations C-5 conversions in Pacer Snow programme; aircraft completed August and November 1990; equipped with Tracor AN/ALE-40 flare dispensers and Honeywell AN/AAR-47 missile warning system; trials at Eglin and Holloman AFBs.

VERIFIED

Lockheed C-5A Galaxy after re-wing programme *(Peter J. Cooper)* *1994*

LOCKHEED 182/282/382 HERCULES

US Air Force designations: C-130, AC-130, DC-130, EC-130, HC-130, JC-130, LC-130, MC-130, NC-130, RC-130 and WC-130
US Navy designations: C-130, DC-130, EC-130, LC-130 and TC-130
US Marine Corps designation: KC-130
US Coast Guard designations: C-130, EC-130 and HC-130
Canadian Forces designations: CC-130 and CC-130T
RAF designations: Hercules C. Mk 1K, C. Mk 1P, W. Mk 2 and C. Mk 3P
Spanish designations: T.10, TK.10 and TL.10
Swedish designation: Tp 84
Export designations: C-130H, C-130H-30, KC-130H, C-130H-MP, VC-130H and C-130J
Israeli Defence Force names: C/K-130 Karnaf (Rhinocerous), EC-130 Aya (Condor)/Sapeer (Saphire)

TYPE: Tactical transport and multimission aircraft.

PROGRAMME: US Air Force specification issued 1951; first production contract for C-130A to Lockheed September 1952; two prototypes, 231 C-130As, 230 C-130Bs and 491 C-130Es manufactured. Lockheed delivered 2,000th Hercules (C-130H to Kentucky ANG) in April 1992.

VERSIONS: **C-130A:** Initial version with Allison T56-A-1A or -9 turboprops, driving Aeroproducts three-blade constant-speed reversible-pitch propellers, diameter 4.60 m (15 ft 0 in). Fuel capacity originally 19,865 litres (5,250 US gallons; 4,371 Imp gallons). Can be supplemented by two 1,705 litre (450 US gallon; 375 Imp gallon) underwing pylon tanks. First flown 7 April 1955. In service with the air forces of Bolivia (1), Honduras (3), Mexico (1), Peru (Air Force 3, Navy 2) and the USA (9+).

RC-130A: Photographic version of C-130A. Externally evident features include TV viewfinder blister under nose radome and camera windows in bottom of fuselage. Cabin contains cameras, mapping equipment, dark room and galley. Additional crew stations for photonavigator, photographer, two HIRAN operators and airborne profile recorder operator.

C-130B: Developed version with additional 6,472 litres (1,710 US gallons; 1,424 Imp gallons) of fuel in wings inboard of inner engine nacelles, strengthened landing gear and four Allison T56-A-7 turboprops, driving four-blade Hamilton Standard propellers. First C-130B for the USAF Tactical Air Command flew 20 November 1958, and entered sevice 12 June 1959. In service with the air forces of the following countries Colombia (3), Indonesia (9+), Jordan (2+), Pakistan (4), Singapore (3+), South Africa (6+) and Turkey (2).

HC-130B: Six modified C-130Bs equipped for Discover satellite recovery. Replaced C-119s of 6593rd Test Squadron, USAF at Hackam AFB.

C-130C: Experimental boundary layer control testbed for USAF which made its first flight in February 1960. Two Allison YT56-A-6 turbojet engines, pod-mounted under the outer wings, are used to blow a constant flow of air over the control surfaces and flaps. Stalling speed, take-off distance and landing distance are greatly reduced.

Lockheed AC-130H Spectre upgraded by Lockheed

1994

C-130D: Modified C-130A aircraft modified for service with the USAF in Antartic and elsewhere. Wheel/ski landing gear. Total ski installation weighs approximately 2,495 kg (5,500 lb). The main skis and nose skis have 8° and 15° nose-up and nose-down pitch enabling them to follow uneven terrain. They have a bearing surface of Teflon plastic to reduce friction and resist the adhesiveness of ice. JATO fitted. Fuel capacity increased by addition of two 1,703 litre (450 US gallon; 375 Imp gallon) underwing pylon tanks. Provision for two 1,891 litre (500 US gallon; 416 Imp gallon) tanks in cargo compartment for longer flights such as ferry flights to Antartic.

C-130E: Extended range version of C-130B. Two 5,145 litre (1,360 US gallon; 1,132 Imp gallon) underwing fuel tanks added. Take-off at overload gross weight of 79,380 kg (175,000 lb) allows extension of range and endurance, with certain operating restrictions at this higher weight. First C-130E flew 25 August 1961. Deliveries began in April 1962. In service with the following countries: Argentina (1), Australia (1), Brazil (6), Canada (15+), Iran (5+), Israel (12), Pakistan (7) and Saudi Arabia (8+).

MC-130E Combat Talon: Special operations aircraft based on C-130E. With Fulton STAR recovery yolk and specialist defensive equipment including RWR, chaff/flare dispenser and AN/AAQ-8 IR countermeasures system.

C-130F: Similar to KC-130F, but without underwing pylons and internal refuelling equipment. AUW 61,235 kg (135,000 lb).

KC-130F: Assault transport, basically similar to C-130B. Equipped for in-flight refuelling to service two jet aircraft simultaneously. Entire refuelling equipment can be quickly and easily installed and removed. Two C-130As loaned to USMC Summer 1957 for flight refuelling tests. The production tanker version, first flown 2 January 1960, has a tankage capacity of 13,620 litres (3,600 US gallons; 2,998 Imp gallons) in its cargo compartment. Able to fly 1,600 km (1,000 miles) at cruise ceiling at 547 km/h (340 mph), and transfer 14,060 kg (31,000 lb) of fuel at 7,620 m (25,000 ft) at a refuelling speed of 571 km/h (355 mph) with normal military reserves. Normal crew of five to seven.

LC-130F: Four C-130Bs for the US Navy with wheel/ski gear, for service in Antartic. Range with skis reduced by 5 to 10 per cent. First delivered in August 1960.

HC-130G: Search and rescue version. Modifications include addition of radio operator's station in-flight compartment, accommodation for two search observer seats and clear-vision panel in rear paratroop doors for low-level search. Able to fly 1,600 km (1,000 miles) offshore at 8,200 m (25,000 ft) at 595 km/h (370 mph), then cruise for 8 hours search duty at 233 to 275 km/h (145 to 170 mph) on two engines, re-start engines and return to base at 480 to 595 km/h (300 to 370 mph), with 15 minutes fuel reserve. Endurance at this range can be increased to 10.2 hours at 240 to 257 km/h (150 to 160 mph) if outward flight is made at 540 km/h (335 mph) and return flight at 480 km/h (300 mph). Normal crew of eight. Can carry 24 to 44 passengers.

C-130H: Deliveries started March 1965 to Royal New Zealand Air Force; in service with 64 countries. Features include updated avionics, improved wing, new corrosion protection, and Allison T56-A-15 engines flat rated at 3,362 kW (4,508 shp). Can deliver up to 19,052 kg (42,000 lb) by low-velocity airdrop, or by low-altitude parachute extraction system (LAPES). Total 216 regular transport C-130Hs funded for USAF/ANG/AFRes in FY73-91; further 29 in FY93; from 1994, new aircraft have AlliedSignal TCAS II collision avoidance system; Night Vision Instrument System introduced to USAF aircraft from May 1993; others for special roles, as detailed below. In January 1993, Chrysler Technologies Airborne Systems and AlliedSignal Aerospace won cockpit upgrade competition for USAF Hercules (and 148 C-141s) involving four 150 × 200 mm (6 × 8 in) liquid crystal displays; AlliedSignal digital autopilot, Sundstrand ground collision avoidance and other avionics improvements. Total 672 Hercules to be modified, including C-130Es and Hs.

C-130H-MP: Maritime patrol version; one delivered to Indonesian Air Force, three to Royal Malaysian Air Force. Maximum T-O weight 70,310 kg (155,000 lb), maximum payload 18,630 kg (41,074 lb), and T56-A-15 engines; search time 2 hours 30 minutes at 1,525 m (5,000 ft) at 1,800 n miles (3,333 km; 2,070 miles) radius or 16 hours 50 minutes at 200 n miles (370 km; 230 miles) radius. Optional and standard search features include sea search radar, observer seats and windows, INS/Omega navigation, crew rest and galley slide-in module, flare launcher, loudspeaker, rescue kit airdrop platform, side-looking radar, passive microwave imager, low-light TV, infra-red scanner, camera with data annotation and ramp equipment pallet with observer station.

C-130H-30: Stretched version of current production C-130; fuselage lengthened by 4.57 m (15 ft 0 in); troop and cargo capability increased by 40 per cent.

AC-130H Spectre: Gunship version with sideways-firing 105 mm recoilless howitzer, 40 mm cannon and two 20 mm Vulcan guns; infra-red and low-light TV sensors, and side-looking head-up display for aiming at night while circling target; in-flight refuelling. Conversion by Lockheed Aircraft Service Company. New fire control computers and navigation and sensors under Special Operations/Forces Improvements (SOFI) being installed. Flight testing began September 1989; first upgrade completed mid-1990; last of current nine were due 1993. In service with 16th Special Operations Squadron at Hurlburt Field, Florida.

EC-130H Compass Call: Works with ground-based C³CM to jam enemy command, control and communications. Operated by 41st Electronic Combat Squadron at Davis Monthan AFB, Arizona. (Eight earlier EC-130Es of 7th ACCS being updated by Unisys to ABCCC III standard in $34 million programme, 1990.)

HC-130H: Extended range USAF aircraft for aerial recovery of personnel or equipment and other duties; 43 delivered from October 1964; update announced Spring 1987 includes self-contained navigation, night vision goggles cockpit and new communications equipment; applied to 31 aircraft; 21 of these modified for in-flight refuelling and now designated HC-130P; US Coast Guard ordered 35, final 10 as HC-130H-7 with T56A-7B power plants. Further three HC-130H(N)s funded FY88-90 for 210th ARS, ANG, in Alaska, delivered from 28 November 1990; has helicopter refuelling capability from outset.

JC-130H: Four US Air Force HC-130Hs equipped to recover re-entering space capsules.

DC-130H: Two US Air Force HC-130Hs modified for drone control.

KC-130H: Probe-drogue tanker similar to KC-130R; exported to Argentina (two), Brazil (two), Israel (two), Morocco (two), Saudi Arabia (eight), Spain (five) and Singapore (one).

LC-130H: Similar to LC-130R (which see); four acquired by 139th AS, ANG, at Schenectady.

MC-130H Combat Talon II: Conversion of new-build C-130H for day/night infiltration and exfiltration, resupply of Special Operations Forces, psychological warfare and aerial reconnaissance; terrain-following radar; six-man crew; 25 (including YMC-130H prototype 74-1686, now retired) funded in FY83-90; first flight by E-Systems (see below) at Greenville Spring 1988; flight testing began at Edwards AFB September 1988; MC-130Hs augment MC-130Es at Hurlburt Field; first four delivered 17 October 1991 and temporarily with 8th SOS prior to formation of 15th SOS on 1 October 1992; others to 1st SOS at Kadena, Japan; 7th SOS at Alconbury, UK (beginning with 86-1699 on 10 September 1992); and 1550th ATS (training) at Kirtland AFB, New Mexico (three aircraft).

Equipment includes Emerson Electric AN/APQ-170 precision ground mapping/weather/terrain-following and avoidance radar in enlarged radome, inertial navigation, automatic computed air release point, high-speed low-level release system, ground acquisition receiver/interrogator, Texas Instruments AN/AAQ-15 infra-red detection system, eight multifunction displays, secure voice UHF/VHF-FM radios, retractable FLIR pod, angle of attack probe, AN/ALQ-8 ECM pod under each wing, and in-flight refuelling. Defensive equipment includes Litton AN/ALR-69 RWR, ITT AN/ALQ-172 detector/jammer, Watkins Johnson WJ-1840 signal detector, Cincinnati Electronics AN/AAR-44 launch warning receiver. Northrop QRC-8402 IR jammer and chaff/flare dispensers. IBM Federal Systems Division is prime contractor for systems integration, with E-Systems as subcontractor for avionics installation and modification.

RC-130H: Unofficial designation for two Moroccan aircraft fitted with SLAR by Flight Systems Inc.

VC-130H: VIP transport.

C-130J: Military version of Hercules II; development of advanced Hercules continuing for foreign sales.

C-130K: RAF version of C-130H; much of avionics and instrumentation made in UK; 66 delivered as **Hercules C. Mk 1** beginning September 1966; one modified by Marshall of Cambridge for RAF Meteorological Research Flight as **Hercules W. Mk 2**. Thirty lengthened by 4.57 m

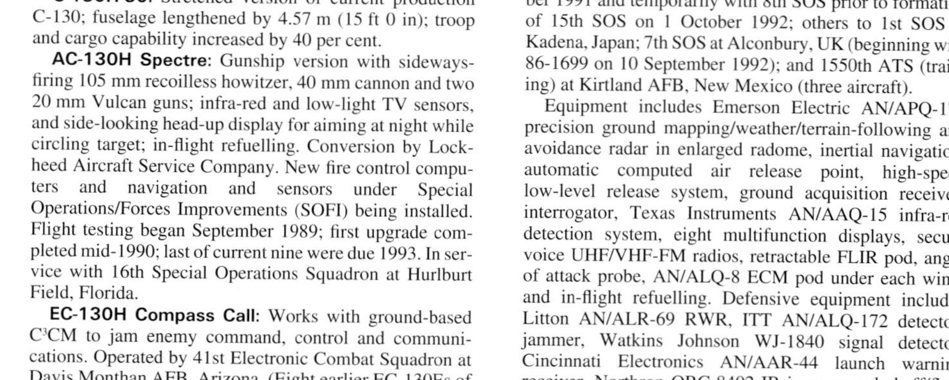

Lockheed HC-130N search and rescue version of C-130N *(Paul Tompkins)*

1995

Lockheed HC-130H Hercules of the US Coast Guard

1992

(15 ft), equivalent to commercial L-100-30, and redesignated **Hercules C. Mk 3**; first aircraft modified at Marietta, remaining 29 by Marshall of Cambridge. Six modified to **C. Mk 1K** tankers, four of which received Racal 'Orange Blossom' ESM, which also fitted to one C. Mk 1; 20 C. Mk 3s received AN/APN-169B station-keeping equipment; some have provision for AN/ALQ-157 IR jammers; AN/ALR-66 RWR ordered; all fitted with refuelling probes 1982-89.

HC-130N: US Air Force search and rescue version of C-130H for recovery of aircrew and space capsules; 15 delivered with helicopter refuelling capability; advanced direction-finding equipment.

HC-130P: C-130H modified for refuelling helicopters in flight and recovering parachute-borne payloads; 20 built for USAF. HC-130N/Ps to be upgraded by Rockwell with SOF Improvement package including self-protection aids for refuelling operations in hostile airspace.

EC-130Q: Similar to earlier EC-130G, but with improved equipment and crew accommodation for TACAMO command communication with submarines; 18 built; HF and VLF SIMOP (simultaneous operation). Last aircraft (161531) retired from VQ-4 26 May 1992; replaced by Boeing E-6A; two EC-130Qs converted to TC-130Q; three to NASA as transports; remainder to storage.

TC-130Q: Two surplus EC-130Qs with trailing wire aerial removed to permit normal cargo loading via rear doors; wingtip pods retained; first (159348) noted 1990. Two TC-130Gs are similar.

KC-130R: Probe-drogue tanker version of C-130H; 14 delivered to US Marine Corps VMGR-252 and -352; changes from KC-130F include 3,362 kW (4,508 shp) engines, higher T-O and landing weights, external fuel tanks for additional 10,296 litres (2,720 US gallons; 2,265 Imp gallons) fuel, and removable 13,627 litre (3,600 US gallon; 2,997 Imp gallon) fuel tank in cargo hold (all fuel can be used to increase tanker's range); single-point refuelling of normal and additional tanks from existing filler; operating weight empty 36,279 kg (79,981 lb); maximum T-O weight 79,378 kg (175,000 lb); can offload up to 20,865 kg (46,000 lb) of fuel, equivalent to 26,790 litres (7,077 US gallons; 5,893 Imp gallons), at radius of 1,000 n miles (1,850 km; 1,150 miles); maximum offload capability 31,750 kg (70,000 lb), equivalent to 40,765 litres (10,769 US gallons; 8,967 Imp gallons).

LC-130R: C-130H with wheel/ski landing gear for US Navy Squadron VXE-6 in Antarctic.

CC-130T: Canadian Forces' designation for five CC-130s converted by Northwest Industries, 1992-93, as tankers with two FRL Mk 32B hose-drum units beneath wings and 13,627 litre (3,600 US gallon; 2,998 Imp gallon) tank in cargo hold.

KC-130T: Tanker for US Marine Corps (Reserve), able to refuel helicopters and fighters; eight delivered to Marine Aerial Refueller Transport Squadron 234 (VMGR-234), starting November 1983; others to VMGR-452; deliveries two per year since 1990; total 28 required by 1995 (14 per squadron). Similar to KC-130R, but with updated avionics including INS, Omega and Tacan, new autopilot and flight director and solid-state search radar; KC-130Ts delivered in 1984 had Bendix/King AN/APS-133 colour radar, flush antennae and orthopaedically designed crew seats.

KC-130T-30: Stretched KC-130T (similar to C-130H-30) for US Navy; first (164597) delivered to VMGR-452 29 October 1991; second in November 1991 to VMGR-234.

C-130T: Transport, with secondary refuelling capability, for US Navy; first of six (164762) to VR-54 at New Orleans 20 August 1991; first of six to VR-48 at NARF Andrews, Maryland, October 1992. Total 22 required. VR-53 next to form, at Martinsburg, West Virginia.

AC-130U Spectre: New gunship version of C-130H; details under Rockwell International entry.

EC-130V: Surveillance conversion of USCG HC-130H (1721) with E-2C Hawkeye's AN/APS-125 radar. Conversion by General Dynamics, Forth Worth; first flight 31 July 1991; to CGAS Clearwater, Florida. Programme terminated, April 1993, due to budget cuts; aircraft transferred to USAF's 514th Test Squadron at Hill AFB, Utah, late 1993. However, USCG HC-130H 1502 to receive AN/APS-137 radar in 1994 after modification at Elizabeth City CGAS.

C-130J: For full details see *Jane's All the World's Aircraft* 1995-1996.

UPGRADES: **Aero Union:** System upgrades. See separate entries.

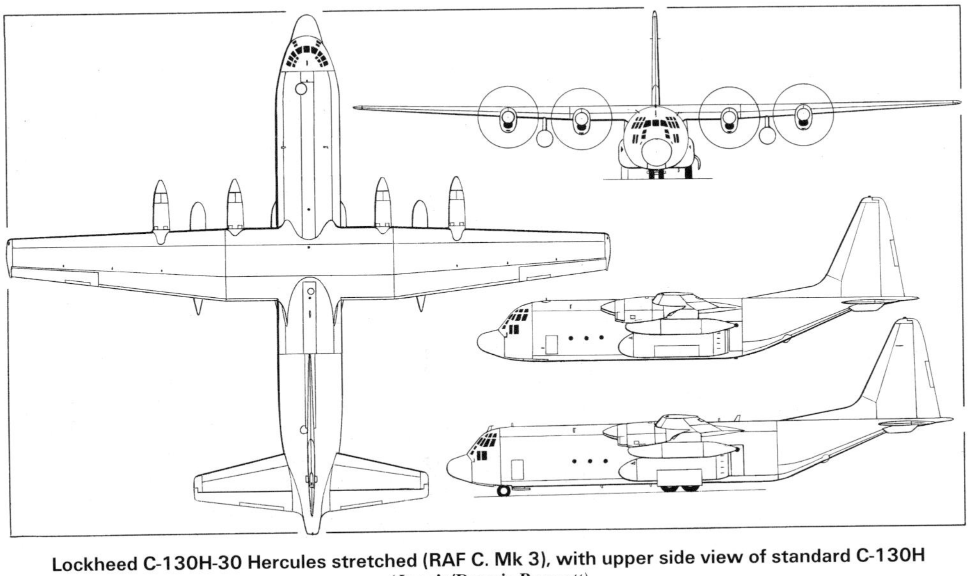

Lockheed C-130H-30 Hercules stretched (RAF C. Mk 3), with upper side view of standard C-130H
(Jane's/Dennis Punnett)

1992

Airwork: Shorts subsidiary reported to be installing Defensive Aids System (DAS) consisting of Loral AN/ALQ-157 fixed-source IR jammer on 23 RAF C-130 aircraft.

Chrysler Technologies Airborne Systems (CTAS): Contract awarded to CTAS as prime contractor with AlliedSignal's Guidance & Control Systems Division as principal subcontractor in early 1993 to upgrade 672 C-130 cockpits and autopilots. Cockpit upgrade entails converting displays from analogue to digital. Cockpit instrumentation to comprise four 150 × 200 mm liquid crystal displays, two primary flight and two subsystem displays. Display processor to be used as mission computer and to control the MIL-STD-1553 databus to link systems. Autopilot to comprise a flight control processor with software based on the digital autopilot used for the Boeing E-3. Contract valued at $200 to 300 million.

E-Systems: Modular modification. See separate entry.

E-Systems/Loral: MC-130H Combat Talon II. See Versions.

Lockheed: AC-130H Spectre. See Versions.

Lockheed Forth Worth (formerly General Dynamics): EC-130V. See Versions.

Marshall Aerospace: Hercules W. Mk 2/C Mk 3/C Mk 1K. See Versions and separate entry.

NorthWest Industries: CC-130TT. See Versions.

Rockwell: AC-130U. See Versions and separate entry.

Rockwell: HC-130P. See Versions.

Sabena Technics: Belgian Air Force upgrade. See separate entry.

Smiths Industries: RAF navigation upgrade: See separate entry.

Snow Aviation: Modern Technology Cockpit (MTC). See separate entry.

Spain: Spanish government gave the go-ahead in July 1995 to extend the lives of the Spanish Air Force's fleet of seven C-130H and five KC-130H aircraft.

Unisys: EC-130H Compass Call. See Versions.

USAF: Reported to have installed selected C-130E/H aircraft with defensive aids system (DAS). Systems said to comprise of AAR-44 and ALQ-156 missile approach warning system (MAWS) together with ALE-40 dispenser.

USMC: Reported to have installed selected KC-130FD/R/T aircraft with AAR-47 passive missile approach warner (MAW).

Westinghouse: Announced June 1995 that it had been awarded a contract to retrofit the Portuguese Air Force fleet of C-130H Hercules with its AN/APN-241 navigation and weather radar with predictive windshear. Deliveries of new radar due to begin in November 1995. Contract value $3 million.

OPERATORS: Versions of the Hercules are in service with the armed forces of the following countries: Algeria (16), Argentina (9), Australia (24), Belgium (12), Bolivia (8), Brazil (13), Cameroon (3), Canada (30), Chad (4), Chile (4), Colombia (5), Denmark (3), Ecuador (2), Egypt (21), France (12), Gabon (3), Greece (1), Honduras (2), Indonesia (24), Iran (50), Israel (28), Italy (12), Japan (20), Jordan (4), South Korea (10), Kuwait (2), Libya (12), Malaysia (9), Mexico (10), Morocco (17), Netherlands (2), New Zealand (5), Niger (2), Nigeria (8), Norway (6), Oman (3), Pakistan (12), Peru (13), Philippines (9), Portugal (6), Saudi Arabia (40), Singapore (10), South Africa (7), Spain (12), Sudan (5), Sweden (8), Taiwan (12), Thailand (8), Tunisia (2), Turkey (9), UAE (6), UK (59), USA (824), Uruguay (3), Venezuela (5), Yemen (2) and Zaire (4).

COSTS: C-130H is $31.5 million (1988) flyaway; $48 million (1987) programme unit cost for AC-130H.

DESIGN FEATURES: Can deliver loads and parachutists over lowered rear ramp and parachutists through side doors; removable external fuel tanks outboard of engines are standard fittings; cargo hold pressurised. Wing section NACA

Lockheed MC-130E Combat Talon with Fulton STAR recovery yolk *(Paul Tompkins)*

1995

64A318 at root and NACA 64A412 at tip; dihedral 2° 30′; incidence 3° at root, 0° at tip. Leading-edges of wing, tailplane and fin anti-iced by engine bleed air.

FLYING CONTROLS: All control surfaces boosted by dual hydraulic units; trim tabs on ailerons, both elevators and rudder, elevator tabs have AC main supply and DC standby; Lockheed-Fowler trailing-edge flaps; provision for two removable afterbody ventral strakes.

STRUCTURE: All-metal two-spar wing with integrally stiffened taper-machined skin panels up to 14.63 m (48 ft 0 in) long.

LANDING GEAR: Hydraulically retractable tricycle type. Each main unit has two wheels in tandem, retracting into fairings built on to the sides of the fuselage. Nose unit has twin wheels and is steerable ±60°. Oleo shock-absorbers. Mainwheel tyres size 56 × 20-20, pressure 6.62 bars (96 lb/sq in). Nosewheel tyres size 39 × 13-16, pressure 4.14 bars (60 lb/sq in). Goodyear air-cooled multiple disc hydraulic brakes with anti-skid units. Retractable combination wheel-skis available. Minimum ground turning radius: C-130H, 11.28 m (37 ft) about nosewheel and 25.91 m (85 ft) about wingtip; C-130H-30, 14.33 m (47 ft) about nosewheel and 27.93 m (90 ft) about wingtip.

POWER PLANT: Four 3,362 kW (4,508 shp) Allison T56-A-15 turboprops, each driving a Hamilton Standard type 54H60 four-blade constant-speed fully feathering reversible-pitch propeller. Fuel in six integral tanks in wings, with total capacity of 26,344 litres (6,960 US gallons; 5,795 Imp gallons) and two optional underwing pylon tanks, each with capacity of 5,146 litres (1,360 US gallons; 1,132 Imp gallons). Total fuel capacity 36,636 litres (9,680 US gallons; 8,060 Imp gallons). Single pressure refuelling point in starboard wheel well. Overwing gravity fuelling. Oil capacity 182 litres (48 US gallons; 40 Imp gallons).

ACCOMMODATION: Crew of four on flight deck, comprising pilot, co-pilot, navigator and systems manager (fully performance qualified flight engineer on USAF aircraft). Provision for fifth man to supervise loading. Sleeping bunks for relief crew and galley. Flight deck and main cabin pressurised and air conditioned. Standard complements for C-130H are as follows: 92 troops, 64 paratroopers, 74 litter patients plus two attendants. Corresponding data for C-130H-30 are 128 troops, 92 paratroopers, and 97 litter patients plus four attendants. Air transport and airdrop loads such as Sheridan light armoured vehicle, 19,051 kg (42,000 lb) when rigged for airdrop, are common to both C-130H and the C-130H-30; light and medium towed artillery weapons, or variety of wheeled and tracked vehicles and multiple 463L supply pallets (five in C-130H and seven in C-130H-30, plus one on rear ramp for each model) are transportable; C-130H-30 is only airlifter which can airdrop entire field artillery section (ammo platform, weapon, prime mover, and eight crew jumping over ramp) in one pass. Hydraulically operated main loading door and ramp at rear of cabin. Paratroop door on each side aft of landing gear fairing. Two emergency exit doors standard; two additional doors optional on C-130H-30.

SYSTEMS: Air conditioning and pressurisation system maximum pressure differential 0.52 bar (7.5 lb/sq in). Three independent hydraulic systems, utility and booster systems operating at a pressure of 207 bars (3,000 lb/sq in), rated at 65.1 litres (17.2 US gallons; 14.3 Imp gallons)/min for utility and booster systems, 30.3 litres (8.0 US gallons; 6.7 Imp gallons)/min for auxiliary system. Reservoirs are unpressurised. Auxiliary system has hand pump for emergencies. Electrical system supplied by four 40 kVA AC alternators, plus one 40 kVA auxiliary alternator driven by APU in port main landing gear fairing. Four transformer-rectifiers for DC power. Current production aircraft incorporate systems and component design changes for increased reliability. There are differences between the installed components for US government and export versions. Babcock Power Ltd High Volume Mine Layer (HVML) system available as an option, using modular roll-on pallets. Leading-edges of wing, tailplane and fin anti-iced by engine bleed air.

AVIONICS: Standard fit specified by US government comprises dual AN/ARC-190 HF com, dual AN/ARC-186 VHC com, dual AN/ARC-164 UHF com, AN/AIC-13 PA system, AN/AIC-18 intercom, AN/APX-100 IFF/AIMS ATC transponder, dual AN/ARN-118 UHF nav, dual AN/ARN-147 VHF nav, self-contained navigation system (SCNS), dual DF-206A ADF, DF-301E UHF direction-finder, emergency locator transmitter (ELT), AN/APN-218 Doppler nav, AN/APN-232 combined altitude radar alt, dual C-12 compass system, dual FD-109 flight director system, either capable of coupling with FD-109 autopilot, Sundstrand ground proximity warning system, Kollsman altitude alerter. Westinghouse low power colour radar (LPCR 130-1) replacing Sperry radar from March 1993, AN/APN-169C station-keeping equipment, A-100A cockpit voice recorder, flight data recorder, AN/AAR-47 missile warning system, provisions for AN/ALE-47 flare and chaff dispensing system, provisions for AN/ALQ-157 infra-red countermeasures system, provisions for KY-58 secure voice, provisions for microwave landing system (Canadian Marconi CMSLA system ordered for C-130 fleet retrofit, 1991), provisions for USTS Satcom system.

ARMAMENT: Fitment of Rockwell Hellfire ASM to AC-130H/U studied 1991-92.

DIMENSIONS, EXTERNAL:
- Wing span: 40.41 m (132 ft 7 in)
- Wing chord: at root: 4.88 m (16 ft 0 in)
  - mean: 4.16 m (13 ft 8½ in)
- Wing aspect ratio: 10.1
- Length overall:
  - all except HC-130H and C-130H-30: 29.79 m (97 ft 9 in)
  - C-130H-30: 34.37 m (112 ft 9 in)
- Height overall: 11.66 m (38 ft 3 in)
- Tailplane span: 16.05 m (52 ft 8 in)
- Wheel track: 4.35 m (14 ft 3 in)
- Wheelbase: 9.77 m (32 ft 0¾ in)
- Propeller diameter: 4.11 m (13 ft 6 in)
- Main cargo door (rear of cabin):
  - Height: 2.77 m (9 ft 1 in)
  - Width: 3.12 m (10 ft 3 in)
  - Height to sill: 1.03 m (3 ft 5 in)
- Paratroop doors (each): Height: 1.83 m (6 ft 0 in)
  - Width: 0.91 m (3 ft 0 in)
  - Height to sill: 1.03 m (3 ft 5 in)
- Emergency exits (each): Height: 1.22 m (4 ft 0 in)
  - Width: 0.71 m (2 ft 4 in)

DIMENSIONS, INTERNAL:
- Cabin, excl flight deck:
  - Length excl ramp: C-130H: 12.50 m (41 ft 0 in)
    - C-130H-30: 17.07 m (56 ft 0 in)
  - Length incl ramp: C-130H: 15.73 m (51 ft 8½ in)
    - C-130H-30: 20.33 m (66 ft 8½ in)
  - Max width: 3.12 m (10 ft 3 in)
  - Max height: 2.81 m (9 ft 2¾ in)
  - Floor area, excl ramp: C-130H: 39.5 m² (425.0 sq ft)
  - Volume, incl ramp: C-130H: 123.2 m³ (4,351.0 cu ft)
    - C-130H-30: 161.3 m³ (5,696.0 cu ft)

AREAS:
- Wings, gross: 162.12 m² (1,745.0 sq ft)
- Ailerons (total): 10.22 m² (110.0 sq ft)
- Trailing-edge flaps (total): 31.77 m² (342.0 sq ft)
- Fin: 20.90 m² (225.0 sq ft)
- Rudder, incl tab: 6.97 m² (75.0 sq ft)
- Tailplane: 35.40 m² (381.0 sq ft)
- Elevators, incl tabs: 14.40 m² (155.0 sq ft)

WEIGHTS AND LOADINGS:
- Operating weight empty:
  - C-130H: 34,686 kg (76,469 lb)
  - C-130H-30: 36,397 kg (80,242 lb)
- Max fuel weight: internal: 20,108 kg (44,330 lb)
  - external: 8,255 kg (18,200 lb)
- Max payload: C-130H: 19,356 kg (42,673 lb)
  - C-130H-30: 17,645 kg (38,900 lb)
- Max normal T-O weight: 70,310 kg (155,000 lb)
- Max overload T-O weight: 79,380 kg (175,000 lb)
- Max normal landing weight: 70,310 kg (155,000 lb)
- Max overload landing weight: 79,380 kg (175,000 lb)
- Max zero-fuel weight, 2.5 *g*: 54,040 kg (119,142 lb)
- Wing loading at max normal T-O weight: 434.5 kg/m² (89 lb/sq ft)
- Power loading at max normal T-O weight: 5.23 kg/kW (8.6 lb/shp)

PERFORMANCE (C-130H at max normal T-O weight, unless indicated otherwise):
- Max cruising speed: 325 kts (602 km/h; 374 mph)
- Econ cruising speed: 300 kts (556 km/h; 345 mph)
- Stalling speed: 100 kts (185 km/h; 115 mph)
- Max rate of climb at S/L: 579 m (1,900 ft)/min
- Service ceiling at 58,970 kg (130,000 lb) AUW: 10,060 m (33,000 ft)
- Service ceiling, OEI, at 58,970 kg (130,000 lb) AUW: 8,075 m (26,500 ft)
- Runway LCN: asphalt: 37
  - concrete: 42
- T-O run: 1,091 m (3,580 ft)
- T-O to 15 m (50 ft): 1,573 m (5,160 ft)
- Landing from 15 m (50 ft):
  - at 45,360 kg (100,000 lb) AUW: 731 m (2,400 ft)
  - at 58,967 kg (130,000 lb) AUW: 838 m (2,750 ft)
- Landing run at 58,970 kg (130,000 lb) AUW: 518 m (1,700 ft)
- Range with max payload, with 5% reserves and allowance for 30 min at S/L: 2,046 n miles (3,791 km; 2,356 miles)
- Range with max fuel, incl external tanks, 7,081 kg (15,611 lb) payload, reserves of 5% initial fuel plus 30 min at S/L: 4,250 n miles (7,876 km; 4,894 miles)

UPDATED

Lockheed Model 1329-25 JetStar II with AiResearch TFE 731-3 turbofan engines

1992

Lockheed Model 1329-25 JetStar II four-turbofan light utility transport aircraft

1992

LOCKHEED MODEL 1329-25 JETSTAR II

US military designation: C-140

TYPE: Four-turbofan utility transport aircraft.

PROGRAMME: The Lockheed JetStar II, which was first announced Summer 1973, has an airframe generally similar to that of the earlier JetStar, but with detail changes in configuration and equipment.

Design of the Model 1329-25 began in October 1972. The major change involved the installation of four AiResearch TFE731-3 turbofan engines, flat rated at 16.5 kN (3,700 lb st) to 76°F (24.4°C), to replace the 14.7 kN (3,300 lb st) Pratt & Whitney JT12 turbojet engines of the JetStar. The new power plant offers significant improvement in both range and noise levels, as well as allowing an increase in maximum take-off weight.

Production of the JetStar II, at Lockheed-Georgia's Marietta plant, began Spring 1975, and the first aircraft off the line flew 18 August 1976. FAA certification of the JetStar II was granted 14 December 1976.

UPGRADES: **AirResearch:** Initiated programme to convert early JetStars to AirResearch TFE 731-3 turbofan engines. First prototype conversion flew in July 1974. First production conversion flew on 18 March 1976.

OPERATORS: The JetStar II is in service with the Iranian armed forces (1).

DESIGN FEATURES: Cantilever low-wing monoplane. Wing section NACA 63A112 at root, NACA 63A309 (modified) at tip. Dihedral 2°. Incidence 1° at tip. Sweepback at quarter-chord 30°. Tail unit variable incidence is achieved by the fin being pivoted, thus allowing an electromechanical dual actuator to move the entire tail unit to rotate the tailplane.

FLYING CONTROLS: Plain aluminium alloy ailerons; trim tab located near the centre of the trailing-edge of the port aileron. An electrically powered dual trim actuator is located within the aileron directly forward of the trim tab. Hydraulically boosted aileron controls are powered by both normal and standby hydraulic systems, either of which is capable of operating the ailerons independently. Manual aileron control is possible in the event of complete hydraulic failure. Aileron booster actuators manufactured by National Water Lift Company. Double-slotted all-metal trailing-edge flaps. Hinged leading-edge flaps. No spoilers. Hydraulically operated speed brake on underside of fuselage aft of pressurised compartment. No trim tabs in elevators. Mechanically operated elevator control system is hydraulically boosted using a National Water Lift Company actuator, sited in the aft fuselage equipment area. The rudder is mechanically controlled, with servo tab assistance. Two pneumatic cylinders, biased by engine bleed air, automatically assist directional control in the event of a power loss from either engine.

STRUCTURE: Conventional fail-safe stressed-skin structure of high-strength aluminium. Bending loads carried by integral skin/stringer extrusion and sheet ribs, shear loads by three beams. The fuselage is a semi-monocoque fail-safe structure of light alloy. The nose section, crew compartment and cabin are pressurised. The aft section, where most of the aircraft's system components are mounted, is unpressurised. The tail unit is a light alloy structure with tailplane mounted part way up fin.

LANDING GEAR: Hydraulically retractable tricycle type, with twin wheels on all main units. Pneumatic emergency extension. Main units retract inward, nosewheels forward. Oleo-pneumatic shock-absorbers. Mainwheels with tubeless tyres size 26 × 6.60, EHP Type VII, 14 ply rating with reinforced tread, pressure 15.17 bars (220 lb/sq in). Nosewheels with tubeless chine tyres size 18 × 4.40, EHP Type VII, 12 ply rating with reinforced tread, pressure 15.17 bars (220 lb/sq in). Hytrol fully modulated anti-skid units.

POWER PLANT: Four AiResearch TFE 731-3 turbofan engines, flat rated at 16.5 kN (3,700 lb st) at 76°F (24.4°C), mounted in lateral pairs on sides of rear fuselage. Thrust reversers fitted. Air intake anti-icing provided by engine bleed air. Fuel in four integral wing tanks and two non-removable external auxiliary tanks glove-mounted on the wings. Capacity of numbers and four internal tanks each 1,420 litres (375 US gallons; 312 Imp gallons); numbers two and three internal tanks each 1,476 litres (390 US gallons; 324 Imp gallons), auxiliary tanks each 2,188 litres (578 US gallons; 481 Imp gallons). Total fuel capacity 10,168 litres (2,686 US gallons; 2,236 Imp gallons). Gravity refuelling point above each tank, or single-point pressure refuelling from starboard wing root. Oil capacity 24.2 litres (6.4 US gallons; 5.3 Imp gallons).

ACCOMMODATION: Normal accommodation for crew of two and 10 passengers, with wardrobe, galley and toilet aft of cabin and baggage compartments fore and aft. Layout and furnishing can be varied to suit customer's requirements. Optional jump-seat available for crew compartment. Door at forward end of fuselage, on port side, opens by moving inward and sliding aft. The fourth window aft on each side of the cabin is a CAR Type IV emergency exit, of plug type and removed inward. Accommodation heated, ventilated, air conditioned and pressurised. High-pressure oxygen system for passengers and crew standard. Integral electric heaters for windscreen anti-icing and demisting.

SYSTEMS: Two independent hydraulic systems with engine-driven pumps, pressure 207 bars (3,000 lb/sq in), to operate landing gear, wheel brakes, nosewheel steering, flight control booster units, flaps, speed-brake and thrust reversers. Separate pneumatic systems installed for emergency extension of the landing gear. Air bottles can be manually discharged into the down ports of the landing gear actuators. Two pneumatic cylinders provided to assist directional control if engine power lost. Four 28 V 300 A engine-driven starter/generators power main DC buses. Two high-discharge 24 V 34 Ah Ni/Cd batteries for engine starting and emergency power. Three 2,500 VA inverters provide AC power for electronics equipment, flight and engine instruments, and windscreen anti-icing, two being on-load and one on standby. A 250 VA rotary inverter is used to power the engine instruments during engine starting. High-pressure oxygen system, 124 bars (1,800 lb/sq in) reduced to 4.83 to 6.21 bars (70 to 90 lb/sq in) at the cylinder, provides selective demand or 100 per cent positive pressure demand for crew. A separate 100 per cent demand system with the safety pressure and manual control for dilution is installed for passengers. An altitude control valve activates the passenger system when the cabin altitude exceeds 4,267 m (14,000 ft), the masks being presented automatically. APU for ground air conditioning and electrical power is optional. Rubber boot de-icers on wing leading-edge. No tail de-icing system.

DIMENSIONS, EXTERNAL:

Wing span	16.60 m (54 ft 5 in)
Wing chord: at root	4.16 m (13 ft 7¾ in)
at tip	1.55 m (5 ft 1 in)
Wing aspect ratio	5.27
Length overall	18.42 m (60 ft 5 in)
Length of fuselage	17.92 m (58 ft 9½ in)
Height overall	6.23 m (20 ft 5 in)
Tailplane span	7.55 m (24 ft 9 in)
Wheel track	3.75 m (12 ft 3½ in)
Wheelbase	6.28 m (20 ft 7 in)
Cabin door: Height	1.50 m (4 ft 11 in)
Width	0.67 m (2 ft 2½ in)
Height to sill	approx 1.37 m (4 ft 6 in)
Servicing door (underfuselage, diameter)	0.61 m (2 ft 0 in)
Emergency exits, each: Height	0.49 m (1 ft 7¼ in)
Width	0.66 m (2 ft 2½ in)

DIMENSIONS, INTERNAL:

Cabin, excl flight deck: Length	8.59 m (28 ft 2½ in)
Max width	1.89 m (6 ft 2½ in)
Max height	1.85 m (6 ft 1 in)
Volume	24.07 m³ (850 cu ft)
Baggage hold volume:	
stbd forward	1.25 m³ (43.1 cu ft)
port forward	0.70 m³ (24.8 cu ft)
centre aft	1.05 m³ (37.0 cu ft)

AREAS:

Wings, gross	50.40 m² (542.5 sq ft)
Ailerons (total)	2.27 m² (24.4 sq ft)
Trailing-edge flaps (extended, total)	5.82 m² (62.6 sq ft)
Leading-edge flaps (total)	3.16 m² (34.0 sq ft)
Speed-brake	0.85 m² (9.2 sq ft)
Fin	8.73 m² (94.0 sq ft)
Rudder, incl tab	1.51 m² (16.2 sq ft)
Tailplane	10.94 m² (117.8 sq ft)
Elevators	2.90 m² (31.2 sq ft)

WEIGHTS AND LOADINGS:

Basic operating weight	10,967 kg (24,178 lb)
Max payload	1,280 kg (2,822 lb)
Max ramp weight	19,958 kg (44,000 lb)
Max T-O weight	19,844 kg (43,750 lb)
Max landing weight	16,329 kg (36,000 lb)
Max zero-fuel weight	12,247 kg (27,000 lb)
Max wing loading	393.5 kg/m² (80.6 lb/sq ft)
Max power loading	300.7 kg/kN (2.96 lb/lb st)

PERFORMANCE (at max T-O weight except where indicated):

Never-exceed speed	Mach 0.87
Max level and cruising speed at 9,145 m (30,000 ft)	475 kts (880 km/h; 547 mph)
Econ cruising speed at 10,670 m (35,000 ft)	441 kts (817 km/h; 508 mph)
Stalling speed, T-O flap setting	123 kts (229 km/h; 142 mph)
Max rate of climb at S/L	1,280 m (4,200 ft)/min
Rate of climb at S/L, one engine out	762 m (2,500 ft)/min
Service ceiling	10,970 m (36,000 ft)
Service ceiling, one engine out	8,685 m (28,500 ft)
T-O to 15 m (50 ft)	1,509 m (4,950 ft)
Landing from 15 m (50 ft) at max landing weight	1,274 m (4,180 ft)
Range with max fuel, 30 min reserves	2,770 n miles (5,132 km; 3,189 miles)
Range with max payload, 30 min reserves	2,600 n miles (4,818 km; 2,994 miles)

OPERATIONAL NOISE CHARACTERISTICS (FAR Pt 36):

T-O noise level	92.7 EPNdB
Approach noise level	97.4 EPNdB
Sideline noise level	87.7 EPNdB

UPDATED

LOCKHEED C-141A/B STARLIFTER

TYPE: Heavy logistics transport aircraft.

PROGRAMME: Operational experience with the Lockheed C-141A StarLifter by the USAF's Military Airlift Command showed that on many occasions the cargo compartment was physically packed to capacity without the aircraft's maximum payload capability reached. Parametric studies carried out by Lockheed showed that lengthening the fuselage by 7.11 m (23 ft 4 in) would provide an optimum relationship between modification cost and payload improvement, while at the same time allowing the existing wings, landing gear and power plant to be retained.

VERSIONS: **NC-141A:** Test aircraft operated by the USAF.

C-141B: Under the USAF contract, Lockheed-Georgia began work on the conversion of a C-141 in 1976. Designated YC-141B, this prototype was rolled out 8 January 1977 and made its first flight 24 March. The fuselage had been lengthened by the insertion of a 4.06 m (13 ft 4 in) plug immediately forward of the wing and by a similar 3.05 m (10 ft 0 in) plug immediately aft of the wing. These two plug sections are designed so that the lengthened aircraft retains the same operational features and mission versatility as the C-141A. As a result of this modification, the floor area of the cargo compartment is increased

Lockheed C-141B StarLifter heavy logistics transport aircraft *(Paul Tompkins)*

1995

by 22.26 m² (239.6 sq ft), and its volume by 61.48 m³ (2,171 cu ft).

To satisfy another proven requirement, for flight refuelling capability, a universal aerial refuelling receptable slipway installation (UARRSI) has been incorporated in the upper surface of the forward fuselage, just aft of the flight deck. The first production C-141B was delivered to the USAF 4 December 1979, and the entire programme was conducted ahead of schedule and below projected cost. Lockheed delivered 86 C-141B conversions during 1980, 125 in 1981, and completed the modification of all 270 C-141 Military Airlift Command StarLifters when the last aircraft was delivered 29 June 1982. There are currently 269 C-141B StarLifters in service with the USAF.

UPGRADES: **Chrysler Technologies Airborne Systems (CTAS):** Contract awarded to CTAS as prime contractor with AlliedSignal's Guidance & Control Systems Division as principal subcontractor in early 1993 to upgrade 148 C-141 cockpits and autopilots. Cockpit upgrade entails converting displays from analogue to digital. Cockpit instrumentation to comprise four 150 × 200 mm liquid crystal displays, two primary flight and two subsystem displays. Display processor to be used as mission computer and to control the MIL-STD-1553 databus to link systems. Autopilot to comprise a flight control processor with software based on the digital autopilot used for the Boeing E-3. Contract valued at $200 to 300 million.

Lockheed: C-141 SLEP proposal. Submitted unsolicited in early 1993 to double the life of 178 C-141s by 40,000 hours, Package includes new centre-wing boxes, new inner and outer wings as well as a fly-by-wire spoiler system.

Lockheed: C-141B. See Versions.

OPERATORS: Versions of the C-141 are in service with the US armed forces (248).

The following description applies to the C-141B:

DESIGN FEATURES: Cantilever high-wing monoplane. Chord NACA 0013.00 (modified) at root. NACA 0010.00 (modified) at tip. Aspect ratio 7.90. Chord 10.11 m (33 ft 2 in) at root. Mean aerodynamic chord 6.77 m (22 ft 2½ in). Thickness/chord ratio 13 per cent at root, 10 per cent at tip. Anhedral 1° 26′ at quarter-chord. Incidence 4° at root, 42′ at tip. Sweepback at quarter-chord 25°. The tail unit is a cantilever all-metal structure with horizontal surfaces mounted at tip of fin.

FLYING CONTROLS: Conventional aluminium alloy ailerons with full power actuation by dual hydraulic units and emergency manual control through geared tabs. Lockheed-Fowler aluminium trailing-edge flaps. Hinged spoilers on upper and lower surfaces of wings are operated by hydraulic systems and are used only for deceleration and landing. No trim tabs. Variable incidence tailplane provides pitch trim, actuated by electric motor or more rapidly by hydraulic motor. Elevators and rudder each controlled by hydraulic units. Third hydraulic system takes over elevator control if one of the normal systems fails. Rudder trim by positioning rudder servo unit control valve with electric trim actuators.

STRUCTURE: Conventional all-metal two-spar box-beam wing structure, built on fail-safe principles. The fuselage is a conventional aluminium alloy semi-monocoque structure built on fail-safe principles. Upswept rear fuselage with built-in ramp.

TAIL UNIT: Cantilever all-metal structure with horizontal surfaces mounted at tip of fin. Variable incidence tailplane provides pitch trim, actuated normally by electric motor or more rapidly by hydraulic motor. Elevators and rudder each controlled by dual hydraulic units. Third hydraulic system takes over elevator control if one of the normal systems fails. Controls can be actuated manually in an emergency. Rudder trim by positioning rudder servo unit control valve with electric trim actuators. Tailplane leading-edge de-iced electrically with metal-clad elements.

LANDING GEAR: Hydraulically retractable tricycle type, with Cleveland Pneumatic twin-wheel nose unit and Bendix Aerospace four-wheel bogie main units. All units retract forward, main bogies into fairings built on to sides of fuselage. Oleo-pneumatic shock-absorbers. Mainwheel tyres size 44 × 16 Type VII, pressure 10.3 to 12.4 bars (150 to 180 lb/sq in). Nosewheel tyres size 36 × 11.0 Type VII, pressure 13.8 bars (200 lb/sq in). Hydraulic multiple-disc brakes with independent braking through Bendix Pacific combined brake metering and anti-skid valve.

POWER PLANT: Four Pratt & Whitney TF33-P-7 turbofan engines, each rated at 9,525 kg (21,000 lb st), mounted in underwing pods and fitted with clamshell door thrust reversers. Fuel in 12 integral tanks in wings, between front and rear spars, with total usable capacity of 87,364 litres (23,080 US gallons; 19,233 Imp gallons). Single-point dual hose adaptor refuelling point in aft section of starboard wheel fairing. Gravity fuelling cap over each tank. Total oil capacity 83.3 litres (22 US gallons; 18.3 Imp gallons).

Lockheed C-141B StarLifter with four Pratt & Whitney TF33-P-7 turbofan engines *(Paul Tompkins)*

1995

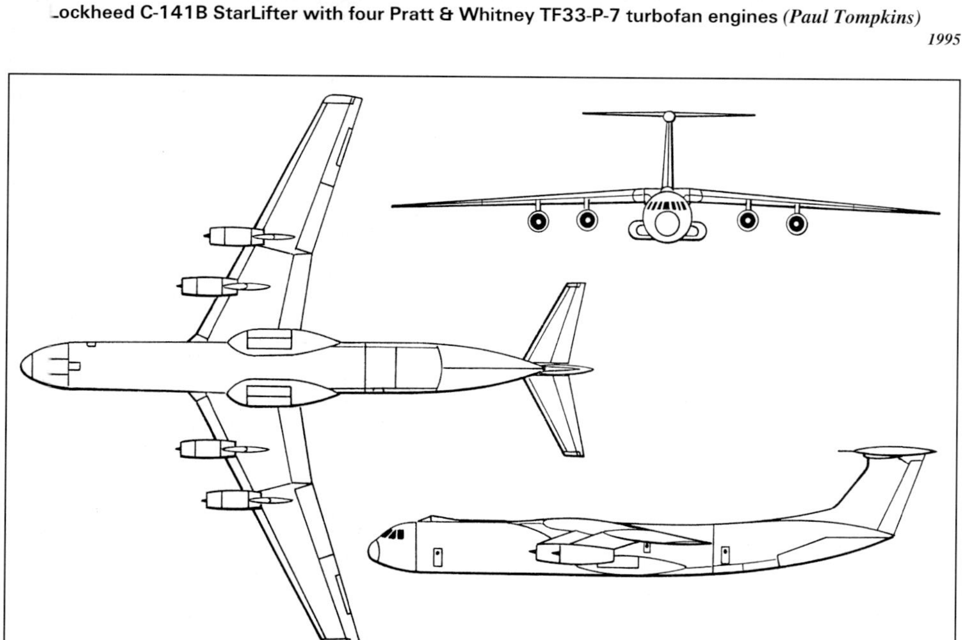

Lockheed C-141B StarLifter of the US Air Force *(Jane's/ Mike Keep)*

1994

DIMENSIONS, EXTERNAL:

Wing span	48.74 m (159 ft 11 in)
Length overall	51.29 m (168 ft 3½ in)
Height overall	11.96 m (39 ft 3 in)

DIMENSIONS, INTERNAL:

Cargo compartment:

Length	28.44 m (93 ft 3½ in)
Max height	2.77 m (9 ft 1 in)
Max width	3.11 m (10 ft 2½ in)
Usable cargo volume	322.71 m³ (11,399 cu ft)

WEIGHTS AND LOADINGS:

Operating weight (MAC)	67,186 kg (148,120 lb)
Max payload (2.5 *g*)	32,025 kg (70,605 lb)
Max payload (2.2 *g*)	41,222 kg (90,880 lb)
Max T-O weight (2.5 *g*)	146,555 kg (323,100 lb)
Max T-O weight (2.25 *g*)	155,580 kg (343,000 lb)
Emergency war planning ramp weight	156,445 kg (344,900 lb)
Max zero-fuel weight (2.5 *g*)	99,210 kg (218,725 lb)
Max zero-fuel weight (2.25 *g*)	108,410 kg (239,000 lb)
Max landing weight	155,580 kg (343,000 lb)

PERFORMANCE (at max 2.5 *g* T-O weight, except where indicated):

Max cruising speed	492 kts (910 km/h; 566 mph)
Long-range cruising speed	430 kts (796 km/h; 495 mph)
Max rate of climb at S/L	890 m (2,920 ft)/min
T-O to 15 m (50 ft)	1,768 m (5,800 ft)
Landing from 15 m (50 ft) at normal landing weight	1,128 m (3,700 ft)
Range with max payload	2,550 n miles (4,725 km; 2,935 miles)
Ferry range	5,550 n miles (10,280 km; 6,390 miles)

UPDATED

LOCKHEED STARFIGHTER

USAF designation: F-104

TYPE: Single-seat multipurpose combat aircraft.

PROGRAMME: Development of the F-104 began in 1951. After production ceased in the USA, Canada, Italy and Japan continued to produce the aircraft under licence.

VERSIONS: **XF-104:** Two single-seat fighter prototypes were built under this designation, with Wright J65-W.6 turbojet engine and afterburner.

F-104A: Single-seat interceptor. Basically similar to the XF-104 but with air intake shock-cones, forward-retracting nosewheel. General Electric J79-GE-3B turbojet engine. The first F-104 flew for the first time 17 February 1956 and deliveries to the USAF commenced 26 January 1958. Withdrawn.

F-104A54A: In service with the Italian Air Force. See Alenia.

NF-104A: An F-104A fitted with a 2,720 kg (6,000 lb st) Rocketdyne AR-2 auxiliary rocket engine in tail above jet pipe. The first of three NF-104s flew in July 1963.

F-104B: Tandem two-seat development of F-104A, for use as both combat aircraft and operational trainer. Has same engine, but considerably greater fin area and fully powered rudder. Prototype flew 7 February 1957. Total of 26 built. Withdrawn.

F-104C: General Electric J79-GE-7A turbojet engine 7,165 kg (15,800 lb st) with the afterburning. Fighter-bomber. Total 76 built; deliveries commenced October 1958. Withdrawn.

F-104D: Two-seat version of F-104C. In service with the Taiwanese Air Force.

F-104DJ: Similar to the F-104D. Lockheed built 20 for Japan Air Self-Defence Force. These were reassembled by Mitsubishi. Withdrawn.

CF-104D: Training version in service with the Turkish Air Force.

F-104F: Basically as F-104D, but some equipment changes. Lockheed built 30 for the Federal German Air Force. These aircraft are no longer in service.

F-104G (Lockheed Model 683-10-19): Single-seat multimission fighter, based on the F-104C. General Electric J79-GE-11A turbojet engine rated at 7,165 kg (15,800 lb st) with afterburning. In service with Taiwanese and Turkish air forces.

RF-104G: F-104G fitted with internal camera pack for reconnaissance duties. In service with the Taiwanese and Turkish air forces.

TF-104G: Two-seat conversion trainers equipped with NASARR and full operational equipment including the same fittings for underwing and wingtip stores as the F-104G, with extra fuselage centreline bomb track. Engine as for F-104G. In service with the Taiwanese Air Force.

F-104J: Similar to the F-104G except for equipment. Built for the Japanese Air Self-Defence Force. Aircraft no longer in service.

F-104N: Astronaut proficiency trainer for NASA. Three delivered in the early 1960s. Withdrawn.

F-104S: Developed from the F-104G Starfighter, primarily as interceptor. Powered by General Electric J79-GE-19 turbojet engine with redesigned afterburner, giving 8,120 kg (17,900 lb st). Nine external attachments for stores, including rockets, bombs and Sidewinder missiles. Built for the Italian Air Force by Fiat.

CF-104: Single-seat strike/reconnaissance aircraft, basically similar to F-104G. 200 built for the Royal Canadian Air Force (RCAF) by Canadair. Aircraft no longer in service.

CF-104D: Two-seat aircraft built for the RCAF.

QF-104: Remotely controlled recoverable target drone conversion of F-104A for missile evaluation. Withdrawn.

UPGRADES: **Alenia:** F-104ASA. See separate entry.

OPERATORS: Versions of the F-104 Starfighter are in service with the armed forces of the following countries: Italy (185), Taiwan (135) and Turkey (179).

The following description applies to the F-104S:

DESIGN FEATURES: Cantilever mid-wing monoplane. Biconvex supersonic wing section with a thickness/chord ratio of 3.36 per cent. Anhedral 10°. No incidence. Sweepback 18° 6′ at quarter-chord. Leading-edge nose radius of 0.41 mm (0.016 in) and razor-sharp trailing-edge. Narrow-chord ventral fin on centreline and two smaller lateral fins under fuselage to improve stability.

FLYING CONTROLS: Full-span electrically actuated drooping leading-edge. Entire trailing-edge hinged, with inboard sections serving as landing flaps and outboard sections as ailerons. Ailerons are of aluminium, each powered by a servo control system which is irreversible and hydraulically powered, and each actuated by 10 small hydraulic cylinders. Trim control is applied to position the aileron relative to the servo control position. An electric actuator positions the aileron trim. Flaps are of aluminium, actuated electrically. Above each flap is the air delivery tube of a boundary layer control system, which ejects air bled from the engine compressor over the entire flap span when the flaps are lowered to the landing position. Rudder is fully powered by a hydraulic servo. Trim control is applied to position the tailplane relative to the servo control position, by means of an electric actuator located in the fin. The rudder itself is trimmed in the same way as the tailplane.

STRUCTURE: All-metal structure with two main spars, 12 spanwise intermediate channels between spars and top and bottom one-piece skin panels, tapering from thickness of 6.3 mm (0.25 in) at root to 3.2 mm (0.125 in) at tip. Each half-wing measures 2.31 m (7 ft 7 in) from root to tip and is a separate structure cantilevered from five forged frames in fuselage. The fuselage is an all-metal monocoque structure. Hydraulically operated aluminium airbrake on each side of rear fuselage. The tail unit is a T-type cantilever unit with 'all-flying' one-piece horizontal tail surface hinged at mid-chord point at top of the vertical fin and powered by a hydraulic servo. Tailplane has a similar profile to wing and is all-metal.

Lockheed F-104S ASA single-seat combat-aircraft with one General Electric J79-GE-19 turbojet engine
1995

Lockheed F-104S ASA Starfighter of the Italian Air Force *(Paul Tompkins)*
1994

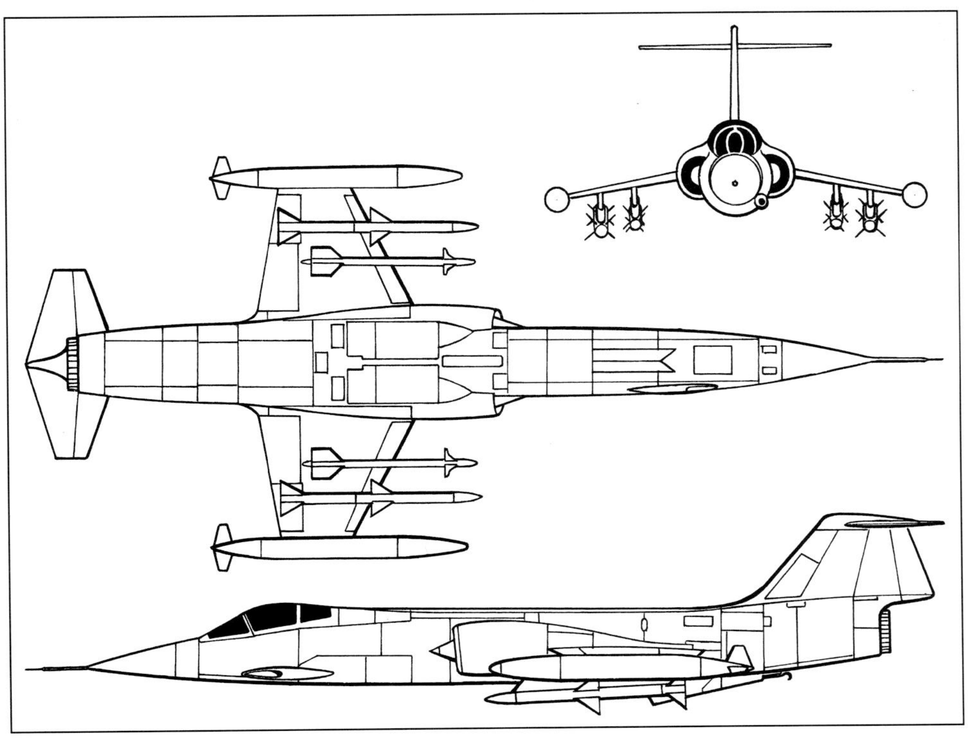

Lockheed F-104 Starfighter single-seat multipurpose combat aircraft *(Jane's/Mike Keep)*
1994

LANDING GEAR: Retractable tricycle type with Dowty patent liquid-spring shock-absorbers on main units, oleo-pneumatic shock-absorbers on nose unit. Hydraulic actuation. Mainwheels raised in and forward. Steerable nosewheel retracts forward into fuselage. Mainwheel legs are hinged on oblique axes so that the wheels lie flush within the fuselage when retracted. Mainwheels size 26 × 8.0, with Goodrich tyres size 26 × 8.0 Type VIII (18 ply rating), pressure 11.93 bars (173 lb/sq in). Nosewheel tyre size 18 × 5.5 Type VII (14 ply rating). Bendix hydraulic disc brakes with Goodyear anti-skid units. Arrester hook under rear of fuselage. Braking parachute in rear fuselage.

POWER PLANT: One General Electric J79-GE-19 turbojet engine, rated at 52.8 kN (11,870 lb st) dry and 79.62 kN (17,900 lb st) with afterburning. Electrical de-icing elements fitted to air intakes. Most of the aircraft's hydraulic equipment mounted inside large engine bay door under fuselage to facilitate servicing. Internal fuel in five bag-type fuselage tanks with total standard capacity of 3,392 litres (896 US gallons; 746 Imp gallons). Provision for external fuel in two 740 litre (195 US gallon; 162 Imp gallon) pylon tanks and two 645 litre (170 US gallon; 142 Imp gallon) wingtip tanks. Pressure refuelling of all internal and external tanks through single point on upper port fuselage just forward of air intake duct. Gravity fuelling point for internal tanks aft of pressure refuelling point, with individual gravity fuelling of external tanks. In-flight refuelling can be provided through Lockheed designed probe-drogue system. Probe, mounted below port sill of cockpit, is removable but when installed is non-retractable. Oil capacity 15 litres (4 US gallons; 3 Imp gallons).

ACCOMMODATION: Pressurised and air conditioned cockpit well forward of wings. Canopy hinged to starboard for access. Martin-Baker IQ-7A zero/zero ejection seat. Air conditioning package by AiResearch using engine bleed air. Pressure differential 0.34 bars (5 lb/sq in). Two completely separate hydraulic systems, using engine-driven pumps operating at 207 bars (3,000 lb/sq in). No. 1 system operates one side of tailplane, rudder and ailerons, also the automatic pitch control actuator and autopilot actuators. No. 2 system operates other half of tailplane, rudder and ailerons, also the landing gear, wheel brakes, airbrakes, nosewheel steering and constant-frequency electrical generator. Emergency ram-air turbine supplies emergency hydraulic pump and 4.5 kVA 115/200 V electric generator. Electrical system supplied by two engine-driven 20 kVA 115/200 V variable frequency (320 to 520 Hz) generators. Constant-speed hydraulic motor drives 2.5 kVA 115/200 V generator to supply fixed-frequency AC. 28 V DC power supplied by two Ni/Cd batteries and a 120 A transformer-rectifier.

ELECTRONICS AND EQUIPMENT: Integrated electronics system in which various communications and navigation components may be self-sustaining units which may be varied to provide for different specific missions. Equipment includes autopilot with 'stick steering', which includes modes for preselecting and holding altitude, speed, heading and constant rate of turn: multipurpose R21G/H radar for air-to-air interception, ground and contour mapping, and terrain-avoidance modes of operation; fixed-reticle gunsight; bombing computer; air data computer; dead reckoning navigation device: Tacan radio air navigation system; provision for datalink-time division set and AN/ARC-552 UHF radio; Litton LN-3-2A lightweight fully automatic inertial navigation system; Sperry C-2G compass system; AN/APN-198 radar altimeter; AIC-18 intercom; and AN/APX-46 IFF/SIF. Provision for fitting a camera pod under the fuselage for reconnaissance duties.

ARMAMENT: Nine external attachment points, at wingtips, under wings and under fuselage, for bombs, rocket pods, auxiliary fuel tanks and air-to-air missiles. Normal primary armament consists of two Raytheon AIM-7 Sparrow III air-to-air missiles under wings and/or two Sidewinders under fuselage and either a Sidewinder or 645 litre (170 US gallon; 142 Imp gallon) fuel tank on each wingtip. Alternatively, an M-61 20 mm multibarrel rotary cannon can be

Lockheed F-104G of the Turkish Air Force

1994

fitted in the port underside of the fuselage instead of the AIM-7 missile control package. Maximum external weapon load 3,402 kg (7,500 lb).

DIMENSIONS, EXTERNAL:
- Wing span without tip tanks: 6.68 m (21 ft 11 in)
- Wing chord (mean): 2.91 m (9 ft 6 7/12 in)
- Wing aspect ratio: 2.45
- Length overall: 16.69 m (54 ft 9 in)
- Length of fuselage: 15.62 m (51 ft 3 in)
- Height overall: 4.11 m (13 ft 6 in)
- Tailplane span: 3.63 m (11 ft 11 in)
- Wheel track: 2.74 m (9 ft 0 in)
- Wheelbase: 4.59 m (15 ft 0½ in)

AREAS:
- Wings, gross: 18.22 m^2 (196.1 sq ft)
- Ailerons (total): 0.85 m^2 (9.2 sq ft)
- Trailing-edge flaps (total): 2.11 m^2 (22.7 sq ft)
- Leading-edge flaps (total): 1.51 m^2 (16.2 sq ft)
- Airbrakes (total): 0.77 m^2 (8.25 sq ft)
- Fin: 3.50 m^2 (37.7 sq ft)
- Ventral fin (centreline): 0.47 m^2 (5.1 sq ft)
- Rear fuselage strakes (total): 0.71 m^2 (7.6 sq ft)
- Rudder, incl tab: 0.51 m^2 (5.5 sq ft)
- Tailplane: 4.48 m^2 (48.2 sq ft)

WEIGHTS AND LOADINGS:
- Weight empty: 6,700 kg (14,900 lb)
- Max internal fuel load: 2,641 kg (5,824 lb)
- Max internal and external fuel load: 5,153 kg (11,362 lb)
- T-O weight 'clean': 9,840 kg (21,690 lb)
- Max T-O weight: 14,060 kg (31,000 lb)
- Max zero-fuel weight 'clean': 6,806 kg (15,006 lb)
- Max zero-fuel weight (fighter-bomber): 7,148 kg (15,760 lb)
- Max wing loading: 540 kg/m^2 (110.7 lb/sq ft)

PERFORMANCE (At 9,840 kg; 21,690 lb AUW except where indicated):
- Never-exceed speed: Mach 2.2
- Max level speed at 11,000 m (36,000 ft): Mach 2.2 (1,259 kts; 2,330 km/h; 1,450 mph)
- Max level speed at S/L: Mach 1.2 (790 kts; 1,464 km/h; 910 mph)
- Max cruising speed at 11,000 m (36,000 ft): 530 kts (981 km/h; 610 mph)
- Econ cruising speed: Mach 0.85
- T-O speed at S/L, Interceptor with two AIM-7 missiles: 189 kts (530 km/h; 217 mph)
- Typical landing speed at S/L: 159 kts (295 km/h; 183 mph)
- Service ceiling: 17,680 m (58,000 ft)
- Zoom altitude: more than 27,400 m (90,000 ft)
- Time to accelerate from Mach 0.92 to 2.0: 2 min
- Time to climb to 10,670 m (35,000 ft): 1 min 20 s
- Time to climb to 17,070 m (56,000 ft): 2 min 40 s
- T-O run at S/L, Interceptor with two AIM-7 missiles: 823 m (2,700 ft)
- Typical landing run at S/L: 762 m (2,500 ft)
- Radius with max fuel: 673 n miles (1,247 km; 775 miles)
- Ferry range (excl flight refuelling): 1,576 n miles (2,920 km; 1,815 miles)

UPDATED

LOCKHEED F-117A

Unofficial name: Nighthawk

TYPE: Precision attack fighter with stealth elements, optimised for radar energy dispersion and low IR emission.

PROGRAMME: Development began with USAF Flight Dynamics Laboratory contract to Lockheed Advanced Development Projects (Skunk Works), funded by DARPA under Have Blue programme; two **XST** (Experimental Stealth Technology) prototypes produced, each powered by two 11.12 to 12.46 kN (2,500 to 2,800 lb st) GE CJ610s; first flight at Groom Lake, Nevada, December 1977 by William C. Park; first prototype crashed 4 May 1978; second XST crashed at Tonopah Test Range 1980; similar to F-117 apart from inward-canted ruddervators. Span 6.71 m (22 ft 0 in), length 11.58 m (38 ft 0 in), maximum T-O weight 5,443 kg (12,000 lb), leading-edge sweep 72° 30′, four-transparency canopy.

Development and manufacture of operational F-117A started November 1978 under Senior Trend programme. First of five preseries aircraft (Article numbers 780 to 784) flew 18 June 1981; one crashed 20 April 1982. Planned production of 100 reduced to 59 (Article numbers 785-843) 785 of which crashed on first flight 21 June 1982; first hand-over to USAF (Article 787) 23 August 1982; funding 13 in FY80 and 11, 10, 11 and 14 in FY82-85; final assembly at Tonopah; first picture and designation released 10 November 1988; first operational deployment in Operation Just Cause over Panama, 21 December 1989, when two F-117As each dropped a 907 kg (2,000 lb) laser-guided bomb on barracks area at Rio Hato; all 56 in-service F-117As participated in 1991 Gulf War against Iraq, flying some 1,270 missions.

VERSIONS: **F-117A:** As described.

UPGRADES: **Lockheed:** Weapon system improvement began with first flight of aircraft modified under Offensive Capability Improvement Program on 1 December 1988; first 'production' redelivery (20805) 27 November 1990; programme to continue to 2005; improvements installed by 1991 include 'four-dimensional' flight management system (time on waypoint ±1 second) and new cockpit instrumentation with full colour MFDs and digital moving map; Lockheed redelivered 73 F-117 upgrades, including 15 second phase, from Palmdale by early 1992; work continues at one aircraft per month. First phase, which began in 1984, involved replacement of Delco M362F computers by IBM AP-102. Second phase known as Offensive Combat Improvement Program; includes entirely revised cockpit with Honeywell MFDs, three-dimensional flight management system, large moving map or digital situation display, autothrottle and pilot-activated recovery system. Third phase flight tested 1992, involves new turret-mounted infra-red acquisition and designation sensor (IRADS) by TI; Honeywell ring laser gyro INS, Collins GPS. Achieved 50,000 flying hours in February 1991.

Lockheed Martin: Commenced conducting a USAF funded Mid-Life Improvement Study (MIPS) in early 1995 to analyse weapons, signature reduction and other upgrades. Work could begin in early 1996 with existing funding. Possible improvements include the installation of a MIL-STD 1760 databus which would allow the F-117A to launch GPS-guided weapons such as Joint Direct Attack Munition (JDAM) and Joint Stand-Off Weapon (JSOW) which are both under development. Other possibilities include a bubble canopy.

OPERATORS: Versions of the F-117 are in service with the US armed forces.

CUSTOMERS: USAF; five preseries plus 59 production; deliveries seven, eight, eight, eight, eight, seven, five, four and three in the calendar years 1982-90 (final delivery, of 88-0843, 12 July 1990).

First F-117A unit, 4450th Tactical Group at Tonopah (122 n miles; 225 km; 140 miles north-west of Las Vegas), Nevada, formed 15 October 1979 and equipped with 18 A-7D Corsair IIs until first F-117A arrived; P-Unit (later 4451st Test Squadron) formed June 1981; Q-Unit (later 4452nd TS) began operations 15 October 1982, flying first 4450th TG training operation (00786); Group transferred from direct control of Tactical Air Command to Tactical Fighter Weapons Center at Nellis AFB 1985; Z-Unit (later 4453rd Test and Evaluation Squadron) formed 1 October 1985; first of several public air display appearances made April 1990; unit renamed 37th Tactical Fighter Wing of 12th Air Force 5 October 1989; strength 40 F-117As divided between 415th and 416th TFS; A-7Ds replaced in training and chase duty by eight T/AT-38 Talons of 417th TFTS, also with balance of 16 remaining F-117As; first assignment away from Tonopah was covert deployment to UK AFB 18 June 1990; wing and squadrons deleted 'Tactical' prefix 1 November 1991; transfer to Holloman AFB, New Mexico, began with delivery of 00791 for maintenance familiarisation, 7 January 1992; movement of remaining aircraft began 9 May 1992, when unit became 49th Fighter Wing (7th, 8th and 9th Fighter Squadrons).

Lockheed F-117A 'Nighthawk' stealth attack aircraft *(Eric Schulzinger and Denny Lombard)*

1995

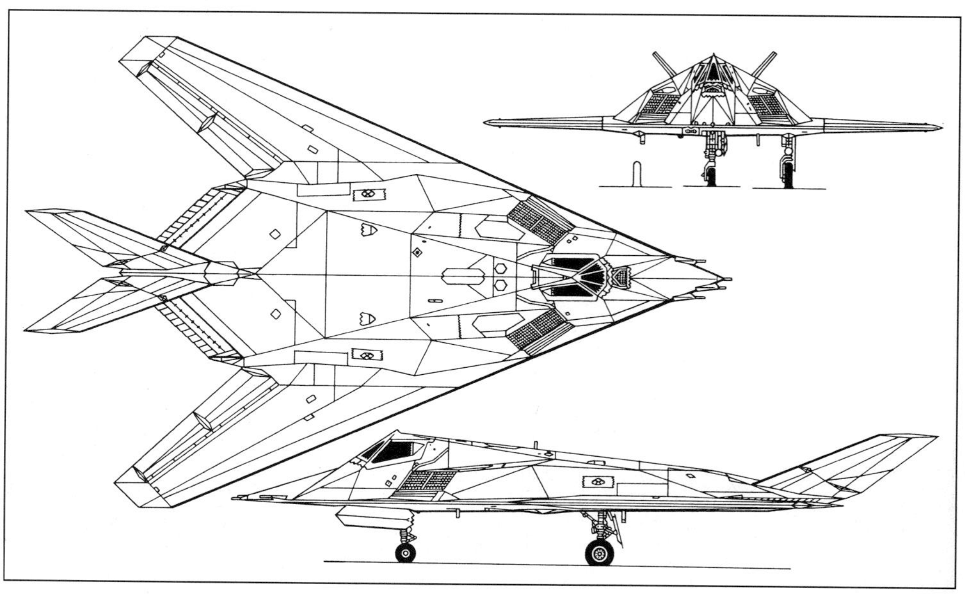

Lockheed F-117A Stealth fighter *(Jane's/Mike Keep)*

1994

COSTS: $6,560 million programme (1990), including $2,000 million R&D, $4,270 million for procurement and $295.4 million for infrastructure. Average unit cost $42.6 million (then-year dollars).

DESIGN FEATURES: Multifaceted airframe designed to reflect radar energy away from originating transmitter, particularly downward-looking AEW aircraft; vortexes from many sharp edges, including leading-edge of wing, designed to form co-ordinated lifting air flow pattern; wings have 67° 30′ sweepback, much greater than needed for subsonic performanes, with aerofoil formed by two flat planes underneath and three on upper surface; forward underwing surface blends with forward fuselage; all door and access panels have serrated edges to suppress radar reflection; internal weapon bay 4.7 m (15 ft 5 in) long and 1.75 m (5 ft 9 in) wide divided longitudinally by two lengthwise doors hinged on centreline; boom refuelling receptacle on port side of top plate, aft of cockpit. Frontal radar cross-section estimated as 0.01 m² (0.1 sq ft).

FLYING CONTROLS: Four omnidirectional air probes at nose indicate GEC Astronics quadruplex fly-by-wire control system, similar to that of F-16, using two-section elevons and all-moving ruddervators together for control and stability; ruddervators swept about 65° and set at 85° to each other.

STRUCTURE: Material principally aluminium; two-spar wings; fuselage has flat facets mounted on skeletal subframe, jointed without contour blending; surfaces coated with various radar-absorbent materials. Ruddervators being replaced by new units of thermoplastic graphite composites construction, removing previous speed restriction due to flutter.

LANDING GEAR: Tricycle type by Menasco, with single wheels all retracting forward. Loral brakes and wheels; Goodyear tyres. All doors have serrated edges to suppress radar reflections. Emergency arrestor hook with explosively jettisoned cover.

POWER PLANT: Two 48.0 kN (10,800 lb st) class General Electric F404-GE-F1D2 non-augmented turbofans. Rectangular overwing air intakes with 2.5 × 1.5 cm (1 × ⅝ in) heated grid for anti-icing and low observability. Auxiliary air intake doors in horizontal surface immediately to the rear. Part of cold air ingested bypasses engine and is mixed with exhaust gases for cooling. Narrow-slot 'platypus' exhausts in rear fuselage, 1.65 m (5 ft 5 in) long and 0.10 m (4 in) high, with extended lower lip, surrounded by heat tiles of type used on Space Shuttle and with 11 vertical, internal guide vanes. Sundstrand air turbine starter. In-flight refuelling receptacle in decking aft of cockpit, illuminated for night refuelling by lamp at apex of cockpit. Optional drop tank on internal weapon pylon.

ACCOMMODATION: Pilot only; McDonnell Douglas ACES II zero/zero ejection seat. Five individually framed flat-plate windows, including single-piece windscreen. Transparencies, gold-coated for radar dissipation, produced by Serracin/Sylmar Corporation. Canopy hinged to open upward and backward.

SYSTEMS: Allied Signal environmental control, auxiliary power and emergency power systems.

AVIONICS: Forwarding-looking infra-red (FLIR) sensor, with dual fields of view, in recessed emplacement, covered by fine mesh screen, below windscreen. Retractable downward-looking DLIR and laser designator beneath forward fuselage to starboard of nosewheel bay; FLIR and DLIR by Texas Instruments (to be replaced by improved equipment during third-phase retrofit in 1994). HUD based on Kaiser AN/AVQ-28; large head-down display for FLIR imagery flanked by two multifunction CRTs. Retractable radio antennae beneath fuselage, ahead of port main landing gear, and on spine. Honeywell radar altimeter, Honeywell SPN-GEANS INS (replaced by Honeywell H-423/E ring laser gyro from August 1991; Rockwell Collins GPS to be added); IBM AP-102 mission computer (replacing original three Delco M362F computers); GEC-Marconi flight control computer/navigation interface and autopilot computer (NIAC) system; SLI Avionic Systems Corporation expanded data transfer system and AHRS. Harris Corporation digital moving map added as retrofit with full-colour MFDS. Proposed adaptation to carry TARPS reconnaissance pod.

Lockheed F-117A Stealth fighter of the US Air Force

1994

ARMAMENT: 'Full range of USAF tactical fighter ordnance', principally two 907 kg (2,000 lb) bombs: BLU-109B low-level laser-guided or GBU-10/GBU-27 laser-guided glide weapons; alternatively, AGM-65 Maverick or AGM-88 HARM ASMs. Internal carriage on two extensible beams in weapon bay. (Only missiles with seeker heads extended below aircraft prior to launch; bombs released from within weapon bay.)

DIMENSIONS, EXTERNAL:

Wing span	13.20 m (43 ft 4 in)
Length overall	20.08 m (65 ft 11 in)
Height overall	3.78 m (12 ft 5 in)

AREAS (estimated):

Wing area	105.9 m² (1,140 sq ft)

WEIGHTS AND LOADINGS:

Weight empty (estimated)	13,608 kg (30,000 lb)
Max T-O weight	23,814 kg (52,500 lb)

PERFORMANCE:

Max level speed	Mach 1+*
Normal max operating speed	Mach 0.9
T-O speed at normal combat weight	165 kts (306 km/h; 190 mph)*
Landing speed	150 kts (227 km/h; 172 mph)*
g limit	+6

(* = not confirmed by USAF)

UPDATED

LOCKHEED 188 ELECTRA

TYPE: Four-engined medium-range airliner.

PROGRAMME: The Electra is a short medium-range airliner powered initially by four 3,750 eshp Allison Model 501 (military designation T56) turboprop engines. Able to take engines developing up to 5,500 eshp each. The first prototype flew 6 December 1957, followed by the second prototype 13 February 1958.

The initial order 'off the drawing board' for 35 Electras was received from American Airlines in 1955 and actual construction of these aircraft began in October 1956. There were 170 built.

VERSIONS: **Lockheed 188A:** Initial version.

Lockheed 188C: Longer range version seating 98 passengers.

Lockheed 188CF: Freighter version.

OPERATORS: Versions of the L-188 Electra are in service with the armed forces of the following countries: Argentina (6), Bolivia (1) and Honduras (1).

DESIGN FEATURES: Cantilever low-wing monoplane. Wing section NACA 0014-1.10 at root, NACA 0012-1.10 at tip. Aspect ratio 7.5. Chord 5.77 m (18 ft 11 in) at root, 2.31 m (7 ft 7 in) at tip. Dihedral 6°. Incidence 3° at root, ½° at tip. Cantilever monoplane tail unit.

FLYING CONTROLS: Conventional aluminium alloy ailerons and high-lift Fowler type flaps. Tailplane elevators.

STRUCTURE: The wing is an extruded and integrally stiffened aluminium box-beam structure. The fuselage is an aluminium alloy semi-monocoque structure. The tail unit is an aluminium alloy cantilever structure.

LANDING GEAR: Retractable tricycle type. Menasco oleo-pneumatic shock-absorbers. Hydraulic retraction. Each main unit has twin wheels with 13.50-16 tyres, pressure 9.5 bars (138 lb/sq in). Steerable nosewheel unit has twin wheels with 7.50-14 tyres. Wheel track 9.50 m (31 ft 2 in). Wheelbase 11.28 m (37 ft).

POWER PLANT: Four 3,750 eshp Allison 501-D13A (later 4,050 eshp Allison 501-D15) turboprop engines, giving 2,060 eshp per engine for cruising at 6,100 m (20,000 ft). Aeroproducts or Hamilton Standard Type 54H60 four-blade reversible-pitch airscrews, 4.11 m (13 ft 6 in) diameter. Total fuel capacity 20,210 litres (5,360 US gallons; 4,467 Imp gallons), in two 4,127 litre (1,090 US gallon; 908 Imp gallons) inboard wing tanks and two 6,018 litre (1,590 US gallon; 1,325 Imp gallons) outboard wing tanks. Single-point fuelling, with underwing fuelling panel at bottom rear of No. 3 nacelle. Oil capacity 91 litres (24 US gallons; 20 Imp gallons).

Lockheed L-188 of Falcon Cargo ***(Peter March)***

1994

Lockheed F-117A precision attack fighter aircraft

1994

ACCOMMODATION: Pressurised cabin for crew of five and passengers maintains 2,440 m (8,000 ft) cabin atmosphere at 9,150 m (30,000 ft). Internal diameter of cabin 3.25 m (10 ft 8 in). Initial version for American Airlines had accommodation for 81 passengers including six in a special lounge; but a high-density version with seats for 85 passengers and a six-seat lounge will also be available. In international configuration, seating can be 44 (siesta arrangement), 65 (mixed custom and standard), 66 (custom) or 88 (standard). The international version has four toilets instead of the normal two. Integral passenger loading steps fold into the fuselage wall when not in use. Main cabin floor area 65.68 m² (707 sq ft). Main cabin luggage space 3.37 m³ (19 cu ft). Lower compartment cargo space 14.47 m³ (511 cu ft). All versions carry up to 1,816 kg (4,000 lb) of air freight, express and mail, in addition to the normal amount of passengers' baggage.

EQUIPMENT: Standard equipment will include autopilot and weather-avoidance radar.

DIMENSIONS, EXTERNAL:

Wing span	30.18 m (99 ft 0 in)
Length	31.81 m (104 ft 6½ in)
Height	9.78 m (32 ft 1 in)

WEIGHTS AND LOADINGS:

Weight empty (domestic)	25,401 kg (56,000 lb)
Weight empty (international)	25,855 kg (57,000 lb)
Normal payload	9,815 kg (21,638 lb)
Max weight limit payload	12,020 kg (26,500 lb)
Gross T-O weight	51,256 kg (113,000 lb)
Max landing weight	43,385 kg (95,650 lb)
Max zero-fuel weight	39,010 kg (86,000 lb)

PERFORMANCE:

Max speed at 3,660 m (12,000 ft) 389 kts (722 km/h; 448 mph)

Cruising speed at 38,782 kg (85,500 lb) AUW at 6,700 m (22,000 ft) 352 kts (652 km/h; 405 mph)

Stalling speed at 43,385 kg (95,650 lb) flaps down 99 kts (172 km/h; 107 mph)

Max rate of climb at S/L at 51,256 kg (113,000 lb) AUW 509 m (1,670 ft)/min

Max rate of climb at S/L at 41,731 kg (92,000 lb) AUW 731 m (2,400 ft)/min

Service ceiling at 45,359 kg (100,000 lb) AUW 8,230 m (27,000 ft)

CAA T-O, runway at 51,256 kg (113,000 lb) AUW 1,775 m (5,820 ft)

CAA T-O runway at 48,217 kg (106,700 lb) AUW 1,600 m (5,250 ft)

CAA landing runway at 38,782 kg (85,500 lb) AUW 1,510 m (4,960 ft)

Range at T-O weight of 51,256 kg (113,000 lb), with 8,165 kg (18,000 lb) payload and 3,220 kg (7,100 lb) fuel reserve 2.407 n miles (4,458 km; 2,770 miles)

UPDATED

LOCKHEED L-1011-1 (MODEL 385) TRISTAR

TYPE: Three-turbofan commercial transport.

PROGRAMME: In January 1966, Lockheed-California began a study of future requirements in the short/medium-haul airliner market. The design which emerged, known as the L-1011 (Lockheed Model 385 TriStar), was influenced by the published requirements of America Airlines, which specified optimum payload/range performance over the Chicago-Los Angeles route, coupled with an ability to take off from comparatively short runways with full payload.

The original design centred around a twin-turbofan configuration. Discussions which followed with American domestic carriers led to the eventual selection of a three-engined configuration, and a Rolls-Royce RB211 high bypass ratio turbofan was chosen as the power plant.

In June 1968 the L-1011 TriStar moved to the production design stage. Construction of the first aircraft began in March 1969, and this was rolled out in September 1970. The first flight was made on 16 November 1970. On 22 December 1971 a Class II provisional type certification was received, permitting delivery of aircraft to customers for route proving and demonstration purposes.

This original version of the TriStar is now known as the L-1011-1. It has been followed by several versions of the same basic airframe.

VERSIONS: **L-1011-1:** Basic TriStar, as described in detail. Initial delivery of the L-1011-1, to Eastern Air Lines for crew training, made 6 April 1972, followed by a similar delivery to TWA. FAA certification was granted in the same month and the first passenger service with the TriStar was flown by Eastern Air Lines 15 April. Scheduled services began 11 days later.

L-1011-100: Longer range version. Outward configuration identical to L-1011-1. Available with RB211-22B engines (each 187 kN; 42,000 lb st) or RB211-22F engines (each rated 193.5 kN; 43,500 lb st). Maximum T-O weight of 204,120 kg (450,000 lb) can be increased to 211,375 kg (466,000 lb) with additional 8,165 kg (18,000 lb) of fuel in new centre-section tanks. Normal range with -22F engines, AUW of 204,120 kg (450,000 lb) and payload of 24,250 kg (55,000 lb) is 3,585 n miles (6,645 km; 4,130 miles). Ordered by Air Canada, Cathay Pacific, Gulf Air and Saudi Arabian Airlines.

Lockheed L-188 medium-range transport aircraft

1994

L-1011-200: Longer range version, with improved take-off and climb performance, offering particular benefits to operators serving 'hot or high' areas. Outward configuration identical with that of L-1011-1. Powered by RB211-524 engines (each rated at 213.5 kN; 48,000 lb st). Optional maximum T-O weights of 204,120 kg (450,000 lb) or 211,375 kg (466,000 lb) according to whether new centre-section tankage is fitted. Nominal range at AUW of 211,375 kg (466,000 lb), with payload of 24,950 kg (55,000 lb st) is 4,020 n miles (7,450 km; 4,630 miles). Ordered by Saudi Arabian Airlines.

L-1011-250: Long-range version, with further increase in maximum T-O weight to 224,980 kg (496,000 lb) and maximum fuel capacity of 96,160 kg (212,000 lb), through added centre-section tankage. Outward configuration identical with that of L-1011-1. Wings, fuselage and fin front spar web reinforced to cater for higher design loads. New nosewheel unit and strengthened main landing gear axles. Larger tyres with increased ply rating on all units. Braking capacity increased. Powered by RB211-524B engines (each 213.5 kN; 48,000 lb st). Galley can be below deck, as on other versions, or dispersed on main deck, which doubles available space in forward cargo hold. Expanded forward hold accommodates 16 LD-3 half-width containers or five pallets, each measuring 2.23 × 3.17 m (88 × 125 in). For pallet loading, the forward cargo door is replaced by a 1.72 × 2.64 m (68 × 104 in) power-operated upward-opening door. Main-deck galleys reduce passenger accommodation from typical 273 to 253 in eight-abreast coach configuration, and from typical 302 to 284 in nine-abreast coach configuration, in each case with 10 per cent first class forward. Nominal range at maximum AUW with 24,950 kg (55,000 lb) payload is 4,303 n miles (7,975 km; 4,955 miles).

L-1011-500: Extended range version, with a maximum T-O weight of 224,980 kg (496,000 lb) and maximum fuel capacity tankage. Fuselage is shortened by 4.11 m (13 ft 6 in); all other external dimensions are the same as the other models. Three RB211-524B engines (each 222.4 kN; 50,000 lb st). Galley located on main deck. Forward cargo hold accommodates 12 LD-3 containers or four pallets each measuring 2.24 × 3.17 m (88 × 125 in). Centre hold takes seven LD-3 containers. In a mixed class configuration, with 24 first class passengers in a six-abreast seating and 222 economy passengers in nine-abreast seating, the aircraft carries 246 passengers. Maximum accommodation for 300 passengers. Maximum range 5,297 n miles (9,815 km; 6,100 miles). Ordered by British Airways.

UPGRADES: **Lockheed Martin Aeronautical Systems Company (LASC):** L-1011 extended range, increase gross weight and re-engine programme. See separate entry.

Marshall Aerospace: Pegasus Space Booster. See separate entry.

Marshall Aerospace: Tanker conversion. See separate entry.

Pemco Aeroplex: L-1011F all-cargo conversion. See separate entry.

OPERATORS: Versions of the L-1011 TriStar Electra are in service with the armed forces of the following countries: Jordan (1), Saudi Arabia (1) and UK (9).

DESIGN FEATURES: Cantilever low-wing monoplane. Special Lockheed aerofoil wing sections. Dihedral at trailing-edge 7° 31′ on inner wings, 5° 30′ outboard. Sweepback at quarter-chord 35°. The tail has a variable-incidence horizontal tailplane-elevator assembly and vertical fin and rudder.

FLYING CONTROLS: Hydraulically powered aluminium ailerons of conventional two-spar box construction, with aluminium honeycomb trailing-edge, inboard and outboard sections on each wing, operate in conjunction with flight spoilers. The low-speed ailerons extend from approximately 80 per cent of semi-span to within 0.25 m (10 in) of the wingtips, the high-speed ailerons extend from approximately WBL 387 to WBL 480 on each wing. Double-slotted Fowler trailing-edge flaps, constructed of aluminium honeycomb. Each flap segment consists of a honeycomb trailing-edge, a front spar, ribs, skin panels, carriages, and tracks mounted on the forward segment to provide for extension and rotation of the aft segment. A sheet metal vane surface, actuated by a linkage system during flap rotation, forms the forward section of the extended flap. Four aluminium leading-edge slats outboard of engine pylon on each wing. Each segment is mounted on two roller-supported tracks and extends in a circular motion down and forward for take-off and landing. Three leading-edge slats inboard of engine on each wing, made of aluminium alloy honeycomb and sheet metal fairings. Six spoilers on the upper surface of each wing, two inboard and four outboard of the inboard aileron, constructed from bonded aluminium tapered honeycomb. No trim tabs. Flight controls fully powered. Each control surface system is controlled by a multiple redundant servo actuator system that is powered by four independent and separate hydraulic sources. The tail elevators are linked mechanically to the tailplane actuation gear, to modify its camber and improve its effectiveness. No trim tabs. Controls are fully powered, the hydraulic servo actuators receiving power from four independent hydraulic sources, under control of electronic flight control system. Control feel is provided, with the force gradient scheduled as a function of flight condition.

STRUCTURE: The wing consists of a centre-section, passing through the lower fuselage, and an outer wing panel on each side. It is of conventional fail-safe construction, with aluminium surfaces, ribs and spars, and integral fuel tanks. The fuselage is a semi-monocoque structure of aluminium alloy. Constant cross-sectional diameter of 5.97 m (19 ft 7 in) for most of the length. Bonding utilised in skin joints, for attaching skin-doublers at joints and around openings to improve fatigue life. Skins and stringers supported by frames spaced at 0.51 m (20 in) intervals, with fail-safe straps midway between frames. These frames, with the exception of main frames and door-edge members, are 0.076 m (3 in) deep at the sides of the cabin increasing progressively to a depth of 0.15 m (6 in) at the top of the fuselage and below the floor. Fuselage length reduced on L-1011-500. The tail unit is a conventional cantilever structure, consisting of variable-incidence horizontal tailplane-elevator assembly and vertical fin and rudder. Primary loads of the fin are carried by a conventional box-beam structure, with ribs spaced at approximately

Lockheed L-1011-200 TriStar three-turbofan commercial transport ***(Peter J. Cooper)***

1994

Lockheed L-1011 TriStar converted to tanker version by Marshall Aerospace

1994

0.51 m (20 in) centres. The rudder comprises forward and aft spars, glassfibre trailing-edges, hinge and actuator back-up ribs, sheet metal formers, box surface panels and leading-edge fairings. Elevators are of similar construction. Truss members for the tailplane centre-section are built-up from forged and extruded sections. Outboard of the centre-section, construction is similar to that of the fin box-beam, leading- and trailing-edges, except that the surface structure is integrally stiffened.

LANDING GEAR: Hydraulically retractable tricycle type, produced by Menasco Manufacturing. Twin-wheel units in tandem on each main gear: twin wheels on nose gear, which is steerable 65° on each side. Nosewheels retract forward into fuselage. Mainwheels retract inward into fuselage wheel-wells. Oleo-pneumatic shock-absorbers on all units. BF Goodrich forged aluminium alloy wheels of split construction. Mainwheels have tubeless tyres size 50 × 20-20, Type VIII, pressure 10.34 to 11.38 bars (150 to 165 lb/sq in) for short- to medium-range operational weights, 12.41 bars (180 lb/sq in) for maximum range weight. Nosewheels have tubeless tyres size 36 × 11-16, Type VII, pressure 12.76 bars (185 lb/sq in). Hydraulically operated brakes, controlled by the rudder pedals. Anti-skid units, with individual wheel skid and modulated control, installed in the normal and alternative braking systems.

POWER PLANT (L-1011-1): Three Rolls-Royce RB211-22B turbofan engines, each rated at 187 kN (42,000 lb st). Two engines mounted in pods on pylons under the wings, the third in the rear fuselage at the base of the fin. Engine bleed air is used to anti-ice the engine inlet lips. Two integral fuel tanks in each wing; inboard tank capacity 30,581 litres (8,828 US gallons; 7,351 Imp gallons). Pressure refuelling points in wing leading-edges. Oil capacity approximately 34 litres (9 US gallons; 7.49 Imp gallons) per engine. A detachable pylon can be fitted between the starboard engine nacelle and fuselage to permit carriage of a replacement engine for another TriStar. Alternative power plants for the -100, -200, -250 and -500 detailed under model listings. These four models each have provision for additional centre-section tankage, raising total fuel capacity to 100,317 litres (26,502 US gallons; 22,067 Imp gallons) in -100 and -200, and 119,774 litres (31,642 US gallons; 26,347 Imp gallons) in -250 and -500.

ACCOMMODATION: Crew of 13. First class and coach mixed accommodation for 256 passengers, with a maximum of 400 in all-economy configuration. Alternative intermediate seating capacities are provided by using eight seat-tracks which permit 6-, 8-, 9- or 10-abreast seating, with two full-length aisles. Underfloor galley. Seven lavatories are provided, two forward and five aft. Three Type A passenger doors of the upward-opening plug type on each side of the fuselage, one pair immediately aft of the flight deck, one pair forward of wing, one pair aft of wing. Two Type I emergency exit doors, one each side of the fuselage, at rear of cabin, replaced by two Type A doors at 10-abreast seating. Baggage and freight compartments beneath the floor able to accommodate 16 containers, totalling 71.58 m³ (2,528 cu ft), and 19.8 m³ (700 cu ft) bulk cargo (19 containers and 14.2 m³; 500 cu ft in -500).

SYSTEMS: Air conditioning and pressurisation system, using engine bleed air and APU air combined with air cycle refrigeration. Pressurisation system maintains equivalent of 2,440 m (8,000 ft) conditions to 12,800 m (42,000 ft). Normal cabin pressure differential 0.582 bars (8.44 lb/sq in). Four indpendent 207 bars (3,000 lb/sq in) hydraulic systems provide power for primary flight control surfaces, normal brake power, landing gear retraction and nose-wheel steering and so on. Electrical system includes four 120/208 V 400 Hz alternators, one on each engine and one driven by the APU, which is sited in the aft fuselage. APU provides ground and in-flight power, to an altitude of 9,145 m (30,000 ft), producing both shaft and pneumatic power for utilisation by the electrical, environmental control and hydraulic systems. Integral electric heaters are used to anti-ice windscreens, pitot masts and total temperature probes. Thermal de-icing of outboard wing leading-edge slats by engine bleed air. No tail de-icing equipment.

ELECTRONICS AND EQUIPMENT: Standard equipment includes two ARINC 546 VHF communication transceivers, two ARINC 547 VHF navigation systems, two ARINC 568 interrogator units, an ARINC 564 weather radar system, two ARINC 572 air traffic control transponders, partial provision for a dual collision system, three vertical gyros, and full blind-flying instrumentation. Space is provided for installation of two ARINC 553A HF transceivers and a dual SATCOM system.

Lockheed L-1011 TriStar 200 with three Rolls-Royce RB211-524 turbofan engines *(Peter J. Cooper)*

1995

DIMENSIONS, EXTERNAL:

Wing span	47.34 m (155 ft 4 in)
Wing chord: at root	10.46 m (34 ft 4 in)
at tip	3.12 m (10 ft 3 in)
Wing aspect ratio	6.95
Length overall:	
-1, -100, -200, -250	54.17 m (177 ft 8½ in)
-500	50.05 m (164 ft 2½ in)
Height overall	16.87 m (55 ft 4 in)
Tailplane span	21.82 m (71 ft 7 in)
Wheel track	10.97 m (36 ft 0 in)
Wheelbase	21.34 m (70 ft 0 in)
Passenger doors (each): Height	1.93 m (6 ft 4 in)
Width	1.07 m (3 ft 6 in)
Height to sill	4.60 m (15 ft 1 in)
Emergency passenger doors (each):	
Height	1.52 m (5 ft 0 in)
Width	0.61 m (2 ft 0 in)
Height to sill	4.60 m (15 ft 1 in)
Baggage and freight compartment doors (forward and centre): Height	1.73 m (5 ft 8 in)
Width	1.78 m (5 ft 10 in)
Height to sill	2.72 m (8 ft 11 in)
Baggage and freight compartment doors (aft):	
Height	1.22 m (4 ft 0 in)
Width	1.12 m (3 ft 8 in)
Height to sill	2.92 m (9 ft 7 in)

DIMENSIONS, INTERNAL:

Cabin, excl flight deck and underfloor galley:	
Length	41.43 m (135 ft 11 in)
Max width	5.77 m (18 ft 11 in)
Max height	2.41 m (7 ft 11 in)
Floor area:	
-1, -100, -200, -250	215.5 m² (2,320 sq ft)
-500	192.6 m² (2,073 sq ft)
Volume	453.0 m³ (16,000 cu ft)
Baggage/cargo holds, bulk capacity:	
-1, -100, -200, -250	110.4 m³ (3,900 cu ft)
-500	118.9 m³ (4,200 cu ft)

AREAS:

Wings, gross	320.0 m² (3,456 sq ft)
Ailerons (total)	14.86 m² (160 sq ft)
Trailing-edge flaps (total)	49.80 m² (536 sq ft)
Leding-edge slats (total):	
inboard slats	11.52 m² (124 sq ft)
outboard slats	21.93 m² (236 sq ft)
Spoilers (total)	19.88 m² (214 sq ft)
Fin	51.10 m² (550 sq ft)
Rudder	11.89 m² (128 sq ft)
Tailplane	119.10 m² (1,282 sq ft)

WEIGHTS AND LOADINGS:

Operating weight empty: -1	109,045 kg (240,400 lb)
-100	110,720 kg (244,100 lb)
-200	111,495 kg (245,800 lb)
-250	112,969 kg (249,054 lb)
-500	108,925 kg (240,139 lb)
Max payload: -1	38,373 kg (84,600 lb)
-100	34,427 kg (75,900 lb)
-200	33,020 kg (72,800 lb)
-250	40,345 kg (88,946 lb)
-500	44,390 kg (97,861 lb)
Max T-O weight: -1	195,045 kg (430,000 lb)
-100, -200	211,375 kg (466,000 lb)
-250, -500	224,980 kg (496,000 lb)
Max zero-fuel weight: -1	147,417 kg (325,000 lb)
-100, -200	145,150 kg (320,000 lb)
-250, -500	153,315 kg (338,000 lb)
Max landing weight: -1	162,385 kg (358,000 lb)
-100, -200, -250, -500	166,920 kg (368,000 lb)

PERFORMANCE (-1, -100, -200 with 273 passengers, -250 with 284 passengers and -500 with 246 passengers):

Cruising speed, all versions	Mach 0.84

T-O field length: -1	2,425 m (7,960 ft)
-100	3,245 m (10,640 ft)
-200	2,460 m (8,070 ft)
-250	2,760 m (9,060 ft)
-500	2,845 m (9,330 ft)
Landing field length: -1	1,735 m (5,690 ft)
-100, -200, -250	1,770 m (5,800 ft)
-500	1,955 m (6,420 ft)
Max range: -1	3,110 n miles (5,760 km; 3,580 miles)
-100	3,820 n miles (7,080 km; 4,400 miles)
-200	3,864 n miles (7,160 km; 4,450 miles)
-250	4,533 n miles (8,400 km; 5,220 miles)
-500	5,297 n miles (9,815 km; 6,100 miles)

OPERATIONAL NOISE CHARACTERISTICS (FAR Pt 36)

T-O noise level	97 EPNdB
Approach noise level	103 EPNdB
Sideline noise level	95 EPNdB

UPDATED

Lockheed P-3C Update II of the Royal Australian Air Force

1994

LOCKHEED 185/285/685/785 ORION

US Navy designation: P-3

CF designations: CP-140 Aurora/CP-140A Arcturus

TYPE: Land-based maritime patrol and ASW aircraft.

PROGRAMME: Lockheed won competition for off-the-shelf ASW aircraft 1958; first flight aerodynamic prototype 19 August 1958; first flight fully equipped YP-3A (YP3V-1) 25 November 1959; P-3C produced from 1969. P-3A/B are Model 185; P-3C is Model 285. Total 642 P-3s built in California (Burbank, then Palmdale); last delivered (to Canada) May 1991; production line reopened at Marietta, Georgia, August 1991; first P-3C from Marietta (for Korea) rolled out 28 June 1994. Following cancellation of P-7, USN, Germany and other considering proposed Orion II.

VERSIONS: **P-3A:** Initial production version for US Navy with Allison T56-A-10W turboprop engines. Maiden flight on 15 April 1961. Deliveries to US Navy began in 1962. Later models known as Deltic as they were fitted with the Deltic system and more sensitive ASW equipment.

EP-3A: Electronic research platforms. In service with the US Navy.

RP-3A: Weapons trials support aircraft. In service with the US Navy.

TP-3A: Conversion training aircraft. In service with the US Navy.

UV-3A: Utility transport aircraft. In service with the US Navy.

VP-3A: VIP transport aircraft. In service with the US Navy.

WP-3A: Weather reconnaissance aircraft. No longer in service.

P-3B: Follow-on production version of P-3A with Allison T56-A-14 turboprop engines which do not need water-alcohol injection. In service with the US Navy, NASA, Chile Norway (P-3N) and Spain.

P-3C: First flight 18 September 1968; in service 1969; introduced A-NEW system based on Univac computer integrating all ASW information for retrieval, display and transmission of tactical data without routine log-keeping; USN received 118 of P-3C Baseline variant. For full details see *Jane's All the World's Aircraft* 1995-96.

P-3C Update I: (Model 285A) Baseline P-3Cs followed from January 1975 by 31 P-3C Update I; new avionics and software included magnetic drum to increase computer memory sevenfold, new versatile computer language, Omega navigation, improved directional acoustic frequency analysis and recording (DIFAR) processing sensitivity. AM/ASA-66 tactical displays for two sensor stations, and improved magnetic tape transport.

Update II: (Model 285A) Applied to 45 aircraft built from August 1977; added infra-red detection system (IRDS) and sonobuoy reference system (SRS); Harpoon missile system incorporated from August 1977. Thirty-six USN P-3Cs received interim **Update II.5** of 1981 including more reliable navigation and communication systems; IACS submarine communications link; MAD compensation group adaptor; standardised wing pylons; and improved fuel tank vents.

Update III: (Model 285G for USN) Deliveries started May 1984; applied to last 36 USN P-3Cs; includes new IBM Proteus acoustic processor (doubling sonobuoy handling capacity), new sonobuoy receiver replacing DIFAR, improved APU, and higher capacity environmental control system. Baseline P-3C to III retrofit kit first installed in P-3C of VP-31 in 1987 (new designation **IIIR**); fitting of 18 more kits started June 1987; USN planned force of 138 Update III/IIIR aircraft; following cancellation of Update IV, further 109 may be added, giving standard fleet of 247 P-3Cs by 2006. Pakistan Orions, basically Update IIIs, have certain systems replaced by export standard equipment and thus known as **II.75** versions. *Detailed description applies to P-3C/Update III except where indicated.*

Update IV: Programme cancelled by US Navy 14 October 1992, but was in full-scale development by Boeing Defense & Space Group for installation from early 1990s; prototype conversion (160292) flew December 1991; equipment stripped out and prototype returned to normal use; originally intended for P-7A LRAACA; all P-3C Update I, II and II.5 were to be retrofitted, equipping VP-8, 10, 11, 23 and 26 at Brunswick, Maine, beginning FY94; one P-3C used for aerodynamic and functional testing of Eaton AIL Division AN/ALR-77 ESM system with 36 antennae mounted in four groups at wingtips; installation abandoned in favour of Litton AN/ALR-66(V)5 in same position; other features included Texas Instruments AN/APS-137(V) inverse synthetic aperture radar, Texas Instruments AN/AAS-36 IR detection system, AT&T AN/UYS-2 signal processor and Texas Instruments AN/ASQ-81 MAD. Planned for 109 retrofits.

Outlaw Hunter/OASIS: Single P-3C (159507) of VP-19 modified for 1991 Gulf War by USN and Tiburon Systems under Outlaw Hunter programme as over-the-horizon (OTH) targeting platform with ability to maintain battle area overall plot for battle group commanders. Two further P-3Cs (157329, 161011), designated OASIS I and II (OTH Airborne Sensor Information System) modified for joint operation.

UP-3C: Kawasaki built service trials aircraft; one funded FY91; one more required; previously designated **NP-3C**.

UP-3D: Kawasaki ECM trainer for JMSDF; planned purchase of two during FY94-95.

WP-3D: Airborne research aircraft for atmospheric research and weather modification experiments.

EP-3: Elint version of P-3C developed by Kawasaki (which see) for JMSDF (first flight October 1990); two aircraft ordered FY87 and FY88; delivered March and November 1991; two more ordered in FY92 and FY93; fifth planned 1995; no relation to USN RP-3D.

EP-3E Aries II: Ten P-3As and two EP-3Bs converted to EP-3E Aries I or similar 'Deepwell' (seven) and 'Batrack' (EP-3Bs) standards; radars in large canoe-shaped fairings above and below fuselage and ventral radome forward of wing; avionics believed to include GTE-Sylvania AN/ALR-60 communications intercept and analysis system, Raytheon AN/ALQ-76 noise jamming pod, Loral AN/ALQ-78 automatic ESM system, Magnavox AN/ALQ-108 IFF jammer, Sanders AN/ALR-132 infra-red jammer, ARGO Systems AN/ALR-52 instantaneous frequency measuring equipment, Texas Instruments AN/APS-115 frequency-agile search radar, Hughes AN/AAR-37 infra-red detector, Loral AN/ASA-66 tactical display, Cardion AN/ASA-69 scan converted and Honeywell AN/ASQ-114 computer. Twelve low-houred P-3Cs selected to replace EP-3E Aries I with USN special reconnaissance squadrons VQ-1 at Agana NAS, Guam, and VQ-2 at Rota, Spain; equipment transferred from original EP-3E but moderate upgrade includes faster processing and standardisation of previous two configurations within Aries I and addition to wingtips of IBM AN/ALR-76 ESM/RWR in pods; first Aries II conversion (156507) completed November 1988, to Patuxent River test centre 21 July 1990; first delivery (157320) to VQ-2 29 June 1991; initial five conversions by Lockheed Aircraft Service Company's Aeromod Center at Greenville, South Carolina, but contract then reassigned to Naval Air Depot, Alameda, 31 July 1991. VQ-2's last aircraft due mid-1995.

CP-140A Arcturus: Three P-3s for Canadian Forces; no ASW equipment; for environmental and fishery patrol replacing CP-121 Trackers; equipment includes Texas Instruments AN/APS-134 Plus radar, Honeywell AN/APN-194 RAWS, Bendix AN/ASW-502 AFCS, Canadian Marconi AN/APN-510 Doppler radar, Litton LN-33 INS and a Leigh AN/ASH-502 flight recorder. Delivered to IMP Group at Halifax, Nova Scotia, for completion; final departure from Palmdale, 30 May 1991; delivery to CF December 1992 to February 1993; based at Greenwood.

P-3 AEW&C: Airborne early warning and control; first flight of prototype (N91LC) converted from Australian P-3B and fitted with empty Randtron AN/APA-171 7.32 m (24 ft) diameter rotodome 14 June 1984; testing of installed General Electric AN/APS-125 radar began 1988; military version would have AN/APS-145 radar as in Grumman E-2C Hawkeye. Other systems would include C^3 system to receive, process and transmit tactical information on HF, UHF, VHF and Satcom channels; AN/ARC-187 satellite communication system; and Collins five-tube colour EFIS-86B flight instruments. General Electric AN/APS-145 radar available from late 1989.

First order from US Customs May 1987 for one plus option for three; first flight US Customs aircraft with AN/APS-125 radar 8 April 1988; aircraft (N145CS ex-N91LC) called *Blue Eagle* delivered to NAS Corpus Christi, Texas, 17 June 1988 and used for anti-narcotics patrol over Caribbean and Gulf of Mexico; retrofitted with AN/APS-138;

Lockheed P-3C Orion of the Netherlands Naval Air Arm shadowing former Soviet 'Akademik Shuleykin' class research ship

1994

Lockheed P-3K of the Royal New Zealand Air Force

1995

second ex-RAAF P-3 AEW&C called *Blue Eagle II*, fitted with improved AN/APS-138 radar, delivered to US Customs June 1989; third and fourth (both ex-USN P-3Bs) delivered 26 June 1992 and late 1993. Additional systems include CDC AN/AYK-14 computer with Honeywell 1601M array processor, dual Sanders Miligraphics touch-sensitive colour display screens for digital target data, Hazeltine AN/TPX-54 IFF, dual AN/ARC-182 VHF/UHF com radios, dual AN/ARC-207 HF and dual Wulfsberg VHF/UHF-FM radios. Type nickname 'P-3 Dome'.

P-3F: P-3C standard version. In service with Iranian Navy.

P-3H Orion II: Not yet funded by US Navy. Proposed P-3C upgrade to replace cancelled P-7A. Update IV avionics plus new wings and engines (Allison T406 or GE T407) and, possibly, HUDs and flat panel cockpit displays having high commonality with Hercules II.

EP-3J Orion: Electronic warfare trainer based on conversion of P-3B with AN/USQ-113 communications intrusion and deception system, plus AN/ALQ-167, AN/ALQ-170 and AN/AST-4/6 pods. Modified by Chrysler Technologies. Service with VAQ-33 squadron at Key West; first delivery 17 March 1992; two aircraft (152719 and 152745); transferred to VP-66 in 1993. Planned Phase 2 modifications to include AN/ALT-40(V) radar jammer.

P-3P: Originally P-3B aircraft of Australian Navy. Upgraded to P-3C standard and transferred to Portugese Navy.

P-3K: Upgraded P-3B aircraft. In service with Royal New Zealand Air Force.

P-3T/UP-3T: Conversions of USN P-3A for Thailand; five airframes purchased 1992; two for modification to P-3T; one to UP-3T; and two for spares breakdown.

P-3W: Australian P-3Cs; have AQS-901 processing system and Barra sonobuoys in place of Proteus and AN/AQA-7 equipment of USN P-3C. Upgrade of 10 with Elta ESM equipment (replacing AN/ALQ-78) begun by AWADI in Australia, 1991; other upgrades planned. P-3W is local designation of P-3C-II.5; earlier P-3C-IIs retain original designation.

UPGRADES: **Aero Union:** P-3A Aerostar. See separate entry.

AWA Defence Industries (AWADI): Contract to install RAAF P-3 aircraft with new electronic support measures (ESM) system reportedly from Israeli Aircraft Industries (IAI).

Boeing: P-3 Update IV. See separate entry.

Canada: Expected RFP from the Aurora Life Extension Project Office for avionics upgrade of the CP-140 aircraft. Upgrade likely to include the Spotlight synthetic aperture radar being developed by Unisys GSG Canada, 99-channel sonobuoy receiver/sonobuoy reference system and global positioning system (GPS).

Canada: Expected RFP from Aurora Life Extension Project Office for contract to extend the Aurora operational life-time to 2010.

Chile: Received eight stripped-down P-3A aircraft, now designated UP-3A, in December 1994. Expected to issue RFP for systems installation.

E-Systems: AP-3. Contract awarded July 1994 by Royal Australian Air Force for the upgrade of 18 P-3C/Ws. Value of contract $360 million. Contract signed 30 January 1995. E-Systems to collaborate with Aerospace Technologies of Australia (ASTA) Defence, AWA Defence Industries (AWADI) and Honeywell of Australia (HLA). Upgraded avionics to include Elta EL/M-202A(V)3 high-resolution, digital, multimode radar with 360° surveillance coverage, Unisys DDC-060 data management subsystem (DMS), Computing Devices of Canada UYS-503 acoustic processor, Honeywell, H-423 ring-laser gyro inertial navigation unit, MAGR 300 500-channel GPS receiver, electronic flight instrumentation system (EFIS), and a redundant digital databus. Replacement radio and intercom system supplied by Magnavox consisting of two MFX-500 HF, two AN/ARC-222 VHF and three AN/ARC-187 UHF comms subsystems.

E-Systems/Northrop Grumman: Sustained Readiness Programme (SRP). Awarded $205 million contract in September 1994 to extend the lives of 98 USN P-3C aircraft into the next century. Programme likely to be extended to cover the entire P-3C fleet.

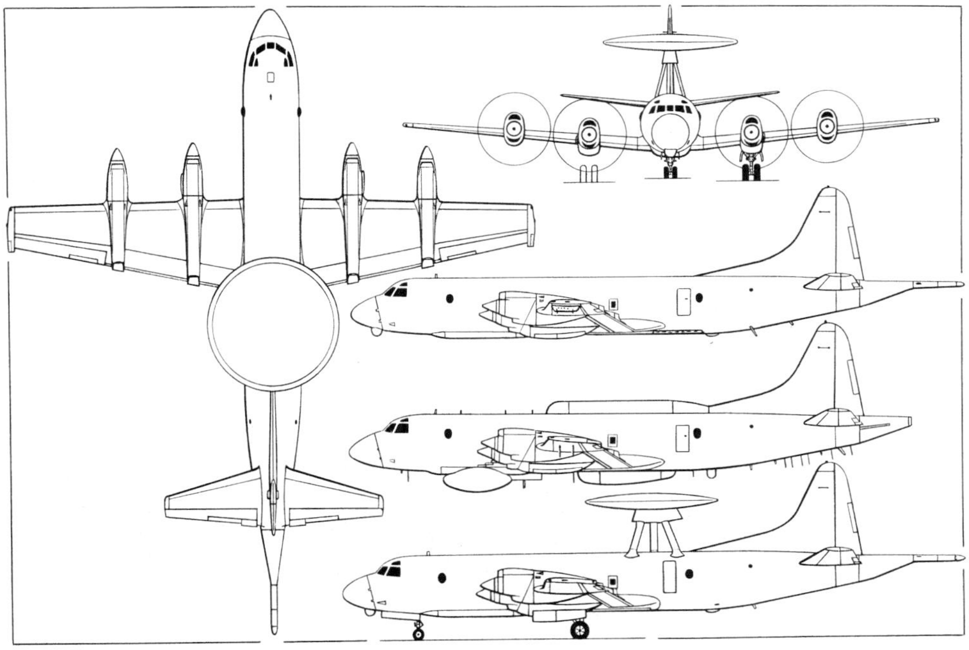

Lockheed P-3 AEW&C, with additional side views of EP-3E Aries II (centre and P-3C Orion (top)
(Jane's/Dennis Punnett)

1994

Japan: Retrofitting Unisys CP-2044 tactical computers in its P-3C Update III Orions.

Kawasaki Heavy Industries: EP-3E. See Versions.

Lockheed: P-3 Updates I/II/III. See Versions.
EP-3E Aries II. See Versions.
EP-3J Orion. See Versions.
P-3 AEW&C. See Versions.
P-3H Orion II. See Versions.
P-3T/UP-3T. See Versions.

Lockheed: Communications Improvement Programme (CIP). Awarded by US Navy to upgrade aircraft radio.

Lockheed Aeronautical Systems Company (LASC): US Customs conversions. See separate entry.

Netherlands: Studies underway by the Royal Netherlands Navy (RNethN) on requirements for a Capabilities Update Programme for their fleet of P-3C Update II.5 aircraft to enhance their anti-surface warfare (ASuW) capability. New systems likely to include: imaging radar, multipurpose high resolution displays, 99-channel acoustic processor and electronic support measures (ESM) system. Delivery expected for 1998.

New Zealand: Project Rigel. See Separate entry in Royal New Zealand Air Force section.

New Zealand: Project Kestrel. See following entry.

New Zealand: Project Sirius. Programme to update the electronic support measures (ESM) and acoustic processing System. Expected to be addressed in 2000.

Norway: Plans to upgrade the Royal Norwegian Air Force (RNoAF) two P-3N aircraft with new imaging radar, the CP-2044 computer, integrated displays, improved defensive aids suite (DAS), Radar/laser warning receivers, missile warner and electronic support measures (ESM) system. Programme also to include installation of GPS and satellite communications.

Portugal: Plans to upgrade the six Portugese Air Force P-3P aircraft with improved radar, electronic support measures (ESM) as well as adding GPS, radar warning receiver (RWR), missile warner, decoy dispenser and AGM-65 Maverick air-to-ground missile capability.

Spain: Plans to upgrade the five Spanish Air Force P-3B aircraft with improved acoustic signal-processor, ALR-66 electronic support measures (ESM) system and a FLIR sensor. Possible programme to install Thorn EMI Searchwater 2000 radar.

Unisys: Undertaking US$115 million contract to improve the anti-surface warfare (ASuW) capabilities of 247 P-3s Orion aircraft of the US Navy. Unisys will provide the information processing and display system, Texas Instruments the improved sensors and the Lockheed Aircraft Service Company will be responsible for installation and integration.

USN/Tiburon Systems: Outlaw Hunter/Oasis. See Versions.

OPERATORS: Versions of the P-3 Orion are in service with the armed forces of the following countries: Australia (19), Chile (11), Greece (on order), Iran (3), Japan (5), South Korea (8), Netherlands (13), New Zealand (6), Norway (6), Pakistan (3), Portugal (6), Spain (7), Thailand (3) and USA (343).

CUSTOMERS: US Navy, 552, comprising one YP-3; 157 P-3As, of which 38 modified to UP-3A, seven to EP-3A, six to RP-3A, five to VP-3A (three via WP-3A), 12 to TP-3A, two to EP-3B and 10 to EP-3E; 124 P-3Bs, one converted to NP-3B and one under conversion in 1991 to RP-3B; 267 P-3Cs, 12 intended for EP-3E-II conversion; one RP-3D; two National Oceanographic and Atmospheric Administration WP-3Ds. Final USN P-3C (163295) delivered 17 April 1990. Last Reserve P-3A ASW mission 22 March 1990 (152158 of VP-64); last regular USN P-3B ASW mission 11 September 1990 (VP-22).

Total 90 P-3s from Californian production for export to Australia (10 Model 185B P-3Bs, one transferred from US Navy; 10 P-3C-IIs and 10 P-3C-II.5s, both Model 285D); six P-3Bs transferred to Portugal as P-3P, one to New

Lockheed P-3C Orion of the Japanese Maritime Self-Defence Force

1994

Zealand and, two to US Customs (after conversion to AEW&C), Canada (18 CP-140 Auroros, three CP-140A Arcturus), Iran (six Model 685A P-3Fs), Japan (three Model 785A P-3C-IIs from Lockheed; Kawasaki produced 66 P-3C-IIs and has 34 P-3C-IIIs, four EP-3s and one UP-3C on order; P-3C-II to be updated), Netherlands (13 Model 285E P-3C-II.5s), New Zealand (five P-3Bs updated to P-3Ks; plus one ex-Australian P-3K), Norway (five Model 185C P-3Bs, two transferred from US Navy, and four P-3C-IIIs; five P-3Bs transferred to Spain and two converted to P-3N; last conversion redelivered 21 June 1992) and Pakistan (three P-3C Update II.75s; crew training completed December 1991; aircraft embargoed by US government and stored). Marietta production initially for South Korea (eight P-3C Update IIIs with AN/APS-134 radar; first roll-out May 1994; last delivery September 1995). Four USN P-3As transferred to US Customs as UP-3As (alternative designation P-3A(CS) 'Slick') fitted with IR detection system and AN/APG-63 look-down radar. Chile obtained two P-3As and six UP-3As from USN early in 1993. Others civilianised for various operators, including N406TP temporarily with Allison GMA 2100 turboprop and Dowty Aerospace R373 composite propeller on port outer nacelle, 1990; and five P-3A 'Aerostar' firefighters of Aero Union Corporation. Greece plans to buy five surplus P-3As and one UP-3A for 1994-95 delivery.

DESIGN FEATURES: Pressurised cabin. Wing section NACA 0014 (modified) at root, NACA 0012 (modified) at tip; dihedral 6°; incidence 3° at root, 0° 30′ at tip.

FLYING CONTROLS: Hydraulically boosted ailerons, elevators and rudders; fixed tailplane; Lockheed-Fowler trailing-edge flaps.

STRUCTURE: Conventional aluminium alloy with fail-safe box beam wing.

LANDING GEAR: Hydraulically retractable tricycle type, with twin wheels on each unit. All units retract forward, mainwheels into inner engine nacelles. Oleo-pneumatic shock-absorbers. Mainwheels have size 40-14 Type VII 26 ply tubeless tyres, pressure 7.58 to 12.41 bars (110 to 180 lb/sq in) at 36,287 kg (80,000 lb) T-O weight, 12.41 bars (180 lb/sq in) at 57,606 kg (127,000 lb) T-O weight, 13.10 bars (190 lb/sq in) at 61,235 kg (135,000 lb) maximum normal T-O weight. Nosewheels have size 28-7.7 Type VII tubeless tyres, pressure 10.34 bars (150 lb/sq in). Hydraulic brakes. No anti-skid units.

POWER PLANT: Four 3,661 kW (4,910 ehp) Allison T56-A-14 turboprops, each driving a Hamilton Standard 54H60-77 four-blade constant-speed propeller. Fuel in one tank in fuselage and four wing integral tanks, with total usable capacity of 34,826 litres (9,200 US gallons; 7,660 Imp gallons). Four overwing gravity fuelling points and central pressure refuelling point. Oil capacity (minimum usable) 111 litres (29.4 US gallons; 24.5 Imp gallons) in four tanks.

ACCOMMODATION: Normal 10-man crew: pilot, co-pilot, flight engineer and nav/com operator on flight deck; tactical co-ordinator, two acoustic sensor operators, MAD operator, ordnance man and flight technician; up to 13 additional relief crew or passengers. Flight deck has wide-vision windows, and circular windows for observers are provided fore and aft in the main cabin, each bulged to give 180° view. Main cabin is fitted out as a five-man tactical compartment (containing advanced electronic, magnetic and sonic detection equipment), an all-electric galley and large crew and rest area.

SYSTEMS: Air conditioning and pressurisation system supplied by two engine-driven compressors. Pressure differential 0.37 bar (5.4 lb/sq in). Hydraulic system, pressure 207 bars (3,000 lb/sq in), for flaps, control surface boosters, landing gear actuation, brakes and bomb bay doors. Three hydraulic pumps, each rated at 30.3 litres (8.0 US gallons; 6.7 Imp gallons)/min at 0 to 152 bars (0 to 2,200 lb/sq in), 22.7 litres (6.0 US gallons; 5.0 Imp gallons)/min at 205 bars (2,975 lb/sq in). Class one non-separated air/oil reservoir. Type B pressurised. Electrical system utilises three 60 kVA generators for 120/208 V 400 Hz AC supply. 24 V DC supply. Integral APU with 60 kVA generator for ground air conditioning, electrical supply and engine starting. Anti-icing by bleed air on wing and electrical heating on tailplane and fin. Electrically de-iced propeller spinners.

AVIONICS: The AN/ASQ-114 general purpose digital computer is the heart of the P-3C system. Together with the AN/AYA-8 data processing equipment and computer-controlled display systems, it permits rapid analysis and utilisation of electronic, magnetic and sonic data. Nav/com system comprises two LTN-72 inertial navigation systems; AN/APN-227 Doppler; AN/ARN-81 Loran A and C; AN/ARN-118 Tacan; two VIR-31A VOR/LOC/GS/MB receivers; AN/ARN-83 LF-ADF; AN/ARA-50 UHF direction finder, AN/AJN-15 flight director indicator for tactical directions; HSI for long-range flight directions; glide slope indicator; on-top position indicator; two AN/ARC-161 HF transceivers; two AN/ARC-143 UHF transceivers; AN/ARC-101 VHF receiver/transmitter; AN/AGC-6 teletype and high-speed printer; HF and UHF secure communication units; AN/ACQ-5 datalink communication set and AN/AIC-22 interphone set; AN/APX-72 IFF transponder and AN/APX-76 SIF interrogator. Electronic computer controlled display equipment includes AN/ASA-70 tactical display, AN/ASA-66 pilot's display, AN/ASA-70 radar display and two auxiliary readout (computer stored data) displays.

ASW equipment includes two AN/ARR-72 sonar receivers, replaced in Update III by AN/ARR-78; two AN/AQA-7(V)8 DIFAR (directional acoustic frequency analysis and recording) sonobuoy indicator sets, replaced in Update III by AN/UYS-1 Proteus; hyperbolic fix unit; acoustic source signal generator; time code generator and AN/AQH-4(V) sonar tape recorder; AN/ASQ-81 magnetic anomaly detector; AN/ASA-64 submarine anomaly detector; AN/ASA-65 magnetic compensator; AN/ALQ-78 electronic countermeasures set; AN/APS-115 radar set (360° coverage); AN/ASA-69 radar scan converter; undernose AN/AAS-36 IRDS; KA-74 forward computer assisted camera; KB-18A automatic strike assessment camera with horizon-to-horizon coverage; RO-308 bathythermograph recorder.

Additional items include AN/APN-194 radar altimeter; two AN/APQ-107 radar altimeter warning systems; A/A24G-9 true airspeed computer; AN/ASW-31 automatic flight control system. P-3Cs delivered from 1975 have the avionics/electronics package updated by addition of an extra 393 kbit memory drum and fourth logic unit. Omega navigation, new magnetic tape transport, and an AN/ASA-66 tactical display for the sonar operators. To accommodate the new systems a new operational software computer programme was written in CMS-2 language. GEC-Marconi AQS-901 acoustic signal processing and display system in RAAF P-3Ws. AN/ALR-66(V)5 passive radar detection system (ESM), to be housed in wingtip pods, is under development for Update IV P-3C by General Instrument, and will also provide targeting data for the aircraft's Harpoon missiles. AN/ALR-66(V)3 installed in Japanese and Norwegian P-3Cs and as retrofit in P-3P and CP-140. Wing span increased by some 0.81 m (2 ft 8 in) to accommodate ESM antennae and receivers. Similar Israeli Elta equipment for Australian retrofit. Loral AN/ALQ-157 IR jammers retrofitted each side of rear fuselage on USN P-3Cs. AN/ALR-66(V)5 replaces Loral AN/ALQ-78A on pod on inboard wing pylon.

EQUIPMENT: Searchlight replaces one wing pylon, starboard. Search stores, such as sonobuoys and sound signals, are launched from inside cabin area in the P-3A/B. In the P-3C sonobuoys are loaded and launched externally and internally. Sonobuoys are ejected from P-3C aircraft with explosive cartridge actuating devices (CAD), eliminating the need for a pneumatic system. Australian P-3Ws use SSQ-801 Barra sonobuoys.

ARMAMENT: Bomb bay 2.03 m wide, 0.88 m deep and 3.91 m long (80 × 34.5 × 154 in), forward of wing, and 10 underwing pylons. Stores can include (weapons bay/underwing, maximum) Mk 46 torpedo 8/0; Mk 50 torpedo 6/0; Mk 54 depth bomb 8/10; B57 nuclear depth charge 3/0; Mk 82 560 lb bomb 8/10; Mk 83 980 lb bomb 3/8; Mk 36 destructor 8/10; Mk 40 destructor 3/8; LAU-68A pod (seven 2.75 in rockets), or LAU-69AS (19 2.75 in rockets), or LAU-10A/C (four 5 in rockets), or SUU-44A (eight flares) 0/4; Mk 52 mine 3/8; Mk 55 or Mk 56 mine 1/6; Mk 60 torpedo 0/6; AGM-85 Harpoon anti-ship missile 0/8. Two AIM-9L Sidewinder AAMs underwing for self-defence. Maximum total weapon load includes six 2,000 lb mines under wings and a 3,290 kg (7,252 lb) internal load made up of two Mk 101 depth bombs, four Mk 44 torpedoes, pyrotechnic pistol and 12 signals, 87 sonobuoys, 100 Mk 50 underwater sound signals (P-3A/B), 18 Mk 3A marine markers (P-3A/B), 42 Mk 7 marine markers, two B. T. buoys, and two Mk 5 parachute flares. Harpoon missiles are standard fit on a proportion of US Navy P-3Cs.

DIMENSIONS, EXTERNAL:

Wing span	30.37 m (99 ft 8 in)
Wing chord: at root	5.77 m (18 ft 11 in)
at tip	2.31 m (7 ft 7 in)
Wing aspect ratio	7.5
Length overall	35.61 m (116 ft 10 in)
Height overall	10.27 m (33 ft 8½ in)
Fuselage diameter	3.45 m (11 ft 4 in)
Tailplane span	13.06 m (42 ft 10 in)
Wheel track (c/l shock-absorbers)	9.50 m (31 ft 2 in)
Wheelbase	9.07 m (29 ft 9 in)
Propeller diameter	4.11 m (13 ft 6 in)
Cabin door: Height	1.83 m (6 ft 0 in)
Width	0.69 m (2 ft 3 in)

DIMENSIONS, INTERNAL:

Cabin, excl flight deck and electrical load centre:	
Length	21.06 m (69 ft 1 in)
Max width	3.30 m (10 ft 10 in)
Max height	2.29 m (7 ft 6 in)
Floor area	61.13 m² (658.0 sq ft)
Volume	120.6 m³ (4,260 cu ft)

AREAS:

Wings, gross	120.77 m² (1,300.0 sq ft)
Ailerons (total)	8.36 m² (90.0 sq ft)
Trailing-edge flaps (total)	19.32 m² (208.0 sq ft)
Fin, incl dorsal fin	10.78 m² (116.0 sq ft)
Rudder, incl tab	5.57 m² (60.0 sq ft)
Tailplane	22.39 m² (241.0 sq ft)
Elevators, incl tabs	7.52 m² (81.0 sq ft)

WEIGHTS AND LOADINGS (P-3B/C):

Weight empty	27,890 kg (61,491 lb)
Max fuel weight	28,350 kg (62,500 lb)
Max expendable load	9,071 kg (20,000 lb)
Max normal T-O weight	61,235 kg (135,000 lb)
Max permissible weight	64,410 kg (142,000 lb)
Design zero-fuel weight	35,017 kg (77,200 lb)
Max landing weight	47,119 kg (103,880 lb)
Max wing loading	507.0 kg/m² (103.8 lb/sq ft)
Max power loading	4.18 kg/kW (6.87 lb/ehp)

PERFORMANCE (P-3B/C, at max T-O weight, except where indicated otherwise):

Max level speed at 4,575 m (15,000 ft) at AUW of 47,625 kg (105,000 lb) 411 kts (761 km/h; 473 mph)

Lockheed P-3B of the Spanish Air Force *(Peter J. Cooper)*

1995

Econ crusing speed at 7,620 m (25,000 ft) at AUW of 49,895 kg (110,000 lb) 328 kts (608 km/h; 378 mph)
Patrol speed at 457 m (1,500 ft) at AUW of 49,895 kg (110,000 lb) 206 kts (381 km/h; 237 mph)
Stalling speed: flaps up 133 kts (248 km/h; 154 mph)
flaps down 112 kts (208 km/h; 129 mph)
Rate of climb at 457 m (1,500 ft) 594 m (1,950 ft)/min
Time to 7,620 m (25,000 ft) 30 min
Service ceiling 8,625 m (28,300 ft)
Service ceiling, OEI 5,790 m (19,000 ft)
T-O run 1,290 m (4,240 ft)
T-O to 15 m (50 ft) 1,673 m (5,490 ft)
Landing from 15 m (50 ft) at design landing weight 845 m (2,770 ft)
Mission radius (3 h on station at 457 m; 1,500 ft) 1,346 n miles (2,494 km; 1,550 miles)
Max mission radius (no time on station) at 61,235 kg (135,000 lb) 2,070 n miles (3,835 km; 2,383 miles)
Ferry range 4,830 n miles (8,950 km; 5,562 miles)
Max endurance at 4,575 m (15,000 ft):
two engines 17 h 12 min
four engines 12 h 20 min

UPDATED

Lockheed S-3B Viking with Sargent-Fletcher refuelling pods *(Paul Tompkins)*

1994

RNZAF/LOCKHEED (LOCKHEED) P-3K ORION KESTREL RELIFE PROGRAMME

TYPE: Maritime patrol aircraft life-extension programme.

PROGRAMME: Royal New Zealand Air Force awarded contract to Lockheed to extend the lives of its six P-3K Orion maritime patrol aircraft by 20 years. Aircraft to undergo rewinging in 1997.

DESIGN FEATURES: Programme consists of rewinging the aircraft by replacing the fatigued structure on existing aircraft with new outer wings, centre-section lower skins and horizontal stabilisers. At the same time, the engine nacelles will be refurbished, new wiring looms installed in the wings and a fuel dump capability added to the five lightweight aircraft which do not currently have the provision.

NEW ENTRY

Lockheed S-3B Viking of the US Navy *(Paul Tompkins)*

1994

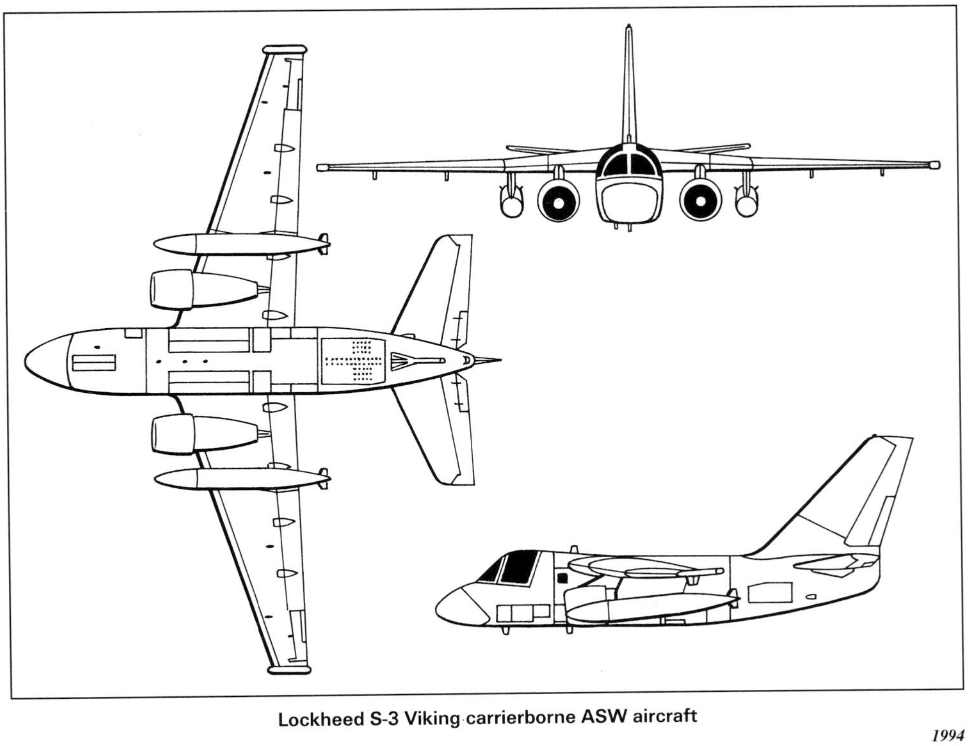

Lockheed S-3 Viking carrierborne ASW aircraft

1994

LOCKHEED S-3 VIKING

US Navy designation: S-3

TYPE: Carrierborne ASW aircraft.

PROGRAMME: Production of 187 S-3As for US Navy ended mid-1978, tooling then stored at Burbank pending further orders; **US-3A** (COD) and **KS-3A** (dedicated tanker) demonstrators evaluated by US Navy early 1980; three earlier S-3As modified to US-3A 1982; KS-3A also converted to US-3A late 1983; S-3A fitted with Sargent-Fletcher buddy pack under port wing (to transfer fuel from internal tanks) and external tank under starboard wing; tested 1984 and adopted (ASW capability not affected).

VERSIONS: **S-3A:** Initial production ASW version.

S-3B: Lockheed defined Weapon System Improvement Program under US Navy contract 1980; full-scale development ordered 18 August 1981 and designated S-3B; first flight of first of two development S-3Bs 13 September 1984; flight testing completed August 1985 and six months' operational evaluation 1986; 22 kits plus spares and support ordered 28 April 1986 for installation at Cecil Field NAS, Florida; two prototype S-3B kits delivered 1987; 24 more kits ordered December 1988 for delivery by August 1992; first S-3B delivery to VS-27 17 December 1987; 42 S-3Bs completed for Atlantic Fleet by December 1991, VS-27, -30 and -32 re-equipped at Cecil Field, Florida, plus VS-28 and -31 in prospect; 38 kits being embodied in Pacific Fleet S-3s at NAS, North Island, California from March 1992.

ES-3A: Contract for development of electronic reconnaissance E-3A awarded March 1988; S-3A for conversion (159401) delivered to Lockheed March 1988. First flight April 1991. Second YS-3A (157993) with 60 aerials and domes completed aerodynamic testing January 1990; original ES-3A development contract allowed for 15 production systems; nine modification kits ordered early 1989 for delivery by August 1991; airframe modification at Burbank and kit installation at Cecil Field, Florida; two new squadrons, VQ-5 and VQ-6, formed at Agana NAS, Guam, 15 April 1991 and Cecil Field, Florida, August 1991; ES-3A will replace, EA-3B Skywarrior in US Navy Battle Group Passive Horizon Extension System for long-range signals monitoring.

Weapons bay used for avionics as well as rear fuselage; starboard dual controls replaced by displays and controls for electronic warfare co-ordinating officer. Operational system, simplified from that of EP-3E Orion, includes AN/ALR-76 ESM, other avionics include three AN/AYK-14 computers replacing one AN/AYK-10, MIL-STD-1553 databus, Navstar GPS and Omega navigation, Link 11 datalink, AN/APX-76 IFF interrogator, AN/APS-137 radar, OR-236 FLIR and avionics cooling system; suppliers include Sanders Associates (cockpit display screens), Lockheed Missiles & Space (wiring and electronics racks).

UPGRADES: **Lockheed:** S-3B. See Versions.

OPERATORS: Versions of the S-3 Viking are in service with the US armed forces (138).

The following dimensions apply to the S-3A:

DESIGN FEATURES: Cantilever shoulder-wing monoplane. Sweepback at quarter-chord 15°. No dihedral. Incidence 3° 15′ at root, −3° 50′ at tip. Cantilever all-metal tail unit structure with swept vertical and horizontal surfaces. Fin and rudder are folded downward by hydraulic servos for carrier stowage. During fin-folding sequence the pedal input to the rudder servo is disconnected to allow the pilot to steer the nosewheel by the rudder pedals.

FLYING CONTROLS: Single-slotted Fowler-type trailing-edge flaps, operated by hydraulic power with an integral electric motor for emergency operation. Electrically operated leading-edge flaps, extending from engine pylons to wingtips, are fully extended after 15° of trailing-edge flap movement. Ailerons augmented by under- and over-wing spoilers for roll control. All primary flight control surfaces are actuated by irreversible servos powered by dual hydraulic systems. Loss of either hydraulic system results in loss of half the available hinge movement, but the remaining system can meet all control requirements. Automatic reversion to manual control in the event of failure of both hydraulic systems. In emergency operation the spoilers are inoperative. Variable incidence tailplane, electrically controlled. Elevator and rudder controlled by hydraulic servos. Trim tabs in elevators and rudder.

STRUCTURE: All-metal fail-safe wing structure. Wings fold upward and inward hydraulically, outboard of engine pylons for carrier stowage. The fuselage is a semi-monocoque all-metal fail-safe structure, incorporating split weapons bays with clamshell doors. Two parallel beams form a keelson from nose gear to tail-hook, strengthening the fuselage and improving cabin structural

Lockheed CT133 Silver Star of the Canadian Forces Air Command

1994

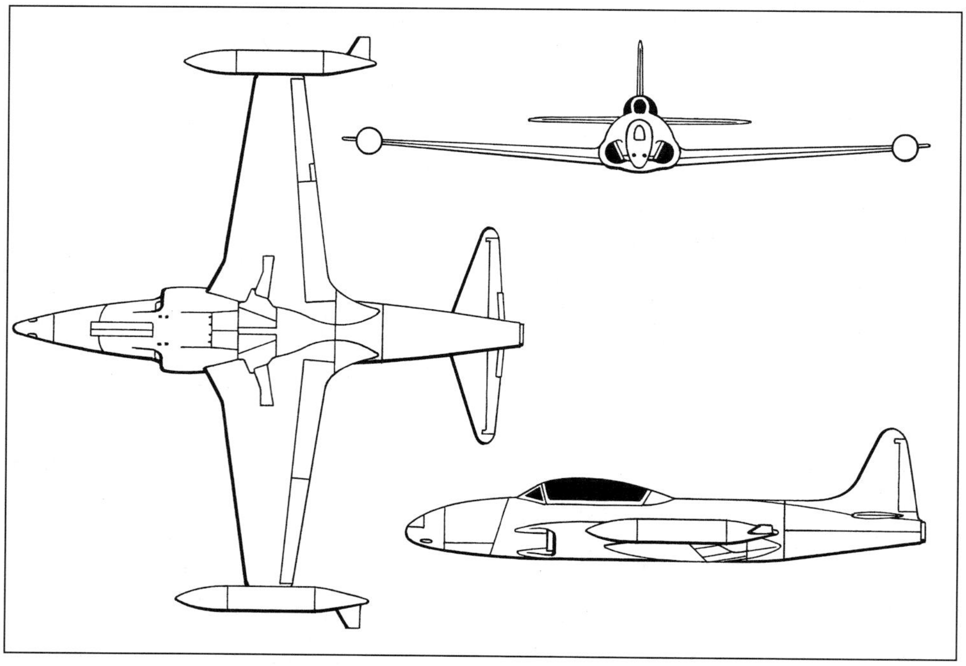

Lockheed T-33A two-seat jet trainer ***(Jane's/Mike Keep)***

1994

integrity by distributing catapult and arrester loads throughout the airframe. Launch tubes for 60 sonobuoys in belly. No provision for in-flight reloading of these launch tubes. Frangible canopies in top of fuselage are so designed that the crew can eject through them in emergency. Electronics bays with external access doors in forward and aft fuselage. An illuminated in-flight refuelling probe, mounted within the fuselage on the top centreline, is operated by an electric drive and protected by a positive-seal door. It can be extended or retracted in emergency by hand crank. MAD boom, extensible in flight, housed in fuselage tail.

POWER PLANT: Two General Electric TF34-GE-2 high-bypass ratio turbofan engines, each rated at 41.25 kN (9,275 lb st), pylon-mounted beneath the wings. Fuel in integral wing tanks, entirely within the wing box beam, one on each side of the fuselage centreline and inboard of the wing fold-line. Usable fuel capacity approximately 7,192 litres (1,900 US gallons; 1,583 Imp gallons). Two 1,136 litre (300 US gallon; 250 Imp gallon) jettisonable fuel tanks can be carried on underwing pylons. Single-point pressure refuelling adapter located on starboard side of fuselage aft of main landing gear door. Internal tanks may also be gravity fuelled through overwing connections. Fuel jettison system. Anti-icing of engine inlet nozzles by engine bleed air.

AVIONICS (S-3B): AN/AYS-1 Proteus acoustic signal processor; modified Sanders AN/OL-320/AYS data processing memory group integrated with IBM AN/UYS-1; updated Honeywell AN/AYK-10A(V) air data computer, interfaced with Harpoon anti-ship missile and other new systems; improved electronic support measures (ESM); Hazeltine AN/ARR-78 sonobuoy receiver system; Precision Echo AN/AQH-7 analog tape recorder, Cubic AN/ARS-4 sonobuoy reference system; Texas Instruments AN/APS-137(V)1 radar, incorporating inverse synthetic aperture radar (ISAR) techniques; modified Goodyear AN/ALE-39 chaff/flare dispensing system; IBM AN/ALR-76 ESM; and provision for future advanced navigation and communications systems including GPS and JTIDS.

ARMAMENT: S-3A/B weapon options include four Mk 46/50 torpedoes, two B57 nuclear depth charges, four Mk 82 560 lb bombs or four Mk 36 destructors all stowed internally; plus underwing armament of six Mk 82/86s, two Mk 52/55/56 mines, two Mk 60 torpedoes, two AGM-84 Harpoon anti-ship missiles, or six LAU-10C/-68A/-69A rocket pods/SUU-44A flare pods.

DIMENSIONS, EXTERNAL:

Wing span	20.93 m (68 ft 8 in)
Wing span, wings folded	8.99 m (29 ft 6 in)
Length overall	16.26 m (53 ft 4 in)
Length overall, tail folded	15.06 m (49 ft 5 in)
Height overall	6.93 m (22 ft 9 in)
Height overall, tail folded	4.65 m (15 ft 3 in)
Tailplane span	8.23 m (27 ft 0 in)
Wheel track	4.19 m (13 ft 9 in)
Wheelbase	5.72 m (18 ft 9 in)

DIMENSIONS, INTERNAL:

Passenger cabin: Max height	2.29 m (7 ft 6 in)
Max width	2.18 m (7 ft 2 in)

AREAS:

Wings, gross	approx 55.74 m² (600 sq ft)
Ailerons (total)	1.23 m² (13.3 sq ft)
Fin	8.51 m² (91.6 sq ft)
Rudder, incl tab	3.48 m² (37.4 sq ft)
Elevators, incl tabs	4.32 m² (46.5 sq ft)

WEIGHTS AND LOADINGS:

Weight empty	10,954 kg (24,150 lb)
Max T-O weight	21,592 kg (47,602 lb)
Max zero-fuel weight	13,290 kg (29,299 lb)
Max carrier landing weight	16,676 kg (36,766 lb)
Max wing loading	388.6 kg/m² (79.6 lb/sq ft)
Max power loading	262 kg/kN (2.56 lb/lb st)

PERFORMANCE (at max T-O weight, unless otherwise indicated):

Max level speed at 6,100 m (20,000 ft)	450 kts (834 km/h; 518 mph)
Service ceiling	12,200 m (40,000 ft)
Operational ceiling (with passengers)	10,670 m (35,000 ft)
T-O to 15 m (50 ft) at T-O weight of 19,251 kg (42,441 lb)	807 m (2,650 ft)
Range with max payload	2,000 n miles (3,706 km; 2,303 miles)
Max ferry range	3,230 n miles (6,085 km; 3,719 miles)

UPDATED

LOCKHEED T-33

TYPE: Two-seat jet trainer.

PROGRAMME: Developed from the Shooting Star jet fighter, the T-33A became the standard jet trainer in service with the USAF and USN and continued in production until late 1959.

The T-33 was produced under licence in Canada by Canadair Ltd powered by Rolls-Royce Nene engines. Kawasaki of Japan also produced the T-33 in Japan from components manufactured by Lockheed.

VERSIONS: **T-33A:** Allison J33-A-35 engine with 2,360 kg (5,200 lb st). Fuselage lengthened 98.1 cm (38.6 in) forward of wings and 30.4 cm (12 in) aft, to accommodate extra seat. Both seats under a continuous canopy hinged at rear edge and raised electrically or manually from either inside or outside aircraft. Smaller fuselage fuel tank of 360 litres (95 US gallons; 79 Imp gallons) and larger wing tanks, which are nylon cells instead of self-sealing type. Instructor in rear seat. Originally, all T-33As had a fixed armament of two .50 in machine guns in their nose. After 1955 the T-33A had all-weather navigational noses, instead of guns many of the pre-1955 T-33As were also upgraded to all-weather capability. The new equipment comprises the first combined radio compass sense antenna and VOR device, and is dual operated with controls in both cockpits.

AT-33: Modified T-33 for strike and counter-insurgency duties.

RT-33A: Single-seat photographic reconnaissance version of the T-33.

TV-2: US Navy version of the T-33A two-seat trainer. Withdrawn from service.

UPGRADES: **Canada:** Canadian DND issued request for proposal (RFP) in mid-1995 to upgrade its fleet of CT-133 Silver Star aircraft. RFP calls for structural refurbishment and avionics upgrade of 45 CT-133s. Some 29 of the CT-133s to be used as standard trainers and seven to be used for maritime target towing. Nine of the aircraft will be further retrofitted with electronic warfare (EW) pods. Avionics upgrade is to include a new ARN-149(V2) automatic direction-finder, an ARN-118(V) Tacan, an ARC-511(V) VHF radio, an APX-100(V) IFF system and an AAN-147(V) VHF Omni-Range System/ILS. The re-engineering package was developed by Canadair of Montreal. Kelowna Flightcraft of British Colombia won $2.5 million contract to complete a prototype aircraft. Refurbishment contract with value of $45 million to be awarded in December 1995.

The nine CT-133s to be modified for EW training are to be retrofitted with Ericsson A100 Erijammer ECM pod, Cross Systems' threat emitter simulator pod and Lundy ALE-43(V)1 chaff pod. Lockheed Canada to act as prime for the separate EW programme will supply pods to CAe Aviation for installation.

Sabreliner: AT-33. See separate entry.

Volpar: T-33V. See separate entry.

OPERATORS: Versions of the T-33 are in service with the armed forces of the following countries: Bolivia (32), Canada (59), Ecuador (23), Greece (47), Iran (9), Japan (40), South Korea (30), Mexico (40), Pakistan (14), Paraguay (5), Philippines (6), Thailand (35), Turkey (75) and Uruguay (7).

The following description applies to the Lockheed T-33A:

DESIGN FEATURES: Cantilever low-wing monoplane. NACA laminar-flow aerofoil section. Dihedral on chord line 3° 50′. Cantilever monoplane type tail unit.

FLYING CONTROLS: All-metal ailerons with hydraulic boost control. Trim tab in port aileron. Electrically operated split trailing-edge flaps inboard of ailerons. Airbrake flaps on underside of fuselage. Controllable inset trim tab in each elevator. Adjustable rudder trim tab.

STRUCTURE: Aluminium alloy wing structure with stressed metal skin. The fuselage is an all-metal semi-monocoque structure. All-metal fin and tailplane, with integral rear section of all-metal construction.

LANDING GEAR: Retractable tricycle type with hydraulic operation.

POWER PLANT: One Allison J-33-A-23 turbojet engine in centre-section of fuselage with air intakes on either side of fuselage forward of wing leading-edge. Aft section of fuselage including tail pipe, removable for servicing and maintenance of jet unit. Self-sealing fuel tanks in fuselage and wings.

DIMENSIONS, EXTERNAL:

Wing span	11.85 m (38 ft 10½ in)
Length	11.48 m (37 ft 8 in)
Height	3.55 m (11 ft 8 in)

WEIGHTS AND LOADINGS:

Weight empty	3,810 kg (8,400 lb)
Weight loaded	5,432 kg (11,965 lb)

PERFORMANCE:

Max speed at S/L	960 km/h (600 mph)
Max speed at 7,620 m (25,000 ft)	874 km/h (543 mph)
Stalling speed: with flaps	88 kts (164 km/h 102 mph)
without flaps	98 kts (182 km/h 113 mph)
Climb to 7,260 m (25,000 ft)	6.5 min
Service ceiling	14,480 m (47,500 ft)
T-O to 15 m (50 ft) with water injection	780 m (2,560 ft)
Landing distance from 15 m (50 ft)	1,061 m (3,480 ft)
Endurance	3.12 h

UPDATED

Lockheed U-2R single-seat reconnaissance aircraft *(Paul Tompkins)*

1994

LOCKHEED U-2

TYPE: High-altitude reconnaissance and research aircraft.

PROGRAMME: The configuration of the U-2 is basically that of a powered sailplane, which explains its unusual 'bicycle' landing gear, combined with underwing units which provide stability during take-off and are then jettisoned. Range can, when necessary, be extended by shutting off the engine and gliding. After touchdown, the aircraft comes to a rest on one of the down-turned wingtips, which serve as skids.

VERIONS: **U-2R:** Single-seat tactical reconnaissance version, described by the Department of Defense as being 'equipped with a variety of electronic sensors to provide continuously available, day or night, high-altitude all-weather standoff surveillance of the battle area in direct support of the US and Allied ground and air forces during peace, crises and war situations.' U-2R fitted with J75-P13B engine. Two additional U-2Rs, fitted with 'super-pods', were ordered for 1989 delivery.

U-2RT: Two-seat trainer version.

U-2S: Re-engined version. See Upgrades.

UPGRADES: **USAF:** U-2S. USAF received full funding mid-1994 to re-engine its U-2R aircraft as well as to implement a series of sensor upgrades. Aircraft to be re-engined with General Electric F118-GE-101 non-afterburning turbofans by 1998. Contract valued at approximately $400 million.

USAF: Congressional approval in FY95 to implement the first $10 million of a $60 million multispectral imagery upgrade to five Senior Year Electro-optical Reconnaissance System (SYERS) cameras.

USAF: Proposal to obtain up to $300 million between FY96-01 to carry out several upgrades including: embedded GPS, replacing analogue processor components in the ASARS-2 with digital systems and improving the self-defence systems.

OPERATORS: Versions of the U-2 are in service with the US armed forces (22).

The following details apply to the U-2R except where indicated:

DESIGN FEATURES: Cantilever mid-wing monoplane, with special Lockheed wing section, and wingtip fittings which serve as skids during landing. Hydraulically variable tailplane incidence, achieved by pivoting entire tail assembly around point at base of fin leading-edge.

FLYING CONTROLS: Small tubular fuel vent fairing between each outermost flap segment and aileron, projecting slightly aft of the trailing-edge. Two small hydraulically actuated plate type roll/lift spoilers on each wing, forward of outboard flap segments. Trim tab in each aileron. All primary flight controls manually operated, with high-density internal mass balance. Manually extended retractable stall strip in leading-edge of each wing to assist in settling aircraft on runway. Balanced rudder and elevators. Trim tab in each elevator.

STRUCTURE: All-metal wing structure. Hydraulically actuated trailing-edge flaps (four segments on each wing), two inboard and two outboard on each underwing pod) occupy approximately 70 per cent of each wing. The fuselage is an all-metal semi-monocoque structure of circular cross-section, with thin-gauge skin. Fineness ratio approximately 10:1. Hydraulically actuated forward-opening door type airbrake on each side of fuselage aft of wings, used mainly as a landing aid. Payload carrying nose sections and mission bay hatches interchangeable. The tail unit is a cantilever all-metal structure.

LANDING GEAR: Retractable bicycle type, with twin mainwheels and twin 8 in diameter steerable tailwheels in tandem, each unit retracting forward into fuselage. Balancer units under outer wings, each with twin small wheels, are jettisoned on take-off. Mainwheel tyre pressure 20.7 bars (300 lb/sq in). Tailwheels and underwing wheels have solid tyres. Brakes on mainwheels. Braking parachute in container under rudder. Provision for an arrester hook (U-2R has provision for foldable outerwing panels and can be operated from aircraft carriers without any change in aircraft structure or operating weights).

POWER PLANT: One 75.6 kN (17,000 lb st) Pratt & Whitney J75-P13B turbojet. All fuel in inboard and outboard main tanks filling each wing except for tip, each with overwing gravity refuelling point. Normal internal fuel capacity approximately 4,448 litres (1,175 US gallons; 978 Imp gallons). General Electric F101-GE-F29 non-afterburning turbofan (derived from B-2's F118-GE-100) began flight trials in TR-1A July 1989; completed mid-1990 after 100 hours. Additional power from 84.5 kN (19,000 lb st) class replacement power plant increases range by 15 per cent and restores operational ceiling to above 24,380 m (80,000 ft); funding decision awaited.

ACCOMMODATION: Pilot only, on ejection seat. Side-hinged transparent canopy, protected internally against ultraviolet radiation. Accommodation is air conditioned and pressurised. Rearview periscope on most aircraft (positions vary), to check that contrails are not being produced in flight. Food warmer, with spaceflight type tubes of food. Second (instructor's) cockpit above and behind standard cockpit in TR-1B.

SYSTEMS: Single hydraulic system, pressure 207 bars (3,000 lb/sq in). Electrical system utilises single AC/DC engine-driven generator, with back-up AC alternator driven from hydraulic system. Liquid oxygen system for pilot.

AVIONICS AND EQUIPMENT: Typical standard avionics include, HF, UHF, and VHF com, INS, Tacan, compass and (for night flying) astro-compass. Equipment includes one vertical and two lateral cameras for training flights, or side-looking airborne radar and T-35 tracking camera for operational missions. Main avionics and equipment compartments are in detachable modular nose section in a 'Q' bay aft of the cockpit (replaced by second cockpit in TR-1B), and in two large pods mounted underwing at approximately one-third span. Each pod is approximately 8.23 m (27 ft) long; has a volume of about 2.55 m³ (90 cu ft), and weighs about 544 kg (1,200 lb) complete with sensors and/or equipment. There is a smaller 'E' bay between the 'Q' bay and mainwheel bay; additional small areas in the bottom of the rear fuselage and in the tailcone can also be used to house mission equipment.

DIMENSIONS, EXTERNAL:

Wing span	31.39 m (103 ft 0 in)
Wing aspect ratio	approx 10.6
Length overall	19.20 m (63 ft 0 in)
Height overall	4.88 m (16 ft 0 in)

AREAS:

Wings, gross	approx 92.9 m² (1,000 sq ft)

WEIGHTS AND LOADINGS:

Weight empty, excl power plant and equipment pods	under 4,535 kg (10,000 lb)
Max T-O weight	18,144 kg (40,000 lb)

PERFORMANCE:

Max cruising speed at normal operational height of 21,650 m (70,000 ft)	more than 373 kts (692 km/h; 430 mph)
Operational ceiling	27,430 m (90,000 ft)
Min ground turning radius	68.6 m (225 ft)
Max range	more than 2,605 n miles (4,830 km; 3,000 miles)
Max endurance	12 h
g limit	+2.5

UPDATED

Lockheed U-2R of the US Air Force *(Paul Tompkins)*

1994

LOCKHEED AERONAUTICAL SYSTEMS COMPANY (LASC) (Division of Lockheed Aeronautical Systems Group)

86 South Cobb Drive, Marietta, Georgia 30063
Telephone: 1 (404) 494 4411
Telex: 542642 LOCKHEED MARA
PRESIDENT: James A Blackwell
DIRECTOR OF COMMUNICATIONSS: Laurie A Tolleson

In April 1991, Lockheed won the competition to produce F-22 with General Dynamics and Boeing Military Airplanes. Lockheed and Vought are studying advanced tactical surveillance (ATS) aircraft, based on combining S-3A airframe with electronically scanned array radar in triangular dorsal radome, to replace E-2C; advanced tactical transport (ATT) for 21st century tactical airlift; and advanced technology tactical transport (ATTT) small, low-cost STOL transport. Other experimental programmes include C-141 with electromechanical instead of hydraulic flying control actuators; and four-year, four-phase investigation of integrated vehicle/propulsion concepts for supersonic cruise.

Other long-term activities at Marietta include production of C-130 Hercules. Re-winging of C-5A completed 1987 and new C-5Bs completed March 1989; new work will consist of P-3C line, transferred from Burbank, and F-22. In December 1991, Marietta was awarded $20 million contract to study potential A-X attack aircraft for US Navy (which see) in consortium also including Boeing and General Dynamics.

VERIFIED

LOCKHEED C-130 CONVERSIONS

TYPE: C-130 upgrades.

PROGRAMME: LAS specialises in design and application of special mission packages for electronic warfare, command, control and communication systems, and signals intelligence; 10 C-130s converted into rapid response mobile hospital systems for Saudi Arabia, including five L-100-30HS; hospital fleet includes aircraft carrying four-wheel drive ambulance and evacuation aircraft for 52 patients; since 1985, LAS updating Special Forces AC-130H Spectre gunships with new navigation and fire control, including new software and improved reliability features; other special mission C-130s produced for USAF and gunships for export; first special operations forces improvement (SOFI Phase I) prototype MC-130E Combat Talon I delivered January 1988; first Phase II redelivered mid-1991, incorporating enhanced electronic warfare capabilities. First SOFI AC-130H (69-6568) redelivered July 1990; received new core avionics, HUD, secure communications, Navstar GPS, FLIR and improved gun mountings.

VERIFIED

LOCKHEED P-3 US CUSTOMS CONVERSIONS

TYPE: P-3 upgrades.

PROGRAMME: LAS modified US Customs anti-narcotics P-3A in 1984, including fittings of Hughes AN/APG-63 radar and infra-red detection; Lockheed Aeromod modified three more P-3As using LAS kits; all have inertial navigation and multistandard com radios; LAS fitted rotodome radar antenna to P-3 AEW&C; two P-3s converted to US Navy VP special mission; order placed 1986 for conversion of EP-3E-II electronic surveillance Orion.

UPDATED

LOCKHEED L-1011 (MODEL 385) TRISTAR CONVERSIONS

TYPE: Three-engined airliner modifications.

PROGRAMME: Total 250 built.

VERSIONS: **L-1011-150:** Four early TriStars being modified with Lockheed kits for First Chicago Leasing Corporation; conversion could be applicable to about 50 L-1011-1s; range increased from 2,800 n miles (5,185 km; 3,220 miles) to 3,600 n miles (6,665 km; 4,140 miles) and maximum T-O weight from 195,045 kg (430,000 lb) to 213,190 kg (470,000 lb).

L-1011-250: Extended range configuration available for 150 TriStars after CG range was extended at c/n 1052; six kits supplied by Lockheed for Delta Air Lines; wings, fuselage and landing gear strengthened to increase gross weight to 231,330 kg (510,000 lb); fuel capacity increased from 71,668 kg (158,000 lb) to 96,905 kg (213,640 lb) to extend range by 2,000 to 5,085 n miles (9,415 km; 5,850 miles); original engines replaced by Rolls-Royce RB211-524B4 under separate arrangement.

L-1011 tanker-freighter for RAF: See under Marshall of Cambridge in UK section.

L-1011 Pegasus Space Booster: See under Marshall Aerospace.

L-1011F: Freighter conversion by Pemco Aeroplex.

VERIFIED

LOCKHEED MARTIN FORT WORTH COMPANY (Division of Lockheed Aeronautical Systems Group)

PO Box 748, Forth Worth, Texas 76101
Telephone: 1 (817) 777 2000
Fax: 1 (817) 763 4797
PRESIDENT: Gordon R. England
VICE-PRESIDENT, F-16 PROGRAMS: Dain M. Hancock
VICE-PRESIDENT, PROGRAM DEVELOPMENT: A. Dwain Mayfield
DIRECTOR OF COMMUNICATIONS: Mike Hatfield
MANAGER, PUBLIC AFFAIRS: Joe W. Stout

General Dynamics' Tactical Military Aircraft Division at Fort Worth sold to Lockheed in December 1992; became Lockheed Forth Worth Company 1 March 1993.

Forth Worth activities include production of F-16 Fighting Falcon, previously shared development of F-22A ATF (with Lockheed and Boeing) and provision of spares, support and modification/update for F-111. Possible reallocation of F-22 work-shares between Fort Worth and original partner at Lockheed Marietta under consideration in 1993. Fort Worth workforce reduced to 5,800 by end 1994.

UPDATED

Lockheed Martin Fort Worth (General Dynamics) F-111 two-seat variable geometry multipurpose fighter aircraft *(Paul Tompkins)*

1995

GENERAL DYNAMICS F-111

TYPE: Two-seat variable geometry multipurpose fighter.

PROGRAMME: Following a detailed evaluation of design proposals submitted by General Dynamics and Boeing, the US Department of Defense announced 24 November 1962 that General Dynamics had been selected prime contractor for development of the F-111 tactical fighter (known originally by the designation TFX), with Grumman Aircraft as an associate. An initial contract was placed for 23 aircraft (18 F-111As for the USAF, five F-111Bs for the US Navy). Since then further orders have been placed, including 24 F-111Cs (similar to F-111As) for the Royal Australian Air Force, and reconnaissance and strategic bomber versions for the USAF. In 1992 the F-111 was still in service with the air forces of Australia (F-111C 13+, RF-111C-4, F-111G 4) and the USA (FB-111A/F 60+, F-111D/E/F225+, EF-111A (40+).

VERSIONS: **F-111A:** USAF two-seat tactical fighter-bomber, with slightly longer nose than the F-111B to accommodate different electronic equipment and TF30-P-3 engines.

EF-111A: ECM jamming version. Converted from F-111As by Grumman.

YF-111A: Strike/reconnaissance fighters completed prior to cancellation of the UK government order for 50 aircraft, under the designation F-111K were assigned to the USAF for research and test evaluation purposes.

F-111B: US Navy version, intended originally for carrier-based fleet defence duties with armament of six Hughes AIM-54 Phoenix air-to-air missiles. Greater wing span and area than F-111A. Powered initially by TF30-P-1 turbofan engines.

F-111C: Strike aircraft. Outwardly similar to F-111B, with Pratt & Whitney TF30-P-3 engines, Mk 1 avionics, cockpit ejection module and eight underwing attachments for stores.

F-111D: Similar to F-111A, but with more advanced Mk II avionics system, offering improvements in navigation and air-to-air as well as air-to-surface weapon delivery.

F-111E: Superseded F-111A from 160th aircraft. Modifed air intakes to permit removal of flight restrictions above Mach 2.2 and 18,300 m (60,000 ft).

F-111F: Fighter-bomber. Generally similar to the F-111D, but with an avionics system that combines the best features of the F-111E and FB-111A systems, to provide effective tactical avionics at the lowest possible cost. Fitted with the more powerful TF3-P100 turbofan engines, producing 25 per cent more thrust than the basic TF30.

F-111G: Converted FB-111A for dual nuclear/conventional bombing.

F-111K: See YF-111A.

FB-111A: Two-seat strategic bomber version for USAF Strategic Air Command. Requirement for 210 announced by US Secretary of Defense 10 December 1965, to replace B-52C/F versions of the Stratofortress and B-58A Hustler. Initial contract for 64 signed Spring 1967, however, the order was increased to 76 aircraft.

RF-111A: Two-seat reconnaissance version of F-111A for USAF. Cameras, radar and infra-red sensors in fuselage weapon bay operate through optical windows and radomes on the outboard sides of the reconnaissance pallet. First flight was made 17 December 1967.

RF-111C: Strike reconnaissance aircraft of Royal Australian Air Force.

UPGRADES: **Northrop Grumman/Lockheed Fort Worth:** EF-111A upgrade. See Northrop Grumman entry.

Rockwell: Pacer Strike and RAAF upgrades. See separate entry.

OPERATORS: Versions of the F-111 are in service with the armed forces of the following countries: Australia (36) and USA (147).

The following description applies to the F-111A except where otherwise indicated:

DESIGN FEATURES: Cantilever shoulder wing. Wing section of NACA 63 series, with conventional washout. Sweepback of outer wing portions variable in flight or on the ground from 16° to 72° 30′.

FLYING CONTROLS: Airbrake/lift dumpers above wing operate as spoilers for lateral control at low speeds. Full-span variable camber leading-edge slats and full-span double-slotted trailing-edge flaps. General Electric flight control system. All-moving horizontal surfaces operate both differentially and symmetrically to provide aileron and elevator functions. Two long narrow ventral stabilising fins.

STRUCTURE: Wing actuating jacks by Jarry Hydraulics. Five-spar wing structure with stressed and sculptured skin panels, each made in one piece between leading- and trailing-edge sections, from root to tip. Leading- and trailing-edge sections of honeycomb sandwich. The fuselage is a semi-monocoque structure mainly of aluminium alloy, with honeycomb sandwich skin. Some steel and titanium. Main structural member is a T-section keel, under the arms of which the engines are hung. The tail unit is of conventional cantilever sweptback surfaces, utilising honeycomb sandwich skin panels, except for tailplane tips and central area of fin on each side.

LANDING GEAR: Hydraulically retractable tricycle type. Single wheel on each main leg. Twin-wheel nose unit retracts forward. Main gear is a triangulated structure with hinged legs which are almost horizontal when the gear is extended. During retraction, the legs pivot downward, the wheels tilt to lie almost flat against them, and the whole gear rotates forward so that the wheels are stowed side by side in fuselage between engine air intake ducts. Low-pressure tyres on mainwheels, size 47-18 in (42-13 in on F-111C and FB-111A). Disc brakes, with anti-skid system. Main landing gear door, in bottom of fuselage, hinges down to act as speed brake in flight.

POWER PLANT: Two Pratt & Whitney TF30-P-3 turbofan engines, each giving 82.3 kN (18,500 lb st) with afterburning. Fuel tanks in wings and fuselage. Pressure fuelling point in port side of fuselage, forward of engine air intake. Gravity refuelling filler/in-flight refuelling receptacle in top of fuselage aft of cockpit. Hamilton Standard hydro-mechanical air intake system with movable shock-cone.

ACCOMMODATION: Crew of two side by side in air conditioned and pressurised cabin. Portion of canopy over each seat is hinged on aircraft centreline and opens upward. Zero-speed, zero-altitude (including underwater) emergency escape module developed by McDonnell Douglas Corporation and utilising a 178 kN (40,000 lb st) Rocket Power Inc rocket motor. Emergency procedure calls for both crew members to remain in capsule cabin section, which is propelled away from aircraft by rocket motor and lowered to ground by parachute. Airbags cushion impact and form flotation gear in water. Entire capsule forms survival shelter.

ARMAMENT: Tactical fighter versions carry one M61 multi-barrel 20 mm gun or two 750 lb bombs in internal weapon bay. External stores are carried on four attachments under each wing. The two inboard pylons on each side pivot as the wings sweep back, to keep the stores parallel with the fuselage. The two outboard pylons on each wing are jettisonable and non-swivelling.

DIMENSIONS, EXTERNAL:

Wing span:	
F-111A, F-111D, F-111E, F-111F:	
spread	19.20 m (63 ft 0 in)
fully swept	9.74 m (31 ft $11\frac{1}{3}$ in)
F-111B, F-111C, FB-111A:	
spread	21.34 m (70 ft 0 in)
fully swept	10.34 m (33 ft 11 in)
Wing chord at root	2.11 m (6 ft 11 in)
Length overall:	
F-111A, F-111C, F-111D, F-111E, F-111F, FB-111A	22.40 m (73 ft 6 in)

Lockheed Fort Worth (General Dynamics) F-111G of the US Air Force *(Paul Jackson)*

1994

Height overall:
F-111A, F-111C, F-111D, F-111E, F-111F, FB-111A 5.22 m (17 ft 1⅔ in)

WEIGHTS AND LOADINGS (F-111A):
Weight empty 20,943 kg (46,172 lb)
Max T-O weight 41,500 kg (91,500 lb)

PERFORMANCE (F-111A):
Max speed at height Mach 2.2
Mach speed at S/L Mach 1.2
Service ceiling over 15,500 m (51,000 ft)
T-O and landing run under 915 m (3,000 ft)
Range with max internal fuel over 2,750 n miles (5,093 km; 3,165 miles)

UPDATED

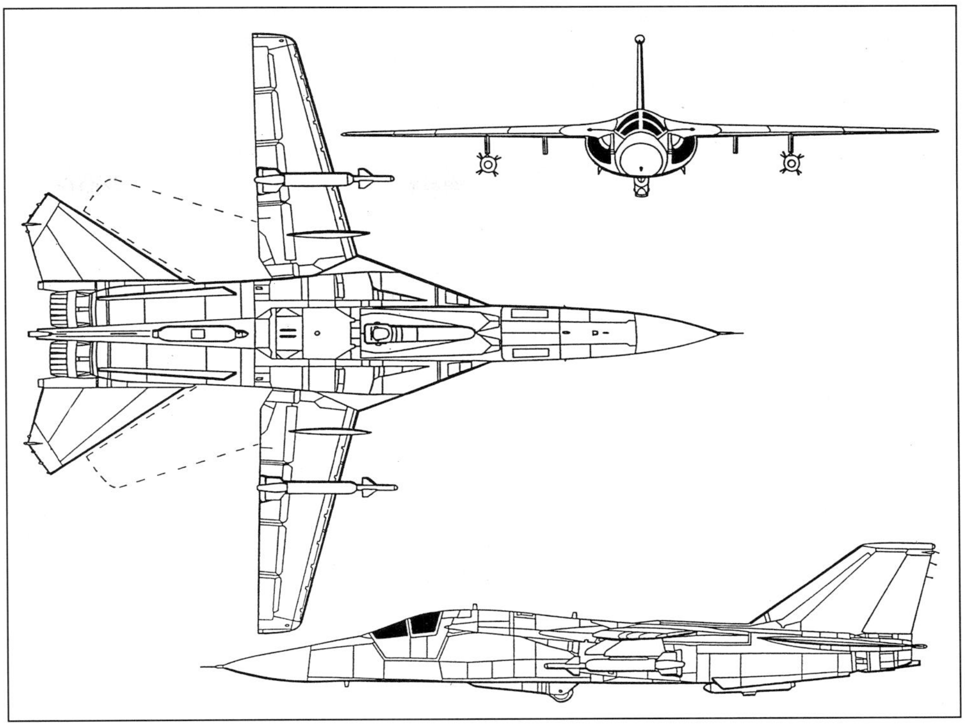

Lockheed Fort Worth (General Dynamics) F-111 *(Jane's/Mike Keep)* ***1994***

LOCKHEED MARTIN (GENERAL DYNAMICS) F-16 FIGHTING FALCON

TYPE: Single/two-seat multirole fighter.

PROGRAMME: Emerged from YF-16 of US Air Force Lightweight Fighter prototype programme 1972 (details under General Dynamics in 1977-78 and 1978-79 *Jane's All the World's Aircraft*); first flight of prototype YF-16 (72-01567) 2 February 1974; first flight of second prototype (72-01568) 9 May 1974; selected for full-scale development 13 January 1975; day fighter requirement extended to add air-to-ground capability with radar and all-weather navigation; production of six single-seat F-16As and two two-seat F-16Bs began July 1975; first flight of full-scale development aircraft 8 December 1976; first flight of F-16B 8 August 1977.

VERSIONS: **F/A-16:** Proposed modification of 300 Block 30/32 aircraft for close air support (CAS)/battlefield air interdiction (BAI) in late 1990s; head-steered FLIR, Pave Penny laser ranger and 30 mm cannon pod. From 1995 200 F-16Cs are to receive CAS/BAI modifications, including DTS, Navstar GPS and improved data modem. Block 30/32 upgrade abandoned January 1992 in favour of CAS/BAI assignment of Block 40/42 aircraft, having LANTIRN capability; these require more simple modification with ground datalink, laser spot-tracker, anti-jam radio, missile approach warner, provision for pilot's night vision goggles and upgrades to LANTIRN pods. Deployment plan envisages 4½ wings of Block 40/42s for night CAS and two wings of Block 30/32s for day CAS. Meanwhile, first dedicated CAS/BAI aircraft are F-16As of ANG's 138th FS at Syracuse, New York; operational 1989, equipment including fixed GPU-5/A 30 mm centreline cannon; first improved data modem installed November 1990; squadron received F-16Cs in 1991.

F-16A: First production version for air-to-air and air-to-ground missions; production for USAF completed March 1985, but still available for other customers; international sales continue; powered since late 1988 (Block 15OCU) by P&W F100-PW-220 turbofan; Westinghouse AN/APG-66 range and angle track radar; first flight of first aircraft (78-0001) 7 August 1978; entered service with 388th TFW at Hill AFB, Utah, 6 January 1979; combat ready October 1980, when named Fighting Falcon; most now serving ANG and AFRES; power plants being upgraded to F100-PW-220E, 1991-1996. Also produced in Europe. Built in Blocks 01, 05, 10 and 15, of which Blocks 01 and 05 retrofitted to Block 10 standard 1982-84; Block 15 retrofitted to OCU standard from late 1987. First **GF-16A** ground trainers relegated to instructural use at 82nd Training Wing, Sheppard AFB.

Operational Capabilities Upgrade (OCU): USAF/NATO co-operative programme to equip F-16A/B for next-generation BVR air-to-air and air-to-surface weapons; radar and software updated, fire control and stores management computers improved, data transfer unit fitted, combined radar-barometric altimeter fitted, and provision for AN/ALQ-131 jamming pods. Ring laser INS and upgrade from P&W F100-PW-200 to F100-PW-220E planned for 1990s. FMS exports since 1988 to Block 15OCU standard with F-16C features including ring laser INS, AN/ALR-69 RWR, F100-PW-220 power plant and AIM-9P-4 Sidewinder AAM capability.

Mid-Life Update (MLU): Development authorised 3 May 1991 (signature of final partner); US government contract to GD 15 June 1991; originally planned to be applied to 533 aircraft of USAF (130), Belgium (110), Denmark (63), Netherlands (172) and Norway (58) from 1996 in co-development/co-production programme. USAF withdrew 1992, but ordered 223 modular computer retrofit kits from MLU to equip Block 50/52 aircraft. European share renegotiated 28 January 1993 to new totals of Belgium 48 plus 24 options; Denmark 61; Netherlands 136; and Norway 56; Letters of Offer and Acceptance finalised mid-1993. Cockpit similar to F-16C/D Block 50 with wide-angle HUD, night vision goggle compatibility, modular mission computer replacing existing three, digital terrain system, AN/APG-66(V2A) fire control radar, Navstar GPS, improved data modem and provision for microwave landing system (MLS). Inlet hardpoints and wiring for FLIR pods will be added to Block 10 aircraft. Options include helmet-mounted display and Hazeltine AN/APX-111 IFF interrogator/transponder; both taken up by Netherlands and Norway. In addition, 150 F-16A/Bs offered to Taiwan in 1992 to be built to Block 15MLU standard. Aircraft for prototype conversion delivered to GD in September 1992, including Danish F-16B ET-204, Netherlands F-16B J-650, Norwegian F-16A 299 and one USAF. Danish Air Force has chosen the AIM-120 AMRAAM medium-range air-to-air missiles for its MLU primary weapon systems. Delivery of the first of 44 missiles ordered under a letter of agreement signed on 6 December 1994 will begin in 1997. Option exists for further 106 missiles.

F-16(ADF) Modification of 279 (actually 272 because of preconversion attrition) Block 15 F-16A/Bs as USAF air defence fighters to replace F-4s and F-106s with 11 Air National Guard squadrons; ordered October 1986. Modifications include upgrade of AN/APG-66 radar to improve small target detection and provide continuous-wave illumination, provision of AMRAAM datalink, improved ECCM, Bendix/King AN/ARC-200 HF/SSB radio (F-16A only), Teledyne/E-Systems Mk XII advanced IFF, provision for Navstar GPS Group A, low-altitude warning, voice message unit, night identification light (port forward fuselage of F-16A only), and ability to carry and guide two AIM-7 Sparrow missiles. First successful guided launch of AIM-7 over Point Mugu range, California, February 1989; F-16(ADF) can carry up to six AIM-120 AMRAAM or AIM-9 Sidewinder or combinations of all three missiles; retains internal M61 20 mm gun. GD converted one prototype, then produced modification kits for installation by USAF Ogden Air Logistics Center, Utah, in conjunction with upgrade to OCU avionics standard; first Ogden aircraft, F-16B 81-0817, completed October 1988. Development completed at Edwards AFB during 1990; operational test and evaluation with 57th Fighter Weapons Wing at Nellis AFB, Nevada; first F-16(ADF), 81-0801, delivered to 114th Fighter Training Squadron at Kingsley Field, Oregon, 1 March 1989; 194th Fighter Interceptor Wing, California ANG, Fresno, achieved IOC in 1989, following receipt of first aircraft (F-16B 82-1048) on 13 April 1989; first AIM-7 launch by ANG (159th FS) June 1991. Programme completed early 1992; includes approximately 30 F-16Bs.

F-16B: Standard tandem two-seat version of F-16A; fully operational both cockpits; fuselage length unaltered; reduced fuel.

MSIP-F-16C/D: Single/two-seat USAF Multinational Staged Improvement Program (**MSIP**) aircraft respectively, implemented February 1980. MSIP expands growth capability to allow for ground attack and beyond-visual-range missiles, and all-weather, night and day missions; **State I** applied to Block 15 F-16A/Bs delivered from November 1981 included wiring and structural changes to accommodate new systems; **Stage II** applied to Block 25 F-16C/Ds from July 1984 includes core avionics, cockpit and airframe changes. **Stage III** includes installation of systems as they became available, beginning 1987 and extending up to Block 50/52, including selected retrofits back to Block 25. Changes include Westinghouse AN/APG-68 multimode radar with better range, resolution, more operating modes and better ECCM than AN/APG-66; advanced cockpit with better interfaces and up-front controls, GEC-Marconi wide-angle HUD, two multifunction displays, Fairchild mission data transfer equipment and radar altimeter; expanded base of fin giving space for proposed later fitment of AN/ALQ-165 Airborne Self Protection Jamming system (since cancelled); increased electrical power and cooling capacity; structural provision for increased take-off weight and manoeuvring limits; and MIL-STD-1760 weapons interface for use of smart weapons such as AIM-120A AMRAAM and AGM-65D IR Maverick. First AIM-120 operational launch (by any

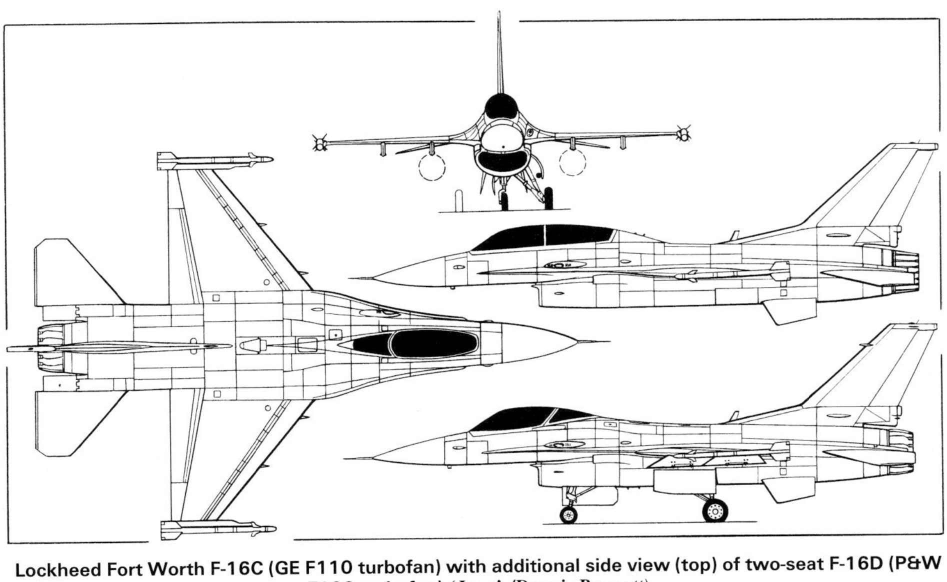

Lockheed Fort Worth F-16C (GE F110 turbofan) with additional side view (top) of two-seat F-16D (P&W F100 turbofan) *(Jane's/Dennis Punnett)* ***1994***

Lockheed Fort Worth F-16A of the Royal Norwegian Air Force with parachute deployed *(Paul Tompkins)*
1994

aircraft), 27 December 1992; F-16D (90-0778) of 33rd FS/363rd FW destroyed Iraqi MiG-25.

Common engine bay introduced at **Block 30/32** (deliveries from July 1986) to allow fitting of either P&W F100 PW-220 (Block 32) or GE F110-GE-100 (Block 30) Alternate Fighter Engine. Other changes include computer memory expansion and seal-bonded fuselage fuel tanks. First USAF wing to use F-16C/Ds with F110 engines was 86th TFW at Ramstein AB, Germany, from October 1986. Additions in 1987 included full Level IV multitarget compatibility with AMRAAM (as Block 30B), voice message unit. Shrike anti-radiation missiles (from August), crash survivable flight data recorder and modular common inlet duct allowing full thrust from F110 at low airspeeds.

Software upgraded for full Level IV multitarget compatibility with AMRAAM early 1988. Industry-sponsored development of radar missile capability for several European air forces resulted in firing of AIM-7F and AIM-7M missiles from F-16C in May 1980; capability introduced mid-1991; missiles guided using pulse Doppler illumination while tracking targets in a high PRF mode of the AN/APG-68 radar.

US Air Force F-16C/Ds of 52nd FW at Spangdahlem AB, Germany, have HARM/Shrike capability and operate alongside A-10A Thunderbolt II in Wild Weasel defence suppression role.

Block 40/42 Night Falcon (deliveries from December 1988) upgrades include AN/APG-68(V) radar allowing 100 hours operation before maintenance, full compatibility with Martin Marietta low altitude navigation and targeting infra-red for night (LANTIRN) pods, four-channel digital flight control system, expanded capacity core computers, diffractive optics HUD, enhanced envelope gunsight. Navstar GPS, improved leading-edge flap drive system, improved cockpit ergonomics, high gross weight landing gear, structural strengthening, and provision for improved EW equipment. LANTIRN gives day/night standoff target identification, automatic target handoff for multiple launch of Mavericks, autonomous laser-guided bomb delivery and precision air-to-ground laser ranging. Combat Edge pressure breathing system installed 1990 for higher pilot *g* tolerance.

First Block 40 F-16C/Ds issued late 1990 to 363rd FW (Shaw AFB, South Carolina); first LANTIRN pods to 36th FS/51st FW as Osan, South Korea, in 1992.

Block 50/52 (deliveries began with F-16C 90-0801 in October 1991 for operational testing) upgrades include F110-GE-129 and F100-PW-229 increased performance engines (IPE). AN/APG-68(V5) radar with advanced programmable signal processor employing VHSIC technology. Have Quick IIA UHF radio. Have Sync VHF anti-jam radio and AN/ALR-56M advanced RWR. Changes planned for 1993 include full integration of HARM/Shrike anti-radiation missiles via Texas Instruments interface, upgraded programmable display generator with digital terrain system (DTS) provisions and scope for digital map capability, ring laser INS (Honeywell H-423 selected 1990) and AN/ALE-47 advanced chaff/flare dispenser. Upgrades being considered for 1994-95 include a digital terrain system with colour map, colour multifunction displays, head-steered FLIR with helmet-mounted display, self-protection jammer and onboard oxygen generating system (OBOGS).

First operational unit is 4th FS of 388th FW at Hill AFB, Utah, from October 1992; others to 480th FS of 52nd FW at Spangdahlem, Germany, replacing Block 30 aircraft from (first delivery) 20 February 1993.

Block 60/62: Projected development to meet possible USAF multirole fighter requirement, employing technology from F-22A programme.

NF-16D: Variable stability in-flight simulator test aircraft (VISTA) modified from Block 30 F-16D (86-0048) ordered December 1988 to replace NT-33A testbed. Features include vertical surface direct force generators above and below wings. Calspan variable stability flight control system, fully programmable cockpit controls and displays, additional computer suite, permanent flight test data recording system, variable feel centrestick and computer, and safety pilot in rear cockpit. Internal gun. RWR and chaff/flare equipment removed, providing space for Phase II and III growth including additional computer, reprogrammable display generator and customer hardware allowance. 'Israeli-type' bulged spine. First flight 9 April 1992; delivery due July 1992 but aircraft stored after five flights because of funding shortage; proposed to fit axisymmetric thrust vectoring engine nozzle (AVEN) for 60-flight trials programme which began May 1993.

F-16N: US Navy supersonic adversary aircraft (SAA) modified from F-16C/D Block 30; selected January 1985; deliveries of 26 aircraft started 1987 and completed 1988; features include AN/APG-66 instead of AN/APG-68 radar, F110-GE-100 engine, deletion of M61 gun, AN/ALR-69 RWR, titanium in lower wing fittings instead of aluminium and cold working of lower wing skin holes to resist greater frequency of high *g*; wingtips fitted only for AIM-9 practice missiles and ACMI AIS pods, but normal tanks and stores on other stations. Four of 26 are two-seat **TF-16N**. F/TF-16Ns serve with 'Top Gun' Fighter Weapons School (eight) and with VF-126 (six) at NAS Miramar, California, VF-45 (six) at NAS Key West, Florida, and VF-43 (six detached from VF-45) at NAS Oceana, Virginia.

FS-X and **TFS-X:** F-16 derivatives selected by Japan Defence Agency for its FS-X requirement 19 October 1987; details under Mitsubishi in Japan section.

F-16 Recce: Four existing European recce pods, including that for Tornado, demonstrated in flight on F-16 fighters with minimum changes; RNethAF reconnaissance **F-16A(R)** operational since 1983 with Orpheus pods.

AFTI/F-16: Modified preseries F-16A (75-0750) used for US Air Force Systems Command Advanced Fighter Technology Integration (AFTI); first flight was 10 July 1982; currently with Block 15 horizontal tail surfaces, Block 25 wing and Block 40 avionics. Trials programmes include automatic target designation and attack (1988), night navigation and map displays (1988-89), digital datalink and two-aircraft operations (1989), autonomous attack (1989-91) and night attack (1989-92); tested automatic ground collision avoidance system and pilot-activated, low-level pilot disorientation recovery system 1991; LANTIRN pod and Falcon Night FLIR trials, 1992. Total 500 sorties November 1991.

F-16XL: Two F-16XL prototypes, in flyable storage since 1985, leased from General Dynamics by NASA; first flight of single-seat No. 1, 9 March 1989; NASA modified this aircraft at Dryden with wing glove having laser-perforated skin to smooth air flow over cranked arrow wing in supersonic flight, reducing drag and turbulence and saving fuel. Two-seat No. 2, with GE F110-GE-129 engine, similarly converted early 1992. F-16XL described in 1985-86 *Jane's All the World's Aircraft*.

F-16B-2: Second prototype F-16B (75-0752) converted to private venture testbed of close air support and night navigation and attack systems; equipment includes F-16C/D HUD, helmet sight or GEC-Marconi Cat's Eyes NVGs, Falcon Eye head-steered FLIR or LANTIRN nav/attack pods, digital terrain system (Terprom), and automatic target handoff system. Alternative nav/attack FLIR pods comprise GEC-Marconi Atlantic and Martin Marietta Pathfinder (LANTIRN derivative). NVG compatible cockpit lighting. Equipment testing continues on AFTI testbed (which see).

F-16ES: Enhanced Strategic two-seat, long-range interdictor F-16 proposal; developed November 1993 in response to Israeli preference for F-15I Eagle; additional fuel in one 2,271 litre (600 US gallon; 500 Imp gallon) centreline tank and two conformal tanks, each 1,893 litres (500 US gallons; 417 Imp gallons), on side of Israeli style enlarged spine. Compbat radius extended to in excess of 1,000 n miles (1,852 km; 1,151 miles). Not purchased, but conformal tanks to be flight tested as retrofit option for existing F-16s.

F-16X: Projected development for 2010 service entry; 1.42 m (5 ft 0 in) fuselage stretch; modified F-22 delta wing with increased leading-edge sweep, but similar taper, section, twist, camber, moving surfaces and structure; some 80 per cent additional internal fuel, obviating drop tanks for most combat missions; conformal AIM-120 AMRAAM carriage; developed version of F100 or F110 engine; cost, two-thirds of F/A-18E Hornet.

UPGRADES: **Base:** F-16 MLU TERPROM. See separate entry.

Fokker Aircraft Services (FAS): Pacer Slip upgrade. See separate entry.

Lockheed: F-16ADF. See Versions.

F-16N. See Versions.

F-16 MLU. See Versions.

F-16 MSIP. See Versions.

F-16 OCU. See Versions.

Netherlands: Certification trials were carried out in late 1994 in the USA for installation of Terma Elektronik/Per Udsen pylon integrated dispenser system (PIDS) on 54 of the Royal Dutch Air Force F-16 aircraft.

OPERATORS: Versions of the F-16 are in service with the armed forces of the following countries: Bahrain (12), Belgium (100), Denmark (63), Egypt (131), Greece (80), Indonesia (12), Israel (159), South Korea (159 on order), Netherlands (184), Norway (59), Pakistan (38), Portugal (20), Singapore (6), Taiwan (150), Thailand (18), Turkey (240), USA (1,653) and Venezuela (24).

DESIGN FEATURES: Cantilever mid-wing monoplane of blended wing-body design and cropped delta planform. The blended wing-body concept is achieved by flaring the wing/body intersection, thus not only providing lift from the body at high angles of attack but also giving less wetted area and increased internal fuel volume. Basic wing is NACA 64A-204 section with 40° sweepback on leading-edges. The tail unit is a cantilever structure with sweptback surfaces. Extension of fin root fairing houses Loral Rapport ECM equipment in Belgian aircraft and a brake parachute in Norwegian and Venezualan aircraft. Ventral fins three-quarters along fuselage.

FLYING CONTROLS: Leading-edge manoeuvring flaps are programmed automatically as a function of Mach number and angle of attack. The increased wing camber maintains lift co-efficients at high angles of attack. These flaps are one-piece bonded aluminium honeycomb sandwich structures actuated by a Garrett drive system using rotary actuators.

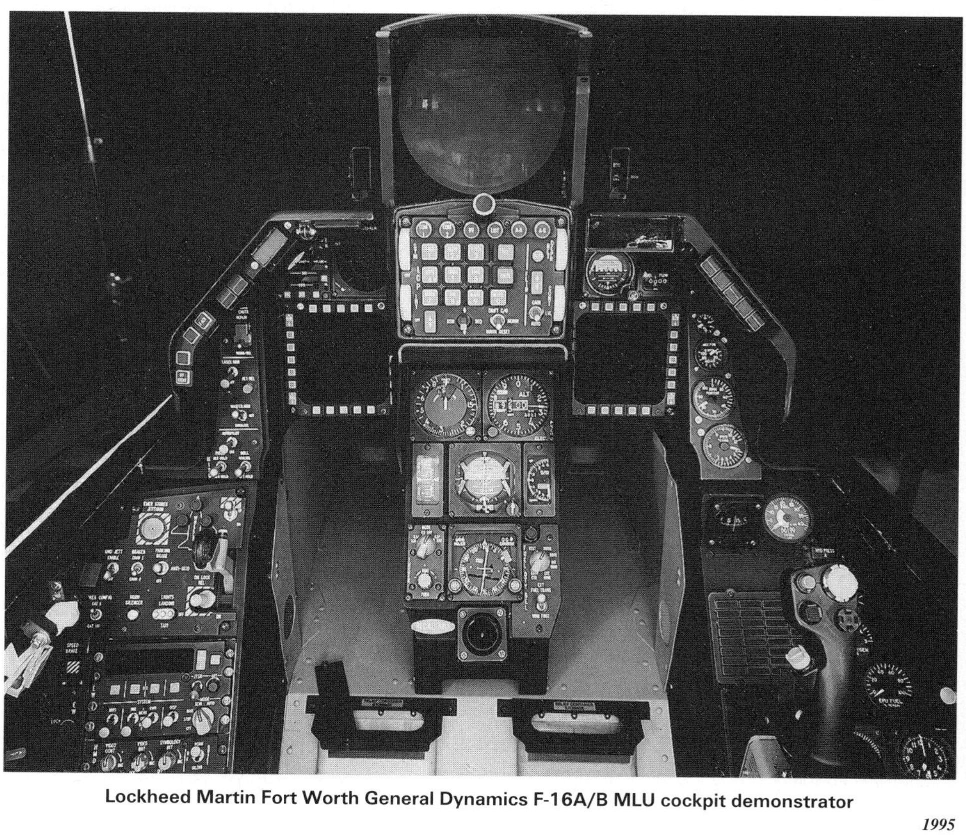
Lockheed Martin Fort Worth General Dynamics F-16A/B MLU cockpit demonstrator
1995

The trailing-edges carry large flaperons (flap/ailerons), which are interchangeable left with right and are actuated by National Water Lift integrated servo-actuators. The maximum rate of flaperon movement is 80°/s. Interchangeable, all-moving tailplane halves. Split speed-brake inboard of rear portion of each horizontal tail surface to each side of nozzle, each deflecting 60° from the closed position. National Water Lift servo-actuators for rudder and tailplane.

STRUCTURE: Wing mainly of aluminium alloy with 11 spars, five ribs and single upper and lower skins is attached to fuselage by machined aluminium fittings. The fuselage is a semi-monocoque all-metal structure of frames and longerons built in three main modules: forward (to just aft of cockpit), centre and aft. Nose radome built by Brunswick corporation. Highly swept vortex control strakes along the fuselage forebody increase lift and improve directional stability at high angles of attack. The tail unit fin is a multi-spar, multirib aluminium structure with graphite epoxy skins, aluminium tip and glassfibre dorsal fin and root fairing. Tailplanes constructed of graphite epoxy composite laminate skins mechanically attached to a corrugated aluminium substructure. Each tailplane half has an aluminium pivot shaft and a removable full depth bonded honeycomb leading-edge. Ventral fins are bonded aluminium skins.

LANDING GEAR: Menaso hydraulically retractable type, nose unit retracting aft and main units forward into fuselage. Nosewheel is located aft of intake, to reduce the risk of foreign objects being thrown into the engine during ground operation, and rotates 90° during retraction to lie horizontally under engine air intake duct. Oleo-pneumatic struts in all units. Goodyear mainwheels and brakes; Goodrich mainwheel tyres, size 25.5 × 8-14, pressure 14.48 to 15.17 bars (210 to 220 lb/sq in) at T-O weight less than 11,340 kg (25,000 lb). Steerable nosewheel with Goodrich tyre, size 18 × 5.5-8, pressure 14.82 to 15.51 bars (215 to 225 lb/sq in) at T-O weights less than 11,340 kg (25,000 lb). All but two main unit components interchangeable. Brake by wire system on main gear, with Goodyear anti-skid units. Runway arrester hook under rear fuselage.

POWER PLANT: One Pratt & Whitney F-100-PW-200 turbofan engine, rated at approximatelty 111.2 kN (25,000 lb st) with afterburning, mounted within the rear fuselage. (General Electric F110-GE-100 to be alternative standard engine in future production aircraft.) Fixed geometry intake, with boundary layer splitter plate, beneath fuselage. Standard fuel contained in wing and five fuselage cells which functions as two tanks; internal fuel weight is 3,162 kg (6,972 lb) in F-16A, and approximately 17 per cent less in F-16B. In-flight refuelling receptacle in top of centre-fuselage, aft of cockpit. Auxiliary fuel can be carried in drop tanks on underwing and underfuselage hardpoints.

ACCOMMODATION: Pilot only in F-16A, in air conditioned cockpit. McDonnell Douglas ACES II zero/zero ejection seat. Transparent bubble canopy made of polycarbonate advanced plastics material. The windscreen and forward

TABLE 2: F-16A/B PRODUCTION

Date	Block	USAF		Belgium		Denmark		Egypt		Indonesia		Israel		Netherlands		Norway		Pakistan		Portugal		Singapore		Taiwan		Thailand		Venezuela	
		A	B	A	B	A	B	A	B	A	B	A	B	A	B	A	B	A	B	A	B	A	B	A	B	A	B	A	B
	—	6	2																										
Aug 78	01	21	22	17	6	3	2				12	6	3	2															
Oct 79	05	89	27	8	4	12	3			18	8	14	2	10	2														
Oct 80	10	26	4	5	—	6	1																						
								10	—	5	1	5	1																
Jan 81	10A	28	5	7	1	3	1			10	—	7	1	4	—														
Apr 81	10B	29	6	6	1	2	1			4	—	4	1	6	1														
Jul 81	10C	27	4	6	—	4	—			16	—	4	2	—	1														
Sep 81	10D	35	6	6	—				9	—																			
Sub-total		**261**	**76**	**55**	**12**	**30**	**8**			**67**	**8**	**46**	**13**	**28**	**7**														
Nov 81	15	31	2		2	1				4	—	3	1																
Feb 82	15A	25	1			3	4																						
Apr 82	15B	26	1	1	1	5	1	2	1			5	2	4	1														
Jun 82	15C	26	1			2	1																						
Aug 82	15D	25	2	4	4	4	—	3	—			6	1	4	1	—	2												
Oct 82	15E	25	2			3	—																						
Dec 82	15F	25	2	5	3	3	1	3	—			6	2	4	1	2	2												
Mar 83	15G	24	2			4	—																						
Apr 83	15H	24	2	6	—	2	—	4	—			6	2	4	1														
Jun 83	15J	25	3			2	—																						
Aug 83	15K	20	5	8	—		3	—			4	2	8	—						1	3								
Oct 83	15L	21	4			3	—										2	—											
Dec 83	15M	21	4	8	—		2	—			4	1	5	—	2	2													
Feb 84	15N	21	3			—	1					1	2																
Apr 84	15P	18	4	5	—		—	1			5	—		3	—														
Jul 84	15Q	22	6		—	1						5	—																
Oct 84	15R	19	2	4	—					10	2		4	—															
Jan 85	15S	11	—								5	—																	
Mar 85	15T							7	2		2	—					3	3											
Jun 85	15U							10	2		4	1					8	—											
Oct 85	15V									—	3					4	—												
—	15W							9	2																				
—	15X							8	—																				
Sub-total		**670**	**122**	**96**	**20**	**46**	**12**	**34**	**8**		**67**	**8**	**130**	**31**	**60**	**12**	**28**	**12**					**18**	**6**					
—	15Y		7	—	—	4				7	1				—	1													
Jan 88	15Z																												
Feb 88	15AA		7	1	4	—				7	1				2	2		1	3										
Jun 88	15AB											2	1		7	1													
—	15AC		7	1	4	—				7	1																		
May 89	15AD								—	2																			
Aug 89	15AE		6	2			—	2		7	2																		
Sep 89	15AF					2	2																						
Mar 90	15AG		8	—			6	—		8	—																		
—	15AH																												
Jan 91	15AJ		5	—					8	—						6	—												
—	15AK																												
—	15AL							3	—																				
Nov 91	15AM									6	5																		
—	15AN																												
—	15AP																												
Apr 93	15AQ									5	1																		
Oct 93	15AR									2	3																		
Jan 94	15AS									—	2																		
Jan 94	15AT										3	3																	
Mar 94	15AU									—	4																		
Jun 94	15AV										10	—																	
Oct 94	15AW										4	—																	
Feb 95	15AX													—	4														
Jun 95	15AY													4	—														
Oct 95	15AZ													8	2														
Sub-total		**670**	**122**	**136**	**24**	**54**	**16**	**34**	**8**	**8**	**4**	**67**	**8**	**177**	**36**	**60**	**14**	**41**	**27**	**17**	**3**	**4**	**4**	**–**	**–**	**26**	**10**	**18**	**6**
	20												120	30															
Totals		**670**	**122**	**136**	**24**	**54**	**16**	**34**	**8**	**8**	**4**	**67**	**8**	**177**	**36**	**60**	**14**	**41**	**27**	**17**	**3**	**4**	**4**	**120**	**30**	**26**	**10**	**18**	**6**
Grand total																**1,744**													

Notes: OCU standard from Block 15Y
Date of manufacture refers to Fort Worth only; overseas production varies slightly.

Lockheed Fort Worth F-16C of the US Air Force

1994

Lockheed Fort Worth F-16A of the Belgian Air Force *(Paul Tompkins)*

1994

canopy are an integral unit without a forward bow frame, and are separated from the aft canopy by a simple support structure which serves also as the breakpoint where the forward section pivots upward and aft to give access to the cockpit. A redundant safety lock feature prevents canopy loss. Windscreen/canopy design provides 360° all-round view, 195° fore and aft, 40° down over the side, and 15° down over the nose. To enable the pilot to sustain high *g* forces, and for pilot comfort, the seat is inclined 30° aft and the heel line is raised. In normal operation the canopy is pivoted upward and aft by electrical power; the pilot is also able to unlatch the canopy manually and open it with a backup handcrank. Emergency jettison is provided by explosive unlatching devices and two rockets. A limited displacement, force sensing control stick is provided on the right hand console, with a suitable armrest, to provide precise control inputs during combat manoeuvres. The F-16B has two cockpits in tandem, equipped with all controls, displays, instruments, avionics and life support systems required to perform both training and combat missions. The layout of the F-16B second station is essentially the same as that of the F-16A, and is fully systems-operational. A single-enclosure polycarbonate transparency, made in two pieces and spliced aft of the forward seat with a metal bow frame and lateral support member, provides outstanding view from both cockpits.

SYSTEMS: Hamilton Standard regenerative bootstrap air cycle environmental control system, using engine bleed air, for pressurisation and cooling. Two separate and independent hydraulic system supply power for operation of the primary flight control surfaces and the utility functions. Electrical system powered by engine-driven Westinghouse 40 kVA and Lear Sielger 5 kVA generators and ground control units, with Sundstrand constant speed drive. Four dedicated, sealed cell batteries provide transient electrical power protection for the fly-by-wire flight control system. Application of the control configured vehicle (CCV) principle of relaxed static stability produces a significant reduction in trim drag, especially at high load factors and supersonic speeds. The aircraft centre of gravity is allowed to move aft, reducing both the tail drag and the change in drag on the wing due to changes in lift required to balance the download on the tail. Relaxed static stability imposes a requirement for a highly reliable, full-time operating, stability augmentation system, including reliable electronic, electrical and hydraulic provisions. The signal paths in the quad-redundant system are used to control the aircraft, replacing the usual mechanical linkages. Direct electrical control is employed from pilot controls to surface actuators. An onboard Sundstrand/Solar jet fuel starter is provided for engine self-start capability. Hamilton Standard turbine compressor and Sundstrand accessory drive gearbox. Simmonds fuel measuring system. Garrett emergency power unit automatically drives a standby generator and pump to provide uninterrupted electrical and hydraulic power for control in the event of the engine or primary power supply becoming inoperative.

AVIONICS AND EQUIPMENT: Westinghouse APG-66 pulse Doppler range and angle track radar, with planar array in nose (APG-68 in F-16C/D). The radar has a look-down range, in ground clutter, of 20 to 30 n miles (37 to 56 km; 23 to 35 miles), and lookup range of 25 to 40 n miles (46 to 74 km; 29 to 46 miles). Forward avionics bay immediatly forward of cockpit, contains radar, air data equipment, inertial navigation system, flight control computer, and space and structural provisions for radar altimeter. Rear avionics bay contains Collins AN/ARN-108 ILS, Tacan and IFF, with space for future equipment. A Dalmo Victor ALR-69 radar warning system is installed. Communications equipment includes Magnavox AN/ARC-164 UHF transceiver; provisions for a Magnavox KY-58 secure voice system; Collins AN/ARC-168 VHF AM/FM transceiver; government furnished AN/AIC-18/25 intercom; and Novatronics interferance blanker. Sperry Flight Systems central air data computer. Singer-Kearfott modified SKN-2400 inertial navigation system (replaced by Litton LN-39 INU on F-16C/D); Collins AN/ARN-108 ILS; Collins AN/ARN-118 Tacan; Teledyne Electronics AN/APX-101 air-to-ground IFF transponder with a government furnished IFF control; government furnished National Security Agency KIT-1A/TSEC cryptographic equipment; Lear Siegler stick force sensors; GEC Avionics electronic head-up display; horizontal situation indicator; Teledyne Avionics angle of attack transmitter; Gull Airborne angle of attack indicator; Clifton Precision attitude director indicator; Delco fire control computer; Photo-Sonics gun camera; Kaiser radar electro-optical display. Landing/taxying light on each main landing gear strut. Essential structure and wiring provision are built into the airframe to allow for easy incorporation of an advanced programmable signal processor radar, and future avionics systems under development for the F-16 by the US Air Force.

ARMAMENT: General Electric M61 A1 20 mm multibarrel cannon in the port side wing/body fairing, equipped with a General Electric ammunition handing system and a 'snap-shoot' gunsight (part of the head-up display system) and 515 rounds of ammunition. There is a mounting for an air-to-air missile at each wingtip, one underfuselage centreline hardpoint, and six underwing hardpoints for a additional stores. For manoeuvring flight at 5.5 *g* the underfuselage station is stressed for a load of up to 1,000 kg (2,200 lb), the two inboard underwing stations for 2,041 kg (4,500 lb) each, the two centre underwing stations for 1,587 kg (3,500 lb) each, the two outboard underwing stations for 318 kg (700 lb) each, and the two wingtip stations for 193 kg (425 lb) each. For manoeuvring flight at 9 *g* the underfuselage station is stressed for a load of up to 544 kg (1,200 lb), the two inboard underwing stations for 1,134 kg (2,500 lb) each, the two-centre underwing stations for 907 kg (2,000 lb) each, the two outboard underwing stations for 204 kg (450 lb) each, and the two wingtip stations for 193 kg (425 lb) each. There are mounting provisions on each side of the inlet shoulder for the specific carriage of sensor pods (electro-optical, FLIR, and so on); each of these stations is stressed for 408 kg (900 lb) at 5.5 *g*, and 250 kg (550 lb) at 9 *g*. Typical stores loads can include two wingtip mounted AIM-9J/L Sidewinders, with up to four more on the outer underwing stations; Sargent-Fletcher 1,400 litre (370 US gallon; 308 Imp gallon) drop tanks on the inboard underwing stations; a 1,136 litre (300 US gallon; 250 Imp gallon) drop tank or a 2,200 lb bomb on the underfuselage station; a Martin Marietta Pave Penny laser tracker pod along the starboard side of the nacelle; and single or cluster bombs, air-to-surface missiles, or flare pods, on the four inner underwing stations. Stores can be launched from Aircraft Hydro-Forming MAU-12C/A bomb ejector racks. Hughes LAU-88 launchers, or Organ triple or multiple ejector racks. Westinghouse AN/ALQ-119 and AN/ALQ-131 ECM (jammer) pods can be carried on the centreline and two underwing stations. Modified Tracor ALE-40 internal pyrotechnic/chaff dispensers have been specified. Current capabilities include air-to-air combat with gun and Sidewinder missiles; and air-to-ground attack with gun, rockets, conventional bombs, special weapons, laser guided and electro-optical weapons. Specific structure and wiring provisions, and system architecture, are built in to ensure the capability to accept future sensor and weapon systems, including electro-optical and FLIR pods, and advanced beyond-visual-range missiles. Weapons already launched successfully from F-16s, in addition to Sidewinders and AMRAAM, include radar guided Sparrow and Sky Flash air-to-air missiles, and TV guided Maverick air-to-surface missiles.

DIMENSIONS, EXTERNAL:

Wing span over missile launchers	9.45 m (31 ft 0 in)
Wing span over missiles	10.01 m (32 ft 10 in)
Wing aspect ratio	3.0
Length overall	15.09 m (49 ft 5.9 m)
Height overall	5.09 m (16 ft 8½ in)
Tailplane span	5.64 m (18 ft 6 in)
Wheel track	2.36 m (7 ft 9 in)
Wheelbase	4.00 m (13 ft 1.44 in)

AREAS:

Wings, gross	27.87 m² (300.0 sq ft)
Flaperons (total)	5.82 m² (62.64 sq ft)
Leading-edge flaps (total)	6.82 m² (73.42 sq ft)

Lockheed Martin Fort Worth (General Dynamics) first MLU F-16A/B trial verification aircraft

1995

Vertical tail surfaces (total)	5.09 m² (54.75 sq ft)
Rudder	1.08 m² (11.65 sq ft)
Horizontal tail surfaces (total):	
Pre-Block 15 aircraft	4.55 m² (49.0 sq ft)
Block 15 and subsequent aircraft	5.92 m² (63.70 sq ft)

WEIGHTS AND LOADINGS:

Weight empty: F-16A	7,070 kg (15,586 lb)
F-16B	7,374 kg (16,258 lb)
Internal fuel load: F-16A	3,162 kg (6,972 lb)
F-16B	2,624 kg (5,785 lb)
Max external load	5,443 kg (12,000 lb)
Structural design gross weight (9 *g*) with full internal fuel	10,205 kg (22,500 lb)
Max symmetrical design load factor with full internal fuel at 10.205 kg (22,500 lb) gross weight: F-16A/B	9.0
Max T-O weight: F-16A, air-to-air, no external tanks	10,800 kg (23,810 lb)
F-16B, air-to-air, no external tanks	10,568 kg (23,298 lb)
F-16A/B with external load	16,057 kg (35,400 lb)
Wing loading: at 9,979 kg (22,000 lb) AUW	356 kg/m² (73 lb/sq ft)
at 14,968 kg (33,000 lb) AUW	537 kg/m² (110 lb/sq ft)
Thrust/weight ratio ('clean')	1.1 to 1

PERFORMANCE: (F-16A):

Max level speed at 12,200 m (40,000 ft)	above Mach 2.0
Service ceiling	more than 15,240 m (50,000 ft)
Radius of action	more than 500 n miles (925 km; 575 miles)
Ferry range, with drop tanks	more than 2,100 n miles (3890 km; 2,415 miles)

UPDATED

MARSH

MARSH AVIATION COMPANY

5060 East Falcon Drive, Mesa, Arizona 85215-2590
Telephone: 1 (602) 832 3770
Fax: 1 (602) 985 2840
Telex: 165028

PRESIDENT: Floyd D. Stilwell
EXECUTIVE VICE-PRESIDENT: Bill Walker Jr

Rockwell Thrush Commander and Schweizer Super Ag-Cat conversions abandoned because of US product liability laws, but would be supplied to foreign governments if ordered.

VERIFIED

MARSH S-2F1T TURBO TRACKER

TYPE: Grumman S-2 re-engined with Garrett TPE331-14GR turboprops.

PROGRAMME: First flight of Marsh S-2 conversion (N426DF) 24 November 1986; certificated 19 February 1990.

CUSTOMERS: Evaluated by California Forestry Department; nine Turbo S-2s in conversion.

DESIGN FEATURES: Basic conversion includes: re-engine with Garrett TPE331-14GR turboprops, Hartzell (HC-B5MP-5X1) five-blade reversing propellers of 2.92 m (9 ft 7 in). Firefighting retardant tank holds 3,785 litres (1,000 US gallons; 833 Imp gallons).

POWER PLANT: Two 1,227 kW (1,645 shp) Garrett TPE331-14GR engines.

ACCOMMODATION: Two seats for minimum crew of one.

DIMENSIONS, EXTERNAL:

Wing span	21.23 m (69 ft 8 in)
Length overall	13.05 m (42 ft 10 in)
Height overall	4.97 m (16 ft 3½ in)
Wheel track	5.63 m (18 ft 6 in)
Wheelbase	4.41 m (14 ft 6 in)

WEIGHTS AND LOADINGS:

Weight empty	5,826 kg (12,840 lb)
Max fuel weight	1,624 kg (3,580 lb)
Max T-O weight	11,340 kg (25,000 lb)
Max landing weight	11,113 kg (24,500 lb)
Max power loading (TPE331-14s)	6.08 kg/kW (10.00 lb/shp)

PERFORMANCE:

Cruising speed:	
at 3,050 m (10,000 ft)	230 kts (426 km/h; 265 mph)
at 7,925 m (26,000 ft)	270 kts (500 km/h; 311 mph)
Retardant drop speed	130 kts (241 km/h; 150 mph)
Max rate of climb at S/L	518 m (1,700 ft)/min
Max range at cruising speed of 210 kts (389 km/h; 242 mph) at 7,620 m (25,000 ft), no reserves	1,468 n miles (2,720 km; 1,690 miles)

VERIFIED

Marsh S-2F1T Turbo Tracker firefighting conversion *1992*

MARSH S-2F3T TURBO TRACKER

TYPE: Grumman S-2 re-engined with Garrett TPE331-15AW turboprops.

PROGRAMME: **TS-2F3 Turbo Tracker:** ASW and maritime patrol version; first flight 26 July 1991; flight testing included carrier FCLP qualification achieved 14 July 1992.

CUSTOMERS: Six military ASW versions delivered to Armada Argentina in 1993.

DESIGN FEATURES: Equipment for military maritime patrol and ASW includes search radar, FLIR, MAD, sonobuoy launchers, pictorial navigation avionics with automatic search and attack modes, six underwing stores points and torpedo bay. Basic conversion includes Hartzell five-blade reversing propellers of 2.92 m (9 ft 7 in) diameter, composite cowlings, maximum usable fuel 1,960 litres (518 US gallons; 431 Imp gallons), modern instrumentation, optional EFIS cockpit, multiple bus electrical with two batteries. Nacelles and fuselage streamlined to achieve 60 knot (111 km/h; 69 mph) increase in cruising speed with 50 per cent lower fuel consumption; take-off and landing runs decreased by 25 per cent and single-engined rate of climb at gross weight increased by 230 m (750 ft)/min.

POWER PLANT: Two Garrett TPE331-15AW turboprop engines, tank holds 3,785 litres (1,000 US gallons; 83 Imp gallons).

ACCOMMODATION: Crew of four.

SYSTEMS: Systron-Donner fire detection and halon extinguishers, optional environmental control system.

ARMAMENT: Six underwing pylons, one torpedo bay, 32 sonobuoy ports and two PDC charge racks.

DIMENSIONS, EXTERNAL:

Wing span	22.12 m (72 ft 7 in)
folded	8.33 m (27 ft 4 in)
folded to prop arc	8.68 m (28 ft 6 in)
Length overall	13.26 m (43 ft 6 in)
Height: overall	5.06 m (16 ft 7½ in)
during wing fold ops	10.08 m (33 ft 1 in)
Tailplane span	8.16 m (26 ft 9½ in)
Wheel track	5.63 m (18 ft 6 in)
Wheelbase	4.41 m (14 ft 6 in)

WEIGHTS AND LOADINGS:

Weight empty	6,282 kg (13,840 lb)
Max T-O weight	13,163 kg (29,000 lb)
Max payload	6,881 kg (15,160 lb)
Wing loading (max) T-O	286.6 kg/m² (58.46 lb/sq ft)

PERFORMANCE:

Max speed at S/L	260 kts (481 km/h; 301 mph)
Max speed at altitude	296 kts (548 km/h; 342 mph)
Ceiling, certified	7,315 m (24,000 ft)
Range, no reserve	647 n miles (1,200 km; 745 miles)

VERIFIED

Marsh S-2F3T Turbo Tracker ASW and maritime patrol aircraft *1994*

MCDONNELL DOUGLAS

MCDONNELL DOUGLAS CORPORATION

PO Box 516, St Louis, Missouri 63166-0516
Telephone: 1 (314) 232 0232
Fax: 1 (314) 234 3826

Corporate Office

CHAIRMAN AND CEO: John F. McDonnell
EXECUTIVE VICE-PRESIDENT: John Capellupo
SENIOR VICE-PRESIDENT, NEW AIRCRAFT AND MISSILE PROGRAMS (NAMP): James M. Sinnett
SENIOR VICE-PRESIDENT, C-17 PROGRAM: Donald Kozlowski
VICE-PRESIDENT AND GENERAL MANAGER, NAMP:
David O. Swain

Formed 28 April 1967 by merger of Douglas Aircraft Company Inc and the McDonnell Company; encompasses their subsidiaries plus former Hughes Helicopter Company acquired in 1984 and renamed McDonnell Douglas Helicopter Company; helicopter company was for sale for a time but then incorporated in main company, now as McDonnell Douglas Helicopter Systems. Employees totalled more than 70,000 worldwide in early 1994, some 17,000 fewer than in 1992.

McDonnell Douglas was the largest US defence contractor in 1993 and, after reorganising to recover from the combined downturn in defence and civil markets, expected to return to profit in 1994. The workforce has been reduced by 20 per cent and non-core businesses sold. This and improved profits generated $1 billion cash in 1993 and reduced debt by $1 billion to $1.8 billion. The C-17 should become profitable in 1994 and the MD-11 should produce positive cash flow. Current military aircraft and missile programmes are continuing and new programmes are at hand. The company is looking for risk-sharing partners for new civil programmes in order to extend its family of airliners and its market share.

Major operating units of McDonnell Douglas Corporation (MDC) Aerospace Group were reorganised in August 1992 and now are:

McDonnell Douglas Aerospace (MDA)

Follows this entry

MDA administratively divided into MDA East and MDA West and comprises the government aerospace business of McDonnell Douglas Corporation including the C-17 Globemaster III, formerly a Douglas Aircraft programme. MDA-W address is 5301 Bolsa Avenue, Huntington Beach, California 92647. *Telephone:* 1 (714) 896 1300 *Fax:* 1 (714) 896 1308

Douglas Aircraft Company (DAC)

Self-contained civil airliner manufacturer. Follows MDA entry

McDonnell Douglas Helicopter Systems

Helicopter Division of McDonnell Douglas. Follows DAC entry

VERIFIED

MCDONNELL DOUGLAS AEROSPACE (Division of McDonnell Douglas Corporation)

PO Box 516, St Louis, Missouri 63166-0516
Telephone: 1 (314) 232 0232
Fax: 1 (314) 234 3826

EXECUTIVE VICE-PRESIDENT: John P. Capellupo
VICE-PRESIDENT/GENERAL MANAGER, F-15:
Peter J. Von Minden
VICE-PRESIDENT/GENERAL MANAGER, F/A-18: Michael M. Sears
DIRECTOR OF COMMUNICATIONS: James R. Reed

VERIFIED

MCDONNELL DOUGLAS PHANTOM II

US Navy and USAF designations: F-4 and RF-4

TYPE: Twin-engined two-seat all-weather fighter aircraft.

PROGRAMME: The Phantom II was developed initially as a twin-engined two-seat long-range all-weather attack fighter for service with the US Navy. A letter of intent to order two prototypes was issued 18 October 1954, at which time the aircraft was designated AH-1. The designation was changed to F4H-1 26 May 1955, with change of mission to missile fighter, and the prototype XF4H-1 flew for the first time 27 May 1958. The first production Phantom II was delivered to US Navy Squadron VF-101 in December 1960. Trials in a ground attack role led to USAF orders, and the basic USN and USAF versions became the F-4B and F-4C respectively.

VERSIONS: **F-4A** (formerly F4H-1F): Basic power plant comprised two General Electric J79-GE-2 turbojet engines, with afterburning. Total of 23 preproduction and 24 production aircraft built. After evaluation of this version, the USAF decided to order land-based versions of the F-4B under the designation F-4C.

F-4B (formerly F4H-1): All-weather fighter for US Navy and Marine Corps, powered by two General Electric J79-GE-8 turbojet engines. Total of 649 built. (See F-4G and F-4N.)

RF-4B (formerly F4H-1P): Multisensor reconnaissance version of the F-4B for US Marine Corps. No dual controls or armament. Reconnaissance system as for RF-4C. J79-GE-8 engines. High frequency single sideband radio. First flown 12 March 1965. Overall length increased to 19.2 m (63 ft). Total 46 built.

F-4C (formerly F-110A): Version of F-4B for USAF, with J79-GE-15 turbojet engines, cartridge starting, wider tread low-pressure tyres size 30 × 11.5, larger brakes, Litton type LN-12A/B (ASN-48) inertial navigation system, APQ-100 radar, APQ-100 PPI scope, LADD timer, Lear Siegler AJB-7 bombing system, GAM-83 controls, dual controls and boom flight refuelling instead of drogue (receptacle in top of fuselage, aft of cockpit). Folding wings and arrester gear retained. For close support and attack duties with Tactical Air Command, PACAF and USAFE, and with the Air National Guard (ANG) from January 1972. Sufficient F-4Cs were modified to equip two squadrons for a defence suppression role under the USAF's **Wild Weasel** programme. These aircraft carry ECM warning sensors, jamming pods, chaff dispensers and anti-radiation missiles. First F-4C flew 27 May 1963; 36 supplied to Spanish Air Force. The last of 583 was delivered to TAC 4 May 1966. Replaced in production by F-4D.

RF-4C (formerly RF-110A): Multisensor reconnaissance version of F-4C for USAF, with radar and photographic systems in modified nose which increases overall length by 0.84 m (2 ft 9 in). Three basic reconnaissance systems, including: side-looking radar to record high-definition radar picture of terrain on each side of flight path on film; infra-red detector to locate enemy forces under cover or at night by detecting exhaust gases and other heat sources; forward- and side-looking cameras, including panoramic models with moving lens element for horizon-to-horizon pictures. Systems are operated from rear seat. HF single sideband radio. YRF-4C flew 9 August 1963; first production RF-4C 18 May 1964. Taken into service with ANG February 1972. Production ended December 1973. Total 505 built.

McDonnell Douglas F-4E equipped with AGCW free flight test vehicle

1994

F-4D: Development of F-4C for USAF, with J79-GE-15 turbojet engines, APQ-109 fire control radar, ASG-22 servoed sight, ASQ-91 weapon release computer, ASG-22 lead computing amplifier, ASG-22 lead computing gyro, 30 kVA generators, and ASN-63 inertial navigation system. First F-4D flew 8 December 1965. Two squadrons of F-4Ds (32 aircraft) delivered to the Imperial Iranian Air Force and 18 to the Republic of Korea. Total of 843 built.

F-4E: Multirole fighter for air superiority, close support and interdiction missions with USAF. Has internally mounted M-61A1 multibarrel gun, improved (AN/APQ-120) fire control system and J79-GE-17 turbojet engines (each 79.6 kN; 17,900 lb st). Additional fuselage fuel cell. First production F-4E delivered to USAF 3 October 1967. Supplied to the Israeli Air Force, Hellenic Air Force, Turkish Air Force, Republic of Korea Air Force and Imperial Iranian Air Force. All F-4Es fitted retrospectively with leading-edge manoeuvring slats.

In early 1973 F-4Es began to be fitted with Northrop's target identification system electro-optical (TISEO). Essentially a vidicon TV camera with a zoom lens, it aids positive visual identification of airborne or ground targets at long range. The ASX-1 TISEO is mounted in a cylindrical housing on the leading-edge of the port wing of the F-4E.

F-4EJ: On 1 November 1968, the Japan Defence Agency selected the F-4E as the main fighter for the JASDF. Except for the first two, these aircraft were built in Japan under a licence agreement, with some components being supplied from St Louis. The first US built F-4EJ flew 14 January 1971. Equipment includes tail warning radar and launchers for Mitsubishi AAM-2 air-to-air missiles. Total 158 built.

RF-4E: Multisensor reconnaissance version developed for the Federal Republic of Germany (which ordered 88), Iran, Israel, and Japan. Generally similar to the RF-4C, it differs by having the J79-GE-17 turbojets of the F-4E and changed reconnaissance equipment.

F-4F: Two-seat fighter with leading-edge slats to improve manoeuvrability and modified electronics. The Federal Republic of Germany ordered 175 for the Luftwaffe. First one rolled out 24 May 1973; last was delivered July 1976.

F-4G: Development of F-4B for US Navy, with AN/ASW-21 datalink communications equipment, first flown 20 March 1963. In service over Vietnam with Squadron VF-213 from USS *Kitty Hawk* Spring 1966. Only 12 built included in the total number of F-4B built. No longer in service.

F-4G (Wild Weasel): The USAF's Wild Weasel programme concerns primarily suppression of hostile weapon radar guidance systems. The provision of airborne equipment able to fulfil such a role, and modification of the necessary aircraft to create an effective force for deployment against such targets, had first priority in tactical air force planning Spring 1975. The requirement for such a weapon system had been appreciated by Tactical Air Command as early as 1968, and feasibility studies were initiated September that year, following which eight sets of equipment were acquired for development, qualification testing and flight testing in two F-4D aircraft. The F-4E however, was eventually chosen as the best aircraft on which to install the required equipment. Modifications include the addition of a torpedo-shaped fairing to the top of the tailfin to carry APR-38 antennae, with other APR-38 antennae installed on the side of the fin and along the upper surface of the fuselage. Other modifications include changes to the LCOSS amplifier in the upper equipment bay, APR-38 CIS installation in the aft cockpit, APR-38 CIS installation in the forward cockpit, removal of the M-61A1 gun system to allow sufficient room for installation of APR-38 subsystems (receiver, HAWC, CIS), and the provision of suitable cockpit displays. The changes give the F-4G Wild Weasel the capability to detect, identify and locate hostile electromagnetic emitters, and to deploy against them suitable weapons for their suppression or destruction. Such aircraft can operate independently in a hunter-killer role.

The USAF sought funding in FY76 for the advanced Wild Weasel concept for provision of an expanded memory of the airborne processor and extended low-frequency emission coverage. The programme provided for the first F-4G operational kit installation Spring 1976 and the second Autumn that year, followed by 15 installations in 1977, 60 in 1978 and 39 in 1979, to provide a force of 116 aircraft. The USAF still has 100 F-4Gs in service.

F-4J: Development of the F-4B for US Navy and Marine Corps, primarily as interceptor but with full ground attack capability. J79-GE-10 turbojets. Use of 16.5° drooping ailerons and slotted tail gives reduced approach speed in spite of increased landing weight. Westinghouse AN/AWG-10 pulse Doppler fire control system. Lear Siegler AJB-7 bombing system; 30 kVA generators. First F-4J demonstrated publicly on 27 May 1966. Production of 518 completed in December 1972.

F-4K: Development of the F-4B for the Royal Navy, with improvements evolved for the F-4J plus other changes. Westinghouse AN/AWG-10 pulse Doppler fire control radar system modified to allow the antenna to swing around with the radome. This 'foldable radome' reduces the length of the aircraft, making it compatible with the deck elevators on British aircraft carriers. Two Rolls-Royce Spey RB168-25R Mk 201 turbofans (each rated at 55.6 kN; 12,500 lb st dry) with 70 per cent afterburning. Air intake ducts 0.15 m (6 in) wider than on US models to cater for more powerful engines. Drooped ailerons. Tailplane has leading-edge fixed slot. Strengthened main landing gear. Nose landing gear strut extends to 1.02 m (40 in), compared to 0.51 m (20 in) on the F-4J, to permit optimum incidence catapulting. Martin-Baker ejection seats. Weapons include Sparrow air-to-air missiles. Initial contracts for two YF-4Ks; ordered as Phantom FG. Mk 1. First flight 27 June 1966. First operational Phantom unit, 892 Squadron, commissioned at RNAS Yeovilton 31 March 1969. Total of 52 built. Withdrawn from service with the Royal Navy.

F-4M: For Royal Air Force. Generally similar to F-4K, but with larger brakes and low-pressure tyres of F-4C, and no tailplane leading-edge slot. Folding wings and arrester gear retained. Up to 50 per cent of the components manufactured in the UK. First F-4M flew 17 February 1967. Deliveries began 23 August 1968. RAF designation is

McDonnell Douglas F-4E Phantom II of the Greek Air Force *(Paul Tompkins)*

1994

Phantom FGR. Mk 2. Total of 118 built some delivered with dual controls for use as conversion trainers. Withdrawn from service with the RAF.

F-4N: The US Navy updated version of the F-4Bs. No longer in service.

F-4S: US Navy upgrade of the F-4Js. No longer in service.

F-4CCV: Experimental Phantom for study of control configured vehicle (CCV) techniques.

UPGRADES: **Deutsche Aerospace:** F-4 ICE programme. See separate entry.

Greece: Issued an interim RFP for structural and avionics upgrade of its fleet of F-4E and RF-4E aircraft. Final RFP document is imminent. Programme to cover new radar (reportedly Hughes APG-65 or Westinghouse APG-68), F-16 controls and displays, new BVR missile. Manufacturers known to be bidding include: Daimler-Benz Aerospace, Israeli Aircraft Industries and Rockwell's North American Aircraft Modification Division.

Israeli Aircraft Industries (IAI): Selected by Turkish Air Force Command to upgrade its fleet of F-4E Phantom II aircraft to Phantom 2000 configuration. If signed the deal will be worth US$800 million. However the Turkish authorities have delayed negotiations in an effort to secure a workshare and technology transfer agreement.

Israeli Aircraft Industries: Bidding for contract to upgrade South Korean F-4D/E and RF-4C Phantom II aircraft.

McDonnell Douglas: F-4G Wild Weasel. See Versions.

Mitsubishi Heavy Industries: F-4EJKai and RF-4EJ. See separate entry.

OPERATORS: Versions of the F-4 Phantom II are in service with the armed forces of the following countries: Egypt (24), Germany (152), Greece (101), Iran (46), Israel (140), Japan (110), South Korea (123), Spain (4), Turkey (174) and USA (270).

The following description applies to the F-4E:

DESIGN FEATURES: Cantilever low-wing monoplane. Wing section NACA 0006.4-64 (mod) at root, NACA 0004-64 (mod) at wing fold line, NACA 0003-64 (mod) at tip. Average thickness/chord ratio 5.1 per cent. Incidence 1°. Dihedral, inner panels 0°, outer panels 12°. Sweepback 45° on leading-edges. Outer panels have extended chord and dihedral of 12°.

FLYING CONTROLS: Trailing-edge is a one-piece aluminium honeycomb structure. Flaps and ailerons of all-metal construction, with aluminium honeycomb trailing-edges. Inset ailerons limited to down movement only, the 'up' function being supplied by hydraulically operated spoilers on upper surface of each wing. Ailerons and spoilers fully powered by two independent hydraulic systems. Hydraulically operated trailing-edge manoeuvring slats. Hydraulically operated airbrake under each wing aft of wheel well. Outer panels fold upward for stowage. Rudder interconnected with ailerons at low speeds.

STRUCTURE: Centre-section and centre wings form one-piece structure from wing fold to wing fold. Portion that passes through fuselage comprises a torsion-box between the front and main spars (at 15 per cent and 40 per cent chord) and is sealed to form two integral fuel tanks. Spars are machined from large forgings. Centre wings also have forged rear spar. Centreline rib, wing-fold ribs, two intermediate ribs forward of main spar and two aft of main spar are also made from forgings. Wing skins machined from aluminium panels 0.635 m (2½ in) thick, with integral stiffening. The fuselage is an all-metal semi-monocoque structure built in forward fuselage fabricated in port and starboard halves, so that most internal wiring and finishing can be done before assembly. Keel and rear sections make use of steel and titanium. Double-wall construction under fuel tanks and for lower section of rear fuselage, with ram-air cooling. The tail unit is a cantilever all-metal structure, with 23° of anhedral on one-piece all-moving tailplane which has slotted leading-edges. Ribs and stringers of tailplane are of steel, skin titanium and trailing-edge of steel honeycomb.

LANDING GEAR: Hydraulically retractable tricycle type, mainwheels retracting inward into wings, nose unit rearward. Single wheel on each main unit, with tyres size 30 × 11.5 Type VIII; twin-wheels on nose unit, which is steerable and self-centring and can be lengthened pneumatically to increase the aircraft's angle of attack for take-off. Brake-chute housed in fuselage tailcone. Mk II anti-skid system.

POWER PLANT: Two General Electric J79-GE-17A turbojet engines (each rated 79.6 kN; 17,900 lb st with afterburning). Variable-area inlet ducts monitored by air data computer. Integral fuel tankage in wings, between front and main spars, and in seven fuselage tanks, with total capacity of 7,022 litres (1,855 US gallons; 1,545 Imp gallons). Provision for one 2,270 litre (600 US gallon; 500 Imp gallon) external tank under fuselage and two 1,400 litre (370 US gallon; 308 Imp gallon) underwing tanks. Equipment for probe-and-drogue and 'buddy tank' flight refuelling, with retractable probe in starboard side of fuselage. Oil capacity 39 litres (10.3 US gallons; 8.6 Imp gallons).

ACCOMMODATION: Crew of two in tandem on Martin-Baker Mk H7 ejection seats, under individual rearward-hinged canopies. Optional dual controls.

SYSTEMS: Three independent hydraulic systems, each of 207 bars (3,000 lb/sq in). Pneumatic system for canopy operation, nosewheel strut extension and ram-air turbine extension. Primary electrical source is AC generator. No battery.

ELECTRONICS AND EQUIPMENT: CPK-92A/A24G-34 central air data computer; AN/ASQ-19(B) com-nav-ident; MS25447/MS25448 counting accelerometer; AN/APN-155 radar altimeter; AN/AJB-7 all-altitude bomb system; AN/ASN-46A navigational computer; AN/ASN-63 INS; AN/ASQ-91 (MOD) weapons release system, AN/ASG-26 (MOD) lead computing optical sight; AN/APR-36, -37 RHAWS; AN/ASA-32 AFCS; AN/APQ-120 fire control system radar AN/ARW-77 AGM-12 control system; TD-709/AJB-7 sequential timer; ID-1755A standby attitude reference system; and KB-25A gunsight camera.

ARMAMENT: Four Falcon, Sparrow, Sidewinder, Shrike or Walleye missiles, or two Bullpup missiles, on four semi-submerged mountings under fuselage and four underwing mountings. Provision for carrying alternative loads of up to 7,250 kg (16,000 lb) of nuclear or conventional bombs and stores on seven attachments under wings and fuselage. Stores which can be carried include B-28, -43, -57, -61 nuclear bombs; M117, M118, M129, MC-1, Mk 36, Mk 81, Mk 82, Mk 83 and Mk 84 bombs; MLU-10 land mine; BLU-1, -27, -52 and -76 fire bombs; cluster bombs; practice bombs; flares; rocket packs; ECM pods; gun pods; spray tanks; tow targets' Pave Knife pod; and AAVSIV camera pod. One M61A-1 nose-mounted gun.

DIMENSIONS, EXTERNAL:

Wing span	11.77 m (38 ft 7½ in)
Wing mean aerodynamic chord	4.89 m (16 ft 0½ in)
Wing aspect ratio	2.82
Width, wings folded	8.41 m (27 ft 7 in)
Length overall	19.20 m (63 ft 0 in)
Height overall	5.02 m (16 ft 5½ in)
Wheel track	5.45 m (17 ft 10½ in)

AREAS:

Wings, gross	49.2 m² (530 sq ft)

WEIGHTS AND LOADINGS:

Weight empty	13,757 kg (30,328 lb)
Weight empty, basic mission	14,448 kg (31,853 lb)
Combat T-O weight	18,818 kg (41,487 lb)
Design T-O weight	26,308 kg (58,000 lb)
Max T-O weight	28,030 kg (61,795 lb)
Max landing weight	20,865 kg (46,000 lb)
Max wing loading	569.2 kg/m² (116.59 lb/sq ft)
Max power loading	176.1 kg/kN (1.73 lb/lb st)

PERFORMANCE (A: at 24,410 kg; 53,814 lb, B: at 24,572 kg; 54,171 lb, C: at 25,397 kg; 55,991 lb, D: at 27,954 kg; 61,629 lb, E: at 28,030 kg; 61,795 lb T-O weight):

Max level speed with external stores	over Mach 2
Average speed: A, B	504 kts (934 km/h; 580 mph)
C	506 kts (938 km/h; 583 mph)
D	502 kts (930 km/h; 578 mph)
E	496 kts (919 km/h; 571 mph)
Stalling speed, approach power with BLC:	
A	148 kts (273.5 km/h; 170 mph)
B	148.5 kts (275 km/h; 171 mph)
C	151 kts (280 km/h; 174 mph)
D	158.4 kts (294 km/h; 182.5 mph)
E	158.6 kts (294.5 km/h; 183 mph)
Max rate of climb at S/L: A	2,847 m (9,340 ft)/min
B	2,816 m (9,240 ft)/min
C	2,621 m (8,600 ft)/min
D	2,003 m (6,570 ft)/min
E	1,881 m (6,170 ft)/min
Rate of climb at S/L, one engine out:	
A	1,731 m (5,680 ft)/min
B	1,713 m (5,620 ft)/min
C	1,591 m (5,220 ft)/min
D	1,167 m (3,830 ft)/min
E	1,067 m (3,500 ft)/min
Service ceiling: A	10,975 m (36,000 ft)
B	10,925 m (35,850 ft)
C	10,620 m (34,850 ft)
D	9,070 m (29,750 ft)
E	8,565 m (28,100 ft)
Service ceiling, one engine out: A	9,905 m (32,500 ft)
B	9,860 m (32,350 ft)
C	9,340 m (30,650 ft)
D	7,055 m (23,150 ft)
E	6,490 m (21,300 ft)
T-O run: A	969 m (3,180 ft)
B	985 m (3,230 ft)
C	1,064 m (3,490 ft)
D	1,329 m (4,360 ft)
E	1,338 m (4,390 ft)
T-O to 15 m (50 ft): A	1,369 m (4,490 ft)
B	1,384 m (4,540 ft)
C	1,478 m (4,850 ft)
D	1,780 m (5,840 ft)
E	1,792 m (5,880 ft)

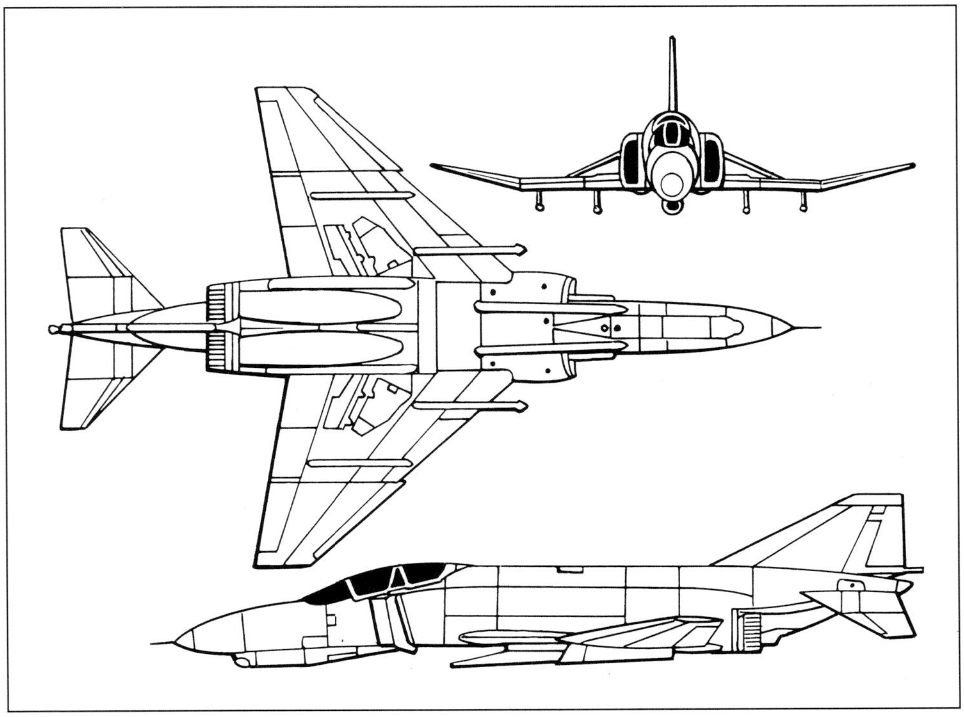

McDonnell Douglas F-4 Phantom II all-weather combat aircraft *(Jane's/Mike Keep)*

1994

Landing run (A: at landing weight of 16,706 kg; 36,831 lb, B and C: at 15,937 kg; 35,134 lb, D: at 17,155 kg; 37,821 lb, E: at 17,211 kg; 37,944 lb):

A	1,122 m (3,680 ft)
B, C	1,073 m (3,520 ft)
D	1,146 m (3,760 ft)
E	1,152 m (3,780 ft)
Landing run, with parabrake: A	927 m (3,040 ft)
B, C	887 m (2,910 ft)
D	948 m (3,110 ft)
E	951 m (3,120 ft)
Landing run from 15 m (50 ft): A	1,704 m (5,590 ft)
B, C	1,655 m (5,430 ft)
D	1,728 m (5,670 ft)
E	1,734 m (5,690 ft)
Landing from 15 m (50 ft), with parabrake:	
A	1,509 m (4,950 ft)
B, C	1,469 m (4,820 ft)
D	1,530 m (5,020 ft)
E	1,533 m (5,030 ft)
Combat radius:	
Area intercept	683 n miles (1,266 km; 786 miles)
Defensive counter-air	429 n miles (795 km; 494 miles)
Interdiction	618 n miles (1,145 km; 712 miles)
Ferry range	1,718 n miles (3,184 km; 1,978 miles)

UPDATED

McDonnell Douglas F-15C Eagle single-seat air superiority fighter ***(Paul Tompkins)***

1995

McDonnell Douglas F-15A Eagle of the US Air Force

1994

MCDONNELL DOUGLAS F-15 EAGLE

Israel Defence Force name: Baz (Falcon)

TYPE: Twin-turbofan air superiority fighter with secondary attack role.

PROGRAMME: First flight of YF-15 27 July 1972; first F-15C (78-468) 26 February 1979; first F-15D 19 June 1979; P&W F100-PW-220 standard since 1985; last of 894 F-15A/B/C/Ds delivered 3 November 1989; production restarted during 1991 to produce five for Israel and 12 for Saudi Arabia; production now concentrated on F-15E.

VERSIONS: **F-15A:** Initial single-seat version.

F-15B: Initial two-seat operational training version; first flight 7 July 1973.

F-15C: Became standard single-seat production version from June 1979.

F-15D: Standard two-seat production version from June 1979.

F-15E: Modified F-15B Strike Eagle. See *Jane's All the World's Aircraft 1994-95.*

F-15J and DJ: Single- and two-seat versions built by Mitsubishi in Japan for JASDF.

MSIP: Multistaged improvement programme, first funded 1983; testing began December 1984; first production MSIP F-15C unveiled 20 June 1985; all other F-15Cs retrofitted. MSIP included AN/APG-70 radar with memory increased to 1,000 kbits and processing speed trebled; central computer capacity multiplied by four and processing speed three times as fast; original armament control system panel replaced by single Honeywell colour video screen which, linked to computer, allows for advanced versions of AIM-7, AIM-9 and AIM-20 AMRAAM. Other improvements include Northrop enhanced AN/ALQ-135 internal countermeasures set, Loral AN/ALR-56C radar warning receiver, Tracor AN/ALE-45 chaff/flare dispenser, and Magnavox electronic warfare system. JTIDS Class 2 terminals to be installed from 1992.

Conformal fuel tanks: All F-15Cs fitted to carry two 2,839 litre (750 US gallon; 624 Imp gallon) conformal fuel tanks (CFT) attached to fuselage sides aft of engine intakes; same load factors as main airframe and removable in 15 minutes; CFT can contain reconnaissance sensors, radar detection and jamming equipment, laser designator and low-light TV as well as fuel; tangential carriage system includes rows of stub pylons carrying up to twelve 1,000 lb, four 2,000 lb class weapons or AIM-7F Sparrows.

Reconnaissance pod: Centreline conformal reconnaissance pod developed as private venture by McDonnell Douglas, tested during Summer 1987; can transmit imagery via datalink to ground stations in near-real-time.

Follow-on Wild Weasel: Proposed replacement for USAF F-4G Wild Weasel.

UPGRADES: **McDonnell Douglas:** F-15 MSIP. See Versions.

McDonnell Douglas/Hughes: Contract awarded October 1994 by US Air Force to upgrade the radar on more than 350 F-15C/D Eagle fighter aircraft. Contract value US$189.8 million. McDonnell Douglas to lead the systems integration, aircraft modification, testing and training. Hughes Aircraft Company's Radar Systems Group will design and develop the upgraded radar system. Flight testing is scheduled to begin in August 1997. Retrofit into fleet set to commence in 1999 at rate of 72 aircraft per year. To reduce costs the radar upgrade programme will use components from the APG-73 radar.

OPERATORS: Versions of the F-15 (not including F-15E/I/S) are in service with the armed forces of the following countries: Israel (67), Japan (180), Saudi Arabia (92) and USA (614).

CUSTOMERS: Including prototypes, 1,017 F-15A/Bs and C/Ds produced up to November 1989, when production temporarily halted; 366 F-15As, 58 F-15Bs, 409 F-15Cs and 61 F-15Ds for US Air Force; 19 F-15As (plus four ex-USAF), two F-15Bs, 24 F-15Cs and two F-16Ds for Israel Defence Force; 47 F-15Cs and 15 F-15Ds for Royal Saudi Air Force; and two F-15Js and 12 F-15DJs for Japan ASDF. Further 17 built in 1991-92, of which nine F-15Cs and three F-15Ds for Saudi Arabia, beginning June 1991 (in addition, USAF transferred 20 F-15Cs and four F-15Ds as emergency aid August 1990); Israeli Air Force operates seven F-15Ds. Total production 1,034 (not including Japanese manufacture), plus F-15Es; including FY91 contracts, 191 JASDF F-15s ordered from Mitsubishi (which see) and McDonnell Douglas (previously noted).

DESIGN FEATURES: NACA 64A aerofoil section with conical camber on leading-edge; sweepback 38° 42′ at quarter-chord; thickness/chord ratio 6.6 per cent at root, 3 per cent at tip; anhedral 1°; incidence 0°. Twin fins positioned to receive vortex off wing and maintain directional stability at high angles of attack. Straight two-dimensional external compression engine air inlet each side of fuselage. Air inlet controllers by Hamilton Standard. Air inlet actuators by National Water Lift.

FLYING CONTROLS: Plain ailerons and all-moving tailplane with dog-tooth extensions, both powered by National Water Lift hydraulic actuators; rudders have Ronson Hydraulic Units actuators; no spoilers or trim tabs; Moog boost and pitch compensator for control column; plain flaps; upward-opening airbrake panel mounted on fuselage between fins and cockpit.

STRUCTURE: Wing based on torque box with integrally machine skins and ribs of light alloy and titanium; aluminium honeycomb wingtips, flaps and ailerons; airbrake panel of titanium, aluminium honeycomb and graphite/epoxy composites skin.

LANDING GEAR: Hydraulically retractable tricycle type, with single wheel on each unit. All units retract forward. Cleveland nose and main units, each incorporating an oleo-pneumatic shock-absorber. Nosewheel and tyre by Goodyear, size 22 × 6.6-10, pressure 17.93 bars (260 lb/sq in). Mainwheels by Bendix, with Goodyear tyres size 34.5 × 9.75-18, pressure 23.44 bars (340 lb/sq in). Bendix carbon heat-sink brakes. Hydro-Aire wheel braking skid control system.

POWER PLANT: Two Pratt & Whitney F100-PW-220 turbofans, each rated at 105.7 kN (23,770 lb st) with afterburning for take-off. Internal fuel in eight Goodyear fuselage tanks, total capacity 7,836 litres (2,070 US gallons; 1,724 Imp gallons). Simmonds fuel gauge system. Optional conformal fuel tanks attached to side of engine air intakes, beneath wing, each containing 2,839 litres (750 US gallons; 624 Imp gallons). Provision for up to three additional 2,309 litre (610 US gallon; 508 Imp gallon) external fuel tanks. Maximum total internal and external fuel capacity 20,441 litres (5,400 US gallons; 4,496 Imp gallons).

ACCOMMODATION: Pilot only, on McDonnell Douglas ACES II ejection seat. Stretched acrylic canopy and windscreen. Windscreen anti-icing valve by Dynasciences Corporation.

SYSTEMS: AiResearch air conditioning system. Three independent hydraulic systems (each 207 bars; 3,000 lb/sq in) powered by Abex engine-driven pumps; modular hydraulic packages by Hydraulic Research and Manufacturing Company. Smiths Industries generating system for electrical power, with Sundstrand 40/50 kVA generator constant-speed drive units and Electro Development Corporation transformer-rectifiers. The oxygen system includes a Simmonds liquid oxygen indicator. Garrett APU for engine starting, and for the provision of electrical or hydraulic power on the ground independently of the main engines.

AVIONICS: General Electric automatic analogue flight control system standard. Hughes Aircraft AN/APG-63 X-band pulse Doppler radar (upgraded to AN/APG-70 under MSIP), equipped since 1980 with a Hughes Aircraft programmable signal processor, provides long-range detection and tracking of small high-speed targets operating at all altitudes to treetop level, and feeds accurate tracking information to the IBM CP-1075 96 kbits (24 kbits on early F-15C/Ds) central computer to ensure effective launch of

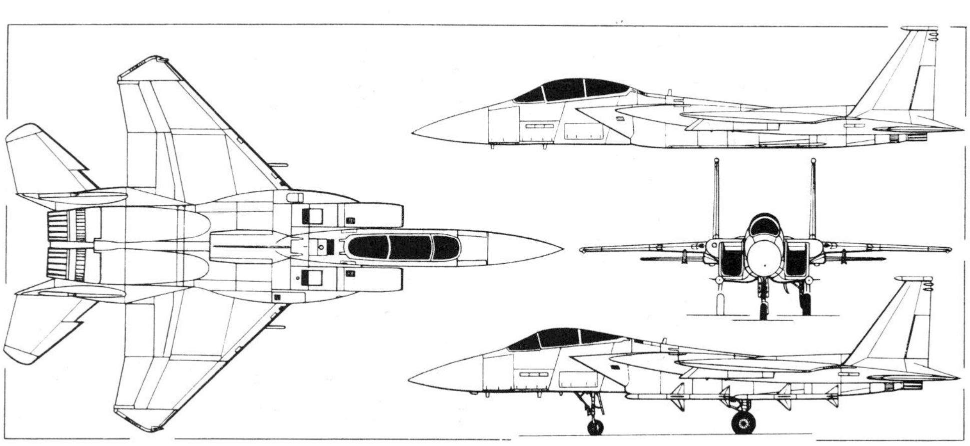

McDonnell Douglas F-15C Eagle single-seat air superiority fighter, with additional side view (top) of two-seat F-15D Eagle ***(Jane's/Dennis Punnett)***

1994

the aircraft's missiles or the firing of its internal gun. For close-in dogfights, the radar acquires the target automatically and the steering/weapon system information is displayed on a McDonnell Douglas Electronics AN/AVQ-20 head-up display. A Teledyne Electronics AN/APX-101 IFF transponder informs ground stations and other suitably equipped aircraft that the F-15 is friendly. It also supplies data on the F-15's range, azimuth, altitude and identification to air traffic controllers. A Hazeltine AN/APX-76 IFF interrogator informs the pilot if an aircraft seen visually or on radar is friendly. A Litton reply evaluator for the IFF system operates with the AN/APX-76. A Honeywell vertical situation display set, using a cathode ray tube to present radar, electro-optical identification and attitude director indicator formats to the pilot, permits inputs received from the aircraft's sensors and the central computer to be visible to the pilot under any light conditions. Honeywell also developed the AN/ASK-6 air data computer and AN/ASN-108 AHRS for the F-15, the latter also serving as a back-up to the Litton AN/ASN-109 INS which provides the basic navigation data and is the aircraft's primary attitude reference. In addition to giving the aircraft's position at all times, the INS provides pitch, roll, heading, acceleration and speed information. Other specialised equipment for flight control, navigation and communications includes a Collins AN/ARN-118 Tacan; Collins HSI to present aircraft navigation information on a symbolic pictorial display; Collins ADF and AN/ARN-112 ILS receivers; Magnavox AN/ARC-164 UHF transceiver and UHF auxiliary transceiver. The communications sets have cryptographic capability. Dorne and Margolin glide slope localiser antenna, and Teledyne Avionics angle of attack sensors. Northrop (Defense Systems Division) Enhanced AN/ALQ-135(V) internal countermeasures set provides automatic jamming of enemy radar signals; Loral AN/ALR-56C radar warning receiver; Magnavox AN/ALQ-128 electronic warfare warning set; and Tracor AN/ALE-45 chaff dispenser. Bendix tachometer, fuel and oil indicators; Plessey feel trim actuators.

McDonnell Douglas F-15B Eagle IFFC/Firefly 3 single-seat air superiority aircraft

1994

ARMAMENT: Provision for carriage and launch of a variety of air-to-air weapons over short and medium ranges, including four AIM-9L/M Sidewinders, four AIM-7F/M Sparrows or eight AIM-120 AMRAAM, and a 20 mm M61A1 six-barrel gun with 940 rounds of ammunition. General Electric lead-computing gyro. A Dynamic Controls Corporation armament control system keeps the pilot informed of weapons status and provides for their management. Three air-to-surface weapon stations (five if configured with conformal fuel tanks) allow for the carriage of up to 10,705 kg (23,600 lb) of bombs, rockets or additional ECM equipment. AN/AWG-20 armament control system.

DIMENSIONS, EXTERNAL:
Wing span 13.05 m (42 ft 9¾ in)
Wing aspect ratio 3.01
Length overall 19.43 m (63 ft 9 in)
Height overall 5.63 m (18 ft 5½ in)
Tailplane span 8.61 m (28 ft 3 in)
Wheel track 2.75 m (9 ft 0¼ in)
Wheelbase 5.42 m (17 ft 9½ in)

AREAS:
Wings, gross 56.5 m² (608.0 sq ft)
Ailerons (total) 2.46 m² (26.48 sq ft)
Flaps (total) 3.33 m² (35.84 sq ft)
Fins (total) 9.78 m² (105.28 sq ft)
Rudders (total) 1.85 m² (19.94 sq ft)
Tailplanes (total) 10.34 m² (111.36 sq ft)

WEIGHTS AND LOADINGS:
Weight empty, equipped (no fuel, ammunition, pylons or external stores 12,973 kg (28,600 lb)
Max fuel load: internal 6,103 kg (13,455 lb)
CFTs (two total) 4,422.5 kg (9,750 lb)
auxiliary tanks (three total) 5,395.5 kg (11,895 lb)
max internal and external 15,921 kg (35,100 lb)
T-O weight (interceptor, full internal fuel and four Sparrows) 20,244 kg (44,630 lb)
T-O weight (incl three 2,309 l; 610 US gallon; 508 Imp gallon drop tanks) 26,521 kg (58,470 lb)
Max T-O weight with CFTs 30,845 kg (68,000 lb)
Max wing loading 546.1 kg/m² (111.8 lb/sq ft)
Max power loading 147.87 kg/kN (1.45 lb/lb st)

PERFORMANCE:
Max level speed more than Mach 2.5
800 kts (1,482 km/h; 921 mph) CAS
Approach speed 125 kts (232 km/h; 144 mph) CAS
Service ceiling 18,300 m (60,000 ft)
T-O run (interceptor) 274 m (900 ft)
Landing run (interceptor), without braking parachute 1,067 m (3,500 ft)
Ferry range: with external tanks, without CFTs more than 2,500 n miles (4,631 km; 2,878 miles)
with CFTs 3,100 n miles (5,745 km; 3,570 miles)
Max endurance: with in-flight refuelling 15 h
unrefuelled, with CFTs 5 h 15 min
Design *g* limits +9/−3

UPDATED

McDonnell Douglas F-15A Eagle equipped under the MSIP programme

1994

DOUGLAS AIRCRAFT COMPANY (Division of McDonnell Douglas Corporation)

HEADQUARTERS: 3855 Lakewood Boulevard, Long Beach, California 90846
Telephone: 1 (310) 593 5511
Fax: 1 (310) 593 2019
Telex: 674357
PRESIDENT: Robert H. Hood Jr
EXECUTIVE VICE-PRESIDENTS:
John J. Van Gels
John D. Wolf
VICE-PRESIDENTS:
Allen C. Haggerty (General Manager, Design and Technology)
James R. Jensen (General Manager, Product Support)
Walter J. Orlowski (General Manager, Marketing and Business Development)
GENERAL MANAGER: Doug Jacobsen (External Relations)

Douglas delivered 976 DC-9s up to 1982 and its 2,000th twin-jet airliner, an MD-80 to American Airlines, on 11 June 1992. Plan to sell up to 49 per cent of Douglas to Taiwan Aerospace during 1992 and jointly develop MD-12 did not materialise; Douglas still willing to consider an airliner manufacturing alliance, but business returning to profitability.

Despite the drastic drop in orders and deliveries, accompanied by a drop of 28 per cent in revenues, and a decline to 4 per cent of the world airliner business, Douglas has made an operating profit in every quarter of the last three years.

Douglas reduced workforce from between 40,000 and 43,000 in 1990, before military C-17 was hived off, to around 11,000 by end of 1993 and has stabilised at that number. Productivity increased so that man-hours to produce an MD-11 reduced from 300,000 in 1991 to 140,000 by end 1993, and for a twin-jet from 100,000 to 50,000; MD-90 modular manufacturing system applied also to MD-80 during 1994; company can therefore remain viable while producing fewer than 30 twin-jets a year, compared with 136 needed in 1989, and around 30 tri-jets a year. Company can fund own development of forthcoming derivatives MD-90-50 and MD-11ER, but development of MD-XX and MD-12 at a standstill in 1994 and awaiting an upturn in market.

UPDATED

MCDONNELL DOUGLAS (DOUGLAS) SKYHAWK

US designation: A-4

TYPE: Single-seat attack bomber aircraft.

PROGRAMME: Designed originally to provide the US Navy and Marine Corps with a simple low-cost lightweight attack and ground support aircraft, the Skyhawk was based on experience gained during the Korean War. Since the initial requirement called for operation by the US Navy, special design consideration was given to providing low-speed control and stability during take-off and landing, added strength for catapult launch and arrested landings, and dimensions that would permit it to negotiate standard aircraft carrier lifts without the complexity of folding wings.

Construction of the XA-4A (originally XA4D-1) prototype Skyhawk began in September 1953 and the first flight of this aircraft, powered by a Wright J65-W-2 engine (32 kN; 7,200 lb st), took place 22 June 1954.

VERSIONS: **A-4A** (formerly A-4D-1): Initial version with Wright J65-W-4 turbojet engine 3,493 kg (7,700 lb st). First A-4A flew 14 August 1954, and this version entered service with the US Atlantic and Pacific Fleets 26 October 1956. 166 built. Uprated engines (3,855 kg; 8,500 lb st) fitted progressively to all aircraft.

A-4B (formerly A4D-2): Similar to A-4A but with improved bomb delivery system, provision for carrying Bullpup missiles, automatic dead reckoning navigation computer, flight refuelling capability (both tanker and receiver), dual hydraulic system, stiffer single surface rudder and powered tail, and Wright J65-W-16A turbojet (3,493 kg; 7,700 lb st). First flight 26 March 1956. 542 built. 50 reconditioned for Argentine Air Force. Uprated engines 3,855 kg (8,500 lb st) fitted progressively to all aircraft.

A-4C (formerly A4D-2N): Similar to A-4B but with longer nose to accommodate additional equipment to improve all-weather capability. New items included advanced autopilot, low-altitude bombing/all-attitude indicating gyro system, terrain-clearance radar and angle of attack indicator. First flight 21 August 1958. Deliveries began in 1959. Production completed in December 1962. 638 built. Uprated engines 3,855 kg (8,500 lb st) fitted to all aircraft progressively.

Douglas A-4M Skyhawk being fitted with the Angle Rate Bombing System (ARBS)

1992

TA-4E: Original designation of prototypes of TA-4F.

A-4F: Attack bomber with J52-P-8A turbojet rated at 41.4 kN (9,300 lb st), new lift-spoilers on wings to shorten landing run by up to 305 m (1,000 ft), nosewheel steering, low-pressure tyres, zero/zero ejection seat, additional bullet- and flak-resistant materials to protect pilot, updated electronics contained in fairing 'hump' aft of cockpit. Prototype flew for the first time 31 August 1966. Deliveries to US Navy began 20 June 1967 and were completed in 1968 with 146 built.

TA-4F: Tandem two-seat dual-control trainer version of A-4F for US Navy. Fuselage extended 0.71 m (2 ft 4 in), fuselage fuel tankage reduced to 379 litres (100 US gallons; 83.3 Imp gallons), Pratt & Whitney J52-P-6 or -8A engines optional, Douglas Escapac rocket ejection seats. Provision to carry full range of weapons available for A-4F. Reduced electronics. First prototype flew 30 June 1965. Deliveries to the US Navy began in 1966.

A-4G: Similar to A-4F for Royal Australian Navy. Equipped to carry Sidewinder air-to-air missiles. First of eight delivered 26 July 1967.

TA-4G: Similar to TA-4F for Royal Australian Navy. First of two delivered 26 July 1967.

A-4H: Designation of version supplied to Israel. Delivery of an initial batch of 48 in 1967-68, followed by 60 more by early 1972. Retrofitted with Rafael MAHAT lightweight analog weapons delivery system.

TA-4H: Tandem two-seat trainer version of the A-4H for Israel, 10 delivered.

TA-4J: Tandem two-seat trainer, basically a simplified version of the TA-4F. Ordered for the US Naval Air Advanced Training Command, under $26,834,000 contract, followed by further contract in mid-1971. Deletion of the following equipment, although provisions retained: radar, dead reckoning navigation system, low-altitude bombing system, air-to-ground missile systems, weapons delivery computer and automatic release, intervalometer, gun pod, standard stores pylons, in-flight refuelling system and spray tank provisions. Addition and relocation of certain instruments. J-52-P-6 engine standard. Provision for J-52-P-8A engine and combat electronics. Prototype flew in May 1969 and the first four were delivered to the US Navy 6 June 1969.

A-4K: Similar to the A-4F, for the Royal New Zealand Air Force. Different radio, and braking parachute. First of 10 delivered to the RNZAF 16 January 1970.

A-4KU: Designation of 30 aircraft similar to the A-4M for Kuwait Air Force. Deliveries began in Spring 1977.

TA-4K: Similar to TA-4F, for Royal New Zealand Air Force. The first of four was handed over 16 January 1970.

TA-4KU: Designation of six aircraft, similar to TA-4F, for Kuwait Air Force.

A-4L: Modification of A-4C with uprated engine, bombing computing system and electronics relocated in fairing 'hump' aft of cockpit as on A-4F. Delivery to US Navy Reserve carrier air wing in December 1969.

A-4M Skyhawk II: Similar to A-4F, but with J52-P-408 turbojet rated at 50 kN (11,200 lb st) and braking parachute standard, making possible combat operation from 1,220 m (4,000 ft) fields and claimed to increase combat effectiveness by 30 per cent. Larger windscreen and canopy; windscreen bullet-resistant. Increased ammunition capacity for 20 mm cannon. More powerful generator, provision of wind-driven back-up generator and self-contained engine starter. First of two prototypes flew for the first time 10 April 1970. About 50 initially ordered for US Marine Corps, the first of which was delivered 3 November 1970. Further order was placed subsequently, and the FY76 budget included $70 million for the procurement of a final 24 aircraft. Funds also allocated for the installation of improved electronic warfare equipment in service aircraft, and for the continued development of an Angle Rate Bombing System (ARBS) for future installation in A-4Ms (see A-4Y).

A-4N Skyhawk II: Light attack version ordered by US Navy for export to Israel. Basically similar to A-4M. First flew 8 June 1972.

A-4P: Revised A-4B for Argentine Air Force (50).

A-4Q: Revised A-4B for Argentine Navy (16).

A-4S: Designation of 40 Skyhawks for service with Singapore Air Defence Command. Conversion from ex-USN A-4Bs began in 1973, carried out by Lockheed Aircraft Service Company.

TA-4S: Three two-seat A-4B conversions for Singapore, by Lockheed Aircraft Service Company.

A-4Y: USMC A-4M with updated HUD, redesigned cockpit and Hughes Angle Rate Bombing System (ARBS). All A-4Ms modified.

UPGRADES: **IAI:** A-4 upgrade. See separate entry.

Lockheed Aircraft Service Company/Allied Signal: Selected by Argentine Air Force to upgrade 36 ex-Marine A-4M Skyhawks received from the US government. Lockheed to act as prime. See also Lockheed aircraft Argentina.

McDonnell Douglas: A-4Y Skyhawk. See Versions.

Royal New Zealand Air Force: Kahu A-4K Skyhawk upgrade. See separate entry.

Singapore Aerospace: A-4S-1 Super Skyhawk upgrade. See separate entry.

OPERATORS: Versions of the A-4 Skyhawk are in service with the armed forces of the following countries: Argentina (65), Indonesia (28), Israel (165), Kuwait (19), Malaysia (36), New Zealand (20), Singapore (70) and USA (127).

The following description applies to the A-4M aircraft:

DESIGN FEATURES: Cantilever low-wing monoplane. Sweepback 33° at wing quarter-chord with all-metal three-spar structure. Spars machined from solid plate in one piece tip-to-tip. One-piece wing skins. Variable incidence tailplane.

FLYING CONTROLS: Hydraulically powered all-metal ailerons, with servo trim tab in port aileron. All-metal split flaps. Automatic leading-edge slats with fences. Hydraulically actuated lift spoilers above flaps. Outward hinged hydraulically actuated airbrake on each side of rear fuselage. Electrically actuated variable incidence tailplane. Hydraulically powered elevators. Powered rudder with unique central skin and external stiffeners.

STRUCTURE: The fuselage is an all-metal semi-monocoque structure in two sections. Rear section removable for engine servicing. Detachable nose over communications and navigation equipment. Integral flak-resistant armour in cockpit area, with internal armour plate below and forward of cockpit. The tail unit is a cantilever all-metal structure.

LANDING GEAR: Hydraulically retractable tricycle type, with single wheel on each unit. All units retract forward. Free-fall emergency extension. Main legs preshorten for retraction and wheels turn through 90° to stow horizontally in wings. Menasco shock-absorbers. Hydraulic nosewheel steering. Ribbon type braking parachute of 4.88 m (16 ft) diameter contained in canister secured in rear fuselage below engine exhaust. Arrester hook for carrier operation.

POWER PLANT: One 50 kN (11,200 lb st) Pratt & Whitney J52-P-408 turbojet engine. Fuel in integral wing tanks and self-sealing fuselage tank aft of cockpit, total capacity 3,028 litres (800 US gallons; 666 Imp gallons). One 568, 1,136 or 1,514 litre (150, 300 or 400 US gallon; 125, 250, 333 Imp gallon) auxiliary tank can be carried on the underfuselage bomb-rack and one 568 or 1,136 litre (150 or 300 US gallon; 125 or 250 Imp gallon) auxiliary tank on each of the inboard underwing racks. Maximum fuel capacity, internal plus auxiliary tanks 6,814 litres (1,800 US gallons; 1,498 Imp gallons). Large refuelling probe on starboard side of nose. Douglas developed self-contained flight refuelling unit can also be carried on the underfuselage standard bomb shackles. Provision for JATO.

ACCOMMODATION: Pilot on Douglas Escapac 1-G3 zero/zero lightweight ejection seat. Enlarged cockpit enclosure to improve pilot's view with rectangular bullet-resistant windscreen.

SYSTEMS: Dual hydraulic system. Oxygen system. Electrical system powered by 20 kVA generator, with wind-driven generator to provide emergency power.

ELECTRONICS: Include Bendix Automatic Flight Control, ARC-159 UHF radio transceiver, ARA-50 UHF direction finder, APX-72 IFF, Marconi-Elliot AVQ-24 head-up display system, Douglas angle of attack indicator, electronic countermeasures, ASN-41 nav computer, APN-153(V) radar nav, ARC-114 VHF/FM radio transceiver, ARR-69 auxiliary radio receiver, ARN-84 Tacan and APN-194 radar altimeter.

ARMAMENT: Provision for several hundred variations of military load, carried externally on one underfuselage rack, capacity 1,588 kg (3,500 lb); two inboard underwing racks, capacity of each 1,020 kg (2,250 lb); and two outboard underwing racks, capacity of each 450 kg

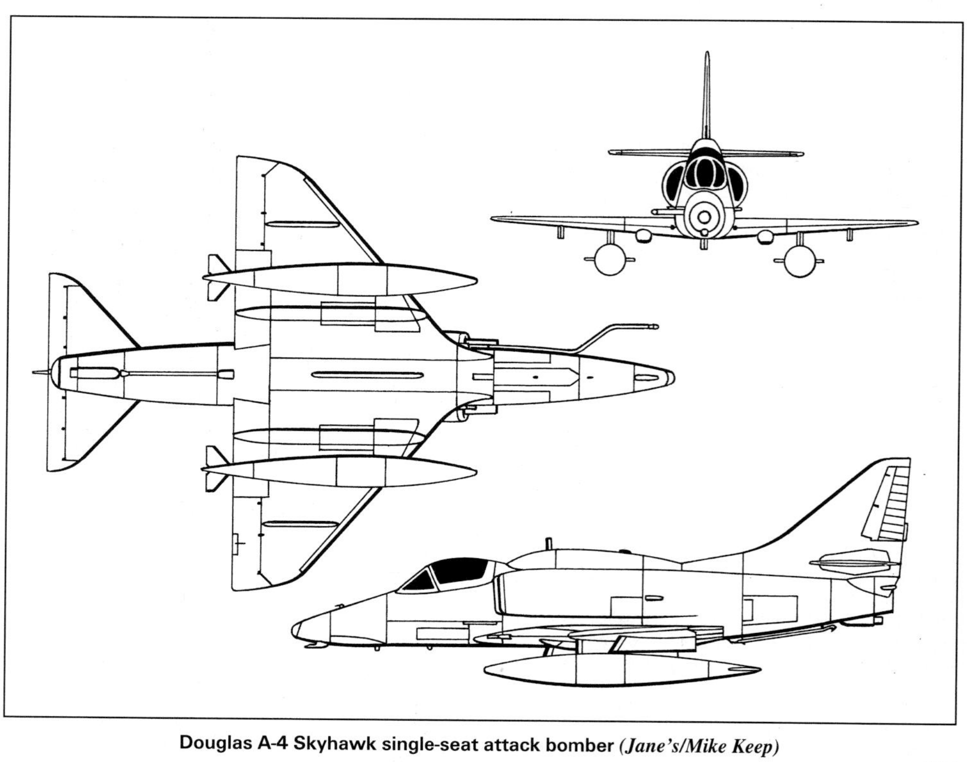

Douglas A-4 Skyhawk single-seat attack bomber ***(Jane's/Mike Keep)***

1994

(1,000 lb). Weapons that can be deployed include nuclear or HE bombs, air-to-air and air-to-surface rockets, Sidewinder infra-red missiles, Bullpup air-to-surface missiles, ground attack gun pods, torpedoes, countermeasures equipment and so on. Two 20 mm Mk 12 cannon in wingroots standard, each with 200 rounds of ammunition. DEFA 30 mm cannon available as optional on international versions, with 150 rounds of ammunition per gun.

DIMENSIONS, EXTERNAL:
Wing span 8.38 m (27 ft 6 in)
Wing chord at root 4.72 m (15 ft 6 in)
Length overall (excl flight refuelling probe):
A-4M 12.29 m (40 ft 4 in)
TA-4F 12.98 m (42 ft 7¼ in)
Height overall: A-4M 4.57 m (15 ft 0 in)
TA-4F 4.66 m (15 ft 3 in)
Tailplane span 3.45 m (11 ft 4 in)
Wheel track 2.37 m (7 ft 9½ in)

AREAS:
Wings, gross 24.16 m² (260 sq ft)
Vertical tail surfaces (total) 4.65 m² (50 sq ft)
Horizontal tail surfaces (total) 4.54 m² (48.85 sq ft)

WEIGHTS AND LOADINGS:
Weight empty: A-4F 4,581 kg (10,100 lb)
TA-4F 4,853 kg (10,700 lb)
A-4M 4,899 kg (10,800 lb)
Normal T-O weight:
A-4F, M, TA-4F 11,113 kg (24,500 lb)
* A-4F from land base 12,437 kg (27,420 lb)
* *export version only: overload condition not authorised by US Navy*

PERFORMANCE (at combat weight):
Max level speed:
TA-4F 568 kts (1,052 km/h; 654 mph)
Max level speed (with 1,814 kg; 4,000 lb bomb load):
A-4F 548 kts (1,015 km/h; 631 mph)
A-4M 561 kts (1,040 km/h; 646 mph)
Max rate of climb (ISA at S/L):
A-4F 2,440 m (8,000 ft)/min
A-4M 3,140 m (10,300 ft)/min
Rate of climb (ISA at 7,620 m; 25,000 ft):
A-4F 1,097 m (3,600 ft)/min
A-4M 1,463 m (4,800 ft)/min
T-O run (at 10,433 kg; 23,000 lb T-O weight):
A-4F 1,030 m (3,380 ft)
A-4M 832 m (2,730 ft)
Max ferry range, A-4M at 11,113 kg (24,500 lb) T-O weight with max fuel, standard reserves
1,740 n miles (3,225 km; 2,000 miles)

UPDATED

Douglas TA-4J Skyhawk two-seat trainer aircraft of the US Navy *(Peter J. Cooper)*

1994

Douglas A-4M Skyhawk assigned to the Naval Weapons Centre, China Lake

1994

DOUGLAS DC-3

UK name: Dakota
Military designation: C-47
Chinese military designation: Li-2

TYPE: Twin-engine transport aircraft.

PROGRAMME: The DC-3 was withdrawn from production after 10,926 had been built as military transports under the designations C-47 and C-53. Over 2000 were also built in Japan and the USSR.

The first DC-3 flew 18 December 1935, and the first transport company to put the DC-3 into commercial service was American Airlines, in June 1936. Before the Second World War the DC-3 was standard aircraft on the major US airlines and on several foreign lines. During the Second World War the DC-3 in military configuration became the standard aircraft in the Transport Commands of the Allied air forces, serving as cargo-carrier, paratroop carrier, personnel transport, glider tug and ambulance. The DC-3 is still in service with airlines worldwide and has undergone a variety of modifications, re-engining and upgrades.

VERSIONS: **DC-3:** Standard 21-passenger aircraft with two Wright Cyclone R-1820 engines.

DC-3A: Same as DC-3 but fitted with two Pratt & Whitney Twin Wasp R-1830 engines.

DC-3B: Convertible day plane or sleeper with two Pratt & Whitney Twin Wasp engines.

DC-3C: Commercial 'executive' conversion by Douglas of the C-47 with two Pratt & Whitney Twin Wasp engines. Two standard interior furnishings, with chairs, settees (convertible to berths) as well as catering facilities.

DC-3D: Converted C117A transport (two Pratt & Whitney Twin Wasp engines), started as a military personnel transport but converted on the production line after termination of military contracts and completed as civil transport.

Li-2: Chinese built version.

UPGRADES: **Aero Services:** DC-3 Interior conversion kits. See separate entry.

AMI: DC-3-65TP. See separate entry.

Basler Turbo: Turbo-67. See separate entry.

Professional Aviation: Jet Prop DC-3 AMI. See separate entry.

Schafer/AMI: DC-3-65TP Cargomaster. See separate entry.

South African Air Force: DC-3 Turbo Dak upgrade. See separate entry.

South African Air Force: Programme to convert eight C-47 Dakota aircraft to undertake a dedicated maritime patrol role. All aircraft are being re-engined with Pratt & Whitney Canada PT6A-65AR turboprops at Zwartkop air base. Aircraft fitted with fuselage plug of 1.02 m (3 ft 4 in) forward of wing centre section, new brakes and hydraulics, two long-range tanks (757 litres; 200 US gallons; 166 Imp gallons) allowing 2,580 n miles (4,800 km; 2,982 miles) range or 12 to 14 hour endurance. Programme also includes new wiring and electrical harnesses. The Avionics fit includes an electro-optical surveillance sensor in the nose and an underfuselage maritime surveillance radar (possibly EL/M-2022) as well as various ESM sensors and sonobuoy dispenser in the rear of the fuselage. Further equipment includes two recce cameras (one vertical and one oblique). The four crew consoles are fitted to the port side. Remaining crew consists of three aircrew and two observers.

OPERATORS: Versions of the DC-3/C-47 are in service with the armed forces of the following countries: Australia (4), Bolivia (3), China (30), Colombia (11), Dominican Republic (3), El Salvador (6), Greece (8), Guatemala (6), Haiti (3), Honduras (6), Indonesia (7), Israel (20), Laos (7), Malawi (2), Mexico (8), Paraguay (7), South Africa (39), Taiwan (9), Thailand (21) and Turkey (40).

DESIGN FEATURES: Low-wing cantilever monoplane. Double cellular multiweb all-metal wing construction.

STRUCTURE: The fuselage is an all-metal structure built up of transverse frames of formed sheet, longitudinal members of extruded bulb angles, with a covering of smooth sheet. The tail unit is a cantilever monoplane type. Tailplane and fin of multicellular construction.

FLYING CONTROLS: Fabric covered ailerons, controllable trim tabs in the starboard aileron. Hydraulically operated all-metal split trailing-edge flaps. Rudder and elevators have aluminium alloy frames and fabric covering and are aerodynamically and statically balanced. Trim tabs in all control surfaces.

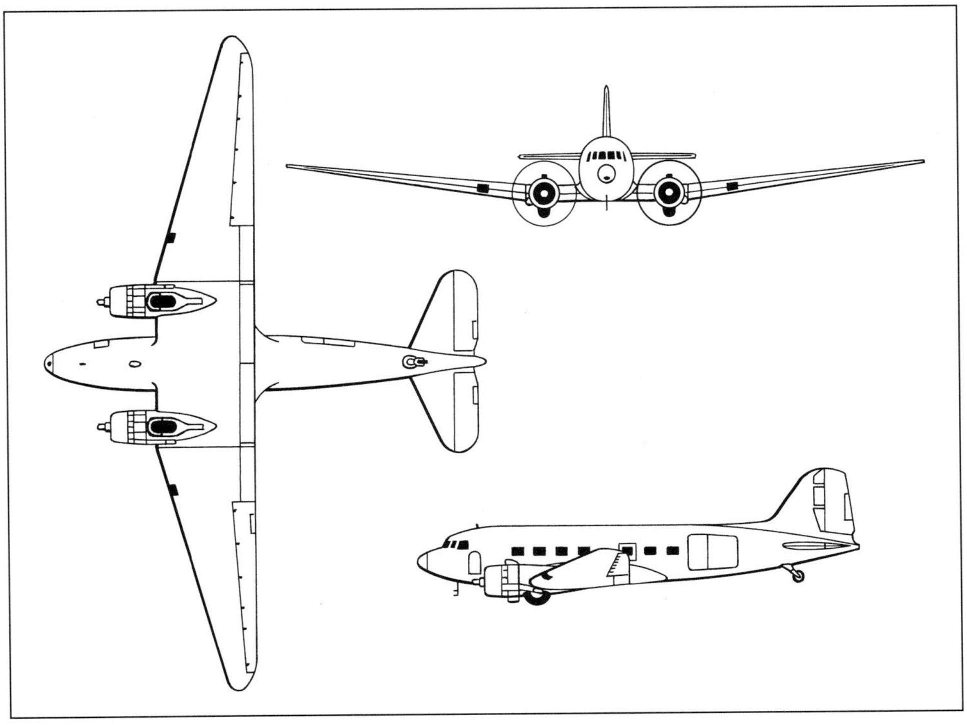

Douglas DC-3 transport aircraft *(Jane's/Mike Keep)*

1994

Douglas C-47 Dakota of the South African Air Force

1994

Douglas EC-47 Dakota of the Italian Air Force *(Peter J. Cooper)*

1994

LANDING GEAR: Retractable type. Each unit comprises two Bendix air-oil shock-absorber legs. Wheels are retracted forward and upward into engine nacelles and can be raised or lowered in 15 seconds, by engine-driven hydraulic system. Bendix wheels and hydraulic wheel brakes. Non-retractable steerable tailwheel.

POWER PLANT: Two Wright Cyclone GR-1820-G102A or GR-1820-G202A nine-cylinder Pratt & Whitney Twin Wasp R-1830-S1C3G 14-cylinder radial air-cooled engines with two-speed superchargers. Three-blade Hamilton Standard constant-speed propellers. Two main fuel tanks 794 litres (210 US gallons; 175 Imp gallons) located forward of centre-section spar. Two auxiliary tanks 760 litres (201 US gallons; 167 Imp gallons) aft of spar. One oil tank 109 litres (29 US gallons; 24 Imp gallons) in each engine nacelle.

ACCOMMODATION: Flight deck is forward of the wing and is reached through a corridor from the passenger cabin. Emergency exit in ceiling of flight deck. Dual controls. The normal cabin accommodates up to 21 passengers and is completely sound-insulated with ventilation and steam heating. There are four mail compartments forward of the main cabin, two on each side of the centre aisle. The right forward compartment has a capacity of 991 m^3 (35 cu ft), and the compartment directly aft 1.42 m^3 (50 cu ft). The left forward compartment has a capacity of 0.38 m^3 (13.5 cu ft), and is connected to a 1.42 m^3 (50 cu ft) compartment directly aft. An outside cargo loading door is located on the left side aft of the pilot's seat. Baggage compartment 2.9 m^3 (103 cu ft) aft of the galley and toilet with loading door on left side of the fuselage.

DIMENSIONS, EXTERNAL:

Wing span	28.9 m (95 ft)
Length overall	19.63 m (64 ft 5½ in)
Height	5.20 m (16 ft 11½ in)

WEIGHTS AND LOADINGS (Pratt & Whitney Twin Wasp R-1830-S1C3G engines):

Weight empty	7,657 kg (16,865 lb)
Disposable load	3,784 kg (8,784 lb)
Weight loaded	11,441 kg (25,200 lb)
Max landing weight	11,080 kg (24,400 lb)
Wing loading	124.4 kg/m^2 (25.5 lb/sq ft)
Power loading	4.76 kg/hp (10.5 lb/hp)

PERFORMANCE (Pratt & Whitney Twin Wasp R-1830-S1C3G engines):

Max speed at 2,590 m (8,500 ft)	199 kts (368 km/h; 229 mph)
Cruising speed	178 kts (331.2 km/h; 207 mph)
Stalling speed	57 kts (107 km/h; 67 mph)
Initial rate of climb	345 m (1,130 ft)/min
Service ceiling	7,076 m (23,200 ft)
Normal range (max fuel)	1,306 n miles (2,420 km; 1,510 miles)

UPDATED

DOUGLAS DC-6

TYPE: Four-engined transport aircraft.

PROGRAMME: The DC-6 is a larger more powerful development of, and successor to the DC-4 it has a longer fuselage and more powerful engines as well as pressurised cabins for crew and passengers; larger cabin seats and berths; reversible-pitch propellers and thermal de-icing for wings, tail and windscreen.

DESIGN FEATURES: Low-wing cantilever monoplane. Wing section NACA 23016-23012. Incidence at root 4°. Dihedral 7°. All-metal structure with smooth Alclad skin.

FLYING CONTROLS: NACA slotted flaps inboard of ailerons.

STRUCTURE: The fuselage is a semi-monocoque all-metal structure with flush-riveted Alclad skin. The tail unit is a cantilever monoplane type. Fin area 8.67 m^2 (93.4 sq ft), rudder area (aft of hinge with tab) 4.55 m^2 (49 sq ft), total vertical area 14.86 m^2 (159.9 sq ft). The tailplane area is 19.59 m^2 (210.9 sq ft), elevator area (aft of hinge, with tab) 10.11 m^2 (108.9 sq ft).

LANDING GEAR: Retractable tricycle type. Hydraulic retraction with emergency manual gear. Hydraulic wheel brakes. Steerable nosewheel. Track 7.52 m (24 ft 8 in). Wheelbase 9.34 m (30 ft 8 in).

POWER PLANT: Four Pratt & Whitney Double-Wasp R-2800-CA15 18-cylinder two-row radial air-cooled engines, each normally rated at 1,343 kW (1,800 hp) at 1,830 m (6,000 ft), 1,194 kW (1,600 hp) at 4,880 m (16,000 ft) and with a take-off output of 1,567 kW (2,100 hp), without or 1,791 kW (2,400 hp) with water injection. Hamilton Standard three-blade constant-speed fully feathering and reversible propellers of 3.98 m (13 ft 1 in) diameter. Fuel capacity from 12,560 to 17,770 litres (3,322 to 4,700 US gallons; 2,766 to 3,913.6 Imp gallons).

ACCOMMODATION: Pressurised accommodation for crew and passengers. Passenger compartment seats 48 by day and for short-range up to a maximum of 58. Entrance vestibule with coat room and galley aft of wings with cabin space fore and aft. Upper and lower berths of 26 to 39 passengers may be fitted in 30 seconds. Upper berths have separate air conditioning controls. All berths have reading lights and storage space for clothing and toilet accessories. Entire cabin space has fibreglass soundproofing and floors covered with foam rubber-backed carpets. Pressurisation ensures cabin pressure altitude of approximately 1,525 m (5,000 ft) at 4,880 m (16,000 ft), or 2,440 m (8,000 ft) at 6,100 m (20,000 ft). Passenger cabin dimensions 19.5 m (64 ft) long, 2.2 m (7 ft 3 in) high. All freight and baggage space below cabin floor with facilities for quick loading and unloading.

SYSTEMS: Thermal de-icing.

DIMENSIONS, EXTERNAL:

Wing span	35.81 m (117 ft 6 in)
Length	30.66 m (100 ft 7 in)
Height overall	8.66 m (28 ft 5 in)

WEIGHTS AND LOADINGS:

Weight empty	23,380 kg (51,495 lb)
Weight loaded	44,130 kg (97,200 lb)
Wing loading	324 kg/m^2 (66.4 lb/sq ft)
Power loading	4.58 kg/hp (10.1 lb/hp)

PERFORMANCE:

Max speed at 5,980 m (19,600 ft)	307 kts (570 km/h; 356 mph)
Cruising speed at 6,220 m (20,400 ft)	270 kts (501 km/h; 313 mph)
Landing speed	79 kts (145.6 km/h; 91 mph)
Rate of climb at S/L	326 m (1,070 ft)/min
Rate of climb at S/L, one engine out	171 m (560 ft)/min
T-O to 15 m (50 ft) at S/L (no wind)	1,133 m (3,795 ft)
Landing from 15 m (50 ft)	872 m (2,860 ft)
Normal range	3,300 n miles (6,112 km; 3,820 miles)
Max range	3,983 n miles (7,376 km; 4,610 miles)

VERIFIED

DOUGLAS DC-6A

TYPE: Four-engine transport aircraft.

PROGRAMME: The DC-6A is a freight-carrying version of the standard DC-6. It uses the wings, tail unit and landing gear of the DC-6 but has a different fuselage.

The fuselage is 1.525 m (5 ft) longer than that of the DC-6, giving the DC-6A a total cargo space of 141.5 m^3 (5,000 cu ft). The main cabin of constant cross-section throughout, is 20.74 m (68 ft) long, 2.36 m (7 ft 9 in) high, and 2.67 m (8 ft 9 in) wide at floor level. Two large doors, one forward and the other aft of the wings, are hinged at their top edges and swing upward to be clear of loading equipment. A self-powered loading elevator, which folds up for storage within the aircraft can be attached to either front or rear cargo door and lifts 1,820 kg (4,000 lb) from truck-bed height to cabin floor level.

The DC-6A has automatically controlled cabin pressurisation and air conditioning systems to permit high-altitude transportation of perishable cargoes.

The military versions, although primarily cargo transports, may be rapidly converted for passenger or troop carrying or air evacuation. In these forms they have accommodation for 60 passengers in rearward-facing seats, 76 on troop benches, or 40 stretcher cases.

POWER PLANT: Four Pratt & Whitney Double-Wasp R-2800-CB17 18-cylinder radial air-cooled engines each developing 1,418 kW (1,900 hp) at maximum continuous cruise and with 1,866 kW (2,500 hp) available for take-off with alcohol-water injection. Hamilton Standard or Curtiss Electric fully feathering and reversible propellers. Standard fuel capacity 15,111 litres (3,992 US gallons; 3,265 Imp gallons), with optional capacities of 20,550 litres (5,406 US gallons; 4,501 Imp gallons) and 20,918 litres (5,512 US gallons; 4,589 Imp gallons).

DIMENSIONS, EXTERNAL: As for DC-6 except:

Length	32.20 m (105 ft 7 in)

WEIGHTS AND LOADINGS:

Weight empty	22,595 kg (49,767 lb)
Gross T-O weight	48,125 kg (106,000 lb)
Wing loading	353.8 kg/m^2 (72.5 lb/sq ft)
Power loading	4.81 kg/hp (10.6 lb/hp)

Douglas DC-6A four-engine transport aircraft *(Peter J. Cooper)*

1995

Douglas DC-6A freighter transport *(D. March)*

1994

PERFORMANCE (at 42,800 kg; 95,000 lb, gross weight):
Max speed at 5,520 m (18,100 ft) 311 kts (576 km/h; 360 mph)
Cruising speed at 7,390 m (22,400 ft) 267 kts (494 km/h; 307 mph)
Landing speed 111 kts (149 km/h; 93 mph)
Initial rate of climb 374 m (1,120 ft)/min
Initial rate of climb, one engine out 203 m (620 ft)/min
T-O to 15 m (50 ft) at max AUW (no wind) 1,492 m (4,500 ft)
Landing distance from 15 m (50 ft) 918 m (3,010 ft)
Normal range 3,334 n miles (6,176 km; 3,860 miles)
Max range 4,242 n miles (7,856 km; 4,910 miles)

UPDATED

DOUGLAS DC-6B

TYPE: Four-engine transport aircraft.

PROGRAMME: The DC-6B is a passenger version of the DC-6A. The larger fuselage offers approximately 7 per cent greater payload capacity and 14 per cent greater passenger capacity than the standard DC-6, with only 4 per cent greater operating costs.

The standard domestic dayplane version of the DC-6B carries 64 passengers, eight more than the standard DC-6, while a transoceanic model with larger galleys, coatrooms and toilet compartments carries 54 passengers. A high-density 92-passenger version with an aircoach type interior also produced.

The DC-6B originally powered by either the Pratt & Whitney R-2800-CB17 or R-2800-CB16 engines. With the CB17 engine, the maximum take-off weight if 48,125 kg (106,000 lb) and the performance is the same as for the DC-A. With the CB16 engine which has 1,791 kW (2,400 hp) available for take-off, the maximum weight is 45,400 kg (100,000 lb) and corresponding take-off to 15 m (50 ft) is 1,153 m (3,780 ft). The cruising and landing performance is the same as for the DC-6A. The empty weight of the standard DC-6B is 24,583 kg (54,148 lb) and the maximum landing weight is 40,043 kg (88,200 lb).

The Super 6, which was certificated for a gross take-off weight of 48,526 kg (107,000 lb), carries a greater payload, has an increased fuel capacity and an improved performance. It has a standard accommodation for 82 passengers for trans-Atlantic tourist class operations.

OPERATORS: One in service with the Guatemalan Air Force.

UPDATED

MCDONNELL DOUGLAS DC-8

TYPE: Four-engined jet transport aircraft.

PROGRAMME: When Douglas decided to proceed with construction of their first jet transport, the four-engined DC-8, on 7 June 1955, they announced that all projected versions would have the same overall dimensions. They adhered to this policy until 1965, and the first five versions of the DC-8 have an identical airframe, with uniform electrical, hydraulic, control and air conditioning systems. The intercontinental versions differ from the domestic models only in having extra fuel capacity and the structural modifications needed to carry the additional fuel. The modifications are limited to the use of thicker skin and stronger material within the wing structure, the aft portion of the fuselage and the tailplane. The landing gear is also more robust in the case of the heavier intercontinental versions.

VERSIONS: **Series 10:** Domestic version with 6,124 kg (13,500 lb st) Pratt & Whitney JT3C-6 turbojet engines. The first Series 10 DC-8 flew 30 May 1958. This version received FAA certification 31 August 1959 and entered service with United Air Lines and Delta Air Lines 18 September 1959.

Series 20: Similar to Series 10, but with 7,167 kg (15,800 lb st) Pratt & Whitney JT4A-3 turbojet engines. The first Series 20 DC-8 flew 29 November 1958 and received FAA certification 19 January 1960.

Series 30: Long-range intercontinental version. Basically similar to Series 20, with 7,620 kg (16,800 lb st) JT4A-9 or 7,945 kg (17,500 lb st) JT4A-11 turbojet engines, and with increased fuel capacity. The first Series 30 flew 21 February 1959 and received certification 1 February 1960.

Series 40: Long-range intercontinental version. Similar to Series 30, but with 7,945 kg (17,500 lb st) Rolls-Royce Conway Rco.12 bypass turbojet engines. First Series 40 DC-8 flew 23 July 1959 and received FAA certification 24 March 1960.

Series 50: Similar to the Series 30, but with turbofan engines. Initial production Srs 50 aircraft powered by 7,718 kg (17,000 lb st) Pratt & Whitney JT3D-1 or 8,172 kg (18,000 lb st) JT3D-3 engines. Subsequent production aircraft, known as Model 55, have improved JT3D-3B engines and aerodynamic refinements. Some Srs 30 aircraft have been converted to Series 50 standard. First Series 50 DC-8 flew 20 December 1960 and received FAA certification 1 May 1961. A major improvement developed for the above versions of the DC-8 involves the installation of a different leading-edge which changes the wing profile and increases the wing chord by 4 per cent, to reduce drag and improve the speed and range of the DC-8. It is standard on all late-production long-range DC-8s and has been fitted retrospectively to some earlier aircraft.

Super 61: First of the Super 60 DC-8s, the Super 61 has the same wing and engine pylon structures as the Series 50, but the fuselage is extended by the insertion of a 6.10 m (20 ft 0 in) cabin section forward of the wing and a 5.08 m (16 ft 8 in) section aft of the wing. Accommodation is provided for up to 259 passengers, and a cargo space under the floor is increased to 70.80 m^3 (2,500 cu ft). Powered by four 8,172 kg (18,000 lb st) Pratt & Whitney JT3D-3B turbofan engines. The first Super 61 was completed 24 January 1966 and made its first flight 14 May 1966. FAA Type Approval was received 2 September 1966. Deliveries began 26 January 1967 and the Super 61 entered scheduled service 25 February 1967.

Super 62: This ultra-long-range version of the Super Sixty Series has an extended wing span and a fuselage 2.03 m (6 ft 8 in) longer than the standard DC-8. A 1.02 m (3 ft 4 in) cabin section is inserted both fore and aft of the wing, giving standard accommodation for up to 189 passengers. Each wingtip is fitted with a 0.91 m (3 ft 0 in) extension which significantly reduces induced drag under cruise conditions and makes possible an increased fuel tankage. Four Pratt & Whitney JT3D-3B turbofan engines. Engine pods of new design augment thrust and reduce drag by ducting bypass air through the entire length of each engine nacelle. Redesigned engine pylons reduce interference drag. This version is capable of carrying a full 18,145 kg (40,000 lb) payload non-stop, against prevailing headwinds, from points in central Europe to the West coast of the USA, with ample fuel reserves. The first Super 62 flew for the first time 29 August 1966. FAA Type Approval was received 27 April 1967, including qualification for automatic landing approach under Cat. II conditions. First delivery, to SAS, was made 3 May 1967, and the Super 62 entered scheduled service 22 May 1967.

Super 63: This developed version combines the long fuselage of the Super 61 with the aerodynamic and power plant improvements of the Super 62, but with the majority of Super 63s having JT3D-7 engines of 8,618 kg (19,000 lb st). Wheel track and tyre size increased. The first Super 63, with JT3D-3B engines, flew for the first time 10 April 1967. FAA Type Approval was received 30 June 1967 and delivery of the first Super 63 was made, to KLM, 15 July 1967. It entered scheduled service 27 July 1967.

Super 71: Re-engined Super 61 with CFM56-2-C1 turbofan engines. First flight 15 August 1981. FAA certification was received in April 1982. Cammacorp of the USA was responsible for the re-engining.

Super 72: Re-engined Super 62 with CFM56-2-C1 turbofan engines. Cammacorp of the USA was responsible for the re-engining.

Super 73: Re-engined Super 63 with CFM56-2-C1 turbofan engines. Cammacorp of the USA was responsible for the re-engining.

UPGRADES: **Alenia:** DC-8 freighter conversion. See separate entry.

Cammacorp: Series 71/72/73, re-engine programme. See Versions.

Dee Howard: DC-8 update. See separate entry.

McDonnell Douglas: DC-8 freighter conversion. See separate entry.

OPERATORS: Versions of the DC-8 are in service with the armed forces of the following countries: France (4), Peru (2) and USA (1).

DESIGN FEATURES: Cantilever low-wing. Dihedral 6° 30′. Sweepback at wing quarter-chord 30°. There is 10° of dihedral on hydraulically operated variable incidence tailplane.

FLYING CONTROLS: Power operated ailerons in two portions, outer portions being operated from inner portions at low speed only, via torsion box springs. Self-coupling manual circuits to tabs on inner ailerons in event of power failure. Two double-slotted flaps on each wing. Spoilers on top-wing surface forward of flaps operate on nosewheel contact during landing. Two slots, 2.03 m (6 ft 8 in) and 0.81 m (2 ft 8 in) long respectively on inboard leading-edge of each wing. Hydraulically operated variable incidence tailplane. Power operated rudder with manual standby. Elevators manually operated through servo tabs.

STRUCTURE: All-metal wing structure, with three plate-web spars inboard, two plate-web spars outboard forming torsion box, and spanwise stringers riveted to top and bottom skins. The fuselage is a double circular section all-metal structure. The tail unit is a cantilever all-metal structure.

LANDING GEAR: Retractable tricycle type. Nosewheel unit retracts forward, main units inward into fuselage. Main units are four-wheel bogies, the rear pair of wheels on each bogie being free to swivel in a sharp turn. Goodyear tyres pressure 9.21 kg/cm^2 (131 lb/sq in). Dual nosewheel steerable through steering wheel and rudder pedals. Goodyear disc brakes.

POWER PLANT: Four turbojet or turbofan engines in separate pods, two under each wing (details under Versions). All engines fitted with noise suppressors and thrust reversers for both ground and in-flight operation. All-fuel in integral wing tanks with total capacity of 66,528 litres (17,600 US gallons; 14,655 Imp gallons) on Srs 10 and 20; 87,360 litres (23,079 US gallons; 19,217 Imp gallons) on Srs 30 and 40; 88,531 litres (23,390 US gallons; 19,476 Imp gallons) on Srs 50 (JT3D-3B) and Super 61; 91,881 litres (24,275 US gallons; 20,213 Imp gallons) on Super 62 and 63. Pressure refuelling. Powered fuel dumping at maximum rate of 6,815 litres (1,800 US gallons; 1,499 Imp gallons)/min.

ACCOMMODATION (Srs 10, 20, 30, 40, 50): Crew of three-five plus cabin attendants. Seats for 105 to 118 persons in first class domestic versions and for 132 in mixed class intercontinental versions, with normal tourist accommodation for 144 and economy class seating for up to 179 in all

Douglas DC-6B four-engine passenger aircraft *(Peter March)*

1994

The first McDonnell Douglas DC-8 Super 61 on its maiden flight

1994

versions. Mixed class arrangements to suit requirements of customer. Windows, size 0.47 × 0.38 m (18.5 × 15 in), correspond with rows of first class seats. Passenger doors at front and rear of cabin on port side. Servicing doors opposite passenger doors on starboard side. Freight and baggage holds under floor forward and aft of wing, each with two doors on starboard side.

ACCOMMODATION (Super Sixty Srs): See under individual descriptions above. Super 61 has coatroom and two galleys at front on starboard side, galley and coatroom at rear on port side and three toilets aft. Doors as for standard DC-8.

SYSTEMS: Air conditioning and pressurisation system supplied by engine-driven turbo-compressors, with closed-circuit Freon system by Carrier Corporation. Maximum pressure differential 0.62 kg/cm² (8.77 lb/sq in). Hydraulic system, pressure 210 kg/cm² (3,000 lb/sq in), for landing gear retraction, nosewheel steering, brakes, flying controls, flaps and spoilers. Electrical system includes four 115/208 V three-phase 400c/s AC alternators and four transformer-rectifiers for DC supply. Cyclic hot-air anti-icing system.

ELECTRONICS AND EQUIPMENT: Radio and radar to customer's specifications. Sperry automatic flight control system, with SP-30 autopilot.

DIMENSIONS, EXTERNAL:
Wing span: except Super 62, 63 — 43.41 m (142 ft 5 in)
Super 62, 63 — 45.23 m (148 ft 5 in)
Wing chord, theoretical on c/l — 9.67 m (31 ft 8⅘ in)
Wing chord at tip, except Srs 60 — 2.22 m (7 ft 3½ in)
Length overall:
except Super Sixty Series — 45.87 m (150 ft 6 in)
Super 61, 63 — 57.12 m (187 ft 5 in)
Super 62 — 47.98 m (157 ft 5 in)
Height overall:
except Super 61, 62, 63 — 12.91 m (42 ft 4 in)
Super 61, 62, 63 — 12.92 m (42 ft 4½ in)
Tailplane span — 14.48 m (47 ft 6 in)
Wheel track, except Super 63 — 6.35 m (20 ft 10 in)
Wheelbase:
except Super Sixty Series — 17.52 m (57 ft 6 in)
Super 61, 63 — 23.65 m (77 ft 7 in)
Super 62 — 18.54 m (60 ft 10 in)
Passenger doors (each): Height — 1.83 m (6 ft 0 in)
Width — 0.88 m (2 ft 10½ in)
Servicing doors (each): Height — 1.63 m (5 ft 4 in)
Width — 0.85 m (2 ft 9½ in)
Freight hold doors (each): Height — 1.12 m (3 ft 8 in)
Width — 0.91 m (3 ft 0 in)

DIMENSIONS, INTERNAL:
Cabin, excl flight deck:
Length: Srs 10, 20, 30, 40, 50 — 31.11 m (102 ft 1 in)
Max width: Srs 10, 20, 30, 40, 50 — 3.50 m (11 ft 6 in)
Max height: Srs 10, 20, 30, 40, 50 — 2.21 m (7 ft 3 in)
Volume: Srs 10, 20, 30, 40, 50 — 215.7 m³ (7,617 cu ft)
Super 61, 63 — 285.4 m³ (10,080 cu ft)
Super 62 — 228.9 m³ (8,084 cu ft)
Freight and baggage holds (under floor, total):
Srs 10, 20, 30, 40, 50 — 39.35 m³ (1,390 cu ft)
Super 61, 63 — 70.80 m³ (2,500 cu ft)
Super 62 — 45.70 m³ (1,615 cu ft)

AREAS:
Wings, gross: Srs 10, 20, 30, 40, 50 early aircraft — 257.6 m² (2,773 sq ft)
Srs 10, 20, 30, 40, 50, extended leading-edge — 266.5 m² (2,868 sq ft)
Super 61 — 267.9 m² (2,884 sq ft)
Super 62, 63 — 271.9 m² (2,927 sq ft)

WEIGHTS AND LOADINGS:
Basic operating weight: Srs 10 — 56,578 kg (124,732 lb)
Srs 20 — 57,632 kg (127,056 lb)
Srs 30 (JT4A-11) — 60,692 kg (133,803 lb)
Srs 40 — 60,068 kg (132,425 lb)
Srs 50 (JT3D-3) — 60,020 kg (132,325 lb)
Super 61 — 67,538 kg (148,897 lb)
Super 62 — 64,366 kg (141,903 lb)
Super 63 — 69,739 kg (153,749 lb)
Capacity payload:
Srs 10, 20, 30, 40, 50 — 15,585 kg (34,360 lb)
Super 61 — 30,240 kg (66,665 lb)
Super 62 — 21,470 kg (47,335 lb)
Super 63 — 30,719 kg (67,735 lb)
Max ramp weight: Srs 10 — 124,740 kg (275,000 lb)
Srs 20 — 126,100 kg (278,000 lb)
Srs 30 (JT4A-11), 40 — 144,240 kg (318,000 lb)
Srs 50 (JT3D-3B), Super 61 — 148,775 kg (328,000 lb)
Super 62 — 153,315 kg (338,000 lb)
Super 63 — 160,000 kg (353,000 lb)
Max T-O weight: Srs 10 — 123,830 kg (273,000 lb)
Srs 20 — 125,190 kg (276,000 lb)
Srs 30, 40 — 142,880 kg (315,000 lb)
Srs 50 (JT3D-3B), Super 61 — 147,415 kg (325,000 lb)
Super 62 — 151,950 kg (335,000 lb)
Super 63 — 158,760 kg (350,000 lb)
Design landing weight: Srs 10 — 87,543 kg (193,000 lb)
Srs 20 — 90,500 kg (199,500 lb)
Srs 30, 40 — 93,900 kg (207,000 lb)
Srs 50 (JT3D-3B) — 98,430 kg (217,000 lb)
Super 61, 62 — 108,860 kg (240,000 lb)
Super 63 — 111,130 kg (245,000 lb)
Max zero-fuel weight: Srs 10 — 75,250 kg (165,900 lb)
Srs 20 — 75,975 kg (167,500 lb)
Srs 30 — 81,420 kg (179,500 lb)
Srs 40 — 80,330 kg (177,100 lb)
Srs 50 (JT3D-3B) — 80,060 kg (176,500 lb)
Max wing loading: Srs 10 — 480.9 kg/m² (98.5 lb/sq ft)
Srs 20 — 486.3 kg/m² (99.6 lb/sq ft)
Srs 30, 40, 50 — 533.2 kg/m² (109.2 lb/sq ft)

PERFORMANCE:
Max recommended cruising speed at 99,800 kg (220,000 lb) AUW at 9,150 m (30,000 ft):
Srs 10 — 471 kts (873 km/h; 542 mph)
Srs 20 — 503 kts (932 km/h; 579 mph)
Srs 30 (JT4A-11) — 514 kts (952 km/h; 592 mph)
Srs 40 — 509 kts (943 km/h; 586 mph)
Srs 50 (JT3D-3B) — 504 kts (933 km/h; 580 mph)
Super 61, 62, 63 — 521 kts (965 km/h; 600 mph)
Landing speed at max landing weight:
Srs 10, 20 — 129 kts (238 km/h; 148 mph)
Srs 30, 40, 50 — 133 kts (246 km/h; 153 mph)
Rate of climb at S/L: Srs 10 — 405 m (1,330 ft)/min
Srs 20 — 808 m (2,650 ft)/min
Super 61 — 692 m (2,270 ft)/min
Super 62 — 683 m (2,240 ft)/min
Super 63 — 660 m (2,165 ft)/min
FAA T-O field length at max AUW:
Srs 10 — 2,860 m (9,380 ft)
Srs 20 — 2,510 m (8,240 ft)
Srs 30 (JT4A-11) — 3,020 m (9,900 ft)
Srs 40 — 2,940 m (9,650 ft)
Srs 50 (JT3D-3B) — 3,220 m (10,560 ft)
Super 61 — 3,042 m (9,980 ft)
Super 62 — 2,980 m (9,780 ft)
Super 63 — 3,505 m (11,500 ft)
FAA landing field length at max landing weight:
Srs 10, 20 — 1,950 m (6,400 ft)
Srs 30, 40 — 2,073 m (6,800 ft)
Srs 50 — 1,713 m (5,620 ft)
Super 61, 62 — 1,870 m (6,140 ft)
Super 63 — 1,801 m (5,910 ft)
Design range with max payload, normal reserves:
Super 61 — 3,256 n miles (6,035 km; 3,750 miles)
Super 62 — 5,210 n miles (9,640 km; 6,000 miles)
Super 63 — 3,907 n miles (7,240 km; 4,500 miles)
Estimated max range in still air, no reserves:
Srs 10 — 3,734 n miles (6,920 km; 4,300 miles)
Srs 20 — 4,159 n miles (7,710 km; 4,790 miles)
Srs 30 (JT4A-11) — 5,184 n miles (9,605 km; 5,970 miles)
Srs 40 — 5,279 n miles (9,817 km; 6,100 miles)
Srs 50 (JT3D-3B) — 6,078 n miles (11,260 km; 7,000 miles)
Super 61, zero payload — 6,209 n miles (11,500 km; 7,150 miles)
Super 62, zero payload — 7,381 n miles (13,675 km; 8,500 miles)
Super 63, zero payload — 6,686 n miles (12,390 km; 7,700 miles)

UPDATED

MCDONNELL DOUGLAS DC-8 FREIGHTER CONVERSION

TYPE: Four-engine cargo aircraft.

PROGRAMME: Under a programme launched at Tulsa works in February 1976, McDonnell Douglas began modifying DC-8 passenger transports into specialised freighters. First to be converted were two DC-8-43 Series airliners, which were also re-engined with Pratt & Whitney JT3D turbofans, under contract from Frederick B. Ayer and Associates. Subsequent orders for a total of eight conversions were received from International Airways, International Air Leases Inc, International Air Service, Overseas National Airways and Transmeridian Air Cargo.

DESIGN FEATURES: Modification includes removal of passenger installations and fitting a production freighter seven-track floor, and a 2.16 × 3.56 m (85 × 140 in) main deck cargo door. Cabin windows are replaced by metal plugs, and a cargo loading system is installed. Conversion to turbofan power is optional for turbojet models.

VERIFIED

MCDONNELL DOUGLAS DC-9

USAF designations: C-9A and VC-9C
US Navy designation: C-9B

TYPE: Twin-turbofan short/medium-range transport aircraft.

PROGRAMME: Design study data on the DC-9, originally known as the Douglas Model 2086, were released in 1962. Preliminary design work began during that year. Production started 6 March 1964. It flew for the first time 25 February 1965 and five DC-9s were flying by the end of June 1965. These aircraft were of the basic version now known as the DC-9 Series 10.

VERSIONS: **Series 10 Model 11:** Initial version, powered by 54.5 kN (12,250 lb st) Pratt & Whitney JT8D-5 turbofan engines. Standard fuel capacity 10,546 litres (2,786 US gallons; 2,320 Imp gallons). Maximum accommodation for 80 passengers at 86 cm (34 in) seat pitch with normal facilities, or 90 passengers with reduced facilities. This version received FAA Type Approval 23 November 1965 and entered scheduled service with Delta Air Lines 8 December 1965.

Series 10 Model 15: Generally similar to Srs 10 Model 11 but with 62.3 kN (14,000 lb st) JT8D-1 turbofan engines, standard fuel capacity of 14,000 litres (3,700 US gallons; 3,080 Imp gallons), and increased AUW.

Series 20: For operation in hot/high-altitude conditions, combining long-span wings of Series 30 with short fuselage of Series 10. Up to 90 passengers. Two 64.5 kN (14,500 lb st) JT8D-9 turbofans. The Series 20 flew for the first time 18 September 1968, and was certificated 11 December 1968. The first Series 20 was delivered to SAS on the same day and entered commercial service 23 January 1969.

Series 30: Developed version, initially with 62.3 kN (14,000 lb st) JT8D-7s, increased wing span, longer fuselage accommodating up to 105 passengers (normal) or 115 (with reduced facilities), and new high-lift devices including full-span leading-edge slats and double-slotted flaps. First Srs 30 flew for the first time 1 August 1966. First delivery, to Eastern Air Lines, was made 27 January 1967 and scheduled services began 1 February 1967. Engine options available include JT8D-9 of 64.5 kN (14,500 lb st); JT8D-11 of 66.7 kN (15,000 lb st); JT8D-15 of 69 kN (15,500 lb st); and JT8D-17 of 71.2 kN (16,000 lb st). All engines have sound-treated nacelles that comply with FAA FAR Pt 36 noise regulations.

Series 40: As Series 30, but with 64.5 kN (14,500 lb st) JT8D-9, 69 kN (15,500 lb st) JT8D-15 or 71.2 kN (16,000 lb st) JT8D-17 turbofans, increased fuel capacity, longer fuselage accommodating up to 125 passengers, and greater AUW. First flight was made 28 November 1967 and FAA certification was received 27 February 1968. The

McDonnell Douglas DC-8F-55 cargo aircraft of Air Charter Systems *(Peter J. Cooper)*

1994

first Series 40 was delivered to SAS 29 February 1968 and entered commercial service with that airline 12 March 1968.

Series 50: 'Stretched' short/medium-range development of the Series 30, announced 5 July 1973. High-density seating for up to 139 passengers made possible by a 4.34 m (14 ft 3 in) fuselage extension. A 'new look' interior features enclosed overhead racks for carry-on baggage, sculptured wall panels, acoustically treated ceiling panels and indirect lighting. Available with either Pratt & Whitney JT8D-15 or -17 turbofan engines, rated at 69 kN (15,500 lb st) and 71.2 kN (16,000 lb st) respectively, and embodying sound absorption materials as developed for the engines and nacelles of the DC-10, the Series 50 meets FAR Pt 36 noise requirements. The engines are smokeless and have thrust reversers rotated 17° from the vertical to reduce the possibility of exhaust gas ingestion. The landing gear is fitted with an improved anti-skid braking system. First flight was made 17 December 1974. First deliveries were made to Swissair, with whom it entered service in August 1975.

Versions offered in passenger, cargo (DC-9F), convertible (DC-9CF) or passenger-cargo (DC-9RC) configurations. The cargo and convertible models have a main cargo door measuring 3.45 m (11 ft 4 in) wide and 2.06 m (6 ft 9 in) high. An executive transport version also offered, with increased fuel, enabling up to 15 persons to be carried non-stop over 2,865 n miles (5,300 km; 3,300 miles) transcontinental or transocean stages. First delivery of an all-cargo model, a DC-9 Series 30F, was made to Alitalia 13 May 1968. This model has 122.1 m³ (4,313 cu ft) of cargo space in main cabin, plus the underfloor hold enabling it to carry eight full cargo pallets and two half-pallets with total weight of nearly 18,144 kg (40,000 lb).

C-9A Nightingale: Aeromedical airlift transport, of which weight ordered in 1967 for operation by the 375th Aeromedical Wing of the USAF Military Airlift Command. Essentially an 'off-the-shelf' DC-9 Srs 30 commercial transport, but with JT8D-9 engines, the C-9A is able to carry 30 to 40 litter patients, more than 40 ambulatory patients or a combination of the two, together with two nurses and three aeromedical technicians. The interior includes a special care compartment with separate atmospheric and ventilation controls. Galleys and toilets are provided fore and aft. There are three entrances, two with hydraulically operated stairways. The third has a forward door 2.06 m (6 ft 9 in) high and 3.45 m (11 ft 4 in) wide, with a hydraulically operated ramp, to facilitate loading of litters. First C-9A was rolled out 17 June 1968 and delivered to the US Air Force at Scott AFB 10 August 1968. In total, 23 in service with the USAF.

C-9B Skytrain II: Fleet logistic support transport. Seventeen in service with the USN.

VC-9C: DC-9-30 type aircraft with special configuration, ordered by the USAF for service in Special Air Missions Wing based at Andrews AFB.

In June 1966, the FAA certificated three Cat. II all-weather landing systems for the DC-9, comprising Collins FD-108 flight director system, Sperry AD-200 flight director system, and coupled approach utilising Sperry SP-50A autopilot.

UPGRADES: **McDonnell Douglas:** DC-9X Programme consists of Phase I and Phase II. Phase I is the baseline airframe upgrade to extend the aircraft life by 15 years; Phase II will consist of a series of options relating to the power plant including re-engining and hush kits. In addition to the Phase I and II modifications, the DC-9X programme also includes optional retrofit of equipment to meet Euronav requirements and operation in Category IIIa weather conditions. Cabin improvements and increased operating weights are also offered.

Pemco: DC-9 cargo conversion. See separate entry.

Rockwell: Rockwell Collins Avionics & Communications Division awarded contract February 1992 to supply and integrate the FMS-800 Flight Management System into the USAF KC-10A, E-4 and C-9 aircraft. FMS-800 includes a control display unit with embedded GPS (CDU-900G). The system automates routine cockpit tasks, integrates GPS and the inertial navigation system (INS). Contract value of US$4.2.

OPERATORS: Versions of the DC-9 are in service with the armed forces of the following countries: Italy (2), Kuwait (1) and USA (69).

The following description applies to the DC-9 Series 30:

DESIGN FEATURES: Cantilever low-wing monoplane. Mean wing thickness/chord ratio 11.6 per cent. Sweepback 24° at quarter-chord. Single boundary-layer fence (vortillon) under each wing. Detachable wingtips. Variable incidence T tailplane.

FLYING CONTROLS: Glassfibre trailing-edges on wings, ailerons and flaps. Hydraulically controlled ailerons, each in two sections, outer sections used at low speed only. Wing-mounted speed brakes. Full-span leading-edge slats (also on Srs 20/40/50). Hydraulically actuated double-slotted trailing-edge flaps over 67 per cent of semi-span. Hydraulically actuated variable incidence tailplane. Manually controlled elevators with servo tabs. Hydraulically controlled rudder with manual override. Glassfibre trailing-edges on control surfaces.

STRUCTURE: All-metal construction, with three spars inboard, two spars outboard and spanwise stringers riveted to skin.

McDonnell Douglas DC-9 Super 80 of Austrian Airlines

1992

McDonnell Douglas DC-9 Series 82 operated by American Airlines

1992

The fuselage is an all-metal semi-monocoque structure. The tail unit is a cantilever all-metal structure.

LANDING GEAR: Retractable tricycle type of Menasco manufacture, with steerable nosewheel. Hydraulic retraction, nose unit forward, main units inward. Twin Goodyear wheels on each unit. Mainwheel tyres size 40 × 14. Nosewheel tyres size 26 × 6.60. Goodyear brakes. Hydro-Aire anti-skid units.

POWER PLANT: Two Pratt & Whitney JT8D turbofan engines (details given in Versions), pod-mounted on each side of rear fuselage. Engines fitted with 40 per cent target-type thrust reversers for ground operation only. Standard fuel capacity 13,925 litres (3,679 US gallons; 3,063 Imp gallons) as well as in Srs 20 and 40. Srs 50 19,073 litres (5,038 US gallons; 4,195 Imp gallons).

ACCOMMODATION: Crew of two on flight deck, plus cabin attendants. Normal accommodation in main cabins for 105 passengers, with seating for maximum of 115 with reduced facilities. Fully pressurised and air conditioned. Toilets at rear of cabin. Provision for galley. Passenger door at front of cabin on port side, with electrically operated built-in airstairs. Optional ventral stairway. Servicing and emergency exit door opposite on starboard side. Underfloor freight and baggage holds, with forward and rear doors on starboard side.

SYSTEMS: Thermal anti-icing of leading-edges.

DIMENSIONS, EXTERNAL:

Wing span: Srs 20, 30, 40, 50	28.47 m (93 ft 5 in)
Wing aspect ratio: Srs 20, 30, 40, 50	8.71
Length overall: Srs 20	31.82 m (104 ft 4¾ in)
Srs 30	36.37 m (119 ft 3½ in)
Srs 40	38.28 m (125 ft 7¼ in)
Srs 50	40.72 m (133 ft 7¼ in)
Height overall: Srs 20, 30	8.38 m (27 ft 6 in)
Srs 40, 50	8.53 m (28 ft 0 in)
Tailplane span:	
Srs 20, 30, 40, 50	11.23 m (36 ft 10¾ in)
Wheel track: Srs 20, 30, 40, 50	5.03 m (16 ft 6 in)
Wheelbase: Srs 20	13.32 m (43 ft 8½ in)
Srs 30	16.22 m (53 ft 2½ in)
Srs 40	17.10 m (56 ft 1¼ in)
Srs 50	18.56 m (60 ft 11 in)
Passenger door (port, fwd): Height	1.83 m (6 ft 0 in)
Width	0.85 m (2 ft 9½ in)
Height to sill	2.13 m (6 ft 11½ in)
Service door (stbd, fwd): Height	1.22 m (4 ft 0 in)
Width	0.69 m (2 ft 3 in)
Height to sill	2.18 m (7 ft 2 in)
Freight and baggage hold doors:	
Height	1.27 m (4 ft 2 in)
Width: fwd	1.35 m (4 ft 5 in)
rear	0.91 m (3 ft 0 in)
Height to sill	1.07 m (3 ft 6 in)

DIMENSIONS, INTERNAL:

Cabin: Max width	3.07 m (10 ft 1 in)
Floor width	2.87 m (9 ft 5 in)
Max height	2.06 m (6 ft 9 in)
Carry-on baggage compartment:	
Srs 20	1.42 m³ (50 cu ft)
Freight hold (underfloor): Srs 20	17.00 m³ (600 cu ft)
Srs 30	25.30 m³ (895 cu ft)
Srs 40	28.90 m³ (1,019 cu ft)
Srs 50	29.30 m³ (1,034 cu ft)

AREAS:

Wings, gross: Series 20, 30, 40, 50	92.97 m² (1,000.7 sq ft)
Ailerons (total): Srs 50	3.53 m² (38.0 sq ft)
Trailing-edge flaps (total): Srs 50	19.58 m² (210.8 sq ft)
Leading-edge slats (total): Srs 50	11.22 m² (120.8 sq ft)
Spoilers (total): Srs 50	3.22 m² (34.7 sq ft)
Fin: Srs 50	14.96 m² (161.0 sq ft)
Rudder: Srs 50	6.07 m² (65.3 sq ft)
Tailplane: Srs 50	25.60 m² (275.6 sq ft)
Elevators, incl tabs: Srs 50	9.83 m² (105.8 sq ft)

WEIGHTS AND LOADINGS:

Manufacturer's empty weight:	
Srs 20	23,985 kg (52,880 lb)
Srs 30	25,940 kg (57,190 lb)
Srs 40	26,612 kg (58,670 lb)
Srs 50	28,068 kg (61,880 lb)
Max space-limited payload: Srs 20	9,925 kg (21,885 lb)
Srs 30	14,118 kg (31,125 lb)
Srs 40	15,610 kg (34,415 lb)
Max weight-limited payload:	
Srs 20	10,565 kg (23,295 lb)
Srs 30	12,743 kg (28,094 lb)
Srs 40	14,363 kg (31,665 lb)
Srs 50	15,617 kg (34,430 lb)
Max T-O weight: Srs 20	44,450 kg (98,000 lb)
Srs 30, 40, 50	54,885 kg (121,000 lb)
Max rate weight:	
Srs 20, 30, 40, 50	55,338 kg (122,000 lb)
Max zero-fuel weight: Srs 20	35,380 kg (78,000 lb)
Srs 30, 40, 50	44,678 kg (98,500 lb)
Max landing weight: Srs 20	42,365 kg (93,400 lb)
Srs 30, 40, 50	49,895 kg (110,000 lb)

McDonnell Douglas C-9A Nightingale aeromedical aircraft of the USAF *(Paul Tompkins)*

1995

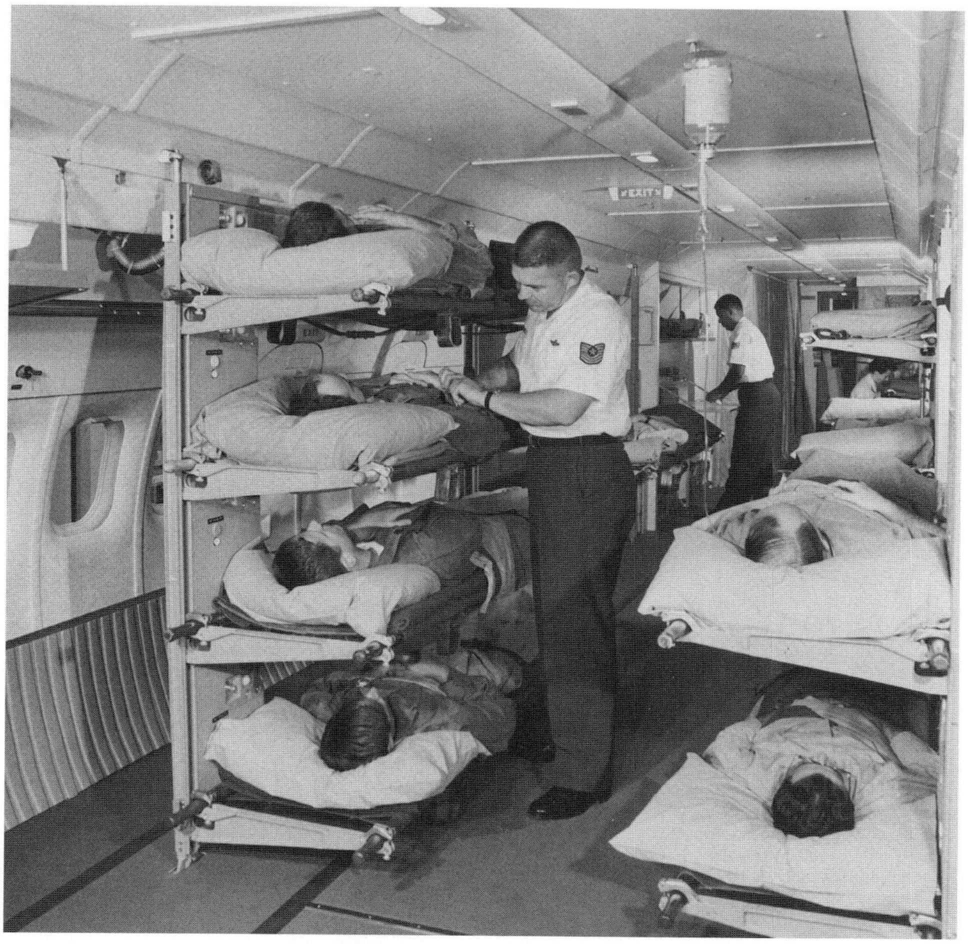

McDonnell Douglas DC-9 C-9A Nightingale interior mock-up

1992

McDonnell Douglas DC-9X artist's impression of the proposed upgrade programme

1994

PERFORMANCE (at max T-O weight, except where indicated):
Never-exceed speed:
Srs 50 537 kts (995 km/h; 618 mph)
Max level speed: Srs 50 500 kts (926 km/h; 575 mph)
Max cruising speed at 7,620 m (25,000 ft):
Srs 20 494 kts (915 km/h; 568 mph)
Srs 30 490 kts (907 km/h; 564 mph)
Srs 40, 50 485 kts (898 km/h; 558 mph)
Average long-range cruising speed at 9,145-10,675 m (30,000-35,000 ft) 443 kts (821 km/h; 510 mph)
Max rate of climb at S/L: Srs 20 1,035 m (3,400 ft)/min
Srs 30 885 m (2,900 ft)/min
Srs 40 865 m (2,850 ft)/min
Srs 50 792 m (2,600 ft)/min
FAA T-O field length: Srs 20 1,555 m (5,100 ft)
Srs 30 1,685 m (5,530 ft)
Srs 40 2,088 m (6,850 ft)
FAR T-O distance to 10.7 m (35 ft):
Srs 20 1,495 m (4,900 ft)
Srs 30, 40 2,255 m (7,400 ft)
Srs 50 2,445 m (8,020 ft)
FAA landing field length: Srs 20 1,355 m (4,450 ft)
Srs 30 1,425 m (4,680 ft)
Srs 40 1,440 m (4,720 ft)
Srs 50 1,485 m (4,880 ft)
FAR landing distance from 15 m (50 ft):
Srs 50 1,440 m (4,720 ft)
Range at Mach 0.8, with reserves for 200 n miles (370 km; 230 miles) flight to alternate and 60 min at 3,050 m (10,000 ft):
Srs 20 at 7,620 m (25,000 ft) with 50 passengers and baggage 1,140 n miles (2,111 km; 1,312 miles)
Srs 30 at 9,150 m (30,000 ft) with 64 passengers and baggage 1,160 n miles (2,148 km; 1,335 miles)
Srs 40 at 7,620 m (25,000 ft) with 70 passengers and baggage 930 n miles (1,723 km; 1,071 miles)
Range at long-range cruising speed at 9,150 m (30,000 ft), reserves for 200 n miles (370 km; 230 miles) flight to alternate and 45 min continued cruise at 9,150 m (30,000 ft):
Srs 20 with 63 passengers and baggage 1,605 n miles (2,974 km; 1,848 miles)
Srs 30 with 80 passengers and baggage 1,670 n miles (3,095 km; 1,923 miles)
Srs 40 with 87 passengers and baggage 1,555 n miles (2,880 km; 1,790 miles)
Srs 50 with 97 passengers and baggage 1,795 n miles (3,326 km; 2,067 miles)
Ferry range, reserves as above:
Srs 20 1,865 n miles (3,455 km; 2,147 miles)
Srs 30 1,980 n miles (3,669 km; 2,280 miles)
Srs 40 1,850 n miles (3,428 km; 2,130 miles)
Srs 50 2,185 n miles (4,049 km; 2,516 miles)

UPDATED

MCDONNELL DOUGLAS C-9B SKYTRAIN II

TYPE: Twin-turbofan passenger/cargo transport aircraft.

PROGRAMME: The US Navy's C-9B Skytrain II is a special convertible passenger/cargo version of the DC-9 Series 30 commercial transport named after long enduring Navy R4D Skytrain, a DC-3 variant of which 624 were procured by that service.

The contract for five (subsequently increased to eight) C-9Bs was signed by Naval Air Systems Command 24 April 1972, and the first of these aircraft made its first flight 7 February 1973, two months ahead of schedule. The first two aircraft were delivered 8 May 1973, to Fleet Tactical Support Squadrons 1 (VR-1) at NAS Norfolk, Virginia, and 30 (VR-30) at NAS Alameda, California. All eight were delivered during 1973. A further six C-9Bs were ordered in late 1974, and delivery was completed by mid-1976. The Skytrain is still in service with the USN (29).

A compromise between the DC-9 Series 30 and 40, the C-9B has the overall dimensions of the former, 64.5 kN (14,500 lb st) Pratt & Whitney JT8D-9 turbofan engines and the optional 3.45 m (11 ft 4 in) by 2.06 m (6 ft 9 in) cargo door at the port forward end of the cabin. This allows loading of standard military pallets measuring 2.24 m (7 ft 4 in) by 2.74 m (9 ft 0 in), and in an all-cargo configuration eight of these can be carried, representing a total cargo load of 14,716 kg (32,444 lb). When loading, each pallet is first lifted to door sill height, and then rolled forward on to a ball transfer system before being positioned finally by means of roller tracks.

Normal flight crew consists of pilot, co-pilot, crew chief and two cabin attendants, and standard accommodation is for 90 passengers on five-abreast seating at 97 cm (38 in) pitch, or up to 107 passengers at 86 cm (34 in) pitch. In a typical passenger/cargo configuration, three pallets are carried in the forward area, with 45 passengers in the rear section. A galley and toilet are located at each end of the cabin. In all-cargo or mixed passenger/cargo configuration, a cargo net can be erected at the forward end of the cabin; in latter configuration a smoke barrier curtain is placed between the cargo section and the passengers.

Normal passenger access is by means of forward port and aft ventral doors, each with hydraulically operated airstairs to make the C-9B independent of ground facilities. The ventral door allows passengers to board while cargo is being loaded in the forward area. Two Type III emergency exits, each 0.91 m (3 ft 0 in) by 0.51 m (1 ft 8 in), are positioned on each side of the fuselage to permit overwing escape, and four 25-man liferafts are carried in stowage racks. To complete the C-9Bs independence of ground facilities, an auxiliary power unit provides both electrical and hydraulic services when the aircraft is on the ground. An environmental control system maintains a sea level cabin altitude to a height of 5,640 m (18,500 ft) and a 2,440 m (8,000 ft) cabin altitude to 10,670 m (35,000 ft).

A maximum fuel capacity of 22,443 litres (5,929 US gallons; 4,937 Imp gallons) provides a ferry range of 2,953 n miles (5,472 km; 3,400 miles), the standard wing fuel tanks being supplemented by a 4,732 litre (1,250 US gallon; 1,040 Imp gallon) tank in the forward underfloor freight hold, and a 3,785 litre (1,000 US gallon; 832 Imp gallon) tank in the aft hold.

Advanced nav/com equipment is installed, including Omega and inertial navigation systems, and FAA certification has been received for both manual and automatic approaches under Cat. II weather conditions.

DIMENSIONS, EXTERNAL:
Wing span: Srs 20, 30, 40, 50 28.47 m (93 ft 5 in)
Wing aspect ratio: Srs 20, 30, 40, 50 8.71
Length overall: Srs 20 31.82 m (104 ft 4¾ in)
Srs 30 36.37 m (119 ft 3½ in)
Srs 40 38.28 m (125 ft 7¼ in)
Srs 50 40.72 m (133 ft 7¼ in)
Height overall: Srs 20, 30 8.38 m (27 ft 6 in)
Srs 40, 50 8.53 m (28 ft 0 in)
Tailplane span:
Srs 20, 30, 40, 50 11.23 m (36 ft 10¾ in)
Wheel track: Srs 20, 30, 40, 50 5.03 m (16 ft 6 in)
Wheelbase: Srs 20 13.32 m (43 ft 8½ in)
Srs 30 16.22 m (53 ft 2½ in)
Srs 40 17.10 m (56 ft 1¼ in)
Srs 50 18.56 m (60 ft 11 in)
Passenger door (port, fwd): Height 1.83 m (6 ft 0 in)
Width 0.85 m (2 ft 9½ in)
Height to sill 2.13 m (6 ft 11½ in)
Service door (stbd, fwd): Height 1.22 m (4 ft 0 in)
Width 0.69 m (2 ft 3 in)
Height to sill 2.18 m (7 ft 2 in)
Freight and baggage hold doors:
Height 1.27 m (4 ft 2 in)
Width: fwd 1.35 m (4 ft 5 in)
rear 0.91 m (3 ft 0 in)
Height to sill 1.07 m (3 ft 6 in)

DIMENSIONS, INTERNAL:
Cabin: Length 20.73 m (68 ft 0 in)
Width 3.05 m (10 ft 0 in)
Volume (cargo) 118.9 m³ (4,200 cu ft)
Baggage holds (underfloor):
forward 8.44 m³ (298 cu ft)
aft 3.82 m³ (135 cu ft)

WEIGHTS AND LOADINGS:
Operating weight, empty:
passenger configuration 29,612 kg (65,283 lb)
cargo configuration 27,082 kg (59,706 lb)
Max ramp weight 50,350 kg (111,000 lb)
Max T-O weight 49,900 kg (110,000 lb)
Max landing weight 44,906 kg (99,000 lb)

PERFORMANCE (at max T-O weight unless otherwise specified):
Max cruising speed 500 kts (927 km/h; 576 mph)
Long-range cruising speed 438 kts (811 km/h; 504 mph)
Military critical field length 2,259 m (7,410 ft)
Landing distance, at max landing weight 786 m (2,580 ft)

McDonnell Douglas C-9B Skytrain II of the US Navy

1994

Range, long-range cruising speed at 9,145 m (30,000 ft) with 4,535 kg (10,000 lb) payload
2,538 n miles (4,704 km; 2,923 miles)

VERIFIED

MCDONNELL DOUGLAS DC-10

TYPE: Three-turbofan medium-range airliner.

PROGRAMME: First flight of original DC-10 Series 10, 29 August 1970 (see also under Versions); certificated 29 July 1971; last delivery Spring 1989.

VERSIONS: DC-10 Series 10: Initial version with maximum T-O weight 185,970 kg (410,000 lb) for US domestic services; three General Electric CF6-6D rated at 178 kN (40,000 lb st) each or -6D1 at 182.4 kN (41,000 lb st) each; later maximum T-O weight 206,385 kg (455,000 lb), with added centre wing fuel.

Series 15: Series 10 with 206,385 kg (455,000 lb) maximum T-O weight and 207 kN (46,500 lb st) GE CF6-50C2 engines; seven built.

Series 30: Intercontinental version; first flight 21 June 1972; certificated 21 November 1972; first deliveries to KLM and Swissair. Initial deliveries at maximum T-O weight 251,745 kg (555,000 lb); engines were CF6-50As rated at 218 kN (49,000 lb st) each or -50C at 227 kN (51,000 lb st) each. Later versions had maximum T-O weight up to 263,085 kg (580,000 lb) and powered by CF6-50C1s or -50C2s rated at 233.5 kN (52,500 lb st) each or CF6-50C2Bs at 236 kW (53,000 lb st) each; other features include more fuel, wing span increased by 3.05 m (10 ft 0 in), and additional two-wheel landing gear leg mounted on centreline.

Series 30ER: Further range extension with fuel tank in rear cargo compartment; powered by 240.2 kW (54,000 lb st) GE CF6-50C2B turbofans; first order (two) by Swissair July 1980 plus conversion of two existing Series 30s.

Series 30CF: Convertible freighter, also basis of USAF KC-10A Extender tanker/transport (which see). First flight 28 February 1973 powered by GE CF6-50A turbofans; first deliveries 1973 to Trans International Airlines and Overseas National Airways; total more than 30 Series 30CFs ordered. Similar to passenger Series 30 and 40 with capacity variable between 380 passengers plus baggage and 64,860 kg (143,000 lb) all-freight on intercontinental routes or up to 70,626 kg (155,700 lb) all-cargo on US transcontinental routes; two later Series 30CF powered by 233.5 kN (52,500 lb st) GE CF6-50C1s delivered to Overseas National 1977. Features include overnight conversion; upward-opening side freight door 2.59 m high × 3.56 m wide (8 ft 6 in × 11 ft 8 in); maximum capacity 30 standard pallets on main deck and 132 m^3 (4,670 cu ft) underfloor bulk cargo capacity.

Series 30F: Cargo-only version, generally similar to Series 30CF. Nine ordered by Federal Express delivered between 24 January 1986 and October 1988. Cargo capacity 80,282 kg (176,992 lb) with intercontinental range; powered by GE CF6-50C2 turbofans; main deck capacity 23 standard pallets; forward and centre underfloor bulk cargo compartments total 117.8 m^3 (4,108 cu ft) volume.

Series 40: Extended range intercontinental version, powered by Pratt & Whitney JT9D turbofans; auxiliary fuel tank in rear of cargo compartment; first flight 28 February 1972; certificated 20 October 1972; Northwest Airlines received 22 early examples powered by 220 kN (49,400 lb st) JT9D-20s with water injection; first flight of later version for Japan Air Lines powered by 236 kN (53,000 lb st) JT9D-59As 25 July 1975. Delco Electronics performance management system (PMS) coupled to autothrottle and autopilot, saving 1 to 3 per cent furl, certificated early 1983; first PMS-equipped DC-10 in service with JAL March 1983.

UPGRADES: **KLM Engineering and Maintenance:** Contract awarded November 1993 to convert two DC-10 civil airliners to military tanker/transports for the Royal Netherlands Air Force. Modification work to include: installation of a 13 m (43 ft) aerial refuelling boom on the aft fuselage and a new remote aerial refuelling operator's system. Engineering for the aircraft modification to be carried out at McDonnell Douglas' Long Beach facilities. Aircraft to enter service in late 1995.

Sabena Technics: DC-10 cargo conversion. See separate entry.

The following description applies to the Series 30, 30ER and 40:

DESIGN FEATURES: Cantilever low-wing monoplane of all-metal fail-safe construction. Several different Douglas wing sections. Thickness/chord ratio varies from slightly more than 12.2 per cent at root to less than 8.4 per cent at tip. Dihedral 5° 14.4′ inboard, 3° 1.8′ outboard. Incidence positive at wingroot, negative at tip. Sweepback at quarter-chord 35°. Variable incidence tailplane.

FLYING CONTROLS: All-metal inboard and outboard ailerons, the former used conventionally, the latter only when the leading-edge slats are extended. Double-slotted all-metal trailing-edge flaps mounted on external hinges, with an inboard and outboard flap panel on each wing. Five all-metal spoiler panels on each wing, at rear edge on each wing of the fixed-wing structure. All spoilers operate in unison as lateral control, speed brake and ground spoilers. Full-span three-position all-metal leading-edge slats. Ailerons are powered by Bertea hydraulic actuators, spoilers by Parker-Hannifin hydraulic actuators. Each aileron is powered by either of two hydraulic systems each spoiler is powered by a single system. All leading-edge slat segments outboard of the engines are anti-iced with engine bleed air. Variable incidence tailplane, actuated by Vickers hydraulic motors. Longitudinal and directional controls are fully powered and comprise inboard and outboard elevators, each segment powered by a Bertea tandem actuator; upper and lower rudder each powered by a Bertea actuator. Rudder standby power supplied by two Abex transfer motor pumps.

STRUCTURE: The fuselage is an aluminium semi-monocoque fail-safe structure of circular cross-section. Except for auxiliary areas, the entire fuselage is pressurised. The tail unit is a cantilever all-metal structure. Graphite epoxy has been used to make upper rudder sections on some DC-10s.

LANDING GEAR: Hydraulically retractable tricycle type, with gravity free-fall for emergency extension. Twin-wheel steerable nose unit (±68°) retracts forward; Menasco main units inward into fuselage. Main gear comprises two four-wheel bogies, with an additional dual-wheel main unit mounted on the fuselage centreline between the bogie units and retracting forward. Oleo-pneumatic shock-absorbers in all units. Goodyear nosewheels and tyres size 40 × 15.5-16, pressure 12.41 bars (180 lb/sq in). Main units and centreline unit have Goodyear wheels and tyres size 52 × 20.5-23. The former have a pressure of 11.38 bars (165 lb/sq in), the latter 9.65 bars (140 lb/sq in). Goodyear disc brakes and anti-skid system, with individual wheel control.

POWER PLANT: Three turbofan (details under Versions listings), two of which are mounted on underwing pylons, the third above the rear fuselage at the base of the fin. All engines are fitted with thrust reversers for ground operation. Engine air inlets have load-carrying acoustically treated for noise attenuation, and each engine fan case and fan exhaust is similarly treated. Four integral wing fuel tanks and an auxiliary tank in the wing centre-section with a connected structural compartment fitted with a bladder cell, giving total capacity of approximately 138,165 litres (36,500 US gallons; 30,392 Imp gallons). Four standard pressure refuelling adaptors, two in each wing outboard of the engine pylons. Lower cargo hold can be used to carry an optional long-range tank of either 5,807 litres (1,534 US gallons; 1,277 Imp gallons) or 12,556 litres (3,317 US gallons; 2,762 Imp gallons). Oil capacity (usable): Series 30, 34.1 litres (9 US gallons; 7.5 Imp gallons); Series 40, 56.8 litres (15 US gallons; 12.5 Imp gallons).

ACCOMMODATION: Crew of three (pilot, first officer, flight engineer), with seating for two observers, plus cabin attendants. Standard seating for 255 or 270 in mixed class versions, with a maximum of 380 passengers in an economy class arrangement. Two aisles run the length of the cabin, which is separated into sections by cloakroom dividers. In the first class section, with three pairs of reclining

McDonnell Douglas DC-10-30 of British Airways *(Peter J. Cooper)*

1994

McDonnell Douglas DC-10-30F freighter of Federal Express *(Peter J. Cooper)*

1994

seats abreast, the aisles are 0.78 m (2 ft 7 in) wide. In the coach class section four pairs of seats, with a table between the centre pairs, also have two aisles being 0.51 m (1 ft 8 in) wide. One pair of seats is exchanged for a three-seat unit in the nine-abreast high-density layout. Up to nine lavatories located throughout the passenger cabin. Cloakrooms of standard and elevating type distributed throughout the cabin. Overhead stowage modules, fully enclosed and providing stowage for passengers' personal effects, are located on the sidewalls and extend the full length of the cabin. Optional centreline overhead baggage racks available. Eight passenger doors, four on each side, open by sliding inward and upward into the above-ceiling area. Containerised or bulk cargo compartments located immediately forward and aft of the wing, with outward-opening doors on the starboard side, maximum capacity 14 LD3 containers. A bulk cargo compartment is located in the lower rear section of the fuselage, with its door on the port side. Entire accommodation is fully air conditioned, with five separate control zones for the below-floor galley configuration. Optional main cabin galley to replace the lower galley, and in this configuration there are four separate control zones for the air conditioning. The lower deck is provided with five to eight high-temperature ovens, and with refrigerators, storage space for linen, china and other accessories. Serving carts are taken to cabin level by two electric elevators, to a buffet service centre, from where stewardesses serve passengers. To permit quick turnaround at terminals, without interference to passenger movement in the main cabin, the kitchen is provisioned through the cargo doors at ground level.

SYSTEMS: Three parallel continuously operating and completely separate hydraulic systems supply the fully powered flight controls and wheel brakes. Normally, one of the systems supplies power for landing gear actuation. Two reversible motor pumps, each sized to deliver power from one of the two systems for standby operation of landing gear, can also power any other hydraulically operated unit. Each hydraulic system is powered by two identical engine-driven pumps, capable of delivering a total of 256 litres (70 US gallons; 58.3 Imp gallons)/min at 207 bars (3,000 lb/sq in) at take-off. All three hydraulic systems are applied to each primary control axis in a manner which ensures maximum control effectiveness in the event of a single or dual hydraulic system failures. A Garrett TSCP-700-4 APU provides ground electrical and pneumatic power, including main engine starting and auxiliary electrical power in flight.

AVIONICS AND EQUIPMENT: A dual fail-operative landing system is installed to meet Cat. IIIA weather minima. Triple inertial navigation system meeting ARINC 561 requirements on Srs 30 and 40, with optional dual area navigation system capability.

DIMENSIONS, EXTERNAL:
Wing span 50.40 m (165 ft 4½ in)
Wing chord: at root 10.71 m (35 ft 1¾ in)
at tip 2.73 m (8 ft 11½ in)
Wing aspect ratio 7.5
Length overall 55.50 m (182 ft 1 in)
Fuselage: Length 51.97 m (170 ft 6 in)
Max width 6.02 m (19 ft 9 in)
Height overall 17.70 m (58 ft 1 in)
Tailplane span 21.69 m (71 ft 2 in)
Wheel track 10.67 m (35 ft 0 in)
Wheelbase: Series 30 22.05 m (72 ft 4 in)
Series 40 22.07 m (72 ft 5 in)

DIMENSIONS, INTERNAL:
Length:
Length, from rear bulkhead of flight deck to rear cabin bulkhead approx 41.45 m (136 ft 0 in)
Max width 5.72 m (18 ft 9 in)
Height (basic) 2.41 m (7 ft 11 in)
Series 30, 40 in lower galley configuration:
Forward baggage and/or freight hold (forward of wing):
containerised volume 27.2 m^3 (960.0 cu ft)
bulk volume 37.9 m^3 (1,339 cu ft)
Centre baggage and/or freight hold (aft of wing):
containerised volume 36.2 m^3 (1,280 cu ft)
bulk volume 43.9 m^3 (1,552 cu ft)
Rear hold: bulk volume 22.8 m^3 (805.0 cu ft)
Series 30, 40 in upper galley configuration:
Forward baggage and/or freight hold (forward of wing):
containerised volume 72.5 m^3 (2,560 cu ft)
bulk volume 86.2 m^3 (3,045 cu ft)
Centre baggage and/or freight hold (aft of wing):
containerised volume 45.3 m^3 (1,600 cu ft)
bulk volume 54.8 m^3 (1,935 cu ft)
Rear hold: bulk volume 14.4 m^3 (510.0 cu ft)

AREAS:
Wings, gross 367.7 m^2 (3,958 sq ft)
Ailerons: inboard (total) 7.68 m^2 (82.7 sq ft)
outboard (total) 9.76 m^2 (105.1 sq ft)
Trailing-edge flaps (total) 62.1 m^2 (668.2 sq ft)
Leading-edge slats (total) 43.84 m^2 (471.9 sq ft)
Spoilers (total) 12.73 m^2 (137.0 sq ft)
Fin 45.92 m^2 (494.29 sq ft)
Rudders (total) 10.29 m^2 (110.71 sq ft)
Tailplane 96.6 m^2 (1,040.2 sq ft)
Elevators (total) 27.7 m^2 (298.1 sq ft)

WEIGHTS AND LOADINGS:
Basic weight empty: Series 30 121,198 kg (267,197 lb)
Series 40 122,951 kg (271,062 lb)
Max payload: Series 30 48,330 kg (106,550 lb)
Series 40 46,243 kg (101,950 lb)
Max fuel weight: standard 111,387 kg (245,566 lb)
small auxiliary tank installed 116,049 kg (255,844 lb)
large auxiliary tank installed 121,467 kg (267,790 lb)
Max T-O weight:
Series 30 259,450-263,085 kg (572,000-580,000 lb)
Series 40 (-20 engines) 251,745 kg (555,000 lb)
Series 40 (-59A engines) 259,450 kg (572,000 lb)
Max ramp weight: Series 30 260,815 kg (575,000 lb)
Series 40 (-20 engines) 253,105 kg (558,000 lb)
Series 40 (-59A engines) 260,815 kg (575,000 lb)
Max zero-fuel weight:
Series 30, 40 166,922 kg (368,000 lb)
Max landing weight:
Series 30, 40 182,798 kg (403,000 lb)
Max wing loading:
Series 30 705.6 kg/m^2 (144.5 lb/sq ft)
Series 40 (-20 engines) 684 kg/m^2 (140.2 lb/sq ft)
Series 40 (-59A engines) 705.6 kg/m^2 (144.5 lb/sq ft)

PERFORMANCE (at max T-O weight except where indicated):
Never-exceed speed Mach 0.95
Max level speed at 7,620 m (25,000 ft)
Mach 0.88 (530 kts; 982 km/h; 610 mph)
Max cruising speed at 9,145 m (30,000 ft)
Series 30 490 kts (908 km/h; 564 mph)
Series 40 (-20 engines) 489 kts (906 km/h; 563 mph)
Series 40 (-59A engines)
498 kts (922 km/h; 573 mph)
Normal cruising speed, all versions Mach 0.82
T-O speed (V_2): Series 30 (-50C engines)
189 kts (351 km/h; 218 mph)
Series 40 (-20 engines) 187 kts (346 km/h; 215 mph)
Series 40 (-59A engines)
178 kts (330 km/h; 205 mph)
Landing speed (with full load of passengers and baggage):
Series 30, 40 138 kts (256 km/h; 159 mph)
Max rate of climb at S/L:
Series 30 884 m (2,900 ft)/min
Series 40 (-20 engines) 829 m (2,720 ft)/min
Series 40 (-59A engines) 762 m (2,500 ft)/min
Service ceiling:
Series 30 at 249,475 kg (550,000 lb) AUW
10,180 m (33,400 ft)
Series 40 (-20 engines) at 242,670 kg (535,000 lb) AUW) 9,660 m (31,700 ft)
Series 40 (-59A engines) 9,965 m (32,700 ft)
En route climb altitude, one engine out:
Series 30 at 251,744 kg (555,000 lb) AUW
4,360 m (14,300 ft)
Series 40 (-20 engines) at 247,205 kg (545,000 lb) AUW 3,565 m (11,700 ft)
Series 40 (-59A engines) at 254,010 kg (560,000 lb) AUW 5,135 m (16,850 ft)
Min ground turning radius:
about nosewheels:
Series 10, 10F 24.02 m (78 ft 9½ in)
Series 30, 30CF, 40, 40CF 24.32 m (79 ft 9½ in)
about wingtip: Series 10, 10F 33.65 m (110 ft 4¾ in)
Series 30, 30CF, 40, 40CF 35.39 m (116 ft 1¼ in)
FAA T-O field length:
Series 30 (-50C2 engines) 3,170 m (10,400 ft)
Series 40 (-59A engines) 3,135 m (10,280 ft)
FAA landing field length:
Series 30, 40 1,630 m (5,350 ft)
Range with max fuel, no payload:
Series 30 6,504 n miles (12,055 km; 7,490 miles)
Series 40 6,305 n miles (11,685 km; 7,260 miles)
Range with max payload at max zero-fuel weight:
Series 30 4,000 n miles (7,413 km; 4,606 miles)
Series 40 (-20 engines)
3,500 n miles (6,485 km; 4,030 miles)
Series 40 (-59A engines)
4,050 n miles (7,505 km; 4,663 miles)

UPDATED

MCDONNELL DOUGLAS EXTENDER

US Air Force designation: KC-10A

TYPE: Long-range military tanker/transport.

PROGRAMME: Selected as Advanced Tanker/Cargo Aircraft for USAF 19 December 1977; 60 ordered in annual lots and multiyear contract between November 1978 and 1987; first flight 12 July 1980; first delivery 17 March 1981; in service with 6th and 9th Air Refuelling Squadron at March AFB, California, 2nd and 32nd ARS at Barksdale AFB and 344th and 911th ARS at Seymour Johnson AFB, North Carolina; some aircraft shared by AF Reserve 77th, 78th and 79th ARS (Associate); final aircraft delivered 4 April 1990 following trials with Flight Refuelling Mk 32B underwing hose-reel pods.

UPGRADES: **Rockwell:** Rockwell Collins Avionics & Communications Division awarded contract February 1992 to supply and integrate the FMS-800 Flight Management System into the USAF KC-10A, E-4 and C-9 aircraft. FMS-800 includes a control display unit with embedded GPS (CDU-900G). The system automates routine cockpit tasks, integrates GPS and the Inertial Navigation System (INS). Contract value of US$4.2.

OPERATORS: Versions of the KC-10 are in service with the armed forces of the following countries: Netherlands (2 on order) and USA (59).

DESIGN FEATURES: Based on DC-10 Series 30CF (which see); can deliver 90,718 kg (200,000 lb) fuel to receivers at radius of 1,910 n miles (3,540 km; 2,200 miles) from base; total onboard fuel, usable for own range or transferable, includes 108,062 kg (238,236 lb) of standard aircraft tankage plus seven fuel cells in underfloor cargo compartment totalling 53,446 kg (117,829 lb) fuel; maximum transfer rate 5,678 litres (1,500 US gallons; 1,249 Imp gallons) through boom and 1,590 litres (420 US gallons; 350 Imp gallons) through hose; Sargent-Fletcher FR600 hose/reel unit with maximum flow 2,271 litres (600 US gallons; 499.6 Imp gallons)/min in fuselage for US Navy/Marine receivers; KC-10A can be refuelled in flight through boom receptacle above flight deck; powered by three 233.53 kN (52,500 lb st) GE CF6-50C2 turbofans.

McDonnell Douglas KC-10A long-range military tanker/transport

1994

DIMENSIONS, EXTERNAL:

Wing span	50.40 m (165 ft 4½ in)
Wing chord: at root	10.71 m (35 ft 1¾ in)
at tip	2.73 m (8 ft 11½ in)
Wing aspect ratio	7.5
Length overall	55.35 m (182 ft 7 in)
Fuselage: Length	51.97 m (170 ft 6 in)
Max width	6.02 m (19 ft 9 in)
Height overall	17.70 m (58 ft 1 in)
Tailplane span	21.69 m (71 ft 2 in)
Wheel track	10.57 m (34 ft 8 in)
Wheelbase	22.05 m (72 ft 4 in)

DIMENSIONS, INTERNAL:

Cabin:

Length, from rear bulkhead of flight deck to rear cabin bulkhead	approx 41.45 m (136 ft 0 in)
Max width	5.72 m (18 ft 9 in)
Height (basic)	2.41 m (7 ft 11 in)

AREAS:

Wings, gross	367.7 m² (3,958 sq ft)
Ailerons: inboard (total)	7.68 m² (82.7 sq ft)
outboard (total)	9.76 m² (105.1 sq ft)
Trailing-edge flaps (total)	62.1 m² (668.2 sq ft)
Leading-edge slats (total)	43.84 m² (471.9 sq ft)
Spoilers (total)	12.73 m² (137.0 sq ft)
Fin	45.92 m² (494.29 sq ft)
Rudders (total)	10.29 m² (110.71 sq ft)
Tailplane	96.6 m² (1,040.2 sq ft)
Elevators (total)	27.7 m² (298.1 sq ft)

WEIGHTS AND LOADINGS:

Operating weight empty: tanker	109,328 kg (241,027 lb)
cargo	110,945 kg (244,591 lb)
Fuel at T-O: tanker	158,291 kg (348,973 lb)
Design fuel capacity	161,508 kg (356,065 lb)
Max cargo capacity	76,843 kg (169,409 lb)
Design max T-O weight	267,620 kg (590,000 lb)
Max wing loading	727.8 kg/m² (149.06 lb/sq ft)
Max power loading	382.2 kg/kN (3.75 lb/lb st)

PERFORMANCE:

Critical field length	3,124 m (10,250 ft)
Max range with max cargo	3,797 n miles (7,032 km; 4,370 miles)
Max ferry range, unrefuelled	9,993 n miles (18,507 km; 11,500 miles)

McDonnell Douglas KC-10A of the US Air Mobility Command *(Paul Tompkins)*

1995

UPDATED

MCDONNELL DOUGLAS HELICOPTER SYSTEMS (Division of McDonnell Douglas Aerospace)

5000 East McDowell Road, Mesa, Arizona 85205-9797
Telephone: 1 (602) 891 3000
Fax: 1 (602) 891 5599
Telex: 3719337 MD HC C MESA
OTHER WORKS: Culver City, California 90230
Telephone: 1 (213) 305 5000
Telex: 182436 HU HELI C CULV
PRESIDENT: Dean C. Borgman
SENIOR VICE-PRESIDENT, OPERATIONS: Patrick R. McGinnis
SENIOR VICE-PRESIDENT, PROGRAMMES: Chuck Suyo
VICE-PRESIDENTS:
Ervin J. Hunter (Military Programmes)
Andrew H. Logan (Commercial Programmes)
Al Winn (Engineering Division)
MANAGER, COMMUNICATIONS: Ken Jensen

Hughes Helicopters Inc became subsidiary of McDonnell Douglas Corporation 6 January 1984; name changed to McDonnell Douglas Helicopter Company 27 August 1985; changed again to McDonnell Douglas Helicopter Systems in 1993. In 1994, alliance announced with Hunting Aviation of Singapore to present McDonnell Douglas helicopters in nations of Pacific Rim and Middle East; teamed with Bell Helicopter Textron for US Army Light Helicopter (LH) competition (unsuccessful); Chain Gun systems manufactured at Culver City.

Main company base at Mesa, Arizona, with 52,955 m² (570,000 sq ft) AH-64 Apache assembly and testing factory and another 123,980 m² (1,334,500 sq ft) completed in 1986; MD 500/530 production line transferred to Mesa 1986-87; workforce 3,800 mid-1993. Single-engined helicopter deliveries totalled 114 in 1991-92 (57 in each year); 1992 total included 38 MD 520Ns.

Model 300 helicopter design rights sold to Schweizer Aircraft Corporation at Elmira, New York, 1986, following licence production by Schweizer since 1983; McDonnell Douglas helicopters produced under licence by RACA, Argentina (civil 500D and 500E), Kawasaki, Japan (civil and military 500D), Korean Air (civil and military 500D and 500E, excluding TOW variants, and fuselages for all MD 500s sold worldwide), Agusta, Italy (500D, 500E and 530F civil variants); Agusta licence extended mid-1993 to include MD 520N (first Italian deliveries mid- to late 1994).

VERIFIED

MCDONNELL DOUGLAS (HUGHES) MODEL 500 (CIVIL VERSIONS)

TYPE: Turbine-powered civil light helicopter.

PROGRAMME: The Model 500, which entered full-scale production in November 1968, originated as a civil development of the OH-6A Cayuse military helicopter. Several new military export versions have been developed from it; these are described separately.

VERSIONS: **Model 500:** Initial basic production version, with Allison 250-C18A turboshaft engine.

Model 500C: Similar to Model 500, but with Allison 250-C20 engine and improved 'hot and high' performance. Licence manufacture also undertaken by RACA (Argentina) and Kawasaki (Japan).

Model 500D: Announced in February 1975, the 500D is similar in size and general appearance to the Model 500C. It differs in having a 313 kW (420 shp) Allison 250-C20B engine; a five-blade main rotor; engine exhaust muffler; sound blanketing of the complete power plant assembly, including the engine air intake; and reshaping of the tips of the main rotor blades. It introduced also a small T tail which gives greater flight stability in both high- and low-speed regimes, as well as better handling characteristics in abnormal manoeuvres. Construction of the prototype and its first flight took place in August 1974, the first flight of a production aircraft was made 9 October 1975. Certificated by FAA 8 December 1976. The 1,000th Model 500D was delivered in July 1981. Licence manufacture by BredaNardi (Italy), Kawasaki (Japan) and KAL (South Korea). Supplied also to air forces of Jordan (8) and Kenya (2) for training.

Model 500E: Introduced at the Helicopter Association International meeting at Las Vegas, Nevada, in early 1982, this new version of the Model 500 has a longer and more streamlined nose, providing increased legroom and improved field of view for front-seat occupants. Rear-seat passengers have 12 per cent more headroom, additional legroom and, as a result of lowering the bulkhead between front and rear seats, a 50 per cent improvement in forward view. First flown 28 January 1982, the Model 500E (N5294A) incorporates improvements including a new auxiliary fuel tank for 28 per cent greater range; better soundproofing around the transmission and cooling fan, a four-blade tail rotor, longer main rotor blade abrasion strips, new endplate fins, new 'T' grouping of flight instruments, and an improved heating system. A more efficient air conditioning system is optional. Initial deliveries of the Model 500E commenced December 1982.

Model 530F Lifter: See *Jane's All the World's Aircraft 1994-95*.

McDonnell Douglas (Hughes) 500D with Allison 250-C20B turboshaft engine *(Peter March)*

1994

OPERATORS: Versions of the McDonnell Douglas (Hughes) 500/530 are in service with the armed forces of the following countries: Argentina (11), Colombia (23), Costa Rica (2), Denmark (13), Dominican Republic (1), El Salvador (6), Finland (2), Indonesia (12), Israel (33), Italy (50), Japan (283), Jordan (8), Kenya (36), North Korea (70), South Korea (25), Malta (2), Mexico (3), Philippines (20), Spain (10), Taiwan (18) and USA (210).

The following description applies to Models 500, 500C and 500D:

DESIGN FEATURES: Four-blade fully articulated main rotor (five blade on 500D), with blades attached to laminated strap retention system by means of quick-disconnect pins for folding. Two-blade tail rotor. Three sets of bevel gears, three driveshafts and one overrunning clutch. Main rotor/engine rpm ratio 1:12.806 on 500/500C; 1:12.594 on 500D. Tail rotor/engine rpm ratio 1:1.987 on 500/500C; 1:1.956 on 500D. The tail unit comprising fixed fin, horizontal stabiliser and ventral fin on Model 500/500C; Model 500D has T tail with horizontal stabiliser at tip of narrow-chord sweptback fin; small auxiliary fin at tip of tailplane on each side; narrow-chord sweptback ventral fin with integral tailskid to protect tail rotor in tail-down attitude near ground.

FLYING CONTROLS: Trim tab outboard on each main blade. Tail rotor brake optional.

STRUCTURE: Each main blade consists of an extruded aluminium spar hot-bonded to one-piece wraparound aluminium skin. Tail rotor blades consists of a swaged steel tube spar and glassfibre skin covering (metal skin on 500D). The fuselage is an aluminium semi-monocoque structure of pod and boom type. Clamshell doors at rear of pod give access to engine and accessories.

LANDING GEAR: Tubular skids carried on Hughes single-acting (oleo-pneumatic on 500D) shock-absorbers. Utility floats, snow skis and emergency inflatable floats optional.

POWER PLANT: The 236.5 kW (317 shp) Allison 250-C18A turboshaft engine installed in the 500 is derated to 207 kW (278 shp) for T-O and has a maximum continuous rating of

181 kW (243 shp). Model 500C is powered by a 298 kW (400 shp) Allison 250-C20 turboshaft engine; this also is derated to 207 kW (278 shp) for T-O and has a maximum continuous rating of 181 kW (243 shp). Model 500D powered by a 313 kW (420 shp) Allison 250-C20B turboshaft engine. Two interconnected bladder fuel tanks with combined usable capacity of 240 litres (63.4 US gallons; 52.8 Imp gallons). Self-sealing fuel tank optional in 500D. Refuelling point on starboard side of fuselage. Auxiliary fuel system with 132 litre (35 US gallon; 29 Imp gallon) crashworthy internal fuel tanks, or two external glassfibre fuel cells with combined capacity of 167 litres (44 US gallons; 37 Imp gallons), optional. Oil capacity 5.7 litres (1.5 US gallons; 1 Imp gallon).

ACCOMMODATION: Pilot and four passengers or equivalent freight in 500/500C. Optional accommodation for seven with litter kit in use or with four in passenger compartment. Model 500D has forward bench seat for pilot and two passengers, with two or four passengers, or two litter patients and two medical attendants, in aft portion of cabin. Baggage space, capacity 0.31 m^3 (11 cu ft), under and behind rear seat in five-seat form. Clear space for 1.19 m^3 (42 cu ft) of cargo or baggage with only three front seats in place. Two doors on each side.

SYSTEMS: Electrical system in 500D includes a 150 A engine-driven generator and a Ni/Cd battery.

AVIONICS AND EQUIPMENT (500D): Optional avionics include dual King KY 195 com, KX 175 nav/com, KR 251 com, VHF-251/351 nav/com, IND-350 nav indicator, ADF-650 ADF, and TDR-950 transponder; SunAir ASB 125; intercom system, headsets, microphones; public address system; stereo tape system; and flight management computer system. Standard equipment includes outside air temperature gauge, eight day clock, engine hour meter, five sets inertia-reel shoulder harness, cargo tiedown fittings, fire extinguisher, first aid kit, passenger steps, external power socket, landing light, skid-tip position light, anti-collision strobe lights, cockpit utility light, aft cabin light, and instrument lights. Optional equipment includes dual controls, blind-flying instrumentation, electric hoist, cargo hook, external baggage pods, cargo racks, underfuselage cargo pod, nylon mesh seats, dual-strap shoulder harnesses, and heating/demisting system.

EQUIPMENT (500/500C): Standard equipment includes engine hour meter, navigation lights, clock, and ground handling wheels. Optional equipment includes shatterproof glass, heating system, radios and intercom, attitude and directional gyros, rate of climb indicator, inertia reels and shoulder harnesses for pilot and co-pilot, dual controls, cargo hook, hoist, auxiliary fuel system, fire extinguisher, heated pitot tube, extended landing gear, blade storage rack, litter kit, emergency inflatable floats, inflated utility floats, seating for four in passenger compartment, and first aid kit.

DIMENSIONS, EXTERNAL:

Main rotor diameter: 500/500C	8.03 m (26 ft 4 in)
500D/E	8.08 m (26 ft 6 in)
530	8.38 m (27 ft 6 in)
Main rotor blade chord	0.171 m (6¾ in)
Tail rotor diameter	1.30 m (4 ft 3 in)
Distance between rotor centres:	
500/500C	4.58 m (15 ft 0¼ in)
500D/E	4.62 m (15 ft 2 in)
Length overall, rotors fore and aft:	
500/500C	9.24 m (30 ft 3¾ in)
500D/E	9.30 m (30 ft 6 in)
Length of fuselage	7.01 m (23 ft 0 in)
Height to top of rotor head:	
500/500C	2.48 m (8 ft 1½ in)
500D/E	2.59 m (8 ft 6 in)
Skid track	2.06 m (6 ft 9 in)
Cabin doors (500/500C, fwd, each):	
Height	1.19 m (3 ft 11 in)
Width	0.89 m (2 ft 11 in)
Cabin doors (500D/E, each):	
Height	1.16 m (3 ft 9½ in)
Max width	0.76 m (2 ft 6 in)
Height to sill	0.76 m (2 ft 6 in)
Cargo compartment doors (each):	
Height	1.04 m (3 ft 5 in)
Width	0.88 m (2 ft 10½ in)
Height to sill	0.57 m (1 ft 10½ in)

DIMENSIONS, INTERNAL:

Cabin: Length	2.44 m (8 ft 0 in)
Max width: 500/500C	1.37 m (4 ft 6 in)
500D	1.31 m (4 ft 3½ in)
Max height: 500/500C	1.31 m (4 ft 3½ in)
500D	1.52 m (5 ft 0 in)

AREAS:

Main rotor blades (each):	
500/500C	0.688 m^2 (7.41 sq ft)
500D/E	0.690 m^2 (7.43 sq ft)
Tail rotor blades (each):	
500/500C	0.079 m^2 (0.85 sq ft)
500D/E	0.09 m^2 (0.94 sq ft)
Main rotor disc: 500/500C	50.60 m^2 (544.63 sq ft)
500D/E	50.89 m^2 (547.81 sq ft)
Tail rotor disc (all versions)	1.32 m^2 (14.19 sq ft)
Fin: 500/500C	0.52 m^2 (5.65 sq ft)
500D/E	0.56 m^2 (6.05 sq ft)
Horizontal stabiliser: 500/500C	0.72 m^2 (7.70 sq ft)
500D/E	0.61 m^2 (6.52 sq ft)

McDonnell Douglas (Hughes) 500C civil light helicopter *(Peter March)*

WEIGHTS AND LOADINGS:

Weight empty: 500	493 kg (1,088 lb)
500C	501 kg (1,105 lb)
500D/E	598 kg (1,320 lb)
530E	1,406 kg (3,100 lb)
Fuel load: 500D	181 kg (400 lb)
Max normal T-O weight:	
500/500C	1,157 kg (2,550 lb)
Max overload T-O weight:	
500/500C	1,360 kg (3,000 lb)
Max T-O and landing weight:	
500D/E	1,360 kg (3,000 lb)
530E	1,406 kg (3,100 lb)
Max disc loading: 500D	26.76 kg/m^2 (5.48 lb/sq ft)
Max power loading: 500D	4.35 kg/kW (7.14 lb/shp)

PERFORMANCE (500/500C at max T-O weight):

Max level speed at 305 m (1,000 ft):	
500	132 kts (244 km/h; 152 mph)
Max cruising speed at S/L:	
500C	125 kts (232 km/h; 144 mph)
Max cruising speed at 1,220 m (4,000 ft):	
500C	126 kts (233 km/h; 145 mph)
Cruising speed for max range at S/L:	
500	117 kts (217 km/h; 135 mph)
500C	124 kts (230 km/h; 143 mph)
Max rate of climb at S/L:	
500, 500C	518 m (1,700 ft)/min
Service ceiling: 500	4,390 m (14,400 ft)
500C	4,420 m (14,500 ft)
Hovering ceiling IGE: 500	2,500 m (8,200 ft)
500C	3,960 m (13,000 ft)
Hovering ceiling OGE: 500	1,615 m (5,300 ft)
500C	2,040 m (6,700 ft)
Range at 1,220 m (4,000 ft):	
500	327 n miles (606 km; 377 miles)
Range at 1,220 m (4,000 ft), 2 min warm-up with max fuel, no reserves:	
500C	325 n miles (603 km; 375 miles)

PERFORMANCE (500D/E at max T-O weight, ISA):

Never-exceed speed	152 kts (282 km/h; 175 mph)
Max cruising speed at S/L	139 kts (258 km/h; 160 mph)
Max cruising speed at 1,525 m (5,000 ft)	135 kts (250 km/h; 155 mph)
Econ cruising speed at S/L	130 kts (241 km/h; 150 mph)
Econ cruising speed at 1,525 m (5,000 ft)	126 kts (233 km/h; 145 mph)
Max rate of climb at S/L	579 m (1,900 ft)/min
Service ceiling	4,570 m (15,000 ft)
Hovering ceiling IGE: ISA	2,590 m (8,500 ft)
ISA + 20°C	1,830 m (6,000 ft)
Hovering ceiling OGE: ISA	2,285 m (7,500 ft)
ISA + 20°C	1,370 m (4,500 ft)
Range, 2 min warm-up, standard fuel, no reserves:	
at S/L	260 n miles (482 km; 300 miles)
at 1,525 m (5,000 ft)	287 n miles (531 km; 330 miles)

PERFORMANCE (530E, provisional, at max T-O weight, ISA):

Never-exceed speed at S/L	152 kts (282 km/h; 175 mph)
Max cruising speed from S/L to 1,525 m (5,000 ft)	135 kts (250 km/h; 155 mph)
Econ cruising speed at 1,525 m (5,000 ft)	130 kts (241 km/h; 150 mph)
Max rate of climb at S/L	543 m (1,780 ft)/min
Vertical rate of climb at S/L	279 m (915 ft)/min
Service ceiling	5,335 m (17,500 ft)
Hovering ceiling IGE: ISA	3,660 m (12,000 ft)
ISA + 20°C	2,985 m (9,800 ft)
Hovering ceiling OGE: ISA	2,925 m (9,600 ft)
ISA + 20°C	2,255 m (7,400 ft)
Range, 2 min warm-up, standard fuel, no reserves:	
at S/L	219 n miles (406 km; 252 miles)
at 1,525 m (5,000 ft)	234 n miles (434 km; 269 miles)

UPDATED

1994

MCDONNELL DOUGLAS (HUGHES) MODEL 500 (MILITARY VERSIONS)

TYPE: Turbine-powered light military helicopter.

VERSIONS: **Model 500M:** Initial uprated version of OH-6A. Power plant as for civil Model 500, but fuel capacity of 227 litres (60 US gallons; 50 Imp gallons). First deliveries to Colombian Air Force in April 1968. Now in service also in Argentina, Bolivia, Denmark, Japan, Mexico, Philippines and Spain. The Model 500Ms delivered to the Spanish Navy for ASW duties have an AN/ASQ-81 magnetic anomaly detector installed on the starboard side of the fuselage, and can carry two Mk 44 torpedoes beneath the fuselage. Control boxes for the MAD equipment are mounted on the instrument panel and centre pedestal, and special instrumentation includes a 6 in attitude indicator and radar altimeter. Licence manufacture also undertaken by RACA (Argentina), and by Kawasaki in Japan as the Models 369HM/OH-6J (now ended) and 369D/OH-6D.

Model 500MD Defender: Multirole military version. Structurally similar to civil Model 500D, from which it differs by having as standard or optional equipment self-sealing fuel cells, engine inlet particle separator, armour protection, Hughes 'Black Hole Ocarina' infra-red suppressor, and provisions for the carriage and deployment of a variety of weapons, including TOW missiles. Its diverse capabilities include training, command and control, scout, light attack, ASW, troop lift and logistical support duties. It can carry up to seven people, including the pilot; or, in ambulance configuration, two stretcher patients with attendants as well as a flight crew of two. Licence manufacture also undertaken by BredaNardi in Italy and (Scout and TOW versions) by KAL in South Korea.

500MD Scout Defender: Basic military version, able to carry a variety of alternative weapons, including 14 2.75 in rockets and either a 7.62 mm Minigun with 2,000 rounds of ammunition, one 40 mm grenade launcher, a 7.62 mm EX-34 Chain Gun machine gun with 2,000 rounds of ammunition, or a 30 mm Chain Gun automatic cannon with 600 rounds of ammunition. Operators include Kenyan Army (15), South Korean Air Force (123), and Royal Moroccan Air Force (24).

500MD Quiet Advanced Scout Defender: Basically similar to Scout, but with added quietening kit and Hughes Aircraft mast-mounted sight (MMS). Quietening kit features a slower turning four-blade tail rotor, which imposes no reduction of performance. The MMS, described in the following Defender II paragraph, is mounted on a static mast 61 cm (2 ft) above the main rotor, and enables the aircraft to hover behind cover, using the small sight as a periscope to scan a large area out to a range of 3,000 m (9,840 ft). If employed to spot enemy armour, it is envisaged that Scouts would call in TOW Defenders to attack the targets.

500MD/TOW Defender: Anti-tank version armed with four TOW air-to-ground missiles. The TOW installation comprises four weapon pods, mounted two each side on a tubular mount carried through the lower aft fuselage, a stabilised telescopic sight mounted on the port side of the nose, sight control and armrest for the gunner, and a steering indicator for the pilot. Maximum T-O weight with four TOW missiles 1,360 kg (3,000 lb). In service with air forces of Israel (30), Kenya (15) and South Korea (50). Available also in **500MD/MMS-TOW** version, with Hughes Aircraft mast-mounted sight and 30 mm Chain Gun automatic cannon.

500MD/ASW Defender: Version for anti-submarine warfare and surface search missions, with two crew, search radar on nose, AN/ASQ-81 towed MAD, smoke marker launchers, hauldown gear, emergency 'popout' floats and armament of two Mk 44 or Mk 46 homing torpedoes. Maximum T-O weight 1,610 kg (3,550 lb). Can remain on station for 1 hour 48 minutes when operated at a typical ASW mission radius of 22 to 87 n miles (40 to 160 km; 25 to 100 miles) from ship or shore base. Using its radar,

McDonnell Douglas (Hughes) 500M of the Danish Air Force *(Peter March)*

1994

500MD/ASW could locate enemy destroyers and gunboats up to 150 n miles (275 km; 172 miles) from its base ship during a 2 hour patrol. Twelve delivered to Taiwanese Navy.

500MD Defender II: Multimission version, introduced Summer 1980; available for delivery within two years. Five-blade main rotor standard; four-blade 'quiet' tail rotor optional: this turns at a rate 25 per cent slower than the standard two-blade rotor and is reported to be 47 per cent quieter in operation. Other options include Hughes Aircraft mast-mounted sight (MMS), two twin-round pods for four TOW anti-tank missiles, Hughes 30 mm Chain Gun automatic cannon (with firing rate reduced to 350 rds/min), Hughes 'Black Hole' infra-red suppression system, pod containing two Stinger or other air-to-air missiles, pilot's FLIR night vision system, AN/APR-39 (V-1) equipment to give warning that the helicopter is being tracked by hostile radar-directed weapon systems, self-sealing fuel tanks, auxiliary fuel tanks, and an advanced avionics/mission equipment package. The MMS uses a video link to TV displays for the crew, and includes laser rangefinder. Use of the MMS enables the Defender II to hover virtually out of sight behind trees or natural terrain, while the crew surveys the battlefield over extended ranges.

Standard lightweight avionics equipment (SLAE) as developed for the OH-6A has been adapted for the 500MD with minimal changes. This equipment comprises AN/ARC-164 UHF/AM, AN/ARC-115 UHF/AM, AN/ARC-114 VHF/FM, ARN-89 ADF, APX-72 IFF transponder, AN/ASN-43 directional gyro, ID-1351 heading and bearing indicator, and C-6533/ARC intercom.

500MG Defender: See *Jane's All the World's Aircraft 1995-96.*

Paramilitary MG Defender: See *Jane's All the World's Aircraft 1995-96.*

TOW Defender: See *Jane's All the World's Aircraft 1995-96.*

DIMENSIONS, EXTERNAL (500MD/TOW):

Main rotor diameter	8.05 m (26 ft 4¾ in)
Tail rotor diameter	1.40 m (4 ft 7 in)
Distance between rotor centres	4.63 m (15 ft 2½ in)
Length overall, rotors turning	9.39 m (30 ft 9½ in)
Length of fuselage	7.01 m (23 ft 0 in)
Height to top of rotor head	2.65 m (8 ft 8½ in)
Height over tail (endplate fins)	2.71 m (8 ft 10¾ in)
Fuselage: Max width	1.40 m (4 ft 7 in)
Width over skids	1.95 m (6 ft 4¾ in)
Width over TOW pods	3.23 m (10 ft 7¼ in)
Tailplane span	1.68 m (5 ft 6 in)
Ventral fin ground clearance	0.67 m (2 ft 2½ in)

WEIGHTS AND LOADINGS (500M):

Weight empty	512 kg (1,130 lb)
Max normal T-O weight	1,157 kg (2,550 lb)
Max overload T-O weight	1,360 kg (3,000 lb)

PERFORMANCE (500M at max normal T-O weight):

Max level speed at 305 m (1,000 ft)	132 kts (244 km/h; 152 mph)
Cruising speed for max range at S/L	117 kts (217 km/h; 135 mph)
Max rate of climb at S/L	518 m (1,700 ft)/min
Service ceiling	4,390 m (14,400 ft)
Hovering ceiling IGE	2,500 m (8,200 ft)
Hovering ceiling OGE	1,615 m (5,300 ft)
Range at 1,220 m (4,000 ft)	318 n miles (589 km; 366 miles)

VERIFIED

MCDONNELL DOUGLAS APACHE UPGRADES

US Army designation: AH-64A, C and D
Israel Defence Force name: Petan (Cobra)

TYPE: Day/night twin-engined attack helicopter upgrade.

PROGRAMME: Original Hughes Model 77 entered for US Army advanced attack helicopter (AAH) competition; first flights of two development prototype YAH-64s 30 September and 22 November 1975; selected by US Army December 1976; named Apache late 1981.

Deliveries started 26 Janauary 1984; 700th delivered December 1991; 720 by April 1992 (including 18 to Israel); 811th and final (plus five flying prototype and pre-production) for delivery December 1994. Self deployment capability shown by 14th Apache with four 871 litre (230 US gallon; 191 Imp gallon) external tanks, which flew 1,020 n miles (1,891 km; 1,175 miles) from Mesa to Santa Barbara with 45 minutes fuel remaining on 4 April 1985; initial operating capability achieved by 3rd Squadron, 6th Cavalry Regiment, July 1986; 29 of 40 planned AH-64A battalions, including seven National Guard and two Army Reserve, combat-ready by December 1992; first combat use (11 AH-64As) in operation Just Cause, Panama, December 1989; used extensively (288) during January/February 1991 Gulf War against Iraq, including first air strike of conflict. First AH-64As issued to Army National Guard in 1987; fourth ArNG unit (I-211 AvRgt in Utah) established 1990; first overseas regiment 2/6 Cavalry Regiment, Illesheim, Germany, September 1987; eighth in Europe (3-4 AvRgt at Finthen) equipped 1990; battalion consists of 18 AH-64As and 13 Bell OH-58 Kiowas; more than 160 AH-64As based in Germany at peak strength, but force now reduced.

Eleven month programme to integrate air-to-air Stinger began October 1987; four missiles mounted in pairs on wingtips, five firings early 1989; air-to-air development programme included firing two AIM-9 Sidewinders in hover and at 80 knots (148 km/h; 92 mph) at White Sands, New Mexico, November 1987; laser ranging and tracking tests on Bell UH-1 and LTV A-7 flown in 1989; M-230 Chain Gun being improved for air-to-air use; Matra Mistral

McDonnell Douglas AH-64D Longbow Apache with Westhinghouse mast-mounted Longbow millimetre-wave radar *(McDonnell Douglas)*

1995

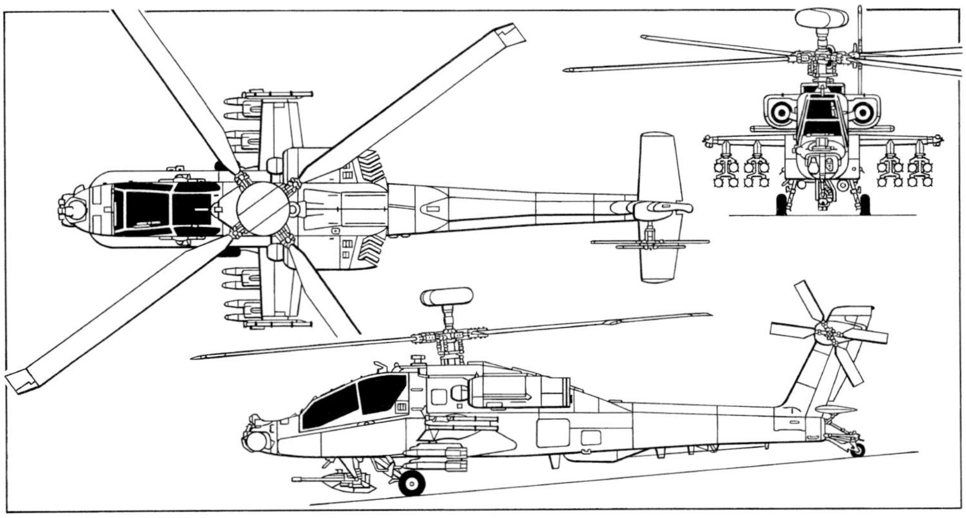

McDonnell Douglas AH-64D Longbow Apache tandem-seat advanced attack helicopter *(Jane's/Mike Keep)*

1995

captive carry tests completed. New missile control system by Base 10 Defense of Trenton, New Jersey, used for two more Stinger firings during 1990; Sidearm anti-radiation missile from AH-64A hit RF emitter on armoured vehicle at US Naval Weapons Test Center 25 April 1988.

VERSIONS: **AH-64A:** Production for US Army and export. All to be upgraded to AH-64D; last in 2010. Retrofit from 1993 with Sincgars secure radios and GPS; first installed in Apaches of 5-501 Aviation Regiment on deployment to Camp Eagle, South Korea, from March 1994, as first AH-64s in Korea.

AH-64B: Cancelled in 1992. Was planned near-term upgrade of 254 AH-64s with improvements derived from operating experience in 1991 Gulf War, including GPS, Sincgars radios, target handover capability, better navigation, and improved reliability including new rotor blades.

AH-64C: Previous designation for upgrade of AH-64As to near AH-64D standard, apart from omission of Longbow radar retention of -701 engines; provisions for optional fitment of both; Army requested draft proposal, August 1991; funding for two prototype conversions awarded in September 1992. With exception of AH-64Ds and re-sales, all remaining US Army AH-64As (approximately 540) to be modified. Designation abandoned late 1993; all Apaches to become AH-64D, including those not fitted with radar.

AH-64D Longbow Apache: Current improvement programme based on Westinghouse mast-mounted Longbow millimetre-wave radar and Martin Marietta Hellfire with RF seeker; includes more powerful GE T700-GE-701C engines, larger generators for 70 kVA peak loads, Plessey AN/ASN-157 Doppler navigation, MIL-STD-1553B databus allied to dual 1750 A processors, and a vapour cycle cooling system for avionics; early user tests completed April 1990.

Full-scale development programme, lasting 51 months, authorised by Defense Acquisition Board August 1990, but airframe work extended in December 1990 to 70 months to coincide with missile development; supporting modifications being incorporated progressively; first flight of AH-64A (82-23356) with dummy Longbow radome 11 March 1991; first (89-0192) of four AH-64D prototypes flown 15 April 1992; second (89-0228) flew 13 November 1992; fitted with radar in mid-1993 and flown 20 August 1993; No.3 (90-0324) flown 30 June 1993; No.4 (85-25410) on 4 October 1993; No.5 (formerly AH-64C No.1) 19 January 1994 (first Apache with new Hamilton Standard lightweight flight management computer); No.6 4 March 1994; last two mentioned converted from 85-25408 and 85-25477. Six AH-64Ds to fly 3,300 hour test programme; production deliveries to start 1997 from planned first batch of 24. Following redesignation of AH-64C in late 1993, entire Army inventory to be known as AH-64D Longbow Apache, although only 227 (original AH-64D total) to carry Longbow radar. AH-64D to equip 26 battalions; company strength to be three with radar plus five without; three companies per battalion. Longbow can track flying targets and see through rain, fog, and smoke that defeat FLIR and TV; RF Hellfire can operate at shorter ranges; it can lock-on before launch or launch on co-ordinates and lock-on in flight; Longbow scans through 360° for aerial targets or scans over 270° in 90° sectors for ground targets; mast-mounted rotating antenna weighs 113 kg (250 lb).

Further modification include 'manprint' cockpit with large displays, air-to-air missiles, digital autostabiliser, integrated GPS/Doppler/INS/air data/laser/radar altimeter navigation system, digitial communications, faster target handoff system, and enhanced fault detection with data transfer and recording. AH-64D No.1 made first Hellfire launch on 21 May 1993; first demonstration of digital air-to-ground data communications with Synmetrics Industries improved data modem, 8 December 1993.

GAH-64A: At least 17 AH-64As grounded for technical instruction.

JAH-64A: Seven AH-64As for special testing, of which one reverted to standard.

For full details of AH-64 see *Jane's All the World's Aircraft* 95-96.

OPERATORS: Versions of the AH-64D are in service with the armed forces of the following countries: Netherlands (30 on order), UK (67 on order) and USA (710).

NEW ENTRY

MID-CONTINENT

MID-CONTINENT AIRCRAFT CORPORATION

Drawer L, Highway 84 East, Hayti, Missouri 63851
Telephone: 1 (314) 359 0500
Fax: 1 (314) 359 0538
Telex: 447183
CHAIRMAN: Richard Reade
PRESIDENT: Ken Mauk

Operator and distributor of Schweizer (Grumman) Ag-Cats and Ayres Thrushes.

VERIFIED

MID-CONTINENT KING CAT

Mid-Continent King Cat, re-engined Schweizer Ag-Cat

1992

TYPE: Re-engined Schweizer Ag-Cat.

PROGRAMME: Also certificated in Canada. Offered ready converted or as retrofit kit. Engineering work for STC by Serv-Aero Engineering Inc.

DESIGN FEATURES: Uses airframe of Super Ag-Cat C; hopper holds 1,893 litres (500 US gallons; 416 Imp gallons) liquid or 1,814 kg (4,000 lb) powder; powered by 895 kW (1,200 hp) Wright R-1820-202A radial engine driving three-blade metal propeller; improved hot and high performance. Options include upper wing installation height increased 20.3 cm (8 in), Serv-O ailerons, fuel capacity increased to 431.5 litres (114 US gallons; 95 Imp gallons), Collins cockpit air conditioning, and a 1,893 litre (500 US gallon; 416 Imp gallon) water bombing system.

FLYING CONTROLS: As for Schweizer Ag-Cat.

STRUCTURE: As for Schweizer Ag-Cat.

WEIGHTS AND LOADINGS:

Weight empty: basic	2,184 kg (4,816 lb)
spray equipped	2,257 kg (4,976 lb)
dust equipped	2,225 kg (4,906 lb)
Max T-O weight: FAR 23	2,857 kg (6,300 lb)
CAM 8	3,855 kg (8,500 lb)
Max wing loading: FAR 23	78.5 kg/m² (16.07 lb/sq ft)
CAM 8	105.9 kg/m² (21.68 lb/sq ft)
Max power loading: FAR 23	3.19 kg/kW (5.25 lb/hp)
CAM 8	4.31 kg/kW (7.08 lb/hp)

PERFORMANCE:

Ferry speed	117 kts (217 km/h; 135 mph)
Typical working speed	87-113 kts (161-209 km/h; 100-130 mph)
Stalling speed, power off, at AUW of 2,857 kg (6,300 lb)	60 kts (111 km/h; 69 mph) CAS
T-O run with 907 kg (2,000 lb) hopper load	293 m (960 ft)
T-O to 15 m (50 ft) with 907 kg (2,000 lb) hopper load	427 m (1,400 ft)
Landing from 15 m (50 ft) at weight of 2,257 kg (4,976 lb)	353 m (1,190 ft)
Landing run at weight of 2,257 kg (4,976 lb)	179 m (588 ft)

VERIFIED

NOISES OFF

NOISES OFF

10 New King Street, White Plains, New York 10604
Telephone: 1 (914) 686 1100
Fax: 1 (914) 686 8223
PRESIDENT: Robert E. Wagenfeld
VICE-PRESIDENT: Walter H. Johnson

Company formed 1992 to produce an active noise cancellation system for turbine powered aircraft. The system cancels random jet noise at its source by generating an identical new noise that is 180° out of phase with the original noise. The system is applicable to all types of jet engine aircraft.

VERIFIED

NORDAM

NORDAM AIRCRAFT MODIFICATION DIVISION

520 South Lansing, Tulsa, Oklahoma 74120
Telephone: 1 (918) 5560/8636
Fax: 1 (918) 560 8528
Telex: 158105
GENERAL MANAGER: J. Roger Collins
DIRECTOR, SALES AND MARKETING: Jack M. Arehart

Repairs, manufactures and modifies aircraft components, and fits interiors.

Received FAA STC 26 June 1992 to modify Boeing 737-200s to meet FAA Stage 3/ICAO Chapter 3 noise levels; reduces 75 dBA footprint by 69 per cent at 52,390 kg (115,500 lb) and allows most 737-200s to use 40° flaps during approach. Aircraft began commercial service with Lufthansa and Air New Zealand July 1992; over 157 firm orders by January 1994; cost $3 million per aircraft.

VERIFIED

NORDAM (BOEING) 737-200 HIGH GROSS WEIGHT HUSH KIT

TYPE: Hush kit for Boeing 737-200.

PROGRAMME: Awarded FAA Supplemental Type Certificate June 1992. First hush-kitted Boeing 737-200 entered service with German Cargo on 19 August 1992 equipped with Nordam/Pratt & Whitney JT8DS.

DESIGN FEATURES: High gross weight ejector suppressor hush kit offers Stage III compliance for JT8D-9/9A, JT8D-15/15A and JT8D-17/17A engines on aircraft with gross weights up to 127,200 MBRGW. System consists of: Hush kit mixer assembly, 12-lobe suppressor, respaced inlet guide vanes, acoustical treatment, enlarged tailpipe and modified reverser.

WEIGHTS AND LOADINGS:

Max T-O weight:	
JT8D-9/9A	56,473 kg (124,500 lb)
JT8D-15/15A	57,697 kg (127,200 lb)
JT8D-17/17A	57,607 kg (127,000 lb)

OPERATIONAL NOISE CHARACTERISTICS:

T-O: JT8D-9/9A	91.87 EPNdB
JT8D-15/15A	91.10 EPNdB
JT8D-17/17A	90.17 EPNdB
Approach: JT8D-9/9A	98.60 EPNdB
JT8D-15/15A	98.50 EPNdB
JT8D-17/17A	98.50 EPNdB
Sideline: JT8D-9/9A	94.87 EPNdB
JT8D-15/15A	95.96 EPNdB
JT8D-17/17A	98.86 EPNdB

VERIFIED

NORDAM (BOEING) 737-200 LOW GROSS WEIGHT HUSH KIT

TYPE: Hush Kit for Boeing 737-200.

PROGRAMME: Received FAA STC in November 1994 for aircraft powered with JT8D-9/-9A, -15/15A and initial deliveries have been made.

DESIGN FEATURES: Hush Kit is certified for aircraft with MTOGW up to 122,000 lbs and meets US FAR 36 Stage 3 and ICAO Annex 16 Chapter III compliance. Primary components include: RIGV, mixer, new tailpipe with specially developed acoustic treatment and standard reverser.

COST: $1.8 million in 1995.

NEW ENTRY

Nordam (Boeing) hush-kitted 737-200 of Lufthansa *1994*

Nordam (Boeing) 737-200 Low Gross Weight (LGW) hush kit ***(Nordam)*** *1995*

NORTHROP GRUMMAN

NORTHROP GRUMMAN CORPORATION

1840 Century Park East, Los Angeles, California 90067
Telephone: 1 (310) 553 6262
Fax: 1 (310) 201-3023/553-2076
CHAIRMAN, PRESIDENT AND CEO: Kent Kresa
CORPORATE MANAGEMENT:
John E. Harrison (Senior Vice-President)
Wallace C. Solberg (Corporate Vice-President)
Ralph D. Crosby Jr (Corporate Vice-President)
Herbert W. Anderson (Corporate Vice-President)
Gordon L. Williams (Vice-President)
Paul C. Bavitz (Vice-President)

Acquisition of Grumman by Northrop completed 1 May 1994 and new corporation formed 18 May. Combined Northrop Grumman workforce more than 41,300 at April 1995. Northrop Grumman subsequently completed acquisition of Vought Aircraft Company, with $130 million purchase of Carlyle Group's 51 per cent interest, in August 1994. Vought workforce numbered 5,200 in Dallas area.

Northrop company formed 1939 to produce military aircraft; activities extended to missiles, target drones, electronics, space technology, communications, support services and commercial products; name changed from Northrop Aircraft Inc to Northrop Corporation in 1959. Reorganisation in wake of Grumman purchase resulted in creation of five divisions (B-2, Military Aircraft, Commercial Aircraft, Electronics and Systems Integration, and Data Systems and Services) plus Advanced Technology and Development Center.

Grumman Aircraft Engineering Corporation incorporated 6 December 1929; Grumman Corporation formed as small holding company 1969 for Grumman Aerospace Corporation, Grumman Allied Industries Inc and Grumman Data Systems Corporation. Ten operating divisions created February 1985 (see 1990-91 and earlier editions), followed by further reorganisations in 1987-88. Corporate structure consolidated into four operating groups in 1991, falling to three (Aerospace and Electronics, Data Systems and Services, and Allied) on 1 January 1993. Consolidation plan, finalised January 1994, required Grumman Corporation to shed one-third of worldwide capacity and abandon independent design and development of aircraft. Grumman Corporation employees totalled 16,700 at time of takeover.

Northrop Grumman divisions responsibilites allocated as follows:

Advanced Technology and Development Center: intended to generate new military business, new non-traditional business and maintain technology base in support of core competencies.

B-2 Division: is prime contractor for advanced technology stealth bomber and functions as depot for B-2 Spirit, plus centre for special projects.

Military Aircraft Division: is responsible for ongoing military aircraft programmes and their derivatives; intercompany work orders for modification and assembly of aircraft (E-2C Hawkeye and E-8C Joint-STARS); production of fuselage sections for McDonnell Douglas F/A-18 and builds UAVs including BQM-74 and International Chukar III. Also responsible for Northrop Worldwide Aircraft Services Inc (NWASI), Lawton, Oklahoma which provides technical assistance and support to US armed forces and customs service.

Commerical Aircraft Division: produces commercial aircraft structures including major subassemblies and components for Boeing 747, 757, 767, 777, Gulfstream IV and Fokker 100. Also manages residual ground transportation programmes.

Electronics and Systems Integration Division: tasked with continuation of ongoing programmes, derivatives and upgrades to existing programmes (including EF-111A and EA-6B, as well as new electronic, battle management and system integration projects. Current work includes production of strategic and tactical navigation and guidance equipment, passive sensor and tracking systems, electronic countermeasures equipment (including AN/ALQ-135 internal countermeasures for USAF F-15 and AN/ALQ-162 for joint service use), rate and rate-integrating gyros and strapdown guidance systems. Also has overall programme responsibility for E-2C and E-8C.

Data Systems and Services Division: continues existing Grumman activity and will incorporate the Northrop Information Services Center and services business of NWASI.

UPDATED

NORTHROP F-5A and F-5B

Canadian Air Force designation: CF-116
Spanish Air Force designations: A.9, AE.9 and AR.9

TYPE: Light tactical fighter and reconnaissance aircraft.

PROGRAMME: Design of this light tactical fighter started in 1955 and construction of the prototype of the single-seat version (then designated N-156C) began in 1958. It flew for the first time 30 July 1959, exceeding Mach 1 on its

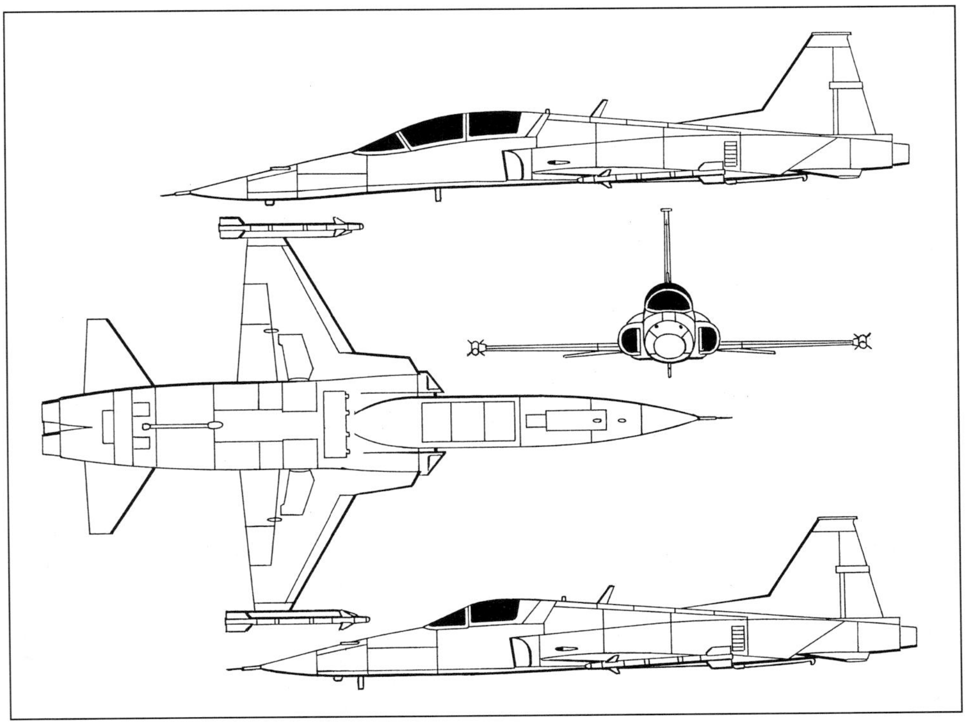
Northrop F-5 single-seat and two-seat verions ***(Jane's/Mike Keep)*** *1994*

Northrop F-5A of the Royal Norwegian Air Force, built with ATO and arrester hook

1994

maiden flight. Two more prototypes were built, followed by several production versions.

VERSIONS: **F-5A:** Basic single-seat fighter. Two General Electric J85-GE-13 afterburning turbojets. First production F-5A flew in October 1963. Norwegian version built with ATO and arrester hook for short field operation. Approximately 14 still in service with the Jordanian Air Force.

F-5B: Generally similar to F-5A, but with two seats in tandem for dual fighter/trainer duties. First F-5B flew 24 February 1964.

CF-5A/D: These are the designations of the versions of the F-5A/B that were produced for the Canadian Armed Forces, the first of them entering service in 1968. Official Canadian designation is CF-116. Several improvements were incorporated in the CF series, including higher thrust engines (J85-CAN-15), and flight refuelling capability. 115 built by Canadair, under licence. Still 80+ in service with the Canadian Air Force.

NF-5A/B: Versions of the F-5 produced for the Royal Netherlands Air Force with a Doppler navigation system, 1,040 litre (275 US gallon; 229 Imp gallon) fuel tanks and manoeuvring flaps. Manufacture and assembly of the 105 aircraft ordered were integrated with CF-5 production by Canadair Ltd. First aircraft entered service in 1969. No longer in service with the Royal Netherlands Air Force.

RF-5A: Reconnaissance version of the F-5; initial deliveries were made in mid-1968. Its four KS-92 cameras, each with a 100 ft film magazine, provides forward oblique, trimetrogen and split vertical coverage, including horizon-to-horizon with overlap. Associated equipment includes four light sensors, defogging and cooling systems, a pitot static nose boom and a computer/'J' box, all housed in a nose compartment with forward hinged clamshell top cover. Still in service with the Greek Air Force (8).

SF-5A/B (C-9/CE-9): Spanish versions of the F-5. Still in service with the Spanish Air Force (45+).

F-5E: See following entry.

F-5F: See following entry.

F-5G: Royal Norwegian Air Force designation for F-5A/B. Approximately 30 still in service.

RF-5G: Royal Norwegian Air Force designation for the RF-5G. No longer in service.

UPGRADES: **Bristol Aerospace:** Awarded contract by Norwegian Air Force for 10 sets of new wings (including the update of its technical order publications) with option for a further five sets. Deliveries to begin in early 1995 and continue through 1996. See separate entry in Canada section.

Northrop Grumman: Structural Update Programme (SUP). See separate entry.

Sierra Technologies: Norwegian F-5A TIGER PAWS upgrade. See separate entry.

OPERATORS: Versions of the F-5A/B/D Freedom Fighter are in service with the armed forces of the following countries: Brazil (4), Canada (46), Greece (88), Jordan (10), Morocco (13), Norway (15), Philippines (5), Saudi Arabia (15), Spain (47), Thailand (12), Turkey (50), Venezuela (18) and Yemen (4).

DESIGN FEATURES: Cantilever low-wing monoplane. Wing section NACA 65A004.8 (modified). No dihedral or incidence. Sweepback at quarter-chord 24°.

FLYING CONTROLS: Hydraulically powered sealed-gap ailerons at approximately mid-span with light alloy single-slotted flaps inboard. Continuous hinge leading-edge flaps of full-depth honeycomb construction. No trim tabs. Two hydraulically actuated airbrakes on underside of fuselage forward of wheel wells. Tail unit has hydraulically powered rudder and one-piece all-moving tailplane. No tail trim tabs. Longitudinal and directional stability augmentors installed in series with control system.

STRUCTURE: Multispar light alloy structure with heavy plate machined skins. The fuselage is a semi-monocoque structure of light alloy, with steel, magnesium and titanium used in certain areas. 'Waisted' area rule lines. The tail unit is a cantilever all-metal structure with single spars with full depth light alloy honeycomb secondary structure.

LANDING GEAR: Hydraulically retractable tricycle type with steerable nosewheel. Emergency gravity extension. Main units retract inward into fuselage, nosewheel forward. Oleo-pneumatic shock-absorbers. Mainwheels fitted with tubeless tyres size 22 × 8.5, pressure 5.86 to 14.48 bars (85 to 210 lb/sq in). Nosewheel fitted with tubeless tyre size 18 × 6.5, pressure 4.14-12.41 bars (60-180 lb/sq in). Multiple-disc hydraulic brakes.

POWER PLANT: Two General Electric J85-GE-13 turbojets (each with maximum rating of 18.15 kN; 4,080 lb st with afterburning). Two internal fuel tanks composed of integral cells with total usable capacity of 2,207 litres (583 US gallons; 485 Imp gallons). Provision for one 568 litre (150 US gallon; 125 Imp gallon) jettisonable tank on fuselage centreline pylon; two 568 litre (150 US gallon; 125 Imp gallon) jettisonable tanks on underwing pylons and two 189 litre (50 US gallon; 41.6 Imp gallon) wingtip tanks. Total fuel capacity, with external tanks, 4,289 litres (1,133 US gallons; 943 Imp gallons). Single pressure point on lower fuselage. Oil capacity 4.5 litres each engine.

ACCOMMODATION (F-5A): Pilot only, on rocket-powered ejection seat in pressurised and air conditioned cockpit. (F-5B): Pupil and instructor in tandem on rocket-powered ejection seats in pressurised and air conditioned cockpits separated by windscreen. Separate manually operated rearward hinged jettisonable canopies. Instructor's seat at rear raised 0.25 m (10 in) higher than that of pupil to give improved forward view.

SYSTEMS: Electrical system includes two 8 kVA engine-driven generators, providing 115 V 400 Hz AC power, and 24 V battery. No de-icing system.

ELECTRONICS AND EQUIPMENT: Standard equipment includes AN/ARC-34C UHF radio, PP-2024 SWIA-Missile AVX, AN/AIC-18 interphone, J-4 compass, Norsight optical sight, AN/APX-46 IFF, and AN/ARN-65 Tacan. Space provision for AN/ARW-77 Bullpup AUX. Blind-flying instrumentation not standard.

ARMAMENT: Basic interception weapons are two Sidewinder missiles on wingtip launchers and two 20 mm guns in the fuselage. Five pylons, one under the fuselage and two under each wing, permit the carriage of a wide variety of other operational loads. A bomb of more than 910 kg (2,000 lb) or high rate of fire gun pack can be suspended from the centre pylon. Underwing loads can include four air-to-air missiles, Bullpup air-to-surface missiles, bombs, up to 20 air-to-surface rockets, gun packs or external fuel tanks. The reconnaissance nose does not eliminate the 20 mm nose gun capability.

DIMENSIONS, EXTERNAL:
- Wing span: 7.70 m (25 ft 3 in)
- Wing span over tip tanks: 7.87 m (25 ft 10 in)
- Wing chord: at root: 3.43 m (11 ft 3 in)
  - at tip: 0.69 m (2 ft 3 in)
- Length overall: F-5A: 14.38 m (47 ft 2 in)
  - F-5B: 14.12 m (46 ft 4 in)
- Height overall: F-5A: 4.01 m (13 ft 2 in)
  - F-5B: 3.99 m (13 ft 1 in)
- Tailplane span: 4.28 m (14 ft 1 in)
- Wheel track: 3.35 m (11 ft 0 in)
- Wheelbase: F-5A: 4.67 m (15 ft 4 in)
  - F-5B: 5.94 m (19 ft 6 in)

AREAS:
- Wings, gross: 15.79 m^2 (170 sq ft)
- Ailerons (total): 0.86 m^2 (9.24 sq ft)
- Trailing-edge flaps (total): 1.77 m^2 (19.0 sq ft)
- Leading-edge flaps (total): 1.14 m^2 (12.3 sq ft)
- Fin: 3.85 m^2 (41.42 sq ft)
- Rudder: 0.57 m^2 (6.1 sq ft)
- Tailplane: 5.48 m^2 (59.0 sq ft)

WEIGHTS AND LOADINGS:
- Weight empty, equipped: F-5A: 3,667 kg (8,085 lb)
  - F-5B: 3,792 kg (8,361 lb)
- Max military load: 2,812 kg (6,200 lb)
- Max T-O weight: F-5A: 9,379 kg (20,677 lb)
  - F-5B: 9,298 kg (20,500 lb)
- Max design landing weight: 9,006 kg (19,857 lb)
- Max zero-fuel weight: F-5A: 6,446 kg (14,212 lb)
  - F-5B: 6,237 kg (13,752 lb)
- Max wing loading: F-5A: 590.8 kg/m^2 (121 lb/sq ft)
  - F-5B: 576 kg/m^2 (118 lb/sq ft)

PERFORMANCE: (F-5A at AUW of 5,193 kg; 11,450 lb, F-5B at AUW of 4,916 kg; 10,840 lb, unless indicated otherwise):
- Never-exceed speed: 710 kts (1,315 km/h; 818 mph) IAS
- Max level speed at 11,000 m (36,000 ft):
  - F-5A: Mach 1.4
  - F-5B: Mach 1.34
- Max cruising speed without afterburning, at 11,000 m (36,000 ft): Mach 0.97
- Econ cruising speed: Mach 0.87
- Stalling speed, 50% fuel flaps extended:
  - F-5A: 128 kts (237 km/h; 147 mph)
  - F-5B: 120 kts (223 km/h; 138 mph)
- Max rate of climb at S/L:
  - F-5A: 8,750 m (28,700 ft)/min
  - F-5B: 9,265 m (30,400 ft)/min
- Service ceiling: F-5A: 15,390 m (50,500 ft)
  - F-5B: 15,850 m (52,000 ft)
- Service ceiling, one engine out:
  - F-5A, F-5B: over 10,365 m (34,000 ft)
- T-O run (with two Sidewinder missiles):
  - F-5A at AUW of 6,203 kg (13,677 lb): 808 m (2,650 ft)
  - F-5B at AUW of 5,924 kg (13,061 lb): 671 m (2,200 ft)
- T-O run to 15 m (50 ft) (with two Sidewinder missiles):
  - F-5A at AUW of 6,203 kg (13,677 lb): 1,113 m (3,650 ft)
  - F-5B at AUW of 5,924 kg (13,061 lb): 960 m (3,150 ft)
- Landing run from 15 m (50 ft), with brake-chute:
  - F-5A at AUW of 4,504 kg (9,931 lb): 1,189 m (3,900 ft)
  - F-5B at AUW of 4,363 kg (9,619 lb): 1,158 m (3,800 ft)
- Landing run, with brake-chute:
  - F-5A at AUW of 4,504 kg (9,931 lb): 701 m (2,300 ft)
  - F-5B at AUW of 4,363 kg (9,619 lb): 671 m (2,200 ft)
- Range with max fuel, with reserve fuel for 20 min max endurance at S/L:
  - F-5A, tanks retained: 1,205 n miles (2,232 km; 1,387 miles)
  - F-5B, tanks retained: 1,210 n miles (2,241 km; 1,393 miles)
  - F-5A, tanks dropped: 1,400 n miles (2,594 km; 1,612 miles)
  - F-5B, tanks dropped: 1,405 n miles (2,602 km; 1,617 miles)

Northrop NF-5A of the Royal Netherlands Air Force

1994

Combat radius with max payload, allowances as above and 5 min combat at S/L:
F-5A 170 n miles (314 km; 195 miles)
F-5B 175 n miles (323 km; 201 miles)
Combat radius with max fuel, two 530 lb bombs, allowances as above and 5 min combat at S/L:
F-5A 485 n miles (898 km; 558 miles)
F-5B 495 n miles (917 km; 570 miles)
Operational hi-lo-lo-hi reconnaissance radius with max fuel, 50 n miles (93 km; 58 mile) S/L dash to and from target and allowances as for combat radius with max fuel: RF-5A 560 n miles (1,036 km; 644 miles)

UPDATED

NORTHROP TIGER II

USAF designations: F-5E and F-5F

TYPE: Single-seat light tactical fighter.

PROGRAMME: Production ended 1987, but five F-5E and three F-5F assembled from spares for Singapore Air Force and delivered by July 1989; total 3,805 of F-5/T-38 family produced; includes 617 F-5As, 89 RF-5As, 183 F-5Bs, 792 F-5Es, 12 RF-5Es, 140 F-5Fs and 1,187 T-38As. Canadair assembled 164 F-5As and 76 F-5B/Ds; CASA (Spain) 18 F-5As, 18 RF-5As and 34 F-5Bs; KA (South Korea) 48 F-5Es and 20 F-05Fs; F+W (Switzerland) 84 F-5Es and six F-5Fs, AIDC (Taiwan) 242 F-5Es and 66 F-5Es. Separate are three prototypes each of T-38 and F-20 Tigershark; unmodified F-5 prototype (N.156F) and two F-5E fuselages used for Grumman X-29 forward sweep test aircraft.

Total 500 F-5A/Bs and 1,080 F-5E/Fs remain in service with 31 air forces; several upgrade and life-extension programmes in hand or planned. Canadian Forces' CF-5s undergoing modernisation by Bristol Aerospace; Chilean F-5Es by Bedek (Israel); Jordanian and Thai F-5Es by Smiths Industries; Norwegian F-5As by Sierra (formerly LTV); Venezuelan VF-5As by Singapore Aerospace.

Other avionics and rebuild contractors offer similar programmes.

The F-5E was selected in November 1970 by the US government as the winner of a competition to determine the single-seat International Fighter Aircraft (IFA) which was to succeed Northrop's F-5A. The two-seat F-5F was developed subsequently.

The F-5E design places particular emphasis on manoeuvrability by the incorporation of auto-manoeuvring flaps. Full-span leading-edge flaps work in conjunction with conventional trailing-edge flaps, and are operated automatically in response to airspeed and angle of attack. The flaps may also be pilot controlled to full down and full up positions. Wing loading is maintained at approximately the same value as the F-5A, as the result of an increase in wing area to 17.30 m² (186 sq ft). This is due principally to the widened fuselage, which also increases wing span. The tapered wing leading-edge extension, between the inboard leading-edge and fuselage, was refined to increase the wing area and maximise the lift coefficient of the wing.

The F-5E incorporated other features developed for the Canadian, Dutch and Norwegian F-5s, including two-position nosewheel gear, which increases wing angle of attack on the ground by 3° 22′. In conjunction with the more powerful engines, this improves F-5E take-off performance by some 30 per cent compared with earlier F-5s. The aircraft is qualified to carry three 1,040 litre (275 US gallon; 228 Imp gallon) drop tanks, and up to nine 500 lb Mk 82 bombs, following the addition of a multiple ejector rack (MER) on the centreline stores station.

The first F-5E made its first flight on 11 August 1972. USAF Tactical Air Command, with assistance from Air Training Command, was assigned responsibility for training pilots and technicians of user countries. First deliveries, to the USAF's 425th Tactical Fighter Squadron, were made in the Spring of 1973. In total, 20 were supplied for the USAF training programme by the end of September 1973, and deliveries to foreign countries began in early 1974. In addition to their use as tactical fighters, F-5Es are operated by the USN in the 'agressor' role, to simulate enemy aircraft.

In March 1985 the USAF announced plans to upgrade its 74 F-5Es and F-5Fs. The programme included new Emerson AN/APQ-159(V)5 radar in the F-5Es and AN/APQ-159(V)6 in the F-5Fs, replacing the current APQ-153 radar. The new radars double detection range and incorporate off-boresight target acquisition and track-while-scan. A new radar warning receiver and radar jammer also installed. Programme also included the modification of the control column grips on the USAFs 70 F-5Es to standardise them with those on the F-5F, and the transfer of radar control switches from panel mounts to the control column handgrip.

VERSIONS: **F-5E:** Standard production version, to which the detailed description applies.

F-5F: Tandem two-seat version of the F-5E, with fuselage lengthened by 1.22 m (4 ft 0 in). Fire control system and one M39 cannon retained, enabling aircraft to be used for both training and combat duties. First flight was made 25 September 1974. Two F-5Fs completed flight test and qualification in early 1976.

RF-5E: Reconnaissance version described separately.

UPGRADES: **Bristol Aerospace:** See separate entry.

Northrop F-5E (bottom) and F-5F (top) Tiger IIs light tactical fighter aircraft

1994

British Aerospace Systems & Equipment Limited (BASE): Inertial navigation upgrade. See separate entry.

Elbit: See separate entry.

Israel Aircraft Industries (IAI): See separate entry.

Israeli Aircraft Industries (IAI): Bidding for contract to upgrade South Korean F-5 aircraft.

Northrop Grumman: See separate entries.

Swiss Federal Aircraft Factory: See separate entry.

SABCA: Selected by Indonesia to act as systems integrator for the upgrade of 12 F-5E/F fighters of the Indonesian Air Force. Value of contract US$40 million. Preselected avionics suite to include a GEC-Marconi Avionics (GMAv) HUD/WAC, Litton LN-93 Inertial Navigation System (INS) and GEC-Marconi Sky Guardian Radar Warning Receiver (RWR). Also included in the upgrade are a MIL-STD-1553 databus, new air data computer, stores management system, mission camera and hands-on throttle and stick (HOTAS). Existing ESCO (formerly) Emerson APG-59(V)3 to be upgraded to (V)5 standard. After initial engineering and integration work on prototype F-5E and F-5F aircraft, the remaining seven F-5Es and three F-5Fs will be upgraded by Indonesian Air Force from kits supplied by SABCA. Programme also includes four-year spares support package and structural repair work.

Sierra Technologies: See separate entry.

Singapore Aerospace: See separate entry.

OPERATORS: Versions of the F-5E/F Tiger II are in service with the armed forces of the following countries: Bahrain (12), Brazil (51), Chile (16), Honduras (12), Indonesia (9), Iran (155), Jordan (59), Kenya (11), South Korea (195), Malaysia (17), Mexico (11), Morocco (24), Saudi Arabia (80), Singapore (43), Switzerland (103), Taiwan (295), Thailand (40), Tunisia (19), USA (36) and Yemen (10).

The following description applies to the F-5E, but is generally applicable to the F-5F also, except for details noted under model listings:

DESIGN FEATURES: Cantilever low-wing monoplane. Wing section NACA 65A004.8 (modified). No dihedral. No incidence. Sweepback at quarter-chord 24°. Tailplane incidence varied by hydraulic actuators of Northrop design for control of rudder and tailplane.

FLYING CONTROLS: Hydraulically powered sealed gap ailerons approximately mid-span. Electrically operated light alloy single-slotted trailing-edge flaps inboard of ailerons. Electrically operated leading-edge flaps. Two hydraulically actuated airbrakes of magnesium alloy, mounted on underside of mainwheel wells. Tail unit has hydraulically powered rudder and one-piece all-moving tailplane.

STRUCTURE: Multispar light alloy structure with heavy plate machined skins. The fuselage is a light alloy semi-monocoque basic structure, with steel, magnesium and titanium used in certain areas. Rear avionics bay and cockpit pressurised; fail-safe structure in pressurised sections. The tail unit is a cantilever all-metal structure.

LANDING GEAR: Hydraulically retractable tricycle type, main units retracting inward into fuselage, noswheel forward. Oleo-pneumatic shock-absorbers of Northrop design in all units. Two-position extending nose unit increases static angle of attack by 3° 22′ to reduce T-O distance, and is shortened automatically during the retraction cycle. Gravity operated emergency extension. Mainwheels and tyres size 24 × 8.00-13, pressure 14.48 bars (210 lb/sq in). Steerable nose unit with wheel and tyre size 18 × 6.50-8, pressure 8.27 bars (120 lb/sq in). All-metal multiple-disc brakes of Northrop design.

POWER PLANT: Two General Electric J85-GE-21B turbojet engines, each rated at 22.24 kN (5,000 lb st) with afterburning. Two independent fuel systems, one for each engine. Fuel for starboard engine supplied from two rubber impregnated nylon fabric bladder cells, comprising a centre-fuselage cell of 803 litre (212 US gallon; 176 Imp gallon) capacity, and a rear fuselage cell of 640 litre (169 US gallon; 140 Imp gallon) capacity. Port engine supplied from a forward fuselage cell of 1,120 litre (296 US gallon; 246 Imp gallon) capacity. Total internal fuel capacity 2,563 litres (677 US gallons; 563 Imp gallons) in F-5E, 2,555 litres (675 US gallons; 562 Imp gallons) in F-5F. No fuel is carried in the wings. Fuel crossfeed system allows fuel from either or both to be fed to either or both engines. Auxiliary jettisonable fuel tanks of 568 or 1,041 litres (150 or 275 US gallons; 125 or 229 Imp gallons) can be carried on the fuselage centreline pylon and the inboard underwing pylons. Single refuelling point on lower fuselage for fuel cell and external tank installation. Oil capacity 4.5 litres (1.2 US gallons; 1 Imp gallon) per engine.

ENGINE INTAKES: Intakes are supplemented with auxiliary air inlet doors for use during T-O and low-speed flight, to improve compressor face pressure recovery and to decrease distortion. Each door consists of a set of six pivot-mounted louvres in removable panels on each side of the fuselage. The doors are actuated by the pilot at T-O, and controlled automatically in flight by Mach sensor switches, and are maintained in the open position at airspeeds below Mach 0.35-0.4.

ACCOMMODATION: Pilot only, in pressurised, heated and air conditioned cockpit, on rocket powered ejection seat. Upward-opening canopy, hinged at rear.

SYSTEMS: Cockpit and rear avionics bay pressurised, heated and air conditioned by engine bleed air, maximum pressure differential 0.34 bars (5 lb/sq in). Hydraulic power supplied by two independent systems at a pressure of 207 bars (3,000 lb/sq in), maximum flow rate 32.9 litres (8.7 US gallons; 7.24 Imp gallons)/min, with non-separated type reservoirs, engine bleed air pressurised at 1.10 bars (16.0 lb/sq in). Flight control system provides power solely for operation of primary flight control surfaces. Utility system provides hydraulic back-up power for primary flight control surfaces and operating power for the landing gear doors, airbrakes, wheel brakes, nosewheel steering, gun bay purge doors, gun gas deflectors and stability augmentation system. Electrical power supplied by two 13/15 kVA 115/200 V three-phase 320 to 480 Hz non-paralleled engine-driven alternators. Each alternator has the capacity to accept full aircraft power load via an automatic transfer function. 250 VA 115 V 400 Hz single-phase solid-state inverter provides secondary AC source for engine starting. Two 33 A 26-32 transformer-rectifiers and a 24 V 13 Ah Ni/Cd battery provide DC power. Liquid oxygen system with capacity of 5 litres. No de-icing system.

AVIONICS AND EQUIPMENT (F-5E): AN/ARC-164 UHF command radio, 7,000-channel with 24 kHz spacing. Emerson Electric AN/APQ-159 lightweight micro-miniature pulse radar for the air-to-air search for target detection with range and angle tracking; target information, at a range of up to 20 n miles (37 km; 23 miles), is displayed on a 0.13 m (5 in) Direct View Storage Tube (DVST) in cockpit. AN/ASG-31 lead computing optical sight; AN/ARA-50 UHF ADF; AN/AIC-25 intercom; AN/APX-101 IFF; AN/ARN-118 Tacan; attitude and heading reference system; angle of attack system; and central air data computer. Full blind-flying instrumentation. Optional avionics include Litton LN-33 inertial navigation system; AN/ARN-108 instrument landing system; CPU-129/A flight director computer; VHF; VOR/ILS with DME; LF ADF; CRT with scan converter for radar or electro-optical weapon (AGM-65 Maverick); AN/ALE-40 countermeasures dispenser system; and Itek AN/ALR-46 digital or analogue radar warning receiver.

AVIONICS AND EQUIPMENT (F-5F): As detailed in F-5E. Optional equipment includes the Northrop AN/AVQ-27 laser target designation set.

ARMAMENT (F-5E): Two AIM-9 Sidewinder missiles on wingtip launchers. Two M39A2 20 mm cannon in fuselage nose, with 280 rds/gun. Up to 3,175 kg (7,000 lb) of mixed ordnance can be carried on one underfuselage and four underwing stations, including M129 leaflet bombs; Mk 84 2,000 lb bombs; Matra Durandal air-to-surface missiles; LAU-68 (7) 2.75 in rockets; LAU-3 (19) 2.75 in rockets; CBU-24, -49, -52 cluster bomb units; SUU bomb and rocket packs; SUU-25 flare dispensers; TDU-10 tow targets (DART); and RMU-10 reel (Dart). Lead computing optical gunsight uses inputs from airborne radar for air-to-air missiles and cannon, and provides a roll-stabilised manually depressible reticle aiming reference for air-to-ground delivery. A 'snap-shoot' capability is included for sttack on violently manoeuvring and fleeting targets. The gunsight incorporates also a detachable 16 mm reticle camera with 15 m (50 ft) film magazine. Optional ordnance capability includes the AGM-65 Maverick; centre-line multiple ejector rack; and laser-guided bombs.

ARMAMENT (F-5F): Two AIM-9 Sidewinder missiles on wing-tip launchers. One M39 20 mm cannon in port side of nose with 140 rounds. Underfuselage and underwing stores as for the F-5E.

DIMENSIONS, EXTERNAL:

Wing span	8.13 m (26 ft 8 in)
Wing span over missiles	8.53 m (27 ft 11⅞ in)
Wing chord: at root	3.57 m (11 ft 8⅝ in)
at tip	0.68 m (2 ft 2⅞ in)
Wing aspect ratio	3.88
Length overall (incl nose probe):	
F-5E	14.45 m (47 ft 4¾ in)
F-5F	15.65 m (51 ft 4 in)
Height overall: F-5E	4.07 m (13 ft 4¼ in)
F-5F	4.13 m (13 ft 2 in)
Tailplane span	4.31 m (14 ft 1½ in)
Wheel track	3.80 m (12 ft 5½ in)
Wheelbase: F-5E	5.17 m (16 ft 11½ in)
F-5F	6.52 m (21 ft 1½ in)

AREAS:

Wings, basic	17.3 m² (186 sq ft)
Ailerons (total)	0.86 m² (9.24 sq ft)
Trailing-edge flaps (total)	1.95 m (21.0 sq ft)
Leading-edge flaps (total)	1.14 m² (12.3 sq ft)
Fin	3.85 m² (41.42 sq ft)
Rudder	0.57 m² (6.10 sq ft)
Tailplane	5.48 m² (59.0 sq ft)

WEIGHTS AND LOADINGS:

Weight empty: F-5E	4,410 kg (9,723 lb)
F-5F	4,797 kg (10,576 lb)
Max internal fuel weight: F-5E	1,996 kg (4,400 lb)
F-5F	1,991 kg (4,390 lb)
Max internal fuel weight:	
F-5E/F-5F	2,415 kg (5,324 lb)
Max T-O weight: F-5E	11,214 kg (24,722 lb)
F-5F	11,409 kg (25,152 lb)
Max landing weight: F-5E	11,406 kg (25,147 lb)
Max zero-fuel weight: F-5E	7,953 kg (17,534 lb)
Max wing loading: F-5E	649.4 kg/m² (133 lb/sq ft)
Max power loading: F-5E	251.6 kg/kN (2.5 lb/lb st)

PERFORMANCE (F-5E at combat weight of 6,055 kg; 13,350 lb, F-5F at combat weight of 6,375 kg; 14,055 lb unless otherwise stated):

Never-exceed speed
710 kts (1,314 km/h; 817 mph) EAS

Max level speed at 10,975 m (36,000 ft):	
F-5E	Mach 1.64
F-5F	Mach 1.56
Max cruising speed:	
F-5E at 10,975 m (36,000 ft)	Mach 0.98
Econ cruising speed	Mach 0.80
Stalling speed, flaps down, power off:	
F-5E	124 kts (230 km/h; 143 mph)
F-5F	136 kts (253 km/h; 157 mph)
Max rate of climb at S/L:	
F-5E	10,516 m (34,500 ft)/min
F-5F	10,030 m (32,900 ft)/min
Service ceiling: F-5E	15,790 m (51,800 ft)
F-5F	15,485 m (50,800 ft)
Service ceiling, one engine out:	
F-5E	over 12,495 m (41,000 ft)
F-5F	12,285 m (40,300 ft)
T-O run:	
F-5E at 7,053 kg (15,550 lb)	610 m (2,000 ft)
F-5F at 7,371 kg (16,250 lb)	701 m (2,300 ft)
T-O run at max T-O weight: F-5E	1,737 m (5,700 ft)
F-5F	1,829 m (6,000 ft)
T-O to 15 m (50 ft):	
F-5E at 7,053 kg (15,550 lb)	884 m (2,900 ft)
F-5F at 7,371 m (16,250 lb)	975 m (3,200 ft)

Landing from 15 m (50 ft):
F-5E at 5,230 kg (11,530 lb), without brake-chute:
1,417 m (4,650 ft)
F-5F at 5,554 kg (12,245 lb), without brake-chute
1,524 m (5,000 ft)

Landing run with brake-chute:	
F-5E at 5,230 kg (11,530 lb)	762 m (2,500 ft)
F-5F at 5,554 kg (12,245 lb)	792 m (2,600 ft)

Combat radius, F-5E:
with max fuel, two Sidewinder missiles, reserves for 20 min max endurance at S/L and 5 min combat with max afterburning power at 4,575 m (150,000 ft)
570 n miles (1,056 km; 656 miles)
with 2,358 kg (5,200 lb) ordnance load, two Sidewinder missiles, max fuel, allowances as above and 5 min combat at military power at S/L, lo-lo-lo mission
120 n miles (222 km; 138 miles)
with max fuel, two Sidewinder missiles and two 530 lb bombs, allowances as above and 5 min combat at military power at S/L, hi-lo-hi mission
480 n miles (890 km; 553 miles)

Combat radius, F-5F:
with max internal fuel, and allowances comprising

Northrop F-5F Tiger II of the Swiss Air Force with ACMI pod *(Paul Tompkins)*

1994

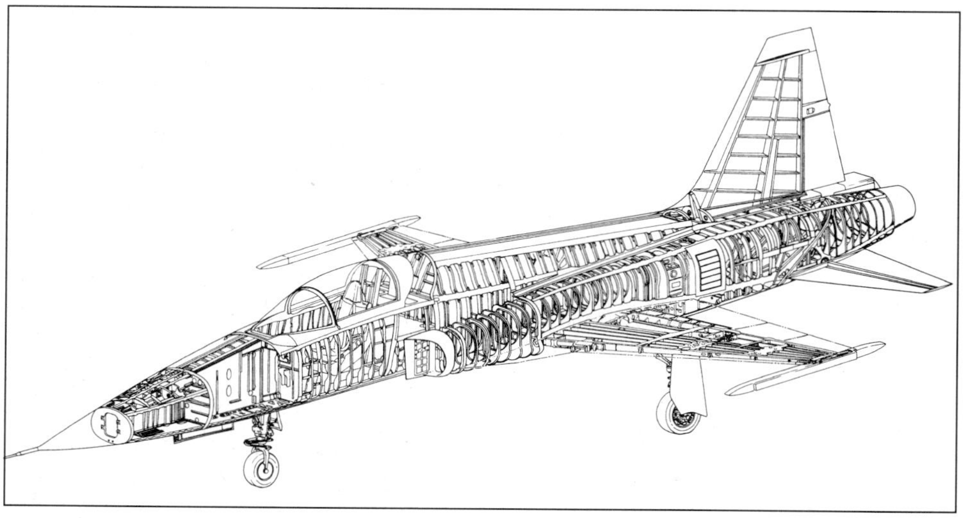

Northrop Grumman F-5 Structural Update Programme diagram *(Northrop Grumman)*

1995

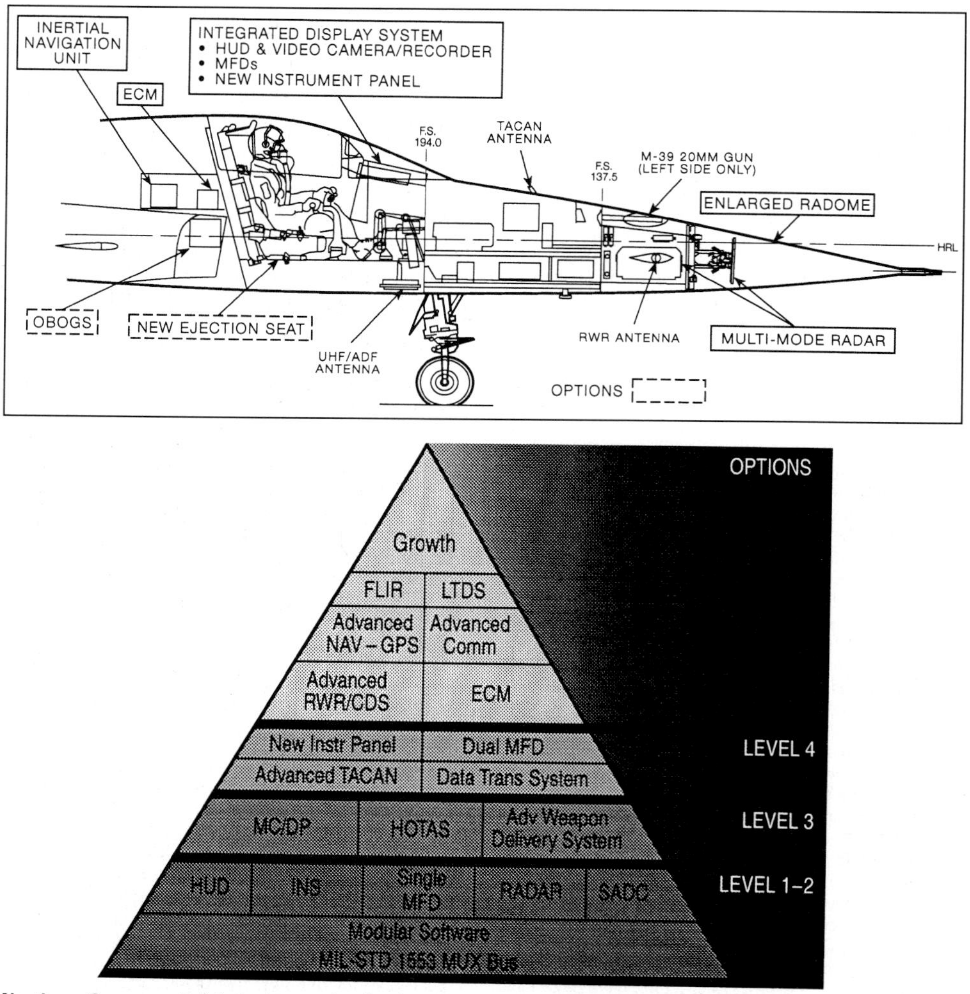

Northrop Grumman F-5 Avionics modernisation diagram (top) and modular architecture (bottom) *(Northrop Grumman)*

1995

Northrop Grumman F-5E Tiger IV during test flights *(Northrop Grumman)*

1995

2 min at normal thrust, 1 min at max thrust, 5 min max thrust for combat at 4,575 m (15,000 ft), 20 min loiter at S/L, plus reserves of 5% of initial fuel
520 n miles (964 km; 599 miles)
with max fuel, two Sidewinder missiles and two 530 lb bombs, allowances as above, and 5 min combat at military power at S/L, hi-lo-hi
450 n miles (834 km; 518 miles)
Range, F-5E:
with max fuel and reserves for 20 min max endurance at S/L: tanks retained
1,340 n miles (2,483 km; 1,543 miles)
tanks dropped
1,545 n miles (2,863 km; 1,779 miles)
Range, F-5F:
ferry range with max fuel, allowances comprising 5 min at normal thrust, 1 min at max thrust, 20 min loiter at S/L; plus reserve of 5% of initial fuel:
crew of two 1,270 n miles (2,353 km; 1,462 miles)

UPDATED

NORTHROP F-5 STRUCTURAL UPDATE PROGRAMME (SUP)

TYPE: Combat aircraft upgrade.

PROGRAMME: Selected by the USAF's San Antonio Air Logistics Center (SA-ALC) to offer F-5 structural upgrades worldwide. RFP issued January 1994. Formal contract award expected December 1994. Programme will combine redesign, damage tolerance analysis, prototype, kitproofing, developing/updating data packages, production kits and installation services into one contract.

DESIGN FEATURES: Upgrade packages consists of: new wing for F-5E/F models with ability to wire for AGM-65 Maverick air-to-ground missile, redesigned forward fuselage lower longerons for all models, horizontal stabilisers for all models, bulkheads (F.S.332/362/388/487) for the F-5A/B. CF-5 and F-5, dorsal longerons for all models and inlet duct skins for the F-5A/B, CF-5 and NF-5.

NEW ENTRY

NORTHROP GRUMMAN F-5E TIGER IV ADVANCED AVIONICS DEMONSTRATOR

TYPE: Combat aircraft avionics upgrade demonstrator.

PROGRAMME: Complete package to upgrade aircraft in stages or all at once including airframe life extension and avionics suite. The demonstrator aircraft is an F-5E on loan from the US Air Force. Following its first flight, 20 April 1995 the F-5E Tiger IV began six months flight testing of its avionics.

DESIGN FEATURES: Use of aluminium alloy and composite materials for structural enhancements including replacement of control surfaces and fuselage.

AVIONICS: The new suite consists of MIL-STD 1553 databus architecture which integrates multimode radar, mission computer, display processor, cockpit multifunction displays (MFD), and head-up display (HUD), ring laser gyro inertial navigation system (with or without GPS capability), weapons management system, and new hands-on stick and throttle (HOTAS) controls. Operational flight program (OFP) software is based on flight proven F-20 OFP. Upgrade also includes new ejection seat and onboard oxygen generating system. Options include advanced electronics warfare (EW) suite and beyond visual range (BVR) missile.

NEW ENTRY

NORTHROP RF-5E TIGEREYE

TYPE: Single-seat tactical reconnaissance aircraft.

PROGRAMME: In March 1978, Northrop announced receipt of US government approval for a company development and flight demonstration programme of an RF-5E having a modified forward fuselage with quick-change capabilities to accommodate a wide variety of reconnaissance equipment. Both day and night photo missions were demonstrated during the subsequent test programme. Modification of a production F-5E made possible the first flight of the RF-5E prototype in January 1979, and this aircraft made its international debut at the 1979 Paris Air Show. The first production RF-5E, one of two delivered to the Royal Malaysian Air Force in 1983, made its first flight on 15 December 1982. Saudi Arabia ordered 10, the first two of which were delivered Spring 1985.

Basicaly similar to the F-5E Tiger II, the RF-5E TigerEye differs by having a modified forward fuselage, and specialised equipment to enable it to fulfil a highly efficient reconnaissance role. The modified forward fuselage extends the overall length by 0.20 m (8 in), and provides 0.74 m^3 (26 cu ft) of space to accommodate reconnaissance equipment. To allow maximum flexibility for differing reconnaissance roles, Northrop decided to group the various combinations of proposed cameras/sensors on portable pallets, any one of which could be loaded easily and quickly into this forward fuselage compartment. In addition to the selected pallet, a KS-87D1 oblique frame camera is mounted in a forward nose compartment and provided with lenses of 6 in and 12 in focal length.

Various pallets have been developed to date for the RF-5E, the first comprising a KA-95B medium-altitude panoramic camera, KA-56E low-altitude panoramic camera, and an RS-710E infra-red linescanner. Pallet 2 also has the KA-56E panoramic camera, with a KA-93B6 panoramic camera with a 145° angle for heights of 3,050 to 15,240 m (10,000 to 50,000 ft). A third pallet configured for long-range oblique photo (LOROP) missions carries a single KS-147A LOROP camera of 66 in focal length.

Northrop Grumman F-5E Tiger IV demonstrator cockpit

1995

Northrop RF-5E TigerEye with AIM-9L Sidewinder air-to-air missiles

1994

Other configurations are being studied, including pallets with mapping cameras, electro-optical, infra-red and elint equipment. Forward-looking (FLIR) or downward-looking infra-red linescanners, sideways-looking airborne radar (SLAR), and sideways-looking airborne multimode radar (SLAMMR) are also potentially usable in the RF-5E. The video viewfinder has future growth possibilities for low-light operations and datalinks to another aircraft or to ground stations for real-time viewing.

The pilot has available advanced nav/com systems to complement the reconnaissance equipment, plus a video viewfinder system which enables him to view the terrain below the aircraft on a cathode ray tube display in the cockpit. Using this system, the pilot can monitor and correct his line of flight during mapping runs, and can also update the INS when passing over recognisable terrain features. In addition, there is a photographic sensor control system (PSCS) which handles many operations automatically.

The RF-5E TigerEye retains the external stores stations of the F-5E, permitting the carriage of up to three external fuel tanks (each 1,041 litres; 275 US gallons; 229 Imp gallons) for maximum range performance. It has essentially the same weights, performance and armament capabilities as the F-5E tactical fighter, and on all missions is able to carry one M39 20 mm gun with 280 rounds, plus two AIM-9 Sidewinder missiles.

UPGRADES: **Singapore Aerospace:** See separate entry.

OPERATORS: Versions of the RF-5E TigerEye are in service with the armed forces of the following countries: Malaysia (2), Saudi Arabia (10) and Singapore (6).

PERFORMANCE: As for F-5E except:

Mission radius (A, with one drop tank and two AIM-9s; B, with three drop tanks and two AIM-9s):

low altitude throughout:

A	245 n miles (454 km; 282 miles)
B	350 n miles (684 km; 403 miles)
hi-lo-hi: A	390 n miles (723 km; 449 miles)
B	530 n miles (982 km; 610 miles)
hi-lo-lo-hi: A	345 n miles (639 km; 397 miles)
B	485 n miles (898 km; 558 miles)
high altitude throughout:	
A	475 n miles (830 km; 547 miles)
B	595 n miles (1,102 km; 685 miles)

UPDATED

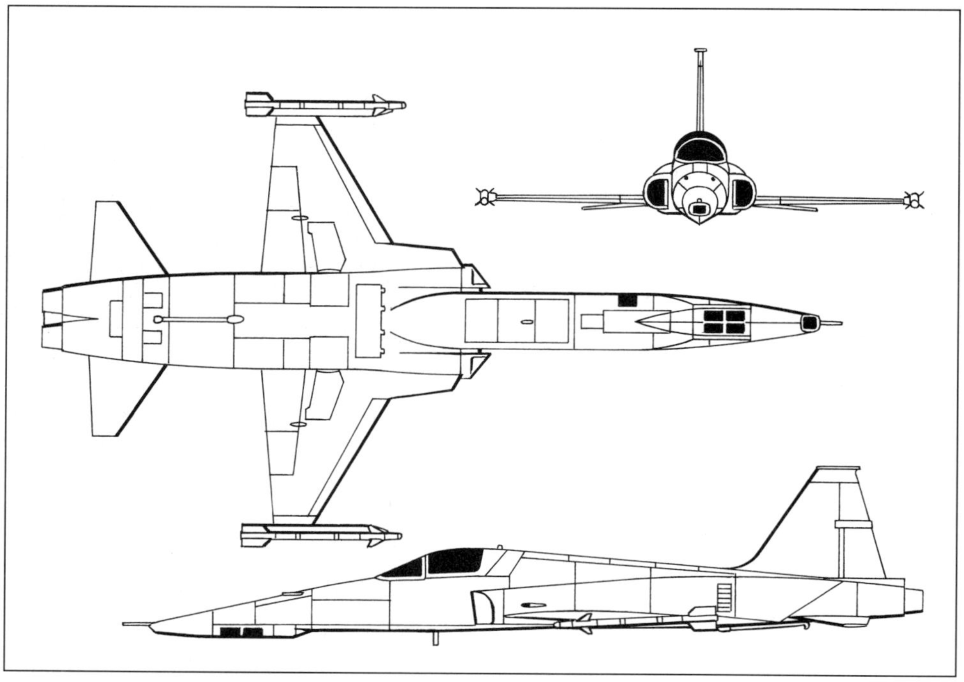

Northrop RF-5E TigerEye light tactical reconnaissance aircraft ***(Jane's/Mike Keep)***

1994

NORTHROP TALON

USAF Designation: T-38

TYPE: Two-seat supersonic basic trainer aircraft.

PROGRAMME: Developed for two years as a private venture the T-38 supersonic lightweight twin-jet trainer went into production in 1956, powered by two General Electric J85-GE-5 turbojets, and reproduces the flying characteristics of a supersonic operational fighter aircraft.

The first T-38 flew for the first time 10 April 1959 and production T-38As became operational 17 March 1961. NASA procured T-38s as flight readiness trainers for astronauts. In service with the air forces of the following countries: Germany (40+; US based), Turkey (15+) and the USA (860+).

UPGRADES: **McDonnell Douglas/Israeli Aircraft Industries:** Signed Memorandum of Understanding in March 1995 to collaborate in competition to upgrade 425 USAF T-38 aircraft with digital avionics. McDonnell Douglas to act as prime with IAI Lahav to act as main subcontractor. RFP expected in September 1995.

Northrop Grumman: Set to bid for contract to upgrade 425 USAF T-38 aircraft with digital avionics. RFP expected in September 1995.

OPERATORS: Versions of the T-38 Talon are in service with the armed forces of the following countries: Germany (41), Taiwan (40), Turkey (26) and USA (619).

DESIGN FEATURES: Cantilever low-wing monoplane. Wing section NACA 65A004.8 (modified). Thickness/chord ratio 4.8 per cent. No dihedral. No incidence. Sweepback at quater-chord 24°.

FLYING CONTROLS: Hydraulically powered sealed-gap ailerons at approximately mid-span with aluminium alloy single-slotted flaps inboard. No trim tabs. Designed to be flown and landed safely using only one aileron. Two hydraulically actuated airbrakes on underside of fuselage forward of wheel wells. No trim tabs on tail unit. Longitudinal and directional stability augomentors installed in series with control system.

STRUCTURE: Multispar aluminium alloy wing structure, with heavy plate machined skins. The fuselage is an aluminium semi-monocoque basic structure with steel, magnesium and titanium used in certain areas. 'Waisted' area-rule lines. The tail unit is a cantilever all-metal structure, with hydraulically powered rudder and one-piece 'all-moving' tailplane. Single spars with full-depth aluminium honeycomb secondary structure.

LANDING GEAR: Hydraulically retractable tricycle type with steerable nosewheel. Emergency gravity extension. Main units retract inward into fuselage, nosewheel forward, oleo-pneumatic shock-absorbers, mainwheel tyres size 20 × 4.4, pressure 16.6 kg/cm^2 (236 lb/sq in). Nosewheel tyre size 18 × 4.4, pressure 5.27 kg/cm^2 (75 lb/sq in). Multiple disc hydraulic brakes.

POWER PLANT: Two General Electric J85-GE-5 turbojet with afterburners (each rated 1,216 kg (2,680 lb st dry, 1,748 kg; 3,850 lb st with afterburner). Two independent fuel systems, one for each engine. Fuel for starboard engine provided by forward fuselage tank and dorsal tank just aft of rear cockpit. Fuel for port engine provided by centre and aft fuselage tanks. All tanks of bladder type.

Northrop T-38 Talon with General Electric J85-GE-5 turbojet engines

1992

Total usable capacity 2,206 litres (583 US gallons; 485 Imp gallons). No external tanks. Single refuelling point on lower fuselage. Oil capacity 5.3 litres (4.7 US quarts) each engine. Aircraft supplied to NASA have an engine inlet lip anti-ice system.

ACCOMMODATION: Pupil and instructor in tandem on rocket-powered ejection seats in pressurised and air conditioned cockpits, separate by windshield. Separate manually operated, rearward-hinged jettisonable canopies. Instructor's seat at rear raised 0.25 m (10 in) higher than that of pupil to give improved forward view.

SYSTEMS: Pressure differential 0.35 kg/cm² (5 lb/sq in). Two separate 210 kg/cm² (3,000 lb/sq in) hydraulic systems, one for flying controls, other for flying controls, airbrakes, landing gear, and nosewheel steering. No pneumatic system. Two 8.5 kVA generators, wide frequency 320 to 480 c/s, manufactured by Westinghouse. Two 25 A transformer-rectifiers for DC supply. No de-icing system.

ELECTRONICS: Magnavox AN/ARC-34X UHF radio, Hoffman AN/ARN-65 Tacan, Hazeltine AN/APX-64 IFF, Andrea AN/AIC-18 intercom, Collins AN/ARN-58 ILS, Bendix compass system. Integrated instrument panel.

DIMENSIONS, EXTERNAL:

Wing span	7.70 m (25 ft 3 in)
Wing chord (mean aerodynamic)	2.36 m (7 ft 9 in)
Wing aspect ratio	3.75
Length overall	14.13 m (46 ft 4½ in)
Height overall	3.92 m (12 ft 10½ in)
Tailplane	4.32 m (14 ft 2 in)
Wheel track	3.28 m (10 ft 9 in)
Wheelbase	5.93 m (19 ft 5½ in)

AREAS:

Wings, gross	15.80 m² (170 sq ft)
Ailerons (total)	0.86 m² (9.24 sq ft)
Trailing-edge flaps (total)	1.77 m² (19.00 sq ft)
Fin	3.85 m² (41.42 sq ft)
Rudder, incl tab	0.59 m² (6.37 sq ft)
Tailplane	5.48 m² (59.00 sq ft)

WEIGHTS AND LOADINGS:

Max T-O and landing weight	5,465 kg (12,050 lb)
Max zero-fuel weight	3,475 kg (7,663 lb)
Max wing loading	346.2 kg/m² (70.9 lb/sq ft)
Max power loading	1.56 kg/kg st (1.56 lb/lb st)

PERFORMANCE (at max weight except where indicated):

Max level speed (50% fuel) at 11,000 m (36,000 ft)	above Mach 1.23
Max permissible diving speed	710 kts (1,316 km/h; 818 mph) IAS
Max cruising speed at 11,000 m (360,000 ft)	above Mach 0.95
Econ cruising speed	Mach 0.88
Stalling speed flaps extended (50% fuel)	136 kts (252 km/h; 156 mph) IAS
Rate of climb at S/L (50% fuel)	9,145 m (30,000 ft)/min
Service ceiling (50% fuel)	16,335 m (53,600 ft)
Service ceiling, one engine out (50% fuel)	12,200 m (40,000 ft)
T-O run	762 m (2,500 ft)
T-O run to 15 m (50 ft)	1,128 m (3,700 ft)
Landing run from 15 m (50 ft) at AUW of 4,014 kg (8,850 lb)	1,372 m (4,500 ft)
Landing run	914 m (3,000 ft)
Range with max fuel, with 272 kg (600 lb) reserve fuel	955 n miles (1,700 km; 1,100 miles)

UPDATED

GRUMMAN G-64/G-111 ALBATROSS

TYPE: Twin-engined general utility triphibian or amphibian flying-boat.

PROGRAMME: The prototype of this general utility amphibian flew in October 1947, and the Albatross entered military service in July 1949. More than 450 built. Production has ended, but conversion of existing airframes for new duties continues. It is estimated that there are currently 200 Albatrosses still in use or in storage.

VERSIONS: **HU-16A** (formerly SA-16A) (**G-64**): Amphibian used by the USAF for sea rescue duties. 305 built. A few fitted with triphibious sprung skis to make possible operation from ice and snow as well as land and water. Withdrawn from service.

HU-16B (formerly SA-16B) (**G-111**): Improved version, with higher speed, increased range and improved single-engine performance. Span increased by 5.03 m (16 ft 6 in). Larger tail surfaces. Wing slots replaced by cambered leading-edges. Antenna housings modified to reduce drag. New high-pressure de-icing boots on wings and tail. Prototype first flown 16 January 1956. First HU-16A aircraft modified to HU-16B standard flew 25 January 1957. In service with Greek Air Force.

There were 16 special models produced for the Royal Norwegian Air Force, for anti-submarine warfare duties. These aircraft had a large nose radome, retractable MAD tail 'sting' wingtip ECM antennae, sonobuoy stowage and launcher in rear fuselage, depth charge and marine marker stowage in centre-fuselage, searchlight under starboard wing, and four underwing attachments for Mk 42-1 torpedoes, 5 in HVAR rockets, depth bombs, Zuni rocket packs or fuel tanks. Withdrawn from service with Norwegian Air Force.

Northrop T-38 Talon trainer of the US Air Force *(Northrop Grumman)*

1995

Northrop T-38 Talon two-seat Supersonic, basic trainer

1994

CSR-110: Ten specially modified HU-16Bs for SAR duties with the RCAF. Each has two 1,525 hp Canadian built Wright R-1820-82 engines, and a new retractable tricycle landing gear designed to facilitate beaching. Withdrawn from service.

HU-16C (formerly UF-1): Amphibian for US Navy and Coast Guard, similar to HU-16A. Modified to HU-16D standard as received for inspection and repair. Withdrawn from service.

HU-16D (formerly UF-2): US Navy utility amphibian similar to HU-16B. Six supplied to Japanese Maritime Self-Defence Force Spring 1961. Withdrawn from service.

HU-16E (formerly UF-2G): Similar to UH-16D, for US Coast Guard. Withdrawn from service.

UPGRADES: **Chalk's International**: G-111T. See separate entry.

Grumman/Resorts International: See separate entry.

Grumman: Upgraded Greek Air Force HU-16B Albatross search and rescue SAR aircraft with MEL Super Searcher nose-mounted radar, GEC AQS-920 lightweight acoustic processing and display system (LAPADS) and IFF.

OPERATORS: Versions of the HU-16B are in service with the armed forces of Greece (8).

The following details apply specifically to the HU-16B and D versions of the Albatross:

DESIGN FEATURES: Cantilever high-wing monoplane. Wing section NACA 23017 with extended leading- and trailing-edges. Aspect ratio 9.0. Chord 4.33 m (14 ft 2¾ in) at root, 2.13 m (6 ft 11¾ in) at tip. Dihedral 1° 10′. Incidence 5° at root. The tail unit is a cantilever structure with dihedral tailplane. Fin integral with hull.

FLYING CONTROLS: Frise ailerons with aluminium alloy structure and fabric covering. Aluminium alloy split flaps. Trim tab in each aileron. Trim tabs in rudder and each elevator.

STRUCTURE: The wing is an aluminium alloy box-beam structure in three sections; a centre-section permanently attached to hull and two outer sections. The hull is a two-step aluminium alloy semi-monocoque structure. The tail unit is an all-metal structure.

LANDING GEAR: Retractable tricycle type. Mainwheels raised into sides of hull, twin nosewheels into nose of hull. Hydraulic retraction. Bendix oleo-pneumatic shock-absorbers. Mainwheels and tyres size 40 × 12, pressure 7.4 bars (107 lb/sq in). Nosewheels size 26 × 6, pressure 4.2 bars (62 lb/sq in). Disc brakes.

POWER PLANT: Two 1,425 hp Wright R-1820-76A or B nine-cylinder radial air-cooled engines, each driving a three-blade Hamilton Standard constant-speed and reversing propeller. Internal fuel in two tanks in centre-section outboard of hull with total capacity of 2,550 litres (675 US gallons; 562 Imp gallons). Drop tanks of 378, 567 or 1,135 litres (100, 150 or 300 US gallons; 83, 125 or 250 Imp gallons) each may be carried on bomb-racks under centre-section, one on each side of hull. Each wingtip float can carry a further 756 litres (200 US gallons; 166 Imp gallons). Oil capacity 200 litres (58 US gallons; 48 Imp gallons).

ACCOMMODATION: Crew of five. Cabin may be adapted for various missions. As an ambulance 12 litters can be carried, and for transport work 10 to 22 passengers in addition to crew. With seats removed cargo or special equipment can be accommodated. A 'dutch-type' door is provided on port side for sea rescue operations or for loading stretchers. Door is split horizontally, the lower half being left in place to give higher freeboard in rough weather. A rescue platform may be attached to bottom door sill from which a crewman secured by safety belt may haul persons aboard from water. Smaller emergency door on starboard side. Both doors may be used for oblique photography and for handling sea anchor. Hatch in roof to facilitate loading of freight. Three liferafts carried, two in cabin, and one of automatically inflatable type in compartment in top of hull aft of wing. Lavatory and tail compartment for stowing equipment and gear.

SYSTEMS AND EQUIPMENT: Stewart-Warner air conditioning equipment. Auxiliary power plant in compartment aft of main cabin. Separate oxygen supplied for crew and passengers with 20 outlets for oxygen masks in cabin. Racks under each wing can carry bombs, auxiliary fuel tanks, rescue boat or other packaged equipment. Radome on nose. Provision for Jato units to be attached to cabin doors from inside hull. Stowage for four Jato units below cabin floor between wheel wells. Stowage for two parachute flares in tail compartment. Pneumatic de-icing boot on leading-edges.

DIMENSIONS, EXTERNAL (HU-16B):

Wing span	29.46 m (96 ft 8 in)
Length overall	19.18 m (62 ft 10 in)
Height overall	7.87 m (25 ft 10 in)
Draught	1.00 m (3 ft 3½ in)
Tailplane span	9.45 m (31 ft 0 in)
Wheel track	5.38 m (17 ft 8 in)
Wheelbase	5.58 m (18 ft 3¾ in)

DIMENSIONS, INTERNAL (HU-16B):

Cabin: Length	7.95 m (26 ft 1 in)
Max width	2.26 m (7 ft 5 in)
Max height	1.93 m (6 ft 4 in)
Floor area	13.5 m² (145 sq ft)
Volume	16.08 m³ (568 cu ft)

AREAS (HU-16B):

Wings, gross	96.2 m² (1,035 sq ft)
Ailerons (total)	8.40 m² (90.4 sq ft)
Flaps (total)	12.58 m² (135.4 sq ft)
Fin	8.83 m² (95.0 sq ft)
Rudder	4.23 m² (45.5 sq ft)
Tailplane	12.98 m² (139.8 sq ft)
Elevators	8.16 m² (87.8 sq ft)

WEIGHTS AND LOADINGS (HU-16B):

Weight empty equipped	10,380 kg (22,883 lb)
Normal T-O weight	13,768 kg (30,353 lb)
Max T-O weight	17,010 kg (37,500 lb)
Max zero-fuel weight	12,467 kg (27,486 lb)
Max wing loading	143.2 kg/m² (29.33 lb/sq ft)
Max power loading	7.06 kg/hp (15.56 lb/hp)

PERFORMANCE (HU-16B at max T-O weight):

Max level speed at S/L	108 kts (379 km/h; 236 mph)
Max cruising speed	195 kts (362 km/h; 224 mph)

Grumman G-111 (HU-16B) Albatross amphibious aircraft no longer in service with the US Coast Guard

1994

Cruising speed for max endurance
107 kts (200 km/h; 124 mph)
Stalling speed 64 kts (119 km/h; 74 mph)
Rate of climb at S/L 442 m (1,450 ft)/min
Service ceiling 6,550 m (21,500 ft)
Service ceiling, one engine out 2,835 m (9,300 ft)
T-O run 640 m (2,100 ft)
T-O run with JATO 213 m (700 ft)
T-O run to 15 m (50 ft) 1,356 m (4,450 ft)
T-O to 15 m (50 ft) with JATO 457 m (1,500 ft)
Landing from 15 m (50 ft) 670 m (2,200 ft)
Landing run 335 m (1,100 ft)
Range with max fuel 5% reserve, 30 min hold-off
2,477 n miles (4,587 km; 2,850 miles)

UPDATED

GRUMMAN/RESORTS INTERNATIONAL G-111 ALBATROSS

TYPE: Amphibious aircraft passenger conversion.

PROGRAMME: By January 1984, Grumman had completed prototype conversion of a UF-2 Albatross amphibian to G-111 configuration and had delivered 12 'production' aircraft to Resorts International, the originator of the redesign.

The first stage of work on each Albatross conversion involves the inspection and replacement of parts to produce the equivalent of a zero-time airframe, followed by modernisation of the flight deck, and the provision of two additional passenger cabin doors and a 28-seat interior. At the same time, the 1,100 kW (1,475 hp) Wright R-1820-982C9HE3 engines are removed and overhauled, and new fire detection and autofeathering systems installed. The resulting aircraft meets FAR Pt 36 noise level standards.

During inspection of the first airframe to be modified, Grumman found that a rear spar capstrip of 7075-T6 light alloy had suffered deterioration, and it was replaced by a titanium capstrip. Subsequent conversions have all four centre-section capstrips replaced by titanium to give the airframe an unlimited service life. Other modifications include conversion of the port main passenger door and the starboard emergency hatch to open outward, and the lower portion of the main door now incorporates a drop-down ladder.

There are 28 non-reclinable seats provided at 81 cm (32 in) pitch, with facilities for a flight attendant, and there is a toilet at the rear of the cabin. New lightweight solid-state avionics include two Collins VHF-20A com, two VIR-30 nav, ADF-30 ADF, and RCA WeatherScout 2 radar. Equipment removed included the autopilot, JATO and drop fuel tank provisions.

The prototype conversion, of a UF-2 Albatross, flew for the first time 13 February 1979, and FAA certification was received 29 April 1980.

Of the initial aircraft acquired by Resorts International, its commuter carrier, Chalks International, has five, and another has been sold to Pelita Air Service of Indonesia. Grumman has purchased 57 HU-16As and Bs for civil conversion, and believes that there is a potential market for conversions of approximately 200 Albatross aircraft with remain in worldwide use. A water-bomber version, and a version with Garrett TPE331-15 turboprop engines and Dowty Rotol four-blade propellers, are under consideration.

DIMENSIONS, EXTERNAL:
Wing span 29.46 m (96 ft 8 in)
Length overall 18.67 m (61 ft 3 in)
Height overall 7.87 m (25 ft 10 in)

DIMENSIONS, INTERNAL:
Cabin: Length 7.95 m (26 ft 1 in)
Max height 1.88 m (6 ft 2 in)
Max width 2.26 m (7 ft 5 in)
Floor area 13.47 m² (145 sq ft)
Baggage compartment 7.93 m³ (280 cu ft)

AREAS:
Wings, gross 96.15 m² (1,035 sq ft)

WEIGHTS AND LOADINGS:
Operating weight empty 10,659 kg (23,500 lb)
Max fuel load 2,920 kg (6,438 lb)
Max T-O weight: land 13,970 kg (30,800 lb)
water 14,129 kg (31,150 lb)
Max landing weight: land 13,226 kg (29,160 lb)
water 14,129 kg (31,150 lb)
Max wing loading 146.9 kg/m² (30 lb/sq ft)
Max power loading 14.15 kg/kW (10.5 lb/hp)

PERFORMANCE (at max T-O weight, ISA at S/L except where indicated):
Never-exceed speed 229 kts (424 km/h; 263 mph) IAS
Normal operating speed
206 kts (382 km/h; 237 mph) IAS
Stalling speed, flaps and landing gear down, power off
72 kts (134 km/h; 83 mph) IAS
Max rate of climb, METO power 381 m (1,250 ft)/min
Max operating altitude 2,440 m (8,000 ft)
T-O to 15 m (50 ft): land 1,341 m (4,400 ft)
water 1,349 m (4,425 ft)
Landing from 15 m (50 ft): land 924 m (3,030 ft)
water 966 m (3,170 ft)
Accelerate/stop distance: land 1,825 m (5,990 ft)
water 1,832 m (6,010 ft)
Range at 1,525 m (5,000 ft), cruising speed of 162 kts (300 km/h; 186 mph) TAS, crew of three plus 28 passengers, 45 min reserves:
land 273 n miles (506 km; 314 miles)
water 405 n miles (750 km; 466 miles)
Max ferry range, height and speed as above, no reserves:
land, water 1,480 n miles (2.742 km; 1,704 miles)

UPDATED

GRUMMAN G-134 MOHAWK

US Army designation: OV-1 (formerly AO-1)

TYPE: Two-seat army observation aircraft.

PROGRAMME: The Mohawk is a high-performance two-seat observation aircraft which Grumman developed for the US Army. The following versions have been built:

VERSIONS: **YOV-1A:** Initial batch of nine service test aircraft, of which the first flew 14 April 1959. Equipment includes ARC-52 UHF, ARC-44 VHF, AIC-12 ICS, ARN-59 ADF, MA-1 compass, APX-6B IFF, ARN-21 VOR/Tacan, ARN-32 marker beacon, KA-30 high-resolution optical photographic system, ARC-39 HF and APA-89 IFF coder.

OV-1A: First 18 production aircraft similar to YOV-1A, but with addition of FD-105 integrated flight system, ARC-55 UHF instead of ARC-52, ARN-30 VOR/Tacan instead of ARN-21, ARN-68 marker beacon instead of ARN-32 and provision for ARC-73 VHF, APX-44 IFF, radar altimeter, ILS, emergency VHF, autopilot, ground track beacon, Doppler and UAS-4 IR. The 19th and subsequent production aircraft have the radar altimeter, ILS, emergency VHF, autopilot and ground track beacon installed and have duplicated VOR/Tacan. All versions can carry 52 flares in each of two removable upward-firing pods mounted above the wingroots for night photography. Withdrawn from service.

OV-1B: Different equipment. Similar to second series of OV-1A, but with APS-94 SLAR (side-looking airborne radar) in underfuselage container and AKT-16 VHF data-link. SLAR provides a permanent radar photographic map of terrain on either side of the flight path, on either 10 × 12.7 cm (4 × 5 in) cut film or 70 mm film strip. An in-flight processor enables the observer to see a developed photograph seconds after the film has been exposed. This version has increased wing span and area. Withdrawn from service.

OV-1C: Different equipment. Similar to second series of OV-1A, but with UAS-4 infra-red surveillance equipment installed. Withdrawn from service.

OV-1D: This final version can be converted from infra-red to SLAR surveillance capability, and vice versa, in an hour, so combining the duties of the OV-1B and OV-1C. Four preproduction prototypes for the US Army were followed by production OV-1Ds, the last of which was completed in December 1970.

RV-1D: See separate entry.

More than 375 Mohawks of all versions were delivered.

UPGRADES: **OV-1D/RV-1D:** See separate entry.

OPERATORS: Versions of the OV-1D Mohawk are in service with the armed forces of the following countries: Argentina (20) and South Korea (quantity unknown).

The description which follows applies in particular to the OV-1D:

DESIGN FEATURES: Cantilever mid-wing monoplane. Wing section NACA 2412. Dihedral 6° 30′. Incidence 1° 30′. The tail unit is a cantilever structure with central and two endplate fins and rudders.

FLYING CONTROLS: Aluminium alloy inboard and outboard ailerons. Servo-tabs and manually adjustable trim tabs in outboard ailerons. Hydraulically operated trailing-edge flaps. Forward-opening hydraulically operated airbrake on each side of fuselage aft of wing. Manually adjustable trim tabs in elevators and central rudder.

STRUCTURE: The wing is an aluminium box-beam structure. The fuselage is an aluminium alloy semi-monocoque structure. The tail unit is an aluminium alloy structure.

LANDING GEAR: Hydraulically retractable tricycle type, with a single wheel on each unit. High-pressure pneumatic system for emergency extension. Bendix oleo-pneumatic shock-absorbers. Main units, which retract outward, each have a mechanical shrink-rod to compress the oleo-pneumatic strut so that it can be housed within the engine nacelle. The steerable nosewheel retracts rearward into the fuselage. Mainwheels and tyres size 8.50 × 10, pressure 6.2 bars (90 lb/sq in). Nosewheel tyre size 6.50 × 8, pressure 4.5 bars (65 lb/sq in). Goodyear hydraulically operated disc brakes. Provision for wheel/ski gear.

POWER PLANT: Two 1,400 shp Lycoming T53-L-701 turboprop engines, driving Hamilton Standard Type 53C51-27 three-blade constant-speed fully feathering reversible-pitch metal propellers, diameter 3.05 m (10 ft 0 in). Fuel in self-sealing fuselage tank above wing, with capacity of 1,045 litres (276 US gallons; 230 Imp gallons). Provision

Grumman OV-1D Mohawk converted to SLAR surveillance duties *(Peter J. Cooper)*

1994

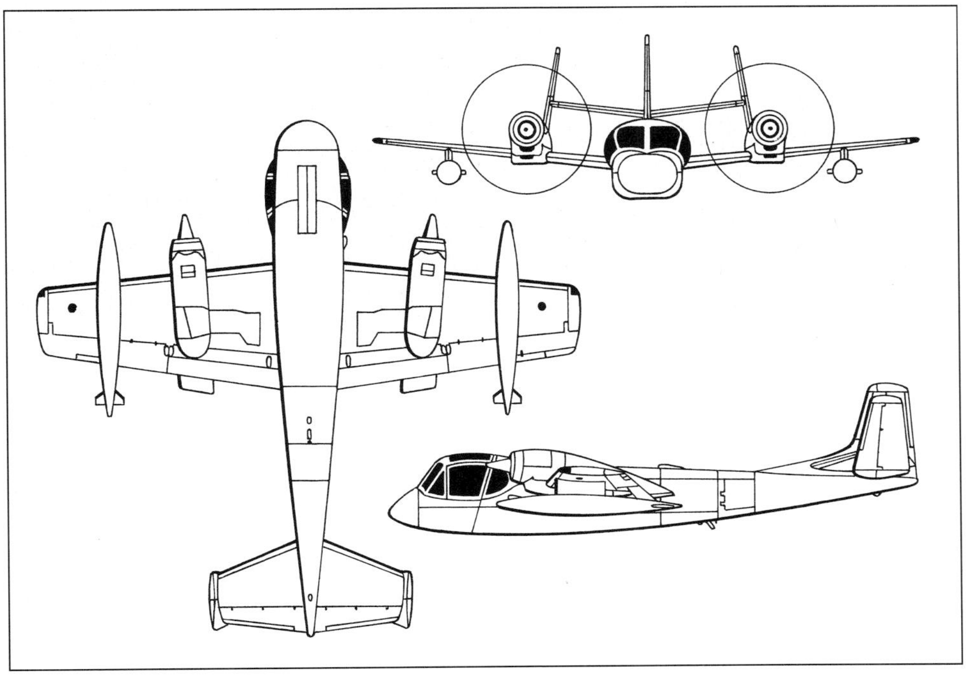

Grumman OV-1 Mohawk two-seat observation aircraft *(Jane's/Mike Keep)*

1994

for one 567 litre (150 US gallon; 125 Imp gallon) Aero 1C jettisonable tank under each wing, outboard of engine. Oil capacity 19 litres (5 US gallons; 4.1 Imp gallons). Electrical de-icing of propeller blades, spinners and air inlet cowlings.

ACCOMMODATION: Flight compartment seating two side by side in nose on Martin-Baker Mk J5 ejection seats. Dual controls, except when electronic surveillance equipment is fitted. Sides of canopy bulged to improve downward visibility. Armouring includes 0.64 cm (¼ in) aluminium alloy cockpit floor, bullet-resistant windshields and removable flak curtains on fore and aft cockpit bulkheads.

SYSTEMS: Hydraulic system supplied by two engine-driven variable-volume hydraulic pumps, pressure 210 kg/cm² (3,000 lb/sq in), to operate inboard ailerons, flaps, airbrakes, landing gear, wheel brakes, nosewheel steering and windshield wipers. Electrical power supplied by two engine-driven starter-generators providing up to 400 A at 28 V DC. Two engine-driven 6.5 kVA alternators for de-icing. Three rotary inverters powered from the DC busbar generate 115/200 V AC three-phase 40 Hz current: 5 kVA unit powers inertial navigation and surveillance systems; a 2.5 kVA unit powers instruments and navigation aids; and a 750 VA unit provides emergency power, 24 V 38 Ah Ni/Cd battery provides power for engine starting and emergency power. An air cycle system supplies a mixture of heated and cooled engine bleed air for automatic or manual control of cockpit temperature and window defogging. Ram-air system for cockpit cooling and window defogging. A separate fully automatic system supplies engine bleed air to all three camera compartments. Engine inlet struts de-iced by engine bleed air. Wing and empennage de-icing boots receive pulses of engine bleed air. Fire extinguishing system for both engines. Pneumatic de-icing boots on leading-edges.

EQUIPMENT: Photo surveillance system consists of two KA-60C 180° panoramic camera systems and one KA-76 serial frame camera. Infra-red AN/AAS-24 surveillance system. Alternative AN/APS-94D side-looking airborne radar (SLAR) system. ADR-6 radiac system. AN/AYA-10 data annotation system. Nav/com systems include ARN-82 VOR, ARC-102 HF-SSB, ARC-114 VHF-FM primary and auxiliary, ARN-52 Tacan, ARN-89 loop, APX-72 IFF, APN-171(V) radar altimeter, ARC-114 VHF-FM homing, ARC-115 and -116 VHF/UHF, ARN-58 glide slope and marker beacon, ARN-89 LF-ADF, AN-ASN-33 flight director, PT489/ASQ-104 map readout unit, AN/ASW-12 automatic flight control, C-6533/ARC intercom, ASH-19 aural reproducer and continuous in-flight performance recorder (CIPR). ECM pods and an LS-59A photo-flash unit can be carried on underwing stations.

DIMENSIONS, EXTERNAL:
- Wing span: OV-1A, C — 12.80 m (42 ft 0 in)
  - OV-1B, OV-1D — 14.63 m (48 ft 0 in)
- Wing chord: at root — 3.20 m (10 ft 6 in)
  - at tip — 1.60 m (5 ft 3 in)
- Wing aspect ratio — 5.35
- Length overall — 12.50 m (41 ft 0 in)
- Height overall — 3.86 m (12 ft 8 in)
- Tailplane span — 4.85 m (15 ft 11 in)
- Wheel track — 2.79 m (9 ft 2 in)
- Wheelbase — 3.56 m (11 ft 8¼ in)

AREAS:
- Wings, gross: OV-1A, C — 30.65 m² (330 sq ft)
  - OV-1B, OV-1D — 33.45 m² (360 sq ft)
- Ailerons (total) — 2.11 m² (22.7 sq ft)
- Flaps (total) — 4.05 m² (43.6 sq ft)
- Fins (total) — 3.84 m² (41.3 sq ft)
- Rudders (total) — 2.55 m² (27.5 sq ft)
- Tailplane — 6.13 m² (66.0 sq ft)
- Elevators — 1.77 m² (19.0 sq ft)

WEIGHTS AND LOADINGS:
- Weight empty, equipped: OV-1A — 4,507 kg (9,937 lb)
  - OV-1B — 5,020 kg (11,067 lb)
  - OV-1C — 4,717 kg (10,400 lb)
  - OV-1D — 5,467 kg (12,054 lb)
- Normal T-O weight: OV-1A — 5,748 kg (12,672 lb)
  - OV-1B — 6,197 kg (13,650 lb)
  - OV-1C — 5,915 kg (13,040 lb)
  - OV-1D SLAR — 7,140 kg (15,741 lb)
  - OV-1D IR — 7,051 kg (15,544 lb)
- Max T-O weight: OV-1A — 6,818 kg (15,031 lb)
  - OV-1B, C — 8,722 kg (19,230 lb)
  - OV-1D SLAR — 8,214 kg (18,109 lb)
  - OV-1D IR — 8,124 kg (17,912 lb)

PERFORMANCE:
- Max level speed at 1,520 m (5,000 ft):
  - OV-1A, 1C — 267 kts (496 km/h; 308 mph)
  - OV-1B — 258 kts (478 km/h; 297 mph)
  - OV-1D, max rated power at 3,050 m (10,000 ft), 40% fuel: SLAR mission — 251 kts (465 km/h; 289 mph)
    - IR mission — 265 kts (491 km/h; 305 mph)
- Max permissible diving speed — 390 kts (724 km/h; 450 mph)
- Max cruising speed:
  - OV-1A — 264 kts (489 km/h; 304 mph)
  - OV-1B — 239 kts (443 km/h; 275 mph)
  - OV-1C — 258 kts (478 km/h; 297 mph)
- Econ cruising speed:
  - All versions — 180 kts (334 km/h; 207 mph)
- Stalling speed (landing configuration):
  - OV-1A — 59 kts (109 km/h; 68 mph)
  - OV-1B — 64 kts (117 km/h; 73 mph)
  - OV-1C — 66 kts (123 km/h; 76 mph)
  - OV-1D, landing configuration, 10% normal rated power, 60% fuel:
    - SLAR mission — 73 kts (135 km/h; 84 mph)
    - IR mission — 72 kts (133.5 km/h; 83 mph)
- Rate of climb at S/L: OV-1A — 900 m (2,950 ft)/min
  - OV-1B — 716 m (2,350 ft)/min
  - OV-1C — 814 m (2,670 ft)/min
  - OV-1D, max rated power:
    - SLAR mission — 1,056 m (3,466 ft)/min
    - IR mission — 1,102 m (3,618 ft)/min
- Service ceiling: OV-1B — 9,235 m (30,300 ft)
  - OV-1C — 9,000 m (29,500 ft)
  - OV-1D — 7,620 m (25,000 ft)
- Service ceiling, one engine out:
  - OV-1A — 4,152 m (13,625 ft)
- Min T-O run: OV-1A — 145 m (475 ft)
  - OV-1B — 177 m (580 ft)
  - OV-1C — 192 m (630 ft)
- T-O to 15 m (50 ft): OV-1A — 281 m (922 ft)
  - OV-1B — 268 m (880 ft)
  - OV-1C — 335 m (1,100 ft)
  - OV-1D SLAR mission — 358 m (1,175 ft)
  - OV-1D IR mission — 349 m (1,145 ft)
- Landing from 15 m (50 ft): OV-1A, 1C — 240 m (787 ft)
  - OV-1B — 264 m (866 ft)
  - OV-1D SLAR mission — 323 m (1,060 ft)
  - OV-1D IR mission — 320 m (1,050 ft)
- Landing run: OV-1A — 138 m (454 ft)
  - OV-1B — 165 m (540 ft)
  - OV-1C — 153 m (502 ft)
- Max range with external tanks, 10% reserves:
  - OV-1A — 1,220 n miles (2,270 km; 1,410 miles)
  - OV-1B — 1,065 n miles (1,980 km; 1,230 miles)
  - OV-1C — 1,155 n miles (2,140 km; 1,330 miles)
- Max range with external tanks at 6,100 m (20,000 ft):
  - OV-1D SLAR mission — 820 n miles (1,520 km; 944 miles)
  - OV-1D IR mission — 878 n miles (1,627 km; 1,011 miles)
- Max endurance at 140 kts (259 km/h; 161 mph) at 4,560 m (15,000 ft): OV-1D SLAR mission — 4.35 h
  - OV-1D IR mission — 4.54 h

UPDATED

GRUMMAN OV-1/RV-1 MOHAWK UPGRADES

TYPE: Two-seat army observation aircraft upgrade.

PROGRAMME: Although production of the OV-1 Mohawk surveillance aircraft ended in the early 1970s, those still in US Army service have undergone a succession of conversion and modernisation programmes. The most recent of these to be completed, in FY87, was the conversion of 78

Grumman RV-1D Mohawk converted under the Quick Look II programme

1994

OV-1Bs (SLAR reconnaissance) and OV-1Cs (infra-red and photographic reconnaissance) to **OV-1D** (interchangeable SLAR/IR) configuration, to augment 37 aircraft built from the outset as OV-1Ds. Specific mission avionics and equipment installed in the OV-1D include AN/AAS-24 infra-red detection system AN/ALQ-147A(V)1 infra-red jammer, AN/APN-171 radar altimeter, AN/APR-39(V)2 or AN/APR-44 radar warning receiver, AN/APS-94F side-looking airborne radar, AN/APX-72 IFF transponder, AN/ARN-103 Tacan, AN/ASN-86 inertial navigation system, AN/AYA-10 signal processor, and a KS-113A photographic survey system with KA-60C and KA-76 cameras.

In an earlier programme known as Quick Look II, completed in 1983, a total of 28 other OV-1Bs were converted as **RV-1D** electronic warfare aircraft equipped with AN/ALQ-133 jamming ECM, AN/USQ-61 digital data set, AN/USM-393 simulation set, AN/ALM-153/154 test sets, and an AN/MSA-34 antenna group. Aircraft withdrawn from service.

The US Army has flight tested in recent years an OV-1D fitted with more powerful engines, and Grumman has tested improved stall warning and electrical power/anti-icing systems for the Mohawk.

UPDATED

Grumman S-2E Tracker of the Royal Australian Air Force *(Peter March)*

1994

Grumman S-2E Tracker with provision for AS.12 air-to-surface missiles *(Peter J. Cooper)*

1994

GRUMMAN S-2 TRACKER

Brazilian military designation: P-16

TYPE: Twin-engined naval anti-submarine aircraft.

PROGRAMME: The Tracker is a twin-engined anti-submarine search and attack aircraft. The prototype, designated XS2F-1, made its first flight 4 December 1952, since when Grumman have delivered more than 1,000 S-2s.

VERSIONS: **S-2A** (formerly S2F-1): Two 1,525 hp Wright R-1820-82 engines. First production model. Supplied also to Argentina, Brazil, Italy, Japan, Netherlands, Taiwan, Thailand and Uruguay. About 500 built. Production completed.

TS-2A: Training version of S-2A.

S-2B: Redesignation of S-2As modified to carry Jezebel acoustic search equipment and Julie acoustic echo-ranging system.

S-2C (formerly S2F-2): Developed version to carry larger anti-submarine weapons. An enlarged torpedo bay with asymmetrical extension on port side of fuselage accommodates homing torpedoes. To compensate for increased weight the tail surfaces were increased in area. Sixty built. Most converted into **US-2C** utility aircraft.

S-2D (formerly S2F-3): Developed version of S-2A ordered in 1958. Increased span. More roomy crew accommodation. Much improved operational equipment and armament. Twice the endurance of S-2A at 200 n miles (370 km; 230 mile) radius. Prototype flew 20 May 1959. Original contracts for 167, for delivery by end of 1963, were increased by a further order for 48 announced in January 1962. Production completed.

S-2E (formerly S2F-3S): Similar to S-2D, but with more advanced ASW electronic equipment, and provision for new types of armament, including nuclear depth charges, AS.12 air-to-surface missiles and Mini-guns. Deliveries to VS-41 began in October 1962. Production completed.

CS2F-1: Canadian production version of S-2A built under licence for Royal Canadian Navy by de Havilland Aircraft of Canada Ltd. 100 built. Seventeen supplied to Royal Netherlands Navy. Withdrawn from service.

CS2F-2/3: New designations for CS2F-1 aircraft fitted with improved operational equipment. Withdrawn from service.

US-2C: Utility version.

UPGRADES: **Conair:** Firecat and Turbo Firecat. See separate entry.

IAI: S-2 and ST-UP. See separate entry.

IMP: S-2E. See separate entry.

Marsh: S-2FT Turbo Tracker. See separate entry.

Northrop Grumman: S-2T. See following entry.

OPERATORS: Versions of the S-2 Tracker are in service with the armed forces of the following countries: Argentina (9), Brazil (13), South Korea (24), Peru (11), Taiwan (32), Thailand (6), Turkey (33) and Uruguay (6).

The following data apply generally to all versions of the S-2:

DESIGN FEATURES: Cantilever low-wing monoplane. Wings fold upward and inward hydraulically from outboard of the engine nacelles. Rudder split vertically into two sections; forward section actuated hydraulically during take-off, landing and single engine operation to increase rudder area. Cantilever tail unit.

FLYING CONTROLS: Fixed leading-edge slots on outer wings. Small ailerons supplemented by wide-span spoilers on upper surfaces. Long-span slotted flaps.

STRUCTURE: All-metal multispar wing structure. The fuselage is an all-metal semi-monocoque structure. The tail unit is an all-metal structure.

LANDING GEAR: Retractable tricycle type, all units retracting rearward. Twin-wheels on nose unit only. Small aft wheel-bumper is extendable but not fully retractable.

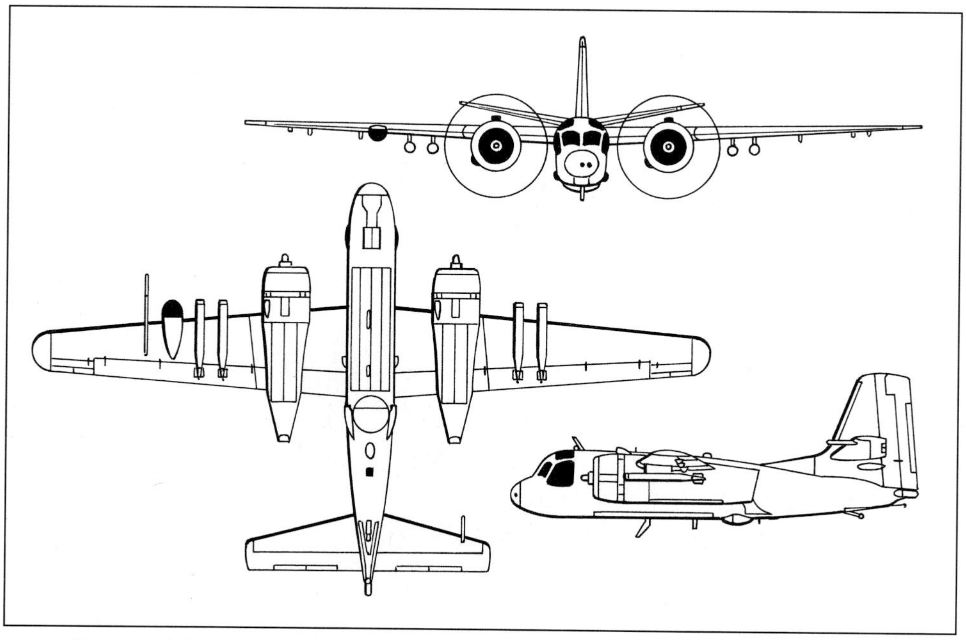

Grumman S-2 Tracker twin-engine anti-submarine warfare (ASW) aircraft *(Jane's/Mike Keep)*

1994

POWER PLANT: Two 1,525 hp Wright R-1820-82 WA nine-cylinder air-cooled radial engines, driving three-blade constant-speed metal propellers.

ACCOMMODATION: Crew of four consisting of pilot, co-pilot (who serves as navigator, radio-operator and searchlight operator), radar operator and MAD operator. Dual controls.

ELECTRONICS AND EQUIPMENT: Equipment of S-2D includes UHF direction-finder, LF direction-finder, Tacan, APN-122 Doppler radar, APN-117 low-altitude radar altimeter, UHF transmitter/receiver, HF transmitter/receiver, APX 6B and APA 89 IFF, ASA-13 ground position indicator, and autopilot. S-2D also has ground track plotter on instrument panel, giving aircraft position and ground track, including Doppler correction for drift and groundspeed, target position from Julie computer, sonobuoy location, radar position, MAD mark on ground attack and exhaust trail mark on ground track.

OPERATIONAL EQUIPMENT (S-2D): AQA-3 Jezebel passive long-range acoustic search equipment, using sonobuoys; ECM instantaneous electronic countermeasures direction finder; Sniffer passive submarine exhaust trail detector; Julie active acoustic echo ranging by means of explosive charges, automatic target computer and automatic target plotting; retractable ASQ-10 MAD (magnetic anomaly detector) tail 'sting'; retractable 75 kW X-band 42 × 20 in search radar under fuselage; 85 Mcd remotely controlled searchlight under starboard wing.

ARMAMENT: Two homing torpedoes or one Mk 101 depth bomb or four 385 lb depth charges in bomb bay. Six underwing attachments for torpedoes, 5 in rockets, Zuni rockets or 250 lb bombs. Housing for sonobuoys and marine markers in rear of engine nacelles. Fuselage dispenser for 60 underwater sounding charges for echo ranging.

DIMENSIONS, EXTERNAL (S-2E):

Wing span	22.13 m (72 ft 7 in)
Width folded	8.33 m (27 ft 4 in)
Length overall	13.26 m (43 ft 6 in)
Height overall	5.06 m (16 ft 7 in)
Wheel track	5.64 m (18 ft 6 in)

AREAS:

Wings, gross:	
S-2A, S-2C, CS2F-1 and 2	45.10 m² (485 sq ft)
S-2D, S-2E	46.08 m² (496 sq ft)

WEIGHTS AND LOADINGS (S-2E):

Weight empty	8,505 kg (18,750 lb)
Max payload	2,182 kg (4,810 lb)
Max internal fuel	1,981 kg (4,368 lb)
Max T-O weight	13,222 kg (29,150 lb)

PERFORMANCE (S-2E at max T-O weight):

Max level speed at S/L	over 230 kts (426 km/h; 265 mph)
Patrol speed at 450 m (1,500 ft)	130.2 kts (241 km/h; 150 mph)
Stalling speed (landing configuration)	65 kts (119 km/h; 74 mph)
Service ceiling	6,400 m (21,000 ft)
Min T-O run	396 m (1,300 ft)
T-O to 15 ft (15 m)	572 m (1,875 ft)
Ferry range	1,128 n miles (2,095 km; 1,300 miles)
Endurance with max fuel, 10% reserves	9 h

UPDATED

GRUMMAN S-2T TURBO TRACKER

TYPE: Twin-engine, naval anti-submarine warfare (ASW) aircraft turboprop upgrade.

PROGRAMME: US Navy $260 million foreign military sales contract placed to fit S-2 Tracker with turboprops and fit new avionics, including GEC Avionics MAPADS 902F acoustic processor, AN/ASQ-504(V) magnetic anomaly detector, AN/APS-509 radar, AN/ARR-84 acoustic receivers, AN/ASN-150 tactical navigation system coupled with the 72R inertial navigation system, and Rockwell Collins radios. Tracor Aviation replaced 1,141 kW (1,530 hp) Curtiss-Wright R1820-82 piston engines with 1,227 kW (1,645 shp) Garrett TPE331-15AW turboprops driving Dowty advanced technology four-blade propellers; Turbo Tracker has 270 knot (500 km/h; 311 mph) maximum speed at 1,525 m (5,000 ft) and payload is increased by 500 kg (1,102 lb); cruising speed, field lengths and single-engined performance and engine TBO all improved.

Taiwan ordered 32 S-2Ts under US Navy FMS contract; Grumman converted two; remainder modified in Taiwan with kits. Refer to IMP Group (Canadian section) and Bedek Aviation (Israel) for details of other military turboprop Tracker conversions.

VERIFIED

GRUMMAN A-6E INTRUDER

US Navy designations: A-6E and KA-6D

TYPE: Carrier-based all-weather attack aircraft, with tanker variant.

PROGRAMME: 708 new A-6 and EA-6s built; about 375 still in service with US Navy and Marine squadrons and three readiness training squadrons. First A-6A entered service in February 1963 with training squadron VA-42; first operational deployment with VA-75, May 1965.

VERSIONS: **A-6A/B/C:** Only those converted to other variants still in service.

EA-6B Prowler: Advanced electronics development of EA-6A; described separately.

KA-6D: Dedicated tanker; 78 converted from A-6A and 12 from A-6E; latest configuration has no weapon capability and can carry 1,514 litre (400 US gallon; 333 Imp gallon) drop tanks; fitted as probe and drogue receiver as well as tanker; drogue retracts into tunnel under rear fuselage.

A-6E Intruder: Began as advanced conversion of A-6A (240 converted) with multimode radar and IBM computer first fitted to EA-6B; first flight 10 November 1970; first deployed September 1972; approved for service November 1972. Total 205 new-build A-6Es funded by end FY88, of which 21 have composites wings (see below).

A-6E/TRAM: Version equipped with Target Recognition Attack MultiSensor; first flight October 1974; delivery of fully equipped aircraft began 14 December 1978; first carrier deployment completed May 1980; all older A-6Es converted to TRAM by 1988; eight Grumman TC-4C (modified Gulfstream I) used for training.

US Navy A-6E squadrons are VA-52, 95, 115, 128 (training), 145, 155, 165, 196 at Whidbey Island, Washington, with two rotated to Atsugi, Japan; VA-34, 35, 36, 42 (training), 55, 65, 75 and 85, at Oceana, Virginia; Reserves VA-205 at Atlanta, Georgia. Marine squadrons are VMA(AW)-224, 332 and 533 at Cherry Point, North Carolina, including one rotated to Atsugi, Japan.

A-6E Harpoon: 50 A-6Es each fitted to carry four McDonnell Douglas Harpoon anti-ship missiles; deployment began 1981; all subsequent new-build and converted aircraft also equipped.

Re-winged A-6E: Last metal-winged A-6E delivered September 1988; final 21 production aircraft being fitted with composites wings. Boeing Military Airplanes won competition to produce graphite/epoxy wing for A-6F (new build and retrofit) to overcome fatigue problems in service and give 8,800 hour service life; first flight 3 April 1989. Naval trials completed June 1990. Initial contract for 179 wing sets plus options for 327. First five wings fitted to A-6Fs; options not exercised and remaining wings installed on A-6E after A-6F cancelled.

A-6E SWIP: Systems Weapon Integration Program partially applied to last 33 production aircraft. Features of Block I system, currently being applied also to earlier aircraft, include provision for AGM-88 HARM, AGM-84 Harpoon and AGM-65 Maverick, and various airframe improvements. SWIP Block 1A covers GPS, AN/ASN-139 INS, GEC Avionics standard central air data computer on MIL-STD-1553B digital databus, upgrade of Tacan from AN/ARN-84 to AN/ARN-118, HUD, multi-function display wing fillet modifications to reduce approach speed and thus increase maximum arresting weight and IDAP (Integrated Defence Avionics Program) combining chaff dispensers and radar warning receiver. Fillet trials A-6 flew August 1991; two avionics trials aircraft were to fly September 1992 and May 1993; upgrade of 290 A-6Es to follow.

A-6F and A-6G: Abandoned Intruder developments. Five preseries A-6Fs employed for trials.

UPGRADES: **Boeing:** Re-winged A-6E. See Versions.

Grumman: KA-6D, A-6E TRAM, A-6E Harpoon and A-6E SWIP. See Versions.

OPERATORS: Versions of the A-6 Intruder are in service with the US armed forces (136).

Grumman A-6E Intruder all-weather attack aircraft

1994

COSTS: $30.9 million (1987) system unit cost.

DESIGN FEATURES: Wing sweepback 25° at quarter-chord; outer wing panels fold upward more than 90° for stowage; TRAM ball under nose, aft of radome. Survivability improvements incorporated; fire extinguishing system (halon) inerting fuel tanks; fire suppression in areas around fuselage fuel tanks wing dry foam blocks; and self-sealing fuel lines in engine cavities.

FLYING CONTROLS: Hydraulic fully powered controls; slab tailplane; lateral control spoilers ahead of near-full-span Fowler flaperons; split surface airbrakes in trailing-edge outboard of flaperons; sharp-edged leading-edge stall strip next to fuselage with fixed drop section outboard, then long-span leading-edge slats. Fences near wingroot and near tip.

STRUCTURE: Conventional all-metal; graphite/epoxy wing being retrofitted; aluminium alloy control surfaces and titanium high strength fittings, such as wing-fold.

LANDING GEAR: Hydraulically retractable tricycle type. Twin-wheel nose unit retracts rearward. Single-wheel main units retract forward and inward into air intake fairings. A-frame arrester hook under rear fuselage; nose-tow catapult fitting.

POWER PLANT: Two 41.4 kN (9,300 lb st) Pratt & Whitney J52-P-408 turbojets. Replacement by two 53.4 kN (12,000 lb st) J52-P-409s under study. Maximum internal fuel capacity 8,873 litres (2,344 US gallons; 1,952 Imp gallons). Provision for up to five external fuel tanks under wing and centreline stations, each of 1,135 litres (300 US gallons; 250 Imp gallons) or 1,514 litres (400 US gallons; 333 Imp gallons) capacity. Removable flight refuelling probe projects upward immediately forward of windscreen.

ACCOMMODATION: Crew of two on Martin-Baker GRU7 ejection seats, which can be reclined to reduce fatigue during low-level operations. Bombardier/navigator slightly behind and below pilot to starboard. Hydraulically operated rearward-sliding canopy.

SYSTEMS: AiResearch environmental control system for cockpit and avionics bay. Dual hydraulic systems for operation of flight controls, leading-edge flaps, wingtip speed-brakes, landing gear brakes and cockpit canopy, each rated at 119.4 litres (31.5 US gallons; 26.2 Imp gallons)/min, with air/oil separated reservoir, pressurised at 2.76 bars (40 lb/sq in). One electrically driven hydraulic pump provides restricted flight capability by supplying the tailplane and rudder actuators only, each rated at 11.4 litres (3 US gallons; 2.5 Imp gallons)/min, with internally pressurised reservoir, pressurised a 1.08 bars (15.7 lb/sq in). Electrical system powered by two Garrett constant-speed drive units that combine engine starting and electric power generation, each delivering 30 kVA; retrofit of Sundstrand generators (as in EA-6B) under way. A Garrett ram-air turbine, mounted so that it can be projected into the airstream above the port wingroot, provides in-flight emergency electric power for essential equipment.

AVIONICS: Single Norden AN/APQ-148 or AN/APQ-156 simultaneous multimode nav/attack radar. IBM and Fairchild nav/attack computer system and interfacing data converter. Conrac Corporation armament control unit. RCA video tape recorder for post-strike assessment of attacks; replacement unit by Precision Echo to be retrofitted. Litton AN/ALR-67 radar warning receiver. Radar provides simultaneous ground-mapping; identification, tracking, and rangefinding of fixed or moving targets; and terrain-clearance or terrain-following manoeuvres. During 1981-83, it was updated by an improved AMTI (airborne moving target indication) to enhance its ability to detect moving targets. IBM AN/ASQ-133 or AN/ASQ-155 solid-state digital computer is coupled to A-6E's radar, inertial and Doppler navigational equipment, communications and AFCS. Fairchild signal data converter accepts analog input data from up to 60 sensors, converting data to a digital output that is fed into nav/attack system computer. Conrac armament control unit (ACU) provides all inputs and outputs necessary to select and release weapons. Kaiser AN/AVA-1 multimode display serves as a primary flight aid for navigation approach landing and weapons delivery. TRAM package includes undernose precision-stabilised turret, with a sensor package containing both infra-red and laser equipment; INS updated with Litton AN/ASN-92 CAINS; new communications-navigation-identification (CNI) equipment including AN/APX-72 IFF transponder; and automatic carrier landing capability. Sensor package is integrated with multimode radar, providing capability to detect, identify and attack a wide range of targets (as well as view the terrain) under adverse weather conditions, and

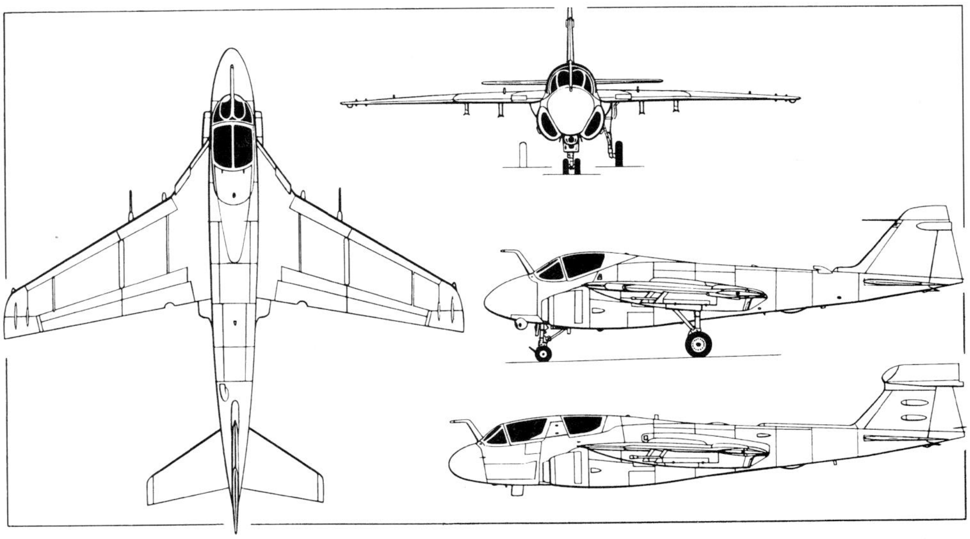

Grumman A-6E/TRAM Intruder, with additional side view of EA-6B (bottom) ***(Jane's/Dennis Punnett)***

1994

Grumman A-6E Intruder landing aboard the USS *John F. Kennedy*. Lockheed S-3 Viking in background
1994

with improved accuracy, using either conventional or laser-guided weapons. Bombardier/navigator operates TRAM system by first acquiring target on his radar screen. He then switches to FLIR (forward-looking infra-red) system, using an optical zoom to enlarge target's image. After identifying and selecting the targets, Bombardier uses a laser designator to mark target with a laser spot, on which the laser-guided weapons, or those from another aircraft, will home. Using TRAM's laser spot detector, A-6E can also acquire a target being illuminated from another aircraft or designated by a forward air controller on the ground. Some A-6s modified for compatibility with crew night vision goggles.

ARMAMENT: Five weapon attachment points, each with a 1,633 kg (3,600 lb) capacity (maximum external stores load 8,165 kg; 18,000 lb). Typical weapon loads are 28 500 lb bombs in clusters of six, or three 2,000 lb general purpose bombs plus two 1,135 litre (300 US gallon; 250 Imp gallon) drop tanks. AIM-9 Sidewinder can be carried for air-to-air use. Harpoon missile capability added to weapons complement of A-6E/TRAM. The HARM missile has been test flown on the A-6E. Up to 20 Brunswick Defense AN/ADM-141 TALD (Tactical Air-Launched Decoy) gliders, or two in addition to normal bomb load. Flight and firing tests have been carried out with the AGM-123A Skipper II, also on an A-6E.

DIMENSIONS, EXTERNAL:

Wing span	16.15 m (53 ft 0 in)
Wing mean aerodynamic chord	3.32 m (10 ft 10¾ in)
Width, wings folded	7.72 m (25 ft 4 in)
Length overall	16.69 m (54 ft 9 in)
Height overall	4.93 m (16 ft 2 in)
Tailplane span	6.21 m (20 ft 4½ in)
Wheel track	3.32 m (10 ft 10¾ in)
Wheelbase	5.24 m (17 ft 2¼ in)

AREAS:

Wings, gross	49.1 m² (528.9 sq ft)
Flaperons (total)	3.81 m² (41.0 sq ft)
Trailing-edge flaps (total)	9.66 m² (104.0 sq ft)
Leading-edge slats (total)	4.63 m² (49.8 sq ft)
Fin	5.85 m² (62.93 sq ft)
Rudder	1.52 m² (16.32 sq ft)
Tailplane	10.87 m² (117.0 sq ft)

WEIGHTS AND LOADINGS:

Weight empty	12,525 kg (27,613 lb)
Fuel load: internal	7,230 kg (15,939 lb)
external (five tanks)	4,558 kg (10,050 lb)
Max external load	8,165 kg (18,000 lb)
Max T-O weight: catapult	26,580 kg (58,600 lb)
field	27,397 kg (60,400 lb)
Max zero-stores weight	20,521 kg (45,242 lb)
Max landing weight: carrier	16,329 kg (36,000 lb)
field	20,411 kg (45,000 lb)
Max wing loading	557. kg/m² (114.2 lb/sq ft)
Max power loading	330.88 kg/kN (3.25 lb/lb st)

PERFORMANCE (no stores, except were stated):

Never-exceed speed (V_{NE})	700 kts (1,297 km/h; 806 mph)
Max level speed at S/L	560 kts (1,037 km/h; 644 mph)
Cruising speed at optimum altitude	412 kts (763 km/h; 474 mph)
Approach speed	110 kts (204 km/h; 127 mph)
Stalling speed: flaps up	142 kts (264 km/h; 164 mph)
flaps down	98 kts (182 km/h; 113 mph)
Max rate of climb at S/L	2,323 m (7,620 ft)/min
Rate of climb at S/L, one engine out	646 m (2,120 ft)/min
Service ceiling	12,925 m (42,400 ft)
Service ceiling, one engine out	6,400 m (21,000 ft)
Min T-O run	1,185 m (3,890 ft)
T-O to 15 m (50 ft)	1,390 m (4,560 ft)
Landing from 15 m (50 ft)	774 m (2,540 ft)
Min landing run	521 m (1,710 ft)
Range with max military load	878 n miles (1,627 km; 1,011 miles)
Ferry range with max external fuel:	
tanks retained	2,380 n miles (4,410 km; 2,740 miles)
tanks jettisoned when empty	2,818 n miles (5,222 km; 3,245 miles)

UPDATED

GRUMMAN TOMCAT

US Navy designation: F-14

TYPE: Two-seat carrier-based interceptor with attack capability.

PROGRAMME: Won US Navy VFX fighter competition 15 January 1969; first flight of 12 development aircraft 21 December 1970; original programme was for 497 Tomcats including 12 development aircraft; programme since extended into 1990s.

Initial F-14A deployed with USN squadrons VF-1 and VF-2 October 1972; total 557 including 12 development aircraft, delivered to US Navy by April 1987, when production ended; final 102 aircraft (beginning 161597) delivered from FY83 powered by improved TF30-P-414A turbofans, having same rating as original 93 kN (20,900 lb st) TF30-P-412A.

VERSIONS: **F-14A:** Initial and main version, supplied to US Navy and Iran. Total of 557 built (last aircraft 162711), ending 13 March 1987; 32 reworked to F-14B and 18 to F-14D(R); 385 remained in 1993. Retrofit with Tape 115B (later Tape 116) started May 1991 (see under Avionics) permitting conventional bombing. First two F-15A **'Bombcat'** squadrons, VF-14 and VF-32, began cruise on USS *Kennedy*, 7 October 1992.

F-14MMCAP (Formerly F-14A++): MultiMission Capability Avionics Program; US Navy proposal, 1992, to upgrade 250 F-14As with improved avionics.

F-14B: Known as F-14A(Plus) until 1 May 1991; second use of F-14B designation; interim improved version re-engined with GE F110-GE-400 turbofans pending introduction of F-14D (see below). Development began July 1984; Grumman prime contractor with General Electric as subcontractor. F110 has 82 per cent parts commonality with F110-GE-100 in USAF F-15s and F-16s; 1.27 m (4 ft 2 in) plug inserted in afterburner section to match engine to F-14A inlet position and airframe contours; only secondary structure requires modification; new engine allows unrestricted throttle handling throughout flight envelope and fewer compressor stalls. NASA scoops for Vulcan cannon; glove vanes eliminated; cockpit modified. Full-scale development used two aircraft, including (first use) F-14B prototype (157986), which made first flight with definitive F110-GE-400 engines 29 September 1986; one prototype was to be upgraded to full F-14D.

F-14B followed F-14A into production; first flight production F-14B (162910) 14 November 1987; 18, 15 and five funded in FY86-88; first delivery, to VF-101, 11 April 1988; IOC with two US Navy squadrons achieved early 1989; F-14B production deliveries (last aircraft 163411) completed May 1990. Additionally, 32 F-14As upgraded to F-14B. Further five-seven conversions funded ($143 million) in FY92; Grumman to build conversion kits, then compete with Norfolk naval depot for installation contract; FY93 funds of $175 million for approximately 12 more conversions. B variant issued to VF-24, 74, 101, 103, 142, 143 and 211.

F-14A/B Upgrade: Total 196 aircraft (all extant F-14Bs plus balance in late production F-14As) retrofitted from FY94 (30 aircraft; then 25 per cent) with some F-14D features, including AN/ALR-67 RWR (to A models), BOL chaff dispenser, AN/AYK-14 mission computer (plus interface to analog 5400 computer), new tactical information display to permit addition of future weapon systems without re-wiring aircraft, MIL-STD-1553B digital databus, and airframe time compliance requirements (TCRs) for 7,500 hour life.

F-14D Super Tomcat: Improved version with AN/APG-71 radar, NACES seats in NVG-compatible cockpits, twin IRST/TV pod (ordered January 1993 for installation before August 1996), JTIDS link (from 1993), AN/ALR-67 RWR (from 1993), plus enhanced missile capability (AIM-120 AMRAAM, AIM-54C Phoenix). Total 37 out of 127 planned new F-14Ds funded (seven, 12 and 18 in FY88-90) before programme cancelled as economy measure in 1989; plans continued to 400 F-14A and F-14B to be remanufactured to F-14D; six **F-14D(R)** conversions funded FY90; Grumman working on four conversions, USN at Norfolk NADEP on two; FY91-95 funding plans for 12, 18, 20, 24 and 24 all cancelled February 1991, but FY91 dozen restored April 1991 (Grumman eight, NADEP four); F-14D programme thus 37 new, 18 rebuilt. Last Grumman redelivery April 1993; last from Norfolk in November 1993. First flight of first of three development aircraft 24 November 1987; 36 months flight testing included a TA-3B Skywarrior. Prototype 161867 delivered trials squadron VX-4 at Point Mugu, California, May 1990; first production F-14D (163412) rolled out 23 March 1990; last (16404) delivered 20 July 1992. First F-14D(R) (161659) delivered September 1991; two production aircraft to **NF-14D** for permanent test use by VX-4 at Point Mugu, California. Training squadron, VF-124 (first user) October 1990; official acceptance 16 November 1990; first embarked squadrons, VF-11 and VF-31; VF-1 and VF-2 to convert from F-14A in 1993; all F-14D based at Miramar, California. Last F-14D(R) was due for redelivery in Spring 1993.

Future upgrades: F-14Ds and F-14A/B Upgrades provide USN with core force of 251 modernised Tomcats for which future planning includes digital flight control system from FY96; GPS from FY95 (to F-14Ds first); AN/ARC-210 radios from FY98; and improved mission

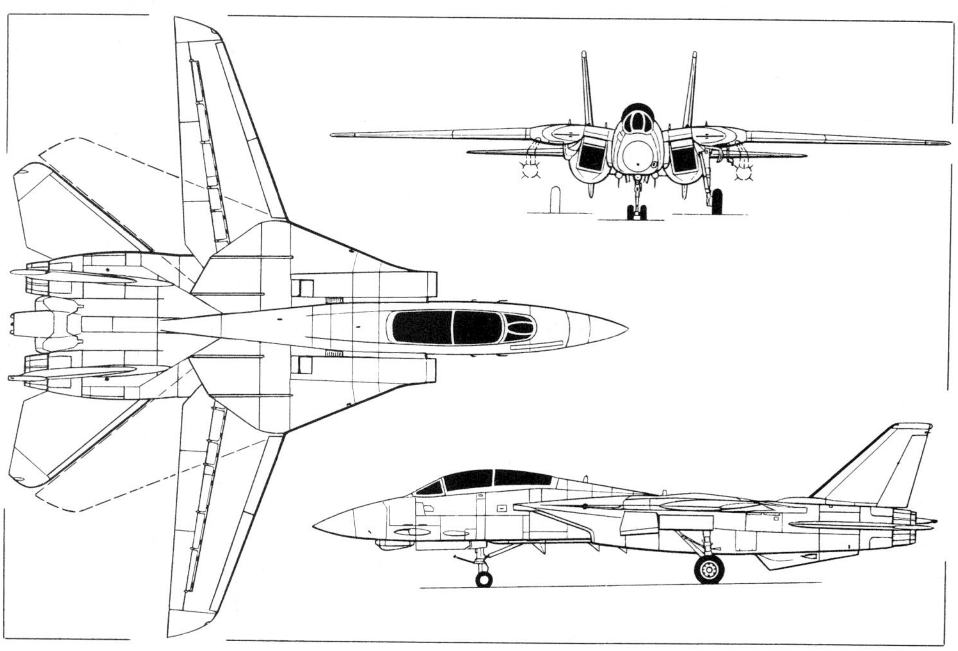

Grumman F-14B Tomcat, carrier-based multirole fighter aircraft *(Jane's/Dennis Punnett)*
1994

Grumman F-14D Tomcat of the multirole fighter aircraft of the US Navy

1994

recorders from FY99. Navy also considering further bombing upgrades (laser designator, FLIR and AN/ALE-50 towed decoy) to offset possible withdrawal of A-6 Intruder by 2000.

Other upgrades recently proposed are:

Quickstrike: Suggested first-stage development of F-14D for extended ground attack potential, providing F-15E Eagle-like capability to the fleet, including standoff weapons; potentially available (132-aircraft proposal) from 1994.

Super Tomcat-21: Company funded study for US Navy multirole fighter for next century as economic alternative to Navy version of USAF advanced tactical fighter (ATF). Quickstrike improvements plus new slotted flaps, extended-chord leading-edge slats, enlarged wing glove fairings containing extra fuel, new frameless windscreen. Grumman claims Tomcat-21 can have 90 per cent of ATF capability for 60 per cent of cost. Conversion possible from F-14A.

Attack Super Tomcat-21 (AST-21): Suggested interim replacement for A-12 Avenger II with low-level penetration capability and nuclear as well as conventional armament. Terrain-avoidance radar; two extra weapon stations.

ASF-14: Grumman proposal for alternative to Navalised ATF as F-14 evolutionary development including avionics, and possibly power plants, developed for ATF. Could be available by year 2000.

UPGRADES: **Grumman:** F-14A/B, F-14D, F-14MMCAP, F-14 Quickstrike, F-14 Super Tomcat-21, F-14 Attack Super Tomcat-21 (AST-21) and F-14 ASF-14. See Versions.

OPERATORS: Versions of the F-14 Tomcat are in service with the armed forces of the following countries: Iran (25) and USA (383).

CUSTOMERS: US Navy (see Versions). Present F-14 squadrons are (variants in parentheses) are VF-1 (A to D 1993), VF-2 (A to D 1993), VF-11 (D), VF-21 (A), VF-24 (B to A 1992), VF-31 (D), VF-51 (A), VF-111 (A), VF-124 (D, training), VF-154 (A), VF-211 (B to A 1992), VF-213 (A), 301 (Reserve) and 302 (Reserve) at Miramar, California; VF-14 (A), VF-32 (A), VF-33 (A), VF-41 (A), VF-74 (B), VF-84 (A), VF-101 (A/B training), VF-102 (A), VF-103 (B), VF-142 (B) and VF-143 (B) at Oceana, Virginia; VF-201 (A) and VF-202 (A) of the Reserve at Dallas, Texas. In August 1991, 'Top Gun' Fighter Weapons School at Miramar received 'aggressor' F-14As in Sukhoi Su-27 markings.

Iran also acquired 79 F-14As in 1976-78: retained Phoenix missile system, but had slightly different ECM equipment.

COSTS: $984 million fixed-price contract in July 1984 for development of E-14B and E-14D Super Tomcats.

DESIGN FEATURES: Wing sweepback variable from 20° leading-edge to 68°; oversweep of 75° used for carrier stowage without wing fold; wing pivot point 2.72 m (8 ft 11 in) from aircraft centreline; fixed glove has dihedral to minimise cross-sectional area and reduce wave drag; small canards on F-14A known as glove vanes extend forward progressively to 15° from inboard leading-edge to balance supersonic trim change and unload tail surfaces.

FLYING CONTROLS: Lateral control by long-span spoilers, ahead of flaps, and tailerons; automatic leading-edge slats assist manoeuvring; strakes emerge from wing glove leading-edge at high airspeeds; automatic wing sweep has manual override; automatic scheduling of control with airspeed; autostabilisation and angle of attack protection; autopilot and automatic carrier landing system (ALCS). Airbrake panel above and below tail, between fins. Twin fins and rudders.

STRUCTURE: Wing carry-through is one-piece electron beam-welded structure of Ti-6A1-4V titanium alloy with 6.71 m (22 ft) span. Fuselage has machined frames, titanium main longerons and light alloy stressed skin; centre-fuselage is fuel-carrying box; radome hinges upwards for access to radar; fuel dump pipe at extreme tail; fins and rudders of light alloy honeycomb sandwich; tailplanes have multiple spars, honeycomb trailing-edges and boron/epoxy composites skins.

LANDING GEAR: Retractable tricycle type. Twin-wheel nose unit and single-wheel main units retract forward, main units inward into bottom of engine air intake trunks. Original beryllium brakes were replaced with Goodyear lightweight carbon barkes from Spring 1981. Arrester hook under rear fuselage, housed in small ventral fairing. Nose-tow catapult attachment on nose unit.

POWER PLANT (F-14B/D): Two General Electric F110-GE-400 turbofans rated at 71.56 kN (16,088 lb st) dry and 120.1 kN (27,000 lb st) with afterburning. Garrett ATS200-50 air turbine starter. F110 engine has 43 per cent more reheated thrust and 37 per cent more military thrust (without afterburning) than TF30-P-414A in F-14A; results in 20 per cent more specific excess energy, 30 per cent lower specific fuel consumption in afterburner, 62 per cent greater deck launch intercept radius and 34 per cent more combat air patrol time; can be launched without afterburner; time to 10,670 m (35,000 ft) reduced by 61 per cent and acceleration time by 43 per cent. Integral fuel fanks in outer wings, each with capacity of 1,117 litres (295 US gallons; 246 Imp gallons); between engines in rear fuselage, with capacity of 2,453 litres (648 US gallons; 539 Imp gallons); and forward of wing carry-through structure, capacity 2,616 litres (691 US gallons; 575 Imp gallons); plus two feeder tanks with combined capacity of 1,726 litres (456 US gallons; 380 Imp gallons). Total internal fuel capacity 9,029 litres (2,385 US gallons; 1,986 Imp gallons). An external auxiliary fuel tank can be carried beneath each intake trunk, each containing 1,011 litres (267 US gallons; 222 Imp gallons). Retractable flight refuelling probe on starboard side of fuselage near front cockpit.

ACCOMMODATION: Pilot and naval flight officer seated in tandem on Martin-Baker NACES (or GRU7A in F-14A/B) rocket-assisted zero/zero ejection seats, under a one-piece bubble canopy, hinges at the rear and offering all-round view.

AVIONICS: In F-14A, Hughes AN/AWG-9 weapons control system, with ability to detect airborne targets at ranges of more than 65 to 170 n miles (120 to 315 km; 75 to 195 miles) according to their size, and ability to track 24 enemy targets and attack six of them simultaneously at varied altitudes and distances. Fairchild AN/AWG-15F fire control set; CP-1066/A central air data computer; CP-1050/A computer signal data converter; AN/ASW-27B digital datalink; AN/APX-76(V) IFF interrogator; AN/APX-72 IFF transponder; AN/ASA-79 multiple display indicator group; Kaiser Aerospace AN/AVG-12 vertical and head-up display system. AN/ARC-51 and AN/ARC-159 UHF com; AN/ARR-69 UHF auxiliary receiver; KY-28 cryptographic system; LS-460/B intercom; AN/ASN-92(V) INS; A/A24G39 AHRS; AN/APN-154 beacon augmentor; AN/APN-194(V) radar altimeter; AN/ARA-63A receiver-decoder; AN/ARN-84 micro Tacan; AN/ARA-50 UHF ADF; AN/APR-27/50 radar receiver; AN/APR-25/45 radar warning set, TV optical unit in undernose pod. Northrop Corporation television camera set (TCS) mounted beneath nose is closed-circuit TV system, offering both wide-angle (acquisition) and telescopic (identification) fields of view. TCS automatically searches for, acquires and locks on to distant targets, displaying them on monitors for the pilot and flight officer. Small undernose pod for Sanders AN/ALQ-100/126 deception jamming system, relocated under camera package of aircraft with Northrop TCS. During 1980-81, 49 F-14As were allocated to carry TARPS (tactical air reconnaissance pod system), containing a KS-87B frame camera, KA-99 low-altitude panoramic camera, and AN/AAD-5 infra-red reconnaissance equipment, on underbelly attachment; by 1993 TARPS capability in 23 F-14As, 11 F-14Bs and 55 F-14Ds.

In F-14D, some 60 per cent of analog avionics made digital, giving new weapons management, navigation, displays and control functions. MIL-STD-1553B digital bus interconnects Litton AN/ALR-67 threat warning and recognition system, joint tactical information distribution system (JTIDS), GE Aerospace Electronic Systems infra-red search and track sensor (IRST) and television camera set (TCS); emphasis on commonality with F/A-18 and latest A-6. (Westinghouse/ITT AN/ALQ-165 Airborne

Grumman F-14A (Plus) Tomcat with General Electric F110-GE-400 turbofan engines

1994

Self-Protection Jammer cancelled 1992.) New Hughes AN/APG-71 replaces AN/AWG-9 radar with improved ECM, monopulse angle tracking, digital scan control, target identification and raid assessment. AN/APG-71 features non-co-operative target identification, and ECCM using low-sidelobe antenna and sidelobe blanking guard channel, frequency agility, new high-speed digital signal processor based on AN/APG-70 radar in US Air Force multistaged improvement programme for F-15. Litton AN/ASN-139 INS; Smiths AN/AYQ-15 stores management system. ECM equipment includes Goodyear AN/ALE-29 and AN/ALE-39 chaff and flare dispensers, with integral jammers.

All in-service F-14A Tomcats given Tape 116 computer software addition in Summer 1992 to allow full ground attack with conventional bombs. F-14Ds of VF-11 and VF-31 have Tape G-6 software changes.

ARMAMENT: One General Electric M61A-1 Vulcan 20 mm gun mounted in the port side of forward fuselage, with 675 rounds of ammunition. Four AIM-7 Sparrow air-to-air missiles mounted partially submerged in the underfuselage, or four AIM-54 Phoenix missiles carried on special pallets which attach to the bottom of the fuselage. Two wing pylons, one under each fixed-wing section, can carry four AIM-9 Sidewinder missiles or two additional Sparrow or Phoenix missiles with two Sidewinders. F-14D has bombing capability; Rockeye and CBU-59 cluster bombs validated for F-14 December 1992; GBU-16 LGB and Gator mine to follow; AGM-88 HARM ARM and SLAM ASMs planned, but not yet funded.

DIMENSIONS, EXTERNAL:

Wing span: unswept	19.54 m (64 ft 1½ in)
swept	11.65 m (38 ft 2½ in)
overswept	10.15 m (33 ft 3½ in)
Wing aspect ratio	7.28
Length overall	19.10 m (62 ft 8 in)
Height overall	4.88 m (16 ft 0 in)
Tailplane span	9.97 m (32 ft 8½ in)
Distance between fin tips	3.25 m (10 ft 8 in)
Wheel track	5.00 m (16 ft 5 in)
Wheelbase	7.02 m (23 ft 0½ in)

AREAS:

Wings, gross	52.49 m² (565.0 sq ft)
Leading-edge slats (total)	4.29 m² (46.2 sq ft)
Trailing-edge flaps (total)	9.87 m² (106.3 sq ft)
Spoilers (total)	1.97 m² (21.2 sq ft)
Fins (total)	7.90 m² (85.0 sq ft)
Rudders (total)	3.06 m² (33.0 sq ft)
Horizontal tail surfaces (total)	13.01 m² (140.0 sq ft)

WEIGHTS AND LOADINGS (F-14D with F110-GE-400):

Weight empty	18,951 kg (41,780 lb)
Fuel (usable): internal	7,348 kg (16,200 lb)
external	1,724 kg (3,800 lb)
Max external weapon load	6,577 kg (14,500 lb)
T-O weight: fighter/escort mission	29,072 kg (64,093 lb)
fleet air defence mission	33,157 kg (73,098 lb)
max	33,724 kg (74,349 lb)
Max wing loading	642.5 kg/m² (131.59 lb/sq ft)
Max power loading	140.4 kg/kN (1.38 lb/lb st)

PERFORMANCE (F110 engines):

Max level speed	Mach 1.88
Max cruising speed	Mach 0.72
Carrier approach speed	125 kts (232 km/h; 144 mph)
Service ceiling	above 16,150 m (53,000 ft)
Field T-O distance	762 m (2,500 ft)
Field landing distance	732 m (2,400 ft)
Field range with external fuel	approx 1,600 n miles (2,965 km; 1,842 miles)

UPDATED

ELECTRONICS AND SYSTEMS INTEGRATION DIVISION

1111 Stewart Avenue, Bethpage, New York 11714
Telephone: 1 (516) 575 0574
Fax: 1 (516) 575 2164

CORPORATE VICE-PRESIDENT AND GENERAL MANAGER:
John E. Harrison

CORPORATE VICE-PRESIDENT AND DEPUTY GENERAL MANAGER:
Max T. Weiss

This Division combines the activities of the former Northrop Electronics Systems Division with the Electronics and Space divisional activities of Grumman. Aircraft programme responsibilities include E-2C Hawkeye, E-8C Joint-STARS, and support for the EA-6B Prowler and EF-111 Raven.

VERIFIED

Grumman EA-6B Prowler carrierborne ECM aircraft in service with the US Navy and Marine Corps

1994

GRUMMAN PROWLER

US Navy designation: EA-6B

TYPE: Four-seat carrierborne ECM aircraft.

PROGRAMME: Development contract issued Autumn 1966; externally similar to basic A-6 except longer nose enclosing four-seat cockpit and large pod on fin; first flight 25 May 1968; delivery of first 12 production aircraft started January 1971; last of 170 (164403) delivered 29 July 1991.

VERSIONS: **Prototypes:** Five aircraft.

Standard: Following 23 aircraft.

EXCAP: Expanded capability; 25 aircraft.

ICAP-1: Improved capability variant; increased jamming capacity; all aircraft updated to ICAP-2; 45 new-build; 21 production EA-6B modified to ICAP-1 and all production aircraft to 1983 built to this standard; modifications include seven-band onboard tactical jamming system, reduced response time and multiformat display, automatic all-weather carrier landing system (ACLS) new defensive ECM, new communications-navigation-identification (CNI) system.

ICAP-2: Introduced further improved jamming capacity; now standard; prototype first flight 24 June 1980; first of 72 production aircraft (161776) delivered 3 January 1984; earlier aircraft modified to same standard; exciter in each of five external jamming pods can generate signals in one of seven different frequency bands instead of one, and each pod can jam in two frequency bands simultaneously.

Subvariants of ICAP-2 are Block 82 and Block 86 (funded FY82 and FY86). Following 12 baseline aircraft, 162223 delivered 21 January 1986 as first of 23 Block 82s with HARM missile capability; Block 86 began 29 July 1988 at 163049, covering final 37 with expanded communications system and enhanced signal processing.

ADVCAP (advanced capability): Programme cancelled early 1994. However since the US Navy plans to operate the EA-6B until 2020, an upgrade package is still required. Initiated 1983; Litton Industries Amecon Division with Texas Instruments and ITT contracted to produce new receiver/processor group for tactical jamming system; ADVCAP EA-6B can carry HARM (flight tested during 1989) or tanks on four inboard underwing pylons; additional outboard station under each wing for ECM pods (total six underwing stations); centreline station for tank or ECM pod; first flight of prototype (156482) rebuilt to ADVCAP 29 October 1989; retrofit of ADVCAP to 102 ICAP-2 Prowlers began 1991 with contract for one; funding for three planned in FY93. Prime feature of ADVCAP is Lockheed-Sanders AN/ALQ-149 communications jammer with underfuselage antenna group; last of seven AN/ALQ-149 development models delivered for flight trials early 1989; AN/ALQ-149 contains eight assemblies in aircraft equipment bay including all antennae, receivers, signal recognisers, computers and controls needed to detect, identify, evaluate and jam hostile communications signals and long-range early warning radar systems; Sanders has options to supply 95 production AN/ALQ-149s. Other ADVCAP changes include MFD/HUD display system, updated AN/ARN-118 Tacan, increased emergency RAT capacity, third AN/ARC-182 radio, Band 2/3 transmitter, improved stall/manoeuvring capability, two additional (making four) AN/ALE-39 chaff dispensers, Navstar GPS, disc-based recorder/onboard programme loader, fitting of 53.38 kN (12,000 lb st) J52-P-409 turbojets to boost maximum landing weight to 21,546 kg (47,500 lb), fin height increased by 0.5 m (1 ft 7¾ in), recontoured slats and flaps, drooped wing leading-edges, new forward fuselage strakes and wingtip speed-brake modification to operate as ailerons in conjunction with digital flight control system. Aerodynamic changes developed as Vehicle Enhancement Program (VEP, formerly VIP); first flight of VEP June 1992. Full ADVCAP changes constitute Block 91; 'export standard' Block 90 identical apart from lower (Block 86) level of jamming suite.

Block 2000: Proposed enhancement of ADVCAP with JTIDS, satellite communications and survivability measures.

UPGRADES: **Grumman:** ICAP-1, ICAP-2. See Versions.

Litton/Texas Instruments: ADVCAP. See Versions.

OPERATORS: Versions of the EA-6B Prowler are in service with the US armed forces (128).

CUSTOMERS: US Navy and Marine Corps, total 170; 12 USN squadrons (VAQ-129 for training, 130, 131, 132, 134, 135, 136, 137, 138, 139, 140 and 141, all at Whidbey Island, Washington, including one rotated to Atsugi, Japan) equipped with Prowler by 1989. First detachment of US Marine Corps Prowler squadron VMAQ-2 began training on EA-6B at Cherry Point, North Carolina, in September 1977 and deployed in late 1978. Two additional detachments became full squadrons on 1 July 1992—VMAQ-1 at Iwakuni, Japan and VMAQ-3 at Cherry Point—while on 1 October 1992, VMAQ-4 elevated from Reserve to full-time squadron at Whidbey Island, Washington; at least one is deployed at all times. Deployment with USN reserve units began in June 1989 with conversion of VAQ-309 at Whidbey, from EA-6As, VAQ-209 followed at Andrews AFB, 1991.

COSTS: $55.7 system unit cost, 1987.

DESIGN FEATURES: Generally as for A-6E; four crew and fintip antenna.

FLYING CONTROLS: As A-6E, but modified wingtip speed brakes in ADVCAP (see above).

STRUCTURE: Wings as for A-6E, but reinforced to allow for greater gross weight, fatigue life and 5.5 *g* load factor. Fuselage as for A-6E, but lengthened by 1.37 m (4 ft 6 in).

LANDING GEAR: As for A-6E, except for reinforcement of attachments. A-frame arrester hook, and upgrading of structure to cater for increased gross weight.

POWER PLANT: Two Pratt & Whitney J52-P-408 turbojets, each rated at 49.8 kN (1,200 lb st). ADVCAP retrofit includes 53.38 kN (12,000 lb st) J52-P-409s.

ACCOMMODATION: Crew of four under two separate upward-opening canopies. Martin-Baker GRUEA 7 ejection seats for crew. The two additional crewmen are ECM Officers to operate the ALQ-99 equipment from the rear cockpit. Either ECMO can independently detect, assign, adjust and monitor the jammers. The ECMO in the starboard front seat is responsible for communications, navigation, defensive ECM and chaff dispensing.

SYSTEMS: Generally as for A-6E.

AVIONICS: AN/ALQ-99F tactical jamming system, in five integrally powered pods, with a total of 10 jamming transmitters. Each pod covers one of seven frequency bands. Sensitive surveillance receivers in the fintip pod for long-range detection of radars; emitter information is fed to a central digital computer (AN/AYK-14 in ICAP-2 aircraft) that processes the signals for display and recording. Detection, identification, direction-finding and jammer set-on sequence can be performed automatically or with manual assistance from crew. PRB Associates AN/TSQ-142 tactical mission support system. Teledyne Systems AN/ASN-123 navigation system with digital dislay group.

ARMAMENT: Originally unarmed, but currently capable of carrying Texas Instruments AGM-88A HARM anti-radar missiles underwing. Four underwing hardpoints on ICAP-2 aircraft, six on ADVCAP EA-6B.

DIMENSIONS, EXTERNAL: As for A-6E, except:

Width, wings folded	7.87 m (25 ft 10 in)
Length overall	18.24 m (59 ft 10 in)
Height overall	4.95 m (16 ft 3 in)
Wheelbase	5.23 m (17 ft 2 in)

WEIGHTS AND LOADINGS:

Weight empty	14,321 kg (31,572 lb)
Internal fuel load	6,995 kg (15,422 lb)
Max external fuel load	4,547 kg (10,025 lb)
T-O weight: from carrier in standoff jamming configuration (5 ECM pods)	24,668 kg (54,383 lb)
from field in ferry range configuration (max internal and external fuel)	27,236 kg (60,045 lb)
Max T-O weight, catapult or field	29,483 kg (65,000 lb)
Max zero-fuel weight	17,672 kg (38,961 lb)
Max landing weight, carrier or field	20,638 kg (45,500 lb)
Max wing loading	600.5 kg/m² (123 lb/sq ft)
Max power loading	296.0 kg/kN (2.90 lb/lb st)

PERFORMANCE (A: clean, B: 5 ECM pods):

Never-exceed speed (VNE)	710 kts (1,315 km/h; 817 mph)
Max level speed at S/L:	
A	566 kts (1,048 km/h; 651 mph)
B	530 kts (982 km/h; 610 mph)

First production Northrop Grumman E-8C Joint-STARS aircraft, to be used as a permanent testbed

1995

Cruising speed at optimum altitude:
A, B — 418 kts (774 km/h; 481 mph)
Stalling speed:
flaps up, max power:
A — 124 kts (230 km/h; 143 mph)
flaps down, max power:
A — 84 kts (156 km/h; 97 mph)
Max rate of climb at S/L: A — 3,932 m (12,900 ft)/min
B — 3,057 m (10,030 ft)/min
Rate of climb at S/L, OEI: A — 1,189 m (3,900 ft)/min
Service ceiling: A — 12,550 m (41,200 ft)
B — 11,580 m (38,000 ft)
Service ceiling, OEI: A — 8,930 m (29,300 ft)
T-O run: B — 814 m (2,670 ft)
T-O to 15 m (50 ft): A — 823 m (2,700 ft)
Landing run: A — 579 m (1,900 ft)
B — 655 m (2,150 ft)
Range with max external load, 5% reserves plus 20 min at S/L: B — 955 n miles (1,769 km; 1,099 miles)
Ferry range with max external fuel:
tanks retained — 1,756 n miles (3,254 km; 2,022 miles)
tanks jettisoned when empty — 2,085 n miles (3,861 km; 2,399 miles)

UPDATED

NORTHROP GRUMMAN (BOEING) E-8 JOINT-STARS

TYPE: Long-range radar reconnaissance aircraft; programme name, Joint Surveillance Target Attack Radar System (Joint-STARS).

PROGRAMME: Full-scale development contract to Grumman, 27 September 1985; two Boeing 707-328Cs purchased from American Airlines and Qantas as EC-18C testbeds, later redesignated E-8A; first aircraft for modification reached Boeing Military Airplanes, Wichita, January 1986; delivered to Grumman Melbourne Systems Division, Florida 31 July 1987; second aircraft to Wichita June 1986, for delivery to Grumman Autumn 1988; first flight in full Joint-STARS configuration (86-0416/N770JS) 22 December 1988; first flight of second aircraft (86-0417/N8411) 31 August 1989; first instantaneous transmission to ground station August 1989; European trials February-March 1990 (N770JS) and September 1990 (86-0417); original requirement for 22 new-build aircraft and 100 ground stations; amended to 20 second-hand Boeing 707s; IOC planned in 1997 with five aircraft, but emergency interim capability achieved for Gulf War against Iraq 1991. Low-rate initial production advanced procurement contract signed by USAF 24 April 1992; progress review, May 1993, authorised first six production aircraft at two per year; further review then to consider outstanding 12.

Airframe modifications now at Northrop Grumman Lake Charles following Boeing's withdrawal from programme; avionics installation at Melbourne; first E-8C (90-0175/N526SJ) completed December 1993 and made first flight 25 March 1994, at which time four more with Grumman and three unmodified airframes in storage with USAF; further 10 conversion subjects to be sought for modification to begin in 1996.

NATO considering integrated Joint-STARS force; planning group established December 1992; programme office established November 1994; short European tour by second E-8A in October 1994 for demonstration to Belgium, France, Germany and UK; requirement for up to 12 Joint-STARS to be based at Geilenkirchen, alongside E-3A Sentry AWACS unit; may use Airbus or similar aircraft as avionics platform. Further interest expressed by Japan (possibly four aircraft), South Korea (four) and Saudi Arabia (up to six). Grumman teamed with BAe in 1993 to offer Joint-STARS to UK in satisfaction of SR(L/A) 925 (ASTOR) requirement.

MultiStage Improvement Program (MSIP) formulated by 1994 and first phase funded; involves TADIL-J datalink with dedicated operator's position for monitoring of air threats, followed by 'End Game' radar- and IR-countermeasures; Phase 2 options include satellite communications, improved data modem and automatic target recognition, to be launched in 1996.

CURRENT VERSIONS: **E-8A**: Two development aircraft, as above. Fitted with consoles for 10 operators; Pratt & Whitney JT3D-3B engines. First operational mission, 14 January 1991 with 4411th Joint-STARS Squadron (Provisional), USAF, from Saudi Arabia; six ground station modules also deployed; flew 49 missions/534.6 hours up to end of conflict. Returned to Grumman for completion of performance testing in 100 sorties and several thousand hours of ground testing. Both accepted by USAF, December 1993, for trials at Edwards AFB to be eventually upgraded as E-8Cs.

E-8B: Originally proposed production version, based on new-built airframe; F108 turbofan engines; 15 operator consoles. One prototype, 88-0322, flown 12 June 1990 in 'green' state; avionics not installed; overtaken by decision to use secondhand Boeing 707 airframes; delivered to USAF 3 October 1991 for storage at Davis Monthan AFB, Arizona; bartered with Omega Air, 1993, for five used 707s and currently in civilian use as N707UM.

E-8C: Production version; 18 operator consoles. Initial two aircraft to Lake Charles for conversion, May-June 1992. First E-8C, completed in March 1994, is permanent testbed, not for delivery to USAF. *Data apply to this version.* Further details in *Jane's Avionics 1995-96.*

CUSTOMERS: US Air Force and US Army (aircraft operated by USAF). First E-8C funded in FY90; second in FY94; two in FY95 and request for two in FY96; total 20 expected, including two upgraded E-8As and one E-8C permanent testbed. To be based at Robins AFB, Georgia.

JOINT-STARS FUNDING

FY	Type	Qty	First aircraft
86	E-8A	2	86-0416
88	E-8B*	(1)	88-0322
90	E-8C	1	90-0175
94	E-8C	1	
95	E-8C	2	
Total		**6 (+1*)**	

*New-build; not delivered

COSTS: $657 million (1985) full-scale development; $523 million award to Grumman November 1990 for conversion of first E-8C. Estimated programme cost (1994) $8,560 million, including ground stations. First (two aircraft) advance procurement contract $125 million (April 1992). Northrop Grumman secured $254 million contract in July 1994 for completion of second pair, with $440 million appropriation for two more in FY95 budget.

DESIGN FEATURES: Boeing 707-320C airliner converted with 7.32 m (24 ft) antenna covered by a 'canoe' fairing under forward fuselage, offset to starboard, for phased-array SLAR.

STRUCTURE, FLYING CONTROLS, LANDING GEAR: As, or similar to, commercial Boeing 707.

POWER PLANT: Four 80.1 kN (18,000 lb st) Pratt & Whitney JT3D-3B turbojets. See Weights and Loadings for fuel.

SYSTEMS: As Boeing 707; additional electronic generating capacity. World Auxiliary Power Company APU.

AVIONICS: *Comms:* Telephonics multiple intercom net control system.

Radar: Norden AN/APY-3 multimode side-looking I-band radar. Synthetic aperture radar (SAR) mode used to detect stationary objects, such as parked tanks, up to 81 to 95 n miles (150 to 175 km; 93 to 109 miles) behind front line. Can interleaf Doppler mode to detect moving targets. Coverage, 1 million km² (386,100 sq miles) in 8 hour sortie, cruising at 9,145 to 12,200 m (30,000 to 40,000 ft).

Flight: Litton INS, Collins flight management system.

Mission: Five Raytheon Model 920/866 supermini computers per aircraft, Ceridian Data Inc programmable signal processors (three), Interstate Electronics graphic displays, Raytheon Model AXP-3000/500 workstations, Orbit International workstation keyboards, Miltope message printers, Cubic Defense Systems surveillance and control datalink, JTIDS for Tacan operation and TADIL-J generation and processing. Satellite communications link, Magnavox encrypted UHF radios (12), two encrypted HF radios, three encrypted VHF radios with SINCGARS (Single Channel Ground and Airborne Radio System) provision. Radar data can be transmitted instantaneously to ground stations; or attacks by aircraft and ground forces directed via JTIDS datalink.

ARMAMENT: Nil.

DIMENSIONS, EXTERNAL (abbreviated):
Wing span — 44.42 m (145 ft 9 in)
Length overall — 46.61 m (152 ft 11 in)
Height overall — 12.95 m (42 ft 6 in)

WEIGHTS AND LOADINGS:
Weight empty — 77,564 kg (171,000 lb)
Max fuel load — 70,307 kg (155,000 lb)
Max T-O weight — 152,407 kg (336,000 lb)

Grumman/Boeing E-8 (Joint-STARS) Joint Surveillance and Target Attack Radar System aircraft

1994

PERFORMANCE:

Max operating Mach number (MMO)	0.84
Service ceiling	12,800 m (42,000 ft)
Endurance: internal fuel	11 h
with one in-flight refuelling	20 h

UPDATED

NORTHROP GRUMMAN (LOCKHEED FORT WORTH) RAVEN UPGRADE

US Air Force designation: EF-111A

TYPE: Electronic warfare aircraft upgrade.

PROGRAMME: Conversion of 42 F-111As. Avionics modernisation programme initiated with January 1987 contract, new $155.8 million contract March 1991 for FSD system improvement programme (SIP) to upgrade hardware and software to improve radar jamming capabilities; Northrop Grumman Electronics & Systems Integration Division (with team members Eaton/AIL, Astronautics Corporation, IBM, Comptech and Smith Industries) will design and install prototype upgrade kit at its Calverton facility; contract to be completed by January 1996. First flight of the SIP test aircraft occured on 22 March 1995 at Eglin AFB Florida.

UPDATED

Northrop Grumman (Lockheed Fort Worth) EF-111A Raven electronic warfare aircraft *1992*

PEMCO

PEMCO AEROPLEX INC Subsidiary of Precision Standard Inc (PCSN)

PO Box 2287, Birmingham, Alabama 35201-2287
Telephone: 1 (205) 591 7870
Fax: 1 (205) 592 6306
TWX: (810) 733 3687
PRESIDENT: Charles Perry
CEO: Matthew L. Gold
VICE-PRESIDENT, CORPORATE DEVELOPMENT: Fredrick Groth
VICE-PRESIDENT, BUSINESS DEVELOPMENT: David Mitchell

Former Hayes International Corporation acquired and renamed by Precision Standard Inc of Birmingham, Alabama, September 1988; aircraft maintenance, modification, repair and cargo conversions; besides those below, company has converted Convair 240, 340, 580, 640, Douglas DC-6, DC-8, Gulfstream I, Boeing 727-100/200. FAA, CAA-JAA approved.

In 1994 Precision Standard (Pemco's parent company) agreed a new maintenance and modification facility in Copenhagen, Denmark since renamed, Pemco World Air Service.

In February 1995 Pemco announced that it had been awarded a contract from First Air of Ottowa, Ontario for a combi conversion of a Boeing 727-200.

UPDATED

Pemco Aeroplex BAe 146-200 undergoing cargo conversion *1994*

PEMCO AEROPLEX BAe 146-200 CARGO CONVERSION

TYPE: Transport aircraft cargo conversion.

PROGRAMME: Appointed by BAe to make first conversion of 146-200 (see UK section) to all-cargo configuration; installed port side upward-opening 3.30 × 1.98 m (10 ft 10 in × 6 ft 6 in) cargo door aft of wing; applied strengthening and fitted out cabin. Converted aircraft can operate as all-cargo or passenger carrier. Pemco Aeroplex obtained FAA supplemental type certificate and makes all conversions, but BAe markets; company has installed Mode-S, TCAS and windshear customer-ordered modifications to BAe 146; 146-200 and 146-300 also converted to QC (quick-change).

VERIFIED

PEMCO AEROPLEX BOEING 737-200/-300 FREIGHTER/QC

TYPE: Transport aircraft cargo conversion.

PROGRAMME: Only post-delivery quick-change and freighter conversions of Boeing 737-300 offered worldwide. Programme certificated in France, Germany, Singapore, Sweden and USA. Stronger floor-beam anchorages; additional fuselage doubles to accept new 3.55 × 2.28 m (11 ft 8 in × 7 ft 6 in) cargo door. Since the launch of the 737 conversion programme, Pemco has contracted for 50 passenger to freighter/QC conversions of which 26 have been redelivered. In April 1994 Pemco completed its first conversion for Polaris Aircraft Leasing (operator: Can Air).

CUSTOMERS: Customers include L'Aeropostale, Falcon Aviation, Malaysia Airlines, Singapore Airlines, Lufthansa German Airlines, Aloha Airlines, Polynesia Airlines and TACA International Airlines.

VERIFIED

PEMCO AEROPLEX BOEING 747-100 CARGO CONVERSION

TYPE: Transport aircraft cargo conversion.

PROGRAMME: Ex-TWA Boeing 747-100 converted to all-cargo configuration; 3.12 × 3.48 m (10 ft 3 in × 11 ft 5 in) cargo door port side aft of wing; 9 *g* bulkhead aft of flight deck; cargo compartment floor beams replaced; upper deck modified; windows blanked off; new interior cabin liner; provision for electric cargo handling system. Main deck can carry 29 cargo containers, with another 13 on lower deck; total payload 91,625 kg (202,000 lb). FAA supplemental type certificate gained April 1988; two modified in 1989; three more modified in 1990; sixth completed in April 1991.

VERIFIED

PEMCO AEROPLEX BOEING 757QC

TYPE: Transport aircraft cargo conversion.

PROGRAMME: Pemco Aeroplex presently developing Boeing 757 passenger/freighter/QC conversion programme through Data Licensing Agreement with Boeing Company. Conversion available from Spring 1993, and Pemco Aeroplex expected to initially certificate modification in Europe and USA.

VERIFIED

Pemco Aeroplex (Boeing) 737-300 converted freighter for L'Aeropostale *(Pemco Aeroplex)* *1995*

PEMCO AEROPLEX LOCKHEED L-1011F TRISTAR

TYPE: Transport aircraft cargo conversion.

PROGRAMME: Ex-Eastern Air Line L-1011-1 converted to all-cargo configuration for Aircraft Sales Company; 2.84 × 4.32 m (9 ft 4 in × 14 ft 2 in) cargo door fitted port side forward of wing; 9 *g* bulkhead aft of flight deck; windows blanked off; tracks and rollers on upper and lower decks; maximum payload 54,430 kg (120,000 lb); first flight 7 May 1987; FAA Type Approved; further conversion to order.

VERIFIED

PEMCO AEROPLEX MCDONNELL DOUGLAS MD-80/DC-9 CARGO CONVERSION

TYPE: Transport aircraft cargo conversion.

PROGRAMME: All-cargo conversion of DC-9; 2.06 × 3.45 m (6 ft 9 in × 11 ft 4 in) cargo door port side aft of wing; 9 *g* net barrier; cockpit emergency exit; smoke detection system; overhead storage rack/bin modification; altered interior panels and modifications to air conditioning and oxygen systems; 15,876 kg (35,000 lb) cargo capacity in main cabin without operating restrictions. Signed a licensing agreement with McDonnell Douglas for passenger/freighter/QC conversion of MD-80/DC-9 aircraft in 1993.

VERIFIED

PEMCO AEROPLEX TUPOLEV Tu-204 CARGO CONVERSION

TYPE: Passenger aircraft cargo conversion.

PROGRAMME: In June 1993, Pemco Aeroplex announced its intention to form a joint stock company with British Russian Aviation Corporation (Bravia) for the conversion of Tu-204 passenger aircraft to freighter, quick-change and combi configurations.

VERIFIED

PEN TURBO

PEN TURBO AVIATION INC

1340 East Road, Cape May County Airport, Rio Grande, New Jersey 08242
Telephone: 1 (609) 886 7631
Fax: 1 (609) 886 7411
PRESIDENT: R. Gobalian
VICE-PRESIDENT: Joseph Wolf

NEW ENTRY

PEN TURBO (DE HAVILLAND CANADA) DHC-4T CARIBOU

TYPE: Turboprop conversion of DHC-4 STOL utility transport.

PROGRAMME: Prototype conversion made first flight 16 November 1991; crashed 27 August 1992 on take-off from NewCal's Gimli, Manitoba, Canada base; programme continuing. Test flight of second aircraft planned Summer 1995.

COSTS: Approximately $3.7 million conversion cost (excluding cost of airframe) in 1995.

DESIGN FEATURES: Replaces 1,081 kW (1,450 hp) P&W R-2000 piston radials with two flat rated 1,062 kW (1,424 shp) P&WC PT6A-67R turboprops and new five-blade propellers; MTOGW unchanged at 12,927 kg (28,500 lb), but basic empty weight reduced by 794 kg (1,750 lb) to 7,484 kg (16,500 lb), resulting in 1,474 kg (3,250 lb) increase in maximum payload to 5,443 kg (12,000 lb); service ceiling raised to more than 8,840 m (29,000 ft), and speeds improved by 10 per cent. A more powerful PT6 turboprop engine is in development and will be available for this conversion.

SYSTEMS: Conversion to include optional installation of a removable aerial dispersant system for combating oil slicks and removable tank for waterbombing fires. The company is also working on a long-range coastal patrol version.

UPDATED

PIPER

PIPER AIRCRAFT CORPORATION

2926 Piper Drive, Vero Beach, Florida 32960
Telephone: 1 (407) 567 4361
Fax: 1 (407) 562 0299
CHAIRMAN: A. Stone Douglass
PRESIDENT AND CEO: Chuck Suma
DIRECTOR, MARKETING: Werner Hartlieb
DIRECTOR, MANUFACTURING: Buddy Watson
DIRECTOR OF ENGINEERING: Elliot Nichols
MARKETING MANAGER: Kimberley Wheeler

Piper purchased by Lear Siegler Inc effective 1 March 1984; Lock Haven and Piper facilities in Pennsylvania closed second half 1984; Lakeland, Florida, plant phased out October 1985. Activities concentrated at Vero Beach and new 12,077 m² (130,000 sq ft) plant completed October 1986 for Cheyenne IIIA and 400 production (transferred from Lakeland).

Company acquired by M. Stuart Millar 12 May 1987 and became subsidiary of Romeo Charlie Inc; LoPresti Piper Advanced Engineering Group set up as subsidiary December 1987; Piper North Corporation subsidiary formed November 1989 at Lock Haven to re-establish aircraft manufacturing at former headquarters under loan guarantee agreement with Commonwealth of Pennsylvania. Increasing cash flow problems in second half of 1990 resulted in closure of both subsidiaries and reduced production. Sale of type certificates and manufacturing rights for number of out of production models were negotiated August 1990 and March 1991. Aerospatiale (Socata) negotiations to buy Piper finally broken off 22 March 1991, reportedly because of product liability uncertainties. Voluntarily filed for protection under Chapter 11 of US Bankruptcy Code 1991. Principal creditor, Congress Financial Corporation, made further loan early 1992 to complete approximately 60 aircraft from outstanding order book. Loan repaid by October 1992.

Piper designs are manufactured in Argentina by Chincul SA (which see) and in Brazil by Embraer subsidiary Neiva; PZL Mielec in Poland produces Seneca II as M-20 Mewa.

According to figures released 10 February 1995 by the General Aviation Manufacturers Association (GAMA), Piper Aircraft Corporation sold the most single-engined aircraft (108) in the USA in 1994 as well as ranking second for the number of twin-engined aircraft sold (24). In total Piper sold 132 aircraft of all types in 1994 and plans to sell 177 in 1995. Piper currently manufactures the Warrior III, Arrow, Seminole and Seneca IV trainer aircraft, the Archer III and Saratoga II HP personal aircraft and Malibu Mirage business aircraft.

In March 1995, Piper announced that together with Dimeling, Schreiber & Park and Teledyne Industries, it had filed a joint plan of re-organisation with the US Bankruptcy Court. The joint plan provides for the sale of substantially all of Piper's assets to Newco Pac, Inc, a new entity owned by Dimeling, Schreiber & Park and Teledyne Industries. The value of the transaction is approximately $95 million. A trust formed for the benefit of Piper's creditors and shareholders will own 25 per cent of the new firm.

UPDATED

PIPER PA-18-150 SUPER CUB

TYPE: Two-seat light cabin monoplane.

PROGRAMME: The original Piper PA-18 with a 67 kW (90 hp) Continental C90-12F engine received FAA Type Approval 18 November 1949. The PA-18-150, PA-18A-150 agricultural aircraft, PA-18S and PA-18AS seaplanes were all approved 1 October 1954.

Piper restored the PA-18-150 Super Cub to production during 1988. Later versions differ by having stamped aluminium wing ribs, Ceconite 7600 Flexi-gloss paint, electric fuel gauges, stainless steel control cables, and heavy-duty wing strut fork ends. Piper intends to build 24 Super Cubs in 1994. All have been sold to Muncie Aviation, the master distributor of the aircraft.

DESIGN FEATURES: Braced high-wing monoplane, with steel tube V bracking struts each side. Wing section USA 35B. Thickness/chord ratio 12 per cent. Dihedral 1°. No incidence at mean aerodynamic chord. Total washout of 3° 18′. Variable incidence tailplane.

FLYING CONTROLS: Plain ailerons and trailing-edge flaps of light alloy construction and skinning. Tailplane incidence variable for trimming. Balanced rudder and elevators.

STRUCTURE: Aluminium spars and stamped aluminium ribs, aluminium sheet leading-edge and aileron false spar, wingtip bow of ash, with Ceconite 7600 covering, pop-riveted to structure. The fuselage is a rectangular welded steel tube structure covered with Ceconite 7600. The tail unit is a wire-braced structure of welded steel tubes and channels, covered with Ceconite 7600.

LANDING GEAR: Non-retractable tailwheel type. Two side Vs and half axles hinged to bottom of fuselage. Rubber chord shock-absorption. Mainwheel tyres size 6.00-6 (4 ply). Scott 20 cm (8 in) steerable leaf spring tailwheel. Dual brakes and parking brake standard. Floats optional.

POWER PLANT: One 112 kW (150 hp) Textron Lycoming O-320 flat-four engine, driving a Sensenich two-blade fixed-pitch metal propeller with spinner. Steel tube engine

Piper Super Cub two-seat light cabin monoplane

1992

Piper Warrior II four-seat light aircraft

1992

mount is hinged at firewall, allowing it to be swung to port for access to rear of engine. One 68 litre (18 US gallon; 15 Imp gallon) metal fuel tank in each wing. Total fuel capacity 136 litres (36 US gallons; 30 Imp gallons), of which 135.5 litres (35.8 US gallons; 29.8 Imp gallons) are usable. Refuelling points on top of wing.

ACCOMMODATION: Enclosed cabin, seating two in tandem with dual controls. Adjustable front seat. Shoulder harness and seat belts standard for front and rear seats. Heater and adjustable fresh air control. Downward hinged door on starboard side, and upward hinged window above, can be opened in flight. Sliding window on port side. Baggage compartment aft of rear seat, capacity 22 kg (50 lb).

SYSTEMS: Electrical system comprising 14 V 55 A alternator, 12 V engine starter and 35 Ah battery, standard.

AVIONICS AND EQUIPMENT: Standard package includes an altimeter, airspeed indicator, magnetic compass, tachometer, oil pressure gauge, ELT 910, oversize tyre (8.00 × 6.00 4 ply), night lighting (three navigation lights, red tail strobe light, instrument panel lights, dome light auxiliary cockpit light and landing/taxi light as well as provisions for one nav/com. No options available.

DIMENSIONS, EXTERNAL:

Wing span	10.76 m (35 ft 3½ in)
Length overall	6.86 m (22 ft 6 in)
Height overall	2.05 m (6 ft 8½ in)
Propeller diameter	1.88 m (6 ft 2 in)

AREAS:

Wings, gross	16.58 m² (178.5 sq ft)

WEIGHTS AND LOADINGS:

Weight empty	482 kg (1,062 lb)
Max T-O weight	794 kg (1,750 lb)
Max wing loading	47.9 kg/m² (9.8 lb/sq ft)
Max power loading	7.09 kg/kW (11.7 lb/hp)

PERFORMANCE (at max T-O weight, ISA except where indicated):

Max level speed at S/L	113 kts (210 km/h; 130 mph)
Cruising speed, 75% power at 1,525 m (5,000 ft)	100 kts (185 km/h; 115 mph)
Stalling speed, flaps down	37 kts (69 km/h; 43 mph)
Max rate of climb at S/L	292 m (960 ft)/min
Service ceiling	5,790 m (19,000 ft)
Absolute ceiling	6,490 m (21,300 ft)
T-O run	61 m (200 ft)
T-O run to 15 m (50 ft)	153 m (500 ft)
Landing from 15 m (50 ft)	270 m (885 ft)
Landing run	107 m (350 ft)
Range with max fuel, 75% power	400 n miles (741 km; 460 miles)

VERIFIED

PIPER PA-28-161 WARRIOR II

TYPE: Four-seat cabin monoplane.

PROGRAMME: Design of the Warrior II began in June 1972, an important feature being replacement of the earlier constant-chord wings of the Cherokee series by a longer span wing with tapered outer panels. As a result of its introduction, the Warrior, which at the time had essentially the same 112 kW (150 hp) engine as the discontinued Cherokee Cruiser, was certificated at a maximum T-O weight of 79 kg (157 lb) greater. First flight of a prototype was made on 17 October 1972, and FAA certification of the original Model PA-28-151 was granted 9 August 1973. Later versions of the Warrior II have 119 kW (160 hp) engines operating at 100 octane low-lead fuel. Over 3,000 Warriors have been sold.

OPERATORS: Versions of the PA-28 (Archer, Arrow, Dakota and Warrior II) are in service with the armed forces of the following countries: Argentina (4), Chile (14), Colombia (4) and Finland (9).

DESIGN FEATURES: Cantilever low-wing monoplane. Wing section NACA 65_2-415 on inboard panels; outboard leading-edges incorporate modification No. 5 of NACA TN 228. Dihedral 7°. Incidence 2° at root, −1° at tip. Sweepback at quarter-chord 5°.

FLYING CONTROLS: Four-position manually actuated trailing-edge flaps of light alloy with ribbed skins. Fin and rudder have ribbed light alloy skins. One-piece all-moving tailplane with combined anti-servo and trim tab. Rudder trimmable, but no trim tab in rudder.

STRUCTURE: Light alloy single-spar structure with glassfibre wingtips. Plain ailerons of light alloy construction. The fuselage is a light alloy semi-monocoque structure with glassfibre nose cowl and tailcone. The tail unit is a cantilever structure of light alloy, except for glassfibre tips on fin and tailplane.

LANDING GEAR: Non-retractable tricycle type. Steerable nosewheel. Piper oleo-pneumatic shock-absorbers; single wheel on each unit. Cleveland wheels with 4 ply tyres size 6.00-6 on main units, pressure 1.65 bars (24 lb/sq in). Cleveland nosewheel and 4 ply tyre size 5.00-5, pressure 1.65 bars (24 lb/sq in). Cleveland disc brakes. Parking brake. Glassfibre wheel fairings standard.

POWER PLANT: One 119 kW (160 hp) Textron Lycoming O-320-D3G flat-four engine driving a Sensenich two-blade fixed-pitch metal propeller type 7DM6-0-60 with spinner. Fuel in two wing tanks, with total capacity of 189 litres (50 US gallons; 41.6 Imp gallons), of which 181.5 litres (48 US gallons; 40 Imp gallons) are usable. Refuelling point on upper surface of each wing. Oil capacity 7.5 litres (2 US gallons; 1.7 Imp gallons).

ACCOMMODATION: Four persons in pairs in enclosed cabin. Individual adjustable front seats with seat belts and shoulder harnesses. Dual controls and brakes standard. Large door on starboard side. Baggage compartment at rear of cabin, with volume of 0.68 m³ (24 cu ft) and capacity of 91 kg (200 lb). External baggage door on starboard side. Heating, ventilation and windscreen defrosting standard.

SYSTEMS: Hydraulic system for brakes only. Electrical system powered by 14 V 60 A engine-driven alternator. 12 V 35 Ah battery standard. Dual vacuum system for blind-flying instrumentation standard, complete with vacuum gauge, regulator, filter and annunciator light. Piper Aire air conditioning system optional.

AVIONICS AND EQUIPMENT: Single Bendix/King nav/com transponder and Narco AR-850 altitude reporter digitiser standard. Choice of three optional factory installed Bendix/King avionics groups including various combinations of nav/com, ADE, DME RNAV, transponder, glide slope and marker beacon receivers, and Bendix/King autopilots, and electric pitch trim, with substitute equipment options in each group. Additional avionics options for all groups include II Morrow 612 and Northstar M1 Loran C, Argus 5000 moving map display, 3M Stormscope WX-1000 and 1000+, Bendix/King KMA 24-H-70 audio control panel with intercom, KAS 297B altitude/vertical speed preselect system. KRA 10A radio altimeter, KI229 RMI, KI 204-02 or KI 203-00 nav indicator, ground clearance energy saving switch, United UI-5035-P23 encoding altimeter, transponder ident and press-to-talk buttons on pilot's control wheel, pilot's and co-pilot's boom microphones with control wheel buttons, and static discharge wicks. Standard equipment includes advanced instrument panel with 3 in gyro horizon and directional gyro, pictorial rate of turn indicator, rate of climb indicator and OAT gauge; electric clock; engine-driven vacuum pump with indicator, regulator, filter, annunciator light and auxiliary electric vacuum pump with selector switch; EGT gauge; heated pitot head; Piper true airspeed indicator; magnetic compass; sensitive altimeter; ammeter; annunciator panel with press-to-test recording tachometer; fuel quantity gauges; fuel pressure gauge; oil temperature and pressure gauges; vacuum gauge; wool carpet and vinyl sidewalls and headliner fabric and vinyl pilot's and co-pilot's fore and aft adjustable and tilting seats with inertia reel shoulder harnesses, seatbelts, armrests and ashtrays; rear bench seat with inertia reel shoulder harnesses, seatbelts and side panel ashtrays; pilot's storm window; map pockets; pilot's and co-pilot's scuff pads and sun visors; door closing strap; soundproofing; ignition lock; carpeted luggage compartment with security straps; compass card with holder; tiedown rings; jack pads; cabin and rear baggage door locks; emergency locator beacon; zinc chromate corrosion protection; stainless steel control cables; exterior styling with single- or two-tone base colour in Piper acrylic enamel with a choice of two trim colours in three designs; instrument panel white backlighting and overhead blue lighting, cabin dome light, two navigation lights, landing/taxi light, shielded white wingtip and red tail strobe lights and avionics dimming; true airspeed indicator; alternate static source; 35Ah battery; entrance step; wheel fairings; engine primer system; and towbar. Optional Deluxe avionics and equipment packages include dual Bendix/King KX 165 nav/com; glide slope receiver; KR 87 digital ADF. KAP 100 autopilot with electric pitch trim; deluxe interior with headrests, window curtains and Sofsuede sidewalls; super soundproofing; ventilation fan; vertically adjustable pilot's and co-pilot's seats; cigarette lighter, crew cup holders; engine hour recorder; external power receptacle, wingtip recognition lights; and fin strobe light. Optional equipment in addition to the Deluxe package includes carburettor ice detector, digital clock in pilot's control wheel, approach plate holder (not with digital clock), lockable fuel caps, Piper Aire air conditioning system, hand fire extinguisher, leather trimmed or all-leather seats, entry/exit assist strap. Du Pont IMRON polyurethane enamel exterior finish, third trim colour, and shadowed registration numbers.

DIMENSIONS, EXTERNAL:

Wing span	10.67 m (35 ft 0 in)
Wing chord: at root	1.60 m (5 ft 3 in)
at tip	1.07 m (3 ft 6¼ in)
Wing aspect ratio	7.2
Length overall	7.25 m (23 ft 9½ in)
Height overall	2.22 m (7 ft 3½ in)
Tailplane span	3.96 m (12 ft 11¾ in)
Wheel track	3.05 m (10 ft 0 in)
Wheelbase	2.03 m (6 ft 8 in)
Propeller	1.88 m (6 ft 2 in)
Propeller ground clearance	0.21 m (8¼ in)
Cabin door: Height	0.89 m (2 ft 11 in)
Width	0.91 m (3 ft 0 in)
Baggage door: Height	0.51 m (1 ft 8 in)
Max width	0.56 m (1 ft 10 in)
Height to sill	0.71 m (2 ft 4 in)

DIMENSIONS, INTERNAL:

Cabin: Length (instrument panel to rear bulkhead)	2.49 m (8 ft 2 in)
Max width	1.05 m (3 ft 5¼ in)
Max height	1.14 m (3 ft 8¾ in)
Floor area	2.28 m² (24.5 sq ft)
Volume (incl baggage area)	3.00 m³ (106 cu ft)

AREAS:
Wings, gross 15.8 m^2 (170.0 sq ft)
Ailerons (total) 1.23 m^2 (13.2 sq ft)
Trailing-edge flaps 1.36 m (14.6 sq ft)
Fin 0.69 m^2 (7.4 sq ft)
Rudder 0.38 m^2 (4.1 sq ft)
Tailplane, incl tab 2.46 m^2 (26.5 sq ft)

WEIGHTS AND LOADINGS:
Weight empty, equipped 653 kg (1,439 lb)
Max T-O and landing weight 1,106 kg (2,440 lb)
Max ramp weight 1,110 kg (2,447 lb)
Max wing loading 70.06 kg/m^2 (14.35 lb/sq ft)
Max power loading 9.33 kg/kW (15.25 lb/hp)

PERFORMANCE (at max T-O weight):
Never-exceed speed 153 kts (282 km/h; 176 mph)
Max level speed at S/L 127 kts (235 km/h; 146 mph)
Best power cruising speed:
75% power at 2,745 m (9,000 ft) 126 kts (233 km/h; 145 mph)
65% power at 3,810 m (12,500 ft) 118 kts (219 km/h; 136 mph)
55% power at 3,810 m (12,500 ft) 107 kts (198 km/h; 123 mph)
Best econ cruising speed:
75% power at 2,745 m (9,000 ft) 122 kts (225 km/h; 140 mph)
65% power at 3,810 m (12,500 ft) 116 kts (215 km/h; 134 mph)
55% power at 3,810 m (12,500 ft) 105 kts (195 km/h; 121 mph)
Stalling speed: flaps up 56 kts (104 km/h; 65 mph)
flaps down 50 kts (93 km/h; 58 mph) CAS
Max rate of climb at S/L 196 m (644 ft)/min
Service ceiling 3,335 m (11,000 ft)
T-O run 320 m (1,050 ft)
T-O to 15 m (50 ft) 503 m (1,650 ft)
Landing from 15 m (50 ft) 354 m (1,160 ft)
Landing run 191 m (625 ft)
Range with max fuel with allowances for taxi, T-O, climb and descent, and 45 min reserves at max range power:
at best settings: 75% power at 2,745 m (9,000 ft) 525 n miles (972 km; 604 miles)
65% power at 3,810 m (12,500 ft) 553 n miles (1,025 km; 637 miles)
55% power at 3,810 m (12,500 ft) 565 n miles (1,047 km; 651 miles)
at best econ settings: 75% power at 2,745 m (9,000 ft) 590 n miles (1,092 km; 679 miles)
65% power at 3,810 m (12,500 ft) 633 n miles (1,173 km; 729 miles)
55% power at 3,810 m (12,500 ft) 640 n miles (1,186 km; 737 miles)

UPDATED

PIPER PA-28-181 ARCHER II

Piper Archer II four-seat cabin monoplane

1992

TYPE: Four-seat cabin monoplane.

PROGRAMME: On 9 October 1972 Piper introduced the Cherokee Challenger as successor to the Cherokee 180. In 1974 this was superseded by the Cherokee Archer, with the same basic airframe and power plant, but with many additional equipment and avionics options. In 1976 this aircraft was redesignated PA-28-181 Cherokee Archer II, and in 1978 introduced the tapered wings of the Warrior II.

A total of 9,919 Cherokee 180s/Archer IIs had been delivered by December 1993.

DESIGN FEATURES: Cantilever low-wing monoplane. Wing section NACA 65_2-415 on inboard panels; outboard leading-edges have modification No. 5 of NACA TN 2228. Dihedral 7°. Incidence 2° at root, −1° at tip. Sweepback at quarter-chord 5°.

FLYING CONTROLS: Plain ailerons of light alloy construction. Light alloy trailing-edge flaps with ribbed skins. One-piece all-moving horizontal surface with combined anti-servo and trim tab. Trim tab in rudder.

STRUCTURE: Light alloy single-spar structure with glassfibre wingtips. The fuselage is an aluminium semi-monocoque structure with glassfibre engine cowling. The tail unit is a cantilever structure of aluminium alloy, except for glassfibre tips on fin and tailplane. Fin and rudder have corrugated metal skin.

LANDING GEAR: Non-retractable tricycle type. Steerable nosewheel. Piper oleo-pneumatic shock-absorbers. Cleveland wheels and Schenuit tyres, size 6.00-6, 4 ply rating, on all three wheels. Mainwheel tyre pressure 1.665 bars (24 lb/sq in), nosewheel 1.24 bars (18 lb/sq in). Cleveland high-capacity disc brakes. Parking brake. Wheel speed fairings standard.

POWER PLANT: One 134 kW (180 hp) Textron Lycoming O-360-A4M flat-four engine, driving a Sensenich two-blade fixed-pitch metal propeller with spinner. Fuel in two tanks in wing leading-edges, with total capacity of 189 litres (50 US gallons; 41.6 Imp gallons), of which 181.5 litres (48 US gallons; 40 Imp gallons) are usable. Oil capacity 7.5 litres (2 US gallons; 1.7 Imp gallons).

ACCOMMODATION: Four persons in pairs in enclosed cabin. Individual adjustable front seats, with dual controls; individual rear seats. Large door on starboard side. Baggage compartment at rear of cabin, with volume of 0.74 m^3 (26 cu ft) and capacity of 90 kg (200 lb); door on starboard side. Rear seats removable to provide 1.25 m^3 (44 cu ft) cargo space. Accommodation heated and ventilated. Windscreen defrosting.

SYSTEMS: Optional Piper Aire air conditioning system. Electrical system includes 14 V 60 A alternator and 12 V 35 Ah battery. Hydraulic systems for brakes only. Vacuum system standard.

AVIONICS AND EQUIPMENT: Single Bendix/King nav/com transponder and Narco AR-850 altitude reporter digitiser standard. Choice of three optional factory installed Bendix/King avionics groups including various combinations of nav/com, ADE, DME RNAV, transponder, glide slope and marker beacon receivers, and Bendix/King autopilots, and electric pitch trim, with substitute equipment options in each group. Additional avionics options for all groups include II Morrow 612 and Northstar M1 Loran C, Argus 5000 moving map display, 3M Stormscope WX-1000 and 1000+, Bendix/King KMA 24-H-70 audio control panel with intercom, KAS 297B altitude/vertical speed preselect system. KRA 10A radio altimeter, KI229 RMI, KI 204-02 or KI 203-00 nav indicator, ground clearance energy saving switch, United UI-5035-P23 encoding altimeter, transponder ident and press-to-talk buttons on pilot's control wheel, pilot's and co-pilot's boom microphones with control wheel buttons, and static discharge wicks. Standard equipment includes advanced instrument panel with 3 in gyro horizon and directional gyro, pictorial rate of turn indicator, rate of climb indicator and OAT gauge; electric clock; engine-driven vacuum pump with indicator, regulator, filter, annunciator light and auxiliary electric vacuum pump with selector switch; EGT gauge; heated pitot head; Piper true airspeed indicator; magnetic compass; sensitive altimeter; ammeter; annunciator panel with press-to-test recording tachometer; fuel quantity gauges; fuel pressure gauge; oil temperature and pressure gauges; vacuum gauge; wool carpet and vinyl sidewalls and headliner fabric and vinyl pilot's and co-pilot's fore and aft adjustable and tilting seats with inertia reel shoulder harnesses, seatbelts, armrests and ashtrays; rear bench seat with inertia reel shoulder harnesses, seatbelts and side panel ashtrays; pilot's storm window; map pockets; pilot's and co-pilot's scuff pads and sun visors; door closing strap; soundproofing; ignition lock; carpeted luggage compartment with security straps; compass card with holder; tiedown rings; jack pads; cabin and rear baggage door locks; emergency locator beacon; zinc chromate corrosion protection; stainless steel control cables; exterior styling with single- or two-tone base colour in Piper acrylic enamel with a choice of two trim colours in three designs; instrument panel white backlighting and overhead blue lighting, cabin dome light, two navigation lights, landing/taxi light, shielded white wingtip and red tail strobe lights and avionics dimming; true airspeed indicator; alternate static source; 35Ah battery; entrance step; wheel fairings; engine primer system; and towbar. Optional Deluxe avionics and equipment packages include dual Bendix/King KX 165 nav/com; glide slope receiver; KR 87 digital ADF. KAP 100 autopilot with electric pitch trim; deluxe interior with headrests, window curtains and Sofsuede sidewalls; super soundproofing; ventilation fan; vertically adjustable pilot's and co-pilot's seats; cigarette lighter, crew cup holders; engine hour recorder; external power receptacle, wingtip recognition lights; and fin strobe light. Optional equipment in addition to the Deluxe package includes carburettor ice detector, digital clock in pilot's control wheel, approach plate holder (not with digital clock), lockable fuel caps, Piper Aire air conditioning system, hand fire extinguisher, leather trimmed or all-leather seats, entry/exit assist strap. Du Pont IMRON polyurethane enamel exterior finish, third trim colour, and shadowed registration numbers.

DIMENSIONS, EXTERNAL:
Wing span 10.67 m (35 ft 0 in)
Wing chord: at root 1.60 m (5 ft 3 in)
at tip 1.07 m (3 ft 6¼ in)
Wing aspect ratio 7.2
Length overall 7.25 m (23 ft 9½ in)
Height overall 2.22 m (7 ft 3½ in)
Tailplane span 3.92 m (12 ft 10½ in)
Wheel track 3.05 m (10 ft 0 in)
Wheelbase 2.00 m (6 ft 7 in)
Propeller diameter 1.93 m (6 ft 4 in)
Propeller ground clearance 0.21 m (8¼ in)
Cabin: Height 1.14 m (3 ft 9 in)
Width 1.06 m (3 ft 5¾ in)
Baggage compartment volume, incl hatshelf 0.74 m^3 (26.0 cu ft)

DIMENSIONS, INTERNAL:
Cabin: Length (instrument panel to rear bulkhead) 2.49 m (8 ft 2 in)
Max width 1.05 m (3 ft 5¼ in)
Max height 1.14 m (3 ft 8¾ in)
Floor area 2.28 m^2 (24.5 sq ft)
Volume (incl baggage area) 3.00 m^3 (106 cu ft)

AREAS:
Wings, gross 15.8 m^2 (170.0 sq ft)
Ailerons (total) 1.23 m^2 (13.2 sq ft)
Trailing-edge flaps 1.36 m^2 (14.6 sq ft)
Fin 0.69 m^2 (7.4 sq ft)
Rudder 0.38 m^2 (4.1 sq ft)
Tailplane, incl tab 2.46 m^2 (26.5 sq ft)

WEIGHTS AND LOADINGS:
Weight emtpy, equipped (standard) 683.5 kg (1,507 lb)
Max T-O and landing weight 1,156 kg (2,550 lb)
Max ramp weight 1,160 kg (2,558 lb)
Max wing loading 73.2 kg/m^2 (15.0 lb/sq ft)
Max power loading 8.63 kg/kW (14.17 lb/hp)

PERFORMANCE (at max T-O weight):
Never-exceed speed 148 kts (275 km/h; 171 mph) CAS
Max level speed at S/L 129 kts (239 km/h; 148 mph)
Best power cruising speed:
75% power at 2,440 m (8,000 ft) 129 kts (239 km/h; 148 mph)
65% power at 3,660 m (12,000 ft) 125 kts (231 km/h; 144 mph)
55% power at 3,810 m (12,500 ft) 111 kts (206 km/h; 128 mph)
Best econ cruising speed:
75% power at 2,440 m (8,000 ft) 126 kts (233 km/h; 145 mph)
65% power at 3,660 m (12,000 ft) 121 kts (224 km/h; 140 mph)
55% power at 3,810 m (12,500 ft) 106 kts (196 km/h; 122 mph)
Stalling speed: flaps up 59 kts (109 km/h; 68 mph) CAS
flaps down 53 kts (98 km/h; 61 mph) CAS
Max rate of climb at S/L 224 m (725 ft)/min
Service ceiling 4,160 m (13,650 ft)
Absolute ceiling 4,800 m (15,750 ft)
T-O run 265 m (870 ft)
T-O to 15 m (50 ft) 506 m (1,660 ft)
Landing from 15 m (50 ft) 424 m (1,390 ft)
Landing run 282 m (925 ft)
Range with max fuel, for taxi, T-O, climb and descent, and 45 min reserves at max range power:
at best settings: 75% power at 2,440 m (8,000 ft) 520 n miles (963 km; 599 miles)
65% power at 3,660 m (12,000 ft) 565 n miles (1,047 km; 650 miles)
55% power at 3,180 m (12,500 ft) 580 n miles (1,075 km; 668 miles)
at best econ power settings:
75% power at 2,440 m (8,000 ft) 600 n miles (1,112 km; 691 miles)
65% power at 3,660 m (12,000 ft) 645 n miles (1,196 km; 743 miles)
55% power at 3,810 m (12,500 ft) 670 n miles (1,242 km; 772 miles)

VERIFIED

PIPER PA-28R-201 ARROW III and PA-28RT-201T TURBO ARROW III

TYPE: Four-seat cabin monoplane.

PROGRAMME: The Piper Arrow was derived from the Cherokee Arrow II, which was generally similar to the Cherokee Archer II but had a retractable landing gear, more powerful engine, and untapered wings of the 1975 PA-28-180 Archer. In 1977, Piper updated this model by fitting long-span tapered wings identical with those of the Archer II, but with increased fuel capacity, giving improved performance. The 1978 version of this aircraft was named Arrow III, the prototype of which flew for the first time on 16 September 1975, followed by the first production aircraft on

Piper Arrow four-seat cabin monoplane

1992

7 January 1977. Piper designation is PA-28R-201. The Turbo Arrow III differed by having a turbocharged engine and the first production example of this version flew 1 December 1976.

In 1979 Piper introduced new models with all-moving T tailplanes, but later recommenced production of the earlier low-tail Arrow and Turbo Arrow.

DESIGN FEATURES: Cantilever low-wing monoplane. Wing section NACA 65_2-415 on inboard panels; outboard leading-edges have modification No. 5 of NACA TN 2228. Dihedral 7°. Incidence 2° at root, −1° at tip. Sweepback at quarter-chord 5°.

FLYING CONTROLS: Plain ailerons of light alloy construction. Light alloy trailing-edge flaps with ribbed skins. One-piece all-moving horizontal surface with combined anti-servo and trim tab. Trim tab in rudder.

STRUCTURE: Light alloy single-spar structure with glassfibre wingtips. The fuselage is an aluminium semi-monocoque structure with glassfibre engine cowling. The tail unit is a cantilever structure of aluminium alloy, except for glassfibre tips on fin and tailplane. Fin and rudder have corrugated metal skin.

LANDING GEAR: Retractable tricycle type, hydraulically retracted with an electrically operated pump supplying the hydraulic pressure. Main units retract inward into wings, nose unit rearward. All units with oleo-pneumatic shock-absorbers. Mainwheels and tyres size 6.00-6, 6 ply rating, pressure 2.07 bars (30 lb/sq in). Nosewheel and tyre size 5.00-5, 4 ply rating, pressure 1.86 bars (27 lb/sq in). High-capacity dual hydraulic disc brakes and parking brake.

POWER PLANT: Arrow: One 149 kW (200 hp) Textron Lycoming IO-360-C1C6 flat-four engine, driving a McCauley two-blade constant-speed metal propeller with spinner. Turbo Arrow: One 149 kW (200 hp) Teledyne Continental TSIO-360-FB, driving a Hartzell two-blade constant-speed metal propeller with spinner. Fuel tanks in wing leading-edges with total capacity of 291 litres (77 US gallons; 64 Imp gallons), of which 273 litres (72 US gallons; 60 Imp gallons) are usable. Oil capacity 7.5 litres (2 US gallons; 1.7 Imp gallons).

ACCOMMODATION: Four persons in pairs in enclosed cabin. Individual adjustable front seats, with dual controls; individual rear seats. Large door on starboard side. Baggage compartment at rear of cabin, with volume of 0.74 m³ (26 cu ft) and capacity of 90 kg (200 lb); door on starboard side. Rear seats removable to provide 1.25 m³ (44 cu ft) cargo space. Accommodation heated and ventilated. Windscreen defrosting.

SYSTEMS: Generally as for Archer II and Warrior II except for electrohydraulic system for landing gear actuation. An oxygen system of 1.37 m³ (48.3 cu ft) capacity is available optionally.

AVIONICS AND EQUIPMENT: Single Bendix/King nav/com transponder and Narco AR-850 altitude reporter digitiser standard. Choice of three optional factory installed Bendix/King avionics groups including various combinations of nav/com, ADE, DME RNAV, transponder, glide slope and marker beacon receivers, and Bendix/King autopilots, and electric pitch trim, with substitute equipment options in each group. Additional avionics options for all groups include II Morrow 612 and Northstar M1 Loran C, Argus 5000 moving map display, 3M Stormscope WX-1000 and 1000+, Bendix/King KMA 24-H-70 audio control panel with intercom, KAS 297B altitude/vertical speed preselect system. KRA 10A radio altimeter, KI229 RMI, KI 204-02 or KI 203-00 nav indicator, ground clearance energy saving switch, United UI-5035-P23 encoding altimeter, transponder ident and press-to-talk buttons on pilot's control wheel, pilot's and co-pilot's boom microphones with control wheel buttons, and static discharge wicks. Standard equipment includes advanced instrument panel with 3 in gyro horizon and directional gyro, pictorial rate of turn indicator, rate of climb indicator and OAT gauge; electric clock; engine-driven vacuum pump with indicator, regulator, filter, annunciator light and auxiliary electric vacuum pump with selector switch; EGT gauge; heated pitot head; Piper true airspeed indicator; magnetic compass; sensitive altimeter; ammeter; annunciator panel with press-to-test recording tachometer; fuel quantity gauges; fuel pressure gauge; oil temperature and pressure gauges; vacuum gauge; wool carpet and vinyl sidewalls and headliner fabric and vinyl pilot's and co-pilot's fore and aft adjustable and tilting seats with inertia reel shoulder harnesses, seatbelts, armrests and ashtrays; rear bench seat with inertia reel shoulder harnesses, seatbelts and side panel ashtrays; pilot's storm window; map pockets; pilot's and co-pilot's scuff pads and sun visors; door closing strap; soundproofing; ignition lock; carpeted luggage compartment with security straps; compass card with holder; tiedown rings; jack pads; cabin and rear baggage door locks; emergency locator beacon; zinc chromate corrosion protection; stainless steel control cables; exterior styling with single- or two-tone base colour in Piper acrylic enamel with a choice of two trim colours in three designs; instrument panel white backlighting and overhead blue lighting, cabin dome light, two navigation lights, landing/taxi light, shielded white wingtip and red tail strobe lights and avionics dimming; true airspeed indicator; alternate static source; 35 Ah battery; entrance step; wheel fairings; engine primer system; and towbar. Optional equipment as for Archer II, plus cold weather start kit and integral or portable oxygen systems (Turbo models only).

DIMENSIONS, EXTERNAL:

Wing span	10.80 m (35 ft 5 in)
Wing chord: at root	1.60 m (5 ft 3 in)
at tip	1.07 m (3 ft 6 in)
Length overall	7.52 m (24 ft 8 in)
Height overall	2.39 m (7 ft 10¼ in)
Tailplane span	3.92 m (12 ft 10½ in)
Wheel track	3.19 m (10 ft 5½ in)
Wheelbase	2.39 m (7 ft 10¼ in)
Propeller diameter	1.93 m (6 ft 4 in)
Cabin door (stbd): Width	0.91 m (3 ft 0 in)
Height	0.89 m (2 ft 11 in)
Baggage door (stbd): Width	0.56 m (1 ft 10 in)
Height	0.51 m (1 ft 8 in)

DIMENSIONS, INTERNAL:

Cabin:	
Length, panel to rear bulkhead	2.42 m (7 ft 11¼ in)
Max width	1.06 m (3 ft 5¾ in)
Max height	1.14 m (3 ft 8¾ in)
Volume (incl baggage area)	3.00 m³ (106.0 cu ft)
Baggage compartment volume, incl hatshelf	0.74 m³ (26.0 cu ft)

AREAS:

Wings, gross	15.79 m² (170.0 sq ft)

WEIGHTS AND LOADINGS (A: Arrow, B: Turbo Arrow):

Weight, empty: A	731 kg (1,612 lb)
B	756 kg (1,667 lb)
Max T-O weight: A	1,247 kg (2,750 lb)
B	1,315 kg (2,900 lb)
Max ramp weight: A	1,252 kg (2,760 lb)
B	1,320 kg (2,912 lb)
Max wing loading: A	79.0 kg/m² (16.2 lb/sq ft)
B	83.3 kg/m² (17.0 lb/sq ft)
Max power loading: A	8.37 kg/kW (13.7 lb/hp)
B	8.83 kg/kW (14.5 lb/hp)

PERFORMANCE (at max T-O weight):

Max level speed: A at S/L	152 kts (281 km/h; 175 mph)
B	178 kts (330 km/h; 205 mph)
Best power cruising speed at optimum altitude:	
A, 75% power	143 kts (265 km/h; 165 mph)
A, 65% power	138 kts (255 km/h; 159 mph)
A, 55% power	131 kts (242 km/h; 151 mph)
B, 75% power	172 kts (318 km/h; 198 mph)
B, 65% power	167 kts (310 km/h; 192 mph)
B, 55% power	157 kts (291 km/h; 181 mph)
Best econ cruising at optimum altitude:	
A, 75% power	135 kts (250 km/h; 155 mph)
A, 65% power	129 kts (230 km/h; 148 mph)
A, 55% power	122 kts (226 km/h; 140 mph)
A, 75% power	168 kts (311 km/h; 193 mph)
B, 65% power	164 kts (304 km/h; 189 mph)
B, 55% power	154 kts (285 km/h; 177 mph)
Stalling speed: A, flaps up	60 kts (111 km/h; 69 mph)
B, flaps up	63 kts (117 km/h; 73 mph)
A, flaps down	55 kts (102 km/h; 64 mph)
B, flaps down	56 kts (104 km/h; 65 mph)
Max rate of climb at S/L: A	253 m (831 ft)/min
B	286 m (940 ft)/min
Service ceiling: A	4,935 m (16,200 ft)
B	6,100 m (20,000 ft)
T-O run: A	313 m (1,025 ft)
B	339 m (1,110 ft)
T-O to 15 m(50 ft): A	488 m (1,600 ft)
B	494 m (1,620 ft)
Landing from 15 m (50 ft): A	465 m (1,525 ft)
B	476 m (1,560 ft)
Landing run: A	188 m (615 ft)
B	197 m (645 ft)

Range with max fuel, allowances for start, T-O, climb and descent, plus 45 min reserves:

at best power settings:	
A, 75% power	725 n miles (1,343 km; 835 miles)
A, 65% power	770 n miles (1,427 km; 886 miles)
A, 55% power	795 n miles (1,473 km; 915 miles)
B, 75% power	675 n miles (1,250 km; 777 miles)
B, 65% power	705 n miles (1,306 km; 811 miles)
B, 55% power	740 n miles (1,371 km; 852 miles)
at best econ power settings:	
A, 75% power	810 n miles (1,501 km; 932 miles)
A, 65% power	845 n miles (1,566 km; 973 miles)
A, 55% power	875 n miles (1,621 km; 1,007 miles)
B, 75% power	785 n miles (1,455 km; 904 miles)
B, 65% power	830 n miles (1,538 km; 955 miles)
B, 55% power	860 n miles (1,593 km; 990 miles)

VERIFIED

PIPER PA-28-236 DAKOTA

TYPE: Four-seat cabin monoplane.

PROGRAMME: Piper introduced in 1978 an addition to the Warrior, Archer, Arrow line known as the PA-28-236 Dakota, which differs primarily by having a 175 kW (235 hp) Textron Lycoming engine to provide increased performance, and increased capacity fuel tanks to cater for this power plant.

The Dakota introduced the same standard items as listed for the Arrow, plus high-performance wheel fairings. A Deluxe package as detailed for the Warrior II is also available, and adds 38 kg (83.8 lb) to aircraft basic weight. A range of Bendix/King and Narco avionics packages is also available optionally.

Licence assembly of the Dakota was undertaken by the Chilean Aircraft Industry (ENAER); a total of 27 was completed by ENAER. A total of 747 was sold by December 1993.

DESIGN FEATURES: Cantilever low-wing monoplane. Wing section NACA 65_2-215 on inboard panels; outboard leading-edges have modification No. 5 of NACA TN 2228. Dihedral 7°. Incidence 2° at root, −1° at tip. Sweepback at quarter-chord 5°.

FLYING CONTROLS: Plain ailerons of light alloy construction. Light alloy trailing-edge flaps with ribbed skins. The tail has a one-piece all-moving horizontal surface with combined anti-servo and trim tab. Trim tab in rudder.

STRUCTURE: Light alloy single-spar structure with glassfibre wingtips. The fuselage is an aluminium semi-monocoque structure with glassfibre engine cowling. The tail unit is a centilever structure of aluminium alloy, except for glassfibre tips on fin and tailplane. Fin and rudder have corrugated metal skin.

LANDING GEAR: Non-retractable tricycle type. Steerable nosewheel. Piper oleo-pneumatic shock-absorbers. Cleveland wheels and Schenuit tyres, size 6.00-6, 4 ply rating, on all three wheels. Mainwheel tyre pressure 1.665 bars (24 lb/sq in), nosewheel 1.24 bars (18 lb/sq in). Cleveland high-capacity disc brakes. Parking brake. Wheel speed fairings standard.

POWER PLANT: One 175 kW (235 hp) Textron Lycoming O-540-J345D flat-six engine driving a Hartzell two-blade constant-speed metal propeller with spinner. Two integral fuel tanks in wing, with total capacity of 291.5 litres (77 US gallons; 64 Imp gallons), of which 272.5 litres (72 US gallons; 60 Imp gallons) are usable. Refuelling point on upper surface of each wing. Oil capacity 11.5 litres (3 US gallons; 2.5 Imp gallons).

AVIONICS AND EQUIPMENT: Avionics groups, optional avionics, standard equipment, and equipment options generally as for Arrow, except oxygen systems not available and heavy-duty brakes and tyres and polished propeller spinner optional.

DIMENSIONS, EXTERNAL AND INTERNAL: As for Archer II except:

Length overall	7.54 m (24 ft 8¾ in)
Height overall	2.18 m (7 ft 2 in)
Wheelbase	1.98 m (6 ft 6 in)
Cabin: Height	1.14 m (3 ft 8¾ in)
Weight empty	730 kg (1,610 lb)
Max T-O weight	1,361 kg (3,000 lb)
Max ramp weight	1,366 kg (3,011 lb)
Max wing loading	85.93 kg/m² (17.6 lb/sq ft)
Max power loading	7.78 kg/kW (12.8 lb/hp)

Piper Dakota four-seat cabin monoplane

1992

PERFORMANCE (at max T-O weight):

Max level speed at S/L	148 kts (274 km/h; 170 mph)
Best power cruising speed at optimum altitude:	
75% power	144 kts (267 km/h; 166 mph)
65% power	139 kts (258 km/h; 160 mph)
55% power	130 kts (241 km/h; 150 mph)
Best econ cruising speed at optimum altitude:	
75% power	139 kts (258 km/h; 160 mph)
65% power	134 kts (248 km/h; 154 mph)
55% power	126 kts (234 km/h; 145 mph)
Stalling speed: flaps up	65 kts (120 km/h; 75 mph) CAS
flaps down	58 kts (108 km/h; 67 mph) CAS
Max rate of climb at S/L	338 m (1,110 ft)/min
Service ceiling	5,335 m (17,500 ft)
Absolute ceiling	5,945 m (19,500 ft)
T-O run	270 m (886 ft)
T-O to 15 m (50 ft)	371 m (1,216 ft)
Landing from 15 m (50 ft):	
standard brakes	526 m (1,725 ft)
heavy-duty brakes	466 m (1,530 ft)
Landing run: standard brakes	252 m (825 ft)
heavy-duty brakes	195 m (640 ft)
Range with max fuel, allowances for taxi, T-O, climb, cruise, descent, and 45 min reserves at max range power:	
at best power settings at optimum altitude:	
75% power	650 n miles (1,205 km; 748 miles)
65% power	710 n miles (1,315 km; 817 miles)
55% power	750 n miles (1,390 km; 863 miles)
at best econ power settings at optimum altitude:	
75% power	720 n miles (1,334 km; 829 miles)
65% power	770 n miles (1,427 km; 886 miles)
55% power	810 n miles (1,501 km; 933 miles)

VERIFIED

PIPER PA-31P-350 MOJAVE

TYPE: Seven-seat pressurised light transport.

PROGRAMME: Announced by Piper on 20 November 1982, the Mojave is a piston-engined counterpart of the turboprop Cheyenne II, of much the same size but with wings of slightly increased span and no tip tanks. It is pressurised, and its direct drive counter-rotating and turbocharged Avco Lycoming engines are similar to those which power the PA-31-350 Chieftain. A prototype is belived to have flown for the first time in late 1979. Certification was gained 9 June 1983.

DESIGN FEATURES: Cantilever low-wing monoplane. Wing dihedral 5° from roots. Wingroot leading-edge extended forward between nacelle and fuselage on each side. Conventional tail unit with swept vertical surfaces and an extended dorsal fin blended into the upper surface of the fuselage.

FLYING CONTROLS: Electrically operated flaps. Trim tab in starboard aileron. The tail unit has horn-balanced control surfaces. Trim tab in rudder and starboard elevator.

STRUCTURE: The fuselage is a semi-monocoque structure primarily of light alloy with fail-safe structure in pressurised section.

LANDING GEAR: Retractable tricycle type with single wheel on each unit; main units retract inward into wingroots, nosewheel rearward. Mainwheels and tyres size 6.50-10, 8 ply rating. Steerable nosewheel size 6.00-6 with tyre size 17.5 × 6.25-6, 10 ply rating. Oleo-pneumatic shock-absorber in each unit. Hydraulic disc brakes on mainwheels. Parking brake.

POWER PLANT: Two 261 kW (350 hp) Avco Lycoming flat-six turbocharged counter-rotating engines, one TIO-540-V2AD and one LTIO-540-V2AD, each driving a Hartzell three-blade constant-speed and fully feathering propeller with spinner. Fuel system usable capacity 900 litres (238 US gallons; 198 Imp gallons). Oil capaicty 24.6 litres (6.5 US gallons; 5.4 Imp gallons).

ACCOMMODATION: Pilot, co-pilot and five passengers, or pilot and six passengers, in pressurised cabin. Four contoured reclining seats for passengers in facing pairs, with side-facing seat at rear on starboard side. Foldaway writing tables, a hot/cold refreshment centre, and stereo system, optional. Door with built-in airstair behind wing on port side. Baggage space in nose, wing lockers and rear cabin, with combined capacity of 308 kg (680 lb).

SYSTEM: Pressurisation system with maximum differential of 0.35 bars (5.0 lb/sq in), to provide a cabin altitude of 2,775 m (9,100 ft) to a height of 7,620 m (25,000 ft).

AVIONICS: Wide range of avionics, including colour weather radar, available from manufacturers that include Bendix, Collins, King and Sperry. Full IFR equipment and autopilot included in basic price.

DIMENSIONS, EXTERNAL:

Wing span	13.56 m (44 ft 6 in)
Length overall	10.52 m (34 ft 6 in)
Height overall	3.96 m (13 ft 0 in)
Tailplane span	5.52 m (18 ft 1½ in)
Wheel track	4.19 m (13 ft 9 in)
Wheelbase	2.64 m (8 ft 8 in)
Propeller diameter	2.03 m (6 ft 8 in)
Distance between propeller centres	3.86 m (12 ft 8 in)
Passenger door (port, rear): Height	1.17 m (3 ft 10 in)
Width	0.71 m (2 ft 4 in)
Baggage door (port, nose): Height	0.53 m (1 ft 9 in)
Width	0.66 m (2 ft 2 in)
Nacelle locker doors (each): Length	1.02 m (3 ft 4 in)
Width	0.51 m (1 ft 8 in)

Piper Mojave seven-seat pressurised light transport aircraft

1992

Piper Cheyenne II six/eight-seat cabin monoplane

1992

DIMENSIONS, INTERNAL:

Cabin: Length, instrument panel to rear bulkhead	4.41 m (14 ft 5¾ in)
Max width	1.27 m (4 ft 2 in)
Max height	1.31 m (4 ft 3½ in)
Baggage compartment volume:	
Nose	0.64 m³ (22.5 cu ft)
Nacelle lockers (total)	0.51 m³ (18 cu ft)
Rear cabin	0.62 m³ (22 cu ft)

AREAS:

Wings, gross	22.02 m² (237.0 sq ft)

WEIGHTS AND LOADINGS:

Weight empty, standard	2,297 kg (5,064 lb)
Max T-O weight	3,266 kg (7,200 lb)
Max ramp weight	3,286 kg (7,245 lb)
Max zero-fuel weight	3,039 kg (6,700 lb)
Max landing weight	3,175 kg (7,000 lb)
Max wing loading	148.3 kg/m² (30.38 lb/sq ft)
Max power loading	6.26 kg/kW (10.29 lb/hp)

PERFORMANCE (at max T-O weight, S/L, ISA, except where indicated):

Max level speed at mid-cruise weight	242 kts (447 km/h; 278 mph)
Cruising speed (mid-cruise weight at optimum altitude):	
75% power	235 kts (435 km/h; 270 mph)
65% power	221 kts (409 km/h; 254 mph)
60% power	195 kts (361 km/h; 224 mph)
Stalling speed: flaps up	78 kts (145 km/h; 90 mph)
flaps down	72 kts (134 km/h; 83 mph)
Max rate of climb at S/L	372 m (1,220 ft)/min
Rate of climb at S/L, one engine out	78 m (225 ft)/min
Service ceiling	9,265 m (30,400 ft)
Service ceiling, one engine out	4,360 m (14,300 ft)
T-O run	495 m (1,625 ft)
T-O to 15 m (50 ft)	753 m (2,469 ft)
Landing from 15 m (50 ft)	700 m (2,300 ft)
Landing run	424 m (1,390 ft)
Range with max fuel at optimum altitude with allowances for engine starting, taxi, T-O, climb, and 45 min reserves at max range power:	
75% power	1,113 n miles (2,061 km; 1,280 miles)
65% power	1,143 n miles (2,116 km; 1,315 miles)
60% power	1,221 n miles (2,261 km; 1,405 miles)

VERIFIED

PIPER PA-31T CHEYENNE II

TYPE: Six/eight-seat cabin monoplane.

PROGRAMME: Design of the PA-31T began at the end of 1965. First flight of the prototype was made 20 August 1969, and FAA certification was granted 3 May 1972. The first production aircraft flew for the first time 22 October 1973. Following the introduction of the low-cost Cheyenne I and six/11-seat Cheyenne III the original version was redesigned Cheyenne II.

OPERATORS: Versions of the PA-31T are in service with the armed forces of the following countries: Argentina (2), Colombia (1), Mauritania (2), Peru (1) and Thailand (1).

VERSIONS: **Standard:** Six individual seats in pairs, with headrests and armrests. Pilot/co-pilot seats four-way adjustable with shoulder harness and inertia reels; third and fourth cabin seats aft-facing; and all cabin seats with seat belts; window curtains and wall to wall carpet. Rear cabin divider with clothes bar and baggage security net. Forward cabin divider curtain. 'No smoking - fasten seat belt' sign. Oxygen outlets and masks at each seat position. Options available include a cabin instrumentation panel comprising digital readouts of altitude, outside air temperature, time, and true airspeed; pneumatic door extender; forward cabin combination unit; storage cabinets; folding tables; aft cabin combination unit, which includes side-facing seventh seat/toilet; seventh and eighth seats; tinted cabin windows; cabin fire extinguisher; stereo system; and all-leather seat covering.

Executive: Six individual seats, comprising two crew seats and four reclining chairs in the Standard arrangement. Other standard equipment as described, plus forward cabin combination unit which includes cabin dividers and curtain, electrically heated Thermos unit, cup dispenser, storage for ice, beverages and manuals; two folding tables; pneumatic door extender and aft cabin combination unit which includes side-facing seventh seat/toilet, cabin divider with mirror, privacy curtain, refreshment centre; AC power outlet for electric razor; and baggage security net. Options are the same as for the Standard interior, unless included in the Executive package. Adds 55.3 kg (121.9 lb) to basic empty weight.

De-icing Group: Pneumatic de-icing boots for wing and tail unit leading-edges, and wing ice inspection light; adding 10.1 kg (22.3 lb) to basic empty weight.

Co-pilot Flight Group: Airspeed and rate of climb indicator, altimeter, electric turn rate indicator, attitude and directional gyros, clock, heated pitot, static system with alternate source, co-pilot's toe-brakes and windscreen wiper; adding 8.7 kg (19.2 lb) to basic empty weight.

Five optional factory-installed avionics packages are available for the Cheyenne II.

In addition to the six/eight-seat passenger versions of the Cheyenne II, as described, Piper developed a special multipurpose model known as the **Maritime Surveillance Cheyenne II**. The prototype (N431PC) has a radar pod under the port wing. It can, however, be equipped for a wide range of special missions, including photo mapping, airways calibration, geological survey, rainmaking and hail suppression. A quick-change interior is standard, permitting conversion for corporate use in 30 minutes. Two aircraft of this type, each fitted with a Global GNS-500A navigation system, two 70 mm reconnaissance cameras (pod-mounted underwing) and a search radar, were delivered to the Mauritanian Islamic Air Force in April 1981.

DESIGN FEATURES: Cantilever low-wing monoplane. Wing section NACA 63_2A415 at root, NACA 63_1A212 at tip. Dihedral 5°. Incidence 1° 30′ at root, −1° at tip. Sweepback 0° at 30 per cent chord. The tail unit is a cantilever structure with sweptback vertical surfaces. Fixed incidence tailplane.

FLYING CONTROLS: Balanced ailerons and single-slotted trailing-edge flaps of 2024ST light alloy. Trim tab in starboard aileron. Trim tabs in elevators and rudder.

STRUCTURE: The wing is a three-spar structure of 2024ST light alloy. The fuselage is a semi-monocoque structure of 2024ST light alloy with fail-safe in the pressurised areas.

LANDING GEAR: Hydraulically retractable tricycle type with single wheel on each unit, main units retracting inward and nosewheel aft. Nosewheel safety mirror. Piper oleo-pneumatic shock-absorbers. Cleveland mainwheels with tyres size 6.50-10, 10 ply rating. Steerable nosewheel with Type VII tyre size 18 × 4.4, 6 ply rating. Cleveland disc-type hydraulic brakes. Goodyear wheels and brakes optional. Parking brake.

POWER PLANT: Two 462 kW (620 ehp) Pratt & Whitney Aircraft of Canada PT6A-28 turboprop engines, each driving a Hartzell HC-BTN-3B three-blade constant-speed reversible-pitch fully feathering metal propeller with spinner. Propeller synchrophasers optional. Each wing has three interconnected fuel cells and a tip tank, giving total fuel capacity of 1,476 litres (390 US gallons; 325 imp gallons), of which 1,446 litres (382 US gallons; 318 Imp gallons) are usable. Refuelling points in engine nacelles and on upper surface of each tip tank. NACA type anti-icing non-siphoning fuel tank vents with flame arresters. Oil capacity 24.6 litres (6.5 US gallons; 5.4 Imp gallons). Electrically heated air intake anti-icing boot, air intake ice deflection and air bypass doors. Electrical propeller de-icing.

ACCOMMODATION: Pilot and co-pilot on two individual adjustable seats. Dual controls standard. Pilot's storm window. Heated windscreen. Windscreen wiper standard for pilot, optional for co-pilot. Cabin seating for four to six passengers on individual seats. Door with built-in airstair on port side, which has seven locking pins and inflatable pressurisation seal. Dual-pane windows. Emergency exit window on starboard side. Cabin heated and air conditioned. Forward and aft cabin dividers. A wide range of options for cabin includes folding tables, beverage dispensers, pneumatic door extender, storage cabinets and tinted windows. Baggage compartments in nose, capacity 136 kg (300 lb), and rear of cabin, capacity 91 kg (200 lb). External access door to nose compartment.

SYSTEMS: Air conditioning and pressurisation, with pressure differential of 0.38 bars (5.5 lb/sq in). Freon-type air conditioner of 23,000 BTU capacity. Janitrol combustion heater of 35,000 BTU capacity with automatic windscreen defroster. Hydraulic system supplied by dual engine-driven pumps for landing gear retraction and brakes. Pneumatic system and vacuum system provided by engine bleed air. Electrical system supplied by two 28 V 200 A starter/generators and 24 V 43 Ah Ni/Cd battery. External power socket standard. Oxygen system of 0.62 m³ (22 cu ft) capacity. De-icing system comprises electric anti-icing boots for air intakes, heated pitot and electric propeller de-icing. Fire detection system with six sensors; engine fire extinguishing system optional.

AVIONICS: Four optional factory-installed avionics packages are available, these include nav/coms, ADF, AFCS, DME, glide slope and marker beacon receivers, R/Nav, transponder, and weather radar by Bendix, Collins, King and Piper. Options include alternatives to or duplication of the above equipment, plus altitude alerter, encoding altimeter, radar altimeter, radio altimeter, radio telephone, global navigation system, and HF transceivers by manufacturers which include the above, plus Aeronetics, RCA, Smith, Sperry and SunAir.

EQUIPMENT: Installed standard equipment is extensive, and optional items include co-pilot's flight instrument group, digital clock, toe brakes and windscreen wiper, cargo door, emergency locator transmitter, wing and tail pneumatic de-icing boots, lockable fuel caps, engine fire extinguisher system, engine wash rings, fuselage ice protection plates, ice inspection lights and propeller synchrophaser.

DIMENSIONS, EXTERNAL:

Wing span over tip tanks	13.01 m (42 ft 8¼ in)
Length overall	10.57 m (34 ft 8 in)
Height overall	3.89 m (12 ft 9 in)
Tailplane span	6.05 m (19 ft 10 in)
Wheel track	4.19 m (13 ft 9 in)
Wheelbase	2.64 m (8 ft 8 in)
Propeller diameter	2.36 m (7 ft 9 in)
Distance between propeller centres	3.85 m (12 ft 7½ in)
Propeller ground clearance	0.27 m (10½ in)
Passenger door (port, rear): Height	1.17 m (3 ft 10 in)
Width	0.71 m (2 ft 4 in)
Height to sill	0.94 m (3 ft 1 in)
Baggage door (fwd): Height	0.53 m (1 ft 9 in)
Width	0.66 m (2 ft 2 in)
Emergency exit (stbd, fwd): Height	0.64 m (2 ft 1 in)
Width	0.48 m (1 ft 7 in)

DIMENSIONS, INTERNAL:

Cabin: Length, excl flight deck	2.57 m (8 ft 5 in)
Max width	1.27 m (4 ft 2 in)
Max height	1.31 m (4 ft 3½ in)
Floor area	4.37 m² (47 sq ft)
Volume	6.29 m³ (222 cu ft)
Forward baggage compartment	0.57 m³ (20 cu ft)
Rear baggage compartment	0.62 m³ (22 cu ft)

AREAS:

Wings, gross	21.3 m² (229 sq ft)
Ailerons (total)	1.21 m² (13 sq ft)
Trailing-edge flaps (total)	3.12 m² (33.6 sq ft)
Fin	1.48 m² (15.9 sq ft)
Rudder, incl tab	0.98 m² (10.6 sq ft)
Tailplane	3.92 m² (42.2 sq ft)
Elevators, incl tab	2.63 m² (28.3 sq ft)

WEIGHTS AND LOADINGS:

Weight empty, standard, equipped	2,276 kg (5,018 lb)
Max T-O and landing weight	4,082 kg (9,000 lb)
Max ramp weight	4,105 kg (9,050 lb)
Max zero-fuel weight	3,265 kg (7,200 lb)
Max wing loading	191.9 kg/m² (39.3 lb/sq ft)
Max power loading	4.42 kg/kW (7.26 lb/ehp)

PERFORMANCE (at max T-O weight, unless specified otherwise):

Cruising speed, max cruise power, average cruise weight of 3,493 kg (7,700 lb) at:

3,350 m (11,000 ft)	283 kts (524 km/h; 325 mph)
4,875 m (16,000 ft)	278 kts (515 km/h; 320 mph)
6,400 m (21,000 ft)	270 kts (500 km/h; 311 mph)
8,840 m (29,000 ft)	253 kts (469 km/h; 291 mph)

Stalling speed, wheels and flaps up, engines idling	86 kts (160 km/h; 99 mph) IAS
Stalling speed, wheels and flaps down, engines idling	75 kts (139 km/h; 86.5 mph) IAS
Rotation speed	91 kts (167 km/h; 105 mph) IAS
Approach speed	98 kts (182 km/h; 113 mph)
Max rate of climb at S/L	826 m (2,710 ft)/min
Rate of climb at S/L, one engine out	201 m (660 ft)/min
Service ceiling	9,630 m (31,600 ft)
Service ceiling, one engine out	4,450 m (14,600 ft)
T-O run	430 m (1,410 ft)
T-O to 15 m (50 ft)	604 m (1,980 ft)
Landing from 15 m (50 ft)	756 m (2,480 ft)
Landing from 15 m (50 ft) with propeller reversal	567 m (1,860 ft)
Landing run	436 m (1,430 ft)
Landing run with propeller reversal	291 m (955 ft)
Accelerate/stop distance	1,006 m (3,300 ft)

Range with max fuel at max cruise power, with allowances for start, taxi, T-O, climb, descent, and 45 min reserves at long-range cruise power at: 3,660 m (12,000 ft)

3,660 m (12,000 ft)	905 n miles (1,676 km; 1,041 miles)
4,875 m (16,000 ft)	1,020 n miles (1,889 km; 1,173 miles)
6,400 m (21,000 ft)	1,155 n miles (2,139 km; 1,329 miles)
8,840 m (29,000 ft)	1,380 n miles (2,555 km; 1,588 miles)

Piper Navajo six/eight-seat corporate and commuter aircraft *1992*

Range at max range power, fuel and allowances as above, at: 3,660 m (12,000 ft)

3,660 m (12,000 ft)	1,090 n miles (2,018 km; 1,254 miles)
4,875 m (16,000 ft)	1,195 n miles (2,213 km; 1,375 miles)
6,400 m (21,000 ft)	1,330 n miles (2,463 km; 1,530 miles)
8,840 m (29,000 ft)	1,510 n miles (2,796 km; 1,737 miles)

UPDATED

PIPER PA-31-310 NAVAJO C

TYPE: Six/eight-seat corporate and commuter airline transport.

PROGRAMME: On 30 September 1964 Piper flew the first of what it described as a new series of larger executive aircraft for corporate and commuter airline service. Named Navajo, it was then available with normally aspirated or turbocharged engines, the latter being known as the Turbo Navajo. Subsequently, three additional versions were introduced, the Navajo C/R, the pressurised Navajo and Navajo Chieftain.

The basic 1977 version known as the Navajo C is turbocharged and introduced new interior colour schemes, improved pilot's door, revised circuit breaker panel, instrument panel changes and relocated engine hour recorder. New cabinets, cabin dividers, Collins Micro-line electronics, and the Bendix FC5-810 autopilot/flight director available as options.

OPERATORS: Versions of the PA-31-310 are in service with the armed forces of the following countries: Argentina (1), Chile (3) and France (8).

DESIGN FEATURES: Cantilever low-wing monoplane. Wing section NACA 63_2415 at root, NACA 63_1212 at tip. 1° aerodynamic twist. 2° 30′ geometric twist. Variable incidence tailplane.

FLYING CONTROLS: Balanced ailerons are interconnected with rudder. Trim tab in starboard aileron. Electrically operated flaps.Trim tabs in rudder and starboard elevator.

STRUCTURE: All-metal wing structure, with heavy stepped-down main spar, front and rear spars, lateral stringers, ribs and stressed skin. Wings spliced on centreline with heavy steel plates. Flush riveted forward of main spar. Wingroot leading-edge extended between nacelle and fuselage. Glassfibre wingtips. The fuselage is a conventional all-metal semi-monocoque structure. The tail unit is a cantilever all-metal structure, with sweptback vertical surfaces.

LANDING GEAR: Hydraulically actuated retractable tricycle type, with single wheel on each unit. Manual hydraulic emergency extension. Mainwheels and tyres size 6.50-10, 8 ply rating, pressure 4.14 bars (60 lb/sq in). Steerable nosewheel and tyre size 6.00-6, 6 ply rating, pressure 2.90 bars (42 lb/sq in). Toe-controlled hydrauic disc brakes. Mainwheel doors close when gear is fully extended.

POWER PLANT: Two 231 kW (310 hp) Lycoming TIO-540-A2C flat-six turbocharged engines. Hartzell three-blade fully feathering metal propellers. Propeller de-icing optional. Four rubber fuel cells in wings; inboard cells each contain 212 litres (55 US gallons; 46.6 Imp gallons), outboard cells 151.5 litres (40 US gallons; 33.3 Imp gallons) each. Total fuel capacity 727 litres (192 US gallons; 160 Imp gallons), of which 708 litres (187 US gallons; 155.7 Imp gallons) are usable. Fuel cells equipped with NACA-type anti-icing non-siphoning fuel vents. Two-piece glassfibre engine nacelles.

ACCOMMODATION: Six individual seats, with headrests and armrests, in pairs with centre aisle. Seventh and eighth seats optional. Dual controls standard. Thermostatically controlled Janitrol 35,000 BTU combustion heater, windscreen defrosters and fresh air system standard. Double-glazed windows. Electrical de-icing and windscreen wiper for port side of windscreen optional. 'Dutch' door at rear of cabin on port side. Top half hinges upward; lower half hinges down and has built-in steps. Baggage compartments in nose, capacity 68 kg (150 lb), and in rear of cabin, capacity 91 kg (200 lb). Cargo door and cockpit door available as optional items.

SYSTEMS: Hydraulic system utilises two engine-driven pumps. 24 V electrical system supplied by two engine-driven 28 V 70 A alternators and 24 V 17 Ah battery; 25 Ah battery optional. External power socket standard. Oxygen system optional. Wing has pneumatic de-icing boots optional. Tail unit has optional pneumatic de-icing boots.

ELECTRONICS AND EQUIPMENT: Blind-flying instrumentation standard, with optional dual installation for co-pilot. Standard equipment includes fore, aft, and vertically adjustable and tilting seats for pilot and co-pilot, with headrests, folding armrests, shoulder safety belts and inertia reels, and oxygen mask storage beneath each seat. Four adjustable and reclining passenger seats in club arrangement, with headrests and folding armrests, seat belts, oxygen mask storage beneath each seat, magazine storage pockets, 'No smoking - fasten seat belt' sign curtain divider and a choice of interior colour schemes. Optional equipment includes cabin ground ventilation fan. Piper Aire air conditioning system, cabin fire extinguisher, cold-weather heater for rear cabin, propeller synchroniser, aft cabin divider with curtain and shelf, fuselage ice protection shields, beverage dispensers, folding tables, forward cabin divider with curtain and magazine racks. Piper automatic locator, ice inspection light, toilet, utility door, tinted windows, toe-brakes for co-pilot, and pilot's windscreen wiper. Standard electrical equipment includes navigation, landing, taxying, cockpit, cabin dome and passenger reading lights, two rotating beacons, stall warning light, courtesy lights, cabin and cockpit speakers and heated pitot tube.

DIMENSIONS, EXTERNAL:

Wing span	12.40 m (40 ft 8 in)
Length overall	9.94 m (32 ft 7½ in)
Height overall	3.96 m (13 ft 0 in)
Tailplane span	5.52 m (18 ft 1½ in)
Wheel track	4.19 m (13 ft 9 in)
Wheelbase	2.64 m (8 ft 8 in)
Propeller diameter	2.03 m (6 ft 8 in)

DIMENSIONS, INTERNAL:

Cabin: Length	4.88 m (16 ft 0 in)
Height	1.31 m (4 ft 3½ in)
Baggage compartments: Nose	0.40 m³ (14 cu ft)
Aft	0.62 m³ (22 cu ft)

AREAS:

Wings, gross	21.3 m² (229 sq ft)

WEIGHTS AND LOADINGS:

Weight empty	1,810 kg (3,991 lb)
Max T-O and landing weight	2,948 kg (6,500 lb)
Max zero-fuel weight	2,812 kg (6,200 lb)
Max wing loading	138.7 kg/m² (28.4 lb/sq ft)
Max power loading	6.38 kg/kW (10.5 lb/hp)

PERFORMANCE (at max T-O weight):

Max level speed at 4,570 m (15,000 ft)	227 kts (420 km/h; 261 mph)
Max cruising speed (75% power) at 6,705 m (22,000 ft)	215 kts (399 km/h; 248 mph)
Max cruising speed (75% power) at 3,660 m (12,000 ft)	195 kts (362 km/h; 225 mph)
Econ cruising speed at 6,100 m (20,000 ft)	207 kts (383 km/h; 238 mph)
Econ cruising speed at 3,660 m (12,000 ft)	192 kts (356 km/h; 221 mph)
Stalling speed, flaps down	63.5 kts (118 km/h; 73 mph)
Max rate of climb at S/L	440 m (1,445 ft)/min
Rate of climb at S/L, one engine out	75 m (245 ft)/min
Service ceiling	8,015 m (26,300 ft)
Service ceiling, one engine out	4,635 m (15,200 ft)
Normal T-O run	314 m (1,030 ft)
Short-field T-O run	262 m (860 ft)
Normal T-O to 15 m (50 ft)	668 m (2,190 ft)
Short-field T-O to 15 m (50 ft)	518 m (1,700 ft)
Normal landing from 15 m (50 ft) at max landing weight	713 m (2,340 ft)
Short-field landing from 15 m (50 ft)	552 m (1,810 ft)
Normal landing run	584 m (1,915 ft)
Short-field landing run	376 m (1,235 ft)
Accelerate/stop distance	741 m (2,430 ft)

Range with max fuel, 45 min reserves:
at max cruising speed at 6,705 m (22,000 ft) 875 n miles (1,620 km; 1,007 miles)
at max cruising speed at 3,660 m (12,000 ft) 830 n miles (1,537 km; 955 miles)
at econ cruising speed at 6,100 m (20,000 ft) 1,065 n miles (1,973 km; 1,226 miles)
at econ cruising speed at 3,660 m (12,000 ft) 995 n miles (1,601 km; 1,145 miles)

UPDATED

PIPER PA-32-301 SARATOGA

TYPE: Six-seat cabin monoplane.

PROGRAMME: In December 1979, Piper announced that it had begun production of a new family of six-seat single-engined aircraft known at Saratoga, to replace the PA-32 SIX 300 and T tail Lance series. Saratogas have a common airframe, with a conventional low-mounted tailplane and a semi-tapered wing of longer span than the wing of the aircraft they supersede.

VERSIONS: **PA-32-301 Saratoga:** Basic version, as described in detail.

PA-32-301 Saratoga SP: Retractable landing gear version of the Saratoga.

PA-32-301 Saratoga II HP: Introduced in June 1992. Redesigned cowling with axisymmetric inlets. Speed modifications include blister wheel fairings and hinge fairings.

A total of 832 Saratogas had been delivered by December 1993.

DESIGN FEATURES: Cantilever low-wing monoplane. Light alloy single-spar structure with glassfibre wingtips.

FLYING CONTROLS: Plain ailerons of light alloy construction. Electrically operated trailing-edge flaps of light alloy construction with ribbed skins. The tail unit has a one-piece all-moving horizontal surface with combined anti-servo and trim tab. Trimmable rudder.

STRUCTURE: The fuselage is a conventional semi-monocoque structure of light alloy. Glassfibre engine cowling. The tail unit is a cantilever structure of light alloy, except for glassfibre tips on fin and tailplane. Fin and rudder have ribbed metal skins.

LANDING GEAR: Non-retractable tricycle-type. Steerable nosewheel. Piper oleo-pneumatic shock-absorbers. Single wheel on each unit. Mainwheel tyres size 6.00-6, 8 ply rating, pressure 3.79 bars (55 lb/sq in). Nosewheel tyre size 5.00-5, 6 ply rating, pressure 2.41 bars (35 lb/sq in). Nosewheel tyre size 6.00-6 optional. High-capacity disc brakes. Parking brake. Wheel fairings optional. Heavy-duty brakes and tyres optional.

POWER PLANT: One 224 kW (300 hp) Textron Lycoming IO-540-K1G5 flat-six engine, driving a Hartzell two-blade constant-speed metal propeller with spinner. Three-blade propeller optional. Polished spinner optional (three-blade propeller only). Two fuel tanks in each wing with combined capacity of 405 litres (107 US gallons; 89 Imp gallons), of which 386 litres (102 US gallons; 85 Imp gallons) are usable. Refuelling points on wing upper surface. Oil capacity 11.5 litres (3 US gallons; 2.5 Imp gallons).

ACCOMMODATION: Enclosed cabin, seating six people in pairs. Dual controls and toe brakes standard. Two forward-hinged doors, one on starboard side forward, overwing, and one on port side at rear end of cabin. Space for 45 kg (100 lb) baggage at rear of cabin, with external baggage/utility door on port side. Additional baggage space, capacity 45 kg (100 lb), between engine fireproof bulkhead and instrument panel, with external door on starboard side. Pilot's storm window. Sun visors. Accommodation heated and ventilated. Windscreen defroster standard.

SYSTEMS: Piper Aire air conditioning, vacuum and oxygen systems optional, including a built-in oxygen system of 1.81 m³ (64 cu ft) capacity. Hydraulic system for brakes only. Electrical system includes a 14 V 60 A engine-driven alternator and 28 V 35 Ah battery. Vacuum system standard.

AVIONICS AND EQUIPMENT: Single Bendix/King KX 155 nav/com, KAP 100 autopilot, Narco AR-850 altitude reporter/digitiser, and transponder standard. Three optional factory installed avionics groups including various single and dual combinations of nav/com, ADF, DME, RNAV, transponder, glide slope and marker beacon receivers, and electric pitch trim, with substitute equipment options in each group. Additional avionics options for all groups Bendix/King KAP 150 autopilot, KFC 150 flight control system, Bendix/King KLN 88, II Morrow 612 and Northstar M1 Loran C Systems, Argus 5000 moving map display, 3M Stormscope WX-1000 and 1000+, Bendix/King KMA 24H-70 audio control panel with intercom, KHF 950 HF, KRA 10A radio altimeter, KI 229 RMI, KI 204-02 or KI 203-00 nav indicator, ground clearance energy saving switch, Wulfsberg Flitefone VI, press-to-talk buttons on pilot's control wheel, pilot's and co-pilot's boom microphones with control wheel buttons, and static discharge wicks. Standard equipment includes Piper true airspeed indicator; magnetic compass; sensitive altimeter; ammeter, annunciator panel with press-to-test; recording tachometer; fuel quantity gauges; fuel pressure gauge; manifold/fuel flow gauge, oil temperature and pressure gauges; engine-driven vacuum system with indicator and annunciator light; electric standby vacuum pump; instrument panel white backlighting and overhead blue lighting, baggage door ajar warning light, four cabin reading lights, two map lights, navigation lights, landing/taxying light, wingtip strobe lights and avionics dimming; advanced instrument panel with 3 in gyro horizon, turn rate indicator, rate of climb indicator, OAT gauge and electric clock; heated pitot head; EGT gauge; 35 Ah heavy-duty battery; emergency locator beacon; entrance step; wheel fairings; wool carpet and Sofsuede sidewalls and headliner; fabric and vinyl pilot's and co-pilot's fore and aft adjustable reclining seats with inertia reel shoulder harnesses, seatbelts, armrests and ashtrays; four rear seats with inertia reel shoulder harnesses, seatbelts and side panel ashtrays; pilot's storm window; map pockets; pilot's and co-pilot's scuff pads and sun visors; conference interior with club seating; headrests for all seats; window curtains; assist strap; tinted windscreen and side windows; vertically adjustable crew seats; ventilation fan; alternate static source; external power receptable; crew cup holders; and courtesey lighting package; door closing straps; soundproofing; ignition lock; carpeted luggage compartment with security straps; compass card with holder; stowable towbar; tiedown rings; jack pads; cabin and rear baggage door locks; zinc chromate corrosion protection; stainless steel control cables; and exterior styling in choice of three styles with single- or two-tone base colour in Piper acrylic enamel. Optional Deluxe avionics and equipment package comprises dual Bendix/King KX 165 nav/com; KAP 150 autopilot with electric pitch trim VOR/LOC/GS coupling; KCS 55A HSI/slaved compass system. Lounge interior with refreshment console and fold-down armrest between fifth and sixth seats; improved cabin soundproofing; cigarette lighter; three-blade propeller; polished spinner; engine hour recorder; wingtip recognition lights; heavy-duty brakes and tyres on main gear, and heavy duty tyre on nosewheel. Optional equipment in addition to the Deluxe package includes engine hour recorder, digital clock in pilot's control wheel, approach plate holder (not with digital clock), lockable fuel caps, strobe light on fin. Piper Aire air conditioning system. Deluxe lounge interior, executive writing table (with club seating only), fire extinguisher, Du Pont IMRON polyurethane enamel exterior finish, third trim colour, and shadowed registration numbers.

DIMENSIONS, EXTERNAL:

Wing span	11.02 m (36 ft 2 in)
Length overall	8.44 m (27 ft 8½ in)
Height overall	2.49 m (8 ft 2 in)
Tailplane span	3.94 m (12 ft 11 in)
Wheel track	3.23 m (10 ft 7 in)
Wheelbase	2.36 m (7 ft 9 in)
Cabin door (fwd, stbd): Height	0.89 m (2 ft 11 in)
Width	0.91 m (3 ft 0 in)
Cabin door (rear, port): Height	0.72 m (2 ft 4½ in)
Width	0.71 m (2 ft 4 in)
Baggage door (fwd): Height	0.41 m (1 ft 4 in)
Width	0.56 m (1 ft 10 in)
Baggage/utility door (rear): Height	0.52 m (1 ft 8½ in)
Width	0.66 m (2 ft 2 in)

DIMENSIONS, INTERNAL:

Cabin: Length (instrument panel to rear bulkhead)	3.15 m (10 ft 4¼ in)
Max width	1.22 m (4 ft 0¼ in)
Max height	1.07 m (3 ft 6 in)
Volume (incl rear baggage area)	5.53 m³ (195.3 cu ft)
Baggage compartment volume: forward	0.23 m³ (8.0 cu ft)
rear	0.49 m³ (17.3 cu ft)

AREAS:

Wings, gross	16.56 m² (178.3 sq ft)

WEIGHTS AND LOADINGS:

Weight, empty, equipped	953 kg (2,101 lb)
Max T-O weight	1,633 kg (3,600 lb)
Max ramp weight	1,640 kg (3,615 lb)
Max wing loading	98.6 kg/m² (20.2 lb/sq ft)
Max power loading	7.30 kg/kW (12.0 lb/hp)

PERFORMANCE (at max T-O weight):

Max level speed at optimum altitude	152 kts (282 km/h; 175 mph)
Best econ cruising speed at optimum altitude: at 75% power	150 kts (278 km/h; 173 mph)
at 65% power	146 kts (270 km/h; 168 mph)
at 55% power	136 kts (252 km/h; 156 mph)
Best econ cruising speed at optimum altitude: at 75% power	148 kts (274 km/h; 170 mph)
at 65% power	144 kts (267 km/h; 166 mph)
at 55% power	133 kts (246 km/h; 153 mph)
Stalling speed: flaps up	66 kts (122 km/h; 76 mph) CAS
flaps down	60 kts (111 km/h; 69 mph) CAS
Max rate of climb at S/L	302 m (990 ft)/min
Service ceiling	4,875 m (16,000 ft)
T-O run: two-blade propeller	361 m (1,183 ft)
three-blade propeller	309 m (1,013 ft)
T-O run to 15 m (50 ft): two-blade propeller	536 m (1,759 ft)
three-blade propeller	479 m (1,573 ft)
Landing from 15 m (50 ft): standard brakes	491 m (1,612 ft)
heavy-duty brakes	466 m (1,530 ft)
Landing run: standard brakes	223 m (732 ft)
heavy duty brakes	198 m (650 ft)

Range with max fuel, allowances for taxi, T-O, climb, descent, and 45 min reserves at max range power:

best power settings at optimum altitude: 75% power	745 n miles (1,381 km; 858 miles)
65% power	805 n miles (1,492 km; 927 miles)
55% power	849 n miles (1,573 km; 978 miles)
best econ power settings at optimum altitude: 75% power	823 n miles (1,525 km; 948 miles)
65% power	911 n miles (1,688 km; 1,049 miles)
55% power	960 n miles (1,778 km; 1,105 miles)

VERIFIED

Piper Saratoga six-seat cabin monoplane

1992

PIPER PA-34-220T SENECA III

TYPE: Six-seat twin-engined light aircraft.

PROGRAMME: On 23 September 1971, Piper announced a twin-engined light aircraft which had the company designation PA-34 and the name Seneca. Built at Piper's Vero Beach factory in Florida, the aircraft was redesignated Seneca II from 1975. On 15 February 1981 Piper introduced the improved PA-34-220T Seneca II, with more powerful engines.

The Seneca III has counter-rotating (C/R) engine and propeller installations. The retractable landing gear is operated by an electrohydraulic system and includes an emergency extension system which allows the wheels to free-fall into the down and locked position. A dual-vane stall warning system provides warning by horn well in advance of the stall in either clean or gear/flaps-down configuration.

It was announced 3 January 1977 that Piper had signed an agreement with Pezetel, the Polish foreign trade organisation, enabling PZL Mielec to assemble, manufacture and distribute the Seneca in Eastern Europe. These aircraft (several hundred are involved in the agreement) are powered by 164 kW (220 hp) PZL-F engines and are named **M-20 Mewa** (Gull). In June 1987 Pezetel awarded Western world marketing rights for the M-20 to the British company Aircraft International.

Well over 4,517 Senecas have been delivered. The newest model, the Seneca IV, was introduced in November 1993.

DESIGN FEATURES: Cantilever low-wing monoplane. Dihedral 7° with single-spar wings.

FLYING CONTROLS: Frise ailerons, and wide span electrically operated slotted flaps, of light alloy construction. The tail unit has a one-piece all-moving horizontal surface with combined anti-balance and trim tab. Anti-servo tab in rudder.

STRUCTURE: Glassfibre wingtips. The fuselage is a light-alloy semi-monocoque structure. The tail unit is a cantilever structure of light alloy.

LANDING GEAR: Hydraulically retractable tricycle type. Main units retract inward, nose unit forward. Oleo-pneumatic shock-absorbers. Steerable nosewheel. Emergency free-fall extension system. Mainwheels and tyres size 6.00-6, 8 ply rating, pressure 3.79 bars (55 lb/sq in); nosewheel and tyre size 6.00-6, 6 ply rating, pressure 2.76 bars (40 lb/sq in). Nosewheel safety mirror. High-capacity disc brakes. Parking brake. Heavy duty tyres and brakes optional.

POWER PLANT: One 164 kW (220 hp) Teledyne Continental TSIO-360-KB and one 164 kW (220 hp) LTSIO-360-KB flat-six turbocharged counter-rotating engine, each driving a Hartzell two-blade constant-speed fully feathering metal propeller with spinner. Three-blade propellers, propeller de-icing, automatic unfeathering and propeller synchrophasers optional. Fuel in four tanks in wings, with a total capacity of 485 litres (128 US gallons; 106.6 Imp gallons), of which 466 litres (123 US gallons; 102.4 Imp gallons) are usable. Oil capacity 7.5 litres (2 US gallons; 1.7 Imp gallons). Glassfibre engine cowlings.

ACCOMMODATION: Enclosed cabin, seating six people in pairs on individual seats with 0.25 m (10 in) centre aisle. Dual controls standard. Pilot's storm window. Two forward-hinged doors, one on starboard side at front, the other on port side at rear. Large utility door adjacent rear cabin door provides an extra-wide opening for loading bulky items. Passenger seats easily removable to provide different seating/baggage/cargo combinations. Space for 45 kg (100 lb) baggage at rear of cabin, and for 45 kg (100 lb) in nose compartment with external door on port side, cabin heated and ventilated. Windscreen defrosters standard. Electrically de-iced windscreen for pilot, and ice inspection light optional.

SYSTEMS: Electrohydraulic system for landing gear actuation. Electrical system powered by two 28 V 60 A alternators. 24 V 35 Ah battery. Oxygen system with six outlets, or built-in oxygen system of 1.81 m³ (64 cu ft), optional. Dual engine-driven pneumatic pumps for flight instruments standard. Piper Aire air conditioning system of 14,500 BTU capacity optional. Janitrol 45,000 BTU combustion heater standard. Pneumatic de-icing boots for wing leading-edges optional. Pneumatic de-icing boots for fin and tailplane leading-edges optional.

AVIONICS AND EQUIPMENT: Standard Bendix/King avionics package comprises dual KX 155 nav/com; KAP 150 autopilot; transponder; and Narco AR-850 altitude reporter/digitiser. Choice of three optional factory installed Bendix/King avionics groups including various single and dual combinations of nav/com, ADF, DME, RNAV, transponder, glide slope and marker beacon receivers, and electric pitch trim, with substitute equipment options in each group. Additional avionics options for all groups include Bendix/King EHI-40, King KAP 150 autopilot, KFC 150 flight control system, Bendix/King KLN 88 Foster LNS 616B, II Morrow 612 and Northstar M1 Loran C systems, Argus 5000 map display, 3M Stormscope WX-1000 and 1000+, Bendix/King RDS-81 or Narco KWX-56 colour weather radar with nose radome; Bendix/King KMA 24H-70 audio control panel with intercom, KHF 950 HF, KRA 10A radio altimeter, KI 229 RMI, KI 204-02 or KI 203-00 nav indicator, ground clearance energy saving switch, Wulfsberg Flitefone VI, press-to-talk buttons on pilot's control wheel, pilot's and co-pilot's boom microphones with control wheel buttons, and static discharge wicks. Standard equipment includes advanced instrument panel with gyro horizon, directional gyro, turn rate indicator, rate of climb indicator, OAT gauge and electric clock; Piper true airspeed indicator; illuminated magnetic compass; sensitive altimeter; dual reading ammeter; annunciator panel with press-to-test; dual electric tachometers; dual fuel quantity gauges; dual fuel flow gauges; dual manifold pressure gauges; dual oil temperature and pressure gauges; dual CHT gauges; dual EGT gauges; pilot's and co-pilot's toe brakes; basic lighting package comprising instrument panel white backlighting and overhead blue lighting, four cabin reading lights. Two map lights, navigation lights, landing lights, and avionics dimming; wingtip strobe lights; heated pitot head; dual engine-driven pneumatic pumps with indicator and annunciator light; emergency locator beacon; wool carpet, Sofsuede sidewalls and white vinyl headliner; fabric and vinyl pilot's and co-pilot's fore and aft adjustable reclining seats with inertia reel shoulder harnesses, seatbelts, armrests and ashtrays; fabric and vinyl passenger seats with quick release feature; inertia reel shoulder harnesses on forward-facing seatbelts and side panel ashtrays; four rear seats with inertia reel shoulder harnesses, seatbelts and side panel ashtrays; pilot's storm window; map pockets; cigarette lighter; glove compartment with rollaway door; pilot's and co-pilot's scuff pads and sun visors; door closing straps; soundproofing; Janitrol 45,000 BTU combustion heater with six floor air/heat outlets, regulating levers and cabin air exhaust vents; six cabin fresh air vents; windscreen defrosters; luggage compartments with security straps; compass card with holder; power setting table and T-O/landing checklists on sun visor; tiedown rings; jack pads and forward baggage door locks; zinc chromate corrosion protection; stainless steel control cables; exterior styling in choice of three styles with single- or two-tone base colour in Piper acrylic enamel; conference interior with club seating, headrests for all seats, window curtains, tinted windscreen and side windows; vertically adjustable crew seats; ventilation fan; external power receptacle; crew cup holders; courtesy lighting package; nose baggage door ajar warning light; and stowage towbar. Optional Deluxe avionics and equipment package comprises dual Bendix/King KX 165 nav/com; KAP 150 autopilot with electric pitch trim, VOR/LOC/GS coupling; KCS 55A HSI/slaved compass system. Lounge interior with refreshment console and fold-down armrest between fifth and sixth seats; improved cabin soundproofing; three-blade propellers and synchrophasers; polished spinners; wingtip recognition lights; heavy-duty brakes and tyres on main gear, and heavy duty tyre on nosewheel. Optional

Piper Seneca III six-seat twin-engine aircraft

1992

equipment to the Deluxe package includes cold weather start kit, heater hour recorder, digital clock in pilot's control wheel, approach plate holder (not with digital clock), co-pilot's instrument panel (deletes glove compartment and not available with some avionics groups or weather radar), de-icing group comprising wing and tail leading-edge pneumatic boots, electric propeller de-icing, pilot's windscreen de-icing plate, ice inspection light and luminous OAT gauge, lockable fuel caps, wingtip recognition lights, Piper Aire air conditioning system, executive writing table (with club seating only), six-outlet portable oxygen system (club seating only) mounted between third and fourth seats, built-in oxygen system, capacity 1.81 m³ (64 cu ft) with remote filler, ventilation fan, fire extinguisher, Du Pont IMRON polyurethane enamel exterior finish, third trim colour, and shadowed registration numbers.

DIMENSIONS, EXTERNAL:

Wing span	11.85 m (38 ft 10¾ in)
Wing chord constant	1.60 m (5 ft 3 in)
Length overall	8.72 m (28 ft 7½ in)
Height overall	3.02 m (9 ft 10¾ in)
Tailplane span	4.14 m (13 ft 6¾ in)
Wheel track	3.38 m (11 ft 1 in)
Wheelbase	2.13 m (7 ft 0 in)
Propeller diameter	1.93 m (6 ft 4 in)
Distance between propeller centres	3.80 m (12 ft 5½ in)
Cabin door (stbd, fwd): Height	0.89 m (2 ft 11 in)
Width	0.91 m (3 ft 0 in)
Baggage door (stbd, rear): Height	0.52 m (1 ft 8½ in)
Width	0.66 m (2 ft 2 in)
Baggage door (port, fwd): Height	0.53 m (1 ft 9 in)
Width	0.61 m (2 ft 0 in)

DIMENSIONS, INTERNAL:

Cabin (incl flight deck): Length	3.15 m (10 ft 4½ in)
Max width	1.24 m (4 ft 0¾ in)
Max height	1.07 m (3 ft 6 in)
Volume	5.53 m³ (195.3 cu ft)
Forward baggage compartment	0.43 m³ (15.3 cu ft)
Rear baggage compartment	0.49 m³ (17.3 cu ft)

AREAS:

Wings, gross	19.39 m² (208.7 sq ft)
Ailerons, incl tab (total)	1.17 m² (12.60 sq ft)
Trailing-edge flaps (total)	1.94 m² (20.84 sq ft)
Fin	1.14 m² (12.32 sq ft)
Rudder, incl tab	0.71 m² (7.62 sq ft)
Horizontal tail surfaces (total)	3.60 m² (38.74 sq ft)

WEIGHTS AND LOADINGS:

Weight empty, standard	1,362 kg (3,004 lb)
Max usable fuel weight	385 kg (738 lb)
Max T-O weight	2,154 kg (4,750 lb)
Max ramp weight	2,165 kg (4,773 lb)
Max zero-fuel weight	2,027 kg (4,470 lb)
Max landing weight	2,047 kg (4,513 lb)
Max wing loading	111.1 kg/m² (22.76 lb/sq ft)
Max power loading	6.57 kg/kW (10.8 lb/hp)

PERFORMANCE (at max T-O weight except where indicated):

Max level speed at optimum altitude, mid-cruise weight	196 kts (363 km/h; 226 mph)
Cruising speed at optimum altitude, mid-cruise weight:	
75% power	193 kts (357 km/h; 222 mph)
65% power	191 kts (354 km/h; 220 mph)
45% power	168 kts (311 km/h; 193 mph)
Cruising speed at 3,050 m (10,000 ft), mid-cruise weight:	
75% power	179 kts (332 km/h; 206 mph)
65% power	175 kts (324 km/h; 202 mph)
45% power	143 kts (265 km/h; 165 mph)
Min single-engine control speed (VMC)	65 kts (120 km/h; 75 mph)
Stalling speed: flaps and landing gear up	66 kts (122 km/h; 76 mph) CAS
flaps and landing gear down	62 kts (115 km/h; 72 mph) CAS
Max rate of climb at S/L	427 m (1,400 ft)/min
Rate of climb at S/L, one engine out	73 m (240 ft)/min
Max certificated ceiling	7,620 m (25,000 ft)
Service ceiling, one engine out	3,750 m (12,300 ft)
T-O run	280 m (920 ft)
T-O run to 15 m (50 ft)	369 m (1,210 ft)
Landing run from 15 m (50 ft):	
standard brakes	658 m (2,160 ft)
heavy-duty brakes	603 m (1,978 ft)
Landing run: standard brakes	427 m (1,400 ft)
heavy-duty brakes	371 m (1,218 ft)
Accelerator/stop distance:	
standard brakes	732 m (2,400 ft)
heavy-duty brakes	636 m (2,088 ft)

Range with standard fuel, allowances for taxi, T-O, climb, descent, and 45 min reserves at max range power:

at optimum altitude:	
75% power	463 n miles (858 km; 533 miles)
65% power	550 n miles (1,018 km; 633 miles)
45% power	670 n miles (1,240 km; 771 miles)
at 3,050 m (10,000 ft):	
75% power	450 n miles (833 km; 517 miles)
65% power	535 n miles (990 km; 615 miles)
45% power	632 n miles (1,170 km; 727 miles)

Range with max optional fuel, allowances as above:

at optimum altitude:	
75% power	665 n miles (1,232 km; 765 miles)
65% power	785 n miles (1,454 km; 904 miles)
45% power	990 n miles (1,834 km; 1,140 miles)
at 3,050 m (10,000 ft):	
75% power	640 n miles (1,186 km; 737 miles)
65% power	758 n miles (1,405 km; 873 miles)
45% power	903 n miles (1,673 km; 1,040 miles)

VERIFIED

PIPER PA-44-180 SEMINOLE

TYPE: Twin-engined lightweight four-seat cabin monoplane.

PROGRAMME: The prototype Piper PA-144 Seminole lightweight twin-engined four-seat cabin monoplane first flew in May 1976, and the production version was announced 21 February 1978. Two versions were produced: the basic Seminole, powered by two counter-rotating 134 kW (180 hp) Textron Lycoming O/LO-360-E1A6D engines, and the PA-44-180T Turbo Seminole with TO/LTO-360-E1A6D engines. Production of both versions was terminated in 1982 after a total of 361 Seminoles and 87 Turbo Seminoles had been built. The normally aspirated Seminole was restored to production during 1988, and a further 29 were built by January 1991. The Seminole production was halted until 1994 when it was re-introduced as a training model.

DESIGN FEATURES: Cantilever low-wing monoplane.

FLYING CONTROLS: Plain ailerons and manually operated four-position trailing-edge flaps of light alloy construction. All-moving tailplane with full-span tab. Rudder anti-servo tab.

STRUCTURE: Conventional structure of light alloy. The fuselage is a conventional semi-monocoque structure, primarily of light alloy. The tail unit is a cantilever T tail of light alloy construction.

LANDING GEAR: Hydraulically retractable tricycle type. Free-fall emergency extension system. Piper oleo-pneumatic shock-absorbers. Mainwheels and tyres size 6.00-6, 8 ply, with tubes. Steerable nosewheel with tyre size 5.00-5, 6 ply, with tube. Dual toe-operated high-capacity disc brakes. Heavy-duty brakes and tyres optional.

POWER PLANT: Two 134 kW (180 hp) Textron Lycoming flat-four counter-rotating engines (one O-360-A1H6 and one LO-360-A1H6), each driving a Hartzell two-blade constant-speed fully feathering metal propeller with spinner. One bladder-type fuel tank in each engine nacelle, with total capacity of 416 litres (110 US gallons; 91.5 Imp

Piper Seminole twin-engine four-seat light aircraft

1992

gallons), of which 409 litres (108 US gallons; 90 Imp gallons) are usable. Refuelling point on upper surface of each nacelle. Oil capacity 11.5 litres (3 US gallons; 2.5 Imp gallons).

ACCOMMODATION: Cabin seats four in two pairs of individual seats. Dual controls standard. Emergency exit on port side. Pilot's storm window. Baggage compartment at rear of cabin, capacity 91 kg (200 lb). Accommodation heated and ventilated. Windscreen defrosters.

SYSTEMS: Electrohydraulic system for landing gear actuation and brakes. Electrical system includes two engine-driven 14 V 70 A alternators and 12 V 35 Ah battery. Janitrol combustion heater of 45,000 BTU capacity. Dual vacuum systems standard.

AVIONICS AND EQUIPMENT: Standard Bendix/King avionics fit comprising KY 197 com transceiver, KX 155 nav/com, KNS 80 integrated nav/RNAV/DME/ILS, KI 206 VOR/LOCS/GS indicator, KI 208 VOR/LOC indicator, KR 87 ADF with KA 44B loop/sense antenna, KT 76A transponder, KR 21 marker beacon receiver and lights, AR-850 encoder, KFC 150 autopilot/flight director, KCS 55A slaved compass, KI 525 HSI, KI 256 attitude indicator, KMA 24H audio selector amplifier with integral intercom, avionics master switch, cabin speaker, pilot's and co-pilot's control wheel microphone buttons and headsets with boom microphones, and microphone/headphone jack plugs for pilot, co-pilot and observer. Standard equipment includes advanced black finish metal instrument panel with post lights, dual vacuum systems with indicator; Piper true airspeed indicator; illuminated magnetic compass; sensitive altimeter; dual ammeters; annunciator panel with press-to-test; dual tachometers; engine hour meter; dual fuel quantity gauges; dual fuel pressure gauges; dual manifold pressure gauges; dual oil temperature gauges; dual CHT gauges; dual EGT gauges; heated pitot head; alternate instrument static source; instrument panel lights and overhead blue lighting; cabin dome light; navigation lights; two landing/taxying lights; wingtip strobe lights; avionics dimming; Janitrol 45,000 BTU heater with four floor air/heat outlets; heater hour recorder; four overhead cabin fresh air vents with vent fan; wool carpet and vinyl sidewalls and headliner in choice of four colours; fabric and vinyl pilot's and co-pilot's fore and aft and vertically adjustable reclining seats with inertia reel shoulder harnesses, seatbelts, armrests, armrest ashtrays, magazine pockets and map pockets; quick-release fabric and vinyl rear passenger seats with inertia reel shoulder harnesses, seatbelts and side panel ashtrays; pilot's storm window; pilot's and co-pilot's scuff pads and sun visors with power setting and T-O/landing checklists; compass card with holder; tinted windscreen and side windows; emergency exit window; assist strap; door closing strap; soundproofing; luggage compartment with security straps and hatshelf area; external power receptacle; static discharge wicks; tie-down rings; jack pads; nose gear safety mirror; stowable towbar; entrance step; cabin and baggage door locks; emergency locator beacon; full corrosion protection; stainless steel control cables; and exterior styling with single- or two-tone base colour (choice of seven) in Piper acrylic enamel with two trim colours (choice of 16) in two designs. **Training Equipment Group** comprises two propeller unfeathering accumulators, heavy-duty brakes and tyres, flight time meter, recognition lights. Quietised cabin soundproofing, two control approach plate holders, work table for third crew member, and hand-held fire extinguisher.

DIMENSIONS, EXTERNAL:

Wing span	11.7 m (38 ft 7¼ in)
Length overall	8.41 m (27 ft 7¼ in)
Height overall	2.59 m (8 ft 6 in)
Wheel track	3.20 m (10 ft 6 in)
Wheelbase	2.56 m (8 ft 4¾ in)
Propeller diameter	1.88 m (6 ft 2 in)
Cabin door (stbd): Height	0.89 m (2 ft 11 in)
Width	0.91 m (3 ft 0 in)
Baggage door: Height	0.51 m (1 ft 8 in)
Width	0.56 m (1 ft 10 in)

DIMENSIONS, INTERNAL:

Cabin, instrument panel to rear bulkhead:	
Length	2.46 m (8 ft 1 in)
Max width	1.05 m (3 ft 5½ in)
Max height	1.25 m (4 ft 1 in)
Volume	3.00 m³ (106.0 cu ft)
Baggage compartment volume	0.74 m³ (26.0 cu ft)

AREAS:

Wings, gross	17.08 m² (183.8 sq ft)

WEIGHTS AND LOADINGS:

Weight empty	1,068 kg (2,354 lb)
Max T-O, landing and zero-fuel weight	1,723 kg (3,800 lb)
Max ramp weight	1,731 kg (3,815 lb)
Max wing loading	100.9 kg/m² (20.67 lb/sq ft)
Max power loading	6.43 kg/kW (10.55 lb/hp)

PERFORMANCE (at mid-cruise weight):

Max level speed	168 kts (311 km/h; 193 mph)
Best power cruising speed at optimum altitude:	
75% power	167 kts (309 km/h; 192 mph)
65% power	164 kts (304 km/h; 189 mph)
55% power	156 kts (289 km/h; 180 mph)
Econ cruising speed at optimum altitude:	
75% power	162 kts (300 km/h; 187 mph)
65% power	159 kts (295 km/h; 183 mph)
55% power	151 kts (280 km/h; 174 mph)
Stalling speed:	
flaps and wheels up	57 kts (106 km/h; 66 mph) IAS
flaps and wheels down	55 kts (102 km/h; 64 mph) IAS
Max rate of climb at S/L	408 m (1,340 ft)/min
Rate of climb at S/L, one engine out	66 m (217 ft)/min
Service ceiling	5,210 m (17,100 ft)
Service ceiling, one engine out	1,250 m (4,100 ft)
T-O run	268 m (880 ft)
T-O to 15 m (50 ft)	427 m (1,400 ft)
Landing from 15 m (50 ft):	
standard brakes	427 m (1,400 ft)
heavy-duty brakes	363 m (1,190 ft)
Landing run: standard brakes	180 m (590 ft)
heavy-duty brakes	117 m (383 ft)
Range at optimum altitude with max fuel, allowances for start, T-O, climb and descent, plus 45 min reserves:	
at best power settings:	
75% power	690 n miles (1,278 km; 794 miles)
65% power	725 n miles (1,343 km; 834 miles)
55% power	765 n miles (1,417 km; 881 miles)
at best econ power settings:	
75% power	750 n miles (1,445 km; 898 miles)
65% power	850 n miles (1,575 km; 979 miles)
55% power	915 n miles (1,695 km; 1,053 miles)

VERIFIED

PIPER PA-60 AEROSTAR 600A

TYPE: Twin-engined light transport aircraft.

PROGRAMME: Design work on this series was started by the Ted Smith Aircraft Company in November 1964 and the first Model 600/601 prototype made its first flight in October 1967. FAA Type Approval of the Model 600 was awarded in March 1968, and of the Model 601 in November 1968. Piper suspended production of the Aerostar 601P for 1982.

VERSIONS: **Model 600A:** Powered by two 216 kW (290 hp) Avco Lycoming IO-540-K1J5 flat-six engines. European models designated 600AE.

Model 601P: As Model 600A, but with increased wing span and two 216 kW (290 hp) Avco Lycoming IO-540-S1A5 flat-six engines, fitted with high flow-rate turbochargers to supply bleed air for cabin pressurisation. European models designated 601PE.

Model 602P: New model introduced in 1981. Described separately.

UPGRADES: **AAC:** Aerostar Super 700. See separate entry.

AAC: Aerostar 3000. See separate entry.

DESIGN FEATURES: Cantilever mid-wing monoplane. Wing section NACA 61_1A212. Dihedral 2°. Incidence 1°. No sweepback. Cantilever tail unit with swept vertical and horizontal surfaces. Both fixed and control surfaces are interchangeable.

FLYING CONTROLS: Ailerons and hydraulically actuated Fowler-type trailing-edge flaps each comprise a spar, ribs, nose skin and one-piece wraparound light alloy skin aft of spar. No trim tabs. Electrically operated trim tab in rudder and each elevator.

STRUCTURE: The wing is an all-metal structure using heavy gauge skins attached to three spars, several bulkheads and stringers. Entire wing assembly, excluding attachments for ailerons and flaps, contains fewer than 50 detail parts. The fuselage is an all-metal fail-safe monocoque structure. Skin composed of large segments of light alloy sheet over stringers and frames. Entire fuselage contains fewer than 100 parts, including skin panels. All fuselage assemblies designed basically for pressurisation.

LANDING GEAR: Hydraulically retractable tricycle type. Main units retract inward, nosewheel forward. Hydraulically powered nosewheel steering. Hydraulically operated heavy-duty dual caliper brakes. Parking brake.

POWER PLANT: Two Avco Lycoming engines, as detailed in model listings, each driving a Hartzell three-blade constant-speed and fully feathering metal propeller with spinner. Fuel in integral wing tanks and fuselage tank with total capacity of 657 litres (173.5 US gallons; 144.6 Imp gallons), of which 626 litres (165.5 US gallons; 138 Imp gallons) are usable. Oil capacity 22.7 litres (6 US gallons; 5 Imp gallons). Propeller synchrophasers and propeller de-icing optional. Engine fire detection system optional.

ACCOMMODATION: Cabin seats six people on track-mounted individual reclining seats, in pairs. Dual controls standard. Door on port side by pilot's seat; top half hinges upward, bottom half downward. Emergency escape window at rear of cabin. Tinted windscreen and dual-pane tinted cabin windows. Large utility shelf in rear of cabin. Baggage compartment, capacity 109 kg (240 lb), aft of cabin, with external access on port side. Individual air vents and reading lights for each seat. Cabin heated/ventilated. Control locks. Windscreen defrosting; alcohol de-icing system, or electrically heated windscreen plate, optional.

SYSTEMS: Heating/ventilation system includes a Janitrol 35,000 BTU heater. Hydraulic system for landing gear

Piper Aerostar 600A twin-engine light transport aircraft

1994

actuation, wheel brakes and flaps, powered by an engine-driven pump, pressure 69 bars (1,000 lb/sq in); auxiliary hydraulic system with electrically driven pump optional. Dual pneumatic systems for instrument gyros and optional de-icing boots. Electrical system powered by two 28 V 70 A engine-driven alternators with failure warning lights. Two 12 V 24 Ah batteries. External power socket. De-icing for leading-edges optional.

AVIONICS AND EQUIPMENT: A wide range of optional avionics is available for both versions of the Aerostar, including nav/coms, ADF, autopilots, DME, encoding altimeter, glide slope and marker beacon receivers, HF transceiver, R/Nav, radio altimeter, transponder and weather radar, by manufacturers which include Aeronetics, Aerosonics, Bendix, Bonzer, Collins, King, Piper and SunAir. Standard equipment includes blind-flying instrumentation with 3 in attitude and directional gyros, clock, flight hour meter, dual manifold pressure gauges, outside air temperature gauge, sensitive altimeter, turn co-ordinator, and vertical speed indicator; control locks; inertia reel shoulder harness for pilot and co-pilot seats; storm windows; sun visors; alternator failure, heater failure, and low fuel warning lights; emergency locator transmitter; heated pitot, cabin air ventilators, dual-pane tinted windows, carpeted floor, window curtains, super soundproofing map, instrument panel red and white, reading, landing, navigation, taxi, and strobe lights; full-flow oil filters; jack pads and towbar. An extensive range of optional equipment is also available.

DIMENSIONS, EXTERNAL:

Wing span	10.41 m (34 ft 2 in)
Wing chord: at root	2.18 m (7 ft 2 in)
at tip	0.87 m (2 ft 10⅜ in)
Wing aspect ratio	6.83
Length overall	10.61 m (34 ft 9¾ in)
Height overall	3.70 m (12 ft 1½ in)
Tailplane span	4.37 m (14 ft 4 in)
Wheel track	3.11 m (10 ft 2½ in)
Propeller diameter	1.98 m (6 ft 6 in)
Passenger door: Height	1.14 m (3 ft 9 in)
Width	0.71 m (2 ft 4 in)
Baggage compartment door: Height	0.61 m (2 ft 0 in)
Width	0.56 m (1 ft 10 in)

DIMENSIONS, INTERNAL:

Cabin: Length	3.81 m (12 ft 6 in)
Width	1.17 m (3 ft 10 in)
Height	1.22 m (4 ft 0 in)
Baggage space	0.85 m³ (30 cu ft)

AREAS:

Wings, gross	15.79 m² (170 sq ft)
Tailplane	4.20 m² (45.2 sq ft)

WEIGHTS AND LOADINGS:

Weight empty, equipped	1,695 kg (3,737 lb)
Max T-O and landing weight	2,495 kg (5,500 lb)
Max ramp weight	2,506 kg (5,525 lb)
Max wing loading	158.2 kg/m² (32.4 lb/sq ft)
Max power loading	5.77 kg/kW (9.5 lb/hp)

PERFORMANCE (at max T-O weight):

Cruising speed at average cruise weight:
75% power at 2,285 m (7,500 ft) — 220 kts (408 km/h; 253 mph)
65% power at 3,050 m (10,000 ft) — 213 kts (394 km/h; 245 mph)
65% power at 2,285 m (7,500 ft) — 208 kts (385 km/h; 239 mph)
55% power at 3,050 m (10,000 ft) — 197 kts (365 km/h; 227 mph)
55% power at 2,285 m (7,500 ft) — 193 kts (357 km/h; 222 mph)

Stalling speed, flaps down — 74 kts (137 km/h; 85 mph) IAS

Max rate of climb at S/L	549 m (1,800 ft)/min
Rate of climb at S/L, one engine out	110 m (360 ft)/min
Service ceiling	6,460 m (21,200 ft)
Service ceiling, one engine out	1,920 m (6,300 ft)
T-O run	472 m (1,550 ft)
T-O to 15 m (50 ft)	594 m (1,950 ft)
Landing from 15 m (50 ft)	561 m (1,840 ft)
Landing run	317 m (1,040 ft)

Range with max fuel, allowances for start, taxi, T-O, climb, and 45 min reserves at long-range cruise power:
75% power at 1,830 m (6,000 ft) — 983 n miles (1,821 km; 1,132 miles)
65% power at 2,745 m (9,000 ft) — 1,079 n miles (2,000 km; 1,242 miles)
55% power at 3,050 m (10,000 ft) — 1,201 n miles (2,225 km; 1,383 miles)

VERIFIED

PIPER AEROSTAR 602P

TYPE: Twin-engined light transport aircraft.

PROGRAMME: On 14 February 1981, Piper announced an addition to the Aerostar family of twin piston-engined aircraft, designated Model 602P. Generally similar to the Aerostar 601P, which it superseded, the Model 602P is pressurised and has low-compression engines with integral turbochargers, providing significantly improved performance, particularly at high altitudes. The 1982 version introduced four optional factory-installed avionics packages, plus a wide variety of other optional avionics items from manufacturers which include Aeronetics, Bendix, Bonzer, Collins, Edo-Aire, Mitchell, King, Piper and SunAir.

An optional group of factory-installed de-icing equipment is available which has received FAA approval for flight into icing conditions. It comprises de-icing boots for wing and tail unit leading-edges, electrically heated propeller boots, alcohol de-icing system for windscreen, wing ice inspection light, and a manual alternative engine air control; adding 31.5 kg (69.5 lb) to basic empty weight.

UPGRADES: **AAC:** Aerostar Super 700. See separate entry.

AAC: Aerostar 3000. See separate entry.

The description of the Aerostar 600A applies generally to the 602P, except as follows:

POWER PLANT: Two 216 kW (290 hp) Avco Lycoming IO-540-AA1A5 flat-six turbocharged engines, each driving a Hartzell three-blade constant-speed fully feathering metal propeller with spinner. Fuel system and propeller options as for the Aerostar 600A.

SYSTEMS: Pressurisation by engine bleed air; pressure differential 0.29 bars (4.25 lb/sq in). Heating/ventilation system includes a Janitrol 35,000 BTU heater, thermostatic control of cabin temperature, individual cabin air ventilators and fan. Electrical air conditioning system optional. Hydraulic system for landing gear actuation, wheel brakes and flaps, powered by an engine-driven pump, pressure 69 bars (1,000 lb/sq in). Auxiliary hydraulic system with electrically driven pump optional. Dual engine-driven pneumatic pumps for instrument gyros and optional wing and tail unit de-icing boots. Electrical system includes two 28 V 70 A engine-driven alternators and a 24 V 24 Ah storage battery. Scott Ambassador oxygen system with outlet at each seat, capacity 0.31 m³ (11 cu ft); system capacity of 3.26 m³ (115 cu ft) optional. Engine fire detection system optional.

AVIONICS AND EQUIPMENT: A wide range of optional avionics incudes nav/coms, ADF, autopilots, DME, encoding altimeter, glide slope and marker beacon receivers, HF transceiver, R/Nav, radio altimeter, transponder and weather radar, by manufacturers that include Aeronetics, Bendix, Collins, Edo-Aire, Mitchell, King, Piper and SunAir. Standard equipment is as detailed for the Model 600A, plus instrument panel glareshield; cabin altitude, rate of climb and differential pressure indicators; alternate static source; baggage compartment lights and individual passenger reading lights; fuel quick drains; external power socket; and polyurethane external paint. Optional equipment includes co-pilot's blind-flying instrumentation, fuel flow indicator and totaliser, cabin fire extinguisher, swivel seat four and writing table installed next to seat three, reclining bench seat in lieu of seats five and six, inertia-reel shoulder harnesses for individual passenger seats, pilot and co-pilot vertically adjustable seats, headrests, reinforced baggage floor, rotating beacons, engine rpm limiter, nosecap erosion protective cover, and wheel jack pads.

DIMENSIONS, EXTERNAL AND INTERNAL: As for Aerostar 600A except:

Wing span	11.19 m (36 ft 8½ in)
Wing chord at tip	0.76 m (2 ft 6 in)
Wing aspect ratio	7.55

WEIGHTS AND LOADINGS:

Weight empty, equipped	1,871 kg (4,125 lb)
Max T-O and landing weight	2,721 kg (6,000 lb)
Max ramp weight	2,735 kg (6,029 lb)
Max wing loading	164.6 kg/m² (33.71 lb/sq ft)
Max power loading	6.30 kg/kW (10.34 lb/hp)

PERFORMANCE (at max T-O weight, except where indicated):

Max level speed at average cruise weight of 2,495 kg (5,500 lb) — 262 kts (486 km/h; 302 mph)

Cruising speed, best power, at average cruise weight of 2,495 kg (5,500 lb): 75% power at 7,620 m (25,000 ft) — 247 kts (458 km/h; 284 mph)
75% power at 7,010 m (23,000 ft) — 242 kts (448 km/h; 279 mph)
75% power at 4,570 m (15,000 ft) — 226 kts (419 km/h; 260 mph)
65% power at 7,620 m (25,000 ft) — 228 kts (423 km/h; 263 mph)
65% power at 7,010 m (23,000 ft) — 224 kts (415 km/h; 258 mph)
65% power at 4,570 m (15,000 ft) — 210 kts (389 km/h; 242 mph)

Piper Aerostar 601P with Avco Lycoming IO-540-S1A5 flat-six engines

1994

Piper Aerostar 700P six-seat light transport aircraft

1994

55% power at 7,620 m (25,000 ft) 203 kts (376 km/h; 234 mph)
55% power at 7,010 m (23,000 ft) 201 kts (372 km/h; 231 mph)
55% power at 4,570 m (15,000 ft) 191 kts (354 km/h; 220 mph)
Stalling speed, flaps and landing gear down, power off 77 kts (143 km/h; 89 mph) IAS
Max rate of climb at S/L 535 m (1,755 ft)/min
Rate of climb at S/L, one engine out 92 m (302 ft)/min
Service ceiling 8,525 m (28,000 ft)
Service ceiling, one engine out 3,930 m (12,900 ft)
T-O run 549 m (1,800 ft)
T-O to 15 m (50 ft) 686 m (2,250 ft)
Landing from 15 m (50 ft) 633 m (2,076 ft)
Landing run 371 m (1,217 ft)
Range with max fuel, allowances for start, taxi, T-O, climb, descent and 45 min reserves at long-range cruise power: 65% power at 4,570 m (15,000 ft) 1,094 n miles (2,027 km; 1,260 miles)
65% power at 7,010 m (23,000 ft) 1,098 n miles (2,035 km; 1,264 miles)
65% power at 7,620 m (25,000 ft) 1,078 n miles (1,998 km; 1,241 miles)
55% power at 4,570 m (15,000 ft) 1,143 n miles (2,118 km; 1,316 miles)
55% power at 7,010 m (23,000 ft) 1,137 n miles (2,107 km; 1,309 miles)
55% power at 7,620 m (25,000 ft) 1,112 n miles (2,060 km; 1,280 miles)

VERIFIED

PIPER PA-60-700P AEROSTAR 700P

TYPE: Six-seat light transport aircraft.

PROGRAMME: On 20 November 1982 Piper announced the addition of the pressurised PA-60-700P Aerostar 700P to the Aerostar range. Design had been initiated in January 1981, and construction of a prototype had started six months later. The first flight had followed during September 1981, and the construction of production aircraft began in December 1982. Certification under FAR Pt 23 Amendment 6 was received in May 1983, with initial customer deliveries in December 1983. Certifications for autopilot and flight into known icing conditions were obtained in September and October 1983 respectively.

Introducing more powerful Avco Lycoming turbocharged and intercooled flat-six engines, the 700P demonstrated during flight tests improved speed and range by comparison with the Aerostar 602P. It retains the same high standard of equipment and interior furnishing.

UPGRADES: **AAC**: Aerostar Super 700. See separate entry.
AAC: Aerostar 3000. See separate entry.

DESIGN FEATURES: Cantilever mid-wing monoplane. Wing section NACA 61_1A212. Dihedral 2°. Incidence 1° constant. Wings swept forward 3° 36′ at quarter-chord. The tail unit has sweptback horizontal and vertical surfaces.

FLYING CONTROLS: Frise balanced ailerons and single-slotted trailing-edge flaps, all of aluminium alloy construction. Electrically operated trim tabs in each elevator and rudder.

STRUCTURE: The wing is an aluminium alloy structure with three spars and flush riveted heavy gauge skins. The fuselage is a pressurised semi-monocoque fail-safe structure of aluninium alloy. Cabin windows have safe life of 16,000 hours, the windscreen 4,860 hours. The tail unit is a conventional riveted all-metal structure of aluminium alloy.

LANDING GEAR: Hydraulically retractable tricycle type, with single wheel on each unit. Main units retract inward, nosewheel forward. Piper oleo-pneumatic shock-absorbers. Goodyear wheels and tyres. Mainwheel tyres size 18.50 × 7-7.50, pressure 4.83 bars (70 lb/sq in); nosewheel tyre size 17.00 × 6-7.25, pressure 2.76 bars (40 lb/sq in). Goodyear air-cooled multiple disc hydraulic brakes. Parking brake.

POWER PLANT: Two 261 kW (350 hp) Avco Lycoming flat-six counter-rotating engines with intercoolers, one TIO-540-U2A and one LTIO-540-U2A, each driving a Hartzell three-blade constant-speed fully feathering propeller with spinner. Fuel in integral wing tanks and fuselage bladder tank with combined capacity of 656.75 litres (173.5 US gallons; 144.6 Imp gallons), of which 626.5 litres (165.5 US gallons; 138 Imp gallons) are usable. Optional auxiliary fuel tank, capacity 151.5 litres (40 US gallons; 33 Imp gallons), aft of bladder tank. Refuelling point adjacent each wingtip and on port side of fuselage over wing. Oil capacity 11.4 litres (3 US gallons; 2.5 Imp gallons) per engine. Electric propeller de-icing boots and propeller synchrophaser optional.

ACCOMMODATION: Pilot and five passengers on track-mounted reclining seats, in pairs, with centre aisle. Separate baggage compartment, capacity 109 kg (240 lb), to rear of cabin, with door on port side. Two-piece airstair door on port side, by pilot's seat. Overwing emergency exit on starboard side. Cabin is pressurised, air conditioned, heated and ventilated. Heated plates and alcohol de-icing for windscreen optional.

SYSTEMS: Piper 16,000 BTU air conditioning system, and pressurisation system with maximum differential of 0.29 bars (4.25 lb/sq in). Hydraulic system, pressure 89.65 bars (1,300 lb/sq in), with single engine-driven pump to actuate flaps, landing gear and nosewheel steering; second engine-driven pump optional. Dual engine-driven pneumatic pumps. Electrical system includes two 70 A alternators. Six-outlet constant flow oxygen system of 0.31 m³ (11 cu ft) capacity optional. Engine fire detection system optional.

AVIONICS AND EQUIPMENT: Standard factory-installed avionics package includes dual King KX-165 nav/com transceivers, KI-525A HSI, KI-202 indicator, VOR/LOC/glide slope receivers, KR-87 digital ADF, KN-63 DME, KAP-150 autopilot, transponder and encoding altimeter. A wide range of optional avionics, including HF transceivers, R/Nav systems, Loran C and weather radar from Aeronetics, Bendix, Bonzer, Collins, Century, King, Piper, Sperry, SunAir and Texas Instruments, is also available. Other optional items include 3 in horizon and directional gyros for co-pilot, fuel consumed totaliser, shoulder harness for passenger seats, cabin fire extinguisher, rotating beacon, heated pitot and stall warning transducer, and wing ice light.

DIMENSIONS, EXTERNAL:
Wing span 11.18 m (36 ft 8 in)
Wing chord: at root 2.18 m (7 ft 2 in)
at tip 0.77 m (2 ft 6½ in)
Wing aspect ratio 7.55
Length overall 10.61 m (34 ft 9¾ in)
Height overall 3.68 m (12 ft 1 in)
Tailplane span 4.37 m (14 ft 4 in)
Wheel track 3.10 m (10 ft 2 in)
Wheelbase 3.55 m (11 ft 7¾ in)
Propeller diameter 1.93 m (6 ft 4 in)
Propeller ground clearance 0.25 m (10 in)
Distance between propeller centres 3.91 m (12 ft 10 in)
Passenger door (port, fwd): Height 1.14 m (3 ft 9 in)
Width 0.71 m (2 ft 4 in)
Height to sill 0.71 m (2 ft 4 in)
Baggage door (port, rear): Height 0.56 m (1 ft 10 in)
Width 0.61 m (2 ft 0 in)
Height to sill 1.30 m (4 ft 3 in)
Emergency exit (stbd, overwing):
Height 0.56 m (1 ft 10 in)
Width 0.69 m (2 ft 3¼ in)

DIMENSIONS, INTERNAL:
Cabin: Length (instrument panel to rear bulkhead) 3.03 m (9 ft 11½ in)
Max width 1.16 m (3 ft 9½ in)
Max height 1.18 m (3 ft 10½ in)
Baggage compartment 0.85 m³ (30 cu ft)

AREAS:
Wings, gross 16.56 m² (178.2 sq ft)
Ailerons (total) 1.275 m² (13.72 sq ft)
Trailing-edge flaps (total) 2.20 m² (23.68 sq ft)
Fin 1.15 m² (12.41 sq ft)
Rudder (incl tab) 0.59 m² (6.33 sq ft)
Tailplane 3.02 m² (32.54 sq ft)
Elevators (incl tab) 1.18 m² (12.66 sq ft)

WEIGHTS AND LOADINGS:
Weight empty 1,939 kg (4,275 lb)
Max payload 544 kg (1,200 lb)
Max standard fuel weight 472 kg (1,041 lb)
Max T-O weight 2,864 kg (6,315 lb)
Max ramp weight 2,883 kg (6,356 lb)
Max zero-fuel weight 2,744 kg (6,050 lb)
Max landing weight 2,721 kg (6,000 lb)
Max wing loading 172.8 kg/m² (35.4 lb/sq ft)
Max power loading 5.49 kg/kW (9.02 lb/hp)

PERFORMANCE (at max T-O weight, ISA, except where indicated:
Never-exceed speed 245 kts (454 km/h; 282 mph) CAS
Max level speed at 5,490 m (18,000 ft) at mid-cruise weight 266 kts (492 km/h; 306 mph)
Cruising speed at mid-cruise weight, at optimum altitude:
81% power 261 kts (483 km/h; 300 mph)
65% power 230 kts (426 km/h; 264 mph)
55% power 211 kts (390 km/h; 242 mph)
Stalling speed, engine idling:
flaps up 82 kts (152 km/h; 95 mph) CAS
flaps down 76 kts (141 km/h; 88 mph) CAS
Max rate of climb at S/L 561 m (1,840 ft)/min
Rate of climb at S/L, one engine out 99 m (325 ft)/min
Certificated ceiling 7,620 m (25,000 ft)
Service ceiling, one engine out 5,025 m (16,500 ft)
T-O run 594 m (1,950 ft)
T-O to 15 m (50 ft) 939 m (3,080 ft)
Landing from 15 m (50 ft) 652 m (2,140 ft)
Landing run 439 m (1,440 ft)
Range with max standard fuel, incl allowances for taxi, T-O, climb and descent, plus 45 min reserves at long-range cruise power:
81% power 675 n miles (1,250 km; 776 miles)
65% power 820 n miles (1,518 km; 943 miles)
55% power 890 n miles (1,648 km; 1,024 miles)
Range with max optional fuel, allowances as above:
81% power 875 n miles (1,620 km; 1,007 miles)
65% power 1,075 n miles (1,990 km; 1,237 miles)
55% power 1,160 n miles (2,148 km; 1,335 miles)

VERIFIED

RAISBECK

RAISBECK ENGINEERING INC

4411 South Ryan Way, Seattle, Washington 98178
Telephone: 1 (206) 723 2000
Fax: 1 (206) 767 7726
CEO: James D. Raisbeck
VICE-PRESIDENT, SERVICE AND SALES: Robert P. Steinbach
DIRECTOR, TECHNICAL SERVICES: Michael P. Tougher

Develops, certificates and sells general aviation advanced technology modifications to improve performance, safety and productivity of business aircraft. In December 1993 Raisbeck completed delivery of a block of seven complete mini-kits to Beechcraft Scandinavia, for installation on factory-fresh Beechcraft Super King Air B200s.

UPDATED

HARTZELL/RAISBECK QUIET TURBOFAN PROPELLER SYSTEM FOR BEECHCRAFT KING AIRS

TYPE: Light aircraft propeller conversion.

PROGRAMME: Four-blade propeller conversion known as Quiet Turbofan; first flight on Mark VI-equipped B200 June 1984; certificated on Super King Air 200 February 1985; also certificated for all King Air 90s. Flight deck and cabin noise reduced between 7 and 10 dBA; lower vibration; shorter take-off and landing; better climb. Uses advanced technology Hartzell lightweight turbofan propellers of 2.39 m (7 ft 10 in) diameter, new timers, slip rings and Goodrich hot propeller de-icing system.

VERIFIED

HARTZELL/RAISBECK QUIET TURBOFAN PROPELLER SYSTEM FOR DHC-6 TWIN OTTER

TYPE: Light aircraft propeller conversion.

PROGRAMME: Similar to above applied to de Havilland Canada Twin Otter Series 300 of Scenic Airlines at Las Vegas, Nevada; cabin and flight deck noise reduced by 10 to 13 dBA; lower external noise, better engine inlet ram air recovery and lower engine temperatures; certificated May 1986; entire Scenic Airlines fleet converted; several European operators also converted to meet ICAO Annex 16 noise limits.

VERIFIED

RAISBECK SHORT-FIELD ENHANCEMENT SYSTEM

TYPE: Light aircraft short-field modification.

PROGRAMME: New composites constructed inboard wing leading-edges, intercooler ducting, wing-to-fuselage fairings, and flush-mounted Goodrich de-icing boots. Announced June 1987 for Beech Super King Air 200/B200; when combined with Hartzell Quiet Turbofan propellers, gives 18 knot (33 km/h; 21 mph) reduction in flaps-up V_2 take-off speed, 35 per cent reduction in FAA T-O distance, 50 per cent reduction in FAA accelerate/go distance, better engine-out climb rate, reduced FAA reference and approach speeds, and 56 per cent reduction in stopping distance with maximum reverse thrust. Other benefits include better outer wing fatigue life, more efficient air conditioning, higher cruising speeds, reduced stalling speed at all flap settings, and better low-speed handling and stall characteristcs.

Raisbeck short-field enhancement system modifications for Beech 200/B200 Super King Airs

1992

PERFORMANCE (at AUW of 5,670 kg; 12,500 S/L, ISA. Super King Air 200/B200):

T-O speed (V_2)	103 kts (191 km/h; 119 mph)
Approach speed	97 kts (180 km/h; 112 mph)
T-O to 15 m (50 ft)	686 m (2,250 ft)
Accelerate/go distance	1,061 m (3,480 ft)
Accelerate/stop distance	988 m (3,240 ft)
Landing from 15 m (50 ft)	500 m (1,970 ft)

VERIFIED

RAISBECK DUAL AFT-BODY STRAKE SYSTEM

TYPE: Light aircraft dual strake modification.

PROGRAMME: Dual ventral strakes to replace single strake on all Beechcraft King Air and Super King Air models; slightly increases cruising speeds and reduces rear cabin area vibration; with dual strakes, yaw damper no longer a mandatory dispatch item on King Air F90 and Super King Air 200.

VERIFIED

RAISBECK WING LOCKERS

TYPE: Light aircraft wing locker installation.

PROGRAMME: Engine nacelle wing lockers on Beechcraft King Air 90/100 and Super King Air 200/300; certificated to FAA FAR Pt 135. Lockers can carry 272 kg (600 lb), increase baggage volume by 0.45 m³ (16 cu ft), improve climb and cruise performance, reduce stalling speed, and reduce pitch-axis trim changes in landing configuration.

VERIFIED

RAISBECK FULLY ENCLOSED LANDING GEAR DOORS

TYPE: Light aircraft landing gear doors installation.

PROGRAMME: Doors applied to Beechcraft King Air 90/100 and Super King Air 200s with high flotation landing gear; made of composites; increase high-altitude cruising speed by 8 to 12 knots (15 to 22 km/h; 9 to 14 mph), increase climb rates, keep wheel wells clean and extend wheel, tyre and brake life.

VERIFIED

RAISBECK SOOTLESS EXHAUST STACK FAIRINGS

TYPE: Light aircraft sootless exhaust stack fairings installation.

PROGRAMME: Small fairings over fronts of exhaust stacks on all P&WC PT6A-powered Beechcraft King Air and Super King Air models; prevent power loss from hot air entering forward cowling and engine inlet, and reduce drag by streamlining propeller wake; fairing held in place by camlocks; two 'delta-wing' vortex generators on fairings divert high-energy gases between exhaust and cowling, keeping exhaust soot from accumulating on nacelle and wing. Stainless steel or composites fairings available.

VERIFIED

RAISBECK RAM AIR RECOVERY SYSTEM

TYPE: Light aircraft air recovery system installation.

PROGRAMME: RARS intended for Super King Air 200; developed in association with P&WC and certificated; more complete sealing of engine nacelle air inlet section, new fixed turning vane, addition of Coanda effect curved surface on rear portion of movable inertial separator vane, and

Raisbeck dual aft strakes modification

1994

Raisbeck wing lockers

1994

new highly porous ice shedder screen; increases cruising speed, rates of climb, power plant flat rating and engine output with ice vanes deployed; reduces block fuel consumption and engine ITT at equal torque.

VERIFIED

RAISBECK BEECHCRAFT KING AIR C90/C90A

TYPE: Light aircraft increase gross weight modification.

PROGRAMME: FAA approval for operation at increased gross weight of early Beechcraft C90/C90As (Serial Nos. LJ-527 to LJ-1137) announced 1 October 1990; maximum T-O weight raised from 4,377 kg (9,650 lb) to 4,581 kg (10,100 lb) permits legal increase of three additional passengers with full tanks; maximum landing weight increased to 4,400 kg (9,700 lb). Approval requires Hartzell Quiet Turbofan propellers, Raisbeck Dual Aft Body Strakes and 10 ply tyres. Similar programme in progress to raise E90 gross weight to 4,762 kg (10,500 lb).

VERIFIED

RAISBECK LEARJET AFT FUSELAGE LOCKERS

Raisbeck Learjet aft fuselage lockers

1994

TYPE: Business aircraft aft fuselage locker modification.

PROGRAMME: Under development for two years has now completed its FAA certification flight tests. Production deliveries began in 1994. Initially for the Learjet 35/36 Series. Subsequent applications for the Learjet will include the Models 55, 60 and the forthcoming 45. The Model 20 series may also be included according to demand.

CUSTOMERS: Jet America took delivery of first production aft fuselage locker for its Lear 36A (N54JA) in March 1995.

DESIGN FEATURES: The aft fuselage fairing is entirely external to the existing fuselage and begins as a rearward extension of the horizontal fuselage keel beam. It then rolls into a deep ventral fin replacing the aircraft's standalone fin. Contained within the fairing is an 2.44 m (8 ft) long drawer which deploys sideways for loading and unloading. The drawer is certified to carry 136 kg (300 lb) of cargo. Installation time is 125 man-hours.

UPDATED

RAM

RAM AIRCRAFT CORPORATION

7505 Airport Drive, PO Box 5219, Waco, Texas 76708-0219
Telephone: 1 (817) 752 8381
Fax: 1 (817) 752 3307
PRESIDENT: Jack M. Riley Jr (Engineering)
INTERNATIONAL SALES:
SALES MANAGER: Doug Mackay (South America)
SALES: Bob Neal (Europe and Pacific)
PRESS RELATIONS: Chuck Morrow

RAM specialises in modification of selected single- and twin-engined general aviation aircraft for improved performance and efficiency. All modifications FAA approved by award of STC. Export STC modification kits available with new engines or RAM-remanufactured 100 per cent balanced versions of Teledyne Continental TSIO-520 engine. RAM Cessna 310, 320, 340, 414 and 414A conversions designated Series II (231 kW; 310 hp) and Series IV (242 kW; 325 hp) with RAM economy camshafts developed by a leading automotive race engine cam designer, Florida; these give easier starting, increase manifold pressure and smoother idling, together with claimed 3 to 5 per cent reduction in fuel consumption at cruise power. Refurbishing, repainting and maintenance also offered.

UPDATED

RAM/Cessna T310 uprated to 224 kW (300 hp)

1994

RAM/CESSNA 172

TYPE: Light aircraft re-engine programme.

PROGRAMME: D to N Models fitted with 119 kW (160 hp) Textron Lycoming O-320-D2G or Textron Lycoming O-320-E2D engine with power increased from 112 kW (150 hp) to 119 kW (160 hp) when overhauled under RAM STC SE3692SW and installed under RAM STC SA2375SW.

VERIFIED

RAM/PIPER PA-28-140/PA-28-151

TYPE: Light aircraft re-engine programme.

PROGRAMME: Re-engined with 119 kW (160 hp) Textron Lycoming O-320-D3G under RAM STC SA2706SW for -140 and SA2969SW for -151.

VERIFIED

RAM/CESSNA T206/210

TYPE: Light aircraft re-engine programme.

PROGRAMME: Replacing TSIO-520-C/H in early models with 231 kW (310 hp) RAM-remanufactured TSIO-520-M/R; performance generally same as later models with 200 hours more overhaul life; optional Hartzell Q-tip or McCauley wide-chord propellers.

VERIFIED

RAM/CESSNA T310

TYPE: Light aircraft re-engine programme.

PROGRAMME: Replaces 212 kW (285 hp) Teledyne Continental TSIO-520-Bs with either 224 kW (300 hp) TSIO-520-Es or RAM-remanufactured 242 kW (325 hp) TSIO-520-NBRs; with 242 kW (325 hp) engines, aircraft known as RAM 310 Series IV; T-O weight of 310P and Q models increased to 2,572 kg (5,670 lb), giving 122 kg (270 lb) greater useful load in 310P and 77 kg (170 lb) in 310Q; dual aerial mapping/reconnaissance ports available for cameras as large as Wild RC10 and Zeiss RMKA 15/23, but RAM will not install (will sell STC per aircraft).

PERFORMANCE (A: 224 kW; 300 hp, B: 242 kW; 325 hp engines, at maximum T-O weight of 2,572 kg; 5,670 lb):

Cruising speed, 75% power at 5,485 m (18,000 ft):
- A: 217 kts (402 km/h; 250 mph)
- B: 243 kts (450 km/h; 280 mph)

Max rate of climb at S/L:
- A: 579 m (1,900 ft)/min
- B: 457 m (1,500 ft)/min

Rate of climb at S/L, one engine out:
- A: 149 m (490 ft)/min
- B: 195 m (640 ft)/min

UPDATED

RAM/CESSNA 340/340A

TYPE: Light aircraft re-engine programme.

PROGRAMME: Replaces 212 kW (285 hp) TSIO-520-Ks with RAM 242 kW (325 hp) TSIO-520-NBRs or 250 kW (335 hp) with RAM intercooler scoops added to the 242 kW (325 hp). Series IV modification has the 242 kW (325 hp) engines, 136 kg (300 lb) increase in useful load and maximum T-O weight increase of 68 kg (150 lb); Series VI has the 250 kW (335 hp) engines and 188 kg (415 lb) increases in useful load and maximum T-O weight; Series IV package also includes Hartzell Q-tip propellers. Series VI includes new 'Sabre Tip' McCauley C-515 propellers.

PERFORMANCE (A: 242 kW; 325 hp engines with Q-tip propellers, B: 250 kW; 335 hp engines with C-515 propellers):

Cruising speed, 75% power at 6,100 m (20,000 ft):
- A: 225 kts (417 km/h; 259 mph)
- B: 228 kts (423 km/h; 262 mph)

Max rate of climb at S/L:
- A: 553 m (1,815 ft)/min
- B: 614 m (2,015 ft)/min

Rate of climb at S/L, one engine out:
- A: 103 m (340 ft)/min
- B: 108 m (355 ft)/min

Time to 5,485 m (18,000 ft):
- A: 16 min
- B: 15 min

Acceleration to 87 kts (161 km/h; 100 mph):
- A: 16 s
- B: 15 s

UPDATED

RAM/CESSNA 401/402A/402B/402C

TYPE: Light aircraft re-engine programme.

PROGRAMME: Remanufactured 224 kW (300 hp) Teledyne Continental TSIO-520-E or 242 kW (325 hp) TSIO-520-VB engines installed in Cessna 402C; engine accessories replaced by new or overhauled parts; Slick 6320 pressurised magnetos; red silicone rubber baffle seal kit; Alcor 46158 exhaust gas temperature gauge and combustion analyser; options include electronic fuel flow management, McCauley MC-I propeller synchrophaser, RAM super soundproofing, Cleveland heavy-duty brakes, polished and balanced propellers and three-colour polyurethane external paint finish. STC (SA8634SW) for 402C to increase gross weight from 3,106 to 3,269 kg (6,850 to 7,210 lb). Zero-fuel weight remains at 2,954 kg (6,515 lb) allowing maximum use of more than 1,523 kg (3,360 lb) to be used to extra payload. STC to install new McCauley C-515 Sabre Tip propellers. C-515 has a TBO of 2,000 hours. STC to install RAM Vortex Generators SA8624SW. This reduces stall and VMC speeds.

UPDATED

RAM/CESSNA 414

TYPE: Light aircraft re-engine programme.

PROGRAMME: Replaces 231 kW (310 hp) TSIO-520-J engines in 414s built between 1970-76 with 250 kW (335 hp) TSIO-520-NBRs; known as RAM 414 Series VI; same options as Cessna 340A conversions; Hartzell Q-tip available.

PERFORMANCE (A: 242 kW; 325 hp, B: 250 kW; 335 hp engines, at max T-O weight of 2,881 kg; 6,350 lb):

Cruising speed, 75% power at 6,100 m (20,000 ft):
- A: 215 kts (398 km/h; 247 mph)
- B: 218 kts (404 km/h; 250 mph)

Max rate of climb at S/L:
- A: 545 m (1,790 ft)/min
- B: 579 m (1,900 ft)/min

Rate of climb at S/L, one engine out:
- A: 83 m (275 ft)/min
- B: 94 m (310 ft)/min

Time to 5,485 m (18,000 ft): A, B: 18 min

Acceleration to 87 kts (161 km/h; 100 mph): A, B: 19 s

UPDATED

RAM/Cessna 414 Series VI uprated to 250 kW (335 hp)

1994

RAM/CESSNA 414AW

TYPE: Light aircraft re-engine programme.

PROGRAMME: Replaces 231 kW (310 hp) TSIO-250-Ns with remanufactured with 242 kW (325 hp) TSIO-520-NBRs, with McCauley 'Saber Tip' C-515 propellers (Q-tip Hartzell propellers optional); certificated by FAA, with 0.89 m (2 ft 11 in) high winglets, 1 March 1983, and then by Australia; RAM/Cessna 414AW gross weight increased by 161 kg (355 lb) and useful load by 144 kg (318 lb).

PERFORMANCE (at max T-O weight with 242 kW; 325 hp engines):

Cruising speed at 6,100 m (20,000 ft):

75% power	215 kts (398 km/h; 247 mph)
55% power	190 kts (351 km/h; 218 mph)
Max rate of climb at S/L	493 m (1,620 ft)/min
Rate of climb at S/L, one engine out	104 m (340 ft)/min
Time to 5,485 m (18,000 ft)	19 min
Acceleration to 87 kts (161 km/h; 100 mph)	21 s

UPDATED

RAM/CESSNA 414AW SERIES V

TYPE: Light aircraft re-engine programme.

PROGRAMME: Replaces 231 kW (310 hp) Teledyne Continental TSIO-520-N/NB engines with new 261 kW (350 hp) liquid-cooled Teledyne Continental TSIO-550-A Voyager engines, driving new McCauley C-514 three-blade propellers. Modification includes installation of coolant radiator in extended rear of each nacelle baggage locker, each with flush non-icing NACA air intake scoop; new nacelle nose caps of graphite/glassfibre with flush-mounted pulse light recognition/landing lights; and optional winglets. Cockpit gauges provided for coolant temperature and pressure; coolant temperature thermostatically controlled. Same standard and optional items as Series IV; maximum T-O weight increased to 3,214 kg (7,087 lb) with winglets, 3,222 kg (7,105 lb) without winglets and useful load by 53.6 kg (118 lb). 414AW/V certificated 10 August 1989; deliveries at one per month; 34 delivered by March 1994.

PERFORMANCE:

Cruising speed at 75% power with winglets:

at 7,010 m (23,000 ft)	227 kts (420 km/h; 260 mph)
at 9,145 m (30,000 ft)	240 kts (445 km/h; 276 mph)
Rate of climb at S/L, one engine out	99 m (325 ft)/min
Time to 9,145 m (30,000 ft)	33 min

UPDATED

RAM/CESSNA 421C and 421CW

TYPE: Light aircraft re-engine programme.

PROGRAMME: Cessna 421B and 421C Golden Eagle powered by RAM or Teledyne Continental GTSIO-520-L/N engines; 421C useful load increased by 50 kg (110 lb). Standard cooling baffles, exhaust risers, slip joints and magnetos replaced by RAM Reliability Package including new red silicone/glassfibre cooling baffle seal material, remanufactured exhaust system and slip joints, Bendix S6RN-1250 pressurised magnetos; also includes digital electronic computerised fuel management system and Alcor direct-reading exhaust gas temperature system.

Winglets for 421Cs up to factory serial number 0799 (becoming 421CW) approved by FAA November 1984, including flight into known icing; Australian certification 1987; winglets, incorporating wing extensions and winglet assembly and made of bidirectional carbon graphite cloth/epoxy with outer layer which includes interwoven aluminium cloth for electrical conductivity and lightning protection, increase wing span by 0.94 m (3 ft 1 in). First 421CW delivery October 1984; 45 conversions completed by January 1994. 421CW useful load increased by 23 kg (50 lb); winglets increase rate of climb at 15 to 20 per cent.

DIMENSIONS, EXTERNAL: As for Cessna 421C except:

Cessna factory basic wing span	12.46 m (41 ft)
Basic wing span	13.47 m (44 ft 2½ in)
Span, winglet tip to winglet tip	13.83 m (45 ft 4½ in)
Wing aspect ratio (basic)	8.64
Effective wing aspect ratio, incl winglets	9.74
Winglet cant angle	10°
Winglet toe-out angle	2°

AREAS:

Wings, gross (basic)	21.0 m² (226.0 sq ft)

WEIGHTS AND LOADINGS:

Max T-O weight	3,429 kg (7,560 lb)
Max zero-fuel weight	2,963 kg (6,533 lb)
Max wing loading	163.3 kg/m² (33.45 lb/sq ft)
Max power loading	6.12 kg/kW (10.06 lb/hp)

PERFORMANCE:

Cruising speed: at 75% power at 6,100 m (20,000 ft) 228 kts (422 km/h; 262 mph)

at 65% power at 6,100 m (20,000 ft) 223 kts (413 km/h; 257 mph)

Rate of climb at S/L, ISA, one engine out 96 m (315 ft)/min

Time to 6,096 m (20,000 ft) 17 min

UPDATED

RAM/BEECH BARON 58

TYPE: Light aircraft re-engine programme.

PROGRAMME: Baron 58P and 58TC re-engined with 231 kW (310 hp) Teledyne Continental TSIO-520-L/LB or 242 kW (325 hp) TSIO-520-WBs; standard equipment includes Alcor exhaust gas temperature gauge and combustion analyser, airborne dry vacuum pumps, electronic fuel management and red silicone rubber baffle seals; options include Teflon fuel and oil hoses, balanced propellers and three-colour polyurethane exterior paint. Optional McCauley C-521 'Sabre Tip' propellers available for aftermarket sale to RAM or factory original status airplanes.

PERFORMANCE (Baron 58P. A: 231 kW; 310 hp, B: 242 kW; 325 hp):

Cruising speed, 75% power at 6,100 m (20,000 ft):

A	223 kts (413 km/h; 257 mph)
B	229 kts (424 km/h; 264 mph)
Max rate of climb at S/L: A	463 m (1,520 ft)/min
B	469 m (1,540 ft)/min

Rate of climb at S/L, one engine out:

A	67 m (220 ft)/min
B	97 m (320 ft)/min

UPDATED

RAM/Cessna 414AW Series V with liquid-cooled Voyager engines

1994

RAM/Cessna 421CW with winglets

1994

RAYTHEON

RAYTHEON AIRCRAFT COMPANY (Subsidiary of Raytheon Company)

9709 East Central, Wichita, Kansas 67201-0085
Telephone: 1 (316) 676 7111
Fax: 1 (316) 676 8286
BRANCH DIVISIONS: Salina, Kansas, and Little Rock, Arkansas
PRESIDENT: Roy H. Norris III
CHAIRMAN AND CEO: Arthur E. Wegner
VICE-PRESIDENT, AEROSPACE: Dr William A. Edgington
VICE-PRESIDENT, ENGINEERING: Steve Hanney
VICE-PRESIDENT, MANUFACTURING: John Booth
VICE-PRESIDENT, MARKETING: Doug Mahin
VICE-PRESIDENT, CUSTOMER SUPPORT: A. John Diebold
VICE-PRESIDENT, JPATS: David Riemer
VICE-PRESIDENT, DOMESTIC BUSINESS JET SALES: Karl Childs
VICE-PRESIDENT, DOMESTIC BEECH SALES: Keith Nadolski
VICE-PRESIDENT, PROGRAMME MANAGEMENT: Robert Brill
VICE-PRESIDENT, QUALITY ASSURANCE: Bill Wise
DIRECTOR, AIRLINE MARKETING: Mike Scheidt
DIRECTOR, CORPORATE AFFAIRS: James M. Gregory
MANAGER, PUBLIC RELATIONS: Pat Zerbe

Raytheon Aircraft Company (RAC) formed 15 September 1994, combining Raytheon Company subsidiaries Beech Aircraft Corporation and Raytheon Corporate Jets Inc. Beech Aircraft Corporation founded 1932 by Walter and Olive Ann Beech; became wholly owned subsidiary of Raytheon 8 February 1980. Raytheon acquired British Aerospace's Corporate Jets division 6 August 1993; founded Raytheon Corporate Jets, headquartered in Little Rock, Arkansas, with responsibility for design, development, production, marketing and support of Hawker family of corporate jets. RAC builds civil and military aircraft, missile targets, and components for aircraft and missiles; Salina division supplies all wings, nonmetallic interior components, ventral fins, nosecones and tailcones used in Wichita piston-engined and turboprop Beechcraft production and builds major subassemblies for the Beechjet. Additional Wichita products include subcontracted composites and metal winglets and composite landing gear doors for McDonnell Douglas C-17.

Completion work on British Aerospace Airbus manufactured Hawker 800 and 1000 business jets performed by Little Rock division but being transferred to Wichita; final assembly of these models also due to be transferred from UK to Wichita beginning last quarter of 1995 and finalised by 1997; Little Rock will continue to provide custom interiors, avionics, sales and customer support services for Hawker models, aircraft painting services for Hawkers and Beechcraft 1900Ds and Beechjets, and refurbishment and depot service work on military aircraft formerly performed by Beech Aerospace Services Inc in Selma, Alabama.

Wholly owned subsidiaries include Beech Aerospace Services Inc (BASI) of Madison, Mississippi (worldwide logistic support for Army/Air Force/Navy C-12s, Army U-21s and Beech MQM-107 targets, and of US Navy T-34C and T-44 trainers in the United States); Beech Acceptance Corporation Inc (business aircraft retail financing and leasing); Travel Air Insurance Company Ltd, Bermuda (aircraft liability insurance), and United Beechcraft Inc (marketing support to parent company).

In early 1995 RAC had 10,645 employees worldwide and occupied 380,902 m² (4,100,000 sq ft) of plant area at its two major facilities in Wichita; production rate adjustments and consolidation of facilities, announced in late 1994, will result in personnel reduction of 940 at RAC divisions in the UK and USA during 1995.

Deliveries during 1994 totalled 361, valued at $1.72 billion, comprising 103 Bonanzas, 31 Barons, 35 King Airs, 59 Super King Airs (including RC-12s), 50 1900Ds, three Starships, 22 Beechjet 400As, 37 Jayhawks, 16 Hawker 800s and five Hawker 1000s.

NEW ENTRY

RAYTHEON HAWKER (DH/HS/BAe) 125

RAF designations: CC. Mk 2, T. Mk 1/2

TYPE: Twin-jet business transport aircraft.

PROGRAMME: The Raytheon Hawker 125 is a twin-jet business aircraft which is also suitable for use by armed forces in the communications SIGINT, ELINT, and SAR roles; an ambulance aircraft; for airways inspection; and as an economical trainer for pilots, navigators and specialised radio and radar operators. All series of HS 125s can operate from unpaved runways without modification.

The HS 125 was developed as a private venture, and the first of two prototypes flew for the first time 13 August 1962. The BAe 125, originally a Hawker Siddeley programme, is a twin-engined business aircraft which is also suitable for use by armed forces in the communications role, as an ambulance aircraft, for airways inspection, and as an economical trainer for pilots, navigators and specialised radio and radar operators. It can operate from unpaved runways without modifications.

The HS 125 was developed as a private venture, and the first of two prototypes flew for the first time 13 August 1962. The Viper turbojet-engined Series 1/1A/B/2/3/3A/3B/3A-R/3A-RA/3B-RA/400A/400B/600A/600B, which totalled 358, were superseded in 1976 by the introduction of the Series 700.

Raytheon (HS/BAe) 125-3B-RA with additional fuel tank attached to rear fuselage ***(Raytheon)*** *1995*

VERSIONS: **Series 1:** First eight production aircraft, built before end of 1964. Two Viper 520 turbojet engines.

Series 1A: In production from beginning of 1965 with 'North American' airframe and modifications to meet FAA requirements. Total 62 built, 30 with Viper 521 turbojets and 32 with Viper 522 turbojets.

Series 1B: For sale throughout the world except where FAA requirements apply. Total of 15 built, five with Viper 521 turbojets and 10 with Viper 522 turbojets.

Series 2/Dominie T. Mk 1: Navigational trainer version built for the RAF. First Dominie T. Mk 1 flew 30 December 1964. Basically as Series 1A/B but with interior and systems modifications. Viper 520 turbojets. Normally carries one pilot, a supernumerary crew member, two students and an instructor. Students sit in rearward-facing seats opposite table and instructor's console, with periscopic sextant station behind their seats in centre-fuselage. External modifications include: belly fairing forward of wing and a small ventral fin.

Series 2/Dominie T. Mk 2: Upgrade of Dominie T. Mk 1 for RAF navigational training. Modifications include updated equipment to bring the aircraft more in line with equipment on current front-line aircraft. THORN EMI is prime contractor for the programme and Marshall Aerospace is to carry out the refurbishment (new cabin layout) and system's installation work. System include Super Searcher radar, tactical processing and improved multifunction, multicolour displays, which enable other advanced sensors data such as navigation aids, EMS, IFF, IRDS and acoustics, to be correlated for display in cabin and cockpit installations.

Series 3: Two aircraft built with Viper 522 turbojets.

Series 3A: Produced for North American market with two Viper 522 turbojets. Improved performance and APU for ground power and cabin air conditioning.

Series 3B: Similar to Series 3A, but intended for world markets other than North America. Two Viper 522 turbojets.

Series 3A-RA and 3B-RA: Versions with increased AUW and fuel capacity. Additional fuel carried in tank attached to the underside of the rear fuselage.

Series 400A and 400B/CC. Mk 1: Version with integral airstair door and improvements to flight deck cabin, vestibule and exterior appearance. Two Viper 522 turbojets. In 1983 the CC. Mk 1s were retrofitted with Garrett TFE731 turbofans and APU units were installed. The CC. Mk 1s were withdrawn from service April 1994.

Series 600A/B/CC. Mk 2: Larger, faster development of the Series 400 with 20 per cent more payload. Series 600A developed for the North American market; Series 600B for the rest of the world. Both versions with Viper 601 engines. In 1983 the CC. Mk 2s were retrofitted with Garrett TFE731 turbofan engines and fitted with APU units. The designation was not changed.

Series 700A/B/CC. Mk 3: Improved version with Garrett TFE731-3-1RH turbofan engines. See separate entry.

Series 700 II: Update of Series 700 aircraft by Arkansas Aerospace. See separate entry in this section.

Raytheon Hawker 800/C-29/U125A: See *Jane's All The World's Aircraft 1995-96.*

Raytheon Hawker 1000: See *Jane's All The World's Aircraft 1995-96.*

UPGRADES: **Arkansas Aerospace:** (HS) 125 700 II. See separate entry in this section.

British Aerospace: CC. Mk 1/2. In 1983 British Aerospace retrofitted Garrett TFE731 engines into both the CC. Mk 1 and CC. Mk 2 aircraft of the RAF. APU units were also installed. Neither designations were changed.

FLS: Dominie overhaul and conversion. See separate entry.

Garrett: Garrett has retrofitted TFE731 turbofan engines in more than 60 Hawker 125 aircraft.

Kanematsu: The Kanematsu Corporation of Japan is prime contractor with Fuji as systems integrator in a programme to upgrade the aircraft for the JASDF. See separate entry.

Racal Radar Defence Systems/Marshall Aerospace: Dominie T. Mk 2: See Versions.

The following details refer specifically to Raytheon (HS) 125-600:

OPERATORS: Versions of the Hawker 125 are in service with the armed forces of the following countries: Botswana (1), Brazil (12), Japan (5), Malawi (1), Saudi Arabia (4), South Africa (4), and the United Kingdom (20).

DESIGN FEATURES: Cantilever low-wing monoplane. Thickness/chord ratio 14 per cent at root, 11 per cent at tip. Dihedral 2°. Incidence 2° 6′ at root, –0° 24′ at tip. Sweepback 20° at quarter-chord. Fence on each upper surface at approximately two-thirds span. Cantilever tail unit has fixed incidence tailplane mounted on fin. Small fairings on tailplane undersurface to eliminate turbulence around elevator hinge cutouts. Triangular ventral fin and extended dorsal fin.

FLYING CONTROLS: Mass balanced ailerons operated by cable linkage. Trim tab and geared tab in port aileron, two geared tabs in starboard aileron. Aileron fences to improve lateral stability. Large four-position double-slotted flaps (45° travel compared with 50° on Series 400), actuated hydraulically via a screwjack on each flap. Mechanically operated hydraulic cutout prevents asymmetric operation of the flaps. Flat-plate spoilers above and below each wing forming part of flap shrouds provide lift dumping facility during landing, and have interconnected controls to prevent asymmetric operation. Tail unit control surfaces operated manually via cable linkage. Tabs in rudder and each elevator.

STRUCTURE: Wings built in one piece and dished to pass under fuselage, to which they are attached by four vertical links, a side link and a drag spigot. The wing is an all-metal two-spar structure with partial centre spar approximately two-thirds span, sealed to form integral fuel tankage which is divided into two compartments by centreline rib. Skins are single one-piece units on each of the upper and lower

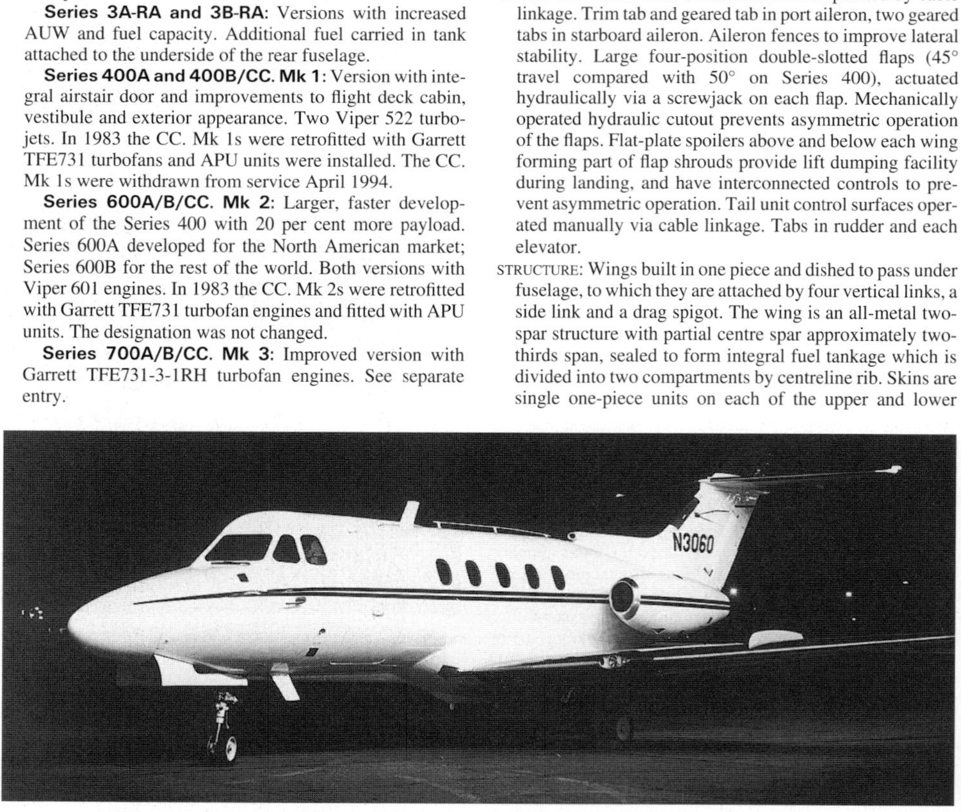

Raytheon Hawker (HS/BAe) 125-1A twin-jet business aircraft ***(Raytheon)*** *1995*

semi-spans. Detachable leading-edges. The fuselage is an all-metal semi-monocoque fail-safe structure making extensive use of Redux bonding. Constant circular cross-section over much of its length. Compared with Srs 400, the Series 600 has an extra 0.61 m (2 ft 0 in) cabin section added forward of the wings, and 12 cabin windows instead of 10; the nose radome is redesigned and is 152 mm (6 in) longer. The tail unit is a cantilever all-metal structure.

LANDING GEAR: Retractable tricycle type, with twin-wheels on each unit. Hydraulic retraction on all units; nosewheels forward, mainwheels inward, into wings. Oleo-pneumatic shock-absorbers. Fully castoring nose unit, steerable 45° to left or right. Dunlop mainwheels and 10 ply tyres, size 23 × 7-12, pressure 8.27 bars (120 lb/sq in). Dunlop nosewheels and 6 ply tyres, size 18 × 4.25-10, pressure 5.17 bars (75 lb/sq in). Dunlop double-disc hydraulic brakes with Maxaret anti-skid units on all mainwheels.

POWER PLANT: Two Rolls-Royce Bristol Viper 601-22 turbojet engines, each rated at 16.7 kN (3,750 lb st), pod-mounted on sides of rear fuselage. Hot-air anti-icing of intake lips and bullets. Integral fuel tanks in wings, with total capacity of 4,673 litres (1,233 US gallons; 1,028 Imp gallons). Overwing refuelling point near each wingtip. Rear underfuselage tank of 509 litres (61 US gallons; 112 Imp gallons) capacity, with refuelling point on starboard side, and 231 litres (51 Imp gallon) dorsal fin tank, raising overall total capacity to 5,414 litres (1,430 US gallons; 1,191 Imp gallons) of which 5,368 litres (1,418 gallons; 1,181 Imp gallons) are usable. Self-contained engine re-oiling system, capacity 15.5 litres (4.1 US gallons; 3.4 Imp gallons).

ACCOMMODATION: Crew of two on flight deck, which is fully soundproofed, insulated and air conditioned. Dual controls standard. Seat provided for third crew member. Standard executive layout has seating for eight passengers, with forward baggage compartment, refreshment bar and coat compartment (forward) and toilet (aft). Compared with Srs 400, there are smoother-line roof panels, with individual recessed lights and air louvres. Cabin restyling offers the operator a choice of interchangeable furnishing units to suit individual requirements. The new wider seats, which on the Srs 600A swivel through 180°, are adjustable fore and aft and sideways, have adjustable lumbar support, and can be reclined hydraulically up to 40°. Typical executive furnishing includes a couch for three persons and five individual seats, foldaway conference table and individual foldaway wall tables. Alternative high-density layout is available, seating up to 14 passengers. Outward-opening door at front on port side, with integral airstairs. Emergency exit over wing on starboard side. Windscreen demisting by engine bleed air; electrical windscreen anti-icing, with methanol spray back-up.

SYSTEMS: AiResearch air conditioning and pressurisation system. Maximum cabin differential 0.58 bars (8.35 lb/sq in), maintaining S/L cabin pressure up to 6,550 m (21,500 ft). Oxygen system standard, with dropout masks for passengers. Hydraulic system, pressure 159 to 207 bars (2,300 to 3,000 lb/sq in), for operation of landing gear, mainwheel doors, flaps, spoilers, nosewheel steering, mainwheel brakes and anti-skid units. Two accumulators provide emergency hydraulic power for wheel brakes in case of a main system failure. Independent auxiliary system for lowering landing gear and flaps in the event of a main system failure. DC electrical system utilises two 300 A 9 kW engine-driven starter/generators and two 24 V 25 Ah batteries. A 24 V 3.5 Ah battery provides separate power for igniter and starter control circuits. AC electrical system includes two 115 V 2.5 kVA 400 Hz three-phase rotary inverters and one 250VA solid-state standby inverter for electronics, and one engine-driven 115 V 3 kVA frequency-wild alternator for windscreen anti-icing. Ground power receptacle on starboard side at rear of fuselage for 28 V external DC supply. AiResearch GTCP-30-92 auxiliary power unit is standard on Srs 600B; Solar T-62T-39 is optional on Srs 600A. Engine ice protection system supplied by engine bleed air. Graviner triple FD Firewire fire warning system and two BCF engine fire extinguishers. TKS liquid system using porous stainless steel leading-edge panels for de-icing or anti-icing on wing and tail unit leading-edges.

ELECTRONICS AND EQUIPMENT: Standard equipment includes full blind-flying instrumentation, complete ice protection system, stick-shaker stall warning, and electrically heated rudder auto-bias to apply corrective rudder during asymmetric engine power conditions. A spring and *g* weight are included in the elevator circuit to reduce variations in stick force to a minimum over a wide CG range. Compared with the Srs 400, the layout of flight deck instrumentation has been completely redesigned, all systems (including the electrical and ice protection systems) have been refined, and a new central warning system is incorporated. A combined slot/stereo tape unit and FM/AM self-seeking radio are fitted as standard in Srs 600B, together with storage for additional tape cartridges, magazines and stationery. Comprehensive electronics, available to customer's requirements, include an automatic flight system comprising autopilot (typically, Sperry SP40C or Bendix PB60 for Srs 600A, Collins AP104 for Srs 600A and 600B), flight director and compass; dual VHF nav/com; HF com; dual ADF; marker; ATC transponder; DME; and weather radar. Doppler, Decca Navigator, flight data recorder and passenger address system may also be installed. Equipment for ICAO Cat. 2 low weather minima operation will be available as an option. A feature console is provided for fitting customer-specified optional items such as digital readouts and a telephone.

Raytheon (HS/BAe) 125-600 CC Mk 2 of the RAF ***(Raytheon)***

1995

DIMENSIONS, EXTERNAL:
Wing span 14.33 m (47 ft 0 in)
Wing chord (mean) 2.29 m (7 ft 6¼ in)
Wing aspect ratio 6.25
Length overall 15.39 m (50 ft 6 in)
Height overall 5.26 m (17 ft 3 in)
Fuselage: Max diameter 1.93 m (6 ft 4 in)
Tailplane span 6.10 m (20 ft 0 in)
Wheel track (c/l of shock-absorbers) 2.79 m (9 ft 2 in)
Wheelbase 6.34 m (20 ft 9½ in)
Passenger door (fwd, port): Height 1.30 m (4 ft 3 in)
Width 0.69 m (2 ft 3 in)
Height to sill 1.07 m (3 ft 6 in)
Emergency exit (overwing, stbd):
Height 0.91 m (3 ft 0 in)
Width 0.51 m (1 ft 8 in)

DIMENSIONS, INTERNAL:
Cabin (excl flight deck): Length 6.50 m (21 ft 4 in)
Max width 1.80 m (5 ft 10¾ in)
Max height 1.75 m (5 ft 9 in)
Floor area 5.11 m² (55.0 sq ft)
Volume 17.8 m³ (628.0 cu ft)
Baggage compartment 0.84 m³ (29.6 cu ft)

AREAS:
Wings, gross 32.8 m² (353.0 sq ft)
Ailerons (total) 2.76 m² (29.76 sq ft)
Trailing-edge flaps (total) 5.21 m² (56.06 sq ft)
Fin, incl dorsal fin 5.31 m² (57.15 sq ft)
Ventral fin 0.61 m² (6.61 sq ft)
Horizontal tail surfaces (total) 9.29 m² (100 sq ft)

WEIGHTS AND LOADINGS:
Weight empty 5,683 kg (12,530 lb)
Typical operating weight empty 6,148 kg (13,555 lb)
Max payload 907 kg (2,000 lb)
Max T-O and ramp weight 11,340 kg (25,000 lb)
Max zero-fuel weight 7,053 kg (15,550 lb)
Max landing weight 9,979 kg (22,000 lb)
Max wing loading 346 kg/m² (70.8 lb/sq ft)
Max power loading 339.5 kg/kN (3.33 lb/lb st)

PERFORMANCE (initial certification, at max T-O weight except where indicated):
Never-exceed speed 375 kts (695 km/h; 432 mph) IAS
Max design Mach number in dive 0.85
Max operating speed: fuselage fuel tanks empty
320 kts (592 km/h; 368 mph) IAS
fuel in fuselage fuel tanks
280 kts (519 km/h; 322 mph) IAS
Max operating Mach number 0.78
Max cruising speed at 8,534 m (28,000 ft)
454 kts (840 km/h; 522 mph)
Econ cruising speed at 11,890 m (39,000 ft)
403 kts (747 km/h; 464 mph)
Rough-air speed 230 kts (426 km/h; 265 mph) IAS
Landing gear operation speed
220 kts (407 km/h; 253 mph) IAS
Flap operating speed:
T-O 220 kts (407 km/h; 253 mph) IAS
approach 175 kts (324 km/h; 201.5 mph) IAS
landing 160 kts (296.5 km/h; 184 mph) IAS
Stalling speed, flaps down
83 kts (155 km/h; 96 mph) EAS
Max rate of climb at S/L 1,493 m (4,900 ft)/min
Rate of climb at S/L, one engine out
420 m (1,380 ft)/min
Servicing ceiling 12,500 m (41,000 ft)
T-O run 1,341 m (4,400 ft)
T-O balanced field length 1,631 m (5,350 ft)
Landing from 15 m (50 ft) at typical landing weight, unfactored 649 m (2,130 ft)
Landing run (scheduled performance):
Srs 600A at typical landing weight 1,036 m (3,400 ft)
Srs 600A at max landing weight 1,295 m (4,250 ft)
Srs 600B at 7,167 kg (15,800 lb) landing weight
1,137 m (3,730 ft)
Min ground turning radius (inside wheel)
4.70 m (15 ft 5 in)
Runway LCN requirement at max T-O weight 10
Typical range with 454 kg (1,000 lb) payload, 45 min reserves plus allowances for T-O, approach, landing and taxying 1,650 n miles (3,057 km; 1,900 miles)
Range with max fuel and max payload, reserves as above 1,560 nm (2,891 km; 1,796 miles)

UPDATED

RAYTHEON (HS/BAe) 125 SERIES 700

TYPE: Twin-turbofan business transport aircraft.

PROGRAMME: Introduced in 1976 with Garrett TFE731 turbofan engines. The prototype was produced by the conversion of a Series 600 airframe and flew for the first time 28 June 1976.

The use of turbofan engines gives an improved specific fuel consumption in comparision to turbojets in earlier versions of the HS 125. The Series 700 also meets all existing and proposed international noise regulations. Turbofan conversions of existing turbojet-powered HS 125s are available.

New production Series 700s embody many other refinements in addition to the change of power plant. As in the case of earlier versions, the intended market is indicated by a suffix: the Series 700A is for the North American market, Series B for the rest of the world.

Improvements to the airframe, to reduce drag and enhance its appearance, include redesign of the wing keel skid, redesign of the ventral fin and adjacent fairings in glassfibre, enlargement of the area of the ventral fin to improve directional stability, deletion of the NACA cooling air intake introduced in the nose of the Series 600, and addition of fairings over the windscreen wiper blades and two ADF loop aerials.

New interior equipment and furnishings include the use of figured walnut veneer on cabin tables and toilet consoles, leather trim, provision of a Blaupunkt Bamberg combined radio/cassette stereo player and recorder, a luxury toilet compartment, digital cabin clock, slide-out portable bar box, full harness on sideways-facing seats, improved lifejacket stowage under seats, improved plug-in meal tray for divan occupants, and a new range of interior colour schemes.

The first flight of a production HS 125 Series 700 was made 8 November 1976. UK certification was received 7 April 1977. The first sale had been made before completion of the prototype. More than 188 orders were received by 1982, 130 of which were from customers in North America. Military operators of the HS 125 include the Irish Air Corps (1) and the RAF (12).

The following description applies specifically to the Series 700:

DESIGN FEATURES: Cantilever low-wing monoplane. Wing thickness/chord ratio 14 per cent at root, 11 per cent at tip. Dihedral 2°. Incidence 2° 6′ at root, −0° 24′ at tip. Sweepback 20° at quarter-chord. Wings built in one piece and dished to pass under fuselage, to which they are attached by four vertical links, a side link and a drag spot. The tail unit is a cantilever all-metal structure with fixed incidence tailplane on fin. Small fairings on tailplane underside to eliminate turbulance around elevator hinge cutouts. Triangular ventral fin, and extended dorsal fin.

FLYING CONTROLS: Mass balanced ailerons, operated manually by cable linkage. Trim tab and geared tab in port aileron, two geared tabs in starboard aileron. Aileron fences to improve lateral stability. Large, four-position double-slotted flaps, actuated hydraulically via a screwjack on each flap. Mechanically operated hydraulic cutout prevents asymmetric operation of flaps. Airbrakes above and below each wing, forming part of flap shrouds, provide lift dumping facility during landing, and have interconnected controls to prevent asymmetric operation. Tail unit control surfaces operated manually via cable linkage. Tabs in rudder and each elevator.

Raytheon Hawker (HS) 125 Series 700 powered by Garrett AiResearch TFE731 turbofan engines

1992

STRUCTURE: The wing is an all-metal two-spar fail-safe structure with partial centre spar of approximately two-thirds span, sealed to form integral fuel tankage which is divided into two compartments by centreline rib. Skins are single-piece units on each of the upper and lower semi-spans. Detachable leading-edges. Fence on each upper surface at approximately two-thirds span. The fuselage is an all-metal fail-safe structure, making extensive use of Redox bonding. Constant circular cross-section over much of its length.

LANDING GEAR: Retractable tricycle type, with twin-wheels on each unit. Hydraulic retraction of all units; nosewheels forward, mainwheels inward into wings. Oleo-pneumatic shock-absorbers. Fully castoring nose unit, steerable 45° to left or right. Dunlop mainwheels and 10 ply tyres, size 23 × 7-12, pressure 8.75 bars (127 lb/sq in). Dunlop nosewheels and 6 ply tyres, size 18 × 4.25-10, pressure 5.51 bars (80 lb/sq in). Dunlop double-disc hydraulic brakes with Maxaret anti-skid units on all mainwheels.

POWER PLANT: Two 16.46 kN (3,700 lb st) Garrett TFE731-3R-1H turbofan engines, pod-mounted on sides of rear fuselage, in pods designed and manufactured by Grumman Aerospace. Engine intake anti-icing by engine bleed air. Integral fuel tanks in wings, with total capacity of 4,628 litres (1,222 US gallons; 1,018 Imp gallons). Single pressure refuelling point in lower starboard side of rear fuselage. Overwing refuelling point near each wingtip. Rear underfuselage tank of 509 litres (134 US gallons; 112 Imp gallons) capacity, with refuelling point on starboard side, and 232 litres (61 US gallons; 51 Imp gallons) dorsal fin tank, raising overall total capacity to 5,369 litres (1,418 US gallons; 1,181 Imp gallons) of which 5,323 litres (1,406 US gallons; 1,171 Imp gallons) are usable.

ACCOMMODATION: Crew of two on flight deck, which is fully soundproofed, insulated and air conditioned. Dual controls. Seat provided for third crew member. Standard executive layout has seating for eight passengers, with forward baggage compartment, refreshment bar and coat compartment (forward) and toilet (aft). There are individual recessed lights and air louvres. Cabin styling offers the operator a choice of interchangeable furnishing units to suit individual requirements. The wide seats, which on the Srs 700A swivel through 180°, are adjustable fore and aft and sideways, have adjustable lumbar support, and can be reclined hydraulically up to 40°. Typical executive furnishing includes a couch for three persons, five individual seats, and individual foldaway wall tables. Alternative high-density layout is available, seating up to 14 passengers. Outward-opening door at front on port side, with integral airstairs. Emergency exit over wing on starboard side. Edge-heating for windscreen. Electrical windscreen anti-icing, with methanol spray back-up.

SYSTEMS: Garrett air conditioning and pressurisation system. Maximum cabin differential 0.58 bars (8.35 lb/sq in), maintaining S/L cabin pressure up to 6,550 m (21,500 ft). Oxygen system standard, with dropout masks for passengers. Hydraulic system, pressure 186 to 207 bars (2,700 to 3,000 lb/sq in), for operation of landing gear, mainwheel doors, flaps, spoilers, nosewheel steering, mainwheel brakes and anti-skid units. Two accumulators provide emergency hydraulic power for wheel brakes in case of a main system failure. Independent auxiliary system for lowering landing and flaps in the event of a main system failure. DC electrical system utilises two 12 kW engine-driven starter/generators and two 24 V 25 Ah Ni/Cd batteries. A 24 V 3.5 Ah battery provides separate power for igniter and starter control circuits. AC electrical system includes two 115 V 2.5 kVA 400 Hz three-phase static inverters and one 250 VA solid-state standby inverter for avionics and one engine-driven 120 V 4.4 VA frequency-wild alternator for windscreen anti-icing. Ground power receptacle on starboard side at rear of fuselage for 28 V external DC supply. Garrett GTCP-30-92 auxiliary power unit is standard of Srs 700B. Engine ice protection system supplied by engine bleed air. Graviner triple FD Firewire fire warning system and two BCF engine fire extinguishers. TKS liquid system using porous stainless steel leading-edge system for de-icing or anti-icing.

AVIONICS AND EQUIPMENT: Standard avionics include dual Collins VHF-20A com transceivers, Collins 718U-5HF com transceiver, Collins VIR-30 VHF nav receivers with dual marker beacon indicators, dual Collins DF-206 ADF, dual Collins MC-103 compasses, dual Collins DME, Marconi AD1550 audio control and passenger address system, RCA Primus 400 weather radar, dual Collins TDR-90 ATC transponder and Blaupunkt Bamberg stereo tape and AM/FM radio. Collins APS-80 autopilot and FDS-80 flight director system standard, to provide altitude hold, altitude preselect airspeed hold, Mach number hold, vertical speed hold, aircraft heading, VOR/LOC, ILS approach and pitch with electric trim.

DIMENSIONS, EXTERNAL:
Wing span 14.33 m (47 ft 0 in)
Wing chord (mean) 2.29 m (7 ft 6¼ in)
Wing aspect ratio 6.25
Length overall 15.46 m (50 ft 8½ in)
Height overall 5.36 m (17 ft 7 in)
Fuselage: Max diameter 1.93 m (6 ft 4 in)
Tailplane span 6.10 m (20 ft 0 in)
Wheel track (c/l of shock-absorbers) 2.79 m (9 ft 2 in)
Wheelbase 6.34 m (20 ft 9½ in)
Passenger door (fwd, port): Height 1.30 m (4 ft 3 in)
Width 0.69 m (2 ft 3 in)
Height to sill 1.07 m (3 ft 6 in)
Emergency exit (overwing, stbd):
Height 0.91 m (3 ft 0 in)
Width 0.51 m (1 ft 8 in)

DIMENSIONS, INTERNAL:
Cabin (excl flight deck): Length 6.50 m (21 ft 4 in)
Max width 1.80 m (5 ft 10¾ in)
Max height 1.75 m (5 ft 9 in)
Floor area 5.11 m² (55.0 sq ft)
Volume 17.10 m³ (604.0 cu ft)
Baggage compartment volume 0.84 m³ (29.6 cu ft)

AREAS:
Wings, gross 32.8 m² (353.0 sq ft)
Ailerons (total) 2.76 m² (29.76 sq ft)
Trailing-edge flaps (total) 5.21 m² (56.06 sq ft)
Fin, incl dorsal fin 5.31 m² (57.15 sq ft)
Ventral fin 0.61 m² (6.61 sq ft)
Horizontal tail surfaces (total) 9.29 m² (100 sq ft)

WEIGHTS AND LOADINGS:
Weight empty 5,826 kg (12,845 lb)
Typical operating weight empty 6,270 kg (13,822 lb)
Max payload 1,124 kg (2,478 lb)
Max ramp and T-O weight 11,566 kg (25,500 lb)
Max zero-fuel weight 7,392 kg (16,300 lb)
Max landing weight 9,979 kg (22,000 lb)
Max wing loading 352.5 kg/m² (72.2 lb/sq ft)
Max power loading 351 kg/kN (3.45 lb/lb st)

PERFORMANCE (at max T-O weight except where indicated):
Never-exceed speed Mach 0.85
Max level speed at S/L 320 kts (592 km/h; 368 mph) IAS
Max cruising speed at 8,380 m (27,500 ft) 436 kts (808 km/h; 502 mph)
Econ cruising speed at 11,275-12,500 m (37,000-41,000 ft) 390 kts (723 km/h; 449 mph)
Stalling speed, flaps down 83 kts (155 km/h; 96 mph) EAS
Service ceiling 12,500 m (41,000 ft)
T-O run 1,367 m (4,484 ft)
T-O to 10.7 m (35 ft), unfactored 1,448 m (4,750 ft)
T-O balanced field length 2,042 m (6,700 ft)

Raytheon (HS/BAe) 125-700 CC Mk 3 aircraft of the Royal Air Force *(Raytheon)*

1995

Landing from 15 m (50 ft) at landing weight of 7,167 kg (15,800 lb), unfactored 619 m (2,030 ft)
Landing run at landing weight of 6,804 kg (15,000 lb) 1,143 m (3,750 ft)
Range with max fuel and max payload, allowances for T-O, approach, landing, taxying and 45 min reserves 2,420 n miles (4,482 km; 2,785 miles)

OPERATIONAL NOISE LEVELS (FAR Pt 36):

T-O	87.6 EPNdB
Approach	96.3 EPNdB
Sideline	88.9 EPNdB

UPDATED

Beechcraft T-34C Turbo Mentor of the US Navy ***(Raytheon)***

1995

BEECHCRAFT T-34C

US Navy designation: T-34C

TYPE: Two-seat turbine-powered primary trainer and light strike aircraft.

PROGRAMME: Although production ended April 1990, numbers of T-34Cs remain in service. In addition to two prototypes, 353 were built for US Navy and 139 T-34C-1s for export to Argentine Navy, Republic of China (Taiwan) Air Force, Ecuador Air Force and Navy, Gabon Presidential Guard, Indonesian Air Force, Moroccan Air Force, Peruvian Navy, Uruguayan Navy; six civil Turbine Mentor 34Cs supplied to Algerian national pilot training school in 1979.

The original piston-engined Beechcraft Model 45 was built as the T-34A Mentor for the USAF (450) and as the similar T-34B for the US Navy (423). Design of the upgraded T-34C with a PT6A-25 turboprop and updated avionics, began in March 1973, and the first of two YT-34C prototypes flew 21 September that year. Under successive US Navy contracts, Beech delivered 334 new production T-34Cs between November 1977 and April 1984. Six T-34s have been transferred to the US Army's, Airborne Special Operations Test Board at Fort Bragg, North Carolina, where they serve as chase and photographic aircraft. In May 1987 the US Navy ordered a further 19 T-34s for delivery between June 1989 and April 1990. By mid-1987 the US Navy's T-34C fleet had logged nearly 1 million flight hours and has established the lowest accident rate for aircraft in the service's current inventory. An export civil version, known as the **Turbine Mentor-34C**, is in service with the Algerian national pilot training school.

VERSIONS: **T-34C-1**: Armament systems trainer version. In addition to its basic role, it is capable of carrying out forward air control (FAC) and tactical strike training missions.

OPERATORS: Versions of the T-34C are in service with the armed forces of the following countries: Argentina (15), Ecuador (17), Gabon (4), Indonesia (24), Morocco (12), Peru (6), Taiwan (43), USA (320) and Uruguay (2).

DESIGN FEATURES: Cantilever low-wing monoplane. Wing section NACA 23016.5 (modified) at root, NACA 23012 at tip. Dihedral 7°. Incidence 4° at root, 1° at tip. No sweepback. Fixed incidence tailplane.

Beechcraft T-34C Turbo-Mentor with HUTTS, TRX-3, TGX-7 and LTC-2 launchers

1994

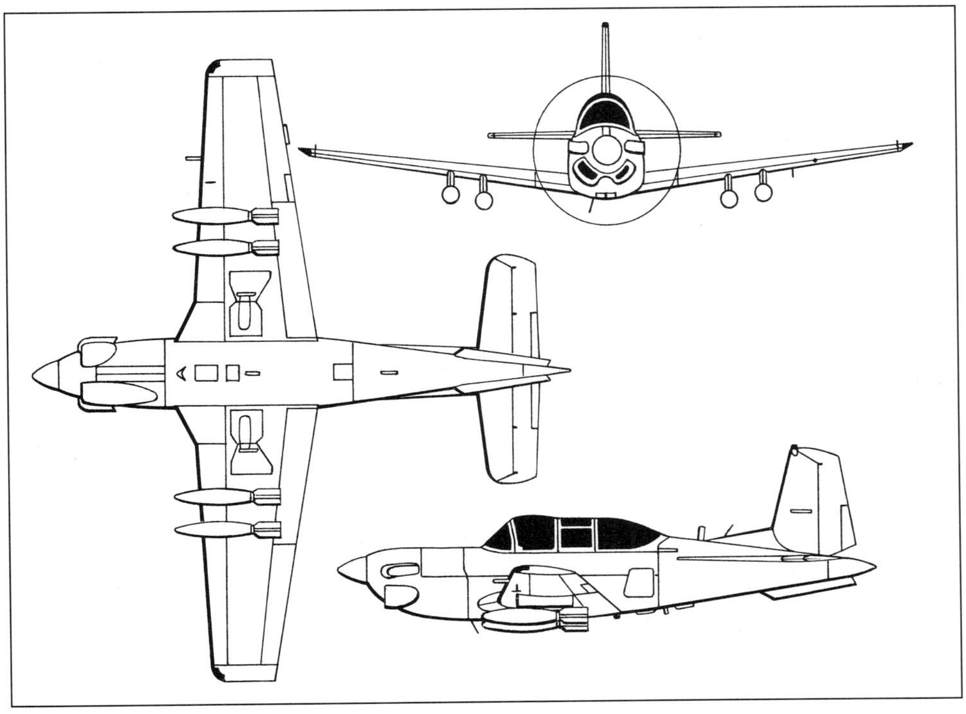

Beechcraft T-34C Turbo-Mentor ***(Jane's/Mike Keep)***

1994

STRUCTURE: Conventional box beam structure of light alloy. Ailerons of light alloy construction. Single-slotted trailing-edge flaps of light alloy. The fuselage is a semi-monocoque light alloy structure. The tail unit is a cantilever structure of light alloy. Twin ventral fins under rear fuselage.

FLYING CONTROLS: Manually operated trim tab in port aileron. Servo tabs in both ailerons. Manually operated trim tabs in elevators and rudder.

LANDING GEAR: Electrically retractable tricycle type. Main units retract inward, nosewheel aft. Beech oleo-pneumatic shock-absorbers. Single wheel on each unit. Mainwheels size 7.00-8, pressure 6.20 bars (90.0 lb/sq in). Nosewheel and tyre size 5.00-5 pressure 4.83 bars (70.0 lb/sq in). Goodyear multiple-disc hydraulic brakes.

POWER PLANT: One 533 kW (715 shp) Pratt & Whitney Canada PT6A-25 turboprop, torque limited to 298 kW (400 shp), driving a Hartzell three-blade constant-speed fully feathering metal propeller with spinner. Version of same engine derated to 410 kW (550 shp) available optionally. Two bladder fuel cells in each wing, in inboard leading-edge and aft of main spar outboard of landing gear; total usuable capacity 492 litres (130 US gallons; 108 Imp gallons). Oil capacity 15 litres (4 US gallons; 3.33 Imp gallons).

ACCOMMODATION: Instructor and pupil in tandem beneath rearward sliding cockpit canopy. Cockpit ventilated, heated by engine bleed air and air conditioned. Dual controls standard. All armament controls in forward cockpit of T-34C-1.

SYSTEMS: Hydraulic system for brakes only. Pneumatic system for emergency opening of cockpit canopy. Diluter demand gaseous oxygen system, pressure 103.5 bars (1,500 lb/sq in). Electrical power supplied by 250 A starter/generator. Freon air conditioner for cockpit cooling.

AVIONICS AND EQUIPMENT: Standard avionics can include UHF or VHF com, VOR or Tacan nav, DME transponder, angle of attack indicator, ADF, marker beacon receiver, compass and intercom system. R/nav, Loran HF and specialised tactical systems available to customer's requirements. US Navy T-34C has ARC-159V UHF com, VIR-30A VOR/Omni, dual 255Y-1 ICS/audio, Tacan and PN-101 remote compass, all by Collins; Two TDR transponders and a CIR-11-2 emergency locator transmitter. Blind-flying instrumentation standard. Electrically heated pitot.

ARMAMENT (T-34C-1): CA-513 fixed-reticle reflector gunsight. Four underwing hardpoints are provided for the carriage of stores. The inboard stations are rated at 272 kg (600 lb) each, the outboard stations at 136 kg (300 lb) each, with maximum load of 272 kg (600 lb) each side and 544 kg (1,200 lb) total. Weapons which can be carried on MA-4 racks include AF/B37K-1 bomb containers with practice bombs or flares, LAU-32 or LAU-59 rocket pods, Mk 81 bombs, SUU-11 Minigun pods, BLU-10/B incendiary bombs, AGM-22A wire-guided anti-tank missiles and TA8X towed target equipment.

DIMENSIONS, EXTERNAL:

Wing span	10.16 m (33 ft 3⅞ in)
Wing chord: at root	2.55 m (8 ft 4½ in)
at tip	1.05 m (3 ft 5¼ in)
Wing aspect ratio	6.2
Length overall	8.75 m (28 ft 8½ in)
Height overall	2.92 m (9 ft 7 in)
Tailplane span	3.71 m (12 ft 2⅛ in)
Wheel track	2.95 m (9 ft 8 in)
Wheelbase	2.41 m (7 ft 11 in)
Propeller diameter	2.29 m (7 ft 6 in)
Propeller ground clearance	0.29 m (11½ in)

DIMENSIONS, INTERNAL:

Cabin: Length	2.74 m (9 ft 0 in)
Max width	0.86 m (2 ft 10 in)
Max height	1.22 m (4 ft 0 in)

AREAS:

Wings, gross	16.69 m² (179.69 sq ft)
Ailerons (total)	1.06 m² (11.4 sq ft)
Trailing-edge flaps (total)	1.98 m² (21.3 sq ft)
Fin	1.20 m² (12.9 sq ft)
Rudder, incl tab	0.64 m² (6.9 sq ft)
Tailplane	3.46 m² (37.2 sq ft)
Elevators, incl tabs	1.26 m² (13.6 sq ft)

Beechcraft V35B Bonanza V-tail monoplane ***(Raytheon)***

1995

WEIGHTS AND LOADINGS:

Weight empty: T-34C	1,342 kg (2,960 lb)
T-34C-1	1,356 kg (2,990 lb)
Max T-O and landing weight:	
T-34C	1,950 kg (4,300 lb)
T-34-1, strike role	2,494 kg (5,500 lb)
Max ramp weight: T-34C	1,962 kg (4,325 lb)
Max wing loading: T-34C	108.3 kg/m² (22.2 lb/sq ft)

PERFORMANCE (T-34C, at max T-O weight of 1,910 kg; 4,210 lb, except where indicated):

Never-exceed speed	280 kts (518 km/h; 322 mph)
Max cruising speed at 5,180 m (17,000 ft)	214 kts (396 km/h; 246 mph)
Stalling speed, flaps down, power off, at typical landing weight of 1,588 kg (3,501 lb)	53 kts (98 km/h; 61 mph)
Max rate of climb at S/L	451 m (1,480 ft)/min
Service ceiling	over 9,145 m (30,000 ft)
T-O run	352 m (1,155 ft)
T-O run to 15 m (50 ft)	586 m (1,920 ft)
Landing run from 15 m (50 ft)	547 m (1,795 ft)
Landing run	226 m (740 ft)
Range with fuel:	
at 181 kts (335 km/h; 208 mph) at 305 m (1,000 ft)	427 n miles (790 km; 491 miles)
at 202 kts (374 km/h; 232 mph) at 3,050 m (10,000 ft)	523 n miles (968 km; 601 miles)
at 180 kts (333 km/h; 207 mph) at 6,100 m (20,000 ft)	708 n miles (1,311 km; 814 miles)
g limits	+6/−3

PERFORMANCE (T-34C-1 with 410 kW; 550 shp engine, estimated A: with two stores at AUW of 2,222 kg; 4,900 lb, B: with four stores at AUW of 2,494 kg; 5,500 lb except where indicated):

Max level speed at 5,500 m (18,000 ft):	
A	209 kts (387 km/h; 241 mph)
B	206 kts (382 km/h; 237 mph)
Stalling speed, flaps down, idle power:	
A	65 kts (120 km/h; 75 mph) CAS
B	69 kts (128 km/h; 80 mph) CAS
Max rate of climb at S/L: A	540 m (1,771 ft)/min
B	436 m (1,431 ft)/min
Typical combat radius:	
FAC mission at AUW of 2,429 kg (5,355 lb), with four stores and optional max fuel; incl 2.6 h loiter over target and 20 min +5% reserves	100 n miles (185 km; 115 miles)
Strike mission at AUW of 2,473 kg (5,452 lb), with four stores and optional max fuel, incl 20 min +5% reseves	300 n miles (555 km; 345 miles)

UPDATED

1995

BEECHCRAFT BONANZA MODEL V35B

TYPE: Four/five-seat light cabin monoplane.

PROGRAMME: The prototype V-tail Bonanza flew for the first time 22 December 1945. The type went into production 1947 and was certificated 25 March of that year. Production of the V35 ended in 1985 after 10,390 had been built.

Bonanzas are equipped with a dual-duct fresh air system to increase cabin air flow. Safety features include single diagonal strap shoulder harness with inertia reel for all occupants as standard equipment. Three optional factory installed IFR avionics packages include dual communication, dual navigation, ADF marker beacon receiver, glide slope, DME, and transponder. The packages meet FAA Technical Standard Order (TSO). Beech was, in 1972, the first general aviation manufacturer to acquire approval on production aircraft with IFR equipment.

DESIGN FEATURES: Cantilever low-wing monoplane. Wing section Beech modified NACA 23012 at tip. Dihedral 6°. Incidence 4° at root, 1° at tip. Sweepback 0° at quarter-chord. The tail unit is a cantilever V tail, with tailplane and elevators set at 33° dihedral angle.

STRUCTURE: Each wing is a two-spar semi-monocoque box-beam of conventional aluminium construction. Symmetrical section ailerons and single-slotted three-position flaps of aluminium alloy. The fuselage is a conventional aluminium alloy semi-monocoque structure. Hat section longerons and channel type keels extend forward from cabin section, making the support structure for the engine and nosewheel an integral part of the fuselage. Semi-monocoque construction. Fixed surfaces have aluminium alloy structure and skin. Control surfaces, aft of the light alloy spar, are primarily of magnesium alloy construction.

FLYING CONTROLS: Ground adjustable trim tab in each aileron. Large controllable trim in each tail control surface. Tail surfaces are interchangeable port and starboard, except for tabs and actuator horns. Electrically operated elevator trimming optional.

LANDING GEAR: Electrically retractable tricycle type, with steerable nosewheel. Mainwheels retract inward into wings, nosewheel aft. Beech oleo-pneumatic shock-absorbers in all units. Cleveland mainwheels, size 6.00-6 and tyres size 7.00-6, pressure 2.28 to 2.76 bars (33 to 40 lb/sq in). Cleveland nosewheel and tyre size 5.00-5, pressure 2.76 bars (40 lb/sq in). Cleveland ring-disc hydraulic brakes. Parking brake, 'Magic Hand' landing gear system optional.

POWER PLANT: One 212.5 kW (285 hp) Continental IO-520-BB flat-six engine, driving a McCauley two-blade constant-speed metal propeller with spinner. Three-blade propeller optional. Manually adjustable engine cowl flaps. Two standard fuel tanks in wing leading-edges, with total fuel capacity of 280 litres (74 US gallons; 61.7 Imp gallons). Refuelling points above tanks. Oil capacity 11.5 litres (3 US gallons; 2.5 Imp gallons).

ACCOMMODATION: Enclosed cabin seating four or five persons on individual seats. Centre windows open for ventilation on ground and have release handles to permit their use as emergency exits. Pilot's storm window, port side. Cabin structure reinforced for protection in turnover. Space for up to 122.5 kg (270 lb) of baggage aft of seats. Passenger door and baggage access door both on starboard side, cabin heated and ventilated.

SYSTEMS: Optional 12,000 BTU refrigeration type air conditioning system comprises evaporator located beneath pilot's seat, condenser on lower fuselage and engine-mounted compressor. Air outlets on centre console, with two-speed blower. Electrical system supplied by 28 V 60 A alternator, 24 V 15.5 Ah battery; a 100 A alternator is available as an option, as is a standby generator. Hydraulic system for brakes only. Pneumatic system for instrument gyros and refrigeration type air conditioning system optional. Oxygen system optional.

AVIONICS AND EQUIPMENT: Standard avionics comprise King KX-155-09 720-channel com transceiver, 200-channel nav receiver with KI 208 VOR/LOC converter-indicator, microphone, headset and cabin speaker. A wide range of optional avionics available by Collins and King. King and S-Tec autopilots, Sperry WeatherScout radar and 3M/Ryan Stormscope optional. Standard equipment includes electric clock, outside air temperature gauge, rate of climb indicator, sensitive altimeter, turn co-ordinator, 3 in horizon and directional gyros, four fore and aft adjustable and reclining seats, armrests, headrests, shoulder harness and lapbelts, pilot's storm window, sun visors, ultraviolet-proof windscreen and windows, emergency locator transmitter, stall warning device, carpeted floor, coat hooks, glove compartment, in-flight storage pockets, approach plate holder, utility shelf, cabin dome light, reading lights, instrument panel floodlights, electroluminescent sub-panel lighting, landing light, taxi light, full-flow oil filter, polyurethane exterior paint, external power socket and towbar. Optional equipment includes control wheel clock, exhaust gas temperature gauge, dual controls, co-pilot's wheel brakes, alternate static source heated pitot, fifth seat, large cargo door, super soundproofing control wheel map lights, entrance door courtesy light, instrument post lights, internally lit instruments, fresh air vent blower, rotating beacons, three light strobe system, and static wicks.

Also available is a Beech designed 'Magic Hand' landing gear safety system. Designed to eliminate the possibility of wheels-up landing or inadvertent retraction of the landing gear on the ground, it lowers the gear automatically on approach when the engine manifold pressure falls below approximately 508 mm (20 in) and airspeed has been reduced to 104 knots (193 km/h; 120 mph). On take-off, it keeps the gear down until the aircraft is airborne and has accelerated to 78 knots (145 km/h; 90 mph) IAS. The system can be switched off by pilot at will.

DIMENSIONS, EXTERNAL:

Wing span	10.21 m (33 ft 6 in)
Wing chord: at root	2.13 m (7 ft 0 in)
at tip	1.07 m (3 ft 6 in)
Wing aspect ratio	6.2
Length overall	8.05 m (26 ft 5 in)
Height overall	2.31 m (7 ft 7 in)
Tailplane span	3.10 m (10 ft 2 in)
Wheel track	2.92 m (9 ft 7 in)
Wheelbase	2.13 m (7 ft 0 in)
Propeller diameter: two-blade	2.13 m (7 ft 0 in)
three-blade	2.03 m (6 ft 8 in)
Passenger door: Height	0.91 m (3 ft 0 in)
Width	0.94 m (3 ft 1 in)
Baggage compartment door:	
Height	0.57 m (1 ft 10½ in)
Width	0.47 m (1 ft 6½ in)

DIMENSIONS, INTERNAL:

Cabin, aft of firewall: Length	3.07 m (10 ft 1 in)
Max width	1.07 m (3 ft 6 in)
Max height	1.27 m (4 ft 2 in)
Volume	3.31 m³ (117 cu ft)
Baggage space	0.99 m³ (35 cu ft)

AREAS:

Wings, gross	16.80 m² (181 sq ft)
Ailerons (total)	1.06 m² (11.4 sq ft)
Trailing-edge flaps (total)	1.98 m² (21.3 sq ft)
Fixed tail surfaces	2.20 m² (23.8 sq ft)
Movable tail surfaces, incl tabs	1.34 m² (14.4 sq ft)

WEIGHTS AND LOADINGS:

Weight empty, standard	955 kg (2,106 lb)
Max T-O and landing weight	1.542 kg (3,400 lb)
Max ramp weight	1,547 kg (3,412 lb)
Max wing loading	91.8 kg/m² (18.80 lb/sq ft)
Max power loading	7.26 kg/kW (11.93 lb/hp)

PERFORMANCE (at max T-O weight, except cruising speeds at mid-cruise weight):

Max level speed at S/L	182 kts (338 km/h; 209 mph)
Cruising speed: 75% power at 1,830 m (6,000 ft)	172 kts (319 km/h; 198 mph)
65% power at 3,050 m (10,000 ft)	168 kts (311 km/h; 193 mph)
55% power at 3,660 m (12,00 ft)	157 kts (291 km/h; 181 mph)

Beechcraft V35B Bonanza four/five-seat light monoplane

1992

45% power at 2,440 m (8,000 ft)
136 kts (253 km/h; 157 mph)

Stalling speed, power off:

flaps up	64 kts (118 km/h; 74 mph) IAS
30% flap	51 kts (94 km/h; 59 mph) IAS
Max rate of climb at S/L	356 m (1,167 ft)/min
Service ceiling	5,443 m (17,858 ft)
T-O run	305 m (1,002 ft)
T-O run to 15 m (50 ft)	539 m (1,769 ft)
Landing from 15 m (50 ft)	404 m (1,324 ft)
Landing run	233 m (763 ft)

Range with 280 l (74 US gallons; 61.7 Imp gallons) usable fuel, allowances for engine start, taxi, T-O climb and 45 min reserves at 45% power:
75% power at 1,830 m (6,000 ft)
716 n miles (1,326 km; 824 miles)
65% power at 3,050 m (10,000 ft)
777 n miles (1,440 km; 894 miles)
55% power at 3,660 m (12 000 ft)
838 n miles (1,553 km; 964 miles)
45% power at 2,440 m (8,000 ft)
889 n miles (1,648 km; 1,023 miles)

VERIFIED

BEECHCRAFT BARON MODEL E55

TYPE: Four/six-seat cabin monoplane.

PROGRAMME: The Baron E55 had its origins in the Baron 95-C55 when that model was added to the Baron series of twin-engined aircraft August 1965. Compared to the B55, the 95-C55 had Continental IO-520-C engines, a pneumatic vacuum system for instrument gyros and the optional wing and tail unit de-icing system, two 24 V 50 A engine-driven alternators, increased tailplane span, swept vertical surfaces and an extended nose baggage compartment. It was followed by the D55 in October 1967, this model introducing a pneumatic pressure system in place of the pneumatic vacuum system. The subsequent Model E55, which has an improved interior and systems accessory refinements, was licenced in the Normal category on 12 November 1969.

Beech had delivered over 1,000 of this Baron series by the end of 1984.

UPGRADES: **Colemill Enterprises:** Foxstar Baron. See separate entry.

DESIGN FEATURES: Cantilever low-wing monoplane. Wing section NACA 23016.5 at root, NACA 23010.5 at tip. Dihedral 6°. Incidence 4° at root, 0° at tip. No sweepback.

FLYING CONTROLS: Electrically operated single-slotted light alloy trailing-edge flaps, with beaded skins. Manually operated trim tab in port aileron. Tail has manually operated trim tab in each elevator and in rudder.

STRUCTURE: Each wing is a two-spar semi-monocoque box beam of conventional aluminium alloy construction. Symmetrical section ailerons of light alloy construction, with beaded skins. The fuselage is a semi-monocoque aluminium alloy structure. Hat section longerons and channel type keels extend forward from the cabin section, making the support structure for the forward nose section and nosewheel gear an integral part of the fuselage. The tail unit is a cantilever all-metal structure. Elevators have smooth magnesium alloy skins.

LANDING GEAR: Electrically retractable tricycle type. Main units retract inward into wings, nosewheel aft. Beech oleo-pneumatic shock-absorbers in all units. Steerable nosewheel with shimmy damper. Cleveland wheels, with mainwheel tyres size 6.50-8, pressure 3.59 to 3.86 bars (52 to 56 lb/sq in). Nosewheel tyre size 5.00-5, pressure 3.79 to 4.14 bars (55 to 60 lb/sq in). Cleveland ring-disc hydraulic brakes. Heavy duty brakes optional. Parking brake.

POWER PLANT: Two 212.5 kW (285 hp) Continental IO-520-CB flat-six engines, each driving a Hartzell two-blade constant-speed fully feathering metal propeller with spinner. Hartzell three-blade propellers optional. Standard fuel system comprises two interconnected tanks in each wing leading-edge, with total usable capacity of 378 litres (100 US gallons; 83.25 Imp gallons). Optional interconnected fuel tanks may be added in each wing to provide a total usable capacity of 628 litres (166 US gallons; 138 Imp gallons). Single refuelling point in each wing for the standard or optional fuel systems. Optional fuel system includes a mechanical sight gauge in each wing leading-edge to give partial fuelling information. Oil capacity 23 litres (6 US gallons; 5 Imp gallons). Full flow oil filters standard. Propeller de-icing, unfeathering accumulator and propeller synchrophaser optional.

ACCOMMODATION: Standard model has four individual seats in pairs in enclosed cabin, with door on starboard side. Single diagonal strap shoulder harness with inertia reel standard on all seats. Optional wider door for cargo. Folding airline style fifth and sixth seats optional, complete with shoulder harness and inertia reel. Baggage compartments aft of cabin and in nose, both with external doors on starboard side and with capacity of 181 kg (400 lb) and 136 kg (300 lb) respectively. Rear compartment extension for an additional 54 kg (120 lb) of baggage standard. Pilot's storm window. Openable windows adjacent the third and fourth seats are used for ground ventilation and as emergency exits. Cabin heated and ventilated. Windscreen defrosting standard. Alcohol de-icing for port side of windscreen optional.

Beechcraft E55 in background, with Model B55 foreground

1992

SYSTEMS: Cabin heated by Janitrol 50,000 BTU heater, which serves also for windscreen defrosting. Oxygen system of 1.41 m³ (49.8 cu ft) or 1.87 m³ (66 cu ft) capacity optional. Electrical system includes two 28 V 60 A engine-driven alternators with alternator failure lights and two 12 V 25 Ah batteries. Two 100 A alternators optional. Hydraulic system for brakes only. Pneumatic pressure system for air-driven instruments, and optional wing and tail unit de-icing system. Oxygen system optional. Cabin air conditioning and windscreen electric anti-icing system optional. Pneumatic rubber de-icing boots optional for wings and rudder.

AVIONICS AND EQUIPMENT: Standard avionics comprise King KX 155-09 720-channel com transceiver and 200-channel nav receiver with KI VOR/LOC converter indicator, microphone, headset and cabin speaker. King KR 87 ADF and KI 227 indicator standard. Optional weather radars are available, by Bendix, King and Sperry, and a wide range of optional avionics by Bendix, Collins, King Edo-Aire Mitchell and Narco. Standard equipment includes blind-flying instrumentation, clock, outside air temperature gauge, sensitive altimeter, turn co-ordinator, pilot's storm window, sun visors, ultraviolet-proof windscreen and windows, armrests, adjustable and retractable starboard side rudder pedals, emergency locator transmitter, heated pitot, carpeted floor, glove compartment, hatshelf, headrests for passenger seats, cabin door courtesy light, cabin dome light instrument panel floodlights, map light, trim tab position indicator lights, navigation lights, position lights, dual landing lights, taxi light, soundproofing, heated fuel vents, towbar, external power socket, and engine winterisation kit. Optional equipment includes control wheel clock or chronograph, approach plate holder, engine and flight hour recorders, instantaneous vertical speed indicator, exhaust gas temperature gauge, dual tachometers with synchroscope, dual alternate static source, cabin fire extinguisher, cabin ventilation blower, super soundproofing, instrument post light lights, internally illuminated instruments, rotating beacon strobe lights, wing ice lights and static wicks.

DIMENSIONS, EXTERNAL:

Wing span	11.53 m (37 ft 10 in)
Wing chord: at root	2.13 m (7 ft 0 in)
at tip	0.90 m (2 ft 11⅜ in)
Wing aspect ratio	7.16
Length overall	8.84 m (29 ft 0 in)
Height overall	2.79 m (9 ft 2 in)
Tailplane span	4.85 m (15 ft 11 in)
Wheel track	2.92 m (9 ft 7 in)
Wheelbase	2.46 m (8 ft 1 in)
Propeller diameter: two-blade	1.98 m (6 ft 6 in)
three-blade	1.93 m (6 ft 4 in)
Passenger door: Height	0.91 m (3 ft 0 in)
Width	0.94 m (3 ft 1 in)
Height to step	0.41 m (1 ft 4 in)
Baggage door (fwd): Height	0.56 m (1 ft 10 in)
Width	0.64 m (2 ft 1 in)
Baggage door (rear):	
Standard: Height	0.57 m (1 ft 10½ in)
Width	0.47 m (1 ft 6½ in)
Height to sill	0.71 m (2 ft 4 in)
Optional: Height	0.57 m (1 ft 10½ in)
Width	0.97 m (3 ft 2 in)

DIMENSIONS, INTERNAL:

Cabin: Length (incl rear baggage compartment extension)	3.58 m (11 ft 9 in)
Max width	1.07 m (3 ft 6 in)
Max height	1.27 m (4 ft 2 in)
Baggage compartment: fwd	0.51 m³ (18 cu ft)
rear	0.99 m³ (35 cu ft)
Extension to rear baggage compartment	0.28 m³ (10 cu ft)

AREAS:

Wings, gross	17.47 m² (188.1 sq ft)
Ailerons (total)	1.06 m² (11.40 sq ft)
Trailing-edge flaps (total)	1.98 m² (21.3 sq ft)
Fin	1.46 m² (15.67 sq ft)
Rudder, incl tab	0.81 m² (8.75 sq ft)
Tailplane	4.95 m² (53.30 sq ft)
Elevators, incl tabs	1.84 m² (19.80 sq ft)

WEIGHTS AND LOADINGS:

Weight empty	1.493 kg (3,291 lb)
Max T-O and landing weight	2,405 kg (5,300 lb)
Max ramp weight	2,415 kg (5,324 lb)
Max wing loading	130.0 kg/m² (26.6 lb/sq ft)
Max power loading	5.66 kg/kW (9.3 lb/hp)

PERFORMANCE (at max weight, except cruising speeds at average cruise weight):

Max level speed at S/L 208 kts (386 km/h; 239 mph)
Max cruising speed 77% power at 1,830 m (6,000 ft)
200 kts (370 km/h; 230 mph)
Cruising speed, 66% power at 3,050 m (10,000 ft)
195 kts (362 km/h; 224 mph)
Econ cruising speed, 56% power at 3,660 m (12,000 ft)
184 kts (341 km/h; 212 mph)

Stalling speed, power off:

flaps up	83 kts (154 km/h; 96 mph) IAS
flaps down	73 kts (135 km/h; 84 mph) IAS
Max rate of climb at S/L:	
two-blade	481 m (1,577 ft)/min
three-blade	503 m (1,650 ft)/min
Rate of climb at S/L, one engine out	118 m (388 ft)/min
Service ceiling	5,820 m (19,100 ft)
Service ceiling, one engine out	1,990 m (6,530 ft)
T-O run	401 m (1,315 ft)
T-O to 15 m (50 ft)	625 m (2,050 ft)
Landing from 15 m (50 ft)	671 m (2,202 ft)
Landing run	377 m (1,237 ft)

Range with 628 l (166 US gallons; 138 Imp gallons) usable fuel, with allowances for engine start, taxi, T-O, climb and 45 min reserves at econ cruise power:
max cruising speed at 1,830 m (6,000 ft)
933 n miles (1,728 km; 1,074 miles)
cruising speed at 3,050 m (10,000 ft)
1,032 n miles (1,912 km; 1,188 miles)
econ cruising speed at 3,660 m (12,000 ft)
1,135 n miles (2,103 km; 1,306 miles)

VERIFIED

BEECHCRAFT BARON MODEL 58P

TYPE: Four/six-seat cabin monoplane.

PROGRAMME: Design of this pressurised version of the Model 58 Baron started June 1972; the first flight of the prototype was made in August 1973. Certification under FAR Pt 23 was received in May 1974; the first production aircraft flew later that year. Examples of the Model 58P produced prior to 1979 are powered by two 231 kW (310 hp) Continental TSIO-520-L (or -LB) engines. Later models were fitted with more powerful TSIO-520-WB engines, and introduced propeller synchrophasers as standard equipment in 1981.

Deliveries of production aircraft began in 1975, and a total of 492 58P were produced by the end of 1987. This total includes 18 for the US Forest Service by 1982, for use

Beechcraft Baron Model 58P four/six-seat monoplane

1992

as lead aircraft in smoke jumping operations, as well as for reconnaissance, administration and cargo missions.

UPGRADES: **Colemill Enterprises:** Foxstar Baron. See separate entry.

DESIGN FEATURES: Cantilever low-wing monoplane. Wing section NACA 23016.5 at root, NACA 23010.5 at tip. Dihedral 6°. Incidence 4° at root, 0° at tip. No sweepback.

FLYING CONTROLS: Electrically operated single-slotted light alloy trailing-edge flaps, with beaded skins. Manually operated trim tab in port aileron.

STRUCTURE: Each wing is a two-spar semi-monocoque box beam of conventional aluminium alloy construction. Symmetrical section ailerons of light alloy construction, with beaded skins. The fuselage is as for Model 58 except structural reinforcement to cater for pressurisation. The tail unit is as for the Model 58.

LANDING GEAR: Electrically retractable tricycle type. Main units retract inward, nosewheel rearward; all three units have fairing doors. Beechcraft oleo-pneumatic shock-absorbers. Goodrich mainwheels and tyres size 19.50 × 6.75-8 (10 ply rating), pressure 5.24 to 5.66 bars (76 to 82 lb/sq in). Steerable nosewheel with shimmy damper, with tyre size 5.0-5, pressure 3.79 to 4.14 bars (55 to 60 lb/sq in). Goodrich single-disc hydraulic brakes. Parking brake.

POWER PLANT; Two 242 kW (325 hp) Continental TSIO-520-WB turbocharged flat-six engines, each driving a McCauley three-blade constant-speed fully feathering metal propeller with spinner. Propeller synchrophasers standard, unfeathering accumulators optional. Electrically operated engine cowl flaps. Integral fuel tanks in wings, with standard capacity of 651 litres (172 US gallons; 143 Imp gallons) of which 628 litres (166 US gallons, 138 Imp gallons) are usable. Optional maximum capacity of 724 litres (196 US gallons; 163 Imp gallons) of which 719 litres (190 US gallons; 158 Imp gallons) are usable. Refuelling points in outboard leading-edge of wings and, for optional maximum fuel, in wingtips. Oil capacity 22.7 litres (6 US gallons; 5 Imp gallons). Electric anti-icing for propellers optional.

ACCOMMODATION: Standard accommodation has four individual seats in pairs, facing forward, with shoulder harness and inertia reel belts. Fifth and sixth seats optional, as is club layout. Doors on starboard side, adjacent co-pilot, and at trailing-edge of wing on port side. Baggage space in aft cabin and in fuselage nose, with door on starboard side of nose. Openable storm window for pilot on port side. Cabin heated and pressurised. Air conditioning optional. Windscreen electric anti-icing optional.

SYSTEMS: Garrett pressurisation system with maximum differential of 0.27 bars (3.9 lb/sq in), giving a 3.050 m (10,000 ft) cabin environment to a height of 6,705 m (22,000 ft). Beechcraft 14,000 BTU air conditioning optional. Janitrol 35,000 BTU heater. Engine-driven compressors supply air for flight instruments, pressurisation control and optional pneumatic de-icing boots. Electrical system powered by two 28 V 60 A alternators, with two 12 V 25 Ah storage batteries. Two 24 V 100 A alternators optional. Hydraulic system for brakes only. Oxygen system 0.42 m³ (15 cu ft) optional. Pneumatic rubber de-icing boots optional.

AVIONICS AND EQUIPMENT: Standard avionics package comprises King KX 155 nav/com (720-channel com transceiver and 200-channel nav receiver) with KI 208 VOR/LOC converter-indicator, KR ADF with KI 227 indicator, microphone, headset and cabin speaker. Optional avionics by Bendix, Collins, King, Narco. Edo-Aire Mitchell, Sperry and RCA. Standard equipment as for Model E55, plus dual controls, emergency locator transmitter, heated stall warning transmitter, nose baggage compartment light, door-ajar warning light, step light, dual rotating beacons and exterior urethane paint. Optional equipment includes engine and flight recorders, control wheel chronometer, electrically operated elevator trim, cabin fire extinguisher, executive writing desk, internally illuminated instruments, strobe lights, wing ice lights, and static wicks.

DIMENSIONS, EXTERNAL:
Wing span 11.53 m (37 ft 10 in)
Wing chord: at root 2.13 m (7 ft 0 in)
at tip 0.90 m (2 ft 11½ in)
Length overall 9.12 m (29 ft 11 in)
Height overall 2.79 m (9 ft 2 in)
Tailplane span 4.85 m (15 ft 11 in)
Wheel track 2.92 m (9 ft 7 in)
Wheelbase 2.72 m (8 ft 11 in)
Propeller diameter 1.98 m (6 ft 6 in)
Propeller ground clearance 0.28 m (10¾ in)
Passenger door (stbd, fwd): Height 0.91 m (3 ft 0 in)
Width 0.94 m (3 ft 1 in)
Height to sill 0.51 m (1 ft 8 in)
Passenger door (port, rear): Height 0.89 m (2 ft 11 in)
Width 0.58 m (1 ft 11 in)
Height to sill 0.79 m (2 ft 7 in)
Baggage door (nose, stbd): Height 0.38 m (1 ft 3 in)
Width 0.64 m (2 ft 1 in)

DIMENSIONS, INTERNAL:
Cabin, incl rear baggage area:
Length 3.84 m (12 ft 7 in)
Max width 1.07 m (3 ft 6 in)
Max height 1.27 m (4 ft 2 in)
Floor area 3.72 m² (40 sq ft)
Volume 3.85 m³ (135.9 cu ft)
Baggage compartment: fwd 0.49 m³ (17.2 cu ft)

AREAS:
Wings, gross 17.47 m² (188.1 sq ft)
Ailerons (total incl tabs) 1.06 m² (11.4 sq ft)
Trailing-edge flaps (total) 1.98 m² (21.3 sq ft)
Fin 1.46 m² (15.67 sq ft)
Rudder, incl tab 0.81 m² (8.75 sq ft)
Tailplane 4.95 m² (53.3 sq ft)
Elevators, incl tabs 1.84 m² (19.8 sq ft)

WEIGHTS AND LOADINGS:
Weight empty, equipped 1,826 kg (4,026 lb)
Max T-O and landing weight 2,812 kg (6,200 lb)
Max ramp weight 2,830 kg (6,240 lb)
Max zero-fuel weight 2,585 kg (5,700 lb)
Max wing loading 161.1 kg/m² (33 lb/sq ft)
Max power loading 5.81 kg/kW (9.54 lb/hp)

PERFORMANCE (at max T-O weight, except cruising speeds at average cruise weight):
Max level speed 261 kts (483 km/h; 300 mph)
Max cruising speed at approx 77% power:
at 4,575 m (15,000 ft) 222 kts (412 km/h; 256 mph)
at 6,100 m (20,000 ft) 232 kts (430 km/h; 267 mph)
at 7,620 m (25,000 ft) 241 kts (447 km/h; 277 mph)
Cruising speed at approx 75% power:
at 4,575 m (15,000 ft) 220 kts (407 km/h; 253 mph)
at 6,100 m (20,000 ft) 229 kts (425 km/h; 264 mph)
at 7,620 m (25,000 ft) 237 kts (439 km/h; 273 mph)
Cruising speed at approx 62% power:
at 4,575 m (15,000 ft) 201 kts (372 km/h; 231 mph)
at 6,100 m (20,000 ft) 210 kts (389 km/h; 242 mph)
at 7,620 m (25,000 ft) 218 kts (404 km/h; 251 mph)
Econ cruising speed at approx 53% power:
at 4,575 m (15,000 ft) 185 kts (343 km/h; 213 mph)
at 6,100 m (20,000 ft) 194 kts (359 km/h; 223 mph)
at 7,620 m (25,000 ft) 202 kts (375 km/h; 233 mph)
Stalling speed, power off:
flaps up 84 kts (156 km/h; 97 mph)
flaps down 78 kts (145 km/h; 90 mph)
Max rate of climb at S/L 450 m (1,475 ft)/min
Rate of climb at S/L, one engine out 68 m (223 ft)/min
Service ceiling above 7,620 m (25,000 ft)
Service ceiling, one engine out 3,725 m (12,220 ft)
T-O run 474 m (1,555 ft)
T-O run to 15 m (50 ft) 806 m (2,643 ft)
Landing from 15 m (50 ft) 740 m (2,427 ft)
Landing run 420 m (1,378 ft)
Range with 719 l (190 US gallons; 158 Imp gallons) usable fuel, and allowances for engine start, taxi, T-O, climb and 45 min reserves at econ cruising speed:
at approx 77% power: at 4,575 m (15,000 ft)
917 n miles (1,699 km; 1,056 miles)
at 6,100 m (20,000 ft)
960 n miles (1,779 km; 1,105 miles)
at 7,620 m (25,000 ft)
1,013 n miles (1,877 km; 1,166 miles)
at approx 75% power: at 4,575 m (15,000 ft)
930 n miles (1,723 km; 1,071 miles)
at 6,100 m (20,000 ft)
975 n miles (1,807 km; 1,122 miles)
at 7,620 m (25,000 ft)
1,030 n miles (1,908 km; 1,186 miles)
at approx 65% power: at 4,575 m (15,000 ft)
1,098 n miles (2,035 km; 1,264 miles)
at 6,100 m (20,000 ft)
1,122 n miles (2,079 km; 1,292 miles)
at 7,620 m (25,000 ft)
1,160 n miles (2,150 km; 1,335 miles)
at approx 53% power: at 4,575 m (15,000 ft)
1,189 n miles (2,182 km; 1,356 miles)
at 6,100 m (20,000 ft)
1,220 n miles (2,261 km; 1,405 miles)
at 7,620 m (25,000 ft)
1,235 n miles (2,288 km; 1,422 miles)

VERIFIED

BEECHCRAFT BARON MODEL 58TC

TYPE: Four/six-seat cabin monoplane.

PROGRAMME: This turbocharged version of the Baron Model 58 is generally similar to the Model 58P, sharing the same power plant, but is unpressurised and has detail differences in the airframe and equipment. The design originated in July 1974, and construction of a prototype to production standard began February 1975. The first flight of this aircraft was made 31 October 1975 and FAA certification in the Normal category was granted 23 January 1976. Deliveries began in June 1976. A total of 149 Baron Model 58TCs were built.

UPGRADES: **Colemill Enterprises:** Foxstar Baron. See separate entry.

DESIGN FEATURES: Generally similar to Model 58P, which is as follows: Cantilever low-wing monoplane. Wing section NACA 23016.5 at root, NACA 23010.5 at tip. Dihedral 6°. Incidence 4° at root, 0° at tip. No sweepback.

FLYING CONTROLS: Symmetrical section ailerons of light alloy construction, with beaded skins. Electrically operated single-slotted light alloy trailing-edge flaps, with beaded skins. Manually operated trim tab in port aileron.

STRUCTURE: Each wing is a two-spar semi-monocoque box beam of conventional aluminium alloy construction. The fuselage is as for Model 58.

LANDING GEAR: Electrically retractable tricycle type. Main units retract inward, nosewheel rearward; all three units have fairing doors. Beechcraft oleo-pneumatic shock-absorbers. Goodrich mainwheels and tyres size 19.50 × 6.75-8 (10 ply rating), pressure 5.24 to 5.66 bars (76 to 82 lb/sq in). Steerable nosewheel with shimmy damper, with tyre size 5.0-5, pressure. 3.79 to 414 bars (55-60 lb/sq in). Goodrich single-disc hydraulic brakes. Parking brake.

POWER PLANT: Two 242 kW (325 hp) Continental TSIO-520-WB turbocharged flat-six engines each driving a McCauley three-blade constant-speed fully feathering metal propeller with spinner. Propeller synchrophasers standard, unfeathering accumulators optional. Electrically operated engine cowl flaps. Integral fuel tanks in wings, with standard capacity of 651 litres (172 US gallons; 143 Imp gallons) of which 628 litres (166 US gallons; 138 Imp gallons) are usable. Optional maximum capacity of 724 litres (196 US gallons; 163 Imp gallons) of which 719 litres (190 US gallons; 158 Imp gallons) are usable. Refuelling points in outboard leading-edge of wings and, for optional maximum fuel, in wingtips. Oil capacity 22.7 litres (6 US gallons; 5 Imp gallons). Electric anti-icing for propellers optional.

ACCOMMODATION: Generally similar to the Model 58 which is as follows: Standard model has four individual seats in pairs in enclosed cabin, with door on starboard side. Single diagonal strap shoulder harness with inertia reel standard on all seats. Vertically adjusting pilot's seat is standard. Vertically adjusting co-pilot's seat, folding fifth and sixth seats, or club seating comprising folding fifth and sixth seats and aft-facing third and fourth seats, are optional. Executive writing desk available as option with club seating. Baggage compartment in nose, capacity 136 kg (300 lb). Double passenger/cargo doors on starboard side of cabin provide access to space for 181 kg (400 lb) of baggage or cargo behind the third and fourth seats. Pilot's storm window. Openable windows adjacent the third and fourth seats are used for ground ventilation and as emergency exits. Cabin heated and ventilated. Windscreen defrosting standard.

SYSTEMS: Generally similar to Model 58P, which is as follows: Garrett pressurisation system with maximum differential of 0.27 bars (3.9 lb/sq in), giving a 3,050 m (10,000 ft) cabin environment to a height of 6,705 m (22,000 ft). Beechcraft 14,000 BTU air conditioned optional. Janitrol 35,000 BTU heater. Engine-driven compressors supply air for flight instruments, pressurisation control and optional pneumatic de-icing boots. Electrical system powered by two 28 V 60 A alternators, with two 12 V 25 Ah storage batteries. Two 24 V 100 A alternators optional. Hydraulic system for brakes only. Oxygen system of 0.4 m³ (15 cu ft) optional. The 58TC, however has Beech freon air conditioning system optional. 50,000 BTU heater standard. Hydraulic system for brakes and propeller unfeathering only. Electrical power from two 24 V 60 A alternators, with two 12 V 25 Ah storage batteries. Two 100 A alternators are optional. Oxygen system to supply crew and passengers is optional. Pneumatic rubber de-icing boots optional.

AVIONICS AND EQUIPMENT: Standard avionics package comprises King KX 155 nav/com (720-channel com transceiver and 200-channel nav receiver) with KI 208 VOR/LOC converter-indicator, KR ADF with KI 227 indicator, microphone, headset and cabin speaker. Optional avionics by Bendix, Collins, King, Narco, Edo-Aire Mitchell, Sperry and RCA. Standard equipment as for Model E55, plus dual controls, emergency locator transmitter, heated stall warning transmitter, nose baggage compartment light, door-ajar warning light, step light, dual rotating beacons and exterior urethane paint. Optional equipment includes engine and flight recorders, control wheel chronometer, electrically operated elevator trim, cabin fire extinguisher, executive writing desk, internally illuminated instruments, strobe lights, wing ice lights, and static wicks.

DIMENSIONS, EXTERNAL:
- Wing span: 11.53 m (37 ft 10 in)
- Wing chord: at root: 2.13 m (7 ft 0 in)
  - at tip: 0.90 m (2 ft 11½ in)
- Length overall: 9.12 m (29 ft 11 in)
- Height overall: 2.79 m (9 ft 2 in)
- Tailplane span: 4.85 m (15 ft 11 in)
- Wheel track: 2.92 m (9 ft 7 in)
- Wheelbase: 2.72 m (8 ft 11 in)
- Propeller diameter: 1.98 m (6 ft 6 in)
- Propeller ground clearance: 0.28 m (10¾ in)
- Passenger door (stbd, fwd): Height: 0.91 m (3 ft 0 in)
  - Width: 0.94 m (3 ft 1 in)
  - Height to sill: 0.51 m (1 ft 8 in)
- Passenger door (port, rear): Height: 0.89 m (2 ft 11 in)
  - Width: 0.58 m (1 ft 11 in)
  - Height to sill: 0.79 m (2 ft 7 in)
- Baggage door (nose, stbd): Height: 0.56 m (1 ft 10 in)
  - Width: 0.64 m (2 ft 1 in)
- Emergency exit window (port and starboard):
  - Height: 0.53 m (1 ft 9 in)
  - Width: 0.61 m (2 ft 0 in)

DIMENSIONS, INTERNAL:
- Cabin, incl rear baggage area:
  - Length: 3.84 m (12 ft 7 in)
  - Max width: 1.07 m (3 ft 6 in)
  - Max height: 1.27 m (4 ft 2 in)
  - Floor area: 3.72 m² (40.0 sq ft)
  - Volume: 3.85 m³ (135.9 cu ft)
- Baggage compartment: fwd: 0.49 m³ (17.2 cu ft)

AREAS:
- Wings, gross: 17.47 m² (188.1 sq ft)
- Ailerons (total, incl tabs): 1.06 m² (11.4 sq ft)
- Trailing-edge flaps (total): 1.98 m² (21.3 sq ft)
- Fin: 1.46 m² (15.67 sq ft)
- Rudder, incl tab: 0.81 m² (8.75 sq ft)
- Tailplane: 4.95 m² (53.3 sq ft)
- Elevators, incl tabs: 1.84 m² (19.8 sq ft)

WEIGHTS AND LOADINGS:
- Weight empty, equipped: 1,720 kg (3,793 lb)
- Max T-O and landing weight: 2,812 kg (6,200 lb)
- Max ramp weight: 2,830 kg (6,240 lb)
- Max zero-fuel weight: 2,585 kg (5,700 lb)
- Max wing loading: 160.96 kg/m² (32.96 lb/sq ft)
- Max power loading: 5.81 kg/kW (9.55 lb/hp)

PERFORMANCE (at max T-O weight, except cruising speeds at average cruise weight):
- Max level speed: 261 kts (483 km/h; 300 mph)
- Max cruising speed at approx 77% power:
  - at 4,575 m (15,000 ft): 222 kts (412 km/h; 256 mph)
  - at 6,100 m (20,000 ft): 232 kts (430 km/h; 267 mph)
  - at 7,620 m (25,000 ft): 241 kts (447 km/h; 277 mph)
- Cruising speed at approx 75% power:
  - at 4,575 m (15,000 ft): 220 kts (407 km/h; 253 mph)
  - at 6,100 m (20,000 ft): 229 kts (425 km/h; 264 mph)
  - at 7,620 m (25,000 ft): 237 kts (439 km/h; 273 mph)
- Cruising speed at approx 62% power:
  - at 4,575 m (15,000 ft): 201 kts (372 km/h; 231 mph)
  - at 6,100 m (20,000 ft): 210 kts (389 km/h; 242 mph)
  - at 7,620 m (25,000 ft): 218 kts (404 km/h; 251 mph)
- Econ cruising speed at approx 53% power:
  - at 4,575 m (15,000 ft): 185 kts (343 km/h; 213 mph)
  - at 6,100 m (20,000 ft): 194 kts (359 km/h; 223 mph)
  - at 7,620 m (25,000 ft): 202 kts (375 km/h; 233 mph)
- Stalling speed, power off:
  - flaps up: 84 kts (156 km/h; 97 mph)
  - flaps down: 78 kts (145 km/h; 90 mph)
- Max rate of climb at S/L: 450 m (1,475 ft)/min
- Rate of climb at S/L, one engine out: 68 m (223 ft)/min

Beechcraft Duchess 76 four-seat monoplane *(Raytheon)*

1995

- Service ceiling: above 7,620 m (25,000 ft)
- Service ceiling, one engine out: 3,725 m (12,220 ft)
- T-O run: 474 m (1,555 ft)
- T-O run to 15 m (50 ft): 806 m (2,643 ft)
- Landing from 15 m (50 ft): 740 m (2,427 ft)
- Landing run: 420 m (1,378 ft)
- Range with 719 l (190 US gallons; 158 Imp gallons) usable fuel, and allowances for engine start, taxi, T-O climb and 45 min reserves at econ cruising speed:
  - at approx 77% power: at 4,575 m (15,000 ft): 917 n miles (1,699 km; 1,056 miles)
    - at 6,100 m (20,000 ft): 960 n miles (1,779 km; 1,105 miles)
    - at 7,620 m (25,000 ft): 1,013 n miles (1,877 km; 1,166 miles)
  - at approx 75% power: at 4,575 m (15,000 ft): 930 n miles (1,723 km; 1,071 miles)
    - at 6,100 m (20,000 ft): 975 n miles (1,807 km; 1,122 miles)
    - at 7,620 m (25,000 ft): 1,030 n miles (1,908 km; 1,186 miles)
  - at approx 65% power: at 4,575 m (15,000 ft): 1,098 n miles (2,035 km; 1,264 miles)
    - at 6,100 m (20,000 ft): 1,122 n miles (2,079 km; 1,292 miles)
    - at 7,620 m (25,000 ft): 1,160 n miles (2,150 km; 1,335 miles)
  - at approx 53% power: at 4,575 m (15,000 ft): 1,189 n miles (2,182 km; 1,356 miles)
    - at 6,100 m (20,000 ft): 1,220 n miles (2,261 km; 1,405 miles)
    - at 7,620 m (25,000 ft): 1,235 n miles (2,288 km; 1,422 miles)

VERIFIED

BEECHCRAFT DUCHESS 76

TYPE: Four-seat cabin monoplane.

PROGRAMME: The Duchess 76, a four-seat twin-engined light aircraft, flew for the first time 24 May 1977, and received FAA certification on 24 January 1978. A testbed version of this aircraft, designated PD 289, has been undergoing a comprehensive flight test programme since 1974. Following certification in the Normal category for day and night VFR and IFR, the first production deliveries were made May 1978. A total of 429 were built.

The Duchess 76 was planned for use by Beech Aero Centers, and is designed for the personal light twin, light charter and multi-engine flight trainer markets. Emphasis has been placed on good low-speed flight and single-engine handling characteristics, and counter-rotating propellers are fitted.

Factory installed optional equipment packages include:

Weekender: Comprising sun visors; tinted windscreen and windows; landing navigation strobe, cabin dome, map, overhead instrument and instrument post lights; cabin boarding steps; propeller unfeathering accumulators; and acrylic enamel paint; adding 16.8 kg (37 lb) to basic empty weight.

Holiday: As above, plus coat hook and garment hanger in baggage compartment; instrument group comprising electric clock, 3 in horizon and directional gyros and pressure system, outside air temperature gauge, rate of climb indicator, turn co-ordinator, and two 12 V 25 Ah batteries; adding 31.7 kg (70 lb) to basic empty weight.

Professional: As above, plus true airspeed indicator; two-seat headrests; heated pitot tube; and wing-mounted taxi lights; adding 34.9 kg (77 lb) to basic empty weight.

Nine factory installed avionics packages also available include: com transceiver; nav receiver-converter, with VOR/LOC indicator, ADF; transponder; and audio panel; from Collins and King. An extensive selection of additional avionics, including autopilot, DME, glide slope and marker beacon receiver systems, available optionally from these manufacturers, and also from Bendix.

Beechcraft Duchess 76 on grass runway

1992

DESIGN FEATURES: Cantilever low-wing monoplane. Wing section NACA 63-2A415 with modified root section. Dihedral 6° 30′. Incidence 3° at root, 0° 38′ 39″ at tip. Wings and modified Frise ailerons of light alloy bonded honeycomb construction. Fixed incidence tailplane.

FLYING CONTROLS: Electrically operated single-slotted trailing-edge flaps of light alloy. Trim tab in rudder and each elevator.

STRUCTURE: The fuselage is a semi-monocoque structure of light alloy. The tail unit is a conventional cantilever T tail structure of light alloy with swept vertical surfaces.

LANDING GEAR: Hydraulically retractable tricycle type, with single wheel on each unit. Self-centring steerable nosewheel retracts forward, main units inward. Oleo-pneumatic shock-absorbers. Mainwheels with tubed tyres size 17.5 x 6.00-6. Nosewheel with tubed tyre size 5.00-5. Hydraulic brakes.

POWER PLANT: Two 134.2 kW (180 hp). Avco Lycoming O-360-A1G6D counter-rotating flat-four engines, each driving a Hartzell Type HC-M2YR-2C(L)EUF/F(J)C 7666A two-blade constant-speed fully feathering metal propeller with spinner. One fuel tank in each wing, giving a combined usable capacity of 378.5 litres (100 US gallons; 83 Imp gallons). Refuelling point on upper surface of each wing. Oil capacity 15 litres (4 US gallons; 3.3 Imp gallons).

ACCOMMODATION: Two individual front seats, adjustable fore and aft, with reclining seatbacks. Rear bench seat for two passengers. Shoulder harness and lap belt for each seat. Door on each side of cabin. Baggage compartment (capacity 90 kg; 200 lb) with external door on port side. Pilot's storm window. Accommodation heated and ventilated. Windscreen defroster.

SYSTEMS: 28 V electrical system supplied by two 55 A engine-driven alternators, 24 V 15.5 Ah battery. Electrically driven hydraulic pump for landing gear retraction. Separate hydraulic system for brakes. 45,000 BTU heater.

AVIONICS AND EQUIPMENT: Optional avionics as noted earlier. Standard equipment includes instrument panel glareshield, sensitive altimeter, control lock, map and storage pockets, emergency locator transmitter, stall warning device, armrests, carpeted floor, super soundproofing, utility shelf, tie-down rings and towbar. Optional equipment includes Hobbs hour meter, ventilation, internal corrosion proofing, and external power socket.

DIMENSIONS, EXTERNAL:
- Wing span: 11.58 m (38 ft 0 in)
- Wing chord: at root: 1.71 m (5 ft 7½ in)
  - at tip: 1.36 m (4 ft 5¾ in)
- Wing aspect ratio: 7.973
- Length overall: 8.86 m (29 ft 0½ in)
- Height overall: 2.90 m (9 ft 6 in)
- Tailplane span: 3.81 m (12 ft 6 in)
- Wheel track: 3.25 m (10 ft 8 in)
- Wheelbase: 2.13 m (7 ft 0 in)
- Propeller diameter: 1.93 m (6 ft 4 in)

Propeller ground clearance 0.25 m (10 in)
Cabin doors (port stbd): Height 0.86 m (2 ft 10 in)
Width 0.97 m (3 ft 2 in)
Baggage door (port): Height 0.84 m (2 ft 9 in)
Width 0.56 m (1 ft 10 in)

DIMENSIONS, INTERNAL:
Cabin: Length 2.41 m (7 ft 11 in)
Max width 1.12 m (3 ft 8 in)
Max height 1.22 m (4 ft 0 in)
Baggage compartment 0.55 m³ (19.5 cu ft)

AREAS:
Wings, gross 16.81 m² (181 sq ft)
Ailerons (total) 0.99 m² (10.7 sq ft)
Trailing-edge flaps (total) 2.29 m² (24.7 sq ft)
Fin 1.70 m² (18.27 sq ft)
Rudder (incl tab) 0.68 m² (7.3 sq ft)
Tailplane 3.66 m² (39.4 sq ft)
Elevators (incl tabs) 1.28 m² (13.77 sq ft)

WEIGHTS AND LOADINGS:
Weight empty 1,119 kg (2,466 lb)
Max T-O and landing weight 1,769 kg (3,900 lb)
Max ramp weight 1,776 kg (3,916 lb)
Max zero-fuel weight 1,587 kg (3,500 lb)
Max wing loading 105.2 kg/m² (21,5 lb/sq ft)
Max power loading 6.59 kg/kW (10.8 lb/hp)

PERFORMANCE (at max T-O weight):
Never-exceed speed 194 kts (359 km/h; 223 mph)
Max level speed 171 kts (317 km/h; 197 mph)
Max cruising speed at 1,830 m (6,000 ft) 166 kts (308 km/h; 191 mph)
Recommended cruising speed at 3,050 m (10,000 ft) 158 kts (293 km/h; 182 mph)
Econ cruising speed at 3,660 m (12,000 ft) 151 kts (280 km/h; 174 mph)
Stalling speed, power off:
flaps up 70 kts (130 km/h; 81 mph) IAS
flaps down 60 kts (111 km/h; 69 mph) IAS
Max rate of climb at S/L 380 m (1,248 ft)/min
Rate of climb at S/L, one engine out 72 m (235 ft)/min
Service ceiling 5,990 m (19,650 ft)
Service ceiling, one engine out 1,880 m (6,170 ft)
T-O run 310 m (1,017 ft)
T-O to 15 m (50 ft) 573 m (1,881 ft)
Landing run 305 m (1,000 ft)
Range with max fuel, incl allowances for start, taxi, T-O, climb, and 45 min reserves at econ cruise power:
max cruising speed at 1,830 m (6,000 ft) 623 n miles (1,155 km; 717 miles)
recommended cruising speed at 3,050 m (10,000 ft) 711 n miles (1,317 km; 818 miles)
econ cruising speed at 4,265 m (14,000 ft) 843 n miles (1,562 km; 970 miles)

VERIFIED

Beechcraft Sierra 200 four/six-seat light aircraft *(Raytheon)*

1995

BEECHCRAFT SIERRA 200 AND SUNDOWNER 180

TYPE: Four/six-seat light aircraft.

PROGRAMME: In December 1971, Beech introduced a new light aircraft marketing programme centred around three models, which were given individual exterior paint schemes and renamed from their previous Musketeer designations.

In 1974 these designations were changed again to indicate the engine horsepower rating, so becoming Beechcraft Sierra 200 (formerly A24R Musketeer Super R), Sundowner 180 (Model C23, formerly Musketeer Custom).

The Sierra 200 was certificated in 1974 and redesignated Model B24R, due to the installation of a new engine, improved cowling and redesign of control features. Designated Model C24R in 1977, improvements included a propeller of increased diameter to enhance performance, and wheel well fairings to reduce drag when the wheels are retracted.

VERSIONS: **Sundowner 180:** Basic four-seat version with 134 kW (180 hp) Avco Lycoming O-360-A4K engine, driving a Sensenich Type 76EM8S5-0-60 two-blade fixed-pitch metal propeller with spinner, and non-retractable landing gear. Aerobatic version is approved for rolls, Immelman turns, loops, spins, chandelles and other manoeuvres, carrying two persons. Three windows standard on each side of cabin.

Sierra 200: Generally similar to the Sundowner but with accommodation for four to six persons, 149 kW (200 hp) Avco Lycoming IO-360-A1B6 engine, driving a Hartzell Type HC-M2YR-1BF/F7666A two-blade constant-speed metal propeller with spinner, and retractable tricycle landing gear. Electricallly actuated hydraulic system based on a self-contained unit in the rear fuselage, comprising electrically driven hydraulic pump, fluid reservoir and valves. An emergency valve, sited adjacent the pilot's feet, allows selection of the landing gear to free-fall within three seconds. Mainwheels retract outward into wings; nosewheel turns through 90° as it retracts rearwards. Four windows standard on each side of cabin.

Factory installed packages are as follows:

Weekender: Includes sun visors; tinted windscreen and windows; dual controls and pedal operated brakes for co-pilot; lighting group comprising rotating beacon, navigation cabin dome, overhead instrument, glareshield and map lights; cabin boarding steps; and acrylic enamel paint; adding 12.2 kg (27 lb) to basic empty weight.

Holiday: As above, plus instrument group comprising 3 in horizon and direction gyros, with vacuum system, electric clock, outside air temperature gauge, rate of climb indicator and turn co-ordinator; coat hook and garment hanger in baggage compartment; wing-mounted landing light; and two 12 V 25 Ah batteries; adding 26.5 kg (58.4 lb) to basic empty weight.

Professional: As above, plus true airspeed indicator; two headrests; instrument post lights; wing-mounted taxi light; and heated pitot tube; adding 30.2 kg (66.6 lb) to basic empty weight.

Six factory installed avionics packages are available optionally, and include com transceiver, nav receiver-converter, ADF transponder and audio panel from Collins, King or Edo-Aire Mitchell, plus cabin speaker and microphone with jack. An extensive selection of additional avionics also available optionally.

A total of 2,384 Sundowners and 1,146 Sierras were produced.

The following details apply to both models:

DESIGN FEATURES: Cantilever low-wing monoplane. Wing section NACA 63-2A415. Dihedral 6° 30′. Incidence 3° at root, 1° at tip. Single extruded main spar at 50 per cent chord.

FLYING CONTROLS: Slotted all-metal riveted ailerons and mechanically controlled (optionally electrically actuated) flaps with corrugated skin. No trim tabs. One-piece all-moving tailplane with full-span anti-servo tab. Optional electric tailplane trim. Rudder and aileron controls interconnected for easy cross-country flying.

STRUCTURE: Aluminium skin and stringers are bonded to honeycomb Trussgrid ribs on forward 50 per cent of wing; rear 50 per cent of wing is riveted. Plastic wingtips. The fuselage cabin section has basic keel formed by floor and lower skin, with rolled skin side panels, stringers, a minimum number of bulkheads and structural top. Conventional semi-monocoque rear fuselage. The tail unit is a cantilever all-metal structure, with swept vertical surfaces.

LANDING GEAR: (Sundowner): Non-retractable tricycle type. Beech rubber tyre size 15 × 6.00-6, pressure 2.76 bars (40 lb/sq in). Mainwheels have extruded tyres size 17.5 × 6.00-6, pressure 1.52 bars (22 lb/sq in). Cleveland hydraulic disc brakes with toe operated control. Steerable nosewheel. Parking brake.

POWER PLANT: One flat-four engine (details given under model listings). Two fuel tanks in inboard wing leading-edges, with usable capacity of 216 litres (57 US gallons; 47 Imp gallons). Refuelling points above tanks. Oil capacity 7.5 litres (2 US gallons; 1.67 Imp gallons).

ACCOMMODATION: Pilot and three to five passengers in Sierra; pilot and three passengers in Sundowner; in pairs, in enclosed cabin with forward hinged door on each side. Compartment for 122 kg (270 lb) baggage, with external door on port side. In-flight-adjustable seats; pilot's storm window, windscreen defroster, instrument panel glareshield, air vents map stowage, wall to wall carpeting. Optional aerobatic kit for Sundowner includes *g* meter and quick-release door.

SYSTEMS: Electrical system includes a 28 V 60 A alternator; 24 V 15.5 Ah battery standard, two 12 V 25 Ah batteries optional. Hydraulic system for brakes only, except on Sierra which has electrohydraulic actuation system for landing gear. Vacuum system for instruments optional.

AVIONICS AND EQUIPMENT: Optional avionics as listed earlier. Standard equipment includes sensitive altimeter, fore and aft adjustable front seats with reclining backs, a two-place rear bench seat, shoulder harness and lap belts, control locks, map stowage, pilot's storm window, emergency locator transmitter, stall warning device and towbar. Optional equipment includes exhaust gas temperature gauge, Hobbs hour meter, outside air temperature gauge true airspeed indicator, headrests, tinted windscreen and windows, alternate static source, heated pilot, central split seats for two persons, and a two-place rear bench seat (Sierra only), internal corrosion-proofing, instrument post lights, two-light strobe system, wing-mounted taxi light, external power socket and acrylic enamel exterior finish. An aerobatic kit is available for the Sundowner, and a 'Magic Hand' landing gear safety system for the Sierra.

DIMENSIONS, EXTERNAL:
Wing span 9.98 m (32 ft 9 in)
Wing chord, constant 1.34 m (4 ft 4¾ in)
Wing aspect ratio 7.4
Length overall 7.85 m (25 ft 9 in)
Height overall: Sundowner 2.51 m (8 ft 3 in)
Sierra 2.46 m (8 ft 1 in)
Tailplane span 3.30 m (10 ft 10 in)
Wheel track: Sundowner 3.61 m (11 ft 10 in)
Sierra 3.86 m (12 ft 8 in)
Wheelbase: Sundowner 1.93 m (6 ft 4 in)
Sierra 1.83 m (6 ft 0¼ in)
Propeller diameter 1.93 m (6 ft 4 in)
Propeller ground clearance:
Sundowner 0.36 m (1 ft 2 in)
Sierra 0.33 m (1 ft 1 in)
Cabin doors: Height 0.86 m (2 ft 10 in)
Width 0.96 m (3 ft 2 in)
Baggage compartment door:
Sundowner: Height 0.47 m (1 ft 6½ in)
Width 0.60 m (1 ft 11¾ in)

Beechcraft Sundowner 180. This aircraft acting as experimental model for liquid methane powered engine *(Raytheon)*

1995

Sierra: Height 0.84 m (2 ft 9 in)
Width 0.56 m (1 ft 10 in)

DIMENSIONS, INTERNAL:
Cabin, aft of instrument panel:
Length 2.41 m (7 ft 11 in)
Max width 1.18 m (3 ft 8 in)
Max height 1.22 m (4 ft 0 in)
Floor area 2.40 m² (25.84 sq ft)
Volume 2.92 m³ (103.2 cu ft)
Baggage compartment 0.55 m³ (19.5 cu ft)

AREAS:
Wings, gross 13.57 m² (146.0 sq ft)
Ailerons (total) 1.27 m² (13.65 sq ft)
Flaps (total) 1.74 m² (18.76 sq ft)
Fin 0.64 m² (6.88 sq ft)
Rudder 0.45 m² (4.82 sq ft)
Tailplane, incl anti-servo tab 2.52 m² (27.08 sq ft)

WEIGHTS AND LOADINGS:
Weight empty (incl oil and unusable fuel):
Sundowner 678 kg (1,494 lb)
Sierra 768 kg (1,694 lb)
T-O weight, Utility category:
Sundowner 920 kg (2,030 lb)
Max T-O and landing weight:
Sundowner 1,111 kg (2,450 lb)
Sierra 1,247 kg (2,750 lb)
Max wing loading:
Sundowner 81.9 kg/m² (16.78 lb/sq ft)
Sierra 91.9 kg/m² (18.84 lb/sq ft)
Max power loading:
Sundowner 8.29 kg/kW (13.61 lb/hp)
Sierra 8.37 kg/kW (13.75 lb/hp)

PERFORMANCE (at max T-O weight):
Max level speed at S/L:
Sundowner 128 kts (237 km/h; 147 mph)
Sierra 145 kts (269 km/h; 167 mph)
Cruising speed: Sundowner, 84% power at 1,370 m (4,500 ft)
126 kts (233 km/h; 145 mph)
Sundowner, 59% power at 1,370 m (4,500 ft)
105 kts (194 km/h; 121 mph)
Sierra 75% power at 3,050 m (10,000 ft)
137 kts (254 km/h; 158 mph)
Sierra, 55% at 3,050 m (10,000 ft)
115 kts (213 km/h; 132 mph)
Stalling speed, flaps down, power off:
Sundowner 51 kts (95 km/h; 59 mph) IAS
Sierra 60 kts (111 km/h; 69 mph) IAS
Max rate of climb at S/L:
Sundowner 241 m (792 ft)/min
Sierra 283 m (927 ft)/min
Service ceiling: Sundowner 3,840 m (12,600 ft)
Sierra 4,690 m (15,385 ft)
Absolute ceiling: Sundowner 4,390 m (14,400 ft)
Sierra 5,315 m (17,430 ft)
Min ground turning radius:
Sundowner 7.29 m (23 ft 11 in)
Sierra 8.18 m (26 ft 10 in)
T-O run: Sundowner 344 m (1,130 ft)
Sierra 324 m (1,063 ft)
T-O run to 15 m (50 ft): Sundowner 596 m (1,955 ft)
Sierra 476 m (1,561 ft)
Landing from 15 m (50 ft):
Sundowner 452 m (1,484 ft)
Sierra 446 m (1,462 ft)
Landing run: Sundowner 214 m (703 ft)
Sierra 249 m (816 ft)
Range with max fuel, allowances for warm-up, T-O, climb and 45 min reserves:
Sundowner, 2,300 rpm at 1,370 m (4,500 ft)
641 n miles (1,187 km; 737 miles)
Sierra, 2,400 rpm at 3,050 m (10,000 ft)
686 n miles (1,271 km; 790 miles)

VERIFIED

BEECHCRAFT QUEEN AIR B80 and QUEEN AIRLINER B80

TYPE: Six/eleven-seat business aircraft, commuter airliner and utility aircraft.

PROGRAMME: The prototype of the original Queen Air 80, which introduced more powerful engines than those of the A65, flew for the first time 22 June 1961 and received its FAA Type Certificate 20 February 1962. It was followed in January 1964 by the Queen Air A80, with increased wing span and AUW, new interior styling, increased fuel capacity and redesigned nose compartment, giving more space for radio. The A80 was followed in turn by the improved B80 and 11-seat Queen Airliner B80, to which the details apply.

The B80 has two optional equipment packages for factory installation:

Executive Package: Comprising quickly removable forward partition with magazine rack and accordion door separating cockpit from cabin; quickly removable aft cabin partition and door; private lavatory, aft installation with relief tube; four executive cabin chairs with inward folding armrests, or two such chairs and one four-seat couch (exchange for five commuter chairs with fixed armrests); and aft coat hanger rod installation; adding 19 kg (42 lb) to basic empty weight.

Beechcraft B80 Queen Air commuter airliner ***(Raytheon)***

1995

Airliner Package: Comprising seven floor-mounted chairs with inward folding armrests and removable upholstery two aft folding chairs with removable upholstery, extended seat tracks, four additional fresh air outlets and reading lights (exchange for five standard chairs); extended aft baggage compartment with baggage restraints; aft baggage door with safety light; fourth cabin window in starboard side; 30 Ah Ni/Cd battery (exchange for 13 Ah battery); metal map case under pilot's seat; cockpit fire extinguisher; cockpit separation half-curtain; second outboard fuel filler caps with dipstick (one each side); coat hanger bar; and upholstery for nose baggage compartment; adding 141 kg (311 lb) to basic empty weight.

By January 1977 Beech had built over 500 Queen Air 80s, A80s, B80s and Queen Airliner B80s.

UPGRADES: **Excalibur Aviation Company:** Excalibur Queenaire 800 and 8000. See separate entry.

DESIGN FEATURES: Cantilever low-wing monoplane. Wing section NACA 23020 at root, NACA 23012 outboard of joint between outer panel and wingtip. Dihedral 7°. Incidence 3° 55′ at root, 0° 1′ at tip. Tailplane dihedral 7°.

FLYING CONTROLS: All-metal ailerons of magnesium. Trim tab in port aileron. Single-slotted aluminium alloy flaps. Trim tabs in rudder and elevators.

STRUCTURE: Two-spar all-metal structure of aluminium alloy. The fuselage is an aluminium alloy semi-monocoque structure. The tail unit is a cantilever all-metal structure of aluminium alloy, with sweptback vertical surfaces.

LANDING GEAR: Electrically retractable tricycle type. Main units retract forward, nosewheel aft. Beechcraft oleo-pneumatic shock-absorbers. Goodyear mainwheels and tyres, size 8.50-10, 8 ply rating, pressure 3.31 bars (48 lb/sq in). Goodyear steerable nosewheel with tyre size 6.50-10, 6 ply rating, pressure 2.90 bars (42 lb/sq in). Goodyear heat sink and air-cooled single-disc hydraulic brakes.

POWER PLANT: Two 283 kW (380 hp) Lycoming IGSO-540-A1D supercharged flat-six engines, each driving a Hartzell three-blade fully feathering constant-speed propeller synchrophasers optional. Fuel in two inboard wing tanks, each with capacity of 166 litres (44 US gallons; 36.6 Imp gallons) and two outboard tanks; each 238.5 litres (63 US gallons); 52.5 Imp gallons). Total standard fuel capacity 809 litres (214 US gallons; 178 Imp gallons). Provision for two optional auxiliary tanks in wings to bring total capacity to 1,000 litres (264 US gallons; 219.8 Imp gallons). Refuelling points above wings. Oil capacity total 30 litres (8 US gallons; 6.6 Imp gallons).

ACCOMMODATION: Crew of one or two on flight deck and four to nine passengers in cabin. Basic layout has five commuter passenger seats; executive layout has four chairs, fore and aft partitions, and lavatory; airliner layout has seven commuter and two folding passenger seats, extended aft baggage compartment and door, map case, cabin fire extinguisher and fourth cabin window. Door on port side of cabin at rear; optionally, double-width cargo doors. Optional toilet baggage compartment opposite door, capacity 160 kg (350 lb). Other optional items include sofa, tables, refreshment cabinets and external cargo pod.

SYSTEMS: Optional electrically driven vapour-cycle air conditioning system with combustion heater. Standard model has 100,000 BTU heater and ventilation system. Hydraulic system for brakes only. Two 28 V 150 A DC engine-driven generators and 24 V 13 Ah Ni/Cd battery for electrical system. 39 Ah Ni/Cd battery optional. Oxygen system of 1.81 m³ (64 cu ft) capacity optional. Pneumatic rubber de-icing boots optional.

ELECTRONICS AND EQUIPMENT: Standard electronics comprise Narco com 11B 720-channel transceiver with nav 12 300-channel nav receiver/converter, VOR/ILS indicator, B3 com antenna and B17 NAV/gs antenna; Narco com 11B 720-channel transceiver with nav 11 200-channel nav receiver/converter, VOR/LOC indicator and B3 com antenna; Collins 356F-3 speaker amplifier and Collins 356C-4 isolation amplifier; Bendix T-12d ADF with 551RL indicator, voice range filter and sense antenna; Narco MBT-24R marker beacon with B16 marker antenna and panel-mounted marker lights; Narco UGR-2A glide slope with B35 glide slope antenna; Beech metal radio panel, accessories and static wicks; edge-lighted audio switch panel; white lighting; dual microphones, headsets and single cockpit speaker; emergency locator transmitter; and electronics master switch. A wide range of optional electronics, by Bendix, Collins, Edo-Aire Mitchell, Narco, RCA and Sunair, is available to customer's requirements. Standard equipment includes dual controls, blind-flying instrumentation, outside air temperature gauge, eight-day clock, flap position indicator, map light, primary and secondary instrument light systems, dual storm windows, sun visors, four-way adjustable pilot and co-pilot seats, map pockets, control locks, landing gear warning system, cabin door 'unlocked' warning light, windscreen defroster, 'no smoking-fasten seat belt' sign, carpeted floor, provisions for removable cabin partitions and toilet, window curtains, emergency exit, coat rack, cabin door and indirect overhead lighting, reading light for each passenger, aft compartment dome light external power socket, two landing lights, navigation lights dual rotating beacons, dynamic brake on landing gear system, heated stall warning device, dual heated pitot heads, heated fuel vents and towbar. Optional equipment includes De Luxe instrument panel carrying duplicate blind-flying instruments, exhaust temperature gauge, flight hour recorder, control-wheel chronograph, cockpit and cabin fire extinguishers dual windscreen wipers, cargo door, photographic installation, super soundproofing, nosewheel taxi light, three-light strobe system, wing ice lights, alcohol or electrical propeller and windscreen anti-icing, propeller unfeathering accumulators, and external power socket.

DIMENSIONS, EXTERNAL:
Wing span 15.32 m (50 ft 3 in)
Wing chord at root 2.15 m (7 ft 0½ in)
Wing chord at tip 1.07 m (3 ft 6 in)
Wing aspect ratio 7.51
Length overall 10.82 m (35 ft 6 in)
Height overall 4.33 m (14 ft 2½ in)
Tailplane span 5.25 m (17 ft 2¾ in)
Wheel track 3.89 m (12 ft 9 in)
Wheelbase 3.75 m (12 ft 3½ in)
Propeller diameter 2.36 m (7 ft 9 in)
Standard passenger door: Height 1.31 m (4 ft 3¾ in)
Width 0.69 m (2 ft 3 in)
Height to sill 1.17 m (3 ft 10 in)
Optional cargo door: Height 1.31 m (4 ft 3¾ in)
Width 1.37 m (4 ft 6 in)
Height to sill 1.17 m (3 ft 10 in)

DIMENSIONS, INTERNAL:
Cabin, incl flight deck and baggage area:
Length 6.97 m (19 ft 7 in)
Max width 1.37 m (4 ft 6 in)
Max height 1.45 m (4 ft 9 in)
Volume 9.5 m³ (335.4 cu ft)
Nose baggage compartment 0.68 m³ (24 cu ft)
Standard aft baggage compartment 1.50 m³ (53 cu ft)
Optional extension to aft baggage compartment:
0.48 m³ (17 cu ft)

AREAS:
Wings, gross 27.3 m² (293.9 sq ft)
Ailerons (total) 1.29 m² (13.9 sq ft)
Trailing-edge flaps (total) 2.72 m² (29.30 sq ft)
Fin 2.20 m² (23.67 sq ft)
Rudder, incl tab 1.30 m² (14.00 sq ft)
Tailplane 4.39 m² (47.25 sq ft)
Elevators, incl tabs 1.66 m² (47.25 sq ft)

WEIGHTS AND LOADINGS:
Weight empty, equipped 2,393 kg (5,277 lb)
Max T-O and landing weight 3,992 kg (8,800 lb)
Max ramp weight 4,016 kg (8,855 lb)
Max wing loading 146.0 kg/m² (29.9 lb/sq ft)
Max power loading 7.05 kg/kW (11.6 lb/hp)

PERFORMANCE (at max T-O weight, except where indicated):
Max level speed at 3,500 m (11,500 ft) at average AUW
215 kts (400 km/h; 248 mph)
Cruising speed, 70% power at 4,570 m (15,000 ft) at average AUW 195 kts (362 km/h; 225 mph)
Econ cruising speed, 45% power at 4,570 m (15,000 ft) at average AUW 159 kts (294 km/h; 183 mph)
Stalling speed, wheels and flaps up
85 kts (157 km/h; 97 mph) IAS

Stalling speed, wheels and flaps down
71 kts (131 km/h; 81 mph) IAS
Max rate of climb at S/L 388 m (1,275 ft)/min
Rate of climb at S/L, one engine out 64 m (210 ft)/min
Service ceiling 8,168 m (26,800 ft)
Service ceiling, one engine out 3,596 m (11,800 ft)
Min ground turning radius 11.58 m (38 ft 0 in)
Runway LCN 3.5
T-O run 612 m (2,007 ft)
T-O run to 15 m (50 ft) 779 m (2,556 ft)
Landing from 15 m (50 ft) at max landing weight
784 m (2,572 ft)
Landing run at max landing weight 494 m (1,620 ft)
Range with max optional fuel, allowances for warm-up, taxi, T-O run and climb to altitude, with 45 min reserves:
70% power at 4,570 m (15,000 ft)
957 n miles (1,773 km; 1,102 miles)
65% power at 5,180 m (17,000 ft)
1,076 n miles (1,994 km; 1,239 miles)
45% power at 4,570 m (15,000 ft)
1,317 n miles (2,410 km; 1,517 miles)

VERIFIED

Beechcraft King Air E90 six/10-seat twin-turboprop business aircraft

1994

BEECHCRAFT KING AIR E90

TYPE: Six/10-seat twin-turboprop business aircraft.

PROGRAMME: On 1 May 1972 Beech announced an addition to the King Air range of business aircraft. Designated King Air E90, this combines the airframe of the C90 with the 507 kW (680 ehp) Pratt & Whitney Aircraft of Canada PT6A-28 turboprop engines that power the King Air A100, each flat rated to 410 kW (550 ehp). The US Navy ordered 15 King Air 90s in 1976 as **T-44A** advanced trainers to meet its VTAM (X) requirement, and this version is described separately.

UPGRADES: **Raisbeck Engineering:** Quiet turbofan propeller system, fully enclosed landing gear doors and sootless exhaust stack fairings modifications. See separate entries. King Air C90/C90A. See separate entry.

OPERATORS: Versions of the King Air E90 are in service with the armed forces of the following countries: Ecuador (1), Thailand (2) and Venezuela (1).

DESIGN FEATURES: Cantilever low-wing monoplane. Wing section NACA 23014.1 (modified) at root, NACA 23016.22 (modified) at outer end of centre-section, NACA 23012 at tip. Dihedral 7°. Incidence 4° 48′ at root, 0° at tip. No sweepback at quarter-chord. Fixed incidence tailplane, with 7° dihedral.

FLYING CONTROLS: All-metal ailerons of magnesium, with adjustable trim tab on port aileron. Single-slotted aluminium alloy flaps. Trim tabs in rudder and each elevator.

STRUCTURE: Two-spar aluminium alloy structure. The fuselage is an aluminium alloy semi-monocoque structure. The tail unit is a cantilever all-metal structure with sweptback vertical surfaces.

LANDING GEAR: Electrically retractable tricycle type. Nosewheel retracts rearward, mainwheels forward into engine nacelles. Mainwheels protrude slightly beneath nacelles, when retracted, for safety in a wheels-up emergency landing. Steerable nosewheel with shimmy damper. Beech oleo-pneumatic shock-absorbers. B.F. Goodrich mainwheels with tyres size 8.50-10, pressure 3.93 bars (52 lb/sq in). B.F. Goodrich nosewheel with tyre size 6.50-10, pressure 3.59 bars (59 lb/sq in). Goodrich heat-sink and air-cooled multidisc hydraulic brakes. Parking brakes.

POWER PLANT: Two 507 kW (680 ehp) Pratt & Whitney Aircraft of Canada PT6A-28 turboprop engines, flat rated to 410 kW (550 ehp), each driving a Hartzell three-blade metal fully feathering and reversible-pitch constant-speed propeller. Standard fuel capacity 1,794 litres (474 US gallons; 394.7 Imp gallons).

ACCOMMODATION: Two seats side by side in cockpit with dual controls standard. Normally, four reclining seats are provided in the main cabin, in pairs facing each other fore and aft. Standard furnishings include cabin forward partition, with fore and aft partition curtain and coat rack, hinged nose baggage compartment door, seat belts and inertia-reel shoulder harness for all seats. Optional arrangements seat up to eight persons, some with two- or three-place couch, lateral tracking chairs, and refreshment cabinets. Baggage racks at rear of cabin on starboard side, with optional toilet on port side. Door on port side aft of wing, with built-in airstairs. Emergency exit on starboard side of cabin. Entire accommodation pressurised and air conditioned. Electrically heated windscreen standard.

SYSTEMS: Pressurisation by dual engine bleed air system with pressure differential of 0.32 bars (4.6 lb/sq in). Cabin heated by 45,000 BTU dual engine bleed air system and auxiliary electrical heating system. Electrical system utilises two 28 V 250 A starter/generators, 24 V 45 Ah air-cooled Ni/Cd battery with failure detector. Complete de-icing and anti-icing equipment. Oxygen system 0.62 m^3 (22 cu ft), 1.39 m^3 (49 cu ft) or 1.81 m^3 (64 cu ft) capacity, optional. Vacuum system for flight instruments. Automatic pneumatic de-icing boots on leading-edges standard.

AVIONICS AND EQUIPMENT: Standard avionics include dual Collins VHF-20A VHF transceivers with Gables controls and B3 antennae; Collins VIR-30 MGM Omni No.1 receiver, with 331A-3G indicator, Gables control and B17 antenna; Collins VIR-30M Omni No.2 receiver with 331H-3G indicator and Gables control; Collins 356C-4 isolation amplifier and 356F-3 speaker amplifier with single set of audio switches; Collins ADF-650 ADF with IND-650 indicator, Gables control, voice range filter and ANT-650 antenna; Collins marker beacon integral with VIR-30 No.1, with marker lights and B16 antenna; Collins glidescope receiver integral with VIR-30 No.1, with B35 antenna; dual flight instrumentation; Collins DME-40 with 339F-12 indicator, Nav 1/Nav 2 switching, DME hold and 237Z-1 antenna; Collins TDR-950 transponder with 237Z-1 antenna; Collins PN-101 compass system (pilot); Standard Electric gyro horizon (pilot); CF gyro horizon and directional gyro (co-pilot); dual FliteTronics PC-14B 125VA inverters with failure light; avionics transient protection; Beech edge-lighted radio panel, radio accessories and static wicks; white lighting, microphone key button in pilot and co-pilot control wheels; dual microphones, headsets and cockpit speakers; and avionics master switch. Optional avionics include a wide range of equipment by Bendix, Collins, King Edo-Aire Mitchell, RCA and Sperry.

DIMENSIONS, EXTERNAL:
Wing span 15.32 m (50 ft 3 in)
Wing chord: at root 2.15 m (7 ft 0½ in)
at tip 1.07 m (3 ft 6 in)
Wing aspect ratio 8.57
Length overall 10.82 m (35 ft 6 in)
Height overall 4.33 m (14 ft 2½ in)
Tailplane span 5.25 m (17 ft 2½ in)
Wheel track 3.89 m (12 ft 9 in)
Wheelbase 3.75 m (12 ft 3½ in)
Propeller diameter 2.36 m (7 ft 9 in)
Passenger door: Height 1.31 m (4 ft 3¾ in)
Width 0.69 m (2 ft 3 in)
Height to sill 1.17 m (3 ft 10 in)

DIMENSIONS, INTERNAL:
Total pressurised length 5.43 m (17 ft 10 in)
Cabin: Length 3.86 m (12 ft 8 in)
Max width 1.37 m (4 ft 6 in)
Max height 1.45 m (4 ft 9 in)
Floor area 6.50 m^2 (70 sq ft)
Volume 8.89 m^3 (314 cu ft)
Baggage compartment aft 1.51 m^3 (53.5 cu ft)

AREAS:
Wings, gross 27.31 m^2 (293.94 sq ft)
Ailerons, (total) 1.29 m^2 (13.90 sq ft)
Trailing-edge flaps (total) 2.72 m^2 (29.30 sq ft)
Fin 2.20 m^2 (23.67 sq ft)
Rudder, incl tab 1.30 m^2 (14.00 sq ft)
Tailplane 4.39 m^2 (47.25 sq ft)
Elevators, incl tabs 1.66 m^2 (17.87 sq ft)

WEIGHTS AND LOADINGS:
Weight empty 2,745 kg (6,052 lb)
Max T-O weight 4,581 kg (10,100 lb)
Max ramp weight 4,608 kg (10,160 lb)
Max landing weight 4,400 kg (9,700 lb)
Max wing loading 168.0 kg/m^2 (34.4 lb/sq ft)
Max power loading 5.59 kg/kW (9.18 lb/ehp)

PERFORMANCE (at max T-O weight, except where indicated):
Max cruising speed at 3,660 m (12,000 ft)
249 kts (462 km/h; 287 mph)
Cruising speed at max recommended cruise power:
at 4,875 m (16,000 ft) 247 kts (459 km/h; 285 mph)
at 6,400 m (21,000 ft) 245 kts (454 km/h; 282 mph)
Cruising speed for max range
197 kts (365 km/h; 227 mph)
Stalling speed, power off, wheels and flaps up
86 kts (159 km/h; 99 mph) IAS
Stalling speed, power off, wheels and flaps down
77 kts (143 km/h; 89 mph) IAS
Max rate of climb at S/L 570 m (1,870 ft)/min
Rate of climb at S/L, one engine out 143 m (470 ft)/min
Service ceiling 8,419 m (27,620 ft)
Service ceiling, at 3,629 kg (8,000 lb) AUW
9,421 m (30,910 ft)
Service ceiling, one engine out 4,386 m (14,390 ft)
Service ceiling, one engine out at, 3,629 kg (8,000 lb) AUW
6,218 m (20,400 ft)
Min ground turning radius 11.58 m (38 ft 0 in)
Runway LCN 4.5
T-O run 473 m (1,553 ft)
T-O run to 15 m (950 ft) 617 m (2,024 ft)
Landing distance, 5° approach, full flap at max landing weight: landing from 15 m (50 ft) 643 m (2,110 ft)
landing run 314 m (1,030 ft)
Accelerate/stop distance, incl 2 s failure recognition time
1,139 m (3,736 ft)
Cruising range at max recommended cruise power:
at 4,875 m (16,000 ft)
1,125 n miles (2,084 km; 1,295 miles)
at 6,400 m (21,000 ft)
1,309 n miles (2,425 km; 1,507 miles)
Cruising range at max range power:
at 4,975 m (16,000 ft)
1,479 n miles (2,741 km; 1,703 miles)
at 6,400 m (21,000 ft)
1,624 n miles (3,009 km; 1,870 miles)
Range at max T-O weight, max recommended power at 6,400 m (21,000 ft), 45 min reserves, five occupants, 73 kg (160 lb) baggage and 1,440 kg (3,176 lb) fuel before engine start
1,309 n miles (2,425 km; 1,507 miles)

UPDATED

BEECHCRAFT T-44A (KING AIR 90)

TYPE: Advanced pilot training aircraft.

PROGRAMME: In 1976 Beech Aircraft won an industry-wide competition for a twin-turboprop advanced pilot training aircraft to meet the US Navy's VTAM (X) requirement. The aircraft selected was a version of the King Air 90, incorporating features of both the Model C90 and E90, with modifications to meet special naval requirements.

The first T-44A was delivered to USN Training Command at NAS Corpus Christi, Texas 5 April 1977, and student pilot training began in 1977. T-44As replaced TS-2As and TS-2Bs in the USN's training aircraft inventory.

The US Navy's T-44As differ primarily from other King Air 90s by their power plant, consisting of two 560 kW (750 ehp) Pratt & Whitney Aircraft of Canada PT6A-34B turboprop engines, flat rated to 410 kW (550 ehp). They have also an engine compressor wash system. The fuel capacity is 1,454 litres (384 US gallons; 319 Imp gallons).

OPERATORS: The T-44A is in service with the US Navy (56).

DESIGN FEATURES: Cantilever low-wing monoplane. Wing section NACA 23014.1 (modified) at root, NACA 23016.22 (modified) at outer end of centre-section. NACA 23012 at tip. Dihedral 7°. Incidence 4° 48′ at root, 0° at tip. No sweepback at quarter-chord. Fixed incidence tailplane with 7° dihedral.

FLYING CONTROLS: All-metal ailerons of magnesium, with adjustable trim tab on port aileron. Single-slotted aluminium alloy flaps. Trim tabs in rudder and each elevator.

STRUCTURE: Two-spar aluminium alloy structure. The fuselage is an aluminium alloy semi-monocoque structure. The tail unit is a cantilever all-metal structure with sweptback vertical surfaces.

ACCOMMODATION: Two seats side by side in cockpit with dual controls standard. Normally, four reclining seats are provided in the main cabin, in pairs facing each other fore and aft. Standard furnishings include cabin forward partition, with fore and aft partition curtain and coat rack, hinged nose baggage compartment door, seat belts and inertia-reel shoulder harness for all seats. Optional arrangements seat up to eight persons, some with two- or three-place couch, lateral tracking chairs, and refreshment cabinets. Baggage racks at rear of cabin on starboard side, with optional toilet on port side. Door on port side aft of wing, with built-in airstairs. Emergency exit on starboard side of cabin. Entire accommodation pressurised and air conditioned. Electrically heated windscreen standard.

SYSTEMS: Pressurisation by dual engine bleed air system with pressure differential of 0.32 bars (4.6 lb/sq in). Cabin heated by 45,000 BTU dual engine bleed air system and auxiliary electrical heating system. Electrical system utilises two 28 V 250 A starter/generators, 24 V 45 Ah air-cooled Ni/Cd battery with failure detector. Complete de-icing and anti-icing equipment. Oxygen system 0.62 m^3 (22 cu ft), 1.39 m^3 (49 cu ft) or 1.81 m^3 (64 cu ft) capacity, optional. Vacuum system for flight instruments. Automatic pneumatic de-icing boots on leading-edges standard.

AVIONICS: Complete commercial package plus Tacan, UHF, and UHF-DF 'failure mode selector box' is also provided, enabling the instructor to initiate any of 10 avionics/instruments failures on the student's instrument panel.

DIMENSIONS, EXTERNAL:

Wing span	15.32 m (50 ft 3 in)
Wing chord: at root	2.15 m (7 ft 0½ in)
at tip	1.07 m (3 ft 6 in)
Wing aspect ratio	8.57
Length overall	10.82 m (35 ft 6 in)
Height overall	4.33 m (14 ft 2½ in)
Tailplane span	5.25 m (17 ft 2½ in)
Wheel track	3.89 m (12 ft 9 in)
Wheelbase	3.75 m (12 ft 3½ in)
Propeller diameter	2.36 m (7 ft 9 in)
Passenger door: Height	1.31 m (4 ft 3¾ in)
Width	0.69 m (2 ft 3 in)
Height to sill	1.17 m (3 ft 10 in)

DIMENSIONS, INTERNAL:

Total pressurised length	5.43 m (17 ft 10 in)
Cabin: Length	3.86 m (12 ft 8 in)
Max width	1.37 m (4 ft 6 in)
Max height	1.45 m (4 ft 9 in)
Floor area	6.50 m² (70 sq ft)
Volume	8.89 m³ (314 cu ft)
Baggage compartment aft	1.51 m³ (53.5 cu ft)

AREAS:

Wings, gross	27.31 m² (293.94 sq ft)
Ailerons (total)	1.29 m² (13.90 sq ft)
Trailing-edge flaps (total)	2.72 m² (29.30 sq ft)
Fin	2.20 m² (23.67 sq ft)
Rudder, incl tab	1.30 m² (14.00 sq ft)
Tailplane	4.39 m² (47.25 sq ft)
Elevators, incl tabs	1.66 m² (17.87 sq ft)

WEIGHTS AND LOADINGS:

Max T-O weight	4,377 kg (9,650 lb)

PERFORMANCE (at max T-O weight):

Cruising speed at 4,570 m (15,000 ft)	240 kts (445 km/h; 276 mph)
Max rate of climb at S/L	597 m (1,960 ft)/min
Service ceiling	8,990 m (29,500 ft)
T-O run to 15 m (50 ft)	625 m (2,050 ft)
Landing run from 15 m (50 ft)	619 m (2,030 ft)
Range with max fuel at 7,620 m (25,000 ft)	1,265 n miles (2,344 km; 1,456 miles)

UPDATED

BEECHCRAFT MODEL 1300 COMMUTER

TYPE: Twin-turboprop pressurised commuter aircraft.

PROGRAMME: In January 1988 Beech announced development of this regional airliner variant of the Super King Air B200, designed for operations on long routes with light passenger loads. The Model 1300 provides accommodation for 13 passengers with individual air, light and oxygen outlets at every seat, and is equipped with dual overwing emergency exits. The adoption of panel-mounted avionics as standard has enabled a baggage compartment to be provided in the aircraft's nose, volume 0.37 m³ (13 cu ft). Three-blade propellers with autofeathering; rudder boost for engine-out operations, hydraulically actuated landing gear, airframe anti- and de-icing systems, air conditioning, cabin paging system, front-mounted post lit instruments interchangeable with those of the Beech 99 Airliner and Beech 1900, and radiant heat cabin warming, are standard. A belly cargo pod, volume 1.25 m³ (44 cu ft), engine fire extinguishing system, and landing gear and brake de-icing systems, are optional. Dual outward canted ventral fins are installed for increased stability at low airspeeds and high angles of attack.

Launch customer for the Beechcraft 1300 Commuter was Mesa Airlines of Farmington, New Mexico, which ordered 10, the first of which was delivered 30 September 1988. All had been delivered by 1 January 1990.

DESIGN FEATURES: Pratt & Whitney Canada PT6A-41 turboprops of Super King Air 200 replaced by 634 kW (850 shp) P&WC PT6A-42s for better cruise and altitude performance: maximum zero-fuel weight raised by 272 kg (600 lb); cabin pressure differential increased from 0.41 bar (6.0 lb/sq in) to 0.44 bar (6.5 lb/sq in). Wing aerofoil NACA 23018 to 23016.5 over inner wing, 23012 at tip; dihedral 6°: incidence 3° 48′ at root. –7′ at tip; swept vertical and horizontal tail.

FLYING CONTROLS: Trim tabs in port aileron and both elevators; anti-servo tab in rudder; single-slotted trailing-edge flaps; fixed tailplane.

STRUCTURE: Two-spar light alloy wing: safe-life semi-monocoque fuselage.

LANDING GEAR: Hydraulically retractable tricycle type, with twin-wheels on each main unit. Single wheel on steerable nose unit, with shimmy damper. Main units retract forward, nosewheel rearward. Beech oleo-pneumatic shock-absorbers. Goodrich mainwheels and tyres size 18 × 5.5, pressure 7.25 bars (105 lb/sq in). Oversize and/or 10 ply mainwheel tyres optional. Goodrich nosewheel size 6.50 × 10 with tyre size 22 × 6.75-10, pressure 3.93 bars (57 lb/sq in). Goodrich hydraulic multidisc brakes. Parking brake.

POWER PLANT: Two 634 kW (850 shp) Pratt & Whitney Canada PT6A-42 turboprops, each driving a Hartzell three-blade constant-speed reversible-pitch metal propeller with autofeathering and synchrophasing. Bladder fuel cells in each wing, with main system capacity of 1,461 litres (386 US gallons; 321.5 Imp gallons) and auxiliary system capacity of 598 litres (158 Imp gallons; 131.5 Imp gallons). Total usable fuel capacity 2.059 litres (544 US gallons; 453 Imp gallons). Two refuelling points in upper surface of each wing. Wingtip tanks optional, providing an additional 401 litres (106 US gallons; 88.25 Imp gallons) and raising maximum usable capacity to 2,460 litres (650 US gallons; 541 Imp gallons). Oil capacity 29.5 litres (7.8 US gallons; 6.5 Imp gallons).

SYSTEMS: Cabin pressurisation by engine bleed air, with a maximum differential of 0.44 bar (6.5 lb/sq in). Cabin air conditioner of 34,000 BTU capacity. Auxiliary cabin heating by radiant panels standard. Oxygen system for flight deck, and 0.62 m³ (22 cu ft) oxygen system for cabin, with automatic drop-down face masks; standard system of 1.39 m³ (49 cu ft); 1.81 m³ (64 cu ft) or 2.15 m³ (76 cu ft) optional. Dual vacuum system for instruments. Hydraulic system for landing gear retraction and extension, pressurised to 171 to 191 bars (2,475 to 2,775 lb/sq in). Separate hydraulic system for brakes. Electrical system has two 250 A 28 V starter/generators and a 24 V 45 Ah air-cooled Ni/Cd battery with failure detector. AC power provided by dual 250 VA inverters. Engine fire detection system standard; engine fire extinguishing system optional. Pneumatic de-icing of wings and tailplane standard. Anti-icing of engine air intakes by hot air from engine exhaust, electrothermal anti-icing for propellers.

AVIONICS: Standard Collins Pro Line II avionics and cockpit equipment generally as for King Air C90A except: dual DME-42, dual RMI-30; pilot's ALT-80A encoding altimeter; cockpit-to-cabin paging; dual maximum allowable airspeed indicators; flight director indicator, co-pilot's 24 hour clock; oxygen pressure indicator and blue-white cockpit lighting standard. Optional avionics include Collins Pro Line II with 5 in FCS; EFIS-85B(-14); Honeywell SPZ-4000 autopilot with 4 or 5 in flight director systems or EDZ-605 three-tube EFIS; Bendix/King Gold Crown avionics packages; Bendix/King RDS-84VP colour weather radar; Fairchild 17M-700-274 flight data recorder; Fairchild A-100A cockpit voice recorder; Bendix-King KHF-950 or Collins HF-230 HF transceiver, Bendix/King CC-2024C or Collins DCP-270 radar checklist; Foster LNS-616B RNAV/Loran C; Bendix/King KNS-660 RNAV/VLF/Omega; Foster WX-1000+ Stormscope; and Wulfsberg Flitefone VII.

EQUIPMENT: Standard/optional equipment generally as for King Air C90A except fluroescent cabin lighting, one place couch with storage drawers, flushing toilet (B200) or chemical toilet (B200C), cabin radiant heating cockpit/cabin partition with sliding doors, and airstair door with hydraulic snubber and courtesy light, standard. FAR Pt 135 operational configuration includes cockpit fire extinguisher and 2.15 m³ (76 cu ft) oxygen bottle as standard. A range of optional cabin seating and cabinetry configuration is available, including quick-removable fold-up seats.

Beechcraft T-44A (King Air 90) USN Advanced pilot training aircraft

1994

ACCOMMODATION: Crew of two and maximum of 13 passengers. Nose baggage compartment, capacity 159 kg (350 lb), and aft baggage compartment, capacity 231 kg (510 lb), standard (aft compartment not usable when two rear seats installed for maximum passenger capacity). Belly cargo pod, capacity 206 kg (455 lb) optional.

WEIGHT:

Empty weight, typical	3,573 kg (7,877 lb)

PERFORMANCE (at max T-O weight, ISA, except where indicated):

Max level speed at S/L	272 kts (504 km/h; 313 mph)
Max operating speed	260 kts (482 km/h; 299 mph) IAS
Max operating speed	260 kts (482 km/h; 299 mph) IAS
Max cruising speed, weight 4,990 kg (11,000 lb):	
at 3,050 m (10,000 ft)	252 kts (467 km/h; 290 mph)
at 7,620 m (25,000 ft)	269 kts (499 km/h; 310 mph)
Stalling speed: flaps up	96 kts (178 km/h; 111 mph)
flaps down	74 kts (137 km/h; 86 mph)
Max rate of climb at S/L	732 m (2,400 ft)/min
Rate of climb at S/L, one engine out	152 m (500 ft)/min
Max certificated ceiling	7,620 m (25,000 ft)
Service ceiling, one engine out:	
at max T-O weight	5,067 m (16,625 ft)
at 4,990 kg (11,000 lb)	6,325 m (20,750 ft)
Landing from 15 m (50 ft)	784 m (2,572 ft)
Landing run	471 m (1,544 ft)
Range with max payload, ISA, zero wind, 1,800 rpm at 7,620 m (25,000 ft)	381 n miles (706 km; 438 miles)
Ferry range	1,235 n miles (2,288 km; 1,422 miles)

VERIFIED

BEECHCRAFT MODEL 1900C AIRLINER and 1900 EXEC-LINER

US Air Force designation: C-12J

TYPE: Twin-turboprop commuter cargo airliner and executive transport.

PROGRAMME: Beech began design of the basic 1900 commuter airliner during 1979, and the first flight of the performance prototype (UA-1) was made 3 September 1982, followed by the systems prototype (UA-2) 30 November 1982. The third prototype (UA-3) was used for function and reliability testing, equipment certification, and demonstration; it is now in operational service. FAA certification under SFAR Pt 41C, obtained 22 November 1983, included single pilot approval under FAR Pt 135 Appendix A.

The Beech 1900C is offered in two variants: **Model 1900C Airliner** with cargo door, the first of which was delivered February 1984; and **1900 Exec-Liner,** which is the corporate version. By May 1990 nearly 200 Model 1900C Airliners and 1900 Exec-Liners had been delivered. They were in service with 17 regional airlines in the USA and seven international operators. The first delivery of an Exec-Liner (N34GT) was made to General Telephone Company of Illinois Summer 1985. Exec-Liners are operating in Africa, Australia, Europe and the USA.

Beechcraft Model 1300 Commuter airliner twin-turboprop pressurised commuter aircraft

1994

Beechcraft 1900C with rear cargo door *(Raytheon)*

1995

In March 1986 the US Air Force ordered six Model 1900Cs for delivery commencing September 1987. These aircraft, designated **C-12J**, serve as Air National Guard mission support aircraft, and replaced Convair C-131s.

Six Model 1900CS (of eight ordered) have been delivered to the Egyptian Air Force. Four are configured for electronic surveillance missions, and two as maritime patrol aircraft. For this latter role the aircraft are equipped with Litton search radar. Motorola sideways-looking airborne multimode radar (SLAMMR) and Singer S-3075 ESM systems. Deliveries of 12 Model 1900Cs for the Chinese Nationalist Air Force began January 1988.

From aircraft c/n UC-1 onwards, all Beechcraft 1900 series aircraft (including those for the US and Egyptian air forces) are equipped with a wet wing providing a maximum fuel capacity of 2,593 litres (685 US gallons; 570 Imp gallons).

In March 1989 Beech announced development of a new **Model 1900D**. See *Jane's All the Worlds Aircraft 1995-96*.

OPERATORS: Versions of the 1900C (C-12J) are in service with the armed forces of the following countries: Egypt (6), Taiwan (11) and the USA (6).

The description which follows applies to the commercial Models 1900C and 1900 Exec-Liner:

DESIGN FEATURES: Cantilever low-wing monoplane. Wing section NACA 23018 modified at root, NACA 23102 modified at tip. Dihedral 6°. Incidence 3° 29′ at root, −1° 4′ at tip. No sweepback at quarter-chord. Cantilever T tail with sweptback vertical and horizontal surfaces. Small fin (tail-let) beneath each side of tailplane near tip, as well as fixed horizontal tail surface (stabilon) on each side of rear fuselage.

FLYING CONTROLS: Single-slotted trailing-edge flaps in two sections on each wing of aluminium alloy construction, symmetrical ailerons of similar construction. Trim tab at inboard end of port aileron. Trim tabs in elevators and rudder.

STRUCTURE: The wing is a semi-monocoque fail-safe structure of aluminium alloy, riveted and bonded with a continuous main spar. The fuselage is a semi-monocoque fail-safe pressurised structure of aluminium alloy, mainly of bonded construction but including some riveting. Small vortex generator on each side of fuselage immediately forward of wing leading-edge. The tail unit is of aluminium structure.

LANDING GEAR: Hydraulically retractable tricycle type. Main units retract forward and nose unit rearward. Beech oleo-pneumatic shock-absorber in each unit. Twin Goodyear wheels on each main unit, size 6.50 × 10, with Goodyear tyres size 22 × 6.75-10, pressure 6.07 bars (88 lb/sq in). Multiple-disc hydraulic brakes. Beech/Hydro-Aire anti-skid units and power steering optional.

POWER PLANT: Two Pratt & Whitney Canada PT6A-65B turboprops, each flat rated at 820 kW (1,100 shp) and driving a Hartzell four-blade constant-speed fully feathering reversible-pitch composite propeller with spinner. 'Wet wing' fuel storage, with a total capacity of 2,593 litres (685 US gallons; 570 Imp gallons), of which 2,525 litres (667 US gallons; 555 Imp gallons) are usable. Refuelling point in each wing leading-edge, inboard of engine nacelle. Oil capacity (total) 27.2 litres (7.2 US gallons; 6 Imp gallons).

ACCOMMODATION: Crew of one (FAR Pt 91) or two (FAR Pt 135) on flight deck, with standard accommodation in cabin of commuter version for 19 passengers, in single seats on each side of centre aisle. Forward and rear carry-on baggage lockers, underseat baggage stowage, rear baggage compartment and nose baggage compartment. Forward and rear doors, incorporating airstairs, on port side. Upward-hinged cargo door instead of rear passenger door on Model 1900C. Two emergency exits over wing on starboard side, plus one on port side (1900C only). Accommodation is air conditioned, heated, ventilated and pressurised. Exec-Liner has 12/18-passenger cabin with forward and rear compartments, combination lavatory/passenger seat and two beverage bars at cabin compartment division. Club and double club seating optional. Customised interiors to customer choice.

SYSTEMS: Bleed air cabin heating and pressurisation, maximum differential 0.33 bars (4.8 lb/sq in). Air cycle and vapour cycle air conditoning. Hydraulic system, pressure 207 bars (3,000 lb/sq in) for landing gear actuation. Electrical system includes two 300 A engine starter/generators and one 22 Ah Ni/Cd battery. Constant flow oxygen system of 4.33 m³ (153 cu ft) capacity standard.

AVIONICS: Duplicated Bendix/King com/nav, glide slope receiver, transponder, audio, ADF, DME, marker beacon receiver and Bendix RDR-160 weather radar. Honeywell EFIS, and Collins autopilot and Pro Line II equipment, optional.

DIMENSIONS, EXTERNAL:
- Wing span: 16.60 m (54 ft 5¾ in)
- Wing chord: at root: 2.18 m (7 ft 1¾ in)
  - at tip: 0.91 m (3 ft 0 in)
- Wing aspect ratio: 9.8
- Length overall: 17.63 m (57 ft 10 in)
- Height overall: 4.54 m (14 ft 10¾ in)
- Tailplane span: 5.63 m (18 ft 5¾ in)
- Wheel track: 5.23 m (17 ft 2 in)
- Wheelbase: 7.26 m (23 ft 10 in)
- Propeller diameter: 2.78 m (9 ft 1½ in)
- Propeller ground clearance: 0.36 m (1 ft 2 in)
- Distance between propeller centres: 5.23 m (17 ft 2 in)
- Passenger doors (fwd and rear, port, each):
  - Height: 1.32 m (4 ft 4 in)
  - Width: 0.68 m (2 ft 2¾ in)
  - Height to sill: fwd: 1.28 m (4 ft 2½ in)
  - rear: 1.15 m (3 ft 9¼ in)
- Cargo door (rear, port): Height: 1.32 m (4 ft 4 in)
  - Width: 1.32 m (4 ft 4 in)
  - Height to sill: 1.15 m (3 ft 9¼ in)
- Baggage door (nose, port):
  - Max height: 0.56 m (1 ft 10 in)
  - Width: 0.66 m (2 ft 2 in)
  - Height to sill: 1.45 m (4 ft 9 in)
- Emergency exits (two stbd; plus one port on 1900C only; all overwing): Height: 0.70 m (2 ft 3½ in)
  - Width: 0.51 m (1 ft 8 in)

DIMENSIONS, INTERNAL:
- Cabin, incl flight deck and rear baggage compartment:
  - Length: 12.02 m (39 ft 5½ in)
  - Max width: 1.37 m (4 ft 6 in)
  - Max height: 1.45 m (4 ft 9 in)
  - Floor area: 15.28 m² (164.5 sq ft)
  - Pressurised volume: 20.00 m³ (706 cu ft)
- Volume of passenger cabin: 14.10 m³ (498 cu ft)
- Baggage space: Cabin:
  - 1900C, fwd: standard: 0.42 m³ (15.0 cu ft)
    - optional: 1.19 m³ (41.9 cu ft)
  - 1900C, rear (both): 4.36 m³ (154.0 cu ft)
- Nose compartment: 0.38 m³ (13.5 cu ft)

AREAS:
- Wings, gross: 28.15 m² (303.0 sq ft)
- Ailerons (total): 1.67 m² (18.0 sq ft)
- Trailing-edge flaps (total): 4.17 m² (44.9 sq ft)
- Fin: 3.42 m² (36.85 sq ft)
- Rudder (incl tab): 1.106 m² (11.9 sq ft)
- Tail-lets (total): 0.305 m² (3.28 sq ft)
- Tailplane: 4.52 m² (48.7 sq ft)
- Elevator (incl tab): 1.79 m² (19.3 sq ft)
- Stabilons (total, exposed): 1.44 m² (15.46 sq ft)

WEIGHTS AND LOADINGS:
- Weight empty (typical): 4.327 kg (9,540 lb)
- Max fuel (usable): 2.027 kg (4,469 lb)
- Max baggage: 866 kg (1.910 lb)
- Max T-O weight: 7,530 kg (16,600 lb)
- Max ramp weight: 7,580 kg (16,710 lb)
- Max landing weight: 7,320 kg (16,100 lb)
- Max zero-fuel weight: 6,350 kg (14,000 lb)
- Max wing loading: 267.5 kg/m² (54.8 lb/sq ft)
- Max power loading: 4.59 kg/kW (7.55 lb/shp)

PERFORMANCE (at max T-O weight except where indicated):
- Max crusing speed at AUW of 6,350 kg (14,000 lb):
  - at 2,440 m (8,000 ft): 267 kts (495 km/h; 307 mph)
  - at 4,875 m (16,000 ft): 267 kts (495 km/h; 307 mph)
  - at 7,620 m (25,000 ft): 254 kts (471 km/h; 292 mph)
- T-O speed, 20° flap: 105 kts (194 km/h; 121 mph) CAS
- Approach speed at max landing weight: 113 kts (209 km/h; 130 mph) CAS
- Stalling speed at max T-O weight: wheels and flaps up: 104 kts (193 km/h; 120 mph) IAS
  - wheels down and approach flaps: 91 kts (169 km/h; 105 mph) IAS
- Stalling speed at max landing weight, wheels and flaps down: 88 kts (163 km/h; 102 mph) IAS
- Max rate of climb at S/L: 719 m (2,360 ft)/min
- Rate of climb at S/L, one engine out: 149 m (490 ft)/min
- Service ceiling: exceeds certificated ceiling of 7,620 m (25,000 ft)
- Service ceiling, one engine out: 3,960 m (13,000 ft)
- Turning circle based on nosewheel: 8.17 m (26 ft 9½ in)
- T-O run, approach flap: 671 m (2,200 ft)
- T-O to 15 m (50 ft), approach flap: 994 m (3,260 ft)
- Landing from 15 m (50 ft) at max landing weight: 780 m (2,560 ft)
- Landing run at max landing weight: 466 m (1,530 ft)
- Accelerate/stop distance 20° flap: 1,158 m (3,800 ft)
- Range with 10 passengers, at long-range cruise power, with allowances for starting, taxi, T-O, climb and descent: with VFR reserves: 1,569 n miles (2,907 km; 1,806 miles)

VERIFIED

Beechcraft 1900C C-12J of the US Air Force

1994

RILEY

RILEY INTERNATIONAL CORPORATION

2206 Palomar Airport Road, Suite A-1, Carlsbad, California 92008
Telephone: 1 (619) 438 9089
Fax: 1 (619) 438 0578
PRESIDENT: Jack M. Riley

Company formed to continue marketing Riley conversions of production aircraft. Two types available in 1992 detailed below. Riley holds total of 120 STCs for conversions developed over 39 years. Future products will include unconventional power plants and turbine conversion of existing aircraft.

UPDATED

RILEY ROCKET CESSNA P-210

TYPE: Light aircraft re-engine programme.

PROGRAMME: Riley conversion of Cessna P-210 Pressurised Centurion, announced 1983; FAA STC issued late 1984; 31 Rocket P-210s delivered by early 1991; current production two a month. Includes fitting remanufactured 231 kW (310 hp) turbocharged Teledyne Continental TSIO-520-AF with Riley intercooler system and flush NACA intake; induction air 45 per cent cooler and engine efficiency improved. Other equipment includes dynamically balanced, electrically de-iced three-blade McCauley propeller, heavy-duty battery, dual vacuum pumps, alternative electrical system, Riley engine baffle system, new Cleveland wheels and brakes, new tyres, luxury interior, Bendix/King avionics including Loran, WX-1000 Stormscope, flush antennae, S-Tech autopilot system, Horton short take-off and landing modification and Riley custom soundproofing and interior. Options include radar, air conditioning, speedbrakes, Flint wingtip fuel tanks increasing fuel capacity to 462 litres (122 US gallons; 102 Imp gallons), air conditioning, full airframe de-icing and special law enforcement and government communications systems. Intercooler also FAA approved for Cessna Turbo Stationair 6 and 7, Turbo Centurion and Pressurised Super Skymaster, and Piper Turbo Saratoga.

PERFORMANCE:

Cruising speed:	
at 5,335 m (17,500 ft)	204 kts (378 km/h; 235 mph)
at 7,000 m (23,000 ft)	217 kts (402 km/h; 250 mph)
Stalling speed, power off, clean	42 kts (79 km/h; 49 mph)
Max rate of climb at S/L	305 m (1,000 ft)/min
Time to 7,000 m (23,000 ft)	23 min
Service ceiling	7,000 m (23,000 ft)
Range, with IFR reserves	1,129 n miles (2,092 km; 1,300 miles)

VERIFIED

Riley Rocket P-210 conversion of a Cessna Pressurised Centurion

1994

RILEY SKY ROCKET CESSNA P-337

TYPE: Light aircraft re-engine programme.

PROGRAMME: Riley was converting two Cessna pressurised P-337 twins a month in Spring 1991 and had then delivered 34. Basic conversion and optional features same as those listed for Rocket P-210 above. Maximum continuous cruising speed 220 knots (408 km/h; 253 mph); stalling speed, power off, clean, 50 knots (93 km/h; 58 mph).

VERIFIED

ROCKWELL

ROCKWELL INTERNATIONAL CORPORATION

2201 Seal Beach Boulevard, Seal Beach, California 90740-8250
Telephone: 1 (310) 797 3311
CHAIRMAN AND CEO: Donald R. Beall
EXECUTIVE VICE-PRESIDENTS AND COOs:
Kent M. Black
Sam F. Iacobellis

VERIFIED

NORTH AMERICAN AIRCRAFT MODIFICATION DIVISION

3370 Miraloma Avenue, Anaheim, California 92803
Telephone: 1 (714) 762 8111
SENIOR VICE-PRESIDENT AND PRESIDENT, DEFENSE SYSTEMS:
John A. McLuckey
VICE-PRESIDENT AND GENERAL MANAGER (NAAMD):
Gene P. Burbey

North American Aviation, founded in 1928, manufactured aircraft from 1934 the company merged with Rockwell-Standard Corporation of Pittsburgh, Pennsylvania (which manufactured Aero Commander aircraft) on 22 September 1967 to form the North American Rockwell Corporation. The present name was adopted in 1973. In 1993 the North American Aircraft Modification Division was created out of the North American Aircraft and integrated into Rockwell's Defense Systems business to handle aircraft modifications. The renamed North American Aircraft Division designs and builds military aircraft and commercial aerostructures.

Rockwell is a diversified, high-technology company active in the fields of automation, avionics, aerospace, defense electronics, telecommunications, automotive components and graphic systems, with annual worldwide sales of $11 billion. Automation includes production of programmable controllers, man-machine interface devices, communications networks and programming and application software. Avionics includes navigation systems, flight instrumentation and radar systems. Aerospace includes spacecraft, rocket propulsion systems, military aircraft modifications and aerostructures. Defense electronics includes communication and navigation systems for air, land and sea applications, global positioning system (GPS) receivers and flight management systems. Telecommunications includes fax and data modems, modem engines, circuit boards and integrated call processing systems. Automotive components includes roof, door and access control systems and heavy duty drivetrain components. Graphic systems includes newspaper and commercial web offset presses.

VERIFIED

Rockwell International B-1B Lancer strategic Bomber of the US Air Force

1994

ROCKWELL INTERNATIONAL B-1B

TYPE: Long-range multirole strategic bomber.

PROGRAMME: The decision to order 100 derivative B-1Bs for the US Air Force was announced by President Reagan in October 1981, and the first production B-1B flew for the first time 18 October 1984. The remaining 99 were ordered under the FY83-86 defence budgets (7, 10, 34 and 48 respectively). Initial delivery, of the second production aircraft (first flown 4 May 1985), to the 96th Bomb Wing at Dyess AFB, Texas, took place 7 July 1985. This base achieved IOC in September 1986 and had received all 29 of its aircraft by the end of the year. Deliveries continued throughout 1987 and early 1988 at a rate of approximately four per month to the 28th Bomb Wing at Ellsworth AFB, South Dakota (35 aircraft), the 319th Bomb Wing at Grand Forks AFB, North Dakota (17), and the 384th Bomb Wing at McConnell AFB, Kansas (17). Each base also deploys supporting tankers. The 100th and final B-1B was rolled out 20 January 1988, at which time the air force had accepted 80 aircraft. Deliveries were completed 30 April 1988. In November 1988 two B-1Bs were lost in crashes, the first near Abilene, Texas and the second while landing at Ellsworth AFB, South Dakota. All crew members ejected safely. Three B-1Bs are allocated for test and development flying.

Operational B-1Bs are able to carry, in three weapons bays, varying combinations of nuclear air-to-ground missiles, conventional or nuclear free-fall bombs, and auxiliary fuel. Using electronic jamming equipment, infra-red countermeasures, radar location and warning systems, other advanced avionics and 'low observable' technology to defeat hostile defensive systems, the B-1B will be able to penetrate predicted sophisticated enemy defences well into the 1990s and to operate within less heavily defended areas into the next century. It will also be suitable for deployment in a variety of roles now flown by the Boeing B-52, including anti-submarine patrol or maritime surveillance at long ranges, and aerial minelaying.

The first launch of an AGM-69 short-range attack missile (SRAM) from a B-1B was made successfully on 16 January 1987, while the bomber was flying at Mach 0.9 at a height of 150 m (500 ft). First live launch of an AGM-86B ALCM, with an instrument package replacing the warhead, was made 24 November 1987.

Outwardly the B-1B is generally similar to the B-1 prototype No. 4, but has major airframe improvements that include a strengthened landing gear; a movable bulkhead in the forward weapons bay to allow for the carriage of a wide range of different sized weapons; optional weapons bay fuel tanks to give extended range; and external stores stations beneath the fuselage to accommodate additional fuel or weapons. The variable engine inlets of the original B-1 are replaced by fixed inlets, with new engine nacelles and simplified overwing fairings. These modifications are designed to provide optimum performance for the high-subsonic low-altitude penetration role.

The B-1B has, through the application of 'low observable' technology, a radar signature only 1 per cent that of a B-52. Offensive avionics include advanced forward-looking and terrain-following radar, an extremely accurate inertial navigation system, a link to the Air Force Satellite Communications (AFSATCOM) system, a new computer-driven avionics system, and a strategic Doppler radar altimeter. The defensive avionics are built around the AN/ALQ-161 ECM system, with extended frequency coverage, and include tail warning radar and expendable decoys such as chaff and flares. Development of the full potential of this system will not be achieved for several more years.

The structure of the B-1B is hardened to withstand nuclear blast and overpressure. More than 60 per cent of the structure and equipment was subcontracted, with some 3,000 subcontractors and suppliers being involved in the programme in addition to AIL/Eaton (defensive avionics), Boeing (offensive avionics) and General Electric (power plant).

During Summer 1987, a series of international speed and distance with payload records was set by the B-1B. On 4 July, a 1,080 n miles (2,000 km; 1,244 miles) closed

Rockwell International B-1B Lancer with four General Electric F101-GE-102 augmented turbofans *1995*

circuit was covered at a speed of 582.18 knots (1,078.2 km/h; 669.96 mph) with a payload of 30,000 kg (66,140 lb). On 17 September a similar payload was carried around a 2,700 n miles (5,000 km; 3,110 miles) circuit at 569.225 knots (1,054.206 km/h; 655.05 mph).

UPGRADES: **Boeing:** $46 million contract from the US Air Force to upgrade the B-1B's offensive avionics software. Three stage contract included developing the software, operating the USAF software integration facility at Tinker AFB, Oklahoma, and integration of software for the B-1B's weapons. Programme commenced 28 September 1991.

Rockwell/Boeing: Weapon systems upgrade. See separate entry in this section.

Rockwell: B-1B ECM upgrade. See separate entry.

OPERATORS: The B-1B is in service with the US Air Force (95).

DESIGN FEATURES: Cantilever low-wing fail-safe blended wing/body structure with variable geometry on outer panels. Wing sweep is actuated by screwjacks driven by four hydraulic motors; it can be powered by any two of the aircraft's four hydraulic systems, asymmetric movement being prevented by a torque shaft between the two screwjacks. Sweep actuators are covered by a leading-edge 'knuckle' fairing which prevents a gap from opening when outer panels are swept back. Aft of the wing pivot on each side are overwing fairings which blend the wing trailing-edge and engine nacelles. Each of outer wing panels, have 15° of leading-edge sweep when fully forward and 67° 30′ when fully swept. Cantilever tail unit with sweepback on all surfaces.

FLYING CONTROLS: Full-span seven-segment leading-edge slats on each outer panel can be drooped 20° for take-off and landing. Six-segment single-slotted trailing-edge flaps on each outer panel, with maximum downward deflection of 40°. No ailerons; lateral control is provided by four segment airbrake/spoilers on each outer wing, forward of outer four flap segments, with maximum upward deflection of 70°. All control surfaces are operated electrohydraulically by rods, cables, pulleys and bellcrank levers, except for two outboard spoilers on each wing which are controlled by a fly-by-wire system. Small sweptback movable vane of composite material with 30° anhedral, on each side of nose actuated by structural mode control system accelerometers in fuselage. These sense lateral and vertical motion of forward fuselage in turbulent conditions and compensate for it by relaying electrical signals to move vanes, providing both yaw and pitch damping. Fin panels have 25° of travel each side. Two-section all-moving tailplane is operated collectively for control in pitch (between 10° up and 25° down) and indifferentially (±20°) for roll, the two halves moving independently on the steel spindle. Rudder and tailplane actuated hydraulically, with fly-by-wire back-up system for use in the event of a mechanical system failure.

STRUCTURE: Wing carry-through structure, sealed as an integral fuel tank, is mainly of diffusion bonded 6AL-4V titanium. Wing pivot mechanism is of the same material with a pin made from a single 6AL-4V forging on each side, in spherical steel bearings, above and below which are integrally stiffened double cover plates of machined titanium. Each outer wing panel is a conventional two-spar aluminium alloy torsion box structure, with machined spars, ribs and one-piece integrally stiffened top and bottom skin panels. Wingtips, wing/body fairings, and some outer wing skin panels are of GRP. The fuselage is a conventional area-ruled fail-safe stressed-skin structure of closely spaced frames and longerons, built mainly of 2025 and 7075 aluminium alloys. Built in five main sections comprising forward, forward intermediate, wing carry-through, rear intermediate and rear fuselage; last two of these being manufactured by Vought. Titanium used for engine bays and firewalls, tail support, rear fuselage skins and other high load or high heat areas. Dorsal spine of steel/boron filled titanium sandwich construction. Nose radome of polyimide; dielectric panels of GFRP. Fin is conventional titanium and aluminium alloy torsion box structure, secured to rear fuselage by a double shear attachment, bolts on tailplane spindle; a vertical shear pin in tailplane spindle fitting, and a shear bolt on front beam of box. Aluminium rudder is in three sections.

LANDING GEAR: Hydraulically retractable tricycle type. Each main unit, which retracts inward and rearward, has two pairs of wheels in tandem. Steerable nose unit has twin wheels and retracts forward. Oleo-pneumatic shock-absorber in each unit. Goodyear wheels and carbon brakes, with anti-skid system. Goodrich tyres. Mainwheel diameter 60 cm (23½ in), tyre size 46 × 16-325, 30 ply rating, pressure 15.2 to 19.0 bars (220 to 275 lb/sq in). Nosewheel diameter 41 cm (16 in), tyre size 35 × 11.5-16, 22 ply rating, pressure 14.5 bars (210 lb/sq in).

POWER PLANT: Four General Electric F101-GE-102 augmented turbofans, each rated at 136.9 kN (30,780 lb st), mounted in pairs beneath fixed centre-section of wing, close to CG, to provide optimum stability in low-altitude turbulence conditions. Fixed geometry inlets. Integral fuel tanks in fuselage and outer wings; provision for auxiliary fuel tanks to be carried in two forward weapons bays and beneath fuselage. Fuel capacity increased considerably over that of original B-1. Simmonds Precision fuel management system maintains CG trim automatically as fuel is consumed. Receptacle in upper nose section, forward of windscreen, for in-flight refuelling; aircraft is compatible with KC-10 and KC-135 tankers.

ACCOMMODATION: Four-man operational crew comprising pilot, co-pilot and two systems operators (defensive and offensive) on Weber ACES II ejection seats in a pressurised crew compartment. Pilots have control sticks rather than conventional bomber/transport yokes. Radiation glareshield standard. Crew access via downward-opening door and electrically retractable ladder under fuselage, aft of nosewheel unit.

SYSTEMS: All systems and subsystems are either fail-operative or fail-safe, to ensure that no single system failure prevents accomplishment of primary mission, and that no second failure in same system prevents a safe return to base. Hamilton Standard air conditioning and pressurisation systems. Four independent hydraulic systems, each 276 bars (4,000 lb/sq in) pressure, for actuation of wing sweep, control surfaces, landing gear and weapons bay doors. Hydraulic system maximum flow rate 238.5 litres (63 US gallons; 52.4 Imp gallons)/min each. Gas/oil reservoirs, pressurised to approximately 11.03 bars (160 lb/sq in). No pneumatic system. Main electrical system has three 115 kVA integrated engine-driven constant-speed generators, supplying 230/400 V three-phase AC power at 400 Hz through four main buses. Harris Corporation self-testing electrical multiplex system (EMUX), using mini-computers, controls major subsystems: it collects and conditions signals at remote terminals and transmits them from point to point over a common databus and also supervises all signal data, using a centralised controller/processor. Requiring only two two-wire cables for its operations, EMUX is designed to control such functions as electrical power distribution to subsystems and avionics equipment, engine instruments, environmental control system, fuel system, landing gear, lights and weapons system operations. Two Garrett APUs, installed between pairs of turbofans, provide self-start capability for operation from advance airfields and drive an emergency generator to power the essential bus. Quadruplex automatic flight control system (AFCS) controls flight path, roll attitude, altitude, airspeed, autothrottle and terrain-following. Flight director panel has heading hold, navigation and automatic approach modes. Central air data computer; gyro-stabilisation systems; stability control augmentation system; and structural mode control subsystem (SMCS). Engine fire extinguishing system.

AVIONICS: Standard GFE (government furnished equipment) includes communications, IFF, ILS, intercom, some navigation equipment, Honeywell ASN-131 SPN/GEANS radar altimeter (similar to that in B-52) and altimeter indicator, rescue beacon and transponder. Boeing Military Airplanes is responsible for the offensive avionics system (OAS). This includes a Singer Kearfott high accuracy inertial navigation system (developed from that used in the F-16); a Teledyne Ryan AN/APN-218 Doppler velocity sensor, comprising a single antenna/receiver/transmitter unit; Westinghouse AN/APQ-164 multimode offensive radar system (ORS), derived from the AN/APG-66 in the F-16, which includes a low-observable phased-array antenna to provide low-altitude terrain-following and precise navigational functions; IBM avionics control units (ACUs), including two for terrain-following based on those used in B-52 plus a mass storage device (MSD), using AP-101C computers initially (1750As later) to provide programme instructions for navigation, weapons delivery, bomb damage assessment, defensive system computation, and central integrated test; Honeywell offensive display sets, similar to those in B-52, comprising three multifunction displays (two at offensive systems operator's station and one for defensive systems operator), an electronics display unit, and a video recorder similar to that used in B-52; Sanders Associates electronic CRT display units, modified from those developed for original B-1, to allow defensive systems operator to analyse threat situations and assign appropriate countermeasures; and Sundstrand data transfer units (similar to those in B-52) to gather and store mission and flight data.

The defensive avionics system, which is the responsibility of Eaton Corporation's AIL Division, is based on that company's AN/ALQ-161 system, which comprises an AN/ALQ-161A radio frequency surveillance/ECM system (RFS/ECMS), tail warning function (TWF), AN/ASQ-184 defensive management system and an expendable countermeasures system, totalling 108 separate elements. Developed to support the original B-1 over a broad spectrum of missions, including deep solo penetration of hostile airspace, the system subsequently received additions to extend both the frequency coverage and repertoire of electronic jamming techniques of the original design. The current AN/ALQ-161 system, which will enable the B-1B to penetrate present and predicted enemy defences well into the 1990s, is controlled by a network of digital computers which can be reprogrammed easily; in addition, all electronic systems boxes 'plug in' to a dedicated databus network, enabling system to be upgraded continuously to adapt to future threats until well into the next century. To protect the B-1B, the system must counter a very dense

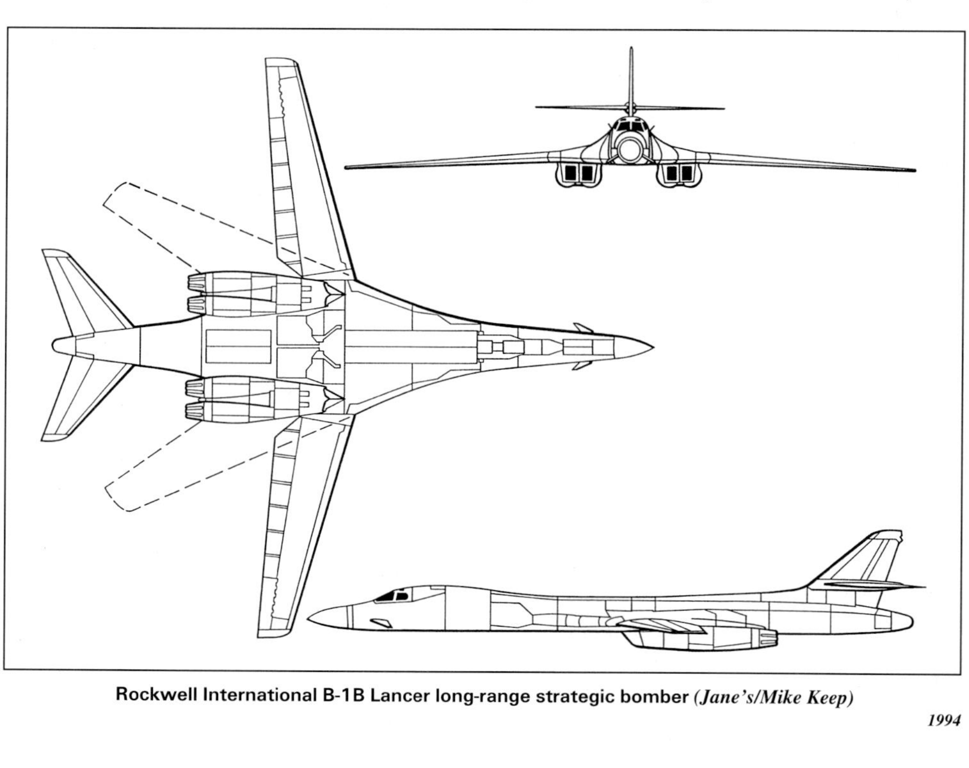

Rockwell International B-1B Lancer long-range strategic bomber *(Jane's/Mike Keep)* *1994*

Rockwell International B-1B Lancer variable geometry strategic bomber

1994

environment of signals from increasingly hostile radar networks. A single AN/ALQ-161 system contains and controls a large number of Northrop (Defense Systems Division) jamming transmitters and Raytheon phased-array antennae. In addition to jamming hardware, a sophisticated control system, managed by a network of special digital computers, is employed. This network can control the jamming chains so rapidly that each can jam signals from many radars simultaneously. The numerous jamming chains are deployed around periphery of B-1B to jam signals in any frequency band coming from any direction. Integrated with jamming control subsystem is an equally sophisticated network of separate receiving antennae, receivers and processors which act as 'ears' of system. By means of these new signals can be picked up, identified and jammed, with optimised jamming techniques, in a fraction of a second. One of the advantages of having the receiving function completely integrated with the jamming function, is that it allows receiving system to detect new signals and continue to monitor old signals while jamming in same frequency band. A special subsystem allows this to be accomplished by monitoring output of jamming transmitters and adjusting receivers continuously. All main systems computers on B-1B, including AN/ALQ-161's main computer, are identical, and communicate over a MIL-STD-1553B databus. Via this, the AN/ALQ-161 communicates with controls and displays used by defensive systems operator. It also uses this bus to send status reports to a central integrated test system (CITS), which records all in-flight failures and battle damage for later diagnosis and repair. Within the AN/ALQ-161 itself is a local status monitoring network called SEAT (status evaluation and test), which reports to CITS and allows system automatically to route electronic signals around failed components and maintain full jamming response against highest priority threat signals. Exclusive of cabling, displays and controls, the current AN/ALQ-161 system weighs approximately 2,360 kg (5,200 lb) and consumes about 120 kW (160 hp) of power in 'all-out' jamming mode. Other defensive equipment includes expendable decoys such as chaff and flares.

ARMAMENT: Three internal weapons bays, comprising 9.53 m (31 ft 3 in) double bay forward of wing carry-through structure and a single 4.57 m (15 ft) long bay aft, with hydraulically actuated doors. Forward bay incorporates a movable bulkhead permitting accommodation of a wide variety of weapons, of various sizes, and mixed loads. Internal capacity in a nuclear role for up to eight AGM-86B air-launched cruise missiles (ALCMs), 24 AGM-69 short-range attack missiles (SRAMs), 12 B-28 or 24 B-61 or B-83 free-fall nuclear bombs; or, in a non-nuclear role, for up to 84 500lb Mk 82 bombs or 500 lb Mk 36 mines. Six external stores stations beneath fuselage, on which can be carried an additional 12 ALCMs. The forward and aft bays can be combined to carry eight ALCMs on a common strategic rotary launcher.

DIMENSIONS, EXTERNAL:

Wing span: fully spread	41.67 m (136 ft 8½ in)
fully swept	23.84 m (78 ft 2½ in)
Length overall	44.81 m (147 ft 0 in)
Height overall	10.36 m (34 ft 0 in)
Tailplane span	13.67 m (44 ft 10 in)
Wheel track (c/l of shock-absorbers)	4.42 m (14 ft 6 in)
Wheelbase	17.53 m (57 ft 6 in)

AREAS:

Wings, gross	approx 181.2 m² (1,950 sq ft)

WEIGHTS AND LOADINGS:

Weight empty, equipped	87,090 kg (192,000 lb)
Max weapons load: internal	34,019 kg (75,000 lb)
external	26,762 kg (59,000 lb)
Max fuel load	88,450 kg (195,000 lb)
Typical conventional weapon load	29,030 kg (64,000 lb)
Max T-O weight	216,365 kg (477,000 lb)
Max wing loading	approx 1,194 kg/m² (244.6 lb/sq ft)

PERFORMANCE (design):

Max level speed approx Mach 1.25

Low-level penetration speed at approx 62 m (200 ft) more than 521 kts (965 km/h; 600 mph)

Max unrefuelled range approx 6,475 n miles (12,000 km; 7,455 miles)

UPDATED

ROCKWELL INTERNATIONAL OV-10 BRONCO

US military designation: OV-10

TYPE: Two-seat multipurpose counter-insurgency aircraft.

PROGRAMME: This aircraft was Rockwell International's North American Aircraft Operation's entry for the US Navy's design competition for a light armed reconnaissance aircraft (LARA) specifically suited for counter-insurgency missions. Nine US airframe manufacturers entered for the competition and the NA-300 was declared the winning design in August 1964. Seven prototypes were then built by the company's Colombus Division, under the designation YOV-10A Bronco. The first of these flew on 16 July 1965, followed by the second in December 1965.

A number of modifications were made as a result of flight experience with the prototypes. In particular, the wing span was increased by 3.05 m (10 ft 0 in), the T76 turboprop engines were uprated from 492 kW (660 shp) to 534 kW (716 shp), and the engine nacelles were moved outboard approximately 0.15 m (6 in) to reduce noise in the cockpit.

A prototype with lengthened span flew for the first time on 15 August 1966. The seventh prototype had Pratt & Whitney (Canada) T74 (PT6A) turboprops for comparative testing.

VERSIONS: **OV-10A**: Initial production version, first flown by the US Marine Corps for light armed reconnaissance, helicopter escort and forward air control duties, and by the US Air Force in the forward air control role, and for limited quick-response ground support pending the arrival of tactical fighters. Total 271 built six transferred to the Royal Moroccan Air Force of which three are still in service. Withdrawn from US service by March 1994.

OV-10B and OV-10B(Z): Versions of OV-10A for Federal Republic of Germany (6 and 12 respectively). Withdrawn from service.

OV-10C: Version of the OV-10A for the Royal Thai Air Force, 40 delivered, of which 35+ still in service.

OV-10D: Designation of 17 USMC OV-10As converted for night observation surveillance (NOS) role. In addition to the NOS systems and the retention of basic OV-10A fuselage stores and external fuel capability, the OV-10D NOS has uprated engines, wing pylons capable of carrying rocket pods, flare pods, free-fall stores, including 113 kg (250 lb) laser-guided bombs, and external fuel tanks when extended radius/loiter time is required. A Texas Instruments AN/AAS-37 FLIR sensor and laser target designator installed in a rotating ball-turret in the nose. The turret can be linked to a turret-mounted General Electric M97 20 mm cannon, mounted beneath the fuselage, in lieu of normal operation with standard OV-10A, armament sponsons and centreline station. Equipment includes optional APR-39 radar homing and warning system, ALE-39 chaff/flare dispensers and IR-suppressant engine exhaust system. Withdrawn from US service by March 1994. Possible transfer to foreign air force (South Korea).

OV-10E and OV-10F: Versions of OV-10A for the Venezuelan and Indonesian air forces. Still in service with the Venezuelan Air Force (12+) and with the Indonesian Air Force (13).

UPGRADES: **Rockwell**: OV-10D-Plus Bronco. See separate entry.

OPERATORS: Versions of the OV-10 Bronco are in service with the armed forces of the following countries: Colombia (10), Indonesia (12), Morocco (4), Philippines (22), Thailand (27) and Venezuela (22).

The following description applies to the OV-10A, except where indicated:

DESIGN FEATURES: Cantilever shoulder-wing monoplane. Constant-chord wing without dihedral or sweep. Tailplane mounted near tips of fins.

FLYING CONTROLS: Manually operated ailerons, supplemented by four small manually operated spoilers forward of the outer flap on each wing, for lateral control at all speeds. Hydraulically operated double-slotted flaps in two sections on each wing, separated by tail booms. Manually operated rudders and elevator.

STRUCTURE: Conventional aluminium alloy two-spar structure. The fuselage is a short pod-type conventional aluminium semi-monocoque construction, suspended from wing. Glassfibre nosecone. The tail unit is a cantilever all-metal structure carried on twin booms of semi-monocoque construction.

LANDING GEAR: Retractable tricycle type, with single wheel on each unit, developed by Cleveland Pneumatic Tool Company. Hydraulic actuation, nosewheel retracting forward, main units rearward into tailbooms. Two-stage oleo-pneumatic shock-absorbers. Forged aluminium wheels with tyres size 29 × 11-10, pressure 4.48 bars (65 lb/sq in). Nosewheel tyre size 7.50-10, pressure 5.52 bars (80 lb/sq in). Cleveland hydraulic disc brakes.

POWER PLANT: Two 533 kW (715 ehp) Garrett T76-G-416/417 turboprop engines, each driving a Hamilton Standard three-blade constant-speed reversible-pitch and fully feathering metal propeller. (OV-10D has 775.5 kW; 1,040 shp T76-G-420/421 engines, each driving a similar

Rockwell International OV-10NOS Night Observation Surveillance aircraft formerly of the USMC. All OV-10s withdrawn from US service

1995

Rockwell International OV-10A Bronco twin-boom counter-insurgency aircraft

1994

Rockwell International OV-10A Bronco of the US Air Force *(Peter J. Cooper)*

1994

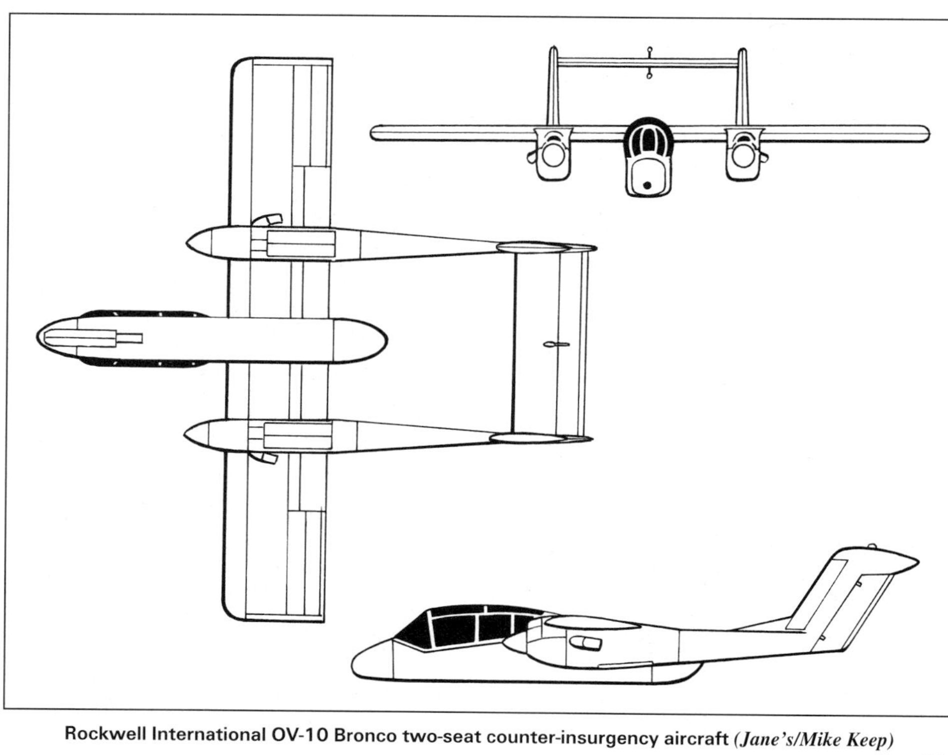

Rockwell International OV-10 Bronco two-seat counter-insurgency aircraft *(Jane's/Mike Keep)*

1994

Hamilton Standard propeller but with glassfibre blades.) Five self-sealing bladder-type fuel tanks in wings, with combined capacity of 954 litres (252 US gallons; 210 Imp gallons). Gravity refuelling point above each tank on wing upper surface. Provision for carrying one 568 litre (150 US gallon; 125 Imp gallon) drop tank on underfuselage pylon; OV-10D also has provisions for carrying one 378 litre (100 US gallon; 83 Imp gallon) drop tank on each wing pylon. Oil capacity 11.5 litres (3 US gallons; 2.5 Imp gallons).

ACCOMMODATION: Crew of two in tandem, on LW-3B zero/zero ejection seats, under canopy with two large upward-opening transparent door panels on each side. Dual controls optional. Cargo compartment aft of rear seat, with rear-loading door at end of fuselage pod. Rear seat removable to provide increased space for up to 1,452 kg (3,200 lb) of freight, or for carriage of five paratroops, or two stretcher patients and attendant.

SYSTEMS: Heating and ventilation system combines engine bleed air and cold ram air to provide temperature controlled conditions. Engine bleed air is used also for windscreen defrosting and to supply crew's anti-*g* suits. Hydraulic system of intermittent-duty powered by an electrically driven hydraulic pump at a pressure of 103.5 bars (1,500 lb/sq in), for actuation of trailing-edge flaps, landing gear and nosewheel steering. Wheel brakes which have two independent manually driven brake units, are fed directly from the hydraulic system reservoir. Electrical system powered by two 30 V 300 A starter/generators and two 24 V 22 Ah Ni/Cd batteries. (OV-10D has two 30 V 400 A starter/generators and two 24 V 30 Ah air-cooled batteries.) AC power derived from two 750 VA inverters which supply 115 V at 400 Hz three-phase; single-phase AC of 115 V or 26 V at 400 Hz can be tapped from the bus system. (An additional 3,000 VA inverter is installed in the OV-10D.) External power sockets for engine starting and utility services; the latter can be used to provide 28 V DC to other aircraft for engine starting or servicing. Demand-regulated oxygen system supplied from two 0.008 m^3 (0.3 cu ft) oxygen cylinders at a pressure of 124 bars (1,800 lb/sq in). Independent fire warning system for each engine, comprising control unit, sensing elements and warning lights. US Air Force aircraft (only) have an electrically fired fire extinguishing system installed in each engine nacelle. No pneumatic system.

AVIONICS: US Air Force OV-10As are equipped with AN/AIC-18 intercom; AN/ARC-51BX UHF, Wilcox 807A VHF, dual FM-622A VHF, and HF-103 HF com radios; nav system includes AN/ASN-75 compass, AN/ARN-52(V) Tacan, AN/ARA-50 UHF-ADF, AN/ARN-83 LF-ADF, 51R-6 VOR, and 51V-4A ILS glide slope; identification system includes AN/APX-64(V) IFF/SIF, and SST-181-X radar beacon. US Marine Corps aircraft are equipped with AN/AIC-18 intercom; AN/ARC-51 AX UHF, AN/ARC-54 VHF, and AN/ARC-120 HF com radios; nav system includes AN/ASN-75 compass, AN/APN-171 radar altimeter, AN/ARN-52(V) Tacan, and AN/ARA-50 UHF-ADF; AN/APX-64(V) IFF/SIF for identification. (OV-10D NOS aircraft have an AN/AAS-37 FLIR sensor system package, comprising FLIR, a laser target designator, and an automatic video tracker.)

ARMAMENT: Four weapon attachment points, each with capacity of 272 kg (600 lb), under short sponsons extending from bottom of fuselage on each side, under wings. Fifth attachment point, capacity 544 kg (1,200 lb) under centre-fuselage. Two 7.62 mm M60C machine guns, each with 500 rounds of ammunition, carried in each sponson. USMC OV-10A has provision also for carrying one AIM-9D Sidewinder missile under each wing. Stores which can be carried on the underfuselage and sponson stations include Mk 81, 82 and 83 GP bombs, Mk 81 and 82 GP (Snakeye) bombs; Mk 77 Mod 2 and Mod 4 fire bombs; LAU-3/4, LAU-10/A, LAU-32/A, LAU-59/A, LAU-60/A, LAU-61/A, LAU-68/A and LAU-69/A rocket packages; SUU-11A/A (7.62 mm Minigun), Mk 4 Mod 0 (20 mm), and GPU-2/A (20 mm) gun pods; SUU-40/A and SUU-44/A with Mk 24 and Mk 45 flares; Mk 12 Mod 0 (Popeye) smoke tank; Mk 86/AA37B-3 MBR with Mk 76 and Mk 106 practice bombs; CBU-55/B cluster bomb. Maximum weapon load on fuselage stations 1,633 kg (3,600 lb). (OV-10D aircraft have wing pylons with stores capacity of 272 kg; 600 lb each, which can carry CBU-55/B cluster bombs; LAU-10/A, LAU-68/A and LAU-69/A rocket packages; SUU-40/A and SUU-44/A with Mk 24 and Mk 45 flares. In lieu of sponsons and fuselage centreline load, the OV-10D can accommodate a 20 mm gun turret kit installed on centreline station hardpoint.)

DIMENSIONS, EXTERNAL:

Wing span	12.19 m (40 ft 0 in)
Length overall: OV-10A	12.67 m (41 ft 7 in)
OV-10D	13.41 m (44 ft 0 in)
Height overall	4.62 m (15 ft 2 in)
Tailplane span	4.45 m (14 ft 7 in)
Wheel track	4.52 m (14 ft 10 in)
Wheelbase	3.56 m (11 ft 8 in)
Propeller diameter	2.59 m (8 ft 6 in)
Rear-loading door: Height	0.99 m (3 ft 3 in)
Width	0.76 m (2 ft 6 in)

AREAS:

Wings, gross	27.03 m^2 (291 sq ft)

WEIGHTS AND LOADINGS:

Weight empty	3,127 kg (6,893 lb)
Normal T-O weight	4,494 kg (9,908 lb)
Overload T-O weight	6,552 kg (14,444 lb)
Max wing loading	242.4 kg/m² (49.6 lb/sq ft)

PERFORMANCE (A: OV-10A/C/E/F, B: OV-10D; at weights stated):

Max level speed at S/L, without weapons:	
A	244 kts (452 km/h; 281 mph)
B	250 kts (463 km/h; 288 mph)
Max rate of climb at S/L at normal T-O weight:	
A	790 m (2,600 ft)/min
B	920 m (3,020 ft)/min
Rate of climb at S/L, one engine out, without weapons:	
A	58 m (190 ft)/min
B	168 m (550 ft)/min
Service ceiling at normal T-O weight:	
A	7,315 m (24,000 ft)
B	9,150 m (30,000 ft)
Service ceiling, one engine out, without weapons:	
A	3,505 m (11,500 ft)
B	3,810 m (12,500 ft)
T-O run at normal T-O weight	226 m (740 ft)
T-O run to 15 m (50 ft):	
at normal T-O weight	341 m (1,120 ft)
at max T-O weight	853 m (2,800 ft)
Landing from 15 m (50 ft) at normal T-O weight	372 m (1,220 ft)
Landing run: at normal T-O weight	226 m (740 ft)
at max T-O weight	381 m (1,250 ft)
Combat radius with max weapon load, no loiter	198 n miles (367 km; 228 miles)
Ferry range with auxiliary fuel	1,200 n miles (2,224 km; 1,382 miles)

UPDATED

Rockwell North American (Lockheed/General Dynamics) F-111F Pacer Strike aircraft taking off from Cannon AFB

1995

Rockwell (General Dynamics) F-111C upgraded for the Royal Australian Air Force

1994

ROCKWELL (NORTH AMERICAN) T-2 BUCKEYE

TYPE: Two-seat general-purpose jet trainer.

PROGRAMME: After a design competition among several leading US manufacturers, what was then the North American's Columbus Division was awarded a contract in 1956 to develop and build a jet training aircraft for the US Navy. The first T2J-1 flew on 31 January 1958. Five versions of the aircraft were built.

VERSIONS: **T-2A** (formerly T2J-1): Initial version with single Westinghouse J34-WE-48 turbojet engine. Withdrawn from service in 1973.

T-2B (formerly T2J-2): Powered by two Pratt & Whitney J60-P-6 turbojet engines. US Navy ordered 97. Deliveries completed 1969. Withdrawn from service.

T-2C: Powered by two General Electric J85-GE-4 turbojet engines. First production aircraft flew on 10 December 1968. Deliveries completed 1975. In service with the US Navy.

T-2D: Basically similar to T-2C but with different electronics and deletion of carrier capability. In service with the VenezuelanAir Force. Deliveries completed in 1977.

T-2E: Generally similar to T-2C but with different electronics, provision of an accessory kit for attack role. In service with Greek Air Force. Deliveries completed 1977.

DESIGN FEATURES: Cantilever mid-wing monoplane. Wing section NACA 64A212 (modified). Thickness/chord ratio 12 per cent. Cantilever tail unit.

STRUCTURE: The wing is an all-metal two-spar structure. The fuselage is an all-metal semi-monocoque structure in three main sections: forward fuselage containing equipment bay and cockpit, centre fuselage housing power plant, fuel and wing carry through structure; and rear fuselage, carrying the arrester hook and airbrakes. The tail unit is an all-metal structure. Each half of tailplane interchangeable.

FLYING CONTROLS: Interchangeable all-metal ailerons, with hydraulic boost. Large all-metal trailing-flaps. Hydraulically actuated airbrake on each side of the fuselage. Elevators boosted hydraulically. Rudder manually controlled. Trim tabs in elevators and rudder.

LANDING GEAR: Retractable tricycle type. Oleo-pneumatic shock-absorbers. Hydraulic retraction. Main units retract inward into wings. Nosewheel retracts forward into fuselage. Mainwheels size 24 × 5.50. Nosewheel size 20 × 4.40. Mainwheel tyre pressure 10.34 bars (150 lb/sq in), nosewheel tyre pressure 5.17 bars (75 lb/sq in). Goodyear air-cooled single-disc hydraulic brakes. Retractable sting-type, universal joint mounted, arrester hook.

POWER PLANT: Two 13.12 kN (2,950 lb st) General Electric J85-GE-4 turbojet engines, with jet outlets under rear fuselage. Fuel in main tanks over engines with capacity of 1,465 litres (387 US gallons; 322 Imp gallons), two wing-tip tanks each of 386 litres (102 US gallons; 85 Imp gallons) capacity, and two tanks in the inboard sections of the wings. Total fuel capacity 2,616 litres (691 US gallons; 575 Imp gallons).

ACCOMMODATION: Pupil and instructor in tandem in enclosed cabin, on rocket-powered LS-1 ejection seats, under clamshell canopy. Instructor is raised 0.25 m (10 in) above level of pupil.

ARMAMENT: Optional packaged installations of guns, target-towing gear, ?? kg (100 lb) practice bombs, M-5 or MK76 practice bomb clusters, Aero 4B practice bomb containers, 2.25 in rocket launchers or seven 2.75 in rockets in Aero 6A-1 rocket containers, can be carried on two store stations, one beneath each wing, with a combined capacity of 290 kg (640 lb). Alternative option includes above package, plus four additional wing store stations, to provide a total underwing ordnance capacity if 1,588 kg (3,500 lb), and protection against small arms fire for the fuel tanks.

DIMENSIONS, EXTERNAL:

Wing span over tip-tanks	11.62 m (38 ft 1½ in)
Length overall	11.67 m (38 ft 3½ in)
Height overall	4.51 m (14 ft 9½ in)
Tailplane span	5.46 m (17 ft 11 in)
Wheel track	5.61 m (18 ft 4¾ in)

AREAS:

Wings, gross	23.69 m² (255 sq ft)
Trailing-edge flaps (total)	4.23 m² (45.56 sq ft)
Fin	2.54 m² (27.29 sq ft)
Rudder, incl tab	0.84 m² (9.01 sq ft)
Tailplane	3.95 m² (42.55 sq ft)
Elevators, incl tabs	1.95 m² (21.00 sq ft)

Rockwell International (North American) T-2C Buckeye in service with the US Navy *(Peter J. Cooper)*

1995

WEIGHTS AND LOADINGS:

Weight empty	3,680 kg (8,115 lb)
Max T-O weight	5,983 kg (13,191 lb)

PERFORMANCE (at max T-O weight):

Max level speed at 7,620 m (25,000 ft)	460 kts (852 km/h; 530 mph)
Stalling speed	87 kts (161 km/h; 100 mph)
Max rate of climb at S/L	1,800 m (5,900 ft)
Service ceiling	13,870 m (45,500 ft)
Max range	930 n miles (1,722 km; 1,070 miles)

NEW ENTRY

ROCKWELL OV-10D-PLUS BRONCO

TYPE: Combat aircraft upgrade.

PROGRAMME: US Navy ordered modification (at Cherry Point MCAS, with Rockwell kits) of 23 US Marine Corps OV-10As and 14 OV-10Ds to OV-10D-Plus July 1988; modifications include fitting 775.5 kW (1,040 shp) Garrett T76-G-420/421 turboprops and composites propellers, new solid-state avionics with Rockwell-Collins cockpit management system, nose-mounted Texas Instruments AN/AAS-37 FLIR turret (already fitted to OV-10Ds), Sanders AN/ALQ-144 IR jammer, General Instrument AN/APR-39 RWR, complete re-wiring, quick-disconnect cargo door and structural strengthening for 4.28 m (14 ft)/s carrier landings; first carrier trial (by prototype OV-10D-Plus, 155468) on USS *Saratoga*, 7 June 1990; first production conversion (155499) completed June 1990 and redelivered to VMO-2 Squadron at Camp Pendleton. All OV-10s withdrawn from US service by March 1994. Surplus aircraft may be transferred to foreign air force (possibly South Korea).

Eighteen ex-USAF OV-10A Broncos were bought by Venezuela in 1991.

UPDATED

LOCKHEED FORT WORTH (GENERAL DYNAMICS) F-111 PACER STRIKE AND OTHER UPGRADES

TYPE: Combat aircraft upgrade.

PROGRAMME: Pacer Strike update of 79 F-111Ds and 84 F-111Fs began September 1989, extending life to 2010; intended to bring F-111 fleet to essentially common avionics standard (F-111A, EF-111A and F-111E upgraded earlier by Grumman) but F-111D component cancelled 30 March 1990, and aircraft to retire. F model upgrade kits reduced to 58, delivery of which was completed by July 1995. US Air Force however, has decided to upgrade only 29 F-111F aircraft. Estimated value of programme is $120 million. Modifications include integrated cockpit displays and complete re-wiring of avionics bay added to ring laser gyro INS and Navstar GPS; Rockwell won Australian contract to update 22 Australian F-111s; announced 23 August 1990, with value of $160 million to Rockwell from total $320 million; AUP (avionics update programme) concerns 18 F-111Cs and four RF-111Cs; gaining digital avionics system and MIL-STD-1553B databus; prototype flown in USA, August 1990; Australian industry to manufacture 21 further kits to complete modification by late 1995.

UPDATED

ROCKWELL B-1B LANCER ECM UPGRADE PROGRAMME

TYPE: Long-range, multirole strategic bomber ECM upgrade programme.

PROGRAMME: The ECM upgrade programme estimated to cost between $750 million to $1 billion forms part of a $1.5 billion survivability upgrade programme plus a $2.5 billion programme to deliver conventional precision munitions. The purpose of the programme is to modify the avionics to give the B-1B Lancer capability to address various regional air defence threats at high and low altitudes and varying speeds.

There are six teams bidding for the ECM contract which will have Rockwell as the source selection authority and prime contractor. Teams include: ITT/Loral with the ALQ-172 jammer, the ALR-56C radar warning receiver and the ALQ-153 tail warning set; AIL/Litton with modified ALQ-161 and parts of ALR-621 RWR; Westinghouse/Hughes with a derivative of the cancelled ALQ-165 airborne self-protection jammer (ASPJ) and ALR-67(V)3 warning control system; Raytheon/TRW with the ALQ-184; Northrop/(unknown partner) with the ALQ-135 and parts of the original ALQ-161; Lockheed-Sanders with a hybrid system combining parts of the ALQ-161 with units from the U-2R ECM system.

A single contractor will be chosen for a three-year engineering and manufacturing development (EMD) phase followed by a four-year production period.

VERIFIED

ROCKWELL/BOEING B-1B LANCER CONVENTIONAL WEAPONS UPGRADE

TYPE: Long-range bomber weapons upgrade.

PROGRAMME: Contract awarded April 1995 by USAF. Rockwell International North American Division to act as prime. Contract value $232 over four years. Boeing Military Airplanes Division received subcontract worth approximately $115 million. Programme to continue Engineering and Manufacturing Development (EMD) of weapons systems upgrade.

DESIGN FEATURES: Programme entails the installation and integration of a NAVSTAR Global Positioning System (GPS), an MIL-STD-1760 databus, GPS-guided Joint Direct Attack Munition (J-DAM) and secure communications. The B-1B will also be able to launch the Wind Corrected Munitions Dispenser (WCMD) and Joint StandOff Weapon (JSOW).

NEW ENTRY

ROCKWELL (LOCKHEED) AC-130U SPECTRE (C-130H UPGRADE)

TYPE: C-130 aerial gunship upgrade.

PROGRAMME: Development of new gunship version of Lockheed C-130 Hercules launched 6 July 1987 with $155,233,489 contract to North American Aircraft Operations; prototype (87-0128) funded in FY86, six production aircraft in FY89, five in FY90 and one in FY92 - last mentioned contract awarded 31 December 1992, increasing total to 13.

AC-130U FUNDING

Batch	FY	Quantity	First aircraft
—	1987	1	87-0128
Lot I	1989	6	89-0509
Lot II	1990	5	90-0163
Lot III	1992	1	92-

Prototype delivered to Rockwell as standard C-130 transport, 28 July 1988; first post-conversion flight 20 December 1990 was also ferry to Edwards AFB, California, for trials with 6510th Test Wing; work on remaining AC-130Us began January 1991; by January 1993, six aircraft complete, of which three with 418th Test Squadron/412th TW (ex-6510th TW) at Edwards AFB; operational unit to be 16th Special Operations Squadron at Hurlburt Field, Florida; deliveries from 1994. Armament, firing to port, consists of (front to rear) General Electric GAU-12/U 25 mm six-barrel Gatling gun with 3,000 rounds, Bofors 40 mm gun, and a 105 mm gun based on US Army howitzer; addition of Rockwell Hellfire ASMs under consideration 1992; guns can be slaved to Hughes AN/APQ-180 (modified AN/APG-70) digital fire control radar, Texas Instruments AN/AAQ-117 FLIR or GEC-Marconi all-light-level television (ALLTV), for night and adverse weather attack on ground targets; sideways-facing HUD for visual aiming. Attack method is to circle target at altitude firing into apex of turn on ground, but guns can now be trained, relieving pilot of absolute precision flying; flight path is also less predictable; can fire on two targets simultaneously. AC-130U can refuel in flight and fly escort, surveillance, search, rescue and armed reconnaissance/interdiction missions.

Prone observer's position on rear ramp; starboard side observer's window aft of flight deck, and battle management centre in cabin with seven positions at monitoring consoles and four IBM AP-102 computers; crew totals 13, including flight crew and loaders. Defensive aids believed similar to those in MC-130H; modified fuel tank pylons contain IR countermeasures; total of 300 chaff bundles and 90 MJU7 or 180 M206 flares in three launchers under fuselage; ITT Avionics AN/ALQ-172 jammer in base of fin; Loral AN/ALR-56M RWR; other equipment includes combined INS and Navstar GPS, triple MIL-STD-1553B digital databuses and Spectra ceramic armour protection.

VERIFIED

Rockwell (Lockheed) AC-130U Spectre aerial gunship

1994

ROCKWELL C-160 AVIONICS AND FLIGHT MANAGEMENT SYSTEM UPGRADE

TYPE: Twin-engine turboprop transport aircraft avionics upgrade.

PROGRAMME: Contract to provide the German Air Force C-160 Transall aircraft with an integrated cockpit. Systems include the FMS-800 flight management system and global positioning system (GPS). Purpose of the contract is to automate routine tasks in order to reduce the crew complement and to enhance navigation accuracy. The enhanced systems will allow German Air Force C-160 Transall aircraft to operate at increased efficiency on military, peacekeeping and humanitarian missions.

COSTS: $21 million production contract.

VERIFIED

ROCKWELL ROTARY-WING UPGRADES

TYPE: Combat rotary-wing aircraft upgrades.

PROGRAMME: Responsible for the installation and integration of: AN/ASC-15 battlefield C2 console in the UH-60 Blackhawk for the US Army; all-weather capable SOF mission equipment and weather radar/electronic flight displays for the MH-47D; and automatic target handover and global positioning systems (GPS) in AH-64 for the US Army.

VERIFIED

Transall C-160 upgraded with Rockwell installed integrated cockpit

1994

SABRELINER

SABRELINER CORPORATION

Pierre Laclede Centre, Suite 1500, 7733 Forsyth Boulevard, St Louis, Missouri 63105-1821
Telephone: 1 (314) 863 6880
Fax: 1 (314) 863 6887
Telex: 44 7227

OTHER WORKS: Perryville, Missouri (Maintenance and Modification Center)
Neosho, Missouri (Engine Overhaul and Repair Center)
Independence, Kansas (Engine Accessory Overhaul Center)
Cahokia, Illinois (Midcoast Aviation Maintenance, Modification and Refurbishment Center)
Little Rock, Arkansas (Midcoast Aviation Maintenance Center and FBO)
St Louis, Missouri (Midcoast Aviation FBO and Service Center)
St Louis, Missouri (FBO at Sabreliner)

CHAIRMAN AND CEO: F. Holmes Lamoreux

EXECUTIVE VICE-PRESIDENT AND COO - SABRELINER: Jerry L. Leah

PRESIDENT AND COO - MIDCOAST AVIATION: Jerry L. Leah

SENIOR VICE-PRESIDENT - FINANCE/CORPORATE DEVELOPMENT AND CFO: Rodney E. Olson

VICE-PRESIDENT - GOVERNMENT MARKETING AND DEVELOPMENT: Gene L. Harbula

VICE-PRESIDENT - SALES AND MARKETING: Robert J. Walker

VICE-PRESIDENT - OPERATIONS: Jackson L. Forney

VICE-PRESIDENT - ENGINEERING QUALITY ASSURANCE AND EPA: Bob D. Hanks

Sabreliner Corporation is a diversified aviation service, modification and manufacting firm that supports a variety of corporate, military and government aircraft, engines, systems and components. The company is the Original Equipment Manufacturer (OEM) of the Sabreliner series of business jets, which it supports through engineering, certification, manufacture and installation of proprietary parts. There are approximately 370 Sabreliner aircraft currently in service.

The Corporation was formed July 1983 after private investors bought the Sabreliner Division of Rockwell International. Today, the Sabreliner Corporation manages sophisticated aviation products, services and training programs; repairs, modifies and remanufactures aircraft; maintains and overhauls aircraft engines and specialised industrial engines; and sells aircraft and aircraft parts.

With its subsidiary, Midcoast Aviation, Sabreliner provides airframe modification, maintenance and repair, component and accessory overhaul and repair; and engine maintenance and repair for business, private and government aircraft of all sizes.

Sabreliner aircraft-specific programmes include the Excalibur, which extends the flying hours of Sabreliner Model 40 and 60s from 10,000 hours to 15,000 hours, changing them to the 40EX and 60EX designation. For the US military, Sabreliner Corporation provides the T-39N, a modified Sabreliner 40 used for Undergraduate Naval Flight Officer (UNFO) training at the Naval Air Station, Pensacola, Florida. Sabreliner Corporation assembled and manages a team of five major subcontractors to provide the training services programme for air-to-ground navigation and air-to-air intercept techniques. In addition, the Corporation supports Sabreliner CT-39 E/G aircraft and their engines at five Navy and Marine Corps sites around the world.

Non-Sabreliner aircraft programmes by the Corporation include the US Air Force and non-US militaries' J85 engine overhaul and repair; US Air Force and non-US militaries' J69 engine overhaul and repair; US Air Force and non-US militaries' T-37 Structural Life Extension Program (SLEP); US

Air Force J85 engine Variable Exhaust Nozzle (VEN) power unit; US Army and non-US militaries' T53 helicopter engine rebuild; US Army T53 fuel control modification/overhaul; US Army T700 engine overhaul and repair; US Army UH-60 hydraulic pump; US Navy J52 engine compressor stators; and B.F. Goodrich C-17 main landing gear test fixture manufacture.

UPDATED

Sabreliner 60 twin-engine business jet transport aircraft *(Sabreliner)*

1995

ROCKWELL SABRELINER

US Air Force and Navy designation: T-39
Swedish Air Force designation: Tp 86

TYPE: Twin-engined business jet transport.

PROGRAMME: To meet the USAF's 'UTX' requirements for a combat readiness trainer and utility aircraft, North American (now Sabreliner) built as a private venture the prototype of a small sweptwing twin-jet monoplane named the Sabreliner. Design work began 30 March 1956 and the prototype, powered by two General Electric J85 turbojet engines, flew for the first time 16 September 1958.

VERSIONS: **Series 40:** Basic version to carry a crew of two and up to nine passengers. More powerful engines than its predecessors. New brakes with longer life. Three windows instead of two on each side of passenger cabin. Early model Sabreliners can be upgraded to Series 40 standard. Series 40 production ended.

T-39A: Pilot proficiency/administrative support aircraft. 143 delivered to the US Air Force.

T-39B: Radar trainer for the US Air Force. Six built.

T-39D: Radar interception officer trainer for the US Navy. 42 built.

CT-39E: Rapid response airlift aircraft for the US Navy.

CT-39G: Fleet tactical support aircraft. 12 built.

T-39N: Modified Sabreliner Model 40 for US Navy navigation training. 17 built.

Sabreliner 60: Generally similar to the Sabreliner 40, but fuselage lengthened by 0.97 m (3 ft 2 in). Accommodation for crew of two and up to 10 passengers. Five windows on each side of passenger cabin.

Sabreliner 60A: Improved version of the Series 60. Described separately.

Sabreliner 65: With TFE731 engines. Described separately.

Sabreliner 75A: With CF 700 turbofan engines. Described separately.

Sabreliner 80A: Improved version of the 75A, with aerodynamic refinements similar to those of Model 60A.

The following details refer to the Sabreliner 60:

UPGRADES: **Sabreliner:** Series 60SC/80SC and Series 40EX/60EX. See separate entries.

OPERATORS: Versions of the Sabreliner are in service with the armed forces of the following countries: Argentina (2), Bolivia (2), Croatia (1), Ecuador (4), Mexico (2), Sweden (2) and USA (25).

DESIGN FEATURES: Cantilever low-wing monoplane. Sweepback 28° 33′. Variable incidence tail unit with moderate sweepback on all surfaces.

FLYING CONTROLS: Electrically operated trim tab in each aileron. Electrically operated trailing-edge flaps. Aerodynamically operated leading-edge slats in five sections on each wing. Large hydraulically operated airbrake under centre-fuselage. Direct mechanical flight controls with electrically operated variable incidence tailplane. Electrically operated trim tab in rudder.

STRUCTURE: All-metal two-spar milled skin structure. The fuselage is an all-metal semi-monocoque structure. The tail unit is a cantilever all-metal structure, with flush antennae forming tip of fin and inset in dorsal fin.

LANDING GEAR: Retractable tricycle type. Twin-wheel nose unit retracts forward. Single wheel on each main unit, retracting inward into fuselage. Mainwheel tyres size 26 × 6.60-14, pressure 12.77 bars (185 lb/sq in). Nosewheel tyres 18 × 4.40-10, pressure 5.18 bars (75 lb/sq in). Hydraulic brakes with anti-skid units. Optional kit for operation from gravel runways.

POWER PLANT: Two Pratt & Whitney JT12A-8 turbojet engines (each 14.68 kN; 3,300 lb st) in pods on sides of rear fuselage. Thrust reversers optional. Integral fuel tanks in wings, with total capacity of 3,418 litres (903 US gallons; 751 Imp gallons). Fuselage tank, capacity 606 litres (160 US gallons; 133 Imp gallons). Total fuel capacity 4,024 litres (1,063 US gallons; 885 Imp gallons). Single-point refuelling.

ACCOMMODATION: Crew of two and up to 10 passengers in pressurised air conditioned cabin. Downward-hinged plug-type door, with built-in steps, forward of wing on port side. Emergency exits on both sides of cabin. Baggage space at front of cabin opposite door, with adjacent coat rack specified in many interior configurations. Beverage galley at forward end of cabin, two folding tables, door between cabin and flight deck, cold/hot air outlets and reading lights at each seat, two magazine racks. Toilets at rear of cabin. With seats removed there is room for 1,135 kg (2,500 lb) of freight.

SYSTEMS: AiResearch air conditioning. Hydraulic system with audible failure warning system. Electrical system powered by engine-driven generators with 34 Ah battery. Automatic oxygen system. APU optional. Optional full-span pneumatically operated de-icing boots on wing. Optional full-span pneumatically operated leading-edge de-icing boots on tail.

ELECTRONICS AND EQUIPMENT: Standard electronics include dual Collins VHF-20A com transceivers, dual Collins VIR-30A nav receivers with VOR/LOC/GS, Collins AP-105 autopilot, Collins FD-108Z pilot's flight director, Collins FD-108Y co-pilot's flight director, Collins NCS-31 navigation control system, dual Collins MC-103 compass systems, dual Collins 332C-10 radio magnetic indicators, Collins TDR-90 ATC transponder, Collins DME-40 DME, Collins ADF-60A ADF, dual Collins 346B-3 audio system control centres, and Bendix RDR-1200C weather radar with digital data display. Standard equipment includes white instrument lights, encoding altimeter and altitude alerting, pilot's and co-pilot clocks, synchroscope and cockpit floor and sidewall heating pads.

DIMENSIONS, EXTERNAL:

Wing span	13.61 m (44 ft 8 in)
Length overall	14.30 m (46 ft 11 in)
Height overall	4.88 m (16 ft 0 in)
Tailplane span	5.35 m (17 ft 6½ in)
Wheel track	2.20 m (7 ft 2½ in)
Wheelbase	4.85 m (15 ft 10¾ in)
Cabin door: Height	1.19 m (3 ft 11 in)
Width	0.71 m (2 ft 4 in)

DIMENSIONS, INTERNAL:

Cabin (excl flight deck): Length	5.79 m (19 ft 0 in)
Max width	1.60 m (5 ft 3 in)
Volume	13.59 m³ (480 cu ft)

AREAS:

Wings, gross	31.78 m² (342.05 sq ft)
Ailerons (total)	1.52 m² (16.42 sq ft)
Flaps (total)	3.74 m² (40.26 sq ft)
Slats (total)	3.38 m² (36.34 sq ft)
Fin	3.86 m² (41.58 sq ft)
Rudder	0.83 m² (8.95 sq ft)
Tailplane	7.15 m² (77.0 sq ft)
Elevators	1.53 m² (16.52 sq ft)

Sabreliner 40 twin-engine business, jet transport aircraft *(Sabreliner)*

1995

WEIGHTS AND LOADINGS:

Weight empty, equipped	5,103 kg (11,250 lb)
Max payload, incl crew	1,156 kg (2,550 lb)
T-O weight with four passengers, baggage and max fuel	8,877 kg (19,572 lb)
Max T-O weight	9,150 kg (20,172 lb)
Max ramp weight	9,221 kg (20,372 lb)
Max zero-fuel weight	6,259 kg (13,800 lb)
Max landing weight	7,938 kg (17,500 lb)
Landing weight with four passengers, baggage and 1 h reserve fuel	6,094 kg (13,435 lb)
Max wing loading	285.5 kg/m² (58.7 lb/sq ft)
Max power loading	311.65 kg/kN (3.03 lb/lb st)

PERFORMANCE (at max T-O weight, except where indicated):

Max diving speed	Mach 0.85
Max level speed at 6,550 m (21,500 ft)	Mach 0.8 (489 kts; 906 km/h; 563 mph)
Max cruising speed	Mach 0.8
Econ cruising speed at 11,900-13,715 m (39,000-45,000 ft)	Mach 0.75
Stalling speed, landing configuration at 6,094 kg (13,435 lb) AUW	83.5 kts (156 km/h; 96.5 mph)
Max rate of climb at S/L	1,433 m (4,700 ft)/min
Max certificated operating altitude	13,715 m (45,000 ft)
Service ceiling, one engine out at AUW of 7,257 kg (16,000 lb)	7,925 m (26,000 ft)
Min ground turning radius	8.69 m (28 ft 6 in)
T-O balanced field length	1,554 m (5,100 ft)
Landing distance at landing weight with four passengers, baggage and 1 h reserve fuel	693 m (2,275 ft)
Max range with four passengers, baggage, max fuel and 45 min reserves	1,748 n miles (3,239 km; 2,013 miles)

UPDATED

ROCKWELL SABRELINER 60A

TYPE: Twin-engine business jet aircraft.

DESIGN FEATURES: The Sabreliner 60A is powered by two Pratt & Whitney JT12A-8 turbojets as used on the Sabreliner 60. Aerodynamic improvements include increased wing chord, leading-edge slats replaced by a blunt cambered leading-edge, provision of Fowler-type trailing-edge flaps, wingtip extensions and a redesigned rudder. These changes significantly reduce the take-off, landing and minimum control speeds. This improved Sabreliner also has an integrated wing thermal anti-icing system and a fully modulated anti-skid system. These improvements were incorporated onto new production Sabreliner 60s. Estimated range is 2,100 n miles (3,892 km; 2,418 miles)

VERIFIED

ROCKWELL SABRELINER 65

TYPE: Twin-engine business jet aircraft.

PROGRAMME: The Sabreliner 65 is powered by Garrett AiResearch TFE731-3 geared turbofan engine, and incorporates the new wing and increased tailplane developed for the Sabreliner 60A. This version flew for the first time 29 June 1977 and was certificated 28 November 1979. Deliveries began in the following month.

DESIGN FEATURES: Cantilever low-wing monoplane. Supercritical wing with dihedral. Incidence 1.68° at root, 4.64° at tip. Sweepback at quarter-chord 32.79°. Variable incidence tailplane.

FLYING CONTROLS: Conventional aileron with single-spar and ribbed skins of light alloy. Electrically operated Fowler-type trailing-edge flaps of light alloy construction. Hydraulically operated spoilers of light alloy construction, two on upper surface of each wing, forward of flaps, are operable in flight and on the ground. Electrically operated trim tab in port aileron. Electrically operated variable incidence tailplane. Electrically operated trim tab in rudder.

STRUCTURE: Integrally stiffened milled skin light alloy structure. The fuselage is a conventional semi-monocoque safe-life structure of light alloy. The tail unit is a cantilever structure of light alloy.

LANDING GEAR: Hydraulically retractable tricycle type, with single wheel on each main unit and twin nosewheels. Main units retract outward, nose unit forward; all wheels enclosed by fairings when retracted. Oleo-pneumatic shock-absorbers in all units. Mainwheels have tyres size 26

Sabreliner 65 twin-engine business aircraft

1994

× 6.75, 16 ply rating, pressure 17.86 bars (259 lb/sq in). Nosewheel tyres size 18 × 4.4 Type VII, 10 ply rating, pressure 5.17 bars (75 lb/sq in). Nose unit steerable. Hydraulically operated disc brakes. Parking brake. Electrically controlled hydraulically actuated fully modulating anti-skid units.

POWER PLANT: Two Garrett AiResearch TFE731-3-1D turbofan engines, each 16.46 kN (3,700 lb st), pod-mounted on sides of rear fuselage. Hydraulically actuated target type thrust reversers. Integral fuel tanks in wings, with capacity of 4,063 litres (1,073.4 US gallons; 893.5 Imp gallons), one aft fuselage bladder tank, capacity of 606 litres (160.2 US gallons; 133.4 Imp gallons), and one forward fuselage bladder tank, beneath cabin floor, with capacity of 213 litres (56.3 US gallons; 46.9 Imp gallons), providing total fuel capacity of 4,882 litres (1,289.9 US gallons; 1,074 Imp gallons). Single-point refuelling standard. Gravity refuelling points on wing upper surface, and on starboard side of fuselage, above engine.

ACCOMMODATION: Crew of two and eight passengers in pressurised and air conditioned cabin with a variety of seating layouts, all with galley, toilet and baggage compartment. Dual controls standard. Bottom-hinged downward-opening door with built-in airstairs forward of wing on port side. Overwing emergency exit on each side of cabin, FAA Type IV, removable from inside or outside. Baggage compartment at forward end of cabin on starboard side; baggage space in aft toilet compartment; total baggage capacity 247 kg (545 lb).

SYSTEMS: Air conditioning and pressurisation by engine bleed air, maximum differential 0.61 bars (8.8 lb/sq in) to provide cabin altitude of 2,440 m (8,000 ft) to a height of 13,715 m (45,000 ft); separate ducting and temperature controls for cabin and flight deck. Hydraulic system supplied by a single electrically driven pump, pressure 207 bars (3,000 lb/sq in). Auxiliary hydraulic accumulator for emergency use provides through separate lines a limited number of cycles for anti-skid units, wheel brakes, nosewheel steering and spoilers. Electrical system includes two 400 A engine-driven generators. Oxygen system comprising one 2.10 m³ (74 cu ft) cylinder in aft cabin, with quick-donning crew masks and automatic dropout masks for passengers. Anti-icing of wing leading-edges and entire air intakes by engine bleed air; electrical de-icing pitots, stall warning vanes and windscreen. APU optional for air conditioning and electrical/hydraulic power on the ground and when airborne.

AVIONICS: Include dual Collins VHF com, dual VHF nav with VOR/ILS, dual FD 109 flight directors, APS-80 autopilot, WXR-250 weather radar, DME-40, ADF-60A, TRD-90 transponder, and dual MC-103 compasses. Full blind-flying instrumentation is standard.

DIMENSIONS, EXTERNAL:

Wing span	15.37 m (50 ft 5⅛ in)
Length overall	14.30 m (46 ft 11 in)
Height overall	4.88 m (16 ft 0 in)
Tailplane span	5.91 m (19 ft 4⅝ in)
Wheel track	2.20 m (7 ft 2½ in)
Wheelbase	4.85 m (15 ft 10¾ in)
Cabin door (port, fwd): Height	1.19 m (3 ft 11 in)
Width	0.71 m (2 ft 4 in)
Emergency exits (port, stbd, each): Height	0.66 m (2 ft 2 in)
Width	0.51 m (1 ft 8 in)

DIMENSIONS, INTERNAL:

Cabin (excl flight deck): Length	5.79 m (19 ft 0 in)
Max width	1.60 m (5 ft 3 in)
Max height	1.60 m (5 ft 3 in)
Volume	13.59 m³ (480 cu ft)
Baggage compartments (fore and aft): Volume	1.22 m³ (43 cu ft)

AREAS:

Wings, gross	35.30 m² (380 sq ft)
Ailerons (total)	1.53 m² (16.42 sq ft)
Trailing-edge flaps (total)	3.87 m² (41.67 sq ft)
Spoilers (total)	1.11 m² (11.92 sq ft)
Fin	3.86 m² (41.58 sq ft)
Rudder, incl tab	0.83 m² (8.95 sq ft)
Tailplane	8.37 m² (90.08 sq ft)
Elevators	1.80 m² (19.43 sq ft)

WEIGHTS AND LOADINGS:

Weight empty, basic (incl two crew)	6,420 kg (14,154 lb)
Max T-O weight	10,886 kg (24,000 lb)
Max zero-fuel weight	7,371 kg (16,250 lb)
Max landing weight	9,868 kg (21,755 lb)
Max wing loading	308.4 kg/m² (63.2 lb/sq ft)
Max power loading	330.7 kg/kN (3.24 lb/lb st)

PERFORMANCE (at max T-O weight, unless specified otherwise):

Max operating speed	Mach 0.83 IAS
High-speed cruise	Mach 0.81
Recommended cruising speed	Mach 0.77
Long-range cruising speed	Mach 0.73
Stalling speed: flaps up	115 kts (212 km/h; 132 mph) IAS
full flap, operating weight empty, plus four passengers and reserve fuel	82 kts (152 km/h; 95 mph) IAS
Max rate of climb at S/L	1,051 m (3,450 ft)/min
Rate of climb at S/L, one engine out	272 m (893 ft)/min
Service ceiling	13,715 m (45,000 ft)
Service ceiling, one engine out	6,100 m (20,000 ft)
FAA T-O field length, ISA	1,628 m (5,340 ft)
FAA (FAR Pt 19) landing field length at max landing weight	957 m (3,140 ft)
Range with 907 kg (2,000 lb) payload, NBAA VFR reserves	2,413 n miles (4,469 km; 2,777 miles)
Range with max fuel and 395 kg (870 lb) payload: NBAA VFR reserves	2,800 n miles (5,185 km; 3,222 miles)
NBAA IFR reserves	2,400 n miles (4,447 km; 2,763 miles)

VERIFIED

ROCKWELL SABRELINER 75

TYPE: Twin-engine business aircraft.

DESIGN FEATURES: First demonstrated publicly at the Reading Air Show on 8 June 1971, the Sabre 75 is generally similar to the Sabreliner Series 60, utilising the same wings, tail unit and power plant. It differs principally in having a fuselage of deeper section, to offer increased headroom in the cabin. It also had square windows and dual wheels on each unit of its tricycle-type landing gear.

DIMENSIONS, EXTERNAL:

Wing span	13.54 m (44 ft 5¼ in)
Length overall	14.33 m (47 ft 0 in)
Height overall	5.35 m (17 ft 6 in)
Tailplane span	5.35 m (17 ft 6 in)
Wheel track	2.56 m (8 ft 4¾ in)

DIMENSIONS, INTERNAL:

Cabin: Max height	1.87 m (6 ft 1½ in)
Volume	18.69 m³ (660 cu ft)

WEIGHTS AND LOADINGS:

Max ramp weight	9,615 kg (21,200 lb)
Max landing weight	8,392 kg (18,500 lb)

VERIFIED

ROCKWELL SABRELINER 75A/80

TYPE: Twin-engined jet business transport.

PROGRAMME: The Sabreliner 75A differs from the earlier 75 by having General Electric turbofan engines instead of Pratt & Whitney JT12A-8 turbojet engines; increased tailplane span; a new landing gear anti-skid system; improved galley, seating and toilet; and a new air conditioning system. From 1978 it was available with a Mark Five improvement kit developed by Raisbeck.

DESIGN FEATURES: Cantilever low-wing monoplane. Wing section NACA 64_1A212 (modified) at wing station 62.90, NACA 64_1A012 (modified) at wing station 254.94. Dihedral 3° 9′. Incidence 0° at root, 2° 54′ at construction tip. Sweepback at quarter-chord 28° 33′. Variable incidence tailplane.

FLYING CONTROLS: Conventional ailerons of light alloy construction with electrically operated trim tab in port aileron. Aerodynamically operated leading-edge slats of light alloy construction. Electrically operated slotted trailing-edge flaps. Large hydraulically operated airbrake under centre-fuselage. Electrically operated variable incidence tailplane. Electrically operated trim tab in rudder. Elevator has electrically operated trim tab, mechanically interconnected.

STRUCTURE: Two-spar milled-skin light alloy structure. The fuselage is a light alloy semi-monocoque structure. The tail unit is a cantilever light alloy structure.

LANDING GEAR: Hydraulically retractable tricycle type with twin-wheels on each unit. Nose unit retracts forward into fuselage nose, main units inward into under surface of wings. Loud oleo-pneumatic shock-absorbers. Twin mainwheels with Goodrich 10 ply tyres size 22 × 5.75-12, pressure 12.41 bars (180 lb/sq in). Steerable nose unit with twin-wheels and Goodrich Type VII tyres size 18 × 4.40-10, pressure 6.90 bars (100 lb/sq in). Goodyear multiple-disc brakes. Fully modulating anti-skid units. Optional kit for operation from gravel runways.

POWER PLANT: Two General Electric CF700-2D-2 turbofan engines, each 20.2 kN (4,500 lb st), mounted in pod on each side of rear fuselage. Cascade-type vertically orientated thrust reversers optional. Integral fuel tanks in wings, with capacity of 3,418 litres (903 US gallons; 751 Imp gallons). Bladder-type fuel tank in aft fuselage with capacity of 753 litres (199 US gallons; 165.7 Imp gallons). Total fuel capacity of 4,171 litres (1,102 US gallons; 917.6 Imp gallons). Single pressure refuelling point in lower surface of starboard inboard wing leading-edge. Alternative gravity refuelling points at each wingtip and on top of aft fuselage tank. Oil capacity 3.8 litres (1 US gallon; 0.83 Imp gallons).

ACCOMMODATION: Crew of two and 8 to 10 passengers in pressurised air conditioned cabin, with a variety of optional seating layouts. Improved seating, galley and toilet. Dual controls standard. Downward-hinged door, with built-in steps, forward of wing on port side. Emergency exit on each side of cabin, over wing. Baggage compartment at forward end of cabin, opposite door. Electrically operated windscreen wipers.

SYSTEMS: Cabin pressurisation and heating by bleed air from both engines. Emergency pressurisation provided by starboard engine bleed air. Air conditioning system incorporates a three-wheel bootstrap refrigeration unit, with separate ducting and temperature controls for cabin and flight deck. Hydraulic system powered by a single electrically driven hydraulic pump, pressure 207 bars (3,000 lb/sq in). Auxiliary hydraulic accumulator for use in event of pump failure. Electrical system of 28 V DC 110 V 400 Hz constant frequency AC. Primary DC power supplied by engine-driven starter/generators, with 34 Ah batteries interconnected in the system. Emergency cabin and exit lighting by standby battery. Oxygen system supplied from 2.10 m³ (74 cu ft) cylinder with quick-donning masks for crew and automatic dropout masks for passengers. Pneumatic system optional for optional wing and tail unit de-icing boots. Solar APU for ground heating and cooling and electrical power generation is optional.

ELECTRONICS AND EQUIPMENT: Standard cabin equipment includes folding tables, cold/hot air outlets and reading lights at each seat position, fluorescent cabin lights, two magazine racks, and doors to isolate flight deck and toilet

Sabreliner 75A also known as Sabreliner 80

1994

from cabin. Equipped to Cat. II IFR requirements with Collins electronics comprising FD-109Y and FD-109Z flight directors; AP-105 autopilot; NCS-31 nav/com control/computer system; dual VHF-20A com transceivers; dual VIR-30A VHF nav receivers with VOR/LOC/GS; ADF-60A ADF; dual 346B-3 audio systems, public address and intercom, DME-40 DME; dual TDR-90 ATC transponders; Rosemont airspeed/Mach pitot system; Bendix RDR-1200C weather avoidance radar; dual MC-103 compass systems; dual 332C-10 radio magnetic indicators; IDC-28007 encoding altimeter and altitude alerting system; plus dual Teledyne SLZ-9123 instant vertical speed indicators; and dual Mach airspeed indicators.

DIMENSIONS, EXTERNAL:	
Wing span	13.61 m (44 ft 8 in)
Wing aspect ratio	5.77
Length overall	14.38 m (47 ft 2 in)
Height overall	5.26 m (17 ft 3 in)
Tailplane span	5.91 m (19 ft 4½ in)
Wheel track	2.54 m (8 ft 4 in)
Wheelbase	4.85 m (15 ft 11 in)
Cabin door: Height	1.19 m (3 ft 11 in)
Width	0.71 m (2 ft 4 in)
Height to sill	0.30 m (1 ft 0 in)
Emergency exits (each): Height	0.74 m (2 ft 5 in)
Width	0.51 m (1 ft 8 in)
DIMENSIONS, INTERNAL:	
Cabin (excl flight deck): Length	5.79 m (19 ft 0 in)
Max width	1.60 m (5 ft 3 in)
Max height	1.83 m (6 ft 0 in)
Volume	15.57 m³ (550 cu ft)
AREAS:	
Wings, gross	31.8 m² (342.05 sq ft)
Ailerons (total)	1.53 m² (16.42 sq ft)
Trailing-edge flaps (total)	3.74 m² (40.26 sq ft)
Leading-edge slats (total)	3.38 m² (36.34 sq ft)
Airbrake	0.70 m² (7.54 sq ft)
Fin	4.85 m² (52.24 sq ft)
Rudder, incl tab	0.91 m² (9.75 sq ft)
Tailplane	8.37 m² (90.08 sq ft)
Elevators, incl tab	1.81 m² (19.43 sq ft)
WEIGHTS AND LOADINGS:	
Weight empty	5,987 kg (13,200 lb)
T-O weight with four passengers, baggage and max fuel	9,698 kg (21,380 lb)
Max T-O and ramp weight	10,432 kg (23,000 lb)
Max zero-fuel weight	7,085 kg (15,620 lb)
Max landing weight	9,979 kg (22,000 lb)
Landing weight with four passengers, baggage and 1 h reserve fuel	6,804 kg (15,000 lb)
Max wing loading	328.3 kg/m² (67.25 lb/sq ft)
Max power loading	260.54 kg/kN (2.66 lb/lb st)
PERFORMANCE (at max T-O weight, except where indicated):	
Max diving speed	Mach 0.85
Max level and cruising speed	Mach 0.80
Econ cruising speed	Mach 0.74
Stalling speed, landing configuration, at 7,008 kg (15,450 lb) AUW	86 kts (160 km/h; 99 mph)
Max rate of climb at S/L	1,372 m (4,500 ft)/min
Max certificated operating altitude	13,715 m (45,000 ft)
Min ground turning radius	8.69 m (28 ft 6 in)
T-O balanced field length	1,326 m (4,350 ft)
Landing distance at 6,804 kg (15,000 lb) landing weight	753 m (2,470 ft)
Max range, with four passengers, baggage, max fuel and 45 min reserves	1,712 n miles (3,173 km; 1,972 miles)

VEREIFIED

ROCKWELL SABRELINER 60SC UPGRADE

TYPE: Twin-engine business jet upgrade.

DESIGN FEATURES: Sabreliner retrofitted 14 Sabreliner Model 60s with the supercritical wing developed for the Sabreliner Model 65. The retrofitted aircraft were redesignated Model 60SC.

VERIFIED

Sabreliner 60SC retrofitted with supercritical wing

1994

Sabreliner 80SC retrofitted with supercritical wing

1994

ROCKWELL SABRELINER 80SC UPGRADE

TYPE: Twin-engine business jet upgrade.

DESIGN FEATURES: Sabreliner retrofitted nine Sabreliner Model 80s with the supercritical wing developed for the Sabreliner 65. The retrofitted aircraft were redesignated Model 80SC.

VERIFIED

ROCKWELL SABRELINER 40EX AND 60EX EXCALIBUR UPGRADE PROGRAMME

TYPE: Twin-engine business jet upgrade.

DESIGN FEATURES: Excalibur programme for Sabreliner 40 and 60 with 10,000 flying hours, introduced early 1989. Airframe hours increased by 5,000 hours. Alternative modifications gave 30,000 h/15,000 mission and 30,000 h/30,000 mission life extension.

VERIFIED

SABRELINER (LOCKHEED) AT-33 UPGRADE

TYPE: Low-wing jet trainer and attack aircraft upgrade.

DESIGN FEATURES: Sabreliner has developed an Attack Trainer AT-33 version of the T-33 as a prime contractor for the USAF. Sabreliner incorporates existing TCTOs and newly developed engineering for installation of updated avionics, gunsight, armaments and structural upgrades of the existing airframe. Programme also includes engine inspections as well as technical and parts support.

VERIFIED

Sabreliner (Lockheed) AT-33A Upgraded T-33 of the Ecuadorean Air Force *(Sabreliner)*

1995

SABRELINER (CESSNA) T-37 SLEP UPGRADE KITS

TYPE: Two-seat light trainer aircraft upgrade.

DESIGN FEATURES: Service Life Extension Programme (SLEP) contract for 644 Cessna T-37Bs of the USAF Air Training Command awarded August 1989. Modification included replacement of wing carry-through structure, lower front wing spars, tail mounting structure and tailplane. SLEP kits installed by USAF.

VERIFIED

SABRELINER T-39N UPGRADE

TYPE: Navigation training aircraft upgrade.

DESIGN FEATURES: Sabreliner contracted to deliver 17 T-39N radar training aircraft for the US Navy Undergraduate Naval Flight Officer (UNFO) training system which also includes: ground-based air-to-air radar interceptor simulators and all logistics, maintenance and training support services. Conversion is based on the Sabreliner Model 40 aircraft. STC approval June 1991. First aircraft flew in May 1991 and was delivered 28 June 1991. All 17 now in service.

VERIFIED

Sabreliner (Cessna) T-37B aircraft of the USAF after undergoing the Service Life Extension Programme (SLEP) *(Sabreliner)*

1995

Sabreliner T-39N radar training aircraft of the US Navy based on the Sabreliner 40 *(Sabreliner)*

1995

SCHAFER

SCHAFER AIRCRAFT MODIFICATIONS INC

Route 10, Box 301, Waco Regional Airport, Waco, Texas 76708
Telephone: 1 (817) 753 1551
Fax: 1 (817) 753 8416
Telex: 795902 SCHAFER CFTO

PRESIDENT: Earl Schafer

EXECUTIVE VICE-PRESIDENT AND DIRECTOR OF MARKETING: R. B. Stevens

Company formed 1977; originally manufactured auxiliary fuel tanks and modifications for Cessna 300 and 400 series aircraft; since 1979, has developed own modification programmes.

VERIFIED

Schafer Comanchero 500 turboprop-powered Piper Chieftain

1994

SCHAFER COMANCHERO

TYPE: Turbine-powered Piper Pressurised Navajo.

PROGRAMME: Original piston engines replaced by turboprops; FAA STC received January 1981; Aerospatiale subsidiary SECA of France appointed sole modification and service centre for Europe, Middle East and Africa.

DESIGN FEATURES: Useful load increased by 345 kg (760 lb); optional dual camera port installation for high-altitude photography.

FLYING CONTROLS: As for Navajo.

STRUCTURE: As for Navajo.

POWER PLANT: Two 559 kW (750 shp) P&WC PT6A-135 turboprops, flat rated at 462 kW (620 shp). Fuel capacity increased to 1,363 litres (360 US gallons; 300 Imp gallons).

WEIGHTS AND LOADINGS:

Weight empty	2,177 kg (4,800 lb)
Max baggage weight	218 kg (480 lb)
Max T-O weight	3,538 kg (7,800 lb)
Max wing loading	176.9 kg/m² (36.24 lb/sq ft)
Max power loading	4.07 kg/kW (6.7 lb/shp)

PERFORMANCE: As for Pressurised Navajo except:

Max cruising speed at 6,100 m (20,000 ft)	282 kts (522 km/h; 325 mph)
Max rate of climb at S/L	1,067 m (3,500 ft)/min
Rate of climb at S/L, one engine out	250 m (820 ft)/min
Service ceiling	more than 11,280 m (37,000 ft)
Service ceiling, one engine out	6,860 m (22,500 ft)
T-O to 15 m (50 ft)	533 m (1,750 ft)
Landing from 15 m (50 ft)	564 m (1,850 ft)
Max range at 8,840 m (29,000 ft), 45 min reserves	1,530 n miles (2,835 km; 1,761 miles)

VERIFIED

SCHAFER COMANCHERO 500

TYPE: Turboprop-powered Piper Chieftain.

PROGRAMME: Original piston engines replaced by turboprops; programme launched mid-1980; flight testing started August 1981; licence to manufacture 50 Comanchero 500B sold to Neiva Industria Aeronautica of Brazil early 1984.

VERSIONS: **Comanchero 500A:** Powered by two 410 kW (550 shp) P&WC PT6A-20 turboprops; 132 litre (35 US gallon; 29 Imp gallon) extra fuel in each engine nacelle; maximum T-O weight increased; airframe inspected, bushings and hydraulic components replaced as necessary; options include special interiors, custom avionics, detachable underfuselage pod with capacity of 0.44 m³ (15.5 cu ft), and larger 341 litre (90 US gallon; 75 Imp gallon) nacelle tanks.

Comanchero 500B: Powered by two more expensive 533 ekW (715 ehp) P&WC PT6A-27 turboprops, flat rated at 431 ekW (578 ehp) for hot and high performance; other features and options as for Comanchero 500A.

Neiva Comanchero/NE-821 Carajá: Powered by 410 kW (550 shp) P&WC PT6A-34 turboprops; otherwise as Comanchero 500B; first two kits supplied by Schafer, remainder made by Neiva.

FLYING CONTROLS: As for Chieftain.

STRUCTURE: As for Chieftain.

WEIGHTS AND LOADINGS:

Max T-O weight	3,629 kg (8,000 lb)
Max power loading (500A)	4.43 kg/kW (7.27 lb/shp)

PERFORMANCE (A: Comanchero 500A, B: Comanchero 500B):

Max cruising speed: A, B at 3,660 m (12,000 ft)	240 kts (445 km/h; 276 mph)
Max rate of climb at S/L: A	853 m (2,800 ft)/min
B	732 m (2,400 ft)/min
Rate of climb at S/L, one engine out: A, B	259 m (850 ft)/min
Service ceiling: A	8,840 m (29,000 ft)
Service ceiling, one engine out: A, B	4,725 m (15,500 ft)
T-O to 15 m (50 ft): A	759 m (2,490 ft)
Landing from 15 m (50 ft): A	724 m (2,375 ft)
Range with pilot and 10 passengers, 13.6 kg (30 lb) baggage allowance per passenger, and 45 min reserves, with standard fuel: A, B	500 n miles (926 km; 575 miles)
Range with pilot and 181 kg (400 lb) payload: B	1,300 n miles (2,410 km; 1,500 miles)

VERIFIED

SCHAFER COMANCHERO 750

TYPE: Re-engined Piper Cheyenne II.

PROGRAMME: FAA STC granted May 1981.

DESIGN FEATURES: Standard 462 kW (620 shp) P&WC PT6A-28 turboprops replaced by PT6A-135s flat rated at 559 kW (750 shp); optional dual camera port installation. Otherwise as for Cheyenne II.

PERFORMANCE:

Max cruising speed at 7,620 m (25,000 ft)	278 kts (515 km/h; 320 mph)
Max rate of climb at S/L	869 m (2,850 ft)/min
Rate of climb at S/L, one engine out	198 m (650 ft)/min
Max range, econ cruising power, at 8,840 m (29,000 ft)	1,630 n miles (3,020 km; 1,877 miles)

VERIFIED

SCHAFER/AMI DOUGLAS DC-3-65TP CARGOMASTER

TYPE: Turboprop Douglas DC-3.

PROGRAMME: Original piston engines replaced by turboprops; design began January 1985; first flight 1 August 1986; first production conversion started September 1986; by early 1989, Cargomasters operating in South America, South Africa and USA; demonstrated to US Air Force and Navy and government agencies early 1989.

CUSTOMERS: Market estimated at 100 conversions in three years. Marketed by Aero Modifications International.

DESIGN FEATURES: Conversion in association with Aero Modifications International of Fort Worth, Texas; re-engined with P&WC PT6A-65AR turboprops; fuselage stretched 1.02 m (3 ft 4 in) forward of wing, with or without extra window each side, to retain CG range; fuel capacity increased and hydraulics modified.

FLYING CONTROLS: Minor changes to horizontal tail to improve stability.

POWER PLANT: Two 1,062 kW (1,424 shp) Pratt & Whitney Canada PT6A-65AR turboprops, each flat rated at 917 kW (1,230 shp) and driving a Hartzell five-blade feathering and reversing propeller with de-icing. Engines have inertial separators and hot lip intake de-icing. One additional fuel tank, usable capacity 447 litres (118 US gallons; 98.25 Imp gallons), in each nacelle. One forward auxiliary fuel tank, usable capacity 765 litres (202 US gallons; 168 Imp gallons), and one rear auxiliary fuel tank, usable capacity 753 litres (199 US gallons; 166 Imp gallons), in each inboard wing, giving total usable fuel capacity of 3,930 litres (1,038 US gallons; 864 Imp gallons). Oil capacity 9.5 litres (2.5 US gallons; 2.1 Imp gallons).

Schafer/AMI Douglas DC-3-65TP Cargomaster conversion with Pratt & Whitney Canada PT6A-65AR turboprops *1994*

SYSTEMS: Two engine-driven 250 A starter/generators. Redesigned electrical system with dual batteries providing simultaneous engine start capability.

DIMENSIONS, EXTERNAL: As for standard DC-3, except:

Length overall	20.34 m (66 ft 9 in)
Propeller diameter	2.79 m (9 ft 2 in)
Baggage door: Height	1.42 m (4 ft 8 in)
Width	2.13 m (7 ft 0 in)

DIMENSIONS, INTERNAL:

Cabin: Length	12.50 m (41 ft 0 in)
Max width	2.18 m (7 ft 2 in)
Max height	1.95 m (6 ft 5 in)
Floor area	26.72 m² (287.6 sq ft)
Volume	48.85 m³ (1,725 cu ft)

WEIGHTS AND LOADINGS:

Basic operating weight	7,167 kg (15,800 lb)
Max fuel weight	3,154 kg (6,954 lb)
Max T-O and landing weight	12,202 kg (26,900 lb)
Max zero-fuel weight	11,793 kg (26,000 lb)
Max wing loading	132.8 kg/m² (27.2 lb/sq ft)
Max power loading	6.65 kg/kW (10.9 lb/shp)

PERFORMANCE:

Never-exceed speed (V_{NE})	187 kts (346 km/h; 215 mph) IAS
Max level speed at 3,050 m (10,000 ft)	217 kts (402 km/h; 250 mph)
Max cruising speed at 3,050 m (10,000 ft)	196 kts (363 km/h; 226 mph)
Econ cruising speed at 3,050 m (10,000 ft)	185 kts (343 km/h; 213 mph)
Stalling speed, flight idle power	52 kts (96 km/h; 60 mph) IAS
Max rate of climb at S/L	610 m (2,000 ft)/min
Rate of climb at S/L, one engine out	137 m (450 ft)/min
Service ceiling	7,620 m (25,000 ft)
Service ceiling, one engine out	3,685 m (12,100 ft)
T-O run	1,051 m (3,450 ft)
T-O to 15 m (50 ft)	1,122 m (3,680 ft)
Landing from 15 m (50 ft)	670 m (2,200 ft)
Range: with max fuel and 2,268 kg (5,000 lb) payload	1,300 n miles (2,409 km; 1,497 miles)
with 4,536 kg (10,000 lb) payload	100 n miles (185 km; 115 miles)

VERIFIED

SIERRA

SIERRA INDUSTRIES

Garner Municipal Airport, PO Box 5184, Uvalde, Texas 78801

Telephone: 1 (210) 278 4381

Fax: 1 (210) 278 7649

Telex: 910 240 2612

PRESIDENT: Mark Huffstutler

Sierra Industries acquired assets of R/STOL Systems Inc 24 September 1986; assets included about 100 supplemental type certificates for STOL and performance modifications to 40 types of Beech, Cessna and Piper light and business aircraft.

VERIFIED

SIERRA IN-WING WEATHER RADAR

TYPE: Light aircraft weather radar upgrade.

PROGRAMME: RCA Weather Scout I weather radar with truncated aerial and limited mapping capability installed in reinforced wing leading-edge bay behind dielectric leading-edge panel in Cessna 182, various 206 models and 210 series; total system weighs 10.5 kg (23.1 lb).

VERIFIED

SIERRA INDUSTRIES CESSNA AUXILIARY FUEL SYSTEMS

TYPE: Light aircraft auxiliary fuel tank upgrade.

PROGRAMME: Extra tankage for 204 litres (54 US gallons; 45 Imp gallons) of fuel fitted to Cessna 180, 182, 185, 206 and 207; three rubber fuel bladders installed outboard of each standard fuel tank; modification takes about 50 man-hours and is certificated to FAR Pt 23.

VERIFIED

SIERRA INDUSTRIES CESSNA 210 LANDING GEAR DOOR MODIFICATION

TYPE: Light aircraft landing gear door modification.

PROGRAMME: FAA approved modification kit removes mainwheel doors and replaces them with metal fairings to cover gaps; advantages include 8.6 kg (19 lb) increase in useful load, faster gear retraction and extension, elimination of possible door-fouling problems, lower maintenance and improved appearance comparable to that of 1979 and later model Centurions; modification applicable to Cessna Model 210, T210 and P210 Centurions manufactured between 1970 and 1978 (c/n 21059200-21062954 inclusive).

VERIFIED

SIERRA/BEECH, SIERRA/CESSNA and SIERRA/PIPER SAFETY and PERFORMANCE CONVERSIONS

TYPE: Light aircraft performance upgrades.

PROGRAMME: Hi-Lift full-span wing leading- and trailing-edge modification, which greatly improves landing and take-off distances; applicable to single- and smaller twin-engined Beech, Cessna and Piper aircraft; first applied to Cessna 182; ailerons linked to flaps to droop when flaps are lowered, virtually doubling lift at low airspeeds; ailerons continue to function for roll control; full-span distributed camber leading-edge provides optimum spanwise lift distribution for maximum cruise efficiency; also reduces leading-edge pressure peak at high angle of attack, giving maximum resistance to stall and high-manoeuvrability at low airspeeds; some Cessna singles built after 1972 have this leading-edge as standard.

Full-span flap system, conical cambered wingtips, dorsal and ventral fins, belly mounted vortex generators and flap/elevator trim linkage can be applied to various models. Stall fences applied between flap and aileron to improve controllability at low airspeeds, and aileron gap sealed with aluminium or rubberised canvas strip; gives safe STOL take-offs and landings, even with novice pilots; cruising speed and range increased by 2 to 4 per cent.

Increased climb performance and slower take-off and landing speeds have allowed gross weight increases for small twin conversions such as Cessna Super Skymaster and Piper Twin Comanche.

Hi-Lift modification for Cessna 400 series involved complete redesign of wing aft of rear spar to accommodate 100 per cent Fowler flaps and flap-actuated aileron droop system; with flap/elevator trim linkage and double-hinged rudder, this modification led to FAA certificated 40 per cent reduction in landing and take-off field lengths.

Sierra equipped Piper Seneca I was FAA certificated in 1974 with full-span slotted flaps, upper surface spoilers for roll control, cambered wingtips and anti-servo rudder tab; roll response increased at all airspeeds, minimum control speed reduced by 16 per cent, best rate of climb speed lowered by 19 per cent and single-engined service ceiling increased by 213 m (700 ft). Similar modification for Seneca II certificated 1976.

Full-span flaps and lateral control spoilers fitted to Piper Cherokee Six; high differential aileron droop Hi-Lift system developed for high-performance singles, such as Cessna P210 Pressurised Centurion; new mechanism droops ailerons 15° when flaps lowered to 20° for take-off; with aileron wheel fully to starboard, starboard aileron moves up 47° (27° up from normal faired position); at same time, port aileron moves down another 12° to a total 27°; aileron droop with 30° flap remains 15°; benefits from Hi-Lift system include greatly reduced adverse yaw tendency in turns, high roll rate for more precise control when flaps are down and speed low, and control wheel forces reduced by aerodynamic effects on up-moving aileron and friction-reducing crossover control cable.

Four aluminium stall strips (rubber when fitted over de-icing boots) retain good stalling characteristics on all Hi-Lift systems; inboard strip produces warning and outboard strip initiates stall of inner and centre wing, leaving outboard section in good lift; aileron control continues into stall.

Sierra Hi-Lift systems can be applied to almost entire range of Cessna and Piper aircraft and to Beechcraft Bonanzas.

VERIFIED

SIERRA/CESSNA EAGLE SERIES

TYPE: R/STOL modifications to Cessna Citations.

PROGRAMME: See below.

VERSIONS: **Eagle:** Formerly Astec Eagle and R/STOL Eagle modifications applied to Cessna Citation 500 and Citation I; inner wing thickness increased and tips extended, as detailed below; fuel capacity increased by 392 kg (865 lb), accounting for about 75 per cent of range increase of 600 n miles (1,112 km; 691 miles); remainder produced by improved aerodynamic efficiency.

Eagle SP: Modifications as above to Cessna Citation 501, which is certificated to FAR Pt 23 for single-pilot operation.

Longwing: Eagle wingtip extension without wingroot thickening; introduced 1983; increases fuel capacity by about 54 kg (120 lb) on aircraft c/n 001-213, giving 10 per cent range increase, better climb rate and lower approach speed.

Eagle 400B: Eagle modification plus 11.12 kN (2,500 lb st) JT15D-4Bs replacing JT15D-1 or -1As.

DESIGN FEATURES: Thickness/chord ratio of inboard wing increased from 14 to 19 per cent; supercritical technology applied to improve wing/fuselage flow; span increased by 0.51 m (1 ft 8 in) wingtip extensions.

FLYING CONTROLS: As for Citation.

STRUCTURE: As for Citation.

POWER PLANT: As for Citation 500, except increased fuel capacity provided in the thicker wing centre-section and wingtip extensions.

DIMENSIONS, EXTERNAL:

Wing span	14.35 m (47 ft 0½ in)
Wing chord: at root	2.95 m (9 ft 8⅛ in)
at tip	0.93 m (3 ft 0½ in)
Wing aspect ratio	7.92
Length overall	13.26 m (43 ft 6 in)
Height overall	4.36 m (14 ft 3¾ in)
Wheel track	3.84 m (12 ft 7⅛ in)
Wheelbase	4.78 m (15 ft 8¼ in)

WEIGHTS AND LOADINGS (A: JT15D-1, B: JT15D-1A, C: JT15D-4B):

Weight empty: A, B	2,971 kg (6,550 lb)
C	3,180 kg (7,011 lb)
Max fuel weight: A, B, C	2,045 kg (4,510 lb)
Max T-O weight: A, B, C	5,670 kg (12,500 lb)
Max zero-fuel weight: A, B, C	4,309 kg (9,500 lb)
Max landing weight: A, B, C	5,148 kg (11,350 lb)
Max power loading: C	254.95 kg/kN (2.50 lb/lb st)

PERFORMANCE (at max T-O weight except where indicated):

Max cruising speed, AUW of 4,309 kg (9,500 lb), at 10,670 m (35,000 ft): A 341 kts (631 km/h; 392 mph)
B 357 kts (661 km/h; 411 mph)
C (at 8,840 m; 29,000 ft) 400 kts (741 km/h; 460 mph)

Cruising speed, AUW of 4,309 kg (9,500 lb), at 12,500 m (41,000 ft): A 317 kts (587 km/h; 365 mph)
B 336 kts (623 km/h; 387 mph)
C (at 13,100 m; 43,000 ft) 375 kts (695 km/h; 432 mph)

Stalling speed at max landing weight: A, B, C 78 kts (145 km/h; 90 mph)

Max rate of climb at S/L: C 1,220 m (4,000 ft)/min

Rate of climb at S/L, one engine out: C 442 m (1,450 ft)/min

Max certificated altitude: A 10,670 m (35,000 ft)
B 12,500 m (41,000 ft)
C 13,100 m (43,000 ft)

Balanced T-O field length: A, B 907 m (2,975 ft)
C 892 m (2,925 ft)

FAA landing field length at max landing weight: A, B, C 722 m (2,370 ft)

Max range, 45 min fuel reserves:
A 1,800 n miles (3,336 km; 2,073 miles)
C 1,997 n miles (3,700 km; 2,300 miles)

Sierra/Cessna Century V Eagle converted Citation 500

1994

VERIFIED

SIERRA

SIERRA TECHNOLOGIES INC

Sierra Research Division

485 Cayuga Road, PO Box 222, Buffalo, New York 14225-0222

Telephone: 1 (716) 631 6200

Fax: 1 (716) 631 7849

VICE-PRESIDENT, PROGRAMS: Joseph A Tomasino

DIRECTOR, PUBLIC RELATIONS: Judith A. Rucki

The Sierra Research Divison of Sierra Technologies is active in the fields of aircraft systems upgrades and modifications. Sierra produces TACAN, transport aircraft station keeping equipment (SKE), navigation aids flight inspection systems (FIS), mobile EW air defence threat simulation systems, and other advanced radar and electronics for civil and military applications. Total number of employees is 500.

UPDATED

SIERRA TECHNOLOGIES F-5A/B/E/F, RF-5A TIGER-PAWS SYSTEM UPGRADE

TYPE: Light supersonic tactical fighter and reconnaissance aircraft upgrade.

PROGRAMME: Previous digital avionics modernisation experience with USAF T-38A and NASA T-38 prototype led to winning a 1991 contact to update 15 Royal Norwegian Air Force (RNoAF) F-5A/B 'Tigers' by creating an F-16 similar, lead-in conbat trainer aircraft. The RNoAF avionics integration, known as the Program for Avionics and Weapon Systems improvements (TIGER-PAWS), was completed in 42 months (July 1994) and featured automatic bombing systems and other F-16A/B cockpit/system similarities.

DESIGN FEATURES: TIGER-PAWS involves replacing old analog electronics with digital core avionics, installing a MIL-STD-1553B databus, head-up display/weapons aiming computer (HUDWAC), ring laser gyro (RLG) inertial navigation system, global positioning system (GPS), standard central air data computer (SCADC), and video camera/recorder. Original F-5E/F pulsed radars can be retrofitted with new Pulse Doppler fire control radars. In addition to the TIGER-PAWS digital core avionics, numerous options are available, including data transfer, radar warning receivers (RWR), chaff/flares, head-down multifunction displays, laser rangefinder and designator, hands-on throttle and stick (HOTAS), ECM systems, and forward-looking infra-red (FLIR). Structural repairs/enhancements are also offered.

UPDATED

SIERRA TECHNOLOGIES (BRITISH AEROSPACE) BAe 125-800 FLIGHT INSPECTION UPGRADE

TYPE: Low-wing, twin-turbofan aircraft flight inspection system upgrade.

PROGRAMME: Sierra Technologies is responsible for the systems integration programme for the USAF of the BAe 125-800 flight inspection aircraft. Six aircraft were modified and are now in service with the FAA's fleet of flight inspection aircraft.

DESIGN FEATURES: Systems include: aircraft modification, integration of advanced automatic flight inspection system, as well as software development and integration.

VERIFIED

Sierra Technologies Royal Norwegian Air Force (RNoAF) upgraded Northrop F-5s

1994

Sierra Technologies (British Aerospace) BAe 125-800 upgraded for flight inspection duties

1994

Sierra Technologies (de Havilland) Dash 8M-100 (E-9A) upgraded for telemetry duties

1994

SIERRA TECHNOLOGIES (DE HAVILLAND) E-9A AIRBORNE PLATFORM TELEMETRY RELAY SYSTEM UPGRADE

US Air Force designation: E-9A

TYPE: Twin-turboprop, short-range transport aircraft telemetry system upgrade.

PROGRAMME: Upgrade package to modify standard de Havilland Dash 8M-100 for use as a missile range control aircraft. The aircraft relays telemetry, voice, incorporates drone and fighter control data and simultaneously observes range with radar.

DESIGN FEATURES: Systems include: large electronically steered phased-array radar in fuselage side and AN/APS-128D surveillance radar in ventral dome.

CUSTOMERS: USAF two.

VERIFIED

SIKORSKY

SIKORSKY AIRCRAFT (Division of United Technologies Corporation)

6900 Main Street, Stratford, Connecticut 06601-1381
Telephone: 1 (203) 386 4000
Fax: 1 (203) 386 7300
Telex: 96 4372

OTHER WORKS: Troy, Alabama; South Avenue, Bridgeport, Connecticut; Shelton, Connecticut; West Haven, Connecticut; Development Flight Test Center, West Palm Beach, Florida

PRESIDENT: Eugene Buckley

SENIOR VICE-PRESIDENTS:
G. C. Kay (Administration)
Robert R. Moore (Production Operations)

VICE-PRESIDENT, PROGRAMMES: Merrick W. Hellyar

VICE-PRESIDENT, INTERNATIONAL BUSINESS:
James J. Satterwhite

VICE-PRESIDENT, RESEARCH AND ENGINEERING:
Dr Kenneth M. Rosen

VICE-PRESIDENT, GOVERNMENT BUSINESS DEVELOPMENT:
Gary F. Rast

MANAGER OF PUBLIC RELATIONS: William S. Tuttle

Founded as Sikorsky Aero Engineering Corporation by late Igor I. Sikorsky; has been division of United Technologies Corporation since 1929; began helicopter production in 1940s; has produced more than 6,000 rotating wing aircraft.

Headquarters and main plant at Stratford, Connecticut; employment at all Sikorsky facilities totalled 10,000 in 1994; main current programmes include UH-60 Black Hawk and derivatives, CH-53E Super Stallion and S-76 series. Sikorsky and Boeing Helicopters, associated in First Team, won US Army LH/RAH-66 Comanche light helicopter demonstration/validation order on 5 April 1991.

Sikorsky licensees include Westland of the UK, Agusta of Italy, Aerospatiale of France, MBB of Germany, Mitsubishi of Japan and Pratt & Whitney Canada Ltd. Sikorsky and Embraer of Brazil signed agreement in Summer 1983 to transfer technology covering design and manufacture of composites components; Sikorsky and CASA of Spain signed memorandum of understanding in June 1984 covering long-term helicopter industrial co-operation programme. CASA builds tail rotor pylon, tailcone and stabiliser components for H-60 and S-70, with first CASA S-70 components delivered to Sikorsky January 1986.

Westland plc shareholders approved joint Fiat/Sikorsky plan involving financial and technical support and minor equity participation in Westland 12 February 1986; Westland then licensed to manufacture S-70 series. Licence to manufacture S-70 in Japan granted to Mitsubishi Heavy Industries in 1988.

UPDATED

Sikorsky CH-53A twin-turbine helicopter *(Sikorsky)*

1995

SIKORSKY S-65A

US Navy designation: CH-53A Sea Stallion
USMC designations: CH-53A/D
USAF designations: HH-53B/C

TYPE: Twin-turbine heavy assault transport helicopter.

PROGRAMME: On 27 August 1962, it was announced that Sikorsky had been selected by the US Navy to produce a heavy assault transport helicopter for use by the Marine Corps. First flight was made 14 October 1964, and deliveries began in mid-1966.

VERSIONS: **CH-53A:** Initial version utilising many components based on those of the S-64A Skycrane, but powered by two General Electric T64 turboshaft engines and has a watertight hull. A full-size rear opening, with built-in ramp, permits easy loading and unloading, with the aid of a special hydraulically operated internal cargo loading system and floor rollers.

Typical cargo loads include two Jeeps, or two Hawk missiles with cable reels and control console, or a 105 mm howitzer and carriage. An external cargo system permits in-flight pickup and release without ground assistance.

The CH-53A is able to operate under all-weather and climatic conditions. Its main rotor blades and tail pylon fold hydraulically for stowage on board ship.

On 17 February 1968, a CH-53A, with two General Electric T64-6 (modified) engines, flew at a gross weight of 23,541 kg (51,900 lb) carrying 12,927 kg (28,500 lb) of payload and fuel, establishing new unofficial payload and gross weight records for a production helicopter built outside the Soviet Union.

On 26 April 1968, a Marine Corps CH-53A made the first automatic terrain-clearance flight in helicopter history and subsequently concluded flight tests of an Integrated Helicopter Avionics System (IHAS). Prime contractor for the IHAS programme was Teledyne Systems Company. Norden Division of United Aircraft Corporation provided the terrain-clearance radar and vertical structure display.

A Marine Corps CH-53A performed a series of loops and rolls, on 23 October 1968, as part of a joint Naval Air Systems Command and Sikorsky flight test programme aimed at investigating the CH-53A's rotor system dynamics and manoeuvrability characteristics.

CH-53D: Improved CH-53A for US Marine Corps, the first of which was delivered on 3 March 1969. Two T64-GE-413 engines, each with a maximum rating of 2,927 kW (3,925 shp). A total of 55 troops can be carried in a high-density arrangement. An integral cargo handling system makes it possible for one man to load or unload 1 ton of palletised cargo a minute. Main rotor and tail pylon fold automatically for carrier stowage.

Last CH-53D (the 26th built) was delivered on 31 January 1972. All but the first 34 CH-53s were provided with hardpoints for supporting towing equipment and transferring tow loads to the airframe, so that the US Marine Corps could utilise the aircraft as airborne minesweepers, giving an assault commander the capability of clearing enemy mines from harbours and off beaches without having to wait for surface minesweepers. Tow kits installed in the 15 CH-53Ds operated by the US Navy Squadron HM-12 included automatic flight control system interconnections to provide automatic cable yaw angle retention and aircraft attitude and heading hold; rearview mirrors for pilot and co-pilot; tow cable tension and yaw angle indicator; automatic emergency cable release; towboom and hook system with 6,803 kg (15,000 lb) load capacity when cable was locked to internal towboom; dam to prevent cabin flooding in emergency water landing with lower ramp open; dual hydraulically powered cable winches; racks and cradles for stowage of minesweeping equipment; auxiliary fuel tanks in cabin to increase endurance.

CH-53G: Version of the CH-53 for the German armed forces, with T64-GE-7 engines. A total of 112 were produced, the first of two built by Sikorsky being delivered 31 March 1969. The next 20 were assembled in Germany from American built components. The remainder embody some 50 per cent components of German manufacture. Prime contractor in Germany was VFW-Fokker, whose first CH-53G flew for the first time 11 October 1971. Deliveries completed in 1975.

HH-53B: Eight ordered by USAF in September 1966 for Aerospace Rescue and Recovery Service. The first of these flew 15 March 1967. The HH-53B is generally similar to the CH-53A, but powered by 2,297 kW (3,080 shp) T64-GE-3 turboshaft engines. It has the same general equipment as the HH-3E, including a retractable flight refuelling probe, jettisonable auxiliary fuel tanks, rescue hoist, all-weather electronics and armament. Withdrawn from service.

HH-53C: Improved version of the HH-53B, with 2,927 kW (3,925 shp) T64-GE-7 engines, auxiliary jettisonable fuel tanks each of 1,703 litres (450 US gallons; 374.7 Imp gallons) capacity on new cantilever mounts, flight refuelling probe, and rescue hoist with 76 m (250 ft) of cable. External cargo hook of 9,070 kg (20,000 lb) capacity. First HH-53C was delivered to the USAF 30 August 1968. A total of 72 HH-53B/Cs were built. Withdrawn from service.

HH-53H Pave Low III: Special operations version for combat rescue and recovery. Eight converted to MH-53J Pave Low III.

RH-53D: Specially equipped minesweeping version for the US Navy and Iran.

YCH-53E: Three-engined development of the CH-53D.

VH-53D: Presidential helicopter of US Marin Corps.

S-65-Oe: Two ordered in 1969 by Austrian Air Force and delivered in 1970. Used for rescue duties in the Alps, they have the same rescue hoist as the HH-53B/C, fittings for auxiliary fuel tanks and accommodation for 38 passengers. Withdrawn from service.

S-65C: Commercial intercity helicopter proposal based on military CH-53.

TH-53A: Training version in service with the USAF.

UPGRADES: **Israel Aircraft Industries (IAI):** CH-53-2000. See separate entry.

Sikorsky: CH-53D. See Versions.

Sikorsky: MH-53J Pave Low III enhanced. See Separate entry.

OPERATORS: Versions of the CH-53 Sea Stallion are in service with the armed forces of the following countries: Germany (108), Iran (2), Israel (36), and USA (125).

DESIGN FEATURES: Rotor system and transmission generally similar to those of S-64A Skycrane, but main rotor head is of titanium and steel, and has folding blades.

STRUCTURE: Fuselage is a conventional semi-monocoque structure of aluminium, steel and titanium. Folding tail pylon. Large horizontal stabiliser on starboard side of tail rotor pylon.

LANDING GEAR: Retractable tricycle type, with twin wheels on each unit. Main units retract into rear of sponsons on each side of fuselage. Fully castoring nose unit. Mainwheels and nosewheels have tyres size 25.65 × 8.50-10, pressure 6.55 bars (95 lb/sq in).

POWER PLANT: Normally two 2,125 kW (2,850 shp) General Electric T64-GE-6 turboshaft engines, mounted in pod on each side of main rotor pylon. The CH-53A can also utilise, without airframe modification, the T64-GE-1 engine of 2,297 kW (3,080 shp) or the later T64-GE-16 (mod) engine of 2,561.5 kW (3,435 shp). Two self-sealing bladder fuel tanks, each with capacity of 1,192 litres (315 US gallons; 262 Imp gallons), housed in forward part of sponsons. Total fuel capacity 2,384 litres (630 US gallons; 524.6 Imp gallons).

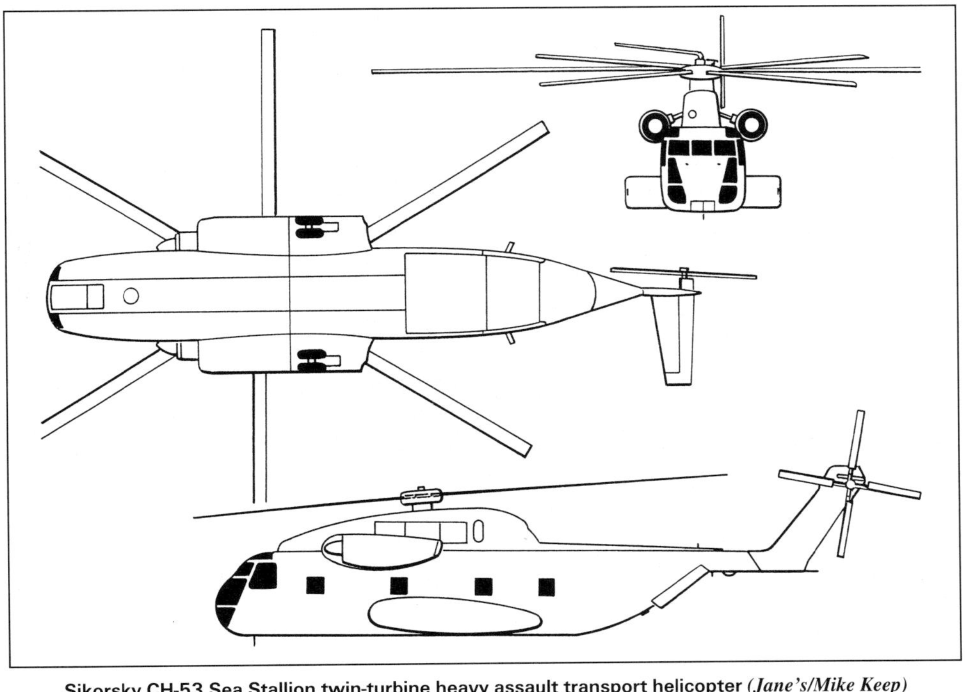

Sikorsky CH-53 Sea Stallion twin-turbine heavy assault transport helicopter *(Jane's/Mike Keep)*

1994

Sikorsky CH-53G of the German Army *(Sikorsky)*

1995

Sikorsky CH-53D Sea Stallion of the US Marine Corps *(Sikorsky)*

1995

ACCOMMODATION: Crew of three. Main cabin accommodates 37 combat-equipped troops on inward-facing seats. Provision for carrying 24 stretchers and four attendants. Roller-skid track combination in floor for handling heavy freight. Door on starboard side of cabin at front. Rear-loading ramp.

DIMENSIONS, EXTERNAL:

Diameter of main rotor	22.02 m (72 ft 3 in)
Diameter of tail rotor	4.88 m (16 ft 0 in)
Length overall, rotors turning	26.90 m (88 ft 3 in)
Length of fuselage, excl refuelling probe	20.47 m (67 ft 2 in)
Width overall, rotors folded	4.72 m (15 ft 6 in)
Width of fuselage	2.69 m (8 ft 10 in)
Height to top of rotor hub	5.22 m (17 ft 1½ in)
Height overall	7.60 m (24 ft 11 in)
Wheel track	3.96 m (13 ft 0 in)
Wheelbase	8.23 m (27 ft 0 in)

DIMENSIONS, INTERNAL:

Cabin: Length	9.14 m (30 ft 0 in)
Max width	2.29 m (7 ft 6 in)
Max height	1.98 m (6 ft 6 in)

AREAS:

Main rotor disc	378.1 m^2 (4,070 sq ft)
Tail rotor disc	18.67 m^2 (201 sq ft)

WEIGHTS AND LOADINGS:

Weight empty: CH-53A	10,180 kg (22,444 lb)
HH-53B	10,490 kg (23,125 lb)
HH-53C	10,690 kg (23,569 lb)
CH-53D	10,653 kg (23,485 lb)
Normal T-O weight: CH-53A	15,875 kg (35,000 lb)
Mission T-O weight: HH-53B	16,964 kg (37,400 lb)
HH-53C	17,344 kg (38,238 lb)
CH-53D	16,510 kg (36,400 lb)
Max T-O weight: HH-53B/C, CH-53D	19,050 kg (42,000 lb)

PERFORMANCE:

Max level speed at S/L:	
HH-53B	162 kts (299 km/h; 186 mph)
HH-53C, CH-53D	170 kts (315 km/h; 196 mph)
Cruising speed: HH-53B/C, CH-53D	150 kts (278 km/h; 173 mph)
Max rate of climb at S/L:	
HH-53B	440 m (1,440 ft)/min
HH-53C	631 m (2,070 ft)/min
CH-53D	664 m (2,180 ft)/min
Service ceiling: HH-53B	5,610 m (18,400 ft)
HH-53C	6,220 m (20,400 ft)
CH-53D	6,400 m (21,000 ft)
Hovering ceiling in ground effect:	
HH-53B	2,470 m (8,100 ft)
HH-53C	3,565 m (11,700 ft)
CH-53D	4,080 m (13,400 ft)
Hovering ceiling out of ground effect:	
HH-53B	490 m (1,600 ft)
HH-53C	1,310 m (4,300 ft)
CH-53D	1,980 m (6,500 ft)
Min ground turning radius	13.46 m (44 ft 2 in)
Runway LCN at max T-O weight	7.1

Range:

HH-53B/C, with 4,502 kg (9,926 lb) fuel (two 1,703 litre; 450 US gallon; 375 Imp gallon auxiliary tanks), including 10% reserves and 2 min warm-up
468 n miles (869 km; 540 miles)

CH-53D, with 1,849 kg (4,076 lb) fuel, 10% reserves at cruising speed and 2 min warm-up
223 n miles (413 km; 257 miles)

UPDATED

SIKORSKY S-65/MH-53J PAVE LOW III ENHANCED

TYPE: Combat helicopter upgrade.

PROGRAMME: US Air Force upgrade of Special Operations Forces combat rescue and recovery fleet; 31 HH-53Bs, HH-53Cs and CH-53Cs converted at NAS Pensacola, Florida, beginning 1986, to MH-53J Pave Low III Enhanced; similar to 11 HH-53H Pave Low III produced earlier, eight survivors of which also converted to MH-53Js; programme completed in 1990.

UPGRADES: **IBM/Loral:** Contract awarded late 1993 to evaluate incorporation of Defensive Aids System (DAS). System includes integration of MIL-STD-1553 databus, AAR-47 missile plume detector, ALW-136 missile jammer as well as a multimission advanced tactical terminal.

US Navy: Block 1. The majority of the MH-53J aircraft have already undergone Service Life Extension Programmes, but a further programme known as the Block 1 programme is intended to keep the aircraft in service until the year 2010. Programme includes basic reworking of structure, hydraulics and wiring. Further work is being carried in order to improve shipboard stowage. Work includes installing fully automatic main and tail rotor folding.

DESIGN FEATURES: Modifications include Texas Instruments AN/AAQ-10 nose-mounted FLIR, inertial navigation, Doppler, computer-projected map display, Navstar GPS, Texas Instruments AN/APQ-158 terrain-following/avoidance radar in offset nose radome, chaff/flare dispensers, Loral AN/ALQ-157 IR jammer on each outrigger pylon, 454 kg (1,000 lb) of extra titanium armour plating and Collins AN/AIC-3 intercom; armament includes three 7.62 mm or 0.50 in machine-guns firing through windows on each side and from open rear ramp. Power plant is two 3,266 kW (4,380 shp) General Electric T64-GE-415 turboshafts; maximum T-O weight increased to 22,680 kg (50,000 lb).

Delivery of 15 MH-53J began mid-1987 to 20th SOS at Hurlburt Field, Florida; six delivered to 21st SOS at RAF Woodbridge, UK; four more issued to 1550th Combat Crew Training Wing at Kirtland AFB, New Mexico, in 1989 (unit also received four TH-53As - with two more in prospect - from US Marines; modified at Kirtland with T64-GE-416 power plant and some standard USAF avionics for commonality with MH-53J).

UPDATED

SIKORSKY S-64 SKYCRANE

US military designation: CH-54 Tarhe

TYPE: Twin-turbine heavy flying crane helicopter.

PROGRAMME: The S-64 flying crane was designed initially for military transport duties. Equipped with interchangeable pods, it is suitable for use as a troop transport, and for minesweeping, cargo and missile transport, anti-submarine or field hospital operations. Equipment includes a removable 9,072 kg (20,000 lb) hoist, a sling attachment and a load stabiliser to prevent undue sway in cargo winch operations. Attachment points are provided on the fuselage and landing gear to facilitate securing of bulky loads.

VERSIONS: **S-64A:** Under this designation the first of three prototypes flew for the first time 9 May 1962 and was used by the US Army at Fort Benning, Georgia, for testing and demonstration. The second and third prototypes were evaluated by the German armed forces.

CH-54A: Six ordered by US Army in 1963 to investigate the heavy lift concept, with emphasis on increasing mobility in the battlefield. Delivery of five CH-54As (originally YCH-54As) to the US Army took place in late 1964 and early 1965. A sixth CH-54A remained at Stratford, with a company owned S-64, for a programme leading toward a restricted FAA certification, which was awarded 30 July 1965. Further US Army orders followed.

Sikorsky MH-53J Pave Low III helicopter with nose-mounted FLIR and GPS *(Sikorsky)*

1995

The CH-54As were assigned to the US Army's 478th Aviation Company, and performed outstanding service in support of the Army's First Cavalry Division, Airmobile, in Vietnam. On 29 April 1965, a CH-54A of this unit lifted 90 persons, including 87 combat-equipped troops in a detachable van. This is believed to be the largest number of people ever carried by a helicopter at one time. Other Skycranes in Vietnam transported bulldozers and road graders weighing up to 7,937 kg (17,500 lb), 9,072 kg (20,000 lb) armoured vehicles and a large variety of heavy hardware. They retrieved more than 380 damaged aircraft, involving savings estimated at $210 million.

Sikorsky Aircraft developed an all-purpose van, known as the Universal Military Pod, for carriage by the US Army's CH-54As, and received an order, worth $2.9 million, to supply 22 to the army. The pods were delivered complete with communications, ventilation and lighting systems, and with wheels to simplify ground handling. They superseded earlier pods which were not approved for the carriage of personnel. The first pod was accepted by the US Army 28 June 1986, following approval for personnel transport.

Internal dimensions of the pod are length 8.36 m (27 ft 5 in), width 2.69 m (8 ft 10 in) and height 1.98 m (6 ft 6 in). Doors are provided on each side of the forward area of the pod, and a double-panelled ramp is located aft. With a maximum loaded weight of 9,072 kg (20,000 lb), each pod accommodates 45 combat-equipped troops, or 24 litters, and in the field may be adapted for a variety of uses, such as surgical unit, field command post and communications post.

On 18 April 1969, two commercial Skycranes were delivered to Rowan Drilling Company Inc of Houston, Texas, for operation in support of oil exploration and drilling operations in Alaska. 115 in service.

S-64E: FAA certification of the improved S-64E for civil use was announced in 1969, for the transportation of external cargo weighing up to 9,072 kg (20,000 lb).

In January 1972 Erickson Air-Crane Company of Marysville, California, purchased the first S-64E, for logging and other heavy-lift tasks. On 1 November 1972 this company ordered three additional S-64Es, the first of which was delivered in the following month, to extend its operations on a worldwide basis, offering heavy-lift capability to the logging, petroleum, power line, shipping and general construction industries.

CH-54B: On 4 November 1968 Sikorsky announced that it had received a US Army contract to increase the payload capacity of the CH-54 from 10 to 12½ tons. The contract called for a number of design improvements to the engine, gearbox, rotor head and structure; altitude performance and hot weather operating capability were also to be improved. Two of the improved flying cranes, designated CH-54B, were accepted by the US Army during 1969.

The original JFTD12-4A engines were replaced by two Pratt & Whitney JFTD12-5As, each rated at 3,579 kW (4,800 shp), and a gearbox capable of receiving 5,891 kW (7,900 hp) from the two engines was introduced. Single-engine performance was increased, since the new gearbox receives 3,579 kW (4,800 hp) from one engine, compared with 3,020 kW (4,050 hp) on the CH-54A.

A new rotor system was also introduced, utilising a high-lift rotor blade with a chord some 0.064 m (2.5 in) greater than that of the blades used formerly.

Sikorsky CH-54A Skycrane (Tarhe) twin-turbine heavy-lift helicopter *(Peter March)* *1994*

Sikorsky CH-54A Skycrane (Tarhe) of the US Army *1994*

Other changes included the provision of twin wheels on the main landing gear, an improved automatic flight control system and some general structural strengthening throughout the aircraft. Gross weight was increased from 19,050 kg (42,000 lb) to 21,318 kg (47,000 lb).

In October 1970, two US Army CH-54Bs lifted an 18,488 kg (40,760 lb) load during a series of tests being conducted to evaluate the technical feasibility and cost of a twin-lift system for potential application to military requirements for greater helicopter external load capacity. Later in the same month, a single US Army CH-54B lifted an 18,497 kg (40,780 lb) load during tests being conducted to evaluate maximum hover lift capability.

S-64F: Designation of a commercial version of the S-64.

UPGRADES: **Sikorsky:** CH-54B. See Versions.

OPERATORS: Versions of the CH-54 Sea Stallion are in service with the armed forces of the USA (45 being phased out).

DESIGN FEATURES: Six-blade, fully articulated main rotor. Four-blade tail rotor. Steel driveshafts. Main gearbox below main rotor, intermediate gearbox at base of tail pylon. Tail gearbox at top of pylon. Main gearbox rated at 4,922 kW (6,600 shp) on CH-54A.

FLYING CONTROLS: Rotor brake standard.

STRUCTURE: Aluminium main blades with aluminium and steel head. Aluminium tail rotor blades with titanium head. Pod and beam type of aluminium and steel semi-monocoque construction.

LANDING GEAR: Non-retractable tricycle type, with single wheel on each unit of CH-54A/S-64E, twin wheels on main units of S-64F. CH-54A/S-64E mainwheel tyres size 38.45 × 12.50-16, pressure 6.55 bars (95 lb/sq in). S-64F mainwheel tyres size 25.65 × 8.50-10, pressure 6.90 bars (100 lb/sq in). Nosewheels and tyres of all versions size 25.65 × 8.50-10, pressure 6.90 bars (100 lb/sq in).

POWER PLANT (CH-54A/S-64E): Two Pratt & Whitney JFTD12-4A (military T73-P-1) turboshaft engines, each rated at 3,356 kW (4,500 shp) for take-off and with maximum continuous rating of 2,983 kW (4,000 shp). Two fuel tanks in fuselage, forward and aft of transmission, each with capacity of 1,664 litres (440 US gallons; 367 Imp gallons). Total standard fuel capacity 3,328 litres (880 US gallons; 733 Imp gallons). Provision for auxiliary fuel tank of 1,664 litres (440 US gallons; 367 Imp gallons) capacity, raising total fuel capacity to 4,992 litres (1,320 US gallons; 1,100 Imp gallons).

POWER PLANT (CH-54B/S-64F): Two Pratt & Whitney JFTD12-5A turboshaft engines, each rated at 3,579 kW (4,800 shp) for take-off and with maximum continuous rating of 3,303.5 kW (4,430 shp). Fuel tanks as for CH-54A/S-64E.

ACCOMMODATION: Pilot and co-pilot side by side at front of cabin. Aft-facing seat for third pilot at rear of cabin, with flying controls. The occupant of this third seat is able to take over control of the aircraft during loading and unloading. Two additional jump seats available in cabin. Payload in interchangeable pods.

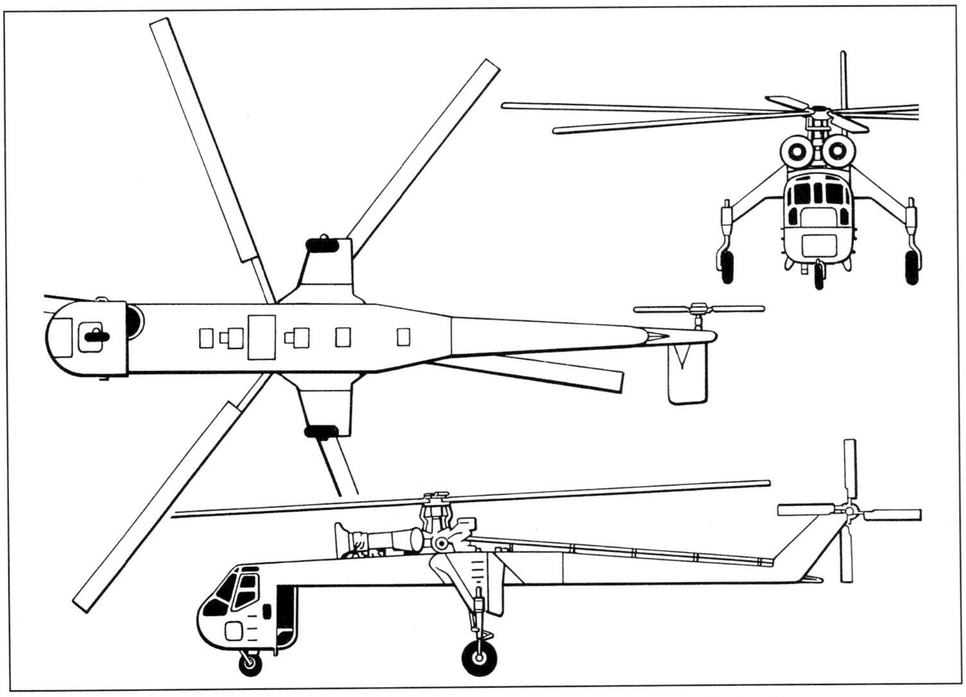

Sikorsky CH-54A Skycrane (Tarhe) flying crane *(Jane's/Mike Keep)* *1994*

DIMENSIONS, EXTERNAL:

Main rotor diameter	21.95 m (72 ft 0 in)
Tail rotor diameter	4.88 m (16 ft 0 in)
Distance between rotor centres	13.56 m (44 ft 6 in)
Length overall	26.97 m (88 ft 6 in)
Length of fuselage	21.41 m (70 ft 3 in)
Width, rotors folded	6.65 m (21 ft 10 in)
Height to top of rotor hub	5.67 m (18 ft 7 in)
Height overall	7.75 m (25 ft 5 in)
Ground clearance under fuselage boom	2.84 m (9 ft 4 in)
Wheel track	6.02 m (19 ft 9 in)
Wheelbase	7.44 m (24 ft 5 in)

AREAS:

Main rotor disc	378.1 m² (4,070 sq ft)
Tail rotor disc	18.67 m² (201 sq ft)

WEIGHTS AND LOADINGS (CH-54A/S-64E):

Weight empty	8,724 kg (19,234 lb)
Max T-O weight	19,050 kg (42,000 lb)

PERFORMANCE (CH-54A/S-64E at normal T-O weight of 17,237 kg; 38,000 lb):

Max level speed at S/L	109 kts (203 km/h; 126 mph)
Max cruising speed	91 kts (169 km/h; 105 mph)
Max rate of climb at S/L	405 m (1,330 ft)/min
Service ceiling	2,475 m (9,000 ft)
Hovering ceiling in ground effect	3,230 m (10,600 ft)
Hovering ceiling out of ground effect	2,100 m (6,900 ft)
Min ground turning radius: CH-54A, S-64E, S-64F	16.4 m (54 ft 0 in)

Runway LCN:
CH-54A, S-64E at max T-O weight of 19,050 kg (42,000 lb) 7.1
S-64F at max T-O weight of 21,318 kg (47,000 lb) 7.7

Range with max fuel, 10% reserves
200 n miles (370 km; 230 miles)

UPDATED

SNOW

SNOW AVIATION INTERNATIONAL, INC

7201 Paul Tibbets Street, Rickenbacker Airport, Columbus, Ohio 43217
Telephone: 1 (614) 492 7669
Fax: 1 (614) 492 7679

PRESIDENT: Harry T. Snow

VICE-PRESIDENT GOVERNMENT AND INTERNATIONAL AFFAIRS: Leonard Ekman

VICE-PRESIDENT SALES: Emil Kluever

PROGRAM MANAGER, C-130: Edwin G. Moitoza

Company formed 24 November 1987 by Harry Snow to design, develop and manufacture the Modern Technology Cockpit (MTC) for retrofit in large passenger/freight aircraft. Company also designing retrofit programme for Boeing 727-200.

NEW ENTRY

SNOW AVIATION MODERN TECHNOLOGY COCKPIT MTC

TYPE: Large turbine-powered passenger, freighter or tanker two-man cockpit upgrade.

PROGRAMME: Design of the MTC (also known as SAI C-130 TMC) began in December 1991. In March 1992 the company acquired a C-130A. This aircraft was put through a comprehensive Programmed Depot Maintenance. Retrofit programme was then initiated. Retrofit kit suitable for a range of large passenger, freight and tanker aircraft including military C-130 Hercules, Commercial L-100 Hercules and Boeing 727-200.

DESIGN FEATURES: Snow Aviation plan to install the cockpits themselves or to supply to the customer with MTC in kit form. MTC consists of four modules each of which will be FAA certified with its own Supplemental Type Certificate (STC). The four modules are: **STC 1:** Electrical system upgrade. This has been carried out on the prototype aircraft and has successfully completed all test procedures required for MIL-SPEC 704E and FAA certification.

STC 2: Communication, navigation and identification (CNI) system. Installed on prototype aircraft.

STC 3: Engine indication and crew alterting system (EICAS).

STC 4: Flat glass displays and HOTAY system.

Additional C-130 modifications are to include: re-engining with four Allison AE2100 turboprop engines, a fuselage stretch (4 ft; 1.2 m) as well as the removal of underwing pylon fuel tanks, installation of centre wing glove and wing tip fuel tanks.

COSTS: $4 million installed (estimate). $3.4 million (estimate) in kit form.

NEW ENTRY

SNOW AVIATION (BOEING) 727-200 RE-ENGINE PROGRAMME

TYPE: Three-engine passenger and freighter aircraft re-engine programme.

PROGRAMME: Proposal to re-engine the three-engined Boeing 727-200.

DESIGN FEATURES: Programme consists of replacing existing Pratt & Whitney JT8-D series turbofan engines with two IAE V2500 turbofan engines as well as the installation of the Modern Technology Cockpit (which see).

COSTS: Estimated at $20 million.

NEW ENTRY

SOLOY

SOLOY CORPORATION

450 Pat Kennedy Way SW, Olympia, Washington 98501
Telephone: 1 (206) 754 7000
Fax: 1 (206) 943 7659

PRESIDENT: Joe I. Soloy

EXECUTIVE VICE-PRESIDENT/GENERAL MANAGER: Scott Smith

Well known for re-engining of Hiller 12E and Bell 47G helicopters with the Allison 250-C20 turboshaft; Turbine Pac turboprop conversion for many fixed-wing aircraft developed in association with Allison. Also provides combiner gearbox for twin Allison installation in Tridair 260L-3ST Gemini ST conversion of Bell LongRanger; this gearbox being adapted for re-engining single-engined business aircraft with coupled twin turbines driving single propeller for FAA certification as twins.

VERIFIED

SOLOY (ENAER) T-35DT TURBO PILLÁN

TYPE: Two-seat turboprop military trainer (converted T-35D).

PROGRAMME: Original **T-35TX Aucán** turboprop prototype (CC-PZC: see 1988-89 *Jane's*) made first flight February1986; flight trials suspended 1987 after some 500 flight hours: subsequently modified to have new one-piece canopy opening sideways to starboard, oxygen system, and partial avionics upgrade. Soloy awarded 1990 contract to develop production-ready modification kit for existing T-35s, based on Allison 250-B17D engine; marketing effort began early 1991. First flight of this **T-35DT** 'prototype' (CC-PZG) March 1991; evaluation continuing in early 1994.

COSTS: Approximately $950,000 (1994).

DESIGN FEATURES: Based on Piper Cherokee series (utilising many components of PA-28 Dakota and PA-32 Saratoga); cleared to FAR Pt 23 (Aerobatic category) and military standards for basic, intermediate and instrument flying training. Low-wing tandem-seat design; sweptback vertical tail, non-swept horizontal surfaces.

Wing section NACA 65_2-415 on constant chord inboard panels, NACA 65_2-415 (modified) at tips; incidence 2° at root –0° 30′ at tip; dihedral 7° from roots.

FLYING CONTROLS: Manual/mechanical actuation of primary control surfaces (ailerons slotted, elevator/rudder mass balanced): single-slotted wing flaps, aileron trim tab (port) and rudder/tailplane/elevator trim all electrically actuated; variable incidence tailplane; one-piece elevator.

STRUCTURE: Main structure of aluminium alloy and steel, with riveted skins, except for glassfibre engine cowling, wingtips and tailplane tips. Single-spar fail-safe wings, with components mainly from PA-28-236 Dakota (leading-edges) and PA-32R-301 Saratoga (trailing-edges), modified for shorter span; vertical tail virtually identical with Dakota; tailplane uses some standard components from Dakota and PA-31 (Navajo/Cheyenne); tailcone from Cherokee components, modified for narrower fuselage.

LANDING GEAR: Hydraulically retractable tricycle type, with single wheel on each unit. Main units retract inward, steerable nosewheel (±25°) rearward. Piper oleo-pneumatic shock-absorber in each unit. Emergency free-fall extension. Cleveland mainwheels and McCreary tyres size 6.00-6 (8 ply), nosewheel and tyre size 5.00-5 (6 ply). Tyre pressures: 2.62 bars (38 lb/sq in) on mainwheels, 2.41 bars (35 lb/sq in) on nosewheel. Single-disc air-cooled hydraulic brake on each mainwheel. Parking brake. Minimum ground turning radius 6.20 m (20 ft 4 in).

POWER PLANT: One 313 kW (420 shp) Allison 250-B17D turboprop in Soloy inverted installation with Soloy reduction gearbox; Hartzell HC-B3TF-7A/T9212K-2B three-blade constant-speed reversible-pitch propeller. Total usable fuel 277.8 litres (73.4 US gallons; 61.1 Imp gallons), in two integral wing tanks; inverted-flight fuel feed system.

ACCOMMODATION: Two vertically adjustable seats, with seat belts and shoulder harnesses, in tandem beneath one-piece transparent jettisonable canopy which opens sideways to starboard. One-piece acrylic windscreen, and one-piece window in glassfibre fairing aft of canopy. Rear (instructor's) seat 22 cm (8.7 in) higher. Dual controls standard. Baggage compartment aft of rear cockpit, with external access on port side. Cockpits ventilated; cockpit heating and canopy demisting by engine bleed air.

SYSTEMS: Electrically operated hydraulic system, at 124 bars (1,800 lb/sq in) pressure for landing gear retraction and 44.8 bars (650 lb/sq in) for gear extension: separate system at 20.7 bars (300 lb/sq in) for wheel brakes. Electrical system is 28 V DC, powered by 28 V 70 A engine-driven Prestolite alternator and 24 V 10 Ah battery, with inverter for AC power at 400 Hz to operate RMIs and attitude indicators. External power socket. Oxygen system.

AVIONICS: Nav/com system comprises dual VHF/VOR transceivers. ATC transponder, DME with dual indicators, dual HSI, ADF, dual RMI, marker beacon receiver, dual audio selector panels, dual magnetic compasses, dual directional gyros and slave meter. Blind-flying instrumentation standard.

DIMENSIONS:

Wing span	8.84 m (29 ft 0 in)
Wing chord: at root	1.88 m (6 ft 2 in)
constant portion	1.60 m (5 ft 3 in)
at tip	1.26 m (4 ft 1½ in)
Wing apsect ratio	5.71
Length overall	8.60 m (28 ft 2½ in)
Height overall	2.64 m (8 ft 8 in)

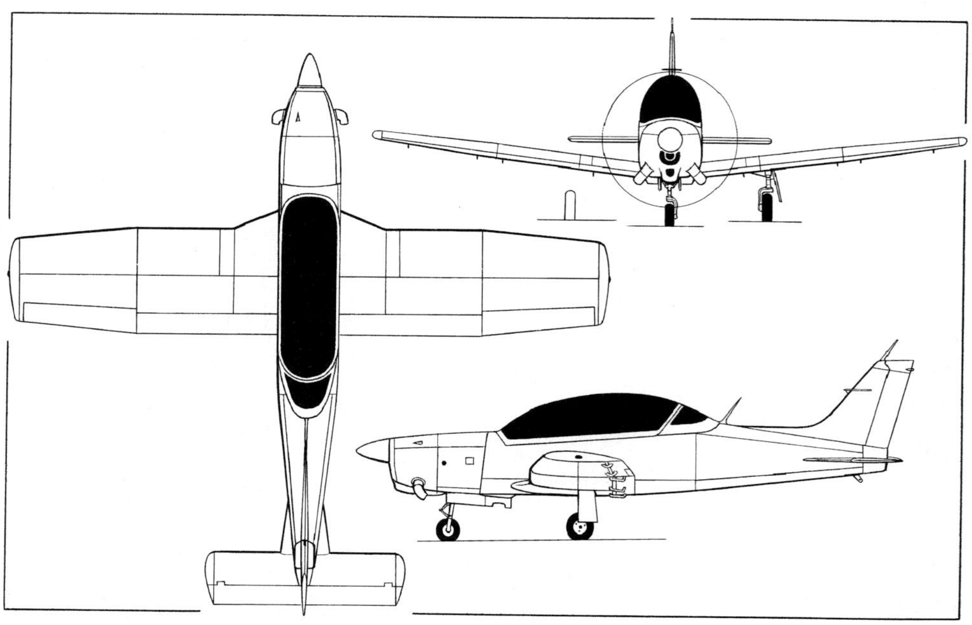

Soloy (Enaer) T-35DT Turbo Pillán tandem seat turboprop trainer *(Jane's/Dennis Punnett)*

1995

Tailplane span	3.05 m (10 ft 0 in)
Wheel track	3.01 m (9 ft 10½ in)
Wheelbase	2.00 m (6 ft 6¼ in)
Propeller diameter	1.93 m (6 ft 4 in)

DIMENSIONS, INTERNAL:

Cockpit: Length	3.24 m (10 ft 7½ in)
Max width	1.04 m (3 ft 5 in)
Max height	1.48 m (4 ft 10¼ in)

AREAS:

Wings, gross	13.69 m² (147.36 sq ft)
Ailerons (total)	1.14 m² (12.27 sq ft)
Trailing-edge flaps (total)	1.36 m² (14.64 sq ft)
Fin	0.69 m² (7.43 sq ft)
Rudder	0.38 m² (4.09 sq ft)
Tailplane	1.57 m² (16.90 sq ft)
Elevator	0.77 m² (8.29 sq ft)

WEIGHTS AND LOADINGS:

Weight empty	943 kg (2,080 lb)
Fuel weight: usable	224 kg (494 lb)
Max	238 kg (524 lb)
T-O and landing weight: Aerobatic	1,315 kg (2,900 lb)
max	1,338 kg (2,950 lb)
Max wing loading	97.73 kg/m² (20.03 lb/sq ft)
Max power loading	4.27 kg/kW (7.02 lb/shp)

PERFORMANCE: (at max T-O weight):

Never-exceed speed (VNE)	241 kts (446 km/h; 277 mph) IAS
Max manoeuvring speed (VA)	183 kts (339 km/h; 211 mph) IAS
Max level speed at S/L	230 kts (426 km/h; 264 mph)
Cruising speed: 75% power at 2,320 m (7,610 ft)	182 kts (337 km/h; 209 mph)
65% power at 3,450 m (11,320 ft)	176 kts (326 km/h; 203 mph)
55% power at 4,630 m (15,190 ft)	169 kts (313 km/h; 194 mph)
Stalling speed, power off: flaps and landing gear up	66 kts (123 km/h; 76 mph)
flaps and landing gear down	62 kts (115 km/h; 71 mph)
Max rate of climb at S/L	869 m (2,850 ft)/min
Service ceiling	7,620 m (25,000 ft)
Absolute ceiling	7,826 m (25,675 ft)
Time to 2,000 m (6,560 ft)	3 min 42 s
Time to 3,000 m (9,850 ft)	5 min 36 s
T-O run	195 m (640 ft)
T-O to 15 m (50 ft)	357 m (1,175 ft)
Landing from 15 m (50 ft)	555 m (1,820 ft)
Landing run	128 m (420 ft)
Range, 45 min reserves: 75% power at 2,500 m (8,200 ft)	350 n miles (648 km; 403 miles)
55% power at 4,000 m (13,125 ft)	410 n miles (759 km; 472 miles)
g limits	+6/–3

NEW ENTRY

SOLOY TURBINE PAC and DUAL PAC CONVERSIONS

TYPE: Light aircraft turboprop upgrades.

PROGRAMME: **Turbine Pac** based on 313 kW (420 shp) Allison 250-C20S, 373 kW (500 shp) 250-C28C or 485 kW (650 shp) 250-C30 turboprop, driving a 522 kW (700 shp) reduction gearbox with output; first flight 23 November 1981 in Cessna 185 (N5010Y) with Allison 250-C20; Cessna 206 and 207 and Beech A36 Bonanza are STC approved; Soloy Turbine Pac Cessna 185 available for government agencies able to operate without certification.

Royal Thai Army evaluated a Soloy Turbine Pac Cessna 185/U-17; Soloy Turbine Pac Cessnas used by US Drug Enforcement Administration and Government of Costa Rica; several Jet-Prop Beech A36 Bonanzas sold for private and business use (see under Tradewind); other installations include Allison Series IV being developed for single- and twin-engined aircraft and 485 kW (650 shp) Allison 250-C30M certificated in Aerospatiale AS 350D Astar in early 1986; Soloy installed Allison 250-B17 turboprop in Chilean Enaer Aucan prototype, first flight 12 February 1986.

Soloy started development in early 1989 of **Dual Pac** twin-engine combining gearbox for both helicopter and fixed-wing applications, using either Allison 250-C20s or 250-C30s; each engine has separate freewheel so that Soloy expects FAA Pt 135 IFR certification of Dual Pac (granted 1991 for Cessna Caravan) as twin-engined power plant; both engines driving on aircraft centreline is important safety and performance factor. Tridair Helicopters uses Soloy Dual Pac for its twin-engined version of Bell 206L LongRanger. Dual Pac PT6B-35F for fixed-wing aircraft under joint development with Pratt & Whitney Canada.

Soloy has also installed two 380 kW (510 hp) Teledyne Continental TP-500E turboprops in Piper Chieftain; first flight in 1989; activity terminated prior to certification.

VERIFIED

Soloy Turbine Pac installation in a Cessna 206 with PK amphibious floats *1994*

Soloy (Cessna) 207 upgraded with Soloy Turbine Pac *(Soloy)* *1995*

SOLOY CESSNA 206 TURBINE PAC

TYPE: Turbine-powered Cessna 206.

PROGRAMME: FAA certificated 22 May 1984.

VERSIONS: **Civil:** Wheel or float landing gear; as detailed below.

Military: Additional fuel and 'super flotation' landing gear with oversize wheels for operation from unprepared terrain. Primarily for liaison and observation.

CUSTOMERS: Operating in Australia, Bolivia, Switzerland and USA; total of 54 Model 206 Turbine Pacs delivered by November 1994.

DESIGN FEATURES: Engine installed inverted and back to front to place output drive above, intake at rear top of nacelle and exhaust pipes under front of cowling. Approved for Wipline and PK floats and Edo 3500 amphibious floats.

FLYING CONTROLS: As for Cessna 206.

STRUCTURE: As for Cessna 206.

Main differences from standard Cessna 206 listed below:

POWER PLANT: One 313 kW (420 shp) Soloy Model 763 Turbine Pac/Allison 250-C20S turboprop, driving a Hartzell three-blade constant-speed fully feathering propeller. Fuel capacity 230 to 348 litres (61 to 92 US gallons; 51 to 77 Imp gallons), depending on airframe model. Extended range fuel tanks, total usable capacity 189 litres (50 US gallons; 42 Imp gallons), optional on military model. Oil capacity 7.1 litres (1.9 US gallons; 1.6 Imp gallons).

WEIGHTS AND LOADINGS:

Weight empty	866 kg (1,910 lb)
Max T-O weight	1,633 kg (3,600 lb)
Max power loading	5.22 kg/kW (8.57 lb/shp)

PERFORMANCE (at max T-O weight):

Never-exceed speed (VNE)	148 kts (274 km/h; 170 mph) IAS
Max level speed and max cruising speed at 3,050 m (10,000 ft)	163 kts (302 km/h; 188 mph)
Econ cruising speed at 6,100 m (20,000 ft)	142 kts (263 km/h; 164 mph)
Stalling speed, power off: flaps up	58 kts (107 km/h; 67 mph) CAS
flaps down	52 kts (96 km/h; 60 mph) CAS
Max rate of climb at S/L	594 m (1,950 ft)/min
Service ceiling	6,100 m (20,000 ft)
T-O run	174 m (570 ft)
T-O to 15 m (50 ft)	324 m (1,063 ft)
Landing from 15 m (50 ft)	334 m (1,094 ft)
Landing run	168 m (553 ft)
Range with 333 litres (88 US gallons; 73 Imp gallons) usable fuel, max range power at 6,100 m (20,000 ft)	882 n miles (1,635 km; 1,016 miles)

VERIFIED

Soloy (Enaer) T-35DT Turbo Pillán *(Charles Bickers)* *1995*

SOLOY CESSNA 207 TURBINE PAC

TYPE: Light aircraft turboprop upgrade.

PROGRAMME: First flight of Cessna 207 (N21190) powered by 313 kW (420 shp) Allison 250-C20S 14 January 1984; second aircraft joined development Spring 1985; certificated early 1986; total of 15 Model 206 Turbine Pacs delivered by November 1994.

Main differences from standard Cessna 207 listed below:

WEIGHTS AND LOADINGS:

Weight empty	986 kg (2,175 lb)
Max T-O weight	1,814 kg (4,000 lb)
Max power loading	5.80 kg/kW (9.52 lb/shp)

PERFORMANCE:

Max level speed: at 1,525 m (5,000 ft)	157 kts (291 km/h; 181 mph)
at 3,050 m (10,000 ft)	164 kts (304 km/h; 189 mph)
Stalling speed, power off: flaps up	59 kts (109 km/h; 68 mph) CAS
flaps down	53 kts (98 km/h; 61 mph) CAS
Max rate of climb at S/L	500 m (1,640 ft)/min
Service ceiling	7,620 m (25,000 ft)

T-O run 224 m (736 ft)
Range with 303 litres (80 US gallons; 67 Imp gallons) usable fuel, max range power at 3,050 m (10,000 ft)
558 n miles (1,034 km; 642 miles)

VERIFIED

SOLOY DUAL PAC® (CESSNA) 208B GRAND CARVAN

TYPE: Single to twin-engined light aircraft conversion.

PROGRAMME: Origins of this conversion can be traced back to an agreement signed in 1990 with Pratt & Whitney Canada to develop Dual Pac power plant using two PT6D-114A engines and a combining gearbox developed by Soloy. This development work is nearing completion and the power plant is scheduled for certification in mid-/late 1995. The Soloy Dual Pac® ran for the first time on 20 September 1993 and was first flown on September 29 1994 in Soloy's DHC-3 Otter flight test vehicle.

After certification Pratt & Whitney Canada will market the power plant to aircraft manufacturers and Soloy will have the right and responsibility to market the power plant for the retrofit of existing aircraft. The Soloy Dual Pac® will be used as an integral part of the Cessna 208B conversion.

DESIGN FEATURES: Programme includes replacing existing engine with Soloy Dual Pac®, strengthening of the forward fuselage, aft fuselage stretch of 1.82 m (72 in), movement of cargo door further to the rear as well as modifications to the landing gear, firewall, engine frame and support, engine compartment, engine controls, cowling and associated structures and instrumentation. Modifications are also carried out on the air induction, exhaust, oil, fuel and electrical systems. The gross weight is increased to 4,762 kg (10,500 lb)

POWER PLANT: Soloy Dual Pac Model 785-1000 utilising Two Pratt & Whitney PT6D-114A engines driving five-bladed, constant-speed full feathering reversible propeller. Fuel capacity 1,257 litres (332 US gallons 276.6 Imp gallons).

DIMENSIONS, INTERNAL:
Cabin (aft of Pilot area)
Length 6.9 m (22 ft)
Height 1.3 m (4.3 ft)
Width 1.6 m (5.2 ft)
Volume 13.4 m³ (474 cu ft)

DIMENSIONS, EXTERNAL:
Length 14.9 m (48.75 ft)
Height 4.5 m (14.8 ft)
Wing span 15.9 m (52.1 ft)

PERFORMANCE (standard day max T-O weight):
T-O distance 335.3 m (1,100 ft)
T-O to 15 m (50 ft) 548.6 m (1,800 ft)
FAR 25 T-O field length 929.6 m (3,050 ft)
Landing distance (Max landing weight):
Ground roll 1,100 m (3,353 ft)
Obstacle clearing at 15 m (50 ft) distance
579.1 m (1,900 ft)
Service ceiling 5,639 m (18,500 ft)
Max cruise speed at 4.572 m (15,000 ft)
218 kts (392 km/h; 272 mph)
Max range at 3,048 m (10,000 ft) 45 min reserve, standard fuel 585 n miles (1,084 km; 673 miles)

NEW ENTRY

Soloy Dual Pac (de Havilland DHC-3) experimental aircraft *(Soloy)*

1995

SUPER 580

SUPER 580 AIRCRAFT COMPANY (Division of Flight Trails Inc)

2192 Palomar Airport Road, Carlsbad, California 92008
Telephone: 1 (619) 438 3600
Fax: 1 (619) 753 1531
Telex: 140414
PRESIDENT: Ted Vallas
SENIOR VICE-PRESIDENT: James Coleman

Licensed by Allison Gas Turbine Division of General Motors for turboprop conversion and remanufacturing of Convair 340/440/580 series transports into Super 580s.

VERIFIED

SUPER 580 AIRCRAFT SUPER 580

TYPE: Re-engined Convair 340/440/580.

PROGRAMME: First Super 580 conversion by Hamilton Aviation, Tucson, Arizona; first flight 21 March 1984; FAA approval to CAR 4b by means of STC; delivered to The Way International June 1984. Super 580 ST stretch for 78 passengers or nine LD3 containers investigated 1984.

DESIGN FEATURES: Replaces piston engines of 340/440 or Allison 501-D13 of 580 with Allison 501-D22G turboprops, each flat rated at 2,983 kW (4,000 shp) and driving Hamilton Standard four-blade, constant-speed, feathering and reversing propeller; compared with 580, Super 580 has 2.5 per cent better sfc, improved engine-out performance, 13 per cent longer range, 40 per cent greater hot day payload, 40 per cent lower operating costs, and cruising speeds up to 325 knots (602 km/h; 374 mph).

Super 580 Aircraft Company Super 580 re-engined Convair

1994

AVIONICS: Basic avionics suite includes: King Gold Crown radio series III with dual Sperry Flight director system. Optional suite includes Rockwell Collins flight director system with Electronic Flight Instrumentation System (EFIS) Horizontal Situation Indicators (HSIs) Pro Line II with dual digital flight director system.

PERFORMANCE:
Cruise speed at 300 kts (557 km/h; 345 mph)
Max T-O weight 24,545 kg (54,000 lbs)

UPDATED

TRADEWIND

TRADEWIND TURBINES CORPORATION

PO Box 31930, 4105 Tradewind Road, Amarillo, Texas 79120
Telephone: 1 (806) 376 5203
Fax: 1 (806) 376 9725
CEO: Joe C. Boyd

TRADEWIND PROP-JET BONANZA

TYPE: Turbine-powered Beechcraft A36 Bonanza.

PROGRAMME: Allison Gas Turbine Division in conjunction with Soloy Corporation (which see) obtained STC for re-engining of Beechcraft A36 Bonanza in 1986; initially known as Allison Prop-Jet Bonanza; STC sold to Tradewind Turbines October 1989.

CUSTOMERS: Allison, DuPage Aviation and Tradewind had completed 43 conversions by April 1995. Production rate six a year.

COSTS: Fittings, engine and installation cost $393,000 (1995).

DESIGN FEATURES: Conversion available for post 1979 A36 and A36TC, but not for F33 Bonanzas; Teledyne Continental 212.5 kW (285 hp) piston engine replaced by Allison 250-B17F/2 turboprop; two support spars added to lower fuselage and engine installed 0.53 m (1 ft 9 in) further forward to maintain centre of gravity; vacated space holds 54 kg (120 lb) of baggage. Other features include 24 V electrical system with 150 A starter/generator; dual actuators for elevator trim tabs; audible VMO overspeed warning; Goodrich propeller de-icing.

POWER PLANT: One 313 kW (420 shp) Allison 250-B17D turboprop driving a metal three-blade Hartzell feathering

Tradewind Prop-Jet Bonanza conversion of a Beechcraft Model A36 Bonanza

1994

and reversing propeller running at 2,030 rpm. Total 424 litres (112 US gallons; 93 Imp gallons) fuel in two tanks in wings and two Osborne wingtip tanks.

DIMENSIONS, EXTERNAL:

Wing span, between tips of winglets 10.20 m (33 ft 5½ in)
Length overall 8.89 m (29 ft 2 in)
Height overall 2.62 m (8 ft 7 in)
Tailplane span 3.71 m (12 ft 2 in)
Wheel track 2.92 m (9 ft 7 in)
Propeller diameter 2.29 m (7 ft 6 in)
Propeller ground clearance 0.18 m (7¼ in)

WEIGHTS AND LOADINGS:

Weight empty 1,089 kg (2,400 lb)
Max ramp weight 1,746 kg (3,849 lb)
Max T-O weight 1,739 kg (3,833 lb)
Max landing weight 1,656 kg (3,650 lb)
Max power loading 5.56 kg/kW (9.13 lb/shp)

PERFORMANCE:

Max level speed 210 kts (389 km/h; 242 mph)
Cruising speed:
at 4,575 m (15,000 ft) 200 kts (370 km/h; 230 mph)
at 6,100 m (20,000 ft) 195 kts (361 km/h; 224 mph)
Stalling speed, flight idle power:
flaps up 65 kts (120 km/h; 75 mph)
flaps down 57 kts (106 km/h; 66 mph)
Max rate of climb at S/L, at max T-O weight 579 m (1,900 ft)/min
Service ceiling above 7,620 m (25,000 ft)
T-O run, 15° flap 177 m (580 ft)
T-O to 15 m (50 ft), 15° flap 244 m (800 ft)
Landing from 15 m (50 ft), 30° flap 160 m (525 ft)
Landing run, 30° flap with reverse pitch 99 m (325 ft)
Range, with max fuel, allowances for start, taxi, T-O, climb and 45 min reserves at max cruising power:
at 3,050 m (10,000 ft) 820 n miles (1,520 km; 944 miles)
at 4,575 m (15,000 ft) 985 n miles (1,825 km; 1,134 miles)
at 6,100 m (20,000 ft) 1,065 n miles (1,974 km; 1,226 miles)

UPDATED

TRIDAIR

TRIDAIR HELICOPTERS INC

3000 Airway Avenue, Costa Mesa, California 92626
Telephone: 1 (714) 540 3000
Fax: 1 (714) 540 1042
PRESIDENT: Douglas Daigle

VERIFIED

TRIDAIR (BELL) 206L-3G AND 206L-4G GEMINI ST

TYPE: Twin-engined conversion of Bell 206L-3 LongRanger.

PROGRAMME: Announced at Helicopter Association International (HAI) show 1989; first flight of prototype N700TH 16 January 1991; two more prototype/demonstrators completed in 1991; final cowling profile decided April 1991; certification granted early 1993. FAA granted single engine operational on 12 September 1994 making the Gemini ST the world's first single/twin helicopter. FAA approved the single engine flight manual supplement for 206L-4 Gemini ST on 15 May 1995. FAA also approved a gross weight increase on the twin 206L-4 to a maximum approved weight of 4,550 lb (2,063 kg).

VERSIONS: **206L-3G:** Based on Bell 206L-1 and 206L-3.
206L-4G: Based on Bell 206L-4 with higher gross weight of 4,450 lb (2,018 kg).

CUSTOMERS: Tridair ordered 25 new airframes from Bell Helicopter Canada for delivery at rate of two per month from November 1990; 18 delivery positions reserved by February 1991; 50 conversions were expected by end of 1993; approved installation centres being established. Bell Helicopter announced in March 1992 that it will build new 206L-35Ts in its Canadian factory; Tridair to specialise in conversion of existing Long Rangers.

COSTS: Modification kit $645,000; new converted aircraft $1,550,000.

DESIGN FEATURES: Replaces single 485 kW (650 shp) Allison 250-C30P with two 335 kW (450 shp) 250-C20Rs with max continuous rating of 276 kW (370 shp) each; fuel capacity increased to 416 litres (110 US gallons; 91.6 Imp gallons); Soloy Dual Pac combining gearbox with individual free-wheels concentrates engine outputs into original single input drive to transmission; transmission rating 325 kW (435 shp); gross weight increased to 4,205 lb (1,928 kg). Category A approval with no weight restriction.

Tridair Helicopters (Bell) 206-3ST Gemini ST helicopter ***(Tridair)***

1995

WEIGHTS AND LOADINGS:

Weight empty 1,249 kg (2,748 lb)
Max fuel weight 340 kg (750 lb)
Max external load 907 kg (2,000 lb)
Max T-O weight (internal load) 1,928 kg (4,250 lb)

PERFORMANCE (estimated):

Never-exceed speed (VNE) 130 kts (241 km/h; 150 mph)
Max cruising speed 117 kts (217 km/h; 135 mph)
Max rate of climb at S/L 472 m (1,550 ft)/min
Service ceiling 6,100 m (20,000 ft)
Hovering ceiling: IGE 4,815 m (15,800 ft)
OGE 1,675 m (5,500 ft)
Max range, with max payload and max internal fuel 300 n miles (555 km; 345 miles)

UPDATED

USH

US HELICOPTER

Blackwell Airport, PO Box 1088, Ozark, Alabama 38361-1088
Telephone: 1 (334) 774 2529
Fax: 1 (334) 774 6756
PRESIDENT: Dick Joyce
CHIELF EXECUTIVE OFFICER: David Thaxton

US Helicopter was formed on 1 January 1995 when Dick Joyce and David Thaxton acquired UNC Helicopter from UNC Incorporated. Main responsibilities are in the fields of repair, overhaul and refurbishment of military and commercial Bell UH-1 helicopters. The company employs 70 personnel and three covered facilites including a 5,576 m² (60,000 sq ft) free span hanger. In August 1990 the company was awarded a US Army Bell UH-1 retirement programme contract and is currently completing the overhaul of 250 aircraft. A follow-on contract will be negotiated in 1996. USH has carried out work on helicopters for the armed fores of Argentina, Bolivia, Chile, Colombia, Greece, Jordan, Lebanon, Peru, Philippines, Singapore, Thailand, Tunisia and USA. USH operates under an FAA Repair Station Certificate and currently holds Type Certificates (TC) and Supplemental Type Cetificates (STC) to register Bell UH-1 helicopters in the restricted category. Further military activities include contracts to strip US Navy training helicopters and refurbish US National Guard helicopters. Commercial activity includes a contract for maintenance support services with a public use operator and airworthiness checks, partial overhaul or complete refurbishment for public use UH-1 aircraft including environmental survey, firefighting and law enforcement.

NEW ENTRY

VAT

VERTICAL AVIATION TECHNOLOGIES INC

PO Box 2527, Sanford, Florida 32773
Telephone: 1 (407) 322 9488
Fax: 1 (407) 330 2647
PRESIDENT: Bradley G. Clark
CONSULTANT ENGINEER: Ralph P. Alex

Vertical Aviation Technologies Inc is FAA approved repair facility for various Sikorsky helicopters. In 1984 development began to modify four-seat Sikorsky S-52-3 former production helicopter (of which 81 originally produced) into kit for assembly by individuals, corporations or military, with all tooling and fixtures completed by 1988. In 1990 development began for new version of Sikorsky S-55 with Garrett turbine engine. In 1991, VAT purchased assets of Orlando Helicopter Airways and is continuing Orlando's activity in remanufactured Sikorsky helicopters.

UPDATED

VAT/ORLANDO/SIKORSKY S-55/H-19

TYPE: Remanufactured Sikorsky S-55/H-19.

VERSIONS: **Vistaplane:** Passenger/air ambulance seating eight passengers or 10 in high-density version; six stretchers and two attendants; optional cabin floor viewing window.

Vertical Aviation Technologies (VAT) Aggressor conversion of the Sikorsky H-19/S-55

1994

Vertical Aviation Technologies (VAT) SS-55 Defender

1994

Vertical Aviation Technologies (VAT) S-55 Bearcat agricultural helicopter

1994

SS-55 Heli-Camper: VIP model sleeping four; fully carpeted and soundproofed; hot and cold water, refrigerator, two-burner stove, shower, wash basin, toilet, air conditioning, AM/FM stereo radio and tape deck, roll-up awning, tinted windows, bar and storage cabinets, dual flight controls, full night lighting and dual landing lights, interphone system, com/nav radio and emergency locator beacon. Options include cargo sling, hydraulic hoist, floats, rotating spotlight and radiotelephone. Heli-Camper sold with remanufactured airframe, new instruments, new or overhauled dynamics, overhauled 596 kW (800 hp) Wright Cyclone R-1300-3D engine. Fuel capacity 715 litres (189 US gallons; 157.5 Imp gallons); maximum T-O weight 3,266 kg (7,200 lb); cruising speed 78 knots (145 km/h; 90 mph); range 304 n miles (563 km; 350 miles).

S-55 Nite-Writer: Aerial advertising model with 12.2 × 2.4 m (40 × 8 ft) Sky Sign In computerised electronic billboard, mounted 40° to helicopter side; displays running messages, logos and graphics; legible at up to 1.7 n miles (3.2 km; 2 miles).

S-55 Bearcat: Carries Orlando-developed dry and spray dispensing systems, with quick conversion to fertilisers and seeds; 946 litre (250 US gallon; 208 Imp gallon) glassfibre chemical tank; dry material can be carried internally; adaptable for firefighting. All Orlando modified S-55s powered by P&W R-1340 engines FAA approved to use automotive fuel; 'super-quiet' engine exhaust; optional Kysor cockpit air conditioning. Certification of civil Bearcat II planned for July 1991.

S-55 Heavy Lift: External load model for logging, firefighting and construction; useful load 1,361 kg (3,000 lb).

AS-55/QS-55 Aggressor: Orlando has modified 15 S-55s to look like Mil Mi-24 'Hind-Es' for US Army Missile Command; used for training and as targets; dummy pilots in Mil cockpits when droned, or with pilot (when not droned) sitting high up and looking out through windows in Mi-24 type oil cooler fairings; retains S-55 four-wheel landing gear; new five-blade rotor being supplied by Rotaire; nose designed and supplied by 3D Industries; Honeywell drone flight control system; noise simulator will emulate sound of Mi-24; stub-wings each stressed to carry 227 kg (500 lb) of stores or ordnance; QS-55 fitted with defensive systems including flare/chaff dispensers. FAA Experimental certification May 1989.

AT-55 Defender: Low-cost multirole military helicopter, for use in troop transport, armoured assault or medevac duties; powered by 626 kW (840 shp) Garrett TPE331-1D. Incorporates much of technology developed for AS-55 Aggressor; five-blade main rotor; stub-wings with pylons; maximum T-O weight 3,583 kg (7,900 lb); maximum payload 952 kg (2,100 lb); maximum level speed 125 knots (231 km/h; 144 mph); maximum cruising speed 115 knots (213 km/h; 132 mph); maximum payload range 325 n miles (602 km; 374 miles). Carries up to 10 fully equipped troops; entrances/exits on either side of cabin; medevac version carries up to six litters and two attendants; assault version with wing pylons each carrying up to 227 kg (500 lb) of missiles, armament or defensive systems.

UPDATED

ORLANDO/SIKORSKY S-58/H-34

TYPE: Remanufactured Sikorsky S-58/H-34.

VERSIONS: **Agricultural and Heavy Lift:** Powered by 1,137 kW (1,525 hp) Wright Cyclone R-1820-84; equipped as equivalent S-55 versions.

VAT Heli-Camper: Equipped as S-55 Heli-Camper; additional features include 8.73 m^2 (94 sq ft) living area, sleeping for six, entertainment centre, separate bar, full-size four-burner stove, 0.11 m^3 (3.9 cu ft) refrigerator, 3,500 W generator, super soundproofing, tinted glass and wraparound windscreen; powered by 1,137 kW (1,525 hp) Wright Cyclone R-1820-84 radial engine; 992 litres (262 US gallons; 218 Imp gallons) fuel; maximum T-O weight 5,670 kg (12,500 lb); cruising speed 96 knots (177 km/h; 110 mph); range 304 n miles (563 km; 350 miles)

VAT Airliner: High-density 18-passenger version of Sikorsky twin-turbine S-58T (see California Helicopters entry in this section); standard equipment includes dual controls, toe brakes, hydraulic rotor brake, two independent powered control systems, full night lighting, rotating beacon, 24 V battery, 1,071 litre (283 US gallon; 235.6 Imp gallon) fuel system, 2,268 kg (5,000 lb) cargo hook, extended passenger cabin, 18 additional cabin windows (tinted), Flightex soundproofing and ram-air cabin cooling; options include 272 kg (600 lb) rescue hoist, pop-out floats, 568 litre (150 US gallon; 125 Imp gallon) external auxiliary fuel tank, stereo system, toilet, air conditioning, one-piece windscreen and choice of avionics; useful load is more than 2,268 kg (5,000 lb); maximum speed at S/L 120 knots (222 km/h; 138 mph); hover ceiling OGE 2,440 m (8,000 ft); range, with auxiliary fuel and 20 minutes reserve, 373 n miles (692 km; 430 miles). 24 conversions completed.

UPDATED

Vertical Aviation Technologies (VAT) S-58T Orlando Airliner

1994

VAZAR

VAZAR AEROSPACE

3025 Eldridge Avenue, Bellingham, Washington 98225
Telephone: 1 (360) 671 7817
Fax: 1 (360) 671 7820
PRESIDENT: Dara Wilder

VERIFIED

VAZAR (DE HAVILLAND) DASH 3

TYPE: STOL utility transport engine conversion.

PROGRAMME: DHC-3 Otter re-engined with 559 kW (750 shp) P&WC PT6A-135 turboprop, giving improved performance and increasing useful load by 331 kg (730 lb); prototype (N9707B) completed 1986. In February 1994 Vazar had completed 18 Dash 3 conversions and was carrying out

Vazar Aerospace Dash 3 conversion of a de Havilland DHC-3 Otter *(Vazar)*

1995

work on a further four aircraft. Conversion takes six weeks; can operate on wheels, skis or choice of Edo 7170 or 7490 amphibious floats. Options also include bubbled double windows and an extra fuel tank to increase endurance for a further hour. The Dash 3 is approved for 15 passengers plus pilot.

DIMENSIONS, EXTERNAL: As for standard DHC-3, except:
Length overall 14.02 m (46 ft 0 in)

WEIGHTS AND LOADINGS:
Weight empty 3,770 kg (8,201 lb)
Max T-O weight 3,630 kg (8,000 lb)
Max power loading 6.49 kg/kW (10.67 lb/shp)

PERFORMANCE (at max T-O weight):
Max cruising speed at 3,050 m (10,000 ft) 144 kts (267 km/h; 166 mph)
Stalling speed: flaps up 63 kts (117 km/h; 73 mph)
flaps down 50 kts (93 km/h; 58 mph)
Max rate of climb at S/L 365 m (1,200 ft)/min
Service ceiling, as tested by July 1988 4,875 m (16,000 ft)
T-O run 171 m (560 ft)
T-O to 15 m (50 ft) 278 m (910 ft)
Landing from 15 m (50 ft) 183 m (600 ft)
Landing run 95 m (310 ft)
Endurance 5 h 18 min
Range 649 n miles (1,203 km; 748 miles)

VERIFIED

VOLPAR

VOLPAR AIRCRAFT CORPORATION

7701 Woodley Avenue, Van Nuys, California 91406
Telephone: 1 (818) 994 5023
Fax: 1 (818) 988 8324
Telex: 651482 VOLPAR B VAN

PRESIDENT: Andy Savva

VICE-PRESIDENTS:
Robert C. Dunigan (Maintenance and Engineering)
Joel Fogelson (Marketing)
Frank V. Nixon (Development)

Volpar Aircraft Corporation acquired by Gaylord Holdings of Switzerland in early 1990. Following Chapter 11 bankruptcy protection, a cash infusion of $7 million was agreed in early 1993. This facilitated work to resume on re-engining the Dassault Falcon 20 business jet. Company not responding to correspondence in 1995.

UPDATED

VOLPAR (DASSAULT) FALCON PW300-F20

TYPE: Business jet re-engine programme.

PROGRAMME: Announced at Paris Air Show 1989; replaces GE CF700 engines on Falcon 20s with 23.24 kN (5,225 lb st) P&WC PW305 turbofans with variable inlet guide vanes for high-altitude performance; Rohr Industries thrust reversers standard. Modified Falcons will meet FAR Pt 36 Stage 3 noise requirements and cruise 50 knots (93 km/h; 58 mph) faster at 12,500 m (41,000 ft); estimated maximum range with NBAA IFR reserves 2,600 n miles (4,818 km; 2,994 miles). Baseline testing with CF700-2D2 engines completed Van Nuys August 1990; prototype then grounded for PW305 installation (first flight 12 February 1991). To be marketed by Advanced Falcon Aircraft Partnership of Greenwich, Connecticut. Cost $3.8 million, including thrust reversers and new paint.

VERIFIED

VOLPAR T-33V

TYPE: Combat training aircraft re-engine programme.

PROGRAMME: T-33V is upgraded and re-engined Lockheed T-33, being developed during 1990 in collaboration with William F. Chana Associates; Allison J33 would be replaced by P&WC PW300 turbofan, flat rated at 21.13 kN (4,750 lb st); aircraft weight reduced by 499 kg (1,000 lb) and fuel consumption reduced by up to two-thirds.

Volpar estimates market for 250 modified T-33Vs and will also supply modification kits; about 1,000 T-33s thought to be in service worldwide.

VERIFIED

VOUGHT

VOUGHT AIRCRAFT COMPANY

9314 West Jefferson Boulevard, PO Box 655907, M/S 49L-06 Dallas, Texas 75265-5907
Telephone: 1 (214) 266 2543
Fax: 1 (214) 266 4982

PRESIDENT AND CEO: Gordon L. Williams

VICE-PRESIDENT, CORPORATE COMMUNICATIONS:
Lynn J. Farris

Original Lewis & Vought Corporation founded 1917; became Chance Vought Corporation, 31 December 1960; merged with Ling-Temco Electronics, 31 August 1961 to form Ling-Temco-Vought Inc (later The LTV Corporation); LTV aerospace/defence operations reorganised September 1986 as LTV Aircraft Products Group and LTV Missiles and Electronics Group; latter included Missiles Division at Grand Prairie, Texas, AM General Division at South Bend, Indiana, and Sierra Research Division at Buffalo, New York. AM General and Sierra Research subsequently sold, and LTV Aircraft Products Group redesignated Aircraft Division.

Early in 1992, LTV Corporation reached agreement with Martin Marietta Corporation and Lockheed Corporation to sell Aircraft and Missile Divisions; this followed by $214 million sale of Aircraft Division to The Carlyle Group and its 49 per cent minority partner, Northrop Corporation, and sale of the Missiles Division for $261 million to Loral Corporation, 13 August 1992; sale finalised on 31 August, at which time Aircraft Division became Vought Aircraft Company and Missiles Division became Loral Vought Systems Corporation; now no formal relationship between Vought Aircraft and Loral Vought Systems.

Current output includes about one-third of Northrop Grumman B-2 airframe by weight, aft fuselage and tail surfaces of Boeing 747, tailplane of Boeing 767, tail surfaces of Boeing 757, engine nacelles and tail sections of McDonnell Douglas C-17A and engine nacelles for Canadair Challenger and Regional Jet; teaming agreement with American Eurocopter Corporation and LHTEC for possible production of Panther 800, a T800-powered version of AS 565 Panther for US Army; Vought and FMA of Argentina teamed to promote IA 63 Pampa 2000 for USAF/USN JPATS programme.

VERIFIED

VOUGHT A-7 CORSAIR II

TYPE: Carrierborne and land-based subsonic single-seat tactical fighter.

PROGRAMME: On 11 February 1964 the US Navy named the former LTV Aerospace Corporation winner of a design competition for a single-seat light attack aircraft. The requirement was for a subsonic aircraft able to carry a greater load of non-nuclear weapons than the A-4E Skyhawk. To keep costs to a minimum and speed delivery it was stipulated by the USN that the new aircraft should be based on an existing design; the LTV design study was based therefore, on the F-8 Crusader. An initial contract to develop and build three aircraft, under the designation A-7A was made on 27 September 1965.

VERSIONS: **A-7A:** Initial attack version for the US Navy, powered by a non-afterburning Pratt & Whitney TF30-P-6 turbofan engine, rated at 50.5 kN (11,350 lb st). The first four were delivered to US Navy Air Test Center on 13-15 September 1966. Deliveries to user squadrons began on 14 October 1966. The A-7A went into combat for the first time with Squadron VA-147 in December 1967, off the USS *Ranger* in the gulf of Tonkin. Delivery of 199 A-7As to the US Navy was completed in 1968. No longer in service.

A-7B: Developed version of the US Navy with non-afterburning TF30-P-8 engine, rated at 54.3 kN (12,200 lb st). First production aircraft flew on 6 February 1968 and A-7Bs entered combat in Vietnam on 4 March 1969. Last of 196 A-7Bs was delivered to the US Navy on 7 May 1969. No longer in service.

A-7C: Designation applied in late 1971 to the first 67 A-7Es to eliminate confusion with subsequent Allison-powered A-7Es. No longer in service.

TA-7C: Sixty-five A-7Bs and A-7Cs converted into tandem two-seat trainers, with operational capability, under this designation. The first of them (154477) flew for the first time 17 December 1976. Flight refuelling capability, gun and weapon pylons retained. Configuration similar to YA-7E, but powered by the non-afterburning 54.3 kN (12,200 lb st) Pratt & Whitney TF30-P-8 engine used in early model Corsair IIs. In service with the USN (18+).

A-7D: Tactical fighter version for the USAF, with a continuous-solution navigation and weapon delivery system, including the capability of all-weather radar bomb delivery. First two powered by TF30-P-8 engine. Subsequent aircraft installed with non-afterburning Allison TF41-1 (Spey) turbofan engine. First flight of an A-7D was made 5 April 1968, and first flight with the TF41 engine 26 September 1968. First A-7D was accepted by the USAF 23 December 1968. First unit equipped with A-7D was the 54th Tactical Fighter Wing at Luke AFB, Arizona. The A-7D entered combat in Southeast Asia in October 1972 with the 354th Tactical Fighter Wing, deployed from Myrtle Beach AFB, South Carolina. Delivery of A-7Ds to Air National Guard units in Arizona, Colorado, Iowa, New Mexico, Ohio, Pennsylvania, Puerto Rico, South Carolina and South Dakota began in 1973. Production of 459 A-7Ds completed in December 1976.

In early 1976, an A-7D was test-flown with a port wing manufactured by Vought Corporation almost entirely from composite materials. The basic material consists of reinforcing graphite and boron fibres supported in an epoxy resin matrix. No longer in service.

A-7E: Developed version for the US Navy, equipped as a light attack/close air support/interdiction aircraft. All except first 67 aircraft (since redesignated A-7C) powered by Allison TF41-A-2 (Spey) non-afterburning turbofan engine, which provides 66.8 kN (15,000 lb st). First flight 25 November 1968; deliveries began 14 July 1969. The A-7E entered combat service in Southeast Asia with

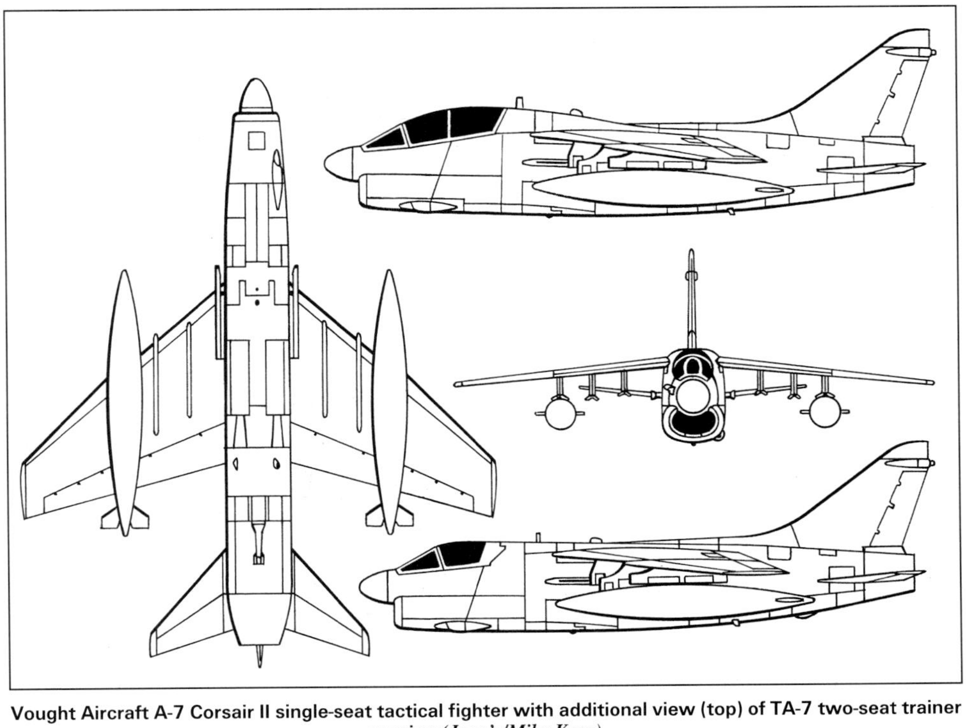

Vought Aircraft A-7 Corsair II single-seat tactical fighter with additional view (top) of TA-7 two-seat trainer version *(Jane's/Mike Keep)*

1994

Vought Aircraft A-7P Corsair II aircraft of the Portuguese Air Force

1995

Vought Aircraft A-7H Corsair II of the Hellenic Air Force ***(Vought)***

1995

Attack Squadrons 146 and 147 in May 1970, and equipped 26 Navy squadrons in 1980. Last of 596 A-7Es was delivered in March 1981. In early 1977 production began of an **A-7E FLIR** version called formerly TRAM (target recognition and attack multisensor system). This comprises a 327 kg (720 lb) pod under the starboard wing to house equipment which includes a Texas Instruments FLIR gimballed sensor, and a Marconi raster-HUD cockpit display, to provide improved night capability. Deliveries of new-production FLIR equipped A-7Es to the Navy began on 15 September 1978. By the beginning of 1981, orders covered the production of 85 FLIR pods and 144 FLIR aircraft provisions. Still in service with the Hellenic Air Force.

A-7H: Land-based version of A-7E, retaining the folding wings. First A-7H flew for the first time on 6 May 1975. Total of 60 delivered to the Hellenic Air Force. Still in service (40+).

TA-7H: Two-seat version for the Hellenic Air Force, with Allison TF41-A-400 engine. Configuration similar to TA-7C, but no in-flight refuelling capability. First TA-7H flew for the first time on 4 March 1980. Five delivered between July and September 1980. Still in service (5).

A-7K: Two-seat version of the US Air Force's A-7D, with fuselage lengthened by 0.86 m (2 ft 10 in). Withdrawn from service.

A-7P: Designation of 20 refurbished A-7As supplied to the Portuguese Air Force from 1981. Re-engined with TF30-P-408 turbofan engines and A-7E-standard avionics. Deliveries began 18 August 1981, following a first flight during the previous month. Still in service (30+).

OPERATORS: Versions of the A-7 Corsair II are in service with the armed forces of the following countries: Greece (85), Portugal (37), Thailand (awaiting delivery) and USA (9).

The following description is applicable to the A-7E but is also generally applicable to other versions of the A-7 except as detailed under the individual model listings.

DESIGN FEATURES: Cantilever high-wing monoplane. Wing section NACA 65A007. Anhedral 5°. Incidence –1°. Wing sweepback at quarter-chord 35°. Outer wing sections fold upward for carrier parking and, in the A-7H, to allow best utilisation of revetments at combat airfields. One-piece all-moving tailplane, swept back 45° at quarter-chord and set at dihedral angle of 5° 25′.

FLYING CONTROLS: Plain sealed inset aluminium ailerons, outboard of wing fold, are actuated by fully triplicated hydraulic system. Leading-edge flaps. Large single-slotted trailing-edge flaps. Spoiler above each wing forward of flaps. Tailplane is operated by triplicated hydraulic systems, and the rudder powered by two systems.

STRUCTURE: All-metal multispar structure with integrally stiffened aluminium alloy upper and lower skins. The fuselage is an all-metal semi-monocoque structure. Large door-type ventral speed-brake under centre-fuselage. The tail unit consists of a large vertical fin and rudder, swept back 44.28° at quarter-chord.

LANDING GEAR: Hydraulically retractable tricycle type, with single wheel on each main unit and twin-wheel nose unit. Mainwheels retract forward into fuselage, nosewheels aft. Mainwheels and tyres size 28 × 9-12; nosewheels and tyres size 22 × 5.50. Nose gear launch bar for carrier catapulting. Sting-type arrester hook under rear fuselage for carrier landings, emergency landings or aborted take-offs. Anti-skid brake system.

POWER PLANT: One Allison TF41-A-2 (Rolls-Royce Spey) non-afterburning turbofan engine, rated at 66.7 kN (15,000 lb st). The A-7E has a pneumatic starter requiring ground air supply; A-7H, TA-7H and A-7K engines have self-start capability through the medium of battery-powered electric motor that actuates a small gas-turbine engine (jet fuel starter), which in turn, starts the main engine through the gearbox. The engine has self-contained ignition for start/airstart, automatic relight and selective ignition. Integral fuel in tanks in wings and additional fuselage tanks. Maximum internal fuel 5,678 litres (1,500 US gallons; 1,249 Imp gallons). Maximum external fuel 4,542 litres (1,200 US gallons; 999 Imp gallons). The A-7E and A-7H have the fuselage sump tank filled with polyurethane fire-suppressing foam. Some fuselage tanks and fuel lines self-selaing. Flight refuelling capability of A-7E provided by a probe and drogue system; A-7K has boom receptacle above fuselage on port side in line with wing leading-edge. The A-7H and TA-7H do not have an air refuelling capability. Boron carbide (HFC) engine armour.

ACCOMMODATION: Pilot on McDonnell Douglas Escapac rocket-powered ejection system. Complete with US Navy life support system on the A-7E/H. Escape system provides a fully inflated parachute three seconds after sequence initiation; positive seat/man separation and stabilisation of the ejected seat and pilot. Boron carbide (HFC) cockpit armour.

SYSTEMS: Triple-redundant hydraulic system for flight controls double-redundant system for flaps, brakes and landing gear retraction. Liquid oxygen system. An air conditioning unit using air provides pressurisation and cooling for the cockpit and cooling for certain avionics components. Automatic flight control system provides control-stick, altitude hold, heading hold, heading preselect and attitude hold, which is coupled for automatic carrier landings. Ram-air turbine provides emergency hydraulic pressure and electrical power down to airspeeds below those used in normal landing approaches.

AVIONICS AND EQUIPMENT: The navigation/weapon delivery system is the heart of the A-7E/H light attack aircraft. It performs continuously the computations needed for greatly increased delivery accuracy, and for manoeuvring freedom during navigation to a target and the attack, weapon release, pull up, and safe return phases of the mission. The system not only provides the pilot with a number of options during the navigation and weapon delivery, but also relieves him of much of his workload. The AN/ASN-91(V) navigation/weapon delivery computer is the primary element of the system, in constant 'conversation' with basic electronic sensors, and computes and displays continuously present position, using computed position and stored data to calculate navigation and weapon delivery solutions, and monitors the reliability of data inputs and outputs. An AN/ASN-90(V) inertial measurement set is the basic three-axis reference system for navigation and weapon delivery. AN/APN-190(V) Doppler measures groundspeed and drift angle. AN/APQ-126(V) forward-looking radar provides pilot with 10 modes of operation; air-to-ground ranging; terrain-following; terrain-avoidance; ground-mapping, shaped beam; ground-mapping, pencil beam; beacon cross-scan terrain-avoidance; cross-scan ground-mapping, pencil; TV; and Shrike integrated display system. AN/AVQ-7(V) HUD receives and displays computed attack, navigation and landing data from the tactical computer; aircraft performance data from flight sensors; and discrete signals from various aircraft systems. CP-953A/AJQ air data computer is a solid-state servo-mechanical analogue computer which measures and computes continuously required altitude and airspeed information. The armament station control unit integrates anc controls the weapon release system; it supplies electrical signals to arm and release or jettison external stores; controls and fires the Vulcan cannon; furnishes store-type information to the tactical computer; supplies weapon status information to the pilot; determines weapon release according to priority of stations; and determines compatibility of selected release mode with the stores on selected stations. Standard aeronautical charts reproduced on 35 mm film in full colour are stored in an AN/ASN-99 projected map display set which, as a subsystem of the tactical computer, provides a continuous display of the aircraft's geographical position. Other avionics include AN/ASN-54 approach power compensator; AN/ASW-30 AFCS; ARA-63 ACLS; dual AN/ARC-159 UHF com; AN/ARN-84 Tacan; AN/APX-72 IFF transponder; AN/APN-154 radar beacon; AN/ASW-25 datalink; AN/ARA-50 ADF; and AN/AIC-25 audio system. ECM equipment includes ALR-45/50 internal homing and warning systems; ALQ-126 active ECM; chaff/flare dispensers; and external pod-mounted systems compatible with the aircraft's internal systems.

ARMAMENT: A wide range of stores, to a total weight of more than 6,805 kg (15,000 lb), can be carried on six underwing pylons and two fuselage weapon stations, the latter suitable for Sidewinder air-to-air missiles. Two outboard pylons on each wing can accommodate a load of 1,587 kg (3,500 lb).

Vought Aircraft F-8E (FN) Crusader of the French Navy *(Peter J. Cooper)*

1994

Inboard pylon on each wing can carry 1,134 kg (2,500 lb). Two fuselage weapon stations, one in each side, can each carry a load of 227 kg (500 lb). Weapons include air-to-air and air-to-ground (anti-tank and anti-radar missiles); electro-optical (TV) and laser-guided weapons; general purpose bombs; bomblet dispensers; rockets; gun pods; and auxiliary fuel tanks. In addition, an M61A-1 Vulcan 20 mm cannon is mounted in the port side of the fuselage. This has a 1,000-round ammunition storage and selected firing rates of 4,000 or 6,000 rds/min. Strike camera in lower rear fuselage for damage assessment.

DIMENSIONS, EXTERNAL:

Wing span	11.80 m (38 ft 9 in)
Width, wings folded	7.24 m (23 ft 9 in)
Wing chord: at root	4.72 m (15 ft 6 in)
at tip	1.18 m (3 ft 10¼ in)
Wing aspect ratio	4
Length overall	14.06 m (46 ft 1½ in)
Height overall	4.90 m (16 ft 0¾ in)
Tailplane span	5.52 m (18 ft 1½ in)
Wheel track	2.90 m (9 ft 6 in)

AREAS:

Wings, gross	34.83 m² (375 sq ft)
Ailerons (total)	1.85 m² (19.94 sq ft)
Trailing-edge flaps (total)	4.04 m² (43.48 sq ft)
Leading-edge flaps (total)	3.46 m² (37.24 sq ft)
Spoilers (total)	0.43 m² (4.60 sq ft)
Deflector	0.32 m² (3.44 sq ft)
Fin	10.70 m² (115.20 sq ft)
Rudder	1.40 m² (15.04 sq ft)
Tailplane	5.22 m² (56.19 sq ft)
Speed-brake	2.32 m² (25.00 sq ft)

WEIGHTS AND LOADINGS:

Weight empty	8,668 kg (19,111 lb)
Max T-O weight	19,050 kg (42,000 lb)

PERFORMANCE:

Max level speed:
at S/L 600 kts (1,112 km/h; 691 mph)
at 1,525 m (5,000 ft): with 12 Mk 82 bombs 562 kts (1,040 km/h; 646 mph)
after dropping bombs 595 kts (1,102 km/h; 685 mph)

Sustained manoeuvring performance at 1,525 m (5,000 ft), at AUW of 13,047 kg (28,765 lb) with six pylons and two Sidewinder missiles 1,770 m (5,800 ft) turning radius at 4*g* and 500 kts (925 km/h; 575 mph)

T-O run at max T-O weight 1,830 m (6,000 ft)

Ferry range:
max internal fuel 1,981 n miles (3,671 km; 2,281 miles)
max internal and external fuel 2,485 n miles (4,604 km; 2,861 miles)

UPDATED

VOUGHT F-8 CRUSADER

French Navy designation: F-8E (FN)

TYPE: Supersonic single-seat carrier fighter.

PROGRAMME: Chance Vought (now Vought Aircraft Company) was given a development contract for the F-8 in May 1953 after winning a design competition in which eight airframe manufacturers had participated. The prototype XF-8A Crusader flew for the first time 25 March 1955, exceeding the speed of sound in level flight. The first production F-8A flew 20 September 1955, and this version began reaching US Naval operational squadrons in March 1957.

On 21 August 1956 an F-8A set up the first US national speed record of over 864 knots (1,600 km/h; 1,000 mph). Operating under restrictions, it recorded a speed of 882 knots (1,634.17 km/h; 1,015.428 mph). On 16 July 1057 an RF-8A photo-reconnaissance version of the Crusader set up the first supersonic US transcontinental record by flying the 2,125 n miles (3,936 km; 2,445.9 miles) from Los Angeles to New York in 3 hours 22 minutes 50 seconds, at an average speed of 628 knots (1,164.39 km/h; 723.52 mph).

An outstanding feature of the F-8 is its two-position variable incidence wing. This provides a high angle of attack for take-off and landing, while permitting the fuselage to remain almost parallel to a flight deck or runway for good pilot visibility.

VERSIONS: **F-8E (FN):** Version of F-8E to equip two French Navy squadrons for service on carriers *Clemenceau* and *Foch*. Main modifications are the incorporation of a boundary layer control system to provide 'blowing' of the flaps and ailerons, the introduction of two-stage leading-edge flaps to reduce landing speed on the comparatively small ships, and provision for carrying Matra R.530 air-to-air missiles in addition to Sidewinders. Pratt & Whitney J57-P-20A turbojet engine. Contract for 42 placed in August 1963, through US Navy. Prototype, a converted F-8D, flew 27 February 1964, and was lost in an accident 11 April 1964. First production F-8E (FN) flew 26 June 1964. Last one was delivered in January 1965, bringing to an end the production of new Crusaders.

UPGRADES: **Dassault:** F-8 life extension programme. See separate entry.

LTV Aerospace Corporation (Vought): On 24 March 1969, LTV Aerospace Corporation announced a $1.7 million contract from the French Navy to manufacture new wings and outer panels for an initial batch of 10 F-8E (FN) Crusaders. The new wings, designed for 4,000 flight hours, were supplied minus droops, flaps, ailerons and actuators. They have titanium wing fold ribs and a thicker skin than standard.

OPERATORS: The F-8 Crusader is in service with the French Navy (19).

DESIGN FEATURES: Cantilever high-wing monoplane. Wings have thin laminar-flow section. Anhedral 5°. Sweepback 35°. Wing is adjustable to two incidence positions by a hydraulic self-locking actuator. Outer wing sections fold upward for carrier stowage. The tail unit consists of a large, swept vertical fin and rudder and one-piece horizontal 'slab' tail.

FLYING CONTROLS: Inset aluminium ailerons inboard of wing fold are actuated by fully duplicated hydraulic system and function also as flaps. Small magnesium alloy trailing-edge flaps. When wing is raised the ailerons, flaps and dogtooth leading-edge are all drooped automatically. Ventral speed brake under centre-fuselage. Tailplane and rudder are actuated by fully duplicated hydraulic system.

STRUCTURE: The wing is an all-metal multispar structure with integrally stiffened aluminium alloy upper and lower skins. The fuselage is an all-metal structure in three main assemblies. Both magnesium alloy and titanium are used in the structure, the aft section and portion of the mid-section are titanium.

LANDING GEAR: Hydraulically retractable tricycle type. Mainwheels retract forward into fuselage, nosewheel aft. Sting-type arrester hook under rear fuselage.

POWER PLANT: One Pratt & Whitney J57 turbojet engine with afterburner. Integral fuel tanks in wings inboard of wing fold. Other tankage in fuselage. Total internal fuel capacity approximately 5,300 litres (1,398 US gallons; 1,165 Imp gallons). Provision for in-flight refuelling, with retractable probe housed in removable pack on port side of fuselage of F-8A and inside flush panel on RF-8A.

ACCOMMODATION: Pilot on Martin-Baker Mk F7 lightweight ejection seat in pressurised cockpit. Liquid oxygen equipment.

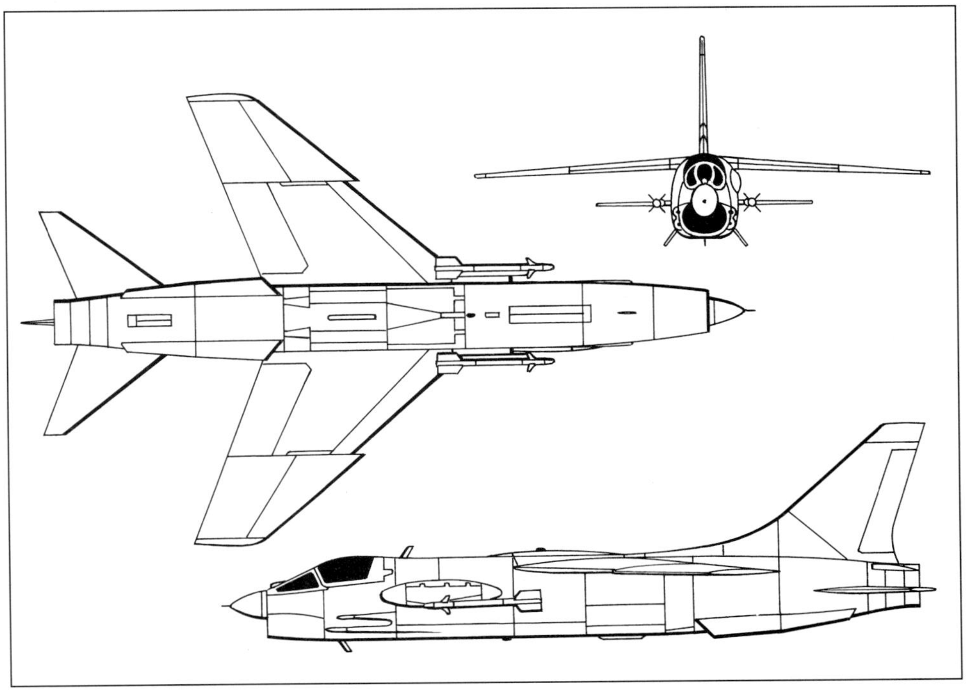

Vought Aircraft F-8 Crusader supersonic single-seat carrier fighter *(Jane's/Mike Keep)*

1994

ARMAMENT: Four 20 mm Colt cannon in fuselage nose, with 84 rounds per gun (average) for F-8C/K, F-8H and F-8E/J, and 144 rounds per gun for NTF-8A and F-8F/L. Two Sidewinder missiles (four in F-8C/K, F-8D/H amd F-8E/J) mounted externally on sides of fuselage.

DIMENSIONS, EXTERNAL:

Wing span	10.87 m (35 ft 8 in)
Length overall: except F-8E/J	16.54 m (54 ft 3 in)
F-8E/J	16.61 m (54 ft 6 in)
Height overall	4.80 m (15 ft 9 in)
Width folded	6.86 m (22 ft 6 in)
Tailplane span:	
except F-8E (FN), F-8J	5.54 m (18 ft 2 in)
F-8E (FN), F-8J	5.88 m (19 ft 3½ in)
Wheel track	2.94 m (9 ft 8 in)

AREAS:

Wings, gross	34.84 m² (375 sq ft)

WEIGHTS AND LOADINGS:

Normal T-O weight: F-8C	12,500 kg (27,550 lb)
Max T-O weight: F-8E/J	15,420 kg (34,000 lb)

PERFORMANCE:

Max level speed:	
F-8A, F-8C	over 868 kts (1,600 km/h; 1,000 mph)
F-8D, F-8E	nearly Mach 2
Combat radius: F-8A	521 n miles (965 km; 600 miles)

UPDATED

WIPAIRE

WIPAIRE INC

8520 River Road, Inver Grove Heights, Minnesota 55076
Telephone: 1 (612) 451 1205
Fax: 1 (612) 451 1786
CHAIRMAN: Robert Wiplinger
PRESIDENT: Robert Nelson
MANAGER, SALES AND MARKETING: Jim Jacobsen

UPDATED

WIPAIRE (DE HAVILLAND) BEAVER CONVERSIONS

TYPE: Light aircraft structural modification.

PROGRAMME: Wipaire, well known for Wipline floats, offers conversion of customers' DHC-2 Beavers and produces Super Beaver rebuilds from surplus military DHC-2 or L-20 airframes.

Modifications to customers' aircraft, applied together or individually, include rearward extension of cabin by 0.71 m (2 ft 4 in), fitting a 0.85 × 0.27 m (33½ × 10½ in) baggage door, two extra Panaview windows each side, tinted forward skylight windows, articulating Cessna or Piper pilot's and co-pilot's seats with inertia reel shoulder harness, forward- or rearward-facing three-seat centre bench with underseat stowage, 3M cabin soundproofing, and hold-open door catches. Wipaire also offers electrically actuated flaps, customised IFR instrument panel, Digiflow fuel metering system, 3M Stormscope, S-Tech Series 50 autopilot with electric trim, altitude hold and flight director, Jasco alternator replacing generator, new battery location, Cessna electric fuel pump/primer, and Hartzell three-blade constant-speed propeller.

Super Beaver represents ex-military airframe completely dismantled, inspected and refurbished; new or overhauled 335.5 kW (450 hp) P&W R-985 engine and accessories; new oil, fuel tanks and instruments; also custom interiors and external paint; floats available. Modification also applicable to DHC Turbo Beavers.

Wipaire (de Havilland) DHC-2 Super Beaver *(Wipaire)*

1995

VERIFIED

YUGOSLAVIA

New state of Yugoslavia proclaimed 27 April 1992, following secession of Bosnia-Hercegovina (which see for details of Soko company; but Soko aircraft described under Utva entry in this section).

UTVA

UTVA—SOUR METALNE INDUSTRIJE, RO FABRIKA AVIONA

Jabučki Put B.B., 26000 Pančevo
Telephone: 381 (13) 512584/515383
Fax: 381 (13) 519859
Telex: 13250
GENERAL MANAGER: Ivan Dundjerski
MANAGER, MARKETING: Borisav Grabovac

Utva (Sheldrake) Aircraft Industry formed at Zemun, 5 June 1937; to Pančevo 1939; Second World War production of Trojka, 212, 213, Aero 3, Utva 56, 60, 65P and 66; currently 85,500 m² (920,320 sq ft) of covered floor area. Subcontract work previously included flaps and weapon pylons for Soko/Avioane Orao/IAR-93; centre and rear fuselage section, surfaces and gun pod for Super Galeb; components for FLS Aerospace Optica; Boeing 737 rib assemblies; Boeing 747 assemblies; Boeing 757 wingtips and floor supports; tools for Israel Aircraft Industries. McDonnell Douglas and CIS aircraft industry; external trade prohibited by UN embargo 1992.

Following termination of aircraft production at the Soko (Falcon) works in newly proclaimed independent Bosnia-Hercegovina, on 1 May 1992, some machinery and jigs from G-4 Super Galeb and Gazelle production lines removed to 'new' Yugoslavia; despite UN trade embargoes, Utva reportedly has assumed development responsibilites for G-4M version of Super Galeb, G-5 derivative and Soko/Avioane Orao (see International section) as well as returning to production the Utva 75 lightplane; Utva also involved with aircraft maintenance and munitions production during civil war.

UPDATED

Utva 66 four-seat utility aircraft in the colours of the former Federal Yugoslav Air Force (JRV)

1994

UTVA 66

TYPE: Four-seat general utility monoplane.

PROGRAMME: When the Utva 66 was announced in early 1968, several prototypes had been under flight test for more than one year. Structure testing of the airframe was completed in mid-1968. In service with the Yugoslav Air Force prior to the break-up in late 1991 (40+).

VERSIONS: **Utva 66:** Basic four-seat utility version. Can be used for glider towing.

Utva 66AM: Ambulance version, able to accommodate two stretchers, which are loaded through upward-hinged rear cabin canopy. Seat for attendant behind the pilot, with drawer for medical equipment.

Utva 66H: Water-based version on BIN-1600 floats designed by Dipl Ing Nikolic. Standard fuel capacity 450 litres (118.9 US gallons; 99 Imp gallons). First flown in September 1968.

OPERATORS: Versions of the Utva 66 are in service with the armed forces of Yugoslavia (45).

DESIGN FEATURES: Strut-braced high-wing monoplane, with single streamline-section light alloy bracing strut each side. Modified NACA 4412 wing section. Dihedral 2° 30′.

FLYING CONTROLS: Flaps and ailerons of pure monocoque construction. Flaps are hydraulically operated and are linked to ailerons, so that ailerons are dropped 15° when flaps are 40° down. Fixed leading-edge slots of same span as ailerons. Servo tab and controllable trim tab in elevator.

STRUCTURE: All-metal single-spar structure. The fuselage is an all-metal construction. Nose and cabin of stressed skin construction. Rear fuselage is a semi-monocoque structure. The tail unit is a cantilever all-metal structure.

LANDING GEAR: Non-retractable tailwheel type. Cantilever steel tube main legs with rubber-in-compression shock-absorption. Wheels and hydraulic disc brakes manufactured by Prva Petoletka. Mainwheels size 500 × 180. Tailwheel size 260 × 85. Tyre pressures: main 2.44 bars (35.5 lb/sq in), tail 1.96 bars (28.5 lb/sq in). Provision for alternative fitment of floats.

POWER PLANT: One 202 kW (270 hp) Lycoming GSO-480-B1J6 six-cylinder horizontally opposed air-cooled engine, driving a Hartzell two-blade constant-speed metal propeller type HC-B3Z20-1/10151 C-5. Two metal fuel tanks in wings, total capacity 250 litres (66 US gallons; 55 Imp gallons). Two optional (standard in 66H) integral fuel tanks, total capacity 200 litres (52.8 US gallons; 44 Imp gallons). Maximum available fuel capacity 450 litres (118.9 US gallons; 99 Imp gallons). Oil capacity 12 litres (3.1 US gallons; 2.6 Imp gallons).

ACCOMMODATION: Enclosed cabin for pilot and three passengers, or pilot, two stretchers and one attendant in ambulance version. Rear canopy hinges upward to form third door, for loading stretchers or freight. Ventilation and heating standard.

ELECTRONICS AND EQUIPMENT: Standard equipment includes blind-flying instrumentation, Narco Mk 5 two-way radio and Narco ADF-30 radio compass. Other navigation, communications and automatic flight control equipment optional. Ambulance version can carry parachute pack of food and medical supplies under fuselage.

DIMENSIONS, EXTERNAL:

Wing span	11.40 m (37 ft 5 in)
Wing chord: at root	1.73 m (5 ft 8 in)
at tip	1.21 m (3 ft 11½ in)
Wing aspect ratio	7.19
Length overall	8.38 m (27 ft 6 in)
Height overall	3.20 m (10 ft 6 in)
Tailplane span	4.08 m (13 ft 4½ in)

Wheel track 2.55 m (8 ft 4½ in)
Cabin doors (each): Height 0.98 m (3 ft 2½ in)
Width 0.93 m (3 ft 0½ in)
Canopy door: Height 0.60 m (1 ft 11½ in)
Width 1.00 m (3 ft 3½ in)

DIMENSIONS, INTERNAL:
Cabin: Length 1.50 m (4 ft 11 in)
Width 1.05 m (3 ft 5 in)
Height 1.20 m (3 ft 11 in)

AREAS:
Wings, gross 18.08 m² (194.50 sq ft)
Ailerons (total) 2.06 m² (22.18 sq ft)
Trailing-edge flaps (total) 2.32 m² (24.97 sq ft)
Fin 1.18 m² (12.70 sq ft)
Dorsal fin 0.44 m² (4.74 sq ft)
Rudder 0.95 m² (10.22 sq ft)
Tailplane 2.42 m² (26.00 sq ft)
Elevators 2.22 m² (23.90 sq ft)

WEIGHTS AND LOADINGS:
Weight empty, equipped: 66 1,250 kg (2,756 lb)
66H 1,464 kg (3,228 lb)
Max T-O weight: 66 1,814 kg (4,000 lb)
66H 2,010 kg (4,431 lb)
Max wing loading: 66 100.5 kg/m² (20.58 lb/sq ft)
Max power loading: 66 5.35 kg/hp (11.79 lb/hp)

PERFORMANCE (at max T-O weight):
Max never-exceed speed 172 kts (320 km/h; 198 mph)
Max level speed:
at S/L: 66 124 kts (230 km/h; 143 mph)
66H 112 kts (208 km/h; 129 mph)
at optimum height:
66 135 kts (250 km/h; 155 mph)
66H 124 kts (230 km/h; 143 mph)
Max cruising speed at optimum height:
66 124 kts (230 km/h; 143 mph)
66H 113 kts (210 km/h; 130 mph)
Stalling speed: 66 43.5 kts (80 km/h; 50 mph)
Max rate of climb at S/L: 66 270 m (885 ft)/min
66H 156 m (512 ft)/min
Service ceiling: 66 6,700 m (22,000 ft)
T-O run: 66 187 m (614 ft)
T-O to 15 m (50 ft): 66 352 m (1,155 ft)
Landing from 15 m (50 ft): 66 274 m (899 ft)
Landing run: 66 181 m (594 ft)
Range with standard fuel:
66 404 n miles (750 km; 466 miles)
66H 593 n miles (1,100 km; 683 miles)

VERIFIED

UTVA 75A

TYPE: Four-seat light aircraft.

PROGRAMME: The **Utva 75** (originally 75A21) side by side two-seat training, glider towing and utility lightplane was projected, designed and built in partnership by Utva-Pančevo, Prva Petoletka-Trstenik, Vazduhoplovno Tehnicki Institut and Institut Masinskog Fakulteta of Belgrade. Design was started in 1974, to the requirements of FAR Pt 23 (Utility category). Construction of two prototypes was undertaken in 1975; the first of these flew for the first time 19 May 1976 and the second 18 December 1976. Over 150 were built. Latest version (flown 1986) is the four-seat **Utva 75A** (previously 75A41), which has larger cabin doors, eliminating the rear quarter-lights. This variant has not yet entered production.

OPERATORS: The Utva 75A is in service with the armed forces of Yugoslavia (70).

DESIGN FEATURES: Cantilever low-wing monoplane, with short span centre-section and two constant chord outer panels. Wing section NACA 65^2-415. Dihedral 0° on centre-section, 6° on outer panels.

FLYING CONTROLS: Ailerons and flaps, with fluted skin, along entire trailing-edge of outer panels, except for tips. Flettner trim tab on each aileron. Rudder and elevator horn balanced. Controllable tab on elevator; ground adjustable tab on rudder.

Utva 75A four-seat tourer/trainer

1994

STRUCTURE: The wing is a conventional all-metal two-spar structure. The fuselage is a conventional all-metal semi-monocoque structure. The tail unit is a cantilever all-metal structure, with sweptback vertical surfaces. Fluted skin on fin, rudder and elevator.

LANDING GEAR: Non-retractable tricycle type, with single wheel on each unit, and small tail bumper. Prva Petoletka-Trstenik oleo-pneumatic shock-absorbers. Dunlop tyres, size 6.00-6, pressure 2.2 bars (32 lb/sq in) on mainwheels; size 5.00-5, pressure 2.0 bars (29 lb/sq in) on nosewheel. Prva Petoletka-Trstenik hydraulic brakes.

POWER PLANT: One 134 kW (180 hp) Textron Lycoming IO-360-B1F flat-four engine, driving a Hartzell HC-C2YK-1BF/F7666A two-blade variable-pitch metal propeller. Two integral fuel tanks in wings, total capacity 150 litres (39.6 US gallons; 33 Imp gallons). Provision for carrying two 100 litre (26.4 US gallon; 22 Imp gallon) drop tanks under wings, raising maximum total capacity to 350 litres (92 US gallons; 77 Imp gallons). Oil capacity 10 litres (2.6 US gallons; 2.2 Imp gallons).

ACCOMMODATION: Four seats in enclosed cabin, with large upward-opening jettisonable canopy door over and to the rear of each seat, hinged on centreline. Dual stick type controls standard. Cabin heated and ventilated.

SYSTEMS: Dual hydraulic systems for brakes. 14 V DC electrical system, with 35 Ah battery, navigation lights, rotating beacon and landing lights as standard equipment.

AVIONICS: Bendix/King equipment standard, including dual KY197 720-channel VHF com transceivers; KR 87 digitally tuned ADF with integral electronic flight timer and push-button elapsed timer; panel-mounted R/Nav system comprising a KNS 81 200-channel nav, 40-channel glide slope indicator and nine-waypoint digital R/Nav computer, combined with a KI 525A pictorial nav indicator; KI 229 RMI; KN 53 200-channel VHF nav with integral 40-channel glide slope indicator and KI 525A indicator; KN 62A 200-channel DME with digital distance, ground speed and time-to-station; KT 79 all-solid-state digital transponder featuring cross-check readout of encoded altitude and automatic VFR code selection; and KMA 24 audio control console with integral marker beacon receiver.

ARMAMENT: Two fittings for light weapon loads underwing on military Utva 75s. Each can carry a bomb, 100 kg (220 lb) cargo container, two-round rocket launcher or machine gun pod.

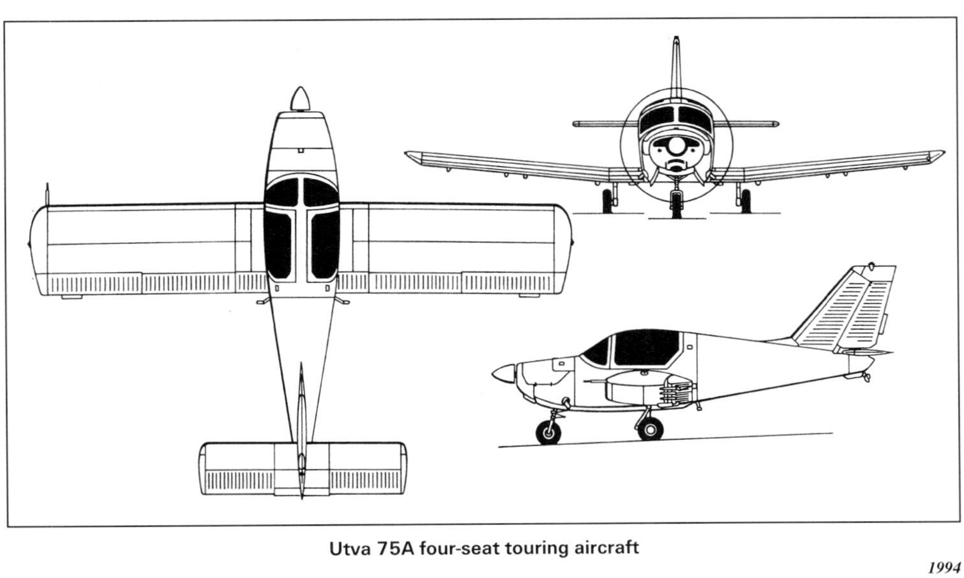

Utva 75A four-seat touring aircraft

1994

DIMENSIONS, EXTERNAL:
Wing span 9.73 m (31 ft 11 in)
Wing chord, constant 1.55 m (5 ft 1 in)
Wing aspect ratio 6.5
Length overall 7.11 m (23 ft 4 in)
Height overall 3.15 m (10 ft 4 in)
Tailplane span 3.80 m (12 ft 5½ in)
Wheel track 2.58 m (8 ft 5½ in)
Wheelbase 1.99 m (6 ft 6¼ in)
Propeller diameter 1.93 m (6 ft 4 in)
Propeller ground clearance 0.295 m (11¾ in)

AREAS:
Wings, gross 14.63 m² (157.5 sq ft)
Ailerons (total) 1.38 m² (14.85 sq ft)
Flaps (total) 1.61 m² (17.33 sq ft)
Vertical tail surfaces (total) 1.78 m² (19.16 sq ft)
Horizontal tail surfaces (total) 3.34 m² (35.95 sq ft)

WEIGHTS AND LOADINGS:
Weight empty, equipped 720 kg (1,587 lb)
Luggage provision 20 kg (44 lb)
Max fuel: standard 108 kg (238 lb)
with drop tanks 251 kg (554 lb)
Max T-O weight 1,140 kg (2,513 lb)
Max wing loading 77.92 kg/m² (15.96 sq ft)
Max power loading 8.50 kg/kW (13.96 lb/hp)

PERFORMANCE (at max T-O weight):
Max level speed 103 kts (190 km/h; 118 mph)
Max cruising speed 94 kts (175 km/h; 109 mph)
Stalling speed 54 kts (100 km/h; 62 mph)
Max rate of climb at S/L 240 m (787 ft)/min
Service ceiling 4,000 m (13,125 ft)
T-O run 200 m (656 ft)
T-O to 15 m (50 ft) 400 m (1,312 ft)
Landing from 15 m (50 ft) 340 m (1,115 ft)
Landing run 180 m (591 ft)
Range:
with max standard fuel
324 n miles (600 km; 373 miles)
with drop tanks, no reserves
755 n miles (1,400 km; 870 miles)

VERIFIED

(SOKO) G-2A GALEB (SEAGULL)

TYPE: Two-seat armed jet basic trainer.

PROGRAMME: Design of the Galeb was started in 1957. Construction of two prototypes began in 1959, and the first of these flew for the first time in May 1961. Production began in 1963 and continued until the early 1980s. First overseas operator was the Zambian Air Force in early 1971.

VERSIONS: **G-2A**: Standard version for Yugoslav Air Force. Progressive design improvements included optional cockpit air conditioning system.

G-2AE: Export version, with updated equipment. First flown in 1974. Series production began in 1975 for the Libyan Air Force.

OPERATORS: Versions of the G-2 Galeb are in service with the armed forces of the following countries: Bosnia-Herzegovina (7), Libya (80), Yugoslavia (200) and Zambia (12).

DESIGN FEATURES: Cantilever low-wing monoplane. Wing section NACA 64A213.5 at root, NACA 64A212.0 at tip. Dihedral 1° 30′. No incidence. Sweepback at quarter-chord. 4° 19′.

FLYING CONTROLS: Manually operated light alloy ailerons. Trim tab on port aileron. Hydraulically actuated Fowler flaps. Two hydraulically actuated door-type airbrakes under centre-fuselage. Rudder and elevators statically and dynamically balanced and manually operated trim tab in each elevator.

STRUCTURE: Conventional light alloy two-spar stressed skin structure, consisting of a centre-section, integral with the fuselage, and two outer panels which can be removed easily. The fuselage is a light alloy semi-monocoque structure in two portions, joined together by four bolts at frame aft of wing trailing-edge. Rear portion removable for engine ser-

Soko G-2A Galeb in the colours of the former Federal Yugoslav Air Force (JRV)

1994

vicing. The tail unit is a cantilever light alloy stressed skin structure. VHF radio aerial forms tip of fin.

LANDING GEAR: Hydraulically retractable tricycle type, with single wheel on each unit. Nosewheel retracts forward, main units inward into wings. Oleo-pneumatic shock-absorbers manufactured by Prva Petoletka of Trstenick. Dunlop mainwheels and tyres size 23 × 7.25-10, pressure 4.41 bars (64 lb/sq in). Dunlop nosewheel and tyre size 6.50-5.5 TC, pressure 3.43 bars (49.8 lb/sq in). Prva Petoletka hydraulic differential disc brakes, toe-operated from both cockpits.

POWER PLANT: One Rolls-Royce Viper Mk 22-6 turbojet engine rated at 11.12 kN (2,500 lb st). Two flexible fuel tanks aft of cockpits with total capacity of 780 kg (1,720 lb). Two jettisonable wingtip tanks, each with capacity of 170 kg (375 lb). Refuelling point on upper part of fuselage aft of cockpits. Fuel system designed to permit up to 15 seconds inverted flight. Oil capacity 6.25 litres (1.68 US gallons; 1.4 Imp gallons).

ACCOMMODATION: Crew of two in tandem on BAe (Folland) Type 1-B fully automatic lightweight ejection seats. Separate sideways-hinged (to starboard) jettisonable canopy over each cockpit. Cockpit air conditioning to special order only.

SYSTEMS: Hydraulic system, pressure 58.5 to 69 bars (850 to 1,000 lb/sq in), for landing gear, airbrakes and flaps. Separate system for wheel brakes. Pneumatic system for armament cocking. Electrical system includes 6 kW 24 V generator, 24 V battery, and inverter to provide 115 V 400 Hz AC supply for instruments. G-2A has low-pressure oxygen system, capacity 1,450 litres (383 US gallons; 319 Imp gallons). G-2AE has high-pressure oxygen system.

AVIONICS AND EQUIPMENT (G-2A): Blind-flying instrumentation. Marconi radio compass (licence-built by Rudi Čajavec), intercom and STR-9Z1 VHF radio transceiver standard. Standard electrical equipment includes navigation lights. 250 W landing light in nose, and 50 W taxying light on nose landing gear. Camera, with focal length of 178 mm (7 in) and 125-exposure magazine, can be fitted in fuselage, under rear cockpit floor. Flares can be carried on the underwing bomb racks for night photography. Target towing hook under centre-fuselage.

AVIONICS AND EQUIPMENT (G-2AE): Full IFR instrumentation. Electronique Aerospatiale (EAS) Type TVU-740 VHF/UHF com rad transceiver. Marconi AD 370B radio compass, EAS RNA-720 VOR/LOC and ILS, Iskra 75R4 marker beacon receiver, and intercom. Vinten Type 360/140A camera with 3 in automatic exposure-control lens. Otherwise as G-2A.

ARMAMENT: All production aircraft have two 0.50 in machine guns in nose (with 80 rds/gun); and underwing pylons for two 50 or 100 kg bombs and four 57 mm rockets; or clusters of small bombs and expendable bomblet containers of up to 150 kg (330 lb) weight (300 kg; 660 lb total).

DIMENSIONS, EXTERNAL:

Wing span	10.47 m (34 ft 4½ in)
Wing span over tip tanks	11.62 m (38 ft 1½ in)
Wing chord: at root	2.36 m (7 ft 9 in)
at tip	1.40 m (4 ft 7 in)
Wing aspect ratio	5.55
Length overall	10.34 m (33 ft 11 in)
Height overall	3.28 m (10 ft 9 in)
Tailplane span	4.27 m (14 ft 0 in)
Wheel track	3.89 m (12 ft 9 in)
Wheelbase	3.59 m (11 ft 9½ in)

AREAS:

Wings, gross	19.43 m² (209.14 sq ft)
Ailerons (total)	2.36 m² (25.40 sq ft)
Trailing-edge flaps (total)	2.02 m² (21.75 sq ft)
Airbrakes (total)	0.34 m² (3.66 sq ft)
Fin	1.34 m² (14.42 sq ft)
Rudder, incl tab	0.56 m² (6.03 sq ft)
Tailplane	3.66 m² (39.40 sq ft)
Elevators, incl tabs	0.83 m² (8.93 sq ft)

WEIGHTS AND LOADINGS:

Weight empty, equipped	2,620 kg (5,775 lb)
Max T-O weight:	
fully aerobatic trainer 'clean'	3,374 kg (7,438 lb)
basic trainer (no tip tanks)	3,488 kg (7,690 lb)
navigational trainer (with tip tanks)	3,828 kg (8,439 lb)
weapons trainer	3,988 kg (8,792 lb)
strike version	4,300 kg (9,480 lb)

PERFORMANCE (at normal T-O weight):

Max level speed: at S/L	408 kts (756 km/h; 470 mph)
at 6,200 m (20,350 ft)	438 kts (812 km/h; 505 mph)
Max cruising speed at 6,000 m (19,685 ft)	394 kts (730 km/h; 453 mph)
Stalling speed:	
flaps and airbrakes down	85 kts (158 km/h; 98 mph)
flaps and airbrakes up	97 kts (180 km/h; 112 mph)
Max rate of climb at S/L	1,370 m (4,500 ft)/min
Time: to 3,000 m (9,840 ft)	2 min 24 s
to 6,000 m (19,685 ft)	5 min 30 s
to 9,000 m (29,520 ft)	10 min 12 s
Service ceiling	12,000 m (39,375 ft)
T-O run on grass	490 m (1,610 ft)
T-O to 15 m (50 ft)	640 m (2,100 ft)
Landing from 15 m (50 ft)	710 m (2,330 ft)
Landing run on grass	400 m (1,310 ft)
Max range at 9,000 m (29,520 ft), with tip tanks full	669 n miles (1,240 km; 770 miles)
Max endurance at 7,000 m (23,000 ft)	2 h 30 min
Design *g* limits	+8/−4

VERIFIED

(SOKO) G-4 SUPER GALEB (SEAGULL)

TYPE: Two-seat basic/advanced trainer and ground attack.

PROGRAMME: Replacement for G-2 Galeb and Lockheed T-33; designed by VTI; programme launched October

Soko G-2A Galeb with a Rolls-Royce Viper Mk 22-6 turbojet engine

1992

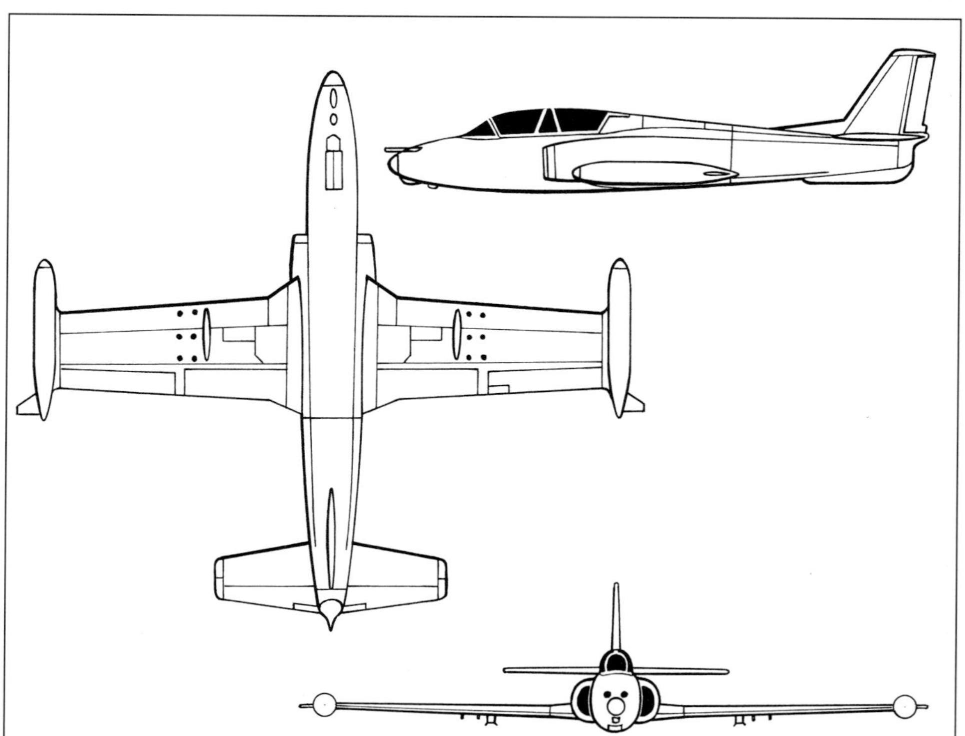

Soko G-2A Galeb two-seat armed jet basic trainer *(Jane's/Mike Keep)*

1994

1973; construction of first prototype begun May 1975; first flight (23004) 17 July 1978; second prototype (23005) 17 December 1979; third airframe for ground testing; first of six preproduction aircraft flew 17 December 1980; first full production, 1983. G-4M with new avionics; planned first flight July 1992, but apparently abandoned. In Spring 1992 the Soko facilities were evacuated to Serbia and all production ceased. Production may have restarted in 1993.

VERSIONS: **G-4 PPP**: First prototype and six preseries aircraft (23601-23606); fixed tailplane with inset elevators and no anhedral.

G-4: Second prototype and all full production aircraft; all-moving tailplane with anhedral. Retrofit planned with provision for AAMs (probably AA-8 and AA-11) and ASMs (probably AS-7 and AS-9), improved electronic equipment and better power plant de-icing. *Main description applies to G-4 except where indicated.*

G-4M: Revised avionics and nav/attack systems, including gyro platform, IFF, flight data recorder, indigenous electronic sight. HUD, multi-function displays and wingtip missile rails. Payload increased by 405 kg (893 lb). Intended to undertake portion of weapon training syllabus previously performed by front-line aircraft.

OPERATORS: Versions of the G-4 Super Galeb are in service with the armed forces of the following countries: Bosnia-Herzegovina (4), Burma (Myanmar) (20) and Yugoslavia (140).

DESIGN FEATURES: Optimised for easy transition of pupils to more advanced combat aircraft; reduces training costs; good manoeuvring characteristics at high Mach numbers and high altitude. Sweptback wing, 26° at leading-edge, 45° at root, 22° at quarter-chord; wing section NACA 64A211; dihedral 1°; no incidence or twist. Sweptback vertical fin; all-moving horizontal tail surfaces have 10° anhedral for optimum rudder efficiency when spinning.

FLYING CONTROLS: Mechanical control with non-linear transmission; dual hydraulic fully powered controls for tailplane and ailerons directional control fully mechanical with direct linkage. Ailerons, without aerodynamic balance (except G-4 PPP); elevator with internal aerodynamic balance on G-4 PPP; rudder has aerodynamic balance and trim tab. Small wedge-type stall strips on wings; fences at two-thirds of wing span; single ventral and dorsal fins. Directional trim by electromechanical tab in rudder (tab in elevator on G-4 PPP). Artificial feel provided by loading simulators within tabs for longitudinal and lateral planes (built-in spring for additional stick loading in longitudinal plane on G-4 PPP). Door-type airbrake under rear fuselage. Single-slotted trailing-edge flaps (double-slotted on G-4 PPP) operated by electrically controlled hydraulic actuators.

STRUCTURE: Single-piece, two-spar, all-metal wing with integrally machined skin panels inboard and chemically milled skin outboard; shallow boundary layer fence above each wing, forward of inboard end of aileron; wings attached to fuselage at six points; entire trailing-edge of conventional all-metal sealed ailerons and flaps; integral fuel tank in wingroot. All-metal, semi-monocoque fuselage; rear detachable for engine access. Optional composites tail unit.

LANDING GEAR: Prva Petoletka-Trstenik (PPT) hydraulically retractable tricycle type, with single wheel on each unit. Nosewheel retracts forward, main units inward into wings and fuselage; maximum steering angle ±90°. Oleo-pneumatic shock-absorber in each leg. Hydraulically steerable nose unit optional (not on current aircraft). Trailing link main units. Mainwheels fitted with Miloje Zakić tyres size 615 × 225-10PR, pressure 4.5 bars (65 lb/sq in), and PPT hydraulic brakes. Nosewheel has tyre size 6.50-5.5 TC, pressure 4.2 bars (61 lb/sq in). Brake parachute container at base of rudder. Provision for attaching two assisted take-off rockets under centre-fuselage. Operable from grass at normal T-O weight. Minimum ground turning radius 4.70 m (15 ft 5 in).

Soko G-4 Super Galeb in the colours of the former Federal Yugoslav Air Force (JRV) *(Peter J. Cooper)* *1994*

POWER PLANT: One licence-built Rolls-Royce Viper Mk 632-46 turbojet, rated at 17.8 kN (4,000 lb st). Fuel in two rubber fuselage tanks, each 506 litres (134 US gallons; 111.3 Imp gallons); collector tank, 57 litres (15.0 US gallons; 12.5 Imp gallons); and two integral wing tanks, each 348 litres (92 US gallons; 76.5 Imp gallons); total internal fuel 1,764 litres (466 US gallons; 388 Imp gallons). Provision for two underwing auxiliary tanks, on inboard pylons, total capacity 737 litres (195 US gallons; 162 Imp gallons); G-4M also has centreline tank of 449 litres (119 US gallons; 99 Imp gallons). Maximum fuel capacity: G-4, 2,501 litres (660 US gallons; 550 Imp gallons); G-4M, 2,950 litres (779 US gallons; 649 Imp gallons). Oil capacity 7.38 litres (1.95 US gallons; 1.6 Imp gallons). Gravity refuelling system standard; pressure system optional.

ACCOMMODATION: Crew of two in tandem on Martin-Baker zero/zero Mk 10Y ejection seats (zero height/90 knot (166 km/h; 103 mph) Mk J8 optional, but not on current aircraft), with ejection through the individual sideways-hinged (to starboard) canopy over each seat. Rear seat raised by 25 cm (10 in). Cockpit pressurised and air conditioned.

SYSTEMS: Engine compressor bleed air used for pressurisation (maximum differential, 0.21 bar; 3.0 lb/sq in), air conditioning, anti-*g* suit and windscreen anti-misting and anti-icing systems, and to pressurise fuel tanks. Engine face and (optional) intake lip de-icing. Dual hydraulic systems, pressure 206 bars (2,988 lb/sq in), for flying control servos, flap and airbrake actuators, landing gear retraction and extension, and wheel brakes. Hydraulic system flow rate 45 litres (12 US gallons; 10 Imp gallons)/min for main system, 16 litres (4.2 US gallons; 3.5 Imp gallons)/min for flight control system. Emergency Abex electric pump providing 4.6 litres (1.2 US gallons; 1.0 Imp gallons)/min to one aileron. Electrical system supplied by 9 kW 28 V DC generator, with 36 Ah Ni/Cd battery for ground/emergency power and self-contained engine starting. Two static inverters, total output 600 VA, provide 115 V/26 V 400 Hz AC power; G-4M total output 900VA. Gaseous oxygen system adequate for two crew for 2 hours 30 minutes comprises two bottles, each 7.3 litres (1.9 US gallons; 1.6 Imp gallons) at nominal 150 bars (2,175 lb/sq in).

AVIONICS: Dual controls and full blind-flying instrumentation in each cockpit. Standard nav/com equipment comprises EAS type ER4.671D or Rudi Čajavec RCE 163 Kondor VHF com radio, GEC Avionics AD 370B or Iskra VARK-01 ADF, Collins VIR-30 VOR/ILS, Iskra 75R4 VOR marker beacon receiver, Collins DME 40, TRT AHV-6 radio altimeter. SFIM CG-90 gyro compass, SFENA H-140 artificial horizon, GEC Ferranti ISIS type D-282 weapon sight and Iskra SD-1 radar warning receiver. Optional M. Kobac UHF. G-4M has RCE 163 VHF, M. Kobac UHF, VARK-01, VIR-30, 75R4, DME-40 and AHV-6 plus IFF, SFIM SHARP gyro platform, Electronic Industry RPL flight data recorder, Zrak ENP-MG4 HUD, Rudi Čajavec ENS-MG4 sight, SD-1 RWR and optional chaff/flare dispensers. Development of a photo-reconnaissance/infra-red linescan pod and night illumination system, and selection of an alternative off-the-shelf reconnaissance pod, deferred pending identification of customer requirement.

ARMAMENT: Removable ventral gun pod containing 23 mm GSh-23L twin-barrel rapid fire cannon with 200 rds; optional replacement on G-4M by 'wet' pylon of 400 kg (882 lb) capacity. Two attachments under each wing: inboard pylons (all 'wet') stressed to 350 kg (772 lb) on G-4 and 500 kg (1,102 lb) on G-4M; outboard pylons to 250 kg (551 lb) on G-4 and 350 kg (772 lb) on G-4M. G-4M wingtip launch rails for R-60 (AA-8 'Aphid') IR AAMs; provision for Hughes AGM-65B Maverick ASM on outboard pylons. Total weapon load capability, with centreline gun pod, 1,280 kg (2,822 lb), or 1,800 kg (3,968 lb) for G-4M. In addition to standard high explosive bombs and napalm pods, typical Yugoslav stores include S-8-16 cluster bombs, each with eight 16 kg fragmentation munitions; KPT-150 expendable containers, each with up to 40 anti-personnel or 54 anti-tank bomblets; L-57-16MD pods, each with 16 57 mm rockets; L-128-04 pods, each with four 128 mm rockets; adaptors for twin 5 in HVAR rockets, single 57 mm VRZ-57 training rockets; SN-3-050 triple carriers for 50 kg bombs; SN-3-100 triple carriers for 100 kg bombs; KM-3 pods each containing a single 12.7 mm (0.50 in) gun; SAM Z-80 towed target system; and auxiliary fuel tanks on the inboard attachments. Provision to be added for air-to-air and air-to-surface missiles.

DIMENSIONS, EXTERNAL:

Wing span: G-4	9.88 m (32 ft 5 in)
G-4M over launchers	10.05 m (33 ft 0 in)
G-4M over missiles	10.31 m (33 ft 9⅞ in)
Wing chord: at root, with apex	3.69 m (12 ft 1¼ in)
at root, without apex, at centreline	2.86 m (9 ft 4⅝ in)
at tip	1.20 m (3 ft 11¼ in)
Wing aspect ratio: G-4	4.73
Length: overall	12.25 m (40 ft 2¼ in)
without pitot	11.35 m (37 ft 2⅞ in)
fuselage	11.02 m (36 ft 1⅞ in)
Height overall	4.30 m (14 ft 1¼ in)
Tailplane span	3.97 m (13 ft 0¼ in)
Wheel track	3.49 m (11 ft 5½ in)
Wheelbase	4.15 m (13 ft 7½ in)

AREAS:

Wings, gross	19.5 m^2 (209.9 sq ft)
Ailerons (total)	1.358 m^2 (14.62 sq ft)
Trailing-edge flaps (total)	3.34 m^2 (35.95 sq ft)
Airbrake	0.438 m^2 (4.71 sq ft)
Fin	2.39 m^2 (25.73 sq ft)
Rudder	0.57 m^2 (6.14 sq ft)
Horizontal tail surfaces (total)	4.669 m^2 (50.26 sq ft)
Elevators, incl tab (G-4 PPP only)	1.00 m^2 (10.76 sq ft)

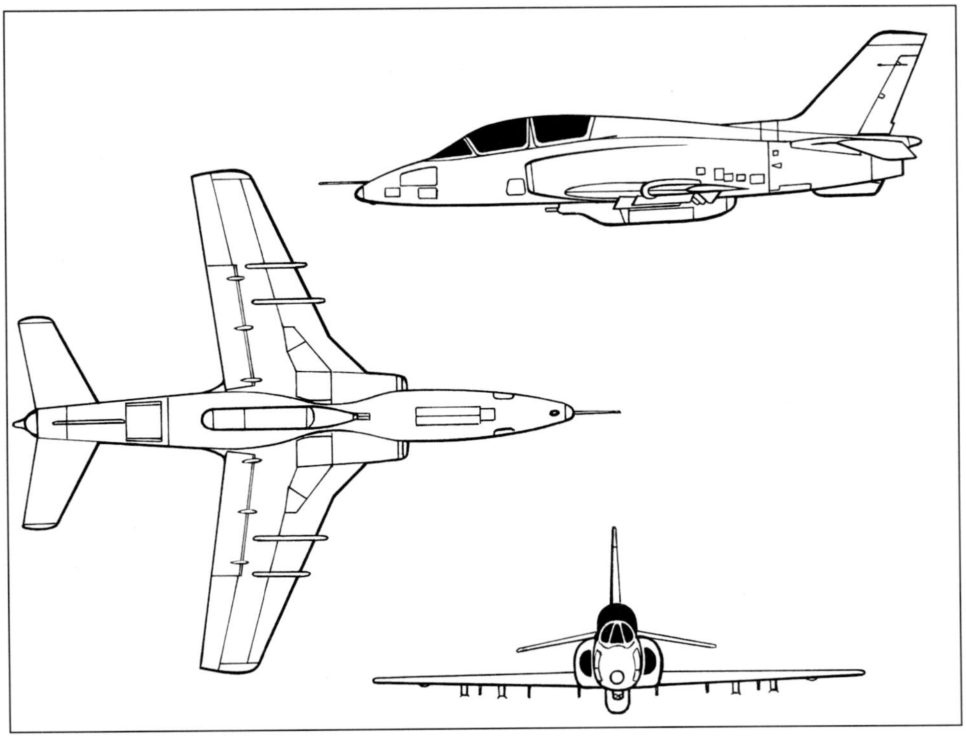

Soko G-4 Super Galeb two-seat operational trainer *(Jane's/Mike Keep)* *1994*

Soko G-4 Super Galeb with Rolls-Royce Viper Mk 632-46 turbojet engine

1992

WEIGHTS AND LOADINGS (G-4 PPP as G-4 unless otherwise stated):

Weight empty, equipped: G-4	3,172 kg (6,993 lb)
G-4M 'clean'	3,403 kg (7,502 lb)
G-4M with missile rails	3,435 kg (7,573 lb)
Max payload: G-4	2,053 kg (4,526 lb)
G-4M	2,458 kg (5,419 lb)
Max fuel weight: internal: G-4	1,307 kg (2,881 lb)
G-4M	1,376 kg (3,034 lb)
external: G-4	575 kg (1,268 lb)
G-4M	925 kg (2,039 lb)
T-O weight, training mission:	
G-4 PPP	4,639 kg (10,227 lb)
G-4	4,708 kg (10,379 lb)
G-4M	4,971 kg (10,959 lb)
Max T-O weight: G-4	6,300 kg (13,889 lb)
G-4M	6,400 kg (14,110 lb)
Max landing weight: G-4 PPP	4,639 kg (10,227 lb)
G-4	4,708 kg (10,379 lb)
G-4M	4,971 kg (10,959 lb)
Max wing loading: training mission:	
G-4 PPP	238.0 kg/m^2 (48.75 lb/sq ft)
G-4	241.0 kg/m^2 (49.36 lb/sq ft)
G-4M	255.0 kg/m^2 (52.23 lb/sq ft)
combat mission:	
G-4 PPP/G-4	323.0 kg/m^2 (66.15 lb/sq ft)
G-4M	328.0 kg/m^2 (67.18 lb/sq ft)
Max power loading: training mission:	
G-4 PPP	260.6 kg/kN (2.56 lb/lb st)
G-4	264.5 kg/kN (2.59 lb/lb st)
G-4M	279.3 kg/kN (2.74 lb/lb st)
combat mission:	
G-4 PPP/G-4	353.9 kg/kN (3.47 lb/lb st)
G-4M	359.6 kg/kN (3.52 lb/lb st)

PERFORMANCE (G-4 at 4,708 kg; 10,379 lb, G-4M at 4,971 kg; 10,959 lb):

Never-exceed speed (VNE)	Mach 0.9
Max level speed:	
at 10,000 m (32,800 ft): G-4	Mach 0.81
G-4M	Mach 0.80
at 4,000 m (13,120 ft):	
G-4	491 kts (910 km/h; 565 mph)
G-4M	486 kts (900 km/h; 559 mph)
Max cruising speed at 6,000 m (19,700 ft) and 95% rpm	456 kts (845 km/h; 525 mph)
Econ cruising speed at 6,000 m (19,700 ft)	297 kts (550 km/h; 342 mph)
Stalling speed, power on:	
flaps up: G-4	119 kts (220 km/h; 137 mph)
G-4M	121 kts (225 km/h; 140 mph)
flaps down: G-4	97 kts (180 km/h; 112 mph)
G-4M	100 kts (185 km/h; 115 mph)
Max rate of climb at S/L: G-4	1,860 m (6,100 ft)/min
G-4M	1,800 m (5,900 ft)/min
Service ceiling: G-4	12,850 m (42,160 ft)
G-4M	12,500 m (41,010 ft)
T-O run: G-4	572 m (1,877 ft)
G-4M	600 m (1,969 ft)
T-O to 15 m (50 ft): G-4	900 m (2,953 ft)
G-4M	950 m (3,117 ft)
Landing from 15 m (50 ft):	
without drag-chute: G-4	1,065 m (3,494 ft)
G-4M	1,130 m (3,707 ft)
with drag-chute: G-4	690 m (2,264 ft)
G-4M	750 m (2,461 ft)
Landing run: G-4	815 m (2,674 ft)
G-4M	860 m (2,822 ft)
Range at 11,000 m (36,090 ft), 10% reserves:	
max internal fuel:	
G-4	1,025 n miles (1,900 km; 1,180 miles)
G-4M	971 n miles (1,800 km; 1,118 miles)
max internal and external fuel:	
G-4	1,349 n miles (2,500 km; 1,553 miles)
G-4M	1,565 n miles (2,900 km; 1,802 miles)
Range at 11,000 m (36,090 ft) with cannon and four 277 kg (610 lb) Hunting BL-755 CBUs:	
G-4	701 n miles (1,300 km; 807 miles)
G-4M, plus two AA-8s	647 n miles (1,200 km; 745 miles)

VERIFIED

(SOKO) J-1/RJ-1 JASTREB (HAWK)

TYPE: Single-seat light attack and tactical reconnaissance aircraft.

PROGRAMME: The basic J-1 Jastreb is a single-seat light attack version of the G-2A Galeb, developed and produced for service with the Yugoslav Air Force. An export version was produced and the first overseas operator was the Zambian Air Force which received them in 1971 and still operates four of the aircraft.

In the J-1 Jastreb, the front cockpit of the G-2A Galeb trainer, with sideways-hinged (to starboard) canopy, is retained, a metal fairing replacing the rear canopy. The engine is the more powerful Rolls-Royce Bristol Viper 531. Other changes include the installation of improved day and night reconnaissance equipment, navigation and communications equipment, and self-contained engine starting. In other respects the airframe and power plant remain essentially unchanged except for some local strengthening and the provision of strongpoints for heavier underwing stores.

VERSIONS: **J-1**: Standard attack version for the Yugoslav Air Force (prior to the break-up of late 1991).

J-1-E: Export attack version with updated equipment. In service with the air forces of Libya (24), Yugoslavia (75+, prior to the break-up in late 1991. Now in storage) and Zambia (4).

RJ-1: Tactical reconnaissance version for the former Yugoslav Air Force. In service (20+. Now in storage).

RJ-1-E: Export reconnaissance version with updated equipment.

TJ-1: Two-seat operational conversion and pilot proficiency training version.

OPERATORS: Versions of the J-1 Jastreb are in service with the armed forces of the following countries: Libya (10), Bosnia-Herzegovina (9), Yugoslavia (150), and Zambia (6).

The details given for the G-2A Galeb apply equally to the J-1, J-1-E, RJ-1 and RJ-1-E Jastreb, with the following exceptions:

POWER PLANT: One Rolls-Royce Viper 531 turbojet engine, rated at 13.32 kN (3,000 lb st). Capacity of each wingtip tank 220 kg (485 lb). Provision for attaching two 4.44 kN (1,000 lb st) JATO rockets under fuselage for use at take-off or in flight.

ACCOMMODATION: Pilot only, on HSA (Folland) Type 1-B fully automatic lightweight ejection seat. Cockpit air conditioning to special order only.

SYSTEMS: Electrical system includes 6 kW 24 V generator and second battery, permitting independent engine starting without ground electrical supply. Two oxygen bottles supply high-pressure oxygen system of nominal 1,900 litres (501 US gallons; 418 Imp gallons) capacity, at a pressure of 138 bars (2,000 lb/sq in).

AVIONICS AND EQUIPMENT (J-1 and RJ-1): Full IFR instrumentation. Standard Telephones & Cables STR-9Z1 VHF com transceiver and Marconi AD 370B radio compass. The fuselage camera of the RJ-1 is supplemented by two further cameras in nose of tip tanks, which are also available for the J-1 and J-1-E attack versions. An aerial target can be towed from a hook under the centre-fuselage. Brake parachute housed in fairing above jet nozzle.

AVIONICS AND EQUIPMENT (J-1-E and RJ-1-E): Nav/com equipment same as for the G-2AE Galeb export version. Photo-reconnaissance equipment of the RJ-1-E consists of a daylight system comprising two Vinten 360/140A cameras with 3 in AEC lenses in nose of tip tanks (also available on J-1-E attack version), and a third camera of the same type in the fuselage, with interchangeable lens units; and a night reconnaissance system comprising one Vinten 1025/527 camera at the fuselage station.

ARMAMENT (J-1 and J-1-E): Three 0.50 in Colt-Browning machine guns in nose (with 135 rds/gun). Total of eight underwing weapon attachments. Two inboard attachments can carry two bombs of up to 250 kg each, two clusters of small bombs, two 200 litre napalm tanks, two pods each with 12 or 16 57 mm (four 128 mm rockets, two multiple carriers each with three 50 kg bombs, two bomblet containers, or two 45 kg photo flares. Other attachments can each carry a 127 mm rocket. Semi-automatic gyro gunsight and camera gun standard.

ARMAMENT (RJ-1 and RJ-1-E): Four underwing attachments, intended basically for carrying flash bombs for night photography, can also be used for carrying high-explosive or other types of bombs. The inboard pylons can each carry a single bomb of up to 250 kg, the outboard pylons up to 150 kg. No rocket armament. Otherwise same as for J-1 and J-1-E.

DIMENSIONS, EXTERNAL: As for Galeb except:

Wing span over tip tanks	11.68 m (38 ft 4 in)
Length overall	10.88 m (35 ft 8½ in)
Height overall	3.64 m (11 ft 11½ in)
Wheelbase	3.61 m (11 ft 10 in)

AREAS:

Wings, gross	19.43 m^2 (209.14 sq ft)
Ailerons (total)	2.36 m^2 (25.40 sq ft)
Trailing-edge flaps (total)	2.02 m^2 (21.75 sq ft)
Airbrakes (total)	0.34 m^2 (3.66 sq ft)
Fin	1.34 m^2 (14.42 sq ft)
Rudder, incl tab	0.56 m^2 (6.03 sq ft)
Tailplane	3.66 m^2 (39.40 sq ft)
Elevators, incl tabs	0.83 m^2 (8.93 sq ft)

WEIGHTS AND LOADINGS:

Weight empty, equipped	2,820 kg (6,217 lb)

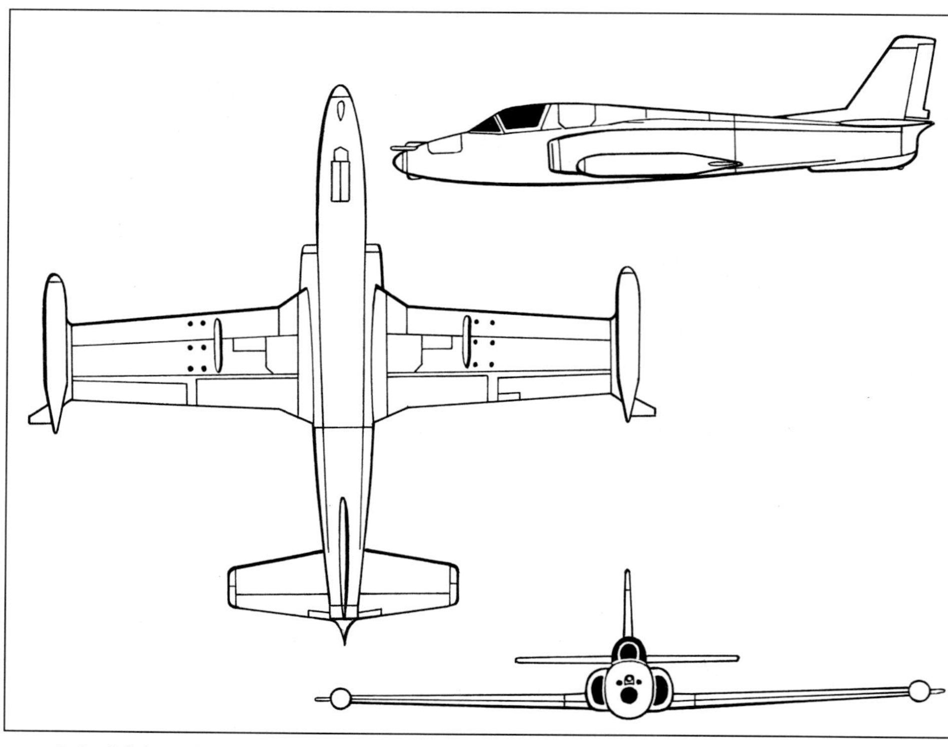

Soko J-1 Jastreb single-seat light attack and tactical reconnaissance aircraft ***(Jane's/Mike Keep)***

1994

Soko J-1 Jastreb in the colours of the former Federal Yugoslav Air Force (JRV)

1994

Max ramp weight 5,287 kg (11,655 lb)
Max T-O weight 5,100 kg (11,243 lb)
Max landing weight 3,750 kg (8,267 lb)

PERFORMANCE (T-O and landing runs on concrete):
Max level speed at 6,000 m (19,680 ft) at AUW of 3,968 kg (8,748 lb) 442 kts (820 km/h; 510 mph)
Max cruising speed at 5,000 m (16,400 ft) at AUW of 3,968 kg (8,748 lb) 399 kts (740 km/h; 460 mph)
Stalling speed, wheels down:
flaps and airbrakes down 82 kts (152 km/h; 95 mph)
flaps and airbrakes up 94 kts (174 km/h; 108 mph)
Max rate of climb at S/L, at AUW of 3,968 kg (8,748 lb) 1,260 m (4,135 ft)/min
Service ceiling at AUW of 3,968 kg (8,748 lb) 12,000 m (39,375 ft)
T-O run at AUW of 3,968 kg (8,748 lb) 700 m (2,300 ft)
T-O run, rocket-assisted, at max T-O weight 404 m (1,325 ft)
T-O to 15 m (50 ft) at AUW of 3,968 kg (8,748 lb) 960 m (3,150 ft)
T-O to 15 m (50 ft), rocket-assisted, at max T-O weight 593 m (1,945 ft)
Landing from 15 m (50 ft) 1,100 m (3,610 ft)
Landing run 600 m (1,970 ft)
Max range at 9,000 m (19,520 ft), with tip tanks full 820 n miles (1,520 km; 945 miles)

VERIFIED

ADDENDA

AIRCRAFT

BELGIUM

SABENA TECHNICS (page 6)

Swissair announced 22 June 1995 that it had acquired a 49.5 stake in Sabena Technic's parent company, Sabena.

NEW ENTRY

SABENA TECHNICS (LOCKHEED) C-130 AVIONICS UPGRADE PROGRAMME AUP

PROGRAMME: Maiden flight of first upgraded C-130 took place on 4 May 1995. The aircraft (CH-01) took off from Melsbroek Air Base at 18:20 local time and landed at 20:45. The Pilot in command was Colonel Jacques Lousberg.

NEW ENTRY

Sabena Technics (Lockheed) C-130 AUP maiden flight (4-5-95)
1995

CANADA

BRISTOL AEROSPACE (page 15)

BRISTOL AEROSPACE (NORTHROP) F-5 UPGRADE

PROGRAMME: Bristol Aerospace announced in June 1995 that it had signed a marketing agreement with the Canadian Government to market and sell the Canadian military's fleet of CF-5 aircraft. In all,72 aircraft are available for sale. These are, 19 aircraft (eight CF-5A and 11 CF-5D) that have had a depot level inspection (DLIR), an avionics upgrade and the replacement of their FS 325/362 bulkheads; 16 aircraft (four CF-5A and 12 CF-5D) that have had the DLIR and an avionics upgrade; eight CF-5Ds which have had a DLIR only; and 29 (24 CF-5A and five CF-5D) that are in their original configuration. In June 1995 24 of the 35 aircraft which were to be upgraded had been completed.

NEW ENTRY

BRISTOL AEROSPACE HELICOPTER WIRE STRIKE PROTECTION SYSTEM

TYPE: Helicopter wire strike protection system retrofit.
PROGRAMME: Developed in collaboration with the Canadian Forces for retrofitting into their helicopter fleet.
DESIGN FEATURES: Basic kit weighs approximately 7.3 kg (15 lbs) and consists of a system of deflectors and cutters which protect the windshield, landing gear, rotors and control linkages. Kit is supplied with all components and fittings ready for installation. Systems is FAA certified for the following helicopter types: Agusta A 109, Bell 204/205/206/212/222/412, Eurocopter (Aerospatiale) AS 350 Ecureuil Astar and AS 355 Ecureuil Twinstar, Eurocopter (Aerospatiale) SA 341/342 Gazelle, Eurocopter (Aerospatiale) AS 330 Puma and AS 332 Super Puma, Eurocopter (MBB) Bk 117 Bo 105 and McDonnell Douglas 500/530.
CUSTOMERS: The system has been installed on helicopters of the following armed services: Canada, France (Ecureuil, Gazelle, Puma and Super Puma), USA (AH-1, AH-64, OH-6, OH-58, UH-1 and UH-60)

NEW ENTRY

INTERNATIONAL

SNECMA - FRANCE
ITP - ITALY
SIMERA - SOUTH AFRICA

COMBAT AIRCRAFT ENGINE UPGRADE

TYPE: Combat aircraft engine upgrade.
PROGRAMME: Snecma of France, ITP of Italy and Simera of South Africa announced in mid-1995 that they will co-operate to develop an upgrade package for the ATAR 8/9K50 turbojet engine used on Atlas Cheetah, Dassault Mirage 50, Dassault Mirage F1, and Dassault Super Etendard aircraft. Upgrade kit will improve aircraft performance to counter threat from new generation infra-red missiles as well as extend their service life to 2015. Package is due to be ready by mid-1997.
DESIGN FEATURES: Main elements of the upgrade will consist of a modification to the engine core to decrease acceleration time by one third; improving maintainability of the engine by increasing the time between overhaul from 900 to 1,200 hours; and reducing maintenance by increasing the time between periodic inspection from 300 to 400 hours. Simera will develop and produce a new high pressure inlet guide vane (IGV) to achieve this. ITP will develop and produce a new compressor outlet guide vane to meet the reduced time between overhaul goal.
COST: System will cost $14.2 million to develop.

NEW ENTRY

IRELAND

SHANNON

SHANNON AEROSPACE LIMITED

Shannon Airport, Shannon, County Clare
Telephone: 353 (61) 370006
Fax: 353 (61) 3700 20
CHIEF EXECUTIVE: Wolfgang F.W. Gohòe
MANAGING DIRECTOR: Dr Herbert Groeger
HEAD, PRODUCTION: Willy Hallauer
HEAD, FINANCE: Pat Shine
HEAD OF MARKETING AND SALES: Peter Ellison

Shannon Aerospace commenced operations in September 1992 and is jointly owned by Swissair and Lufthansa. GPA's Shareholding is now held in trust. Main responsibilities are in the fields of maintenance and overhaul of McDonnell Douglas DC-9 and MD-80 series aircraft as well as Boeing 737, 757 and 767 aircraft. The company is also active in passenger aircraft interior refurbishment, exterior repainting. By the end of 1998, Shannon Aerospace will commence overhauling Airbus A320 passenger aircraft.

NEW ENTRY

ISRAEL

ELISRA

ELISRA ELECTRONIC SYSTEMS LTD (A subsidiary of Tadiran Ltd)

48 Mivtza Kadesh Street, IL-51203 Bene Beraq
Telephone: 972 (3) 754 5111
Fax: 972 (3) 754 5468
PRESIDENT: Avner Raz
VICE-PRESIDENT, MARKETING AND SALES: Eitah Livneh
MANAGER, PRESS RELATIONS: Gil Goldfine

Company founded in 1966 to design, develop and manufacture electronic equipment for airborne, naval and land-based applications. Main responsibilities are in the fields of electonic warfare (EW) equipment for combat aircraft (fixed- and rotary-wing), naval ESM/ECM suites, mobile and fixed ground-based communications and intelligence equipment. The company is also involved in the design and manufacture of microwave and RF components, high power transmitters, antennas, hybrid microelectronics, digital design and computer hardware, software processing development and programming, receivers, RF measurement systems, digital signal processing, image processing technology, high rate communication links and pulse Doppler radar. In June 1995 Elisra announced that it had adopted a strategy to become active in the airborne electronic warfare (EW) systems retrofit/upgrade market. In 1995 the company employed 1,000 personnel and had covered floor space of 2,600 m2 (28,000 sq m). For detail description of equipment see *Jane's Avionics 1995-96*.

NEW ENTRY

ISRAELI AIRCRAFT INDUSTRIES (page 121)

IAI (MIKOYAN) MiG-21-2000

PROGRAMME: First flight of IAI Lahav Division's MiG-21-2000 took place 24 May 1995.

NEW ENTRY

IAI (MIKOYAN) MiG-21-2000 during maiden flight 24 May 1995.
1995

SOUTH AFRICA

SIMERA (A division of Denel (Pty) Ltd)

Astro Park, Atlas Road, PO Box 117, 1620 Kempton Park
Telephone: 27 (11) 927 9222
Fax: 27 (11) 927 2400
GENERAL MANAGER: J J Eksteen
MANAGER, MARKETING: Bob Watson
MANAGER, PUBLIC RELATIONS: Gero Eckermann

Company active in the fields of aircraft repair and overhaul and modification of a wide range of civil aircraft, helicopters, engines and components as well as the manufacture of aircraft components and parts.

NEW ENTRY

SIMERA (AIRBUS) A300/A310 FREIGHTER CONVERSION

TYPE: Stage III freighter conversion.

PROGRAMME: Based on a technical exchange agreement between Simera and Daimler-Benz Aerospace Airbus. Conversions to be carried out at Simera facilities at Jan Smuts International Airport using a complete kit and tooling supplied to Simera by Daimler-Benz Aerospace Airbus. Conversion will meet Stage III requirements.

DESIGN FEATURES: Standard conversion consists of: 3.58 m × 2.56 m (11 ft 7 in × 8 ft 4 in) main deck cargo door, floor structure reinforcement, 9 *g* safety barrier net, fire detection system, smoke curtain, window blanking and system adaptation. Optional features include additional loading places, extended loading height, provisions for AMA container transportation, provisions for engine transport capability as well as weight reduction, door blocking/deactivation and replacement. Cargo loading systems (CLS) can be single row, side by side or integrated.

WEIGHTS AND LOADINGS: (A A300 B4-200F, B A310-200F, C A310-300F)

Max T-O weight	
A	165,001 kg (363,760 lb)
B	138,597 kg (305,550 lb)
B	142,022 kg (313,100 lb)
C	150,005 kg (330,700 lb)
C	152,999 kg (337,300 lb)
C	156,990 kg (346,100 lb)
Max landing weight	
A	134,002 kg (295,420 lb)
A	135,998 kg (299 820 lb)
B	121,995 kg (268,950 lb)
B	123,016 kg (271,200 lb)
C	122,993 kg (271,150 lb)
C	124,014 kg (273,400 lb)
Max zero fuel weight	
A	124,000 kg (273,370 lb)
A	126,001 kg (277,780 lb)
B	111,993 kg (246,900 lb)
B	112,991 kg (249,100 lb)
C	112,991 kg (249,100 lb)
C	113,989 kg (251,300 lb)
Max operating weight	
A	81,761 kg (180,250 lb)
A	81,784 kg (180,300 lb)
B	73,710 kg (162,500 lb)
B	73,846 kg (162,800 lb)
C	73,846 kg (162,800 lb)
C	73,891 kg (162,900 lb)
Max payload	
A	42,239 kg (93,120 lb)
A	44,216 kg (97,480 lb)
B	38,283 kg (84,400 lb)
B	39,145 kg (86,300 lb)
C	39,145 kg (86,300 lb)
C	40,098 kg (88,400 lb)

NEW ENTRY

SPAIN

DASSAULT MIRAGE F1 UPGRADE

The Spanish Government announced in July 1995 that it had given the go-ahead for upgrading the air force's 55 Mirage F1 aircraft. The contract is expected to be awarded to CASA in collaboration with Thomson-CSF of France. Upgrade is to be completed by 1999 and is aimed at extending the aircraft's operational life until 2010 when EF2000 enters service.

NEW ENTRY

UNITED KINGDOM

BRITISH AIRWAYS
BRITISH AIRWAYS ENGINEERING

Room A208 TBA (S316), PO Box 10, London Heathrow Airport, Hounslow, Middlesex TW6 2JA
Telephone: 44 (181) 562 8762
Fax: 44 (181) 562 3581
DIRECTOR, ENGINEERING: Alistair Cumming
HEAD OF ENGINEERING SALES: Bill Coyle

British Airways Engineering operates a global network of sites as well as covered facilities at Cardiff, Heathrow and Gatwick airports and has a total staff of over 10,000 personnel. The company is active in transport aircraft maintenance, modification, component and engine overhaul/repair, exterior painting, interior refurbishing and redesign, avionics repair and modification as well as undercarriage repair and loan. Further activities are in the fields of re-engining, hush-kitting and cargo conversions. The company has ISO 9000/BS 5750 accreditiation as well as CAA, JAR 145, FAA and MoD approvals. Maintenance responsibilities include A, B and C checks on Airbus A320, Boeing 737, 747, 757, 767, Lockheed L-1011 Tristar and McDonnell Douglas DC-10 aircraft.

NEW ENTRY

DHSA
DEFENCE HELICOPTER SUPPORT AUTHORITY

Defence Helicopter Support Authority Gazelle House, Yeovilton, Yeovil, Somerset BA22 8HJ
Telephone: +44 (1935) 453657
Fax: +44 (1935) 453699

The DHSA was formed on 1 April 1994 to bring together the engineers, suppliers, contracts, finance and administrative staffs from the military helicopter community into a tri-service multi-disciplinary group (MDG). The DHSA's main responsibility is to provide support authority services for all eight helicopter types in service with the UK armed forces. As such the DHSA will arrange for the implementation of all in-service helicopter retrofit/upgrade programmes.

NEW ENTRY

HUNTING
HUNTING BUSINESS AVIATION LTD

Biggin Hill, Hangar 446, Biggin Hill Airport, Biggin Hill, Kent TN16 3BN
Telephone: +44 (1959) 570700
Fax: +44 (1959) 540668
MANAGER, CORPORATE COMPLETIONS: Steve Brenchley

Company active in the design, manufacture and installation of interiors for Corporate/VIP/VVIP aircraft. Paint spray booth facility capable of taking aircraft up to Gulfstream IV. Company also acts as integrator of Honeywell avionic systems and design/installation of avionic modifications.

NEW ENTRY

MARSHALL AEROSPACE (page 259)

MARSHALL (LOCKHEED) L-1011 TRISTAR CONVERSION

TYPE: Passenger aircraft freighter conversion.

PROGRAMME: Contract awarded by American International Airways for freighter conversions of nine of its Lockheed L-1011 TriStar aircraft. First aircraft flew 16 August 1995. Aircraft delivered 18 August 1995.

DESIGN FEATURES: Programme includes removal of passenger interior, strengthening of floor, installation of 2.94 × 3.94 m (155 × 125 in) cargo door, fixed crash bulkhead, repainting and certification with US FAA STC. Main deck has capacity for twenty-three 2.23 × 3.17 m (88 × 125 in) pallets, or twenty-six 2.23 × 2.74 m (88 × 108 in) pallets. The three lower deck holds have capacity for various combinations of pallets and up to 24 LD3 containers and bulk cargo.

CUSTOMERS: American International Airways (eight aircraft).

NEW ENTRY

Hand over ceremony of Lockheed L-1011 TriStar converted to freighter by Marshall Aerospace *(Marshall Aerospace)*

1995

RACAL AVIONICS LIMITED (page 264)

RACAL/WESTLAND COMBAT HELICOPTER NAVIGATION UPGRADE

TYPE: Combat helicopter avionics upgrade.

PROGRAMME: Contract awarded 22 July 1995 to Racal Avionics (prime) and Westland Helicopters Ltd (aircraft design authority and installer). Programme to be completed by 1999. Contract includes integrated logistic support (ILS) package to run until 2003. Aircraft to be upgraded are Lynx HMA 8 and Sea King Mk 4/5/6. The Lynx HMA 8 will have the CDNU interfaced with a central tactical system (CTS).

DESIGN FEATURES: Programme covers the installation of a navigation/mission computer known as the control/display and navigation unit (CDNU) which hosts a P(Y) code global positioning system (GPS) module.

CUSTOMERS: UK Royal Navy (150 aircraft).

COSTS: Approximately £25 million.

NEW ENTRY

ROYAL NAVAL AIRCRAFT

Fleetlands, Gosport, Hampshire PO13 OAW
Telephone: 44 (1705) 822351
CHIEF EXECUTIVE: Capt D E Symonds

NEW ENTRY

WESTLAND GAZELLE AH.1 UPGRADE

TYPE: Attack helicopter avionics upgrade.

PROGRAMME: Deliveries commenced in June 1995 of AH.1 Gazelle helicopters fitted with new Radar Warning Receivers. Trials took place in 1993 on six Lynx helicopters and six Gazelles.

DESIGN FEATURS: Upgrade consists of installation of ARI.23491/1 Rewarder variant of the GEC Marconi Defence Systems AWARE-3 radar warning receiver (RWR).

CUSTOMER: UK Army Air Corps.

NEW ENTRY

UNITED STATES OF AMERICA

GREENWICH
GREENWICH AIR SERVICES

PO Box 522187 Miami, Florida 33152-2187
Telephone: 1 (305) 526 7090
Fax: 1 (305) 526 7005
SENIOR VICE-PRESIDENT, INDUSTRIAL AND GOVERNMENT BUSINESS: Moti Volovelsky
VICE-PRESIDENT, MARKETING AND SALES: William Madison

NEW ENTRY

GREENWICH (DE HAVILLAND) DHC-5 BUFFALO UPGRADE

TYPE: Twin-turboprop STOL aircraft upgrade.

PROGRAMME: Greenwich Air Service awarded contract in September 1992 to upgrade nine DHC-5 Buffalo aircraft for the Egyptian Air Force. Fifth aircraft delivered in July 1995. Programme is in collaboration with LAB of France. Work being carried out at LAB's facilities in Dinard (France).

DESIGN FEATURES: Upgrade involves stripping aircraft down, overhauling all components installing new avionics and refurbishing interior.

CUSTOMERS: Egyptian Air Force (9). Upgrade sponsored by US Army under the Foreign Military Sales agreement.

NEW ENTRY

Greenwich (de Havilland) DHC-5 Buffalo STOL aircraft upgraded for the Egyptian Air Force (Greenwich)

1995

PEMCO AEROPLEX (page 410)

(BOEING) 727-200 WEIGHT INCREASE

Pemco received Supplemental Type Certificate (STC) approval (July 1995) from the Federal Aviation Authority (FAA) for its maximum zero fuel weight and maximum landing weight increased for Boeing 727-200 aircraft. This upgrade was incorporated in three Boeing 727-200 combi quick-change aircraft for First air. This aircraft was the first 727-200 to receive FAA and Transport Canada certification.

NEW ENTRY

Pemco (Boeing) 727-200 fully convertible combi-quick change aircraft belonging to First Air
1995

QNC
QUIET NACELLE CORPORATION

8000 Northwest 56th Street, Miami, Florida 33166
Telephone: 1 (305) 593 0731
Fax: 1 (305) 592 78265

Company founded in 1984 to develop Stage 2 hushkits for the Douglas DC-8-50 series 61 aircraft. This was followed by the development of Stage 2 kits for Boeing 707-100 and -300 series aircraft. The company is currently developing Stage 3 hushkits for Boeing 707 aircraft and Douglas DC-8-50 series 61 aircraft. DC-8-50 hushkit is to commence flight testing in mid-1995. Proof of concept flight testing of a Stage 3 hushkit for the BAC One-Eleven has been completed and is expected to achieve certification in 1996.

NEW ENTRY

QNC (BOEING) OC-135 OPEN SKIES HUSHKIT

TYPE: Treaty observation aircraft hushkit.

PROGRAMME: Acoustic and performance testing successfully completed in January 1995. Production of first shipset of components to be used in the certification programme commenced July 1995. First kit will be installed on USAF Boeing 707 (WC-135B) test aircraft by October 1995. Delivery of first OC-135 Open Skies aircraft with QNC hushkit is due in December 1995. Supplemental Type Certificate (STC) of the QNC hushkit for commerical Boeing 707 is expected in 1996.

DESIGN FEATURES: Construction is of acoustically treated carbon graphite honeycomb.

CUSTOMER: USAF (3).

NEW ENTRY

QNC Hushkit installed on Boeing 707 showing detail of inner acoustic ring and ejector *(QNC)*
1995

RAYTHEON (page 426)

RAYTHEON (BEECH) 1900C CARGO CONVERSION

TYPE: Twin-turboprop commuter aircraft cargo conversion.

PROGRAMME: First of four converted aircraft delivered to Falcon Express Cargo Airlines of Dubai in August 1995. Options exist for two more.

DESIGN FEATURES: Conversion consists of installation of: aluminium liners on main cabin ceilings and sidewalls, seven 54 × 57 in web nets to divide cabin area into cargo compartments, special lighting to facilitate night operations as well as a removable seat for extra crew member and plugs in the windows for ease of maintenance.

CUSTOMERS: Falcon Express Cargo Airlines (4 with options for further 2).

NEW ENTRY

Raytheon (Beech) 1900C converted for cargo operations for Falcon Express Cargo Airlines *(Raytheon)*
1995

INDEX

A

B

C

D

E

F

G

H

I

N

O

P

Q

R

S

T

U

V

W

X

Y

Z